Volkswagen Corrado

Repair Manual
1990, 1991, 1992, 1993, 1994
including G60, VR6 and SLC

Selected Books and Repair Information From Bentley Publishers

Motorsports

Alex Zanardi: My Sweetest Victory *Alex Zanardi and Gianluca Gasparini*
ISBN 978-0-8376-1249-2

The Unfair Advantage *Mark Donohue and Paul van Valkenburgh*
ISBN 978-0-8376-0069-7

Equations of Motion - Adventure, Risk and Innovation *William F. Milliken*
ISBN 978-0-8376-1570-7

Engineering

Bosch Fuel Injection and Engine Management *Charles O. Probst, SAE*
ISBN 978-0-8376-0300-1

Maximum Boost: Designing, Testing, and Installing Turbocharger Systems *Corky Bell* ISBN 978-0-8376-0160-1

Supercharged! Design, Testing and Installation of Supercharger Systems *Corky Bell* ISBN 978-0-8376-0168-7

Race Car Aerodynamics *Joseph Katz*
ISBN 978-0-8376-0142-7

Scientific Design of Exhaust and Intake Systems *Phillip H. Smith & John C. Morrison* ISBN 978-0-8376-0309-4

Audi

Audi A4 (B6, B7) Service Manual: 2002-2008, 1.8L Turbo, 2.0L Turbo, 3.0L, 3.2L including Avant and Cabriolet *Bentley Publishers*
ISBN 978-0-8376-1574-5

Audi A4 Service Manual: 1996-2001, 1.8L Turbo, 2.8L, including Avant and quattro *Bentley Publishers*
ISBN 978-0-8376-1675-9

Audi A6 Service Manual: 1998-2004 *Bentley Publishers*
ISBN 978-0-8376-1670-4

Audi TT Service Manual: 2000-2006, 1.8L turbo, 3.2 L, including Roadster and quattro *Bentley Publishers*
ISBN 978-0-8376-1625-4

Audi A6, S6: 2005-2007 Repair Manual on DVD-ROM *Audi of America*
ISBN 978-0-8376-1362-8

Audi A6 Electrical Wiring Manual *Bentley Publishers*
ISBN 978-0-8376-0166-3

BMW

BMW 5 Series (E60, E61) Service Manual: 2004-2010 *Bentley Publishers*
ISBN 978-0-8376-1689-6

BMW X5 (53) Service Manual: 2000-2006 *Bentley Publishers*
ISBN 978-0-8376-1643-8

BMW Z3 (E36/7) Service Manual: 1996-2002 *Bentley Publishers*
ISBN 978-0-8376-1617-9

BMW 3 Series (E30) Service Manual: 1984-1990 *Bentley Publishers*
ISBN 978-0-8376-1647-6

BMW 3 Series (E36) Service Manual: 1992-1998 *Bentley Publishers*
ISBN 978-0-8376-0326-1

BMW 3 Series (E46) Service Manual: 1999-2005 *Bentley Publishers*
ISBN 978-0-8376-1657-5

BMW 3 Series (E90, E91, E92, E93) Service Manual: 2006-2010 *Bentley Publishers*
ISBN 978-0-8376-1685-8

BMW 5 Series (E39) Service Manual: 1997-2003 *Bentley Publishers*
ISBN 978-0-8376-1672-8

BMW 7 Series (E38) Service Manual: 1995-2001 *Bentley Publishers*
ISBN 978-0-8376-1618-6

Porsche

Porsche Boxster Service Manual: 1997-2004 *Bentley Publishers*
ISBN 978-0-8376-1645-2

Porsche 911 Carrera Service Manual: 1984-1989 *Bentley Publishers*
ISBN 978-0-8376-1696-4

Porsche 911 SC Service Manual: 1978-1983 *Bentley Publishers*
ISBN 978-0-8376-0290-5

Porsche: Excellence Was Expected *Karl Ludvigsen* ISBN 978-0-8376-1769-5

Ferdinand Porsche – Genesis of Genius *Karl Ludvigsen*
ISBN 978-0-8376-1557-8

Volkswagen

Battle for the Beetle *Karl Ludvigsen*
ISBN 978-0-8376-1695-7

Volkswagen Rabbit, GTI Service Manual: 2006-2009 *Bentley Publishers*
ISBN 978-0-8376-1664-3

Volkswagen Jetta, Golf, GTI Service Manual: 1999-2005 *Bentley Publishers*
ISBN 978-0-8376-1678-0

Volkswagen Jetta, Golf, GTI: 1993-1999, Cabrio: 1995-2002 Service Manual *Bentley Publishers*
ISBN 978-0-8376-0366-7

Volkswagen GTI, Golf, Jetta Service Manual: 1985-1992 *Bentley Publishers*
ISBN 978-0-8376-1637-7

Volkswagen Corrado Repair Manual: 1990-1994 *Bentley Publishers*
ISBN 978-0-8376-1699-5

Volkswagen Jetta: 2005-2009 Repair Manual on DVD-ROM *Volkswagen of America*
ISBN 978-0-8376-1360-4

Volkswagen Passat, Passat Wagon: 1998-2005 Service Manual *Bentley Publishers*
ISBN 978-0-8376-1669-8

MINI Repair Manuals

MINI Cooper Service Manual: 2007-2011 *Bentley Publishers*
ISBN 978-0-8376-1671-1

MINI Cooper Service Manual: 2002-2006 *Bentley Publishers*
ISBN 978-0-8376-1639-1

MINI Cooper Diagnosis Without Guesswork: 2002-2006 *Bentley Publishers*
ISBN 978-0-8376-1571-4

Mercedes-Benz

Mercedes-Benz C-Class (W202) Service Manual 1994-2000 *Bentley Publishers*
ISBN 978-0-8376-1692-6

Mercedes Benz E-Class (W124) Owner's Bible: 1986-1995 *Bentley Publishers*
ISBN 978-0-8376-0230-1

Mercedes-Benz Technical Companion *Staff of The Star and members of Mercedes-Benz Club of America*
ISBN 978-0-8376-1033-7

B BentleyPublishers™ .com

Automotive Reference

Bentley Publishers has published service manuals and automobile books since 1950. For more information, please contact Bentley Publishers at 1734 Massachusetts Avenue, Cambridge, MA 02138, visit our web site at **www.BentleyPublishers.com**

Volkswagen Corrado

Repair Manual
1990, 1991, 1992, 1993, 1994

including G60, VR6 and SLC

BentleyPublishers®
.com

BENTLEY PUBLISHERS® | Automotive Reference™

Bentley Publishers, a division of Robert Bentley, Inc.
1734 Massachusetts Avenue
Cambridge, MA 02138 USA
800-423-4595 / 617-547-4170

Information that makes the difference®

BentleyPublishers®.com

Technical contact information

We welcome your feedback. Please submit corrections and additions to our technical discussion forum at:

`http://www.BentleyPublishers.com`

Updates and corrections

We will evaluate submissions and post appropriate editorial changes online as updates or tech discussion. Appropriate updates and corrections will be added to the book in future printings. Check for updates and corrections for this book before beginning work on your vehicle. See the following web address for additional information:

`http://www.BentleyPublishers.com/updates/`

WARNING—Important Safety Notice

Do not use this manual for repairs unless you are familiar with basic automotive repair procedures and safe workshop practices. This manual illustrates the workshop procedures for some maintenance and service work. It is not a substitute for full and up-to-date information from the vehicle manufacturer or for proper training as an automotive technician. Note that it is not possible for us to anticipate all of the ways or conditions under which vehicles may be serviced or to provide cautions as to all of the possible hazards that may result.

We have endeavored to ensure the accuracy of the information in this manual. Please note, however, that considering the vast quantity and the complexity of the service information involved, we cannot warrant the accuracy or completeness of the information contained in this manual.

FOR THESE REASONS, NEITHER THE PUBLISHER NOR THE AUTHOR MAKES ANY WARRANTIES, EXPRESS OR IMPLIED, THAT THE INFORMATION IN THIS MANUAL IS FREE OF ERRORS OR OMISSIONS, AND WE EXPRESSLY DISCLAIM THE IMPLIED WARRANTIES OF MERCHANTABILITY AND OF FITNESS FOR A PARTICULAR PURPOSE, EVEN IF THE PUBLISHER OR AUTHOR HAVE BEEN ADVISED OF A PARTICULAR PURPOSE, AND EVEN IF A PARTICULAR PURPOSE IS INDICATED IN THE MANUAL. THE PUBLISHER AND AUTHOR ALSO DISCLAIM ALL LIABILITY FOR DIRECT, INDIRECT, INCIDENTAL OR CONSEQUENTIAL DAMAGES THAT RESULT FROM ANY USE OF THE EXAMPLES, INSTRUCTIONS OR OTHER INFORMATION IN THIS MANUAL. IN NO EVENT SHALL OUR LIABILITY WHETHER IN TORT, CONTRACT OR OTHERWISE EXCEED THE COST OF THIS MANUAL.

Before attempting any work on your Volkswagen, read **X** and **XI Warnings and Cautions** and any **WARNING** or **CAUTION** that accompanies a procedure in the manual. Review the **WARNINGS** and **CAUTIONS** each time you prepare to work on your Volkswagen.

Your common sense and good judgment are crucial to safe and successful service work. Read procedures through before starting them. Think about whether the condition of your car, your level of mechanical skill, or your level of reading comprehension might result in or contribute in some way to an occurrence which might cause you injury, damage your car, or result in an unsafe repair. If you have doubts for these or other reasons about your ability to perform safe repair work on your car, have the work done at an authorized Volkswagen dealer or other qualified shop.

Part numbers listed in this manual are for identification purposes only, not for ordering. Always check with your authorized Volkswagen dealer to verify part numbers and availability before beginning service work that may require new parts.

Special tools required to perform certain service operations are identified in the manual and are recommended for use. Use of improper tools may be detrimental to the car's safe operation as well as the safety of the person servicing the car.

The vehicle manufacturer will continue to issue service information updates and parts retrofits after the editorial closing of this manual. Some of these updates and retrofits will apply to procedures and specifications in this manual. We regret that we cannot supply updates to purchasers of this manual.

This manual is prepared, published and distributed by Bentley Publishers, 1734 Massachusetts Avenue, Cambridge, Massachusetts 02138. All information contained in this manual is based on information available to the publisher from Volkswagen at the time of editorial closing. Volkswagen has not reviewed and does not vouch for the accuracy or completeness of the technical specifications and work procedures described and given in this manual.

Library of Congress Catalog Card No. 93-73889
Volkswagen Part No. LPV 800 300

ISBN 978-0-8376-1699-5 **Bentley Stock No. VC94** **Editorial closing 10/93, Supplement 7/94** Mfg. code: VC94-12

Manufactured in the United States of America

Contents

Important Safety Notice ... IV
Foreword and Disclaimers ... VII
Vehicle Identification and VIN Decoder ... VIII
Warnings and Cautions ... X
Index ... XII

Owner Information

Vehicle Care ... Page 1
Do-it-yourself Service ... 21
Technical Description ... 31
Gas Station Information ... 40
Maintenance/Lubrication Schedules ... 41

General
Engine

General ... Repair Group 00
Engine—Assembly ... 10
—Crankshaft/Crankcase ... 13
—Cylinder Head/Valve Drive ... 15
—Lubrication System ... 17
—Cooling System ... 19

Fuel Systems
Supercharger
Exhaust
Engine Electrical

Fuel Supply ... 20
Supercharger ... 21
Fuel Injection, AFC System ... 24
Exhaust System—Emission Controls ... 26
Electrical—Battery, Starter, Generator ... 27
Ignition System ... 28

Clutch
Transmission

Clutch ... 30
Torque Converter ... 32
Manual Transmission—Controls, Assembly ... 34
Manual Transmission—Case, Gears, Shafts ... 35
Automatic Transmission—Controls, Housing ... 37
Automatic Transmission—Case, Gears, Shafts ... 38
Differential—Manual Transmission ... 39
—Automatic Transmission ... 39

Suspension

Front Wheel Suspension—Shafts & Axle ... 40
Rear Wheel Suspension—Shafts & Axle ... 42
Wheels, Tires, Wheel Alignment ... 44

Brakes
Steering

Anti-Lock Brake System ... 45
Brake—Mechanical Components ... 46
Brake—Hydraulic Components, Regulator, Booster ... 47
Steering ... 48

Body

Body ... 50
Hoods/Lids ... 55
Body—Front Doors ... 57
Sunroof ... 60
Bumpers ... 63

Glass, Window Regulators ... 64
Body Accessories—Exterior ... 66
Body Accessories—Interior ... 68
Trim—Interior ... 70
Body—Seats ... 72

Heater
Air Conditioner

Heater and Ventilation ... 80
Air Conditioner ... 87

Electrical

Electrical System—Instruments ... 90
—Radio, Multi-function Indicator ... 91
—Windshield Wiper & Washer ... 92
—Lights ... 94
—Wiring ... 97

Wiring Diagrams

Index ... Page 1
General Information ... 3

Appendix

On-Board Diagnostic ... Repair Group 01 (D2)

Supplement

Motronic Ignition system—VR6 engine code AAA ... 28
Automatic Transmission 096 Shiftlock III ... 37
Air Conditioning—Systems using R134a refrigerant ... 87
Wiring Diagrams

Foreword and Disclaimers

Service to Volkswagen owners is a top priority of the Volkswagen organization and has always included the continuing development and introduction of new and expanded services. In line with this purpose, Robert Bentley, Inc., in cooperation with Volkswagen of America, Inc., has introduced this Volkswagen Corrado Official Factory Repair Manual.

This manual covers Volkswagen Corrados for the model years 1990 through 1994 based on Volkswagen's A3 platform. Please see the Vehicle Identification & VIN decoder remarks on the following page for more information. This manual was created specifically to cover only those models built for sale in the United States and Canada.

For the Volkswagen owner with basic mechanical skills and for independent auto service professionals, this manual includes the specifications and procedures that were available in an authorized Volkswagen dealer service department as this manual went to press. The Volkswagen owner with no intention of working on his or her car will find that owning and referring to this manual will make it possible to be better informed and to more knowledgeably discuss repairs with a professional automotive technician.

The Volkswagen owner intending to do maintenance and repair should have screwdrivers, a set of metric wrenches and sockets, and metric hex wrenches, since these basic hand tools are needed for most of the work described in this manual. Most procedures will also require a torque wrench to ensure that fasteners are tightened properly and in accordance with specifications. In some cases, the text refers to special tools that are recommended or required to accomplish adjustments or repairs. These tools are usually identified by their Volkswagen special tool number and illustrated.

Disclaimer

We have endeavored to insure the accuracy of the information in this manual. When the vast array of data presented in the manual is taken into account, however, no claim to infallibility can be made. We therefore cannot be responsible for the result of any errors that may have crept into the text. Please also read the **Important Safety Notice** on the copyright page at the beginning of this book.

A thorough pre-reading of each procedure, and the **WARNINGS** and **CAUTIONS** at the front of the book and those that accompany the procedure is essential. Reading a procedure before beginning work will help you determine in advance the need for specific skills, identify hazards, prepare for appropriate capture and handling of hazardous materials, and the need for particular tools and replacement parts such as gaskets.

Bentley Publishers encourages comments from the readers of this manual with regard to errors, and/or suggestions for improvement of our product. These communications have been and will be carefully considered in the preparation of this and other manuals. If you identify inconsistencies in the manual, you may have found an error. Please contact the publisher and we will endeavor to post applicable corrections on our website. Posted corrections (errata) should be reviewed before beginning work. Please see the following web address:

`http://www.BentleyPublishers.com/errata/`

Volkswagen offers extensive warranties, especially on components of the fuel delivery and emission control systems. Therefore, before deciding to repair a Volkswagen that may be covered wholly or in part by any warranties issued by Volkswagen of America, Inc., consult your authorized Volkswagen dealer. You may find that the dealer can make the repair either free or at minimum cost. Regardless of its age, or whether it is under warranty, your Volkswagen is both an easy car to service and an easy car to get serviced. So if at any time a repair is needed that you feel is too difficult to do yourself, a trained Volkswagen technician is ready to do the job for you.

Bentley Publishers

Vehicle Identification & VIN Decoder

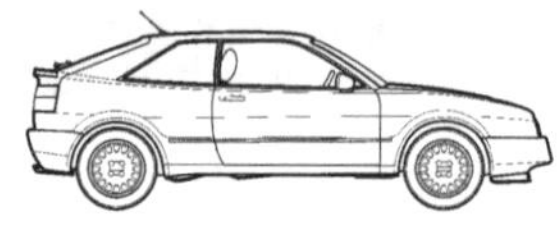

Some of the information in this manual applies only to cars of a particular model year or range of years. For example, "1994 m.y." refers to the 1994 model year. The model year does not necessarily match the calendar year in which the car was manufactured or sold. To be sure of the model year of a particular car, check the vehicle identification number (VIN) on the car. Technical changes made in production within a model year are identified in this manual by listing the VIN for the first car produced with this change.

The VIN is a unique sequence of 17 characters assigned by Volkswagen to identify each individual car. WVWDB4502NK005678 is an example. Each of the 17 letters and numbers indicates certain facts about the car and its manufacture. VINs used to distinguish information in this manual may refer only to the last eleven digits—the characters 502NK005678 in the example above.

Your Volkswagen's VIN can be found on a plate mounted on the top of the instrument panel, on the driver's side where the number can be seen through the windshield. The 10th character is the model year code. The letters "I", "O", "Q" and "U" are not used for model year designation, for example, "L" for 1990 m.y., "M" for 1991 m.y., "N" for 1992, "P" for 1993 m.y., etc. This manual covers Volkswagen Corrados for the model years 1990 through 1994. The table below explains some of the various codes in the VIN numbers of models covered by this manual.

position	1	2	3	4	5	6	7	8	9	10	11	12	13	14	15	16	17
Corrado:							5	0				0	0	0	0	0	1
to							5	0				9	9	9	9	9	9

Position	Description	Decoding Information	
1	Country of origin	W	Europe
2	Manufacturer	V	Volkswagen
3	Vehicle Type	W	Passenger Car
4	Series		Varies with model year See BentleyPublishers.com for more information
5	Engine		
6	Restraint system		
7	Model	50	Corrado
8			
9	Check Digit		0-9 or X, calculated by NHTSA
10	Model Year	L M N P R	1990 1991 1992 1993 1994
11	Assembly Plant	K	Osnabrück, Germany
12-17	Serial Number		Sequential production number of the specific vehicle

Before attempting any work on your Volkswagen, please turn the page to read the warnings and cautions. Also read any warning and caution that accompanies a procedure in the manual. Review the warnings and cautions on the following pages each time you prepare to work on your Volkswagen.

Please read these WARNINGS and CAUTIONS before proceeding with maintenance and repair work.

WARNING—

- Read the **Important Safety Notice** on the copyright page at the beginning of the book.

- Some repairs may be beyond your capability. If you lack the skills, tools and equipment, or a suitable workplace for any procedure described in this manual, we suggest you leave such repairs to an authorized Volkswagen dealer service department, or other qualified shop.

- A thorough pre-reading of each procedure, and the **WARNINGS** and **CAUTIONS** that accompany the procedure is essential. Posted corrections (errata) should also be reviewed before beginning work. Please see **www.BentleyPublishers.com/errata/**.

- Volkswagen is constantly improving its cars. Sometimes these changes, both in parts and specifications, are made applicable to earlier models. Therefore, before starting any major jobs or repairs to components on which passenger safety may depend, consult your authorized Volkswagen dealer about Technical Bulletins that may have been issued since the editorial closing of this manual.

- Do not re-use any fasteners that are worn or deformed in normal use. Many fasteners are designed to be used only once and become unreliable and may fail when used a second time. This includes, but is not limited to, nuts, bolts, washers, self-locking nuts or bolts, circlips and cotter pins. Always replace these fasteners with new parts.

- Never work under a lifted car unless it is solidly supported on stands designed for the purpose. Do not support a car on cinder blocks, hollow tiles or other props that may crumble under continuous load. Never work under a car that is supported solely by a jack. Never work under the car while the engine is running.

- If you are going to work under a car on the ground, make sure that the ground is level. Block the wheels to keep the car from rolling. Disconnect the battery negative (–) terminal (Ground strap) to prevent others from starting the car while you are under it.

- Never run the engine unless the work area is well ventilated. Carbon monoxide kills.

- Finger rings, bracelets and other jewelry should be removed so that they cannot cause electrical shorts, get caught in running machinery, or be crushed by heavy parts.

- Tie long hair behind your head. Do not wear a necktie, a scarf, loose clothing, or a necklace when you work near machine tools or running engines. If your hair, clothing, or jewelry were to get caught in the machinery, severe injury could result.

- Do not attempt to work on your car if you do not feel well. You increase the danger of injury to yourself and others if you are tired, upset or have taken medication or any other substance that may keep you from being fully alert.

- Illuminate your work area adequately but safely. Use a portable safety light for working inside or under the car. Make sure the bulb is enclosed by a wire cage. The hot filament of an accidentally broken bulb can ignite spilled fuel or oil.

- Catch draining fuel, oil, or brake fluid in suitable containers. Do not use food or beverage containers that might mislead someone into drinking from them. Store flammable fluids away from fire hazards. Wipe up spills at once, but do not store the oily rags, which can ignite and burn spontaneously.

- Always observe good workshop practices. Wear goggles when you operate machine tools or work with battery acid. Gloves or other protective clothing should be worn whenever the job requires working with harmful substances.

- Friction materials such as brake or clutch discs may contain asbestos fibers. Do not create dust by grinding, sanding, or by cleaning with compressed air. Avoid breathing asbestos fibers and asbestos dust. Breathing asbestos can cause serious diseases such as asbestosis or cancer, and may result in death.

- Disconnect the battery negative (–) terminal (ground strap) whenever you work on the fuel system or the electrical system. Do not smoke or work near heaters or other fire hazards. Keep an approved fire extinguisher handy.

- Batteries give off explosive hydrogen gas during charging. Keep sparks, lighted matches and open flame away from the top of the battery. If hydrogen gas escaping from the cap vents is ignited, it will ignite gas trapped in the cells and cause the battery to explode.

- Connect and disconnect battery cables, jumper cables or a battery charger only with the ignition switched off, to prevent sparks. Do not disconnect the battery while the engine is running.

- Do not quick-charge the battery (for boost starting) for longer than one minute. Wait at least one minute before boosting the battery a second time.

- Do not allow battery charging voltage to exceed 16.5 volts. If the battery begins producing gas or boiling violently, reduce the charging rate. Boosting a sulfated battery at a high charging rate can cause an explosion.

- The air-conditioning system is filled with chemical refrigerant, which is hazardous. The A/C system should be serviced only by trained technicians using approved refrigerant recovery/recycling equipment, trained in related safety precautions, and familiar with regulations governing the discharging and disposal of automotive chemical refrigerants.

- Do not expose any part of the A/C system to high temperatures such as open flame. Excessive heat will increase system pressure and may cause the system to burst.

- Some aerosol tire inflators are highly flammable. Be extremely cautious when repairing a tire that may have been inflated using an aerosol tire inflator. Keep sparks, open flame or other sources of ignition away from the tire repair area. Inflate and deflate the tire at least four times before breaking the bead from the rim. Completely remove the tire from the rim before attempting any repair.

- Most cars covered by this manual are equipped with a supplemental restraint system (SRS), that automatically deploys an airbag in the event of a frontal impact. The airbag is inflated by an explosive device. Handled improperly or without adequate safeguards, it can be accidently activated and cause serious injury.

- To prevent personal injury or airbag system failure, **only factory trained Volkswagen service technicians** should test, disassemble or service the airbag system.

Please read these WARNINGS and CAUTIONS before proceeding with maintenance and repair work.

WARNING (continued) —

- Disconnect the power supply before working on the airbag system, or when doing repairs that require removing airbag system components. Disconnect the battery negative (–) terminal and cover the battery.

- On airbag-equipped cars, never apply stickers or any other type of covering on the steering wheel. Do not let chemical cleaners, oil or grease come into contact with vinyl covering of the airbag unit.

- Never open or otherwise attempt to repair airbag system parts. Always use new parts. Never leave airbag parts or the partially disassembled airbag system unattended.

- Never use a test light to conduct electrical tests on the airbag system. The system must only be tested by trained Volkswagen Service technicians using the Volkswagen VAG 1551/1552 Scan Tool (ST) or an approved equivalent. The airbag unit must never be electrically tested while it is not installed in the car.

- Do not expose the airbag unit to temperatures above 194°F (90°C), even for brief periods. Keep clear of heat sources such as hot plates, soldering irons, heat lamps and welding equipment.

- When driving or riding in an airbag-equipped vehicle, never hold test equipment in your hands or lap while the vehicle is in motion. Objects between you and the airbag can increase the risk of injury in an accident.

- Greases, lubricants and other automotive chemicals contain toxic substances, many of which are absorbed directly through the skin. Read manufacturer's instructions and warnings carefully. Use hand and eye protection. Avoid direct skin contact.

CAUTION—

- If you lack the skills, tools and equipment, or a suitable workshop for any procedure described in this manual, we suggest you leave such repairs to an authorized Volkswagen dealer or other qualified shop. We especially urge you to consult an authorized Volkswagen dealer before beginning repairs on any car that may still be covered wholly or in part by any of the extensive warranties issued by Volkswagen of America.

- Volkswagen offers extensive warranties, especially on components of fuel delivery and emission control systems. Therefore, before deciding to repair a Volkswagen that may still be covered wholly or in part by any warranties issued by Volkswagen United States, Inc., consult your authorized Volkswagen dealer. You may find that he can make the repair for free, or at minimal cost.

- Volkswagen part numbers listed in this manual are for identification purposes only, not for ordering. Always check with your authorized Volkswagen dealer to verify part numbers and availability before beginning service work that may require new parts.

- Before starting a job, make certain that you have all the necessary tools and parts on hand. Read all the instructions thoroughly, do not attempt shortcuts. Use tools appropriate to the work and use only replacement parts meeting Volkswagen specifications. Makeshift tools, parts and procedures will not make good repairs.

- Use pneumatic and electric tools only to loosen threaded parts and fasteners. Never use these tools to tighten fasteners, especially on light alloy parts. Always use a torque wrench to tighten fasteners to the tightening torque specification listed.

- Be mindful of the environment and ecology. Before you drain the crankcase, find out the proper way to dispose of the oil. Do not pour oil onto the ground, down a drain, or into a stream, pond or lake. Consult local ordinances that govern the disposal of wastes.

- On cars equipped with anti-theft radios, make sure you know the correct radio activation code before disconnecting the battery or removing the radio. If the wrong code is entered into the radio when power is restored, that radio may lock up and be rendered inoperable, even if the correct code is then entered.

- Connect and disconnect a battery charger only with the battery charger switched off.

- Do not quick-charge the battery (for boost starting) for longer than one minute. Wait at least one minute before boosting the battery a second time.

- Sealed or "maintenance free" batteries should be slow-charged only, at an amperage rate that is approximately 10% of the battery's ampere-hour (Ah) rating.

- Do not allow battery charging voltage to exceed 16.5 volts. If the battery begins producing gas or boiling violently, reduce the charging rate. Boosting a sulfated battery at a high charging rate can cause an explosion.

Index

WARNING

- Automotive service and repair is serious business. You must be alert, use common sense, and exercise good judgement to prevent personal injury and complete the work safely.
- Before beginning any work on your vehicle, thoroughly read the safety notice, disclaimer, and the Cautions and Warnings listed near the front of this manual.
- Always read the complete procedure before you begin the work. Pay special attention to any Cautions and Warnings that accompany that procedure, or other information on a specific topic.

ABS
see Anti-Lock Brake System or On Board Diagnostic (ABS)

Air filter
see Owner Information

Alignment
see Wheels, Tires, Wheel Alignment

Alternator
see Electrical—Battery, Starter, Generator

Anti-freeze
see Engine—Cooling System

Air Conditioner
Systems using R-134a refrigerant
see Supplement
Systems using R12 refrigerant
—A/C compressor
assembly 87.22
clutch
checking 87.9
removing/installing 87.23
lines
installing 87.9a
shaft seal
removing/installing 87.28
valve plate, head and gasket
removing/installing 87.35
—A/C pressure switch
checking 87.7
—A/C system
fuses 87.7
relays
location 87.8
temperatures and pressures
checking 87.16
—Air distribution and evaporator case
assembly 87.10
—Belt and pulleys
assembly 87.21
—Belt tensioning damper
assembly 87.20
—Engine components
mounting 87.18
—Evaporator water drain valve
identifying/removing 87.9
expansion valve insulator
checking 87.9

Air Conditioner (cont'd)
—Refrigerant system
components 87.5
discharging 87.3
oil specifications 87.4a
safety measures 87.2
—Safety relief plug
identifying 87.8
—Sight glass
checking 87.8
—Troubleshooting
air conditioner
does not cool 87.46
compressor clutch does not engage 87.48
cooling output diminishes after several minutes 87.51
Insufficient cooling output 87.50
defroster
does not clear windshield 87.44
fresh air
cannot be shut off 87.39
no air flow from dash board or foot well vents 87.42
heater
insufficient or no heat output 87.40
—Vacuum system
checking 87.13
hose layout 87.12
reservoir
removing/installing 87.15
servos
checking 87.14

Anti-Lock Brake System (ABS)
—Hydraulic modulator
assembly
with EDL 45.7
without EDL 45.3
removing/installing 45.6
—Pressure warning switch
removing/installing 45.5
—Repair work
general notes 45.2
repair rules 45.2

ATF (automatic transmission fluid)
see Automatic Transmission—Controls, Housing

Automatic Transmission—Case, Gears, Shaft
—1st to 3rd gear clutch (K1) with turbine shaft
assembly 38-130
—2nd/4th gear brake (B2)
adjusting 38-90
—3rd/4th gear clutch (K3) With pump shaft
assembly 38-170
circlip for clutch spring support
replacing 38-160
—ATF pump
assembly 38-100
—Clutches K1/K2
adjusting/checking freeplay 38-80
disassembling/assembling 38-140
—One-way clutch with B1 piston
assembly 38-110
disassembling/assembling 38-120

Automatic Transmission—Case, Gears, Shaft (cont'd)
—Parking lock
assembly 38-210
installing 38-220
—Planet carrier
adjusting/checking 38-60
—Reverse brake (B1)
adjusting/checking 38-70
—Reverse clutch (K2)
assembly 38-150
circlip for clutch spring support
replacing 38-180
—Transmission
assembling 38-50
assembly 38-30
clutches/brakes
installed positions 38-10
disassembling 38-40
selector/clutch/brake engagement
chart 38-20
—Valve body
assembly 38-190
disassembling/assembling 38-200

Automatic Transmission—Controls, Housing
096 automatic transmission
—ATF cooler/ATF filler tube
assembly 37-70
—ATF filter
replacing 37-80
—Automatic shiftlock
troubleshooting 37-110
—Identification 00.3a, 00.3b
—Removing/installing 37-60
—Shift points
checking 37-90
—Stall speed
checking 37-100
—Selector lever/detent segment/solenoid switch
adjusting 37-40
assembly 37-30
—Selector mechanism
assembly 37-20
support
removing/installing 37-50
—Technical data
chart 37-10

Battery
see Electrical—Battery, Starter, Generator or Owner Information

Belts
see Electrical—Battery, Starter, Generator

Body
—Acid rain
repairing 50.11
—Body dimensions
door opening 50.4
floor section 50.2
front section 50.4
rear section 50.5
—Front fender
adjusting 50.7
mounting 50.6

Body (cont'd)

—Head lamp
 removing 50.9
—Lock support
 assembly 50.8
 removing/installing 50.10
—Shroud
 removing/installing 50.9

Body Accessories—Exterior

—Mirror glass
 removing/installing 66.8
—Radiator grille
 removing/installing 66.5
—Rear spoiler
 installing 66.5
 removing 66.4
—Rearview mirror
 assembly 66.7
 removing/installing 66.8
—Rocker panel
 protective molding
  replacing 66.9
—Roof trim molding
 removing /installing 66.3
—Wheel house liner
 assembly 66.6
 installing
  1991 m.y. 66.6a

Body Accessories—Interior

—Automatic shoulder belt system (USA only)
 assembly 68.2
 door frame track trim
  removing/installing 68.4
 drive unit
  installing 68.4
  removing 68.3
 knee bar
  removing/installing 68.4
 knee bar support bracket
  removing/installing 68.5
 seat switch release magnet
  removing/installing 68.6
 shoulder belt release magnet
  installing 68.6
  removing 68.5
—Automatic shoulder belt troubleshooting (USA only)
 belts do not move to A-pillar
  after removing ignition key
  (driver or passenger side) 68.17
 belts only move to A-pillar
  door opened with key removed
  (driver or passenger) 68.16
 both belts do not move to B-pillar
  doors shut, ignition ON 68.9
 belt does NOT move to A-pillar
  when door opened 68.13
 belt does NOT move to B-pillar 68.11
 driver belt moves to A-pillar
  when vehicle shifted into reverse 68.18
 seat belt warning light comes on
  with belts connected and ignition on 68.20
 seat belt warning light does not light
  belts at B-pillar, ignition ON 68.19
 wiring diagrams 68.21

Body Accessories—Interior (cont'd)

—Child seat
 installing on front passenger seat 68.31
—Interior mirror
 removing /installing 68.8
—Rear seat belt
 assembly 68.7
 buckle
  removing/installing 68.8
 mounting
  removing /installing 68.8
—Seat belts
 inspecting 68.24

Body—Front Doors

—Central locking
 assembly 57.9
 bi-pressure pump
  removing 57.10
  checking 57.13
 electrical troubleshooting 57.14
 gas flap activator,
  adjusting 57.11a
  removing/installing 57.11
 master activators
  removing/installing 57.10
 pump/control module
  location 57.17
 rear lid activator
  removing/installing 57.12
—Door lock
 checking 57.6
 door lock
  removing/installing 57.8
 inner handle
  removing 57.7
 installing 57.7
 locks
  re-keying 57.18
—Front door
 assembly 57.2
 impact bar
  removing 57.5
 removing/adjusting 57.3
 seal
  removing/installing 57.8
—Window regulator
 bordering frame
  removing/installing 57.4
 carrier to glass
  bonding 57.3
 drilling out rivet 57.3
 removing/installing 57.3

Body—Seats

—Center armrest
 removing/installing 72.3
—Front seat
 backrest
  removing/installing 72.2
 removing/installing 72.2
—Rear seat
 backrest
  removing/installing 72.4
 removing/installing 72.3

Brake fluid

 see Brake—Hydraulic Components

Brake—Hydraulic Components, Regulator, Booster

—Brake booster/master cylinder
 assembly 47.13
—Brake caliper piston
 pre-bleeding 47.7
 removing/installing 47.3, 47.6, 47.7
 seal
  removing/installing 47.3, 47.6
—Brake fluid
 replacing
  with ABS 47.11
  without ABS 47.15
—Brake pressure regulator
 checking/adjusting 47.8
—Brake system
 bleeding
  with ABS 47.9
  without ABS 47.14
—Front brake caliper
 component layout 47.2
—Rear brake caliper
 component layout 47.5

Brake—Mechanical Components

—Brake caliper
 removing 46.4
—Front brake
 assembly 46.3
 pads
  removing/installing 46.4
—Brake pedal
 assembly 46.9
—Parking brake
 adjusting 46.8
 assembly 46.7
—Rear brake
 assembly 46.5
 pads
  replacing 46.6
—Technical data
 chart 46.2

Bumpers

—Front bumper
 assembly 63.2
 gap dimension 63.4
 removing/installing 63. 3
—Rear bumper
 assembly 63.5
 gap dimension 63.6
 removing/installing 63.6

Central locking

 see Body—Front Doors

Clutch

02A 5-speed manual transmission

—Clutch
 assembly 30.7
 removing/installing 30.8
—Clutch diaphragm spring ends
 checking 30.8
—Clutch pedal
 assembly 30.2
—Clutch release
 assembly 30.5

WARNING

- Automotive service and repair is serious business. You must be alert, use common sense, and exercise good judgement to prevent personal injury and complete the work safely.
- Before beginning any work on your vehicle, thoroughly read the safety notice, disclaimer, and the Cautions and Warnings listed near the front of this manual.
- Always read the complete procedure before you begin the work. Pay special attention to any Cautions and Warnings that accompany that procedure, or other information on a specific topic.

Clutch (cont'd)
—Flywheel
bolt torque 30.7
—Hydraulic components
assembly 30.4
—Main shaft seal
driving in 30.6
replacing 30.9
—Pressure plate
replacing 30.9
—Return spring
removing/installing 30.3
—Troubleshooting
chart 30.10

Compression pressure
See Engine—Cylinder Head Valve Drive

Constant velocity (CV) joints
see Front Wheel Suspension—Shafts and Axle

Coolant fan
See Engine—Cooling System

Cruise control
see Electrical—Battery, Starter, Generator or Electrical System—Lights

Cylinder block
See Engine—Crankshaft/Crankcase

Cylinder head
See Engine—Crankshaft/Crankcase

Differential—Automatic Transmission
—Axle flange oil seal
removing/installing 39-10
—Differential
adjusting 39-130
assembly 39-80
disassembling 39-90
modification 39-85
—Differential pinion
assembly 39-60
removing/installing 39-70
—Final drive/differential
assembly 39-60
removing/installing 39-70
—Output gear
adjusting 39-120
assembly 39-20
removing/installing 39-30

Differential—Automatic Transmission (cont'd)
—Pinion
adjusting 39-110
assembly 39-40
removing/installing 39-50
—Shims/bearings
installed positions 39-100

Differential—Manual Transmission
02A 5-speed
—Bearings
inner race
removing/installing 39.5
outer race
removing/installing 39.5
—Differential
adjusting 39.8
assembly 39.4
gears
installing 39.7
—Drive flange oil seal
installing 39.5
replacing 39.2
—Final drive gear
installing 39. 7
removing 39.6
—Gear
installing 39.7
—Rivet heads
boring 39.6
—Roll pins
installing 39.7
—Turning torque
checking 39.10

Electrical—Battery, Starter, Generator
—Battery
charging 27.4
checking 27.3
jump starting 27.5
load testing 27.6
removing/installing 27.2
troubleshooting 27.7
voltage test 27.6
—Belt
routing
vehicles with AC 27.13
vehicles without AC 27.12
tensioning 27.12
—Cruise control system
checking/adjusting 27.15
component layout 27.14
vacuum system
leak testing 27.17
—Generator components
layout 27.10
removing/installing 27.11
—Generator (GEN) V-belt pulley
removing/installing 27.12
—Starter motor
removing/installing 27.8

Electrical System—Instruments
1990 m.y.
—Digital clock
connector wiring 90.7

Electrical System—Instruments (cont'd)
—Engine Coolant Temperature (ECT) gauge
checking 90.14
connector wiring 90.6
troubleshooting 90.15
—Fuel gauge
checking 90.13
connector wiring 90.6
troubleshooting 90.13
—Instrument cluster
assembly 90.2
multi-point connector
identification 90.4
printed circuit
removing 90.8
removing 90.3
testing 90.10
—Instruments
removing/installing 90.5
—Light emitting diode (LED)
installing 90.16
—Printed circuit
removing 90.8
—Tachometer/oil pressure warning control unit
connector wiring 90.7
—Voltage stabilizer
checking 90.9
removing/installing 90.7
—Warning light (left and right)
installation position 90.6
From 1991 m.y
—Engine Coolant Temperature (ECT)
checking 90.14
connector wiring 90.6
troubleshooting 90.15
—Fuel gauge
checking 90.13
connector wiring 90.6
troubleshooting 90.13
—Instrument cluster
assembly 90.17
multi-point connector
identification 90.20
removing 90.19
lights
identification 90.24
removing/installing 90.24
testing 90.28
—Instruments
removing/installing 90.21
—Light emitting diode(LED)
installing (90.23
—MFA vacuum sensor
removing/installing 90.27
—Printed circuit
component identification 90.26
—Turn signal indicator lights
installing 90.23
—Vehicle Speed Sensor (VSS), speedometer (G22)
removing/installing 90.27
—Voltage stabilizer
checking 90.25
—Warning lights (left and right)
installation position 90.22
removing/installing 90.23

Electrical System—Instruments (cont'd)

Anti—theft alarm system 1991—1992 m.y

—General information 90.31
—Troubleshooting
Alarm horn sounds but lights don't flash with alarm activated 90.40
Alarm system cannot be switched off 90.38
Alarm system cannot be switched on 90.33
Engine does not crank with alarm system off 90.41

From 1993 m,y

—General information 90.31
—Troubleshooting
activated alarm system does not trigger alarm horn 90.50
activated alarm system does not trigger flasher lights 90.50
alarm cannot be switched off 90.49
alarm cannot be switched on 90.46
starter does not crank with alarm system switched off 90.52

Electrical System—Lights

—Additional brake light
removing 94.18
—Back up light housing
removing/installing 94.18
—Cruise control system
connector wiring 94.11
—Fog lights
adjusting 94.16
removing 94.16
—Front turn signal light
removing/installing 94.16
—Headlights
adjusting 94.15
aiming 94.14
assembly 94.13
removing/installing 94.15
—Horn contact ring
lubricating 94.3, 94.7
—Horn pad
removing 94.3, 94.7
—Horn plate
removing 94.3, 94.7
—License plate lights
removing 94.19
—Lock cylinder
removing/installing 94.9
—Multi-function indicator
connector wiring 94.11
—Side marker lights
removing 94.15
—Steering column
assembly 94.2, 94.5
—Steering column switches
connector wiring 94.10
wire routing (USA only) 94.12
—Taillight
assembly 94.17
—Turn signal switch
removing/installing 94.4, 94.8
—Windshield wiper switch
removing/installing 94.4, 94.8

Electrical System—Radio, MFA

Radio

—Heidelberg VI fixed coded radio
coding a new or remanufactured radio 91.2
reactivating radio after reconnecting power 91.4
unlocking an electronically locked radio 91.5

Multi-function Indicator (MFA)

—Hall Vehicle Speed Sensor (VSS)
checking (up to Aug. 1990) 91.24
contact plate
removing/installing 91.7
location 91.6
—Multi-function Indicator (MFA)
cancelling the memories
(from 1993 m.y.) 91.53
connector wiring 91.6
driving data memory
(from 1993 m.y.) 91.52
general notes 91.6
instrument cluster
checking (from 1993 m.y.) 91.55
MFA check mode test
(from 1993 m.y.) 91.54
Multi-function Indicator Mode Select Switch (E 86) and Memory Switch (E 109)
(from 1993 m.y.) 91.53
operation and function
(from 1993 m.y.) 91.52
printed circuit
identifying 91.8
removing 91.8
setting the time
(from 1993 m.y.) 91.53
vacuum sensor
removing 91.7
wiring
(up to Aug.1990) 91.25
(from Aug. 1990) 91.47
—Multi-function Indicator (MFA) self test (up to Aug. 1990)
checking 91.9
check mode test 91.9
measuring mode test 91.11
—Multi-function Indicator (MFA) self test (Aug. 1990 through 1992 m.y.)
checking 91.31
check mode test 91.31
measuring mode test 91.33
—Multi-function Indicator (MFA) troubleshooting (up to Aug. 1990)
ambient (outside) temperature display inaccurate 91.19
average driving speed indicator is not accurate or displays "— — —" or "0" 91.14
average fuel consumption display not accurate 91.15
clock inaccurate 91.13
clock operates in incorrect mode 91.13
display errors or no display 91.13
driving distance display not accurate 91.14
driving distance display remains at 0 91.14
driving time remains at 0:00 91.14

Electrical System—Radio, MFA (cont'd)

—Multi-function Indicator (MFA) troubleshooting (up to Aug. 1990) (cont'd)
memory cannot be canceled/memory does not store information 91.21
no engine RPM display or engine RPM display faulty, (while performing measuring mode test) 91.17
oil temperature display inaccurate 91.20
"signal from speed sensor" display does not change (during measuring mode test) 91.24
—Multi-function Indicator (MFA) troubleshooting (Aug. 1990 through 1992 m.y.)
ambient (outside) temperature display inaccurate 91.42
average driving speed indicator is not accurate or displays "— — —" or "0" 91.36
average fuel consumption display not accurate 91.37
clock inaccurate 91.35
clock operates in incorrect mode 91.35
display errors or no display 91.35
driving distance display not accurate 91.36
driving distance display remains at 0 91.36
driving time remains at 0:00 91.36
electronic speedometer and odometer
checking 91.44
memory cannot be canceled, memory does not store information. 91.39
oil temperature display inaccurate 91.43
tachometer inaccurate or does not work 91.45
Vehicle Speed Sensor (VSS), Speedometer (G 22)
checking 91.46
—Multi-Function Indicator (MFA) troubleshooting (from 1993 m.y.)
average fuel consumption display not accurate 91.57
average speed display remains at 0 or display not accurate 91.57
clock inaccurate 91.56
driving distance not accurate 91.57
driving distance remains at 0 91.56
driving time remains at 0:00 91.56
Multi-Function Indicator Mode Select Switch (E86), Memory Switch (E109) and wiring
checking 91.60
oil temperature display
checking 91.58
oil temperature display inaccurate 91.57
outside (ambient) temperature display
checking 91.59
outside (ambient) temperature display inaccurate 91.57
total or single trip memory of MFA not OK 91.56
Vehicle Speed Sensor (VSS), Speedometer
checking 91.63

> **WARNING**
> • Automotive service and repair is serious business. You must be alert, use common sense, and exercise good judgement to prevent personal injury and complete the work safely.
> • Before beginning any work on your vehicle, thoroughly read the safety notice, disclaimer, and the Cautions and Warnings listed near the front of this manual.
> • Always read the complete procedure before you begin the work. Pay special attention to any Cautions and Warnings that accompany that procedure, or other information on a specific topic.

Electrical System—Windshield Wiper & Washer
—Crank
installing 92.3
—Rear window wiper
assembly
(1989) 92.7
(from 1990) 27.7a
—Rear wiper arm
adjusting 92.8
—Rubber wiper inserts
removing/installing 92.4
—Spray jets
adjusting 92.6
—Windshield washer
system 92.5
—Windshield washer pump
hose connections 92.6
—Windshield wiper
assembly 92.2
troubleshooting 92.9
—Wiper arm
adjusting 92.3
angle, adjusting 92.8

Electrical System—Wiring
—Central locking system pump and control module
location/terminal identification 97.7
—Coolant fan control module
location/terminal identification 97.8
—Connector terminal rubber grommet
modifying 97.5
—Fuse/relay panel
connections/plugs 97.3
identification 97.2
removing/installing 97.4
—Multi-point connector
removing/installing 97.4
—Power window control module
location/terminal identification 97.6

Engine—Assembly
—Engine code PG
Engine/Transmission
adjusting 10.5
installing 10.3
mounting points 10.4
mounting points, hydraulic 10.4
removing 10.2
tightening torques 10.4

Engine—Assembly (cont'd)
—Engine code-AAA
Engine/Transmission
installing 10.8
mounting points 10.10
removing 10.6
tightening torques 10.11
— Identification
type and model 00.2
engine /vehicle number location 00.2a
—Technical data
charts 00.7

Engine—Cooling System
Engine code-PG
—Coolant
draining 19.5
filling 19.6
hydrometer
checking 19.6
mixture ratios 19.5
—Pump
assembly 19.3
pulleys 19.4
—Radiator
assembly 19.2
—Thermoswitch
checking 19.4
switching temperature 19.4
Engine code-AAA
—Coolant
draining/filling 19.11
hydrometer
checking 19.6
mixture ratio 19.11
—Pump
assembly 19.9
auxiliary electric coolant pump
checking 19.19
fan control module
location 19.20
removing/installing 19.14
—Radiator/fan
assembly 19.7
removing/installing 19.13
—Thermostat
assembly 19.17

Engine—Crankshaft/Crankcase
Engine code-PG
—Clutch flywheel
assembly 13.10
—Crankshaft, crankcase
assembly 13.9
dimensions 13.12
ignition mark 13.12
oil seal
removing/installing 13.11
—Cylinder block
assembly 13.4
—Cylinder head
assembly 13.3
—Drive belt
installing 13.6
—G-charger, A/C brackets
assembly 13.5

Engine—Crankshaft/Crankcase (cont'd)
—Piston, connecting rod
assembly 13.13
cylinder bore
checking 13.15
dimensions 13.15
piston rings
checking 13.14
pistons
checking 13.14
identification 13.14
Engine code-AAA
—Clutch, flywheel
assembly 13.29
—Crankshaft
assembly 13.27
dimensions 13.31
oil seal, drive belt side
removing/installing 13.26
—Cylinder block
assembly 13.21
bore
checking 13.37
—Cylinder head
assembly 13.19
—Drive chains
assembly sprockets 13.16
lower tensioner
installing 13.24
marking direction of travel 13.18
—Pistons
assembly with connecting rod 13.32
checking 13.36
dimensions 13.38
rings
end gap, checking 13.34
side clearance 13.35
—Poly-ribbed belt
removing/installing 13.25
—Valve timing
adjusting 13.23

Engine—Cylinder Head, Valve Drive
Engine Code-PG
—Camshaft
assembly 15.5
identification 15.6
oil seal
removing/installing 15.9
play
checking 15.6
removing/installing 15.10
—Cylinder head
assembly 15.2
compression
checking 15.4
distortion
checking 15.3
installing 15.3
—Valve
assembly 15.5
dimensions 15.6
exhaust valves
discarding 15.8a

Engine—Cylinder Head, Valve Drive (cont'd)
—Valve (cont'd)
guides
checking 15.11
replacing 15.11
hydraulic valve lifters
checking 15.8
seats
refacing 15.7
stem seal
removing/installing 15.12
Engine Code-AAA
—Camshaft
assembly 15.19
axial play
checking 15.22
bearing caps
installed position 15.21
removing/installing 15.27
—Compression pressure
checking values 15.18
—Cylinder head
assembly 15.13
distortion
checking 15.16
gaskets
replacing 15.15
installing 15.17
—Valves
assembly 15.14
dimensions 15.23
guides
checking 15.29
replacing 15.31
hydraulic valve lifters
checking 15.26
seats
replacing 15.24
stem seals
removing/installing 15.32

Engine—Lubrication System
Engine Code-PG
—Dynamic oil pressure system
troubleshooting 17.6
—Lubrication system
assembly 17.2
—Oil
capacity, specifications 17.4
pan 17.3
pump
assembly 17.3
checking 17.4
gears 17.3
switches
checking 17.5
Engine Code-AAA
—Lubrication system
assembly 17.11
—Oil
capacity, specifications 17.18
pump
assembly 17.13
checking 17.15
relief valve
checking 17.17

Engine—Lubrication System (cont'd)
—Oil (cont'd)
spray jet
removing/installing 17.16
warning system 17.18

Exhaust System—Emission Controls
Engine code-PG
—Evaporative system leakage
troubleshooting 26.5
—Exhaust system
component layout 26.2
—Heated crankcase breather
checking 26.4
Engine code-AAA
—EGR system
component layout 26.8
valve checking 26.9
—Exhaust system
component
removal/installation 26.7

Flywheel
See Clutch

Fuel Supply
Engine Code-PG
—Accelerator cable
checking/adjusting 20.9
—Carbon canister
component layout 20.16
shutoff valve
checking 20.17
—Fuel delivery rate
checking 20.7
—Fuel pump
after run
checking 20.13
checking 20.10
check valve
checking 20.15
relay
checking 20.12
removing/installing 20.11
—Fuel supply system
component layout 20.2
—Rules of cleanliness
chart 20.6
Engine Code-AAA
—Accelerator cable
checking 20.22
adjusting 20.22
—Carbon canister
component layout 20.32
—Fuel delivery rate
checking 20.27
—Fuel pump
check valve
checking 20.29
checking 20.25
relay
checking 20.30
—Fuel supply/fuel pump assembly
installation position 20.20
removing/installing 20.23
—Fuel supply system
component layout 20.18
—Rules of cleanliness
chart 20.6

Fuel Supply (cont'd)
—Vent valve
checking 20.21

Front Wheel Suspension—Shafts & Axle
Engine code-PG
—Axle shaft
assembly 40.11
length 40.13
—Boot
installing 40.13
—Coil spring
removing/installing 40.10
—Control arm bushing
installing 40.4
pressing out/in 40.3
—CV Joint, inner
assembling/checking 40.16
checking/disassembling 40.15
pressing off 40.12
pressing on 40.13
—CV joint, outer
assembling 40.15
checking/disassembling 40.14
removing 40.12
—Engine/transmission
supporting 40.4
—Front suspension
assembly 40.1, 40.6
—Steering knuckle/suspension strut
separating 40.7
—Subframe mounting
repairing 40.5
—Suspension strut
assembly 40.9
strut nut
removing/installing 40.7
—Vibration damper
removing/installing 40.12
—Wheel bearing
pressing out/in 40.8
—Wheel hub
pressing in 40.8
pressing out 40.7
Engine code-AAA
—Axle shaft
assembly 40.28
length 40.32
removing/installing 40.24
—Ball joint
checking 40.19a
removing/installing 40.19b
—Coil spring
removing/installing 40.27
—Control arm bushings
removing/installing 40.19
—CV joint, inner
greasing 40.33
removing/installing 40.30
—CV joint, outer
greasing 40.33
removing/installing 40.30
—Front subframe
assembly 40.17
—Front suspension
assembly 40.20

WARNING

- Automotive service and repair is serious business. You must be alert, use common sense, and exercise good judgement to prevent personal injury and complete the work safely.
- Before beginning any work on your vehicle, thoroughly read the safety notice, disclaimer, and the Cautions and Warnings listed near the front of this manual.
- Always read the complete procedure before you begin the work. Pay special attention to any Cautions and Warnings that accompany that procedure, or other information on a specific topic.

Front Wheel Suspension—Shafts & Axle (cont'd)
—Suspension strut
assembly 40.26
nut
removing/installing 40.22
—Wheel bearing/hub
installing 40.23
removing 40.22

Fuel Injection, AFC System
Engine Code—PG
49 States and Canada, 1989 m.y. to present
California m.y. 1989-1990
—Basic throttle adjustment
checking/adjusting 24-180
—Bypass valve adjustment
checking 24-150
—Control unit
component layout 24-40
—Deceleration fuel shut-off
checking 24-130
—Digifant system
under hood component layout 24-30
—Fuel injectors
checking 24-170
—Fuel pressure regulator
checking 24-160
—Full throttle enrichment
checking 24-120
—Idle and full throttle switches
checking/adjusting 24-140
—Idle speed and CO content
checking/adjusting 24-90
—Idle stabilization
checking 24-100
—Intake air
component layout 24-20
—Oxygen sensor
control
checking 24-110
general information 24-60
—Residual pressure
checking 24-160
—Safety precautions
rules of cleanliness 24-80
safety measures 24-80
—Technical data
chart 24-70
—Throttle switches
component layout 24-50

Fuel Injection, AFC System (cont'd)
—Vacuum connections
checking 24-190
Engine Code—PG
California starting m.y. 1991
—Acceleration full throttle enrichment and deceleration fuel cut-off
checking 24-240
—Basic throttle adjustment
checking 24-320
—Bypass valve adjustment
checking 24-310
—CO potentiometer
checking 24-270
—Coolant temperature sensor
checking 24-250
—Digifant System
under hood component layout 24-200
—Fuel injectors
checking 24-300
—Fuel pressure
checking 24-160
—Idle speed and CO content
checking/adjusting 24-220
—Intake air temperature sensor
checking 24-260
—Oxygen sensor and oxygen sensor control
checking 24-230
—Residual pressure
checking 24-160
—Throttle body
component layout 24-210
—Throttle potentiometer
checking 24-290
—Vacuum connections
checking 24-190
—Vacuum sensor
checking 24-280
Engine Code—AAA
—Acceleration, full throttle enrichment and fuel deceleration shut-off
checking 24-430
—Dashpot
checking 24-420
adjusting 24-420
—Engine Coolant Temperature (ECT) sensor
checking 24-440
—Fuel Injectors
checking 24-470
—Fuel injection system
component layout 24-370
repairing 24-340
—Fuel pressure regulator and residual pressure
checking 24-460
—Idle speed
checking 24-400
control valve (ISC)
checking 24-410
—Intake Air Temperature (IAT) sensor
checking 24-480
—Motronic system
component layout 24-330
engine control module
component layout 24-350
intake air system
component layout 24-360
throttle assembly
component layout 24-380

Fuel Injection, AFC System (cont'd)
—Safety precautions
chart 24-390
—Throttle position (TP) potentiometer
checking 24-450

Fuel tank
see Owner Information

Fuses
see Electrical System—Wiring

Glass, Window Regulators
—Power windows
troubleshooting 64.20
control module
location 64.26
—Rear window
assembly 64.11
installing 64.14
removing 64.12
—Side window
assembly 64.16
installing 64.18
removing 64.17
—Window replacing
general information 64.2
—Windshield
assembly 64.5
installing 64.8
removing 64.6

Heater and Ventilation
—Air intake duct
identifying 80.4
—Air outlet grill/housing
removing 80.4
—Control cables
installing/adjusting 80.5
—Fresh air blower
removing 80.5
series resistance
removing 80.5
—Heater
assembly 80.7
ventilation air ducts
assembly 80.2
—Heater/fresh air controls
faceplate/assembly
removing 80.4
—Side window air vent
removing 80.4

Hoods, Lids
—Gas-filled strut
assembly 55.13
discarding 55.14
—Hood
assembly 55.4
bowden cable
assembly 55.3
hinge
assembly 55.6
gap
adjusting 55.7
radiator support 55.3
rubber stop
adjusting 55.5
sound deadener
installing 55.5

Hoods, Lids (cont'd)
—Hood (cont'd)
striker
adjusting 55.5
support
replacing 55.14
—Rear lid
adjusting 55.9
assembly 55.8
lid release
replacing 55.12
lock assembly 55.10
lock cylinder
replacing 55.11
lock repairing 55.11
rubber stop. adjusting 55.9
spring pin
removing 55.11

Hydraulic fluid
see Owner Information

Idle speed
see Fuel Injection AFC System

Ignition System
Engine Code—PG
49 States and Canada
m.y. 1989 to present
California m.y. 1989—1990
—Digifant Ignition system
component layout 28.2
—Hall sender unit
checking 28.9
function
checking 28.10
—Ignition distributor
installing 28.6
—Ignition timing
checking/adjusting 28.7
map
checking 28.8
trigger function
checking 28.9
—Knock sensor
checking 28.11
—Safety precautions chart 28.5
—Technical data, spark plugs 28.4
Engine Code—PG
California starting m.y. 1991
—Digifant Ignition system
component layout 28.12
—Control unit
component layout 28.13
technical data 28.15
—Hall sender
checking 28.20
—Ignition coil, checking 28.19
—Ignition distributor, installing 28.16
—Ignition timing
advance, checking 28.18
checking/adjusting 28.17
—Safety precautions
chart 28.14

Ignition System (cont'd)
Engine Code—AAA
production up to January, 1993
—Hall sender, checking 28.26
—Ignition coil, checking 28.25
—Ignition distributor
disassembling/assembling 28.22
—Ignition and spark plugs
technical data 28.23
—Motronic System
component layout 28.21
—Safety precautions
chart 28.24
Engine Code—AAA
production from January, 1993
see Supplement

Lifting vehicle
see Owner Information

Locks
see Body—Front Doors

Manual Transmission— Case, Gears, Shafts
02A 5-speed
—1st gear
operating sleeve and hub
assembling 35.39
installing 35.40
shift forks, ball bearings
changes 35.53
synchronizer ring
checking 35.38
marking 35.39
—2nd gear
operating sleeve and hub
assembling 35.39
installing 35.40
synchronizer ring
checking 35.40
marking 35.40
—3rd gear
modifications 35.49, 35.50
needle bearing sleeve
modified 35.50
operating sleeve/synchronizer hub
disassembling/assembling 35.41
removing/installing 35.42
removing/installing 35.30
synchronizer ring
checking 35.38
synchronizer sleeve
modified 35.49
thrust washer
modified 35.49
—4th gear
installing 35.29
modifications 35.49
needle bearing sleeve
installing 35.42
operating sleeve/synchronizer hub
disassembling/assembling 35.41
removing/installing 35.42
removing
with bearing and bushing 35.28

Manual Transmission— Case, Gears, Shafts (cont'd)
—4th gear (cont'd)
—shift forks
ball bearings
changes 35.53
synchronizer ring
checking 35.38
synchronizer sleeve
modified 35.49
thrust washer
modified 35.49
—5th gear assembly
adjusting 35.16
adjusting, checking
(transmissions from 31 10 1) 35.55
gear
installing 35.15
removing 35.7
gear/synchronizer hub bolts
modified 35.51
modified synchronizer
installed position 35.55
modified synchronizer ring and 5th gear synchronizer
assembling 35.54
operating sleeve
installing 35.16
removing 35.7
shift fork
changes 35.54
installing 35.16
removing 35.7
synchronizer
changes 35.53
synchronizer hub
modified 35.51
installing with 5th gear 35.15
removing with 5th gear 35.7
synchronizer ring
changes 35.53
checking 35.30
installing 35.15
synchronizer sleeve
modified 35.52
—Backup light switch/socket
pre-assembly 35.22
—Ball bearing inner race
removing/installing 35.24
—Bearing sleeve
removing/installing 35.21
—Differential
assembly 35.5
installing 35.11
removing 35.10
—Drive flange shaft
circlip
replacing 35.18
seal
replacing 35.17
shaft
installing 35.18
removing 35.6
—Gear carrier housing
assembly 35.5

WARNING

- Automotive service and repair is serious business. You must be alert, use common sense, and exercise good judgement to prevent personal injury and complete the work safely.
- Before beginning any work on your vehicle, thoroughly read the safety notice, disclaimer, and the Cautions and Warnings listed near the front of this manual.
- Always read the complete procedure before you begin the work. Pay special attention to any Cautions and Warnings that accompany that procedure, or other information on a specific topic.

Manual Transmission— Case, Gears, Shafts (cont'd)
—Guide sleeve assembly
sealing ring
removing/installing 35.18
sleeve
installing 35.19
removing 35.6
—Inner race
testing for wear 35.40
—Large pinion shaft bearing
inner race
removing/installing 35.38
—Lock ring
installing 35.24
—Main shaft
adjusting 35.31
adjusting shim
chart 35.32
assembly 35.5
checking 35.34
disassembling/assembling 35.26
installing 35.11
removing 35.10
—Main shaft bearing
inner race
installing 35.28
outer race
installing 35.28, 35.30
removing 35.28
—Needle bearing
bushing
installing 35.29
installing 35.43, 35.47
removing 35.42, 35.47
revised
installing 35.52
sleeve
installing 35.41, 35.43
—Pinion shaft
adjusting 35.44
assembly 35.5
disassembling/assembling 35.35
installing 35.11
removing 35.10
—Reverse gear assembly
disassembling/assembling 35.46
idler shaft
revised 35.52

Manual Transmission— Case, Gears, Shafts (cont'd)
—Reverse gear assembly (cont'd)
idler shaft support
installing 35.11, 35.13
removing 35.8
modified 35.52
shift fork
installing 35.11
removing 35.9
sliding gear
installing 35.11
removing 35.9
—Sealing ring
installing 35.21
—Selector shaft
changes 35.56
disassembling/assembling 35.20
end cover
changes 35.57
shaft assembly
installing 35.14
removing 35.8
support, bearing sleeve
removing/installing 35.14
—Shift fork assembly
assembly 35.5
installing 35.12
removing 35.9
disassembling/assembling 35.23
pivot pins
installing 35.13
removing 35.8
—Shift segments
identification 35.24
installing 35.25
—Small pinion shaft bearing
inner race
removing/installing 35.37
outer race
removing/installing 35.37
—Synchronizer
ring
checking 35.38
sleeve w/hub and 1st gear
removing 35.39
—Transmission
adjustment overview 35.48
assembly 35.2
assembly stand
clamping in 35.6
disassembling/assembling 35.6
housing
installing 35.12
removing 35.8
housing cover
installing 35.17
modified 35.51
removing 35.6
identification 00.3
modifications 35.53
—Turning torque
measuring 35.45

Manual Transmission—Controls, Assembly
02A 5-speed
—Gearshift cable/lever
installing 34.6
—Gearshift housing
repairing 34.7
—Gearshift lever
repairing 34.8
rubber cover 34.9
—Gearshift mechanism
adjusting/checking 34.10
modifications
vehicles w/ 6-cyl. engine 34.6a
repairing 34.4
—Selector housing cap
installing 34.9
—Technical data
chart 34.2
—Transmission
controls
overview 34.3
removing/installing 34.12

Multi-Function Indicator (MFA)
See Electrical System— Radio, MFA

Oil
—Engine
see Engine—Lubrication System or Owner Information
—Manual transmission
see Owner Information

Oil filter
see Engine—Lubrication System or Owner Information

Oil pump
see Engine—Lubrication System or Owner Information

On Board Diagnostic
ABS
—Teves 02 ABS 01-1
on board diagnostic
ABS 01-3
ABS/EDL 01-53
electrical test
ABS 01-24
ABS/EDL 01-100
096 Automatic Transmission
—On Board Diagnostic 01-1
Fuel Injection and Ignition
—Engine code PG–Digifant II 01-1
—Engine Code PG–Digifant I 01-9
—Engine Code AAA 01-65

Owner Information
—Do-it-yourself service
changing a wheel 22
compact spare wheel 21
emergency starting 28
emergency towing with commercial towtruck 29
lifting vehicle
lifting points hoist/floor jack
using hoist/floor jack 00.6
fuses 23
headlight adjustment 27

Owner Information (cont'd)
—Do-it-yourself service (cont'd)
jack and tools 21
lifting vehicle 30
replacing bulbs 25
replacing the radio 28
—Gas station information
servicing points 40
—Maintenance/lubrication schedule
tables 41
—Technical description
body/chassis 34
brakes 33
capacities/dimensions 37
emission control system 32
engine (4 cylinder) 31
engine (6 cylinder) 31
engine data/spark plugs
(4 cylinder) 34
engine data/spark plugs
(6 cylinder) 35
steering, suspension 33
transmission 33
v-belts (4 cylinder) 36
vehicle identification 38
weights 38
—Vehicle care
additional accessories, modifications and parts replacement 20
air cleaner 11
battery 14
brake fluid 13
cleaning and protecting 2
cooling system 11
corrosion protection 5
difficult operating conditions 19
engine compartment 7
engine hood 7
engine oil filter 10
engine oil 8
fuel supply 1
fuel tank 1
maintenance 6
power steering 11
tires/wheels 16
transmission fluids 10
v–belt 13
windshield/rear window
washer container 15
winter driving 20

Oxygen sensor (OXS)
see Fuel Injection—AFC System or Fuel Injection and Ignition—On Board Diagnostic (OBD)

Radio
See Electrical System—Radio, MFA

Rear Wheel Suspension, Shafts & Axle
—Axle beam bushings
removing/installing 42.3
—Axle beam mounting bracket
installing 42.3
—Axle mounting
repairing 42.4

Rear Wheel Suspension, Shafts & Axle (cont'd)
—Cover ring
installed position 42.9
—Grease cap
removing/installing 42.8
—Rear axle
assembly 42.2
installing 42.4
—Seal
installing 42.10
—Stub axle/disc brake
assembly 42.7
—Support
installed position 42.6
—Suspension strut
assembly 42.5
removing/installing 42.4
—Wheel bearing
outer races
installing 42.9
—Wheel bearing play
adjusting 42.9

Seat belts
see Body Accessories—Interior

Spark plugs
see Ignition System

Starter
see Electrical— Battery, Starter, Generator

Steering
—Adaptor sleeve
removing/installing 48.5
—Auxiliary cooling line
routing 48.20
—Lock cylinder
removing/installing 48.5
—Power steering gear
adjusting
TRW 48.9
ZF 48.15
assembly
TRW 48.7
ZF 48.13
removing/installing 48.11
universal joint shaft and pinion
modifications 48.5a
—Power steering pump
assembly 48.17
pressure
checking (4-cyl. engine) 48.10
checking (6-cyl. engine) 48.16
—Power steering system
assembly 48.18
cleaning 48.6
pressure
checking 48.10
troubleshooting 48.21
—Steering column
assembly 48.2, 48.4
length
checking 48.3
—Tie rod
removing/installing 48.12

Struts
see Front Wheel Suspension—Shafts and Axle

Supercharger
—G-charger
boost pressure 21-30
component layout 21-10
damage
repairing 21-40
—Safety precautions
chart 21-20

Sunroof
—Electrical cable drive
checking/adjusting 60.6
—Parallel alignment
checking/adjusting 60.4
—Slide/tilt sunroof
assembly 60.2
assembly frame
installing 60.9
assembly frame
removing 60.7
installing 60.4
removing 60.3
—Sunroof panel
adjusting 60.6
headliner trim frame
removing 60.5
—Water drain hoses
cleaning 60.10

Suspension
see Front Wheel Suspension—Shafts and Axle and Rear Wheel Suspension—Shafts and Axle

Thermostat
see Engine—Cooling System

Tires
see Wheels, Tires, Wheel Alignment

Torque Converter
096 automatic transmission
—Torque converter
identification 32.10
removing/installing 32.20
seal, replacing 32.30

Trim—Interior
—Cassette storage box
removing 70.8
—Door trim panel
assembly 70.12
grab handle outer cover
removing/installing 70.13
—Front storage trays
removing 70.7
—Glove box
assembly 70.9
lid
removing/installing 70.10
lock cylinder 70.10
removing/installing 70.10
—Instrument panel
installing 70.8
removing 70.7
trim plate, removing 70.8

WARNING

• Automotive service and repair is serious business. You must be alert, use common sense, and exercise good judgement to prevent personal injury and complete the work safely.

• Before beginning any work on your vehicle, thoroughly read the safety notice, disclaimer, and the Cautions and Warnings listed near the front of this manual.

• Always read the complete procedure before you begin the work. Pay special attention to any Cautions and Warnings that accompany that procedure, or other information on a specific topic.

Trim—Interior (cont'd)
—Molded headliner
installing 70.5
removing 70.3
—Plastic moisture barrier
replacing/installing 70.11
—Rear side trim panel
assembly 70.6

Valves
See Engine—Cylinder Head, Valve Drive

Windshield
See Glass, Window Regulators

Wheels, Tires, Wheel Alignment
—Front wheel camber
adjusting 44.4, 44.4a
—Rear wheel toe
determining 44.3
—Snow tires
size 44.9
—Vibrations
eliminating 44.5
—Wheel alignment
data (4 cylinder) 44.2
data (6 cylinder) 44.2a

Water pump
See Engine—Cooling System

Wiring Diagrams
General information
—Cautions and warnings 3
—Handling components and connectors 15
—How to read wiring diagrams 5
—Test equipment 9
computer memory saver **14**
LED test light **9**
multimeter or volt/ohm meter **11**
test kits **13**
—Troubleshooting 17
basic electricity **17**
checking for short circuit to ground **27**
checking for voltage **20**
checking ground connections **21**
checking wiring and components **23**
Main wiring diagrams
—1990 **29**
—1991 **57**
—1992
engine code PG **57**
engine code AAA **83**

Wiring Diagrams (cont'd)
Main wiring diagrams (cont'd)
—1993
production up to Jan. 1993 **83**
production from Jan.1993 to Sept. 1993 **113**
—1994
production from Sept. 1993 to Mar. 1994
see Supplement
production from Mar. 1994 **145**
Auxiliary wiring diagrams
—Air conditioning
1990, 1991 **177**
1992 (engine code PG) **177**
1992 (engine code AAA) **181**
1993 (production up to Aug. 1992) **181**
1993 (production from Aug. 1992) **185**
1994 (production from Aug. 1993)
see Supplement
—Anti-lock brake system (ABS)
1990 **191**
1991 **195**
1992 (engine code PG) **195**
1992 (engine code AAA) **199**
1993 **199**
1994 with electronic differential lock (EDL) **327**
—Anti-theft alarm system
1991 **205**
1992 **205**
1993 (production up to Aug. 1992) **205**
1993 (production from Aug. 1992) **209**
1994 **209**
—Automatic shoulder belt system
1990 **213**
1991 **217**
1992 (engine code PG) **217**
1992 (engine code AAA) **221**
1993, 1994 **221**
—Automatic transmission
1991 **225**
1992 (engine code PG) **225**
1992 (engine code AAA) **231**
1993 (production up to Jan. 1993) **231**
1993 (production from Jan. 1993) **237**
1994 (production up to Mar. 1994) **237**
1994 (production from Mar. 1994) **243**
—Central locking
1990 **291**
1991 **297**
1992 (engine code PG) **297**
1992 (engine code AAA) **303**
1993 (production up to Aug. 1992) **303**
1993 (production from Aug. 1992) **307**
1994 **307**
—Cruise control
1990 **249**
1991 **251**
1992 (engine code PG) **251**
1992 (engine code AAA) **253**
1993, 1994 **253**
—Daytime driving lights (Canada only)
1990 **255**
1991 and later **259**

Wiring Diagrams (cont'd)
Auxiliary wiring diagrams (cont'd)
—Electronic engine controls
(Digifant I, California only)
1991, 1992 (engine code PG) 263
—Fuel pump
1990 (production from Sept. 1989) **269**
—Heated seats
1990, 1991 **271**
1992 (engine code PG) **271**
1992 (engine code AAA) **273**
1993, 1994 **273**
—Headlight washers (Canada only)
1993, 1994 **275**
—Multi-function indicator
1991–1994
(production from Aug. 1990) **277**
—Power mirrors
1992–1994 (engine code AAA) **283**
—Power sunroof
1990 **325**
1991 **285**
1992 (engine code PG) **285**
1992 (engine code AAA) **287**
1993
(production up to Aug. 1992) **287**
1993
(production from Aug. 1992) **289**
1994 **289**
—Power windows
1990 **291**
1991 **297**
1992 (engine code PG) **297**
1992 (engine code AAA) **303**
1993 (production up to Aug. 1992) **303**
1993 (production from Aug. 1992) **307**
1994 **307**
—Radio, stereo
1990 **315**
1991, 1992 **319**
1993 **323**
1994 *see Supplement*
—Radio with CD changer
see Supplement
—Recaro power seats (Canada only)
1992–1994 (engine code AAA) **311**
—Seatbelt warning system (Canada only)
1990–1994 **313**

Index

Do-it-yourself service

- changing a wheel 22
- compact spare wheel 21
- emergency starting 28
- emergency towing with commercial towtruck 29
- fuses 23
- gas station information 40
- headlight adjustment 27
- jack and tools 21
- lifting vehicle 30
- replacing bulbs 25
- replacing the radio 28

Maintenance/lubrication schedule

- tables 41

Technical description

- body/chassis 34
- brakes 33
- capacities/dimensions 37
- emission control system 32
- engine (4 cylinder) 31
- engine (6 cylinder) 31
- engine data/spark plugs (4 cylinder) 34
- engine data/spark plugs (6 cylinder) 35
- steering, suspension 33
- transmission 33
- v-belts (4 cylinder) 36
- vehicle identification 38
- weights 38

Vehicle care

- additional accessories, modifications and parts replacement 20
- air cleaner 11
- battery 14
- brake fluid 13
- cleaning and protecting 2
- cooling system 11
- corrosion protection 5
- difficult operating conditions 19
- engine compartment 7
- engine hood 7
- engine oil filter 10
- engine oil 8
- fuel supply 1
- fuel tank 1
- maintenance 6
- power steering 11
- tires/wheels 16
- transmission fluids 10
- v–belt 13
- windshield/rear window washer container 15
- winter driving 20

Gas station information

- servicing points 40

FUEL TANK

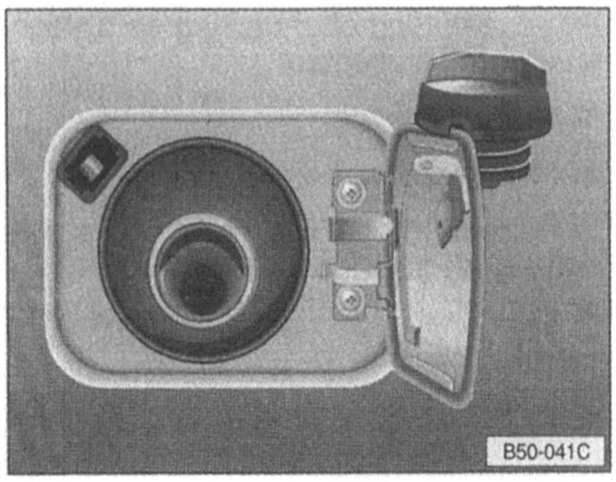

The fuel filler neck is located on the right rear side panel behind the fuel tank flap. The central locking system automatically locks and unlocks the fuel tank flap. If the central locking system should fail, you can still open the flap manually. See page 8 for further details.

The capacity of the tank is approx. 18.5 gal/55 litres (Reserve is approx 2.6 gal/10 lit. of total capacity).

When refuelling, hang the fuel filler cap on the fuel tank flap as illustrated above.

Trouble-free refueling depends on the correct use of the filler nozzle.

Always insert the nozzle all the way into the fuel filler neck, to completely open the spring-loaded flap located just below the filler neck opening. Make sure the filler nozzle is not tilted.

Do not fill the fuel tank too quickly, otherwise the fuel may foam up and cause the nozzle to switch off too soon.

❀ ***As soon as the correct operated nozzle switches off automatically for the first time, the tank is full. Do not try to add more fuel, because the expansion space in the fuel tank will be filled – the fuel can then overflow when it becomes warm and pollute the environment.***

To avoid fuel spilling or evaporating from the fuel tank always close fuel cap properly.

WARNING

Never carry additional fuel containers in your vehicle. Such containers, full or empty, may leak, cause an explosion, or result in fire in case of a collision.

FUEL SUPPLY

❀ ***Use only unleaded fuel in your vehicle.***

For maximum engine performance and under high load operating conditions, unleaded premium gasoline with a minimum octane rating of 95 RON (91 AKI) is recommended.

You may also use unleaded regular gasoline with a minimum octane rating of 91 RON (87 AKI), if the above mentioned conditions no longer apply, or if premium is not readily available.
Do not use any fuel with octane ratings lower than 87 AKI or 91 RON.

Do not use Leaded Gasoline.
The use of leaded gasoline will severely damage your vehicle's catalytic converter and it's ability to control exhaust emissions.
Unleaded gasolines are available throughout the continental USA, Canada, and in most European countries. We recommend you do not take your vehicle to areas or countries where unleaded gasoline may not be available.

Octane rating

Octane rating indicates a gasoline's ability to resist engine damaging "knock" caused by pre-ignition and detonation. Therefore, buying the correct grade of gasoline is very important to prevent possible engine damage and a loss of engine performance.

The recommended gasoline octane rating for your engine can be found on a label on the fuel filler flap. This rating may be specified as AKI (CLC) or RON (ROZ).

Explanation of the abbreviations:

ROZ = „**R**esearch **O**ktan **Z**ahl"
RON = **R**esearch **O**ctane **N**umber
AKI = **A**nti **K**nock **I**ndex
CLC = **C**ost of **L**iving **C**ouncil

The AKI (Anti-knock Index) or CLC (Cost of Living Council) octane ratings are normally displayed on gas station pumps in the USA and Canada and are calculated as follows:

RON (Research Octane Number) plus MON (Motor Octane Number), divided by 2.

$$\text{AKI (CLC)} = \frac{\text{RON} + \text{MON}}{2}$$

According to this calculation, gasoline most commonly used in the USA and Canada have the following octane ratings:
Premium Grade: 91 – 96 AKI
Regular Grade: 87 – 90 AKI

Use of Gasoline Containing Alcohol or MTBE (methyl tertiary butyl ether)

You may use unleaded gasoline blended with alcohol or MTBE (commonly referred to as oxygenates), if the blended mixture meets the following criteria:

■ **Blend of gasoline methanol (wood alcohol or methyl alcohol)**

- Anti-knock index must be 87 or higher.
- Blend must contain no more than 3 % methanol.
- Blend must contain more than 2 % CO-solvents.

Note

Methanol fuels which do not meet these requirements, may cause corrosion and damage to plastic and rubber components in the fuel system.

■ **Blend of gasoline and ethanol (grain alcohol or ethyl alcohol)**

- Anti-knock index must be 87 or higher.
- Blend must not contain more than 10 % ethanol.

■ **Blend of gasoline and MTBE**

- Anti-knock index must be 87 or higher.
- Blend must contain not more than 15 % MTBE.

General notes

■ **Do not use fuels that fail to meet the criteria specified above.**

■ **Do not use fuel for which the contents cannot be identified.**

■ **Fuel system damage and performance problems resulting from the use of fuels different from those specified above are not the responsibility of Volkswagen and are not covered under the New Vehicle of the Emission Control System Warranties.**

■ **If you experience a loss of fuel economy or driveability and performance problems due to the use of one of these fuel blends, we recommend that you switch to unblended fuel.**

■ If you are unable to determine whether or not a particular fuel blend meets the specification above, ask your service station or its fuel supplier.

Seasonally adjusted gasoline

Many gasolines are blended to perform especially well for winter or summer driving. During seasonal change-over, we suggest that you fill up at busy gas stations where the seasonal adjustment is more likely to be made in time.

Gasoline additives

A major concern among many auto manufacturers is carbon deposit build-up caused by the type of gasoline you use.

Although gasolines differ from one oil company to another, they do have one thing in common. All gasolines contain properties that can cause deposits to collect on fuel injectors and intake valves. Although most gasoline brands include additives to keep engine and fuel systems clean, they are not equally effective.

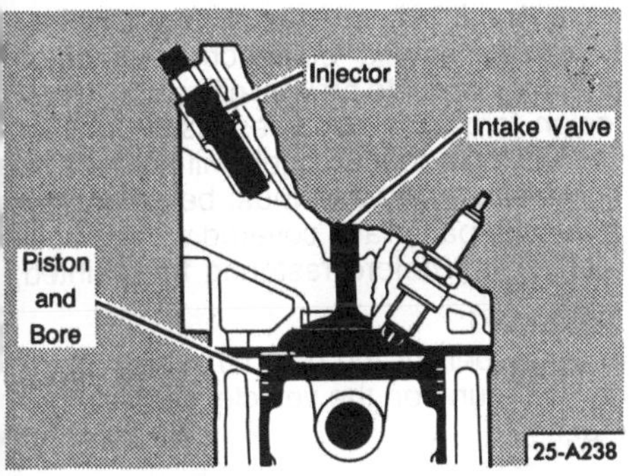

25-A238

After an extended period of using inadequate fuels, built-up carbon deposits can rob your engine of peak performance. And carbon deposits like those in the illustration can lead to other engine performance problems such as:

- unstable idling
- surging
- misfiring
- power loss
- engine run-on
- engine pinging or knocking

If these problems continue over a long period of time, engine damage can be a result.

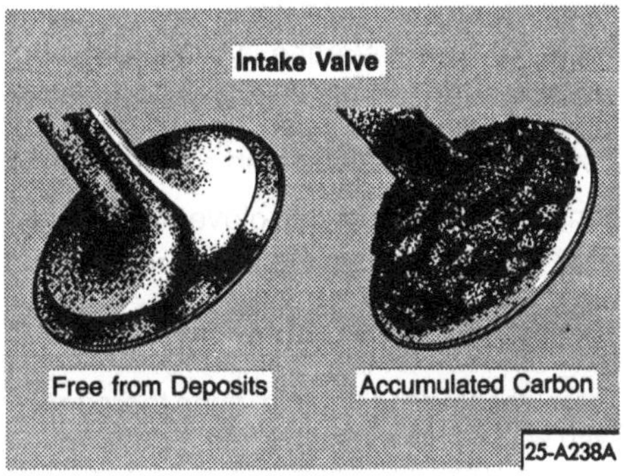

25-A238A

Damage or malfunction due to poor fuel quality is not covered by the Volkswagen New Vehicle Limited Warranty.

Recent fuel developments

Over the past few years, we have seen a great increase in gasoline manufacturer's advertising the use of detergent additives. Many additives are primarily intended to keep fuel injectors clean.

However, they are not all equally effective in reducing carbon deposit build-up on intake valves. We are aware that, as of the time of printing, some manufacturers advertise and sell gasoline that contains the proper deposit control additives that will keep intake valves and injectors clean.

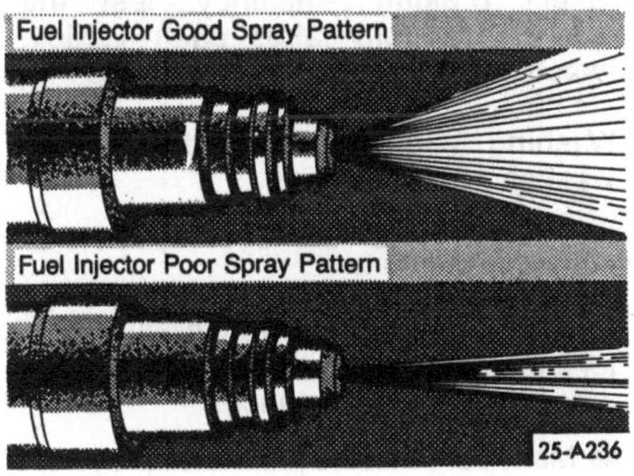

25-A236

To assure the long term performance of your car's engine, you should use gasoline brands that include these deposit control additives.

If gasolines with additives are not available, contact your Volkswagen dealer about proper fuel additives.

CLEANING AND PROTECTING

Any automobile is subjected to abuse from industrial fumes, corrosive road salt, abandoned lollipops, muddy dog feet, etc., to name just a few. A well cared for Volkswagen can look like new many years later. Regular and correct care will contribute to maintaining the beauty and value of your Volkswagen.

Furthermore, this may be a condition for upholding a warranty claim should corrosion damage or paint defects occur.

Your Volkswagen dealer has a number of vehicle-care products and can advise you which ones to use for cleaning the exterior and interior of your vehicle.

Whether you use Volkswagen recommended products or other commercially available cleaning agents, make sure first of their correct application.

Select only environmentally friendly cleaning products. Any remaining cleaning products do not belong in the household wastes.

WARNING

- **Cleaning agents may be poisonous. Keep them out of the reach of children.**
- **Observe all caution labels.**
- **Always read directions on the container before using any product.**
- **Most chemical cleaners are concentrates which require dilution.**
- **Only use spot removing fluids in well ventilated areas.**
- **Do not use gasoline, kerosene, naphtha, nail polish remover or other volatile fluids. They may be toxic, flammable or hazardous in other ways.**

Care of exterior

Washing

The best protection against environmental influences is frequent washing and waxing.

How often this is required depends on how much the vehicle is used, where it is parked (garage, in the open under trees), the seasons, weather conditions and environmental influences. The longer bird droppings, insects, tree resin, road and industrial grime, tar, soot, road salt and other materials remain on the vehicle paint, the more lasting their destructive effect will be. High temperatures, e. g. from strong sun light intensifies the corrosive effect.

Under certain circumstances weekly washing may be necessary. Under other conditions, a monthly washing and waxing may be adequate.

After the winter the underside of the vehicle should be thoroughly washed.

Automatic car wash

The vehicle paint is so durable that the vehicle can normally be washed without problems in an automatic car wash. However, the effect on the paint depends to a large extent on the design of the facility, the filtering of the wash water, the type of wash and care material, etc. If the paint has a dull appearance after going through the car wash or is scratched, bring this to attention of the operator immediately. If necessary use a different car wash.

Notes

■ Before going through a car wash be sure to take the usual precautions such as closing the windows and sliding roof and retracting the radio antenna. The Volkswagen roof antenna does not have to be removed.

■ If you have installed additional accessories on the vehicle – such as spoilers, roof rack, etc. – it is best to ask the car wash operator if these should be removed.

Washing the vehicle by hand

❀ ***In the interest of the environment the vehicle should only be washed in special wash bays.***

First soften the dirt with plenty of water and rinse off as well as possible.

Then clean the car with a soft sponge, glove or brush starting on the roof and working down using only slight pressure. Special car wash soap should only be used for very persistent dirt.

Rinse the sponge or glove thoroughly often.

Clean wheels and sill panels last, using a different sponge if possible.

After washing the vehicle, rinse thoroughly with water and wipe it off.

Notes

■ Do not wash the vehicle in direct sunlight.

■ If the vehicle is rinsed with a hose, do not direct the water into the lock cylinders – otherwise they may freeze up in the winter.

> WARNING
>
> **■ Do not wash, wax or dry the vehicle with the engine running.**
>
> **■ Do not clean the underside of chassis, fenders, wheel covers, etc., without protecting your hands and arms. You may cut yourself on sharp-edged metal parts.**
>
> **■ Moisture and ice on brakes may affect braking efficiency. Test the brakes carefully after each vehicle wash.**

Washing your vehicle with a power washer

If you use a power washer, please note the following:

■ Always follow the operating instructions for the power washer – especially for pressure and spraying distance.

■ Make sure that the jet on the spray hose produces a "fan shape" spray. Do not use a jet which sprays the water out in a direct stream or one that has a rotating jet.

■ The water temperature should not exceed 140° F (60° C).

> WARNING
>
> **Never wash tires with a jet that sprays the water out in a direct stream. This could damage the tires – even if they are sprayed from a relatively large distance for a very short amount of time.**

Waxing

■ A good wax coating protects the vehicle paint to a large extent against the environmental influences listed under "Washing" and even against light scratches.

■ When water does not form small drops and roll off clean paint, apply a coat of good hard wax. Even if a wax solution is used regularly when washing your vehicle, it is advisable to protect the paint with a coat of hard wax at least twice a year.

Polishing

Polish your vehicle only if the paint has lost its shine and the gloss cannot be brought back with wax. If the polish used does not contain preservative compounds, the paint must be waxed afterwards.

Note

Do not treat matte painted and plastic parts with polish.

Touch-up paint

Your Volkswagen dealer has touch-up paint for minor scratches and stone chips. Scratches should be touched up soon after they occur to prevent corrosion. If corrosion formation becomes visible, however, a simple touch-up job will not suffice. The affected surface must be smoothed with sand paper and covered with an anti-rust primer, before restoring the painted finish.

The number for the original vehicle paint can be found on the vehicle identification label.

Windows

Clean all windows regularly to remove road film and car-wash wax buildup. Use a lukewarm soap water solution or an alcohol base commercial window cleaning agent. If a chamois is used for polishing the glass, it should exclusively be used for that purpose.

Also, be sure to clean all windows regularly on the inside.

Use a plastic scraper to remove snow and ice from windows and mirrors. To prevent dirt from scratching the window, always scrape in a forward direction – never back and forth.

In order not to damage the **wires of the rear defogger,** do not place any adhesive stickers in the inside of the rear window over the wires.

Weatherstrips

Keep silicone sprays off the windshield to avoid wiper smear in rain.

To seal properly, weatherstrips around hood, rear lid, doors, etc., must be pliable. Spray with silicone or coat with talcum powder or glycerine to retain flexibility of the rubber and to protect against freezing in the winter.

Sliding roof guide rails

To assure that the sliding roof* functions properly, clean the guide rails and lubricate with silicone spray at least once a year. See your Volkswagen dealer for correct procedures and lubricant.

Dull finishes and plastics

Plastic parts, such as light bulb lenses, decorative stripes, panels, bumpers, etc., will come clean with regular washing. Should additional cleaning or spot removal be necessary, use a soft brush or cloth soaked with a mild detergent solution. Then rinse thoroughly and immediately with clear water.

Do not use anything which could mar the finish of the plastic or dull finished surfaces, such as wax, polish, abrasive detergents or chemical cleaning solvents.

Bright metal trim

Bright or black anodized trim will come clean with regular washing. To protect metal trim, use car wax. Remove spots or dirt from chrome and stainless steel with a chrome cleaner. Apply a chrome polish for continued luster and protection.

Steel wheels

Always include the hub caps and wheel rims when washing your vehicle to remove road dirt, salt sprays and brake dust. If necessary, use a commercial wheel cleaner to remove accumulated brake dust. Paint scratches should be touched up to prevent corrosion.

Light alloy wheels

To preserve the decorative appearance of the light aluminium cast, some special care is necessary. In addition to road dirt and salt sprays, brake dust is also corrosive. If left on too long, brake dust can cause pitting. Wash the wheels with a sponge or hose brush every other week. Road salt should be removed weekly with an acid free cleaning solution. Every three months (after regular cleaning) the wheels should be coated with petroleum jelly or car wax. Rub it in firmly with a soft cloth. Never use abrasive or metal polishing cleaning agents.

WARNING

When washing wheels, remember that moisture, ice and road salt on brakes may affect braking efficiency. Test the brakes carefully after each washing.

Care of interior

Glass

Use the same cleaning agents as for exterior and polish dry.

Fabric

Use a vacuum cleaner or a soft bristle brush to remove dust and loose dirt from carpeting, upholstery, headliner and other trim. Dirt stains can usually be removed with lukewarm soapy water or all purpose cleaner solution, or a dry foam cleaner. For greasy, oily and other stubborn stains, use a spot remover. Do not pour the liquid on the fabric. Dampen a clean cloth and rub carefully, starting at the edge and working inwards.

Plastic and vinyl

Use a clean, damp cloth or sponge to remove dust. For other soil, use a lukewarm all purpose cleaning solution or a mild saddle soap for vinyl trim. Remove water spots and soap traces with a clean, damp cloth or sponge. Use a clean, soft cloth to rub dry.
Grease, tar or oil stains can be removed with a clean cloth or sponge soaked with all purpose cleaner or with a solvent type vinyl cleaning agent.
Occasionally apply a colorless vinyl or leather preservative to retain the material's luster and pliability.

Cleaning and care of leather upholstery

Volkswagen does everything possible to preserve the special qualities of natural leather upholstery. When treating the hides, care is taken to avoid all processes which could impair the natural advantages of the material which make for extra comfort. Because of the exclusive nature of the hides selected and the natural properties of the leather, this kind of upholstery requires a certain amount of care in everyday use since the leather is sensitive to oil, grease and dirt or wet clothing etc. If it is given regular attention the leather upholstery will retain its high quality for many years.

Dust and small particles of dirt lodging in the pores, grain and seams can have an abrasive effect and damage the surface. If the car is left standing for a long time in the sun, cover the upholstery or use the sun blinds to protect the leather from direct sunlight and prevent fading or discoloration.

Depending on the amount of use, clean leather occasionally according to the following instructions. **Never use solvents, wax polish, shoe cream, spot removers or similar materials.**

For normal cleaning of leather-covered steering wheels, leather seat upholstery, etc., slightly moisten a cotton or woollen cloth with water and wipe over the areas affected.

More stubborn dirt can be removed using a mild soap solution (pure liquid soap: two teaspoons dissolved in 1 quart of water). **It is very important not to let the water soak through the leather or penetrate into the seams.** Then wipe off with a soft, dry cloth.

In addition to this regular cleaning it is worthwhile treating the leather with a suitable leather-care product. This should be done about twice a year for upholstery in normal use. Apply these products very sparingly indeed, and wipe off with a soft cloth after leaving for some time to take effect.

Conolly leather

Only use a slightly moistened cotton or woollen cloth to clean the leather. **Do not let water soak into the surface at any point.** Then wipe off with a soft, dry cloth.

The upholstery should be treated with Connolly leather balsam, which was specially developed for this kind of leather. This product nourishes the fibers and keeps the leather supple. Apply the leather balsam very sparingly with a soft cloth, and wipe off with a soft cloth after leaving for about one hour to take effect.

Safety belts

Keep belts clean. Very dirty belts may not retract properly.

For cleaning, use a mild soap water solution. Let belts dry thoroughly and away from direct sunlight.

Do not allow inertia reel safety belts to retract before they are completely dry.

WARNING

- **Do not use chemical cleaning agents, bleach or dyes. The contain corrosive properties which weaken the webbing.**
- **When cleaning your safety belts, inspect them at the same time for damage. If you discover damage, see your dealer.**
- **Do not remove belts from the vehicle to clean them.**

Engine Compartment

WARNING

Before working in the engine compartment, be sure to read the information on page 7.

Remove leaves occasionally from the plenum panel in front of the windshield under the engine hood. This prevents the water drain holes from becoming blocked, and also prevents debris from entering the vehicle interior through the heating and ventilation ducts.

WARNING

Before reaching into the front plenum panel, always remove the ignition key. Otherwise, the windshield wiper system could unintentionally be switched on possibly causing personal injury by the moving wiper linkage.

CORROSION PROTECTION

The engine compartment, as well as all engine, transmission, front and rear axle assembly surfaces have been treated at the factory with a wax based coating for protection against corrosion. However, we recommend that you have the engine compartment and the underside inspected twice a year for any damage to the protective coating preferably before and after the winter season. Have necessary repairs done as soon as possible. See your dealer.

WARNING

Do not apply additional undercoating or rustproofing on or near the exhaust manifold, exhaust pipes, catalytic converter or heat shields. During driving, the substance used for undercoating could overheat and cause a fire.

Engine Compartment

If it becomes necessary to steam clean or wash the engine compartment, the wax based protective coating is usually lost. It is therefore important to reapply this protective coating to all engine compartment panels, flanges, cavities, seams and engine assembly surfaces.

When washing the engine compartment, only use commercially available grease cutting solvents made especially for this purpose. Carefully follow directions printed on the container by the manufacturer.

Always switch off the ignition before washing the engine.

WARNING

Never use gasoline, Diesel fuel or solvents which could damage rubber parts or painted surfaces and could cause a fire.

Chassis

The lower body shell of your Volkswagen is thoroughly protected against corrosion. Any damage to the undercoating, due to road hazards, should be repaired promptly. Oil based protective sprays must not be applied. Only tar or wax based anti-corrosion protectors are compatible with the factory applied undercoating. Before application, road dirt, salt spray deposits and oily substances must be removed.

Whenever the lower body shell, axle, transmission or engine assemblies have been repaired, the lost anti-corrosion coating of the affected surfaces should be reapplied.

Your Volkswagen dealer has the appropriate materials, the necessary equipment and is familiar with the application procedure. Therefore, any additional corrosion protection work should be performed by an authorized Volkswagen dealer.

Body cavity sealing

All body cavities which could be affected by corrosion have been given thorough protection at the factory.

This sealing does not require any inspection or additional treatments. If any wax should seep out of the cavity when the ambient temperature is high, it can be removed with a plastic scraper and a suitable solvent. Be sure to observe all safety and environmental regulations.

❀ ***If any wax should seep out of the cavity when the outside temperature is very high, it can be removed with a plastic scraper. If you use benzine, keep sparks, flame and lighted cigarettes away. Never dump benzine on the ground, into open streams or down sewage drains.***

MAINTENANCE

Your vehicle has been designed to help keep maintenance requirements to a minimum.

However, a certain amount of regular maintenance is still necessary to assure your vehicle's safety, economy and reliability.

❀ ***By regularly maintaining your vehicle, you assure that emission standards are maintained thus minimizing adverse effects on the environment.***

Maintenance Service is required every 12 months or every 15,000 miles (24,000 kilometers), whichever occurs first.

In addition, it is necessary to perform an oil change 6 months after each Maintenance Service, or after each 7,500 miles (12,000 kilometers) whichever occurs first.

The first oil change must be performed 6 months after the delivery of the vehicle or after 7,500 miles (12,000 kilometers).

Under difficult operating conditions, for example at extreme low outside temperatures, in very dusty regions, towing a trailer very frequently, etc., some service work should be performed between the intervals specified.

This applies particularly to:

- oil changes, and
- cleaning or replacing the air filter.

Important Considerations for You and Your Vehicle:

The increasing use of electronics, sophisticated fuel injection and emission control systems together with the generally increasing technical complexity of today's automobiles has steadily reduced the scope of maintenance and repairs which can be carried out by vehicle owners. **Also, safety and environmental** concerns place very strict limits on the nature of repairs and adjustment on engine and transmission parts which an owner can perform.

Maintenance, adjustments and repairs usually require special tools, testing devices and other equipment by specially trained workshop personnel in order to assure proper performance, reliability and safety of the vehicle and its many systems.

Improper maintenance, adjustments and repairs can impair the operation and reliability of your vehicle and even void your vehicle warranty. Therefore, proof of servicing in accordance with the maintenance schedule may be a condition for upholding a possible warranty claim made within the warranty period.

Above all, operational safety can be adversely affected creating unnecessary risks for you and your passengers.

If in doubt about any servicing, have it done by your Volkswagen dealer or any other properly equipped and qualified workshop. However, we strongly urge you to give your authorized Volkswagen dealer the opportunity to perform all scheduled maintenance and necessary repairs. He has the facilities, original parts and trained specialists to keep your vehicle running properly.

The following pages describe a limited number of procedures which can be performed on your vehicle with ordinary tools, should the need arise and trained personnel be unavailable. Before performing any of these procedures, always thoroughly read all of the applicable text and carefully follow the instructions given. Always rigorously observe the WARNINGS provided.

WARNING

- **Serious personal injury may occur as a result of improperly performed maintenance, adjustments or repairs.**
- **The engine compartment of any motor vehicle is a potentially hazardous area.**
- **Do not attempt any of the maintenance, checks or repairs described on the following pages if you are not fully familiar with these or other procedures with respect to the vehicle, or are uncertain as to how to proceed. Have the necessary work done by your Volkswagen dealer or any other properly equipped and qualified workshop.**
- **Always support your vehicle with safety stands if it is necessary to work underneath the vehicle. The jack supplied with the vehicle is not adequate for this purpose.**
- **If you must work underneath the vehicle without safety stands but with the wheels on the ground, always make sure the vehicle in on level ground, that the wheels are always securely blocked and that the engine cannot be started. Always remove the ignition key.**
- **Always be extremely careful when working on the vehicle. Always follow commonly accepted safety practices and general common sense. Never risk personal injury.**

❀ ***If you change the engine settings, this will adversely affect the emission values. This is detrimental to the environment and increases fuel consumption.***
Always observe environmental regulations when disposing of old engine oil, used brake fluid, dirty engine coolant, spent batteries or worn out tires.

ENGINE HOOD

■ To unlock engine hood, pull the release lever on left under dashboard. Hood springs up slightly under spring pressure.

■ In front of vehicle, open hood by depressing the safety catch under hood to the side (arrow) to disengage the catch.

■ Raise hood by hand.

Note

Before opening the engine hood, make sure that the windshield wipers are folded flat against the windshield. Otherwise, they could damage the paint on the hood.

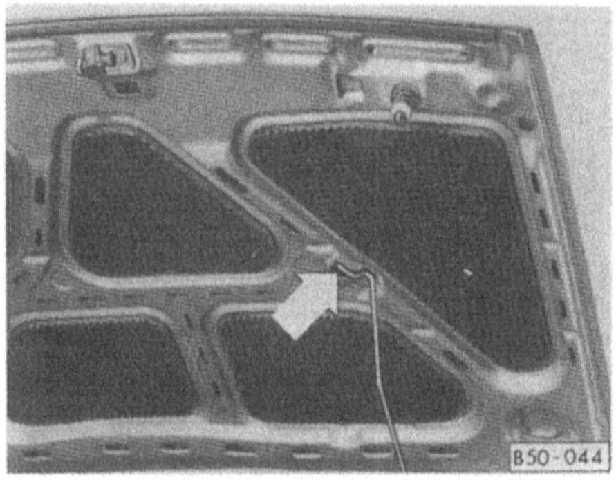

■ Take rod out of retainer on front cross panel and insert in hole provided in hood reinforcement (arrow).

■ To close hood, lift hood slightly and unhook rod.

■ Press rod into retainer on front cross panel.

■ Let hood fall into locks from a height of about 1 foot (30 cm). **Do not press hood down.**

■ Try to pull hood up to check that it is securely latched.

WARNING

Should you notice at any time while driving that the hood is not secured properly, stop at once and close it.

ENGINE COMPARTMENT

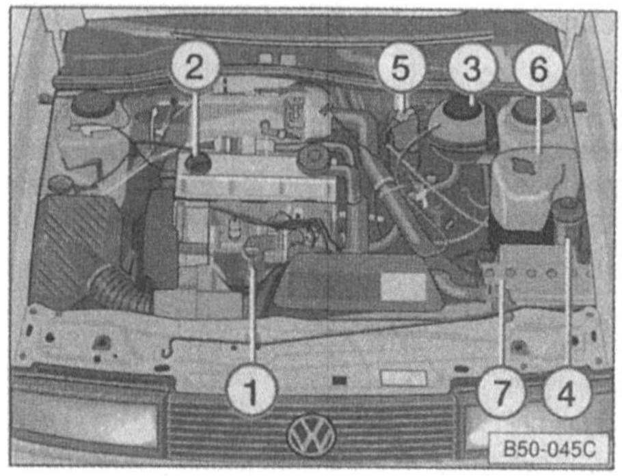

	Page
1 – Engine oil dipstick	8
2 – Engine oil filler cap	9
3 – Coolant expansion tank	11
4 – Power steering reservoir	11
5 – Brake fluid reservoir	13
6 – Windshield washer container	15
7 – Battery	14

Always exercise extreme caution when working under the engine hood

WARNING

Before you check anything under the hood:

■ Switch off engine – remove ignition key.

■ Always fully set the parking brake, move selector lever to P (automatic transmission) or Neutral (manual transmission).

■ Always let the engine cool down. Hot components will burn skin on contact.

■ Never spill fluids on hot engine components. These fluids can cause a fire.

■ Never touch the radiator fan, it is temperature controlled and can switch on suddenly – even when the ignition is off.

■ If work on the fuel system or the electrical system is necessary:
– Always disconnect the battery
– Never smoke or work near heaters or open flames.
– Always keep an approved fire extinguisher immediately available.

ENGINE OIL

■ Be especially careful if the engine is running.
– Always fully set the parking brake, move selector lever to P (Automatic transmission) or Neutral (manual transmission)
– Always exercise extreme caution to prevent neckties, jewelry, long hair or loose clothing from getting caught in the radiator fan or other moving parts.

■ To avoid electrical shock and personal injury while the engine is running or being started never touch
– ignition cables
– other components of the high voltage electronic ignition system.

When adding fluids always make sure that they are poured into the proper container or filler opening, otherwise serious damage to vehicle systems will occur.

Specification and viscosity

Your engine was factory filled with an all-season high quality engine oil. If you need to add oil between oil changes use any high quality petroleum or synthetic based oil with correct specifications.

The following term must appear on the oil container singly or in combination with other designations: "API Service SF or SG".

Engine oils are graded according to their viscosity. The proper grade to be used in your engine depends on existing climatic or seasonal conditions.

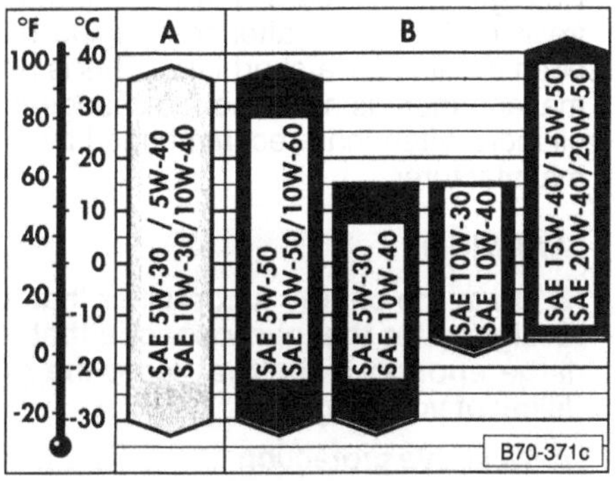

A = Energy conserving oil
B = Multi grade oil

Refer to the temperature chart when selecting engine oil. As temperature ranges of the different oil grades overlap, brief variations in outside temperatures are no cause for alarm. It is also permissible to mix oil of different viscosities if you find it necessary to add oil.

When using SAE 5 W-30 engine oil, avoid high speed long distance driving if outside temperature rises above the indicated limits.

Engine Oil Identification Symbol

Select an oil for your vehicle which conforms to the standards of the American Petroleum Institute (API). A symbol has been added to the top of some oil containers to help you select the correct oil.

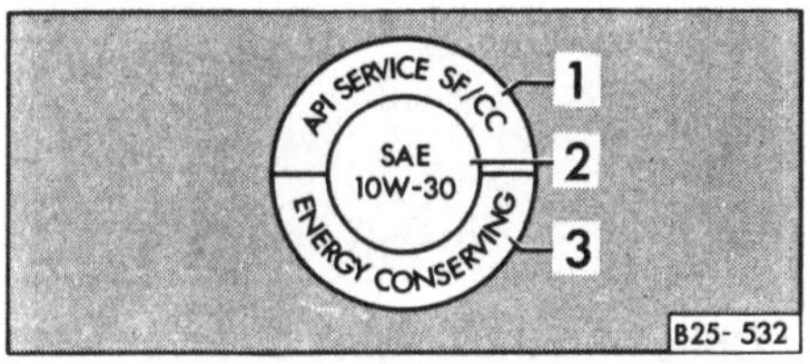

(1) The top portion indicates the oil quality by API designations.
(2) The center portion shows the SAE oil viscosity grade.
(3) The lower portion indicates that the oil has fuel saving capabilities.

Lubricant additives

Volkswagen does not recommend the use of oil additives. They may adversely affect your warranty.

Engine oil consumption

It is normal for your engine to consume oil. The rate of oil consumption depends on the quality and viscosity of oil, the speed at which the engine is operated, the climate, road conditions as well as the amount of dilution and oxidation of the lubricant.

Because of these variables, no standard rate of oil consumption can be established, but drivers should expect an increased oil consumption at high speeds and when the engine is new.

■ The engine in your vehicle depends on oil to lubricate and cool all of its moving parts. Therefore the engine oil should be checked regularly and kept at the required level.

■ Make it a habit to have the engine oil level checked with every fuel filling.

■ Lack of sufficient engine oil may lead to severe engine damage.

■ The oil pressure warning light is not an oil level indicator.

Always heed WARNINGS on page 7.

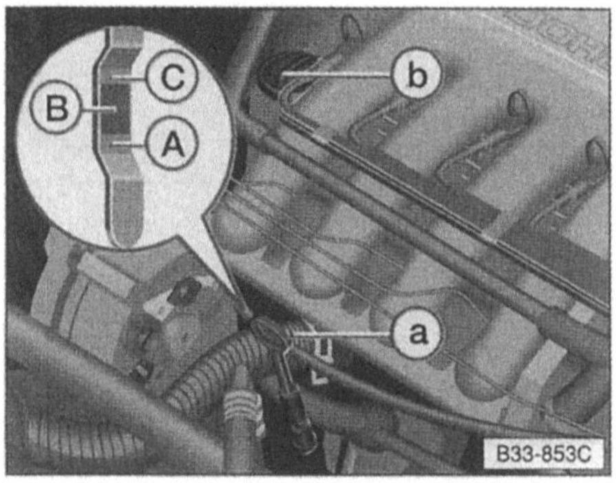

Checking the engine oil level

The best time to check the engine oil level is when the oil is warm.

■ To get a true reading, the vehicle should be on level ground.

■ Turn off the engine, wait a few minutes for the oil to return to the oil pan.

■ Pull out dipstick (a) and wipe it clean with a rag.

■ Reinsert dipstick; push it all the way in.

■ Pull out the dipstick and read the level.

A – You **must** fill the oil. After filling the oil, make sure that the oil level is somewhere within the b range.

B – You **can** fill the oil. It can happen that the oil level goes into the C range.

C – You **must not** fill the oil.

■ Make sure that the dipstick is pushed in all the way.

Adding engine oil

■ Turn off the engine.

■ Unscrew cap from cylinder head cover.

■ Add oil by half quarts (liter) at a time.

■ After doing this, check the oil level using the dipstick.
The oil level must never be above the C range. Otherwise, the excess oil may be drawn in through the crankcase breather and finally get into the exhaust system. The oil would then burn in the catalytic converter and cause damage.

Always select a quality oil with the correct specification. See page 8.

■ Replace cap and hand tighten securely.

WARNING

The oil filler cap must be secure to avoid oil spill which is a fire hazard.

Changing the engine oil

Have your engine oil changed regularly, but at least twice a year (see Maintenance booklet).

This is very important because the lubricating properties of oil diminish gradually during normal vehicle operation.

If you drive mostly short distances, or if you operate the vehicle in dusty areas, or under predominantly stop-and-go traffic conditions, or when temperatures remain below freezing for extended periods, the engine oil should be changed more frequently.

Detergent additives in the oil, will make fresh oil look dark after the engine has been running for a short time. This is normal and there is no reason to change the oil more often than recommended by the manufacturer.

WARNING

■ Drain the oil into a container that is designed for this purpose, one that is large enough to hold at least the oil filling of your engine.

■ Wear eye protection.

■ To reduce the risk of burns from hot engine oil let the engine cool down to the touch.

■ When removing the oil drain plug with your fingers, stay as far away as possible. Always keep your forearm parallel to the ground to help prevent hot oil from running down your arm.

■ Engine oil is poisonous. Keep it well out of the reach of children. Continuous contact with used engine oil is harmful to your skin. Always protect your skin by washing thoroughly with soap and water.

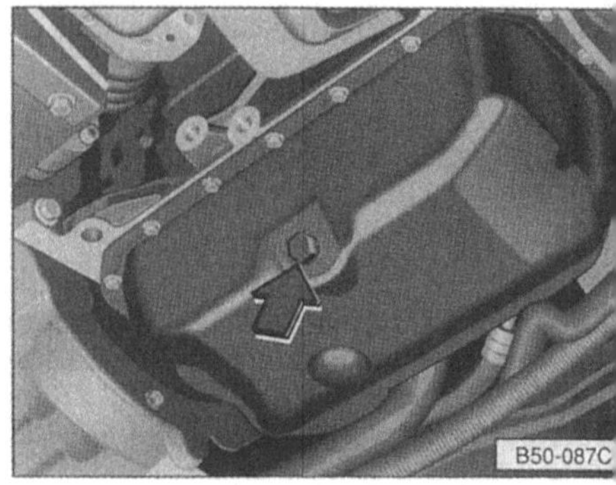

■ Turn off the engine.

■ Remove the oil drain plug (arrow) when the engine is still warm and allow the oil to drain.

■ Always use a new gasket when reinstalling the plug. Do not overtighten.

■ Fill the engine with oil. **Do not overfill.** Always check engine oil level with dipstick as described on previous page.

Engine oil capacity is listed under "Capacities", page 37.

❀ ***Before changing your oil, first make sure you know where you can properly dispose of the old oil.***
Always dispose of used engine oil properly. Do not dump it on garden soil, in wooded areas, into open streams or down sewage drains.
Recycle used engine oil by taking it to a used engine oil collection facility in your area, or contact a service station.
Because of the problem of proper disposal, along with the special tools required and the necessary expertise, we strongly recommend that you have your oil changed by a Volkswagen dealer or a qualified workshop.

Always heed WARNINGS on page 7.

ENGINE OIL FILTER

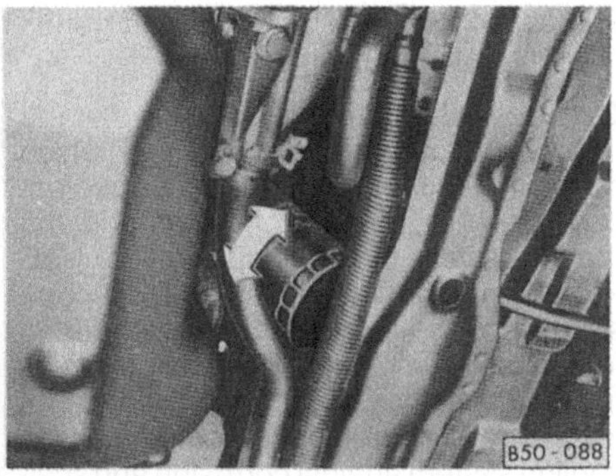

Changing the oil filter

The oil filter should be changed at the intervals listed in your Maintenance booklet.

■ Remove old oil filter element and discard properly.

■ Lightly coat seal of new filter element with oil.

■ Screw on filter element and hand-tighten according to manufacturer's instructions on the carton or on the filter element. Unless installed very carefully according to the instructions, the filter may leak which could seriously damage the engine and pollute the environment.

■ Fill engine with appropriate amount of engine oil.

■ Run engine at various speeds for three to five minutes and check for leaks.

■ Check dipstick for correct oil level, top up if necessary.

❀ ***Because of the problem of proper disposal, along with the special tools required and the necessary expertise, we recommend that you have your oil filter changed by a Volkswagen dealer or a qualified workshop.***

Always heed WARNINGS on page 7.

TRANSMISSION FLUIDS

Specifications

Manual transmission with final drive

■ Transmission oil SAE 80 (Mil-L-2105 or API/GL4) or

■ Synthetic transmission oil G 50, SAE 75 W-90

Automatic transmission

■ Final drive: Transmission oil G 50, SAE 75 W-90

■ Torque converter: ATF Dexron® or Dexron® II.

Checking oil level

On Manual transmission with final drive, the oil does not need to be checked.

On the Automatic transmission, the oil level in the final drive also does not need to be checked.

However, you must check the ATF level in the torque converter between the maintenance intervals specified in the Maintenance booklet.

To obtain an accurate reading, the vehicle must be on level ground. Check the level only when the ATF is **warm.** (The ATF is warm after driving the vehicle for approximately 6 miles or 10 kilometers). If the fluid is cold or too hot, the reading will be incorrect.

The engine must be running at idle, with the parking brake firmly applied and the selector lever in P (Park).

WARNING

■ **Before checking the ATF, the selector lever must be in the Park position and the parking brake firmly applied.**

■ **Checking the ATF level is potentially hazardous because the hot engine has to be running during the check. Do not attempt to check the ATF before reading all WARNINGS on page 7.**

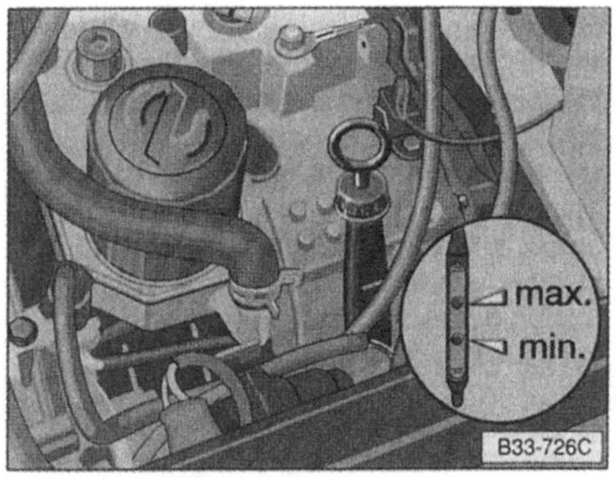

To check, pull the dipstick out (see illustration), wipe it off with a clean lint-free rag, and reinsert it completely.

Then, pull the dipstick out once again and check the ATF level. **The level must be between the two marks.** If not, then contact your Volkswagen dealer immediately to determine the cause. It is not sufficient to simple add more ATF.

Changing transmission oils

On Manual transmission with final drive, the oil does not need to be changed.

On the Automatic transmission, the oil level in the final drive also does not need to be changed. However, the ATF in the transmission must be changed according to the intervals specified in the Maintenance booklet.

❀ ***Because of the problem of proper disposal with the special tools required and the necessary expertise, we strongly recommend that you have your ATF fluid changed by your Volkswagen dealer.***

Notes

■ **Do not tow the vehicle or run the engine without lubricant in the Manual or Automatic transmission.**

■ **Do not use any transmission fluid additives.**

AIR CLEANER

A dirty filter element reduces engine output, increases fuel consumption and can cause premature engine wear.

Normally, it is not necessary to service the air cleaner more often than recommended in the Maintenance booklet. If the vehicle is driven on very dusty roads, the air cleaner must be serviced more frequently, even daily. We recommend that you have the air cleaner serviced by your Volkswagen dealer or a qualified workshop.

If your air filter must be changed more frequently, we recommend that you have your Volkswagen dealer show you the procedure before you attempt doing it yourself.

The paper filter element must never be cleaned or soaked with gasoline, cleaning solvents or oil.

POWER STEERING

The fluid reservoir is located in the engine compartment near the battery.

The power steering is filled with VW Hydraulik oil No. G 002000.

A correct fluid level in the reservoir is important for the proper functioning of power steering.

With the engine running and the wheels in straight-ahead position, the fluid level in the power steering system must be checked at regular intervals. The fluid level should always be between the "max" and "min" markings. If the level should go below the "min" marking, have the power steering system inspected by your Volkswagen dealer. It is not sufficient to just add hydraulic fluid.

- **Be sure not to confuse brake fluid and hydraulic fluid. If the wrong fluid is used, the systems may fail.**
- **If the power steering system should fail, or if the engine is not running (for example while being towed), you will still be able to steer the vehicle, however, more effort will be required.**
- **Since the hydraulic fluid also functions as a lubricant, serious damage to the pump and to other components of the system could result if there is insufficient hydraulic fluid in the system.**

COOLING SYSTEM

The cooling system is sealed and generally requires little attention.

The cooling system has been filled at the factory with a permanent coolant which does not need to be changed. The coolant consists of a mixture of water and the manufacturer's coolant additive G 11 V8B – antifreeze on glycol basis with anticorrision additives (40% for USA models; 50% for Canadian models). This mixture assures the necessary frost protection and protects above all the alloy components in the cooling system against corrosion and scaling, and raises the boiling point of the coolant.

Do not reduce the concentration of the coolant in the summer by adding plain water. **The coolant additive proportion must be at least 40% but not more than 60%,** to maintain the antifreeze protection and the cooling efficiency.

For year round driving, antifreeze is added at the factory for temperatures down to:

−13° F/−25° C (USA)
−31° F/−35° C (Canada)

Use only the approved type of antifreeze, which is available from your Volkswagen dealer.

Other types of anti-freeze can significantly reduce corrosion protection. The resulting corrosion can cause a loss of coolant and therefore serious engine damage.

Checking the coolant level

The coolant expansion tank is located in the engine compartment next to the front plenum panel.

WARNING

To reduce the risk of being burned, never open the hood if you see or hear steam or coolant escaping from the engine compartment. Wait until no steam or coolant can be seen or heard before carefully opening the hood.

The correct coolant level is important for proper functioning of the cooling system.

Therefore, check the coolant level regularly – preferably at every fuel filling.

To obtain an accurate reading, the engine must be switched off.

Always heed WARNINGS on page 7.

Since the expansion tank is transparent, the cap need not be removed to check the coolant level.

The coolant level must be between the min. and max. marks when engine is cold and can be slightly above the max. mark if the engine is warm.

When the coolant level is too low, the warning light in the coolant temperature gauge will flash until the coolant level has been restored to normal.

Coolant losses

Coolant loss may indicate a leak in the cooling system. In the event of coolant losses, the cooling system should be inspected immediately by your Volkswagen dealer. It is not sufficient merely to add coolant.

In a sealed system, losses can only occur if the boiling point of the coolant is exceeded as a result of overheating.

Overheating can occur if:

- the flow of air to the radiator is obstructed by leaves, dust, insects or by additional lights installed in front of the radiator grille;
- the boiling point of the coolant has been reduced due to an incorrect mixture of coolant and water
- the radiator fan is not working – see "Radiator fan", or;
- the vehicle is being driven up a long hill in a low gear with engine speed very high ant at very high ambient temperature – see "Radiator fan".

If the cause for the overheating cannot be determined and corrected, contact your Volkswagen dealer immediately, otherwise serious engine damage could result.

Adding coolant

Turn off the engine and allow it to cool down.

WARNING

Reduce the risk of scalding from hot coolant by following these steps.

- **Turn off the engine and allow it to cool down. The cooling system is under pressure.**
- **Protect face, hands and arms by covering the cap with a large, thick rag to protect against escaping fluid and steam.**
- **Turn the cap slowly and very carefully in a counter-clockwise direction while applying light, downward pressure on the top of the cap.**

If, in an emergency, only water can be added, the correct ratio between water and antifreeze (see previous page) must be restored as soon as possible.
If you have lost a considerable amount of coolant, then you should add cold antifreeze and cold water only when the engine is cold.

Do not fill above the max. mark.

Excess coolant will be forced out through the pressure relief valve in the cap when the engine becomes hot.

Screw cap on again tightly.

WARNING

Antifreeze is poisonous. Always store antifreeze in its original container and well out of the reach of children. If you drain the coolant, it must be caught and sefely stored.

Drained coolant should normally not be reused. Always dispose of used coolant observing all environment regulations.

Always heed WARNINGS on page 7.

Radiator fan

The electric radiator fan is controlled by thermoswitches from the coolant and engine compartment temperature.

WARNING

Never touch the radiator fan, it is temperature controlled and can switch on suddenly – even when the ignition is off.

The radiator fan switches on automatically when the coolant reaches a certain temperature and will continue to run until the coolant temperature drops. This may last up to about 10 minutes.

Notes

- If the radiator fan does not come on when the coolant temperature is very high, check the fuse and replace it if necessary – see page 23.
- The speed of the radiator fan does not depend on the engine speed, therefore cooling effect cannot be increased by downshifting.
- As long as the engine runs smoothly and driving an uphill road does not considerably reduce speed, there is no need to downshift.

Winter operation

At the beginning of the winter season, have the coolant checked for antifreeze concentration. If you have to add, only use quality phosphate-free antifreeze containing ethylene glycol which is available at your Volkswagen dealer.

The ratio between water and antifreeze depends on the anticipated outside temperatures.

V–BELT

Tension checking

Correct V-belt tension is important for overall vehicle performance.

Tensioning and replacing of V-belts should be performed by your Volkswagen dealer or a qualified workshop.

See page 36 for V-belt designation.

WARNING

■ To prevent serious personal injury stay well clear of the V-belt when the engine is running. If you must check the V-belt tension turn the engine off and let it cool down. Hot components can burn skin on contact.

■ Never touch the radiator fan blades. They will rotate spontaneously when the thermoswitch turns the fan on. The radiator fan switches on automatically when the coolant or engine compartment reaches a certain temperature and continues to run (even with ignition off) until the coolant temperature drops. This may last up to about 10 minutes.

BRAKE FLUID

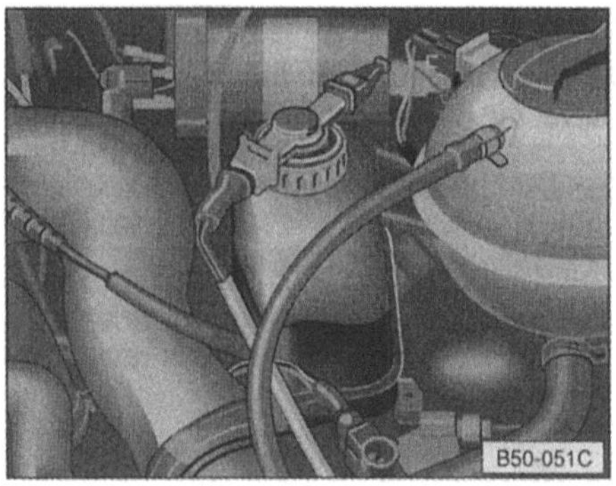

B50-051C

The brake fluid reservoir is located at the rear partition of the engine compartment on the left side as seen in the driving direction.

Checking fluid level

The correct fluid level is important for the proper functioning of the brake system. The fluid level in the brake fluid reservoir should always be between the MAX and MIN markings.

The fluid level may drop slightly after some time due to the automatic adjustment of the brake pads. This is no cause for alarm.

If the brake fluid level falls considerably below the MIN mark, the brake warning light will come on. Do not continue to operate the vehicle. The complete brake system should be thoroughly checked by a

Volkswagen dealer or any other qualified workshop and the cause corrected.

The brake warning lamp lights up to indicate a low fluid in the reservoir. Contact a Volkswagen dear immediately

Changing brake fluid

Brake fluid absorbs moisture from the atmosphere. If the water content in the brake fluid is too high, corrosion in the brake system may result after a period of time. The boiling point of the brake fluid will also decrease considerably. **Therefore, the brake fluid must be changed every 2 years.**

WARNING

■ If the brake fluid becomes too old and the brakes are under constant and extreme use, a vapor lock may result and impair the effectiveness of the brake as well as vehicle safety.

■ Use only DOT 4 brake fluid that meets SAE specification J 1703 and conforms to Federal Motor Vehicle Standard 116.
The brake fluid must be new.

■ Do not add or mix DOT 5 silicone type brake fluid with the brake fluid in your vehicle as severe component corrosion may result. Such corrosion could lead to brake system failure.

■ Brake fluid is poisonous. Therefore, it must only be stored in the closed original container out of the reach of children.

Remember also that brake fluid is harmful to the paint of your vehicle.

Because of the problem of proper disposal of brake fluid along with the special tools required and the necessary expertise, we recommend that you have your brake fluid changed by your Volkswagen dealer.

It is advisable to have the fluid changed during a Maintenance Service.

Always heed WARNINGS on page 7.

BATTERY

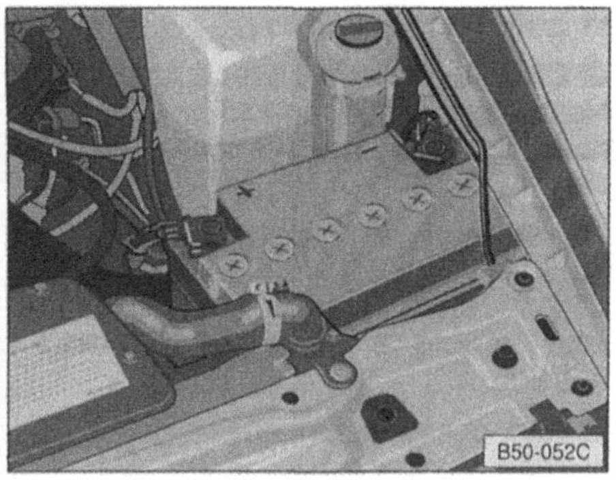

The battery is located under the front hood.

Battery acid level

Under normal operating conditions, the battery in your Volkswagen is maintenance-free.

At high outside temperatures it is advisable, however, to check the fluid level at regular intervals through the transparent battery housing. The fluid level should always be between the "min" and "max" marks in each cell.

If the fluid level is below the "min" mark, let your Volkswagen dealer correct the condition.

WARNING

- **Always shield your eyes and avoid leaning over the battery whenever possible.**
- **Do not let battery acid come in contact with skin, eyes, fabric, or painted surfaces.**
- **If you get battery acid in your eyes or on your skin, immediately rinse with cold water for several minutes and call a doctor.**
- **Do not expose the battery to an open flame or electric spark. Hydrogen gas generated by the battery can explode and cause personal injury.**

Winter operation

During the winter months, battery capacity tends to decrease as temperatures drop. More power is also consumed while starting, and the headlights, blower fan, rear window defogger, etc., are used more frequently. Avoid unnecessary power consumption, particularly in city traffic or when travelling only short distances. Let your Volkswagen dealer test the battery's capacity before winter sets in. A well charged battery will not only prevent starting problems but will also last longer.

If your vehicle is left standing for several weeks at extremely low temperatures, the battery should be removed and stored where it will not freeze. This will prevent it from being damaged.

When removing the battery, first disconnect both cables (see Charging of Battery), then unscrew the battery mounting.

WARNING

Always keep the battery well out of the reach of children.

Charging of battery

WARNING

- **The battery contains sulfuric acid. Therefore, keep battery out of reach of children.**
- **Charge battery in a well ventilated area. Keep away from open flame or electrical spark. Do not smoke. Hydrogen gas generated by the battery is explosive.**
- **Always shield your eyes and avoid leaning over the battery whenever possible.**
- **If you get battery acid in your eyes or on your skin, immediately rinse with cold water for several minutes and call a doctor.**
- **Never charge a frozen battery. It may explode because of gas trapped in the ice. Allow a frozen battery to thaw out first.**
- **Fast charging a battery is dangerous and should only be attempted by a competent mechanic with the proper equipment.**

Always heed WARNINGS on page 7.

- **Never use a fast charger as a booster to start the engine. This will seriously damage sensitive electronic components, such as relays, radio, etc., as well as the battery charger.**
- **Battery acid that may spill during charging should be washed off with a solution of warm water and baking soda to neutralize the acid.**

Slow battery charging

WARNING

- **Heed all WARNINGS and follow instructions that come with your battery charger.**

- It is not necessary to remove the battery from the engine compartment, and it is also not necessary to disconnect the cables.
- Make sure the electrolyte level in each cell is between the "min" and "max" marks. If the fluid level is below the "min" mark, let your VW dealer correct the condition.
- Connect charger cables.

Charger cables must be connected POSITIVE (+) to POSITIVE (+) and NEGATIVE (–) to NEGATIVE (–).

Do not connect or disconnect charger cables while charger is operating to reduce the danger of explosion.

- Switch on charger.
- **Charging rate not over 6 Amp.** Normally, a battery should be charged at no more than 10 percent of its rated capacity.

For example, a charging current of 4.5 Amp. would be used on a battery having 45 Ah. Rated capacity of the battery in your vehicle is listed on the battery housing.

- After charging, turn off charger and disconnect charger cables.

Always heed WARNINGS on page 7.

To remove battery from vehicle

■ Disconnect negative ground cable.

■ Disconnect positive cable.

■ Unscrew bolt of holding plate with open end wrench.

To reinstall battery in vehicle

■ Place battery in car and tighten bolt of holding plate.

■ Reconnect positive cable.

■ Reconnect negative ground cable.

WARNING

Do not reverse polarity.

Cleaning terminals and connections

WARNING

■ Before work is done on the electrical system, disconnect the negative ground cable.

■ When working on the battery, be sure not to short circuit the terminals with tools or other metal objects. This would cause the battery to heat up very quickly, which could lead to damage or explosion and personal injury.

The terminals and connections should be kept clean and greased with silicone spray or petroleum jelly. Make sure the ground connection to the body is tight and free of corrosion.

With a disconnected battery the engine must not run because the electrical system will be damaged.

Replacing battery

A replacement battery must have the same specifications and dimensions as the original equipment battery. Specifications are listed on the battery housing. Make sure the replacement battery is installed correctly and securely.

Because of the problem of proper disposal of a battery, we recommend that you have your Volkswagen dealer change the battery for you. Batteries contain sulphuric acid and lead and must always be disposed of properly observing all environmental regulations.

WINDSHIELD/WINDOW WASHER

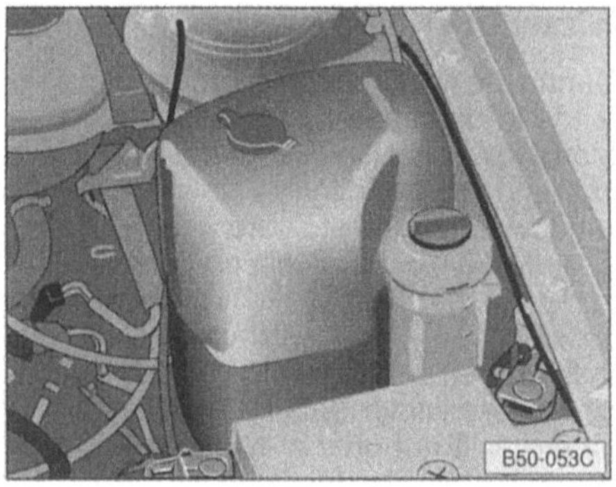

The container for the windshield and rear window washer is located in the engine compartment next to the left wheel housing, as illustrated above.

The capacity of the container is 4.2 qt/4.0 liters.

Filling the container

To add washer fluid just lift the filler cap by the small tongue. The container can be filled to the top.

Since clear water is usually not adequate for cleaning the glass, add a cleaning solution to the water.

Use winterized windshield washer solvent during the cold season although the vehicle is equipped with **heated washer jets.** It helps to keep your windshield clean and prevents the fluid from freezing in the winter.

Do not use engine coolant antifreeze or any other solution that can damage the vehicle paint.

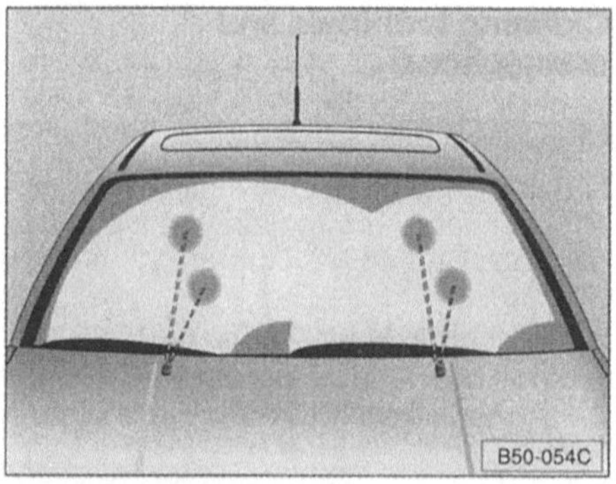

Adjusting washer jets

When vehicle is stationary, the water should hit the windshield as illustrated.

The washer jet for the rear window should be adjusted so that the fluid hits the glass in the center of the wiped area.

Adjust the jets with a needle.

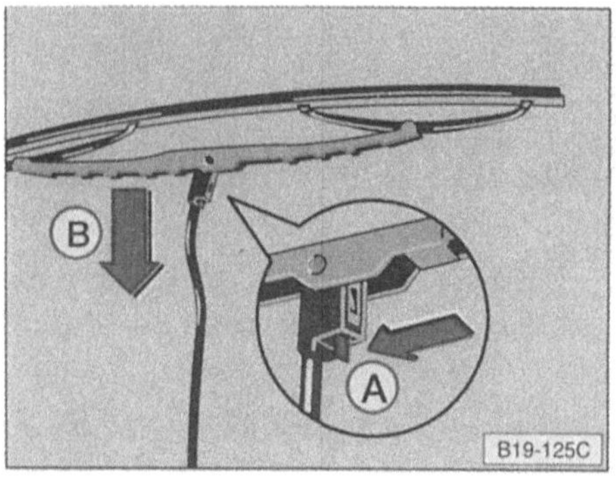

Removing wiper blades

■ Raise the wiper arm and hold wiper blade horizontally.

■ Depress locking spring (A) and remove blade by pulling it towards the windshield (B).

Installing new wiper blades

■ The locking spring must click into place on the wiper arm.

When installing wiper blades with moulded wind deflectors, always make sure the deflector points down.

Wiper blades must be in good condition to provide clear vision.

Clean your wiper blades regularly with a windshield washer solution to prevent streaking. If the blades are very dirty, for example with insects, carefully clean the blades with a sponge or a soft brush.

Replace your wiper blades once or twice a year. See your Volkswagen dealer for replacement blades.

Notes

■ Commercial hot waxes applied by automatic car washes have been known to affect the cleanability of the windshield.

■ To prevent damage to wiper blades, do not use gasoline, kerosene, paint thinner, or other solvents on or near the wiper blades.

■ To prevent damage to the wiper arms or other components, do not attempt to move the wipers manually.

Always heed WARNINGS on page 7.

TIRES/WHEELS

Tires are important but often abused parts of a car, they not only influence comfort and ride, but perform vital safety functions. This section is intended to provide you with important information regarding their proper use, care and replacement.

■ Be sure to inspect your tires at least every 2000 miles (3000 kilometers) for wear and damage.

■ Damage to wheels and tires is not always easy to see. If you believe that a tire or wheel has been damaged, it is wiser to have it replaced as soon as possible. Internal tire damage can also lead to tire failure. See your authorized Volkswagen dealer for advice and assistance.

■ Mark tires before removing them. Remount tires on the same vehicle side because the rotation direction must stay the same.

■ Store removed tires in a cool, dry and preferably dark place. Tires which are not on wheels should be stored standing up.

WARNING

■ Tires age even if they are not being used. Tires which are older than 6 years should only be used in an emergency and then with caution.

■ Avoid damaging tires and wheel rims. If you must drive over a curb or other obstacle, drive slowly and as nearly as possible at a right angle. Frequently check tires for uneven wear and damage.

■ Remove imbedded material.

■ Replace worn or damaged tires immediately.

■ Replace missing valve dust caps.

■ Keep oil, fuel, brake fluid, etc. away from tires.

■ Keep tires inflated correctly.

New Tires

New tires tend to be slippery and cannot offer full traction until they have been properly broken in.

WARNING

To help avoid loss of control, always operate a vehicle equipped with new tires at lower speeds and with special caution for the first 100 miles (160 kilometers).

Tire service life

The service life of your tires depends for the most part on the following factors:

Tire pressures

Required cold tire inflation pressures are listed on a sticker on the passenger's door lock jamb.

WARNING

■ Incorrect tire pressures cause increased tire wear and adversely affect road holding of the vehicle, leading to loss of control and personal injury.

■ Incorrect tire pressure can also lead to sudden deflation, resulting in an accident and personal injury.

❀ ***Low tire pressures increase fuel consumption and are detrimental to the environment.***

The tire pressures should be checked at least once a month and always before a long trip.
Tire pressures are very important particularly when driving at higher speeds.
Always check tire pressures when the tires are cold. When the tires are warm, the pressure will be higher. Do not reduce the pressure of warm tires.

Do not forget to check the spare wheel.

The inflation pressure is marked on the tire sidewall.

Use an accurate tire pressure gauge when checking inflation pressures. Do not exceed the maximum tire inflation pressure listed on the tire sidewall. Cold tire inflation pressure means: when a vehicle has been standing for at least 3 hours or driven for less than 1 mile.

Don't forget to replace the valve stem caps after checking tire pressure.

Driving habits

Fast cornering, heavy acceleration and hard braking all increase tire wear.

Wheel balancing

The wheels on new vehicles are balanced. When driving, however, various conditions can cause a wheel to become unbalanced. This may be noticed as vibrations in the steering.

Since tire imbalance can cause wear on the steering, suspension and tires, you should have your wheels rebalanced. A wheel should always be balanced if a new tire has been mounted or a tire was repaired.

Incorrect wheel alignment

Incorrect wheel alignment causes excessive and uneven tire wear impairing the safety of the vehicle. If you notice excessive tire wear, contact your Volkswagen dealer.

Tire wear

The original tires on your vehicle have built-in wear indicators. They are molded into the bottom of the tread grooves and will appear as approximately ½ inch (12 mm) bands when the tire tread depth wears down to 1/16 inches (1.6 mm). Depending on the tire manufacturer, there are six to eight wear indicators evenly spaced around the circumference of the tire. Markings on the sides of the tires (e.g. the letters "TWI" or a triangle) show the locations of the wear indicators.
When the indicators appear in two or more adjacent grooves, it is time to replace the tires. We recommend, however, that you do not let the tires wear down to this extent. Worn tires cannot grip the road

surface properly, and are even less effective on wet roads.

WARNING

■ Do not drive with worn tires or tires showing cuts, bruises or other damage because they may lead to sudden deflation causing loss of vehicle control and personal injury.

■ Since worn tires do not grip the road surface properly when driving on wet roads, the vehicle may tand to aquaplane sooner. We strongly urge you to replace your tires when the remaining tread depth is .12 inches (3 mm).

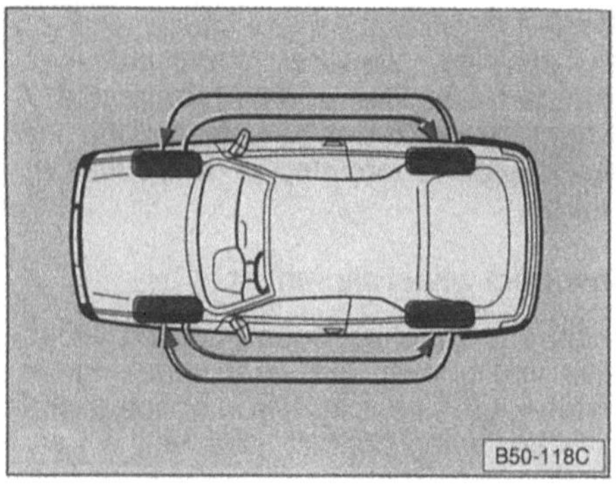

Tire rotation

If the front tires are worn more than rear tires, we recommend that you rotate the front tires with the rear tires as shown in the illustration. By doing this, all tires will have approximately the same service life.

Tires should always remain on same side of vehicle.

Only when tires show unusual wear such as feather-edging should they be rotated diagonally. If in doubt, discuss any unusual tire wear with your Volkswagen dealer's service department.

After rotation adjust tire pressure and torque wheel bolts diagonally to 80 ft lb/ 110 Nm. Refer to "Changing a wheel" on page 22 for details.

Wheels and tire replacement

Wheels and tires approved by the manufacturer have been specially matched to your vehicle and contribute largely to the roadholding and driving characteristics of the vehicle.

■ Fitting and repairing tires requires expert knowledge and special tools. This work should only be performed by a specialist.

■ We recommend that you have your tires changed by a Volkswagen dealer, because they have the special tools and the necessary expertise.

Furthermore, your Volkswagen dealer stocks a range of tires and wheels.

■ For safety reasons, tires should be replaced in pairs and not individually. The tires with the deepest tread should always be mounted on the front wheels.

■ Never mount used tires if you are not sure of their previous history.

■ Whenever replacing a tubeless tire, always install a new valve stem. Tire repair should only be performed by a specialist.

■ In the interest of maximum safety and best all-around vehicle handling, always buy replacement radial tires that have the same specifications with regard to tire size, design, load carrying capacity, speed rating, tread pattern, tread depth etc. This also applies to Volkswagen recommended alternate replacement tires.

Make sure that the new tires also meet the specifications listed on the label located on the inside of the fuel tank flap or on the driver's door lock jamb.

Tire specifications

> WARNING
>
> ■ **Never mix tires of different design such as steel belted radials with radial bias belted or bias ply tires etc. Mixing tire types will adversely affect road holding and can lead to loss of vehicle control and personal injury.**

The knowledge of tire designations makes it easier to choose the correct tires. Radial ply tires have the following designations:

195 / 50 R 15 82 V

- 195 – Tire width in mm
- 50 – Height/width ratio in %
- R – Tire construction: Radial
- 15 – Rim diameter in inches
- 82 – Load rating code
- V – Speed rating code letter

The speed rating letter code indicates the maximum permissible road speeds:

Summer tires:
S – up to 180 km/h (110 mph)
T – up to 190 km/h (118 mph)
H – up to 210 km/h (130 mph)
V – up to 240 km/h (150 mph)
Z – over 240 km/h (150 mph)

"Z" tires carry the speed rating code letter in combination with the tire construction letter "R".

Winter tires (see page 18)

> WARNING
>
> ■ **Never operate a vehicle at speeds greater than the maximum speed rating of the tires with which it is equipped.**
>
> ■ **Exceeding the maximum speed for which tires are rated and approved will lead to sudden tire failure which can cause loss of vehicle control and personal injury.**
>
> ■ **Always observe speed limits and adjust your vehicle to prevailing road and traffic conditions.**

The manufacturing date is also to be seen on the tire wall (possibly only on the inside of the wheel):

DOT ... 121 ... means that the tire was produced in the 12th week of 1991.

Tire strength and performance decline with age – even if they are not used.

Tire manufacturers caution that a tire which is more than six years old cannot be expected to perform reliably in normal use.

> WARNING
>
> **Old tires can fail in use, causing loss of vehicle control and personal injury. Replace tires after six years regardless of tread wear. Always reduce speed and drive cautiously if you must use an old tire in an emergency. Replace the tire as soon as possible.**

Treadwear

The treadwear grade is a comparative rating based on the wear rate of the tire when tested on a government test course. For example, a tire graded 150 would wear one and a half (1½) times as well as a tire graded 100. The relative performance of tires depends on actual conditions of use, however, and may depart significantly from the norm due to variations in driving habits, service practices and differences in road characteristics and climate.

Traction A, B, C

The traction grades, from highest to lowest, are A, B and C and represent the tire's ability to stop on wet pavement as measured on government test surfaces of asphalt and concrete. A tire marked C may have poor traction performance.

The traction grade is based on braking (straightahead) traction tests and does not include cornering (turning) traction.

Temperature A, B, C

The temperature grades A (the highest), B and C represent the tire's resistance to the generation of heat. Sustained high temperatures can reduce tire life, and lead to sudden tire failure. Grade C corresponds to a performance which all passenger car tires must meet under the Federal Motor Vehicle Safety Standard No. 109. Grades B and A represent higher levels than the minimum required by law.

WARNING

The temperature grade for this tire is established for a tire that is properly inflated and not overloaded. Excessive speed, underinflation, or excessive loading, either separately or in combination, can cause heat buildup, possible tire failure, loss of control and personal injury.

If you wish to equip your vehicle with tires or wheels other than those installed at the factory, please note the following:

■ For technical reasons it is not possible in every case to use wheels from other vehicles – under certain conditions not even wheels from the same vehicle model.

■ Wheel rims and wheel bolts are matched to fit your Volkswagen.

■ When installing different wheels (for example, wheels with winter tires), the correct wheel bolts with the proper length and conical shape must be used. The secure fit of the wheels and the proper functioning of the brake system are dependent upon this.

■ Before you plan on exchanging steel wheels, light alloy wheels, or winter tires already mounted on wheel rims, consult your Volkswagen dealer. He has the technical information necessary to advise you which wheel rims and wheel bolts are compatible with the original factory installations.

WARNING

■ The use of tires, wheel rims and wheel bolts that do not meet specifications of the original factory installed equipment will adversely affect the safe operation of your vehicle and may cause an accident and personal injury.

■ If your vehicle is equipped with BBS-light alloy wheels, never loosen the bolts which hold the two parts of the wheel together, as this could cause leakage.
This can load to loss of vehicle control and personal injury!

■ If wheel trim discs or a front spoiler are installed, make sure the air flow for cooling the brakes is not obstructed.

Winter tires

The tires for your vehicle were selected for optimal performance under normal driving conditions.

For winter driving, the roadability of your vehicle can be improved by installing all season tires or radial winter tires (M + S) with or without studs[1].

All season tires are performance tires advertised by the tire manufacturer as suitable for all weather use or with special mud and snow (M + S) capability.

Winter tires, sometimes also called snow tires, are designed for maximum traction in mud and snow.

If your Volkswagen is equipped with high performance tires (identified by the code letter H or V on the tire flank) and you drive frequently on ice or snow, you should consider the installation of all season or winter tires. Let your Volkswagen dealer advise and assist you.

When installing winter tires, please note the following:

■ Only radial ply winter tires must bve installed. Ask your Volkswagen dealer for the recommended tire size.

[1] Check with your local Motor Vehicle Bureau for possible restrictions.

■ Winter tires **should** be mounted on all four wheels.

■ The inflation pressure for winter tires is the same as required for the regular radial ply tires. However, do not exceed the maximum tire inflation pressure listed on the tire sidewall.

■ Winter tires do not fulfill their purpose, if the tread depth is less than 5⁄32 inch (4 mm).

WARNING

■ Tires with badly worn treads and studs are very dangerous. Make sure they are replaced immediately.

■ Never mix tires of different design such as steel belted radials with radial bias belted or bias ply tires etc. Mixing tire types will adversely affect road holding and can lead to loss of vehicle control and personal injury.

■ Winter tires with studs should be run at moderate speeds when new in order to give the studs time to settle.

Do not drive a vehicle equipped with winter tires at prolonged high speed.

Winter tires use a softer rubber compound and a deeper tread pattern and therefore have a lower speed rating and reduced traction and durability when used on dry roads.

■ When winter tires or all season tires are mounted, make sure you observe the speed limitations for which the tire is rated. The following speed rating code letters are applicable:
Q – up to 100 mph (160 km/h)
T – up to 118 mph (190 km/h)
H – up to 130 mph (210 km/h)

WARNING

■ Winter tires available for your car may have a maximum speed rating which is less than the maximum speed of your vehicle.

■ Exceeding the maximum speed for which tires are rated and approved will lead to sudden tire failure which can cause loss of vehicle control and personal injury.

■ Never operate a vehicle at speeds greater than the maximum speed rating of the tires with which it is equipped. Always observe speed limits and adjust your vehicle to prevailing road and traffic conditions.

DIFFCULT OPERATING CONDITIONS

Snow chains

Snow chains can be used on the front wheels only.

Only use chains with fine pitch links protruding no more than ½ in/15 mm from tire tread and side walls, including tensioner. Wheels must rotate freely in all steering positions with chains mounted to prevent damage to body, axle or brake components. **Drive slowly and follow the chain manufacturer's instructions.**

Snow chains may not be used on the compact spare wheel.

Remove chains when roads are free of snow. Otherwise, they can be damaged and impair vehicle handling.

Driving under difficult conditions

Your Corrado was constructed and equiped for normal operating conditions. This also applies to the frequency and the extent of service requirements as stated in the Maintenance brochure.

If you are planning to drive your vehicle under difficult operating conditions (for example, continuous trailer towing, very hot or cold weather, very dusty conditions, poor fuel quality etc.) you may want to make special preparations such as changing to an appropriate oil viscosity, having your car thoroughly inspected, etc. Furthermore, the maintenance should be matched to the operating conditions (see page 6).

Operating your vehicle outside the U.S.A. or Canada

Government regulations in the United States and Canada require that automobiles meet specific emission regulations and safety standards. Therefore, vehicles built for the U.S.A. and Canada differ from vehicles sold in other countries.

If you plan to take your vehicle outside the continental limits of the United States or Canada, there is the possibility that

- unleaded fuels for vehicles with catalytic converter may not be available;
- fuel may have a considerably lower octane rating. Improper fuel may cause engine damage;
- service may be inadequate due to lack of proper service facilities, tools or testing equipment;
- replacement parts may not be readily available.

Volkswagen cannot be responsible for mechanical damage that could result because of inadequate fuel, service or parts availability.

European delivery program

Certain Volkswagen models are available for delivery in Europe under our tourist delivery and return shipment program.

For details consult your Volkswagen dealer or write to:

U.S.A. Volkswagen United States, Inc.
Tourist Delivery
P.O. Box 3951
Troy, MI 48007-3951

Canada Volkswagen Canada, Inc.
Tourist Delivery
1940 Eglinton Avenue East
Scarborough, Ontario
M1L 2M2

If you bought your vehicle abroad and want to bring it back home, be sure to inquire about shipping and forwarding requirements, as well as current import and customs regulations first.

Make sure the vehicle meets United States emission regulations and safety standards. Otherwise it may prove too costly or even impossible to add the equipment necessary to bring the vehicle into compliance with these regulations and standards.

Canadian regulations require that any car not complying to standards has to be shipped back to the car's country of origin at the owner's expense.

WINTER DRIVING

Remember the following points when driving in the winter:

■ Winter weather is particularly hard on the battery. We recommend having the battery inspected, prefer ably by a Volkswagen dealer, before the cold weather starts – see page 14.

■ If the vehicle is left standing for several weeks at extremely low temperatures, the battery should be removed – see page 14 for further details.

■ Have the antifreeze concentration in the cooling system checked before the cold weather starts – see page 12.

■ The engine oil must have the correct viscosity grade for the outside temperatures to be expected – see page 8.

■ The best way to protect the bodywork is to wash and wax it frequently, especially in winter – see page 3.

■ Always use a windshield washer fluid with antifreeze for the windshield washer system in winter – see page 15.

■ Use a plastic scraper to remove snow and ice from the windows and mirrors – see page 4.

■ On winter roads the use of winter tires or all-weather tires will improve handling – see page 18.

■ When driving in the mountains in winter it is best to take along a set of snow chains. Snow chains may be compulsory for some mountain roads – see page 19.

ADDITIONAL ACCESSORIES, MODIFICATIONS AND PARTS REPLACEMENT

The Volkswagen Corrado incorporates the latest safety design features ensuring a high standard of active and passive safety. This safety could be impaired by any non-approved changes to the original new-vehicle condition. For this reason, please observe the following points when installing additional accessories, if parts have to be replaced or if any modifications are made to the vehicle:

■ **Always consult** an authorized Volkswagen dealer **before** purchasing accessories and **before** any modification are carried out.

■ **In your own interest, we advise you to use only expressly approved Volkswagen accessories and genuine Volkswagen spare parts.**

These parts and accessories have been specially designed to be used on the Corrado.

■ Approved Volkswagen accessories and genuine Volkswagen parts are available from authorized Volkswagen dealers. Volkswagen dealers also have the necessary facilities to install the parts properly.

Mobile telephones and CB Radios

Mobile telephones and CB radios may not be used in the vehicle without having a separate outside antenna.

These devices emit a high frequency energy when they are being used. This energy can not penetrate the steel body and will be reflected inside the vehicle.

WARNING

Using a mobile telephone or a CB radio inside the vehicle without having a separate outside antenna could be dangerous to your health.

Note

The high frequency energy emitted inside the vehicle can also cause malfunctions in the vehicle's electronic system.

For these reasons, and in order not to restrict the range of the mobile telephone or CB radio, use a special outside antenna on the vehicle.

JACK AND TOOLS

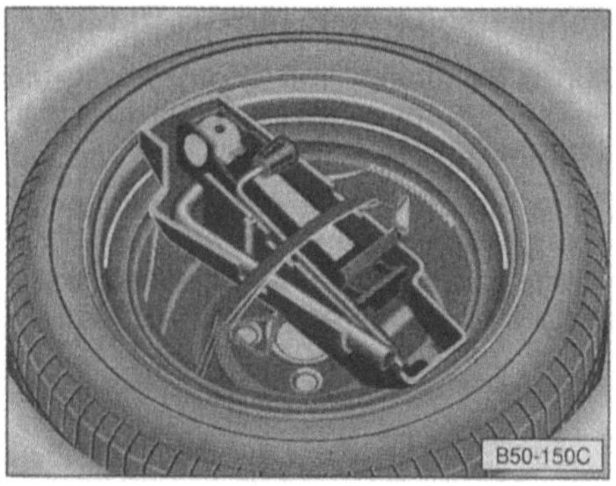

B50-150C

Jack and tools are attached to the rim of the spare wheel by a rubber strap.

A screwdriver with a reversible blade (flat blade on one end and Phillips head on the other) is inserted in the lug wrench.

Before returning the jack to its storage space, it must be fully wound down.

Always securely restow spare tire (or flat tire) and all jacking equipment and the hub cap.

WARNING

■ Always make sure jack and tools are secured in place and not loose, otherwise they could fly forward causing personal injury to passengers in the vehicle in an accident.

■ Use the jack only for changing a wheel. Never use the jack to lift other vehicles or other loads as this may lead to accidents and personal injury.

■ The jack must never be used as a support to work underneath the vehicle. If the jack is accidentally dislodged, you could be seriously injured.

■ Do not raise the vehicle using a bumper jack. The bumper system would be damaged. Also the jack may slip which could cause personal injury.

■ Do not support your car on cinder blocks, bricks or other props. These may crumble under continuous load.

■ Do not start or run the engine while the vehicle is supported by the jack.

■ If you must work under the vehicle, always use safety stands specifically designed for this purpose.

COMPACT SPARE WHEEL

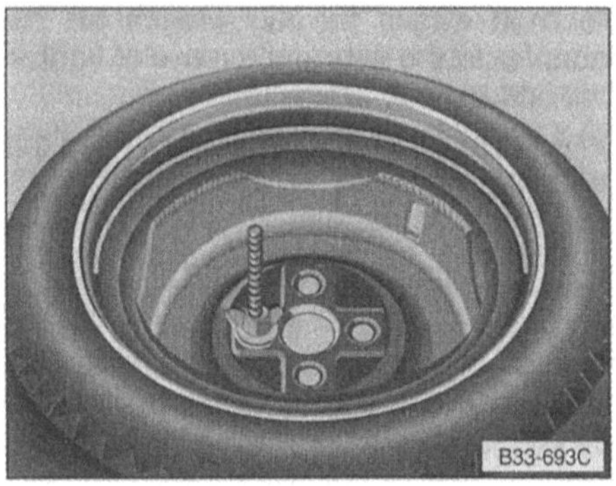

B33-693C

The compact spare wheel lies flat in a recess underneath the floor covering in the luggage compartment.

The wheel is secured with a plastic nut.

WARNING

Always tighten mounting bolt for the spare wheel securely. Unter no circumstances should you store the spare wheel unsecured in the luggage compartment. This could cause serious personal injury to passengers in the vehicle in case of a collision.

Check the inflation pressure of the compact spare (60 psi/4.2 bar) periodically to keep the tire ready for use.

When using the compact spare wheel, note the following:

■ After installing the compact spare wheel, the tire inflation pressure should be checked and, if necessary, increased to 4.2 bar.

■ The load carrying capacity is marked on the tire sidewalls and must not be exceeded.

■ The compact spare should be properly remounted in its storage area as soon as the standard road tire can be repaired or replaced.

■ Because the compact spare was specifically designed for your vehicle, it must not be used on any other vehicle. Likewise, compact spare wheels from other vehicles must not be used on your vehicle.

■ The compact spare tire must not be mounted on any other wheels, nor must standard tires, snow tires, wheel covers or trim rings be used with the compact spare wheel. If such use is attempted, damage to these items or other vehicle components may occur.

WARNING

■ The compact spare tire is for temporary use only and is so marked on the sidewalls.

■ Do not drive at speeds exceeding 50 mph (80 km/h) with the compact spare wheel mounted. Also rapid acceleration, hard braking and hard cornering must be avoided to help prevent loss of vehicle control!

■ Any continuous road use of the compact spare tire may result in tire failure, loss of vehicle control and injury to vehicle occupants.

■ Do not put a snow chain on your compact spare tire. Due to the smaller tire size, the chain will not fit properly and can thus damage your vehicle. You could also lose the chain, as well as control of the vehicle.

In the event of a flat front tire in road conditions which require snow chains, remove the undamaged rear road wheel on the same side first and mount the compact spare wheel there. Then remove the damaged front wheel, take off the snow chains and attach them to the rear road wheel prior to mounting on the front. Check and adjust tire pressure promptly.

■ Never use two or more compact spare tires at the same time.

CHANGING A WHEEL

WARNING

■ If you have a flat tire, move a safe distance off the road. Turn off the engine, turn the emergency flasher on and use other warning devices to alert other motorists.

■ Passengers must not remain in the vehicle when jacked up.

■ Make sure that passengers wait in a safe place away from the vehicle and well away from the roadway and traffic.

■ Before you change a wheel, be sure the ground is level and firm. If necessary, use a board under the jack.

■ Vehicle movement is not prevented when one driving wheel is lifted off the ground by just engaging a gear.

■ To help prevent the vehicle from moving suddenly and possibly slipping off the jack, always fully set the parking brake and block the wheel diagonally opposite the wheel being changed.

Step 1

■ Take jack, tools and spare wheel out of the luggage compartment.

B50-120C

Step 2

■ The alloy wheels have a plastic trim-piece which covers the wheel bolts. Lever the trim piece off using the flat blade the screwdriver (from the vehicle tools).

Step 3

Your vehicle may be equipped with theft-proof hub caps*. On each wheel the theft-proof hubcap has one wheel bolt which is not hexagon-shaped. Included in your vehicle tools, there is a special adapter to loosen and tighten this bolt.

■ First, place the adapter over the theft-proof bolt and turn slowly until the adapter completely engages into the coded slot.

■ Then mount the lug wrench on the adapter all the way and remove or tighten the bolt as described.

We recommend that you always keep the adapter with your vehicle tools. This is especially important if you take your vehicle to a workshop. Always keep the code card which comes with your adapter in a secure place – if possible not in the vehicle. If you should loose your adapter, see your authorized Volkswagen dealer.

■ Mount the lug wrench over the wheel bolt all the way and turn to the left – see illustration. When doing this, hold the lug wrench at the end.
If the wheel bolts do not come loose, you can push the end of the lug wrench, if necessary, using your foot. Make sure you are standing firmly on the ground, and hold on to the car for support.

■ Loosen all wheel bolts about one turn.

Do not yet remove the bolts.

WARNING

If your vehicle is equipped with BBS – light alloy wheels, never loosen the bolts which hold the two parts of the wheel together, as this could cause air leakage from the tire.
This can lead to loss of vehicle control and personal injury!

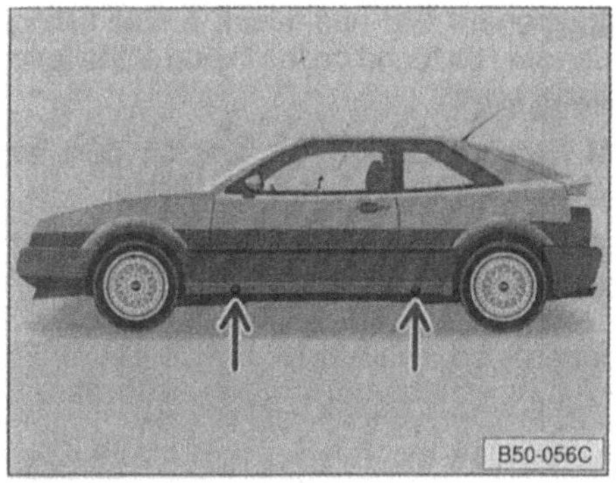

B50-056C

Step 4

■ To position the jack under the vehicle: There are two jack points on each side under the body marked with a triangular impression. See left illustration.

The front impression is located approx. 8 inches (20 cm) behind the front wheel house. The rear impression is located approx. 9 inches (23 cm) in front of the rear wheel house.

WARNING

■ Jacking at any other place may damage the vehicle or may result in personal injuries.

■ An unstable surface under the jack may cause the vehicle to slip off the jack. Always provide a firm base for the jack on the ground. If necessary, use a board under the jack.

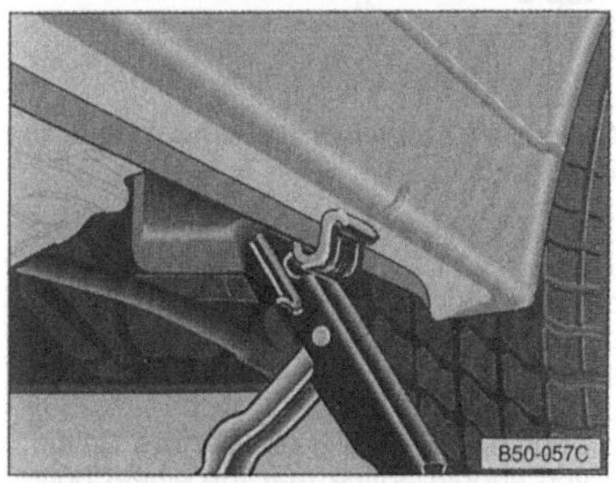

B50-057C

■ Turning the "T" piece on the jack by hand, raise the jack until the claw securely cradles the vertical rib underneath the vehicle so that the jack cannot slip when the vehicle is lifted. See illustration above.

■ Making sure the jack is securely in place, attach the jack handle (from the vehicle tools) on to the "T" piece of the jack.

Step 5

WARNING

■ **Do not raise the vehicle until you are sure the jack is securely engaged.**

■ **Passengers must not remain in the vehicle when jacked up.**

■ **Make sure that passengers wait in a safe place away from the vehicle and well away from the roadway and traffic.**

■ **To raise** the vehicle, turn the handle clockwise. Only raise the vehicle as much as is needed to change a wheel.

Step 6

■ Fully unscrew the wheel bolts and remove wheel. Remove any dirt or corrosion present on mounting surface of wheel or vehicle before wheel replacement. Place the spare wheel against the wheel hub so that the bolt holes in the wheel are in line with the threaded holes in the wheel hub.

■ Insert the wheel bolts and tighten them slightly crosswise before lowering the vehicle.

Wheel bolts must be clean and easy to turn – **never lubricate or oil wheel bolts.**

Step 7

■ **To lower** the vehicle, turn the handle counterclockwise until the jack is fully released.

Step 8

■ Then go crosswise from one bolt to another tightening them firmly with the lug wrench.

Correct tightness of the wheel bolts is important.

Correctly tightened bolts should have a torque of 80 ft lb (110 Nm). This torque can be obtained with the lug wrench by any person of average strength. If in doubt about the correct tightness of the wheel bolts, have it checked with a torque wrench by your dealer or a service station.

If you should notice that the wheel bolts are corroded and difficult to turn, replace them.

Drive carefully with special care until the bolts are replaced.

Step 9

■ Reinstall the protective caps for the wheel bolts.

Step 10

■ Fully lower the vehicle and remove jack.

■ Correct the air pressure of the wheel you have just put on. The cold tire inflation pressure is 60 psi/4.2 bar.

WARNING

Always store securely damaged wheel (or spare wheel), jack and tools in luggage compartment.

■ Have flat tire repaired and replaced promptly.

Remember that the compact spare is for temporary use only!

FUSES

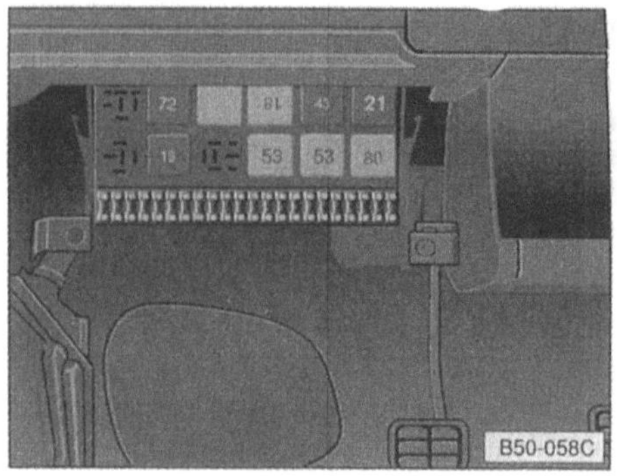

A problem in the electrical system may be caused by a blown fuse.

Fuses are arranged in one centralized unit. The unit is located on the left side below the dashboard, just behind the storage bin or behind the knee bar*.

It is a good idea to keep a supply of spare fuses on hand. They are available at your Volkswagen dealer.

Replacing a fuse

To avoid damaging the electrical system turn off all lights and accessories and remove the ignition key before replacing a fuse.

■ Consult the fuse listing on the following page to find out which fuse belongs to the component that has failed. A fuse listing can also be found on the inside of the fuse panel cover.

■ Remove blown fuse, which can be recognised by the burnt metal strip.

■ Replace the blown fuse with a fuse of equal ampere rating.

WARNING

Do not use a fuse with a higher amperage. This could damage the electrical part and fire could result.

If a fuse blows repeatedly, do not keep on replacing it. The cause of the short circuit or overload must be found. On no account should fuses be patched up with tin foil or wire as this may cause serious damage elsewhere in the electrical circuit or cause a fire.

FUSES

Fuse arrangement

from left to right:

	Amps
1 – Low beam, left	10
2 – Low beam, right	10
3 – Instruments and license plate lights	10
4 – Rear window wiper and sliding/prop-up roof	15
5 – Windshield wiper, windshield and rear window washer	15
6 – Fresh air fan, air conditioner	20
7 – Tail and side marker right	10
8 – Tail and side marker left	10
9 – Rear window and mirror heating	20
10 – Fog lights	15
11 – High beam left, high beam indicator light	10
12 – High beam right	10
13 – Horn and radiator fan run-on	10
14 – Back up lights , electric outside mirrors, heated windshield washer jets	15
15 – Engine electronic, additional electric coolant pump	10
16 – Warning/indicator lights, multi-function indicator, glove box light, cassette storage light, rear spoiler	15
17 – Turn signals	10
18 – Electric fuel pump, Oxygen sensor	20
19 – Radiator fan, air conditioner	30
20 – Brake lights, cruise control system	10
21 – Dome and luggage compartment light, cigarette lighter, central locking, multi-function indicator	15
22 – Radio	10

Separate fuses

located above the relays:

	Amps
Fuse for ABS hydraulic pump relay	30
Fuse for ABS main relay	30
Fuse for air conditioner	30
Fuse for power windows	20

Color codes

Red:	10 Amp fuses
Blue:	15 Amp fuses
Yellow:	20 Amp fuses
Green:	30 Amp fuses

The number on the face of the fuse is the amperage rating.

REPLACING BULBS

Before replacing a bulb, switch off the respective circuit.

Do not touch the glass part of the new bulb with bare fingers. Finger prints left on the glass evaporate when the bulb gets hot, the vapor settles on the reflector surface and dulls it.

Always use the same type of bulb. The designation is marked on the bulb.

It is advisable to always carry a box of spare bulbs in the vehicle. Bulbs can be obtained from any Volkswagen dealer.

The box should contain, at least, the following bulbs which are important for traffic safety:

12 V/70/50 W	– Headlight
12 V/4 W	– Parking light
12 V/21 W	– Front and rear turn signal
12 V/21/5 W	– Brake and tail light
12 V/5 W	– Side marker light, center stop light and license plate light

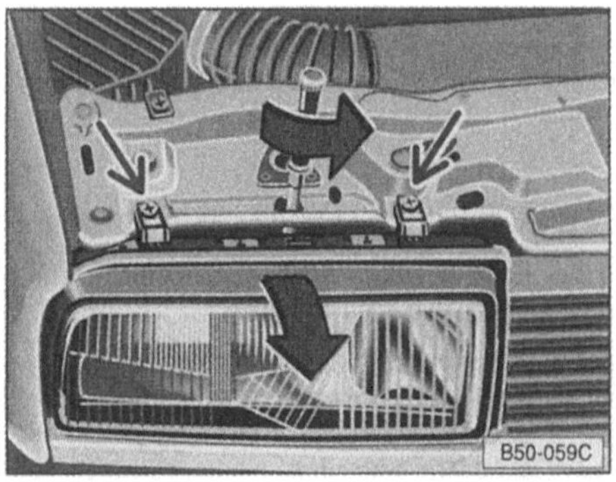
B50-059C

Headlights and Parking lights

You must remove the headlight to change a bulb.

■ Open engine hood and prop up.

■ Remove the two outer Phillips screws (see illustration above) and loosen the center screw a few turns.

■ Pull headlight forward slightly at the top and lift it out.

After changing the defective bulb (see next column) reinstall headlight and have the headlight setting checked – see also page 27.

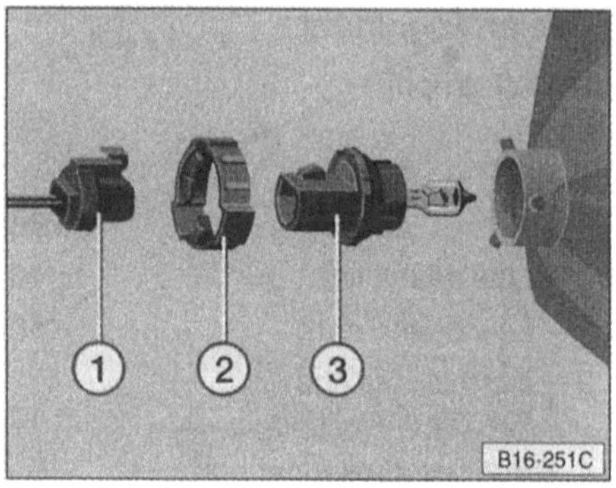

B16-251C

Headlight bulb

■ Disconnect wire connector (1)

■ Twist lock ring (2) counterclockwise, pull bulb (3) out of headlight housing and discard.

■ Install new bulb so that the locating lugs on the headlight housing engage the recesses on the bulb holder.

■ Reinstall lock ring so that its recesses engage the locating lugs on the headlight housing.

■ Twist lock ring clockwise until firmly seated.

■ Reconnect wire connector.

B50-094C

Parking lights

■ Turn bulb holder counterclockwise and pull out.

■ Press bulb gently into holder, turn and take out.

■ Install new bulb.

■ Push bulb holder into light housing and turn clockwise until firmly seated.

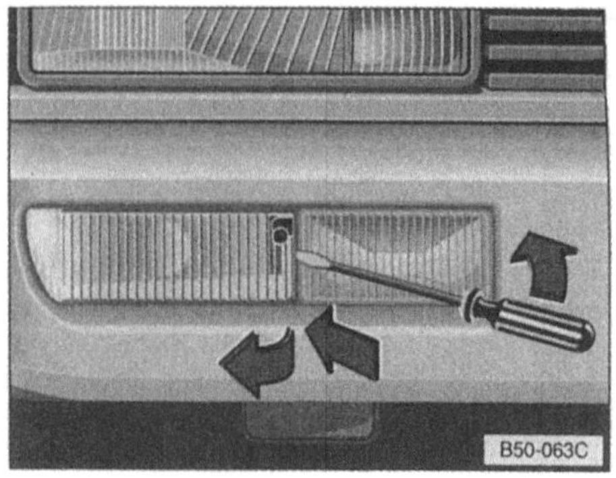
B50-063C

Front turn signals

■ Carefully lever light housing out using the flat blade screwdriver (from vehicle tools) as shown in the illustration above.

■ Turn the bulb holder slightly to the left and remove.

■ Press the defective bulb into socket, turn the left and remove.

■ Install the new bulb into socket and turn to the right until it stops.

■ Reinstall the bulb holder into light housing and turn slightly to the right until it stops.

■ Reinstall light housing by first fitting the outer edge into place, then pressing the inner edge in firmly until both detent lugs engage.

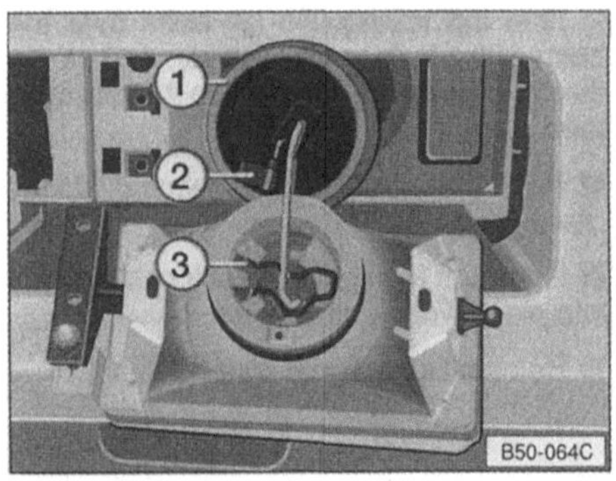

B50-064C

Fog lights

■ First remove turn signal light housing (see left colums).

■ Remove both screws are now visible and pull fog light out.

■ Remove plastic cover (1).

■ Disconnect plug (2) from the wire connector.

■ Unhook spring clip (3) on the bulb holder and fold it over to the side.

■ Remove the Halogen bulb. Install new bulb so that the locating lug in the reflector matches the recess in the bulb base.

■ Fold the spring clip (3) back over the base of the bulb, press the clip ends together and let them engage into the retaining tabs.

■ Reconnect plug (2) and reinstall plastic cover (1).

■ Remount fog light so that it securely fits into place. Reinstall screws.

■ Reinstall turn signal light housing.

Side marker light bulbs

■ Remove Phillips screw.

■ Pull light housing away from bumper sliding toward rear (arrow), until tongue on housing disengages from slit in bumper.

■ Turn socket counterclockwise and pull out of housing.

■ Remove defective bulb and replace. Turn socket clockwise until seated securely.

■ When reinstalling the light housing, slide tongue into slit in fender first, then fasten the screw.

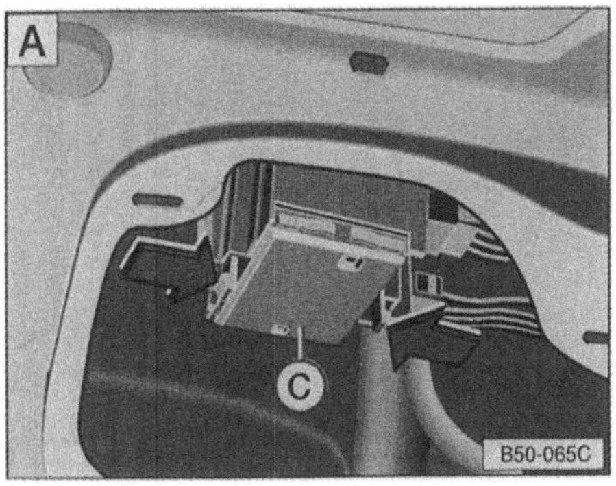

Taillights

The taillights are arranged as follows:

The back-up lights are located **in the rear lid (A).**

The rear turn signals, brake lights and taillights are located **in the vehicle body (B).**

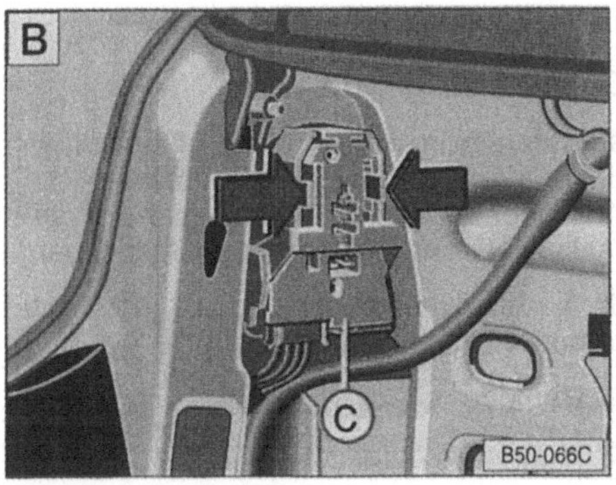

Accessing bulbs

■ Open rear lid.

■ Turn the six fasteners located at the bottom and side of the rear lid trim panel one quarter turn counterclockwise. Remove trim panel.
The taillights located in the body are not covered by a trim panel.

■ Press both spring tags (arrows) and remove the bulb holder.

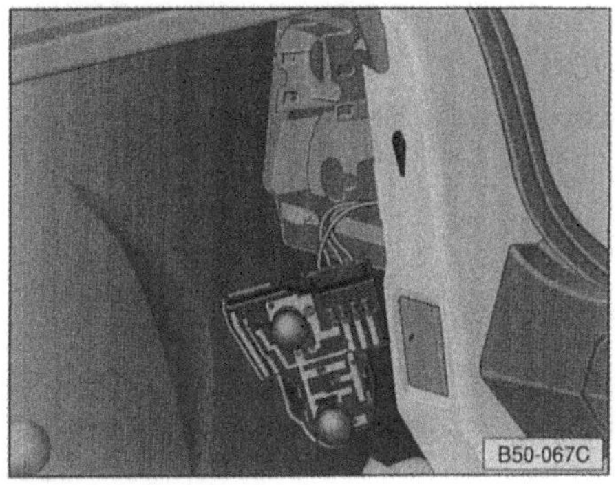

Replacing bulbs

■ Press defective bulb into holder, turn to the left and remove.

■ Install new bulb and turn to the right until it stops.

■ Remount bulb holder making sure the spring tabs are securely engaged.

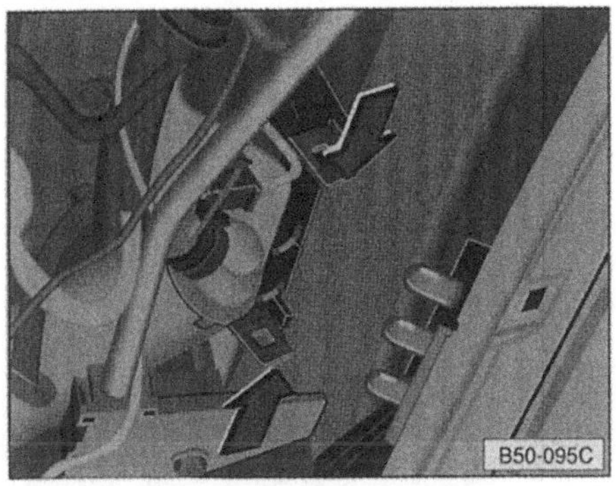

Center stop light in rear lid*

■ Remove trim panel as described before.

■ Squeeze clips (arrows) on both sides of stop light housing inwards and take housing off.

■ Pull defective bulb out and replace.

■ Reinstall housing and trim panel.

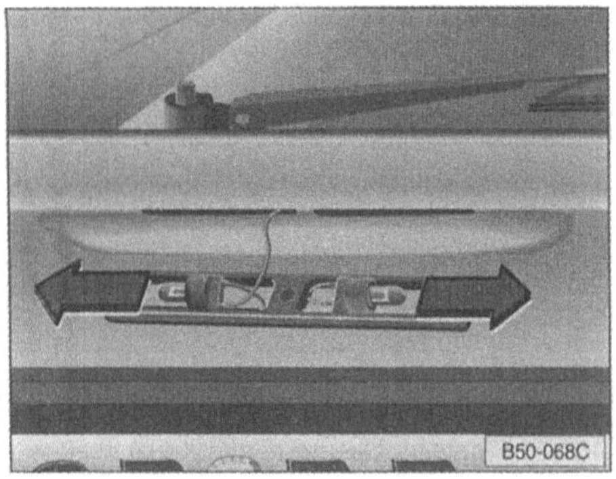

License plate light bulbs

- Remove cover.
- Pull complete light out and remove defective bulb.
- Insert new bulb.
- Install light.
- Attach cover with screws.

Luggage compartment light

- Insert screwdriver in recess at side of light and lever light out carefully.
- Change bulb.
- Insert light at the cable connecting end first.

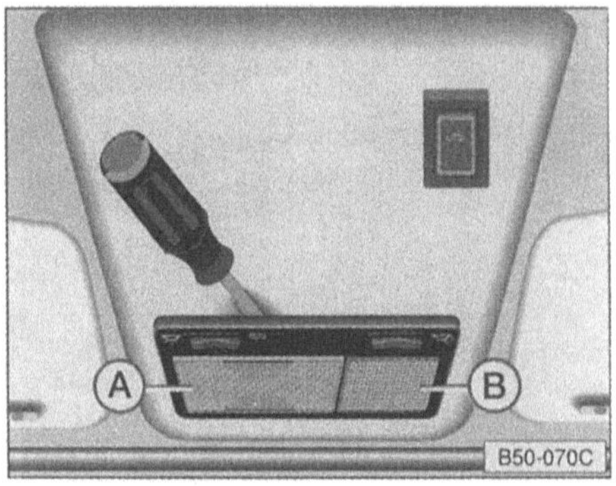

A – Dome light

- Insert flat blade screwdriver (from vehicle tools) between the light housing and the lens and carefully lens pry off.
- Replace defective bulb.
- Press lens onto the housing. The tab on the lens must fit into the recess on the left of the housing.

B – Reading light

- Carefully insert first blade of the screwdriver (from vehicle tools) between the headliner and the light housing and pry off making sure not to damage the headliner.
- Turn the white bulb holder to the left and remove from light housing.
- Remove defective bulb.
- Insert new bulb.
- Install bulb holder into light housing and turn to the right until it stops.
- Press light housing into headliner until it securely fits into place.

HEADLIGHT ADJUSTMENT

The proper headlight adjustment is very important for traffic safety.

Adjust the headlight using the Phillips screwdriver from the vehicle tools.

Both illustrations show the adjustment on the right headlight and fog light. On the left headlight and fog light the adjustments are symmetrically opposite, however, there is no cover behind the headlight.

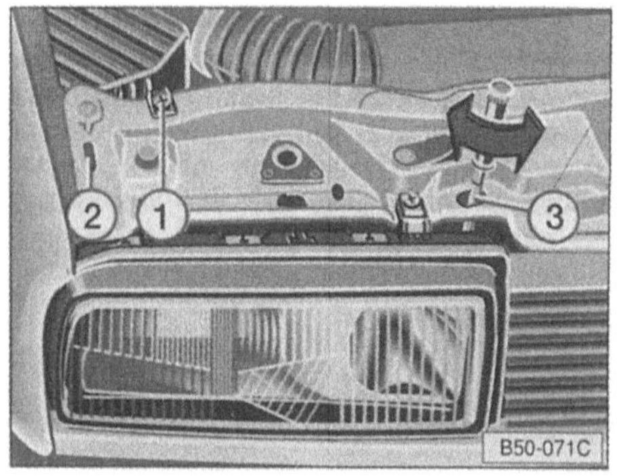

Headlights – height adjustment

- Remove screw (1).
- Remove cover.
- To reduce the beam range, turn the white plastic knob (can be seen through opening 2) counterclockwise.
- Reinstall cover and screw (1).

Headlights – lateral adjustment

- Insert screwdriver into hole (3) and turn white plastic knob.

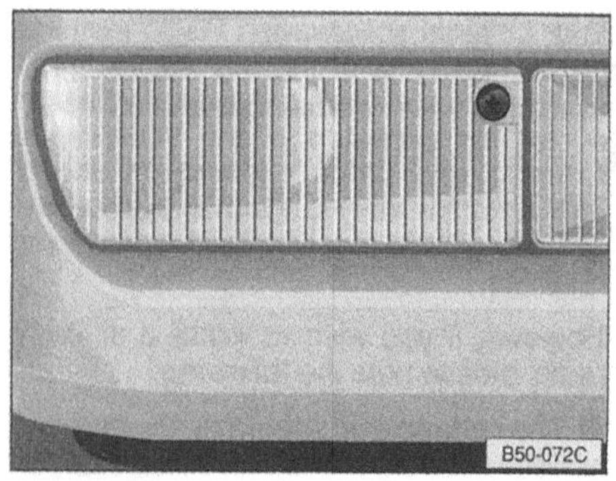

Fog lights

- To reduce beam range, turn screw counterclockwise.

You cannot adjust the fog lights laterally.

REPLACING THE RADIO

Your Original Volkswagen Radio installed at the factory is of the highest technical quality. The speakers are specially designed for the acoustics of the Volkswagen Corrado and produce excellent sound. The system is easy to operate and possesses exceptional receiving and reproducing qualities.

However, if you wish to install a different radio, please note the following:

■ The connectors, speakers, the antenna and the suppressor are designed for Original Volkswagen Radios. When installing a different radio, difficulties could arise such as:
– the radio may not fit into the space provided
– the electrical connections may not be compatible
– different connector terminals may be needed

■ Therefore, we recommend that you have your authorized Volkswagen dealer remove and install the radio. There are the most familiar with the technical features of your vehicle. They also offer Original Volkswagen Radios with the necessary installation components and instructions.

WARNING

Improperly installing a radio could cause a short circuit. This could result in an electrical fire.

EMERGENCY STARTING

Starting by pushing or towing

Vehicles with Automatic Transmission cannot be started by pushing or towing.

Vehicles with Manual Transmission must not be started by pushing or towing. Damage to the catalytic converter and/or other parts of the vehicle may result.

Starting with jumper cables

WARNING

■ Always shield your eyes and avoid leaning over the battery whenever possible.

■ Do not allow battery acid to contact eyes or skin. Flush any contacted area with water immediately.

■ Improper use of a booster battery to start a vehicle may cause an explosion.

■ Vehicle batteries generate explosive gases. Keep sparks, flame and lighted cigarettes away from batteries.

■ Do not try to jump start any vehicle with a low acid level in the battery.

■ The voltage of the booster battery must also have a 12-volt rating. The capacity (Ah) of the booster battery should not be lower than that of the discharged battery. Use of batteries of different voltage or substantially different Ah rating may cause an explosion and personal injury.

■ Do not charge a frozen battery; allow it to thaw out first. Gas trapped in the ice may cause an explosion.

■ Applying a higher voltage booster battery will additionally cause expensive damage to sensitive electronic components, such as relays, radio, etc.

■ Vehicle with discharged battery: turn off lights and accessories, move lever to **N** or **P** (Automatic transmission) or into Neutral (Manual transmission) and set parking brake.

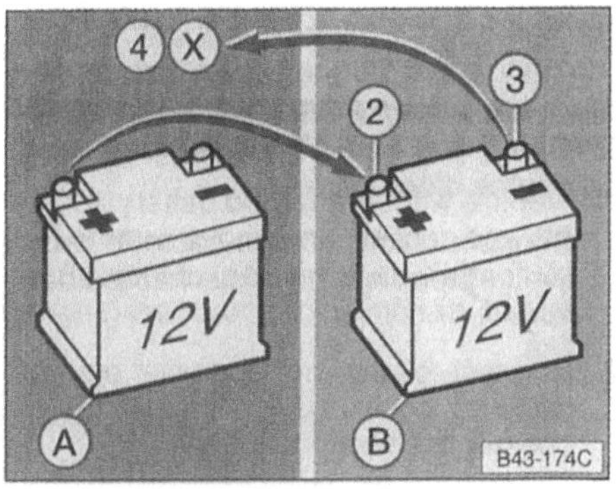

A – Discharged battery
B – Booster battery
X – To ground on engine block

Use of jumper cables

Warning

■ To avoid serious personal injury and damage to the vehicle, heed all warnings and instructions of the jumper cable manufacturer. If in doubt, call for road service.

■ The jumper cables must be long enough so that the vehicles do not touch.

■ When connecting jumper cables, make sure that they cannot get caught in any moving parts in the engine compartment.

Always heed WARNINGS on page 7.

Improper hook-up of jumper cables can ruin the alternator.

Always connect POSITIVE (+) to POSITIVE (+), and NEGATIVE (–) to ground on engine block (x).

1. Connect clamp of plus-cable to positive (+) terminal (1) of discharged battery.
2. Connect clamp on opposite end of cable to positive (+) terminal (2) of booster battery.
3. Connect clamp of minus-cable to negative (–) terminal (3) of booster battery.
4. Connect clamp on opposite end of cable (4) to a bare metal part bolted directly to the engine block or to the engine block itself (x) of vehicle with discharged battery. Connect clamp as far away from battery as possible!
5. Start the engine of the vehicle with the booster battery. Run the engine at a moderate speed.
6. Start engine with discharged battery in the usual manner. If engine fails to start, do not continue to crank but contact nearest workshop.
7. With engine running, remove jumper cables from both vehicles in exact reverse order: Steps 4 through 1.

EMERGENCY TOWING WITH COMMERCIAL TOWTRUCK

The following information is to be used by commercial tow truck operators who know how to operate their equipment safely.

Detailed towing instructions can be found in a special towing instruction brochure at your Volkswagen dealer.

General hints

■ Whenever possible, tow with front wheels off ground.

■ Vehicle may be lifted in rear and moved to position for front hook-up.

■ If the vehicle must be towed with the front wheels on the ground, make sure that transmission oil has not leaked or been drained.

Do not exceed 50 mph or 80 km/h nor a distance of 50 miles or 80 kilometers.

Note

The Volkswagen Corrado **cannot** be towed with conventional sling-type equipment or wheel dollies. Towing with this type of equipment will cause bumper and body panel damage.

WARNING

Never allow passengers to ride in a towed vehicle for any reason.

Front hook-up

■ Attach wheel lift equipment to wheels.

■ Attach safety straps to wheels.

■ Attach safety chains to lower control arms.

■ Towing clearance: 6–12 inches or 15–30 cm.

Rear hook-up

■ Attach wheel lift equipment to wheels.

■ Attach safety straps to wheels.

■ Attach safety chains to axle beam.

■ Towing clearance: 6–12 inches or 15–30 cm.

■ Towing speed: 50 mph or 80 km/h
Towing distance: 50 miles or 80 km.

Note

Vehicles with automatic transmission **must** be transported on a flat bed!

LIFTING VEHICLE

The vehicle should never be lifted or jacked up from underneath the engine oil pan, the transmission housing, or the front or rear axle. This could lead to serious damage.

The same lifting points as illustrated for the hoist also apply when using a floor jack. To avoid damage to the underbody or chassis frame, it is necessary to insert a rubber pad between the floor jack and the lift points.

Lifting with workshop hoist and floor jack

Make sure there is sufficient clearance between pads and vehicle before driving vehicle on to hoist, especially if the vehicle has a large front panel or spoiler.

The vehicle must be lifted only at the lift points illustrated.

WARNING

■ When removing components such as engine block, transmission housing, fuel tank, wheels, front or rear axle, anchor vehicle to hoist or add corresponding weights to maintain the center of gravity. Otherwise the vehicle might tilt or slip off the hoist, causing serious damage or personal injury.

■ To reduce the risk of serious personal injury and vehicle damage, lift vehicle only at the special workshop

hoist and floor jack lift points illustrated. Failure to lift vehicle at these points could cause the vehicle to tilt or fall from a lift when, for example, heavy components such as the engine block or transmission are removed resulting in a change in vehicle weight and balance.

■ If you must lift your vehicle with a floor jack to work underneath, be sure the vehicle is safely supported on stands intended for this purpose.

Lifting points

When using a workshop hoist or a floor jack, make sure you lift your vehicle at the points shown in the illustrations above. The center of the workshop hoist or jack must be aligned with the center of the depression in the side member. (See illustrations).

Front lifting point

On the floor panel reinforcement and the vertical ridge of the side member (left illustrations).

Rear lifting point

At the floor panel reinforcement (right illustration).

Be careful not to damage the parking brake cables with the workshop hoist or jack when lifting the vehicle.

Lifting with vehicle jack

Refer to "Changing a wheel" on page 22.

ENGINE (4 CYLINDER)

■ Four stroke gasoline engine, transversely mounted.

■ Four cylinder in line.

■ Crankshaft with five main bearings.

■ Light alloy cylinder head.

■ Spur-belt driven overhead camshaft.

■ Self-adjusting hydraulic valve lifters.

■ Liquid cooling system, thermostatically controlled.

■ Thermoswitch operated electric fan.

■ Pressure oil feed with gear-type pump and full flow filter.

■ Digital electronic ignition with knock sensor.

■ Mechanically driven G-Charger with charge air cooling.

■ Digital controlled fuel injection system (Digifant).

■ Long-life, low maintenance ribbed V-belts.

■ Paper element air cleaner.

■ Exhaust emission control system with activated charcoal filter in the fuel system.

Digital electronic ignition system

Your engine is equipped with a digital electronic ignition system with knock sensor.

The correct timing for optimum efficiency under all engine operating conditions is calculated by a "timing map" stored in the system's memory. The timing map contains the ignition timing value for any particular engine rpm and engine load.

The combustion process in the engine is monitored by knock sensor. If engine knock (pinging) occurs, the timing for the appropriate cylinder is adjusted briefly by a corresponding amount (knock sensor).

With knock sensor adjustment, the electronic ignition, together with the fuel injection system, provides maximum engine power, low exhaust emissions and reduced fuel consumption.

G-Charger

The engine of your Volkswagen Corrado G60 is equipped with a newly developed, mechanically driven G-Charger.

The name "G-Charger" was derived from the shape of the compressor charger with resembles the letter "G". The number 60 stands for the width of the compressor channel.

The G-Charger increases engine output by approximately 50 %.

A special advantage is a noticeable increase in engine torque that even at low engine speeds.

Since the compressed air from the working chamber of the G-Charger overlaps as it travels to the pressure side, there is never a sudden interruption in the air flow. The charging is therefore very steady and the noise level is low.

You will note the spontaneous power boost when accelerating along with reduced fuel consumption.

The drive train, brakes, clutch, cooling systems etc. have all been matched with the high engine output.

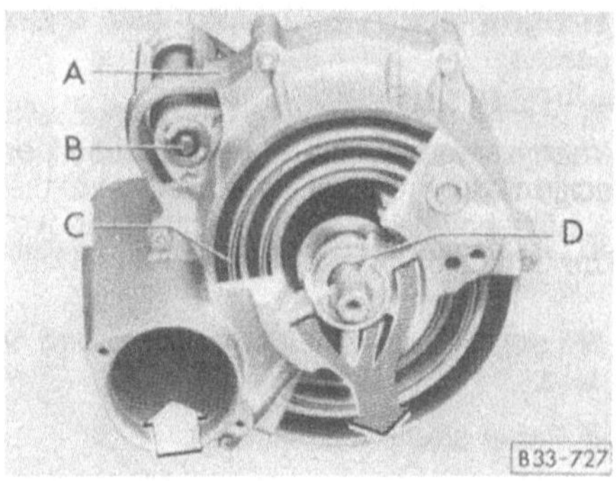

A – Channel housing
B – Secondary shaft
C – Compressor
D – Eccentric shaft

Technical description of the G-Charger

The G-Charger consists essentially of four main parts:

A two-piece channel housing with two spiral shaped channels in which a double-acting compressor is moved on an eccentric shaft via a secondary shaft.

The eccentric shaft is driven by a ribbed V-belt from the engine crankshaft.

The circular movement of the compressor reduced the size of the compression space. The air is forced uniformly into the housing center and then via a charge air cooler to the cylinders.

ENGINE (6 CYLINDER)

■ Four stroke gasoline engine, transversely mounted.

■ Six cylinder V-configuration (cylinder angle 15° in a compact row)

■ Cast iron cylinder block.

■ Seven main bearing crankshaft

■ Steel oil pan

■ Alloy cylinder head

■ Two valves per cylinder

■ Chain driven double overhead camshafts

■ Maintenance-free self adjusting hydraulic valve lifters

■ Radiator with separate reservoir

■ Cooling system with an additional electric coolant pump

■ Thermoswitch operated electric fan

■ Dynamic engine oil pressure indicator

■ Electronic engine control (dual control of fuel and ignition systems)

■ Ignition system with
– stored "timing map"
– dual knock sensors (if engine knock occurs, the timing for the appropriate cylinder is adjusted briefly by a corresponding amount

■ Long-life spark plugs

■ Long-life, low maintenance ribbed drive belts

■ Low maintenance battery

■ Paper element air cleaner

■ Long life exhaust system

■ Exhaust emission control system with oxygen sensor

EMISSION CONTROL SYSTEM

❀ ***The emission control system effectively reduces the amount of pollutant in the exhaust gas.***

In the interest of clean air

Pollution of our environment has become a problem that is of increasing concern to all of us. We urge you to join us in our efforts for cleaner air in controlling the pollutants emitted from the automobile.

You can make a significant contribution to keeping our air clean by:

- Always using unleaded gasoline.
- Not letting your vehicle idle unnecessarily.
- Following the recommended preventive maintenance schedule contained in the maintenance booklet.
- Not removing or altering the emission control system.
- Taking the precautions described below to prevent damage to the emission control system in your vehicle.

Volkswagen warrants the Emission Control System in your new vehicle under the terms and conditions set forth in the Warranty booklet.

Your Volkswagen is equipped with an emission control system, which contains the following major components:

Catalytic converter

The catalytic converter is an efficient "clean-up" device built into the exhaust system of the vehicle. The catalytic converter burns the undesirable pollutants in the exhaust gas before it is released into the atmosphere.

The exclusive use of unleaded fuel is critically important for the life of the catalytic converter and proper functioning of the engine.

The catalytic converter will be permanently damaged by

- **exceeding the correct engine oil level (see page 9)**
- **push or tow starting your vehicle**
- **misfiring of the engine**
- **turning off the ignition while the vehicle is moving or**
- **other unusual operating conditions.**

Do not continue to operate your vehicle under these conditions, as otherwise fuel can reach the catalytic converter. This could result in overheating of the converter, requiring its replacement.

Oxygen Sensor (OXS)

The oxygen sensor, installed in the exhaust manifold, continuously senses the oxygen content of the exhaust and signals the information to an electronic control unit.

Carbon canister

❀ ***The vehicle is equipped with a carbon canister to prevent fuel vapor from escaping into the atmosphere.***

The vapor is passed into a container filled with activated charcoal, where it is retained when the engine is turned off. When the engine is running, a valve allows air into the canister, and the fuel vapor is passed to the engine for combustion.

The system does not require any maintenance.

WARNING

Parking

Do not park or operate the vehicle in areas where the hot exhaust system may come in contact with dry grass, brush, fuel spill or other material, which can cause a fire.

Undercoating

Do not apply additional undercoating or rustproofing on or near the exhaust manifold, exhaust pipes, catalytic converter or heat shields. During driving, the substance used for undercoating could overheat and cause a fire.

To assure efficient operation of the Emission Control System:

- **Have your vehicle maintained properly and in accordance with the service recommendations as described in your Maintenance booklet. Lack of proper maintenance as well as improper use of the vehicle will impair the function of the emission control system and could lead to damage.**
- **Do not alter or remove any component of the Emission Control System unless approved by the manufacturer.**
- **Do not alter or remove any device, such as heat shields, switches, ignition wires, valves, which are designed to protect your vehicle's emission control system.**

Note

Do not leave engine idling unattended after starting. If warning lights should come on to indicate improper operation, they would go unheeded. Extended idling also produces heat, which could result in overheating or other damage to the vehicle or other property.

TRANSMISSION

Manual Transmission

■ Hydraulic operated single plate dry clutch.

■ Synchronized five-speed transmission and differential in one housing with common lubrication.

■ Front wheel drive, with two constant velocity joints per drive shaft.

Automatic Transmission

■ Hydraulic torque converter with elimination of converter slip in 3rd gear (under certain conditions) and always in 4th gear.

■ Planetary gear with four forwards gears.

■ Two electronically controlled driving programs.

■ Final drive in separate housing.

■ Front wheel drive, with two constant velocity joints per drive shaft.

STEERING, SUSPENSION

Steering

■ Power assisted rack and pinion steering with energy absorbing steering wheel and column.

■ Steering gear and steering column connected by an angled shaft with two universal joints.

■ Tilt steering wheel

Front wheel suspension

■ Independent coil spring/shock absorber struts.

■ Stabilizer bar.

■ Negative steering roll radius.

Rear wheel suspension

■ Torsion beam axle, coil springs and shock absorbers.

■ Stabilizer.

■ Track correcting axle mountings to improve handling when cornering.

BRAKES

■ Hydraulic dual diagonal brake circuit on vehicles with ABS hydraulic three circuit brake system (front wheels individually, rear wheels together).

■ Electronically controlled anti-locking brake system (ABS)

■ Front: Ventilated disc brakes
Rear: Disc brakes

■ Power braker

■ Hydraulic power assisted brakes with electric pump on ABS system

■ Load-sensitive brake pressure regulator for rear axle

■ Mechanical parking brake effective on rear brake discs

BODY/CHASSIS

- All steel unitized body/chassis.
- Bolt-on front fenders.
- Front and rear ends designed to help absorb impact energy.

Speed sensitive rear spoiler

The air flowing around a vehicle traveling at high speeds has a lifting effect which tends to reduce the pressure of the wheels on the road surface. On vehicles with modern streamlined bodies, this undesirable lift is offset by aerodynamically designed front and rear ends, or by the installation of additional spoilers.

A large spoiler or high rear end with an air cut-off edge can interfere with the vision to the rear.

On the Corrado, this problem is solved by the speed sensitive rear spoiler. At high speeds, the rear spoiler automatically extends. This increases the pressure on the rear wheels. When driving slowly, the spoiler automatically retracts. This improves the view to the rear when maneuvering and parking.

ENGINE DATA/SPARK PLUGS (4 CYLINDER)

Engine

Maximum output SAE net	158 hp at 5600 rpm
Maximum torque SAE net	166 ft. lbs at 4000 rpm
No. of cylinders	4
Displacement	109 CID/1781 cm^3
Stroke	3.40 in (86.4 mm)
Bore	3.19 in (81.0 mm)
Compression ratio	8.0 : 1
Fuel[1)]	Premium[2)] or Regular unleaded
Spark plugs	
Bosch	W6 DPO

Notes

Spark plugs are replaced during the Volkswagen scheduled Maintenance service. If you replace the spark plugs between the Volkswagen Maintenance services, the following should be noted:

- Engine, spark plugs and the ignition system are matched. To avoid faulty operation or engine damage, use only Original Volkswagen spark plugs. It is especially important to note the number of electrodes and heat value.

[1)]Minimum octane rating and further details see page 1.
[2)]For maximum engine performance unleaded Premium is recommended.

- Since spark plug specifications may change for technical reasons during a current model year, we recommend that you obtain your spark plugs from your Volkswagen dealer who has the latest information.

ENGINE DATA/SPARK PLUGS (6 CYLINDER)

Engine

Maximum output SAE net	178 hp at 5800 rpm
Maximum torque SAE net	173 ft. lbs at 4200 rpm
No. of cylinders	6
Displacement	170 CID/2792 cm^3
Stroke	3.56 in (90.3 mm)
Bore	3.19 in (81.0 mm)
Compression ratio	10.0 : 1
Fuel[1)]	Premium[2)] or Regular unleaded
Spark plugs	
Bosch	C 9 MCC

Notes

Spark plugs are replaced during the Volkswagen scheduled Maintenance service. If you replace the spark plugs between the Volkswagen Maintenance services, the following should be noted:

■ Engine, spark plugs and the ignition system are matched. To avoid faulty operation or engine damage, use only Original Volkswagen spark plugs. It is especially important to note the number of electrodes and heat value.

[1)]Minimum octane rating and further details see page 1.
[2)]For maximum engine performance unleaded Premium is recommended.

■ Since spark plug specifications may change for technical reasons during a current model year, we recommend that you obtain your spark plugs from your Volkswagen dealer who has the latest information.

V-BELTS (4 CYLINDER)

Power steering/waterpump[1)]	9.5 × 735
G-charger, alternator (Multiple V-belt)	6 PK 1755

[1)] This belt is essential to engine operation. It is advisable therefore to always carry a spare belt in the vehicle to allow prompt replacement.

Note

The V-belts are among the most severely stressed parts of a vehicle. The belts therefore have very high quality requirements.

When replacing a belt, it is not sufficient to use just any belt of the same size. For safe operation, use only Original Volkswagen V-belts specially designed for your vehicle. The correct belts can be obtained at your Volkswagen dealership.

CAPACITIES/DIMENSIONS (4 CYLINDER)

Capacities

Fuel tank	approx. 14.5 gal/55 liters
Reserve (of total capacity)	2.6 gal/10 liters
Cooling system	1.7 gal/6.5 liters
Windshield/rear window washer container	4.4 qt/4.2 liters
Engine oil	
with filter change	4.3 qt/4.0 liters
without filter change	3.7 qt/3.5 liters

Do not fill above "max" mark.
The oil level should be checked while topping up.

Difference between "min" and "max" marks on the dipstick is . . . approx. 0.75 qt/0.75 liter

Dimensions

Dimensions	in	mm
Length	159.4	4048
Width	65.9	1674
Height (unloaden)	51.9	1318
Overhang, front	34.5	876
Overhang, rear	27.5	697
Wheelbase	97.5	2475
Front track	56.4	1433
Rear track	56.2	1428
Ground clearance[1)] with gross vehicle weight	4.6	116

Turning circle diameter approx. 34.4 ft/10.5 m (wall to wall).

[1)] When driving up steep ramps, on rough roads, over curbs, etc. it is important to remember that some parts of your car are close to the ground, such as spoilers or exhaust system components. Be sure to avoid damage.

CAPACITIES/DIMENSIONS (6 CYLINDER)

Capacities

Fuel tank	approx. 18 gal/70 liters
Reserve (of total capacity)	2.6 gal/10 liters
Cooling system	9.1 qt/8.6 liters
Windshield/rear window washer container	4.4 qt/4.2 liters
Engine oil	
with filter change	6.3 qt/6.0 liters
without filter change	5.3 qt/5.0 liters

The oil level should be checked while topping up.

Do not overfill (see page 9).

Dimensions

Dimensions	in	mm
Length	159.4	4050
Width	66.5	1690
Height (unloaden)	52.0	1320
Overhang, front	34.6	880
Overhang, rear	27.4	695
Wheelbase	97.5	2475
Front track	57.1	1450
Rear track	56.4	1432
Ground clearance[1)] with gross vehicle weight	5.1	129

Turning circle diameter approx. 36.0 ft/11.0 m (wall to wall).

[1)] When driving up steep ramps, on rough roads, over curbs, etc. it is important to remember that some parts of your car are close to the ground, such as spoilers or exhaust system components. Be sure to avoid damage.

WEIGHTS

The vehicle capacity weight, the Gross Vehicle Weight Rating (GVWR), and the Gross Axle Weight Ratings (GAWR) for front and rear, are listed on a sticker on the left door jamb.

The Gross Vehicle Weight Rating includes the weight of the basic vehicle plus full fuel tank, oil and coolant, plus max. load which combines passenger (150 pounds/68 kg per designated seating position) and the luggage weight.

Luggage weight is not increased by the use of a roof rack, unless the passenger capacity is reduced accordingly.

The Gross Axle Weight Rating is the maximum load that can be applied at each axle of the vehicle.

WARNING

- **The actual Gross Axle Weight Rating at the front and rear axles should not exceed the permissible weights, and their combination must not exceed the Gross Vehicle Weight Rating.**
- **Exceeding permissible weight ratings can result in vehicle damage, accidents and personal injury.**

Roof weight

The maximum permissible roof weight is 110 lb/50 kg. Only use roof racks which have been specifically designed for this vehicle.

Only use the roof rack system tested and approved by Volkswagen.

Distribute the load evenly and do not exceed the permissible roof weight (including the weight of the roof rack) or the gross vehicle weight.

Maximum Permissible Trailer Weights

(braked or unbraked trailer)	450 kg	(1.000 lb.)
Tongue load	50 kg	(110 lb.)

Do not exceed the weight limitations of your trailer hitch.

Notes

The vehicle capacity weight figures apply when the load is distributed evenly in the vehicle (passengers and luggage). When transporting a heavy load in the luggage compartment, the load should be carried as near to the rear axle as possible so that the vehicle's handling is not impaired. Do not exceed the maximum permissible axle loads or the maximum gross vehicle weight. Always remember that the vehicle's handling will be affected by the extra load, therefore, adjust your speed accordingly.

VEHICLE IDENTIFICATION

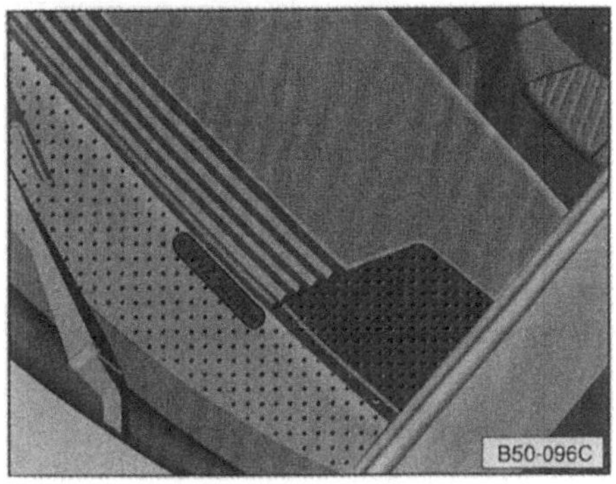

The Vehicle Identification Number (VIN)

is located on the instrument panel on the driver's side so that it is visible from the outside through the windshield.

By law, labels are attached to some vehicle parts, which also bear the vehicle identification number.

Do not remove the labels.

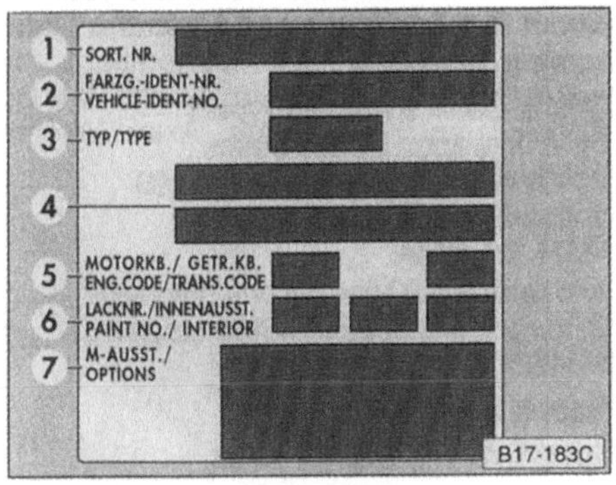

The identification label

is located on the left at the rear panel of the luggage compartment. The label contains the following information:

1. Production control no.
2. Vehicle identification no.
3. Type code number
4. Type designation
5. Engine and transmission code letter
6. Paint no./Interior
7. Optional equipment no.'s.

Vehicle data 2 to 7 are also found in your Maintenance booklet.

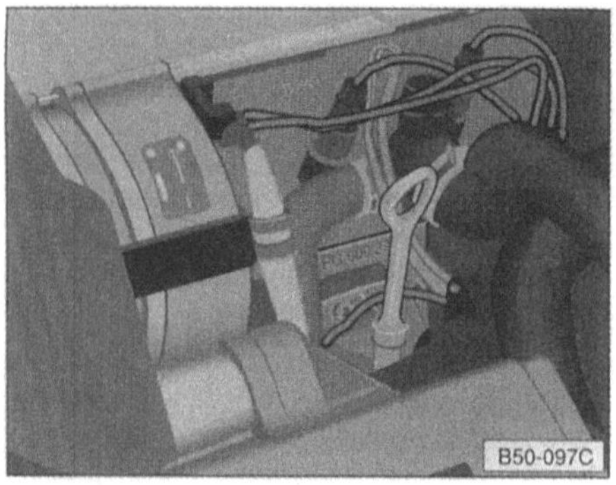
B50-097C

The engine number

is located on the front face of the engine block, just below the cylinder head.

The safety compliance sticker

is your assurance that your new vehicle complies with all applicable Federal Motor Vehicle Safety Standards which were in effect at the time the vehicle was manufactured. You can find this sticker on the left doorjamb. It shows the month and year of production and the vehicle identification number of your vehicle (perforation) as well as the Gross Vehicle Weight Rating (GVWR) and the Gross Axle Weight Rating (GAWR).

GAS STATION INFORMATION

In order to assure that your vehicle remains ready for the road at all times you should carry out a few checks regularly – preferably when filing the tank.

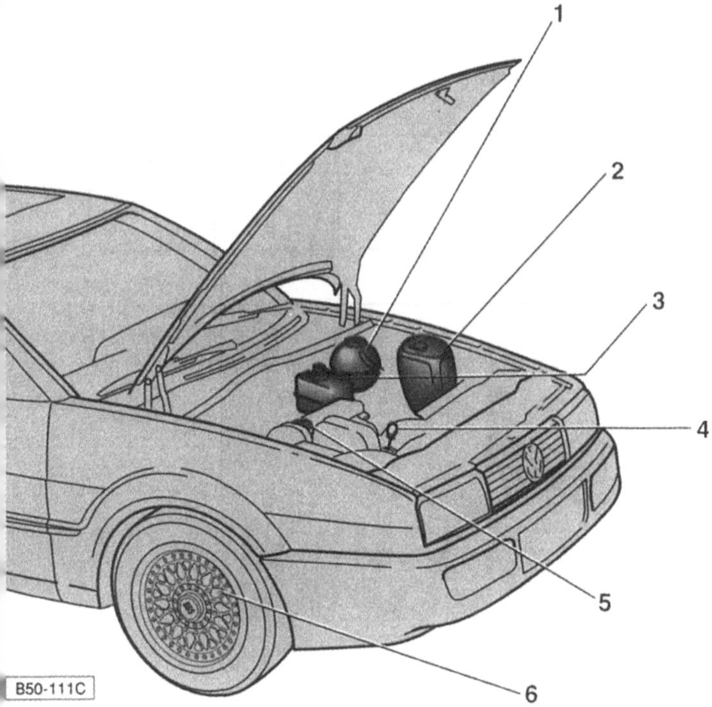

We have listed the most important points to be checked.

How to open the engine hood is explained on page 7.

WARNING
Take particular care when working in the engine compartment – see page 7.

1 – Coolant expansion tank

WARNING
Do not open the coolant reservoir when the engine is hot. Danger of scalding!

If the level is below the Min. mark, top it up with coolant additive and water – see page 11. If unavoidable, plain water may be used temporarily.

2 – Windshield washer container
Always fill with water and a glass cleaning solution. Remember to use a winterized washer solvent during the cold season.

3 – Brake fluid reservoir
The fluid level must always be between the min. and max. marks – see page 13.

4 – Engine oil level
Every engine uses a certain amount of oil. If the oil level is too low, however, this can cause serious engine damage. For this reason, check the level regularly. The oil level must be between the min. and max. marks on the dipstick – see page 9.

5 – Engine oil filler cap
Oil specifications are given on page 8, oil capacities are given on page 37.

6 – Tire pressure
Check the pressures at least once a month. Adjust the tire pressures, if necessary, to suit different vehicle loads. The pressures are given on a sticker on the passenger's door lock jamb.

Every 7,500 miles/12, 000 km	4 cylinder	6 cylinder
Engine: Change oil	✔	✔
Engine: Replace oil filter (replace at the first 7,500 miles/12,000 km and 15,000 miles/24,000 km maintenance intervals and at every 15,000 miles/24,000 km intervals thereafter)	✔	✔
Fluid levels: Check coolant, hydraulic, brake, windshield washer levels; add if necessary	✔	✔
Automatic Shift Lock: Check operation including park/neutral switch	✔	✔

15,000 mile/24,000 km and 45,000 mile/72,000 km Service Interval	4 cylinder	6 cylinder
Engine: Change oil, replace oil filter	✔	✔
Cooling System: Check coolant level and add if necessary	✔	✔
Fluid levels: Check coolant, hydraulic, brake, windshield washer levels; add if necessary	✔	✔
Automatic Shift Lock: Check operation including park/neutral switch	✔	✔
Manual Transmission: Check for leaks	✔	✔
Brake System: Check for damage/leaks, thickness of pads and brake fluid level	✔	✔
Drive Shafts: Check boots	✔	✔
Automatic Shoulder Belt: Check function of belt warning system, seat adjustment lever, belt transport (also in first position), locking hook, acceleration sensor	✔	✔
On-Board Diagnosis System: Check fault memory with VAG 1551 and purge if necessary	✔	✔
Tires/Spare: Check condition and pressure	✔	✔
During road test		
Check Kickdown, Braking, Steering, Heating, Ventilation, Air Conditioning, Power Accessories and Electrical Systems	✔	✔
After Road Test		
Engine Idle Speed: Check (except California vehicles)	✔	✔
Engine: Check for leaks	✔	✔
Automatic Transmission: Check ATF*	✔	✔
Automatic Transmission Final Drive: Check for leaks	✔	✔
Power Steering: Check fluid level (where applicable)	✔	✔
Exhaust System: Check for damage/leaks	✔	✔

*In the case of severe conditions, such as extremely high temperatures, continuous mountain driving, trailer towing and predominantly stop-and-go traffic conditions, it is necessary to change the ATF every 30,000 miles/48,000 km.

During this time, the pan and strainer should be cleaned and gasket replaced.

30,000 mile/48,000 km and 60,000 mile/96,000 km Service Interval	4 cylinder	6 cylinder
Engine: Change oil, replace oil filter	✔	✔
Air Cleaner: Replace filter element	✔	✔
Cooling System: Check coolant level and add if necessary	✔	✔
V–Belt(s): Check tension and condition	✔	✔
Spark Plugs: Replace every 30,000 miles/48,000 km	✔	
Spark Plugs: Replace every 30,000 miles/48,000 km	✔	✔
Oxygen Sensor: Replace every 60,000 miles/96,000 km regardless of time of delivery	✔	
Automatic Shift Lock: Check operation including park/neutral switch	✔	✔
Manual Transmission: Check for leaks	✔	✔
Brake System: Check for damage/leaks, thickness of pads and brake fluid level	✔	✔
Brake Fluid: Replace every two years regardless of mileage	✔	✔
Drive Shafts: Check boots	✔	✔
Front Axle: Check dust seals on ball joints, tie rod ends and check tie rods	✔	✔
Automatic Shoulder Belt: Check function of belt warning system, seat adjustment lever, belt transport (also in first position), locking hook, acceleration sensor	✔	✔
On-Board Diagnosis System: Check fault memory with VAG 1551 and purge if necessary	✔	✔
Tires/Spare: Check condition and pressure	✔	✔
During road test		
Check Kickdown, Braking, Steering, Heating, Ventilation, Air Conditioning, Power Accessories and Electrical Systems	✔	✔

(continued)

30,000 mile/48,000 km and 60,000 mile/96,000 km Service Interval	4 cylinder	6 cylinder
After Road Test	✔	✔
Engine Idle Speed: Check (except California vehicles)	✔	✔
Engine: Check for leaks	✔	✔
Automatic Transmission: Check ATF*	✔	✔
Automatic Transmission Final Drive: Check for leaks	✔	✔
Power Steering: Check fluid level (where applicable)	✔	✔
Exhaust System: Check for damage/leaks	✔	✔

*In the case of severe conditions, such as extremely high temperatures, continuous mountain driving, trailer towing and predominantly stop-and-go traffi c conditions, it is necessary to change the ATF every 30,000 miles/48,000 km.

During this time, the pan and strainer should be cleaned and gasket replaced.

THIS PAGE INTENTIONALLY LEFT BLANK

Index

Identification
- type and model **00.2**
- engine/vehicle number location **00.2a**
- manual transmission identification **00.3**
- “096” automatic transmission identification **00.3a**
- SLC “096” automatic transmission identification **00.3b**

Lifting vehicle
- lifting points hoist/floor jack **00.5**
- using hoist/floor jack **00.6**

Technical data
- chart **00.7, 00.7a**

Towing
- general information **00.4**

Type and model identifications

Model code*	Model years	Designation USA	Designation CANADA	Body style	Trans	Engine code**
509014	1990-1992	Corrado	Corrado G-60	2-Door Sport	5-Spd	PG
509018	1990-1992	Corrado	Corrado G-60		Auto.	PG
5090Q4	1992 (USA) 1993 (Canada)	Corrado SLC	Corrado VR-6		5-Spd	AAA
5090Q8	1992 (USA) 1993 (Canada)	Corrado SLC	Corrado VR-6		Auto.	AAA

*The 5th digit of the Vehicle Identification Number indicates if the vehicle engine is equipped for either 49-States or California.

Engine code KW @ RPM (BHP)

PG	118 @ 5600 (158)
AAA	132 @ 5800 (178)

Assemblies	Type	Code letters	BHP net (SAE)	RPM
Engine 1.8L G-Charger (Digifant)	–	PG	158	5600
Engine 2.8L VR-6 Motronic	–	AAA	178	5800
5-Speed Manual Transmission	02A	CDM AYL ATA	–	–
4-Speed Automatic	096	APB APC	–	–

Engine number

G-60

Location

- stamped on front of cylinder block (**arrow**)

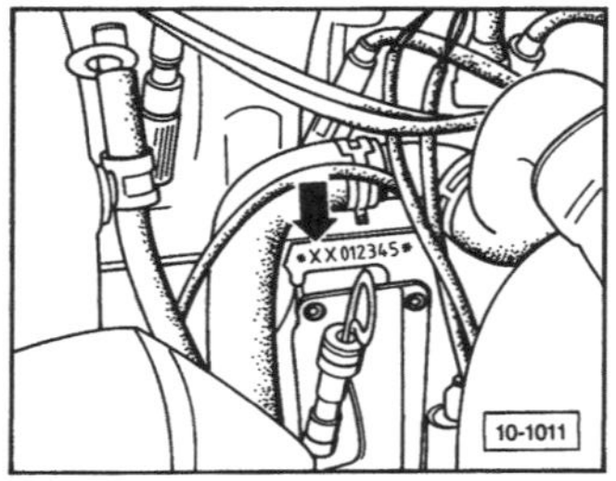

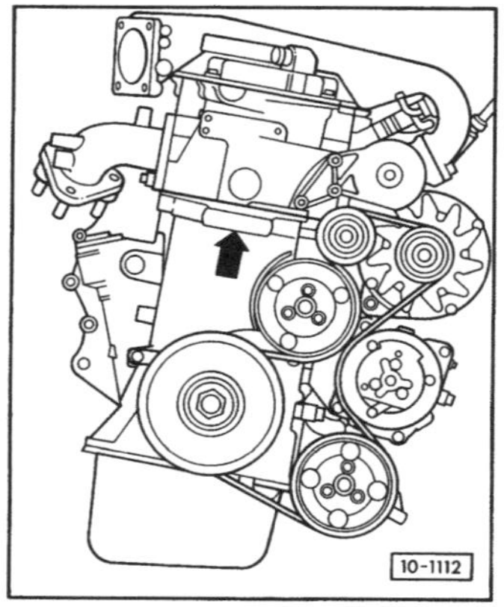

SLC (VR-6)

Location

- stamped on front of cylinder block (**arrow**)
- sticker affixed on valve cover (pulley side)
- vehicle identification label in luggage compartment

Vehicle identification

- on top of instrument panel on driver's side, visible from the outside through the windshield
- on identification label in luggage compartment

▶ Manual transmission identification

1 – transmission code letters and date of production
2 – transmission type no.

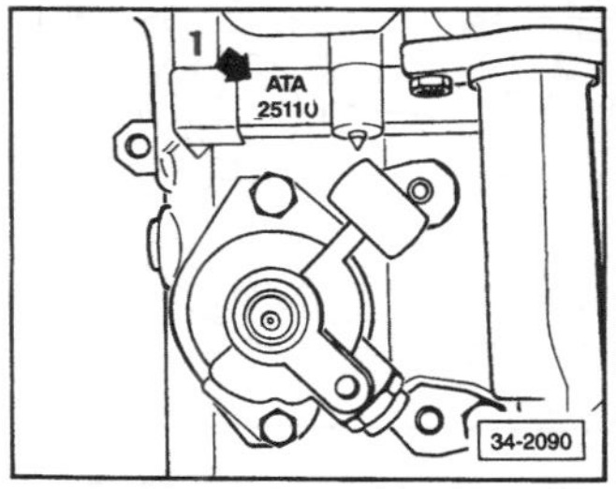

▶ Transmission code letters and date of production (arrow)

Example:

ATA	25	11	0
code	day	month	year (90)

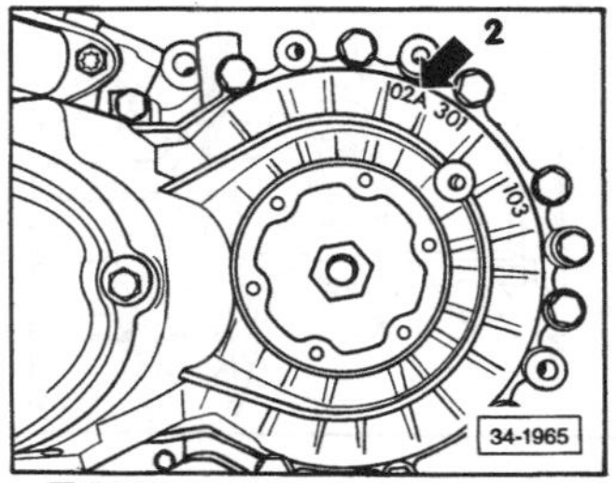

▶ Transmission type no. (arrow)

Note
Transmission code also listed on Vehicle Identification Label.

Transmission code letters		ATA	AYL	CDM
Manufactured	from: to:	9/90 07/90	08/90	10/91 –
Allocation model		Corrado		
Engine		1.8L/118 kW (158 HP)	1.8L/118 kW (158 HP)	2.8L 132 kW (178 HP)
Ratios	1st gear	3.778 : 1	3.778 : 1	3.300 : 1
	2nd gear	2.105 : 1	2.118 : 1	1.944 : 1
	3rd gear	1.345 : 1	1.429 : 1	1.308 : 1
	4th gear	0.971 : 1	1.029 : 1	1.034 : 1
	5th gear	0.795 : 1	0.837 : 1	0.838 : 1
	final drive	3.684 : 1	3.684 : 1	3.647 : 1
	reverse gear	3.800 : 1	3.800 : 1	3.060 : 1

Automatic transmission 096 identification

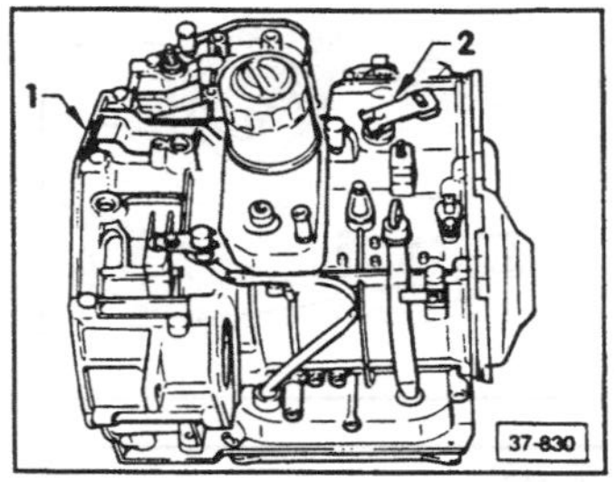

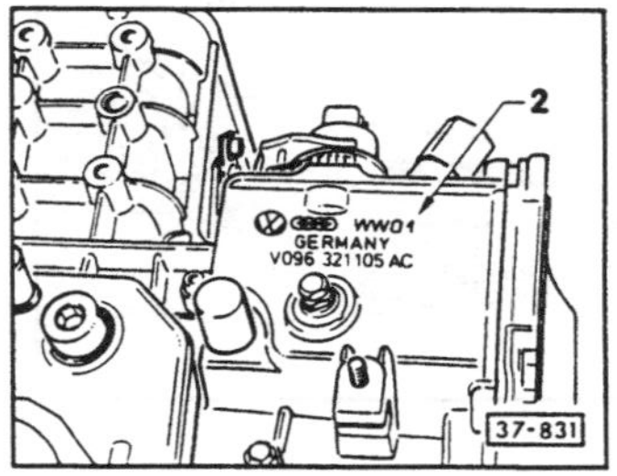

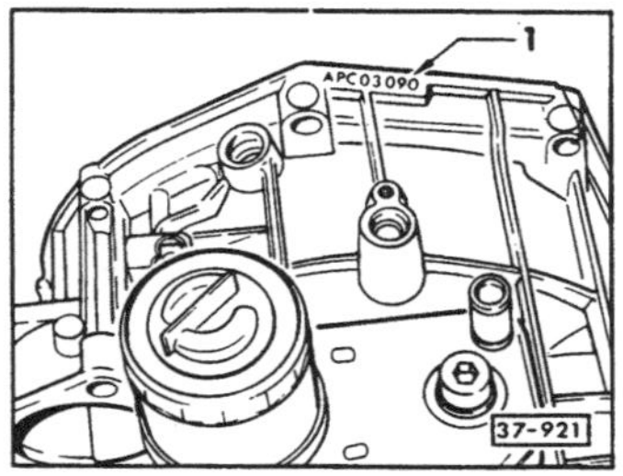

Location on Transmission

1 — Code letters
2 — Type No.

Code letters and date of production (1)

Example:

APC	03	09	0
code letters	day	month	year (90)

Note

The transmission code letters are also given on the vehicle data plates.

Transmission code letters		APC
Manufactured	from: to:	9/90 —
Torque converter code letters		QBCA
Manufactured	from: to:	9/90 —
Allocation model		Corrado
Engine		1.8 ltr./118 kW (158 HP)
Ratios	1st gear	2.714 : 1
	2nd gear	1.551 : 1
	3rd gear	1.000 : 1
	4th gear	0.679 : 1
	Reverse gear	2.111 : 1

Automatic transmission 096 identification

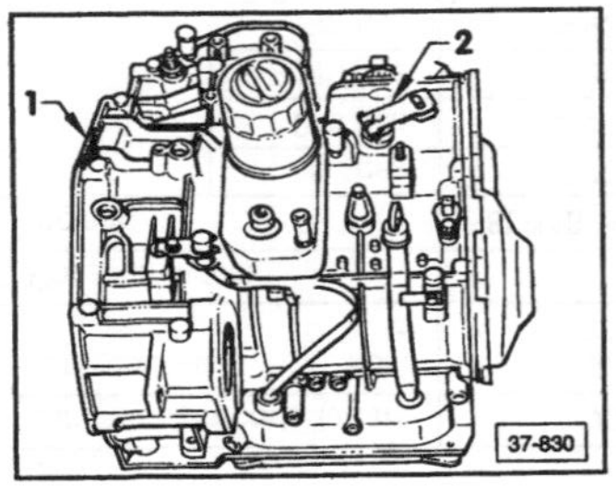

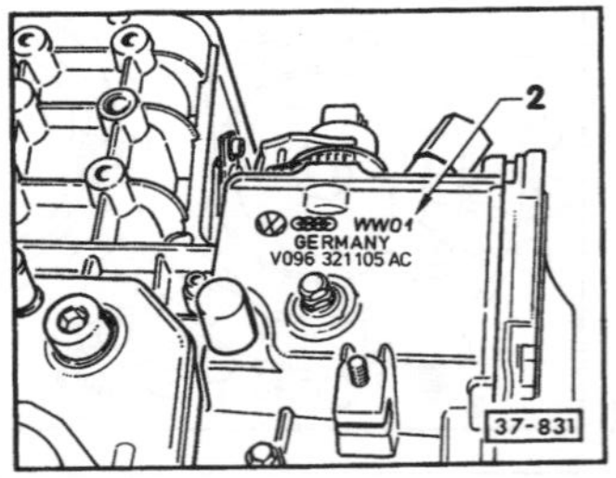

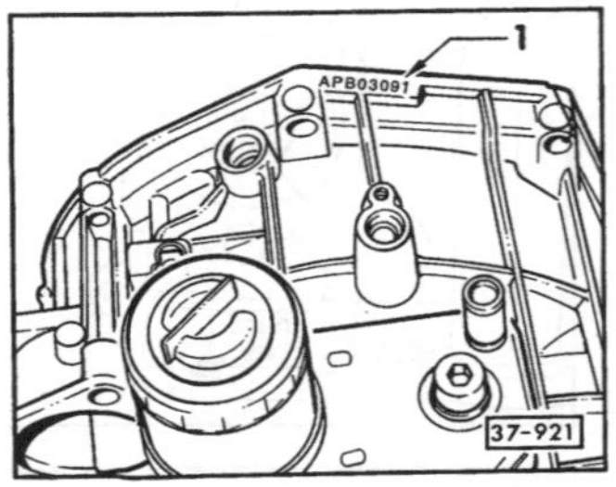

Location on Transmission

1 – Code letters
2 – Type No.

Code letters and date of production (1)

Example:

APB	03	10	1
code letters	day	month	year (91)

Note
The transmission code letters are also given on the vehicle data plates.

Transmission code letters		APB	CFA	CFF
Allocation model		Corrado		
Engine		2.8L 132 kW (178 HP)		
Ratios	1st gear	2.714:1		2.714:1
	2nd gear	1.551:1		1.441:1
	3rd gear	1.000:1		1.000:1
	4th gear	0.679:1		0.742:1
	Reverse gear	2.111:1		2.884:1

Towing

Note

Detailed towing instructions can be found in a special towing instruction brochure at your Volkswagen dealer, and in your owner's manual.

> **WARNING**
>
> Never allow passengers to ride in a towed vehicle for any reason.

> **CAUTION**
>
> - the Corrado **cannot** be towed with conventional sling-type equipment, or wheel dollies
> - towing with this type of equipment will cause bumper and body panel damage
> - do not wrap safety chains around brake lines

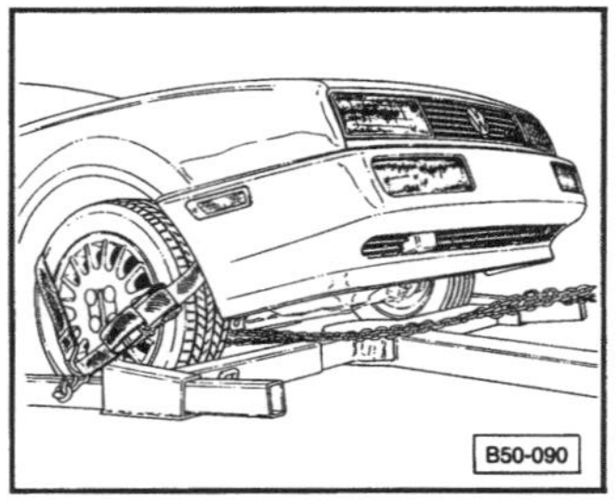

Front hook-up

- attach wheel lift equipment to wheels
- attach safety straps to wheels
- attach safety chains to lower control arms
- towing clearance: 6-12 in. (15-30 cm) between tires and ground

Note

Whenever possible, tow with front wheels off ground. If the vehicle must be towed with the front wheels on the ground, make sure that transmission oil has not leaked or been drained.

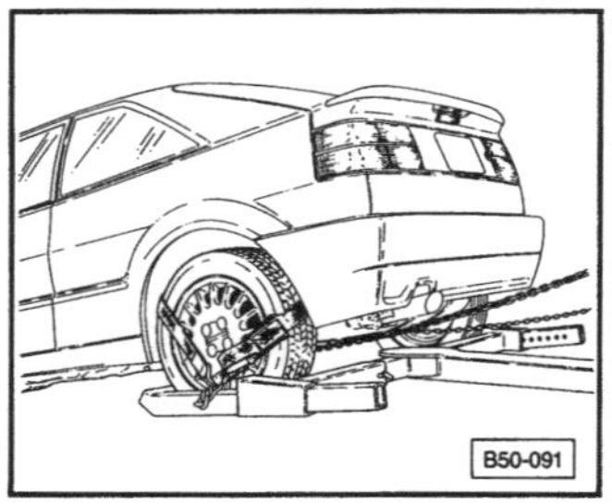

Rear hook-up

- attach wheel lift equipment to wheels
- attach safety straps to wheels
- attach safety chains to axle beam
- towing clearance: 6-12 in. (15-30 cm) between tires and ground

Towing speed/distance

50 mph/50 mi (80 km/h/80 km)

Lifting points, hoist/floor jack

In the contact area for the lift arms, insert a rubber pad between the floor jack and the lift points (**arrows**).

> **CAUTION**
>
> Only lift vehicle at these locations.

Front lifting point

Note

Ensure to position the contact point of the hoist or floor jack so that the center of the workshop hoist or jack must be aligned with the center of the depression **A** in the side member .

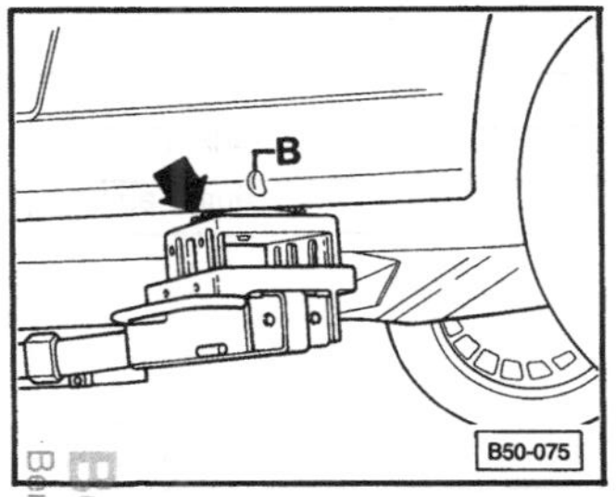

Rear lifting point

Note

Ensure to position the contact point of the hoist or floor jack so that the center of the workshop hoist or jack must be aligned with the center of the depression **B** in the side member .

Lifting vehicle with hoist/floor jack

> **CAUTION**
>
> The vehicle should never be lifted or jacked up from underneath the engine oil pan, the transmission housing, or the front or rear axle. This could lead to serious damage.

The same lifting points as illustrated for the hoist also apply when using a floor jack. To avoid damage to the underbody or chassis frame, it is necessary to insert a rubber pad between the floor jack and the lift points.

Ensure there is sufficient clearance between pads and vehicle before driving the vehicle on a hoist, especially if the vehicle has a large front panel or spoiler.

> **WARNING**
>
> - when removing components such as engine block, transmission housing, fuel tank, wheels front, or rear axle, anchor vehicle to hoist or add corresponding weights to maintain the center of gravity. Otherwise the vehicle might tilt or slip off the hoist, causing serious damage or personal injury
> - to reduce the risk of serious personal injury and vehicle damage, lift vehicle only at the **special workshop hoist and floor jack lift** points illustrated. Failure to lift vehicle at these points could cause the vehicle to tilt or fall from a lift when, for example, heavy components such as the engine block or transmission are removed resulting in a change in vehicle weight distribution and balance
> - if you must lift your vehicle with a floor jack to work underneath, be sure the vehicle is safely supported on stands intended for this purpose

Technical data

Corrado-Digifant

Engine code letters		PG
Start of production		Oct. 1988
Displacement	Liter, (cu. in.)	1.8, (109.8 cu. in.)
Power output	KW @ RPM, (SAE HP @ RPM)	118/5600 (158 @ 5600
Torque	Nm @ RPM, (ft lb @ RPM)	225/4000 (166 @ 4000)
Cylinder bore diameter	mm, (in.)	81, (3.16)
Stroke	mm, (in.)	86.4, (3.40)
Compression ratio		8.0:1
Valve timing at 1mm valve lift	Inlet opens at TDC Inlet closes after BDC Outlet opens before BDC Outlet closes at TDC	0° 40° 40° 0°
Octane rating	RON minimum	91 (leadfree)*
Injection type		Digifant
Firing order		1-3-4-2
Catalytic converter		Yes
Oxygen sensor		Yes
Knock sensor		Yes
G-Charger		Yes

*Use of 95 RON results in maximum performance

Technical data

Corrado VR-6 – Motronic

Engine code letters		AAA
Start of production		Oct. 1990
Displacement	Liter, (cu. in.)	2.8, (170 cu. in.)
Power output	KW @ RPM, (SAE HP @ RPM)	132/5800 (178 @ 5800)
Torque	Nm @ RPM, (ft lb @ RPM)	245/4200 (177 @ 4200)
Cylinder bore diameter	mm, (in.)	81.0, (3.16)
Stroke	mm, (in.)	90.3, (3.56)
Compression ratio		10.0:1
Valve timing at 1mm valve lift	Inlet opens after TDC Inlet closes after TDC Outlet opens before TDC Outlet closes after TDC	8° 228° 12′ 217° 3° 12′
Octane rating	RON minimum	91 (leadfree)*
Injection type		Motronic
Firing order		1-5-3-6-2-4
Catalytic converter		Yes
Oxygen sensor		Yes
Knock sensor		Yes

*Use of 95 RON results in maximum performance

Index

Engine code - PG

- engine/transmission, adjusting **10.5**
- engine/transmission, mounting points **10.4**
- hydraulic engine/transmission, mounting points **10.4**
- installing **10.3**
- removing **10.2**
- tightening torques **10.4**

Engine code - AAA

- engine/transmission mounting points **10.10**
- installing **10.8**
- removing **10.6**
- tightening torques **10.11**

NOTE

Remove engine from top with the transmission attached.

Do not disconnect

- hoses or brackets from power steering pump
- A/C hoses

CAUTION

Coolant/antifreeze **must not** be reused when replacing engine, cylinder head, cylinder head gasket, radiator and heater core.

Engine, removing

- disconnect battery ground strap
- remove power steering pump and attach to body
- remove compressor mounting bolts together with mount and move to one side
- unscrew condenser from radiator, and move to one side

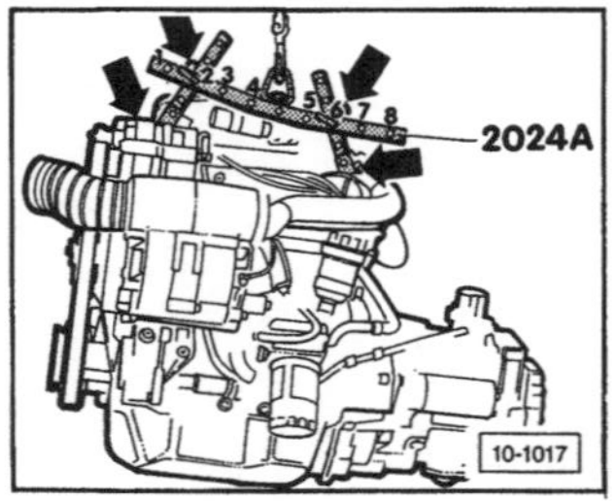

▶

- attach engine sling as shown and take weight with crane

Pulley side lower (**left arrow**)
second hole in vertical at position **2**.

Flywheel side lower (**right arrow**)
second hole in vertical bar at position **6**.

The positions marked 1-4 on the bar must be towards the pulley side. The holes in the hooks are counted **upwards** from the hook.

CAUTION

Use cotter pins on hooks and holding pins (**arrows**) to secure tool.

CAUTION

Guide engine carefully when lifting to avoid damage to body.

- mount engine onto work bench with engine mount **VW 540** if it is to be disassembled

Engine, installing

Install in reverse order of removal. Note the following:

- check clutch release bearing for wear, replace if necessary
- lubricate clutch release bearing, driveshaft lightly with MoS_2 grease
- check and ensure that the dowels for centering engine/transmission are in cylinder block
- replace self-locking nuts for engine mounts
- lower engine onto mounts

CAUTION

Guide engine carefully when lowering to avoid damage to driveshaft, body.

- connect rear engine mount first, then connect transmission mount
- loosen front engine mount and mount bracket
- align all engine supports with mount bushing before tightening mount bolts
- adjust accelerator cable (see repair group 20)
- reconnect all electrical connections (see repair group 97)
- check and adjust shift linkage (see repair group 34)
- fill coolant (see repair group 19)
- check and adjust engine oil level (see repair group 17)
- adjust engine timing/idle (see repair group 28)
- check and adjust headlights, if necessary

Tightening torques

Engine to transmission	
M10	45 Nm (33 ft lb)
M12	55 Nm (41 ft lb)
Driveshaft to flange	45 Nm (33 ft lb)
Exhaust pipe to exhaust manifold	40 Nm (30 ft lb)
Power steering pump to mounting bracket	25 Nm (18 ft lb)

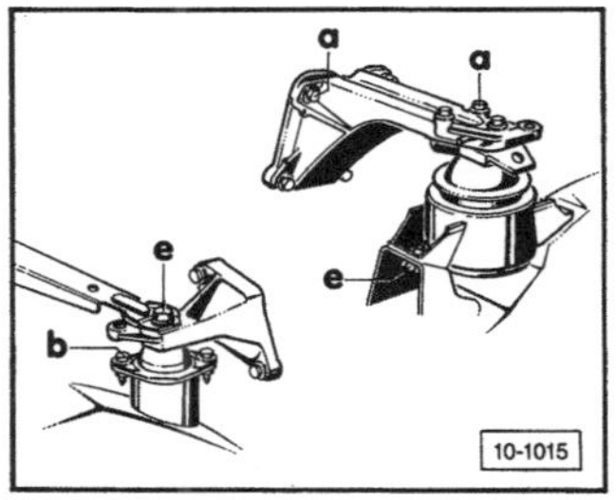

▶ **Engine/transmission mounting points**

a = 25 Nm (18 ft lb)
b = 30 Nm (22 ft lb)
e = 60 Nm (44 ft lb)

▶ **Hydraulic engine/transmission mounting points**

1 = hydraulic mount
b = 30 Nm (22 ft lb)
e = 60 Nm (44 ft lb)
f = 70 Nm (53 ft lb)

Engine/transmission, adjusting

Adjustment is necessary only if engine/transmission is removed, or mounts are loosened from engine/transmission.

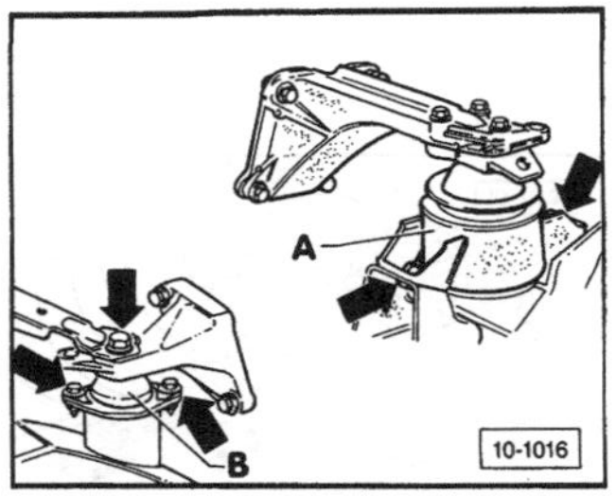

▶
- loosen bolts (**arrows**) from engine mount **A**
- loosen bushing **B** on transmission mount (**arrows**)

▶
- loosen front engine mount and mount bracket (**arrows**)
- lightly rock or shake engine to adjust position
- re-tighten bolts in same sequence

Engine, removing/installing

Removing

Engine code AAA

Note

Remove engine from top with transmission attached.

> **CAUTION**
>
> Wait at least 20 seconds **after** switching off ignition before disconnecting battery. This will allow the platinum hot-wire in Air Mass Sensor (**G 70**) to burn itself clean.

> **CAUTION**
>
> Always switch off ignition when disconnecting/reconnecting battery, or any part of Motronic ignition wiring system to prevent damage to Motronic control unit (**J 220**).

> **CAUTION**
>
> Always obtain radio security code before disconnecting battery ground strap.

- switch off ignition
- disconnect battery ground strap

Note

Ensure all removed electrical wiring, tie wraps, are reinstalled and positioned as originally placed.

- remove air cleaner housing
- remove front lock support, see Repair Group 50
- remove front bumper, see Repair Group 63

> **CAUTION**
>
> Coolant/antifreeze must not be reused when replacing engine, cylinder head, cylinder head gasket, radiator and heater core.

- remove radiator
  - save coolant for reuse, when applicable

Manual transmission

- unbolt and disconnect hydraulic clutch cylinder
- disconnect shift cable and support bracket

Automatic transmission

- unclip and disconnect selector lever cable

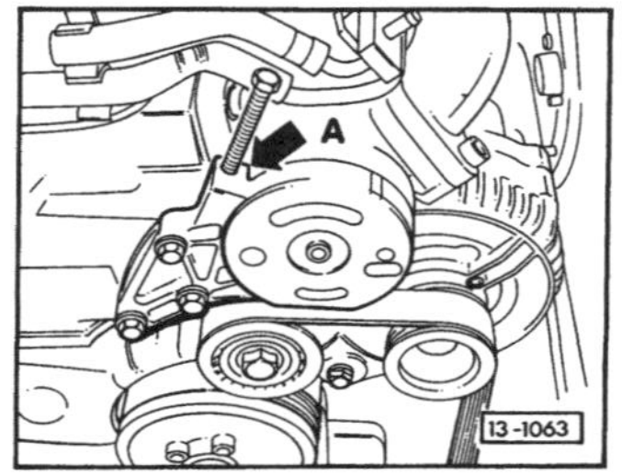

- screw in a long 8.0 mm bolt into a threaded hole **A** on tensioner to release poly-ribbed belt tension

Note

In order to prevent damage to tensioner, only screw in bolt sufficiently to release belt tension.

- remove poly-ribbed belt from power steering pump, A/C compressor

> **CAUTION**
>
> Do not kink or disconnect A/C refrigerant lines.
>
> If A/C refrigerant system is opened or requires service refer to repair group 87 for **CAUTIONS, WARNINGS** and procedures.

- remove A/C compressor and secure to body or frame

> **CAUTION**
>
> Do not kink power steering hoses.

- remove power steering pump from bracket and secure to body or frame, see Repair Group 48

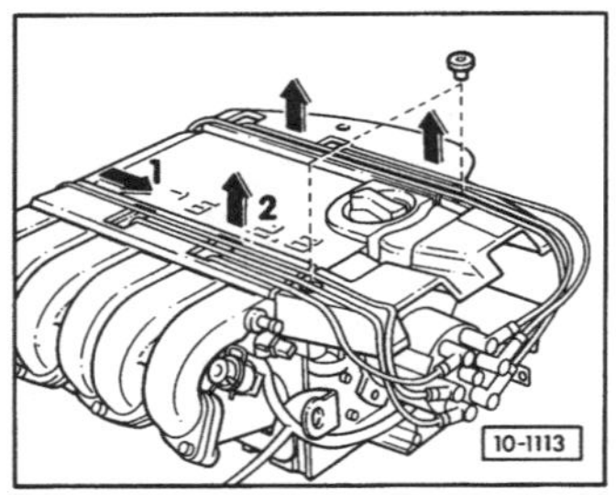

- remove ignition cable guides and ignition cables (**arrows**)
- remove ignition distributor cap
- remove trim cover over cylinder head covers, and disconnect accelerator cable at throttle body and cable guide
- disconnect 42-point connector from engine wiring harness
- remove and disconnect all other alternator, starter and transmission electrical wiring, and lay to one side
- pull off all coolant, vacuum and intake hoses from engine
- disconnect inlet and return fuel lines at fuel distributor

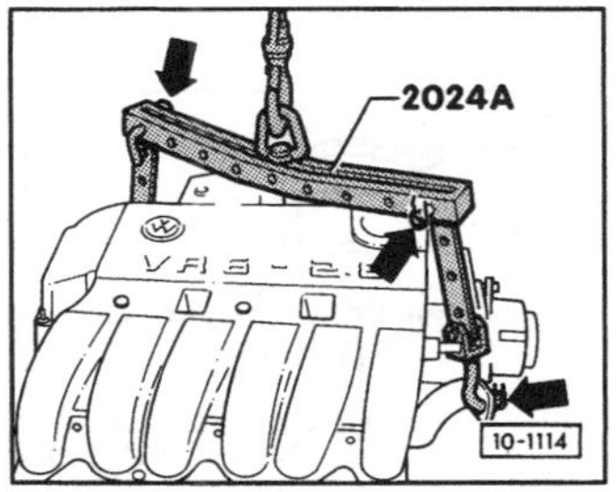

Note

Carefully cover any open fuel lines to prevent spillage and contamination.

- disconnect oxygen sensor and front exhaust pipe, see Repair Group 26
- disconnect and secure axle shafts
- attach engine sling as shown and take weight with crane

CAUTION
Use cotter pins on hooks and holding pins (**arrows**) to secure tool.

Pulley side (**left arrow**)

fourth hole in vertical bar at position **1** (**center arrow**)

Flywheel side (**right arrow**)

fourth hole in vertical bar at position **8**

The positions marked 1-4 on the bar must be towards the pulley side. The holes in the vertical bar are counted **upwards** from the hook.

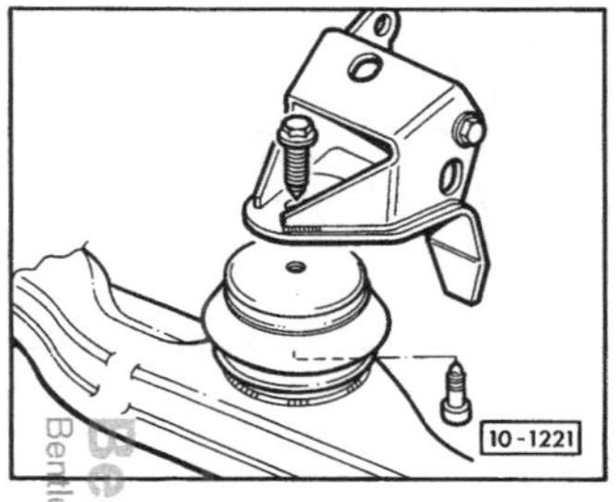

- unbolt engine from engine carrier at front

CAUTION
Guide engine carefully when lifting to avoid damage to body.

- mount engine onto work bench with engine mount **3269** if it is to be disassembled

Installing

CAUTION
Part numbers are for reference only. Always check with your Parts Department for latest information.

Reinstall engine in reverse order of removal, noting the following:

- check clutch release bearing for wear, replace if necessary
- lubricate clutch release bearing, driveshaft lightly with grease **G 000 100**
  - do not lubricate release bearing guide sleeve

- check and ensure that the dowels for centering engine/transmission are in cylinder block

WARNING
DO NOT re-use any fasteners that are worn or deformed in normal use. Many fasteners are designed to be used only once and become unreliable and may fail when used a second time. This includes, but is not limited to, nuts, bolts, washers, self-locking nuts or bolts, circlips, cotter pins. Always follow recommendations given in this publication. For replacements always use new parts.

CAUTION
Guide engine carefully when lowering to avoid damage to driveshaft, body.

- lower engine onto mounts

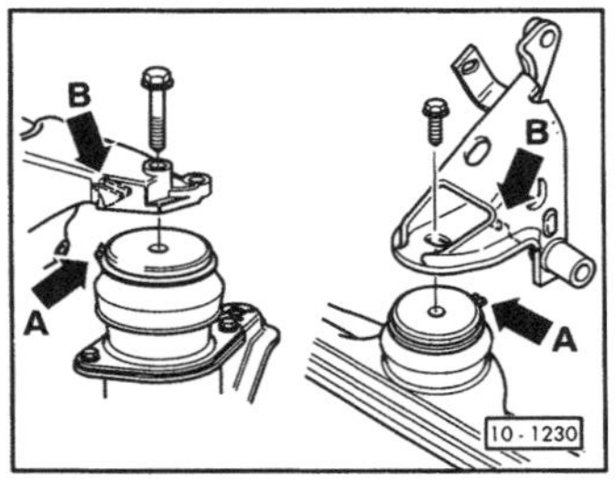

- ensure tabs (**arrows A**) fit into opening (**arrows B**) on engine mounts
- handtighten engine and transmission carrier bolts approximately 5 to **6** turns
- lightly rock or shake engine to adjust position
- fully tighten engine and transmission carrier bolts
- install V-belt, see Repair Group 13
- reconnect all electrical connections, see appropriate wiring diagram
- adjust accelerator cable, see Repair Group 20
- refill coolant, see Repair Group 19
- check engine timing, see Repair Group 28
  - not adjustable
- check and adjust engine oil level
- check and adjust shift linkage, see Repair Group 34
- check and adjust headlights, if necessary

Tightening torques

Engine support to body	50 Nm (37 ft lb)
Engine to transmission	
M 10	60 Nm (44 ft lb)
M 12	80 Nm (59 ft lb)
Driveshaft to flange	45 Nm (33 ft lb)
Front exhaust pipe to exhaust manifold	40 Nm (30 ft lb)
Front exhaust pipe to catalytic convertor	25 Nm (18 ft lb)
Power steering pump to mounting bracket	25 Nm (18 ft lb)

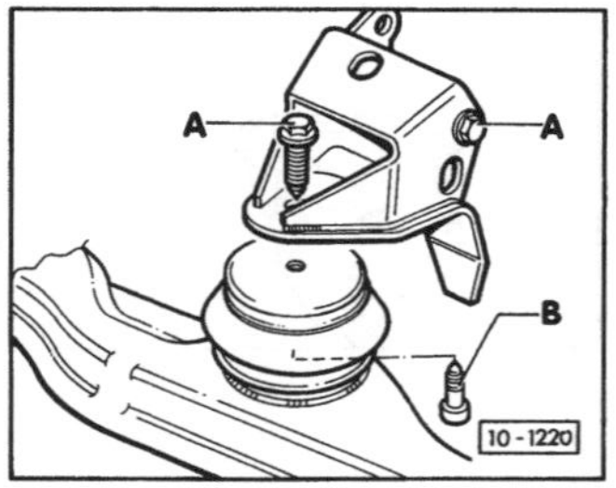

Engine to engine carrier, front

A = 60 Nm (44 ft lb)
B = 55 Nm (41 ft lb)

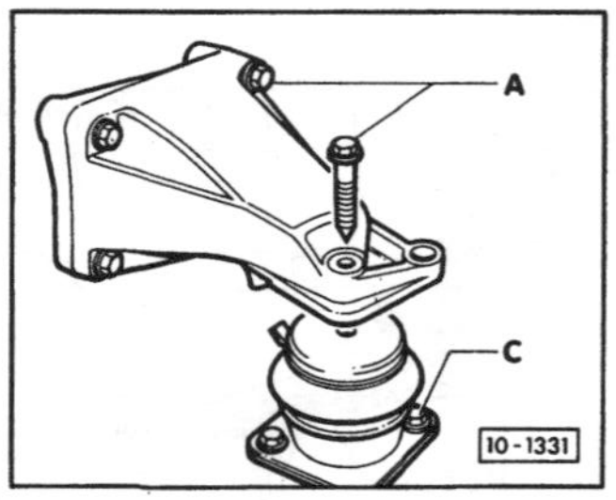

Engine to engine carrier, right

A = 60 Nm (44 ft lb)
C = 25 Nm (18 ft lb)

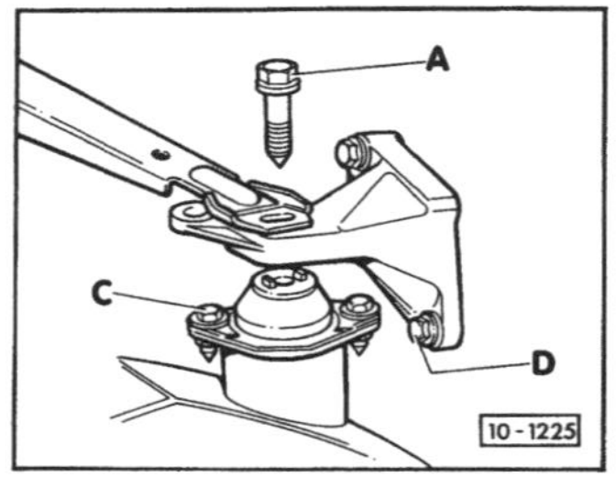

Transmission to transmission carrier

A = 60 Nm (44 ft lb)
C = 30 Nm (22 ft lb)
D = 25 Nm (18 ft lb)

Index

Engine Code – PG

Clutch flywheel
- assembly 13.10

Crankshaft,crankcase
- assembly 13.9
- dimensions 13.12
- ignition mark 13.12
- oil seal, removing/installing 13.11

Cylinder block
- assembly 13.4

Cylinder head
- assembly 13.3

Drive belt
- installing 13.6

G–charger, A/C brackets
- assembly 13.5

Piston, connecting rod
- assembly 13.13
- cylinder bore,checking 13.15
- dimensions 13.15
- piston rings, checking 13.14

Engine Code – AAA

Clutch, flywheel
- assembly 13.29

Crankshaft
- assembly 13.27
- dimensions 13.31
- oil seal, drive belt side removing/installing 13.26

Cylinder block
- assembly 13.21
- bore, checking 13.37

Cylinder head
- assembly 13.19

Drive chains
- assembly sprockets 13.16
- installing lower tensioner 13.24
- marking direction of travel 13.18

Drive plate
- removing/installing 13.31a

Pistons
- assembly with connecting rod rod 13.32
- checking 13.36
- dimensions 13.38
- rings,end gap checking 13.35
- ring, side clearance 13.34

Poly–ribbed belt
- removing/installing 13.25

Valve timing
- adjusting 13.23

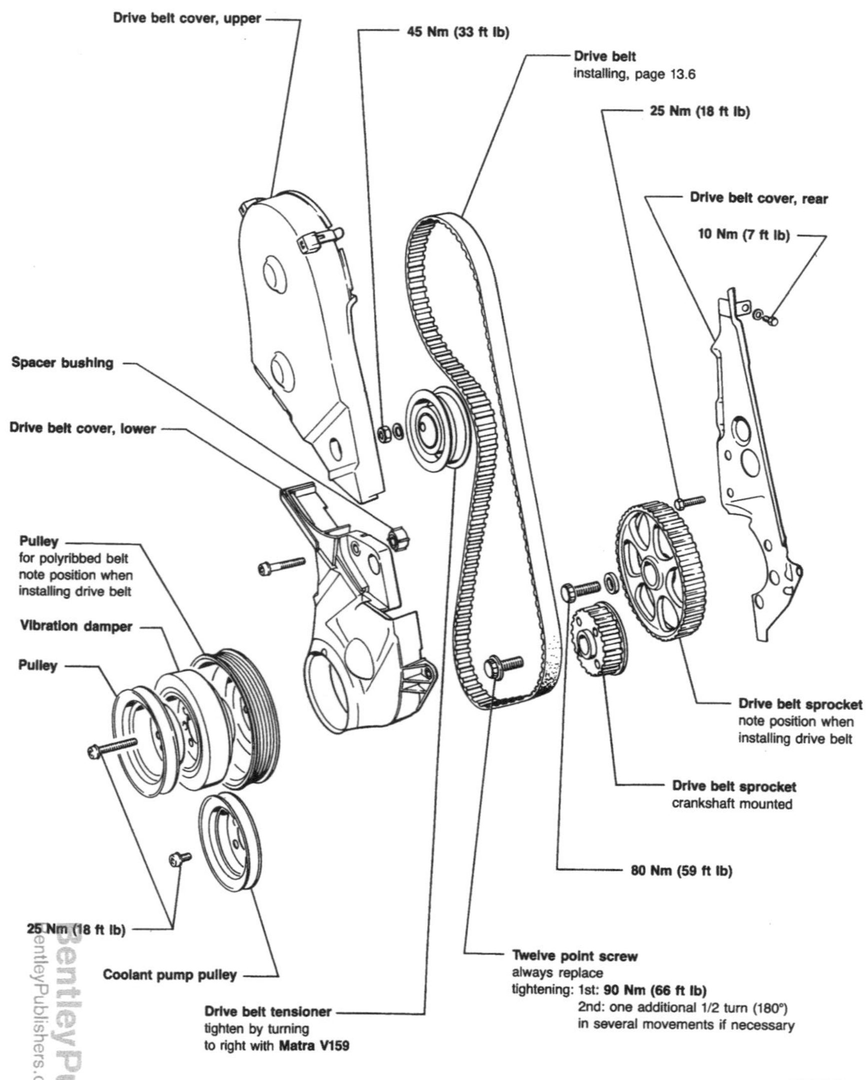

CAUTION

Coolant/anti-freeze **must not** be reused when replacing engine, cylinder head, cylinder head gasket, radiator and heater core.

CAUTION

Always replace gaskets and seals.

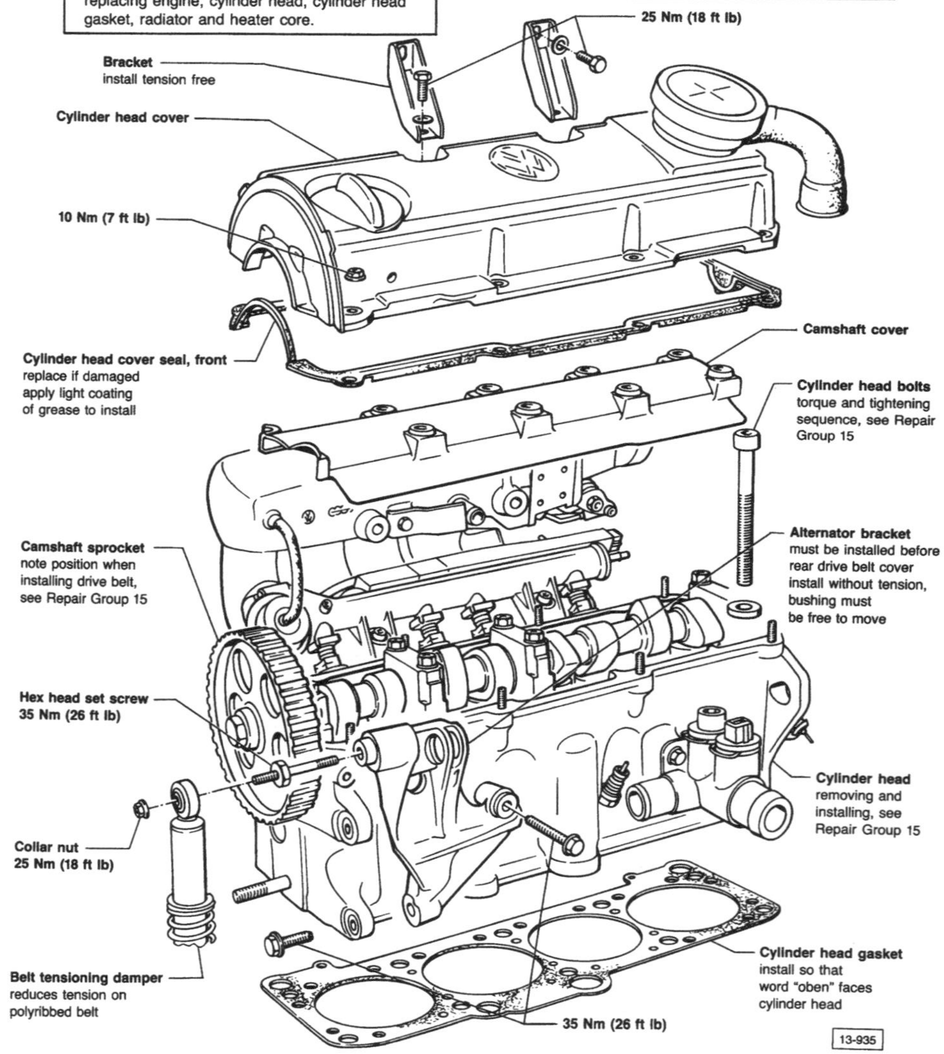

CAUTION

If you find metal shavings in the engine oil as a result of engine damage; clean oil passages thoroughly, then replace oil cooler and oil filter.

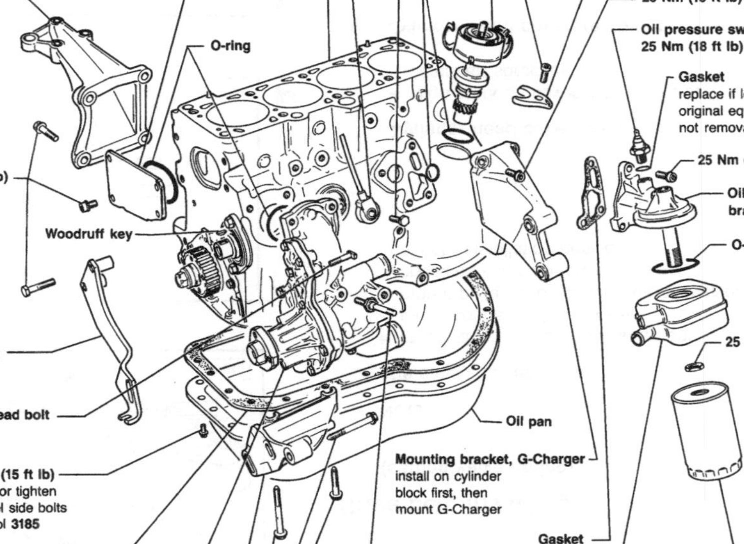

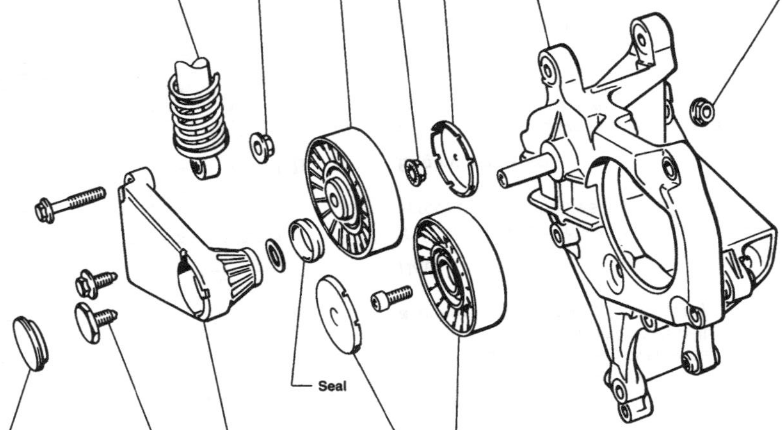

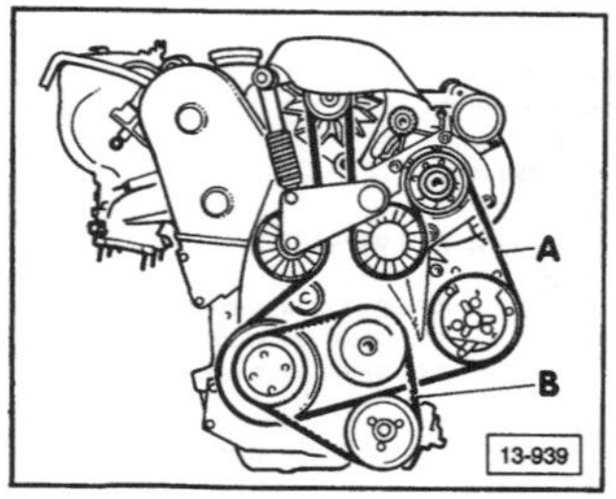

▶ Fig. 1 **Belt routing**

A — Polyribbed belt
B — Drive belt
- Tension drive belt, see Repair Group 48

▶ Fig. 2 **Tensioning polyribbed belt**

- tension using tool **3191**

Drive belt, installing

(setting valve timing)

Note

The crankshaft must **not** be at TDC.

- mount drive belt on crankshaft and intermediate shaft sprockets
- mount pulley and vibration dampener on crankshaft with all four fasteners
  - note installation position

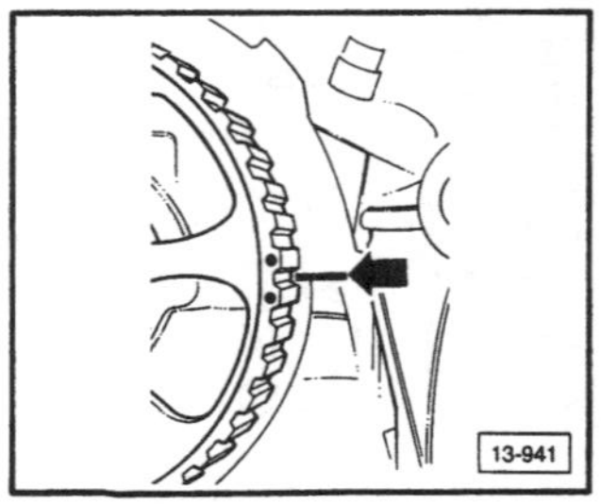

▶ **Cylinder head cover, installing**

- align mark on cover with marks on camshaft sprocket

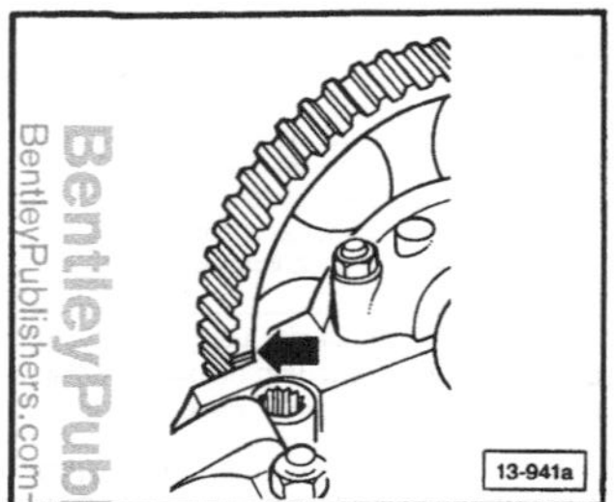

▶ **Cylinder head cover, removing**

- align mark on camshaft sprocket with cylinder head cover

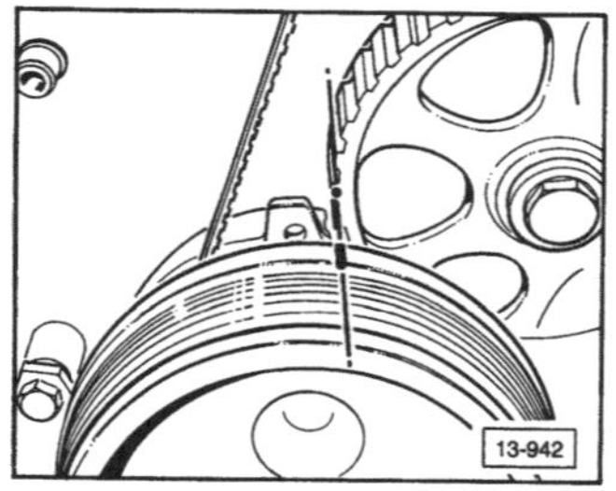

▶
- align mark on crankshaft pulley/vibration dampener with mark on intermediate sprocket (TDC cylinder **1**)
- mount drive belt on camshaft sprocket
- tighten drive belt by turning tensioner in direction of **arrow**
  - it must be possible to twist drive 90° in center between camshaft and intermediate sprockets

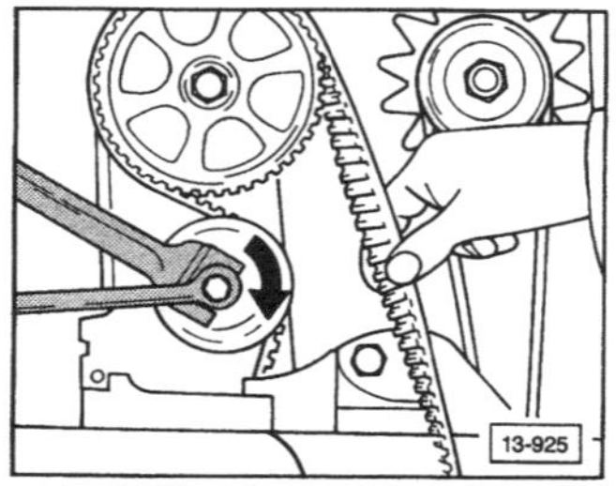

▶
- tighten lock nut on tensioner
  - torque 45 Nm (33 ft lb)
- turn crankshaft pulley twice and recheck drive belt tension
- remove crankcase pulley and vibration dampener
- install lower drive belt cover
- install upper drive belt cover, v-belt pulley, vibration dampener and drive belt
- check ignition timing, if necessary, adjust

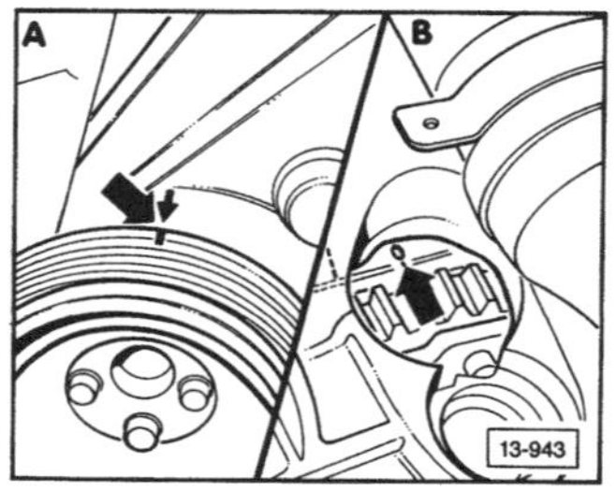

Note

If drive belt was removed from the camshaft sprocket, adjust belt as follows:

▶ **Engine removed**

- align mark on drive belt tensioner and drive belt cover (**left arrow**)

▶ **Engine installed**

- align TDC mark **0** on flywheel with mark on housing (**right arrow**)

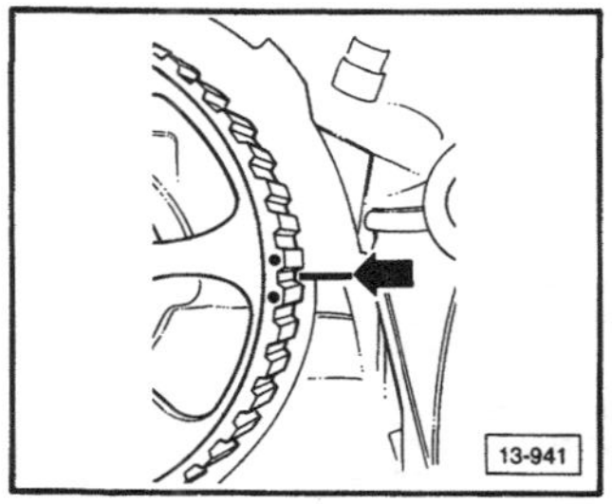

Cylinder head cover, installing

- align mark on cover with marks on camshaft sprocket

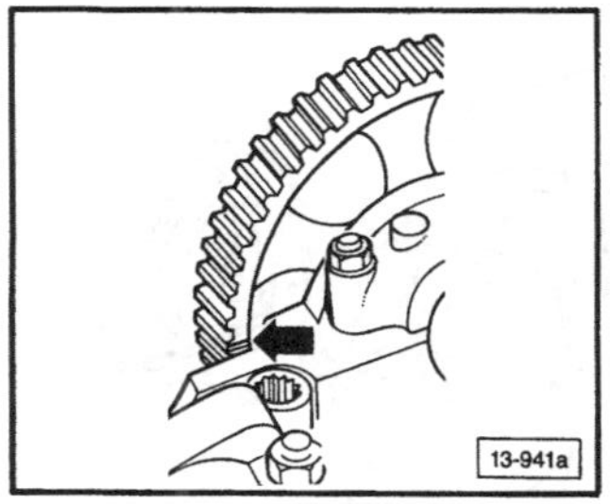

Cylinder head cover, removing

- align mark on camshaft sprocket with cylinder head cover
- mount drive belt and tighten
- check if ignition distributor rotor is pointing to marking for cylinder **1** on distributor housing. If not, turn distributor until mark and rotor align. If necessary install a new distributor
- turn crankshaft twice and check that crankshaft and camshaft marks are aligned with proper reference points
- check ignition timing and if necessary, adjust

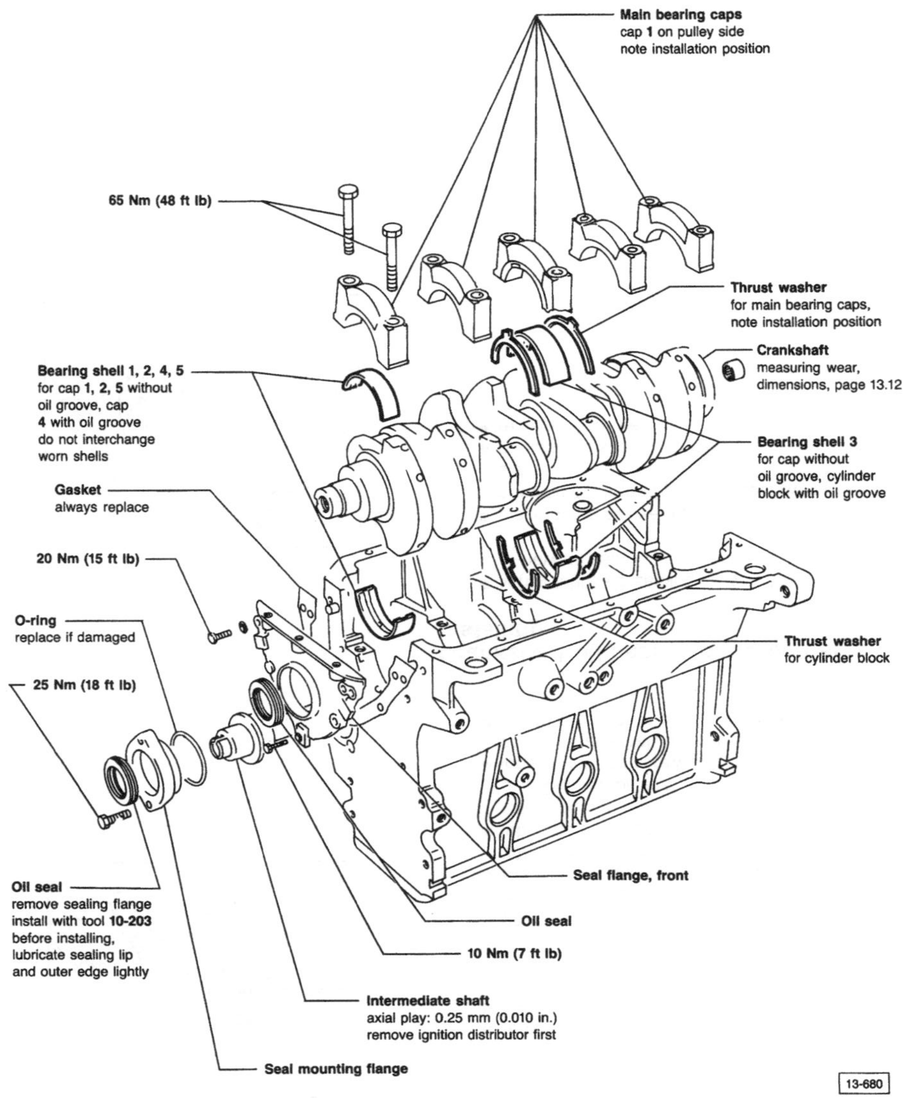

> **WARNING**
>
> Friction materials such as brake and clutch linings, or brake pads may contain asbestos fibers.
>
> Do not create dust by grinding, sanding or by cleaning with compressed air.
>
> Avoid breathing asbestos fibers and asbestos dust.
>
> Breathing asbestos may result in serious diseases, such as asbestosis or cancer.
>
> It may cause severe injury and death.

Repairing clutch, see Repair Group 30

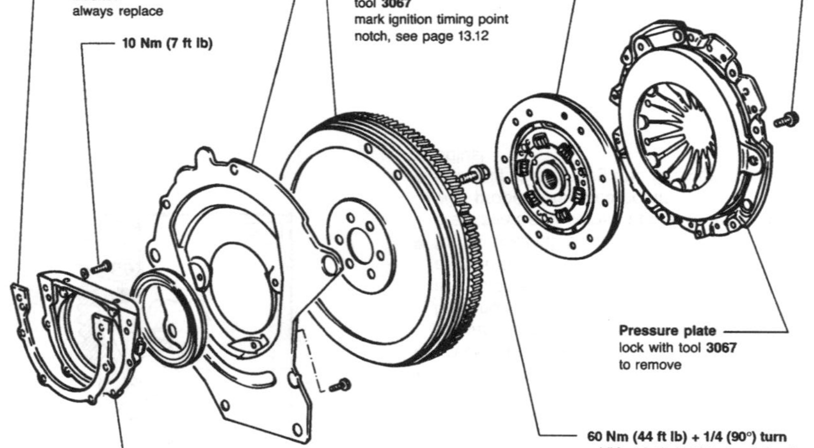

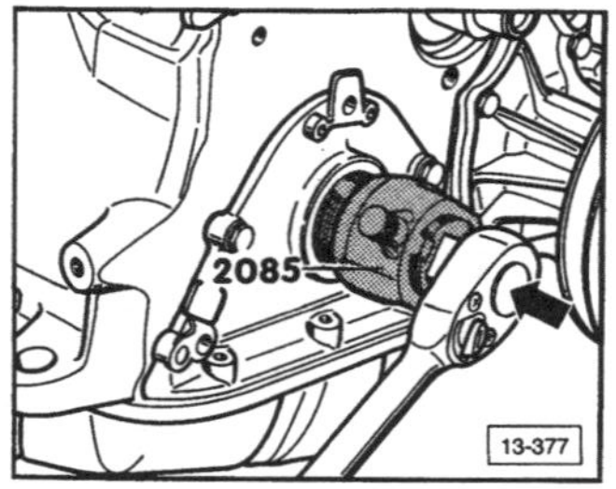

Crankshaft oil seal — drive belt side, removing

- remove drive belt
- remove drive belt sprocket (use socket **3099** to loosen mounting bolts) (**arrow**)

- unscrew inner part of oil seal extractor **2085** two turns (approximately 3 mm) out of outer part and lock with knurled screw
- to guide extractor, insert cylinder bolt from **3083** into crankshaft until it stops
- lubricate threaded head of oil seal extractor, set in position and push as far as possible into oil seal
- loosen knurled screw and turn inner part against crankshaft until oil seal is pulled out
- clamp extractor in a vise and remove oil seal using pliers

Crankshaft oil seal — drive belt side, installing

- lightly lubricate sealing lip and outer edge of new oil seal

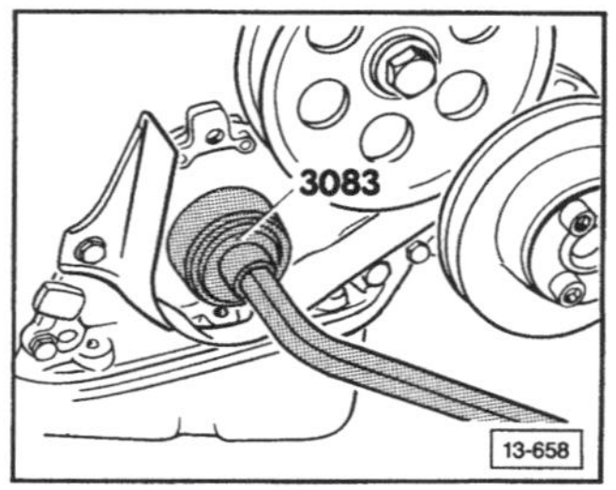

- place guide sleeve from **3083** onto crankshaft pin and push oil seal over guide sleeve
- press in oil seal up to stop
- reinstall drive belt

Crankshaft dimensions (mm)

	Main bearing journal (mm)	Connecting rod journal (mm)
Basic dimension	53.958-53.978	47.758-47.778
1st undersize	53.608-53.728	47.758-47.778
2nd undersize	53.458-53.478	47.258-47.278
3rd undersize	53.208-53.228	47.008-47.028

Making ignition timing mark

Note

If you replace the flywheel/drive plate you will have to inscribe the ignition timing mark. Replacement flywheel/drive plates have the 0 degree TDC mark only.

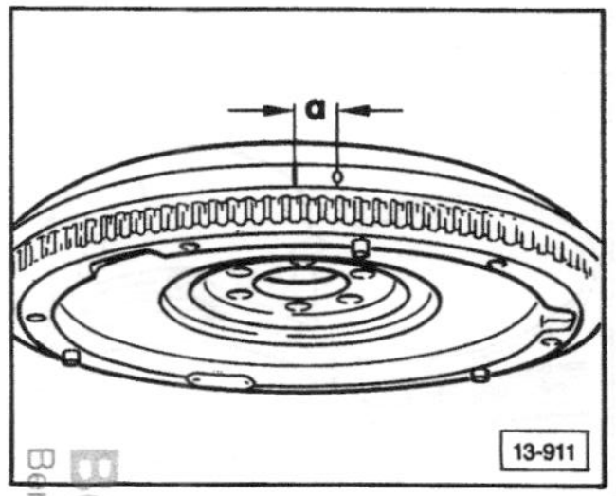

▶ Make ignition timing mark at a point in an arc left from center of TDC marking.

- **a** = 14.5 mm (37/64 in.) along arc

CAUTION

Do not turn crankshaft when measuring radial play.

CAUTION

Connecting rod bolts may be reused **only once** when checking radial play with engine installed. After checking radial play mark each bolt on thread side with center punch to show bolt was used once already.

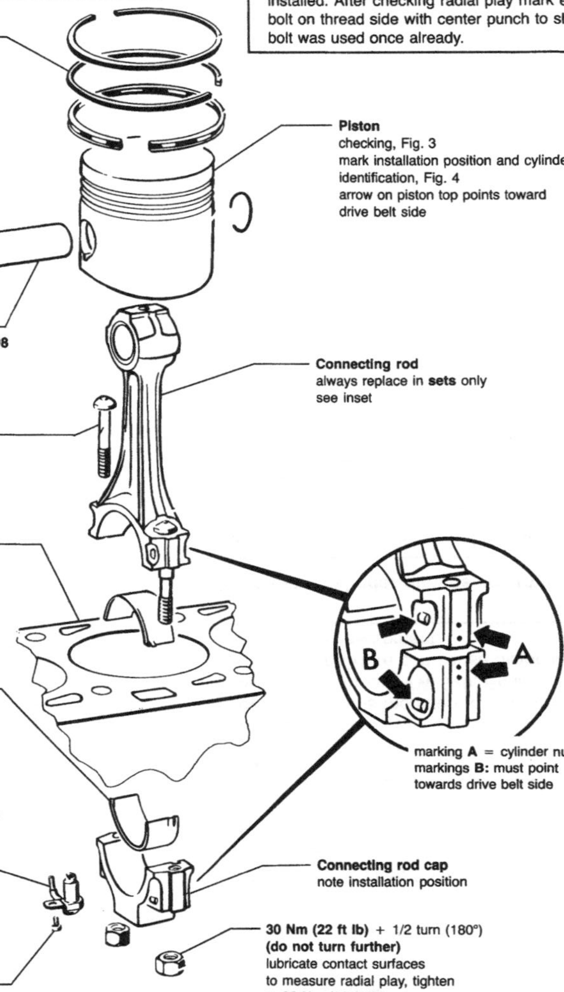

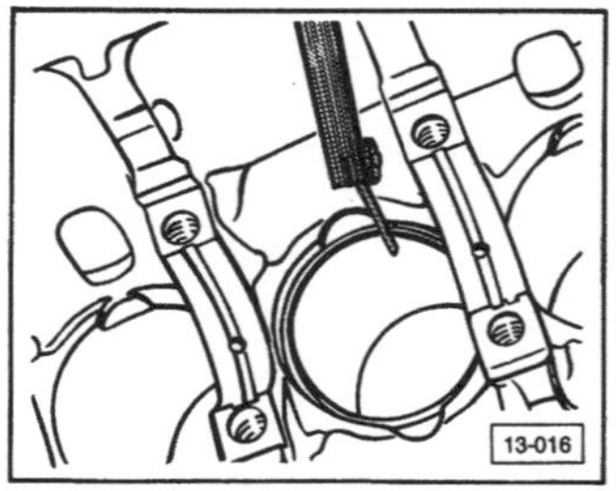

▶ Fig. 1 **Piston rings, checking end gap**

- insert piston ring squarely into cylinder until it is approximately 15 mm (19/32 in.) from bottom edge of cylinder

New
- compression ring: 0.15-0.35 mm (0.006-0.014 in.)
- oil scraper ring: 0.25-0.50 mm (0.01-0.020 in.)

Wear limit:
- 1.0 mm (0.04 in.)

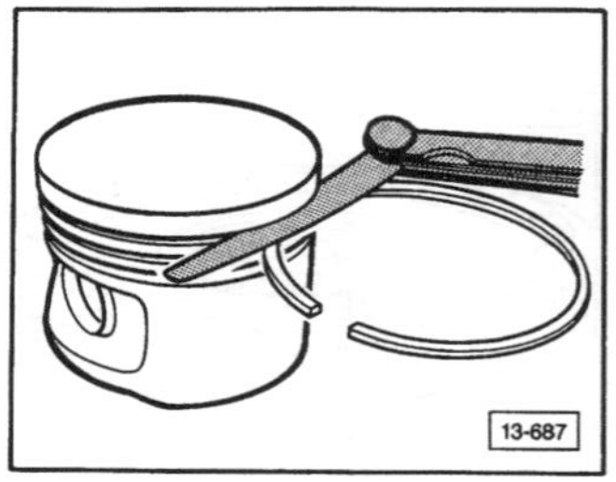

▶ Fig. 2 **Piston ring side clearance, checking**

New:
- compression ring: 0.02-0.07 mm (0.001-0.003 in.)
- oil scraper ring: 0.02-0.06 mm (0.001-0.002 in.)

Wear limit:
- 0.15 mm (0.006 in.)

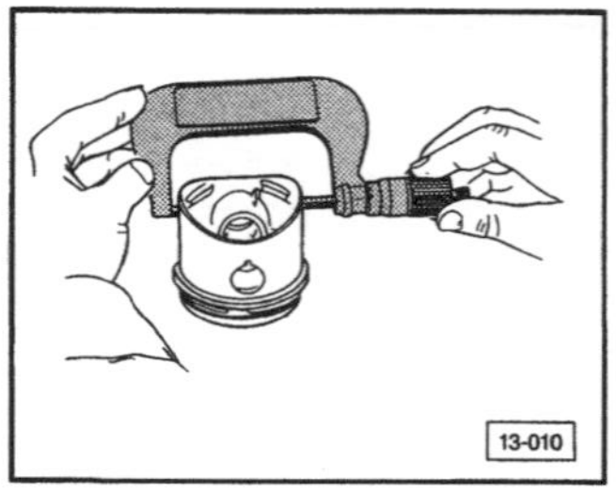

▶ Fig. 3 **Piston, checking**

- measure approximately 10 mm (25/64 in.) from lower edge of skirt at 90° angle to piston pin axis

Nominal dimension tolerance:
- maximum 0.04 mm (0.0016 in.)

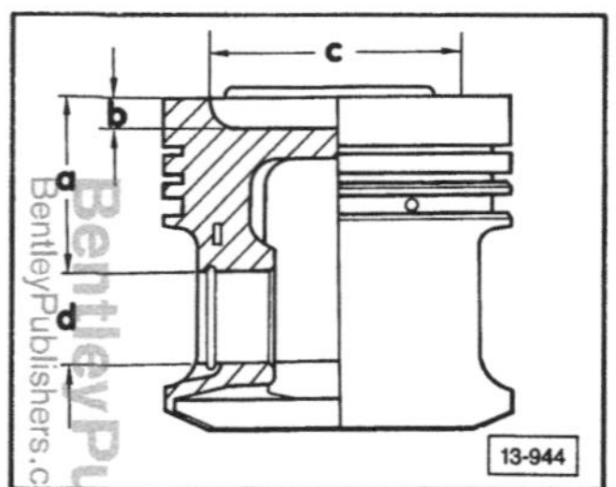

▶ Fig. 4 **Piston identification (mm)**

a = 28.5
b = 8.6
c = 55.0
d = 22.0

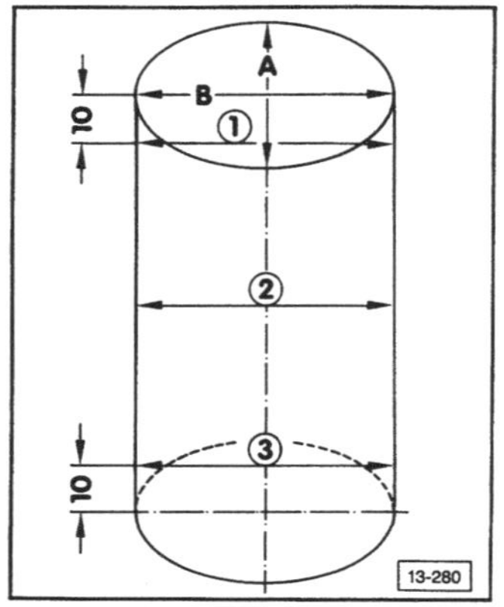

▶ Fig. 5 **Cylinder bore, checking**

- measure at three points in cross direction **A** and longitudinal direction **B**

Use inside micrometer 50-100 mm (2 to 4 inches).

Maximum deviation from nominal dimension:
- 0.08 mm (0.003 in.)

CAUTION

Do not measure cylinder bore when cylinder block is mounted to work bench with engine mount **VW 540**. Engine mounted in fixture can cause distortion affecting measurement.

Piston and cylinder dimensions (mm)

Size	Piston	Cylinder bore
Standard	80.985 mm	81.01 mm
1st oversize	81.23 mm	81.26 mm
2nd oversize	81.48 mm	81.51 mm

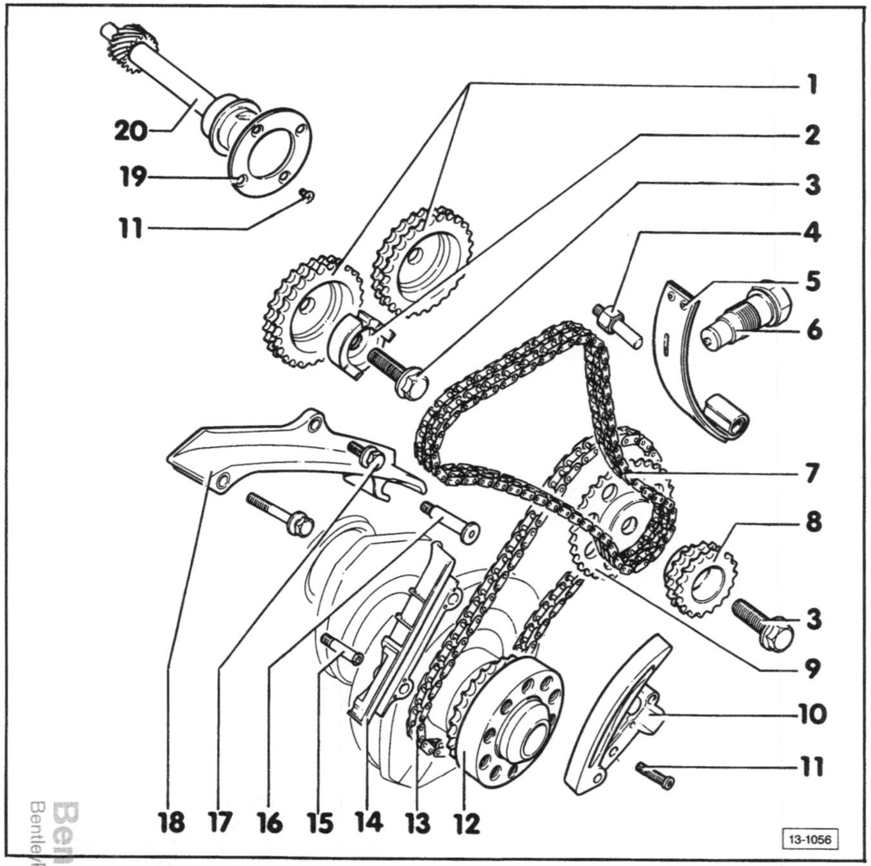

CAUTION
Coolant/anti-freeze must not be reused when replacing engine, cylinder head, cylinder head gasket, radiator and heater core.

CAUTION
Always replace gaskets and seals, with exception of camshaft cover or sealing flange, see page 13.19.

WARNING
DO NOT re-use any fasteners that are worn or deformed in normal use. Many fasteners are designed to be used only once and become unreliable and may fail when used a second time. This includes, but is not limited to, nuts, bolts, washers, self-locking nuts or bolts, circlips, cotter pins. Always follow recommendations given in this publication. For replacements always use new parts.

1 – **Camshaft sprockets**
Removing, see Repair Group 15

2 – **Distributor drive**

3 – **Bolt — 100 Nm (74 ft lb)**

4 – **Pivot pin — 25 Nm (18 ft lb)**

5 – **Tensioning plate**

6 – **Tensioner — 20 Nm (15 ft lb)**

7 – **Double drive chain**
note direction of travel, see page 13.18

8 – **Sprocket, double drive chain**

9 – **Sprocket, single drive chain**

10 – **Chain tensioner with plate**

11 – **Bolt — 10 Nm (7 ft lb)**

12 – **Drive sprocket**
- installed on crankshaft
- ground down tooth on bearing groove points toward **TDC** cylinder 1
- installing chain tensioner with plate, see page 13.24a

13 – **Single drive chain**
note direction of travel, see page 13.18

14 – **Guide rail**

15 – **Retaining bolt — 25 Nm (18 ft lb)**

16 – **Locating pin w/collar — 25 Nm (18 ft lb)**

17 – **Bolt — 20 Nm (15 ft lb)**

18 – **Guide rail**

19 – **Guide ring**

20 – **Intermediate shaft**

Drive chains, marking direction of travel

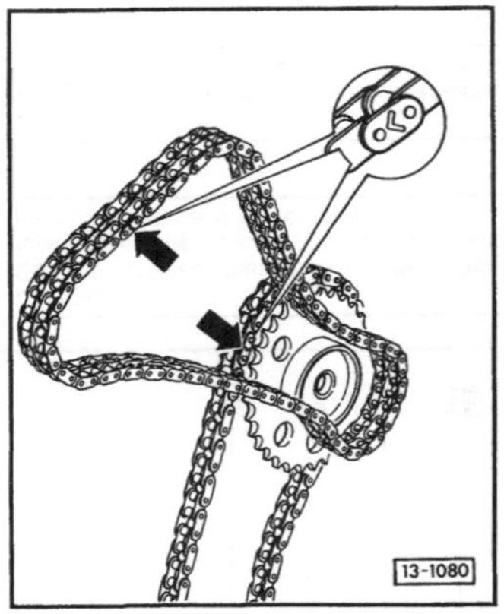
13-1080

- using touch-up paint, mark direction of travel of drive chains before removing (**arrow**)

THIS PAGE INTENTIONALLY LEFT BLANK

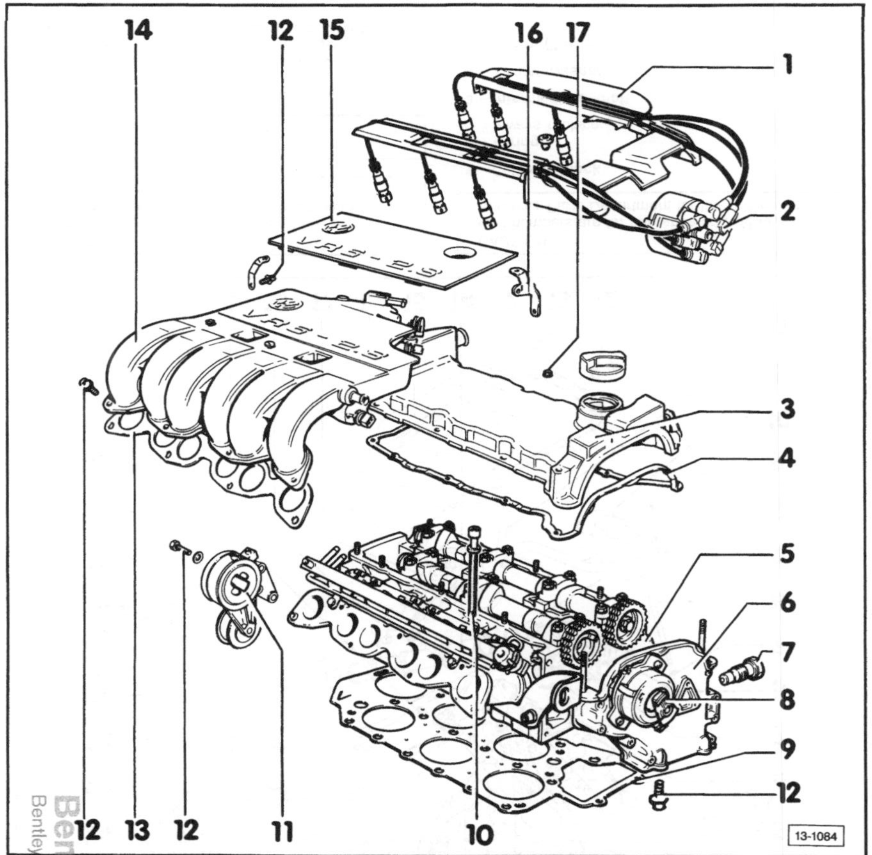

CAUTION
Coolant/anti-freeze must not be reused when replacing engine, cylinder head, cylinder head gasket, radiator and heater core.

CAUTION
Always replace gaskets, O-rings, and seals.

CAUTION
Part numbers are for reference only. Always check with your Parts Department for latest information.

1 – **Ignition wire guide**

2 – **Ignition distributor cap, with ignition wires**

3 – **Cylinder head cover**

4 – **Cylinder head cover seal**
- note position when reinstalling
- always replace

5 – **Cylinder head**
removing/installing, see Repair Group 15

6 – **Camshaft sprocket cover**
coat sealing surface with **AMV 188 001 02**

7 – **Tensioner bolt for double drive chain – 20 Nm (15 ft lb)**

8 – **Ignition distributor**
replacing rotor, see Repair Group 28

9 – **Cylinder head gasket**
- install so that word "TOP" faces cylinder head
- always replace

10 – **Cylinder head bolts**
- torque and tightening sequence, see Repair Group 15
- always replace

11 – **Tension roller, poly-ribbed belt**
removing/installing poly-ribbed belt, see page 13.25

12 – **Bolt – 25 Nm (18 ft lb)**

13 – **Upper intake manifold gasket**
always replace

14 – **Intake manifold, upper**
installing: tighten lower intake manifold, then intake manifold support **16**

15 – **Trim cover**

16 – **Intake manifold support**

17 – **10 Nm (7 ft lb)**

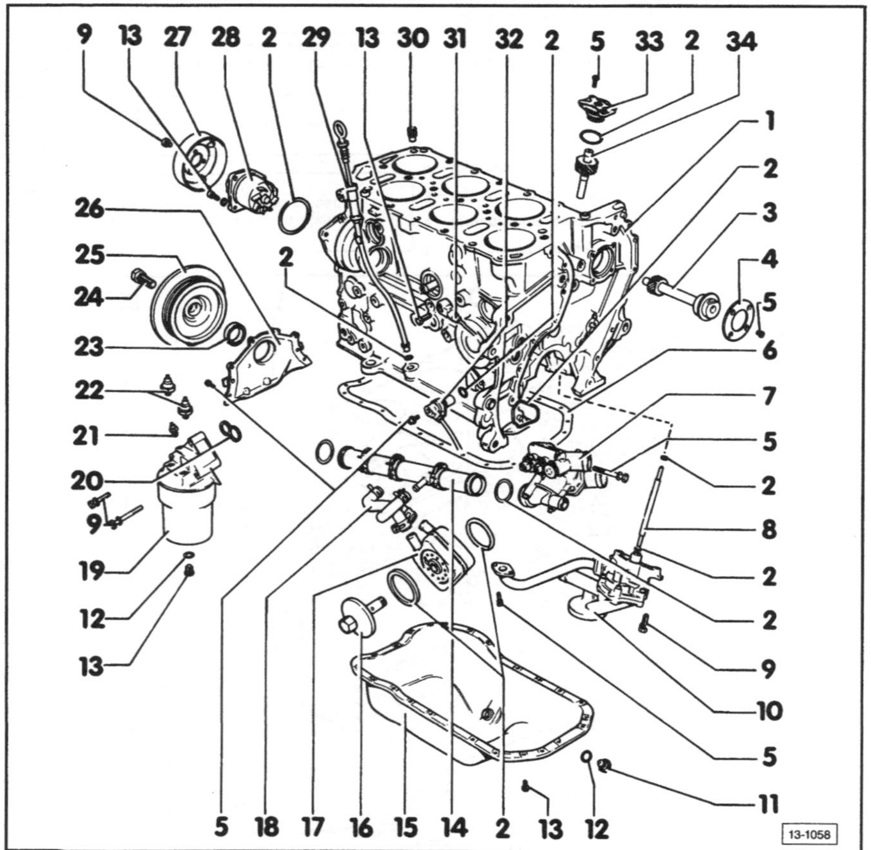

CAUTION
Coolant/anti-freeze must not be reused when replacing engine, cylinder head, cylinder head gasket, radiator and heater core.

CAUTION
Always replace gaskets and seals.

CAUTION
If metal shavings are found in the engine oil as a result of engine damage; clean oil passages thoroughly, then replace all oil spray jets, oil pressure relief valve, oil cooler and oil filter.

CAUTION
Part numbers are for reference only. Always check with your Parts Department for latest information.

1 – **Cylinder block**

2 – **O-ring**
always replace

3 – **Intermediate shaft**

4 – **Thrust washer**

5 – **10 Nm (7 ft lb)**

6 – **Oil pan gasket**
- before installing, coat sealing flange /cylinder block with **AMV 188 001 02**
- always replace

7 – **Thermostat housing**

8 – **Oil pump driveshaft**

9 – **Bolt – 25 Nm (18 ft lb)**
remove and install pulley **27** with tool **VAG 1590**

10 – **Oil pump**
assembling/disassembling, see Repair Group 17

11 – **Oil drain plug – 50 Nm (37 ft lb)**

12 – **Gasket**
always replace

13 – **Bolt – 20 Nm (15 ft lb)**

14 – **Coolant pipe w/drain plug**

15 – **Oil pan**

16 – **Cover, oil cooler – 25 Nm (18 ft lb)**

17 – **Oil cooler**
- coat seal surface with **AMV 188 001 02**
- ensure adequate side clearance

18 – **Coolant hose to crankcase**

19 – **Oil filter housing – 30 Nm (22 ft lb)**
removing /installing, see Repair Group 17

20 – **O-ring**
- note installation position
- always replace

21 – **Oil temperature sender – 10 Nm (7 ft lb)**

22 – **22 – Oil pressure switch – 25 Nm (18 ft lb)**
checking, see Repair Group 17

23 – **Crankshaft oil seal, drive belt side**
removing/installing, see page 13-26

24 – **100 Nm (74 ft lb) +1/4 (90°) turn**
- use support bracket **3273** when loosening /tightening vibration dampener
- always replace

25 – **Vibration dampener**

26 – **Sealing flange**
coat sealing surface with **AMV 188 001 02**

27 – **Coolant pump pulley**

28 – **Coolant pump**
removing/installing, see Repair Group 19

29 – **Dipstick w/guide tube**

30 – **Oil pressure relief valve – 5 Nm (44 in. lb)**
- note installation position
- replace if badly contaminated

31 – **Knock sensor 2 (G66) – 20 Nm (15 ft lb)**
- torque has influence on operation of knock sensor
- torque specifications must be strictly adhered to
- do **NOT** install with spring washer or equivalent

32 – **Speed/reference Sender (G28)**

33 – **Oil pump drive cover**
coat O-ring with engine oil before installing

34 – **Oil pump drive**

Valve timing, checking and adjusting

Vehicles with manual transmission

With engine installed:

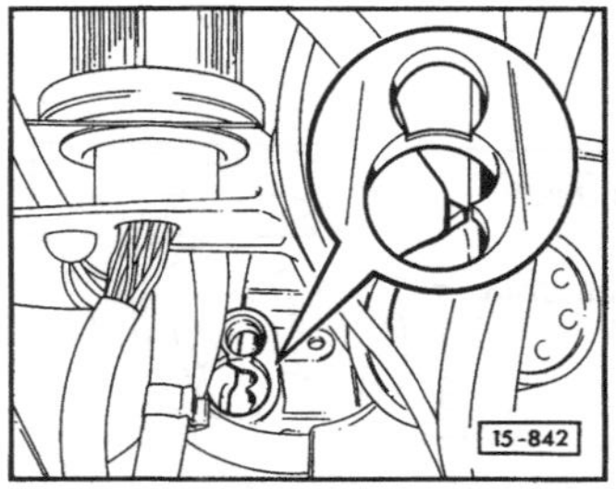

- rotate engine until piston for cylinder **1** is at **TDC**
  - starting with 11-91 production, note 3 cornered tooth (formerly had a rounded tooth)

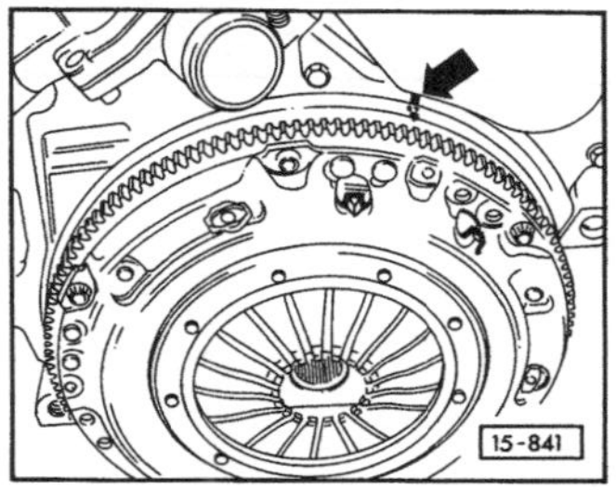

With engine removed

- align casting on sealing flange with **TDC** marking on flywheel
- remove ignition wire guides
- remove intake manifold upper section
- remove valve cover

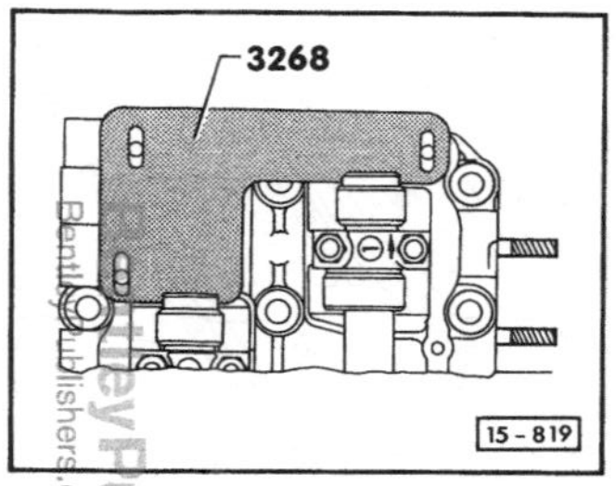

- install and align **3268** camshaft guide onto cylinder head studs with both camshafts snugly positioned in the guide indentations (as shown)
  - this alignment is only possible every 2nd engine revolution

Vehicles with automatic transmission

With engine installed:

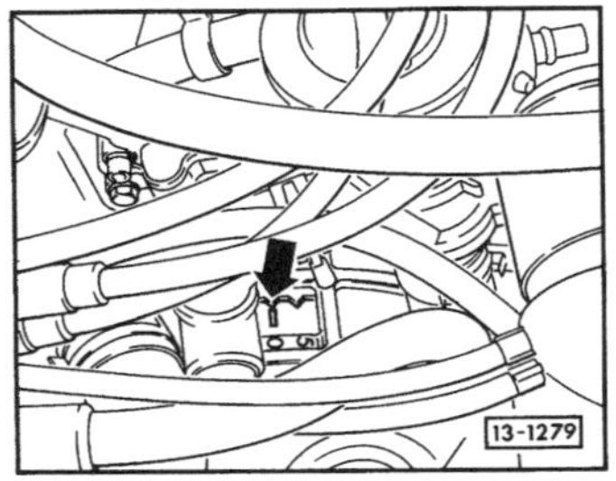

- rotate engine until piston for cylinder **1** is at **TDC**

> **CAUTION**
> There are three marks for cylinder **1** **TDC** on the torque convertor spaced in 120° intervals. To be sure the engine is actually at **TDC** it is mandatory that you verify this condition via the cylinder **1** spark plug hole.

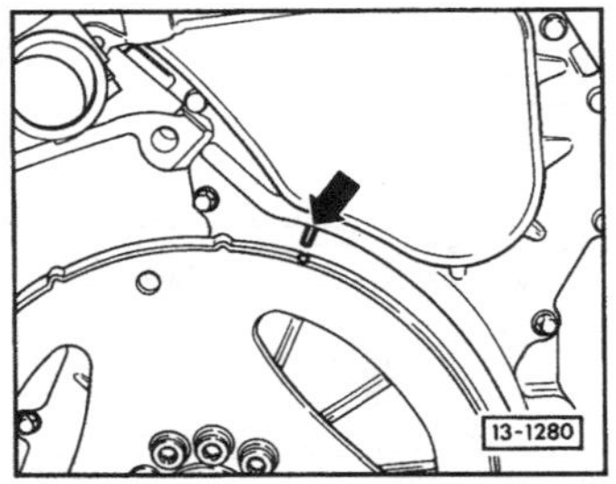

With engine removed:

- align the **TDC** marking on the convertor mounting plate with the mark on the flange
- remove ignition wire guides
- remove intake manifold upper section
- remove valve cover

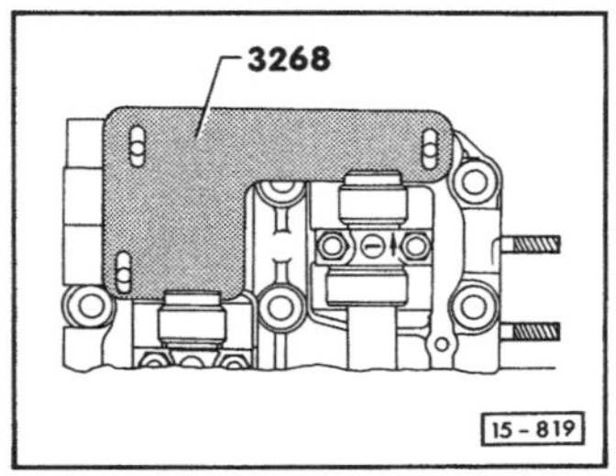

- install and align **3268** camshaft guide onto cylinder head studs with both camshafts snugly positioned in the guide indentations (as shown)
  - this alignment is only possible every 2nd engine revolution

Timing chains, installing

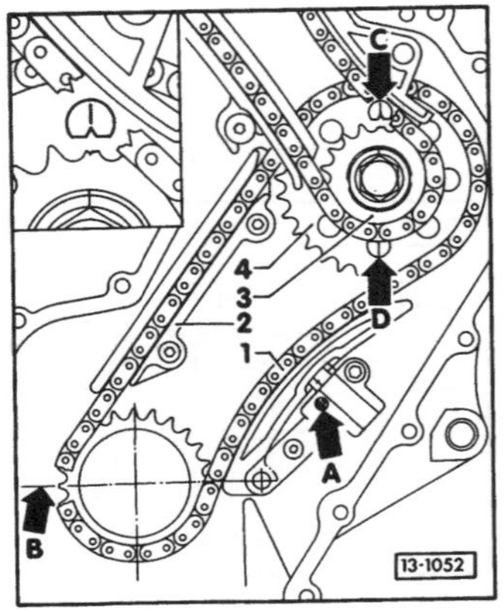

- check position of crankshaft with reference to intermediate shaft
  - ground tooth of drive gear **B** must align with bearing split, re-position if necessary

- install guide rail **2**
- install chain **1** and intermediate shaft gears **3** and **4**
  - note direction of chain rotation, see page 15.28
  - marking on intermediate shaft **4** must align with notch **C** or **D** on the thrust washer

- unlatch lock gear in chain tensioner using a small screwdriver **A**
  - this relaxes the chain tensioner providing additional play for chain routing

- tighten intermediate shaft gears **3** and **4**
  - tighten to 100 Nm (74 ft lb)

To prevent the engine from turning while tightening the shaft gears (Manual transmission):

- fasten flywheel/driving plate to the crankshaft with at least 3 bolts and lock in place with holding tool **VW 558**

To prevent the engine from turning while tightening the shaft gears (Automatic transmission):

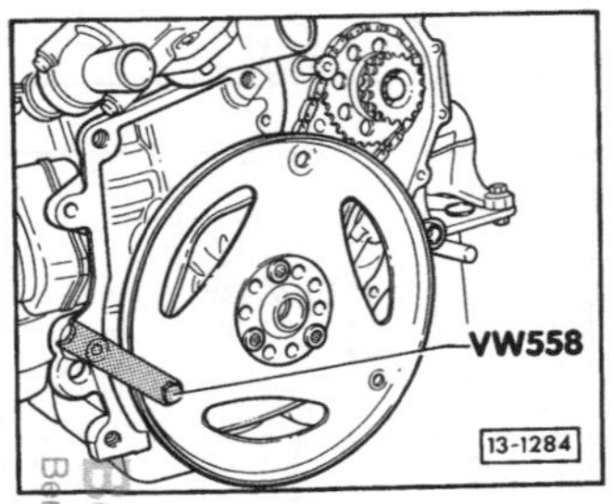

- fasten drive plate to engine using holding tool **VW 558**
  - use an M8 x 45 bolt
  - place 2 M10 hex nuts between the holding tool and the drive plate

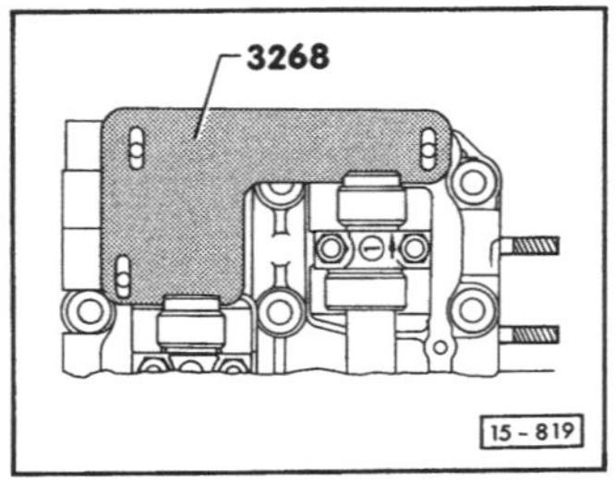

- position the camshafts in the cylinder head using the **3268** camshaft guide
  - with the cylinder head removed

- install cylinder head, see Repair Group 15
- install tensioning rail for chain tensioner
  - dual roller chain

- position rail on both bearing bolt and cylinder head
- position dual roller chain on intermediate shaft gear
  - note previously marked run direction

- attach camshaft chain gear and dual roller chain to the shorter camshaft (cylinders **2**, **4** and **6**) using distributor connector and hand tighten
- attach camshaft chain gear and dual roller chain to the longer camshaft (cylinders **1**, **3** and **5**) and hand tighten
- remove **3268** camshaft guide
- tighten camshaft chain gears
  - tightening torque: 100 Nm (74 ft lb)

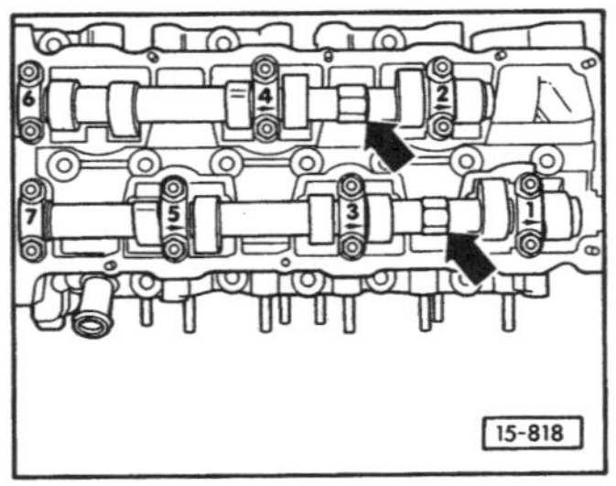

- use a 24mm wrench on hexes (**arrows**) to prevent camshaft rotation while torquing drive gears

CAUTION
DO NOT have the **3268** camshaft guide installed while tightening the camshaft gears.

- coat flange sealing surface with **AMV 18800102** and install
- prepare cylinder head gasket for installation, see Repair Group 15
- coat valve cover sealing surface with **AMV 18800102** and insert sealing ring
- install front camshaft cover
- tighten chain tensioner
- rotate crankshaft two full turns in direction of engine rotation and stop when in cylinder **1 TDC** position

CAUTION
DO NOT rotate the engine unless the chain tensioner is tightened; otherwise, the control chain will jump out.

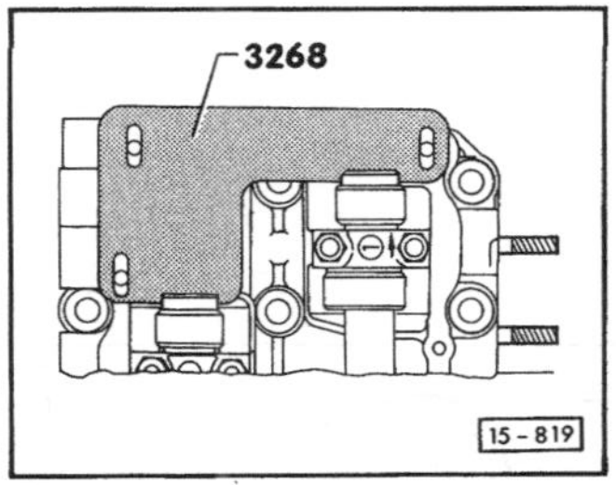

- install and align **3268** camshaft guide onto cylinder head studs with both camshafts snugly positioned in the guide indentations (as shown)
  - this alignment is only possible every 2nd engine revolution

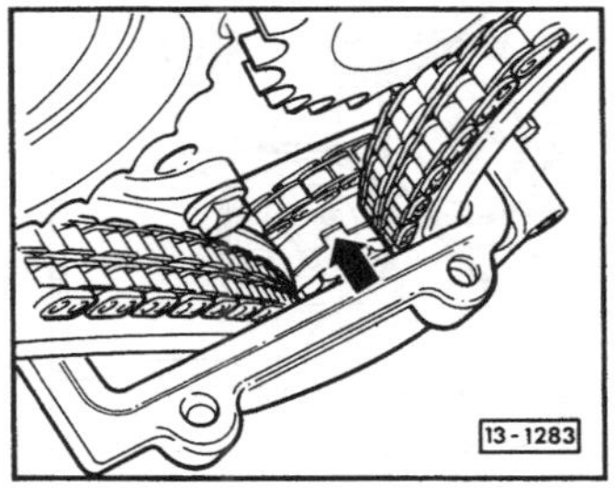

With engine installed and cylinder head removed

- rotate engine until piston for cylinder **1** is in **TDC** position
- when piston is at **TDC**, a notch **(arrow)** should be visible on the intermediate shaft chain gear

If the notch is not visible

- turn crankshaft one complete revolution in direction of normal running rotation

Note

If the engine is rotated with the cylinder head removed, a second mechanic is required to guide the dual roller chain to prevent it from hanging up.

- position the camshaft in the cylinder head using the **3268** camshaft guide
- fill the 3 mm holes in the cylinder head gasket with **AMV 18800102** sealant
- cylinder head assembling, see Repair Group 15
- insert rail on bearing bolt and bolt to cylinder head
- attach camshaft chain gear and dual roller chain to the shorter camshaft (cylinders **2**, **4** and **6**) using distributor connector and hand tighten
- attach camshaft chain gear and dual roller chain to the longer camshaft (cylinders **1**, **3** and **5**) and hand tighten
- remove camshaft **3268** camshaft guide
- tighten camshaft chain gear
  - tightening torque: 100 Nm (74 ft lb)

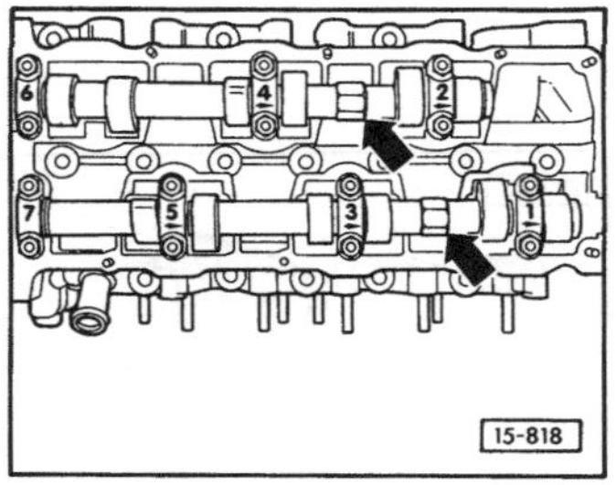

- use a 24mm wrench on hexes **(arrows)** to prevent camshaft rotation while torquing drive gears

> **CAUTION**
> **DO NOT** have the **3268** camshaft guide installed while tightening the camshaft gears.

- prepare cylinder head gasket for installation, see Repair Group 15
- coat valve cover sealing surface with **AMV 18800102** and insert sealing ring
- install front camshaft cover
- tighten chain tensioner
- rotate crankshaft two full turns in direction of engine rotation and stop when in cylinder **1 TDC** position

> **CAUTION**
> **DO NOT** rotate the engine unless the chain tensioner is tightened; otherwise, the control chain will jump out.

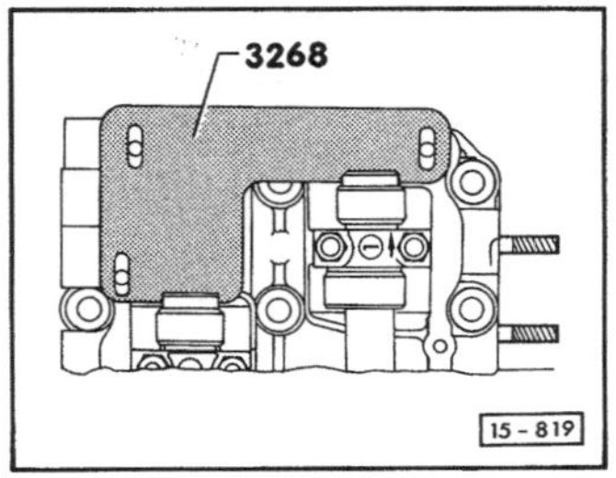

- install and align **3268** camshaft guide onto cylinder head studs with both camshafts snugly positioned in the camshaft guide indentations (as shown)
  - this alignment is only possible every 2nd engine revolution

Poly-ribbed belt, removing/installing

Removing

- remove air cleaner housing

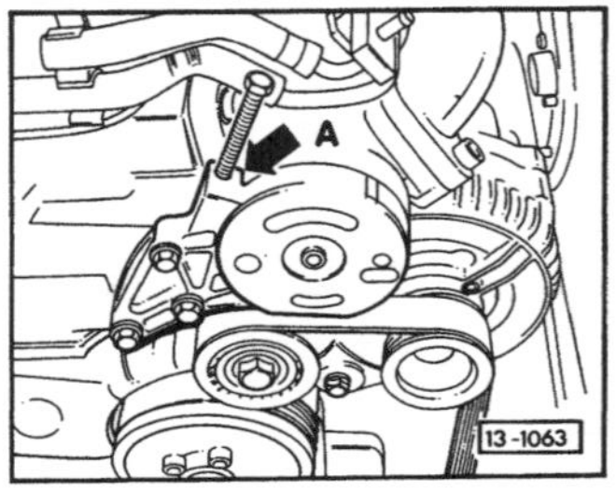

- insert a long (M 8 x 80) bolt into threaded hole **A** on tensioner
- screw in bolt until poly-ribbed belt is loose

Note

Only screw in bolt until belt is loosened. Screwing bolt further will damage tensioner.

- remove poly-ribbed belt

Installing

Note

Before reinstalling belt, ensure all belt driven components, such as alternator, A/C compressor and power steering pump are installed.
Ensure correct mounting of poly-ribbed belt on all pulleys and rollers.

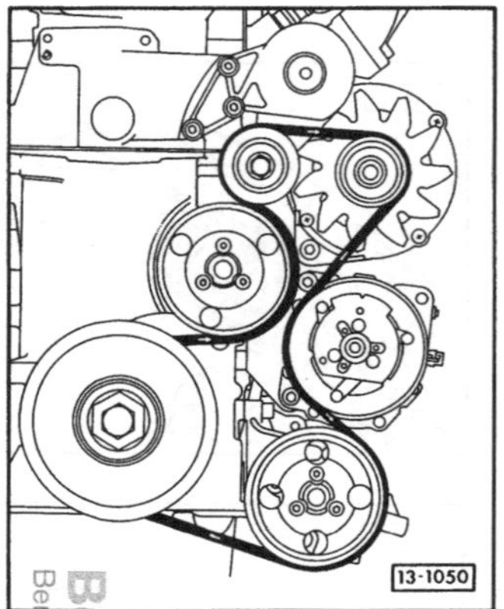

- reinstall and route poly-ribbed belt as shown
- remove bolt from tensioner
- start engine
- ensure belt runs properly

Note

Vehicles with air conditioning are equipped with a double ribbed belt.

Crankshaft oil seal – drive belt side, removing/installing

Removing

- remove poly-ribbed belt
- using support bracket **3273**, remove vibration dampener
- unscrew inner part of oil seal extractor **3203** three turns (approximately 4.0 mm) out of outer part and lock in position with knurled screw
- lubricate threaded head of oil seal extractor, set in position and push as far as possible into oil seal
- loosen knurled screw and turn inner part against crankshaft until oil seal is pulled out
- clamp extractor in a vise and remove oil seal using pliers

Installing

- lightly lubricate sealing lip and outer edge of new oil seal

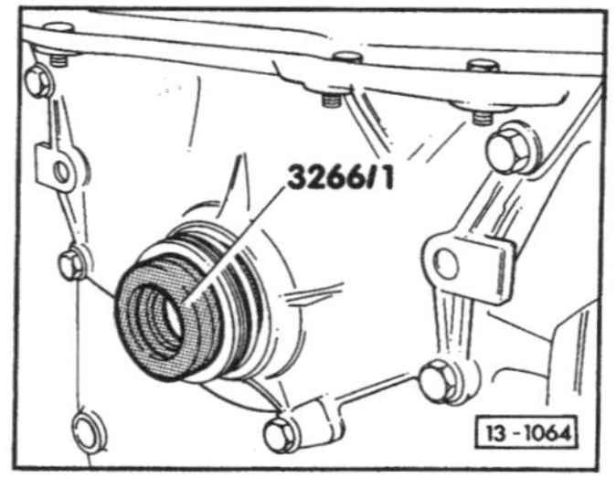

- place guide sleeve from **3266/1** onto crankshaft pin and push oil seal over guide sleeve

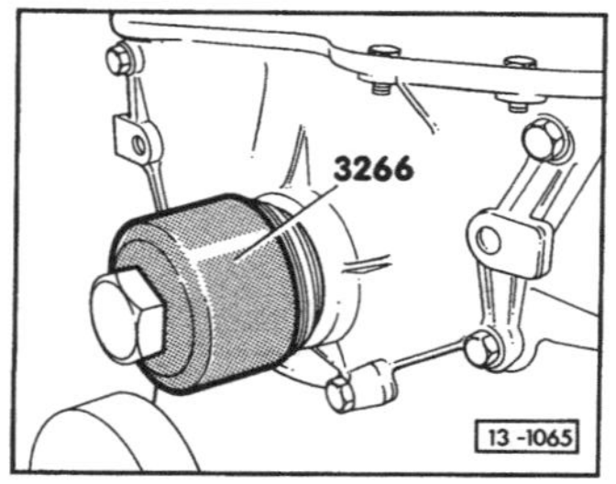

- using vibration dampener securing bolt, press in oil seal with sleeve **3266** up to stop
- using support bracket **3273**, reinstall vibration dampener
  - torque 450 Nm (328 ft lb)
- reinstall poly-ribbed belt

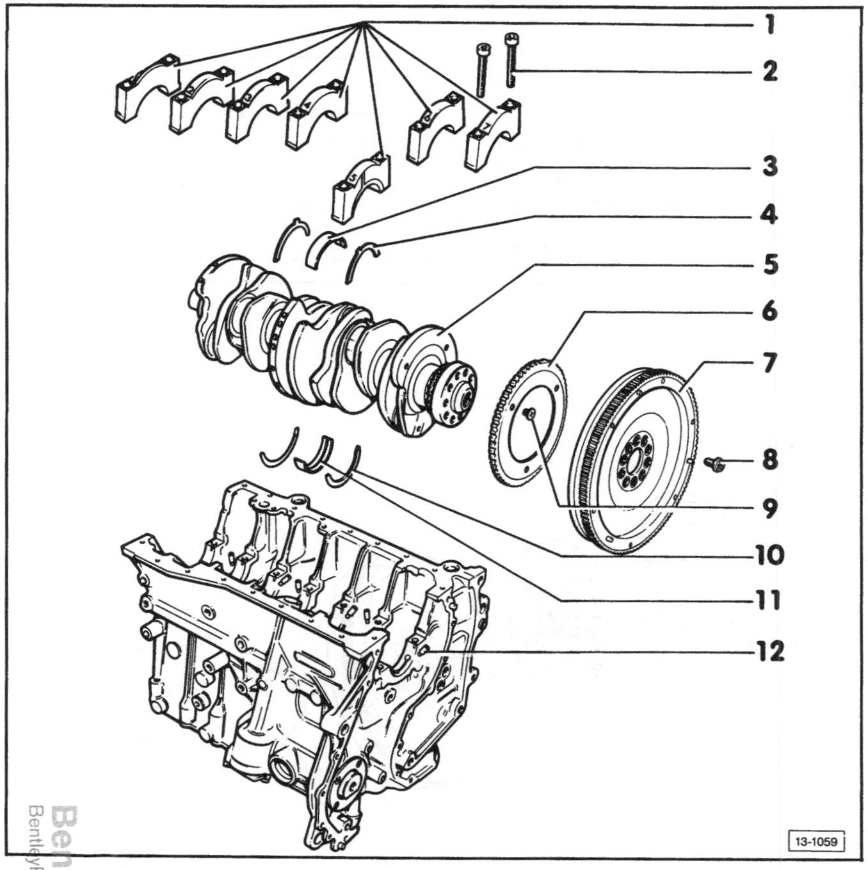

1 – **Main bearing caps**
- cap **1** on drive belt side
- note installation position

2 – **30 Nm (22 ft lb) + 1/2 (180°) turn**
Additional 1/2 (180°) turn may be done in two 90° steps

3 – **Bearing shells 1-7**
for main bearing caps

4 – **Thrust washer**
- for bearing **5**
- note installation position in bearing cap

5 – **Crankshaft**
- end play:
  New: 0.07-0.17 mm (0.003-0.007 in.)
  Wear limit: 0.25 mm (0.01 in.)
- radial play: check with Plastigage©
  New: 0.02-0.06 mm (0.001-0.002 in.)
  Wear limit: 0.10 mm (0.004 in.)

6 – **Toothed wheel**
- for Engine Speed/Reference Sensor (**G 28**)
- always replace

7 – **Flywheel**
hold flywheel with tool **VW 558** when removing

8 – **60 Nm (44 ft lb) + 1/4 (90°) turn**
- always replace

9 – **10 Nm (7 ft lb) + 1/4 (90°) turn**
always replace

10 – **Thrust washer for bearing 5**
- for cylinder block
- note installation position

11 – **Bearing shells 1-7**
- for cap with oil groove
- do **NOT** interchange worn shells
- for cylinder block

12 – **Oil spray jet**
- removing/installing, see Repair Group 17
- see **CAUTION** page 13.21

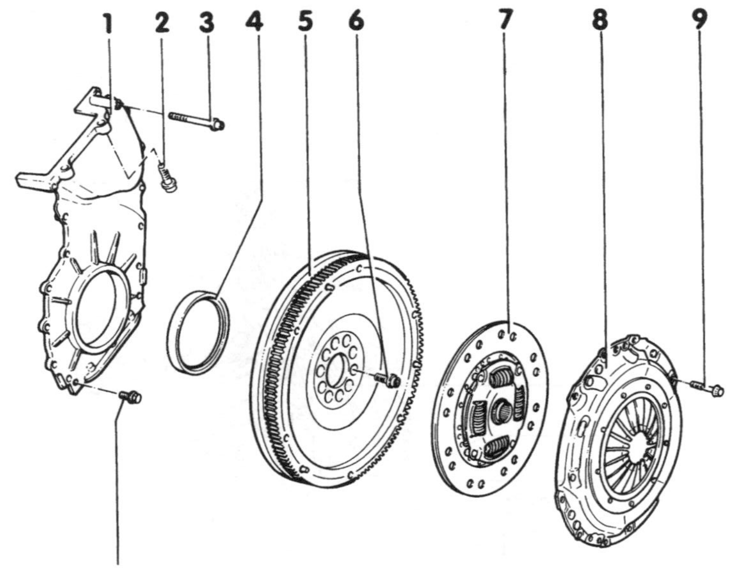

CAUTION
Friction materials such as brake and clutch linings, or brake pads may contain asbestos fibers.

Do **NOT** create dust by grinding, sanding or by cleaning with compressed air.

Avoid breathing asbestos fibers and asbestos dust.

Breathing asbestos may result in serious diseases, such as asbestosis or cancer.

It may cause severe injury and death.

CAUTION
Part numbers are for reference only. Always check with your Parts Department for latest information.

1 – **Seal flange**
- coat sealing surface with **AMV 188 001 02**
- if flange has been removed, check and prepare cylinder head seal, see Repair Group 15

2 – **25 Nm (18 ft lb)**

Engine – Crankshaft, Crankcase

3 – **10 Nm (7 ft lb)**

4 – **Oil seal**
- remove with tool **2086**
- installing: lightly lubricate sealing lip and outer edge
  install with pull sleeves from tool **2003/2A**
  press in seal to stop with tool **2003/1**

5 – **Flywheel**
- can be replaced with engine installed
- removing 5-speed transmission, see Repair Group 34
- hold flywheel with tool **558** when removing

6 – **60 Nm (44 ft lb) + 1/4 (90°) turn**
- always replace

7 – **Clutch disc**
center with tool **3190 A**

8 – **Pressure plate**
- when removing, hold flywheel with tool **558**
- note installed position with **TDC** marking for No. **1** cylinder (rounded, ground down tooth)

9 – **20 Nm (15 ft lb)**
tightly diagonally

10 – **10 Nm (7 ft lb)**

Crankshaft dimensions (in mm)

Wear limit	Crankshaft bearing journal diameter	Connecting rod journals diameter
Standard	54.00 −0.022 / −0.042	47.80 −0.022 / −0.042
1st undersize	53.75 −0.022 / −0.042	47.55 −0.022 / −0.042
2nd undersize	53.50 −0.022 / −0.042	47.30 −0.022 / −0.042
3rd undersize	53.25 −0.022 / −0.042	47.05 −0.022 / −0.042

Drive plate, removing and installing

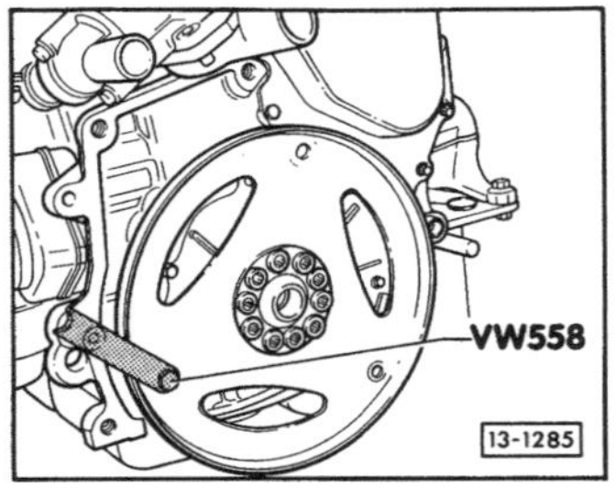

Loosening and tightening

- fasten holding tool **VW 558** to the drive plate using hex head bolt M8 x 45
  - insert 2 M10 hex nuts between holding tool and drive plate

Installing

- position drive plate to crankshaft and tighten using 3 used bolts
  - tighten to 30 Nm (22 ft lb)

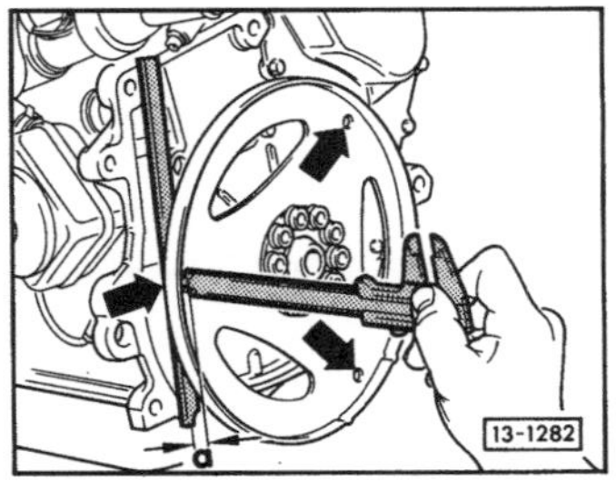

- measure "**a**" through the 3 torque converter mounting holes using a ruler and calculate the average dimension
  - the average dimension (measured distance + ruler thickness) must be between 15.7 and 16.5 mm

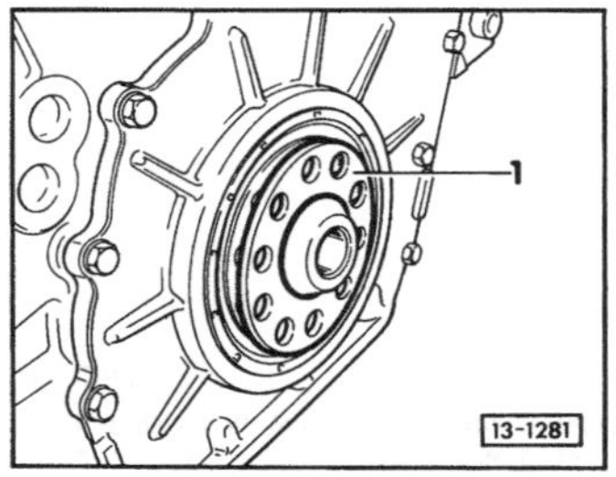

If average value is greater or less than specification

- remove drive plate and remount using used bolts again with the appropriate shims 1 to obtain the specification dimension

If OK

- remove used bolts
- fasten drive plate using new bolts
  - tighten to 60 Nm (44 ft lb) + 90° (1/4 additional turn)

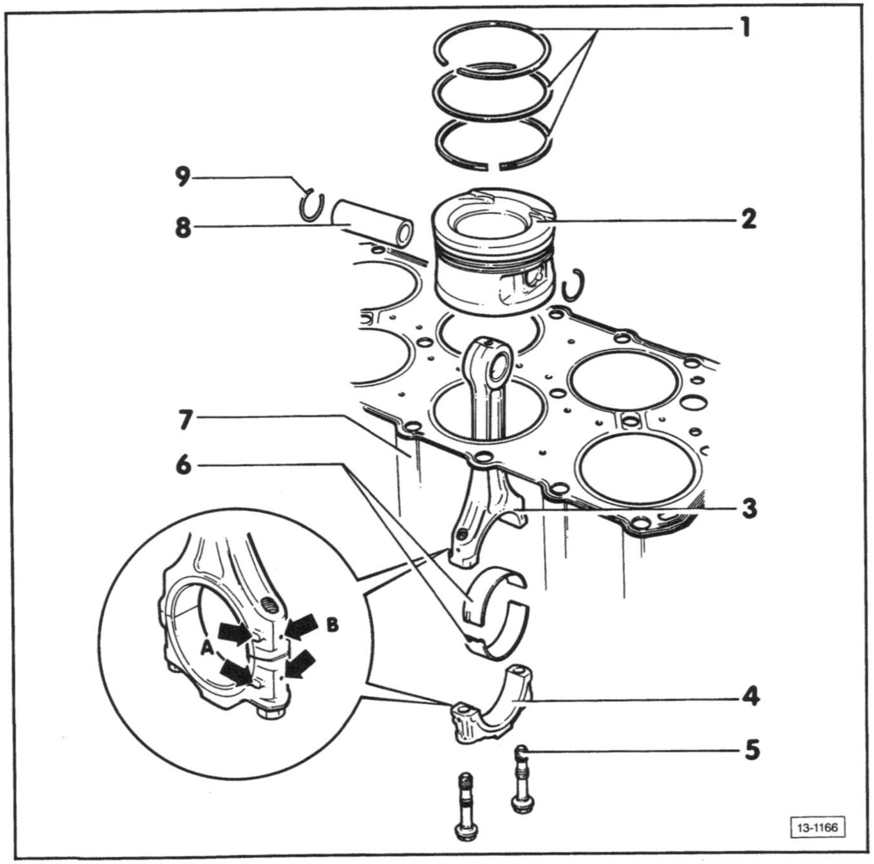

CAUTION
Do **NOT** turn crankshaft when measuring radial play.

1 – **Piston rings**
- offset gaps 120
- remove/install using piston ring pliers
- "**TOP**" marking faces top of piston
- checking gap, see page 13.34
- checking side clearance, see page 13.35

2 – **Piston**
- checking, see page 13.36
- mark installation position and cylinder number
- highest side of piston crown is toward center of cylinder block
- install with tool **3278**

3 – **Connecting rod**
- replace in **sets** only
- markings **B (inset)** indicate cylinder number
- markings **A (inset)** must be on top of each other

4 – **Connecting rod cap**
note installation position

5 – **Connecting rod bolt**
- Use old bolts to check radial play, tighten to **30 Nm (22 ft lb)**, do **NOT** tighten further
- always replace
- new bolts tighten to **30 Nm (22 ft lb) + 1/4 (90°)** turn

6 – **Bearing shell**
- note installation position
- do **NOT** interchange used bearing shells
- retaining tabs must fit securely
- end play,
  New: 0.05-0.31 mm (0.002-0.012 in.)
  Wear limit: 0.4 mm (0.016 in.)
  radial clearance, with Plastigage©,
  New: 0.01-0.06 mm (0.0004-0.002 in.)
  Wear limit: 0.10 mm (0.004 in.)

7 – **Cylinder block**
- checking cylinder bore, see page 13.37
- piston and cylinder dimensions, see page 13.38

8 – **Piston pin**
- remove and install with tool **VW 222a**
- if difficult to remove, heat piston to 60°C (140°F)

9 – **Circlip**

Piston rings, checking end gap

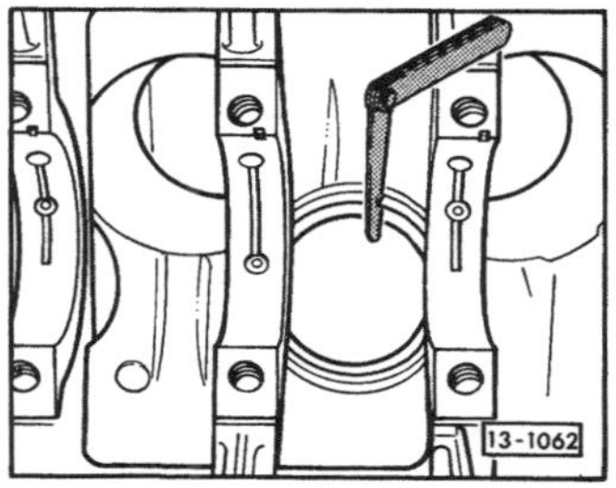

- insert piston ring squarely into cylinder
  - use a piston **without** rings to push
- push ring down until it is approximately 15 mm (19/32 in.) from bottom edge of cylinder

New:
- compression ring: 0.20-0.40 mm (0.008-0.056 in.)
- oil scraper ring: 0.25-0.50 mm (0.010-0.020 in.)

Wear limit:
- 1.0 mm (0.04 in.)

Piston ring side clearance, checking

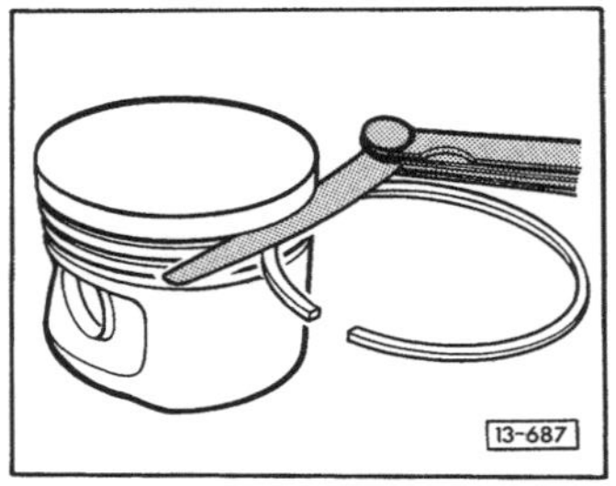

- clean groove before checking

New:
- compression ring: 0.02-0.07 mm (0.001-0.003 in.)
- oil scraper ring: 0.02-0.06 mm (0.001-0.002 in.)

Wear limit:
- 0.15 mm (0.006 in.)

Piston, checking

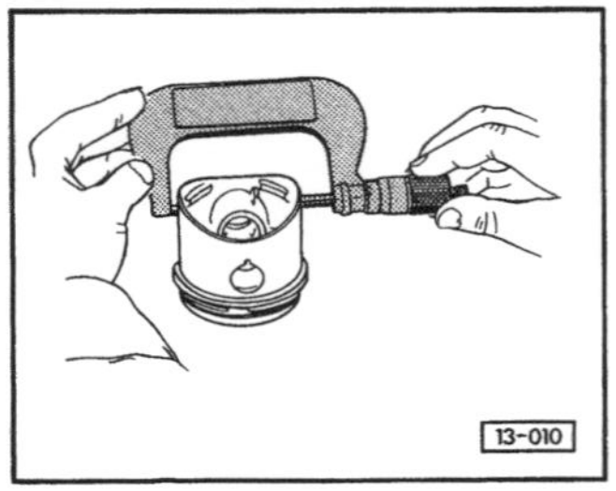
13-010

- measure approximately 6.0 mm (15/64 in.) from lower edge of skirt at 90° angle to piston pin axis

Nominal dimension tolerance:
- maximum 0.04 mm (0.0016 in.)

Cylinder bore, checking

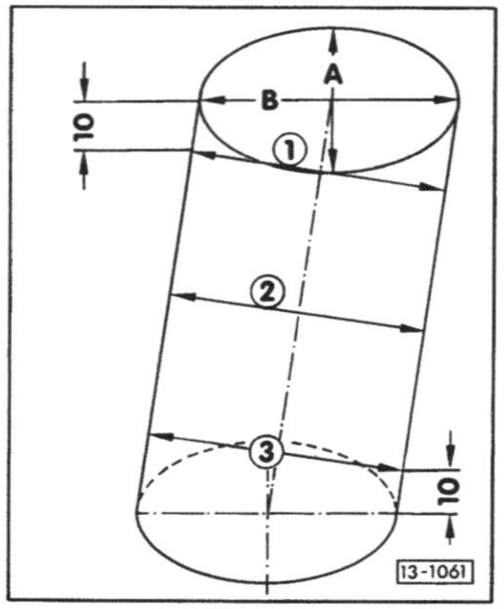

13-1061

- measure at three points in cross direction **A** and longitudinal direction **B**

Use inside micrometer 50 - 100 mm (2 to 4 in.).

Maximum deviation from nominal dimension:
- 0.08 mm (0.003 in.)

> **CAUTION**
> Do not measure cylinder bore when cylinder block is mounted to work bench with engine mount **3269**. Engine mounted in fixture can cause distortion affecting measurement.

Piston and cylinder dimensions (in mm)

Size	Piston	Cylinder bore
Standard	80.985	81.01
1st oversize	81.485	81.51
2nd oversize	81.985	82.01

THIS PAGE INTENTIONALLY LEFT BLANK

THIS PAGE INTENTIONALLY LEFT BLANK

Index

Engine Code – PG

Camshaft
- assembly **15.5**
- identification **15.6**
- oil seal removing/installing **15.9**
- play, checking **15.6**
- removing/installing **15.10**

Cylinder head
- assembly **15.2**
- compressor, checking **15.4**
- distortion, checking **15.3**
- installing **15.3**

Valve
- assembly **15.5**
- dimensions **15.6**
- exhaust valves, discarding **15.8a**
- guides, checking **15.11**
- guides, replacing **15.11**
- hydraulic valve lifters, checking **15.8**
- seats, refacing **15.7**
- stem seal, removing/installing **15.12**

Engine Code – AAA

Camshaft
- assembly **15.19**
- axial play, checking **15.22**
- bearing caps, installed position **15.21**
- removing/installing **15.27**

Compression pressure
- checking values **15.18**

Cylinder head
- assembly **15.13**
- distortion checking **15.16**
- gaskets, replacing **15.15**
- installing **15.17**

Valves
- assembly **15.14**
- dimensions **15.23**
- guides, checking **15.29**
- guides, replacing **15.31**
- hydraulic valve lifters, checking **15.26**
- seats, replacing **15.24**
- stem seals, removing/installing 15.32

CAUTION

Coolant/antifreeze **must not** be reused when replacing engine, cylinder head, cylinder head gasket, radiator and heater core.

Note

The cylinder head may be removed with engine installed.

When replacing cylinder head, lubricate hydraulic valve and camshaft contact surface.

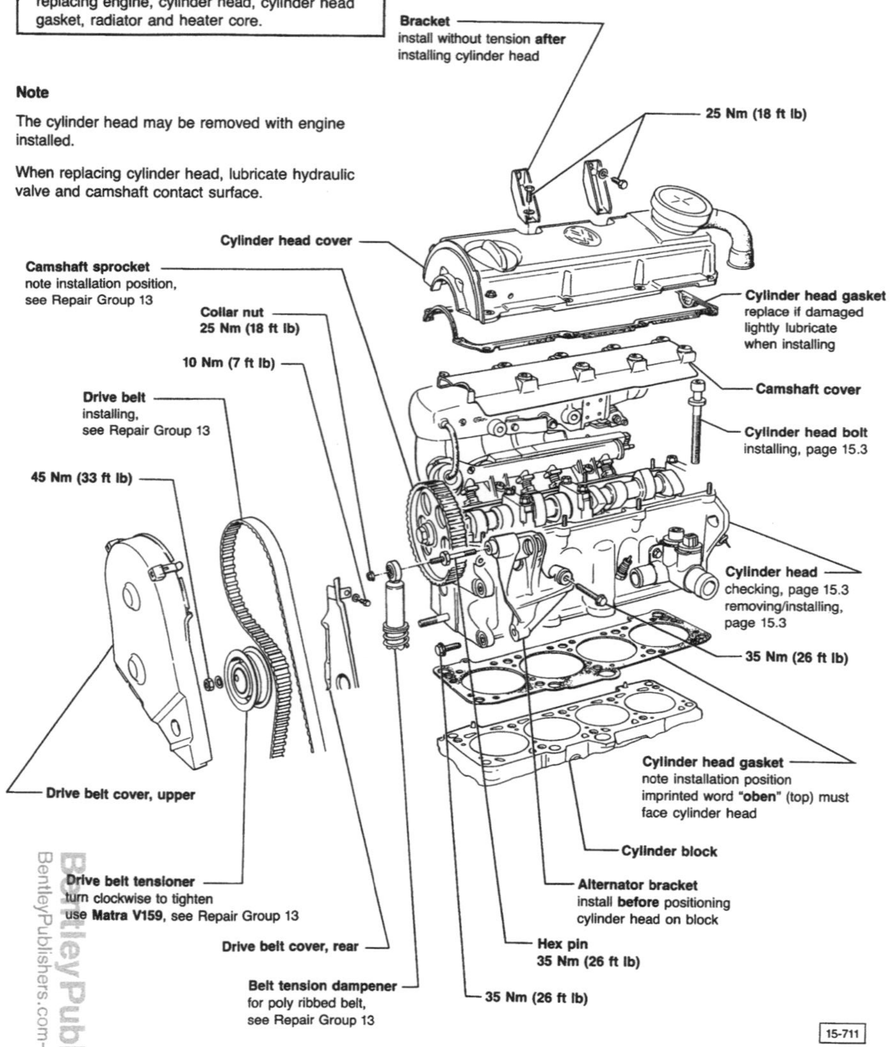

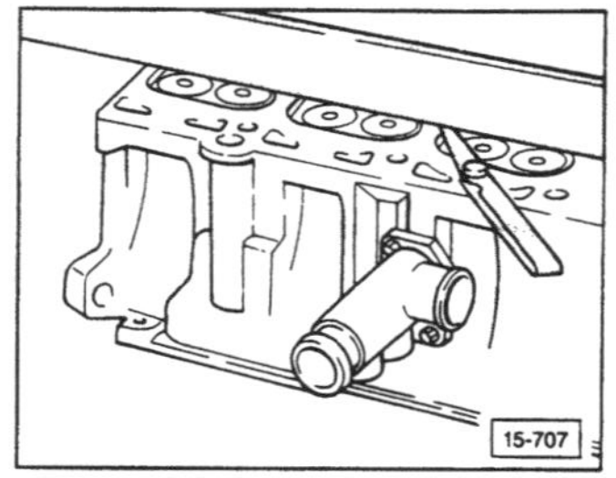

▶ Fig. 1 **Cylinder head, distortion checking**

Note

Cylinder head should be clean and free of gasket material before performing this check.

- use straight edge **US 1030** and feeler gauge to check for distortion
  - maximum permissible distortion allowed: 0.1 mm (0.004 in.)

Cylinder head, installing

- install cylinder head gasket
  - note correct position
- screw in guide bolt from tool **3070** in threaded holes **8** and **10**
- install cylinder head
- fit all cylinder head bolts
- remove guide bolt with tool **3070** and replace with cylinder head bolt
- **hand tighten only**

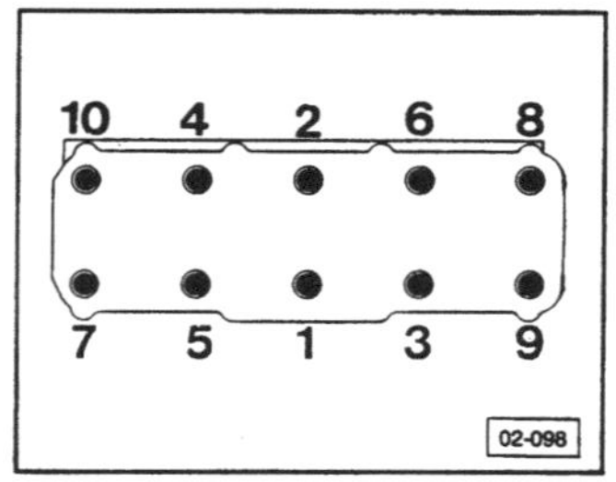

▶ **Tightening torque: engine cold**

- tighten in sequence to 40 Nm (29 ft lb)
- tighten in sequence to 60 Nm (43 ft lb)
- tighten in sequence additional 1/2 turn (180°) further in one movement (two 90° turns are permissible)
- remove cylinder head bolts in reverse sequence

Note

It is not necessary to re-tighten cylinder head bolts during scheduled maintenance or after performing repairs.

Compression pressure, checking

- engine oil temperature minimum 30°C (86°F)
- throttle plate completely open
- ignition coil lead removed from ignition distributor and connected to ground
- check compression pressure using pressure recorder **US 1120**
- follow operating instructions for recorder
- operate starter until recorder unit indicates no further pressure increase

Compression pressure values bar (psi)

Engine code letters	New	Wear limit
PG	8-12 (116-174)	6 (87)

- permissible difference between all four cylinders: 3 bar (43.5 psi)

Note

Heads with fine cracks between valve seats or a valve seat ring and plug threads can be installed again without reducing engine service life provided that cracks are small and not more than 0.5 mm wide or when only the first coils of the plug threads are cracked.

CAUTION

Coolant/antifreeze **must not** be reused when replacing engine, cylinder head, cylinder head gasket, radiator and heater core.

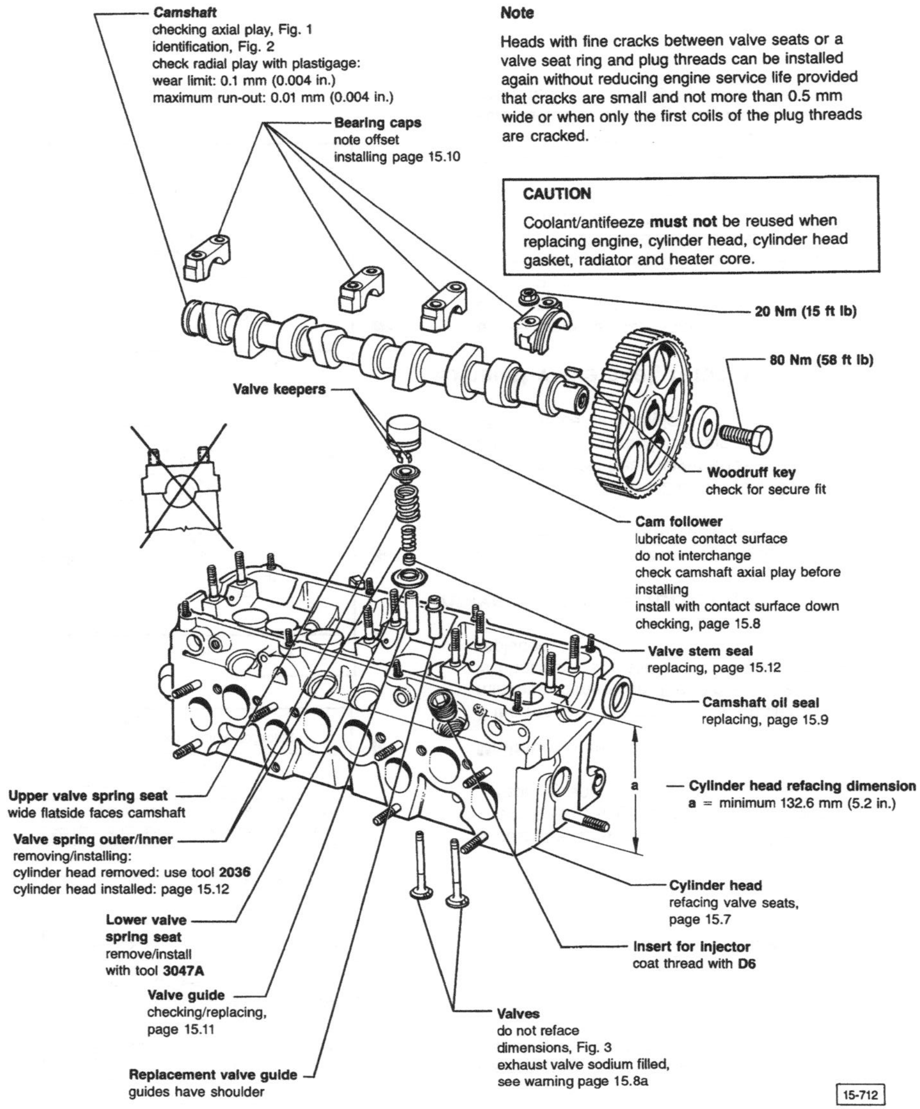

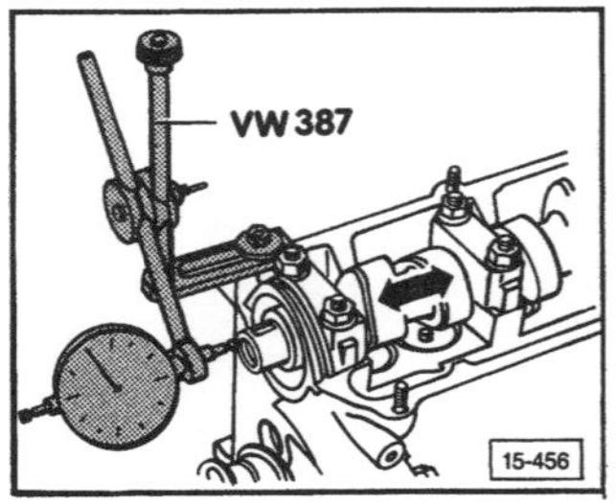

Fig. 1 **Camshaft, axial play, checking**

- measure with valve lifters removed, and bearing caps **1** and **5** removed
  - wear limit: 0.15 mm (0.006 in.)

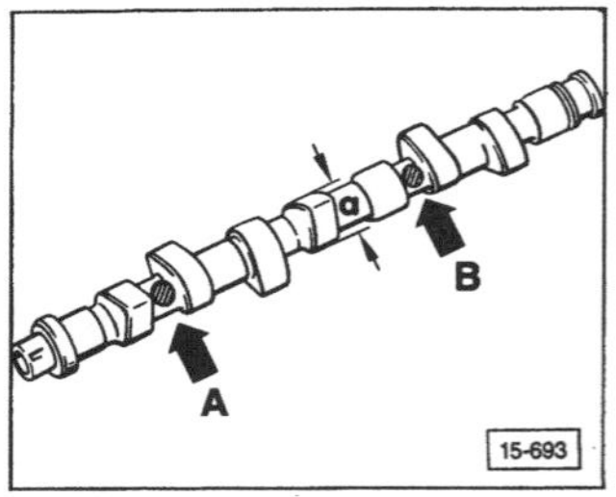

Fig. 2 **Camshaft identification**

Identification between cams

Engine code	Cylinder one arrow A	Cylinder three arrow B
PG	A	026

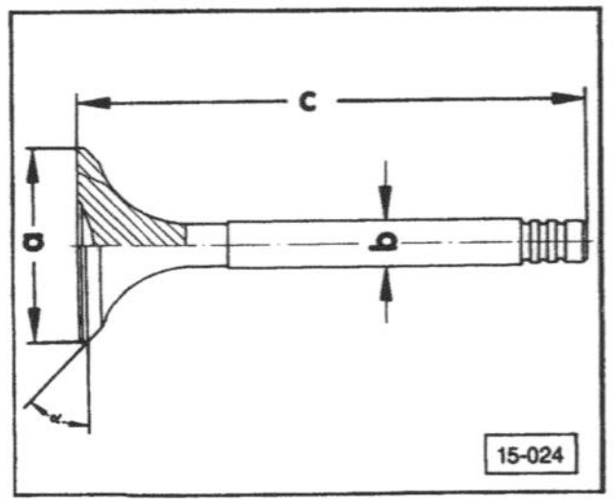

Fig. 3 **Valve dimensions**

	Intake	Exhaust
a	40.00 mm (1.575 in.)	33.15 mm (1.305 in.)
b	7.97 mm (0.314 in.)	7.94 mm (0.313 in.)
c	91.40 mm (3.598 in.)	91.45 mm (3.6 in.)
α	45	45

CAUTION

Do not reface valves, only reface seats.

Valve seats, refacing

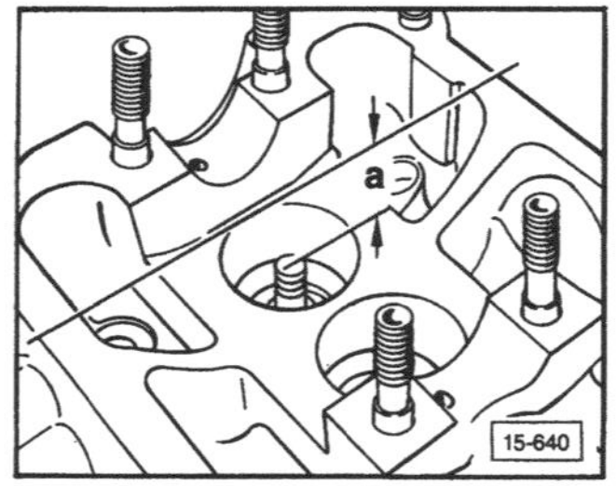

Calculating the maximum permissible refacing dimension

- insert valve and hold firmly against valve seat
- measure distance **a** between end of valve shaft and upper edge of cylinder head

Minimum dimension — intake valve = 33.8 mm
Minimum dimension — exhaust valve = 34.1 mm

measured distance **a**
— minimum dimension

= maximum permissible refacing dimension **b**

- calculate refacing dimension **b** (see example) using the measured distance **a** (above) and:

Minimum dimension

If the measured distance **a** is smaller than the minimum dimension, the hydraulic valve lifter will not work and the cylinder head must be replaced.

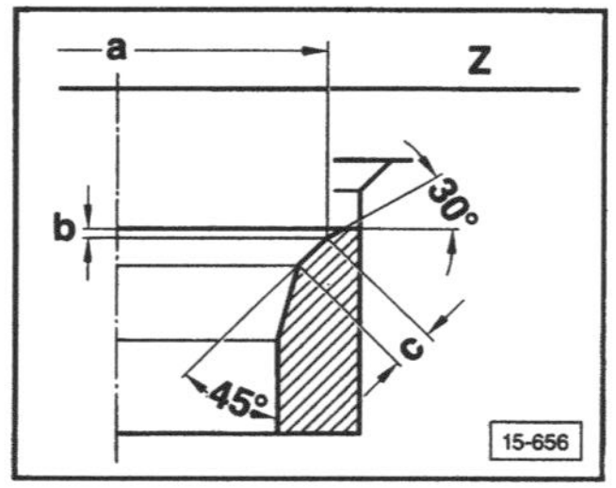

Intake valve seat, refacing

- **a** — diameter 39.2 mm (1.543 in.)
- **b** — maximum permissible refacing dimension
- **c** — approximately 2.0 mm (0.079 in.)
- **Z** — cylinder head lower edge
- **30°** — upper correction angle
- **45°** — valve seat angle

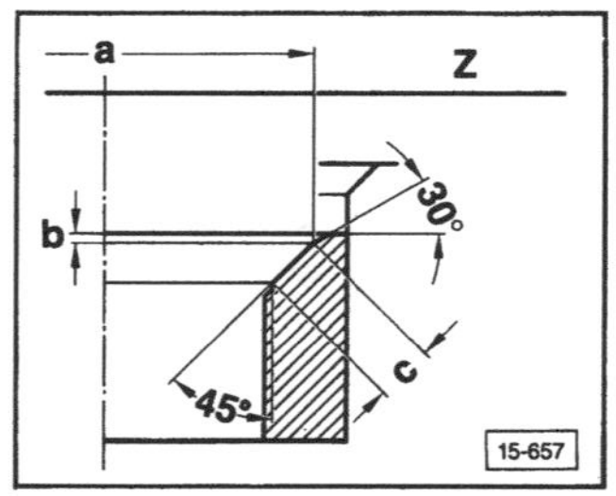

Exhaust valve seat, refacing

- **a** — diameter 32.4 mm (1.276 in.)
- **b** — maximum permissible refacing dimension
- **c** — approximately 2.0 mm (0.079 in.)
- **Z** — cylinder head lower edge
- **30°** — upper correction angle
- **45°** — valve seat angle

Exhaust valve seat rings are also made with a narrowed diameter. When refacing make sure radius of the narrowed diameter is not damaged.

> **CAUTION**
>
> After replacing lifters, do not start engine for approximately 30 minutes, otherwise valves may hit piston. Lifter must be allowed to bleed down to proper adjustment.

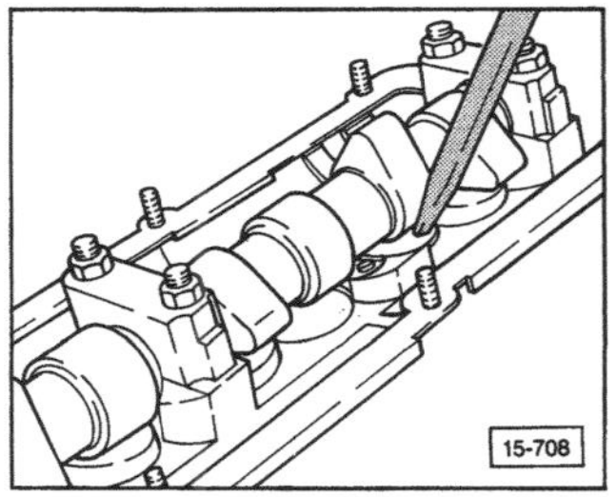

Hydraulic valve lifters, checking

Always place removed valve lifter on a clean surface with the contact surface (camshaft side) facing downward.

Always replace complete valve lifter. Lifter cannot be adjusted or repaired.

Unusual noise from valve lifter while starting is normal.

- run engine until radiator fan comes on at least once
- bring engine speed to approximately 2500 RPM for two minutes

If valve lifter is still noisy, replace as follows:
- remove cylinder head cover
- turn crankshaft pulley bolt clockwise until cam lobes of cylinder to be checked point upward
- push down against valve lifter with light pressure using a suitable wood stick (as shown). If valve lifter can be pushed down more than 0.1 mm (0.004 in.) replace lifter

Exhaust valves, discarding

Before discarding exhaust valves read this warning:

> **WARNING**
>
> Exhaust valves are sodium filled. Before discarding these valves neutralize the sodium to avoid possible injury to people in the recycling chain.

Before discarding observe the following:

- always wear protective goggles or glasses
- note that sodium reacts violently with water
- when cutting valves, valves must not come into contact with water
- use grinding equipment which does not use water or other liquids for lubrication and/or cooling
- to neutralize sodium, drop cut-off valve stems into a bucket of water and move quickly away from water bucket
- do not drop more than 10 valve stems at a time into water bucket

Discarding procedure

- use grinding stone or grinding disc to cut valve stem off near head of valve
- drop cut-off valve stems into bucket of water (sodium will now react with water!)
- once reaction has stopped (water stops bubbling) scrap valve stems
- water is biodegradable and may be discarded accordingly

Camshaft oil seal, removing

- remove upper drive belt cover
- set crankshaft to TDC on cylinder no. **1**
- slacken drive belt and remove
- remove camshaft sprocket
- remove woodruff key from camshaft
- install mounting bolt for camshaft sprocket with washer up to stop

- remove inner part of oil seal extractor **2085** two turns (approximately 3.0 mm) from outer part of tool and lock with screw
- lubricate threaded head of oil seal extractor. Mount and push in as far as possible into oil ring
- loosen knurled screw and turn inner part of puller against camshaft until oil seal is pulled out
- clamp extractor in vise and remove oil seal with pliers

Camshaft oil seal, installing

- coat oil seal seat and seal lip lightly with oil
- press oil seal into cylinder head until **flush** using **10-203**, with special hex head bolt **10-203/1**
- reinstall drive belt, per Repair Group 13

Camshaft, removing

- remove upper drive belt cover
- remove cylinder head
- turn crankshaft to TDC on cylinder **1**
- slacken and remove drive belt
- remove camshaft sprocket
- remove woodruff key from camshaft

- remove bearing caps **1, 3**
- loosen bearing caps **2** and **5** alternately and diagonally

Camshaft, installing

When installing the camshaft, the cams for number **1** cylinder must be pointing upwards. When installing bearing caps, note center offset of bore. Before installing caps, place in position and determine installed position.

- tighten bearing caps **2** and **5** alternately and diagonally
  - tightening torque: 20 Nm (15 ft lb)

- install bearing caps **1** and **3**
  - tightening torque: 20 Nm (15 ft lb)

- position camshaft sprocket and tighten
  - tightening torque: 80 Nm (59 ft lb)

- reinstall drive belt, see Repair Group 13

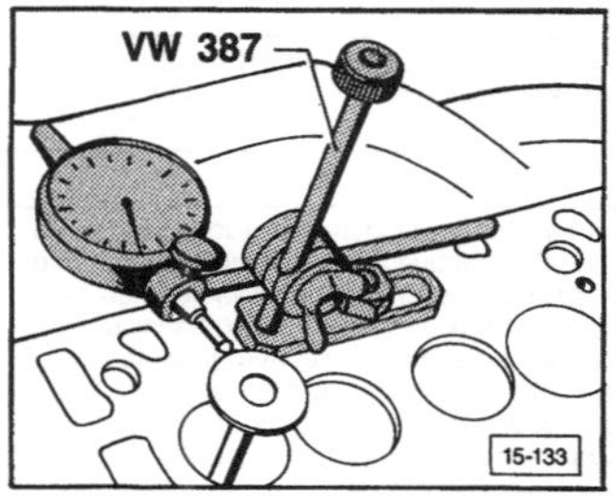

Valve guides, checking for wear

Due to slight difference in stem dimensions, **only** use intake valve in intake guide, and exhaust valve in exhaust guide.

- remove carbon
- install **VW 387** and dial indicator
- insert new valve into valve guide
- valve stem end must be flush with valve guide end
- rock valve back and forth (**arrow**) against dial indicator. Dial indicator reading shows valve guide wear

	Intake valve	Exhaust valve
Maximum dial indicator reading	1.0 mm (0.039 in.)	1.3 mm (0.051 in.)

Valve guides, replacing

CAUTION

Once guide shoulder is seated do not use more than 1 ton pressure or guide shoulder may break.

Cylinder heads on which the valve seat surface can no longer be refaced or cylinder heads that have been refaced down to the minimum dimension cannot be repaired further.

- press worn valve guides out from camshaft side with tool **10-206**
  - valve guides with shoulder (service version) must be pressed out from **combustion** side
- coat new guides with oil and press into cold cylinder head to the shoulder from camshaft side with tool **10-206**
- ream valve guides by hand with reamer tool **10-215**
  - use proper cutting lubricant
- reface valve seats

CAUTION

Once guide shoulder is seated do not use more than 1 ton pressure or guide shoulder may break.

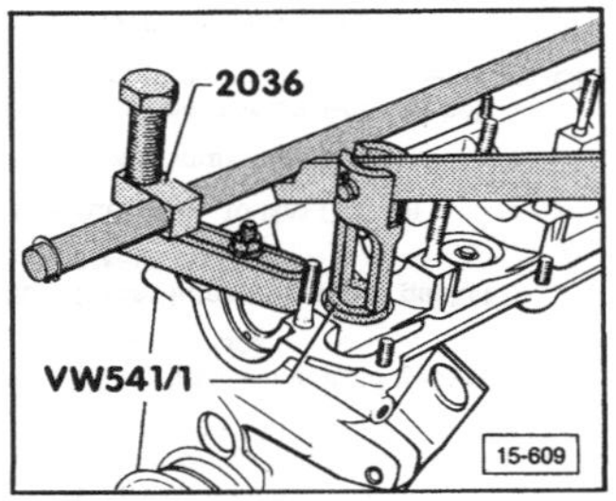

Valve stem seal, removing

(with cylinder head installed)

- remove spark plugs
- remove camshaft and hydraulic lifters
- turn crankshaft until piston of cylinder concerned is in BTDC position
- install tool **2036** and adjust to stud height
- insert pressure hose **US 1106** or **VW 653/3** into spark plug threads and apply steady pressure of at least 6 bar (87 psi)
- remove valve springs with lever **VW 541/1**

If valve keepers are very hard to come out, they can be loosened by tapping **lightly** on assembly lever with hammer.

- remove valve stem seal with tool **3047 A**

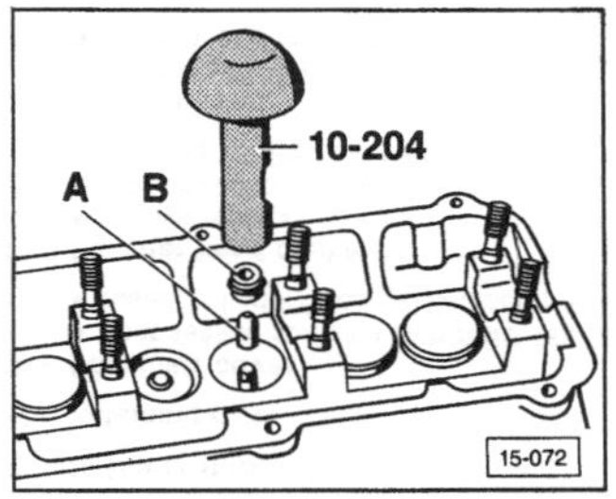

Valve stem seal, installing

- slide plastic sleeve **A** onto valve stem
- lubricate valve stem seal **B**
- push seal carefully onto valve guide using plastic seal protector

CAUTION

When installing valve stem seals, always use plastic sleeve to prevent damage.

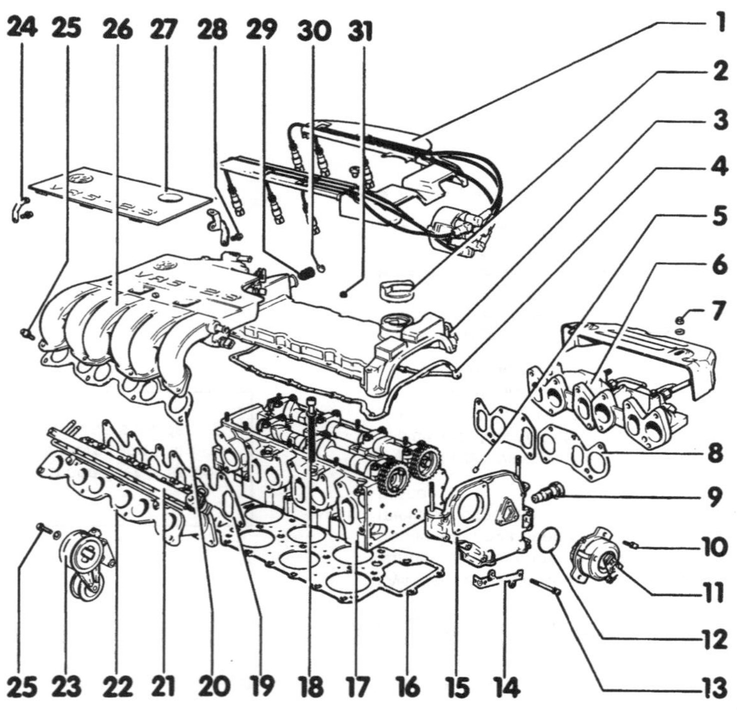

WARNING
DO NOT re-use any fasteners that are worn or deformed in normal use. Many fasteners are designed to be used only once and become unreliable and may fail when used a second time. This includes but is not limited to, nuts bolts, washers, self-locking nuts or bolts, circlips, cotter pins. Always follow recommendations given in this publication. For replacements always use new parts.

CAUTION
Part numbers are for reference only. Always check with your Parts Department for latest information.

CAUTION
Coolant/antifreeze **MUST NOT** be reused when replacing engine, cylinder head, cylinder head gasket, radiator and heater core.

Note
When replacing cylinder head, lubricate hydraulic valve and camshaft contact surface.

1 – **Ignition cable guide**

2 – **Cap**

3 – **Cylinder head cover**

Engine – Cylinder Head, Valve Drive

4 – **Cylinder head cover gasket**
- always replace
- note installation position

5 – **O-ring**
always replace

6 – **Exhaust manifold**
- comes in 2 parts
- with CO take-off pipe attached
- removing and installing, see Repair Group 26

7 – **25 Nm (18 ft lb)**

8 – **Exhaust manifold gasket**
- two-piece
- always replace
- note installation position

9 – **Chain tensioner for double drive chain 20 Nm (15 ft lb)**

10 – **10 Nm (7 ft lb)**

11 – **Ignition distributor with rotor arm**
replacing rotor, see Repair Group 28

12 – **O-ring**
always replace

13 – **10 Nm (7 ft lb)**

14 – **Retaining plate**

15 – **Camshaft cover**
- coat sealing surfaces with **AMV 188 001 02**
- reseal cylinder head gasket, see page 15.15

16 – **Cylinder head gasket**
- always replace
- note installation position
- marking **"TOP"** must face cylinder head

17 – **Cylinder head**
- checking for distortion, see page 15.16
- installing, page 15.17

18 – **Cylinder head bolt**
- always replace
- note torque and tightening sequence, see page 15.17

19 – **Lower intake manifold gasket**
always replace

20 – **Upper intake manifold gasket**
always replace

21 – **Fuel rail – 10 Nm (7 ft lb)**

22 – **Lower intake manifold**

23 – **Tensioner for poly-ribbed belt**
removing and installing, see Repair Group 13

24 – **Intake manifold support**

25 – **25 Nm (18 ft lb)**

26 – **Upper intake manifold**
installing, first install and tighten lower intake manifold **22**, then intake manifold support **24**

27 – **Trim cover**

28 – **25 Nm (18 ft lb)**

29 – **Flame trap coil**

30 – **Circlip**

31 – **10 Nm (7 ft lb)**

Cylinder head gasket, replacing

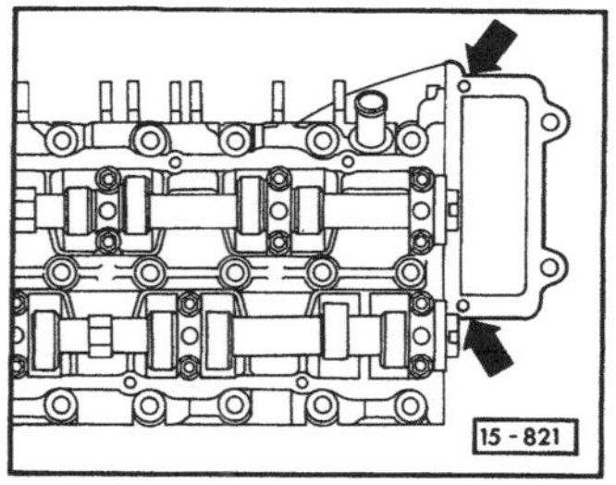

CAUTION

Part numbers are for reference only. Always check with your Parts Department for latest information.

- remove old sealant from 3.0 mm holes in cylinder head gasket **(arrows)**
- fill holes with sealant **AMV 188 001 02**
- position cylinder head gasket on cylinder head
  - imprinted word **"TOP"** must face cylinder head

Cylinder head, checking distortion

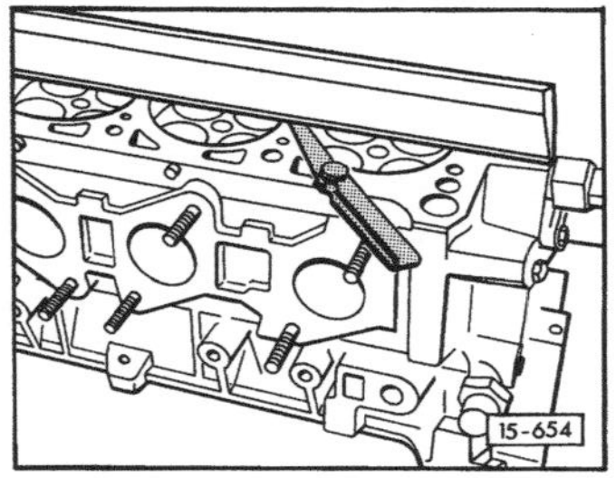

Note

Cylinder head should be clean and free of gasket material before performing this check.

- use straight edge and feeler gauge to check for distortion
  - maximum permissible distortion:

    0.1 mm (0.004 in.)

Cylinder head, installing

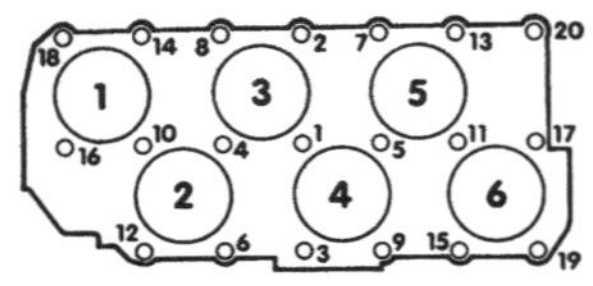

- install cylinder head gasket
  - note installation position, check centering dowels at holes **12** and **20**
- install cylinder head
- fit all cylinder head bolts
  - **hand tighten only**
  - **always replace**

Tightening torque, engine cold

Tighten bolts in sequence:

Stage I = 40 Nm (29 ft lb)

Stage II = 60 Nm (43 ft lb)

- tighten in sequence additional 1/2 (180°) turn further in one movement
  - two additional 1/4 (90°) turns are permissible
- reverse sequence to remove bolts

Note

It is not necessary to re-tighten cylinder head bolts during scheduled maintenance or after performing repairs.

Compression pressure, checking/values

Test conditions
- engine oil temperature minimum 30°C (86°F)
- throttle plate completely open
- fuse **18** removed
- ignition coil lead removed from ignition distributor

- check compression with pressure recorder **US 1120**
  - follow recorder instructions
- operate starter until recorder unit indicates no further pressure increase

Pressure values, bar (psi)

Engine code	New	Wear limit
AAA	11-13 (160-189)	7.5 (109)

- maximum permissible difference between all four cylinders: 3 bar (43.5 psi)

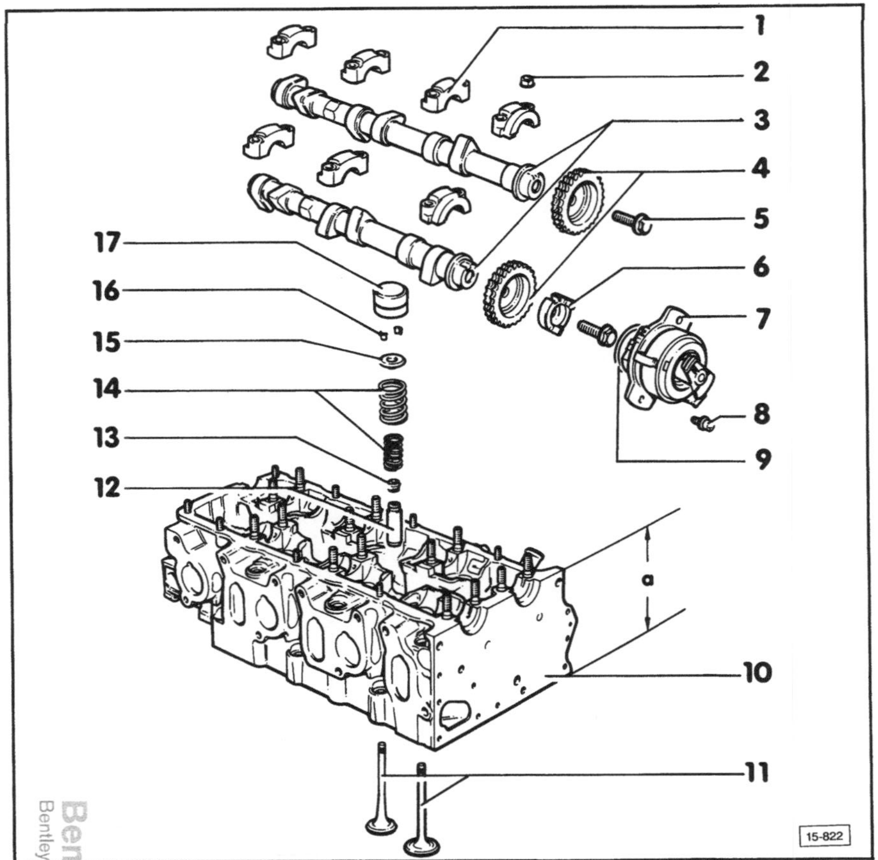

CAUTION
Part numbers are for reference only. Always check with your Parts Department for latest information.

CAUTION
Coolant/antifreeze **MUST NOT** be reused when replacing engine, cylinder head, cylinder head gasket, radiator and heater core.

Note
Heads with fine cracks between valve seats or a valve seat ring and plug threads can be installed again without reducing engine service life provided that cracks are small and not more than 0.5 mm wide or when only the first coils of the plug threads are cracked

1 – **Bearing caps**
installation position, see page 15.21

2 – **20 Nm (15 ft lb)**

3 – **Camshaft**
- checking axial play, see page 15.22
- removing/installing, see page 15.27
- checking radial clearance with plastigage:
  wear limit: 0.1 mm (0.004 in.)
  run out: max. 0.01 mm (0.0004 in.)
- identification: 021 109 101 M cylinders **1**, **3**, **5**
  021 109 102 A cylinders **2**, **4**, **6**

4 – **Camshaft sprockets**
removing/installing/camshaft, page 15.27

5 – **100 Nm (74 ft lb)**

6 – **Distributor drive**

7 – **Distributor**
replacing rotor, see Repair Group 28

8 – **10 Nm (7 ft lb)**

9 – **Seal**
always replace

10 – **Cylinder head**
- 140 ± 0.1 mm (5.5 ± 0.004 in.)
- minimum refacing dimension
  a = 139.5 mm (5.4 in.)

11 – **Valves**
dimensions, page 15.23

12 – **Valve guide**
- checking wear, see page 15.30
- replacing, see page 15.31

13 – **Valve stem seal**
removing/installing see page 15.32

14 – **Valve springs, inner/outer**
removing/installing with tool **2036** or **VW 541/5** and **VW 541/1A**

15 – **Upper valve spring seat**

16 – **Valve keepers**

17 – **Hydraulic valve lifter**
- **DO NOT** interchange
- self-adjusting hydraulic valve clearance
- checking, see page 15.26
- store with cam contact surface down
- before installing, check camshaft axial play, see page 15.22
- lubricate contact surface

Camshaft bearing caps, installed position

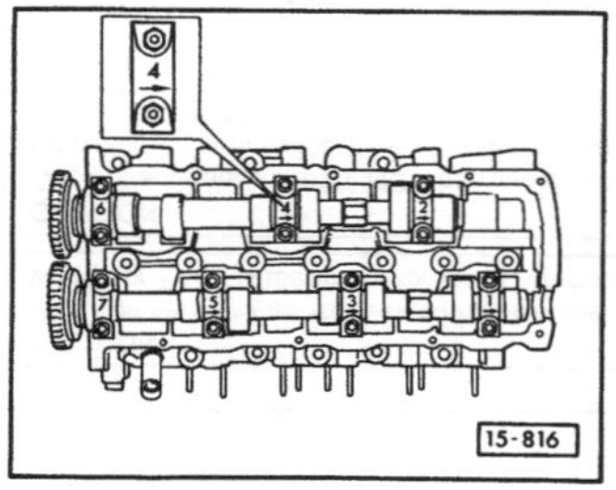

Note

When installing bearing cap, ensure marking on cap is readable from exhaust manifold side and arrow points to vibration dampener side (**inset**).

Camshaft axial play, checking

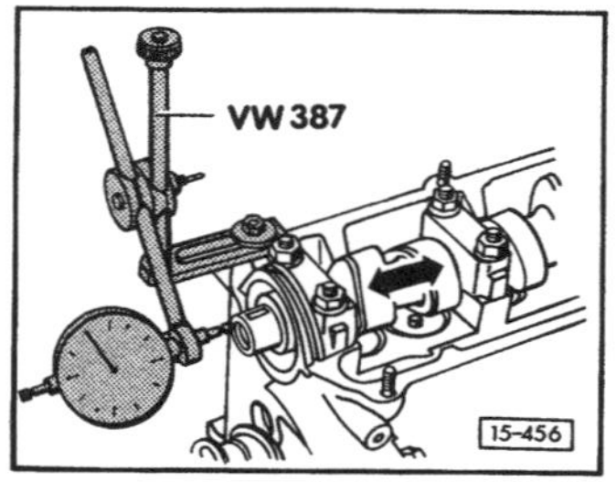

- measure with valve lifters removed, and bearing caps **1** and **7** or **2** and **6** installed
  - wear limit: 0.15 mm (0.006 in.)

Valve dimensions (in mm)

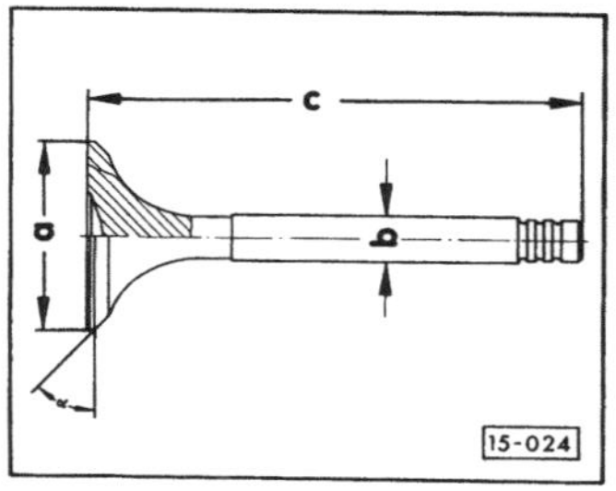

CAUTION
Do not reface valves, only reface seats.

	Intake	Exhaust
a	39.00	34.20
b	6.97	6.95
c	105.95	106.95
α	45°	45°

Valve seats, refacing (in mm)

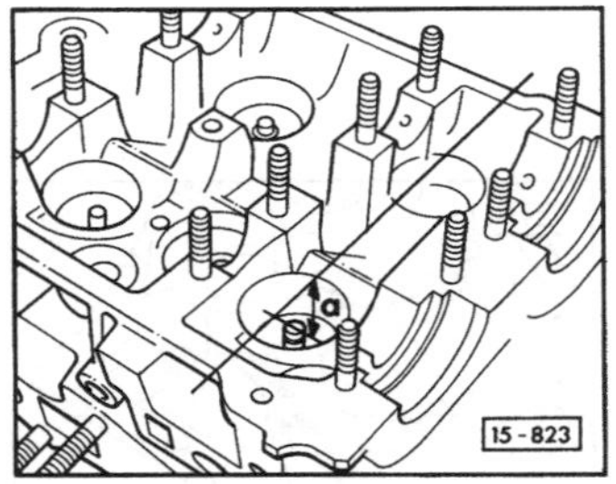

Calculating the maximum permissible refacing dimension

- insert valve and hold firmly against valve seat
- measure distance **a** between end of valve shaft and upper edge of cylinder head
- calculate refacing dimension **b** (see example) using the measured distance **a** (above) and:

Minimum dimension

Minimum dimension - intake valve = 33.9 mm

Minimum dimension - exhaust valve = 34.1 mm

Example:

measured distance **a**
– minimum dimension

= maximum permissible refacing dimension **b**

Note

If the measured distance **a** is smaller than the minimum dimension, the hydraulic valve lifter will not work and the cylinder head must be replaced.

Intake valve seat, refacing

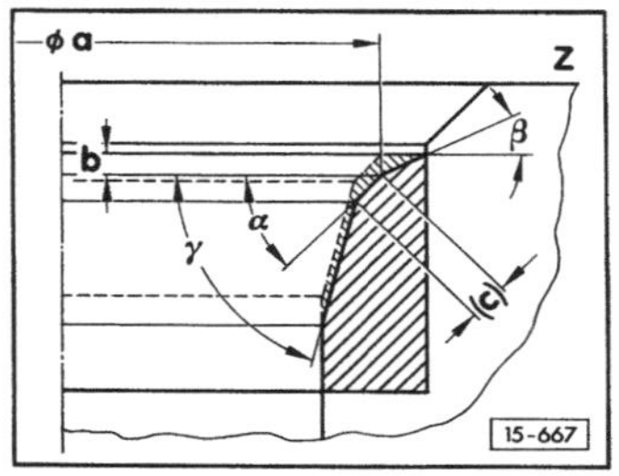

a = diameter 38.3 mm

b = maximum permissible refacing dimension

c = maximum 1.4 mm
- if necessary rework valve seat insert with 75° correction cutter

z = cylinder head lower edge

α = 45° valve seat angle

β = 30° upper correction angle

γ = 75° lower correction angle

Exhaust valve seat, refacing

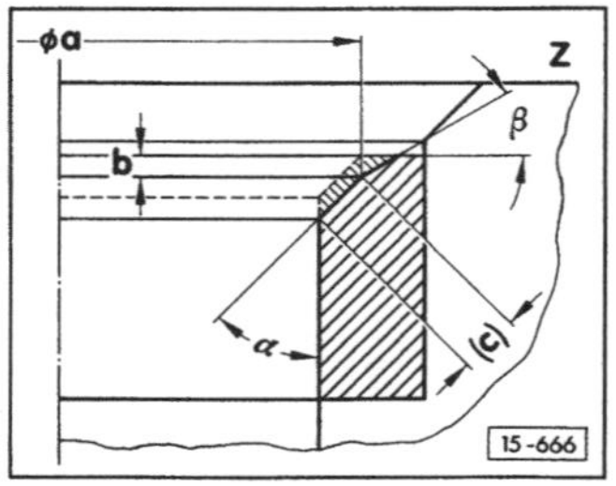

a = diameter 33.5 mm

b = maximum permissible refacing dimension

c = maximum 2.0 mm

- if necessary rework valve seat insert with 75° correction cutter

z = cylinder head lower edge

α = 45° valve seat angle

β = 30° upper correction angle

Hydraulic valve lifter, checking

CAUTION

After replacing lifters, do not start engine for approximately 30 minutes, otherwise valves may hit piston. Lifter must be allowed to bleed down to proper adjustment.

Note

- always place removed valve lifter on a clean surface with the contact surface (camshaft side) facing downward
- always replace complete valve lifter. Lifter cannot be adjusted or repaired
- unusual noise from valve lifter while starting is normal

- run engine until radiator fan comes on at least once
- bring engine speed to approximately 2500 RPM for two minutes

If valve lifter is still noisy, replace as follows:

- remove cylinder head cover
- remove upper intake manifold
- turn crankshaft pulley bolt clockwise until cam lobes of cylinder to be checked point upward

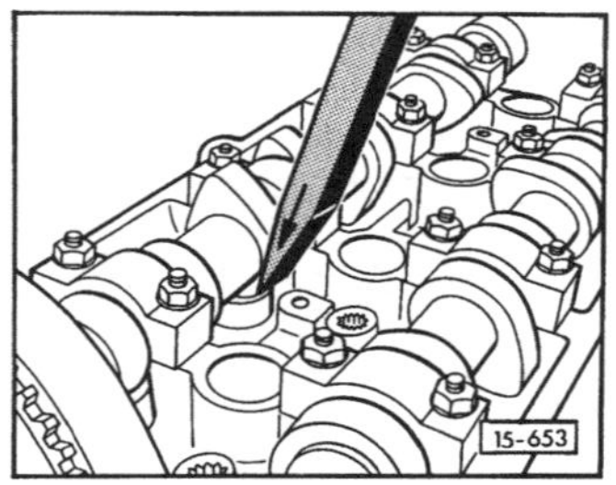

- push down against valve lifter with light pressure using a suitable wood stick (as shown)
- if valve lifter can be pushed down more than 0.1 mm (0.004 in.), replace lifter

Camshafts, removing/installing

Removing

- remove ignition cable guide
- remove upper intake manifold

Note
Cover intake channels with a clean cloth to prevent becoming contaminated.

With engine installed

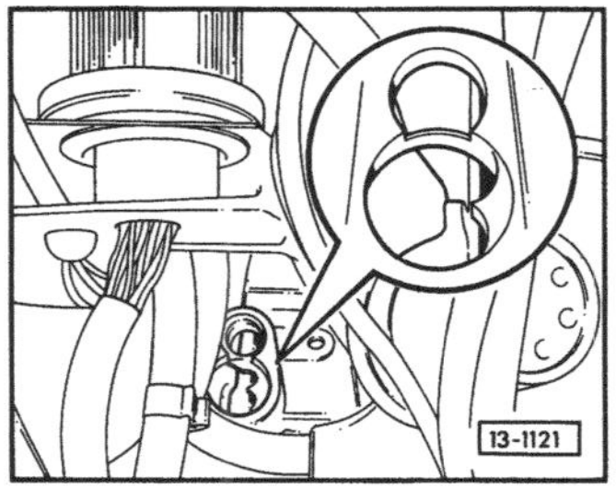

- set crankshaft to **TDC** (cyl. number **1**) (**inset**)
- remove distributor
- remove cylinder head cover
- unscrew chain tensioner
- remove camshaft sprocket cover

Vehicles built up to 10.91
- note rounded teeth

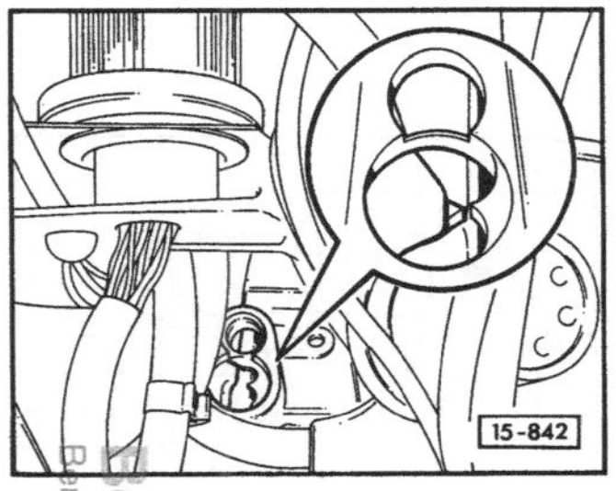

Vehicles built after 11.91
- note pointed teeth

With engine removed

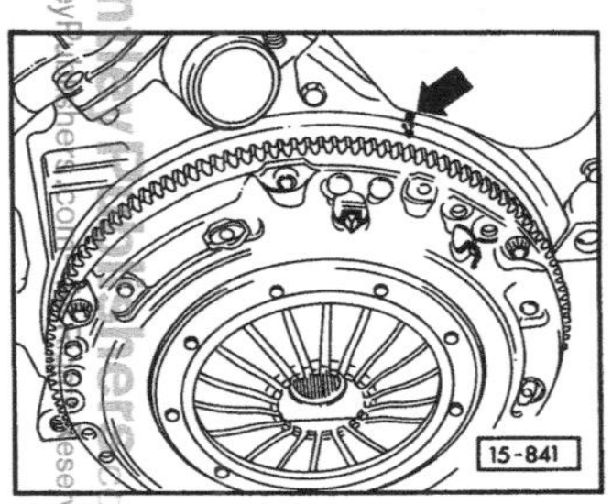

- set flywheel marking to **TDC** (cyl. number **1**)
  - lug on flange must align with mark
- remove ignition distributor
- remove valve cover
- remove camshaft sprocket cover
- unfasten chain tensioner

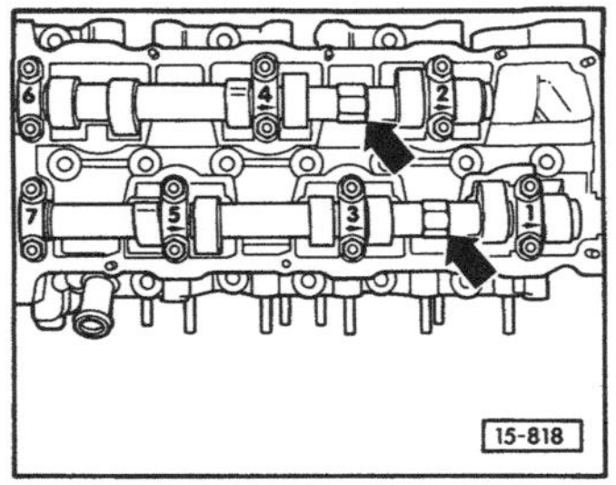

- loosen camshaft sprocket bolts

Note
Only hold camshafts with 24 mm spanner wrench at points shown **(arrows)**. The camshaft guide **3268** must not be fitted when tightening/loosening the sprockets.

- remove camshaft sprockets

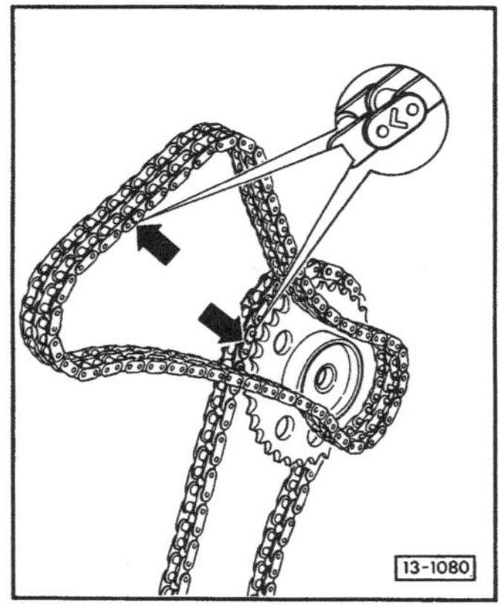

Note
Before removing double chain mark direction of rotation with touch-up paint (**arrows**)

Camshaft cylinders, 1, 3, 5

- remove bearing caps **1** and **7**
- loosen bearing caps **3** and **5** alternately and diagonally

Camshaft cylinders 2, 4, 6

- remove bearing cap **4**
- loosen bearing caps **2** and **6** alternately and diagonally

Installing

> **CAUTION**
> Part numbers are for reference only. Always check with your Parts Department for latest information.

- note marking on the chain

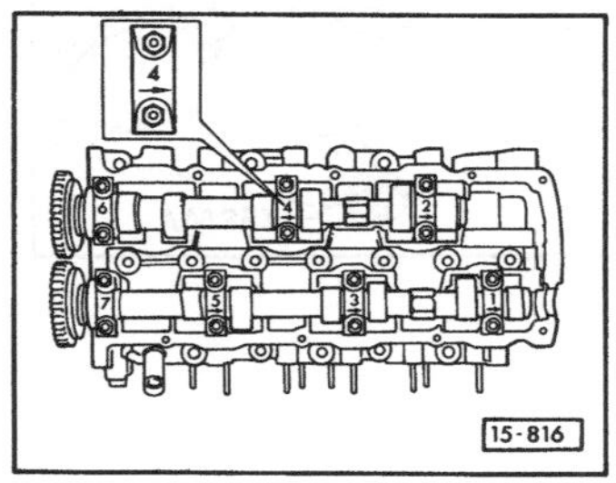

Note
When installing the bearing caps ensure that the marking can be read from the exhaust manifold side and the arrow points to vibration dampener end.

Camshaft cylinders 1, 3, 5

- tighten bearing caps **3** and **5** alternately and diagonally
  - torque 20 Nm (15 ft lb)
- install bearing caps **1** and **7**
  - torque 20 Nm (15 ft lb)

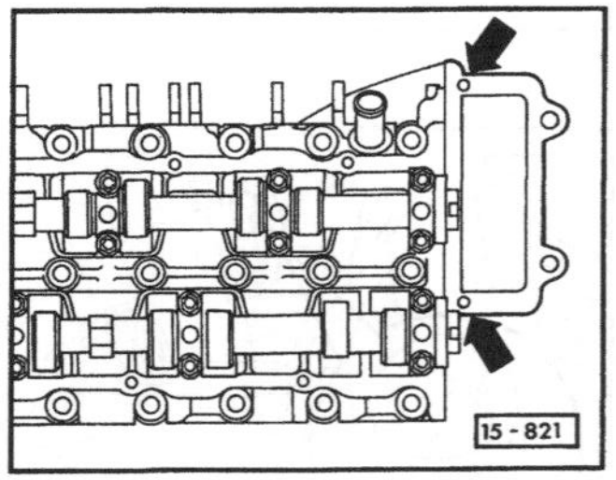

Camshaft cylinders 2, 4, 6

- tighten bearing caps **2** and **6** alternately and diagonally
  - torque 20 Nm (15 ft lb)
- install bearing cap **4**
  - torque 20 Nm (15 ft lb)
- adjust valve timing, see Repair Group 13
- clean camshaft sprocket cover sealing surfaces
- clean old sealant from 3.0 mm holes in cylinder head gasket **(arrow)**
- fill holes with sealant **AMV 188 001 02**
- clean sealing surfaces of cover and coat with **AMV 188 001 02**
  - replace O-ring
- reinstall cover, hand tighten all bolts
- retighten all bolts
  - torque 10 Nm (7 ft lb)
- install chain tensioner
  - torque 20 Nm (15 ft lb)

Valve guides, checking

Note

When repairing engines with leaky valves, it is also necessary to check valve guides for wear. This is particularly important on engines with high mileage.

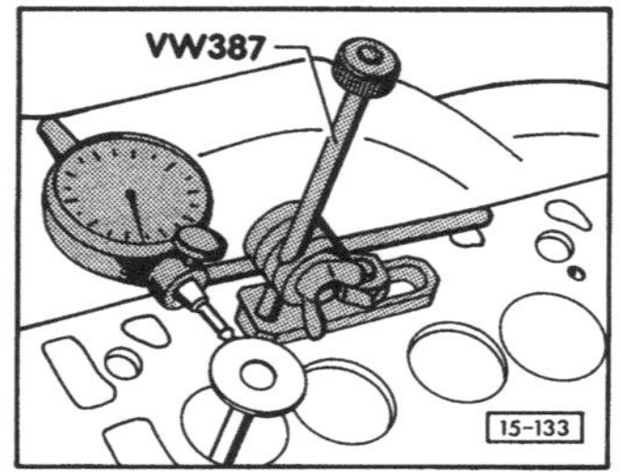

Due to slight difference in stem dimensions, **only** use intake valve in intake guide, and exhaust valve in exhaust guide.

- remove carbon
- install **VW 387** and dial indicator
- insert new valve into valve guide
- valve stem end must be flush with valve guide end
- rock valve back and forth **(arrow)** against dial indicator. Dial indicator reading shows valve guide wear

	Intake valve	Exhaust valve
Maximum dial indicator reading	1.0 mm (0.039 in.)	1.3 mm (0.051 in.)

Valve guides, replacing

Note

Clean and check cylinder head.

Cylinder heads on which the valve seat surface can no longer be refaced, valve guides have been previously installed, or cylinder heads that have been refaced down to the minimum dimension cannot be repaired further.

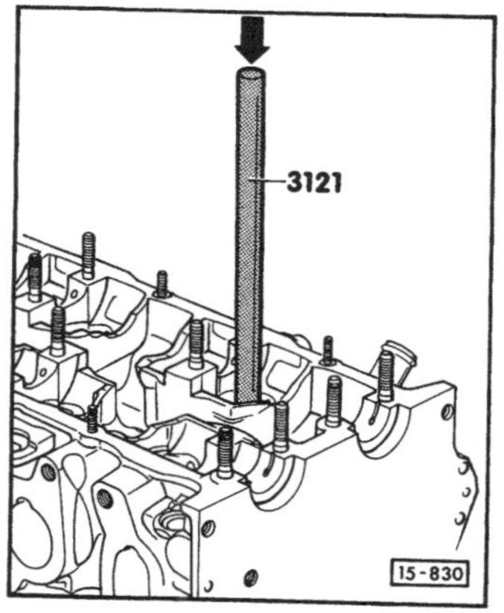

- press worn valve guides with tool **3121** from camshaft side
- lay cylinder head flat on workbench

> **CAUTION**
>
> Once guide shoulder is seated do not use more than 1 ton pressure or guide shoulder may break off.

- coat new guides with oil and press into cold cylinder head to the shoulder from camshaft side with tool **3121**
- ream valve guides by hand with reamer **3120**
  - use proper cutting lubricant
- reface valve seats

Valve stem seals, removing/installing

(with cylinder head installed)

Removing

- remove spark plugs
- remove camshaft and hydraulic lifters
- turn crankshaft until piston of respective cylinder is in **BTDC** position

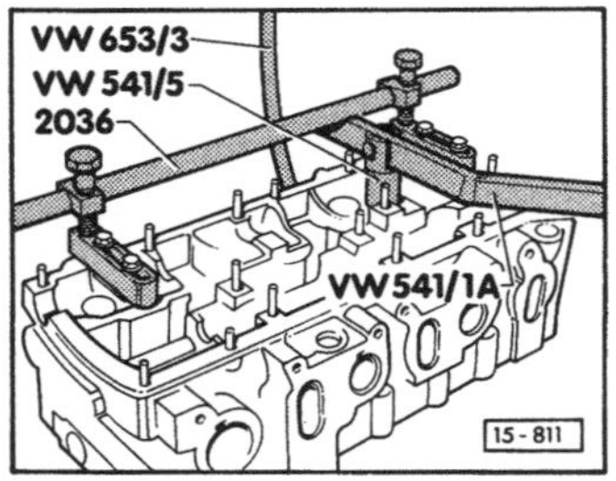

- install tool **2036** and adjust to stud height
- insert pressure hose **VW 653/3** into spark plug threads and apply steady pressure of at least 6 bar (87 psi)
- remove valve springs with lever **VW 541/1A** and **VW 541/5**

Note

If valve keepers are very hard to come out, they can be loosened by tapping **lightly** on assembly lever with hammer.

- remove valve stem seal with tool **3047A**

Installing

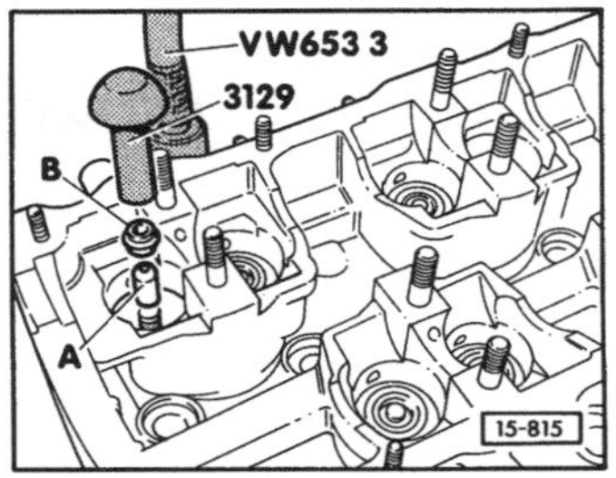

> **CAUTION**
>
> Always use plastic sleeve **A** when installing valve stem seals, to prevent damage to the seal.

- slide plastic sleeve **A** onto the valve stem
- lubricate valve stem seal **B**
- push seal carefully onto valve guide using tool **3129**

THIS PAGE INTENTIONALLY LEFT BLANK

Index

Engine Code – PG

Dynamic oil pressure system
- troubleshooting 17.6

Lubrication system
- assembly 17.2

Oil
- capacity, specifications 17.4
- pan 17.3
- pump 17.3
- pump, checking 17.4
- pump, gears 17.3
- switches, checking 17.5

Engine Code – AAA

Lubrication system
- assembly 17.11

Oil
- capacity, specifications 17.18
- pump, assembly 17.13
- pump, checking 17.15
- relief valve, checking 17.17
- spray jet, removing/installing 17.16
- warning system 17.18

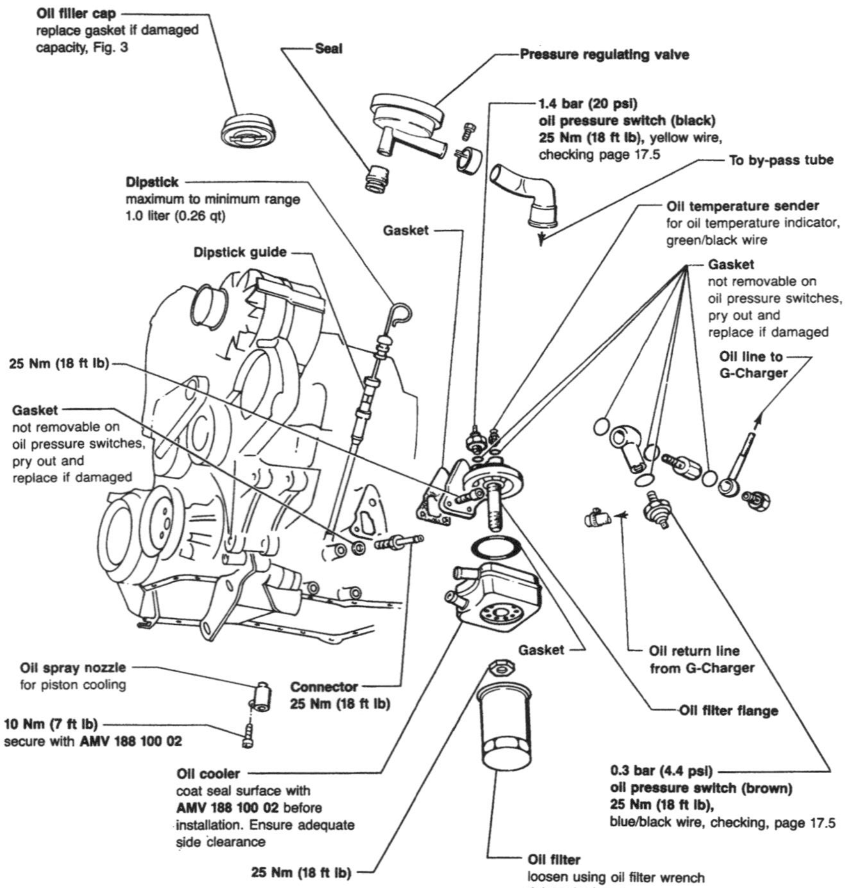

Note

Components shown can be removed or installed with engine in vehicle.

Always replace seals and gaskets.

> **CAUTION**
>
> If, as a result of engine damage, large quantities of metal shavings are found in the engine oil, oil passages should be cleaned and the oil cooler replaced.

Note

Ensure that no coolant hoses or lines touch the knock sensor. This could dampen the sensor and diminish the signal.

17-475

Note

Components shown can be removed or installed with engine in vehicle.

Always replace seals and gaskets.

> **CAUTION**
>
> If, as a result of engine damage, large quantities of metal shavings are found in the engine oil, oil passages should be cleaned and the oil cooler replaced.

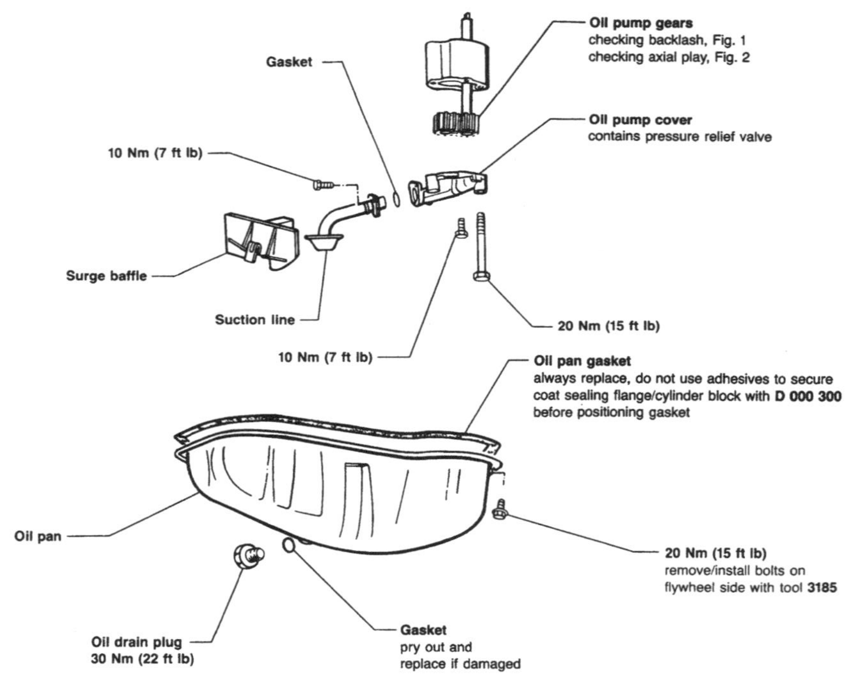

17-475

▶ **Fig. 1 Oil pump, checking backlash**

New: 0.05 mm (0.002 in.)
Wear limit: 0.20 mm (0.008 in.)

▶ **Fig. 2 Oil pump, checking axial play**

Wear limit: 0.15 mm (0.006 in.)

Oil capacity, specifications

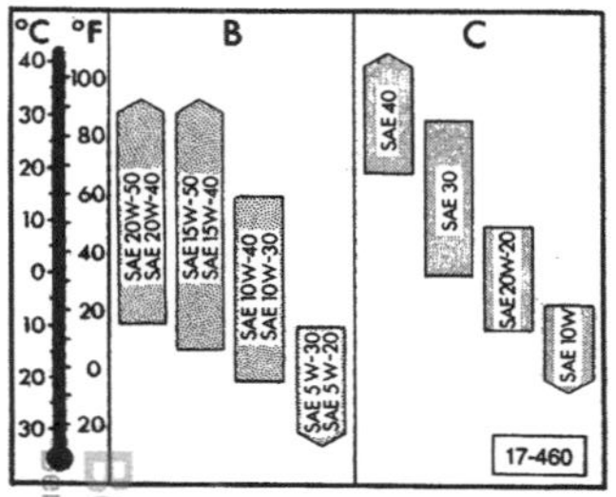

▶ **Fig. 3 Oil specification**

Use only oil marked **SF** or **SG.**

B = Multi-grade oil
C = Single-grade oil

Engine oil capacity

- With oil filter 4.0L (4.3 qts)
- Without oil filter 3.5L (3.7 qts)

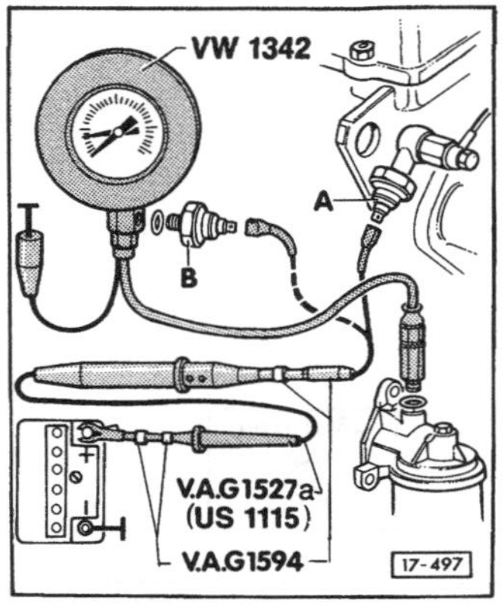

Oil pressure warning system, checking

Note

- Oil temperature must not exceed 85°C (205°F)

■ remove 1.4 bar (20 psi) oil pressure switch (black insulation)
■ insert into test unit **VW 1342** as shown
■ connect test unit to oil filter bracket in place of switch, as shown
■ connect brown wire of test unit **VW 1342** to ground
■ connect LED test lamp **V.A.G1527a** (**US 1115**) with lead from **VW 1594** to 0.3 bar (4.4 psi) oil pressure switch **A**
- test lamp must light

■ start engine and slowly increase RPM
- test lamp must **go out** at 0.15-0.45 bar (2.2-6.5 psi)

If **NO**, replace switch

■ connect LED test lamp to 1.4 bar (20 psi) oil pressure switch **B** (black insulation), as shown
- test lamp must light at 1.2-1.6 bar (17-23 psi)

If **NO**, replace switch

■ increase and hold engine speed at a steady 2000 RPM
- oil pressure must be minimum of 2.0 bar (29 psi) at 80°C (176°F)

Dynamic oil pressure system, troubleshooting

Note

Always refer to appropriate wiring diagram.

Test conditions

- engine oil level **OK**
- tachometer **OK**
- oil pressure switches **OK**

Tools required

- Multimeter **US 1119**
- test box **V.A.G 1598** with adapter **1598/8**

System operation

Ignition ON, engine OFF

- LED light flashes **IF** oil pressure switches **OK**

Ignition ON, engine at idle

- LED flashes when oil pressure drops below 0.3 bar (4.4 psi)

Ignition ON, engine speed over 2100 RPM

- LED flashes at oil pressure **below** 1.4 bar (20.0 psi)
- audio warning sounds after approximately 3.0 second delay
- audio/visual warning continues until oil pressure increases to 1.4 bar (20.0 psi) or ignition is turned off
- **RPM** tolerance ± 100

Wiring between oil pressure switches and instrument cluster, checking continuity

- Switch ignition **OFF**
- disconnect 0.5 yellow wire from 1.8 bar oil pressure switch and 0.5 blue/black wires from oil pressure switch
- remove 28-point instrument cluster connector (**T28**) and connect test box **VAG 1598** using test harness **VAG 1598/8**
- set **US 1119** to Ohm scale
- connect test leads of **US 1119** as follows:

Red (+) lead to:	Black (−) lead to:	Specified readings
0.5 yellow wire	test box **V.A.G 1598** terminal **8**	must be continuity (0 Ohms)
0.5 yellow wire	test box **V.A.G 1598** terminal **3**	must be no continuity (∞ Ohms)
0.5 blue/black wire	test box **V.A.G. 1598** terminal **9**	must be continuity (0 Ohms)
0.5 blue black wire	test box **V.A.G. 1598** terminal **3**	must be no continuity (∞ Ohms)
test box **V.A.G 1598** terminal **8**	test box **V.A.G 1598** terminal **9**	must be no continuity (∞ Ohms)

If specified values obtained →

- reconnect wires to oil pressure switches

If **all specified valves NOT** obtained →

- repair open in wiring to instrument cluster according to wiring diagram

→ End

Oil pressure warning control unit (**J114**), checking

- switch ignition **ON**

LED oil pressure warning light (**K3**) does **NOT** flash → Go to Ⓐ next page

LED oil pressure warning light (**K3**) flashes → Go to Ⓑ two pages following

(A)

Instrument cluster voltage supply, checking

- set **US 1119** to 20 V scale
- connect red (+) test lead of **US 1119** to text box **V.A.G. 1598** terminal **13**
- connect black (–) test lead of **US 1119** to test box **V.A.G. 1598** terminal **3**

U.S. 1119 registers approximately 12.5 V	**U.S. 1119** does **NOT** register voltage
→ LED for oil pressure warning, checking	→ check wiring for open circuit and repair voltage supply to instrument cluster according to wiring diagram → End

LED for oil pressure warning, checking

- switch ignition **OFF**
- remove instrument panel trim
- remove and check LED

LED **OK**	LED **NOT OK**
■ reinstall LED	■ replace LED
Go to (B) next page	End

(B)

LED contacts on printed circuit, checking continuity

- remove tachometer
- set **US 1119** to Ohm scale
- connect test leads of **US 1119** as follows:

Red (+) lead to:	**Black (–) lead to:**
right LED (**K3**) connector	test box **V.A.G. 1598** terminal **13**
left LED (**K3**) connector	terminal **3** of engine oil pressure control unit
US 1119 registers ∞ Ohms (no continuity)	**US 1119** registers 0 Ohms (continuity)
■ replace printed circuit	■ replace LED
End	Go to (C) next page

Ⓒ

Oil pressure control unit (**J114**), checking terminal continuity

■ connect test leads of **US 1119** as follows:

Red (+) lead to:	**Black (−) lead to:**
test box **V.A.G. 1598**, terminal **10**	control unit terminal **1**
test box **V.A.G. 1598**, terminal **13**	control unit terminal **4**
test box **V.A.G. 1598**, terminal **3**	control unit terminal **2**
test box **V.A.G. 1598**, terminal **8**	control unit terminal **7**
test box **V.A.G. 1598**, terminal **9**	control unit terminal **8**
US 1119 registers 0 Ohms (continuity)	**US 1119** registers ∞ Ohms (no continuity)

- **US 1119** registers 0 Ohms (continuity) → ■ replace control unit (**J114**) → End
- **US 1119** registers ∞ Ohms (no continuity) → ■ replace printed circuit → End

THIS PAGE INTENTIONALLY LEFT BLANK

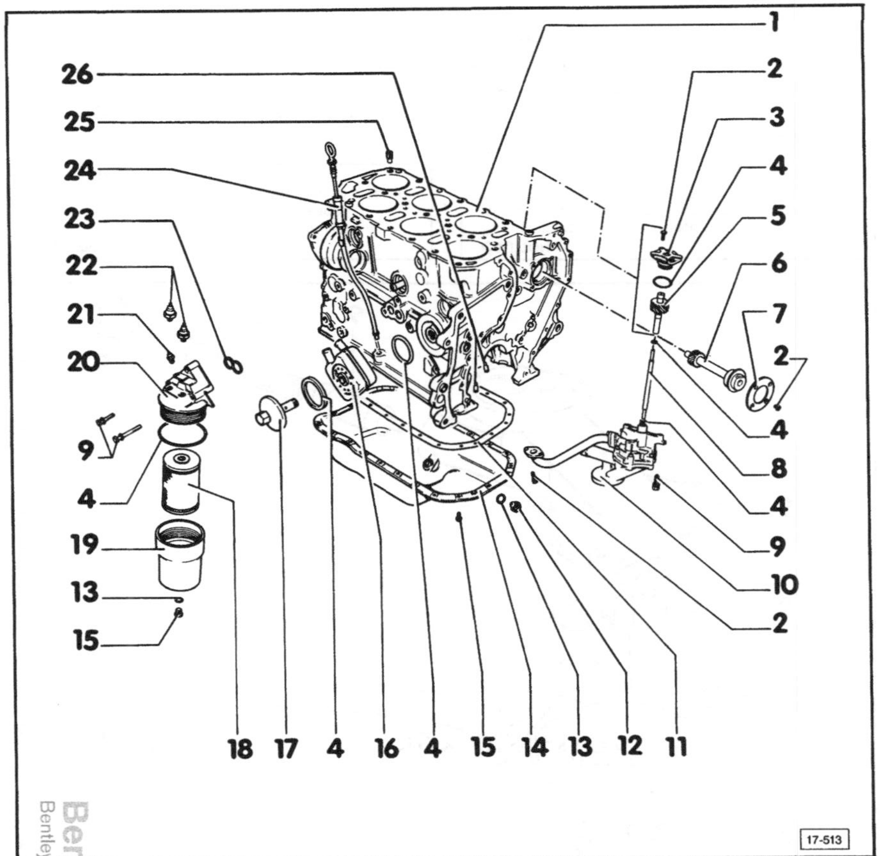

WARNING
DO NOT re-use any fasteners that are worn or deformed in normal use. Many fasteners are designed to be used only once and become unreliable and may fail when used a second time. This includes, but is not limited to, nuts, bolts, washers, self-locking nuts or bolts, circlips, cotter pins. Always follow recommendations given in this publication. For replacements always use new parts.

CAUTION
Always replace gaskets and seals.

CAUTION
Part numbers are for reference only. Always check with your Parts Department for latest information.

CAUTION
If metal shavings are found in the engine oil as a result of engine damage, clean oil passages thoroughly, then replace all oil spray jets, oil pressure relief valve, oil cooler and oil filter.

Note
All components shown, with exception of oil spray jets and the oil pressure relief valve, can be replaced with engine installed.

Note
Engine oil capacity
- with oil filter 6.0L (6.3 qts.)
- without oil filter 5.0L (5.3 qts.)

1 – **Cylinder block**

2 – **10 Nm (7 ft lb)**

3 – **Oil pump drive cover**
installing, coat seal with engine oil

4 – **O-ring**
- replace
- installing, coat O-ring with engine oil

5 – **Oil pump drive**

6 – **Intermediate shaft**

7 – **Thrust washer**

8 – **Drive shaft**
for oil pump drive

9 – **25 Nm (18 ft lb)**

10 – **Oil pump**
- see page 17.13
- assembly checking, page 17.15

11 – **Oil pan gasket**
- seal flange area cylinder block/seal flange front and rear, coat both sides with **AMV 188 001 02**
- always replace

12 – **Oil drain plug – 50 Nm (37 ft lb)**

13 – **Gasket**
replace

14 – **Oil pan**

15 – **Oil filter housing drain plug – 20 Nm (15 ft lb)**

16 – **Oil cooler**
- coat seal surface with **AMV 188 001 02**
- do not allow sealant to get onto oil cooler gasket
- ensure adequate side clearance

17 – **Oil cooler cover – 25 Nm (18 ft lb)**

18 – **Oil filter**
note change intervals

19 – **Oil filter housing, lower – 30 Nm (22 ft lb)**
drain before removing

20 – **Oil filter housing, upper – with by-pass valve**

21 – **Oil temperature sender**

22 – **Oil pressure switches – 25 Nm (18 ft lb)**
- 1.4 bar (20 psi) switch, yellow wire
- 0.3 bar (4.4 psi), blue/black wire
- checking, see page 17.19

23 – **O-ring**
- replace
- note installation position

24 – **Dipstick**

25 – **Oil pressure relief valve – 5 Nm (44 in. lb)**
- note installation position
- replace if badly contaminated

26 – **Oil spray jet**
removing and installing, page 17.16

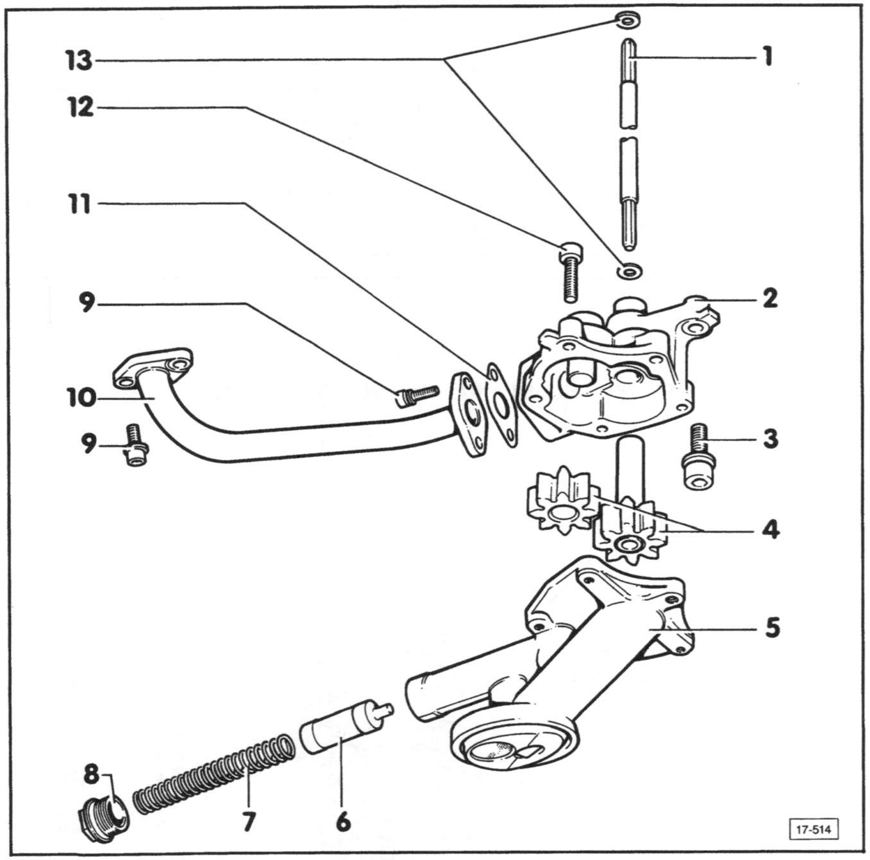

CAUTION
If metal shaving are found in the engine oil as a result of engine damage, clean oil passages thoroughly, then replace all oil spray jets, oil pressure relief valve, oil cooler and oil filter.

CAUTION
Always replace gaskets and seals.

CAUTION
Part numbers are for reference only. Always check with your Parts Department for latest information.

Note
All components shown may be replaced with engine installed.

1 – **Oil pump drive shaft**

2 – **Oil pump housing**

3 – **25 Nm (18 ft lb)**

4 – **Oil pump gears**
- checking backlash, page 17.15
- checking axial play, page 17.15

5 – **Oil pump cover with pressure relief valve**
opening pressure: 5.5 ± 0.2 bar (80 ± 3.0 psi)

6 – **Piston**
note installation position

7 – **Spring**

8 – **40 Nm (30 ft lb)**

9 – **10 Nm (7 ft lb)**
replace

10 – **Oil pressure pipe**
secure to cylinder block with **AMV 188 001 02**

11 – **Gasket**
replace

12 – **10 Nm (7 ft lb)**

13 – **O-ring**
- replace
- coat O-ring with engine oil when installing

Oil pump, checking

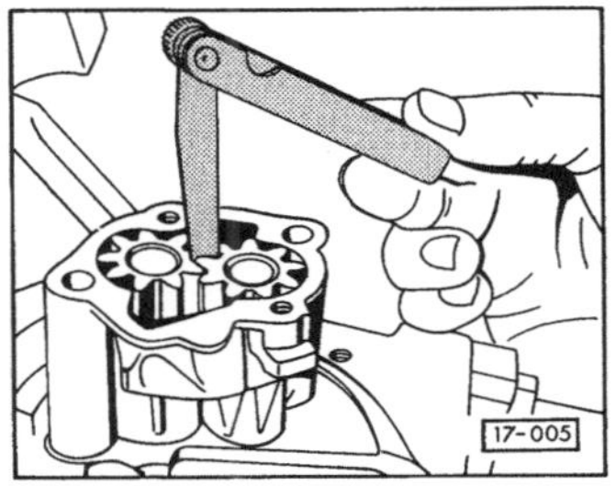

Backlash, checking

Wear limit: 0.20 mm (0.008 in.)

Axial play, checking

Wear limit: 0.10 mm (0.004 in.)

Oil spray jet, removing/installing

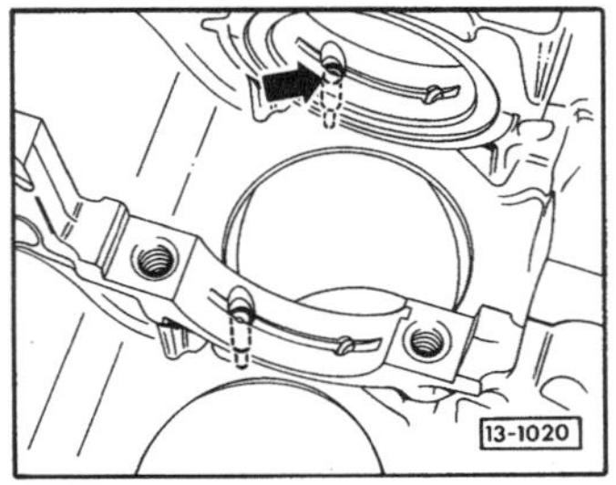

Removing

- opening pressure 2.0 bar (29 psi)
- remove with 4.0 mm drift

Installing

- install by hand with 6.0 mm mandrel (**arrow**)

Oil pressure relief valve, checking

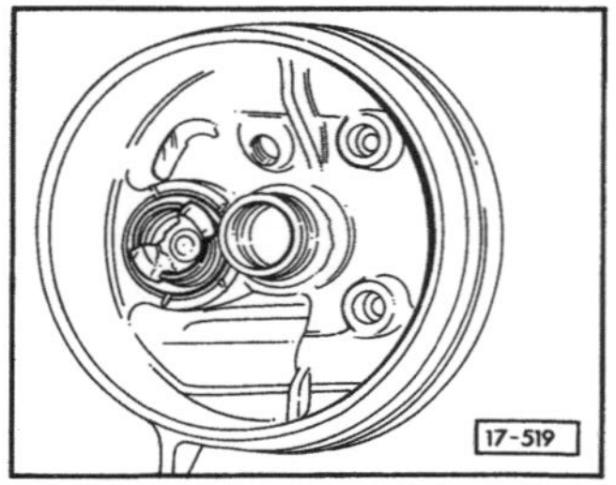
17-519

- opening pressure 2.5 bar (37 psi)
- clean if oil filter is excessively contaminated

Oil capacity, specifications

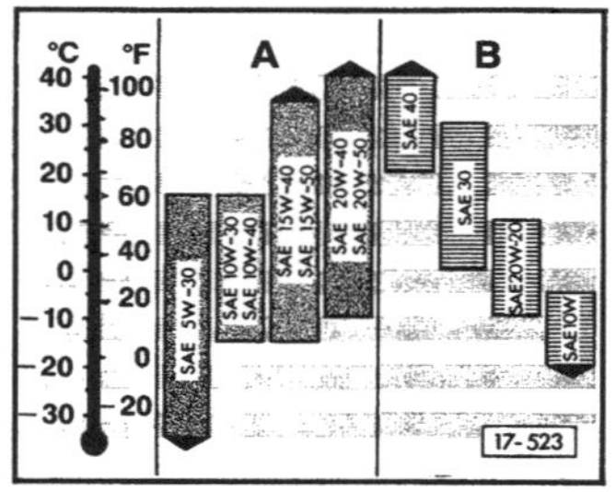

17-523

Use only oil marked API Service **SF** or **SG**

A = Multi-grade oil

B = Single-grade oil

Engine oil capacity
- with oil filter 6.0 L (6.3 qts.)
- without oil filter 5.0 L (5.3 qts.)

Oil pressure warning system

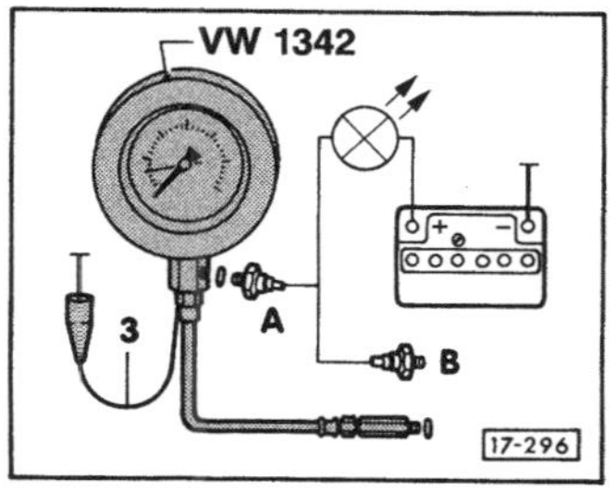

- remove 1.4 bar (20 psi) oil pressure switch **B** **(F1-black insulation)**, and screw into test unit **VW 1342**
- screw test unit **VW 1342** into oil filter bracket in place of switch
- connect brown wire of test unit to ground (-)

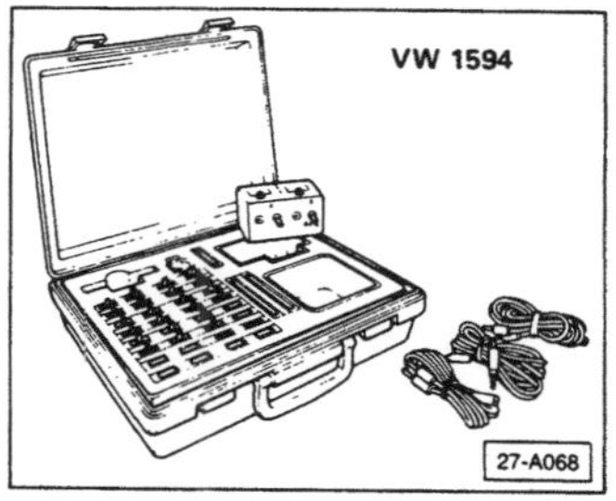

- connect LED test lamp **VAG 1527a** (**US 1115**) with lead from Connector Test Kit **VW 1594** to 0.3 bar (4.4 psi) oil pressure switch **A** **(F22-brown insulation)**
  - test lamp must light
- start engine and slowly increase RPM
  - test lamp must **go out** at 0.15 - 0.45 bar (2.2 - 6.5 psi)

If **NO**, replace switch

- connect LED test lamp to 1.4 bar oil pressure switch **B** (**F1-black insulation**)
  - test lamp must light at 1.6 - 2.0 bar (23 - 29 psi)

If **NO**, replace switch

- increase and hold engine speed at a steady 2000 RPM
  - oil pressure must be minimum of 2.0 bar (29 psi)

THIS PAGE INTENTIONALLY LEFT BLANK

THIS PAGE INTENTIONALLY LEFT BLANK

Index

Engine code - PG

Coolant
- draining **19.5**
- filling **19.6**
- hydrometer, checking **19.6**
- mixture ratios **19.5**

Pump
- assembly **19.3**
- pulleys **19.4**

Radiator
- assembly **19.2**

Thermoswitch
- checking **19.4**
- switching temperature **19.4**

Engine code - AAA

Coolant
- draining/filling **19.11**
- hydrometer, checking **19.6**
- mixture ratio **19.11**

Pump
- assembly **19.9**
- removing/installing **19.14**
- auxiliary electric coolant pump, checking **19.19**
- fan control modual location **19.20**

Radiator/fan
- assembly **19.7**
- removing/installing **19.13**

Thermostat
- assembly **19.17**

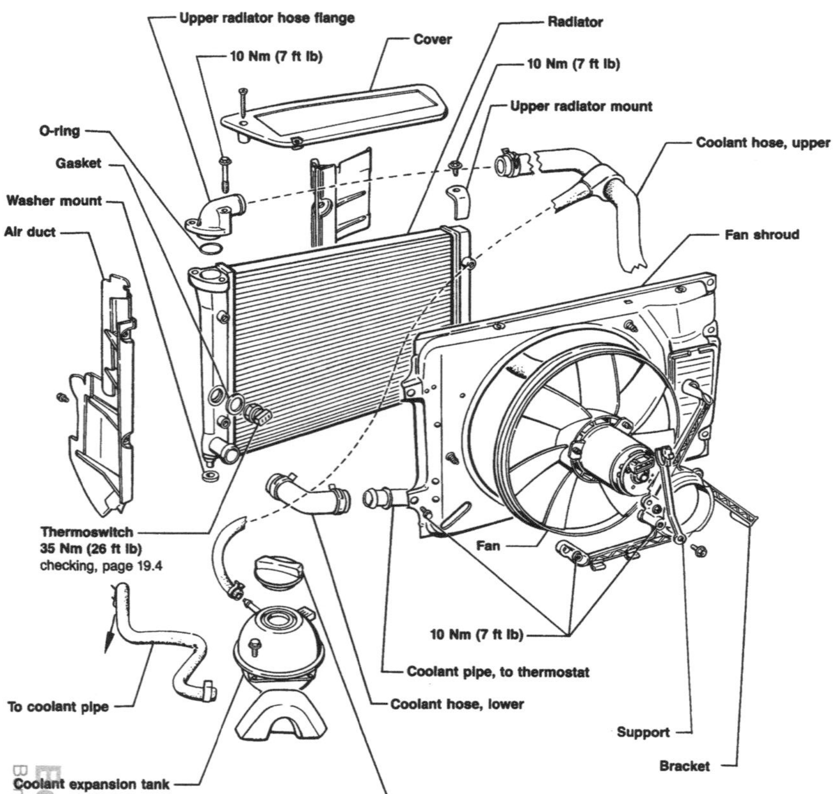

Note

Ensure that no coolant hoses or lines touch the knock sensor. This could dampen the sensor and diminish the signal.

> **CAUTION**
>
> Coolant/antifreeze **must not** be reused when replacing engine, cylinder head, cylinder head gasket, radiator and heater core.

Note

Always replace all gaskets and O-rings. All components shown can be removed/installed with the engine in vehicle.

During repairs, the original spring type hose clamps may be replaced by the screw type.

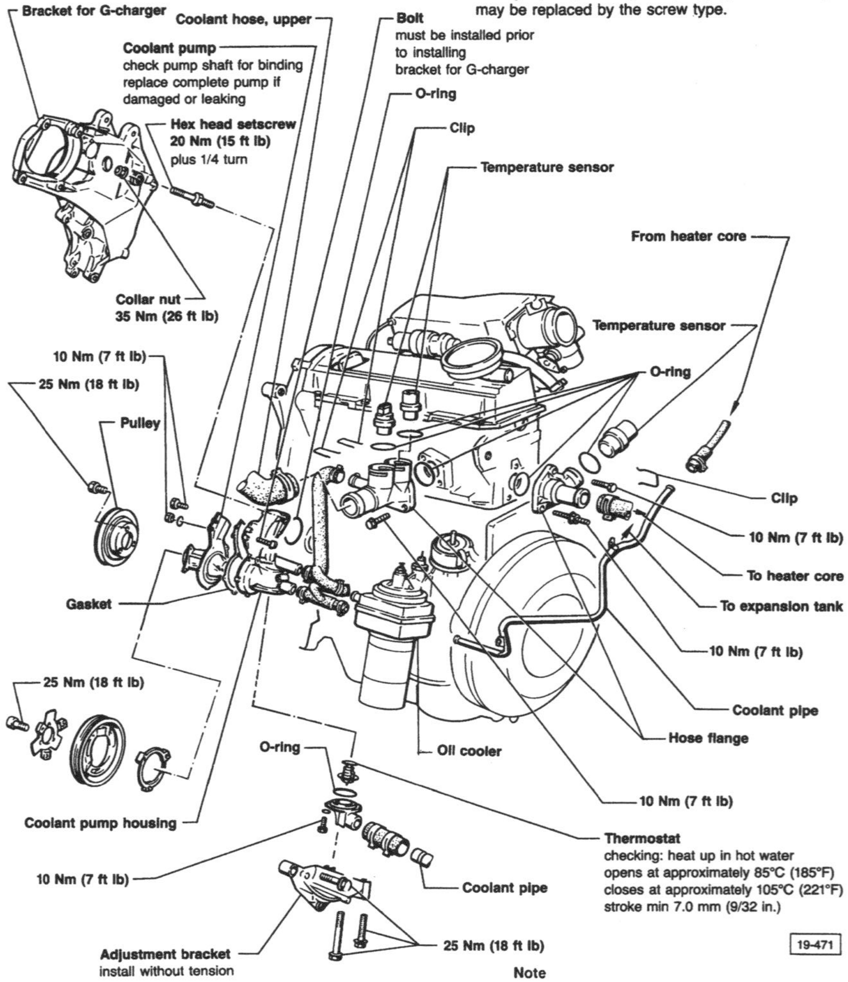

Note

Ensure that no coolant hoses or lines touch the knock sensor. This could dampen the sensor and diminish the signal.

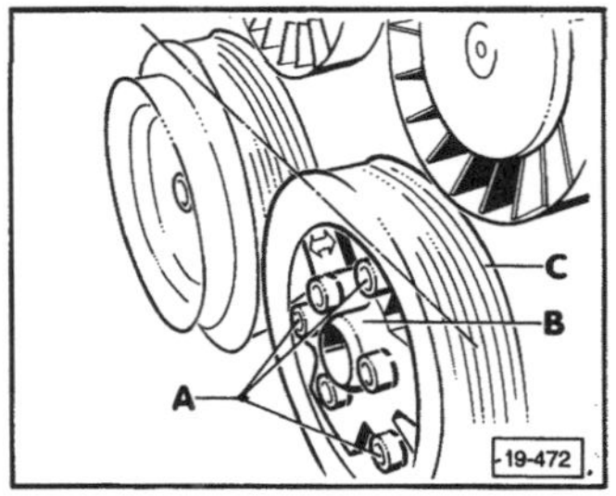

▶ Fig. 1 **Coolant pump pulley, adjusting**

- loosen bolts **A**
- turn inner part **B** against pulley **C** until pulleys are aligned
- re-tighten bolts **A**

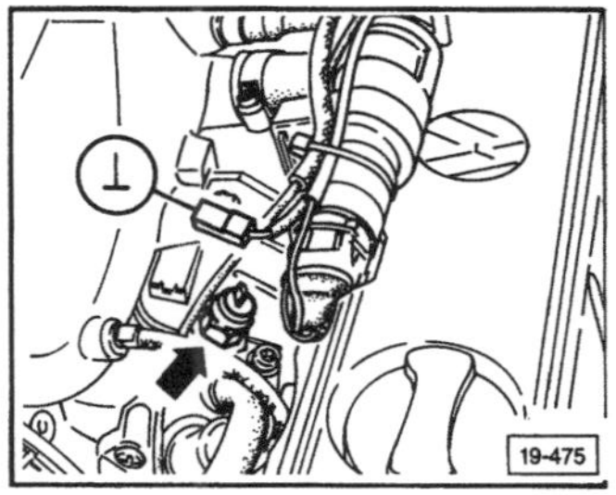

▶ Fig. 2 **Thermoswitch, radiator fan after-run**

Thermoswitch comes on: approximately 90°C (194°F)
Thermoswitch cuts off: approximately 80°C (176°F)

Thermoswitch, checking

- turn ignition key on/off
- disconnect thermoswitch and touch to ground
  - fan must come on

If **NO**, replace switch

Thermoswitch, specifications

Switching temperatures		
Switch	**On**	**Off**
Low speed	92-97°C (198-207°F)	84-91°C (183-196°F)
High speed	99-105°C (210-221°F)	91-98°C (196-208°F)

CAUTION

Coolant/antifreeze **must not** be reused when replacing engine, cylinder head, cylinder head gasket, radiator and heater core.

Cooling system, draining

The cooling systems of **all** Volkswagen vehicles are filled at the factory with a mixture of water and an antifreeze solution (**G 11**) with corrosion inhibitors. This coolant mixture should be used year-round.

Phosphate-free anti-freeze prevents frost and corrosion damage, the formation of chalk and in addition, it raises the boiling point of water. Due to the higher boiling point the coolant is an aid to operational efficiency, when the engine is operating under full load, particularly in tropical climates.

When replacing coolant/antifreeze solution in all Volkswagen models, all model years, use phosphate-free coolant/antifreeze **ZVW237 102**.

Recommended mixtures

Frost Protection to:	G 11	Water
– 13°F/ – 25°C	2.3L (2.4 qt.)	3.3L (3.59 qt.)
– 31°F/ – 35°C*	2.75L (2.9 qt.)	2.75L (2.9 qt.)

* for countries with cold climate

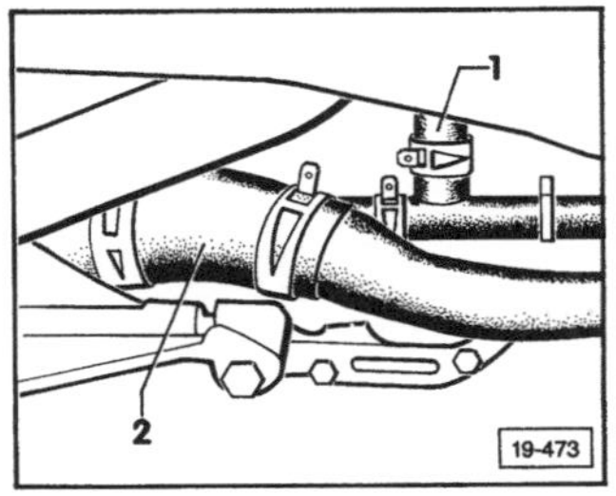

▶
- open cap on coolant expansion tank
- drain coolant through loosening either coolant hose **1** or **2**

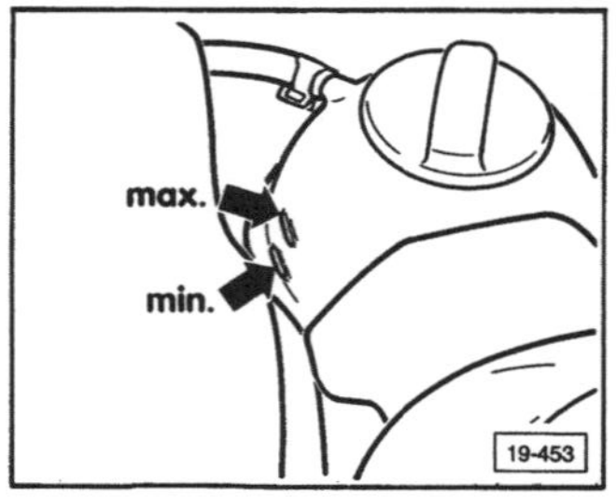

Cooling system, filling

- retighten loosened coolant hoses
- fill with coolant up to **max** marking on coolant expansion tank
- close coolant expansion tank
- let engine run until radiator fan comes on
- check coolant level. If necessary, top off up to the marking. With engine at operating temperature, coolant level should be at **max** mark. With the engine cold, the coolant level should be between the **min** and **max** marking.

Antifreeze hydrometer, checking

Calibration of hydrometer must be checked frequently to ensure accuracy.

- mix a 50/50 mixture of antifreeze and water together in a small container
- hydrometer should read −35°C (−30°F)
  - if not, put a mark (paint dot) to indicate where −35°C (−30°F) should be.

THIS PAGE INTENTIONALLY LEFT BLANK

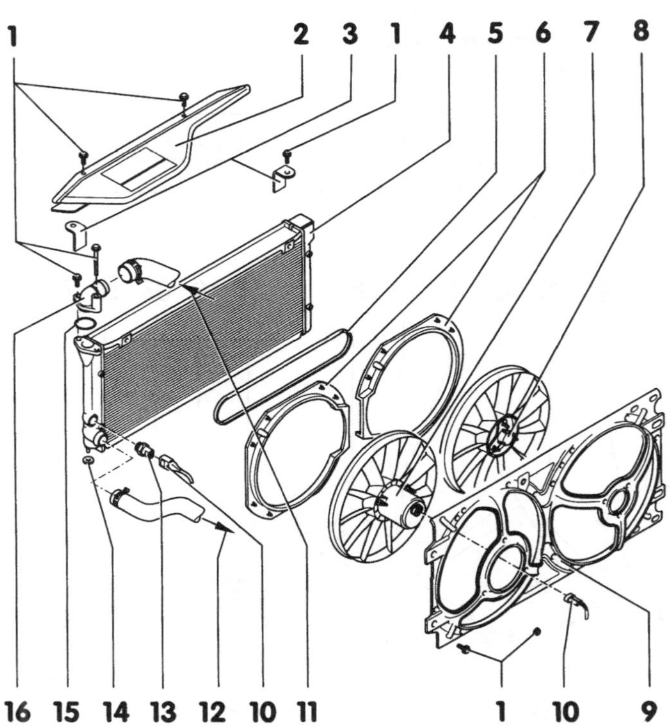

CAUTION
Coolant/antifreeze **must not** be reused when replacing engine, cylinder head, cylinder head gasket, radiator and heater core.

WARNING
DO NOT re-use any fasteners that are worn or deformed in normal use. Many fasteners are designed to be used only once and become unreliable and may fail when used a second time. This includes, but is not limited to, nuts, bolts, washers, self-locking nuts or bolts, circlips, cotter pins. Always follow recommendations given in this publication. For replacements always use new parts.

Note
- always replace all gaskets and O-rings
- all components shown can be removed/installed with the engine in vehicle.
- during repairs, the original spring type hose clamps may be replaced by the screw type
- install spring clamps using pliers **Hazlet 798-5** or a commercial plier equivalent

1 – **10 Nm (7 ft lb)**

2 – **Cover**

3 – **Upper radiator mount**
note installation position

4 – **Radiator**
- removing and installing, see page 19.13
- see replacement of coolant/antifreeze **CAUTION** on page 19.7

5 – **V-belt**

6 – **Guide rings**
- note installation position
- secured to fan shroud **9** with clips

7 – **Fan (V7)**
removing and installing, see page 19.13

8 – **Additional fan for cooling**
removing and installing, see page 19.13

9 – **Fan shroud**
note installation position

10 – **Connector**

11 – **Coolant hose, upper**
from upper connector on thermostat housing, see page 19.9

12 – **Coolant hose, lower**
to regulator on thermostat housing, see page 19.9

13 – **Thermo-switch (F18) – 35 Nm (26 ft lb)**
- for coolant fan
- switch temperatures:
  1st stage:
  on: 92-97°C (198-208°F)
  off: 84-91°C (143-164°F)

  2nd stage:
  on: 99-105°C (178-189°F)
  off: 91-98°C (164-176°F)

14 – **Washer mount**

15 – **O-ring**
replace

16 – **Upper radiator hose flange**

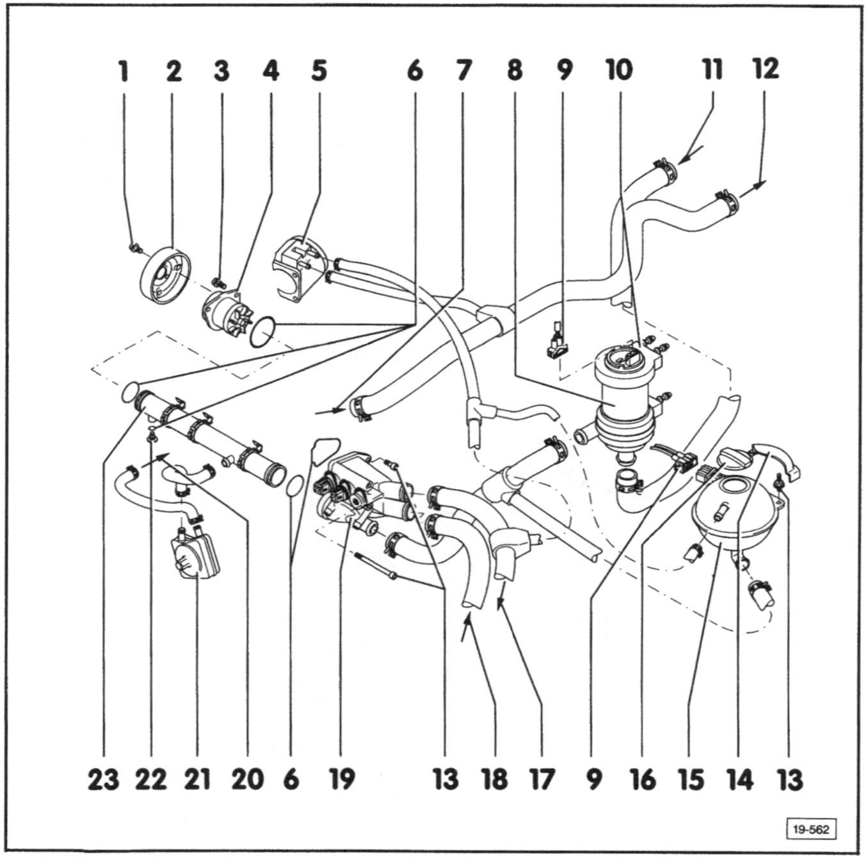

CAUTION
Coolant/antifreeze must not be reused when replacing engine, cylinder head, cylinder head gasket, radiator and heater core.

Note
- always replace all gasket and O-rings
- all components shown can be removed/installed with the engine in vehicle
- during repairs, the original spring type hose clamps may be replaced by the screw type
- install spring clamps using pliers **Hazlet 798-5** or a commercial plier equivalent

1 – **25 Nm (18 ft lb)**
use coolant pump wrench **VAG 1590** to loosen/tighten

2 – **V-belt pulley**

3 – **20 Nm (15 ft lb)**

4 – **Coolant pump**
removing and installing, see page 19.14

5 – **Throttle body housing**

6 – **O-ring**
replace

7 – **From cylinder head**

8 – **Auxiliary electric coolant pump (V51)**
checking, page 19.19

9 – **Connector**

10 – **Rubber retainer**

11 – **From heat exchanger**

12 – **To heat exchanger**

13 – **10 Nm (7 ft lb)**

14 – **Trim**

15 – **Coolant expansion tank**
with coolant low level switch

16 – **Pressure cap**
- check with cooling system tester **VAG 1274**, and adapters **VAG 1274/3**, **VAG 1274/4**
- relief valve opens: 1.2-1.5 bar (17.4-21.8 psi)

17 – **Coolant hose, upper**
to radiator, upper, see page 19.7

18 – **Coolant hose, lower**
from radiator, lower, see page 19.7

19 – **Thermostat housing**
disassembling/assembling, see page 19.17

20 – **To cylinder block**

21 – **Oil cooler**

22 – **Drain plug**
draining and refilling coolant, see page 19.11

23 – **Coolant tube**

Cooling system, draining/filling

Draining

CAUTION

Coolant/antifreeze **must not** be reused when replacing engine, cylinder head, cylinder head gasket, radiator and heater core.

- remove cap from expansion tank
- remove drain plug from coolant pipe, see page 19.9

Note

Direct the coolant when draining by slightly twisting the coolant pipe.

Filling

The cooling systems of **all** Volkswagen vehicles are filled at the factory with a mixture of water and an antifreeze solution (**G 11**) with corrosion inhibitors. This coolant mixture should be used year-round.

Phospate-free anti-freeze prevents frost and corrosion damage, the formation of chalk, and in addition, it raises the boiling point of water. Due to the higher boiling point the coolant is an aid to operational efficiency, when the engine is operating under full load, particularly in tropical climates.

When replacing coolant/antifreeze solution in all Volkswagen models, all model years, use phosphate-free coolant/antifreeze **ZVW 237-104**.

Recommended mixtures

Frost protection to:	G 11	Water
−25°C (−13°F)	3.5L (3.7 qt.)	6.5L (6.8 qt.)
−35°C (−31°F)*	5.0L (5.3 qt.)	5.0L (5.3 qt.)

*for countries with cold climates

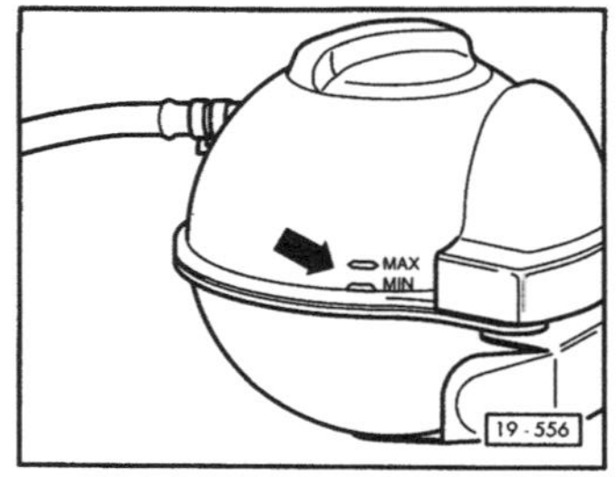

- install coolant pipe drain plug
- slowly fill coolant expansion tank with coolant up to **max** marking (**arrow**)
- install coolant expansion tank cap
- run engine until cooling fan (**V7**) comes on
- check coolant level. If necessary, top off up to the marking. With engine at operating temperature, coolant level should be slightly above **max** mark. With engine cold, the coolant level should be between the **min** and **max** marks.

Radiator/fan, removing/installing

Removing

CAUTION

Wait at least 20 seconds **after** switching off ignition before disconnecting battery. This will allow the platinum hot-wire in Air Mass Sensor (**G70**) to burn itself clean.

CAUTION

Always switch off ignition when disconnecting/reconnecting battery, or any part of Motronic ignition wiring system to prevent damage to Motronic control unit (**J 220**).

CAUTION

Always obtain security code before disconnecting battery ground strap.

- switch off ignition
- disconnect and remove battery
- drain coolant, see page 19.11
- disconnect coolant hoses from radiator
- disconnect thermo-switch, and coolant fan
- unscrew radiator mounting bolts (retaining bracket on lock support)
- remove front bumper, see Repair Group 63
- remove lock support, see Repair Group 50
- remove radiator from the top

Vehicle with A/C

CAUTION

Do not kink or disconnect A/C refrigerant lines.

If A/C refrigerant system is opened or requires service, refer to Repair Group 87 for **CAUTIONS**, **WARNINGS** and procedures.

- unscrew retaining clamps for freon lines
- separate A/C condenser from radiator, and pull forward as far as possible
- remove radiator from the top between condenser and engine

Installing

Reinstall all components in reverse order, noting the following:

CAUTION

Coolant/antifreeze **must not** be reused when replacing engine, cylinder head, cylinder head gasket, radiator and heater core.

- refill coolant, see page 19.11
- check headlight adjustment, adjust if necessary

Coolant pump, removing/installing

Removing

> **CAUTION**
>
> Wait at least 20 seconds **after** switching off ignition before disconnecting battery. This will allow the platinum hot-wire in Air Mass Sensor (**G70**) to burn itself clean.

> **CAUTION**
>
> Always switch off ignition when disconnecting/reconnecting battery, or any part of Motronic ignition wiring system to prevent damage to Motronic control unit (**J 220**).

> **CAUTION**
>
> Always obtain security code before disconnecting battery ground strap.

- switch off ignition
- disconnect battery ground strap
- drain coolant, see page 19.11
- disconnect front exhaust pipe at flange to catalyst, see Repair Group 26
  - replace fastening nuts
- remove poly ribbed belt, see Repair Group 13

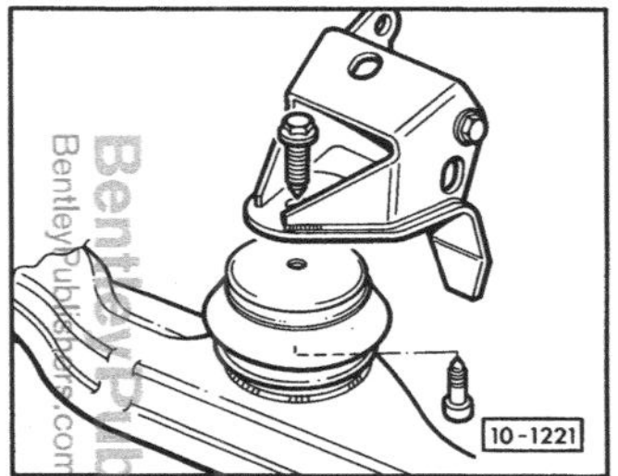

- disconnect engine with bracket from front engine rubber mount, from top (as shown)

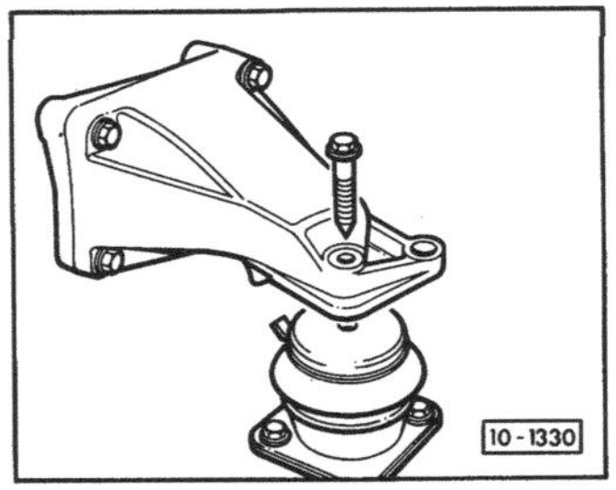

- disconnect engine with bracket from rear engine rubber mount (as shown)

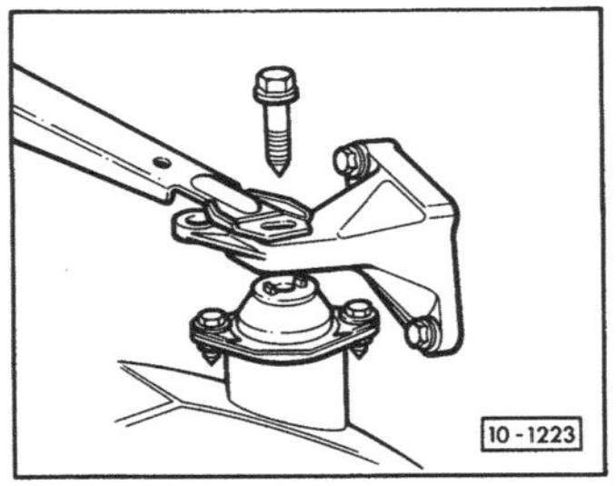

- unscrew transmission with bracket on rear transmission rubber mount (as shown)

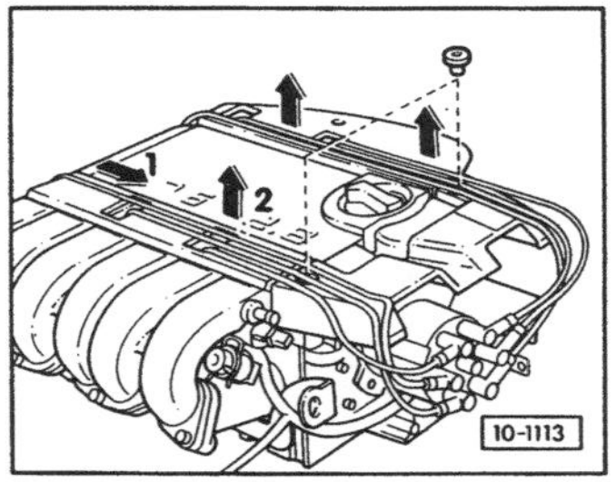

- remove ignition cable guide

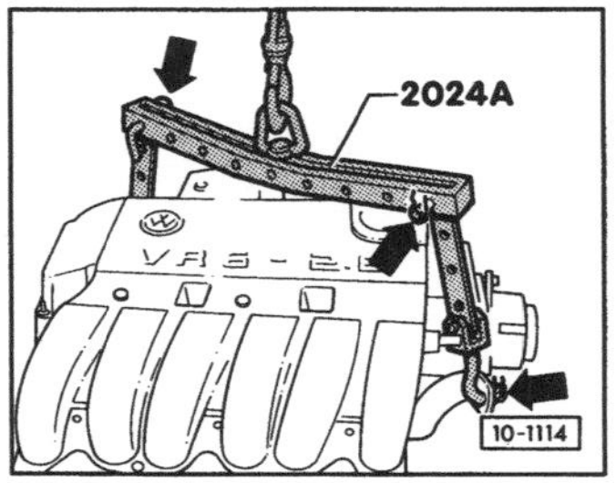

> **CAUTION**
>
> Use cotter pins on hooks and holding pins (**arrows**) to secure engine sling **2024A**.

- attach engine sling, as shown and take weight with crane

Pulley side (**left arrow**)

fourth hole in vertical bar at position **1**

Flywheel side (**right arrow**)

fourth hole in vertical bar at position **8**

The positions marked 1-4 on the bar must be towards the pulley side. The holes in the vertical bar are counted **upwards** from the hook.

- lift engine with transmission sufficiently to gain access to coolant pump
- using water pump wrench **VAG 1590**, remove coolant pump pulley
- unscrew coolant pump
- remove pump by pushing engine/transmission slightly by hand to the left for clearance

Installing

Reinstall all components in reverse order, noting the following:

- when installing engine brackets ensure that the recess on the right rear engine bracket and the front engine bracket fits into the mounting tab of the bonded rubber bushings (**arrow**)
- hand tighten engine and transmission carrier bolts approximately 5 - 6 turns
- lightly rock or shake engine to adjust position
- fully tighten engine, then transmission carrier bolts
- reinstall poly-ribbed belt, see Repair Group 13
- refill coolant, see page 19.11

Tightening torques

Pulley to pump	25 Nm (18 ft lb)
Coolant pump to cylinder block	20 Nm (15 ft lb)
Exhaust pipe to catalytic convertor	25 Nm (18 ft lb)

Engine/transmission carrier, brackets, see Repair Group 10

THIS PAGE INTENTIONALLY LEFT BLANK

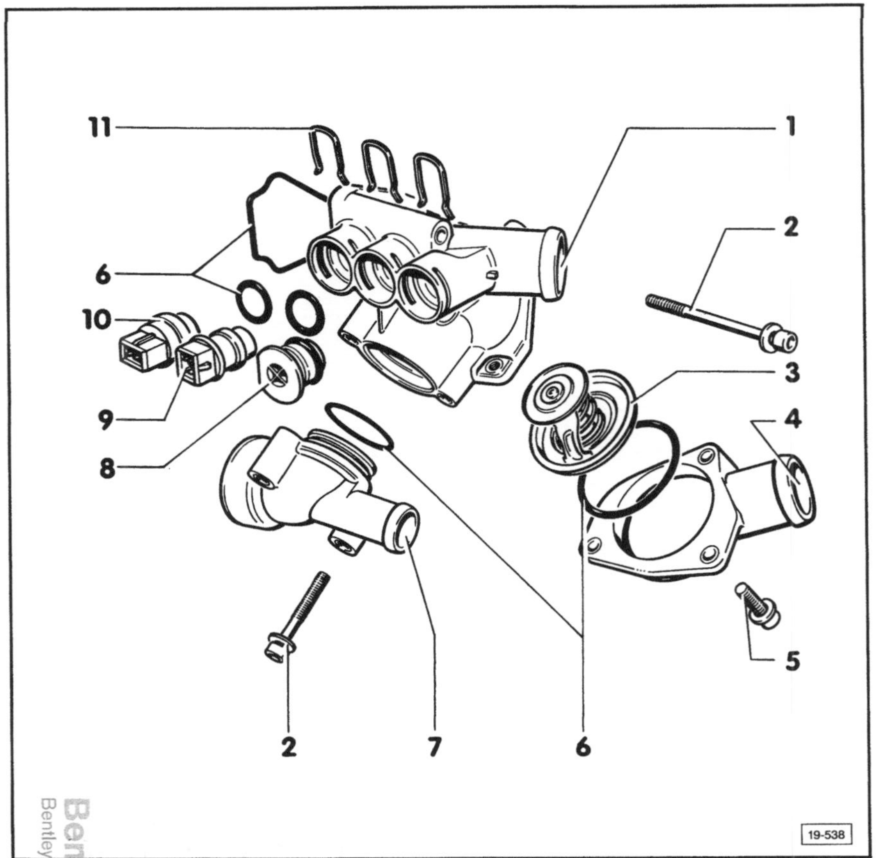

1 – **Thermostat housing**
with hose connection to upper radiator, see page 19.8

2 – **10 Nm (7 ft lb)**

3 – **Thermostat**
- checking: heat thermostat in water
  - opens at approximately 80°C (176°F)
  - closes at approximately 105°C (221°F)
  - stroke min. 7.0mm (9/32 in.)

4 – **Cover**
with hose connection from lower radiator, see page 19.9

5 – **10 Nm (7 ft lb)**

6 – **O-ring**
replace

7 – **Cover**
with hose connection to pump for coolant run-on, see pages 19.9

8 – **Plug**
- for vehicles without air conditioning
- plug is replaced with temperature sender for vehicles with air conditioning

9 – **Temperature sender switch (G2/F87)**
- yellow
- 4-point
- for coolant run-on
- checking, see page 19.19
- switch temperatures:
  on: 101-107°C (214-225°F)
  off: 94-100°C (201-212°F)

10 – **Sender for coolant temperature (G62)**
- blue
- 2-point
- for motronic
- checking, see Repair Group D1

11 – **Mounting clip**
- ensure secure fit

Auxiliary electric coolant pump (V51), radiator fan (V7) run-on, checking

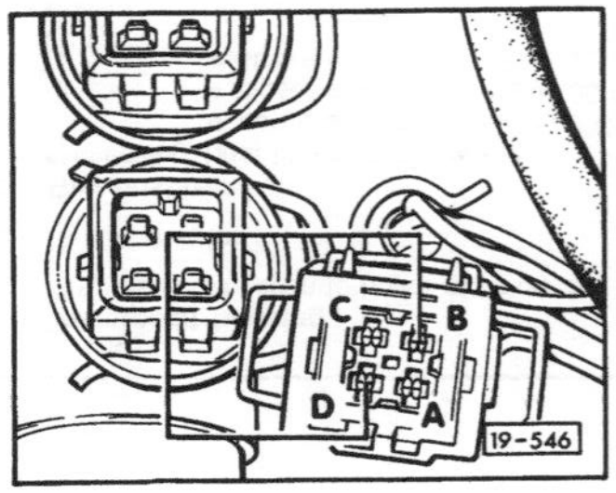

- turn ignition key on/off
- disconnect 4- point connector from temperature sender switch (**G2/F87**)
- bridge contacts **B, D** (brown/red and brown wire) using Connector Test Kit **VAG 1594**
  - radiator fan (**V7**), auxiliary coolant pump (**V51**) must come on

If **NO**, check with **VAG 1466** or equivalent.

Coolant fan control module, location/terminal identification

Location

The coolant fan control module is located in the engine compartment, on the left.

Connector terminal identification

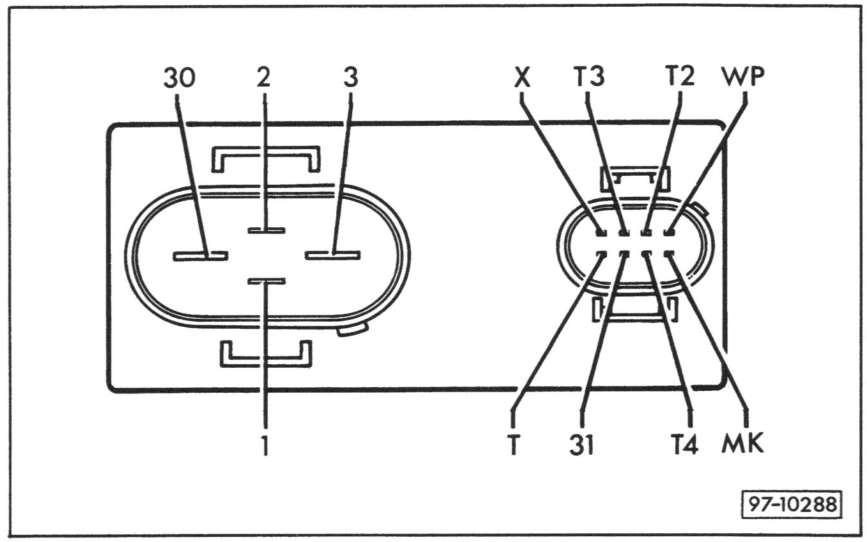

T 4/1 = 3 T 4/2 = 2 T 4/3 = 1 T 4/4 = 30	T 8e/1 = WP T 8e/2 = T2 T 8e/3 = T3 T 8e/4 = X	T 8e/5 = MK T 8e/6 = T4 T 8e/7 = 31 T 8e/8 = T

Note

See wiring diagram for wire routing to **T 4** and **T 8** connectors. Always refer to appropriate wiring diagram when performing any electrical tests or repairs.

Index

Engine code – PG

Accelerator cable
- checking/adjusting **20.9**

Carbon canister
- component layout **20.16**
- shutoff valve, checking **20.17**

Fuel delivery rate
- checking **20.7**

Fuel pump
- afterrun, checking **20.13**
- checking **20.10**
- check valve, checking **20.15**
- relay, checking **20.12**
- removing/installing **20.11**

Fuel supply system
- component layout **20.2**

Rules of cleanliness
- chart **20.6**

Index

Engine code AAA

Accelerator cable
- checking adjusting **20.22**

Carbon canister
- component layout **20.32**

Fuel delivery rate
- checking **20.27**

Fuel pump
- check valve checking **20.29**
- checking **20.25**
- relay checking **20.30**

Fuel supply/fuel pump assembly
- installation position **20.20**
- removing/installing **20.23**

Fuel supply system
- component layout **20.18**

Rules of cleanliness
- chart **20.6**

Vent valve
- checking **20.21**

Accelerator cable, adjusting, page 20.9

Fuel pump relay, checking, page 20.12

Fuel pump afterrun, checking, page 20.13

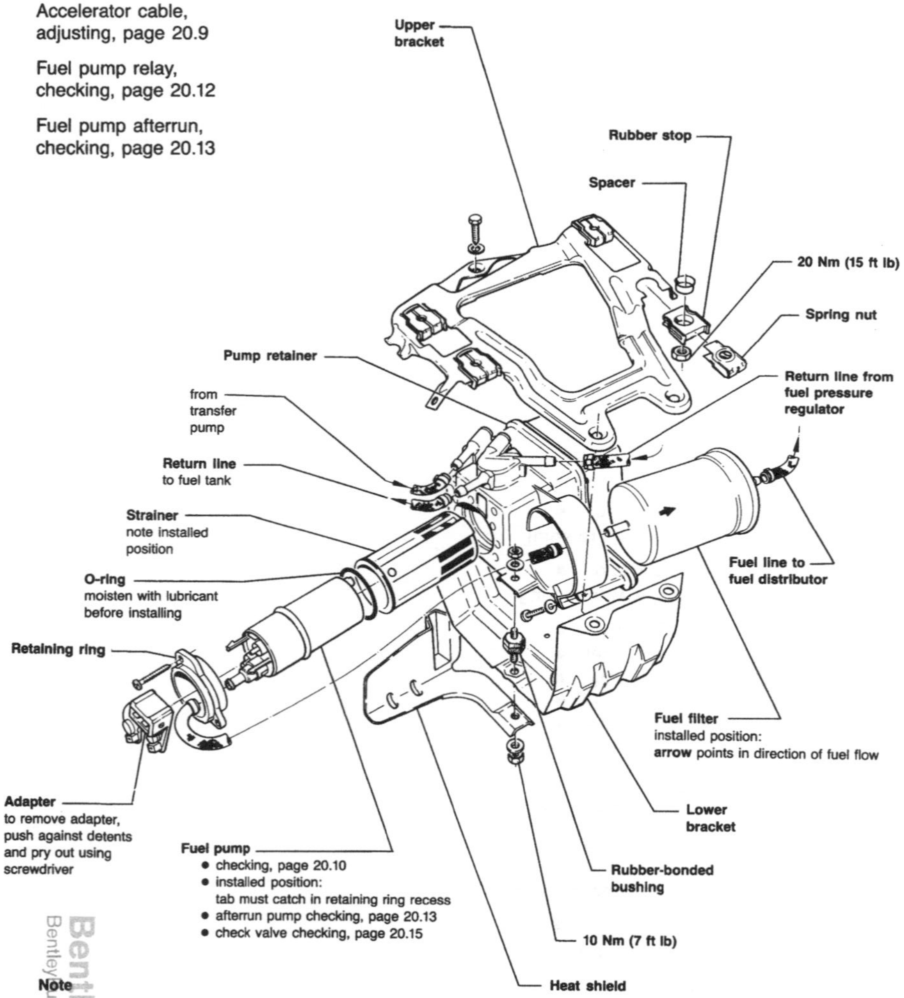

Note

- always replace sealing rings when doing repair
- hose connections are secured with either screw clamps or spring clamps. However; when replacing always use screw clamps
- observe Rules of Cleanliness, page 20.6

Note

Replace spring clamps with screw clamps.

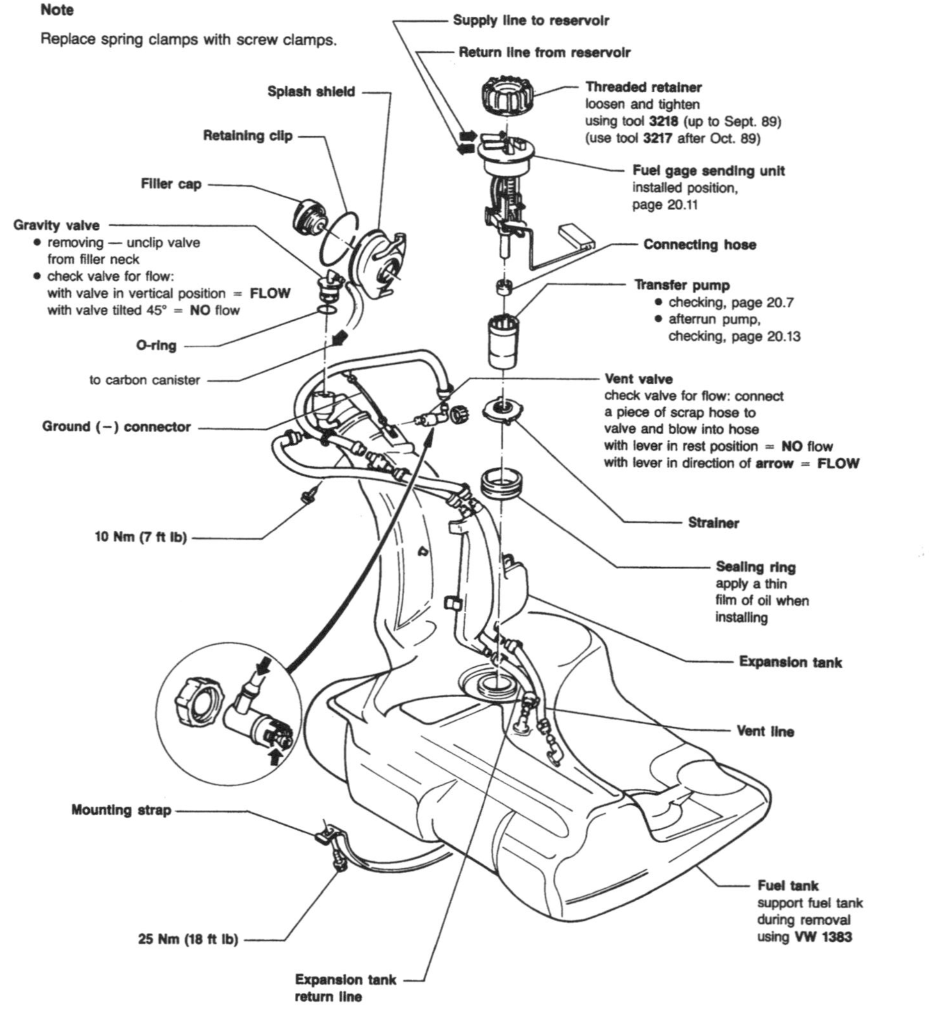

Note

Sender with transfer pump is an assembly.

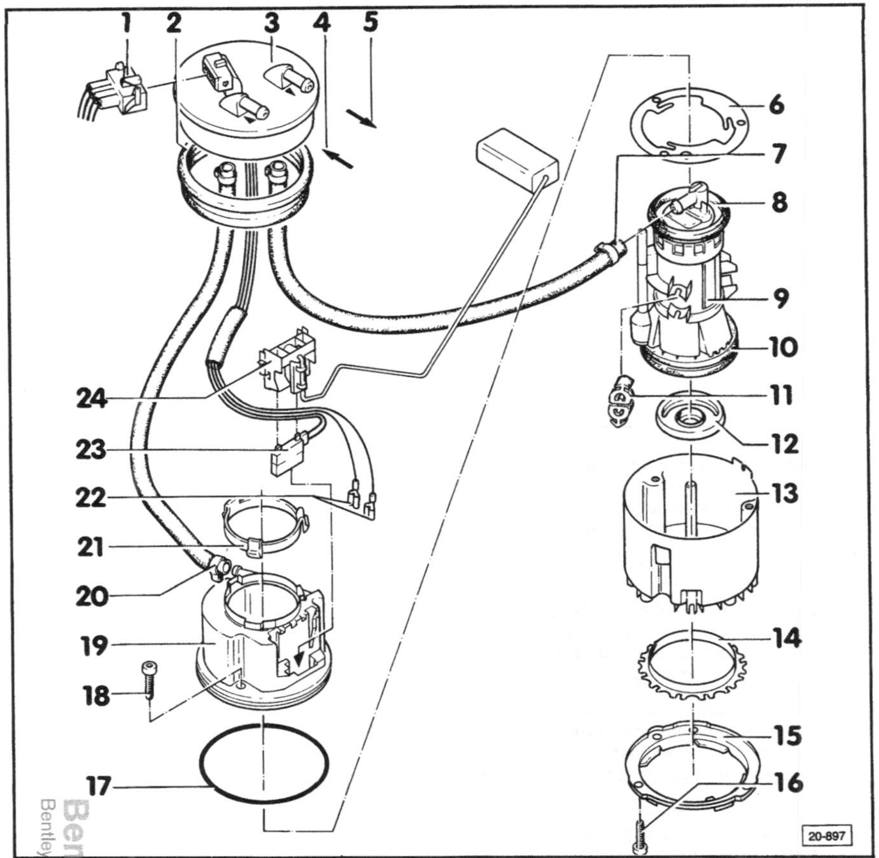

Note
Threaded retainer (not shown)
— loosen and tighten using tool **3217**

Starting with October 1989 production the fuel supply system was revised by installing the fuel pump into the fuel tank and eliminating the reservoir assembly and transfer pump.

WARNING
When removing or installing the fuel pump provide adequate ventilation in the trunk area by drawing off any gasoline fumes with an approved exhaust vent or equivalent.

Avoid the possibility of fuel contacting your hands by using disposable gloves.
- secure all hose connections using screw type clamps
- observe the Rules of Cleanliness, see page 20.6

1 — **Fuel gauge sending unit harness connector**

2 — **Sealing ring**

3 — **Flange assembly**

4 — **Return port from fuel pressure regulator**

5 — **Supply port to fuel filter**

6 — **Retaining ring**

7 — **Pre-supply hose**

8 — **Sealing ring, upper**
- pull in through upper housing (**19**) and fasten using lock ring (**21**)

9 — **Fuel pump**
- removing/installing page 20.11
- delivery quantity, checking page 20.7

10 — **O-ring, lower**
- pull into lower housing and attach using rings (**14** and **15**)

11 — **Deadener rubber**
- between fuel pump and retaining ring (**6**)

12 — **Filter/strainer**

13 — **Housing, lower section**

14 — **Retaining ring, lower**

15 — **Bayonet ring**

16 — **Screw**

17 — **O-ring**

18 — **Screw**

19 — **Housing, upper section**

20 — **Return hose**

21 — **Retaining ring, upper**
- unclip using screwdriver

22 — **Electrical terminals for fuel pump**

23 — **Harness connector for fuel gauge sending unit**

24 — **Fuel gauge sending unit**

Rules of Cleanliness

CAUTION

When working on the fuel supply/ injection system, carefully observe the following rules:

1 — Thoroughly clean connection and surrounding areas before loosening connection.

2 — After removing components, place in clean area and cover with foil or paper. Avoid using rags!

3 — Components which have been opened or disassembled must be carefully covered or sealed if repair cannot be carried out immediately.

4 — Install clean parts only.
- remove replacement parts from package just before installing
- do **NOT** use spare parts that were stored loose or unpackaged (e.g. in tool boxes, etc.)

5 — When fuel system is open:
- avoid using compressed air whenever possible
- avoid moving the vehicle whenever possible

Fuel delivery rate, checking

Applies to both single and two pump systems.

WARNING

Fire hazard, Do not have anything in area that can ignite fuel.

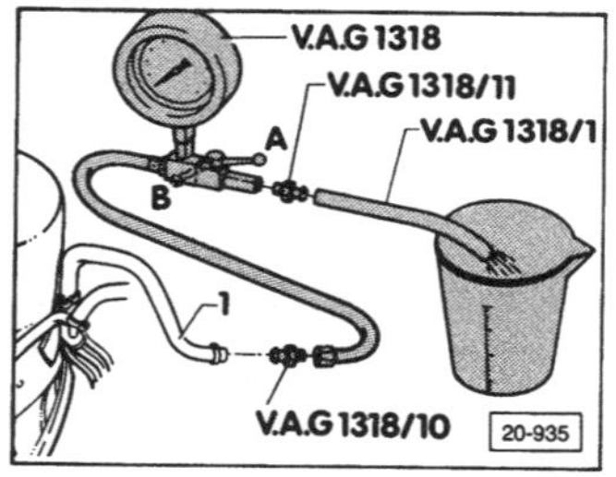

▶

Requirements
- supply voltage **OK**
- remote control **VAG 1348/3A** connected

- remove fuel filler cap
- connect **VAG 1318** pressure tester to pre-supply line 1 using a length of scrap hose and adaptor **VAG 1318/10**
- insert hose **VAG 1318/1** onto adaptor **VAG 1318/11** of pressure tester and hold in measuring glass
- open shutoff valve of tester **VAG 1318** (lever points in direction of flow) position **A**
- operate remote control **VAG 1348/3A** while slowly moving tester shutoff lever in direction **B** until the following pressure is indicated on the tester
  - two pump system (prior to 9-89 production): 3 bar (43.5 psi)
  - single pump system (from 10-89 production): 4 bar (58 psi)

When the indicated pressure is obtained stop moving the valve lever
- shut off remote control
- drain measuring glass
- operate remote control for 30 seconds then shut off
- compare quantity of accumulated fuel in measuring glass with appropriate chart

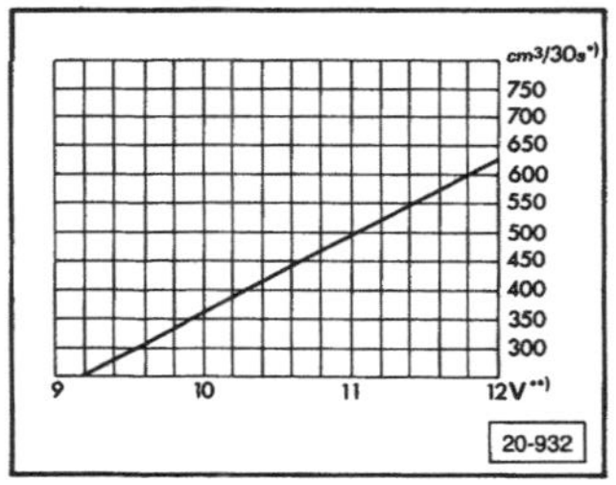

- up to 9-89, system with two pumps
  - *minimum supply per 30 seconds
    voltage of running fuel pump with engine **NOT running (should be approximately 2 volts less than battery voltage)

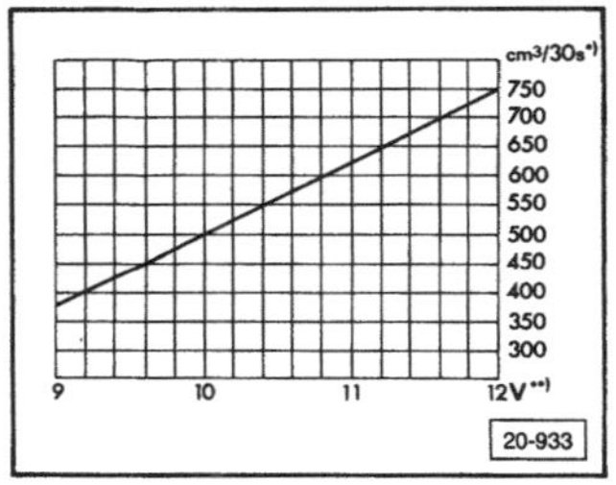

- from 10-89, system with one pump
  - *minimum supply per 30 seconds
    voltage of running fuel pump with engine **NOT running (should be approximately 2 volts less than battery voltage)

If minimum delivery value NOT obtained
- check if fuel lines are bent, pinched or plugged; repair as necessary
- check if fuel filter is plugged, replace if necessary

Note

The chart is expressed in units of cubic centimeters (cc's). It is also acceptable to substitute milliliters (ml) for cc's.

This will allow you to use containers calibrated in cc's or ml's with equal results.

- 1 milliliter = 1 cubic centimeter

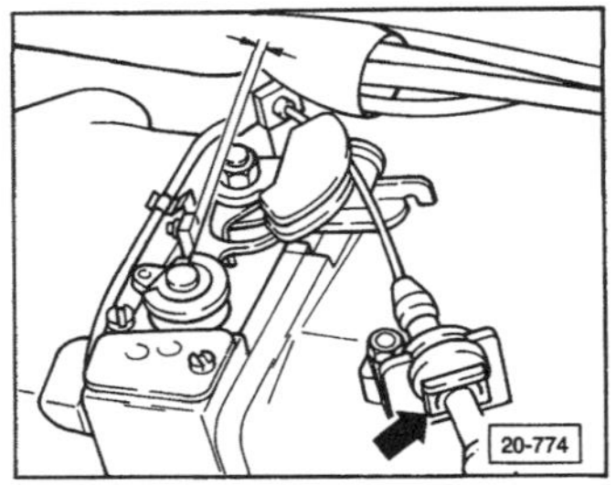

Accelerator cable, checking/ adjusting

Checking

- depress accelerator pedal fully (to floor stop)
  - throttle valve must just reach full throttle position (maximum of 1 mm of play at throttle valve lever)

If **NO**
- adjust as follows

Adjusting

- remove retaining clip (**arrow**) and reposition slotted portion of cable
  - (until checking condition is met)
- re-insert positioning clip
- verify that checking condition is met

Fuel pump, checking

Supply voltage, checking

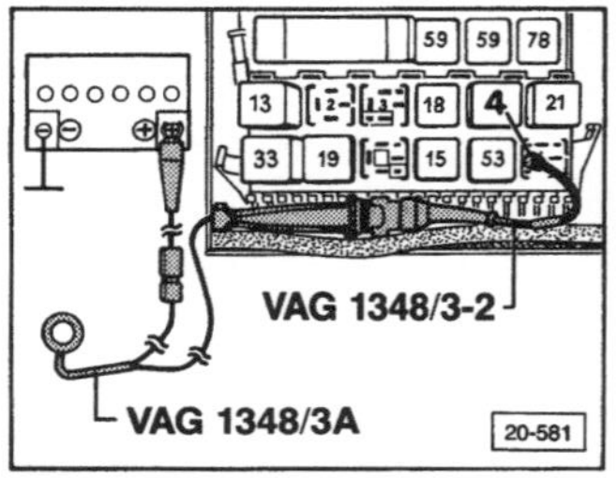

Check these first:
- fuse **18 OK**
- ignition coil wire removed from distributor and connected to ground using jumper wire/clip

- ■ crank starter for 3-4 seconds
  - fuel pumps **MUST** be audible for short time

If **NO**
- ■ remove fuel pump relay from fuse relay panel **(arrow)**
- ■ connect remote control adaptor **VAG 1348/3-2** to terminal **4** of fuel pump relay socket
- ■ connect adaptor lead from remote control **VAG 1348/3A** to battery (+)

- ■ activate remote control
  - fuel pumps should run audibly

If **YES**
- ■ replace fuel pump relay

If **NO** or only one pump runs:
- ■ disconnect transfer pump harness connector
- ■ connect **US 1115** LED tester to middle wire and outer brown wire
- ■ operate remote control
  - LED must light up

If **NO**
- ■ check wiring using wiring diagram, repair as necessary

If **OK**
- ■ replace fuel pump(s)

Fuel supply assembly/fuel pump, removing and installing

Note

This page refers to single pump systems installed since 10-89.

Numbers refer to exploded view **20-897** on page 20.4.

WARNING

Fire hazard, Do not have anything in area that can ignite fuel.

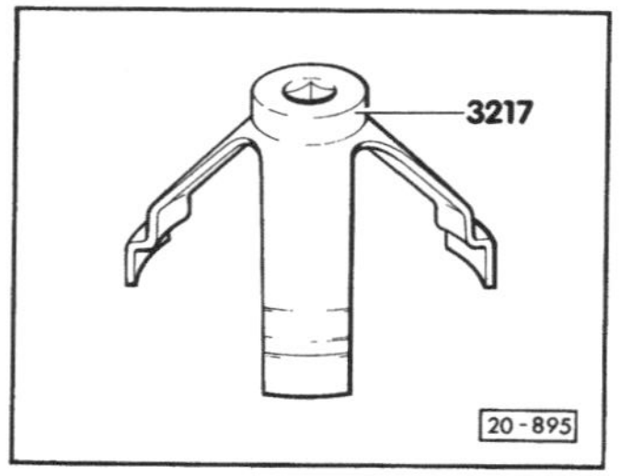

To remove fuel supply assembly

- ■ remove trunk floor cover
- ■ disconnect harness connector and fuel lines from flange **3**
- ■ unscrew flange nut using special tool **3217**
- ■ withdraw flange **3** and O-ring **2** from opening on fuel tank
- ■ loosen and remove supply unit from bayonet lock by turning to the left

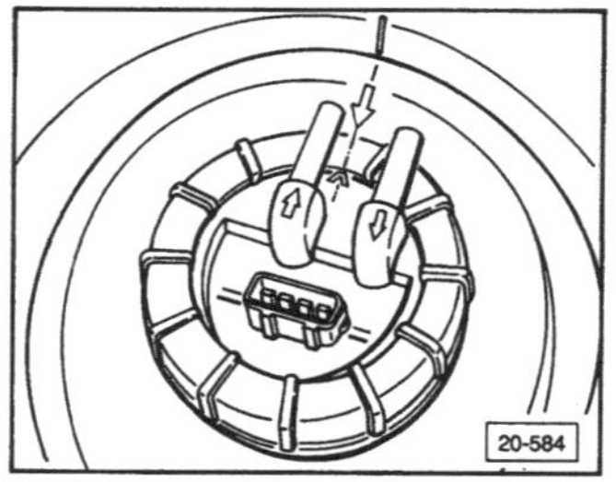

Note

When installing be sure that sender **24** is neither bent nor pointing towards the front of the tank.

The correct position is obtained by aligning the arrows as shown in the illustration.

To remove fuel pump

Requirement
- remove fuel supply assembly

- ■ remove fuel gauge sending unit **24**, connector **22**, connector **23**, hoses **7** and **20** from supply assembly
- ■ unclip upper retaining ring **21** using screwdriver
- ■ unscrew bolts **18** and remove upper housing **19** from lower housing **13**
- ■ installation is reverse of removal procedure

CAUTION

Carefully install O-rings **8** and **10** in the upper and lower housings. Do not pinch, roll or misalign before tightening.

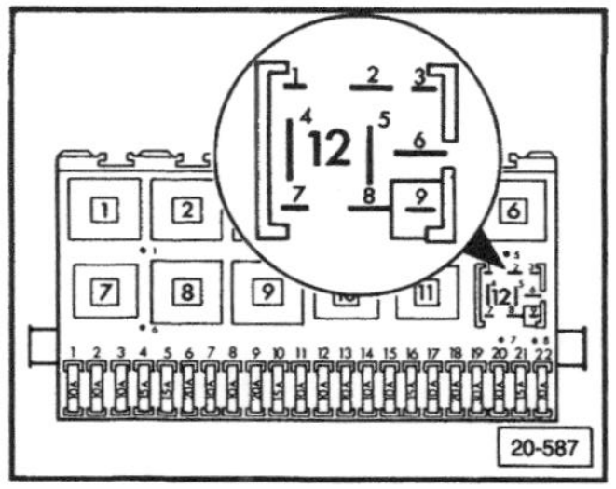

Fuel pump relay, checking

Symptom: Fuel pumps **NOT** running when ignition is switched **ON**.

Check this first
- fuse **18 OK**

▶

- remove fuel pump relay from fuse relay panel **(arrow)**
- switch ignition **ON**
- connect **US 1115** LED tester as follows:
  - between terminal **6** and ground (−)
  - between terminal **2** and ground (−)
  - between terminals **6** and **4**
  LED must light up

If **NO**
- check wiring using wiring diagram, repair as necessary

- switch ignition **OFF**
- connect **US 1115** LED tester between terminals **2** and **3**
- switch ignition **ON**
  - LED must light up for approximately 1 second

If **NO**
- check for open circuit in wiring to control unit using wiring diagram

If circuit **OK**
- replace fuel injection control unit

Fuel pump afterrun, checking

Check this first:
- battery voltage **OK**
- fuel tank must be 1/4 full, minimum

WARNING

Fire hazard, Do not have anything in area that can ignite fuel.

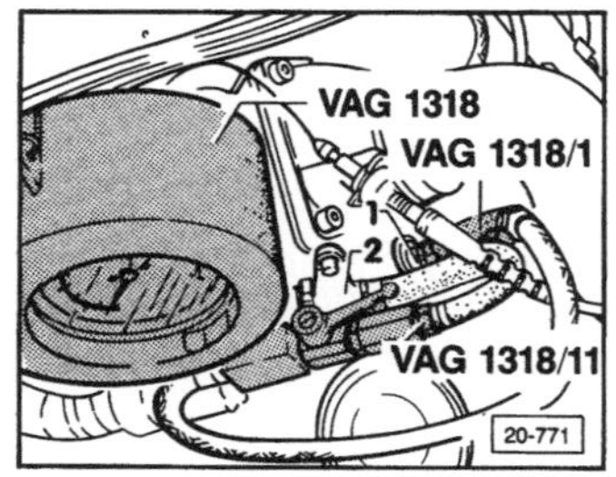

▶

- connect pressure tester **VAG 1318** using adaptors **1318/1** and **1318/11** between fuel distributor **1** and pressure switch **2**

Note

The shutoff valve must be open (lever in direction of fuel flow).

CAUTION

When loosening the line at the fuel distributor, place a rag around the connection to avoid the risk of being sprayed.

- start engine and run at idle
- observe fuel pressure
  - approximately 2.5 bar (36.3 psi)
- switch ignition **OFF**

▶

Remove harness connector from pressure switch **1** and measure resistance between pressure switch terminals.
- 0 Ohms approximately

- remove harness connector from ambient temperature switch for fan afterrun **2** and connect to ground (−) using adaptor wire from **VW 1594** kit
  - radiator fan **MUST** run
- re-connect pressure switch harness connector

- switch ignition **ON** then switch **OFF**
  - after about 2-3 minutes fuel pumps **MUST** run
- re-connect ambient temperature switch harness connector
  - fuel pump must **NOT** run
- remove harness connector from pressure switch and measure resistance between terminals
  - 0 Ohms approximately

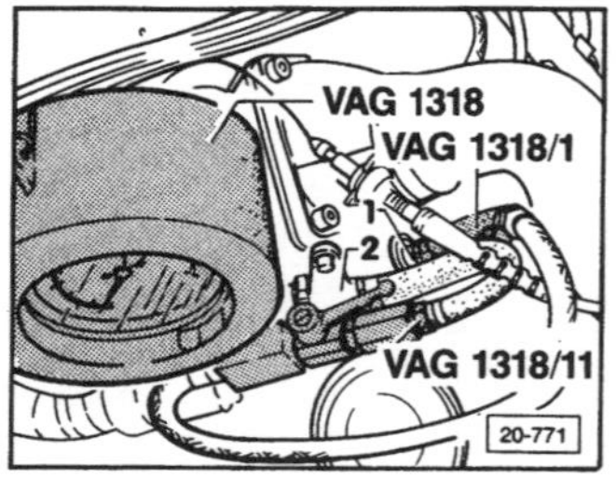

- reduce system pressure by carefully loosening pressure gage/adaptor **VAG 1318/1** at connection

> **CAUTION**
>
> When loosening the adaptor place a rag around the connection to avoid the risk of being sprayed.

The switch must open (infinite ∞) when pressure drops below 1 bar (14 psi).

Fuel pump check valve, checking

Note

This leak test includes the entire fuel supply line from the fuel pump to the **VAG 1318** tester.

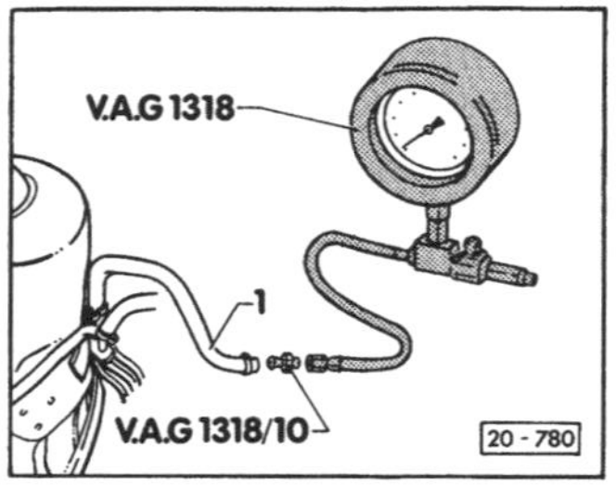

- remove (**black**) fuel supply line **1** on fuel distributor and connect to **VAG 1318** pressure gage using adaptor **VAG 1318/10**

> **CAUTION**
>
> Pressure gage shutoff lever **MUST** be closed (lever at right angles to direction of fuel flow).

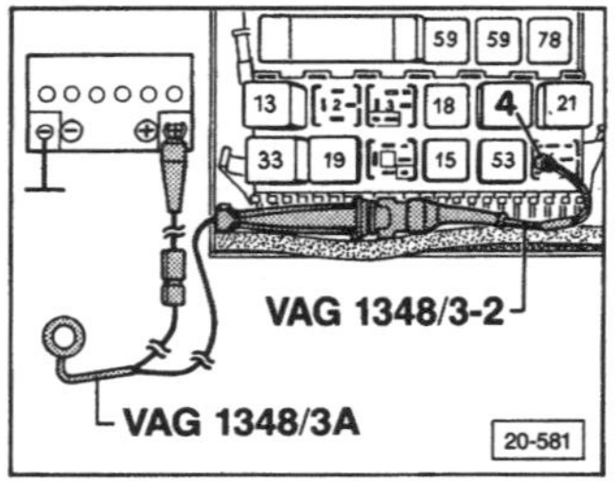

- remove fuel pump relay from fuse relay panel (**arrow**)
- connect remote control adaptor **VAG 1348/3-2** to terminal **4** of fuel pump relay socket
- connect adaptor lead from remote control **VAG 1348/3A** to battery (+)
- activate remote control for brief intervals until a pressure of approximately 3 bar (43.5 psi) has accumulated
- reduce pressure over 3 bar to approximately 2 bar (29 psi) by **CAREFULLY** opening the shutoff valve a small amount

> **CAUTION**
>
> There is a risk of being sprayed when opening the shutoff valve. Hold a rag in front of the pressure gage connector.

- note pressure drop, pressure must **NOT** fall below 2.0 bar (29 psi) after 10 minutes

If **YES**
- check line connectors for leakage and replace fuel pump, if necessary

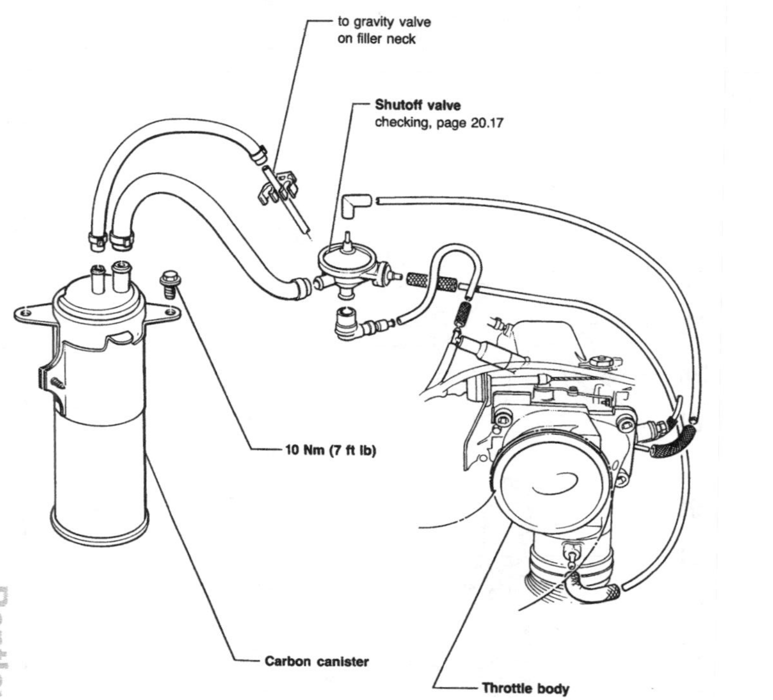

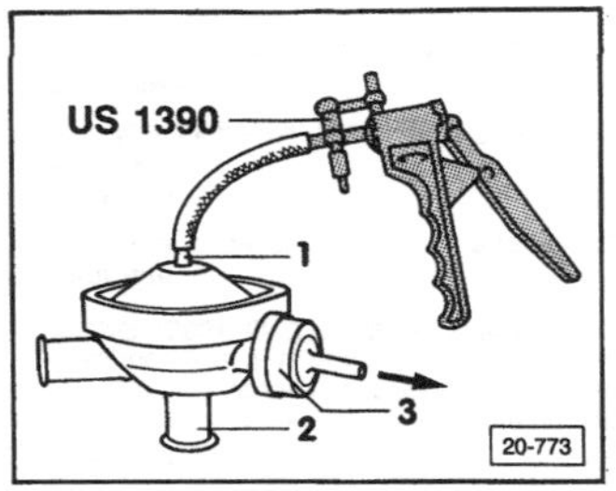

Carbon canister shutoff valve, checking

▶

- connect hand vacuum pump **US 1390** to small diameter connector **1**
- with suction at connector **2**
  - valve must close
- operate vacuum pump
  - valve **MUST** open
- test check valve **3**. It must have flow in the direction of the **arrow**, and be closed in the opposite direction

If **NO**

- replace shutoff valve

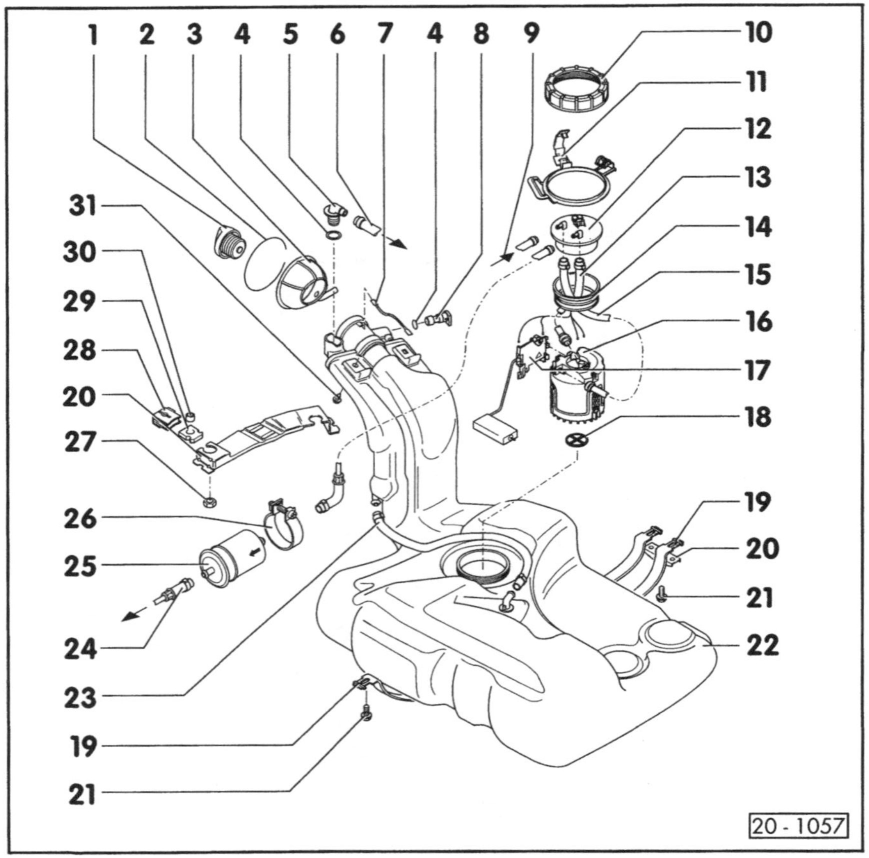

WARNING
When removing or installing the fuel pump provide adequate ventilation in the trunk area by drawing off any gasoline fumes with an approved exhaust vent or equivalent.

Wear disposable gloves to avoid the possibility of fuel contacting your hands.

WARNING
DO NOT re-use any fasteners that are worn or deformed in normal use. Many fasteners are designed to be used only once and become unreliable and may fail when used a second time. This includes, but is not limited to, nuts, bolts, washers, self-locking nuts or bolts, circlips, cotterpins. Always follow recommendation given in this publication. For replacements always use new parts.

Note
- observe the Rules of Cleanliness, see page 20.6
- secure all hose connections using screw type clamps
- adjust accelerator cable, see page 20.22
- check fuel pump relay, see page 20.30
- remove/install carbon canister components, see page 20.32

1 – **Filler cap**

2 – **Retaining clip**

3 – **Splash shield**

4 – **O-ring**
replace

5 – **Gravity valve**
removing, unclip valve from filler neck and pull out from filler neck

6 – **Vent line**
to carbon canister, see page 20.32

7 – **Ground (–) connector**

8 – **Vent valve**
checking, see page 20.21

9 – **Return line**
- blue
- from fuel distributor, see Repair Group 24

10 – **Threaded retainer**
loosen and tighten with tool **3217**

11 – **Hose guide**

12 – **Flange**
note installation position on fuel tank, see page 20.24

13 – **Supply hose**

14 – **Sealing ring**
coat with fuel before installing

15 – **Return hose**

16 – **Fuel pump**
removing/installing, see page 20.23

17 – **Fuel gauge sending unit**

18 – **Filter/strainer**

19 – **Mounting strap**

20 – **Bracket**

21 – **25 Nm (18 ft lb)**

22 – **Fuel tank**
- removing: remove right rear wheel to remove tank, remove vent hose **23** at expansion tank
- support fuel tank during removal using support tool **VAG 1383A** or **OTC Model 1785A**

23 – **Vent line**
remove at expansion tank **before** removing fuel tank

24 – **Fuel line**
- black
- to fuel distributor, see Repair Group 24

25 – **Fuel filter**
installed position: **arrow** points in direction of flow

26 – **Clamp**

27 – **20 Nm (15 ft lb)**

28 – **Rubber stop**

29 – **Spring nut**

30 – **Spacer**

31 – **10 Nm (7 ft lb)**

Fuel supply/fuel pump assembly, installation position

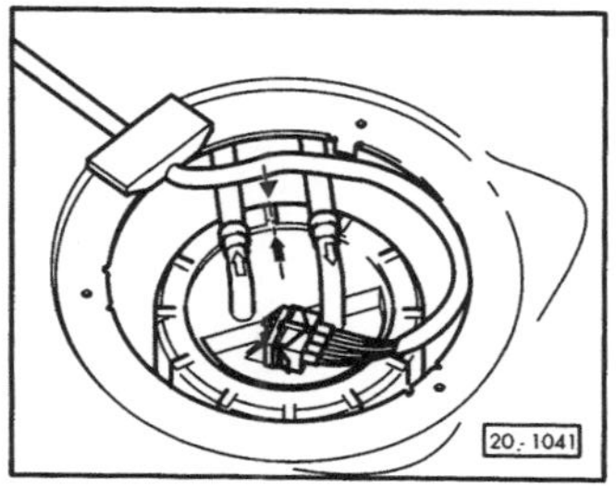

- mark on flange (**arrow**) must align with mark on fuel tank

Vent valve, checking

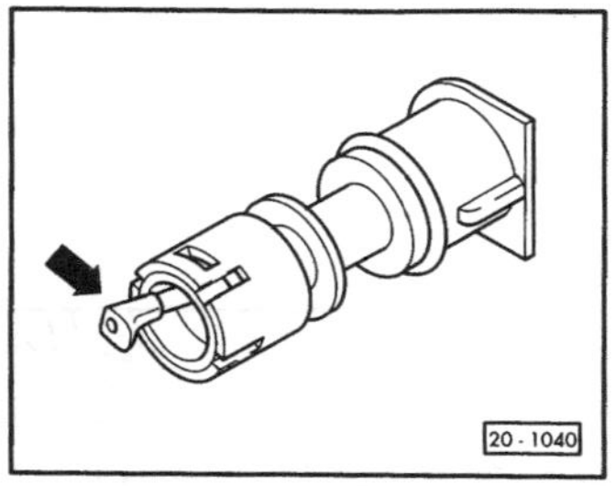

- lever in rest position: valve **closed**
- lever in direction of arrow: valve **open**

Accelerator cable, checking/adjusting

Manual transmission

Checking

- depress accelerator pedal fully (to floor stop)
  - throttle valve must just reach the wide open throttle position

Adjusting

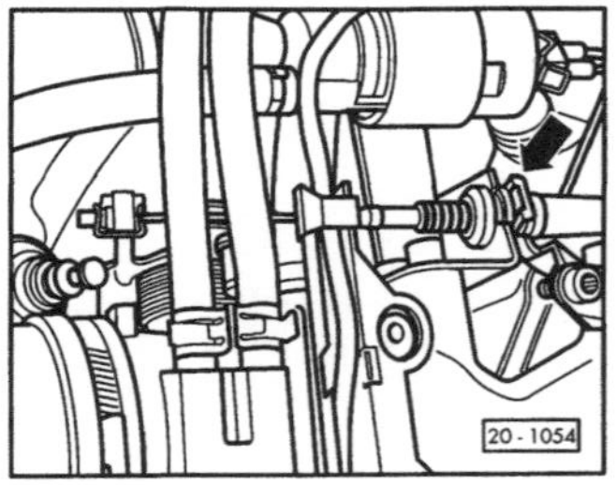

- remove retaining clip (**arrow**) and reposition slotted portion of cable
  - (until checking condition is met)
- re-insert positioning clip
- verify that checking condition is met

Throttle cable, checking and adjusting

Vehicles with automatic transmission

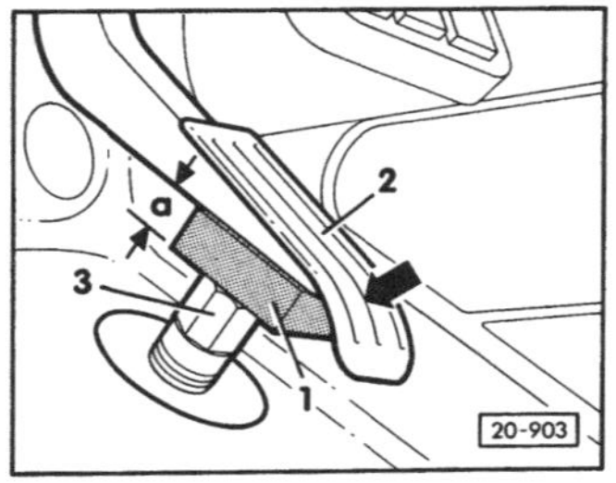

- position spacer **1** (which provides a space of **a** = 15 mm) between accelerator pedal **2** and pedal stop **3**
- fully depress accelerator pedal until it contacts spacer and hold pedal in this position during the following adjustment

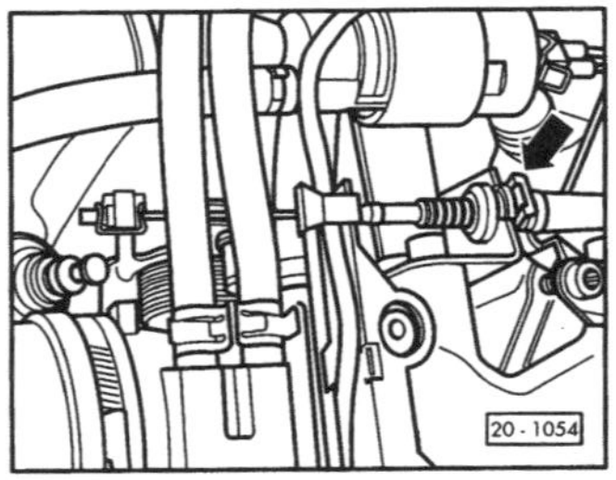

- fully open the throttle by pulling the outer cable and locate at this position using the retaining clip (arrow)
- release the accelerator pedal and remove the spacer

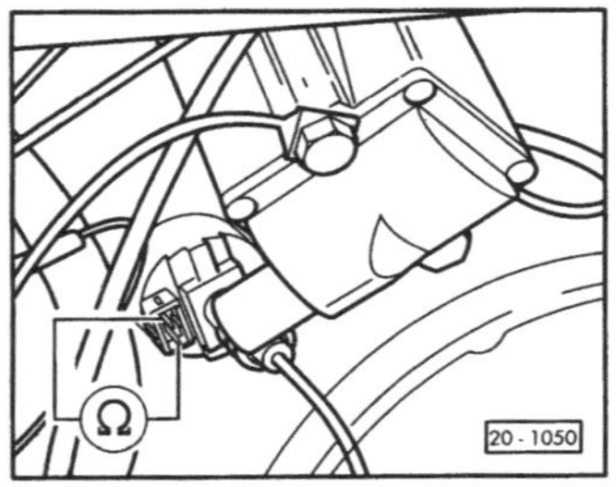

- connect Fluke 83 multimeter betweeen terminals of kickdown switch using jumper wires from **VW 1594** adaptor kit
  - must be infinite Ohms (open)
- slowly depress throttle toward wide open throttle
  - shortly after the kickdown pressure point, resistance must fall to 0 Ohms (continuity)
  - when continuity is obtained, the accelerator pedal should be closely approaching contact with the pedal stop

Fuel pump assembly, removing/installing

Removing

> **WARNING**
> Fire hazard. Do not have anything in area that can ignite fuel.

- switch ignition **OFF**
- disconnect battery ground strap

> **CAUTION**
> Wait at least 20 seconds **after** switching off ignition before disconnecting battery. This will allow the platinum hot-wire in the Mass Air Flow (MAF) Sensor to burn itself clean.

> **CAUTION**
> Always switch off ignition when disconnecting or reconnecting battery, or any part of Motronic system to prevent damage to Motronic engine control module (**ECM**).

> **CAUTION**
> Always obtain radio security code before disconnecting battery ground strap.

- remove cover plate in luggage compartment floor
- disconnect harness connector and fuel lines from flange

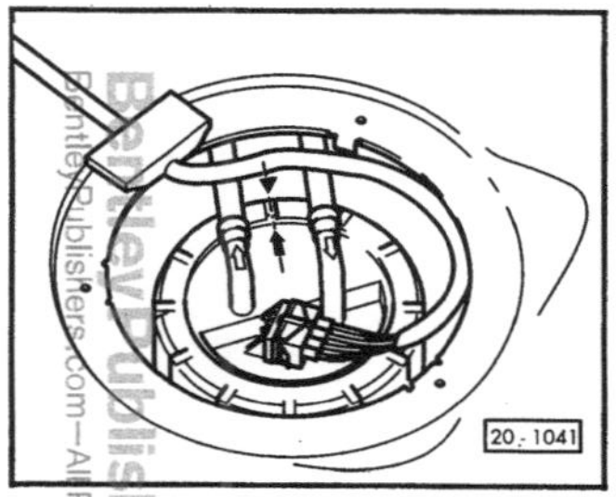

- unscrew flange nut using special tool **3217**
- withdraw flange and O-ring from opening on fuel tank
- loosen and remove supply unit from bayonet lock by turning to the left

Installing

Reinstall all components in reverse order, noting the following:

> **CAUTION**
> Ensure that the sender is not bent when reinstalling.

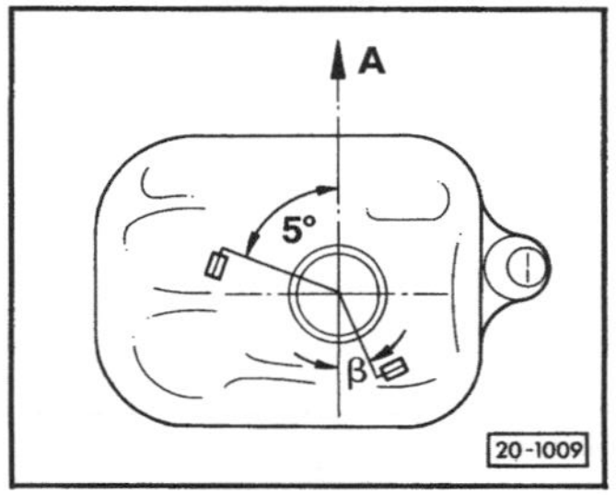

- fuel pump assembly is correctly installed when sender float angle (from direction of travel **A**) is 5°
  - coat flange sealing ring with fuel before installing

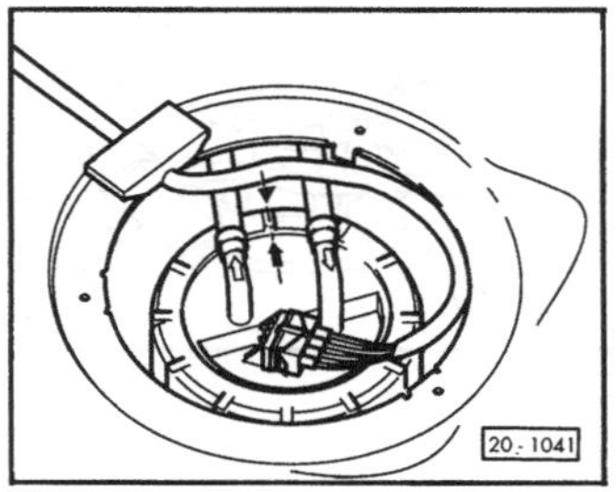

- align mark on flange with mark on fuel tank, as shown

Fuel pump, checking

Supply voltage, checking

Note
Always refer to appropriate wiring diagram.

Requirements
- fuse **18 OK**
- battery **OK**

■ switch ON ignition
- fuel pumps must be audible for approximately **1** second

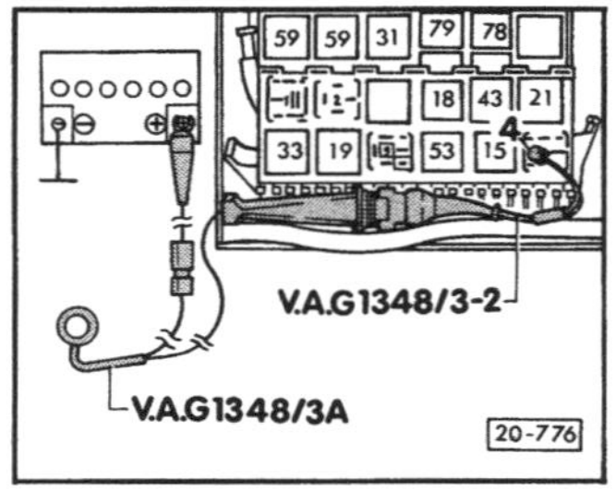

If **NO**

■ remove fuel pump relay (**J 17**) from fuse/relay panel, position **12**
■ connect remote control adaptor **VAG 1348/3-2** to terminal **4** of fuel pump relay socket
■ connect adaptor lead from remote control **VAG 1348/3A** to battery positive (+)
■ activate remote control switch
- fuel pump must run

If **YES**

■ check and replace fuel pump relay, see page 20-30

If **NO**

■ remove cover plate in luggage compartment floor
■ disconnect connector from flange on fuel tank

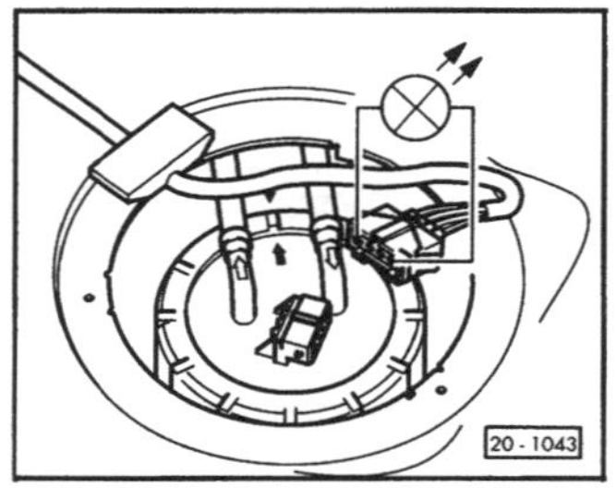

■ connect **US 1115** LED Tester to the outer wires of the connector using jumper wires from **VW 1594** Adaptor Kit
■ operate remote control
- LED must light up

If **NO**

■ check wiring, using appropriate wiring diagram, repair or replace as necessary

If **YES**

■ remove ring nut with **3217**
■ check wiring between flange and fuel pump

If **OK**

■ replace fuel pump, see page 20-23

Fuel delivery rate, checking

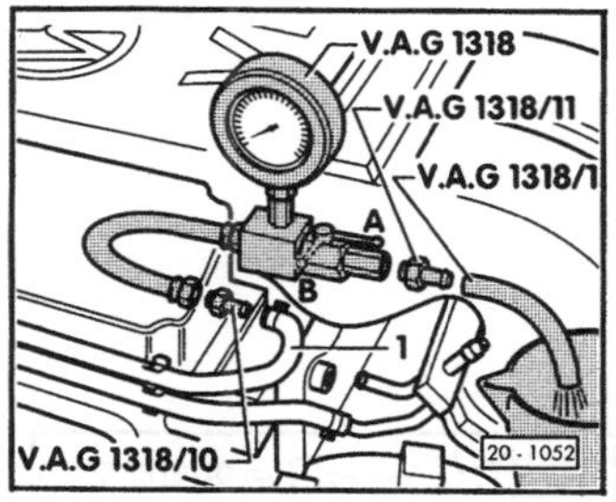

> **WARNING**
> Fire hazard. Do not have anything in area that can ignite fuel.

Requirements
- supply voltage **OK**
- remote control adaptor lead **VAG 1348/3A** connected
- fuel filter **OK**

■ remove fuel tank cap
■ remove fuel supply line **1** on fuel rail

Note
Cover connection with cloth to protect from fuel spray.

■ connect **VAG 1318** pressure tester to the fuel supply line **1** with adaptor **VAG 1318/1A**
■ insert hose **VAG 1318/1A** onto adaptor **VAG 1318/11** and hold in measuring glass
■ open shutoff valve of tester **VAG 1318** (lever points in direction of flow) position **A**
■ operate remote control **VAG 1348/3A** while slowly closing valve of tester until pressure tester indicates 4 bar (58 psi)

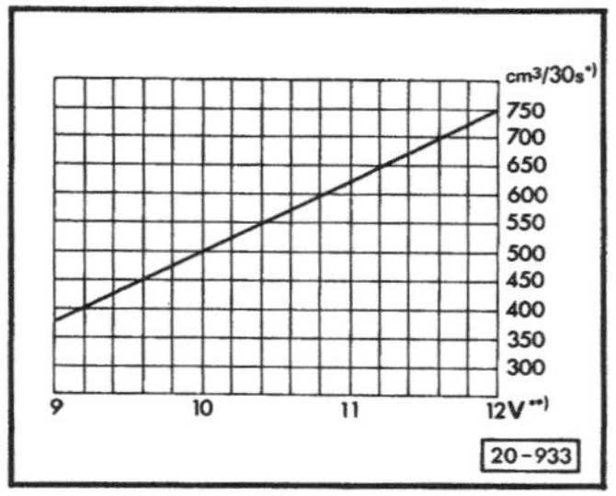

When the indicated pressure is obtained, stop moving the valve lever.

■ switch **OFF** remote control
■ empty measuring glass
■ operate remote control for 30 seconds then shut off

Note
The chart is expressed in units of cubic centimeters (cc's). It is also acceptable to substitute milliliters (ml) for cc's.

■ compare quantity of accumulated fuel in measuring glass with appropriate chart

* minimum supply per 30 seconds

** voltage of running fuel pump with engine **NOT** running (should be approximately 2 Volts less than battery voltage)

If minimum delivery **NOT OK**
- fuel lines bent, pitched or plugged; repair as necessary
- check if fuel filter is plugged, replace as necessary
- fuel pump faulty, replace if necessary, see page 20-23

Fuel pump check valve, checking

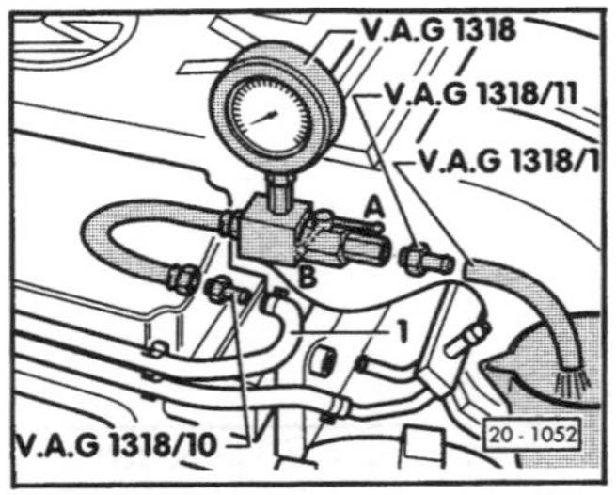

> **WARNING**
>
> Fire hazard. Do not have anything in area that can ignite fuel.

Note

This leak test includes the entire fuel supply line from the fuel pump to the **VAG 1318** tester.

- connect pressure tester **VAG 1318**
- connect remote control adaptor **VAG 1348/3A**

> **CAUTION**
>
> Pressure gauge shutoff lever **MUST** be closed (lever at position **B**).

- activate remote control for brief intervals until a pressure of approximately 4 bar (58 psi) has accumulated
- reduce pressure if over 4 bar (58 psi) by **CAREFULLY** opening the shutoff valve a small amount

> **CAUTION**
>
> There is a risk of being sprayed with fuel when opening the shutoff valve. Hold a measuring glass in front of the pressure gauge connector.

- note pressure drop, pressure must not drop below 3 bar (43.5 psi) within 10 minutes

If **YES**

- check line connectors for leakage and replace fuel pump, if necessary

Fuel pump relay, checking

Symptom: Fuel pump **NOT** running when ignition is switched **ON**.

Note
Always refer to the appropriate wiring diagram.

Requirement
- fuse **18 OK**

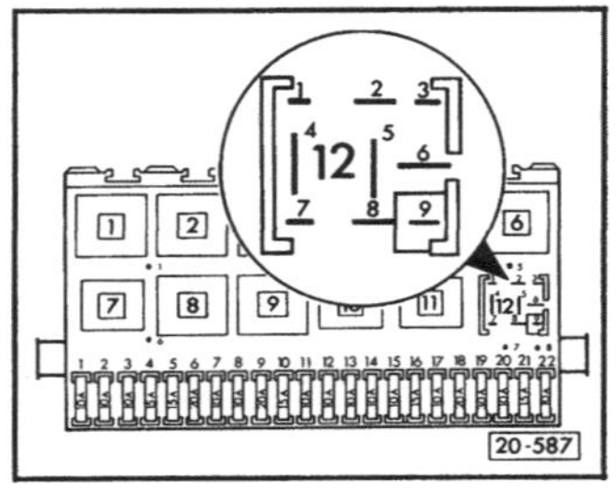

- remove fuel pump relay (**J 17**) from fuse relay panel (**arrow**) position **12**
- switch ignition **ON**
- using Adaptor Test Kit **VW 1594** connect **US 1115** LED Tester as follows:
  - between terminal **6** and ground (-)
  - between terminal **2** and ground (-)
  - between terminals **6** and **4**

LED must light up

If **NO**

- check wiring using wiring diagram, repair or replace as necessary

If **YES**

- switch ignition **OFF**
- using Adaptor Test Kit **VW 1594**, connect LED Tester between terminals **2** and **3**
- switch ignition **ON**
  - LED must light up for **1** second

If **NO**

- check triggering of fuel pump relay, see Repair Group D2 for testing using **VAG 1598** Test Box

If **YES**

- switch ignition **OFF**
- using Adaptor Test Kit **VW 1594**, connect LED Tester between terminals **2** and **3**
- switch ignition **ON**
  - LED must light up for **1** second

If **NO**

- check triggering of fuel pump relay, see Repair Group 01 for testing with Test Box **VAG 1598**

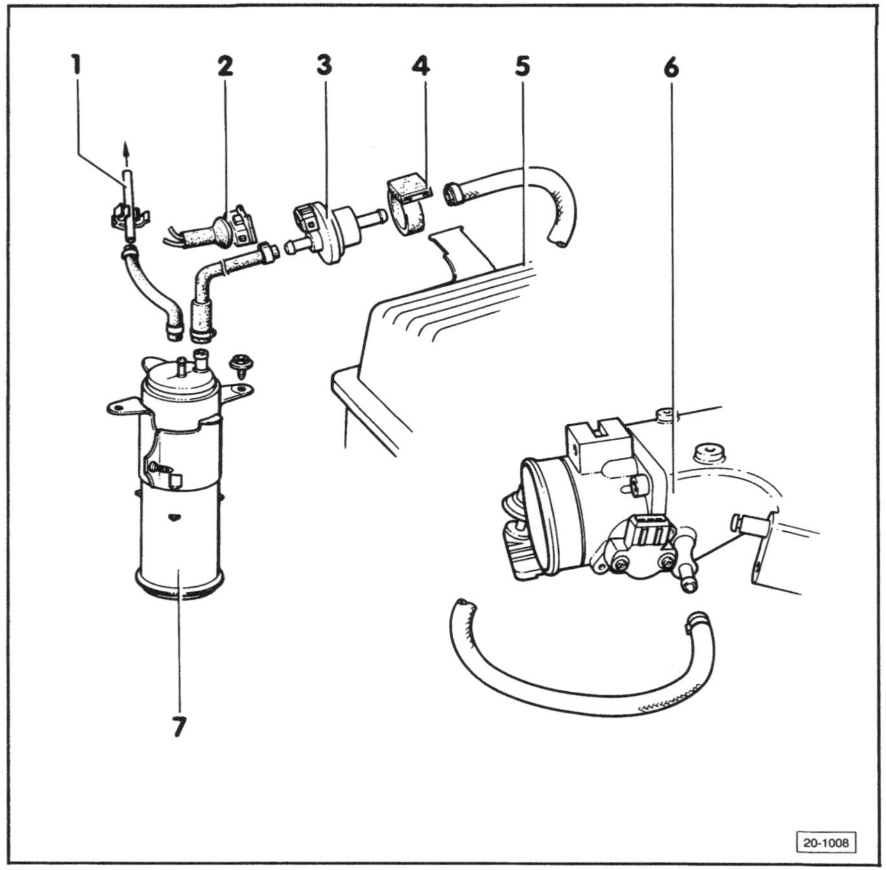

1 – **Vent pipe to gravity valve**
see page, 20.18

2 – **Harness connector**

3 – **EVAP solonoid valve 1 (N80)**
- with ignition switched off, valve to throttle valve housing open
- the valve is activated from Motronic engine control module when engine warm (pulsed)
- checking, see Repair Group 01 (Fuel Injection and Ignition)

4 – **Retaining clamp**

5 – **Air cleaner housing**

6 – **Throttle valve housing**

7 – **Carbon canister**
- location: right front wheel housing
- removing, take off air filter housing completely

THIS PAGE INTENTIONALLY LEFT BLANK

Index

G–charger
- boost pressure **21-30**
- component layout **21-10**
- damage, repairing **21-40**

Safety precautions
- chart **21-20**

Before starting repairs, see
Safety Precautions, section 21-20.

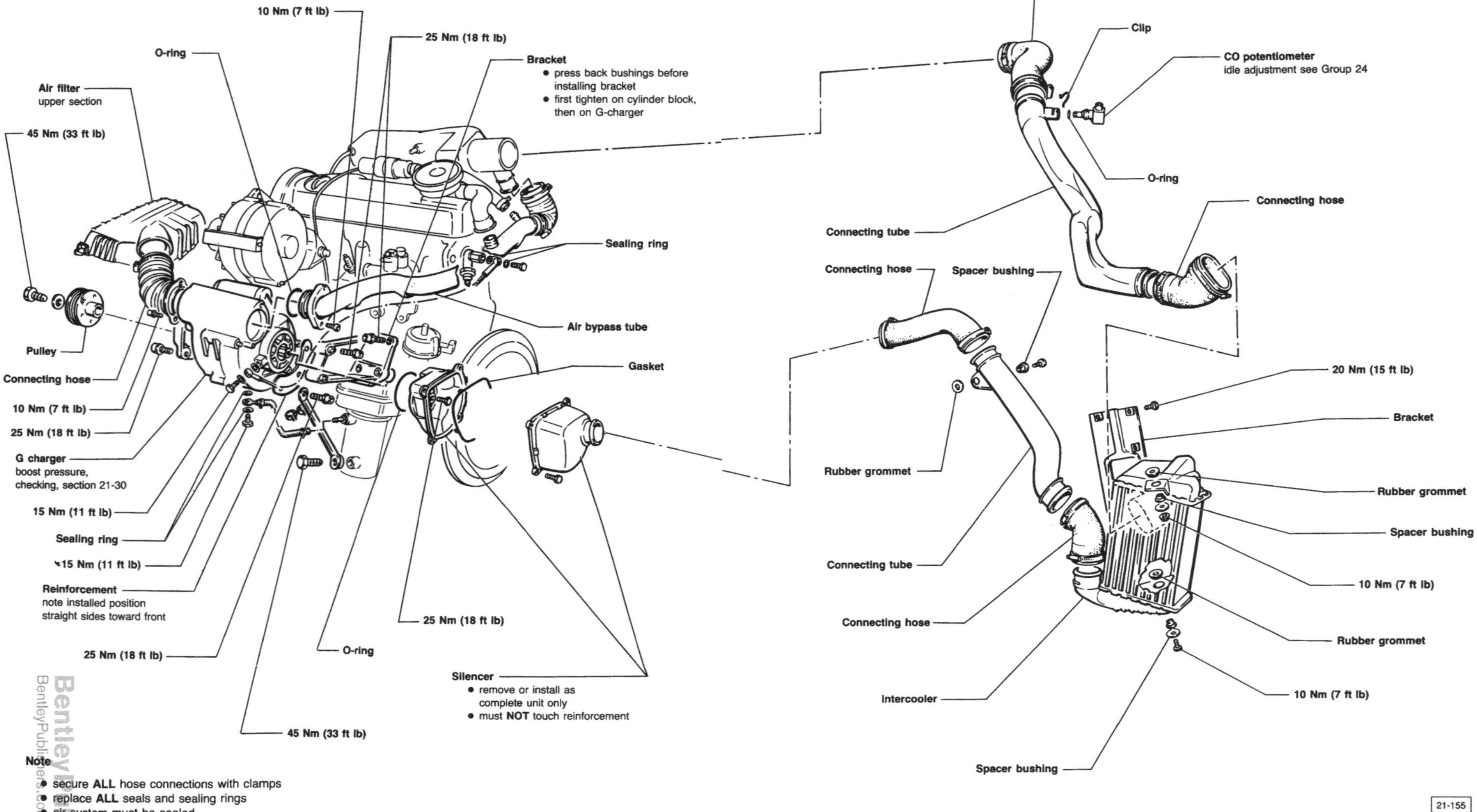

Note
- secure **ALL** hose connections with clamps
- replace **ALL** seals and sealing rings
- air system must be sealed

Safety Precautions

Rules of Cleanliness

When working on fuel system, always follow precautions on cleanliness below:

- thoroughly clean all unions and area near connections **before** disconnecting
- place removed parts on a **clean** surface and cover over. Use paper or plastic sheet. Do **NOT** use fluffy cloths
- components which have been opened or disassembled must be covered or sealed carefully if repair cannot be carried out immediately
- only install **clean** components
- only unpack replacement parts immediately before they are installed
- do **NOT** use parts that have been stored loose (for instance, in toolboxes)

- when fuel system is open, do **NOT** work with compressed air if this can be avoided or move car unless absolutely necessary

System Precautions

CAUTION

When working on vehicles with Digifant fuel/ ignition system, observe the following precautions to prevent personal injury and/or damage to the ignition system:

- connect/disconnect ignition system wires (including high tension wires and test instrument wires) only when the ignition is switched **OFF**
- do **NOT** crank engine before high tension wire of ignition distributor (terminal **4**) is connected to ground with jumper wire
- do **NOT** use battery booster longer than 1 minute or exceed 16.5 Volts with booster
- do **NOT** wash engine unless the ignition is switched **OFF**
- disconnect battery completely when arc or spot welding
- do **NOT** connect condenser to terminal **1**

- do **NOT** substitute rotor of ignition distributor or radio noise suppressor with one of different type
- when installing suppressor, use only 1000 Ohms for high tension wires and 5000 Ohms for spark plug connectors
- do **NOT** allow terminal **1** of coil to contact ground. Possible destruction of the control unit could result

Boost pressure, checking

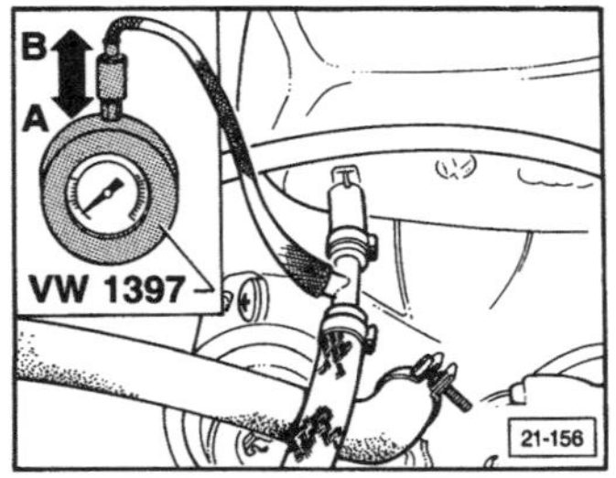

Check these first

- idle setting **OK** (checking, see repair Group 24)
- knock sensor **OK** (checking, see repair Group 28)
- engine oil temperature 80°C (176°F) minimum

▶

- connect pressure gage **VW 1397** to pressure regulator using **T** adaptor
- open pressure gage check valve to position **A**
- start engine and run at idle

- remove harness connectors from CO potentiometer and temperature sensor (in front flange on cylinder head, color code blue)
- accelerate to full throttle
  - engine speed must increase periodically and then drop (surges)

- observe pressure gage
  - boost pressure must reach 0.6 bar (8.7 psi) at high engine speed

If **NO**

- check air system for leakage, check bypass setting, see Repair Group 24 and/or replace G-charger

G-charger damage, repairing

CAUTION

Strictly observe Rules of Cleanliness when working on or assembling components in the G-charger system.

If the G-charger is damaged by foreign material entering the inlet air system, perform the following steps **BEFORE** replacing the G-charger
- thoroughly clean air filter housing
- replace air filter element
- replace intercooler
- replace silencer
- clean G-charger connecting hoses, if cleaning not possible, replace
- replace seals and sealing rings

- replace air bypass tube
- check for filter screen in air bypass tube

If no filter screen
- repair is complete

If filter screen installed
- check air bypass tube for foreign material

If present
- thoroughly clean intake manifold (inside)
- remove cylinder head, see Repair Group 15 for additional information on this topic
- carefully inspect cylinder head for obvious damage and stress induced damage resulting from foreign material, repair or replace as necessary
- inspect piston and cylinder walls for foreign material damage, repair or replace as necessary

Boost pressure system, leak checking

Note

By using leak tester tool **VAG 1687**, **Test A** will completely check the inlet air system for leakage. The following tests are performed with the engine **not** running.

Test A

If there is still leakage after performing **Test A**

- perform **Test B**

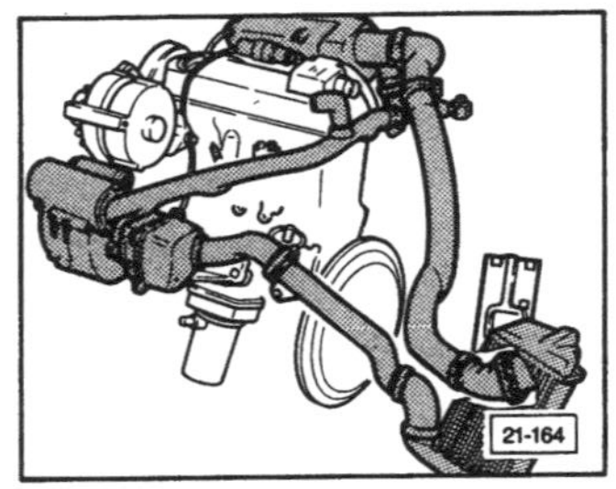

Test A leak-checks the following:

- G-charger including silencer
- connecting hoses between silencer and intercooler
- connecting hoses between intercooler (with CO potentiometer **G 74**) and by-pass
- by-pass tube between G-charger and by-pass
- connecting hose to carbon canister shut-off valve
- throttle
- intake manifold
- fuel injector sealing at seat

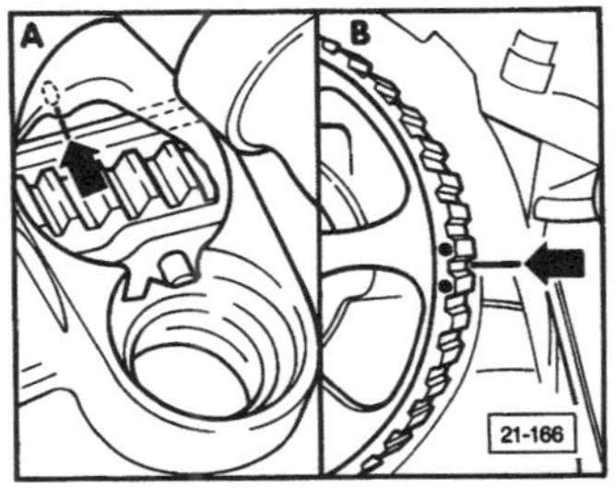

- remove camshaft drive belt cover from above engine

A: position crankshaft at TDC for cylinder **1**.
B: camshaft cover marking must be between dots on cam gear.

- note position of camshaft gear dots
- turn back camshaft gear approx. 3 teeth by turning crankshaft counterclockwise

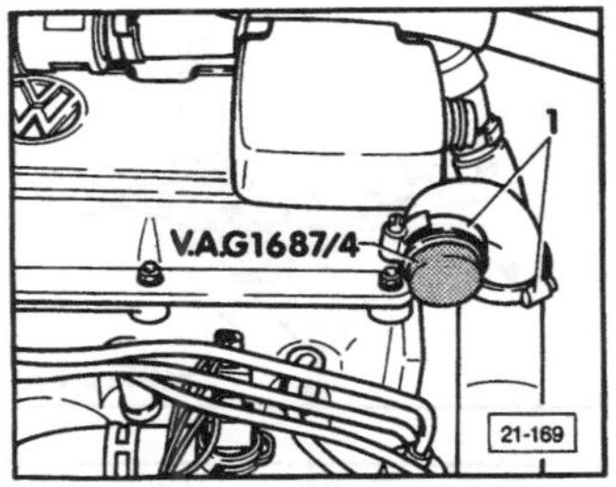

- remove crankcase ventilation hose from pressure regulator valve and seal off using adaptor **VAG 1687/4**
- tighten hose to adaptor using two hose clamps **1**

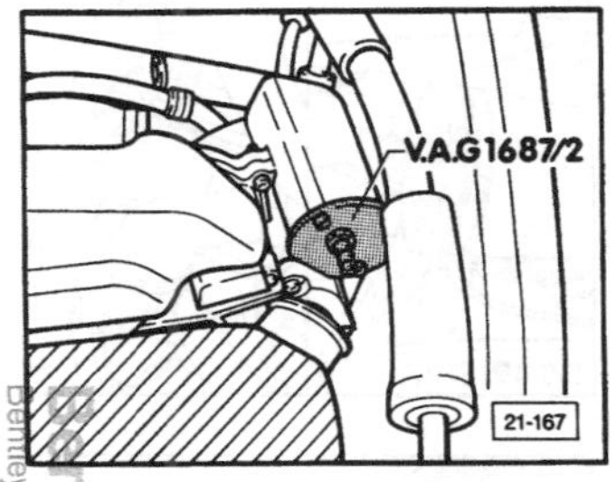

- remove intake air hose flange from G-charger
- install adaptor **VAG 1687/2**

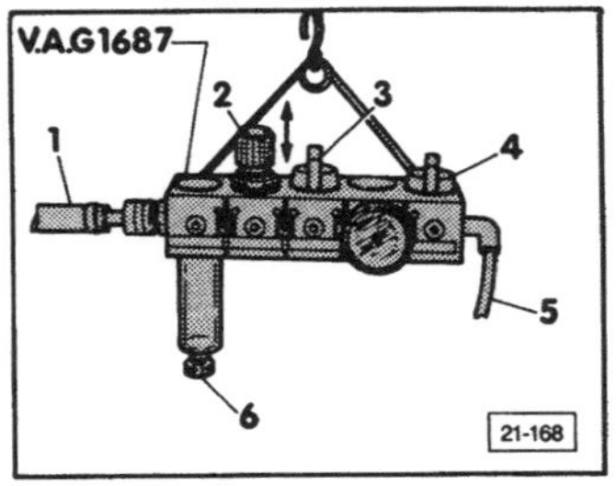

- close valves **2, 3, 4** on **VAG 1687** leak tester, drain viewing glass if necessary using drain screw **6**
- connect shop air **1** to **VAG 1687** leak tester
- connect pressure hose **5** to tester using adaptor **VAG 1687/2**
- open valve **3**
- adjust pressure on valve **2** to 0.9 bar
  - cap pushed upwards

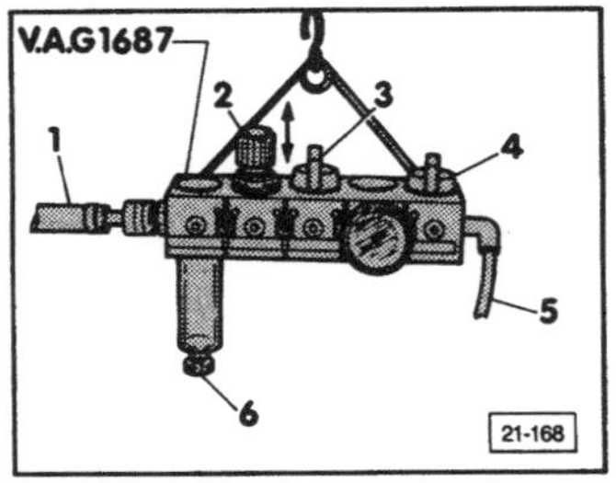

- open valve **4** until the boost pressure circuit (air path) is pressurized
  - regulate using valve **2** until 0.9 bar pressure is obtained
- close valve **3** and observe pressure drop
  - pressure build-up must not exceed 0.9 bar
- observe pressure after 30 seconds
  - pressure must not be less than 0.3 bar

If **YES**

- determine source of leak by listening, feeling or by applying a commercial leak detector spray

Note

A slight air leak near the top toothed gear of the G-charger does not affect boost pressure. This area is not connected to the boost system when the engine is running.

- release the applied pressure before removing the adaptor by removing the coupling from adaptor **VAG 1687/2**

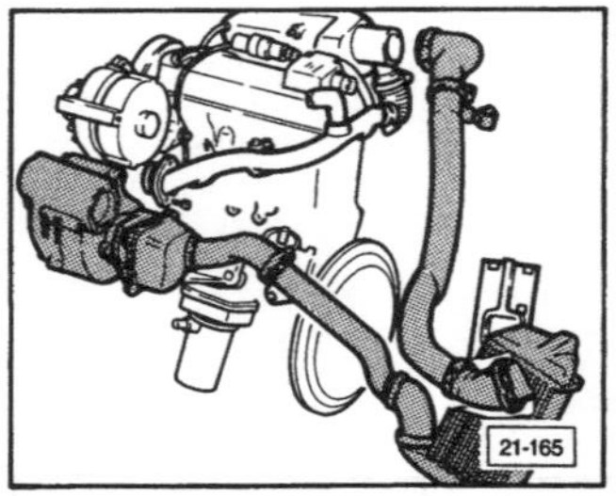

Test B

Test B leak-checks the following:
- G-charger with silencer
- connecting hoses between silencer and intercooler
- connecting hoses between intercooler (with CO potentiometer G 74) and throttle body

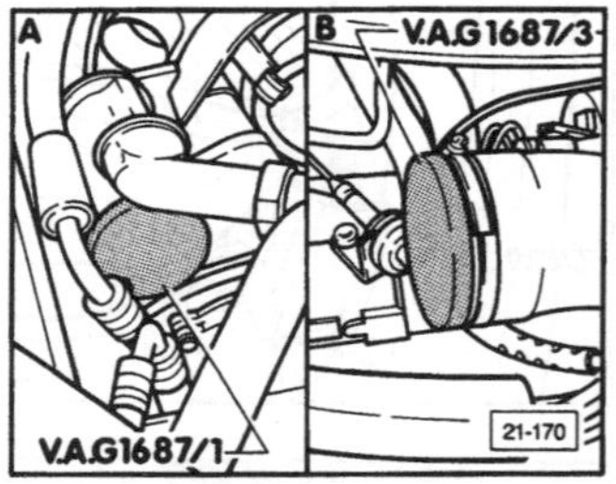

- **A:** disconnect by-pass tube at G-charger and install adaptor VAG 1687/1
- **B:** disconnect intercooler hose from throttle body

- plug off hose opening using adaptor **VAG 1687/3**
- tighten hose to adaptor using hose clamp

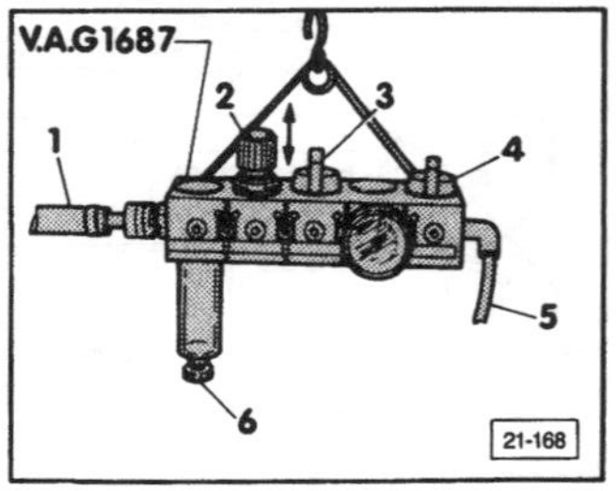

- close valves **2, 3,** and **4** on leak tester **VAG 1687**
- connect air pressure hose **5** to adaptor **VAG 1687/2**
- open valve **3**
- adjust pressure on valve **2** to 0.9 bar
  - with cap pushed upwards

Boost Pressure System, Leak Checking

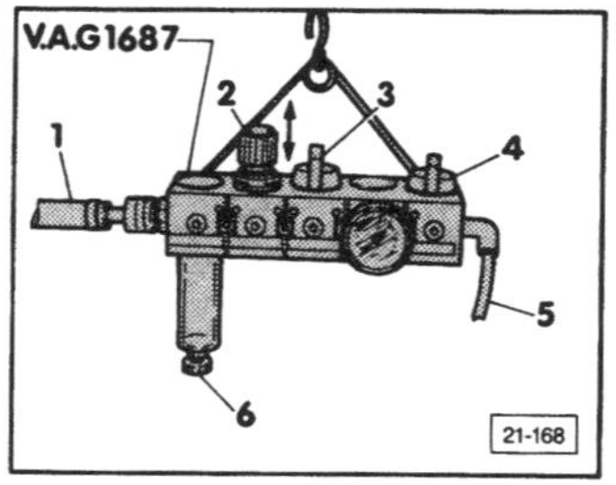

- open valve **4** until the boost pressure circuit (air path) is pressurized
  - regulate using valve **2** until 0.9 bar pressure is obtained
- close valve **3** and observe pressure drop
  - pressure build-up must not exceed 0.9 bar
- observe pressure after 40 seconds
  - pressure must not be less than 0.3 bar

If **YES**

- determine source of leak by listening, feeling or by applying a commercial leak detector spray
- release the applied pressure before removing the adaptor by removing the coupling from adaptor **VAG 1687/2**
- perform **Test A** to check for leakage

THIS PAGE INTENTIONALLY LEFT BLANK

Index

Engine Code – PG, 49 States and Canada, m.y. 1989 to present

California m.y. 1989 - 1990

Basic throttle adjustment
- checking/adjusting **24-180**

Bypass valve adjustment
- checking **24-150**

Control unit
- component layout **24-40**

Deceleration fuel shut-off
- checking **24-130**

Digifant system
- underhood component layout **24-30**

Fuel injectors
- checking **24-170**

Fuel pressure regulator and residual pressure
- checking **24-160**

Full throttle enrichment
- checking **24-120**

Idle and full throttle switches
- checking/adjusting **24-140**

Idle speed and CO content
- checking/adjusting **24-90**

Idle stabilization
- checking **24-100**

Intake air
- component layout **24-20**

Oxygen sensor
- control checking **24-110**
- general information **24-60**

Safety precautions
- rules of cleanliness **24-80**
- safety measures **24-80**

Technical data
- chart **24-70**

Throttle switches
- component layout **24-50**

Vacuum connections
- checking **24-190**

Index

Engine Code – PG
California starting m.y. 1991

Acceleration/full throttle enrichment and deceleration fuel cut-off
- checking 24-240

Basic throttle adjustment
- checking 24-320

Bypass valve adjustment
- checking 24-310

CO potentiometer
- checking 24-270

Coolant temperature sensor
- checking 24-250

Digifant System
- underhood component layout 24-200

Fuel injectors
- checking 24-300

Fuel pressure regulator and residual pressure
- checking 24-160

Idle speed and CO content
- checking/adjusting 24-220

Intake air temperature sensor
- checking 24-260

Oxygen sensor and oxygen sensor control
- checking 24-230

Throttle body
- component layout 24-210

Throttle potentiometer
- checking 24-290

Vacuum connections
- checking 24-190

Vacuum sensor
- checking 24-280

Index

Engine Code AAA

Acceleration, full throttle enrichment and fuel deceleration shut-off
- checking 24-430

Dashpot
- checking and adjusting 24-420

Engine Coolant Temperature (ECT) sensor
- checking 24-440

Fuel injectors
- checking 24-470

Fuel injection system
- component layout 24-370
- repairing 24-340

Fuel pressure regulator and residual pressure
- checking 24-460

Idle speed
- checking 24-400

Idle air control valve (N71)
- checking 24-410

Intake Air Temperature (IAT) sensor
- checking 24-480

Motronic system
- component layout 24-330
- Engine Control Module, component layout 24-350
- Intake air system, component layout 24-360
- throttle assembly component layout 24-380

Safety precautions
- chart 24-390

Throttle position (TP) potentiometer
- checking 24-450

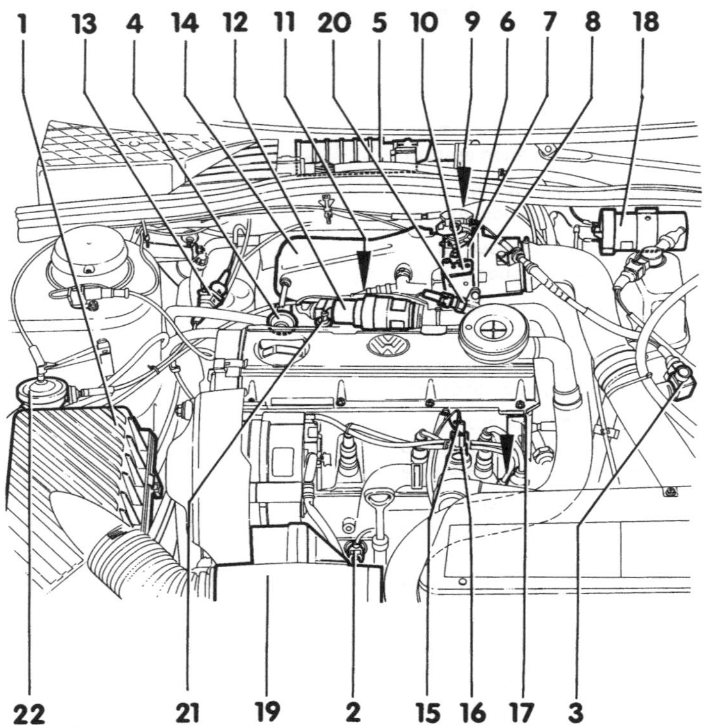

1 — **Air filter**

2 — **Knock sensor (G 61)**

3 — **CO potentiometer (G 74)**
with CO adjustment screw and intake air temperature sensor (G 42)

4 — **Fuel pressure regulator**

5 — **Digifant control unit (J 169)**

6 — **Idle speed adjustment screw**

7 — **Throttle valve housing**

8 — **Boost control valve housing**

9 — **Idle switch (F 60)**

10 — **Full throttle switch (F 81)**

11 — **Fuel rail assembly including fuel injectors**

12 — **Idle stabilizer valve and boost pressure regulator (N 71)**

13 — **Oxygen sensor heater/probe harness connector**

14 — **Intake manifold**

15 — **Coolant temperature sensor (blue)**
supplies signal to control unit

16 — **Coolant temperature sensor (black)**
supplies signal to temperature gauge

17 — **Ignition distributor with Hall sender (G 40)**

18 — **Ignition coil (N 152)**

19 — **G-charger**
boost pressure checking, see Repair Group 21

20 — **Pressure switch (F 130)**
- for fuel pump after-run
- checking, Repair Group 20

21 — **Thermoswitch (F 87)**
for radiator cooling fan after-run

22 — **Carbon canister control valve**
checking, Repair Group 20

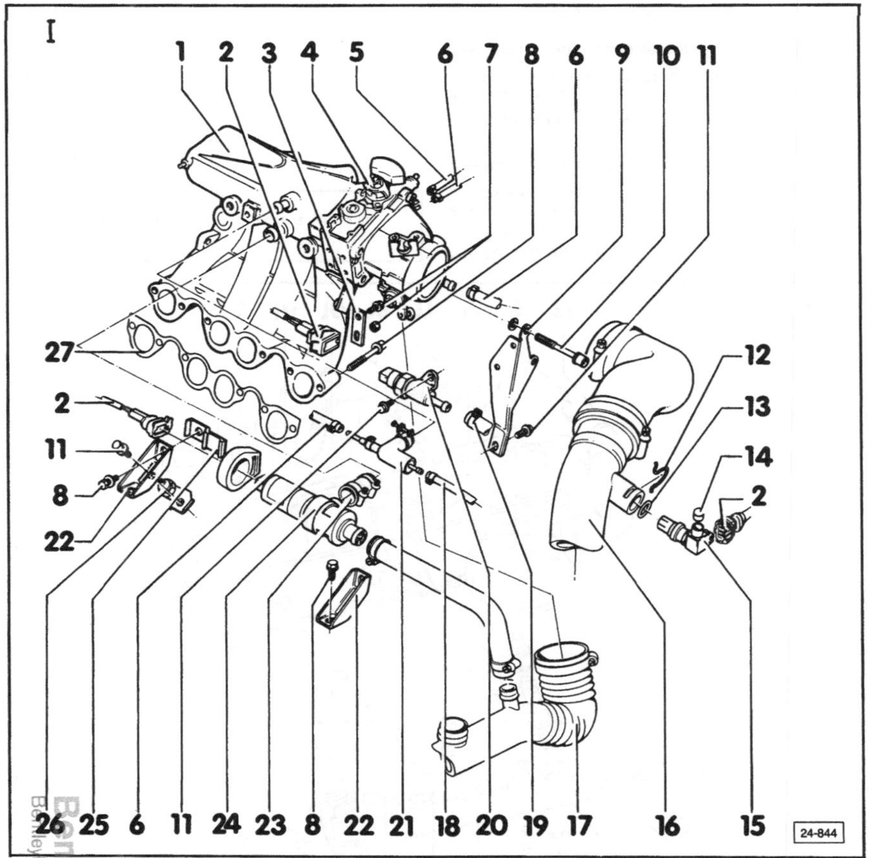

1 — **Intake manifold**

2 — **Harness connector**
for pressure switch

3 — **Support bracket**
between intake and exhaust manifolds

4 — **Throttle valve housing**
basic throttle adjustment, section 24-180

5 — **Vacuum hose**
- to Digifant control unit
- critical length of 1 meter; do not lengthen or shorten

6 — **Vacuum hose**
- to carbon canister control valve
- vacuum hose layout and connections, section 24-190

7 — **15 Nm (11 ft lb)**

8 — **25 Nm (18 ft lb)**

9 — **Intake manifold support**
between bypass valve housing and cylinder head

10 — **Socket head bolt**
20 Nm (15 ft lb)

11 — **10 Nm (7 ft lb)**

12 — **Locking clip**

Fuel Injection, AFC System 24

13 — **O-ring**

14 — **Tamper proof plug**
- for CO adjustment screw
- remove to adjust CO
- idle speed adjustment, section 24-90

15 — **Potentiometer**
- for 49 States and Canada starting m.y. 1990
- for California m.y. 1990 only
  - — measure potentiometer resistance between terminals **1** and **3**
    - must be between 0 and 2000 Ohms

Note

Resistance changes during idle speed adjustment.

16 — **Connecting hose**
for intercooler

17 — **Bypass tube**
between bypass housing and G-charger

18 — **Hose**
to multi-function indicator and A/C

19 — **Hose**
to fuel rail

20 — **Pressure switch**
- for fuel pump after-run
- checking, see Repair Group 20

21 — **Connector tee**

22 — **Mounting brackets**
- for idle stabilizer valve
- mounting must be free of stress

23 — **Hose**

24 — **Idle stabilizer valve**
idle speed stabilization checking, section 24-100

25 — **Support tab**

26 — **Thermo-switch**
for fan after-run

27 — **Gasket**
between intake manifold and cylinder head

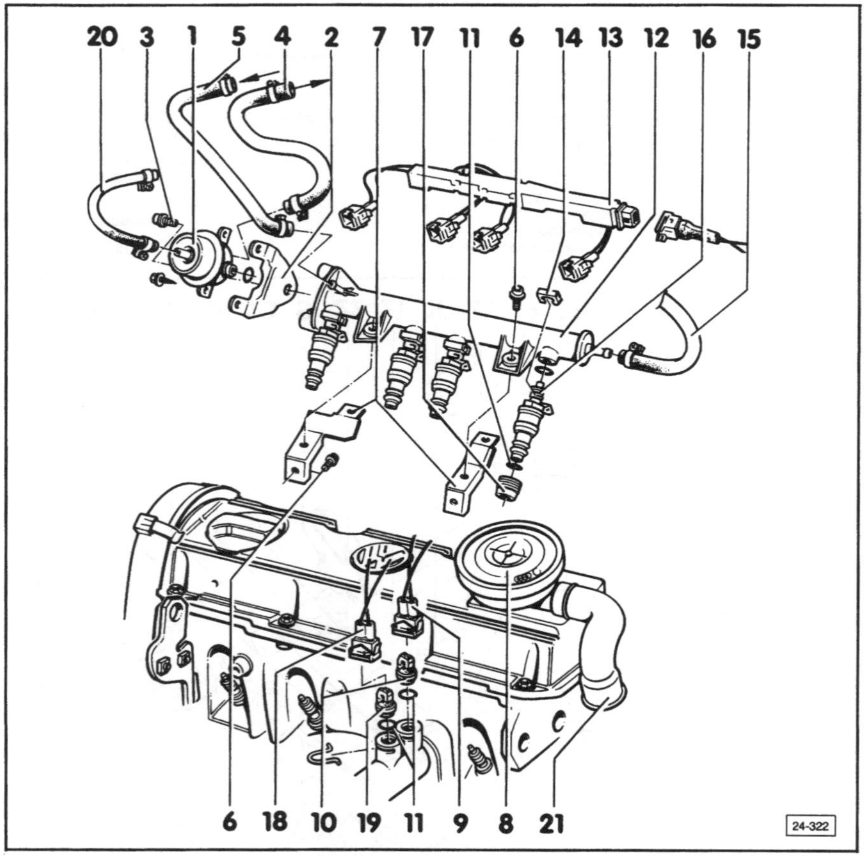

WARNING

Fire Hazard. Do **NOT** smoke or have anything in area that can ignite fuel!

1 — **Fuel pressure regulator**
checking, section 24-160

2 — **Mounting bracket**
for fuel pressure regulator

3 — **Socket head bolt**
15 Nm (11 ft lb)

4 — **Fuel return line**
blue

5 — **Fuel supply line**
black

6 — **Socket head bolt**
10 Nm (7 ft lb)

7 — **Mounting bracket**
for fuel rail

8 — **Emission control valve**

9 — **Harness connector**
for blue coolant temperature sensor

10 — **Coolant temperature sensor**
- blue
- provides signal to Digifant control unit

11 — **O-ring**
always replace

12 — **Fuel rail**

13 — **Wiring guide**

14 — **Retaining clip**
ensure that injector is properly seated (top and bottom) before installing clip

15 — **Hose**
- for fuel pump run-on pressure switch
- connecting pressure gauge **VAG 1318**, section 24-160

16 — **Fuel injector**
checking, section 24-170

17 — **Fuel injector insert**
- use **D6** to install
- **20 Nm (15 ft lb)**

18 — **Harness connector**
- black
- for temperature gauge

19 — **Harness connector**
for black coolant temperature sensor

20 — **Hose**
to intake manifold

21 — **Hose**
to bypass tube

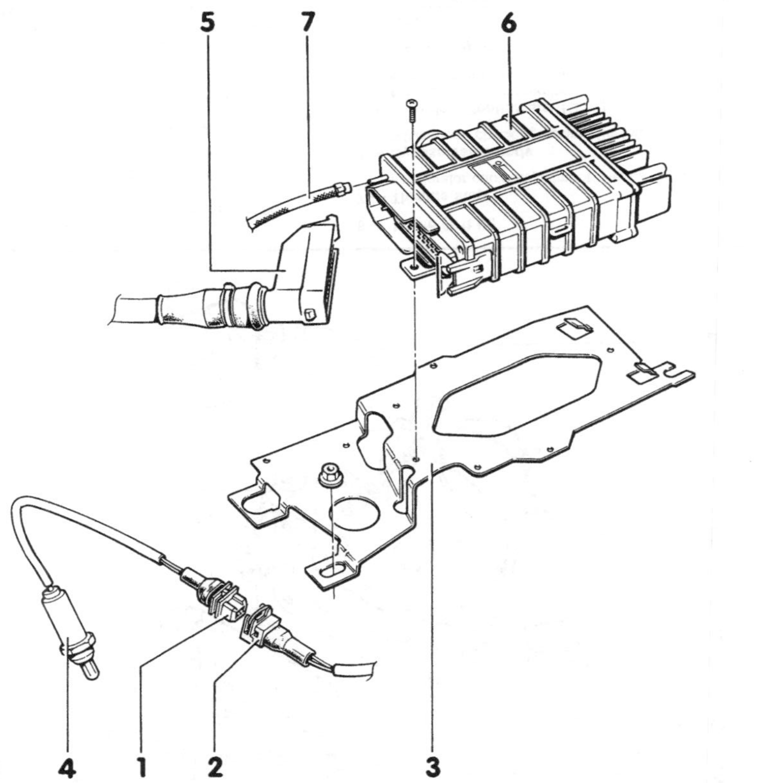

1 — **Connector**
- for Oxygen sensor probe (signal) and Oxygen sensor heater
  - check heater resistance across white wires
    - must be 3 to 4 Ohms

2 — **Harness connector**
- for Oxygen sensor/Oxygen sensor heater
  - check supply voltage at white wires, with engine running
    - must be approx. battery voltage

3 — **Mounting plate**

4 — **Oxygen sensor**
- **20 Nm (15 ft lb)**
- checking, section 24-110
- install using **G5** anti-sieze compound but do not allow compound to enter probe slots

5 — **Harness connector**

> **CAUTION**
>
> Ignition **must** be switched off before connecting or disconnecting!

- for Digifant control unit (**J 169**)
- checking, see Repair Group 01 (Fuel Injection and Ignition)
- to disconnect: disengage spring clip on control unit

6 — ***Digifant control unit (J 169)***
- controls: Fuel injection, Ignition, Boost pressure, Oxygen sensor and idle stabilization
- location: in right side of plenum
- remove fresh air intake before removing control unit

7 — **Vacuum line**
- critical length 1 meter, do **NOT** lengthen or shorten
- only use correct replacement part, black with yellow stripe

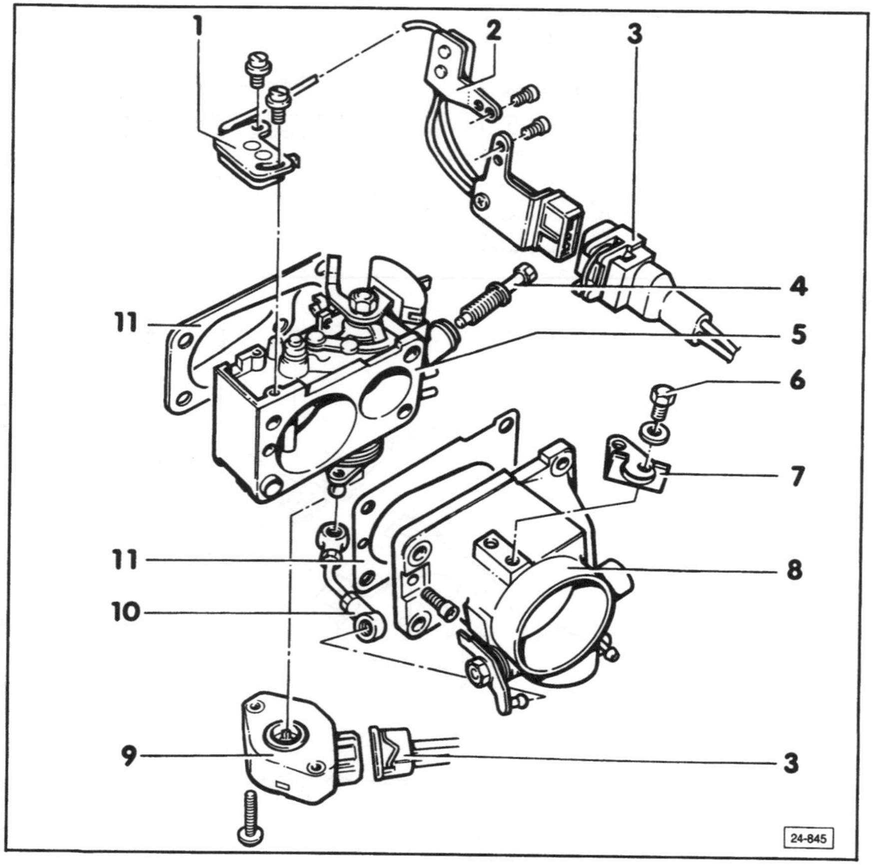

1 — **Full throttle switch (F 81)**
- for 49 states and Canada starting m.y. 1990
- for California m.y. 1990 only
- checking/adjusting, see section 24-140

2 — **Idle switch (F 60)**
- for 49 states and Canada starting m.y. 1990
- for California m.y. 1990 only
- checking/adjusting, see section 24-140

3 — **Harness connector**

4 — **Idle speed adjustment screw**
- idle speed checking/adjusting, section 24-90
- check O-ring for damage, replace if necessary

5 — **Throttle body**
basic throttle adjustment, section 24-180

6 — **10 Nm (7 ft lb)**

7 — **Throttle stop**
accelerator cable adjustment, Repair Group 20

8 — **Bypass valve body**
adjustment, checking section 24-150

9 — **Throttle potentiometer (G 69)**
- for automatic transmission only
- for 49 states and Canada starting m.y. 1991

10 — **Connecting rod**

11 — **Gasket**
always replace

Oxygen sensor, general information

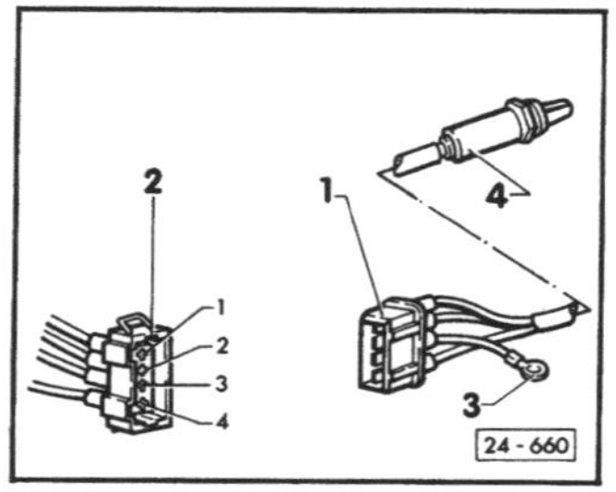

- Oxygen sensor with 4-pin harness connector
- Connector (**1**) for Oxygen sensor and ground wire **3** are attached to right engine mount

Terminal description of harness connector 2
- **1** (+) plus wire to OXS heater
- **2** (−) ground wire to OXS heater
- **3** shield for OXS signal wire
- **4** OXS signal wire

Note

Terminal **3** of connector **2** does not apply to California vehicles, starting m.y. 1991.

> **CAUTION**
> Do Not re-use any sprays or compounds containing silicone on engines equipped with Oxygen sensors. Do Not use these compounds on or near intake air system or near the Oxygen sensor. Silicone drawn into the intake air system is not burned during combustion and will lead to contamination and malfunctioning of the Oxygen sensor.

Technical data

Engine code	PG
Engine RPM limit	6350 ± 150 RPM
Idle speed	800 ± 50 RPM
CO content	0.7 ± 0.4 volume %

THIS PAGE INTENTIONALLY LEFT BLANK

Safety precautions

Rules of Cleanliness

CAUTION

When working on the fuel supply/injection system, carefully observe the following rules:

1 – Thoroughly clean connection and surrounding areas before loosening connection.

2 – After removing components, place in clean area and cover with foil or paper. Avoid using rags!

3 – Components which have been opened or disassembled must be carefully covered or sealed if repair cannot be carried out immediately.

4 – Install clean parts only.

- remove replacement parts from package just before installing
- do **not** use spare parts that were stored loose or unpackaged (e.g. in tool boxes, etc.)

5 – When fuel system is open:

- avoid using compressed air whenever possible
- avoid moving the vehicle whenever possible

Safety measures

CAUTION

Observe the following precautions to prevent personal injury as well as possible damage to sensitive electrical components.

- switch **OFF** the ignition before connecting or disconnecting components or test equipment
- connect and disconnect battery **ONLY** with ignition switched **OFF** otherwise the control unit could be damaged
- if the engine must be cranked but not started (for compression testing etc.) disconnect power output stage of ignition coil and
- do **not** use battery booster longer than one minute nor should 16.5 Volts be exceeded
- do **not** wash engine unless ignition is switched **OFF**
- disconnect **BOTH** battery terminals whenever arc or spot welding
- do **not** connect a condenser of any kind to terminal **1** of the ignition coil
- when installing noise suppressors, **ONLY** use 1000 Ohms for high tension wires and 5000 Ohms for spark plug connectors
- do **not** replace distributor rotor (marked **R1**) with a different type
- if the vehicle is heated up (e.g. in a painting booth) do **not** start the engine until it has had sufficient time to return to room temperature

Note

A variety of electrical connectors are used on this vehicle, **ALWAYS** use the **VW 1594** adaptor kit to connect test instruments to these connectors.

CAUTION

Before disconnecting a customer's battery; **ALWAYS** ask for the radio code (if equipped with an anti-theft radio).

Idle speed and CO content, checking/adjusting

> **CAUTION**
>
> Ignition timing, CO content and idle speed RPM's are inter-related and must be checked/adjusted together.

Note

The Oxygen sensor remains connected during checking/adjusting.

Requirements
- engine oil temperature 80°C (176°F) minimum
- all electrical consumers switched **OFF** (do not take measurements while radiator cooling fan is running)
- A/C switched **OFF**
- ignition timing **OK**
- idle stabilizer valve **OK**, (when ignition is switched on the valve must buzz and hum)
- exhaust system must be leak tight

Checking/adjusting

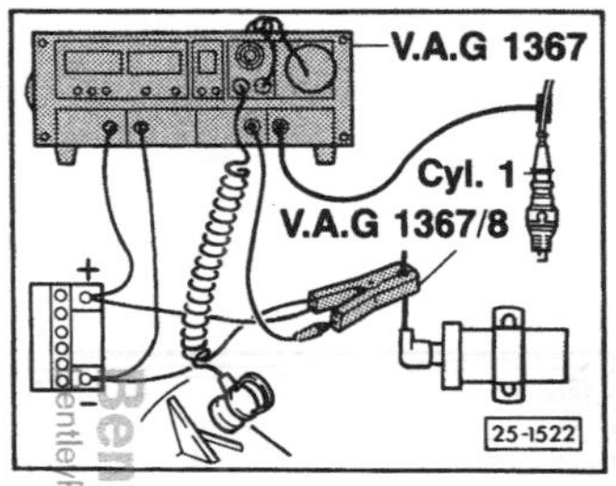

- switch **OFF** ignition
- connect **VAG 1367** engine tester using inductive pickup **VAG 1367/8** on ignition coil high tension lead
- connect **SUN 105** CO tester (or EPA approved equivalent) to CO tap tube using hose adaptor **VAG 1363/3**

> **CAUTION**
>
> Hose must fit securely on CO tap tube to avoid measuring inaccuracies due to dilution or leakage.

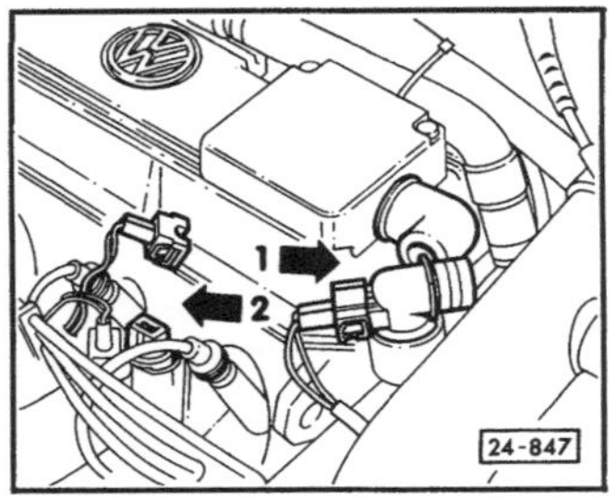

- disconnect crankcase ventilation hose from emission control valve (**arrow 1**) and route such that only fresh air is drawn in
- start engine and let idle
  - will run from 1 to 4 minutes at increased engine RPM

When idle speed is reduced approx. 100 RPM

- disconnect blue coolant temperature sensor harness connector

> **CAUTION**
>
> Do **not** start engine with coolant temperature sensor disconnected. The control unit will go into open loop making accurate adjustment impossible.

- raise engine speed in excess of 3000 RPM 3 times then let idle

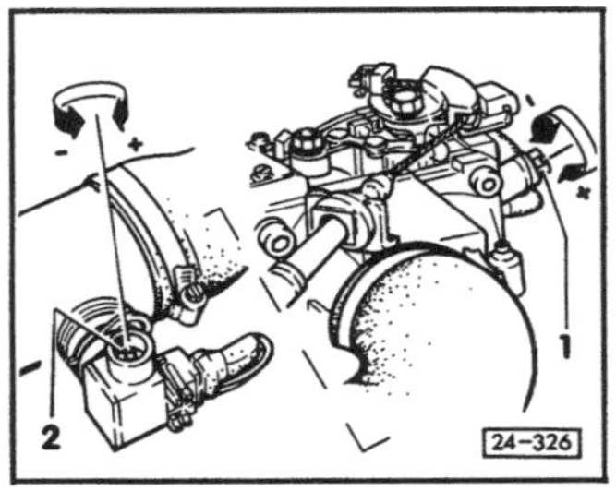

- check idle speed and CO content
  - idle speed must be 800 ± 50 RPM
  - CO content must be 0.7 ± 0.4 volume %

If **NO**

- adjust by alternately turning adjustment screws (**arrows**) until both specifications are obtained
- reconnect blue coolant temperature sensor
- raise engine speed 3 times then let idle
- check CO content
  - must be 0.7 ± 0.4 volume %

If **NO**

- check OXS control, see section 24-110
- after repairs or replacements have been performed, install new plug over CO adjustment screw

Idle stabilization, checking

Function, checking

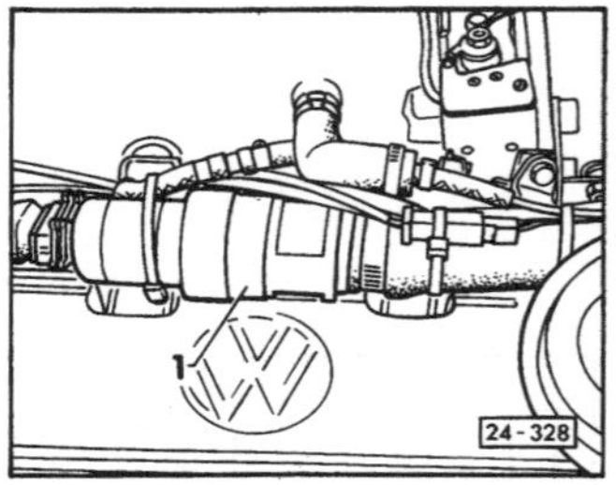

- switch **ON** ignition
  - idle stabilizer valve **1** must vibrate and hum

If **NO**

- switch **Fluke 83** multimeter (**US 1119** or equivalent) to resistance range
- disconnect harness connector from idle stabilizer valve
- connect multimeter across stabilizer valve terminals
  - must be between 2 and 10 Ohms

If **NO**

- replace idle stabilizer valve

If **YES**

- reconnect stabilizer valve harness connector
- check stabilizer valve wiring for open circuit using wiring diagram, also see Repair Group D2, **Test A**, step **11**

If wiring **OK**

- replace Digifant control unit (**J 169**)

Control, checking

Requirements
- engine oil temperature 80°C (176°F) minimum
- coolant temperature sensor (blue) **OK**
- idle speed **OK**
- intake air system **OK** (leak tight)

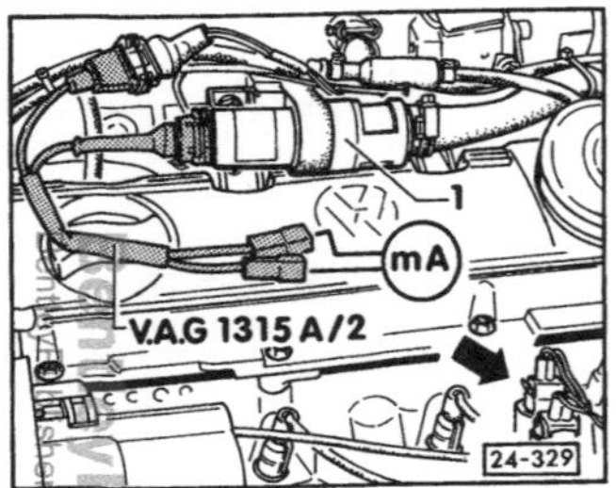

- switch multimeter to mA range
- connect test adaptor **VAG 1315A/2** between idle stabilizer valve and its harness connector
- connect multimeter to terminals of **VAG 1315A/2** adaptor using jumpers from **VW 1594** adaptor kit
- start engine and let idle
  - will run from 1 to 4 minutes at increased engine RPM

When idle speed is reduced approx. 100 RPM

- raise engine speed in excess of 3000 RPM 3 times then let idle
  - control current must be 460 ± 60 mA and **fluctuating**

- disconnect blue coolant temperature sensor harness connector (**arrow**)
  - control current must be 460 ± 60 and **constant**

If test conditions are maintained but specifications are not obtained

- replace Digifant control unit (**J 169**)

Note

Idle speed stabilization control current is dependent upon engine load during idle. Due to the following loads the control current can vary between 400 and 1200 mA:
- engine cold
- A/C switched **ON**
- electrical consumers switched **ON**
- power steering at full lock

CAUTION
Do Not re-use any sprays or compounds containing silicone on engines equipped with Oxygen sensors. Do Not use these compounds on or near intake air system or near the Oxygen sensor. Silicone drawn into the intake air system is not burned during combustion and will lead to contamination and malfunctioning of the Oxygen sensor.

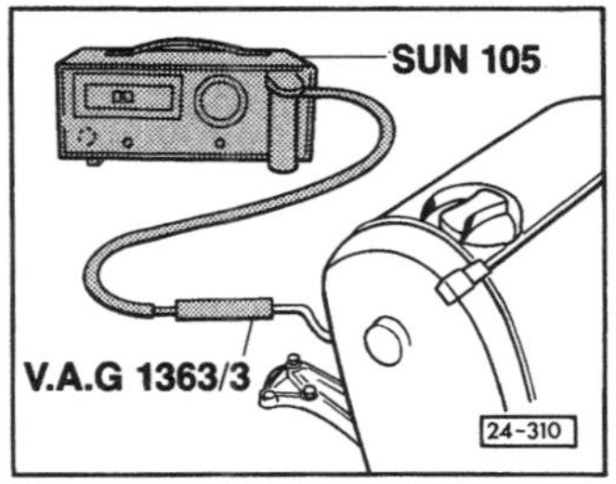

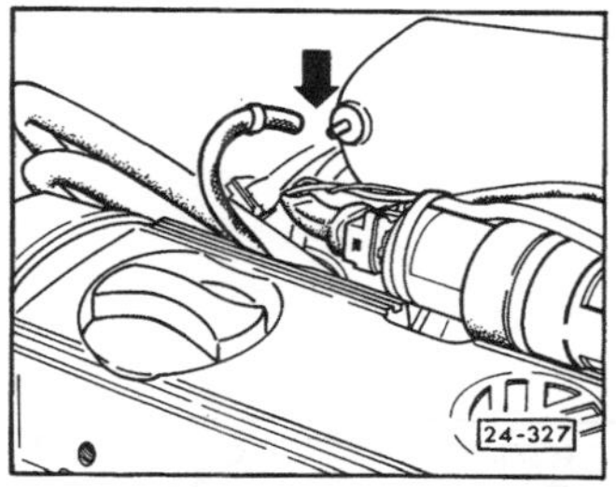

Oxygen sensor control, checking

Requirements
- engine oil temperature 80°C (176°F) minimum
- idle speed **OK**
- exhaust system between catalyst and cylinder head must be leak tight
- coolant temperature sensor (blue) connected
- voltage supply to OXS heater **OK**

■ connect **SUN 105** CO tester (or EPA approved equivalent) to CO tap tube using hose adaptor **VAG 1363/3**

CAUTION
Hose must fit securely on CO tap tube to avoid measuring inaccuracies due to dilution or leakage.

■ start engine and let idle for at least 2 minutes
■ record CO content

■ disconnect pressure regulator vacuum hose from intake manifold (**arrow**) and plug off
- CO content must increase briefly then return to recorded value

If CO content does not drop back

■ disconnect OXS heater harness connector (**arrow**)
■ connect jumper wire from **VW 1594** adaptor kit to OXS harness connector signal wire (**arrow**)
■ alternately touch the free end of the adaptor wire to ground and battery positive
- CO content must alternately increase and decrease

If **YES**

■ replace Oxygen sensor

If **NO**

■ check wiring between Oxygen sensor and Digifant control unit using wiring diagram, repair or replace as necessary

If wiring **OK**

■ replace Digifant control unit (**J 169**)

Full throttle enrichment, checking

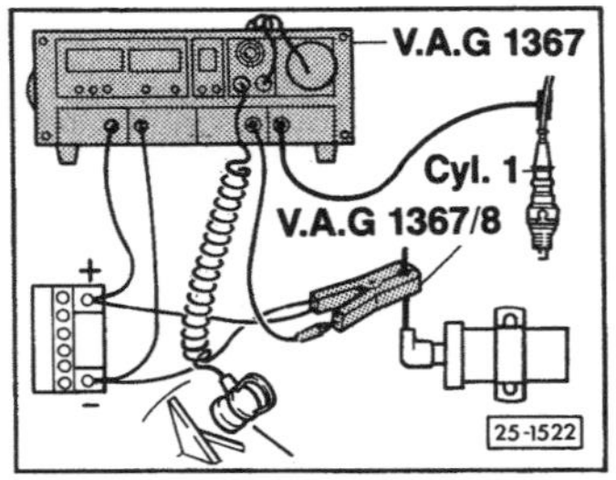

■ switch **OFF** ignition
■ connect **VAG 1367** engine tester using **VAG 1367/8** inductive pickup
■ start engine and let idle

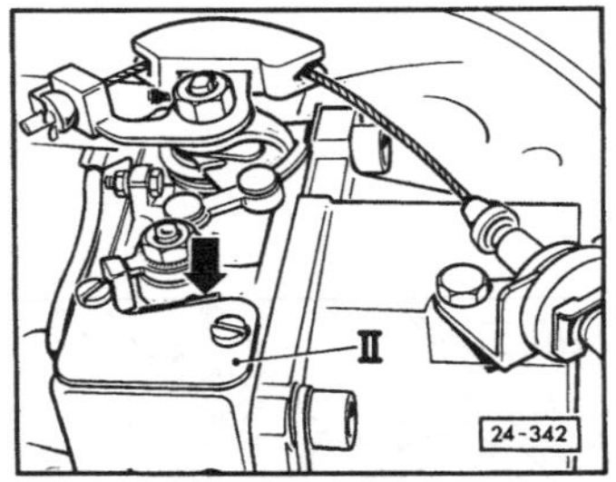

■ operate full throttle switch by hand (**arrow**)
- engine speed must increase and decrease (hunts as deceleration fuel shut off kicks in)

If **NO**

■ check idle and full throttle switch adjustment, checking section 24-140

Deceleration fuel shut-off, checking

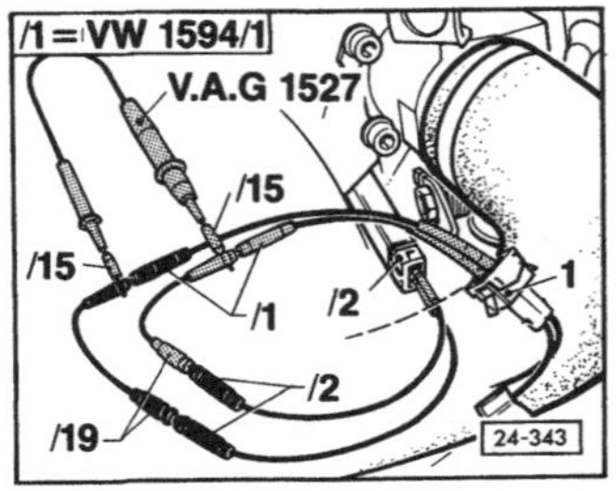

Requirements
- engine oil temperature 80°C (176°F) minimum
- idle switch **OK**

- disconnect main injector harness connector **1** from fuel rail wiring guide connector **2**
- bridge connectors **1** and **2** using adaptor wires **VW 1594/1**, **2**, **15** and **19**
- connect **US 1115** LED tester to adaptors as shown
- start engine and let idle
  - LED tester must light up

- raise engine speed to approx. 3000 RPM and then release throttle abruptly
  - LED must go out briefly (indicating deceleration fuel shut-off has been activated)

Idle and full throttle switches, checking and adjusting

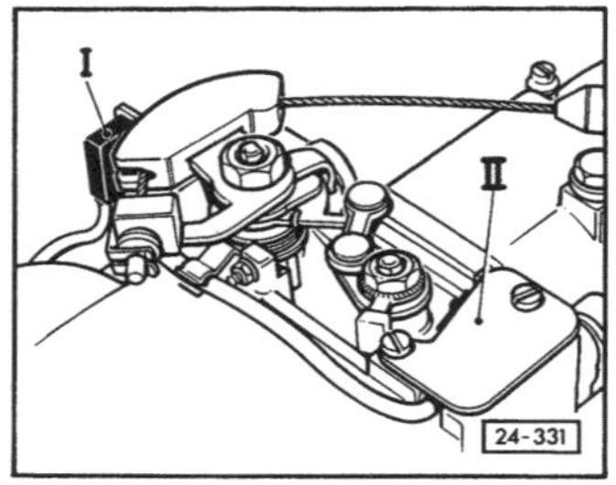

The idle switch **I** signals the Digifant control unit that the throttle is fully closed for the following functions:
- fuel cut-off during deceleration
- idle speed stabilization
- ignition timing control during idle (DIS function) and deceleration

The full throttle switch **II** signals the Digifant control unit that the throttle is fully opened for the following function:
- injection quantity during full throttle enrichment

Voltage supply, checking

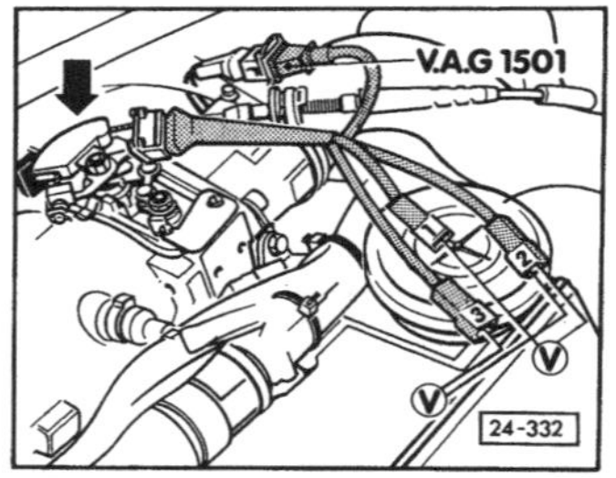

- disconnect throttle switch harness connector (**arrow**)
- connect test adaptor **VAG 1501** to harness connector
  - throttle switch connector remains open

- switch **Fluke 83** multimeter to 20 Volt range
- connect multimeter to terminals **1** and **2** of **VAG 1501** test adaptor using jumpers from **VW 1594** adaptor kit
- switch **ON** ignition
  - must be approx. 5 Volts

- switch **OFF** ignition
- connect multimeter to terminals **2** and **3** of **VAG 1501** test adaptor using jumpers from **VW 1594** adaptor kit
- switch **ON** ignition
  - must be approx. 5 Volts

If voltage values are **not** obtained

- switch **OFF** ignition
- remove test adaptors and re-connect harness connector to throttle switch
- determine break in wiring between throttle switch connector and ECU, see Repair Group D2, **Test A**, steps **11** and **12**, repair or replace as necessary

If wiring **OK**

- replace Digifant control unit (**J 169**)

Idle switch (F 60), checking and adjusting

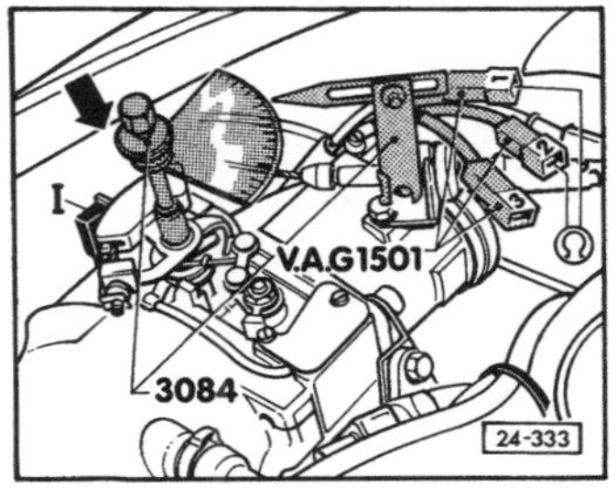

- disconnect throttle switch harness connector **(arrow)**
- connect test adaptor **VAG 1501** between throttle switch harness connector and throttle switch
- switch **Fluke 83** multimeter to resistance range
- connect multimeter to terminals **1** and **2** of **VAG 1501** test adaptor using jumpers from **VW 1594** adaptor kit
- check idle switch **I** for continuity
  - throttle closed = 0 Ohms (continuity)
    throttle open = Infinite resistance (open)

- attach protractor **3084** to throttle shaft using magnetic adaptor **3084/1**
- fasten protractor pointer to throttle cable sleeve using circlip
- note protractor position via pointer
- open throttle valve approx. 10° then slowly and carefully close throttle until idle switch is activated (continuity)
  - must be 1.0 ± 0.5° before idle speed stop

If **NO**

- adjust switch-on point by loosening throttle switch I and moving switch until throttle switch bracket is just resting against the throttle lever
- tighten switch and re-check

Full throttle switch (F 81), checking/adjusting

Requirement
- bypass valve adjustment **OK**, checking, section 24-150

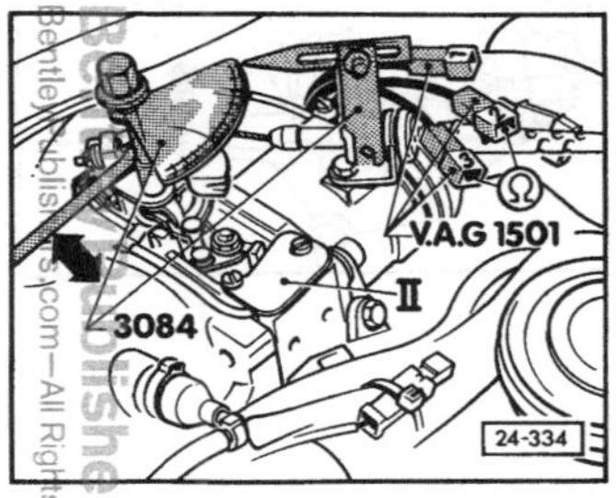

- disconnect throttle switch harness connector **(arrow)**
- connect test adaptor **VAG 1501** between throttle switch harness connector and throttle switch
- switch **Fluke 83** multimeter to resistance range
- connect multimeter to terminals **2** and **3** of **VAG 1501** test adaptor using jumpers from **VW 1594** adaptor kit
- check full throttle switch **II** for continuity
  - throttle open = 0 Ohms (continuity)
    throttle closed = Infinite resistance (open)

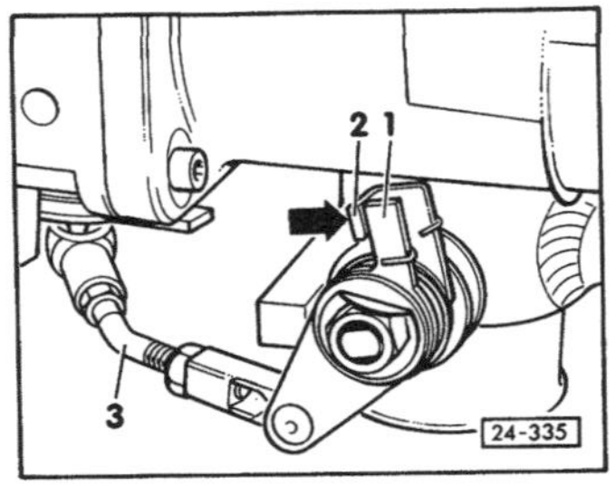

- *attach protractor* **3084** *to throttle shaft using* magnetic adaptor **3084/1**
- fasten protractor pointer to throttle cable sleeve using circlip
- using screwdriver, push throttle lever in direction of full throttle until lever **1** just starts to lift off by-pass valve lever **2** **(arrow)**
- with throttle in this position, set protractor disc to 0°
- close throttle approx. 10° then slowly open again until full throttle switch **II** just begins to activate (continuity)
  - must occur at 4 ± 1°

If **NO**

- adjust switch-on point by loosening full throttle switch **II** and moving switch until specification is obtained
- tighten switch and re-check
- adjust throttle cable such that with accelerator pedal fully depressed, full throttle is just attained (1 mm maximum play at throttle lever)

By-pass valve adjustment, checking

Note

The by-pass valve controls the boost pressure. It is adjusted at the factory and must **not** be changed. Use the following check to determine if the by-pass valve adjustment has been inadvertently altered.

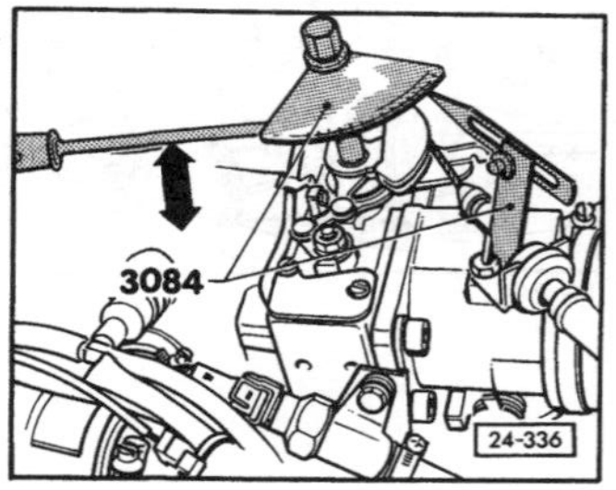

- attach protractor **3084** to throttle shaft using magnetic adaptor **3084/1**
- fasten protractor pointer to throttle cable sleeve using circlip
- using screwdriver, push throttle lever to full throttle stop
- with throttle in this position, set protractor to 10°

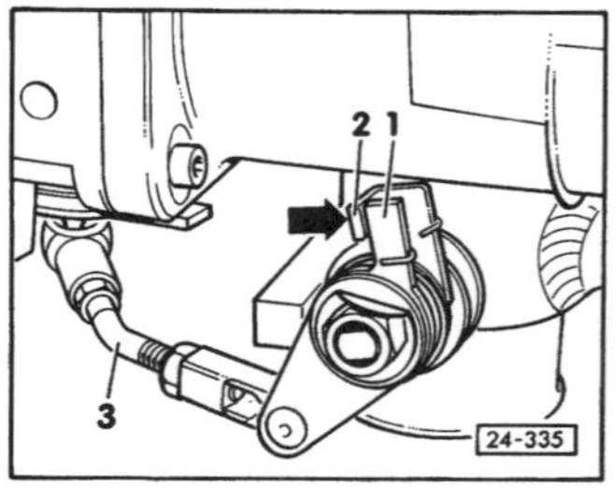

- slowly close throttle until lever **1** just makes contact with lever **2** (**arrow**)
- read value on protractor
  - must be 5 ± 3° before full throttle stop

If **NO**

- adjust by-pass valve on connecting rod **3**

Note

For full throttle switch checking/adjusting see section 24-140.

THIS PAGE INTENTIONALLY LEFT BLANK

Fuel pressure regulator and residual pressure, checking

> **WARNING**
>
> Fire Hazard. Do **not** smoke or have anything in area that can ignite fuel!

The fuel pressure regulator regulates the fuel pressure in proportion to intake manifold vacuum.

- wrap a lint free cloth around pressurized fuel hose to prevent fuel overspray
  - located between fuel rail and fuel pump after-run pressure switch

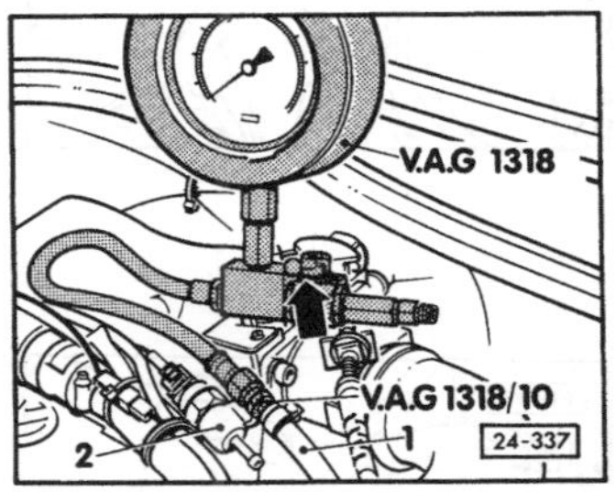

- disconnect hose **1** from pressure switch **2** and catch any fuel that runs out
- connect pressure tester **VAG 1318** to disconnected hose **1** using adaptor **VAG 1318/10**

Note

The valve on the pressure gauge must be in the closed position, that is; the lever must be at a right angle to the direction of flow.

- start engine and let run at idle
- observe fuel pressure
  - must be approx. 2.5 bar

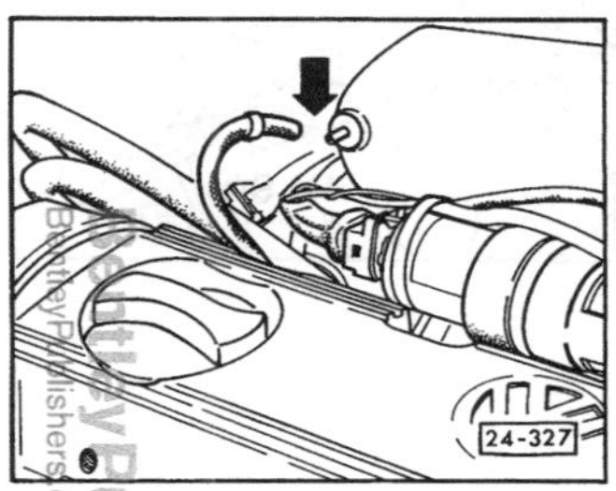

- disconnect fuel pressure regulator vacuum hose from intake manifold (**arrow**)
  - fuel pressure must increase to approx. 3.0 bar

- switch **OFF** ignition
- check for leaks and residual pressure by watching pressure drop on gauge
  - after 10 minutes there must be 2 bar minimum pressure

If the residual pressure drops below 2 bar

- start engine until fuel pressure has built up
- switch **OFF** ignition, at the same time securely clamp fuel return hose (blue) and observe pressure drop on gauge
  - if pressure does **not** drop; pressure regulator is defective

If pressure still drops; check the following
- hose connections are leaking
- fuel rail O-rings are leaking
- fuel injectors are leaking
- fuel pump check valve is leaking
- pressure gauge shut-off valve is leaking

- repair or replace as necessary
- drain fuel from **VAG 1318** before removing the pressure gauge

Fuel injectors, checking

> **WARNING**
>
> Fire Hazard. Do **not** smoke or have anything in area that can ignite fuel!

Voltage supply and resistance, checking

Requirement
- Hall sender **OK**, checking, see Repair Group 28

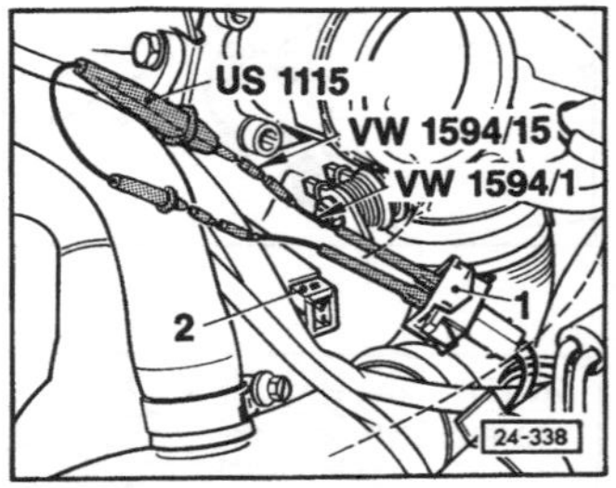

- ■ disconnect harness connector **1** from connector **2**
- ■ check fuel injector supply voltage at harness connector **1** using **US 1115** LED tester and jumper wires from **VW 1594** adaptor kit
- ■ operate starter
  - LED tester must flicker

If LED does not flicker and does not light up

- ■ re-connect harness connector **1** with connector **2**
- ■ locate break in wiring for injectors (**N 30, N 31, N 32, N 33**) and eliminate see Repair Group D2, **Test A**, step **4**

If wiring **OK**

- ■ replace Digifant control unit (**J 169**)

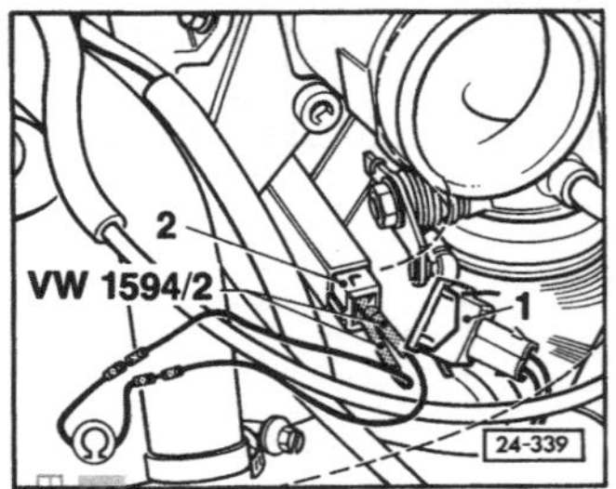

- ■ switch **Fluke 83** multimeter to resistance range
- ■ check fuel injector resistance at connector **2** using multimeter and jumper wires from **VW 1594** adaptor kit

If 3.7 to 5.0 Ohms
- all 4 injectors **OK**

If 5.0 to 6.7 Ohms
- 3 injectors **OK**

If 7.5 to 10.0 Ohms
- 2 injectors **OK**

If 15.0 to 20.0 Ohms
- 1 injector **OK**

If resistance is greater than 5 Ohms

- ■ remove wiring from fuel rail brackets and carry out individual injector resistance checks at connectors and injectors

- must be 15 to 20 Ohms maximum per injector
- must be 0.5 Ohms maximum (continuity) across each injector harness connector

- ■ repair or replace wiring or injectors as necessary

Injectors, removing

> **WARNING**
>
> Fire Hazard. Do **not** smoke or have anything in area that can ignite fuel!

- ■ remove idle stabilizer valve and place to one side
- ■ remove valve cover and intake manifold support

- ■ disconnect harness connector **1** from injector wiring guide **2**
- ■ carefully pry wiring guide **2** from fuel rail brackets (**arrows**)
- ■ remove fuel rail mounting bolts **3** and **4**
- ■ pull injectors complete with fuel rail and wire guide out of cylinder head

Spray pattern and sealing, checking

> **WARNING**
>
> Fire Hazard. Do **not** smoke or have anything in area that can ignite fuel!

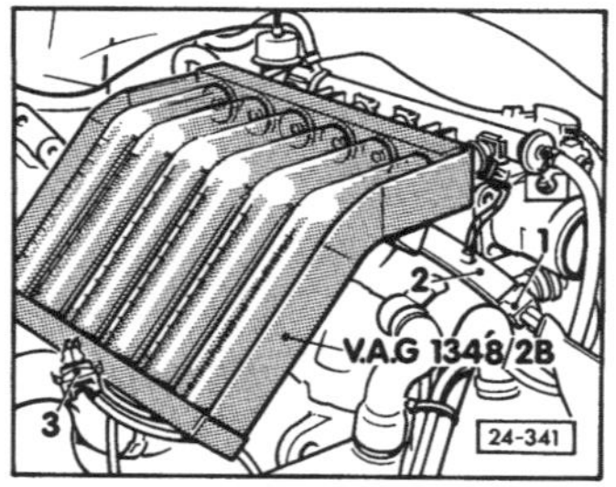

- ■ remove fuel rail assembly (injectors/wire guide/fuel rail)
- ■ insert injector assembly into openings of fuel analyzer **VAG 1348/2B**
- ■ disconnect blue coolant temperature sender harness connector **3**
- ■ re-connect harness connector **1** to connector **2**
- ■ briefly operate starter and observe spray pattern
  - must be uniform among all injectors in the set

- ■ disconnect harness connector **1**
- ■ switch **ON** ignition for approx. 5 seconds and check injectors for leaks
  - not more than 2 drops per minute of leakage may occur

If **YES**

- replace faulty injector(s)

Note

When reinstalling fuel injectors, make sure O-rings are not damaged.

Basic throttle adjustment, checking/adjusting

The stop screw is set at the factory and should NOT be changed. If the screw position should accidentally change, adjust as follows.

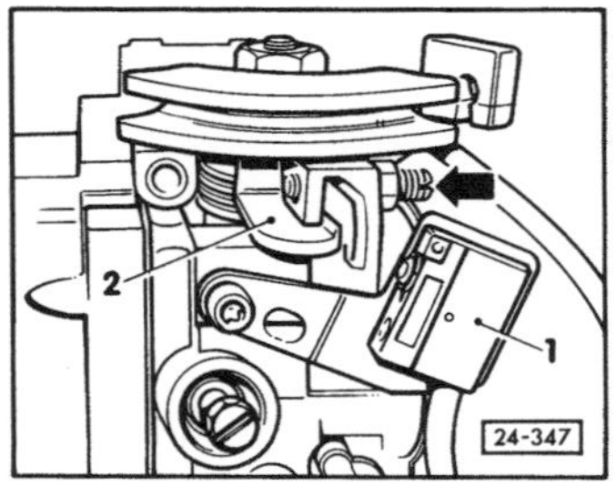

- unscrew throttle stop screw (**arrow**) until there is a gap between throttle lever **2** and throttle stop screw
- turn screw in until lever **2** just makes contact

To determine stop point of adjustment screw

- place a thin piece of cellophane between throttle lever **2** and throttle stop screw (**arrow**)
- pull up on the cellophane while turning in the adjustment screw, until cellophane begins to bind
- turn screw clockwise an additional 1/2 turn
- check idle switch and adjust if necessary, see section 24-140
- check idle speed and CO content and adjust if necessary, see section 24-90

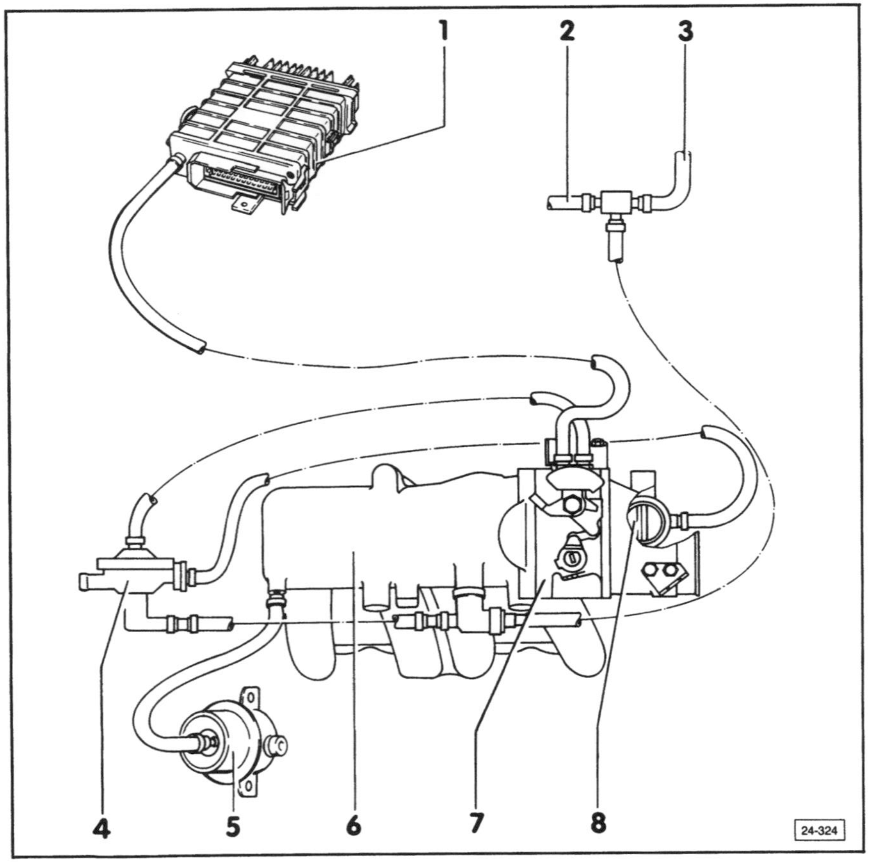

1 — **Digifant control unit (J 169)**
- vacuum hose length between ECU and throttle body critical; 1 meter, do not lengthen or shorten
- only use correct replacement part; black with yellow stripe

2 — **to A/C**

3 — **to multi function indicator**

4 — **Carbon canister control valve**

5 — **Fuel pressure regulator**

6 — **Intake manifold**

7 — **Throttle valve housing**

8 — **By-pass valve**

THIS PAGE INTENTIONALLY LEFT BLANK

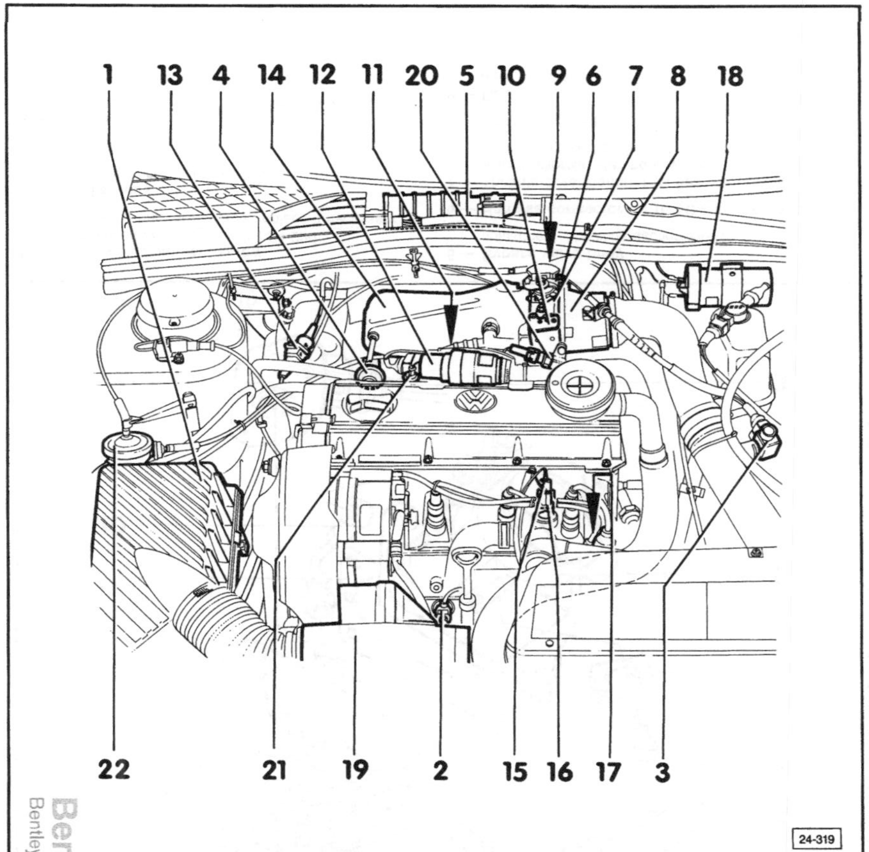

1 — **Air filter**

2 — **Knock sensor (G 61)**

3 — **CO potentiometer (G 74)**
with CO adjustment screw and intake air temperature sensor (G 42)

4 — **Fuel pressure regulator**

5 — **Digifant control unit (J 169)**

6 — **Idle speed adjustment screw**

7 — **Throttle valve housing**

8 — **Boost control valve housing**

9 — **Throttle potentiometer (G 69)**
eliminates former idle and full throttle switches

10 — **Full throttle switches**

11 — **Fuel rail assembly including fuel injectors**

12 — **Idle stabilizer valve and boost pressure regulator**
Idle air control valve (N71)

13 — **Oxygen sensor heater/probe harness connector**

14 — **Intake manifold**

15 — **Coolant temperature sensor (blue)**
supplies signal to control unit

16 — **Coolant temperature sensor (black)**
supplies signal to temperature gauge

17 — **Ignition distributor with Hall sender (G 40)**

18 — **Ignition coil (N 152)**

19 — **G-charger**
boost pressure checking, see Repair Group 21

20 — **Pressure switch (F 130) for fuel pump**
after-run checking, Repair Group 20

21 — **Thermoswitch (F 87) for radiator**
cooling fan after-run

22 — **Carbon canister control valve**
checking, Repair Group 20

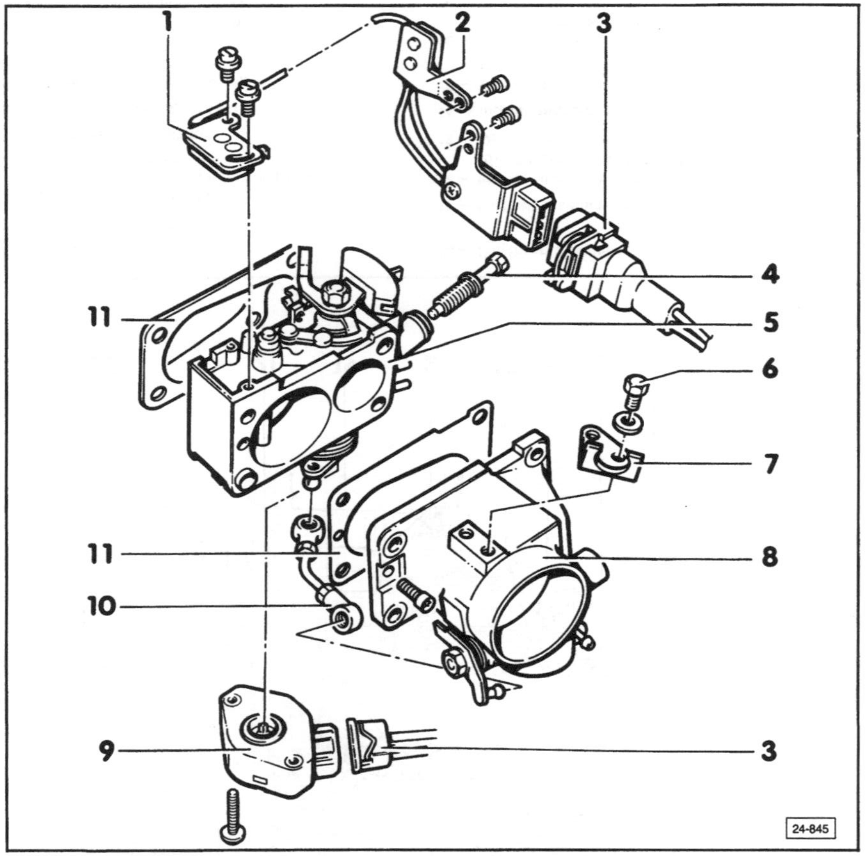

1 — **Idle speed adjustment screw**
- idle speed adjusting, section 24-220
- inspect O-ring for damage, replace if necessary

2 — **Throttle body**
basic adjustment, section 24-320

3 — **10 Nm (7 ft lb)**

4 — **Throttle stop**
accelerator cable adjustment, Repair Group 20

5 — **Bypass valve body**
adjustment, checking section 24-310

6 — **Harness connector for throttle potentiometer**

7 — **Throttle potentiometer (G 69)**
- for California, starting m.y. 1991
- checking, section 24-290

8 — **Connecting rod**

9 — **Gasket**
always replace

THIS PAGE INTENTIONALLY LEFT BLANK

Idle speed and CO content, checking/adjusting

> **CAUTION**
>
> Ignition timing, CO content and idle speed RPM's are inter-related and must be checked/adjusted together.

Note

The Oxygen sensor remains connected during checking/adjusting.

Requirements
- engine oil temperature 80°C (176°F) minimum
- all electrical consumers switched **OFF** (do not take measurements while radiator cooling fan is running)
- A/C switched **OFF**
- ignition timing **OK**
- idle stabilizer valve **OK**, (when ignition is switched on the valve must buzz and hum)
- exhaust system must be leak tight

Checking/adjusting

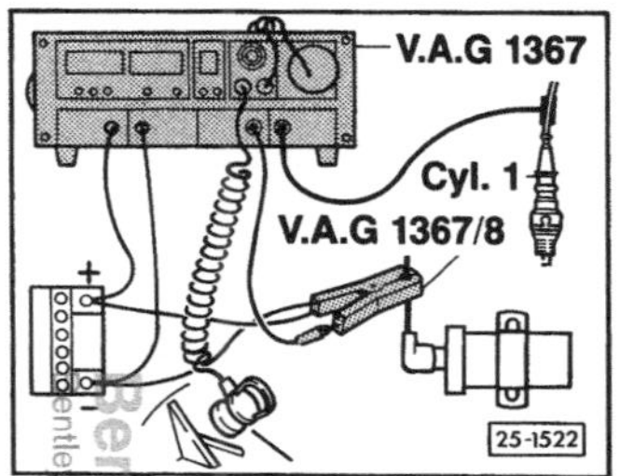

- switch **OFF** ignition
- connect **VAG 1367** engine tester using inductive pickup **VAG 1367/8** on ignition coil high tension lead
- connect **SUN 105** CO tester (or EPA approved equivalent) to CO tap tube using hose adaptor **VAG 1363/3**

> **CAUTION**
>
> Hose must fit securely on CO tap tube to avoid measuring inaccuracies due to dilution or leakage.

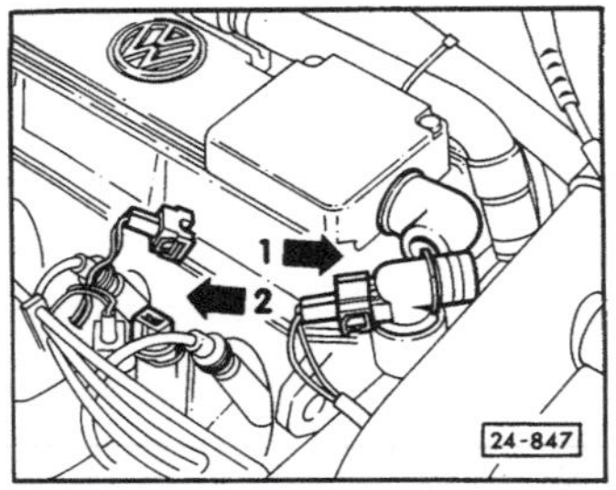

- disconnect crankcase ventilation hose from emission control valve (**arrow 1**) and route such that only fresh air is drawn in
- start engine and let idle
  - will run from 1 to 4 minutes at increased engine RPM

When idle speed is reduced approx. 100 RPM

- disconnect blue coolant temperature sensor harness connector

> **CAUTION**
>
> Do **not** start engine with coolant temperature sensor disconnected. The control unit will go into open loop making accurate adjustment impossible.

- raise engine speed in excess of 3000 RPM 3 times then let idle

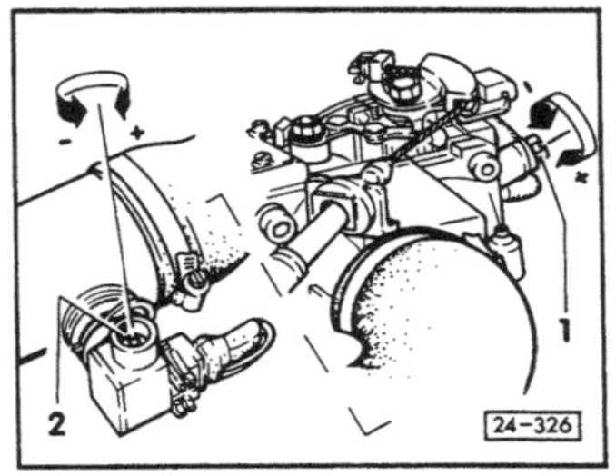

- check idle speed and CO content
  - idle speed must be 800 ± 50 RPM
  - CO content must be 0.7 ± 0.4 volume %

If **NO**

- adjust by alternately turning adjustment screws (**arrows**) until both specifications are obtained
- reconnect blue coolant temperature sensor
- raise engine speed 3 times then let idle
- check CO content
  - must be 0.7 ± 0.4 volume %

If **NO**

- check OXS control, see section 24-110
- after repairs or replacements have been performed, install new plug over CO adjustment screw

Note

After adjusting the idle speed, reconnect the crankcase ventilation hose. If the CO content then changes, this is not due to improper adjustment, but rather to oil dilution caused by short distance driving and insufficient warmup. With longer distance driving, the amount of fuel in the oil will be reduced and the CO will normalize again. A short term solution would be an oil change.

CAUTION
Do Not re-use any sprays or compounds containing silicone on engines equipped with Oxygen sensors. Do Not use these compounds on or near intake air system or near the Oxygen sensor. Silicone drawn into the intake air system is not burned during combustion and will lead to contamination and malfunctioning of the Oxygen sensor.

Oxygen Sensor and Oxygen Sensor Control, Checking

Requirements
- no leaks in exhaust system between catalytic converter and cylinder head
- voltage supply for Oxygen sensor heater **OK**

- connect **VAG 1551** Diagnostic tester, see Repair Group D2 for additional information
- start engine and let idle
- select Function **"Read measuring value block"**, see Repair Group D2 for additional information

or

- while observing **VAG 1551** display, select the following functions in sequence

  1– Rapid data transfer

  01–Engine electronics

  08–Read measuring value block

- press **0** key twice to select display group number **00**
  - display will appear as shown

Read measuring value block Q
Display group number input 00

▶
- press **Q** key to enter input
  - following display appears

Measuring value block reading
1 2 3 4 5 6 7 8 9 10

▶
- continue with checking **only** after:
  - the displayed value in channel **3** is less than **11** (which corresponds to a coolant temperature of at least 80°C or 176°F)
- let engine idle for 2 more minutes
  - read display value in channel **5** (Oxygen sensor signal)

Display value	Possible cause of fault	Fault correction
Varies more than **14**	No fault, Oxygen sensor control **OK**	Checking complete
0 – **5**	Short circuit to positive (+) in wiring between Oxygen sensor and control unit	Perform Test **A**
164 – **168**	Open circuit in wiring between Oxygen sensor and control unit	Perform Test **B**
230 minimum	Short circuit to ground (–) in wiring between Oxygen sensor and control unit	Perform Test **C**

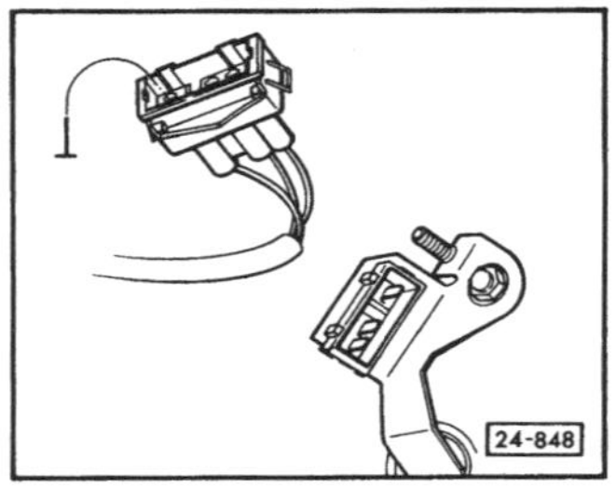

Test A

- disconnect Oxygen sensor/heater harness connector (on valve cover)
  - displayed value must be **164 - 168**

If **YES**
- Oxygen sensor control **OK**

- switch **OFF** ignition
- replace Oxygen sensor (**G 39**)
- activate Fault memory and then erase, see Repair Group D2 for additional information
  - displayed value must be **0** to **5**

If **NO**
- Oxygen sensor control **not OK**

- switch **OFF** ignition
- check wiring for short circuit to control unit positive (+), repair or replace as necessary, also see Repair Group D2: Digifant system electrical testing, **Test A**, step **16**

If wiring **OK**

- replace Digifant control unit (**J 169**)
- activate Fault memory and then erase, see Repair Group D2 for additional information

Test B

- disconnect Oxygen sensor/heater harness connector (on valve cover)
- touch violet wire (from Digifant control unit) to ground (-) using jumper from **VW 1594** adaptor kit
  - displayed value must be **230** minimum

If **YES**
- Oxygen sensor control **OK**

- switch **OFF** ignition
- replace Oxygen sensor (**G 39**)
- activate Fault memory and then erase, see Repair Group D2 for additional information
  - displayed value must be **164 - 168**

If **NO**
- Oxygen sensor control **NOT OK**

- switch **OFF** ignition
- check wiring for open circuit to control unit using wiring diagram, repair or replace as necessary, also see Repair Group D2, **Test A**, step **16**

If wiring **OK**

- replace Digifant control unit (**J 169**)
- activate Fault memory then erase, see Repair Group D2 for additional information

Test C

- disconnect Oxygen sensor/heater harness connector (on valve cover)
  - displayed value must be **164 - 168**

If **YES**

- Oxygen sensor control **OK**

- switch **OFF** ignition
- replace Oxygen sensor (**G 39**)
- activate Fault memory and then erase, see Repair Group D2 for additional information
  - displayed value must be **230** minimum

If **NO**

- Oxygen sensor control **NOT OK**

- switch **OFF** ignition
- check wiring for short circuit between control unit and ground, repair or replace as necessary, also see Repair Group D2, **Test A**, step **16**

If wiring **OK**

- replace Digifant control unit (**J 169**)
- activate Fault memory and then erase, see Repair Group D2 for additional information

THIS PAGE INTENTIONALLY LEFT BLANK

Acceleration/Full Throttle Enrichment and Deceleration Fuel Cutoff, Checking

Note

Use the following procedure to check the deceleration fuel cutoff. If this check **and** the check for the air flow sensor potentiometer are **OK**, acceleration and full throttle enrichment are also **OK**.

- connect **VAG 1551** Diagnostic tester, see Repair Group D2 for additional information
- start engine and let idle
- select Function **08 "read measuring block value"**, see Repair Group D2 for additional information

or:

- while observing **VAG 1551** display, select the following functions in sequence

  1– Rapid data transfer

  01–Engine electronics

  08–Read measuring block value

- press **0** and **1** key to select display group number **00**
  - display will appear as shown

Read measuring value block Q
Display group number input 01

▶

- press **Q** key to enter input
  - display will appear as shown

Read measuring value block 1
1 2 3 4

▶

Continue with checking **only** if:
- the coolant temperature displayed in channel **3** is 80°C (176°F) minimum

- while observing channel **1** displayed value, raise engine speed in excess of 3000 RPM three times
- let engine idle for 2 minutes
- observe displayed value in channel **4**
- raise engine speed to a minimum of 3000 RPM by stepping on accelerator pedal then releasing
  - the value displayed in channel **4** must increase in value; drop to 0 ms; then return to the previously observed value

If **YES**
- acceleration/full throttle enrichment and deceleration fuel shutoff **OK**

If displayed value does not increase and does not return to 0 ms

- check throttle potentiometer, see section 24-290

Coolant temperature sensor, checking

Requirement
- engine cold

- ■ connect **VAG 1551** Diagnostic tester, see Repair Group D2 for additional information
- ■ start engine and let idle
- ■ select Function **"Read measuring value block"**, see Repair Group D2 for additional information

or

- ■ while observing **VAG 1551** display, select the following functions in sequence

  1– Rapid data transfer

  01–Engine electronics

  08–Read measuring value block

- ■ press keys **0** and **1** to select **"Display group number 01"**
  - display will appear as shown

Read measuring value block Q
Display group number input 01

- ■ press **Q** key to acknowledge input
  - channels will be displayed as shown

Read measuring value block 1
1 2 3 4

- ■ observe display value in channel **3**

Display value	Possible cause of fault	Fault correction
Must increase uniformly without interruption*	Coolant temperature sensor (**G 62**) and control **OK**	Checking complete
–40°C	Break in wiring, sensor wiring, or wiring to control unit	Perform Test **A**
80°C minimum **	Short circuit to negative (–) between coolant temperature sensor, wiring or control unit	Perform Test **B**

*If engine malfunctions occur during certain temperature ranges **and** the displayed value in channel **3** does **NOT** increase uniformly and without interruption, the problem is a faulty coolant temperature sensor.

If **YES**

- ■ replace coolant temperature sensor (G-62)

**If the displayed value, 80°C (176°F) minimum does not coresond to the actual coolant temperature, e.g. actual coolant temperature under 60°C

- ■ perform test **A**

- ■ perform test **A**

Test A

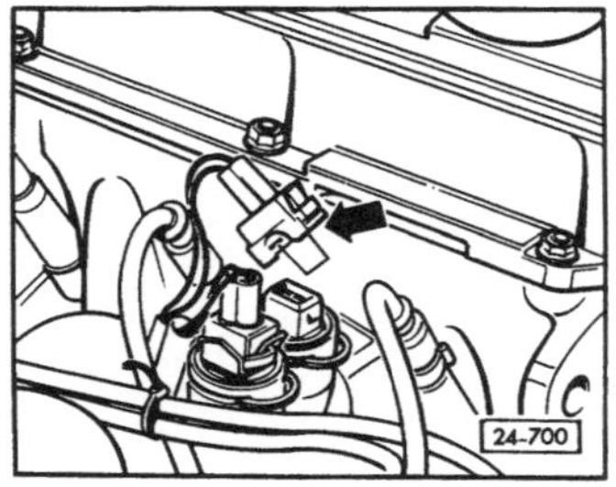

- ■ disconnect coolant temperature sensor harness connector (blue) and bridge the connector terminals using jumpers from **VW 1594** adaptor kit
  - displayed value must be 80°C (176°F) minimum

If **YES**
- coolant temperature sensor control **OK**

- ■ switch **OFF** ignition
- ■ replace coolant temperature sensor (**G 62**)
- ■ activate Fault memory and erase – Repair Group D2, Self Diagnosis

Test B

- ■ switch **OFF** ignition
- ■ disconnect (blue) coolant temperature sensor harness connector
- ■ switch **Fluke 83** (**US 1119**) multimeter to resistance range
- ■ connect multimeter between terminals of temperature sensor using jumpers from **VW 1594** adaptor kit

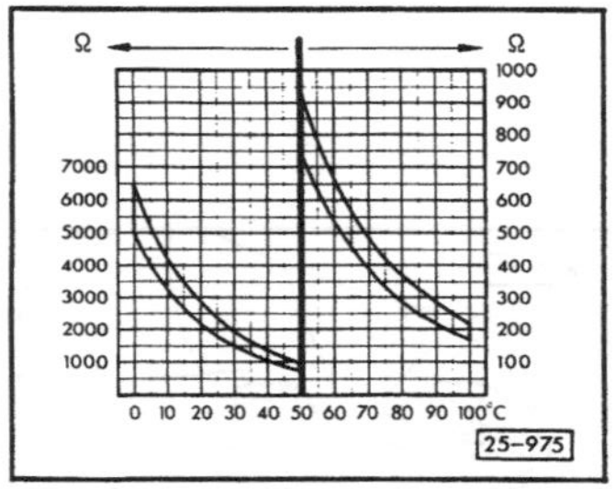

Specified Value: per chart

If **OK**

- ■ re-connect coolant temperature sensor (**G 62**) harness connector
- ■ check wiring for short circuit to control unit negative (–), repair or replace as necessary, also see Repair Group D2, **Test A**, step **9**

If wiring **OK**

- ■ replace Digifant control unit (**J 169**)
- ■ activate Fault memory and then erase, see Repair Group D2 for additional information
- ■ return control unit to basic setting, see Repair Group D3, section D3-70

If measured value is out of range

- ■ replace coolant temperature sensor (**G 62**)

Intake Air Temperature Sensor, checking

- connect **VAG 1551** Diagnostic tester, see Repair Group D2 for additional information
- switch **ON** ignition
- select Function **08 "Read measuring value block"**, see Repair Group D2 for additional information

or

- while observing **VAG 1551** display, select the following functions in sequence

  1 – Rapid data transfer

  01 – Engine electronics

  08 – Read measuring value block

- press **0** keys twice to select **"Display group number 02"**
  - display will appear as shown

Read measuring value block Q
Display group number input 00

- press **Q** key to enter input
  - display will appear as shown

Read measuring value block 2
1 2 3 4 5 6 7 8 9 10

- read displayed value in channel **1**

Display value	Possible cause of fault	Fault correction
10 to **190**	Coolant temperature sensor (**G 62**) and control **OK**	Checking complete
at least **250**	Open circuit in wiring between sensor and control unit	Perform Test **A**
0	Short circuit to positive (+) in wiring between sensor and control unit	Perform Test **B**

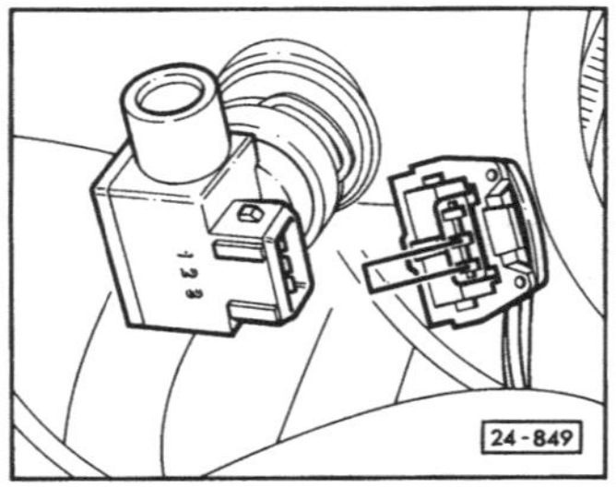

Test A

- disconnect CO potentiometer harness connector (with sensor for intake air temperature and bridge terminals **2** and **3** of the connector using jumpers from **VW 1594** adaptor kit
  - displayed value must be **0**

If **YES**

- Intake air temperature sensor control **OK**

- switch **OFF** ignition
- replace CO potentiometer (**G 74**) and intake air temperature sensor (**G 42**)
- activate Fault memory and erase, Repair Group D2

Test B

- switch **OFF** ignition
- disconnect CO potentiometer harness connector

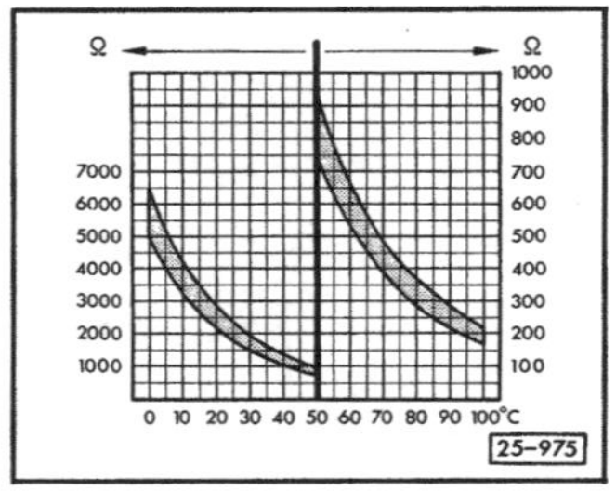

- measure sensor resistance across terminals **2** and **3** using **US 1119** multimeter and jumper wires from **VW 1594** adaptor kit
  - measured value must fall within appropriate area of chart

If specified value obtained

- reconnect CO potentiometer harness connector
- check wiring for short circuit to control unit negative (–) using wiring diagram, repair or replace as necessary, see Repair Group D2, **Test A**, step **10**

If wiring **OK**

- replace Digifant control unit
- activate Fault memory and then erase, see Repair Group D2

If the specified value **not** obtained

- replace CO potentiometer (**G 74**)
  - comes complete with with intake air temperature sensor (**G 72**)

- activate Fault memory and then erase, see Repair Group D2
- check idle speed and CO content, adjust if necessary

CO potentiometer, checking

Requirement
- idle speed **OK**
- coolant temperature 80°C (176°F) minimum

■ connect **VAG 1551** Diagnostic tester, see Repair Group D2 for additional information
■ start engine
■ select Function **08 "Read measuring value block"**, see Repair Group D2 for additional information

or

■ while observing **VAG 1551** display, select the following functions in sequence

1– Rapid data transfer

01–Engine electronics

08–Read measuring value block

■ press **0** keys twice to select **"display group number 00"**
- display will appear as shown

Read measuring value block Q
Display group number input 00

■ press **Q** key to enter input
- channels will be displayed as shown

Read measuring value block 1
1 2 3 4 5 6 7 8 9 10

■ observe displayed value in channel **2**

Display value	Possible cause of fault	Fault correction
40 - 140*	CO potentiometer (**G 74**) and triggering **OK**	Checking complete
0 ... 2	Short negative (–): wire, CO Potentiometer, control unit	Perform Test **A**
at least **250**	Open circuit in wiring between potentiometer and control unit	Perform Test **B**

*The displayed value depends on the CO content adjustment during idle adjustment.

Test A

■ disconnect CO potentiometer harness connector
- display value: **250** minimum

If **YES**
- triggering of CO potentiometer (**G 74**) **OK**

■ switch **OFF** ignition
■ replace CO potentiometer assembly (**G 74**)
- includes intake air temperature sensor (**G 42**)

■ activate Fault memory and then erase, see Repair Group D2 for additional information
■ check idle and CO values, adjust if necessary
- displayed value must be **0** to **2**

If **NO**
- CO potentiometer triggering (**G 74**) **NOT OK**

■ switch **OFF** ignition
■ reconnect CO potentiometer (**G 74**) harness connector
■ eliminate short in wiring between control unit and ground using wiring diagram, see Repair Group D2, **Test A**, step **11**

If wiring **OK**

■ replace control unit (**J 169**)
■ activate Fault memory and then erase, see Repair Group D2

Test B

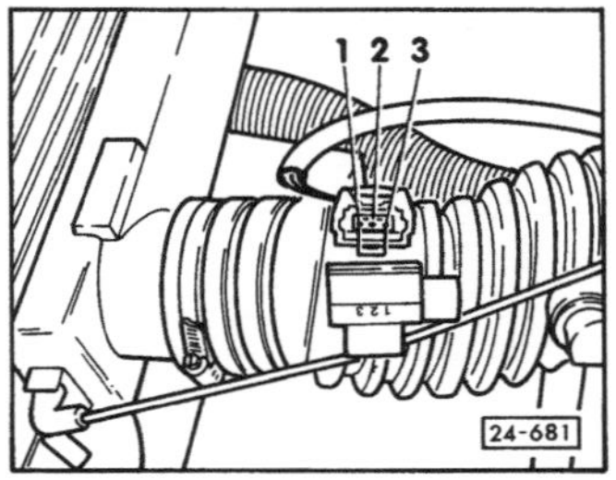

■ disconnect CO potentiometer harness connector (with intake air temperature sensor) and bridge terminals **1** and **3** of connector using jumper wires from **VW 1594** adaptor kit
- displayed value must be **0** to **2**

If **YES**
- control unit triggering **OK**

■ switch **OFF** ignition
■ replace CO potentiometer (**G 74**) assembly
- includes intake air temperature sensor (**G 42**)

■ activate Fault memory and then erase, see Repair Group D2
■ check idle speed and CO content, adjust if necessary
- displayed value must be **250** minimum

If **NO**
- control unit triggering **not OK**

■ switch **OFF** ignition
■ reconnect CO potentiometer (**G 74**) harness connector
■ determine open circuit and eliminate using wiring diagram, see Repair Group D2, **Test A**, step **11**

If wiring **OK**

- replace Digifant control unit (**J 169**)
- activate Fault memory and then erase, see Repair Group D2

Vacuum sensor, checking

Note

The vacuum sensor (**G 71**) is installed inside the Digifant Control Unit (**J 169**).

- connect **VAG 1551** Diagnostic tester, see Repair Group D2
- start engine and let idle
- select Function **08 "Read measuring value block"**, see Repair Group D2 for additional information

or

- while observing **VAG 1551** display, select the following functions in sequence

  1 – Rapid data transfer

  01 – Engine electronics

  08 – Read measuring value block

- press **0** and **1** keys to select **"display group number 01"**
  - display will appear as shown

Read measuring value block Q
Display group number input 01

- press **Q** key to enter input
  - channels will be displayed as shown

Read measuring value block 1
1 2 3 4

- continue with the test only if the coolant temperature in channel **3** is 80°C minimum
- read display value in channel **2**
  - displayed value must be 20 to 40% **fluctuating**

If **YES**
- vacuum line and vacuum sensor (**G 71**) inside Digifant control unit (**J 169**) **OK**

If displayed value 45% (minimum) and **constant**
- vacuum line and sensor (**G 71**) inside Digifant control unit (**J 169**) **NOT OK**

- check vacuum line between throttle body and control unit, replace if necessary

If vacuum line **OK**

- replace Digifant control unit
- activate Fault memory and then erase, see Repair Group D2

Throttle Potentiometer, Checking

Note

The throttle potentiometer provides information to the Digifant control unit (**J 169**) for the following functions:

with throttle closed:
- injector shut off during fuel deceleration cutoff
- idle stabilization
- ignition adjustment during idle using digital idle stabilization function

with throttle partially opened:
- determines injection quantity for acceleration enrichment

with throttle fully opened:
- determines injection quantity during full throttle enrichment

Checking

- connect **VAG 1551** Diagnostic tester, see Repair Group D2 for additional information
- switch **ON** ignition
- select Function **08 "Read measuring value block"**, see Repair Group D2 for additional information

or

- while observing **VAG 1551** display, select the following functions in sequence

  1 – Rapid data transfer

  01 – Engine electronics

  08 – Read measuring value block

- press **0** and **5** keys to select **"Display group number 05"**
  - display will appear as shown

Read measuring value block Q
Display group number input 05

- press **Q** key to enter input
  - channels will be displayed as shown

Read measuring value block 5
1 2 3 4

- read displayed value in channel **2**

Display value	Possible cause of fault	Fault correction
Idle stop **5 - 50** <°	Throttle potentiometer (**G 69**) and control **OK**	Checking complete
full throttle stop* min. **90** >°	—	—
0.0>°	Open circuit in wiring between throttle potentiometer and control unit	Perform Test **A**
min. **90** >° at idle stop	Short circuit to positive (+) in wiring between throttle potentiometer and control unit	Perform Test **A**

- open throttle slowly and uniformly
  - displayed value in channel **2** must uniformly increase without interruption

Test A

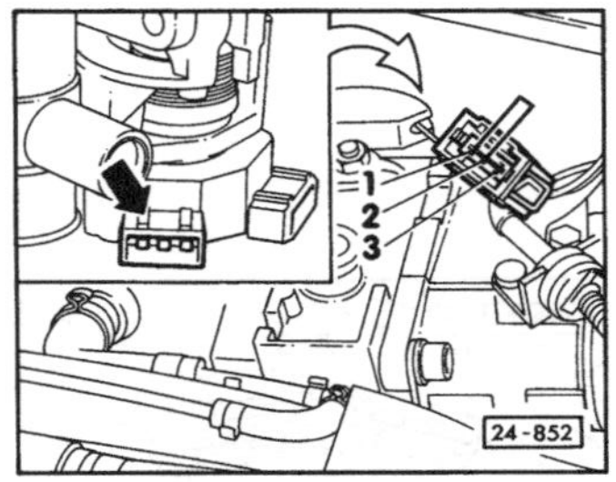

- disconnect 3-pin throttle potentiometer (**G 69**) harness connector (**arrow A**)
  - displayed value must be 0.0 >°
- bridge terminals **1** and **2** of the harness connector (**arrow 2**) using jumpers from **VW 1594** adaptor kit
  - displayed value must be 90 >° minimum

If **YES**
- throttle potentiometer control **OK**

- switch **OFF** ignition
- replace throttle potentiometer (**G 69**)
- activate Fault memory and then erase, see Repair Group D2

Vehicles with automatic transmission only:

- after removal/installation or replacement of throttle potentiometer reset control unit to reference setting, see 01 (Transmission) OBD

If **NO**
- throttle potentiometer control **NOT OK**

- switch **OFF** ignition
- check wiring for open circuit or short to control unit positive (+), repair or replace as necessary, also see Repair Group D2, **Test A**, step **13** and/or **Test B** steps **1** and **2**

If wiring **OK**

- replace Digifant control unit (**J 169**)
- activate Fault memory and then erase, see Repair Group D2

Fuel Injectors, Checking

WARNING

Fire Hazard. Do **not** smoke or have anything in area that can ignite fuel!

Voltage Supply and Resistance, checking

Requirement
- Hall sender **OK**

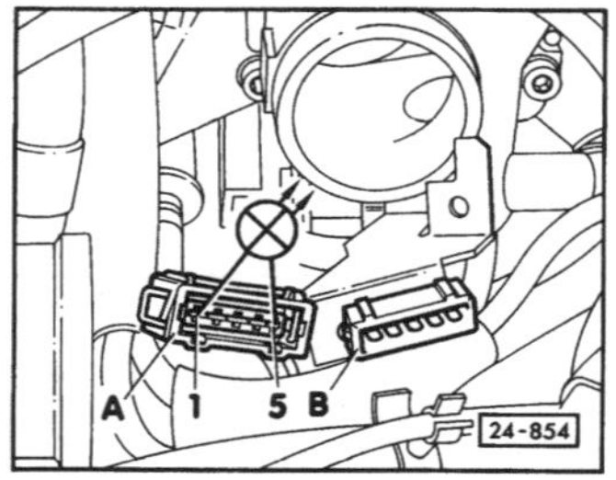

- check voltage supply to all injectors using **US 1115** LED tester and jumper wires from the **VW 1594** adaptor kit in the following sequence of terminals in connector **A**

  1 and **5**

  2 and **5**

  3 and **5**

  4 and **5**

- operate starter
  - LED must flicker

If LED tester lights up continuously

- replace ground wire between battery, transmission and Digifant control unit
- fasten ground wire to transmission using new mounting bolt

If LED tester does not flicker or light up

- reconnect harness connectors **A** and **B**
- check wiring for open circuit to control unit, repair or replace as necessary, also see Repair Group D2: Digifant system electrical testing **Test A** – step **6**

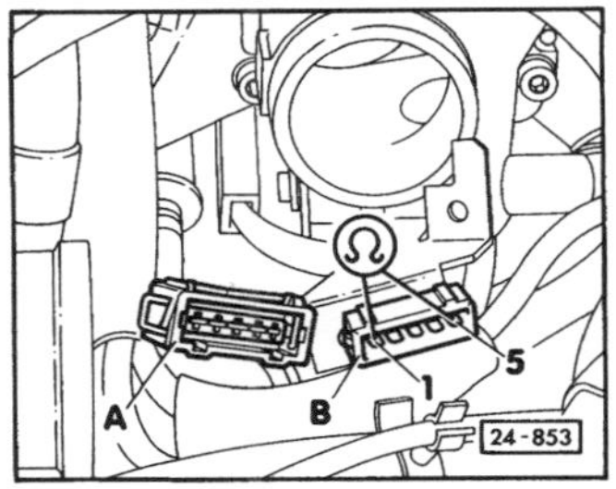

- check resistance for all injectors using **Fluke 83** multimeter
- connect multimeter between the following terminals in connector **B** (in the sequence given):

  1 and **5**

  2 and **5**

  3 and **5**

  4 and **5**

  - must be 15 – 20 Ohms (for each reading)

Fuel injector spray pattern and sealing, checking

> **WARNING**
>
> Fire Hazard. Do **not** smoke or have anything in area that can ignite fuel!

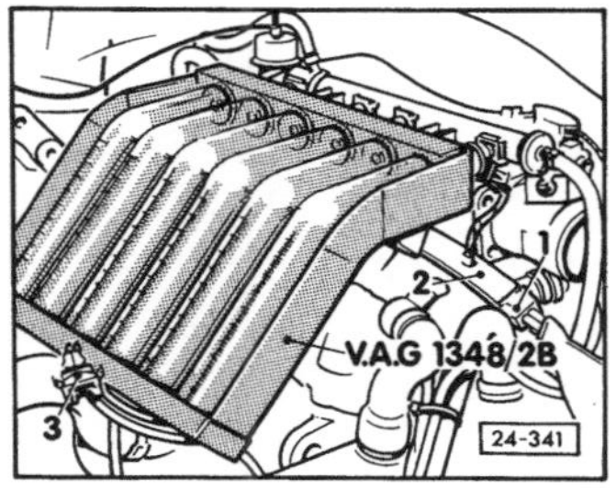

- remove fuel rail assembly (injectors/wire guide/fuel rail)
- insert injector assembly into openings of fuel analyzer **VAG 1348/2B**
- disconnect blue coolant temperature sender harness connector **3**
- re-connect harness connector **1** to connector **2**
- briefly operate starter and observe spray pattern
  - must be uniform among all injectors in the set

- disconnect harness connector **1**
- switch on ignition for approx. 5 seconds and check injectors for leaks
  - not more than 2 drops per minute of leakage may occur

If **YES**

- replace faulty injector(s)

Note

When reinstalling fuel injectors, make sure O-rings are not damaged.

By-pass valve adjustment, checking

Note

The by-pass valve controls the boost pressure. It is adjusted at the factory and must **not** be changed. Use the following check to determine if the by-pass valve adjustment has been inadvertently altered.

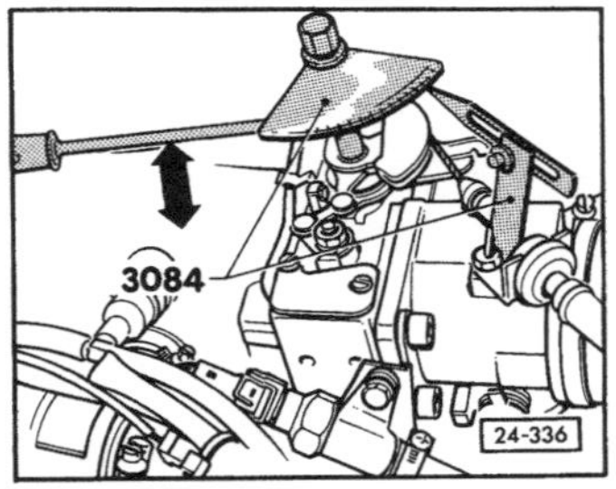

- attach protractor **3084** to throttle shaft using magnetic adaptor **3084/1**
- fasten protractor pointer to throttle cable sleeve using circlip
- using screwdriver, push throttle lever to full throttle stop
- with throttle in this position, set protractor to 10°

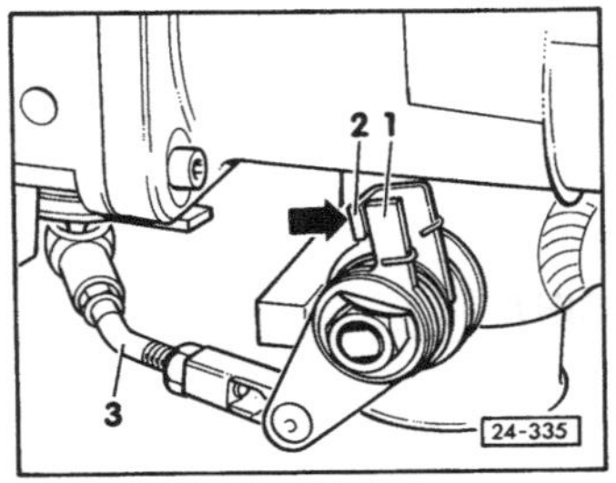

- slowly close throttle until lever **1** just makes contact with lever **2** (**arrow**)
- read value on protractor
  - must be 5 ± 3° before full throttle stop

If **NO**

- adjust by-pass valve on connecting rod **3**

Note

For full throttle switch checking/adjusting see section 24-140.

Basic throttle adjustment, checking/adjusting

The stop screw is set at the factory and should **not** be changed. If the screw position should accidentally change, adjust as follows.

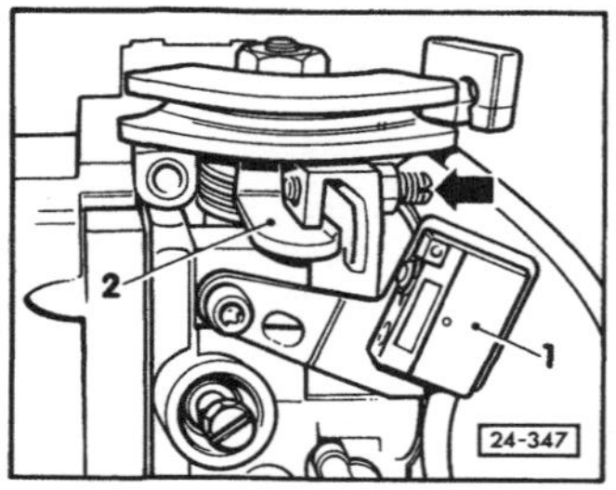

- unscrew throttle stop screw (**arrow**) until there is a gap between throttle lever **2** and throttle stop screw
- turn screw in until lever **2** just makes contact

To determine stop point of adjustment screw

- place a thin piece of cellophane between throttle lever **2** and throttle stop screw (**arrow**)
- pull up on the cellophane while turning in the adjustment screw, until cellophane begins to bind
- turn screw clockwise an additional 1/2 turn
- check idle switch and adjust if necessary, see section 24-140
- check idle speed and CO content and adjust if necessary, see section 24-90

THIS PAGE INTENTIONALLY LEFT BLANK

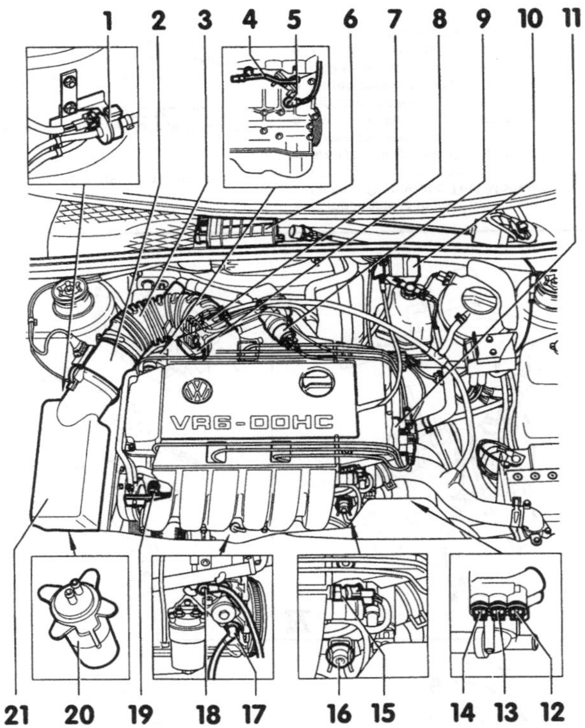

1 – **EVAP solenoid valve I (N 80)**

2 – **Mass air flow (MAF) sensor (G 70)**
before removing; ignition must have been switched off for a minimum of 20 seconds

3 – **Intake air boot**
with connection for idle air control and heater element

4 – **CO tap tube**

5 – **Knock sensor I (G 61)**

6 – **Motronic engine control module (ECM) (J 220)**

7 – **Throttle body**
heated via coolant

8 – **Throttle position (TP) sensor (G 69)**

9 – **Idle speed control (ISC) valve (N 71)**

10 – **Ignition coil (N 152)**

11 – **Ignition distributor**
with Hall sensor

12 – **Temperature sensor**
- brown
- for A/C

13 – **Engine coolant temperature (ECT) sensor (G 62)**
- blue
- 2 pole

14 – **Engine coolant temperature (ECT) sensor**
- supplies signal to instrument panel temperature gauge
- yellow
- 4 pole

15 – **Intake air temperature (IAT) sensor (G 72)**
10 Nm (7 ft lb)

16 – **Fuel pressure regulator**

17 – **Engine speed (RPM) sensor (G 28)**

18 – **Knock sensor II (G 66)**
20 Nm (15 ft lb)

19 – **Fuel injectors (N 30, N 31, N 32, N 33, N 83, N 84)**

20 – **Carbon canister**
- mounted below air cleaner
- for additional information see Repair Group 20

21 – **Air cleaner**

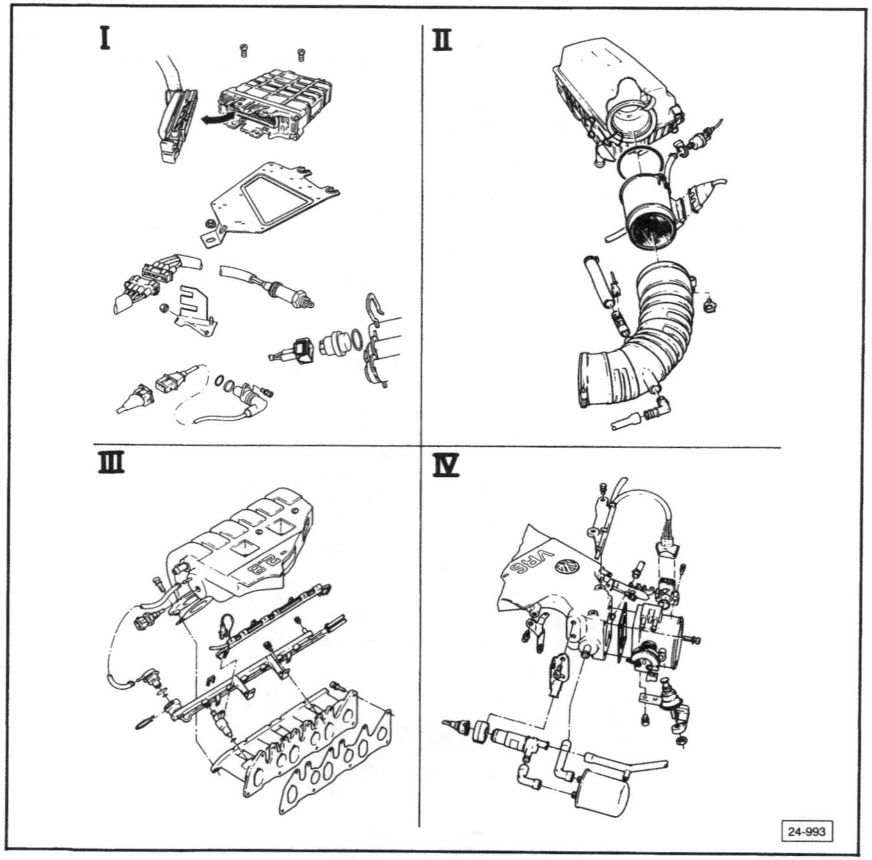

WARNING
Fire Hazard. Do **NOT** smoke or have anything in area that can ignite fuel!

- the engine control module for the fuel injection and ignition system is equipped with **DTC MEMORY** (Fault memory)
- activate **DTC MEMORY** (Fault memory) before repairing, adjusting or troubleshooting the system
- always check all vacuum connections for tightness and condition
- components identified with an * are checked via **On Board Diagnosis**, see Repair Group 01 (Fuel Injection and Ignition)
- components identified with ** are checked using **Output Diagnostic Test Mode (DTM)**, see Repair Group 01(Fuel Injection and Ignition)
- always replace the affected seals and sealing rings during assembly work
- always replace clamps with screw type when replacing hoses
- for repairing ignition portion of this system, see Repair Group 28
- 10.5 Volts minimum is required for proper function of electrical components

I Motronic ECM, component layout, section 24-350
II Intake air system, component layout, section 24-360
III Fuel injection system, section 24-370
IV Throttle assembly, component layout, section 24-380

- Safety precautions see section 24-390
- Rules of Cleanliness see section 24-390
- Idle speed, checking see section 24-400

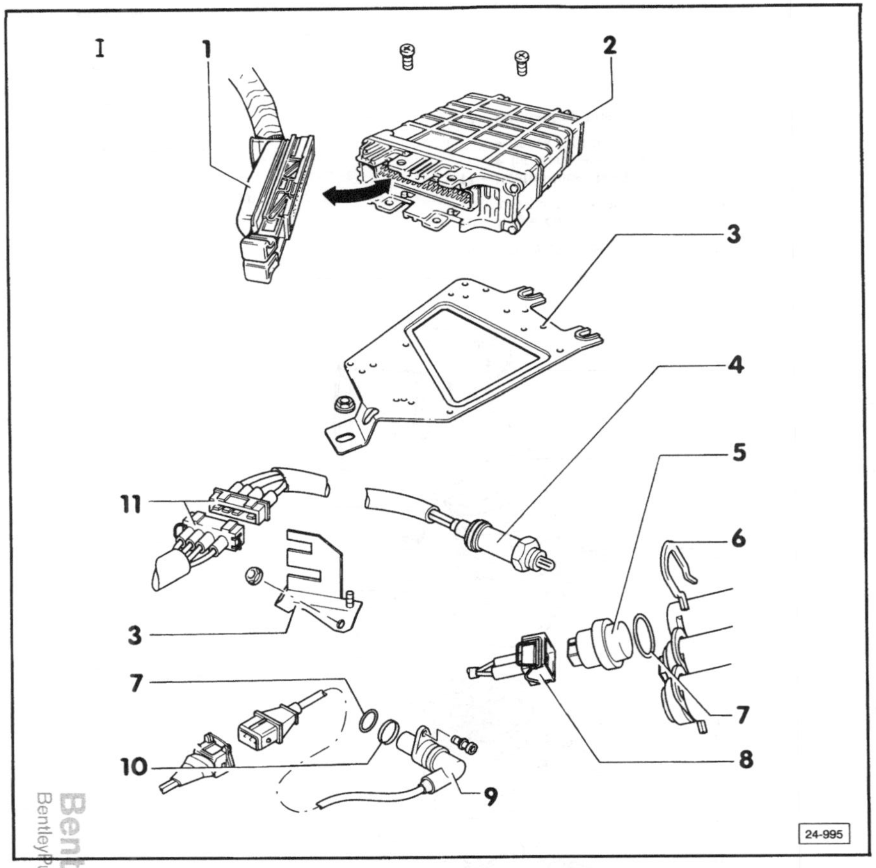

1 – **Motronic ECM harness connector**
- before removing; ignition must have been switched off for a minimum of 20 seconds
- 68-pin
- Motronic System Electrical Testing, see Repair Group 01 (Fuel Injection and Ignition)

2 – **Motronic engine control module (ECM) (J 220)***
- installation location: drivers side of plenum panel in engine compartment
- with **DTC MEMORY** (Fault memory), see Repair Group 01 (Fuel Injection and Ignition) for additional information
- idle speed and ignition timing point checking/adjusting see section 24-400

3 – **ECM mounting plate**

4 – **Heated Oxygen sensor (HO2S) (G 39)***
- **50 Nm**
- installation location: front exhaust pipe
- coat threads with **G5** but do not allow **G5** to contact slotted area of sensor tip.

To check heater voltage supply:
- switch **Fluke 83** multimeter to 20 Volt range
- connect multimeter between terminals **1** (red/white) and **2** (brown)
- start engine and let idle
- must be approx. battery voltage

5 – **Engine coolant temperature (ECT) sensor (G 62)***
- blue
- 2-pin
- checking see section 24-440

Fuel Injection, AFC System 24

6 – **Retaining clip**
check for secure fit

7 – **O-ring**

8 – **Harness connector for blue ECT sensor**

9 – **Engine speed (RPM) sensor (G 28)***

10 – **Spacer**

11 – **Harness connector for HO2S sensor**
- 4-pin
- for **HO2S** and **HO2S** heater
- fastened to rear engine mount facing front of vehicle

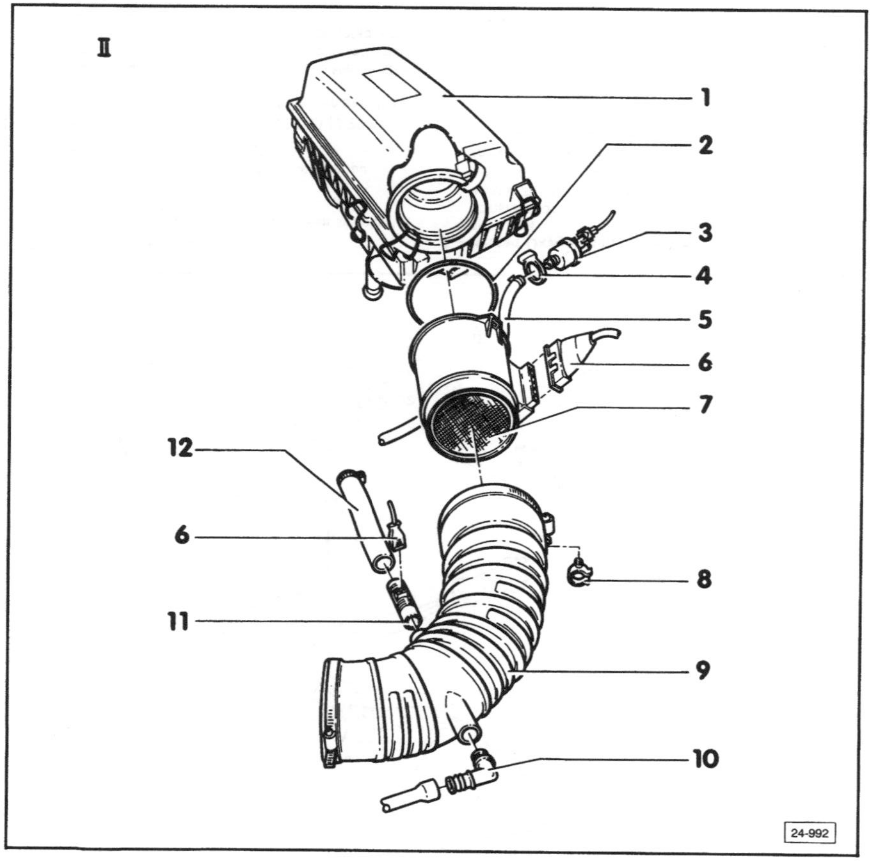

1 – **Air cleaner housing**

2 – **O-ring**

3 – **EVAP frequency valve (N 80)***

4 – **Retaining ring**
for (**N 80**) solenoid valve

5 – **Hose**
to throttle body

6 – **Harness connector**
to (**N 79**) heated tube

7 – **Hot wire Mass Air Flow (MAF) sensor (G 70)***

8 – **Mounting clip**

9 – **Intake air boot**

10 – **Angle connector**
for hose to idle air control (**IAC**) valve

11 – **Heated tube (N 79)**
- for crankcase housing ventilation
- check for continuity

12 – **Connecting hose**

THIS PAGE INTENTIONALLY LEFT BLANK

III

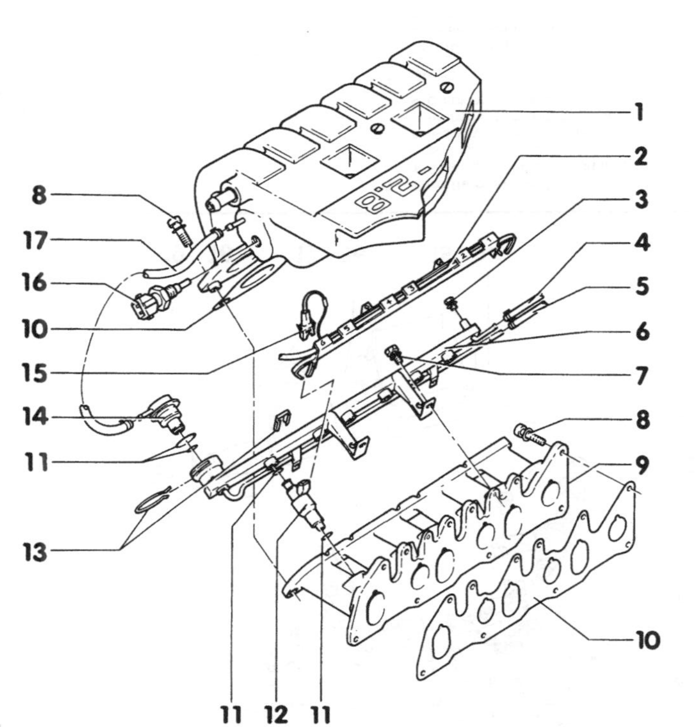

1 – **Intake manifold**
upper section

2 – **Wiring guide**
for fuel injector wiring

3 – **Plug**
pressure measurement connection

4 – **Fuel supply line**
black

5 – **Fuel return line**
blue

6 – **Fuel rail**

7 – **10 Nm (7 ft lb)**

8 – **25 Nm (18 ft lb)**

9 – **Intake manifold**
lower section

10 – **Manifold gasket**
- always replace
- note installed position

11 – **O-ring**
carefully replace if damaged

12 – **Fuel injectors (N 30, N 31, N 32, N 33, N 83, N 84)**
removing and installing, see section 24-470

13 – **Retaining clip**
check for secure fit

14 – **Fuel pressure regulator**
checking, see section 24-460

15 – **Harness connector (6 total)**
for fuel injectors

16 – **Intake air temperature (IAT) sensor (G 72)**
10 Nm (7 ft lb)

17 – **Vacuum line**
- replace if damaged
- check for secure fit

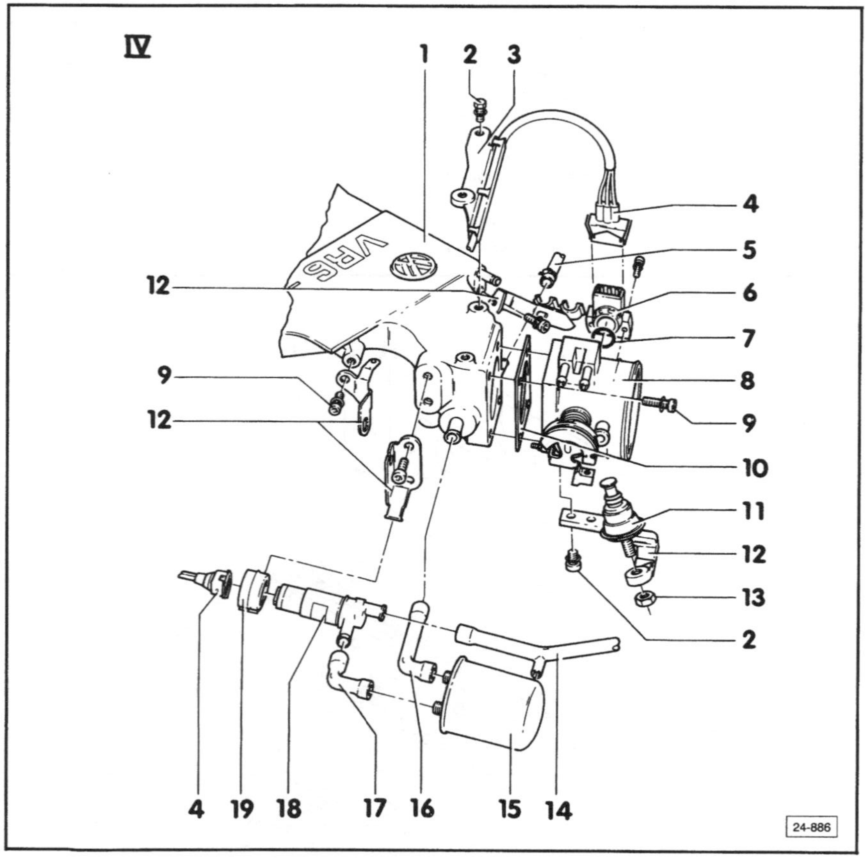

1 – **Intake manifold**
upper section

2 – **10 Nm (7 ft lb)**

3 – **Wiring guide**
for fuel injector wiring

4 – **Harness connector**
for throttle position (TP) potentiometer (**G 69**)

5 – **Hose**
from EVAP frequency valve (**N 80**)

6 – **Throttle position (TP) sensor (G 69)***
- 3-cavity connector (manual trans) for fuel injection and ignition systems
- 4-cavity connection (automatic trans) to **TCM** (transmission control module) (**J 217**).
- checking, see section 24-450

7 – **O-ring**

8 – **Throttle body**
- heated with coolant
- if setscrew for basic adjustment of throttle has been moved, replace throttle body

9 – **25 Nm (18 ft lb)**

10 – **Gasket**
always replace

11 – **Dashpot**
checking and adjusting, see section 24-420

12 – **Mounting bracket for dashpot**

13 – **20 Nm (15 ft lb)**

14 – **Air hose**
to intake air boot

15 – **Noise damper**
for **IAC** (idle air control) valve

16 – **Vacuum hose**
from noise damper to intake manifold upper part

17 – **Connector hose**
from **IAC** (idle air control) valve to noise damper

18 – **Idle air control (IAC) valve (N 71)****
checking, see section 24-410

19 – **Mounting ring**

Safety precautions

Observe the following precautions to prevent personal injury as well as possible damage to sensitive electrical components.

- switch **OFF** the ignition before connecting or disconnecting components or test equipment
- connect and disconnect battery only with ignition switched **OFF**, otherwise the **ECM** (Engine Control Module) could be damaged
- if the engine must be cranked but not started (for compression testing etc.) disconnect the power output stage of the ignition coil and remove fuse **18**
- do not touch or disconnect the ignition wires with the engine running or while cranking

Rules of Cleanliness

CAUTION

When working on the fuel supply/injection system, carefully observe the following rules:

1 – Thoroughly clean connection and surrounding areas before loosening connection.

2 – After removing components, place in clean area and cover with foil or paper. Avoid using rags!

3 – Components which have been opened or disassembled must be carefully covered or sealed if repair cannot be carried out immediately.

4 – Install clean parts only.

- remove replacement parts from package just before installing
- do **NOT** use spare parts that were stored loose or unpackaged (e.g. in tool boxes, etc.)

5 – When fuel system is open:

- avoid using compressed air whenever possible
- avoid moving the vehicle whenever possible

Note
A variety of electrical connectors are used on this vehicle, **ALWAYS** use the **VW 1594** adaptor kit to connect test instruments to these connectors.

CAUTION

Before disconnecting a customers battery; **ALWAYS** ask for the radio code (if equipped with an anti-theft radio).

Idle speed, checking

Note
Idle speed, ignition timing and CO-content are **NOT** adjustable.

Requirements
- engine coolant temperature 80°C (176°F) minimum
- all electrical consumers switched **OFF**
- A/C switched **OFF**
- exhaust system must be leak tight

- start engine and let idle
- activate **DTC MEMORY** (Fault memory) and then erase, see Repair Group D2 for additional information
- connect **EPA 105** CO tester to CO tap tube using **VAG 1363/3** adaptor

After connecting the **VAG 1551** Scan Tool, the display will alternate between the two following operating modes:

1 – Rapid Data Transfer	
2 – Blink Code Output	

- press **1** button to select operating mode **"Rapid Data Transfer"**
  - display will appear as shown

Rapid data transfer	HELP
Insert address word XX	

- press **0** and **1** buttons to select address word **"Engine Electronics"**
  - display will appear as shown

Rapid data transfer	Q
01 – Engine electronics	

- press **Q** button to enter input
- press ➡ button
  - display will appear as shown

Rapid data transfer	HELP
Select function XX	

- press buttons **0** and **4** to select function **04** **"basic setting"**
  - display will appear as shown

Rapid data transfer	Q
04 – Start basic setting	

- press **Q** button to enter input
  - display will appear as shown

Start basic setting	HELP
Input display group number XX	

- press **0** and **1** buttons to select display group **01**
  - display will appear as shown

Perform basic adjustment	Q
Input display group number 01	

- press **Q** button to enter input
  - display fields **1** through **4** will appear as shown

System in Basic Adjustment 1			➡
1	2	3	4

Note
When this mode is selected; the EVAP solenoid valve I (**N 80**) is closed.

- observe display field **2**
  - when the temperature exceeds 80°C (176°F) continue to next step

- raise engine speed briefly then let engine idle for 2 minutes
- read display value for engine RPM and ignition timing point as follows
  - display field **1** = engine RPM

    must be 650 to 750 RPM's
  - display field **4** = ignition timing point

    must be 6° BTDC and fluctuating

- observe CO content on tester
  - checking value must be 0.3 to 1.5 Vol.%

- press ➡ button
- press **0** and **6** buttons to select function **06** **"Ending output"**
  - display will appear as shown

Rapid data transfer	HELP
Select function XX	

- press **Q** button to enter input

Rapid data transfer	Q
06 – Ending output	

If specified RPM value not obtained

- check gas pedal for ease of movement, correct as necessary
- check dashpot, see section 24-420 for checking/adjusting
- check **TP** (Throttle Position Sensor), see section 24-450 for checking/adjusting
- check **IAC** valve (Idle Air Control), see section 24-410 for checking
- check **EGR** valve, see Repair Group 26

If all specified values are obtained

- check intake system for air and vacuum leaks

Idle air control (IAC) valve (N 71), checking

Requirement
- **VAG 1551** Scan Tool connected

Function, checking

- see Repair Group 01 (Fuel Injection and Ignition) for Output Diagnostic Test Mode (**DTM**) checking

Triggering, checking

- switch **OFF** ignition
- disconnect harness connector from **IAC** valve
- connect **US 1115** LED tester to terminals **1** and **2** of harness connector
- switch **ON** ignition and perform Output DTM again until **IAC** valve (**N 71**) is triggered
  - LED tester must flash

If **NO**

- perform Motronic system electrical checks, see Repair Group D2

Resistance, checking

- remove ignition wire guide from cylinders **1**, **3** and **5**
- disconnect harness connector from **IAC** valve (**N 71**)
- switch **Fluke 83** multimeter to resistance range
- connect multimeter between terminals **1** and **2** of **IAC** valve connector using jumper wires from **VW 1594** adaptor kit
  - must be between 10 and 20 Ohms

Dashpot, checking and adjusting

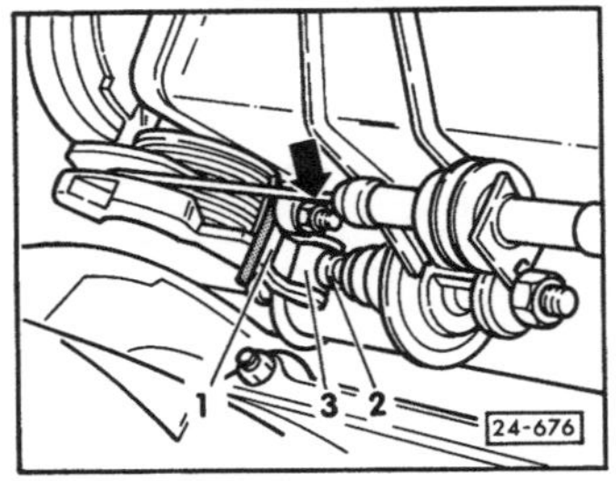

Checking

- open and close throttle until dashpot cam **2** contacts roller **3**
- in this position measure gap between limiter screw (**arrow**) and throttle lever **1** using feeler gauge or drill bit
  - must be 2.5 to 3.5 mm

Adjusting

- place 3.0 mm drill between limiter screw (**arrow**) and throttle lever **1**
- adjust dashpot such that cam **2** just touches roller **3**
- tighten dashpot with jam nut to prevent turning

Acceleration, full throttle enrichment and fuel deceleration shut off, checking

Requirement
- **VAG 1551** Scan Tool connected

- start engine and let idle
- press 1 button to select operating mode **"Rapid Data Transfer"**
  - display will appear as shown

Rapid data transfer	HELP
Enter address word XX	

- press **0** and **1** buttons to select address word **"Engine Electronics"**
  - display will appear as shown

Rapid data transfer	Q
01 – Engine electronics	

- press **Q** button to enter input
- press ➡ button
- press **0** and **8** buttons to select **08** **"Read test value block"**
- press **Q** button to enter input
  - display will appear as shown

Read test value block	HELP
Insert display group number XX	

- press **0** and **4** buttons to select display group **04**
  - display will appear as shown

Read test value block	Q
Insert display group number 04	

- press **Q** button to enter input
  - display will appear as shown

Read test value block 4			
1	2	3	4

- continue with checking only if engine coolant temperature exceeds 80°C (176°F)
- let engine idle for approx. 2 minutes
- observe display value in field **4**
  - display value must be **00010** at idle

- raise engine speed to approx. 3000 RPM using accelerator pedal
  - display value must be **10100** (briefly)

- release accelerator pedal
  - display value must be **00011** (briefly) and then become **00010** at idle

Note
If anything else appears in display, see Repair Group 01 (Fuel Injection and Ignition) for additional information

- press ➡ button
- press **0** and **6** buttons to select function **06** **"Ending output"**
  - display will appear as shown

Rapid data transfer	HELP
Select function XX	

- press **Q** button to enter input

Rapid data transfer	Q
06 – Ending output	

If specified values not obtained

- check TP (Throttle Position Sensor) see section 24-450

Engine coolant temperature (ECT) sensor, checking

Requirements
- engine cold
- **VAG 1551** Scan Tool connected

- ■ start engine and let idle
- ■ press **1** button to select operating mode **"Rapid Data Transfer"**
  - display will appear as shown

Rapid data transfer	HELP
Insert address word XX	

- ■ press **0** and **1** buttons to select address word **"Engine Electronics"**
  - display will appear as shown

Rapid data transfer	Q
01 – Engine electronics	

- ■ press **Q** button to enter input
- ■ press ➡ button
- ■ press **0** and **8** buttons to select function **08** **"Read test value block"**
- ■ press **Q** button to enter input
  - display will appear as shown

Read test value block	HELP
Insert display group number XX	

- ■ press **0** and **1** buttons to select display group **01**
  - display will appear as shown

Read test value block	Q
Insert display group number 01	

- ■ press **Q** button to enter input
  - display will appear as shown

Read test value block 1			➡
1	2	3	4

- ■ observe display value in field **2**
  - engine temperature must be uniform and increase without interruption

- ■ press ➡ button
- ■ press **0** and **6** buttons to select function **06** **"Ending output"**
  - display will appear as shown

Rapid data transfer	HELP
Select function XX	

- ■ press **Q** button to enter input

Rapid data transfer	Q
06 – Ending output	

Note
If engine malfunctions occur during certain temperature ranges and the display value does not increase without interruption, temperature signal has been interrupted and the **ECT** sensor (**G 62**) must be replaced.

If display values do not change

- ■ perform electrical check, see Repair Group D2

Resistance, checking

- ■ switch **OFF** ignition
- ■ disconnect **ECT** harness connector from blue **ECT** sensor
- ■ switch **Fluke 83** multimeter to resistance range
- ■ connect multimeter to terminals of **ECT** sensor using jumper wires from **VW 1594** adaptor kit

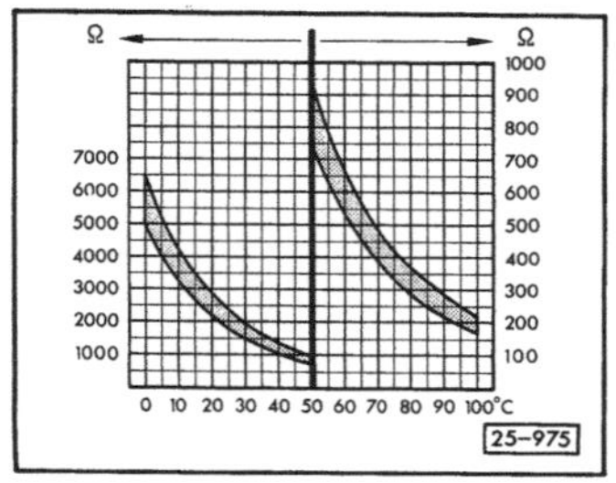

Throttle position (TP) sensor, checking

Vehicles with automatic transmissions:

- the harness with the 4 pin connector leads to the transmission control module (**J 217**)
- if the throttle position (**TP**) sensor is replaced, the transmission control module (**J 217**) must **"learn"** the voltage value of the new potentiometer. See Repair Group 01 (Automatic Transmission) for On Board Diagnosis procedures to return the control module to its basic setting

Requirement
- **VAG 1551** Scan Tool connected

■ switch **ON** ignition
■ press **1** button to select operating mode **"Rapid Data Transfer"**
- display will appear as shown

Rapid data transfer	HELP
Insert address word XX	

▶ ■ press **0** and **1** buttons to select address word **"Engine Electronics"**
- display will appear as shown

Rapid data transfer	Q
01 – Engine electronics	

▶ ■ press **Q** button to enter input
■ press ➡ button
■ press **0** and **8** buttons to select function **08** **"Read test value block"**
■ press **Q** button to enter input
- display will appear as shown

Read test value block	HELP
Insert display group number XX	

▶ ■ press **0** and **3** buttons to select display group **03**
- display will appear as shown

Read test value block	Q
Insert display group number 03	

▶ ■ press **Q** button to enter input
- display will appear as shown

Read test value block 3			
1	2	3	4

▶ ■ observe "throttle angle" in channel **3** of display
- displayed value must be 9 to 19 ∠°

■ open throttle slowly
- displayed value must be uniform and increase without interruption
- with throttle completely open displayed value must be 94 to 101∠°

■ press ➡ button

If **NO**

■ press **0** and **6** buttons to select function **06** **"Ending output"**
- display will appear as shown

Rapid data transfer	HELP
Select function XX	

▶ ■ press **Q** button to enter input

Rapid data transfer	Q
06 – Ending output	

▶ ■ perform electrical checks, see Repair Group 01 (Fuel Injection and Ignition)
■ if necessary replace **TP** (throttle position) sensor (**G 69**)

Fuel pressure regulator and residual pressure, checking

> **WARNING**
>
> Fire hazard. Do **NOT** smoke or have anything in area that can ignite fuel!

> **CAUTION**
>
> Fuel under pressure! Cover the fuel rail test measuring point with a rag to protect against fuel spray when removing the sealing screw.

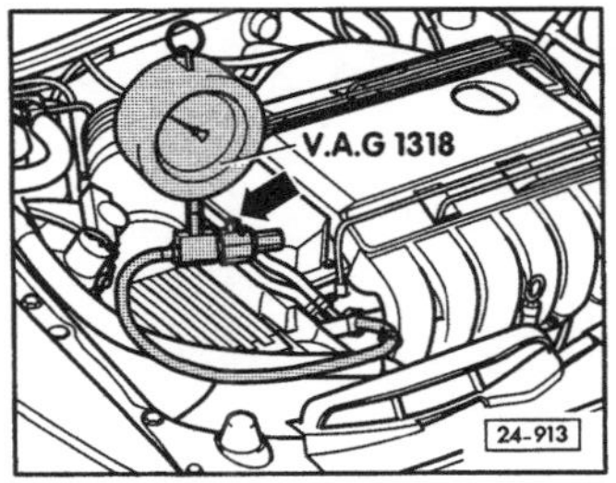

- connect **VAG 1318** pressure tester to fuel rail using adaptor **VAG 1318/10**

> **CAUTION**
>
> Shut-off valve for **VAG 1318** pressure tester must be closed. The lever (**arrow**) must be perpendicular to the direction of flow.

- start engine and let idle
- observe fuel pressure on gauge
  - must be approx. 3.5 bar gauge pressure

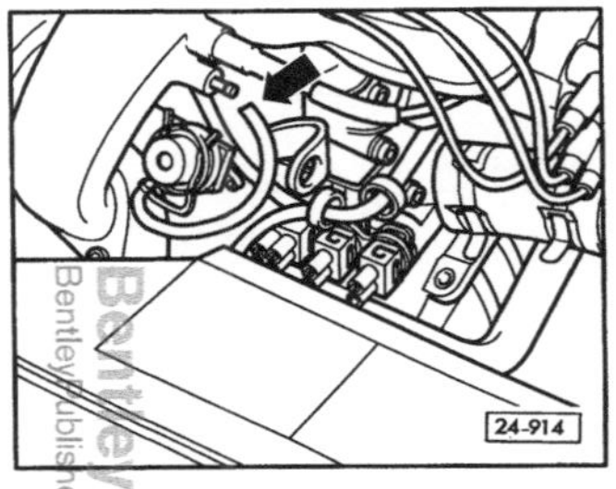

- disconnect hose from fuel pressure regulator (**arrow**)
  - fuel pressure must increase to approx. 4.0 bar gauge pressure, after 10 minutes, gauge pressure must not be less than 2.5 bar

If after 10 minutes gauge pressure is less than 2.5 bar

- check for leaks in system
- start engine and after fuel pressure builds up, switch **OFF** ignition
- at same time, clamp blue return line tightly and observe pressure drop on pressure gauge

If pressure does **NOT** drop

- fuel pressure regulator is defective

If pressure drops again

- check if:
  - line connections are tight
  - O-rings or seals on fuel distributor leak
  - fuel injectors leak
  - fuel pump check valve leaks
  - **VAG 1318** Pressure tester leaks (at shut-off valve)

Note
Drain any residual fuel from the pressure tester before disconnecting it from the fuel rail.

Fuel injectors, checking

> **WARNING**
> Fire hazard. Do **NOT** smoke or have anything in area that can ignite fuel!

Removing

- remove ignition wire guides
- remove suction hose from hot wire Mass Air Flow (**MAF**) sensor
- disconnect hose to EVAP solenoid valve at throttle body

- disconnect harness connector from TP (throttle position) sensor
- disconnect IAC (idle AIR control) valve
- disconnect accelerator cable from throttle
- disconnect coolant hoses from throttle body and plug off
- disconnect connector hose from heated tube

- unclip fuel lines at cylinder head cover and remove from fuel rail
- disconnect vacuum hose from fuel pressure regulator
- remove intake manifold, upper section
- unclip lines to fuel injectors from wiring guide
- remove wiring guide
- remove fuel injectors and fuel rail as an assembly

Note
If the injectors are too tight within the holes of the lower section of the intake manifold; carefully pry under the valve body using wrench **SW 16**.

Installing

- install in reverse order of removal

Note
Use care when installing the fuel injectors to avoid damaging the O-rings. Lightly lubricate the O-rings with oil to ease assembly.

Fuel injectors, checking

- remove upper section of intake manifold (see previous section)

Electrical, checking

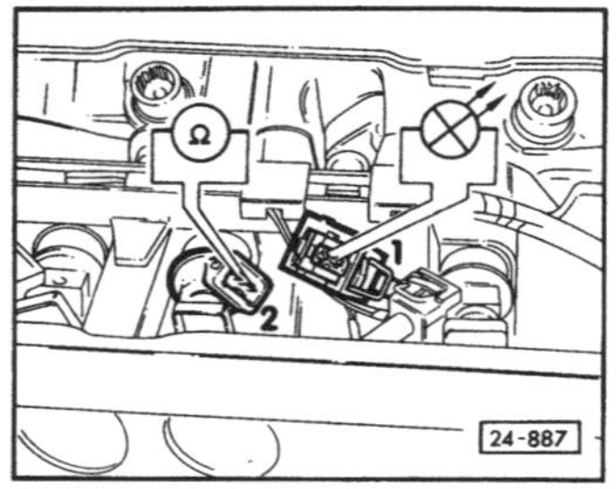

- disconnect harness connectors from all fuel injectors
- switch **Fluke 83** multimeter to resistance range
- connect multimeter to terminals of fuel injector connector **2** using jumper wires from **VW 1594** adaptor kit
  - must be 15 to 21.5 Ohms

If 0 Ohms (continuity)

- check for short circuit in wiring using wiring diagram, repair or replace as necessary

If wiring **OK**

- replace fuel injector(s)
- disconnect harness connector from ignition coil power output stage

- check voltage supply to each injector
- connect **US 1115** LED tester between terminals of harness connector **1** using jumper wires from **VW 1594** adaptor kit
- operate starter
  - LED tester must flicker

If **NO**

- re-connect the harness connector(s) to fuel injector(s)
- check for open circuit in wiring between fuel injectors (**N 30, N 31, N 32, N 33, N 83, N 84**) and engine control module using wiring diagram, repair or replace as necessary, see Repair Group 01 (Fuel Injection and Ignition) Test steps **8** through **13**

Checking spray pattern and for leaks

- remove fuel rail assembly complete with fuel injectors

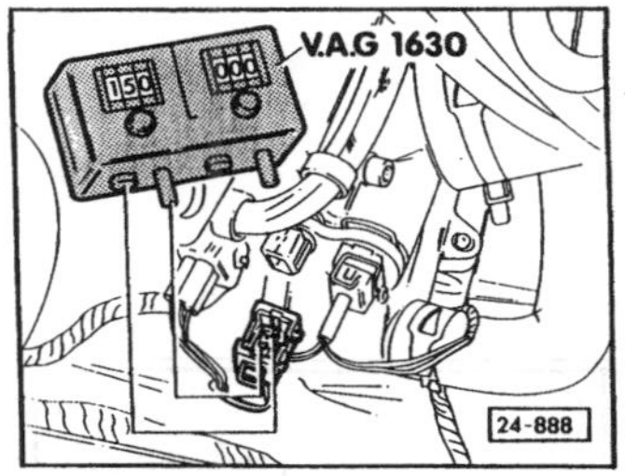

- disconnect harness connector from blue engine coolant temperature (ECT) sensor (**G 62**)
- preset digital potentiometer **VAG 1630** to 15,000 Ohms
- connect the **ECT** harness connector to the **VAG 1630** potentiometer using jumper wires from the **VW 1594** adaptor kit

Note
If the **VAG 1630** potentiometer is not available; use adaptor **VAG 1490** by connecting the 15,000 Ohm side directly to the **ECT** harness connector.

- fuel injectors remain electrically connected
- feed and return line remain connected to fuel rail

Note
There is a tool (**VAG 1602**) that can be used to measure the fuel quantity output of the fuel injectors, simultaneously. If a **VAG 1602** is not available, an acceptable method would be to substitute 6 separate beakers to collect the fuel and a single graduated beaker to measure each of the accumulations individually.

- adjust location of measuring tubes in **VAG 1602** to align with fuel injectors on fuel rail
- insert injectors into each of the **VAG 1602** measuring tubes
- operate starter for several seconds and observe spray pattern
  - spray pattern must be the same for all injectors

- switch **ON** ignition for approx. 5 seconds and check for injector leakage
  - no more than 2 drops per injector is allowed

- empty **VAG 1602** (or equivalent) of any residual fuel

Note
When installing the fuel injectors, be sure that the O-rings in the fuel rail are not damaged. Always replace damaged O-rings.

- activate **DTC MEMORY** (Fault memory) and then erase, see Repair Group 01 (Fuel Injection and Ignition)
  - any faults that are generated by checking the (**G 62**) ECT sensor will then be erased

THIS PAGE INTENTIONALLY LEFT BLANK

Intake air temperature (IAT) sensor, checking

Requirement
- **VAG 1551** Scan Tool connected

- switch **OFF** ignition
- disconnect harness connector from **IAT** sensor
- remove **IAT** sensor from intake manifold
- reconnect sensor to harness connector
- plug hole in manifold where sensor was removed

- start engine and let idle
- press **1** button to select operating mode **"Rapid Data Transfer"**
  - display will appear as shown

Rapid data transfer	HELP
Insert address word XX	

- press **0** and **1** buttons to select address word **"Engine Electronics"**
  - display will appear as shown

Rapid data transfer	Q
01 – Engine electronics	

- press **Q** button to enter input
- press ➡ button
- press **0** and **8** buttons to select function **08** **"Read test value block"**
- press **Q** button to enter input
  - display will appear as shown

Read test value block	HELP
Insert display group number XX	

- press **0** and **3** buttons to select display group **03**
  - display will appear as shown

Read test value block	Q
Insert display group number 03	

- press **Q** button to enter input
  - display will appear as shown

Read test value block 3			
1	2	3	4

- read value displayed in channel **4** "Intake manifold temperature" (approx. ambient temperature)
- spray **IAT** sensor with a commercial cooling spray
  - displayed temperature must decrease

- press ➡ button
- press **0** and **6** buttons to select function **06** **"Ending output"**
  - display will appear as shown

Rapid data transfer	HELP
Select function XX	

- press **Q** button to enter input

Rapid data transfer	Q
06 – Ending output	

If displayed value does not decrease

- perform electrical checks, see Repair Group 01 (Fuel Injection and Ignition),Test step **22**
- if necessary, replace air intake temperature (**IAT**) sensor

Index

Engine Code – PG

Evaporative system leakage
- troubleshooting **26.5**

Exhaust system
- component layout **26.2**

Heated crankcase breather
- checking **26.4**

Engine Code AAA

EGR system
- component layout **26.8**
- valve checking **26.9**

Exhaust system component
- removal and installation **26.7**

Exhaust System – Emission Controls

Note

- after working on the exhaust sytem, it should be free of tension with sufficient clearance between itself and the vehicle body to allow clamp loosening and muffler alignment
- the hanging load should be distributed equally over the entire set of retaining rings
- always replace all self locking nuts

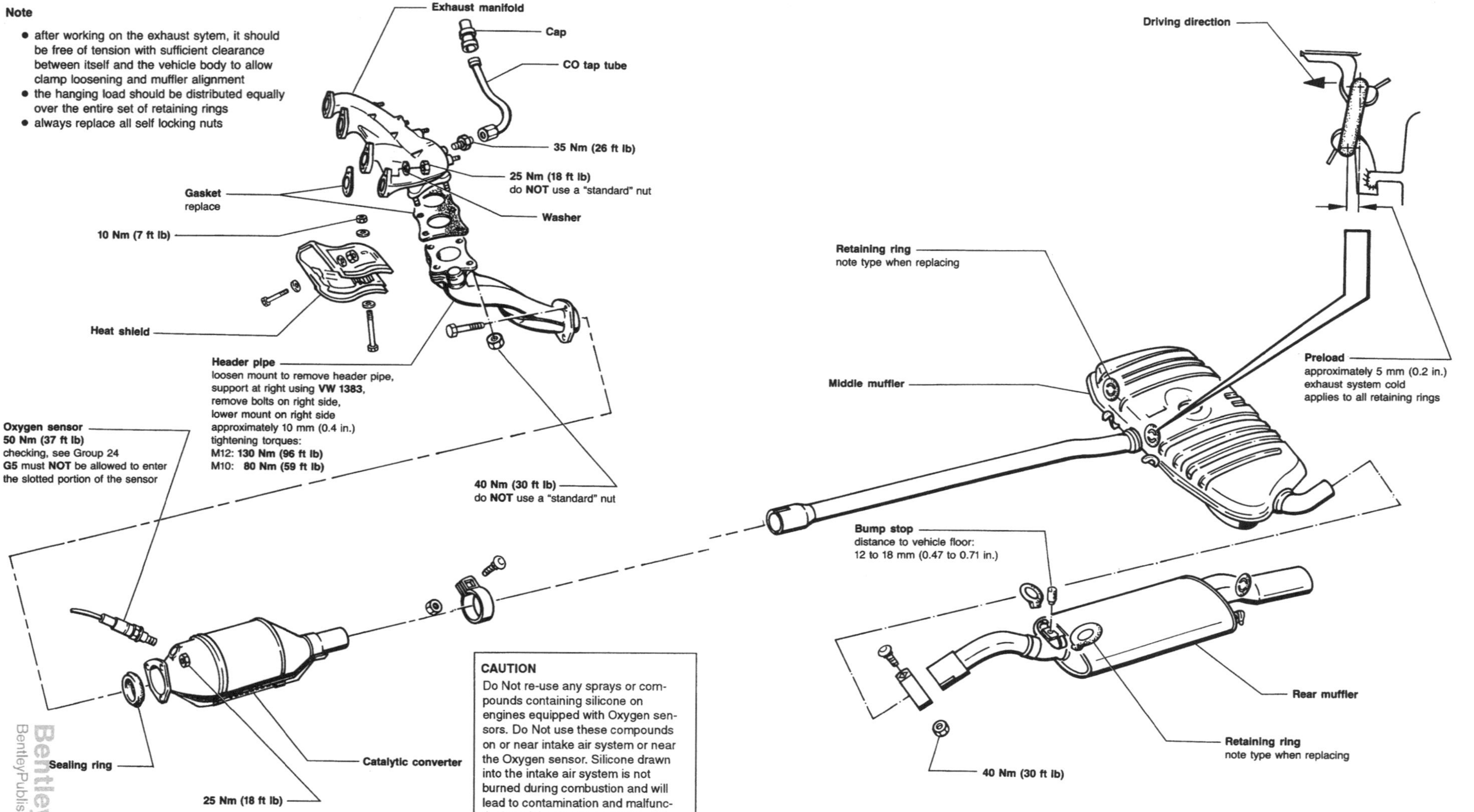

CAUTION
Do Not re-use any sprays or compounds containing silicone on engines equipped with Oxygen sensors. Do Not use these compounds on or near intake air system or near the Oxygen sensor. Silicone drawn into the intake air system is not burned during combustion and will lead to contamination and malfunctioning of the Oxygen sensor.

Heated crankcase breather, checking

- disconnect heated crankcase breather harness connector
- switch multimeter **US 1119** to resistance range
- connect multimeter to terminals of heater unit (**arrow**)
  - must be between 4 and 17 ohms at approximately 25°C (77°F)

If **NO**

- replace heater unit

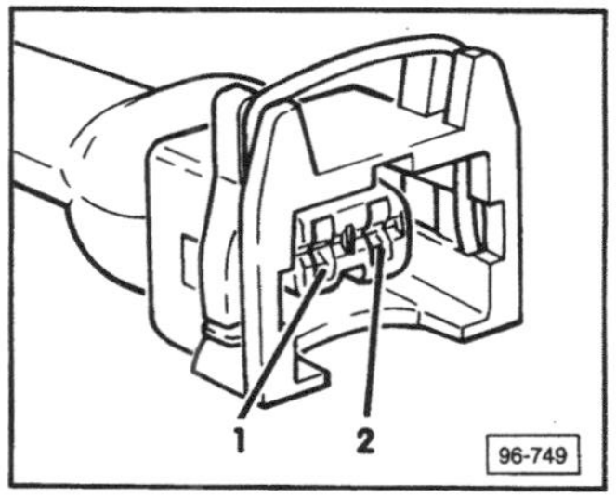

If **OK**

- switch multimeter **US 1119** to 20 volt range
- connect to terminals **1** and **2** of harness connector
- switch **ON** ignition
  - must be approximately 12 volts

If **NO**

- check wiring and connectors from harness connector to battery using wiring diagram, repair as necessary

Note

The heater operates whenever the ignition is switched ON, there are no relays or control units in this circuit.

Evaporative system leakage, troubleshooting

> **WARNING**
>
> These tests involve the use of equipment that contain mercury.
>
> Mercury is a toxic and hazardous material.
> - exercise extreme care when handling
> - do **NOT** allow to come in contact with eyes, nose, skin, etc.
> - when not in use, store carefully in a properly designated area
> - be sure both valves are closed when storing tester

Check these first:

- fuel filter cap securely closed
- fuel level at least 2/3 full (otherwise it will require excessive pumping to pressurize the system)

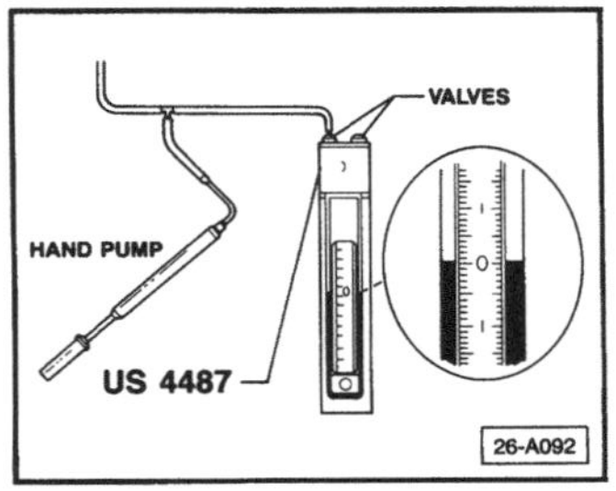

- disconnect small hose from charcoal canister
- vertically connect slack tube tester **US 4487** at disconnected hose from charcoal canister
- open both slack tube valves a 1/2 turn
- move scale on tester such that the zero line is even with the tops of the mercury columns

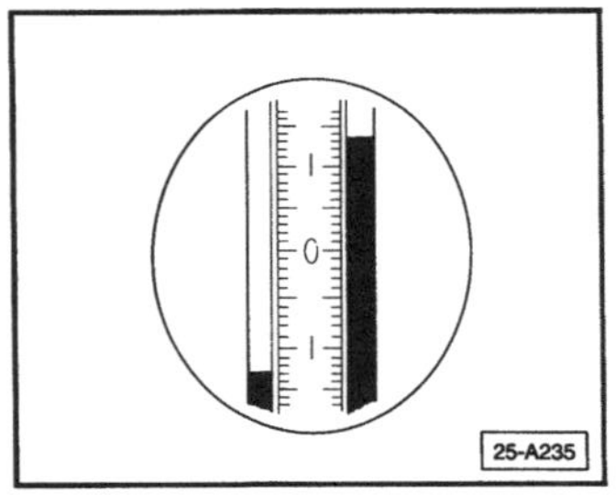

- pressurize the system to 1.3 inches of mercury, using the handpump

Note

If the system reaches the 1.3 inches of mercury soon after you begin pumping, there is reason to suspect that the gravity valve or hoses to it, might be pinched or blocked.

With 2/3 of a tank of fuel it should take considerably more pumping to achieve the 1.3 inches of mercury.

After reaching 1.3 inches of mercury on scale, wait 5 minutes:

- system **OK** if pressure is 1.2 inch of mercury or greater

If pressure drops below 1.2 inch of mercury:

- check fuel filler cap for leakage using soap solution, replace if necessary
- pressurize system to 1.3 inches of mercury, wait 5 minutes

If pressure still drops below 1.2 inch of mercury:

- disconnect hose from top of gravity valve (between gravity valve and charcoal canister) and plug hose
- pressurized system to 1.3 inches of mercury, wait 5 minutes

If pressure still drops below 1.2 inch of mercury:

- the leak is between the gravity valve and charcoal canister

If pressure does **NOT** drop below 1.2 inch of mercury:

- re-connect gravity valve and re-pressurize system
- check hoses/connections at expansion tanks and fuel tank by applying a soap solution
- seal, repair or replace as necessary

After you have repaired the leak(s):

- repeat test to verify that you have fixed all of the leakage
- after tests are completed, close both valves on the tester then properly store

THIS PAGE INTENTIONALLY LEFT BLANK

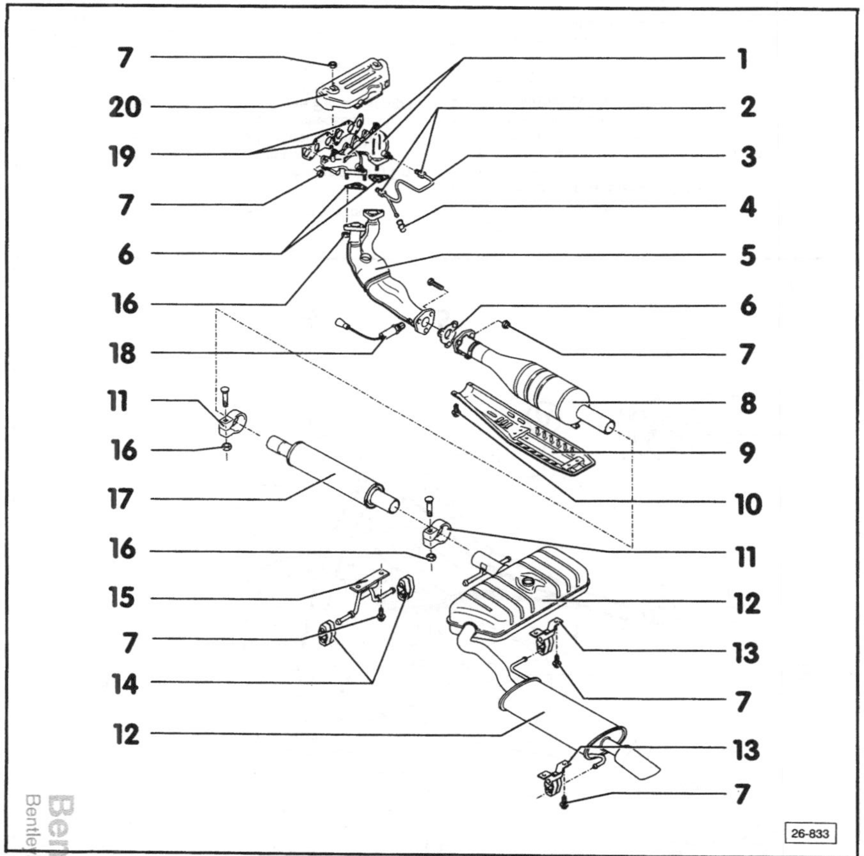

Note
- after working on the exhaust system ensure that the system (cold) is not preloaded
- check for sufficient clearance between the exhaust system and the chassis
- if necessary, loosen the appropriate clamps to allow the mufflers and pipes to be turned and pushed in a longitudinal direction until sufficient clearance is obtained
- the hanging load must be distributed equally over the entire set of retaining rings
- the front muffler assembly is supplied as a single part. For removal and installation; lower the rear axle assembly, see Repair Group 42 for additional information
- always replace self locking nuts

1 – **Exhaust manifold**
in two sections

2 – **30 Nm (22 ft lb)**

3 – **CO tap tube**

4 – **Cap for CO tap tube**
check for secure fit

5 – **Front exhaust pipe**

6 – **Gasket**
always replace

7 – **25 Nm (18 ft lb)**
always replace

8 – **Catalytic converter**

9 – **Heat shield**
note installed position

10 – **10 Nm (7 ft lb)**

11 – **Clamp**

12 – **Center muffler**

13 – **Hanger bracket for rear retaining ring**
note installed position

14 – **Retaining ring**
note style when replacing

15 – **Hanger bracket for mid-exhaust retaining rings**
note installed position

16 – **40 Nm (30 ft lb)**

17 – **Front muffler**
installed position

18 – **Oxygen sensor (G 39)**
- **50 Nm (37 ft lb)**
- for checking see Repair Group 01 (Fuel Injection and Ignition)
- lubricate threads with G5 but do not allow G5 to contact or enter the slotted portion of the sensor tip

19 – **Exhaust manifold gasket**
- in 2 pieces
- always replace
- note installed position

20 – **Shield**

CAUTION
Do Not re-use any sprays or compounds containing silicone on engines equipped with Oxygen sensors. Do Not use these compounds on or near intake air system or near the Oxygen sensor. Silicone drawn into the intake air system is not burned during combustion and will lead to contamination and malfunctioning of the Oxygen sensor.

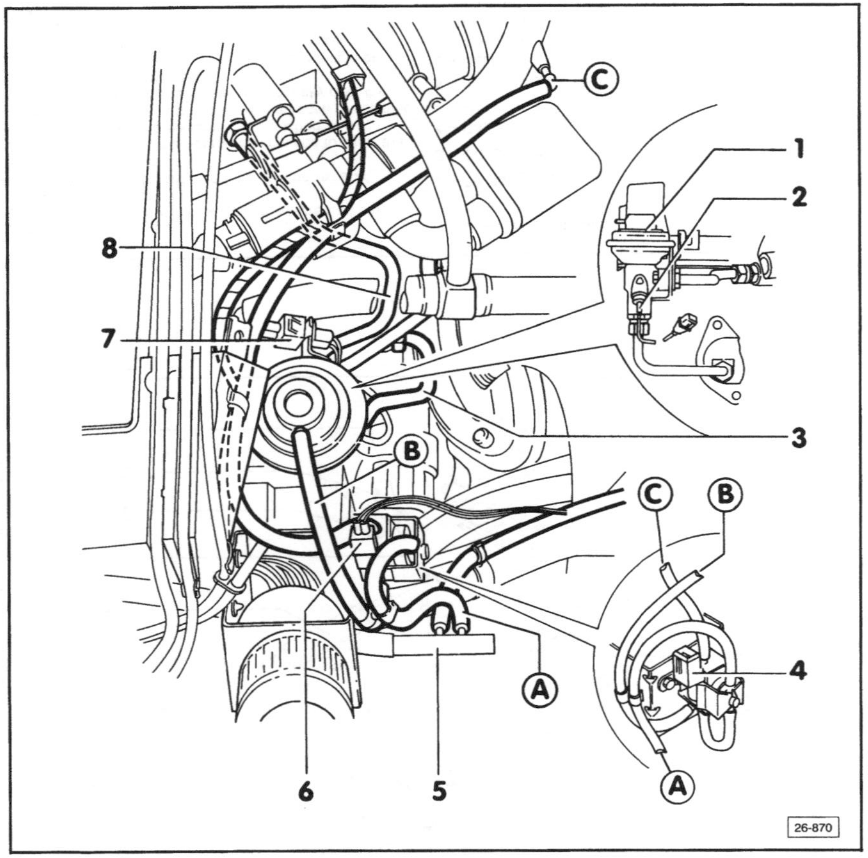

Note
The function of the EGR system is checked via the Motronic On Board Diagnostic system.

- components identified by an * can be checked using On Board Diagnostic, see Repair Group 01 (Fuel Injection and Ignition) for additional information
- components identified by two ** can be checked using Output DTM (Diagnostic Test Mode) see Repair Group 01 (Fuel Injection and Ignition) for additional information
- always replace any seals or O-rings when performing assembly work
- ensure that all vacuum lines and hose connections are tight
- ensure that vacuum lines are not plugged or bent

1 – **EGR valve**
checking see page 26.9

2 – **EGR temperature Sensor (G 98)***
- **20 Nm (15 ft lb)**
- for electrical checking, see Repair Group 01 (Fuel Injection and Ignition)

3 – **Exhaust inlet tube**

4 – **EGR frequency valve (N 18)****
for checking, see Repair Group 01 (Fuel Injection and Ignition) Output DTM and electrical checks

5 – **Connector port**
for intake manifold pressure measurement

6 – **Harness connector for EGR frequency valve (N 018)**

7 – **Harness connector for (G 098) temperature sensor**

8 – **Exhaust outlet tube**

EGR valve, checking

Requirement
- engine oil temperature 50°C (122°F) minimum

■ start engine and let idle

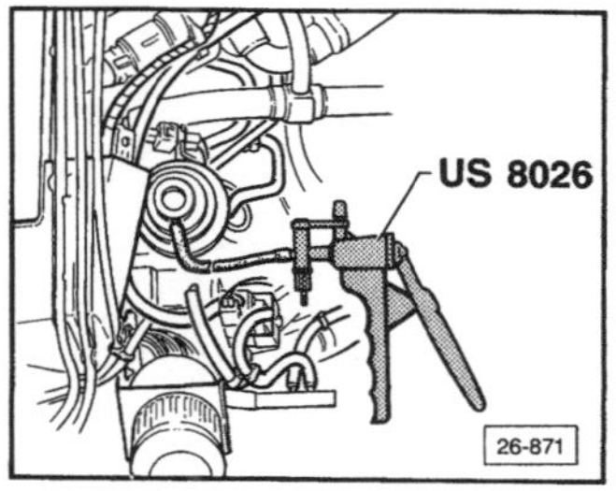

■ disconnect EGR valve vacuum hose (from the connector hose)
■ connect **US 8026** hand vacuum pump to EGR valve
■ operate vacuum pump
- idle behavior must deteriorate

If no change
■ check exhaust inlet and outlet tubes for flow, clean or replace as necessary
■ check exhaust tube connections for tightness, tighten if necessary

If **OK**
■ replace EGR valve

■ activate Fault memory and then erase, see Repair Group 01 (Fuel Injection and Ignition) for additional information
- checking and testing can generate a Diagnostic Trouble Code (DTC) for this component which will then be stored in the DTC memory

THIS PAGE INTENTIONALLY LEFT BLANK

THIS PAGE INTENTIONALLY LEFT BLANK

Index

Battery
- charging **27.4**
- checking **27.3**
- jump starting **27.5**
- load testing **27.6**
- removing/installing **27.2**
- troubleshooting **27.7**
- voltage test **27.6**

Belt
- routing, vehicles w/AC **27.13**
- routing, vehicles w/o AC **27.12**
- tensioning **27.12**

Cruise control system
- checking/adjusting **27.15**
- component layout **27.14**
- vacuum system, leak testing **27.17**

Generator components
- layout **27.10**
- removing/installing **27.11**

Generator (GEN) V – belt pulley
- removing/installing **27.12**

Starter motor
- removing/installing **27.8**

Battery, removing and installing

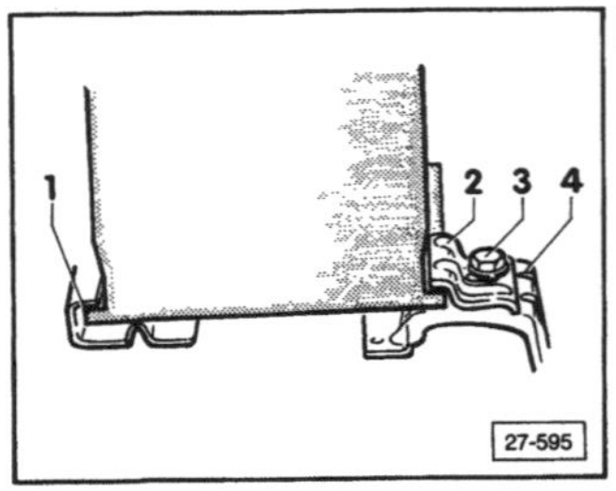

Removing

- first remove battery negative cable from negative terminal
- remove battery positive cable from positive terminal
- remove clamping bracket **2**

Installing

- insert battery into tray, ensuring that lip on battery fits into channel **1**
- install clamping bracket **2** to frame member **4**, ensuring that bracket seats properly on front lip of battery
- install bolt **3**
  - tighten bolt to: 20 Nm (15 ft lb)

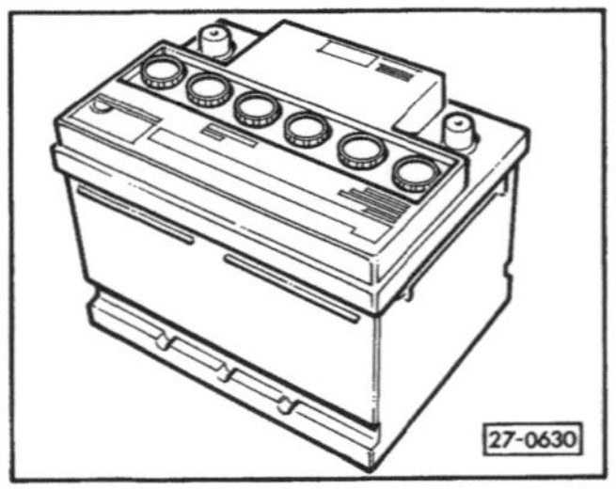

- install a cover on this style of battery
  - normal maintenance style battery with vented caps

Note

Reinstall this cover after removal, and replace it if damaged.

The battery cover is designed to prevent electrolyte vapor and water from accumulating during normal vehicle operation. If left uncovered these liquids can accumulate and spill, damaging components and finish.

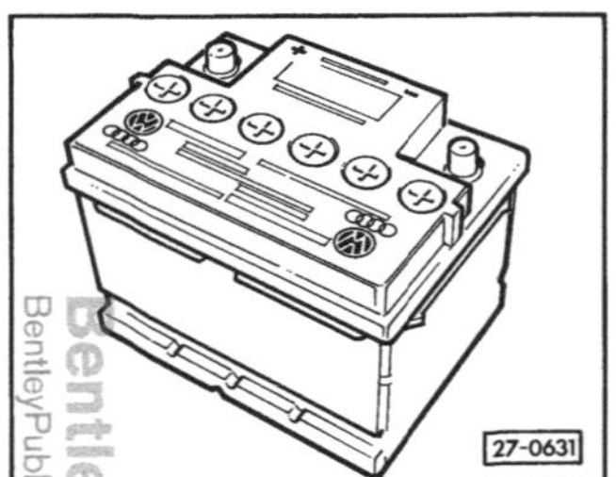

- a cover is **NOT** required on this style of battery
  - low maintenance style battery with sealed caps and side vents

CAUTION

Disconnecting the battery may cause the base setting of Electronic Control Modules having On Board Diagnostic (OBD) (VAG 1551) capabilities to be erased. This may result in vehicle driveability problems.

Perform OBD function "Basic Setting" on Electronic Control Modules having OBD (VAG 1551) capabilities whenever a battery has been disconnected or replaced during repair.

WARNING

Battery acid can cause severe burns. When working with electrolyte always wear goggles, rubber gloves, and apron. If electrolyte is spilled on skin or clothing, flush at once with large quantities of water. If it gets into eyes, immediately flush with large quantities of water for several minutes and call a doctor.

Batteries produce explosive gases. Keep flames and sparks away from batteries. Do not smoke near batteries.

Battery condition, checking

Weak battery can be caused by:
- Generator belt slipping (replace belt when glazed)
- ground straps corroded, loose or broken
- Generator or voltage regulator defective
- Generator warning light bulb burned out
- poor ground connection between warning light socket and circuit board

Battery terminals and cables should be coated with acid resistant grease and cleaned to prevent corrosion and ensure good contact.

Specific gravity	% of charge
1.265	100%
1.225	75%
1.190	50%
1.155	25%
1.120	0%

Visual checks

- check for
  - cracked or leaking case
  - loose or corroded terminals
  - low electrolyte level. If electrolyte is low, refill with distilled water and fast charge battery for several minutes after cells begin gassing (boiling)

CAUTION

Do not expose fully discharged battery to freezing temperature because battery will be damaged by freezing.

Hydrometer test (specific gravity)
(for batteries with *removable caps*)

- test electrolyte in all cells
  - average specific gravity should be at least 1.225
- correct for electrolyte temperature if necessary
  - for every 10°F **above** 80°F **add** .004 to hydrometer reading
  - for every 10°F **below** 80°F **subtract** .004 from hydrometer reading

If specific gravity is **above** 1.225
- load test battery

If specific gravity is **below** 1.225
- recharge battery

If specific gravity varies by more than .050 between cells
- replace battery

Battery, charging

Follow battery charger manufacturer's instructions. Before testing a battery that has been charged, load battery with 15 amps for 1 minute to remove surface charge.

Battery with removable caps

If water is added, fast charge battery for a few minutes after battery begins gassing (boiling). Slow charging current should be approximately 10% of battery capacity. Fast charging should be approximately 80-90% of battery capacity. Remove cell caps while charging.

> **WARNING**
>
> Gases given off during charging are explosive. Do **NOT** smoke or allow sparks or flame near a charging battery.
>
> Battery charger **MUST** be turned off when connecting or disconnecting cables on battery.
>
> "Boosting" a sulfated battery at a high charging rate can cause an explosion!

> **CAUTION**
>
> Do not allow battery charging voltage to exceed 17 volts.
>
> If battery begins gassing (boiling) violently when charging, **REDUCE** charging rate.
>
> Do **NOT** disconnect battery while engine is running.

Specific gravity	Fast charge up to
1.150 or less	1 hour
1.150 to 1.175	3/4 hour
1.175 to 1.200	1/2 hour
1.200 to 1.225	1/4 hour
above 1.225	slow charge **ONLY** to 1.250-1.280

- charge battery according to following table:

Sealed batteries

Only **slow** charge sealed batteries. Sealed batteries, sometimes called "maintenance free," will not accept high rate of charge, making it necessary to charge it for up to twice as long as battery with removable caps. Also, voltage reading will **NOT** increase as rapidly as when charging battery with removable caps.

Battery, jump starting

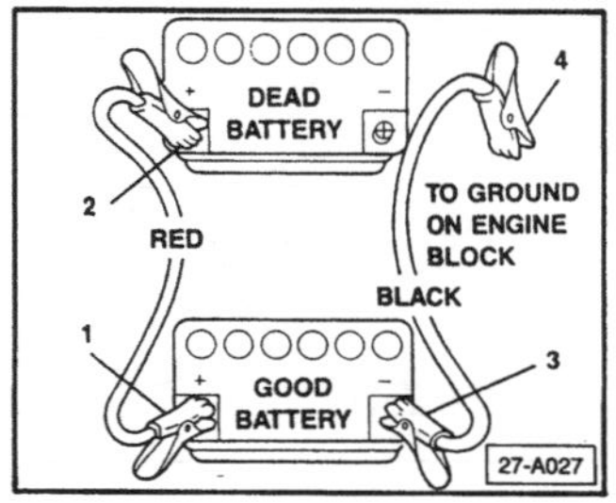

- car with good battery must **NOT** be running when connecting jumper cables

- connect jumper cables in following order:
  - **1** — one end of positive cable to + post of good battery
  - **2** — other end of positive cable to + post of dead battery
  - **3** — one end of negative cable to – post of good battery
  - **4** — other end of negative cable to engine block of car with dead battery

- start car which has good battery first
- next start car which has dead battery
- disconnect jumper cables in reverse order

Voltage	% of charge
12.6 or more	100%
12.4	75%
12.2	50%
12.0	25%
11.7 or less	0%

Battery voltage test — engine not running (for sealed batteries only)

- turn headlights on high beam for 1 minute to remove surface charge
- disconnect battery ground strap
- check battery voltage using multimeter **US 1119**
  - 12.4 volts minimum

If above
- load test battery

If below
- charge until 12.4 volts minimum

Load test

Note

Before load testing, battery must be at least 75% charged.

- connect load tester according to manufacturer's instructions
- load battery to 3 times amp/hour rating **or** 1/2 0°F cold cranking current rating
- wait 15 seconds and read voltage
  - battery voltage at room temperature should be 9.6 volts minimum

If **NO**
- replace battery

Note

If battery is colder than room temperature, voltage under load will be **lower**.

If battery must be tested cold:
- use table

Approximate electrolyte temp. °F (°C)	Minimum acceptable voltage under load
60 (16)	9.5
50 (10)	9.4
40 (4)	9.3
30 (−1)	9.1
20 (−7)	8.9
10(−12)	8.7
0(−18)	8.5

Battery discharges — troubleshooting

Current draw suspected

Check these first:
- fully charged battery
- clock disconnected

Battery discharges

Possible causes
- corroded terminals
- worn insulation
- internal short circuits in electrical equipment

- disconnect battery ground (−) strap
- connect test light between negative (−) battery terminal and ground strap
- turn off all electrical consumers
- close all doors

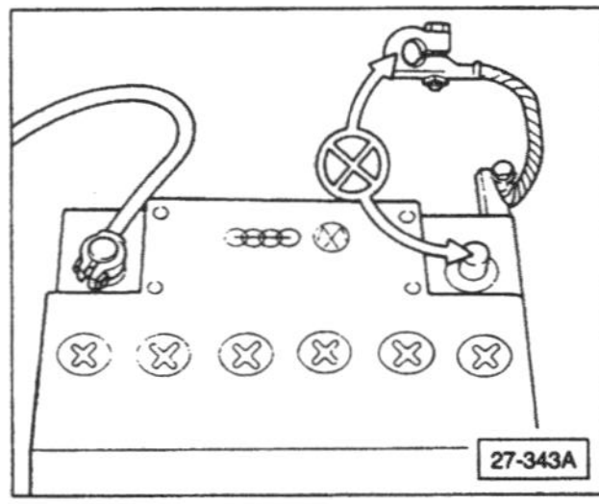

27-343A

Test light lights up	Test light does not light up

Test light lights up:
- eliminate individual circuits by removing fuses
- if test light goes out in one of interrupted circuits, source of fault is to be found there

Test light does not light up:
- in place of test light, connect ammeter (measuring range 0-200 Amp) between negative (−) battery terminal and battery ground strap
- reduce measuring range until a clear reading is shown (1-3 mA is permissible)

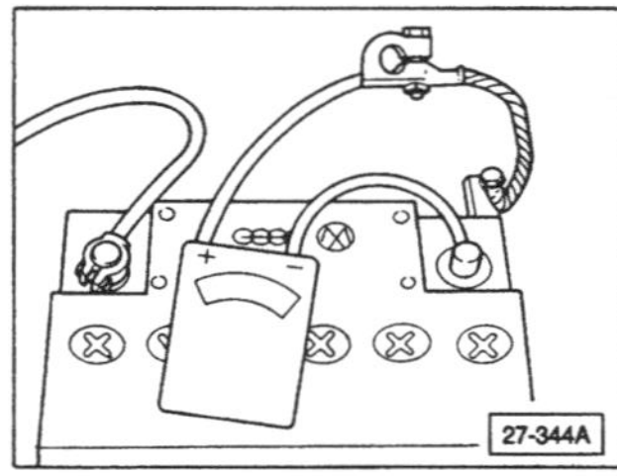

27-344A

- eliminate individual circuits by removing fuses
  if reading drops to zero in one of interrupted circuits, source of fault is to be found there

- if no faults are found in fused circuits, disconnect wiring connectors from nonfused components:
  Generator
  starter
  ignition system
  cluster instruments
  coolant fan and so on until defective circuit or component is discovered

- repair or replace as necessary

End

Starter motor, removing/installing

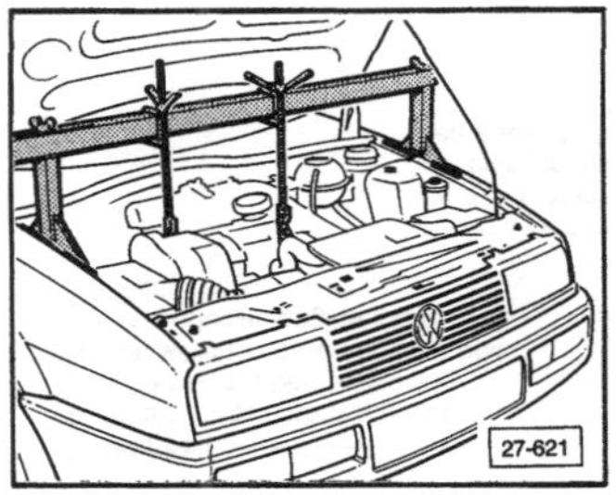

- disconnect battery ground cable
- support engine/gearbox assembly using engine support **10-222A** with leg set **10-222A/1**

Note

Starter motor installation requires precise alignment of the engine, etc. The following removal procedure will cause these components to shift position if the engine support is **NOT** used.

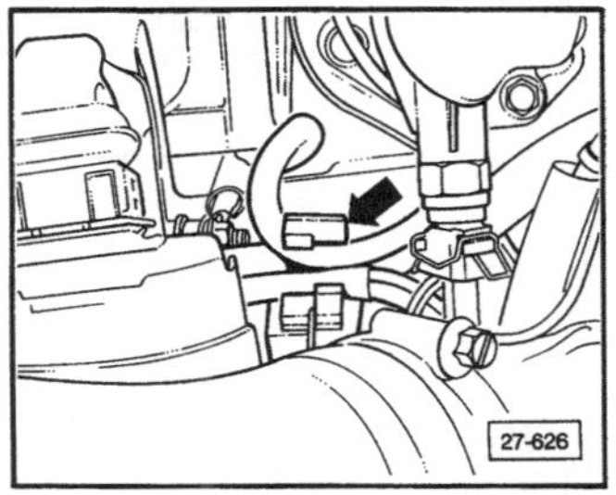

- disconnect terminal **50** from starter switch then pull wire out of retainer (**arrow**)
- when installing; push wires back into retainer clip (**arrow**)

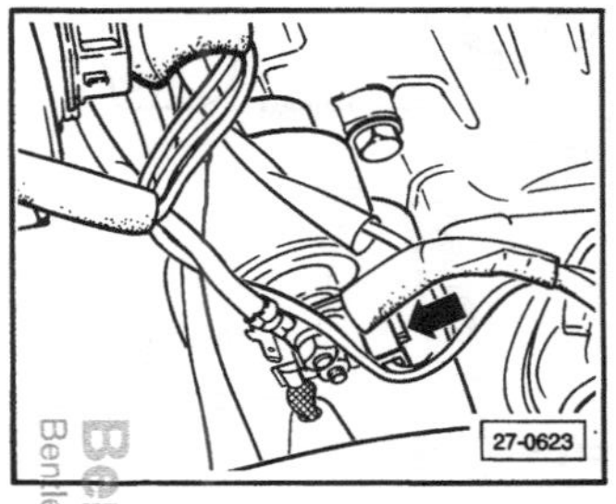

- unclip and disconnect harness connector (**arrow**)

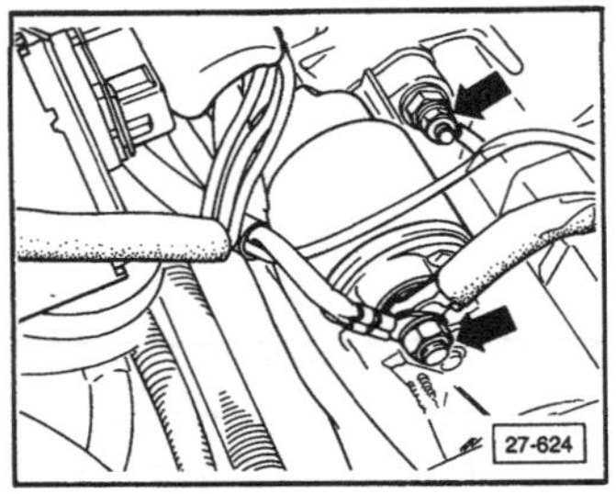

- disconnect wire from Generator B+ and battery positive to starter terminal **30** (**arrows**)
- when installing; tighten starter solenoid nut to 15 Nm (11 ft lb) (**lower arrow**)
- when installing; tighten ground connection nut (**upper arrow**) to 10 Nm (7 ft lb)

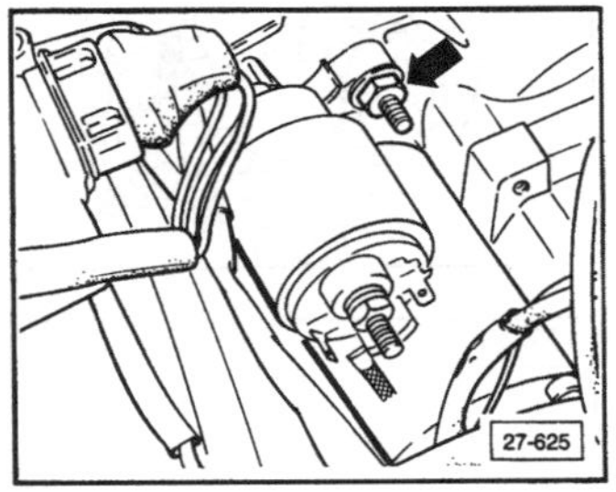

- remove nut from mounting bolt (**arrow**)
- when installing; tighten nut (**arrow**) to: 60 Nm (44 ft lb)

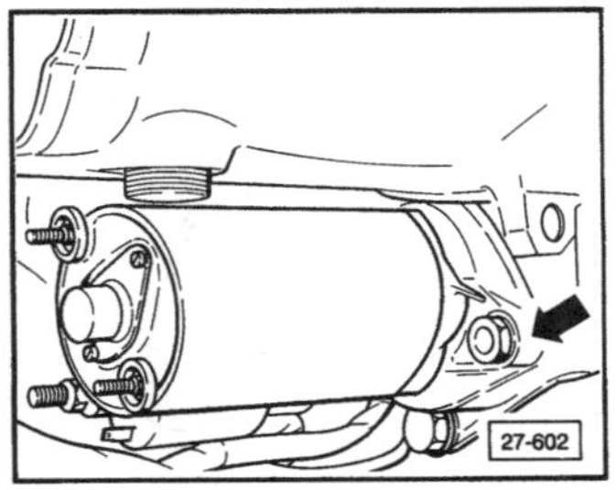

- remove mounting bolt (**arrow**) and remove starter motor
- when installing; tighten mounting bolt (**arrow**) to: 60 Nm (44 ft lb)

Note

Installation is reverse of removal procedure.

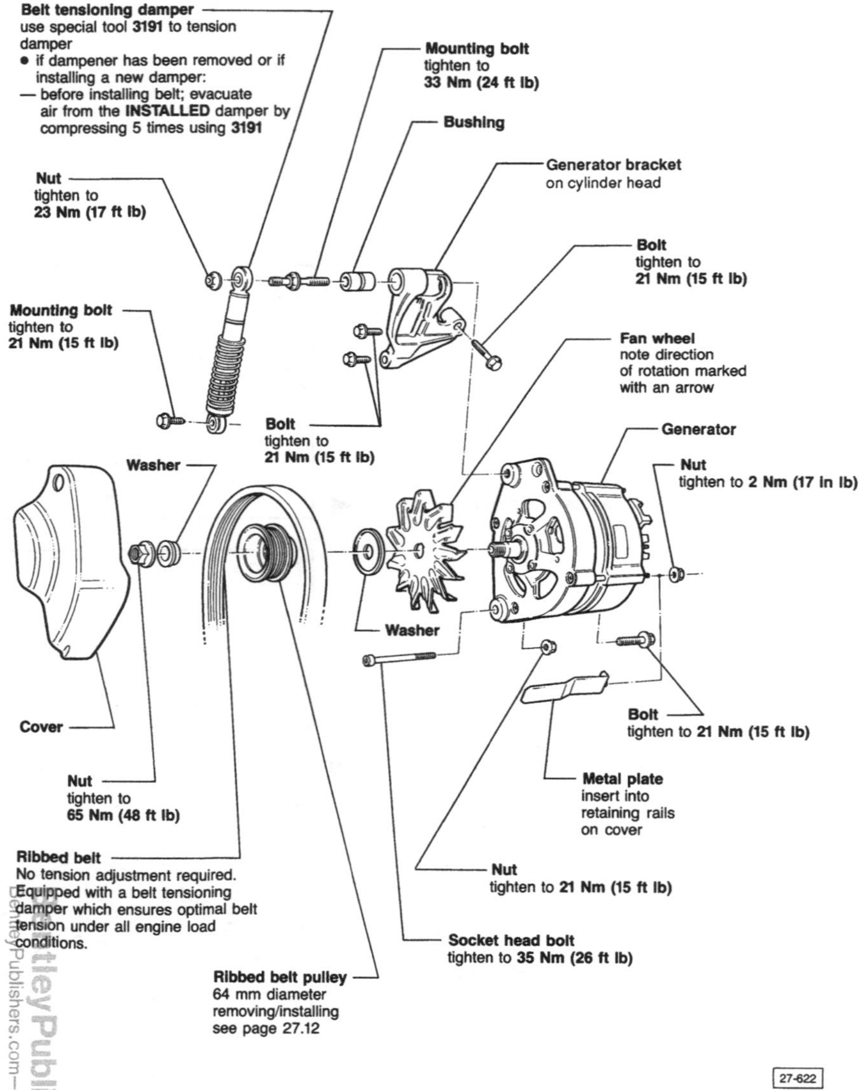

Voltage regulator, removing and installing

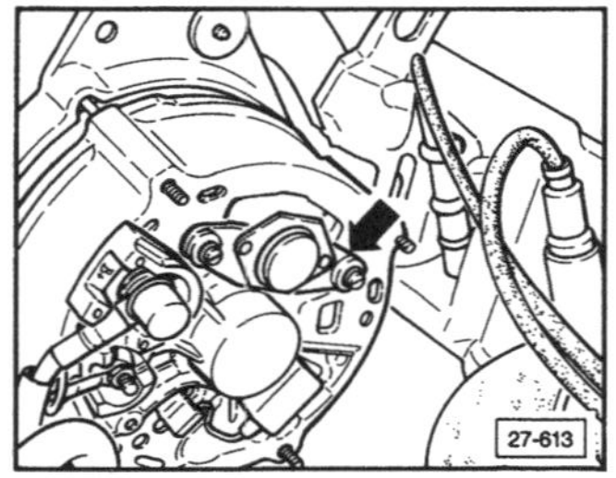

- remove two screws and angle regulator out of Generator

Voltage regulator can be removed with the Generator installed.

Carbon brushes, checking

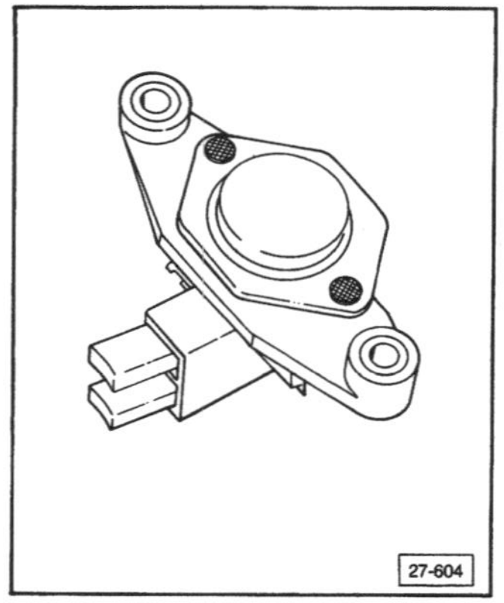

- length of brushes new: 12 mm (0.47 in.)
  wear limit: 5 mm (0.20 in.)
  tolerance: 1 mm (0.04 in.)

Suppression condenser, connecting

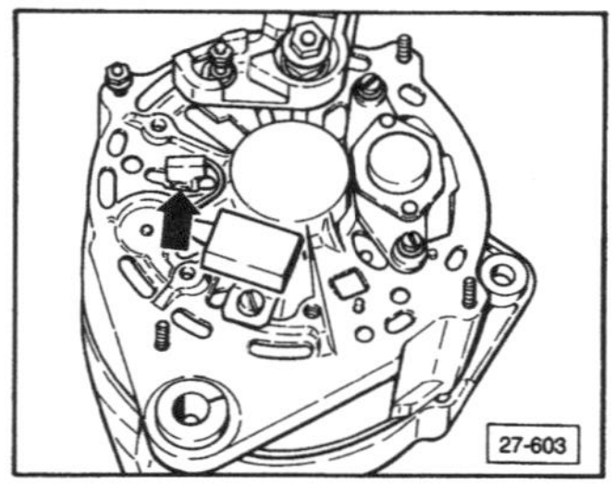

- connect terminal at **arrow**

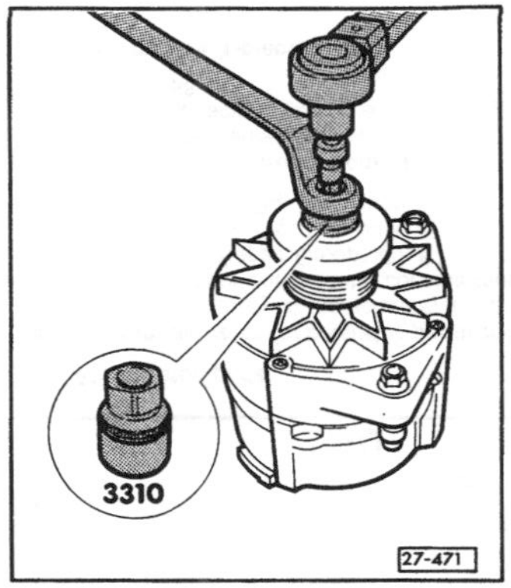

Generator (GEN) V–belt pulley, removing/installing

Use special tool **3310** to remove/install the Generator (**GEN**) V–belt pulley.

- install tool **3310** onto pulley retaining nut
- hold shaft with hex key (through center of tool) while loosening nut with box wrench
- when installing, tighten pulley retaining nut to 65 Nm (48 ft lb)

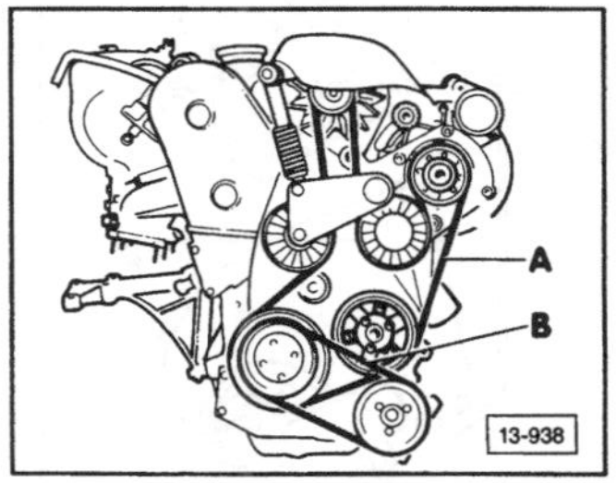

Belt routing, vehicles without Air conditioning

▶

A — Ribbed belt for: Generator, G-charger and water pump
B — Vee belt for: power steering pump

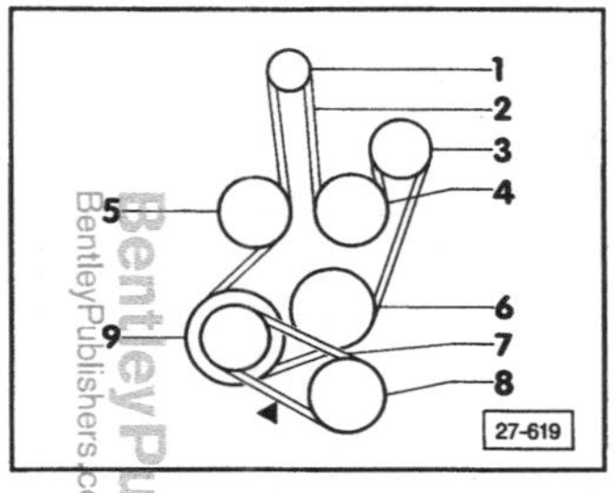

▶

1 — Ribbed belt pulley — Generator
2 — Ribbed belt
- 21.36 mm wide x 1653 mm circumference (0.84 in. wide x 65.0 in. circumference)

3 — Ribbed belt pulley — G-charger
4 — Ribbed belt idler pulley
5 — Ribbed belt tensioning pulley
6 — Ribbed belt pulley — Water pump
7 — Vee belt
- 9.5 mm wide x 663 mm circumference (0.37 in. wide x 26.1 in. circumference)
- arrow = tension checking point

8 — Vee belt pulley — Power steering pump
9 — Combination vee belt and ribbed belt pulley — crankshaft

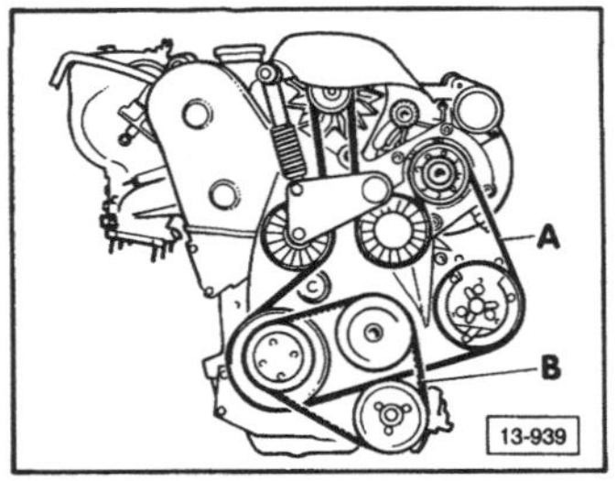

Belt routing, vehicles with Air conditioning

▶

A — Ribbed belt for: Generator, G-charger and air conditioning compressor
B — Vee belt for: water pump and power steering pump

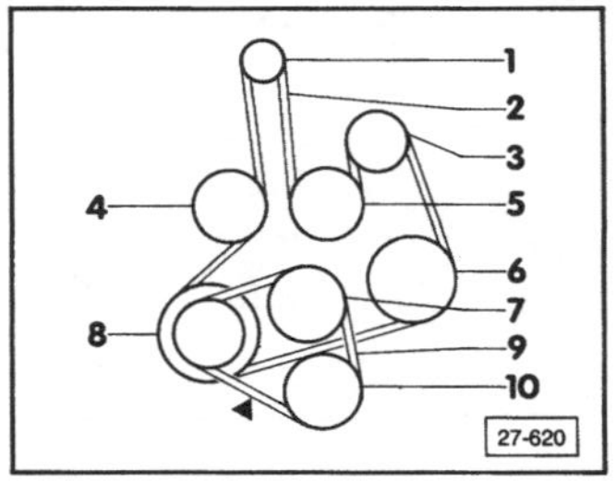

▶

1 — Ribbed belt pulley — Generator
2 — Ribbed belt
- 21.36 mm wide x 1755 mm circumference (0.84 in. wide x 69.0 in. circumference)

3 — Ribbed belt pulley — G-charger
4 — Tensioning pulley
5 — Idler pulley
6 — Ribbed belt pulley — A/C compressor
7 — Vee belt pulley — Water pump
8 — Combination vee belt and ribbed belt pulley — crankshaft
9 — Vee belt
- 9.5 mm wide x 735 mm circumference (0.37 in. wide x 28.9 in. circumference)
- arrow = tension checking point

10 — Vee belt pulley — Power steering pump

Note

Vehicles with VR6 engine (engine code AAA): see Repair Group 13 for information on the poly - ribbed belt and belt routing

CAUTION
Disconnect battery ground strap, before working on the electrical system.

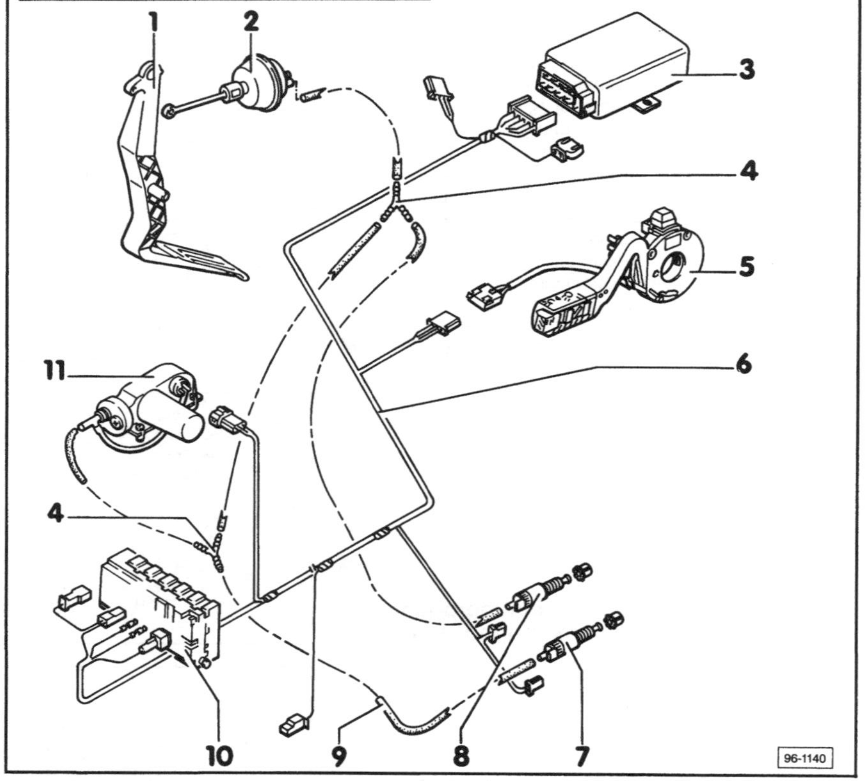

1 – **Accelerator pedal**

2 – **Vacuum actuator (accelerator pedal mounted)**
To remove and adjust the actuator; the drivers side air duct or the floor outlet/center console must be removed see Repair Groups 80, 87.
- removing, see **Fig. 1** on page 27.15
- installing, see **Fig. 2** on page 27.15

3 – **Control unit for cruise control**
- behind the center console
- center console removing and installing see Repair Group 68

4 – **Vacuum line Y-piece**

5 – **Cruise control switch (E 45)**
removing and installing, see Repair Group 94

6 – **Wiring harness for cruise control**

7 – **Ventilation valve with clutch pedal switch (F 36)**
removing and installing, see **Fig. 3** on page 27.15

8 – **Ventilation valve with brake pedal switch (F 47)**
removing and installing, see **Fig. 3** on page 27.15

9 – **Vacuum lines**
leak checking, see page 27.16

10 – **Central electric module**
removing and installing, see Repair Group 97

11 – **Vacuum pump for cruise control (V 18)**
removing, see **Fig. 4** on page 27.15

Cruise control system, checking/adjusting

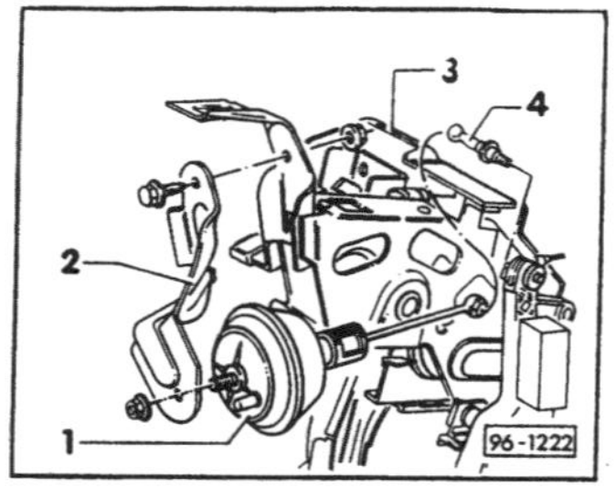

Fig. 1

Removing actuator element

1 – Vacuum actuator
2 – Actuator mounting bracket
3 – Pedal cluster
4 – Linkage

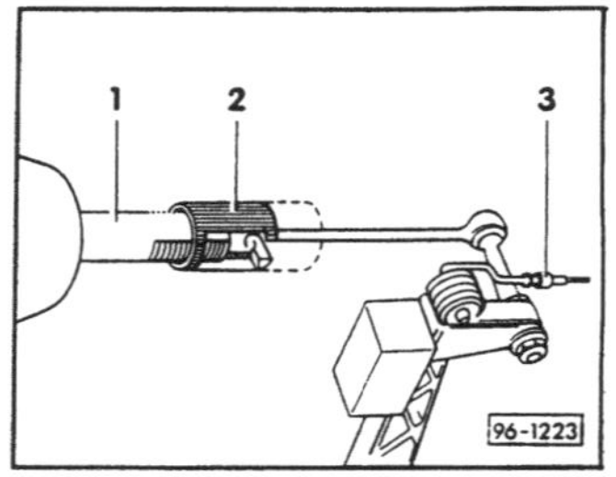

Fig. 2

Vacuum actuator, adjusting

- loosen adjustment sleeve **2** on actuator **1** and push completely forward
- adjust cable **3**
- pull adjustment sleeve to rear until 3.5 mm of play is obtained at the pedal plate
- turn to adjustment sleeve to right to lock into place
- check accelerator pedal clearance

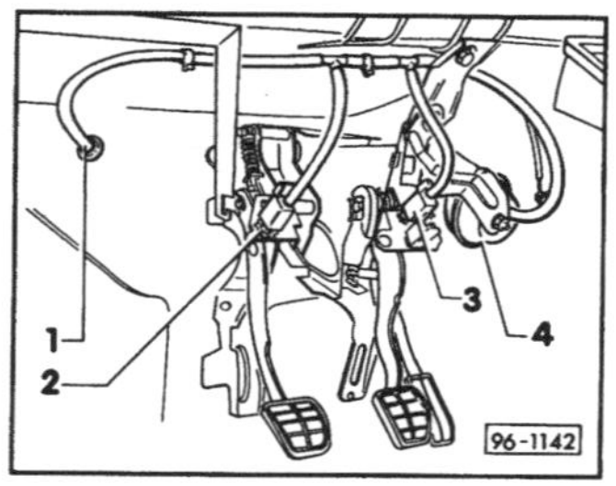

Fig. 3

Ventilation valves, removing and installing

1 – Vacuum lines
2 – Ventilation valve on clutch pedal
3 – Ventilation valve on brake pedal
4 – Vacuum actuator

Removing

- remove driver's side instrument panel trim
- remove connector and vacuum line
- unscrew ventilation valve out of bracket

Installing

- press ventilation valve up to stop in bracket
  - pedal must be completely pulled back

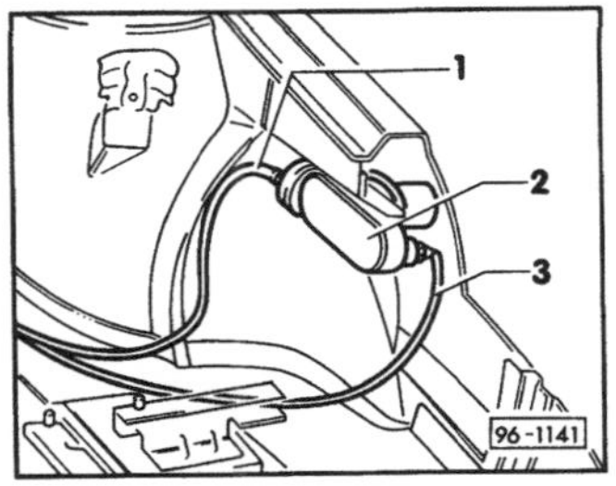

Fig. 4

Vacuum pump (V 18), removing

1 – Vacuum line

2 – Vacuum pump

3 – Wiring harness to vacuum pump

- remove battery
- remove windshield washer mounting bracket
- remove vacuum pump

Vacuum system, leak testing

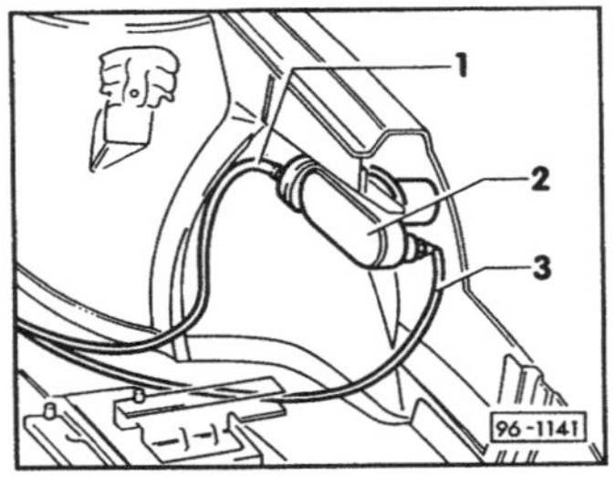

- disconnect vacuum line **1** from vacuum pump **2** and seal tightly
- remove instrument panel trim from driver's side lower

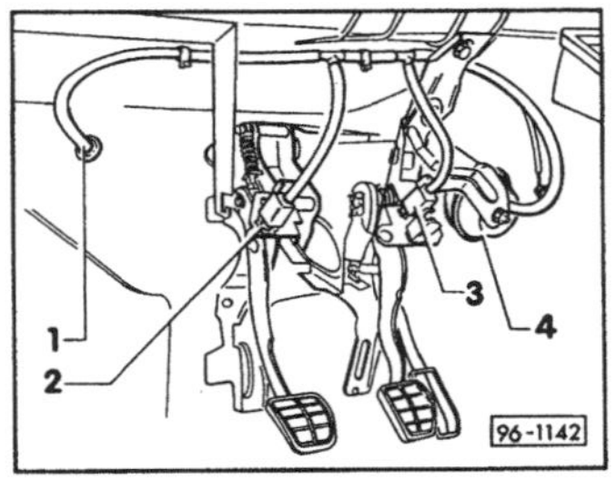

- push in membrane on vacuum actuator **4** then push brake pedal down by hand up to pressure point
- release brake pedal
  - membrane should not move

THIS PAGE INTENTIONALLY LEFT
BLANK

Index

Engine Code – PG, 49 States and Canada, m.y. 1989 to present

California m.y. 1989 – 1990

Digifant Ignition system
- component layout **28.2**

Hall sender unit
- checking **28.9**
- function, checking **28.10**

Ignition distributor
- installing **28.6**

Ignition timing
- checking/adjusting **28.7**
- map, checking **28.8**
- trigger function, checking **28.9**

Knock sensor
- checking **28.11**

Safety measures
- precautions chart **28.5**

Technical data
- spark plugs **28.4**

Engine Code – PG, California starting m.y. 1991

Digifant Ignition system
- component layout **28.12**
- control unit, component layout **28.13**
- technical data **28.15**

Hall sender
- checking **28.20**

Ignition coil
- checking **28.19**

Ignition distributor
- installing **28.16**

Ignition timing
- advance, checking **28.18**
- checking/adjusting **28.17**

Safety precautions
- chart **28.14**

Index

Engine Code AAA

Hall sender
- checking **28.26**

Ignition coil
- checking **28.25**

Ignition distributor
- disassembling/assembling **28.22**

Ignition and spark plugs
- technical data **28.23**

Motronic System
- component layout **28.21**

Safety precautions
- chart **28.24**

Note

Refer to the Supplement at the back of the manual for additional information pertaining to this Repair Group.

Engine Code **PG** applying to:

- 49 States and Canada, model years **1989** to **present.**
- California vehicles model years **1989** and **1990.**

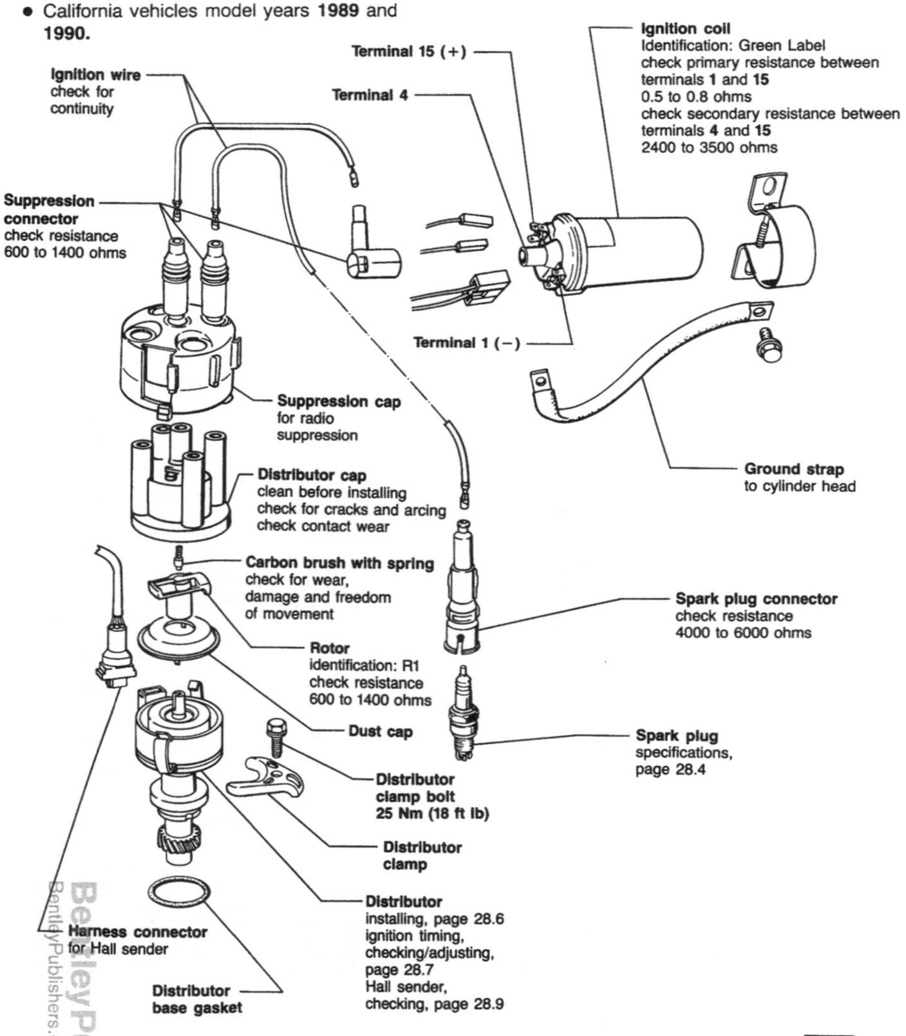

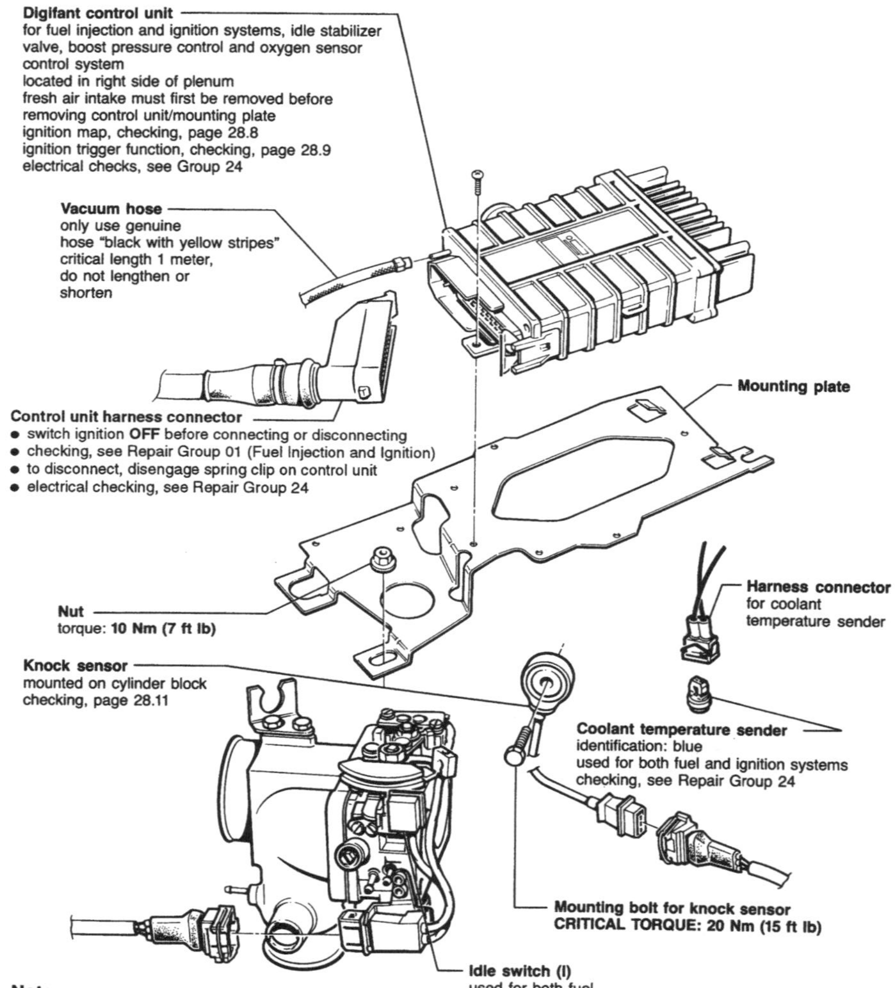

Note

Ensure that no coolant hoses or lines touch the knock sensor. This could dampen the sensor and diminish the signal.

Adjustment data, spark plugs

Engine code letters		PG
Ignition timing*	checking adjusting	4° to 8° BTDC 6° ± 1° BTDC The above values observed at 2000 to 2500 RPM 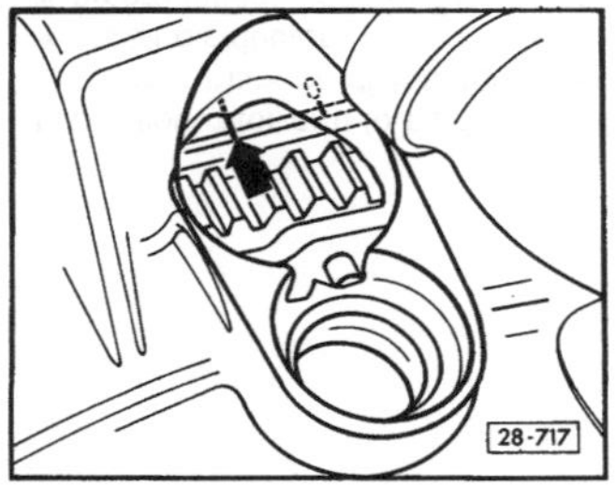
Ignition map**	checking RPM	2500 RPM
	degrees	20° to 30° + ignition timing point
Engine speed limit	RPM	6200 RPM
Firing order	cylinders	1 — 3 — 4 — 2
Spark plugs	Bosch part number	W 6 DPO
	torque	20 Nm (15 ft lb)
	gap	0.6 to 0.7 mm (0.024 to 0.028 in.)

* Engine oil temperature 80°C (176°F) minimum
Coolant temperature sensor **DISCONNECTED**

** Engine oil temperature 80°C (176°F) minimum
Coolant temperature sensor **CONNECTED**

Safety measures

CAUTION

Observe the following precautions to prevent personal injury as well as possible damage to the ignition system components.

Switch **OFF** the ignition before connecting or disconnecting components or test equipment
Do **NOT** crank engine before high tension wire of ignition distributor (terminal **4**) is connected to ground
Do **NOT** use battery booster longer than one minute nor should 16.5 volts be exceeded

Do **NOT** wash engine unless ignition is switched **OFF**
Disconnect **BOTH** battery terminals whenever arc or spot welding
Before towing, vehicles with a defective ignition system (or where this is suspected) must have terminal **1** (green) of the ignition coil disconnected

Do **NOT** connect a condenser of any kind to terminal **1** of the ignition coil
When installing noise suppressors, **ONLY** use 1000 ohms for high tension wires and 5000 ohms for spark plug connectors
Do **NOT** replace distributor rotor (marked **R1**) with a different type

If the vehicle is heated up (e.g. in a painting booth) do **NOT** start the engine until it has had sufficient time to return to room temperature

Ignition distributor, installing

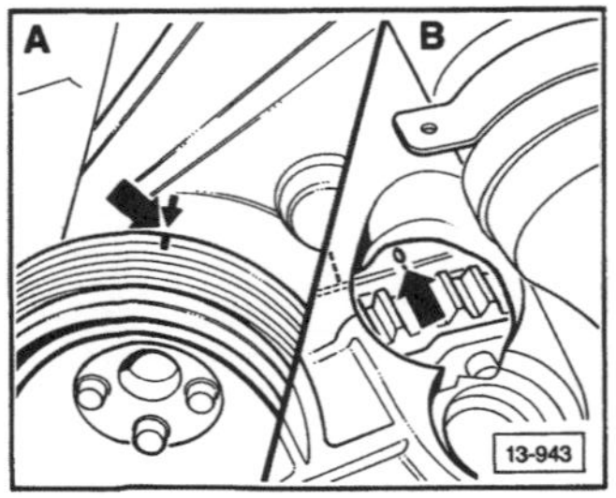

Engine installed

- rotate flywheel to **TDC** for number **1** cylinder (view **B**)

Engine NOT installed

- rotate engine until pulley mark aligns with arrow on toothed belt guard (view **A**)

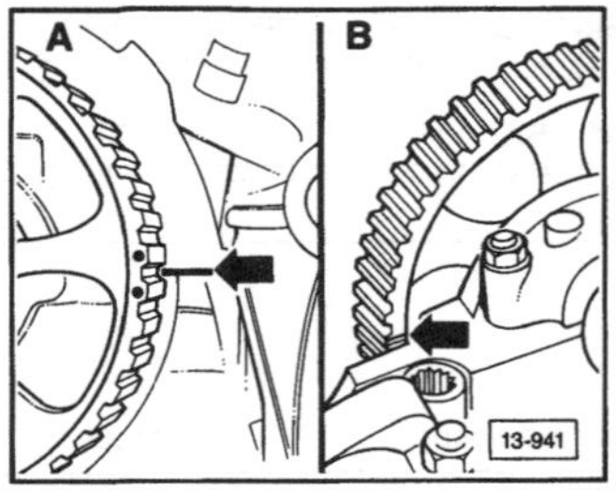

Cylinder head cover INSTALLED

- marking on cylinder head cover (**arrow**) must fall between dots on camshaft sprocket (view **A**)

Cylinder head cover NOT INSTALLED

- marking on camshaft sprocket must align (**arrow**) with edge of cylinder head (view **B**)

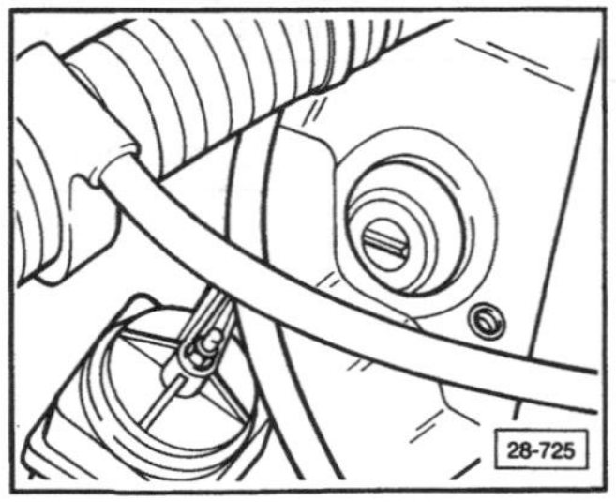

- align oil pump drive pinion lug parallel to crankshaft

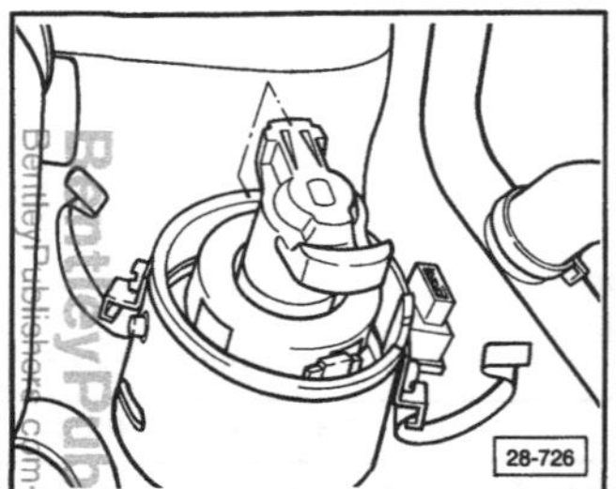

- position rotor arm to point at number **1** cylinder mark on distributor housing
- install distributor
- before fitting; clean distributor and check for cracks and signs of arcing, replace if necessary
- adjust ignition timing (see page 28.7)

Ignition timing, checking and adjusting

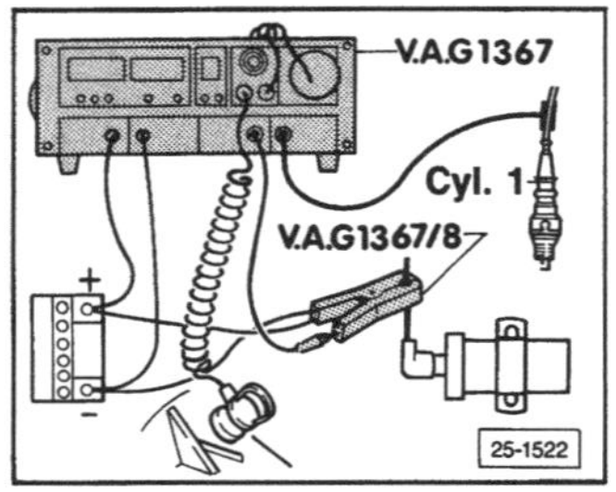

Requirement
- engine oil temperature 80°C (176°F) minimum

- switch ignition **OFF**
- connect **VAG 1367** engine tester using **VAG 1367/8** inductive pickup

- start engine and let idle
- disconnect (blue) coolant temperature sensor (**arrow**)
- check ignition timing at 2000 to 2500 RPM

Ignition timing is displayed directly on **VAG 1367**
- checking 4° to 8° BTDC

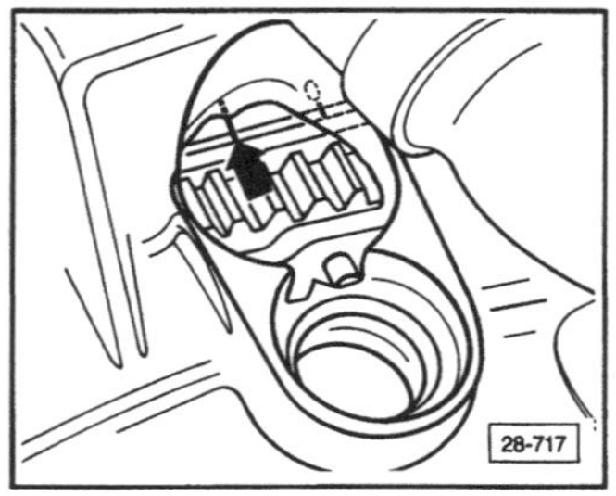

With stroboscope

- aim strobe directly at flywheel opening
  - checking 4° to 8° BTDC

Regardless of checking method, if adjustment is necessary, proceed as follows

- destroy tamper proof cap covering head of distributor base clamp bolt
- loosen distributor base clamp
- rotate distributor as necessary to obtain specification
  - adjust 6° ± 1° BTDC

- re-connect coolant temperature sensor (blue)
- raise engine speed three times then let engine idle
- check idle speed
  - 800 ± 30 RPM, adjust if necessary (see page 24-90-1)

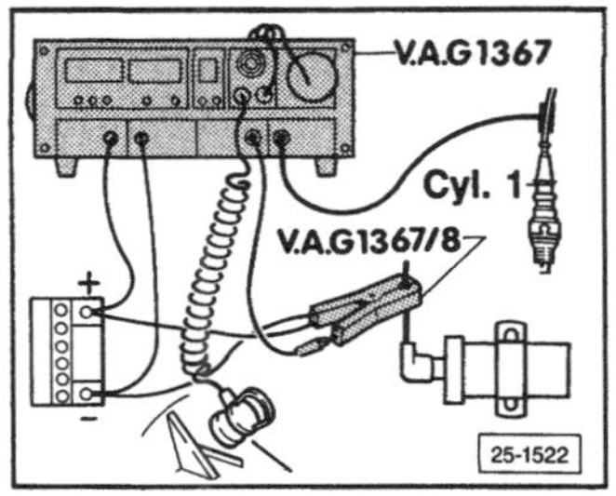

Ignition timing map, checking

Requirements
- engine oil temperature 80°C (176°F) minimum
- coolant temperature sender **OK**

- switch ignition **OFF**
- connect **VAG 1367** engine tester using **VAG 1367/8** inductive pickup

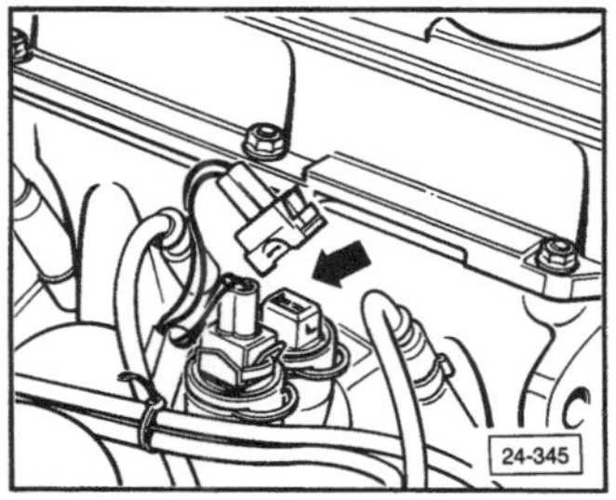

- start engine and let idle
- disconnect (blue) coolant temperature sensor (**arrow**) and record ignition timing at 2500 RPM

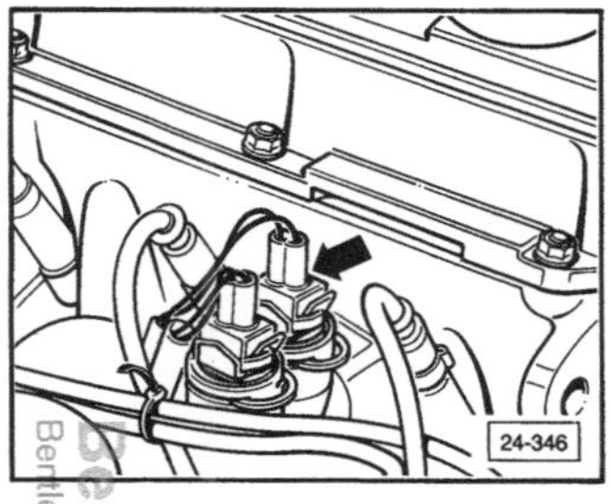

- re-connect coolant temperature sender harness connector (**arrow**) and check ignition timing at 2500 RPM
  - ignition timing value must advance by 20° to 30° in addition to the previously recorded value

If same value is obtained (with re-connected sensor) as previously recorded value
- locate and eliminate open circuit in coolant temperature sensor wiring

If open circuit is **NOT** found
- replace Digifant control unit

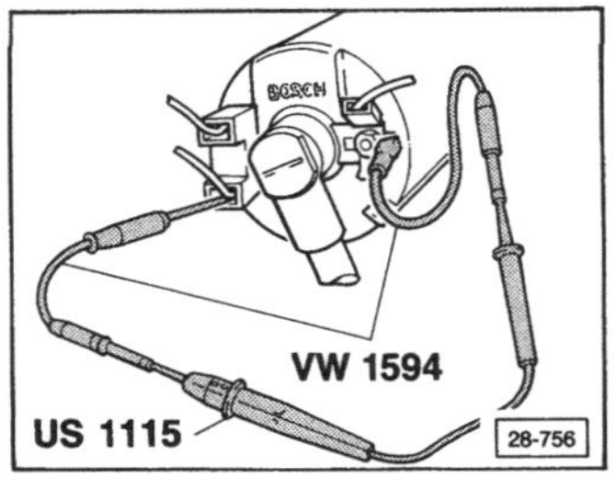

Ignition trigger function, checking

Requirements
- Ignition coil **OK**
- Hall sensor **OK**
- Electrical check of Digifant system **OK**

- connect **US 1115** LED tester to ignition coil terminals **1** and **15** using **VW 1594** adaptor kit
- operate starter
  - LED must flicker

If **NO**
- replace Digifant control unit

Hall sender unit, checking

Requirement
- Electrical check of Digifant system **OK**

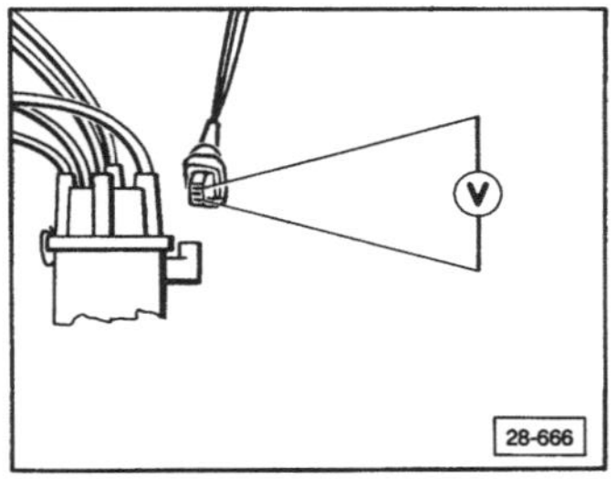

Voltage supply, checking

remove harness connector from Hall sender (on distributor)
- switch multimeter **US 1119** to 20 volt range
- connect multimeter to outer terminals of harness connector

- switch **ON** ignition
  - 10 volts minimum

If less than 10 volts
- replace Digifant control unit

If zero volts
- locate and eliminate open circuit in wiring using wiring diagram

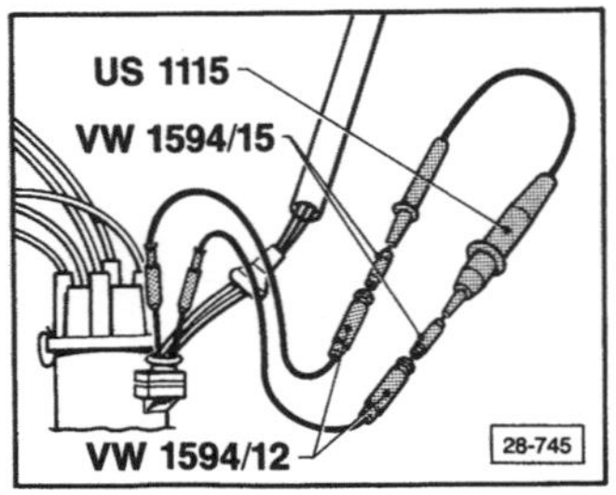

Hall sender function, checking

▶

- disconnect harness connector from fuel injector wiring guide
- loosen rubber boot on Hall sender connector (on ignition distributor)
- slide boot up wiring to expose harness connector terminals (while leaving connector connected)

- connect **US 1115** LED tester to center and outside wires of connector (using **VW 1594** adaptor kit)
- operate starter while watching LED tester
  - LED must flicker

If **NO**
- replace Hall sender (in distributor)

Knock sensor, checking

Vehicles **WITHOUT** On-Board diagnostic capability **CANNOT** have the knock sensor functionally checked.

The complaint **"high speed not attainable"** can be caused by an intermittent or non-functioning knock sensor.

Possible causes:

- disconnected wiring
- corroded terminals (poor or no contact)
- knock sensor shield damaged
- mounting bolt incorrectly torqued; must be 20 Nm (15 ft lb)
- abnormal noises caused by:
  - torn or loose brackets
  - loose wiring
  - loosened components
  - broken, loose or missing bolts
  - mechanical damage

If **NO** fault found

- replace knock sensor

- electrical check of Digifant System, see *Repair Group* 01 (*Fuel Injection* and Ignition)
- fuel injection checking, see Repair Group 24
- ignition Technical data, see page 28.15
- observe Safety precautions, see page 28.14

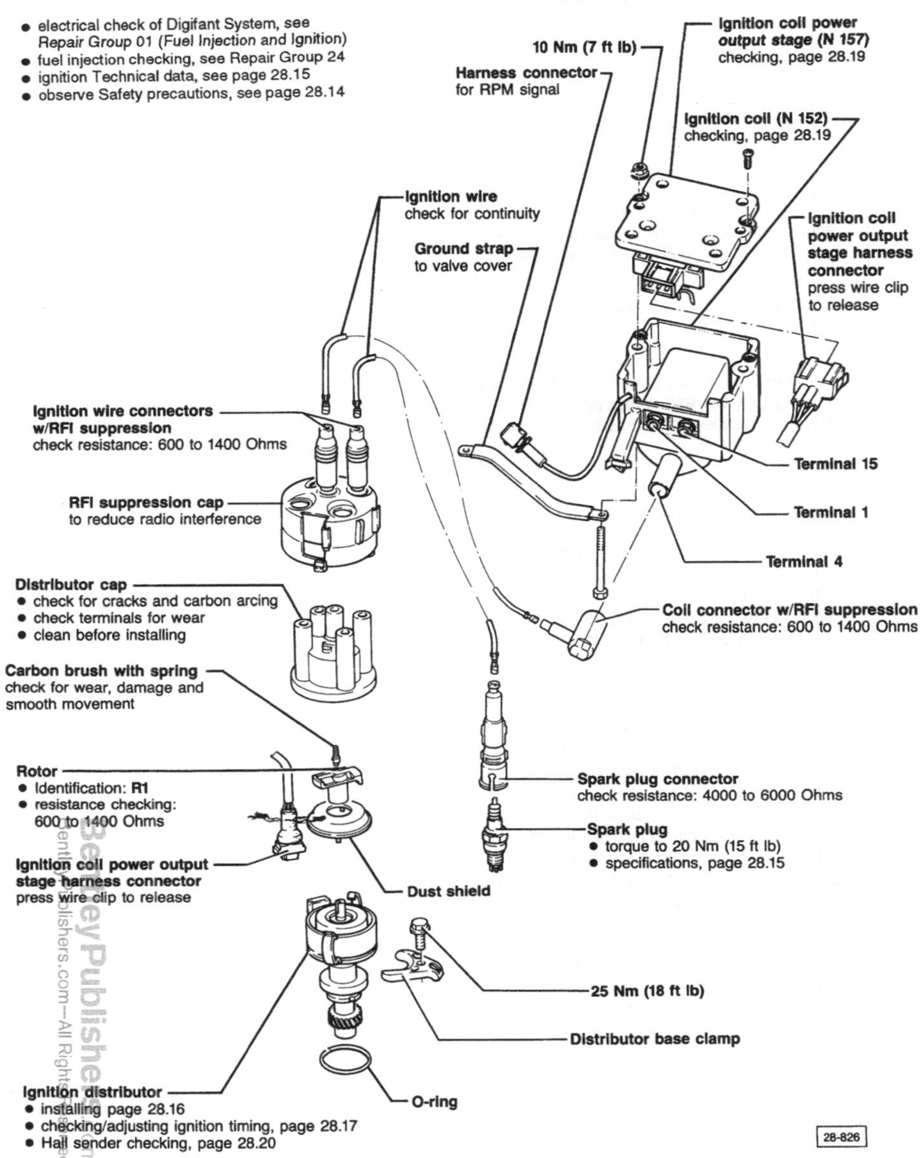

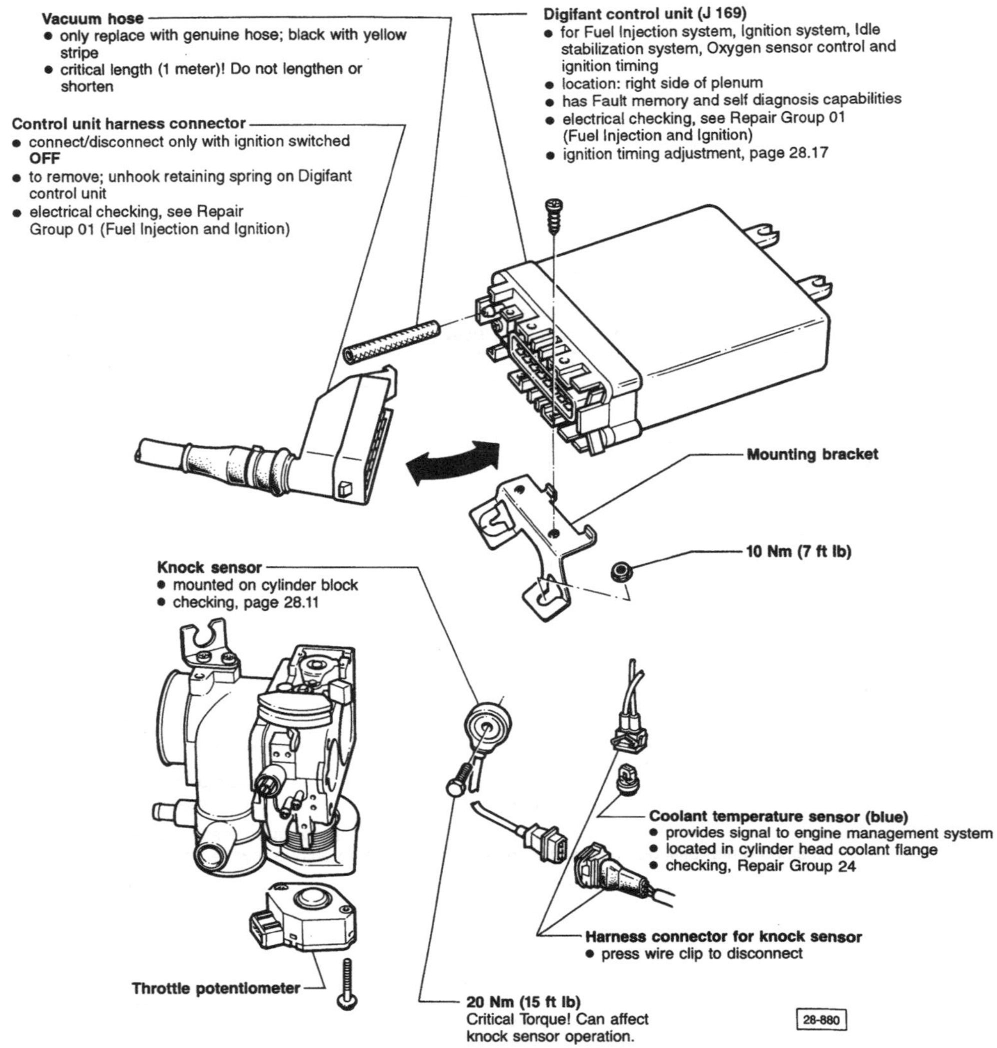

Safety precautions

CAUTION

Observe the following precautions to prevent personal injury as well as possible damage to sensitive electrical components.
- switch **OFF** the ignition before connecting or disconnecting components or test equipment
- connect and disconnect battery **ONLY** with ignition switched **OFF** otherwise the control unit could be damaged
- if the engine must be cranked but not started (for compression testing etc.) disconnect power output stage of ignition coil and
- disconnect **BOTH** battery terminals whenever arc or spot welding
- do **NOT** connect a condenser of any kind to terminal 1 of the ignition coil
- when installing noise suppressors, **ONLY** use 1000 Ohms for high tension wires and 5000 Ohms for spark plug connectors
- do **NOT** replace distributor rotor (marked **R1**) with a different type

Note

A variety of electrical connectors are used on this vehicle, **ALWAYS** use the **VW 1594** adaptor kit to connect test instruments to these connectors.

CAUTION

Before disconnecting a customers battery; **ALWAYS** ask for the radio code (if equipped with an anti-theft radio).

Ignition system, Technical Data

Engine Code	PG
RPM cutoff limit	6200 RPM
Ignition timing point, with engine oil temperature at 80°C minimum and with (blue) coolant temperature sensor **disconnected:**	
• checking value	4 to 8° before TDC
• adjusting value	6 ± 1° before TDC
• engine speed	2000 to 2500 RPM
Ignition advance, with engine oil temperature at 80°C minimum and with (blue) coolant temperature sensor **connected:**	
• engine speed	4500 RPM
• calculating advance	30 ± 3° advance + ignition timing point
Timing mark	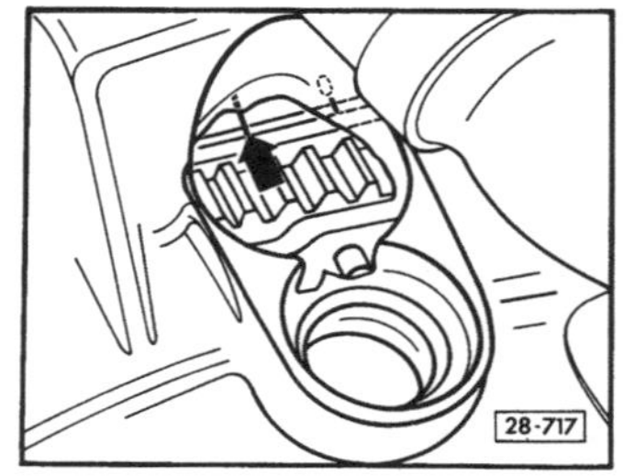
Firing order	1 - 3 - 4 - 2
Spark plugs **Bosch**	W6 DPO or W6 DPI
• torque	25 Nm (18 ft lb)
• gap	0.6 to 0.7 mm (0.024 to 0.028 in.)
Ignition coil identification	Green label

Ignition distributor, installing

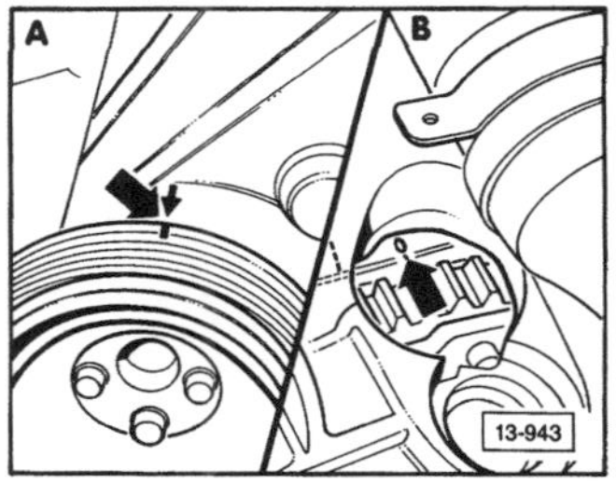

A With engine installed

■ rotate flywheel to TDC for cylinder 1

B With engine removed

■ align arrow on drive belt guard with mark on pulley

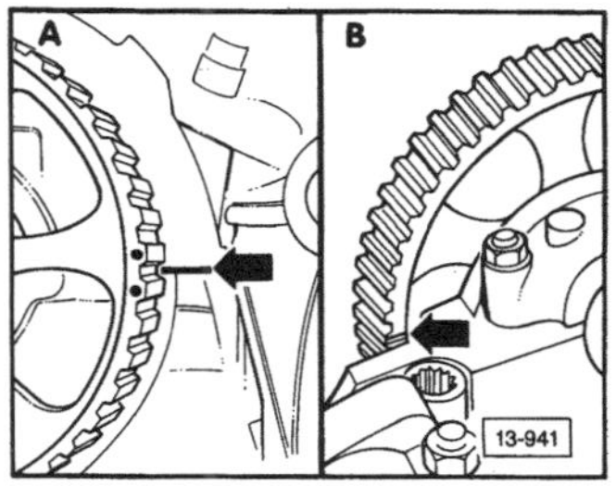

A With valve cover installed

■ align dots on camshaft sprocket with marking on valve cover

B With valve cover removed

■ align marking on camshaft sprocket with edge of cylinder head

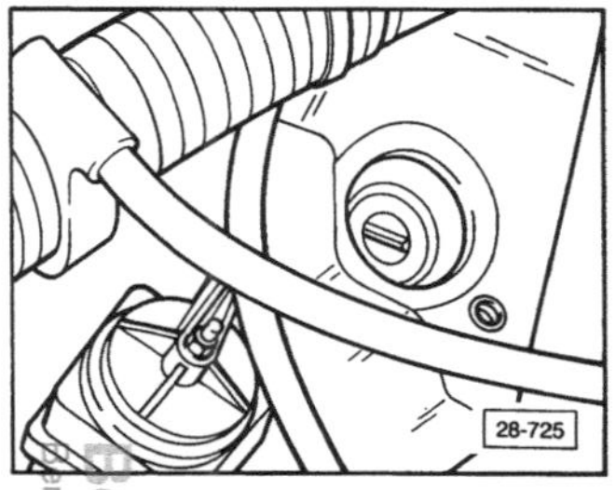

■ align oil pump shaft parallel with crankshaft

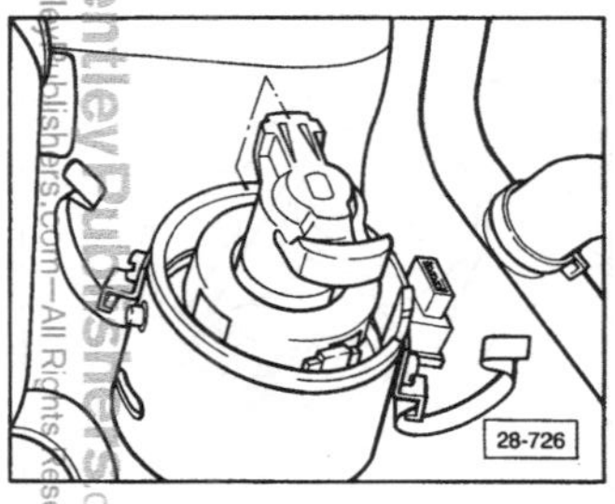

■ position distributor rotor to point at number 1 cylinder on distributor housing
■ inspect distributor cap before installing
  ● check for cracks and carbon arcing
  ● check terminals for wear
  ● clean before installing

■ adjust ignition timing, page 28.17

Ignition timing, checking/adjusting

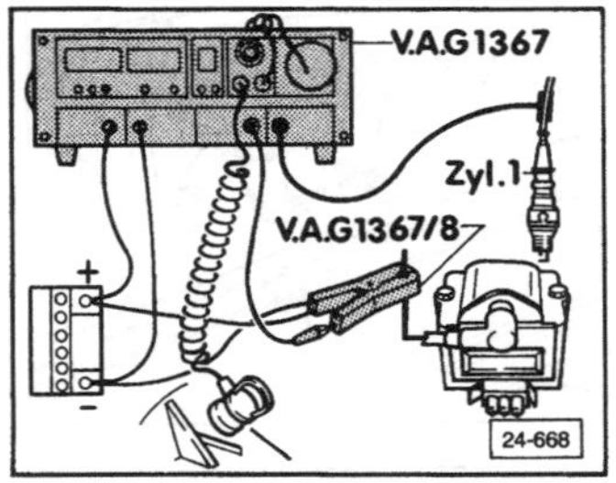

Requirement
  ● engine oil temperature 80°C (176°F) minimum

■ switch **OFF** ignition
■ connect **VAG 1367** engine tester using **VAG 1367/8** inductive pickup
  ● to display engine RPM and ignition timing point

CAUTION

Be certain that TDC sensor is properly seated in recess on flywheel housing.

■ start engine and let idle

With VAG 1367 engine tester

■ disconnect blue coolant temperature sender
■ raise engine speed to a level between 2000 and 2500 RPM and check ignition timing point (displayed directly on **VAG 1367** engine tester)
  ● must be 4 to 8° Before **TDC**

With strobe light

■ disconnect blue coolant temperature sender
■ raise engine speed to a level between 2000 and 2500 RPM and check ignition timing point

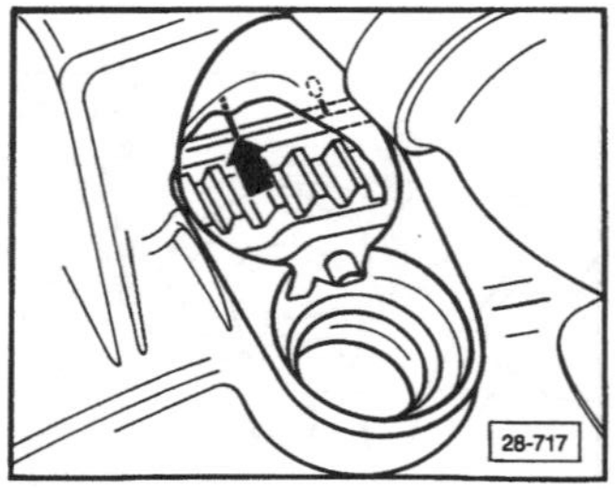

- point strobe at flywheel timing mark
  - must be 4 to 8° Before **TDC**

If **NO** (either method), adjust as follows
- adjust ignition timing by loosening and turning distributor until specification is obtained
  - adjusting value: 6° ± 1° Before **TDC**
- activate fault memory and then erase, see Repair Group 01 (Fuel Injection and Ignition) for additional information

Ignition timing advance, checking

Requirements
- engine oil temperature 80°C (176°F) minimum

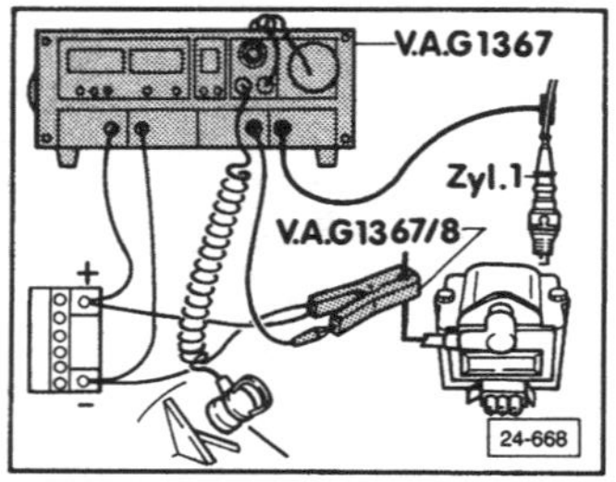

- read Fault memory and check for any stored faults, correct if necessary and then erase Fault memory
- connect **VAG 1367** engine tester using inductive pickup **VAG 1367/8**
  - for engine RPM and ignition timing point display

> **CAUTION**
>
> Be certain that TDC sensor is properly seated in recess on flywheel housing.

- start engine and let idle

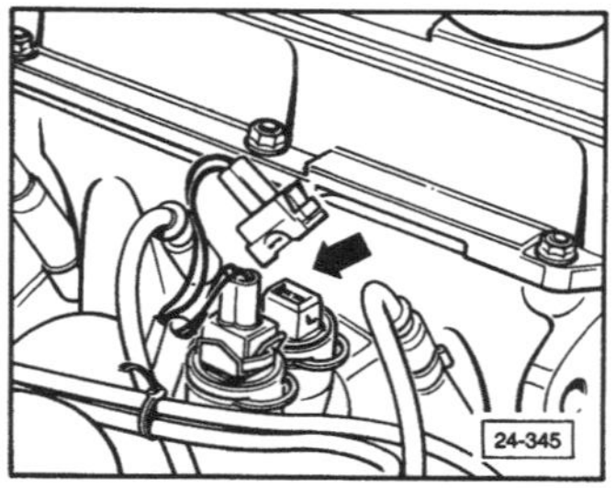

- disconnect blue coolant temperature sender
- raise engine speed to 2300 RPM and record ignition timing point display value

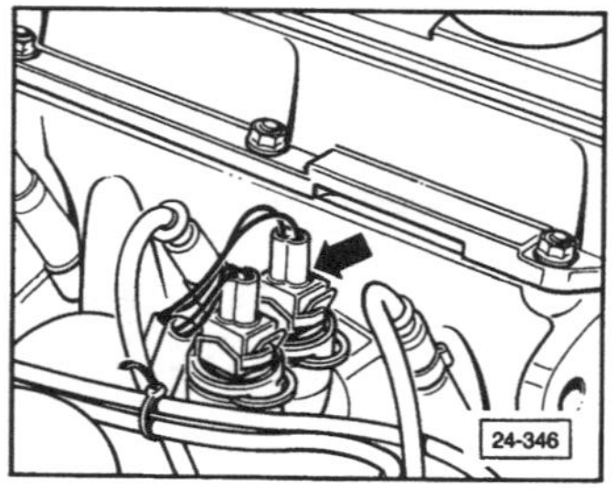

- reconnect blue coolant temperature sender (**arrow**)
- raise engine speed to 4500 RPM and observe ignition timing point display value
  - displayed value must advance at least 30° above the previously recorded value

If **NO**
- replace Digifant control unit (**J 169**)
  - even if no fault is displayed while reading fault memory

Ignition coil (N 152), checking

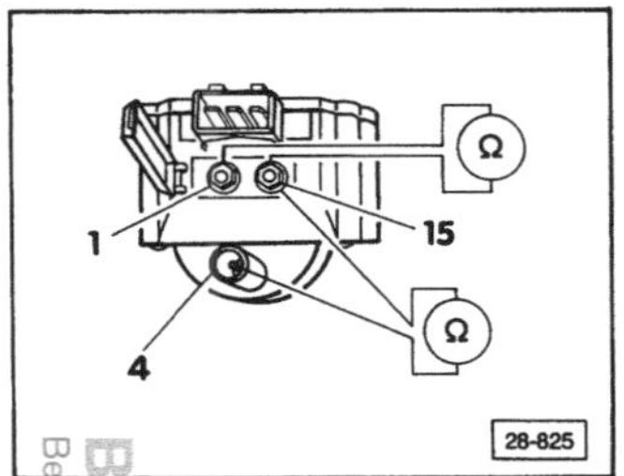

- perform following checks using **Fluke 83** multimeter (**US 1119**)
  - specified values are valid when measured at a temperature between 0 and 40°C (32 and 104°F)

If measured values deviate from specified values

- check for open or short circuits using proper wiring diagram before replacing components
- use the **VW 1594** adaptor kit to make connections between the multimeter and the harness connectors

> **CAUTION**
>
> Switch the multimeter to the appropriate measuring range before making test connections.

- disconnect coil wire and power output stage harness connector from coil
- check primary resistance between terminals **1** and **15**
  - must be 0.5 to 0.7 Ohms
- check secondary resistance between terminals **4** and **15**
  - must be 3000 to 4000 Ohms

If specified values not obtained

- remove ignition coil and disconnect power output stage of ignition coil
- repeat check

If specifications still not obtained

- replace ignition coil (**N 152**)

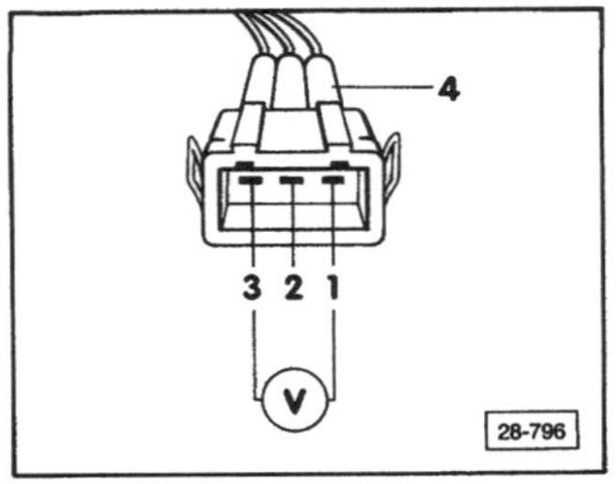

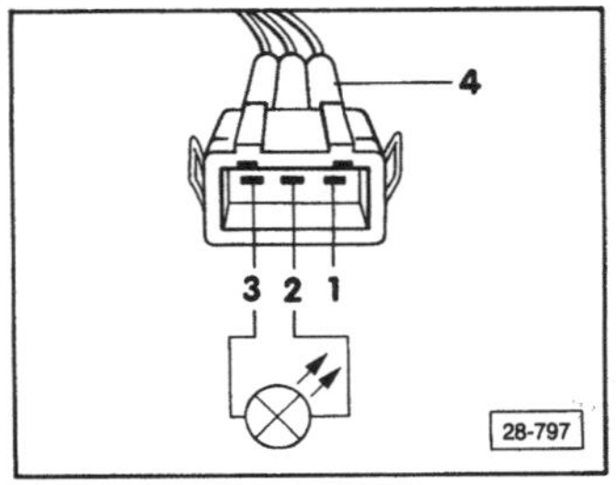

Power output stage, checking

- switch multimeter to 20 Volt range
- connect multimeter between terminals **1** and **3** of disconnected power output stage harness connector **4**
- switch **ON** ignition
  - must be approx. battery voltage

If **NO**

- check for open or short circuit in harness wiring using wiring diagram, repair or replace as necessary
- switch **OFF** ignition
- disconnect multimeter
- disconnect fuel injector main harness connector or remove fuse **18**
- connect **US 1115** LED tester between terminals **2** and **3** of disconnected fuel injector harness connector **4** using jumper leads from **VW 1594** adaptor kit
- operate starter to check signal from the Hall sender and the ignition switching function of the control unit
  - LED tester must flicker

If **NO**

- check Hall sender (**G 40**), page 28.20

If Hall sender **OK**

- **replace Digifant control unit (J 169)**
- reconnect fuel injector main harness connector or reinstall fuse **18**
- reconnect ignition coil wire and power output stage harness connector
- switch **OFF** ignition

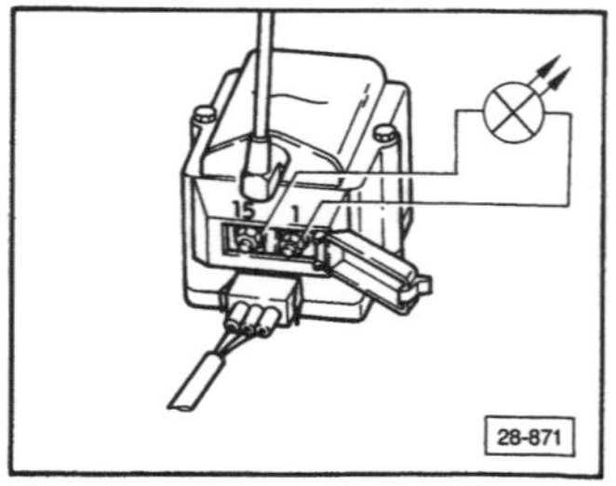

- connect multimeter between terminals **1** and **15** of ignition coil

> **CAUTION**
>
> Do not handle or cause the test connections to become disconnected during the following step.

- switch **ON** ignition
  - must be 2 Volts minimum and then drop to 0 Volts after 1 to 2 seconds
- operate starter
  - LED tester must flicker

If **NO**

- replace ignition coil with power output stage (**N 157**)
- switch **OFF** ignition

Hall sender (G 40), checking

Requirement

- Electrical Test A **OK** from Repair Group 01 (Fuel Injection and Ignition)

Note

The Hall sender can also be checked using the **VAG 1598** Test box. See Repair Group 01 (Fuel Injection and Ignition).

- perform following checks using **Fluke 83** multimeter (**US 1119**)
  - specified values are valid when measured at a temperature between 0 and 40°C (32 and 104°F)

If measured values deviate from specified values

- check for open or short circuits using proper wiring diagram before replacing components
- use the **VW 1594** adaptor kit to make connections between the multimeter and the harness connectors

> **CAUTION**
>
> Switch the multimeter to the appropriate measuring range before making test connections.

Voltage supply, checking

Requirement

- electrical check of Digifant system **OK**, see Repair Group 01 (Fuel Injection and Ignition)

- switch **OFF** ignition
- disconnect harness connector from (**N 41**) Hall control unit
- switch multimeter to 20 Volt range

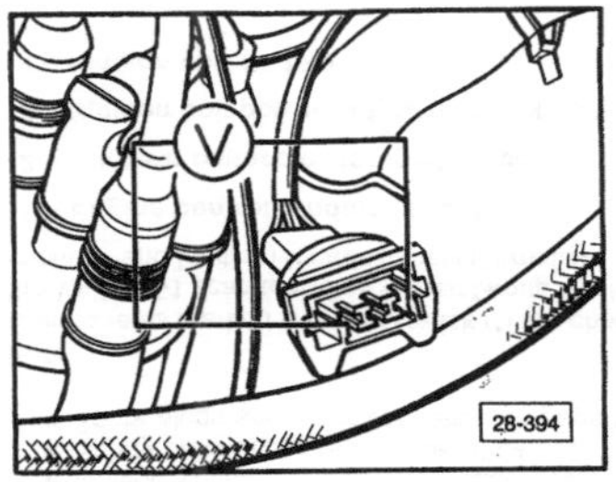

- connect multimeter between outer terminals of harness connector using **VW 1594** adaptor kit
- switch **ON** ignition
  - must be 10 Volts minimum

If **NO**
- check for open or short circuit in harness wiring using wiring diagram, repair or replace as necessary

If wiring **OK**
- replace Digifant control unit (**J 169**)
- switch **OFF** ignition
- reconnect harness connector to Hall control unit

Functional check

- disconnect main harness connector for fuel injectors
- disconnect Hall sender harness connector at distributor
- push up rubber boot (covering connector) to expose terminals and reconnect to distributor
- connect **US 1115** LED tester to center terminal and then either of the outer terminals using **VW 1594** adaptor kit
- operate starter
  - LED tester must flicker

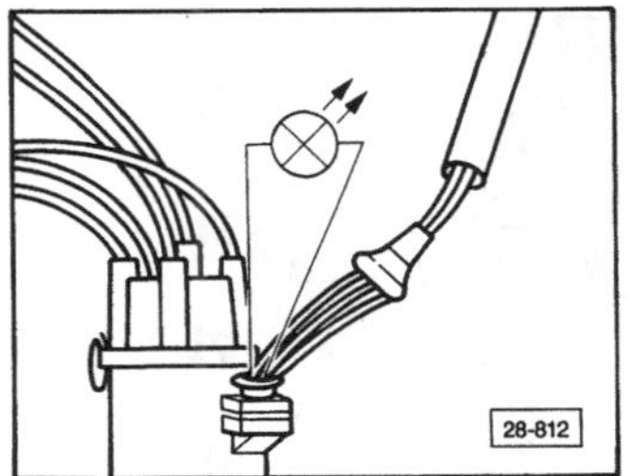

If **NO**
- replace distributor

THIS PAGE INTENTIONALLY LEFT BLANK

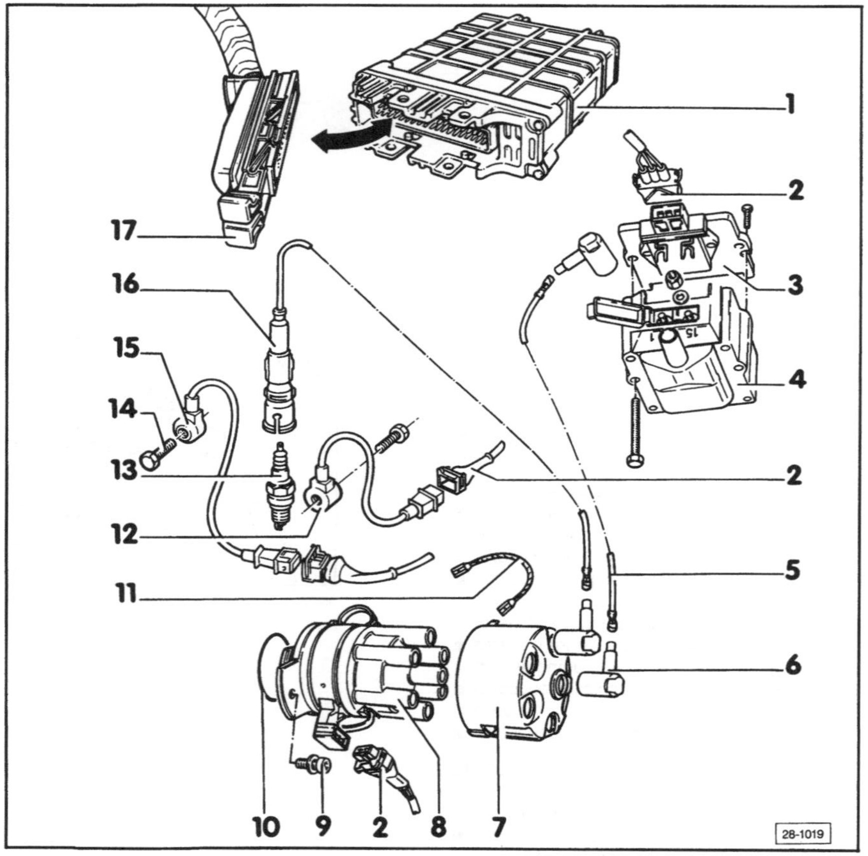

- Ignition system repairing, see page 28.22
- Safety precautions, see page 28.24
- Spark plugs and technical data, see page 28.23

Note

Components marked with an asterisk (*) are checked via the **VAG 1551** scan tool. See Repair Group 01 (Fuel Injection and Ignition) for additional information.

1 – **Engine control module (J 220)***

2 – **Harness connector for ignition coil**

3 – **Ignition coil power output stage (N 157)**

4 – **Ignition coil (N 152)**

5 – **Ignition wire**

6 – **RFI suppression connector**

7 – **RFI suppression cap**

8 – **Distributor cap**

9 – **10 Nm (7 ft lb)**

10 – **O-ring**
replace if damaged

11 – **Ground connector**

12 – **Knock Sensor II (G 66)***

Ignition System

13 – **Spark plug**
- **25 Nm (18 ft lb)**
- use **3122 B** adaptor to remove and install

14 – **20 Nm (15 ft lb)**

15 – **Knock Sensor I (G 61)***

16 – **Spark plug connector**
use **3277** adaptor to remove and insert

17 – **Harness connector for engine control module**

> CAUTION
> Do **NOT** disconnect until ignition has been switched off for at least 20 seconds.

Note

Refer to the Supplement at the back of the manual for additional information pertaining to this Repair Group.

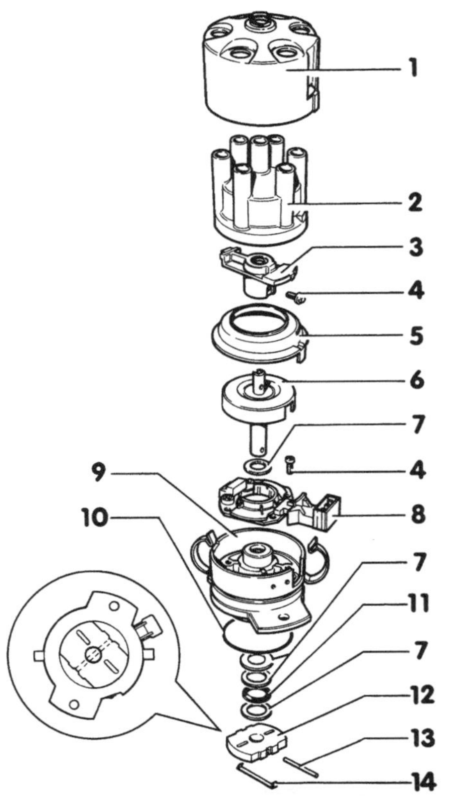

1 – **RFI suppression Cap**

2 – **Distributor cap**

3 – **Distributor rotor**
- 1000 Ohms resistance
- always replace

Note

The rotor is glued to the shaft. If the rotor has to be replaced, remove it by destroying it with a pair of pliers.

Carefully remove any adhesive residue from the shaft and glue on the new rotor using adhesive **AMV 185 100 01** or equivalent.

> **CAUTION**
> **Do NOT** hit the rotor, there is danger of damaging the rotor shaft or bearing.

4 – **5 Nm (44 in. lb)**

5 – **Dust shield**

6 – **Distributor shaft**
- before removing coupling **12**, mark coupling per position to shaft **6**
- after removing shaft, pull out pin **13**

7 – **Shims**

8 – **Hall sensor**
checking see page 28.26

9 – **Distributor housing**

10 – **O-ring**
replace if damaged

11 – **Plastic shim**

12 – **Coupling**
before removing coupling, mark coupling per position to shaft **6**

13 – **Pin**
always replace

14 – **Clip**

Ignition and spark plugs, technical data

Ignition Timing	Check using **VAG 1551** Scan Tool for Idle checking see Repair Group 24
Engine control module Part Number	**021 906 258 P** for manual transmission **021 906 258 S** for automatic transmission
Firing order	1-5-3-6-2-4
Spark Plugs	
To remove or insert the spark plug connectors, use adaptor **3277**.	
Bosch Part Number	C 9 MCC
Electrode Gap	0.7 to 0.8 mm
Tightening Torque	25 Nm (18 ft lb)

Safety precautions

CAUTION
Observe the following precautions to prevent personal injury as well as possible damage to sensitive electrical components.

- switch **OFF** the ignition before connecting or disconnecting components or test equipment
- connect and disconnect battery **ONLY** with ignition switched **OFF** otherwise the engine control module could be damaged
- do not touch or remove ignition wires with engine running or while cranking
- if the engine must be cranked but not started (for compression testing etc.) disconnect harness connector from ignition coil power output stage and remove fuse **18**
- do **NOT** connect a condenser of any kind to terminal **1** of the ignition coil
- when installing noise suppressors, **ONLY** use 1000 Ohms for high tension wires and 5000 Ohms for spark plug connectors
- do **NOT** replace distributor rotor (mark **R1**) with a different type

Ignition coil, checking

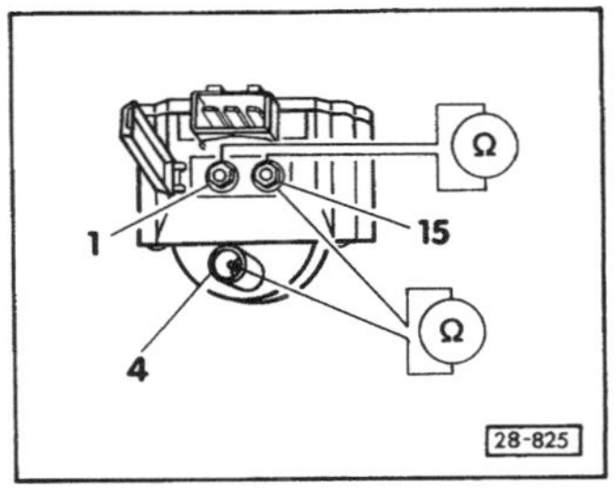

- switch **OFF** ignition
- disconnect harness connector and ignition wire from ignition coil (**N 152**)
- switch **Fluke 83** multimeter to resistance range
- connect multimeter between terminals **1** and **15**
  - primary resistance must be 0.5 to 0.7 Ohms
- connect multimeter between terminals **4** and **15**
  - secondary resistance must be 3000 to 4000 Ohms

If specified values not obtained

- disconnect ground strap from battery
- remove ignition coil and unscrew power output stage (**N 157**)
- repeat check

Ignition coil power output stage

Voltage supply, checking

Requirements
- no DTC's stored in DTC memory
- electrical check of Motronic system **OK**, see Repair Group 01 (Fuel Injection and Ignition) for additional information
- ignition coil **OK**

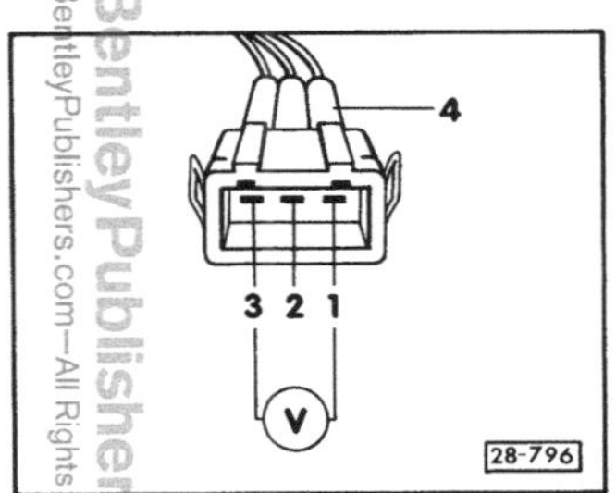

- disconnect harness connector from ignition coil power output stage
- switch **Fluke 83** multimeter to 20 Volt range
- connect multimeter to terminals **1** and **3** of harness connector using adaptor leads from **VW 1594** adaptor kit
- switch **ON** ignition
  - must be approx. battery voltage
- switch **OFF** ignition

Power output stage triggering, checking

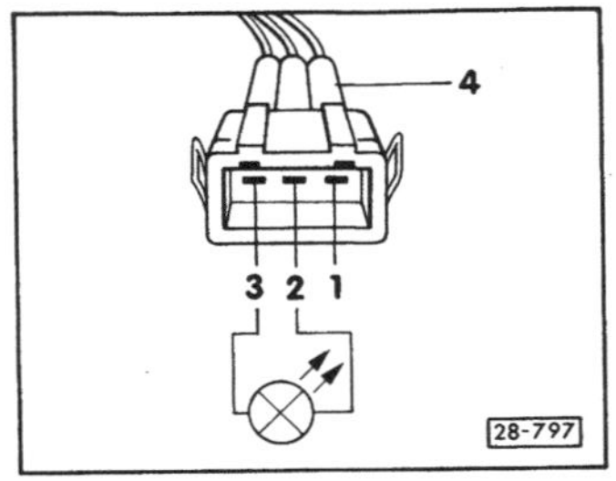

- connect **US 1115** LED tester to terminals **2** and **3** of harness connector using test adaptor **VW 1594/15** from **VW 1594** adaptor kit
- operate starter to check ignition signal from engine control module
  - LED tester must flicker

If **NO**

- replace engine control module (**J 220**)
- re-connect harness connector **4** and coil wire to ignition coil

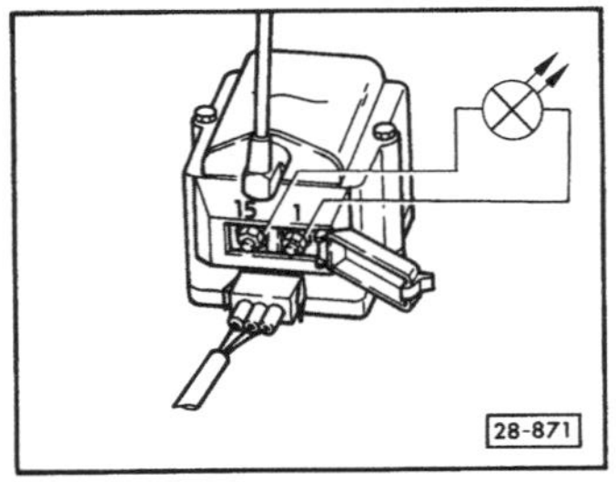

- connect LED tester between terminals **1** and **15** on ignition coil using jumper leads from **VW 1594** adaptor kit

> **CAUTION**
> During the following check, **DO NOT** touch any of the exposed terminals or connectors from the ignition coil or adaptor wires.

- switch **ON** ignition
  - LED tester must light up for 1 to 2 seconds
- operate starter
  - LED tester must flicker

If **NO**

- replace ignition coil power output stage

Hall sensor, checking

Note
The Hall sensor can also be checked using the **VAG 1598** Test Box and **VAG 1598/18** adaptor. See Repair Group 01 (Fuel Injection and Ignition)

If **VAG 1598** Test Box is not available; check Hall sensor as follows:

Voltage supply, checking

- disconnect Hall sensor harness connector at ignition distributor
- switch **Fluke 83** multimeter to 20 Volt range
- connect multimeter between outer terminals of harness connector using jumper wires from **VW 1594** adaptor kit
- switch **ON** ignition
  - must be 4.5 Volts minimum

If no voltage indicated

- switch **OFF** ignition
- check wiring using wiring diagram

If wiring **OK**

- replace engine control module (**J 220**)

Function, checking

- remove fuse **18**
- slip rubber boot for Hall sensor harness connector away from the connector (leaving harness connector connected, but with terminals exposed to allow access for test connections)
- connect **US 1115** LED tester between center terminal of connector and brown/white wire using jumper leads from **VW 1594** adaptor kit
- operate starter
  - LED tester must flicker

If **NO**

- replace Hall sensor

THIS PAGE INTENTIONALLY LEFT BLANK

Index

02A 5–speed

Clutch
- assembly **30.7**
- removing/installing **30.8**

Clutch diaphragm spring ends
- checking **30.8**

Clutch pedal
- assembly **30.2**

Clutch release
- assembly **30.5**

Flywheel
- bolt torque **30.7**

Hydraulic components
- assembly **30.4**

Main shaft seal
- driving in **30.6**
- replacing **30.9**

Pressure plate
- replacing **30.9**

Return spring
- removing/installing **30.3**

Troubleshooting
- chart **30.10**

Note

Lubricate all bearing and contact surfaces with MoS_2 lubricant.

Replace all self-locking nuts and bolts when working on pedal cluster.

> **CAUTION**
>
> Remove ground strap from battery when working on pedal cluster.

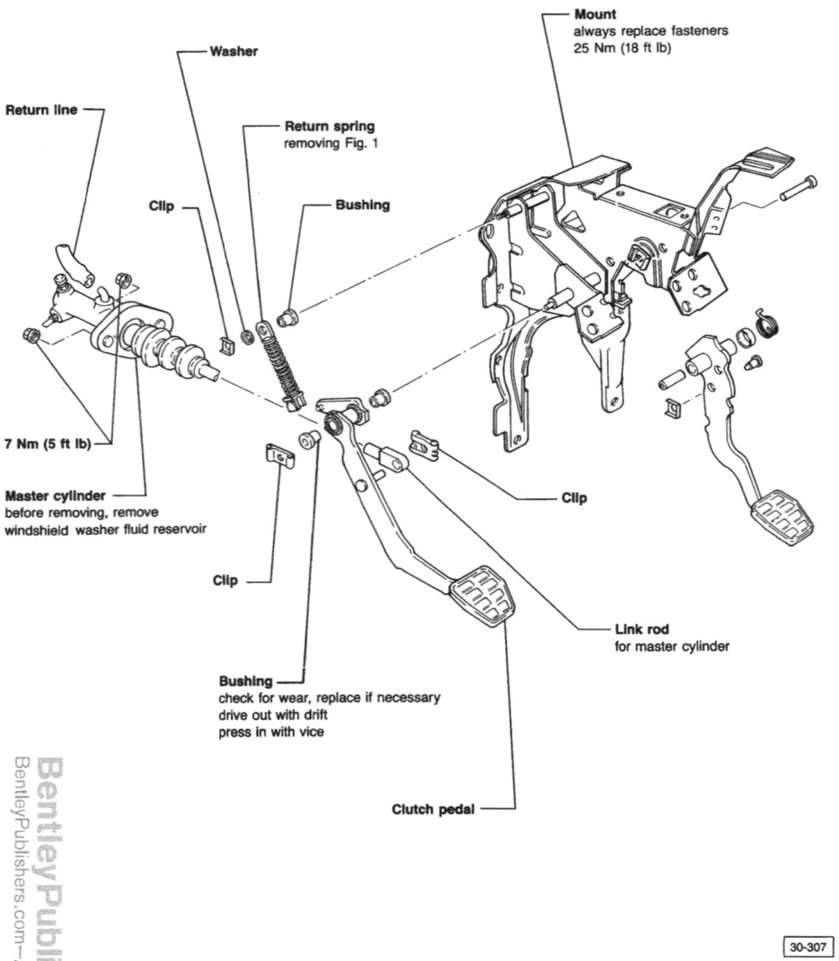

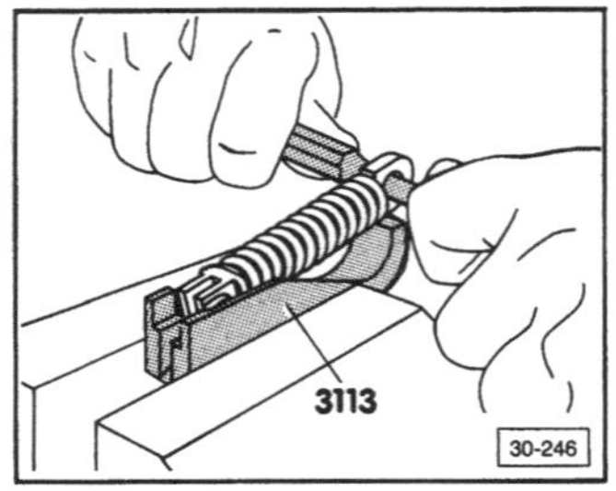

▶ Fig. 1 **Return spring, removing/installing**

- pry off clip
- remove washer
- tighten return spring by moving clutch pedal

- press on holder **3113A** from above
- disconnect linkage rod at pedal
- pull clutch pedal up to stop

- remove return spring together with **3113A**
- install in reverse order

> **CAUTION**
>
> If clutch pedal does not return fully:
> - there could be air in hydraulic system
> - pedal bushing may be tight or return spring could be jammed.

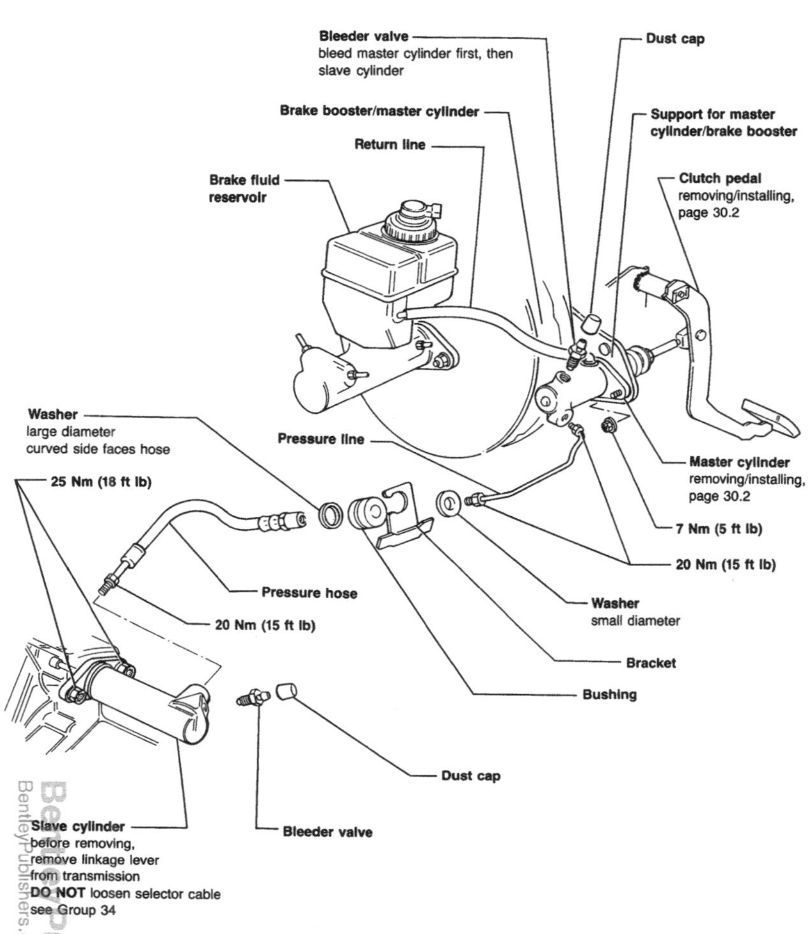

CAUTION
DO NOT depress clutch pedal after slave cylinder is removed.

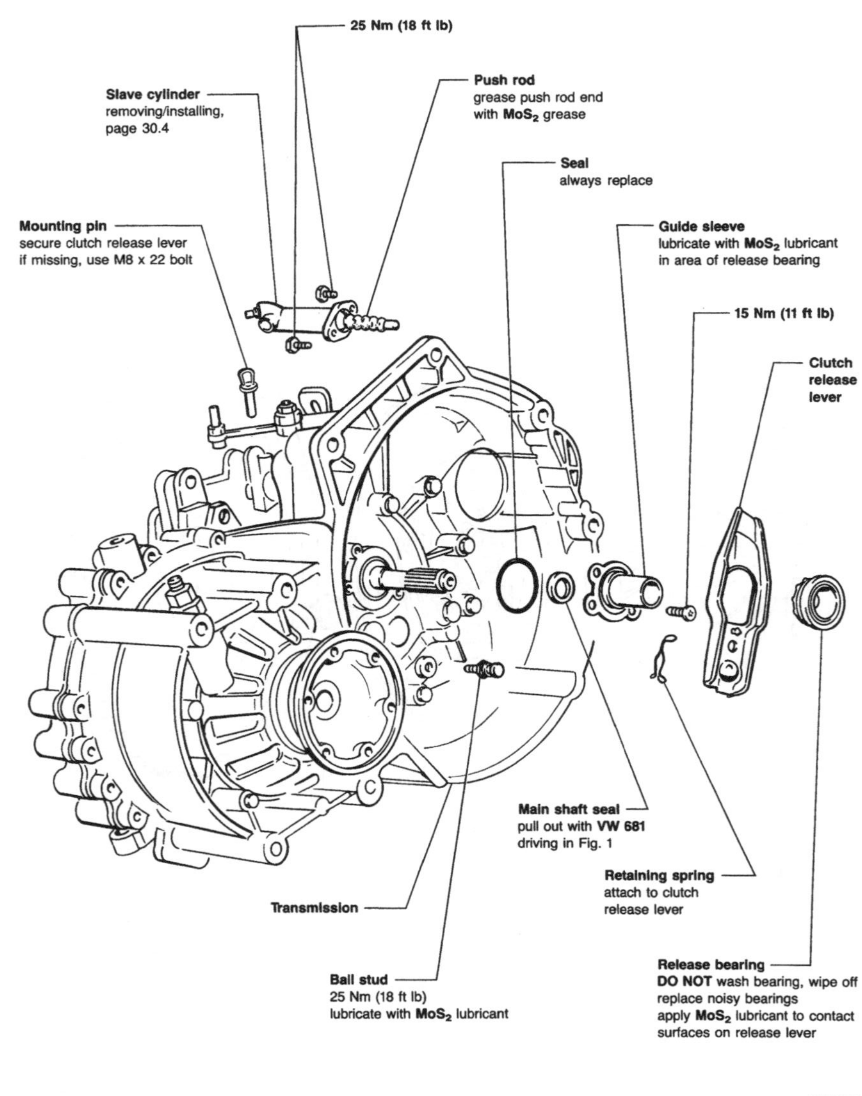

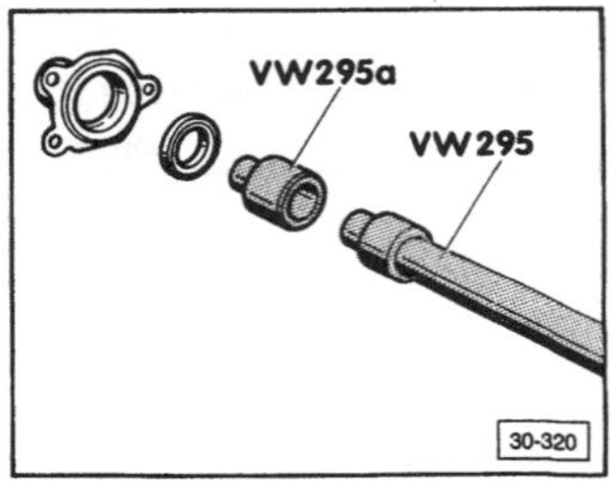

▶ **Fig. 1 Main shaft seal, driving in**

Note

Transmission must be removed for clutch repair.

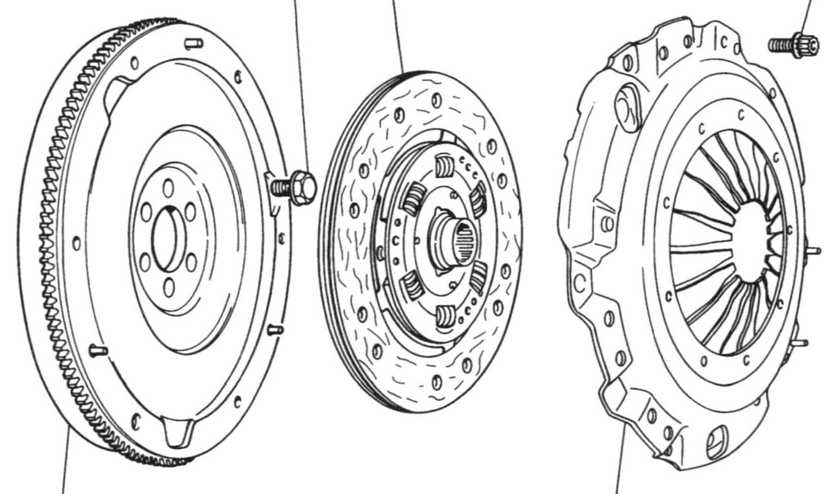

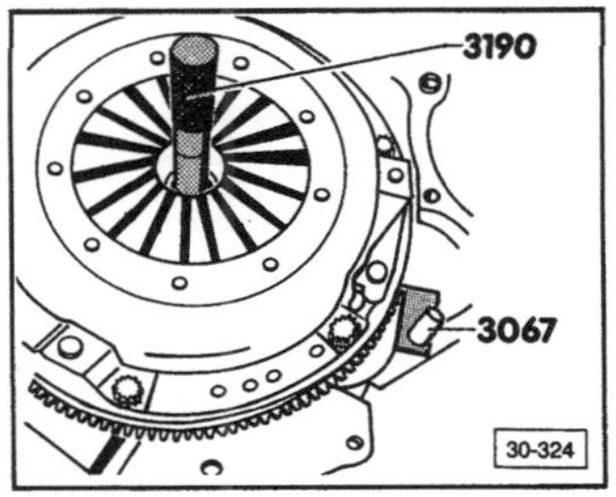

▶ Fig. 1 **Clutch, removing/installing**

- lock flywheel in position
- loosen/tighten bolts diagonally
  - 20 Nm (14 ft lb)

> **CAUTION**
>
> Pressure plate must make complete contact with flywheel before installing the mounting bolts.

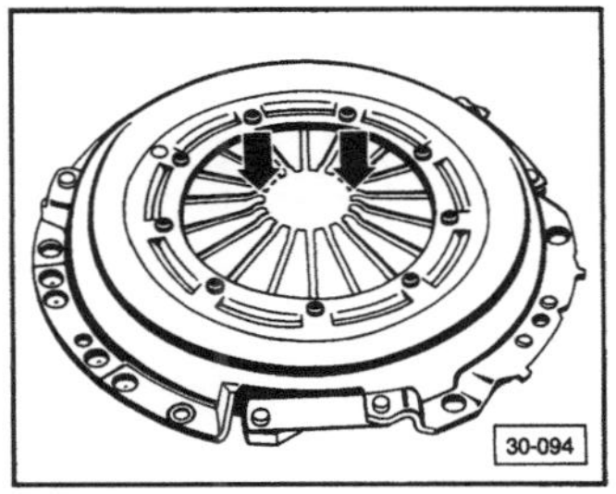

▶ Fig. 2 **Clutch diaphragm spring ends, checking**

- scores up to a depth of 0.3 mm are acceptable for reuse

Note

Remove corrosion and grease residue from the input shaft/splines, and also from the hub teeth if the disc is to be re-used. Grease hub lightly with **G 000 100**, then move clutch disc back and forth on input shaft until hub slides easily.

Excess grease **must** be removed.

Clutch disc, replacing

Replace clutch disc only if one or more of the following has occurred:

Clutch lining

- worn up to rivet head
- oiled/greased
- burned
- split
- loosened
- cracks or grooves
- damaged rivets

Hub

- rusted
- damaged grooves

Spring

- rusted solid
- broken
- cracked
- loose

Torsion damper

- guide for torsion spring broken
- cover damaged
- torsion spring broken or loose
- grooved
- non-functioning

Pressure plate, replacing

Replace the pressure plate only if one or more of the following has occurred:

Contact surface

- burn marks
- cracks
- chatter marks
- grooves
- defective spring connections

Diaphragm spring

- scores due to wear
- pointed tips
- bent or broken

Housing

- distorted
- broken
- defective spring connections
- bores for dowel pins deformed

Riveted connections

- damaged
- loose

> **CAUTION**
>
> Pressure plates are protected against corrosion and greased. Clean only the contact surface or service life of the pressure plate will be shortened.

Complaint	Possible cause	Corrective action
Clutch pedal does not return to original position (hydraulic clutch)	Air in hydraulic system, brake fluid level too low	Replenish brake fluid; bleed hydraulic system
	Piston seizes in master or slave cylinder	Replace defective part; bleed hydraulic system
	Hydraulic system, or master and slave cylinder leaking	
(mechanical clutch)	Clutch cable hard to operate	Replace clutch cable
	Self-adjusting mechanism (if equipped) defective	Replace clutch cable
(all vehicles)	Return spring (if equipped) defective	Replace return spring
	Clutch pedal hard to operate	Clean pivot points, lubricate. If necessary, replace bushing
	Linkage on transmission hard to operate	Clean pivot points, lubricate. Make necessary repairs
	Mechanical components in clutch housing binding or dragging	Clean pivot points, lubricate. Make necessary repairs
	Clutch release bearing twisted on guide sleeve, seized	Replace guide sleeve and clutch release bearing
	Diaphragm spring of pressure plate broken	Replace pressure plate
Excessively hard clutch pedal (mechanical clutch)	Clutch cable binds or drags. Cable corroded in guide. Self-adjusting mechanism defective (if equipped)	Replace clutch cable
	Linkage on transmission binding or dragging	Clean pivot points, lubricate. Replace bushing
(manual transmission — 020 only)	Pushrod and pressure plate bind or drag	Lubricate contact points lightly with **G 000 100** Lightly lubricate clutch pushrod
(all vehicles)	Clutch pedal binding or dragging **Note:** Prior to checking, disconnect master cylinder and/or clutch cable from clutch pedal	Clean pivot points, lubricate. If necessary, replace bushing
	Over-center spring action drags or binds. Over-center spring defective	Clean pivot points, lubricate. If necessary, replace over-center spring
	Return springs (if equipped) too strong/wrong return spring	Replace with correct return spring

Complaint	Possible cause	Corrective action
Excessively hard clutch pedal (all vehicles) — continued	Clutch release force increased due to wear of clutch linings	Inform customer: release force becomes higher with increasing wear Replace clutch disc if lining/rivet distance is below 0.1 mm
	Mechanical components in clutch housing bind or drag	Clean pivot points, lubricate. If necessary, replace pivot bushings
	Release bearing twisted on guide sleeve, seized	Replace defective parts
	Contact surface release bearing/release lever worn	Replace defective parts
	Pressure plate with wrong spring	Replace with correct part no.
	Clutch disc binds or drags on gears	Check gears of hub for defects (burrs). If necessary, replace clutch disc. Clean corrosion and lubricant from gears of hub and input shaft. Lubricate with grease, **G 000 100**. Move clutch disc back and forth; remove surplus grease

Complaint	Possible cause	Corrective action
Noises during clutch operation (mechanical clutch)	Transmission noises enter passenger compartment via clutch cable	Replace or add insulating components as necessary
	Clutch cable creaks, binds or drags	Replace clutch cable
	On full pedal application, diaphragm spring rubs against clutch disc (insufficient clutch free play)	Adjust clutch free play
	Self-adjusting mechanism (if equipped) defective	Replace clutch cable
	Release plate contact points and diaphragm spring need lubricating	Lightly grease contact points with **G 000 100**. Replace worn parts.
(manual transmission — 020 only)	Contact points, pushrod, release plate need lubricating	Lightly grease contact points with **G 000 100**
(all vehicles)	Clutch pedal binds or drags/ pivot points misaligned	Clean pivot points, lubricate. If necessary, replace bushing
	Pivot of over-center spring dry, dragging, misaligned	
	Return spring (if equipped) noisy	Lubricate pivot points
	Release bearing or release bearing guide defective, contact surface worn	Generally, replace noisy release bearing. Replace damaged guide sleeves
	Contact surface (diaphragm spring tips) of pressure plate defective (bent, broken). Release bearing off-center in contact area	Replace pressure plate. Check release bearing and guide sleeve; replace if necessary. Check adaptor sleeves.
	Pilot bearing (if equipped) in crankshaft defective, engine/ transmission offset from center	Replace; lubricate with MoS_2 grease. Check adaptor sleeves
	Clutch disc installed improperly	Correct installation
	Wrong clutch disc installed	Replace with correct clutch disc

Complaint	Possible cause	Corrective action
Grinding noises when engaging a forward or the reverse gear, shift mechanism binds, drags; shifting not possible, clutch inoperative (hydraulic clutch)	Brake fluid level too low	Check system. Replenish brake fluid, bleed system
	Air in system; clutch does not disengage completely	
	Master/slave cylinder leaking, aged, line is too elastic	Replace defective part. Replenish brake fluid, bleed system
	Adjustment of clevis not correct	Correct adjustment
(mechanical clutch)	Mechanical components misaligned	Replace mechanical components
	Clutch free play excessive	Check clutch free play, adjust if necessary
	Wrong clutch cable installed (too long)	Replace with correct clutch cable
	Clutch cable defective; binds or drags	Replace clutch cable
	Self-adjusting mechanism (if equipped) defective	
(manual transmission — 020 only)	Pushrod too short due to wear	Replace pushrod, check release bearing. Lightly lubricate pushrod/release bearing contact points with **G 000 100** grease
(all vehicles)	Clutch pedal travel insufficient (carpet, floor mat beneath pedal). Clutch not being fully depressed	Inform customer
	Only reverse gear grinds when engaged	Inform customer. Depending on clutch diameter, wait approximately 3-6 seconds after depressing clutch before engaging reverse gear. Input shaft with clutch disc must first come to a stop
	Adjustment of shifting mechanism	Check, correct if necessary
	Bearing for shift lever and shift operation not lubricated; misaligned	Lubricate shift mechanism, replace defective parts

Complaint	Possible cause	Corrective action
Grinding noises when engaging a forward or the reverse gear, shift mechanism binds, drags; shifting not possible, clutch inoperative (all vehicles) — continued	Clutch disc binds or drags on gears. Hub corroded, or was damaged during installation. Hub misaligned on one side	Check gears of hub for damage, replace clutch disc if necessary. Remove corrosion and grease residue from hub and shaft. Grease hub lightly with **G 000 100**. Move clutch disc back and forth; remove surplus grease. When hub is misaligned, check position of adaptor sleeves. Check release bearing, guide sleeve, pressure plate and pilot bearing. Replace if necessary
	Pressure plate lift-off too slight (wrong pressure plate installed)	Replace with correct pressure plate
	Pilot bearing on crankshaft defective. Input shaft still driven when clutch disengaged	Replace pilot bearing and lubricate with MoS_2. Check adaptor sleeves. Replace if necessary. Check shaft
	Engine/transmission offset too large (adaptor sleeves missing), thus support plate of clutch disc bent	Install adaptor sleeves prior to transmission installation. Check clutch disc, pressure plate and pilot bearing (if equipped) for damage; replace if necessary
	Pilot bearing in crankshaft defective	
	Lining worn due to excessive rpms: down-shifting at too fast a speed	Replace clutch disc; inform customer
	Lining worn from riding the clutch when accelerating	
	Synchronizing system and/or shifting mechanism in transmission defective	Repair transmission

Complaint	Possible cause	Corrective action
Grinding noises when engaging a forward or the reverse gear, shift mechanism binds, drags; shifting not possible, clutch inoperative (all vehicles) — continued	Pressure plate uneven due to wrong installation. Clutch disc distorted due to improper handling	Check parts, replace if necessary. Observe position of locating pins If grinding occurs thereafter, check splines on clutch disc hub and shaft for ease of operation, check pilot bearing (if equipped) in crankshaft. If necessary, repair transmission
	2nd gear grinds only when cold	Inform customer. If necessary, replace transmission oil with oil of different viscosity (see specs/ procedures in this manual)
	Diaphragm spring tips broken or bent (installation error: release bearing runs off center)	Replace pressure plate. Check guide sleeve; replace if necessary. Check adaptor sleeves
	Clutch disc too thick	Replace with correct clutch disc
	Lining rusted onto flywheel (long periods of disuse, high relative humidity)	Lightly sand friction surfaces. Replace parts when corrosion is severe

Complaint	Possible cause	Corrective action
Clutch slips; little or no clutch action (hydraulic clutch)	Master/slave cylinder piston does not return to rest position.	Replace master/slave cylinder. Change brake fluid, bleed system.
(mechanical clutch)	Clutch cable improperly adjusted (insufficient clutch free play). Wrong clutch cable	Correct adjustment. Replace with correct cable, if necessary
	Self-adjusting mechanism (if equipped) defective	Replace clutch cable, if necessary
	Clutch cable binds or drags	Replace clutch cable, if necessary
(manual transmission — 020 only)	Clutch release pushrod oil seal in input shaft leaks	Replace oil seal, clutch disc. Clean pressure plate and flywheel
(all vehicles)	Wrong clutch disc, wrong pressure plate installed	Replace with correct clutch disc or pressure plate
	Clutch disc worn, burnt pressure plate, overheated grooves, pressure plate distorted due to wrong installation, pressing force of pressure plate too low. Driving errors, normal wear	Replace clutch disc, pressure plate. Instruct customer
	Mechanical components of transmission drag, pedal linkage binds	Clean pivot points, lubricate; repair if necessary
	Clutch disc, pressure plate, flywheel oil-contaminated. Crankshaft oil seal defective. Grease on contact surfaces from over-lubrication of hub	Replace clutch disc. Clean contact surfaces of pressure plate and flywheel. Replace crankshaft oil seal, remove surplus grease from input shaft
	Clutch disc installed from wrong side	Correct installation. Check clutch disc; replace if necessary
	Flywheel too thick; excessive wear on contact surfaces	Replace with correct flywheel. Check disc and pressure plate; replace if necessary

Complaint	Possible cause	Corrective action
Clutch pulls, power train rattles (hydraulic clutch)	Adjustment on clevis not correct	Correct adjustment
	Air in the system; master cylinder/slave cylinder defective	Replace defective part. Check brake fluid level, bleed system, check for leaks
(mechanical clutch)	Master cylinder/slave cylinder pushrod does not return to rest position	Replace defective part. Change brake fluid, if necessary. Bleed system
	Clutch cable binding	Replace clutch cable
	Clutch pedal binding **Note:** Prior to checking, disconnect clutch pedal from clutch cable.	Clean pivot points, lubricate if necessary. Replace bushing if necessary
	Linkage on transmission binds or drags	Clean pivot points, lubricate. Replace bushing if necessary
(all vehicles)	Engine runs unevenly	Check engine adjustment; correct if necessary
	Driving errors: acceleration rpm too low	Instruct customer
	Wrong clutch disc installed	Replace with correct clutch disc
Noises in idle (manual transmission — 020 only)	Clutch lining, contact surface of pressure plate and flywheel oil-contaminated (oil seeps out of clutch housing)	Check clutch release pushrod oil seal; replace if necessary. Replace clutch disc; clean pressure plate and flywheel
(all vehicles)	Torsion spring broken	Replace clutch disc
	Clutch disc installed without spring cage (rattling in idle)	Install clutch disc with spring cage
	Pressure plate distorted, broken, out-of-round	Replace pressure plate
	Engine runs unevenly	Check engine adjustment; correct if necessary

Complaint	Possible cause	Corrective action
Noises in idle (all vehicles) — continued	Engine mounts are too "soft"; misaligned	Check contact points. Replace if necessary, with correct engine mounts
	Shock absorbers defective	Replace shock absorbers
	Clutch lining, contact surface of pressure plate and flywheel oil-contaminated	Locate cause of contamination; repair as necessary. Replace clutch disc; clean pressure plate and flywheel
	Release bearing twisted on guide sleeve, seized (presses from one side on diaphragm spring of the pressure plate)	Replace release bearing and guide sleeve. Check mechanical components and pivot points
	Contact surface of pressure plate has lift on one side only, due to twisted release bearing	Check contact surface for clutch lining on flywheel pressure plate and diaphragm spring; if necessary replace pressure plate. Replace release bearing and guide sleeve.
	Housing of pressure plate warped during assembly. Contact surface of pressure plate has lift on one side only	
	Input shaft greased excessively (traces of grease on clutch disc, pressure plate and flywheel)	Clean grease from pressure plate and flywheel. Replace if damaged (i.e., scoring, signs of overheating, grooves). Remove all lubricant from hub and input shaft, lubricate input shaft lightly with **G 000 100**. Move clutch disc back and forth; remove excess grease.

Index

Automatic Transmission 096

Torque converter
- removing/installing **32–20**
- identification **32–10**

Torque converter seal
- replacing **32–30**

Torque converter, identification

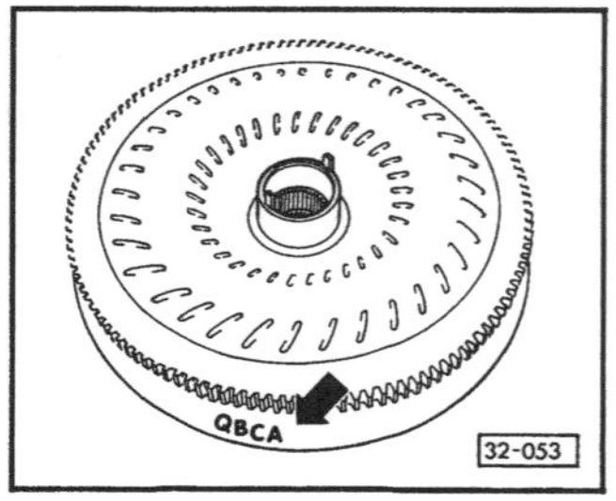

- code letters stamped on converter

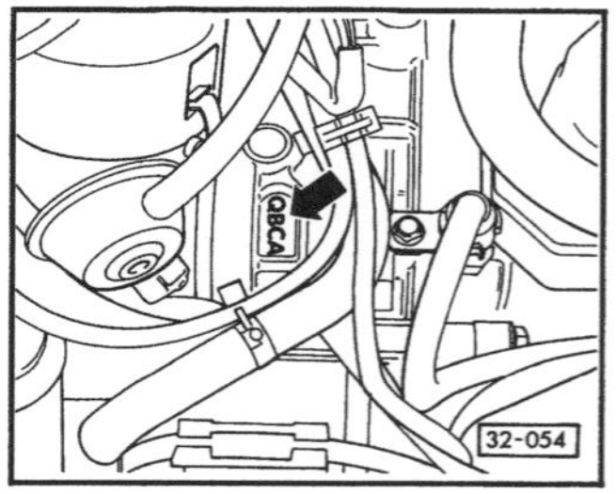

(transmission installed)
- code letters stamped on transmission housing

Torque converter, draining

Note

If the ATF is contaminated, drain the torque converter.

Use a suitable device (e.g. model **EX1**, manufactured by **Fluid Evacuators**) and follow the manufacturer's instructions. Otherwise, drain using the manual method:

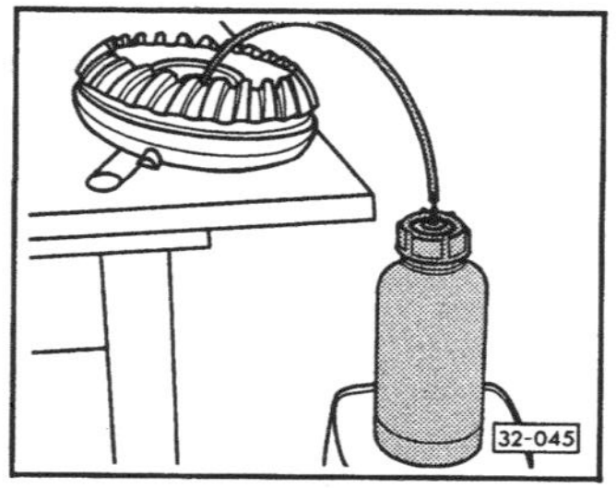

- place converter in sloping position
- attach small inside diameter hose, with maximum outside diameter of 8 mm (5/16 in.), to plastic bottle having 2 liter (2 qt) capacity
  - hose connection and cap must be airtight

- cut converter-end of drain hose on an angle
- squeeze bottle and hold, while inserting free end of hose into converter hub
  - make sure end of hose rests on bottom

- release bottle to allow ATF to flow
- loosen bottle cap at beginning of flow, to allow fluid to siphon

Torque converter seal, replacing

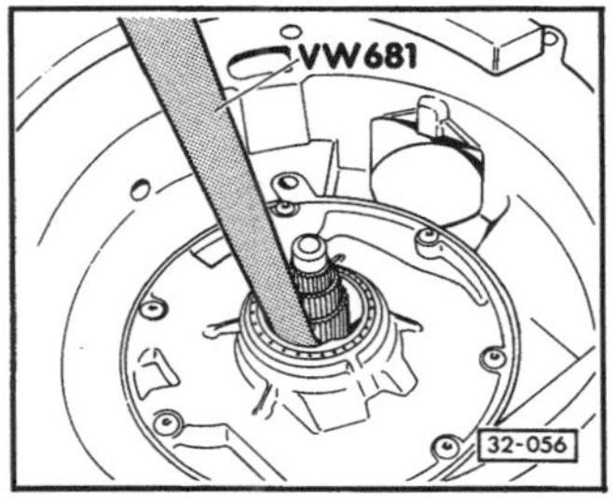

- remove torque converter seal using seal extractor **VW681**

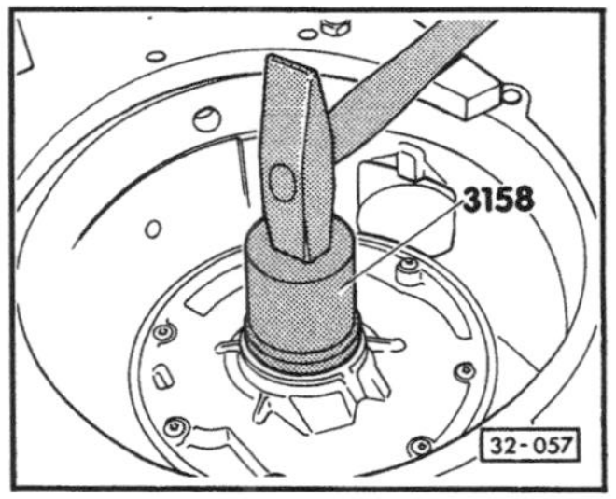

- drive torque converter seal in flush

THIS PAGE INTENTIONALLY LEFT BLANK

THIS PAGE INTENTIONALLY LEFT BLANK

Index

02A– 5 speed

Gearshift cable/lever
- installing **34.6**

Gearshift housing
- repairing **34.7**

Gearshift lever
- repairing **34.8**
- rubber cover **34.9**

Gearshift mechanism
- adjusting/checking **34.10**
- modifications, vehicles with 6 cyl. engines **34.6a**
- repairing **34.4**

Selector housing cap
- installing **34.9**

Technical data
- chart **34.2**

Transmission
- controls, overview **34.3**
- removing/installing **34.12**

Technical data

Code letters		ATA
Production	from	05/89
	to	07/90
Application	type	Corrado
	engine	1.8 Liter 118 KW (160 bhp)
Ratios	final drive	70:19 = 3.684
	1st gear	34:9 = 3.778
	2nd gear	40:19 = 2.105
	3rd gear	39:29 = 1.345
	4th gear	34:35 = 0.971
	5th gear	35:44 = 0.795
	reverse gear	19:9, 36:20 = 3.800
Lubricant	capacity	2.0 Liters (2.1 US qts)
	specification	API GL4, Mil-L 2105, SAE 80 or G50 Synthetic, SAE 75W-90
Clutch actuation		hydraulic
Clutch disc diameter		228 mm
Axle shaft flange diameter		100 mm

Technical data

Code letters		AYL
Production	from	08/90
	to	—
Application	type	Corrado
	engine	1.8 Liter 118 KW (160 bhp)
Ratios	final drive	70:19 = 3.684
	1st gear	34:9 = 3.778
	2nd gear	36:17 = 2.118
	3rd gear	40:28 = 1.429
	4th gear	35:34 = 1.029
	5th gear	36:43 = 0.837
	reverse gear	19:9, 36:20 = 3.800
Lubricant	capacity	2.0 Liters (2.1 US qts)
	specification	API GL4, Mil-L 2105, SAE 80 or G50 Synthetic, SAE 75W-90
Clutch actuation		hydraulic
Clutch disc diameter		228 mm
Axle shaft flange diameter		100 mm

Technical data

Code		CDM
Production	from	10/91
	to	–
Application	type	Corrado
	engine	2.8L – 132 kW
Transmission type		02A
Ratios	final	62:17 = 3.647
	1st gear	33:10 = 3.300
	2nd gear	35:18 = 1.944
	3rd gear	34:26 = 1.308
	4th gear	30:29 = 1.034
	5th gear	31:37 = 0.838
	reverse gear	17:10 x 36:20 = 3.060
	speedometer drive	13:22 = 0.591
Lubricant capacity		2.0 liters (2.1 U.S. qt)
Lubricant type		SAE 75W-90, API GL-4 or Transmission oil G50 (synthetic oil)
Clutch actuation		hydraulic
Clutch disc diameter		228 mm
Drive flange diameter		100 mm
Overall ratio in 5th gear		3.056

THIS PAGE INTENTIONALLY LEFT BLANK

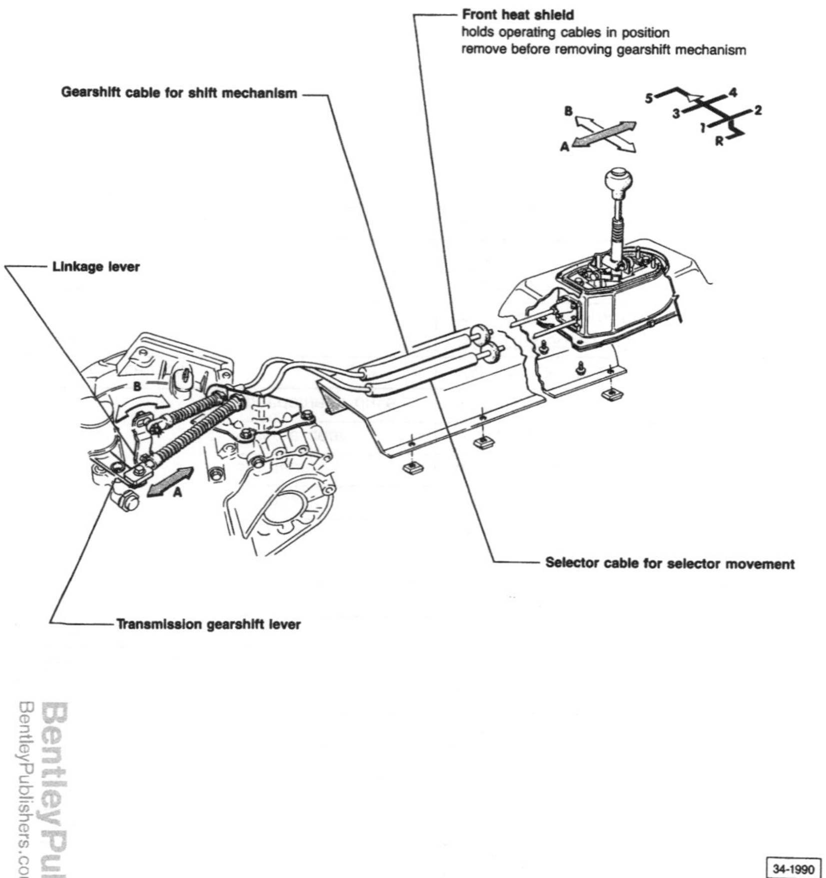

Gearshift mechanism, repairing

Before removing the gearshift mechanism, remove catalytic converter — Repair Group 26, and front heat shield, page 34.3.

Remove gearshift mechanism as a complete unit to change the operating cables.

Do not bend operating cables. Gearshift mechanism, adjusting/checking, page 34.10.

Lubricate bearing locations and sliding contact surfaces with **MoS$_2$** grease.

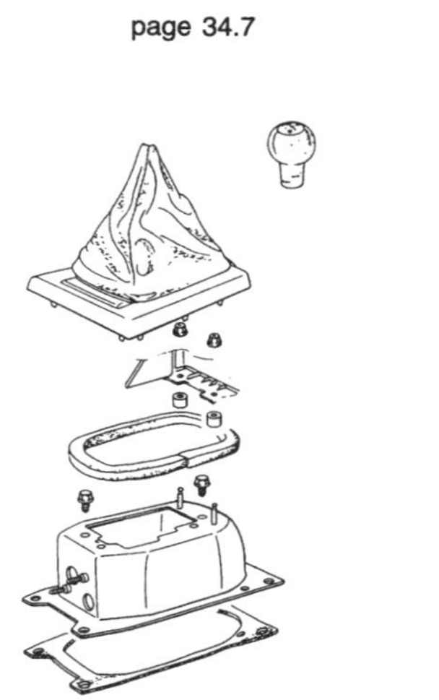

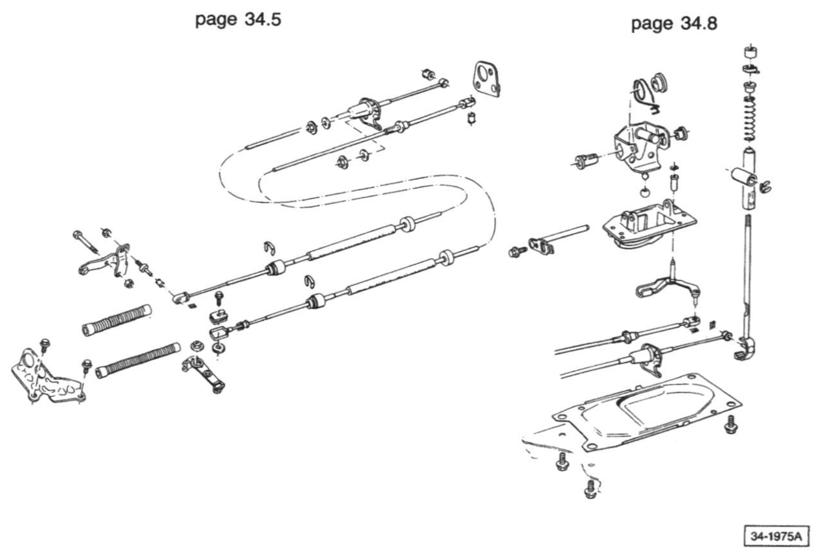

Note

Lubricate bearing locations and sliding contact surfaces with MoS_2 grease.

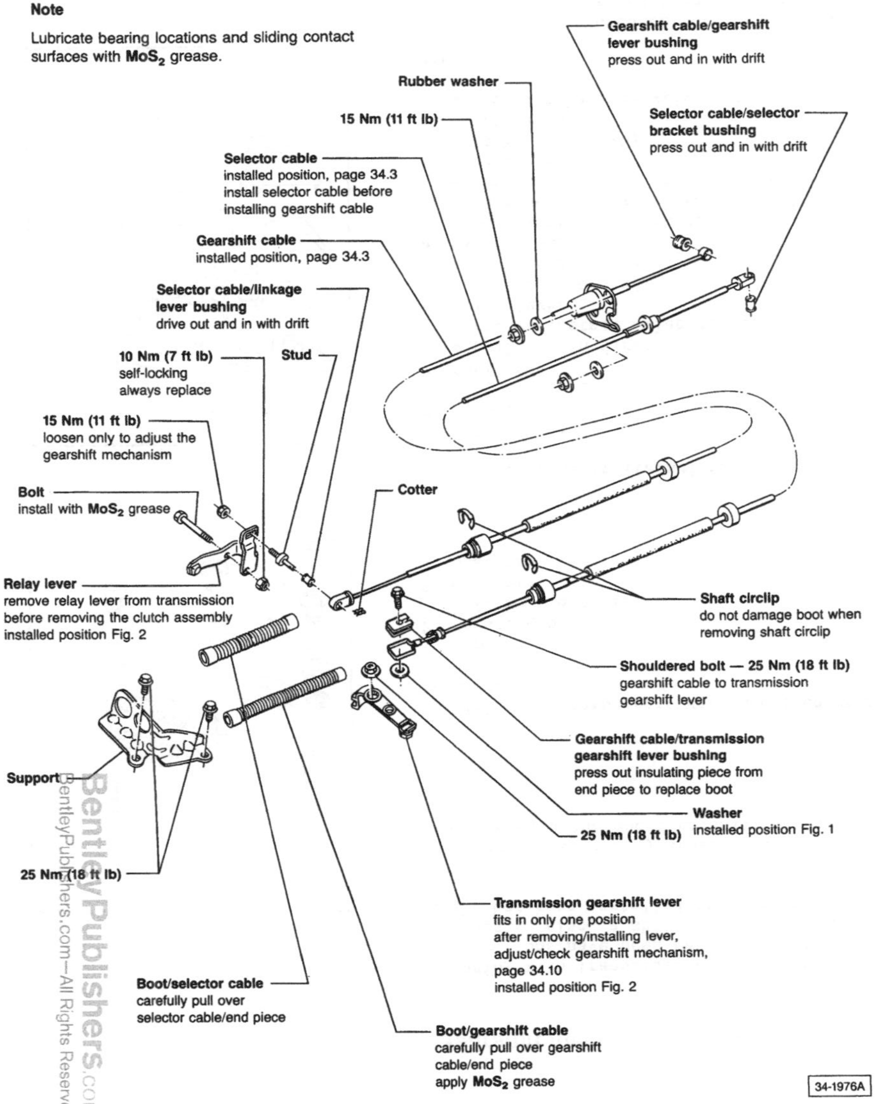

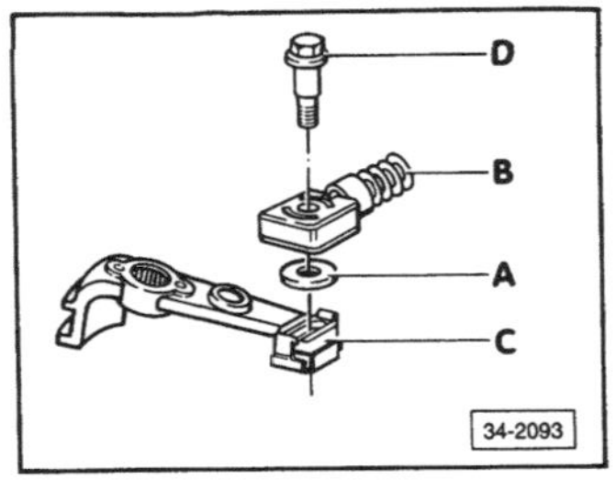

▶ **Fig. 1 Transmission gearshift cable to gearshift lever, installing**

- place washer **A** between shift cable **B** and shift lever **C**, and position with bolt **D**
  - washer prevents jamming during shifting
- tighten bolt **D**
  - tightening torque 25 Nm (18 ft lb)

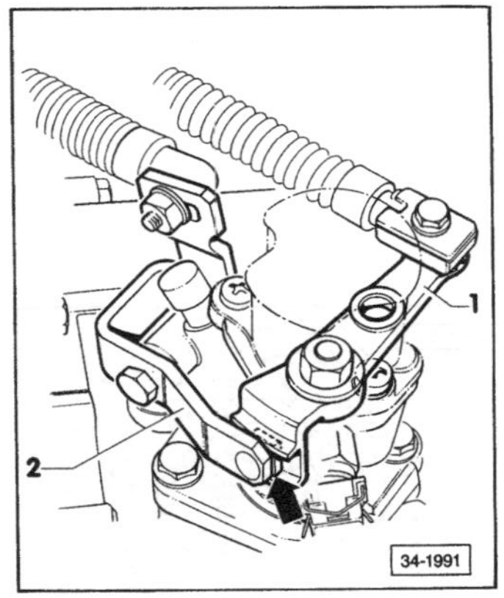

▶ **Fig. 2 Transmission gearshift lever/relay lever, installing**

1 — Transmission gearshift lever
2 — Relay lever

- lubricate with MoS_2 grease
  - meshes into notch of transmission gearshift lever via sliding projection **(arrow)**

Gearshift mechanism, modified

Beginning with transmission build date 03 09 1, modifications have been made to the shift mechanism. Also, a shift mechanism balance weight has been installed in production since 10/91

All vehicles with 6-cylinder engines have been produced with the modified shift mechanism.

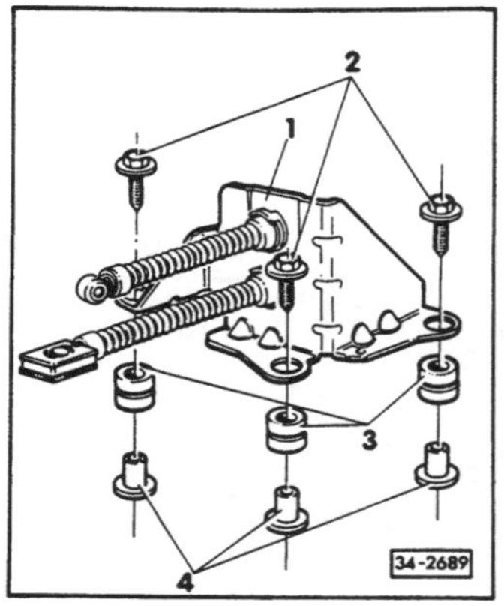

The following components were modified or have been added:

1 Support
2 Mounting bolts
  • 25 Nm (18 ft lb)
3 Grommets
4 Spacers

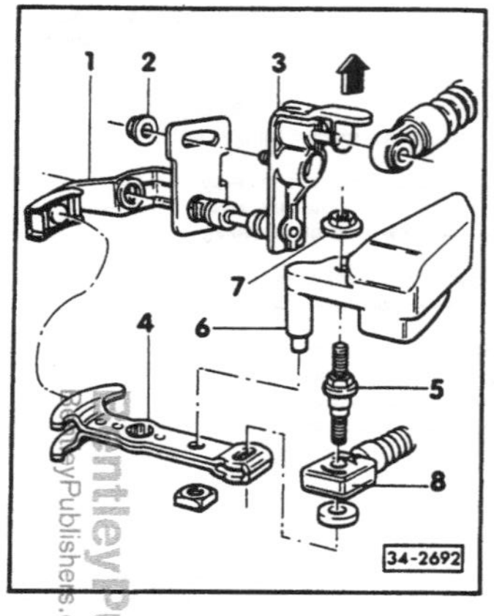

Shift mechanism on transmission

1 Relay lever
2 Mounting nut
  • 15 Nm (11 ft lb)
3 Carrier
  • pull tab in direction of arrow to remove/install selector cable
4 Shift lever
5 Bolt/threaded shaft
  • 25 Nm (18 ft lb)
6 Balance weight
  • only install on vehicles up to 10/91 together with shift cable (8)
  • remove to adjust shift mechanism
7 Mounting nut
  • 25 Nm (18 ft lb)
8 Shift cable
  • use with balance weight (6)

Note

The adjustment procedure for the new shift mechanism is unchanged from the previous version.

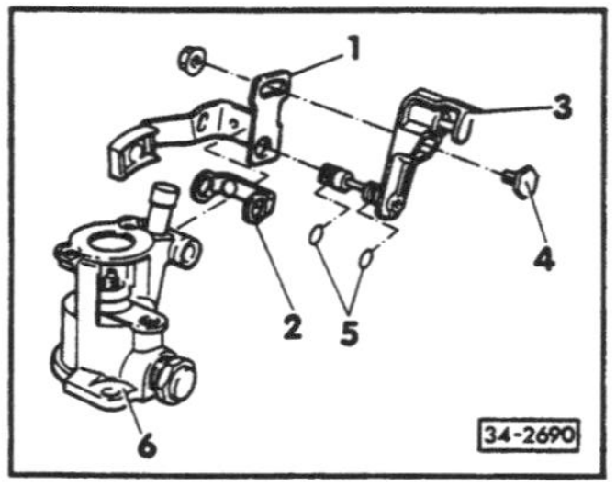

Shift mechanism on transmission, assembly

1 Relay lever
2 Spacer
  • press into relay lever
3 Carrier
4 Bolt
  • press into carrier
5 Bushings
  • install on axle of carrier
6 Gearshift cover

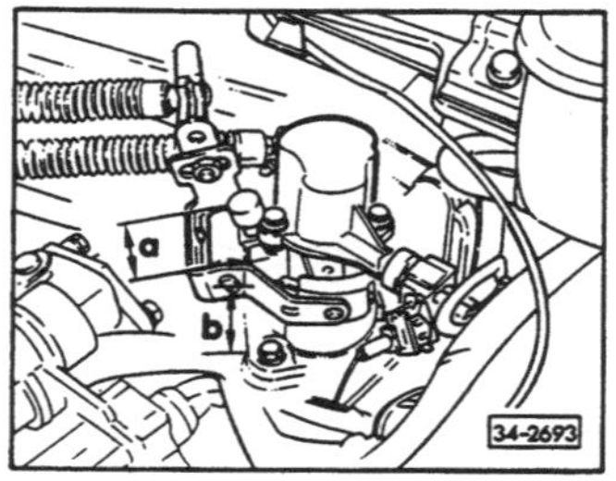

Modified bracket for relay lever

Dimension	New	Previous
a width	40 mm	33 mm
b height from trans housing	85 mm	60 mm

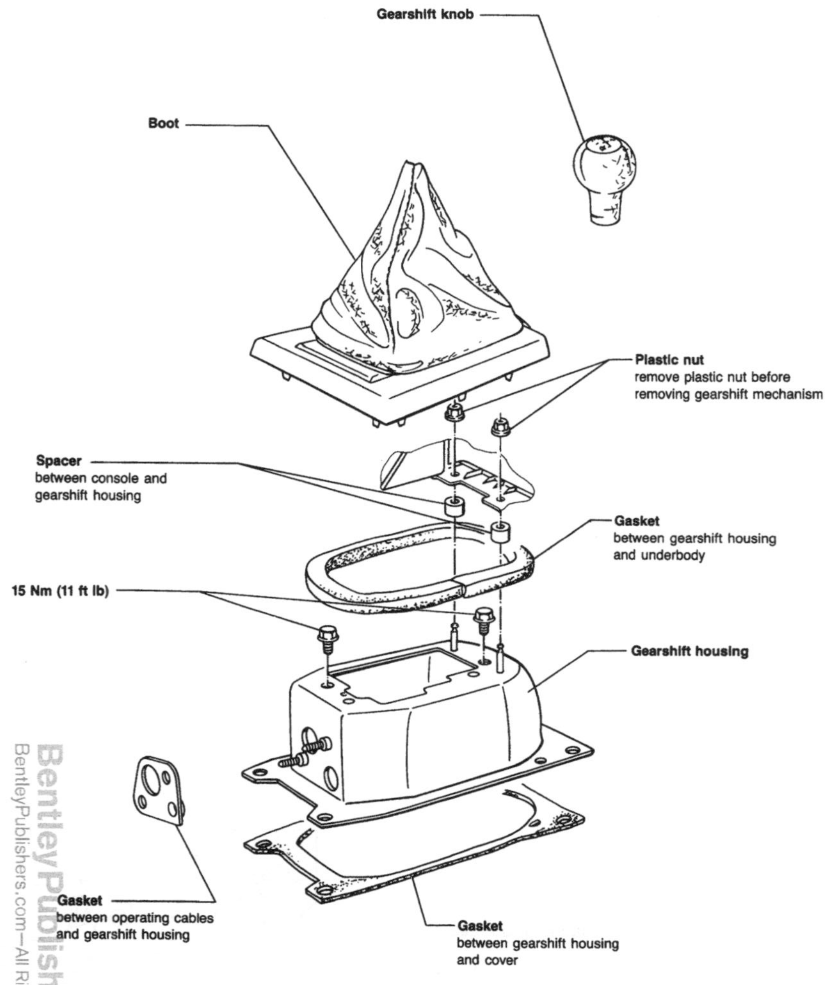

Note

Lubricate bearing locations and sliding contact surfaces with **MoS_2** grease.

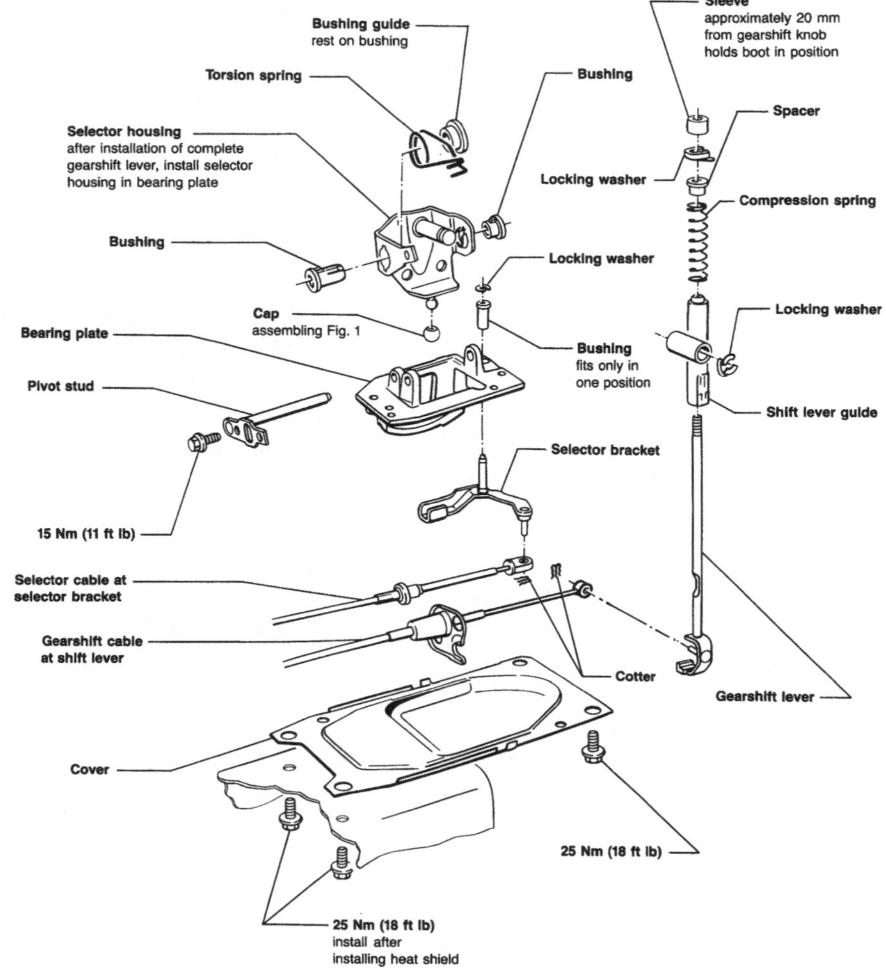

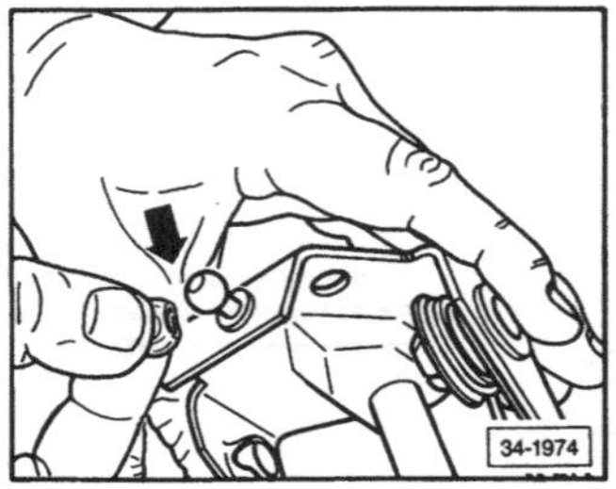

▶ **Fig. 1 Cap, installing**

- press on laterally (**arrow**)
- press opening to form a slit

Note

Heat cap, if necessary (e.g., hot water).

Manual Transmission – Controls, Assembly Replacement

Gearshift mechanism, adjusting/checking

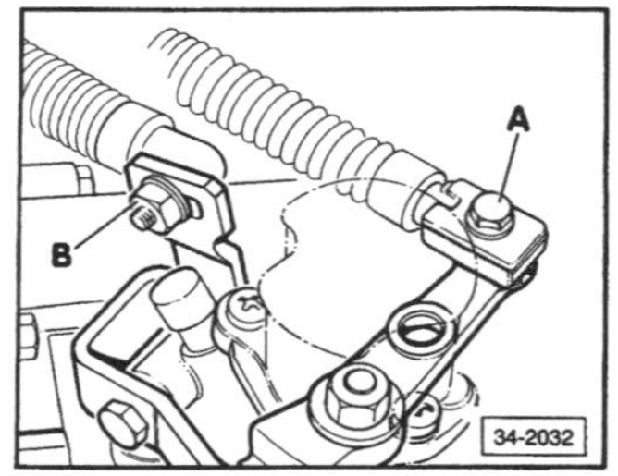

- place gearshift lever in neutral
- remove gearshift lever knob and shift boot

▶
- loosen bolt **A** and nut **B** until operating cables move freely in centering holes

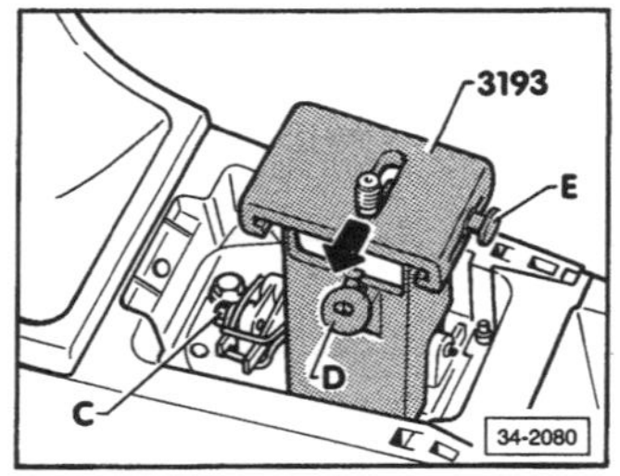

▶
- loosen bolt **C**
- install shift lever gage
- pivot locating pin (for attaching gage) under bearing plate and tighten nut **D**
- press gearshift lever into left detent of slide
- press gearshift lever with slide to left stop (direction of **arrow**) and tighten slide with bolt **E**

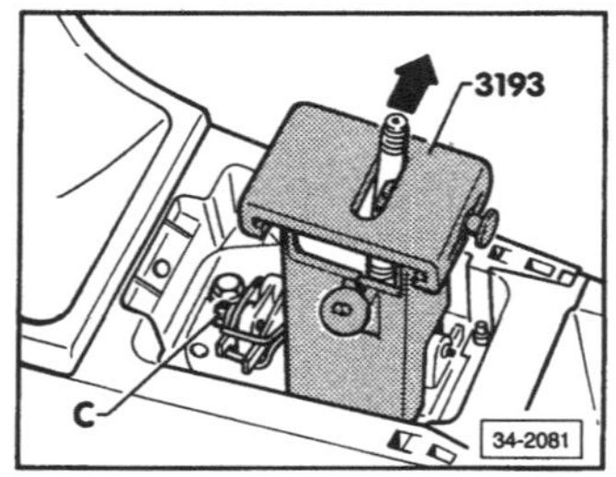

▶
- move gearshift lever to right detent (direction of **arrow**)
- tighten bolt **C**

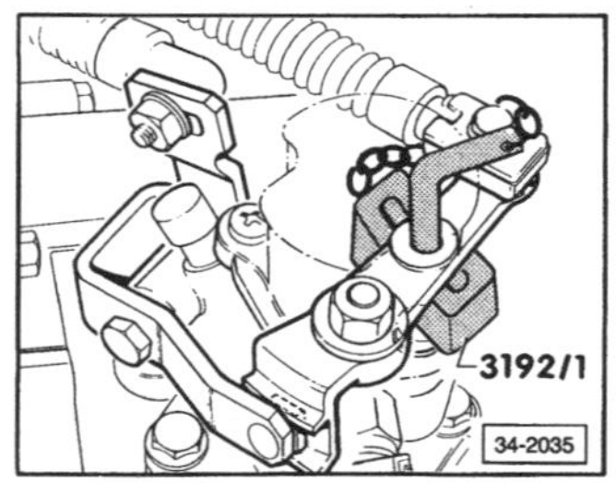

▶
- install wedge and locating pin
- push in wedge between gearshift lever and cover such that there is **no play**

Note

Wedge must not raise transmission gearshift lever.

- attach operating cables in this position; check position of the wedge
- remove wedge and pin
- remove shift lever gage

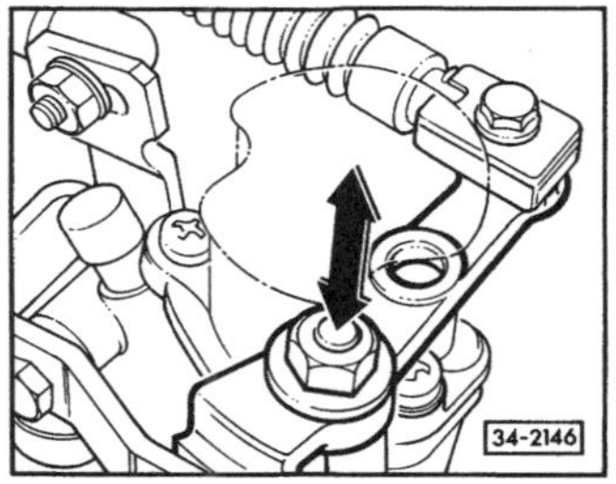

Checking

- press down on clutch pedal
- shift into all gears, several times, noticing effectiveness of reverse gear lock

Note

If any gear binds or drags, check transmission rod travel, as follows:

- shift into first gear
- press shift lever to the left, up to stop; release
- have assistant watch shift rod on transmission
  - rod must travel approximately 1 mm (3/64 in.) in direction of **arrow**, when activating shift lever

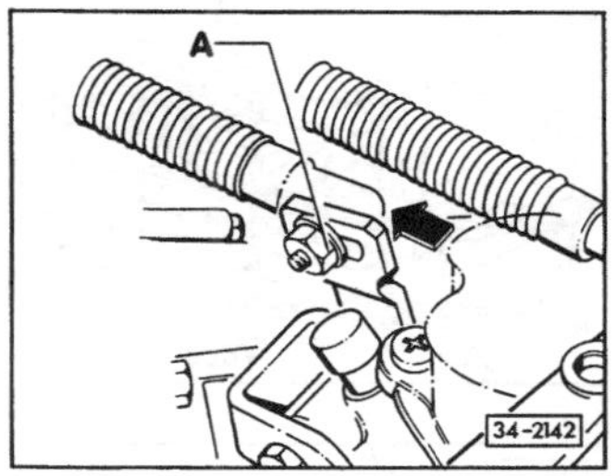

Note

If rod travel is not within specifications:

- shift out of first gear
- loosen nut **A** at selector cable mounting lug
  - minimal cable play at lug is normal
- reposition cable/lug slightly, toward instrument panel
- tighten nut **A**
  - selector cable play will be increased

Note

To obtain a correct gearshift adjustment, the following conditions are required:

1. Shift linkage operating satisfactorily
2. Shift/selector cable mounting components, in satisfactory condition
3. Shift/selector cable boots, in satisfactory condition
4. Shift cables operating smoothly
5. Transmission and clutch, in satisfactory condition

Transmission, removing/installing

Removing

- disconnect ground strap from battery
- disconnect electrical connectors from transmission
- on vehicles with G-charger, remove connecting tubing to intercooler, Repair Group 21
- remove gearshift cable from transmission gearshift lever, page 34.5
- remove selector cable with linkage lever from transmission, page 34.5
- remove cable support from transmission
- remove clutch slave cylinder and place aside
  - secure with wire
  - do not open hydraulic lines
- remove driveshaft for speedometer on transmission
- remove engine/transmission bolts from above
- if necessary, remove intake hose from air-flow sensor

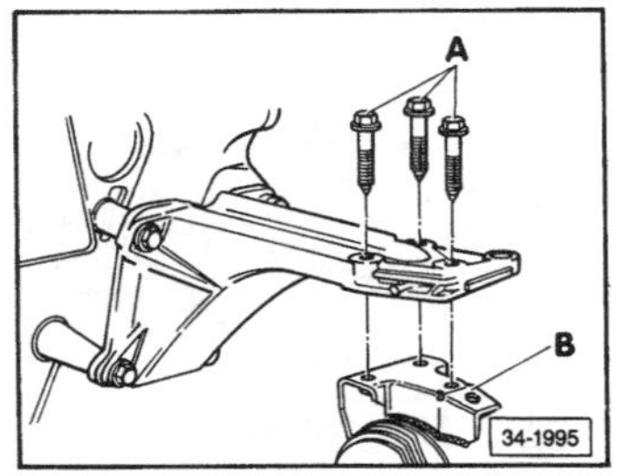

- remove 3 bolts **A** on mount, right, for engine **B**

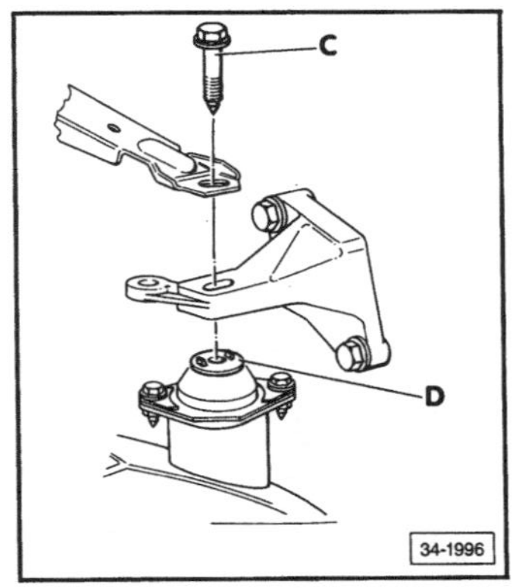

- remove hex head bolt **C** on mount, left, **D** for transmission

Note

For vehicles with ABS, the hex head bolt for the left mount is difficult to access. If necessary, remove overflow reservoir for cooling system, Repair Group 19.

- remove fan with cover and mount

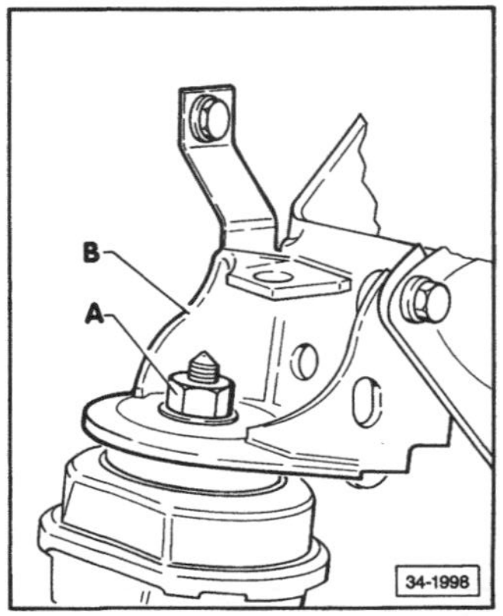

▶
- remove nut **A** for front transmission bracket, **B**

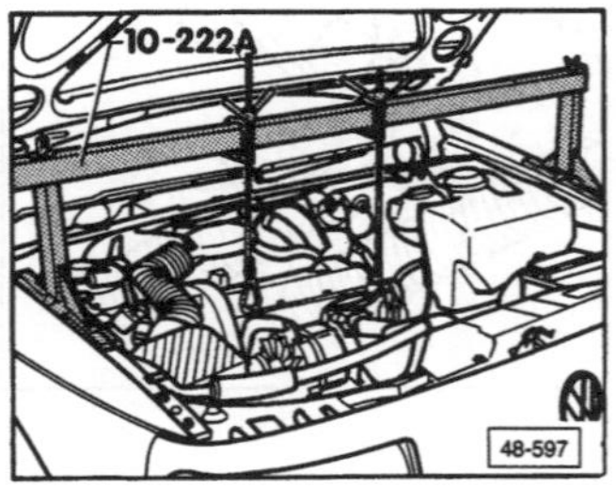

▶
- install device **10-222 A** in conjunction with bases **10-222 A/1**

> **CAUTION**
>
> Bases **10-222 A/1** are adapted to the angle of the body. If the previous bases are used, an accident could occur.

- slightly tighten engine/transmission support via spindles
- remove starter and front transmission bracket

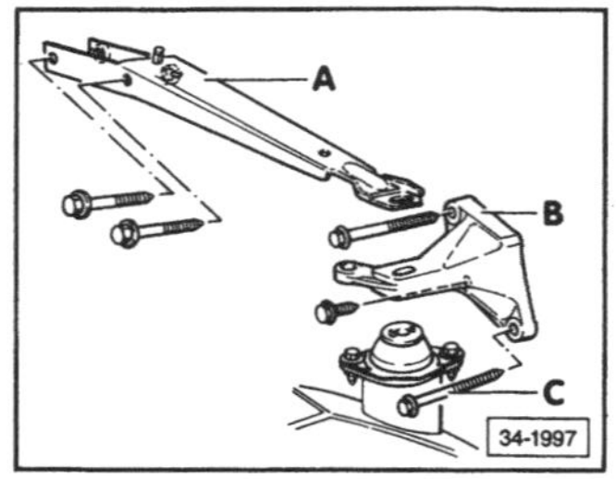

▶
- remove transmission support **A** from transmission
- remove balance weight from mount
- remove heat shield for inner right constant velocity joint
- remove driveshafts from flanges and tie up
- remove transmission bracket **B** from transmission
- press engine/transmission mount rearward and remove lower mounting bolt **C** for transmission bracket

- push engine/transmission mount as far as possible to right and lower slightly via left spindle
- remove small cover for gear carrier housing and bolts for cover
- position a transmission jack under transmission and remove lower engine/ transmission attachment bolt
- pry transmission from bushings and lower carefully

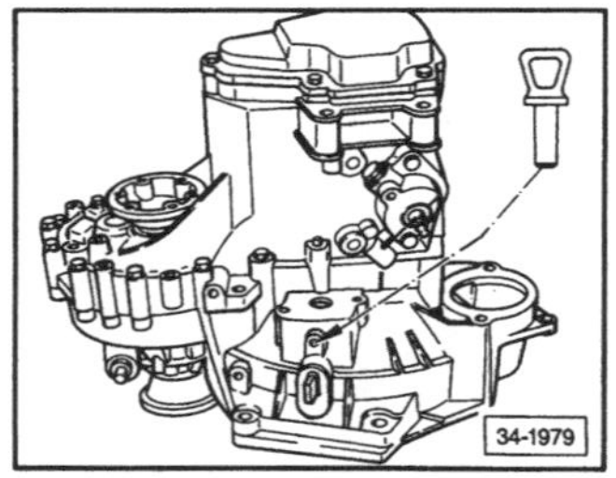

Installing

- install transmission in reverse order

▶
- press clutch release lever toward transmission housing and secure by mounting pin or bolt M 8 x 22
  - remove pin after transmission is installed

- plug bore
- when installing transmission, make sure cover fits properly
- install engine-transmission mount free of tension
- adjust/check gearshift mechanism, page 34.10

Tightening torques

Transmission to engine (M 12)	80 Nm (59 ft lb)
Starter to transmission and engine (M 10)	60 Nm (44 ft lb)
Driveshaft to flange	45 Nm (33 ft lb)
Left transmission bracket and transmission support to mount	60 Nm (44 ft lb)
Left transmission bracket to transmission	25 Nm (18 ft lb)
Transmission support to transmission	60 Nm (44 ft lb)
Right rear transmission bracket to mount	25 Nm (18 ft lb)
Front transmission bracket to rubber bonded bushing	60 Nm (44 ft lb)
Cable support to transmission	25 Nm (18 ft lb)
Gearshift cable to gearshift lever	25 Nm (18 ft lb)
Slave cylinder to transmission	25 Nm (18 ft lb)
Balance weight to mount	30 Nm (22 ft lb)

THIS PAGE INTENTIONALLY LEFT BLANK

Index

02A 5-speed

1st gear

- operating sleeve and hub, assembling **35.39**
- operating sleeve w/hub, installing **35.40**
- shift forks, ball bearings changes **35.53**
- synchronizer ring, checking **35.38**
- synchronizer ring, marking **35.39**

2nd gear

- operating sleeve and hub, assembling **35.39**
- operating sleeve w/hub, installing **35.40**
- synchronizer ring marking/checking **35.40**

3rd gear

- modifications **35.49, 35.50**
- needle bearing sleeve, modified **35.50**
- operating sleeve/synchronizer hub, disassembling/assembling **35.41**
- operating sleeve/synchronizer hub, removing/installing **35.42**
- removing/installing **35.30**
- synchronizer ring, checking **35.38**
- synchronizer sleeve, modified **35.49**
- thrust washer, modified **35.49**

Index

4th gear
- installing **35.29**
- modifications **35.49**
- needle bearing sleeve, installing **35.42**
- operating sleeve/synchronizer hub, disassembling/assembling **35.41**
- operating sleeve/synchronizer hub, removing/installing **35.42**
- removing w/bearing and bushing **35.28**
- shift forks, ball bearings changes **35.53**
- synchronizer ring, checking **35.38**
- synchronizer sleeve, modified **35.49**
- thrust washer, modified **35.49**

5th gear assembly
- adjusting **35.16**
- adjusting, checking (transmissions from 31 10 1) **35.55**
- gear, installing **35.15**
- gear, removing **35.7**
- gear/synchronizer hub bolts modified **35.51**
- modified synchronizer, installed position **35.55**
- modified synchronizer ring and 5th gear synchronizer, assembling **35.54**
- operating sleeve, installing **35.16**
- operating sleeve, removing **35.7**
- shift fork, changes **35.54**
- shift fork, installing **35.16**
- shift fork removing **35.7**
- synchronizer, changes **35.53**
- synchronizer hub, modified **35.51**
- synchronizer hub w/5th gear, installing **35.15**
- synchronizer hub w/5th gear, removing **35.7**
- synchronizer ring, changes **35.53**
- synchronizer ring, checking **35.30**
- synchronizer ring, installing **35.15**
- synchronizer sleeve, modified **35.52**

Backup light switch/socket
- pre-assembly **35.22**

Ball bearing inner race
- removing/installing **35.24**

Bearing sleeve
- removing/installing **35.21**

Differential
- assembly **35.5**
- installing **35.11**
- removing **35.10**

Drive flange shaft
- circlip, replacing **35.18**
- seal, replacing **35.17**
- shaft, installing **35.18**
- shaft, removing **35.6**

Gear carrier housing
- assembly **35.5**

Guide sleeve assembly
- sealing ring, removing/installing **35.18**
- sleeve, installing **35.19**
- sleeve, removing **35.6**

Inner race
- testing for wear **35.40**

Large pinion shaft bearing
- inner race, removing/installing **35.38**

Lock ring
- installing **35.24**

Main shaft
- adjusting **35.31**
- adjusting shim, chart **35.32**
- assembly **35.5**
- checking **35.34**
- disassembling/assembling **35.26**
- installing **35.11**
- removing **35.10**

Index

Main shaft bearing
- inner race, installing **35.28**
- outer race, installing **35.28, 35.30**
- outer race, removing **35.28**

Needle bearing
- bushing, installing **35.29**
- installing **35.43, 35.47**
- removing **35.42, 35.47**
- revised, installing **35.52**
- sleeve, installing **35.41, 35.43**

Pinion shaft
- adjusting **35.44**
- assembly **35.5**
- disassembling/assembling **35.35**
- installing **35.11**
- removing **35.10**

Reverse gear assembly
- disassembling/assembling **35.46**
- idler shaft, revised **35.52**
- idler shaft support, installing **35.11, 35.13**
- idler shaft support, removing **35.8**
- modified **35.52**
- shift fork, installing **35.11**
- shift fork, removing **35.9**
- sliding gear, installing **35.11**
- sliding gear, removing **35.9**

Sealing ring
- installing **35.21**

Selector shaft
- changes **35.56**
- disassembling/assembling **35.20**
- end cover, changes **35.57**
- shaft assembly, installing **35.14**
- shaft assembly, removing **35.8**
- support, bearing sleeve, removing/installing **35.14**

Shift fork assembly
- assembly **35.5**
- assembly, installing **35.12**
- assembly, removing **35.9**
- disassembling/assembling **35.23**
- pivot pins, installing **35.13**
- pivot pins, removing **35.8**

Shift segments
- identification **35.24**
- installing **35.25**

Small pinion shaft bearing
- inner race, removing/installing **35.37**
- outer race, removing/installing **35.37**

Synchronizer
- ring, checking **35.38**
- sleeve w/hub and 1st gear removing **35.39**

Transmission
- adjustment overview **35.48**
- assembly **35.2**
- clamping in assembly stand **35.6**
- disassembling/assembling **35.6**
- housing, installing **35.12**
- housing, removing **35.8**
- housing cover, installing **35.17**
- housing cover, modified **35.51**
- housing cover, removing **35.6**
- modifications **35.53**

Turning torque
- measuring **35.45**

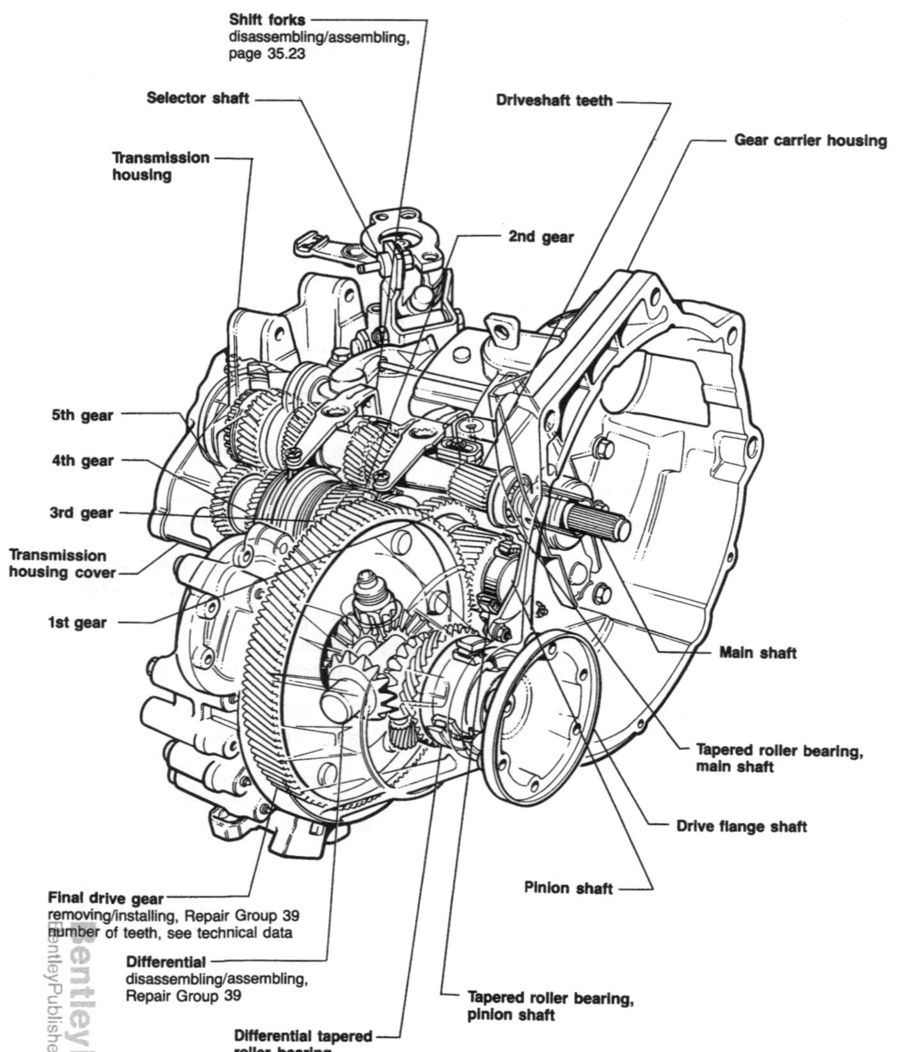

Note

For transmission disassembly/assembly procedures, see page 35.6

THIS PAGE INTENTIONALLY LEFT BLANK

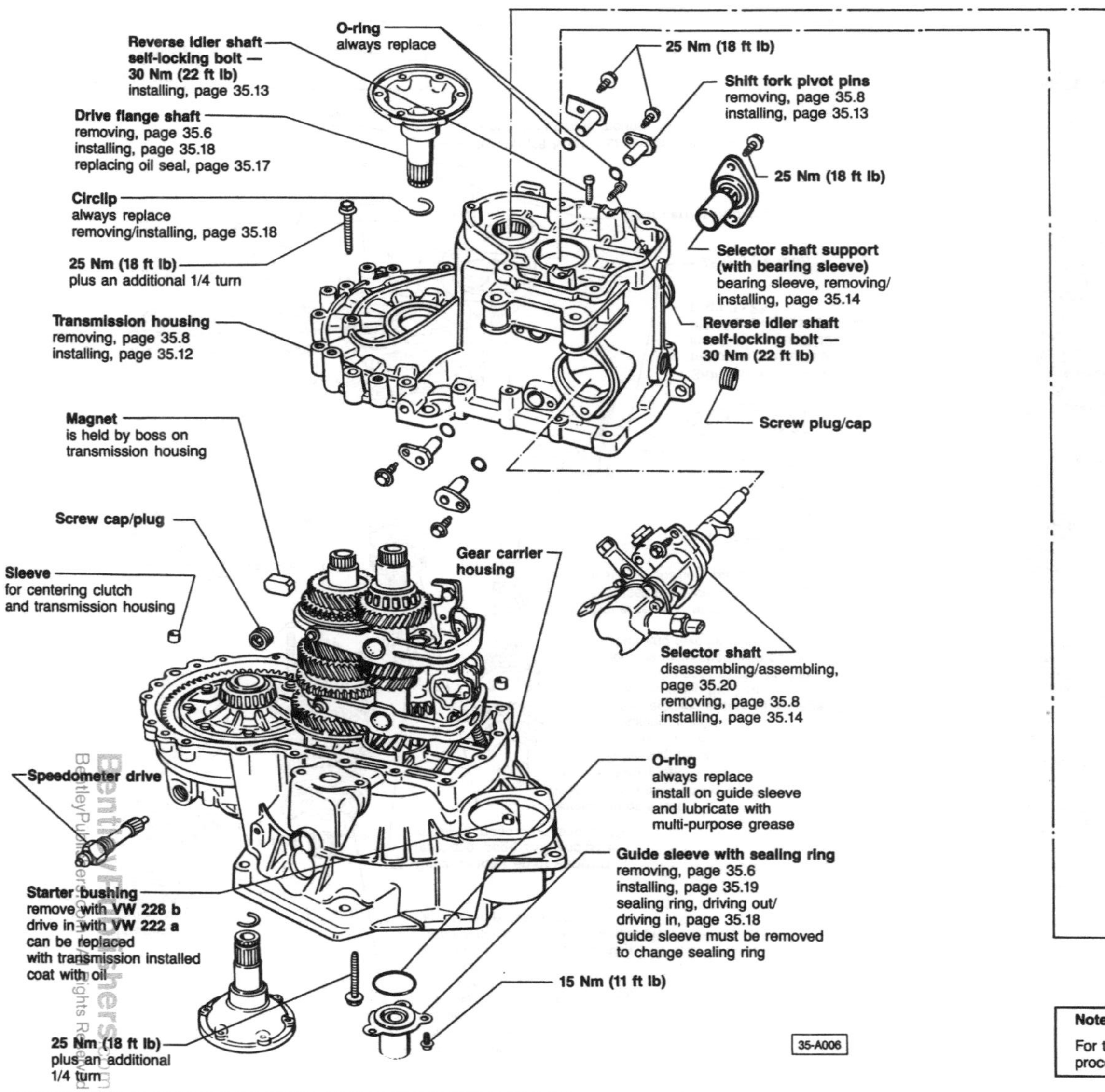

Note

If the transmission housing cover is being installed with the transmission installed, check oil level as follows. Fill oil up to lower edge of filler hole and reinstall plug. Start engine, engage gear and allow transmission to run approximately 2 minutes. Stop engine and fill transmission oil to lower edge of filler hole.

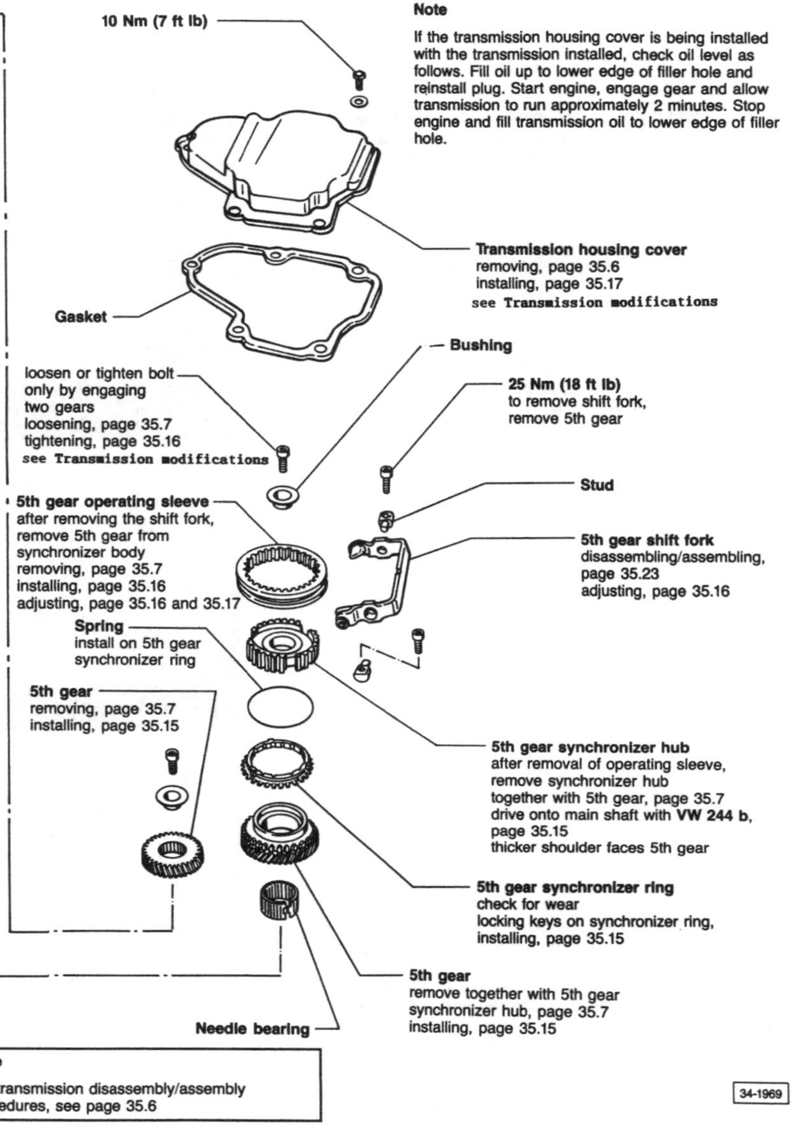

Note

For transmission disassembly/assembly procedures, see page 35.6

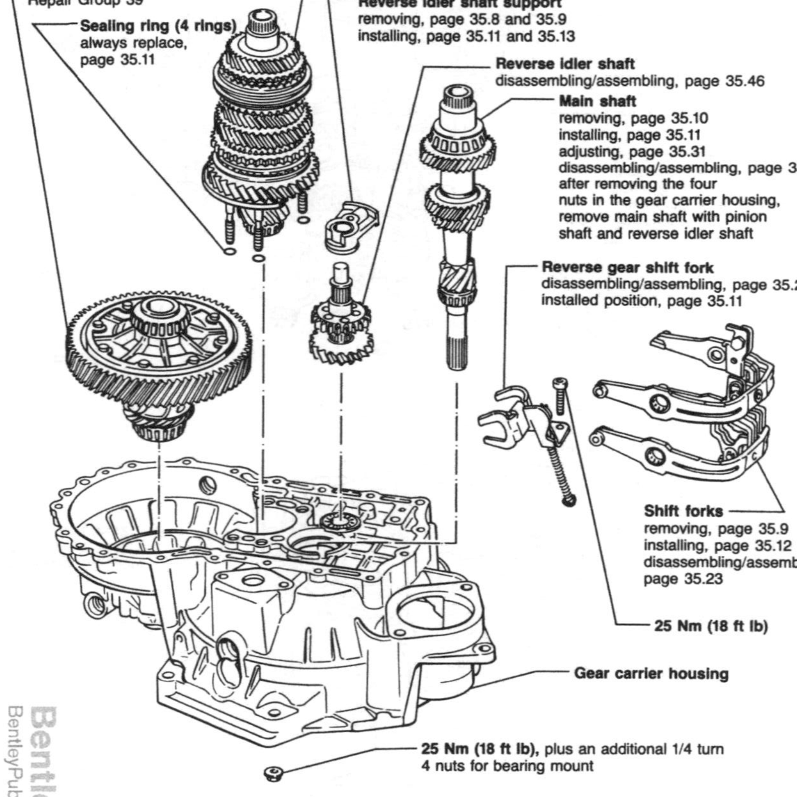

Note

For transmission disassembling/assembling procedures, see page 35.6

34-1970A

Transmission, disassembling/ assembling

Disassembling

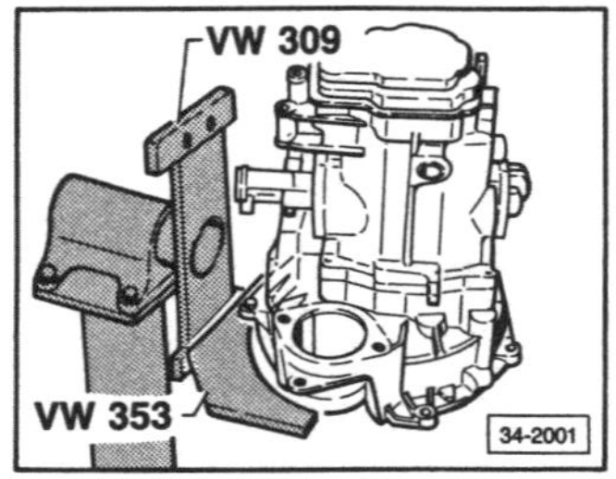

- clamp transmission in assembly stand
  - **VW 353** and **VW 309**
- drain transmission fluid
- remove clutch release lever/bearing
- remove guide sleeve
- remove transmission housing/clutch housing bolts

Note

Loosen but do not remove nuts (**arrow**) for bearing mount.

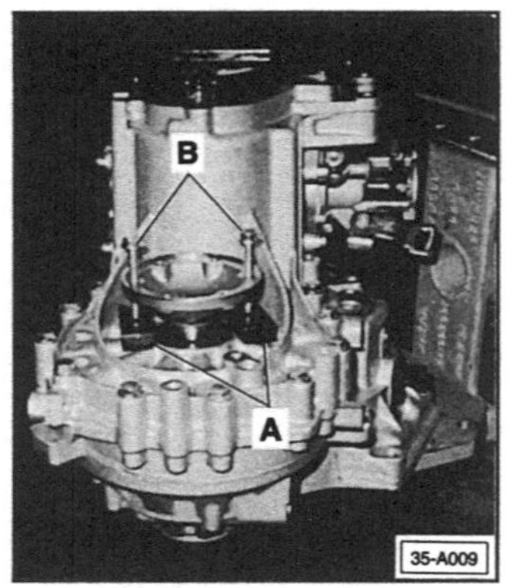

- remove flange shaft
  - position spacers (**A**) below flange, and pull flange shaft out of differential gears by turning bolts (**B**), alternately and evenly

Note

Use **two** of the previously removed housing bolts, to avoid damage to flange shaft or bolts.

- remove opposite flange shaft and speedometer drive
  - if differential is to be removed
- remove transmission cover and gasket

Note

Before loosening synchronizer body and 5th gear bolts (**arrows**), two gears must be engaged, as follows:

- remove 5th gear shift fork
- engage 2nd gear via gearshift lever shaft
- engage 5th gear with operating sleeve
- remove synchronizer body and 5th gear bolts
- remove 5th gear operating sleeve

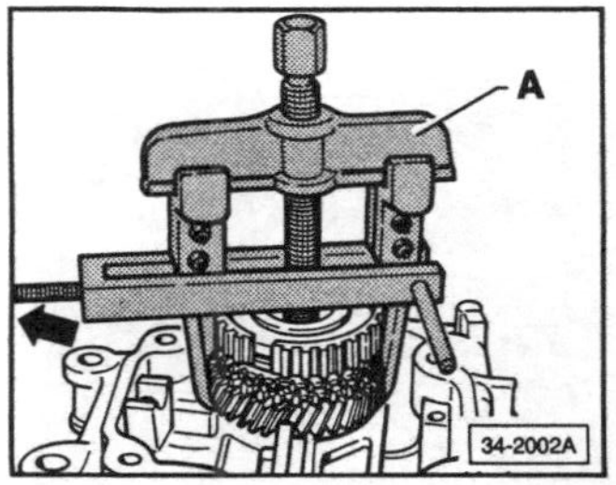

- thread gear puller (**A**) into main shaft, and position puller arms under 5th gear/ synchronizer hub
  - **A** = **US 1078** with **V/170**
- remove 5th gear together with synchronizer hub
- remove puller

Note

When removing the hub/gear assembly, make sure that the locating pin does not bend outward. Retighten screw (**arrow**) if necessary. Check 5th gear for damage after tightening.

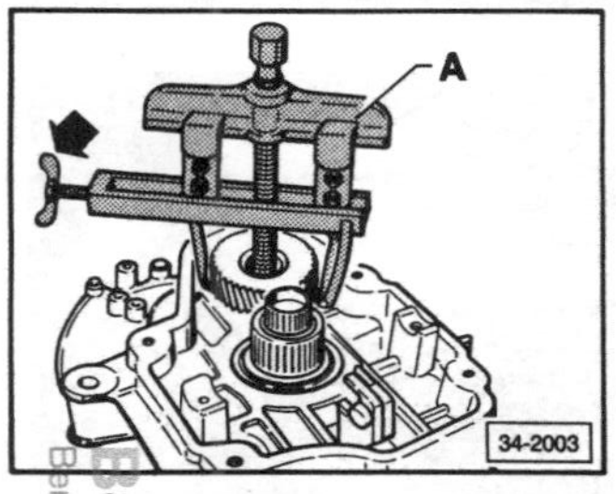

- thread gear puller (**A**) into pinion shaft, and position puller arms under 5th gear
  - **A** = **US 1078** with **V/170**
- remove 5th gear from pinion shaft

Note

When removing the gear, make sure that the locating pin does not bend outward. Retighten screw (**arrow**) if necessary. Check 5th gear for damage after tightening.

- remove puller

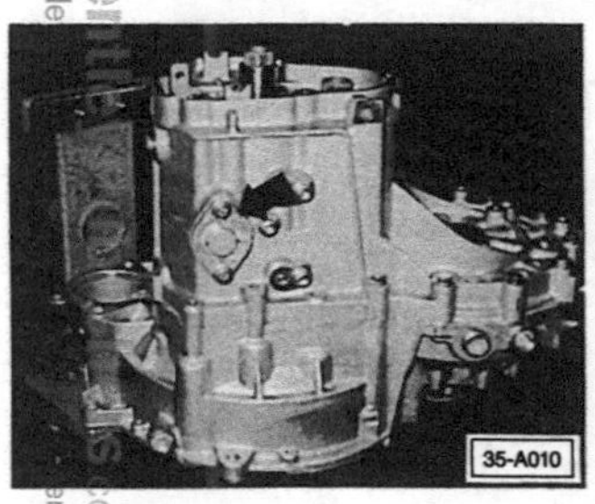

- remove selector shaft support bolts (**arrow**)
- remove support and bearing together

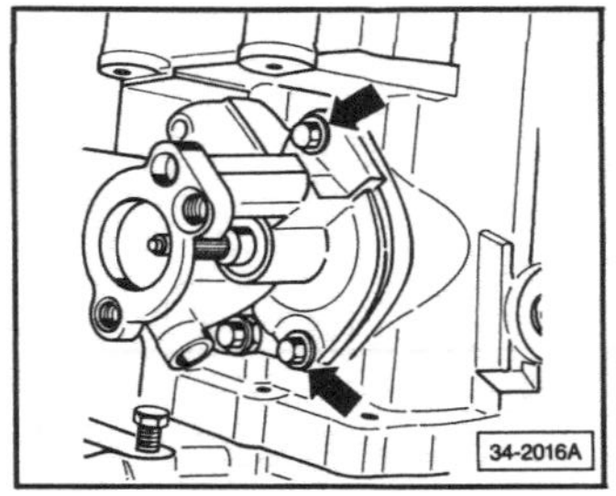

- shift selector shaft to neutral, then remove bolts (**arrows**)
- pull selector shaft assembly out of transmission housing

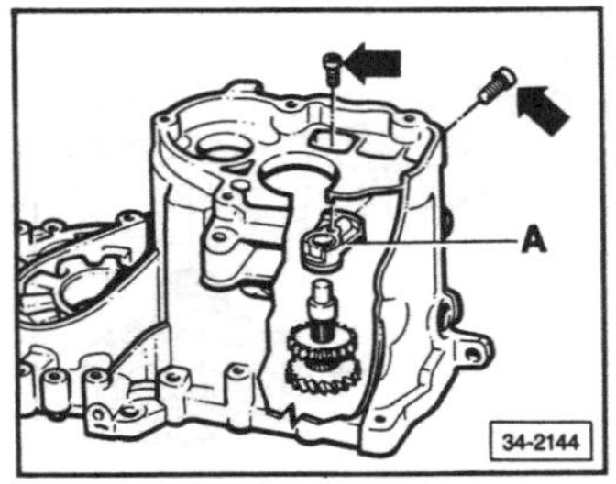

- remove self-locking bolts (**arrows**) retaining reverse idler shaft support (**A**)

> **CAUTION**
>
> Clean threads in shaft support using a thread cutter, to avoid bolt breakage during subsequent disassembly.

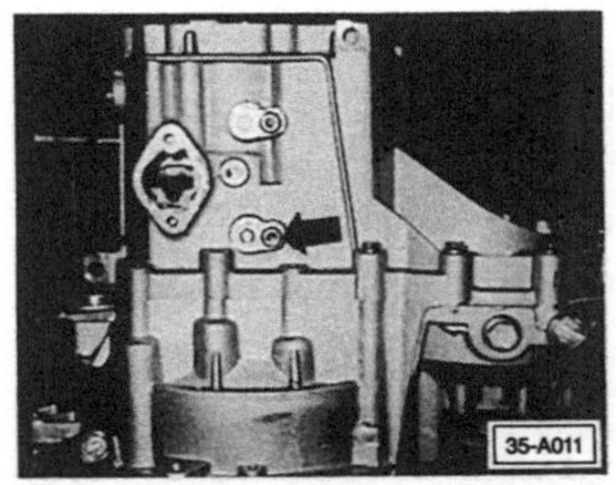

- remove shift fork pivot pins/bolts (**arrow**)
- mark all four pins for re-installation in same position

- remove remaining housing bolts
- tilt transmission 45°
- remove housing

Note

If housing won't separate, tap at mating surfaces with soft-faced mallet, to break seal.

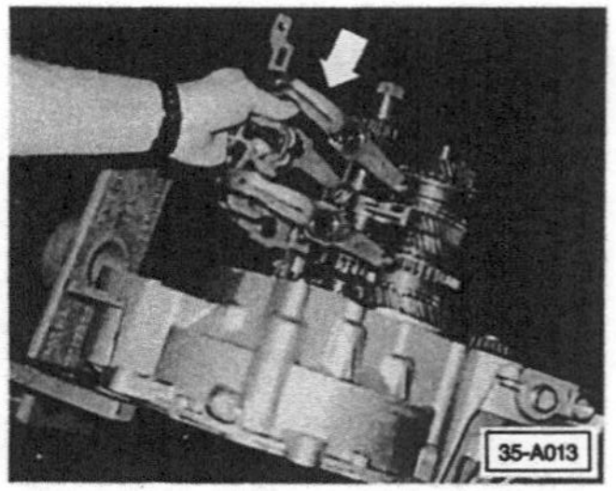

- remove shift forks (**arrow**) as an assembly

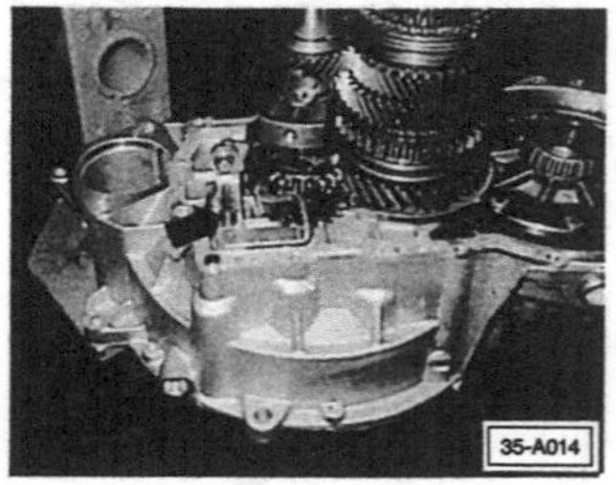

- remove bolt (**arrow**) retaining reverse shift fork
- lift out reverse shift fork assembly

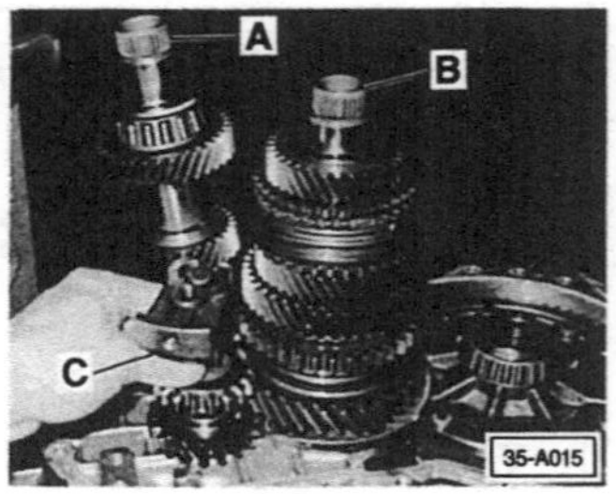

- move main shaft (**A**) away from pinion shaft (**B**)
- lift reverse idler shaft assembly (**C**) up and out

- remove remaining reverse gear (**arrow**)

- remove main shaft (**arrow**) separately
  - tilt away from pinion shaft

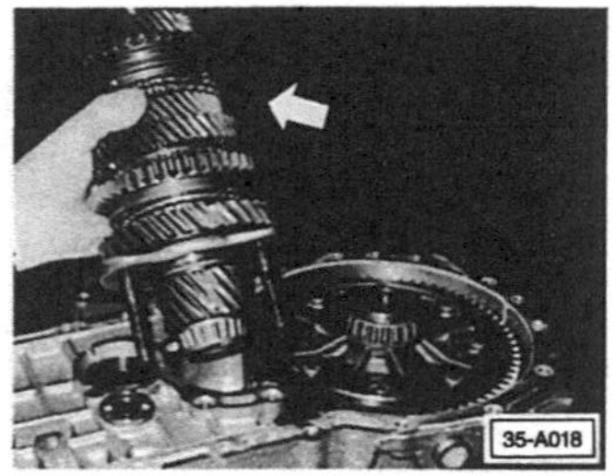

- remove pinion bearing mount nuts (four), on gear carrier housing side
- lift pinion shaft (**arrow**) up and out
- remove speedo drive, if not done previously

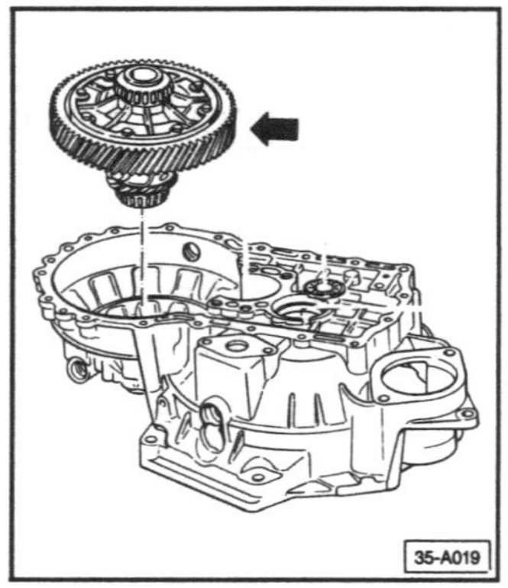

- remove differential assembly (**arrow**)

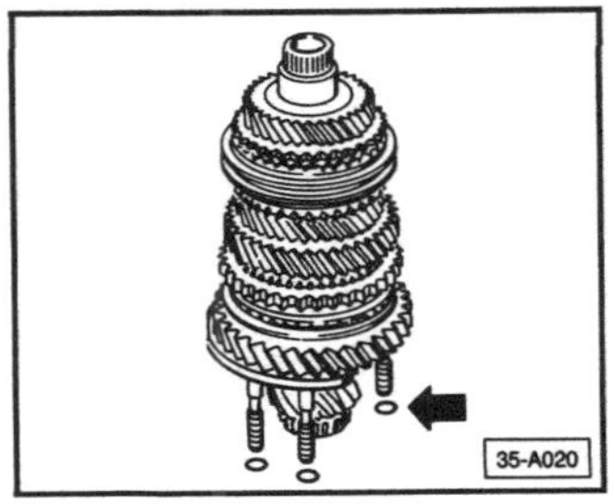

Assembling

- install differential into housing
- install new seals (**arrow**) on pinion bearing support studs
- install main shaft and pinion shaft together
- install pinion bearing mount nuts
  - hand-tighten
- install reverse gear
  - mesh with main shaft
- install sliding reverse gear, reverse shaft and shaft support
  - do not install support bolts at this time
  - move main shaft aside, as in removal
- lift sliding reverse gear into mesh with reverse gears on main shaft and pinion shaft

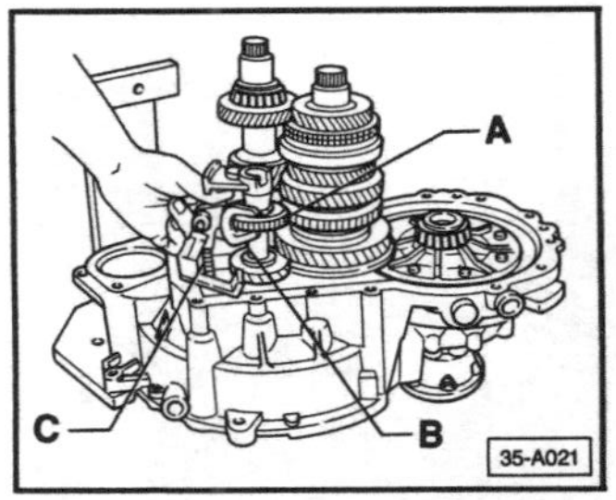

- position shift fork (**B**) on sliding reverse gear (**A**)
- seat over-center spring (**C**) fully in housing
- install shift fork retaining bolt
  - tighten after the following checks have been performed:

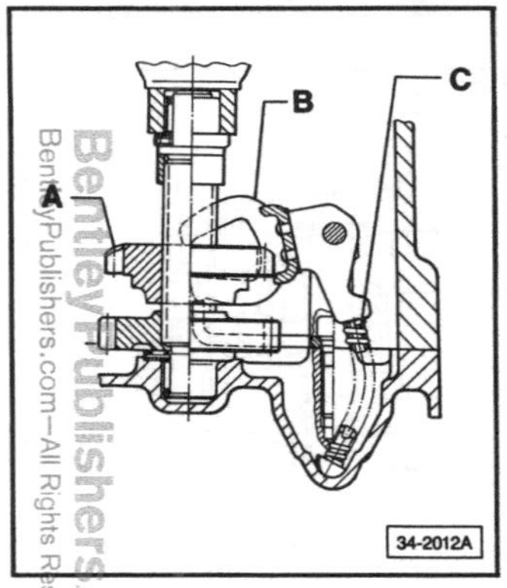

- check installed position of:
  A — Sliding reverse gear
  B — Shift fork
  C — Over-center spring
- check sliding reverse gear for smooth operation
- tighten shift fork retaining bolt to 25 Nm (18 ft lb)

- tilt housing 45°
- *slide shift fork assembly into position*
  - ensure that shift segments (**arrows**) are positioned in grooves of operating sleeve

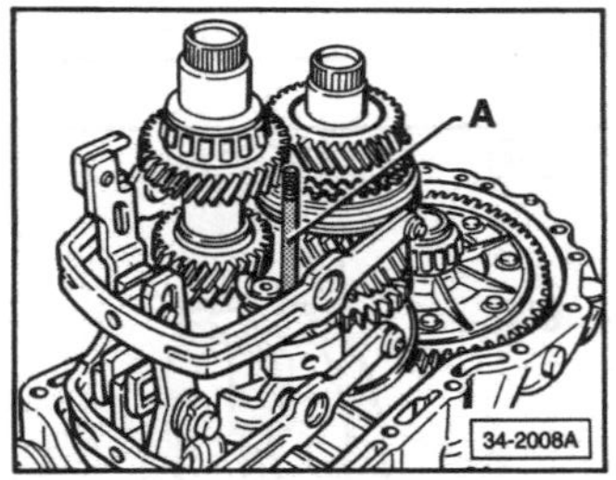

- thread locating stud (**A**) into reverse idler shaft support, and align shift rails
  - **A** = M8 x 100 mm

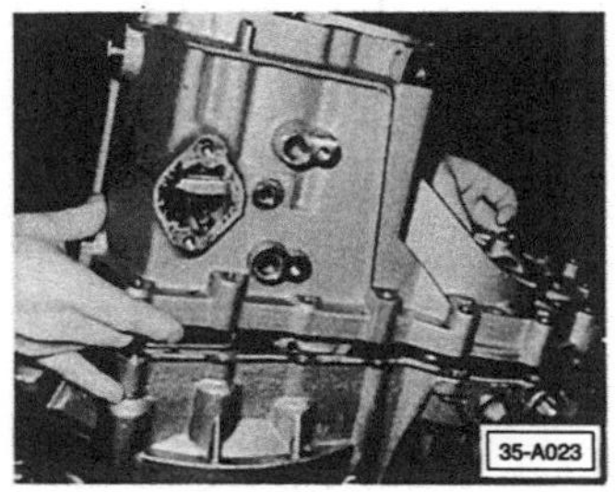

- apply sealant (**AMV188200 03**) to mating surfaces of housing
- guide housing over specially installed locating stud
- install housing bolts
  - tightening torque: 25 Nm (18 ft lb), plus an additional 1/4 turn
- tighten pinion bearing mount nuts
  - tightening torque: 25 Nm (18 ft lb), plus an additional 1/4 turn

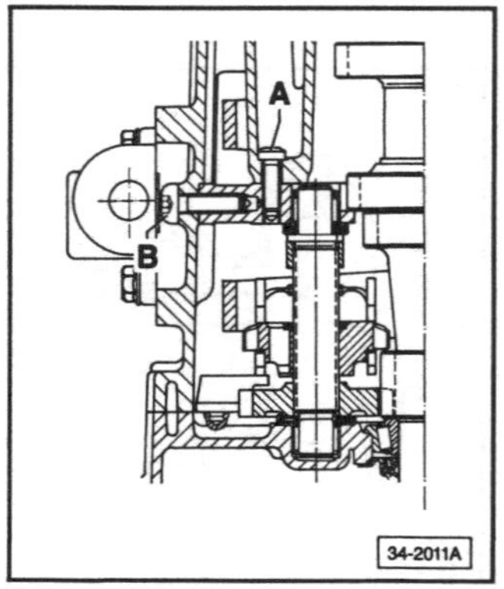

- install bolt (**B**) for reverse idler shaft support
  - hand-tighten
- remove locating stud
- install bolt (**A**) for reverse idler shaft support
  - hand-tighten
- tighten bolts in following sequence:
  - bolt (**B**) — 30 Nm (22 ft lb)
  - bolt (**A**) — 30 Nm (22 ft lb)

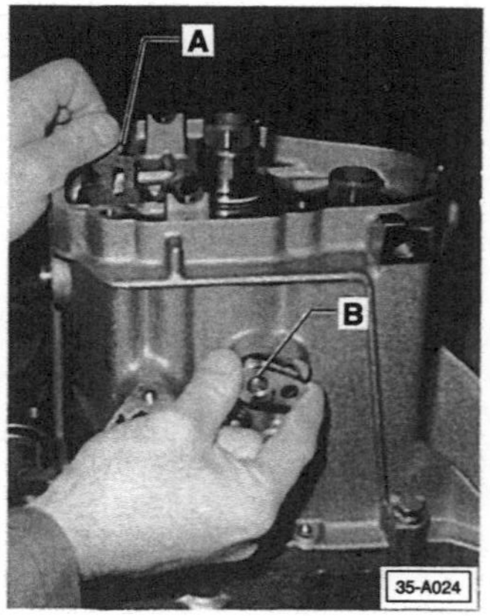

- replace seals on shift fork pivot pins
- align shift fork assembly by lifting lever (**A**) such that pin (**B**) can be installed
- install all four pins
  - each in its original position, as marked during disassembly
  - tightening torque: 25 Nm (18 ft lb)

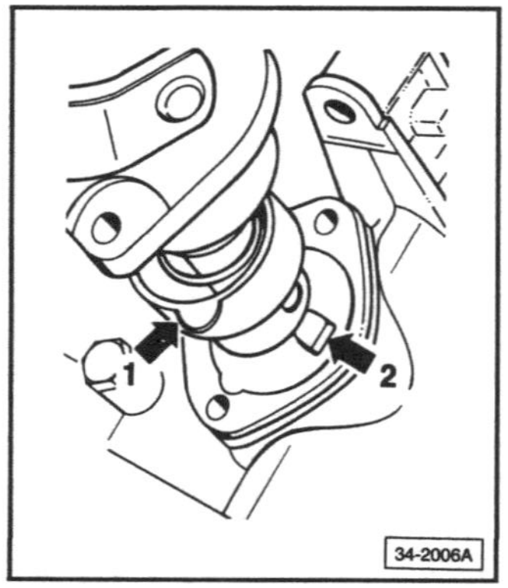

- apply **AMV 188200 03** to sealing surface of selector shaft assembly
- position gearshift rails in neutral
- set projection **(arrow 1)** in recess of transmission housing
- position selector shaft such that shift finger **(arrow 2)** can be inserted in shift rail

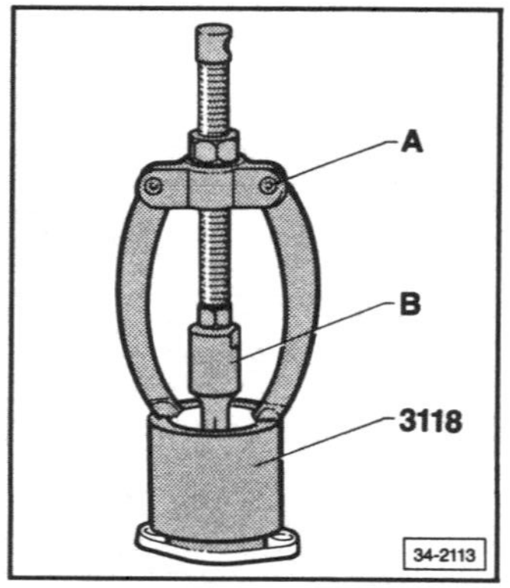

- remove bearing sleeve, if necessary, from selector shaft support
  A — Support **US 1039** or **Kukko 22/2**
  B — Puller **US 1088** (18.5-23.5 mm) or **Kukko 21/3**

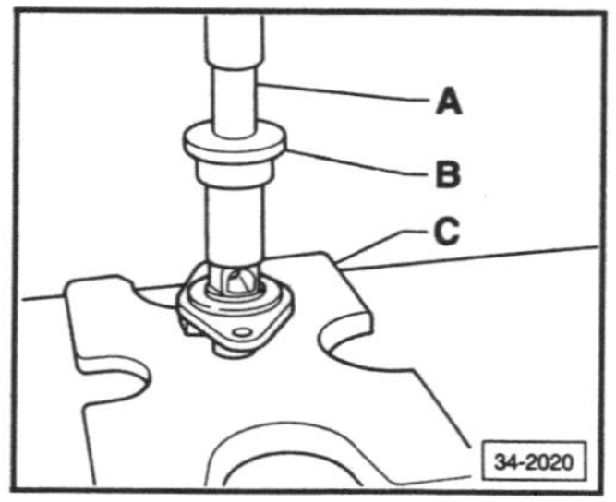

- install bearing sleeve, if removed, in selector shaft support
  A — Punch **VW 408a**
  B — Thrust pad **VW 447h**
  C — Thrust plate **VW 402**

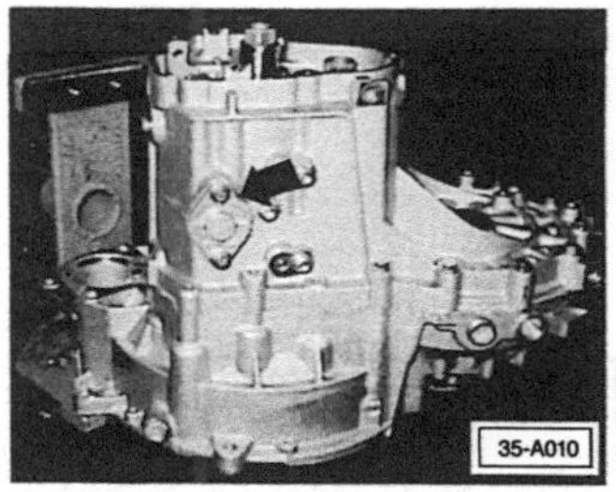

- apply **AMV 188200 03** to sealing surface of selector shaft support
- install selector shaft support
  - tighten bolts (**arrow**) to 25 Nm (18 ft lb)

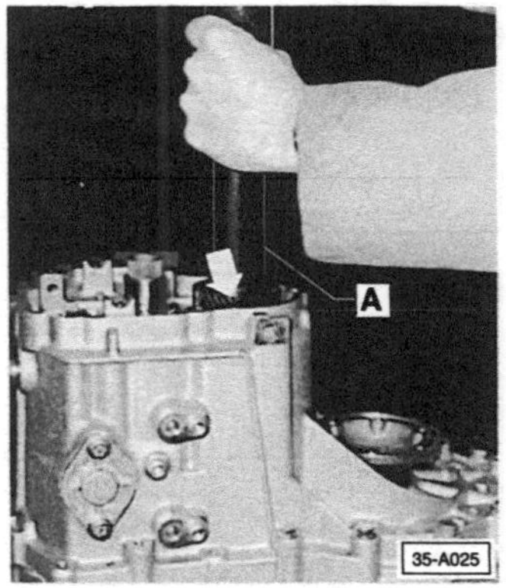

- heat 5th gear to 100°C (212°F)
- drive 5th gear (**arrow**) into position using installer (**A**)
  - shoulder of gear faces housing
  - **A = VW 244b**

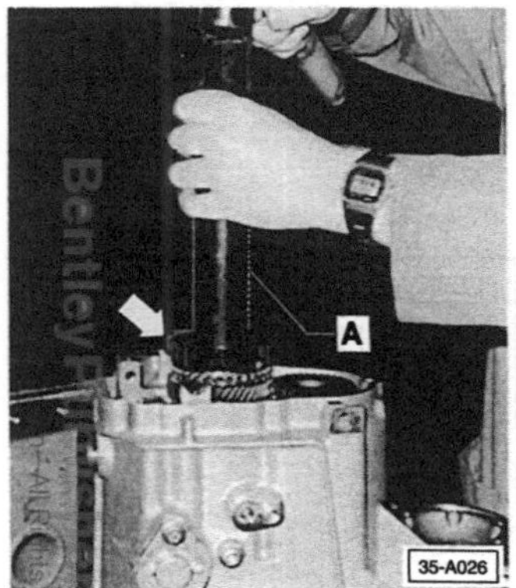

- position 5th gear bearing and synchronizer ring
- heat synchronizer hub to 100°C (212°F)
- drive hub assembly (**arrow**) into position using installer (**A**)
  - **A = VW 244b**

Note

The locking keys on the synchronizer ring must align with the notches in the synchronizer hub.

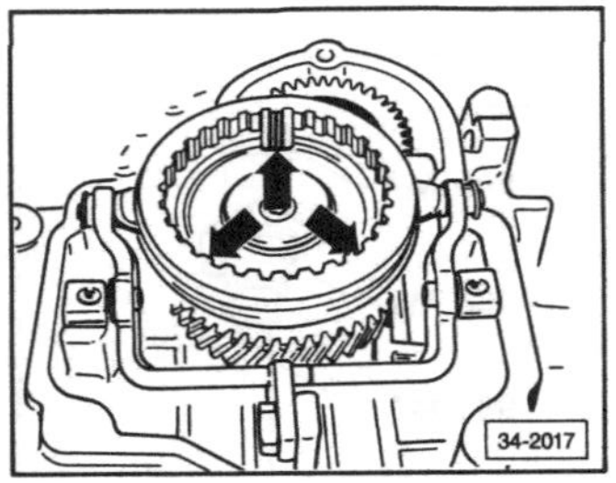

- install operating sleeve
  - pointed parts of teeth face 5th gear
  - slide sleeve over synchronizer hub such that identified teeth (**arrow**) face recesses of hub

Note

Before installing synchronizer body and 5th gear bolts (**arrow**), two gears must be engaged, as follows:

- push operating sleeve toward 5th gear and select another gear with selector lever

up to transmission build date 03 05 0:
- apply D6 locking compound to both bolts, then install
  - tightening torque: **40 Nm (30 ft lb)**

from transmission build date 04 05 0:
- use dished washers on M10 bolts and install
  - tightening torque: **80 Nm (59 ft lb)**
  - see **Transmission modifications**

- install 5th gear shift fork/bolts
  - tightening torque: **25 Nm (18 ft lb)**

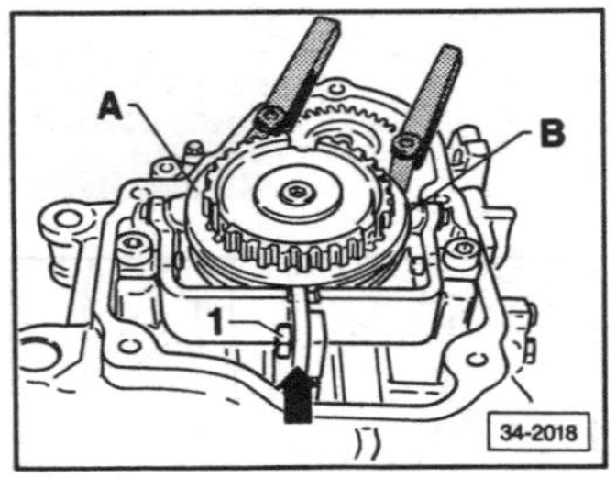

- adjust 5th gear, as follows:
  - engage 5th gear
  - loosen bolt (**1**)
  - push feeler gage (1.2 mm) between operating sleeve and gear, positions (**A**) and (**B**)
  - press transmission gearshift lever in direction of 5th gear and selector arm in direction of **arrow**, with screwdriver
  - tighten bolt (**1**) to **25 Nm (18 ft lb)**
  - disengage 5th gear

Note

Operating sleeve must now be in neutral position.

Spring and synchronizer ring must have clearance.

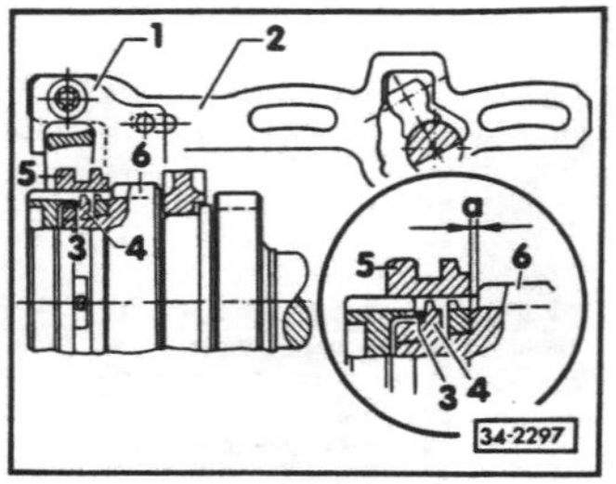

- install 5th gear operating sleeve
  - 1– Selector arm
  - 2– Shift rail
  - 3– Spring
  - 4– Synchronizer hub
  - 5– Synchronizer sleeve
  - 6– Gear
  - dimension **a** = 1.3 mm with 5th gear engaged

- replace transmission cover gasket
- install end cover/bolts **(arrow)**
  - tightening torque 10 Nm (7 ft lb)

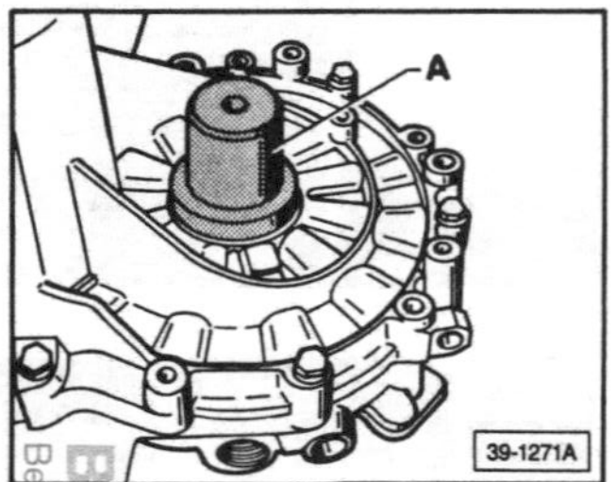

- pull out drive flange oil seal with **VW 681**
- drive in new oil seal with installer (**A**), until flush
  - **A = 3142**
- fill space between sealing and dust lip with multi-purpose grease

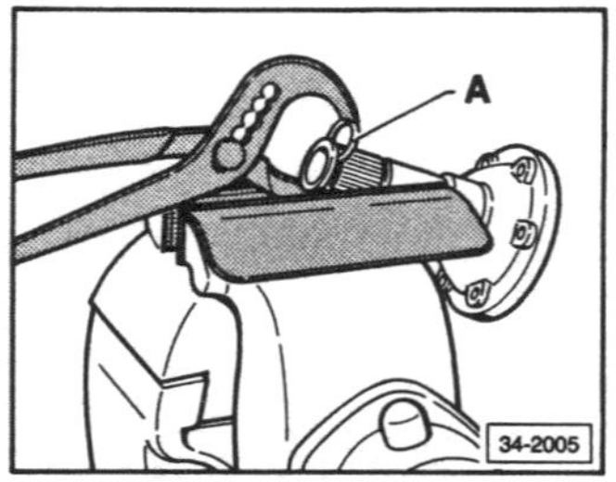

- clamp drive flange shaft in vise equipped with protective jaws
- press circlip out of flange shaft groove with new circlip **A**

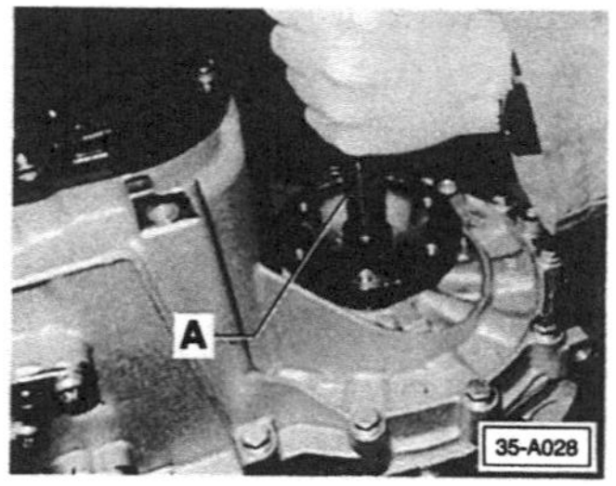

- drive in flange shafts with installer (**A**)
  - **A = VW 295**

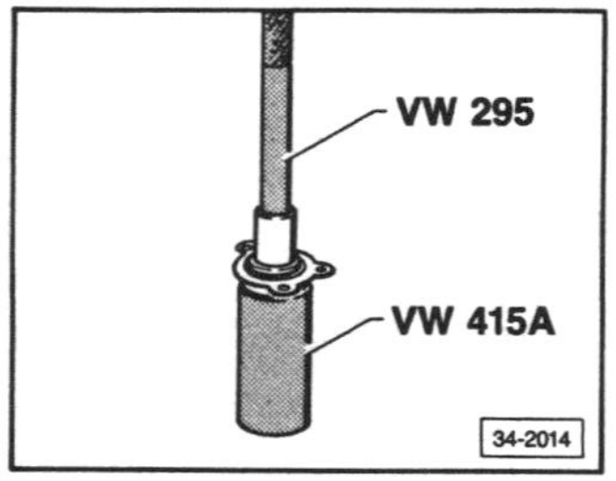

- remove sealing ring from guide sleeve

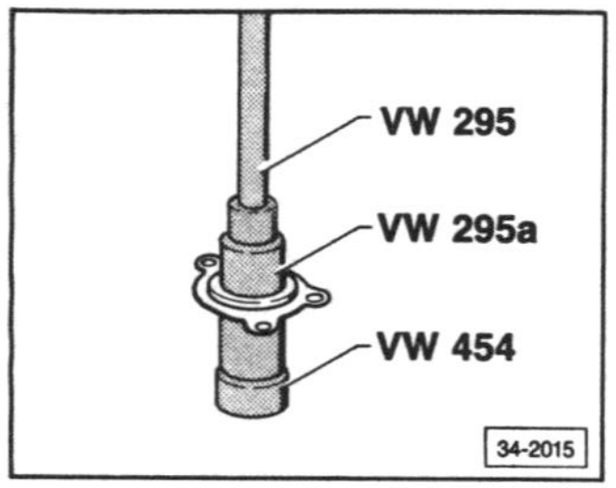

- install sealing ring in guide sleeve
  - drive into sleeve, up to stop

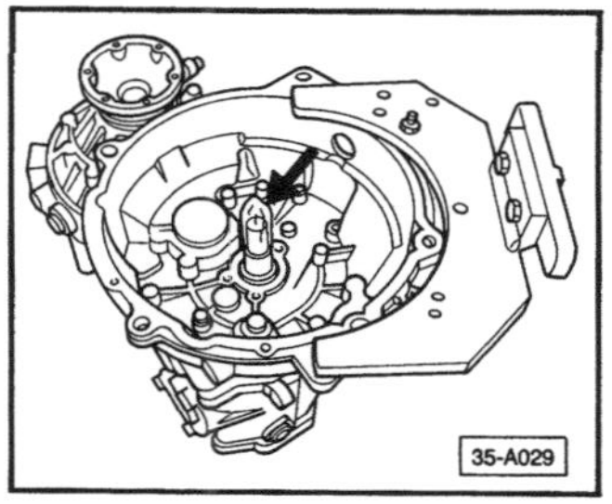
35-A029

- lubricate guide sleeve sealing lips with multi-purpose grease
- install new o-ring
- wrap plastic (**arrow**) or paper around mainshaft, to protect lip seal
- position guide sleeve
  - tighten bolts to 15 Nm (11 ft lb)
- press clutch operating lever and release bearing into place
- lubricate all surfaces with light film of MoS_2 grease

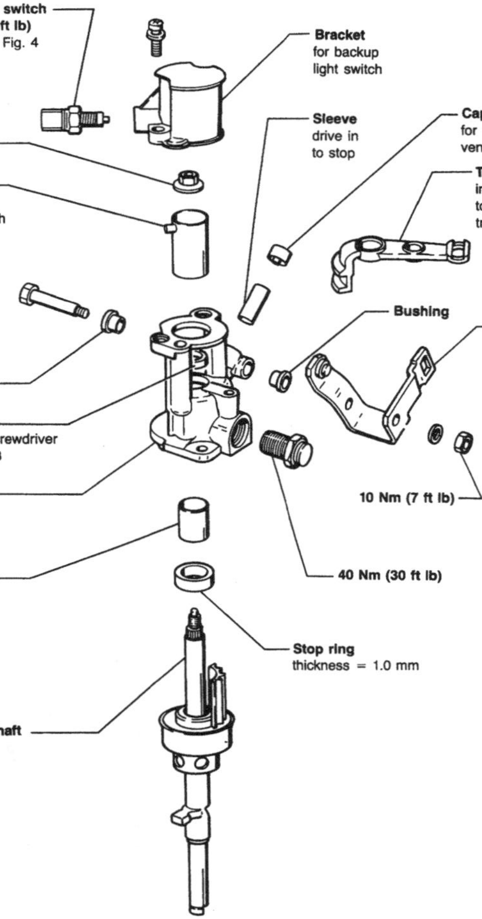

34-1972A

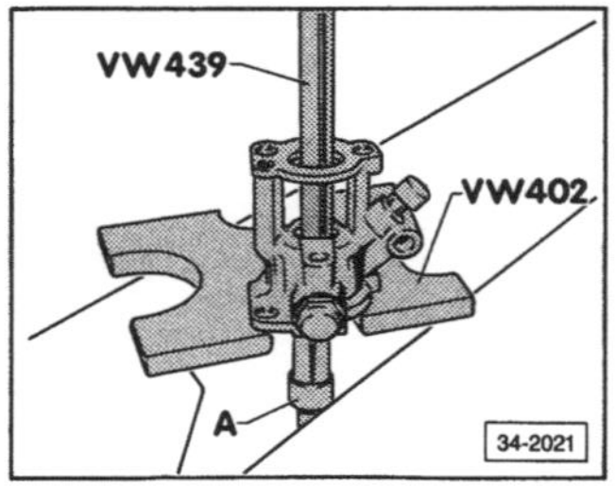

▶ **Fig. 1 Bearing sleeve, removing**

- before removing ball sleeve, destroy plastic cage and remove balls

A — Puller 18.5-23.5 mm, e.g. **US 1088 (Kukko 21/3)**

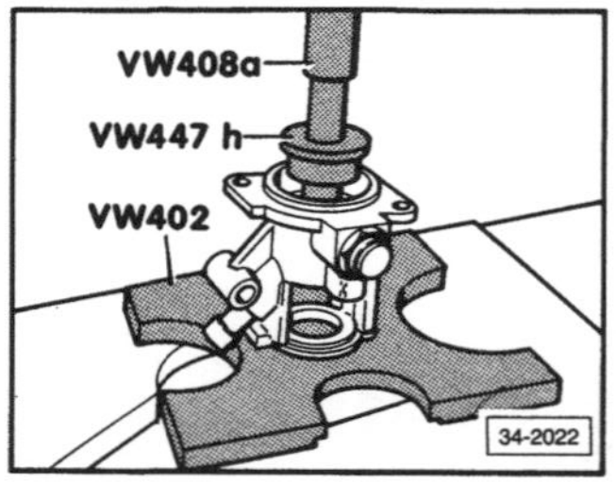

▶ **Fig. 2 Bearing sleeve, installing**

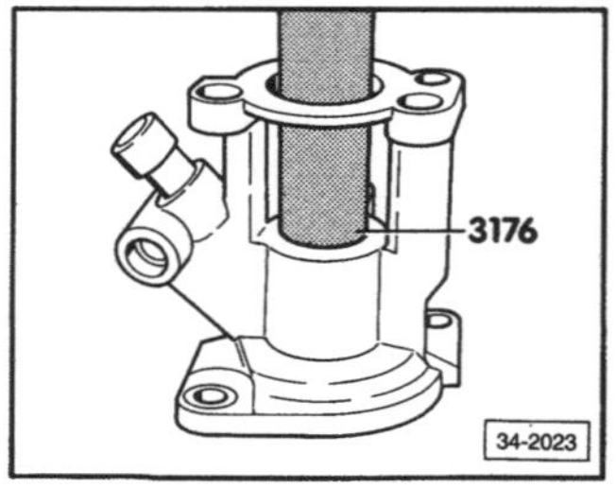

▶ **Fig. 3 Sealing ring, installing**

- press in to stop

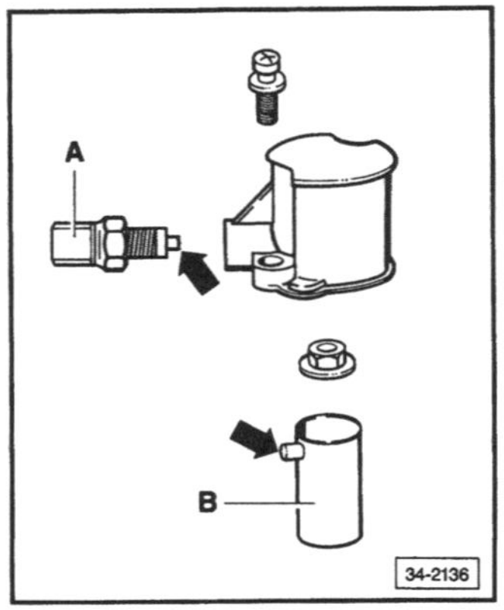

▶ **Fig. 4 Backup light switch and socket, pre-assembly**

- apply light coat of MoS_2 to contact areas (**arrows**) of switch **A** and socket **B**

Note

When removing and installing races, do not bend shift forks.

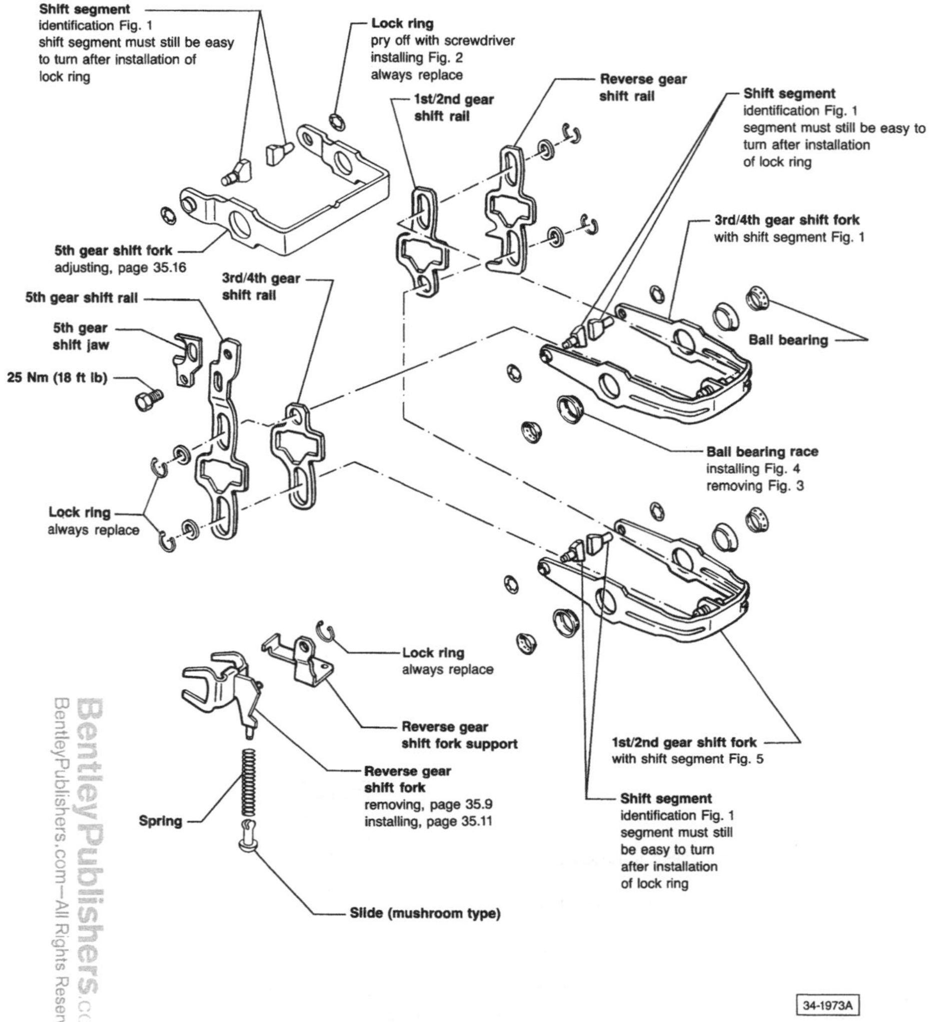

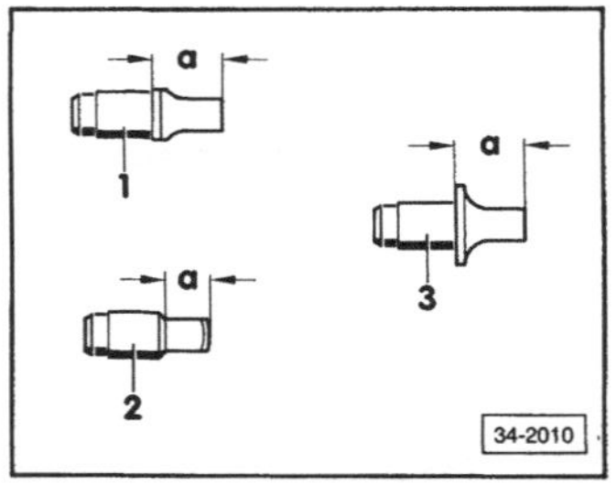

▶ Fig. 1 **Identification of shift segments**

Dimension **a**

1 — Shift segments 1st and 2nd gear = 12.1 mm
2 — Shift segments 3rd and 4th gear = 7.7 mm
3 — Shift segments 5th gear = 12.1 mm

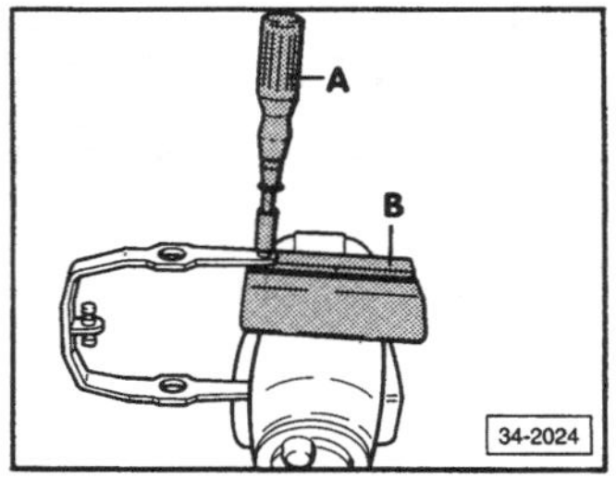

▶ Fig. 2 **Lock ring, installing**

- press new lock ring in groove of shift segment with handle socket wrench

Note

Shift segment must still be easy to turn after installation of the lock ring.

A — Handle socket wrench — width across flats 10 mm

B — Protective jaws

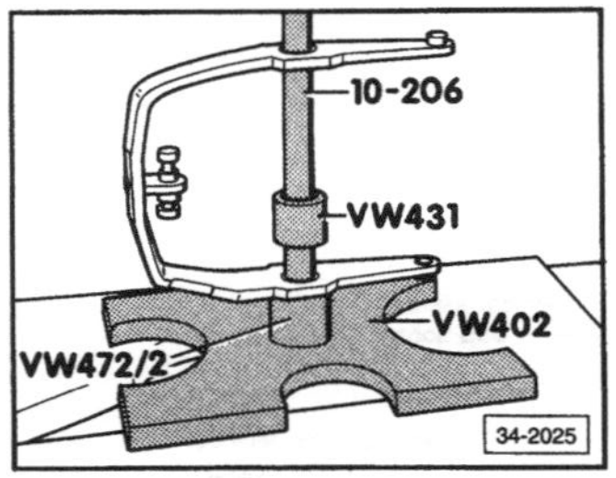

▶ Fig. 3 **Ball bearing race, removing**

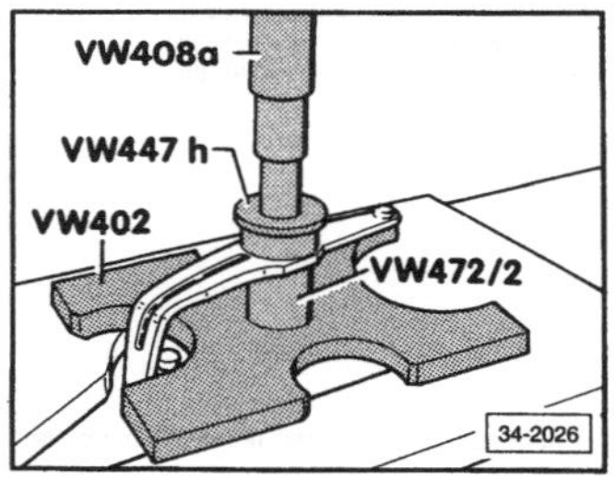

▶ Fig. 4 **Ball bearing race, installing**

- press in to stop

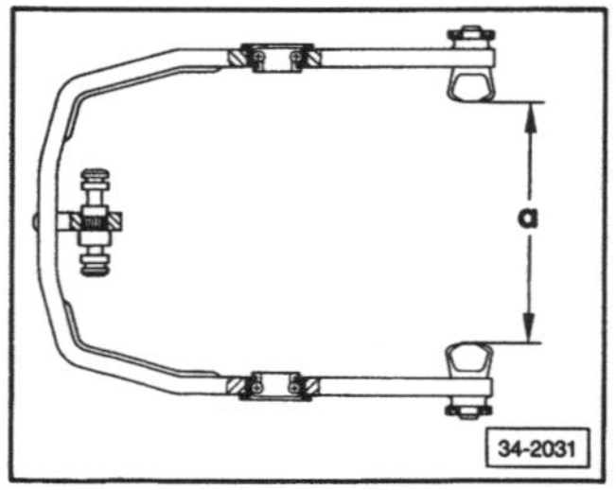

▶ **Fig. 5 Shift fork with installed shift segments**

Dimension **a**

- shift fork 1st/2nd gear = 87.9 to 87.2 mm
- shift fork 3rd/4th gear = 96.7 to 96.0 mm
- shift segment assignment Fig. 1

Note

Pay attention to technical data when installing new gears.

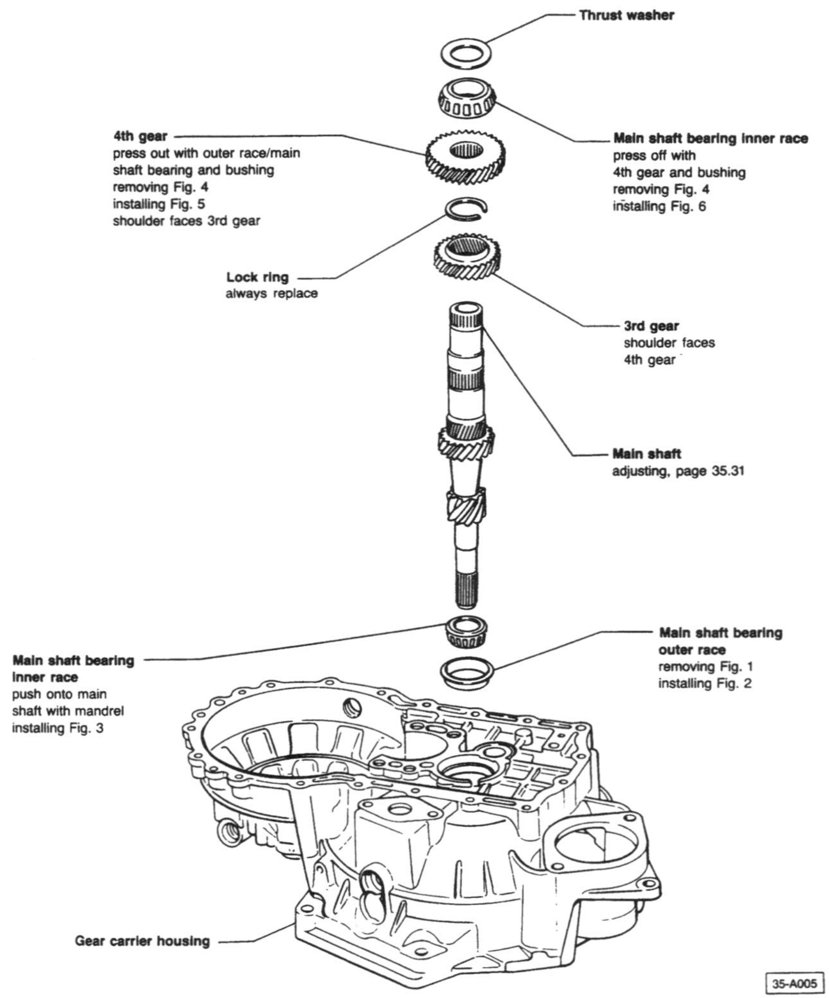

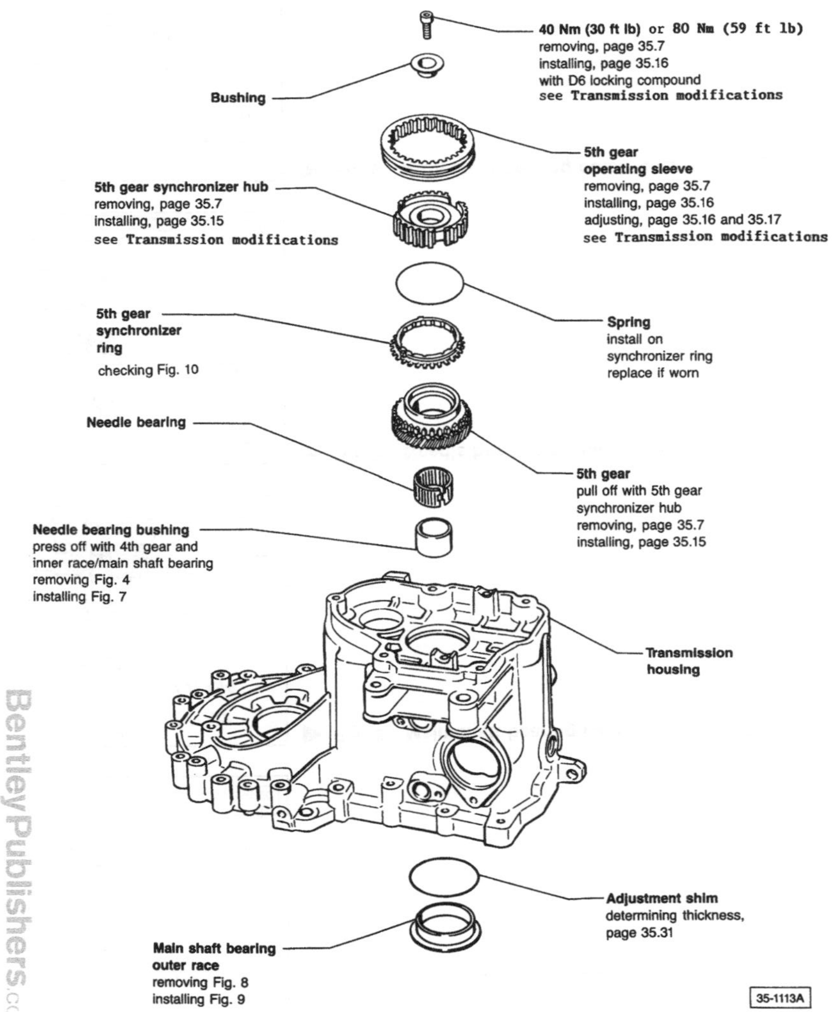

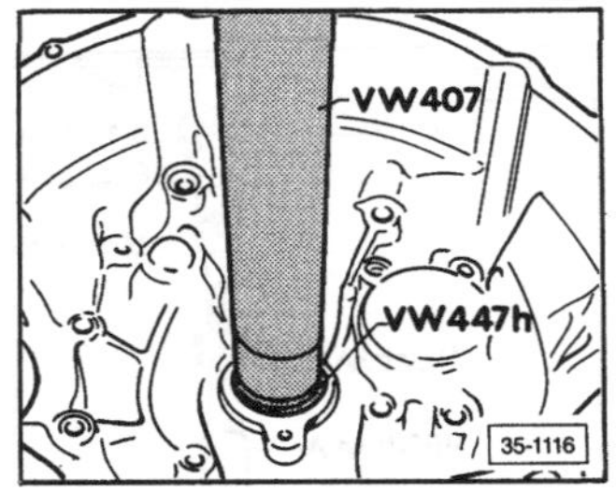

▶ **Fig. 1** **Main shaft bearing outer race, removing**

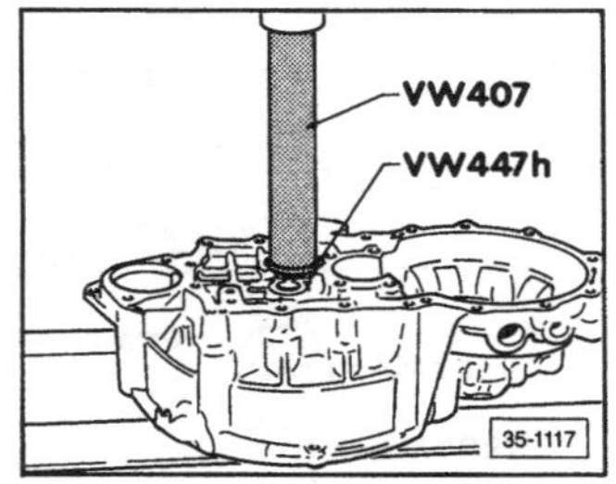

▶ **Fig. 2** **Main shaft bearing outer race, installing**

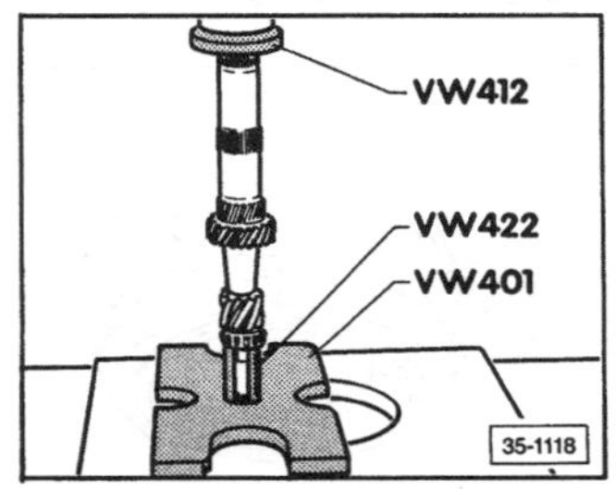

▶ **Fig. 3** **Main shaft bearing inner race, installing**

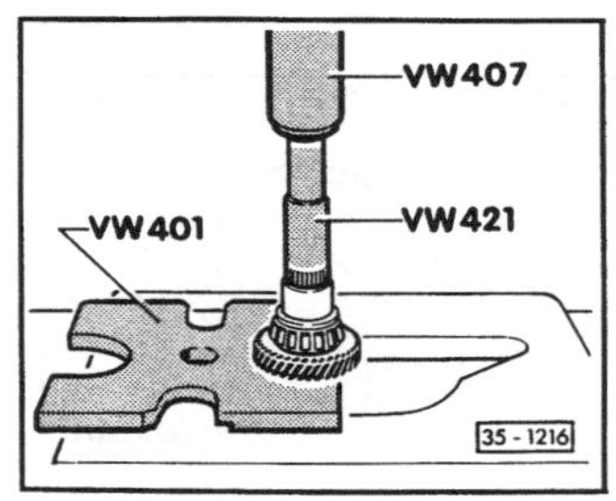

▶ **Fig. 4** **4th gear with main shaft bearing and bushing, removing**

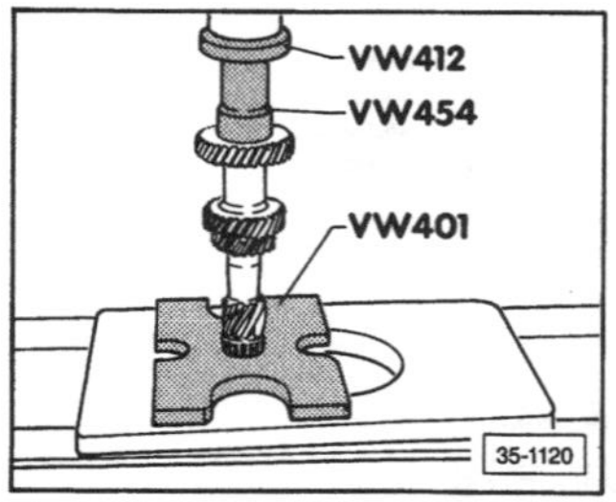

▶ Fig. 5 **4th gear, installing**

- shoulder faces 3rd gear

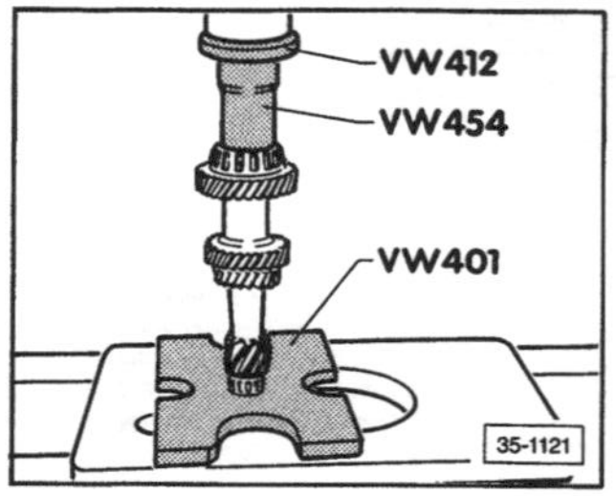

▶ Fig. 6 **Main shaft bearing inner race, installing**

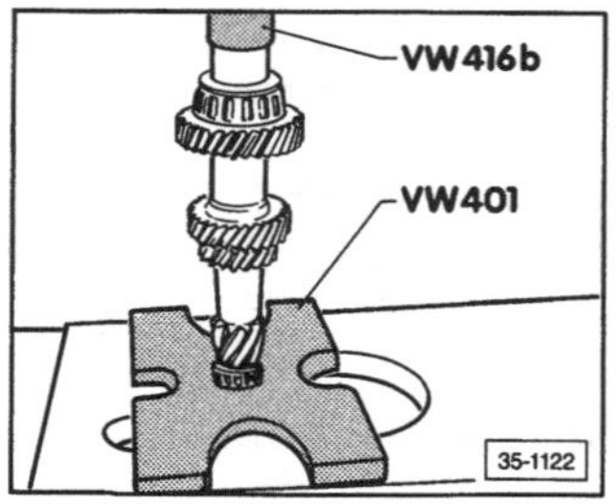

▶ Fig. 7 **Needle bearing bushing, installing**

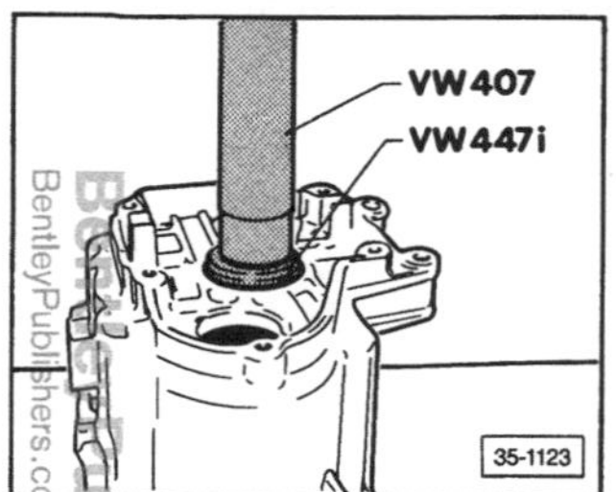

▶ Fig. 8 **Main shaft bearing outer race, removing**

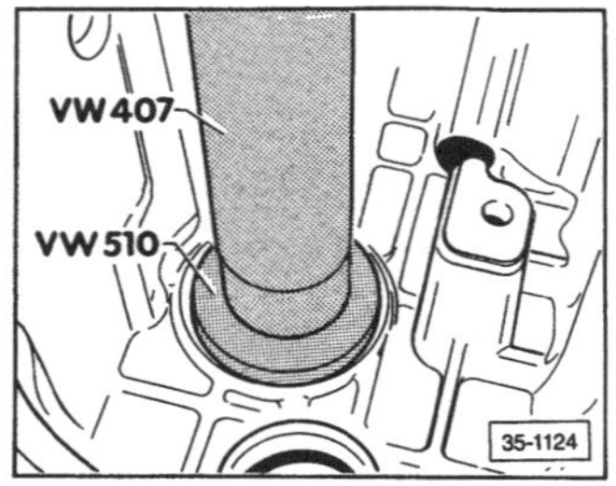

▶ Fig. 9 **Main shaft bearing outer race, installing**

- insert with adjustment shim after adjusting main shaft

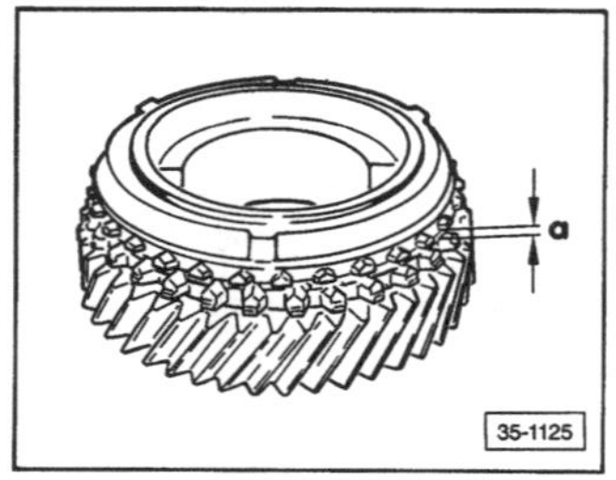

▶ Fig. 10 **5th gear synchronizer ring, checking**

- press synchronizer ring onto cone of gear and measure gap dimension **a** with feeler gage

Gap dimension a	Installation dimension (mm)	Wear limit
5th gear	1.1-1.7 mm	0.5 mm

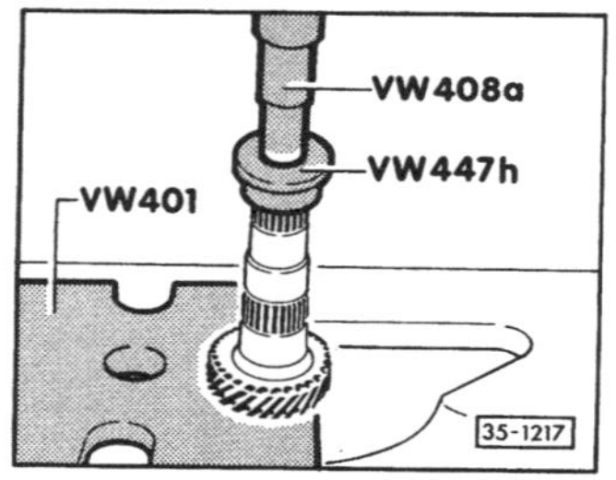

3rd gear, removing

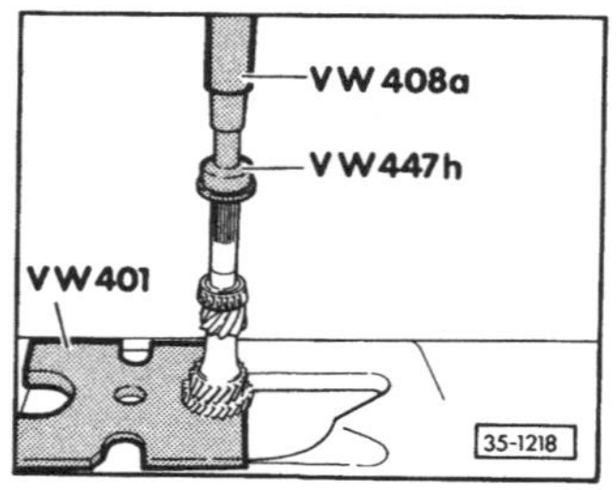

3rd gear, installing

Main shaft, adjusting

Main shaft shim, determining

Adjustment of main shaft is required when any of the following are replaced.

Gear carrier housing
Transmission housing
Main shaft
4th gear
Main shaft bearing

See Adjustment Overview, page 35.48.

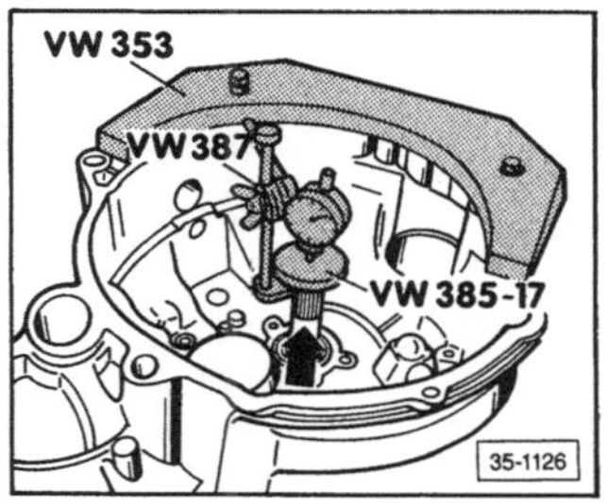

▶

- push main shaft bearing outer race without adjustment shim and with **VW 510** into gear carrier housing
- insert main shaft into gear carrier housing with bearing outer race installed
- tighten bolts to 25 Nm (18 ft lb), plus an additional 1/4 turn

- install measuring device and dial gage in gear carrier housing
- rotate main shaft before measuring to settle bearings
- adjust dial gage with 1 mm preload to zero

CAUTION

This process is to be repeated for each measurement because the dial gage does not return to its original position.

- push main shaft in the direction of the dial gage (**arrow**)
- read tolerance on the dial gage and record (example 1.21)

Note

Dial gage does not return to its original position.

- determine thickness of adjustment shim from table on page 35.32
- remove main shaft and main shaft bearing outer race from gear carrier housing (see page 35.29)
- press main shaft bearing outer race with adjustment washer into the gear carrier housing to the depth determined using table on page 35.32
- attach transmission housing and tighten bolts to 25 Nm (18 ft lb), plus an additional 1/4 turn

Adjustment shims available

Bearing tolerance measured value	Gage (mm)	Adjustment shim replacement part number
0.671-0.699	0.650	02A 311 140
0.700-0.724	0.675	02A 311 140 A
0.725-0.749	0.700	02A 311 140 B
0.750-0.774	0.725	02A 311 140 C
0.775-0.799	0.750	02A 311 140 D
0.800-0.824	0.775	02A 311 140 E
0.825-0.849	0.800	02A 311 140 F
0.850-0.874	0.825	02A 311 140 G
0.875-0.899	0.850	02A 311 140 H
0.900-0.924	0.875	02A 311 140 J
0.925-0.949	0.900	02A 311 140 K
0.950-0.974	0.925	02A 311 140 L
0.975-0.999	0.950	02A 311 140 M
1.000-1.024	0.975	02A 311 140 N
1.025-1.049	1.000	02A 311 140 P
1.050-1.074	1.025	02A 311 140 Q
1.075-1.099	1.050	02A 311 140 R
1.100-1.124	1.075	02A 311 140 S
1.125-1.149	1.100	02A 311 140 T
1.150-1.174	1.125	02A 311 140 AA
1.175-1.199	1.150	02A 311 140 AB
1.200-1.224	1.175	02A 311 140 AC
1.225-1.249	1.200	02A 311 140 AD
1.250-1.274	1.225	02A 311 140 AE
1.275-1.299	1.250	02A 311 140 AF
1.300-1.324	1.275	02A 311 140 AG
1.325-1.349	1.300	02A 311 140 02A 311 140
1.350-1.374	1.325	02A 311 140 02A 311 140 A
1.375-1.399	1.350	02A 311 140 A 02A 311 140 A
1.400-1.424	1.375	02A 311 140 A 02A 311 140 B
1.425-1.449	1.400	02A 311 140 B 02A 311 140 B

Bearing tolerance measured value	Gage (mm)	Adjustment washer replacement part number
1.450-1.474	1.425	02A 311 140 B 02A 311 140 C
1.475-1.499	1.450	02A 311 140 C 02A 311 140 C
1.500-1.524	1.475	02A 311 140 C 02A 311 140 D
1.525-1.549	1.500	02A 311 140 D 02A 311 140 D
1.550-1.574	1.525	02A 311 140 D 02A 311 140 E
1.575-1.599	1.550	02A 311 140 E 02A 311 140 E
1.600-1.624	1.575	02A 311 140 E 02A 311 140 F
1.625-1.649	1.600	02A 311 140 F 02A 311 140 F
1.650-1.674	1.625	02A 311 140 F 02A 311 140 G
1.675-1.699	1.650	02A 311 140 G 02A 311 140 G
1.700-1.724	1.675	02A 311 140 G 02A 311 140 H
1.725-1.749	1.700	02A 311 140 H 02A 311 140 H
1.750-1.774	1.725	02A 311 140 H 02A 311 140 J
1.775-1.791	1.750	02A 311 140 J 02A 311 140 J

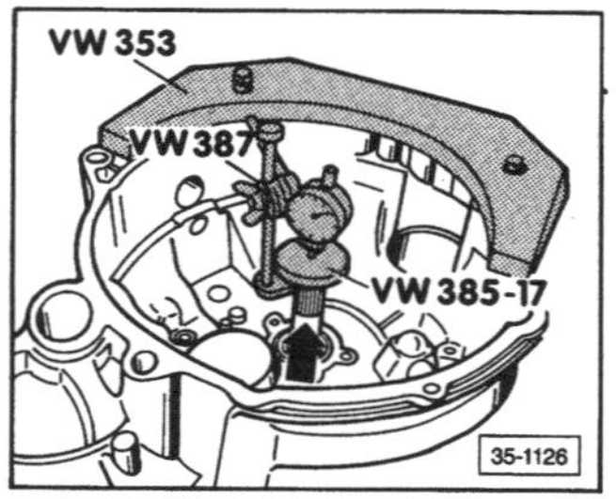

Checking adjustment

- install measuring device and dial gage
- rotate main shaft to settle bearings
  - the bearing clearance must be a minimum of 0.01 and a maximum of 0.09 mm

Note

If no bearing clearance is measurable, but the main shaft can be easily rotated, the adjustment is also satisfactory.

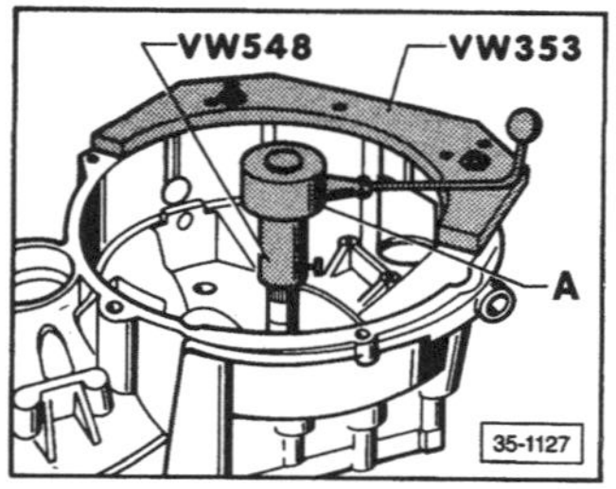

- check that turning torque equals a maximum of 20 Ncm (2 in. lb)

A — Commercial torque gage 0-600 Ncm (0-54 in. lb)

Note

Pay attention to technical data when installing new gears or a new shaft. If replacing both shaft bearings see page 35.44.

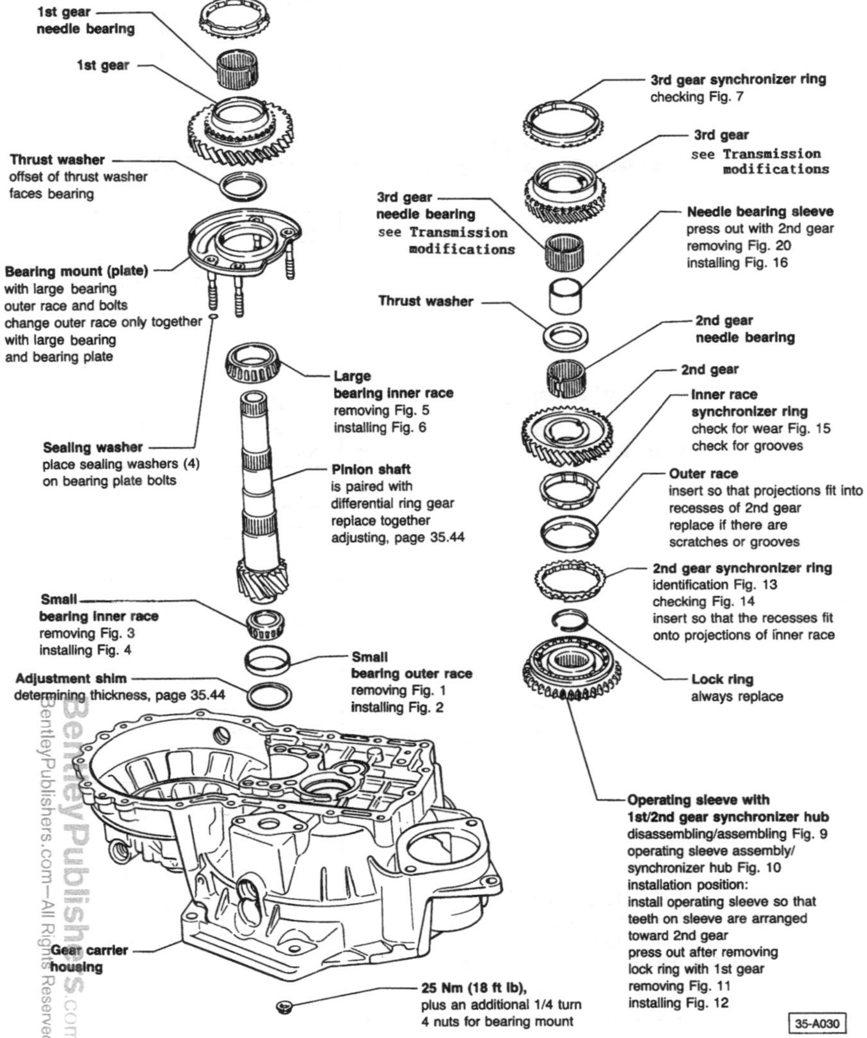

The pinion shaft is to be disassembled as follows:

1 — set separating device under 2nd gear and press off as described in Fig. 20
2 — remove lock ring
3 — set two-arm puller under 1st gear and pull off as described in Fig. 11

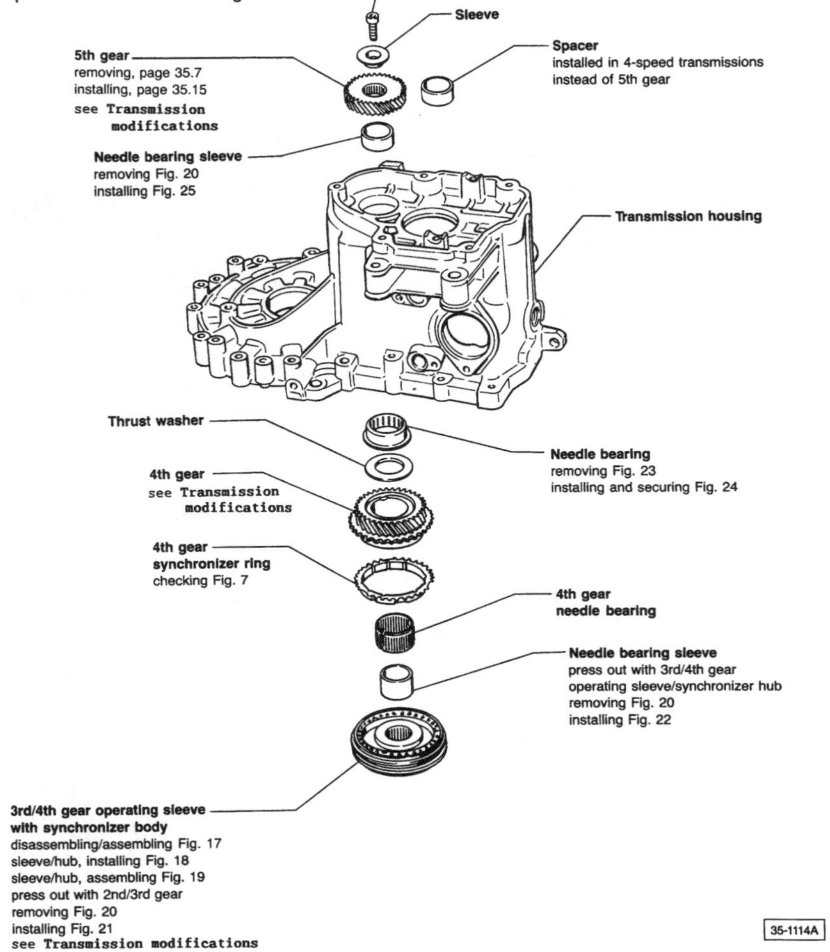

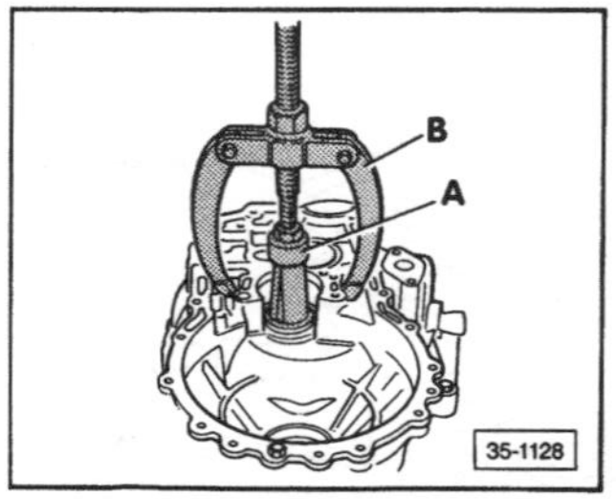

▶ **Fig. 1 Small bearing outer race, removing**

A — Internal extruder, e.g. **US 1099 (Kukko 21/6)**
B — Counter support, e.g. **US 1039 (Kukko 22/2)**

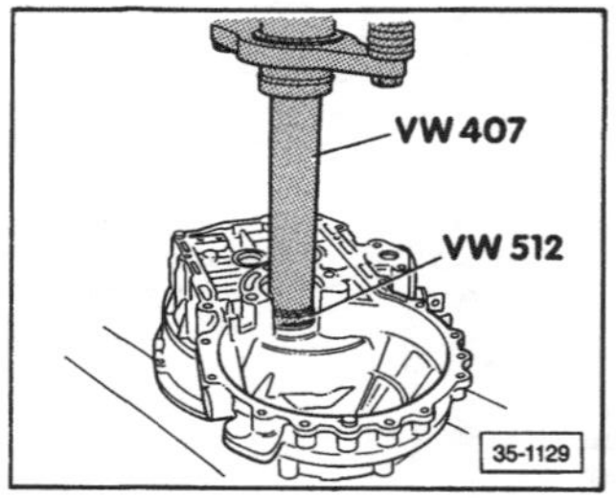

▶ **Fig. 2 Small bearing outer race, installing**

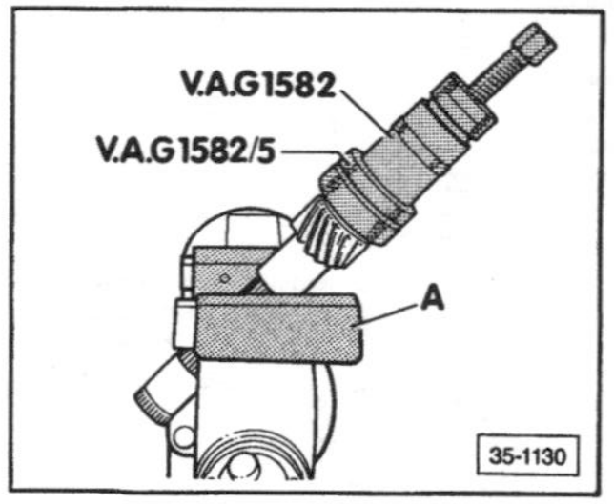

▶ **Fig. 3 Small bearing inner race, removing**

A — Protective jaws

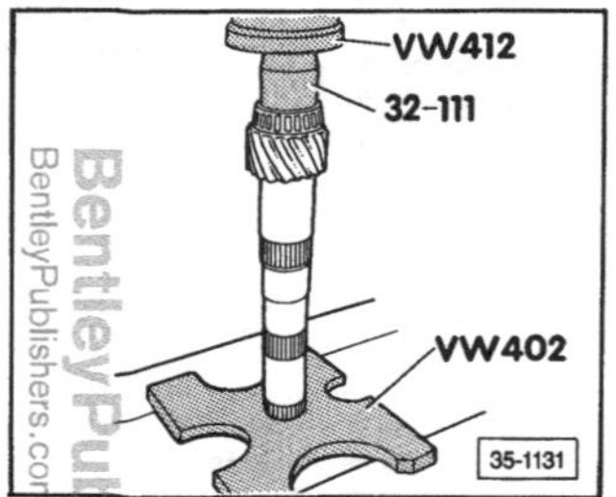

▶ **Fig. 4 Small bearing inner race, installing**

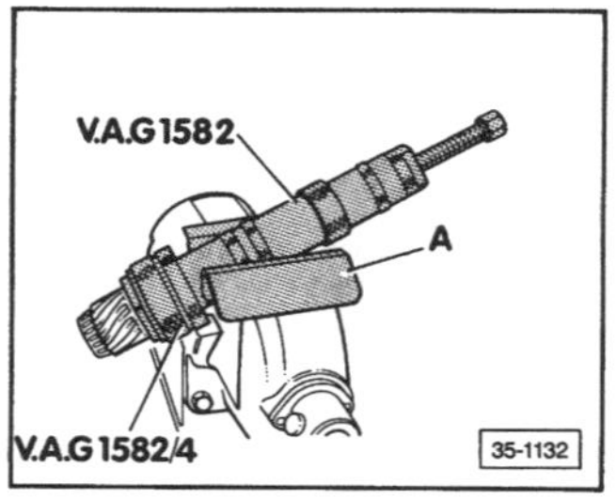

▶ **Fig. 5 Large bearing inner race, removing**

A — Protective jaws

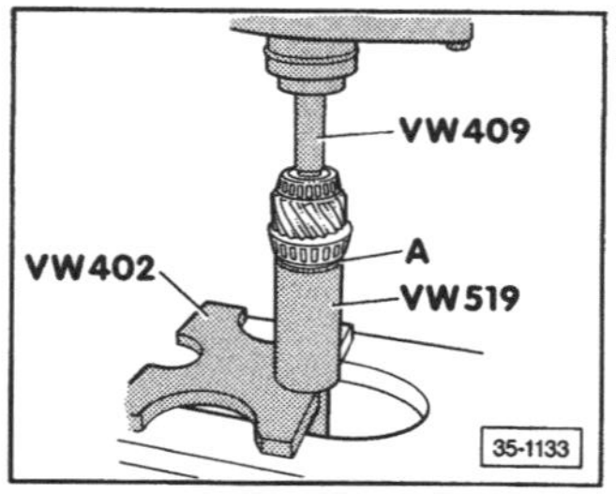

▶ **Fig. 6 Large bearing inner race, installing**

A — Thrust washer
- insert thrust washer prior to pressing on inner race

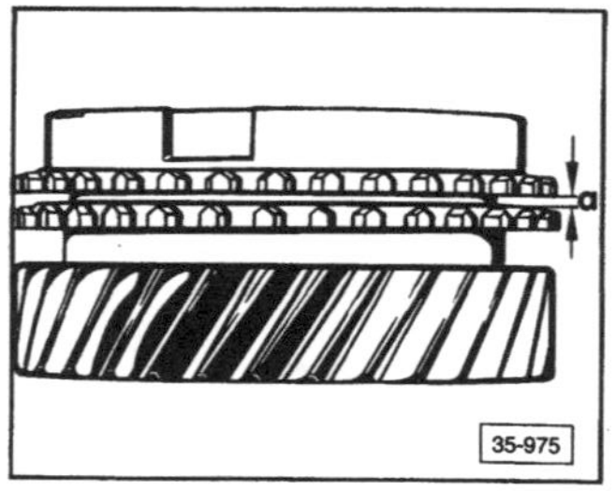

▶ **Fig. 7 Synchronizer ring, checking**

- press synchronizer ring onto cone of gear and measure gap dimension **a** with feeler gage

Gap dimension	Installing dimension (mm)	Wear limit (mm)
1st gear	1.0-1.7	0.5
3rd gear	1.0-1.7	
4th gear	1.0-1.7	

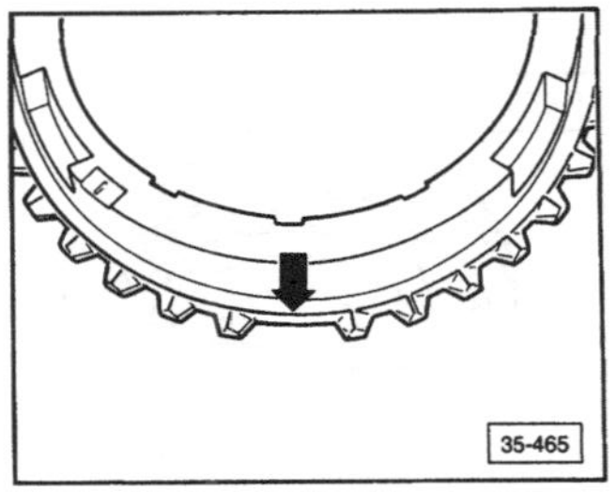

▶ **Fig. 8 1st gear synchronizer ring, marking**

- marking: 3 teeth missing (**arrow**)
- this synchronizer ring is only to be installed on 1st gear

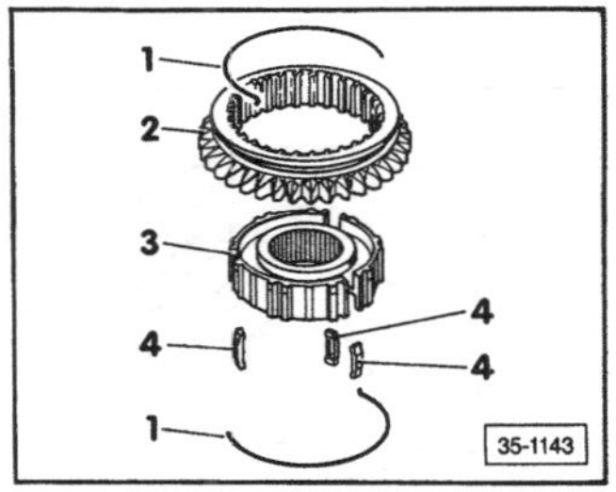

▶ **Fig. 9 1st/2nd gear operating sleeve and hub, assembling**

1 — Spring
2 — Sleeve
3 — Hub
4 — Keys

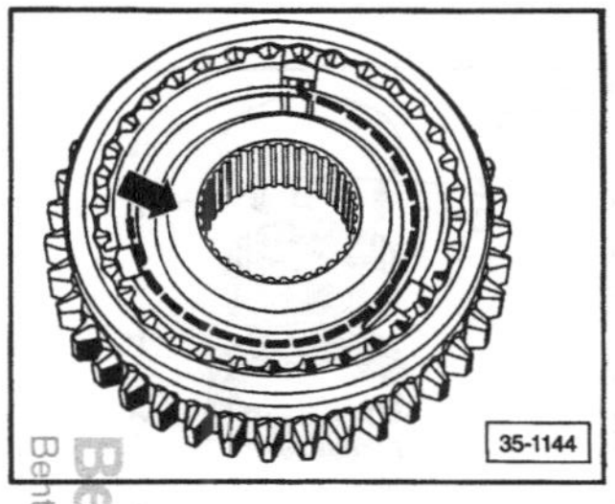

▶ **Fig. 10**

- push sleeve onto synchronizer hub
- insert keys and springs 120° apart
  - the spring bent end must grip into hollow
  - wide shoulder (**arrow**) faces 1st gear

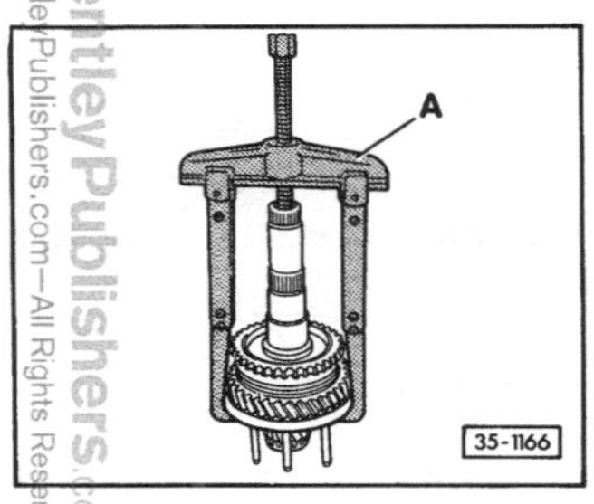

▶ **Fig. 11 Operating sleeve with hub and 1st gear, removing**

A — Two-armed puller, e.g. **US 1078 (Kukko 20/10)** with hooks, 250 mm reach, **US 1079**

Note

Remove lock ring before removing individual parts.

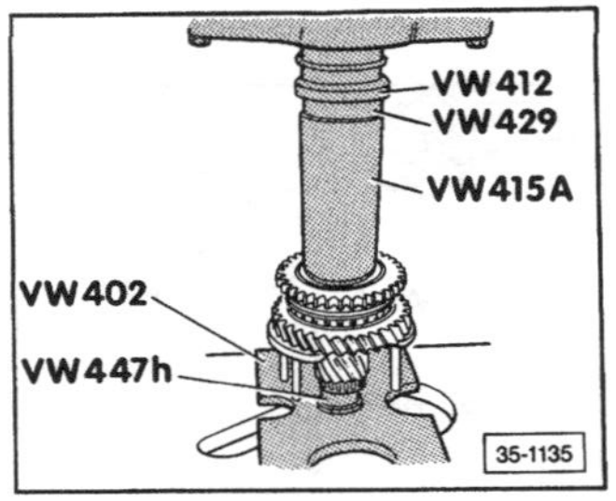

▶ **Fig. 12 1st/2nd gear operating sleeve with hub, installing**

- pointed teeth of sleeve face 1st gear

▶ **Fig. 13 2nd gear synchronizer ring, marking**

- three recesses (**arrows**) for tabs of inner race

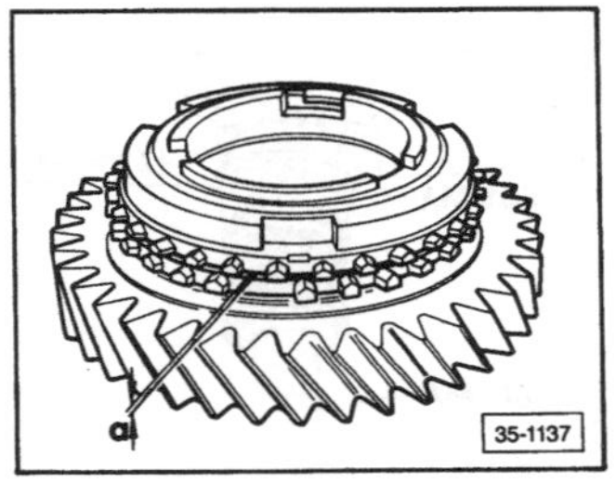

▶ **Fig. 14 2nd gear synchronizer ring, checking**

- press synchronizer ring, outer race and inner race onto cones of 2nd gear and measure gap dimension **a** with feeler gage

Gap dimension	Installing dimension (mm)	Wear limit (mm)
2nd gear	1.2-1.8	0.5

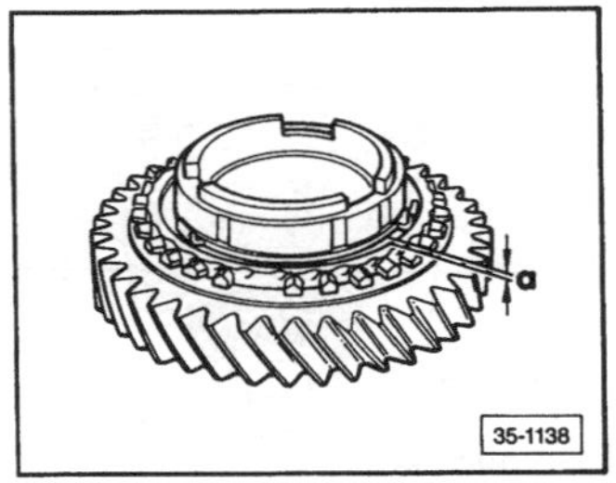

▶ **Fig. 15 Inner race, testing for wear**

- press inner race onto cone of 2nd gear and measure gap dimension **a** with feeler gage

Gap dimension	Installing dimension (mm)	Wear limit (mm)
2nd gear	0.75-1.25	0.3

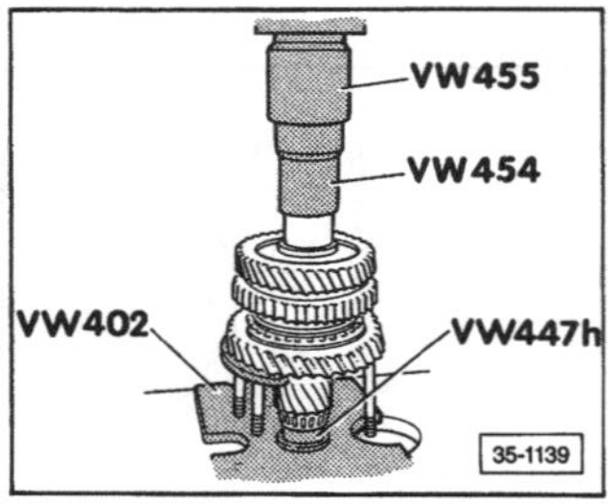

▶ Fig. 16 Needle bearing sleeve, installing

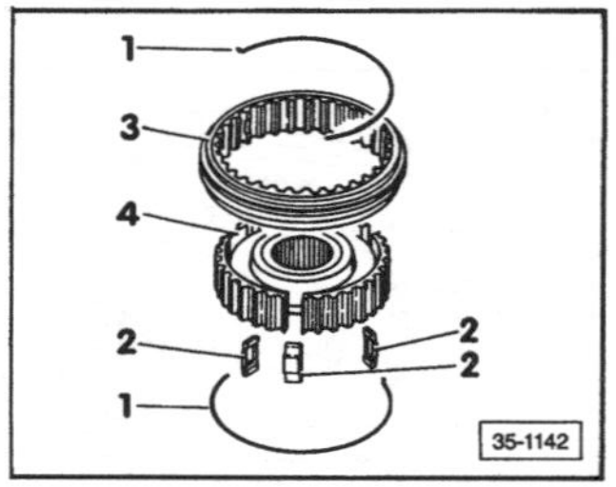

▶ Fig. 17 3rd/4th gear operating sleeve and hub, disassembling/assembling

1 — Spring
2 — Keys
3 — Sleeve
4 — Hub

- push sleeve over synchronizer hub
  - recesses for keys on synchronizer hub and sleeve must align

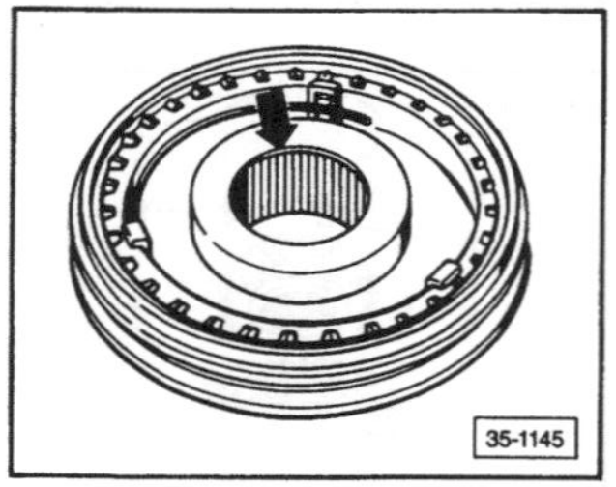

▶ Fig. 18 3rd/4th gear synchronizer hub with sleeve, installation position

- chamfer (**arrow**) faces 4th gear

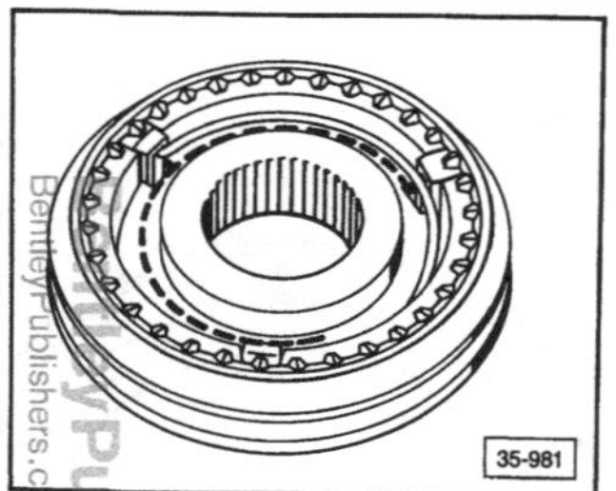

▶ Fig. 19 3rd/4th gear synchronizer with sleeve, assembling

- push sleeve on synchronizer hub
- insert keys and mount springs 120° apart
  - spring bent end must engage in key

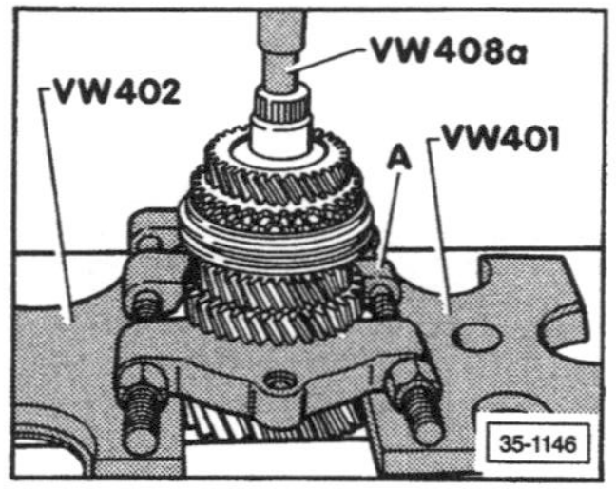

▶ Fig. 20 3rd/4th gear synchronizer hub sleeve, removing

- press off with 2nd, 3rd and 4th gears, and needle bearing sleeve

A — Split bearing puller — 11-115 mm, e.g. **US 1103 (Kukko 17/1)**

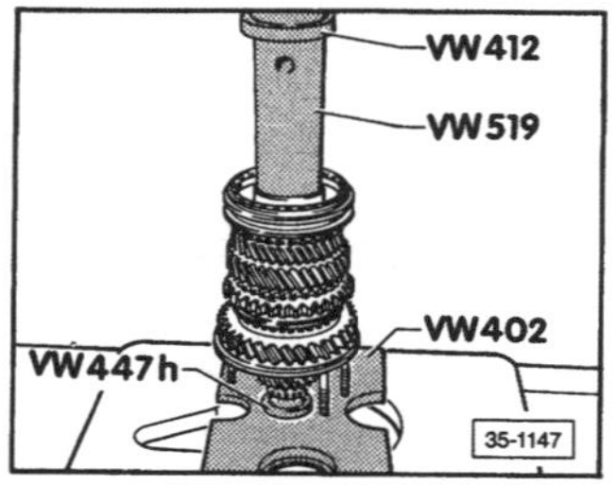

▶ Fig. 21 3rd/4th gear synchronizer hub with sleeve, installing

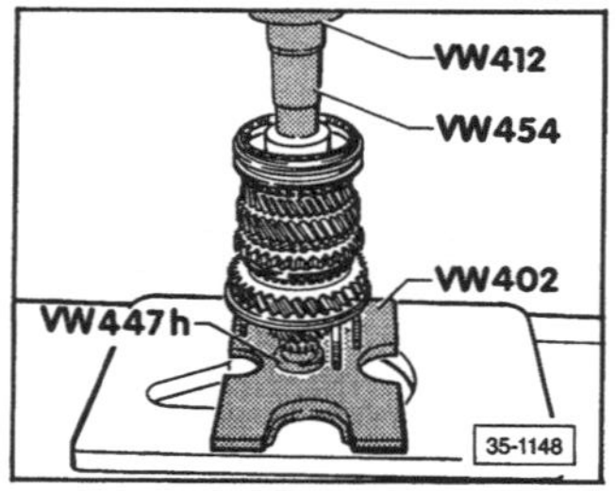

▶ Fig. 22 4th gear needle bearing sleeve, installing

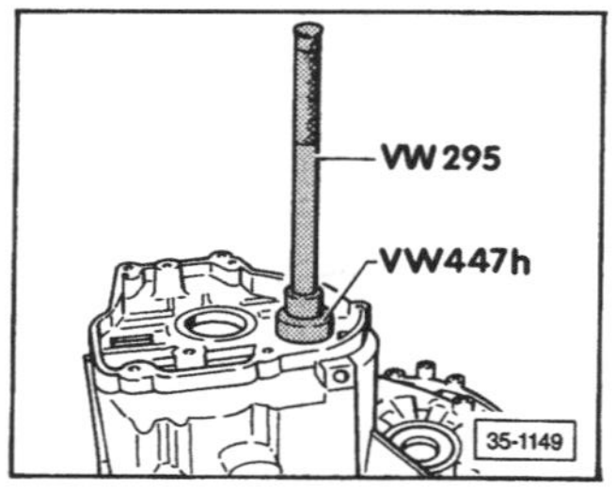

▶ Fig. 23 Needle bearing, removing

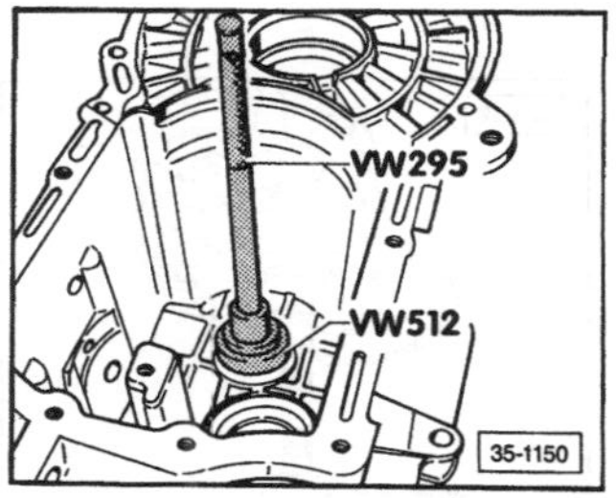

▶ Fig. 24 **Needle bearing, installing**

- secure needle bearing in gear carrier housing by staking at three points 120° apart

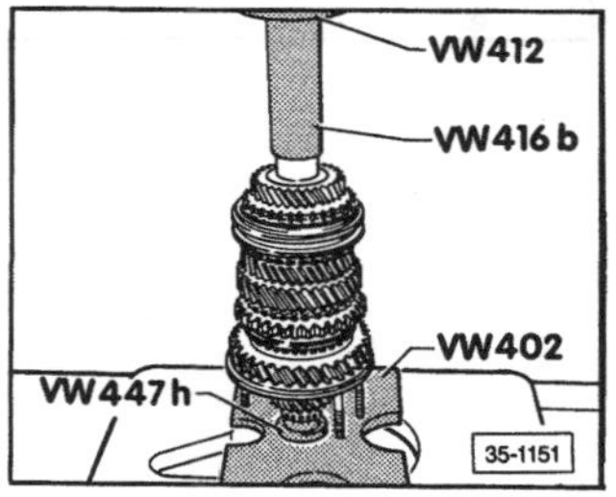

▶ Fig. 25 **Needle bearing sleeve, installing**

Pinion shaft, adjusting

Pinion shaft adjustment shim, determining

Readjustment of the pinion shaft is required when any of the following parts are replaced:

Ring and pinion gear set
Gear carrier housing or
Main shaft bearing
See Adjustment Overview, page 35.48

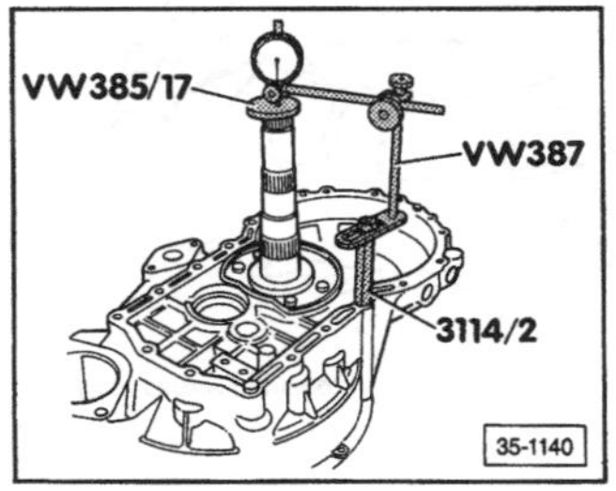

▶

- press in small main shaft bearing outer race with a 0.65 mm thick adjustment washer up to stop (see page 35.37)
- install shaft
- tighten bearing mount nuts to 25 Nm (18 ft lb), plus an additional 1/4 turn
- install dial gage (3 mm measuring range) and set to zero with 1 mm preload
- move pinion shaft up and down
- read tolerance on dial gage and record (example 0.30 mm)

CAUTION

Do not rotate pinion shaft while measuring, since the bearings would settle and result in an inaccurate measurement.

Adjustment shim, determining

The specified preload for the bearing is reached if, to the determined measurement value (0.30 mm) and the installed adjustment shim (0.65 mm), a constant value (0.20 mm) for preload is added.

Example

inserted shim	0.65 mm
measurement value	0.30 mm
preload (constant value)	+ 0.20 mm
Thickness of adjustment shim	= 1.15 mm

- remove pinion shaft and pull out small main shaft bearing outer race, page 35.37

The following adjustment shims are available:

Thickness mm	Replacement part number
0.65	020 311 391 P
0.70	020 311 391 Q
0.75	020 311 391
0.80	020 311 391 A
0.85	020 311 391 B
0.90	020 311 391 C
0.95	020 311 391 D
1.00	020 311 391 E
1.05	020 311 391 F
1.10	020 311 391 G
1.15	020 311 391 H
1.20	020 311 391 J
1.25	020 311 391 K
1.30	020 311 391 L
1.35	020 311 391 M
1.40	020 311 391 N

Note

Varying shim tolerances make it possible to match the thickness of the shim exactly.

- drive in outer race main shaft bearing with determined adjustment shim (in example 1.15 mm) and install driven shaft
- lubricate bearings with transmission oil prior to testing
- tighten nuts for bearing mount in gear carrier housing to 25 Nm (18 ft lb), plus an additional 1/4 turn

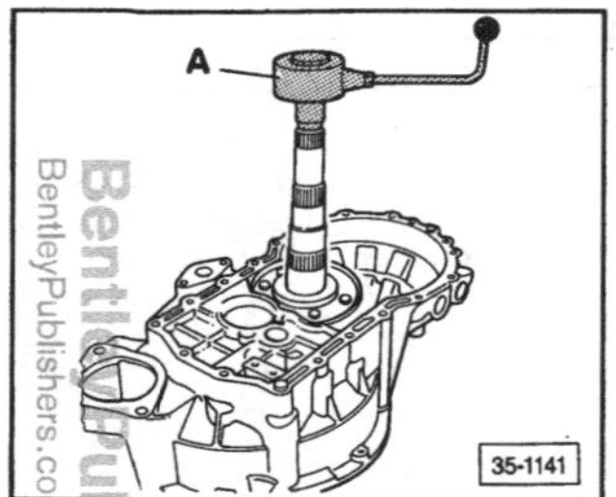

▶

Turning torque, measuring

A — Commercial torque gage 0-600 Ncm (0-54 in. lb)

- for new main shaft bearings, turning torque must be 130-180 Ncm (12-16 in. lb)
- for main shaft bearings that have been run, turning torque must be at least 30 Ncm (3 in. lb)

Note

Needle bearings must always be replaced if removed.

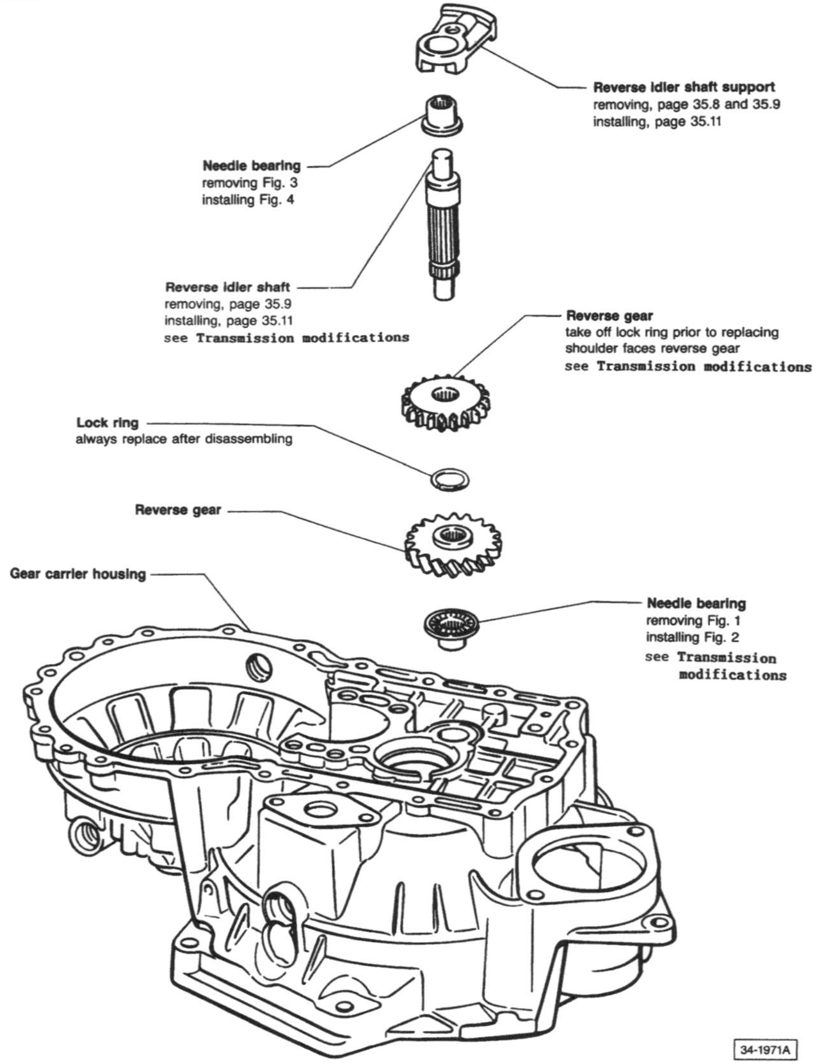

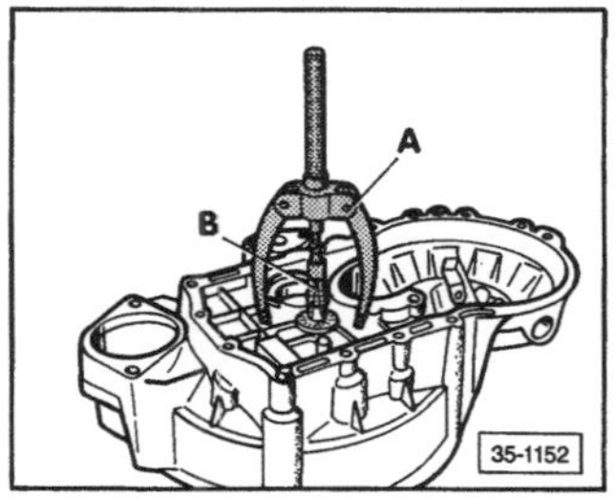

▶ **Fig. 1 *Needle bearing, removing***

A — Counter support, e.g. **Kukko 22/1**
B — Internal puller 14.5-18.5 mm, e.g. **US 8028 (Kukko 21/2)**

Note

Bearing is destroyed when removing and must be replaced.

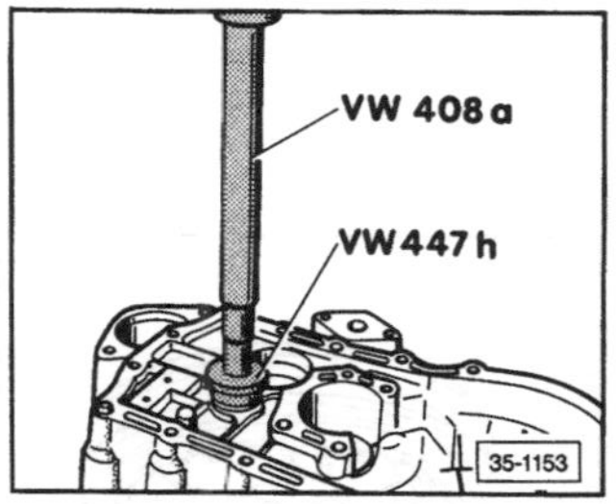

▶ **Fig. 2 Needle bearing, installing**

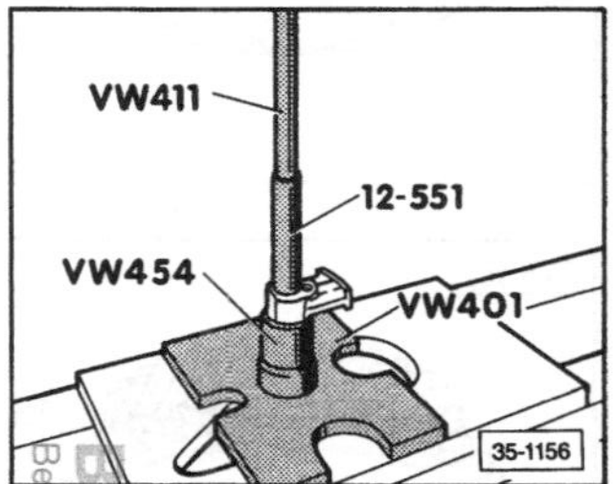

▶ **Fig. 3 Needle bearing, removing**

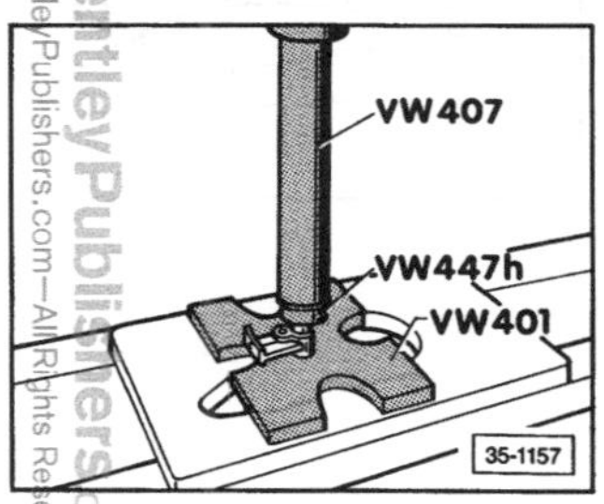

▶ **Fig. 4 Needle bearing, installing on shaft support**

Adjustment Overview

When working on the transmission assembly, adjustment of the main shaft, pinion shaft or differential is only required when parts are replaced that affect the adjustment of the transmission directly. Please pay attention to the following table in order to avoid unnecessary adjusting procedures.

Exchanged part	Main shaft page 35.31	Pinion shaft page 35.44	Differential page 39.8
Gear carrier housing	X		X
Transmission housing	X	X	X
Main shaft	X		
Pinion shaft (gears)		X	
Differential housing			X
Main shaft bearing	X		
Driven shaft bearing		X	
Differential bearing			X

Transmission modifications

The following modifications were made to the 02A transmission beginning on the dates specified for each item.

All parts modified beginning with transmission date 06 07 0 must only be replaced together.

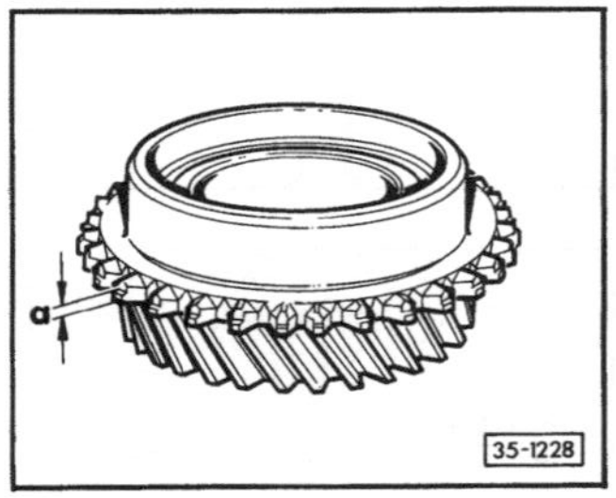

Beginning with transmission date 31 05 0 changes were made to 3rd gear and 4th gear as follows:

- dimension **a** =
  - current 1.45 mm
  - **New** 1.95 mm
  - (illustration shows 4th gear)

Note

Replace 3rd gear and 4th gear with modified parts on transmissions produced prior to 31 05 0.

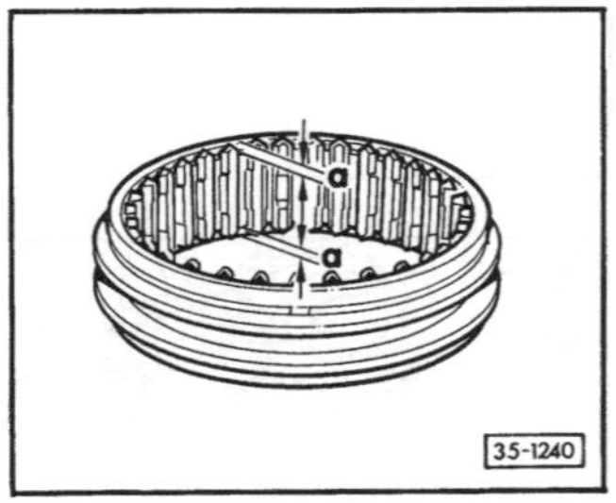

3rd gear/4th gear synchronizer sleeve, modified

Beginning with transmission date 31 05 0:

- dimension **a** =
  - current 2.5 mm
  - **New** 3.0 mm

Note

Replace synchronizer sleeve with modified part on transmissions produced prior to 31 05 0.

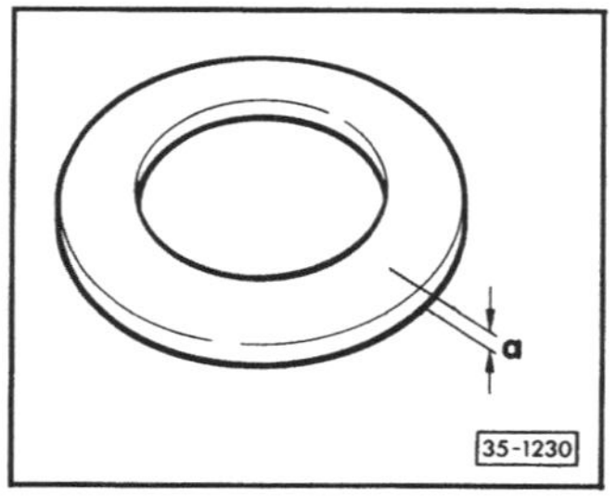

3rd gear thrust washer, modified

Beginning with transmission date 06 07 0:

- dimension **a** =
  - current 2.5 mm
  - **New** 3.5 mm

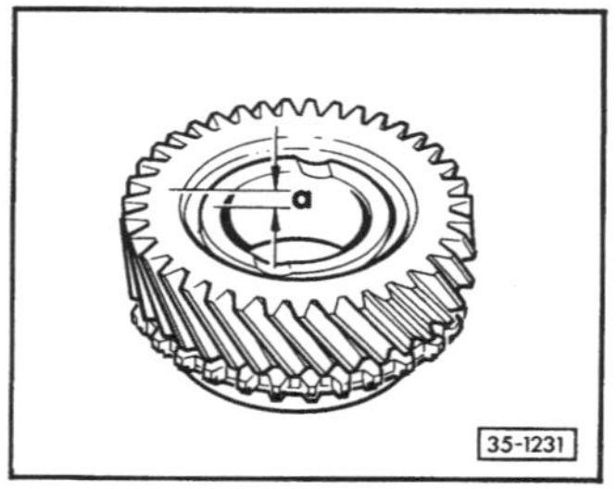

3rd gear, modified

Beginning with transmission date 06 07 0:

- dimension **a** =
  - current 0.3 mm (hub appears flush with outer diameter)
  - **New** 1.3 mm (recess increased due to thicker thrust washer)

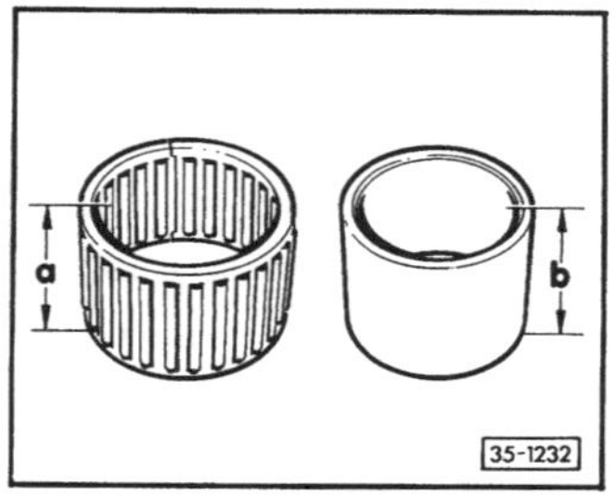

3rd gear needle bearing/sleeve, modified

Beginning with transmission date 06 07 0:

- dimension **a** =
  - current 26.5
  - **New** 25.5 mm
- dimension **b** =
  - current 26.8 mm
  - **New** 25.8 mm

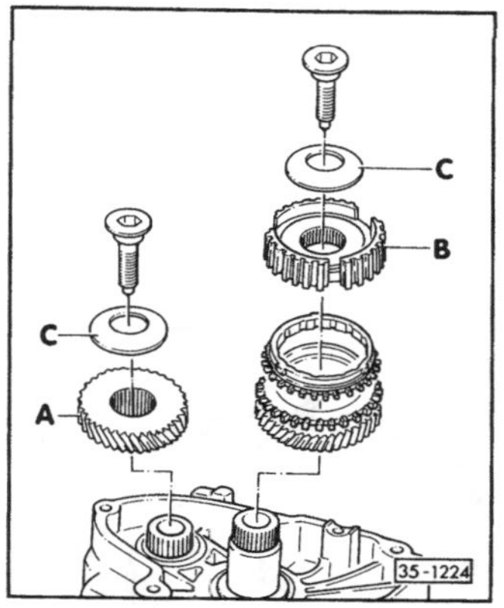

***5th gear (A)/5th gear synchronizer hub (B) bolts*, modified**

Beginning with transmission date 31 05 0 changes were made as follows:

- bolt size and type =
  New M10 with shoulder for centering dished washer **C**
- tightening torque =
  current 40 Nm (30 ft lb)
  New 80 Nm (59 ft lb)
- locking arrangement =
  current **D6** locking compound
  New dished washer
- threaded holes in input/output shafts =
  current **M8**
  New **M10**

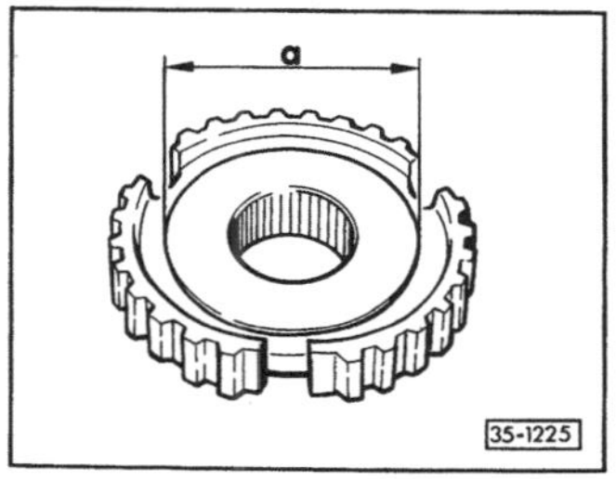

5th gear synchronizer hub, modified

Beginning with transmission date 31 05 0 changes were made as follows to ensure correct positioning of the new dished washer on the synchronizer hub:

- dimension **a** =
  current 39 mm
  New 55 mm

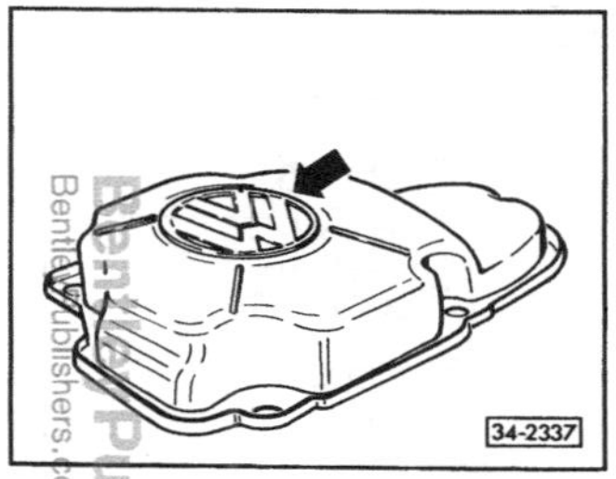

Transmission housing cover, modified

To accommodate changes as of transmission date 31 05 0 the housing cover was provided with a bulge (**arrow**). The cover was previously flat in this area.

Manual Transmission – Case, Gears, Shafts

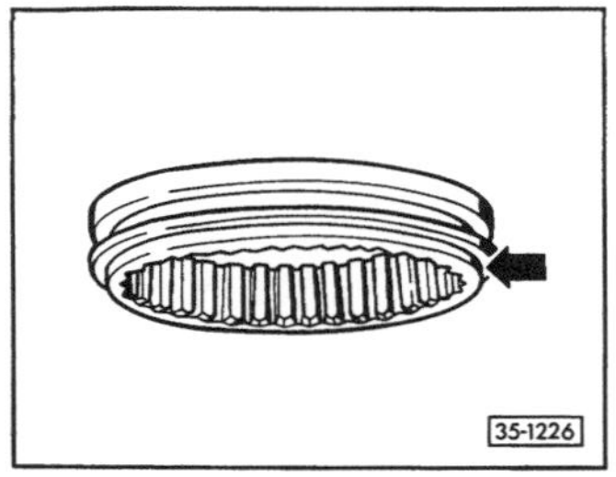

***5th gear synchronizer sleeve*, modified**

To prevent the 5th gear synchronizer sleeve from coming into contact with the 5th gear dished washer, the sleeve has been provided with a collar (**arrow**) on the outer diameter.

Note

The modified synchronizer hub and sleeve can also be installed in earlier transmissions.

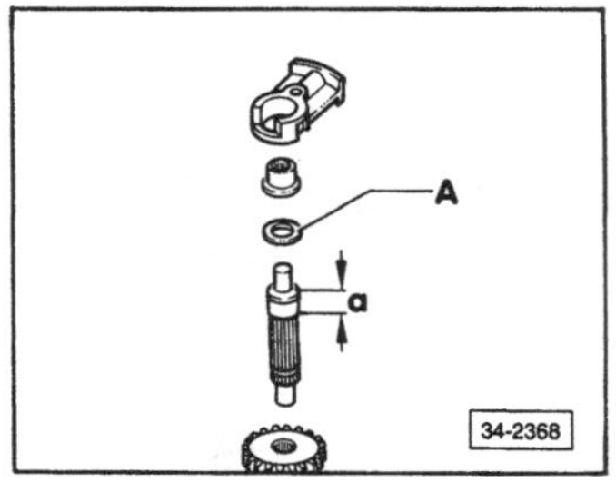

Revised reverse gear idler shaft

Beginning with transmission date 15 08 0, washer **A** between reverse gear shaft and needle bearing, is deleted.

The contact surface of the reverse gear shaft is raised in the area of the needle bearing.

- dimension **a** = current 14.2 mm
  new 14.4 mm

The revised reverse gear shaft can also be installed in current transmissions (before 15 08 0). The washer **A** must then **not** be installed.

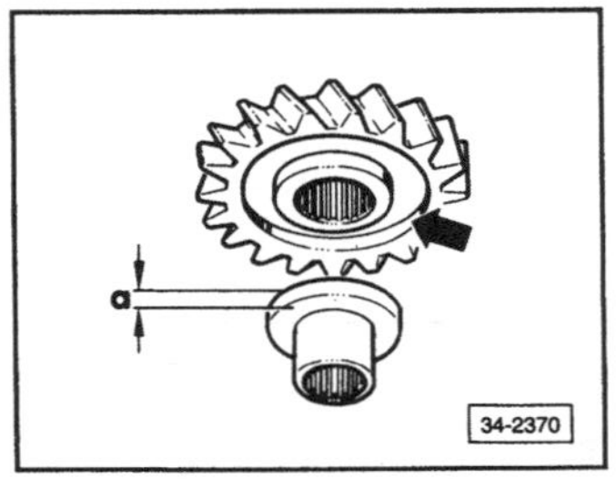

Reverse gear modified

Beginning with transmission date 18 09 0, the groove on the gear carrier housing side of the reverse gear is deleted.

The height of the needle bearing (dimension **a**) is reduced to prevent contact with the gear.

- dimension **a** = current 5.00 mm
  new 3.85 mm

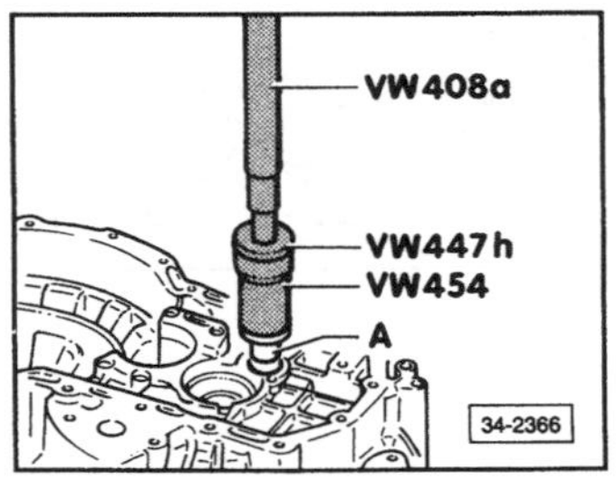

Revised needle bearing, installing

Revised needle bearing **A** , together with the revised reverse gear, can be installed in current transmissions (before 18 09 0).

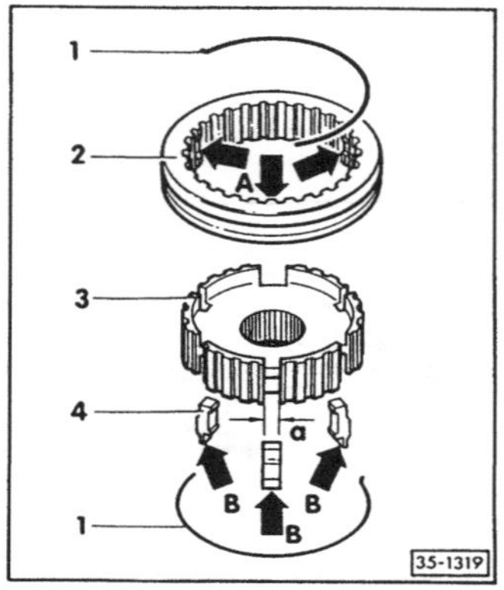

5th gear synchronizer, changes

Transmissions from 31 10 1

Note
The modified assembly has three new locking keys (item **4**) which are additional to the three existing keys cast on the synchronizer ring.

1 — Springs for 5th gear synchronizer ring
- no longer a single-spring design
- installed position: one spring at 120° offset to the other

2 — Synchronizer for 5th gear
- three cutouts (**arrows A**) for new locking keys (item **4**)

3 — Synchronizer hub for 5th gear
- three smaller additional cutouts **a** for new locking keys (item **4**)
- dimension **a** = 8.2 mm

4 — Locking keys
- installed position: smaller end (**arrows B**) toward edge of meshing teeth in synchronizer (item **2**)

5th gear synchronizer ring, changes

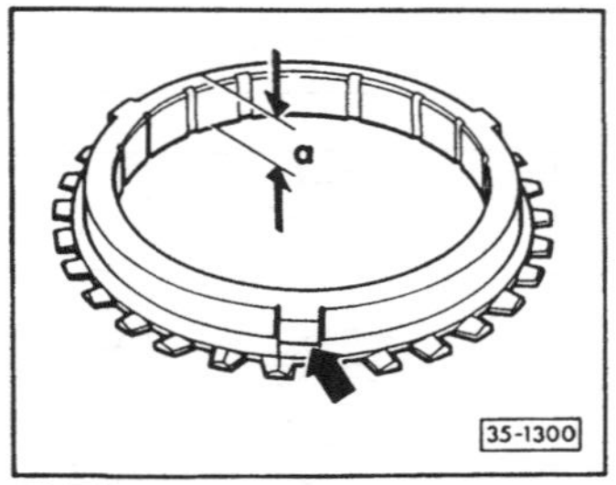

previous (transmissions up to 30 10 1)
- dimension **a** = 9.0 mm
- spring recess (lower **arrow**) is present

new (transmissions from 31 10 1)
- dimension **a** = 8.6 mm
- spring recess (lower **arrow**) has been eliminated

CAUTION
Do not install early version synchronizer rings in later transmissions, nor later version synchronizer rings in earlier transmissions.

Modified synchronizer ring and 5th gear synchronizer, assembling

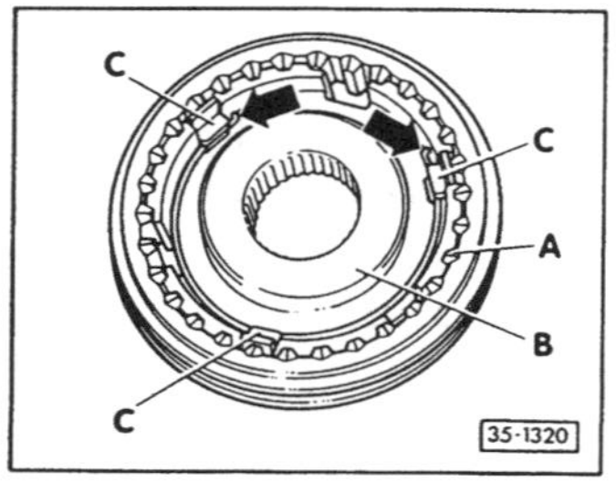

Transmissions from 31 10 1

- ■ slide synchronizer over synchronizer hub (see also, illustration **35-1319**)
  - edges **A** of meshing teeth in synchronizer face in same direction as collar **B** of synchronizer hub
  - cutouts for locking keys in synchronizer and in synchronizer hub must align, one above the other
- ■ insert locking keys, as shown in illustration **35-1319**
- ■ insert springs at a 120° offset, seating each spring under all three lugs **C** of the locking keys
  - be sure ends of spring (**arrows**) are positioned at lugs **C**, as shown

5th gear shift fork, changes

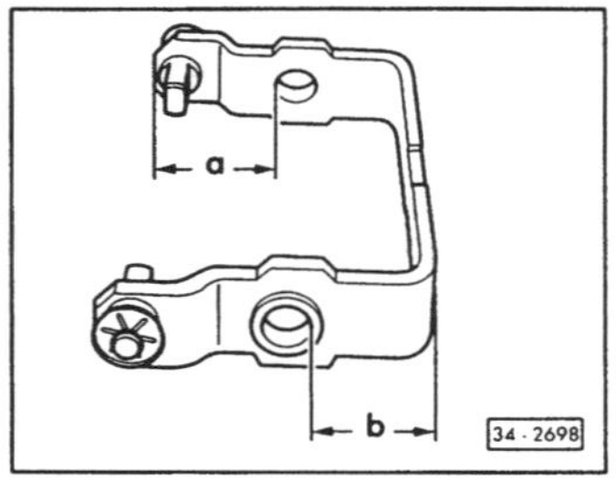

Note
Modifications to the 5th gear shift fork correspond to changes in the 5th gear synchronizer hub.

previous (transmissions up to 30 10 1)
- dimension **a** = 35.0 mm
- dimension **b** = 23.3 mm

new (transmissions from 31 10 1)
- dimension **a** = 33.7 mm
- dimension **b** = 24.3 mm

Note
The new shift fork can be installed in earlier transmissions.

1st/4th gear shift forks/ball bearings, changes

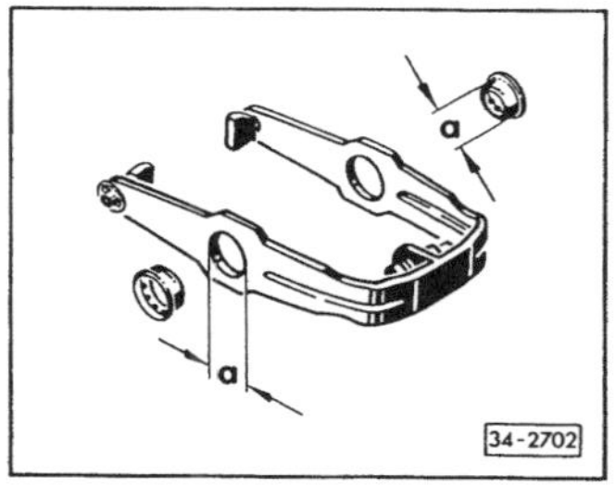

previous (transmissions up to 05 02 2)
- dimension **a** = 20.0 mm

new (transmissions from 06 02 2)
- dimension **a** = 21.0 mm

Note
The new shift fork with larger ball bearings can also be installed in earlier transmissions.

Modified 5th gear synchronizer, installed position

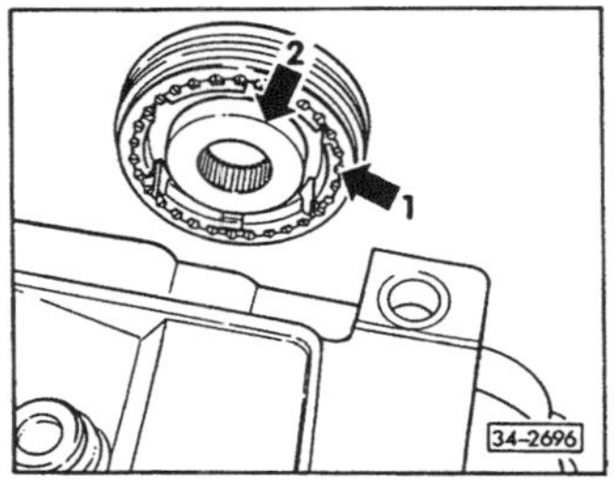

Transmissions from 31 10 1

- ■ assemble hub and synchronizer, as described previously
- ■ install so that teeth (**arrow 1**) in synchronizer, and high collar of synchronizer hub (**arrow 2**) face transmission housing

5th gear, adjusting/checking

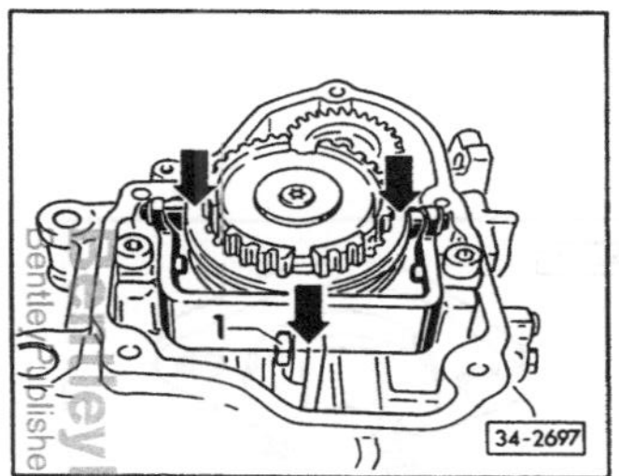

Transmissions from 31 10 1

Adjusting

- ■ engage 5th gear
- ■ loosen bolt **1**
- ■ press synchronizer and selector arm in direction of **arrows**
- ■ tighten bolt **1** to 25 Nm (18 ft lb)

Checking

- ■ check measurement with feeler gage
  - gap between synchronizer and gear must not exceed 0.2 mm
- ■ repeat adjustment if necessary
- ■ remove 5th gear:
  - synchronizer must now be in neutral position
  - synchronizer ring must have enough clearance to move freely

Selector shaft, changes

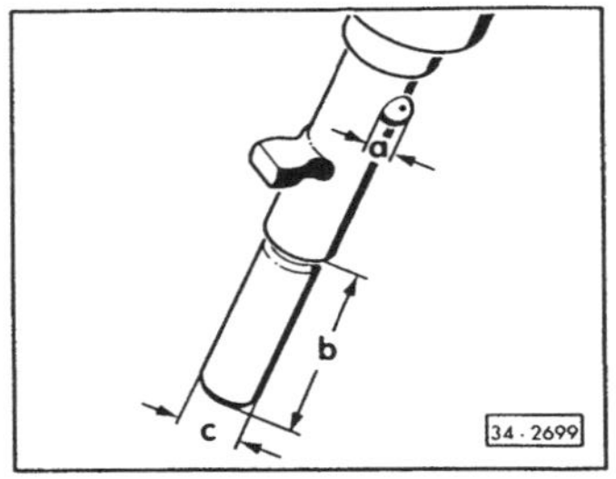

Depth of spring-loaded ball

Note
The spring-loaded ball is now imbedded deeper into the selector shaft. This modified design slows down the main shaft (via the selector shaft and 5th gear synchronizer), and eliminates the noise of shifting into reverse gear.

previous (**all** transmissions up to 30 10 1)

previous (**only** transmissions **NOT** paired with VR6 engines, up to 03 12 1)
- dimension **a** = 5.0 mm

new (**only** transmissions paired with VR6 engines from 31 10 1)

new (**all** transmissions from 04 12 1)
- dimension **a** = 4.2 mm

Length/diameter changes

Note
Instead of a ball bearing for the selector shaft in the end cover, there is now a sleeve bearing. Accordingly, the lower end of the selector shaft is now shorter and has a larger diameter.

previous (transmissions up to 30 10 1)
- dimension **b** = 57.0 mm
- dimension **c** = 15.0 mm

new (transmissions from 31 10 1)
- dimension **b** = 41.0
- dimension **c** = 16.0 mm

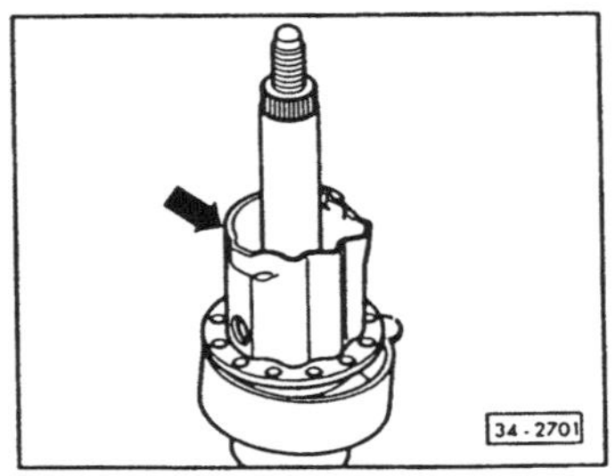

Introduction of detent element

Note
Beginning in April 1992, a selector shaft with a new detent element **(arrow)** was phased into production.

The new detent element can be installed in previous transmissions, with manufacture dates beginning 31 10 1.

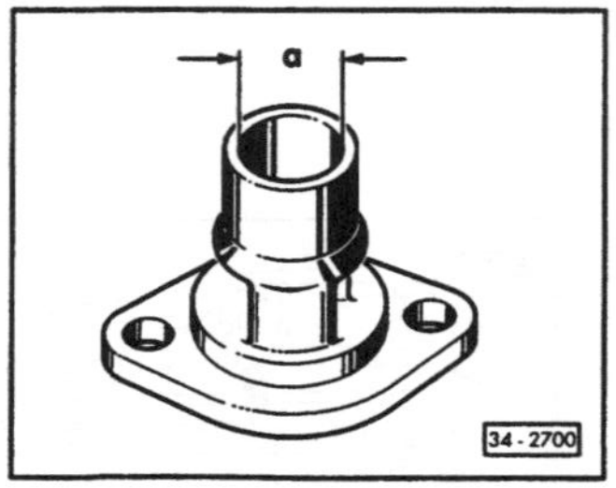

End cover, changes

previous (transmissions up to 30 10 1)
- dimension **a** = 21.0 mm

new (transmissions from 31 10 1)
- dimension **a** = 16.0 mm

Note
The end cover modifications correspond to changes in the selector shaft (i.e. shaft length and diameter changes).

THIS PAGE INTENTIONALLY LEFT BLANK

Index

ATF cooler/ATF filler tube
- assembly **37-70**

ATF filter
- replacing **37-80**

Automatic shiftlock
- troubleshooting **37-110**

Automatic transmission 096
- removing/installing **37-60**

Shift points
- checking **37-90**

Stall speed
- checking **37-100**

Selector lever/detent segment/solenoid switch
- adjusting **37-40**
- assembly **37-30**

Selector mechanism
- assembly **37-20**
- support, removing/installing **37-50**

Technical data
- chart **37-10**

Note
Refer to the Supplement at the back of the manual for additional information pertaining to this Repair Group.

Technical data

- code letters
- transmission applications
- ratios
- equipment

Transmission type		096			
Transmission	Code letters	APC		APB	
	Manufactured from	09/90		10/91	
	to	10/91		02/92	
Torque converter	Code letters	QBCA		QCCR	
Valve body	Code letters	QAC	QAD[1]	QAC	QAD[1]
	Manufactured from	09/90		10/91	
	to	10/91		02/92	
Number of plates		Inner	Outer	Inner	Outer
	Clutch (K1)	5	4	5	4
	Clutch (K2)	5	5	5	5
	Clutch (K3)	5	4	5	4
	Brake (B1)	5	5	5	5
	Brake (B2)	6	7	6	7
Application	Model	Corrado		Corrado	
	Engine	1.8 L 118 kW (158 hp)		2.8 L 132 kW (178 hp)	
Gear ratios	1st gear	2.714		2.714	
	2nd gear	1.551		1.551	
	3rd gear	1.000		1.000	
	4th gear	0.679		0.679	
	Reverse	2.111		2.111	
Intermediate drive	No. of teeth				
	Input gear	74		45	
	Output gear	73		44	
	Ratio	0.983[2]		0.978[2]	
Final drive	No. of teeth				
	Input gear	18		18	
	Final drive gear	76		76	
	Ratio	4.222		4.222	
Axle shaft flange	Diameter (mm)	100		100	
ATF cooler	Equipment	pressure side – 6 row		pressure side – 6 row	

[1] The solenoid valves of valve body **QAD** have flexfoil wiring

[2] When repairing, the intermediate drive ratio 0.984 gearing can also be installed

Transmission type		096			
Transmission	Code letters	CFA		CFF[1]	
	Manufacturedfrom	12/91		01/93	
	to	01/93		–	
Torque converter	Code letters	QCCR		QCCR	
Valve body	Code letters	QAC	QAD[2]	QBA	
	Manufacturedfrom	02/92		11/92	
	to	01/93		–	
Number of plates		Inner	Outer	Inner	Outer
	Clutch (K1)	5	4	5	5
	Clutch (K2)	5	5	5	5
	Clutch (K3)	5	4	6	5
	Brake (B1)	5	5	5	5
	Brake (B2)	6	7	6	7
Application	Model	Corrado		Corrado	
	Engine	2.8 L 132 kW (178 hp)		2.8 L 132 kW (178 hp)	
Gear ratios	1st gear	2.714		2.714	
	2nd gear	1.551		1.441	
	3rd gear	1.000		1.000	
	4th gear	0.679		0.742	
	Reverse	2.111		2.884	
Intermediate drive	No. of teeth				
	Input gear	62		62	
	Output gear	61		61	
	Ratio	0.984		0.984	
Final drive	No. of teeth				
	Input gear	18		20	
	Final drive gear	76		74	
	Ratio	4.222		3.700	
Axle shaft flange	Diameter (mm)	100		100	
ATF cooler	Equipment	pressure side – 6 row		pressure side – 6 row	

[1] Beginning with manufacture date 04 01 3, only the 4th gear (4GR) is mechanically shifted

[2] The solenoid valves of valve body **QAD** have flexfoil wiring

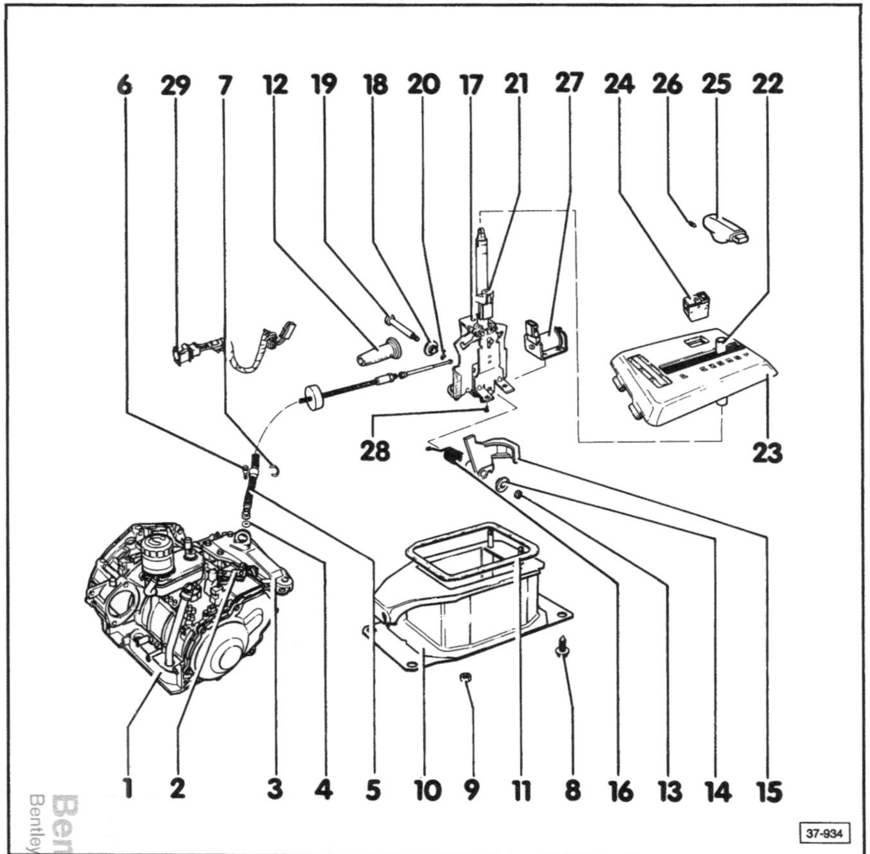

CAUTION

Before working on a running engine, position the selector lever at "P" and apply the parking brake.

Before engaging a selector position from position "P" after starting the engine, the brake pedal must be applied.

1 – **Transmission**

2 – **Lever**
for selector cable

3 – **Bracket**
for selector cable

4 – **Washer**

5 – **Selector lever cable**
- do not bend or kink
- grease cable ends lightly before installing
- check that protective sleeve is located properly
- secure selector cable with clip
- adjusting: place selector lever at "P", loosen screw of gear lever shaft, set gear lever at "P" (detent must engage parking lock), tighten screw to **20 Nm (14 ft lb)**

Note

The front exhaust pipe and deflector must be removed when replacing the selector lever cable

6 – **Shoulder screw — 25 Nm (18 ft lb)**

7 – **Clip**
for selector lever cable

8 – **25 Nm (18 ft lb)**
removing /installing, section 37-50

9 – **15 Nm (11 ft lb)**

10 – **Selector support**
removing/installing, section 37-50

11 – **Gasket**
always replace

12 – **Boot**
install in support before installing cable

13 – **10 Nm (7 ft lb)**

14 – **Bushing**

15 – **Lock lever**
locks selector lever in "N" or "P"

16 – **Spring**

17 – **Selector lever**
- with locking segment and pressure rod
- replace complete assembly only
- assemble before installing

18 – **Bushing**

19 – **Bolt**

20 – **Clip**
install after inserting selector cable

21 – **Bulb holder**

22 – **Sleeve**

23 – **Cover**

24 – **Program switch — E122**
checking in vehicle, Repair Group 01(Automatic Transmission)

25 – **Handle**

26 – **Set screw**
install with locking fluid, Part No. **AMV 200 000**

27 – **Selector lever lock solenoid**
- adjusting, section 37-40
- functional check, section 37-40
- to adjust or replace switch, detent segment must be removed
- checking in-vehicle with Self Diagnosis, Repair Group 01(Automatic Transmission)

28 – **3 Nm (27 in. lb or 31 cm kg)**
install with locking fluid, Part No. **AMV 200 000**

29 – **Wiring harness**
for scale illumination and selector lock solenoid - **N110**

Selector lever, detent segment and solenoid switch, assembling

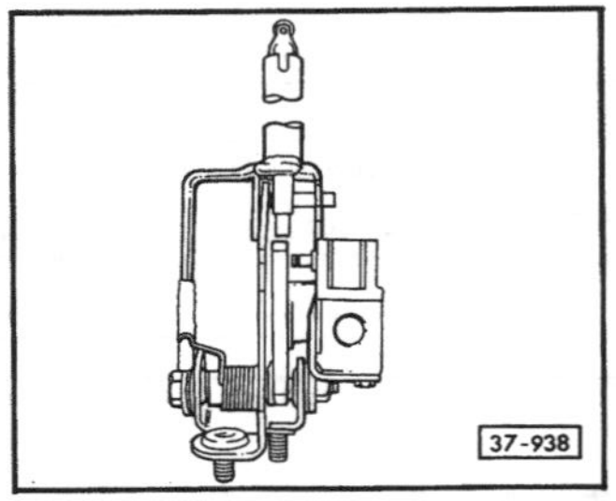

37-938

- adjust the selector lever lock solenoid and perform a functional test, see section 37-40, before installing selector lever into selector support
- install screws for solenoid switch with locking fluid, Part No. **AMV 200 000**

Solenoid switch, adjusting

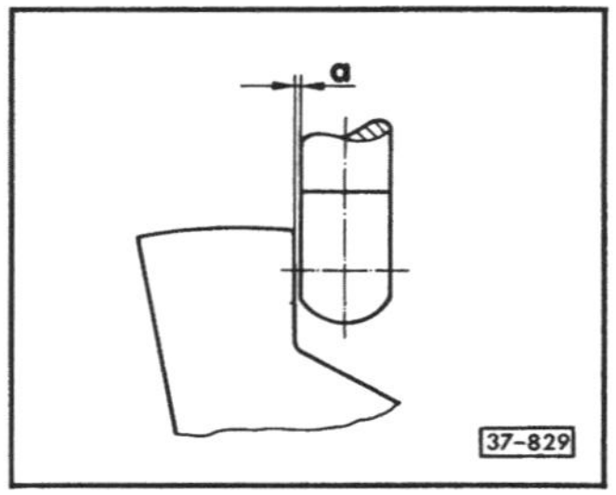

37-829

- measure play between push rod and lever; move switch in elongated holes to adjust to dimension **a**
  - dimension **a** = 0.3 mm

Functional check

- place selector lever at "P" and apply 12V to solenoid
  - must not be possible to move selector lever from "P" position
- stop solenoid operation
  - selector lever can now be moved out of "P" position
- place selector lever at "N" and repeat functional check
- after adjusting play, install complete selector lever assembly

Selector support, removing/installing

Removing

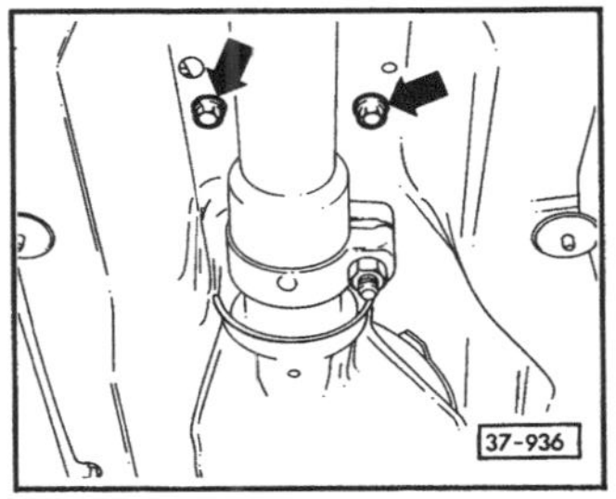

- remove front bolts (**arrows**)
- disconnect front catalytic converter
- disconnect oxygen sensor plug connector
- remove rear exhaust system holding rings
- remove heat shield and rear bolts

Installing

Note

Install in reverse order of removal procedure.

Automatic transmission 096, removing/installing

Removing

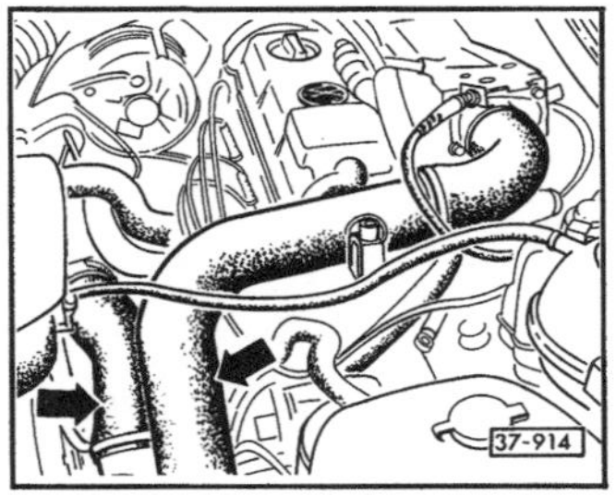

- disconnect battery ground strap
- loosen upper air cooler connecting pipe clamps
- remove brake disc air duct
- loosen air cooler lower clamps and remove upper connector pipes (**arrows**)
- remove speedometer drive from transmission
- remove engine/transmission connecting bolts

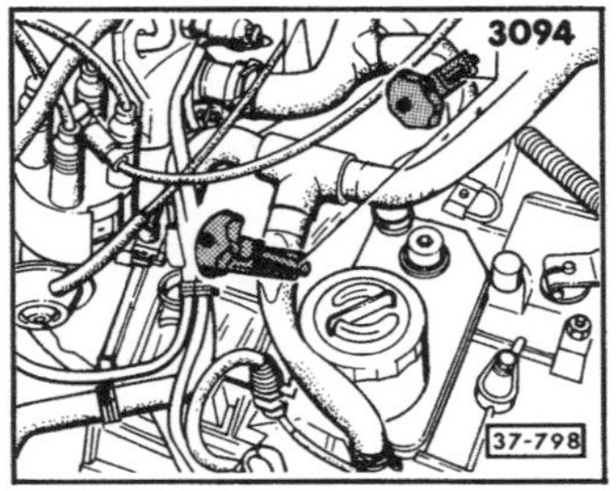

- clamp off coolant hoses at ATF cooler with **3094** and disconnect hoses

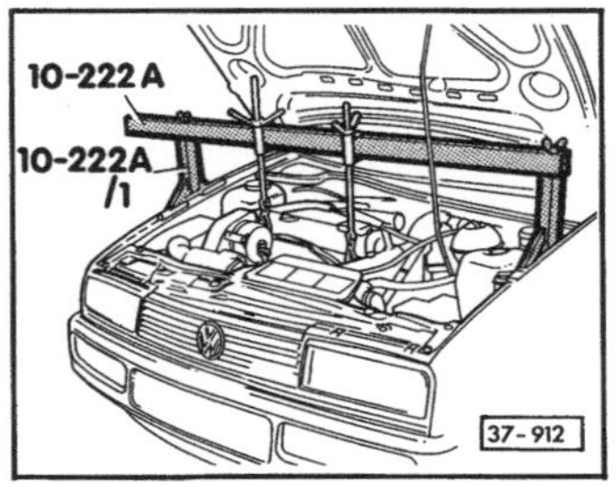

- install engine support bar **10-222** and retainer **3180**
- remove upper starter screw
- place selector lever at "**P**" and remove selector cable from lever
- remove cable bracket clip and remove cable

- remove left transmission mount

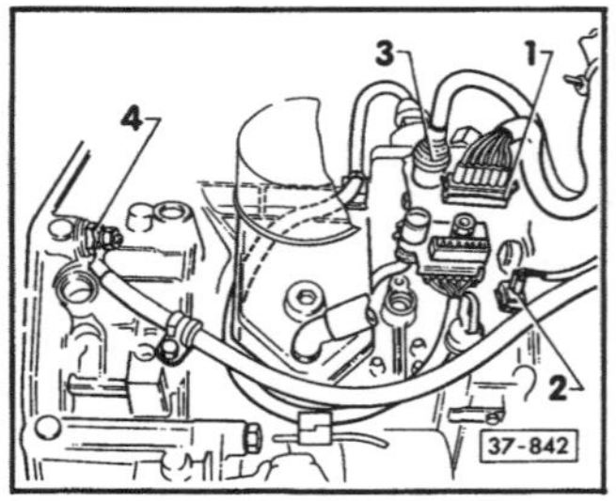

- disconnect electrical connectors **1**, **2**, **3** and **4** from transmission
- remove front assembly mounting
- remove starter
- remove axle shaft protective cap
- disconnect right axle shaft at transmission
- remove left axle shaft from vehicle

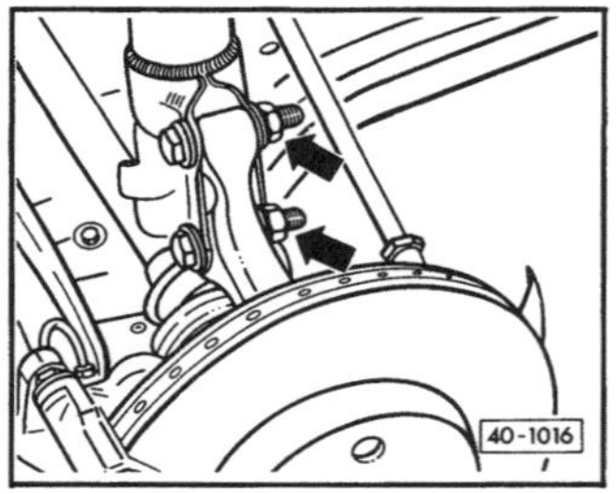

- remove axle shaft nuts
- remove wheel
- mark wheel bearing housing/strut installation position
- remove wheel bearing housing/strut bolts then swing wheel bearing housing to the side and remove axle shaft

> **CAUTION**
>
> Tighten wheel bearing housing/strut back into installation position.

- remove torque converter fasteners
- lower support bar
- slide engine and transmission to right and install transmission jack
- remove lower engine/transmission bolts and separate transmission from engine
- lower transmission carefully
- secure torque converter so it does not fall

Installing

- check that guide sleeves are installed
- install transmission in reverse order of removal procedure
- check and adjust selector cable if necessary (see section 37-20)
- if necessary, check and correct transmission fluid level

Tightening torques

Axle shaft to flange	45 Nm (33 ft lb)
Converter to drive plate	60 Nm (44 ft lb)
M12 bolts	80 Nm (59 ft lb)
M10 bolts	60 Nm (44 ft lb)
Front bracket to mount	60 Nm (44 ft lb)
Left side bracket to transmission	25 Nm (18 ft lb)
Cover plate to transmission	15 Nm (11 ft lb)
Wheel bearing housing to strut	95 Nm (129 ft lb)
Shaft to wheel hub	265 Nm (359 ft lb)

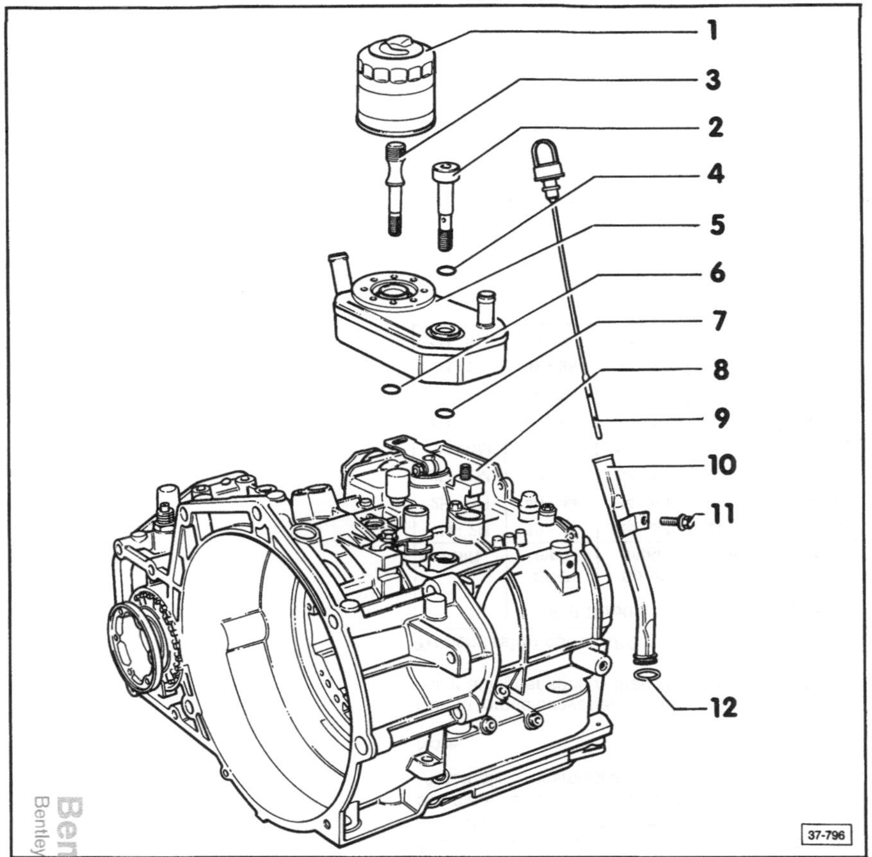

1 — **ATF filter**
discontinued, see next page

2 — **Banjo bolt**
40 Nm (30 ft lb)

3 — **Filter stud**
40 Nm (30 ft lb)

4 — **O-ring**
always replace

5 — **ATF cooler**

6 — **O-ring**
always replace

7 — **O-ring**
always replace

8 — **Transmission housing**

9 — **ATF dipstick**
checking ATF level, Repair Group 32

10 — **ATF filler tube**

11 — **Bolt**

12 — **O-ring**
always replace

ATF filter, replacing

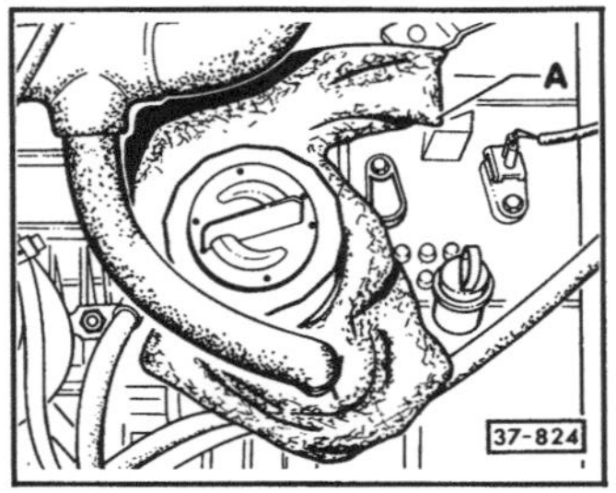

Production Change

Effective with production date of February 1991, the automatic transmission **ATF** filter has been eliminated.

> **CAUTION**
> **Part numbers are for reference only. Always check with your Parts Department for latest information.**

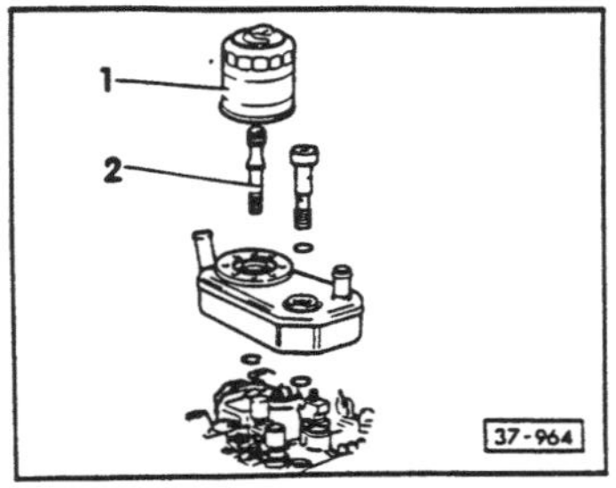

ATF filter, Removing

- wrap workshop cloth (A) around **ATF** filter, loosen filter slightly
- allow **ATF** to flow back into transmission
- remove **ATF** filter, **Part No. 096 409 091** (1) and discard

Note

ATF filters without the turning lugs can be loosened with a strap wrench.

- wipe **ATF** off **ATF** cooler
- remove filter bolt (2) and discard
- install O-ring **Part No. 096 409 069A**, onto new banjo bolt, **Part No. 096 409 067A**
- install new banjo bolt (1), with O-ring (3), in place of **ATF** filter bolt **(arrow)**

Note

Banjo bolt, **Part No. 096 409 067A** and O-ring, **Part No. 096 409 069A**, is also available as a Kit, **Part No. 096 498 067**.

- tighten bolt (1)
  - Torque 40 Nm (29 Ft. Lb.)
- wipe off excess **ATF** from transmission case
- check **ATF** level

Shift points, checking

If the transmission seems to shift to the next gear too early or too late, check the shift points with those specified in the chart.

The observed shift points must fall within the mph range specified in the chart. Speedometers can deviate from the specified value within a certain tolerance,.

H = hydraulic shift

M = mechanical shift

Determine the causes of any deviations in the shifting points by using the transmission self-diagnosis procedures, Repair Group 01 (Automatic Transmission).

Note

Due to torque converter slip, the 3H - 3M shift can occur at higher speeds than shown in the chart.

Transmission code: APC

	Full Throttle		Kickdown	
Shift	ECO mph	Sport mph	ECO mph	Sport mph
1H - 2H	18 - 22	28 - 32	28 - 32	28 - 32
2H - 3H	40 - 44	54 - 58	54 - 58	54 - 58
3H - 3M	40 - 44	54 - 58	54 - 58	54 - 58
3M - 4M	71 - 75	89 - 93	89 - 93	89 - 93
4M - 3M	63 - 60	83 - 79	90 - 86	90 - 86
3M - 3H	36 - 34	36 - 34	53 - 49	53 - 49
3H - 2H	32 - 28	32 - 28	53 - 49	53 - 49
2H - 1H	7 - 3	19 - 15	29 - 25	29 - 25

Transmission code: APB

	Full Throttle		Kickdown	
Shift	ECO mph	Sport mph	ECO mph	Sport mph
1H - 2H	19 - 22	30 - 34	30 - 34	30 - 34
2H - 3H	39 - 43	57 - 60	57 - 60	57 - 60
3H - 3M	57 - 60	82 - 86	57 - 60	82 - 86
3M - 4M	85 - 88	96 - 100	96 - 100	96 - 100
4M - 3M	58 - 55	83 - 79	96 - 93	96 - 93
3M - 3H	37 - 34	41 - 37	56 - 52	56 - 52
3H - 2H	32 - 28	32 - 28	56 - 52	56 - 52
2H - 1H	7 - 4	19 - 15	29 - 33	29 - 33

Transmission code: CFA

	Full Throttle		Kickdown	
Shift	ECO mph	Sport mph	ECO mph	Sport mph
1H - 2H	19 - 22	32 - 36	32 - 36	32 - 36
2H - 3H	40 - 44	57 - 61	57 - 61	57 - 61
3H - 3M	57 - 61	83 - 86	57 - 61	83 - 86
3M - 4M	78 - 82	97 - 101	97 - 101	97 - 101
4M - 3M	64 - 60	82 - 77	96 - 93	96 - 93
3M - 3H	34 - 30	37 - 34	57 - 53	57 - 53
3H - 2H	32 - 28	32 - 28	57 - 53	57 - 53
2H - 1H	10 - 6	19 - 15	29 - 25	29 - 25

Transmission code: CFF

Shift	Kickdown – mph
1H - 2H	61 - 67
2H - 3H	122 -128
3H - 4M	176 -182
4M - 3H	178 -172
3H - 2H	119 -113
2H - 1H	50 - 44

Stall speed, checking

Check stall speed only if vehicle exhibits poor performance or poor acceleration.

Before measuring stall speed, check that the correct torque converter is installed.

Note

The stall speed decreases as height above sea level increases (approx. 125 RPM per 325 feet).

- connect tachometer **VW 1367** or **VAG 1551**
- engage parking brake
- start and warm engine
- hold foot brake firmly, put selector lever in **D** and depress accelerator pedal briefly to full throttle
- check that engine runs within stall speed range

CAUTION

Do not continue stall speed test longer than time required to read tachometer. Maximum stall speed test time is 5 seconds. Wait at least 20 seconds before repeating test.

Transmission code	Torque converter code	Stall speed (RPM)
APC	QBCA	2750 - 2950
APB, CFA, CFF	QCCR	2350 - 2650

- if stall speed is not within range, following are possible causes:
  - stall speed too high = K1 clutch slipping
    or
    one-way clutch slipping
  - stall speed up to 200 rpm too low
    = check engine adjustments
  - stall speed over 200 rpm too low
    = torque converter defective

AUTOMATIC SHIFTLOCK , TROUBLESHOOTING USING TESTER VAG 1466

Test conditions:
- Transmission Control Module (TCM) (J217) OK

Troubleshooting preparations:
- Remove Park/Neutral Position (PNP) Relay (J226), stamped with production control number 150, from relay position 20
- Insert test adaptor VAG 1466/6 into relay position 20
- Connect tester VAG 1466 to test adaptor

Test procedure information:
- Perform test steps in numerical sequence according to table starting on next page.
- If specified value/test result is obtained, continue troubleshooting with next test step (test result OK).
- If specified value/test result is not obtained, locate and eliminate malfunction using table.
- Use valid wiring diagram for checking wiring.
- Remove all test wires from sockets of VAG 1466 after each step.

IF MALFUNCTION HAS NOT BEEN ELIMINATED AFTER PERFORMING TEST STEPS 1 THROUGH 8, REPLACE PARK/NEUTRAL POSITION (PNP) RELAY (J226).

After repairing malfunction, check automatic shiftlock for proper operation.

Note

Refer to the Supplement at the back of the manual for additional information pertaining to this Repair Group.

Test step	1	2	3	4
To be checked:	Voltage supply terminal 30 (B+)	Voltage supply terminal 50	Voltage supply terminal 15 (B+)	Multi-Function Transmission Range (TR) Switch (F125)
Socket assignment VAG 1466: Red test wire	from 2 to S +	from 6 to S +	from 5 to S +	from 5 to S +
Black test wire	from S - to L +	from S - to L +	from S - to L +	from S - to L +
Blue test wire	from L - to 7	from L - to 7	from L - to 7	from L - to 1
Multimeter Fluke 83 (US 1119)	-----	-----	-----	
Additional work:	-----	Switch Ignition/ Starter Switch (D) to "START" position	Switch ignition ON	Move shift lever to position "P" and "1"
Specified value/ test result:	Indicator lamp L (green) in tester lights up	Indicator lamp L (green) in tester lights up	Indicator lamp L (green) in tester lights up	Indicator lamp L (green) in tester lights up

Possible causes if specified value/ test result is not obtained:

Wiring from fuse/relay panel to component, or from component to component	T36 to J226, terminal 2, or: J226, terminal 7 to ground	F/1 to J226, terminal 6	F/6 to J226, terminal 5	F125 , terminal 1 to J226, terminal 1
or: Ignition/Starter Switch (D)		X	X	-----
or: Multi-Function Transmission Range (TR) Switch (F125)		-----	-----	-----
or: Fuse/relay panel	X	X	X	-----

5	6	7	8
Multi-Function Transmission Range (TR) Switch (F125)	Wiring (circuit) to backup lights left (M16) and right (M17)	Wiring to Starter (B), terminal 50	Wiring to Trans-mission Control Module (TCM) (J217)
from 9 to S +	from 5 to 4	from 5 to S +	----
from S - to L +	----	from S - to L +	----
from L - to 7	----	from L - to 8	----
-----	----	----	Connect multimeter between 3 and J217, terminal 38/5
Move shift lever to position "P", "R" and "N"	----	----	----
Indicator light L (green) in tester lights up	Backup lights left (M16) and right (M17) light up	Indicator light L (green) in tester lights up	Must be approx. 0 Ohms (continuity)

F125, terminal 7 to J226, terminal 9	F/7 to J226, terminal 4, or K/8 to M16 and M17	Starter (B), terminal 50 to J226, terminal 8	J217, terminal 38/5 to J226, terminal 3
-----	-----	-----	-----
X	-----	-----	-----
X	X	-----	-----

Index

ATF pump
- assembly **38-100**

Clutches K1/K2
- adjusting/checking freeplay **38-80**

1st to 3rd gear clutch (K1) with turbine shaft
- assembly **38-130**
- disassembling/assembling **38-140**

One-way clutch with B1 piston
- assembly **38-110**
- disassembling/assembling **38-120**

Parking lock
- assembly **38-210**
- installing **38-220**

Planet carrier
- adjusting/checking **38-60**

Reverse brake (B1)
- adjusting/checking **38-70**

Reverse clutch (K2)
- assembly **38-150**
- circlip for clutch spring support, replacing **38-160**

2nd/4th gear brake (B2)
- adjusting **38-90**

3rd/4th gear clutch (K3) with pump shaft
- assembly **38-170**
- circlip for clutch spring support, replacing **38-180**

Transmission
- assembling **38-50**
- assembly **38-30**
- clutches/brakes installed positions **38-10**
- disassembling **38-40**
- selector/clutch/brake engagement chart **38-20**

Valve body
- assembly **38-190**
- disassembling/assembling **38-200**

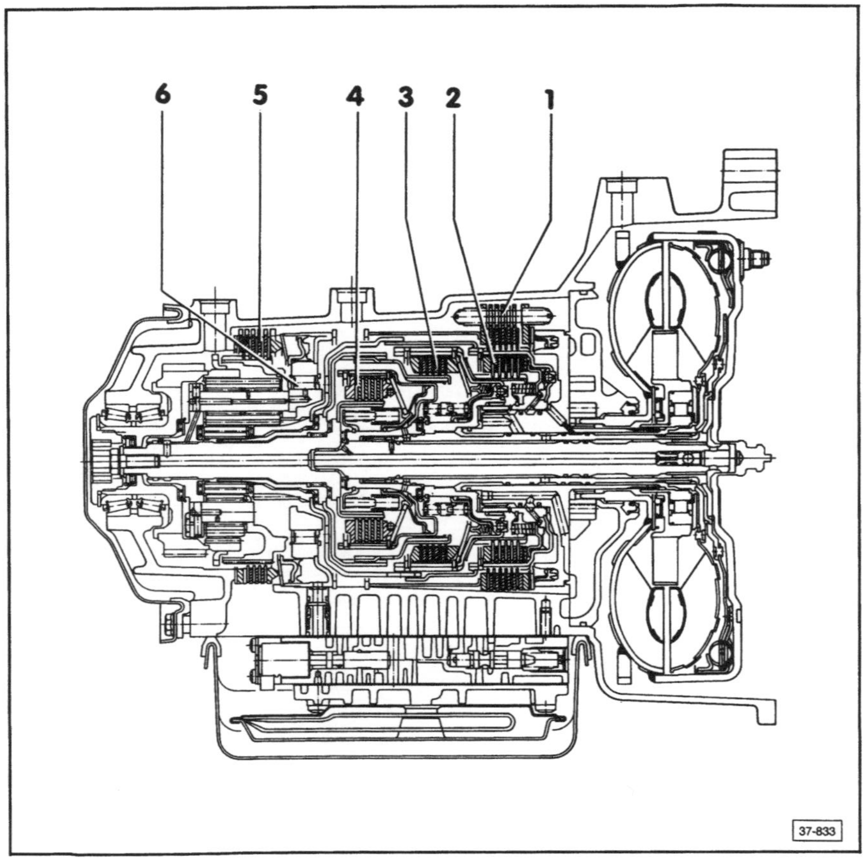

Note

The above illustration identifies shift-related transmission clutches and brakes in their installed positions. Refer also to clutch/brake engagement chart in section 38-20.

1 — One-way clutch (F)
2 — Reverse brake (B1)
3 — 3rd and 4th gear clutch (K3)
4 — 1st to 3rd gear clutch (K1)
5 — Reverse clutch (K2)
6 — 2nd and 4th gear brake (B2)

Selector lever positions and corresponding clutch/brake/gear engagement

Note

The following chart can help isolate transmission shift components which are causing performance problems. Consult the chart to identify the clutches/brakes/gears which engage during the various selector lever positions. These clutches and brakes are shown in their installed positions in section 38-10.

		Brakes		Clutches			
Selector lever position	**Gear[a]**	**Reverse (B1)**	**2nd/4th (B2)**	**1st to 3rd (K1)**	**Reverse (K2)**	**3rd/4th (K3)**	**One-Way (F)**
R	R	X			X		
D	1H			X			X
D,2	2H		X	X			
D,3	3H			X	X		
D	3LTI			X		X	
D	3M			X	X	X	
D	3LT2				X	X	
D	4M		X			X	

[a] H = hydraulic
LT1, LT2 = transitional gear stages between hydraulic and mechanical operation
M = mechanical

Note

Transmission troubleshooting should also include self-diagnosis using testers **VAG 1551** and **VAG 1598**. See Repair Group 01 (Automatic Transmission.)

For additional information on transmission operation and gear systems, refer to Service Training book WSP 521 204 00.

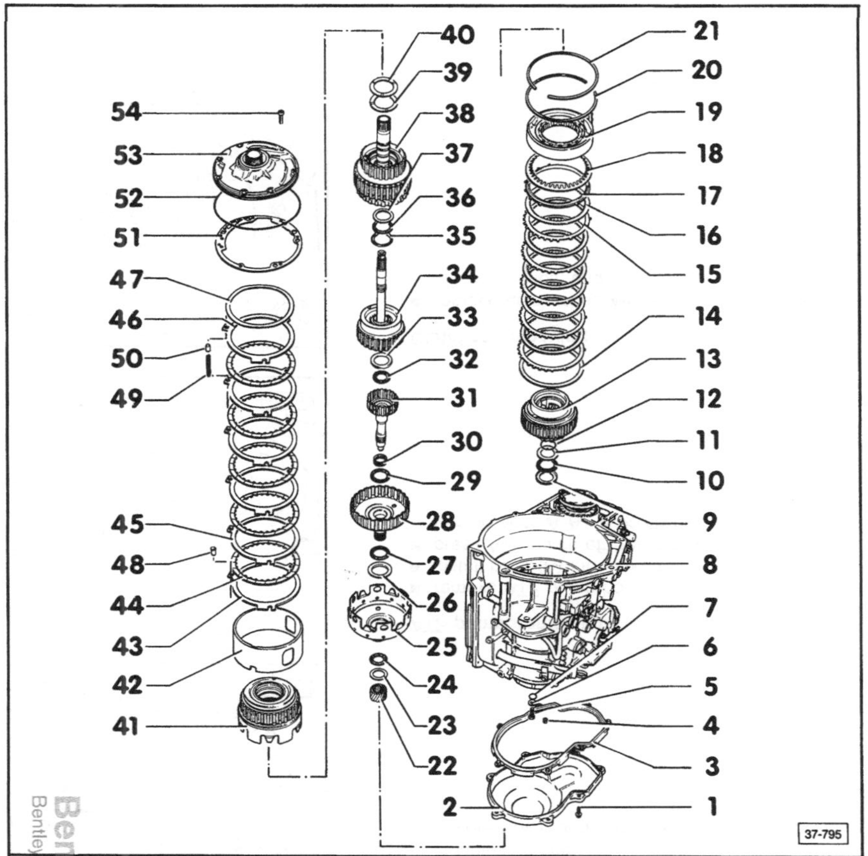

CAUTION
Before disassembling planetary gears remove valve body.

1 — **10 Nm (7 ft lb)**

2 — **End cover**

3 — **Gasket**
always replace

4 — **Spacer**
install into gasket

5 — **30 Nm (22 ft lb)**
- to loosen, engage parking brake

6 — **Washer**

7 — **Shim**
- for planet carrier
- measuring, section 38-50
- adjusting planet carrier: remove drive gear, Repair Group 39

8 — **Transmission housing**
- valve body assembly, section 38-190
- final drive removing/installing, Repair Group 39
- mounting in repair stand, section 38-40

9 — **Washer**
lugs toward axial needle bearing

10 — **Axial bearing**
install in planet carrier with washers

Automatic Transmission – Case, Gears, Shaft

11 — **Washer**

12 — **O-ring**
- always replace
- install in planet carrier

13 — **Planet carrier**
- lightly grease washers and needle bearing
- place washers and needle bearing on planet carrier, then install planet carrier plates

14 — **Shim**
- for reverse gear brake (**B1**)
- determining thickness, section 38-70

15 — **Outer splined plate**
number of plates, see Technical data, Repair Group 37

16 — **Inner splined plate**
- number of plates, see Technical data, Repair Group 37
- before installing new plates, soak in ATF for 15 minutes

17 — **Pressure plate**
flat side toward plates

18 — **Plate spring**
curved side toward one-way clutch

19 — **One-way clutch**
- with **B1** piston
- assembly, section 38-110
- removing section 38-40
- installing, section 38-50
- before removing/installing, remove valve body sealing plug, section 38-190

20 — **Circlip**
- for one-way clutch
- removing, section 38-40
- installing, section 38-50

21 — **Circlip**
- for support tube
- removing, section 38-40
- installing, section 38-50

22 — **Sun gear, small**
- removing, section 38-40
- install in planet carrier, section 38-50

23 — **Washer**
install with rounded side toward small sun gear

24 — **Axial needle bearing**

25 — **Sun gear, large**
- removing, section 38-40
- install in planet carrier, section 38-50

26 — **Washer**
install with chamfered side in large sun gear

27 — **Axial needle bearing**

28 — **Driveshaft, large**
- removing, section 38-40
- installing in planet carrier, section 38-50

29 — **Axial needle bearing**

30 — **Needle bearing**

31 — **Driveshaft, small**
- removing, section 38-40
- installing, section 38-50

32 — **Axial needle bearing**

33 — **Washer**
lugs toward axial needle bearing

34 — **3rd/4th gear clutch (K3)**
- with pump shaft
- removing, section 38-40
- installing, section 38-50
- assembly, section 38-170

35 — **Washer**
lugs toward axial needle bearing

36 — **Axial needle bearing**

37 — **Washer**

38 — **1st to 3rd gear clutch (K1)**
- with turbine shaft
- assembly, section 38-130

39 — **Shim**
- for **K1/K2** freeplay adjustment
- determining thickness, section 38-80

40 — **Shim**
- for **K1/K2** freeplay adjustment (if second shim is needed)
- determining thickness, section 38-80

41 — **Reverse clutch (K2)**
- removing, section 38-40
- installing, section 38-50
- assembly, section 38-140

42 — **Support tube**
- removing, section 38-40
- installing, section 38-50

43 — **Outer splined plate**
- number of plates, see Technical data, Repair Group 37
- always install 3mm thick outer plate, section 38-50

44 — **Inner splined plate**
- number of plates, see Technical data, Repair Group 37
- before installing new plates, soak in ATF for 15 minutes
- removing, section 38-40
- installing, section 38-50

45 — **Outer splined plate**
- number of outer plates, see Technical data, Repair Group 37
- always install 2 mm thick outer plate, section 38-50

46 — **Outer splined plate**
- number of outer plates, see Technical data, Repair Group 37
- determining thickness of outer plates, section 38-90
- removing, section 38-40
- installing, section 38-50

47 — **Spring washer, wavy**

48 — **Spring cap**
- remove all three
- insert, after installing 1st outer splined plate

49 — **Spring**
- remove all three
- insert, before installing the last outer splined plate

50 — **Spring cap**
- remove all three
- insert, before installing the last outer splined plate

51 — **Gasket**
always replace

52 — **O-ring**
- always replace
- install on ATF pump

53 — **ATF pump**
- removing, section 38-40
- installing, section 38-50
- assembly, section 38-100

54 — **10 Nm (7 ft lb)**
- tighten to 10 N•m (7 ft lb) then turn additional 90°

Transmission, disassembling

> **CAUTION**
>
> Before disassembling the transmission, remove the valve body and pull out the sealing plug with O-rings. see section 38-190.

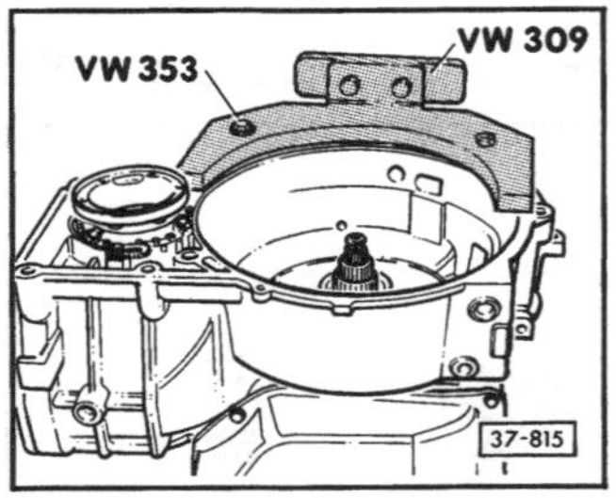

Mounting transmission in assembly stand

- remove ATF oil filter and cooler assembly before mounting in stand
- remove cover and transmission housing gasket

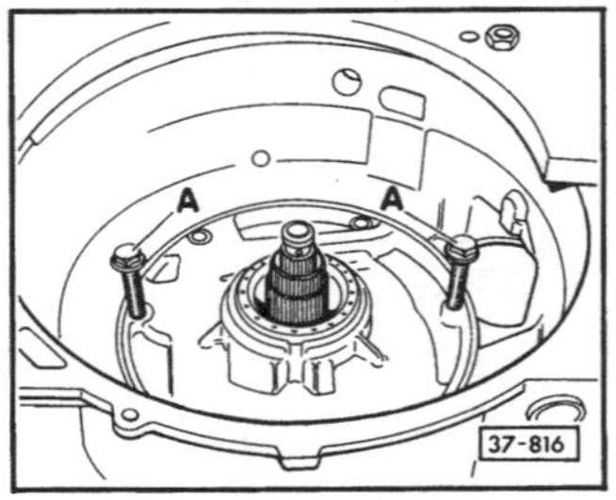

ATF pump, removing

- remove ATF pump bolts
- install bolts (**A**) into threaded holes in ATF pump
- press ATF pump off transmission housing by turning bolts **A A**=8 × 45 bolt

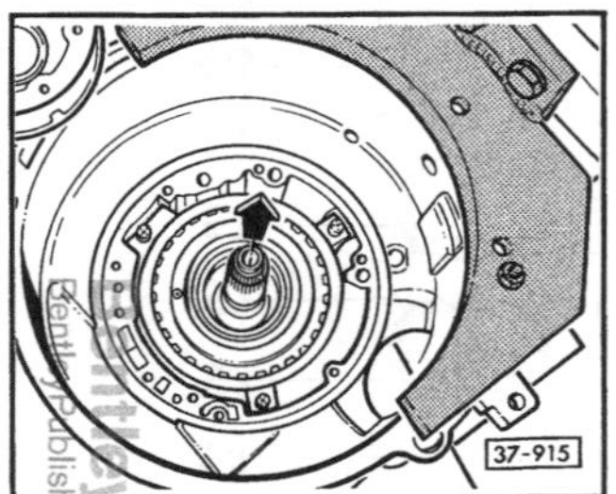

Clutches, removing

- remove all clutches with support pipe, B2 splined plates, springs, and spring caps

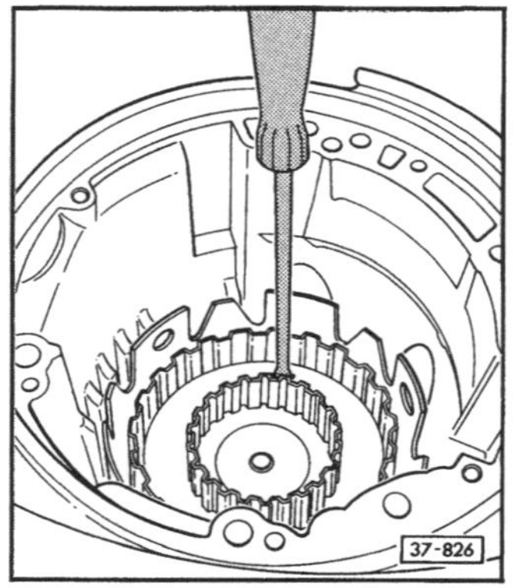

Small driveshaft, removing

- engage parking brake and place transmission in PARK
- prevent driveshaft and sun gear from turning
  - hold with a screwdriver through aligned holes
- remove bolt
- pull out small drive shaft and bearings

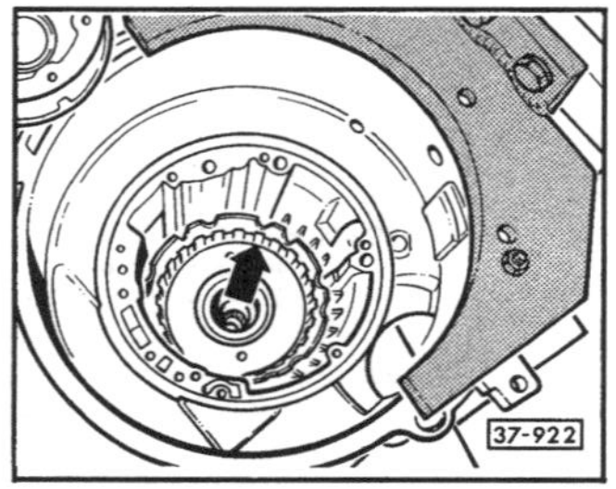

Large driveshaft, removing

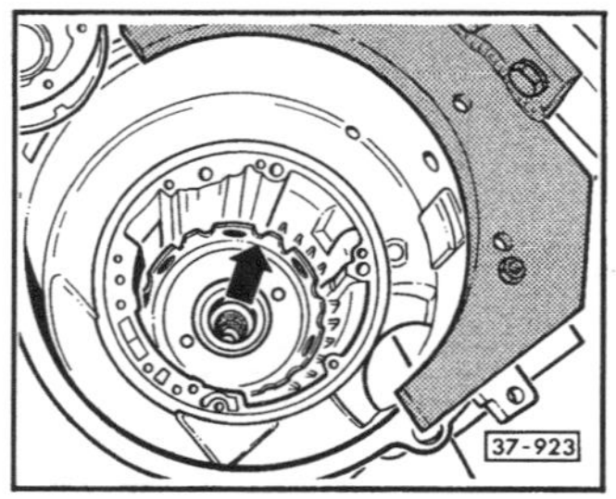

Large sun gear, removing

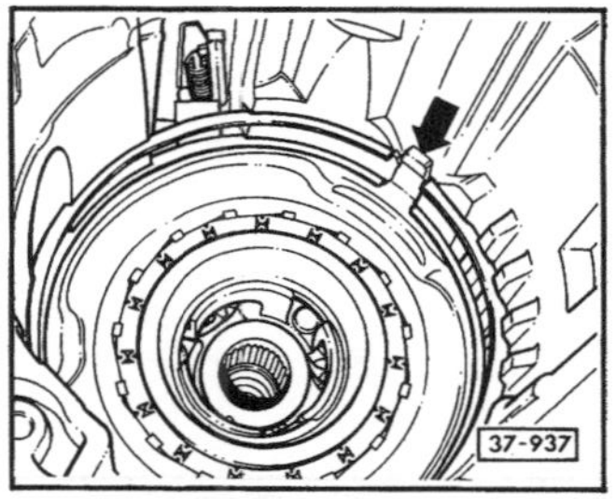

One-way clutch, *removing*

- remove one-way clutch circlip and support pipe circlip
- pull one-way clutch out of transmission housing by placing pliers on holding wedge (**arrow**)

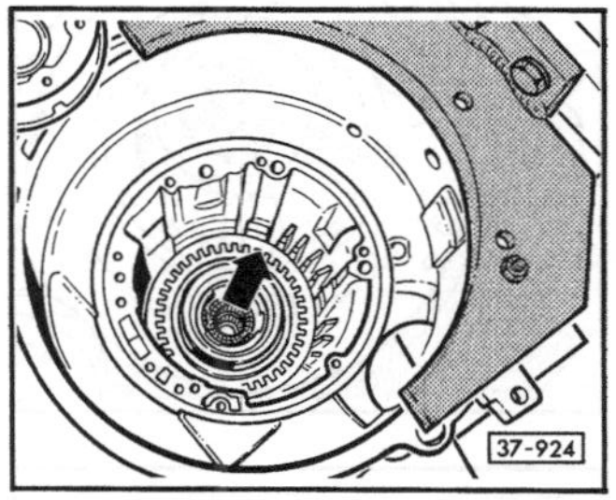

Planet carrier with small sun wheel, removing

- pull out planet carrier with plate spring
- remove B1 splined plates

THIS PAGE INTENTIONALLY LEFT BLANK

Transmission, assembling

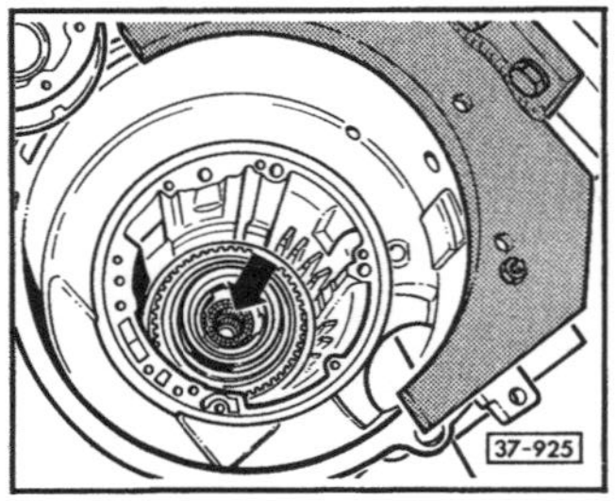

Planet carrier, installing

Note

When replacing planet carrier, adjust planet carrier as shown in section 38-60.

- insert O-rings in planet carrier
- install small sun gear and large sun gear axial needle bearing into planet carrier
- install planet carrier

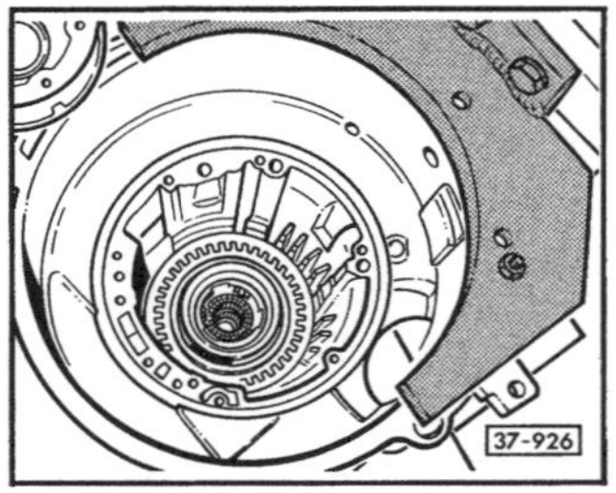

Inner/outer splined plates and B1 plate springs, installing

Note

When replacing the one-way clutch, the B1 piston, and/or the splined plates require adjustment of the reverse gear brake B1, see section 38-70.

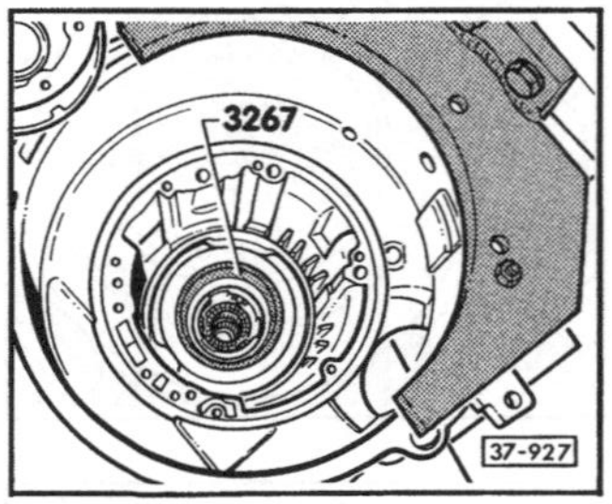

One-way clutch, installing

- place mounting ring **3267** in the one-way clutch and tighten rollers

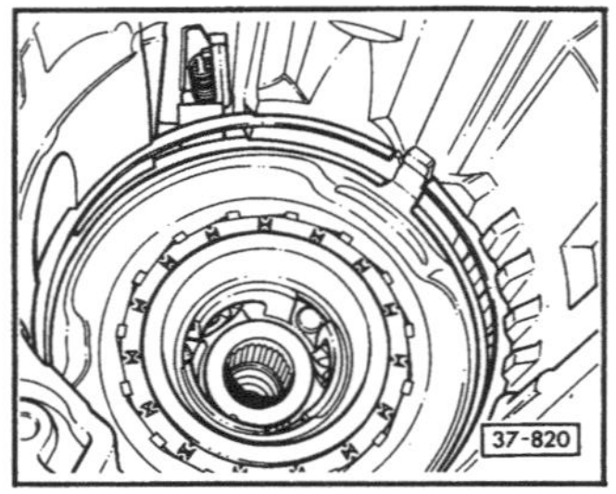

One-way clutch circlip and support pipe circlip, installed positions

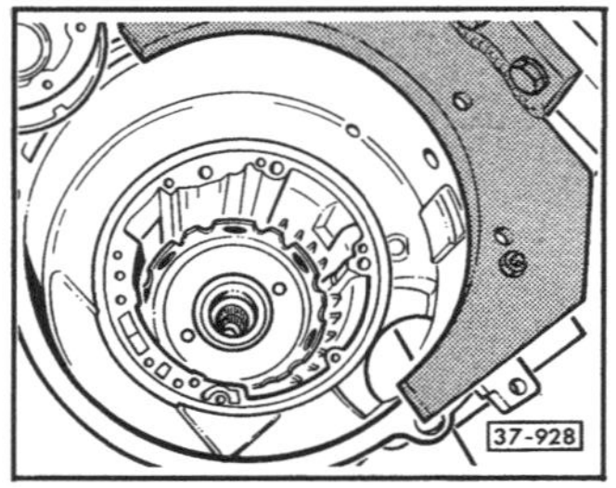

Large sun gear, installed position

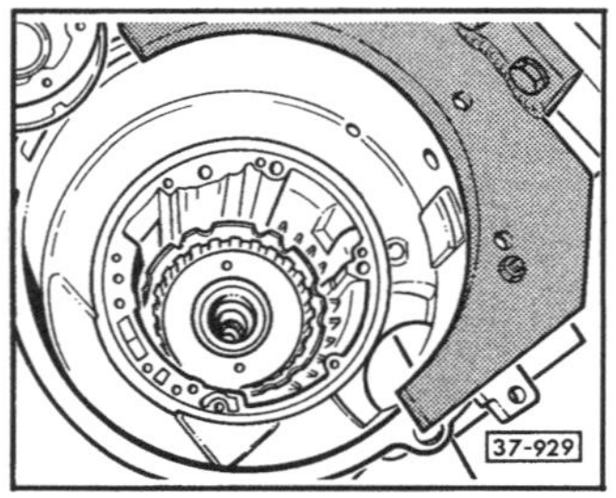

Large driveshaft, installed position

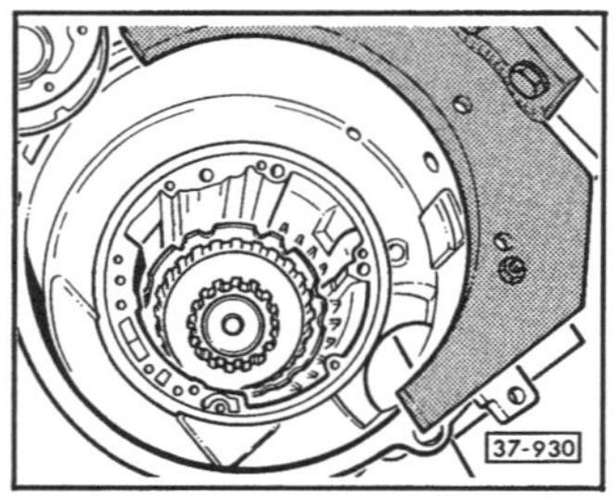

Small driveshaft, installing

- install axial needle bearing and bearing
- install small driveshaft
- tighten bolt (with washer and shim) to 30 Nm (21 ft lb)
- check measurement, section 38-60

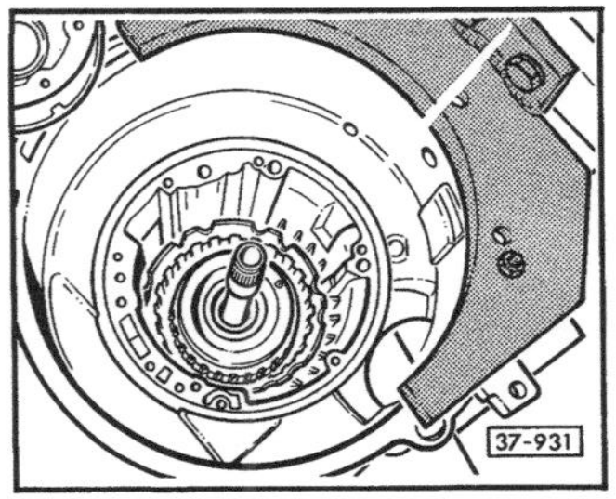

3rd and 4th gear clutch K3, installing

- install axial needle bearing with washer
- install clutch K3

> **CAUTION**
>
> Ensure proper seat of piston rings.

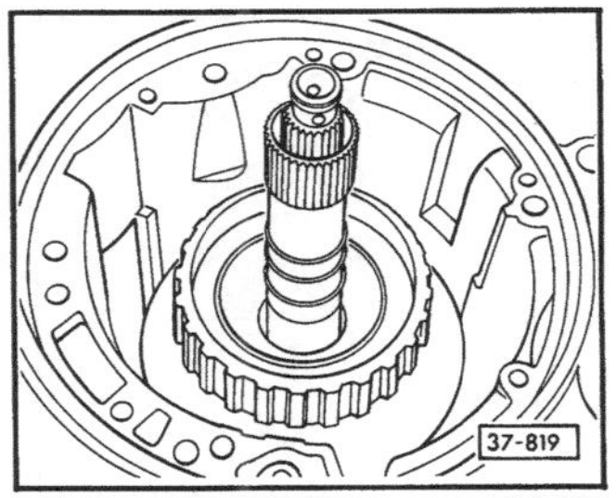

1st and 3rd gear clutch K1, installing

- install axial and needle bearing and washers
- install clutch K1
- install clutch shim

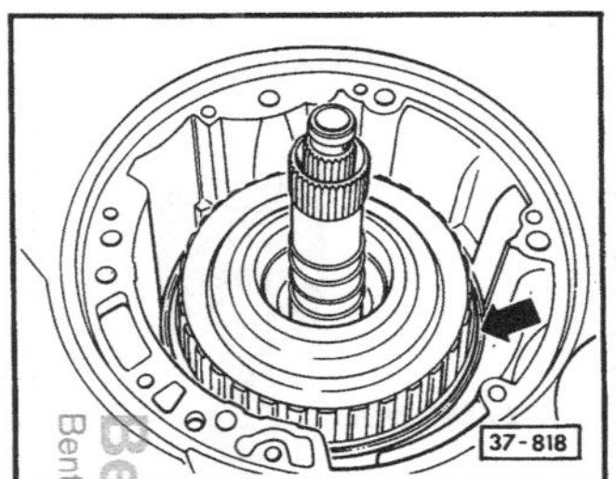

Reverse gear clutch K2 and support tube, installing

- install reverse clutch on shims
- install support tube (**arrow**) so that groove engages in wedge of one-way clutch

Note

When replacing K1, K2, or the ATF pump, determine shim size as shown in section 38-80.

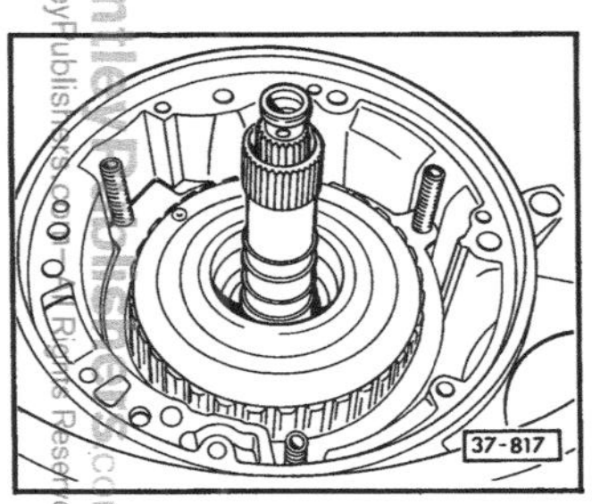

B2 plates, installing

- install 3 mm outer plate
- install spring caps/springs (three each)
- install all inner plates and 2 mm outer plates

Note

Before installing the last measured outer plate, place a spring cap over each of the three springs.

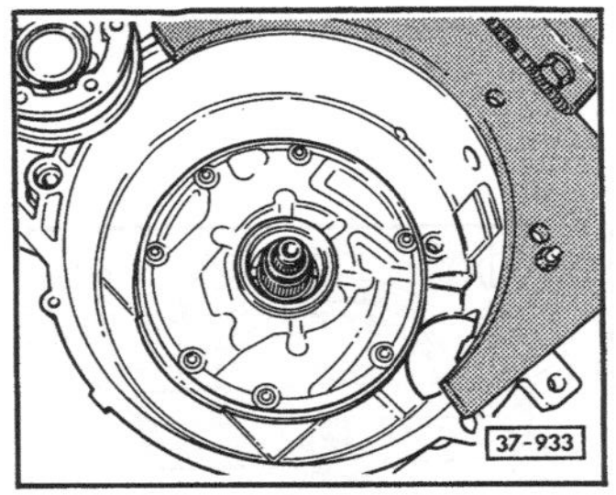

ATF pump, installing

Note

When replacing the support pipe or the ATF pump, the 2nd and 4th gear brake B2 must be adjusted as shown in section 38-90.

- install transmission housing gasket
- install O-ring on ATF pump
- install ATF pump
- tighten bolts uniformly in diagonal sequence
  - make sure that the O-ring is not be damaged

Final installation steps

- measure clutch play, section 38-90
- install valve body, sealing plug, and oil pan, section 38-190
- install cover with seal and spacer bushing

Planet carrier, adjusting/checking

Note

Ensure that the following components are properly assembled to the small driveshaft:
- planet carrier
- small sun gear
- washers
- axial bearing
- small driveshaft

Shim thickness, determining

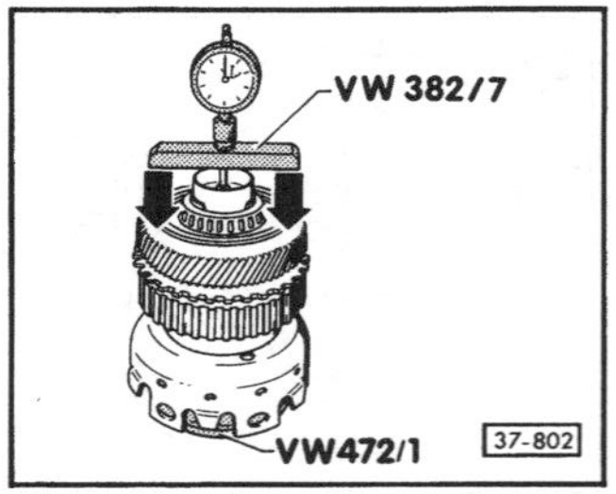

- set up measuring equipment as shown
  - dial indicator sensor must contact small driveshaft
- "zero" dial indicator, with 3 mm preload
- reposition dial indicator sensor onto axial needle bearing

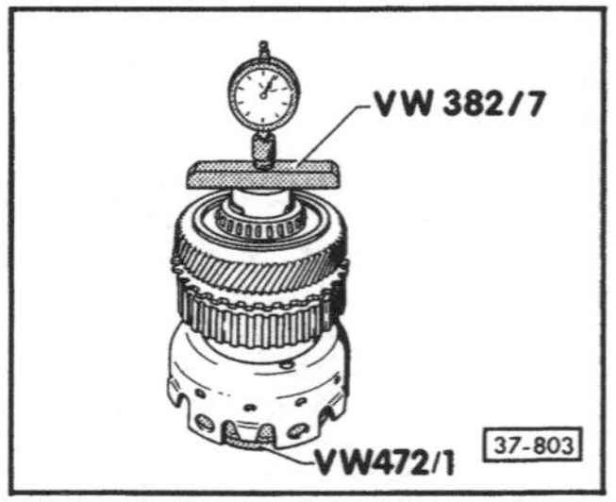

- record measurement
  - example: 2.00 mm
- use recorded measurement to select appropriate shim thickness from chart which follows:

Dial Gauge Measurement	Thickness Shim (mm)
1.26 . . . 1.35	1.0
1.36 . . . 1.45	1.1
1.46 . . . 1.55	1.2
1.56 . . . 1.65	1.3
1.66 . . . 1.75	1.4
1.76 . . . 1.85	1.5
1.86 . . . 1.95	1.6
1.96 . . . 2.05	1.7
2.06 . . . 2.15	1.8
2.16 . . . 2.25	1.9
2.26 . . . 2.35	2
2.36 . . . 2.45	2.1
1.46 . . . 2.55	2.2
2.56 . . . 2.65	2.3
2.66 . . . 2.75	2.4
2.76 . . . 2.85	2.5
2.86 . . . 2.95	2.6
2.96 . . . 3.05	2.7
3.06 . . . 3.15	2.8
3.16 . . . 3.25	2.9

Example

For a measurement of 2.00 mm, select a 1.7 mm shim.

Adjustment, checking

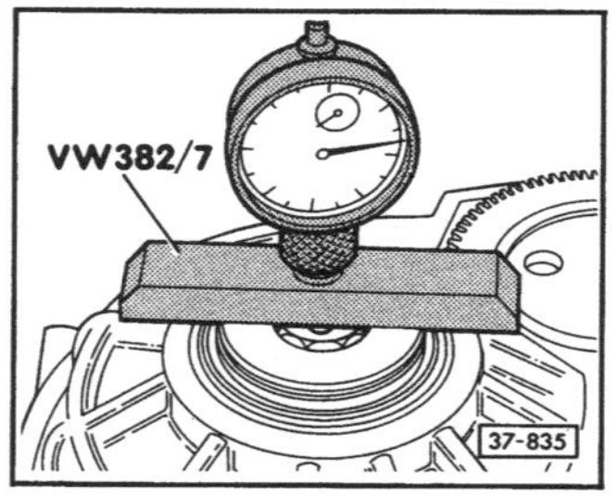

- set up measuring equipment, with dial indicator sensor at small driveshaft
- move small shaft to upper/lower limits of travel
- measure freeplay
  - freeplay minimum: 0.23 mm
  - freeplay maximum: 0.37 mm

Reverse brake (B1), adjusting/checking

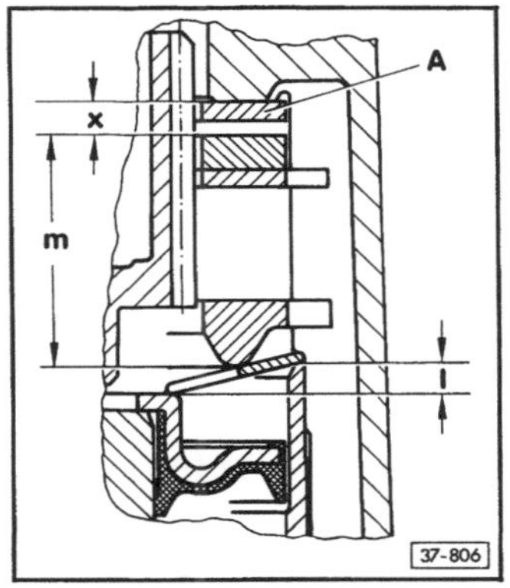

Shim thickness, determining

Note

A shim selection chart appears in this section. Thickness values in the chart are determined by the value of Gap **x**, which must be calculated in the following equation:

Gap **x** = **k** + $\frac{I}{2}$ – **m**

k = constant value = 26.8 mm
A = shim(s)

The remaining variables (**l**,**m**) in the equation are measured, as follows:

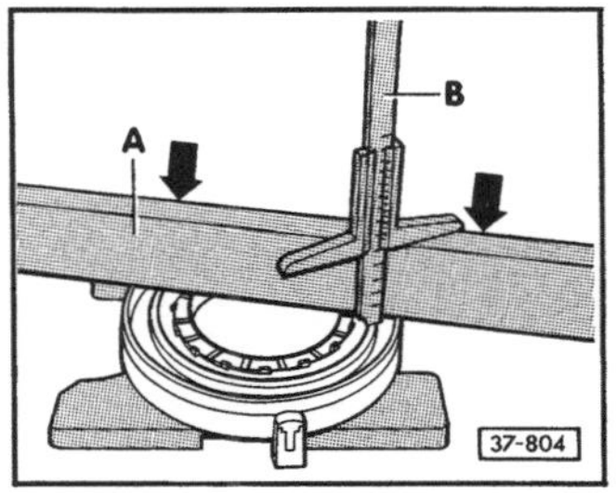

Variable l, determining

- place straightedge **A** on outer race of one-way clutch
- press piston into stop, in direction of **arrows**
- measure inner edge of piston with depth gauge **B**

Example

depth gauge reading	51.8 mm
– straightedge thickness	48.2 mm
	I = 3.6 mm

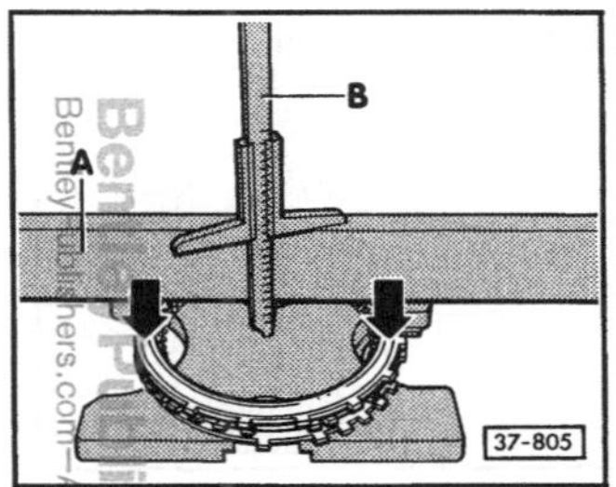

Variable m, determining

- place straightedge **A** on thrust plate
- press plates together in direction of **arrows**
- measure thickness of plate set using depth gauge **B**

Example

depth gauge reading	73.5 mm
– straightedge thickness	48.2 mm
	m = 25.3 mm

Value of Gap x, calculating

Example

$$\text{Gap } \mathbf{x} = \mathbf{k} + \frac{I}{2} - \mathbf{m}$$
$$= 26.8 + \frac{3.6}{2} - 25.3$$
$$= 3.3 \text{ mm}$$

Note

Use calculated value for Gap **x** (derived from actual measurements) to select appropriate shim thickness(es):

Gap x (mm)	Shim (mm)
2.36 . . . 2.45	1.0
2.46 . . . 2.55	1.1
2.56 . . . 2.65	1.2
2.66 . . . 2.75	1.3
2.76 . . . 2.85	1.4
2.86 . . . 2.95	1.5
2.96 . . . 3.05	1.6
3.06 . . . 3.15	1.7
3.16 . . . 3.25	1.8
3.26 . . . 3.35	1.9
3.36 . . . 3.45	1 + 1
3.46 . . . 3.55	1 + 1.1
3.56 . . . 3.65	1.1 + 1.1
3.66 . . . 3.75	1.1 + 1.2
3.76 . . . 3.85	1.2 + 1.2
3.86 . . . 3.95	1.2 + 1.3
3.96 . . . 4.05	1.3 + 1.3
4.06 . . . 4.15	1.3 + 1.4
4.16 . . . 4.25	1.4 + 1.4

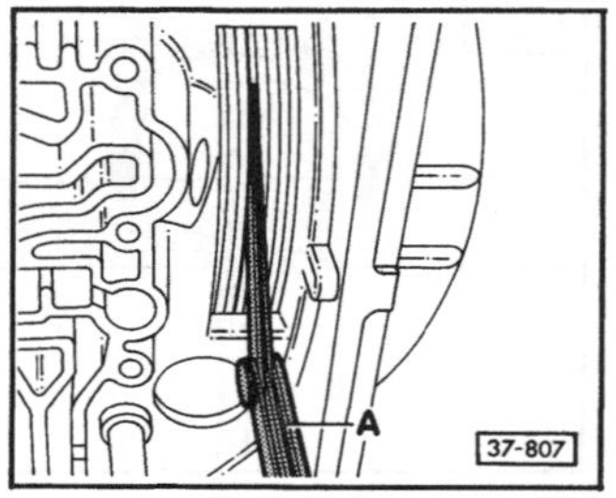

Adjustment, checking

Note

Ensure that components up to one-way clutch are properly assembled and secured with circlip.

- measure freeplay between plates with feeler gauge **A**
  - freeplay minimum 1.20 mm
  - freeplay maximum 1.80 mm

THIS PAGE INTENTIONALLY LEFT BLANK

Clutches K1 and K2, adjusting/checking freeplay

Note

K1 = 1st to 3rd gear clutch

K2 = reverse clutch

Shim thickness, determining

Note

A shim selection chart appears in this section. Thickness values in the chart are determined by the value of Gap **x**, which must be calculated in the following equation:

Gap **x** = **y** – **z**

Variables **y** and **z** in the equation are measured as follows:

Variable y, determining

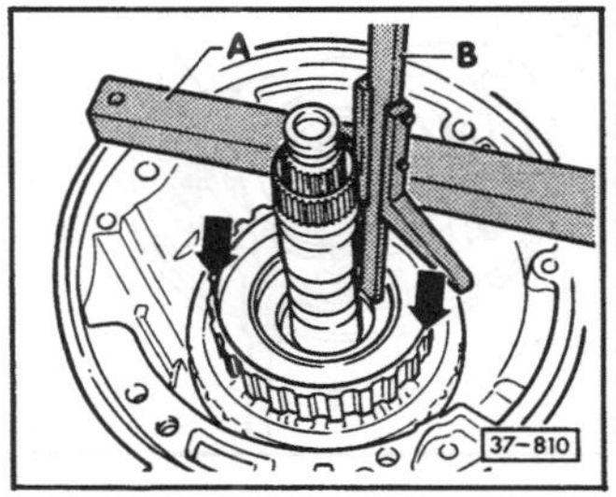

- place straightedge **A** on transmission housing
- press **K1** down, in direction of **arrows**
- measure into **K1** with depth gauge **B**

Example

depth measurement 1 = 88.5 mm

Record value, and take depth measurement 2:

Note

The straightedge thickness must be subtracted to produce the actual measurement.

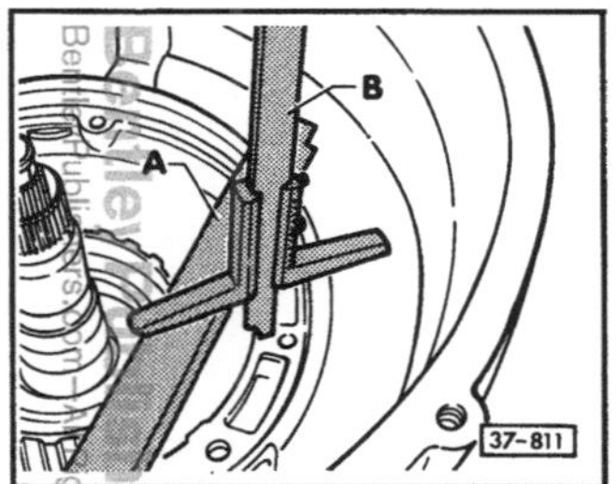

- place straightedge **A** on transmission housing
- position depth gauge **B** on straightedge **A** and measure to ATF pump flange

Example

depth measurement 2 = 34.3 mm

y = measurement 1 – measurement 2
= 88.5 mm – 34.3 mm
= 54.2 mm

measurement from pump flange into **K1**

Variable z, determining

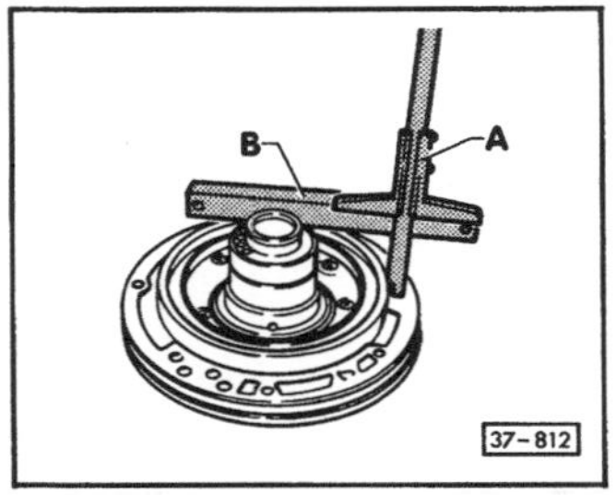

- place straightedge **B** on stator support as shown
- measure to pump flange gasket with depth gauge **A**

Example

depth gauge reading	70.5 mm
– straightedge thickness	19.5 mm
	z = 51.0 mm

Value of Gap x, calculating

Example

Gap **x** = **y** – **z**
= 54.2 mm – 51.0 mm
= 3.2 mm

Note

Use calculated value for Gap **x** (derived from actual measurements) to select appropriate shim thickness(es):

Gap x (mm)	Shims (mm)
. . . . 2.54	1.4
2.55 . . . . 3.09	1 + 1
3.10 . . . . 3.49	1.2 + 1.2
3.50 . . . . 3.89	1.4 + 1.4
3.90 . . . . 4.29	1.6 + 1.6
4.30 . . . . 4.69	1.8 + 1.8
4.70 . . . . 5.04	1.2 + 1.2 + 1.6
5.05 . . . . 5.25	1.2 + 1.2 + 1.8

Adjustment, checking

See Clutch adjustment, checking (section 38-90).

2nd/4th gear brake (B2), adjusting

Last outer splined plate, determining thickness

Note

A shim selection chart appears in this section. Thickness values in the chart are determined by the value of Gap **x**, which must be calculated in the following equation:

Gap **x** = **y** – **z** – **k**

k = constant value = 3.6

The remaining variables (**y, z**) in the equation are measured as follows:

Variable y, determining

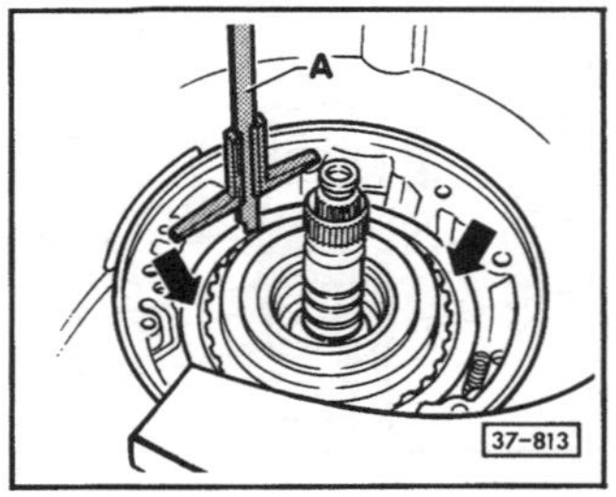

- install outer and inner splined plates except for last outer plate
- press inner splined plate (**arrows**) downward
- measure from pump flange **A** to last inner splined plate, using depth gauge **B**

Example

gauge reading = variable **y** = 30.2 mm

Variable z, determining

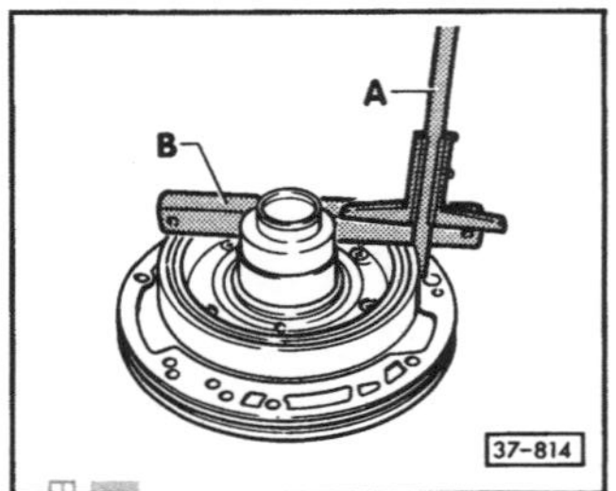

- place straightedge **A** on stator support as shown
- measure to pump flange gasket with depth gauge **B**

Example

depth gauge reading	40.1 mm
– straightedge thickness	19.5 mm
	z = 20.6 mm

Value of Gap x, calculating

Example

Gap **x** = **y** – **z** – **k**
= 30.2 – 20.6 – 3.6
= 6.0 mm

Note

Use calculated value for Gap **x** (derived from actual measurements) to select appropriate plate thickness(es):

Gap x (mm)	Plate (mm)
4.25 . . . . 4.49	2.75
4.50 . . . . 4.74	3.00
4.75 . . . . 4.99	3.25
5.00 . . . . 5.24	3.50
5.25 . . . . 5.49	3.75
5.50 . . . . 5.74	2.00 + 2.00
5.75 . . . . 5.99	2.00 + 2.25
6.00 . . . . 6.24	2.25 + 2.25
6.25 . . . . 6.49	2.25 + 2.50
6.50 . . . . 6.74	2.50 + 2.50
6.75 . . . . 7.00	2.50 + 2.75

Clutch adjustment, checking

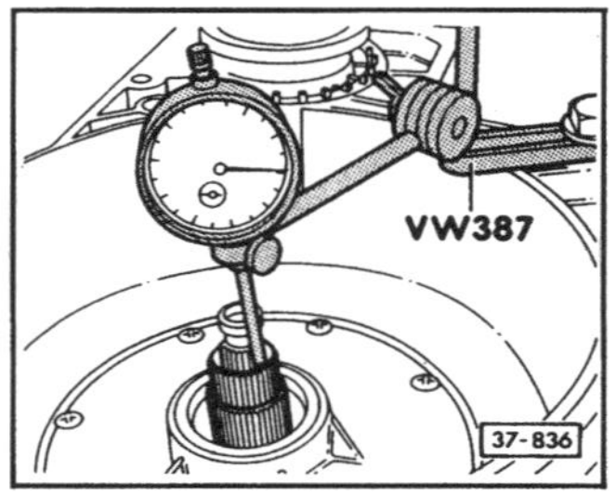

Note

Install ATF pump before measuring clutch freeplay.

- secure dial indicator holder **VW 387** to transmission housing
- position dial indicator sensor on turbine shaft, with 1 mm preload
- move turbine shaft to upper/lower limits of travel
  - freeplay minimum 0.5 mm
  - freeplay maximum 1.2 mm

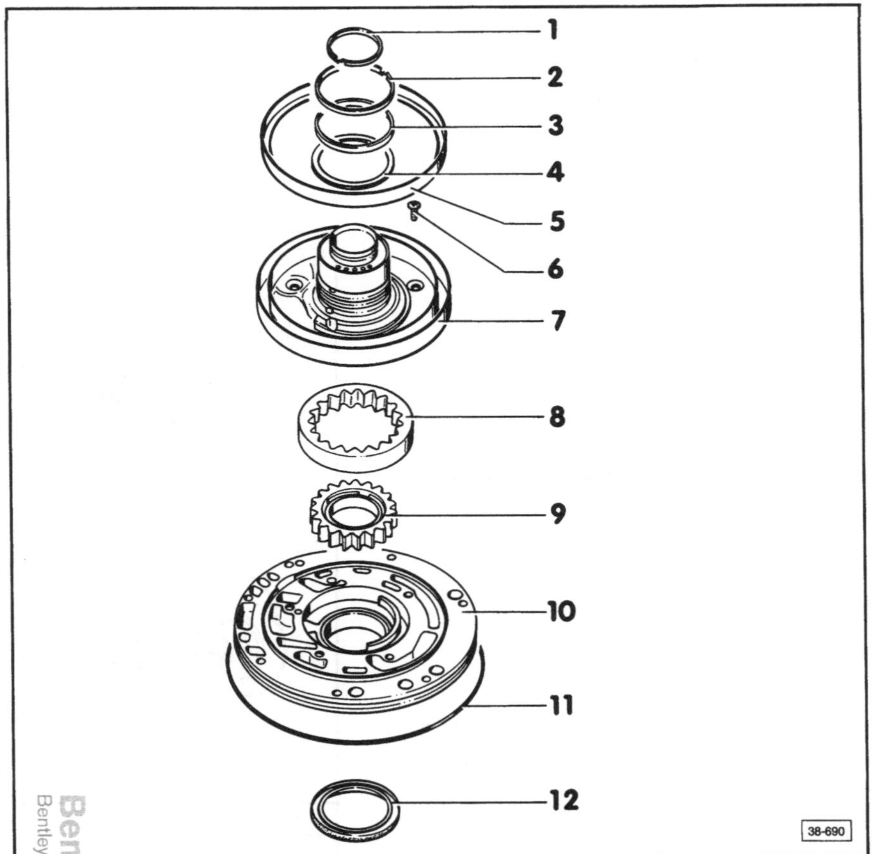

1 — **Piston ring**
check for proper seating

2 — **Piston ring**
check for proper seating

3 — **Piston ring**
check for proper seating

4 — **Thrust washer**

5 — **Piston**
- coat sealing lips with ATF before installing piston
- turn piston slightly while installing

6 — **10 Nm (7 ft lb)**

7 — **Stator support**

8 — **Outer gear**
- coat with ATF before installing
- marked side must face stator support

9 — **Inner gear**
- coat with ATF before installing
- install with same side up as when removed (gear not marked)

10 — **ATF pump housing**

11 — **O-ring**
always replace

12 — **Oil seal for torque converter**
removing/installing, see Repair Group 32

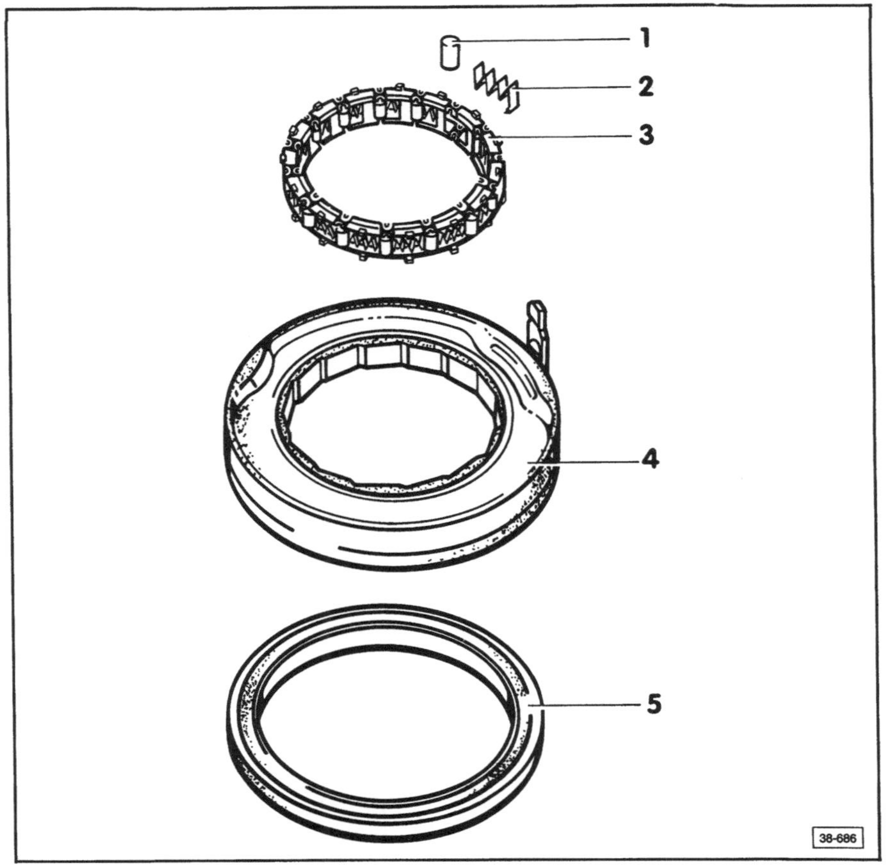

1 — **Rollers**
installing, section 38-120

2 — **Springs**
installing, section 38-120

3 — **Cage**
- installing, section 38-120
- securing, section 38-120

4 — **Outer ring**

5 — **Piston**
- coat sealing lips with ATF before installing piston
- turn piston slightly while installing
- installed position, section 38-120

One-way clutch with B1 piston, disassembling/assembling

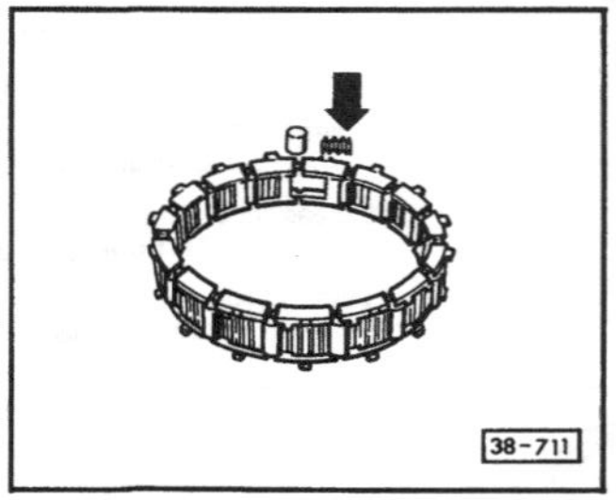

Springs/rollers, installing

- install rollers so that springs (**arrow**) can be pressed into cage

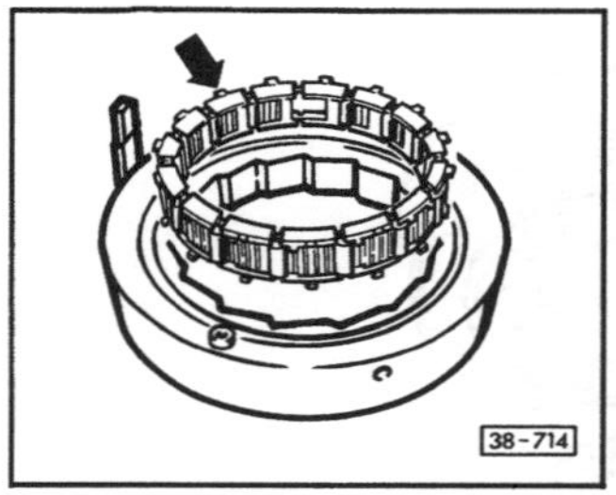

Spring/roller cage assembly, installing

- install with large lugs (**arrow**) upward

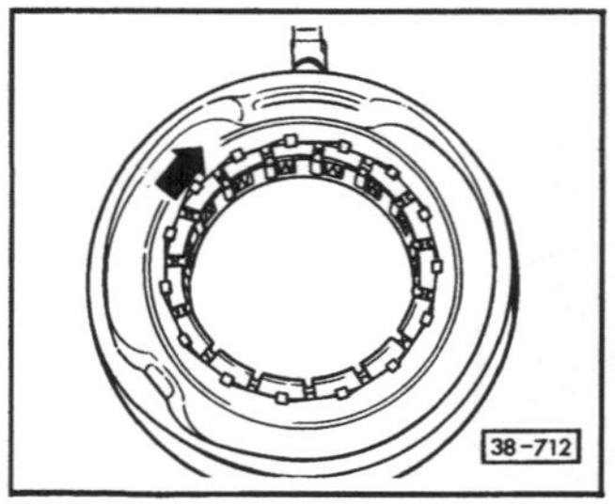

Cage assembly, securing

- turn cage in direction of **arrow**, to stop

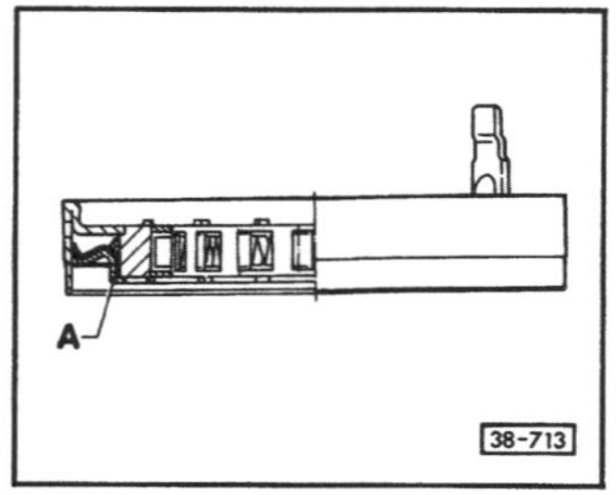

Piston, installed position

- install piston **A** as shown

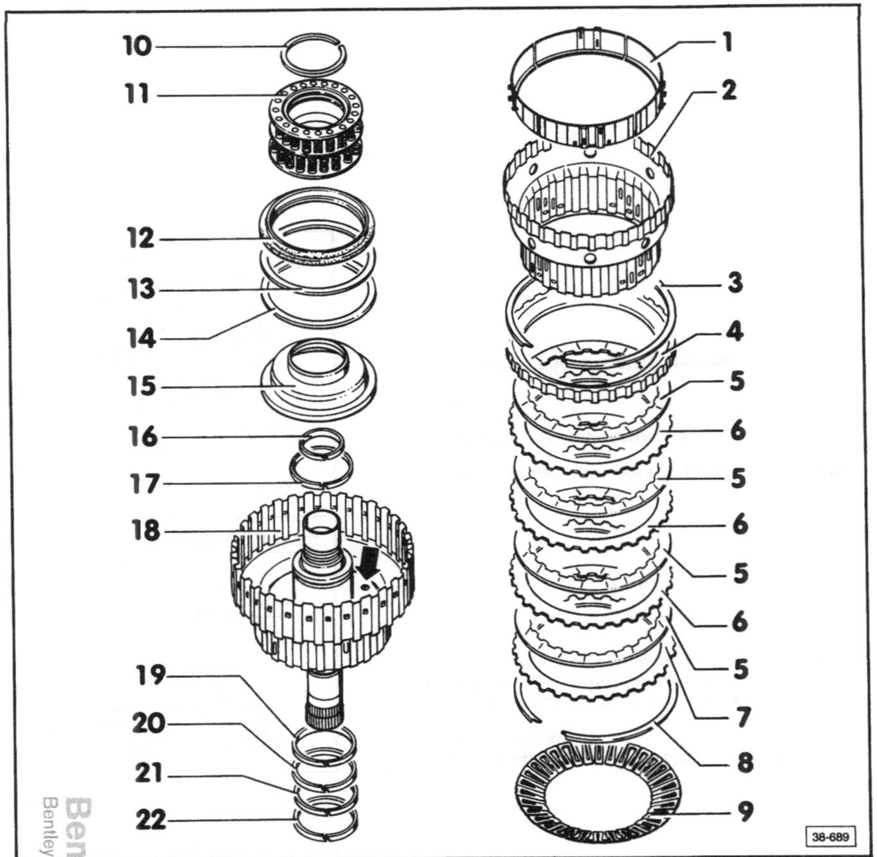

CAUTION

During clutch repair, inspect ball valve (**arrow**) for damage.

1 — **Support ring**
- has four segments
- unclip from inner splined plate carrier
- installing, section 38-140

2 — **Inner splined plate carrier**
- assembling, section 38-140
- installing, section 38-140

3 — **Circlip**
- mark after removal and install in same position
- different thicknesses available

4 — **Thrust plate**
- smooth side faces inner plates
- install together with inner plate carrier

5 — **Inner splined plate**
- number of plates, see Technical data, Repair Group 37
- installing, section 38-140

6 — **Outer splined plate**
- number of plates, see Technical data, Repair Group 37
- installing, section 38-140

7 — **Thrust plate**
curved side faces diaphragm spring

8 — **Circlip**
installing, section 38-140

9 — **Diaphragm spring**
curved side faces piston

10 — **Circlip**
removing/installing, section 38-140

11 — **Spring intermediate ring assembly**
with guide for compression springs

12 — **Operating ring**
rounded surface faces plate spring assembly

13 — **Plate spring**
- curved side faces operating ring
- installing, section 38-140

14 — **Plate spring**
- curved side faces piston
- installing, section 38-140

15 — **Piston**
- coat sealing lips with ATF before installing piston
- turn piston slightly while installing
- cannot be replaced separately

16 — **Piston ring, smaller**
to remove, unhook ends

17 — **Piston ring, larger**
to remove, unhook ends

18 — **Clutch drum with turbine shaft**
- replace drum if bushings are damaged
- before installing inner plate carrier, place an inner and outer plate in clutch drum
- assembling, section 38-140

19 — **Piston ring**
- to remove, unhook ends
- rings are easily broken; when installing shaft push hooked ends into grooves

20 — **Piston ring**
- to remove, unhook ends
- rings are easily broken; when installing shaft push hooked ends into grooves

21 — **Piston ring**
- to remove, unhook ends
- rings are easily broken; when installing shaft push hooked ends into grooves

22 — **Piston ring**
- to remove, unhook ends
- rings are easily broken; when installing shaft push hooked ends into grooves

1st to 3rd gear clutch (K1) with turbine shaft, disassembling/assembling

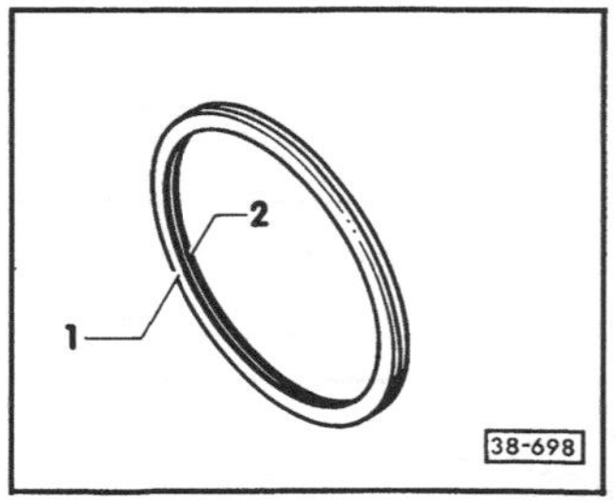

Plate springs, installing

- install plate springs **1** and **2** so that outer edges make contact
  - at inner diameter of plates there is clearance

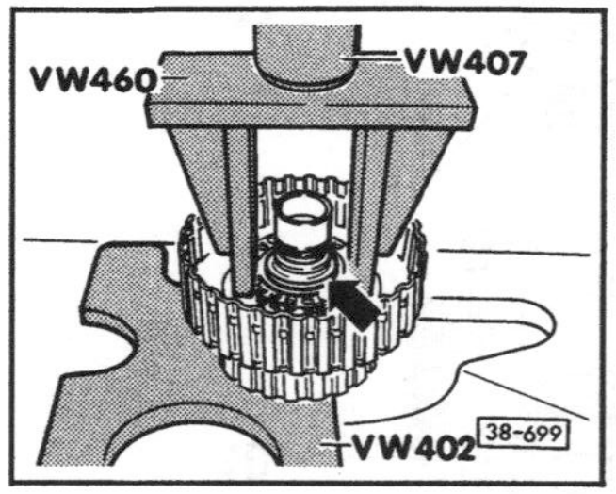

Circlip at spring intermediate ring assembly, removing/installing

- set up pressing equipment as shown
- push spring intermediate ring downward until circlip **(arrow)** can be seated/unseated at groove

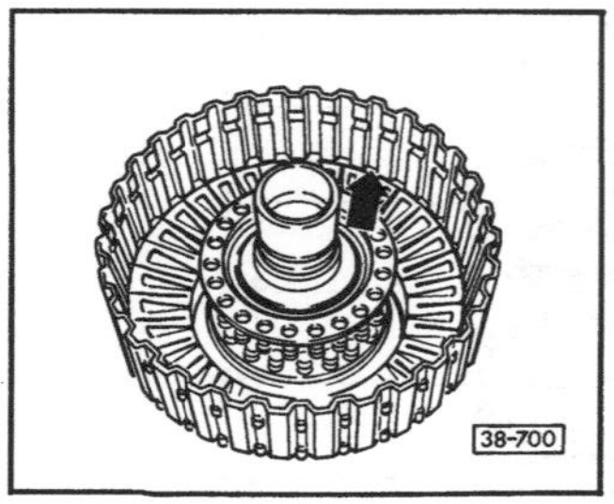

Circlip at diaphragm spring, removing/installing

- seat/unseat circlip **(arrow)** at groove

Note

Diaphragm spring is slightly preloaded.

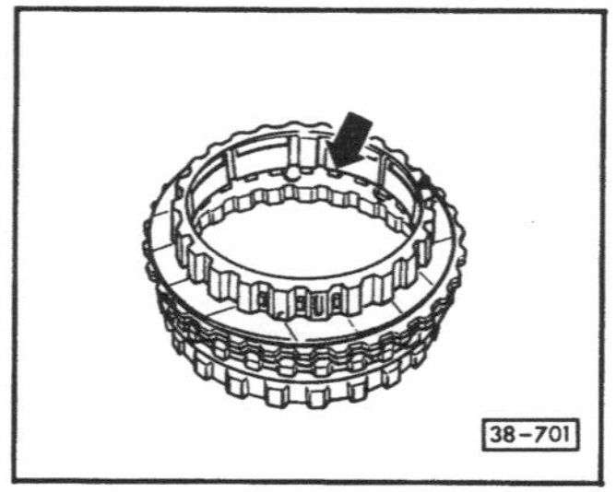

Inner splined plate carrier, assembling

- install thrust plate
- install inner and outer plates
- clip support ring to carrier

Note

One inner and one outer plate must be placed in clutch drum before inner splined plate carrier is installed (see below).

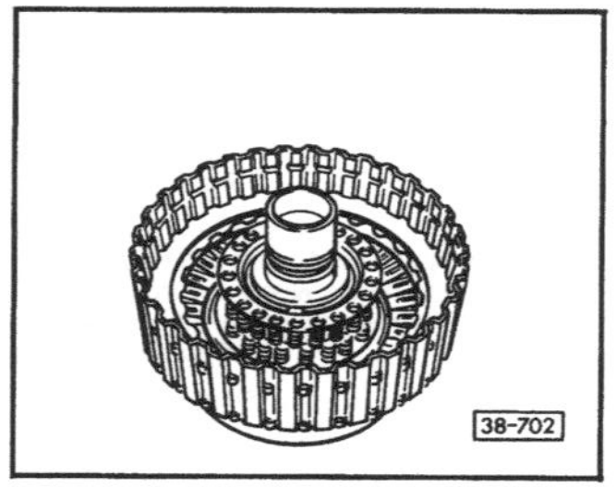

Clutch drum, assembling

- install thrust plate
- install one inner plate
- install one outer plate

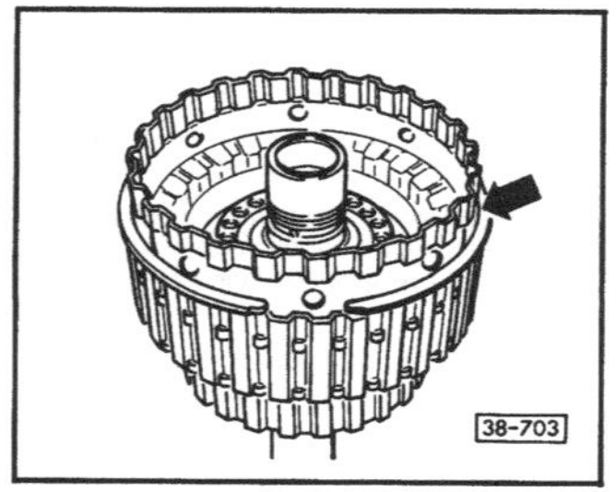

Circlip at inner splined plate carrier, removing/installing

- seat/unseat circlip **(arrow)** at groove

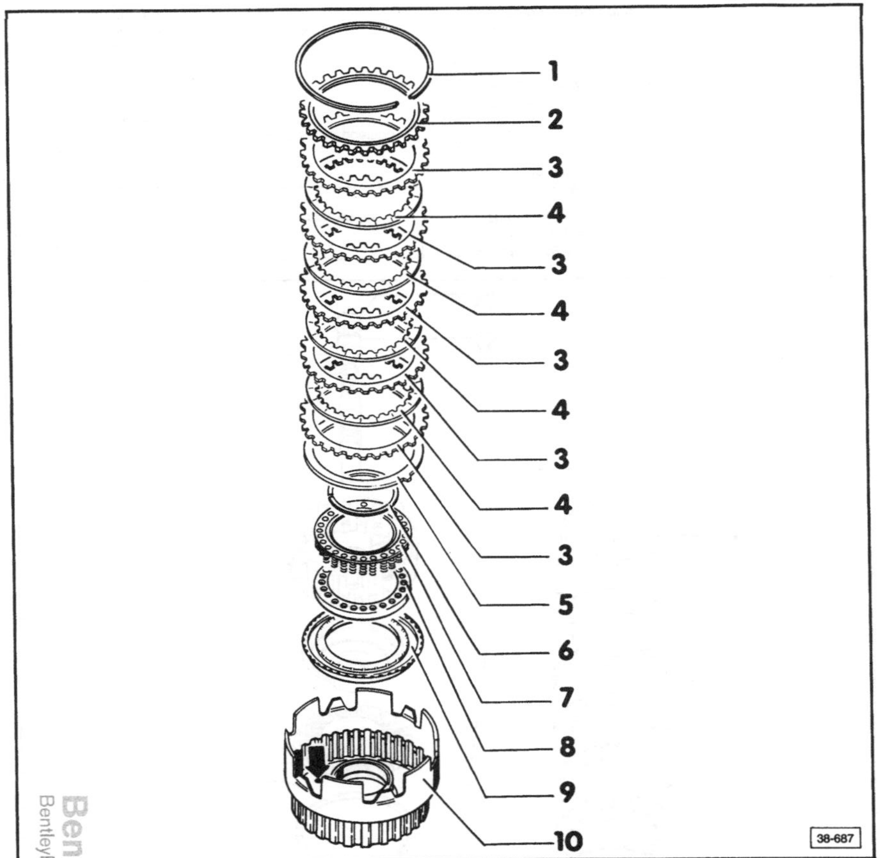

CAUTION

During clutch repair, inspect ball valve (**arrow**) for damage

1 — **Circlip**
- mark after removal and install in same position
- different thicknesses available

2 — **Thrust plate**
shouldered side faces circlip

3 — **Outer plate**
number of plates, see Technical data, Repair Group 37

4 — **Inner plate**
number of plates, see Technical data, Repair Group 37

5 — **Spring washer, wavy**

6 — **Circlip**
removing/installing, section 38-160

7 — **Spring support plate**
with springs

8 — **Spring support ring**
installed position, section 38-160

9 — **Piston**
- coat sealing lips with ATF before installing piston
- turn piston slightly while installing
- cannot be replaced separately

10 — **Clutch drum with clutch tube**

Circlip for reverse clutch (K2) spring support assembly, replacing

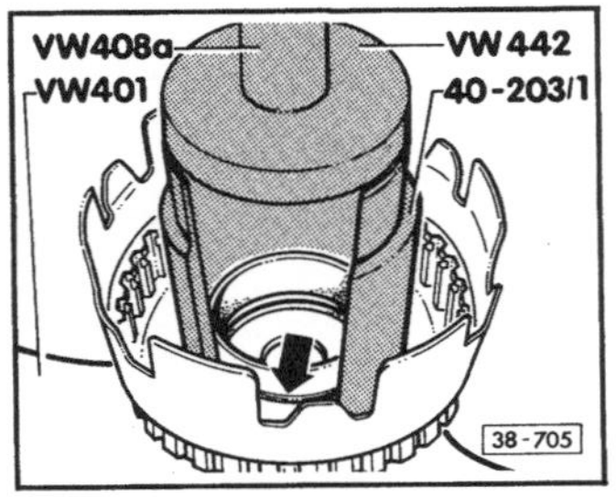

Removing

- set up pressing equipment as shown
- push spring support plate downward

CAUTION

Spring plate bends easily.

- spread/remove circlip

Installing

- position new circlip
- press new circlip down carefully with spring support plate, as in removal
- seat circlip in groove at clutch drum

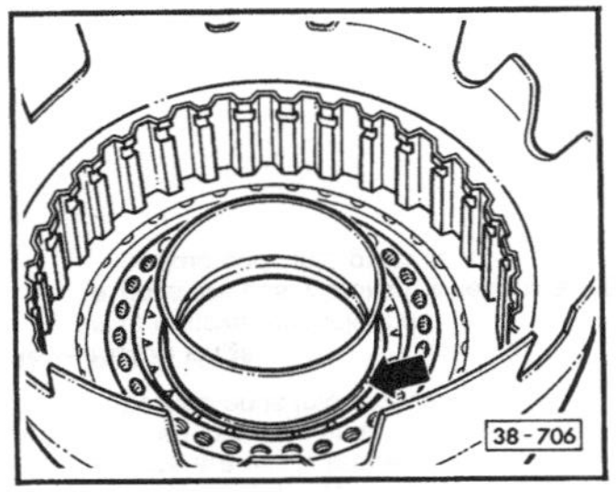

Installed position, checking

- check that circlip (**arrow**) is fully seated at clutch drum
- check that spring support plate is concentric with hub
  - if not, plate will be bent

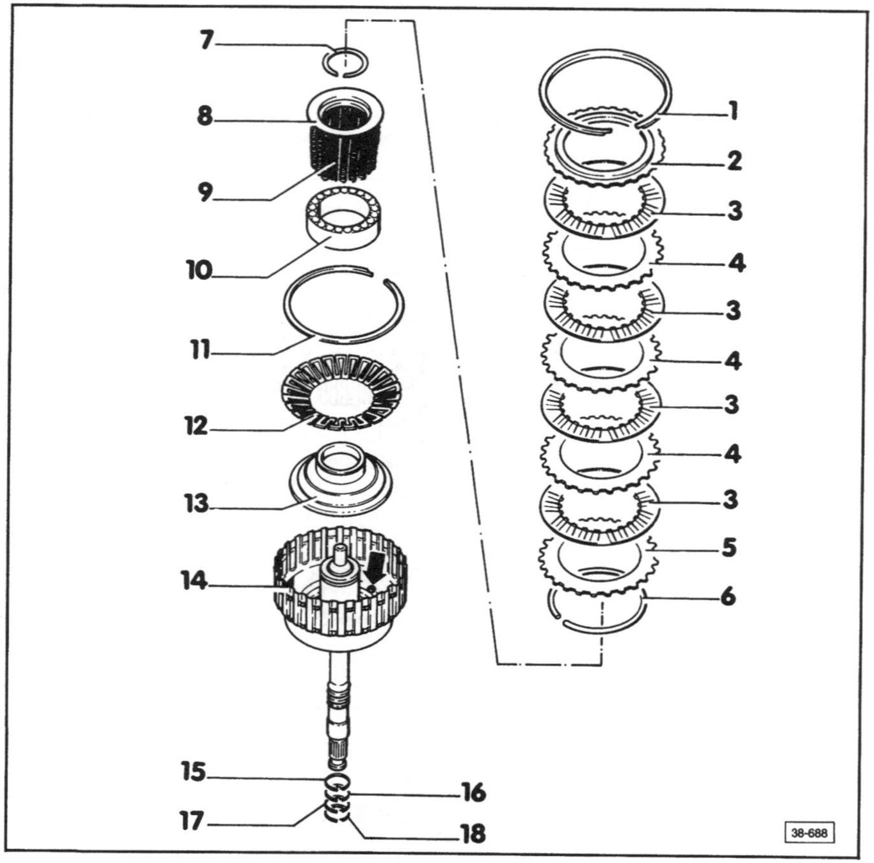

CAUTION

During clutch repair, inspect ball valve (**arrow**) for damage.

1 — **Circlip**
- mark after removal and install in same position
- different thicknesses available

2 — **Pressure plate**
shouldered side faces circlip

3 — **Inner plate**
number of plates, see Technical data, Repair Group 37

4 — **Outer plate**
number of plates, see Technical data, Repair Group 37

5 — **Thrust plate**
rounded side faces ring

6 — **Ring**
install in rounded side of thrust plate

7 — **Circlip**
replacing, section 38-180

8 — **Spring support ring**

9 — **Springs**
place on support ring

10 — **Spring support plate**

11 — **Circlip**
seat in groove at clutch drum

12 — **Diaphragm spring**
curved side faces piston

13 — **Piston**
- coat sealing lips with ATF before installing piston
- turn piston slightly while installing
- cannot be replaced separately

14 — **Clutch drum with pump shaft**

15 — **Piston ring**
- remove/replace only if defective
- rings are destroyed during removal

16 — **Piston rings**
- remove/replace only if defective

17 — **Piston rings**
- to remove, unhook ends
- rings are easily broken; when installing shaft push hooked ends into grooves

18 — **Piston rings**
- to remove, unhook ends
- rings are easily broken; when installing shaft push hooked ends into grooves

Circlip for 3rd/4th gear clutch (K3) spring support assembly, replacing

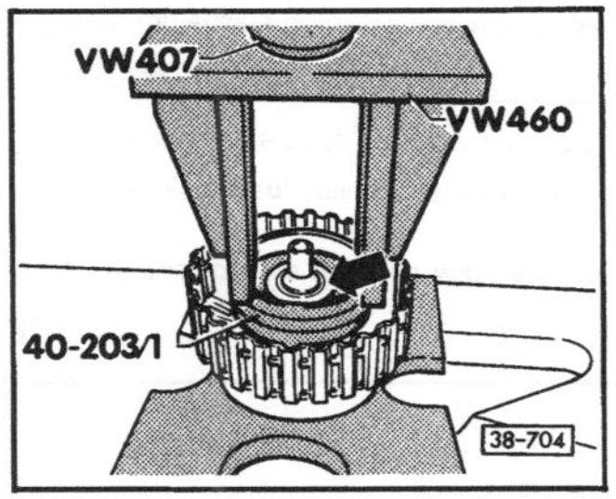

- set up pressing equipment as shown
- seat new circlip at spring support assembly (arrow)

THIS PAGE INTENTIONALLY LEFT BLANK

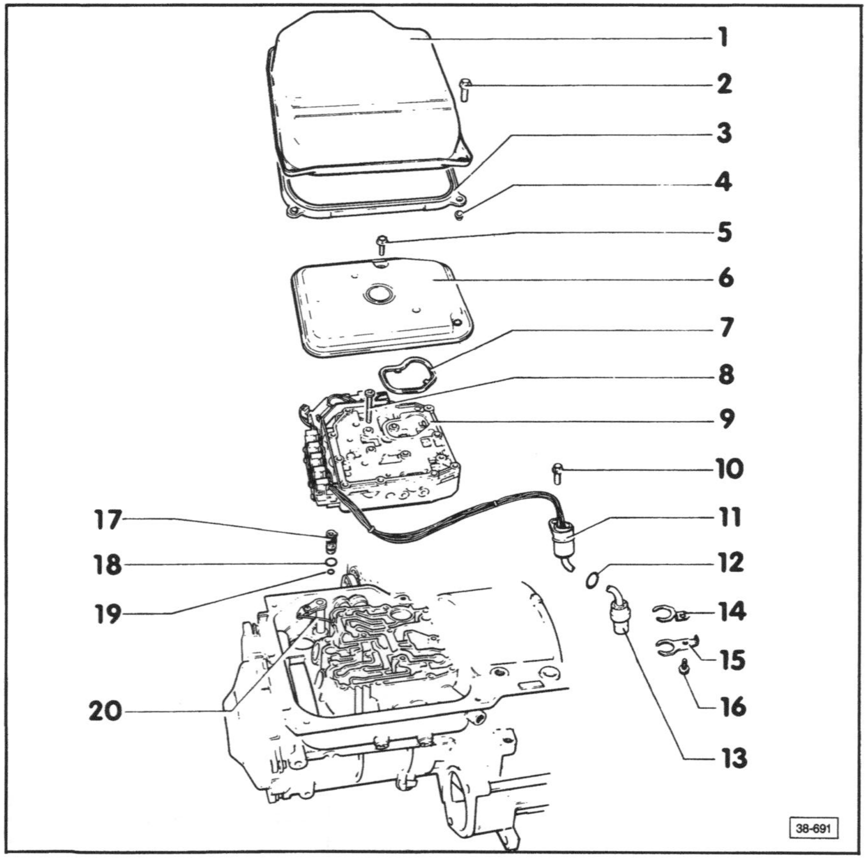

Note

The valve body can be removed with transmission in vehicle.

After installing oil pan, refill with ATF to specifications.

Always replace valve body if dirty or defective.

> **CAUTION**
> Do not run engine or tow vehicle with oil pan removed, or without oil in transmission.

1 — **Oil pan**

2 — **10 Nm (7 ft lb)**

3 — **Gasket**
- always replace
- press spacers into gasket

4 — **Spacer**
press into gasket

5 — **8 Nm (71 in. lb or 82 cm kg)**

6 — **Oil strainer**
- seat seal
- installing strainer, section 38-200

7 — **Seal**
- always replace
- seat in oil strainer

8 — **5 Nm (44 in. lb or 51 cm kg)**

9 — **Valve body**
- installed position, section 38-200
- identifying, section 38-200

10 — **10 Nm (7 ft lb)**

11 — **Wiring sleeve**
install with O-ring

12 — **O-ring**
always replace

13 — **Wiring connector**

14 — **Retainer**

15 — **Retainer**

16 — **10 Nm (7 ft lb)**

17 — **Sealing plug**
- remove plug before removing/installing one-way clutch
- installing, section 38-200

18 — **O-ring**
always replace

19 — **O-ring**
always replace

20 — **Operating rod for manual valve**
- positioning, section 38-200
- shouldered side faces manual valve

Valve body, disassembling/assembling

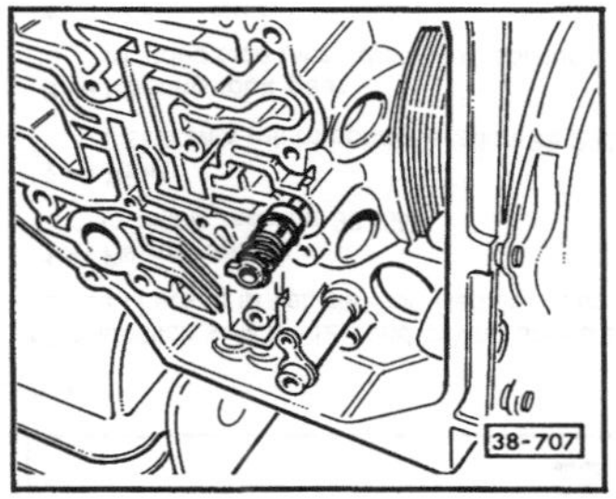

Sealing plug, installing

- install O-rings on plug
- install plug so that lug seats in groove on housing

> **CAUTION**
>
> Before removing/installing one-way clutch, remove sealing plug from housing to avoid damage to plug and O-ring.

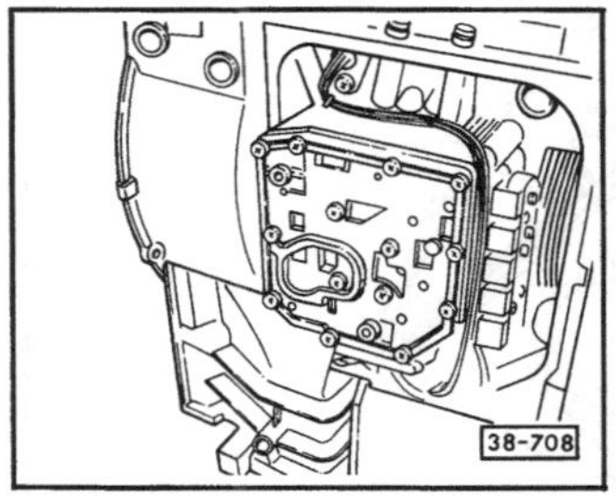

Valve body, installed position

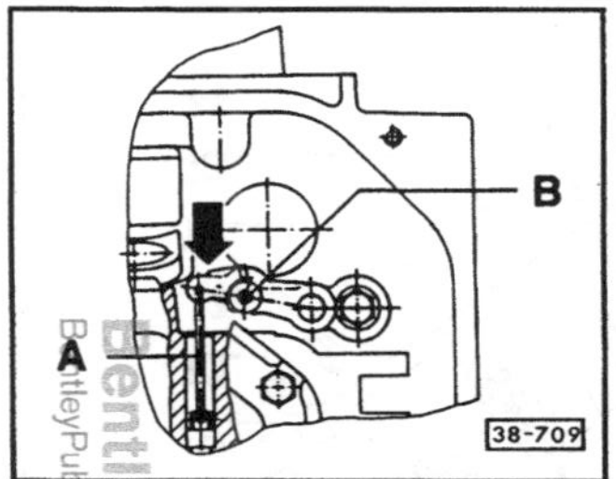

Manual valve assembly; installing/adjusting

- remove transmission oil pan
- move gearshift to "P" position
- loosen tensioning bolt **–B–**
- push operating rod **–A–** with manual shift valve into valve body as far as the stop (direction arrow)

> **CAUTION**
>
> The manual shift valve must always be resting against the stop. Hold tight in direction of arrow when tightening bolt.

- Tighten bolt -**B**- to 3.5 Nm (31 in. lb)
- Install transmission oil pan and refill with ATF to proper level
- Connect VAG 1551 Scan Tool
- Clear all DTCs and return engine/transmission to basic setting
- Road test vehicle to verify repairs
- Recheck for DTCs

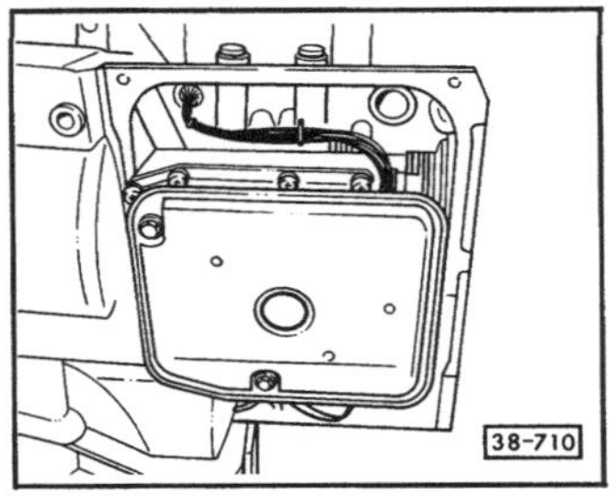

Oil strainer, installed position

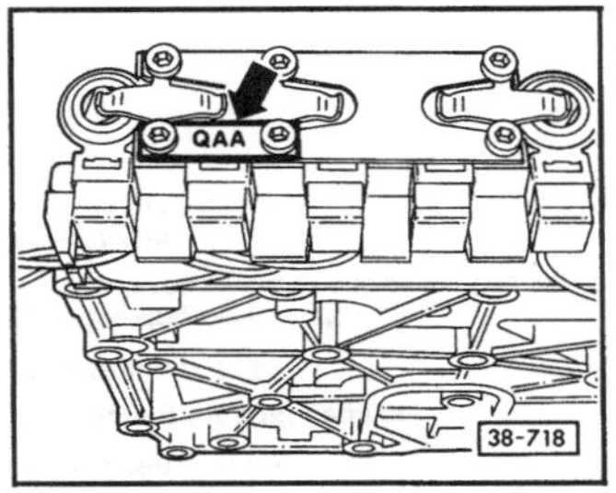

Valve body, identifying

Note

Code letters are stamped on a metal tag which must remain fixed to the valve body.

For valve body application, see Technical data in Repair Group 37.

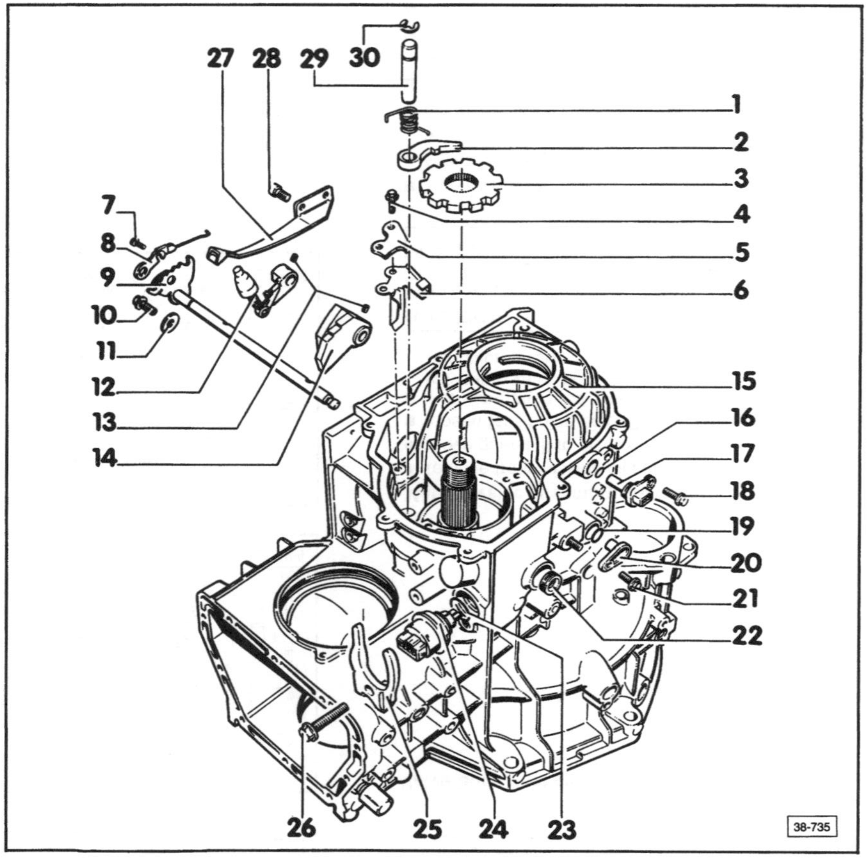

Note

Multi-function switch F125 and driving speed sender G68 can be checked in vehicle, see Repair Group 01.

1 — **Return spring**
installing, section 38-220

2 — **Pawl**
install with return spring, section 38-220

3 — **Parking lock gear**
rounded side faces driveshaft splines

4 — **10 Nm (7 ft lb)**

5 — **Support plate**
installing, section 38-220

6 — **Guide plate**
install before support plate, section 38-220

7 — **10 Nm (7 ft lb)**

8 — **Manual valve lever**
removing/installing, section 38-190

9 — **Gear change shaft**
installing, section 38-210

10 — **10 Nm (7 ft lb)**

11 — **Lock washer**

12 — **Parking lock lever**
install together with shaft and detent segment

13 — **Spring sleeves**
- drive in after installing lock lever and detent segment, section 38-220
- removing, section 38-220

14 — **Detent segment**
install together with shaft and lock lever

15 — **Transmission housing**

16 — **O-ring**
always replace

17 — **Speed sensor**

18 — **10 Nm (7 ft lb)**

19 — **O-ring**
always replace

20 — **Plug**

21 — **10 Nm (7 ft lb)**

22 — **Oil seal**
- pry out with screwdriver
- drive in flush using sleeve **VW 423**

23 — **O-ring**
always replace

24 — **Multi-function switch**

25 — **Retainer**
for multi-function switch

26 — **10 Nm (7 ft lb)**

27 — **Spring for shift segment**

28 — **10 Nm (7 ft lb)**

29 — **Parking pawl shaft**
fastened by circlip and pinion bearing cover

30 — **O-ring**
always replace

Parking lock, installing

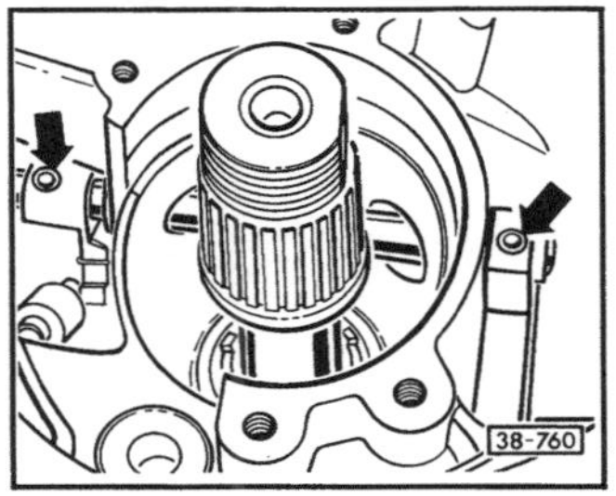

Removing

- drive spring pins (**arrows**) out far enough to allow shaft to be pulled out

Installing

- drive in spring pins after shaft, lock lever and detent segment

> **CAUTION**
>
> Sleeve in lock lever (not pictured) may accidentally fall into transmission.

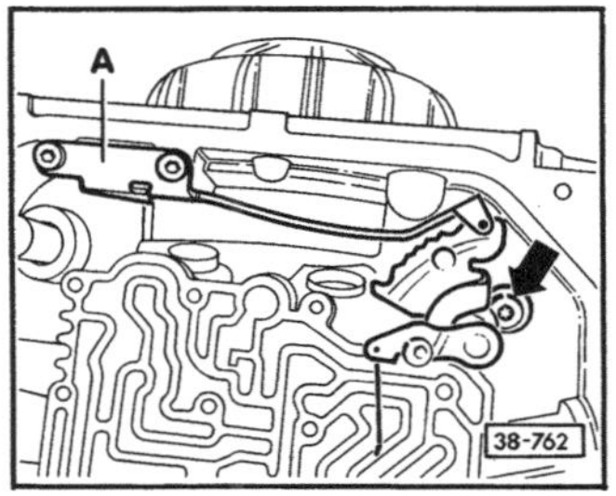

Gear change shaft/shift segment, installed position

- remove bolt (**arrow**) and washer

Note

Prior to installation, check notches for wear.

A–spring for shift segment

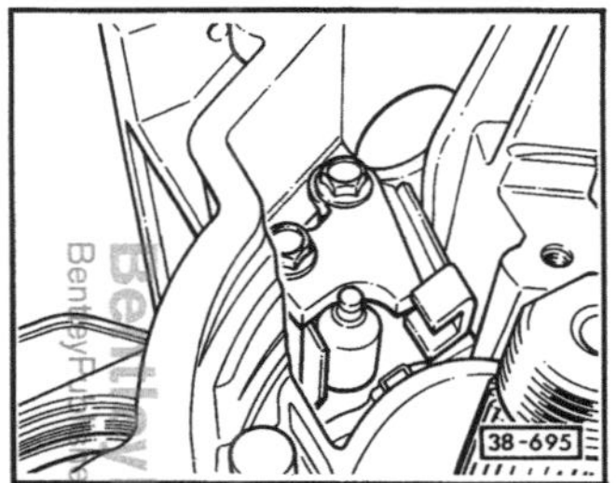

Guide plate/support plate, installed position

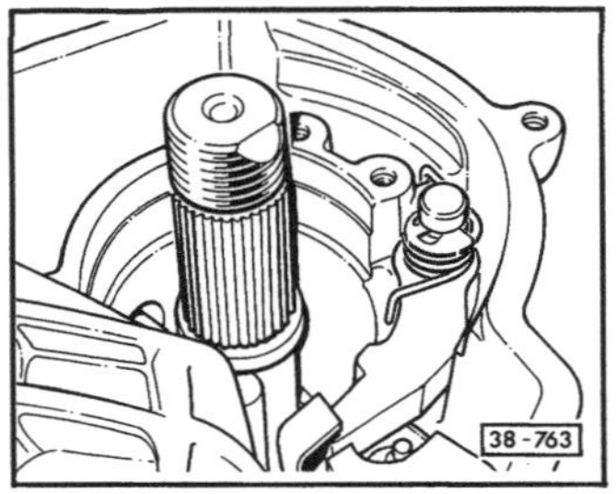

Parking pawl *and return spring, installed* position

THIS PAGE INTENTIONALLY LEFT
BLANK

Index

02A 5-speed

Bearings
- inner race, removing/installing **39.5**
- outer race, removing/installing **39.5**

Differential
- adjusting **39.8**
- assembly **39.4**
- gears, installing **39.7**

Drive flange oil seal
- installing **39.5**
- replacing **39.2**

Final drive gear
- installing **39.7**
- removing **39.6**

Gear
- installing **39.7**

Rivet heads
- boring **39.6**

Roll pins
- installing **39.7**

Turning torque
- checking **39.10**

Drive flange oil seal, replacing

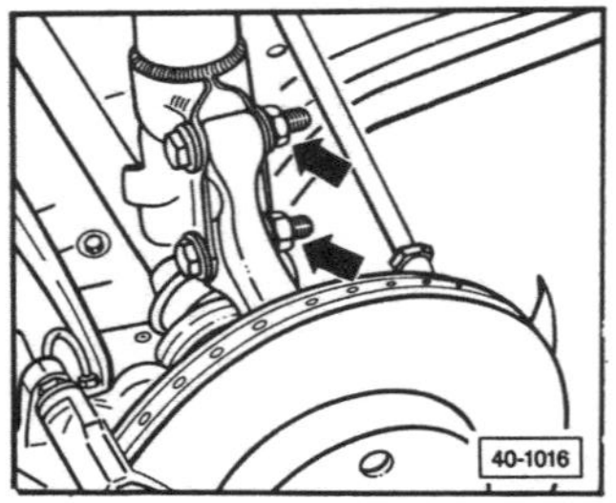

(transmission installed)

- remove axle shaft from drive flange
- press axle shaft toward side and attach with wire
- remove axle shaft when replacing left oil seal

- remove self-locking nut from axle shaft
- remove wheel
- remove wheel bearing housing/suspension strut bolts **(arrows)**
- pivot wheel bearing housing to side
- remove axle shaft

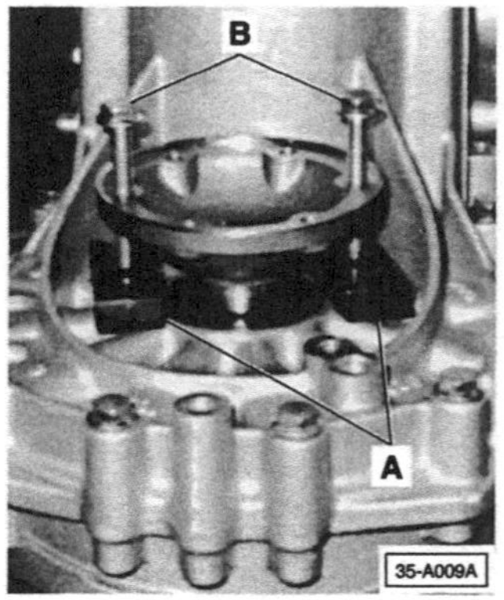

- remove drive flange
  - position spacers (**A**) below flange, and pull flange shaft out of differential gears by turning bolts (**B**), alternately and evenly

Note

Use **two** housing bolts to avoid damage to flange shaft or bolts.

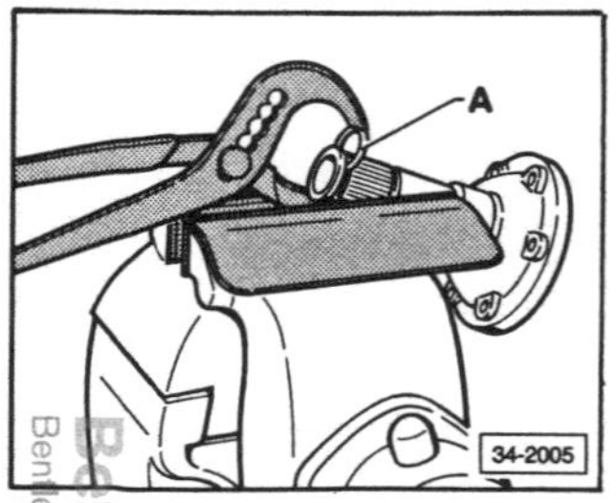

Note

Always replace drive flange lock ring.

- install drive flange in vise with protective jaws
- press lock ring out of flange shaft groove with new lock ring **A**
- pull out drive flange oil seal with **VW 681**

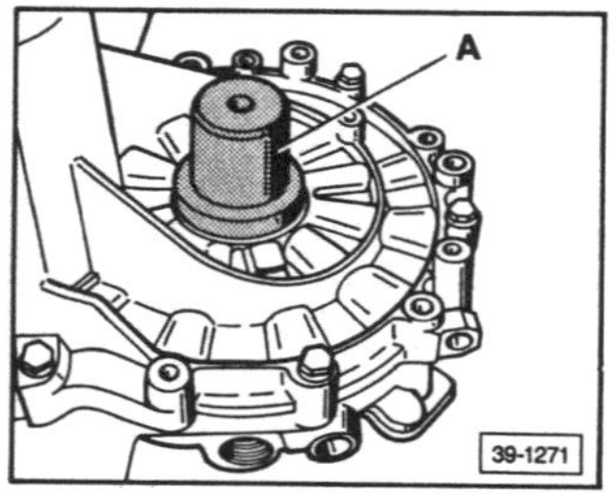

- drive in new drive flange oil seal until flush
  - **A** = 3142
- fill space between sealing and dust lip with multi-purpose grease
- drive in drive flange with **VW 295**
- install axle shaft in reverse order of removal

Tightening torques

axle shaft to drive flange	45 Nm (33 ft lb)
axle shaft to wheel hub	230 Nm (170 ft lb)
wheel bearing housing to suspension strut	95 Nm (70 ft lb)

- check transmission oil
  - fill to edge of filler opening, if necessary

Note

When replacing tapered roller bearings, adjust differential, page 39.8.

Replace both tapered roller bearings together.

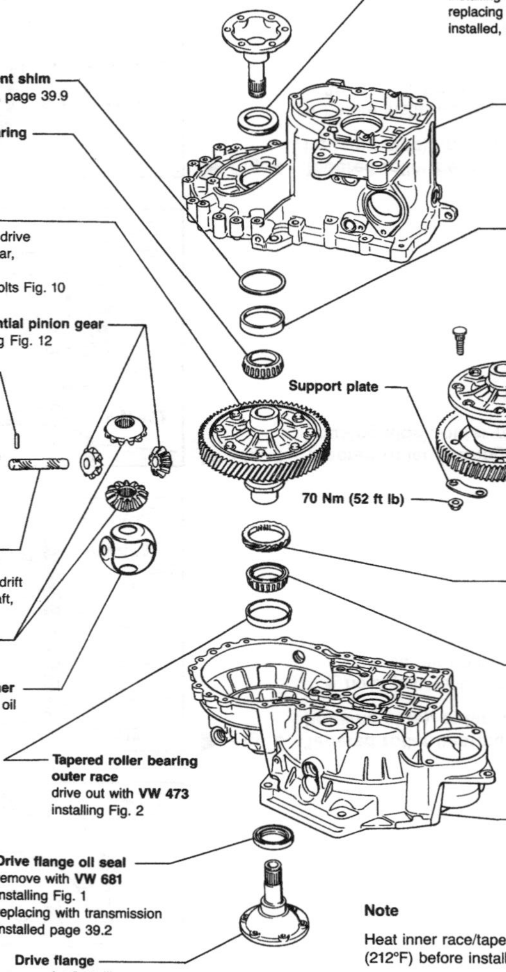

Note

Heat inner race/tapered roller bearings to 100°C (212°F) before installing.

39-1270

Modified differential pinion shaft roll pin

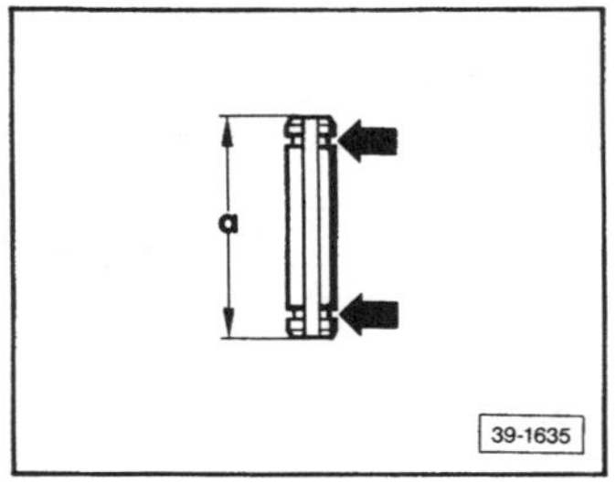

39-1635

Beginning with transmission build date 08 11 0, a longer roll pin is used to secure the differential shaft.

- dimension **a**: old = 14.0 mm
  new = 28.5 mm

The new roll pin has two grooves (**arrows**).

Modified roll pin, removing/installing

Removing

- cover bearing inner race and speedometer drive gear to prevent damage due to entry of particles

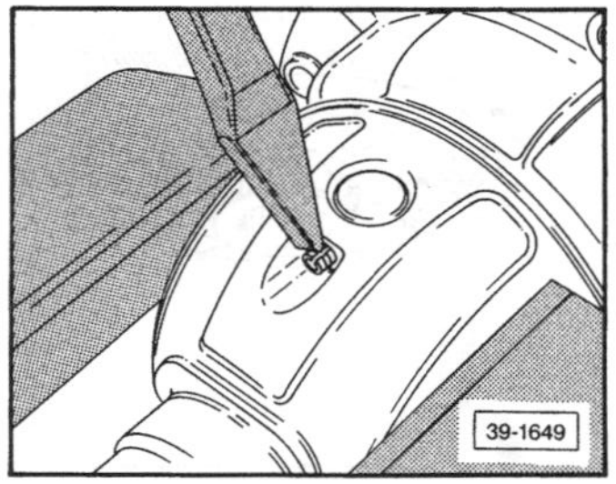

39-1649

- drive roll pin out using a chisel placed in groove of pin

Installing

- drive roll pin in to stop

Note

Only the old version roll pin, Part No. **N 043 302 1**, is available as a replacement part.

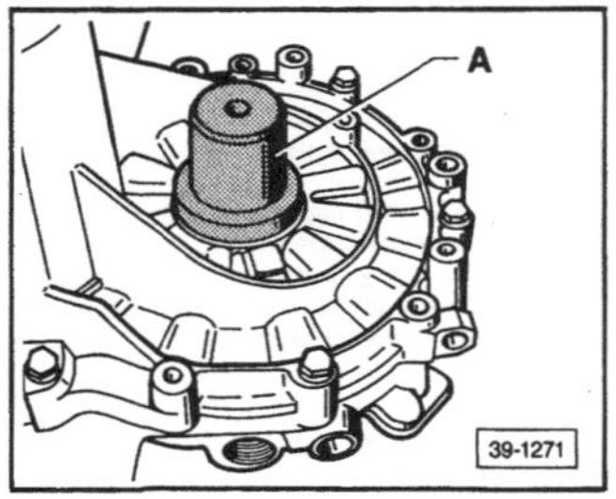

▶ **Fig. 1 Drive flange oil seal, installing**

- drive in flange shaft oil seal until flush
  - **A** = 3142

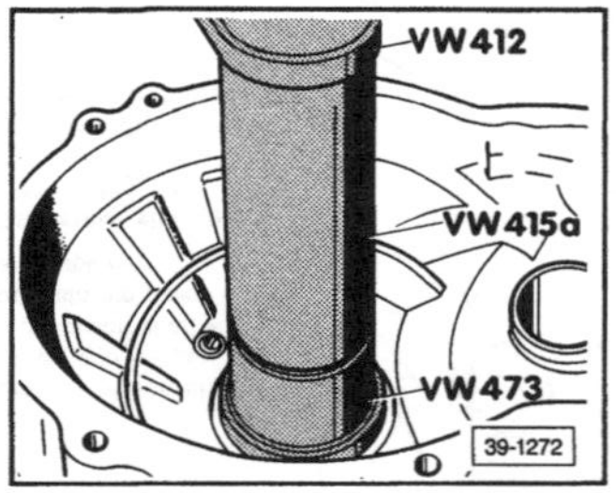

▶ **Fig. 2 Tapered roller bearing (gear carrier housing side), outer race, installing**

- no adjustment shim is installed on the gear carrier housing side

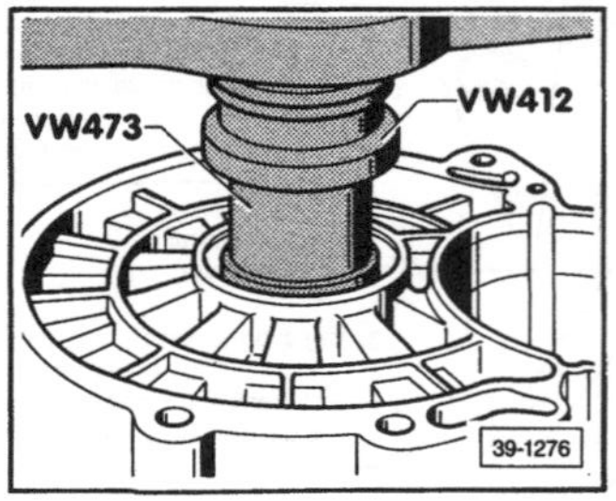

▶ **Fig. 3 Tapered roller bearing (transmission housing side), outer race, installing**

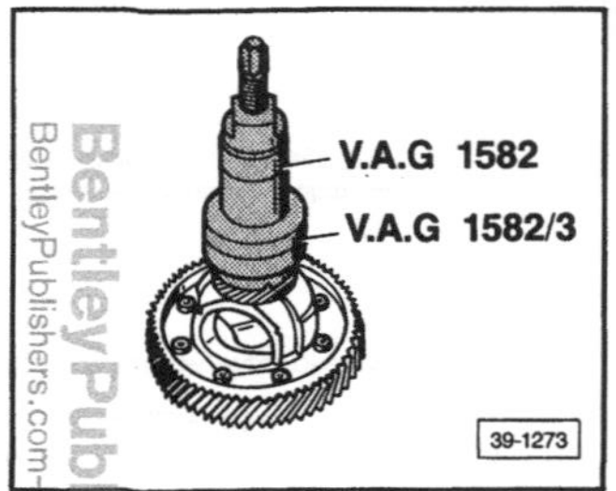

▶ **Fig. 4 Tapered roller bearing (gear carrier housing side), inner race, removing**

- before installing pulling device, install piece **30-11** on differential housing

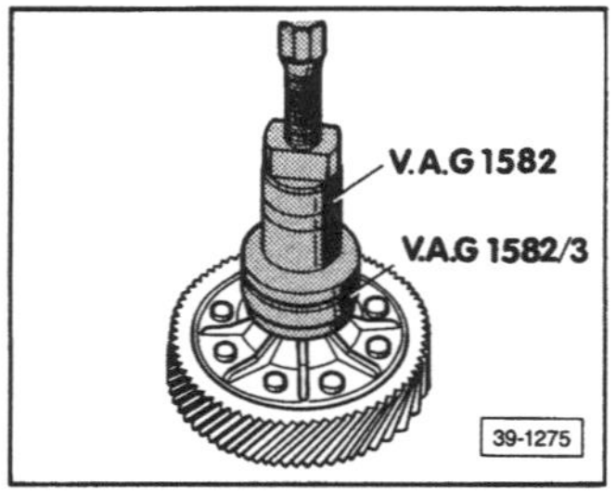

▶ **Fig. 5 Tapered roller bearing (transmission housing side), inner race, removing**

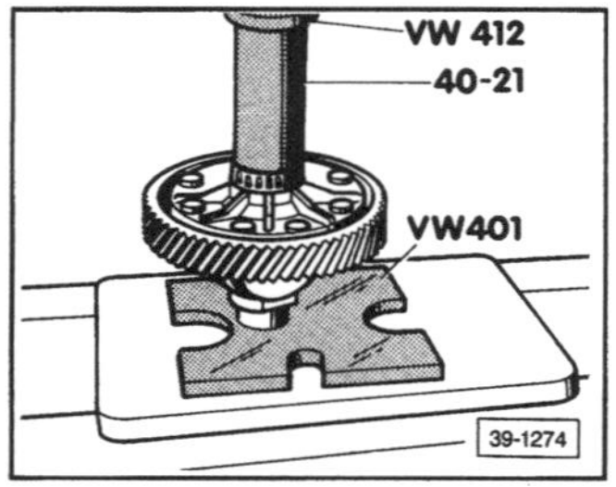

▶ **Fig. 6 Tapered roller bearings (both), inner race, installing**

- heat to approximately 100°C (212°F) before installing
- tapered roller bearing inner races for transmission housing and gear carrier housing are pressed on with same pressing tools

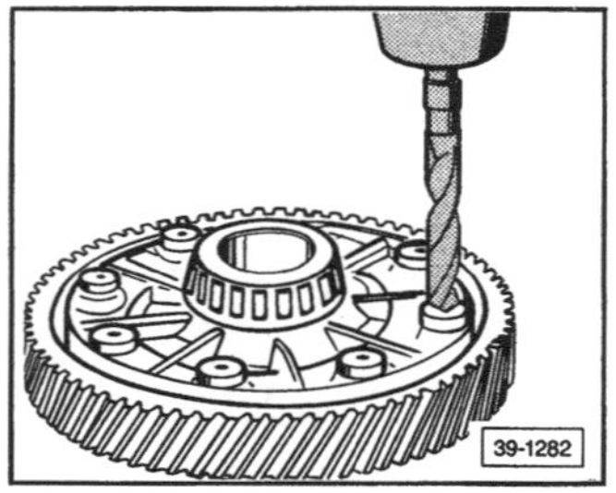

▶ **Fig. 7 Rivet heads, boring**

- bore rivet heads on counterbore side with a 12 mm drill and drive out rivets with a drift

CAUTION

Clean differential before and after the boring procedure. Protect tapered roller bearings from filings.

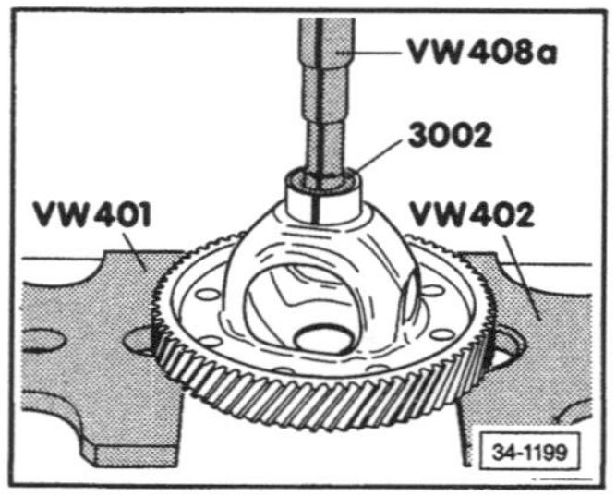

▶ **Fig. 8 Final drive gear, removing**

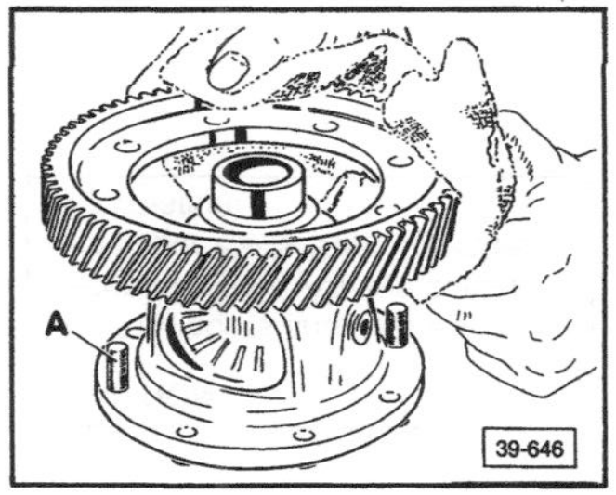

▶ **Fig. 9 Final drive gear, installing**

- heat to approximately 100°C (212°F) before installing
- guide final drive gear with bolts **A** from the repair kit when installing

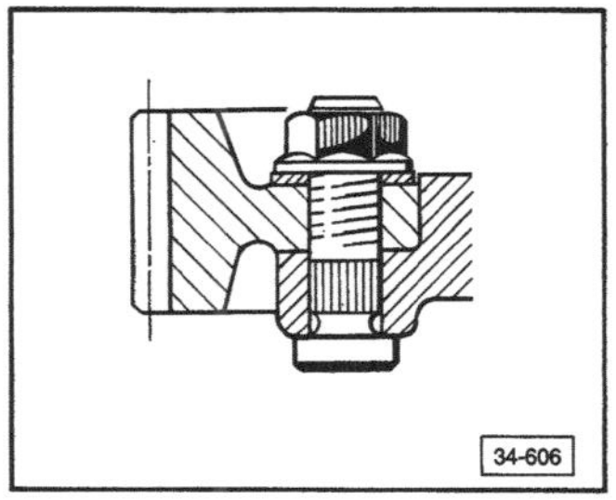

▶ **Fig. 10 Gear, installing**

- use special bolts with support discs and nuts
  - replacement Part No. **02A 498 088**

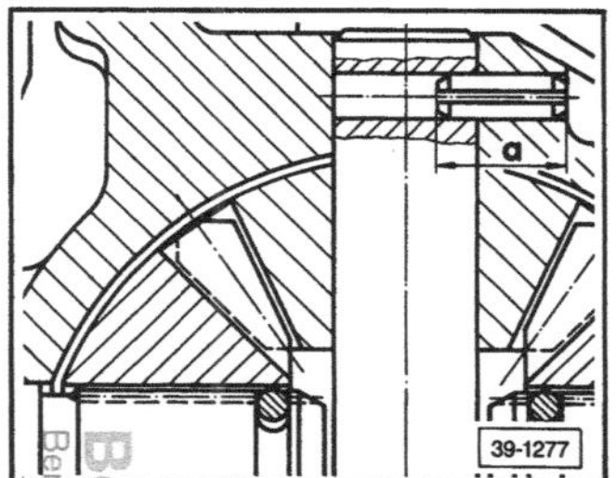

▶ **Fig. 11 Roll pin, installing**

- drive roll pin into differential housing until flush
  - length of roll pin dimension **a** = 14 mm
  - 28.5 mm long pin may be installed in production

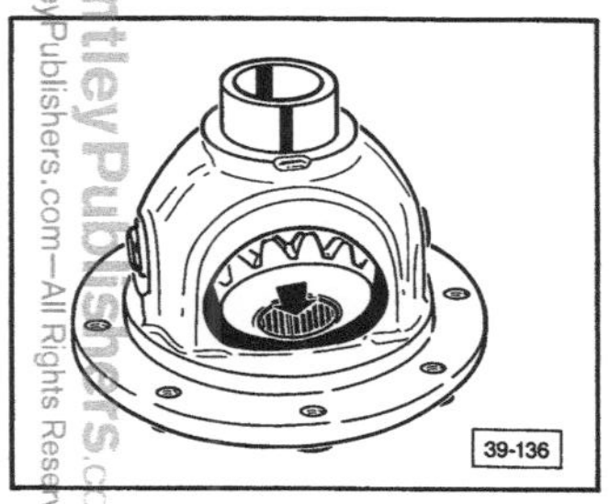

▶ **Fig. 12 Differential gears, installing**

- install one-piece thrust washer with transmission oil
- install differential pinion gears
- drive in differential pinion shaft and secure with roll pin
- install differential side gears turned 180° and pivot in direction of **arrow**

Differential, adjusting

Adjustment of the differential is necessary if any of the following parts are replaced:

transmission housing
gear carrier housing
differential housing
differential tapered roller bearings

See Adjustment Overview, Repair Group 35.

- press tapered roller bearing outer race into gear carrier housing **without** washer with **VW 473**

CAUTION

Tapered roller bearings inner and outer races are paired and must not be interchanged.

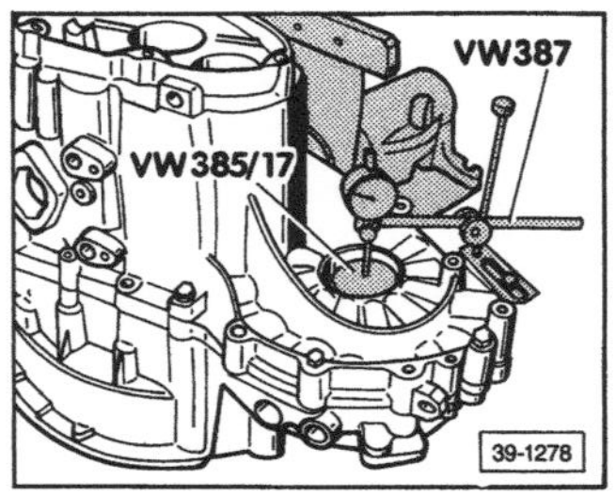

▶

- press tapered roller bearing outer race into transmission housing without adjustment shim using **VW 473**
- install differential in gear carrier housing
- install transmission housing and 5 bolts
  - tighten bolts to 25 Nm (18 ft lb)
- install dial indicator and set to zero with 1 mm pre-load
- note reading on gage while moving differential gear up and down

CAUTION

When taking measurement, do not turn differential, or a false reading will result.

Adjustment shim, determining

- determine specified bearing pre-load by adding a constant value for pre-load (0.40 mm) to the determined measurement value

Example

measurement value = 0.70 mm

measured value	0.70 mm
pre-load (constant value)	0.40 mm
shim thickness	1.10 mm

- remove transmission housing and drive out tapered roller bearing outer race with **VW 473**
- install adjustment shims of the determined thickness
  - install thickest shim first

The following adjustment shims are available:

Thickness (mm)	Part number
0.65	02A 409 210
0.70	02A 409 210 A
0.75	02A 409 210 B
0.80	02A 409 210 C
0.85	02A 409 210 D
0.90	02A 409 210 E
0.95	02A 409 210 F
1.00	02A 409 210 G
1.05	02A 409 210 H
1.10	02A 409 210 J
1.15	02A 409 210 K
1.20	02A 409 210 L
1.25	02A 409 210 M

Note

If the shim thickness measured is greater than that listed in the table, two shims that correspond to the measured value can be installed.

Due to varying tolerances, exact shim thickness can be attained

- press in outer race again and attach transmission housing

CAUTION

Part numbers shown are for reference only. Always consult with your Parts Dept. for latest information.

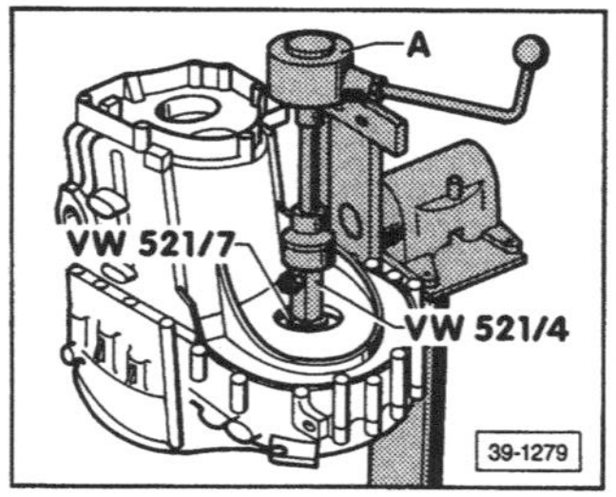

▶

Turning torque, checking

- **A** — commercial torque gage 0-600 Ncm (0-54 in. lb)
- apply hypoid transmission oil to tapered roller bearing

Turning torque

new tapered roller bearing 120-320 Ncm (11-29 in. lb)
used tapered roller bearing at least 30 Ncm (3 in. lb)

Index

Axle flange oil seal
- removing/installing **39–10**

Differential
- adjusting **39–130**
- assembly **39–80**
- disassembling **39–90**
- modification **39–85**

Differential pinion
- assembly **39–60**
- removing/installing **39–70**

Final drive/differential
- assembly **39–60**
- removing/installing **39–70**

Output gear
- adjusting **39–120**
- assembly **39–20**
- removing/installing **39–30**

Pinion
- adjusting **39–110**
- assembly **39–40**
- removing/installing **39–50**

Shims/bearings
- installed positions **39–100**

Axle flange oil seal, removing/installing (transmission installed)

Removing

- remove axle shaft from axle flange
- move axle shaft to one side and secure with wire

Remove left side axle shaft completely when replacing left side axle flange oil seal:

- remove axle shaft nut (vehicle on ground)
- raise vehicle and remove wheel
- mark position of wheel bearing housing on suspension strut

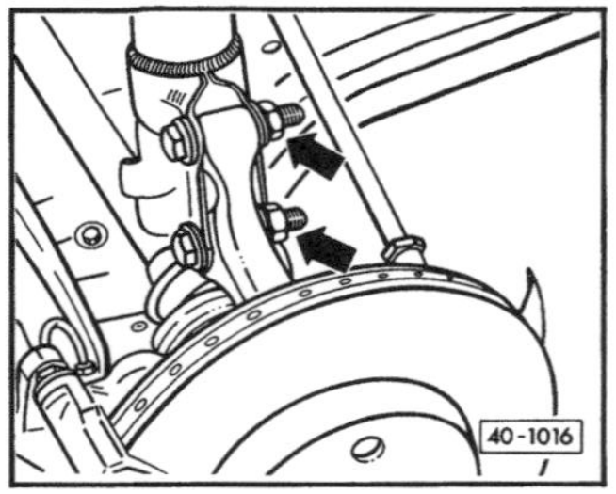

- remove wheel bearing housing/suspension strut bolts **(arrows)**
- swing wheel bearing housing to one side and remove axle shaft
- remove cap and circlip
- remove axle flange with puller **VW 391**
- remove oil seal with extractor lever **VW 681**

Installing

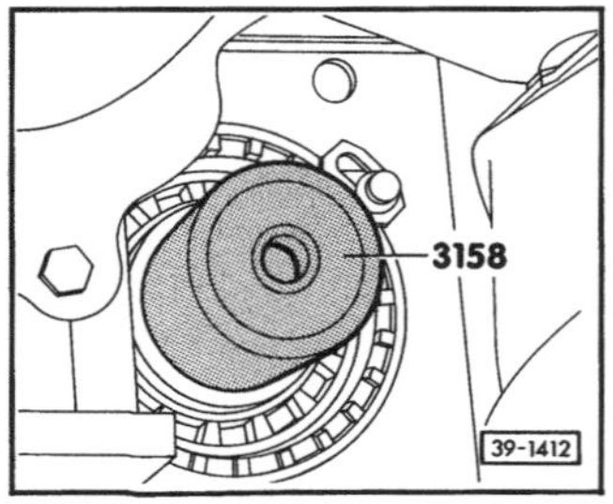

- drive in new oil seal to stop
- fill space between seal lips with multi-purpose grease
- install axle flange

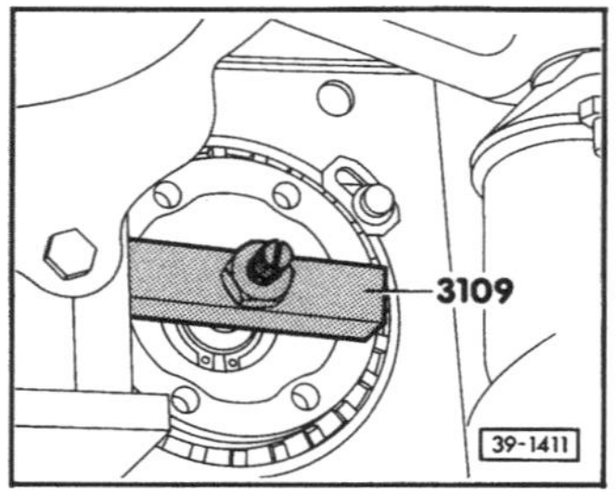

- install puller **3109** and draw flange in
- install circlip and flange cap
- install axle shaft and suspension strut/wheel bearing housing

Note

Ensure that left side suspension strut/wheel bearing housing are aligned to previously made marks. Check wheel alignment if necessary.

Check final drive oil level, see Repair Group 32.

Tightening torques

Axle shaft to flange	45 Nm (33 ft lb)
Axle shaft nut to hub	265 Nm (195 ft lb)
Wheel bearing housing to strut	95 Nm (70 ft lb)

Note

Ensure that left side suspension strut/wheel bearing housing are aligned to previously made marks. Check wheel alignment if necessary.

Check final drive oil level. See Note below.

Tightening torques

Axle shaft to flange	45 Nm (33 ft lb)
Axle shaft nut to hub	265 Nm (195 ft lb)
Wheel bearing housing to strut	95 Nm (70 ft lb)

Note

SAE 75W/90 synthetic oil must be used as replacement oil for the final drive units of automatic transmission 096.

If the final drive contains standard oil, drain the oil completely and refill with synthetic oil. Do not mix synthetic oil with standard gear oil.

Synthetic oil is available as follows:

- 1.0 L size: Part No. G 052 145 A2

CAUTION
Use ONLY the synthetic oil specified above for Volkswagen automatic transmission final drive units.

Part numbers are for reference only. Always check with your Parts Dept. for the latest information.

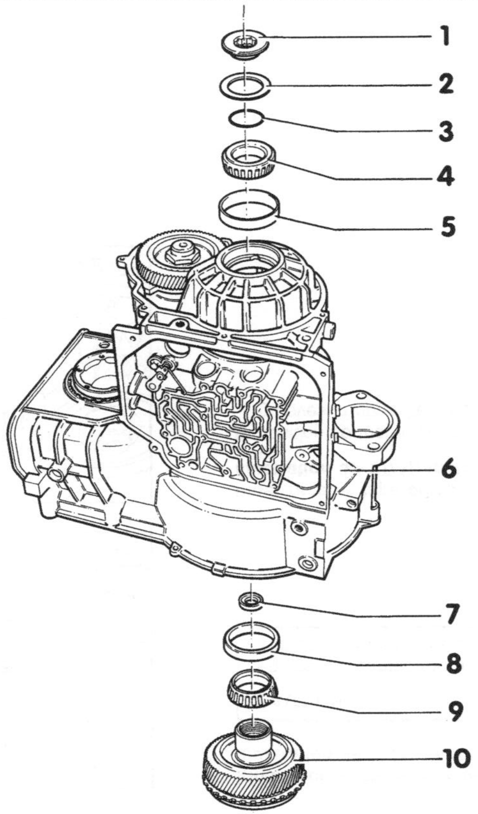

Note

Pinion shaft and differential can remain installed during this procedure.

1 — **Tensioning bolt — 250 Nm (184 ft lb)**
engage parking pawl when removing/installing

2 — **Dished washer**
curved side faces tensioning bolt

3 — **Shim**
determining thickness, section 39-120

4 — **Inner race, tapered roller bearing**
install so that lugs engage in groove of other tapered roller bearing inner race (item **9**). See section 39-30

5 — **Outer race, tapered roller bearing**
- drive out with suitable drift
- drive in using thrust pad **30-205**
- install with locking compound **AMV 185 1001**

6 — **Transmission housing**

7 — **Axial bearing**
- flat side faces driveshaft
- place in output gear before installing adjusting bolt (item 1)

8 — **Outer race, tapered roller bearing**
- drive out with suitable drift
- drive in using thrust pad **30-205** and long drift
- install with locking compound **AMV 185 1001**

9 — **Inner race, tapered roller bearing**
removing/installing, section 39-30

10 — **Output gear**
- number of teeth, see Technicial data, Repair Group 37
- adjusting, section 39-120
- when replacing output gear, adjust planet carrier

Output gear assembly, removing/installing

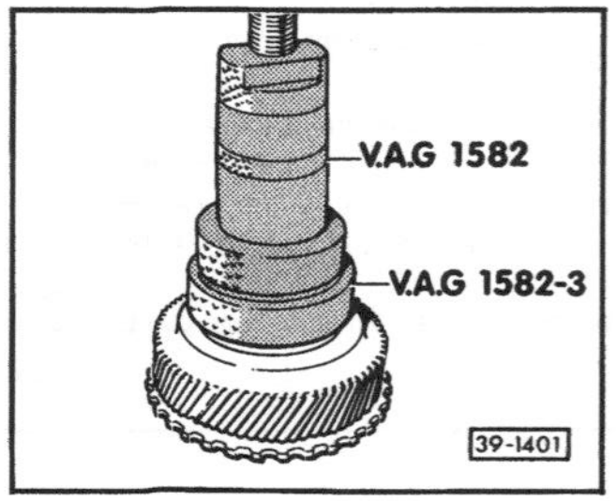

Inner race – tapered roller bearing (item 9, illustration 39-1371), removing

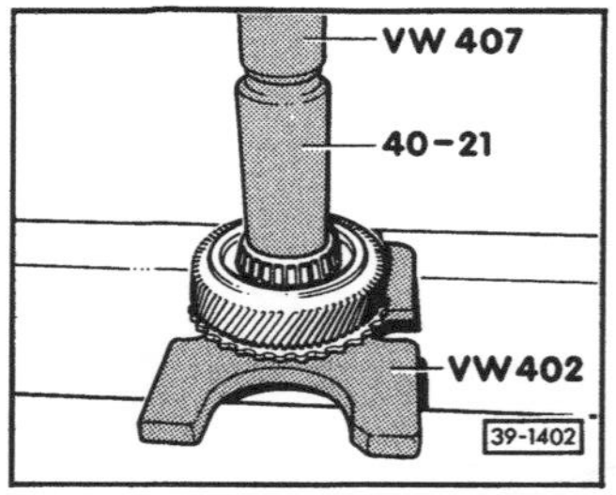

Inner race – tapered roller bearing (item 9, illustration 39-1371), installing

> **CAUTION**
> To prevent damage to the output gear, position a thrust pad under the gear hub.

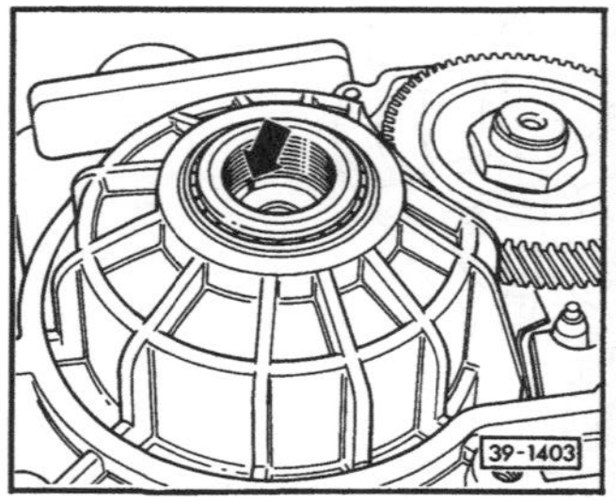

Inner race – tapered roller bearing (item 4, illustration 39-1371), installing

- position so that lugs on inner race engage grooves in opposite inner race (**arrow**)

THIS PAGE INTENTIONALLY LEFT BLANK

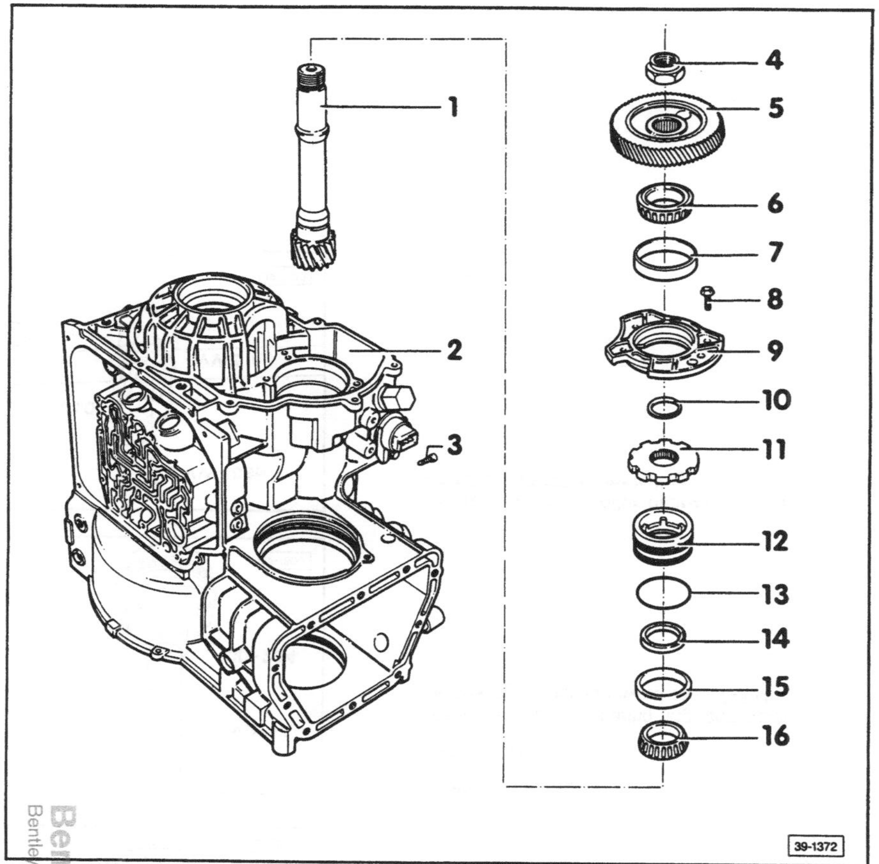

Note:
Beginning Oct. 1991, a stronger tapered roller bearing has been installed.

Therefore items 1, 2, 8, 9, 12, 13, 14, 15 and 16 are changed.

1 — **Pinion shaft**
- number of teeth, see Technical data, Repair Group 37
- before pinion shaft can be pulled, remove items **4-14**, and **16**. After removing item **11**, parking lock gear, remove gear change shaft (see Repair Group 38)
- to remove pinion shaft, reinstall parking lock gear and tensioning nut onto pinion shaft. Install **US 1078 (Kukko 20-10)**, placing hooks under gear/nut. Pull pinion shaft, along with outer race **15** and inner race **16**
- adjusting, section 39-110

2 — **Transmission housing**
- mounting holes for M6 bolts (up to 10/91)
- mounting holes for M8 bolts (from 10/91)

3 — **10 Nm (7 ft lb)**
install using locking compound,
Part No. **AKD 456 000 01**

> CAUTION
> Remove Bolt before removing/installing bearing race.

4 — **Tensioning nut — 250 Nm (184 ft lb)**
- engage parking pawl when removing/installing
- peen against pinion edge to secure

5 — **Pinion drive gear**
- number of teeth, see Technical data, Repair Group 37
- removing, section 39-50
- replace both gears if damaged

6 — **Inner race, tapered roller bearing**
- removing/installing, section 39-50

7 — **Outer race, tapered roller bearing**
- drive out with suitable drift
- installing, section 39-50

8 — **Bolt**
- mounted to transmission housing
- M6 = 14 Nm (10 ft lb)
- M8 = 25 Nm (18 ft lb)

9 — **Pinion cover**
- mounted to transmission housing
- different holes for M6 or M8 bolts

10 — **Shim**
determining thickness, section 39-110

11 — **Parking lock gear**
rounded side faces pinion shaft splines

12 — **Bearing ring — 200 Nm (148 ft lb)**

> CAUTION
> Remove bolt (item **3**) before removing/installing bearing ring.

- removing/installing, section 39-50
- before removing, first remove gear change shaft, Repair Group 38
- do not damage bearing ring, sealing lip or spring when installing

13 — **O-ring**
- always replace
- install on bearing ring

14 — **Oil seal for pinion shaft**
- drive out using screwdriver
- installing, section 39-50
- installed position (from 10/91), see section 39-50

15 — **Outer race, tapered roller bearing**
installing, section 39-50

16 — **Inner race, tapered roller bearing**
- drive out with suitable mandrel
- installing, section 39-50

Pinion assembly, removing/installing

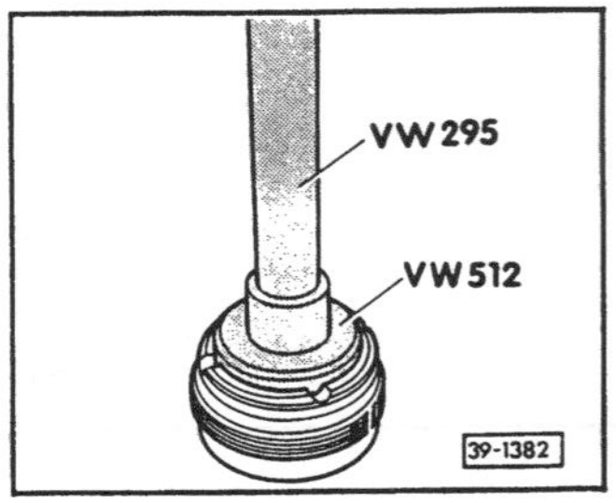

Oil seal, installing

(Up to 10/91)

- drive oil seal into bearing ring, completely

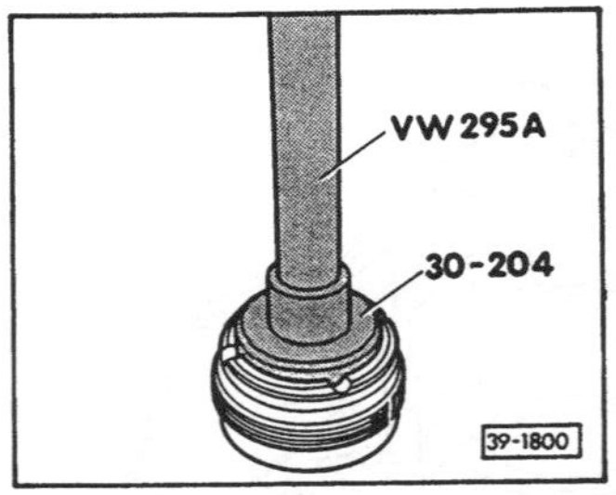

Oil seal, installing

(From 10/91)

- drive oil seal into bearing ring, completely
  - sealing lip points toward tool **30-204**

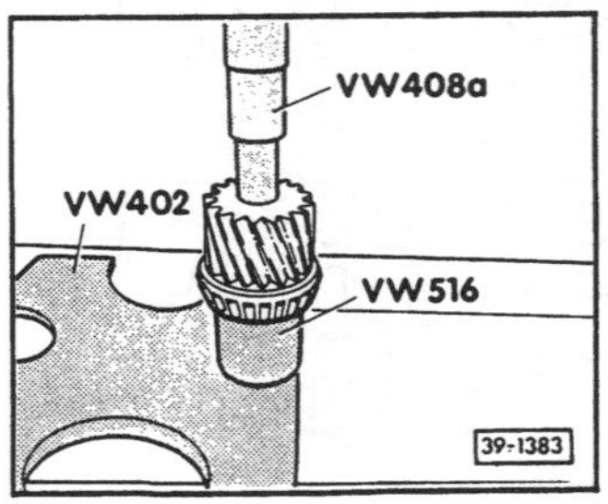

Inner race, tapered roller bearing (item 16, illustration 39-1372), installing

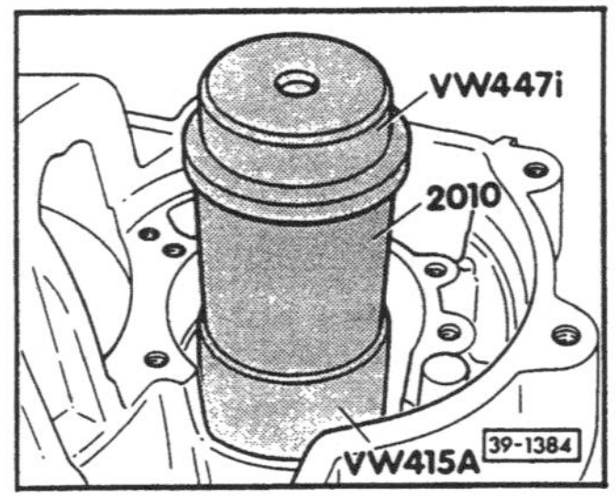

Outer race, tapered roller bearing (item 15, illustration 39-1372), installing

(Up to 10/91)

- insert pinion shaft after pressing inner race of tapered roller bearing on
- drive outer race on

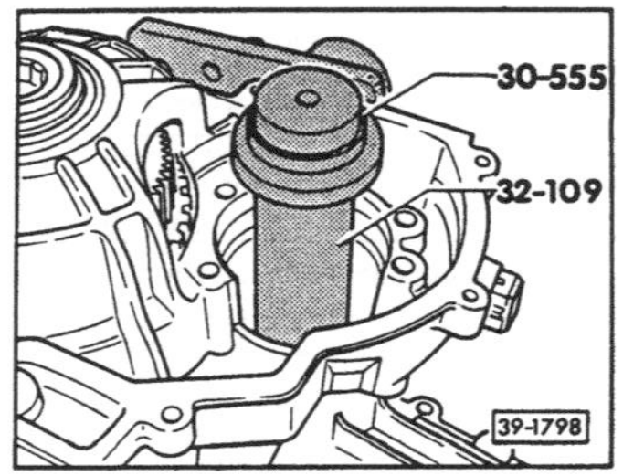

Outer race, tapered roller bearing (item 15, illustration 39-1372), installing

(From 10/91)

- insert pinion shaft after pressing inner race of tapered roller bearing on
- drive outer race on

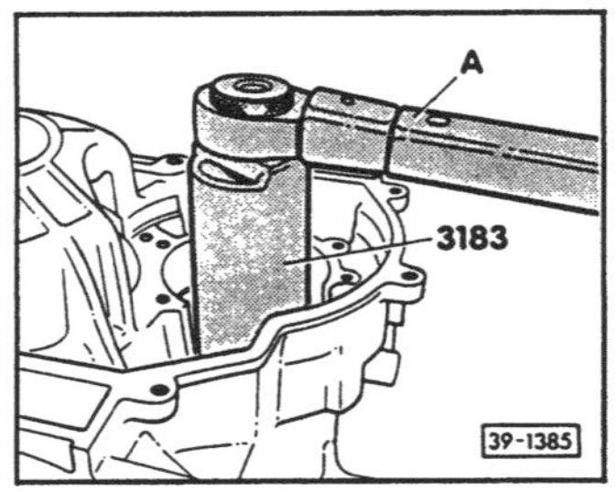

Bearing ring, removing/installing

(Up to 10/91)

A = torque wrench

- tighten bearing ring to 200 Nm (148 ft lb) using **3183**
- secure bearing ring using bolt coated with locking compound, Part No. **AKD 456 000 01**

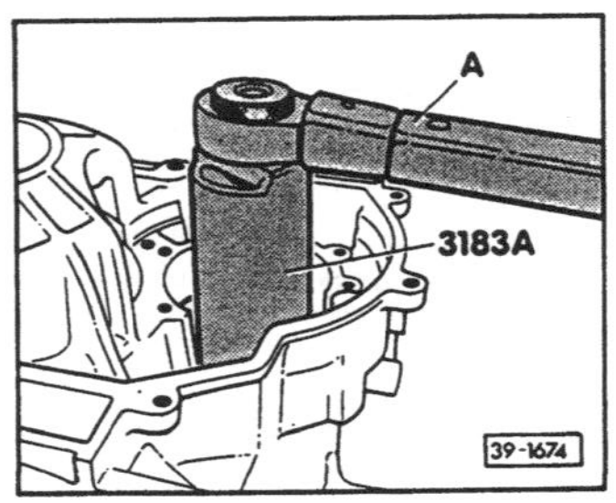

Bearing ring, removing/installing

(From 10/91)

A = torque wrench

- tighten bearing ring to 200 Nm (148 ft lb) using **3183A**
- secure bearing ring using bolt coated with locking compound, Part No. **AKD 456 000 01**

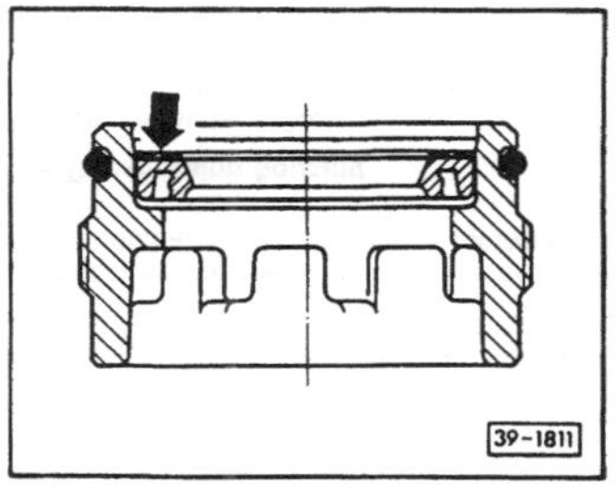

Bearing ring with seal, installed position

(Up to 10/91)
- closed side **(arrow)** points toward outer ring/tapered roller bearing

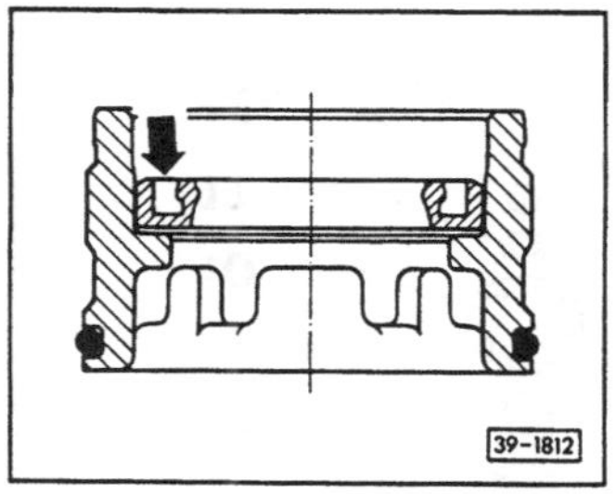

Bearing ring with seal, installed position

(From 10/91)
- open side **(arrow)** points toward outer ring/tapered roller bearing

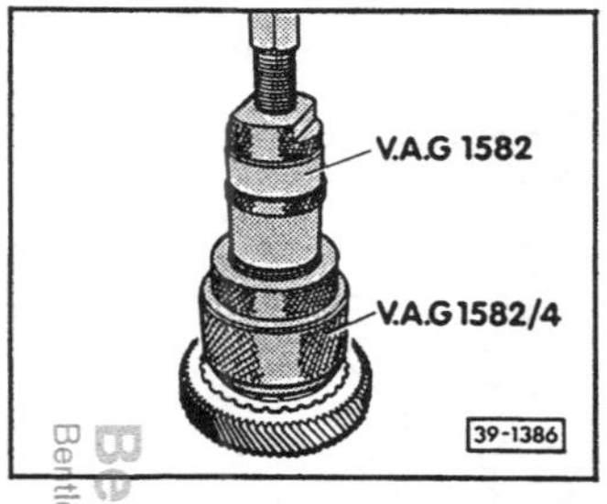

Inner race, tapered roller bearing (item 6, illustration 39-1372), removing

■ position thrust pad on collar of gear

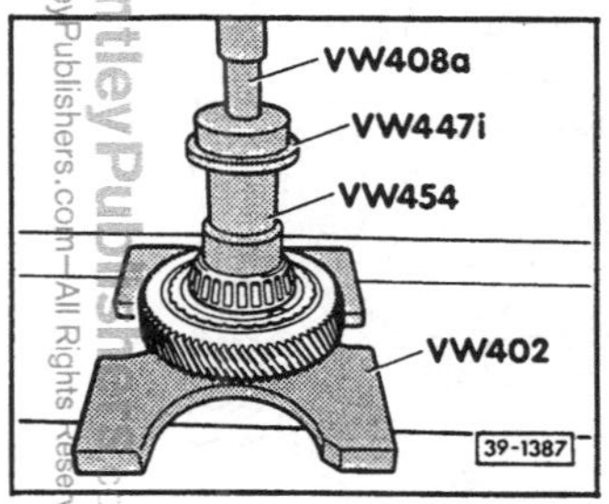

Inner race, tapered roller bearing (item 6, illustration 39-1372), installing

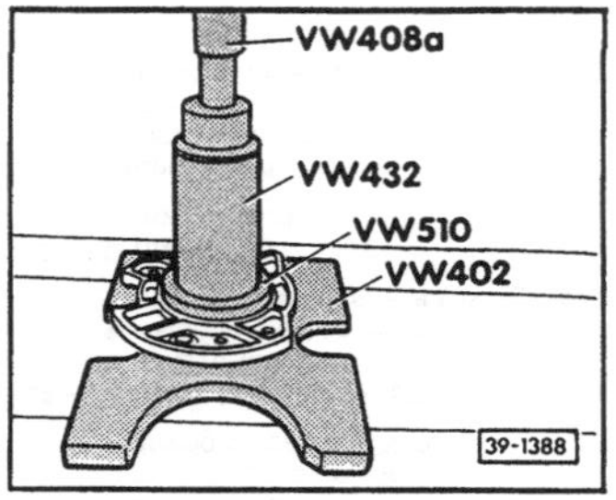

Outer race, tapered roller bearing (item 7, illustration 39-1372), installing

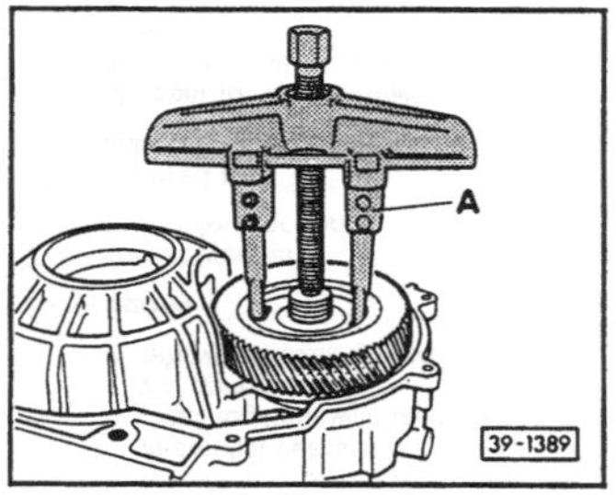

Pinion drive gear, removing

A = two arm puller **US 1078 (Kukko 20-10)** with puller hooks **Matra V172**

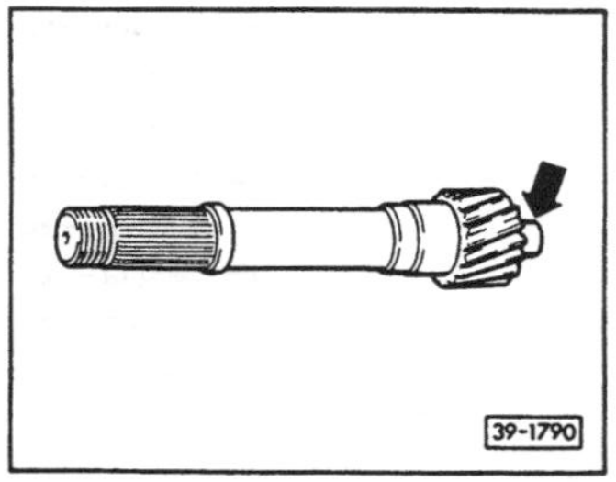

Pinion shaft, identification

(From 10/91)
- note shaft offset **(arrow)** before gear teeth

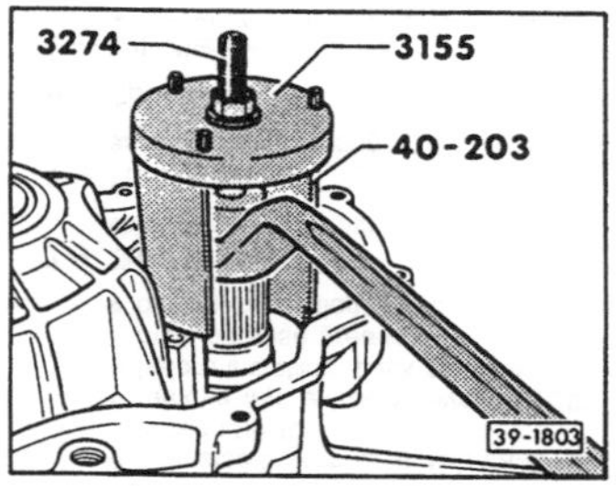

Pinion shaft, removing

(From 10/91)

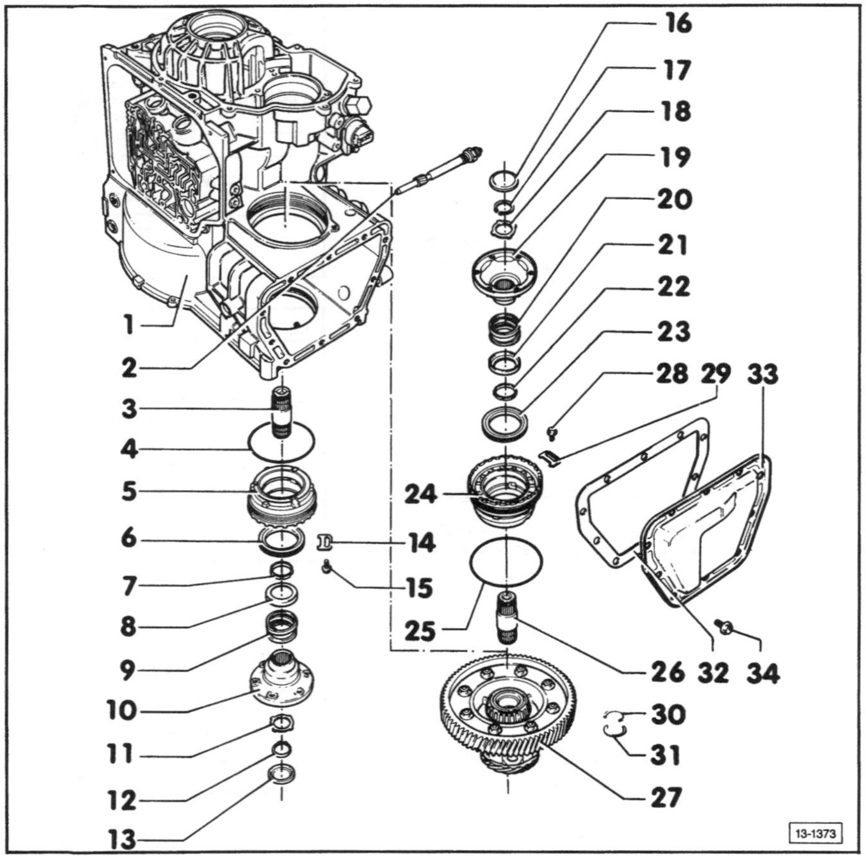

Note

The pinion shaft can remain installed when replacing the differential.

If the tapered roller bearings will be used again, mark the adjusting ring during disassembly.

After adjusting tapered roller bearings, secure bearing housing and adjusting ring with respective bolt/retainer.

1 — **Transmission housing**

2 — **Speedometer drive**
serves also as oil level dipstick, section 39-70

3 — **Output shaft**
- remove, before taking out differential
- install, after adjusting differential
- threaded end faces axle flange

4 — **O-ring**
always replace

5 — **Adjusting ring**
- marking/removing (used bearings), section 39-70
- installing/adjusting (new bearings), section 39-130

6 — **Oil seal for output shaft**
- may also be replaced with transmission installed, section 39-10
- remove with extractor lever **VW 681**
- before installing, pack area between lips with multi-purpose grease
- driving in, section 39-70

7 — **Tapered ring**
shoulder faces thrust washer

8 — **Thrust washer**
assemble over spring

9 — **Spring**

10 — **Axle flange**
- remove first, before removing inner circlips at differential gears
- install together with tapered ring, thrust washer, and spring
- removing/installing, section 39-70

11 — **Dished washer**

12 — **Circlip**

13 — **Rubber cap**

14 — **Retainer**

15 — **10 Nm (7 ft lb)**

16 — **Rubber cap**

17 — **Circlip**

18 — **Dished washer**

19 — **Axle flange**
- remove first, before removing inner circlip at differential gears
- install together with tapered ring, thrust washer and spring
- removing/installing, section 39-70

20 — **Spring**

21 — **Thrust washer**
assemble over spring

22 — **Tapered ring**
shoulder faces thrust washer

23 — **Oil seal for axle flange**
- may also be replaced with transmission installed, section 39-10.
- remove with extractor lever **VW 681**
- before installing, pack area between lips with multi-purpose grease
- driving in, section 39-70

24 — **Bearing housing — 150 Nm (111 ft lb)**
- remove with **3155**
- installing, section 39-70
- with new bearings — adjusting, section 39-130

25 — **O-ring**
always replace

26 — **Output shaft**
- remove, before taking out differential
- threaded end faces axle flange
- install, after adjusting differential

27 — **Differential**
before removing differential, first remove bearing housing, adjusting ring and output shafts

28 — **10 Nm (7 ft lb)**

29 — **Retainer**

30 — **Circlip**
- remove, after taking off axle flange
- removing/installing, section 39-70

31 — **Circlip**
- remove after taking off axle flange
- removing/installing, section 39-

32 — **Gasket**
always replace

33 — **Cover for differential**

34 — **28 Nm (21 ft lb)**
install with locking compound **AMV 200 000**

Final drive/differential assembly, removing/installing

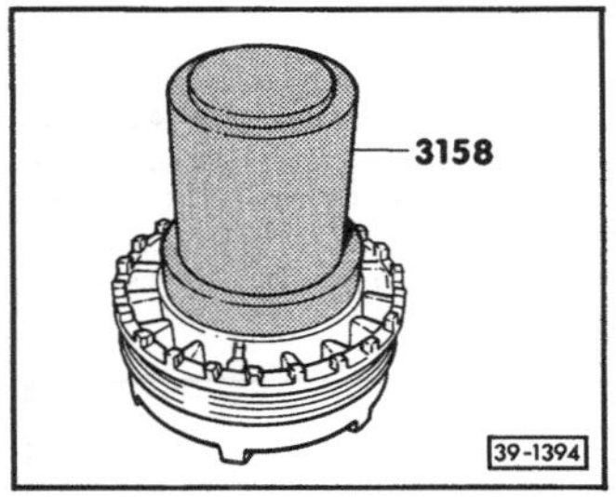

Oil seal for axle flange, installing

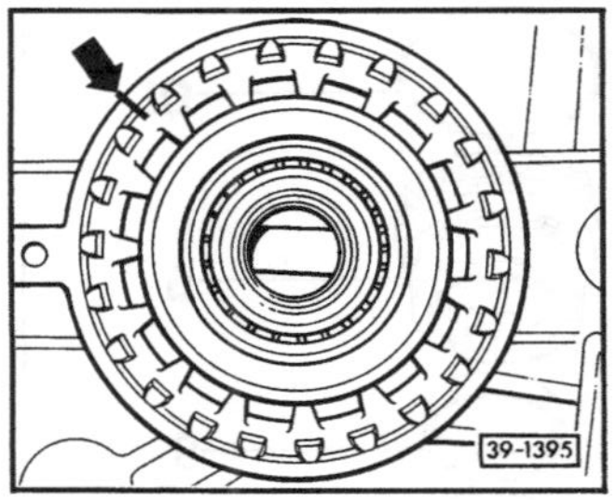

Adjusting ring, marking

Note

If same bearings will be reinstalled, mark at adjusting ring (**arrow**). Realign scribe marks during assembly.

If new bearings are installed, see adjusting procedure, section 39-130.

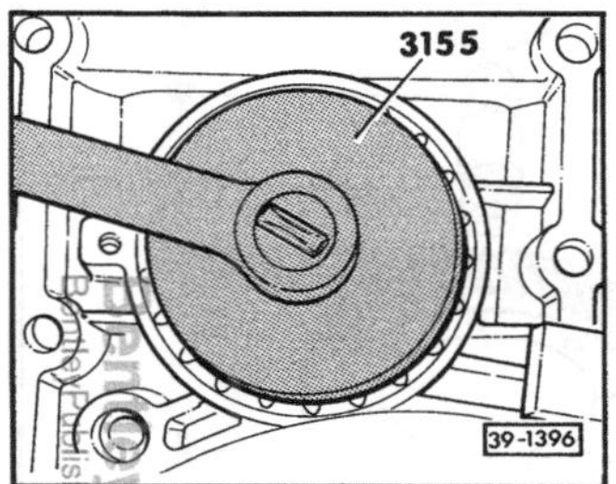

Adjusting ring, removing

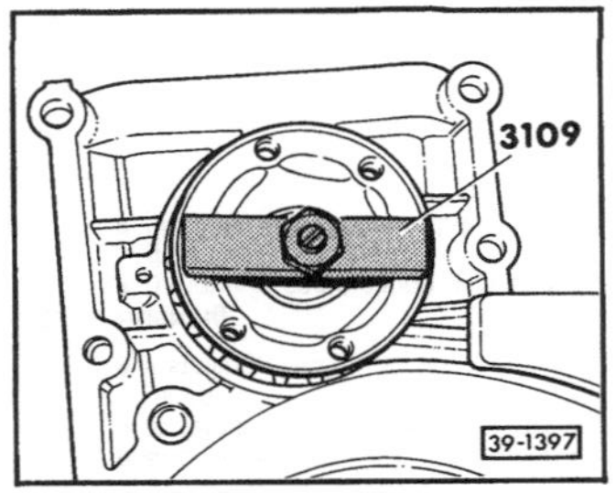

Axle flange, installing

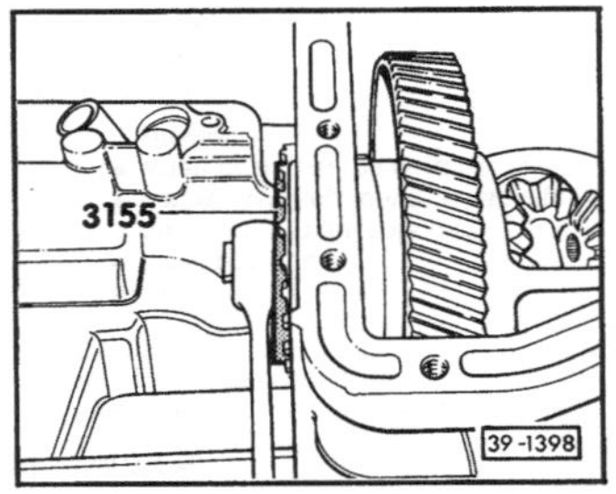

Bearing housing, installing

- install bearing housing up to stop
  - differential must already be installed
  - housing tightening torque = 150 Nm (111 ft lb)

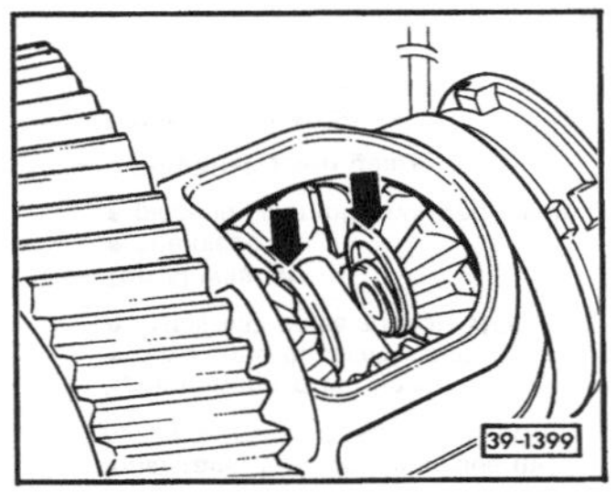

Circlips, removing

- pry out circlips (**arrows**) with two screwdrivers

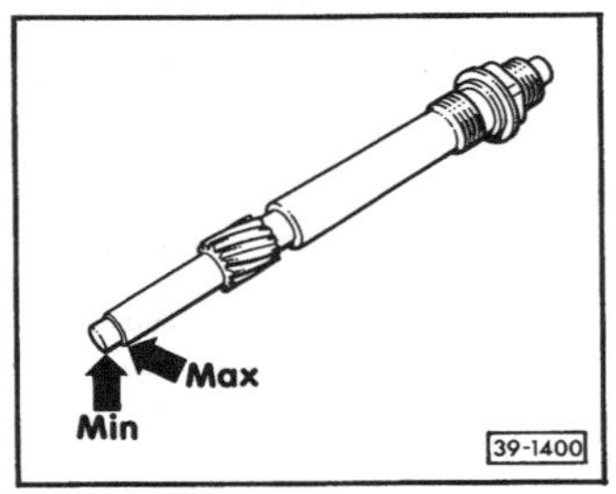

Differential oil level, checking

- oil level must be between **Min** and **Max** points **(arrows)** on speedometer drive.
- use only SAE 75W/90 synthetic oil when replacing differential oil.
- do not mix synthetic oil with standard gear oil.

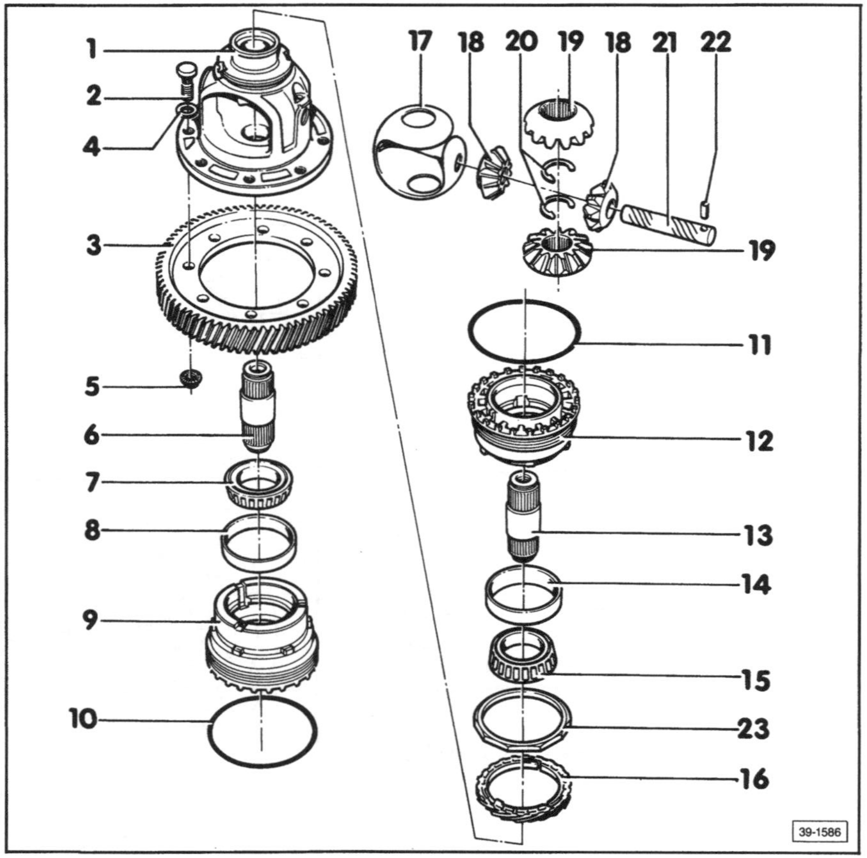

Note

Heat inner race of tapered roller bearings to 100°C (212°F) before pressing on.

Heat adjusting ring for tapered roller bearings to 100°C (212°F) before replacing the outer race.

1 – **Differential housing**
replace complete with ring gear only

2 – **Bolt**

3 – **Ring gear**
- number of teeth, see Technicial data, Repair Group 37
- replace complete with differential housing only.

4 – **Washer**

5 – **Nut**
- tighten to **60 Nm (44 ft lb)**
- then turn 90° further

6 – **Output shaft**
- removing/installing, section 39-60
- tighten to **60 Nm (44 ft lb)**
- then turn 90° further

7 – **Inner race, tapered roller bearing**
removing/installing, section 39-90

8 – **Outer race, tapered roller bearing**
- before removing/installing, heat adjusting ring
- drive out with suitable drift
- installing, section 39-90

9 – **Bearing housing**
removing/installing, section 39-60

10 – **O-ring**
always replace

11 – **O-ring**
always replace

12 – **Adjusting ring**
removing/installing, section 39-90

13 – **Output shaft**
removing/installing, section 39-60

14 – **Outer race, tapered roller bearing**
- before removing/installing, heat adjusting ring
- drive out with suitable drift
- installing, section 39-90

15 – **Inner race, tapered roller bearing**
removing/installing, section 39-90

16 – **Speedometer drive gear**
- drive out with suitable drift
- install with press support **3144**

17 – **Thrust washer**
- one-piece
- place in differential housing before installing gears

18 – **Differential pinon gears**

19 – **Differential side gears**
installing, section 39-90

20 – **Circlips**

CAUTION

After removing axle flanges, pry off circlips, because springs are preloaded.

21 – **Shaft**
- for pinion gears
- drive out with suitable drift
- before installing, drive out spring pin
- when driving in, be sure not to damage thrust washer

22 – **Roll pin**
- locks pinion shaft into position
- pin length = 14 mm or 28.5 mm
- to remove 14 mm pin/shaft, drive pin completely into shaft
- after removing shaft, drive pin out
- to remove 28.5 mm pin, pull out with side-cutting pliers
- installing, section 39-90

23 – **Bushing**
- remove in combination with speedometer drive gear
- drive on with suitable drift

Differential assembly modified

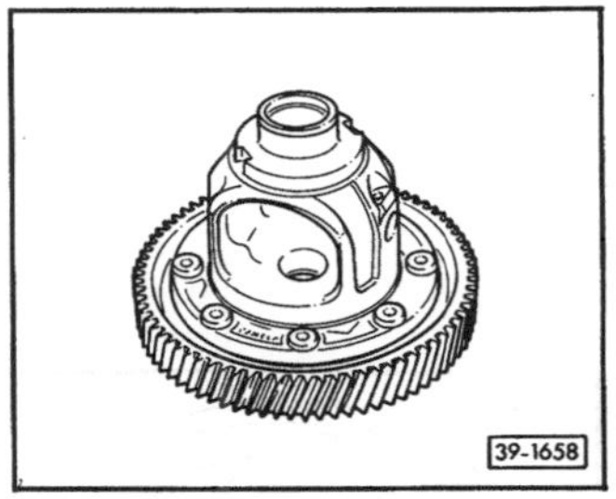

The ring gear on the differential housing can no longer be replaced separately. Ring gears are riveted to the differential housing in production and then machined.

If the differential housing or the ring gear need replacing, the differential housing and ring gear must be replaced as a complete assembly.

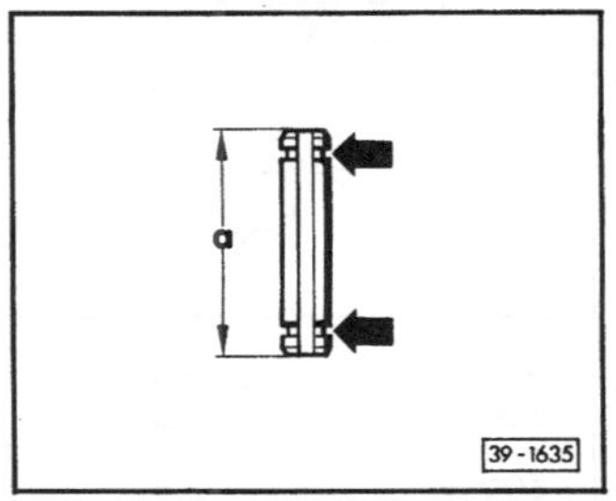

Modified roll pin, removing/installing

A longer roll pin is now installed to secure the differential shaft. The new roll pin has a groove at each end.

- dimension **a**: old = 14.0 mm

  new = 28.5 mm

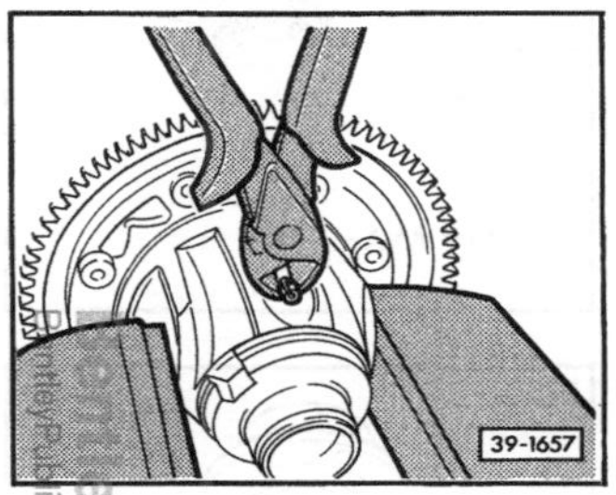

Removing

- pull roll pin out with side cutting pliers

Installing

- drive roll pin in to stop

Note

Only the old version roll pin, Part No. **N 043 302 1**, is available as a replacement part, which can be used for both applications.

THIS PAGE INTENTIONALLY LEFT BLANK

Differential, disassembling/assembling

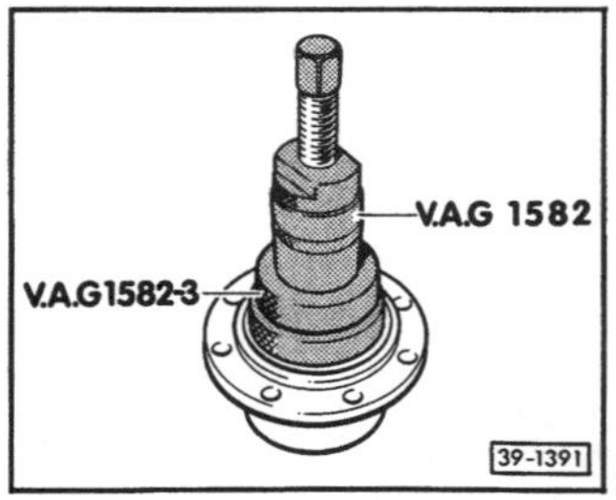

Inner race – tapered roller bearing (item 7, illustration 39-1586), removing

- position thrust pad **30-555** on differential housing

Note

When removing inner race (item **15**, illustration 39-1374), follow this same procedure.

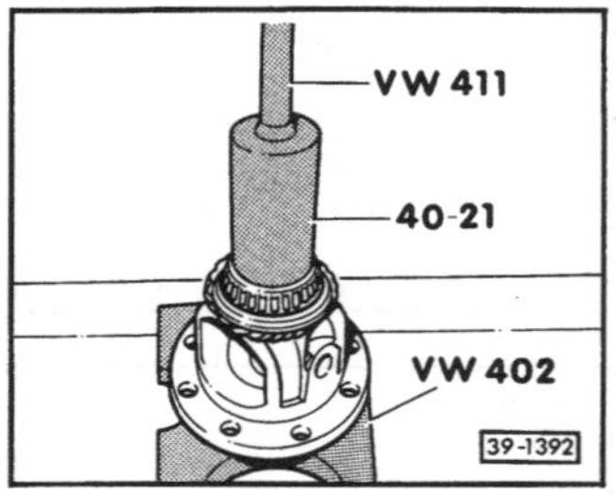

Inner race – tapered roller bearing (item 7, illustration 39-1586), installing

Note

When installing inner race (item **15**, illustration 39-1586), follow this same procedure.

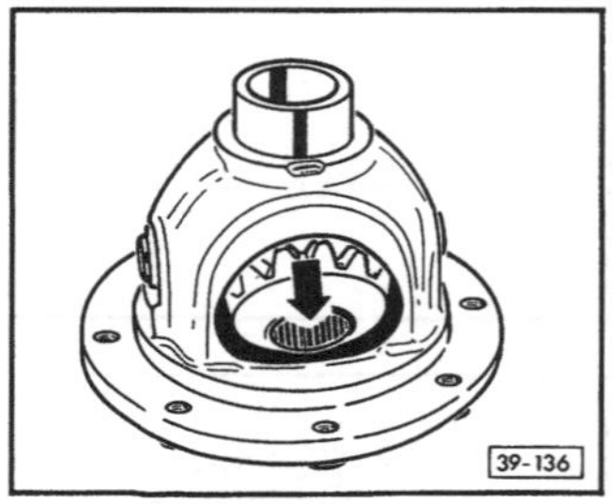

Differential gears, assembling

- position thrust washer in differential housing
  - lubricate washer with gear oil, before inserting
- install pinion gears and secure with shaft and spring pin
- install side gears at 180° angle
  - pivot into position (**arrow**)

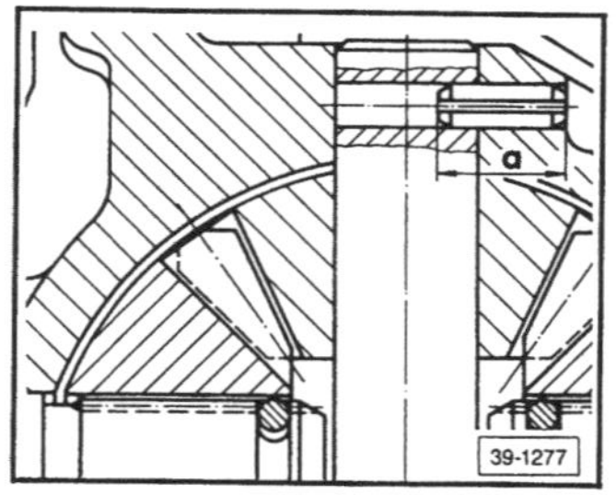

Spring pin for pinion shaft, installing

- drive pin in, flush with housing
  - **a** (length of pin) = 14 mm
  - 28.5 mm long pin may be installed in production

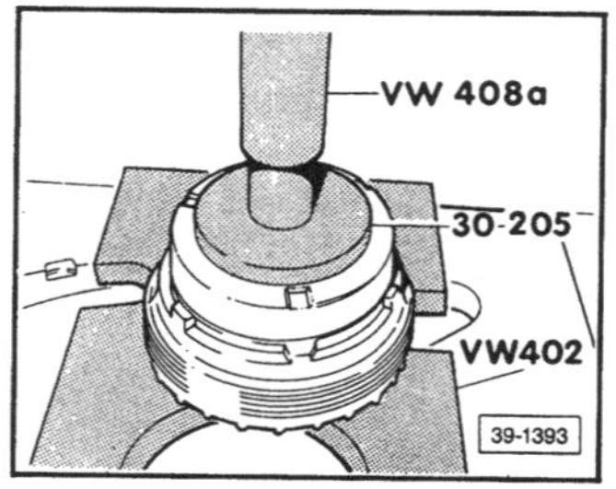

Outer race, tapered roller bearing (items 8 and 14, illustration 39-1586), installing

- press outer race into heated bearing housing or adjusting ring, until race is seated

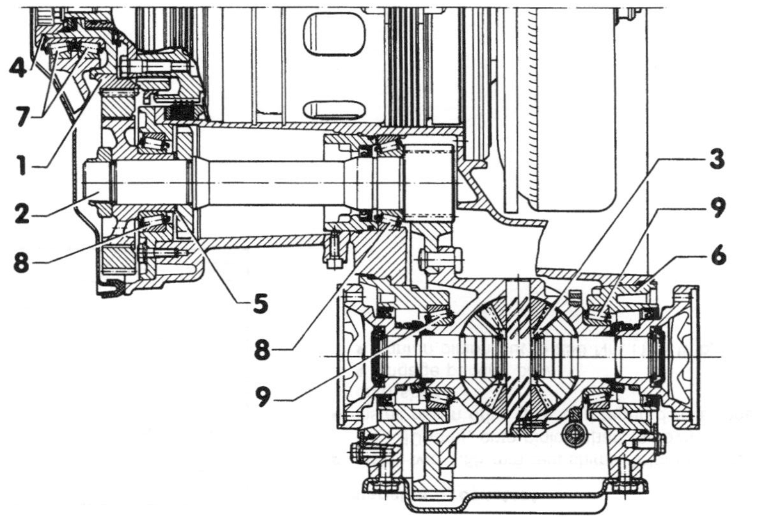

1 — **Output gear**
adjusting, section 39-120

2 — **Pinion shaft**
adjusting, section 39-110

3 — **Differential**
adjusting, section 39-130

4 — **Shim for output gear**
determine thickness from chart, section 39-110

5 — **Shim for pinion shaft**
determine thickness from chart, section 39-110

6 — **Adjusting ring for differential**

7 — **Tapered roller bearing for output gear**
- always replace bearings as a set
- measure turning torque via pinion shaft. For new bearings, torque must be 180-220 Ncm (15.9-19.5 in. lb or 18.3-22.5 cm kg) **above** turning torque for pinion shaft

8 — **Tapered roller bearings for pinion shaft**
- always replace bearings as a set
- once bearings are replaced, redetermine shim thickness

9 — **Tapered roller bearing for differential**
- always replace bearings as a set
- measure turning torque via pinion shaft. For new bearings, torque must be 60-80 Ncm (5.3-7.1 in. lb or 6.1-8.2 cm kg) **above** turning torque for pinion shaft

THIS PAGE INTENTIONALLY LEFT BLANK

Pinion shaft, adjusting

Note

Before adjusting the pinion shaft, the following must be installed:
- bearing ring for pinion shaft bearing (tightened to specifications)
- parking pawl
- parking lock gear

Adjust pinion shaft by shimming it to a specified thickness. Determine the specified thickness from the chart at the end of this section. In order to use the chart, the specified bearing preload must be calculated, using the equation:

Specified bearing preload=**k - x - y - z**.

Values for **k, x, y** and **z** are given or determined, as follows:

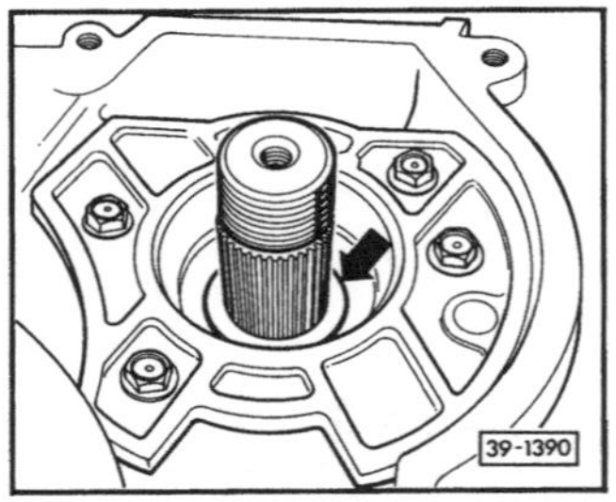

Dimension k, constant

- install two 1.55 mm test shims (total thickness **k**=3.10 mm), onto pinion shaft (**arrow**)
- install pinion drive gear and tapered roller bearing onto shaft
  - engage parking pawl
  - tighten pinion nut to 250 Nm (185 ft lb)

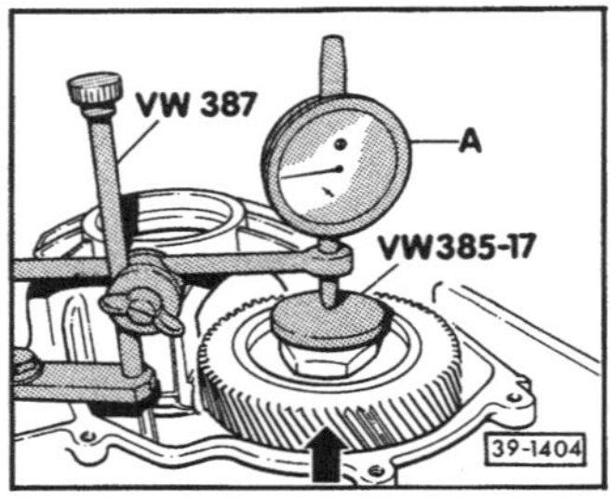

Dimension x, determining

- install dial gauge **A**, as shown
  - set to zero with 1 mm preload

> **CAUTION**
> Do not allow pinion shaft to turn during procedure, or bearings will settle and reading will be incorrect.

Dimension y, constant

Bearing preload **y**=0.12 mm.

Dimension z, constant

Bearing settling amount **z**=0.10 mm.

Specified bearing preload, calculating

Example

k = test shim, thickness	3.00 mm
x = dial gauge reading	−0.93 mm
y = bearing preload	−0.12 mm
z = settling amount	−0.10 mm
	= 1.85 mm

Specified shim value, determining

Note

Using the specified bearing preload just calculated, find the correct adjustment shim value in the chart which follows.

Due to varying tolerances, check all shims with a micrometer so that exact thickness can be installed.

Specified bearing preload (mm)	Shim to be installed (mm)	Specified bearing preload (mm)	Shim to be installed (mm)
0.950-0.974	1.000	1.825-1.849	1.875
0.975-0.999	1.025	1.850-1.874	1.900
1.000-1.024	1.050	1.875-1.899	1.925
1.025-1.049	1.075	1.900-1.924	1.950
1.050-1.074	1.100	1.925-1.949	1.975
1.075-1.099	1.125	1.950-1.974	2.000
1.100-1.124	1.150	1.975-1.999	2.025
1.125-1.149	1.175	2.000-2.024	2.050
1.150-1.174	1.200	2.025-2.049	2.075
1.175-1.199	1.225	2.050-2.074	2.100
1.200-1.224	1.250	2.075-2.099	2.125
1.225-1.249	1.275	2.100-2.124	2.150
1.250-1.274	1.300	2.125-2.149	2.175
1.275-1.299	1.325	2.150-2.174	2.200
1.300-1.324	1.350	2.175-2.199	2.225
1.325-1.349	1.375	2.200-2.224	2.250
1.350-1.374	1.400	2.225-2.249	2.275
1.375-1.399	1.425	2.250-2.274	2.300
1.400-1.424	1.450	2.275-2.299	2.325
1.425-1.449	1.475	2.300-2.324	2.350
1.450-1.474	1.500	2.325-2.349	2.375
1.475-1.499	1.525	2.350-2.374	2.400
1.500-1.524	1.550	2.375-2.399	2.425
1.525-1.549	1.575	2.400-2.424	2.450
1.550-1.574	1.600	2.425-2.449	2.475
1.575-1.599	1.625	2.450-2.474	2.500
1.600-1.624	1.650	2.475-2.499	2.525
1.625-1.649	1.675	2.500-2.524	2.550
1.650-1.674	1.700	2.525-2.549	2.575
1.675-1.699	1.725	2.550-2.574	2.600
1.700-1.724	1.750	2.575-2.599	2.625
1.725-1.749	1.775	2.600-2.624	2.650
1.750-1.774	1.800	2.625-2.649	2.675
1.775-1.799	1.825	2.650-2.674	2.700
1.800-1.824	1.850		

Adjustment shim, installing

- remove drive gear
- select adjustment shim from chart and place on pinion shaft
- install drive gear
- install nut
  - tighten to 150 Nm (111 ft lb)
  - to secure nut, peen to edge of pinion shaft

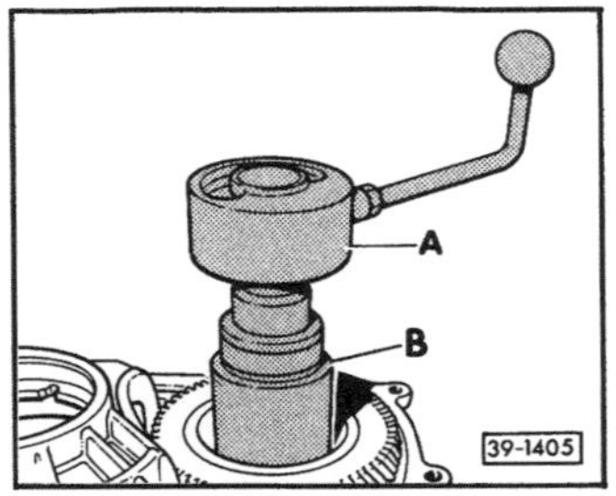

Pinion drive gear bearing, checking turning torque

A = torque wrench, with range of 0-600 Ncm (0-55 in. lb or 0-65 cm kg)

B = 41 mm socket

Note

Lubricate bearings with gear oil before checking.

For **new** tapered roller bearings, the turning torque must be 80-120 Ncm (7.1-10.6 in. lb or 8.2-12.2 cm kg).

Output gear, adjusting

Note

Adjust output gear by shimming it to a specified thickness. Determine the specified thickness from the chart in section 39-110. In order to use the chart, the specified bearing preload must be calculated, using the equation:

specified bearing preload = **c** + **d** - **e**.

Values for **c, d** and **e** are given or determined, as follows:

Dimension c, constant

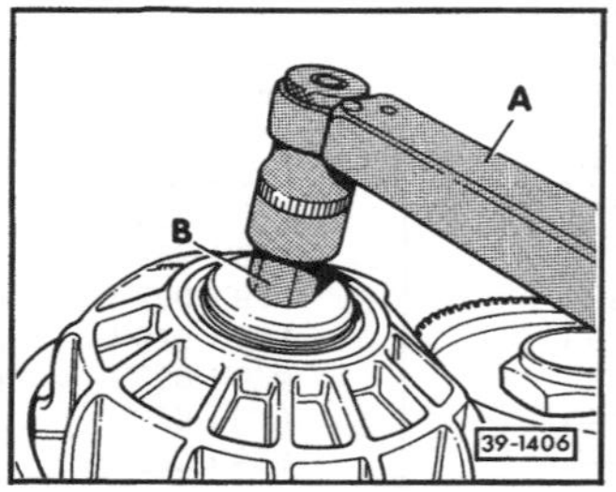

A = torque wrench

B = 22 mm socket

- tighten tensioning bolt to 100 Nm (74 ft lb)
  - **do not** install dished washer (thickness **c**), and shim, at this time
  - parking pawl must be engaged
- remove tensioning bolt

Dimension d, determining

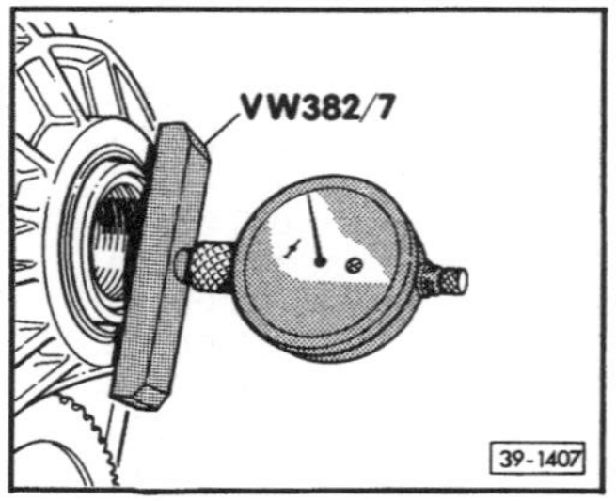

- install dial gauge along with measuring bar **VW 382/7**
  - zero dial gauge with 3 mm preload
- measure **d** = distance between drive gear and inner race of tapered roller bearing
- record dial gauge reading

Example

d = 1.00 mm

Dimension e, constant

Bearing preload **e** = 0.18 mm

Specified bearing preload, calculating

Example

c = thickness of dished washer	1.50 mm
d = dial gauge reading	+1.00 mm
e = bearing preload	−0.18 mm
	= 2.32 mm

Specified shim value, determining

Note

Using the specified bearing preload just calculated, find the correct adjustment shim value in the chart, section 39-110.

Due to varying tolerances, check all shims with a micrometer so that exact thickness can be installed.

Adjustment shim, installing

- select adjustment shim from chart and place on output shaft
- install axial bearing and dished washer
- install tensioning bolt
  - tighten to 250 Nm (184 ft lb)

Output gear bearing, checking turning torque

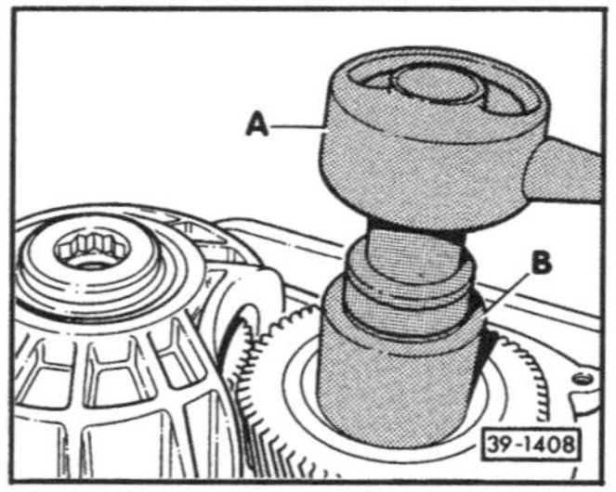

A = torque wrench with range of 0-600 Ncm (0-55 in. lb or 0-65 cm kg)

B = 41 mm socket

Note

Lubricate bearings with gear oil, before checking.

For **new** tapered roller bearings, the turning torque must be 180-220 Ncm (15.9-19.5 in. lb or 18.3-22.5 cm kg) **above** the turning torque for the output shaft.

Example

Output shaft turning torque	100 Ncm (8.6 in. lb or 10.2 cm kg)
Pinion shaft turning torque (mean value)	+200 Ncm (17.7 in. lb or 20.4 cm kg)
	= 300 Ncm (26.6 in. lb or 30.6 cm kg)

Differential, adjusting

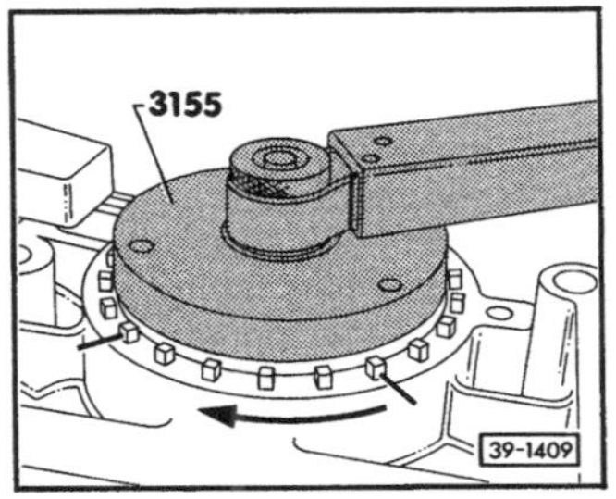

- install bearing housing, up to stop
  - tighten to 150 Nm (111 ft lb) and secure with bolt/retainer
- tighten adjusting ring to 50 Nm (37 ft lb) and mark position

CAUTION

Do not allow differential to turn during procedure, or bearings will settle and reading will be incorrect.

- tighten adjusting ring five (5) locking lugs farther (**arrow**) and secure with bolt/retainer
  - preload is now adjusted

Differential bearings, checking turning torque

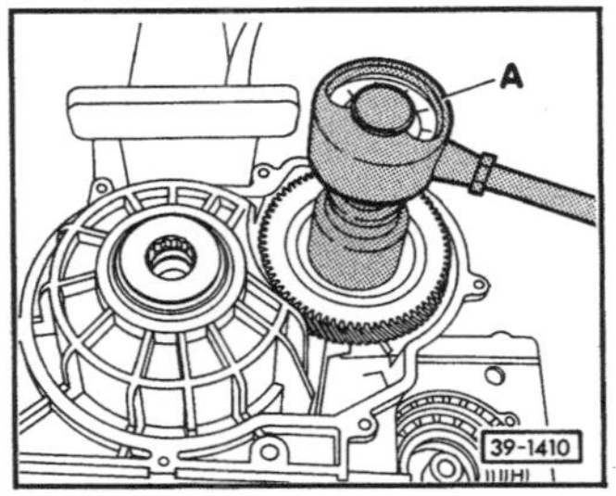

A = torque wrench, with range of 0-600 Ncm (0-55 in. lb or 0-65 cm kg)

Note

Lubricate bearings with gear oil before checking.

For new tapered roller bearings, the turning torque must be 60-80 Ncm (5.3-7.1 in. lb or 6.1-8.2 cm kg) **above** the torque of the pinion shaft.

Example

Pinion shaft turning torque	100 Ncm (8.9 in. lb or 10.2 cm kg)
Differential turning torque (mean value)	+70 Ncm (6.2 in. lb or 7.1 cm kg)
	= 170 Ncm (15.0 in. lb or 17.3 cm kg)

THIS PAGE INTENTIONALLY LEFT BLANK

THIS PAGE INTENTIONALLY LEFT BLANK

Index

Vehicles with 4 cyl. engine

Axle shaft
- assembly 40.11
- length 40.13

Boot
- installing 40.13

Coil spring
- removing/installing 40.10

Control arm bushing
- installing 40.4
- pressing out/in 40.3

Engine/transmission
- supporting 40.4

Front suspension
- assembly 40.1 40.6

Inner CV joint
- assembling/checking 40.16
- checking/disassembling 40.15
- pressing off 40.12
- pressing on 40.13

Outer CV joint
- assembling 40.15
- checking/disassembling 40.14
- removing 40.12

Steering knuckle/suspension strut
- separating 40.7

Subframe mounting
- repairing 40.5

Suspension strut
- assembly 40.9

Suspension strut nut
- removing/installing 40.7

Vibration damper
- removing/installing 40.12

Wheel bearing
- pressing out/in 40.8

Wheel hub
- pressing in 40.8
- pressing out 40.7

Index

Vehicles with VR6 engine

Axle shaft
- assembly **40.28**
- length **40.32**
- removing/installing **40.24**

Ball joint
- checking **40.19a**
- removing/installing **40.19b**

Coil spring
- removing/installing **40.27**

Control arm bushings
- removing/installing **40.19**

Front subframe
- assembly **40.17**

Front suspension
- assembly **40.20**

Inner CV joint
- greasing **40.33**
- removing/installing 40.30

Outer CV joint
- greasing **40.33**
- removing/installing **40.30**

Suspension strut
- assembly **40.26**
- nut, removing/installing **40.22**

Wheel bearing/hub
- installing **40.23**
- removing **40.22**

CAUTION

If vehicles have to be moved after removing an axle shaft, an outer CV joint must be installed to prevent damage to the wheel bearing.

CAUTION

DO NOT attempt to weld or straighten load-carrying components of the front suspension or steering system.

CAUTION

Always replace locking nuts during suspension repairs.

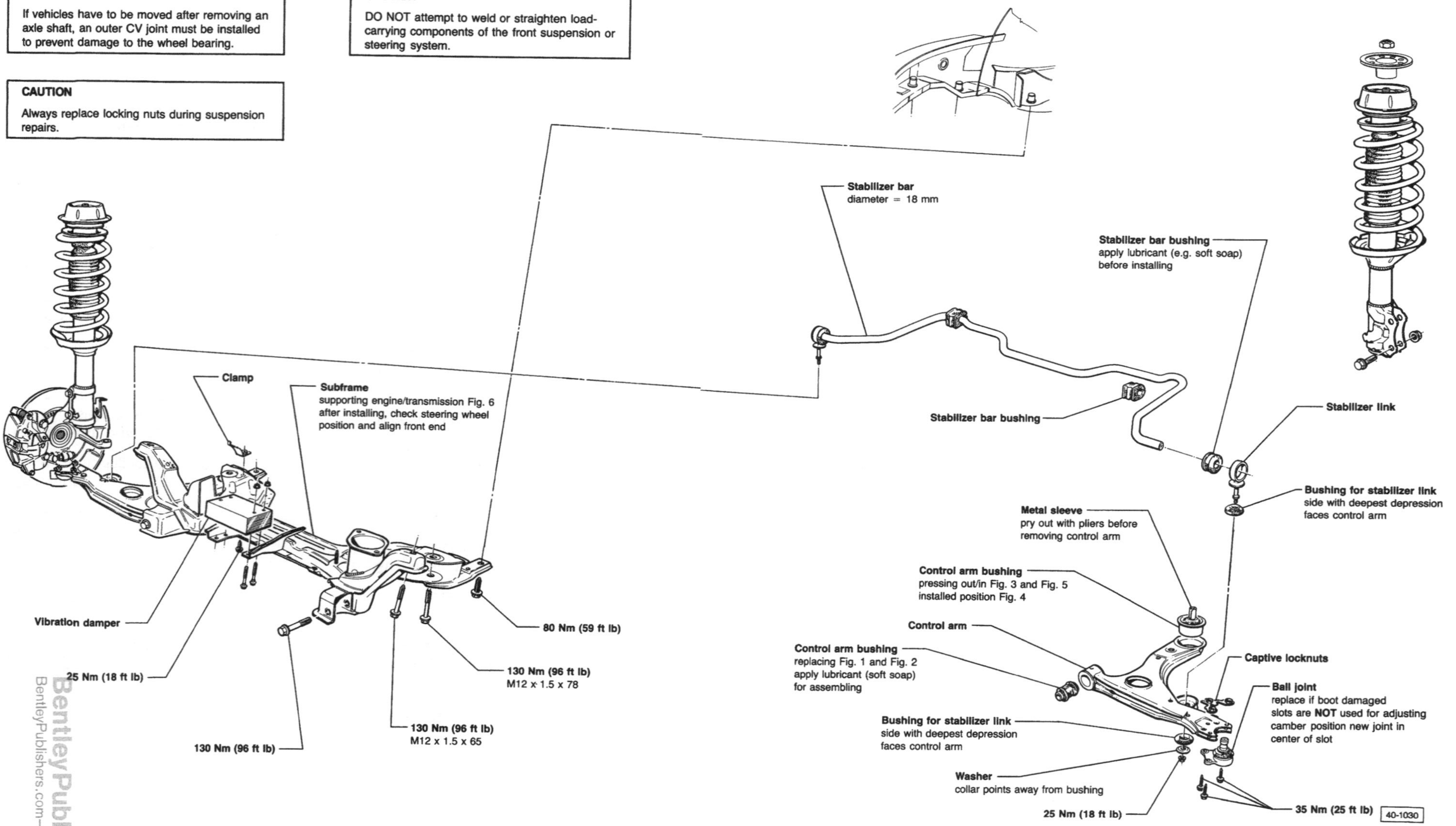

Front suspension assembly

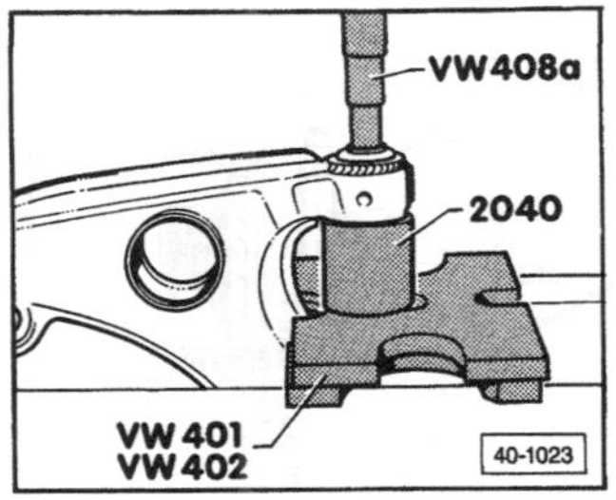

▶ Fig. 1 **Control arm bushing, pressing out**

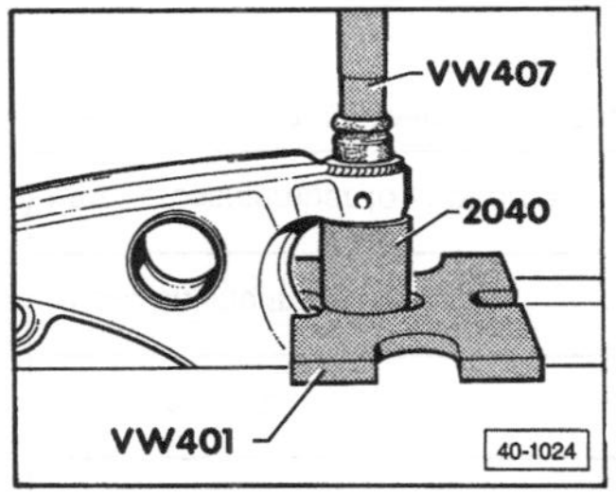

▶ Fig. 2 **Control arm bushing, pressing in**

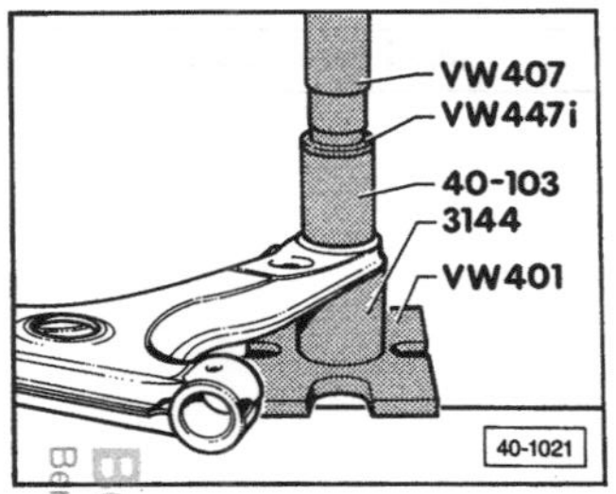

▶ Fig. 3 **Control arm bushing, pressing out**

- if bushing is too corroded to press out, cut out rubber and saw through jacket, then press out

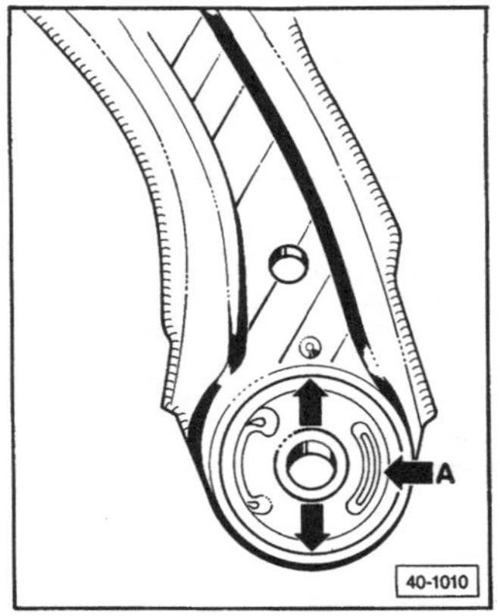

▶ Fig. 4 **Control arm bushing, installed position**

- one imprinted arrow on bushing points to dimple on control arm
- kidney-shaped cavity (**arrow A**) points to center of vehicle

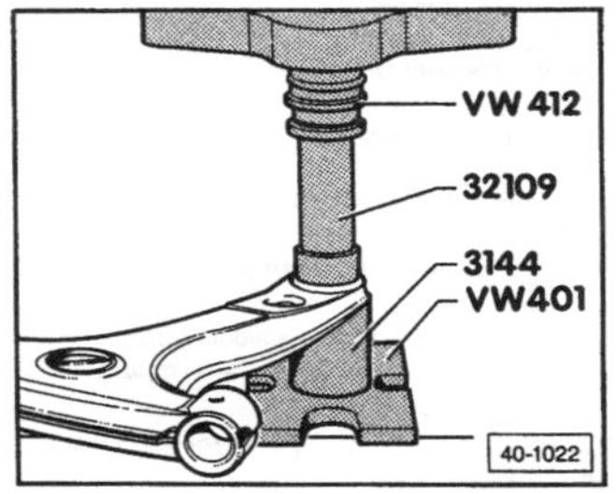

▶ Fig. 5 **Control arm bushing, pressing in**

- tool **30-14** can also be used instead of **3144**

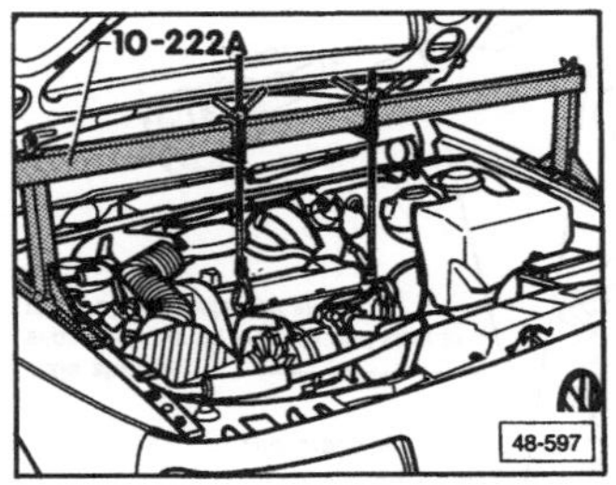

▶ Fig. 6 **Engine/transmission, supporting**

Subframe mounting, repairing

Longer bolts are available for repair of the subframe mountings. Repair damaged threads in the longmember and install new bolts using the following procedure:

- support engine/transmission with **10-222A** and **10-222A/1**
- remove subframe
- drill through cap nut in longmember

Note

Maintain straight alignment of drill during drilling operation.

WARNING

Wear protective goggles.

- cut new threads in cap nut
- install subframe using new bolt

Attachment point	Thread	Bolt	Part number
Subframe, front	M12 x 1.5	M12 x 1.5 x 95 (10.9)	**N 10 161 2.01***
Subframe, rear	M12 x 1.5	M12 x 1.5 x 95 (10.9)	**N 10 161 2.01 ***
Strap, engine/transmission	M10	M10 x 38 (10.9)	**N 01 954 1.1**
Front crossmember to subframe	M10	M10 x 75 (10.9)	**N 90 407 8.01 ***

*these part numbers need to be special ordered

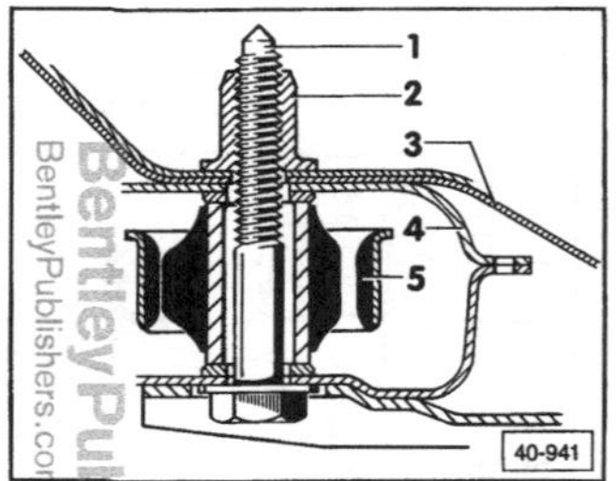

Note

The tightening torque for the replacement bolts remains the same as for the original fasteners.

1 Bolt
2 Cap nut
3 Longmember
4 Subframe
5 Bushing in control arm

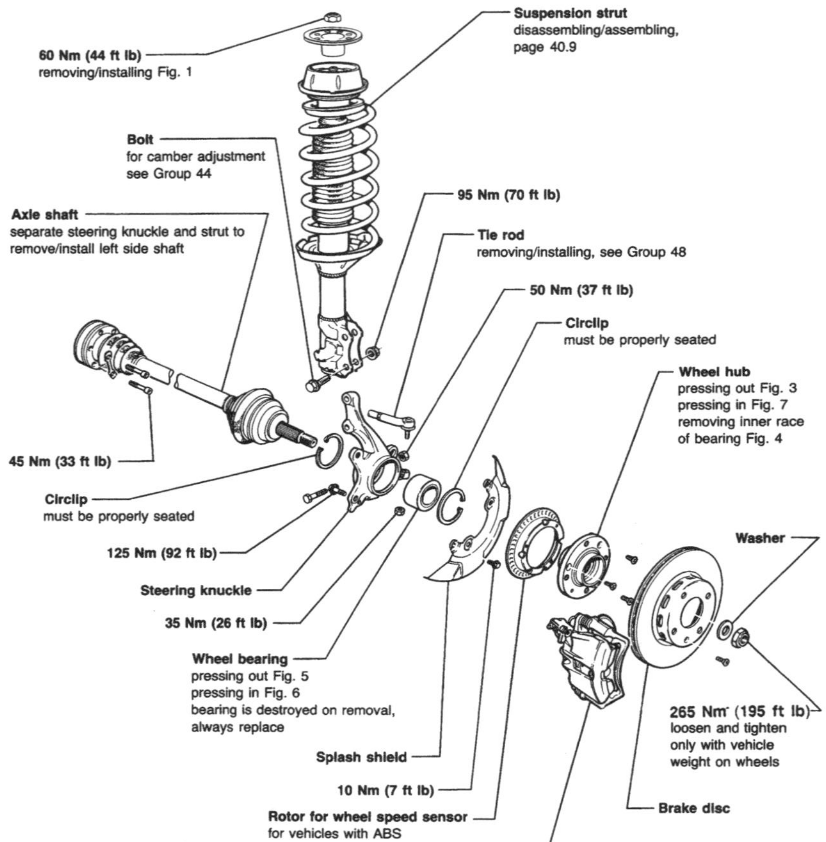

CAUTION

If vehicles have to be moved after removing an axle shaft, an outer CV joint must be installed to prevent damage to the wheel bearing.

CAUTION

Always replace locking nuts during suspension repairs.

CAUTION

DO NOT attempt to weld or straighten load-carrying components of the front suspension or steering system.

40-1031

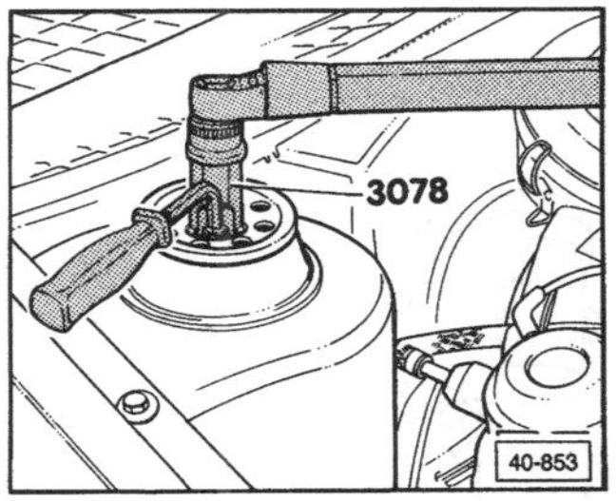

▶ **Fig. 1 Suspension strut nut, removing/installing**

▶ **Fig. 2 Steering knuckle/suspension strut, separating**

- remove bolts **(arrows)** and swing steering knuckle to side

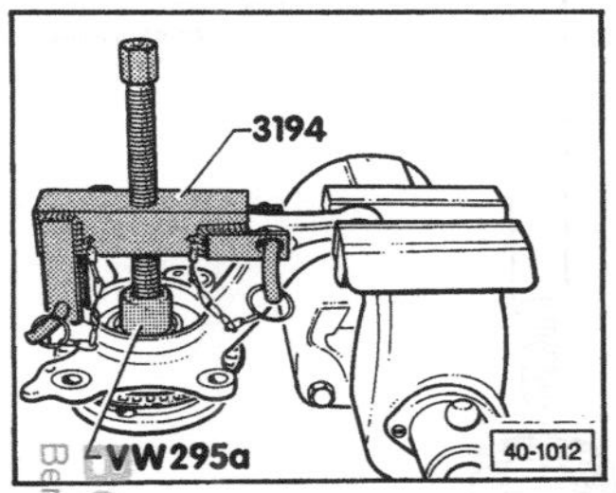

▶ **Fig. 3 Wheel hub, pressing out of steering knuckle**

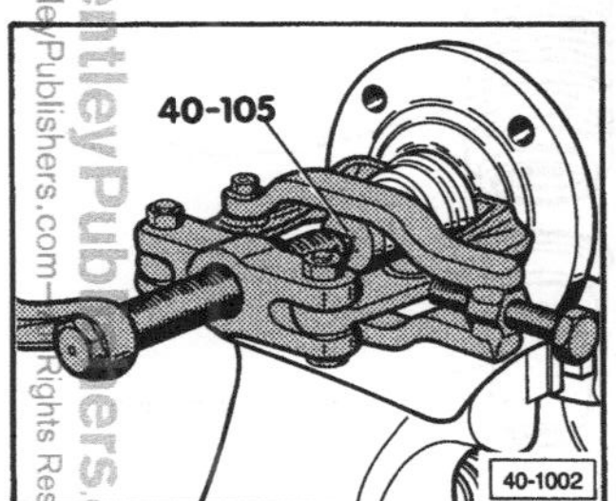

▶ **Fig. 4 Wheel bearing inner race, pulling from wheel hub**

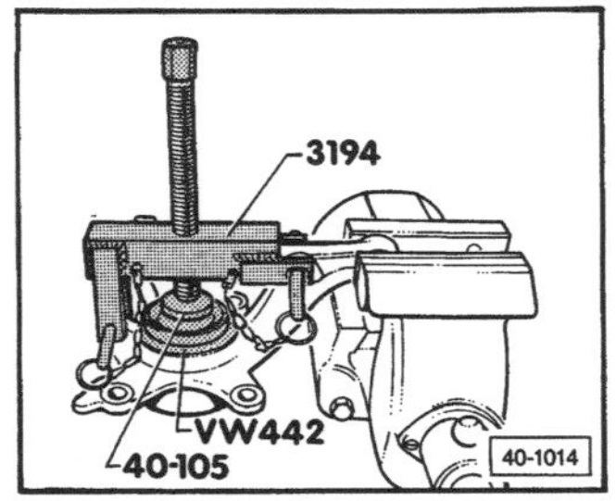

▶ **Fig. 5 Wheel bearing, pressing out**

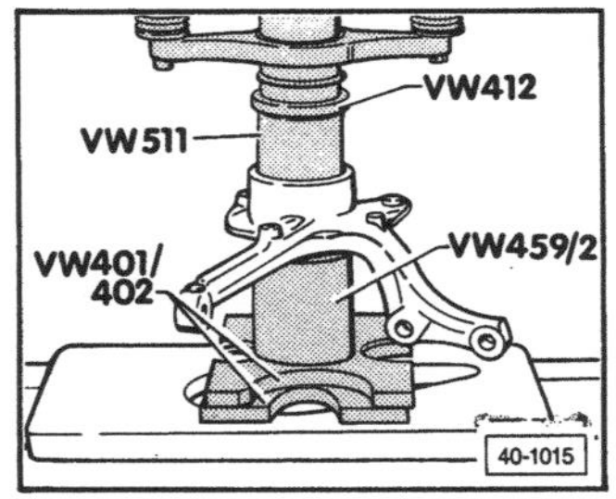

▶ **Fig. 6 Wheel bearing, pressing in**

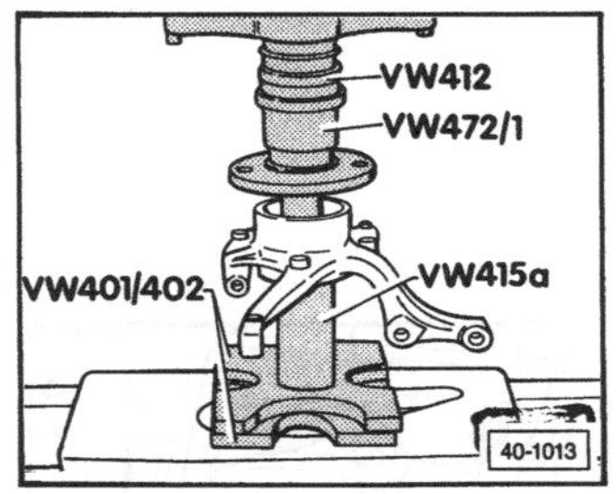

▶ **Fig. 7 Wheel hub, pressing into steering knuckle**

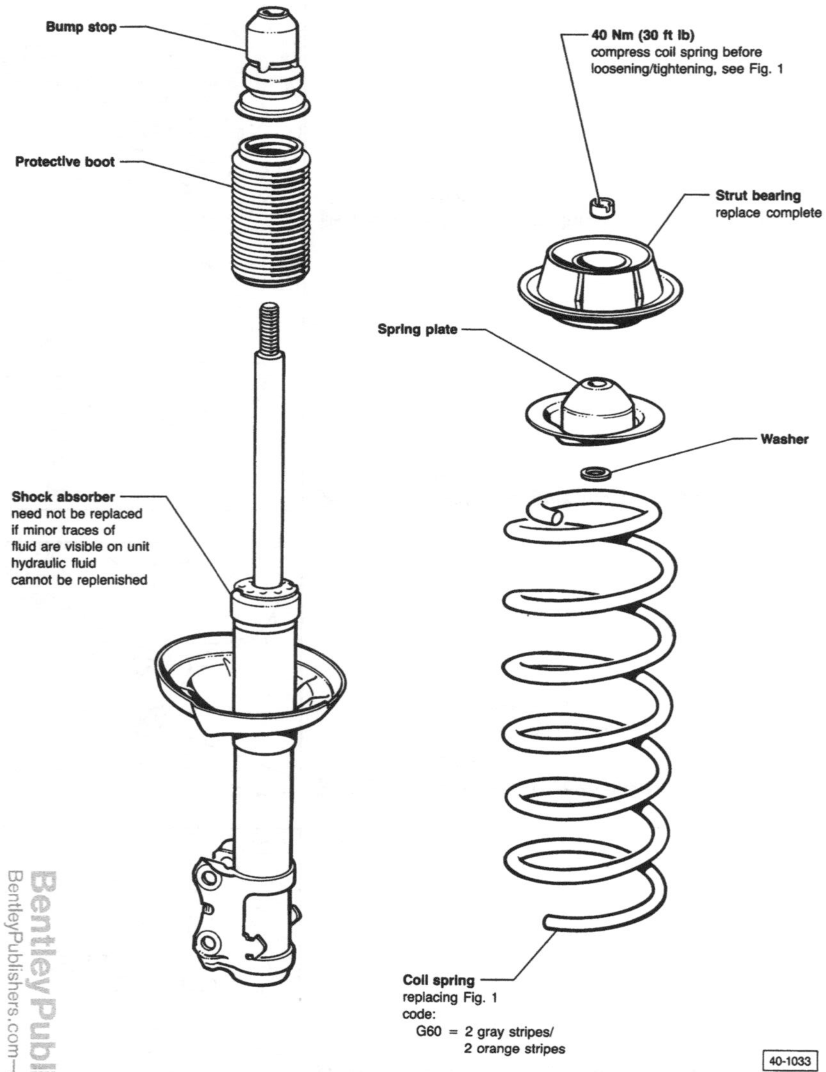

Front Wheel Suspension – Shafts & Axle

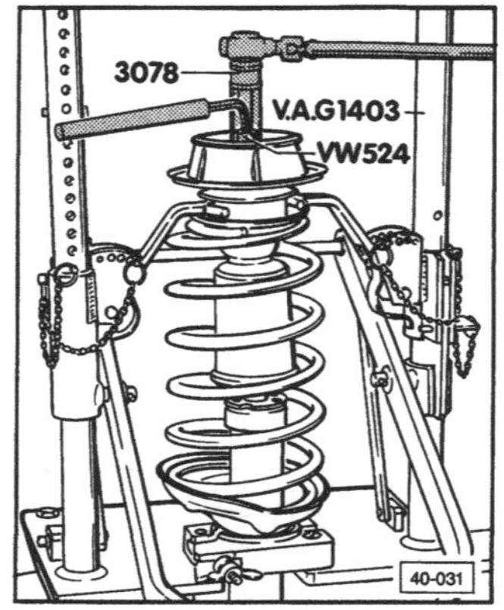

▶ **Fig. 1** **Coil spring, removing/installing**

- tool **VW 340** can also be used to compress spring

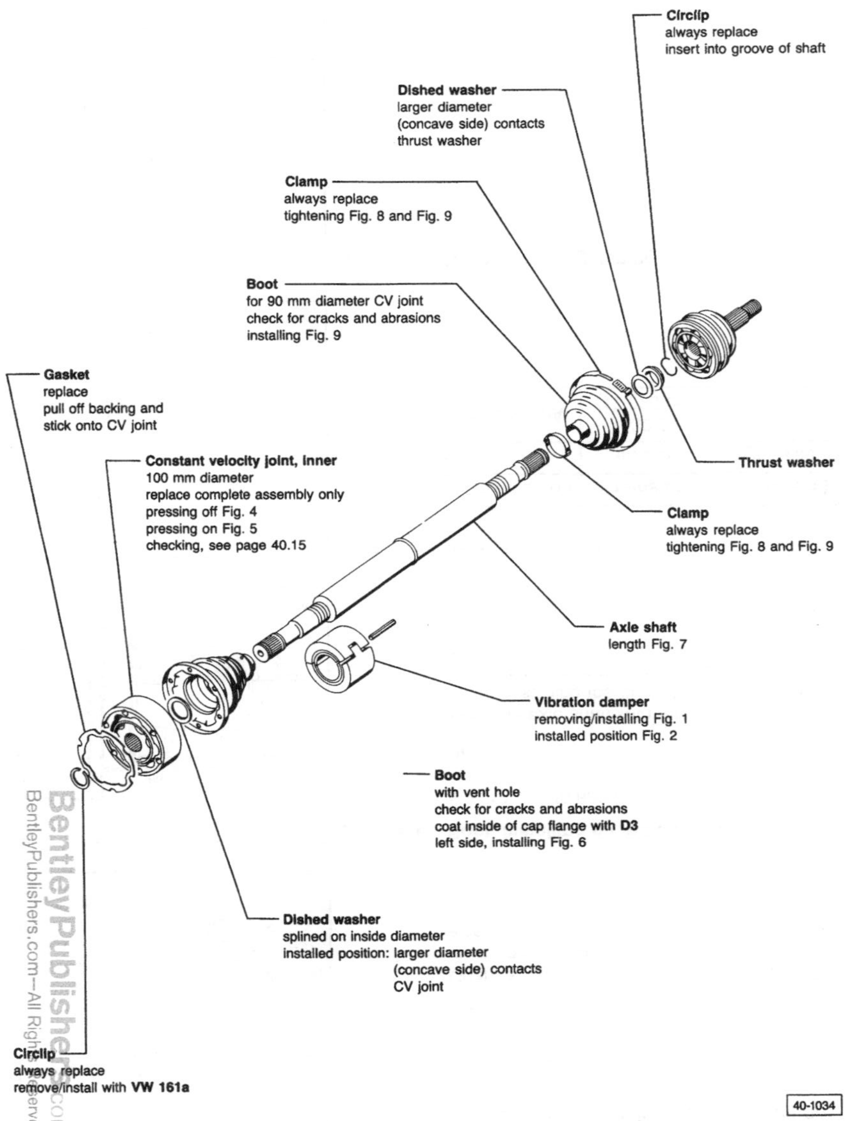

40-1034

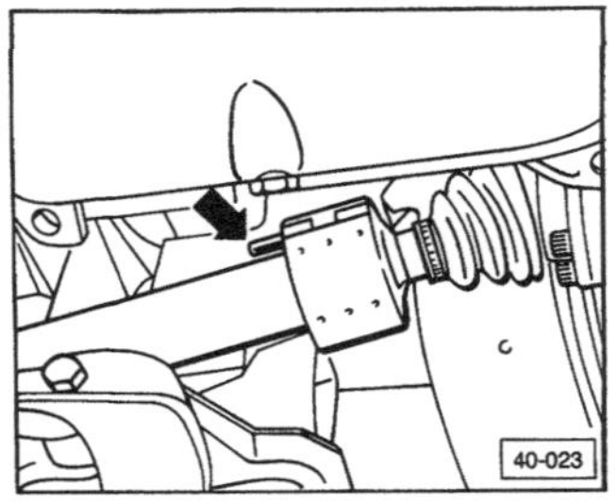
40-023

▶ Fig. 1 **Vibration damper, removing/installing**

- join two halves of damper by driving in spring pins **(arrow)**
  - installed position Fig. 2

Note

An adhesive strip on the inside diameter of the damper prevents it from moving out of position after installation. Check the strength of the adhesive before reinstalling the damper and replace the strip if necessary.

Protect the paint on the driveshaft against damage before driving in the spring pins.

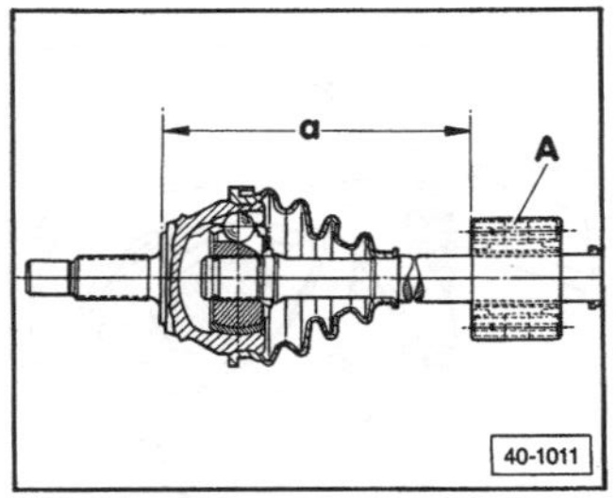

40-1011

▶ Fig. 2 **Vibration damper, installed position**

a = 452.5 + 1 mm (17.815 + 0.039 in.)

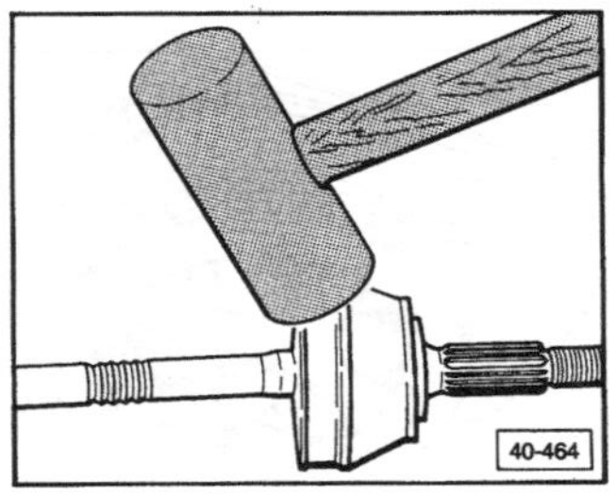
40-464

▶ Fig. 3 **Outer constant velocity joint, removing**

- drive joint off shaft by striking with an aluminum mallet

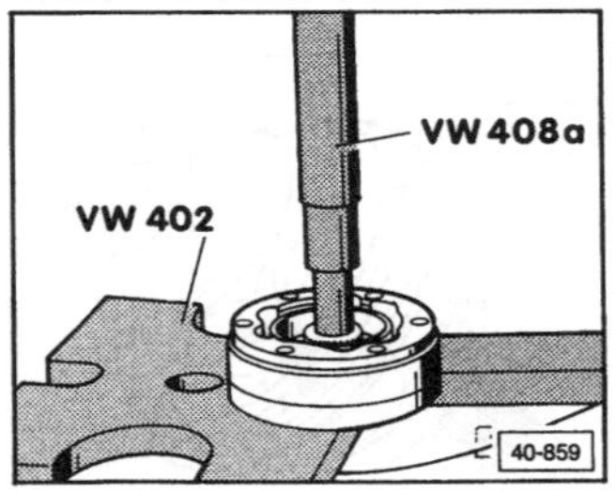

40-859

▶ Fig. 4 **Inner constant velocity joint, pressing off**

- support hub while pressing

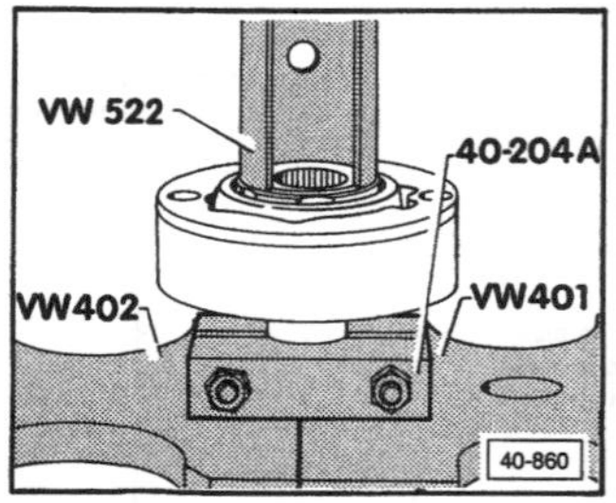

▶ Fig. 5 **Inner constant velocity joint, pressing on**

- press joint until seated
- compress circlip

Note

Chamfer on inside diameter of hub (splines) must point toward the contact shoulder on axle shaft.

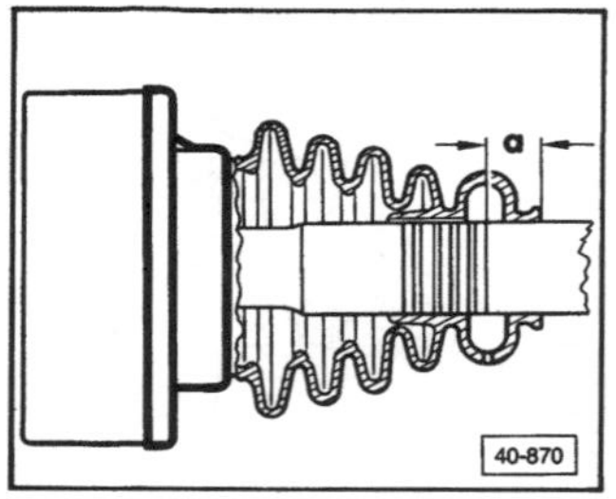

▶ Fig. 6 **Boot, installing on left axle shaft**

a = 17 mm (0.669 in.)

Note

Mark the axle shaft with tape or a crayon before installing the boot. The paint on the axle shaft must not be damaged by the use of a sharp instrument to make the mark.

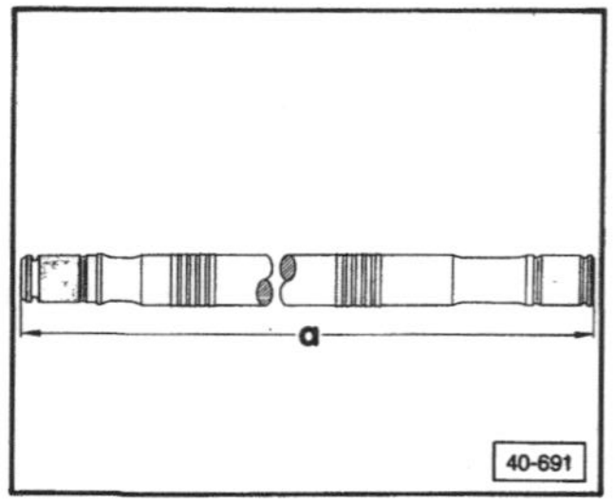

▶ Fig. 7 **Axle shaft, length**

Transmission	Dimension "a" in mm (in.)	
	left side	right side
manual 02A	432.3 (17.020 in.)	666.5 (26.239 in.)

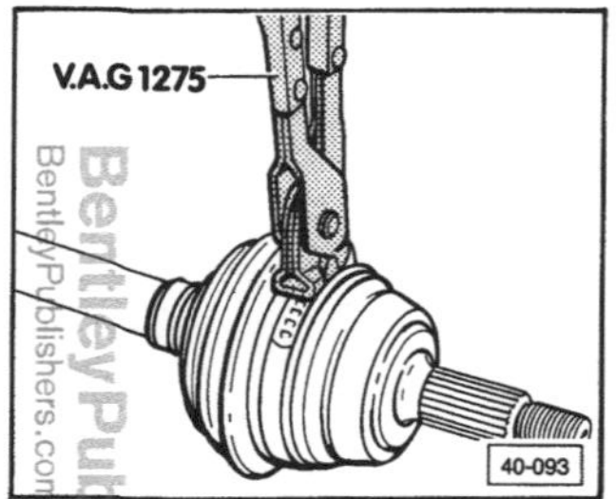

▶ Fig. 8 **Hose clamp, tightening**

▶ Fig. 9 **Boot, installing**

Note

The boot can be pressed in during installation causing an inward fold (**arrow**) during vehicle operation. After installing, pull the small diameter end of the boot open momentarily to allow the pressure to equalize. This venting of the boot will prevent an inward fold.

Outer constant velocity joint, checking

The constant velocity joint must be disassembled to replace heavily contaminated lubricant or to check the race surface and balls for wear.

Disassembling

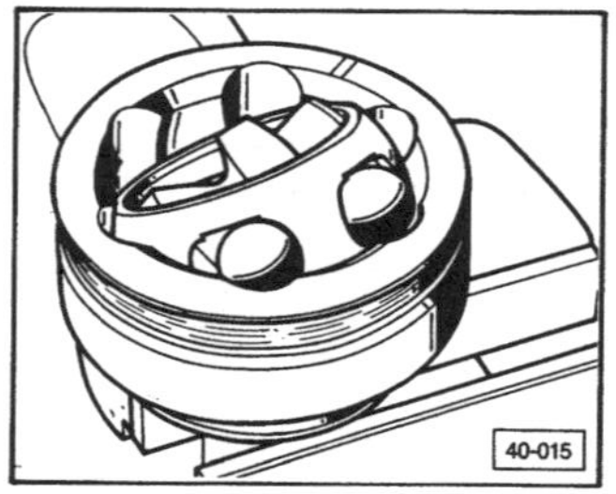

▶

- mark position of ball hub in relation to cage and housing with an engraving tool before disassembling
- pivot ball hub and cage
- remove balls one after another

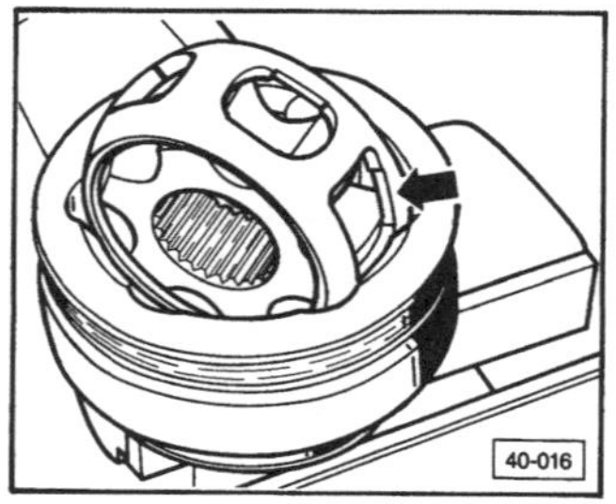

▶

- turn cage until two rectangular openings (**arrow**) are even with joint edge
- remove cage with hub

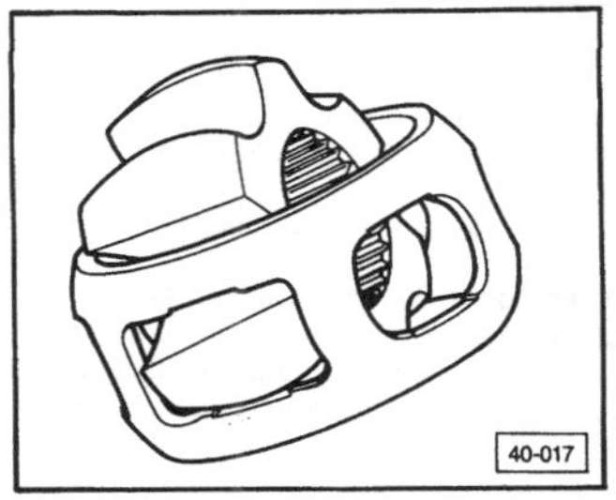

- turn hub until one segment can be pushed into rectangular opening of cage
- tilt hub out of cage

Note

Check housing, hub, cage and balls for pitting. Excessive radial clearance in joint is noticeable when shifting gears. In such cases, replace joint. Do not replace joint because of a polished appearance or because ball track is visible.

Assembling

- fill joint with 45 grams of **G 6.2** grease (half of total amount)
- install cage with hub into housing
  - balls must be installed from opposite side
  - marked position of hub, cage and housing must match
- install new circlip on shaft
- fill joint with remaining **G 6.2** grease (total amount: 90 grams)

Inner constant velocity joint, checking

The inner constant velocity joint must be disassembled to replace heavily contaminated lubricant or to check the race surface and balls for wear.

Disassembling

- pivot ball hub and cage out of outer ring and pull out (**arrow**)
- pull balls out of cage

> **CAUTION**
>
> Six balls of each joint belong to one tolerance group. Do not interchange between joints.

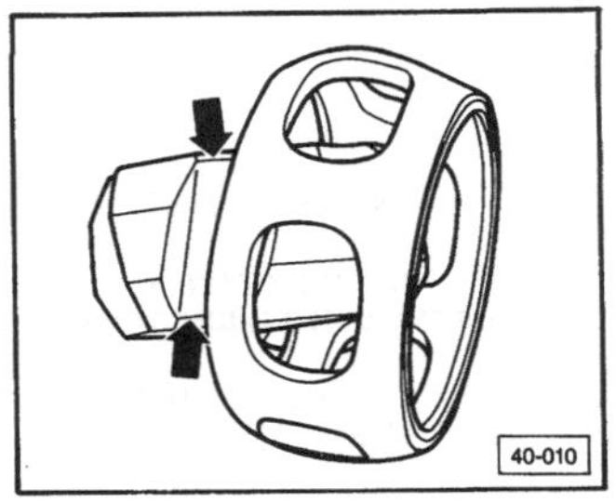

- align two grooves (**arrows**) and take ball hub out of ball cage

Note

Check housing, hub, cage and balls for pitting. Excessive radial clearance in joint is noticeable when shifting gears. In such cases, replace joint. Do not replace joint because of a polished appearance or because ball track is visible.

Assembling

- insert ball hub along both grooves in ball cage
  - installation position not important
- push balls into cage
- insert balls into proper outer ring
  - chamfer on inside diameter of ball hub (splines) must point to contact shoulder on axle shaft and to large diameter of outer ring

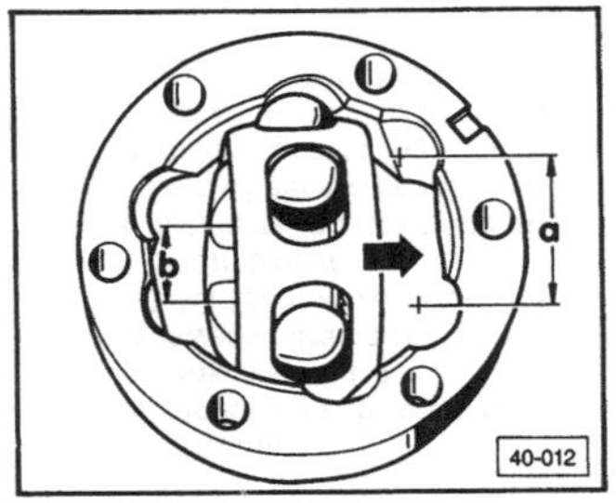

- insert hub with cage and balls into outer ring vertically
  - wide ball groove in outer ring **a**, and a narrow groove in hub **b** are together on one side when hub is swung into outer ring

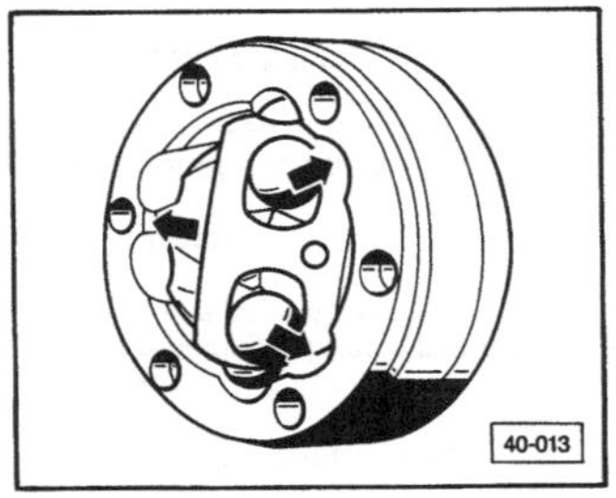

  - when pivoting ball hub with cage and balls in outer ring, hub should be pivoted out of cage (**arrows**) until balls are spaced to fit groove

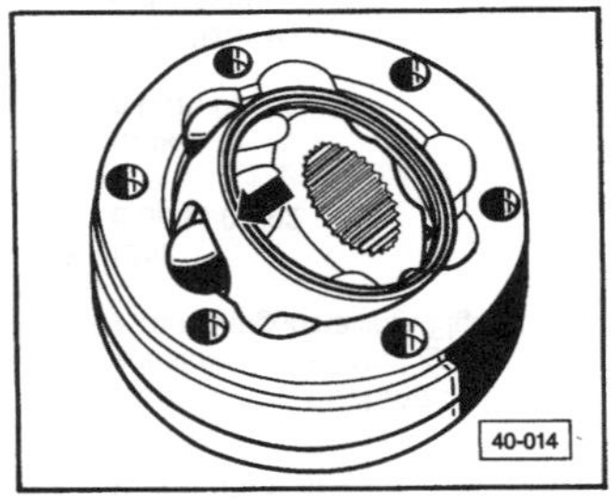

- press cage firmly (**arrow**) until hub swings fully into position

Checking

Joint is correctly assembled when ball hub can be moved in and out over full range of axial movement by hand.

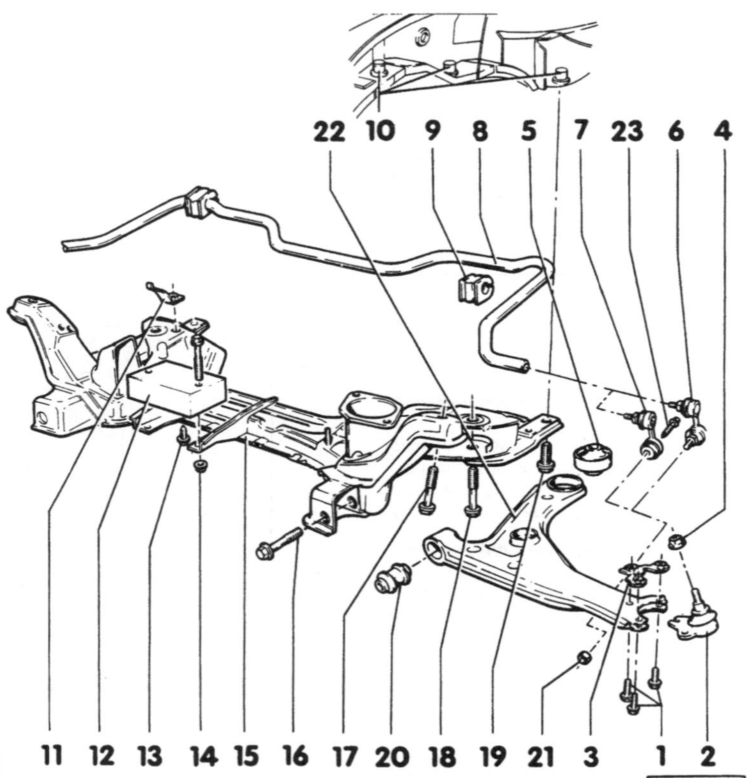

Note

- if the vehicle has to be moved after removing an axle shaft, an outer CV joint must be installed in place of the axle shaft and tightened to 50 Nm (37 ft lb) or wheel bearing will be damaged.
- Wheel alignment specifications, see Group 44
- DO NOT attempt to weld or straighten the load–carrying components of the front suspension
- Always replace self–locking nuts and all corroded nuts and bolts.

1 – **35 Nm (26 ft lb)**

2 – **Ball joint**
- checking, see page 40.19a
- removing/installing, see page 40.19b

3 – **Captive locknuts**

4 – **45 Nm (33 ft lb)**

5 – **Control arm bushing**
- installed position, Fig. 2
- removing/installing, Fig. 3

6 – **Link**
- up to 03/92

7 – **Link**
- from 03/92

8 – **Stabilizer bar**
- diameter: 19 mm

9 – **Stabilizer bar bushing**

10 – **Nut**
- repairing subframe mounting in longmember, see page 40.5

11 – **Clamp**

12 – **Vibration damper**
- not used on vehicles with VR6 engine

13 – **25 Nm (18 ft lb)**

14 – **25 Nm (18 ft lb)**

15 – **Subframe**
- supporting engine/transmission when removing, Fig. 4
- lower subframe with transmission jack VAG 1383 with control arms but without steering gear
- after installing, check steering wheel position and wheel alignment
- removing/installing engine, see Group 10
- not necessary to check wheel alignment if steering gear mounting was not disturbed

16 – **50 Nm (37 ft lb) plus 1/4 turn (90°)**
- M12 x 1.5 x 82

17 – **70 Nm (52 ft lb) plus 1/4 turn (90°)**
- M12 x 1.5 x 65

18 – **70 Nm (52 ft lb) plus 1/4 turn (90°)**
- M12 x 1.5 x 78

19 – **65 Nm (48 ft lb)**

20 – **Control arm bushing**
- removing/installing, Fig.1

21 – **25 Nm (18 ft lb)**

22 – **Control arm**

23 – **Bolt**

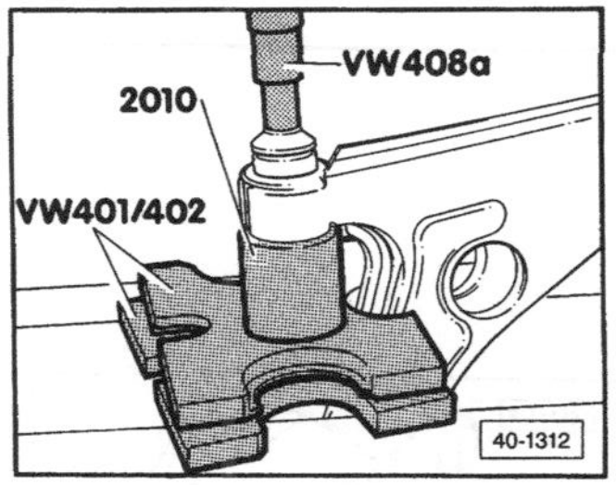

▶ Fig. 1 **Control arm bushing, removing/installing**

- coat bushing with lubricant before pressing in

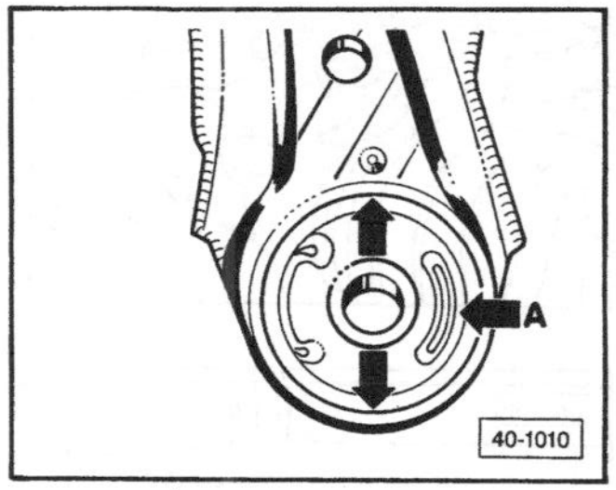

▶ Fig. 2 **Control arm bushing, installed position**

- one arrow on bushing must point toward depression in control arm
- kidney-shaped opening in bushing **(arrow A)** must point toward center of vehicle

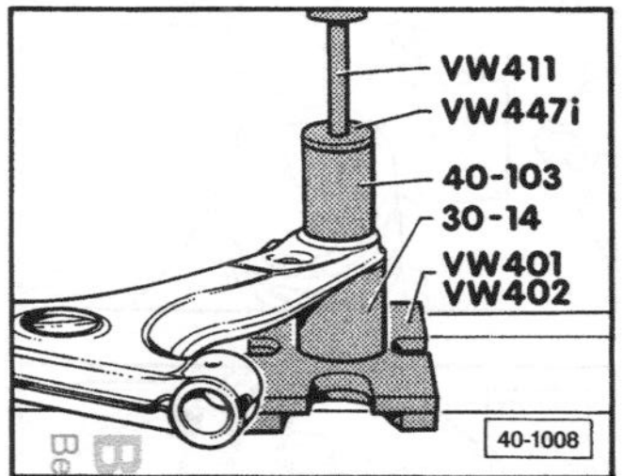

▶ Fig. 3 **Control arm bushing, removing/installing**

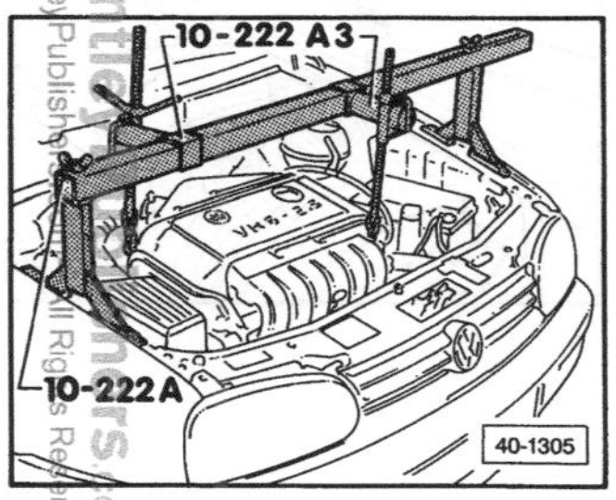

▶ Fig. 4 **Engine/transmission, supporting**

Ball joint, checking

Axial play, checking

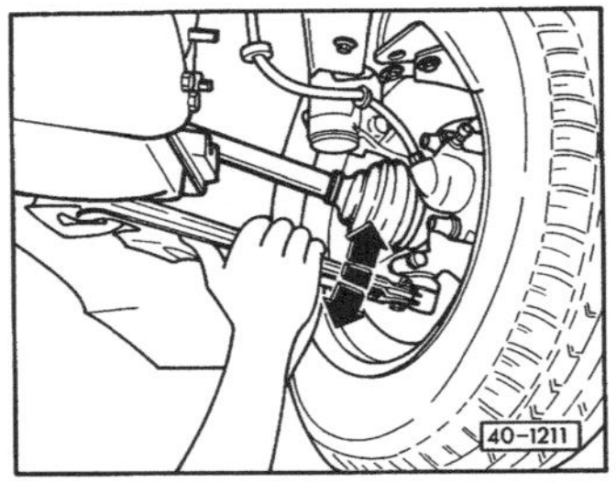

- pull control arm down forcibly and push up again while observing ball joint

Radial play, checking

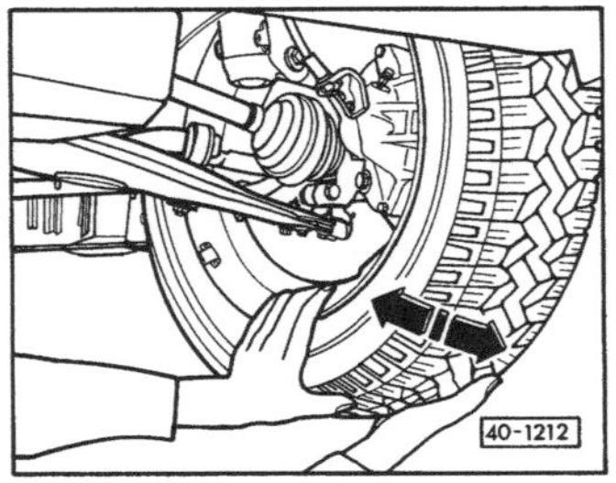

- push wheel inward and outward forcibly at the bottom while observing ball joint
  - no perceptible or visible "play" may exist during either check
  - allow for play of wheel bearings and top strut bushing

Ball joint, removing/installing

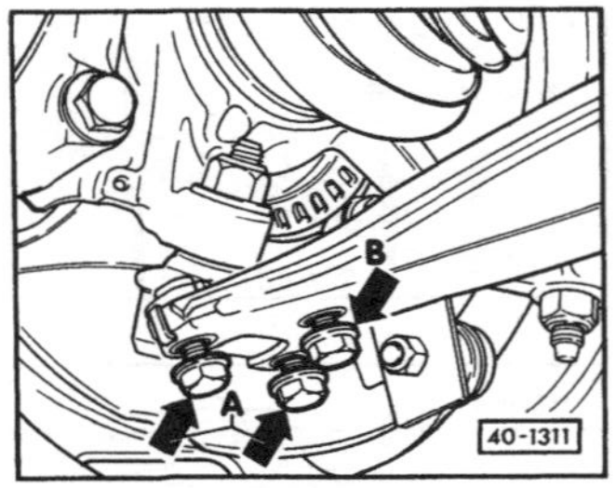

Removing

- remove axle shaft from transmission drive flange
- remove bolts (arrow **A**)
- loosen bolt (arrow **B**)
- mark location of ball joint and separate ball joint from control arm

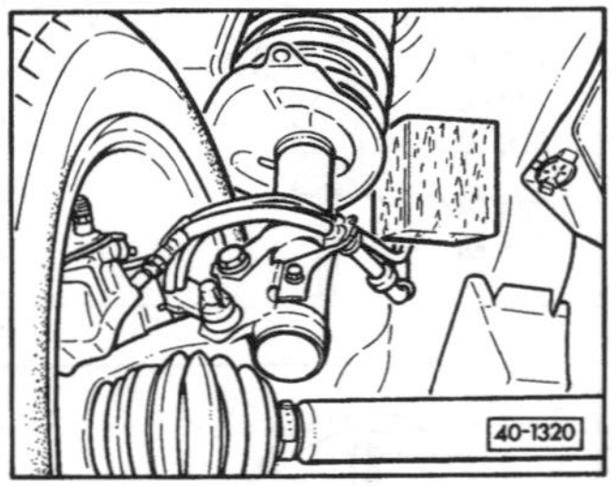

- turn wheel outward and support in position

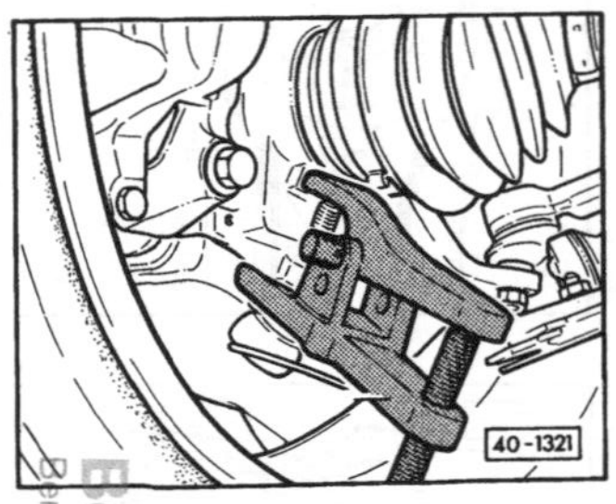

- position extractor tool and press out ball joint
  - leave nut on ball joint a few turns to protect threads

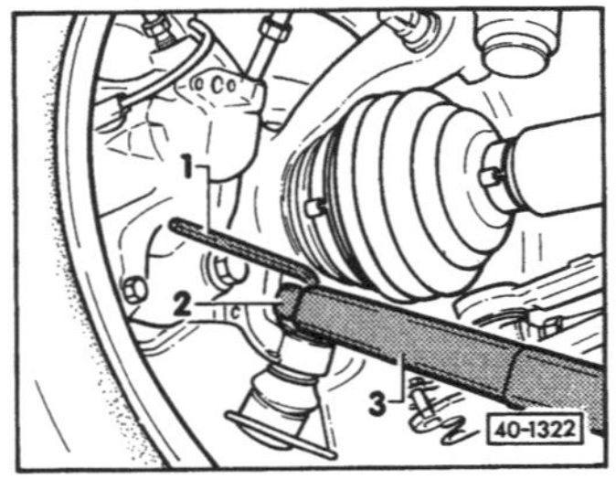

Installing

- insert ball joint into wheel bearing housing
- thread on new self–locking nut and tighten while holding pin of ball joint with 6mm Allen key –1–
- bolt ball joint to control arm after aligning to previously made marks

Tightening torques

Ball joint to wheel bearing housing	45 Nm (33 ft lb)
Ball joint to control arm	35 Nm (26 ft lb)

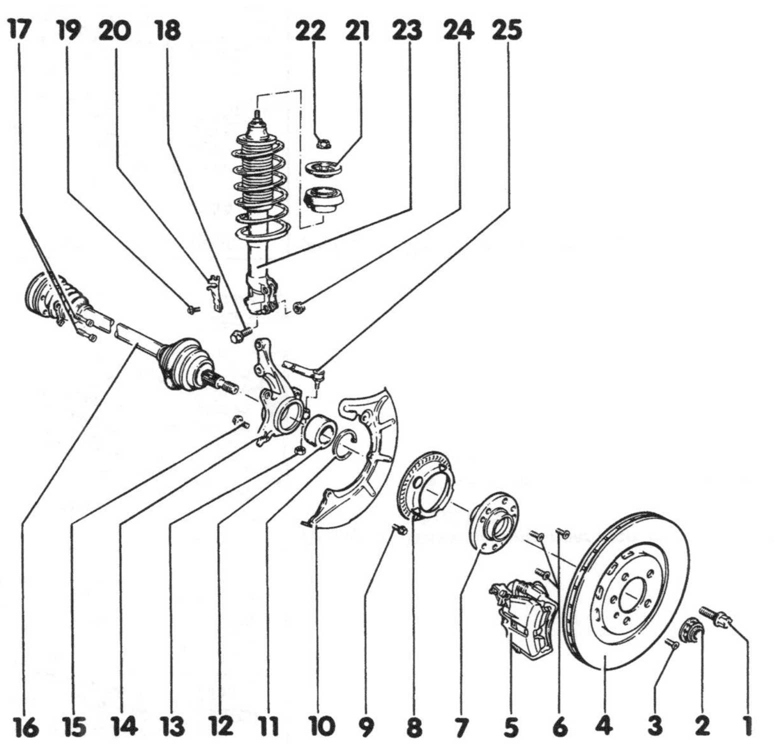

CAUTION
If vehicles have to be moved after removing axle shaft, an outer CV joint must be installed to prevent damage to the wheel bearing.

Always replace self-locking nuts during suspension repairs.

1 – **110 Nm (81 ft lb)**

2 – **90 Nm (66 ft lb) plus 1/8 turn (45°)**
loosen and tighten only with vehicle on ground

3 – **Screw**

4 – **Brake disc**
repair notes, see Group 46

5 – **Brake caliper**
- do not disconnect brake hose when repairing suspension
- do not let caliper hang from brake hose
- replacing brake pads, see Repair Group 46

6 – **Screw**

7 – **Wheel hub**
- pressing out, Fig. 2
- pressing in, Fig. 6
- removing bearing inner race, Fig. 3

8 – **Rotor for wheel speed sensor**

9 – **10 Nm (7 ft lb)**

10 – **Splash shield**

11 – **Circlip**
must be properly seated

12 – **Wheel bearing**
- pressing out, Fig. 4
- bearing is destroyed during removal, always replace

13 – **35 Nm (26 ft lb)**

14 – **Wheel bearing housing**

15 – **125 Nm (92 ft lb)**

16 – **Axle shaft**
- removing/installing, page 40.24
- repairing, page 40.28
- clean threads of outer CV joint before installing nut

17 – **45 Nm (33 ft lb)**

18 – **Bolt**
before loosening, mark installed position

19 – **Bolt**

20 – **Bracket**

21 – **Stop**

22 – **60 Nm (44 ft lb)**
removing/installing, Fig. 1

23 – **Suspension strut**
disassembling/assembling, page 40.26

24 – **95 Nm (70 ft lb)**

25 – **Tie rod**
removing/installing, see Repair Group 48

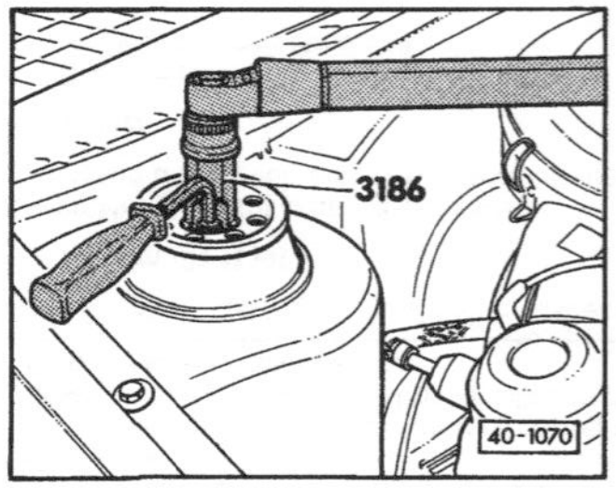

▶ Fig. 1 **Suspension strut nut, removing/installing**

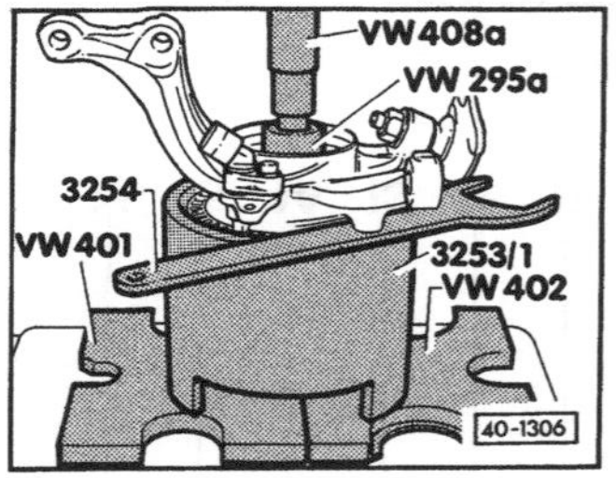

▶ Fig. 2 **Wheel hub, pressing out of wheel bearing housing**

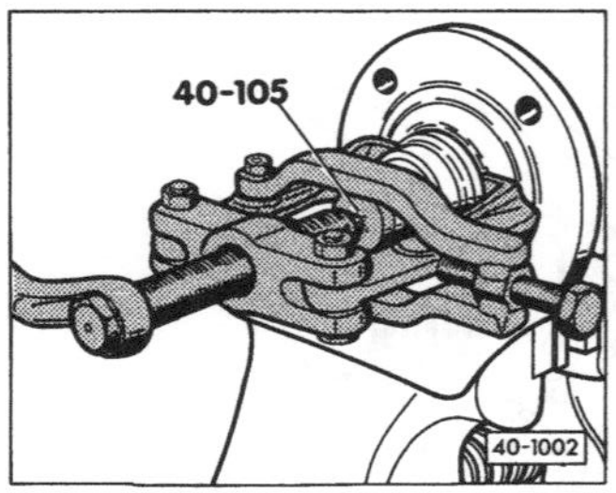

▶ Fig. 3 **Wheel bearing inner race, removing from wheel hub**

- use puller with tensioner, e.g. Kukko 204-2

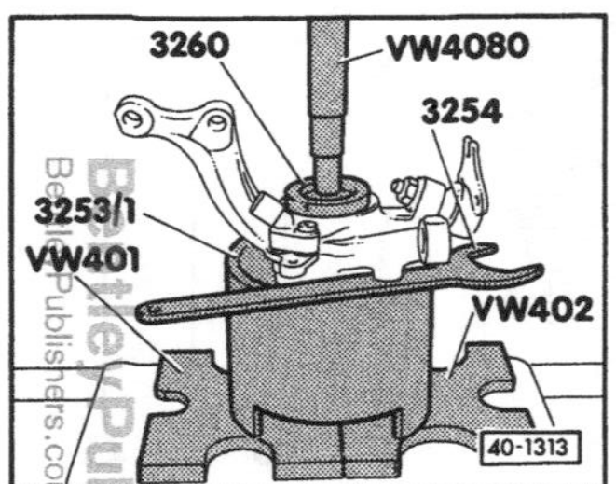

▶ Fig. 4 **Wheel bearing, pressing out**

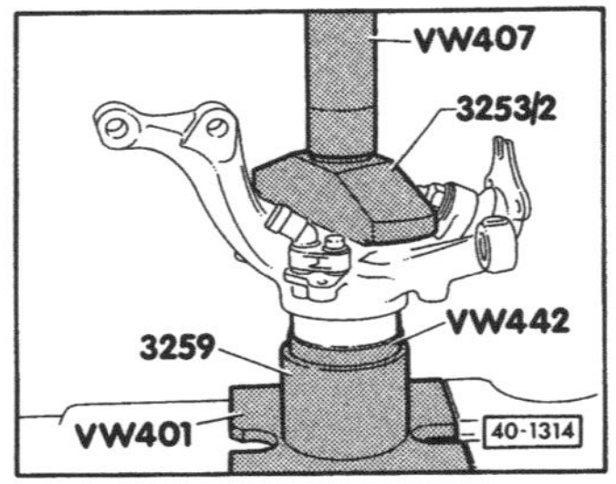

▶ Fig. 5 **Wheel bearing, pressing in**

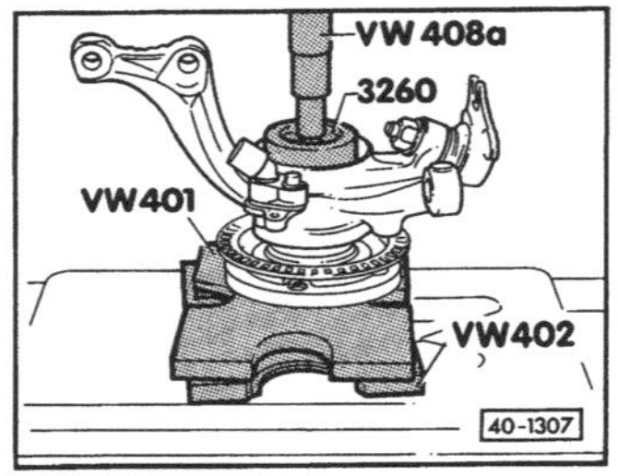

▶ Fig. 6 **Wheel hub, pressing into wheel bearing**

Axle shaft, removing/installing

Removing

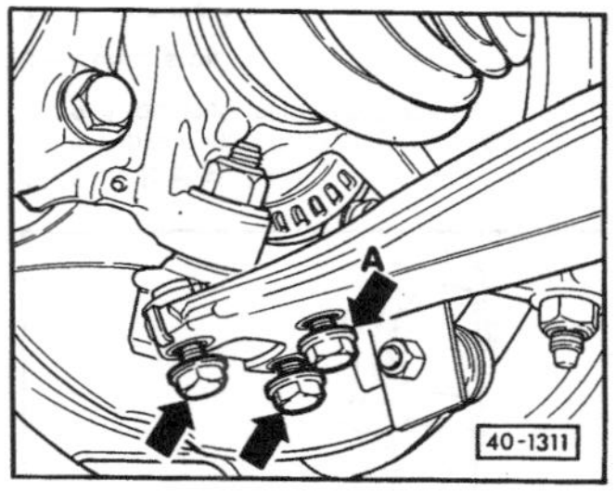

- loosen axle shaft nut with vehicle on ground
- raise vehicle and remove wheel
- mark position of bolt (**arrow A**) for attaching ball joint to control arm
- remove bolts and separate ball joint from control arm

> **CAUTION**
> The camber adjustment will be disturbed if the ball joint is not reinstalled in the exact position in the control arm.

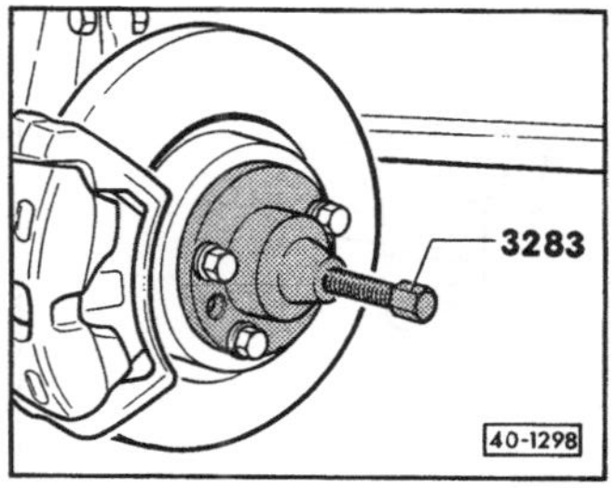

- mount tool and push axle shaft out, checking frequently for sufficient clearance

Installing

- clean splines of outer CV joint and wheel hub thoroughly to remove any lubricant residue

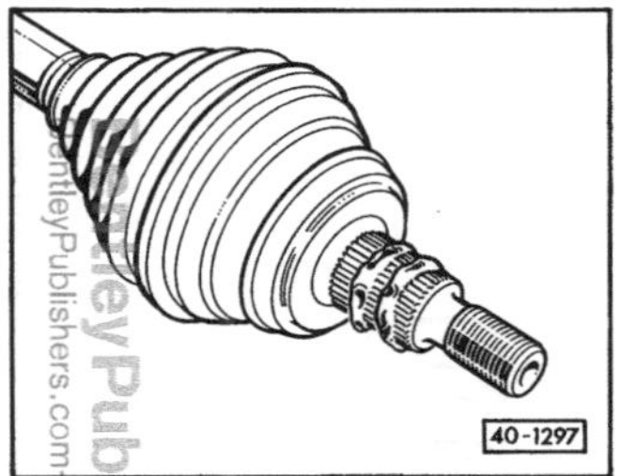

- apply locking compound, Part No. **D 185 400 A2** to splines in two 3.0 mm beads
- install axle shaft
- connect ball joint and control arm
- bolt axle shaft to transmission flange and tighten

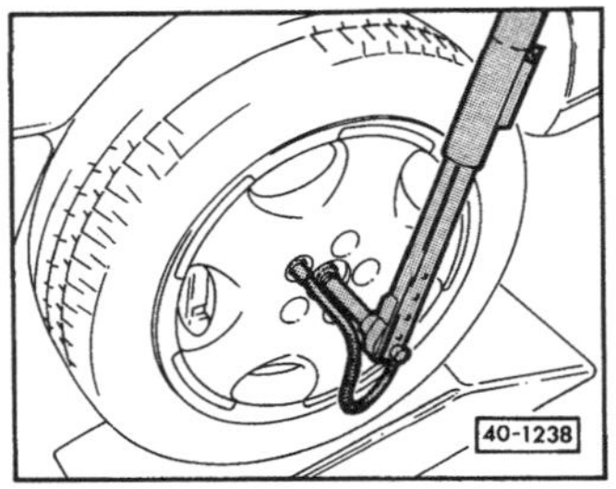

- install axle shaft nut (12 point) and tighten
  - tighten nut only with vehicle on ground

Tightening Torques

Axle shaft to transmission flange **45 Nm (33 ft lb)**

Axle shaft to wheel bearing housing nut **90 Nm (66 ft lb) plus 1/8 turn (45°)**

Ball joint to control arm **35 Nm (26 ft lb)**

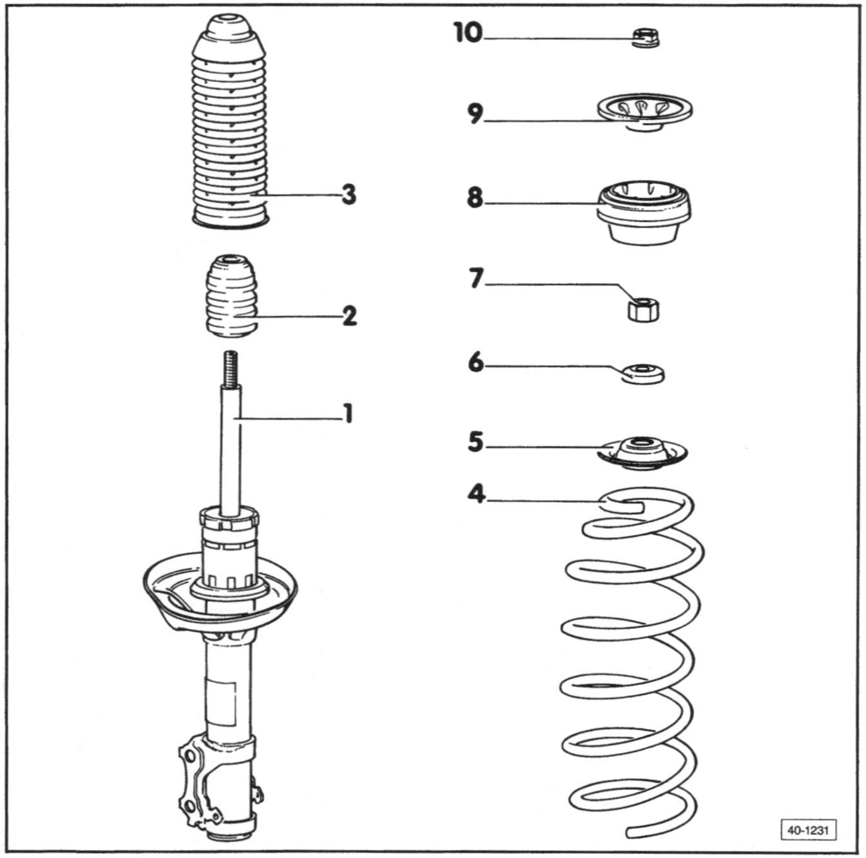

1 – **Shock absorber**
- replace individually if necessary
- if replaced, align front wheels

2 – **Bump stop**

3 – **Protective sleeve**

4 – **Coil spring**
- removing/installing, Fig. 2
- avoid damaging surface of spring

5 – **Spring plate**

6 – **Ball bearing**

7 – **60 Nm (44 ft lb)**

8 – **Strut bearing**

9 – **Stop**

10 – **60 Nm (44 ft lb)**
loosening/tightening, Fig. 1

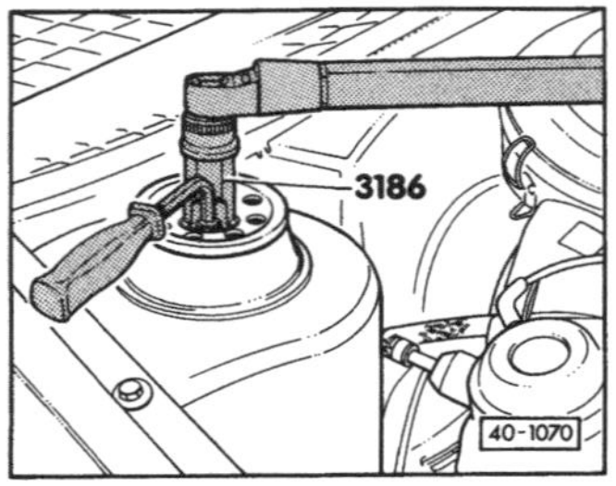

▶ **Fig. 1 Axle nut, loosening/tightening**

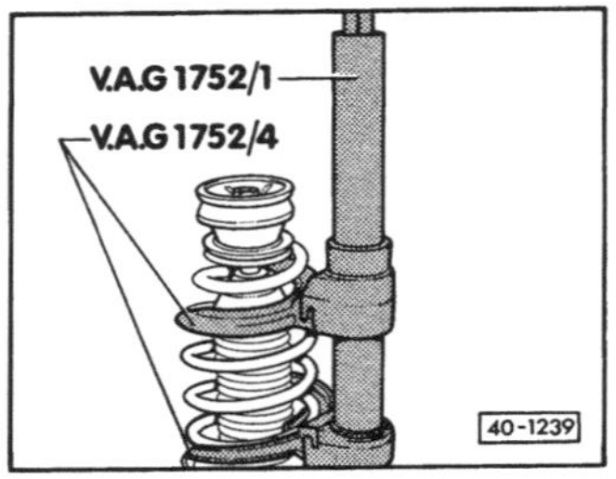

▶ **Fig. 2 Coil spring, removing/installing**

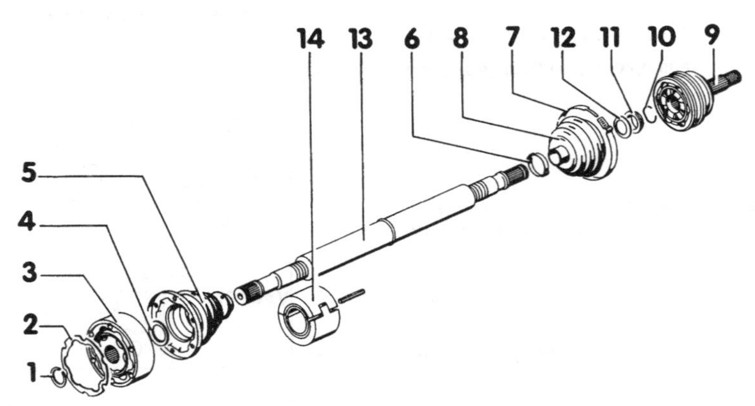

Axle shafts for vehicles with the VR6 engine have different splines for the outer CV joint. Also, the shaft is fastened to the hub/wheel bearing housing with a 12-point nut.

CAUTION
The splines of the outer CV joint and the hub must be bonded with locking compound, Part No. **D 185 400 A2**.

1 – **Circlip**
- always replace
- remove with **VW 161A**

2 – **Gasket**
- always replace
- remove protective film and stick to CV joint

3 – **Constant velocity joint, inner**
- 100 mm diameter
- replace complete assembly only
- pressing off, Fig. 2
- pressing on, Fig. 3
- greasing, see page 40.33

4 – **Dished washer**
- splined on inner diameter
- installed position: concave side contacts CV joint

5 – **Boot**
- with vent hole
- check for cracks and abrasions
- installation position for left side, Fig. 4
- installation position for right side, Fig. 5
- coat inside of cap flange with D3

6 – **Clamp**
- always replace
- tensioning, Fig. 7

7 – **Clamp**
- always replace
- tensioning, Fig. 7

8 – **Boot**
- check for cracks and abrasions
- venting pressure after installation, Fig. 6

9 – **Constant velocity joint, outer**
- replace complete assembly only
- removing, Fig. 1
- installing: drive on with plastic hammer until circlip locks onto shaft
- greasing, see page 40.33

10 – **Circlip**

11 – **Thrust washer**

12 – **Dished washer**
concave side faces thrust washer

13 – **Axle shaft**
- shaft splines 4 mm longer than standard, measured from dished washer to circlip groove
- length of shaft, Fig. 8

14 – **Vibration damper**
- removing/installing, see page 40.12
- installed position, see page 40.12

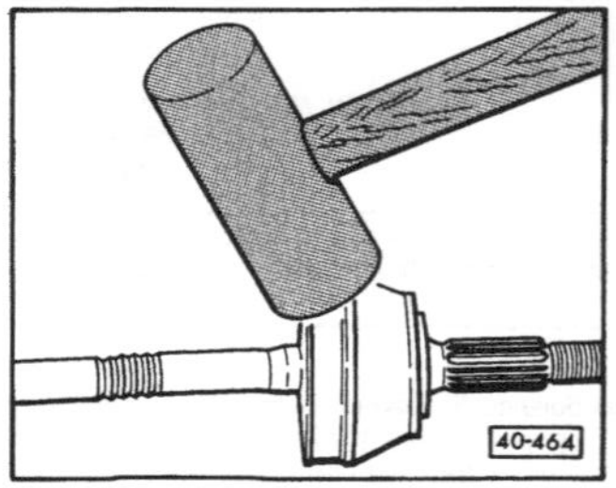

▶ Fig. 1 **Outer CV joint, removing**

- drive joint off shaft by striking with a soft metal mallet

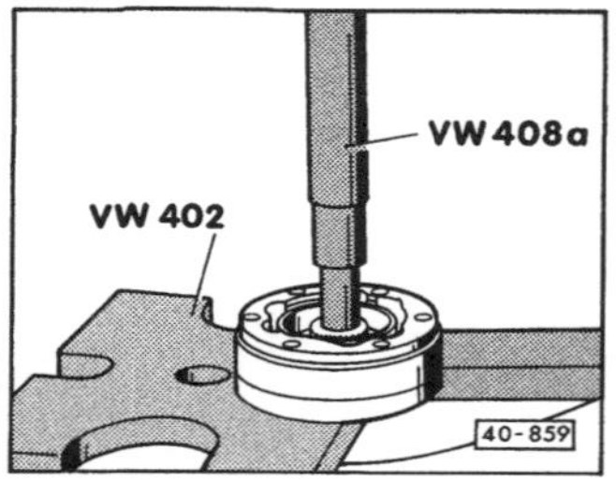

▶ Fig. 2 **Inner CV joint, removing**

- support hub while pressing

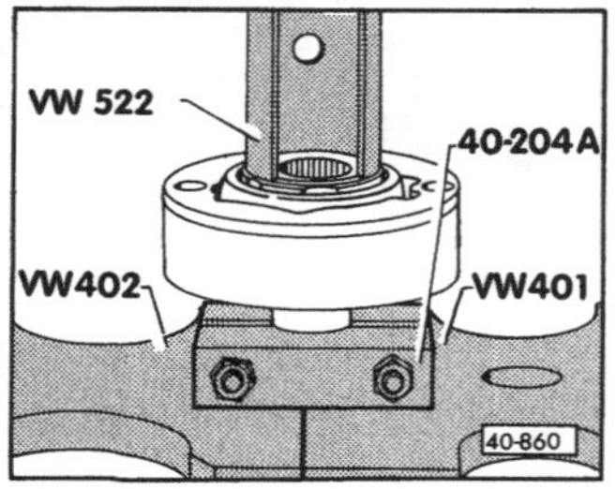

▶ Fig. 3 **Inner CV joint, pressing on**

- press joint on until seated
- compress circlip

Note
Chamfer on inside diameter of hub (splines) must face contact shoulder on axle shaft.

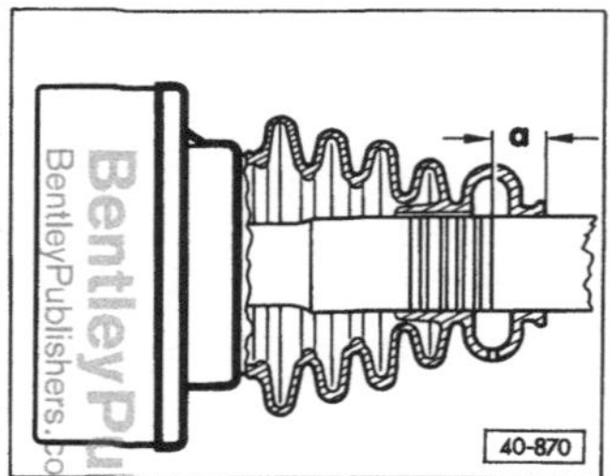

▶ Fig. 4 **Boot, installing on left axle shaft**

- dimension **a** = 17 mm (11/16 in.)

Note
Mark the axle shaft with tape or a crayon before installing the boot. The paint on the axle shaft must not be damaged by the use of a sharp instrument to make the mark.

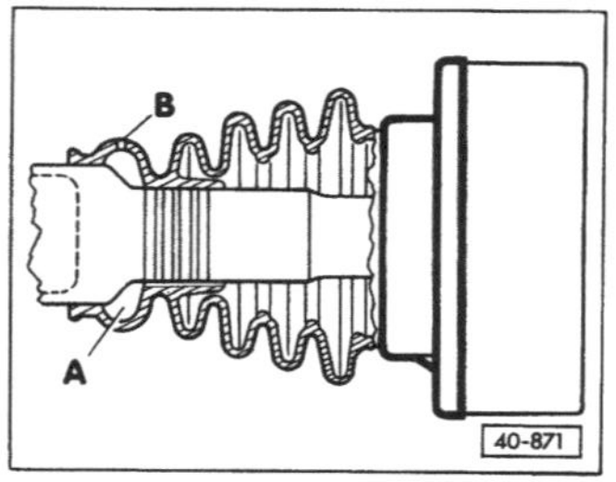

▶ Fig. 5 **Boot, installing on right axle shaft**

- large diameter opening for vent cavity **A** must be installed over large diameter of axle shaft
  - **B** = vent hole

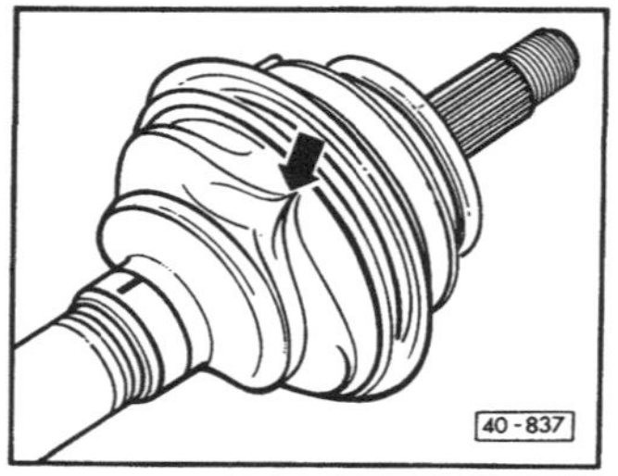

▶ Fig. 6 **Boot, venting pressure after installing**

- pull up boot at small diameter opening to equalize pressure and avoid an inward fold during driving (**arrow**)

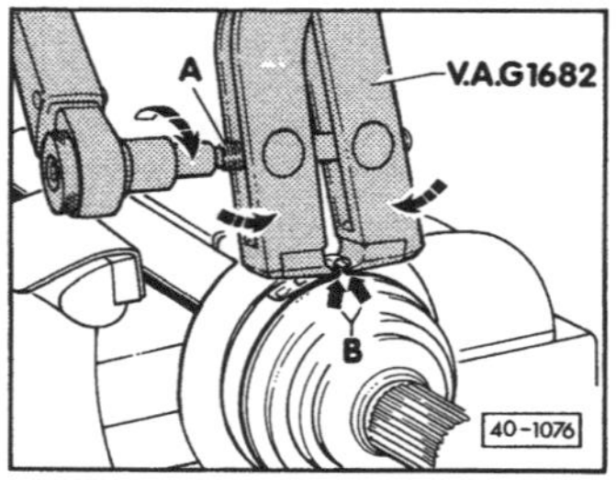

▶ Fig. 7 **CV joint boot clamps, tensioning**

> **CAUTION**
> Hose clamp can only be tensioned using pliers **VAG 1682** or Snap On **YA 3080**.
>
> Ensure that jaws of pliers contact the corners of the hose clamp (**arrows B**).

- tension hose clamp by turning spindle to a torque of 25 N·m (18 ft lb)

Note
Coat the threads of the tool spindle with MoS_2 grease to ensure smooth operation and provide the required torque to the clamp.

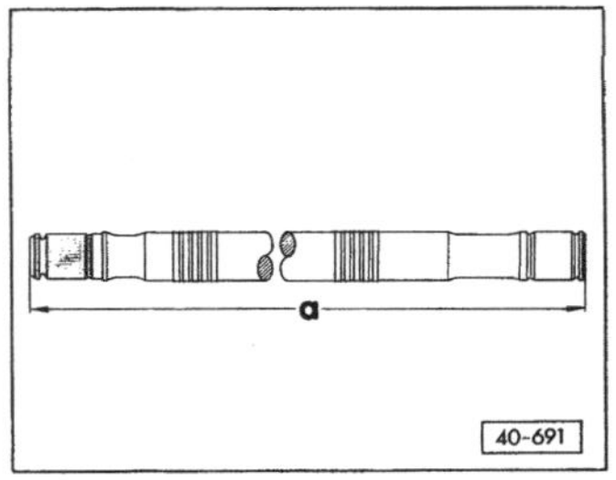

▶ **Fig. 8 Axle shaft, length**

Transmission	Dimension "a" in mm	
	left side	right side
02A	447.4	681.5
096	447.4	681.5

CV joints, greasing

Use Part No. **G 6.3** to grease CV joints.

Outer	Inner
total: 110 grams	total: 120 grams
press 55 grams into joint and spread 55 grams evenly in boot	press 60 grams into joint from both sides and spread 60 grams evenly in boot

Note
When replacing the CV joint boot, regrease the CV joint.

THIS PAGE INTENTIONALLY LEFT
BLANK

Index

Axle beam bushings
- removing/installing **42.3**

Axle beam mounting bracket
- installing **42.3**

Axle mounting
- repairing **42.4**

Cover ring
- installed position **42.9**

Grease cap
- removing/installing **42.8**

Outer/inner wheel bearing
- outer races, installing **42.9**

Rear axle
- assembly **42.2**
- installing **42.4**

Seal
- installing **42.10**

Stub axle/disc brake
- assembly **42.7**

Support
- installed position **42.6**

Suspension strut
- assembly **42.5**
- removing/installing **42.4**

Wheel bearing play
- adjusting **42.9**

CAUTION

Do not attempt to weld or straighten rear axle beam or stub axles.

WARNING

Do not re-use any fasteners that are worn or deformed in normal use. Many fasteners are designed to be used only once and become unreliable and may fail when used a second time. This includes, but is not limited to, nuts, bolts, washers, self-locking nuts or bolts, circlips, cotter pins. Always replace these fasteners with new parts.

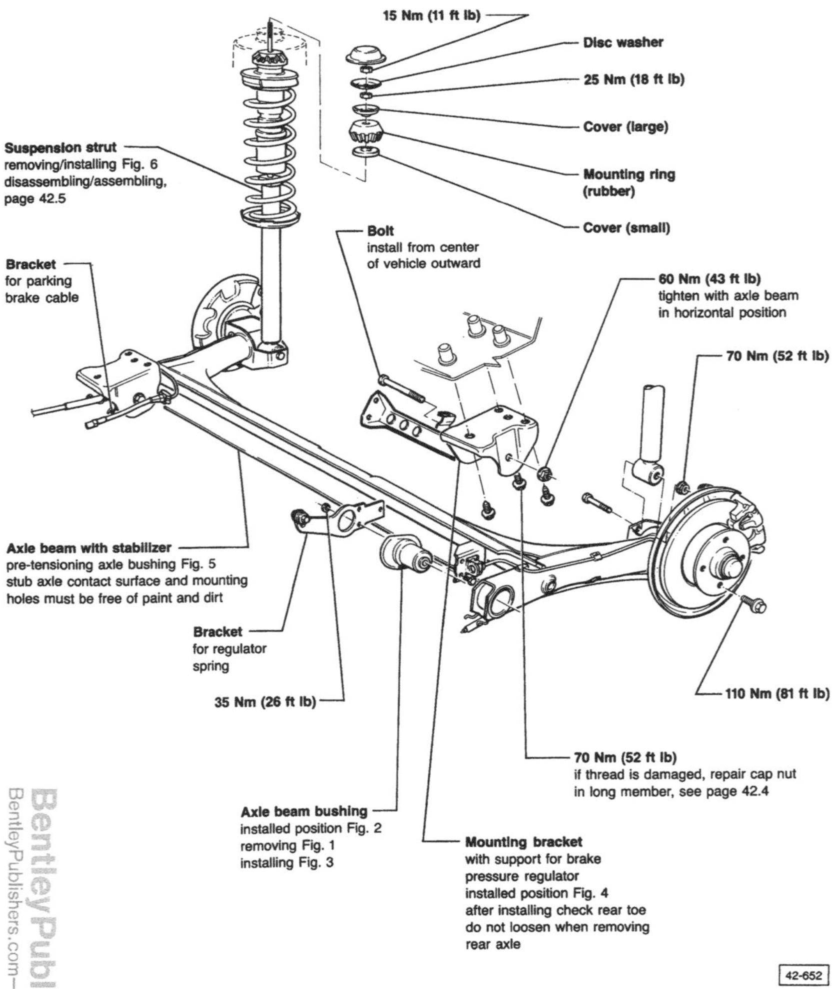

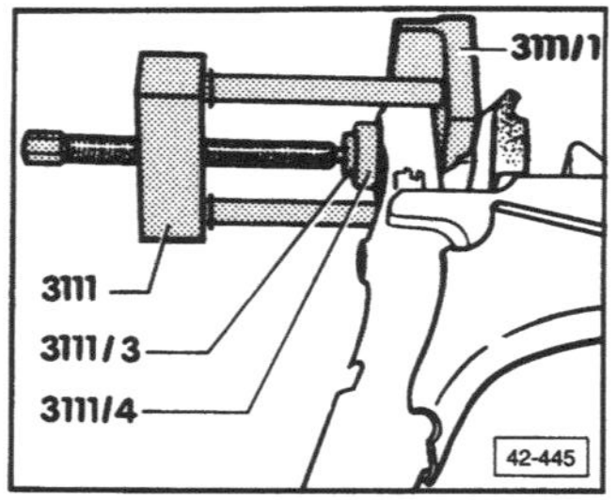

▶ **Fig. 1 Axle beam bushings, removing**

Note

Before using tool **3111**, lubricate spindle with **MoS_2** grease.

Remove any corrosion from area surrounding bushing before pressing out.

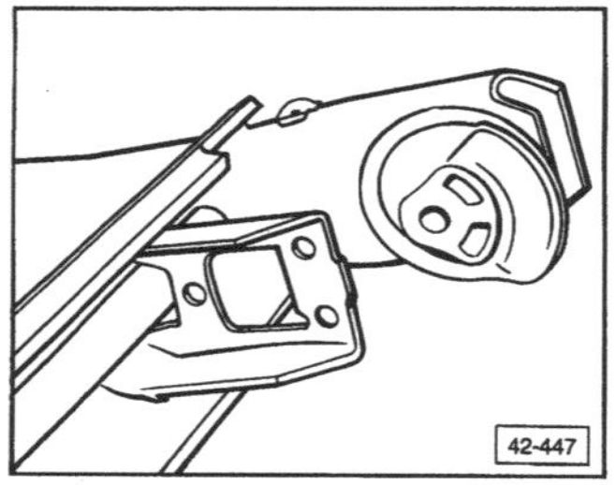

▶ **Fig. 2 Axle beam bushing, installed position**

- protruding rubber/metal segment must point toward front of vehicle

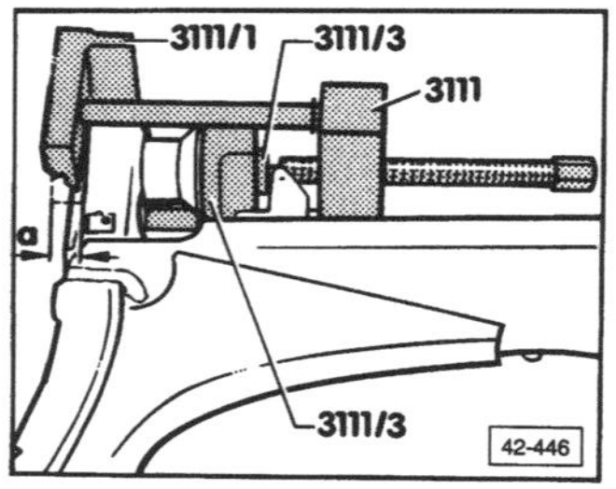

▶ **Fig. 3 Axle beam bushing, installing**

- cylindrical part of bushing must protrude 8 mm
  a = 8 mm (0.315 in.)

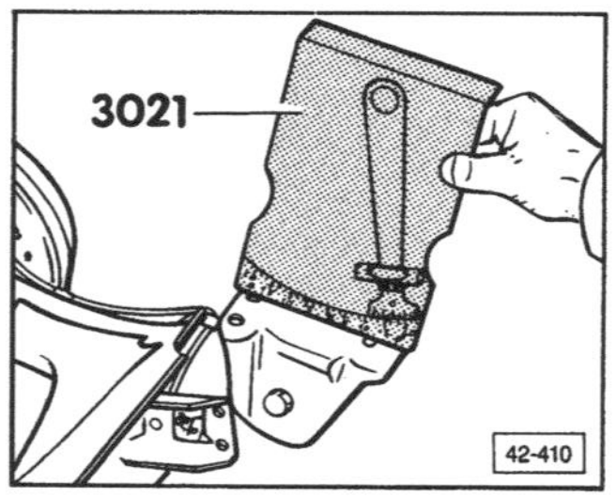

▶ **Fig. 4 Axle beam mounting bracket, installed**

- angle of bracket on axle = 12° 30′ + 1°

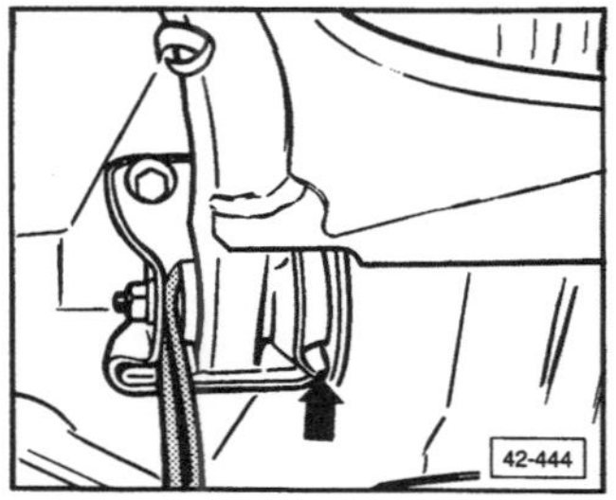

▶ **Fig. 5 Rear axle beam, installing**

The axle bushing must be pre-tensioned in the installation position.

- install all mounting bolts
- align mounting bracket on right side so mounting bolts are in center of slots
- using two prybars, push left mounting bracket against bushing, leaving only a slight gap (**arrow**) on inside of bushing
- tighten mounting bolts to 70 Nm (52 ft lb) in this position

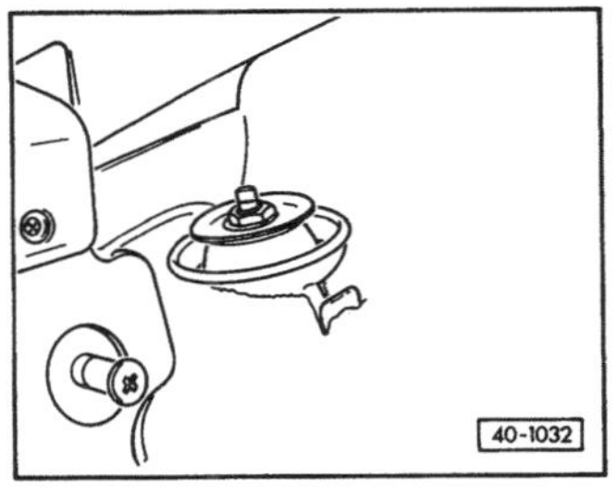

▶ **Fig. 6 Suspension strut, removing/installing**

Axle mounting, repairing

- remove rear axle
- drill through damaged cap nut in long member with 10.2 mm drill bit
- cut new M12 x 1.5 in cap nut
- use new bolt, part number **N 10 127.4** (M12 x 1.5 x 30), and washer, part number **N 11 531.7**, to attach rear axle
- torque bolts to 70 Nm (52 ft lb)

WARNING

Wear safety glasses.

CAUTION

Do not tilt drill.

Note

During repairs check that protective sleeve is properly installed. Missing or broken sleeves will allow dirt and water to enter shock absorber causing premature damage.

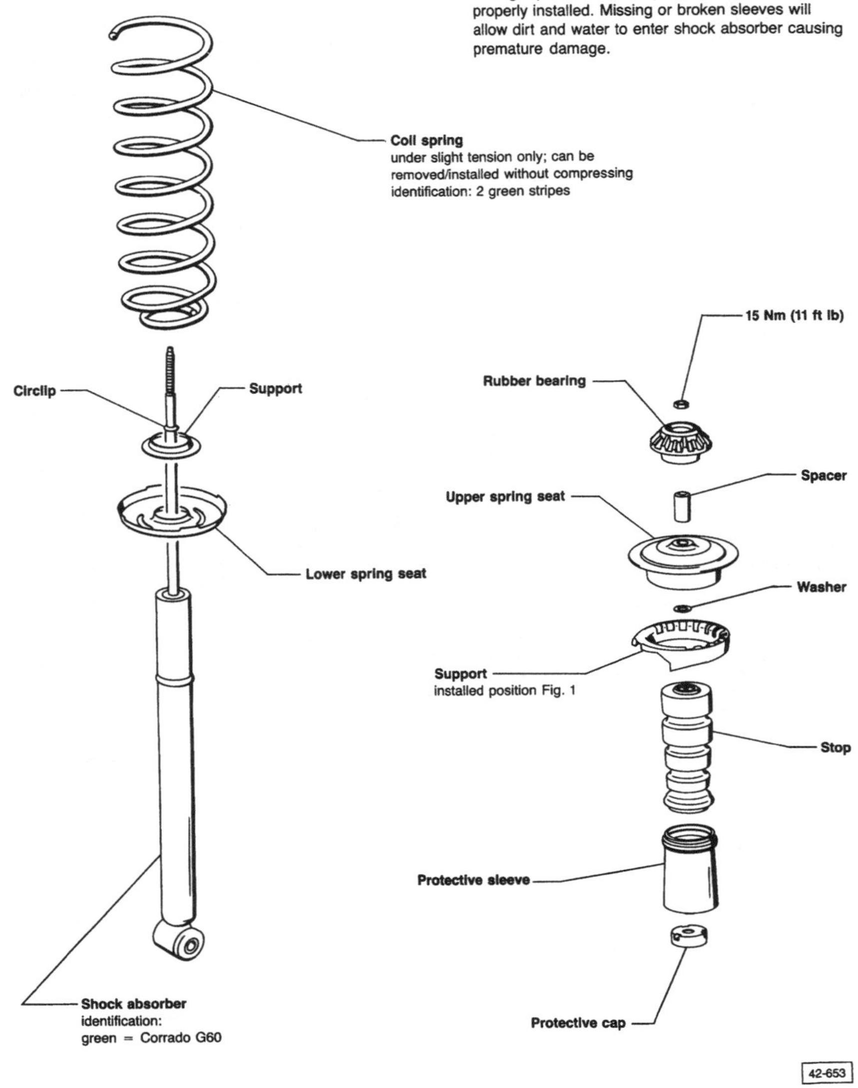

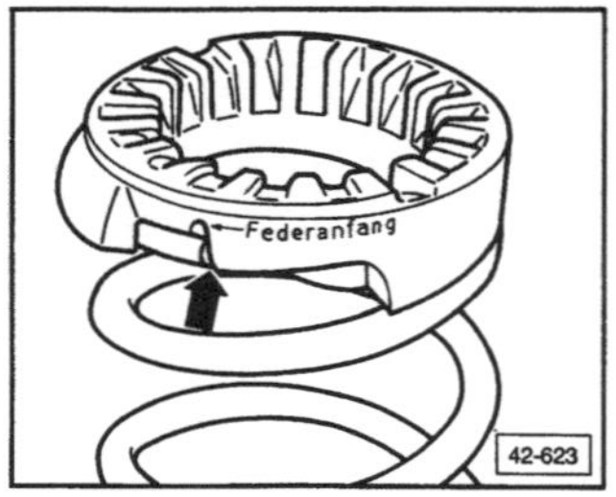

▶ **Fig. 1 Support, installed position**

"Federanfang" = start of spring (**arrow**)

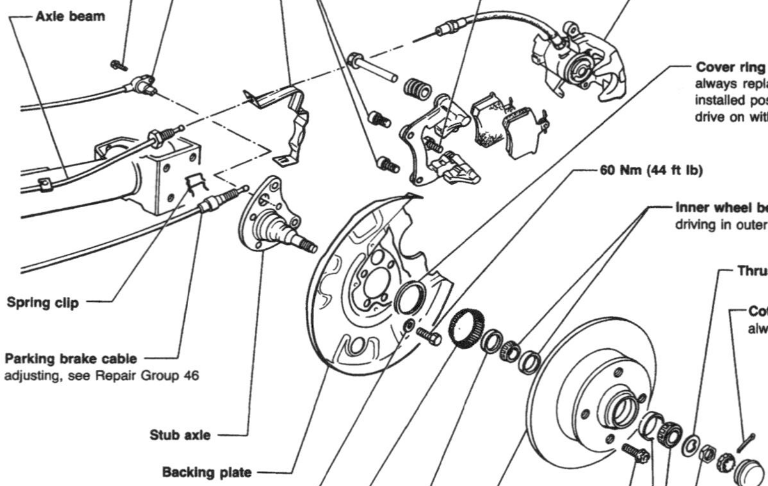

WARNING

Do not re-use any fasteners that are worn or deformed in normal use. Many fasteners are designed to be used only once and become unreliable and may fail when used a second time. This includes, but is not limited to, nuts, bolts, washers, self-locking nuts or bolts, circlips, cotter pins. Always replace these fasteners with new parts.

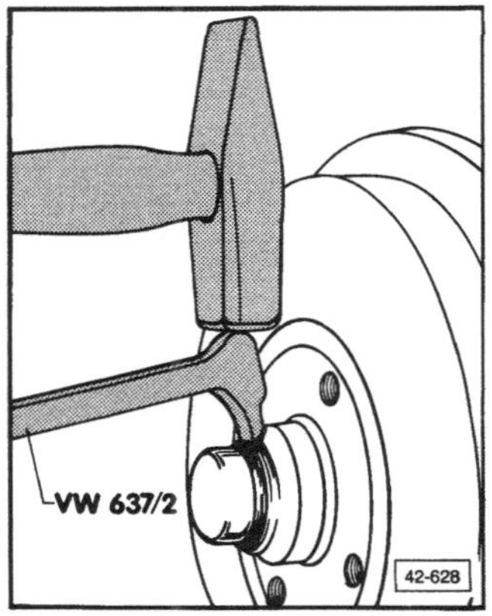

Fig. 1 Grease cap, removing

- loosen cap from seat by lightly striking **VW 637/2**

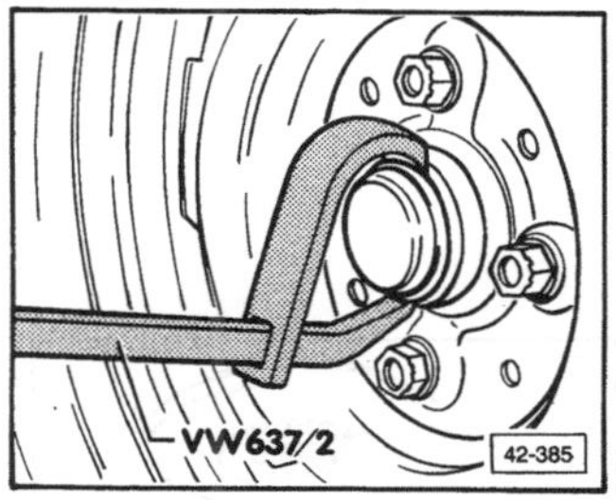

Fig. 2 Grease cap, removing

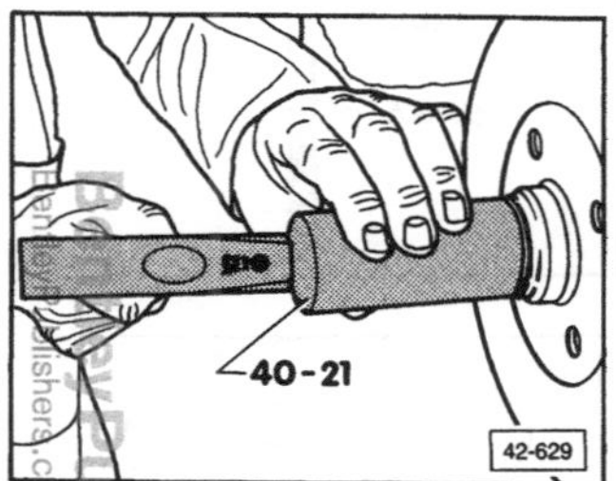

Fig. 3 Grease cap, installing

> **CAUTION**
>
> Damaged (dented) grease caps allow water to enter. Always use **40-21** or equivalent to install grease cap.

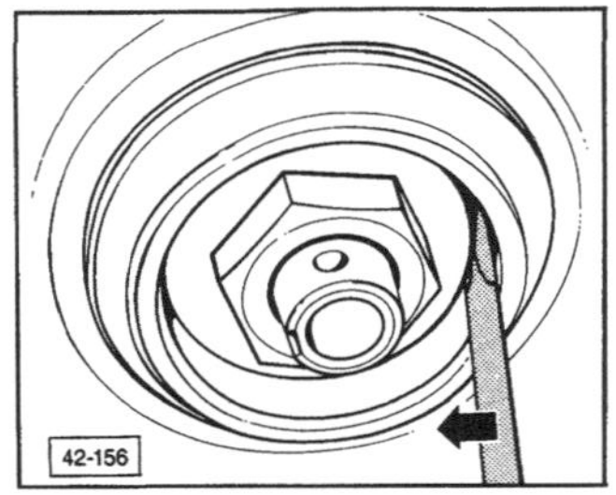

Fig. 4 Wheel bearing play, adjusting

- thrust washer must be movable using finger pressure against a screwdriver without prying

> **CAUTION**
>
> Always use a **new** cotter pin to secure lock nut and bearing.

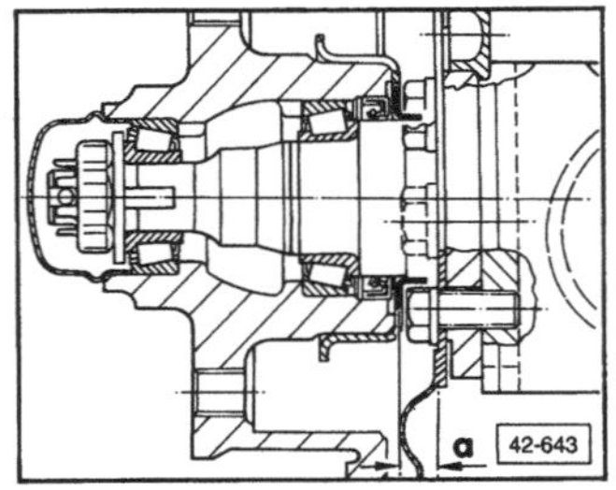

Fig. 5 Cover ring, installed position

- dimension **a** = 9.5 mm (0.373 in.)

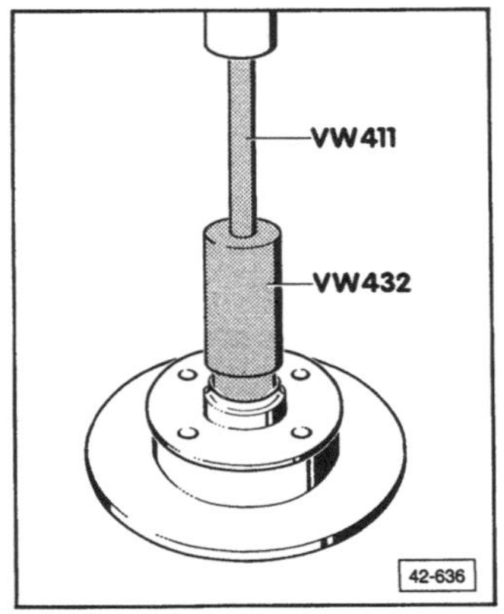

Fig. 6 Outer wheel bearing outer race, installing

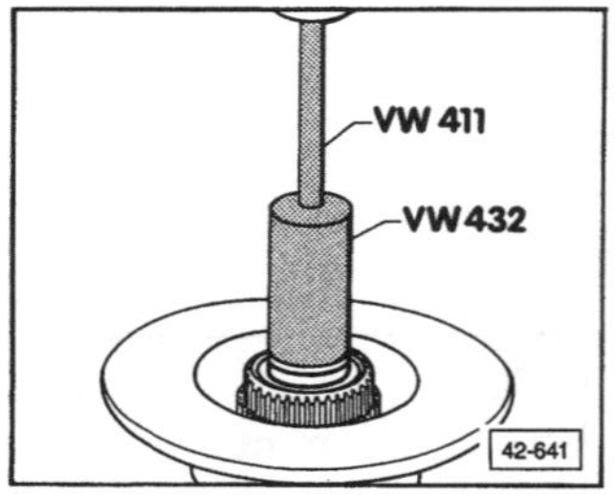

▶ Fig. 7 Inner wheel bearing outer race, installing

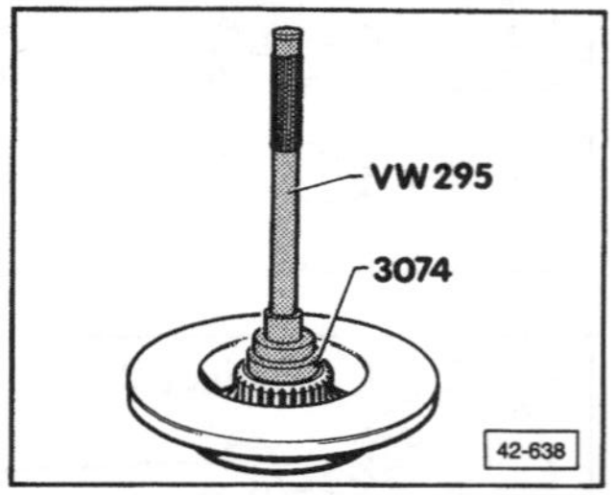

▶ Fig. 8 Seal, installing

THIS PAGE INTENTIONALLY LEFT BLANK

Index

Front wheel camber
- adjusting **44.4, 44.4a**

Rear wheel toe
- determing **44.3**

Snow tires
- size **44.9**

Vibrations
- eliminating **44.5**

Wheel alignment
- data (4 cylinder) **44.2**
- data (6 cylinder) **44.2a**

Wheel alignment data for front and rear axles

Note

It is advisable to measure wheel alignment after at least 1000-2000 km (600-1200 miles) when the coil springs have been given a chance to settle.

Checking requirements

- correct adjustment of measuring equipment
- curb weight of vehicle (w/spare tire, full fuel tank)
- tire pressures correct
- vehicle positioned accurately and suspension bounced
- no excessive play in steering and steering linkage
- check suspension for excessive play and damage

Front axle	
Total toe (wheels not pressed)	0° ± 10′
Camber (at straight-ahead position)	−40′ ± 20′
maximum permissible difference between right and left	maximum 30′
Toe angle difference at 20° lock between left and right	−1° 20′ ± 30′
Caster (not adjustable)	+1° 35′ ± 30′
maximum permissible difference between both sides	maximum 1°

Rear axle	vehicles produced up to 3/90	vehicles produced from 4/90
Camber (not adjustable)	−1° 40′ ± 20′	−1° 30′ ± 10′
maximum permissible difference between both sides	maximum 30′	maximum 20′
Total toe (not adjustable)	+25′ ± 15′	+20′ ± 10′
maximum permissible difference between both sides	maximum 20′	maximum 20′

Wheel alignment data for front and rear axles

Note
It is advisable to measure wheel alignment after at least 1000-2000 km (600-1200 miles) when the coil springs have been given a chance to settle.

Checking requirements

- correct adjustment of measuring equipment
- curb weight of vehicle (w/spare tire, full fuel tank)
- tire pressures correct
- vehicle positioned accurately and suspension bounced
- no excessive play in steering and steering linkage
- check suspension for excessive play and damage

Front axle	
Total toe (wheels not pressed)	0° ± 10′
Camber (at straight-ahead position)	−1° 20′ ± 20′
maximum permissible difference between right and left	maximum 20′
Toe angle difference at 20° lock between left and right	−1° 30′ ± 20′
Caster (not adjustable)	+3° 25′ ± 30′
maximum permissible difference between both sides	maximum 30′

Rear axle	
Camber (not adjustable)	−1° 30′ ± 10′
maximum permissible difference between both sides	maximum 20′
Total toe (not adjustable)	+20′ ± 10′
maximum permissible difference between both sides	maximum 25′

Example

Toe angle of left rear wheel +15′	Toe angle of right rear wheel +5′
15′ − 5′ = 10′ 10′ ÷ 2 = 5′	
Toe (Deviation in alignment) = + 5′	

Example

Toe angle of left rear wheel +15′	Toe angle of right rear wheel −5′
15′ + 5′ = 20′ 20′ ÷ 2 = 10′	
Toe (Deviation in alignment) = + 10′	

Determining total toe/alignment of rear wheels

◀ 1 — if both toe angles are positive (+/+) or negative (−/−), subtract the lower number from the higher number and divide by 2

◀ 2 — if one toe angle is positive and the other negative (+/−), add the two numbers and divide by 2

The result obtained is the actual deviation of the running direction from the logitudinal direction of the vehicle.

THIS PAGE INTENTIONALLY LEFT BLANK

Camber, adjusting

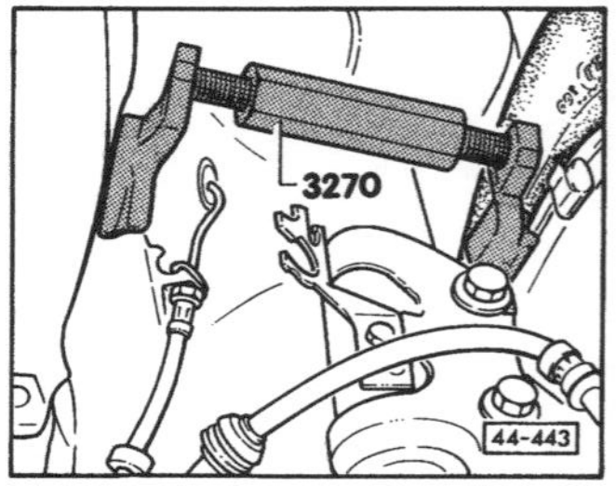

- remove brake hose from bracket on suspension strut
- install tool between wheel and body near suspension strut and tighten slightly
- loosen bolt connection of suspension strut/ wheel bearing housing
- adjust camber by turning tool spindle
- tighten bolt connection of suspension strut/ wheel bearing housing when camber specification is reached
- remove tool and install brake hose in bracket
- check camber and readjust if necessary.

Note

Slight diagonal positioning of tool **3270** may be necessary.

Front wheel camber, adjusting

The Corrado front suspension generally does not require camber adjustment after the installation of suspension struts/wheel bearing housings.

If an alignment check indicates the camber exceeds the recommended tolerance for other reasons, use the following procedure to adjust camber:

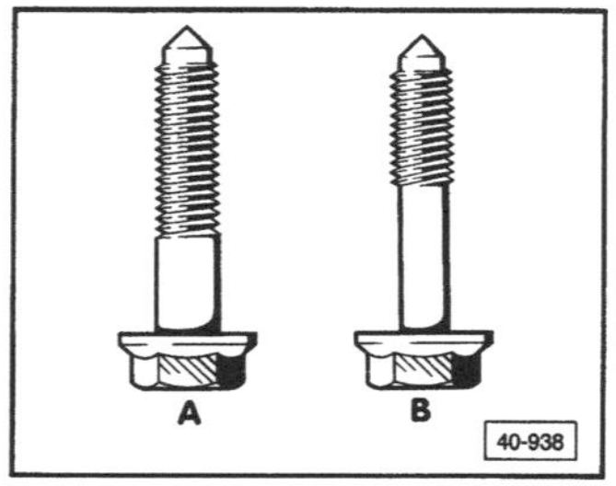

- visually inspect suspension and replace any damaged components
- loosen bolts securing suspension strut to wheel bearing housing
- move top of wheel/tire in or out in direction required to obtain proper camber
- tighten bolts to **80 Nm (59 ft lb)** and recheck camber
- if camber is out of specified range, replace original top bolt **A** with special camber correction bolt **B**
  - bolt Part No. **N 101 740.01** (thinner diameter allows approximately 1° camber adjustment)
- using original bolt **A** in lower position, pivot top of wheel/tire in or out to obtain proper camber
- if more movement is required for specified camber, also replace original lower bolt with special camber adjustment bolt **B**
- tighten bolts to **80 Nm (59 ft lb)** and recheck camber

CAUTION

DO NOT attempt to adjust camber by moving position of ball joint in control arm.

Eliminating Vibrations

Wheel imbalance generates dynamic forces that can cause vehicle components such as steering gear, wheel suspension components and body to resonate. This, in turn, leads to vibrations of the steering wheel and the vehicle itself.

When vibrations of this type occur, use the following procedure to determine the source of the imbalance.

- check tire pressures and correct if necessary
- check surfaces of tires for scuffing, flat spots or other damage
- road test vehicle to determine type of condition and speed range
- check suspension components for damage or wear

Note

If a component is replaced or if tires are worn unevenly, check and align front suspension as necessary. Road test vehicle again after alignment.

- check wheel mountings
    - center boss of hub or brake drum must protrude beyond, or be flush with, center collar of wheel. If not, replace wheel.

Wheel and tire runout, checking

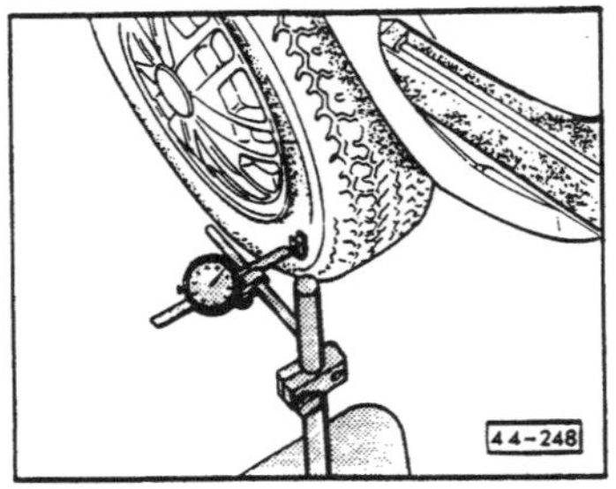

- set up dial indicator so roller is just making contact with center of tire tread or tire wall
- rotate wheel slowly by hand and read needle deflections off meter

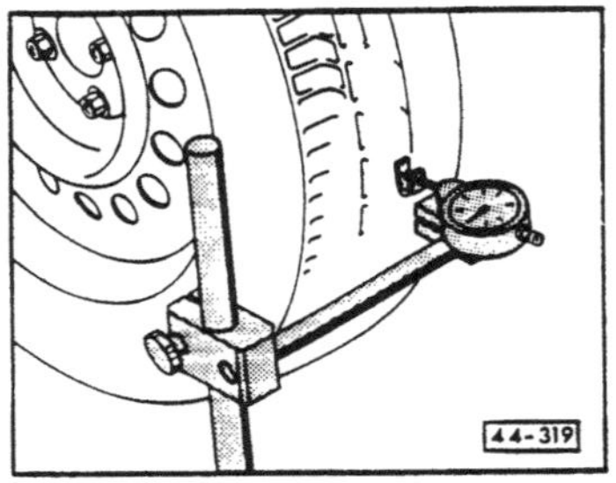

- mark location of maximum radial runout on tire

	Radial	Lateral
Tire with wheel	0.8 mm (0.032 in.)	1.2 mm (0.059 in.)

- if runout figures are within limits, balance wheel/tires (see below Wheels/ tires balancing)
- if runout figures are outside limits, rotate tire on wheel
- deflate tire and push tire beads down into wheel bed
- rotate tire 120° on wheel
- inflate tire and remeasure radial runout
- if maximum figure is still outside limits, rotate tire a further 120° on wheel and remeasure radial runout
- if outside limits, check lateral and radial runout of wheels (see below)

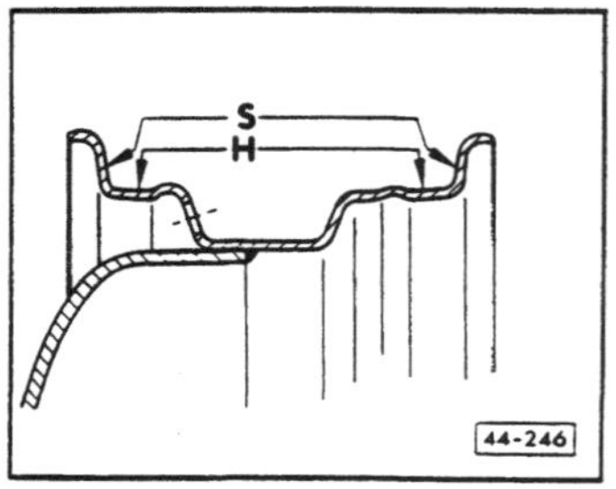

Wheel runout, checking

- dismount tire and mount wheel in balancing machine or on vehicle
- measure radial and lateral runout at points shown above

	Radial (H)	Lateral (S)
Steel wheel	0.6 mm (0.024 in.)	0.8 mm (0.032 in.)
Alloy wheel	0.5 mm (0.020 in.)	0.5 mm (0.020 in.)

Note

Peak readings, up or down, which are caused by small imperfections on the wheel surface, can be ignored.

- if maximum values are exceeded, replace wheel and recheck runout of wheel and tire assembly
- balance tire and wheel assembly

Wheels/tires, balancing

- remove existing balance weights before balancing

> CAUTION
> **Wheels MUST be mounted with same centering method as on vehicle (i.e. bolted to hub/arbor) and centered on a cylindrical (not conical) surface.**

- maximum permissible residual imbalance = 5 grams

Wheels/tires, balancing (continued)

Note

For on-the-vehicle balancing of the driving wheels, the wheels must be driven by the engine so that wheel speeds are synchronized.

Wheels, installing

- install wheels with point of maximum radial runout at top and tighten lug bolts in this position

Note

If tire wear is approximately equal, the wheels with the lowest amount of radial runout and the smallest balance weights should be installed at the front.

- repeat road test of vehicle

If vibrations are still present, the radial and lateral oscillations of one or more tires are too high. These oscillations cannot be measured with normal workshop equipment. For such cases, the front tires, rear tires or all four tires should be replaced.

Snow tires

Corrado vehicles can use 185/55 R15 81T size snow tires with 6J x 15 steel wheels, The Part No. for the steel wheel is **535 601 025 C**.

> CAUTION
> **14 in. wheels cannot be installed on the Corrado because the brake discs and calipers are too large for the wheel.**

> CAUTION
> **Part numbers are for reference only. Always check with your Parts Department for latest information.**

THIS PAGE INTENTIONALLY LEFT BLANK

THIS PAGE INTENTIONALLY LEFT
BLANK

Index

Hydraulic modulator
- assembly **45.3**
- assembly (ABS/EDL) **45.7**
- removing/installing **45.6**

Pressure warning switch
- removing/installing **45.5**

Repair work
- general notes **45.2**
- repair rules **45.2**

When working on brakes

1 — Clean all hydraulic line connections and the surrounding areas thoroughly before loosening

2 — Place removed parts on a clean surface and cover. Always use lint-free cloths

3 — Carefully cover or plug opened components if repair is not going to be done immediately

4 — Install clean parts only
- do not remove replacement parts from packaging until just before installation
- do not use parts that were stored unpacked

5 — When hydraulic brake system is opened:
- do not use compressed air in area
- do not move vehicle unless absolutely necessary

General notes for repair work

Before working on the ABS-System, switch off ignition and disconnect battery ground strap. Also, the pressure in the reservoir (high pressure) must be reduced by pressing the brake pedal frequently (approximately 20 times), until a clearly noticeable increase in pedal effort occurs.

ABS is maintenance free, however when working on vehicles with the ABS System note the following:

- before doing any welding work with an electric welding device, disconnect ABS control unit connector
- the control unit cannot withstand temperatures from paint drying oven or heat lamp greater than 95°C (203°F) even for a short time or a temperature of 85°C (185°F) for more than two hours

ABS hydraulic modulator assembly, repairing

The following components of the ABS hydraulic modulator assembly can be replaced:
- hydraulic modulator
- reservoir
- pressure-warning switch
- pressure accumulator
- pump subassembly

Before the hydraulic modulator assembly is removed, check the following:
- battery voltage
- ABS fuses
- ABS relay
- pump relay
- ABS connectors
- ground connectors on modulator assembly and control unit
- pump assembly and pressure warning switch

to check: reduce pressure in accumulator, turn ignition on, pump must start running and shut off after a maximum of 60 seconds, depress brake pedal 2 or 3 times: pump must start running again

Note

The reservoir, pressure accumulator and pressure warning switch can be replaced with the hydraulic modulator assembly installed. For all other work, the complete modulator assembly must be removed.

WARNING

The ABS modulator assembly is capable of "self-pressurizing" up to 210 bar (3045 p.s.i.). Serious injury may result if you service the unit without properly depressurizing it. Depress brake pedal 25 to 35 times with ignition switch in **OFF** position, to depressurize system.

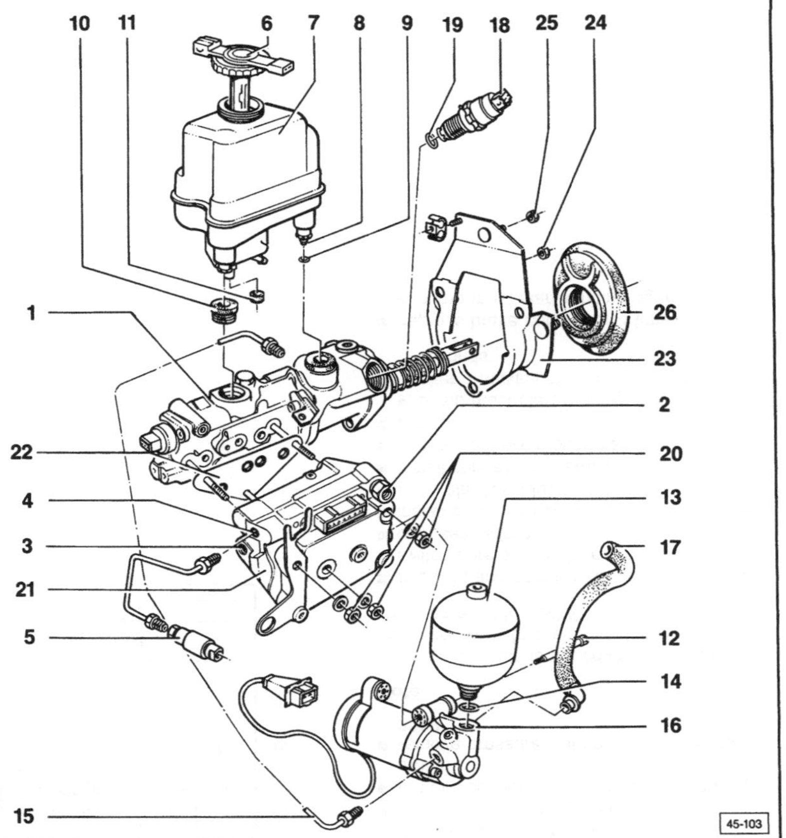

WARNING

The ABS modulator assembly is capable of "self-pressurizing" up to 210 bar (3045 p.s.i.). Serious injury may result if you service the unit without properly depressurizing it. Depress brake pedal 25 to 35 times with ignition switch in **OFF** position, to depressurize system.

1 — **Hydraulic modulator (upper)**
removing/installing page 45.6

2 — **Brake line connection**
for right brake caliper

3 — **Brake line connection**
for left brake caliper

4 — **Brake line connection**
for brake pressure regulator

5 — **Pressure regulator valve**
opens to rear axle starting at 8 bar
cannot be checked in shop

6 — **Cap**

7 — **Brake fluid reservoir**

8 — **Sleeve**
always replace

9 — **O-ring**
always replace

10 — **Sealing plug**
always replace

11 — **Reservoir mounting retainer**
always replace

12 — **10 Nm (7 ft lb)**

13 — **Pressure accumulator**
45 Nm (33 ft lb)
coat O-ring with brake fluid before installing

14 — **O-ring**
coat with brake fluid before installing accumulator

15 — **Brake line**
with rubber damper

16 — **Hydraulic pump**
if pump is replaced, also replace relay

17 — **Connecting hose**

18 — **Pressure warning switch**
25 Nm (18 ft lb)
removing/installing page 45.5
coat O-ring with brake fluid before installing

19 — **Seal**
with rectangular section
check for correct fit
replace if necessary

20 — **25 Nm (18 ft lb)**

21 — **Hydraulic modulator (lower)**
removing/installing page 45.6

22 — **Gasket**
always replace

23 — **Bracket**

24 — **25 Nm (18 ft lb)**

25 — **25 Nm (18 ft lb)**

26 — **Gasket**

Pressure warning switch, removing/installing (with hydraulic modulator assembly installed)

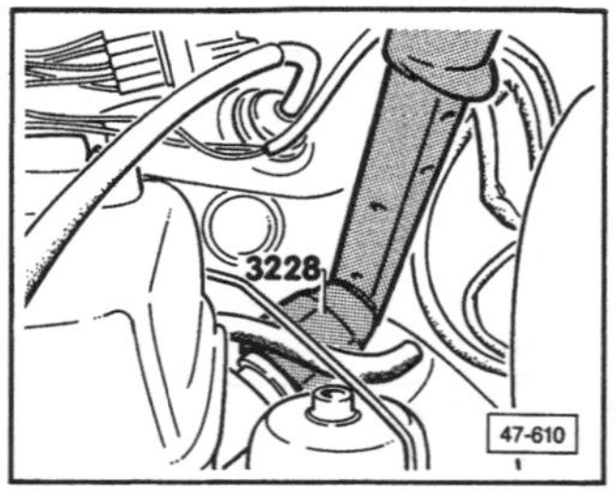

CAUTION

If the pressure warning switch has electrical damage, the pump relay must be replaced also.

- switch ignition off and reduce accumulator pressure by pressing brake pedal approximately 25 times
- disconnect battery ground cable
- remove connector from pressure warning switch
- remove pressure warning switch with tool 3228

Note

If installed switch is not compatible with tool size, remove with pliers. Replacement switches all fit tool 3228.

- check fit and condition of seal in warning switch bore and replace if necessary
- coat seal with brake fluid
- install pressure warning switch and tighten to 25 Nm (18 ft lb)
- install connector to switch
- connect battery ground cable
- turn ignition on
  - pump must start running and shut off after a maximum running time of 60 seconds
- depress brake pedal 2 or 3 times
  - pump must start running again

Hydraulic modulator, removing/installing

Removing

- switch off ignition and press brake pedal approximately 20 times to reduce accumulator pressure
- disconnect battery ground strap
- remove all cable terminals on ABS unit and remove ground connection
- drain brake fluid from the reservoir with a suction bottle
- remove brake lines from ABS unit and seal threaded bores with suitable plug
- remove left shelf under instrument panel
- remove bolts and safety bolt from pushrod clevis
- remove lock nuts on ABS component and remove ABS component

Installing

- installation is conducted in reverse order of removal procedure

Tightening torques

Hydraulic modulator to cross panel (new self locking nuts used)	25 Nm (18 ft lb)
Brake lines to hydraulic modulator	15 Nm (18 ft lb)

- bleeding brake system, Repair Group 47

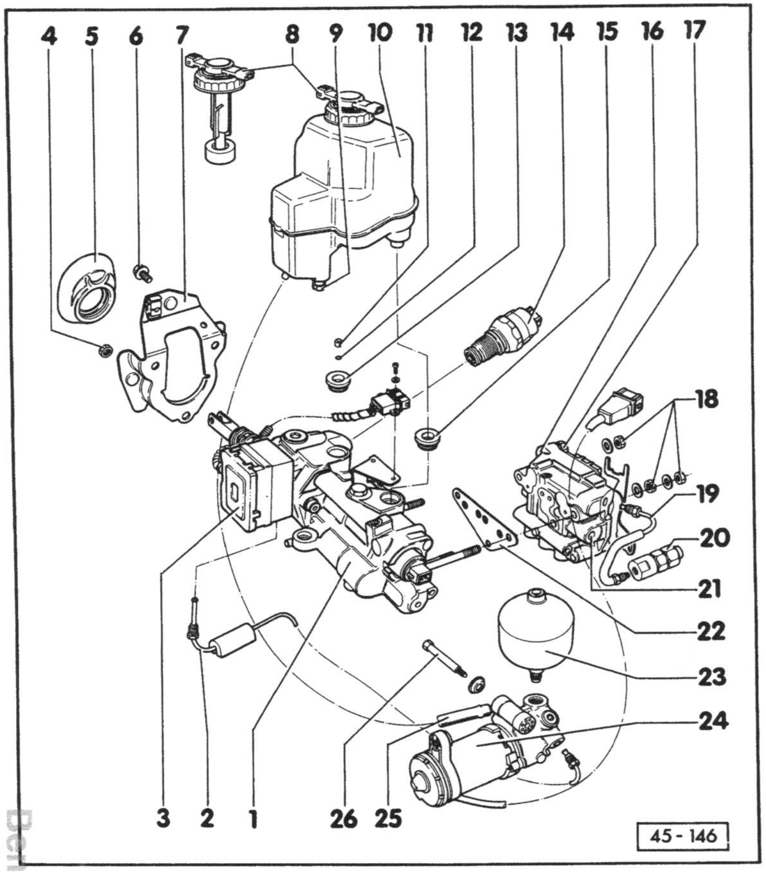

Note
Reservoir (**10**), pressure accumulator (**23**) and pressure warning switch (**14**) can be replaced with the hydraulic unit installed. For all other work the hydraulic unit must be removed.

1 – **Servo unit/brake master cylinder**
- removing/installing hydraulic unit complete, page 45–6

2 – **Brake line**
- with rubber weight
- from hydraulic pump to EDL valve block

3 – **EDL valve block**
- DO NOT separate servo unit with master cylinder and EDL valve block

4 – **25 Nm (18 ft lb)**

5 – **Gasket**

6 – **25 Nm (18 ft lb)**

7 – **Bracket**

8 – **Cap**

9 – **Sleeve**
o always replace

10 – **Reservoir**
- supplied with attachment kit
- can be replaced with unit installed

11 – **Sleeve**
- always replace; remove old sleeve from servo unit

12 – **O–ring**
- always replace; remove old O–ring from servo unit

13 – **Sealing plug**
- always replace; moisten with brake fluid

14 – **Pressure warning switch**
- torque to 25 Nm (18 ft lb)
- can be replaced with hydraulic unit installed
- supplied as replacement part with O–ring and rubber ring (rectanular shape)
- moisten O–ring with brake fluid before installing and ensure correct seating of rubber ring

15 – **Sealing plug**
- always replace; moisten with brake fluid

16 – **ABS valve block**

17 – **Brake line connection**
- for right brake caliper

18 – **25 Nm (18 ft lb)**

19 – **Brake line connection**
- for brake pressure regulator

20 – **Pressure regulator valve**
- on all vehicles with rear disc brakes
- opens to rear axle from approx. 8 bar; cannot be tested with shop equipment

21 – **Brake line connection**
- left brake caliper

22 – **Gasket**
- always replace

23 – **Pressure accumulator**
- torque to 45 Nm (33 ft lb)
- can be replaced with hydraulic unit installed
- supplied with O–ring installed
- moisten O–ring with brake fluid before installing

24 – **Hydraulic pump**
- if faulty, also replace pump relay

25 – **Hose**
- hydraulic pump to reservoir

26 – **10 Nm (7 ft lb)**

THIS PAGE INTENTIONALLY LEFT BLANK

Index

Brake caliper
- removing **46.4**

Front brake
- assembly **46.3**
- pads, removing/installing **46.4**

Brake pedal
- assembly **46.9**

Parking brake
- adjusting **46.8**
- assembly **46.7**

Rear brake
- assembly **46.5**
- pads, replacing **46.6**

Technical data
- chart **46.2**

Technical data

Engine	118 kW (160 bhp) G-Lader 133 kW (178 bhp) VR6	
	without ABS	with ABS
Master cylinder diameter	22.2 mm	20.64 mm
Brake booster diameter	9"	hydraulic

	Front	Rear
Brake caliper piston diameter	54 mm	38 mm
Brake disc diameter	280 mm	226 mm
Brake disc thickness	22 mm	10 mm
Brake pad thickness	14 mm	10 mm

THIS PAGE INTENTIONALLY LEFT BLANK

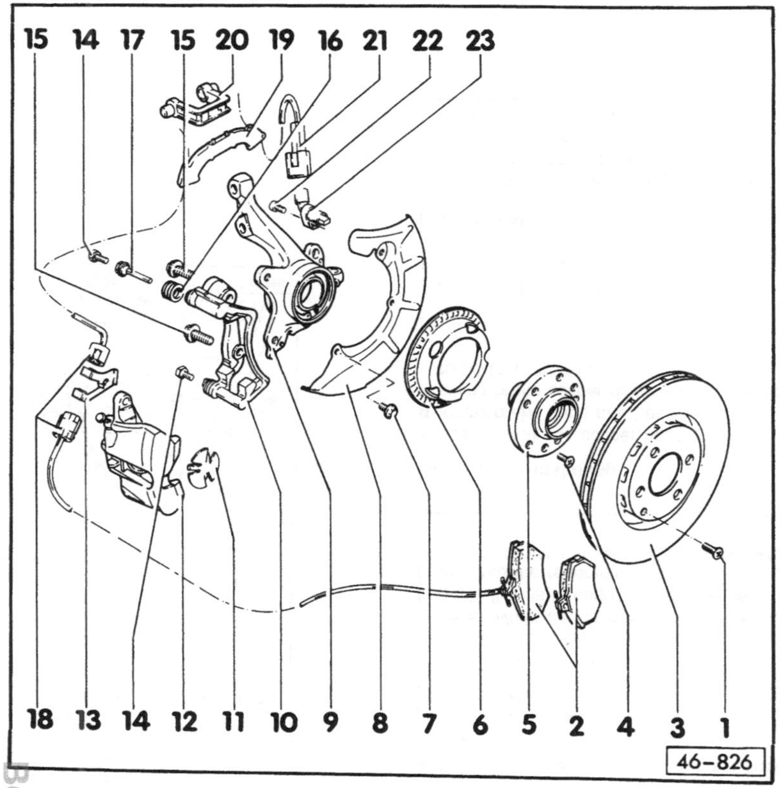

Note
After installing new brake pads, depress brake pedal firmly several times while vehicle is stationary to permit piston and brake pads to adjust to brake disc.

Always remove some brake fluid from the reservoir before installing new brake pads. When the caliper piston is pushed back, fluid is forced out of the caliper and into the reservoir. After pads are installed, refill the reservoir to the MAX mark.

1 – **Screw**

2 – **Brake pads**
- some versions have pad wear indicator
- always replace pads on both sides
- removing/installing, page 46.4
- pad thickness when new: 14 mm
- wear limit: 7 mm (including backing plate)

3 – **Brake disc**
- always replace discs on both sides only
- machine discs on both sides only and ensure adequate margin for wear
- wear limit: 20 mm
- diameter: 280 m
- thickness: 22 mm

4 – **Screw**

5 – **Wheel hub**
- removing/installing, see Repair Group 40

6 – **Rotor**
- for wheel speed sensor (vehicles with ABS)

7 – **10 Nm (7 ft lb)**

8 – **Splash shield**

9 – **Wheel bearing housing**

10 – **Brake pad carrier**
- is supplied as replacement part assembled with grease on guide pins
- if protective caps are damaged, install repair kit and lubricate guide pins with grease pack supplied

11 – **Heat shield**
- install into piston

12 – **Brake caliper housing**
- do not disconnect hydraulic line when replacing brake pads

13 – **Bracket**

14 – **35 Nm (26 ft lb)**
- always replace

15 – **125 Nm (92 ft lb)**
- clean serrations before reusing

16 – **Protective cap**

17 – **Guide pin**

18 – **Plug connection**
- for brake pad wear indicator

19 – **Bracket**

20 – **Bracket**

21 – **Plug connection**
- for ABS wheel speed sensor

22 – **10 Nm (7 ft lb)**

23 – **Wheel speed sensor**
- clean bore and coat with lubricating paste, Part No. G 000 650

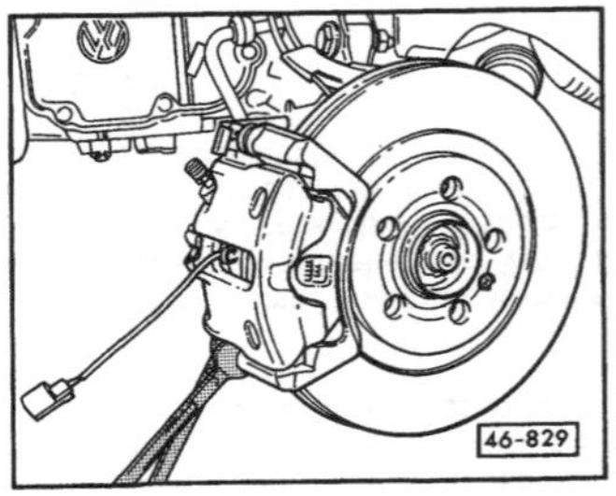

▶ Fig. 1 Brake caliper, removing

- remove lower mounting bolt (hold guide pin with open end wrench while loosening)

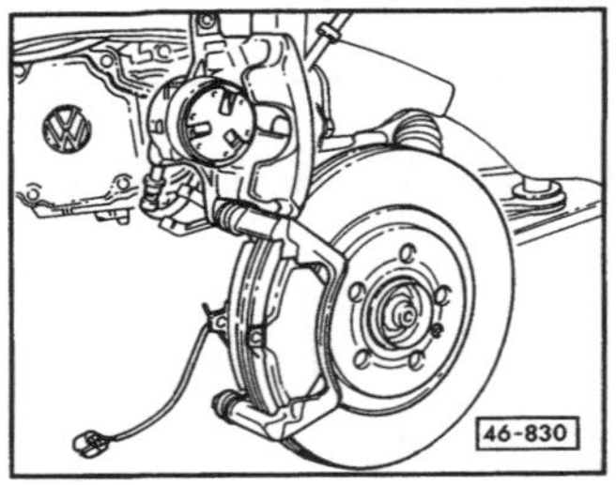

▶ Fig. 2 Brake pads, removing/installing

- swing brake caliper up and remove brake pads

CAUTION

Always remove some brake fluid from the reservoir before installing new brake pads. When the caliper piston is pushed back, fluid is forced out of the caliper and into the reservoir. After pads are installed, refill the reservoir only the **MAX** mark.

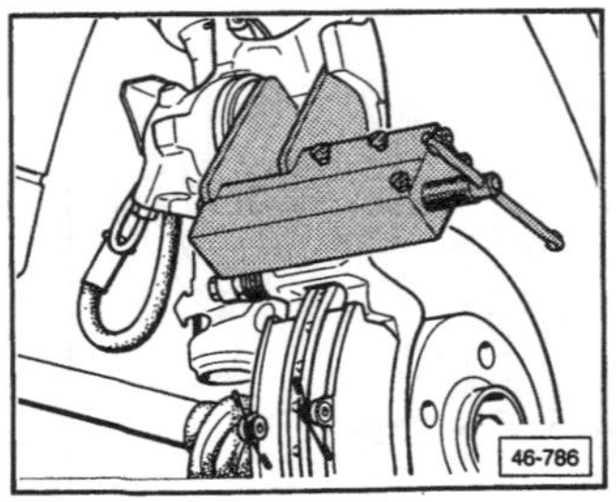

▶ Fig. 3 Brake pads, installing

- push piston into caliper housing
- install brake pads and heat shield
- swing brake caliper down and tighten bolts to 35 Nm (26 ft lb)

Note

The two self-locking bolts in the repair kit must always be installed.

CAUTION

After installing new brake pads, depress brake pedal firmly several times while vehicle is stationary to permit piston and brake pads to adjust to brake disc.

CAUTION

Disc machining must be performed on both sides of disc — never on one side only.

Always check condition and thickness of brake pads.

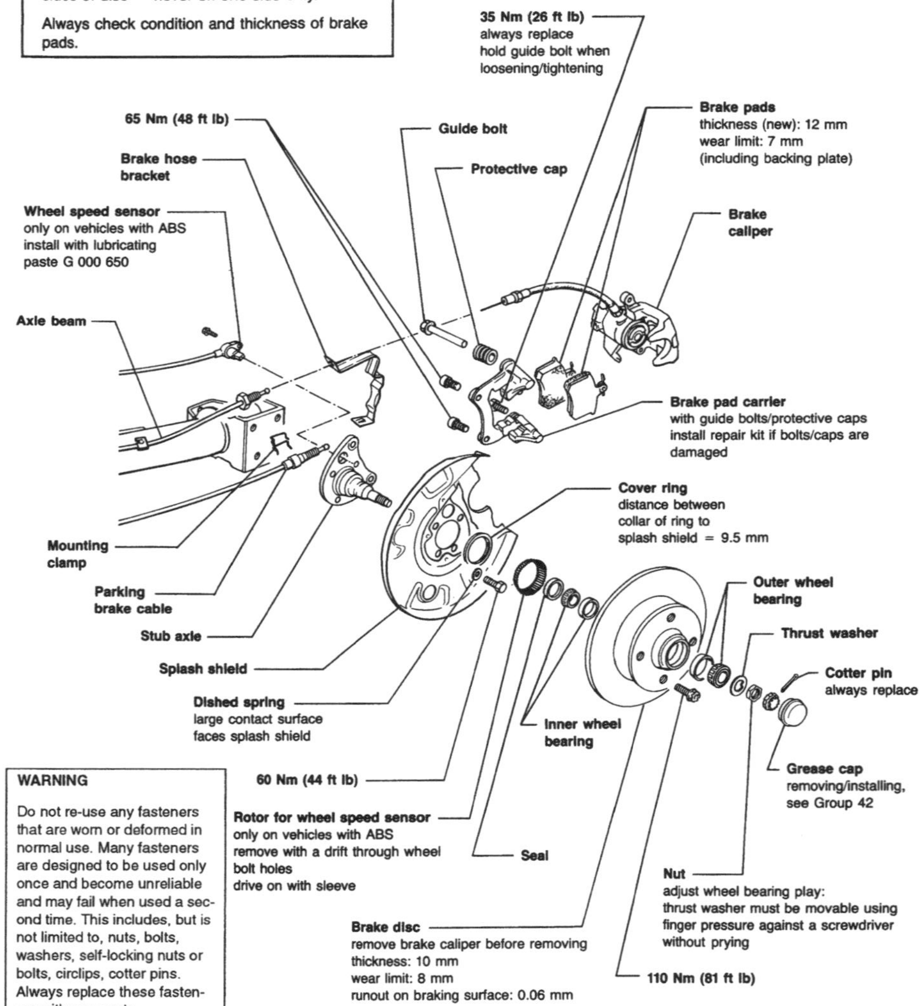

WARNING

Do not re-use any fasteners that are worn or deformed in normal use. Many fasteners are designed to be used only once and become unreliable and may fail when used a second time. This includes, but is not limited to, nuts, bolts, washers, self-locking nuts or bolts, circlips, cotter pins. Always replace these fasteners with new parts.

Rear brake pads, replacing

- remove wheels
- remove parking brake cable from brake caliper

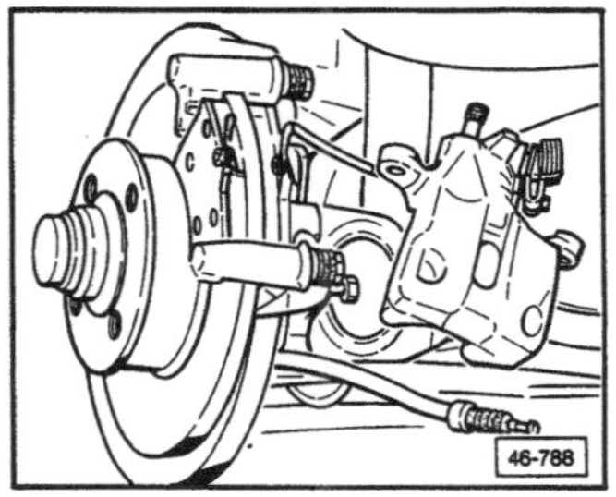

- remove mounting bolts from brake caliper housing
- remove brake pads

> **CAUTION**
> Always remove some brake fluid from the reservoir before installing new brake pads. When the caliper piston is pushed back, fluid is forced out of caliper and into the reservoir. After pads are installed, fill reservoir only to the MAX mark.

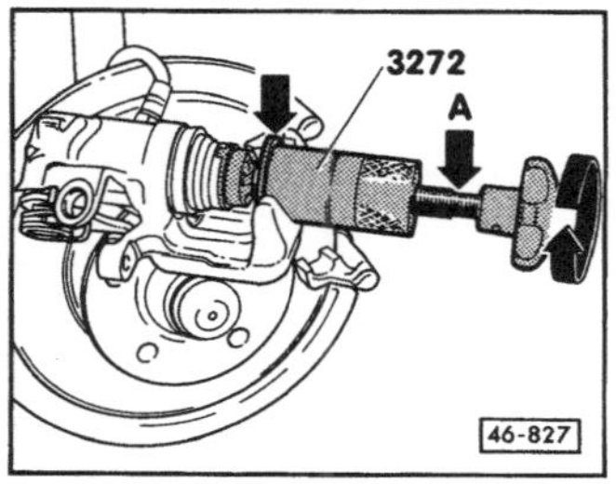

- push piston in by turning knurled wheel of tool clockwise
  - collar of tool (**left arrow**) must make contact with caliper
  - if piston is difficult to move, use 13 mm wrench on tool (**arrow A**)

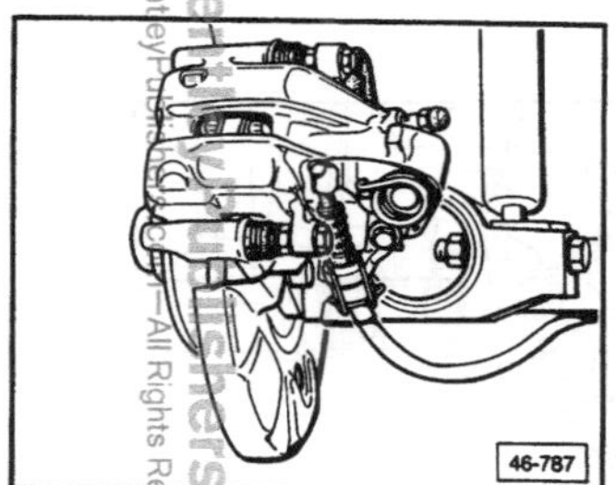

- install brake caliper housing using new self–locking bolts
- attach parking brake cable to brake caliper

> **CAUTION**
> Always pump the brake pedal **before** driving the vehicle to properly seat the pads in their normal operating position.

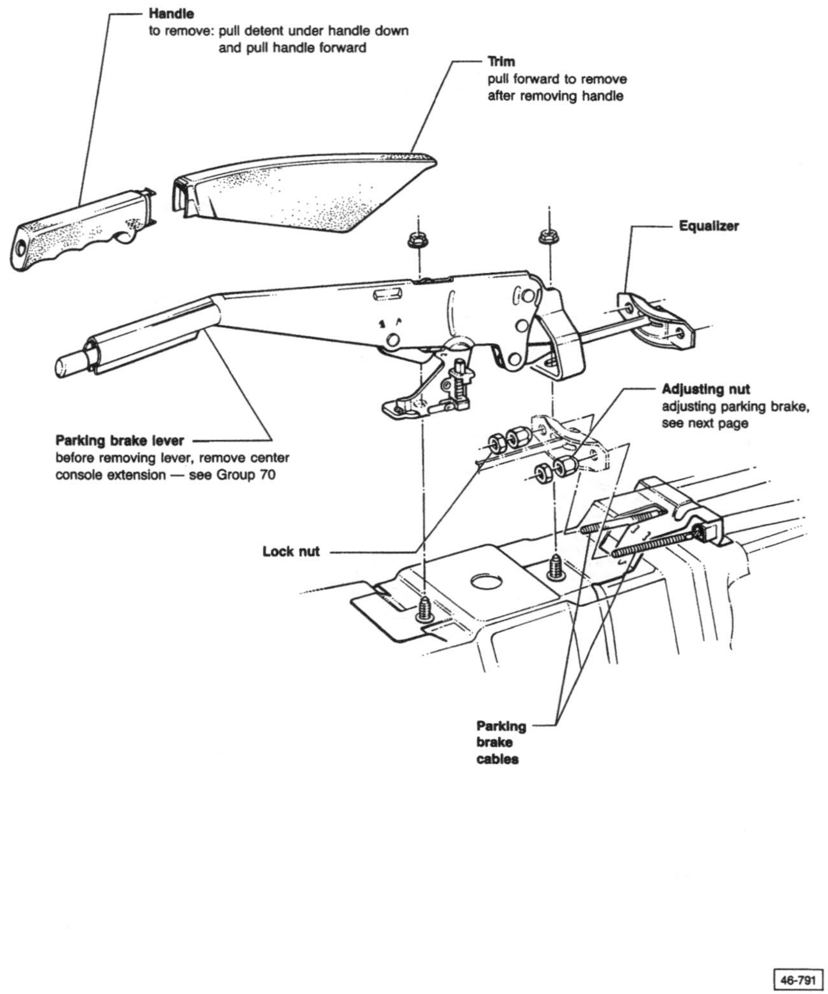

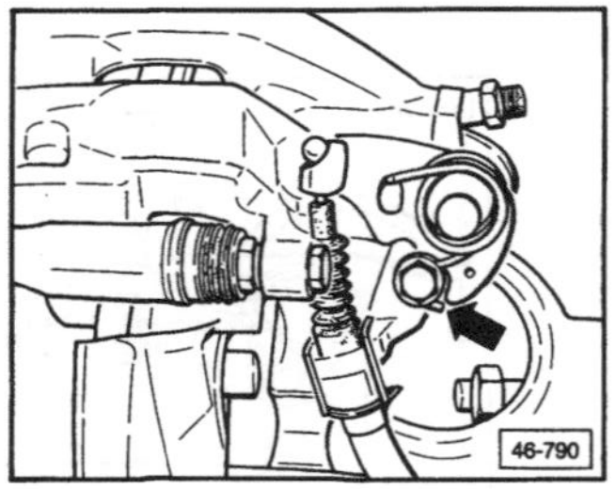

Parking brake, adjusting

Adjustment of the parking brake is only necessary when the parking brake cables, rear brake calipers, brake pads or discs are replaced.

- put parking brake lever into down (disengaged) position
- tighten adjusting nuts until levers on calipers (arrow) just move off their stops
  - maximum persmissible distance from stop = 1.5 mm
- apply and release parking brake
- check that both wheels rotate freely

Note

Lubricate all bushings and pivots before assembly with G 000 602 grease.

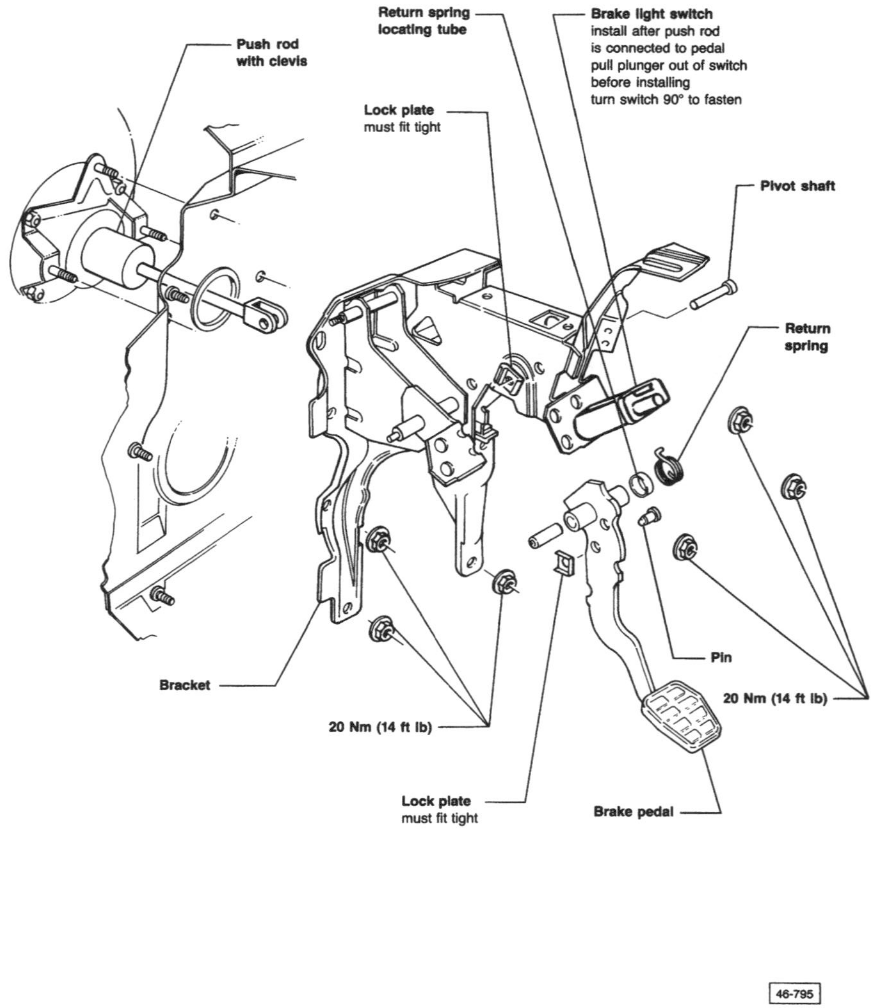

Index

Brake booster/master cylinder
- assembly **47.13**

Brake caliper piston
- pre–bleeding **47.7**
- removing/installing **47.3, 47.6, 47.7**
- seal, removing/installing **47.3, 47.6**

Brake fluid
- replacing **47.11, 47.15**

Brake pressure regulator
- checking/adjusting **47.8**

Brake system
- bleeding **47.9**
- replacing brake fluid **47.9, 47.12, 47.14**

Front brake caliper
- component layout **47.2**

Rear brake caliper
- component layout **47.5**

CAUTION

Brake fluid must not come in contact with painted parts.

Brake fluid absorbs moisture from the air and must therefore be replaced every two years.

Do not use silicone-based brake fluid (DOT 5). Even the smallest traces may cause severe corrosion in the brake system.

Use only new, approved brake fluid which complies with standard **FMVSS 116**, DOT 4.

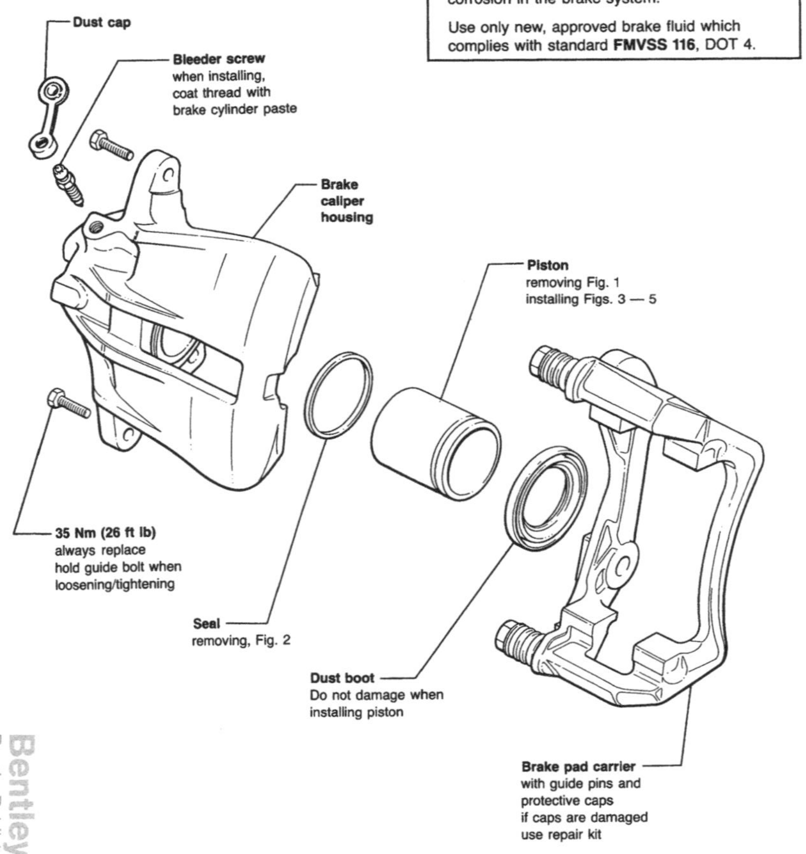

CAUTION

Use all parts supplied in repair kit.

Brake – Hydraulic Components, Regulator, Booster

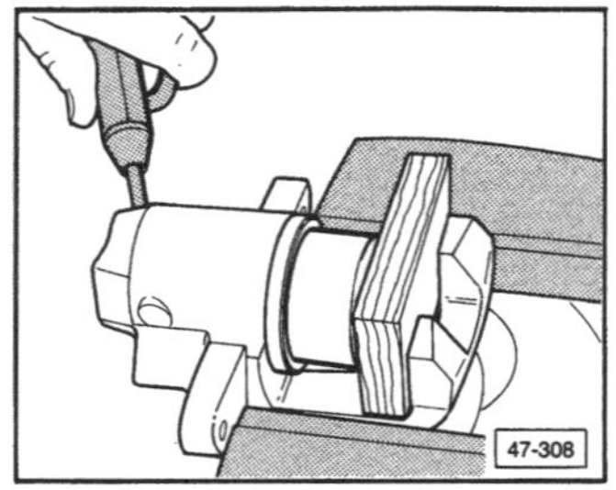

▶ **Fig. 1 Brake caliper piston, removing**

CAUTION

Place a wooden block in the caliper to prevent damage to the piston.

Use only enough air pressure to force the piston out.

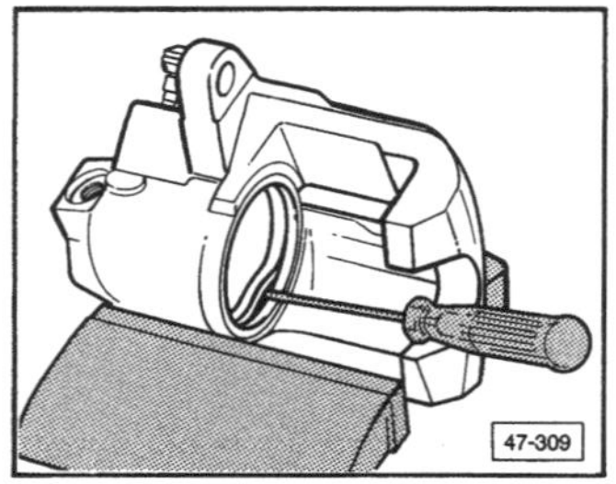

▶ **Fig. 2 Brake caliper piston seal, removing**

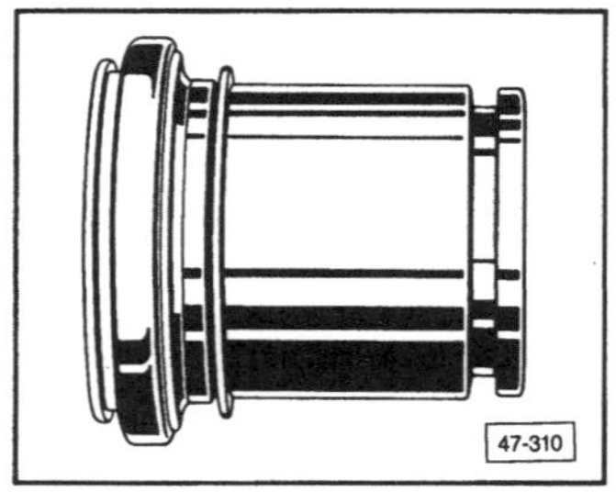

▶ **Fig. 3 Brake caliper piston, installing**

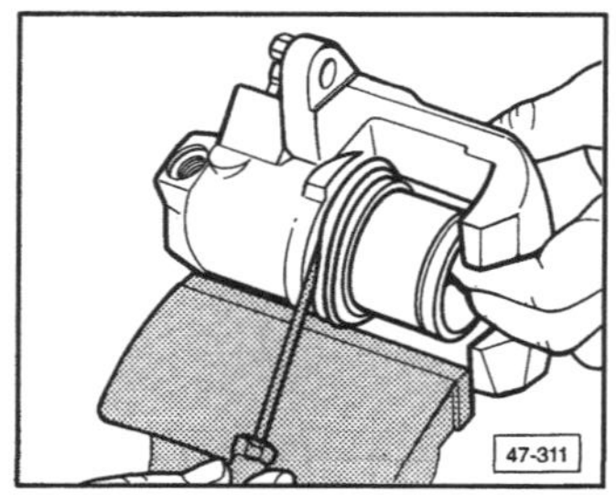

▶ **Fig. 4 Brake caliper piston, installing**

- lubricate piston and cylinder bore lightly with brake cylinder paste
- install piston and insert inner lip of dust boot into groove in brake cylinder

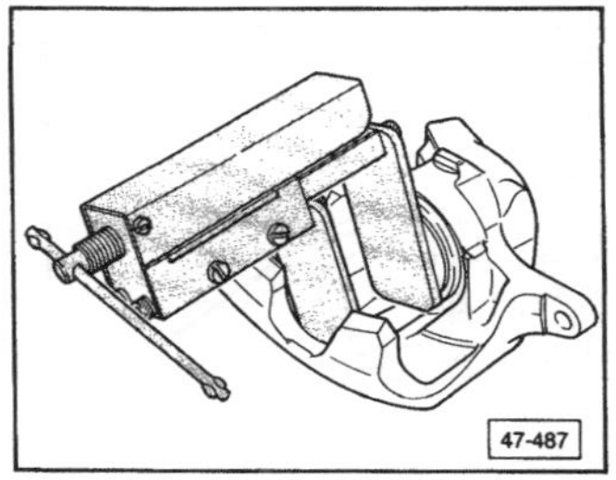

▶ **Fig. 5 Brake caliper piston, installing**

> **CAUTION**
>
> The outer lip of the dust boot must slip into the groove in the piston.

> **CAUTION**
>
> New brake calipers are filled with brake fluid and pre-bled. Repaired calipers must be pre-bled before installation.

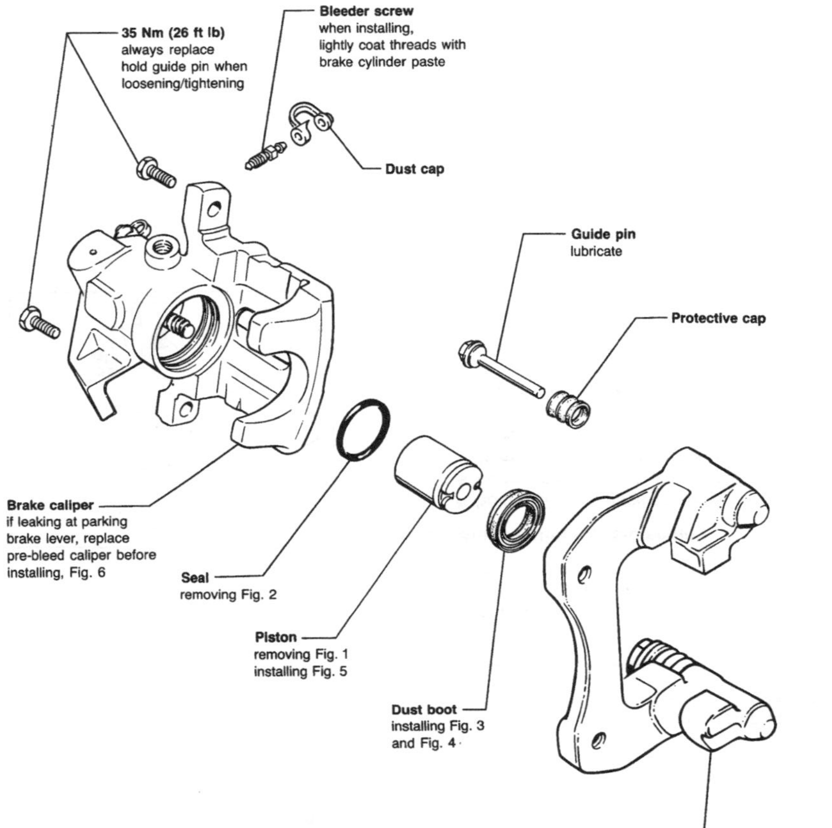

> **CAUTION**
>
> Use all parts in repair kit.
>
> Coat seals, pistons and cylinders lightly with VW brake cylinder paste or equivalent before installing.

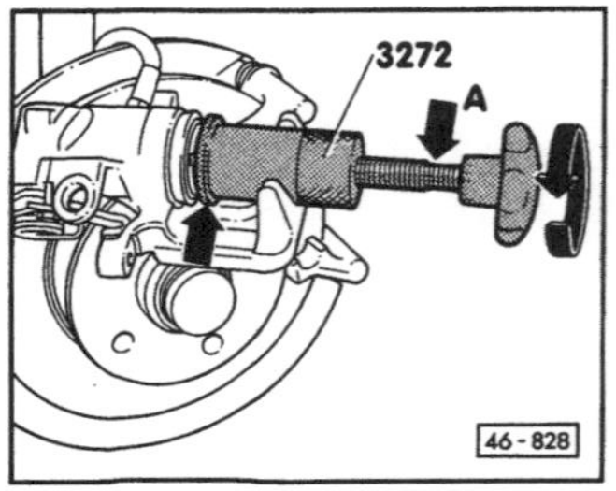

▶ **Fig. 1** **Brake caliper piston, removing**

- turn knurled wheel of tool counterclockwise to remove piston from caliper housing

Note
If the piston is difficult to turn use an open–end wrench (13 mm) on the flats provided (**arrow A**).

> **CAUTION**
> Collar of tool (**arrow**) must be next to caliper piston.

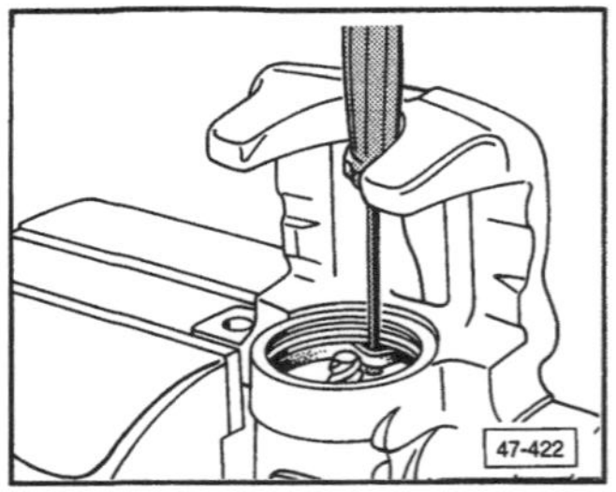

▶ **Fig. 2** **Brake caliper piston seal, removing**

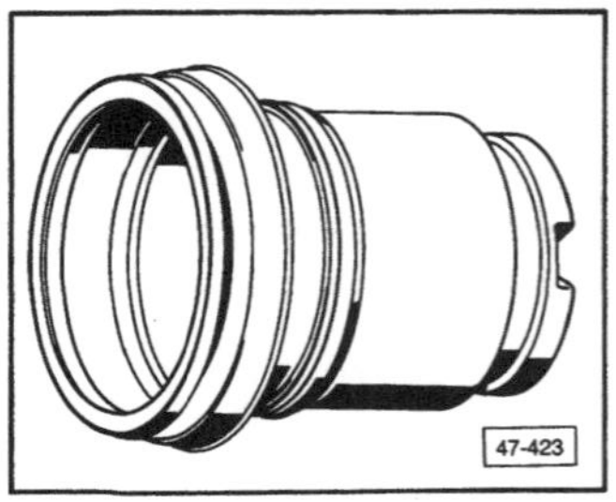

▶ **Fig. 3** **Brake caliper piston, installing**

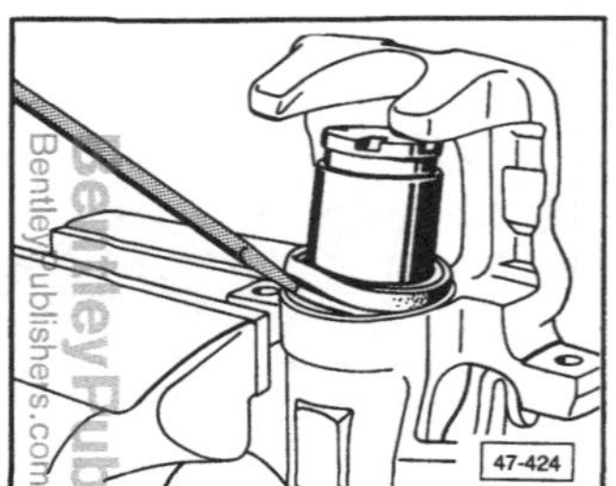

▶ **Fig. 4** **Brake caliper piston, installing**

- install inner lip of dust boot into groove in cylinder with a screwdriver

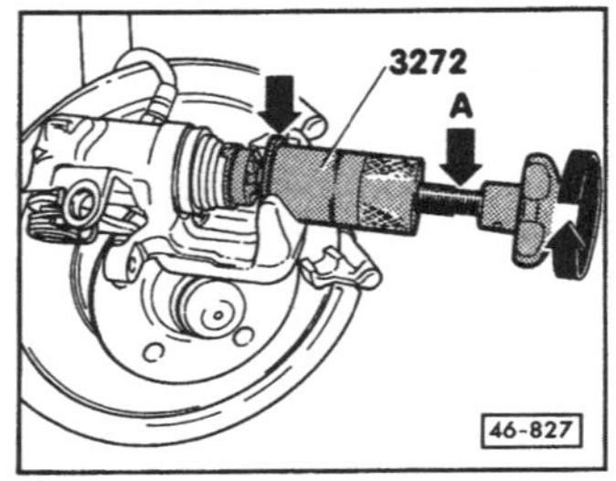

▶ **Fig. 5** **Brake caliper piston, installing**

- turn knurled wheel of tool clockwise to install piston

> **CAUTION**
> Collar of tool (**arrow**) must contact brake caliper.

Note
If the piston is difficult to turn use an open–end wrench (13 mm) on the flats provided (**arrow A**).

> **CAUTION**
> The outer lip of the dust boot must slip into the groove in the piston.

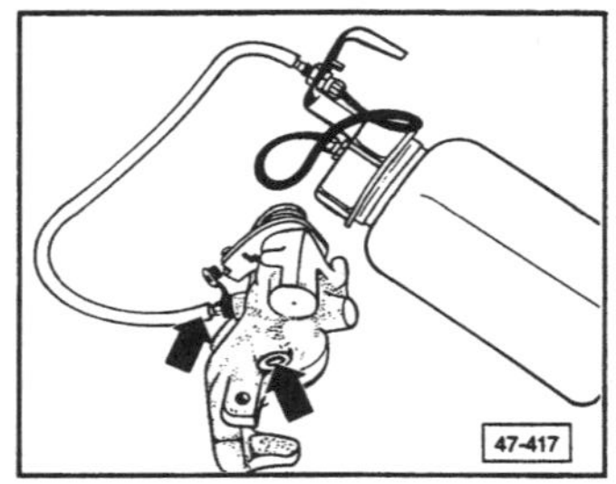

▶ **Fig. 6** **Brake caliper, pre-bleeding**

- hold caliper in position shown
- open bleeder valve and fill with brake fluid from container (**left arrow**) until fluid flows from brake connection (**right arrow**) without bubbles
- close bleeder valve

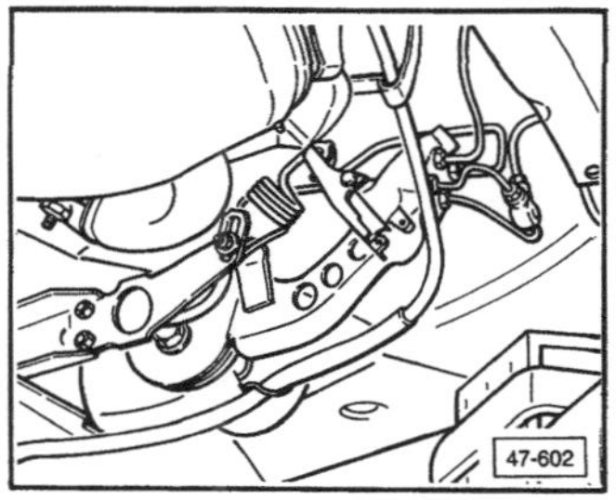

Brake pressure regulator, checking/adjusting

The brake pressure regulator is mounted on a bracket and operated by a spring attached to the rear axle.

Check function

- depress brake pedal firmly once and release suddenly (vehicle on ground)
  - pressure regulator lever must move

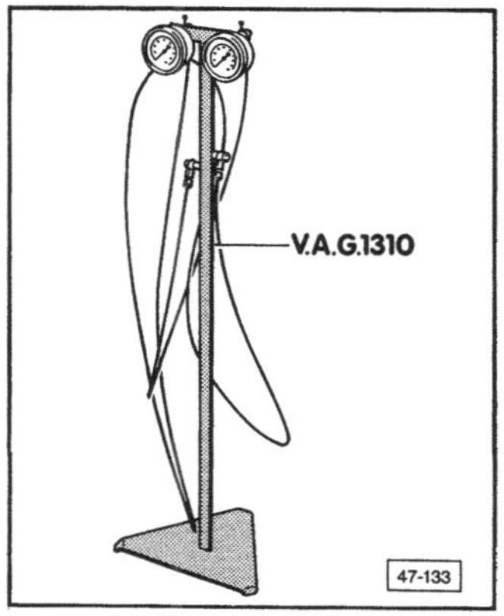

Checking/adjusting pressure (vehicles without ABS)

Check the regulator pressure with the vehicle empty (curb weight — trunk unloaded), fuel tank full and with a driver in the vehicle.

- raise vehicle and connect pressure gauges to left front and right rear brake calipers
- bleed both gauges
- lower vehicle and bounce rear of car several times
- depress brake pedal and obtain specified pressure at front axle and read measured pressure at rear axle:
  - front axle = 50 bar (725 psi)
    rear axle = 27-33 bar (391-478 psi)
  - front axle = 100 bar (1450 psi)
    rear axle = 48-54 bar (696-783 psi)

- release brake pedal
- adjust regulator:
  - if pressure is too high, decrease spring tension
  - if pressure is too low, increase spring tension

- disconnect pressure gauges and bleed brakes

CAUTION

Do not adjust pressure regulator with brake pedal depressed.

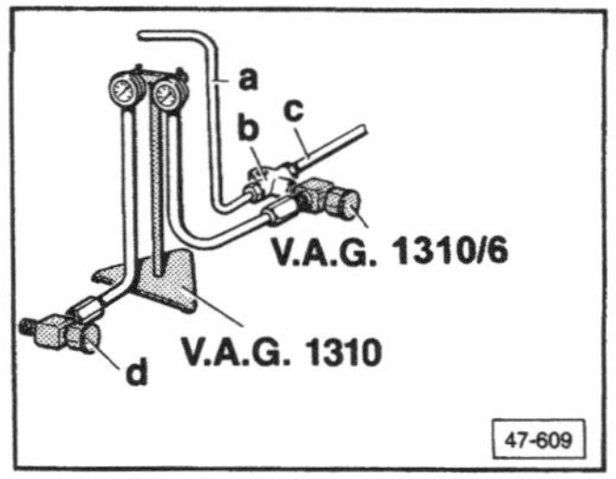

Brake pressure regulator, checking/adjusting (vehicles with ABS)

Check the regulator pressure with the vehicle empty (curb weight — trunk unloaded), fuel tank full and with a driver in the vehicle.

- raise vehicle and connect one pressure gauge in front of pressure regulator and one gauge to rear brake caliper:

  a brake line — to brake pressure regulator (this brake line may contain a distributor in front of the pressure regulator)
  b additional distributor, Part No. **803 611 755**
  c length of brake line, 200 mm long — install between additional distributor and brake pressure regulator (or distributor in front of regulator)
  d pressure gauge connector on rear caliper

- bleed both gauges and rear brakes (see page 47.9)
- lower vehicle and bounce rear suspension several times
- depress brake pedal to obtain specified pressure at front axle, then read measured pressure at rear axle:
  - front axle = 50 bar (725 psi)
    rear axle = 27-33 bar (391-478 psi)
  - front axle = 100 bar (1450 psi)
    rear axle = 48-54 bar (696-783 psi)

- release brake pedal
- adjust regulator:
  - if pressure is too high, decrease spring tension
  - if pressure is too low, increase spring tension

- disconnect pressure gauges and bleed brakes

CAUTION

Do not adjust pressure regulator with brake pedal depressed.

Brake system, bleeding/replacing brake fluid

(vehicles with ABS)

WARNING

Brake fluid is poisonous.

CAUTION

Brake fluid must not come in contact with painted parts.

Brake fluid absorbs moisture from the air and must therefore be replaced every two years.

Do not use silicone-based brake fluid (DOT 5). Even the smallest traces may cause severe corrosion in the brake system.

Use only new, approved brake fluid which complies with **MVSS 116**, DOT 4.

Brake system, bleeding

(with pressure bleeder US1116)

Front axle

- turn ignition off
- depress brake pedal approximately 20 times
- connect pressure bleeder **US1116** to brake fluid reservoir and turn on
- connect hose from bleeder bottle to left front caliper and open bleeder screw
- depress brake pedal slowly until brake fluid flows without bubbles
- close bleeder screw and release pedal
- remove bleeder hose, connect to right front brake caliper and repeat bleeding operation
- remove pressure bleeder **US1116**

Rear axle

- depress brake pedal approximately 20 times (ignition off)
- depress brake pedal and hold
- connect hose of bleeder bottle to right or left rear caliper and open bleeder screw
- turn ignition on
- press lever of pressure regulator in direction of rear axle until brake fluid flows out without bubbles

CAUTION

The fluid level in the brake fluid reservoir must never fall below the **MIN** mark during bleeding.

The pump of the ABS system must not be in operation longer then 120 seconds at a time during this procedure. If the running time of the pump exceeds 120 seconds, a cool-down period of 10 minutes is required.

- bleed second rear brake caliper
- turn on ignition until pump shuts off
- fill brake fluid reservoir up to MAX mark

Brake system, bleeding

(without pressure bleeder US1116)

Front axle

- turn ignition off
- depress brake pedal approximately 20 times
- fill brake fluid reservoir to upper edge

CAUTION

The brake fluid level in the reservoir must not fall below the **MIN** mark during bleeding.

- insert hose of bleeder bottle on bleeder screw of either front caliper
- open bleeder screw
- depress brake pedal until brake fluid flows out without bubbles
- bleed second front brake caliper

Rear axle

Procedure is the same as described for bleeding with pressure bleeder.

Brake fluid, replacing

- turn ignition off
- depress brake pedal approximately 20 times
- connect pressure bleeder **US1116** to brake fluid reservoir and turn on
- insert hose of bleeder bottle on left front caliper and open bleeder screw
- allow 500 cc of brake fluid to flow out by depressing brake pedal
- move bleeder bottle to right side and allow 500 cc of brake fluid to flow out of right front caliper

- remove pressure bleeder **US1116**
- connect hose from bleeder bottle to a rear brake caliper and open bleeder screw
- depress brake pedal and hold down
- turn ignition on
- press lever of pressure regulator in direction of rear axle until new brake fluid flows out without bubbles
- move bleeder bottle to second rear brake caliper and allow 500 cc of brake fluid to flow out

CAUTION

The fluid level in the brake fluid reservoir must never fall below the **MIN** mark during the fluid replacement procedure.

The pump of the ABS system must not be in operation longer than 120 seconds at a time during the fluid replacement procedure. If this running time is exceeded, a cool-down period of 10 minutes is required.

- turn on ignition until pump shuts off
- fill brake fluid reservoir up to **MAX** mark

Note

Brake master cylinders and boosters from different suppliers are installed at the factory. If replacement is necessary, parts are interchangeable.

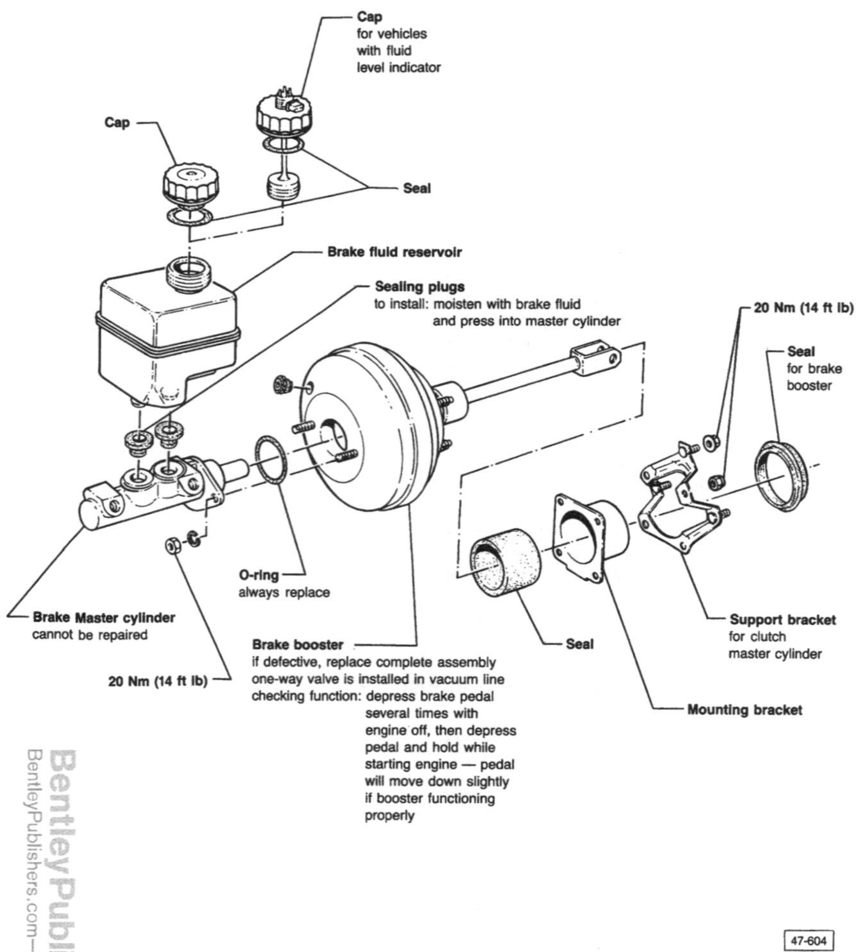

Brake – Hydraulic Components, Regulator, Booster

Brake system, bleeding/replacing brake fluid

(vehicles without ABS)

WARNING

Brake fluid is poisonous.

CAUTION

Brake fluid must not come in contact with painted parts.

Brake fluid absorbs moisture from the air and must therefore be replaced every two years.

Do not use silicone-based brake fluid (DOT 5). Even the smallest traces may cause severe corrosion in the brake system.

Use only new, approved brake fluid which complies with **MVSS 116**, DOT 4.

Brake system, bleeding

(with pressure bleeder US1116)

- connect pressure bleeder **US1116** to brake fluid reservoir and turn on
- connect hose from bleeder bottle to calipers and bleed in following sequence:
  1 — right rear caliper
  2 — left rear caliper
  3 — right front caliper
  4 — left front caliper

Note

Press the regulator lever towards the rear axle during bleeding operation.

Brake system, bleeding

(without using pressure bleeder)

CAUTION

The brake fluid level in the reservoir must not fall below the **MIN** mark during bleeding.

- connect hose from bleeder bottle to each caliper using above sequence
- pump brake pedal several times and then hold pedal down
- open bleed screw
- close bleed screw
- release brake pedal
- repeat operation until brake fluid flows without air bubbles

Brake fluid, replacing

(using pressure bleeder US1116)

- connect pressure bleeder **US1116** to brake fluid reservoir and turn on
- connect hose from bleeder bottle and remove approximately 500 cc of fluid from each caliper following in the following sequence:
  1 — right rear
  2 — left rear
  3 — right front
  4 — left front

Note

To replace the fluid without the pressure bleeder **US1116**, follow the same procedure for bleeding brakes without a pressure bleeder on page 47.14. Remove fluid until the new fluid is visible without air bubbles.

CAUTION

The brake fluid level in the reservoir must not fall below the **MIN** mark during the fluid replacement procedure.

THIS PAGE INTENTIONALLY LEFT BLANK

THIS PAGE INTENTIONALLY LEFT
BLANK

Index

Adaptor sleeve
- removing/installing **48.5**

Auxilliary cooling line
- routing **48.20**

Lock cylinder
- removing/installing **48.5**

Power steering gear
- adjusting (TRW) **48.9**
- adjusting (ZF) **48.15**
- assembly (TRW) **48.7**
- assembly (ZF) **48.13**
- modifications to universal joint shaft and pinion **48.5a**
- removing/installing **48.11**

Power steering pump
- assembly **48.17**
- pressure, checking **48.10**
- pressure, checking (w/VR6 engine) **48.16**

Power steering system
- assembly **48.18**
- cleaning **48.6**
- pressure, checking **48.10**
- troubleshooting **48.21**

Steering column
- assembly **48.2, 48.4**
- length, checking **48.3**

Tie rod
- removing/installing **48.12**

Note

All parts marked with an * can be assembled outside the vehicle.

CAUTION

Always replace self-locking nuts and bolts.

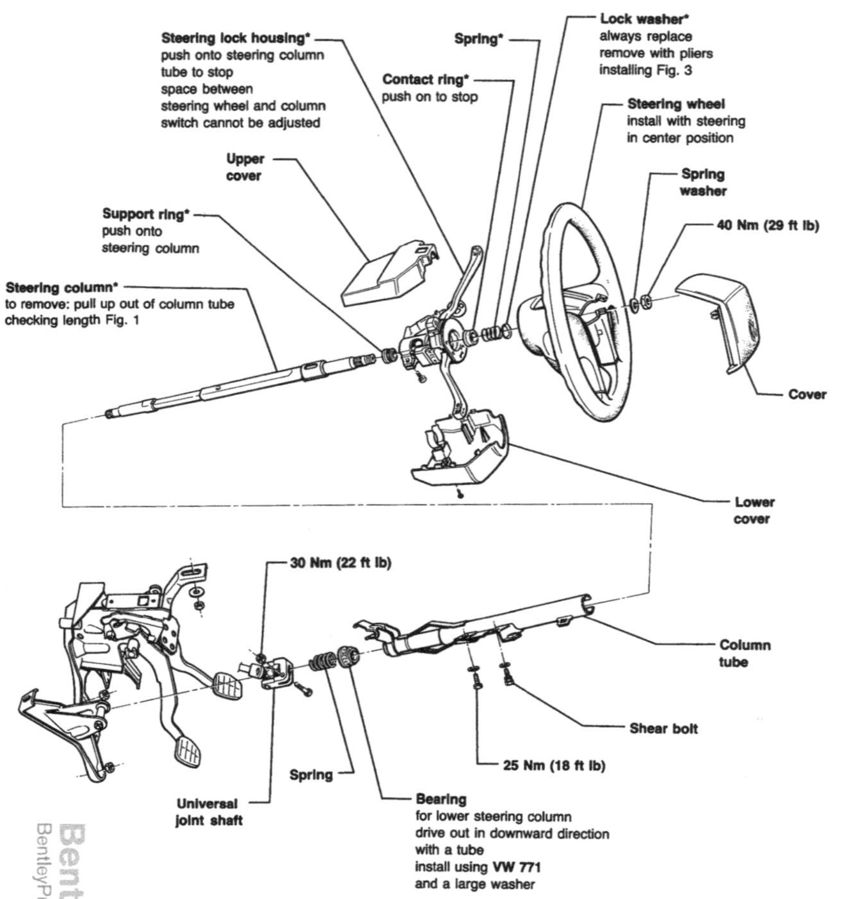

CAUTION

Do not attempt to weld or straighten steering parts.

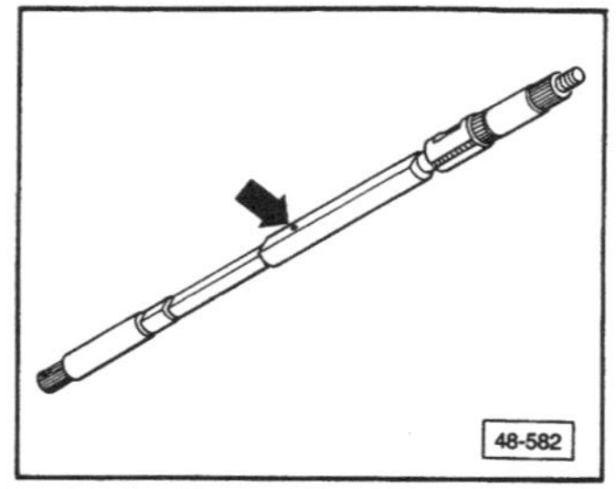

▶ Fig. 1 **Length of steering column, checking**

- pull both upper and lower column halves apart if necessary so that small metal tab on lower column is visible in hole of upper column (**arrow**)

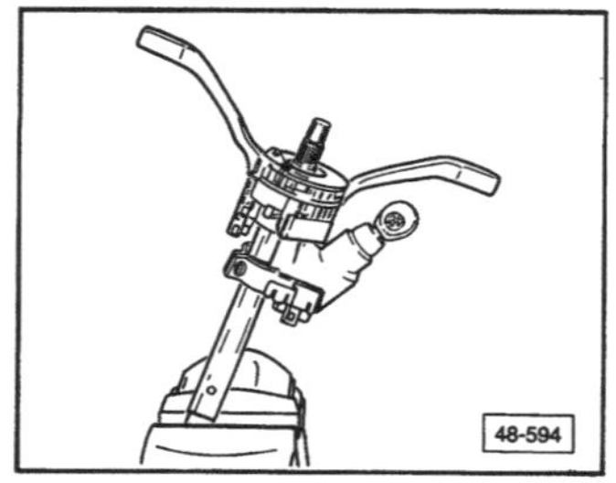

▶ Fig. 2 **Steering column, assembling**

- clamp lower column in vise so upper column is supported and both halves do not telescope

Note

Use protective covers on vise jaws.

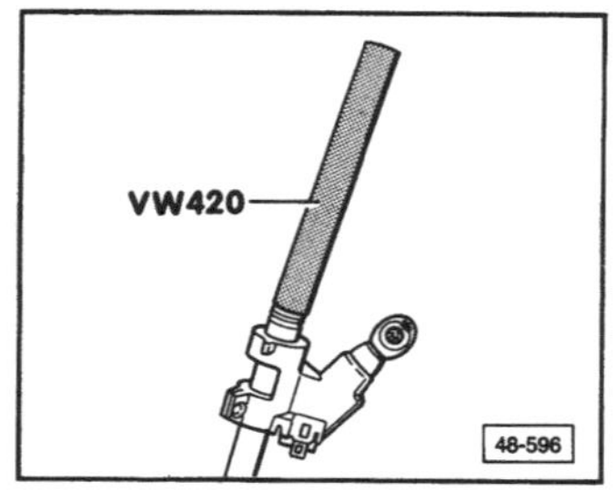

▶ Fig. 3 **Lock washer, installing**

- push lock washer on until spring is completely compressed

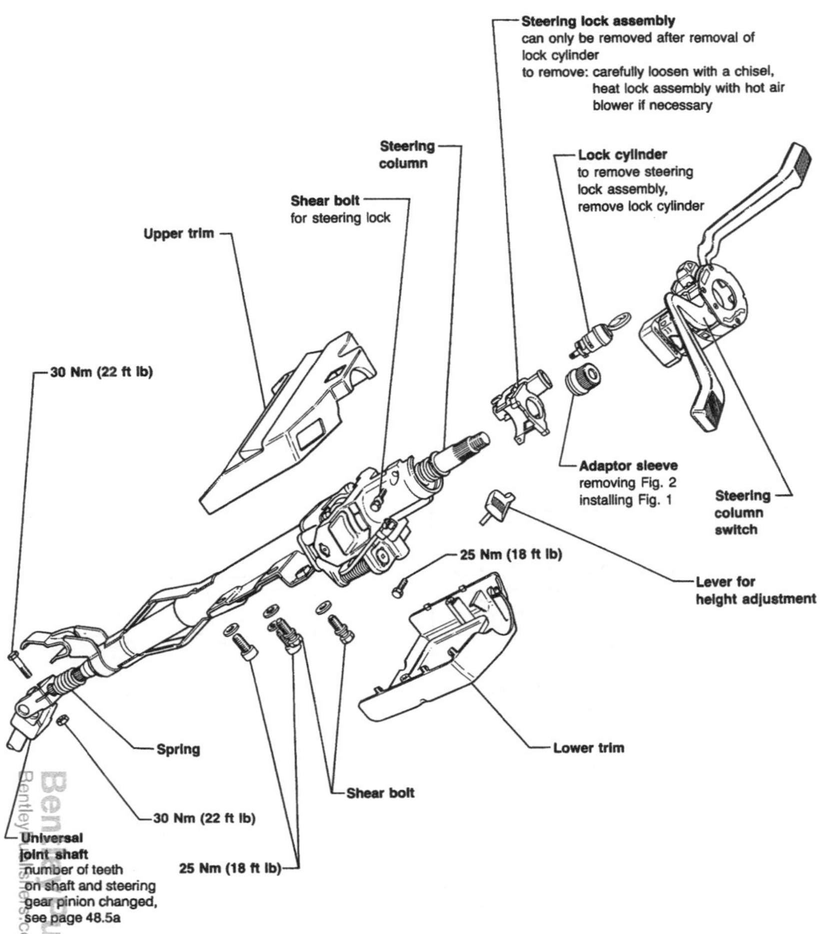

Note

The height-adjustable steering column cannot be disassembled. The steering column is to be replaced complete if there is damage to the column tube, column or universal joint.

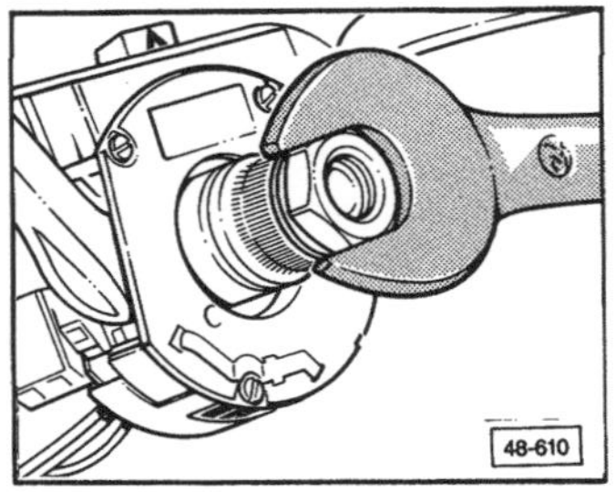

▶ **Fig. 1 Adaptor sleeve, installing**

- place sleeve on column and tighten until firmly seated using a nut and washer

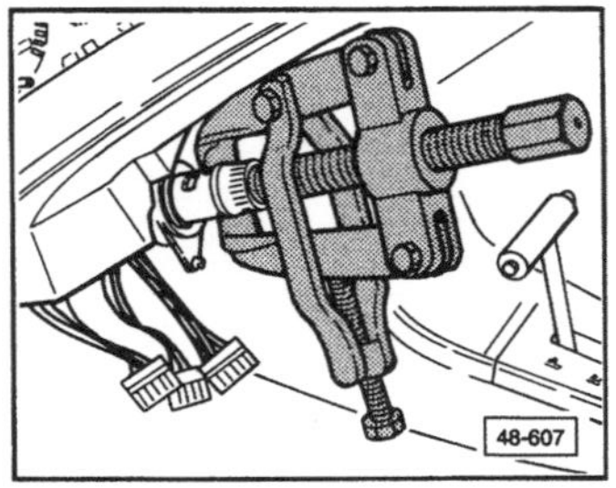

▶ **Fig. 2 Adaptor sleeve, removing**

- use only puller with a clamp, e.g. **Kukko 204-2**

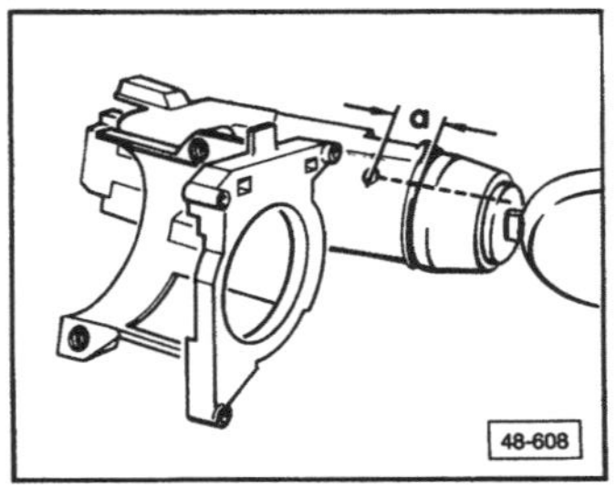

Lock cylinder, removing/installing

▶

Removing

- remove steering column
- remove steering column switch
- insert ignition key in ignition lock
- mark dimension **a** = 12 mm (on line with key)
- bore hole (approximately 3 mm diameter) through steering lock housing until stop spring of lock cylinder is visible (hole depth = approximately 3 mm)
- compress stop spring with a punch and pull out lock cylinder with key

Installing

- press lock cylinder into steering lock housing

Note

The lock cylinder can be pressed in to stop by gently turning the key in the lock cylinder.

Modifications to Universal Joint Shaft, Steering Gear Pinion

From VIN: 50 N 012 001 the number of teeth on the lower universal joint shaft and the steering gear pinion have been changed.

This change allows the universal joint shaft to slide easily onto the steering gear pinion.

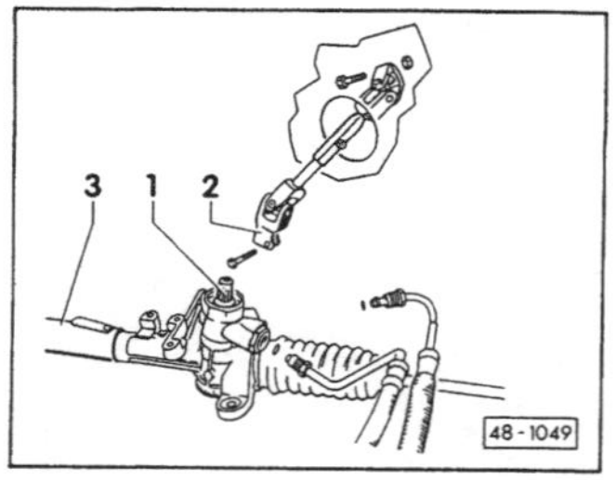

1 — Steering gear pinion
Previously: 36 teeth
New: 22 teeth

2 — Lower universal joint shaft
Previously: 36 teeth
New: 22 teeth

3 — Power steering gear

CAUTION
Old and new universal joint shafts and steering gear pinions are not interchangeable. Always replace as a matched pair.

When working on the power steering system

CAUTION

Always follow these "rules" for cleanliness when repairing the power steering system:

1. Clean connections and surrounding area thoroughly before loosening.
2. Place parts removed on a clean surface and cover them. Use plastic, paper or lint-free cloths.
3. Carefully cover opened components or plug openings if repair is not going to be done immediately.
4. Install clean parts only.
   - do not remove replacement parts from packaging until just before installation
   - do not use parts that were stored unpacked
5. When the power steering hydraulic system is opened:
   - do not use compressed air in the immediate area
   - do not move the vehicle unless absolutely necessary

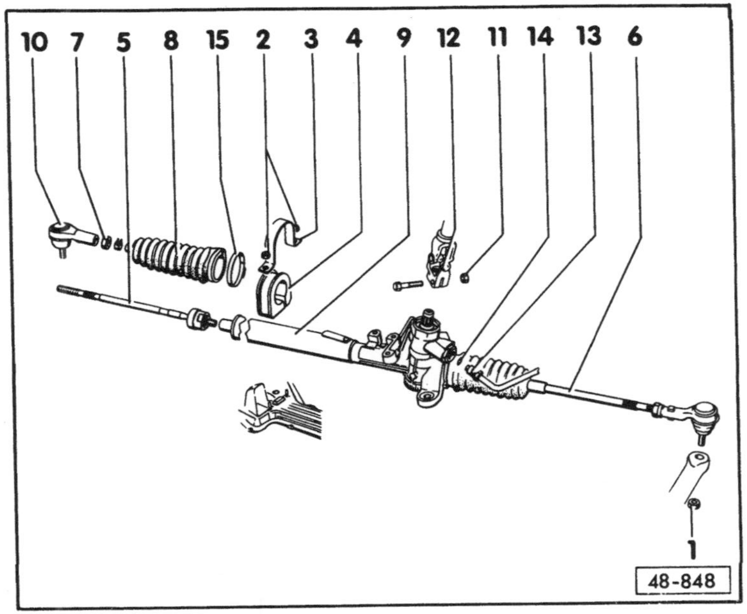

No provision is made for repairing the power steering gear. If a malfunction occurs determine the cause by performing a pressure test and a leak test. Replace the complete power steering gear if it is faulty.

Note
- Always replace self–locking nuts and bolts
- DO NOT attempt to weld or straighten steering components
- Use only steering gear grease for the steering gear rack, Part No. AOF 063 000 04
- Use only hydraulic oil, Part No. G 002 000 in the power steering system
- capacity of power steering system: 0.7 to 0.9 liters

1 – **35 Nm (26 ft lb)**

2 – **30 Nm (22 ft lb)**

3 – **Clamp**

4 – **Rubber bushing**

5 – **Tie rod, right**
- 70 Nm (52 ft lb)
- removing/installing, page 48.12
- adjusting toe, see Repair Group 44

6 – **Tie rod, left**
- removing/installing, page 48.12
- adjust to length, see page 48.12

7 – **50 Nm (37 ft lb)**

8 – **Boot**
- can be replaced with steering gear installed
- pull off and on over tie rod end
- must not be twisted after adjusting toe

9 – **Steering gear**
- removing/installing, page 48.11
- adjusting, page 48.9
- number of teeth on pinion changed as of VIN 50 N 012 001

10 – **Tie rod end**
- o pressing off, page 48.12

11 – **30 Nm (22 ft lb)**

12 – **Universal joint shaft**
- number of teeth in lower joint clevis changed as of VIN 50 N 012 001

13 – **30 Nm (22 ft lb)**

14 – **O–ring**
- always replace

15 – **Clamp**
- tension using V.A.G 1275 or equivalent

TRW power steering gear, adjusting

Note
Two mechanics are required for the steering gear adjustment.

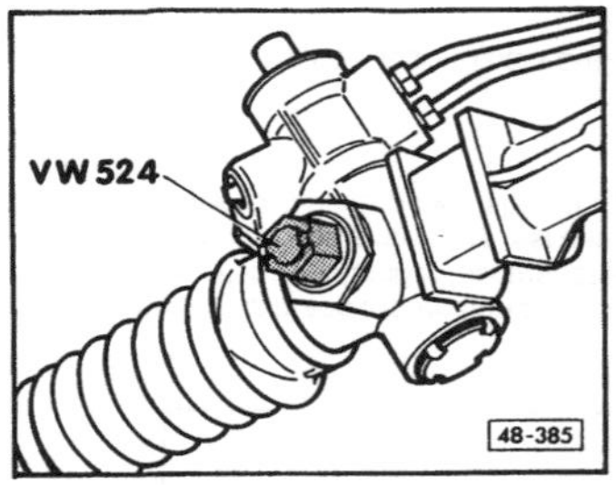

- raise vehicle
- turn wheels approx. 30° left and right of center axis (engine OFF) several times and listen for knocking noises

- turn adjusting bolt (**arrow**) clockwise until knocking noise is no longer heard inside vehicle
- road test vehicle and ensure that steering returns to straight–ahead position by itself without binding after cornering
- correct steering adjustment if necessary

Power steering pump pressure, checking

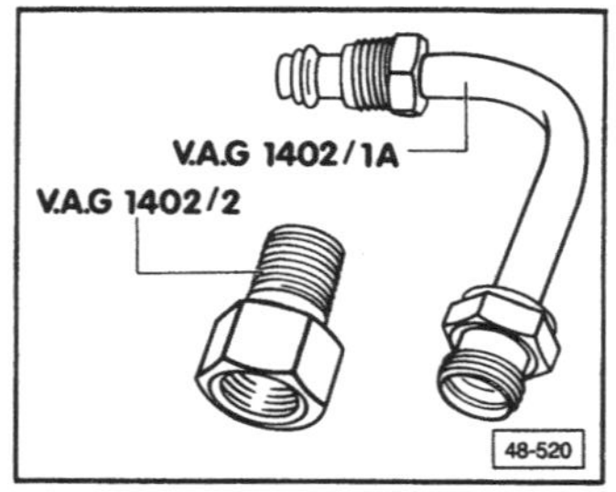

Two adaptors are required to connect pressure gage **US 1074B** to the power steering system.

VW 1402/1A = adaptor for pump

VW 1402/2 = adaptor for pressure gage

- remove pressure line from pump

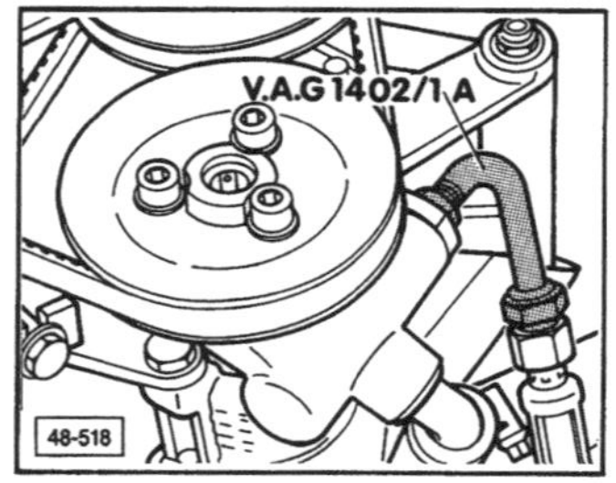

- install adaptor and gage (valve in open position)
- start engine and, if necessary, top up fluid level in reservoir
- close valve with engine running at idle (leave closed no longer than 5 seconds) and note pressure
  - 76-82 bar (1102-1189 psi)
- if pressure is outside specified range, replace pump

System pressure, checking

- with engine idling and valve **US 1074B** open, rotate steering wheel lock to lock and hold (not longer than 5 seconds) to check pressure in each end position
  - 76-82 bar (1102-1189 psi)
- if pressure is outside specified range at each end position, replace power steering gear assembly

CAUTION

If fluid level in reservoir is low, always check power steering system for leaks.

If the leak is at the steering gear pinion or rack seal in the steering gear housing, replace the steering gear assembly.

Power steering gear, removing/ installing

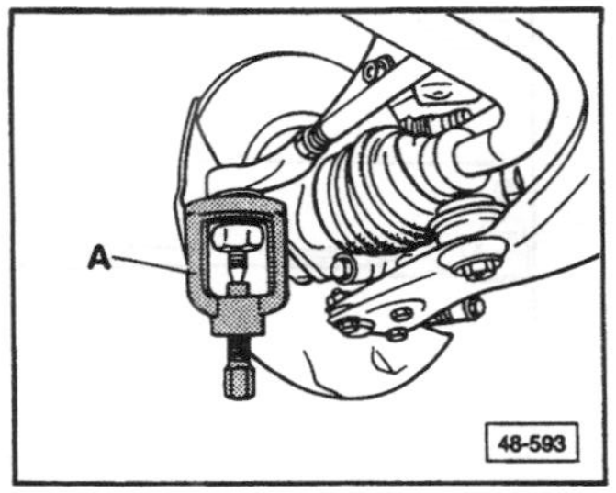

Removing

- remove tie rods
  - **A** = commercial tie rod puller, e.g. **Kukko No. 128/2**

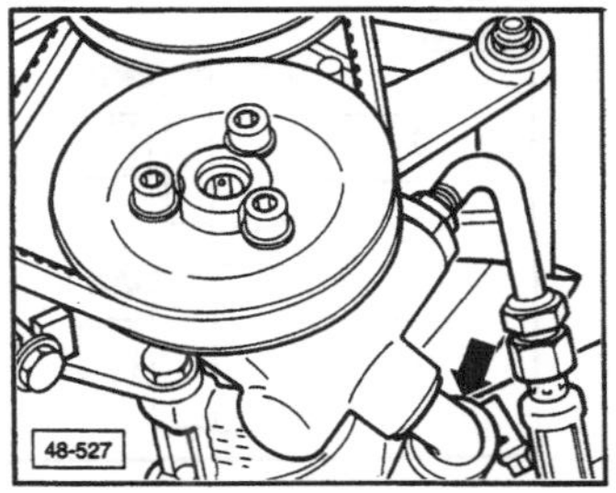

- remove suction hose at pump (**arrow**) and drain
- remove lock bolt for universal joint shaft
- remove nuts and take steering gear assembly off mounting
- disconnect pressure and return lines from steering gear
  - width across flats = 18 mm

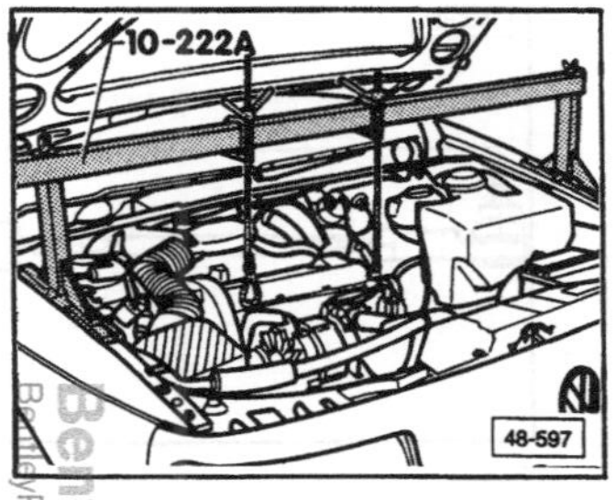

- support engine/transmission with **10-222A** and **10-222A/1**
- remove bolts for subframe and remove steering gear towards rear

Installing

- install in reverse order and tighten subframe bolts to specified torque — see Group 40
- refill system with hydraulic oil, Part No. **G 002 000**

Tie rods, removing/installing

Removing

> **CAUTION**
> To avoid damage to the rack and pinion, DO NOT attempt to remove tie rods from an installed steering gear.

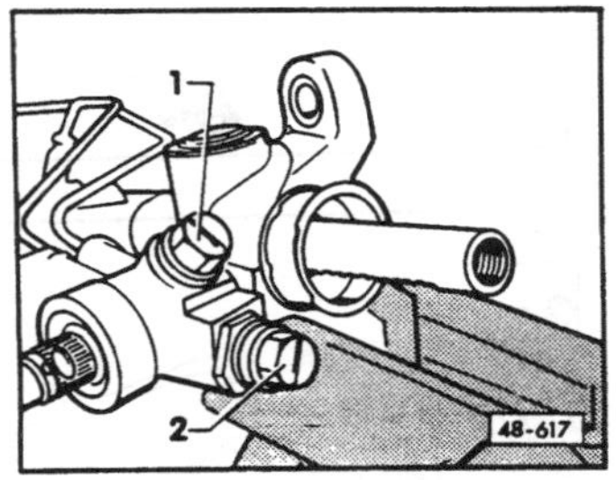

- remove steering gear from vehicle
- seal hydraulic line openings with plastic bags and adhesive tape
- seal hydraulic ports in steering gear, **1** and **2**, with plastic sealing plugs
- clean steering gear
- loosen clamp and push boot back

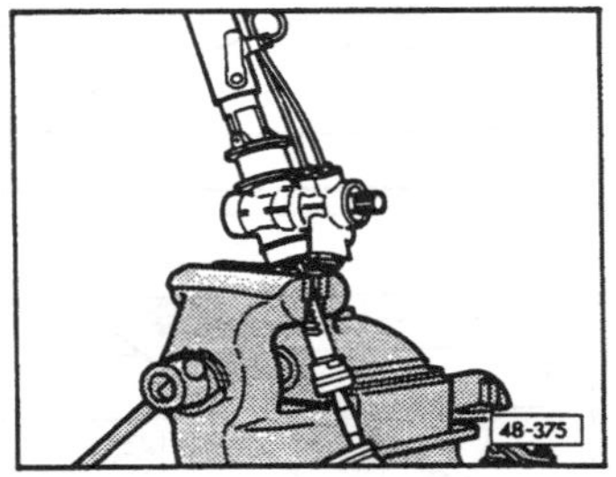

- clamp steering gear in vice and remove tie rods
  - use soft jaw covers in vice to protect rack teeth

Note
Ensure that parts will not be damaged at reassembly. Remove the old locking compound from the tie rod threads and from threaded hole in rack, as follows:

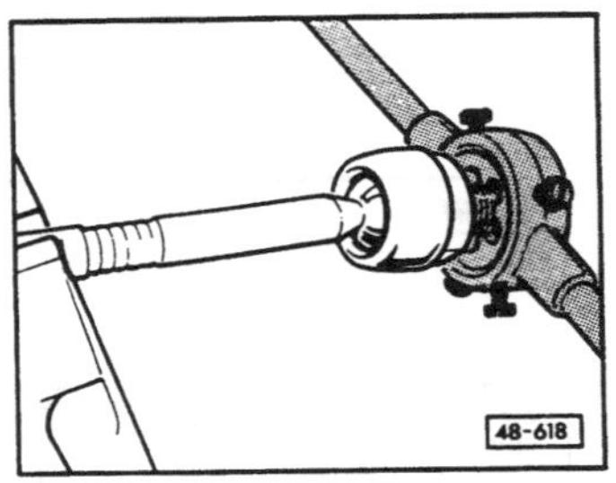

- use die (M14 x 1.5) or wire brush to clean threads at end of tie rod

> **CAUTION**
> When starting the die onto the threads, do not tilt the die. If the die is difficult to turn or if metal shavings form, the threads have been damaged.

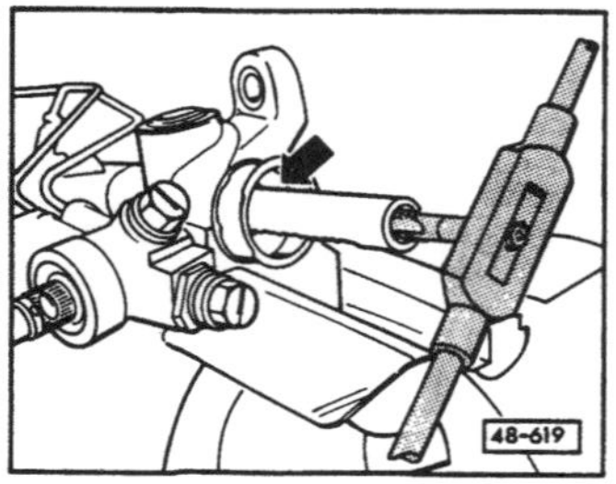

- seal opening in steering gear (arrow) with clean rags
- use tap (M14 x 1.5) to clean threaded hole in rack
- clean oil/grease residue from threads using thinner L 160 or equivalent
- clean threads with compressed air

- remove rags from opening in steering gear

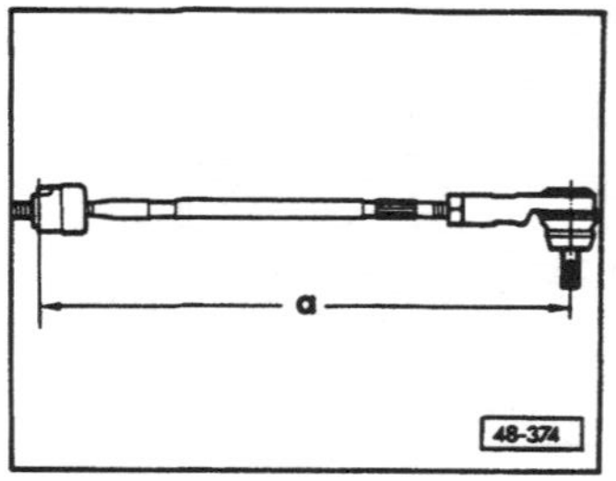

Installing

- check/adjust length of left tie rod
  - vehicles with G60 engine and 185/60 or 195/60 tires:
    a = 379.5 ± 1 mm (14.94 ± 0.039 in.)

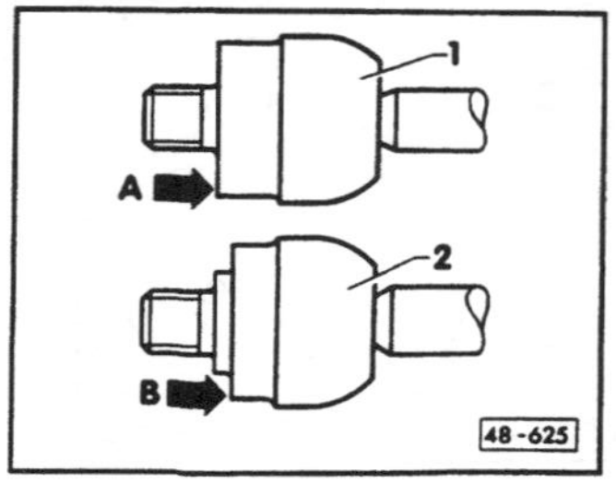

The Corrado G60 was also produced with 205/60 tires. The steering stroke is reduced on these vehicles due to the revised inner tie rod joint, therefore, the turning circle is slightly larger.

1 Inner joint of tie rod for Corrado G60 with 205/60 tires – contact surface **A** without offset

2 Inner joint of tie rod for all other Corrado G60 models – contact surface **A** with offset

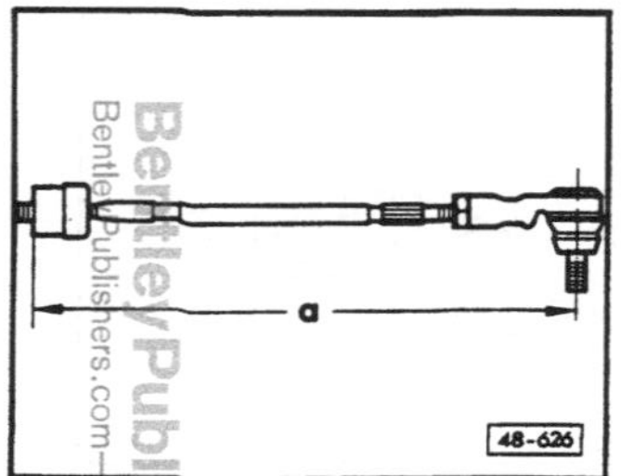

- vehicles with G60 engine and 205/60 tires:
  a = 383 ± 1 mm (15.01 ± 0.039 in.)

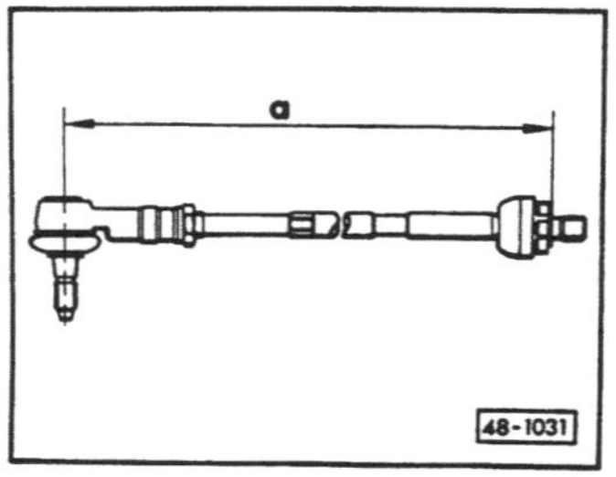

- vehicles with VR6 engine:
  a = 406.0 ± 1 mm (15.99 ± 0.039 in.)

Note
Adjust toe on the right tie rod only.

New left tie rods must always be checked/adjusted for the correct length before installing.

Whenever the length of the left tie rod is disturbed by the removal/replacement of the right tie rod end, it must be readjusted to the correct length specification.

- apply **one drop** of locking compound D6 (or equivalent) to **two** threads in hole of steering rack (arrow)
- thread on tie rod and tighten to 70 Nm (52 ft lb)
- check boot for wear (slits, cracks) and check sealing surfaces for cleanliness
- turn tie rod end to installation position
- install pressure compensating line

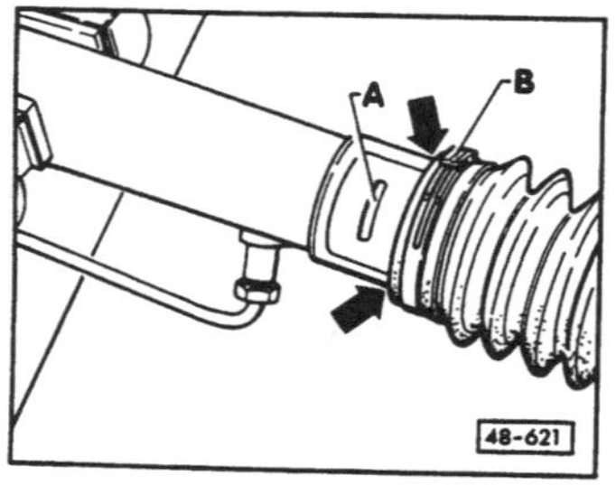

- install boot and seat (arrows) without damaging label **A**
- use only new clamps of original specification and tension with pliers VAG 1275 or KD 424

- install steering gear in vehicle and adjust toe on right side only – see Repair Group 44 in Repair Manual

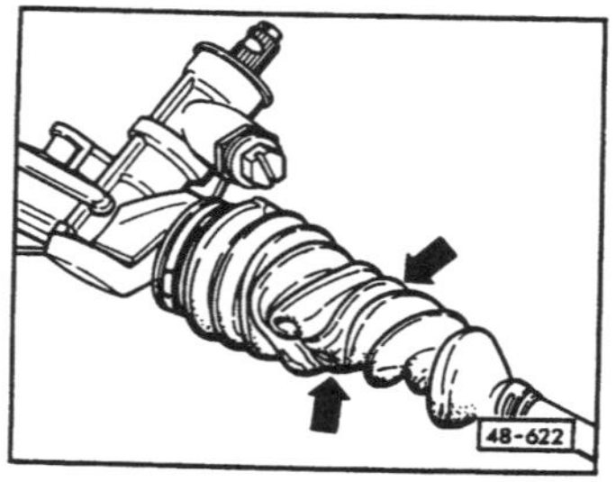

Note
Always check that steering boots are installed without being twisted (arrows).

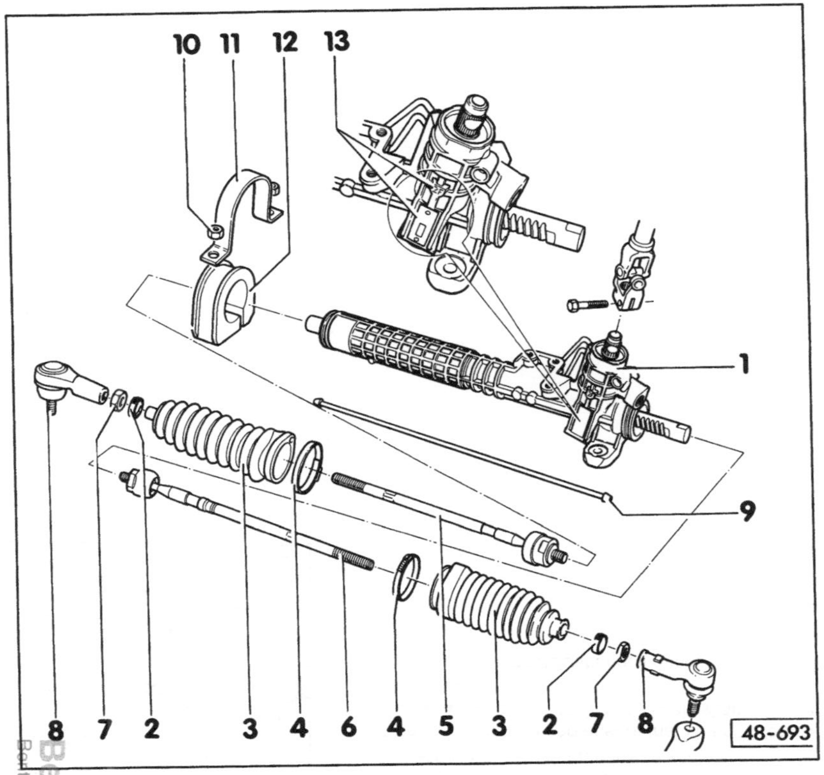

No provision is made for repairing the power steering gear. If a malfunction occurs determine the cause by performing a pressure test and a leak test. Replace the complete power steering gear if it is faulty.

Note
- Always replace self–locking nuts and bolts
- DO NOT attempt to weld or straighten steering components
- Use only steering gear grease, Part No. AOF 063 000 04 for the steering gear rack
- Use only hydraulic oil, Part No. G 002 000 in the power steering system
- capacity of power steering system, 0.7 to 0.9 liters

1 – **Steering gear**
- removing/installing, page 48.11
- adjusting, page 48.15

2 – **Clamp**
- cut open; always replace
- tension using V.A.G 1275 or equivalent
- not interchangeable with TRW clamp

3 – **Boot**
- remove tie rod end before removing
- check for wear or damage
- not interchangeable with TRW steering gear boots

4 – **Clamp**
- tension using V.A.G 1275 or equivalent
- not interchangeable with TRW clamp

5 – **Tie rod, right**
- removing/installing, page 48.12
- adjust toe on right tie rod only
- diameter of tie rod at boot sealing surface: ZF: 15 mm (TRW: 12.7 mm)
- must not be installed on TRW steering gear

6 – **Tie rod, left**
- removing/installing, page 48.12
- adjust toe at right tie rod only
- diameter of tie rod at boot sealing surface: ZF = 15 mm (TRW = 12.7 mm)
- must not be installed on TRW steering gear
- adjusting length, page 48.12

7 – **50 Nm (37 ft lb)**

8 – **Tie rod end**
- pressing off, page 48.11

9 – **Pressure balancing line**

10 – **30 Nm (22 ft lb)**

11 – **Clamp**

12 – **Rubber bushing**
- inner diameter: ZF = 50 mm (TRW) = 44 mm

13 – **ZF data plate**

ZF power steering gear, adjusting

Note
Two mechanics are required for the steering gear adjustment.

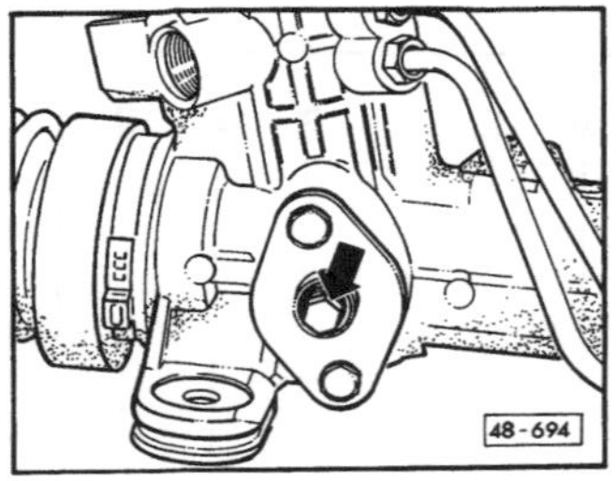

- raise vehicle
- turn wheels approx. 30° left and right of center axis (engine OFF) several times and listen for knocking noises
- turn adjusting bolt (**arrow**) clockwise until knocking noise is no longer heard inside vehicle
- road test vehicle and ensure that steering returns to straight–ahead position by itself without binding after cornering
- correct steering adjustment if necessary

Power steering pump pressure, checking

(vehicles with VR6 engine)

Test requirements

- v–belt and v–belt tension **OK**
- no leaks in system
- hoses/lines not bent or collapsed

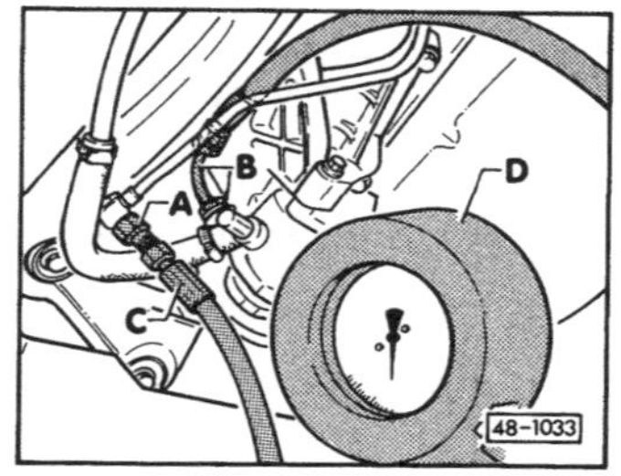

- remove pressure line from pump
- install adapter and gauge (valve in open position)
- start engine and top up fluid level in reservoir, if necessary

A = V.A.G 1402/2

B = V.A.G 1402/1

C = hose from adapter set 1402/6

D = US 1074B

- close valve with engine running at idle (leave closed no longer than 5 seconds) and read pressure
  - 85 – 95 bar (1232 – 1377 psi)
- if pressure is outside range shown, replace pump

CAUTION
If fluid level in reservoir is low, always check power steering stytem for leaks.

If the leak is at the steering gear pinion or rack seal in the steering gear housing, replace the steering gear assembly.

CAUTION
Use only hydraulic oil, Part No. **G 002 000,** in power steering system.

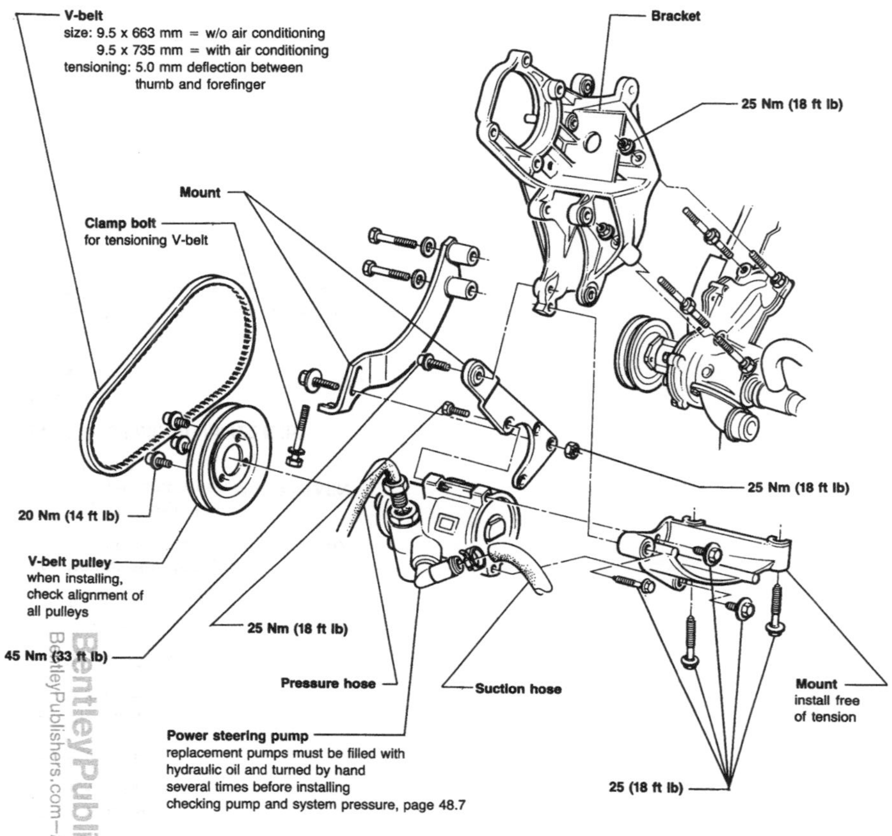

CAUTION
Always replace self-locking nuts and bolts.
Always replace seals.

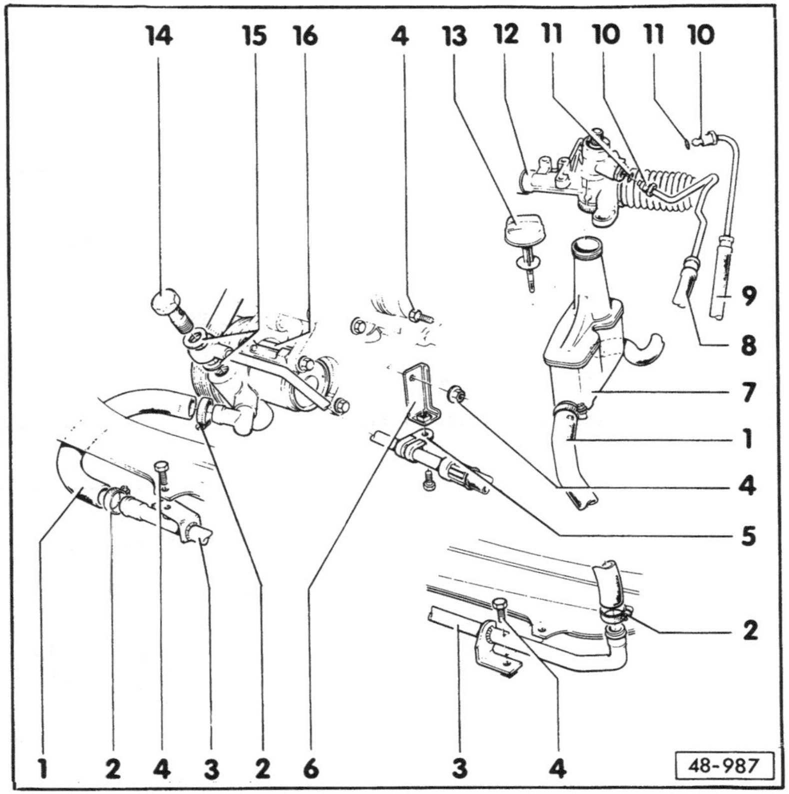

No provision is made for repairing the power steering pump. If a malfunction occurs determine the cause by performing a pressure test and a leak test. Replace the complete power steering pump if it is faulty.

Note
Replacement pumps do not contain oil. Fill the pump with hydraulic oil, Part No. G 002 000, before installing and rotate by hand to prime. Failure to prime the pump before use may cause excessive noise in operation or damage.

Numbered list continued on next page

1 – **Suction hose**

2 – **Hose clamp**

3 – **Cooling line**
- see Fig. 1 and Fig. 2

4 – **10 Nm (7 ft lb)**

5 – **Line clamp**

6 – **Bracket**

Note
Numbered list continued from previous page

7 – Reservoir
- see Fig. 3

8 – Return hose

9 – Pressure line
- tighten to 30 Nm (22 ft lb)

10 – 30 Nm (22 ft lb)

11 – O–ring
- always replace

12 – Power steering gear
- removing/installing, page 48.10

13 – Cap with dipstick
- oil level: between MAX and MIN marks

14 – Banjo bolt, 30 Nm (22 ft lb)

15 – Seal
- always replace

16 – Power steering pump
- removing/installing, page 48.11
- fill with oil before installing, see note on previous page

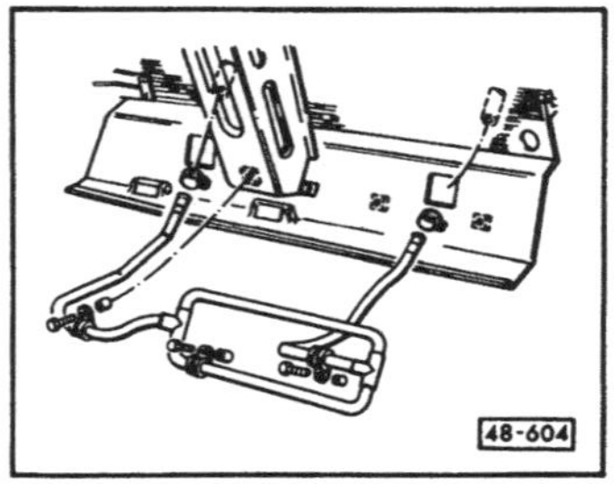

Fig. 1 Auxilliary cooling line routing
- for vehicles with air conditioner

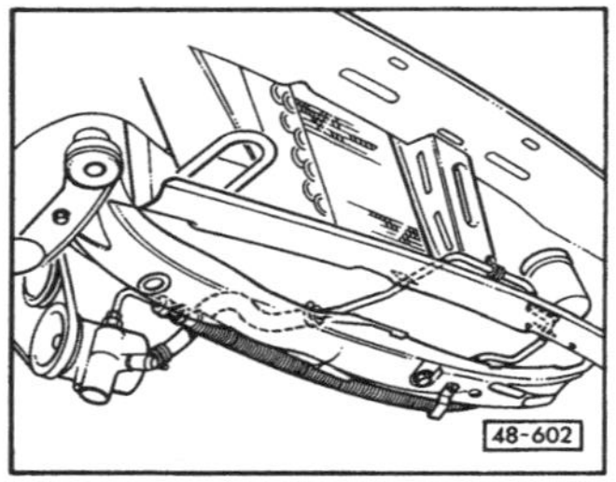

Fig. 2 Additonal cooling line routing

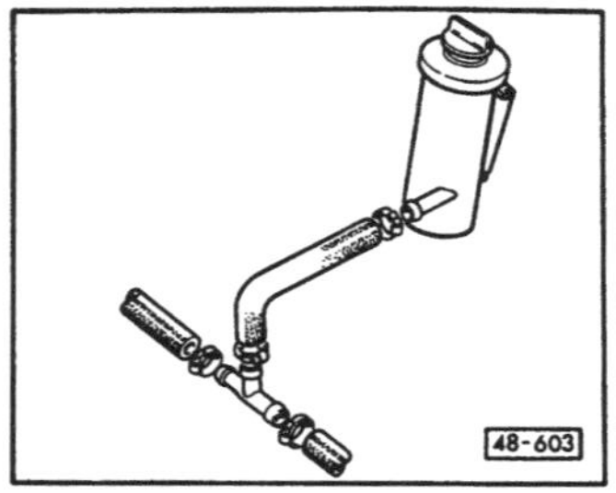

Fig. 3 Power steering reservoir
- for vehicles with auxilliary cooling line

Power steering, troubleshooting

Noises

A properly functioning power steering system will generate certain noises and still be considered in working order.

Normal noises:

Note

In order to evaluate a noise from the power steering system it may be necessary to conduct a "noise comparison test" with an identical new vehicle.

- hissing sounds when vehicle stationary, engine running and steering wheel is slowly turned or turned briefly to full left/right stop
- increased hissing/groaning noises when wheels turned fully to either side during parking

Abnormal noises:

Noises determined not to be normal may be caused by one or more of the following:

- hydraulic oil level too low in fluid reservoir
- steering pump defective
- incorrect adjustment of power steering belt
- hydraulic hose bent or pinched
- hydraulic hoses/lines touching other parts of vehicle
- steering gear mounting loose
- steering pump mounting to bracket/ engine loose
- pulley loose on pump

Note

It may be necessary to run the engine with the steering pump and belt removed to verify if noises are from power steering system.

Leakage

Note

The source of a leak in the power steering system may not always be at the spot where it is seen. Always check hose/line connections for tightness.

CAUTION

If contaminants are found in fluid reservoir, or the hydraulic oil, carefully clean the pump, reservoir, pressure or return hoses. Replace contaminated components as required.

CAUTION

When steering components and connecting lines are replaced, always fill steering system with fresh hydraulic oil and bleed system. Do not reuse hydraulic oil which has been drained. Do not allow reservoir to become dry.

Steering drag checking

Note

Corrado and vehicles with 16V engines require higher steering force to turn wheels.

Test conditions

- vehicle on ground
- tire pressures **OK**
- engine idling
- hydraulic oil level in fluid reservoir **OK**
- wheels in straight ahead position
- power steering belt tension **OK**

■ remove horn pad

■ install Snap-on Torqometer style torque wrench TESI 120 or equivalent
  - set to 10 Nm (7 ft lb)

■ turn wheels slowly to left and right stop

If it takes more than 10 Nm (7 ft lb) to turn wheels then refer to troubleshooting guide on next page.

Power steering system, troubleshooting guide

- verify complaint
- satisfy basic test conditions

Test conditions:

- vehicle on ground
- tire pressures **OK**
- hydraulic oil level in fluid reservoir **OK**
- wheels in straight ahead position
- power steering belt adjustment **OK**

If the problem still exists:

- find "symptom" in top section
- scan the vertical column for ALL possible cause symbols
- read across from the symbols to determine possible cause(s)

OIL SPOTTING UNDER VEHICLE	NO HYDRAULIC OIL IN FLUID RESERVOIR	HYDRAULIC OIL LEVEL IN FLUID RESERVOIR TOO LOW	HYDRAULIC OIL IN FLUID RESERVOIR HAD TO BE REPLACED A SECOND TIME	HYDRAULIC OIL SPRAY IN ENGINE COMPARTMENT	HYDRAULIC OIL IN FLUID RESERVOIR MILKY OR FOAMY	POSSIBLE CAUSE	SOLUTION
▲	▲	●				Excessive leakage at steering gear (oil in boot)	Replace steering gear
●	●	●				Excessive leakage at power steering pump (intake connection)	Replace power steering pump
●	○	○		●		Leak at fluid reservoir	Replace fluid reservoir
○	○	○	○	○		Hydraulic oil hose/line defective	Replace hydraulic oil hose/line
○	○	▲	▲	○	●	Leak at hydraulic oil hose/line	Tighten hose/line connections
					▲	Air leak system	Check and tighten hose connections, add hydraulic oil
○				▲		Fluid reservoir overflowing	Fill reservoir level to between **MIN/MAX** marks

Symbols

- ▲ Generally the cause
- ● Occasionally the cause
- ○ Infrequently the cause

STEERING DIFFICULT/INSUFFICIENT POWER ASSIST	STEERING EFFORT OCCASIONALLY HIGH	STEERING WHEEL JERKS AT FULL LEFT/RIGHT LOCK	STEERING IN CENTER POSITION DIFFICULT/POOR STRAIGHT LINE DRIVING	STEERING WHEEL VIBRATES/SHAKES WHILE DRIVING	Possible cause	Solution
▲	▲	▲			Hydraulic oil level too low/oil loss	Check system for leaks and repair. Fill fluid reservoir to **"MAX"**
●	○	●	●		Power steering pump defective Steering gear defective	Check feed pressure, replace pump if necessary Check system pressure, replace steering gear if necessary
●	○	●	●		Air in hydraulic system, hose connections loose	Tighten hose connection(s)
		●	○		Tire pressure too low	Check and adjust
●			●		Front end alignment incorrect	Align front end to specification
○			○	○	Front wheel/wheel assembly out of balance	Correct imbalance
				▲	Insufficient wheel clearance • wide tires/rims	Locate and correct insufficient wheel clearance
○	○	○			Steering gear mounting loose	Tighten steering gear mounting
		○			Steering gear mounted under tension	Loosen and retighten steering column mounting bolts
○			○		Bearing in steering column switch defective	Replace bearing
●	●	▲			Power steering belt tension incorrect	Adjust drive belt tension, check belt/pulley for damage

Symbols

- ▲ Generally the cause
- ● Occasionally the cause
- ○ Infrequently the cause

Index

Acid rain
- repairing **50.11**

Body dimensions
- door opening **50.4**
- floor section **50.2**
- front section **50.4**
- rear section **50.5**

Front fender
- adjusting **50.7**
- mounting **50.6**

Headlamp
- removing **50.9**

Lock support
- assembly **50.8**
- removing/installing **50.10**

Shroud
- removing/installing **50.9**

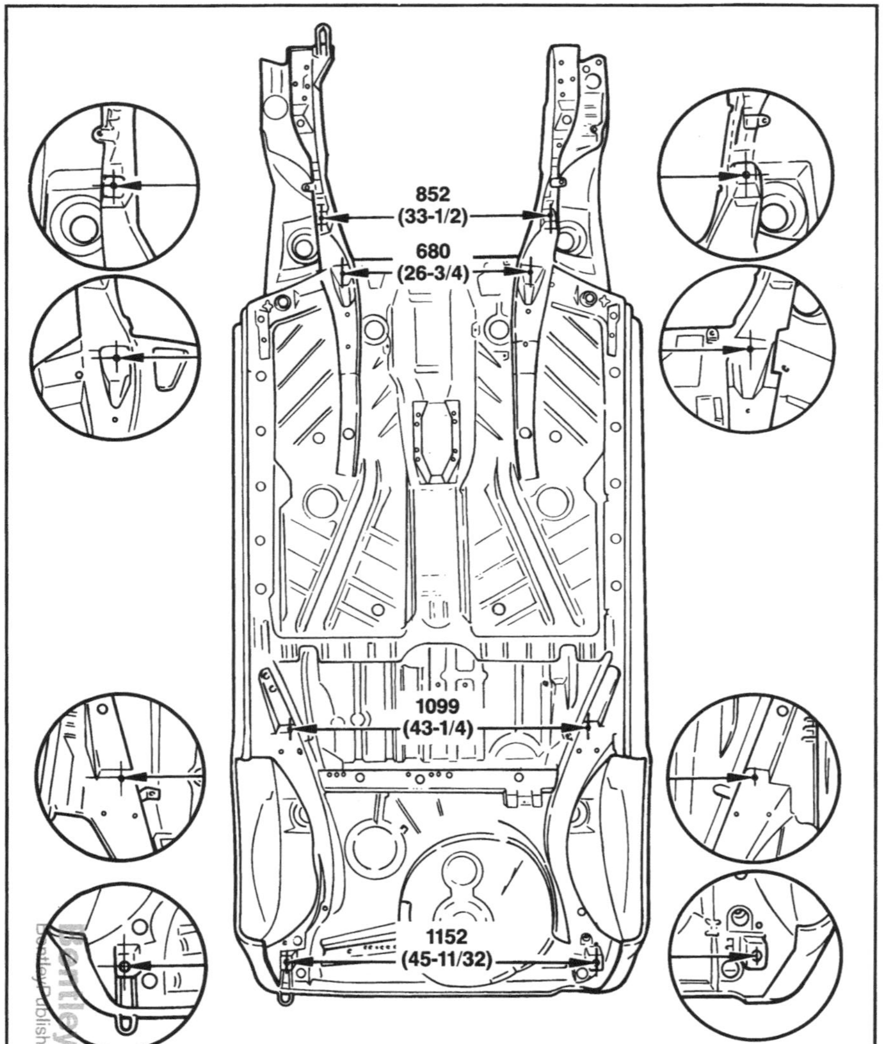

Note

All measurements in mm (in.)
Measurement tolerance = ± 2 mm (3/32 in.)

00-854

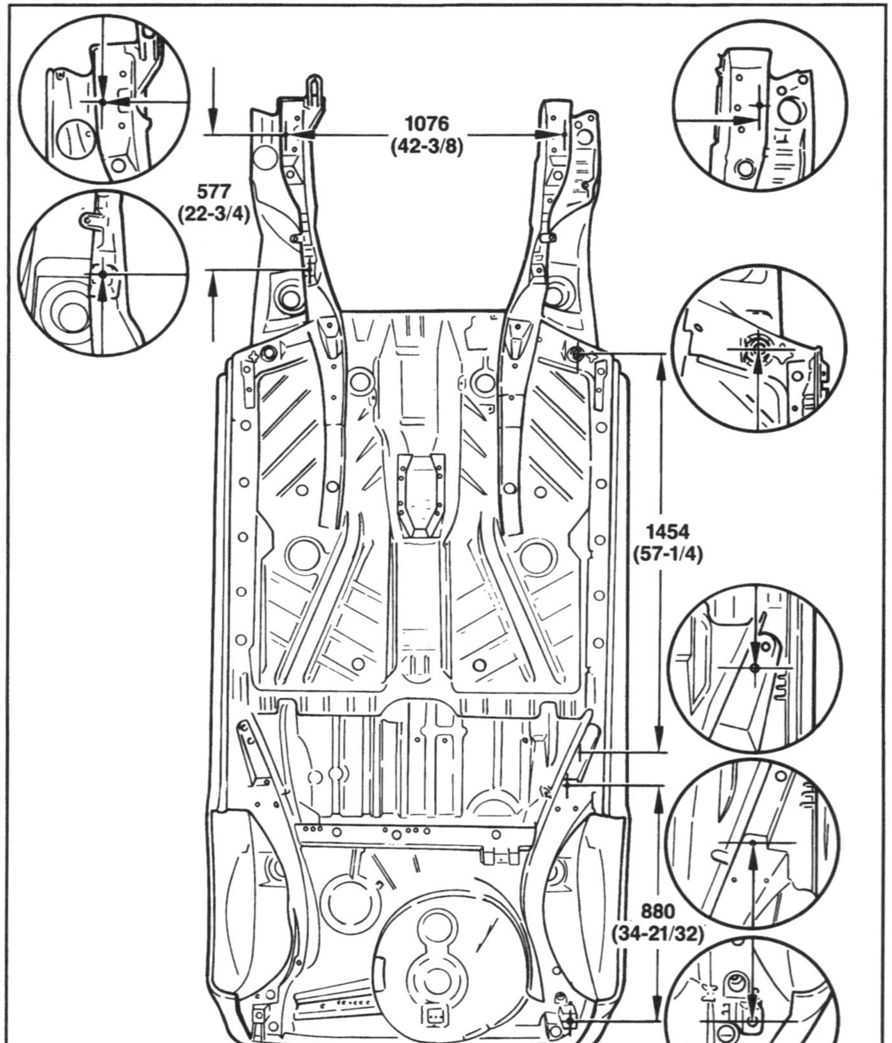

Note

All measurements in mm (in.)
Measurement tolerance = ± 2 mm (3/32 in.)

00-855

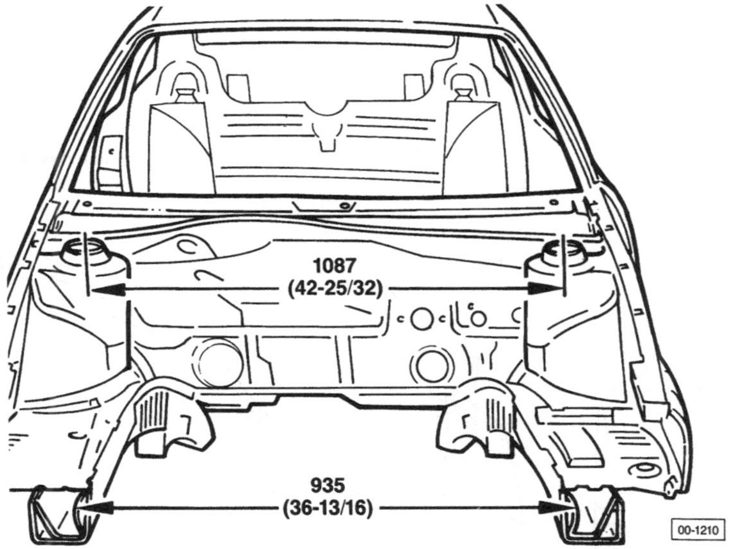

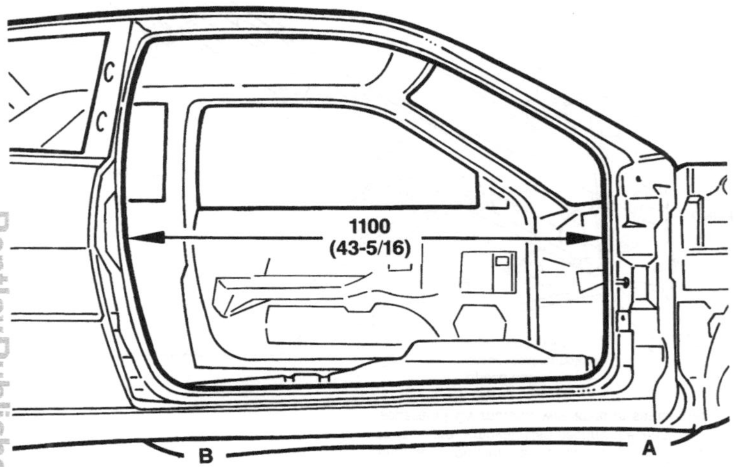

Note

All measurements in mm (in.)
Measurement tolerance = ± 2 mm (3/32 in.)

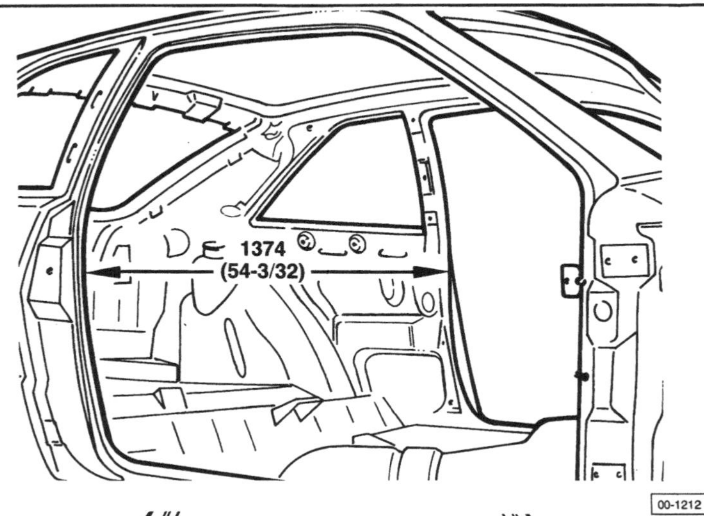

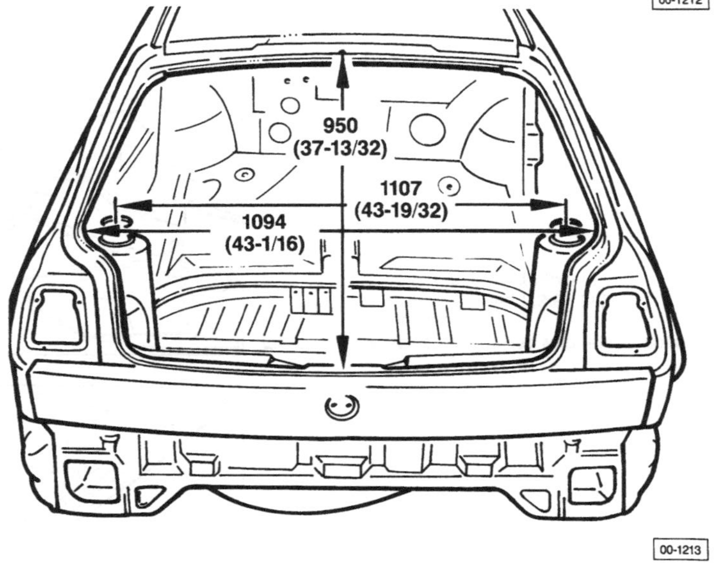

Note

All measurements in mm (in.)
Measurement tolerance = ± 2 mm (3/32 in.)

Note

To remove front bumper, see Repair Group 63. To remove wheelhouse shell, see Repair Group 66.

Note

When removing fender, heat and soften PCV material at low temperature with hot air gun. **DO NOT OVERHEAT**. PCV material must not blister or change color. Heat fender at A-pillar with hot air gun to facilitate removal. Use second mechanic to separate PCV material with knife at same time.

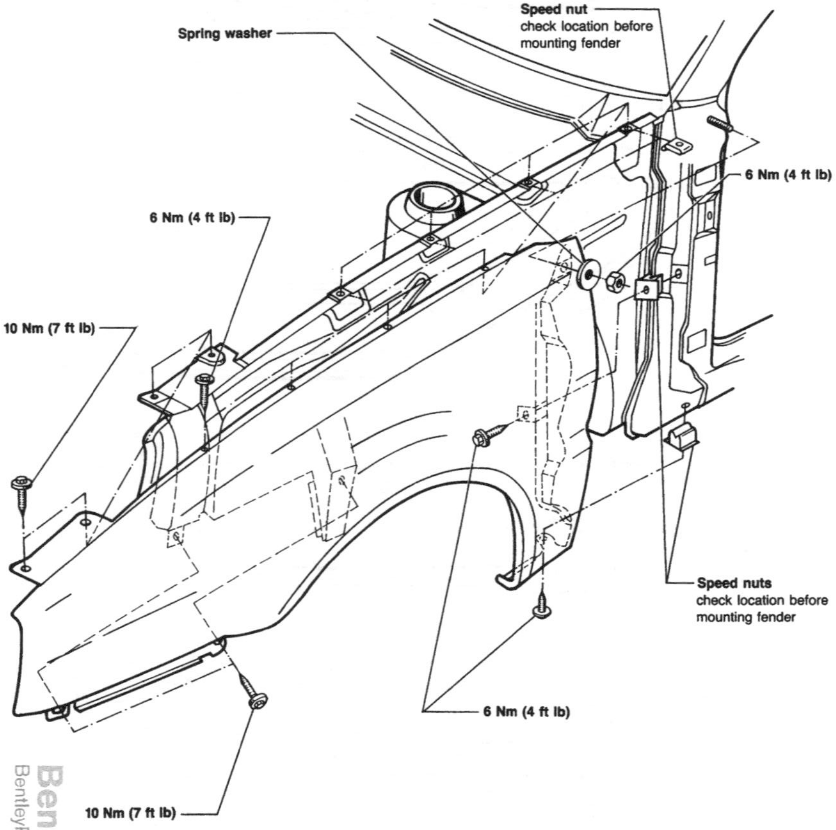

Front fender, adjusting

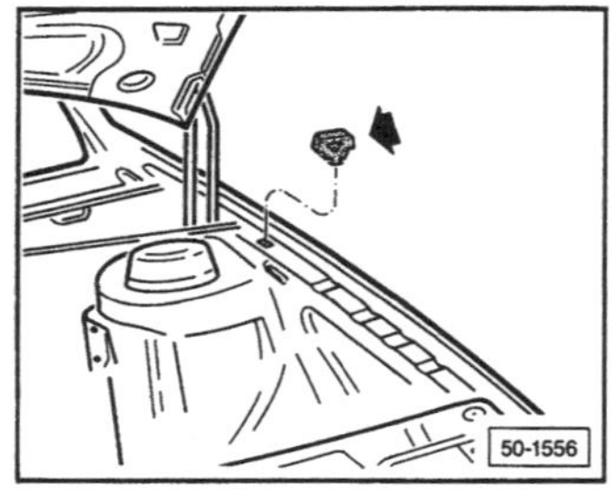

- insert rubber guide piece (**arrow**) before adjusting

- adjust fender gap with fender bolts loosened
  a = 5.0 mm (13/64 in.)
  b = 3.5 mm (9/64 in.)

Note

The lock support can be removed completely with radiator grille, headlamp and horn as well as all attached parts. Before removing the bowden cable, see Repair Group 55. Before removing the bumper, see Repair Group 63.

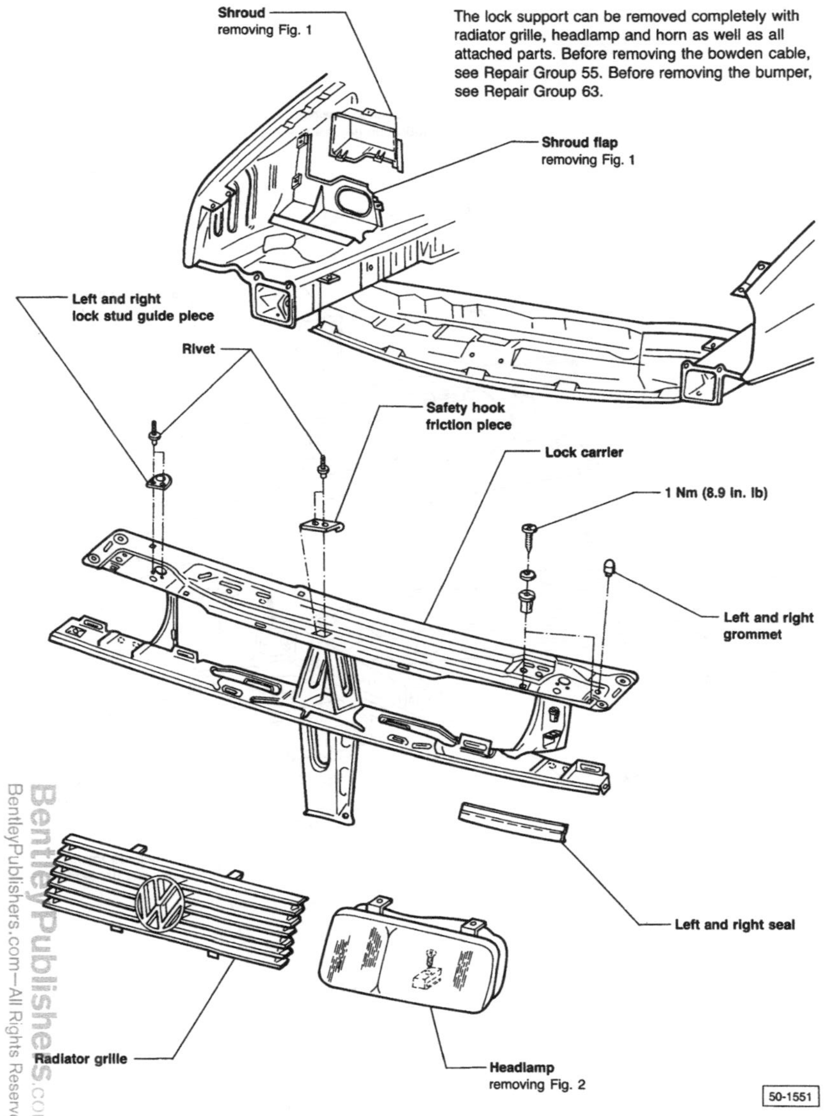

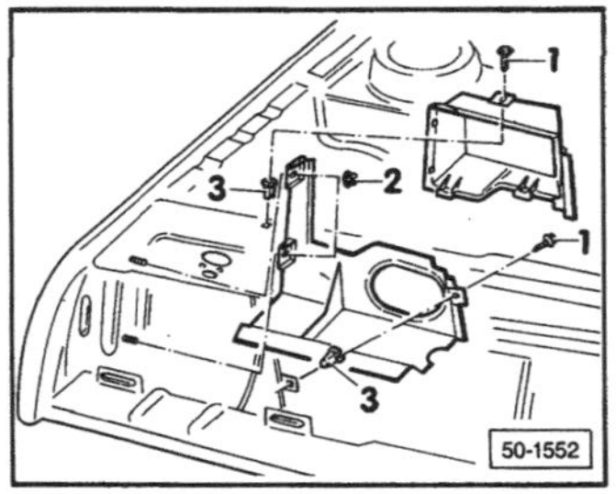

▶ **Fig. 1 Shroud, removing/installing**

- remove hex-head tapping screws **1**
  - torque = 1.5 Nm (1 ft lb)
- remove hex nuts **2**
  - torque = 6 Nm (4 ft lb)
- clip on nuts **3**

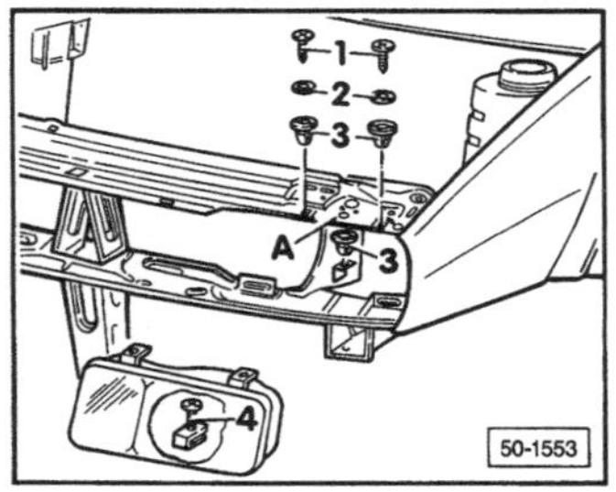

▶ **Fig. 2 Headlamp, removing**

- remove screws **1**
  - torque = 1 Nm (8.9 in. lb)
- loosen screw **4** through assembly hole **A** enough to allow removal of headlamp
  - torque = 1 Nm (8.9 in. lb)
    2 — washer
    3 — nut

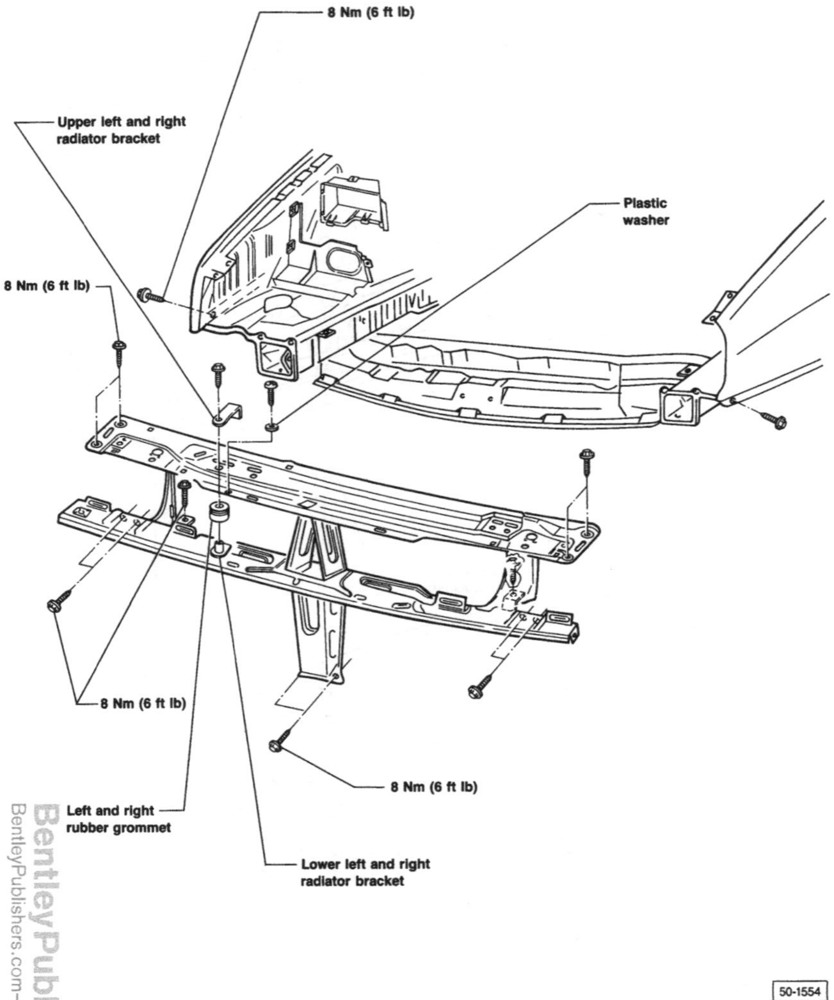

Acid rain spots on paint, removing

Note
Applies to all vehicles without acrylic transport protection

Acid rain spots are most visible under fluorescent light and look like water spots.

In most cases, you can remove these spots by performing the following procedures and thus avoid re-painting.

Do **NOT** confuse the effects of acid rain damage with industrial fallout. Industrial fallout consists of small airborne metallic particles that settle onto the paint and eventually eat their way into the finish. Rubbing your hand across the clean surface will reveal a gritty texture.

If acid rain spot removal is new to you, we recommend that you contact a good body shop or reconditioning center.

WARNING
Before using any chemicals or cleaning compounds, carefully read all of the cautions and medical information on the container. Always follow all of the manufacturer's recommendations.

Most makers of professional automotive reconditioning products recommend the following six steps **IN SEQUENCE** to repair acid rain paint spots:

Step 1: Inspection

- wash vehicle with mild soap and water, then dry completely

CAUTION
Do **NOT** wash or dry vehicle in direct sunlight. Ordinary water spots could form and confuse the inspection.

- *inspect the following surfaces under a fluorescent light*
  - hood, roof, rear deck, tops of fenders, painted surfaces of bumpers and doors from the belt line up

Step 2: Neutralization

The continuing effect of the acid must be stopped by neutralizing.

- mix a solution of baking soda and water (4 tablespoons per gallon) and stir until completely dissolved
- apply solution to all affected areas and let stand for a few minutes before rinsing off with clear water

Commercial neutralizing products such as "Blue Max" by Car Brite® or equivalent work well.

Step 3: Buffing

You can buff by hand or with a low speed orbital buffer or with a high speed buffer.

> **CAUTION**
> It is extremely important that power buffing be done by an experienced person. If done incorrectly, power buffing can quickly and easily damage the paint finish.
> Use clean buffing pads at **ALL** times.

- mask off any trim or glass close to affected areas
- buff small areas at a time, no larger than two feet by two feet
- buff using a light polishing compound such as "Finesse-It" by 3MR, "Acid Rain Creme" by Car Brite®, or equivalent. Follow the manufacturers instructions to achieve the best possible results.

> **CAUTION**
> Do **NOT** use general purpose rubbing or buffing compounds. They are too coarse for this application and could damage the finish.

- buff surface until acid spots are removed

Note

Inspect the work surface frequently during buffing to ensure that only a minimum of original finish is removed. If the color of the vehicle shows up on the buffing pad, you have cut through the clearcoat and you have to repaint the affected area.

Step 4: Cleaning

- wash vehicle thoroughly with a solution of mild soap and water
- rinse vehicle with clear water then dry completely
- inspect vehicle to see if another application is necessary

Step 5: Polishing

Polish the finish to restore the original luster either by hand or with a power buffer. Always follow polishing with an application of wax.

> **CAUTION**
> It is extremely important that power buffing be performed by an experienced person. If done incorrectly, power buffing can quickly and easily damage the paint and finish.
> Use clean buffing pads at **ALL** times.

- buff with polishes such as "Liquid Polish" by 3MR, "Crystal Shine" by Car Brite®, or equivalent. Follow the manufacturers instructions to get the best possible results.

Step 6: Waxing

Waxing (unlike polishing) will seal the finish and prevent acid rain from leaving spots.

- **HAND WAX** using clean cheesecloth and a good quality non-abrasive wax.

If, in extreme cases, you can still see acid rain spots after the previous six steps, or if the clearcoat has been rubbed through, it will be necessary to re-paint the affected areas.

Acid Rain Spot Removal Flowchart
(See detailed instructions on previous pages)

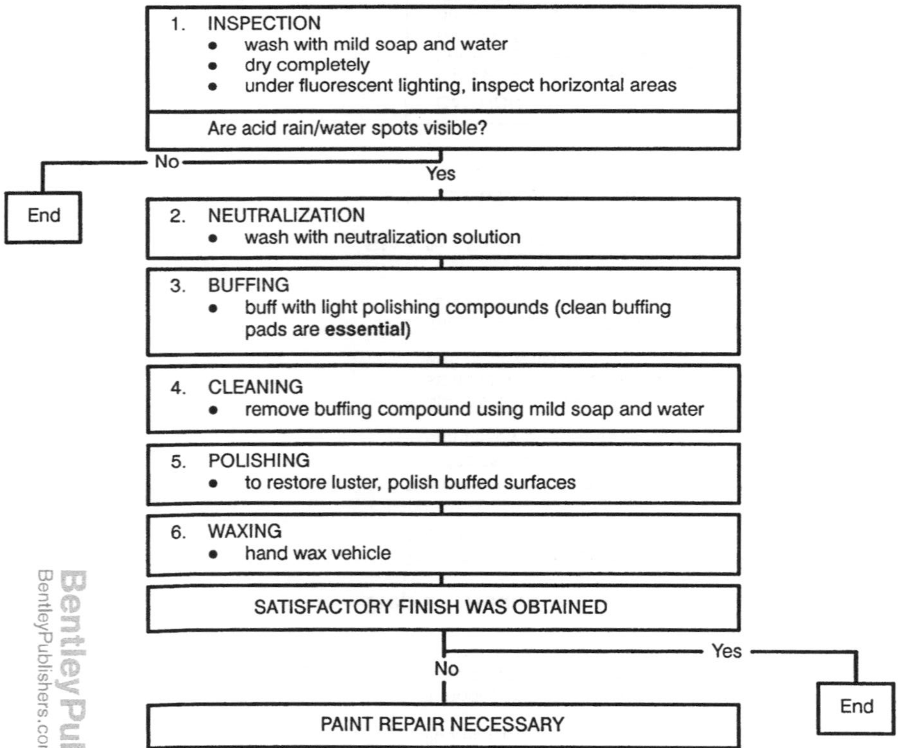

THIS PAGE INTENTIONALLY LEFT BLANK

Index

Gas–filled strut
- assembly **55.13**
- discarding **55.14**

Hood
- assembly **55.4**
- bowden cable, assembly **55.3**
- hinge, assembly **55.6**
- gap, adjusting **55.7**
- radiator support **55.3**
- rubber stop, adjusting **55.5**
- sound deadener, installing **55.5**
- striker, adjusting **55.5**
- support, replacing **55.14**

Rear lid
- adjusting **55.9**
- assembly **55.8**
- lock, assembly **55.10**
- lock cylinder, replacing **55.11**
- lid release, replacing **55.12**
- lock, repairing **55.11**
- rubber stop, adjusting **55.9**
- spring pin, removing **55.11**

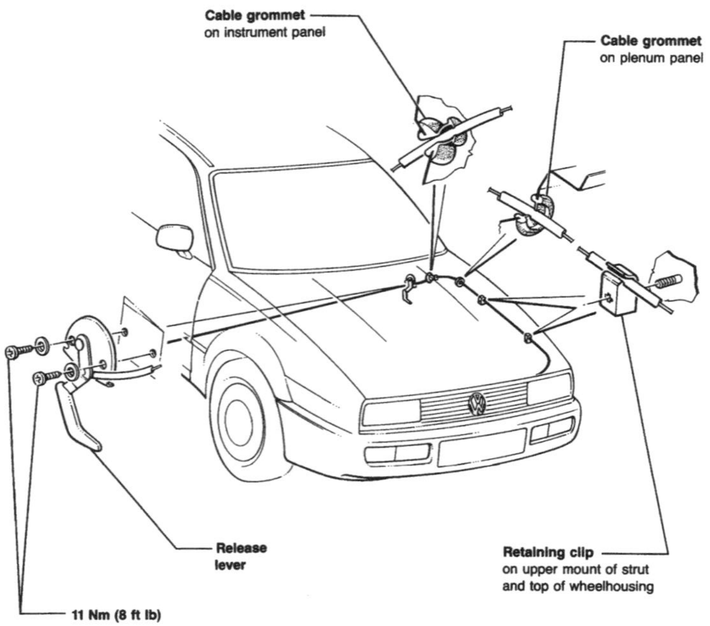
Cable grommet
on instrument panel
Cable grommet
on plenum panel
Release
lever
Retaining clip
on upper mount of strut
and top of wheelhousing
11 Nm (8 ft lb)

55-192

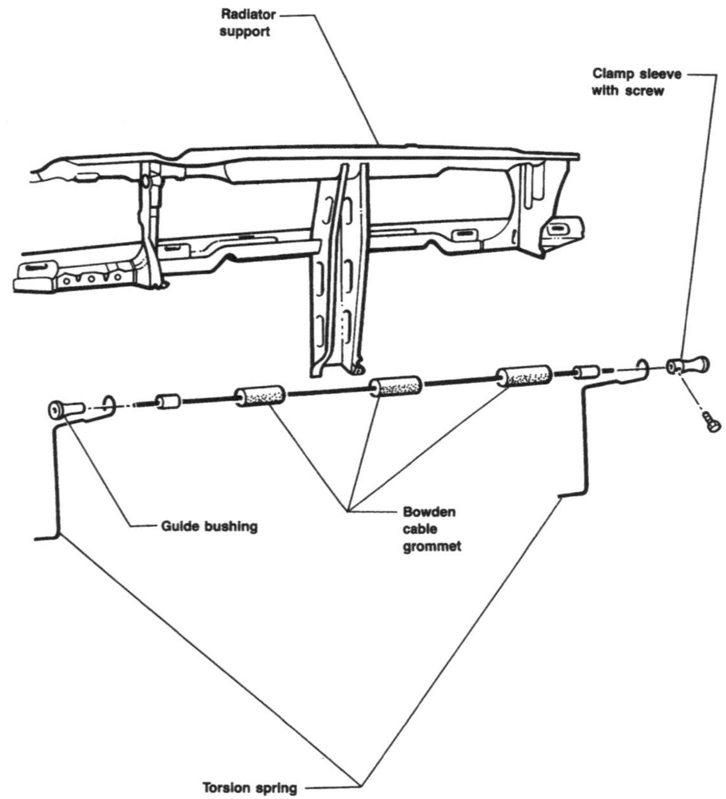
Radiator
support
Clamp sleeve
with screw
Guide bushing
Bowden
cable
grommet
Torsion spring

55-193

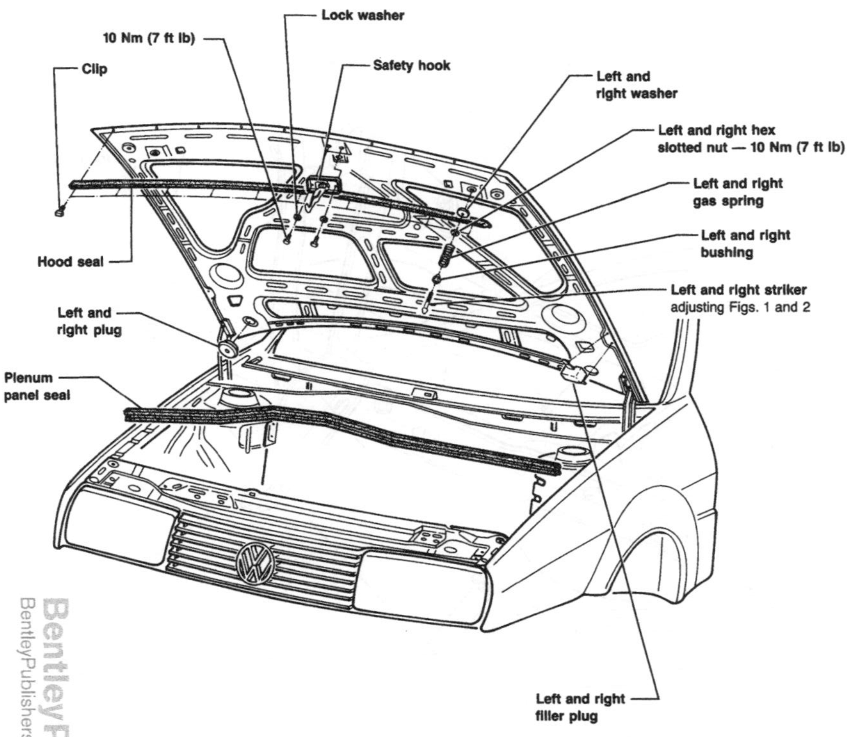

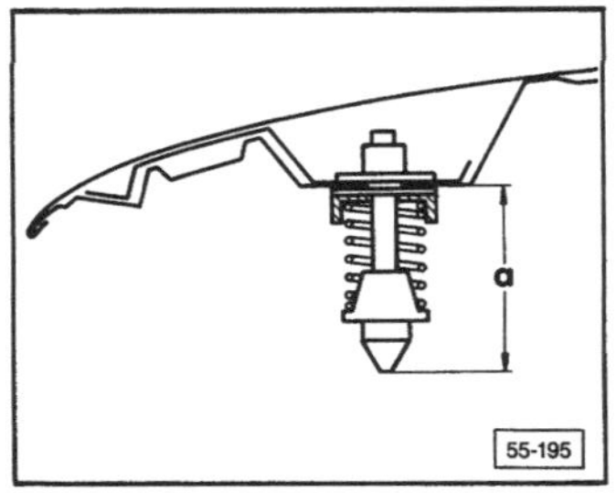

▶ **Fig. 1 Striker, adjusting**

- adjust both strikers to dimension **a = 57 mm (2.1 in.)**
  - rubber stops must be screwed completely in

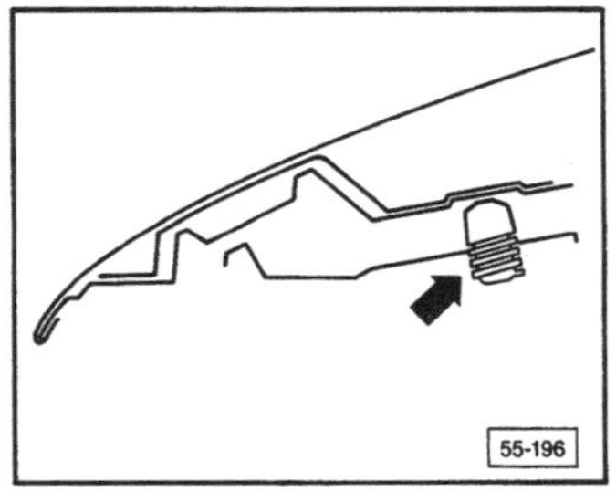

▶ **Fig. 2 Rubber stop, adjusting**

- fully screw in both rubber stops
- adjust hood flush with fenders by turning out both rubber stops (**arrow**)

CAUTION

Part numbers are for reference only. Always check with your Parts Department for latest information.

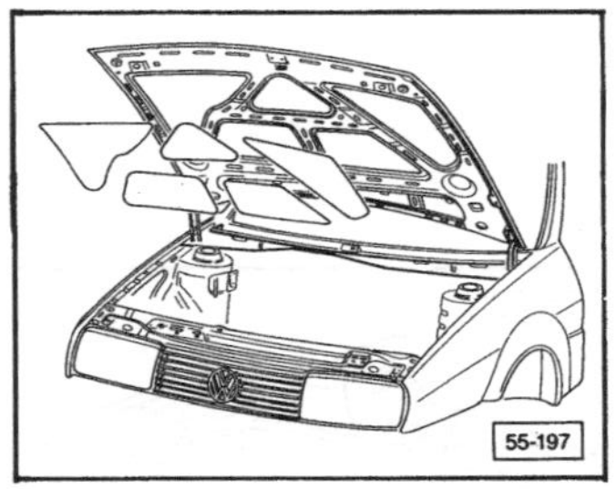

▶ **Hood sound deadener, installing**

- before bonding sound deadener to hood, clean inner panel with cleaning solution, Part No. **D 009 400 01**

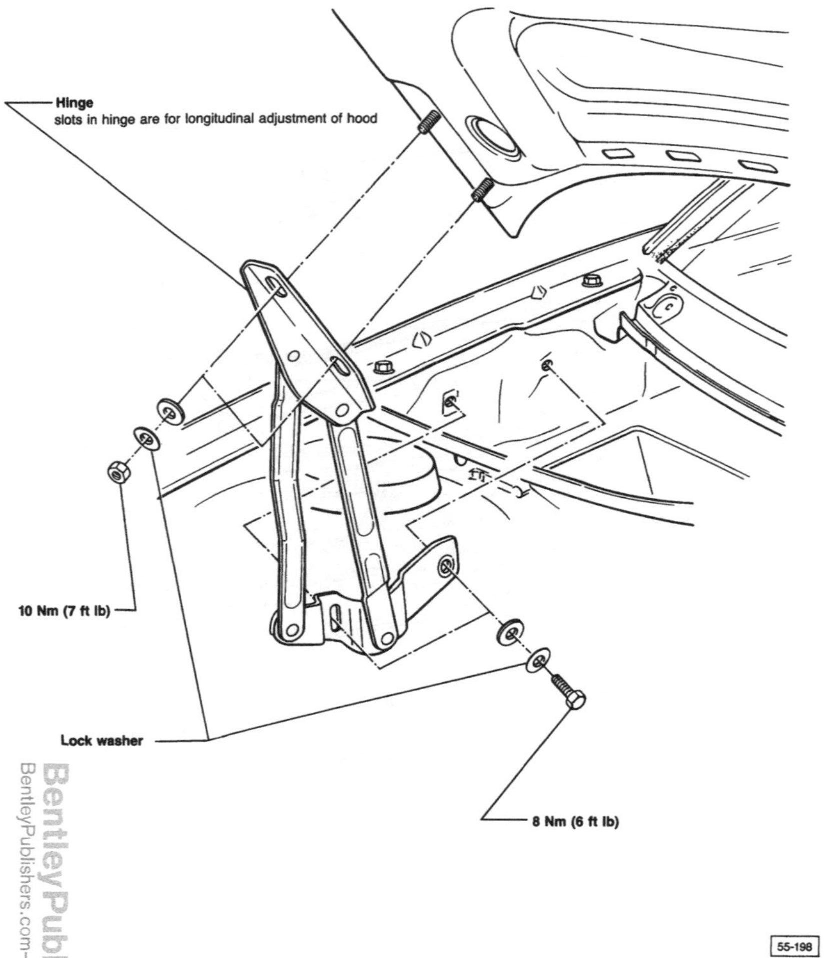

Hoods, Lids

Hood gap, adjusting

- adjust gap by moving fender after loosening fender screws
  - a = 5.0 mm (13/64 in.)

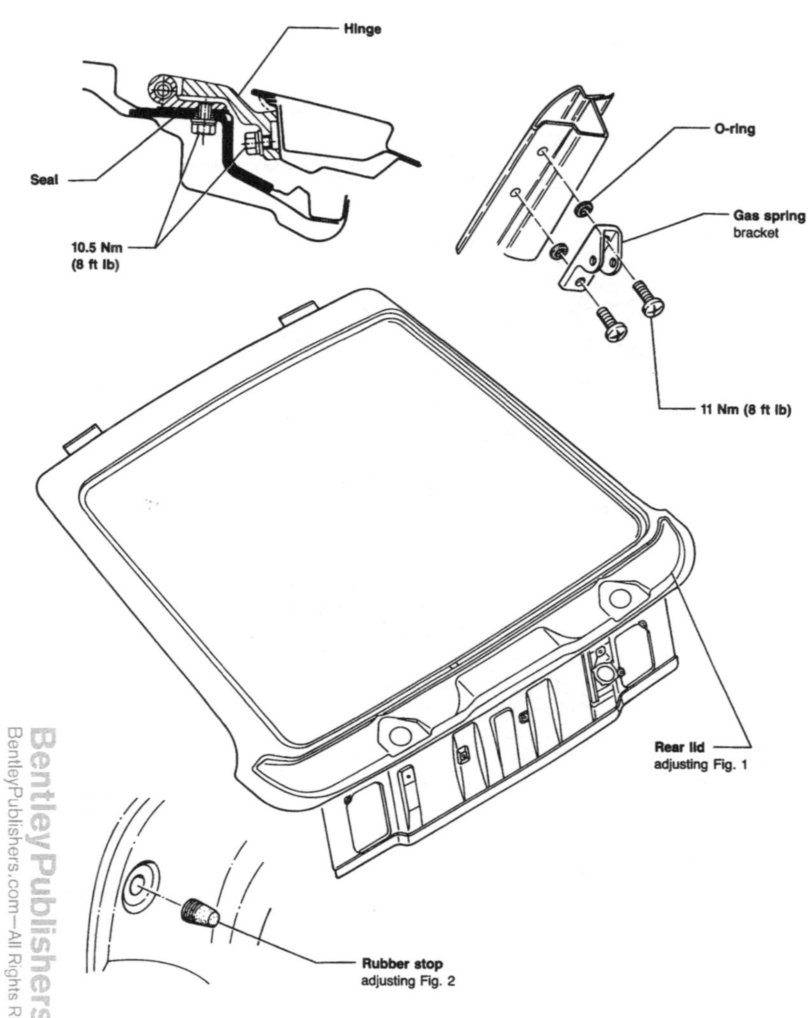

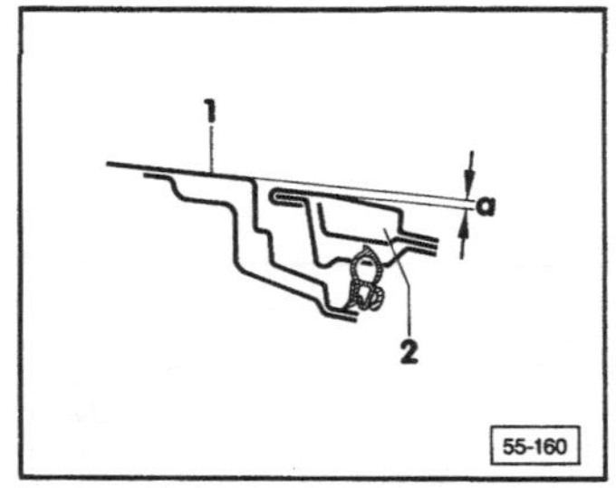

▶ Fig. 1 **Rear lid, adjusting**

- rear lid adjustment is made at hinge (over-size holes)
- adjust lid as shown
  a = 1.7 mm (1/16 in.) (measure at center of roof)
  1 — roof contour
  2 — lid

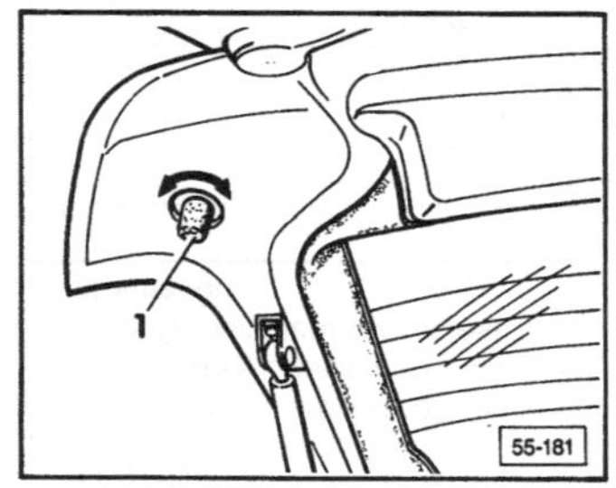

▶ Fig. 2 **Rubber stop, adjusting**

- after adjusting striker, adjust rear lid flush with body by means of rubber stop **1**

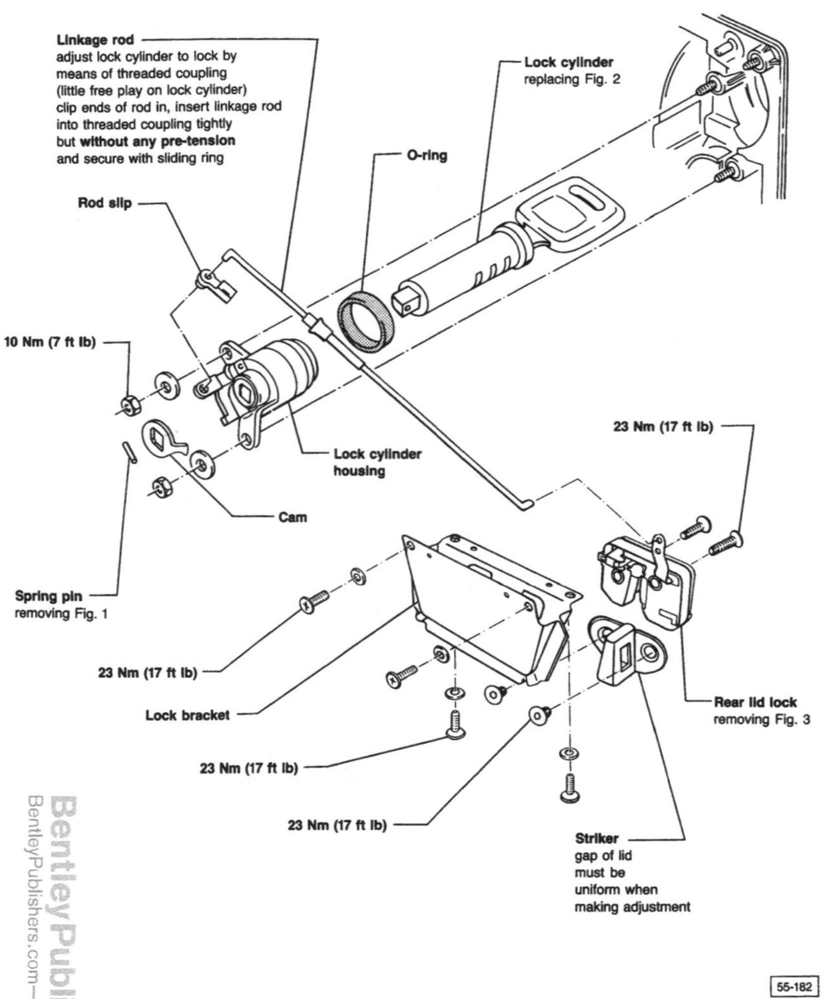

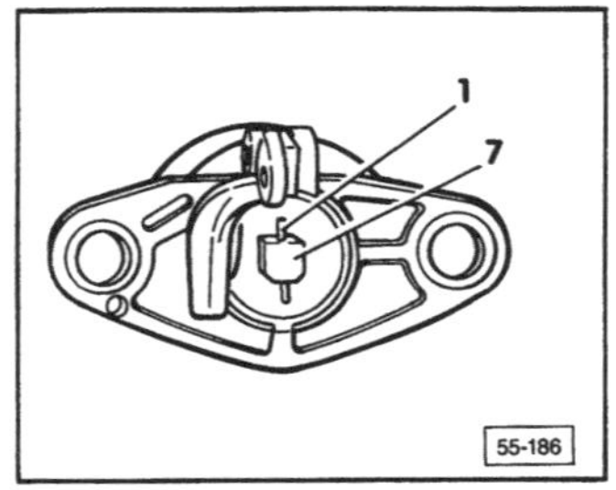

Fig. 1 Spring pin, removing

- to remove lock cylinder, spring pin **1** must first be driven out of lock cylinder **7** with a punch

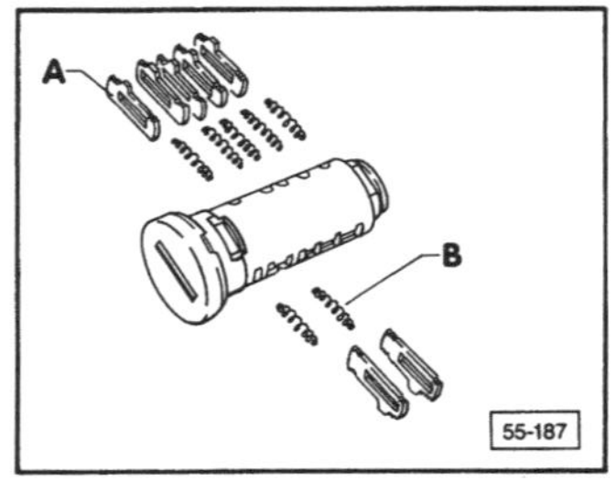

Fig. 2 Lock cylinder, replacing

- to keep same lock setting of existing locks on vehicle, new lock cylinder must be converted to previous setting by re-arranging tumblers
  - for converting of lock setting, tumblers and required springs, check with Parts Department
- lubricate tumblers and springs with grease, Part No. **G 000 400**, before inserting into lock cylinder channels,
- the following tumblers are supplied:
  - Part No. **443 898 041 A** (lock tumbler length 13.5 mm)
  - Part No. **443 898 041** (lock tumbler length 12 mm)
  - **A** — tumblers
  - **B** — tumbler springs

CAUTION

Part numbers are for reference only. Always check with your Parts Department for lastest information.

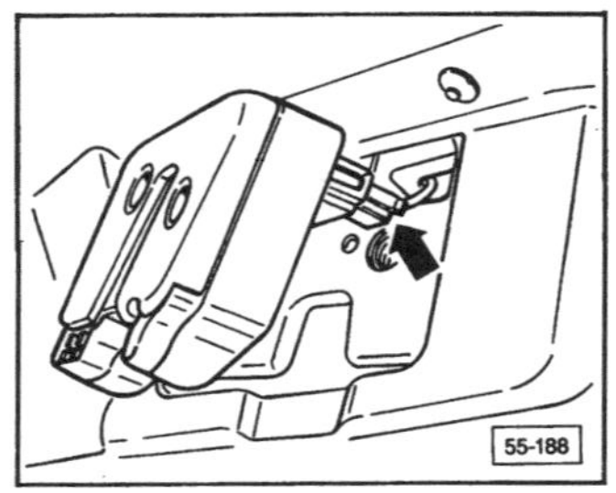

Fig. 3 Rear lid lock, repairing

- remove trim and unclip linkage rod
- remove hex head bolt
  - tightening torque = 23 Nm (17 ft lb)
- pull lock out far enough to separate electrical connection (**arrow**) for luggage compartment lighting

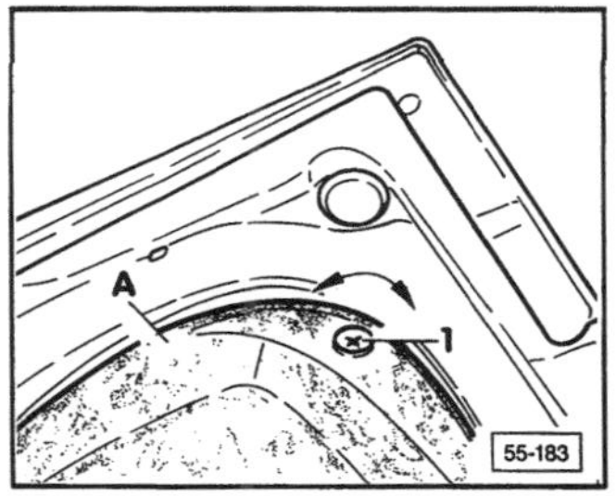

▶ ## Rear lid release, replacing

- rotate twist clip **1** 90° (¼ turn) (unlock) and remove trim **A**

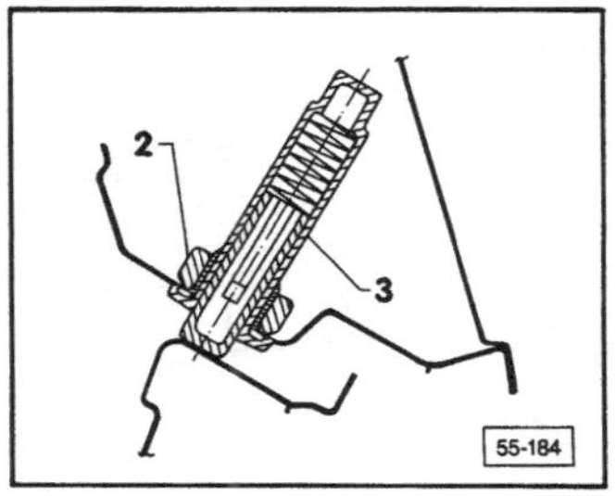

▶

- unscrew threaded plate **2** and pull opener **3** downward and out from lid

▶ **Note**

When installing, make sure that tab on threaded plate engages cutouts (**arrow**) in lid.

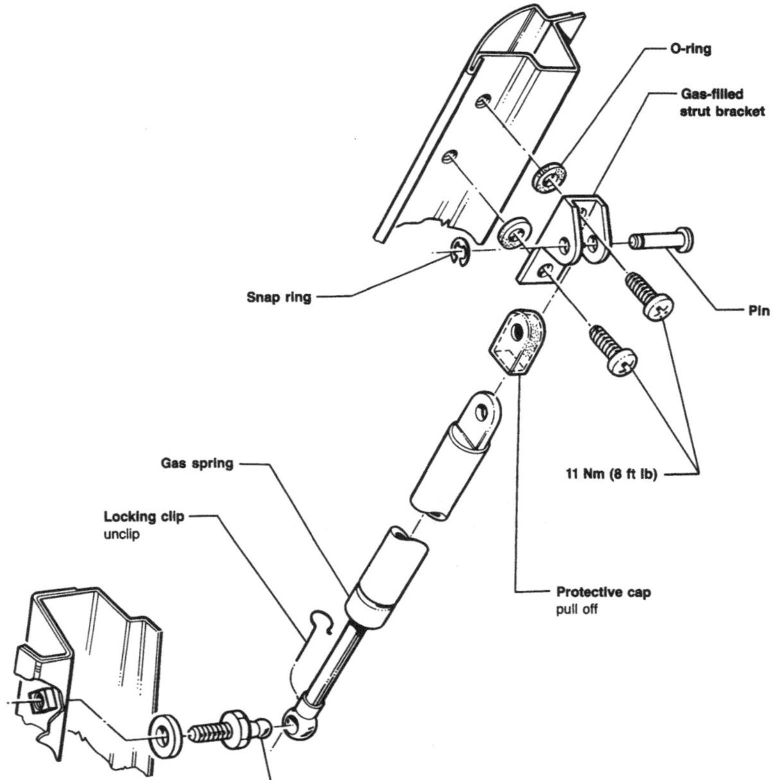

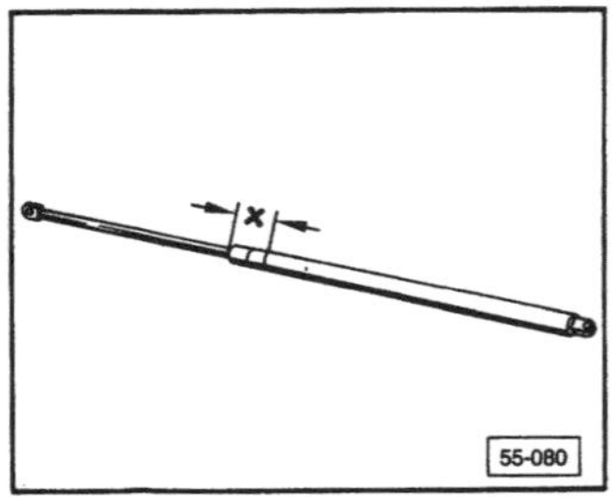

Gas-filled strut, discarding

Note

After replacing gas-filled strut, gas should be released from unserviceable unit before discarding

- clamp area **X** of strut cylinder in vise
  - **area X** = 50 mm (2 in.)
- using hack saw, cut cylinder open immediately to right of vice

CAUTION

Clamp only in the area marked, otherwise there will be risk of accident! Wear eye protection when sawing.

Note

Cover the area being sawed with a cleaning cloth to catch any oil that may spray out.

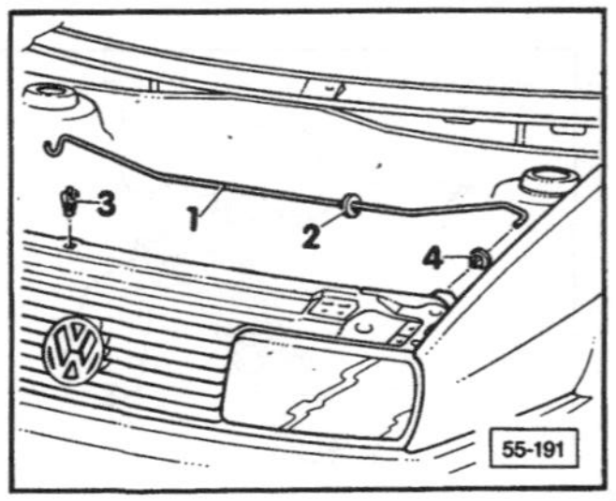

Hood support, replacing

- raise hood support **1** and pull out from bushing **4**
  - **2** — rubber stop
  - **3** — clip for support rod

THIS PAGE INTENTIONALLY LEFT BLANK

Index

Central locking
- assembly **57.9**
- bi-pressure pump, removing **57.10**
- checking **57.13**
- gas flap activator, adjusting **57.11a**
- gas flap activator, removing/installing **57.11**
- electrical troubleshooting **57.14**
- master activators, removing/installing **57.10**
- pump/control module, location **57.17**
- rear lid activator, removing/installing **57.12**

Door lock
- checking **57.6**
- door lock, removing/installing **57.8**
- inner handle, removing **57.7**
- installing **57.7**
- re-keying locks **57.18**

Front door
- assembly **57.2**
- impact bar, removing **57.5**
- removing/adjusting **57.3**
- seal, removing/installing **57.8**

Window regulator
- bordering frame, removing/installing **57.4**
- carrier to glass, bonding **57.4**
- drilling out rivet **57.3**
- removing/installing **57.3**

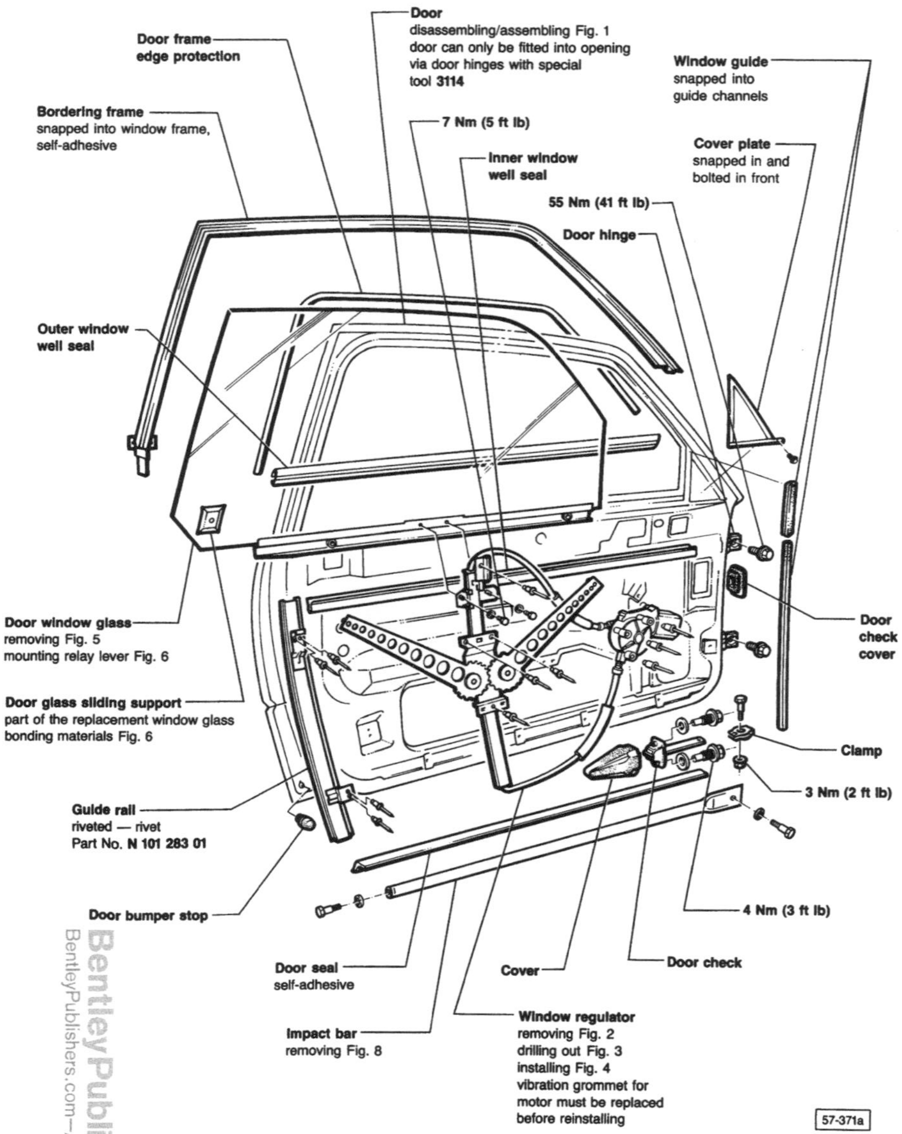

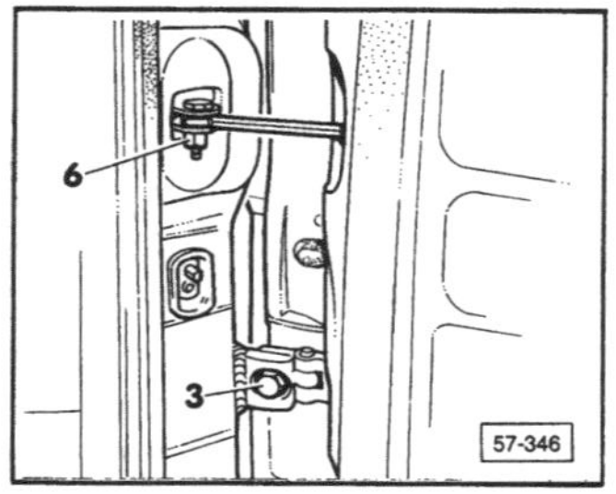

▶ Fig. 1

Door, removing/adjusting

- unscrew nut **6** and pull out bolt
- remove hex head bolt **3** from upper and lower hinge part
- remove door with help of a second mechanic
  - tightening torques:
    nut **6** = 3 Nm (2 ft lb)
    hex head bolt **3** = 55 Nm (41 ft lb)

▶ Fig. 2

Window regulator, removing

- lower door glass fully
- unscrew door glass and pull off guide arms **(arrows)**
- push door glass upward and wedge in place

▶ Fig. 3

Drilling out rivet

- drill out rivets **(arrows)** with a 5 mm (13/64 in.) drill and drive out rivet shafts with a mandrel
- remove window regulator from bottom

Note

Check rivet points for paint damage and apply anti-corrosion primer as required.

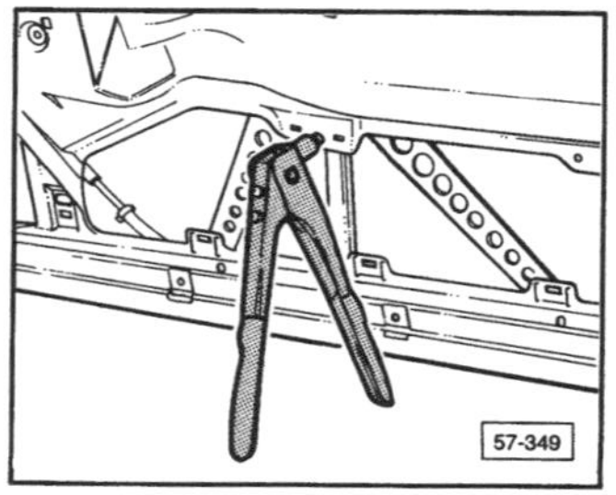

▶ Fig. 4

Window regulator, installing

- insert window regulator in door and align with mounting holes
- rivet window regulator to inner door panel (rivet Part No. **N 101 283.01**)

CAUTION

Part numbers are listed for reference only. Always consult with the Parts Department for latest information.

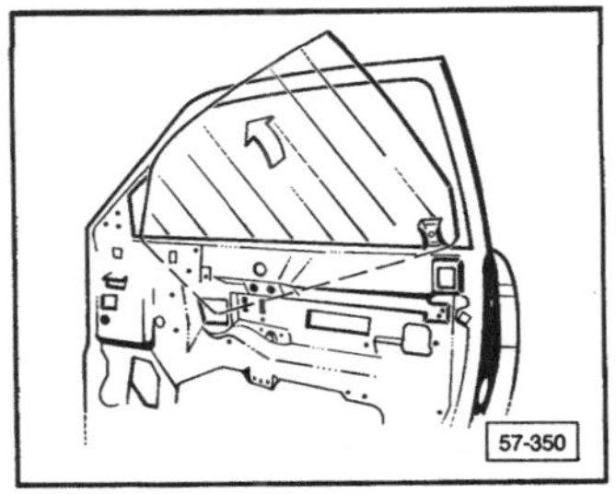

▶ Fig. 5 Door glass, removing/installing/adjusting

- lower door glass fully
- unscrew door glass, pull off guide arms (Fig. 2), tilt door glass toward front and remove upward (window well seal, inner, removed)
  - tightening torque = 7 Nm (5 ft lb)
- when installing door glass, sliding support must engage with guide rail when door glass is installed
- adjust door glass by loosely bolting door glass together with window regulator
- raise door glass to top (glass will center) and tighten bolts
  - tightening torque = 7 Nm (5 ft lb)

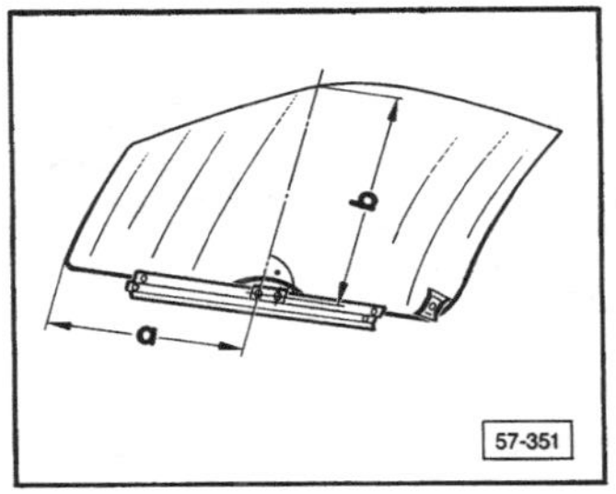

▶ Fig. 6 Window carrier to door glass, bonding

- when repairing, area that is to be covered with adhesive should be cleaned prior to bonding (Part No. **D 009 400 01**) and primed (glass primer Part No. **D 009 200**)
- attach window carrier to door glass with adhesive, Part No. **D 009 100 03** (small cartridge)
- prior to installation, the bonding area (window carrier/door glass) must harden for at least 24 hours
- dimensions for window carrier:
  - **a** = 425 mm (16-39/64 in.)
  - **b** = 475 mm (18-1/2 in.), curvature of glass not included

CAUTION

Part numbers are listed for reference only. Always consult with the Parts Department for latest information.

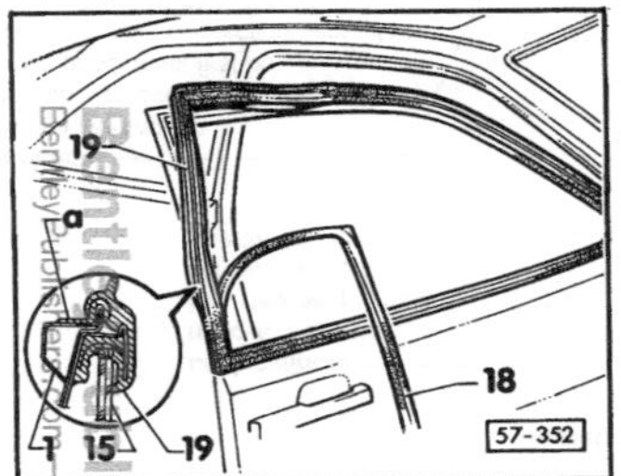

▶ Fig. 7 Trim frame, removing/installing

- remove cover (triangular plates inside and out)
- pull off edge protection **18** from flange
- snap off trim frame **19** from flange trim frame
  - trim frame is bonded with door frame at interior lip

    a — adhesive surface
    1 — door frame
    15 — door glass

- install by inserting trim frame into rear portion of window well
- push trim frame down and press onto flange from upper corner
  - remove protective paper on adhesive tape only after trim frame has been properly seated

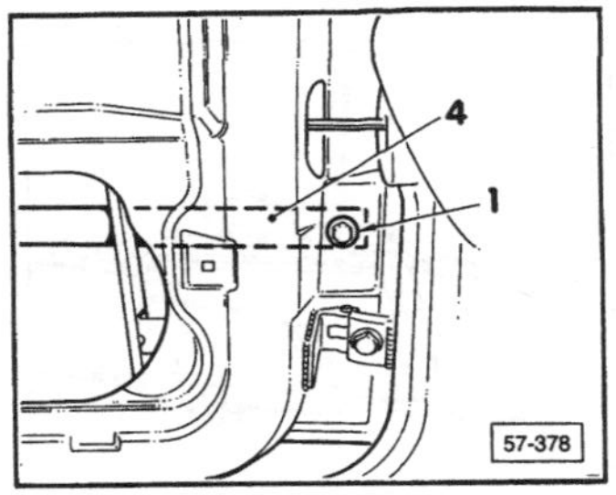

▶ Fig. 8 Impact bar, removing

- remove inner door panel and plastic moisture barrier, Repair Group 70
- remove guide rail
  - **4** — impact bar

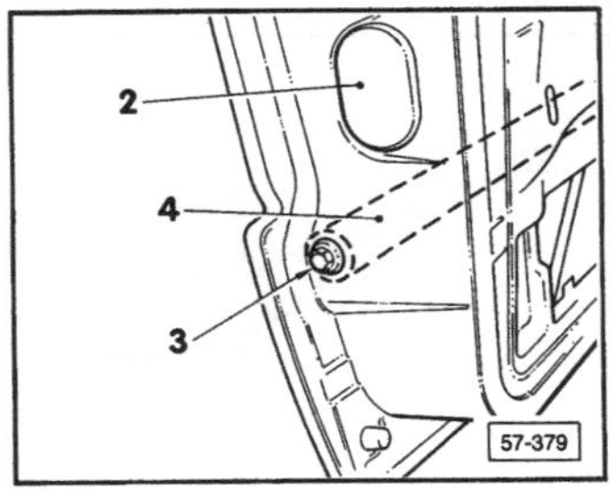

▶

- pry off cover cap **2**, remove hex head bolts **1,3** and take out impact bar from mounting opening
  - tightening torque = 25 Nm (18 ft lb)

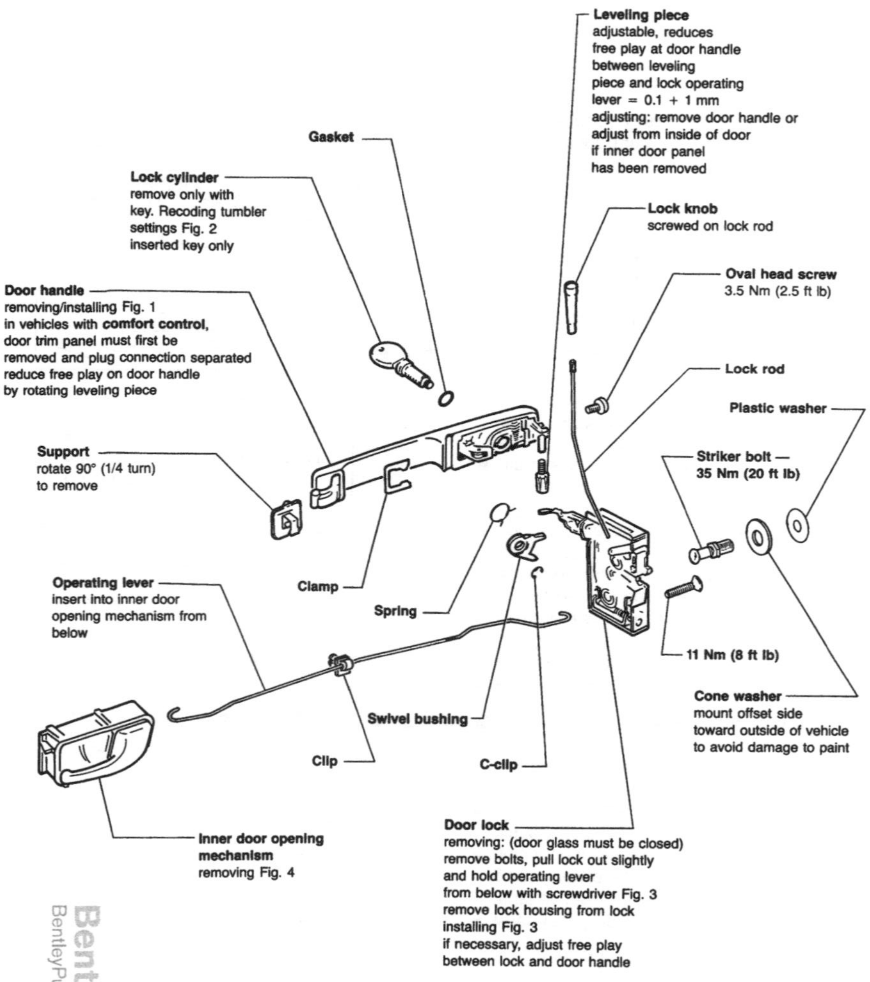

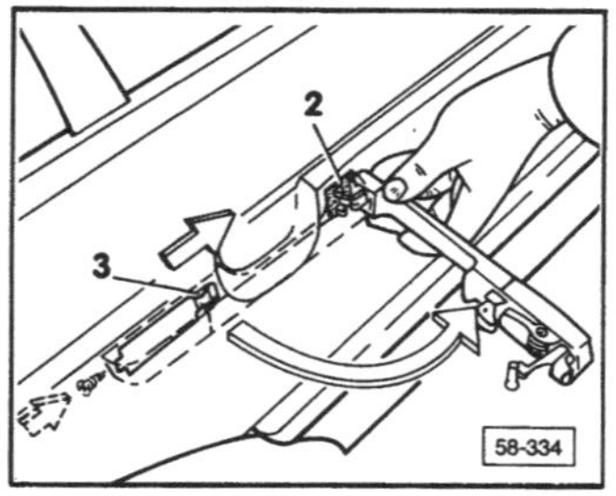

▶ Fig. 1 **Door handle, removing/installing**

- remove bolt and washer assembly
- push door handle in direction of arrow and pivot out of door
- when installing check washer **2** and clip **3** for damage
- check tolerance (free play of door handle) between leveling piece and operating lever. If necessary, adjust through rotating leveling piece
  - 0.1 + 1 mm
  - torque fastening bolt 7 Nm (5 ft lb)

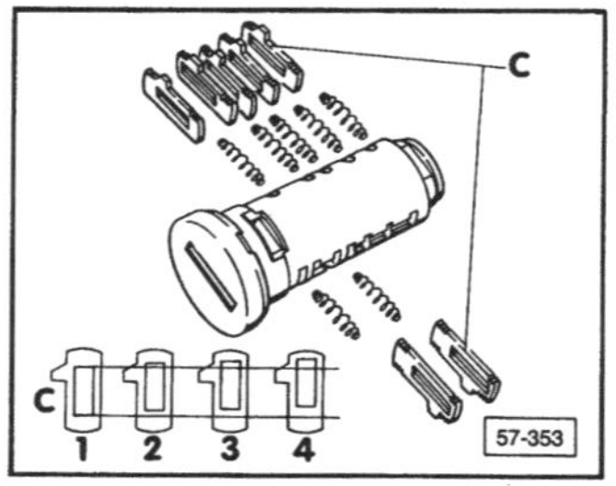

▶ Fig. 2 **Identical tumbler setting, maintaining**

- matching tumbler settings in vehicle can be made by rearranging tumblers **C**
  - possible locking plate sizes = 1 2 3 4

> **CAUTION**
>
> Part numbers are listed for reference only. Always consult with the Parts Department for latest information.

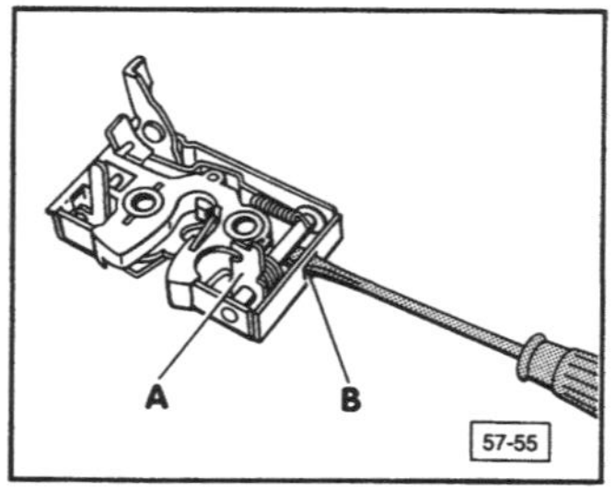

Note

Lubricate tumblers and springs with lubricant, Part No. **G 000 400**, and insert lock cylinder guide channels.

▶ Fig. 3 **Door lock, installing**

- pull operating lever **A** 90° and fasten through opening **B** with screwdriver
- check tolerance (free play of door handle) between leveling piece and door lock operating lever, if necessary, adjust by rotating leveling piece
  - 0.1 + 1 mm, see page 57.6

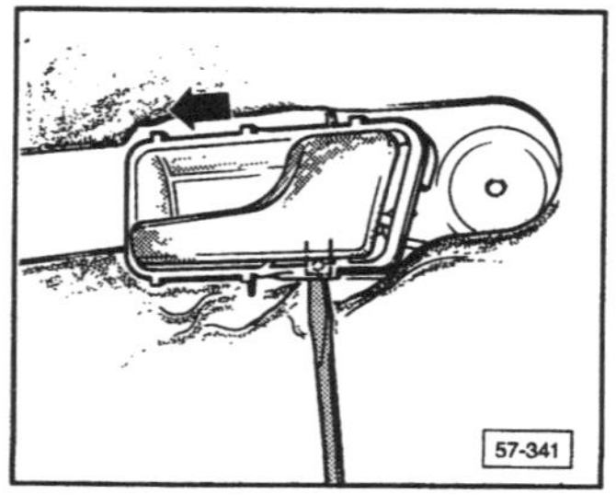

▶ Fig. 4 **Door inner handle, removing**

- push flange out of hole in inner door panel with a screwdriver
- push door handle in direction of arrow

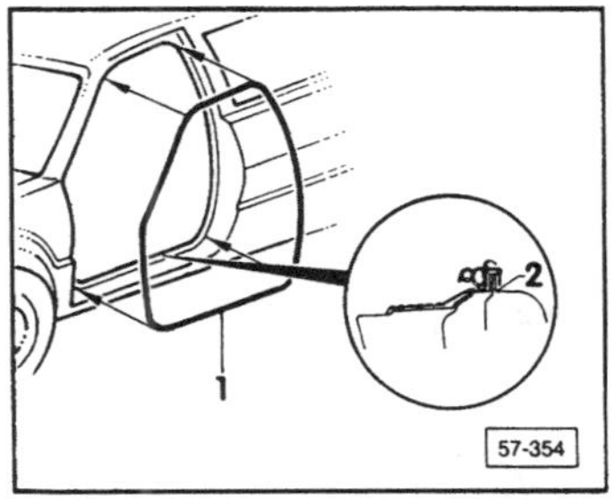

Door inner seal, removing

▶ ■ remove or loosen inner trim, see Repair Group 70
- door inner seal **1** is fitted on door flange **2**

Door inner seal, installing

Reinstall in reverse order of removal.

Door lock with freeplay adjustment as of VIN 50 NK 015 279

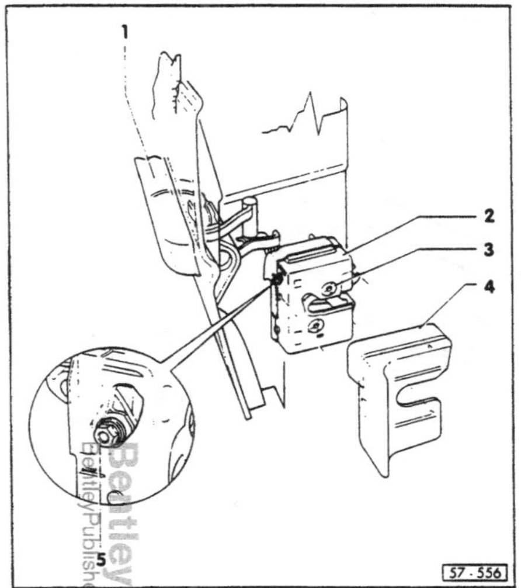

1 – Door handle

2 – Door lock with freeplay adjustment

3 – Bolt
- 20 Nm (15 ft lb)

4 – Cap

5 – Nut
- with freeplay adjustment

Removing

- unclip cap -4-
- remove bolts -3-
- pull away door lock -2- slightly and lock operating lever with 3 mm mandrel
- pull out door lock and unhook pull rod from inner lock mechanism

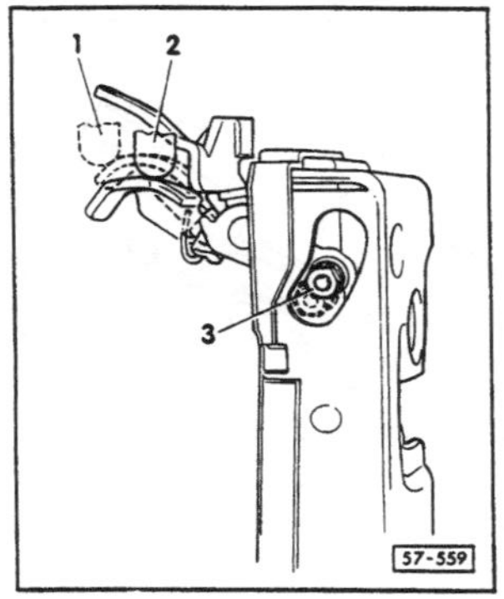

Installing

- loosen nut -**3**- with 5.5 mm wrench
  - release lever will give way
  - right side = left hand thread
  - left side = right hand thread

1 Assembly position

- install door lock
  - freeplay adjustment actuated between release lever and actuating lever

2 Installed position

- tighten nut -**3**-
  - use torque wrench **VAG 1783** and insert **VAG 1784** or equivalent
  - torque to 2.4 Nm (2 in. lb)
- check for proper function

Note
This adjustment must be made whenever the door handle is removed, replaced or reinstalled.

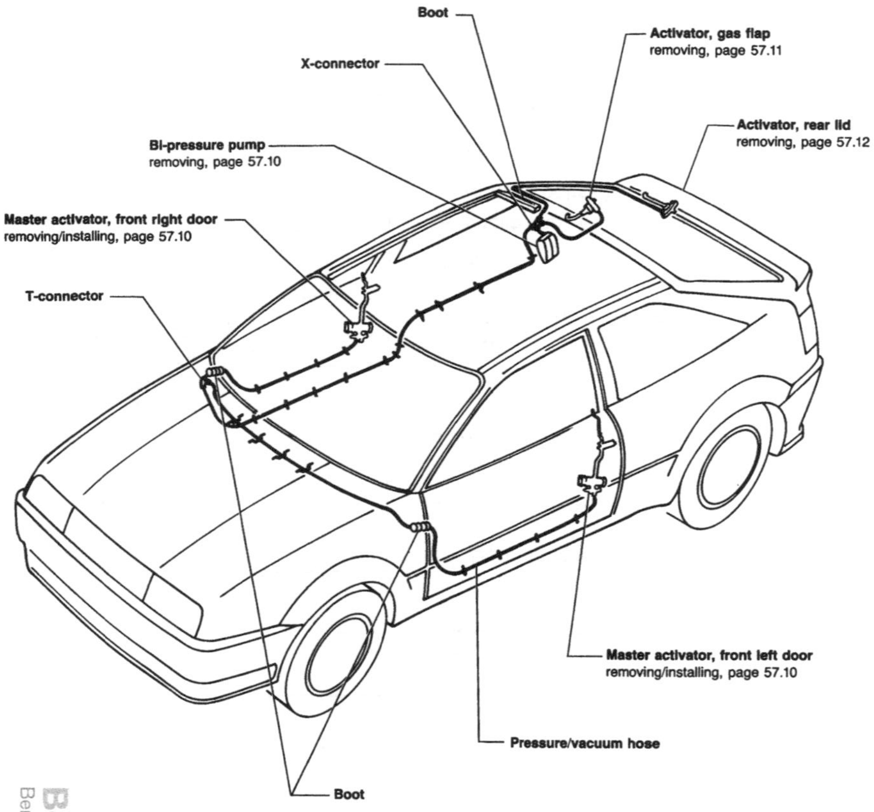

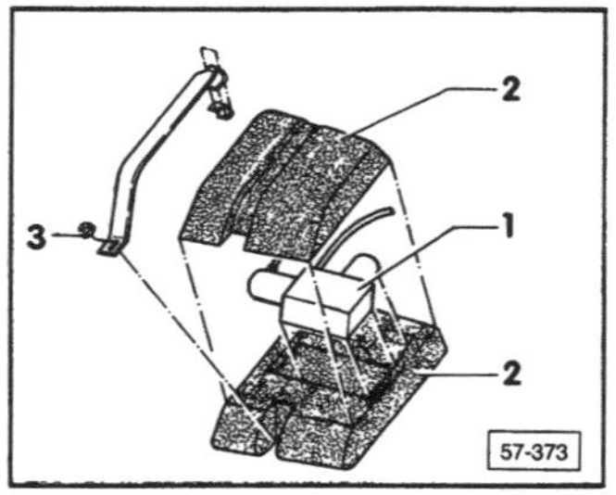

▶ Fig. 1 **Bi-pressure pump, removing**

- loosen right luggage compartment trim in rear area
- remove hex nut **3**
  - tightening torque = 6 Nm (4 ft lb)
- remove bi-pressure pump **1** from deadener **2**

Bi-pressure pump, installing

Install all components in reverse order of removal.

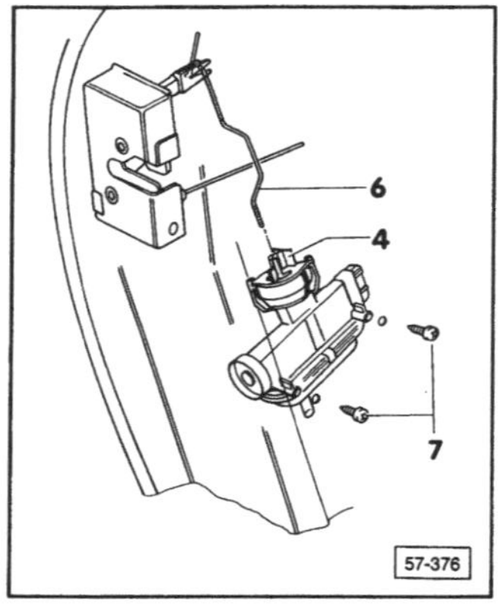

▶ Fig. 2 **Master activators, removing**

- close window
- remove door trim, Repair Group 70
- carefully remove door film
- remove pressure line and connector from switch element
- open hinge and activator **4** and unhook lock rod **6**
- remove screws **7**

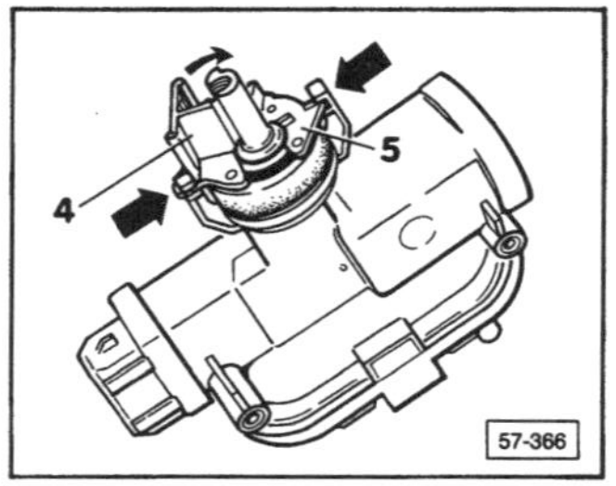

▶ Fig. 3 **Master activators, installing**

Note

For lock rod assembly, activator and door lock must be in locked position.

Door lock on driver's side must be in **closed** position.

- lock washer **5** must be locked
- push in lock rod and close hinge **4**

CAUTION

Do not pull lock rod downward when pushing in.

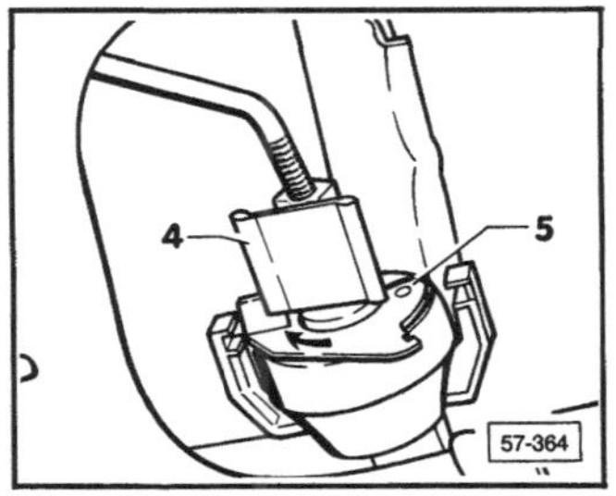

Note

When locking, hinge **4** must turn lock washer **5** and unlock.

- conduct functional test, page 57.13

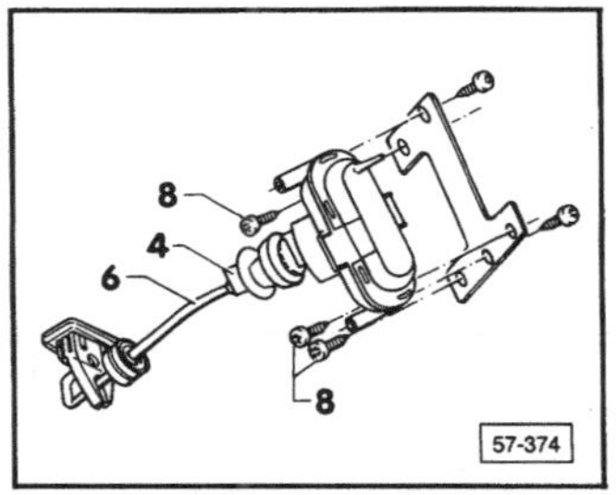

Fig. 4

Gas flap activator, removing

- remove pressure line on activator
- remove screws **8**
- open hinge on activator **4** and unhook lock rod **6**

Gas flap activator, installing

Install all components in reverse order of removal, noting adjustment of activator, page 11a

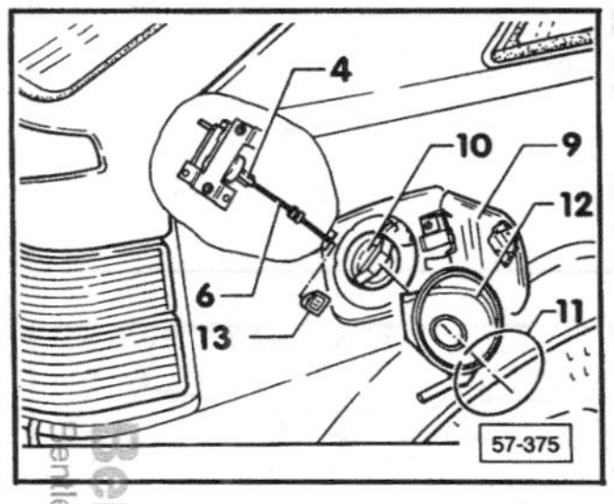

Gas flap lock rod, removing

- open hinge on activator **4** and unhook lock rod **6**
- open gas flap **9**, unscrew tank lock **10**
- unhook lock ring **11** and pry out rubber sleeve **12**
- screw on tank lock **10**
- unclip lock plug **13**
- pull lock rod **6** downward out of tank opening

Gas flap lock rod, installing

Reinstall all components in reverse order of removal.

Gas flap activator, adjusting

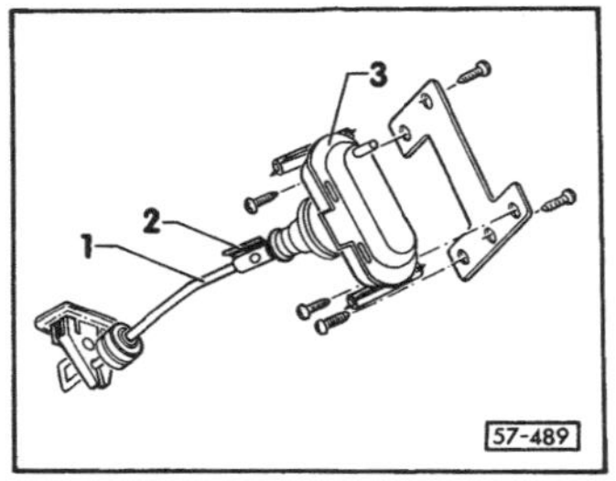

Note

Gas flap activators must be adjusted on vehicles that are not equipped with the 4.0 mm (5/32 in.) distance spacer. The spacer is not available as a replacement part, and may not be retrofitted to earlier vehicles.

- hook lock rod **1**, hinge **2** on activator **3**

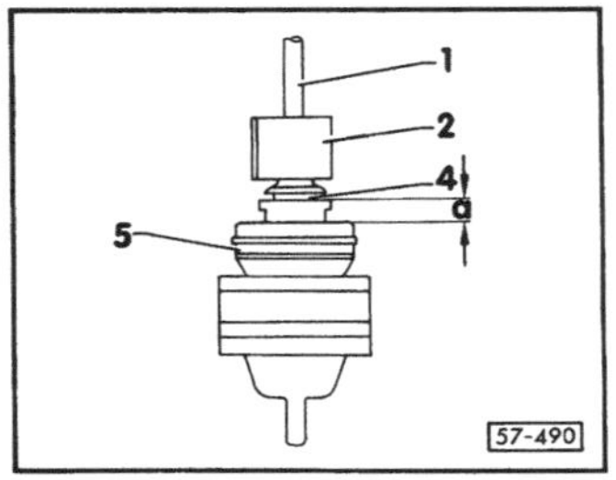

- pull lock rod **1** out of activator until pretension distance **a** between support **4** and rubber boot **5** is attained

  a = 4.0mm (5/32 in.)

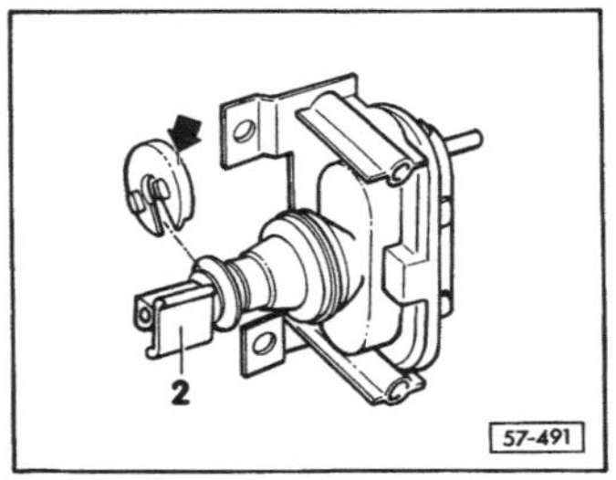

Note

From VIN 50 MK 002 478, a 4.0 mm (5/32 in.) spacer (**arrow**) is installed to maintain pretension distance **a**

- if gas flap activator is replaced, the spacer may be transferred to the replacement activator

- close hinge **2** on activator

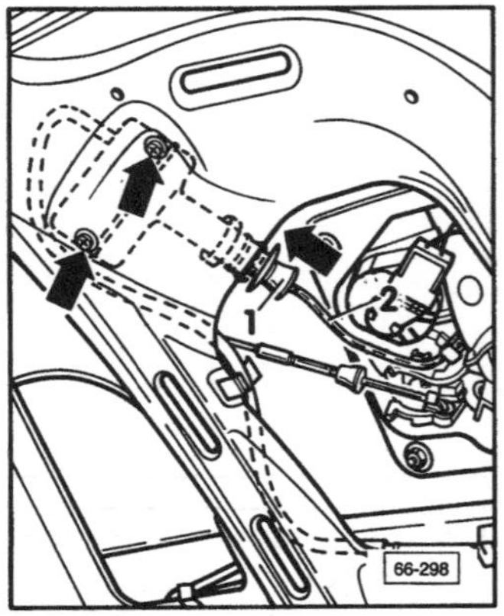

▶ Fig. 5 **Activator for rear lid, removing**

- turn clips (**arrows**) 90° (1/4 turn) (unlock) and unclip inner trim for rear lid

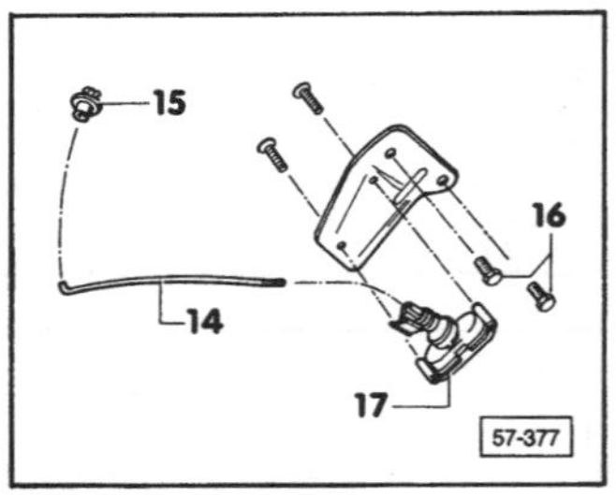

▶

- remove pressure line from activator **17**
- unclip tension rod **14** from clip **15**
- remove hex head bolts **16**
  - tightening torque = 10 Nm (7 ft lb)

Activator for rear lid, installing

Install all components in reverse order of removal.

Central locking, checking

Note

All locks must close within approximately 2 seconds. If the bi-pressure pump runs longer than 5 seconds, either the linkage or the activator is not correctly adjusted or the system has leaks.

The shut-off point in the activator is not attained if the linkage is not adjusted correctly — bi-pressure pump "**pulses**." After 35 seconds the control unit must shut off the bi-pressure pump.

Test conditions

- rear lid and gas flap must be closed
- windows completely open

Locking, checking

- push lock button on either door down
  - both lock buttons must go down, both doors, rear lid and gas flap are locked
- repeat on passenger door

Note

The central locking procedure can be done on the passenger side. Central locking of all activators is possible through the door locks on both doors.

Central locking, electrical troubleshooting

Note
Always refer to appropriate wiring diagram.

Test requirements

- fuses S16, S52 **OK**
- battery voltage **OK**
- vacuum system **OK**
- bi-pressure pump with control module (**V 69**) accessible
- activators properly adjusted, see Index

Tools required

- digital multimeter **Fluke 83 (US 1119)**

Note
Should one or more activators not operate, check if bi-pressure pump operates longer than 5 seconds, then pulses (pumps) again. If yes, eliminate leak in vacuum system.

Voltage supply (terminal 30), checking

- remove 12-pin connector at bi-pressure pump with control module (**V 69**)
- set multimeter **Fluke 83 (US 1119)** to 20-Volt range
- connect multimeter black test lead to terminal **3**, and multimeter red test lead to terminal **1** of 12-pin connector

No voltage:

- check and repair wiring between terminal **30** and Ground (**GND**) **103** according to wiring diagram

End

Voltage:

Go to Ⓐ next page

Ⓐ

Door lock switches (F 59/F 114), checking

- connect multimeter red test lead to terminal **7** of 12-pin connector
  - black test lead remains connected to terminal **3**
- insert key in passenger or driver's door lock
- turn key to close (lock) door

No voltage:

- check switches (**F 59/ F114**) according to wiring diagram, and replace if necessary

End

Voltage:

- connect multimeter red test lead to terminal **10** of 12-pin connector
  - black test lead remains connected to terminal **3**
- insert key in passenger or driver's door lock
- turn key to open (unlock) door

No voltage:

- check switches (**F 59/F 114**) according to wiring diagram, and replace if necessary

End

Voltage:

Go to Ⓑ next page

(B)

Activator (F 131), checking

- reconnect 12-pin connector
- close (lock) doors with key, leaving rear lid raised
  - door buttons down
- disconnect 12-pin connector at bi-pressure pump with control module (**V 69**)
- set multimeter **Fluke 83 (US 1119)** to Ohm range
- connect multimeter black test lead to terminal **3** and multimeter red test lead to terminal **4** of 12-pin connector

No continuity	Continuity

No continuity:

- check and repair wiring according to wiring diagram or,
- replace activator (**F 131**)

End

Continuity:

- reconnect 12-pin connector
- open (unlock) doors with key, leaving rear lid raised
  - door buttons up
- disconnect 12-pin connector
- set multimeter **Fluke 83 (US 1119)** to 20-Volt range
- connect multimeter black test lead to terminal **3** and multimeter red test lead to terminal **4** of 12-pin connector

No voltage	Voltage

No voltage:

- check and repair wiring according to wiring diagram or,
- replace activator (**F 131**)

End

Voltage:

- replace bi-pressure pump with control module (**V 69**)

End

Central locking system pump and control module, location/terminal identification

Location

The central locking system pump and control module are located in the luggage compartment, on the right.

Connector terminal identification

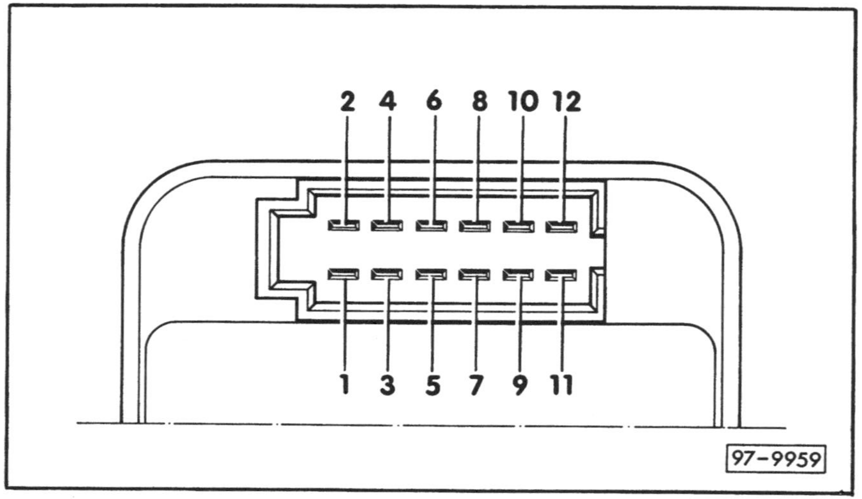

1 = T 12/1 2 = T 12/2 3 = T 12/3 4 = T 12/4	5 = T 12/5 6 = T 12/6 7 = T 12/7 8 = T 12/8	9 = T 12/9 10 = T 12/10 11 = T 12/11 12 = T 12/12

Connector color

T 12 - transparent

Note

See wiring diagram for wire routing to connector **T 12**. Always refer to appropriate wiring diagram when performing any electrical tests or repairs.

Re-keying locks

Lock cylinders cannot be ordered by specified key numbers. On most cars one-key convenience can be retained by changing the appropriate lock tumblers to match the key that is going to be used.

Note

Tumblers in the ignition lock **cannot** be changed.

On certain older vehicles, the glove box key is different than the ignition key. A separate kit is available from the Parts Department to change this lock.

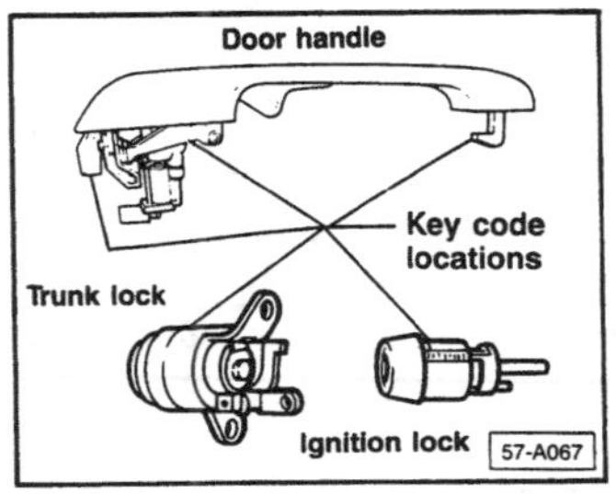

Key code, locating

The key code is a two-letter, four-digit number that represents the contour of the key.

- The key code may be found on:
  - the plastic plate that comes with the original set of keys
  - recorded by owner or selling dealer in vehicle owners manual
  - door handle
  - ignition lock
  - rear lid lock

Lock cylinder tumblers, changing

- select correct tumbler repair kit
  - bags are marked by numbers **1-2-3-4** to identify the tumblers
- insert lock's original key into cylinder lock

- carefully remove lock cylinder
  - do not damage O-ring
- place thumb and forefinger over tumblers to hold them in place
- pull out key

- if tumblers are peened in place, carefully remove peened metal with suitable tool until lock tumblers can be removed
- place lock cylinder on clean flat surface
- release the cylinder and tumblers slowly
  - tumblers will come out as you let go

CAUTION

Do not lose any springs or tumblers.

- remove all tumblers and springs from cylinder

Selecting tumblers

CAUTION

VW Key Code Service Catalog is strictly confidential.

- look up key code of ignition lock in Key Code Service Catalog
  - catalog will tell you what tumblers to use and the sequence they are to be inserted into lock cylinder

Example:

Key code **AH 9129**

The code in catalog is listed in following sequence:

3-1-3-2-3-1-3

- add **1** to the **first** and **third** tumbler size to initially make the lock work. In example use tumbler sizes in this order:

  4-3-4-1-3-4-1

  3 1 3 2 3 1 3 = tumbler arrangement according to code list

  + 1   1
  4 1 4 2 3 1 3 = tumbler arrangement of converted lock cylinder

CAUTION

Part numbers are for reference only. Always consult with the parts department for the latest information.

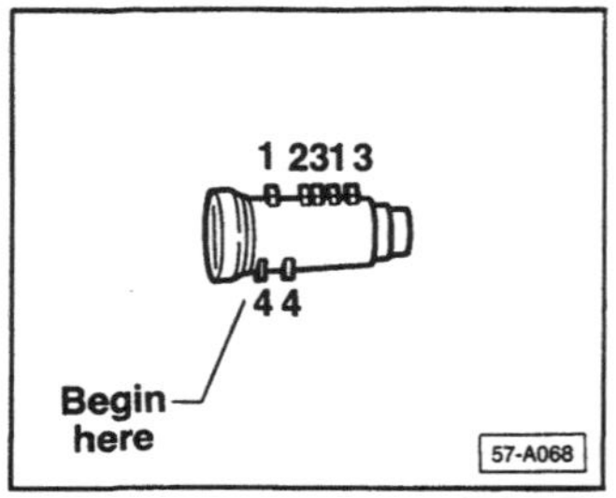

57-A068

Note

Lubricate the tumblers and springs with **G 000 400** grease or equivalent as they are inserted into the lock cylinder.

- reading key code numbers from left to right, insert tumblers in the lock cylinder housing starting at the key slot

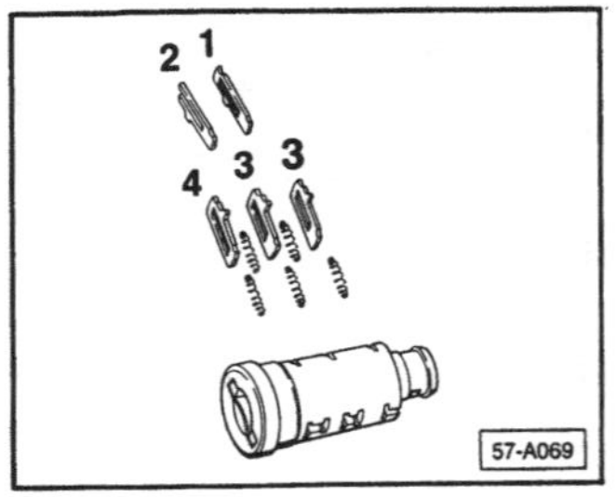

57-A069

Note

When working with a lock cylinder with only 5 tumblers, use tumbler number sequence **minus** the first two tumblers.

Example:

4 1 4 2 3 1 3 = code from catalog

- - 4 2 3 1 3 = are the tumblers to be used

Note

The central locking procedure can be done on the passenger side. Central locking of all activators is possible through the door locks on both doors.

THIS PAGE INTENTIONALLY LEFT BLANK

Index

Electrical cable drive
- checking/adjusting **60.6**

Parallel alignment
- checking/adjusting **60.4**

Slide/tilt sunroof
- assembly **60.2**
- assembly frame, installing **60.9**
- assembly frame, removing **60.7**
- installing **60.4**
- removing **60.3**

Sunroof panel
- adjusting **60.6**

Sunroof panel headliner trim frame
- removing **60.5**

Water drain hoses
- cleaning **60.10**

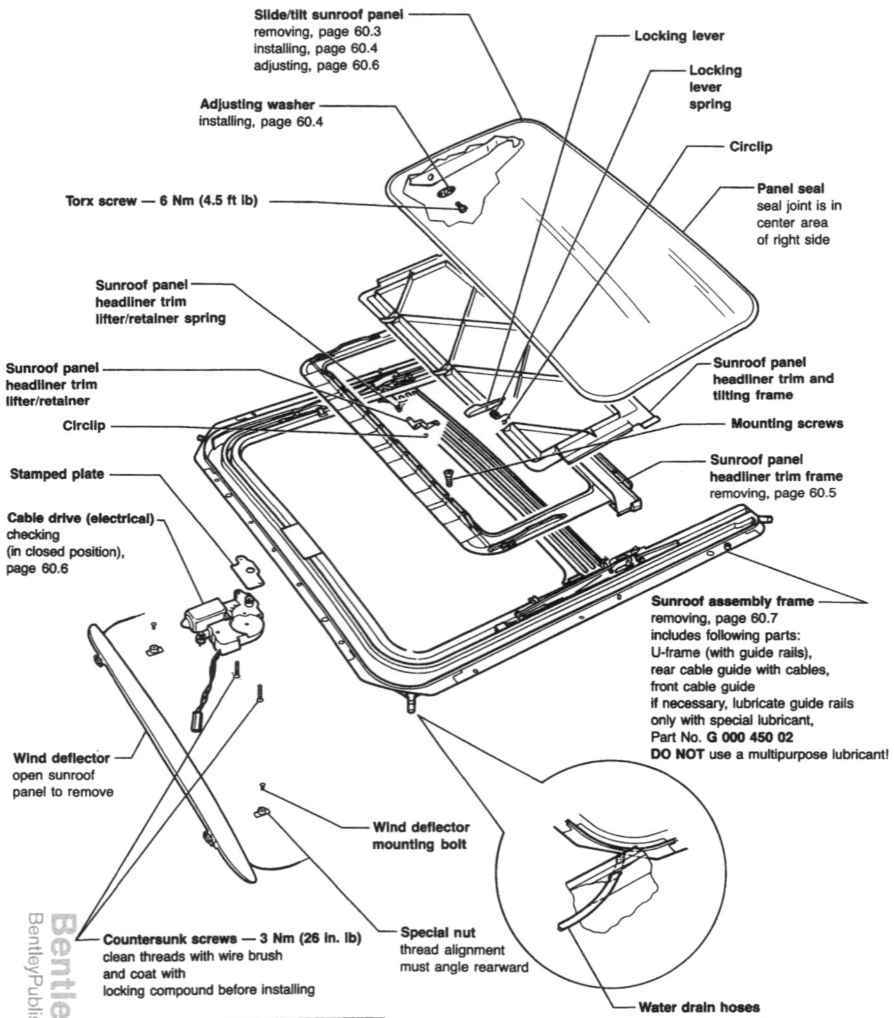

CAUTION

Part numbers are for reference only. Always consult your Parts Department for latest parts information.

Slide/tilt sunroof panel, removing

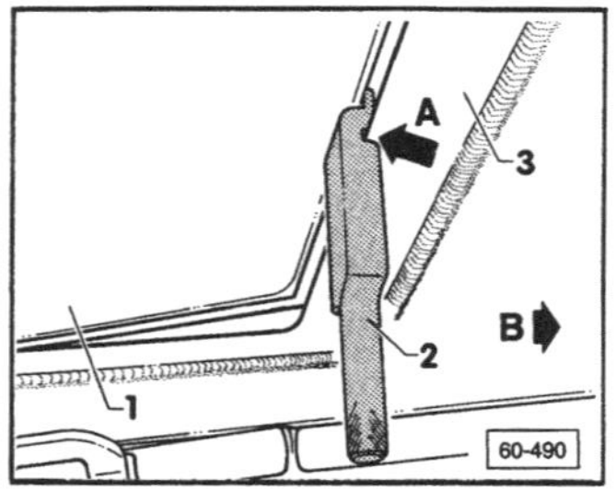

- tilt sunroof panel up
- insert special tool **3195** into sunroof opening
- pull headliner and tilting frame **1** down until special tool **2** engages headliner trim frame **3** (**arrow A**)
- disengage headliner trim frame from guide rails by hitting special tool **3195** backwards with palm of hand (**arrow B**)
- push back headliner trim frame until tool contacts edge of roof cutout

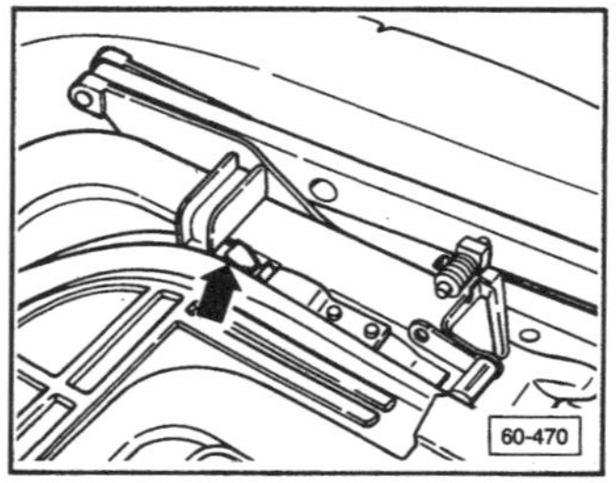

- remove special tool and manually push headliner trim frame to stop (**arrow**)

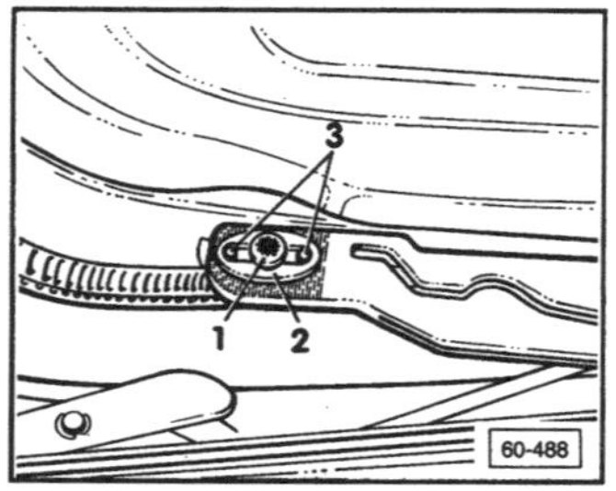

- remove mounting screw **1** (Torx) and adjusting washers **2** from locating pins **3**
- remove slide/tilt sunroof panel from above

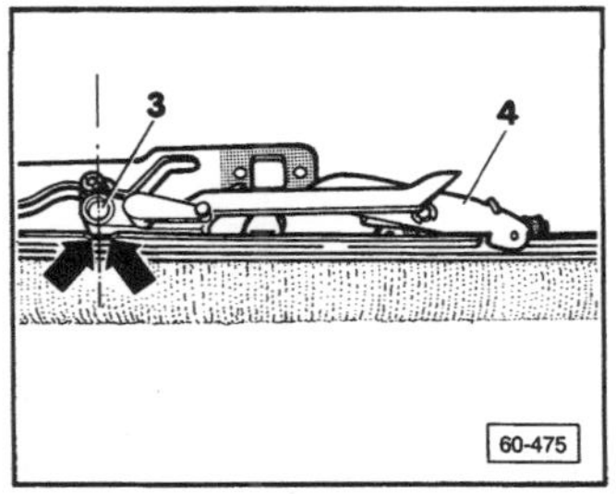

Slide/tilt sunroof panel, installing

Sunroof panel operating mechanism must be in closed position before installing sunroof panel.

To determine closed position:
- pivots **3** must be within the marked area (**arrow**) and locking levers **4** must be engaged in the guide rails

If **NO**,
- adjust sunroof panel to closed position as specified in **Parallel alignment, checking/adjusting** below

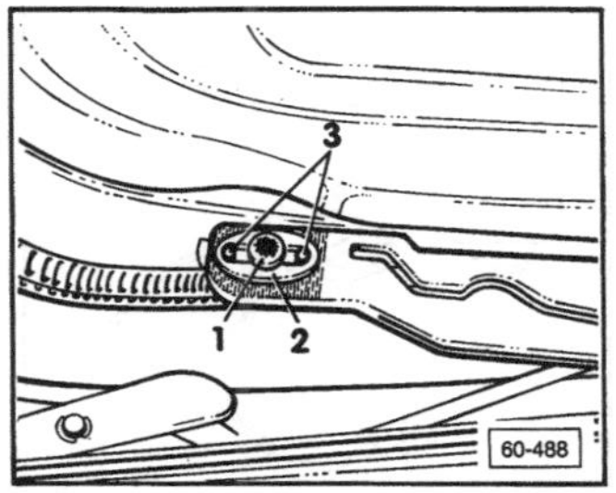

- install slide/tilt sunroof panel from above and insert mounting screws for lever/panel
- position slot of adjusting washer **2** on locating pins **3** before tightening Torx screw **1**
- lightly tighten mounting screws **1** (Torx)
- adjust sunroof panel height (see page 60.6)
- tighten mounting screws fully
  - 6 Nm (4.5 ft lb)
- install headliner frame onto guide rails

Parallel alignment, checking/adjusting

Checking and adjusting parallel alignment can be performed with the sunroof panel either removed or installed, but the headliner trim frame must be shoved back.

- disconnect battery ground strap
- disengage headliner trim frame from guide rails using special tool **3195** and shove back to stop (see page 60.3)
- pry off interior light

- remove screws (**arrows**) and remove sunroof cable drive cover
- remove cable drive mounting screws (countersunk) and lower cable drive
  - drive pinion must no longer mesh with cables

> **CAUTION**
>
> Sunroof drive pinion must not be engaged when checking/adjusting parallel alignment.

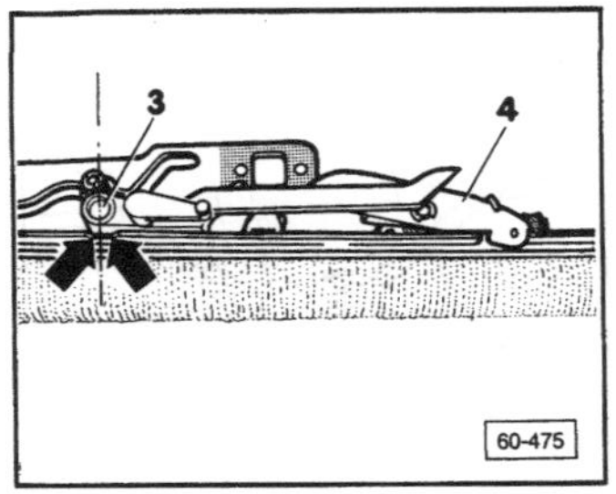

- move rear guides with cables to rear and position pivots **3** within marked area (**arrows**)

> **CAUTION**
>
> When lining up rear guides and cables, always move guides from front of vehicle rearward to marked area. Do not move guides from rear of vehicle forward to marked area.

- engage locking levers **4** into guide rails
- check/adjust cable drive to closed position (see page 60.6)
- clean cable drive mounting screws with wire brush and coat with thread locking compound
- install cable drive
  - 3 Nm (26 in. lb)
- install interior light and reconnect battery ground strap
- install headliner trim frame onto guides

Headliner trim frame, removing

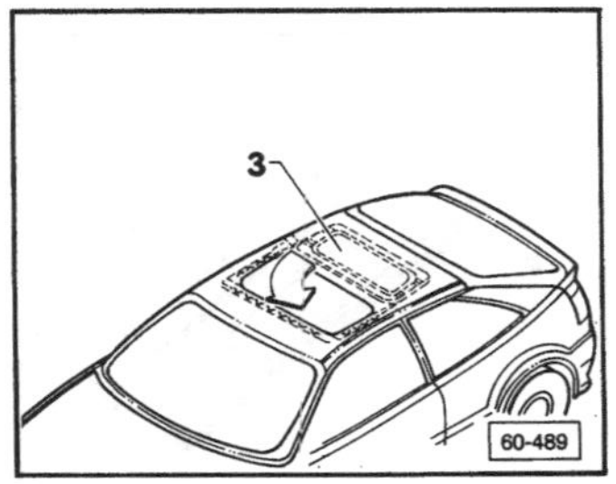

- remove slide/tilt sunroof panel (see page 60.3)
- slightly raise headliner trim frame **3**, rotate 90° (1/4 turn), and remove from roof cutout from front
- install headliner trim frame in reverse order of removal

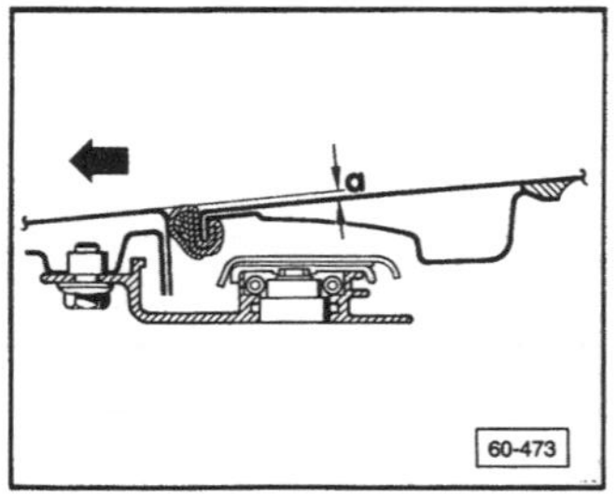

Sunroof panel, adjusting

Adjust height at front and back of sunroof lid to prevent wind noise.

- check parallel alignment of rear guides and cables and adjust if necessary, (see page 60.4)
- always check height adjustment at edge of sunroof opening near center of vehicle

Front height, adjusting

a = 1 mm below roof line
arrow = front of vehicle

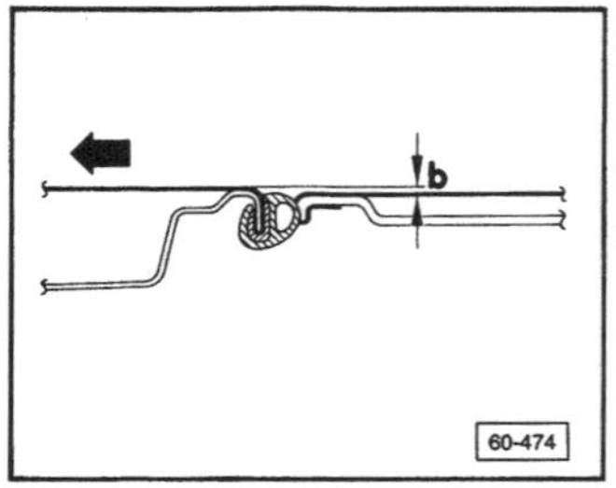

Rear height, adjusting

b = 1 mm above roof line
arrow = front of vehicle

Electrical cable drive, checking/ adjusting

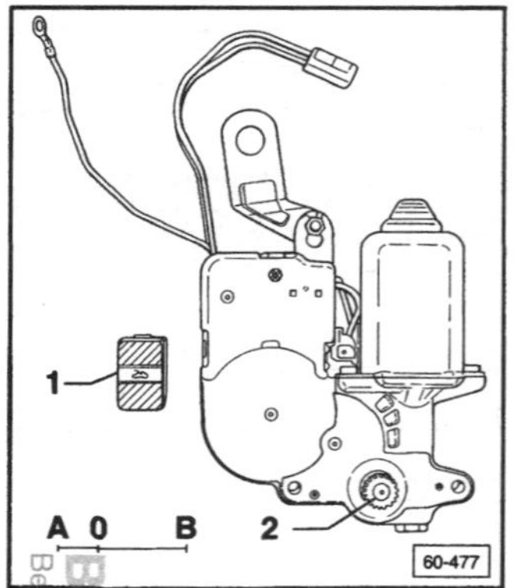

When reinstalling cable drive, it must be installed in **closed position** (cable drive in closed position **and** sunroof panel in closed position)

- adjust slide/tilt sunroof panel to closed position, (see page 60.4)
- actuate switch **1** and count drive pinion **2** revolutions:
  - closed position setting **0** of cable drive is between end stops **A** and **B**
  - end stop **B** is reached by turning drive pinion **2** clockwise
  - end stop **A** is reached by turning drive pinion **2** counter-clockwise

A = end stop, sunroof tilt function; two turns clockwise of drive pinion to closed position **0**
B = end stop, sunroof slide function; nine turns counterclockwise of drive pinion to closed position **0**

- actuate switch **1** and turn drive pinion **2**, two turns clockwise from position **A** or 9 turns counterclockwise from position **B** to obtain closed position **0**

Slide/tilt sunroof assembly frame, removing

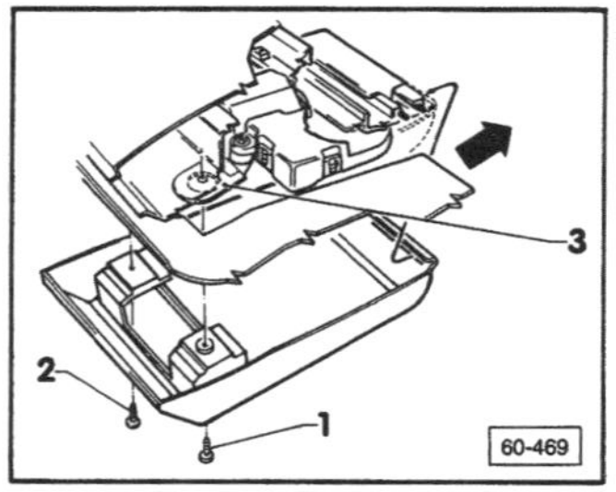

- disconnect battery ground strap
- remove interior light assembly
- remove screws **1** and **2** and pull off electric cable drive cover in direction of (**arrow**)

Note

Screw **1** also secures bracket **3**.

- remove left/right sunvisors

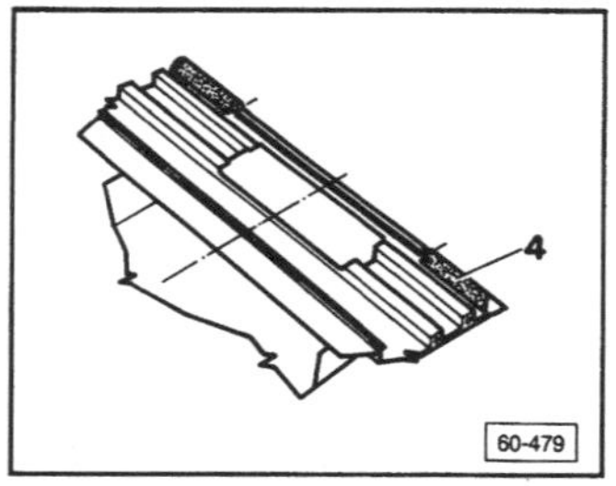

- remove perimeter sealing bead **4** from sunroof opening

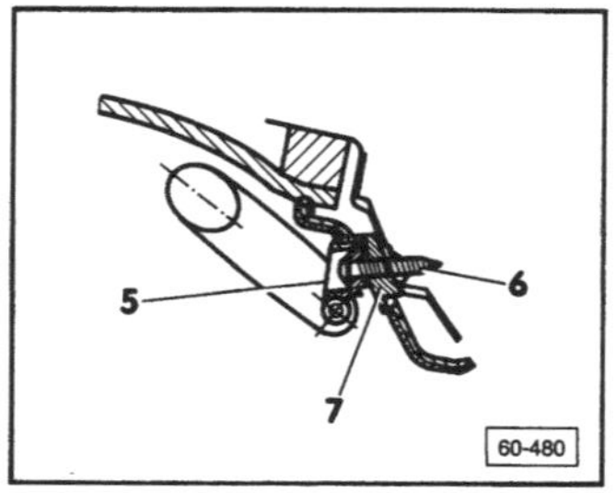

- pry off cap **5**, remove screw **6**, and remove grab handle together with molded mounting base **7**

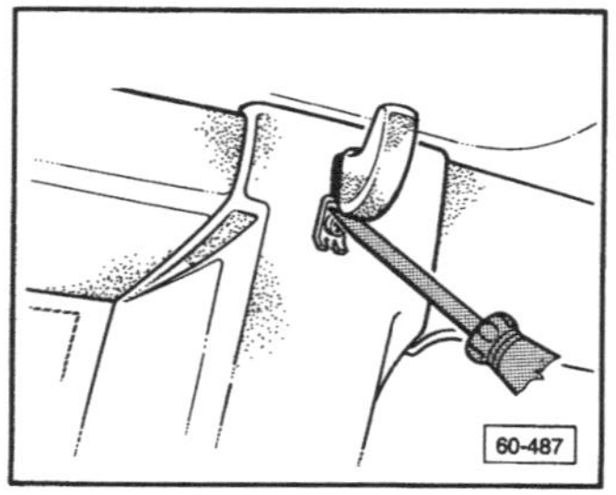

- carefully pry off clothes hook cover with screwdriver
- remove screw, remove clothes hook

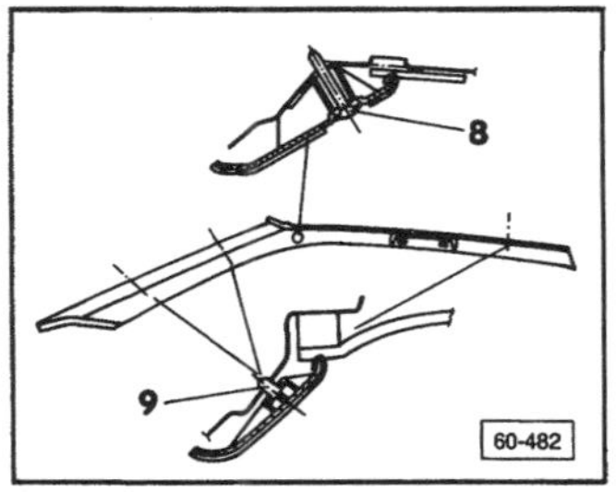

▶ ■ remove screw **8** and pull off A-pillar trim from clips **9**
■ remove seat belt guides
■ pull off seat belt height-adjuster knobs
■ remove rear side panels

■ remove B-pillar trim
■ pull off C-pillar trim in area of contact with headliner

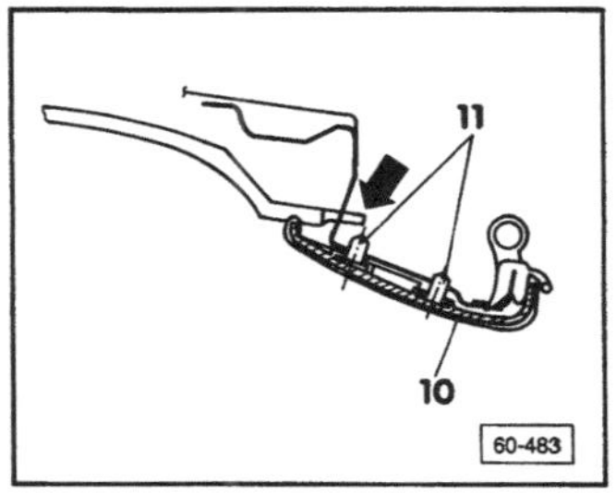

▶ ■ pry off rear header trim **10** from clips **11**

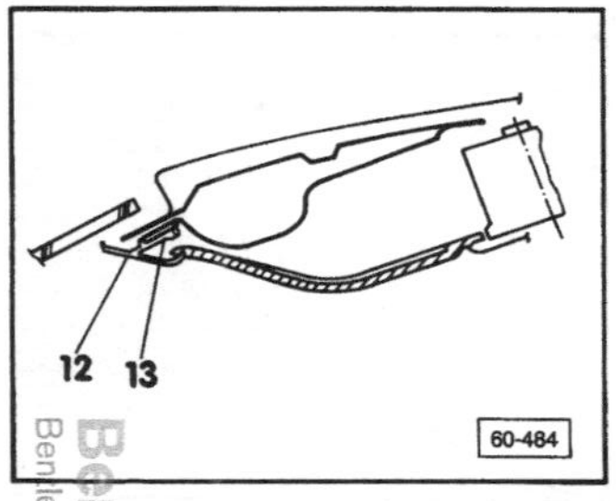

▶ ■ pry off front header trim **12** from U-clips **13**

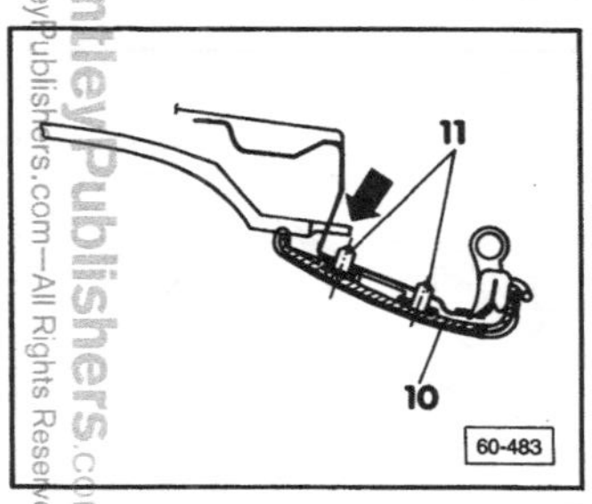

▶ ■ lower headliner and pull forward slightly to pull rear tab (**arrow**) from roof reinforcement
■ remove headliner through rear lid opening

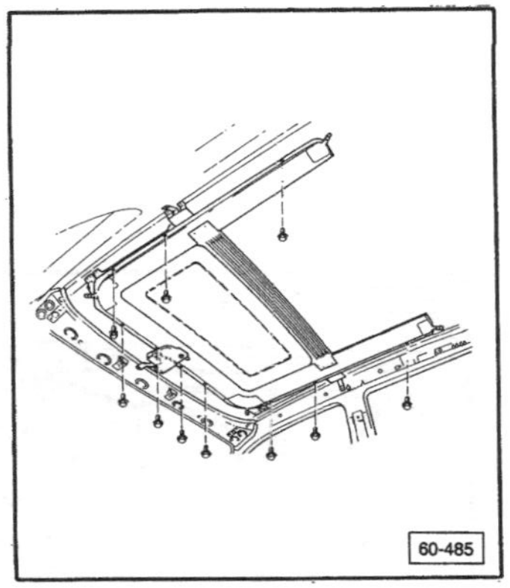

▶ ■ disconnect water drain hoses from sunroof assembly frame
■ remove electric cable drive
■ remove sunroof panel from above (for easier panel frame installation) (see page 60.3)
■ remove hex head bolts and pull out assembly frame with help of an assistant

Slide/tilt sunroof assembly frame, installing

■ install sunroof assembly frame and tighten hex head mounting bolts
- 7.5 Nm (6.0 ft lb)

■ install sunroof lid (see page 60.4)
■ adjust electric cable drive to closed position (see page 60.6)

■ install electric cable drive
■ install water drain hoses to sunroof assembly frame
■ install one-piece molded headliner so that rear tabs engage roof reinforcement

Note

If a functional test of the sunroof is to be performed without the headliner, first install the sunroof perimeter sealing bead, (see below).

■ install front header trim
■ install rear header trim
■ install A-pillar, B-pillar, and C-pillar trim

■ install seat belt guides and height-adjuster knobs
■ install clothes hook
■ install grab handle

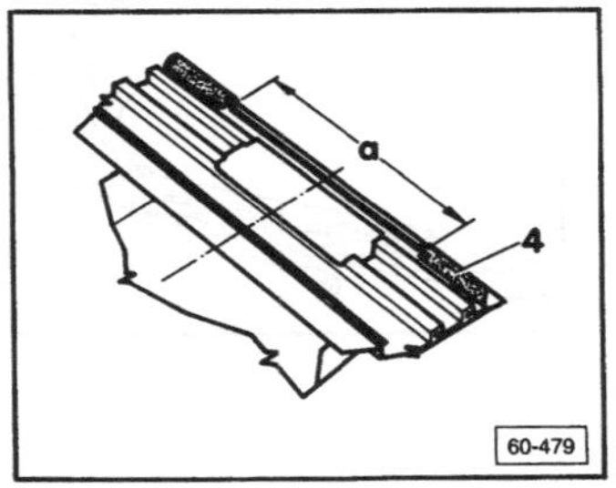

▶ ■ install sunroof opening perimeter sealing bead **4** leaving gap **a** in the area of the electric cable drive
- gap area is denoted by notches in frame
- **a** = 180 mm (7.0 in)

■ install left and right sunvisors
■ install electric cable drive cover

■ install interior light
■ connect battery ground strap

Water drain hoses, cleaning

Front water drain hoses are routed through the A-pillars and discharge between the front wheelhousing and wheelhousing liner. Rear water drain hoses are routed through the C-pillars and discharge behind the rear bumper corners on side of vehicle.

- clean front drain hoses through sunroof opening

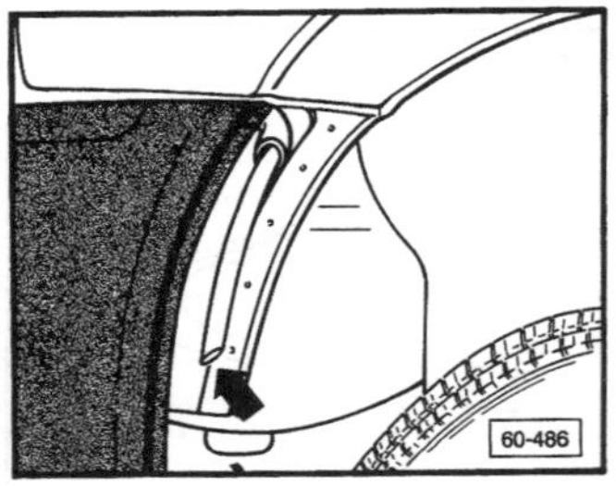

- clean rear drain hoses from lower end of hose (**arrow**)
- remove rear bumper to access rear drain hose

Note

For cleaning, it is recommended that an auxiliary tool be used — for example, a 230 cm (90.5 in) long probe fabricated from the cores of speedometer cables

Sunroof noises, eliminating

Motor cover not seated on assembly, repairing

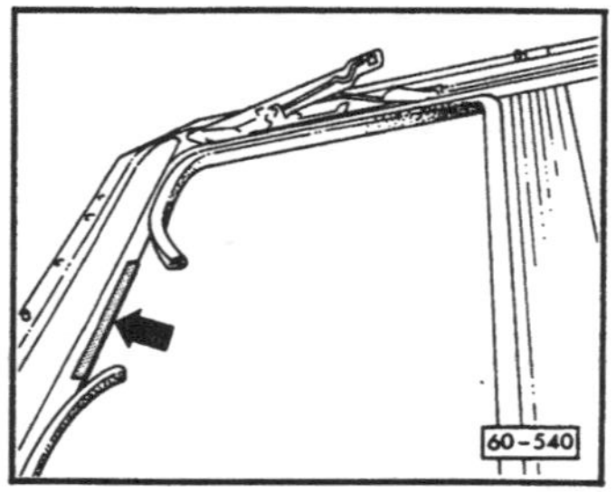

- remove cover, pull assembly unit panel seal off, as shown
- apply textile tape (**arrow**) at cross hatched area as shown
  - if tape already installed, replace

> **CAUTION**
>
> Part numbers are listed for reference only. Always consult with the Parts Department for latest information.

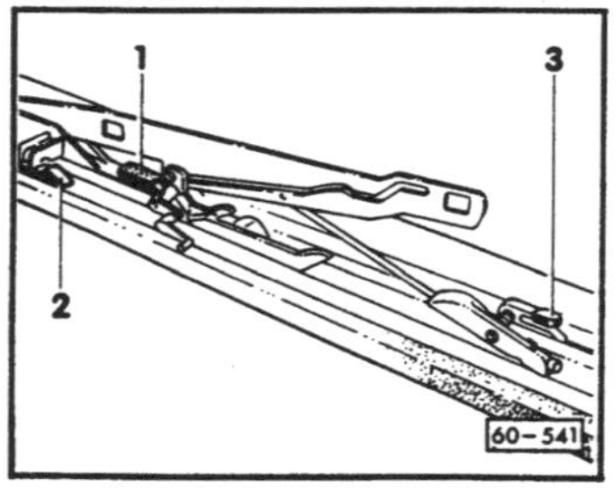

Front guide, guide plates, lifter mechanism hook mounting does not have insulation

- remove frame portion of headlining
- check felt strips **1**, plastic covers **2**, **3** for damage
  - replace as required
- apply felt strip **1** Part No. **533 863 288A**, cut to 10 mm (25/64 in.) x 20 mm (25/32 in.) x 3.0 mm (1/8 in.) in position
- repair plastic covers **2**, **3** by installing covering with cloth or textile tape

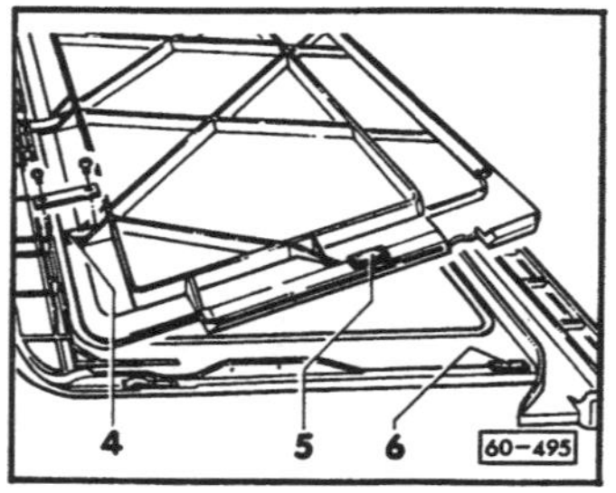

Sunroof panel headliner trim frame insulation, installing

- remove frame position
- check if felt strips **4** (under springs), **5**, **6** are installed

If **NO**:

- install felt strips, Part, No. **533 867 910A** 1.0 mm (3/64 in.) thick

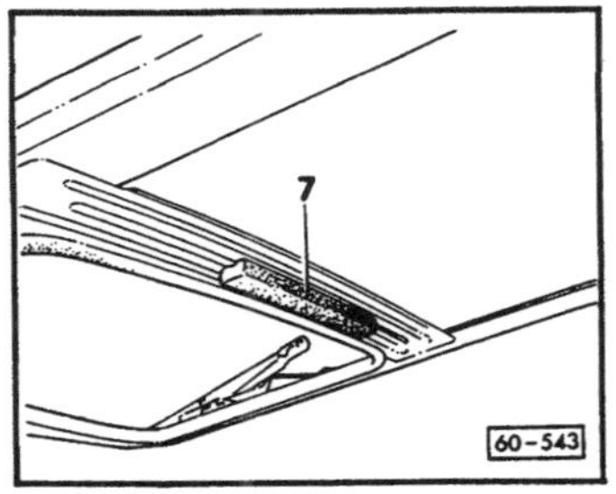

Headliner contacts assembly unit, repairing

- pull off panel seal
- press headlining down at rear
- ensure sealing strip **7**, is positioned between rear assembly unit member and headliner
  - replace with felt of equivalent thickness

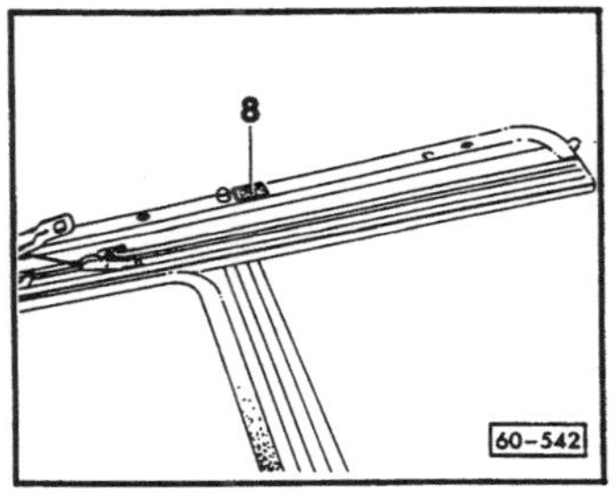

Assembly unit does not contact roof frame, repairing

- remove assembly unit
- check if felt **8** stuck at guide pin areas
- if felt **8** stuck, install new piece of felt 20 mm (25/32 in.) x 20 mm (25/32 in.), 1.0 mm (3/64 in.)

THIS PAGE INTENTIONALLY LEFT BLANK

THIS PAGE INTENTIONALLY LEFT
BLANK

Index

Front bumper
- assembly **63.2**
- gap dimension **63.4**
- removing/installing **63.3**

Rear bumper
- assembly **63.5**
- gap dimension **63.6**
- removing/installing **63.6**

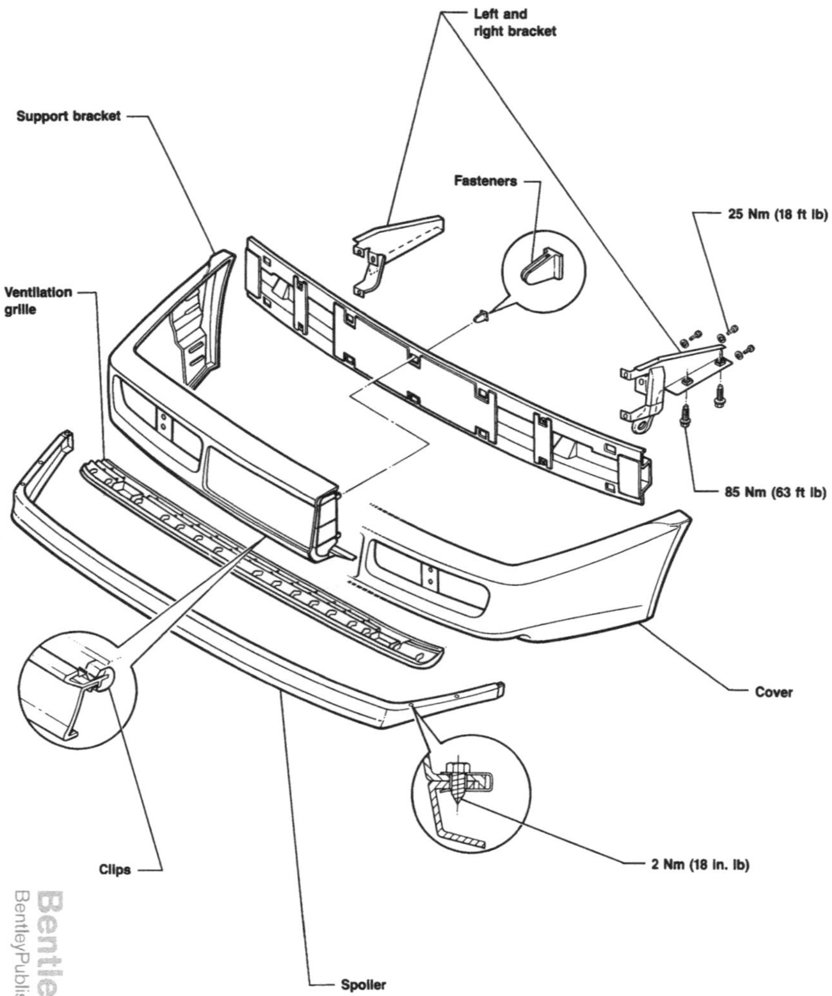

63-150

Front bumper, removing

> **WARNING**
>
> **DO NOT** drive vehicle without the front bumper and brackets installed. Subframe is **NOT** properly secured with bumper removed.

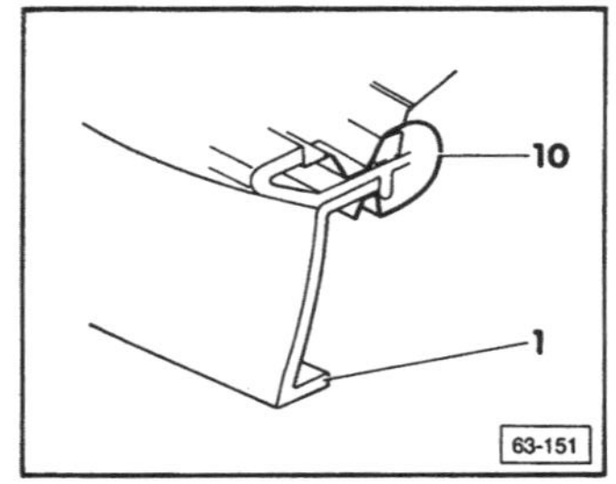

63-151

- remove clips **10** between cover **1** and front engine mount

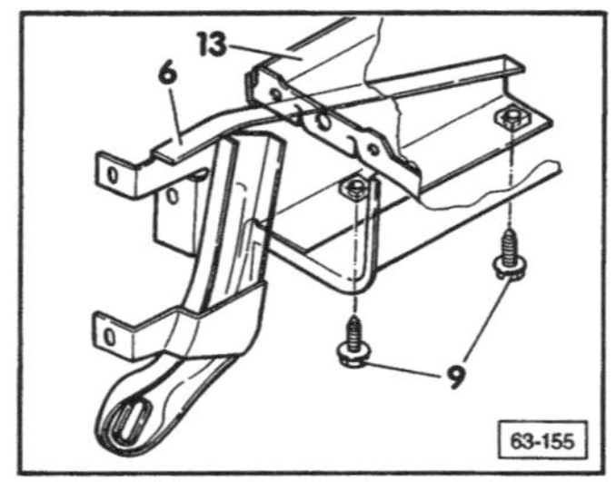

63-155

- remove bolts **9** from engine mount **13** and bracket **6**

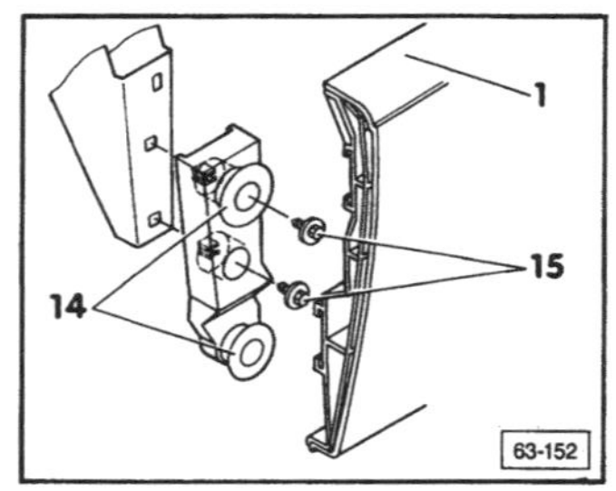

63-152

- pull bumper **1** forward out of plastic retainers **14** (requires two mechanics)
  - bolts **15** tightening torque = 2 Nm (18 in. lb)
- pry fasteners out of plastic clips with screwdriver
- remove support bracket from cover

Front bumper, installing

Note

Install front bumper in reverse order of removal procedure, noting the following:

- press cover and bumper together
  - begin at center of bumper and push both parts together until all plastic clips snap into place
  - bolts **9** (see Fig. 63-155) tightening torque = 85 Nm (63 ft lb)

▶ ■ when replacing bracket **6**, attach it centered on support bracket in order to adhere to the gap dimension **a** = 8 mm (5/16 in.) between the bumper cover and fender

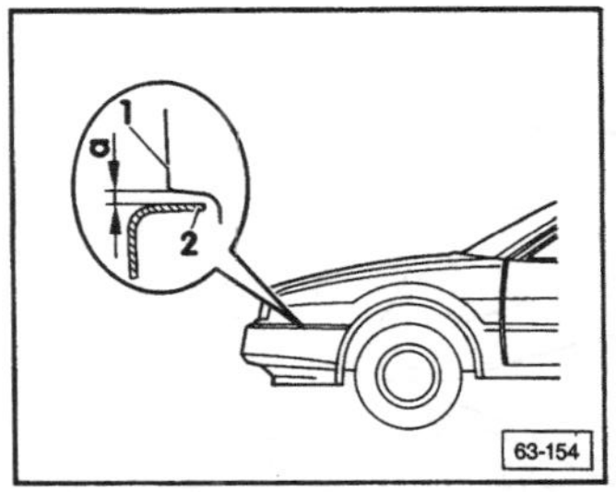

▶ **Gap dimension, fender to bumper**

1 — fender
2 — bumper cover
a = 8 mm (5/16 in.)

THIS PAGE INTENTIONALLY LEFT BLANK

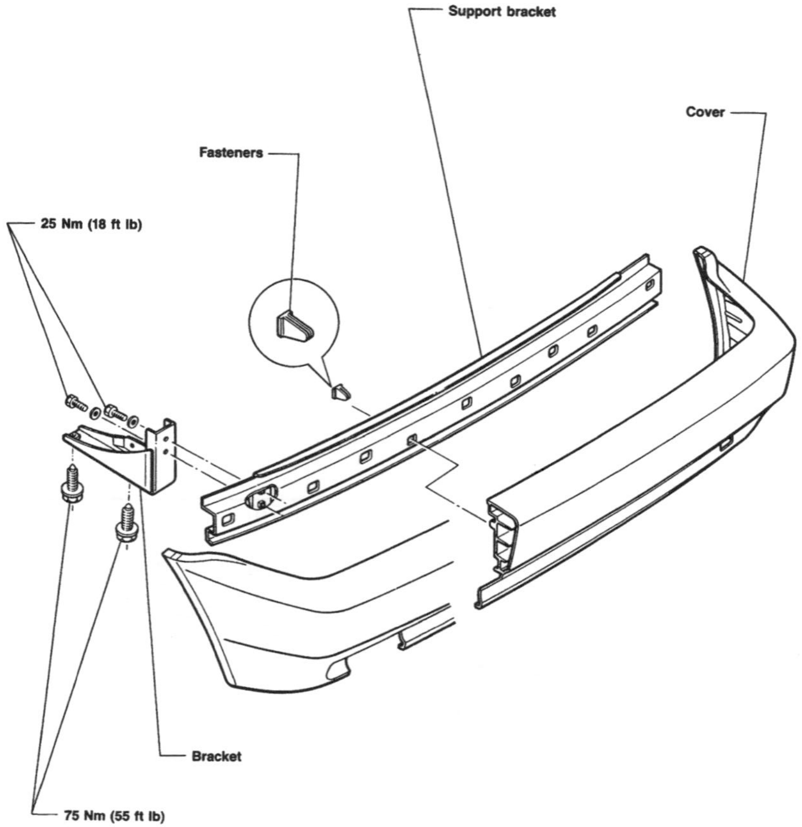

Rear bumper, removing

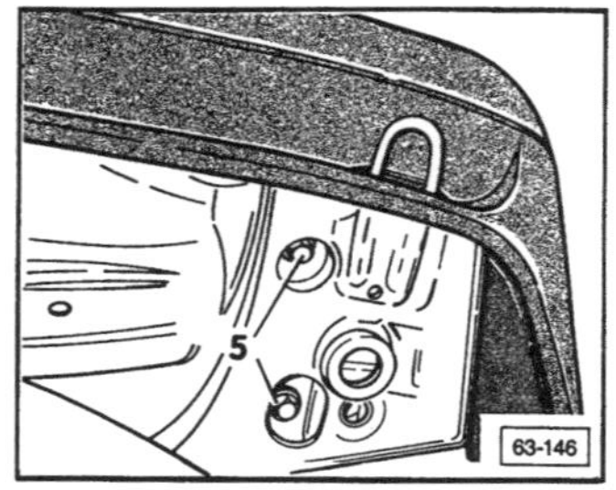

- remove bolts **5** from trunk floor area

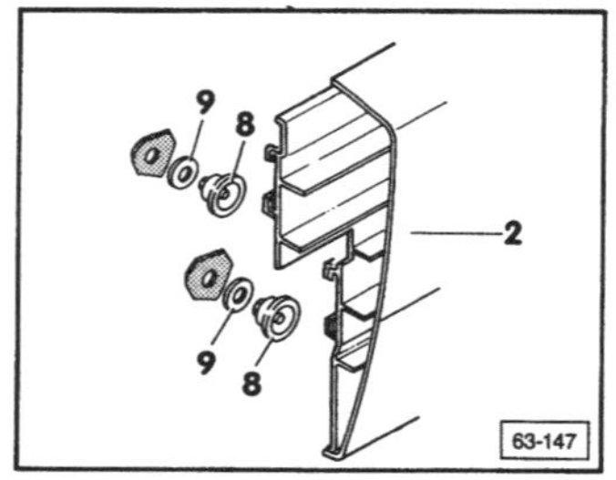

- pull bumper **2** rearward out of plastic retainers **8** (two mechanics required)
  - **8** — plastic retainer
  - **9** — seals
- pry fastener out of plastic retainers with screwdriver and remove support bracket from cover

Rear bumper, installing

Note

Install rear bumper in reverse order of removal procedure, noting the following:

- press cover and bumper together
  - begin at center of bumper and push both parts together until all plastic clips snap into place
  - bolt **5** (see Fig. 63-146) tightening torque = 75 Nm (55 ft lb)

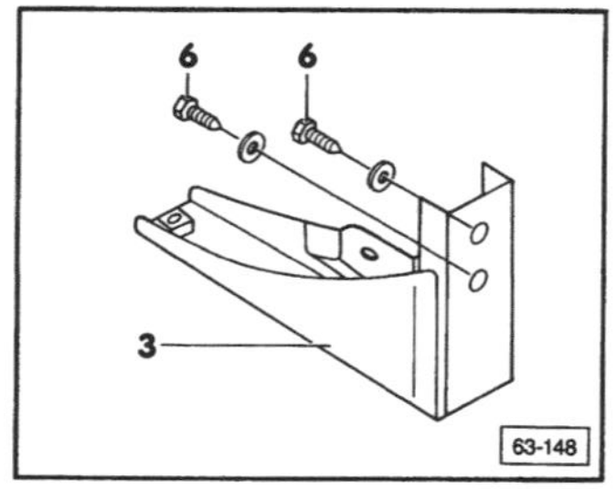

- when replacing bracket **3**, attach it while centered on support bracket in order to adhere to gap dimension **a** = 8 ± 1 mm (5/16 ± 3/64 in.) between cover and quarter panel
  - bolts **6** tightening torque = 25 Nm (18 ft lb)

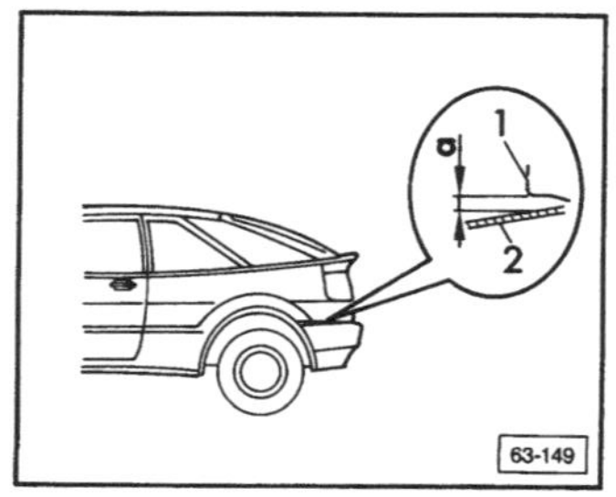

Gap dimension, fender to bumper

1 — quarter panel
2 — bumper cover
a = 8.0 mm (5/16 in.)

Index

Power windows
- troubleshooting **64.20**
- control module, location **64.26**

Rear window
- assembly **64.11**
- installing **64.14**
- removing **64.12**

Side window
- assembly **64.16**
- installing **64.18**
- removing **64.17**

Window replacing
- general information **64.2**

Windshield
- assembly **64.5**
- installing **64.8**
- removing **64.6**

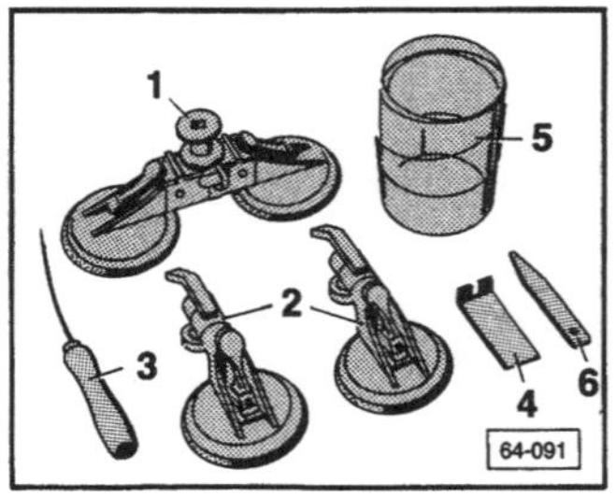

Window replacing, general information

To replace windshield, rear window or side windows, the following tools and materials are required.

Tool kit VW 1474

1 — Reel
2 — Relay roller with holder
3 — Awl
4 — Guide rail
5 — Protective sheeting
6 — Wedge

Use protective sheeting **VW 1474/7** to protect instrument panel.

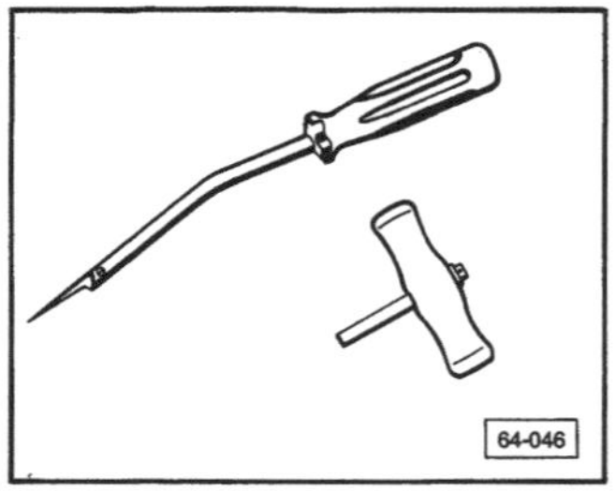

Tool kit VW 1351

- for removing broken windshields, rear windows and broken or unbroken side windows

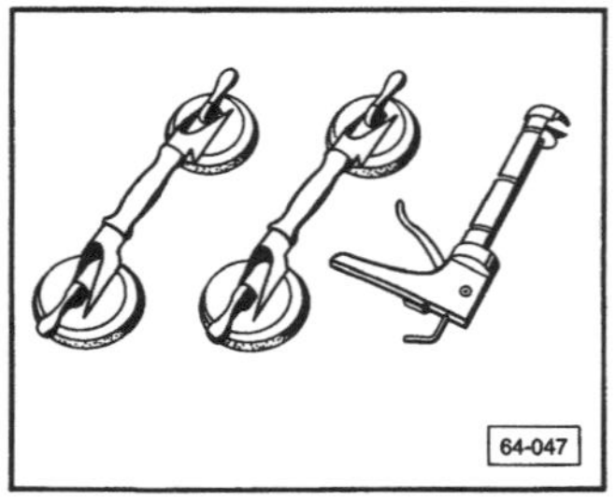

- caulking gun
- double suction pad (only for rear window, windshield)
- protective goggles
- utility knife
- protective gloves (leather)
- cloth or duct tape

CAUTION

Part numbers are listed for reference only. Always consult with the Parts Department for latest information.

Parts required

Description	Part Number	Quantity
cutting wire 50 m (150 ft) roll	443 845 955	1
adjusting wedges for windshield	893 845 631	2
adjusting wedges for rear windows	893 845 955	2
adhesive kit	D 004 300 04	1
mixing rod • reusable	D 009 700	1

Hardening of adhesive sealing compound

- vehicle **must** stand **on an even surface** for at least one hour at a temperature of 20°C (68°F) and a relative humidity of 65-70% before allowing vehicle to be driven
- keeping the sealed bonded area wet promotes hardening process

CAUTION

When using any cleaning agent, follow all cautions and warnings listed on the containers.

Excess bonding/sealing compound, removing

- clean painted areas with dry cloth
- remove excess adhesive sealing compound using cleaning solvent supplied in repair kit or equivalent
  - do not use cleaners which contain alcohol
- to clean plastic trim, let adhesive sealing compound harden three hours, then scrape off excess with plastic wedge

CAUTION

While adhesive sealant is hardening, avoid putting pressure on glass from the inside of the car when cleaning around glass.

Sealing leaks

- water test window area before installing moldings, trim or lip of rubber molding
- dry area around leaks with compressed air and clean
- apply adhesive sealant to leak area between glass and flange, then smooth over
  - windshield can only be re-sealed around area of A-pillar and roof from inside
  - rear window glass can be re-sealed around its perimeter from inside
  - side window glass can be re-sealed around its perimeter from inside
- retest area for water leaks

The 110 ml adhesive sealant cartridge **D 009 100 03** is recommended for resealing

THIS PAGE INTENTIONALLY LEFT BLANK

Note

The windshield is mounted flush. The replacement part is supplied with water repellent frame.

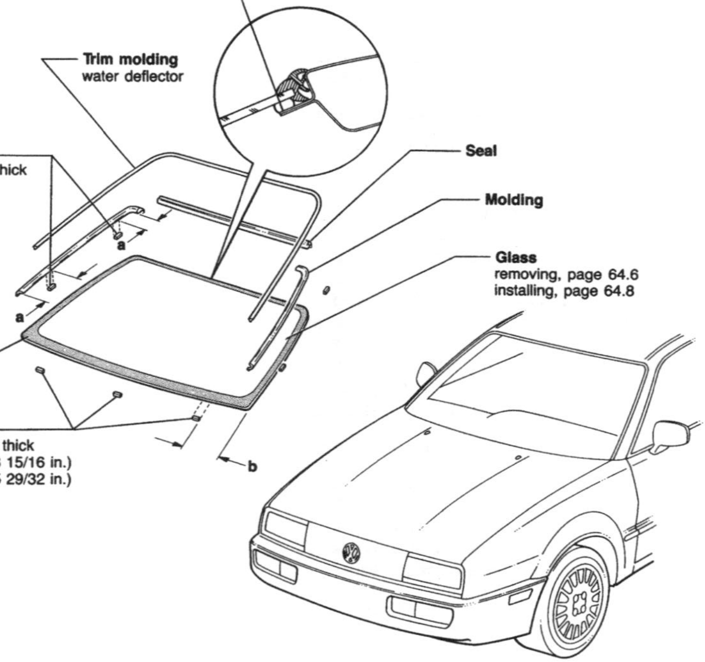

Windshield, removing

> **WARNING**
>
> Always wear protective goggles and leather gloves when removing and installing windows.

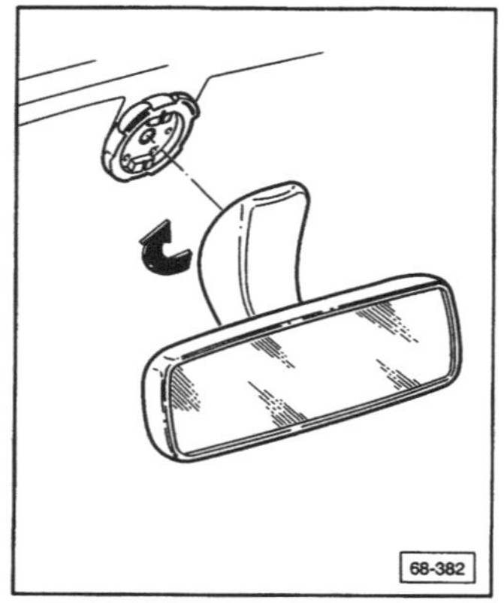

- remove trim from A pillar (see Repair Group 70)
- rotate inner rear view mirror with base 15° - 20° in opposite direction from installation position (**arrow**)

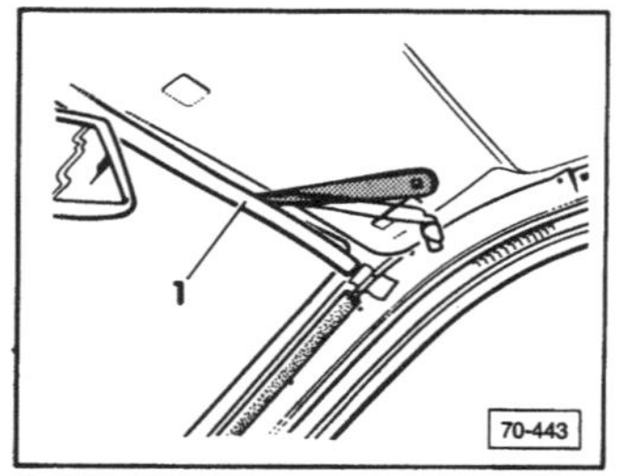

- pry out front push-on trim molding **1** using a plastic wedge
- remove windshield wiper

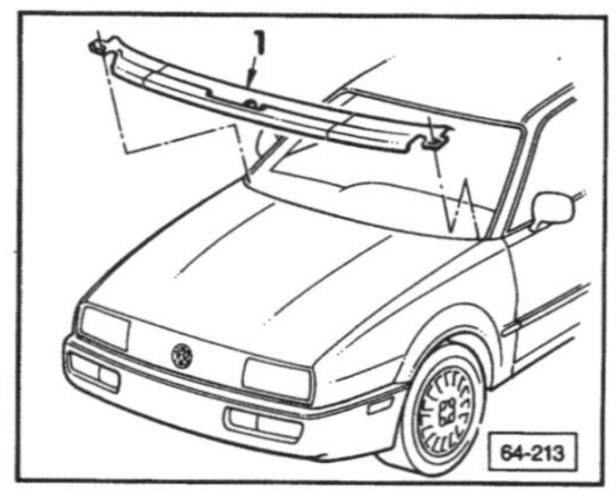

- remove molding **1** (three parts)

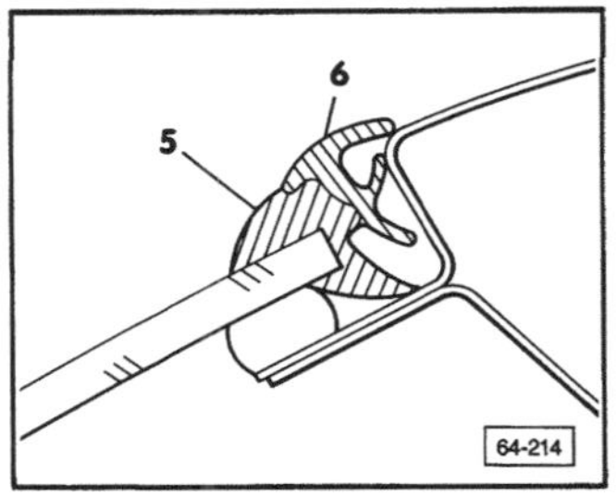

- pull off moldings **5** and **6**

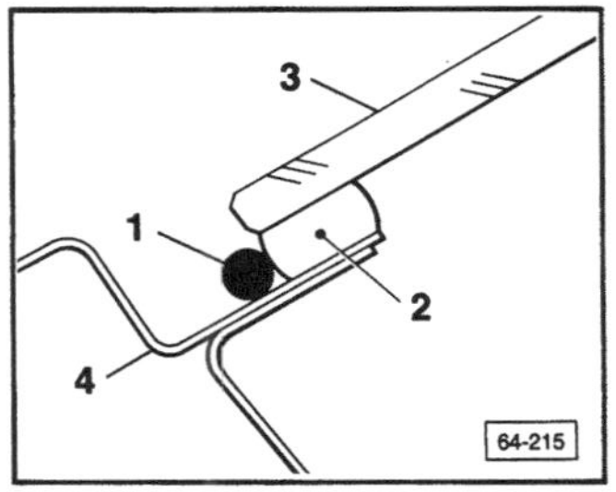

- cut off 7 m (23 ft) of cutting wire
- locate center of cutting wire
- on the outside, starting at top center of windshield, press cutting wire into window channel so that wire runs completely around perimeter of windshield ending at bottom center
  - **1** — cutting wire
  - **2** — adhesive sealant
  - **3** — windshield
  - **4** — body flange

> **CAUTION**
>
> Do not bend or twist wire, otherwise wire will break under tension.

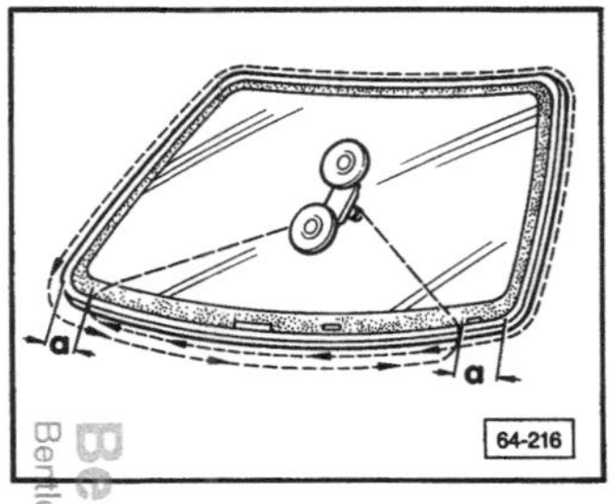

- use the following procedure to insert ends of cutting wire individually in lower area of glass at a distance **a** from each corner
  - **a** = 150 mm (6 in.)
  - cutting wire is doubled up at bottom of glass
- slide plastic foil between windshield and instrument panel
- slide guide rail between windshield and instrument panel to base of windshield
- push awl through adhesive sealant until awl eye is visible from outside
- turn awl slightly and thread in one end of wire
- bend over wire and pull through to inside of vehicle
- remove guide rail
- repeat procedure for other end of wire
- attach reel (as shown) to center of windshield

- thread cutting wire into reel
- place plastic foil between cutting wire and instrument panel

> **WARNING**
>
> Always wear protective goggles and leather gloves when removing and installing windows.

> **CAUTION**
>
> Depending on extent of paint damage in removing windshield, touch-up or repaint as required.

- slightly tension cutting wire in reel
- check whether cutting wire is beneath glass
- execute cutting process evenly
- first separate adhesive sealant at bottom of window from corners toward center
- if wire tears in lower area, move reel until a pointed cutting angle is created and continue cutting with only one wire up to A-pillars
- when approximately 30 mm (1 1/4 in.) adhesive sealant at top center of window is still to be cut, slowly rotate reel until adhesive sealant is completely separated

If cutting wire breaks several times (wire gets caught on windshield flange or on sheet metal lip) and can no longer be fitted on the reel (because the wire is too short), cut remaining area with tool **VW 1351** (pulling handle and mounting).

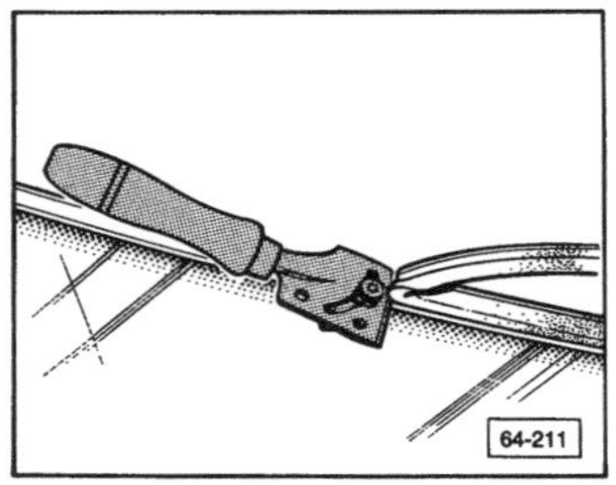

Windshield, installing

If windshield is to be reused, use a commercial glass scraper to remove enough old adhesive sealant from around glass and body flange to provide a smooth uniform surface. Do not damage ceramic black-out band.

Area must be clean and free of grease. It is not necessary to remove all old adhesive sealant.

- clean edge of glass around perimeter 30 mm (1 1/4 in.) wide with cleaning solvent and wipe dry with a lint-free cloth
- cut applicator head
- insert felt in applicator head
- thoroughly shake bottle of primer (at least 30 seconds) and fill applicator bottle
- attach applicator head on bottle

CAUTION

Ceramic black-out band around windshield is not a primer! Before applying adhesive sealant, prime this area with primer from adhesive kit.

- apply primer evenly all around edge of windshield in one continuous stroke. Ensure applicator head is always completely saturated
  - drying time is approximately 10 minutes
- clean body flange with cleaning solvent **D 009 400 01**
- cut back residual material on flange with carpet knife, but do not remove completely
  - residual material is adhesion base of newly applied adhesive sealant compound
- keep surfaces free from dirt and grease

CAUTION

If paint damage at the glass flange is found, it should be properly repaired.

If the glass flange has been machined or partially replaced, the respective area must be cleaned and primed again after painting.

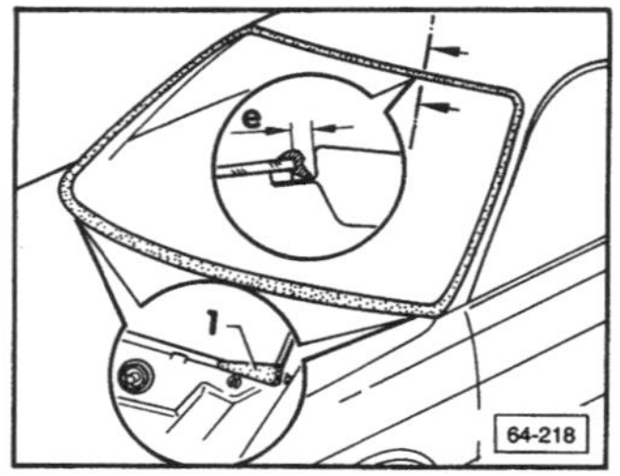

Prior to inserting the glass, new spacers have to be bonded to the windshield flange/vehicle body. See installing position, page 64.5, or use old imprints on the windshield flange/vehicle body.

- apply adhesive sealant compound around perimeter of glass
- bead cross section:
  - width = 8 mm (5/16 in.)
  - height = 13 mm (1/2 in.)

The height of the adhesive bead is to be reduced by the dimension of the residual material remaining on the glass or on the glass flange.

- install glass into window cut-out with aid of two double suction pads
- center glass from side
- with aid of adjustment wedges **1** (Part No. **893 845 631**) position glass so that space **e** 42 mm (2 in.) exists between trim frame and roof edge
  - hardening time is approximately three hours.
- remove adjustment wedges after hardening

Note

The rear window is mounted flush

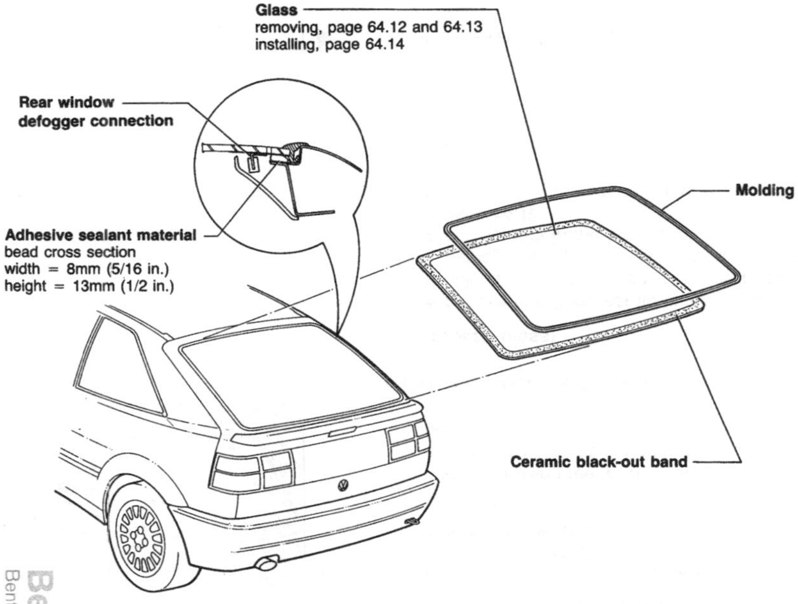

Rear window, removing

(unbroken glass)

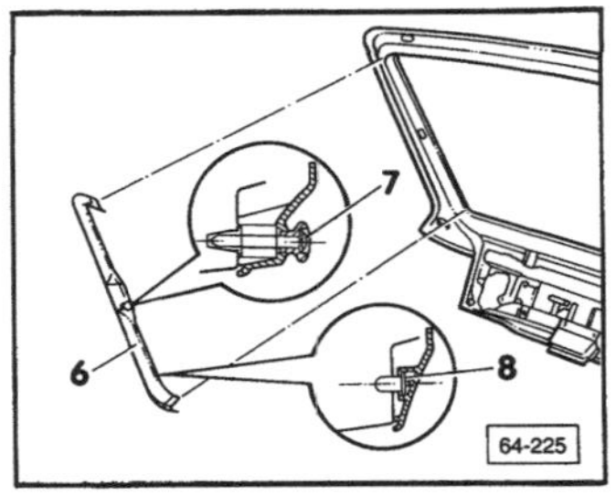

- remove side molding from rear lid
  - **6** — Trim
  - **7** — Phillips screw
  - **8** — Clip
- pull off plug connections for rear window defogger and press tabs flatly against glass

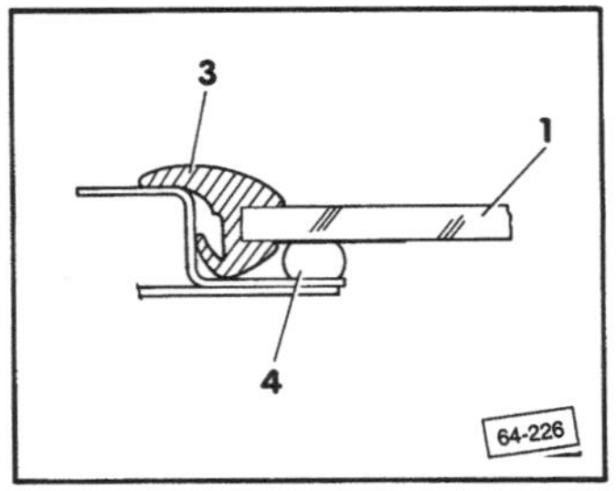

- pry out molding **3** along entire perimeter of window
  - **1** — Rear window glass
  - **3** — Molding
  - **4** — Adhesive sealant
- cut cutting wire to 5 m (16.5 ft) length,
- find center of cutting wire
- starting at top center of glass, press cutting wire into window channel so wire runs completely around perimeter of window ending at bottom center

Do not bend or twist wire, otherwise wire will break under tension.

- slide guide rail between rear window and body to base of window
- push awl through adhesive sealant, awl eye is visible from outside
- turn awl slightly and thread in one end of wire
- bend over wire and pull through to inside of vehicle
- remove guide rail
- repeat procedure for other end of wire

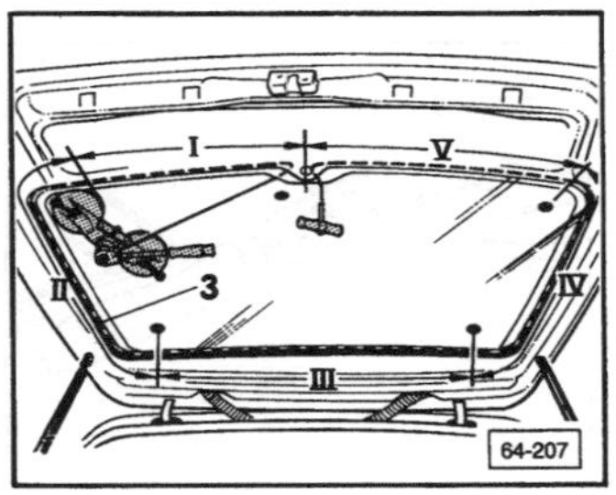

- attach reel as shown
- thread one end of cutting wire in reel and fasten
- fasten other end of cutting wire on pulling handle (from **VW 1351**)
  - pulling handle prevents pulling cutting wire through window

> **CAUTION**
>
> Ends of wire must be in adhesive sealant as close together as possible, but not overlapping, otherwise glass will break or wire will bend and twist.

> **WARNING**
>
> Always wear protective goggles and leather gloves when removing and installing windows.

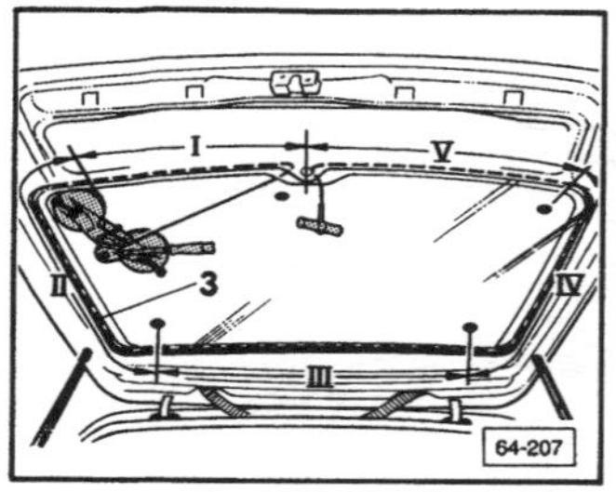

▶

- start cutting process from **position I** evenly up to **position V**
- relocate reel according to individual positions

Instead of using reel device, two pulling handles can also be used for cutting out glass.

Rear window, removing

(broken glass)

- remove trim from rear lid, page 64.12
- pull out plug connection for rear window defogger and press tabs flat against glass
- pull out trim frame around entire perimeter, page 64.12
- protect vehicle body and interior from glass splinters
- remove pieces of glass up to sealant material

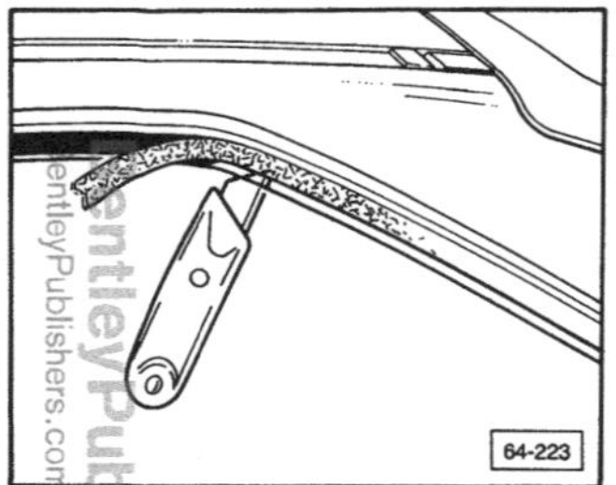

▶

- cut away adhesive sealant compound with knife

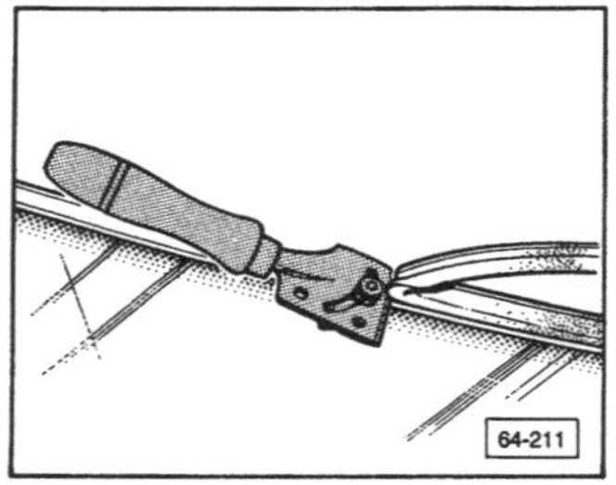

Rear window, installing

If rear window glass is to be reused, use a commercial glass scraper to remove enough old adhesive sealant from around glass and body flange to provide a smooth uniform surface. Do not damage ceramic black-out band.

Area must be clean and free of grease. It is not necessary necessary to remove all old adhesive sealant.

- clean edge of glass around perimeter 30 mm (1 1/4 in.) wide with cleaning solvent and wipe dry with a lint-free cloth
- cut applicator head
- insert felt in applicator head
- thoroughly shake bottle of primer (at least 30 seconds) and pour into applicator bottle
- attach applicator head on bottle

> **CAUTION**
>
> Ceramic black-out band is not a primer! Priming is absolutely required prior to application of adhesive sealant compound.

- apply primer evenly all around edge of windshield in one continuous stroke. Ensure applicator head is always completely saturated
  - drying time is approximately 10 minutes
- clean body flange with cleaning solvent **D 009 400 01**

- cut back residual material on flange with carpet knife, but do not remove completely
  - residual material is adhesion base of newly applied adhesive sealant compound
- keep surfaces free from dirt and grease

> **CAUTION**
>
> Depending on extent of paint damage in removing rear window, touch up or repair as required.

If the glass flange has been machined or partially replaced, the respective area must be cleaned and primed again with primer from adhesive kit.

- apply adhesive sealant compound around perimeter of glass
- bead cross section:
  - width = 8 mm (5/16 in.)
  - height = 13 mm (1/2 in.)

The height of the adhesive bead is to be reduced by the dimension of the residual material remaining on the glass or on the glass flange.

- install glass into window cut-out with aid of two double suction pads
- center glass from side
  - hardening time is approximately three hours.
- mount molding on rear window

THIS PAGE INTENTIONALLY LEFT BLANK

Note

Side windows are mounted flush.

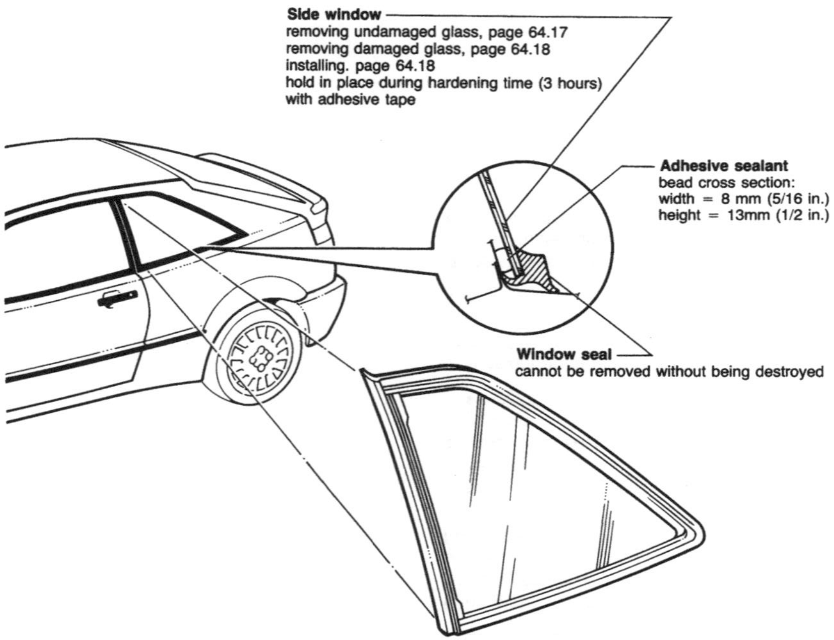

64-220

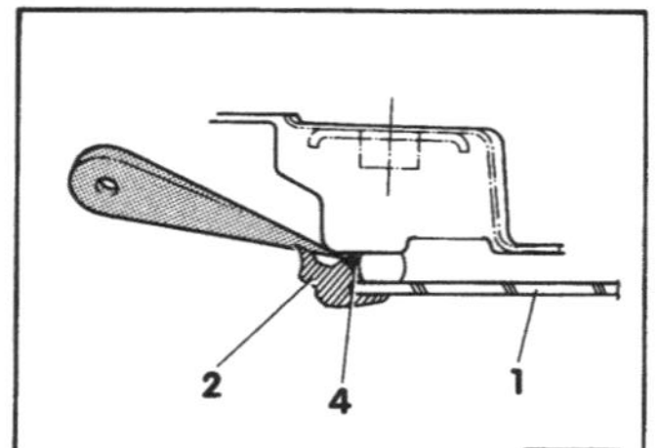

64-221

Side window, removing

(unbroken glass)

- remove B-pillar trim, rear side panels and loosen C-pillar trim, see Repair Group 70
- cut cutting wire **4** to approximately 2.3 m (7.5 ft) length
- find center of cutting wire
- press cutting wire into window channel so wire runs completely around perimeter of window leaving two ends of identical length at B-pillar area

Do not bend or twist wire, otherwise wire will break under tension.

- pry out sealing lip **2** of window seal with mounting wedge
  - **1** — Side window
  - **2** — Molding
  - **3** — Cutting wire
- push cutting wire through to inside in area of B-pillar using awl
  - cutting wire is doubled in B-pillar area

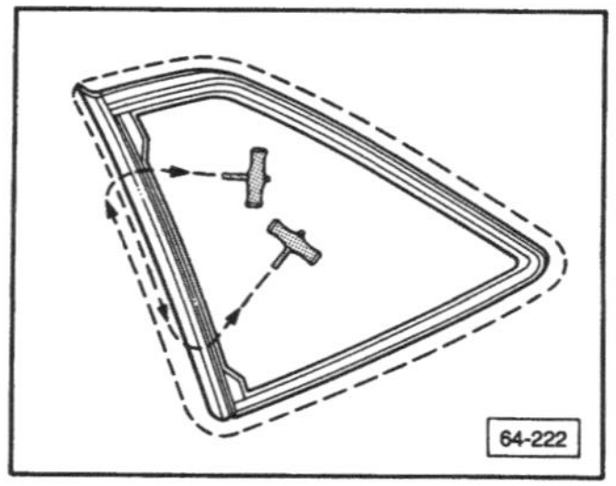

64-222

- fasten ends of cutting wire on pulling handles (from **VW 1351**)
  - second pulling handle is to prevent pulling cutting wire through window

WARNING

Always wear protective goggles and leather gloves when removing and installing windows.

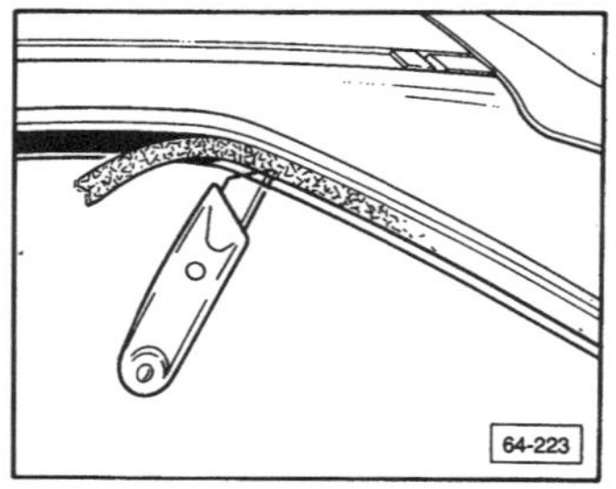

64-223

- start cutting process evenly and at as sharp an angle as possible
- sever adhesive sealant material in rear corner area using a carpet knife if necessary

When cutting out window, do not use sawing action with wire, otherwise wire will heat up very quickly and break.

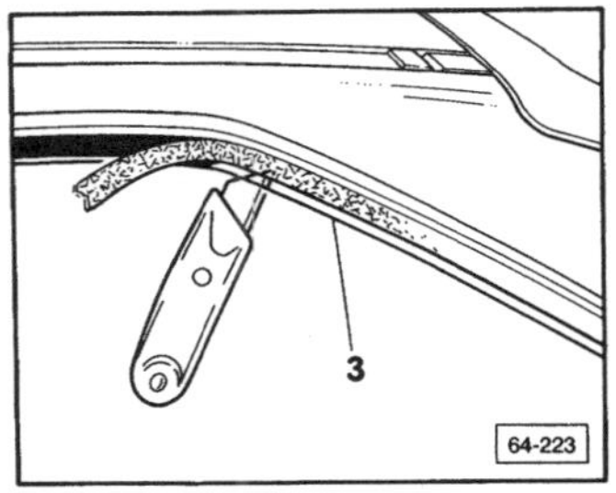

Side window, removing

(broken glass)

- protect vehicle body and interior from glass splinters
- tear off seal
- remove pieces of glass up to sealant material
- sever adhesive sealant **3** (with rest of glass) using knife
- sever adhesive sealant material in rear corner area with carpet knife if necessary

WARNING

Always wear protective goggles and leather gloves when removing and installing windows.

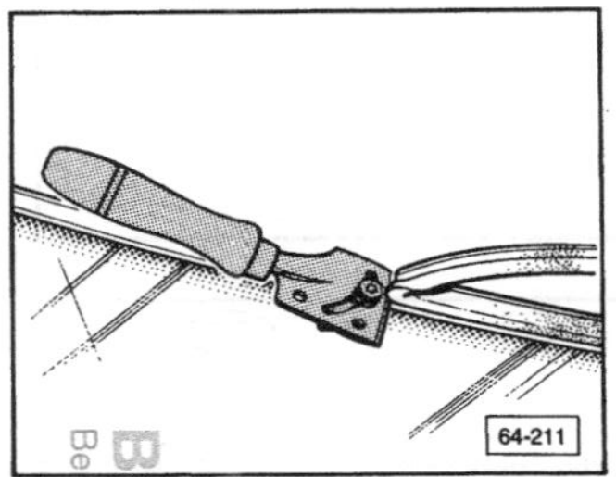

Side window, installing

If side window glass is to be reused, use a commercial glass scraper to remove enough old adhesive sealant from around glass and body flange to provide a smooth uniform surface. Do not damage ceramic black-out band.

Area must be clean and free of grease. It is not necessary to remove all old adhesive sealant.

- clean edge of glass around perimeter 30 mm (1 1/4 in.) wide with cleaning solvent and wipe dry with a lint-free cloth
- cut applicator head
- insert felt in applicator head
- thoroughly shake bottle of primer (at least 30 seconds) and fill applicator bottle
- attach applicator head on bottle

CAUTION

Ceramic black-out band is not a primer! Before applying adhesive sealant, prime this area with primer **D 009 200**

- apply primer evenly all around edge of windshield in one continuous stroke. Ensure applicator head is always completely saturated
  - drying time is approximately 10 minutes
- clean body flange with cleaning solvent **D 009 400 01**
- cut back residual material on flange with carpet knife, but do not remove completely
  - residual material is adhesion base of newly applied adhesive sealant compound
- keep surfaces free from dirt and grease

CAUTION

Depending on extent of paint damage in removing rear window, touch up or repaint as required.

If the glass flange has been machined or partially replaced, the respective area must be cleaned and primed with primer from adhesive kit.

- apply adhesive sealant compound around perimeter of glass
- bead cross section:
  - width = 8 mm (5/16 in.)
  - height = 13 mm (1/2 in.)

The height of the adhesive bead is to be reduced by the dimension of the residual material remaining on the glass or on the glass flange.

- install glass into window cut-out and center (pay attention to door gap)
- hold glass in place during hardening with adhesive tape

Hardening time is approximately three hours.

Power windows, troubleshooting

Note
Always refer to appropriate wiring diagram.

Tools required

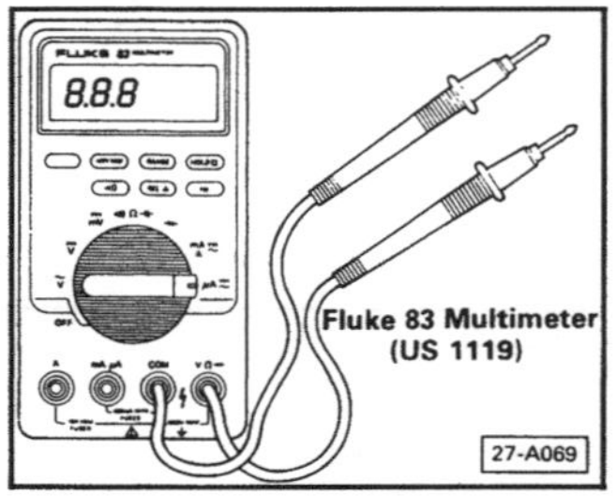

- Multimeter **Fluke 83** (**US 1119**)

- Connector Test Kit **VW 1594**

> **CAUTION**
>
> **DO NOT** damage, enlarge or bend connector terminals or cavities by forcing probes into them when performing electrical checks. Use Connector Test Kit **VW 1594** to make the necessary electrical connections.

> **CAUTION**
> Wait at least 20 seconds after switching **OFF** ignition before disconnecting battery. This will allow the platinum hot-wire in Air Mass Sensor (**G 70**) to burn itself clean.

> **CAUTION**
> Always switch **OFF** ignition when disconnecting/reconnecting battery, or any part of Motronic ignition wiring system to prevent damage to Motronic control unit (**J 220**).

> **CAUTION**
> Always obtain radio security code before disconnecting battery ground strap.

Glass, Window Regulators

No windows operate

Test conditions

- battery **OK**
- fuse **S43 OK**
- fuse **S14 OK**

Window voltage supply, checking

- ■ disconnect battery ground strap
- ■ remove left rear trim panel
- ■ disconnect connectors **T8b** (white), **T8c** (black) from power window control unit
- ■ set multimeter **Fluke 83** (**US 1119**) to 20 Volt range
- ■ first connect black test lead to terminal **1** and red test lead to terminal **7** of connector **T8b** (white)
- ■ reconnect battery ground strap
- ■ switch ignition **ON**
  - must read approximately battery voltage
- ■ now connect red test lead to terminal **6** of connector **T8b** (white)
  - must read approximately battery voltage

One or more specified values **NOT OK** → ■ check and repair wiring according to wiring diagram → End

All specified values **OK** → Central locking switch

Central locking switch (F 59 or F 114) in door lock, checking

- ■ connect red test lead to terminal **6** of connector **T8c** (black)
  - must read approximately battery voltage

Specified value **OK** → ■ using wiring diagram, check central locking switches (**F 59** or **F 114**), and replace as necessary → End

Specified value **NOT OK** → ■ replace power window control unit (**J 139**) → End

Driver's window does not operate

Test condition

- right front window operates

(A)

Driver's side window switch (E 40) in door, checking

- disconnect battery ground strap
- remove left rear trim panel
- disconnect connectors **T8a** (brown), **T8b** (white) from power window control unit (**J 139**)
- set multimeter **Fluke 83** (**US 1119**) to Ohm range
- connect black test lead to terminal **4** of connector **T8a** (brown) and red test lead to terminal **7** of connector **T8b** (white)
- operate window switch to **lower** window

Multimeter reads ∞ Ohms (no continuity)	Multimeter reads 0 Ohms (continuity)
■ check and repair wiring according to wiring diagram, or ■ replace window switch (**E 40**) → End	■ connect red test lead to terminal **6** of connector **T8a** (brown) ■ operate window switch to raise window

Multimeter reads ∞ Ohm (no continuity)	Multimeter reads 0 Ohm (continuity)
■ check and repair wiring according to wiring diagram, or ■ replace window switch (**E 40**) → End	Go to (B), next page

(B)

Driver's door window regulator motor (V 14), checking

- connect black test lead to terminal **5** and red test lead to terminal **4** of connector **T8b** (white)

Multimeter reads ∞ Ohms (not continuity)	Multimeter reads 0 Ohms (continuity)
■ check and repair wiring according to wiring diagram, or ■ replace window regulator motor (**V 14**) → End	■ replace power window control unit (**J 139**) → End

Right front window does not operate

Test condition

- driver's window operates

Right front window switch (E 81) in driver's door, checking

- disconnect battery ground strap
- remove left rear trim panel
- disconnect connectors **T8a** (brown), **T8b** (white), and **T8c** (black) from power window control unit (**J 139**)
- set multimeter **Fluke 83** (**US 1119**) to Ohm range
- connect black test lead to terminal **1** and red test lead to terminal **7** of connector **T8a** (brown)
- operate window switch to **lower** window

Multimeter reads ∞ Ohms (no continuity) →
- check and repair wiring according to wiring diagram, or
- replace window switch (**E 81**)

→ End

Multimeter reads 0 Ohms (continuity) →
- connect red test lead to terminal **6** of connector **T8a** (brown)
- operate switch to **raise** window

  Multimeter reads ∞ Ohms (no continuity) →
  - check and repair wiring according to wiring diagram, or
  - replace window switch (**E 81**)

  → End

  Multimeter reads 0 Ohms (continuity) → Right front door window switch (E 41), checking

Right front door window switch (E 41), checking

- connect black test lead to terminal **4** of connector **T8c** (black)
- connect red test lead to terminal **7** of connector **T8a** (brown)
- operate window switch to **lower** window

Multimeter reads ∞ Ohms (no continuity) → Go to Ⓐ, next page

Multimeter reads 0 Ohms (continuity) → Go to Ⓑ, next page

Ⓐ
- check and repair wiring according to wiring diagram, or
- replace window switch (**E 41**)

→ End

Ⓑ
- connect red test lead to terminal **6** connector **T8a** (brown)
- operate window switch to **raise** window

Multimeter reads ∞ Ohms (no continuity) →
- check and repair wiring according to wiring diagram, or
- replace window switch (**E 41**)

→ End

Multimeter reads 0 Ohms (continuity) → Right front door window regulator motor (V 15), checking

Right front door window regulator motor (V 15), checking

- connect black test lead to terminal **5** and red test lead to terminal **8** of connector **T8b** (white)

Multimeter reads ∞ Ohms (no continuity) →
- check and repair wiring according to wiring diagram, or
- replace window regulator motor (**V 15**)

→ End

Multimeter reads 0 Ohms (continuity) →
- replace power window control unit (**J 139**)

→ End

Power windows do not operate with driver's door open

Test condition

- all other window functions **OK**

Left front door contact switch (F 2), checking

- disconnect battery ground strap
- remove left rear trim panel
- disconnect connectors **T8a** (brown), **T8b** (white) from power window control unit (**J 139**)
- set multimeter **Fluke 83** (**US 1119**) to Ohm range
- connect black test lead to terminal **5** of connector **T8a** (brown) and red test lead to terminal **1** of connector **T8b** (white)
- open driver's door

Multimeter reads ∞ Ohms (no continuity) →

- check and repair wiring according to wiring diagram, or
- replace door contact switch (**F 2**)

→ End

Multimeter reads 0 Ohms (continuity) →

- replace power window control unit (**J 139**)

→ End

Power window control module, location/terminal identification

Location

The power window control module is located in the left side-panel, behind the interior trim.

Connector terminal identification

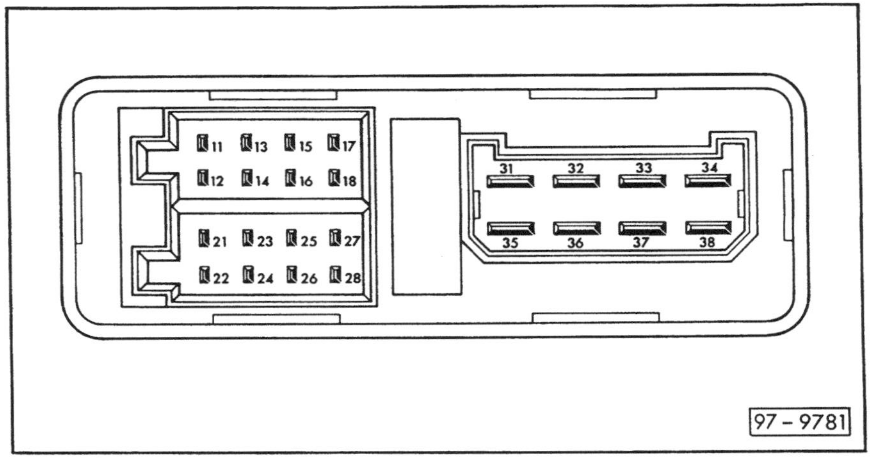

11 = T 8a/1	21 = T 8c/1	31 = T 8b/1
12 = T 8a/2	22 = T 8c/2	32 = T 8b/2
13 = T 8a/3	23 = T 8c/3	33 = T 8b/3
14 = T 8a/4	24 = T 8c/4	34 = T 8b/4
15 = T 8a/5	25 = T 8c/5	35 = T 8b/5
16 = T 8a/6	26 = T 8c/6	36 = T 8b/6
17 = T 8a/7	27 = T 8c/7	37 = T 8b/7
18 = T 8a/8	28 = T 8c/8	38 = T 8b/8

Connector colors

T 8a - Brown

T 8b - White

T 8c - Black

Note
See wiring diagram for wire routing to connectors **T 8a, T 8b and T 8c.** Always refer to appropriate wiring diagram when performing any electrical tests or repairs.

THIS PAGE INTENTIONALLY LEFT
BLANK

Index

Mirror glass
- removing/installing **66.8**

Radiator grille
- removing/installing **66.5**

Rear spoiler
- installing **66.5**
- removing **66.4**

Rearview mirror
- assembly **66.7**
- removing/installing **66.8**

Rocker panel, protective molding
- replacing **66.9**

Roof trim molding
- removing/installing **66.3**

Wheel house liner
- assembly **66.6**
- installing, 1991 m.y. **66.6a**

Note

T-stud or threaded bolt, **See Volkswagen Body Repair Information, LIT NO. W42-701-006-1**

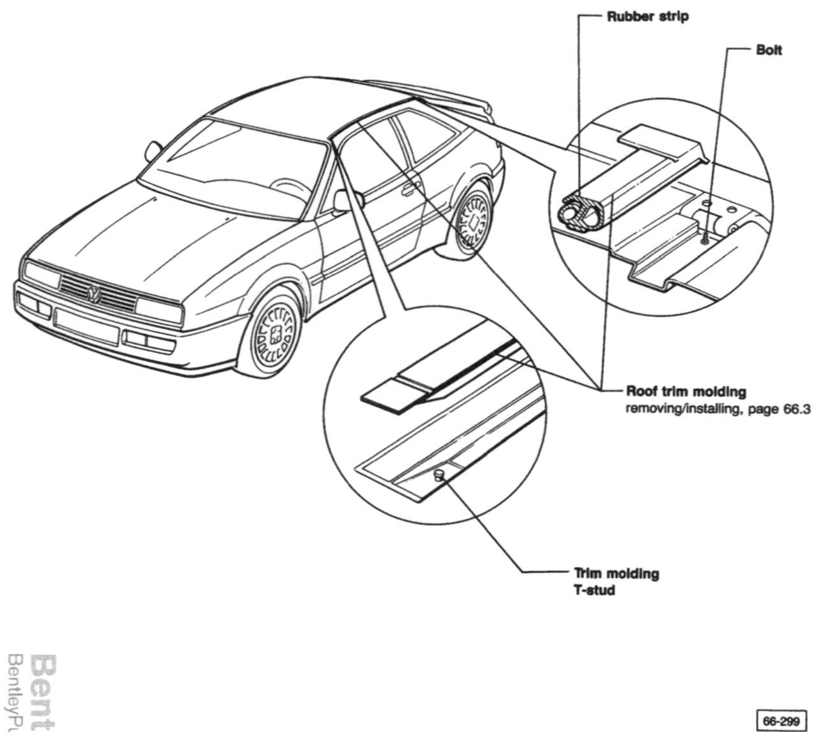

Roof trim molding, removing

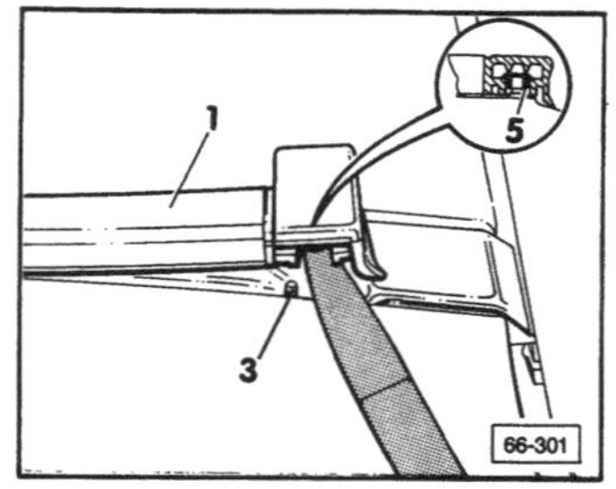

- pry roof molding **1** off of rear bolt **3** with a wedge
  - do not lose plastic clip **5**
- pull roof trim molding out of channel up to front shaped piece
  - shaped piece is bonded to roof trim molding

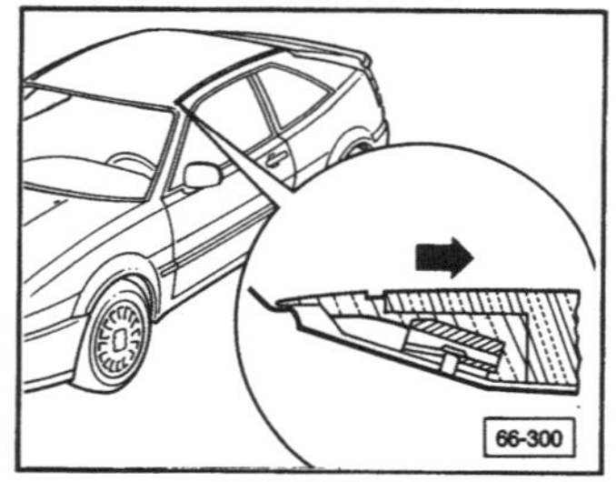

- remove roof trim molding from front T-stud toward the rear

Roof trim molding, installing

- install all components in reverse order of removal, noting the following:
- when reusing an undamaged roof trim molding, check rubber strip for damage and replace if necessary — rubber strip **AKL 450 005 05**
- hand press roof molding in channel (**arrow**)

> **CAUTION**
>
> Part numbers are listed for reference only. Always consult with the Parts Department for latest information.

Rear spoiler, removing

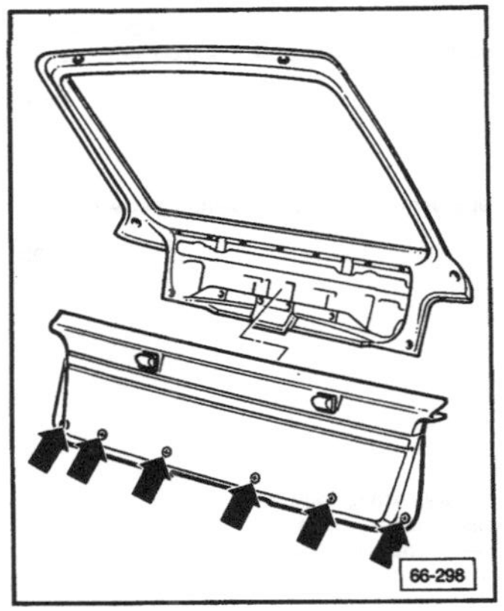

- turn clips **(arrows)** 90° (unlock) and remove inner trim for rear lid

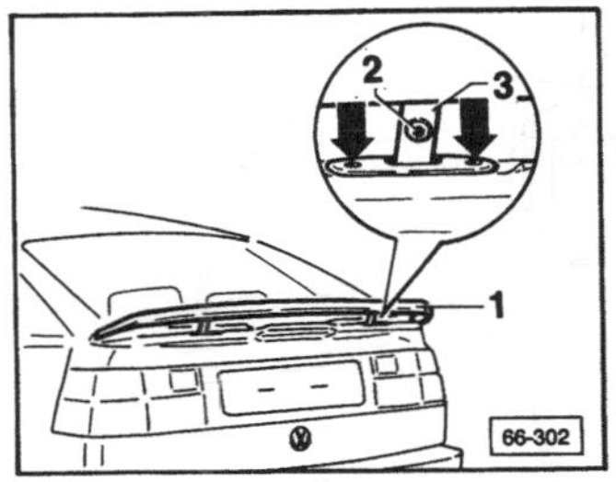

- set rear spoiler **1** to up position, electrically or mechanically
- remove cap screws **2**
- remove rear spoiler **1** from mounting studs **3**
- remove screws **(arrows)**

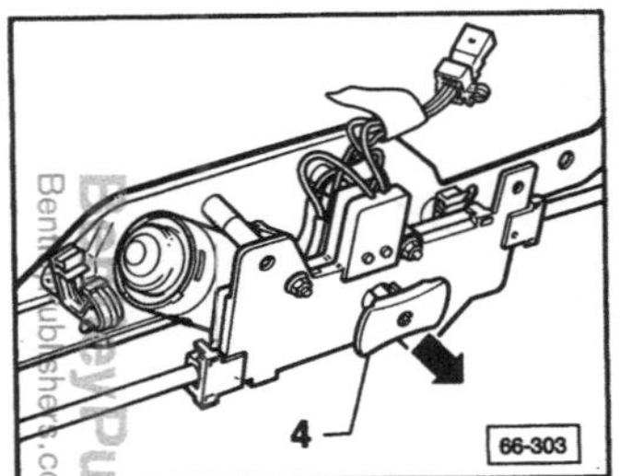

Note

If the rear spoiler cannot be adjusted by the electric spoiler drive, it can be operated mechanically by the hand crank **4**.

- pull hand crank **4** out of detent **(arrow)**
  raise — turn handle **4** clockwise
  lower — turn handle **4** counterclockwise
- press hand crank **4** back into detent
- remove electric leads from retainers on spoiler drive support bracket and separate connectors

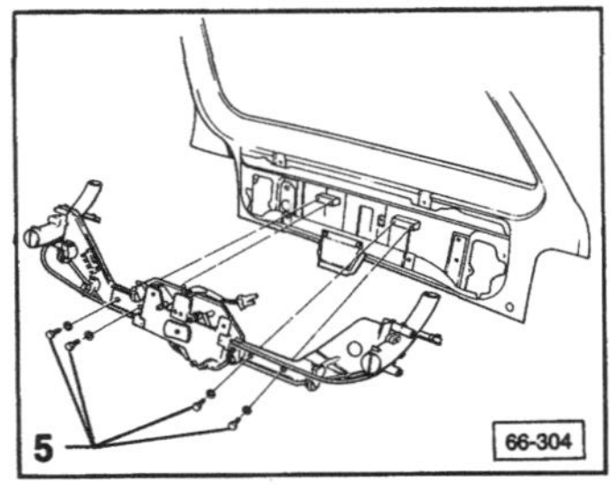

- unscrew four hex head bolts **5** and remove spoiler drive

Rear spoiler, installing

- install all components in reverse order of removal, noting the following:
  - torque cap screws to 8 Nm (6 ft lb)
  - torque hex head bolts to 10 Nm (7 ft lb)
  - torque Phillips screws to 3 Nm (27 in. lb)

Radiator grill, removing

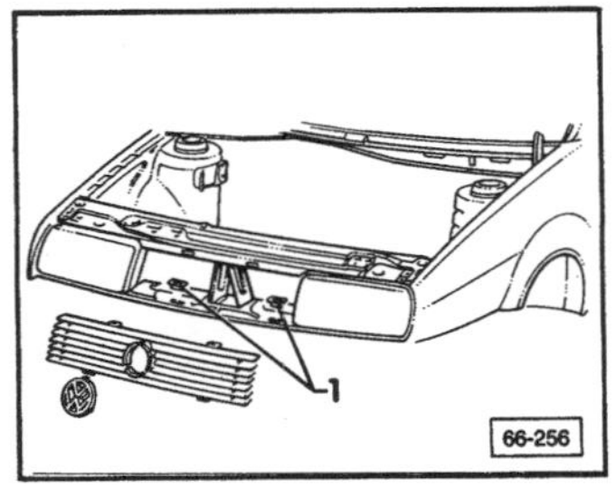

- unclip radiator grill at lock carrier and remove upward
- remove emblem

1 — rubber supports for radiator grill

Radiator grill, installing

Install all components in reverse order of removal.

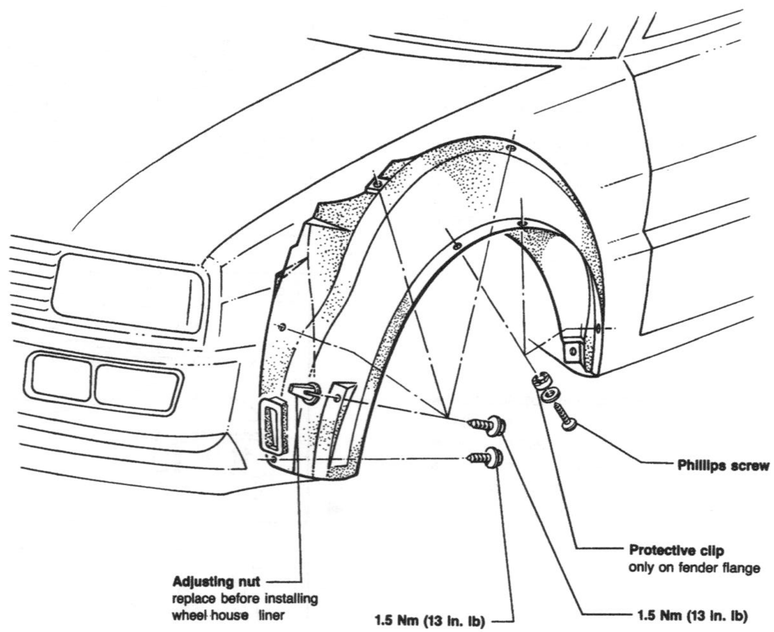

Note

Remove wheel house liner from bumper side.

Note

Install wheel house liner on A-pillar and press over fender flange.

66-257

Wheel house liner — 1991 m.y.

Note

The front fender flange was changed for 1991 m.y. to provide more wheel clearance.

A modified wheel house liner is now used and will be applicable for all years.

Installing

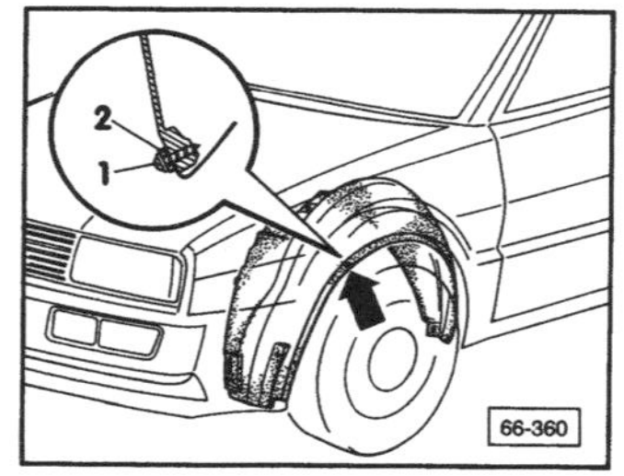

New style wheel house liner with old style fender:

- insert new wheel house liner in position in fender
- drill 7.0 mm (9/32 in.) mounting holes in fender flange using liner as template
- prime drilled holes with anti-corrosion primer and refinish as required
- secure liner with Phillips screw and plastic washer

Old style wheel house liner with new style fender:

The same installation procedures apply as with new style wheel house liner.

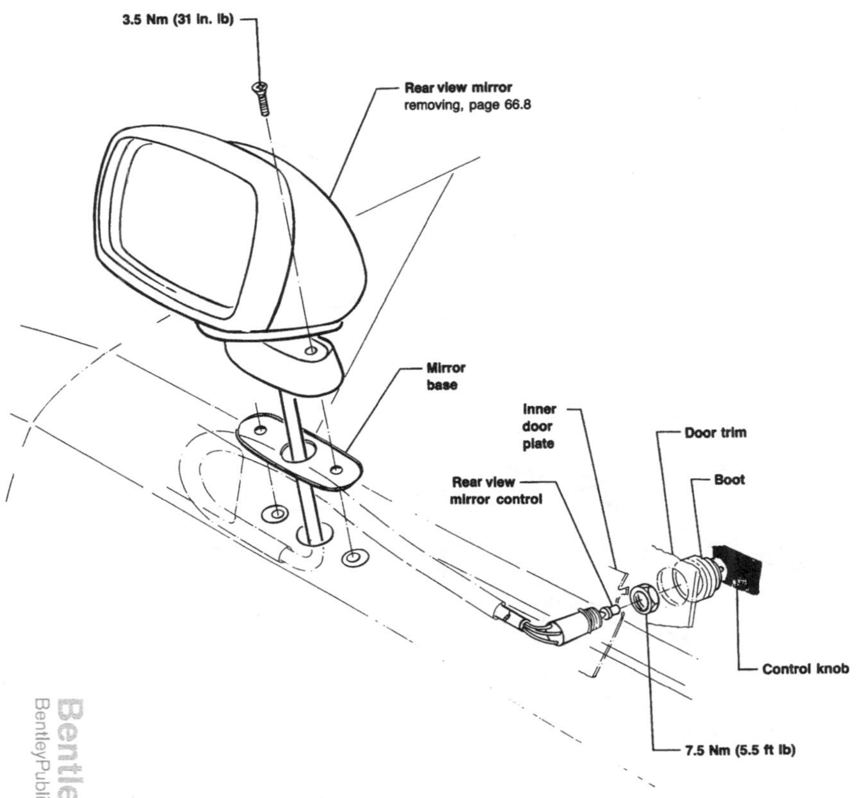

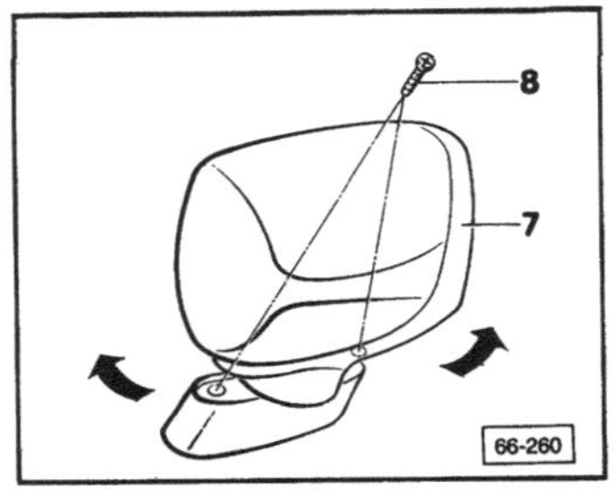

Rearview mirror, removing

- lower window
- remove control knob
- remove boot
- remove door trim (Repair Group 70) and loosen moisture barrier in area of rearview mirror control
- remove lock nut

- pivot rearview mirror **7** towards inside and unscrew oval head screw **8**
- pivot back rearview mirror **7** and remove second oval head screw
- remove rearview mirror **7** with control and mirror base

Rearview mirror, installing

Reinstall all components in reverse order of removal.

Mirror glass, removing

Note

The mirror glass is only clipped on.

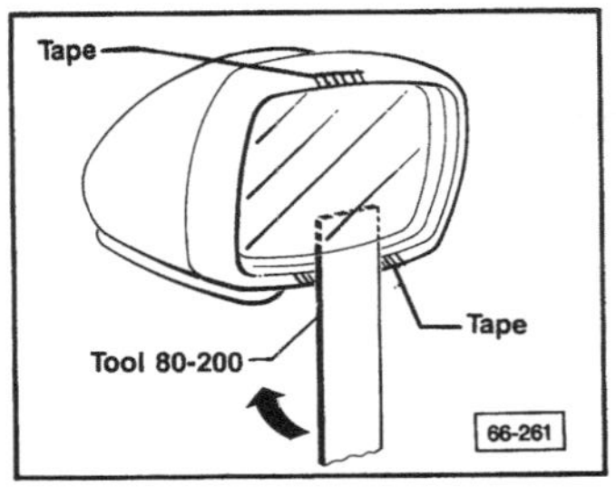

- tape upper/lower edge of mirror housing
- remove mirror glass using special tool **80-200** to pry off at bottom first, then top
- hold mirror, unplug electrical connector

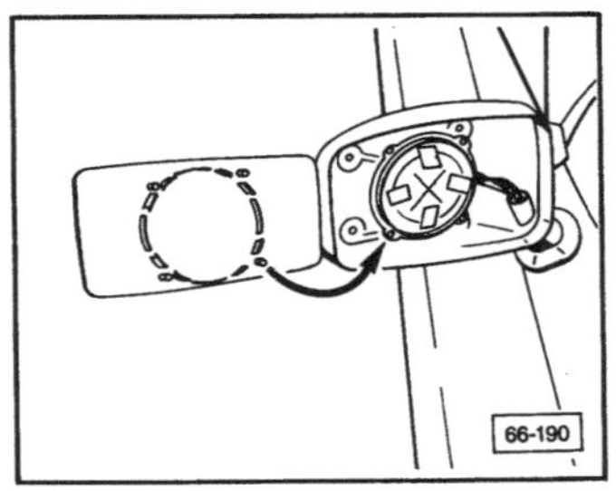

Mirror glass, installing

- connect electric connector
- place mirror glass in notches and press on
- check mirror for proper function

> **CAUTION**
>
> Press only on the center of the glass and always wear protective gloves.

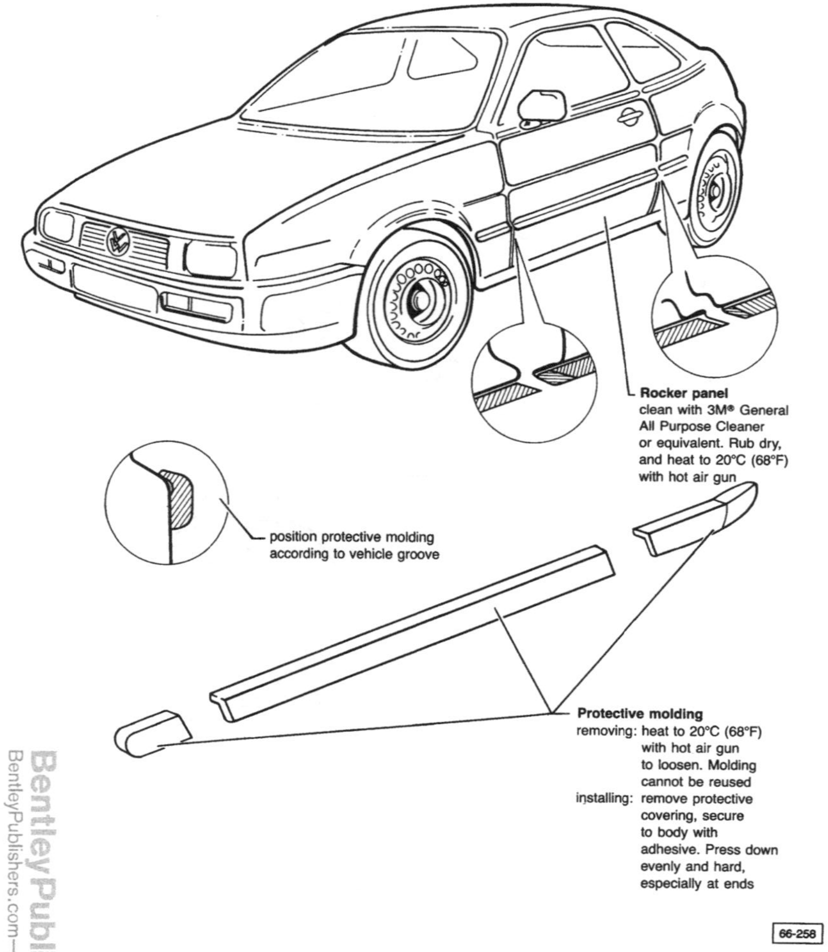

THIS PAGE INTENTIONALLY LEFT BLANK

Index

Automatic shoulder belt system (USA only)
- assembly **68.2**
- door frame track trim, removing/installing **68.4**
- drive unit, installing **68.4**
- drive unit, removing **68.3**
- knee bar, removing/installing **68.4**
- knee bar support bracket, removing/installing **68.5**
- seat switch release magnet, removing/installing **68.6**
- shoulder belt release magnet, installing **68.6**
- shoulder belt release magnet, removing **68.5**

Automatic shoulder belt troubleshooting (USA only)
- belts do not move to A–pillar after removing ignition key (driver or passenger side) **68.17**
- belts only move to A–pillar when door opened with key removed (driver or passenger) **68.16**
- both belts do not move to B–pillar with doors shut, ignition **ON** **68.9**
- driver or passenger belt does **NOT** move to A–pillar when door opened **68.13**

Automatic shoulder belt troubleshooting (USA only) (continued)
- driver or passenger belt does **NOT** move to B–pillar **68.11**
- driver belt moves to A–pillar when vehicle shifted into reverse **68.18**
- seat belt warning light comes on with belts connected and ignition **on** **68.20**
- seat belt warning light does not light belts at B–pillar, ignition **ON** **68.19**
- wiring diagrams **68.21**

Child seat
- installing on front passenger seat **68.31**

Interior mirror
- removing/installing **68.8**

Rear seat belt
- assembly **68.7**
- buckle, removing/installing **68.8**
- mounting, removing/installing **68.8**

Seat belts
- inspecting **68.24**

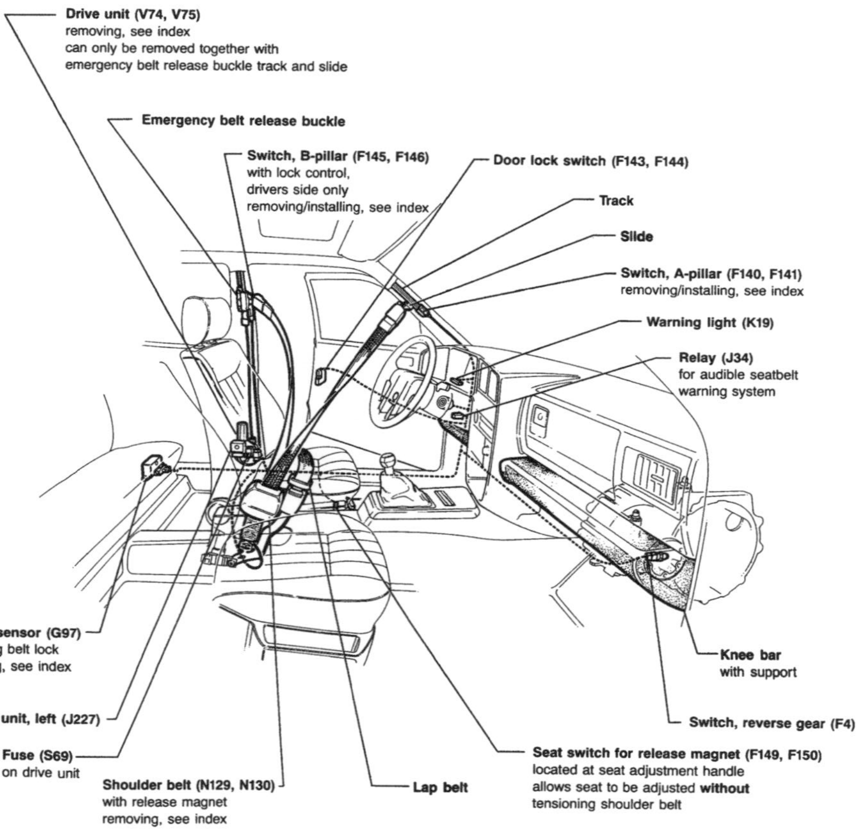

Drive unit, removing

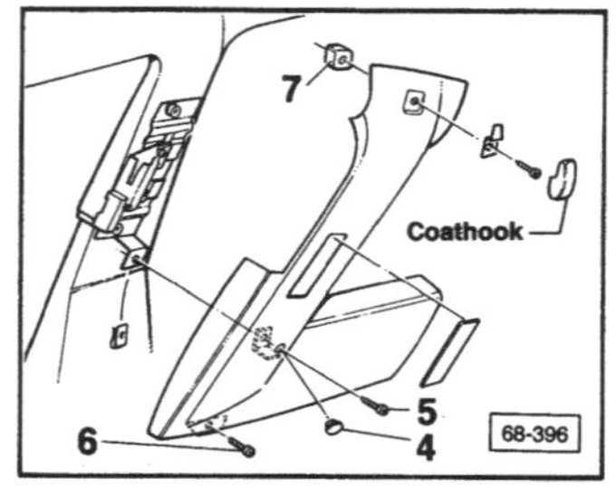

- remove coat hook
- pry off cover cap **4**
- remove Phillips screws **5, 6**
- remove spacer **7**
- take off B-pillar trim
- disconnect battery ground strap

- remove rear seat, see Repair Group 72
- remove rear side trim, see Repair Group 70
- remove passenger assist handle

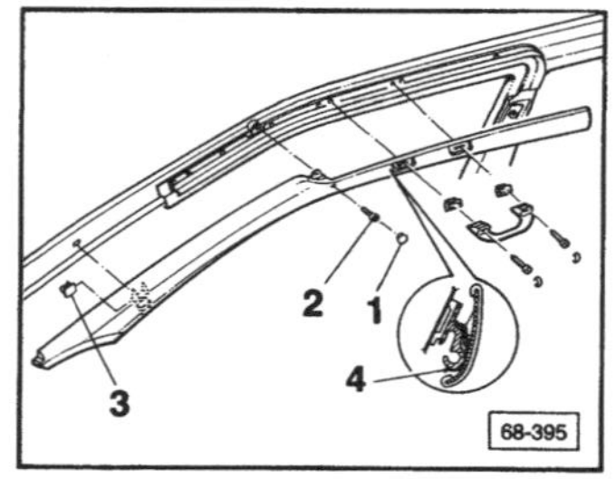

- pry off cover cap **1**, remove Phillips screw **2**
- unclip and remove A-pillar trim
  - **3** is a plastic clip
  - **4** is a sheet metal clip

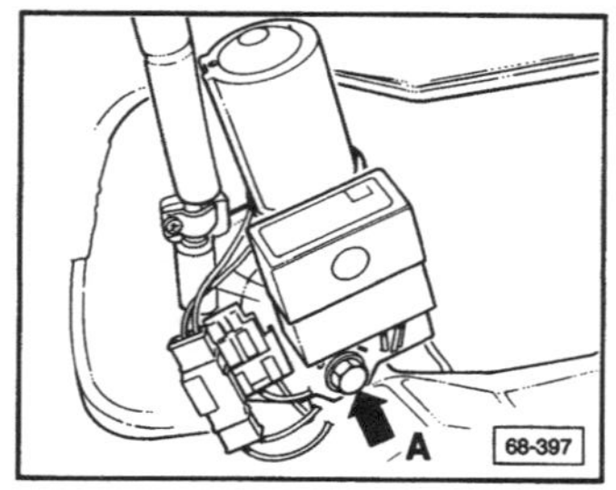

- pull off multi-pin connector from A-pillar switch
- remove control unit
- remove fuse holder from control unit
- disconnect dual connector from motor
- unscrew (**arrow A**) and remove motor

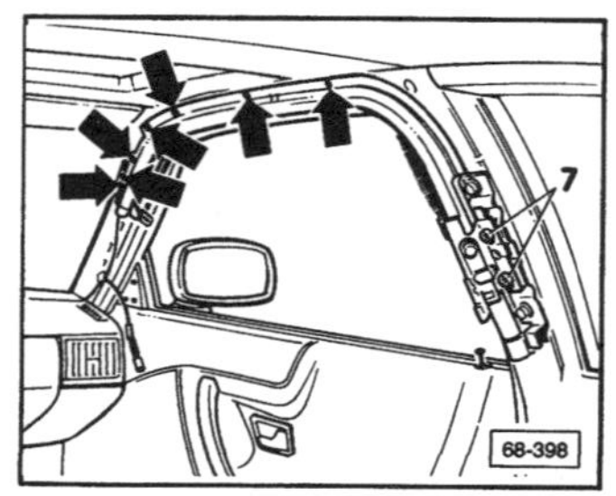

- remove hex head screws **7**
- remove round head track screws (**arrows**)

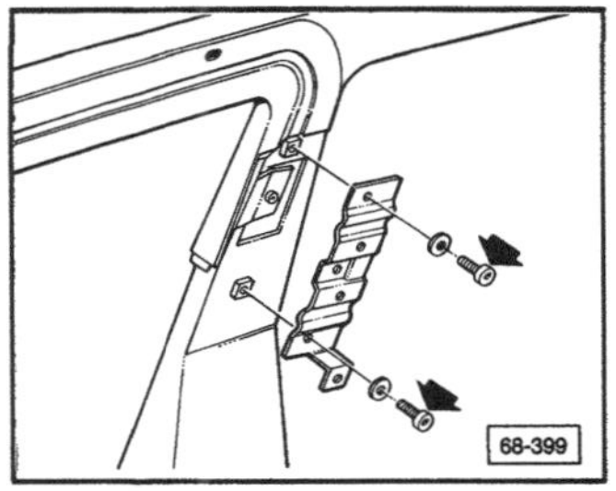

- unscrew hex head socket bolts (**arrows**)
- remove B-pillar trim

Drive unit, installing

Reinstall all components in reverse order of removal, noting the following:
- torque hex head socket bolts to **23 Nm (17 ft lb)**
- torque round head track screws to **2 Nm (1.5 ft lb)**
- torque motor unit to **40 Nm (30 ft lb)**

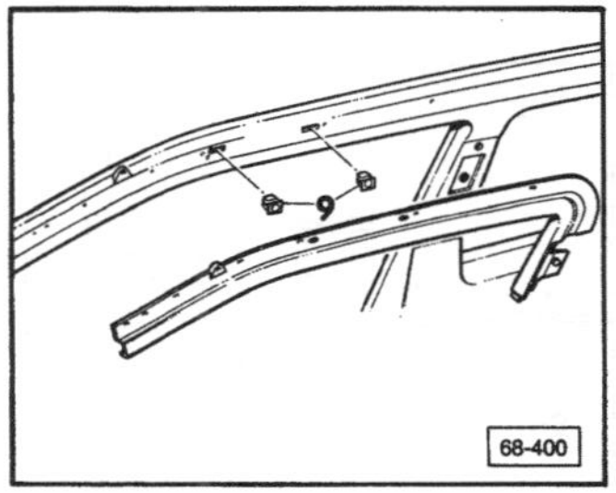

Door frame track trim, removing

Note

B-pillar trim must be removed first.

- remove expanding clips **9**
- remove door seal lip

Door frame track trim, installing

Install all components in reverse order of removal.

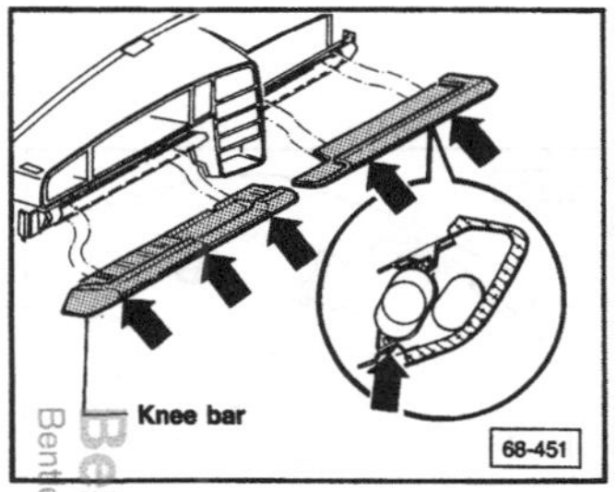

Knee bar, removing

- remove floor outlet covers, see Repair Group 70
- remove screws (**arrows**)
- remove knee bar

Knee bar, installing

Reinstall all components in reverse order of removal.

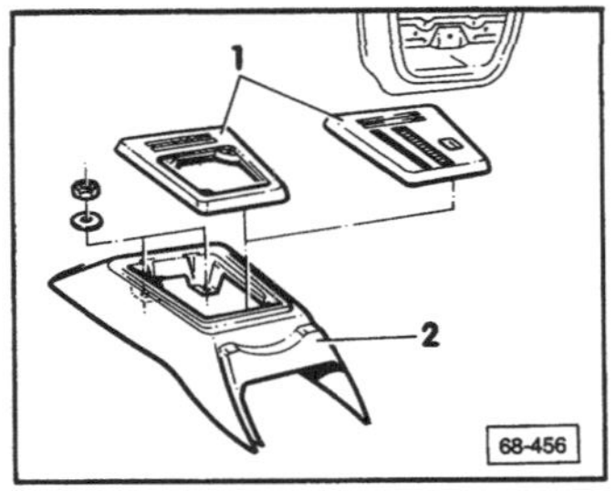

Knee bar support bracket, removing

- pry off and remove trim **1**, cover **2**

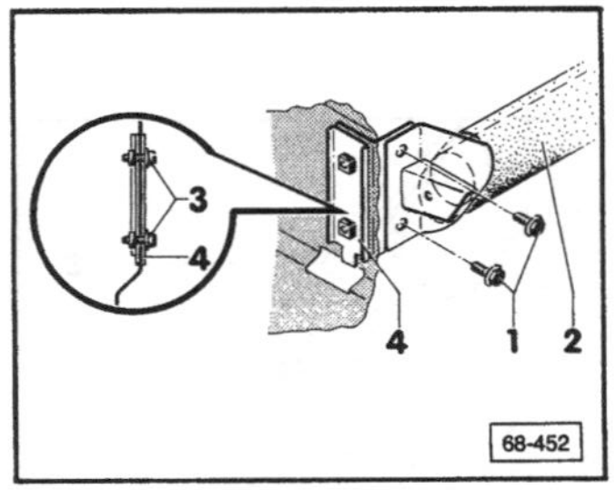

- loosen instrument panel, see Repair Group 70
- remove hex head collar screws **1** from both sides
- remove support **2** toward left side
- remove Phillips screws **3**
- remove bracket **4** from A-pillar

Knee bar support bracket, installing

Reinstall all components in reverse order of removal, noting the following:
- torque hex head collar screws to **6 Nm (4.4 ft lb)**
- torque bracket Phillips screws to **6 Nm (4.4 ft lb)**

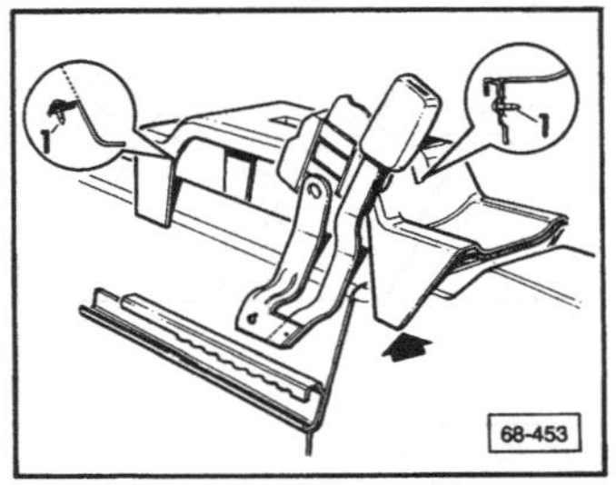

Shoulder belt release magnet, removing

- remove Phillips screws **1**
- remove upper tunnel cover (**arrow**)

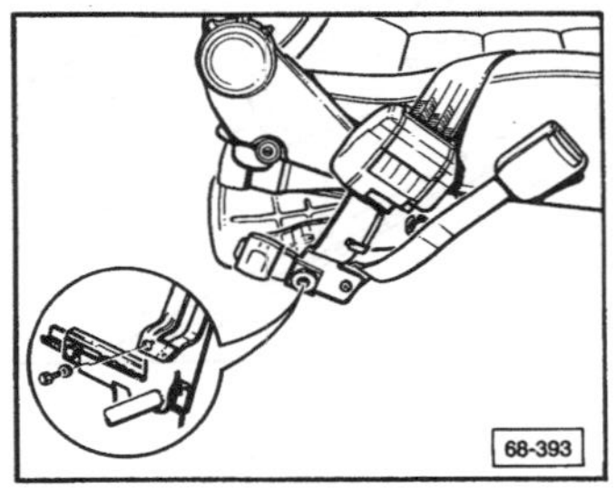

- remove front seat, see Repair Group 72
- disconnect electrical wires to seat
- unscrew and remove shoulder belt with dual bracket (**inset**)

Shoulder belt release magnet, installing

Reinstall all components in reverse order of removal, noting the following:
- torque shoulder belt bracket **60 Nm** (**44 ft lb**)

Motion sensor (G97), removing

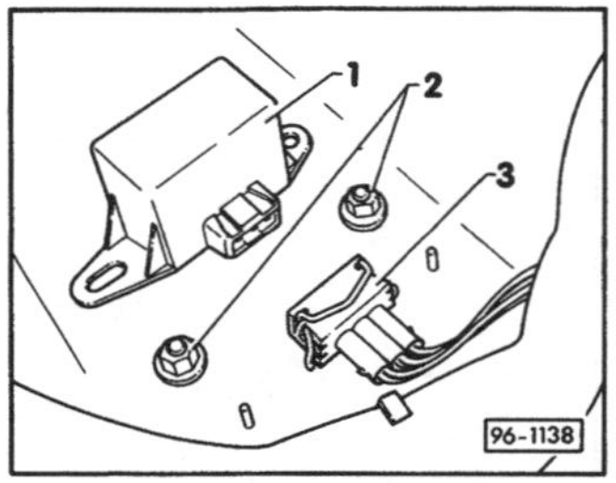

- lift up rear seat bench
- remove sensor

1 – sensor (**G97**)
2 – plastic nuts
3 – plug

Motion sensor (G97), installing

Reinstall all components in reverse order of removal.

Seat switch for reverse magnet (F149, F150), removing

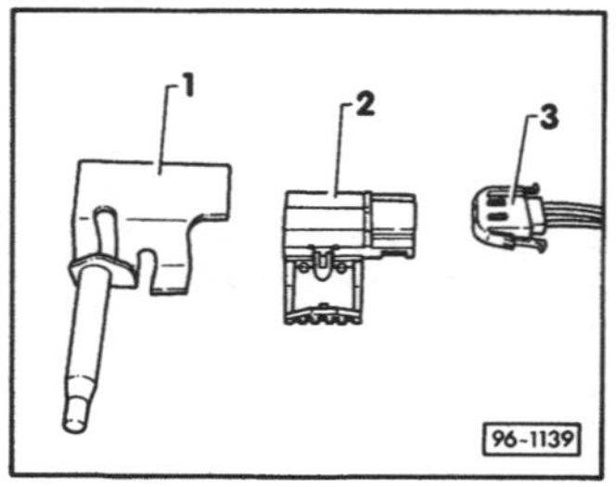

- remove seat
- disconnect plug **3**
- pull latch pin **1** out of front seat mounting
- pry off switch **2** with screwdriver

Seat switch for reverse magnet (F149, F150), installing

Reinstall all components in reverse order of removal.

A-pillar switch (F140, F141), removing

- remove upper A-pillar trim
- using screwdriver, carefully pry switch **2** from drive unit **1**
- disconnect plug **3** and remove switch

A-pillar switch (F140, F141), installing

Reinstall all components in reverse order of removal.

B-pillar switch (F145, F146), removing

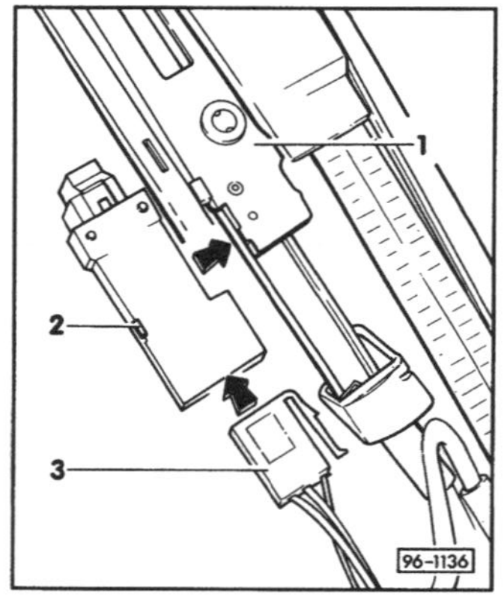

- remove B-pillar trim
- using screwdriver, carefully pry switch **2** from drive unit **1**
- disconnect plug **3** and remove switch

B-pillar switch (F145, F146), installing

Reinstall all components in reverse order of removal.

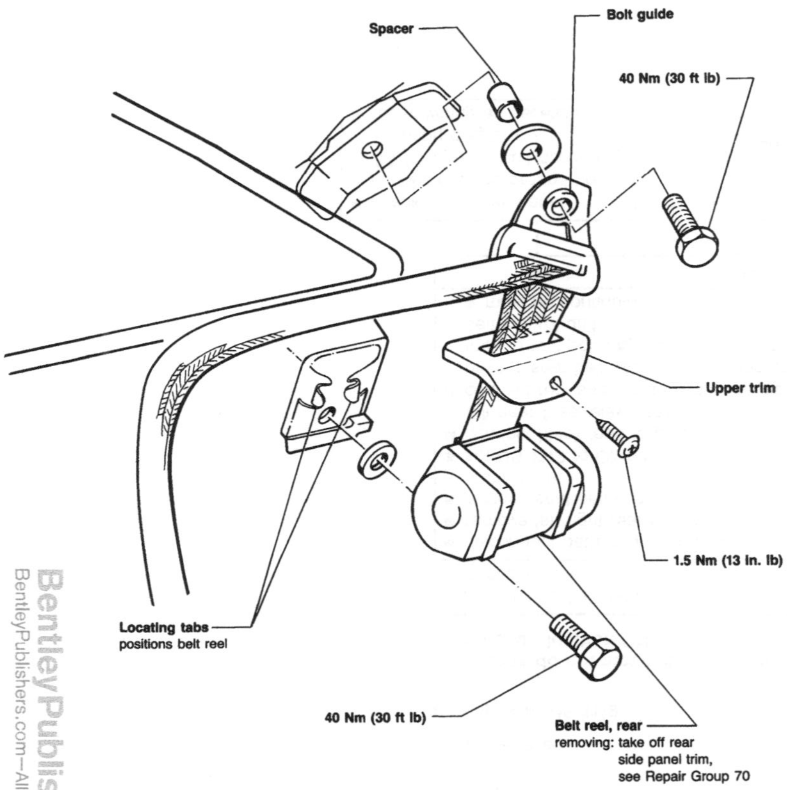

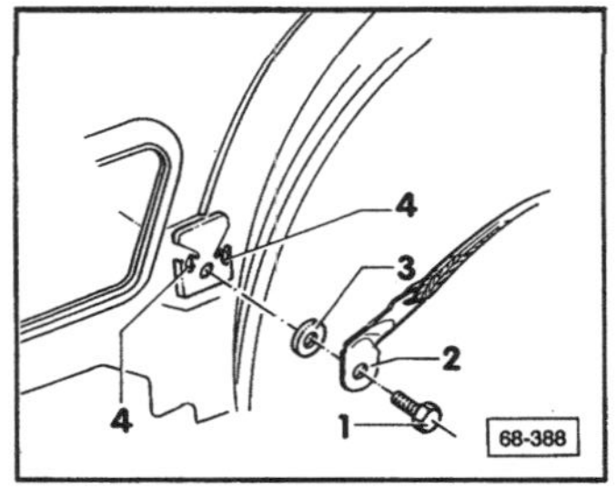

Rear seat belt mounting, removing

- remove rear seat, see Repair Group 72
- remove rear side panel, see Repair Group 70
- remove bolt **1**, belt mounting **2**, washer **3**

Rear seat belt mounting, installing

Reinstall all components in reverse order of removal, noting the following:

- torque bolt **1** to 40 Nm (30 ft lb)
- locating tabs in belt mounting must fit into tabs **4**

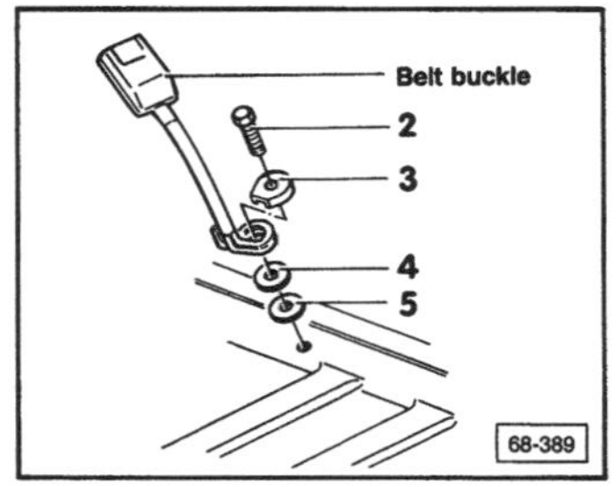

Rear seat belt buckle, removing

- remove rear seat, see Repair Group 72
- remove bolt **2**, belt **1**, cap **3**, washers **4, 5**

Rear seat belt buckle, installing

Reinstall all components in reverse order of removal, noting the following:

- torque bolt **1** to 40 Nm (30 ft lb)

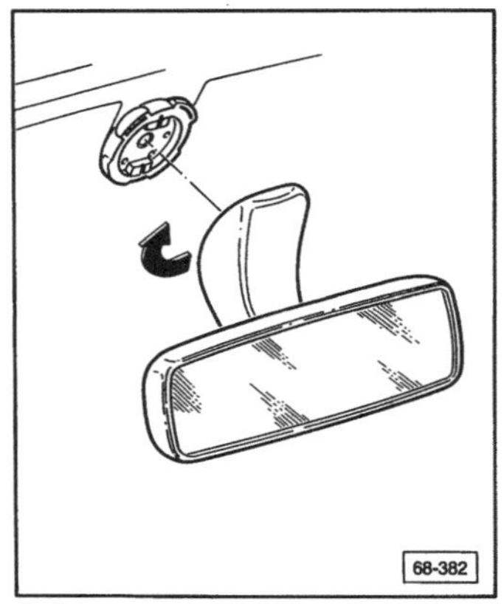

Interior mirror, removing

- rotate mirror 15°-20° from installed position (**arrow**), and remove

Interior mirror, installing

Reinstall in reverse order of removal.

Automatic shoulder belt system, troubleshooting

Test conditions

- battery **OK**

Tools required

- multimeter **US 1119**

Both belts do not move to B-pillar with doors shut, ignition ON

Control unit (**J227/J228**) ground connection, checking

- disconnect battery ground strap
- remove left/right rear trim panel
- remove seat belt control unit (**J227** left or **J228** right)
- set **US 1119** to Ohm scale
- connect black lead to ground point (30) near fuse/relay panel
- connect red lead first to test point (2) on driver's seat belt control unit connector then to test point (4) on passenger seat belt control unit connector
  - must be continuity (0 Ohms)

NO →

- check fuse/relay panel connectors and ground point (30) for open circuit according to wiring diagram
- repair wiring as necessary

→ End

YES → Go to (A) next page

(A)

Power (from ignition switch) to control unit (**J227/J228**), checking

- set **US 1119** to 20V scale
- connect **US 1119** between test points (1) and (2) on drivers seat belt control unit connector, or (8) and (4) on passenger seat belt control unit connector
- reconnect battery ground strap
- insert key into ignition switch, push in and hold
  - must be approximately 12.0V

NO →

- check fuse/relay panel connector **H 1/7** and ignition switch for open circuit according to wiring diagram
- repair wiring as necessary

or

- replace ignition switch

→ End

YES →

- check fuse/relay panel connector **Y/1** (plus connection terminal **30**)
- using **US 1119**, check voltage between test points (9) and (4) on passenger seat belt control unit connector and (3) and (2) on driver's seat belt control unit connector according to wiring diagram
  - must be approximately 12.0V

If **NO**

- check wiring according to wiring diagram
- repair wiring as necessary

→ End

Driver or passenger belt does not move to B-pillar

B-pillar microswitch (F145/F146), checking

- disconnect battery ground strap
- remove appropriate rear side panel, and B-pillar trim
- remove control unit (**J227**, left or **J228**, right)
- separate 4-point connector (**T4n**) (located behind fuse/relay panel)
- set **US 1119** to Ohm scale
- connect test leads between test points ❺ and ❷ on driver's seat belt control unit connector or ❹ and ❼ of passenger seat belt control unit connector
  - must be **NO** continuity (∞ Ohms)

NO → check and replace B-pillar microswitch (**F145 or F146**) → End

YES →

Door latch switches (F143 or F144), checking

- connect **US 1119** between test points ❷ and ❿ on driver's seat belt control unit connector or ❹ and ⓫ on passenger seat belt control unit connector
  - see wiring diagram
- close door(s)
  - must be **NO** continuity (∞ Ohms)

NO → check and replace door latch switch (**F143 or F144**) → End

YES → Go to Ⓐ next page

Ⓐ

Power (from ignition switch) to control unit (J227/J228), checking

- set **US 1119** to 20V scale
- connect **US 1119** between test points ❶ and ❷ on driver's seat belt control unit connector or ❽ and ❹ on passenger seat belt control unit connector
- reconnect battery ground strap
- insert key into ignition switch, push in and hold
  - must be approximately 12.0V

NO →
- check fuse/relay panel multi-connector **H 1/7** for open circuit according to wiring diagram
- repair wiring as necessary

or

- replace ignition switch

→ End

YES →

Belt adjuster motors (V74/75), checking

- disconnect battery ground strap
- set multimeter **US 1119** to Ohm scale and connect between test points ⓬ and ⓭ on driver's seat belt control unit connector and/or ⓮ and ⓯ on passenger seat belt control unit connector according to wiring diagram
  - must be continuity

NO →
- check wiring between control unit (**J227** or **J228**) and belt adjuster motor (**V74** or **V75**) for open circuit according to wiring diagram
- repair wiring as necessary

or

- replace belt adjuster motor (**V74** or **V75**)

→ End

YES → replace control unit (**J227** or **J228**) → End

Driver or passenger belt does NOT move to A-pillar when door opened

Power (from ignition switch) to control unit (**J227** and **J228**), with ignition key removed, checking
- fuse S 51 (10A) on fuse/relay panel **OK**

- set **US 1119** to 20V scale
- connect **US 1119** between test points 1 and 2 on driver's seat belt control unit connector or 8 and 4 on passenger seat belt control unit connector
- reconnect battery ground strap
- remove ignition key
  - Must be 0.0V

NO →
- check wiring according to wiring diagram
- repair wiring as necessary

  or
- replace ignition switch

→ End

YES →

Motion/time delay sensor (**G97**) voltage supply, checking

- remove left rear seat
- disconnect battery ground strap
- remove connector from motion/time delay sensor (**G97**)
- set **US 1119** to 20V scale
- connect red lead to sensor connector terminal **3/30**
- connect black lead to sensor connector terminal **2/ES**
- reconnect battery ground strap
  - must be approximately 12.0V

NO → Go to (B) Two pages forward

YES → Go to (A) next page

(A)

Motion/time delay sensor (**G 97**) defective (always activated), checking

- disconnect battery ground strap
- reconnect motion/time delay sensor (**G 97**) connector
- remove seat belt control unit, driver side (**J227**)
- connect **US 1119** to test points 16 and 17 on control unit connector
- reconnect battery ground strap
  - must be approximately 12.0V

YES → End

NO →
- check for open circuit in wiring to motion/time delay sensor (**G 97**) according to wiring diagram

  or
- check and replace seat belt control unit, driver side (**J 227**)

  or
- replace motion/time delay sensor (**G 97**)

→ End

(B)

- check fuse/relay panel connector **Y/1** (plus connection, terminal 30)
- using **US 1119**, check voltage between test points ❾ and ❹ on passenger seat belt control unit connector and ❸ and ❷ on driver's seat belt control unit connector according to wiring diagram
  - must be approximately 12.0V

NO →
- check wiring from connector **Y/1** to control units according to wiring diagram
- repair wiring as necessary

→ End

YES →
- check wiring between fuse S51 and terminal **3/30** on motion/time delay sensor for open circuit according to wiring diagram
- repair wiring as necessary

→ End

Belts (driver or passenger) only move to A-pillar when door opened with key removed

Door latch switch (**F143/F144**), checking

- disconnect battery ground strap
- remove control unit (**J227 or J228**)
- connect **US 1119** between test points ❷ and ❿ on driver's seat belt control unit connector or ❹ and ⓫ on passenger seat belt control unit connector according to wiring diagram
- adjust **US 1119** to Ohm scale
- open door
  - must be continuity (0 Ohms)

NO →
- check and replace door latch switch (**F143/F144**)

→ End

YES →
- replace seat belt control unit (**J227** or **J228**)

→ End

Belts do not move to A-pillar after removing ignition key, (driver or passenger side)

A-pillar microswitch (**F140**/**F141**), checking

- disconnect battery ground strap
- remove seat belt control unit (**J227 or J228**)
- set **US 1119** to Ohm scale
- connect **US 1119** to test points (2) and (18) on driver's seat belt control unit connector or (4) and (19) on passenger seat belt control unit connector
  - must be continuity (0 Ohms)

NO → check and replace defective A-pillar microswitch (**F140**/**F141**) → End

YES, Driver's side → Backup light switch (**F4**), checking

YES, Passenger side → replace seat belt control unit (**J227 or J228**) → End

Backup light switch (**F4**), checking

- set **US 1119** to 20V scale
- connect **US 1119** to test points (2) and (20) on driver's seat belt control unit connector
- reconnect battery ground strap
- switch ignition **ON**
  - must be 0.0V

NO → replace backup light switch (**F4**) → End

YES → replace seat belt control unit (**J227 or J228**) → End

Driver belt moves to A-pillar when vehicle shifted into reverse

Test condition

- battery **OK**
- place shift into reverse

Backup light switch (**F4**), checking

- disconnect battery ground strap
- set **US 1119** to 20V scale
- connect **US 1119** to test points (2) and (20) on driver's seat belt control unit connector
- reconnect battery ground strap
- switch ignition **ON**
  must be approximately 12.0V

NO → check and repair wiring to backup light switch according to wiring diagram or replace backup light switch (**F4**) → End

YES → replace seat belt control unit (**J227** or **J228**) → End

Seat belt warning light does not light, belts at B-pillar, ignition ON

Seat belt warning relay (**J34**), checking

- remove fuse/relay panel cover
- disconnect battery ground strap
- remove seat belt warning relay (**J34**) from fuse/relay panel position **9**
- set **US 1119** to 20V scale
- connect red lead to terminal **4** at relay position **9**
- connect black lead to terminal **1** at relay position **9**
- reconnect battery ground strap
- switch ignition **ON**
  - must be approximately 12.0V

NO →
- check and repair wiring on fuse/relay panel and/or instrument cluster according to wiring diagram

→ End

YES →
- replace seat belt warning relay (**J34**)

→ End

Seat belt warning light comes on, with belts connected and ignition ON

B-pillar microswitch (**F145/F146**), checking

- disconnect battery ground strap
- remove appropriate rear side panels and B-pillar trim
- remove seat belt control units (**J227/J228**)
- separate 4-point connector **T4n** located behind fuse/relay panel
- set **US 1119** to Ohm scale
- connect leads to test points ❺ and ❷ on driver's seat belt control unit connector and/or ❹ and ❼ on passenger seat belt control unit connector according to wiring diagram
  - must be continuity (0 ohms)

NO →
- check and replace B-pillar microswitch (**F145/F146**)

→ End

YES – Driver side → Magnetic switch in driver's side belt buckle

YES – Passenger side → replace seat belt warning relay (**J34**)

Magnetic switch in driver's side belt buckle (**F157**), checking

- connect tester between test points ❺ and ❻ on driver's seat belt control unit connector according to wiring diagram
  - must be continuity (0 Ohms)

NO →
- check and replace B-pillar microswitch **F 145** and magnetic switch **F 157**

→ End

YES →
- replace seat belt warning relay (**J34**)

→ End

Description	Current track
Door latch switch, left, F143	27
Door latch switch, right, F144	7
Magnetic switch, left rear (B-pillar), F157	24
Microswitch, left front (A-pillar), F140	21
Microswitch, left rear (B-pillar), F145	23
Microswitch, right front (A-pillar), F141	5
Microswitch, right rear (B-pillar), F146	2
Motion/time delay sensor, G97	13-14
Seat belt control unit, left, J227	16-26
Seat belt control unit, right, J228	2-10
Seat belt motor, left, V74	25
Seat belt motor, right, V75	3
Seat belt warning system control unit, J34	4-7
Shoulder belt lock release solenoid, left, N129	15
Shoulder belt lock release solenoid, right, N130	12
Shoulder belt lock release switch, left (under seat), F149	15
Shoulder belt lock release switch, right (under seat), F150	12

Wire connectors

T4n — four-point, behind fuse/relay panel
T18 — eighteen-point, on seat belt control unit
T28 — twenty eight-point, on instrument cluster

Ground connections

(30) — beside fuse/relay panel
(141) — in seat belt wiring harness
(142) — in seat belt wiring harness

Welded wiring harness points

(Q60) — plus connection (30), in seat belt wiring harness

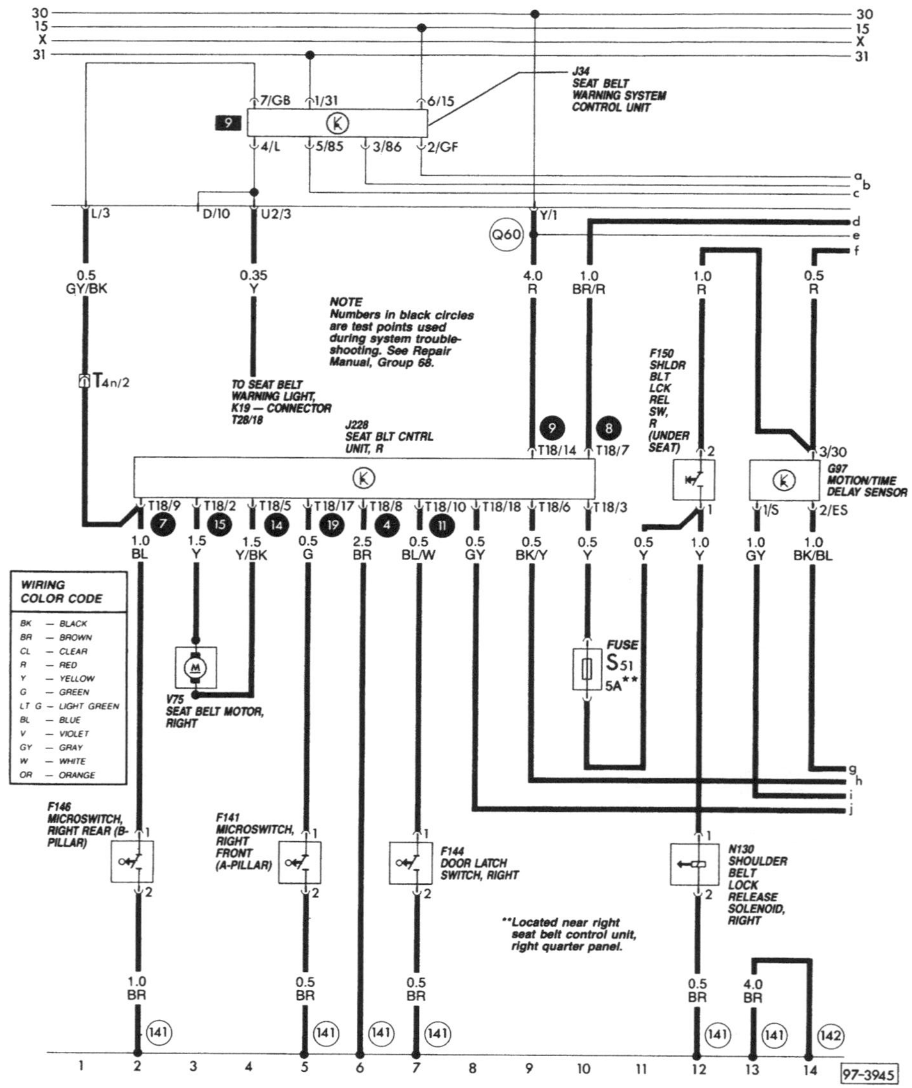

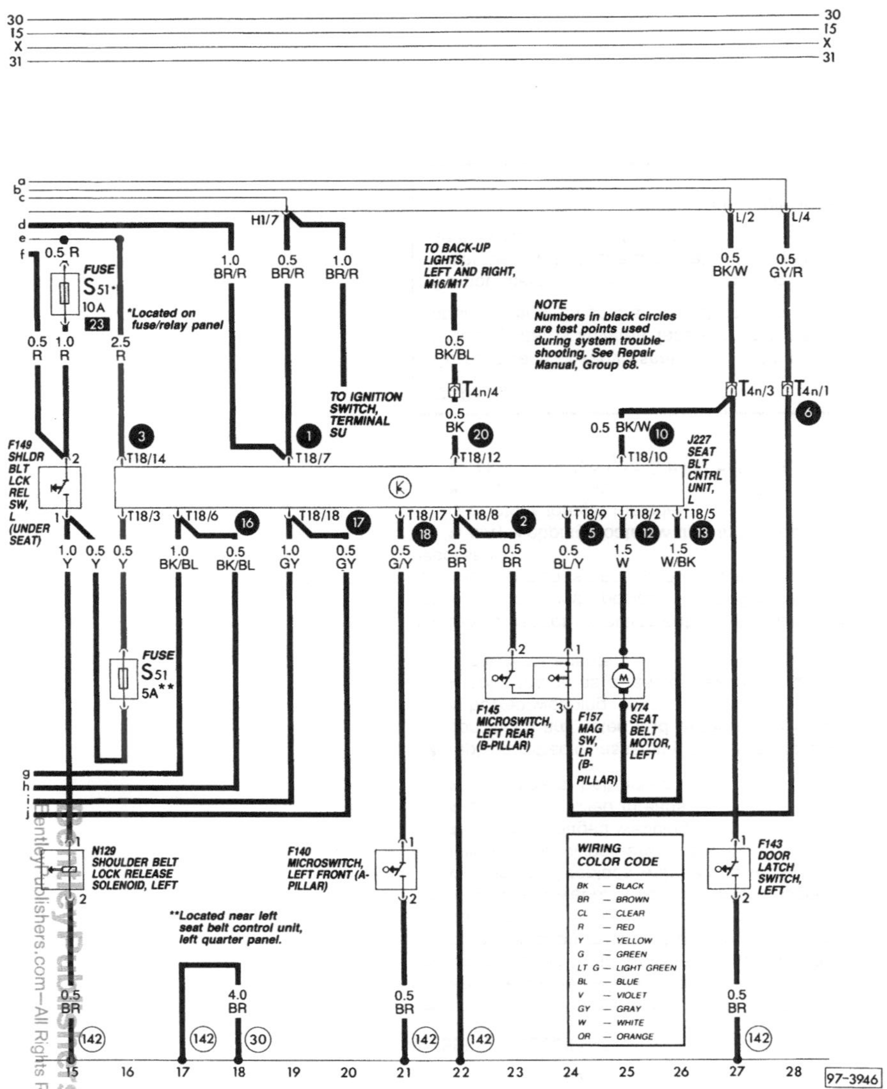

Seat belt system, inspection

A visual and functional inspection of seat belts is recommended to assure maximum protection for vehicle occupants.

> **CAUTION**
> Seat belt assemblies must be replaced after being subjected to loading by occupants in collision. Replace damaged components as necessary.

Front seat

Automatic shoulder belt system, checking for damage

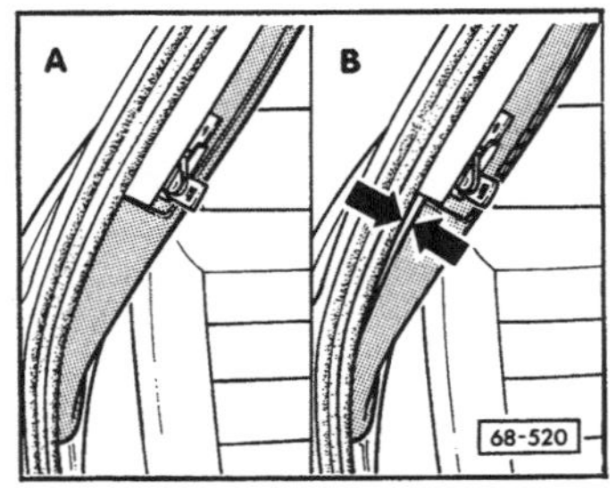

B-pillar trim deformation

A – not damaged

B – damaged

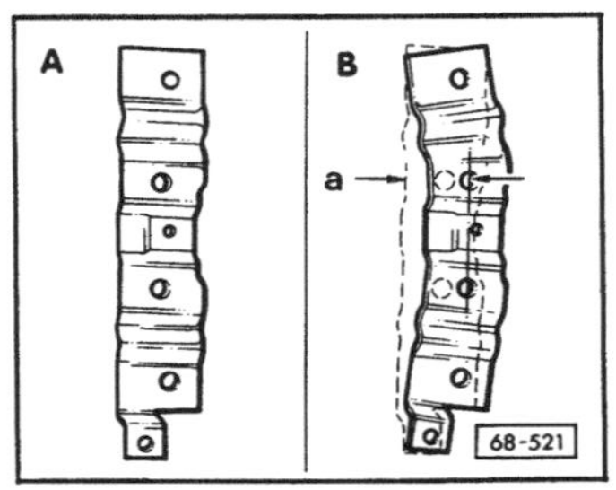

Deformation element

A – not damaged

B – damaged

- place element on flat surface, check for distortion **a**

  if **a** more than 23-31 mm (0.9-1.2 in.) replace element

Knee bar cushion

- check for deformation or indentation, replace as necessary

Webbing inspection

- check for twisted webbing due to improper alignment when connecting buckle
- if seat belt is twisted or binding, pull webbing out far enough to untwist and return to proper position
- fully extend webbing from retractor. Inspect webbing and replace with new assembly if following conditions are noted:
  - cut or damaged webbing
  - broken or pulled threads
  - cut loops at belt edges
- replace retractor assembly only if the following conditions are noted **and** excessive or severe:
  - bowed webbing
  - color fading as a result of exposure to sun or chemical agents

If webbing cannot be pulled out of retractor or will not retract to stowed position, check for the following conditions and clean or correct as necessary:

- dirty webbing coated with gum, syrup, grease or other foreign material
- twisted webbing
- retractor or loop on C-pillar out of position

CAUTION

Do no bleach or redye seat belt webbing. Webbing that is severely faded will not meet strength requirements and must be replaced.

Do not interchange buckle and retractor assemblies with those designated for other seating positions or other car models.

Do not lubricate retractors/buckles to eliminate noises.

Cleaning

CAUTION

Do not use chemical cleaning agents, bleach or dyes. They contain corrosive properties which will weaken the webbing. Do not allow inertia reel safety belts to retract before they are completely dry.

For cleaning, use a mild soap water solution. Let belts dry thoroughly and away from direct sunlight.

Buckle inspection

- insert tongue of seat belt into buckle until audible click is heard. Pull back on webbing quickly to assure buckle is latched properly
- replace seat belt assembly if buckle will not latch
- replace seat belt assembly if buckle will not release
- depress button on buckle to release belt
  - belt should release with a pressure of approximately 2 lbs.
- replace seat belt assembly if buckle cover is cracked, push button is loose or pressure required to release buckle is too high

Retractor inspection

Front retractors are dual sensing and will lock up in two modes. The unit will lock up as the belt is being withdrawn from the retractor if the inertia increases dramatically. Also, the unit will lock up with a change in vehicle motion such as rapid braking or cornering.

Note

Vehicles with Automatic Shoulder Belt System, can only be checked with braking test, at approximately 12 m.p.h.

- grasp seat belt webbing and, while pulling from retractor, give belt a fast jerk

Note

For automatic belt vehicles this mode is difficult to inspect due to a low setting of the inertia unit.

- drive vehicle in an open area away from other vehicles at a speed of approximately 5 to 15 mph and quickly apply footbrake
- if retractor does not lock up under these conditions, remove and replace seat belt assembly

Note

A passenger can be used to test the right hand retractor. The driver alone can test both retractors simultaneously by grasping the right hand webbing, extracting approximately 26 inches from the retractor and holding the belt while braking.

Seat belt warning light/buzzer

A light in the instrument panel displaying the words "fasten seat belts" or "fasten belts" along with an audible signal that lasts for four to eight seconds should be noticed when the ignition switch is moved to the "on" or "start" position and the driver's belt is unlatched. If the driver's belt is latched the buzzer should not activate.

Note

Correct any malfunctions in this system if the customer desires.

D-rings

Check for scoring. If belt worn during accident, plastic covered D-rings will show fine parallel grooves indicating stress. D-rings worn from normal useage will be smooth.

Anchorage inspection

Seat belt mounting bolts are installed at 20 Nm to 60 Nm (15 to 44 ft lb). Retractors must be securely anchored in place. The bolts at the lower belt anchor point (not the retractor) and upper seat belt loop must be tight and still allow the hardware to swivel. Anchorage areas must be sound and uncorroded.

Note

Reinforce bolt mounting area if necessary.

Rear seat

Webbing/buckle inspection

same as for front seat

Lap and shoulder belts (rear seats)

The rear seat is a 3-point system, see section 68-26

Anchorage inspection

Same as for front seat

Webbing defects, examples

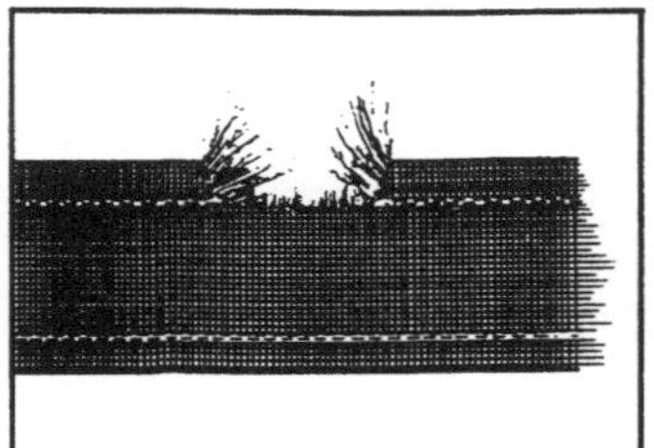
68-A030

Cut or damaged webbing

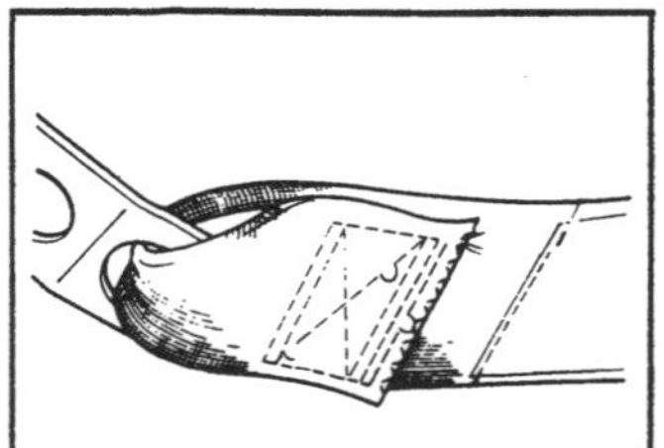
68-A031

Broken or pulled threads

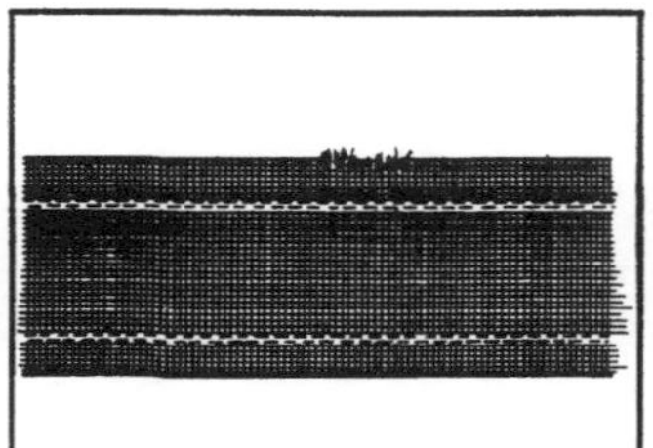
68-A032

Cut loops at one belt edge

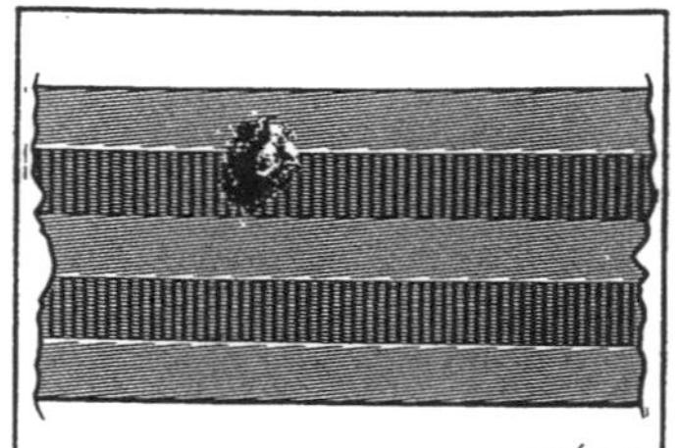
68-460

Cigarette burns

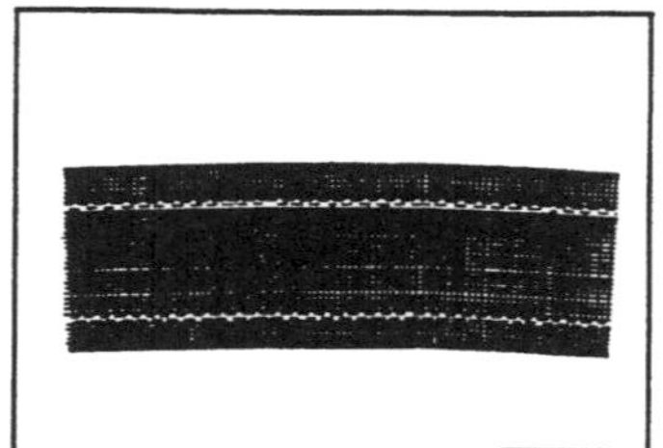
68-A033

Bowed webbing

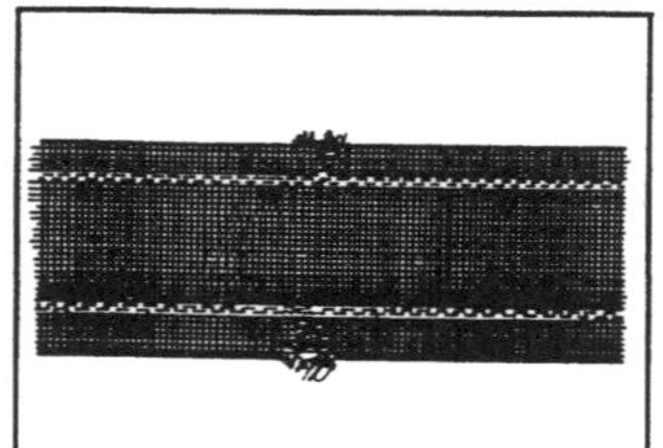
68-A034

Cut loops at both belt edges

- from being caught in door

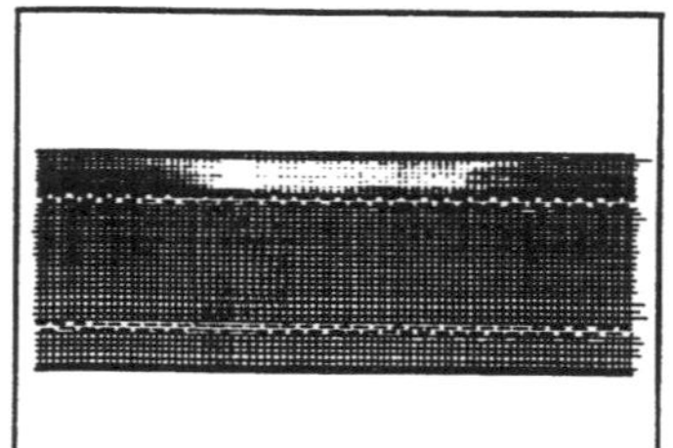
68-A035

Color fading

Installing child's seat on front passenger seat — vehicles w/passive restraint

WARNING

This **child seat restraint belt** is not intended to be used as a lap belt by a child or an adult. Improper use could result in personal injury.

Parts required

Part number	Description	Quantity
176 898 696	Child Seat Restraint Kit	1

CAUTION

Part numbers are for reference only. Always check with your Parts Department for latest information.

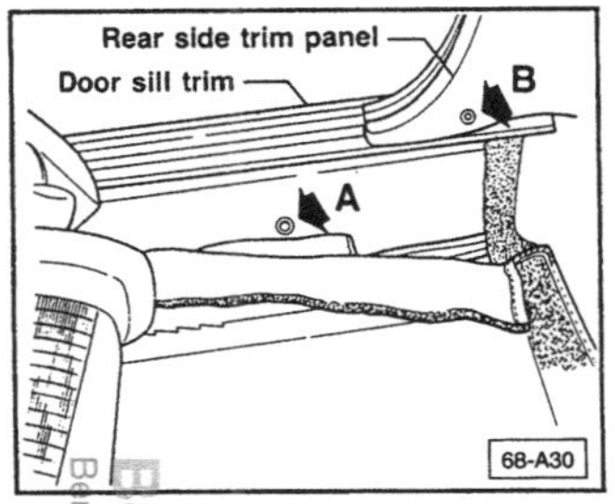

Note

Depending on vehicle, all parts of kit may not be required.

- remove screw (**arrow B**) from rear trim
- pull up and loosen door sill trim
- pull back carpet to expose threaded hole (**arrow A**)
- mark and punch hole in carpet

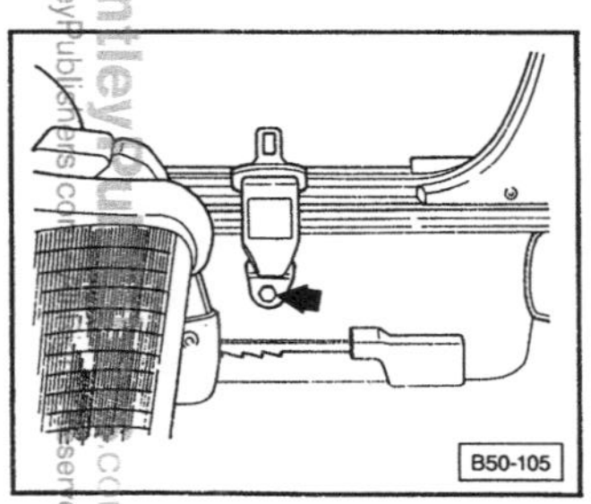

- install tongue and anchor assembly from kit
  - torque mounting bolt (**arrow**) 40 Nm (40 ft lb)
- install cover cap over bolt
- place child's seat on front passenger seat
- buckle both ends of the child seat belt and adjust snugly over child seat by pulling loose end of webbing to tighten seat belt
- unbuckle both ends of belt and store in glove box
- reinstall trim components in reverse order of removal

THIS PAGE INTENTIONALLY LEFT BLANK

THIS PAGE INTENTIONALLY LEFT BLANK

Index

Cassette storage box
- removing **70.8**

Door trim panel
- assembly **70.12**
- grab handle outer cover, removing/installing **70.13**

Front storage trays
- removing **70.7**

Glove box
- assembly **70.9**
- lid, removing/installing **70.10**
- lock cylinder **70,.10**
- removing/installing **70.10**

Instrument panel
- installing **70.8**
- removing **70.7**
- trim plate, removing **70.8**

Molded headliner
- installing **70.5**
- removing **70.3**

Plastic moisture barrier
- replacing/installing **70.11**

Rear side trim panel
- assembly **70.6**

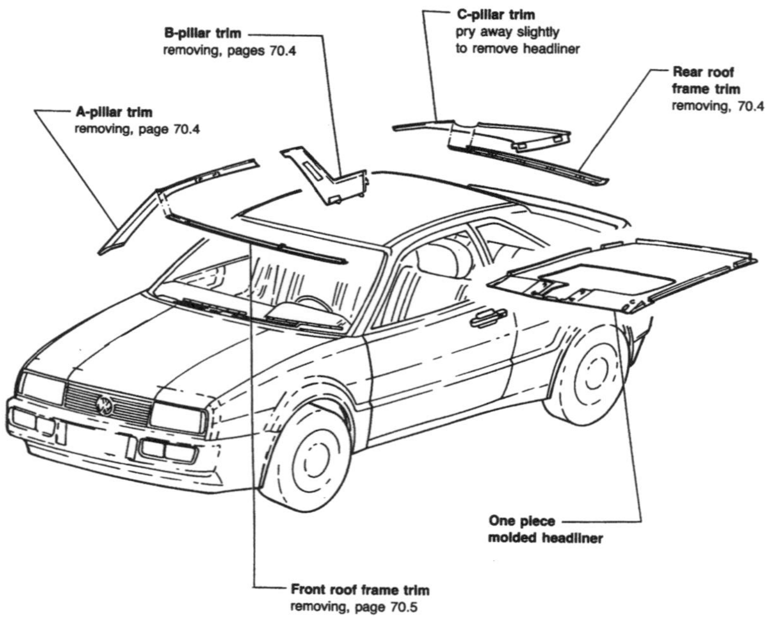

Molded headliner, removing

- disconnect battery-ground strap
- pry off interior light with screwdriver

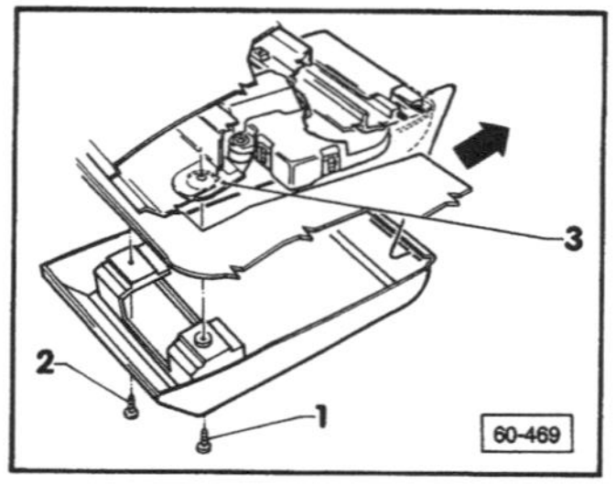

- remove screws **1**, **2** and remove drive cover in direction of **arrow**
  - screw **1** also used for attaching bracket **3**

- remove sun visor bracket **1**, holding bracket **2** (**arrows**)
- disconnect plug connector from passenger side sun visor

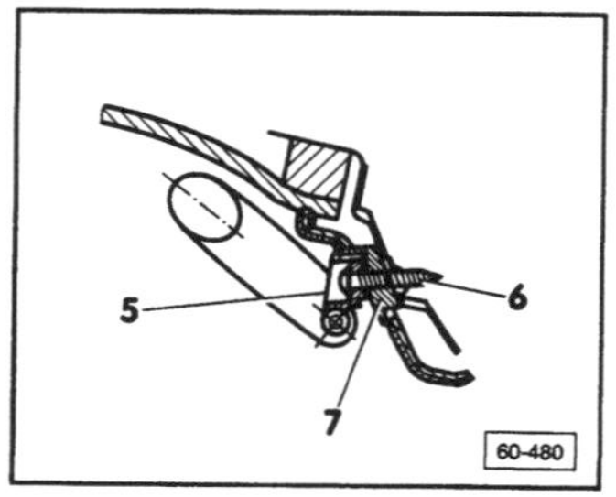

- pry off cap **5**, remove screw **6** and remove grab handle with molded part **7**

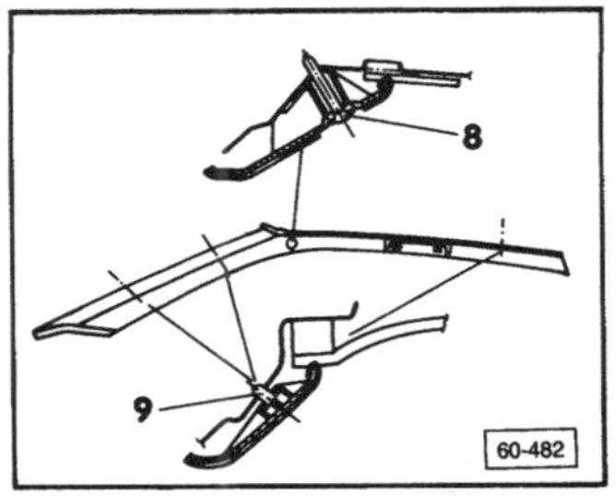

■ remove screw **8**, pull off A-pillar trim (secured with clip **9**)

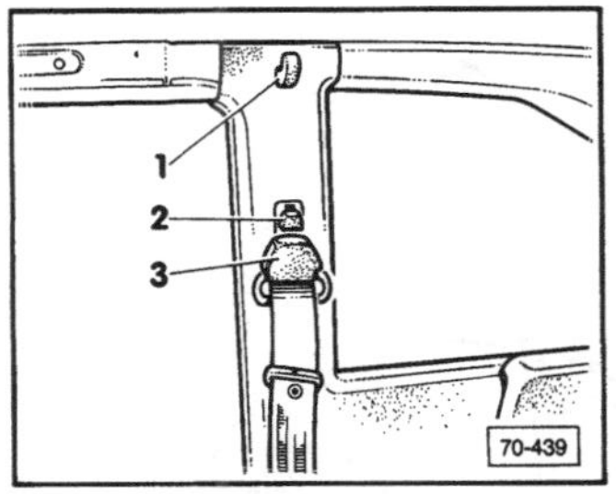

■ remove coat hook **1**
■ press off cover upward
■ pull off adjusting knob **2**
■ pry off routing hardware cap **3** and unscrew routing hardware

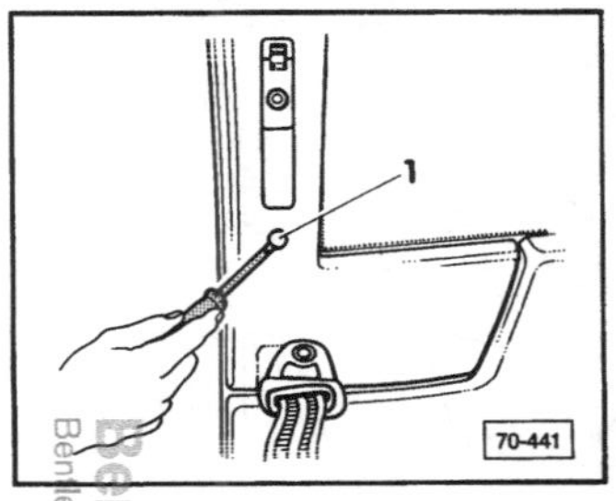

■ pry out cap **1** and remove screw
■ pull off B-pillar trim

Note

The lower attachment screws for the B-pillar trim are not unscrewed.

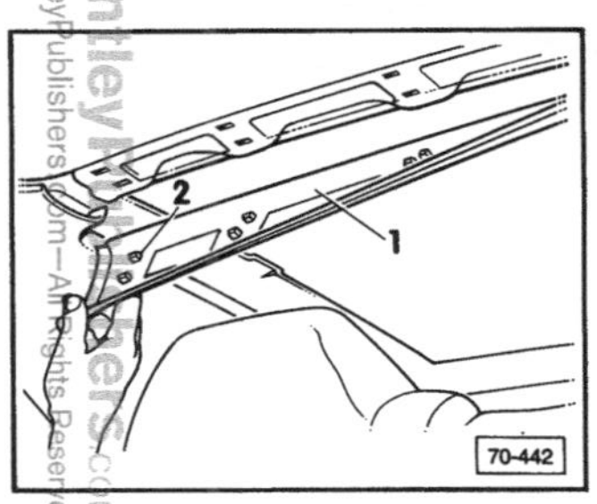

■ pull off rear roof frame cover molding **1**
  ● **2** plastic clips

■ pull off C-pillar trim in contact area with headliner

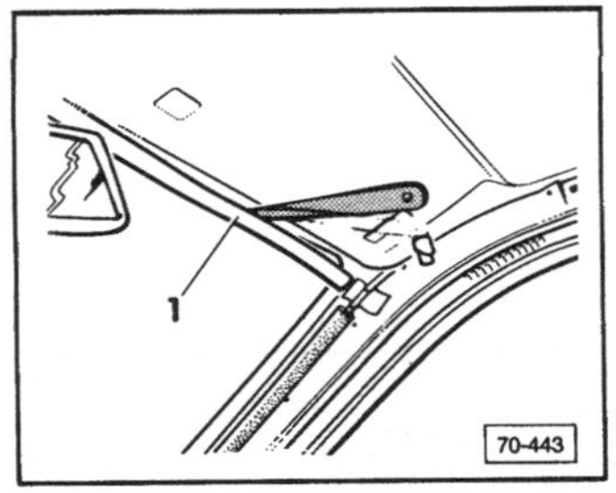

■ pry off front roof frame trim **1** using a plastic wedge

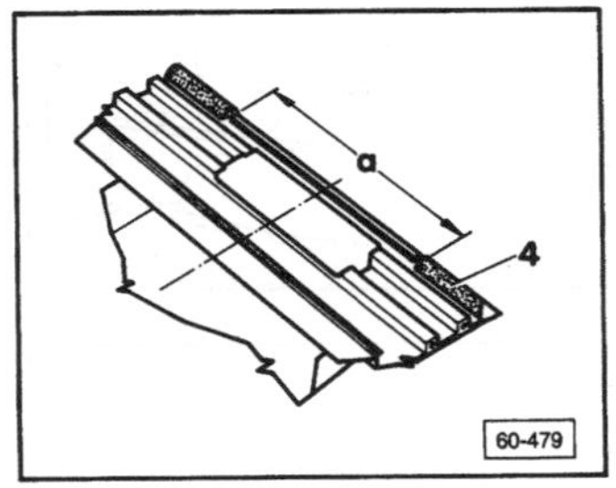

■ in vehicles with sunroof pull off beading **4** from around sunroof opening
■ during installation note cut out **a** in area of roof-drive
  ● **a** = 180 mm (7.0 in.)
  ● cut out is marked with notches in frame

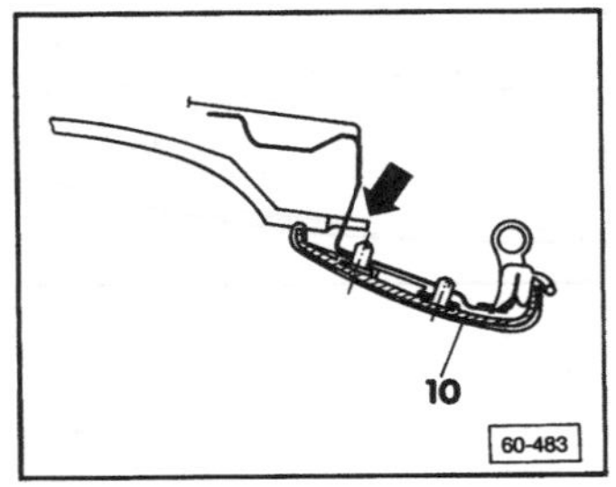

■ lower headliner and pull slightly toward front until tabs (**arrow**) are pulled from roof reinforcement
■ remove headliner through rear lid opening

Molded headliner, installing

Reinstall all components in reverse order of removal.

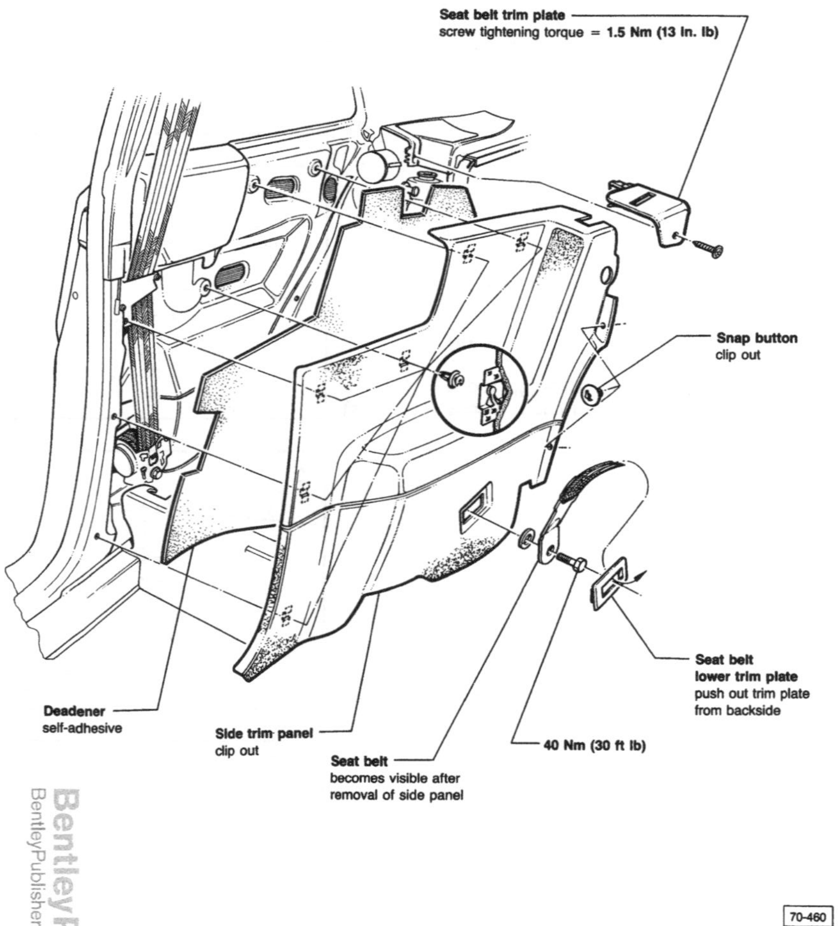

Instrument panel, removing

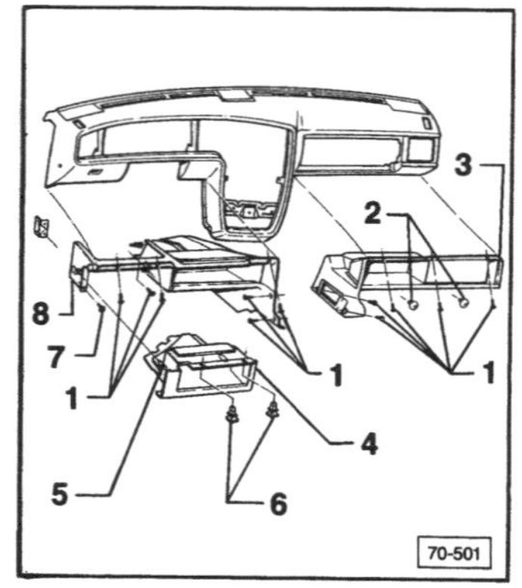

Front storage trays, removing

- disconnect battery ground strap
- remove shelf screws **1**
- remove plastic nuts **2**, pull out tray **3**
- turn rotating clips **4**, pull out storage box **6**
  - when installing, locating pins **5** must be pushed into trim plate retainers **8**
- unscrew oval head screw **7**
- unscrew multi-purpose screws **1** and remove cover **8**
- remove steering wheel, see Repair Group 48

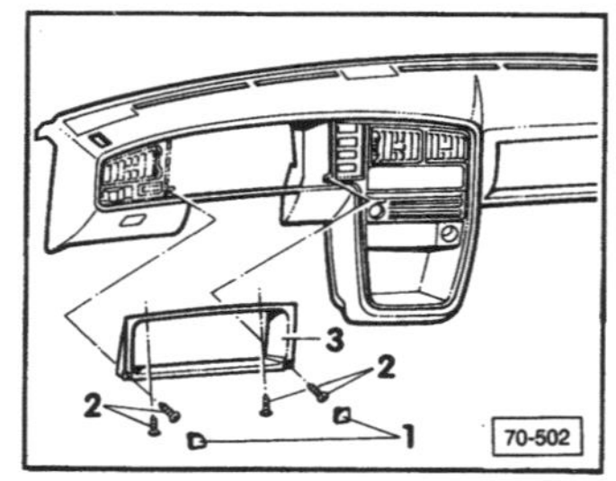

- pry off cover cap **1**
- remove screws **2** and remove trim plate **3**

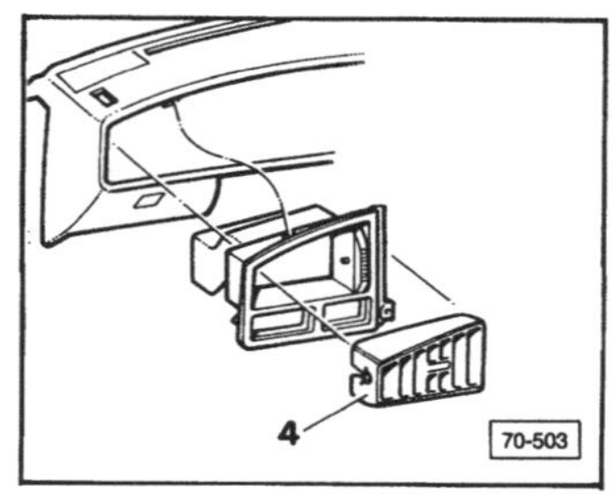

- pry out vents **4** with screwdriver
- remove radio or glove box

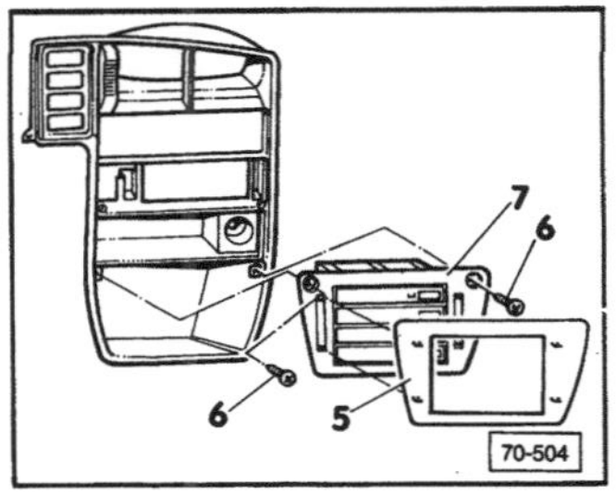

Cassette storage box, removing

- pry off trim plate **5**, remove screws **6**
- pull out cassette storage box **7**

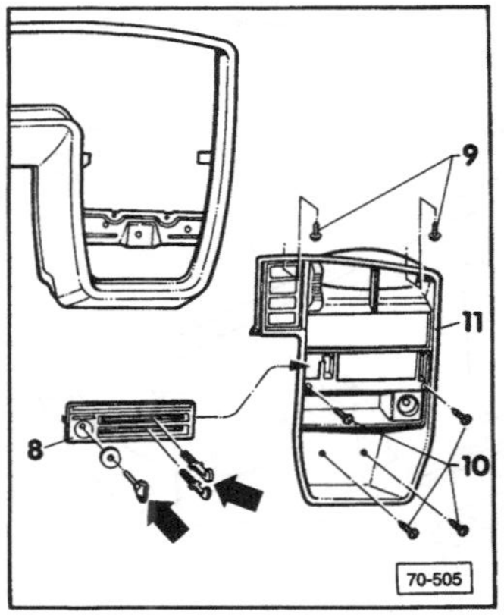

Instrument panel trim plate, removing

- pull off temperature control knobs and sliding levers **(arrows)**
- pry out trim plate **8**, remove heating and ventilation controls
- unscrew multi-purpose screws **9** (operating switch removed)
- unscrew multi-purpose screws **10** and pull out trim plate **11**

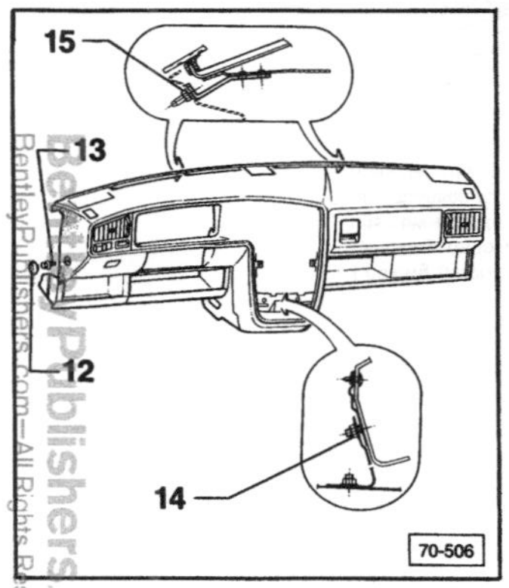

- remove instrument panel cover and steering column switch, see Repair Group 90
- pry off left/right caps **12** and unscrew left/right hex head bolts **13**
- unscrew hex nut **14** (nut is in back of console)
- unscrew hex nuts **15** (nuts are located in plenum panel and can only be seen with aid of mirror)
- slightly pull off instrument panel, loosen cables and remove instrument panel

Instrument panel, installing

- install all components in reverse order of removal procedure, noting the following:
- push locating pins **5** into retainers **8** (see fig. 70-501, page 70.7)

THIS PAGE INTENTIONALLY LEFT BLANK

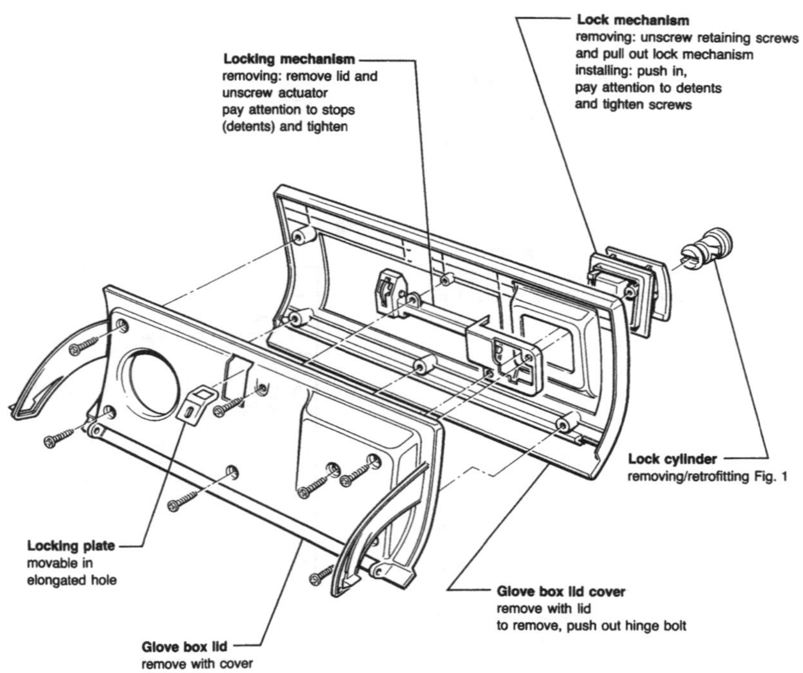

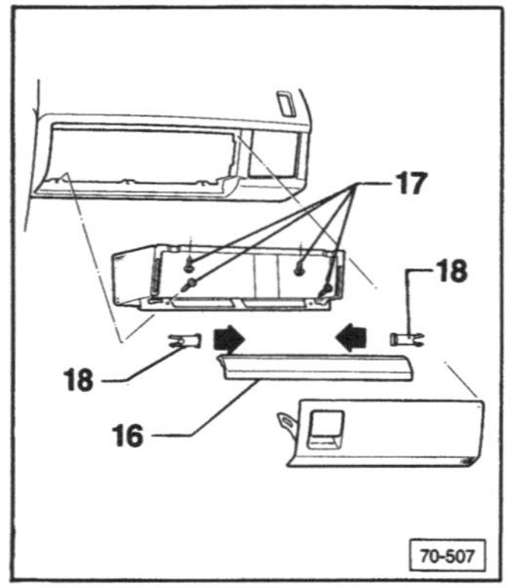

▶ **Glove box, removing**

- clip out light for glove box
- pull off cover molding **16**
- unscrew multi-purpose screws **17** and remove glove box from instrument panel

Glove box, installing

- install all components in reverse order of removal

Glove box lid, removing

- pull out bolts **18** in direction of arrows

Glove box lid, installing

- install all components in reverse order of removal

▶ Fig. 1 **Lock cylinder, removing**

- remove actuator and carefully saw off detent (**arrows**)
  - actuator cannot be used again

Lock cylinder, retrofitting

- pry out cover from operating mechanism and press in lock cylinder
  - original lock combination can be maintained by rearranging of tumblers

CAUTION

Part numbers are for reference only. Always check with your Parts Department for latest information.

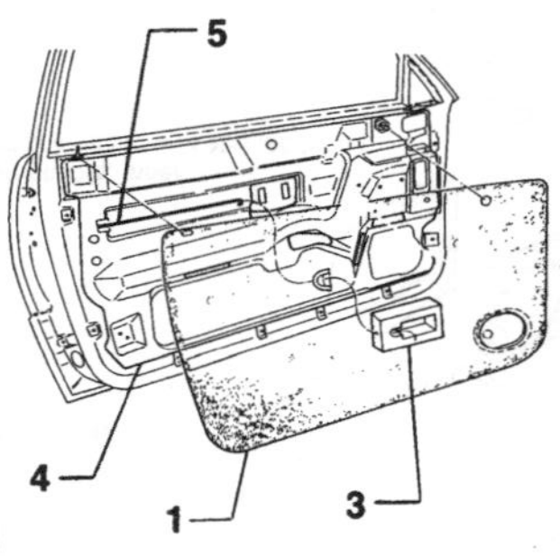

70-459

Plastic moisture barrier, replacing/ installing

- remove door trim panel
- remove plastic moisture barrier
- remove all remaining adhesive from door with 3M© General Purpose Adhesive Cleaner or equivalent

Note

If plastic moisture barrier is to be reused:

- remove all traces of old adhesive material from plastic
- apply double sided adhesive tape **AKL 434 019 025** to door
- apply a 5.0 mm (13/64 in.) bead **4** of sealant **AKL 450 005 05** into prestamped groove in door shell
- install plastic moisture barrier to door shell
- push seal **3** for inside door handle on pull rod **5**

CAUTION

Install plastic moisture barrier from bottom up without wrinkles on seating surface.
Also, no foreign particles (dust, lint etc.) should become stuck to plastic. That may form a capillary action and cause a leak.

- reinstall door trim panel

Note

Take care not to damage plastic moisture barrier when replacing inner door related items.

THIS PAGE INTENTIONALLY LEFT BLANK

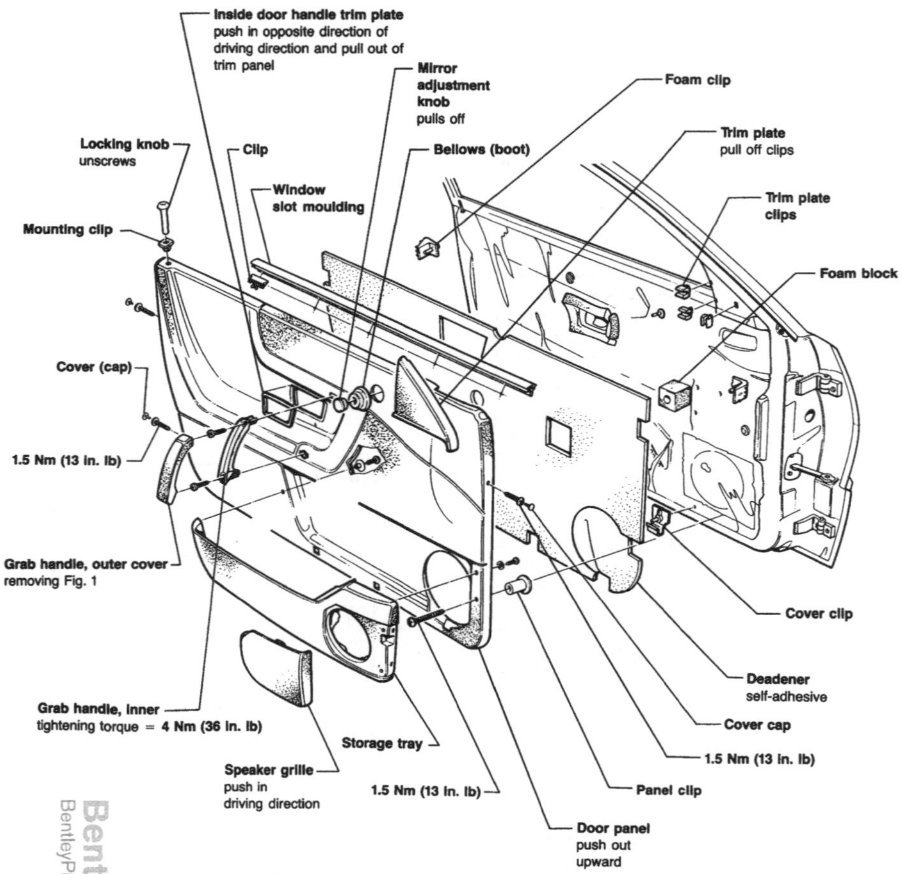

CAUTION

When door panel is removed, the window slot moulding is removed with it. Prior to installing door panel, the window slot moulding must be installed onto door panel.

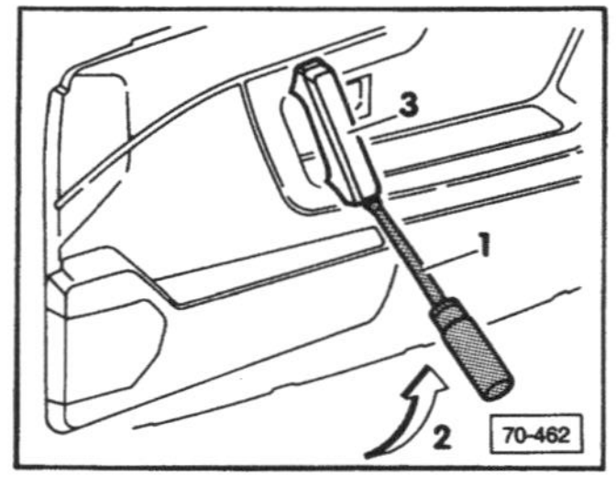

▶ Fig. 1

Grab handle outer cover, removing

- push screw driver **1** with flat side into slot under grab handle
- carefully lift handle upward **2**
- pull off cover **3**

Grab handle outer cover, installing

Reinstall all components in reverse order of removal.

Index

Center armrest
- removing/installing **72.3**

Front seat
- removing/installing **72.2**

Front seat backrest
- removing/installing **72.2**

Rear seat
- removing/installing **72.3**

Rear seat backrest
- removing/installing **72.4**

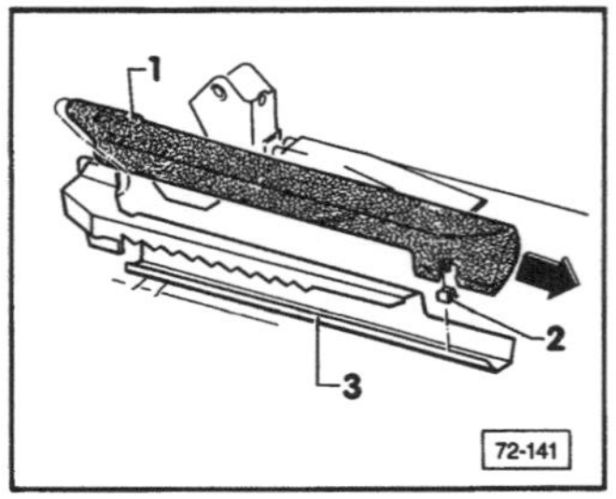

Front seat, removing

- pry inner cover **1** (including clamp **2**) off seat track **3**
- remove (in direction of arrow) inner cover **1** from seat track **3**

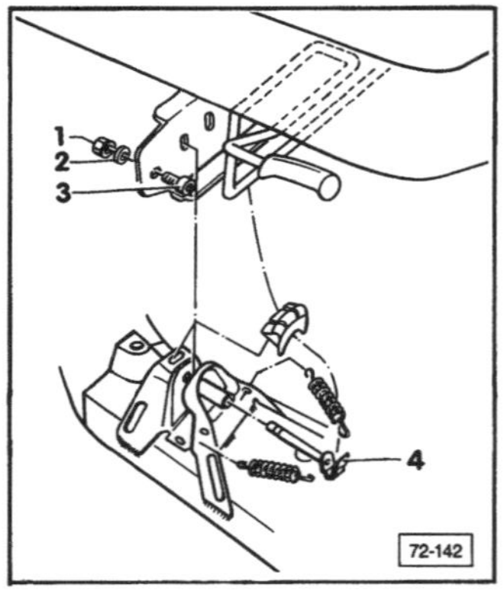

- push seat fully forward
- remove cap nut **1**, lock ring **2**, and bolt **3**
- release retaining rod **4** and push seat to rear and lift out of tracks

Front seat, installing

- install all components in reverse order of removal
  - torque cap nut to 8 Nm (6 ft lb)

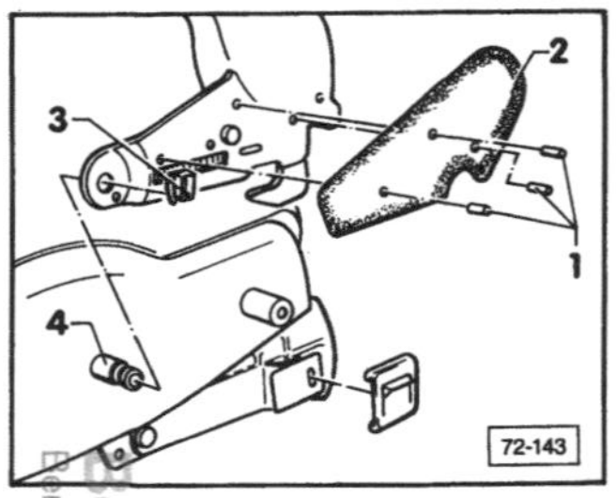

Front seat backrest, removing

- remove front seat
- drive retaining pins **1** left and right through toward inside with drift, and pry off cap **2**
- remove left and right clamps **3**
- remove backrest from supports **4**

Front seat backrest, installing

- install all components in reverse order of removal

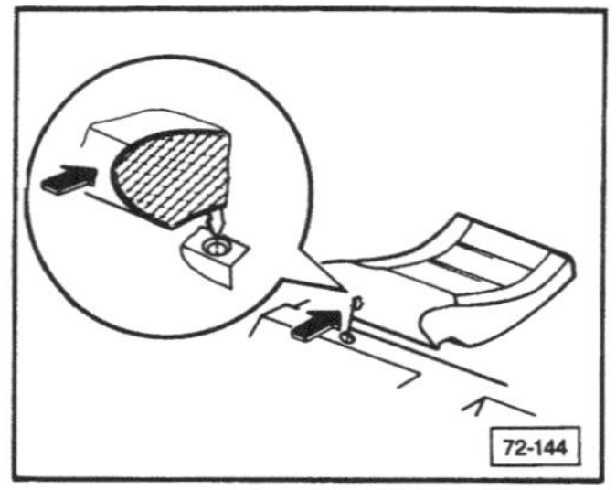

Rear seat, removing

- push rear seat back to unlock at catches **(arrows)**, lift seat up, and pull forward to remove

Rear seat, installing

- install all components in reverse order of removal

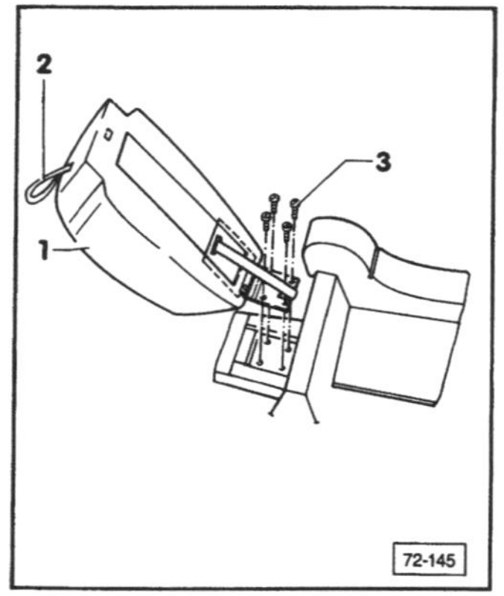

Center armrest, removing

- fold down center armrest **1** forward with pull loop **2**
- remove four screws **3** and remove center armrest **1**

Center armrest, installing

- install all components in reverse order of removal
  - torque screws **3** to 5 Nm (45 in. lb)

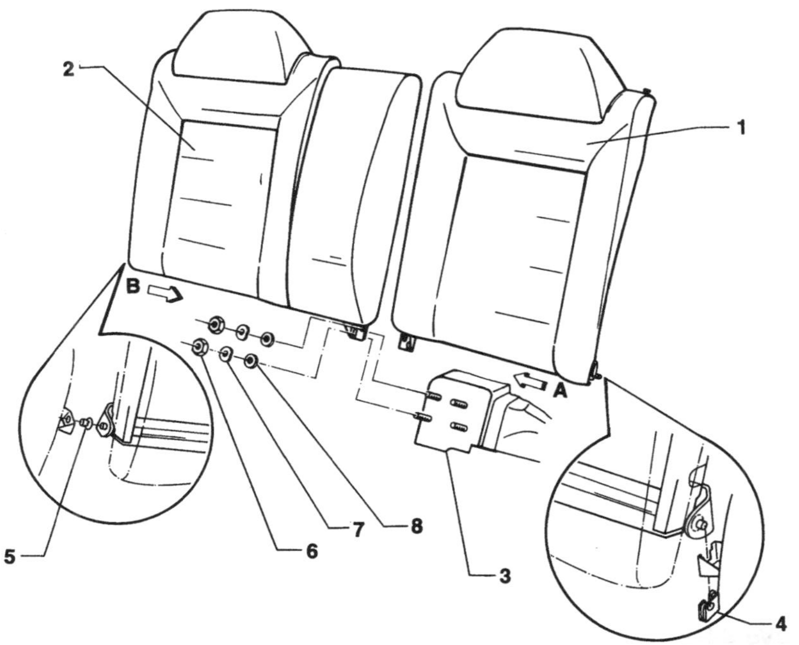

72-146

Rear seat backrest, removing

- fold down center armrest
- remove left and right rear seat
- remove nuts **6**, spring washers **7**, and washers **8**
- unlock left rear seat backrest **1**, pivot from mount **3** toward front, and pull toward middle (**arrow A**) to remove from support **4**
- unlock rear seat backrest, left **1**, pivot from mount **3** toward front, and pull toward middle (**arrow B**) to remove from guide bushing **5**

Rear seat backrest, installing

- install all components in reverse order of removal
  - torque nuts **6** to 4 Nm (36 in. lb)

THIS PAGE INTENTIONALLY LEFT BLANK

THIS PAGE INTENTIONALLY LEFT BLANK

Index

Air intake duct
- identifying **80.4**

Air outlet grill/housing
- removing **80.4**

Control cables
- installing/adjusting **80.5**

Fresh air blower
- removing **80.5**
- series resistance, removing **80.5**

Heater
- assembly **80.7**
- ventilation air ducts, assembly **80.2**

Heater/fresh air controls
- faceplate/assembly, removing **80.4**

Side window air vent
- removing **80.4**

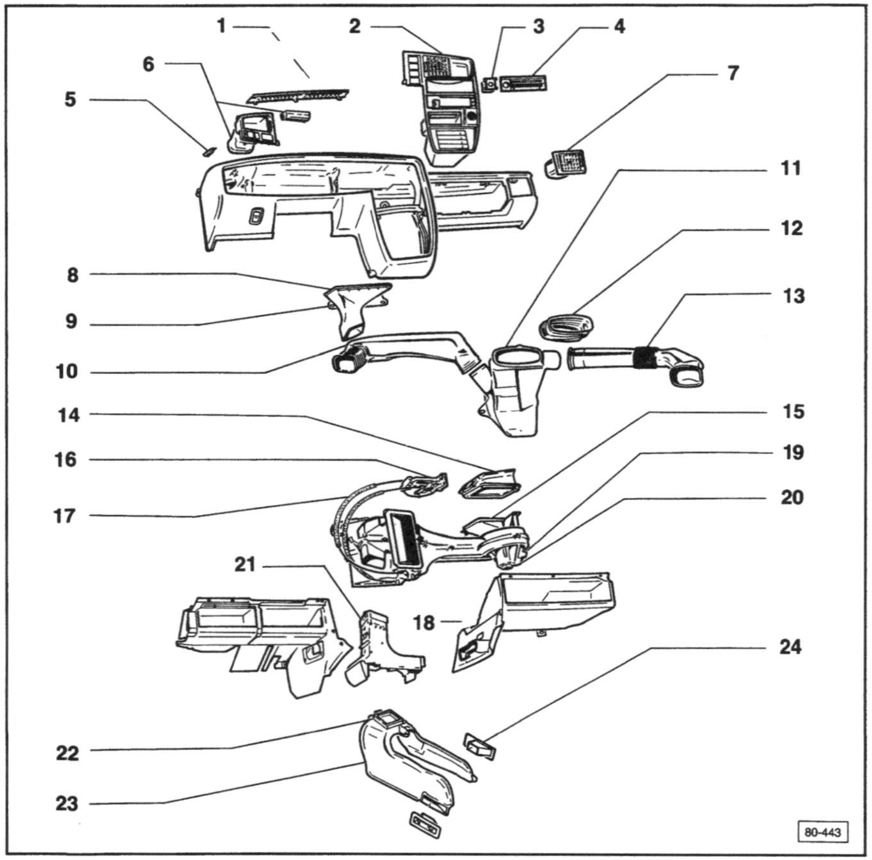

1 — **Windshield defroster vents**
carefully pry up to remove

2 — **Instrument panel center insert**
air outlet housing is integral
air outlet grille, removing,
page 80.4, Fig. 1

3 — **Fresh air blower switch**

4 — **Heater/fresh air control face plate**
removing/installing, page 80.4, Fig. 3

5 — **Side window air vent**
removing, page 80.4, Fig. 2

6 — **Air outlet housing and grille**
removing, page 80.4, Fig. 1

7 — **Air outlet**

8 — **Seal**
remove instrument panel to replace

9 — **Defroster air duct**
remove instrument panel to replace

10 — **Air duct, left**
remove instrument panel to replace

11 — **Air distribution housing**
remove instrument panel to replace

12 — **Center air outlet duct**

13 — **Air duct, right**

14 — **Air intake duct**
identifying, page 80.4, Fig .4

15 — **Seal**

16 — **Heater/fresh air control assembly**
removing, page 80.4, Fig. 3

17 — **Control cables**
installing/adjusting, page 80.5

18 — **Air distribution housing with heater core (heater assembly)**
assembly, page 80.7
remove instrument panel to remove

19 — **Fresh air blower series resistance**
removing, page 80.5, Fig. 6

20 — **Fresh air blower**
removing/installing, page 80.5, Fig. 5

21 — **Footwell air outlets**
remove instrument panel and heater assembly to remove

22 — **Seal**

23 — **Rear footwell air duct**
remove instrument panel and heater assembly to remove

24 — **Rear footwell air outlet**

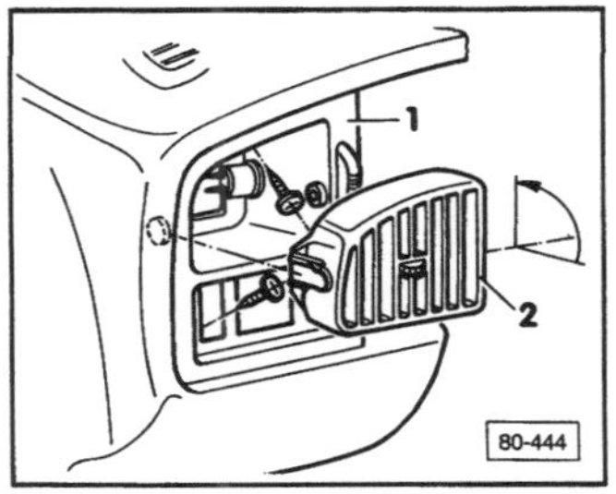

▶ **Fig. 1 Air outlet grille and housing, removing**

- turn grille in direction of arrow and push down to disengage
- remove grille from housing
- remove retaining screws and pull housing from dashboard
- install in reverse order of removal

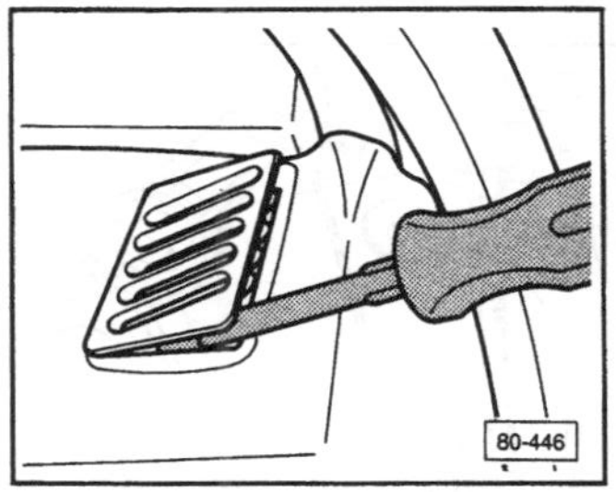

▶ **Fig. 2 Side window air vent, removing**

- carefully pry side window vent off with screwdriver

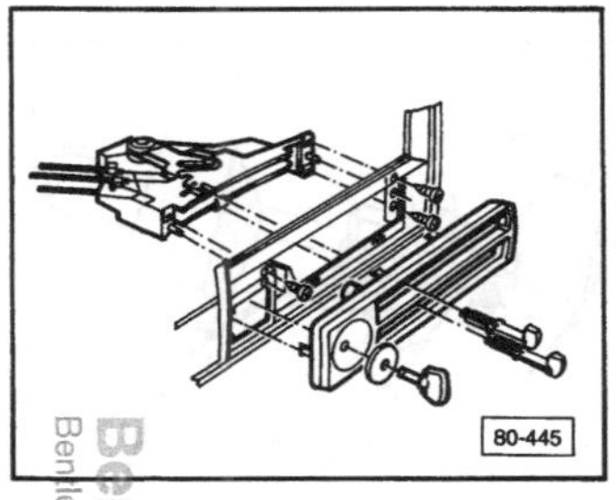

▶ **Fig. 3 Heater/fresh air control face plate and assembly, removing**

- pull control levers off
- pull blower switch knob off
- gently pry off face plate
- remove control assembly retaining screws
- remove control cables and remove assembly from dashboard
- install in reverse order of removal

- adjust all three control cables, 80.5

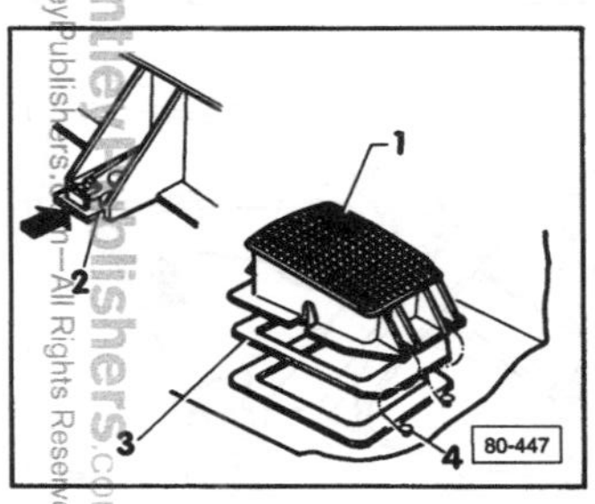

▶ **Fig. 4 Air intake duct, identifying**

1 — Air intake grille
2 — Retaining clips (4)
3 — Seal
4 — Screw with sealing washer
- tighten screw to stop

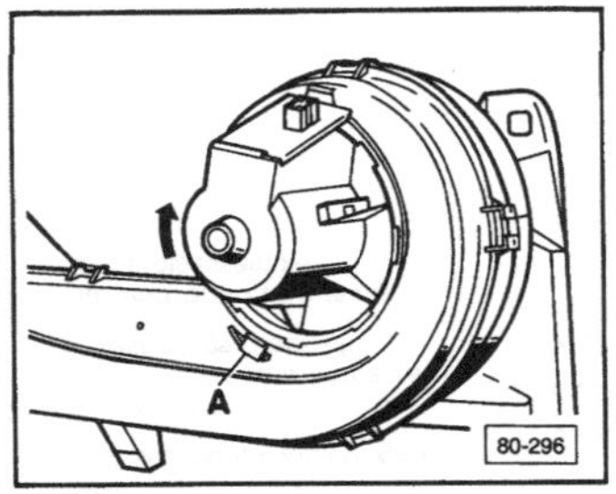

▶ **Fig. 5 Fresh air blower, removing**

- disconnect wire at blower
- disengage retaining lug **A**
- turn fresh air blower in direction of arrow
- pull blower from housing
- install in reverse order of removal

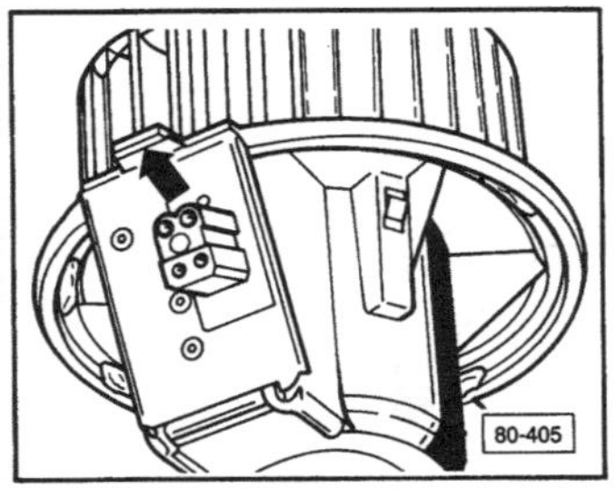

▶ **Fig. 6 Fresh air blower series resistance, removing**

- remove fresh air blower motor
- unclip plate from blower (**arrow**) and remove

Control cables, installing/adjusting

Install control cables to heater control levers before installing on heater box levers.

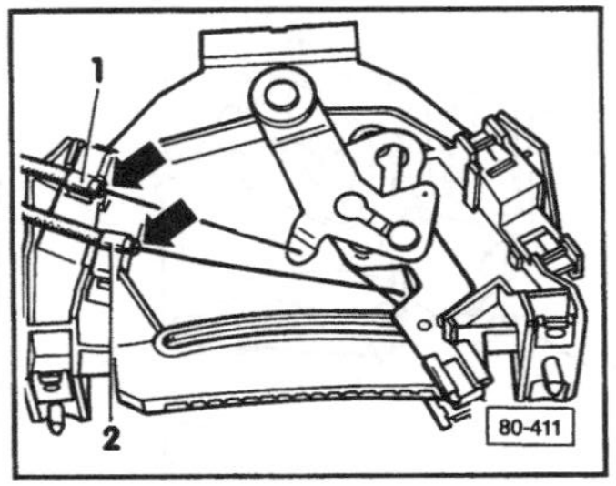

▶

- install cables **1** and **2** on control levers
- install clips onto cables (**arrows**)
  - cable **1** (black, long) — to central flap (mode) door lever
  - cable **2** (black, short) — to heater/defroster control flap

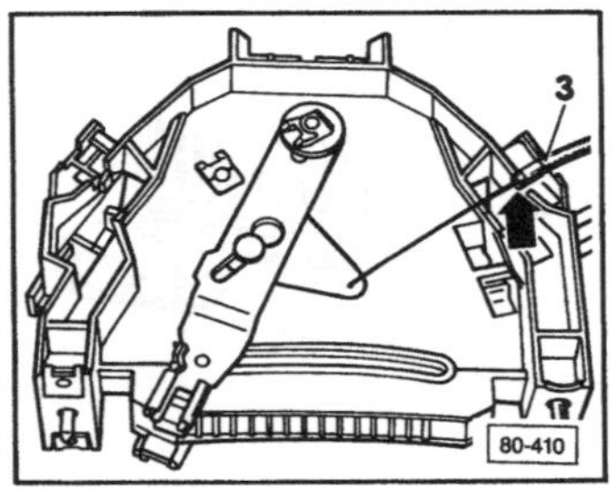

▶

- install cable **3** to heater control lever
- install clip onto cable (**arrow**)
  - cable **3** (blue) — to air blend (temperature) flap
- install fresh air/heater control assembly into instrument panel
- slide both control levers to left stop (off/cool)

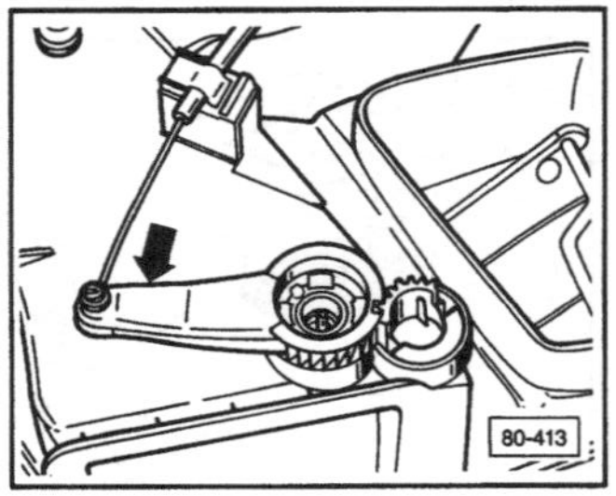

▶
- connect cable **1** (black, long) to mode door (central flap) lever on heater box
- push lever in direction of **arrow** to stop
- hold lever in this position and install cable retaining clip

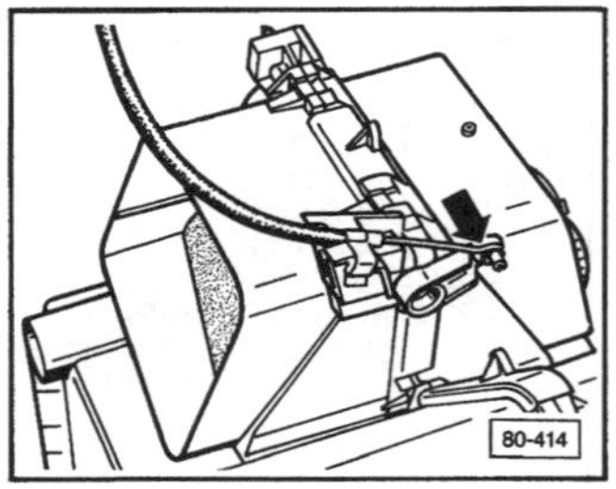

▶
- connect cable **2** (black, short) to heater/defroster control lever
- push lever in direction of **arrow** to stop
- hold lever in this position and install cable retaining clip

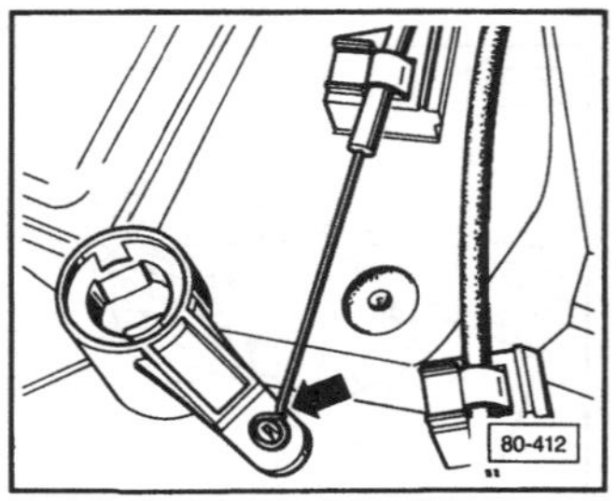

▶
- connect cable **3** (blue) to air blend (temperature) door lever
- push lever in direction of **arrow** to stop
- hold lever in this position and install cable retaining clip

- check operation of heater/fresh air controls by moving control levers back and forth
  - all flaps must audibly contact stops during lever movement

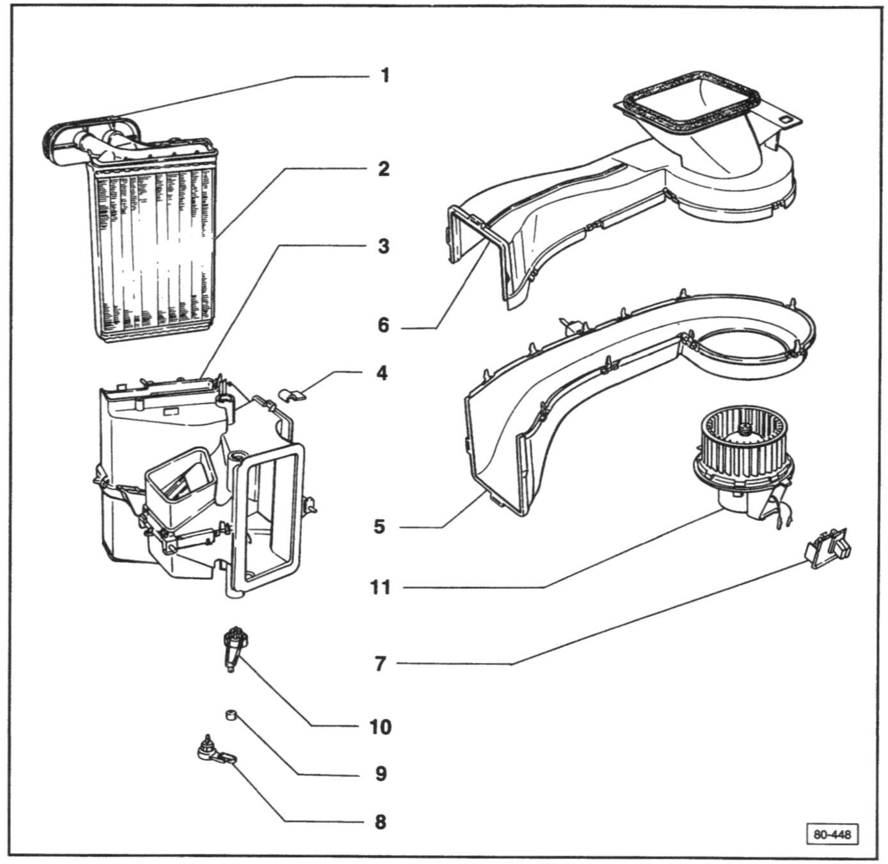

1 — **Seal**
must seal tightly without gaps

2 — **Heater core**
mount heater core with screws if retaining lugs break off

3 — **Air distribution housing**
do not disassemble

4 — **Retaining clip**

5 — **Lower air duct**

6 — **Upper air duct**

7 — **Fresh air blower series resistance**

8 — **Temperature flap lever**

9 — **Lever bushing**

10 — **Central (mode) flap lever**

11 — **Fresh air blower**

Index

A/C compressor
- assembly **87.22**
- clutch, checking **87.9**
- clutch, removing/installing **87.23**
- lines, installing **87.9a**
- shaft seal, removing/installing **87.28**
- valve plate, head and gasket, removing/installing **87.35**

A/C pressure switch
- checking **87.7**

A/C system
- fuses **87.7**
- relays, location **87.8**
- temperatures and pressures, checking **87.16**

Air distribution and evaporator case
- assembly **87.10**

Belts and pulleys
- assembly **87.21**

Belt tensioning damper
- assembly **87.20**

Engine components
- mounting **87.18**

Evaporator water drain valve
- identifying/removing **87.9**
- Expansion valve insulator checking **87.9**

Refrigerant system
- components **87.5**
- discharging **87.3**
- oil specifications **87.4a**
- safety measures **87.2**

Safety relief plug
- identifying **87.8**

Sight glass
- checking **87.8**

Vacuum system
- checking **87.13**
- hose layout **87.12**
- reservoir, removing/installing **87.15**
- servos, checking **87.14**

WARNING

This repair group covers only 1990, 1991 and 1992 vehicles, those with A/C systems using Freon R-12 refrigerant.

The 1993 and 1994 vehicles covered by this manual are equipped with a new A/C system using the non-CFC refrigerant R-134a.

The two refrigerant types are not compatible and the system components, especially those carrying the refrigerant, are not interchangeable. Different equipment is required to service the A/C systems using R-134a refrigerant.

Refrigerant capacities, system pressures and other specifications listed in this repair group apply to **R-12 systems only**.

Refer to the Supplement at the back of this manual for additional information covering A/C systems using R-134a refrigerant.

Index

Troubleshooting

Systems using R-12 refrigerant

Air conditioner
- does not cool **87.46**
- compressor clutch does not engage **87.48**
- cooling output diminishes after several minutes **87.51**
- Insufficient cooling output **87.50**

Defroster
- does not clear windshield **87.44**

Fresh air
- cannot be shut off **87.39**
- no air flow from dash board/foot well vents **87.42**

Heater
- insufficient or no heat output **87.40**

Safety measures for working on vehicles with air conditioning and for handling refrigerant.

The subassemblies and lines of the air conditioner are filled with difluorodichloro methane (CF_2 CL_2) as a refrigerant. This refrigerant is known under the commercial name of Frigen R 12 or Freon R 12.

Replacement A/C compressors supplied by the Parts Department are filled with 135 cc (4.6 oz) of A/C refrigerant oil.

A/C system refrigerant capacity:
1100 g ± 50 g
(38.8 oz ± 1.7 oz)

A/C system refrigerant oil capacity:
135 cc (4.6 oz)

WARNING

- A/C system is filled with R12 refrigerant gas which is under pressure
- always be careful that refrigerant does not come in contact with your skin
- always wear eye protection when working around the system
- if liquid refrigerant has come in contact with your skin or eyes:
  - do not rub skin or eyes
  - immediately flush with cool water for 15 minutes
  - rush to a doctor or hospital
  - do not attempt to treat yourself
- keep refrigerant containers stored below 50°C (122°F) and never drop from high places
- keep refrigerant away from open flames because poisonous gas will be produced if it burns. Do not smoke when refrigerant gases are present for the same reason
- electric welding near refrigerant hoses causes refrigerant to decompose from ultraviolet light. Discharge system before electric welding

WARNING

- do not expose any component of the A/C system to high temperatures (i.e., open flames). Excessive heat will cause a pressure increase which could burst the system
- do not steam clean condensers or evaporators. Use only cold water or compressed air
- automotive refrigerant containing CFC's is hazardous to the earth's atmosphere. To protect our environment, use an Underwriter's Laboratory (UL) approved refrigerant recovery/recycling unit such as **Kent-Moore ACR³**, or equivalent, whenever discharging an A/C system

Refrigerant system, discharging

Follow the safety measures listed above and on previous page when discharging or charging A/C refrigerant system.

WARNING

Automotive refrigerant containing CFC's is hazardous to the earth's atmosphere. To protect our environment, use an Underwriter's Laboratory (UL) approved refrigerant recovery/recycling unit such as **Kent-Moore ACR³**, or equivalent, whenever discharging an A/C system.

WARNING

The A/C system should be serviced only by trained personnel familiar with:
- equipment use
- related safety procedures
- regulations governing the discharging/handling/disposal of automotive refrigerants

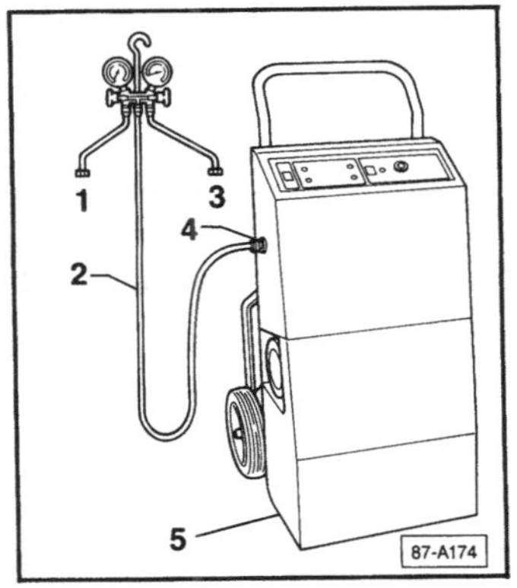

WARNING

Always wear safety goggles when charging or discharging system.

Avoid inhaling fumes when using flame type leak detector. Refrigerant-12 becomes poisonous gas after coming into contact with open flame.

- locate A/C service valves on refrigerant hoses and remove protective caps
- close both valves on A/C manifold gauge set
- connect hose **1** from low pressure gauge to low pressure service valve
- connect hose **3** from high pressure gauge to high pressure service valve
- connect manifold gauge hose **2** to inlet connection **4** on refrigerant recovery/ recycling unit **5**

CAUTION

Always follow manufacturer's instructions when using a refrigerant recovery/ recycling unit.

- following refrigerant recovery/recycling unit manufacturer's instructions, discharge A/C system into refrigerant recovery/recycling unit
- close manifold gauge valves when refrigerant is fully discharged
- disconnect manifold gauge set hoses from service valves
- reinstall protective caps on service valves

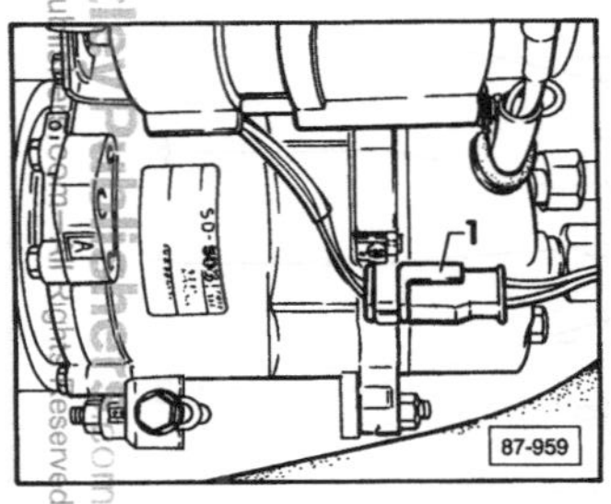

CAUTION

DO NOT leave A/C system open for long periods of time. Prolonged exposure to the air will contaminate the system and necessitate the replacement of the receiver drier.

- disconnect plug **1** from A/C compressor clutch to prevent accidental compressor operation while system is discharged

Refrigerant system, oil specifications

Use A/C refrigerant oil that meets the specifications of the following oils:

Shell	Clavus G 100
Sunoil	Suniso 5 GS
Texaco, DEA	Capella MS 100 or Capella WF 100
Indemitsu	Dens Oil 6
Fuchs	Reniso KES 100

Replacement A/C compressors (Sanden SD-709) supplied by the Parts Department are filled with 135cc (4.6 oz) of A/C refrigerant oil. This is the total system oil capacity.

A/C system refrigerant oil capacity: 135 cc (4.6 oz).

If replacing one of the following components, add specified amount of new refrigerant oil to component before installing:

Condensor	15cc (0.5 oz)
Evaporator	20cc (0.7 oz)
Receiver drier	10cc (0.3 oz)
High pressure hose	10cc (0.3 oz)
Low pressure hose	10cc (0.3 oz)
Compressor	70cc (2.4 oz)
Total oil capacity	135 cc (4.6 oz)

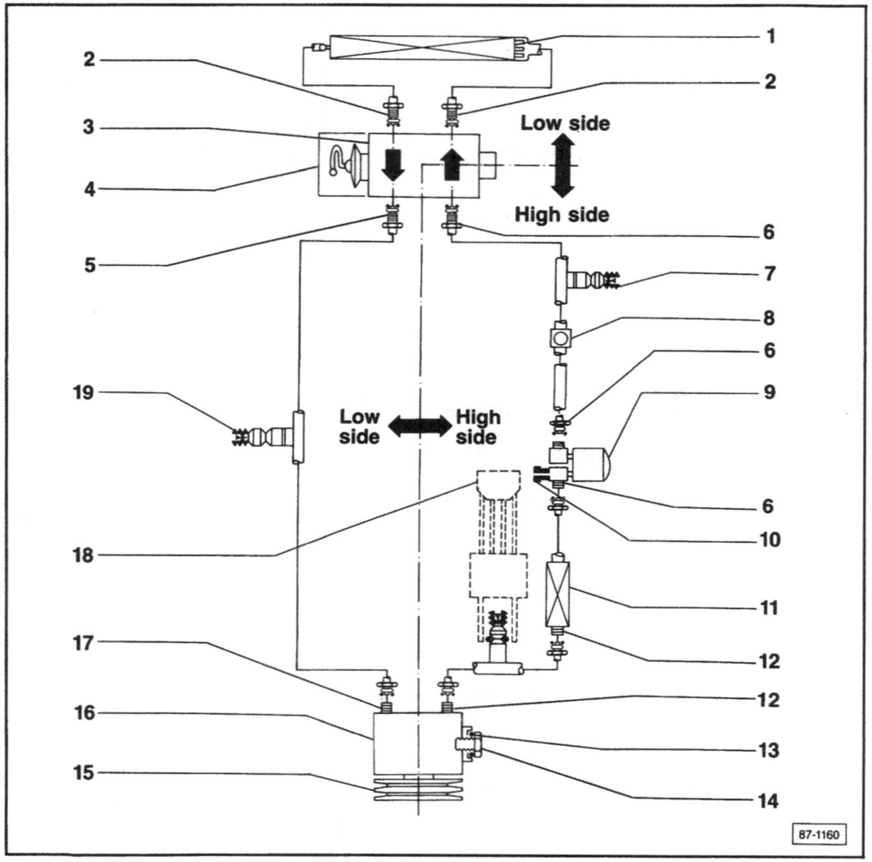

Refrigerant capacity: 1100 g ± 50 g (38.8 oz ± 1.7 oz)

Refrigerant oil capacity:135 cc (4.6 oz)

1 — **Evaporator**

2 — **Threaded connection (¾ in. — 18 UNS)**
- O-ring seal on connection: 10.8 mm inside diameter, 1.8 mm thickness
- 25 Nm (18 ft lb)

3 — **Expansion valve**

4 — **Insulator**
- checking, see Index

5 — **Threaded connection (⅞ in. — 18 UNS)**
- O-ring seal on connection: 14.0 mm inside diameter, 1.8 mm thickness
- 35 Nm (26 ft lb)

6 — **Threaded connection (⅝ in. — 18 UNF)**
- O-ring seal on connection: 7.5 mm inside diameter, 1.8 mm thickness
- 13 Nm (10 ft lb)

7 — **High pressure service valve**
- located between sight glass and expansion valve
- use adapter **VAG 1637** for connection when servicing A/C

8 — **Sight glass**
- checking, see Index

9 — **Receiver drier**
- replace if system is contaminated or if system has been open for several hours

10 — **Safety relief plug**
- identifying, see Index

11 — **Condenser**

12 — **Threaded connection (¾ in. — 16 UNF)**
- O-ring seal on connection: 10.8 mm inside diameter, 1.8 mm thickness
- 32 Nm (24 ft lb)

13 — **O-ring**
- 9.3 mm inside diameter, 1.8 mm thickness

14 — **Oil drain plug**

15 — **Compressor clutch**
- checking, see Index

16 — **A/C Compressor**
- Sanden SD 709
- replacement compressors supplied by the Parts Department are filled with 135 cc (4.6 oz) of oil

17 — **Threaded connection (⅞ in. — 14 UNF)**
- O-ring seal on connection: 14.0 mm inside diameter, 1.8 mm thickness
- 35 Nm (26 ft lb)

18 — **A/C pressure switch**
- switches off A/C compressor clutch if system pressure is too low (low refrigerant charge)
- switches off A/C compressor clutch if system pressure is too high
- switches radiator cooling fan into second speed as a result of A/C system pressure rise
- unscrew switch to remove (schrader valve installed below switch to allow removal without system discharge)
- O-ring seal on connection: 10.8 mm inside diameter, 1.8 mm thickness
- 8 Nm (6 ft lb)
- checking, see Index

19 — **Low pressure service valve**
- 7/16 in. — 20 UNF
- use adapter **VAG 1637** for connection when servicing A/C

WARNING

Automotive refrigerant containing CFC's is hazardous to the earth's atmosphere. To protect our environment, use an Underwriter's Laboratory (UL) approved refrigerant recovery/recycling unit such as **Kent-Moore ACR³**, or equivalent, whenever discharging an A/C system.

A/C pressure switch, checking

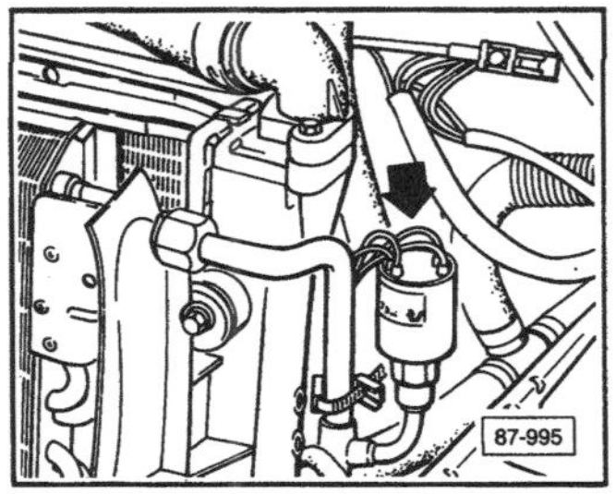

The A/C pressure switch (**arrow**) is located near the left side of the condenser

A/C compressor clutch switching (between two blue wires):
switch closes above:
2.0 bar — 3.0 bar
(29.0 psi — 43.5 psi)
switch opens above:
24.0 bar — 32.0 bar
(348.0 psi — 464.0 psi)

Radiator cooling fan second speed switching (between red and black wires):
switch closes above:
12.5 bar — 16.0 bar
(181.3 psi — 232.0 psi)

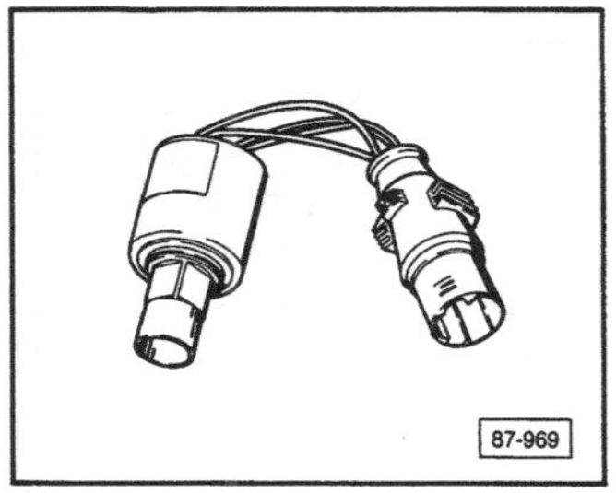

- visually inspect O-ring seal (10.8 mm × 1.8 mm) to ensure proper seat in groove

Note

The A/C pressure switch can be removed without discharging the A/C system.

A/C system, fuses

The following fuses are used in the A/C system:

- fuse S6 (20A) — fresh air fan
- fuse S19 (30A) — radiator cooling fan, A/C compressor clutch
- fuse S23 (30A) — main heater fuse (**arrow**)
- fuse S24 — fresh air fan overheat fuse

A/C system relays, location

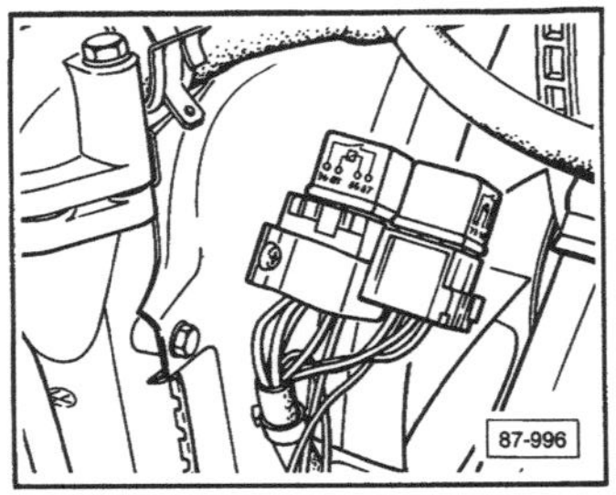

The radiator cooling fan relay (2nd stage) and A/C compressor clutch relay are mounted in the engine compartment on the top left side of the radiator fan shroud.

Safety relief plug, identifying

A safety plug (**arrow**) is mounted on the receiver drier. The plug will release if maximum safe A/C system operating pressure is exceeded.

Sight glass, checking

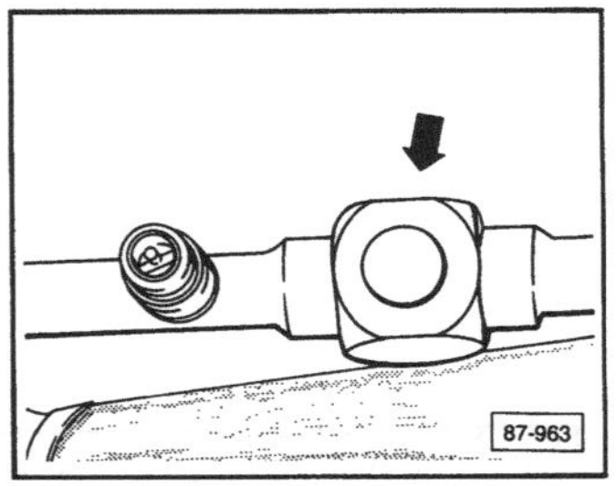

The sight glass (**arrow**) is located in the high pressure tube between the high pressure service valve and receiver drier.

- start engine and switch A/C **ON**
  - during A/C compressor operation, sight glass should remain clear and bubble-free

If bubbles are present all the time, system is low on refrigerant

- check for leaks and recharge A/C system as necessary

WARNING

Automotive refrigerant containing CFC's is hazardous to the earth's atmosphere. To protect our environment, use an Underwriter's Laboratory (UL) approved refrigerant recovery/recycling unit such as **Kent-Moore ACR³**, or equivalent, whenever discharging an A/C system.

A/C compressor clutch, checking

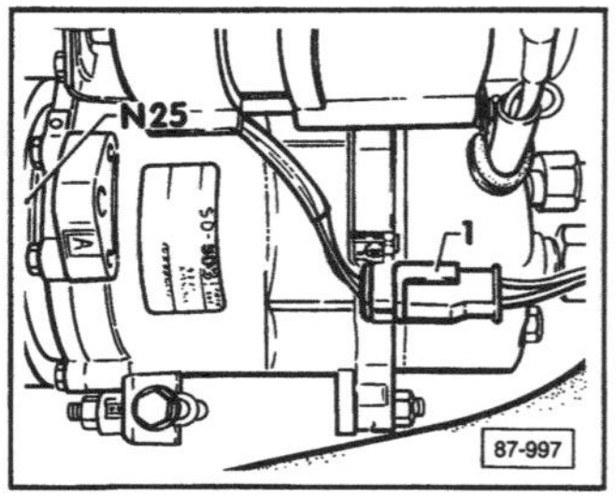

- disconnect plug **1** from A/C compressor clutch
- measure resistance of compressor clutch using **US 1119** set to ohm scale
  - must be approximately 3.6 ohms

Expansion valve insulator, checking

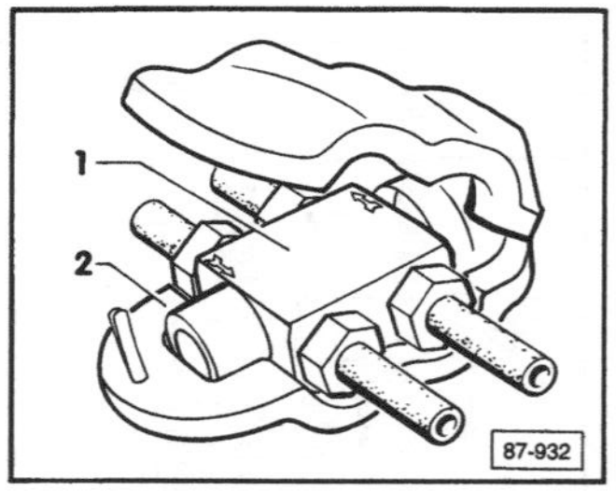

Insulator **2** protects expansion valve **1** from hot engine compartment temperatures

- inspect insulator **2** and replace if damaged

Evaporator housing water drain valve, identifying/removing

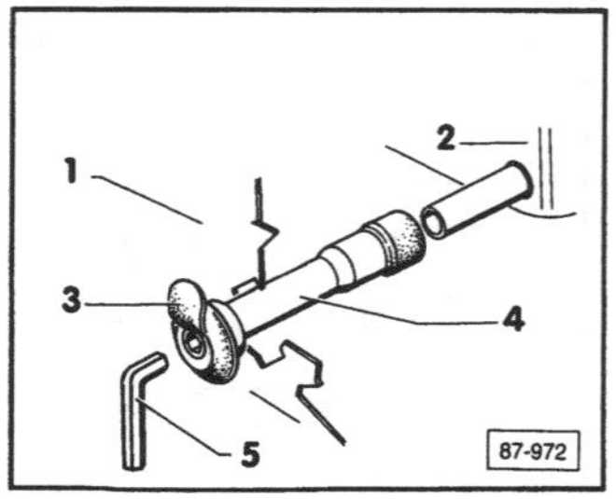

1 — Firewall
2 — Evaporator housing
3 — Valve flap (shown lifted)
4 — Water drain valve
5 — 14 mm allen wrench

Note

If water sprays from the dashboard air vents during A/C operation, check water drain valve for proper operation.

To remove drain valve:

- lift up valve flap **3** and insert 14 mm allen wrench **5** into valve outlet
- rotate valve counter-clockwise 45° (1/8 turn) and pull from evaporator housing
- clean any dirt or foreign matter from valve tube
- clean any dirt, wax, or undercoating from valve flap **3**
- check valve flap for proper operation
- install valve in reverse order of removal

A/C refrigerant compressor lines, installing

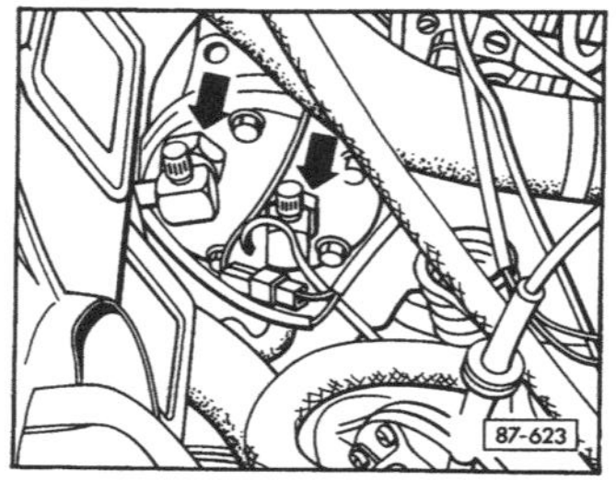

- install A/C lines (**arrows**) on compressor and tighten as follows:
  - suction line: 35 Nm (26 ft lb)
  - discharge line: 25 Nm (18 ft lb)

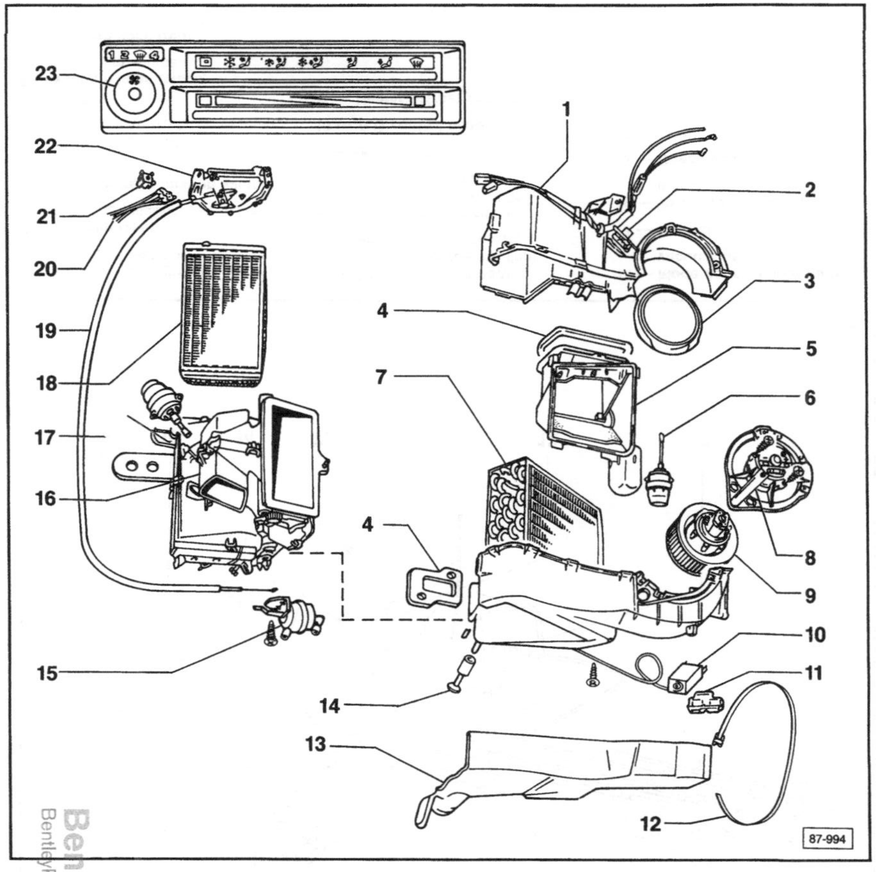

Note

A/C system must be discharged before removing and/or replacing components marked with an asterisk *.

1 — **Evaporator housing***

2 — **Fresh air blower series resistance**
- includes overheat fuse

3 — **Intake air ring***

4 — **Seal (watertight)***

5 — **Intake duct***

6 — **Vacuum servo***
- for fresh air/recirculation door
- insert rod into servo 68 mm (2.7 in.) before installing

7 — **Evaporator***

8 — **Fresh air blower housing**
- coat contact surfaces with silicone sealer before assembly
- note wire polarity code (+/−) on housing

9 — **Fresh air blower**
- to remove, loosen right side of dashboard and remove blower housing

10 — **A/C thermostat**
- *insert sensing tube into evaporator 330 mm* (13.0 in.)

11 — **Cap**

12 — **Tension strap**

13 — **Cover**

14 — **Water drain valve**
- identifying/removing, see Index

15 — **Vacuum servo**
- for mode door (central flap)

16 — **Air distribution case***

17 — **Vacuum servo**
- for footwell/defrost flap

18 — **Heater core***

19 — **Control cable**
- for temperature (air blend) flap

20 — **Vacuum hoses with connector**
- layout, see Index

21 — **Fresh air blower switch**

22 — **A/C control head**

23 — **A/C control head face plate**

> **WARNING**
>
> Automotive refrigerant containing CFC's is hazardous to the earth's atmosphere. To protect our environment, use an Underwriter's Laboratory (UL) approved refrigerant recovery/recycling unit such as **Kent-Moore ACR³**, or equivalent, whenever discharging an A/C system.

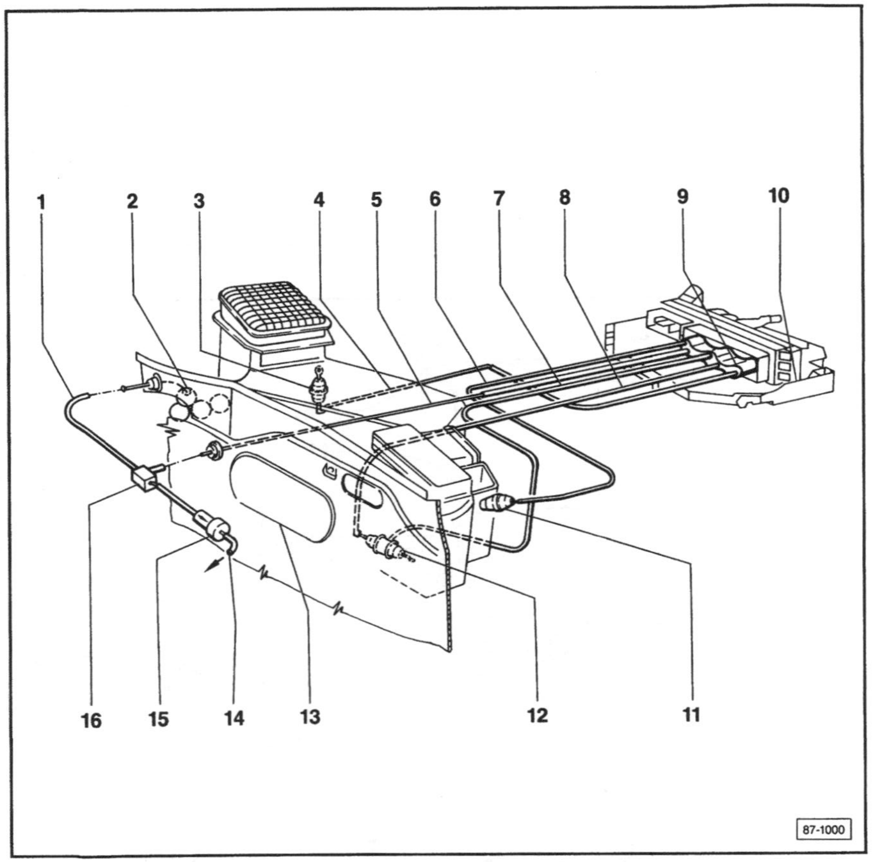

1 — **Vacuum hose (green)** to vacuum reservoir
2 — **Vacuum reservoir** removing/installing, see Index
3 — **Fresh air/recirculation vacuum servo**
4 — **Vacuum hose (white)** to fresh air/recirculation vacuum servo
5 — **Vacuum hose (black)** to vacuum source
6 — **Vacuum hose (red)** to heat/defrost vacuum servo
7 — **Vacuum hose (green)** to mode door (central flap) vacuum servo, side port
8 — **Vacuum hose (yellow)** to mode door (central flap) vacuum servo, top port
9 — **Vacuum hose connector**
10 — **A/C control head**
11 — **Heater/defrost vacuum servo**
12 — **Mode door (central flap) vacuum servo**
13 — **Firewall**
14 — **To vacuum supply (intake manifold)**
15 — **Check valve**
16 — **T-fitting**

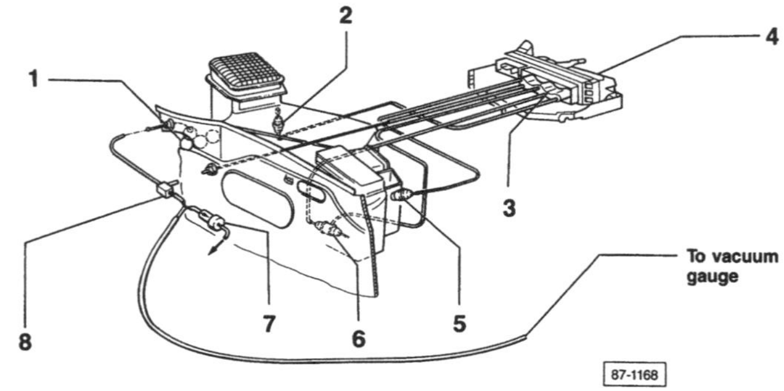

Vacuum system, checking

Refer to illustration 87-1168 above.

- connect vacuum gauge between vacuum T-fitting **8** and check valve **7**
- start engine and accelerate several times until approximately 600 mbar (18 in Hg) vacuum is obtained
- set upper control lever to "Defrost" position (right hand stop)
- shut engine **OFF**
  - vacuum is stored in system
  - vacuum measurement of 600 mbar (18 in. Hg) must be maintained for several minutes

If **NO**,

- check for vacuum leaks at vacuum reservoir **1**, vacuum hose connector **3**, control head **4**, and check valve **7** by pinching off vacuum supply hose leading to each component
- if vacuum leak stops when a hose is pinched, check all hoses, connections, and components supplied by that hose
- repair/replace hoses, connections, or components as necessary
- re-check vacuum system

Note

If upper A/C control lever is moved during test, vacuum in system will decrease due to actuation of one or more vacuum servos.

Vacuum servos, checking

Upper control lever setting	OFF	A/C MAX	A/C NORM	A/C BI-LEV	VENT	HEAT	DEFROST
Vacuum servo for fresh/air recirculation	V	V	NV	NV	NV	NV	NV
Vacuum servo for mode door (central flap)							
1st stage	V	V	V	V	V	NV	NV
2nd stage	V	V	V	NV	V	NV	NV
Vacuum servo for heat/defrost	NV	NV	NV	V	V	V	NV

87-1169

V — vacuum
NV — no vacuum

- remove lower instrument panel trim
- disconnect vacuum hose from fresh air/recirculation servo
- push actuation rod into servo and plug servo vacuum connection with finger
- release rod
  - actuation rod must not move (servo holds vacuum)

If rod moves (extends),
- vacuum leak in servo, replace servo

- repeat test with mode door (central flap) vacuum servo and heat/defrost vacuum servo
- re-connect vacuum hoses to vacuum servos

Note

Perform the following check only after checking vacuum servos.

- move upper A/C control lever to OFF position
- start engine
- move upper A/C control lever to each position one after the other
- compare vacuum servo setting with chart above

Note

A/C control head is defective when only one of specified settings cannot be obtained (vacuum supply and vacuum servo **OK**)

- switch ignition **OFF**
  - vacuum servos must remain in same position for several minutes

If **NO**,
- replace A/C control head

Vacuum reservoir, removing/installing

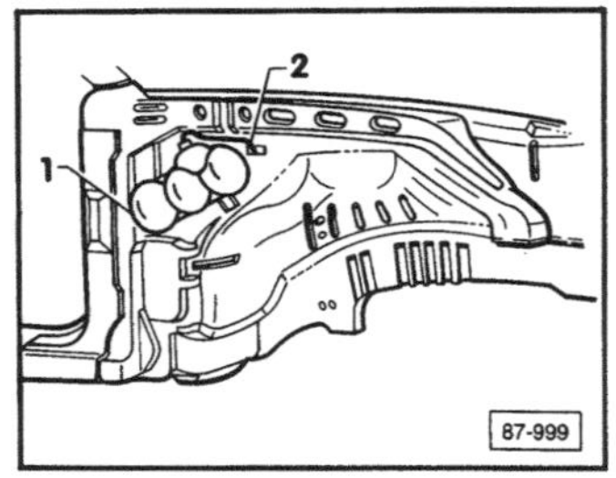

Removing

- remove right hand wheel housing liner
- detach vacuum hose **2** and remove vacuum reservoir **1**

Installing

- insert vacuum hose **2** into vacuum reservoir **1**, 30 mm (1.2 in.)
- install reservoir onto wheel housing
- re-install wheel housing liner

A/C system temperatures and pressures, checking

Observe the following:

- do not park car in sunlight
- condenser and radiator clean and free of obstructions
- ribbed drive belt in good condition and properly tensioned
- air ducts for condenser and radiator properly installed
- temperature (air blend) flap cable is properly adjusted
- A/C compressor clutch engages when A/C is switched on

Test preparations:

- connect pressure gauge set to low pressure and high pressure service valves
- open all air outlets on dashboard
- insert thermometer into left side air outlet
- run engine until warm and set speed to 2000 rpm
- set fresh air blower speed to **4** (high)
- move upper control head lever to MAX A/C
- set temperature control lever to FULL COLD (far left)
- close windows and doors
- check A/C system performance according to the following charts

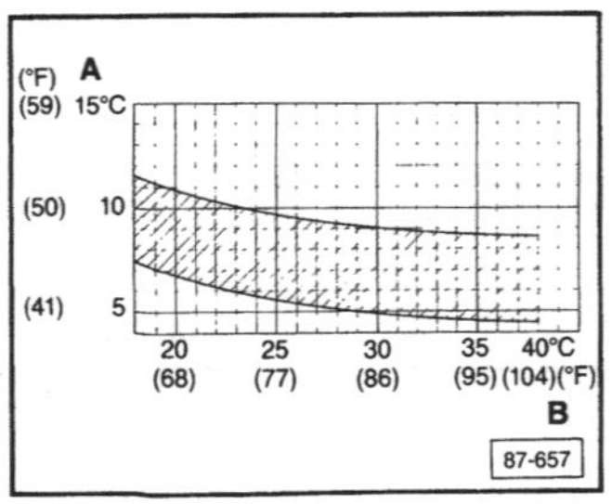

Outlet air temperature at left hand vent vs. ambient temperature

A — Outlet air temperature at left hand vent
B — Ambient temperature

- read temperature at vent after A/C compressor has cycled off for the first time
- compare temperature reading with chart at left

Example:
When outside temperature is 86°F, left hand vent outlet air temperature should be between 41°F - 48°F.

If specified readings are not obtained,
- check A/C system pressures, following

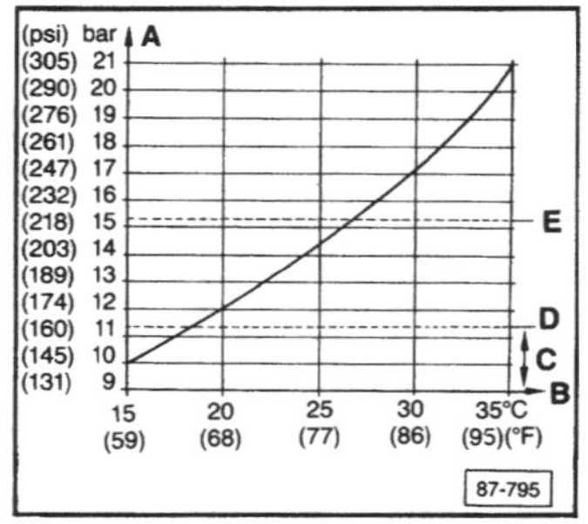

A/C system high pressure vs. ambient temperature

A — Discharge pressure (high side)
B — Ambient (outside) temperature
C — Radiator cooling fan on low speed
D — Radiator cooling fan high speed switches **OFF**
E — Radiator cooling fan high speed switches **ON**

- compare actual A/C system readings with chart at left

Example:
When outside temperature is 86°F, A/C system high side pressure should be approximately 247 psi. Radiator cooling fan on high speed.

If specified readings are not obtained,
- check A/C refrigerant charge

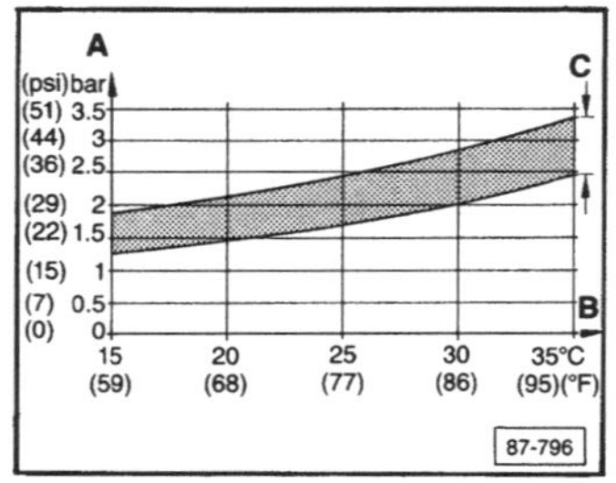

A/C system low pressure vs. ambient temperature

A — Suction pressure (low side)
B — Ambient (outside) temperature
C — A/C thermostat
- switch closes at upper level arrow (A/C compressor **ON**)
- switch opens at lower level arrow (A/C compressor **OFF**)

- compare actual A/C system readings with chart at left

Example:
When outside temperature is 86°F, A/C system low side pressure should be between 29 psi - 41 psi.

If specified readings are not obtained,
- check A/C refrigerant charge

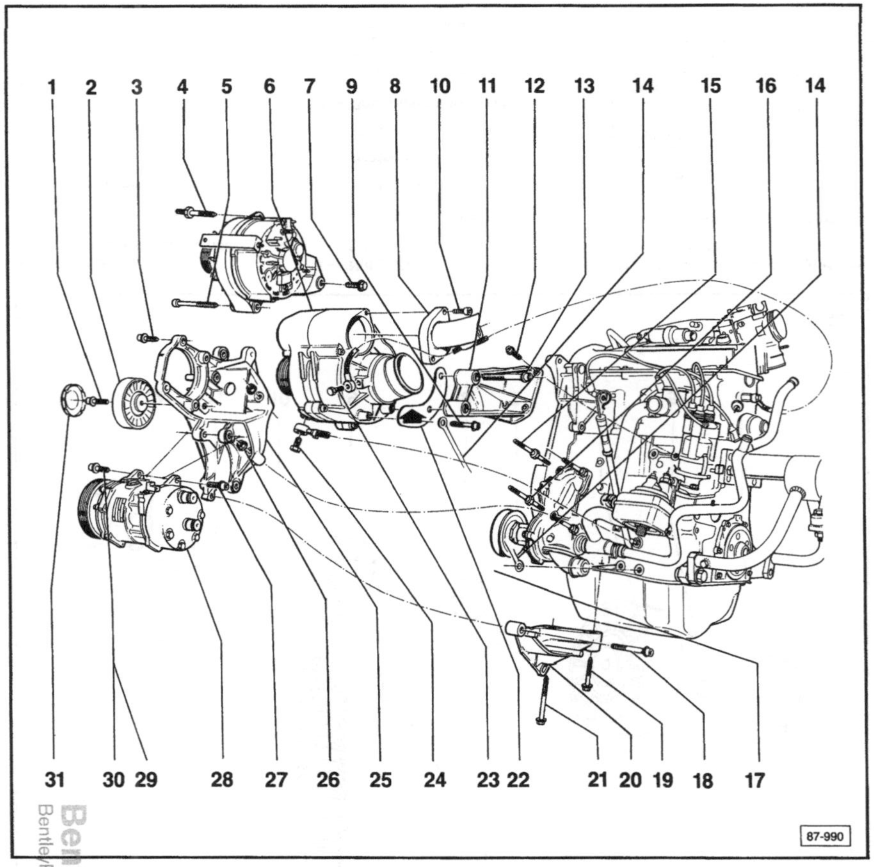

CAUTION

Part numbers are for reference only. Always check with your Parts Department for latest parts information.

1 — **Bolt, socket head** (M8 x 20 mm)
- 25 Nm (18.4 ft lb)

2 — **Idler pulley**

3 — **Bolt, socket head with point** (M8 x 40 mm)
- 25 Nm (18.4 ft lb)

4 — **Stud** (M8)
- 30 Nm (22.1 ft lb)

5 — **Bolt, socket head with point** (M8 x 100 mm)
- 35 Nm (25.8 ft lb)

6 — **G-charger**

7 — **Bolt, with point** (M8 x 38 mm)
- 25 Nm (18.4 ft lb)

8 — **G-charger by-pass with O-ring**

9 — **Bolt, socket head with point** (M8 x 28 mm)
- 25 Nm (18.4 ft lb)

10 — **Bolt, socket head** (M6 x 22 mm)

11 — **Support bracket**
- *tap all bushings flush with rear of bracket before* installing
- install O-ring (28 mm inside diameter x 1.8 mm thickness) between engine and support bracket

12 — **Bolt, socket head with point** (M8 x 28 mm)
- 25 Nm (18.4 ft lb)

13 — **Bolt, socket head with point** (M8 x 65 mm)
- replace with part number N 901 442 01
- 25 Nm (18.4 ft lb)

14 — **Support strut**

15 — **Stud** (M8 x 55 mm/34 mm)
- 20 Nm (14.8 ft lb) + ¼ turn (90°) more

16 — **Stud** (M8 x 55 mm/50 mm)
- 20 Nm (14.8 ft lb) + ¼ turn (90°) more

17 — **Bolt, with point** (M10 x 38 mm) (not shown)
- 45 Nm (33.2 ft lb)

18 — **Bolt** (M8 x 50 mm)
- 25 Nm (18.4 ft lb)

19 — **Bolt, with shoulder and point** (M8 x 50 mm)
- 25 Nm (18.4 ft lb)

20 — **Power steering pump bracket**

21 — **Bolt, with shoulder and point** (M8 x 85 mm)
- 25 Nm (18.4 ft lb)

22 — **Support plate**
- install so that straight edge of plate (**arrow**) mounts against G-charger and curved edge mounts on support bracket toward engine

23 — **Banjo bolt** (M8 x 19 mm)
- use a sealing washer on each side of fitting

24 — **Banjo bolt** (M10 x 19 mm)
- use a sealing washer on each side of fitting

25 — **G-charger bracket**
- push all bushings flush with rear of bracket before installing

26 — **Self-locking nut** (M8)
- replace with part number N 901 838 01
- 35 Nm (25.8 ft lb)

27 — **Bolt, socket head with lock washer** (M10 x 30 mm)
- 45 Nm (33.2 ft lb)

28 — **A/C compressor**

29 — **Bolt, socket head with washer** (M10 x 50 mm)
- 45 Nm (33.2 ft lb)

30 — **Bolt, socket head with washer** (M10 x 38 mm)
- 45 Nm (33.2 ft lb)

31 — **Idler pulley cap**

WARNING

Automotive refrigerant containing CFC's is hazardous to the earth's atmosphere. To protect our environment, use an Underwriter's Laboratory (UL) approved refrigerant recovery/recycling unit such as **Kent-Moore ACR³**, or equivalent, whenever discharging an A/C system.

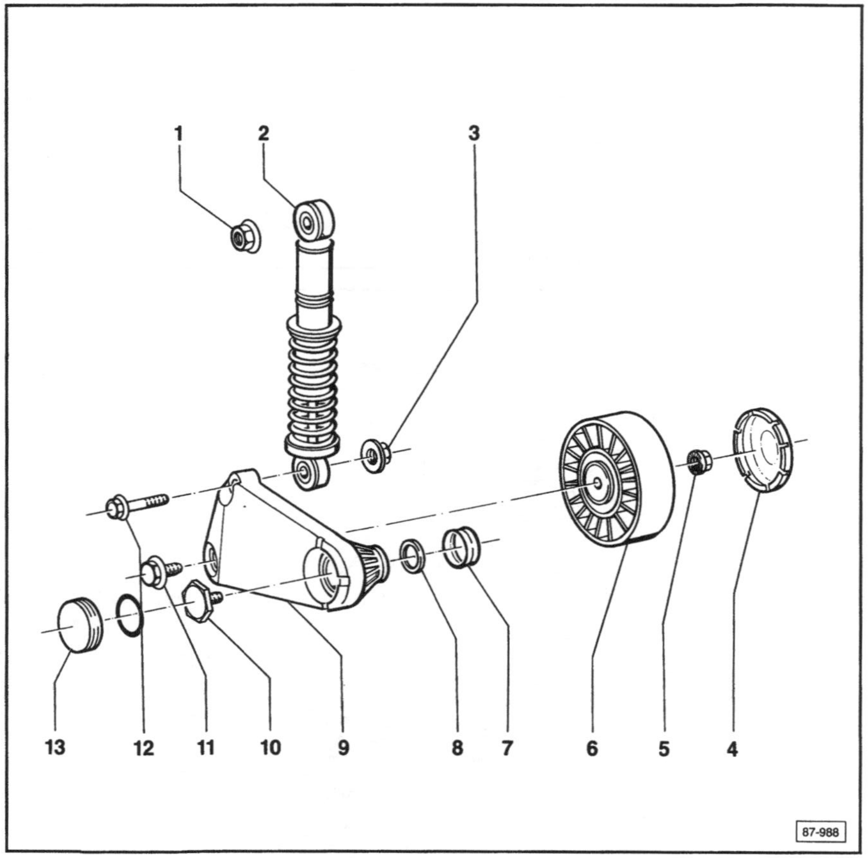

1 — **Nut** (M8)
- replace with part number N 902 920 01
- 25 Nm (18.4 ft lb)

2 — **Belt tensioning damper**

3 — **Nut** (M8)
- replace with part number N 902 920 01
- 25 Nm (18.4 ft lb)

4 — **Idler pulley cover**

5 — **Nut** (M8)
- replace with part number N 901 838 01
- 25 Nm (18.4 ft lb)

6 — **Idler pulley**

7 — **Protective sleeve**

8 — **Washer**

9 — **Belt tensioning bracket**

10 — **Bolt** (M8 x 18 mm)
- replace with part number N 904 056 01
- 25 Nm (18.4 ft lb)

11 — **Bolt, with shoulder and point** (M8 x 20 mm)

12 — **Bolt, with shoulder and point** (M8 x 45 mm)

13 — **Protective cap**

CAUTION

Part numbers are for reference only. Always check with your Parts Department for latest parts information.

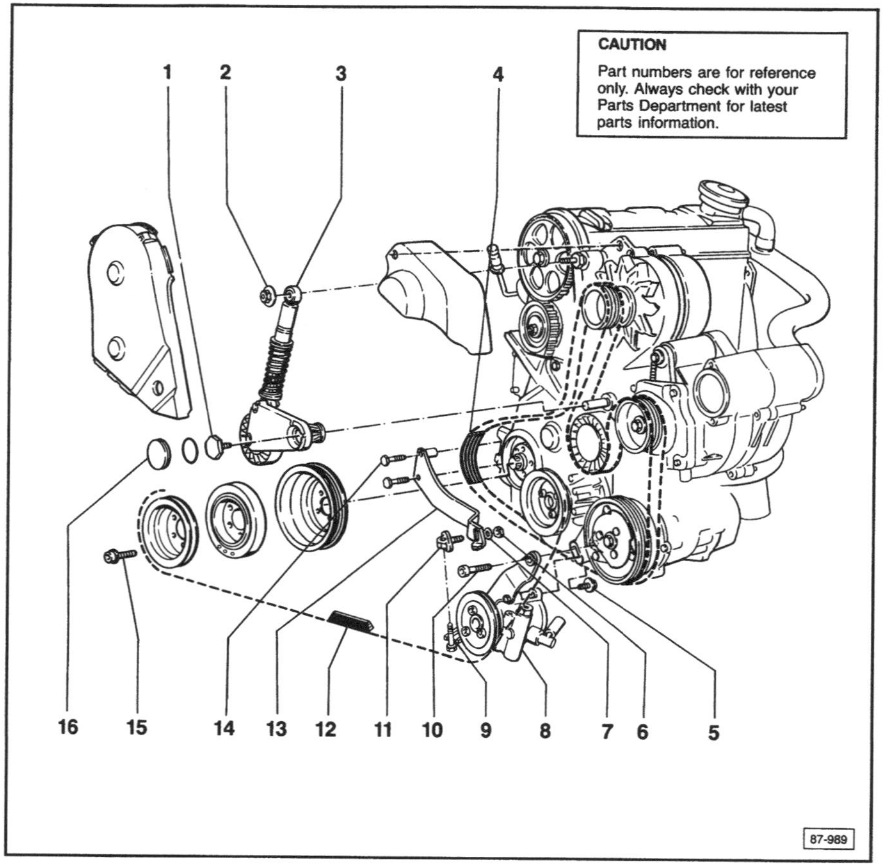

CAUTION

Part numbers are for reference only. Always check with your Parts Department for latest parts information.

1 — **Bolt** (M8 x 18 mm)
- replace with part number N 904 056 01
- 25 Nm (18.4 ft lb)

2 — **Nut** (M8)
- replace with part number N 902 920 01
- 25 Nm (18.4 ft lb)

3 — **Belt tensioning damper**
- use special tool **3191** to adjust tension

4 — **Ribbed belt**
- 21.36 mm x 1755 mm (0.84 in. x 69.1 in.)

5 — **Bolt, with shoulder and point** (M8 x 22 mm)
- 25 Nm (18.4 ft lb)

6 — **Nut** (M8)
- 25 Nm (18.4 ft lb)

7 — **Washer**

8 — **Power steering pump**

9 — **Adjusting bolt**

10 — **Bolt, socket head** (M8 x 45 mm)
- 45 Nm (33.2 ft lb)

11 — **Eye bolt**

12 — **V-belt**
- 9.5 mm x 735 mm (3/8 in. x 29.0 in.)

13 — **G-charger tensioner bracket**

14 — **Bolt** (M8 x 40 mm)
- 25 Nm (18.4 ft lb)

15 — **Bolt, socket head** (M8 x 45 mm)
- 25 Nm (18.4 ft lb)

16 — **Protective cap**

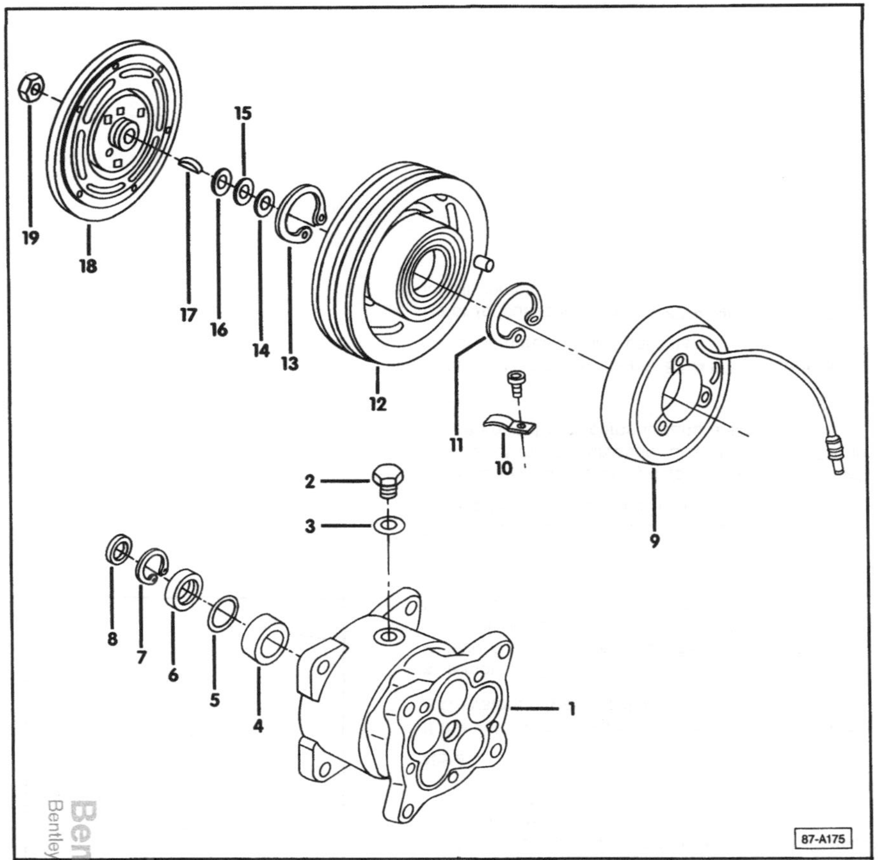

1 — Compressor block
2 — Oil drain plug
3 — Drain plug O-ring
4 — Shaft seal
5 — Shaft seal O-ring
6 — Shaft seal seat
7 — Snap ring
8 — Felt ring
9 — Clutch coil
10 — Wire clamp
11 — Snap ring — coil
12 — Clutch pulley
13 — Pulley snap ring
14 — Shim
15 — Shim
16 — Shim
17 — Shaft key
18 — Clutch plate
19 — 19 mm nut

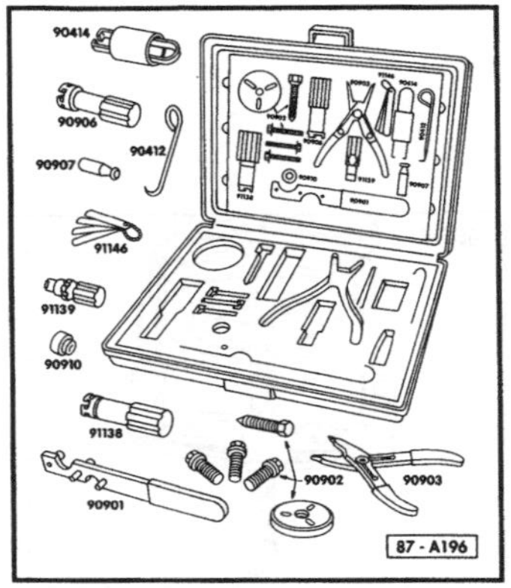

A/C compressor clutch, removing/installing

Tools required

Use Mastercool A/C compressor service kit No. **91296** or equivalent. Kit contains:

Tool No.	Description
90902	Clutch plate puller
90414	Shaft seal seat remover/installer • SD-508 only
90412	O-ring hook
90906	Shaft seal remover/installer • SD-508 only
90910	Compressor shaft protector
91139	Shaft seal seat remover/installer • SD-709 only
91146	Feeler gauges
91138	Shaft seal remover/installer • SD-709 only
90903	Snap ring pliers
90907	Shaft seal protector
90901	Spanner wrench

Removing

WARNING

Follow all recommended safety precautions listed on page 87.2 and 87.3 before performing any A/C system servicing.

Automotive refrigerants containing CFC's are hazardous to the earth's atmosphere. To protect our environment, use an Underwriters Laboratory (UL) approved refrigerant recovery/recycling unit such as Kent-Moore ACR[3] or equivalent, whenever discharging an A/C system.

The A/C system should only be serviced by trained personnel familiar with:
- equipment use,
- related safety procedures,
- regulations governing the discharging, handling and disposal of automotive refrigerants.

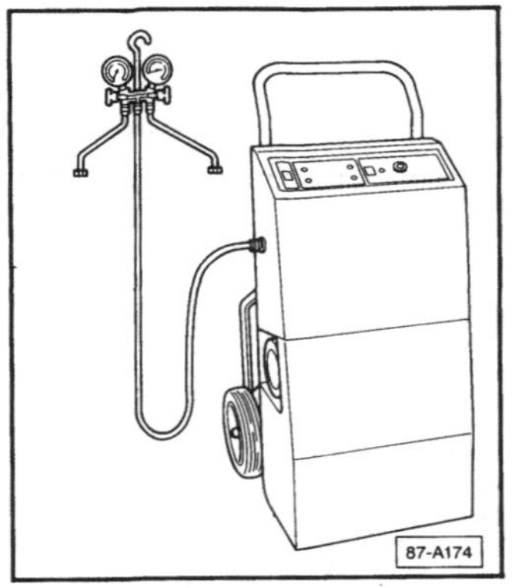

CAUTION

Some A/C compressor replacement parts are not available. Always check with your Parts Department for latest parts information before attempting compressor repairs.

CAUTION

Always follow manufacturer's instructions when using a refrigerant recovery/recycling unit.

- discharge A/C refrigerant system using refrigerant recovery/recycling unit Kent-Moore ACR[3] or equivalent.

Note

If the vehicle is received with no refrigerant in the A/C system, always replace the receiver drier.

- remove A/C lines from compressor
- plug A/C lines with a suitable plug to prevent moisture contamination of receiver drier
- remove compressor from vehicle

Note

Perform all A/C compressor repairs on a clean dry service bench equipped with a vise.

When servicing A/C compressors, verify the model number of the compressor, for example SD-709, to insure correct replacement parts are ordered.

Note

Refer to illustration number **87-A175** on page 87.22 for all compressor part callouts.

Refer to illustration number **87-A196** on page 87.23 for all compressor tool callouts.

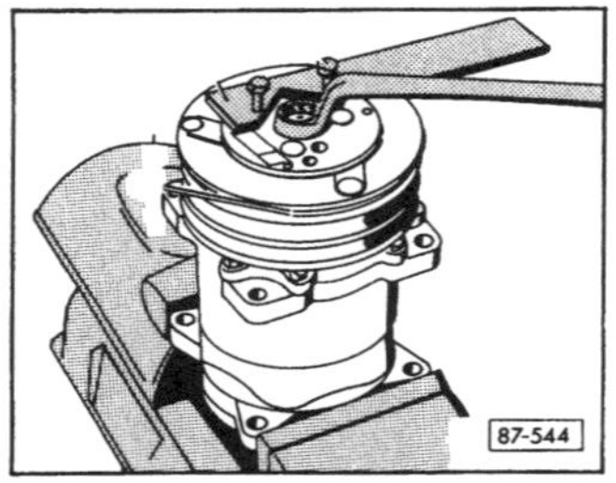

- insert spanner wrench (**Mastercool tool No. 90901** or equivalent) into clutch plate 18 of compressor
- hold plate 18 stationary with spanner and remove 19 mm hex nut **19** from compressor shaft

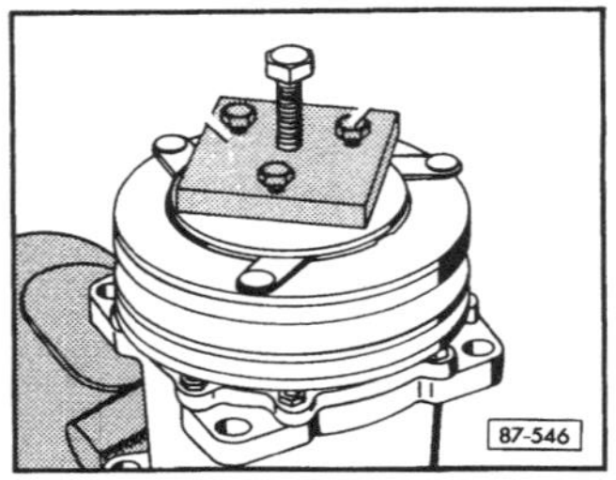

- remove front clutch plate **18** with clutch plate puller (**Mastercool tool No. 90902** or equivalent - similar tool shown)
- if equipped, carefully remove bearing dust cover with screwdriver

CAUTION

Do not to distort dust cover when removing.

- remove shaft key **17** by lightly tapping with slotted screwdriver and small hammer
- remove all clutch shims **14**, **15**, **16** from compressor shaft
- remove bearing snap ring **13** with snap ring pliers (**Mastercool tool No. 90903** or equivalent)

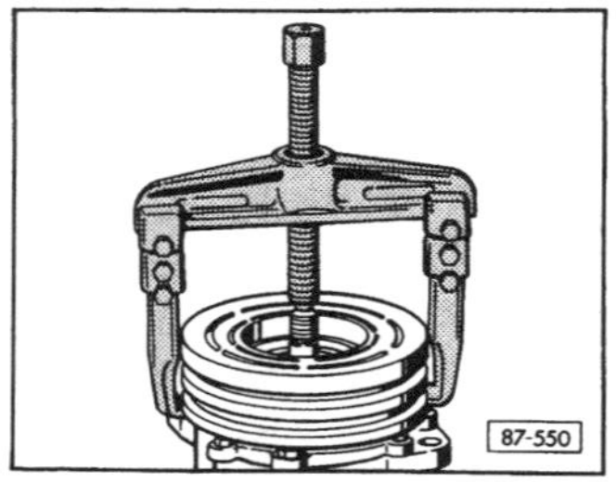

- remove clutch pulley **12** using puller **US 1078**
- loosen clutch field coil wire clamp **10** and connector holder, if equipped
- remove clutch field coil retainer snap ring **11** with snap ring pliers (**Mastercool tool No. 90903** or equivalent)
- remove clutch field coil **9**

Installing

- install clutch field coil **9**
    - clutch field coil flange protrusion must match hole in front of compressor housing to prevent coil movement and to correctly locate power leads for attachment
- install clutch field coil retainer snap ring **11** with snap ring pliers
- attach field coil wire clamp **10** and connector holder, if equipped

- support compressor in bench vise on mounting tabs at base of compressor
- align pulley **12** on front housing hub squarely
- place driver kit **VW 519** on pulley
  - ensure outer edge of driver ring rests firmly on inner race of pulley bearing

- while guiding pulley **12** to prevent binding, tap end of **VW 519** with small hammer until pulley bottoms out against compressor front housing hub

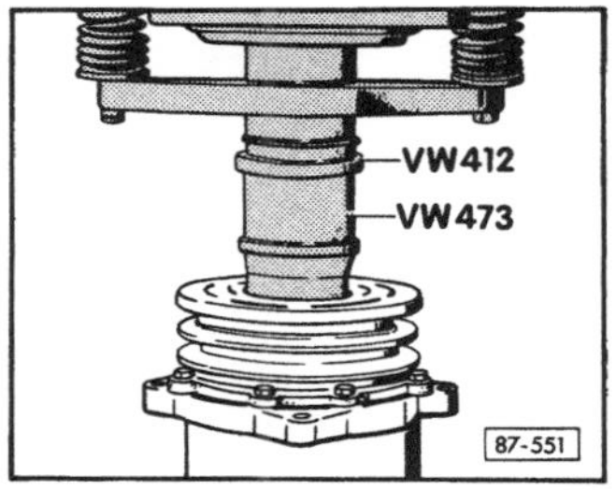

- as an alternate method, press pulley onto compressor shaft using spacers **VW 412** and **VW 473**
- install snap ring **13** with snap ring pliers (**Mastercool tool No. 90903** or equivalent)
- install shims **14, 15, 16**
- install shaft key **17** using small hammer and screwdriver
- install bearing dust cover, if equipped
- align front clutch plate keyway to compressor shaft key **17**
- install front clutch plate **18** onto compressor shaft

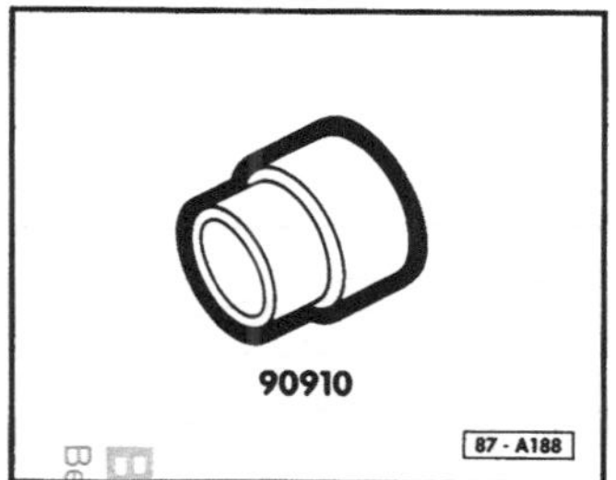

- install shaft protector (**Mastercool tool No. 90910** or equivalent) onto compressor shaft and tap with a small hammer until clutch plate **18** bottoms on shims **14, 15, 16**

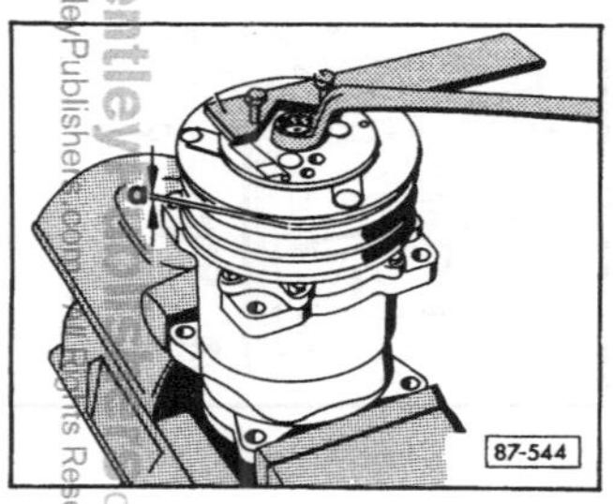

- install 19 mm hex nut **19** onto compressor shaft
- insert spanner wrench (**Mastercool tool No. 90901** or equivalent) into holes of front clutch plate **18** and hold stationary

> **CAUTION**
>
> Use a beam type torque wrench. Damage may result from use of inaccurate "click" wrenches.

- tighten hex nut **19**
  - 30-34 Nm (22-25 ft lb)

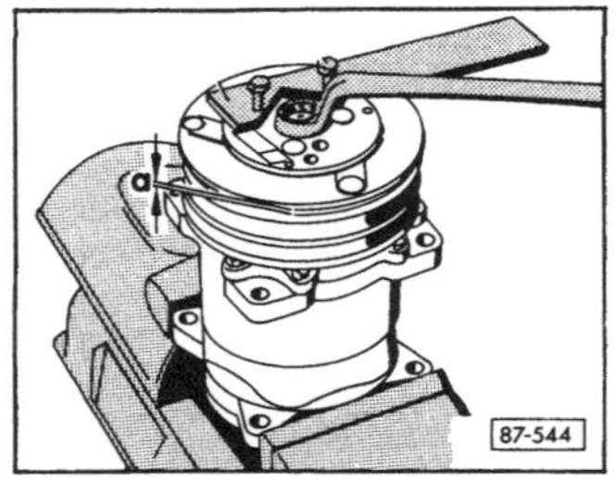

Note

Air gap **a** is determined by spacer shims **14, 15, 16**. When installing a clutch assembly, try the original shims first. When installing a new clutch on a new compressor use the 1 mm, 0.5 mm, and the 0.127 mm shims provided in the clutch installation kit. If the air gap does not meet specification, add or remove shims accordingly.

- check air gap **a** of clutch with feeler gauge (**Mastercool tool No. 91146** or equivalent)
  - **a** = 0.400 mm to 0.800 mm gap (0.016 in. to 0.031 in.)

Note

If air gap **a** is not consistent around the circumference, lightly pry up on the front face plate at the low spots and lightly tap down on the high spots with a small hammer.

- lubricate all O-rings with clean refrigerant oil
- install compressor in vehicle
- remove plugs from A/C lines

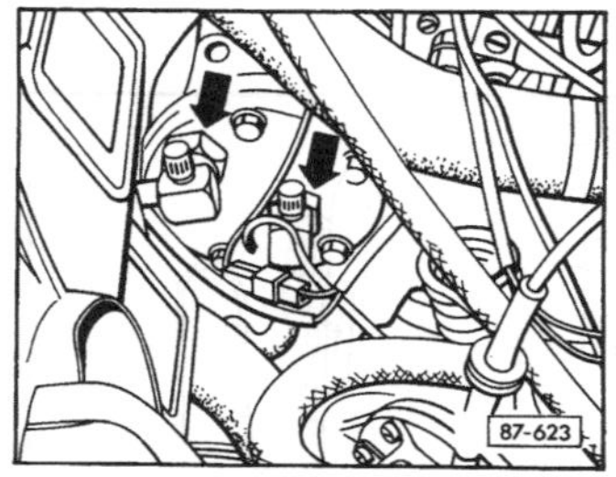

- install A/C lines (**arrows**) on compressor and tighten as follows:
  - suction line: 35 Nm (26 ft lb)
  - discharge line : 25 Nm (18 ft lb)

- install and properly tension compressor drive belt
- evacuate and charge A/C system
- check compressor clutch and A/C system for proper operation

A/C compressor shaft seal, removing/installing

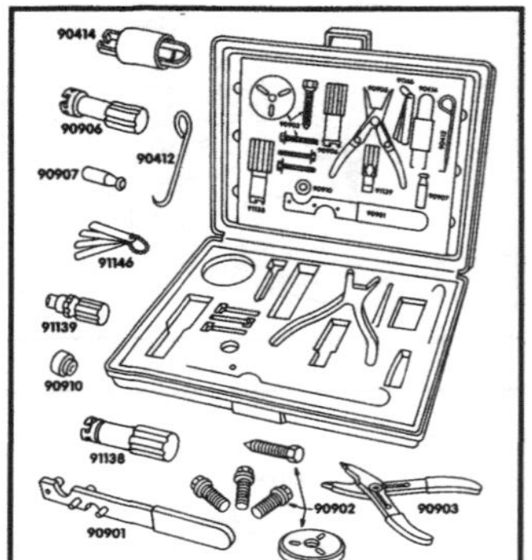
87 - A196

Tools required

Use Mastercool A/C compressor service kit No. **91296** or equivalent. Kit contains:

Tool No.	Description
90902	Clutch plate puller
90414	Shaft seal seat remover/installer • SD-508 only
90412	O-ring hook
90906	Shaft seal remover/installer • SD-508 only
90910	Compressor shaft protector
91139	Shaft seal seat remover/installer • SD-709 only
91146	Feeler gauges
91138	Shaft seal remover/installer • SD-709 only
90903	Snap ring pliers
90907	Shaft seal protector
90901	Spanner wrench

Removing

WARNING

Follow all recommended safety precautions listed on page 87.2 and 87.3 before performing any A/C system servicing.

Automotive refrigerants containing CFC's are hazardous to the earth's atmosphere. To protect our environment, use an Underwriters Laboratory (UL) approved refrigerant recovery/recycling unit such as Kent-Moore ACR³ or equivalent, whenever discharging an A/C system.

The A/C system should only be serviced by trained personnel familiar with:
- equipment use,
- related safety procedures,
- regulations governing the discharging, handling and disposal of automotive refrigerants.

CAUTION

Some A/C compressor replacement parts are not available. Always check with your Parts Department for latest parts information before attempting compressor repairs.

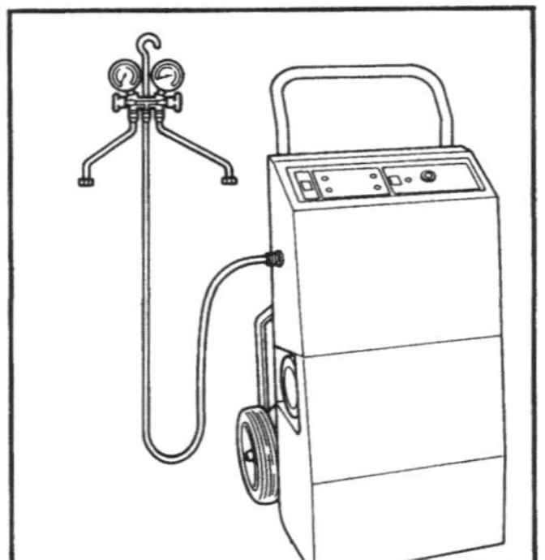
87-A174

CAUTION

Always follow manufacturer's instructions when using a refrigerant recovery/recycling unit.

- discharge A/C refrigerant system using refrigerant recovery/recycling unit Kent-Moore ACR³ or equivalent

Note

If the vehicle is received with no refrigerant in the A/C system, always replace the receiver drier.

- remove A/C lines from compressor
- plug A/C lines with a suitable plug to prevent moisture contamination of receiver drier
- remove compressor from vehicle

Note

Perform all A/C compressor repairs on a clean dry service bench equipped with a vise.

When servicing A/C compressors, verify the model number of the compressor, for example SD-709, to insure correct replacement parts are ordered.

Note

Refer to illustration number **87-A175** on page 87.22 for all compressor part callouts.

Refer to illustration number **87-A196** on page 87.28 for all compressor tool callouts.

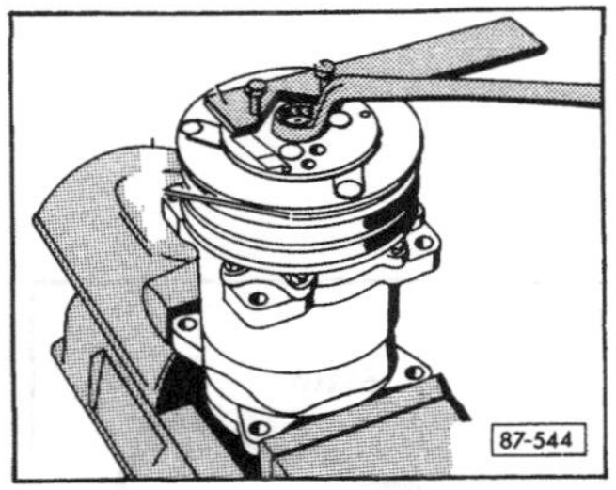

- insert spanner wrench (**Mastercool tool No. 90901** or equivalent) into clutch plate **18** of compressor
- hold plate **18** stationary with spanner and remove 19 mm hex nut **19** from compressor shaft

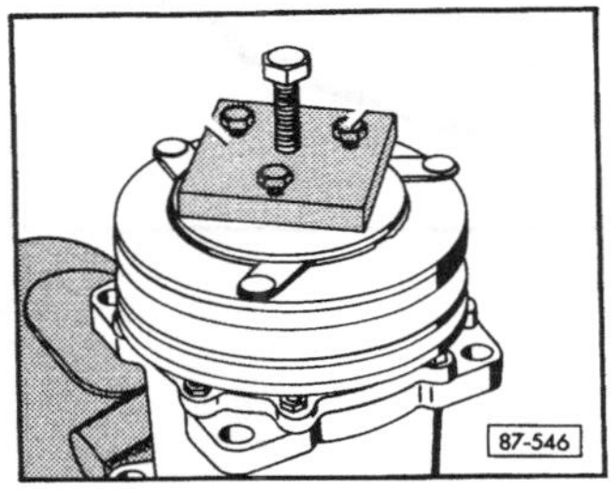

- remove front clutch plate **18** with clutch plate puller (**Mastercool tool No. 90902** or equivalent - similar tool shown)
- if equipped, carefully remove bearing dust cover with screwdriver

> **CAUTION**
>
> Do not distort dust cover when removing.

- remove shaft key **17** by lightly tapping with slotted screwdriver and small hammer
- remove all clutch shims **14**, **15**, **16** from compressor shaft

> **CAUTION**
>
> Shaft seal replacement must be performed on a clean, dry surface. Do not use old parts on the shaft seal assembly. Never touch the shaft seal with your fingers. Oil from skin contact will cause seal deterioration. Always use the metal carrier to handle the seal assembly.

- remove felt ring **8** with small screwdriver or O-ring hook
- remove shaft seal retainer snap ring **7** with snap ring pliers (**Mastercool tool No. 90903** or equivalent)

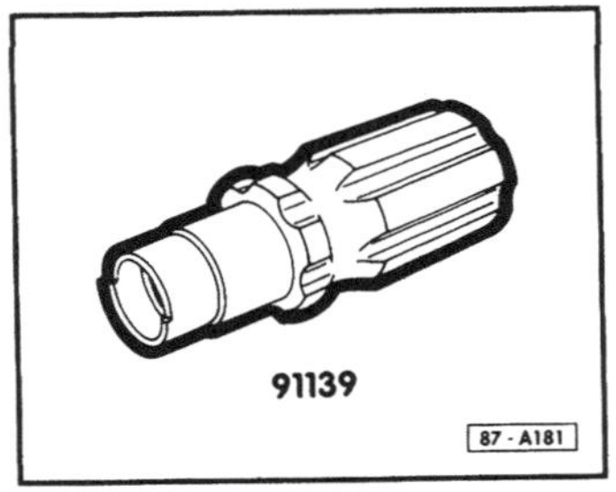

- remove shaft seal seat **6** with seal seat remover/installer tool (**Mastercool tool No. 91139** or equivalent)

> **CAUTION**
>
> Do not scratch the O-ring groove. Scratches will cause refrigerant leaks in the seal area.

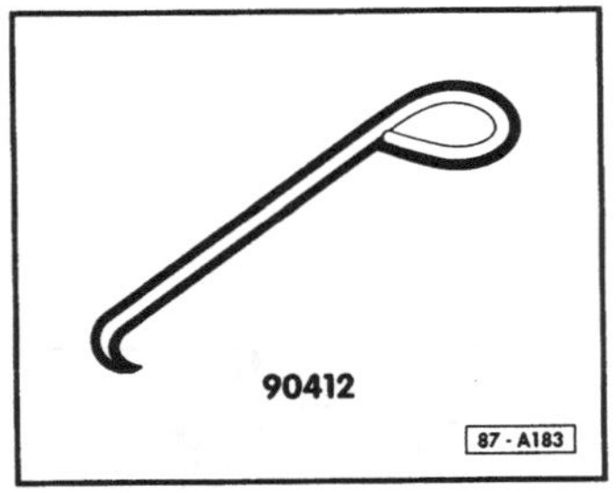

- carefully remove O-ring from seal seat using O-ring hook (**Mastercool tool No. 90412** or equivalent)

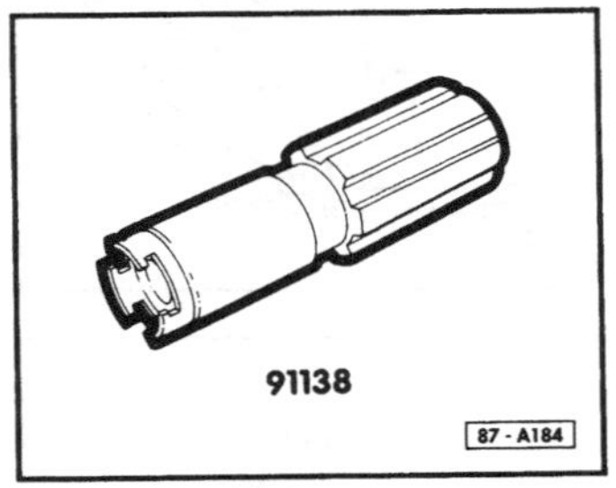

- insert shaft seal remover/installer tool (**Mastercool tool No. 91138** or equivalent) onto compressor shaft and against seal assembly
- press down seal remover/installer tool on seal assembly and twist tool until it engages slots on seal
- pull out shaft seal assembly **4**
- clean seal cavity thoroughly with a suitable cleaning fluid, blow out cavity with dry pressurized air

Note

Insure that dirt and all foreign substances are removed from the seal cavity area.

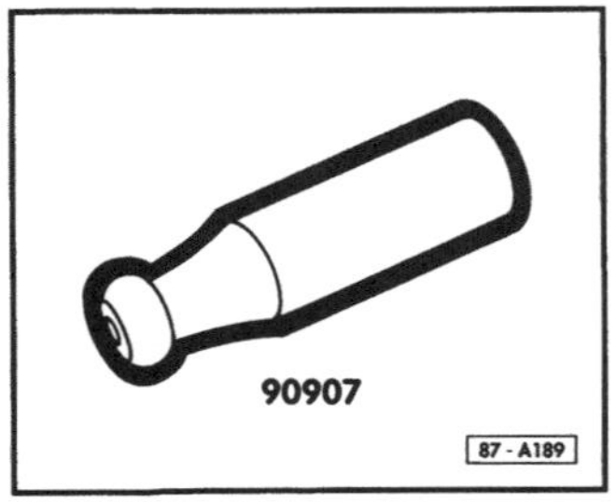

Installing

- install shaft seal protector (**Mastercool tool No. 90907** or equivalent) over compressor shaft
- dip new shaft seal assembly **4** in clean, new refrigerant oil

> **CAUTION**
> Do not touch the sealing surfaces of the shaft seal. Skin oil will contaminate the seal and cause it to leak.

- engage seal remover/installer tool (**Mastercool tool No. 91138**, or equivalent) into slots of shaft seal assembly **4**
- insert shaft seal **4** into compressor seal cavity
- twist installer tool in opposite direction to disengage tool from shaft seal
- remove tool and shaft seal protector from compressor shaft
- coat new O-ring **5** with new clean refrigerant oil

> **CAUTION**
> Do not scratch O-ring groove when installing O-ring.

- install new O-ring on new seal seat and coat with clean refrigerant oil
- coat seal seat **6** with clean, new refrigerant oil

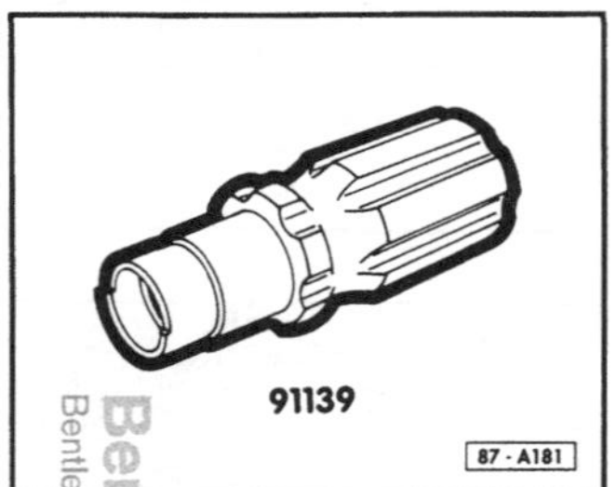

- install seal seat **6** with remover/installer tool (**Mastercool tool No. 91139**, or equivalent)
- lightly press against seal seat **6** with tool to insure proper seating of seal seat
- install snap ring **7** with snap ring pliers (**Mastercool tool No. 90903** or equivalent)
  - beveled edge of snap ring lies outward (toward clutch) and flat side towards compressor
- lightly tap snap ring into groove to ensure snap ring is fully seated
- dip new felt ring **8** into fresh, clean refrigerant oil and install onto compressor shaft
- using a small hammer, lightly tap new felt ring **8** into place
- install shims **14**, **15**, **16**
- install shaft key **17** using small hammer and screwdriver

Air Conditioner

- install bearing dust cover, if equipped
- align front clutch plate keyway to compressor shaft key **17**
- install front clutch plate **18** onto compressor shaft

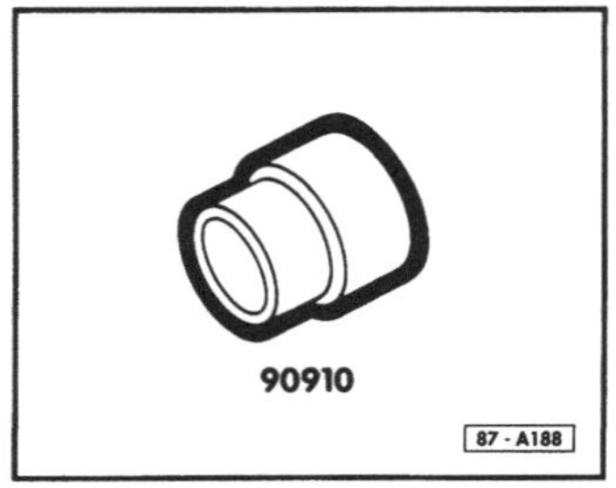

- install shaft protector (**Mastercool tool No. 90910** or equivalent) onto compressor shaft and tap with a small hammer until clutch plate **18** bottoms on shims **14**, **15**, **16**

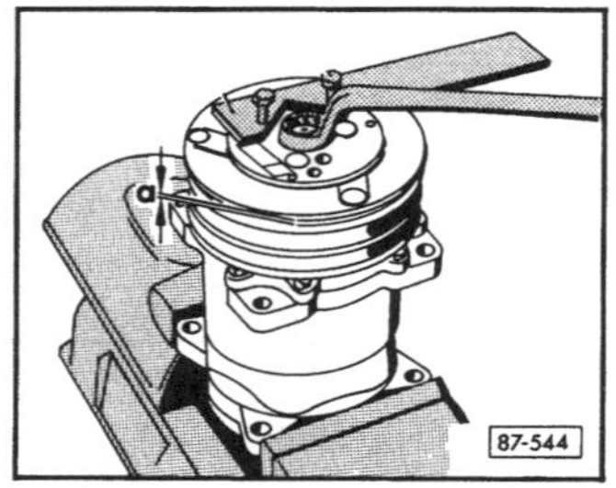

- install 19 mm hex nut **19** onto compressor shaft
- insert spanner wrench (**Mastercool tool No. 90901** or equivalent) into holes of front clutch plate **18** and hold stationary

> **CAUTION**
> Use a beam type torque wrench. Damage may result from use of inaccurate "click" wrenches.

- tighten hex nut **19**
  - 30-34 Nm (22-25 ft lb)

Note

Air gap **a** is determined by spacer shims **14, 15, 16**. When installing a clutch assembly, try the original shims first. When installing a new clutch on a new compressor use the 1 mm, 0.5 mm, and the 0.127 mm shims provided in the clutch installation kit. If the air gap does not meet specification, add or remove shims accordingly.

- check air gap **a** of clutch with feeler gauge (**Mastercool tool No. 91146** or equivalent)
  - **a** = 0.400 mm to 0.800 mm gap (0.016 in. to 0.031 in.)

Note

If air gap **a** is not consistent around the circumference, lightly pry up on the front face plate at the low spots and lightly tap down on the high spots with a small hammer.

- *lubricate all O-rings with clean refrigerant oil*
- wash exterior case of compressor with hot soapy water to remove any CFC residues
- dry compressor with dry pressurized air
- install compressor in vehicle
- remove plugs from A/C lines

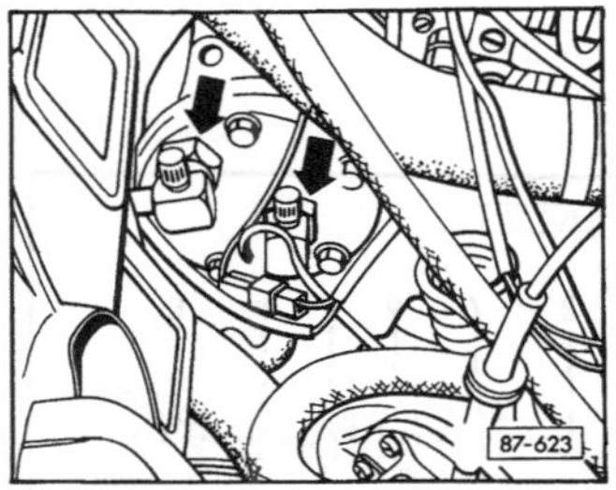

- install A/C lines (**arrows**) on compressor and tighten as follows:
  - suction line: 35 Nm (26 ft lb)
  - discharge line : 25 Nm (18 ft lb)

- install and properly tension compressor drive belt
- evacuate and charge A/C system
- operate A/C system and check front seal for leaks

Note

If the front bearing seal still leaks after a new seal has been installed, replace the compressor. No further repairs are possible to the seal; the mating surfaces are damaged beyond repair.

A/C compressor valve plate, head and gasket, removing/installing

Removing

WARNING

Follow all recommended safety precautions listed on page 87.2 and 87.3 before performing any A/C system servicing.

Automotive refrigerants containing CFC's are hazardous to the earth's atmosphere. To protect our environment, use an Underwriters Laboratory (UL) approved refrigerant recovery/recycling unit such as Kent-Moore ACR[3] or equivalent, whenever discharging an A/C system.

The A/C system should only be serviced by trained personnel familiar with:
- equipment use,
- related safety procedures,
- regulations governing the discharging, handling and disposal of automotive refrigerants.

CAUTION

Some A/C compressor replacement parts are not available. Always check with your Parts Department for latest parts information before attempting compressor repairs.

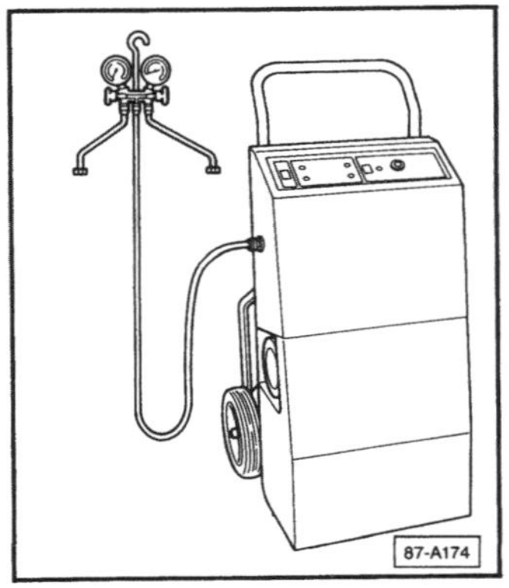

CAUTION

Always follow manufacturer's instructions when using a refrigerant recovery/recycling unit.

- discharge A/C refrigerant system using refrigerant recovery/recycling unit Kent-Moore ACR[3] or equivalent

Note

If the vehicle is received with no refrigerant in the A/C system, always replace the receiver drier.

- remove A/C lines from compressor
- plug A/C lines with a suitable plug to prevent moisture contamination of receiver drier
- remove compressor from vehicle

Note

Perform all A/C compressor repairs on a clean dry service bench equipped with a vise.

When servicing A/C compressors, or filling with new oil, always verify the model number of the compressor (shown on compressor case or block), for example SD-709, to insure the correct quantity of oil and correct tool and parts are used.

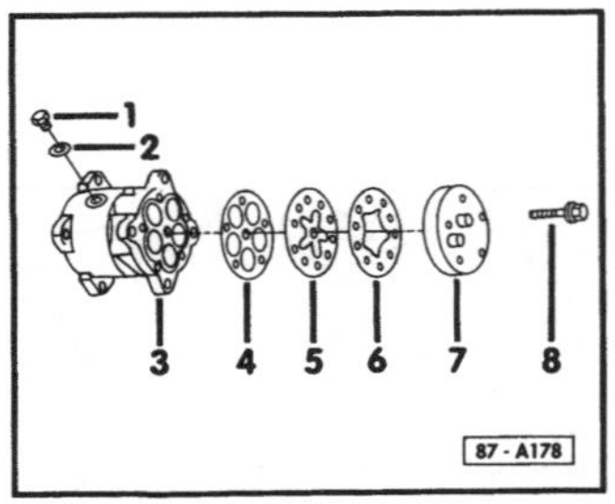

- remove drain plug **1** and O-ring **2** from compressor
- inspect O-ring **2** for damage, replace if necessary
- drain refrigerant oil from compressor
- using measuring cup from recycling equipment, measure quantity of oil removed from compressor and quantity of oil removed during refrigerant discharging with Kent-Moore ACR[3], record the total amount then discard used oil
- remove compressor cylinder head bolts **8** using 13 mm wrench
- remove compressor cylinder head **7**

Note

If necessary, use a small hammer and gasket scraper to lightly tap the outer edge of the cylinder head until it is free.

- inspect cylinder head **7** for damage, replace if necessary

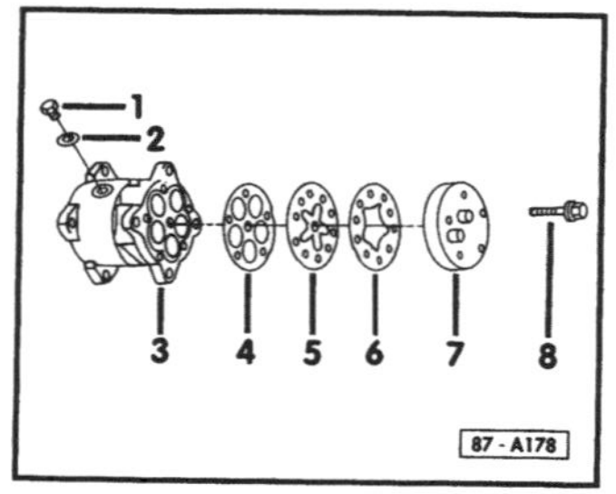

Note

The cylinder head gasket **6** normally sticks to the valve plate **5**.

CAUTION

Do not damage cylinder block, valve plate or head mating surfaces when removing head, valve plate and gaskets from compressor.

- using a gasket scraper, carefully remove any gasket material from cylinder head
- remove valve plate **5** and valve plate gasket **4** from compressor
- inspect compressor cylinder block **3** for damage and **carefully** remove any gasket material remaining on the compressor cylinder block

Installing

- coat cylinder block **3** mating surface with clean refrigerant oil
- install new valve plate gasket **4** insuring that gasket aligns with locator pins on cylinder block **3**
- install new valve plate **5** in same manner
- coat top of valve plate **5** with clean refrigerant oil
- install and align new head gasket **6** on valve plate **5** locator pins
- coat cylinder head **7** mating surface with clean refrigerant oil
- install cylinder head **7** and head bolts **8**

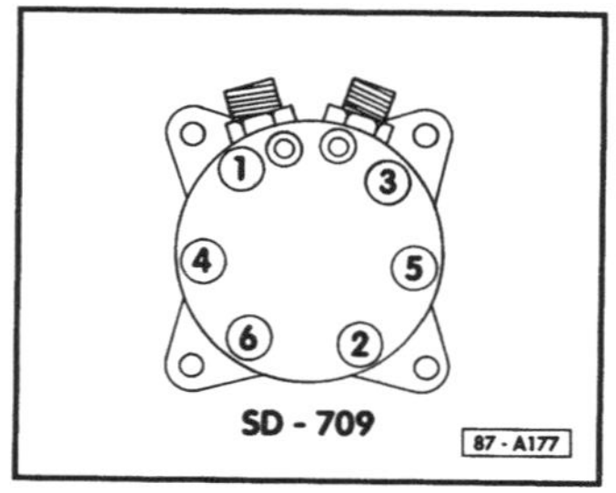

CAUTION

Use a beam type torque wrench. Damage may result from use of inaccurate "click" wrenches.

- torque head bolts according to sequence shown
  - 30-34 Nm (22-25 ft lb)
- refill compressor with new refrigerant oil (same quantity measured previously)

If all or most of the refrigerant oil from the compressor is lost due to refrigerant leaks,

- refill compressor with 70cc (2.36 oz) of new refrigerant oil

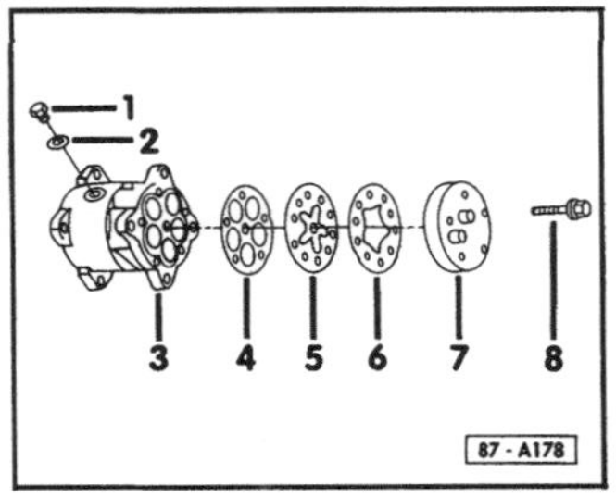

- inspect, lubricate and install O-ring **2** and drain plug **1** in compressor, replace O-ring if damaged
- install compressor in vehicle
- remove plugs from A/C lines
- inspect and lubricate all A/C Hose O-rings with clean refrigerant oil, replace any damaged O-rings

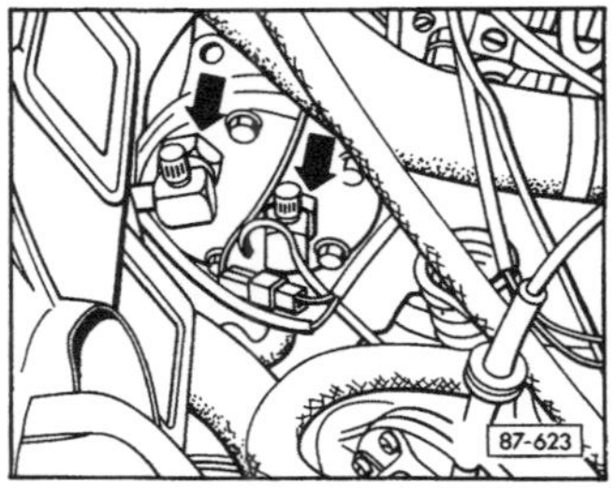

- install A/C lines (**arrows**) on compressor and tighten as follows:
  - suction: 35 Nm (26 ft lb)
  - discharge: 25 Nm (18 ft lb)
- install and properly tension compressor drive belt
- evacuate and charge A/C system
- operate A/C system and check head, valve plate, and gaskets for leaks

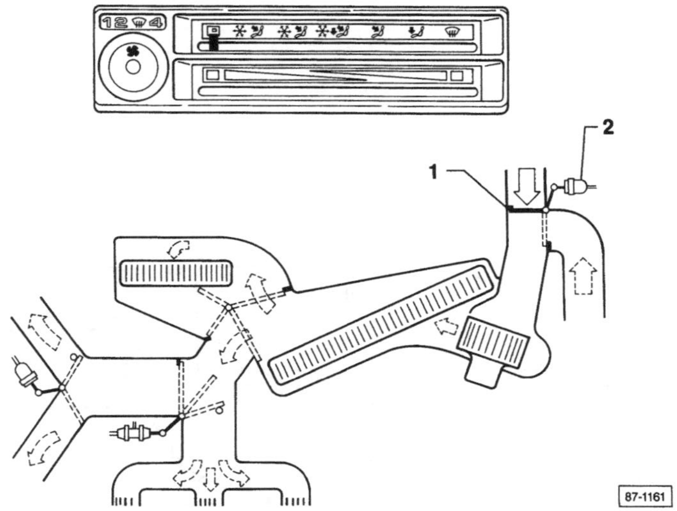

Fresh air cannot be shut OFF

Refer to illustration 87-1161 above.

- start engine
- push upper control lever to **OFF** position (far left)
  - fresh air/recirculation vacuum servo **2** must actuate/retract (vacuum applied)
  - fresh air/recirculation air flap **1** must close to outside air

If **NO**

- check for vacuum supply at servo
- check vacuum system, see Index
- check vacuum servos, see Index

Note

The lower control (temperature) lever and blower switch do not affect air flow distribution.

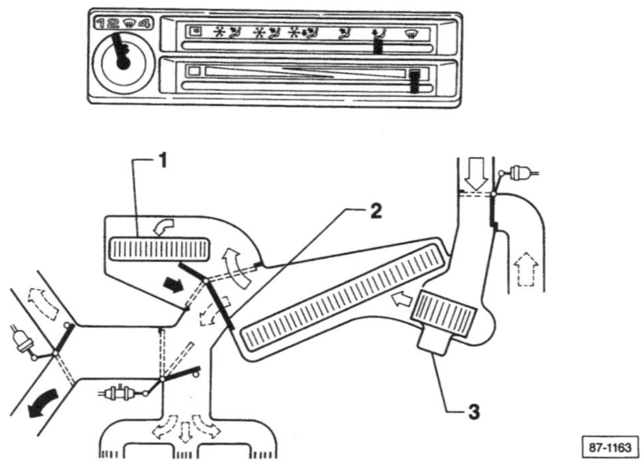

Insufficient or no heat output

Refer to illustration 87-1163 above.

- start and warm engine to normal operating temperature
- test operation of fresh air blower **3**
  - must operate on all four speeds
- check operation of lower control (temperature) lever
  - air blend flap **2** must audibly contact stop when lever is moved to full cold

If **NO**

- adjust air blend (temperature) flap control cable, see Repair Group 80
- adjust lower control lever to full hot (far right) position
- adjust upper control lever to HEAT position (footwell vents)
- adjust fresh air blower switch to second speed
  - warm air must radiate from footwell vents (through heater core **1**)

If **NO**,

- check thermostat, see Repair Group 19
- check vacuum system, see Index
- check vacuum servos, see Index

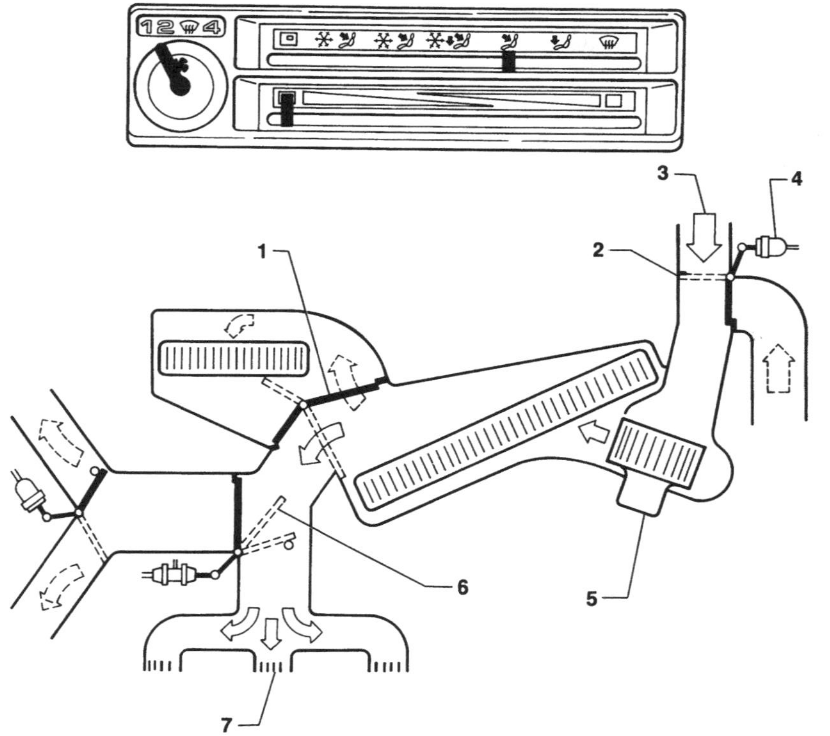

No fresh air from dashboard or footwell vents (fresh air cannot be adjusted)

Refer to illustration 87-1162 on previous page.

- switch ignition **ON**
- test operation of fresh air blower **5**
  - must operate on all four speeds

- adjust lower control (temperature) lever to full cold (far left) position
  - air blend flap **1** must audibly contact stop

If **NO**

- adjust air blend (temperature) flap control cable, see Repair Group 80

- adjust upper control lever to VENT position
- set fresh air blower switch to first speed (low)
- start engine
  - mode (central) flap **6** must operate
  - air must flow from dashboard vents **7**
  - fresh air/recirculation vacuum servo **4** must be extended (no vacuum applied)
  - fresh air/recirculation flap **2** and fresh air inlet (**arrow 3**) must be open

If **NO**

- check for vacuum supply at servos
- check vacuum system, see Index
- check vacuum servos, see Index

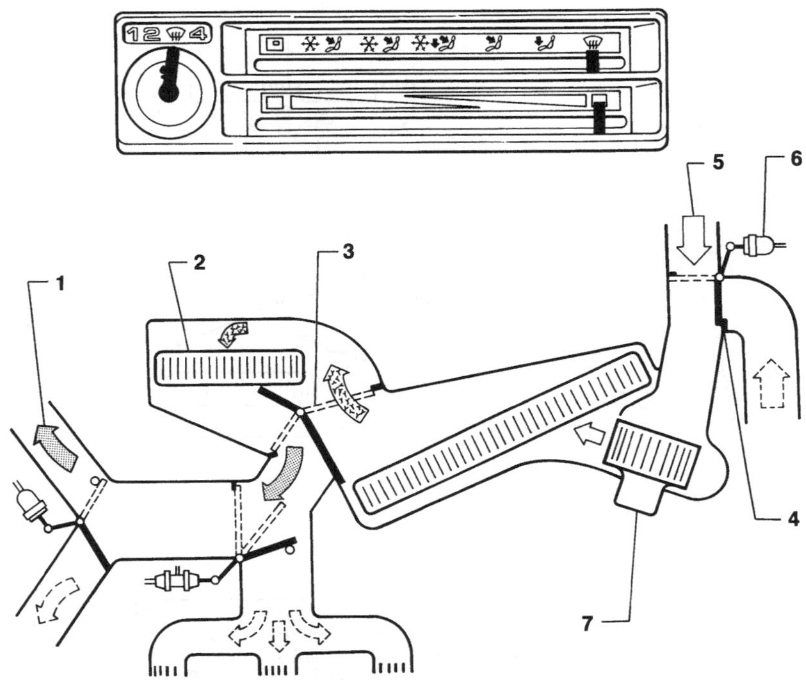

Windshield does not clear in DEFROST mode

Refer to illustration 87-1164 on previous page.

- check A/C operation
  - must cool with full output
- test operation of fresh air blower **7**
  - must operate on all four speeds
- check operation of lower control (temperature) lever
  - air blend flap **3** must audibly contact stop when lever is moved to full cold

If **NO**

- adjust air blend (temperature) flap control cable, see Repair Group 80
- adjust lower control lever (flap **3**) to full hot (far right) position
- adjust upper control lever to DEFROST position
- adjust fresh air blower to third speed (defrost)
- start engine
  - fresh air/recirculation vacuum servo **6** must be extended (no vacuum applied)
  - fresh air/recirculation flap **4** and fresh air inlet (**arrow 5**) must be open
  - cool air flows through heater core **2**
  - luke-warm air flows from defroster vents **1**

If **NO**,

- check thermostat, see Repair Group 19
- check vacuum system, see Index
- check vacuum servos, see Index

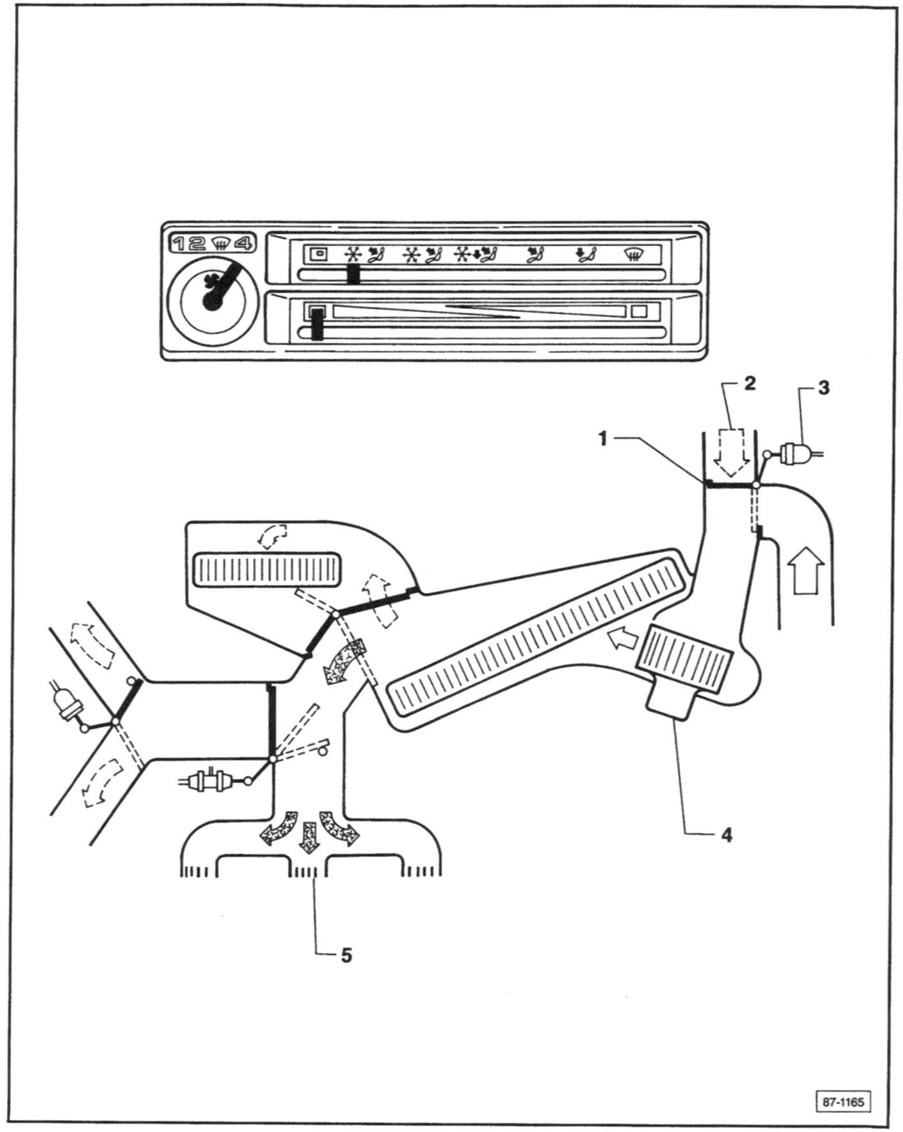

A/C does not cool

Refer to illustration 87-1165 on previous page.

- check A/C system fuses **S6**, **S19**, **S23**, and **S24**, see Index
  - fuses must be **OK**
- check A/C compressor drive belt
  - must be properly tensioned and in good condition

- open dashboard vents **5**
- adjust upper control lever to MAX A/C
- adjust lower control lever to full cold (far left)
- adjust fresh air blower to fourth speed (high)
- start engine
  - fresh air/recirculation vacuum servo **3** must retract (vacuum applied)
  - fresh air flap **1** must close and block outside air inlet (**arrow 2**)
  - air must flow from dashboard vents only

If **NO**
- check vacuum supply to servos
- check vacuum system, see Index
- check vacuum servos, see Index

- switch engine **OFF**
- adjust fresh air blower **4** to first speed (low)
- switch ignition back **ON**, but do not start car

- move upper control lever back and forth between MAX A/C and OFF
  - A/C compressor clutch must audibly click on and off during lever movement

If **NO**
- leave ignition switch **ON** and upper control lever set to MAX A/C

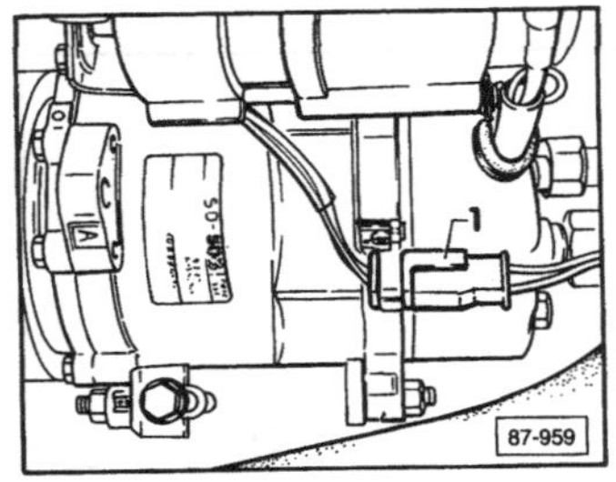

▶
- disconnect A/C *compressor connector* **1**
- measure voltage between each connector terminal and ground using digital multimeter **US 1119** set to voltmeter scale
  - must be approximately battery voltage at one terminal

If no voltage is present,
- repair wiring circuit according to wiring diagram

If voltage is present,
- replace A/C compressor clutch

A/C compressor clutch does not engage (A/C does not cool)

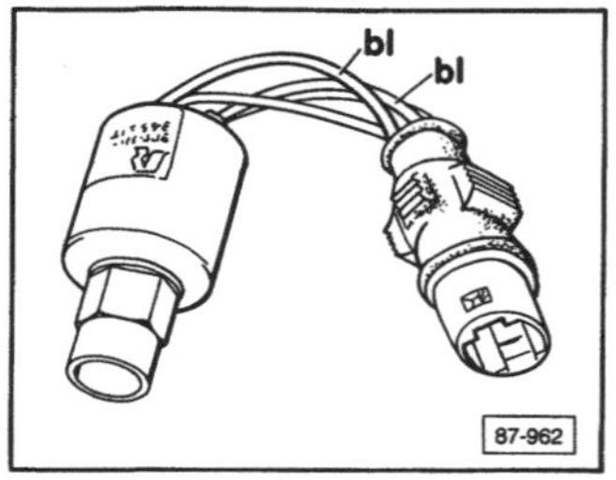

- check A/C system fuses **S6**, **S19**, **S23**, and **S24**, see Index
  - fuses must be **OK**
- check A/C compressor drive belt
  - must be properly tensioned and in good condition
- start engine
- switch A/C **ON**
- briefly connect two blue wires on A/C pressure switch using a jumper wire

If A/C compressor clutch engages, A/C system is low on refrigerant,

- check system for leaks, evacuate and recharge

CAUTION

DO NOT keep the A/C compressor clutch engaged for more than a moment or damage to the compressor and/or system will result.

Replace the receiver drier if the A/C system has been open (exposed to the air) for a long period of time. Operating the A/C with a contaminated system will cause A/C component damage.

WARNING

Automotive refrigerant containing CFC's is hazardous to the earth's atmosphere. To protect our environment, use an Underwriter's Laboratory (UL) approved refrigerant recovery/recycling unit such as **Kent-Moore ACR³**, or equivalent, whenever discharging an A/C system.

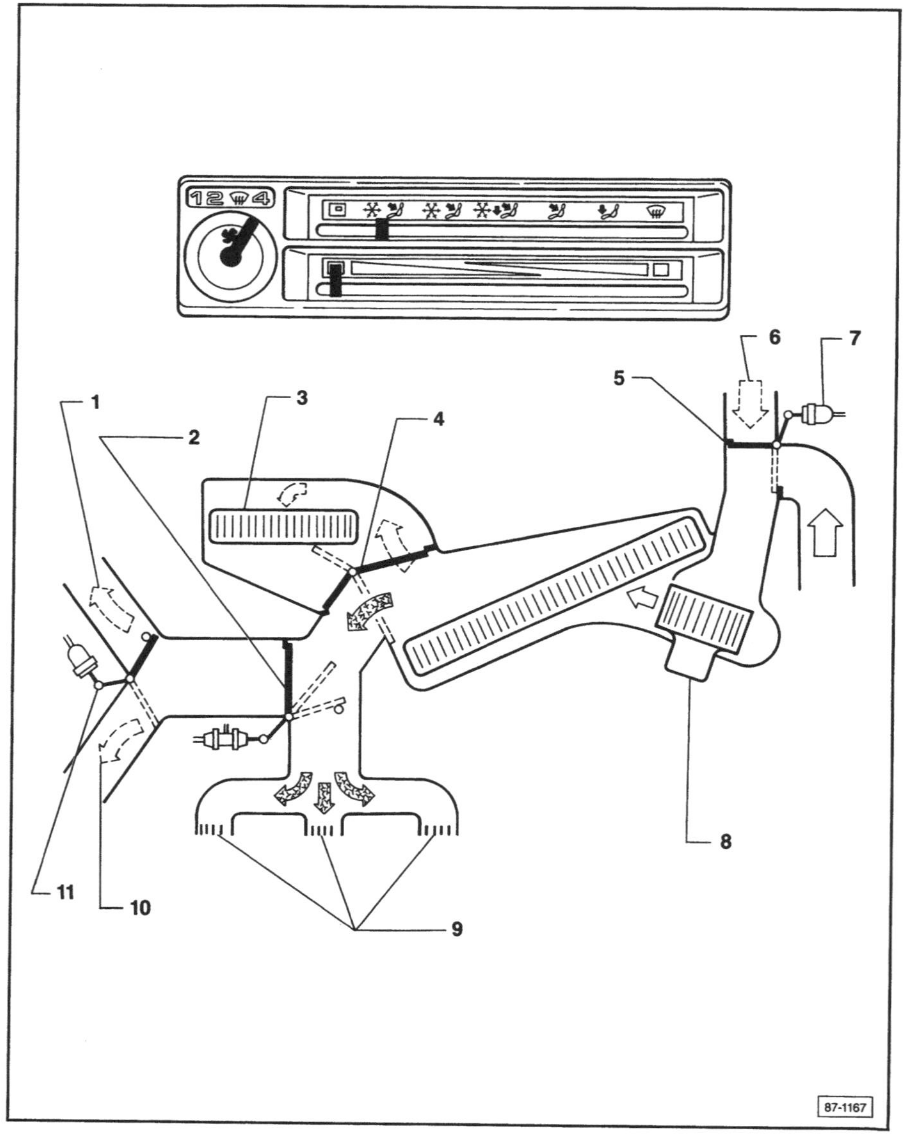

Insufficient cooling output

Refer to illustration 87-1167 on previous page.

Observe the following:
- ■ do not park car in sunlight
- ■ condenser and radiator clean and free of obstructions
- ■ compressor drive belt in good condition and properly tensioned
- ■ air ducts for condenser and radiator properly installed
- ■ fresh air blower **8** operates on all four speeds

- ■ open dashboard air vents **9**
- ■ insert thermometer in left dashboard vent
- ■ adjust fresh air blower to fourth speed (high)
- ■ adjust upper A/C control lever to MAX A/C
- ■ adjust lower A/C control lever to full cold (far left)
  - ● air blend flap **4** must audibly contact stop
  - ● air flow to heater core **3** must be blocked

If **NO**
- ■ adjust air blend (temperature) flap control cable, see Repair Group 80

- ■ start engine
  - ● heat/defrost vacuum servo **11** must be extended (no vacuum applied)
  - ● mode (central) flap **2** must close to defrost/footwell vents
  - ● no air must flow from footwell vents **10** or defroster vents **1**
  - ● fresh air/recirculation servo **7** must retract (vacuum applied)
  - ● fresh air/recirculation flap **5** must close to outside air
  - ● fresh air inlet (**arrow 6**) must be blocked

If air distribution **NOT OK**
- ■ check for vacuum supply to all servos
- ■ check vacuum system, see Index
- ■ check vacuum servos, see Index

If air distribution **OK**
- ■ check A/C system temperatures and pressures (outlet air temperature at left vent), see Index

Cooling output diminishes after running A/C for several minutes

If A/C cooling output and vent air flow diminish after several minutes of A/C operation, the evaporator may be icing up.
- ■ check evaporator for ice accumulation

If ice is present,
- ■ check insertion depth of A/C thermostat sensing tube into evaporator
  - ● must be 330 mm (13.0 in.)
- ■ check sensing tube and replace thermostat if damaged

THIS PAGE INTENTIONALLY LEFT BLANK

Index

1990 m.y.

Engine Coolant Temperature (ECT) gauge
- checking **90.14**
- connector wiring **90.6**
- troubleshooting **90.15**

Digital clock
- connector wiring **90.7**

Fuel gauge
- checking **90.13**
- connector wiring **90.6**
- troubleshooting **90.13**

Instrument cluster
- assembly **90.2**
- multi-point connector, identification **90.4**
- printed circuit, removing **90.8**
- removing **90.3**
- testing **90.10**

Instruments
- removing/installing **90.5**

Light emitting diode (LED)
- installing **90.16**

Printed circuit
- removing **90.8**

Tachometer/oil pressure warning control unit
- connector wiring **90.7**

Voltage stabilizer
- checking **90.9**
- removing/installing **90.7**

Warning lights (left and right)
- installation position **90.6**

From 1991 m.y.

Engine Coolant Temperature (ECT) gauge
- checking **90.14**
- connector wiring **90.6**
- troubleshooting **90.15**

Fuel gauge
- checking **90.13**
- connector wiring **90.6**
- troubleshooting **90.13**

Instrument cluster
- assembly **90.17**
- multi-point connector, identification **90.20**
- removing **90.19**
- lights, identification **90.24**
- lights, removing/installing **90.24**
- testing **90.28**

Instruments
- removing/installing **90.21**

Light emitting diode (LED)
- installing **90.23**

MFA vacuum sensor
- removing/installing **90.27**

Printed circuit
- component identification **90.26**

Turn signal indicator lights
- installing **90.23**

Vehicle Speed Sensor (VSS), speedometer, (G22)
- removing/installing **90.27**

Voltage stabilizer
- checking **90.25**

Warning lights (left and right)
- installation position **90.22**
- removing/installing **90.23**

Index

Anti-theft alarm system

1991-1992 m.y.

Anti-theft alarm system
- general information **90.31**

Anti-theft alarm system, troubleshooting
- alarm horn sounds but lights don't flash with alarm activated **90.40**
- alarm system cannot be switched off **90.38**
- alarm system cannot be switched on **90.33**
- engine does not crank with alarm system off **90.41**
- lights flash but alarm horn does not sound with alarm activated **90.39**
- wiring diagram **90.42**

From 1993 m.y.

Anti-theft alarm system
- general information **90.31**

Anti-theft alarm system, troubleshooting
- activated alarm system does not trigger alarm horn **90.50**
- activated alarm system does not trigger flasher lights **90.51**
- alarm system cannot be switched off **90.49**
- alarm system cannot be switched on **90.46**
- starter does not crank with alarm system switched off **90.52**

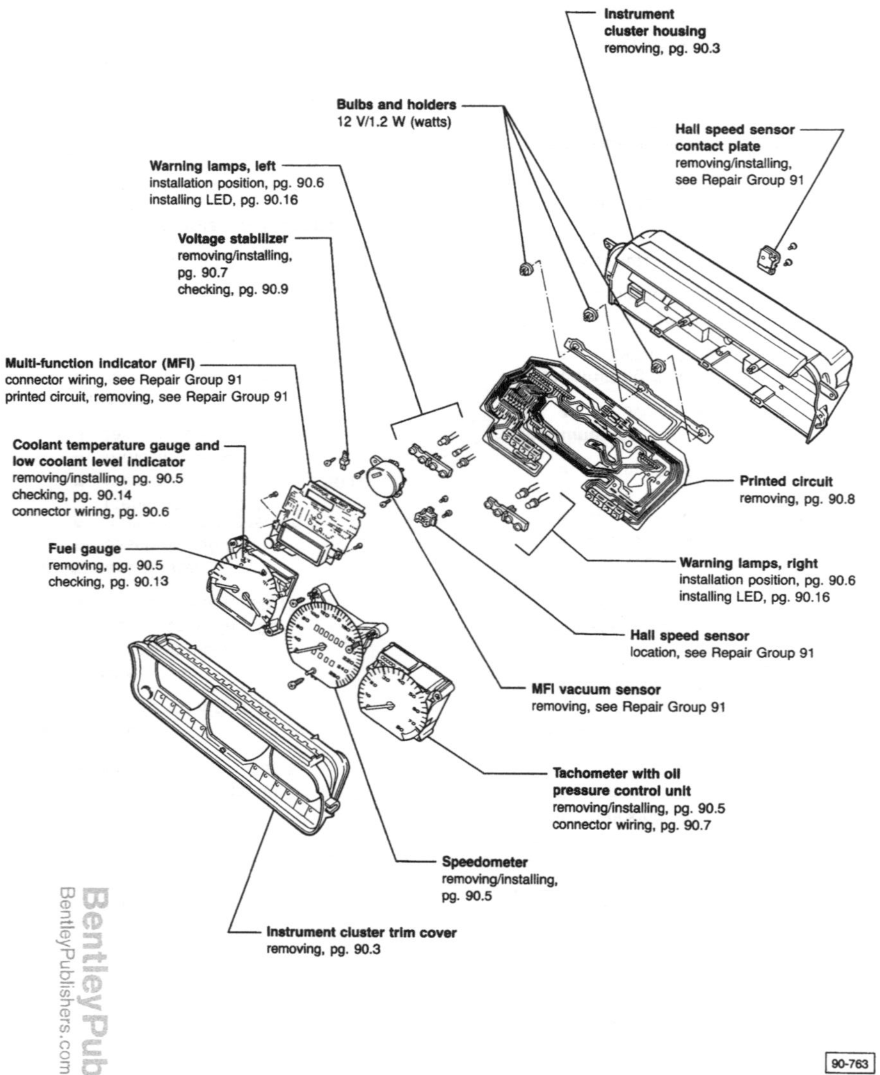

Instrument cluster, removing

CAUTION

Before working on any part of the electrical system, disconnect battery ground strap.

All components in the instrument cluster, except for the printed circuit, can be removed from the front without removing the complete instrument cluster housing.

- remove battery ground strap
- remove steering wheel horn cover (see Repair Group 94)
- remove steering wheel
  - 40 Nm (29.5 ft lb)

Remove steering wheel only when removing the complete instrument cluster housing.

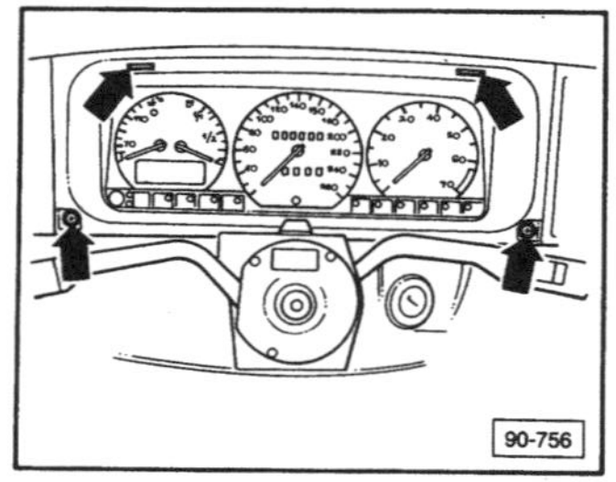

▶

- remove screw trim caps located on lower part of instrument cluster trim
- remove instrument cluster trim retaining screws (**arrows**)
- remove instrument cluster trim

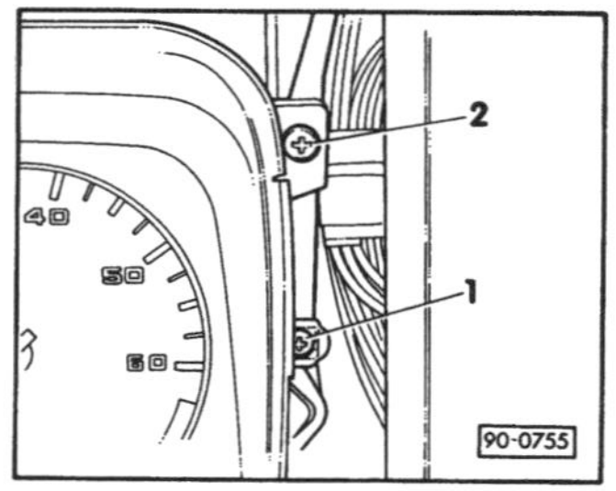

▶

- unscrew trip odometer reset button
- remove trim cover retaining screws **1**
- remove trim cover from instrument cluster
- remove instrument cluster housing mounting screws **2**, left and right
- pull instrument cluster housing from dash panel
- disconnect multi-point connector, MFI vacuum sensor hose, and speedometer drive cable
- remove instrument cluster housing

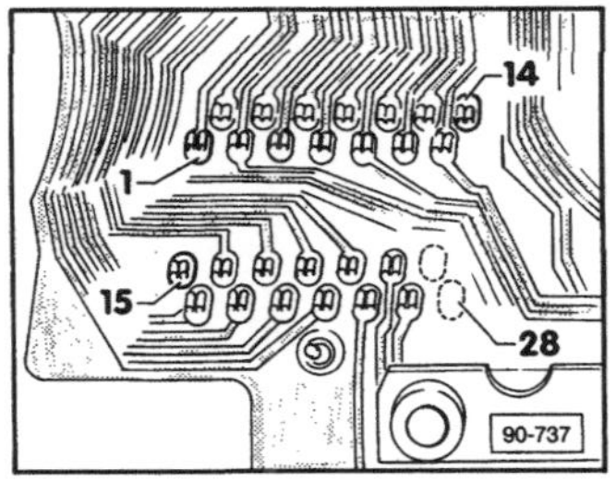

Multi-point connector on instrument cluster, identification

Remove combination fuel, coolant temperature, and low coolant level indicator gauge from instrument cluster to access multi-point connector.

or

Use test box **V.A.G. 1598** with adaptor harness **1598/8** for checking and testing. The numbered sockets on test box **V.A.G. 1598** correspond to the numbered terminals of the 28-point instrument cluster connector.

1 — ground (−) for outside temperature sensor
2 — not used
3 — ground (−), terminal **31**
4 — to multi-function indicator (MFI) reset switch
5 — ground (−), terminal **31**
6 — to MFI memory switch
7 — hall speed sensor signal
8 — from oil pressure switch (1.8 bar)
9 — from oil pressure switch (0.3 bar)
10 — tachometer input signal, terminal **1**
11 — plus (+) connection, terminal **30**
12 — plus (+) connection for instrument lighting, terminal **58b**
13 — plus (+) connection, terminal **15**
14 — not used
15 — to MFI mode switch
16 — to alternator warning light, terminal **61**
17 — from oil temperature sensor
18 — to seat belt warning light/brake warning light
19 — outside temperature sensor signal
20 — not used
21 — to fuel gauge
22 — not used
23 — to coolant temperature gauge
24 — to turn signal indicator light
25 — to high beam indicator light
26 — not used
27 — not connected
28 — not connected

Instruments, removing/installing

> **CAUTION**
>
> Before working on any part of the electrical system, disconnect battery ground strap.

- remove battery ground strap
- remove instrument cluster trim
- remove instrument cluster trim cover (see pg. 90.3)

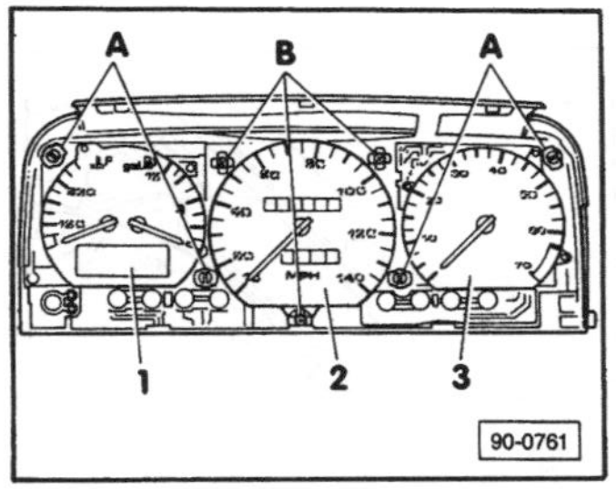

For combination fuel, coolant temperature and low coolant level indicator gauge **1** and tachometer **3**:

- squeeze locating pins **A** and pull gauge from housing and connector

For speedometer **2**:

- disconnect speedometer drive cable
- remove mounting screws **B** and pull speedometer from housing
- install instruments in reverse order of removal

> **CAUTION**
>
> **DO NOT** damage printed circuit when installing the instruments.

- carefully guide plugs on rear of instruments into connector on printed circuit board

Warning lights (left), installation position

1 — not used
2 — oil pressure warning light (red), LED
3 — high beam indicator light (blue), bulb
4 — turn signal indicator light (green), LED

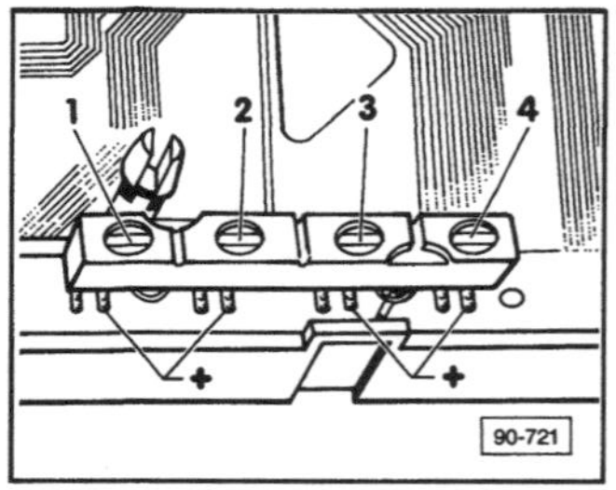

Warning lights (right), installation position

1 — seat belt warning light/brake warning light (red), LED
2 — alternator warning light (red), LED
3 — not used
4 — not used

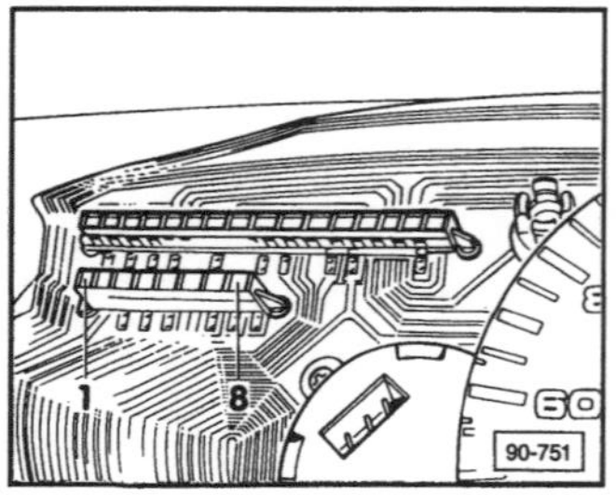

Fuel Gauge and Engine Coolant Temperature Gauge/Engine Coolant Level Indicator, connector wiring (All m.y.)

1 — not used
2 — from Fuel Level Sensor
3 — from Engine Coolant Temperature Sensor
4 — plus connection (B+), terminal **15**
5 — not used
6 — not used
7 — ground (–), terminal **31**
8 — voltage stabilizer output voltage, 9.5 V to 10.5 V

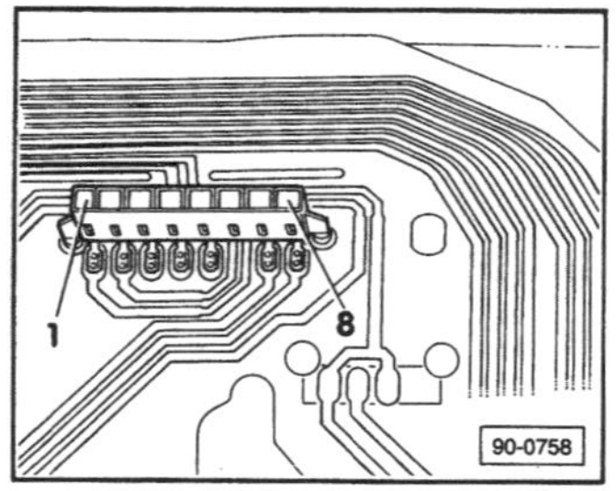

Tachometer and oil pressure warning control unit, connector wiring

1 — tachometer input signal, terminal **1**
2 — ground (–), terminal **31**
3 — to oil pressure warning light
4 — plus (+) connection, terminal **15**
5 — tachometer output signal
6 — not used
7 — from oil pressure switch, 1.8 bar
8 — from oil pressure switch, 0.3 bar

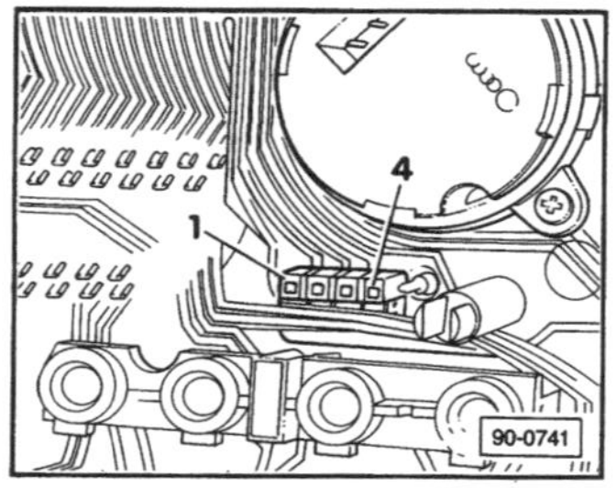

Digital clock, connector wiring

1 — MFI plus (+) connection, terminal **15**
2 — plus (+) connection for instrument lighting, terminal **58b**
3 — plus (+) connection, terminal **30**
4 — ground (–), terminal **31**

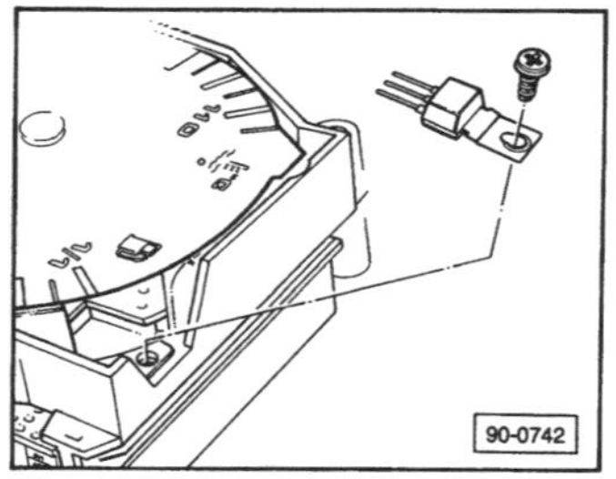

Voltage stabilizer, removing/installing

The voltage stabilizer is mounted on the instrument cluster housing above the fuel gauge.

CAUTION

Before working on any part of the electrical system, disconnect battery ground strap.

- remove mounting screw and slide voltage stabilizer from connector
- install voltage stabilizer into connector making sure terminals engage connector slots properly
- tighten mounting screw securely

CAUTION

Voltage stabilizer must be tight against the mounting base to ensure proper heat transfer.

Printed circuit, removing

> **CAUTION**
>
> Before working on any part of the electrical system, disconnect battery ground strap.

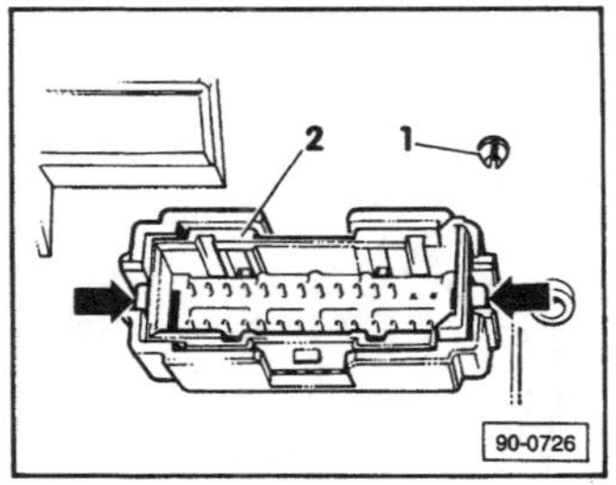

- remove instrument cluster (see page 90.3)
- remove instruments from instrument cluster housing (see page 90.5)
- remove MFI vacuum sensor, see Repair Group 91
- remove Hall speed sensor contact plate, see Repair Group 91
- squeeze spreader pins **1** and carefully remove connector strip with printed circuit

> **CAUTION**
>
> **DO NOT** damage printed circuit when removing.

- push locating lugs (**arrows**) in direction indicated and remove multi-point connector plug **2** from housing
- remove printed circuit

Electrical System – Instruments

Voltage stabilizer, checking

Check voltage stabilizer when fuel gauge and coolant temperature gauge simultaneously indicate false readings.

Leave battery ground strap and multi-point connector on instrument cluster connected for the following tests.

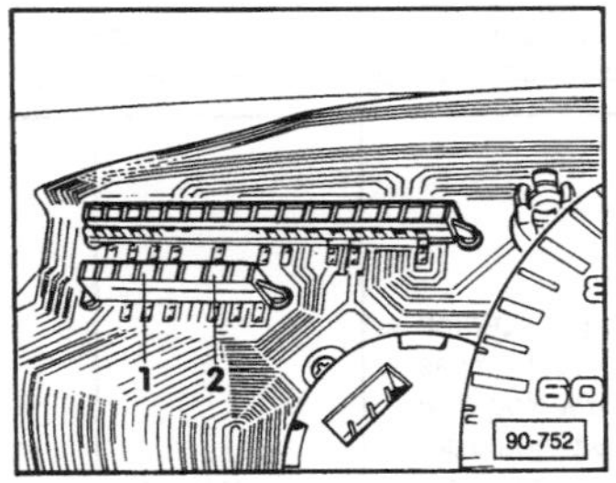

- remove instrument cluster trim
- remove combination fuel, coolant temperature, and low coolant level indicator gauge (see page 90.5)
- connect multimeter **US 1119** (set to voltmeter scale) between positive connection **1** and ground **2**
- switch ignition **ON**
  - approximately battery voltage

If **NO**
- check wiring according to current flow diagram

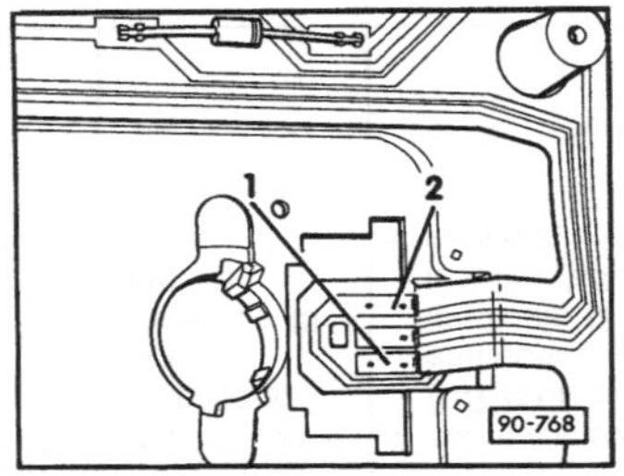

- switch ignition **OFF**
- reinstall combination fuel, coolant temperature, and low coolant level indicator gauge
- remove speedometer (see page 90.5)
- connect multimeter **US 1119** (set to voltmeter scale) between positive connection **1** and ground **2**
- switch ignition **ON**
  - 9.5-10.5 V

If **NO**
- open in printed circuit or voltage stabilizer defective, replace

Instrument cluster, testing

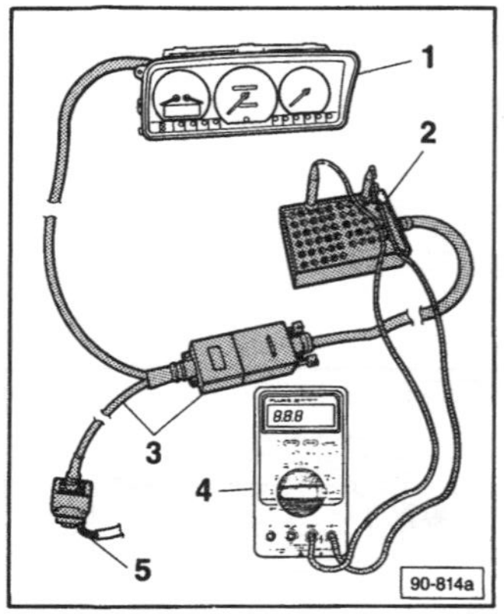

▶

Note

Always refer to appropriate wiring diagram when performing electrical tests.

Test conditions

- battery **OK**
- fuses **OK**
- 28-point connector at instrument cluster disconnected
- fuse/relay cover folded down for access to fuse/relay panel

Tools required

- multimeter **Fluke 83** (**US 1119**)
- test box **V.A.G. 1598** with harness adaptor **1598/8**
  - connect to instrument cluster 28-point connector and wiring harness
- L.E.D. tester **US 1115**

1 — instrument cluster
2 — test box **V.A.G. 1598**
3 — adaptor **1598/8**
4 — multimeter **Fluke 83** (**US 1119**)
5 — instrument cluster wiring harness

Test step	Item tested	Test condition	Specified value	Cause of problem
1	Power supply and ground (terminals **30** and **31**)	■ switch ignition **OFF** ■ set multimeter **Fluke 83** (**US 1119**) to volt range and connect between terminals **11** and **3** of test box **V.A.G. 1598**	Must be approx. 12V	If not **OK**, ■ check wiring for open or short circuits according to wiring diagram, repair as necessary If **OK**, ■ check printed circuit and connections for open or short circuits, repair/replace as necessary
2	Ignition power supply and ground (terminals **15** and **31**)	■ set multimeter **Fluke 83** (**US 1119**) to volt range and connect between terminals **13** and **3** of test box **V.A.G. 1598** ■ switch ignition **ON**	Must be approx. 12V	If not **OK**, ■ check wiring for open or short circuits according to wiring diagram, repair as necessary If **OK**, ■ check printed circuit and connections for open or short circuits, repair/replace as necessary
3	Turn signal indicator light	■ connect L.E.D. tester **US 1115** between terminals **24** and **3** of test box **V.A.G. 1598** ■ switch ignition **ON** ■ actuate left turn signal	L.E.D. tester must flash	If not **OK**, ■ check wiring for open or short circuits according to wiring diagram, repair as necessary If **OK**, ■ check printed circuit and connections for open or short circuits, repair/replace as necessary
4	Instrument cluster illumination	■ set multimeter **Fluke 83** (**US 1119**) to volt range and connect between terminals **12** and **3** of test box **V.A.G. 1598** ■ switch ignition **ON** ■ switch lights **ON**	Must be approx. 2.5V to 12V depending on position of instrument panel light dimmer switch (**E20**)	If not **OK**, ■ check wiring for open or short circuits according to wiring diagram, repair as necessary If **OK**, ■ check printed circuit and connections for open or short circuits, repair/replace as necessary

Test step	Item tested	Test condition	Specified value	Cause of problem
5	High beam indicator light	■ set multimeter **Fluke 83** (**US 1119**) to volt range and connect between terminals **25** and **3** of test box **V.A.G. 1598** ■ switch ignition **ON** ■ switch high beam headlights **ON**	Must be approx. 12V	If not **OK**, ■ check wiring for open or short circuits according to wiring diagram, repair as necessary If **OK**, ■ check printed circuit and connections for open or short circuits, repair/replace as necessary
6	Alternator light	■ set multimeter **Fluke 83** (**US 1119**) to volt range and connect between terminals **16** and **3** of test box **V.A.G. 1598** ■ start engine	Must be approx. 12V	If not **OK**, ■ check wiring for open or short circuits according to wiring diagram, repair as necessary If **OK**, ■ check printed circuit and connections for open or short circuits, repair/replace as necessary

Fuel gauge, checking

Leave battery and electrical connectors connected for the following tests.

■ check these first:
- verify complaint
- battery **OK**
- chassis ground **OK**
- sending unit ground **OK**
- fuse **OK**

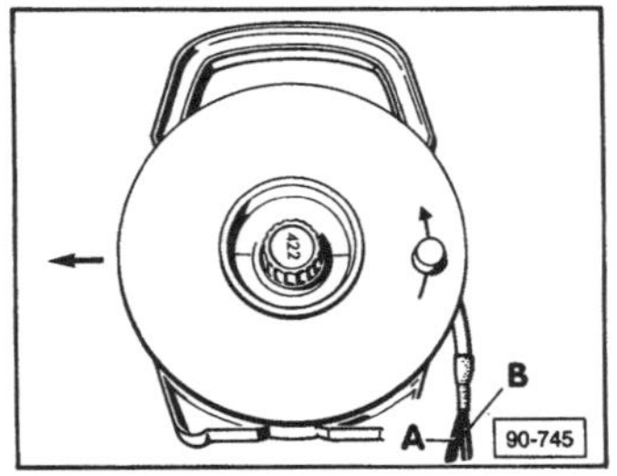

■ remove wire connector from fuel gauge sending unit
■ connect lead **A** of **VW 1301** to ground
■ connect lead **B** of **VW 1301** to wire (purple/black) leading to fuel gauge
■ switch ignition **ON**
■ adjust **VW 1301** as follows:

dial setting	gauge reads	
50	full	
160	half	
350	empty	(reserve area)

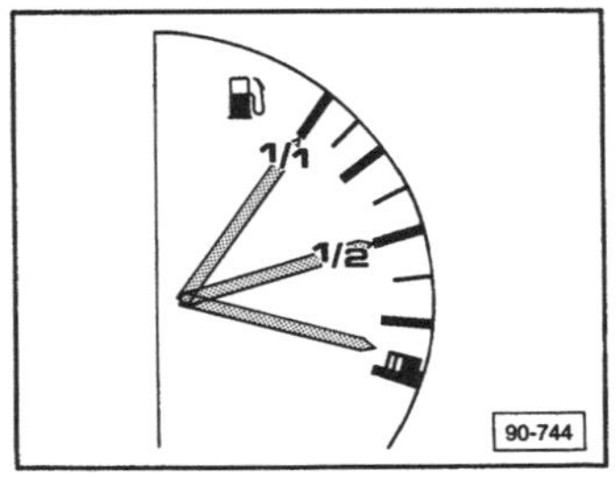

- fuel gauge needle must correspond with specified values
- deviation of one needle width to left or right is permissible

Fuel gauge, troubleshooting

Use the following procedure to determine the cause of false fuel gauge readings.

■ check fuel gauge (see above)

If fuel gauge checks **OK**,
■ replace fuel gauge sender unit

If fuel gauge needle moves but is inaccurate,
■ check voltage stabilizer, see Index
- output voltage must be 9.5-10.5 V

Note

A malfunctioning voltage stabilizer will also cause a false coolant temperature gauge reading.

If voltage stabilizer specified value is obtained,
- replace fuel gauge

If voltage stabilizer specified value is not obtained,
- replace voltage stabilizer

If fuel gauge needle does not move during check,
- check wiring from sender to gauge according to wiring diagram

If wiring is **OK**,
- check voltage stabilizer, see Index
  - output voltage must be 9.5-10.5 V

If voltage stabilizer checks **OK**,
- replace fuel gauge

Engine Coolant Temperature (ECT) Gauge, checking

Leave battery and electrical connectors connected for the following tests.

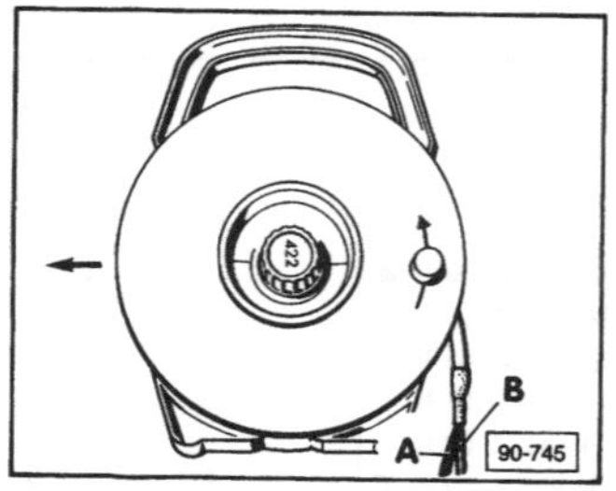

- check these first:
  - verify complaint
  - battery **OK**
  - chassis ground **OK**
  - fuse **OK**

- remove connector from Engine Coolant Temperature (ECT) Sensor
- connect lead **A** of **VW 1301** to ground
- connect lead **B** of **VW 1301** to wire (yellow/red) leading to Engine Coolant Temperature (ECT) Gauge

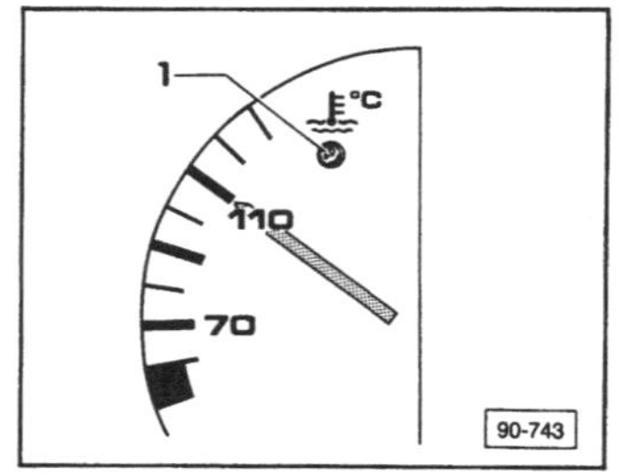

- switch ignition **ON**
  - Engine Coolant Temperature (ECT) warning light **1** flashes for approx. 3 seconds
- adjust **VW 1301** as follows:

dial setting	**gauge reads**
80	engine hot (110°C/230°F)
50	warning zone, engine too hot (warning light **1** flashes)

  - coolant gauge needle must correspond to specified values
  - deviation of one needle width to left or right is permissible

Engine Coolant Temperature (ECT) Gauge, troubleshooting

Use the following procedure to determine the cause of false Engine Coolant Temperature (ECT) Gauge readings.

- check Engine Coolant Temperature Gauge (see above)

If temperature gauge checks **OK**,
- replace Engine Coolant Temperature Sensor

If temperature gauge needle moves during check, but is inaccurate
- check voltage stabilizer, see Index
  - output voltage must be 9.5-10.5 V

A malfunctioning voltage stabilizer will also cause a faulty fuel gauge reading.

If voltage stabilizer specified value is obtained,
- replace Engine Coolant Temperature Gauge

If voltage stabilizer specified value is not obtained,
- replace voltage stabilizer

If temperature gauge needle does not move during check,
- check wiring from ECT Sensor to ECT Gauge according to wiring diagram

If wiring checks **OK**,
- check voltage stabilizer, see Index
  - output voltage must be 9.5-10.5 V

If voltage stabilizer checks **OK**,
- replace ECT Gauge

Light emitting diode (LED), installing

Always check for correct polarity when installing diodes.

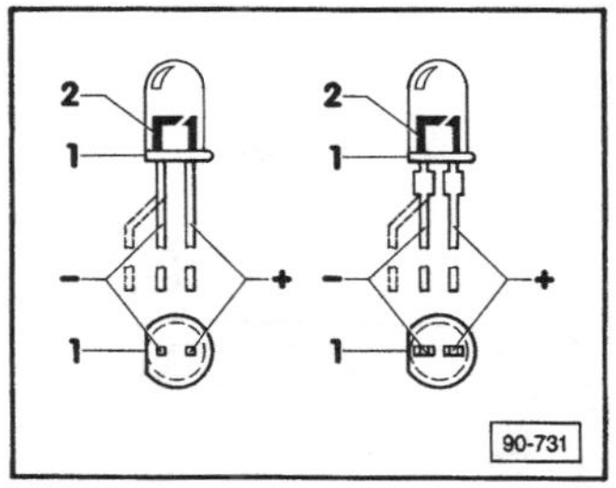

▶ The minus pole is indicated by:

1 — flat on diode housing
2 — larger pole in diode housing

On some diodes, minus pole is also angled.

THIS PAGE INTENTIONALLY LEFT BLANK

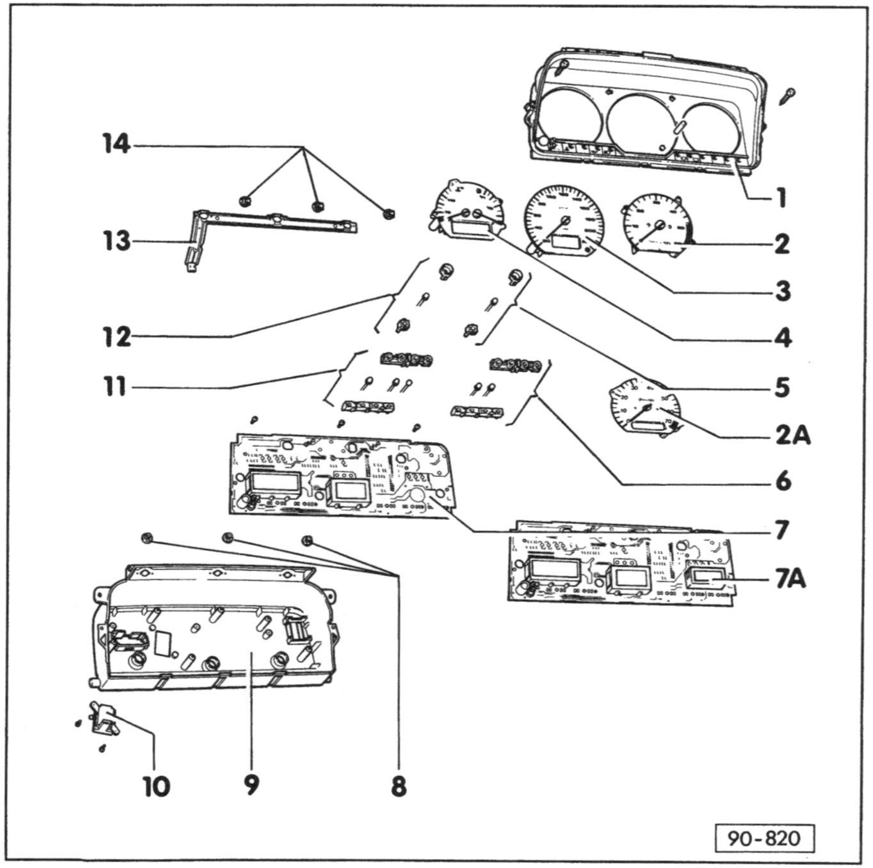

1 – **Instrument cluster trim cover**
- removing, 90.19

2 – **Tachometer (Manual Trans. only)**

2A – **Tachometer (with Transmission Range Selector Lever Display, Auto Trans. only)**

3 – **Speedometer (with electronic odometer display)**
- Vehicle Speed Sensor (VSS), speedometer, (G 22), removing/installing, 90.27

4 – **Engine Coolant Temperature (ECT) Gauge, Fuel Gauge and Multi-function Indicator (MFA)**

5 – **Turn signal indicator light, right**
- LED installing, 90.23

6 – **Warning lamps, right**
- installation position, 90.22
- Led installing 90.23

7 – **Printed circuit (Manual Trans. only)**
- replace with instrument cluster housing as an assembly only
- component identification, 90.26

7A – **Printed circuit (with Transmission Range Selector Lever Display, Auto Trans. only)**
- replace with instrument cluster housing as an assembly only
- component identification, 90.26

8 – **Bulbs and holders (black sockets)**
- for lighting Multi-function indicator (MFA) and trip odometer
- 12V/1.2W
- removing/installing, 90.24

9 – **Instrument cluster housing**
- replace with printed circuit as an assembly only

10 – **MFA vacuum sensor**
- removing/installing, 90.27

11 – **Warning lamps, left**
- installation position, see index
- LED installing, 90.23

12 – **Turn signal indicator light, left**
- installing, 90.23

13 – **Printed circuit for instrument cluster lighting**

14 – **Instrument cluster lighting, bulbs and holder (blue sockets)**
- 12V/1.2W

Instrument cluster, removing

CAUTION

Before working on any part of the electrical system, disconnect battery ground strap.

All components in the instrument cluster, except for the printed circuit, can be removed from the front without removing the complete instrument cluster housing.

- remove battery ground strap

Note

Remove steering wheel only when removing the complete instrument cluster housing.

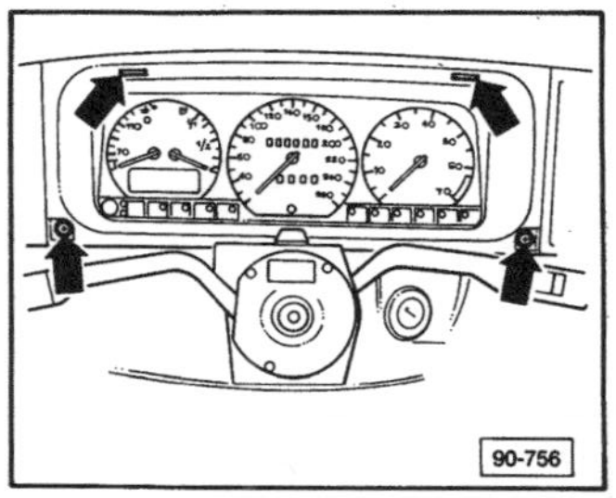

- remove steering wheel horn cover (see Repair Group 94)
- remove steering wheel
  - 40 Nm (29.5 ft lb)
- remove screw trim caps located on lower part of instrument cluster trim
- remove instrument cluster trim retaining screws (**arrows**)
- remove instrument cluster trim

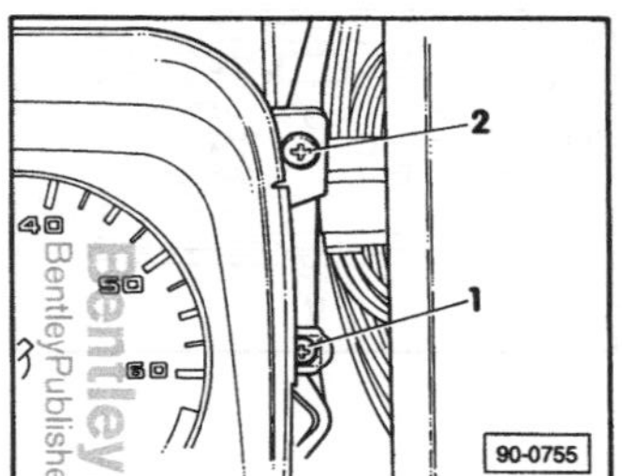

- unscrew trip odometer reset button
- remove trim cover retaining screws **1**
- remove trim cover from instrument cluster
- remove instrument cluster housing mounting screws **2**, left and right
- pull instrument cluster housing from dash panel
- disconnect multi-point connector, MFA vacuum sensor hose, and speedometer drive cable
- remove instrument cluster housing

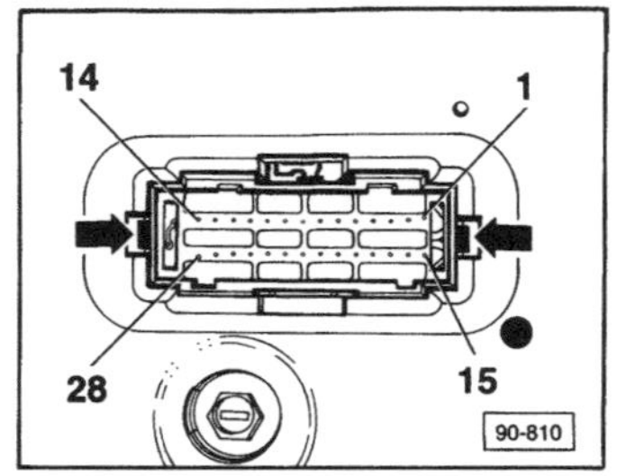

Multi-point connector on instrument cluster, identification

Note

Use test box **V.A.G. 1598** with adaptor harness **1598/8** for checking and testing. The numbered sockets on test box **VAG 1598** correspond to the numbered terminals of the 28-point instrument cluster connector.

1 — ground (–) for outside temperature sensor (MFA)
2 — to Engine Coolant Level (ECL) sensor
3 — ground (–), terminal **31**
4 — to multi-function indicator (MFA) memory switch (reset)
5 — ground (–), terminal **31**
6 — to MFA memory switch (memory)
7 — Vehicle Speed Sensor (VSS), (**G 21**) signal
8 — from oil pressure switch (1.4 bar)
9 — from oil pressure switch (0.3 bar)
10 — tachometer input signal, terminal **1**/terminal **W**
11 — plus (B+) connection, terminal **30**
12 — plus (B+) connection for instrument lighting, terminal **58b**
13 — plus (B+) connection, terminal **15**
14 — not used
15 — to MFA mode switch
16 — to Generator warning light, terminal **61**
17 — from oil temperature sensor (MFA)
18 — to seat belt warning light/brake warning light
19 — outside temperature sensor signal (MFA)
20 — not used
21 — to fuel gauge
22 — to turn signal indicator light, left
23 — Engine Coolant Temperature Sensor (to gauge)
24 — to turn signal indicator light, right
25 — to high beam indicator light
26 — not used
27 — speedometer signal from Vehicle Speed Sensor (VSS), speedometer (**G 22**)
28 — not connected

Instruments, removing/installing

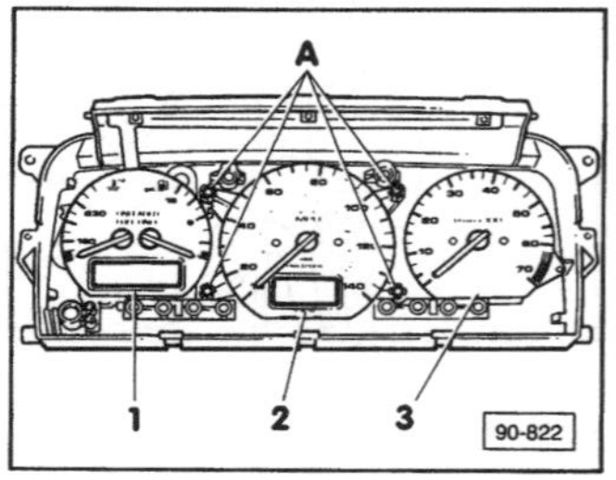

> **CAUTION**
>
> Before working on any part of the electrical system, disconnect battery ground strap.

- remove battery ground strap
- remove instrument cluster trim
- remove instrument cluster trim cover

For combination fuel gauge, Engine Coolant Temperature (ECT) gauge/Engine Coolant Level (ECL) indicator **1** and tachometer **3**:
- carefully pull gauge from housing and connector

For speedometer (with odometer display) **2**:
- carefully pull speedometer from housing and locating pins **A**

- install instruments in reverse order of removal

> **CAUTION**
>
> **DO NOT** damage printed circuit when installing the instruments.

- carefully guide plugs on rear of instruments into connector on printed circuit board

Electrical System – Instruments

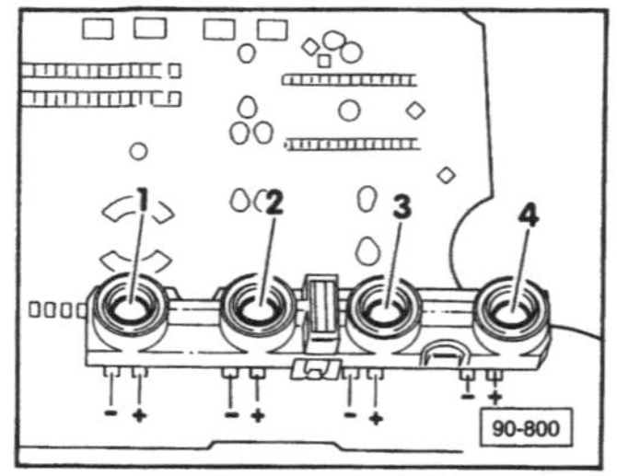

Warning lights (left), installation position

1 — warning light (red) for Engine Coolant Temperature/Engine Coolant Level Indicator LED
2 — not used
3 — oil pressure warning light (red), LED
4 — high beam indicator light (blue), bulb

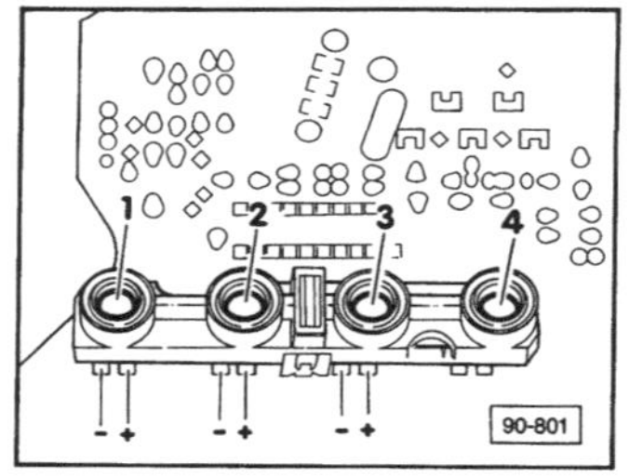

Warning lights (right), installation position

1 — seat belt warning light/brake warning light (red), LED
2 — Generator warning light (red), LED
3 — not used
4 — not used

Light emitting diode (LED), installing

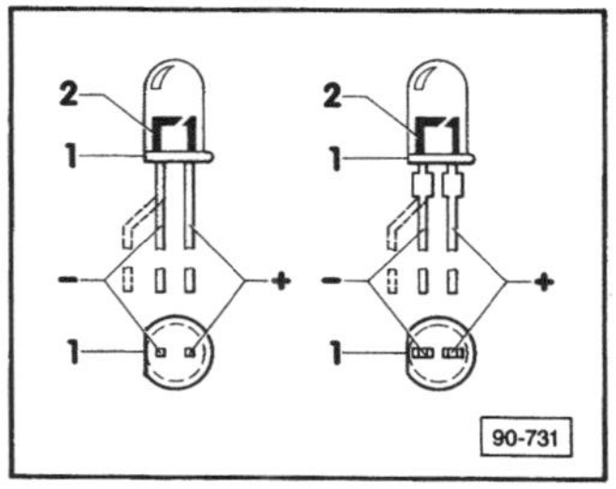

Always check for correct polarity when installing diodes.

The minus (-) pole is indicated by:

1 — flat on diode housing
2 — larger pole in diode housing

On some diodes, minus (-) pole is also angled.

Warning lights, removing/installing

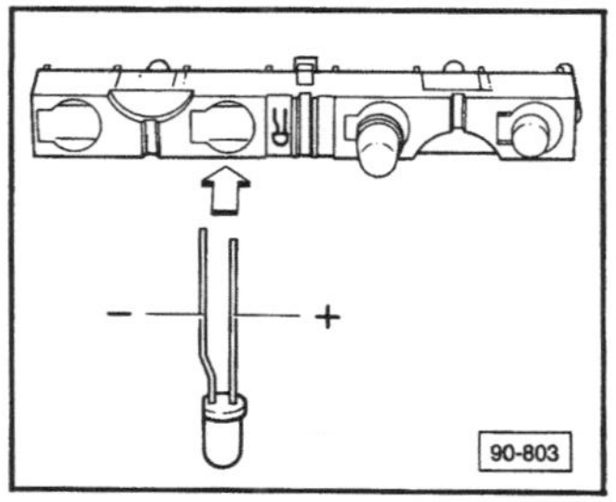

Removing

- unclip and remove upper half of light base (socket)
- pull out and remove LED

Installing

- install in reverse order of removal
  - proper installation position of LED is marked on top ledge of lower light base (socket)

Turn signal indicator lights, installing

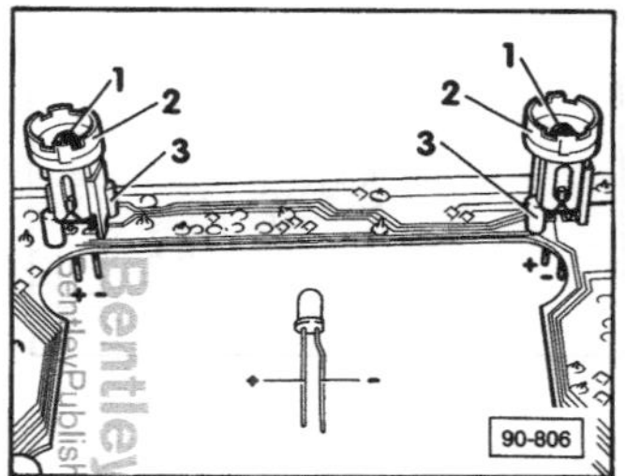

- install LED's as shown

1 — LED
2 — upper light socket (with clip)
3 — lower light socket

Instrument cluster lights, identification

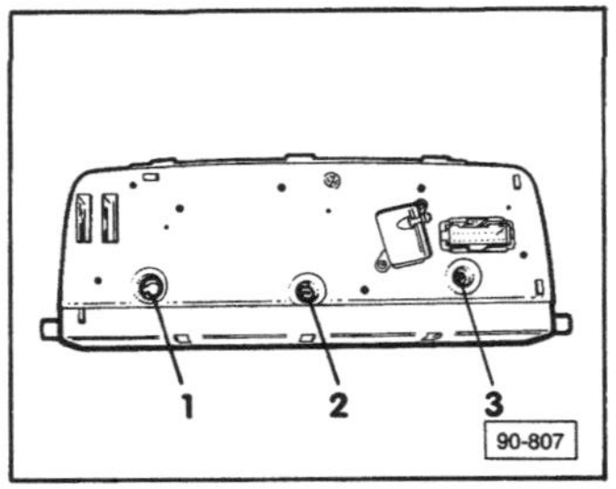

1 — open
2 — bulb for trip odometer display
3 — bulb multi-function indicator (MFA) display

Instrument cluster lights, removing/installing

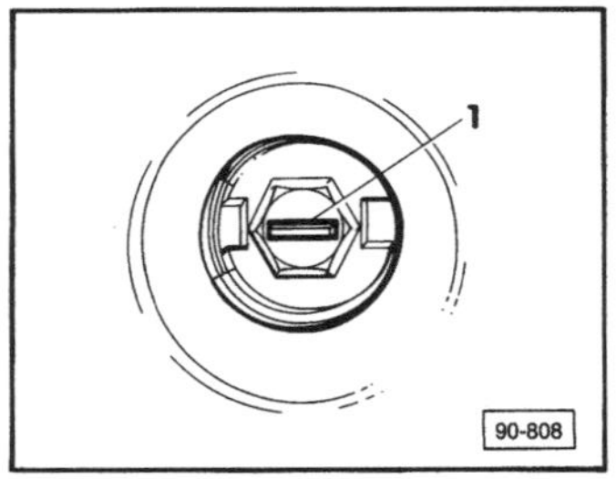

Removing

- turn slot **1** vertical to unlocked position
- remove bulb

Installing

- insert bulb
- turn slot **1** horizontal to locked position

Voltage stabilizer, checking

Check voltage stabilizer when fuel gauge and coolant temperature gauge simultaneously indicate false readings.

Leave battery ground strap and multi-point connector on instrument cluster connected for the following tests.

- remove instrument cluster trim
- remove tachometer by carefully pulling from printed circuit board

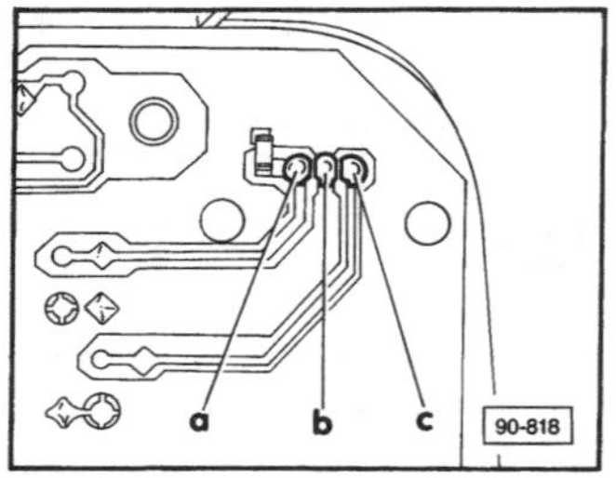

- set multimeter **Fluke 83 (US 1119)** to voltmeter scale and connect between positive connection **a** (terminal **15**) and ground **b**
- switch ignition **ON**
  - approximately battery voltage

If **NO**
- check wiring according to wiring diagram

- switch ignition **OFF**

- connect multimeter **Fluke 83 (US 1119)** between positive connection **c** and ground **b**
- switch ignition **ON**
  - must be 9.5-10.5 V

If **NO**
- replace instrument cluster printed circuit and housing

Note

The voltage stabilizer cannot be replaced separately. If the voltage stabilizer is defective, the entire instrument cluster printed circuit and housing must be replaced.

Printed circuit, component identification

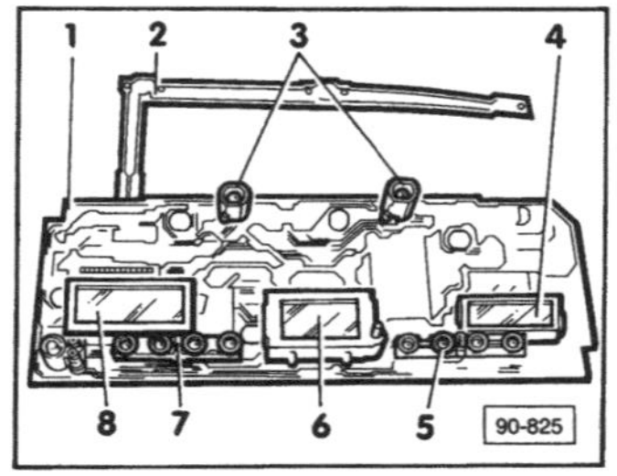

Front

1 — printed circuit
2 — printed circuit for instrument lighting
3 — turn signal indicator lights, left/right
4 — selector lever display window (not used)
5 — warning lights, right
6 — trip odometer display
- replace with printed circuit and housing as an assembly only

7 — warning lights, left
8 — multi-function indicator (MFA)
- replace with printed circuit and housing as an assembly only

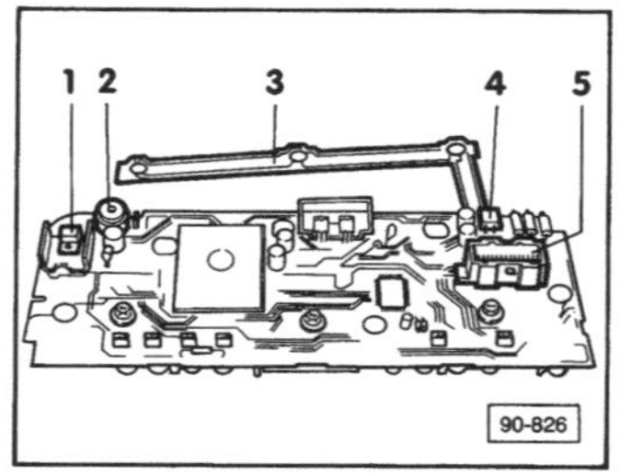

Rear

1 — voltage stabilizer
- replace with printed circuit and housing as an assembly only

2 — dynamic oil pressure warning buzzer
3 — printed circuit for instrument lighting
4 — printed circuit plug connector
5 — 28-point connector to wiring harness

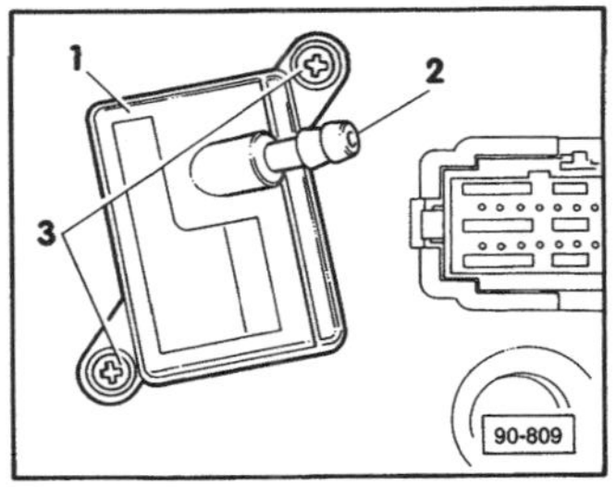

MFA vacuum sensor, removing/ installing

Removing

- remove instrument cluster
- remove vacuum hose from connector **2**
- remove screws **3**
- remove vacuum sensor **1** from instrument cluster

Installing

- install in reverse order of removal

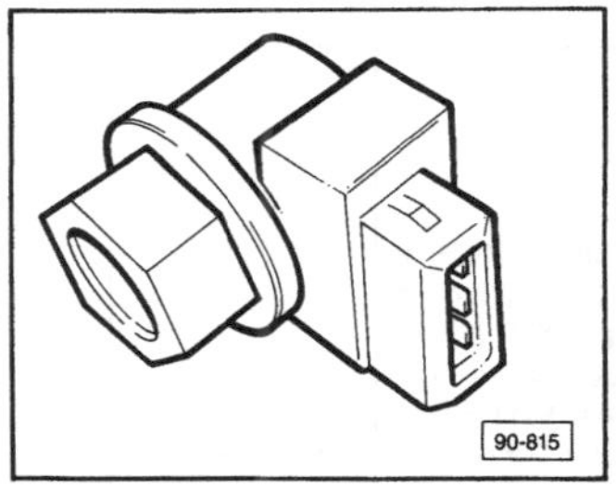

Vehicle Speed Sensor (VSS), speedometer (G 22), removing/installing

The Vehicle Speed Sensor (VSS), speedometer (**G 22**) is mounted on the transmission.

Removing

- remove wire connector
- loosen sensor using a 22 mm wrench
- remove sensor

Installing

- install in reverse order of removal

Instrument cluster, testing

Note

Always refer to appropriate wiring diagram when performing electrical tests.

Test conditions

- battery **OK**
- fuses **OK**
- 28-point connector at instrument cluster disconnected
- fuse/relay cover folded down for access to fuse/relay panel

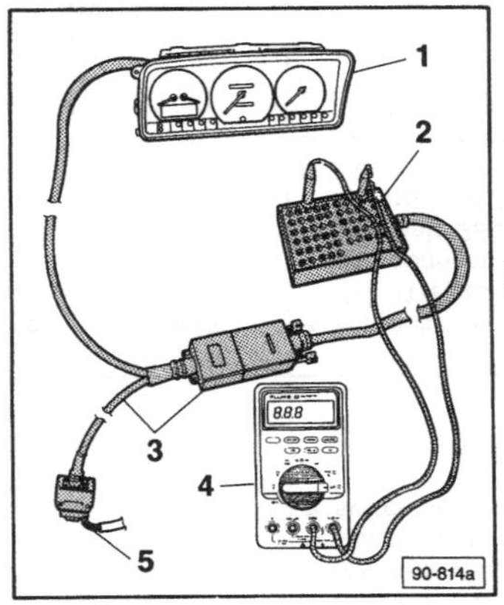

Tools required

- multimeter **Fluke 83 (US 1119)**
- test box **V.A.G. 1598** with harness adaptor **1598/8**
  - connect to instrument cluster 28-point connector and wiring harness
- L.E.D. tester **US 1115**

1 — instrument cluster
2 — test box **V.A.G. 1598**
3 — harness adaptor **1598/8**
4 — multimeter **Fluke 83 (US 1119)**
5 — instrument cluster wiring harness (28-point connector)

Test step	Item tested	Test condition	Specified value	Cause of problem
1	Power supply and ground (terminals **30** and **31**)	■ switch ignition **OFF** ■ set multimeter **Fluke 83** (**US 1119**) to volt range and connect between terminals **11** and **3** of test box **V.A.G. 1598**	Must be approx. 12V	If not **OK**, ■ check wiring for open or short circuits according to wiring diagram, repair as necessary If **OK**, ■ check printed circuit and connections for open or short circuits, repair/replace as necessary
2	Ignition power supply and ground (terminals **15** and **31**)	■ set multimeter **Fluke 83** (**US 1119**) to volt range and connect between terminals **13** and **3** of test box **V.A.G. 1598** ■ switch ignition **ON**	Must be approx. 12V	If not **OK**, ■ check wiring for open or short circuits according to wiring diagram, repair as necessary If **OK**, ■ check printed circuit and connections for open or short circuits, repair/replace as necessary
3	Turn signal, left	■ connect L.E.D. tester **US 1115** between terminals **22** and **3** of test box **V.A.G. 1598** ■ switch ignition **ON** ■ actuate left turn signal	L.E.D. tester must flash	If not **OK**, ■ check wiring for open or short circuits according to wiring diagram, repair as necessary If **OK**, ■ check printed circuit and connections for open or short circuits, repair/replace as necessary

Test step	Item tested	Test condition	Specified value	Cause of problem
4	Turn signal, right	■ connect L.E.D. tester **US 1115** between terminals **24** and **3** of test box **V.A.G. 1598** ■ switch ignition **ON** ■ actuate right turn signal	L.E.D. tester must flash	If not **OK**, ■ check wiring for open or short circuits according to wiring diagram, repair as necessary If **OK**, ■ check printed circuit and connections for open or short circuits, repair/replace as necessary
5	Instrument cluster/digital display illumination	■ set multimeter **Fluke 83** (**US 1119**) to volt range and connect between terminals **12** and **3** of test box **V.A.G. 1598** ■ switch ignition **ON** ■ switch lights **ON**	Must be approx. 2.5V to 12V depending on position of instrument panel light dimmer switch (**E20**)	If not **OK**, ■ check wiring for open or short circuits according to wiring diagram, repair as necessary If **OK**, ■ check printed circuit and connections for open or short circuits, repair/replace as necessary
6	High beam indicator light	■ set multimeter **Fluke 83** (**US 1119**) to volt range and connect between terminals **25** and **3** of test box **V.A.G. 1598** ■ switch ignition **ON** ■ switch high beam headlights **ON**	Must be approx. 12V	If not **OK**, ■ check wiring for open or short circuits according to wiring diagram, repair as necessary If **OK**, ■ check printed circuit and connections for open or short circuits, repair/replace as necessary
7	Generator light	■ set multimeter **Fluke 83** (**US 1119**) to volt range and connect between terminals **16** and **3** of test box **V.A.G. 1598** ■ start engine	Must be approx. 12V	If not **OK**, ■ check wiring for open or short circuits according to wiring diagram, repair as necessary If **OK**, ■ check printed circuit and connections for open or short circuits, repair/replace as necessary

Anti-theft alarm system, general information

The anti-theft alarm system has two types of alarms, visual and audible. If the alarm system is activated (vehicle broken into), the emergency flashers will flash, the alarm horn will sound and the starter will be deactivated.

The alarm system monitors and protects the following:

- engine compartment (hood)
- luggage compartment (rear hatch)
- doors
- ignition (starter lock-out)

The alarm system has a self-diagnostic feature which checks the operation of its components and wiring. When the alarm system is armed and all components are functional, the alarm horn will sound briefly after either front door is locked with the key.

The alarm system can only be armed when the hood, rear hatch and all doors are closed.

Note

The alarm system control module is mounted on the left A-pillar, above the fuse/relay panel.

Anti-theft alarm system, troubleshooting

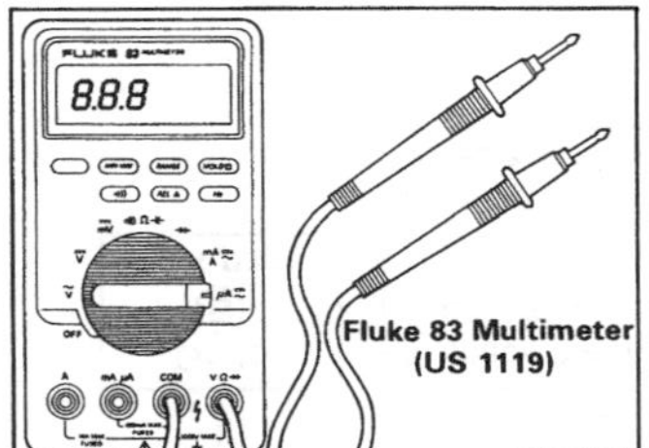

27-A069

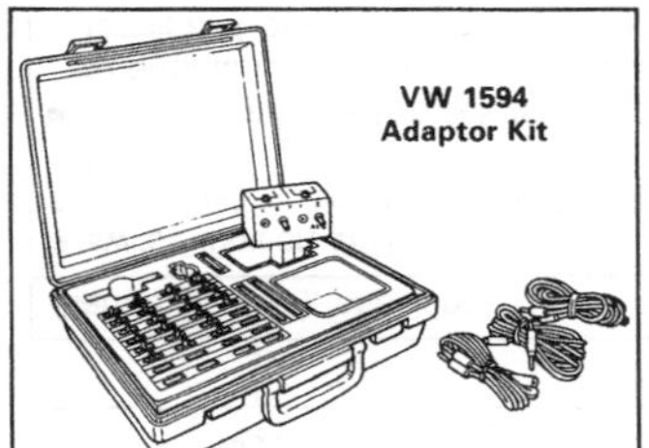

01-A06

Tools required

- use digital multimeter **Fluke 83** (**US 1119**) for all electrical tests
- use connector test kit **VW 1594** for all electrical test connections
- refer to wiring diagram at end of this section when troubleshooting

Alarm system cannot be switched ON (alarm horn does not beep when either front door is locked with key)

Test conditions

- battery **OK**
- power window and central locking system circuit breaker (S52, 20A) **OK** (located above fuse/relay panel)
- hood closed
- doors closed
- rear hatch closed
- obtain radio security code before disconnecting battery ground strap

Ground circuit for alarm system control unit (J284), checking

- disconnect battery ground strap
- remove instrument panel passenger side shelf
- remove ten-point (**T10**) and six-point (**T6**) connectors from anti-theft alarm system control unit, **J284**
- set digital multimeter **Fluke 83** (**US 1119**) to Ohm range
- connect multimeter black test lead to ground (−) and red test lead to terminal **3** of ten-point connector (**T10/3**)
  - must be continuity

Yes → Go to Ⓐ next page

No →
- check wiring from terminal **T10/3** to ground connection ㉜ for open circuits according to wiring diagram, repair as necessary

→ End

Ⓐ

Voltage supply (terminal 30), checking

- reconnect battery ground strap
- set digital multimeter **Fluke 83** (**US 1119**) to volt range
- connect multimeter black test lead to ground (−) and red test lead to terminal **1** (+) of ten-point connector (**T10/1**)
  - must be approximately 12 V
- leave multimeter black test lead connected to ground (−) and connect red test lead to terminal **3** (+) of six-point connector (**T6/3**)
  - must be approximately 12 V

No →
- check wiring for open circuits according to wiring diagram, repair as necessary

→ End

Yes →

Voltage supply (terminal 15), checking

- connect multimeter black test lead to ground (−) and red test lead to terminal **4** (+) of ten-point connector (**T10/4**)
- switch ignition **ON**
  - must be approximately 12 V

No →
- check wiring for open circuits according to wiring diagram, repair as necessary

→ End

Yes →

Alarm horn (H8), checking

- switch ignition **OFF**
- set digital multimeter **Fluke 83** (**US 1119**) to Ohm range
- connect multimeter black test lead to ground (−) and red test lead to terminal **4** of six-point connector (**T6/4**)
  - must be continuity

Yes → Go to Ⓑ next page

No →
- check alarm horn wiring for open circuits according to wiring diagram, repair as necessary

If wiring checks **OK**,
- replace alarm horn

→ End

(B)

Central locking system switches (F59 and F114), checking

- set digital multimeter **Fluke 83** (**US 1119**) to volt range
- connect multimeter black test lead to ground (−) and red test lead to terminal **7** (+) of ten-point connector (**T10/7**)
- insert key into driver's door lock
- turn key to lock position and hold against stop
- repeat for passenger door
  - must be approximately 12 V each time

Yes →
- connect multimeter black test lead to ground (−) and red test lead to terminal **6** (+) of ten-point connector (**T10/6**)
- insert key into driver's door lock
- turn key to open position and hold against stop
- repeat for passenger door
  - must be approximately 12 V each time

  Yes → Go to (C) next page

  No → check central locking system wiring and switches for open circuits according to wiring diagram, repair or replace as necessary → End

No → check central locking system wiring and switches for open circuits according to wiring diagram, repair or replace as necessary → End

(C)

Alarm system switches, checking

Alarm switch, hood (F120)

- set digital multimeter **Fluke 83** (**US 1119**) to Ohm range
- connect multimeter black test lead to ground (−) and red test lead to terminal **8** of ten-point connector (**T10/8**)
  - must be no continuity (∞) with hood closed and continuity (0 Ohms) with hood open

No → check wiring and switch for open or short circuits according to wiring diagram, repair or replace as necessary → End

Yes →

Door contact switch, left front (F2)

- close both doors
- connect multimeter black test lead to ground (−) and red test lead to terminal **10** of ten-point connector (**T10/10**)
  - must be no continuity (∞) with door closed and continuity (0 Ohms) with door open

No → check wiring and switch for open or short circuits according to wiring diagram, repair or replace as necessary → End

Yes →

Door contact switch, right front (F3)

- close both doors
- connect multimeter black test lead to ground (−) and red test lead to terminal **10** of ten-point connector (**T10/10**)
  - must be no continuity (∞) with door closed and continuity (0 Ohms) with door open

Yes → Go to (D) next page

No → check wiring and switch for open or short circuits according to wiring diagram, repair or replace as necessary → End

(D)

Alarm switch, trunk lock (F124)

- connect multimeter black test lead to ground (–) and red test lead to terminal **9** of ten-point connector **(T10/9)**
  - must be no continuity (∞) with trunk (rear hatch) closed and continuity (0 Ohms) with trunk (rear hatch) open

Yes → go to "Alarm trunk latch/luggage compartment light switch (F5)"

No →
- check wiring and switch for open or short circuits according to wiring diagram, repair or replace as necessary

→ End

Alarm trunk latch/luggage compartment light switch (F5)

- connect multimeter black test lead to ground (–) and red test lead to terminal **5** of ten-point connector **(T10/5)**
  - must be no continuity (∞) with trunk (rear hatch) closed and continuity (0 Ohms) with trunk (rear hatch) open

Yes →
- replace anti-theft alarm system control unit, J284

→ End

No →
- check wiring and switch for open or short circuits according to wiring diagram, repair or replace as necessary

→ End

Alarm system cannot be switched OFF

Test conditions

- anti-theft alarm system is switched **ON**
- battery **OK**
- obtain radio security code before disconnecting battery ground strap

Central locking system switches, driver's door (F59) and passenger's door (F114), checking

- disconnect battery ground strap
- remove instrument panel passenger side shelf
- remove ten-point (**T10**) and six-point (**T6**) connectors from anti-theft alarm system control unit, **J284**
- set digital multimeter **Fluke 83** (**US 1119**) to volt range
- connect multimeter black test lead to ground (–) and red test lead to terminal **7** (+) of ten-point connector (**T10/7**)
- reconnect battery ground strap
- insert key into driver's door lock
- turn key to lock position and hold against stop
- repeat for passenger door
  - must be approximately 12 V each time

Yes →
- connect multimeter black test lead to ground (–) and red test lead to terminal **6** (+) of ten-point connector (**T10/6**)
- insert key into driver's door lock
- turn key to open position and hold against stop
- repeat for passenger door
  - must be approximately 12 V each time

  Yes →
  - replace anti-theft alarm system control unit, **J284**

  → End

  No →
  - check central locking system wiring and switches for open circuits according to wiring diagram, repair or replace as necessary

  → End

No →
- check central locking system wiring and switches for open circuits according to wiring diagram, repair or replace as necessary

→ End

Lights flash but alarm horn does not sound when alarm is activated

Test conditions

- anti-theft alarm system is switched **ON**
- battery **OK**
- obtain radio security code before disconnecting battery ground strap

Alarm horn (**H8**), checking

- disconnect battery ground strap
- remove instrument panel passenger side shelf
- remove ten-point (**T10**) and six-point (**T6**) connectors from anti-theft alarm system control unit, **J284**
- set digital multimeter **Fluke 83** (**US 1119**) to Ohm range
- connect multimeter black test lead to ground (−) and red test lead to terminal **4** of six-point connector (**T6/4**)
  - must be continuity

Yes →

- replace anti-theft alarm system control unit, **J284**

→ End

No →

- check alarm horn wiring for open circuits according to wiring diagram, repair as necessary

If wiring checks **OK**,

- replace alarm horn

→ End

Electrical System – Instruments

Alarm horn sounds but lights do not flash when alarm is activated

Test conditions

- anti-theft alarm system is switched **ON**
- battery **OK**
- emergency flasher system **OK**
- obtain radio security code before disconnecting battery ground strap

Emergency flasher wiring to anti-theft alarm system control unit (**J284**), checking

- disconnect battery ground strap
- remove instrument panel passenger side shelf
- remove ten-point (**T10**) and six-point (**T6**) connectors from anti-theft alarm system control unit, **J284**
- set digital multimeter **Fluke 83** (**US 1119**) to volt range
- connect multimeter black test lead to ground (−) and red test lead to terminal **1** (+) of six-point connector (**T6/1**)
- reconnect battery ground strap
- switch emergency flashers **ON**
  - must be approximately battery voltage when lights flash **ON**
- leave multimeter black test lead connected to ground (−) and repeat test with red test lead connected to terminal **2** (+) of six-point connector (**T6/2**)
  - must be approximately battery voltage when lights flash **ON**

Yes →

- replace anti-theft alarm system control unit, **J284**

→ End

No →

- check wiring from control unit to lights for open circuits according to wiring diagram, repair as necessary

→ End

Engine does not crank with alarm system OFF

Test conditions

- anti-theft alarm system switched **OFF**
- battery **OK**
- obtain radio security code before disconnecting battery ground strap

Ignition/start wiring to anti-theft alarm system control unit (**J284**), checking

- disconnect battery ground strap
- remove instrument panel passenger side shelf
- remove ten-point (**T10**) and six-point (**T6**) connectors from anti-theft alarm system control unit, **J284**
- set digital multimeter **Fluke 83** (**US 1119**) to volt range
- connect multimeter black test lead to ground (−) and red test lead to terminal **5** (+) of six-point connector (**T6/5**)
- reconnect battery ground strap
- insert ignition key into ignition switch
- turn ignition switch to "start" position while reading multimeter
  - must be approximately battery voltage

Yes

- remove ignition key
- disconnect battery ground strap
- set digital multimeter **Fluke 83** (**US 1119**) to Ohm range
- connect multimeter black test lead to ground (−) and red test lead to terminal **6** of six-point connector (**T6/6**)
  - must be continuity

Yes

- replace anti-theft alarm system control unit, **J284**

End

No

- check wiring to starter for open circuits according to wiring diagram, repair as necessary

End

No

- check wiring to ignition/start switch for open circuits according to wiring diagram, repair as necessary

End

Description	Current track
Alarm trunk latch/luggage compartment light switch, F5	16
Alarm horn, H8	1
Alarm switch, hood, F120	2
Alarm switch, trunk lock, F124	6
Anti-theft alarm system control unit, J284	1-22
Central locking system switch (Passenger's door), F114	23-24
Door contact switch, left front, F2	9
Door contact switch, right front, F3	14
Fuel pump relay, J17	3

Wire connectors

T1 — single, behind fuse/relay panel
T1a — single, behind fuse/relay panel
T1b — single, on left A-pillar, near fuse/relay panel
T1c — single, on lower right A-pillar
T1d — single, in rear lid
T2a — double, on lower right A-pillar
T2b — double, in luggage compartment, right
T2c — double, behind fuse/relay panel
T2d — double, behind fuse/relay panel
T3 — three-point, on lower right A-pillar
T3c — three-point, on lower right A-pillar
T3d — three-point, on lower left A-pillar, near fuse/relay panel
T3f — three-point, in passenger's door
T3j — three-point, on left front turn signal
T3k — three-point, on right front turn signal
T4 — four-point, behind fuse/relay panel
T6 — six-point, on anti-theft alarm system control unit
T8a — eight-point, on power window control unit
T10 — ten-point, on anti-theft alarm system control unit
T12 — twelve-point, on central locking system pump and control unit
T28 — twenty eight-point, on instrument cluster
T36 — junction connector, terminal 30, behind fuse/relay panel

Ground connections

(32) — behind instrument panel, left
(52) — in rear lid, left
(53) — in rear lid, right
(106) — in anti-theft alarm system wiring harness

Welded wiring harness points

(C15) — wire connection, in headlight wiring harness
(C16) — wire connection, in headlight wiring harness
(Q4) — plus (+) connection (30), in power window/central locking system wiring harness
(Q22) — wire connection, in rear lid wiring harness
(Q51) — plus (+) connection (30), in anti-theft alarm system wiring harness
(Q53) — plus (+) connection (15), in anti-theft alarm system wiring harness
(Q54) — wire connection (left flasher), in anti-theft alarm system wiring harness
(Q55) — wire connection (right flasher), in anti-theft alarm system wiring harness
(Q56) — wire connection (door locks open), in anti-theft alarm system wiring harness
(Q57) — wire connection (door locks locked), in anti-theft alarm system wiring harness
(S1) — plus (+) connection (30), in central locking system wiring harness
(S4) — plus (+) connection (30), in central locking system wiring harness

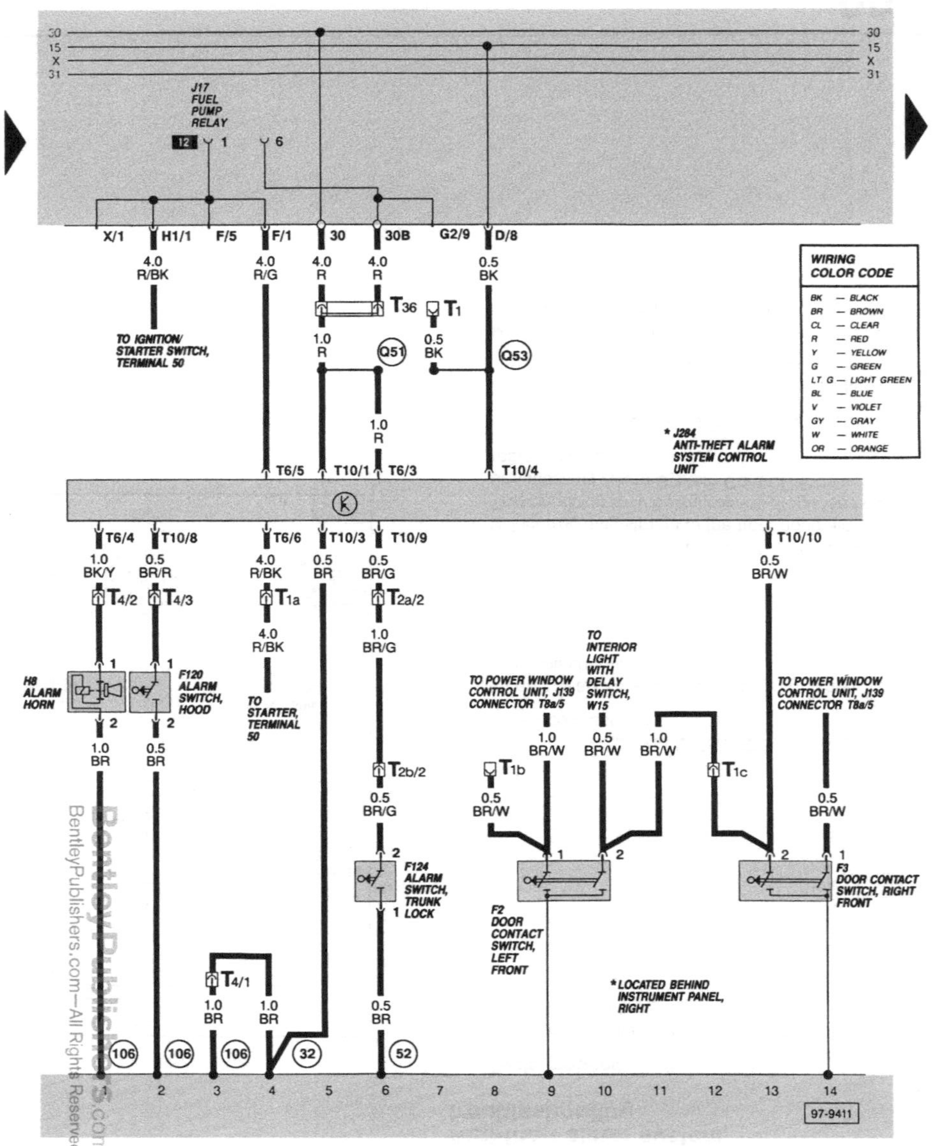

Electrical System – Instruments

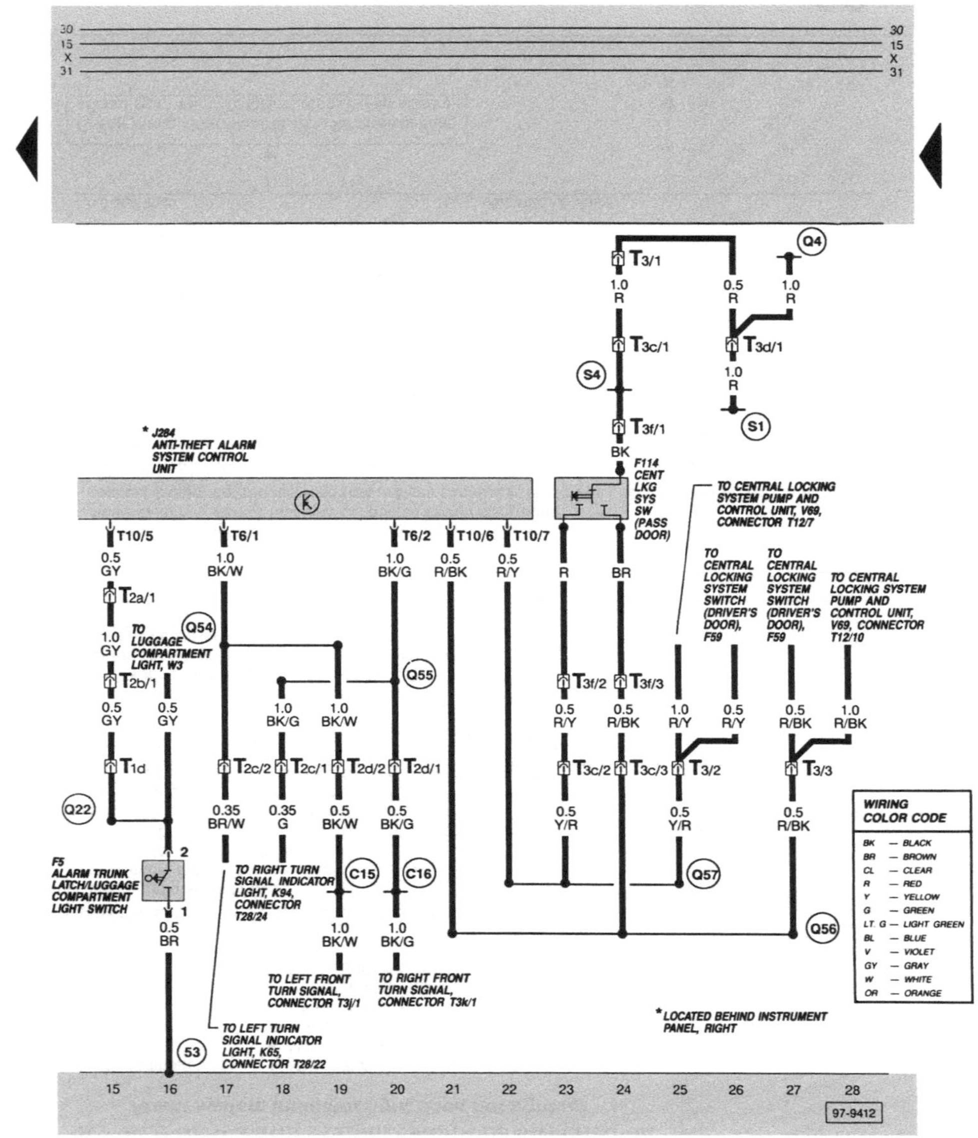

Anti-theft alarm system, troubleshooting

Tools required

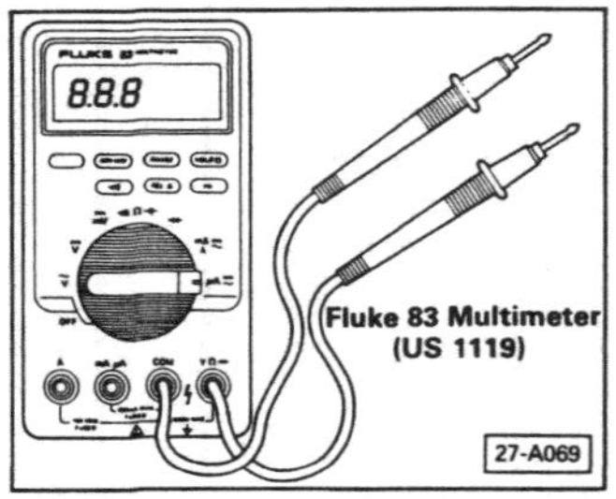

- use digital multimeter **Fluke 83** (**US 1119**) for all electrical tests

- use connector test kit **VW 1594** for all electrical test connections
- refer to appropriate wiring diagram when troubleshooting

Note

The alarm system may be inadvertently activated due to incorrect adjustment of the front hood alarm switch. This condition may be improperly diagnosed causing unnecessary Alarm System Control Module replacements.

If alarm system activates for no apparent reason,

- check system according to troubleshooting information on the following pages
- check adjustment of front hood alarm switch and adjust as necessary by bending switch bracket

A - Anti-theft alarm system cannot be switched on
Alarm system indicator light does not light up

Test conditions:
- Rear hatch closed and locked
- Doors closed
- Hood closed
- Battery voltage OK
- Power window and central locking system fuses OK
- Fuses S90 and S91 OK

Alarm system control module (J284) ground connection, checking

- obtain radio security code
- disconnect battery ground strap
- remove lower right kick panel
- disconnect 10-pin and 6-pin connectors from alarm system control module (J284)
- set multimeter Fluke 83 to Ohm measuring range
- connect multimeter black test lead to vehicle ground (GND)
- connect multimeter red test lead to 10-pin connector, terminal 3 (T10/3)

∞ Ohms (no continuity)	0 Ohms (continuity)
Check wiring from T10/3 to ground connection (32) for open circuits according to wiring diagram, repair as necessary.	→ next step

Voltage supply (B+, terminal 30), checking

- reconnect battery ground strap
- set Fluke 83 multimeter to DC Volt measuring range
- leave multimeter black test lead connected to vehicle ground
- connect multimeter red test lead first to 6-pin connector, terminal 3 (T6/3) then to 10-pin connector, terminal 1 (T10/1)

No voltage at one or both terminals	Approx.12 Volts at both terminals
Check wiring to terminal 30 (B+) for open circuits according to wiring diagram, repair as necessary	→ next step

Voltage supply (B+, terminal 15), checking

- leave multimeter set to Volt range and black test lead connected to ground
- connect multimeter red test lead to 10-pin connector, terminal **4** (T10/4)
- switch ignition on

No voltage	Approx.12 Volts
Check wiring to terminal 15 (B+) for open circuits according to wiring diagram, repair as necessary.	→ A

A

Alarm horn (H8), checking

- leave multimeter black test lead connected to ground
- disconnect battery ground strap
- set multimeter to Ohm measuring range
- connect multimeter red test lead to 6-pin connector, terminal 4 (T6/4)

∞ Ohms (no continuity)	0 Ohms (continuity)

∞ Ohms (no continuity) →

Check wiring to alarm horn for open circuit according to wiring diagram, repair as necessary or

- replace alarm horn (H8)

0 Ohms (continuity) →

Central locking system switches (F59 and F114), checking

- leave multimeter black test lead connected to ground
- connect multimeter red test lead to 10-pin connector, terminal 7 (T10/7)
- set multimeter to DC Volt measuring range
- reconnect battery ground strap
- insert ignition key into driver's door lock cylinder, turn to "LOCKED" position and hold
- repeat for passenger's door

No voltage for one or both doors	Approx.12 volts for both doors

No voltage for one or both doors →

Check central locking system switches (F59/F114) and wiring to switches for open circuit according to wiring diagram, repair wiring or replace switches as necessary

Approx.12 volts for both doors →

- leave multimeter set to Volt range and black test lead connected to ground
- connect multimeter red test lead to 10-pin connector, terminal 6 (T10/6)
- insert ignition key into driver's door lock cylinder, turn to "UNLOCKED" position and hold
- repeat for passenger's door

No voltage for one or both doors	Approx. 12 Volts for both doors

No voltage for one or both doors → Check central locking system switches (see above)

Approx. 12 Volts for both doors → B

B

Check alarm switch, hood (F120), door contact switches (F2 and F3), alarm/central locking switch, trunk lock (F 124) and luggage compartment light switch (F5)

- leave multimeter black test lead connected to ground
- disconnect battery ground strap
- set multimeter to Ohm measuring range
- connect multimeter red test lead to the following terminals, one after the other:
♦ 10-pin connector, terminal 8 (T10/8)
♦ 10-pin connector, terminal 10 (T10/10)
♦ 10-pin connector, terminal 9 (T10/9)
♦ 10-pin connector, terminal 5 (T10/5)

0 Ohms (continuity) at one or more terminals	∞ Ohms (no continuity) at all terminals

0 Ohms (continuity) at one or more terminals →

Check wiring and respective switch for short circuit to ground according to wiring diagram, repair wiring or replace switch as necessary

∞ Ohms (no continuity) at all terminals →

Alarm system indicator light (K95), checking

- leave multimeter black test lead connected to ground
- reconnect battery ground strap
- set multimeter to 10A DC measuring range
- connect multimeter red test lead to 10-pin connector, terminal 2 (T10/2)

Alarm system indicator light not on	Alarm system indicator light on

Alarm system indicator light not on →

Check wiring to indicator light for open circuits according to wiring diagram, repair as necessary

Alarm system indicator light on →

Replace alarm system control module (J284)

B - Anti-theft alarm system cannot be switched off

Test conditions:
- Anti-theft alarm system switched on
- Battery voltage OK

Central locking system switches (F59 and F114), checking

- obtain radio security code
- disconnect battery ground strap
- remove lower right kick panel
- disconnect 10-pin and 6-pin connectors from alarm system control module (J284)
- set multimeter Fluke 83 to DC Volt measuring range
- connect multimeter black test lead to vehicle ground
- connect multimeter red test lead to 10-pin connector, terminal 7 (T10/7)
- reconnect battery ground strap
- insert ignition key into driver's door lock cylinder, turn to "LOCKED" position and hold
- repeat for passenger's door

No voltage for one or both doors	Approx. 12 Volts for both doors

No voltage for one or both doors →

Check central locking system switches (F59/F114) and wiring to switches for open circuit according to wiring diagram, repair wiring or replace switches as necessary.

Approx. 12 Volts for both doors →

- leave multimeter set to Volt range and black test lead connected to ground
- connect multimeter red test lead to 10-pin connector, terminal 6 (T10/6)
- insert ignition key into driver's door lock cylinder, turn to "UNLOCKED" position and hold
- repeat for passenger's door

No voltage for one or both doors	Approx. 12 Volts for both doors

No voltage for one or both doors → Check central locking system switches (see above)

Approx. 12 Volts for both doors →

Replace alarm system control module (J284)

C - Activated anti-theft alarm system does not trigger alarm horn

Test conditions:
- Anti-theft alarm system activated
- Battery voltage OK

Alarm horn (H8), checking

- obtain radio security code
- disconnect battery ground strap
- remove lower right kick panel
- disconnect 10-pin and 6-pin connectors from alarm system control module (J284)
- set multimeter Fluke 83 to Ohm measuring range
- connect multimeter black test lead to vehicle ground
- connect multimeter red test lead to 6-pin connector, terminal 4 (T6/4)

∞ Ohms (no continuity)	0 Ohms (continuity)

∞ Ohms (no continuity) →

Check wiring to alarm horn for open circuits according to wiring diagram, repair as necessary or

→ Replace alarm horn (H8)

0 Ohms (continuity) →

Replace alarm system control module (J284)

D - Activated anti-theft alarm system does not trigger flasher lights

Test conditions:
- Anti-theft alarm system activated
- Battery voltage OK
- Emergency flasher system OK

Connection from emergency flasher system to alarm system control module (J284), checking

- obtain radio security code
- disconnect battery ground strap
- remove lower right kick panel
- disconnect 10-pin and 6-pin connectors from alarm system control module (J284)
- set multimeter Fluke 83 to DC Volt measuring range
- connect multimeter black test lead to vehicle ground
- connect multimeter red test lead to 6-pin connector, terminal 1 or 2 (T6/1 or T6/2)
- reconnect battery ground strap
- switch on emergency flasher system

No voltage at one or both terminals	Approx. 12 Volts at both terminals
↓ Check wiring for open circuits according to wiring diagram, repair as necessary	↓ Replace alarm system control module (J284)

E - Starter does not crank with anti-theft alarm system switched off

Test conditions:
- Anti-theft alarm system switched off
- Battery voltage OK

Alarm system control module (J284), checking

- obtain radio security code
- disconnect battery ground strap
- remove lower right kick panel
- disconnect 10-pin and 6-pin connectors from alarm system control module (J284)
- set multimeter Fluke 83 to DC Volt measuring range
- connect multimeter black test lead to vehicle ground
- connect multimeter red test lead to 6-pin connector, terminal 5 (T6/5)
- reconnect battery ground strap
- insert key into ignition/starter switch, turn to "START" position and hold

No voltage	Approx. 12 Volts
↓ Check wiring to ignition/starter switch for open circuit according to wiring diagram, repair as necessary	↓ (continue below)

- leave multimeter black test lead connected to ground
- remove ignition key from ignition/starter switch
- set multimeter to Ohm measuring range
- connect multimeter red test lead to 6-pin connector, terminal 6 (T6/6)

∞ Ohms (no continuity)	0 Ohms (continuity)
↓ Check wiring to starter for open circuits according to wiring diagram, repair as necessary	↓ Replace alarm system control module (J284)

THIS PAGE INTENTIONALLY LEFT BLANK

Index

Radio

Heidelberg VI fixed coded radio
- coding a new or remanufactured radio **91.2**
- reactivating radio after reconnecting power **91.4**
- unlocking an electronically locked radio **91.5**
- Vehicle Speed Sensor (VSS) -G54- circuit, wire connection at radio **91.5**

Multi-function Indicator (MFA)

Hall Vehicle Speed Sensor (VSS)
- checking (up to Aug. 1990) **91.24**
- contact plate, removing/installing **91.7**
- location **91.6**

Multi-function Indicator (MFA)
- cancelling the memories (from 1993 m.y.) **91.53**
- connector wiring **91.6**
- driving data memory (from 1993 m.y.) **91.52**
- general notes **91.6**
- instrument cluster, checking (from 1993 m.y.) **91.55**
- MFA check mode test (from 1993 m.y.) **91.54**
- Multi-function Indicator Mode Select Switch (E 86) and Memory Switch (E 109) (from 1993 m.y.) **91.53**
- operation and function (from 1993 m.y.) **91.52**
- printed circuit, identifying **91.8**
- printed circuit, removing **91.8**
- setting the time (from 1993 m.y.) **91.53**
- vacuum sensor, removing **91.7**
- wiring (from Aug. 1990) **91.47**
- wiring (up to Aug.1990) **91.25**

Multi-function Indicator (MFA) self test (up to Aug. 1990)
- checking **91.9**
- check mode test **91.9**
- measuring mode test **91.11**

Multi-function Indicator (MFA) self test (Aug. 1990 through 1992 m.y.)
- checking **91.31**
- check mode test **91.31**
- measuring mode test **91.33**

Multi-function Indicator (MFA) troubleshooting (up to Aug. 1990)
- ambient (outside) temperature display inaccurate **91.19**
- average driving speed indicator is not accurate or displays "— — —" or "0" **91.14**
- average fuel consumption display not accurate **91.15**
- clock inaccurate **91.13**
- clock operates in incorrect mode **91.13**
- display errors or no display **91.13**
- driving distance display not accurate **91.14**
- driving distance display remains at 0 **91.14**
- driving time remains at 0:00 **91.14**
- memory cannot be canceled/memory does not store information **91.21**
- no engine RPM display or engine RPM display faulty, (while performing measuring mode test) **91.17**
- oil temperature display inaccurate **91.20**
- "signal from speed sensor" display does not change (during measuring mode test) **91.24**

Index

Multi-function Indicator (MFA) troubleshooting (Aug. 1990 through 1992 m.y.)
- ambient (outside) temperature display inaccurate **91.42**
- average driving speed indicator is not accurate or displays "— — —" or "0" **91.36**
- average fuel consumption display not accurate **91.37**
- clock inaccurate **91.35**
- clock operates in incorrect mode **91.35**
- display errors or no display **91.35**
- driving distance display not accurate **91.36**
- driving distance display remains at 0 **91.36**
- driving time remains at 0:00 **91.36**
- electronic speedometer and odometer checking **91.44**
- memory cannot be canceled/memory does not store information. **91.39**
- oil temperature display inaccurate **91.43**
- tachometer inaccurate or does not work **91.45**
- Vehicle Speed Sensor (VSS), Speedometer, (G 22), checking **91.46**

Multi–Function Indicator (MFA) troubleshooting (from 1993 m.y.)
- average fuel consumption display not accurate **91.57**
- average speed display remains at 0 or display not accurate **91.57**
- clock inaccurate **91.56**
- driving distance not accurate **91.57**
- driving distance remains at 0 **91.56**
- driving time remains at 0:00 **91.56**
- Multi-Function Indicator Mode Select Switch (E86), Memory Switch (E109) and wiring, checking **91.60**
- oil temperature display, checking **91.58**
- oil temperature display inaccurate **91.57**
- outside (ambient) temperature display, checking **91.59**
- outside (ambient) temperature display inaccurate **91.57**
- total or single trip memory of MFA not OK **91.56**
- Vehicle Speed Sensor (VSS), Speedometer, checking **91.63**

Note
As of model year 1993, the Multi–function indicator will be identified in all service information with the new abbreviation MFA. The old abbreviation, MFI, is now used as an emission–related term for Multiport Fuel Injection. MFA will only appear on Multi–function indicator information from model year 1993; Multi–function indicator information up to model year 1993 will not change and will continue to use the MFI abbreviation.

Heidelberg VI fixed coded radio, coding

The Heidelberg VI radio is assigned a fixed four digit security code from the factory. Unlike previous coded radios, **the security code cannot be changed**.

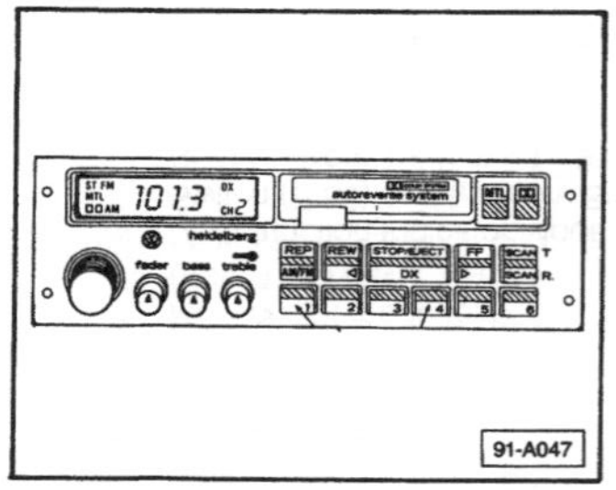

91-A047

The Heidelberg VI radio can be identified by white lettering on the face plate and separate knobs for fader, bass and treble control.

The radios are not coded at the factory. To activate the theft protection, the radio must be coded during the Pre-Delivery Inspection.

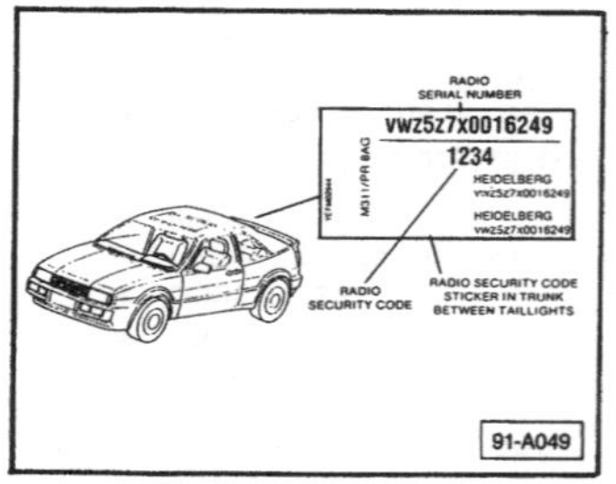

91-A049

The security code and serial number of the radio are printed on a removable sticker located in the trunk near the vehicle identification sticker.

Coding a new or remanufactured radio

- remove radio security code/serial number sticker from trunk and attach to radio card in radio owner's manual

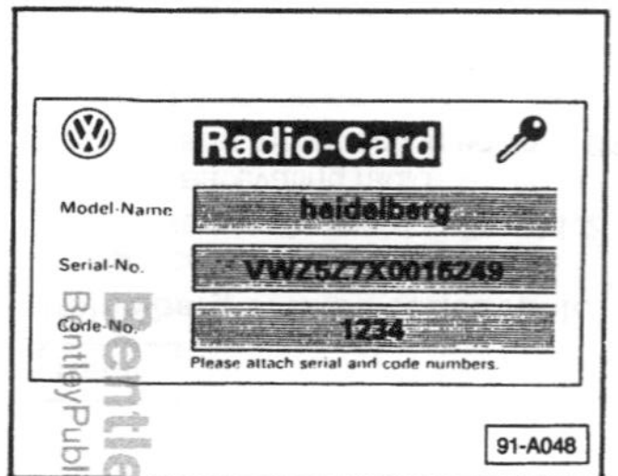

91-A048

> **CAUTION**
>
> **DO NOT** keep the security code or radio card with the vehicle. Always instruct customers to detach radio card from operating instructions manual and keep in a safe place.

Note

The security code/serial number sticker for remanufactured replacement radios comes in the box with the new radio. Place the new sticker over the old sticker on the radio card.

- switch radio **ON**
  - radio plays and radio station frequency is displayed
- push and hold **AM/FM** and **SCAN** buttons to activate coding procedure
  - radio display changes to **CODE** and then **1000**
- release **AM/FM** and **SCAN** buttons
  - **1000** will remain on display
- enter radio's security code using first four program station buttons
  - security code will appear on display

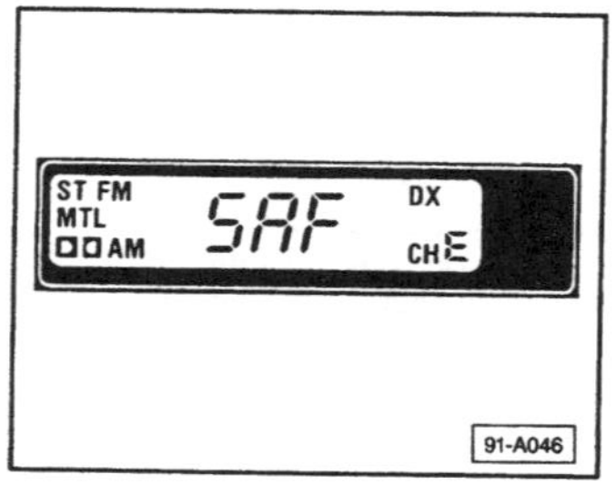

91-A046

- push and hold **AM/FM** and **SCAN** buttons once again until display changes to **SAFE**

Note

A new or remanufactured radio will only accept its factory assigned security code. If an incorrect code is entered, the radio display will not change to **SAFE**.

- release **AM/FM** and **SCAN** buttons
  - radio display changes to a radio station frequency
  - radio is now coded

91-A046

Reactivating radio after reconnecting power

After disconnecting power (fuse, battery, etc.), the radio will lock up electronically.

To reactivate:

- obtain radio's security code
  - located on radio card
- switch radio **ON**
  - radio will not play
  - radio display shows **SAFE**
- push and hold **AM/FM** and **SCAN** buttons
  - display changes to **1000**
- release **AM/FM** and **SCAN** buttons
  - **1000** remains on display

Note

If the **AM/FM** and **SCAN** buttons are held down too long or pushed again, the radio will misinterpret the **1000** as an attempt at coding and one incorrect attempt will be logged.

- enter radio's security code using first four program station buttons
  - security code will appear on display
- push and hold **AM/FM** and **SCAN** buttons once again until display changes to **SAFE**
- release **AM/FM** and **SCAN** buttons
  - radio display changes to a radio station frequency
  - radio is now coded and plays

Unlocking an electronically locked radio

After two incorrect attempts at entering the security code, the radio will lock-up electronically. The display shows **SAFE** and will not change.

To unlock:

- leave radio switched **ON** for approximately one hour
- after one hour, reactivate (code) radio as it would be after a power interruption

Note

The radio will never permanently lock-up no matter how many incorrect coding attempts are made. The reactivation procedure can be repeated indefinitely.

Vehicle Speed Sensor (VSS) -G 54 circuit, wire connection to radio

As of 1991 m.y., the Vehicle Speed Sensor (VSS) -G 54 circuit is supplied to radio connector T8/1 via a blue/white wire.

The purpose of this circuit is to provide Vehicle Speed information to the automatic volume control circuitry of radios used in some markets.

To avoid problems in the engine management system, shifting problems in the electronically controlled automatic transmission and functional problems of the cruise control system, **DO NOT** splice any wires into this circuit or connect this wire to any component (for example, aftermarket radios).

If a problem exists in the Vehicle Speed Sensor (VSS) circuit, do the following first:

1. Inspect vehicle to determine if the radio is a VW supplied radio. If not, ensure that the VSS -G 54 circuit is disconnected from the radio.
2. Determine if a wire is spliced into the VSS -G 54 circuit; if so, remove splice.

Multi-function indicator (MFA), general notes

The multi-function indicator (MFA) provides driving information using the following inputs:

- impulse signal from ignition coil, terminal **1**
- impulse signal from hall speed sensor (**G54**)
- signal from MFA vacuum sensor (**G55**)
- resistance value from oil temperature sensor (**G8**)
- resistance value from outside temperature sensor (**G17**)

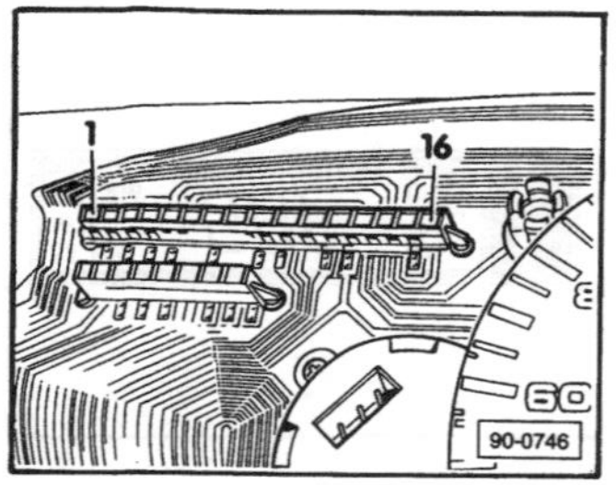

Multi-function indicator (MFA), connector wiring

1 — to MFA mode switch
2 — from oil temperature sensor
3 — outside temperature sensor signal
4 — plus (+) connection, terminal **15**
5 — not used
6 — ground (−) for outside temperature sensor
7 — not used
8 — ground (−) for MFA, terminal **31**
9 — to MFA reset switch
10 — not used
11 — to MFA memory switch
12 — Hall speed sensor signal
13 — not used
14 — not used
15 — tachometer output signal
16 — not used

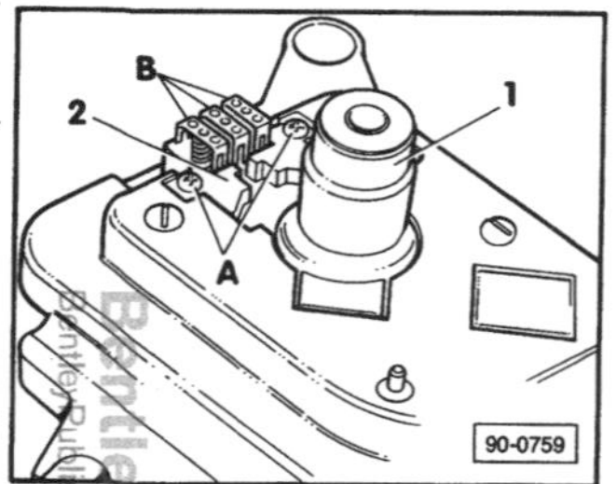

Hall Vehicle Speed Sensor, location

The Hall Vehicle Speed Sensor is used to provide an input signal to the multi–function indicator (MFA), cruise control system, and speed dependent rear spoiler adjustment system.

The Hall Vehicle Speed Sensor is located on the rear of the speedometer housing.

1 — speedometer drive cable connection
2 — Hall Vehicle Speed Sensor
A — mounting screws
B — spring loaded contact

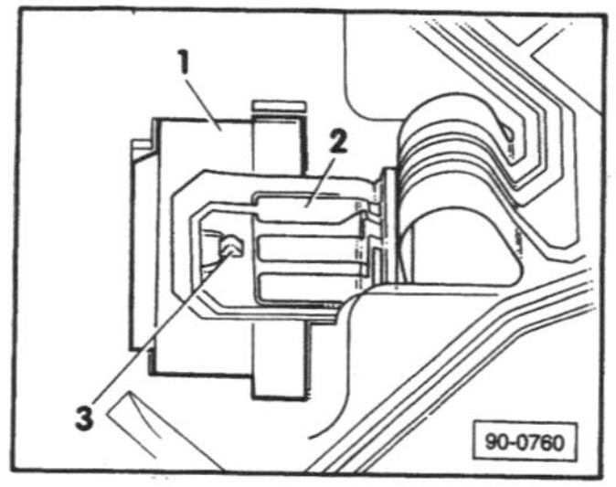

Hall Vehicle Speed Sensor contact plate, removing/installing

Contact plate **1** presses printed circuit **2** against spring loaded contacts to make connection.

CAUTION

Before working on any part of the electrical system, disconnect battery ground strap.

Removing

- remove speedometer, see Repair Group 90
- remove Hall Vehicle Speed Sensor mounting screws
- detach printed circuit from lug **3**
- remove contact plate **1**

Installing

- install (keeping flat) printed circuit **2** into contact plate **1**
- locate printed circuit **2** on lug **3**
- install contact plate **1** and printed circuit **2** into the rear of the speedometer housing
- install mounting screws and tighten

MFA vacuum sensor, removing

Two different MFA vacuum sensors are used depending upon manufacturer of MFA printed circuit. Always replace MFA vacuum sensor with the same make as original.

CAUTION

MFA vacuum sensor and MFA printed circuit are a functional combination. **DO NOT** interchange MFA components between the two types of printed circuits.

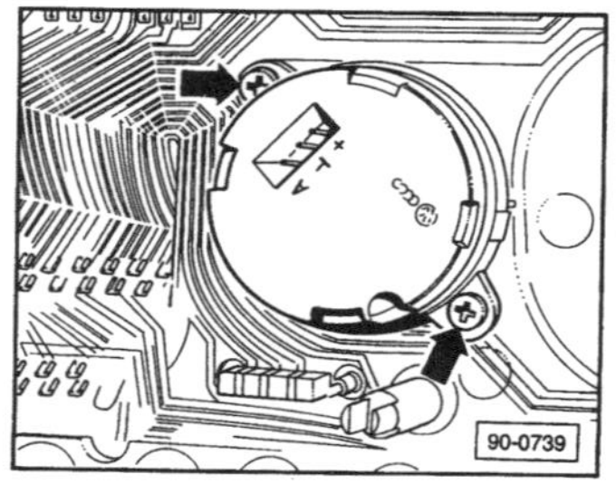

- remove vacuum hoses from back of instrument cluster
- remove combination fuel, coolant temperature, and low coolant level indicator gauge from instrument cluster
- remove MFA vacuum sensor mounting screws (**arrows**)
- remove MFA vacuum sensor

MFA printed circuit, identifying

Two different makes of MFA printed circuits are used for the MFA system. One printed circuit is manufactured by Motometer and the other by VDO. Identification is the same for both types, Fig. 1 and Fig. 2

> **CAUTION**
>
> Before working on any part of the electrical system, disconnect battery ground strap.

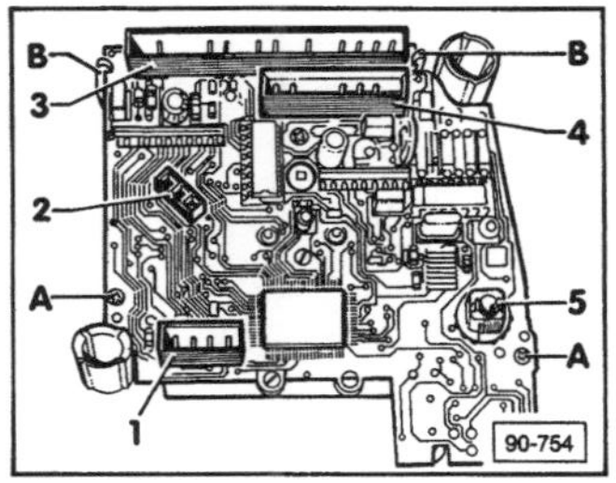

▶ **Fig. 1 MFA printed circuit, manufacturer: Motometer**

1 — 4-point connector, digital clock
2 — 3-point connector, MFA vacuum sensor
3 — 16-point connector, MFA
4 — 8-point connector, combination fuel, coolant temperature, and low coolant level indicator gauge
5 — bulb 12 V/1.2 W (watts)

MFA printed circuit, removing

Repair and removal are the same for both makes of MFA printed circuits, Fig. 1 and Fig. 2.

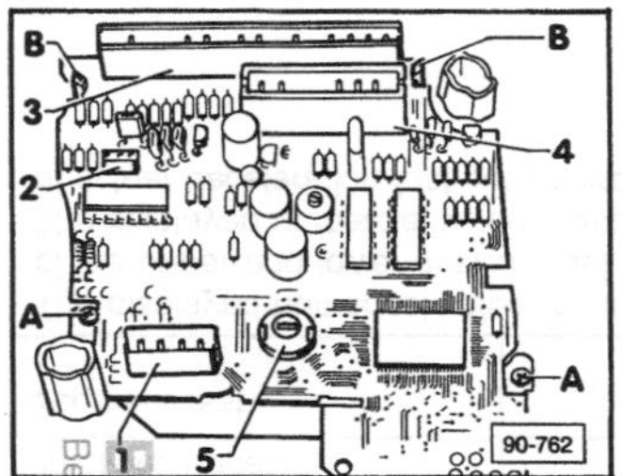

▶ **Fig. 2 MFA printed circuit, manufacturer: VDO**

- remove combination fuel, coolant temperature, and low coolant level indicator gauge, see Repair Group 90
- remove gauge mounting screws **A**
- spread retaining lugs **B** and separate MFA printed circuit from combination fuel, coolant temperature, and low coolant level indicator gauge

> **CAUTION**
>
> The MFA printed circuit and MFA vacuum sensor are a functional combination. **DO NOT** interchange MFA components between the two types of printed circuits.

Multi-function indicator (MFI) self-test, checking

Perform MFI check mode and measuring mode tests for all MFI complaints before doing any repairs.

Check mode test

Check mode test verifies MFI is operating with the correct base vehicle information.

- set MFI memory switch (**E109**) to position **1** (short term/single trip memory)
- switch ignition **ON**
- press and hold MFI mode select button
- switch ignition **OFF**, then back **ON**
- release mode select button
  - engine map code is displayed
- compare engine map code with chart on next page
  - numerical value must match
- press and release mode select button to advance to next mode (country code)
- compare code to chart
  - country code must match
- continue pressing and releasing mode select button to advance sequentially from mode to mode
- compare numerical values to chart
  - all values must match

If any values do not match chart,
- replace MFI printed circuit board

To leave check mode test,
- switch ignition **OFF**

Check mode test functions

Function	Times MFI mode select button is pressed	Mode indicated	Value displayed
Engine map code	0	—	**018**
Country code	1	clock	**4** (USA) **1** (CANADA)
Speedometer drive ratio	2	mpg or L/100km	**955** (USA) **950** (CANADA)
Upper rpm limit*	3	miles or km	**7418**
Lower rpm limit*	4	mph or km/h	**1175**
Deceleration fuel shut-off pressure (mbar)**	5	oil temp.	**57**
Segment test	6	—	all segments of display light up

*The displayed numerical value must match the numerical value in the chart, it is **not** an indication of the upper and lower overrun fuel cut-off rpm.

The displayed numercial value must match the numerical value in the chart, it is **not an indication of actual deceleration fuel cut-off pressure.

Note

As of model year 1993, the Multi–function indicator will be identified in all service information with the new abbreviation MFA. The old abbreviation, MFI, is now used as an emission–related term for Multiport Fuel Injection. MFA will only appear on Multi–function indicator information from model year 1993; Multi–function indicator information up to model year 1993 will not change and will continue to use the MFI abbreviation.

Measuring mode test

Measuring mode test provides a display of measured functions during vehicle operation.

- set MFI memory switch (**E109**) to position **2** (long term/total trip memory)
- switch ignition **ON**
- press and hold MFI mode select button
- switch ignition **OFF**, then back **ON**
- release mode select button
  - first function is displayed, see chart below

- start engine and drive vehicle
- press and release mode select button to advance sequentially from mode to mode
- note functions/displays as indicated on chart

If any display is blank or does not correspond to the guidelines in the chart,
- see troubleshooting section

Measuring mode test functions

Function/display	Times MFI mode select button is pressed	Mode/symbol indicated
Test sum of programmed bytes*	0	—
Intake manifold vacuum (mbar)**	1	clock
Engine rpm***	2	mph or km/h
Momentary fuel consumption	3	mpg or L/100 km
Signal from speed sensor****	4	miles or km
Oil temperature (°F or °C)	5	oil temperature
Ambient (outside) temperature (°F or °C)	6	ambient (outside) temperature

*Displayed numerical value has no importance for the check.

**The VDO multi-function indicator (printed circuit) displays intake manifold vacuum as a decimal numeric value instead of in mbar. See intake manifold vacuum chart on next page.

***If engine rpm is not displayed or is inaccurate, see troubleshooting section.

****Displayed numerical value =

USA: Speedometer drive ratio (from check mode test) × constant (always 4) × miles driven × conversion factor (miles to km)
Example: 955 × 4 × 0.5 × 1.609 = 3073

CANADA: Speedometer drive ratio (from check mode test) × constant (always 4) × kilometers driven
Example: 950 × 4 × 0.5 = 1900

Numeric value will change as vehicle is driven. If displayed value does not change, see troubleshooting section.

Intake manifold vacuum chart

Intake manifold vacuum		Numeric value displayed on MFI	
(mbar)*	(in. Hg)	VDO	Motometer
200	23.7	15	200
400	17.8	55	400
600	11.9	90	600
800	5.9	127	800
1000	0.0	148	1000
1200	2.9 psi	172	1200
1400	5.8 psi	192	1400
1600	8.7 psi	211	1600
1800	11.6 psi	228	1800
2000	14.5 psi	242	2000
2200	17.4 psi	255	2200

*Vacuum readings given in mbar (millibar) do not take into account atmospheric air pressure. The standard reading for atmospheric air pressure is approximately 1000 mbar/1 bar (depending on altitude). Readings less than 1000 mbar indicate vacuum and readings greater than 1000 mbar indicate pressure.

Note

As of model year 1993, the Multi–function indicator will be identified in all service information with the new abbreviation MFA. The old abbreviation, MFI, is now used as an emission–related term for Multiport Fuel Injection. MFA will only appear on Multi–function indicator information from model year 1993; Multi–function indicator information up to model year 1993 will not change and will continue to use the MFI abbreviation.

Multi-function indicator (MFI), troubleshooting

Observe the following:
- obtain radio security code from customer before disconnecting battery. Radio will electronically lock once power is disconnected.
- reactivate (unlock) radio after reconnecting power. See Repair Group 91
- numbers in black circles are electrical test points used during system troubleshooting. See wiring diagram at end of this section for test point references

Display errors or no display

If errors occur in MFI display, or if there is no display do this first:
- ■ disconnect battery ground strap and reconnect
- ■ recheck MFI display

If display still malfunctions,
- ■ continue with troubleshooting

Clock inaccurate

If clock deviates more then 5 seconds every 24 hours,
- ■ replace MFI (printed circuit), **J119**

Clock operates in incorrect mode (12 hour display or 24 hour display)

- ■ disconnect battery ground strap and reconnect
- ■ set time on clock

If clock mode remains incorrect,
- ■ perform check mode test as previously outlined in this Repair Group
- ■ check displayed engine map code and country code
  - engine map must be **018**
  - country code must be **4** (USA) or **1** (CANADA)

If **NO**
- ■ replace multi-function indicator (printed circuit), **J119**

Driving time remains at 0:00

- see troubleshooting section, "Memory cannot be cancelled/Memory does not store information"

Driving distance display remains at 0

- see troubleshooting section, "Memory cannot be cancelled/Memory does not store information"

If memory cancelling function **OK**
- check Hall speed sensor, see Index

Driving distance display not accurate

Maximum difference between actual and calculated driving distance is 5%.

If difference is greater than 5%,
- perform check mode test, as previously outlined in this Repair Group
- check speedometer drive ratio information
  - must be **955** (USA) or **950** (CANADA)

If speedometer drive ratio **OK**
- check Hall speed sensor, see Index

If speedometer drive ratio **NOT OK**
- replace multi-function indicator (printed circuit), **J119**

Average driving speed indicator is not accurate or displays "— — —", or "0"

- see troubleshooting section, "Memory cannot be cancelled/Memory does not store information"

If memory cancelling function **OK**
- check Hall speed sensor, see Index

Average fuel consumption display not accurate

- perform check mode test as previously outlined in this Repair Group
- check displayed engine map code, country code, and speedometer drive ratio
  - engine map code must be **018**
  - country code must be **4** (USA) or **1** (CANADA)
  - speedometer drive ratio must be **955** (USA) or **950** (CANADA)

YES → Intake manifold vacuum display, checking

NO →
- replace multi-function indicator (printed circuit), **J119**

→ End

Intake manifold vacuum display, checking

- verify that driving distance display is **OK**
- verify that average speed display is **OK**
- connect hand vacuum pump **US 8026** to intake manifold vacuum source
- perform measuring mode test as follows:
- set MFI memory switch (**E109**) to position **2** (long term/ total trip memory)
- switch ignition **ON**
- press and hold MFI mode select button
- switch ignition **OFF**, then back **ON**
- release mode select button
  - first function is displayed
- press and release mode select button once to display intake manifold vacuum (numeric value)

Go to Ⓐ next page

(A)

- match numeric value displayed with vacuum chart (see below) to obtain a corresponding vacuum reading
- compare reading with actual vacuum measured from **US 8026**
  - engine **OFF**, display must read approximately 148 (VDO) or 1000 (Motometer) — intake manifold vacuum 960-1040 mbar (0 in. Hg — no vacuum)
  - engine idling, display must read approximately 55-90 (VDO) or 400-600 (Motometer) — intake manifold vacuum 400 - 600 mbar (11.9 - 17.8 in. Hg)
  - engine momentarily at full throttle, display must read approximately 148 (VDO) or 1000 (Motometer) — intake manifold vacuum falls to approximately 960-1040 mbar (0 in. Hg — no vacuum)
  - display values must change with varying engine conditions as noted above

YES:

- check Hall speed sensor, see Index
  - Hall speed sensor **OK** → Go to (B) next page
  - Hall speed sensor **NOT OK** → replace Hall speed sensor, **G54** → End

NO:

- check vacuum hoses for damage/leaks
  - Vacuum hoses **OK** → replace MFI vacuum sensor, **G55** → End
  - Vacuum hoses **NOT OK** → repair as necessary → End

Intake manifold vacuum chart

Intake manifold vacuum		Numeric value displayed on MFI	
(mbar)*	**(in. Hg)**	**VDO**	**Motometer**
200	23.7	15	200
400	17.8	55	400
600	11.9	90	600
800	5.9	127	800
1000	0.0	148	1000
1200	2.9 psi	172	1200
1400	5.8 psi	192	1400
1600	8.7 psi	211	1600
1800	11.6 psi	228	1800
2000	14.5 psi	242	2000
2200	17.4 psi	255	2200

*Vacuum readings given in mbar (millibar) do not take into account atmospheric air pressure. The standard reading for atmospheric air pressure is approximately 1000 mbar/1 bar (depending on altitude). Readings less than 1000 mbar indicate vacuum and readings greater than 1000 mbar indicate pressure.

No engine rpm display or engine rpm display faulty (while performing measuring mode test)

(B)

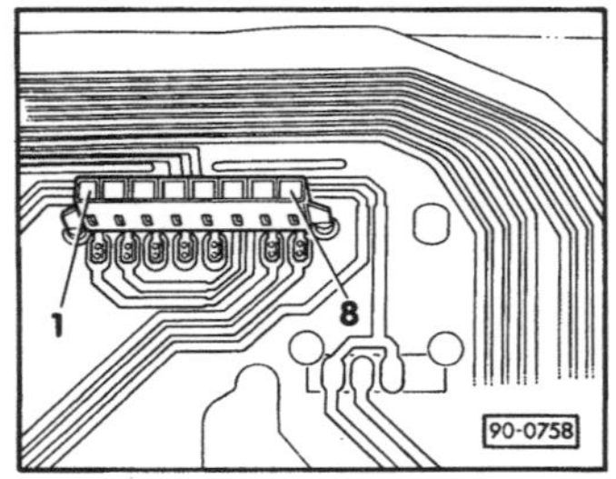

- disconnect battery ground strap
- remove instrument cluster trim and trim cover
- remove tachometer by squeezing locating pins and pulling from housing and connector
- re-connect battery ground strap
- connect multimeter **US 1119** (set to 20 V range) to tachometer connector as follows:
  - red (+) test lead to **4**, terminal **15**
  - black (−) test lead to **2**, terminal **31** (ground)
- switch ignition **ON**
  - must be approximately battery voltage

YES: Go to (C) next page

NO:

- switch ignition **OFF**
- re-install tachometer
  - install test box **VAG 1598** to 28-point instrument cluster connector using test harness **VAG 1598/8**
- connect multimeter **US 1119** (set to 20 V range) to test box as follows:
  - red (+) test lead to test point **13** (T28/13, terminal **15**)
  - black (−) test lead to test point **3** (T28/3, terminal **31**)
- switch ignition **ON**
  - must be approximately battery voltage

YES:

- check printed circuit for open circuit according to wiring diagram, replace as necessary → End

NO:

- check wiring for open circuit according to wiring diagram, repair as necessary → End

(C)

- switch ignition **OFF**
- disconnect battery ground strap
- re-install tachometer
- remove combination fuel, coolant temperature and low coolant level indicator gauge (with MFI) by squeezing locating pins and pulling from housing and connector
- connect multimeter **US 1119** (set to 2 V range) to 16-point MFI connector as follows:
  - red (+) test lead to **15** (connector T16/15)
  - black (–) test lead to **8** (connector T16/8, ground)
- re-connect battery ground strap
- start engine and run at 1000 rpm
  - must be approximately 270 mV
- raise and lower engine rpm
  - voltage must vary with engine speed

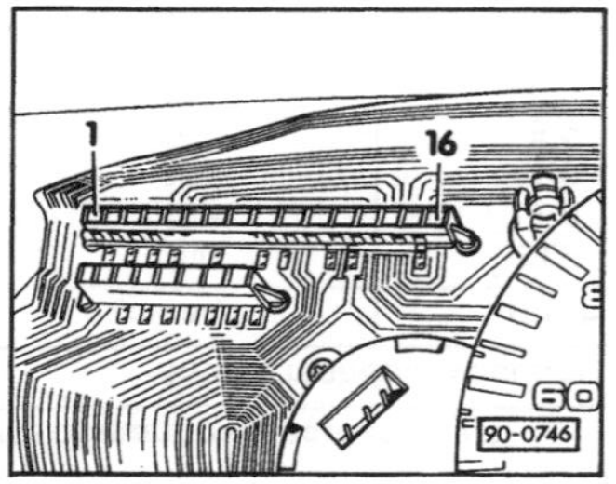

YES → Signal from tachometer to MFI **OK**
- replace MFI (printed circuit), **J119**

→ End

NO →
- check connection between tachometer and printed circuit
  - must be tight
- recheck voltage output test as described above

- Voltage **OK** →
  - replace MFI (printed circuit), **J119**

  → End
- Voltage **NOT OK** →
  - replace tachometer

  → End

Ambient (outside) temperature display inaccurate

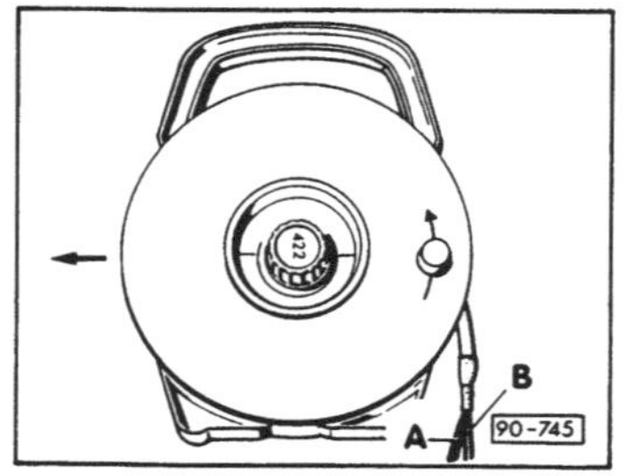

Test conditions

- ignition **ON**

Note

Numbers in black circles are electrical test points. See wiring diagram at end of this section for test point references.

- select ambient (outside) temperature display on MFI
- remove connector from ambient (outside) temperature sensor
- connect leads **A** and **B** of **VW 1301** to test points ❸ and ❹ (sensor connector)
- adjust **VW 1301** to 990
  - display must read 77°F (25°C)

YES →
- replace ambient (outside) temp. sensor, **G17**

→ End

NO →
- check wiring and printed circuit for open or short circuits according to wiring diagram

- Wiring/printed circuit **OK** →
  - replace multi-function indicator (printed circuit), **J119**

  → End
- Wiring/printed circuit **NOT OK** →
  - repair/replace as necessary

  → End

Oil temperature display inaccurate

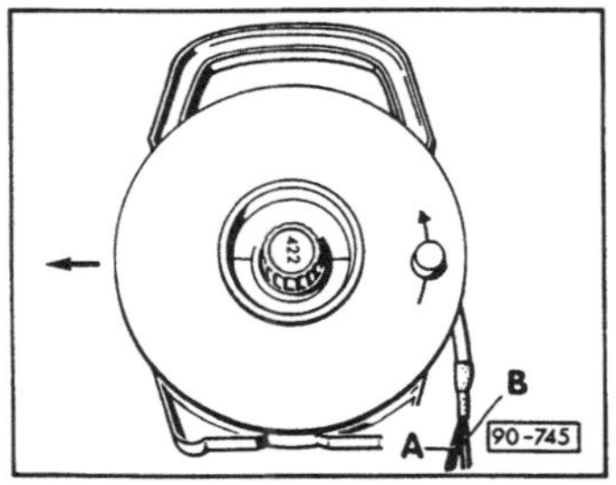

Test conditions
- ignition **ON**

Note

Numbers in black circles are electrical test points. See wiring diagram at end of this section for test point references.

- select oil temperature display on MFI
- remove wire from oil temperature sensor
- connect leads **A** and **B** of **VW 1301** to test points ❶ (ground) and ❷ (sensor connector)
- adjust **VW 1301** to 430
  - display must read 140°F (60°C)

YES → replace oil temp. sensor, **G8** → End

NO → check wiring and printed circuit for open or short circuits according to wiring diagram

- Wiring/printed circuit **OK** → replace multi-function indicator (printed circuit), **J119** → End
- Wiring/printed circuit **NOT OK** → repair/replace as necessary → End

Memory cannot be cancelled/Memory does not store information

- switch ignition **ON**
- set MFI memory switch (**E109**) to position **1** (short term/single trip memory)
- cancel memory by sliding switch fully left (beyond position **1**) and holding for longer than one second
  - short term memory must be cancelled
  - after one minute, driving time display must change from 0:00 to 0:01

NO → Go to Ⓐ next page

YES →

- set MFI memory switch (**E109**) to position **2** (long term/total trip memory)
- cancel memory by sliding switch fully right (beyond position **2**) and holding for longer than one second
  - long term memory must be cancelled
  - after one minute, driving time display must change from 0:00 to 0:01

NO → Go to Ⓐ next page

YES →

- move MFI memory switch (**E109**) back to position **1** (short term/single trip memory)
  - driving time display in memory **1** must be larger than memory **2**

YES → End

NO → Go to Ⓐ next page

(A)

Note

Numbers in black circles are electrical test points used during system troubleshooting. See wiring diagram at end of this section for test point references.

- disconnect battery ground strap
- remove steering column switch trim cover
- disconnect MFI switch 4-point connector on steering column
- set MFI memory switch to position **2** (long term/total trip memory)
- connect multimeter **US 1119** (set to ohm range) between test point ❼ (T4d/1) and test points ❽ (T4d/2), ❾ (T4d/3) and ❿ (T4d/4) one after the other
  - must be ∞ ohms (no continuity) each time

YES → MFI mode select button (E86) and MFI memory switch (E109), checking

NO → replace MFI memory switch (**E109**) → End

MFI mode select button (**E86**) and MFI memory switch (**E109**), checking

- leave MFI switch 4-point connector disconnected
- connect multimeter **US 1119** (set to ohm range) between test points ❼ (T4d/1) and ❽ (T4d/2)
- press MFI mode select button (**E86**) and hold
  - must be 0.0 ohm (continuity)

YES → Go to (B) next page

NO → replace MFI mode select button (**E86**) and MFI memory switch (**E109**) → End

(B)

- connect multimeter **US 1119** (set to ohm range) between test points ❼ (T4d/1) and ❾ (T4d/3)
- set MFI memory switch to position **1** (short term/single trip memory)
  - must be 0.0 ohm (continuity)

YES → next check

NO → replace MFI memory switch **E109** → End

- connect multimeter **US 1119** (set to ohm range) between test points ❼ (T4d/1) and ❿ (T4d/4)
- slide MFI memory switch fully right (reset position, long term memory) and hold
  - must be 0.0 ohm (continuity)
- slide MFI memory switch fully left (reset position, short term memory) and hold
  - must be 0.0 ohm (continuity)

YES → Printed circuit, checking

NO → replace MFI memory switch **E109** → End

Printed circuit, checking

- remove instruments, see Repair Group 90
- visually inspect printed circuit and check for continuity using multimeter **US 1119** (set to ohm range)

• Printed circuit **OK** → replace multi-function indicator (printed circuit), **J119** → End

• Printed circuit **NOT OK** → replace printed circuit → End

"Signal from speed sensor" display does not change (during measuring mode test)

Hall speed sensor (G54), checking

Note

Numbers in black circles are electrical test points. See wiring diagram at end of this section for test point references.

- set MFI memory switch (**E109**) to position **2** (long term/total trip memory)
- switch ignition **ON**
- press and hold MFI mode select button
- switch ignition **OFF**, then back **ON**
- release mode select button
  - first function is displayed
- press mode select button four times to display signal from speed sensor (numeric value)
- drive vehicle while briefly looking at display
  - numeric value must change (increase)

If **NO**
- switch ignition **OFF**
- remove combination fuel, coolant temperature, and low coolant level indicator gauge from instrument cluster
- connect multimeter **US 1119** (set to 20 V range) between test points ❺ (T28/7) and ❻ (T28/5) of instrument cluster multi-point connector
- place transmission in neutral and apply parking brake
- secure vehicle with wheel chocks to prevent rolling
- raise front of vehicle at proper lift points until wheels turn freely
- place jack stands under vehicle for safety
- switch ignition **ON**
- slowly rotate one front wheel
  - voltage must alternate between 0 V and 5 V (one rotation of wheel = four voltage jumps)

YES →
- replace multi-function indicator (printed circuit), **J119**

End

NO →
- check Hall speed sensor for proper contact, replace if necessary
- visually inspect printed circuit and check for continuity using multimeter **US 1119** (set to ohm range), replace if necessary

End

Description	Current track
Alternator warning light, K2	14
Brake and parking brake indicator light, K7	41
Brake fluid level warning contact, F34	55
Cassette storage light, L66	58
Cigarette lighter, U1	68
Cigarette lighter light, L28	67
Clock light, L8	29
Coolant low level control unit, J120	2-3
Coolant low level sensor, G32	2
Coolant overheat warning light, K28	23
Coolant temperature gauge, G3	21
Coolant temperature sending unit, G2	3
Digital clock, Y2	30-33
Engine oil pressure control unit, J114	7-9
Engine oil pressure switch (0.3 bar), F22	4
Engine oil pressure switch (1.8 bar), F1	5
Engine oil pressure warning light, K3	6
Engine oil temperature sensor, G8	49
Fresh air control lever light, L16	65
Fuel gauge, G1	20, 21
Fuse, S16	57
Fuse, S21	68
Glove compartment light, W6	57
Headlight high beam indicator light, K1	19
Instrument panel light, L10	26-28
Memory switch (multi-function ind.), E109	44-47
Multi-function indicator recall button, E86	43
Multi-function indicator vacuum sensor, G55	35-37
Multifunction indicator (MFI), J119	33-39
Outside air temperature sensor, G17	51
Parking brake warning light switch, F9	53
Seat belt warning light, K19	12
Speed sensor, G54	21
Tachometer, G5	16
Turn signal indicator light, K5	18
Voltage stabilizer, J6	20

Wire connectors

T1n — single, behind fuse/relay panel
T1o — single, behind fuse/relay panel
T1p — single, behind fuse/relay panel
T4d — four-point, below steering column trim cover
T16 — sixteen point, on instrument cluster (MFI)
T28 — twenty-eight point, on instrument cluster

Ground connections

(14) — on transmission
(80) — in instrument panel wiring harness

Welded wiring harness points

(A11) — plus connection (58b), in instrument panel wiring harness
(A15) — plus connection (15), in instrument panel wiring harness

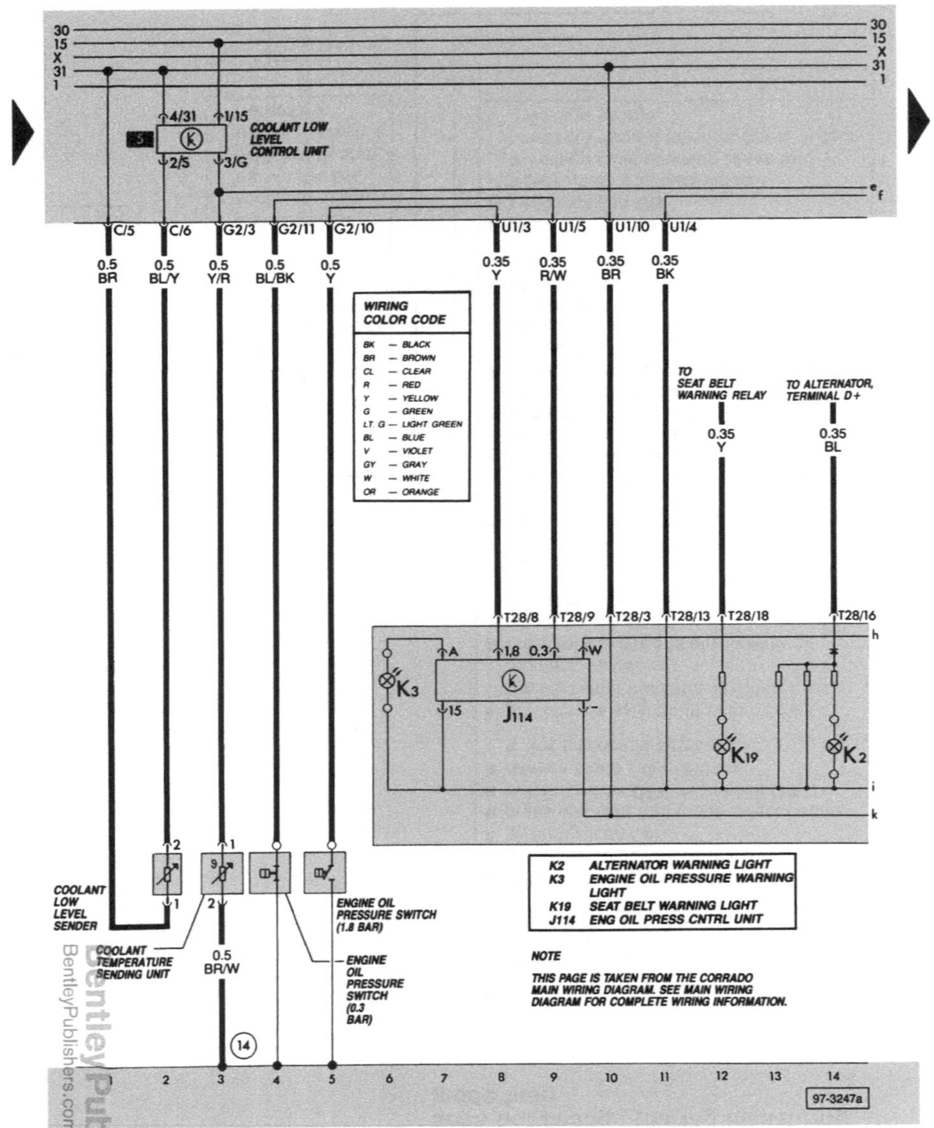

K2	ALTERNATOR WARNING LIGHT
K3	ENGINE OIL PRESSURE WARNING LIGHT
K19	SEAT BELT WARNING LIGHT
J114	ENG OIL PRESS CNTRL UNIT

NOTE

THIS PAGE IS TAKEN FROM THE CORRADO MAIN WIRING DIAGRAM. SEE MAIN WIRING DIAGRAM FOR COMPLETE WIRING INFORMATION.

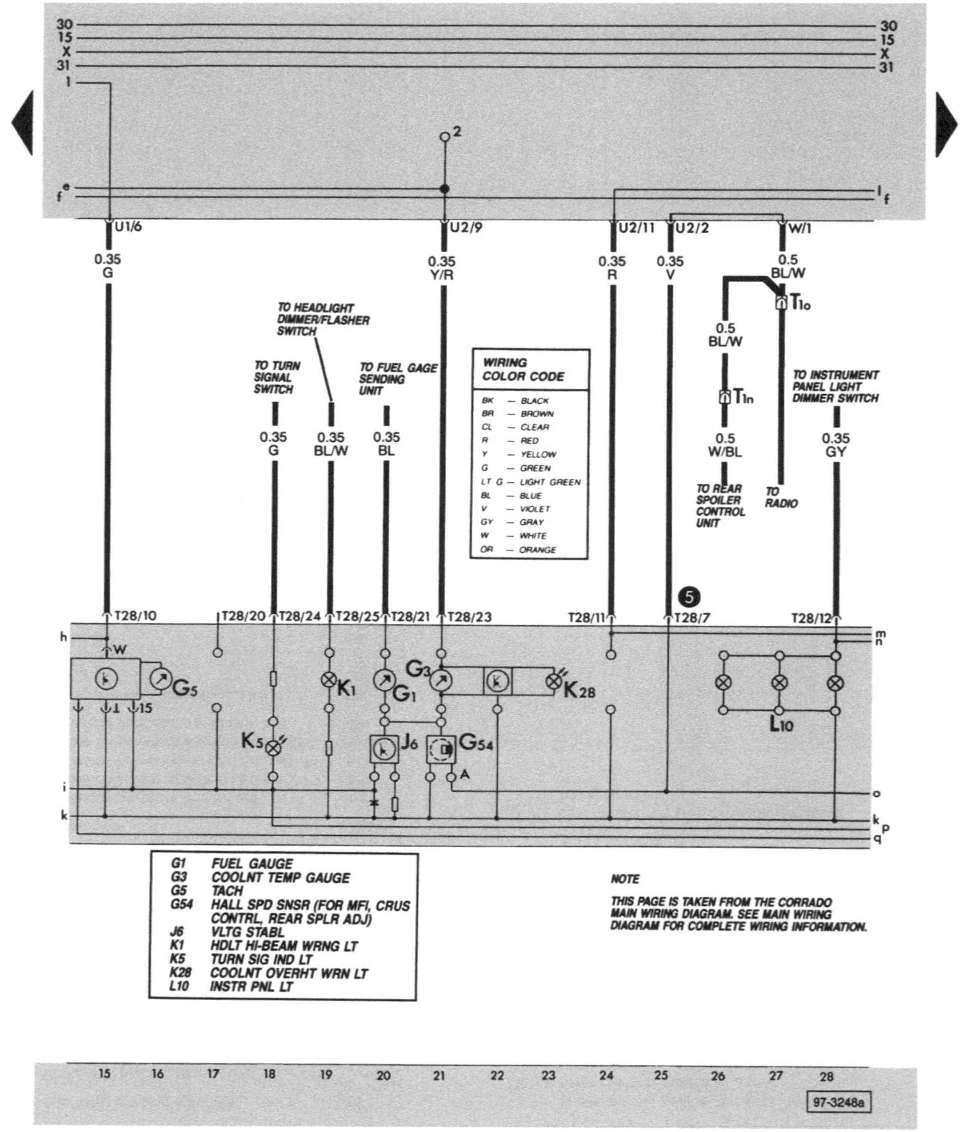

G1	FUEL GAUGE
G3	COOLNT TEMP GAUGE
G5	TACH
G54	HALL SPD SNSR (FOR MFI, CRUS CONTRL, REAR SPLR ADJ)
J6	VLTG STABL
K1	HDLT HI-BEAM WRNG LT
K5	TURN SIG IND LT
K28	COOLNT OVERHT WRN LT
L10	INSTR PNL LT

NOTE

THIS PAGE IS TAKEN FROM THE CORRADO MAIN WIRING DIAGRAM. SEE MAIN WIRING DIAGRAM FOR COMPLETE WIRING INFORMATION.

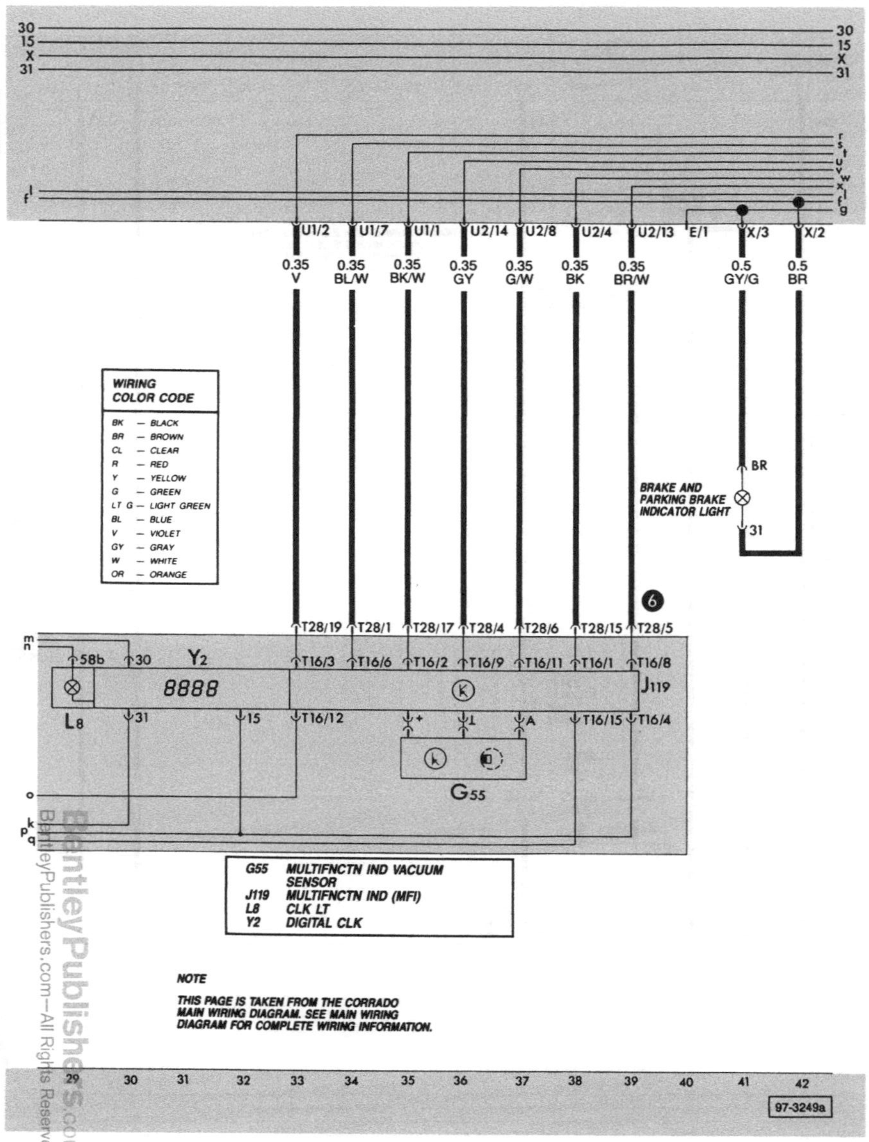

NOTE

THIS PAGE IS TAKEN FROM THE CORRADO MAIN WIRING DIAGRAM. SEE MAIN WIRING DIAGRAM FOR COMPLETE WIRING INFORMATION.

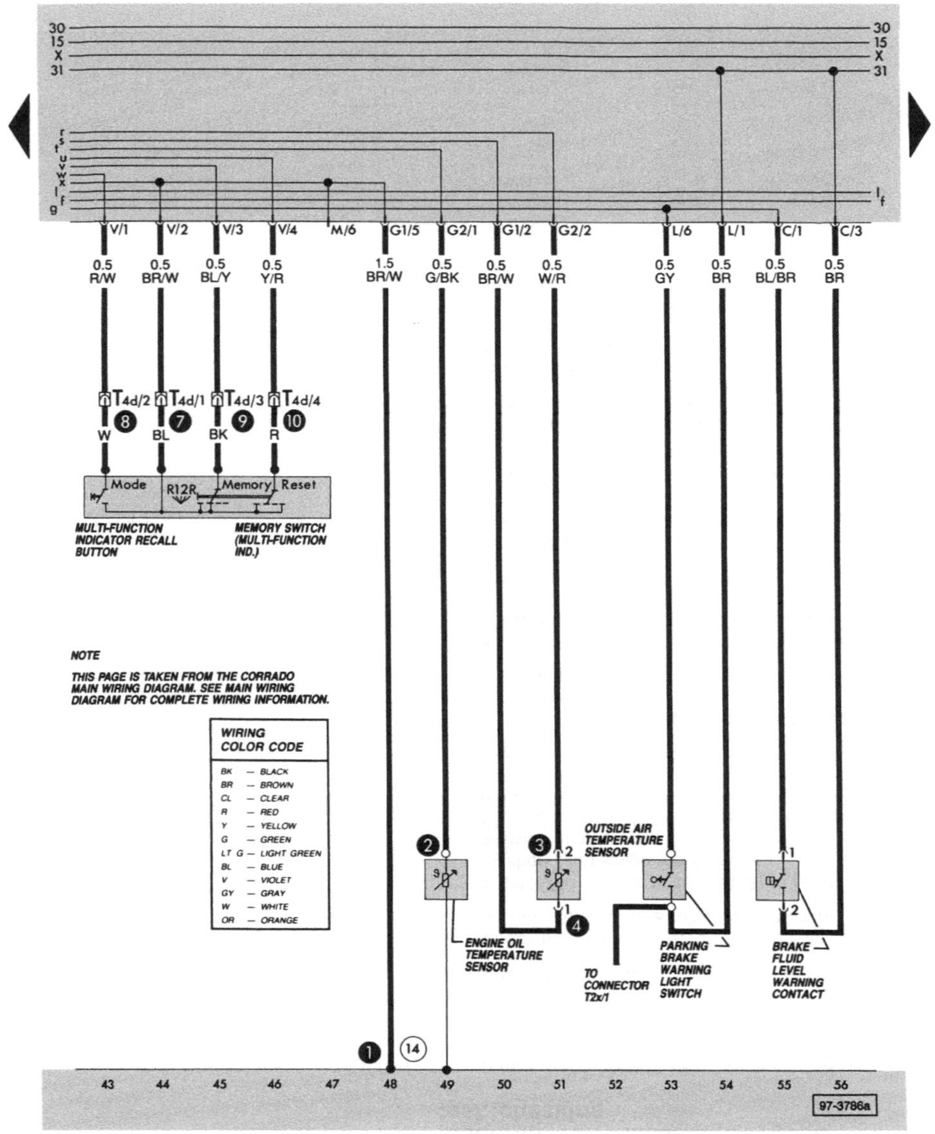

NOTE

THIS PAGE IS TAKEN FROM THE CORRADO MAIN WIRING DIAGRAM. SEE MAIN WIRING DIAGRAM FOR COMPLETE WIRING INFORMATION.

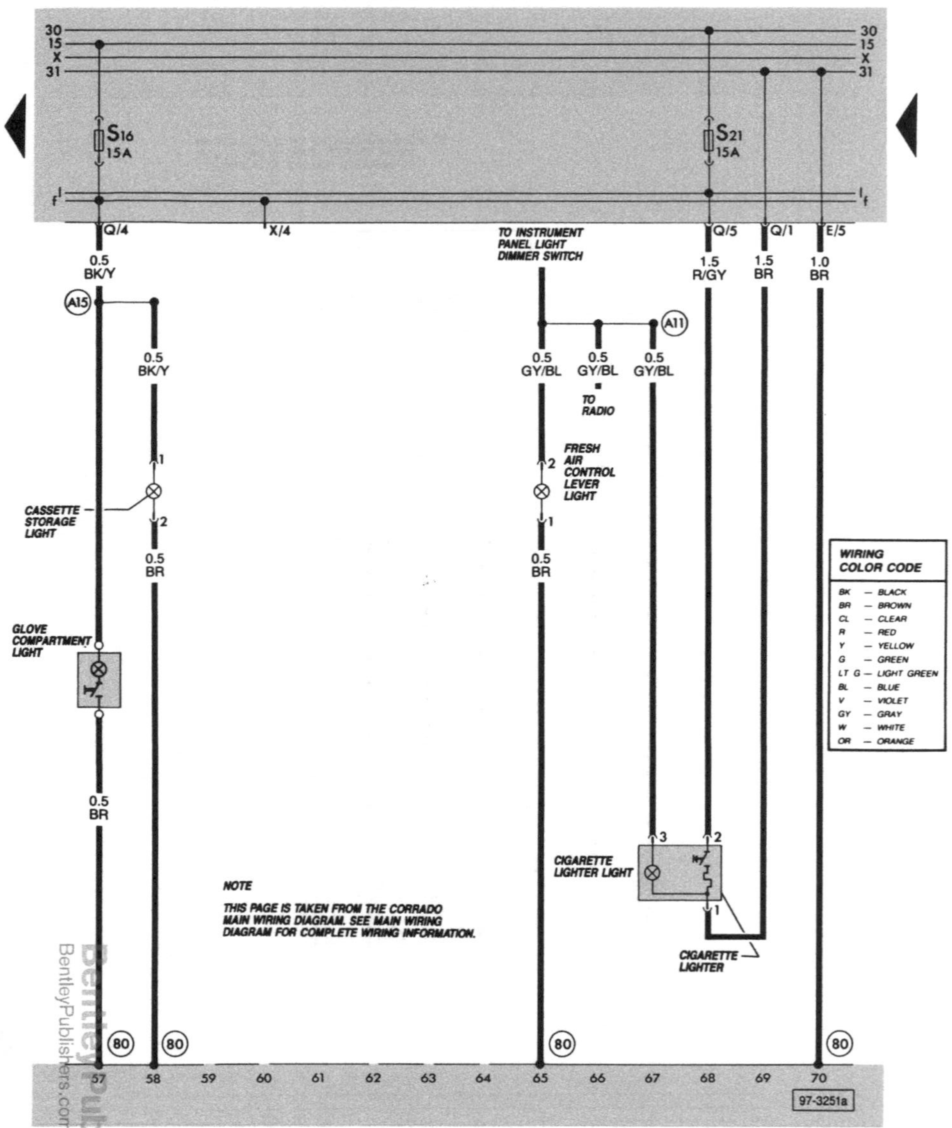

Multi-function indicator (MFA) self-test, checking

Perform MFA check mode and measuring mode tests for all MFA complaints before doing any repairs.

Check mode test

Check mode test verifies MFA is operating with the correct base vehicle information.

- set MFA memory switch (**E109**) to position **1** (short term/single trip memory)
- switch ignition **ON**
- press and hold MFA mode select button
- switch ignition **OFF**, then back **ON**
- release mode select button
  - engine map code is displayed
- compare engine map code with chart on next page
  - numerical value must match
- press and release mode select button to advance to next mode (country code)
- compare code to chart
  - country code must match
- continue pressing and releasing mode select button to advance sequentially from mode to mode
- compare numerical values to chart
  - all values must match

If any values do not match chart,

- replace MFA printed circuit board

To leave check mode test,

- switch ignition **OFF**

Check mode test functions

Function	Times MFA mode select button is pressed	Mode indicated	Values displayed – Engine code letters: PG	Values displayed – Engine code letters: AAA
Engine map code	0	–	**026**	**039**
Country code	1	clock	**4** (USA) **1** (CAN.)	**4** (USA) **1** (CAN.)
Speedometer drive ratio	2	mpg or l/100 km	**6160** (USA) **3800** (CAN.)	**6160** (USA) **3800** (CAN.)
Upper RPM limit*	3	miles or km	**7418**	**7418**
Lower RPM limit*	4	mph or km/h	**1175**	**1175**
Overrun cut-off vacuum (mbar)	5	oil temp.	**320**	**220**
Segment test	6	–	all segments of display light up	

*The displayed numerical values are random numbers and have no meaning for diagnosis.

Note

As of model year 1993, the Multi–function indicator will be identified in all service information with the new abbreviation MFA. The old abbreviation, MFI, is now used as an emission–related term for Multiport Fuel Injection. MFA will only appear on Multi–function indicator information from model year 1993; Multi–function indicator information up to model year 1993 will not change and will continue to use the MFI abbreviation.

Measuring mode test

Measuring mode test provides a display of measured functions during vehicle operation.

- set MFA memory switch (**E109**) to position **2** (long term/total trip memory)
- switch ignition **ON**
- press and hold MFA mode select button
- switch ignition **OFF**, then back **ON**
- release mode select button
  - first function is displayed, see chart below

- start engine and drive vehicle
- press and release mode select button to advance sequentially from mode to mode
- note functions/displays as indicated on chart

If any display is blank or does not correspond to the guidelines in the chart,
- see troubleshooting section

Measuring mode test functions

Function/display	Times MFA mode select button is pressed	Mode/symbol indicated
Test sum of programmed bytes*	0	—
Intake manifold vacuum (mbar)**	1	clock
Engine rpm***	2	mph or km/h
Momentary fuel consumption	3	mpg or L/100 km
Signal from speed sensor****	4	miles or km
Oil temperature (°F or °C)	5	oil temperature
Ambient (outside) temperature (°F or °C)	6	ambient (outside) temperature

*Displayed numerical value has no importance for the check.
**The VDO multi-function indicator (printed circuit) displays intake manifold vacuum as a decimal numeric value instead of in mbar. See intake manifold vacuum chart on next page.
***If engine RPM is not displayed or is inaccurate, see troubleshooting section, "Tachometer inaccurate or does not work".
****Displayed numerical value = speedometer drive ratio (from check mode test).

Intake manifold vacuum chart

Intake manifold vacuum		Numeric value displayed on MFA			
		VDO			
(mbar)*	(in. Hg)	Engine code	PG	AAA	Motometer
100	26.6		7	15	100
200	23.7		28	55	200
300	20.7		45	90	300
400	17.8		60	121	400
500	14.8		74	148	500
600	11.9		85	172	600
700	8.9		95	193	700
800	5.9		105	211	800
900	3.0		114	228	900
1000	0.0		121	242	1000
1100	1.5 psi		127	255	1100

*Vacuum readings given in mbar (millibar) do not take into account atmospheric air pressure. The standard reading for atmospheric air pressure is approximately 1000 mbar (1 bar) depending on altitude. Readings less than 1000 mbar indicate vacuum and readings greater than 1000 mbar indicate pressure.

Note

As of model year 1993, the Multi–function indicator will be identified in all service information with the new abbreviation MFA. The old abbreviation, MFI, is now used as an emission–related term for Multiport Fuel Injection. MFA will only appear on Multi–function indicator information from model year 1993; Multi–function indicator information up to model year 1993 will not change and will continue to use the MFI abbreviation.

Multi-function indicator (MFA), troubleshooting

Observe the following:

- get radio security code from customer before disconnecting battery. Radio will electronically lock once power is disconnected.
- reactivate (unlock) radio after reconnecting power. See Repair Group 91
- numbers in black circles are electrical test points used during system troubleshooting. See wiring diagram at end of this section for test point references

Display errors or no display

If errors occur in MFA display, or if there is no display do this first:

- disconnect battery ground strap and reconnect
- recheck MFA display

If display still malfunctions,

- continue with troubleshooting

Clock inaccurate

If clock deviates more then 5 seconds every 24 hours,

- replace MFA (printed circuit)

Clock operates in incorrect mode (12 hour display or 24 hour display)

- disconnect battery ground strap and reconnect
- set time on clock

If clock mode remains incorrect,

- perform check mode test as previously outlined in this Repair Group
- check displayed engine map code and country code
  - engine map code must be **026** (engine code PG) or **039** (engine code AAA)
  - country code must be **4** (USA) or **1** (CANADA)

If **NO**

- replace multi-function indicator (printed circuit)

Driving time remains at 0:00

- see troubleshooting section, "Memory cannot be cancelled/Memory does not store information"

Driving distance display remains at 0

- see troubleshooting section, "Memory cannot be cancelled/Memory does not store information"

If memory cancelling function **OK**
- check Vehicle Speed Sensor (VSS), speedometer (**G 22**), see Index

Driving distance display not accurate

Maximum difference between actual and calculated driving distance is 5%.

If difference is greater than 5%,
- perform check mode test as previously outlined in this Repair Group
- check speedometer drive ratio information
  - must be **6160** (USA) or **3800** (CANADA)

If speedometer drive ratio **OK**
- check Vehicle Speed Sensor (VSS), speedometer (**G 22**), see Index

If speedometer drive ratio **NOT OK**
- replace multi-function indicator (printed circuit)

Average driving speed indicator is not accurate or displays "– – –", or "0"

- see troubleshooting section, "Memory cannot be cancelled/Memory does not store information"

If memory cancelling function **OK**
- check Vehicle Speed Sensor (VSS), speedometer (**G 22**), see Index

Average fuel consumption display not accurate

- perform check mode test as previously outlined in this Repair Group
- check displayed engine map code, country code, and speedometer drive ratio
  - engine map code must be **026** (engine code PG) or **039** (engine code AAA)
  - country code must be **4** (USA) or **1** (CANADA)
  - speedometer drive ratio must be **6160** (USA) or **3800** (CANADA)

YES → Intake manifold vacuum display, checking

NO →
- replace multi-function indicator (printed circuit)

→ End

Intake manifold vacuum display, checking
- verify that driving distance display is **OK**
- verify that average speed display is **OK**
- connect hand vacuum pump **US 8026** to intake manifold vacuum source
- perform measuring mode test as follows:
- set MFA memory switch (**E109**) to position **2** (long term/ total trip memory)
- switch ignition **ON**
- press and hold MFA mode select button
- switch ignition **OFF**, then back **ON**
- release mode select button
  - first function is displayed
- press and release mode select button once to display intake manifold vacuum (numeric value)

Go to Ⓐ next page

Ⓐ

- match numeric value displayed with vacuum chart (see below) to obtain a corresponding vacuum reading
- compare reading with actual vacuum measured from **US 8026**
  - engine **OFF**, display must read approximately 121 (VDO – engine code PG), 242 (VDO – engine code AAA) or 1000 (Motometer) – intake manifold vacuum 960-1040 mbar (0 in. Hg – no vacuum)
  - engine idling, display must read approximately 60-85 (VDO – engine code PG), 121-172 (VDO – engine code AAA) or 400-600 (Motometer) – intake manifold vacuum 400-600 mbar (11.9 - 17.8 in. Hg)
  - engine momentarily at full throttle, display must read approximately 121 (VDO – engine code PG), 242 (VDO – engine code AAA) or 1000 (Motometer) – intake manifold vacuum falls to approximately 960-1040 mbar (0 in. Hg – no vacuum)
  - display values must change with varying engine conditions as noted above

YES →
- check Vehicle Speed Sensor (VSS), speedometer (**G 22**), see Index

NO →
- check vacuum hoses for damage/leaks

Vacuum hoses **OK** →
- replace MFA vacuum sensor, **G55**

→ End

Vacuum hoses **NOT OK** →
- repair as necessary

→ End

Intake manifold vacuum chart

Intake manifold vacuum		Numeric value displayed on MFA			
		VDO			
(mbar)*	(in. Hg)	Engine code	PG	AAA	Motometer
100	26.6		7	15	100
200	23.7		28	55	200
300	20.7		45	90	300
400	17.8		60	121	400
500	14.8		74	148	500
600	11.9		85	172	600
700	8.9		95	193	700
800	5.9		105	211	800
900	3.0		114	228	900
1000	0.0		121	242	1000
1100	1.5 psi		127	255	1100

*Vacuum readings given in mbar (millibar) do not take into account atmospheric air pressure. The standard reading for atmospheric air pressure is approximately 1000 mbar (1 bar) depending on altitude. Readings less than 1000 mbar indicate vacuum and readings greater than 1000 mbar indicate pressure.

Memory cannot be cancelled/Memory does not store information

- switch ignition **ON**
- set MFA memory switch (**E109**) to position **1** (short term/single trip memory)
- cancel memory by sliding switch fully left (beyond position **1**) and holding for longer than one second
  - short term memory must be cancelled
  - after one minute, driving time display must change from 0:00 to 0:01

NO → Go to Ⓐ next page

YES →
- set MFA memory switch (**E109**) to position **2** (long term/total trip memory)
- cancel memory by sliding switch fully right (beyond position **2**) and holding for longer than one second
  - long term memory must be cancelled
  - after one minute, driving time display must change from 0:00 to 0:01

NO → Go to Ⓐ next page

YES →
- move MFA memory switch (**E109**) back to position **1** (short term/single trip memory)
  - driving time display in memory **1** must be larger than memory **2**

YES → End

NO → Go to Ⓐ next page

(A)

Note

Numbers in black circles are electrical test points used during system troubleshooting. See wiring diagram at end of this section for test point references.

- disconnect battery ground strap
- remove steering column switch trim cover
- disconnect MFA switch 4-point connector on steering column
- set MFA memory switch to position **2** (long term/total trip memory)
- connect multimeter **Fluke 83** (**US 1119**) (set to Ohm range) between test point 7 (T4a/1) and test points 8 (T4a/2), 9 (T4a/3) and 10 (T4a/4) one after the other
  - must be ∞ Ohms (no continuity) each time

YES → MFA mode select button (**E86**) and MFA memory switch (**E109**), checking

NO → replace MFA mode select button (**E86**) memory switch (**E109**) and harness → End

MFA mode select button (**E86**) and MFA memory switch (**E109**), checking

- leave MFA switch 4-point connector disconnected
- connect multimeter **Fluke 83** (**US 1119**) (set to Ohm range) between test points 7 (T4a/1) and 8 (T4a/2)
- press MFA mode select button (**E86**) and hold
  - must be 0.0 Ohm (continuity)

YES → Go to (B) next page

NO → replace MFA mode select button (**E86**), MFA memory switch (**E109**) and harness → End

(B)

- connect multimeter **Fluke 83** (**US 1119**) (set to Ohm range) between test points 7 (T4a/1) and 9 (T4a/3)
- set MFA memory switch to position **1** (short term/single trip memory)
  - must be 0.0 Ohm (continuity)

YES →

- connect multimeter **Fluke 83** (**US 1119**) (set to Ohm range) between test points 7 (T4a/1) and 10 (T4a/4)
- slide MFA memory switch fully right (reset position, long term memory) and hold
  - must be 0.0 Ohm (continuity)
- slide MFA memory switch fully left (reset position, short term memory) and hold
  - must be 0.0 Ohm (continuity)

  YES → replace printed circuit board → End

  NO → replace MFA mode select button (**E86**), MFA memory switch (**E109**) and harness → End

NO → replace MFA mode select button (**E86**), MFA memory switch (**E109**) and harness → End

Ambient (outside) temperature display inaccurate

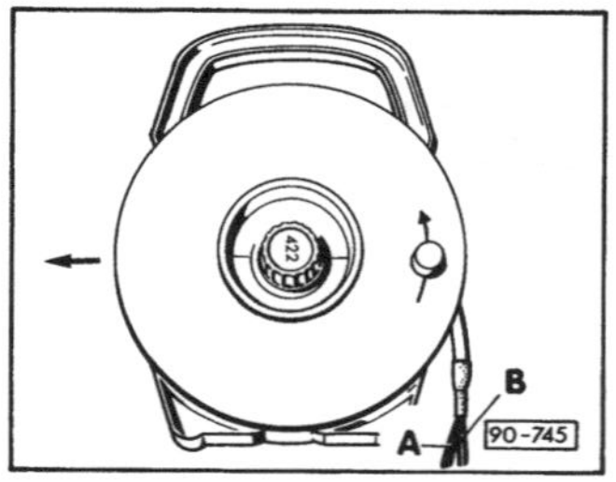

Test conditions

- ignition **ON**

Note

Numbers in black circles are electrical test points. See wiring diagram at end of this section for test point references.

- select ambient (outside) temperature display on MFA
- remove connector from ambient (outside) temperature sensor
- connect leads **A** and **B** of **VW 1301** to test points ❸ and ❹ (sensor connector)
- adjust **VW 1301** to 990
  - display must read 77°F (25°C)

YES →
- replace ambient (outside) temp. sensor, **G17**

→ End

NO →
- check wiring and printed circuit for open or short circuits according to wiring diagram

Wiring/printed circuit **OK** →
- replace multi-function indicator (printed circuit), **J 119**

→ End

Wiring/printed circuit **NOT OK** →
- repair/replace as necessary

→ End

Oil temperature display inaccurate

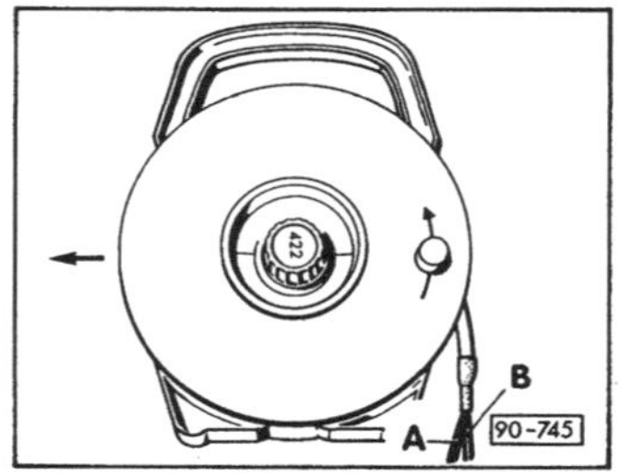

Test conditions
- ignition **ON**

Note

Numbers in black circles are electrical test points. See wiring diagram at end of this section for test point references.

- select oil temperature display on MFA
- remove wire from oil temperature sensor
- connect leads **A** and **B** of **VW 1301** to test points ❶ (ground) and ❷ (sensor connector)
- adjust **VW 1301** to 430
  - display must read 140°F (60°C)

YES →
- replace oil temp. sensor, **G8**

→ End

NO →
- check wiring and printed circuit for open or short circuits according to wiring diagram

Wiring/printed circuit **OK** →
- replace multi-function indicator (printed circuit), **J 119**

→ End

Wiring/printed circuit **NOT OK** →
- repair/replace as necessary

→ End

Electronic speedometer and odometer, checking

If speedometer and/or odometer do not work,

- check Vehicle Speed Sensor (VSS), speedometer (**G 22**), see Index
- check for open/short circuits in wiring from speed sensor (speedometer) to speedometer/odometer according to wiring diagram, repair as necessary

If speed sensor (speedometer) and wiring check **OK**,

- replace speedometer and/or
- replace odometer printed circuit board

Tachometer, inaccurate or does not work

- perform MFA measuring mode test (for engine RPM) as follows:
- set MFA memory switch (**E109**) to position **2** (long term/total trip memory)
- switch ignition **ON**
- press and hold MFA mode select button
- switch ignition **OFF**, then back **ON**
- release mode select button
  - first function is displayed
- press mode select button twice to display "engine RPM"
- start engine
  - engine RPM must be displayed on MFA

YES →

- replace tachometer, see Repair Group 90

→ End

NO →

- switch ignition **OFF**
- remove instrument cluster, see Repair Group 90
- install Test box **VAG 1598** to 28-point instrument cluster connector using test harness **VAG 1598/8**
- set digital multimeter **Fluke 83** (**US 1119**) to Volt range
- connect multimeter black test lead to socket **3** (T28/3) and red test lead to socket **10** (T28/10) of Test box **VAG 1598**
- start engine
  - must be approximately 6.6 V

YES →

- replace instrument cluster printed circuit board

→ End

NO →

- check wiring from engine management control unit (**J 169** or **J 220**) to instrument cluster connector T28/10 for open/short circuit according to wiring diagram, repair as necessary

→ End

Vehicle Speed Sensor (VSS), speedometer, (G 22), checking

Note

Numbers in black circles are electrical test points. See wiring diagram at end of this section for test point references.

If an aftermarket radio is installed, ensure that the blue/white speed signal wire (**T8/1**) at the radio is **NOT** connected to either ground or the antenna amplifier.

- set MFA memory switch (**E109**) to position **2** (long term/total trip memory)
- switch ignition **ON**
- press and hold MFA mode select button
- switch ignition **OFF**, then back **ON**
- release mode select button
  - first function is displayed

- press mode select button four times to display "signal from speed sensor" (numeric value)
- drive vehicle while briefly looking at display
  - numeric value must change (increase)

If **NO**
- switch ignition **OFF**
- remove instrument cluster, see Repair Group 90
- install Test box **VAG 1598** to 28-point instrument cluster connector using test harness **VAG 1598/8**
- set digital multimeter **Fluke 83** (**US 1119**) to volt range
- connect multimeter between sockets **7** (test point ❺, T28/7) and **5** (test point ❻, T28/5) of Test box **VAG 1598**
- place transmission in neutral and apply parking brake
- secure vehicle with wheel chocks to prevent rolling
- raise front of vehicle at proper lift points until wheels turn freely

- place jack stands under vehicle for safety
- switch ignition **ON**
- slowly rotate one front wheel
  - voltage must alternate between 0 V and 5 V (one rotation of wheel = four voltage jumps)

YES	NO
■ replace multi-function indicator (printed circuit), **J119** → End	■ check speed sensor (speedometer), **G22**, for proper contact, replace if necessary ■ visually inspect printed circuit and check for continuity using multimeter **Fluke 83** (**US 1119**) (set to Ohm range), replace if necessary → End

Description	Current track
Alternator warning light, K2	24
Control unit with display unit in instrument cluster, J285	9-46
Coolant low level sensor, G32	8
Coolant overheat warning light, K28	12
Coolant temperature gauge, G3	13
Coolant temperature sending unit, G2	5
Digital display light, L75	41, 43
Engine oil pressure switch (0.3 bar), F22	6
Engine oil pressure switch (1.8 bar), F1	7
Engine oil pressure warning light, K3	10
Engine oil temperature sensor, G8	53
Fuel gauge, G1	14
Fuse, S16	19
Fuse, S21	32
Headlight high beam indicator light, K1	28
Instrument panel light, L10	32-34
Multi-function indicator memory switch, E109	51
Multi-function indicator recall button, E86	49
Multi-function indicator vacuum sensor, G55	29-31
Odometer display, Y4	38-40
Outside air temperature sensor, G17	55
Seat belt warning light, K19	23
Speed sensor (speedometer), G22	48-49
Speedometer, G21	32-35
Tachometer, G5	15-19
Turn signal indicator light, left, K65	26
Turn signal indicator light, right, K94	27
Voltage stabilizer, J6	20-22
Warning buzzer, H3	14
Wire connection, terminal 15, T35	20
Wire connector, speed signal, T43	29

Wire connectors

T2d – double, behind fuse/relay panel
T2h – double, behind fuse/relay panel
T4a – four-point, behind steering column switch cover
T8a – eight point, above fuse/relay panel
T28 – twenty-eight point, on instrument cluster
T35 – wire connection, terminal 15, behind fuse/relay panel
T42 – forty-two point, on left side of cylinder head (6 cyl. only)
T43 – wire connection for speed signal, behind fuse/relay panel

Ground connections

(18) – on engine block
(125) – in headlight wiring harness
(159) – in Digifant/Motronic wiring harness

Welded wiring harness points

(A15) – plus connection (15), in instrument panel wiring harness
(C15) – wire connection, in headlight wiring harness
(C16) – wire connection, in headlight wiring harness
(E6) – plus connection (15), in Digifant/Motronic wiring harness
(W9) – plus connection (30), in rear wiring harness
(W10) – plus connection (15), in rear wiring harness

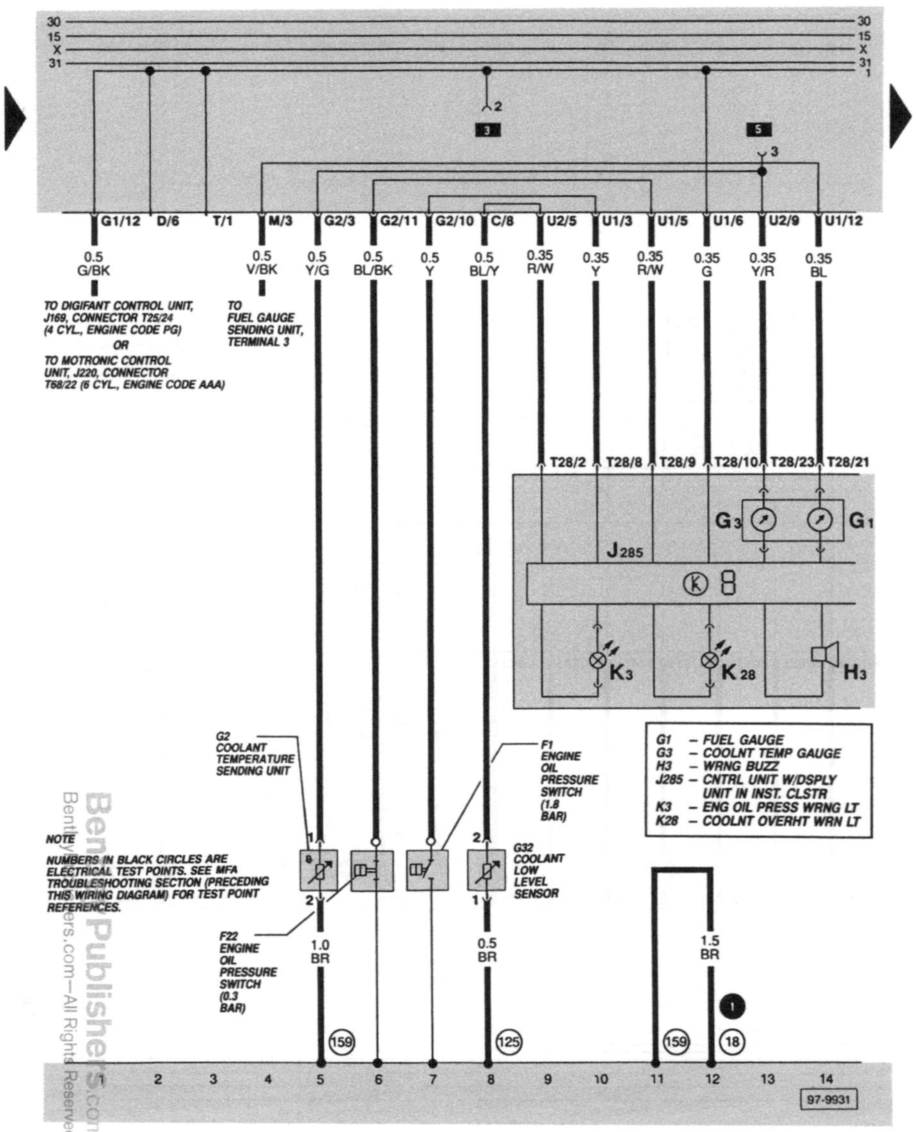
30
15
X
31
G1/12
D/6
T/1
M/3
G2/3
G2/11
G2/10
C/8
U2/5
U1/3
U1/5
U1/6
U2/9
U1/12
0.5 G/BK
0.5 V/BK
0.5 Y/G
0.5 BL/BK
0.5 Y
0.5 BL/Y
0.35 R/W
0.35 Y
0.35 R/W
0.35 G
0.35 Y/R
0.35 BL
TO DIGIFANT CONTROL UNIT, J169, CONNECTOR T25/24 (4 CYL., ENGINE CODE PG) OR TO MOTRONIC CONTROL UNIT, J220, CONNECTOR T68/22 (6 CYL., ENGINE CODE AAA)
TO FUEL GAUGE SENDING UNIT, TERMINAL 3
T28/2
T28/8
T28/9
T28/10
T28/23
T28/21
J285
K3
K28
H3
G3
G1
G1 – FUEL GAUGE
G3 – COOLNT TEMP GAUGE
H3 – WRNG BUZZ
J285 – CNTRL UNIT W/DSPLY UNIT IN INST. CLSTR
K3 – ENG OIL PRESS WRNG LT
K28 – COOLNT OVERHT WRN LT
G2 COOLANT TEMPERATURE SENDING UNIT
F1 ENGINE OIL PRESSURE SWITCH (1.8 BAR)
G32 COOLANT LOW LEVEL SENSOR
F22 ENGINE OIL PRESSURE SWITCH (0.3 BAR)
NOTE
NUMBERS IN BLACK CIRCLES ARE ELECTRICAL TEST POINTS. SEE MFA TROUBLESHOOTING SECTION (PRECEDING THIS WIRING DIAGRAM) FOR TEST POINT REFERENCES.
1.0 BR
0.5 BR
1.5 BR
1 2 3 4 5 6 7 8 9 10 11 12 13 14
97-9931

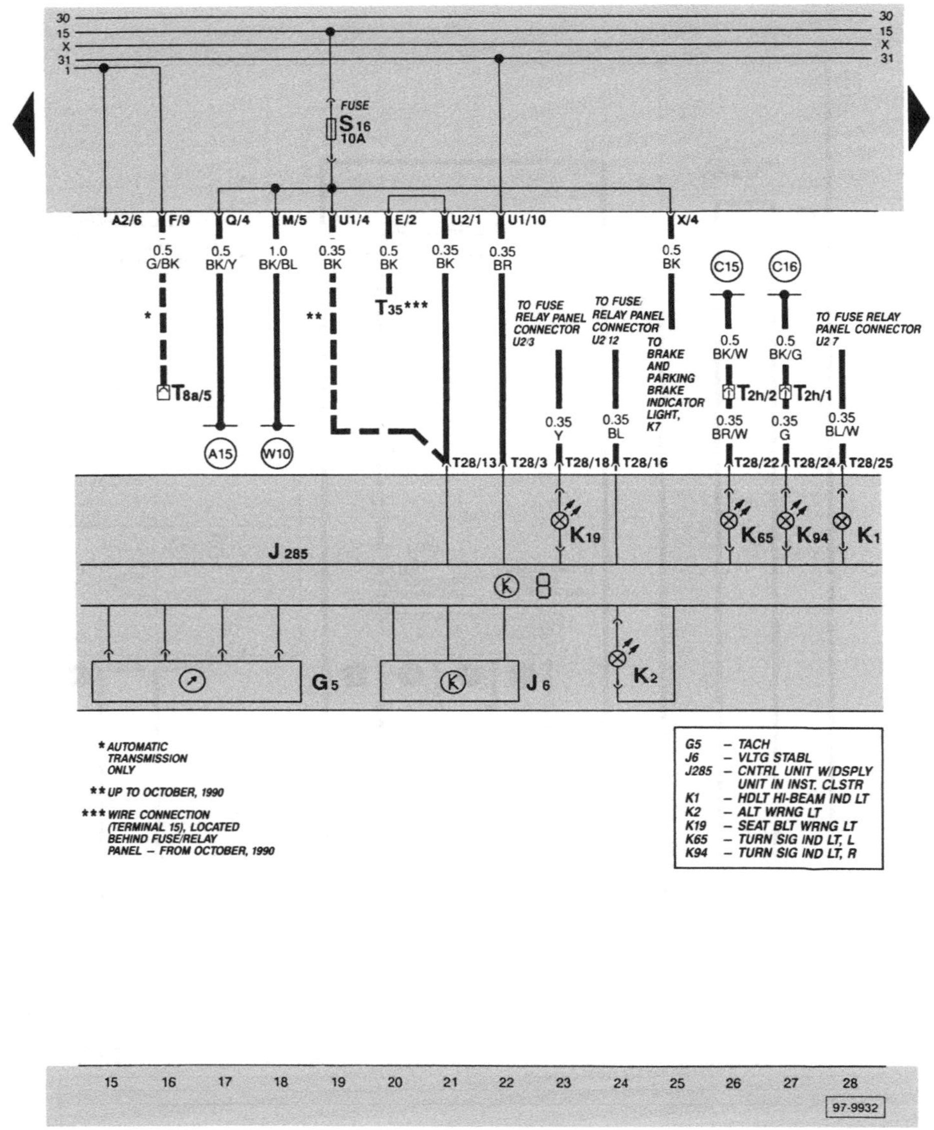
30
15
X
31
FUSE S16 10A
A2/6
F/9
Q/4
M/5
U1/4
E/2
U2/1
U1/10
X/4
0.5 G/BK
0.5 BK/Y
1.0 BK/BL
0.35 BK
0.5 BK
0.35 BK
0.35 BR
0.5 BK
T8a/5
T35***
TO FUSE RELAY PANEL CONNECTOR U2/3
TO FUSE RELAY PANEL CONNECTOR U2 12
TO BRAKE AND PARKING BRAKE INDICATOR LIGHT, K7
TO FUSE RELAY PANEL CONNECTOR U2 7
0.5 BK/W
0.5 BK/G
T2h/2
T2h/1
0.35 Y
0.35 BL
0.35 BR/W
0.35 G
0.35 BL/W
T28/13
T28/3
T28/18
T28/16
T28/22
T28/24
T28/25
J285
K19
K65
K94
K1
G5
J6
K2
* AUTOMATIC TRANSMISSION ONLY
** UP TO OCTOBER, 1990
*** WIRE CONNECTION (TERMINAL 15), LOCATED BEHIND FUSE/RELAY PANEL – FROM OCTOBER, 1990
G5 – TACH
J6 – VLTG STABL
J285 – CNTRL UNIT W/DSPLY UNIT IN INST. CLSTR
K1 – HDLT HI-BEAM IND LT
K2 – ALT WRNG LT
K19 – SEAT BLT WRNG LT
K65 – TURN SIG IND LT, L
K94 – TURN SIG IND LT, R
15 16 17 18 19 20 21 22 23 24 25 26 27 28
97-9932

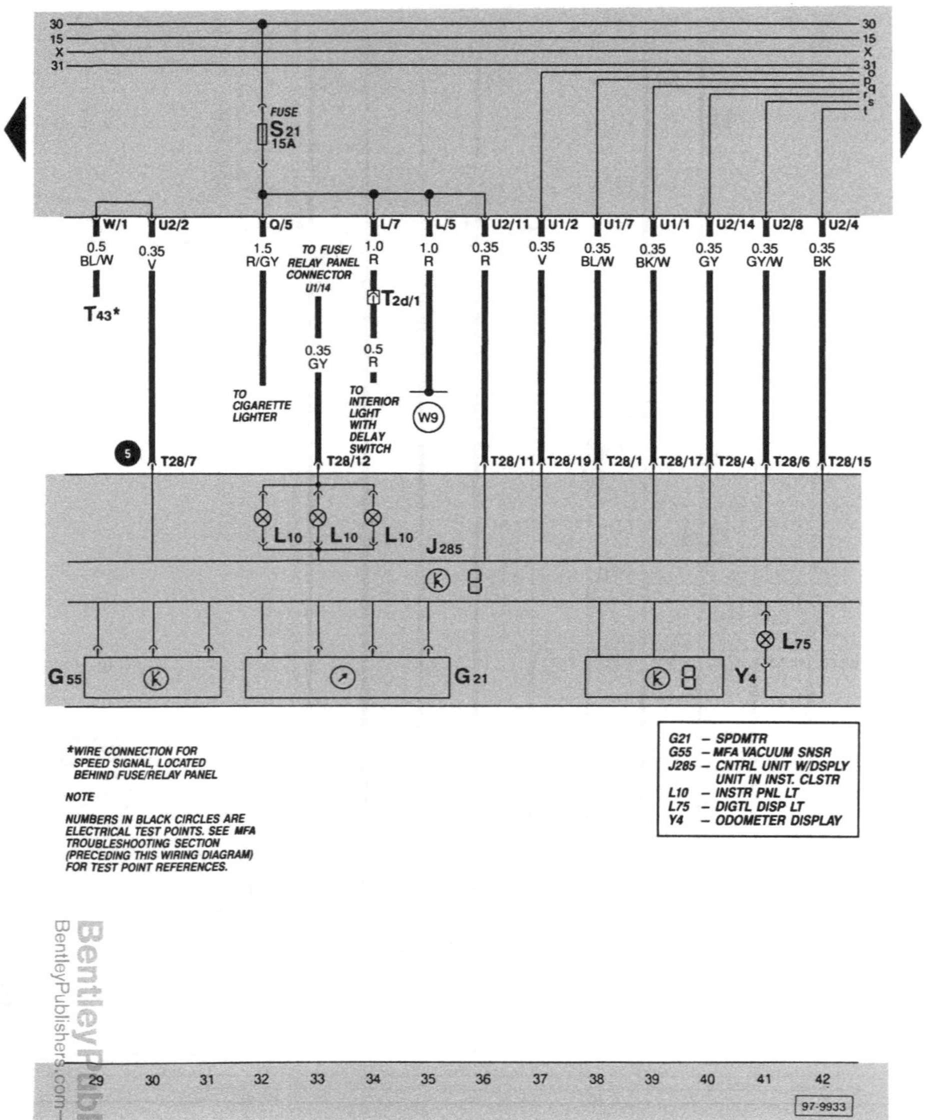

*WIRE CONNECTION FOR SPEED SIGNAL, LOCATED BEHIND FUSE/RELAY PANEL

NOTE

NUMBERS IN BLACK CIRCLES ARE ELECTRICAL TEST POINTS. SEE MFA TROUBLESHOOTING SECTION (PRECEDING THIS WIRING DIAGRAM) FOR TEST POINT REFERENCES.

G21 – SPDMTR
G55 – MFA VACUUM SNSR
J285 – CNTRL UNIT W/DSPLY UNIT IN INST. CLSTR
L10 – INSTR PNL LT
L75 – DIGTL DISP LT
Y4 – ODOMETER DISPLAY

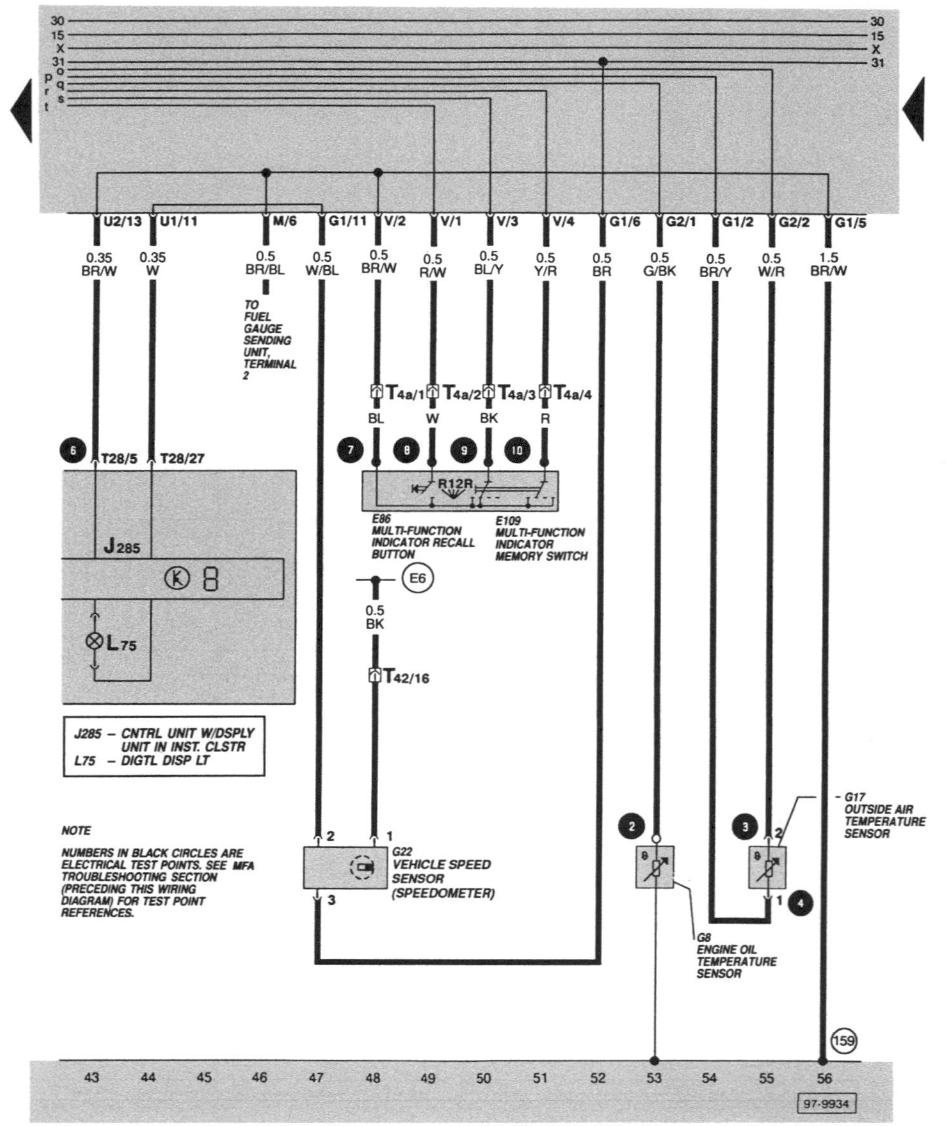

J285 – CNTRL UNIT W/DSPLY UNIT IN INST. CLSTR
L75 – DIGTL DISP LT

NOTE

NUMBERS IN BLACK CIRCLES ARE ELECTRICAL TEST POINTS. SEE MFA TROUBLESHOOTING SECTION (PRECEDING THIS WIRING DIAGRAM) FOR TEST POINT REFERENCES.

Multi-Function Indicator (MFA), Checking

Operation and function

The Multi-function indicator is an integral part of the instrument cluster and consists of an electronic processor unit and a digital display unit. Electrical signals from external components (sensors and control modules) are fed into the electronic unit, processed and stored. The results can be shown as figures in the display.
The Mode Select Switch (E 86) and the Memory Switch (E 109) are located in the wiper switch lever.

The following signals are processed by the Multi-function indicator:

- Pulse from Speedometer Vehicle Speed Sensor (VSS) (G 22)
- Resistance change from Oil Temperature Sensor (G 8)
- Resistance change from Outside Air Temperature Sensor (G 17)
- Fuel consumption signal from engine control module
- Pulse from terminal TD/terminal 1

When the ignition is off, the MFA display shows time. When ignition is switched on, the time remains in the display until the Multi-function Indicator Mode Select Switch (E 86) is pressed. When the Mode Select Switch (E 86) has been operated, the selected MFA function and the stored data appear. If the Mode Select Switch (E 86) is pressed briefly and repeatedly, different functions appear one after the other.

Sequence of functions:
- **Time** (in hours and minutes)
- **Driving time** (max. 100 hours)
- **Average fuel consumption** (liters/100 km or miles/gallon)
  At speeds below 8 km/hr or 8 mph the average consumption display switches to liters/hour or gallons/hour.
- **Distance driven** (max.10,000 km or 10,000 miles)
- **Average speed** (in km/h or mph)
- **Engine oil temperature** (in degrees Celsius or Fahrenheit)
  At temperatures below 122 °F (50°C) dashes appear.
- **Outside (ambient) temperature** (in degrees Celsius or Fahrenheit)

When ignition is on, the sensor values for driving time, distance driven and fuel consumption are recorded continuously and stored. The calculation of the average speed and the average fuel consumption only takes place when the function is called up.

Driving data memory

The Multi-function indicator contains two selectable memories:

- Memory 1: **Single trip memory**
- Memory 2: **Total trip memory**

The data in these memories can be displayed using the Mode Select Switch (E 86).

The two memories collect all driving data as long as the ignition is on. Both memories can be cancelled manually, see "Cancelling the memories". If the capacity in one memory is exceeded, all data in this memory will be cancelled. Memory 1 (single trip memory) will automatically be cancelled when ignition is switched on if the ignition has been off continuously for longer than 120 minutes.

Multi-function Indicator Mode Select Switch (E 86) and Memory Switch (E 109)

Position 1 of Multi-function Indicator Memory Switch (E 109)

In position 1 the contents of the single trip memory can be displayed. With ignition on, and until the ignition is switched off, the following data is collected:

- **Driving time**
- **Distance driven**
- **Fuel consumed**

From this data the average speed and the average fuel consumption are calculated and stored. If the trip is continued within 2 hours of switching off the ignition, the new values are included in the calculation.

Position 2 of Multi-function Indicator Memory Switch (E 109)

In position 2 the contents of the total trip memory can be displayed.
The following data from any number of single trips is collected:

- **Driving time** (up to 100 hours)
- **Distance driven** (up to 10,000 km/miles)
- **Fuel consumed** (1,000 liters/gallons)

From this data the average speed and consumption for all single trips are calculated and stored. If one of the given maximum memory capacities is exceeded, the memory is cancelled and the calculations start over once again.

Unlike the single trip memory, the total trip memory is not cancelled after switching the ignition off for two hours.

Cancelling the memories

Both memories can be cancelled separately with the ignition on. To cancel memories, slide Memory Switch (E 109) fully left (for memory 1) or fully right (for memory 2) into position 0 and hold longer than 1 second.

If the battery is disconnected, then both memories are cancelled automatically.

Setting the time

To set the time there are two buttons on the left below the engine coolant temperature (ECT) gauge. The upper button (h) is for the hours and the lower one (min) for the minutes. Pressing the appropriate button briefly will advance the time one hour or one minute. With button depressed, hours or minutes will advance continuously.

With the minute button the time can be set exactly to the second as follows:

- Press minute button until time displayed is one minute before time to be set.
- Press minute button at the moment when the second hand of an accurate watch shows a full minute or when the time signal is heard on the radio.

Multi-function Indicator (MFA) check mode test

If the display values for the momentary fuel consumption, oil temperature or outside air (ambient) temperature are inaccurate, perform check mode test before attempting repairs.

Calling up Multi-function indicator check mode test

- Set Multi-function indicator memory switch (E 109) to position 1 (single trip memory)
- Switch ignition on
- Press and hold mode select switch (E 86)
- Switch ignition off, then back on
- Release mode select switch (E 86)

The value for momentary fuel consumption appears in the display. Press and release mode select switch (E 86) to call up oil temperature, then outside air (ambient) temperature.
The selected function appears in the display.
The data in the display is aligned on the right hand side.

Multi-function indicator check mode displays:

Momentary fuel consumption	Below 8 km/h or 8 mph, value is in liters/hour or gallons/hour.
Oil temperature	Shown in °F or °C Range from 122°F to 320°F (50°C to 160°C) At temperatures below 122°F (+50°C) dashes appear (---)
Outside (ambient) temperature	Shown in °F or °C Range from -40°F to 122°F (-40°C to +50°C) If temperature sensor is not connected, -40°F (-40°C) will be displayed

Instrument cluster, checking

The electronic components and instruments in the instrument cluster must match the appropriate vehicle equipment.
The electronic unit in each instrument cluster has an integrated test mode which can be checked to determine whether the cluster has the correct electronic components and instruments to ensure that the vehicle specific input signals are processed properly and shown in the display.
This check mode for the instrument cluster can be used for troubleshooting and should be checked in all cases of complaints concerning the functioning of the instrument cluster (MFA).

Calling up instrument cluster check mode test

- Switch ignition on
- Press trip odometer reset button and hold down
- Switch ignition off, then back on again
- Release odometer reset button

The function number and data for the selected function appear in the display window of the speedometer.
Press and release odometer reset button to advance to next function.

Function number	Function displayed	Engine code letters		
		AAA (2.8 L, VR6)		
1	Distance pulse number	3735	6010	6010
2	Country code	1	3	4
3	Speedometer/Tachometer	Not required for troubleshooting; used for production only		
4	No. of cylinders	6	6	6
5	Segment test	All segments of display light up		
6	Pulse from Speedometer Vehicle Speed Sensor (VSS)	4 pulses counted per 1 wheel revolution and added to counter total. Display range from 0 to 9999		
7	Engine Speed (RPM)	Engine Speed (RPM) is displayed		
8	Program version	Not required for troubleshooting; used for production only		

Country codes:

1 - All countries (including Canada) except USA and Great Britain
3 - Great Britain
4 - USA

Distance pulse number:

If the displayed value does not match the value in the table, the distance driven will be recorded incorrectly. The distance pulse number is a constant, engine specific value. The distance pulses for any given driven distance can be calculated using the number in the table.

Example: Distance pulse number (from table) x driven distance (miles or km) = pulses for given distance.
6010 x 0.5 = 3005

Troubleshooting

Note:
If the Multi-function indicator shows signs of malfunctioning such as, for example, unrealistic displays or no reaction to switch operation, disconnect battery ground strap (obtain radio security code first) then reconnect. This will cancel all Multi-function indicator memories and zero the electronic unit.

Clock inaccurate

- Reset clock to proper time
- Time deviation of 5 seconds/24 hours is permissible
- If clock deviates more than 5 seconds/24 hours or loses time (1 minute per day), ensure digital clock counter is properly synchronized as follows (VDO instrument clusters only):
- Obtain radio security code
- Disconnect battery ground strap and wait at least 30 seconds
- Reconnect battery ground strap and set clock to correct local time
- Verify clock keeps accurate time
- Activate radio using security code
- On non-VDO instrument clusters, if clock deviates more than 5 seconds/24 hours, replace instrument cluster printed circuit

Total or single trip memory of Multi-function indicator not OK

Memories cannot be switched over or controlled
See page 60, **"Multi-function Indicator Mode Select Switch (E 86), Memory Switch (E 109) and wiring, checking". If Mode Select Switch (E 86), Memory Switch (E 109) and wiring check OK, replace printed circuit.**

Memories cannot be cancelled or are permanently cancelled (do not store information)
- Switch ignition on
- Set Memory Switch (E 109) to position 1: Time is displayed
- Press mode select switch once: Driving time (stored value) is displayed
- Slide Memory Switch (E 109) fully left from position 1 to position 0 and hold for longer than 1 second: The display goes to 0:00 and changes to 0:01 after 1 minute
- If function is OK, slide Memory Switch (E 109) fully right past position 2 to position 0 and hold for longer than 1 second. The display goes to 0:00 and changes to 0:01 after 1 minute

If one or both memories cannot be cancelled, see "Multi-function Indicator Mode Select Switch (E 86), Memory Switch (E 109) and wiring, checking", **page 60**. If the Mode Select Switch (E 86), Memory Switch (E 109) and wiring are OK, replace printed circuit.

Driving time remains at 0:00

See "Multi-function Indicator Mode Select Switch (E 86), Memory Switch (E 109) and wiring, checking", page 60. If the Mode Select Switch (E 86), Memory Switch (E 109) and wiring are OK, replace printed circuit.

Driving distance remains at 0

If the speedometer (G 21) is also malfunctioning , see "Speedometer Vehicle Speed Sensor (VSS) (G 22), checking", page 63.
If the wiring and pulses from Speedometer Vehicle Speed Sensor (G 22) are OK, see "Multi-function indicator Mode Select Switch (E 86), Memory Switch (E 109) and wiring, checking", page 60.
If the Mode Select Switch (E 86), Memory Switch (E 109) and wiring are OK, replace printed circuit.

Driving distance not accurate

Deviations up to max. ± 4% between actual and calculated driving distance are permissible. If the deviation is more than 4%, see "Speedometer Vehicle Speed Sensor, (G 22), checking", page 63.
If the wiring and pulses from the Vehicle Speed Sensor (G 22) are OK, replace printed circuit.

Average speed display remains at "0" or display not accurate

Check driving distance display and driving time display . If driving time and distance are also inaccurate, troubleshoot these conditions first, see page 56.
If the driving time and distance displays are OK, replace printed circuit.

Average fuel consumption display not accurate

If the average speed display and driving distance display are also inaccurate, troubleshoot these conditions first, see above.
If the average speed and driving distance displays are OK:
Check fuel consumption signal from engine control module (see Diagnostic Trouble Code (DTC), checking using VAG 1551, Repair Groups 01 and 24, Fuel injection and ignition).
If the consumption signal from engine control module is OK: Check wiring from engine control module to instrument cluster.
If the wiring is OK, replace printed circuit.

Oil temperature display inaccurate

See "Oil temperature display, checking", page 58.

Outside (ambient) temperature display inaccurate

See "Outside (ambient) temperature display, checking", page 59.

Oil temperature display, checking

Tools required:

- Instrument tester VW 1301
- Connector test kit VW 1594

Oil Temperature Sensor (G 8), checking

- remove connector from Oil Temperature Sensor (G 8)
- connect tester VW 1301 between detached connector and ground (use connector test kit VW 1594)

Initiate Multi-function indicator check mode test:
- switch ignition on
- press and hold Mode Select Switch (E 86)
- switch ignition off, and then back on again
- select oil temperature display by pressing and releasing Mode Select Switch (E 86)
- set the following values on VW 1301 and check display:

VW 1301	Specified display
220	167°F to 185°F (75°C to 85°C)
110	203°F to 221°F (95°C to 105°C)
40	257°F to 275°F (125°C to 135°C)

- "----" appears in display or specified values not OK → Check wiring from Oil Temperature Sensor (G 8) to instrument cluster for open circuit according to wiring diagram
  - Wiring not OK → Repair wiring
  - Wiring OK → Replace printed circuit
- Specified values OK → Replace oil temperature sensor (G 8)

Outside (ambient) temperature display, checking

Tools required:

- Digital potentiometer VAG 1630
- Connector test kit VW 1594

Note:
Display range for outside (ambient) temperature is from -40°F to 122°F (-40°C to 50°C).
If temperature sensor is not connected "-40°" is displayed.

Outside Air Temperature Sensor (G 17), checking

- remove 2-pin connector from Outside Air Temperature Sensor (G 17), located on right side behind front bumper
- connect digital potentiometer VAG 1630 to 2-pin connector (use connector test kit VW 1594)

Initiate Multi-function indicator check mode test:
- switch ignition on
- press and hold Mode Select Switch (E 86)
- switch ignition off, then back on again
- select outside (ambient) temperature display by pressing and releasing Mode Select Switch (E 86)
- set digital potentiometer VAG 1630 to the following resistance values and compare display with specified values:

Resistance VAG 1630 (kΩ)	Specified display °F (°C)
14.7	14°F (-10°C)
8.6	32°F (0°C)
1.0	120°F (49°C)
less than 0.9	122°F (50°C)

- Deviation more than ± 4°F (2°C) = display not OK → Check wiring to instrument cluster for open circuit according to wiring diagram
  - Wiring not OK → Repair wiring
  - Wiring OK → Replace printed circuit
- Deviation less than ± 4°F (2°C) = display OK → Replace outside (ambient) temperature sensor, (G 17)

Multi-function Indicator Mode Select Switch (E 86), Memory Switch (E 109) and wiring, checking

Tools required:

- Multimeter Fluke 83 (US 1119)
- Connector test kit VW 1594
- Test box VAG 1598 with adapter harness VAG 1598/8

Troubleshooting preparations:

- Separate 4-pin connector located behind steering column switch trim
- Use valid wiring diagram to check wiring

Multi-function Indicator Mode Select Switch (E 86), checking

Continuity check with multimeter Fluke 83:
- Set multimeter to Ohm measuring range
- Connect multimeter black test lead to terminal 1 of 4-pin connector (T4a/1)
- Connect multimeter red test lead to terminal 2 of 4-pin connector (T4a/2)
- Press Mode Select Switch (E 86) and hold

0 Ohms (continuity)	∞ Ohms (no continuity)

∞ Ohms (no continuity) → Replace Mode Select Switch (E 86)

0 Ohms (continuity) →

Multi-function Indicator Memory Switch (E 109) position "1", checking

- Leave multimeter set to Ohm measuring range
- Connect multimeter black test lead to terminal 1 of 4-pin connector (T4a/1)
- Connect multimeter red test lead to terminal 3 of 4-pin connector (T4a/3)
- Slide Memory Switch (E 109) to position "1" (memory 1)

0 Ohms (continuity)	∞ Ohms (no continuity)

∞ Ohms (no continuity) → Replace Memory Switch (E 109)

0 Ohms (continuity) →

- Slide Memory Switch (E 109) fully left from position "1" to position "0" and hold

0 Ohms (continuity)	∞ Ohms (no continuity)

∞ Ohms (no continuity) → Replace Memory Switch (E 109)

0 Ohms (continuity) →

Multi-function Indicator Memory Switch (E 109) position "2" , checking

- Slide Memory Switch (E 109) to position "2" (memory 2)

∞ Ohms (no continuity)	0 Ohms (continuity)

0 Ohms (continuity) → Replace Memory Switch (E 109)

∞ Ohms (no continuity) →

- Slide Memory Switch (E 109) fully right from position "2" to position "0" and hold

∞ Ohms (no continuity)	0 Ohms (continuity)

0 Ohms (continuity) → Replace Memory Switch (E 109)

∞ Ohms (no continuity) → A

A ↓

Multi-function Indicator Memory Switch (E 109) position "0", checking

- Leave multimeter set to Ohm measuring range
- Connect multimeter black test lead to terminal 1 of 4-pin connector (T4a/1)
- Connect multimeter red test lead to terminal 4 of 4-pin connector (T4a/4)
- Slide Memory Switch (E 109) from position "1" to "0", then from "2" to "0"

0 Ohms (continuity) in position "0"	∞ Ohms (no continuity) in position "0"

∞ Ohms (no continuity) in position "0" → Replace Memory Switch (E 109)

0 Ohms (continuity) in position "0" →

Ground wire from fuse/relay panel to 4-pin connector, checking

- Leave multimeter set to Ohm measuring range
- Connect multimeter black test lead to vehicle ground
- Connect multimeter red test lead to terminal 1 of 4-pin connector (T4a/1)

0 Ohms (continuity)	∞ Ohms (no continuity)

∞ Ohms (no continuity) → Check wiring for open circuits according to wiring diagram, repair as necessary

0 Ohms (continuity) →

Wiring from instrument cluster to 4-pin connector, checking

- Leave multimeter set to Ohm measuring range
- Switch ignition off
- Remove instrument cluster, see Electrical System, Repair Group 90

Use test box VAG 1598 to check wiring:
- Remove 28-pin connector from instrument cluster
- Connect instrument cluster wiring harness to test box VAG 1598 using adapter harness VAG 1598/8 (do not connect instrument cluster to test box; leave disconnected)
- Connect multimeter black test lead to test box, socket 15
- Connect multimeter red test lead to terminal 2 of 4-pin connector (brown) (T4a/2)

0 Ohms (continuity)	∞ Ohms (no continuity)

∞ Ohms (no continuity) → Check wiring for open circuits according to wiring diagram, repair as necessary

0 Ohms (continuity) →

- Leave multimeter set to Ohm measuring range
- Connect multimeter black test lead to test box VAG 1598, socket 6
- Connect multimeter red test lead to terminal 3 of 4-pin connector (T4a/3)

0 Ohms (continuity)	∞ Ohms (no continuity)

∞ Ohms (no continuity) → Check wiring for open circuits according to wiring diagram, repair as necessary

0 Ohms (continuity) → B

Socket numbers on test box VAG 1598 are identical to instrument cluster 28-pin terminal numbers.

B

- Leave multimeter set to Ohm measuring range
- Connect multimeter black test lead to test box VAG 1598, socket 4
- Connect multimeter red test lead to terminal 4 of 4-pin connector (T4a/4)

0 Ohms (continuity)	∞ Ohms (no continuity)

∞ Ohms (no continuity) → Check wiring for open circuit according to wiring diagram, repair as necessary

0 Ohms (continuity) →

- Leave multimeter set to Ohm measuring range
- Connect multimeter black test lead to test box VAG 1598, socket 5
- Connect multimeter red test lead to terminal 1 of 4-pin connector (T4a/1)

0 Ohms (continuity)	∞ Ohms (no continuity)

0 Ohms (continuity) → Multi-function Indicator Mode Select Switch (E 86), Memory Switch (E 109) and wiring OK → END

∞ Ohms (no continuity) → Check wiring for open circuit according to wiring diagram, repair as necessary → END

Socket numbers on test box VAG 1598 are identical to instrument cluster 28-pin terminal numbers.

Speedometer Vehicle Speed Sensor (VSS) (G 22), checking

Tools required:

- Multimeter Fluke 83 (US 1119)
- Connector test kit VW 1594
- Test box VAG 1598 with adapter harness VAG 1598/8

Test conditions: Distance pulse number and country code OK, see "Calling up MFA check mode test", page 54.

Speedometer Vehicle Speed Sensor (VSS) (G 22), checking pulse signal

Initiate instrument cluster check mode test:
- switch ignition on
- press trip odometer reset button and hold down
- switch ignition off and then back on:

The function number and data for the selected function appear in the display window of the speedometer.
- compare distance pulse number with specifications in table, see page 55
- press and release odometer reset button several times until function number 6 appears in display
- slowly rotate front wheel and watch display

No change in number or change does not correspond with 4 pulses per rotation of wheel	For every wheel rotation, 4 pulses are added to number in display

For every wheel rotation, 4 pulses are added to number in display → Speedometer Vehicle Speed Sensor (VSS) (G 22) OK

No change in number or change does not correspond with 4 pulses per rotation of wheel → Check wiring between Speedometer Vehicle Speed Sensor (VSS) (G 22) and instrument cluster for open circuit according to wiring diagram

Wiring not OK	Wiring OK

Wiring not OK → Repair wiring

Wiring OK → Replace Speedometer Vehicle Speed Sensor (VSS) (G22)

Index

Crank
- installing **92.3**

Rear window wiper
- assembly (1989) **92.7**
- assembly (from 1990) **92.7a**

Rear wiper arm
- adjusting **92.8**

Rubber wiper inserts
- removing/installing **92.4**

Spray jets
- adjusting **92.6**

Windshield washer
- system **92.5**

Windshield washer pump
- hose connections **92.6**

Windshield wiper
- assembly **92.2**
- troubleshooting **92.9**

Wiper arm
- adjusting **92.3**

Wiper arm angle
- adjusting **92.8**

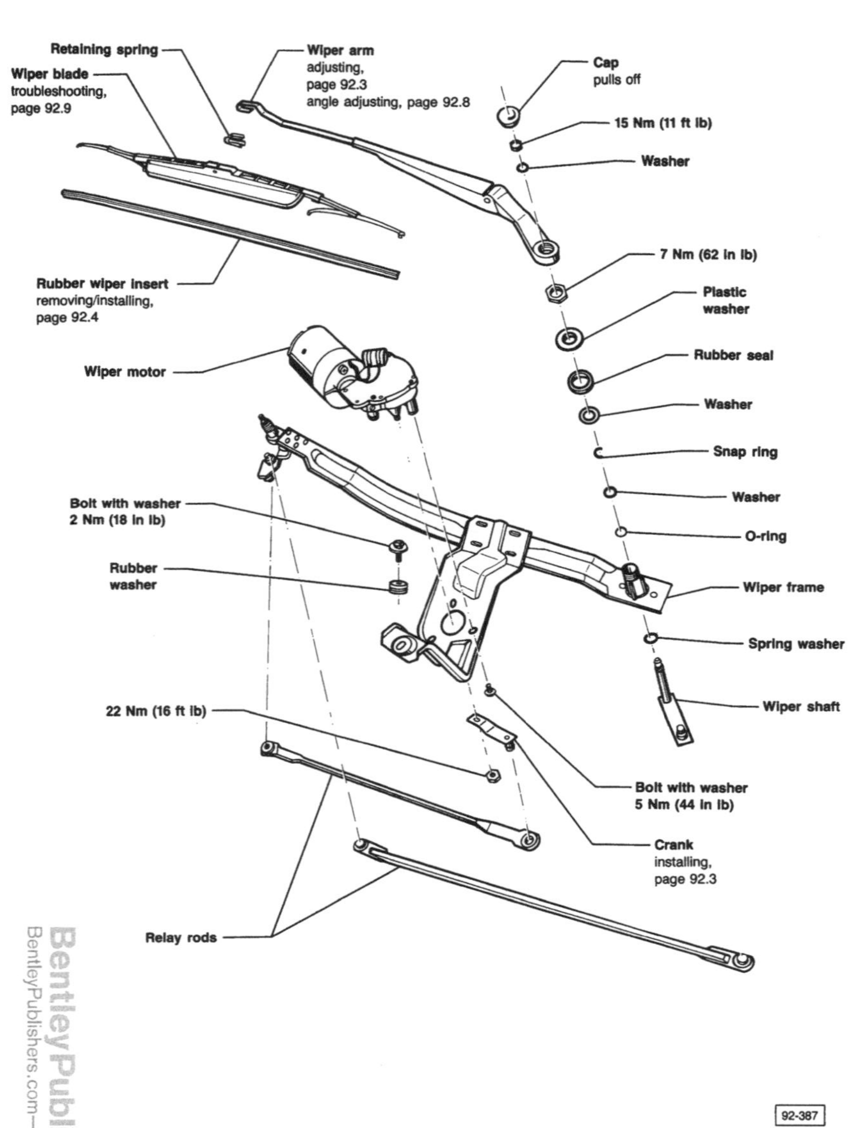

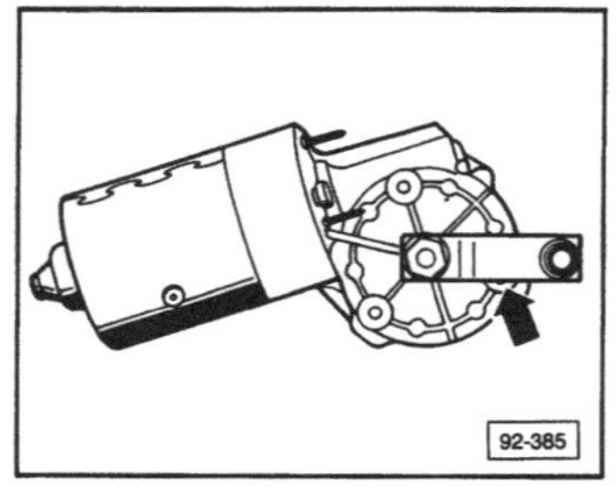

▶ Fig. 1 **Crank, installing**

Windshield wiper motor must be in the park position before installing the crank.

- run wiper motor to stop position
- install crank and align so that threaded hole (**arrow**) is just covered
  - 22 Nm (16 ft lb)

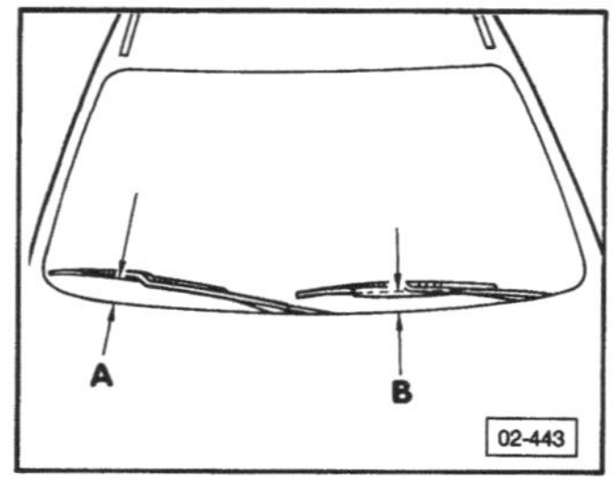

▶ Fig. 2 **Wiper arm, adjusting**

Windshield wiper motor must be in the park position before adjusting the wiper arms.

- run wiper motor to stop position
- install right wiper arm and adjust to position **A**
  - **A** = 25 mm (1.0 in)
  - 15 Nm (11 ft lb)
- install left wiper arm and adjust to position **B**
  - **B** = 45 mm (1.75 in)
  - 15 Nm (11 ft lb)

Rubber wiper inserts, removing and installing

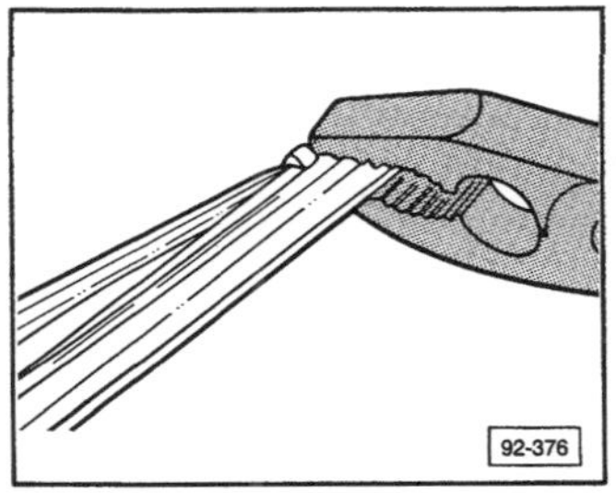

▶ Fig. 3 **Removing**

- squeeze metal inserts at open end of rubber insert using pliers
- remove rubber insert from upper retaining clip
- slide rubber insert from other clips of wiper blade
- remove metal inserts from rubber

▶ Fig. 4 **Installing**

- install metal inserts into upper groove of new rubber insert

Note

Notches in metal inserts must face rubber and engage rubber lugs in groove.

- slide rubber insert into lower clips of wiper blade
- squeeze metal inserts at open end of rubber insert using pliers
- install rubber insert into clip making sure clip engages into rubber retaining slot **(arrow)**

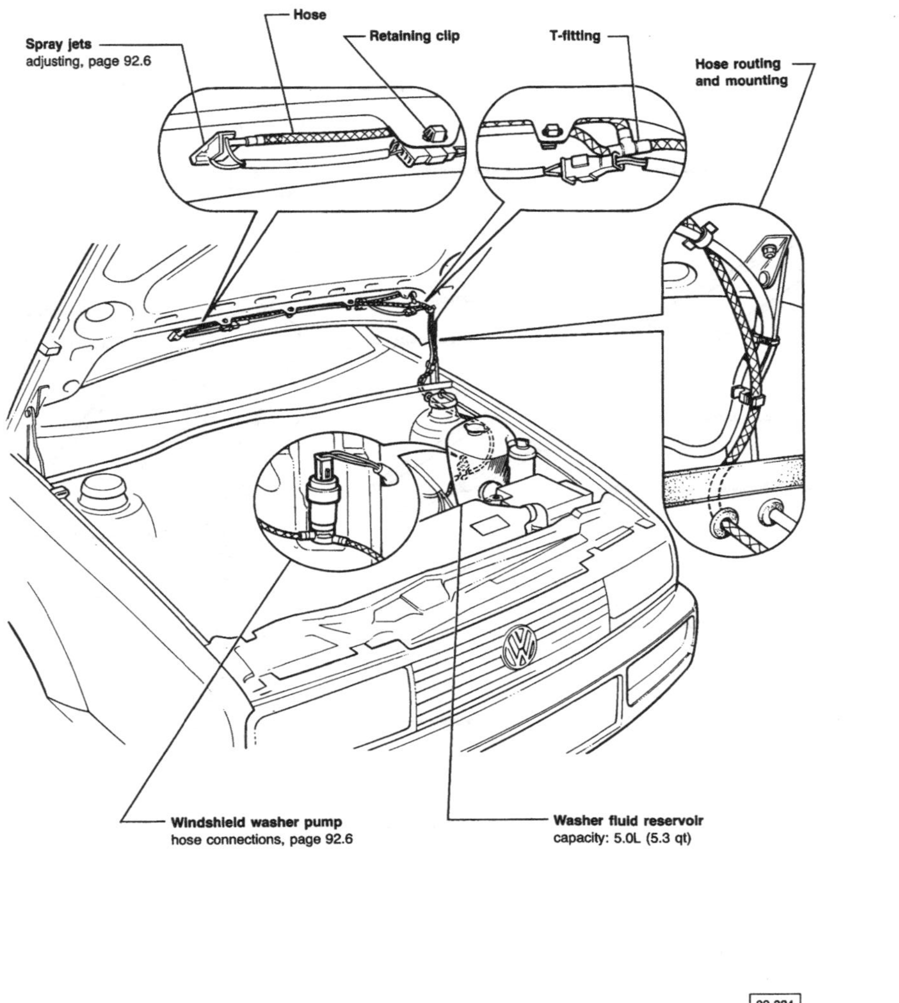

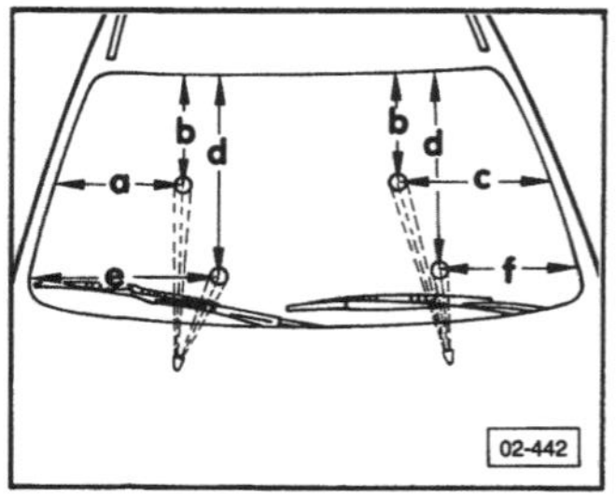

▶ **Fig. 1 Spray jets, adjusting**

a = 320 mm (12.6 in)
b = 310 mm (12.2 in)
c = 380 mm (15.0 in)
d = 560 mm (22.0 in)
e = 460 mm (18.1 in)
f = 340 mm (13.4 in)

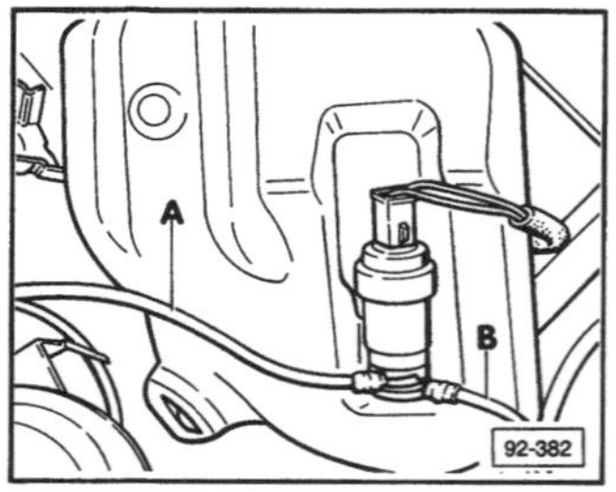

▶ **Fig. 2 Windshield washer pump, hose connections**

A = Hose to windshield washer system
B = Hose to rear window washer system

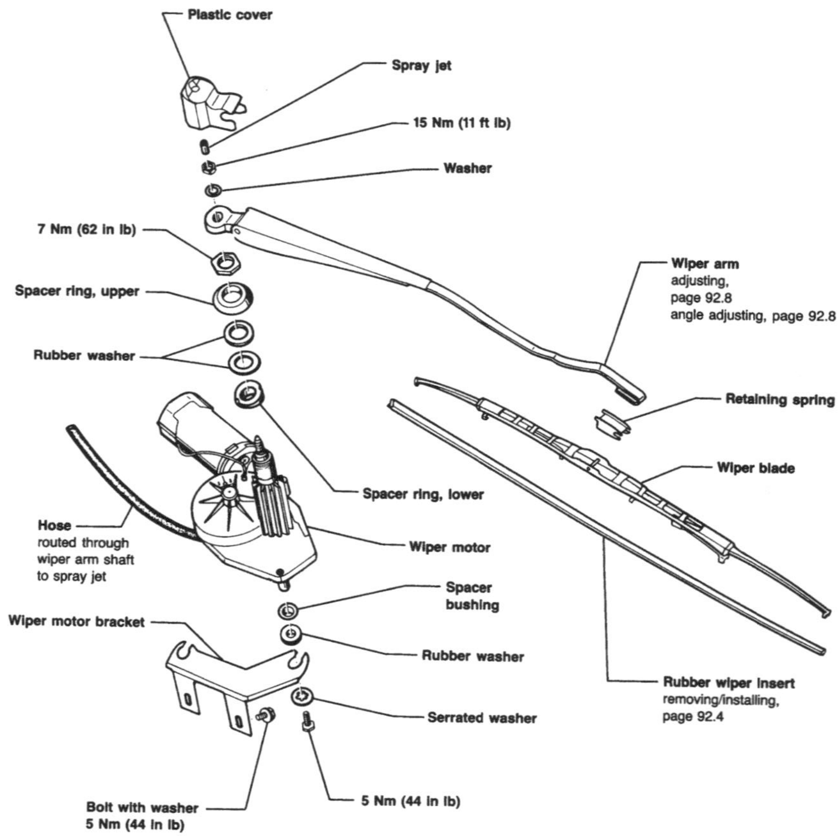

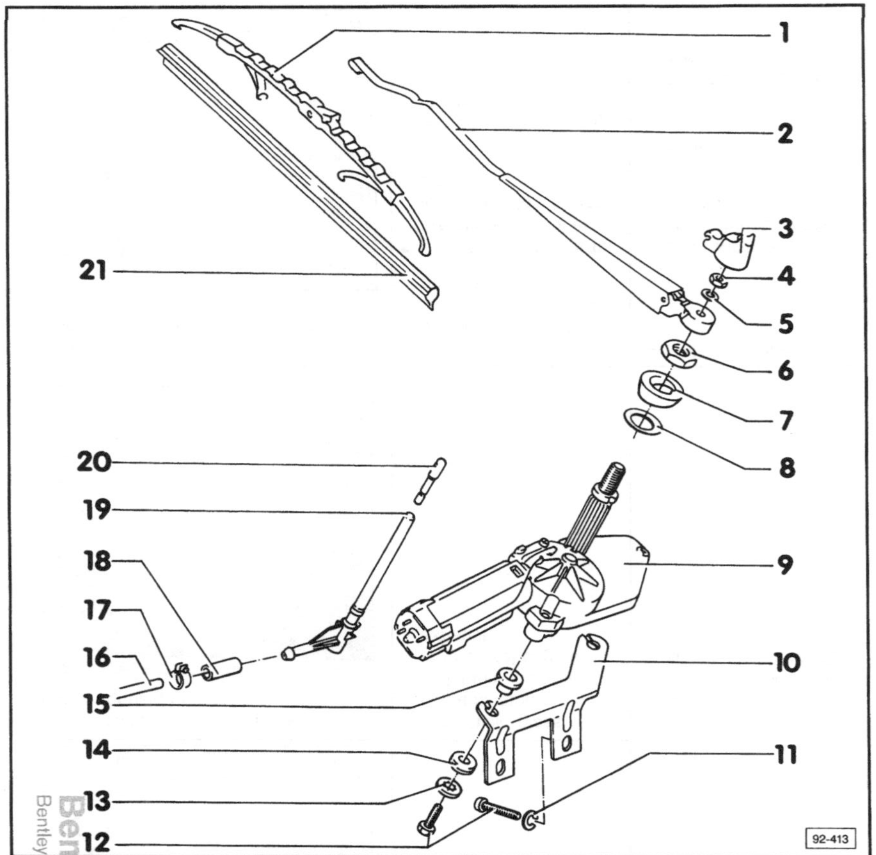

1 – **Wiper blade**
2 – **Wiper arm**
adjusting, see Index
3 – **Plastic cover**
4 – **15 Nm (11 ft lb)**
5 – **Washer**
6 – **7 Nm (62 in. lb)**
7 – **Spacer ring**
8 – **Washer**
9 – **Wiper motor**
10 – **Wiper motor bracket**
11 – **Washer**
12 – **5 Nm (44 in. lb)**
13 – **Serrated washer**
14 – **Rubber washer**
15 – **Spacer bushing**
16 – **Hose**
17 – **Hose clamp**
18 – **Hose connector**
19 – **Connector tube for spray jet**
20 – **Spray jet**
adjust so that washer fluid sprays in center of wiper contact area
21 – **Rubber wiper insert**
removing/installing, see Index

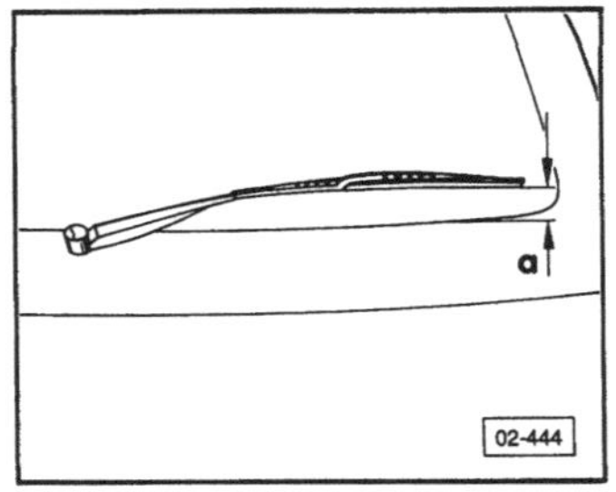

▶ Fig. 1 **Rear wiper arm, adjusting**

- run wiper motor to stop position
- adjust wiper arm to position **a**
  - **a** = 50 mm (2.0 in)

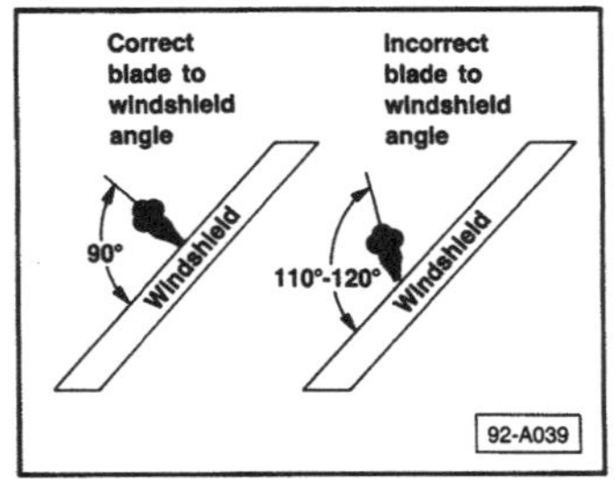

▶ Fig. 2 **Wiper arm angle, adjusting**

Windshield wiper blade chatter and skipping of blade across windshield can be caused by incorrect wiper arm angle.

The incorrect wiper arm angle will not allow the wiper blade to flip and follow the wiper arm.

It is most noticeable on the downward stroke of the wiper.

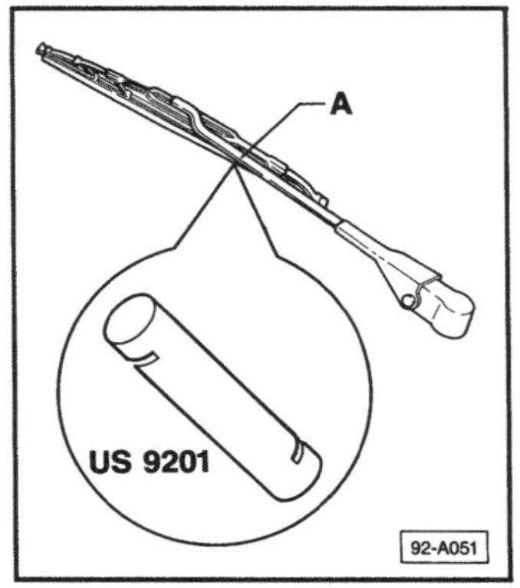

▶ To adjust wiper arm angle,
- insert slot of wiper arm adjusting tool **US 9201** over wiper arm
- bend wiper arm at **A** to obtain correct wiper arm angle

Note

Wiper arm must be bent at location **A**. Bending the wiper arm in the wrong location will not provide a lasting repair.

Wax and dirt build-up on windshield wiper blades can cause streaking and smearing on windshield when using wipers.
- **do not replace** windshield wiper blades that may only need to be cleaned
- clean wax and dirt build-up from wiper blades and entire windshield with alcohol based cleaning solution or rubbing alcohol and water
- see troubleshooting chart on page 92.9
- refer customer to instructions in Owner's Manual

Windshield wiper troubleshooting chart

Symptoms	Possible causes	Solutions
Bad wiping (sheeting) 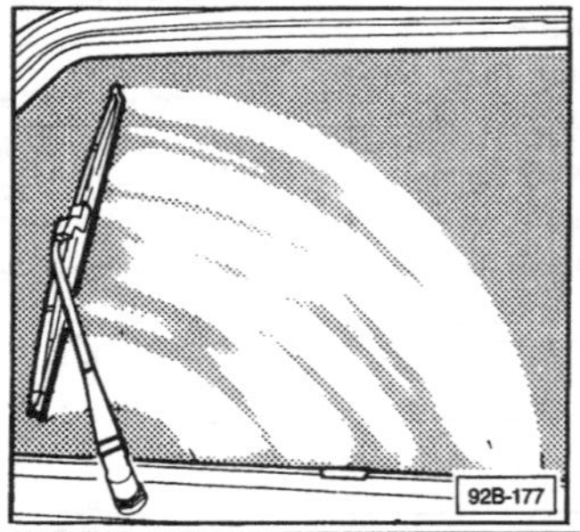	• dirty blade lip • blade lip worn, torn or broken • old blade lip	— clean blade lip with wet cloth and soap — rinse with water — replace blades — replace blades
Some water remains in the wiping area, forming droplets 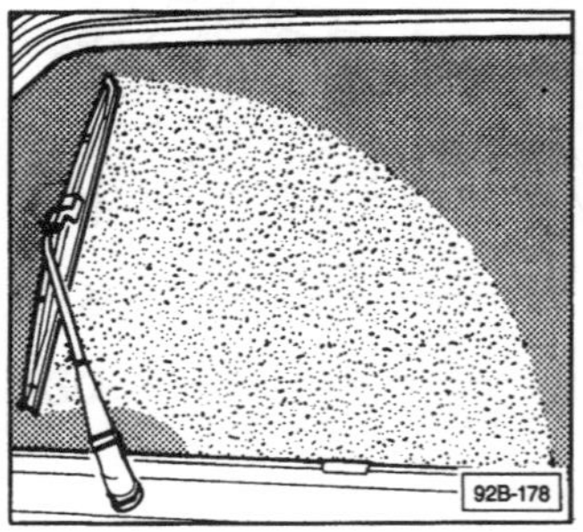	• windshield glass contaminated by pollutants, oil or film from diesel engine exhaust	— clean windshield glass with clean cloth soaked in solution of alcohol (or ammonia) and warm water
Windshield wiper blades clean to one side and are deficient to the other (chatter) 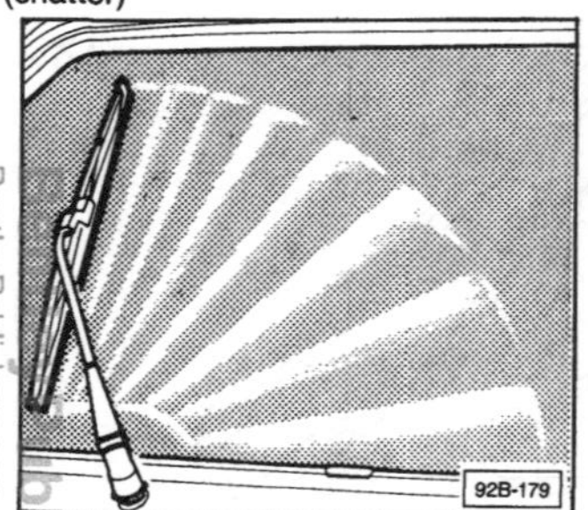	• blade lips are deformed or filled with dirt or foreign matter on one side • windshield wiper arms are twisted. Rubber lips do not touch glass properly • glass is dirty or not wet enough, resulting in varying friction between blade and glass	— replace blades — bend wiper arm to obtain proper blade contact — clean the glass with clean cloth soaked in solution of alcohol (or ammonia) and warm water

THIS PAGE INTENTIONALLY LEFT BLANK

Index

Additional brake light
- removing **94.18**

Back–up light housing
- removing/installing **94.18**

Cruise control system
- connector wiring **94.11**

Fog lights
- adjusting **94.16**
- removing **94.16**

Front turn signal light
- removing/installing **94.16**

Headlights
- adjusting **94.15**
- aiming **94.14**
- assembly **94.13**
- removing/installing **94.15**

Horn contact ring
- lubricating **94.3**
- lubricating (auto. trans. from 1993 m.y.) **94.7**

Horn pad
- removing **94.3**
- removing (auto. trans. from 1993 m.y.) **94.7**

Horn plate
- removing **94.3**
- removing (auto. trans. from 1993 m.y.) **94.7**

License plate lights
- removing **94.19**

Lock cylinder
- removing/installing (auto. trans. from 1993 m.y.) **94.9**

Multi–function indicator
- connector wiring **94.11**

Side marker lights
- removing **94.15**

Steering column
- assembly **94.2**
- assembly (auto. trans. from 1993 m.y) **94.5**

Steering column switches
- connector wiring **94.10**
- wire routing (USA only) **94.12**

Steering lock housing
- removing/installing (auto trans from 1993 m.y.) **94.8**

Taillight
- assembly **94.17**

Turn signal switch
- removing/installing **94.4**
- removing (auto. trans. from 1993 m.y.) **94.8**

Windshield wiper switch
- removing/installing **94.4**
- removing (auto. trans. from 1993 m.y.) **94.8**

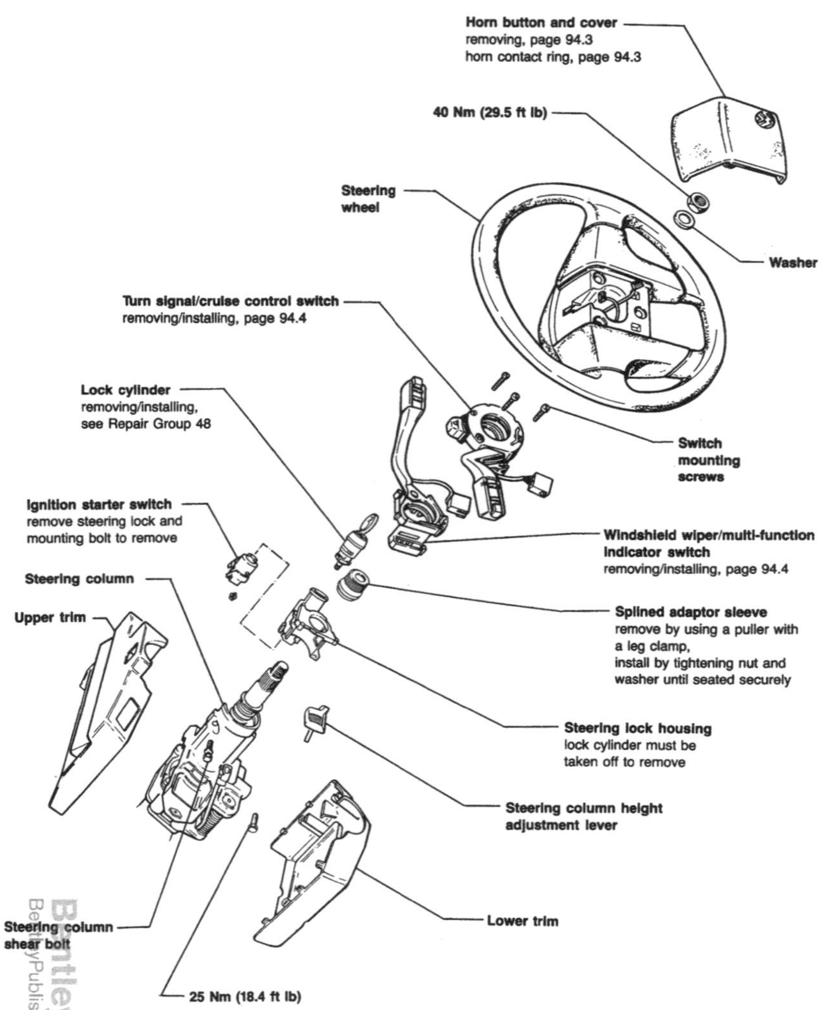

Horn pad, removing

- lift horn cover from bottom
- carefully pull from steering wheel in direction of arrow
- remove horn wire

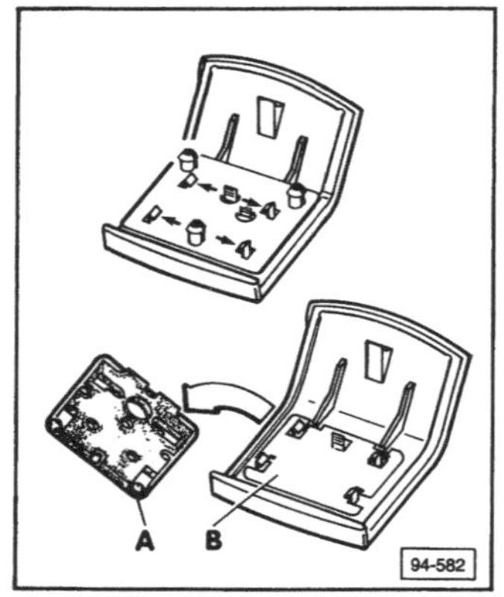

Horn plate, removing

- spread locating lugs (**arrows**) in direction indicated and remove contact plate **A** from plate **B**

Note

Plate **A** and plate **B** are normally separated from each other. Horn operates as plate **A** contacts plate **B** when horn button is pushed.

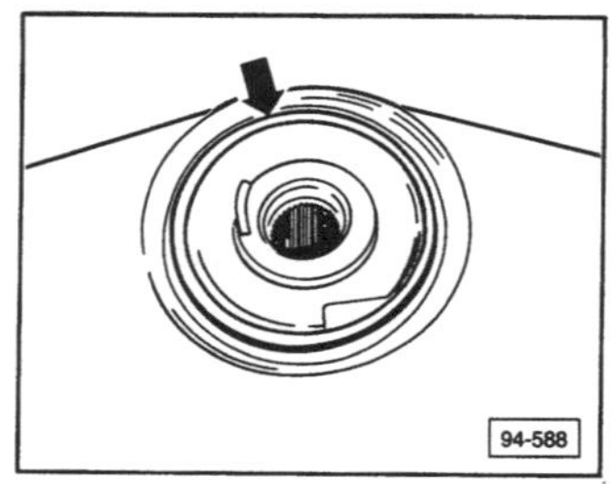

Horn contact ring, lubricating

Electrical connection to the horn button is made through a horn contact ring (**arrow**) under the steering wheel

- lubricate ring with universal grease

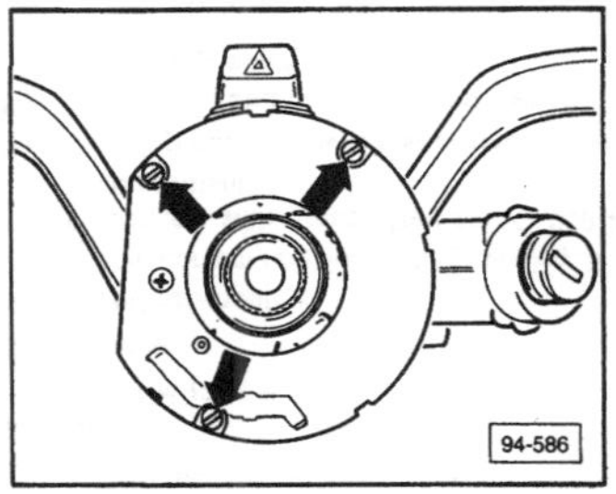

Turn signal and windshield wiper switch, removing/installing

Removing

- remove steering wheel
- remove mounting screws (**arrows**)
- detach electrical connectors
- remove turn signal and windshield wiper switch from steering column

Installing

- install turn signal and windshield wiper switch to steering column
- attach electrical connectors
- install mounting screws (**arrows**) and tighten
- install steering wheel
  - 40 Nm (29.5 ft lb)

THIS PAGE INTENTIONALLY LEFT BLANK

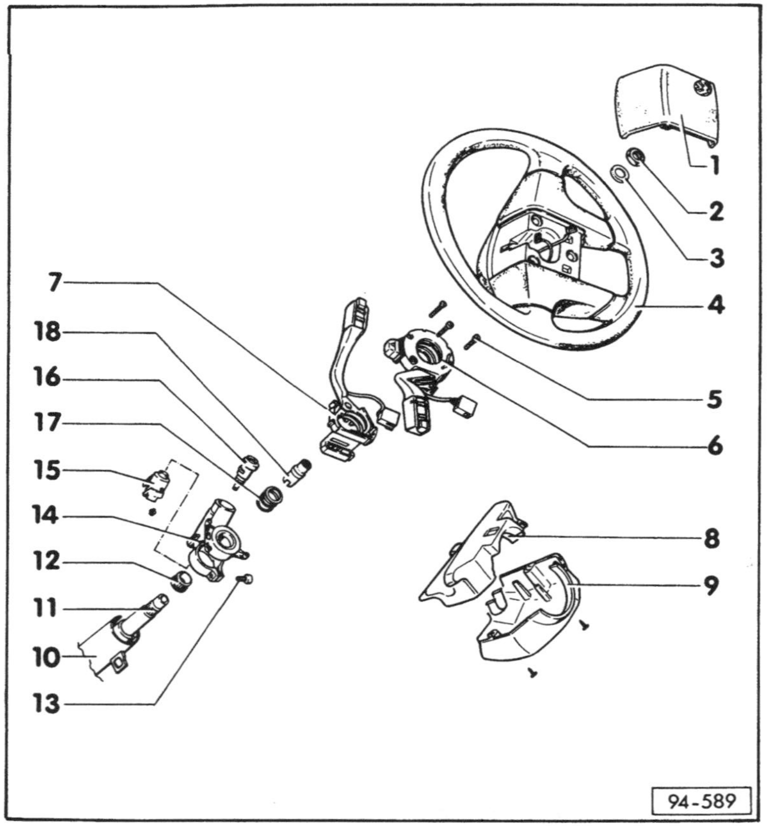

1 — **Horn pad**
- removing, see Fig. 1
- removing horn plate, see Fig. 2

2 — **Nut**
- 40 Nm (29.5 ft lb)

3 — **Washer**

4 — **Steering wheel**
- horn contact ring, lubricating, see Fig. 3
- lubricate with grease Part No. G 000 400 or equivalent

5 — **Steering column switch securing screws**

6 — **Turn signal and cruise control switch**
- removing, see Fig. 4

7 — **Windshield wiper/washer and Multi-function Indicator (MFA) switch**
- removing, see Fig. 4

8 - **Upper steering column trim**

9 - **Lower steering column trim**

10 - **Steering column tube**

11 - **Steering column**

12 - **Support ring**

13 - **Steering lock housing securing bolt**
- 10 Nm (7.4 ft lb)

14 - **Steering lock housing**
- removing/installing, see Fig. 5

15 - **Ignition/starter switch**
- to remove, first remove steering lock housing, then remove securing screws

16 - **Lock cylinder**
- removing/installing, see Fig. 6

17 - **Spring**

18 - **Splined adaptor sleeve**
- remove using puller with leg clamp (i.e. Kukko 204-2 or equivalent)
- install by tightening nut and washer securely

Fig. 1 Horn pad, removing

–lift horn pad at bottom and pull from steering wheel in direction of **arrow**

–remove horn wire

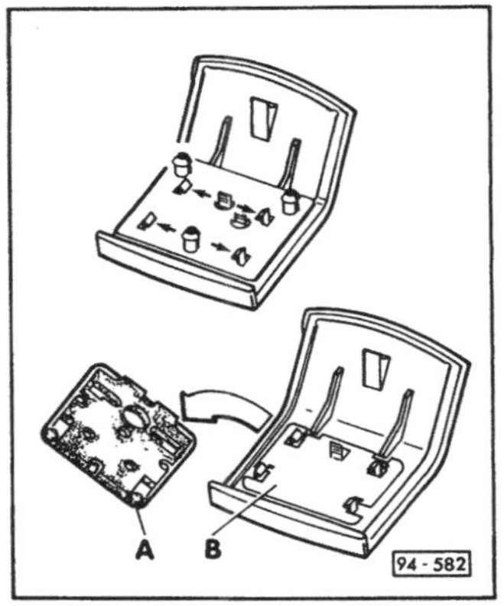

Fig. 2 Horn plate, removing

–remove wire connector from horn plate

–disengage retaining lugs (**arrows**) and remove contact plate **A** from contact plate **B**

Note

The horn plate consists of two uninsulated contact plates. Horn operates when plate **A** contacts plate **B** when the horn pad is pressed.

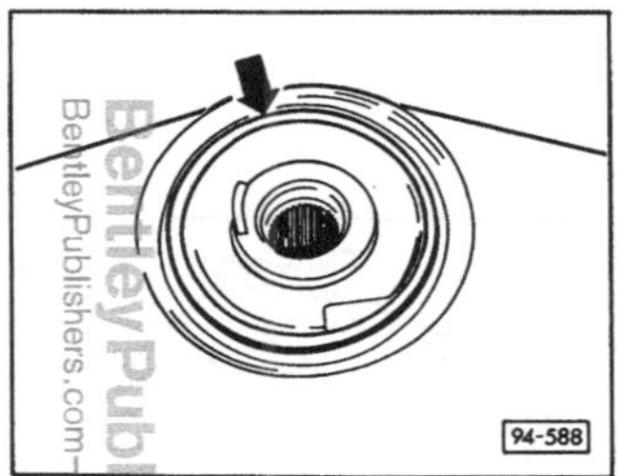

Fig. 3 Horn contact ring, lubricating

The positive voltage (B+) for horn operation is supplied via a spring contact on the turn signal switch and a contact (slip) ring (**arrow**) under the steering wheel.

–lubricate ring with grease Part No. G 000 400 or equivalent

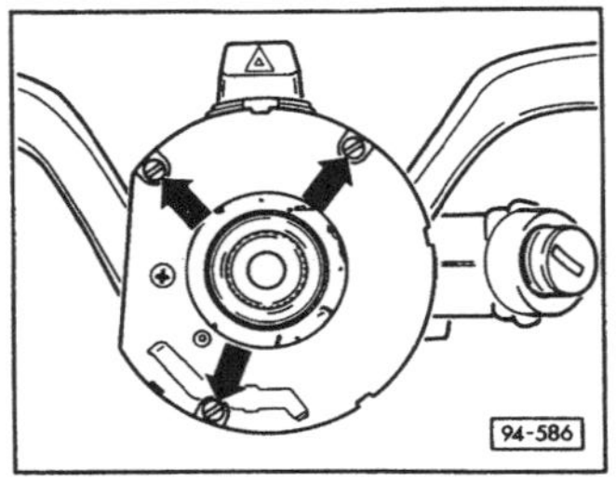

Fig. 4 Turn signal and windshield wiper/washer switch, removing

–remove steering wheel

–remove switch securing screws (**arrows**)

–disconnect switch connectors

–remove turn signal and windshield wiper/washer switch from steering column

Fig. 5 Steering lock housing, removing/ installing

Removing

–remove turn signal and windshield wiper/washer switch, see Fig. 4

–remove washer, spring and support ring from steering column

–remove mounting bolts and remove steering lock housing

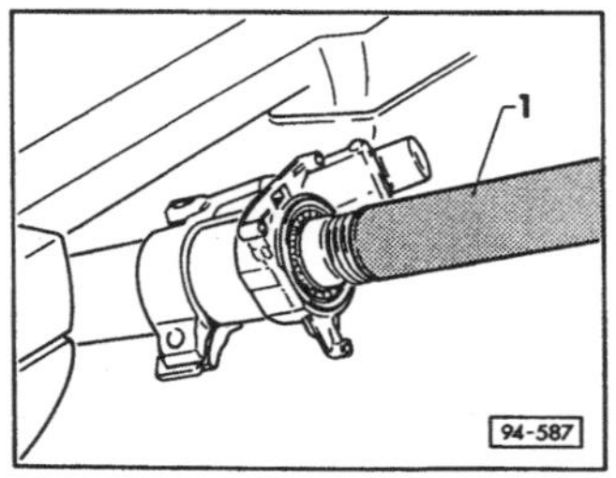

Installing

–slide steering lock housing onto steering column and install mounting bolts

–install support ring, spring and washer onto steering column

– using special tool VW 420 **-1-**, drive support ring, spring and washer onto steering column until fully seated

Note

If steering column shaft recedes into steering column tube, thread nut onto shaft and pull shaft out.

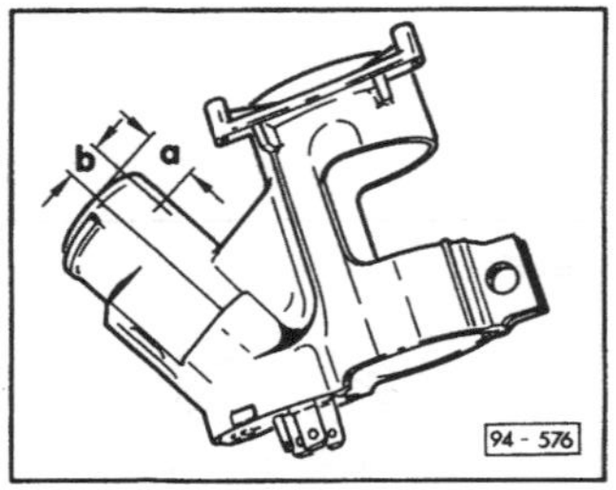

Fig. 6 **Lock cylinder, removing/installing**

Removing

- remove steering lock housing, see Fig. 5
- mark steering lock housing for hole (where **a** and **b** intersect)

a = 12 mm (15/32 in.)

b = 10 mm (25/64 in.)

- drill hole (approx. 3 mm diameter) into steering lock housing at mark until lock cylinder stop spring is visible; hole depth = approx. 3 mm (1/8 in.)
- using opposite end of drill, compress stop spring and pull out lock cylinder

Installing

- insert lock cylinder into steering lock housing
- insert key into lock cylinder
- push lock cylinder fully into housing while gently turning key

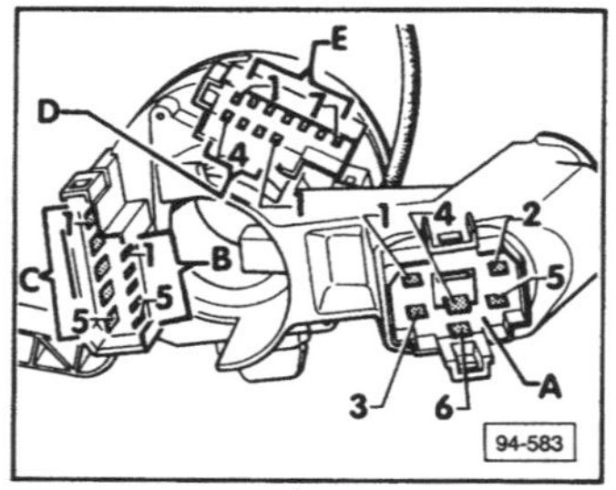

Steering column switches, connector wiring

A. Ignition starter switch
1 — plus (+) connection, terminal **15**
2 — plus (+) connection, terminal **X**
3 — plus (+) connection, terminal **50**
4 — plus (+) connection, terminal **30**
5 — plus (+) connection, terminal **SU**
6 — plus (+) connection, terminal **P**

B. 5-point connector
1 — windshield wiper switch, terminal **53b**
2 — windshield wiper switch, terminal **53a**
3 — windshield wiper switch, terminal **53**
4 — windshield wiper switch, terminal **53e**
5 — windshield wiper switch, terminal **J**

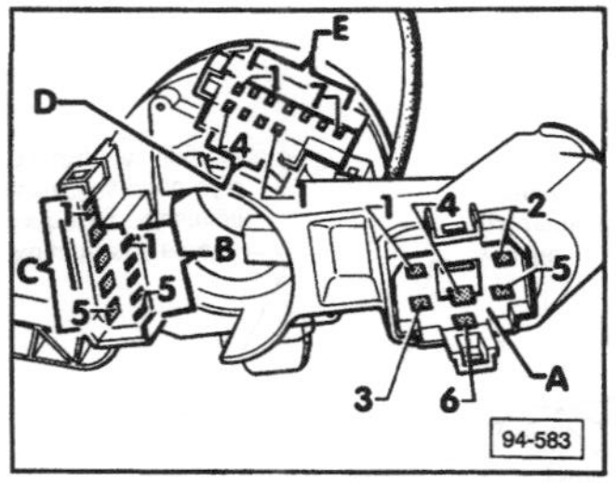

C. 5-point connector
- 1 — emergency flasher switch, terminal **49**
- 2 — headlight dimmer/flasher switch, terminal **56b**
- 3 — headlight dimmer/flasher switch, terminal **56**
- 4 — headlight dimmer/flasher switch, terminal **56a**
- 5 — headlight dimmer/flasher switch, terminal **30**

D. 4-point connector
- 1 — windshield wiper switch, terminal **L**
- 2 — windshield wiper switch, terminal **T**
- 3 — windshield wiper switch, terminal **31**
- 4 — emergency flasher switch, terminal **15**

E. 7-point connector
- 1 — horn button, terminal **71**
- 2 — emergency flasher switch, terminal **49a**
- 3 — turn signal switch, terminal **L**
- 4 — parking light switch, terminal **PL**
- 5 — parking light switch, terminal **P**
- 6 — parking light switch, terminal **PR**
- 7 — turn signal switch, terminal **R**

Multi-function indicator (MFA) and cruise control system (CCS), connector wiring

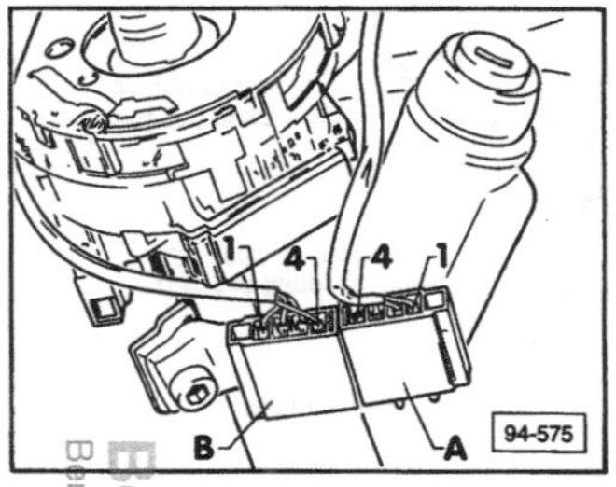

The following connectors are located on the steering lock housing

A. 4-point connector MFA switch
- 1 — ground (−), terminal **31**
- 2 — mode selector switch
- 3 — memory selector switch
- 4 — reset switch

B. 4-point connector, cruise control switch
- 1 — on switch
- 2 — resume switch
- 3 — memory switch
- 4 — plus (+) connection, terminal **30**

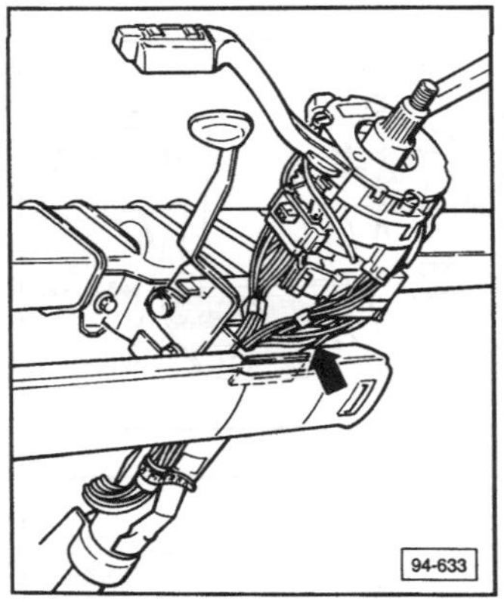

Steering column switches, wire routing

If any of the steering column switches (windshield wiper, turn signal, cruise control, ignition, etc.) are removed/installed, the switch wires must be properly re-routed against the steering column.

> **CAUTION**
>
> Routing wires incorrectly on steering column may cause wire damage and lead to inoperative components, shorts/ grounds and blown fuses.

- route wires against side of steering column
- **DO NOT** route wires between steering column and knee bar (**arrow**)
- if necessary, fasten wires with tie-wrap

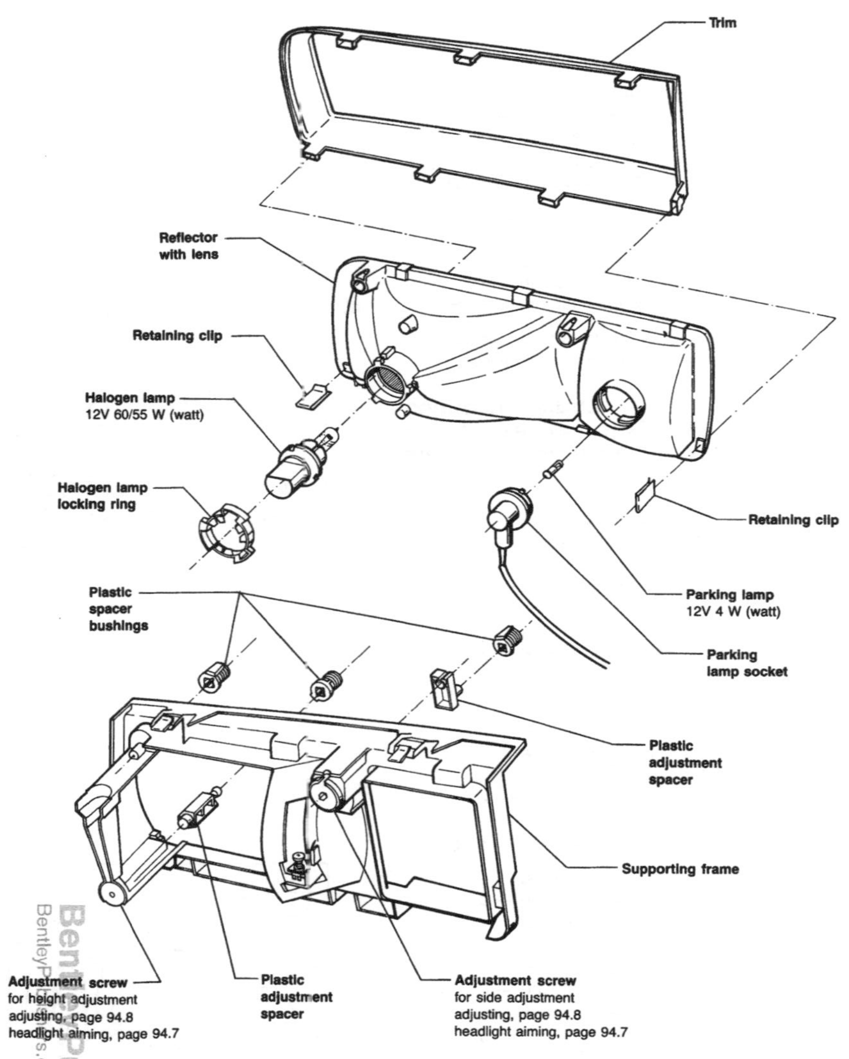

Headlights, aiming

Aiming requirements

Before checking headlight aim

- place vehicle on flat surface

Note

It is not necessary that the surface be exactly level.

- remove any large accumulations of mud or ice from wheel wells, fender undersides, bumpers etc.
- check tire pressure, correct if necessary
- verify that vehicle has no loading other than driver
- rock vehicle to equalize suspension
- clean headlight lenses and aiming pads
- check for bulb burnout, broken aiming pads and proper beam switching. Correct if necessary

Setting up universal adaptor

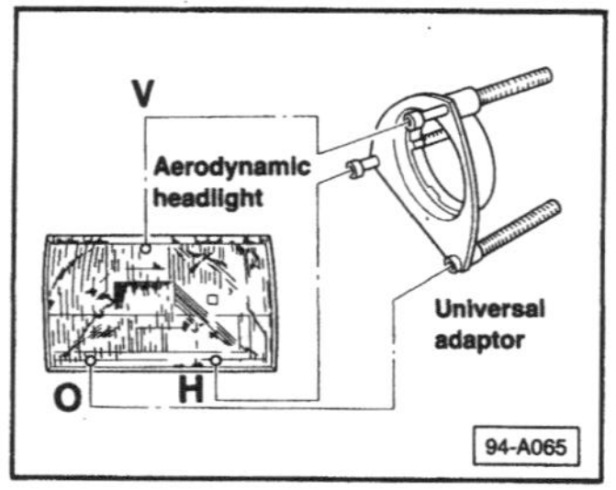

- locate aim settings on face of headlight lens

Note

The settings will be designated by the letters **V** and **H**, each followed by a number. Aiming pad nomenclature places the **V** number at the top aiming pad. This is the vertical setting. The **H** number is the outside aiming pad and is the horizontal setting. The inside pad is always set to **ZERO**

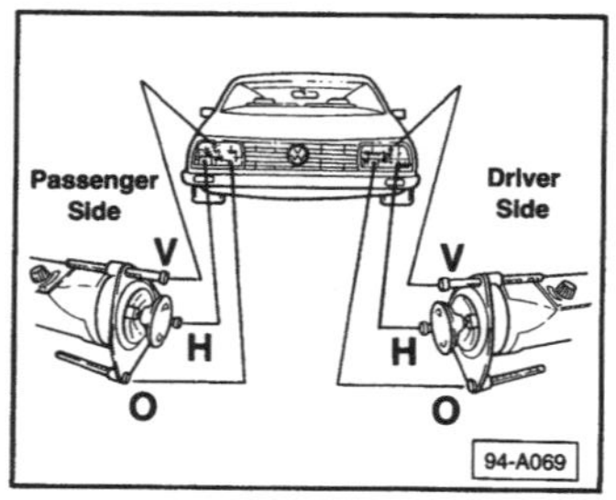

- move adjustable rods of adaptor to correspond to settings found on headlight
- attach aiming device to adaptor
- mount assembly to headlight lens

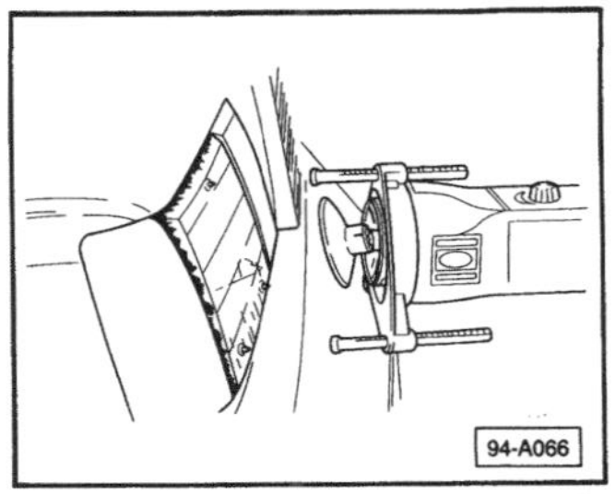

- check headlight aim using mechanical aiming procedure provided by equipment manufacturer
- correct if necessary

Headlights, adjusting

Adjust headlights by turning adjusting screws through openings **A** and **B** with long Phillips head screwdriver (right headlight shown in this illustration)

A — height adjustment
B — side adjustment

Note

Adjustment screws for the left headlight are the same.

Headlights, removing/installing

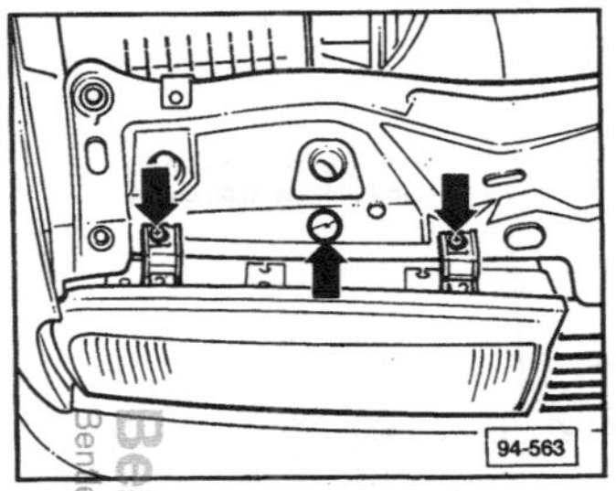

- remove both outer screws (**arrows**)
- loosen center screw (**arrow**) a few turns
- remove headlight by sliding outward and sideways towards center of vehicle
- disconnect wiring connections from headlight

- install in reverse order of removal

Side marker lights, removing

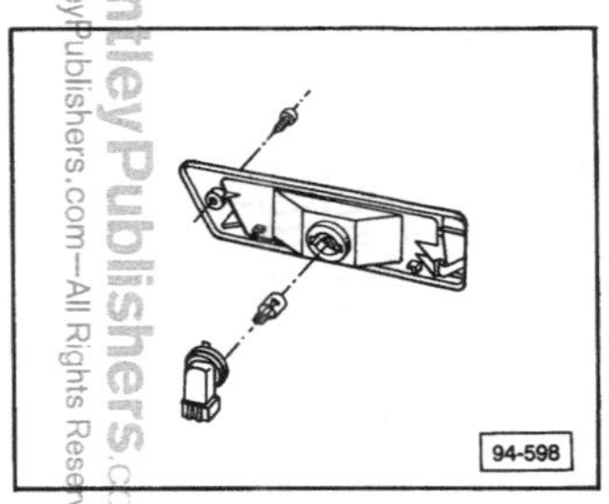

The side marker lights are installed in the left and right front bumper side covers.

- remove mounting screws
- pull out side marker light housing
- remove socket from housing
- remove bulb from socket
  - halogen, 12 V/5 W (watt)

Front turn signal light, removing/installing

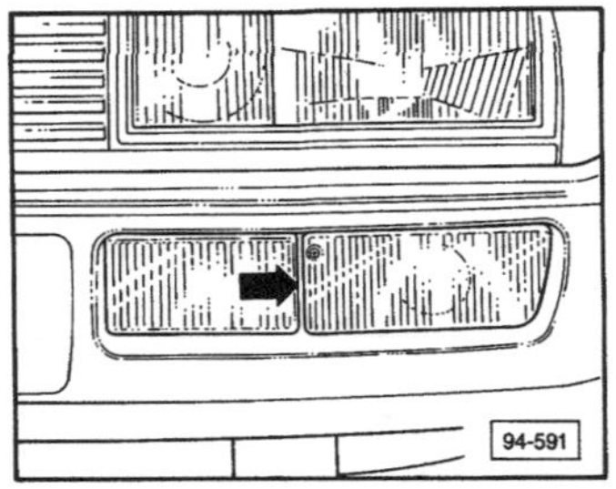

Removing

- carefully pry out turn signal light (**arrow**) with flat blade screwdriver
- pull light from bumper

Installing

- install outside edge of turn signal light into retainer
- push inside edge of turn signal light firmly into retainer until lugs engage

Fog lights, removing

- remove turn signal light
- remove fog light mounting screws (**arrows**)
- remove fog light

Fog lights, adjusting

Fog lights are adjustable for height only. Side adjustment cannot be changed.

- turn adjustment screw (**arrow**) clockwise to raise fog light beam
- turn adjustment screw (**arrow**) counter-clockwise to lower fog light beam

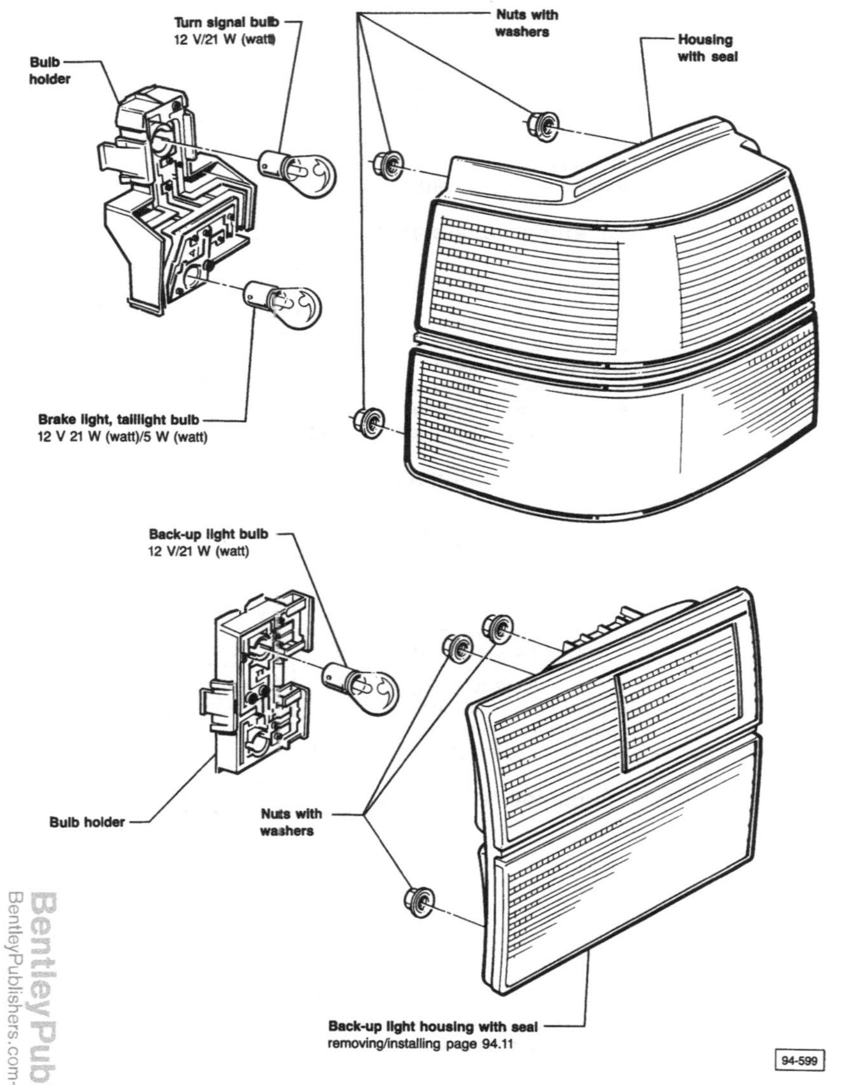

Back-up light housing, removing/ installing

The back-up light and housing are located in the rear lid.

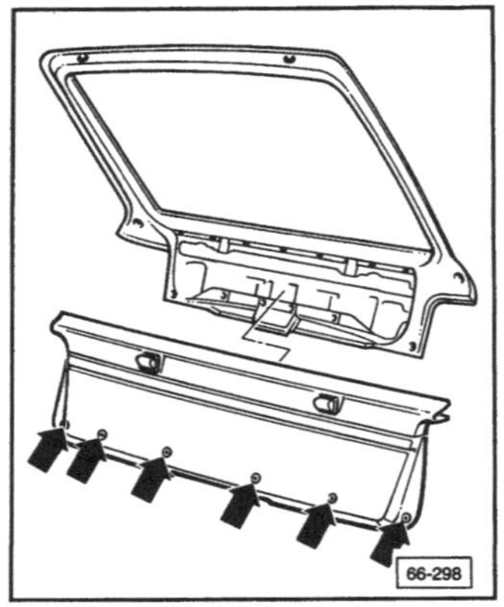

- unfasten interior trim on rear lid by turning retaining clips **(arrows)** 90° (1/4 turn)
- remove interior trim
- remove wiring connector from back-up light housing

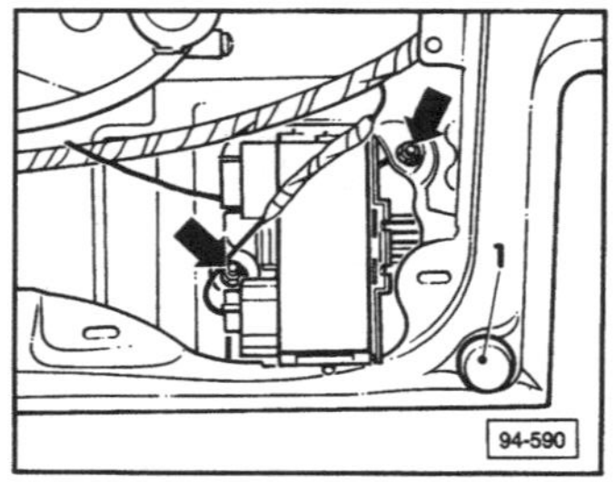

- pull out rubber plug **1** and remove mounting nut behind it
- remove remaining nuts **(arrows)**
- remove back-up light housing
- install in reverse order of removal

Additional brake light, removing

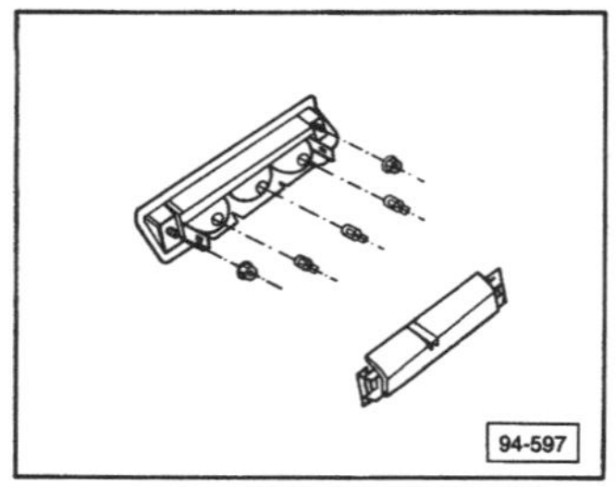

- remove rear lid interior trim by turning retaining clips 90° (1/4 turn)
- remove wiring connector
- remove both outer mounting nuts
- remove lamp socket by spreading tabs on housing
- remove bulbs from socket
  - halogen, 12 V/5 W (watt)

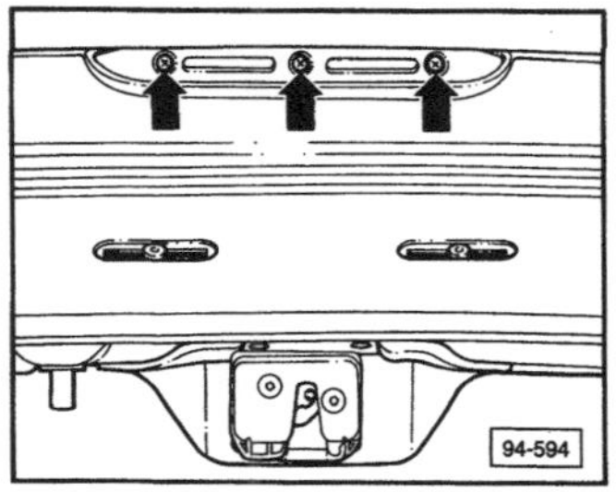

License plate lights, removing

▶

- remove mounting screws (**arrows**)
- remove trim
- remove license plate lights

THIS PAGE INTENTIONALLY LEFT BLANK

THIS PAGE INTENTIONALLY LEFT
BLANK

Index

Central locking system pump and control module
- location/terminal identification **97.7**

Coolant fan control module
- location/terminal identification **97.8**

Connector terminal rubber grommet
- modifying **97.5**

Fuse/relay panel
- connections/plugs **97.3**
- identification **97.2**
- removing/installing **97.4**

Multi–point connector
- removing/installing **97.4**

Power window control module
- location/terminal identification **97.6**

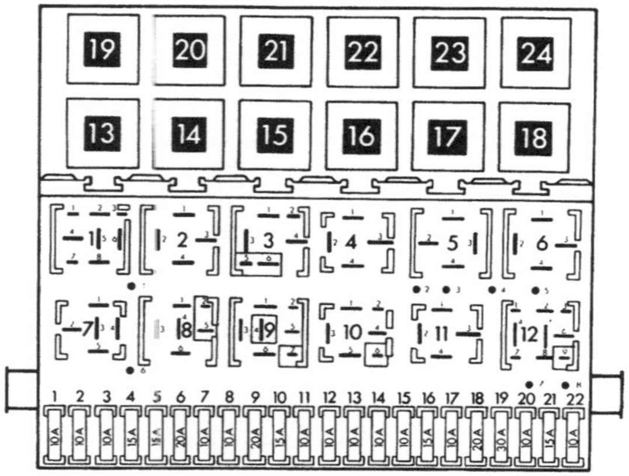

97-3000

Fuse arrangement

according to the numbers on fuse panel from left to right:

		Amp.
1	Headlight, low beam, left	10
2	Headlight, low beam, right	10
3	Instrument panel lights, license plate lights	10
4	Rear window wiper	15
5	Wiper/washer, front	15
6	Fresh air fan	20
7	Tail lights, side marker light, right	10
8	Tail lights, side marker light, left	10
9	Rear window defogger	20
10	Foglights	15
11	Headlight, high beam, left	10
12	Headlight, high beam, right	10
13	Horn, radiator fan run-on	10
14	Back-up lights, washer jet heaters	10
15	Electronic ignition, idle stabilizer, deceleration fuel shut off	10
16	Dash warning lights, glove box light, cassette storage light, rear spoiler adjustment control unit	15
17	Emergency flasher switch, turn signals	10
18	Fuel pump	20
19	Radiator coolant fan, A/C	30
20	Brake lights	10
21	Interior lights, cigarette lighter, digital clock, rear spoiler adjustment	15
22	Radio	10

Relay location on fuse/relay panel

1 — A/C relay
2 — Rear window wiper relay
3 — Digifant control unit
4 — Load reduction relay
5 — Low coolant level
6 — Emergency flasher relay
7 — Open
8 — Intermittent wiper relay
9 — Seat belt warning system
10 — Fog light relay
11 — Horn relay
12 — Fuel pump relay

Separate relays above fuse/relay panel

13 — ABS hydraulic pump relay
14 — ABS relay
15 — Radiator fan after run relay
16 — Fuel pump after run control unit
17 — Heated seat relay, driver
18 — Heated seat relay, passenger
19 — Open
20 — Open
21 — ABS hydraulic pump fuse, power window fuse
22 — ABS valves fuse
23 — Open
24 — Open

Electrical System – Wiring

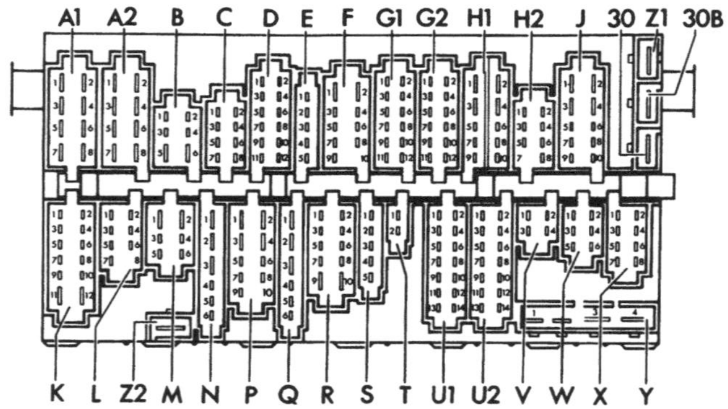

97-3001

Connections and plugs on fuse/relay panel

A1 — Headlight wiring harness (yellow)
A2 — Headlight wiring harness (yellow)
B — Open
C — Headlight wiring harness (yellow)
D — Optional equipment wiring harness (green)
E — Instrument wiring harness (green)
F — Engine compartment wiring harness, right (white)
G1 — Engine compartment wiring harness, right (white)
G2 — Engine compartment wiring harness, right (white)
H1 — Steering column switch wiring harness (red)
H2 — Steering column switch wiring harness (red)
J — Steering column switch wiring harness (red)
K — Rear wiring harness (black)
L — Rear wiring harness (black)
M — Rear wiring harness (black)
N — A/C wiring harness (green)
P — Rear window defroster and fog light wiring harness (blue)
Q — Instrument wiring harness (blue)
R — Light switch wiring harness (blue)
S — Engine compartment wiring harness, right (white)
T — Two-point connector (green)
U1 — Instrument cluster wiring harness (blue)
U2 — Instrument cluster wiring harness (blue)
V — Multi-function indicator wiring harness (green)
W — Six-point connector (green)
X — Warning lamp wiring harness (green)
Y — Single connector, terminal 30
Z1 — Single connector
Z2 — Single connector, terminal 31
30 — Single connector, terminal 30
30B — Single connector
31 — Single connector, terminal 31

Note

Information on pages 97.2 and 97.3 represents a typical Corrado fuse/relay panel. For specific fuse and relay information about a particular car, refer to the appropriate Main Wiring Diagram. See **Wiring Diagrams**.

Fuse/relay panel, removing/installing

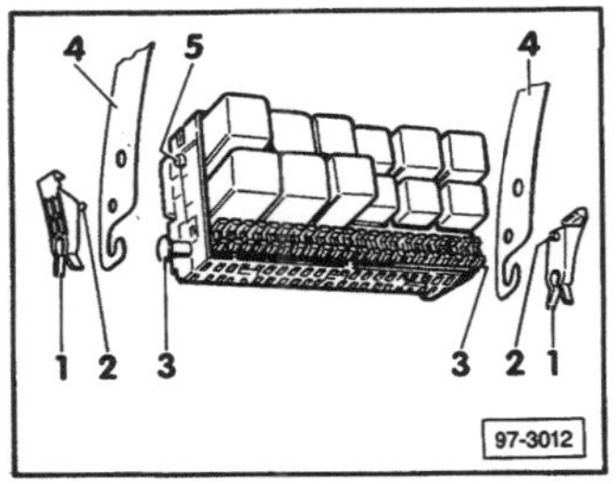

The fuse/relay panel is located underneath the left side of the instrument panel.

- remove lower instrument panel trim, see Repair Group 70
- spread retaining clips **1** to disengage lock pins **2** from brackets **4**
- swing retaining clips **1** forward and pull from fuse/relay panel pins **3**
- spread left retaining bracket **4** and disengage from pin **5**
- remove fuse/relay panel from brackets **4**

- install in reverse order of removal making sure pin **5** and retaining pins **2** engage proper recesses in brackets **4**

Multi-point connector, on fuse/relay panel

Removing

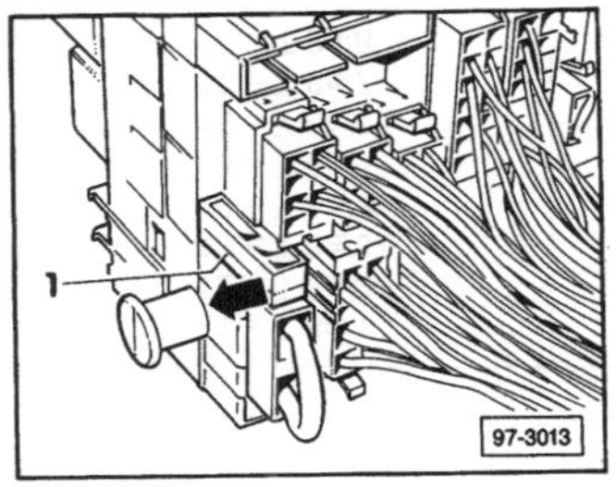

- pull open locking slide **1** on back of fuse/relay panel in direction of (**arrow**)

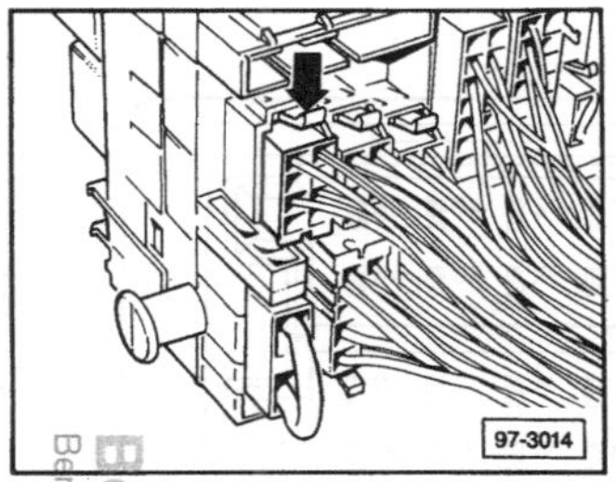

- depress locking clip (**arrow**) and pull multi-point connector from fuse/relay panel

Installing

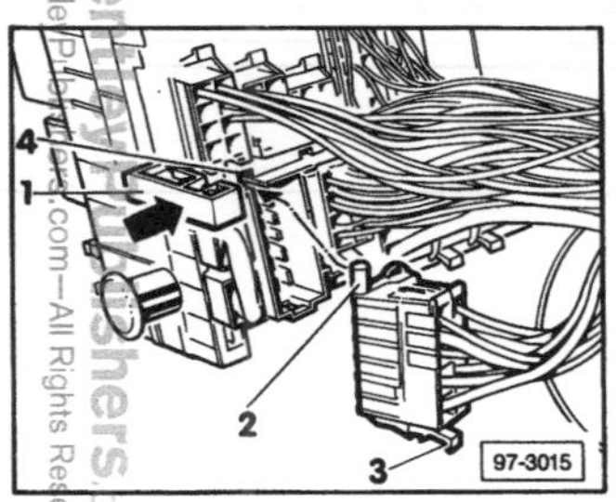

- insert connector so that locking pins **2** face locking slide **1**
- push multi-point connector into appropriate location **4** until locking clip **3** engages
- push locking slide **1** in direction of (**arrow**) until locked

Note

Fully insert all multi-point connectors into the fuse/relay panel before engaging locking slide **1**.

Connector terminal rubber insulating grommet, modifying

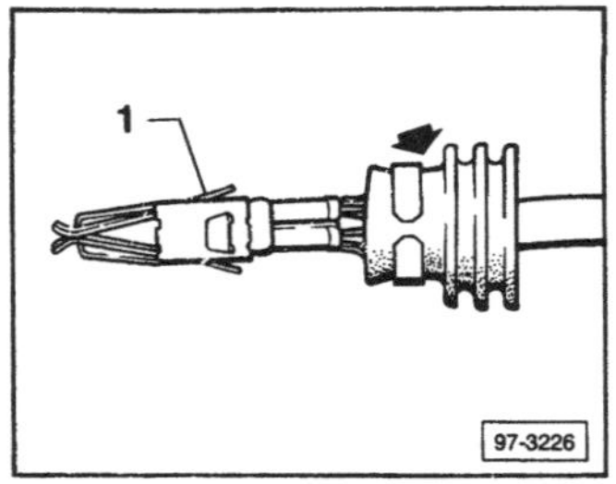

A rubber insulating grommet (**arrow**) is installed on some connector terminals at the factory. During wiring repair, the rubber insulating grommet must be reversed before re-inserting into the connector housing.

- remove and discard terminal **1**
- slide rubber insulating grommet from wire and rotate end-to-end (180°)
- inspect grommet and replace if damaged

> **CAUTION**
>
> New replacement grommet must be the same color as original to ensure proper sealing between wire, grommet and housing.

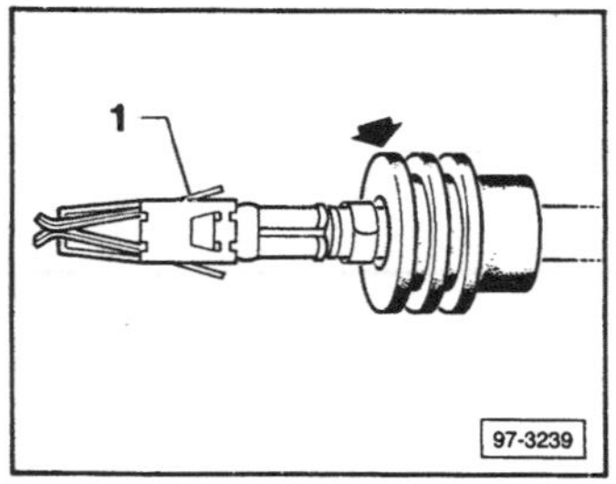

- reinstall grommet onto wire so that collar (**arrow**) faces terminal **1**
- properly crimp new terminal **1** to wire

Power window control module, location/terminal identification

Location

The power window control module is located in the left side-panel, behind the interior trim.

Connector terminal identification

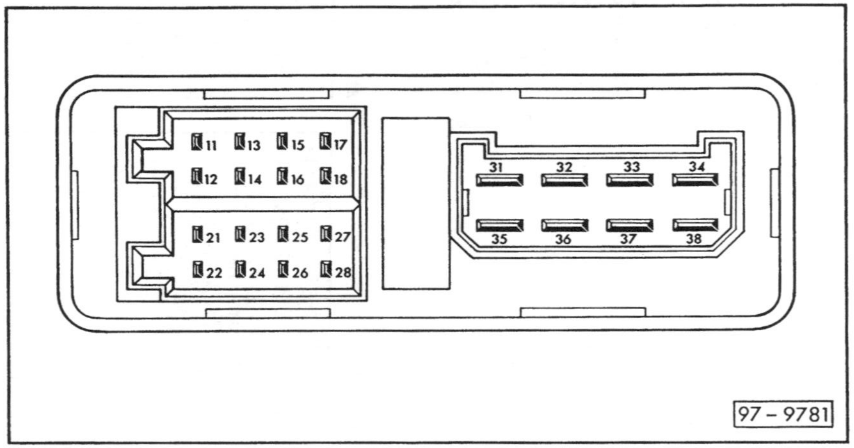

11 = T 8a/1 12 = T 8a/2 13 = T 8a/3 14 = T 8a/4 15 = T 8a/5 16 = T 8a/6 17 = T 8a/7 18 = T 8a/8	21 = T 8c/1 22 = T 8c/2 23 = T 8c/3 24 = T 8c/4 25 = T 8c/5 26 = T 8c/6 27 = T 8c/7 28 = T 8c/8	31 = T 8b/1 32 = T 8b/2 33 = T 8b/3 34 = T 8b/4 35 = T 8b/5 36 = T 8b/6 37 = T 8b/7 38 = T 8b/8

Connector colors

T 8a - Brown

T 8b - White

T 8c - Black

Note
See wiring diagram for wire routing to connectors **T 8a, T 8b and T 8c.** Always refer to appropriate wiring diagram when performing any electrical tests or repairs.

Central locking system pump and control module, location/terminal identification

Location

The central locking system pump and control module are located in the luggage compartment, on the right.

Connector terminal identification

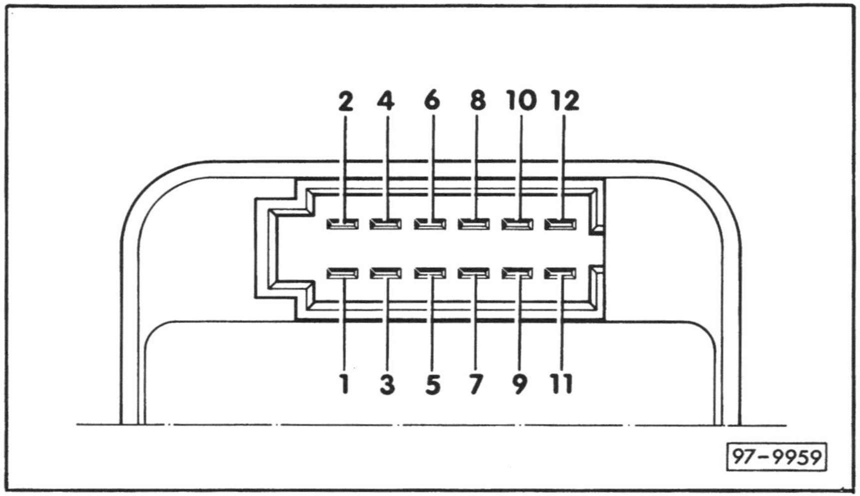

1 = T 12/1 2 = T 12/2 3 = T 12/3 4 = T 12/4	5 = T 12/5 6 = T 12/6 7 = T 12/7 8 = T 12/8	9 = T 12/9 10 = T 12/10 11 = T 12/11 12 = T 12/12

Connector color

T 12 - transparent

Note

See wiring diagram for wire routing to connector **T 12**. Always refer to appropriate wiring diagram when performing any electrical tests or repairs.

Coolant fan control module, location/terminal identification

Location

The coolant fan control module is located in the engine compartment, on the left.

Connector terminal identification

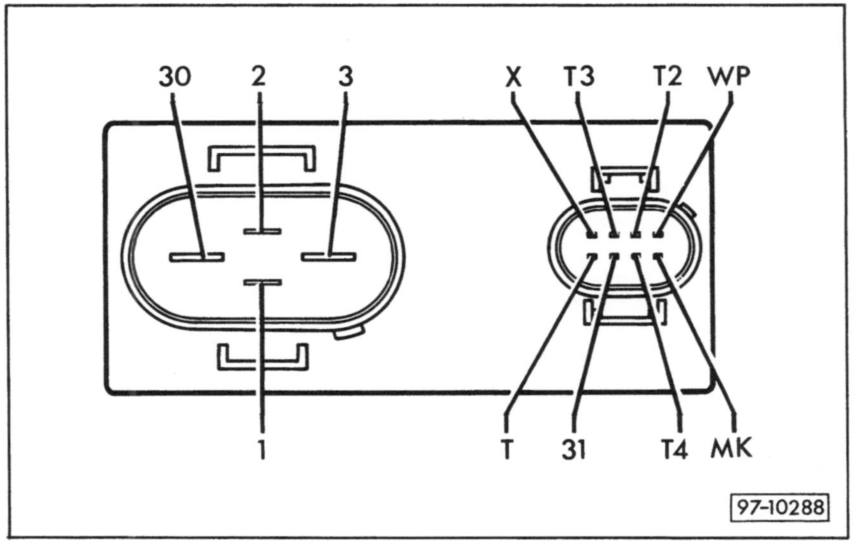

T 4/1 = 3	T 8e/1 = WP	T 8e/5 = MK
T 4/2 = 2	T 8e/2 = T2	T 8e/6 = T4
T 4/3 = 1	T 8e/3 = T3	T 8e/7 = 31
T 4/4 = 30	T 8e/4 = X	T 8e/8 = T

Note

See wiring diagram for wire routing to **T 4** and **T 8** connectors. Always refer to appropriate wiring diagram when performing any electrical tests or repairs.

THIS PAGE INTENTIONALLY LEFT BLANK

THIS PAGE INTENTIONALLY LEFT BLANK

Index

General information

Cautions and warnings 3

Handling components and connectors 15

How to read wiring diagrams 5

Test equipment 9
- computer memory saver 14
- LED test light 9
- multimeter or volt/ohm meter 11
- test kits 13

Troubleshooting 17
- basic electricity 17
- checking for short circuit to ground 27
- checking for voltage 20
- checking ground connections 21
- checking wiring and components 23

Main wiring diagrams

- 1990 **29**
- 1991 **57**
- 1992 (engine code PG) **57**
- 1992 (engine code AAA) **83**
- 1993 (production up to Jan. 1993) **83**
- 1993 (production from Jan.1993) **113**
- 1994 (production up to Mar. 1994) **Supplement 339**
- 1994 (production from Mar. 1994) **145**

Auxiliary wiring diagrams

Air conditioning
- 1990, 1991 **177**
- 1992 (engine code PG) **177**
- 1992 (engine code AAA) **181**
- 1993 (production up to Aug. 1992) **181**

Auxiliary wiring diagrams

Air conditioning (continued)
- 1993 (production from Aug. 1992) **181**
- 1994 **Supplement 333**

Anti-lock brake system (ABS)
- 1990, 1991 **191**
- 1992 (engine code PG) **195**
- 1992 (engine code AAA) **199**
- 1993, 1994 **199**
  ABS w/EDL **327**

Anti-theft alarm system
- 1991 **205**
- 1992 **205**
- 1993 (production up to Aug. 1992) **205**
- 1993 (production from Aug. 1992) **209**
- 1994 **209**

Automatic shoulder belt system
- 1990 **213**
- 1991 **217**
- 1992 (engine code PG) **217**
- 1992 (engine code AAA) **221**
- 1993, 1994 **221**

Automatic transmission
- 1991 **225**
- 1992 (engine code PG) **225**
- 1992 (engine code AAA) **231**
- 1993 (production up to Jan. 1993) **231**
- 1993 (production from Jan. 1993) **237**
- 1994 (production up to Mar. 1994) **237**
- 1994 (production from Mar. 1994) **243**

Index

Auxiliary wiring diagrams

Central locking
- 1990 **291**
- 1991 **297**
- 1992 (engine code PG) **297**
- 1992 (engine code AAA) **303**
- 1993 (production up to Aug. 1992) **303**
- 1993 (production from Aug. 1992) **307**
- 1994 **307**

Cruise control
- 1990 **249**
- 1991 **251**
- 1992 (engine code PG) **251**
- 1992 (engine code AAA) **253**
- 1993, 1994 **253**

Daytime driving lights (Canada only)
- 1990 **255**
- 1991 – 1994 **259**

Electronic engine controls (Digifant I, California only)
- 1991, 1992 (engine code PG) **263**

Fuel pump
- 1990 (production from Sept. 1989) **269**

Heated seats
- 1990, 1991 **271**
- 1992 (engine code PG) **271**
- 1992 (engine code AAA) **273**
- 1993, 1994 **273**

Headlight washers (Canada only)
- 1993, 1994 **275**

Multi-function indicator
- 1991–1994 (production from Aug. 1990) **277**

Power mirrors
- 1992–1994 (engine code AAA) **283**

Power sunroof
- 1990 **325**
- 1991 **285**
- 1992 (engine code PG) **285**
- 1992 (engine code AAA) **287**
- 1993 (production up to Aug. 1992) **287**
- 1993 (production from Aug. 1992) **289**
- 1994 **289**

Power windows
- 1990 **291**
- 1991 **297**
- 1992 (engine code PG) **297**
- 1992 (engine code AAA) **303**
- 1993 (production up to Aug. 1992) **303**
- 1993 (production from Aug. 1992) **307**
- 1994 **307**

Recaro power seats (Canada only)
- 1992–1994 (engine code AAA) **311**

Seatbelt warning system (Canada only)
- 1990–1994 **313**

Stereo radio
- 1990 **315**
- 1991, 1992 **319**
- 1993 **323**
- 1994 **Supplement 371**

Radio with CD Changer
- 1994 **Supplement 373**

Please read these warnings and cautions before proceeding with maintenance and repair work.

WARNING—

- Some repairs may be beyond your capability. If you lack the skills, tools and equipment, or a suitable workplace for any procedure described in this manual, we suggest you leave such repairs to an authorized Volkswagen dealer service department, or other qualified shop.
- Volkswagen is constantly improving its cars. Sometimes these changes, both in parts and specifications, are made applicable to earlier models. Therefore, before starting any major jobs or repairs to components on which passenger safety may depend, consult your authorized Volkswagen dealer about Technical Bulletins that may have been issued since the editorial closing date of this manual.
- Do not re-use any fasteners that are worn or deformed in normal use. Many fasteners are designed to be used only once and become unreliable and may fail when used a second time. This includes, but is not limited to, nuts, bolts, washers, self-locking nuts or bolts, circlips, cotter pins. Always replace these fasteners with new parts.
- Never work under a lifted car unless it is solidly supported on stands designed for the purpose. Do not support a car on cinder blocks, hollow tiles or other props that may crumble under continuous load. Never work under a car that is supported solely by a jack. Never work under the car while the engine is running.
- If you are going to work under a car on the ground, make sure that the ground is level. Block the wheels to keep the car from rolling. Disconnect the battery negative (–) terminal (ground strap) to prevent others from starting the car while you are under it.
- Never run the engine unless the work area is well ventilated. Carbon monoxide kills.
- Finger rings, bracelets and other jewelry should be removed so that they cannot cause electrical shorts, get caught in running machinery, or be crushed by heavy parts.
- Tie long hair behind your head. Do not wear a necktie, a scarf, loose clothing, or a necklace when you work near machine tools or running engines. If your hair, clothing, or jewelry were to get caught in the machinery, severe injury could result.
- Illuminate your work area adequately but safely. Use a portable safety light for working inside or under the car. Make sure the bulb is enclosed by a wire cage. The hot filament of an accidentally broken bulb can ignite spilled fuel or oil.
- Always observe good workshop practices. Wear goggles when you operate machine tools or work with battery acid. Gloves or other protective clothing should be worn whenever the job requires working with harmful substances.
- Disconnect the battery negative (–) terminal (ground strap) whenever you work on the fuel system or the electrical system. Do not smoke or work near heaters or other fire hazards. Keep an approved fire extinguisher handy.
- Batteries give off explosive hydrogen gas during charging. Keep sparks, lighted matches and open flame away from the top of the battery. If hydrogen gas escaping from the cap vents is ignited, it will ignite gas trapped in the cells and cause the battery to explode.
- Connect and disconnect battery cables, jumper cables or a battery charger only with the ignition switched off, to prevent sparks. Do not disconnect the battery while the engine is running.
- Do not quick-charge the battery (for boost starting) for longer than one minute. Wait at least one minute before boosting the battery a second time.
- Do not allow battery charging voltage to exceed 17 volts. If the battery begins producing gas or boiling violently, reduce the charging rate. Boosting a sulfated battery at a high charging rate can cause an explosion.
- The air-conditioning system is filled with chemical refrigerant, which is hazardous. The A/C system should be serviced only by trained technicians using approved refrigerant recovery/recycling equipment, trained in related safety precautions, and familiar with regulations governing the discharging and disposal of automotive chemical refrigerants.
- Do not expose any part of the A/C system to high temperatures such as open flame. Excessive heat will increase system pressure and may cause the system to burst.
- On cars equipped with anti-lock brakes (ABS), the ABS modulator assembly is capable of "self-pressurizing" up to 210 bar (3045 psi). Serious injury may result if the unit is not properly depressurized before servicing the system.

CAUTION—

- If you lack the skills, tools and equipment, or a suitable workshop for any procedure described in this manual, we suggest you leave such repairs to an authorized Volkswagen dealer or other qualified shop.
- Volkswagen offers extensive warranties, especially on components of fuel delivery and emission control systems. Therefore, before deciding to repair a Volkswagen that may still be covered wholly or in part by any warranties issued by Volkswagen United States, Inc., consult your authorized Volkswagen dealer. You may find that he can make the repair for free, or at minimal cost.
- Volkswagen part numbers listed in this manual are for identification purposes only, not for ordering. Always check with your authorized Volkswagen dealer to verify part numbers and availability before beginning service work that may require new parts.
- Before starting a job, make certain that you have all the necessary tools and parts on hand. Read all the instructions thoroughly, do not attempt shortcuts. Use tools appropriate to the work and use only replacement parts meeting Volkswagen specifications. Makeshift tools, parts and procedures will not make good repairs.
- Use pneumatic and electric tools only to loosen threaded parts and fasteners. Never use these tools to tighten fasteners, especially on light alloy parts. Always use a torque wrench to tighten fasteners to the tightening torque specification listed.
- On cars equipped with the anti-lock brakes (ABS), the ABS control unit cannot withstand temperatures from a paint-drying booth or a heat lamp in excess of 203°F (95°C) and should not be subjected to temperatures in excess of 185°F (85°C) for more than two hours.
- Before doing any electrical welding on cars equipped with ABS, disconnect the battery negative (–) terminal (ground strap) and the ABS control unit connector.
- On cars equipped with anti-theft radios, make sure you know the correct radio activation code before disconnecting the battery or removing the radio. If the wrong code is entered into the radio when power is restored, that radio may lock up and be rendered inoperable, even if the correct code is then entered.
- Connect and disconnect a battery charger only with the battery charger switched off.
- Do not quick-charge the battery (for boost starting) for longer than one minute. Wait at least one minute before boosting the battery a second time.
- Sealed or "maintenance free" batteries should be slow-charged only, at an amperage rate that is approximately 10% of the battery's ampere-hour (Ah) rating.
- Do not allow battery charging voltage to exceed 17 volts. If the battery begins producing gas or boiling violently, reduce the charging rate. Boosting a sulfated battery at a high charging rate can cause an explosion.

Notes, Cautions, and Warnings

Throughout this manual are many passages with the headings **NOTE**, **CAUTION**, or **WARNING**. These very important headings have different meanings.

WARNING
A warning is the most serious of the three. It warns of unsafe practices that are very likely to cause injury, either by direct threat to the person(s) doing the work or by increased risk of accident or mechanical failure while driving.

CAUTION
A caution calls attention to important precautions to be observed during the repair work that will help prevent accidently damaging the car.

Note
A note contains helpful information, tips which will help in doing a better job and completing it more easily.

Please read every **NOTE**, **CAUTION**, and **WARNING** at the front of the manual and as they appear throughout the manual. They are very important. Read them before you begin any work. Some **CAUTION**s and **WARNING**s are repeated wherever they apply. Read them all. Do not skip any. These messages are important, even to the owner who never intends to work on the car.

1. How to Read Wiring Diagrams

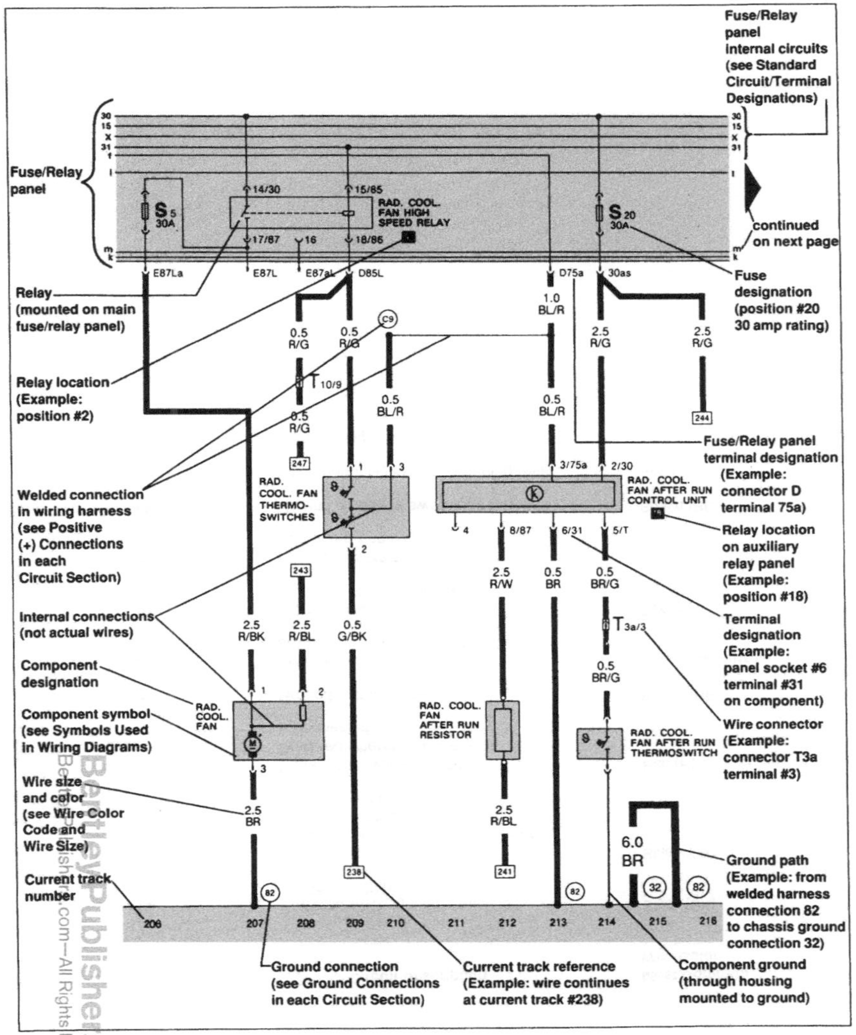

Symbols Used in Wiring Diagrams

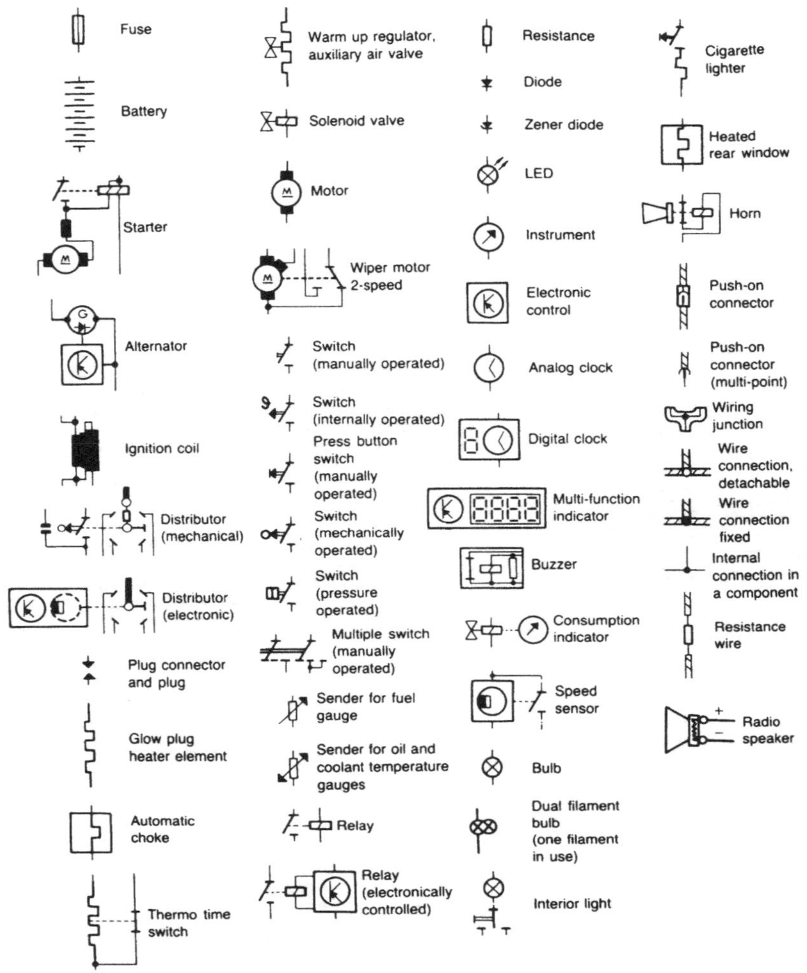

Standard Circuit/Terminal Designations

Circuit Number	Circuit Description	Most Common Wire Color
15	Powered when ignition switch is in "On" or "Start" positions	Black (BK)
X	Load-reduction circuit Powered by load-reduction relay when ignition switch is in "On" position (not powered in "Start" position)	Black/Yellow (BK/Y)
30	Battery positive (+) voltage Powered whenever battery is connected	Red (R)
31	Ground or battery negative (-)	Brown (BR)
50	Powered only when ignition switch is in "Start" position	Red/Black(R/BK)
B+	From alternator Charging voltage to battery	Red (R)
D+	Alternator warning light and field energizing circuit	—
85	Ground (-) side of switching relay	Brown (BR)
86	Power-input (+) side of switching relay	—
87	Relay change-over contact	—

Wire Color-Code

BK	Black
BL	Blue
BR	Brown
CL	Clear
G	Green
GY	Gray
LT.G	Light Green
OR	Orange
R	Red
V	Violet
W	White
Y	Yellow

Combined codes indicate a multi-colored wire.

Example: The code R/G indicates a red wire with a green tracer stripe.

> **Note**
>
> *Sometimes the wire color on the car may be different than the one on the wiring diagram. Don't be concerned, just be sure to confirm that the wires connect to the proper terminals.*

Wire Size

The wiring diagrams in this manual identify wires by their metric wire size. Metric wire sizes indicate cross-sectional area in square millimeters (mm^2). The table below lists metric wire sizes and their equivalents in American Wire Gauge (AWG) sizes.

Metric size (cross-sectional area - mm^2)	American Wire Gauge (AWG)
0.35	22
0.5	20
0.75	18
1.0	16
1.5	14
2.5	12
4.0	10
6.0	8
16.0	4
25.0	2
35.0	2

Connectors

T1 .... Single-point
T2 .... 2-point
T3 .... 3-point
T4 .... 4-point
T6 .... 6-point
T8 .... 8-point
T10 .... 10-point
T12 .... 12-point
T26 .... 26-point

Each connector has a unique number. Example: T6, T6a, and T6b are three different 6-point connectors.

Wiring diagrams usually also identify a particular connector terminal. Example: T3a/2 is terminal two of connector T3a.

T3a

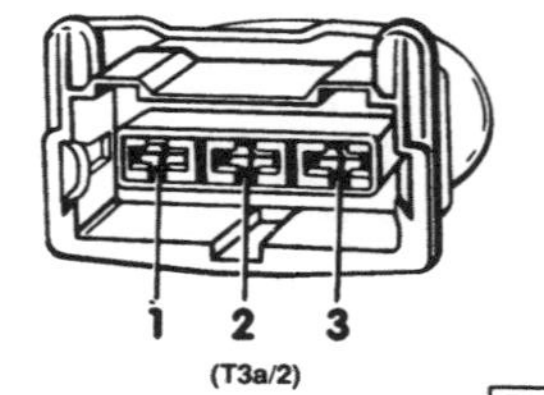

96-902

2. Test Equipment

Special Tools

The most basic electrical troubleshooting calls for specialized electrical test equipment. Dealer Service Department technicians do this work using Special Tools, many of which are designed specifically for working on electrical systems. Proper tools are essential to ensure accurate test results and to prevent damage to sensitive components.

The Special Tools shown in this manual are identified by the order numbers that can be used to special-order the tools through your authorized dealer Parts Department.

2.1 LED Test Light

A test light is an inexpensive tool used to perform many simple electrical tests that would otherwise require a multimeter. It lights to indicate when there is voltage potential between any two test-points in a circuit.

With one probe contacting the battery negative (-) terminal or a known ground connection, the other probe can be used to check for voltage at other points in the circuit. The test light will light when the test probe contacts a voltage source, and current flows from the voltage source to ground.

With one probe contacting a known source of voltage, the other probe can be used to check for continuity to ground at other points in the circuit. Current will flow and the test light will light when the test-point provides the known voltage source with a path to ground.

> **CAUTION**
> *A common test light (with incandescent bulb) may no longer be used. The bulb's high current draw will damage sensitive electronic components.*

LED Test Light (US1115)

Used to check for voltage reaching components or when searching for faults in a circuit.

Voltage range: 3 to 48 VDC

> **CAUTION**
> *A common test light (with incandescent bulb) may no longer be used. The bulb's high current draw will damage sensitive electronic components.*

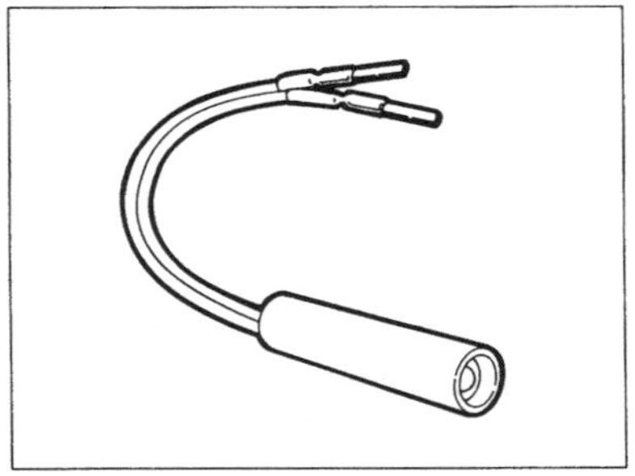

> **Note**
> *This tool has been superseded by Special Tool VAG 1527B, Voltage Tester. (US1115 is still considered an acceptable equivalent to VAG 1527B).*

Voltage Tester (VAG1527B)
Order Number: TAG 152 7B0 28 ZEL

Two-pole voltage tester with two LEDs (light emitting diodes), suitable for measuring from 3 to 48 volts DC or AC.

When used to test DC ciruits, the positive (+) or negative (-) LED will light, depending on polarity. When used to test AC circuits, both LEDs will light.

Supplied with banana clip and detachable probe—can be used directly with VAG1598 Test Box.

For troubleshooting all common electrical and electronic components in automobiles. Safe for testing electronic components and circuits because of extremely low current consumption (1.5 mA maximum).

Voltage range: 3 to 48 volts (AC or DC)

> **CAUTION**
> *A common test light (with incandescent bulb) may no longer be used. The bulb's high current draw will damage sensitive electronic components.*

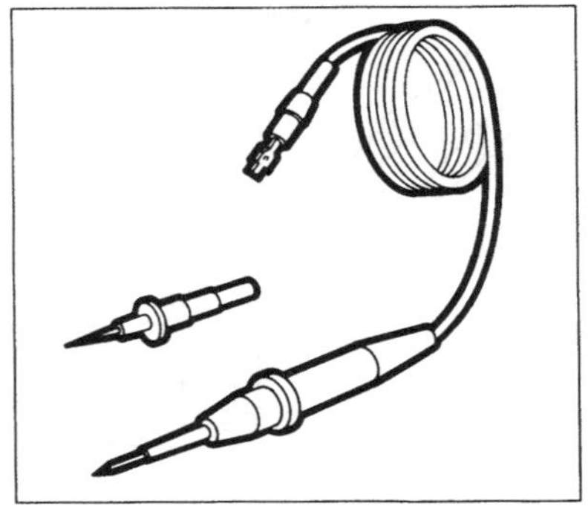

> **Note**
> *This tool supersedes Special Tool US1115 LED Test Light.*

Making An LED Test Light

A lower-cost alternative to buying an LED test light such as US1115 or VAG1527B is to make one, using parts available from most any electronics supply outlet.

Assemble the components as shown. Use a needlenose plier to hold the parts and to act as a heat dam while soldering. Insulate the connections with heat-shrinkable tubing or electrical tape.

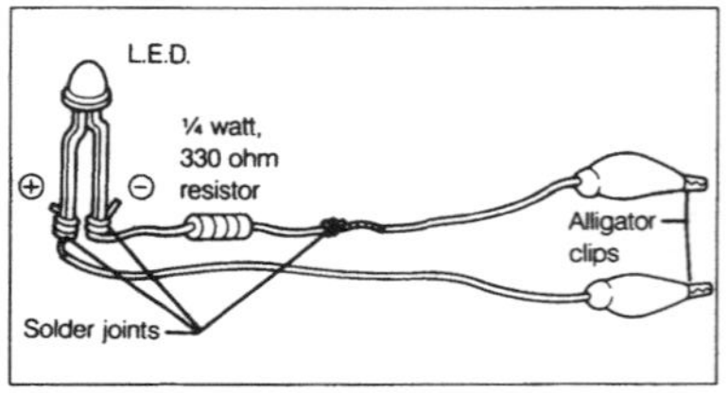

Do-it-yourself LED test light for safely testing sensitive solid-state circuits.

Parts:

1. LED (1)
2. 1/4 watt, 330 Ω resistor (1)
3. Alligator clips (2)
4. Wire, Solder and Soldering Iron
5. Heat-shrinkable tubing or electrical tape

2.2 Multimeter or Volt/Ohm Meter (VOM)

A multimeter or volt-ohm meter (VOM) is used to measure voltage, resistance (ohms) or current (amps or milliamps). Two types of meters are in common use. The analog or swing-needle volt-ohm meter displays test values according to the position of a needle on the meter face. A digital multimeter displays test values as numbers. The meters shown here and recommended are digital multimeters.

Accuracy is an important consideration when choosing a meter. Analog or swing-needle voltmeters are generally rated for accuracy as a percentage of full-scale on the meter face. A typical analog meter may be rated as accurate to ± 3% of full scale. Some analog meters offer 10-volt and 50-volt scales, with no range in between. For 12-volt automotive electrical systems, this means testing on the 50-volt scale, with a corresponding decrease in accuracy. On a 50-volt scale, ± 3% accuracy is equal to ± 1.5 volts. While still useful for testing circuits such as lights and horns, analog meters may not be accurate enough for use where precise measurements are required.

Digital multimeters are preferred for precise measurements and for electronics work because most digital multimeters are more accurate than most analog meters. Another advantage of digital multimeters is that they are less likely to be misread, since there is no needle position to be misinterpreted or distorted by reading at an angle.

Perhaps the most important consideration is input impedance—how much of a load the meter places on the circuit being tested. Meters with low input impedance place a greater load on the circuit because they allow more current flow through the meter. This can exceed design limits and be damaging to sensitive electronic components.

Most digital multimeters have very high input impedance—10 megohms (10,000,000 Ω) or higher—and make their measurements while allowing very little current flow through the meter. The meter itself is not a load in the circuit, and does not induce damaging current flow.

Multimeter Kit - Digital (US1119, VAG1526)
Order Number: TU1 119 000 00 KTM

Used for monitoring electronic engine controls (Example: setting/checking the idle stabilization system) and for general electrical troubleshooting.

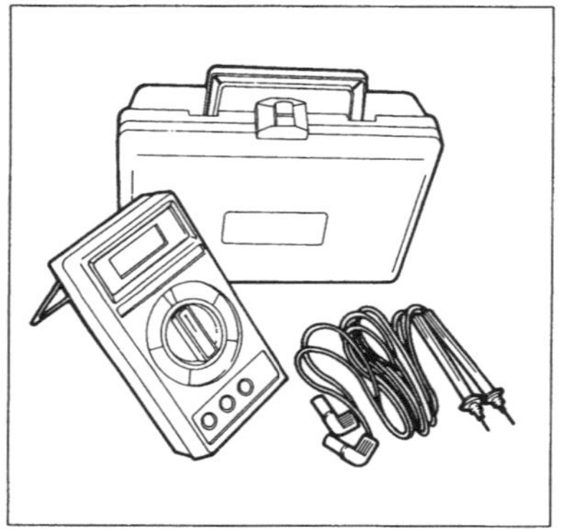

> **Note**
> *This tool has been superseded by the Fluke 83® Digital Multimeter. (US1119 or VAG1526 are still considered acceptable equivalents.)*

Fluke 83® Digital Multimeter
Order Number: TFL UKE 830 00 RSE

Used for monitoring electronic engine controls (Example: setting/checking the idle stabilization system) and for general electrical troubleshooting.

— 10 Megohm input impedance
— ± 0.3% DC voltage accuracy
— fully overload-protected

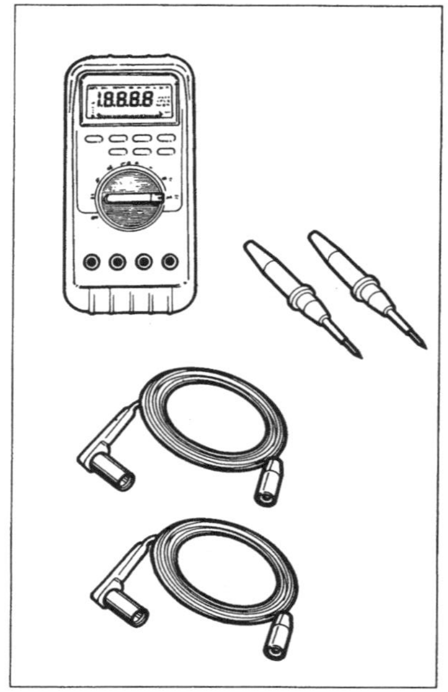

> **Note**
> *This tool supersedes Special Tool US1119 Digital Multimeter Kit.*

2.3 Test Kits

Connector Test Kit (VW1594)
Order Number: TV1 594 400 97 KTM

Used to connect electrical measuring equipment to various harness connectors in the electrical system.

CAUTION
Test connections made carelessly or without proper tools can damage harness connectors, causing poor connections and future problems.

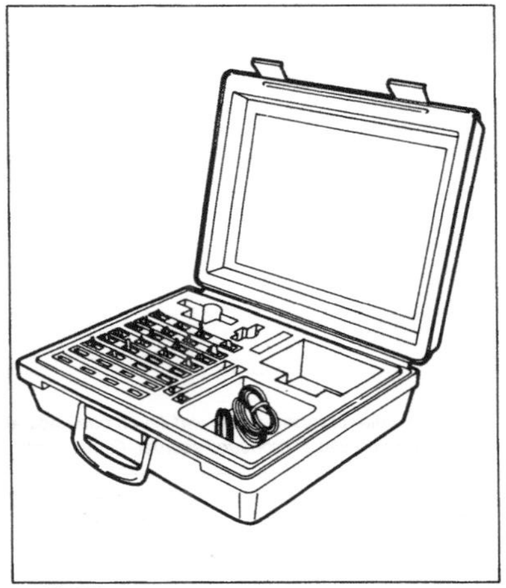

Test Box (VAG1598)
Order Number: WAG 159 800 00 VOA

Used to perform static and dynamic tests of electrical and electronic systems without damaging miniature contacts, especially where limited access to connectors makes testing difficult.

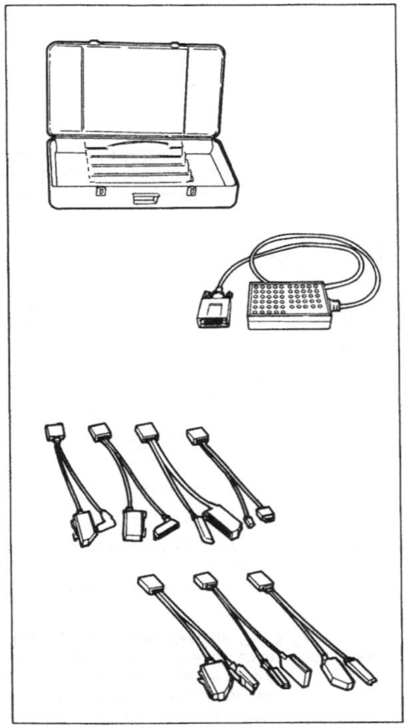

2.4 Computer Memory Saver

Used on cars with theft-protected radios to prevent the radio from electronically locking when battery power is disconnected. Use of this tool eliminates the need to re-activate the radio (reenter the correct code) after reconnecting battery power.

WARNING
Always separate the airbag voltage supply connector before using a computer memory saver. Failure to do so may result in accidental activation of the airbag..

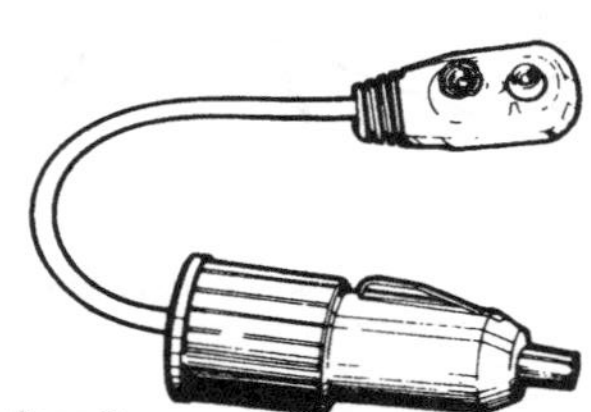

Snap-On
Computer Memory Saver
Part #YA960

91-A051

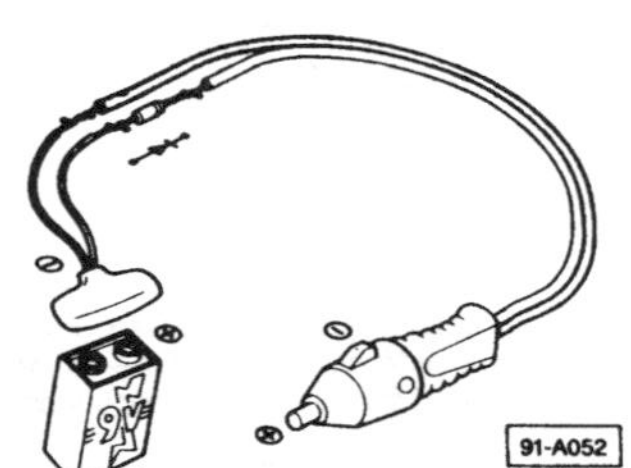

91-A052

Home-made computer memory saver

3. Handling Components and Connectors

Harness Connectors

The harness connectors used throughout the car are designed to positively lock into place. Press on wire lock as shown to release the connector.

> **CAUTION**
> *To disconnect, pull only on the connector body. Never pull on the wires themselves.*

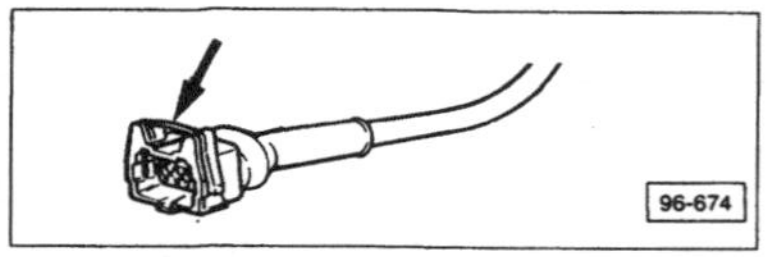

*Typical wiring harness connector. Push on wire lock (**arrow**) and gently pull on connector to release.*

Making Test Connections

Many electrical troubleshooting tests will require hooking up to wiring harness connectors, or socket connectors on electrical components. Test connections made carelessly or without proper tools can easily damage the connectors, causing poor connections and future problems.

To avoid connector damage, test connections to wiring harness connectors must be made using small, flat-blade terminals which will mate properly with the connectors. The Special Tools recommended include special adapters which can be used to make test connections safely and prevent connector damage. See **2. Test Equipment**.

All test connections to harness connectors should be made using the proper adapter or correct size flat connectors that will not deform the connector cavities.

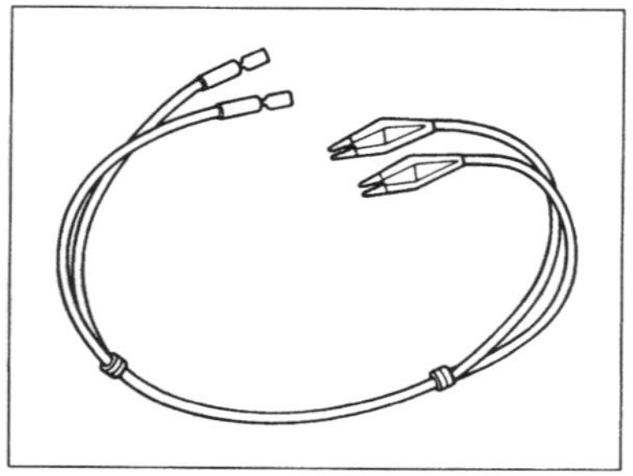

Alligator-clip test leads with flat connectors (Part No. N 17 457 2).

Wiring Harness Repairs

When repairing wiring harnesses, use only high-quality electrical connectors suitable for use with electronic components.

To make connections, use crimping pliers that make a "W"-type crimp. Only this type of crimp provides the necessary mechanical strength.

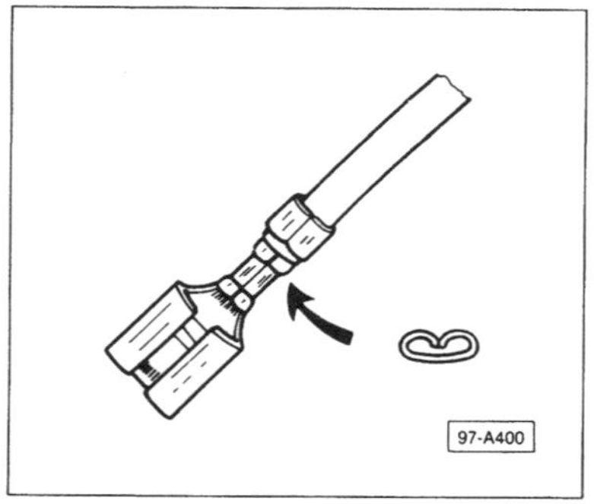

Typical "W"-type crimp connection

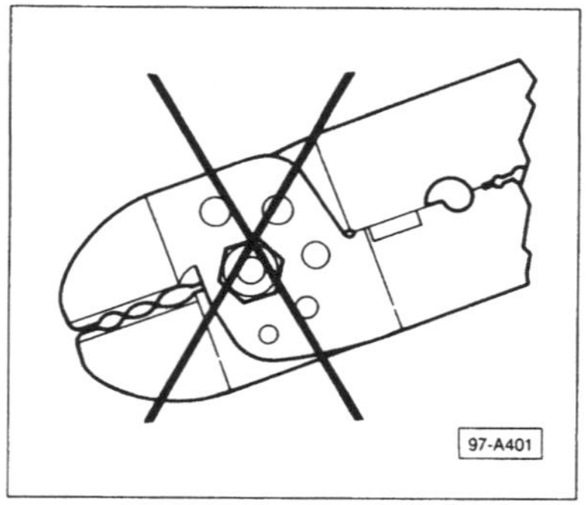

Typical wire crimping pliers, not suitable for making "W"-type crimps.

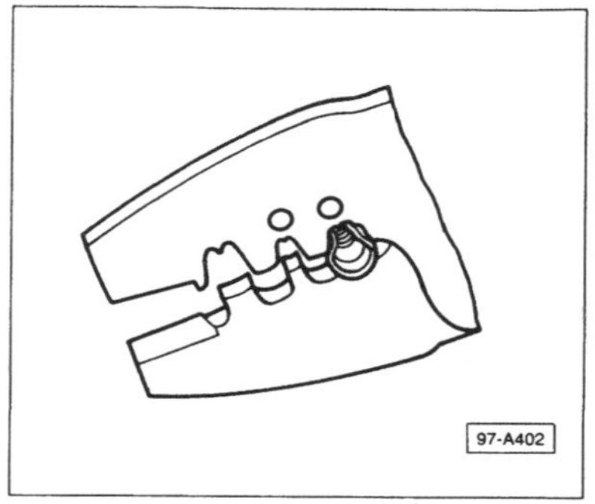

Wire crimping pliers, used to make "W"-type crimps.

Connector repair kits, connectors and the correct crimping pliers are available through your authorized dealer Parts department.

4. Troubleshooting

4.1 Basic Electricity

Electricity is defined by three basic elements: Voltage, Current and Resistance.

Voltage

Voltage is a measure of electromotive force, sometimes referred to as electrical "pressure". It can be described as the difference in potential (potential for the flow of electricity) between any two points in a circuit.

A typical automobile battery, for example, has a difference in potential of about 12 volts between the positive (+) terminal and the negative (-) terminal.

The basic units of electrical potential are **volts** (V). Very low voltages are expressed as **millivolts** (mV).
1 V = 1000 mV; 1 mV = .001 V.

Current

Current is the term describing the flow of electricity through a conductor. In a complete circuit, potential (voltage) will cause current to flow from positive (+) to negative (-).

The basic units of current flow are amperes or **amps** (A). Small amounts of current flow are often measured in **milliamps** (mA).
(1 A = 1000 mA; 1 mA = .001 A)

Resistance

Resistance resists or opposes the flow of electricity. Conductors are made from materials of low resistance that allow electricity to flow easily. Insulators are materials of very high resistance that inhibit the flow of electricity.

The basic unit of resistance is the **ohm Ω.** High resistance values are often expressed as kilohms (KΩ).
1 KΩ = 1000Ω

Resistance vs. Current Flow

The basic rule of electricity (Ohm's Law) states that one unit of force (1 volt) is required for one unit of current (1 amp) to flow against one unit of resistance (1 ohm). From Ohm's Law, we also know that:

Voltage = Current × Resistance

When voltage is approximately constant, as in an automobile electrical system, current and resistance affect each other. As resistance increases, there will be less current flow. And lower resistance will permit higher current flow.

Higher resistance = lower current flow
Example: Corrosion on a headlight connector (higher resistance) causes the light to be dim (lower current flow).

Lower resistance = higher current flow
Example: A damaged wire shorted to ground (lower resistance) overloads circuit capacity (higher current flow) and blows a fuse.

Definition of a Circuit

Four things are required for current to flow in any electrical circuit, and for that circuit to function as intended:

— **Power Source** (voltage)
— **Conductors** (wires, printed circuits, etc.)
— **Load or Consumer** (a user of electrical power)
— **Complete Circuit** (a connection to ground)

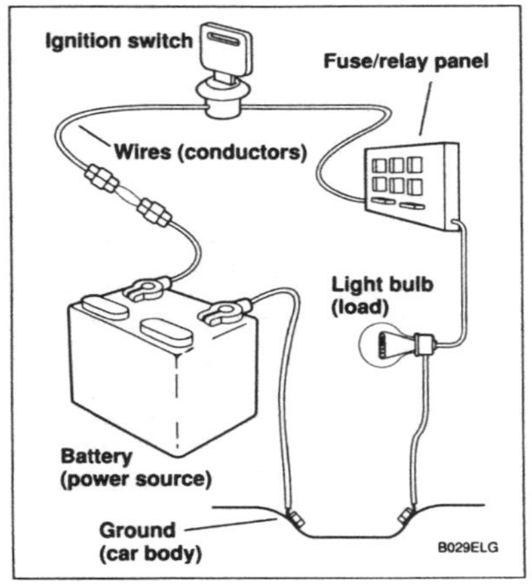

A complete circuit.

Open Circuits

An open circuit is an incomplete circuit. An open circuit occurs when some kind of fault or failure interrupts the circuit path and prevents current flow. Some common causes of open circuits are:

— broken wire
— loose or disconnected connector
— loose or damaged connector terminal
— corrosion
— failed fuse or component

Test for an open circuit by checking continuity using an ohmmeter (multimeter), or by checking for voltage at various points of the circuit using a test light or voltmeter (multimeter). See **4.4 Checking Wiring and Components.**

Short Circuits

A short circuit is an unintended complete circuit. A short circuit occurs when some kind of fault or failure causes current flow to follow the wrong path.

A short circuit to ground (grounded circuit) may prevent voltage from reaching a component. If voltage is shorted directly to ground, bypassing any load, the unrestricted current flow will damage fuses, wires or components. Some common causes of short circuits are:

— damaged wire or wiring harness
— faulty insulation
— internally damaged component
— incorrect connection

Test for a short circuit to ground using a multimeter or a test light to indicate circuit faults and abnormal current flow paths. See **4.5 Checking for Short Circuit to Ground**

Troubleshooting Procedure

Verify the complaint — Check the complaint. Try to understand the problem. If possible, let the driver show you what happens. Check all functions of the system and note the symptoms before starting any testing or disassembly.

Analyze the problem — Identify the part of the electrical system that is most likely to be causing the problem. Find the Circuit Section in the manual that applies to that part of the system. Find the wiring diagram that applies to the car. By following the circuit from ground back to the power source, get an understanding of how the circuit works.

Find the problem — You will find the problem if you follow a simple and logical step-by-step procedure. Test portions of the circuit one at a time, starting with the area or component most likely to be at fault. Test first at points that you can reach most easily.

Repair the problem — When you find the cause of the problem, make the repair. Use appropriate tools and procedures.

Check the results — Be sure it works. Check the functions of all parts of the circuit that you worked on.

Working on the Electrical System

A test light or a digital multimeter can be very helpful for testing circuits. See **2. Test Equipment** for more information on test equipment.

Current flow is logical, always moving from the highest potential at the voltage source (+) toward the lowest potential at ground (-). Using a wiring diagram to trace a circuit, you should start with the ground and then follow the wires back to the source of power.

To troubleshoot a circuit:

1. Inspect all connections, especially grounds. Make sure they are clean, tight and corrosion-free.
2. Check the fuses.

> **Note**
>
> *Repeated fuse failures are the sign of a faulty wire, a failed component, or a short to ground somewhere in the circuit.*

3. Check for voltage reaching particular components or points in the circuit.
4. Check continuity between points to look for breaks in the circuit (open circuit).
5. Check voltage drop at connections, especially ground connections.

4.2 Checking for Voltage

Checking for voltage confirms that the circuit is uninterrupted between the voltage source and the test point. The example illustrates troubleshooting the high-speed circuit for the radiator cooling fan.

Voltage: If the test light or multimeter indicates voltage potential, then the circuit between the voltage source and terminal 1 of the fan connector is OK.

NO Voltage: Power is not reaching the fan connector. The fan is probably OK. Look for a fault somewhere between the voltage source and the fan connector.
(Example: check for voltage reaching terminal E87La of the fuse/relay panel)

> **CAUTION**
>
> *Direct contact with meter probes at the connector terminals can easily damage the small contacts, causing poor connections and risking future intermittent faults.*
>
> *Special Tools include adapters for making test connections safely and preventing connector damage. See* **2. Test Equipment.**

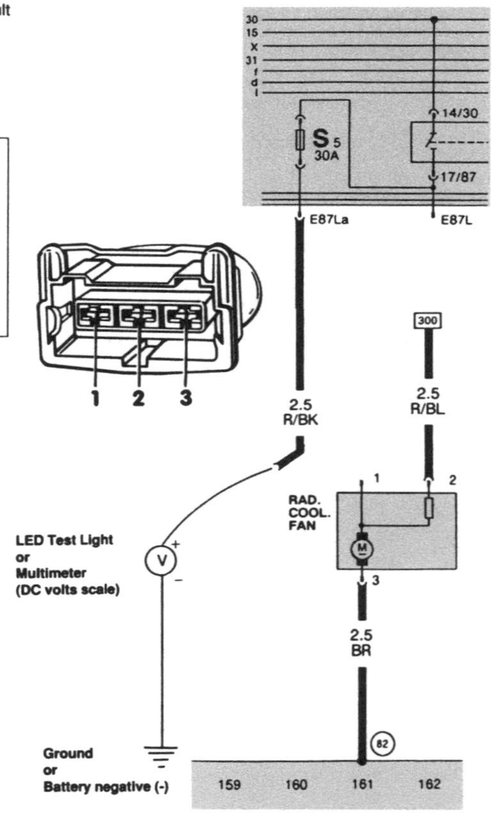

4.3 Checking Ground Connections

Checking ground connections as shown confirms that the circuit is complete—that the necessary path to ground is uninterrupted and current can flow in the circuit. The example illustrates troubleshooting the high-speed circuit for the radiator cooling fan.

Voltage: If the test light or multimeter indicates voltage, then there is potential for current flow between the known voltage source and ground at the test point. The ground side of the circuit, between terminal 3 of the fan connector and battery negative (-), is OK.

NO Voltage: The test point is not providing a path that completes the circuit to ground. The fan is probably OK. Look for a fault somewhere in the wiring between the fan connector and chassis ground. Also check the mechanical ground connection at the chassis (body).

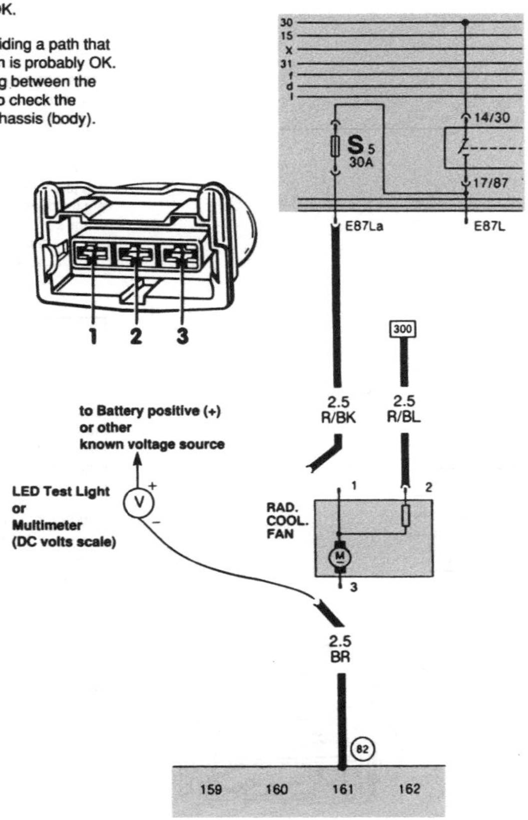

Continuity (approximately 0 Ω): Little or no resistance indicates that there is a continuous conductive path between the two test points—the circuit's ground path between terminal 3 and battery negative (-) is OK.

NO Continuity: There is resistance to current flow in the ground side of the circuit. The fan is probably OK. Look for a fault somewhere in the wiring between the fan connector and chassis ground. Also check the mechanical ground connection at the chassis (body).

> ***CAUTION***
>
> *Direct contact with meter probes at the connector terminals can easily damage the small contacts, causing poor connections and risking future intermittent faults.*
>
> *Special Tools include adapters for making test connections safely and preventing connector damage. See* **2. Test Equipment.**

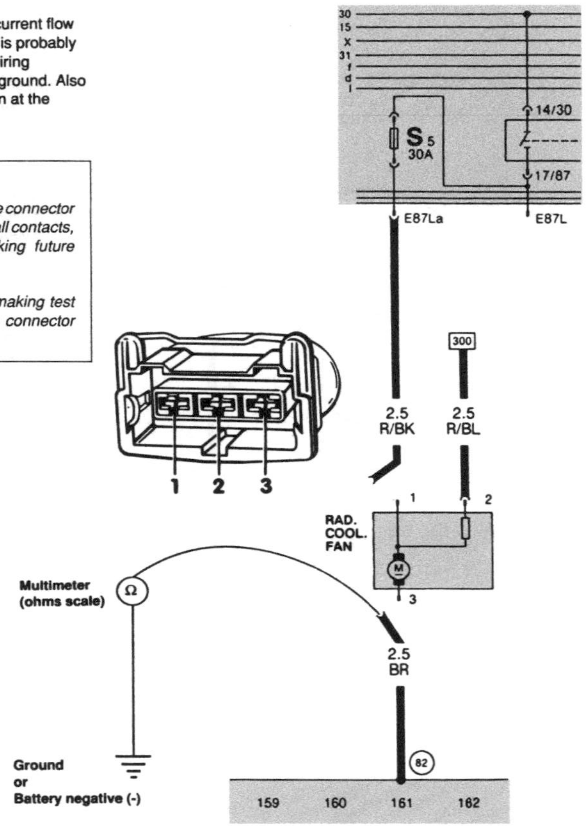

4.4 Checking Wiring and Components

Resistance or Continuity

Checking a portion of the wiring harness or a component as shown indicates whether or not there is a continuous conductive path—whether current can flow between the two test points. The example illustrates troubleshooting the idle and full-throttle switches in the fuel injection system.

> **CAUTION**
>
> *Resistance measurements and continuity checks must always be made with all power to the circuit or component switched* **Off.** *When testing continuity in a circuit that is always powered (fuse/relay panel "30" circuit for example) disconnect the battery before testing.*
>
> *Always use a digital (low current) meter. An ohmmeter, or the ohms scale of a multimeter, measures resistance by passing a small amount of current through the circuit or component being checked.*
>
> *Improper testing may damage sensitive electronic components.*

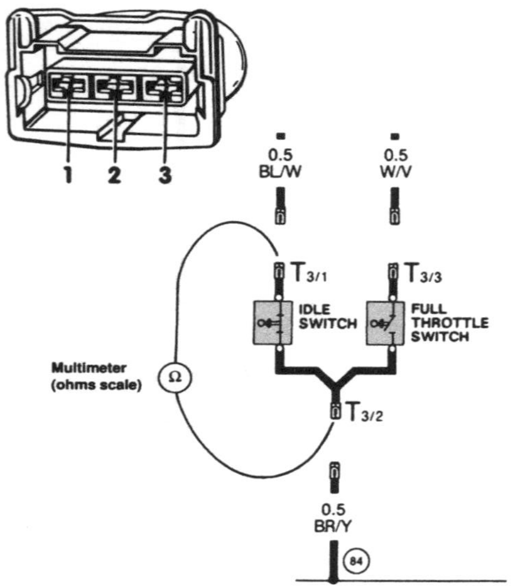

Continuity (approximately 0 Ω): Little or no resistance indicates that there is a continuous conductive path between the two test points. As shown, this is correct for the normally-closed idle switch in the idle position (switch not actuated). Also check that the switch opens (no continuity) when actuated by the throttle.

NO Continuity: There is resistance to current flow through the switch. In the example, this indicates that the normally-closed idle switch or the wires between it and the connector (T3) are faulty.

> **CAUTION**
>
> *Direct contact with meter probes at the connector terminals can easily damage the small contacts, causing poor connections and risking future intermittent faults.*
>
> *Special Tools include adapters for making test connections safely and preventing connector damage. See* **2. Test Equipment.**

Continuity (approximately 0 Ω): Little or no resistance indicates that there is a continuous conductive path between the two test points. In the example, this indicates that the normally-open full-throttle switch is faulty.

NO Continuity: There is no connection—an open circuit. As shown, this is correct for the normally-open full-throttle switch. Also check that the switch closes, completing the circuit, when actuated by the fully-open throttle.

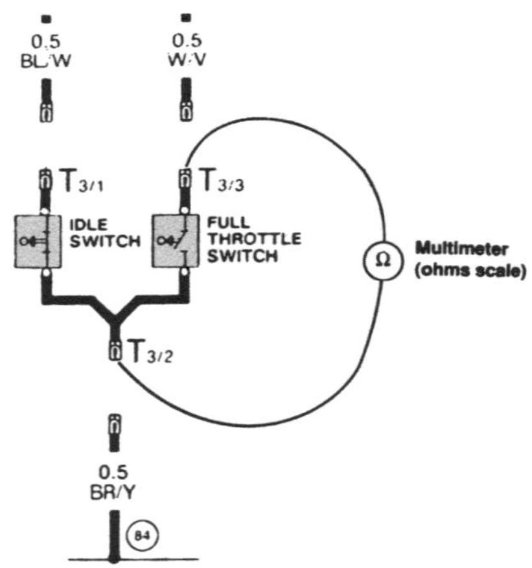

Voltage Drop

Checking voltage drop across connections or components as shown will indicate whether there is abnormal resistance creating an additional load in the circuit—consuming power and dropping the voltage available to other parts of the circuit.

> **Note**
>
> *Voltage drop measurements can only be made when the circuit is powered and there is normal current flow.*

The example illustrates troubleshooting the back-up light switch. The switch is in the circuit to switch power to the back-up lights On and Off. When Reverse gear is selected and the switch is closed, it should have very little resistance and not be a consumer.

If dirt or corrosion on the switch contacts creates resistance, some of the available battery voltage goes into overcoming that resistance. Less voltage is available to light the back-up lights, and they will not be as bright.

> **Note**
>
> *An ohmmeter can measure resistance (or check continuity) only when the circuit is not powered, i.e. when there is almost no load.*

> ***CAUTION***
>
> *Direct contact with meter probes at the connector terminals can easily damage the small contacts, causing poor connections and risking future intermittent faults.*
>
> *Special Tools include adapters for making test connections safely and preventing connector damage. See* **2. Test Equipment.**

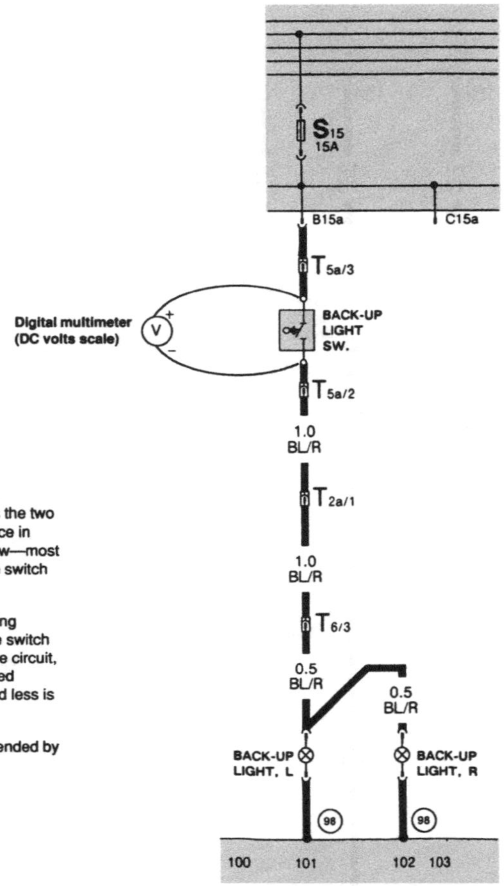

Low voltage: A low voltage reading across the two switch contacts indicates almost no difference in potential. Resistance across the switch is low—most of the battery voltage is passing through the switch and is still available to power the lights.

High voltage: Any significant voltage reading indicates a difference in potential across the switch contacts. Excessive resistance is loading the circuit, causing a voltage drop—voltage is consumed overcoming the resistance of the switch, and less is left to power the lights.

Maximum allowable voltage drops recommended by the Society of Automotive Engineers are:

- — 0.0 volt for small wire connections
- — 0.1 volt for high current connections (example: fuel pump, headlights)
- — 0.1 volt for ground connections
- — 0.2 volt for high-current cables (Example: battery/starter cable)
- — 0.3 volt for switch or relay contacts

4.5 Checking for Short Circuit to Ground

Checking the circuit as shown will detect circuit faults that are providing an unintended current flow path to ground. The examples illustrate two methods of troubleshooting a short to ground that is causing a blown fuse in the circuit powering the license plate lights and the glove compartment light.

Using an Ohmmeter (Multimeter)

Step 1 - Remove the fuse

Step 2 - Disconnect the load (powered components) to eliminate the circuit's normal path to ground

> **CAUTION**
> *Always use a digital (low current) meter. An ohmmeter, or the ohms scale of a multimeter, measures resistance by passing a small amount of current through the circuit or component being checked. Improper testing may damage sensitive electronic components.*

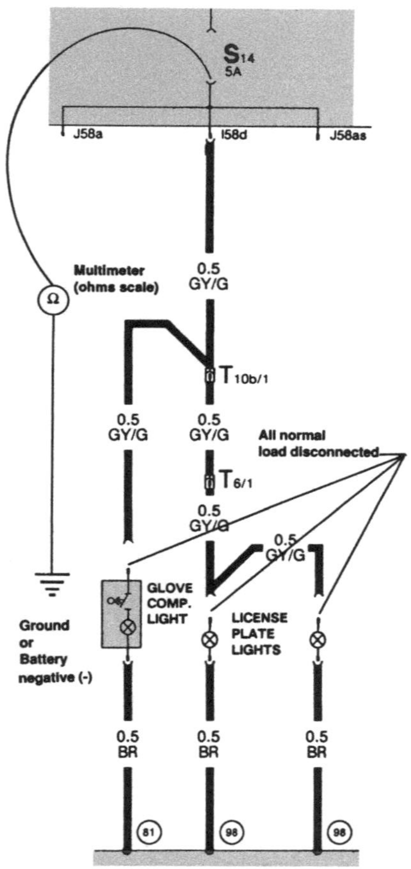

Continuity (approximately 0 Ω): Little or no resistance indicates that there is a continuous conductive path between the isolated circuit and ground, even though all the circuit's normal ground paths are eliminated. There is a short—an unintentional connection to ground—somewhere in the circuit.

NO Continuity: The circuit's normal ground paths have been disconnected, and there is no other connection between the isolated circuit and ground—no short has been detected.

> **CAUTION**
> *Direct contact with meter probes at the connector terminals can easily damage the small contacts, causing poor connections and risking future intermittent faults.*
>
> *Special Tools include adapters for making test connections safely and preventing connector damage. See* **2. Test Equipment.**

Using an LED Test Light or Voltmeter (Multimeter)

Step 1 - Remove the fuse

Step 2 - Disconnect the load (powered components) to eliminate the circuit's normal path to ground

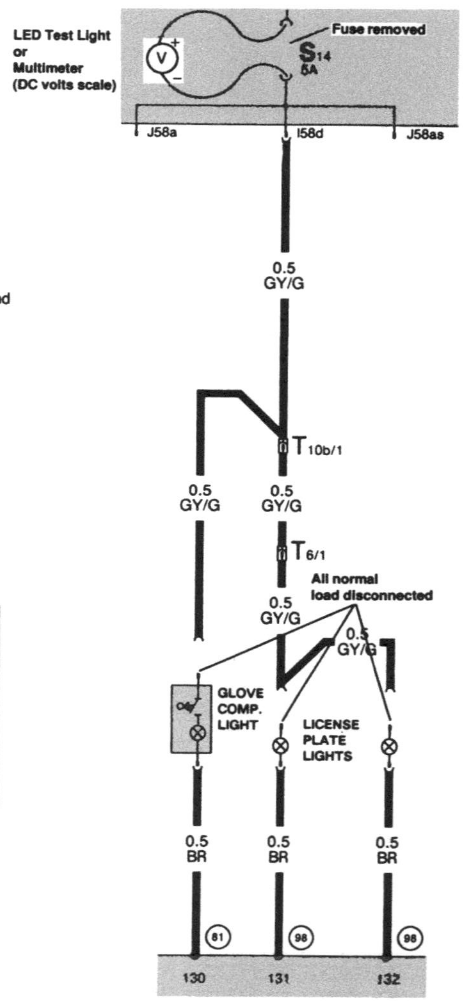

Voltage: If the test light or multimeter indicates voltage, then there is a complete circuit—a connection to ground even though all the circuit's Normal ground paths are eliminated. There is a short—an unintentional connection to ground—somewhere in the circuit.

NO Voltage: There is not a complete circuit. The circuit's normal ground paths have been disconnected, and there is no other connection between the isolated circuit and ground—no short has been detected.

> **CAUTION**
> *Direct contact with meter probes at the connector terminals can easily damage the small contacts, causing poor connections and risking future intermittent faults.*
>
> *Special Tools include adapters for making test connections safely and preventing connector damage. See* **2. Test Equipment.**

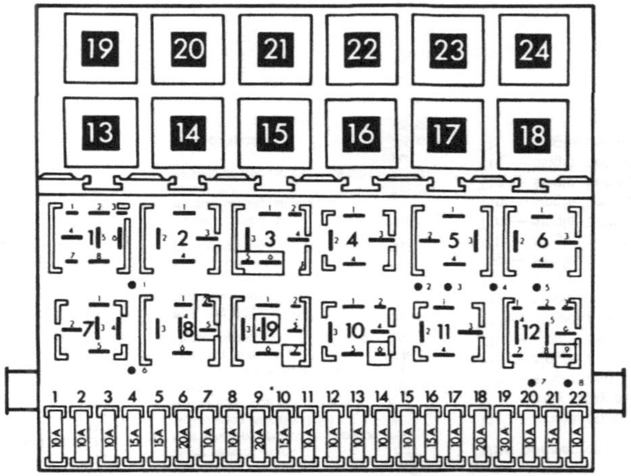
97-3000

Fuse arrangement

according to the numbers on fuse panel from left to right:

		Amp.
1	Headlight, low beam, left	10
2	Headlight, low beam, right	10
3	Instrument panel lights, license plate lights	10
4	Rear window wiper, sunroof	15
5	Wiper/washer, front, rear window washer	15
6	Fresh air fan, A/C	20
7	Tail lights, side marker light, right	10
8	Tail lights, side marker light, left	10
9	Rear window defogger, heated mirrors	20
10	Fog lights	15
11	Headlight, high beam, left	10
12	Headlight, high beam, right	10
13	Horn, radiator fan run-on	10
14	Back-up lights, washer jet heaters	10
15	Electronic ignition, idle stabilizer, deceleration fuel shut off	10
16	Dash warning lights, glove box light, multi-function indicator, cassette storage light, rear spoiler control unit	15
17	Emergency flasher switch, turn signals	10
18	Fuel pump, oxygen sensor	20
19	Radiator cooling fan, A/C	30
20	Brake lights, cruise control	10
21	Interior lights, cigarette lighter, digital clock, rear spoiler adjustment, central locking	15
22	Radio	10

Relay location on fuse/relay panel

1 — A/C relay
2 — Rear window wiper relay
3 — Digifant control unit
4 — Load reduction relay
5 — Low coolant level
6 — Emergency flasher relay
7 — Open
8 — Intermittent wiper relay
9 — Seat belt warning system
10 — Fog light relay
11 — Horn relay
12 — Fuel pump relay

Separate relays above fuse/relay panel

13 — ABS hydraulic pump relay
14 — ABS relay
15 — Radiator fan after run relay
16 — Fuel pump after run control unit
17 — Heated seat relay, driver
18 — Heated seat relay, passenger
19 — ABS hydraulic pump fuse
20 — ABS valves fuse
21 — Power window fuse
22 — ABS valves fuse
23 — Open
24 — Open

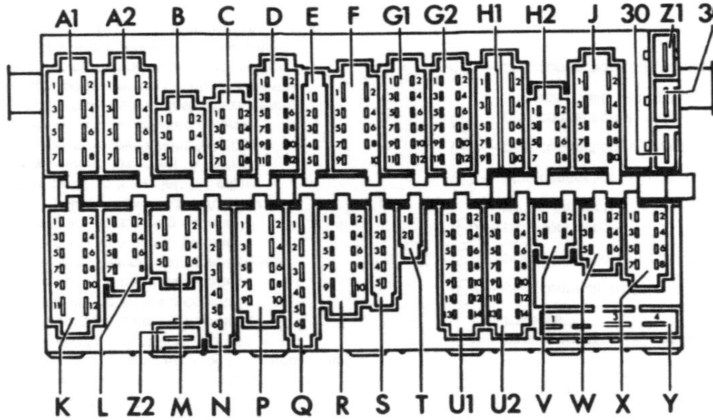

97-3001

Connections and plugs on fuse/relay panel

A1 — Headlight wiring harness (yellow)
A2 — Headlight wiring harness (yellow)
B — Open
C — Headlight wiring harness (yellow)
D — Optional equipment wiring harness (green)
E — Instrument wiring harness (green)
F — Engine compartment wiring harness, right (white)
G1 — Engine compartment wiring harness, right (white)
G2 — Engine compartment wiring harness, right (white)
H1 — Steering column switch wiring harness (red)
H2 — Steering column switch wiring harness (red)
J — Steering column switch wiring harness (red)
K — Rear wiring harness (black)
L — Rear wiring harness (black)
M — Rear wiring harness (black)
N — A/C wiring harness (green)
P — Rear window defroster and fog light wiring harness (blue)
Q — Instrument wiring harness (blue)
R — Light switch wiring harness (blue)
S — Engine compartment wiring harness, right (white)
T — Two-point connector (green)
U1 — Instrument cluster wiring harness (blue)
U2 — Instrument cluster wiring harness (blue)
V — Multi-function indicator wiring harness (green)
W — Six-point connector (green)
X — Warning lamp wiring harness (green)
Y — Single connector, terminal 30
Z1 — Single connector
Z2 — Single connector, terminal 31
30 — Single connector, terminal 30
30B — Single connector
31 — Single connector, terminal 31

Description	Current track
Air inlet temperature sender	40
Alternator	1,2
Alternator warning light	84
Amplifier speaker, left front	312
Amplifier speaker, right front	319
Antenna booster	312
Back-up light switch	224
Back-up light, left	222
Back-up light, right	224
Battery	4
Brake and parking brake indicator light	111
Brake fluid level warning contact	125
Brake light switch	195
Brake light, left	186
Brake light, right	190
Cassette storage light	128
Cigarette lighter	138
Cigarette lighter light	137
Clock light	99
CO potentiometer	41
Coolant low level control unit	72,73
Coolant low level sender	72
Coolant overheat warning light	93
Coolant temperature gauge	90
Coolant temperature sender	37
Coolant temperature sending unit	73
Digifant control unit	29-54
Digifant control unit relay	52-55
Digital clock	101,102
Door contact switch, left front	145
Door contact switch, right front	147
ECS-Malfunction indicator light and switch	20
Emergency flasher relay	171-173
Emergency flasher switch	171
Emergency flasher warning light	174
Engine oil pressure control unit	78
Engine oil pressure sensor	132
Engine oil pressure switch (0.3 bar)	74
Engine oil pressure switch (1.8 bar)	75
Engine oil pressure warning light	76
Engine oil temperature sensor	119
Fog light indicator light	214
Fog light relay	213,214
Fog light switch	216
Fog light switch bulb, front and rear	213
Fog light, left	218
Fog light, right	219
Fresh air blower series resistance	248
Fresh air control lever light	135
Fresh air fan	249
Fresh air fan switch	246-249
Fuel enrichment switch	32
Fuel gauge	90
Fuel gauge sending unit	63
Fuel pump	58
Fuel pump afterrun control unit	69
Fuel pump afterrun pressure switch	67
Fuel pump relay	49,50
Glove compartment light	127
Hall generator	31-33
Headlight dimmer/flasher switch	167,169
Headlight high beam warning light	89
Headlight switch light	198
Headlight, left	160
Headlight, right	164
Heated crankcase breather	19
Hi-mounted brake light	187,188, 189
High-beam headlight, left	159
High-beam headlight, right	163
Horn button	275
Horn relay	261
Horns	263, 265
Idle speed stabilizer valve	53-55
Idle switch	35
Ignition coil	27
Ignition distributor	27
Ignition/starter switch	10-14
Injector, cyl. 1	53
Injector, cyl. 2	54
Injector, cyl. 3	55
Injector, cyl. 4	56
Instrument panel light	97
Instrument panel light dimmer switch	206
Interior light with delay switch	144,145
Knock sensor	43-45
License plate light	206,209
Light switch	202
Load reduction relay	12,13
Luggage compartment light	149
Luggage compartment light switch	149
Make-up mirror light, right	141,142
Make-up mirror light switch, right	141
Memory switch (multi-function ind.)	115,116
Midrange speaker, left front	297
Midrange speaker, right front	299
Mirror adjustment switch	226
Mirror heat element, left	236
Mirror heated element, right	235
Mirror magnetic clutch, left	231
Mirror magnetic clutch, right	233
Mirror motor, left	228
Mirror motor, passenger side	226
Mirror selector switch	226
Multi-function indicator recall button	113
Multifunction indicator (MFI)	109
Multifunction indicator pressure sender	105-107
Outside air temperature sensor	121
Overheat fuse	246
Oxygen sensor	50
Park light indicator light	205
Park light switch	178
Parking brake warning light switch	123
Parking light, left	158
Parking light, right	162
Radiator cooling fan	240
Radiator cooling fan after run control unit	242-244
Radiator cooling fan thermoswitch	239-241, 244
Radio connection	295-308, 309-322
Rear spoiler control unit	151-153
Rear spoiler end switch 1	151,152
Rear spoiler end switch 2	153,154
Rear spoiler motor	151
Rear spoiler switch	150-153
Rear window defogger indicator light	255
Rear window defogger switch	256
Rear window defogger switch light	254
Rear window heater element	258
Rear window wiper motor	289
Rear window wiper/washer relay	285-287
Seat belt warning light	82
Seat belt warning relay	82
Side marker lights, front	157
Side marker lights, front	161
Spark plug connectors	27
Spark plugs	27
Speaker, left rear	304
Speaker, right rear	306
Speed sensor (for MFI, cruise control, and rear spoiler system)	91
Starter	7-9
Tachometer	86
Tail light, left	185
Tail light, right	191
Transfer fuel pump	62
Turn signal indicator light	88
Turn signal light, left front	183
Turn signal light, left rear	184
Turn signal light, right front	196
Turn signal light, right rear	192
Turn signal switch	175,176
Voltage regulator	1,2
Voltage stabilizer	90
Washer nozzle heater, left	237
Washer nozzle heater, right	238
Washer/wiper intermittent relay	278-280
Windshield and rear window washer pump	283
Windshield and/or rear window antenna	312
Windshield wiper motor	268-271
Windshield-wiper intermittent switch	279

Wire connectors

T1 — single, on lower left A-pillar, near relay panel
T1a — single, behind relay panel
T1b — single, behind relay panel
T1c — single, behind relay panel
T1d — single, near battery
T1e — single, behind relay panel
T1f — single, behind relay panel
T1g — single, on left firewall
T1h — single, next to relay panel
T1m — single, behind relay panel
T1n — single, behind relay panel
T1o — single, behind relay panel
T1q — single, behind relay panel
T1r — single, behind relay panel
T1s — single, in rear deck lid
T1t — single, above relay panel
T1u — single, behind relay panel
T1v — single, near left headlight
T1x — single, on center firewall
T2a — double, near left brake light
T2b — double, left front speaker
T2c — double, right front speaker
T2d — double, upper right "A" pillar
T2e — double, in left rear of luggage compartment
T2f — double, near left headlight
T2g — double, near right headlight
T2h — double, in rear deck lid
T2i — double, in luggage compartment, left
T2j — double, on engine hood
T2k — double, in luggage compartment, left
T2l — double, near left headlight
T2m — double, in luggage compartment, left
T2n — double, in right front of luggage compartment
T2p — double, on engine hood
T2r — double, near intake manifold
T2s — double, near right headlight
T2x — double, under gear shift lever housing
T2y — double, under gear shift lever housing
T3 — three-point, near ignition distributor
T3a — three-point, near intake manifold
T3b — three-point, on left firewall
T3c — three-point, left front speaker
T3d — three-point, right front speaker
T3e — three-point, in rear deck lid, left
T3f — three-point, in rear deck lid, right
T3g — three-point, on left firewall
T3m — three-point, in driver door
T3n — three-point, in right plenum
T3p — three-point, in passenger door
T3q — three-point, in passenger door
T3r — three-point, behind dashboard, center
T3s — three-point, in driver door
T4a — four-point, in left rear luggage compartment
T4b — four-point, in right rear luggage compartment
T4e — four-point, in right front footwell
T4c — four-point, in steering column
T4d — four-point, in steering column
T4n — four-point, behind relay panel
T5 — five point, behind instrument cluster
T5b — five-point, in steering column
T5c — five-point, in steering column
T5d — five-point, in lower right "A" pillar
T5e — five-point, in lower left "A" pillar
T6a — six-point, in lower left "A" pillar
T6b — six-point, in lower right "A" pillar
T7a — seven-point, in steering column
T8 — eight-point, behind dashboard, center
T8a — eight-point, behind dashboard, center
T10 — ten-point, behind dashboard, center
T16 — sixteen-point, on instrument cluster
T28 — twenty eight-point, on instrument cluster

Ground connections

(1) — battery to body
(2) — transmission to body
(14) — on transmission
(17) — intake manifold
(29) — near intake manifold
(30) — beside relay panel
(43) — on lower "A" pillar, right
(46) — near relay panel
(50) — in luggage compartment, left
(51) — in luggage compartment, right
(52) — in rear deck lid, left
(53) — in rear deck lid, right
(80) — in instrument panel wiring harness
(94) — in Digifant wiring harness
(119) — in headlight wiring harness, left
(120) — in headlight wiring harness, right
(129) — in dual horn wiring harness

Welded wiring harness points

(A11) — plus connection (58b), in instrument panel wiring harness
(A15) — plus connection (15), in instrument panel wiring harness
(C1) — plus connection (58l), in front wiring harness, left
(C2) — plus connection (58r), in front wiring harness, right
(C10) — plus connection (30), in headlight wiring harness
(C12) — plus connection, in headlight wiring harness
(C13) — plus connection, in dual horn wiring harness
(C14) — plus connection (15), in heated washer jet wiring harness
(G3) — plus connection, in injector wiring harness
(G4) — plus connection, in injector wiring harness
(G5) — plus connection, in ECS malfunction wiring harness (Calif. vehicles only)
(W1) — plus connection (30), in rear wiring harness

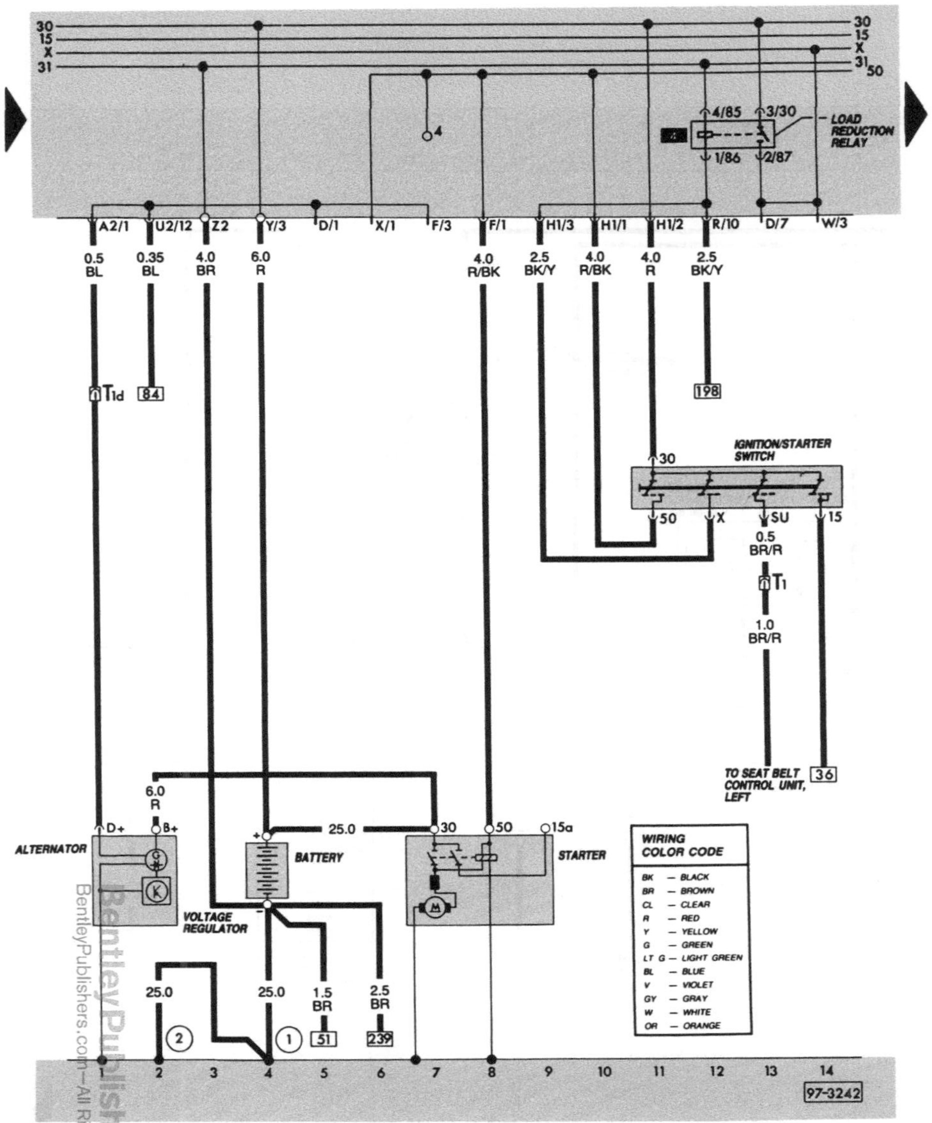
LOAD REDUCTION RELAY
IGNITION/STARTER SWITCH
TO SEAT BELT CONTROL UNIT, LEFT
ALTERNATOR
VOLTAGE REGULATOR
BATTERY
STARTER
WIRING COLOR CODE
BK — BLACK
BR — BROWN
CL — CLEAR
R — RED
Y — YELLOW
G — GREEN
LT G — LIGHT GREEN
BL — BLUE
V — VIOLET
GY — GRAY
W — WHITE
OR — ORANGE
97-3242

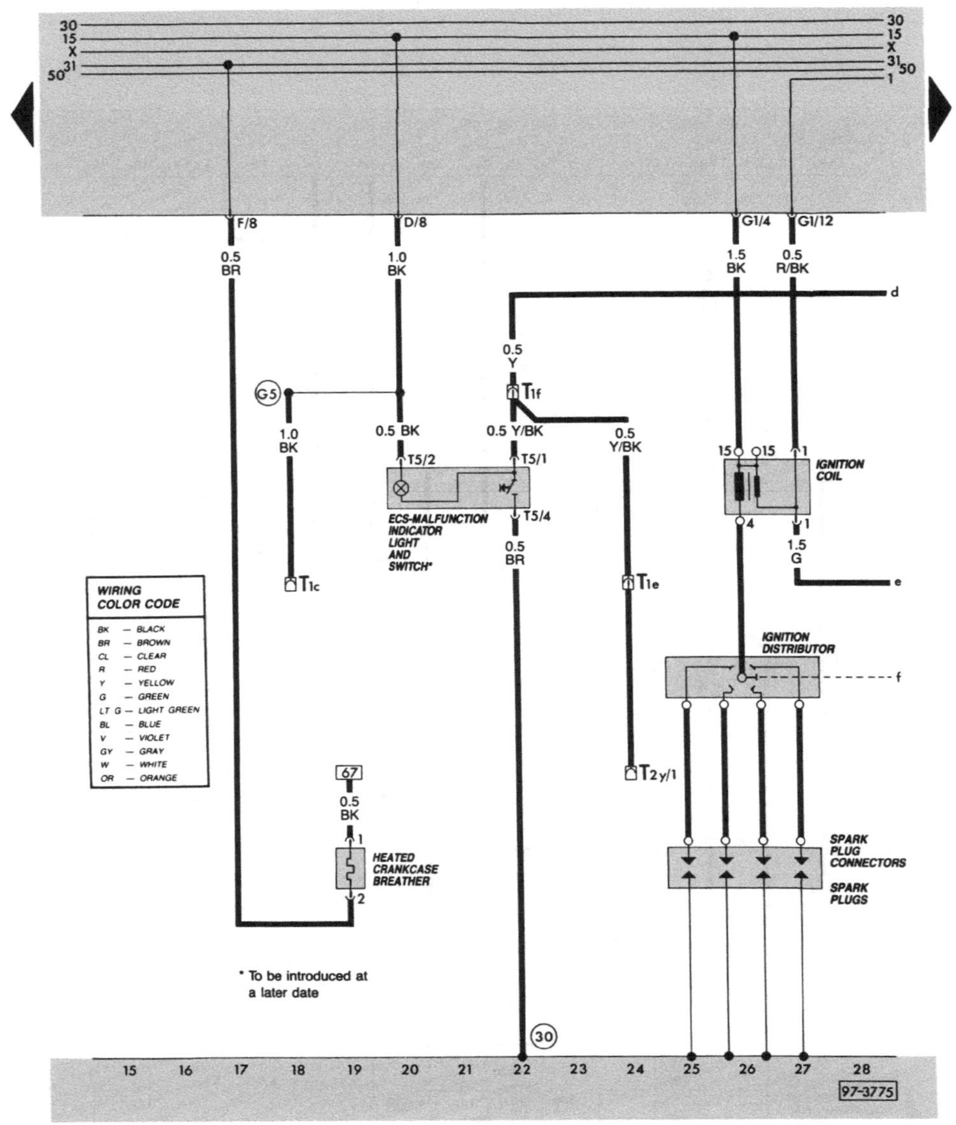
IGNITION COIL
ECS-MALFUNCTION INDICATOR LIGHT AND SWITCH*
IGNITION DISTRIBUTOR
HEATED CRANKCASE BREATHER
SPARK PLUG CONNECTORS
SPARK PLUGS
WIRING COLOR CODE
BK — BLACK
BR — BROWN
CL — CLEAR
R — RED
Y — YELLOW
G — GREEN
LT G — LIGHT GREEN
BL — BLUE
V — VIOLET
GY — GRAY
W — WHITE
OR — ORANGE
* To be introduced at a later date
97-3775

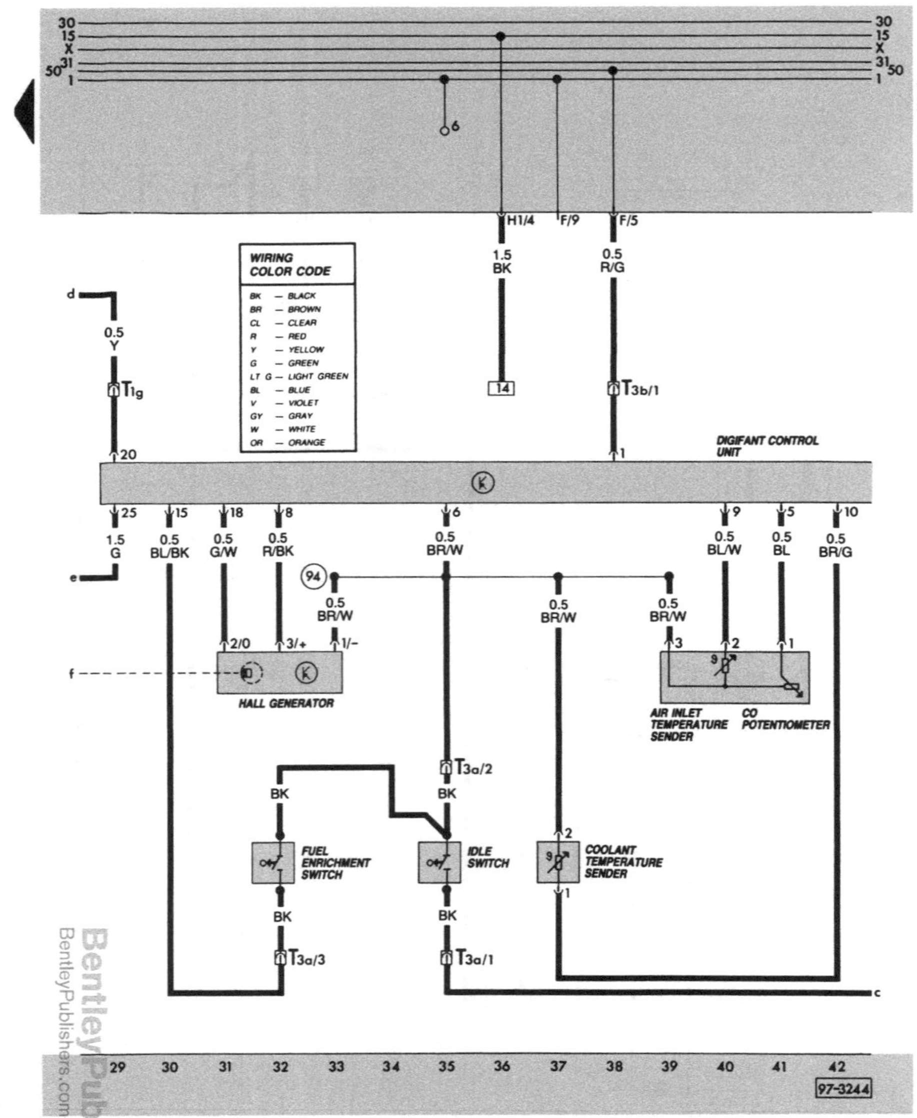

WIRING COLOR CODE
BK — BLACK
BR — BROWN
CL — CLEAR
R — RED
Y — YELLOW
G — GREEN
LT G — LIGHT GREEN
BL — BLUE
V — VIOLET
GY — GRAY
W — WHITE
OR — ORANGE
DIGIFANT CONTROL UNIT
HALL GENERATOR
AIR INLET TEMPERATURE SENDER
CO POTENTIOMETER
FUEL ENRICHMENT SWITCH
IDLE SWITCH
COOLANT TEMPERATURE SENDER
97-3244

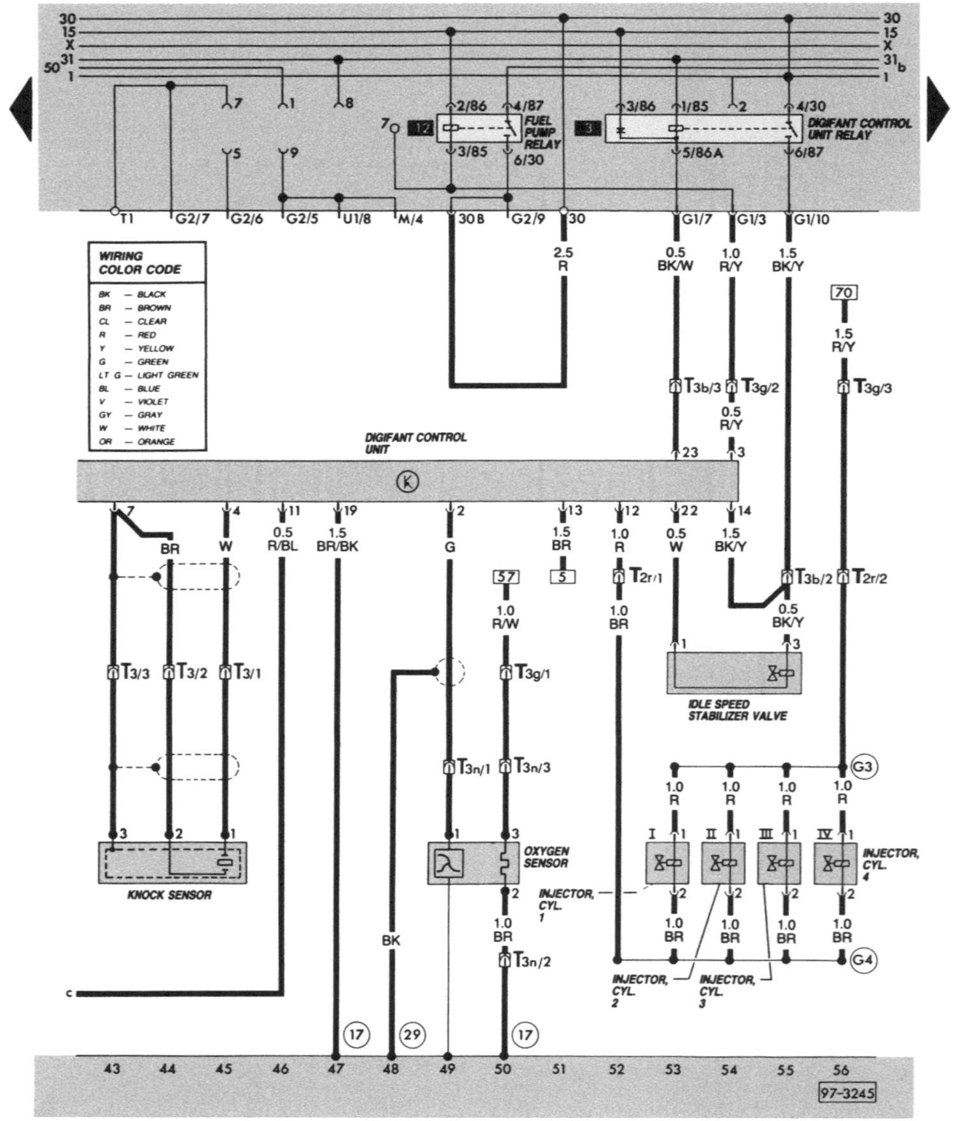

FUEL PUMP RELAY
DIGIFANT CONTROL UNIT RELAY
WIRING COLOR CODE
BK — BLACK
BR — BROWN
CL — CLEAR
R — RED
Y — YELLOW
G — GREEN
LT G — LIGHT GREEN
BL — BLUE
V — VIOLET
GY — GRAY
W — WHITE
OR — ORANGE
DIGIFANT CONTROL UNIT
IDLE SPEED STABILIZER VALVE
KNOCK SENSOR
OXYGEN SENSOR
INJECTOR, CYL. 1
INJECTOR, CYL. 2
INJECTOR, CYL. 3
INJECTOR, CYL. 4
97-3245

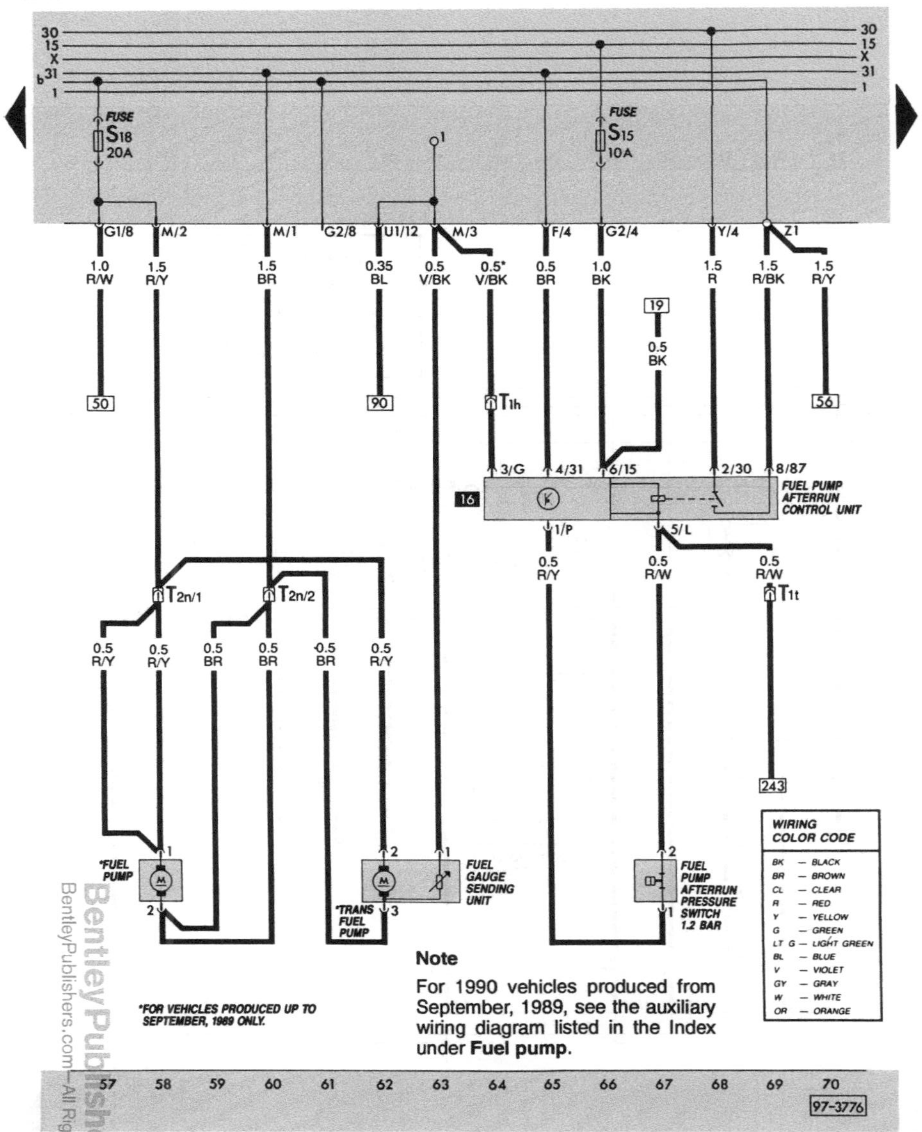

Note

For 1990 vehicles produced from September, 1989, see the auxiliary wiring diagram listed in the Index under **Fuel pump**.

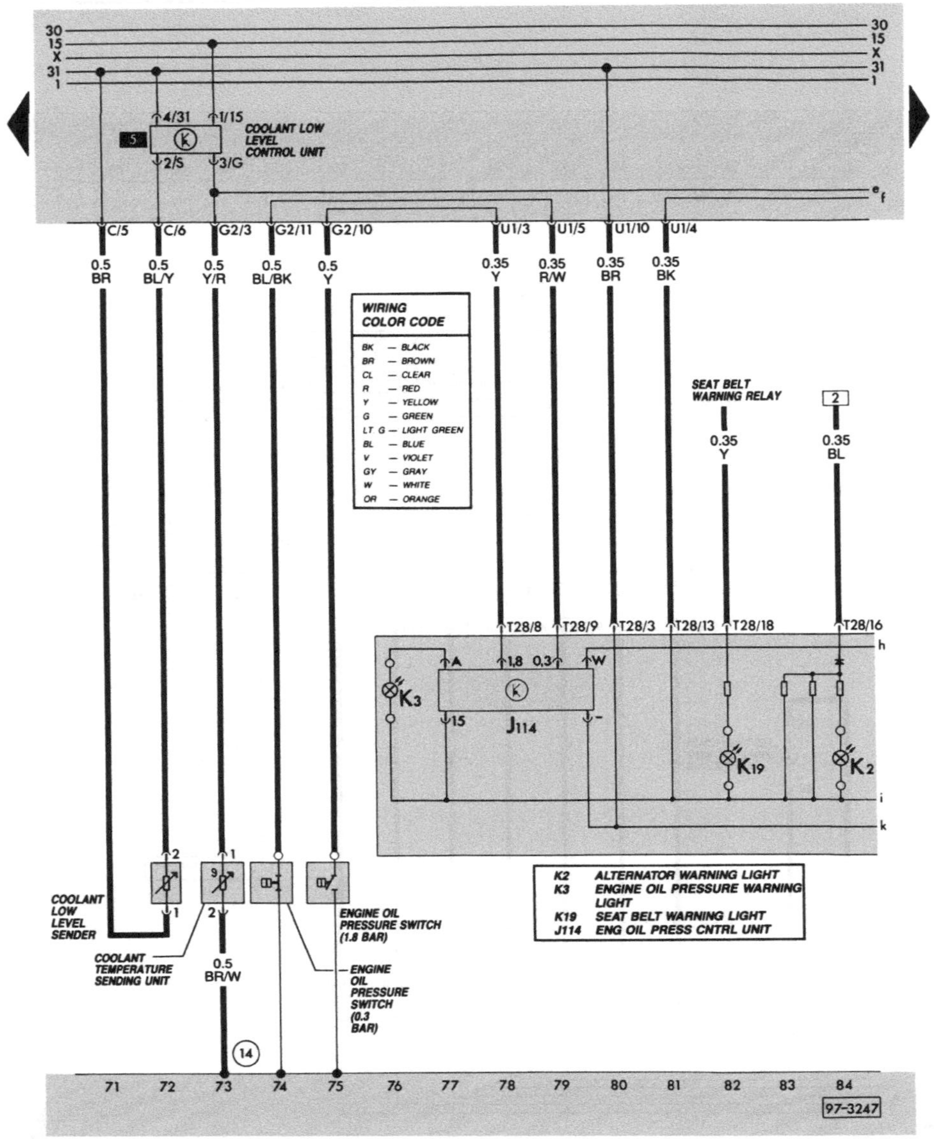

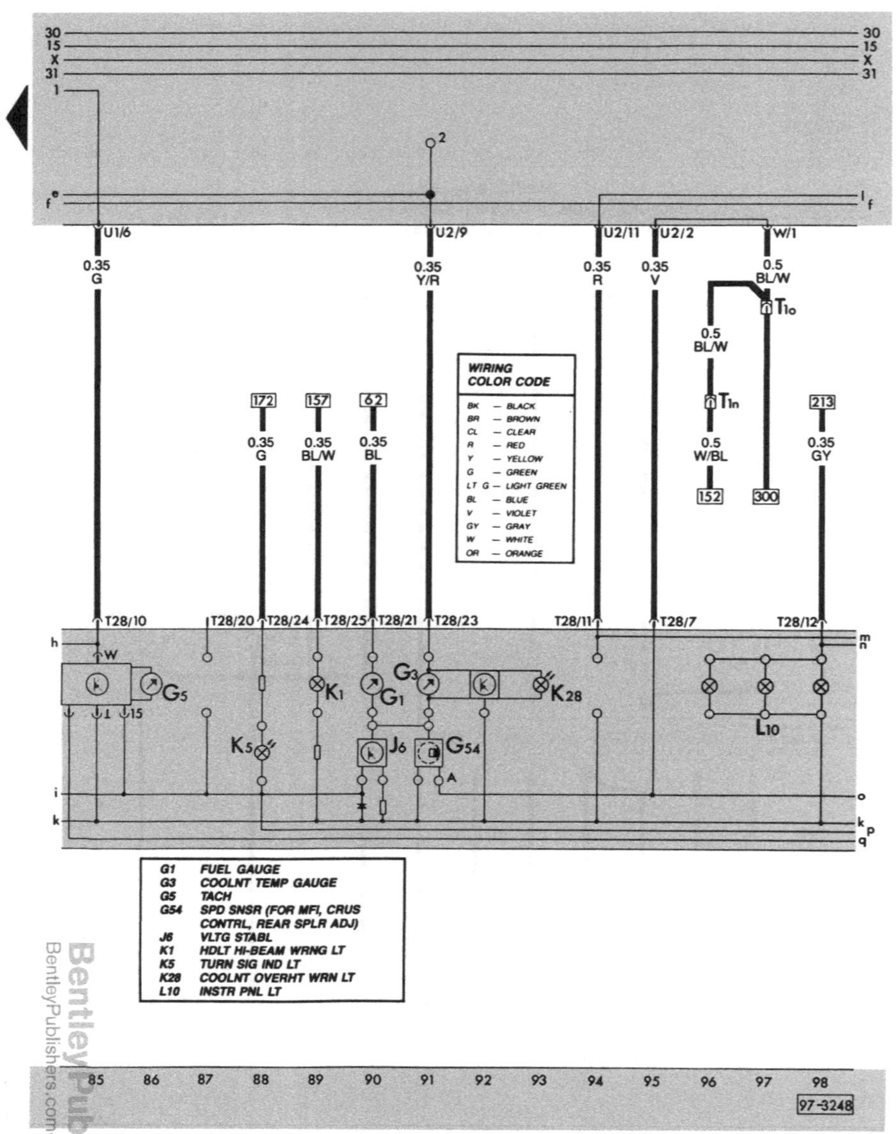
30
15
X
31
U1/6
U2/9
U2/11
U2/2
W/1
0.35 G
0.35 Y/R
0.35 R
0.35 V
0.5 BL/W
T10
T11
0.5 W/BL
172
157
62
213
152
300
0.35 BL/W
0.35 BL
0.35 GY
T28/10
T28/20
T28/24
T28/25
T28/21
T28/23
T28/11
T28/7
T28/12
G5
K1
G1
G3
K28
K5
J6
G54
L10
WIRING COLOR CODE
BK — BLACK
BR — BROWN
CL — CLEAR
R — RED
Y — YELLOW
G — GREEN
LT G — LIGHT GREEN
BL — BLUE
V — VIOLET
GY — GRAY
W — WHITE
OR — ORANGE
G1 FUEL GAUGE
G3 COOLNT TEMP GAUGE
G5 TACH
G54 SPD SNSR (FOR MFI, CRUS CONTRL, REAR SPLR ADJ)
J6 VLTG STABL
K1 HDLT HI-BEAM WRNG LT
K5 TURN SIG IND LT
K28 COOLNT OVERHT WRN LT
L10 INSTR PNL LT
85 86 87 88 89 90 91 92 93 94 95 96 97 98
97-3248

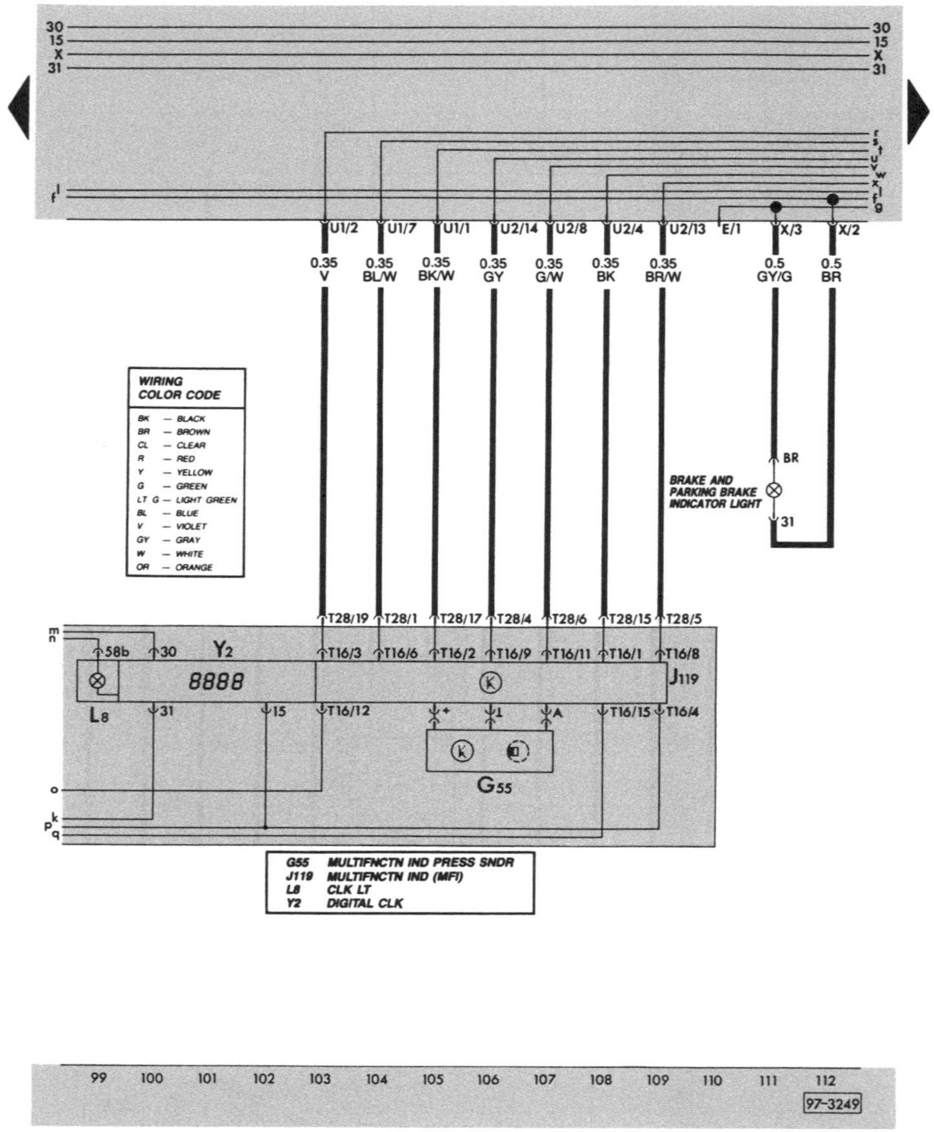
30
15
X
31
U1/2
U1/7
U1/1
U2/14
U2/8
U2/4
U2/13
E/1
X/3
X/2
0.35 V
0.35 BL/W
0.35 BK/W
0.35 GY
0.35 G/W
0.35 BK
0.35 BR/W
0.5 GY/G
0.5 BR
BRAKE AND PARKING BRAKE INDICATOR LIGHT
WIRING COLOR CODE
BK — BLACK
BR — BROWN
CL — CLEAR
R — RED
Y — YELLOW
G — GREEN
LT G — LIGHT GREEN
BL — BLUE
V — VIOLET
GY — GRAY
W — WHITE
OR — ORANGE
T28/19
T28/1
T28/17
T28/4
T28/6
T28/15
T28/5
T16/3
T16/6
T16/2
T16/9
T16/11
T16/1
T16/8
T16/12
T16/15
T16/4
58b
30
31
15
Y2
8888
J119
L8
G55
G55 MULTIFNCTN IND PRESS SNDR
J119 MULTIFNCTN IND (MFI)
L8 CLK LT
Y2 DIGITAL CLK
99 100 101 102 103 104 105 106 107 108 109 110 111 112
97-3249

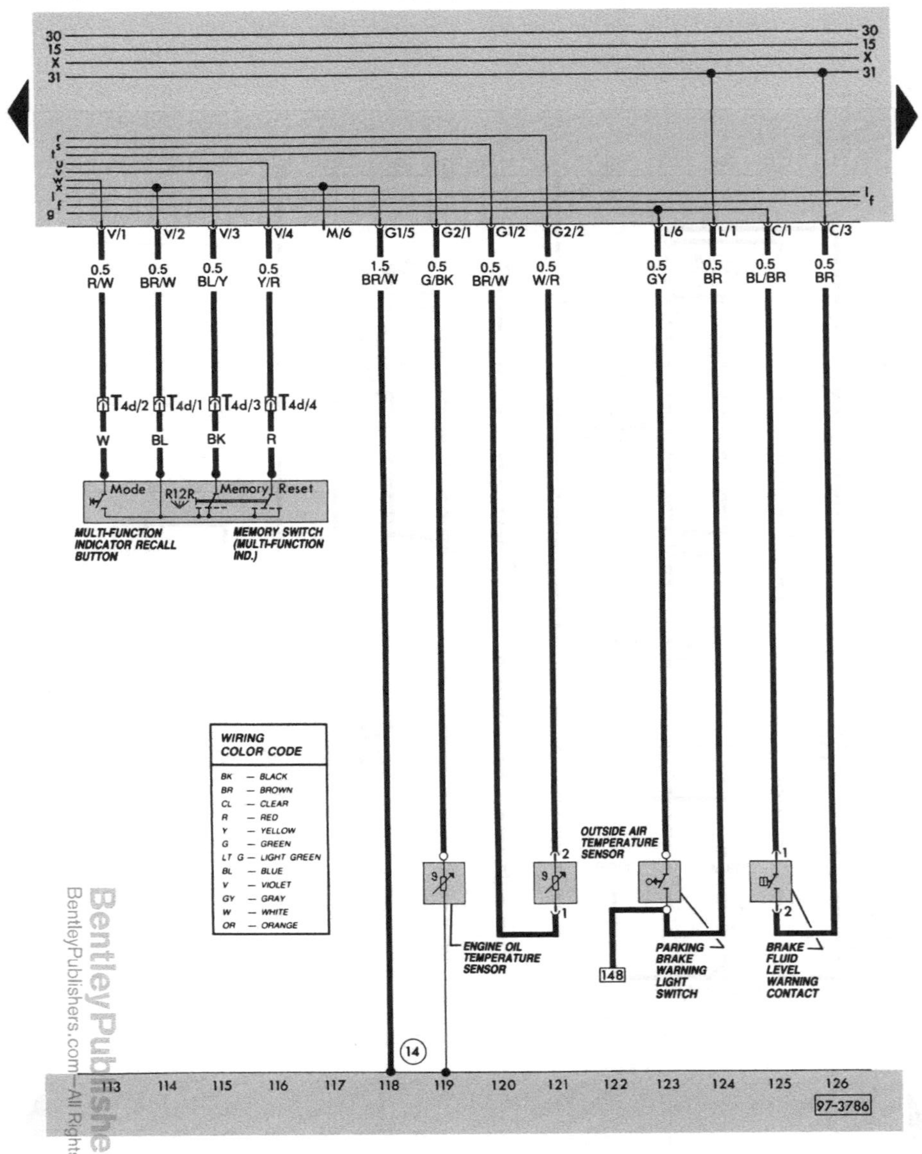
30
15
X
31
V/1
V/2
V/3
V/4
M/6
G1/5
G2/1
G1/2
G2/2
L/6
L/1
C/1
C/3
0.5 R/W
0.5 BR/W
0.5 BL/Y
0.5 Y/R
1.5 BR/W
0.5 G/BK
0.5 BR/W
0.5 W/R
0.5 GY
0.5 BR
0.5 BL/BR
0.5 BR
T4d/2
T4d/1
T4d/3
T4d/4
W
BL
BK
R
Mode
R12R
Memory
Reset
MULTI-FUNCTION INDICATOR RECALL BUTTON
MEMORY SWITCH (MULTI-FUNCTION IND.)
WIRING COLOR CODE
BK — BLACK
BR — BROWN
CL — CLEAR
R — RED
Y — YELLOW
G — GREEN
LT G — LIGHT GREEN
BL — BLUE
V — VIOLET
GY — GRAY
W — WHITE
OR — ORANGE
ENGINE OIL TEMPERATURE SENSOR
OUTSIDE AIR TEMPERATURE SENSOR
148
PARKING BRAKE WARNING LIGHT SWITCH
BRAKE FLUID LEVEL WARNING CONTACT
14
113 114 115 116 117 118 119 120 121 122 123 124 125 126
97-3786

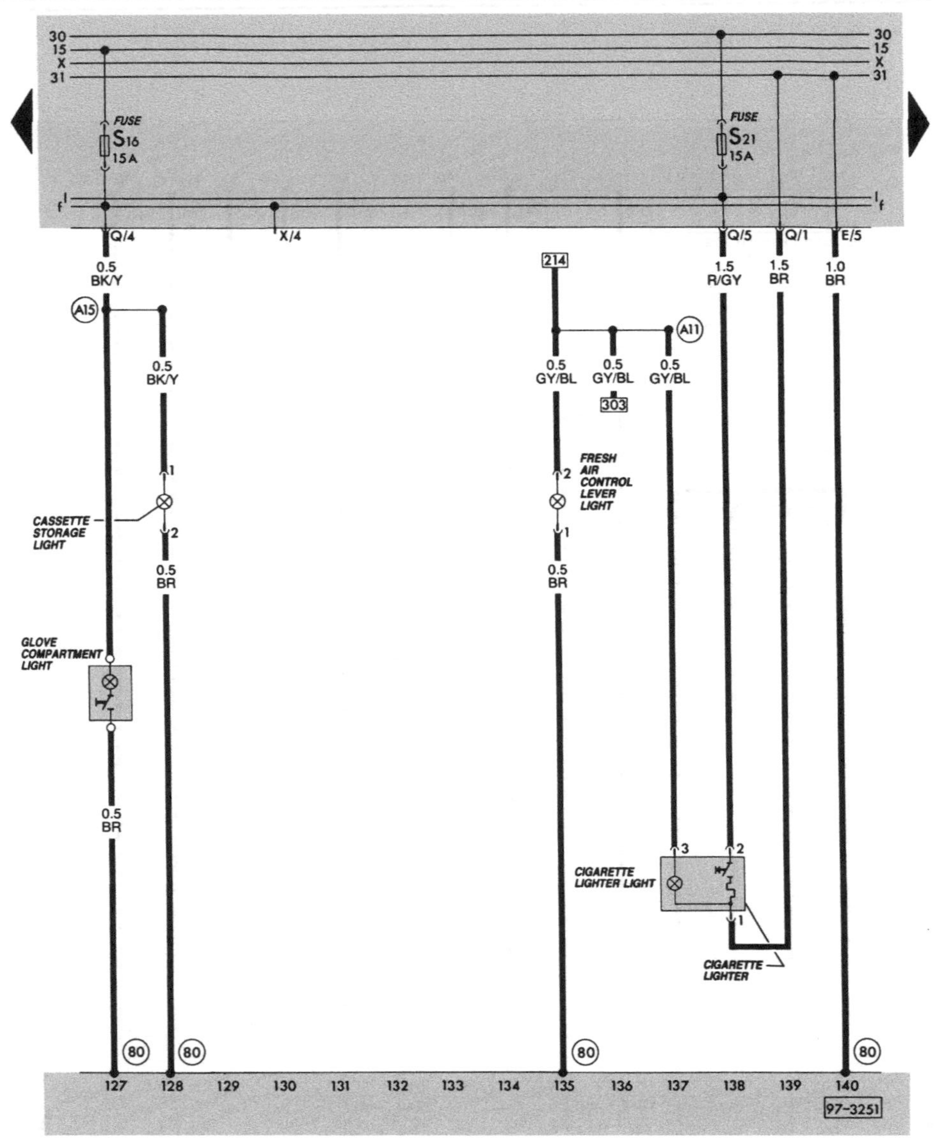
30
15
X
31
FUSE
S16
15A
FUSE
S21
15A
Q/4
X/4
Q/5
Q/1
E/5
0.5 BK/Y
A15
0.5 BK/Y
CASSETTE STORAGE LIGHT
0.5 BR
GLOVE COMPARTMENT LIGHT
0.5 BR
214
0.5 GY/BL
0.5 GY/BL
303
0.5 GY/BL
A11
FRESH AIR CONTROL LEVER LIGHT
0.5 BR
1.5 R/GY
1.5 BR
1.0 BR
CIGARETTE LIGHTER LIGHT
CIGARETTE LIGHTER
80
127 128 129 130 131 132 133 134 135 136 137 138 139 140
97-3251

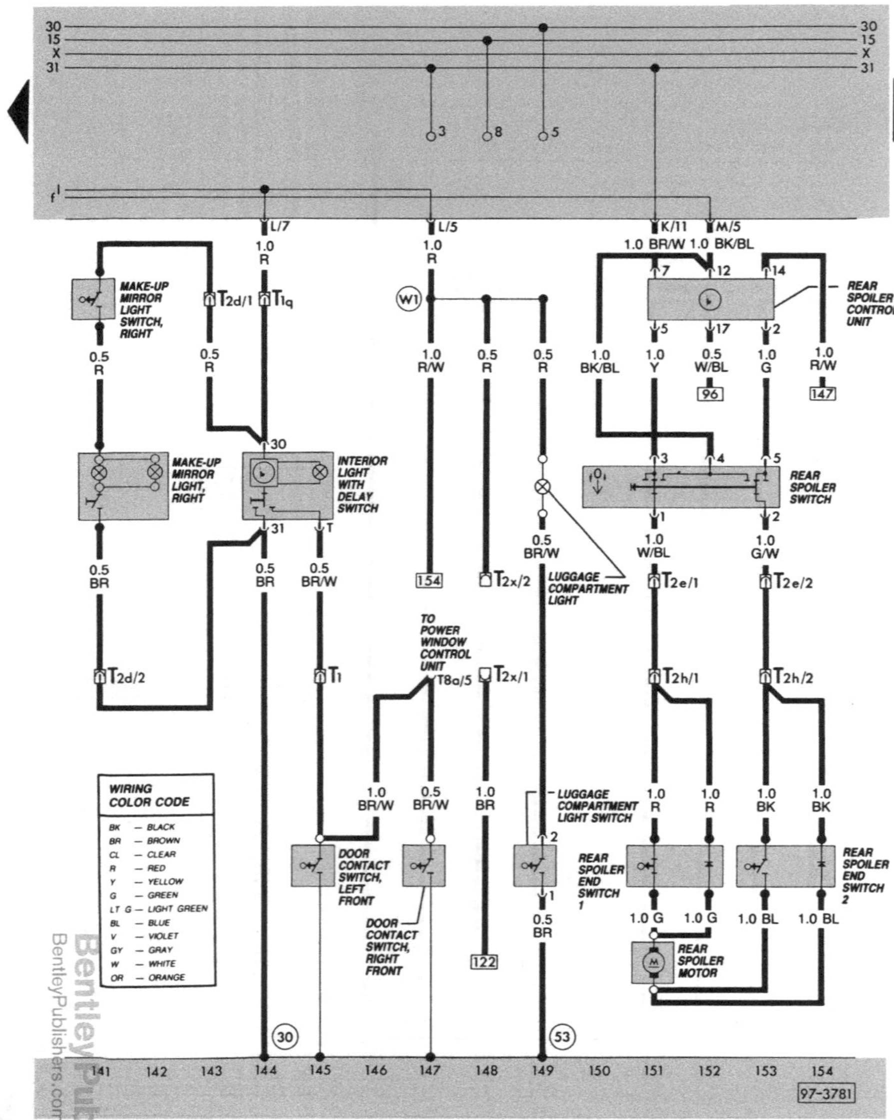

Corrado 1990

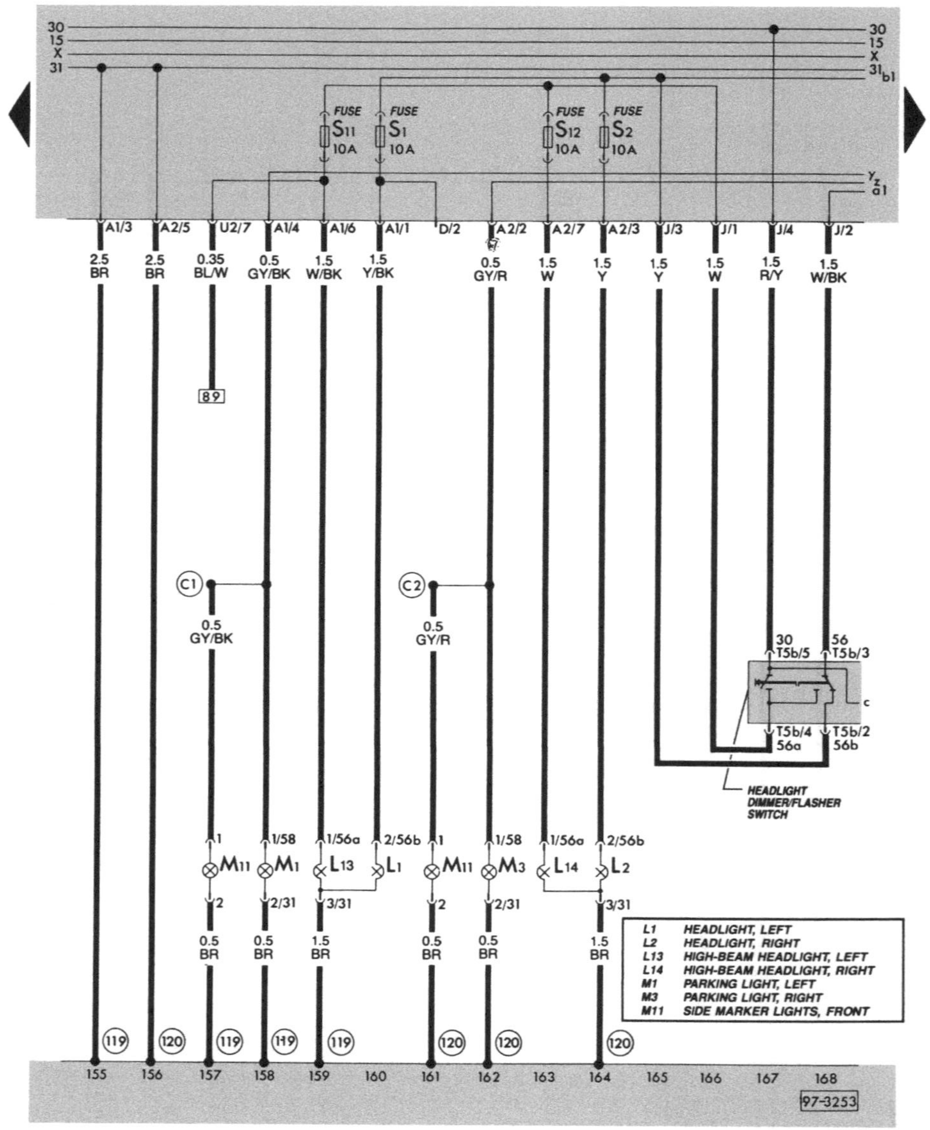

Corrado 1990

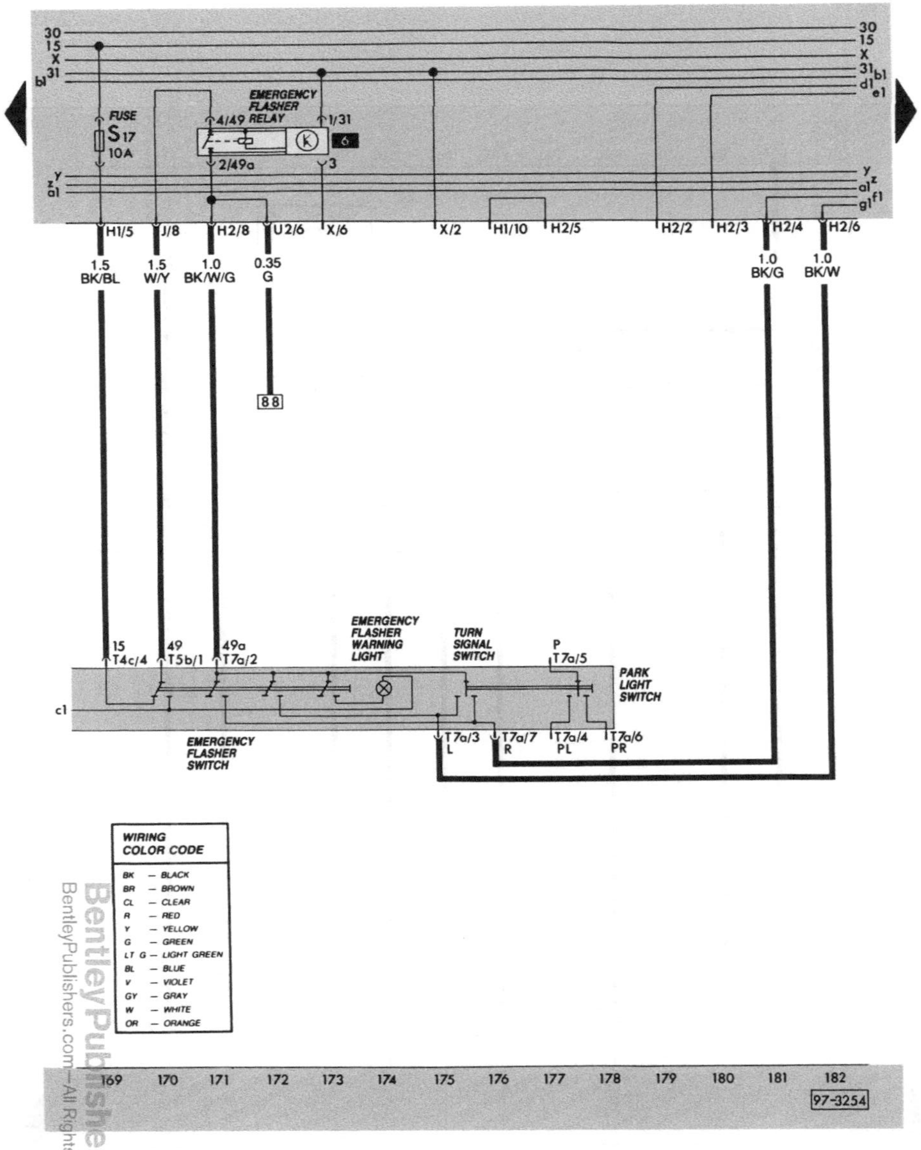
30
15
X
31
bl
d1
e1
y
z
a1
g1
f1
FUSE
S17
10A
EMERGENCY FLASHER RELAY
4/49
2/49a
1/31
3
6
H1/5
J/8
H2/8
U2/6
X/6
X/2
H1/10
H2/5
H2/2
H2/3
H2/4
H2/6
1.5 BK/BL
1.5 W/Y
1.0 BK/W/G
0.35 G
1.0 BK/G
1.0 BK/W
88
EMERGENCY FLASHER WARNING LIGHT
TURN SIGNAL SWITCH
PARK LIGHT SWITCH
EMERGENCY FLASHER SWITCH
15 T4c/4
49 T5b/1
49a T7a/2
P T7a/5
T7a/3 L
T7a/7 R
T7a/4 PL
T7a/6 PR
c1
WIRING COLOR CODE
BK — BLACK
BR — BROWN
CL — CLEAR
R — RED
Y — YELLOW
G — GREEN
LT G — LIGHT GREEN
BL — BLUE
V — VIOLET
GY — GRAY
W — WHITE
OR — ORANGE
169 170 171 172 173 174 175 176 177 178 179 180 181 182
97-3254

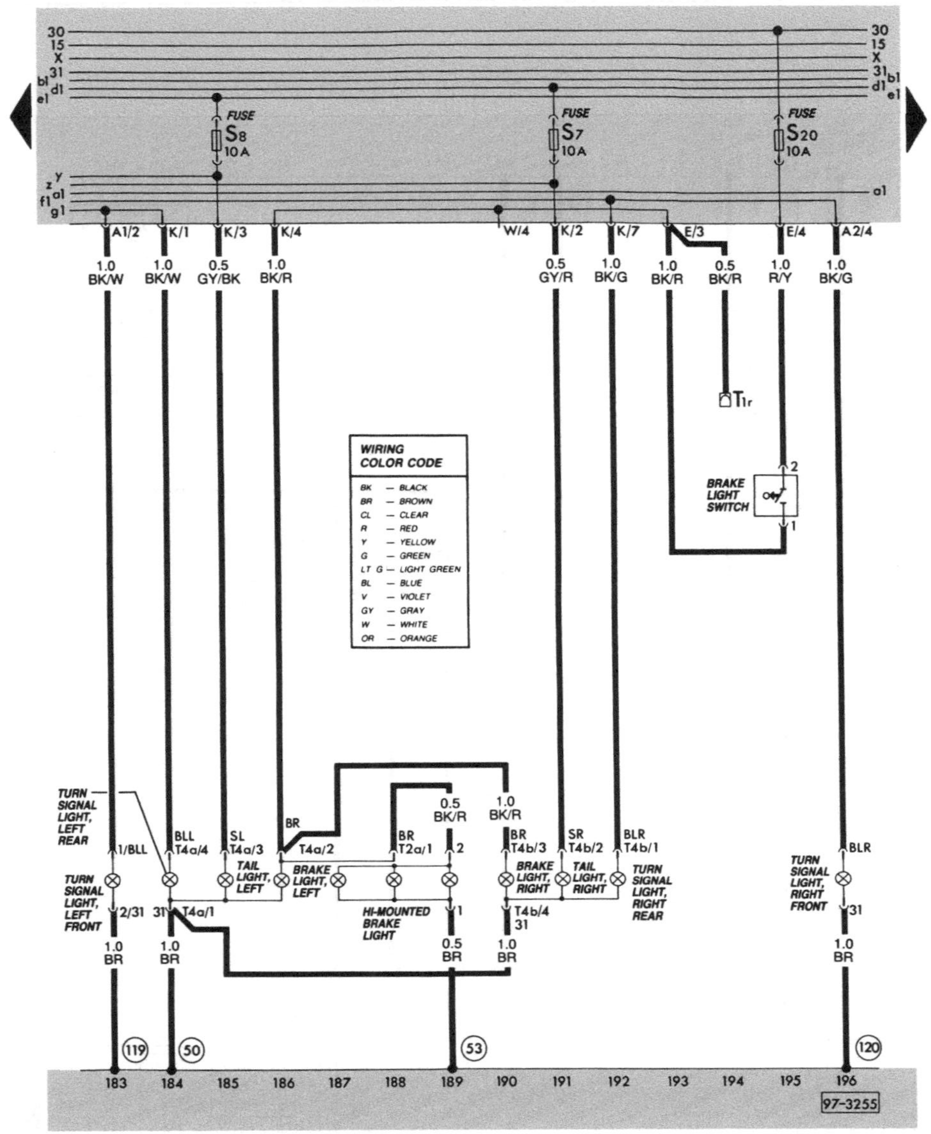
30
15
X
31
bl
d1
e1
y
z
a1
f1
g1
FUSE
S8
10A
FUSE
S7
10A
FUSE
S20
10A
A1/2
K/1
K/3
K/4
W/4
K/2
K/7
E/3
E/4
A2/4
1.0 BK/W
1.0 BK/W
0.5 GY/BK
1.0 BK/R
0.5 GY/R
1.0 BK/G
1.0 BK/R
0.5 BK/R
1.0 R/Y
1.0 BK/G
T1r
BRAKE LIGHT SWITCH
2
1
WIRING COLOR CODE
BK — BLACK
BR — BROWN
CL — CLEAR
R — RED
Y — YELLOW
G — GREEN
LT G — LIGHT GREEN
BL — BLUE
V — VIOLET
GY — GRAY
W — WHITE
OR — ORANGE
TURN SIGNAL LIGHT, LEFT REAR
TURN SIGNAL LIGHT, LEFT FRONT
1/BLL
2/31
BLL T4a/4
31 T4a/1
SL T4a/3
TAIL LIGHT, LEFT
BR T4a/2
BRAKE LIGHT, LEFT
BR T2a/1
0.5 BK/R
1.0 BK/R
HI-MOUNTED BRAKE LIGHT
BR T4b/3
T4b/4 31
BRAKE LIGHT, RIGHT
SR T4b/2
TAIL LIGHT, RIGHT
BLR T4b/1
TURN SIGNAL LIGHT, RIGHT REAR
TURN SIGNAL LIGHT, RIGHT FRONT
BLR
31
1.0 BR
0.5 BR
119
50
53
120
183 184 185 186 187 188 189 190 191 192 193 194 195 196
97-3255

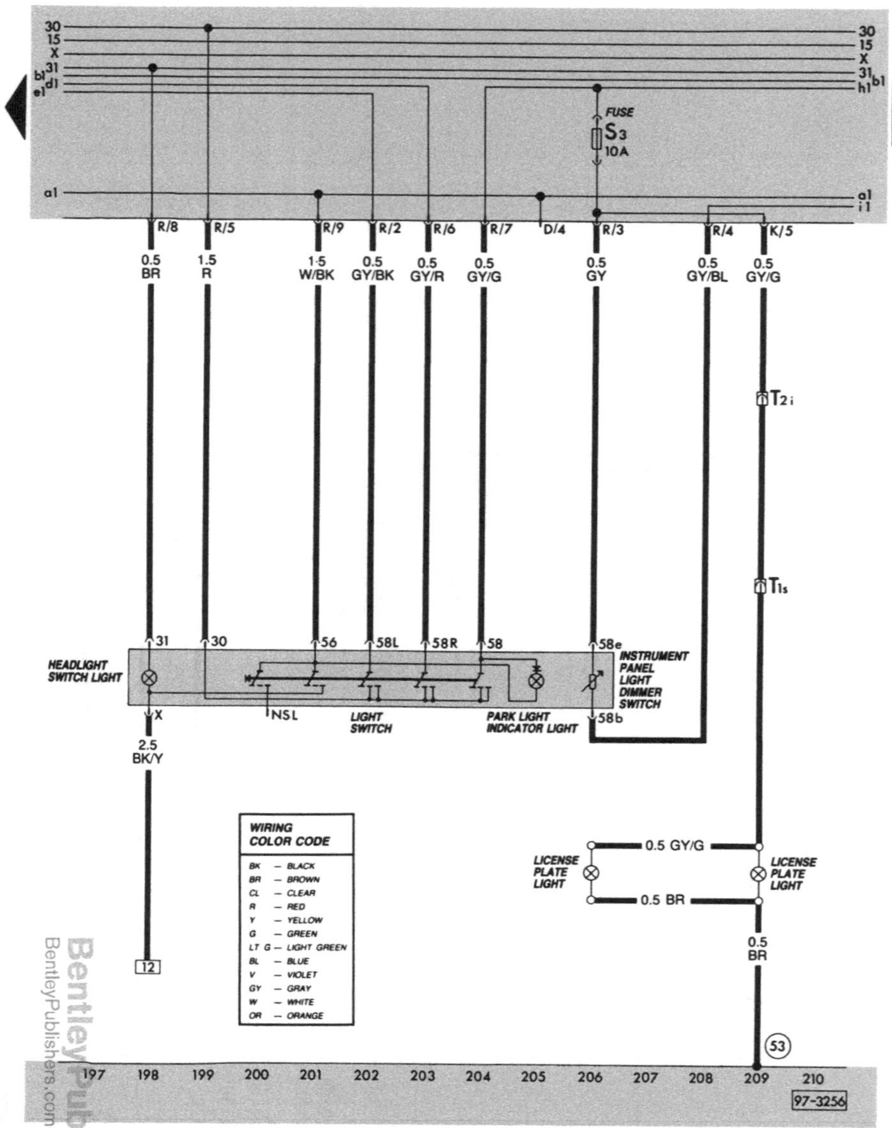
30
15
X
31
FUSE
S3
10A
R/8
R/5
R/9
R/2
R/6
R/7
D/4
R/3
R/4
K/5
0.5 BR
1.5 R
1.5 W/BK
0.5 GY/BK
0.5 GY/R
0.5 GY/G
0.5 GY
0.5 GY/BL
0.5 GY/G
T2i
T1s
HEADLIGHT SWITCH LIGHT
NSL
LIGHT SWITCH
PARK LIGHT INDICATOR LIGHT
INSTRUMENT PANEL LIGHT DIMMER SWITCH
58e
58b
2.5 BK/Y
12
LICENSE PLATE LIGHT
0.5 GY/G
0.5 BR
LICENSE PLATE LIGHT
0.5 BR
53
WIRING COLOR CODE
BK — BLACK
BR — BROWN
CL — CLEAR
R — RED
Y — YELLOW
G — GREEN
LT G — LIGHT GREEN
BL — BLUE
V — VIOLET
GY — GRAY
W — WHITE
OR — ORANGE
197 198 199 200 201 202 203 204 205 206 207 208 209 210
97-3256

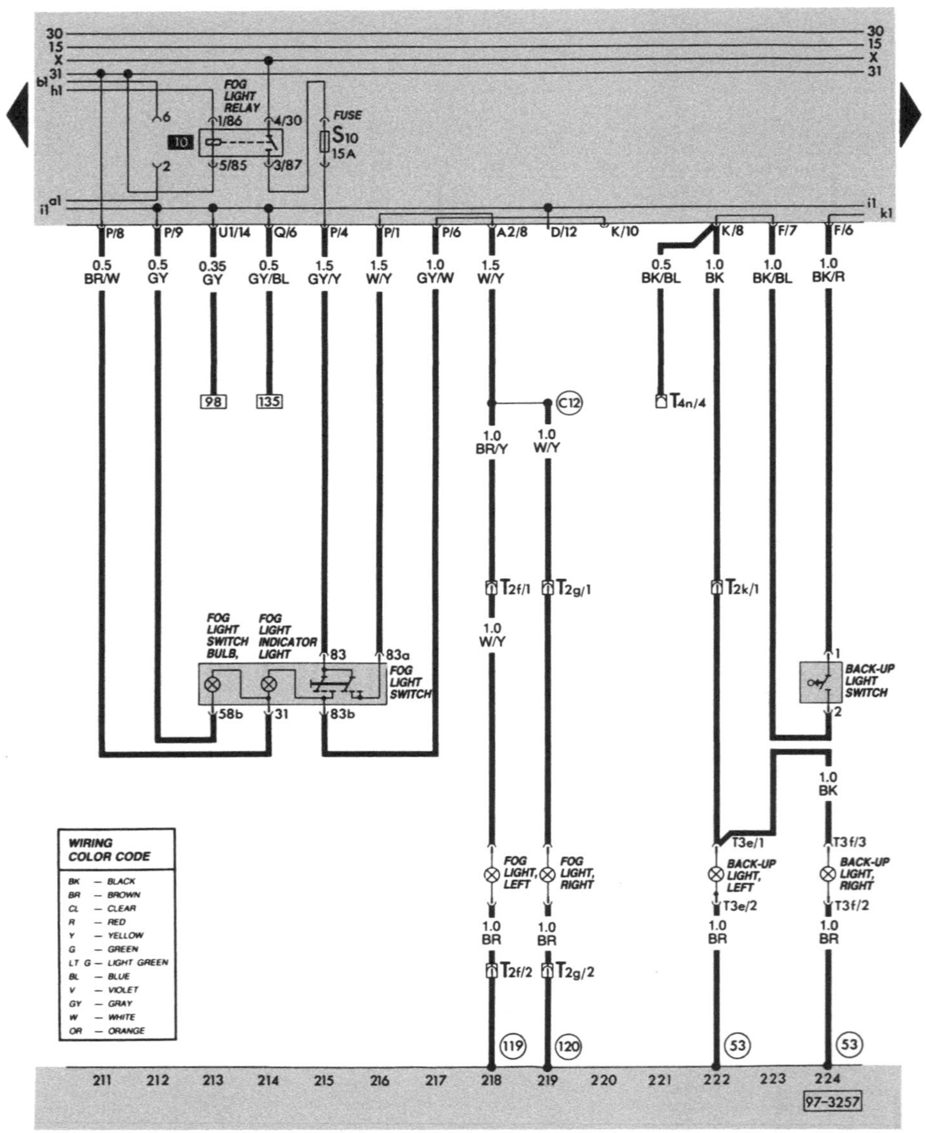
30
15
X
31
FOG LIGHT RELAY
10
1/86
4/30
5/85
3/87
FUSE
S10
15A
P/8
P/9
U1/14
Q/6
P/4
P/1
P/6
A2/8
D/12
K/10
K/8
F/7
F/6
0.5 BR/W
0.5 GY
0.35 GY
0.5 GY/BL
1.5 GY/Y
1.5 W/Y
1.0 GY/W
1.5 W/Y
0.5 BK/BL
1.0 BK
1.0 BK/BL
1.0 BK/R
98
135
C12
1.0 BR/Y
1.0 W/Y
T4n/4
T2f/1
T2g/1
T2k/1
FOG LIGHT SWITCH BULB,
FOG LIGHT INDICATOR LIGHT
FOG LIGHT SWITCH
83
83a
83b
58b
31
BACK-UP LIGHT SWITCH
1.0 BK
T3e/1
T3f/3
FOG LIGHT, LEFT
FOG LIGHT, RIGHT
BACK-UP LIGHT, LEFT
BACK-UP LIGHT, RIGHT
T3e/2
T3f/2
1.0 BR
T2f/2
T2g/2
119
120
53
WIRING COLOR CODE
BK — BLACK
BR — BROWN
CL — CLEAR
R — RED
Y — YELLOW
G — GREEN
LT G — LIGHT GREEN
BL — BLUE
V — VIOLET
GY — GRAY
W — WHITE
OR — ORANGE
211 212 213 214 215 216 217 218 219 220 221 222 223 224
97-3257

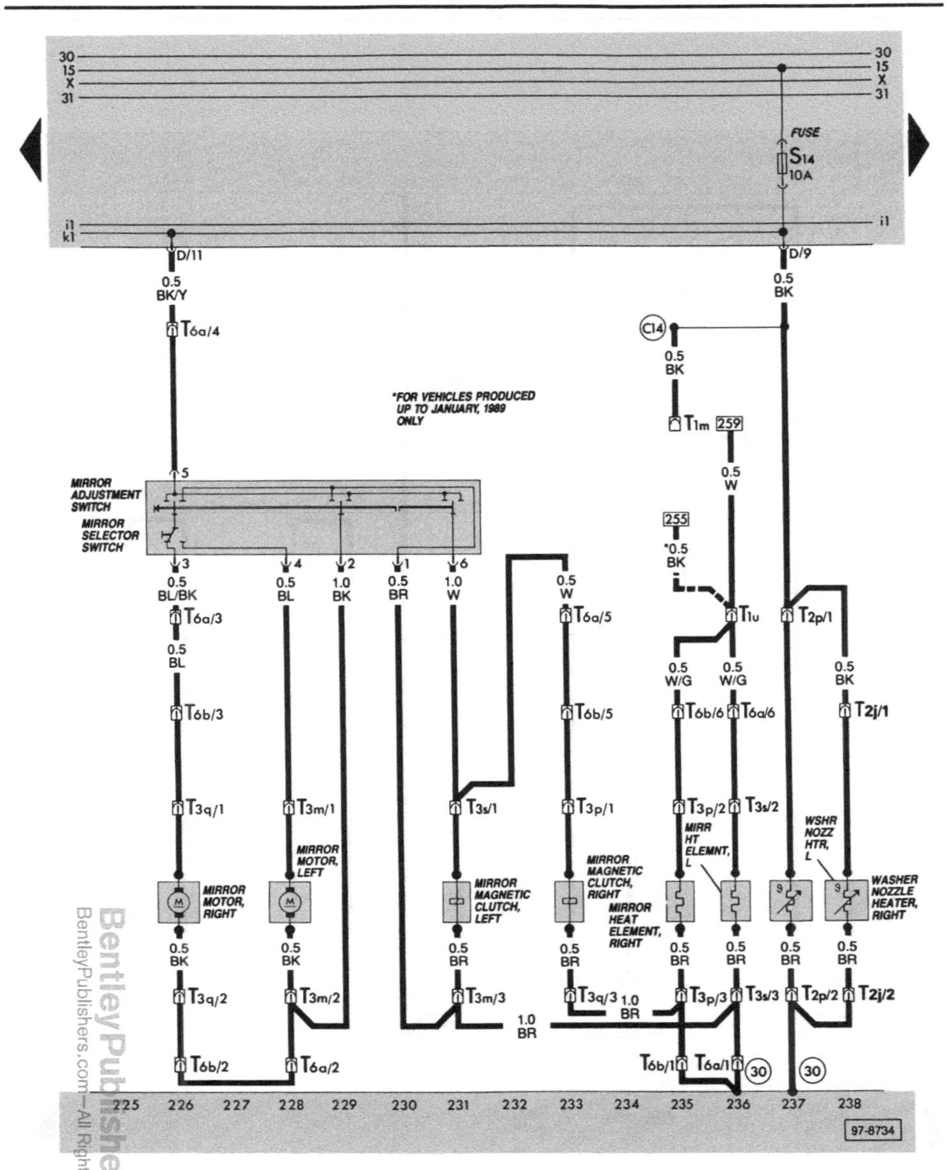

Corrado 1990

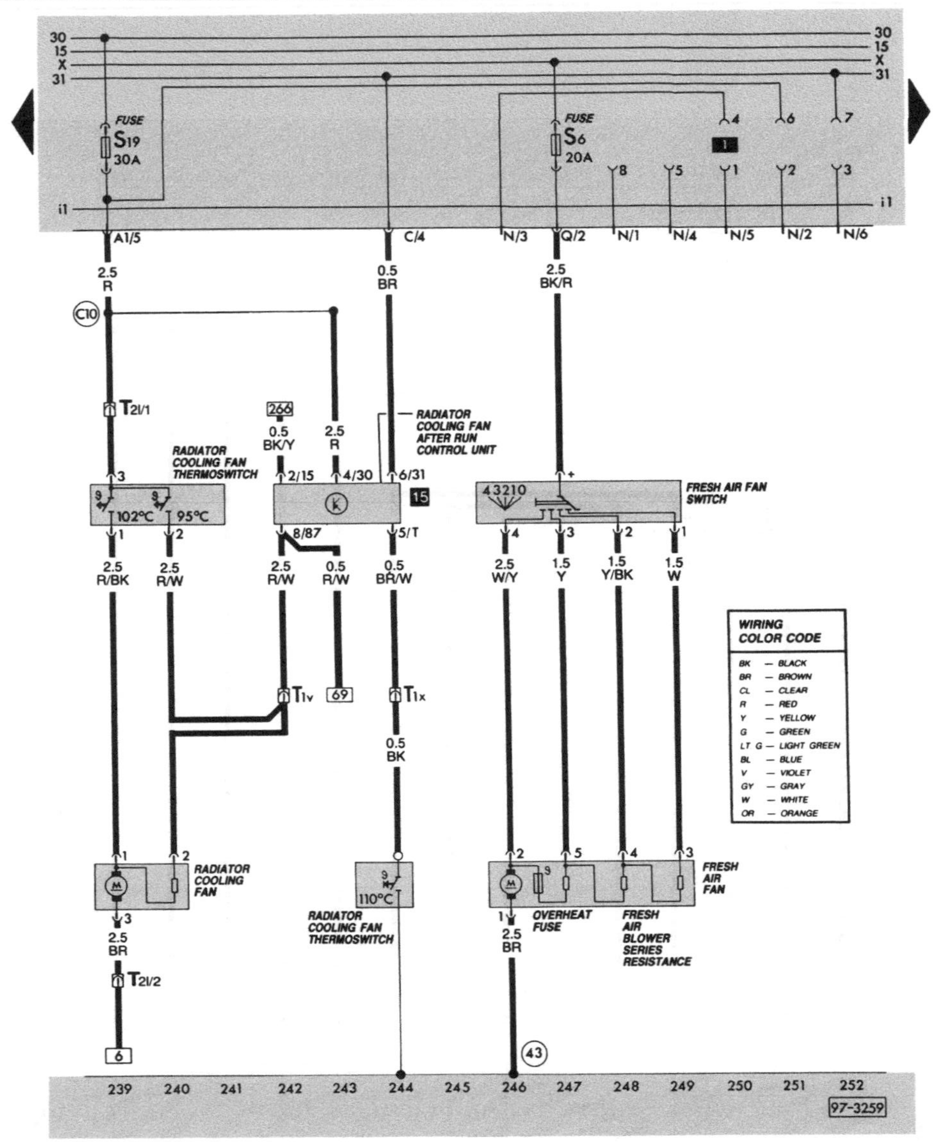

Corrado 1990

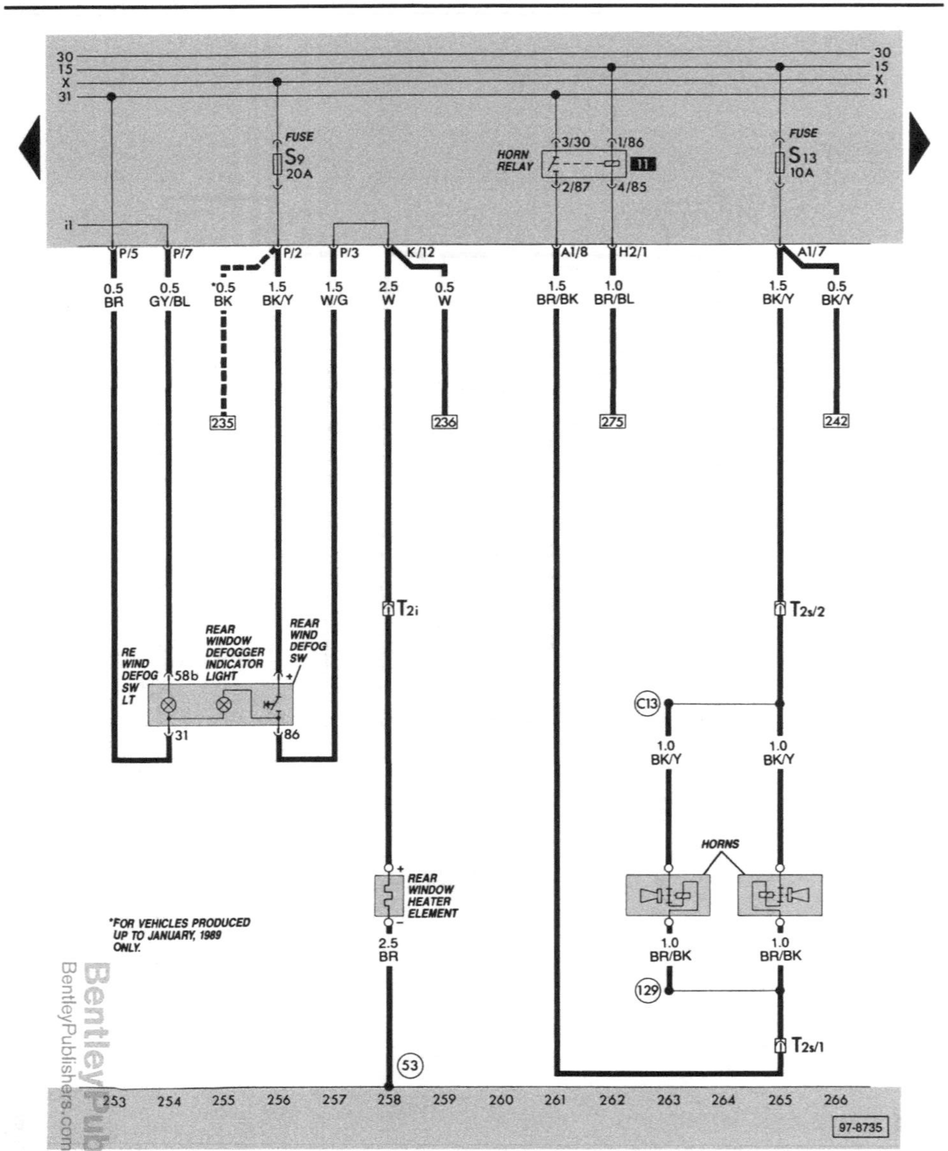

Corrado 1990

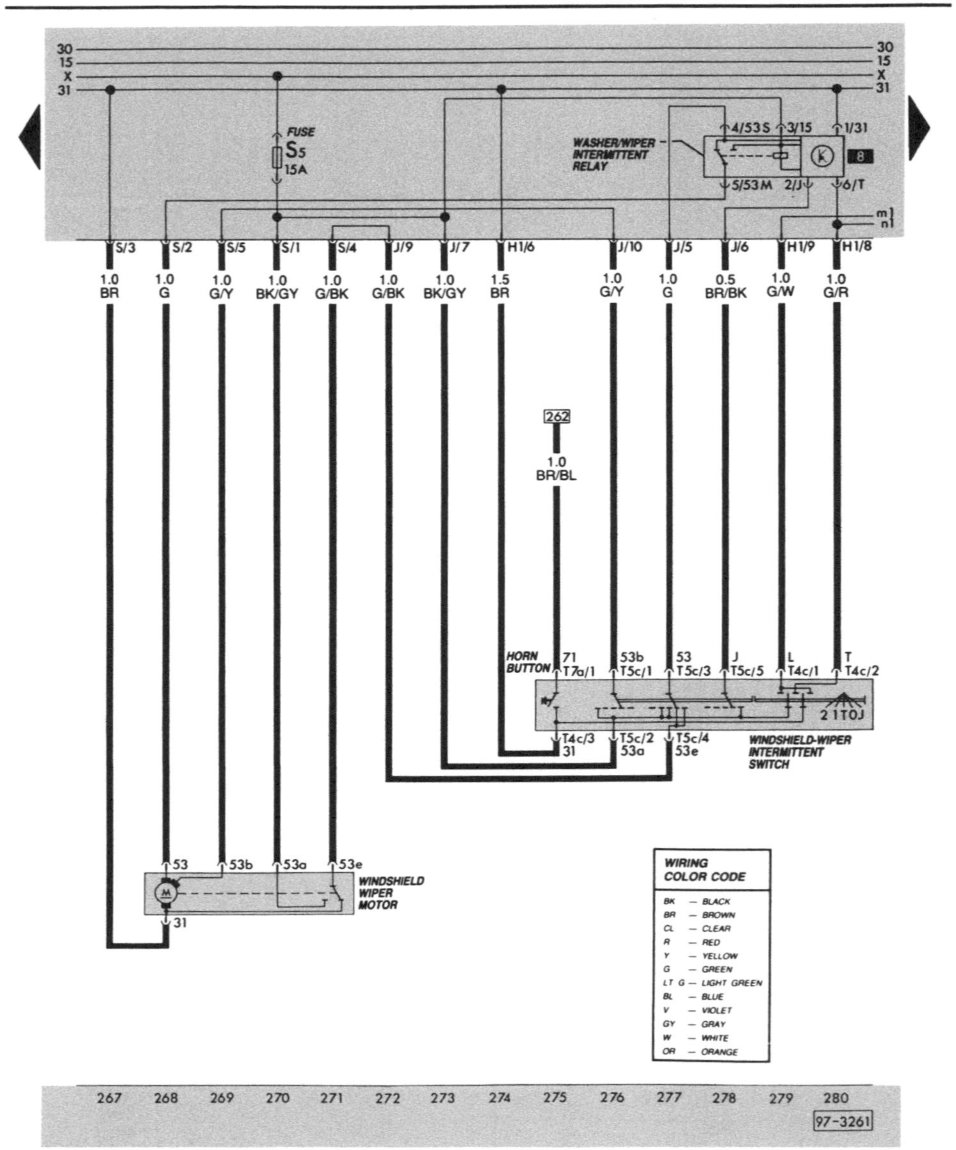

Corrado 1990

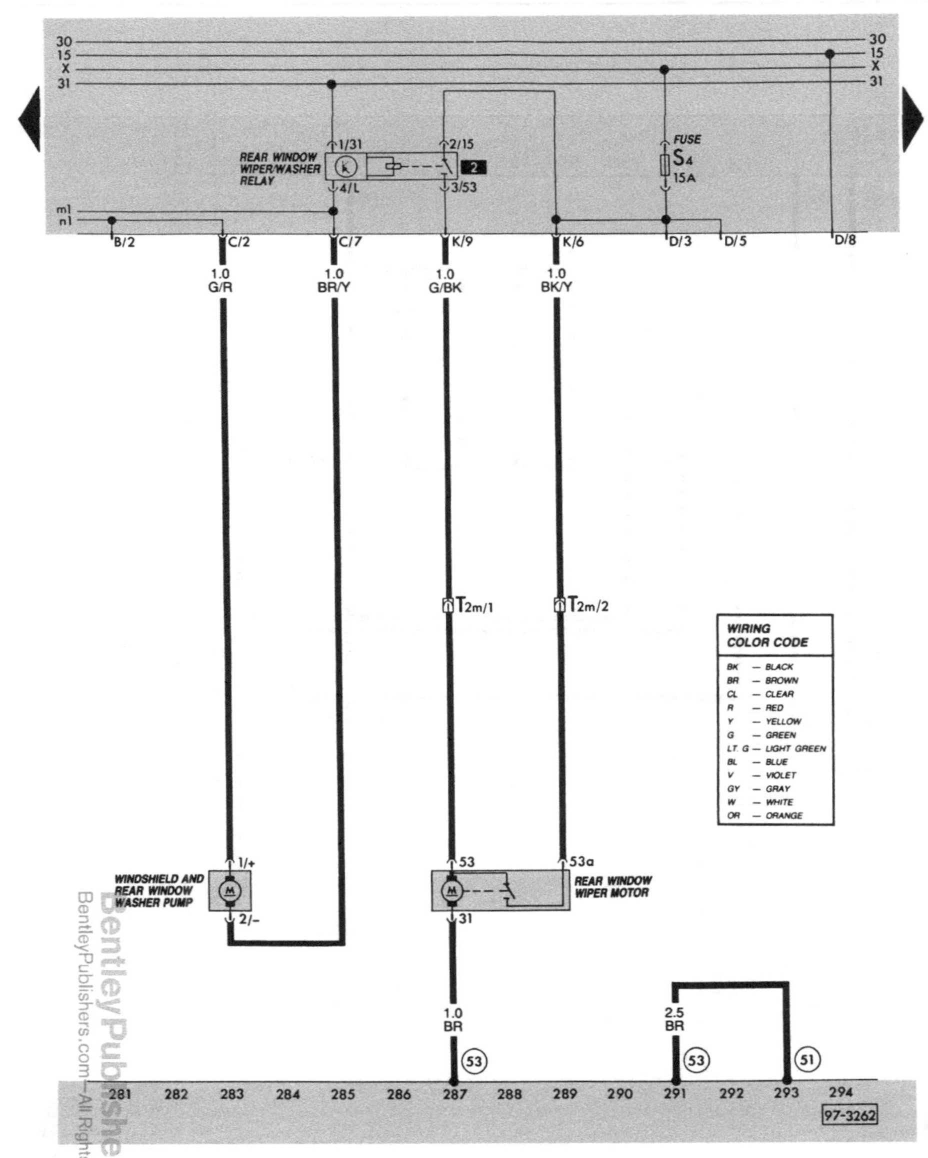
30
15
X
31
REAR WINDOW WIPER/WASHER RELAY
1/31
4/L
2/15
3/53
2
FUSE
S4
15A
m1
n1
B/2
C/2
C/7
K/9
K/6
D/3
D/5
D/8
1.0 G/R
1.0 BR/Y
1.0 G/BK
1.0 BK/Y
T2m/1
T2m/2
WIRING COLOR CODE
BK — BLACK
BR — BROWN
CL — CLEAR
R — RED
Y — YELLOW
G — GREEN
LT. G — LIGHT GREEN
BL — BLUE
V — VIOLET
GY — GRAY
W — WHITE
OR — ORANGE
1/+
2/–
WINDSHIELD AND REAR WINDOW WASHER PUMP
53
53a
31
REAR WINDOW WIPER MOTOR
1.0 BR
2.5 BR
53
53
51
281 282 283 284 285 286 287 288 289 290 291 292 293 294
97-3262

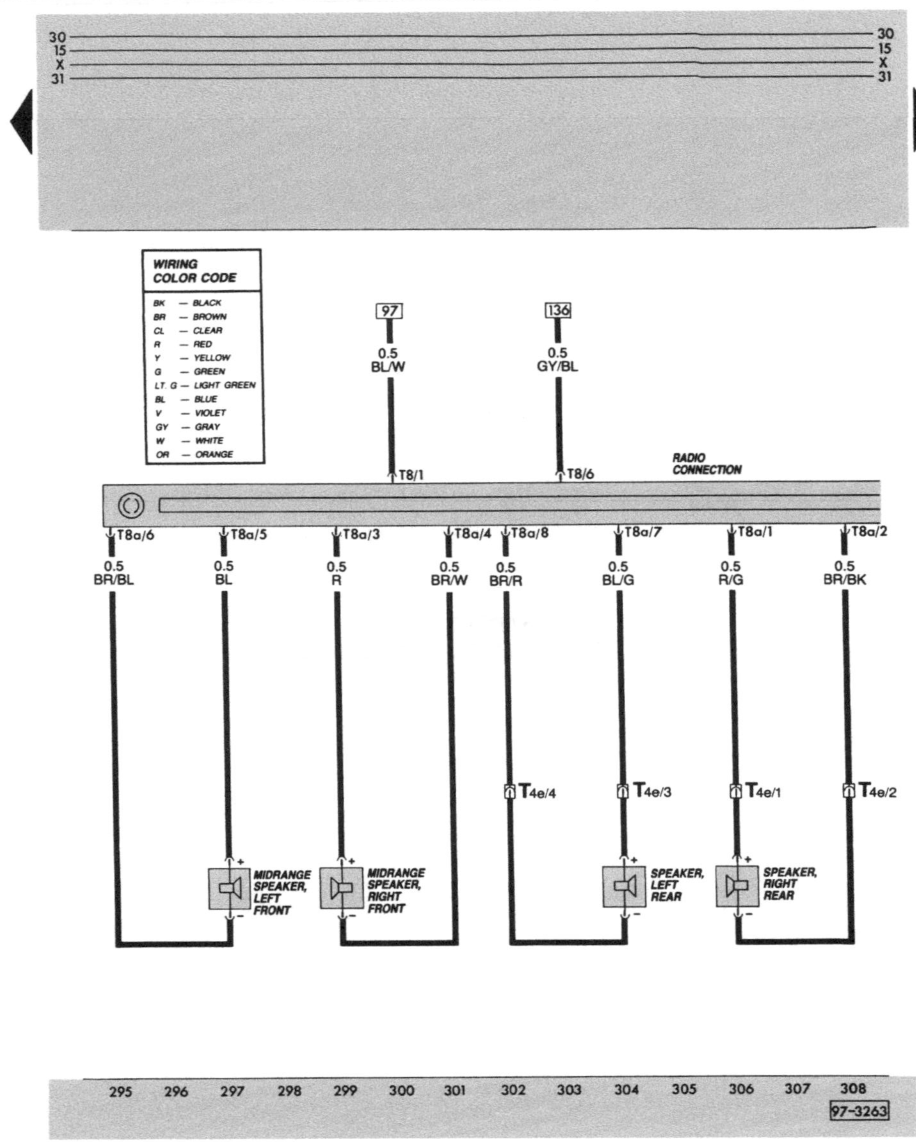
30
15
X
31
WIRING COLOR CODE
BK — BLACK
BR — BROWN
CL — CLEAR
R — RED
Y — YELLOW
G — GREEN
LT. G — LIGHT GREEN
BL — BLUE
V — VIOLET
GY — GRAY
W — WHITE
OR — ORANGE
97
0.5 BL/W
136
0.5 GY/BL
T8/1
T8/6
RADIO CONNECTION
T8a/6
T8a/5
T8a/3
T8a/4
T8a/8
T8a/7
T8a/1
T8a/2
0.5 BR/BL
0.5 BL
0.5 R
0.5 BR/W
0.5 BR/R
0.5 BL/G
0.5 R/G
0.5 BR/BK
T4e/4
T4e/3
T4e/1
T4e/2
MIDRANGE SPEAKER, LEFT FRONT
MIDRANGE SPEAKER, RIGHT FRONT
SPEAKER, LEFT REAR
SPEAKER, RIGHT REAR
295 296 297 298 299 300 301 302 303 304 305 306 307 308
97-3263

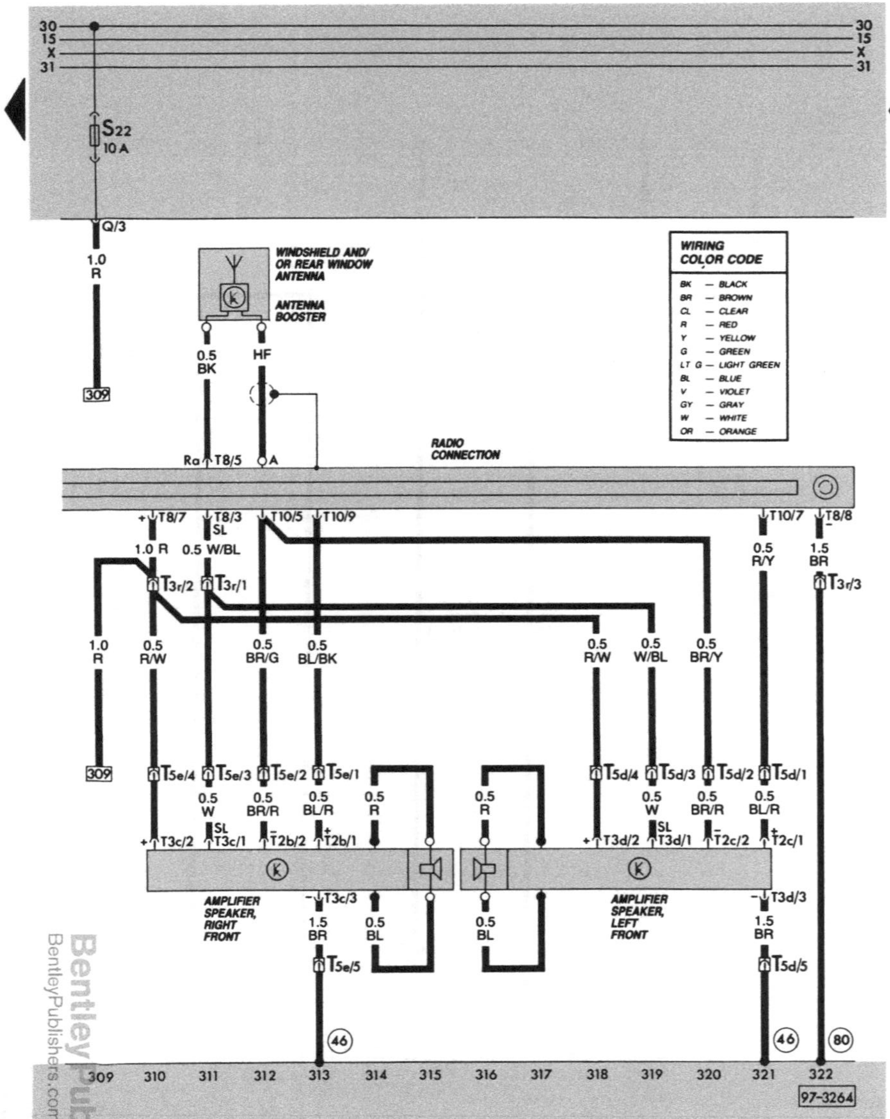

THIS PAGE INTENTIONALLY LEFT BLANK

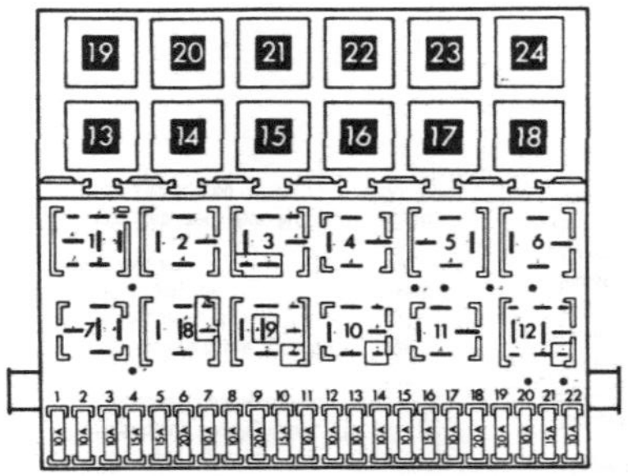

97-3000

Fuse arrangement

according to the numbers on fuse panel from left to right:

		Amp.
1	Headlight, low beam, left	10
2	Headlight, low beam, right	10
3	Instrument panel lights, license plate lights	10
4	Rear window wiper, sunroof	15
5	Wiper/washer, front; rear window washer	15
6	Fresh air blower, A/C	20
7	Tail lights, side marker light, right	10
8	Tail lights, side marker light, left	10
9	Rear window defogger, heated mirrors	20
10	Fog lights	15
11	Headlight, high beam, left, high beam indicator light	10
12	Headlight, high beam, right	10
13	Horn, radiator cooling fan run-on	10
14	Back-up lights, heated windshield washer nozzles, power mirrors	10
15	Electronic ignition, idle stabilizer, deceleration fuel shut off	10
16	Dash warning/indicator lights, glove box light, multi-function indicator, cassette storage light, rear spoiler control unit	15
17	Emergency flasher switch, turn signals	10
18	Fuel pump, oxygen sensor	20
19	Radiator cooling fan, A/C	30
20	Brake lights, cruise control	10
21	Interior lights, luggage compartment light, cigarette lighter, digital clock, rear spoiler adjustment, engine diagnostic, central locking, multi-function indicator	15
22	Radio	10

Relay location on fuse/relay panel

1 — A/C relay
2 — Rear window wiper/washer relay (72)*
3 — Digifant control unit relay (32)*
4 — Load reduction relay (18)*
5 — Open
6 — Emergency flasher/turn signal relay (21)*
7 — Open
8 — Intermittent wiper/washer relay (19)*
9 — Seat belt warning system control unit (4)*
10 — Fog light relay (53)*
11 — Horn relay (53)*
12 — Fuel pump relay (80)*

Separate relays above fuse/relay panel

13 — ABS hydraulic pump relay
14 — ABS relay
15 — Radiator cooling fan after run control unit (31)*
16 — Fuel pump after run control unit (90)*
17 — Heated seat relay, driver
18 — Heated seat relay, passenger
19 — ABS hydraulic pump fuse
20 — ABS valves fuse
21 — Power window fuse
22 — ABS valves fuse
23 — Automatic shoulder belt system fuse, S51 (USA only)
24 — Open

*Numbers in parentheses indicate production control number stamped on relay housing.

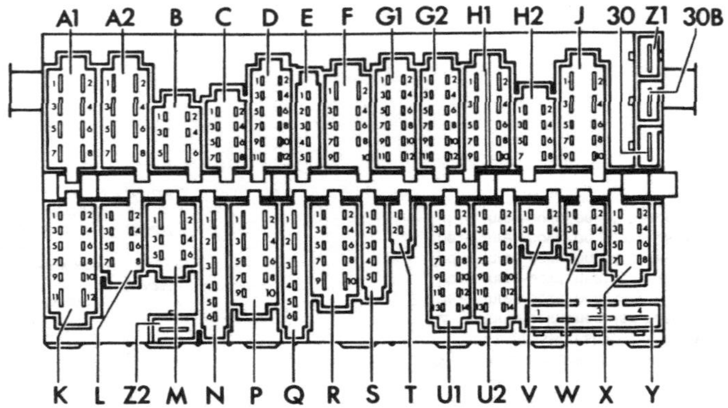

97-3001

Connections and plugs on fuse/relay panel

A1 — Headlight wiring harness (yellow)
A2 — Headlight wiring harness (yellow)
B — Open
C — Headlight wiring harness (yellow)
D — Optional equipment wiring harness (green)
E — Instrument wiring harness (green)
F — Engine compartment wiring harness, right (white)
G1 — Engine compartment wiring harness, right (white)
G2 — Engine compartment wiring harness, right (white)
H1 — Steering column switch wiring harness (red)
H2 — Steering column switch wiring harness (red)
J — Steering column switch wiring harness (red)
K — Rear wiring harness (black)
L — Rear wiring harness (black)
M — Rear wiring harness (black)
N — A/C wiring harness (green)
P — Rear window defroster and fog light wiring harness (blue)
Q — Instrument wiring harness (blue)
R — Light switch wiring harness (blue)
S — Engine compartment wiring harness, right (white)
T — Two-point connector (green)
U1 — Instrument cluster wiring harness (blue)
U2 — Instrument cluster wiring harness (blue)
V — Multi-function indicator wiring harness (green)
W — Six-point connector (green)
X — Warning lamp wiring harness (green)
Y — Single connector, terminal 30
Z1 — Single connector
Z2 — Single connector, terminal 31
30 — Single connector, terminal 30
30B — Single connector
31 — Single connector, terminal 31

Description	Current track
Alternator, C	2
Alternator warning light, K2	94
Back-up light switch, F4	238
Back-up light, left, M16	236
Back-up light, right, M17	238
Battery, A	4
Brake and parking brake indicator light, K7	135-137
Brake fluid level warning contact, F34	154
Brake light switch, F	205
Brake/tail light indicator lights, left, M21	200, 201
Brake/tail light indicator lights, right, M22	206, 207
Cigarette lighter, U1	142-143
Cigarette lighter light, L28	143
CO potentiometer, G74	41
Coolant low level sensor, G32	75
Coolant overheat warning light, K28	80
Coolant temperature gauge, G3	81
Coolant temperature sending unit, G2	72
Coolant temperature sensor, G62	42
Crankcase ventilation heating element, N79	63
Diagnostic connectors	165, 167
Digifant control unit, J169	33-54
Digifant control unit relay, J176	51-53
Digital display light, L75	105, 116
Door contact switch, left front, F2	127
Door contact switch, right front, F3	128
Door latch switch, left, F143	180
Emergency flasher relay, J2	189-191
Emergency flasher switch, E3	185-194
Emergency flasher warning light, K6	193
Engine oil pressure gauge, G11	150-152
Engine oil pressure gauge light, L27	152
Engine oil pressure sensor, G10	150
Engine oil pressure switch (0.3 bar), F22	73
Engine oil pressure switch (1.8 bar), F1	74
Engine oil pressure warning light, K3	76
Engine oil temperature sensor, G8	71
Fog light indicator light, K17	228
Fog light relay, J5	227-228

Description	Current track
Fog light switch, E23	227-230
Fog light switch illumination, L40	227
Fog light, left, L22	233
Fog light, right, L23	234
Fresh air blower, V2	22
Fresh air blower series resistance, N23	23-27
Fresh air blower switch, E9	24-27
Fresh air control lever light, L16	28
Fuel gauge, G1	88
Fuel gauge sending unit, G	68
Fuel pump, G6	69
Fuel pump afterrunning control unit, J206	64-66
Fuel pump afterrunning pressure switch, F130	65
Fuel pump relay, J17	60-61
Full throttle switch, F81	48
Fuse, S19	16
Fuse, S6	27
Fuse, S15	64
Fuse, S18	68
Fuse, S21	142
Fuse, S16	145
Fuse, S11	171
Fuse, S1	172
Fuse, S2	175
Fuse, S12	176
Fuse, S17	187
Fuse, S8	200
Fuse, S20	205
Fuse, S7	207
Fuse, S3	219
Fuse, S10	229
Fuse, S14	234
Fuse, S9	242
Fuse, S13	252
Fuse, S5	256
Fuse, S22	268
Fuse, S4	280
Glove compartment light, W6	148
Hall generator, G40	33-35
Headlight dimmer/flasher switch, E4	185-186
Headlight high beam indicator light, K1	98
Headlight switch light, L9	212
Headlight, left, L1	172
Headlight, right, L2	175
High-beam headlight, left, L13	171
High-beam headlight, right, L14	176
High-mount brake light, M25	204
Horn button, H	261
Horn relay, J4	249-250
Horns, H1	250, 252
Idle stabilizer valve, N71	55-56
Idle switch, F60	50

Description	Current track
Ignition coil, N	29-31
Ignition distributor, O	29-31
Ignition/starter switch, D	9-13
Injector, cyl. 1, N30	56
Injector, cyl. 2, N31	55
Injector, cyl. 3, N32	54
Injector, cyl. 4, N33	53
Instrument panel light, L10	100
Instrument panel light dimmer switch, E20	219
Intake air temperature sensor, G42	40
Interior light with delay switch, W15	127-128
Knock sensor, G61	38-40
License plate light, X	219, 221
Light switch, E1	212-219
Load reduction relay, J59	10-11
Luggage compartment light, W3	166
Luggage compartment light switch, F5	166
Make-up mirror light switch, right, F148	132
Make-up mirror light, right, W14	131-132
Map/reading light, right front, W13	129
Multi-function indicator (MFI), J119	105-116
Multi-function indicator memory switch, E109	118-121
Multi-function indicator recall button, E86	118-121
Multi-function indicator vacuum sensor, G55	107-109
Odometer display, Y4	113-116
Oil pressure warning and coolant temperature/low level indicator control unit, J252	76-80
Outside air temperature sensor, G17	125
Overheat fuse, S24	25
Oxygen sensor, G39	45-46
Park light indicator light, K4	218
Parking brake warning light switch, F9	140
Parking light, left, M1	170
Parking light, right, M3	177
Radiator cooling fan, V7	16-17
Radiator cooling fan after run control unit, J138	18-21
Radiator cooling fan thermoswitch, F18	15-17
Radiator cooling fan thermoswitch, F87	20
Radio, R	269
Rear spoiler control unit, J223	157-164

Description	Current track
Rear spoiler end switch 1, F135	159-160
Rear spoiler end switch 2, F136	163-164
Rear spoiler motor, V52	159
Rear spoiler switch, E127	158-163
Rear window defogger indicator light, K10	241
Rear window defogger switch, E15	240-242
Rear window defogger switch light, L39	240
Rear window heater element, Z1	245
Rear window wiper motor, V12	277-279
Rear window wiper/washer relay, J30	276-277
Seat belt switch, left, E24	180
Seat belt warning control unit, J34	178-182
Seat belt warning light, K19	93
Side marker lights, front, M11	169, 178
Spark plug connectors, P	29-31
Spark plugs, Q	29-31
Speed sensor (speedometer), G22	117-118
Speedometer, G21	110-112
Starter, B	6-7
Tachometer, G5	87
Turn signal indicator light, left, K65	96
Turn signal indicator light, right, K94	97
Turn signal light, left front, M5	197
Turn signal light, left rear, M6	199
Turn signal light, right front, M7	209
Turn signal light, right rear, M8	208
Turn signal switch, E2	194
Voltage regulator, C1	2
Voltage stabilizer, J6	89-90
Voltmeter, G14	145-146
Voltmeter light, L25	146
Warning buzzer, H3	79
Washer nozzle heater, left, Z20	223
Washer nozzle heater, right, Z21	224
Washer/wiper intermittent relay, J31	262-266
Windshield and rear window washer pump, V59	275
Windshield wiper motor, V	254-257
Windshield wiper intermittent switch, E22	261-266

Wire connectors

T1 — single, near starter
T1a — single, behind fuse/relay panel
T1b — single, behind fuse/relay panel (A/C connector)
T1c — single, behind fuse/relay panel
T1d — single, behind fuse/relay panel
T1d1— single, above fuse/relay panel
T1e — single, on left A-pillar, near fuse/relay panel
T1f — single, behind fuse/relay panel
T1g — single, behind fuse/relay panel
T1h — single, behind fuse/relay panel
T1i — single, behind fuse/relay panel
T1j — single, in rear lid
T1k — single, behind fuse/relay panel
T1l — single, behind fuse/relay panel
T1m— single, behind fuse/relay panel
T1n — single, behind fuse/relay panel
T1o — single, on left back-up light
T1p — single, on left back-up light
T1r — single, behind fuse/relay panel
T2 — double, above fuse/relay panel
T2a — double, on fresh air blower series resistance
T2a1— double, below driver's seat
T2b — double, behind instrument panel, center
T2c — double, near intake manifold
T2d — double, on left A-pillar, near fuse/relay panel
T2e — double, on upper right A-pillar
T2f — double, in luggage compartment, left
T2g — double, on rear lid
T2h — double, behind fuse/relay panel
T2i — double, in luggage compartment, left
T2j — double, on engine hood
T2k — double, on engine hood
T2l — double, near left headlight
T2m— double, near right headlight
T2n — double, in luggage compartment, left
T2o — double, near transmission
T2p — double, near right headlight
T2q — double, in luggage compartment, left
T2r — double, on left parking light
T2s — double, on left front side marker light
T2t — double, on right parking light
T2u — double, on right front side marker light
T2v — double, on left taillight
T2w— double, on high mount brake light
T2x — double, black, diagnostic connector, below transmission shift lever cover
T3 — three-point, near left headlight
T3a — three-point, near starter
T3b — three-point, near throttle valve housing
T3d — three-point, on left back-up light
T3e — three-point, on right back-up light
T3f — three-point, on left headlight

Wire connectors

T3g — three-point, on right headlight
T4 — four-point, near exhaust manifold
T4a — four-point, near steering column
T4d — four-point, behind steering column switch cover
T4e — four-point, on left taillight
T4f — four-point, on right taillight
T4i — four-point, behind fuse/relay panel
T5 — five-point, on fresh air blower series resistance
T5a — five-point, above fuse/relay panel
T5b — five-point, behind steering column switch cover
T5c — five-point, behind steering column switch cover
T5d — five-point, on brake/parking brake indicator light
T7 — seven-point, behind steering column switch cover
T8 — eight-point, on radio
T25 — twenty five-point, on Digifant control unit
T28 — twenty eight-point, on instrument cluster
T35 — junction, terminal 15, behind fuse/relay panel
T43 — junction, speed signal connector, behind fuse/relay panel

Ground connections

(1) — battery to body ground strap
(2) — transmission to body ground strap
(18) — on engine block
(29) — near exhaust manifold
(30) — beside fuse/relay panel
(43) — on lower right A - pillar
(50) — in luggage compartment, left
(53) — on rear lid, right
(80) — in instrument panel wiring harness
(86) — in rear wiring harness
(94) — in Digifant wiring harness
(98) — in rear lid wiring harness
(119) — in headlight wiring harness
(120) — in headlight wiring harness
(125) — in headlight wiring harness
(129) — in horn wiring harness
(159) — in Digifant wiring harness

Welded wiring harness points

(A11) — plus connection (58b), in instrument panel wiring harness
(A15) — plus connection (15), in instrument panel wiring harness
(B6) — wire connection, in steering column switch wiring harness
(C1) — plus connection (58l), in left front wiring harness
(C2) — plus connection (58r), in left front wiring harness
(C10) — plus connection (30), in headlight wiring harness
(C11) — wire connection, in radiator cooling fan wiring harness
(C12) — plus connection, in headlight wiring harness
(C13) — plus connection, in horn wiring harness
(C14) — plus connection (15), in heated windshield washer nozzles wiring harness
(C15) — wire connection, in headlight wiring harness
(C16) — wire connection, in headlight wiring harness
(G1) — plus connection, in Digifant wiring harness
(G3) — plus connection, in injector valve wiring harness
(G4) — wire connection, in injector valve wiring harness
(W9) — plus connection (30), in rear wiring harness
(W10) — plus connection (15), in rear wiring harness

Wire Size

Wiring Diagrams identify wires by the metric wire size. Metric wires sizes indicate cross-sectional area in square millimeters (mm²). The chart below lists metric wire sizes and their approximate equivalents in American Wire Gauge (AWG) sizes.

Metric size — cross section (mm²)	American Wire Gauge Size (AWG)
0.35	22
0.5	20
0.75	18
1.0	16
1.5	14
2.5	12
4.0	10
6.0	8
16.0	4
25.0	2
35.0	2

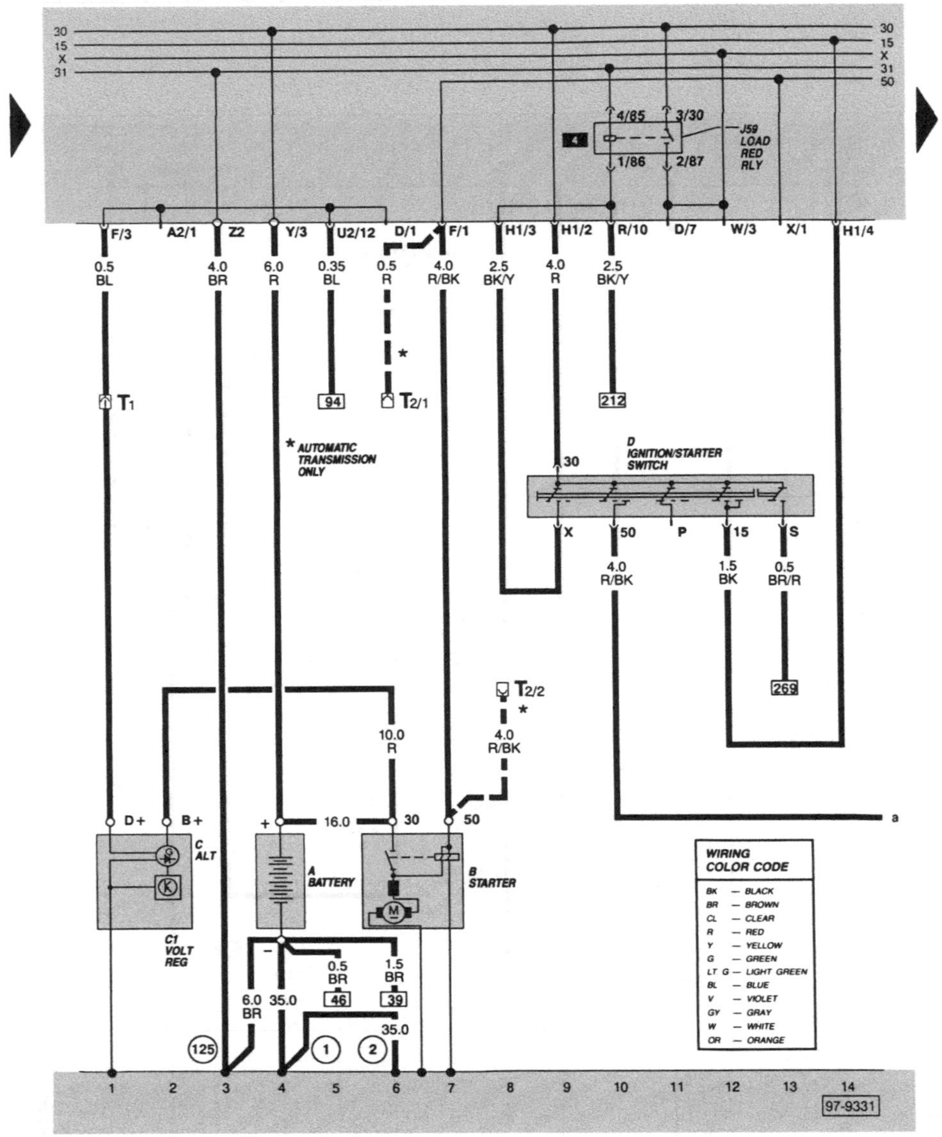

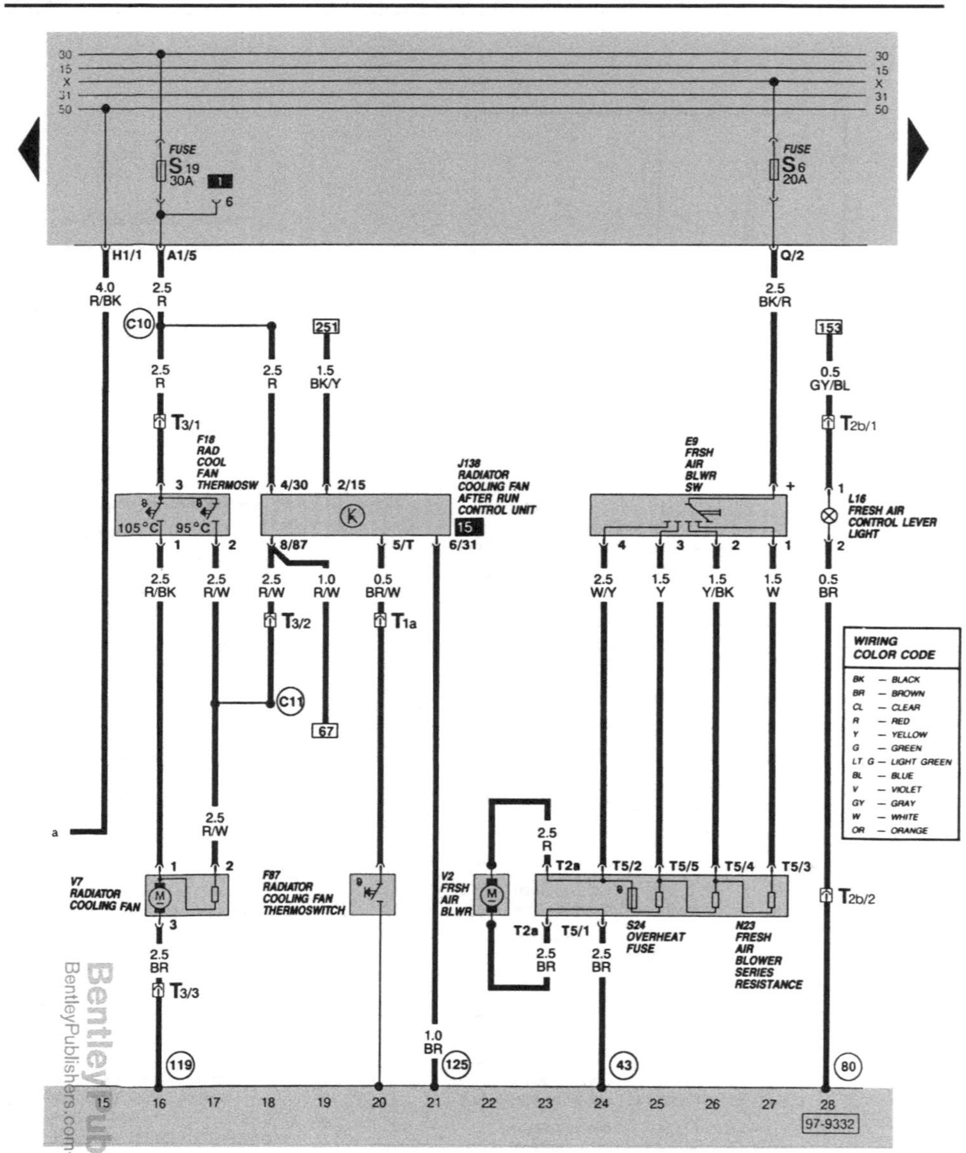
FUSE
S19
30A
FUSE
S6
20A
H1/1
A1/5
Q/2
C10
C11
251
153
67
F18
RAD
COOL
FAN
THERMOSW
105°C
95°C
J138
RADIATOR
COOLING FAN
AFTER RUN
CONTROL UNIT
E9
FRSH
AIR
BLWR
SW
L16
FRESH AIR
CONTROL LEVER
LIGHT
V7
RADIATOR
COOLING FAN
F87
RADIATOR
COOLING FAN
THERMOSWITCH
V2
FRSH
AIR
BLWR
S24
OVERHEAT
FUSE
N23
FRESH
AIR
BLOWER
SERIES
RESISTANCE
WIRING
COLOR CODE
BK – BLACK
BR – BROWN
CL – CLEAR
R – RED
Y – YELLOW
G – GREEN
LT G – LIGHT GREEN
BL – BLUE
V – VIOLET
GY – GRAY
W – WHITE
OR – ORANGE
97-9332

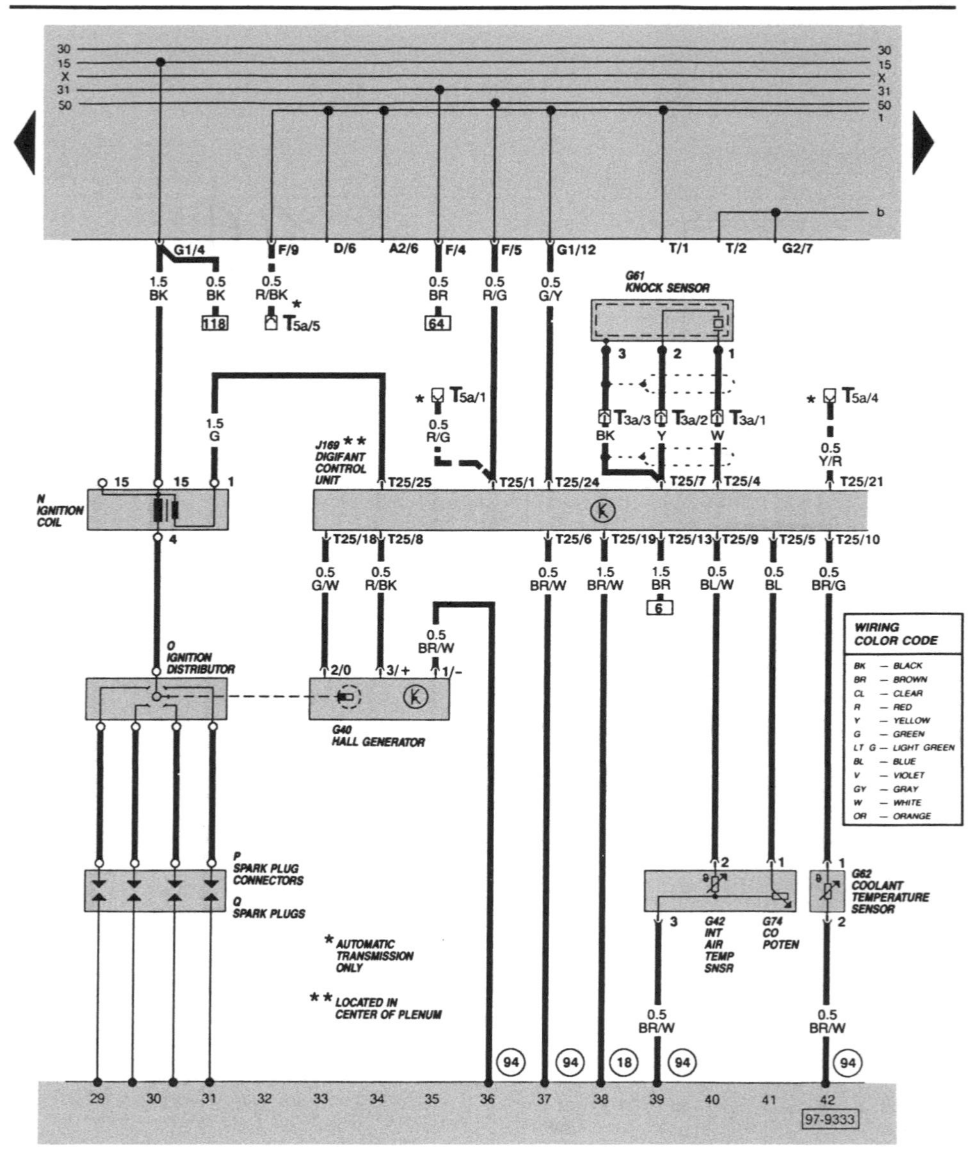
G1/4
F/9
D/6
A2/6
F/4
F/5
G1/12
T/1
T/2
G2/7
G61
KNOCK SENSOR
N
IGNITION
COIL
J169 * *
DIGIFANT
CONTROL
UNIT
O
IGNITION
DISTRIBUTOR
G40
HALL GENERATOR
P
SPARK PLUG
CONNECTORS
Q
SPARK PLUGS
* AUTOMATIC
TRANSMISSION
ONLY
* * LOCATED IN
CENTER OF PLENUM
G42
INT
AIR
TEMP
SNSR
G74
CO
POTEN
G62
COOLANT
TEMPERATURE
SENSOR
WIRING
COLOR CODE
BK – BLACK
BR – BROWN
CL – CLEAR
R – RED
Y – YELLOW
G – GREEN
LT G – LIGHT GREEN
BL – BLUE
V – VIOLET
GY – GRAY
W – WHITE
OR – ORANGE
97-9333

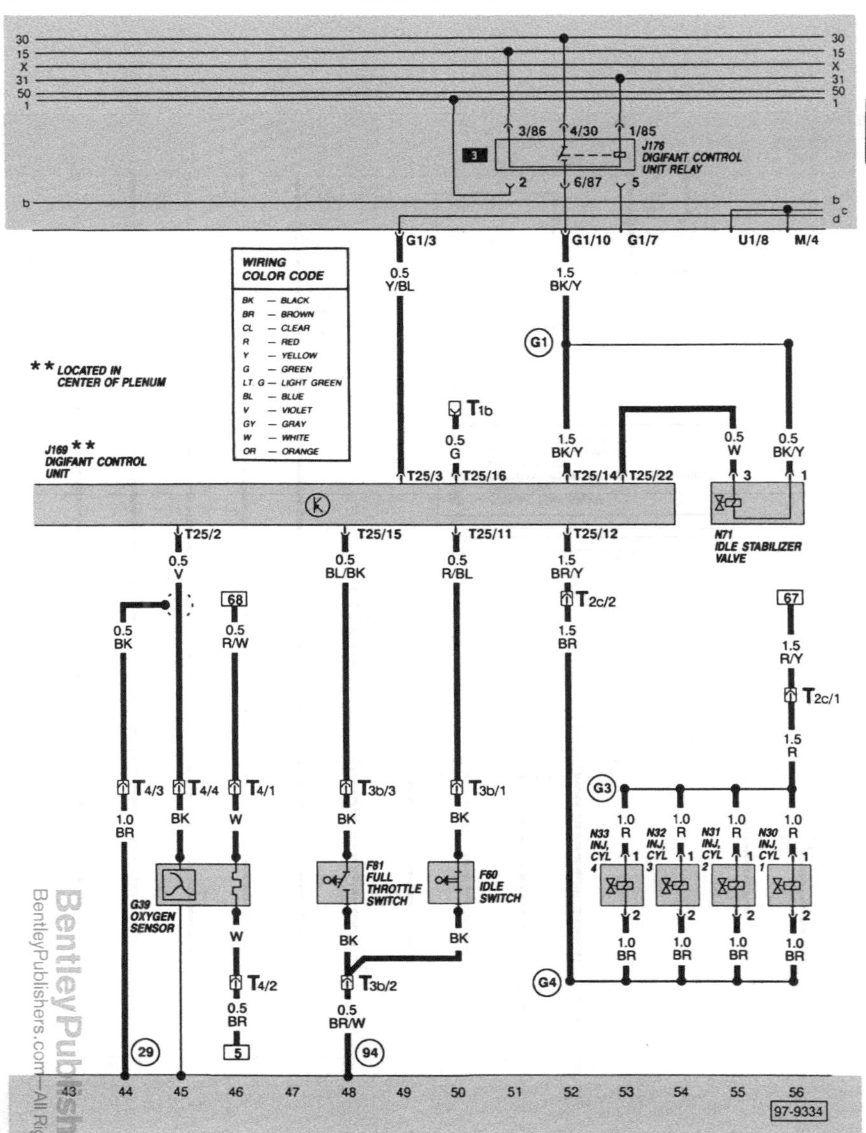

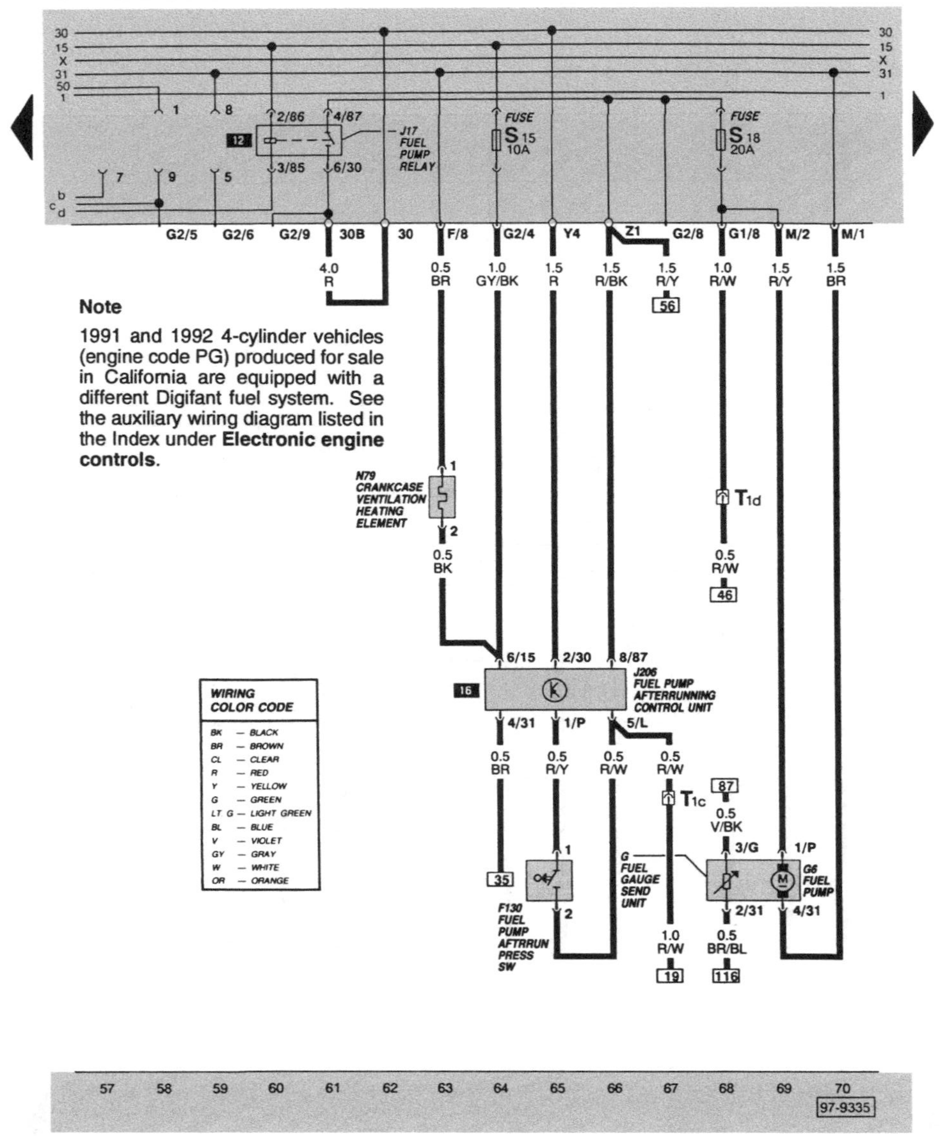

Note

1991 and 1992 4-cylinder vehicles (engine code PG) produced for sale in California are equipped with a different Digifant fuel system. See the auxiliary wiring diagram listed in the Index under **Electronic engine controls.**

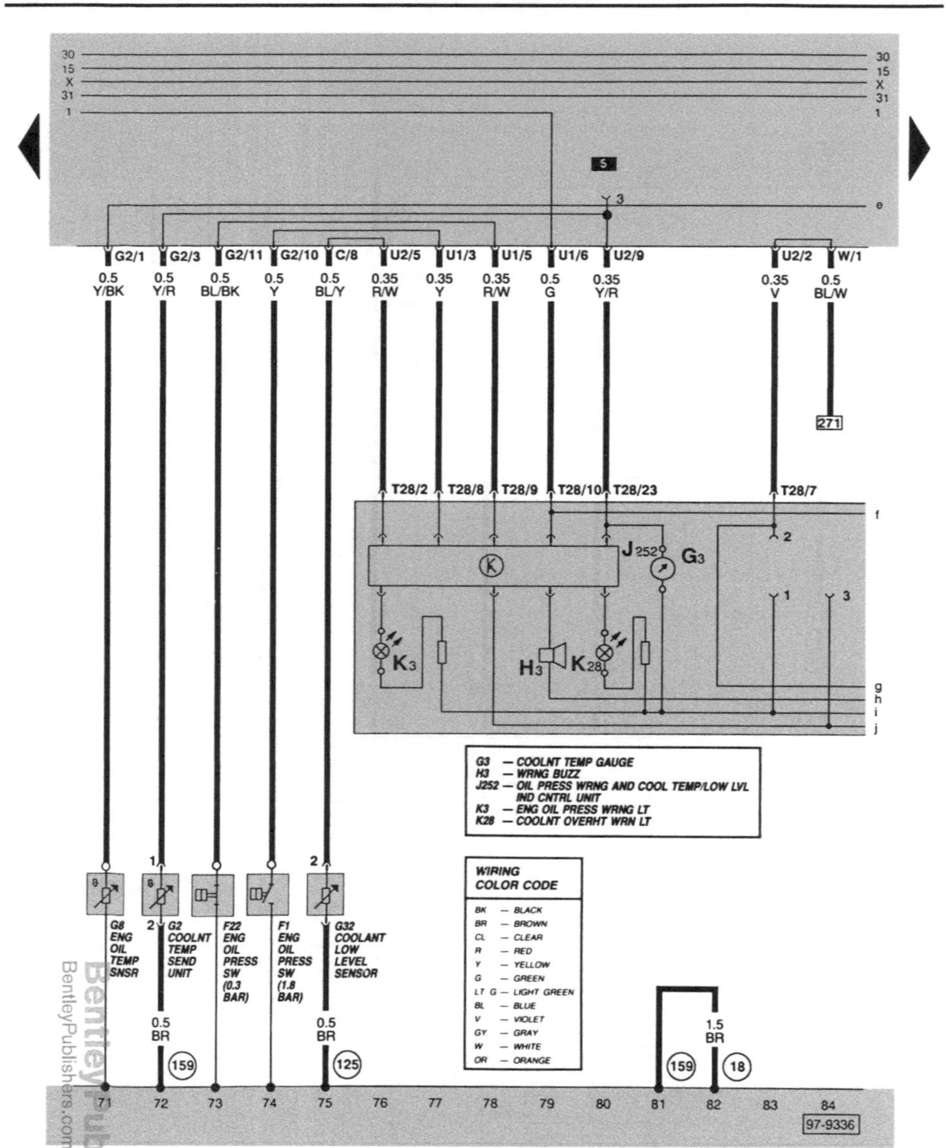

G3 — COOLNT TEMP GAUGE
H3 — WRNG BUZZ
J252 — OIL PRESS WRNG AND COOL TEMP/LOW LVL IND CNTRL UNIT
K3 — ENG OIL PRESS WRNG LT
K28 — COOLNT OVERHT WRN LT

WIRING COLOR CODE

BK — BLACK
BR — BROWN
CL — CLEAR
R — RED
Y — YELLOW
G — GREEN
LT G — LIGHT GREEN
BL — BLUE
V — VIOLET
GY — GRAY
W — WHITE
OR — ORANGE

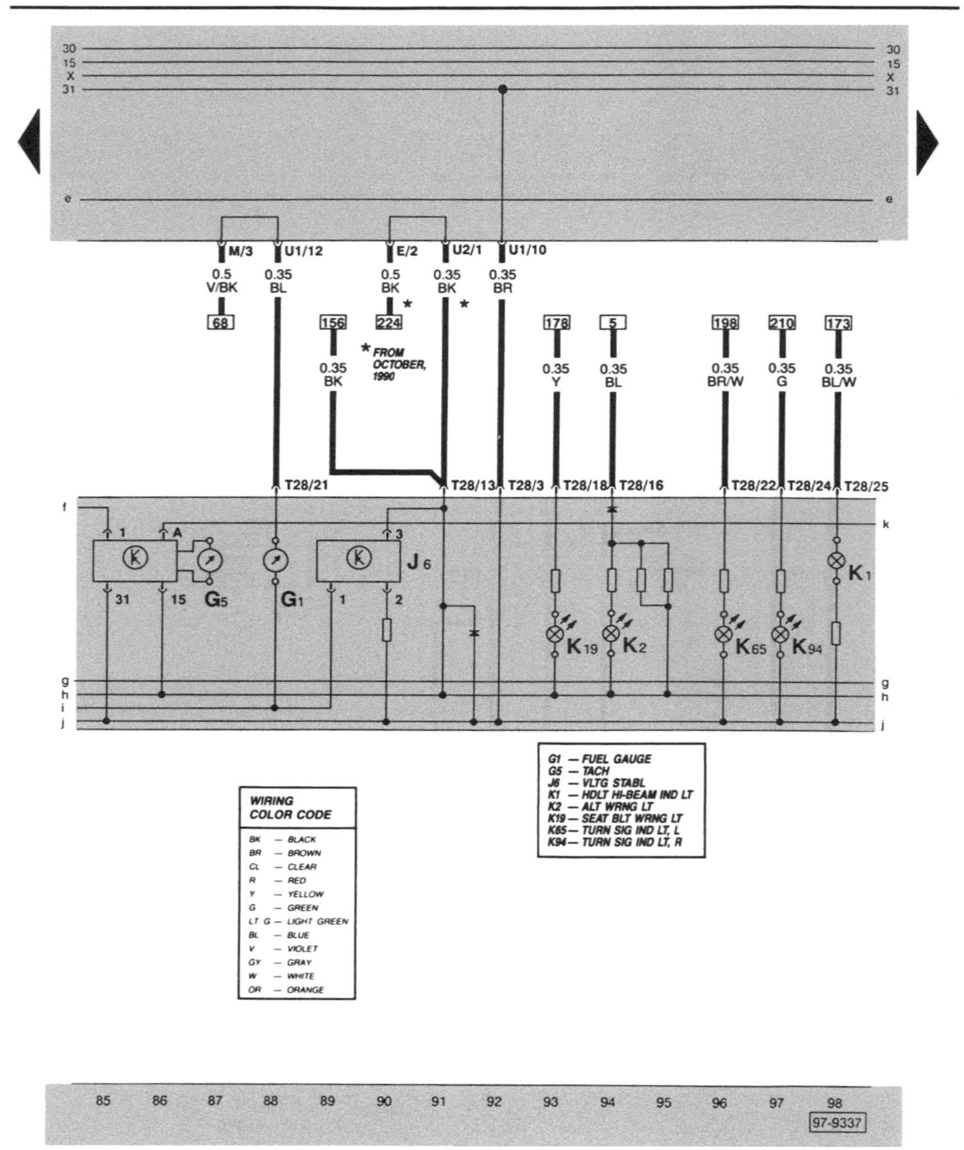

G1 — FUEL GAUGE
G5 — TACH
J6 — VLTG STABL
K1 — HDLT HI-BEAM IND LT
K2 — ALT WRNG LT
K19 — SEAT BLT WRNG LT
K65 — TURN SIG IND LT, L
K94 — TURN SIG IND LT, R

WIRING COLOR CODE

BK — BLACK
BR — BROWN
CL — CLEAR
R — RED
Y — YELLOW
G — GREEN
LT G — LIGHT GREEN
BL — BLUE
V — VIOLET
GY — GRAY
W — WHITE
OR — ORANGE

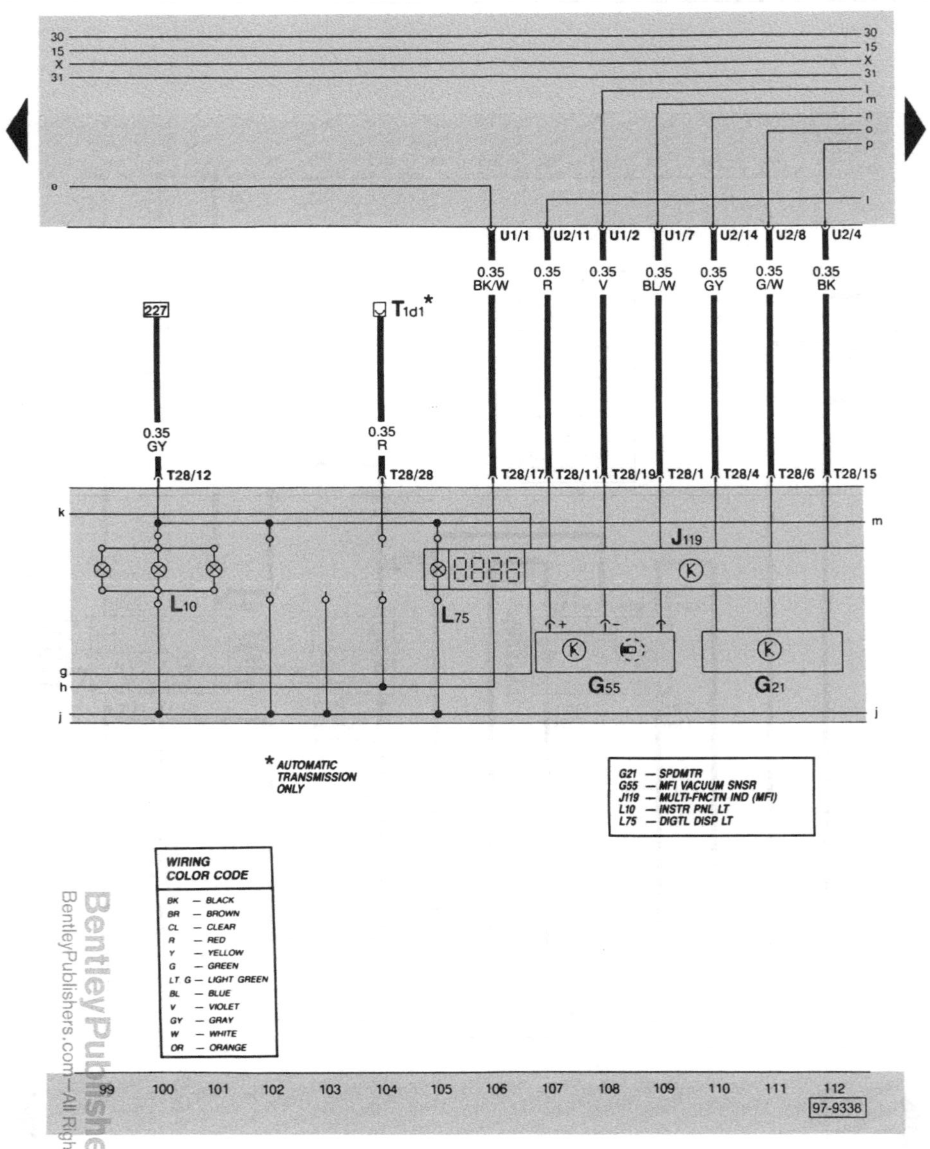

Corrado 1991–1992
Engine code – PG

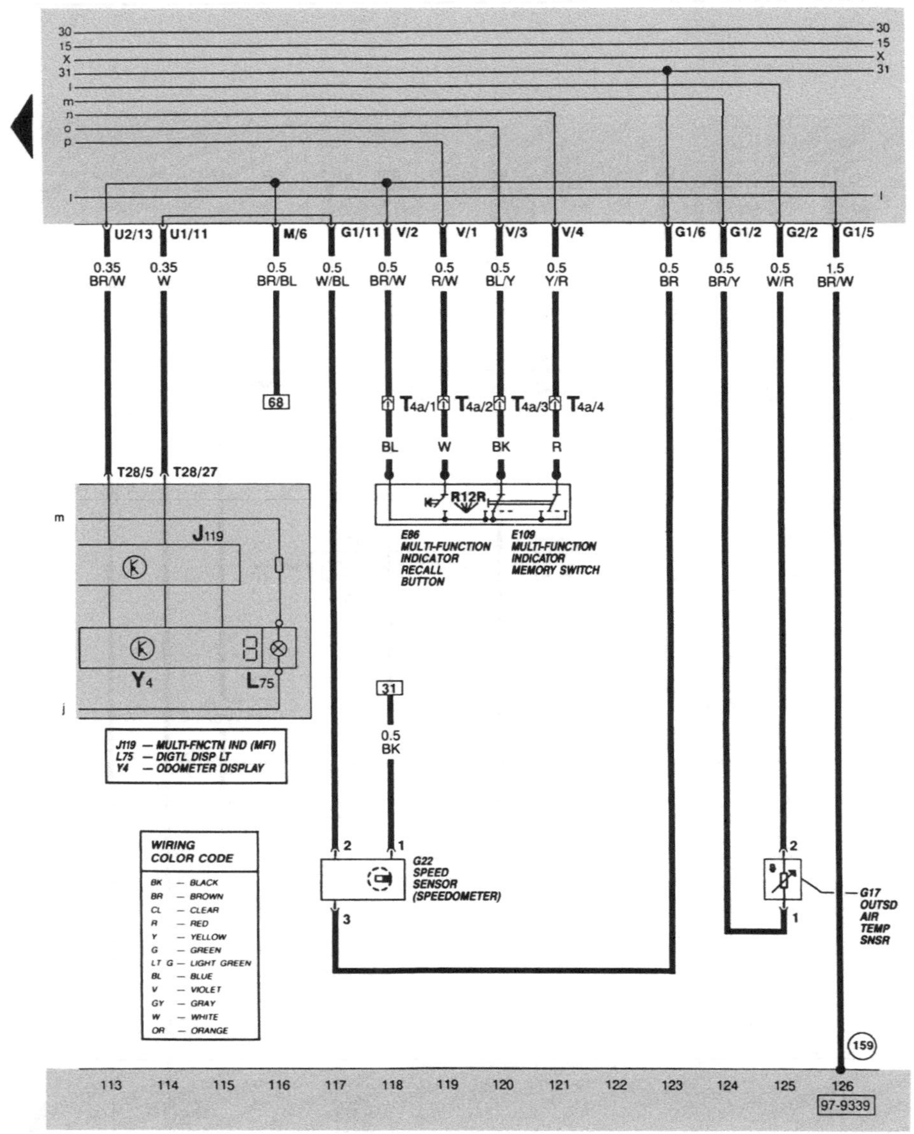

Corrado 1991–1992
Engine code – PG

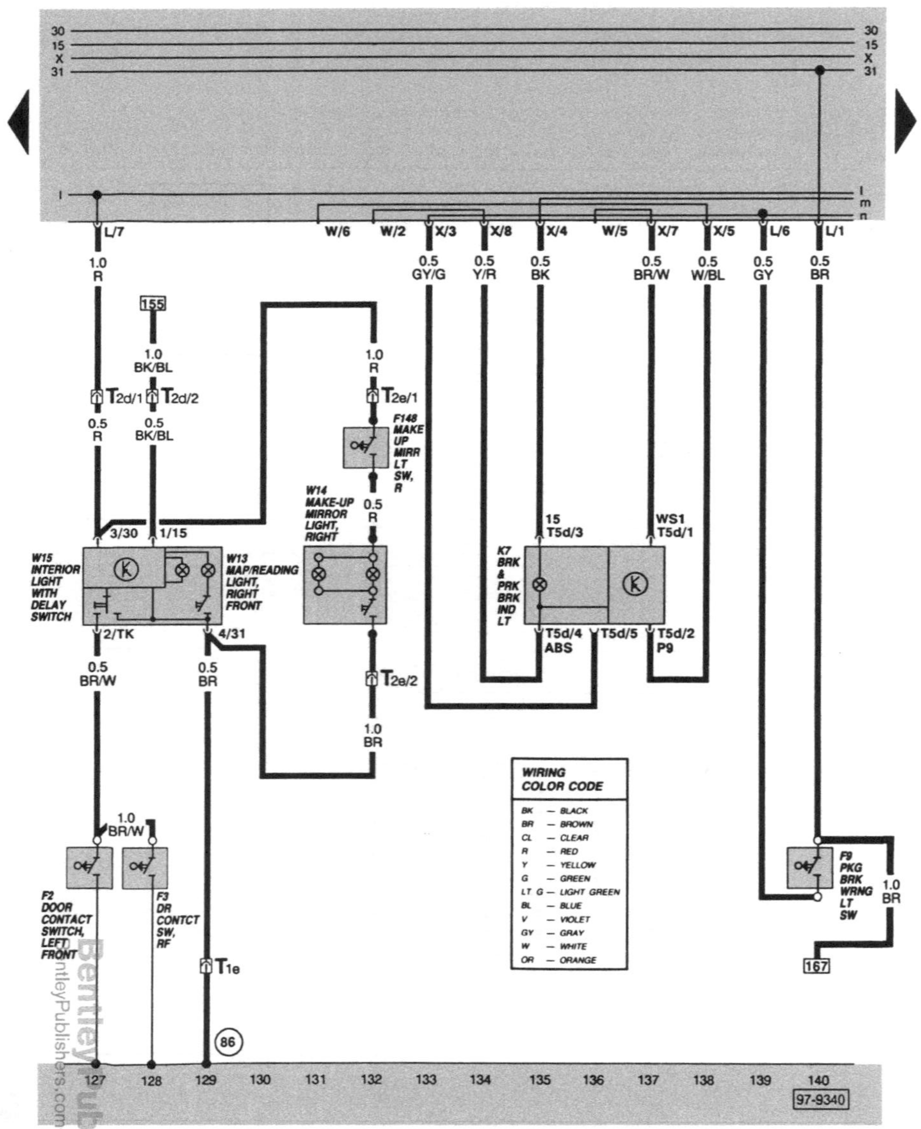

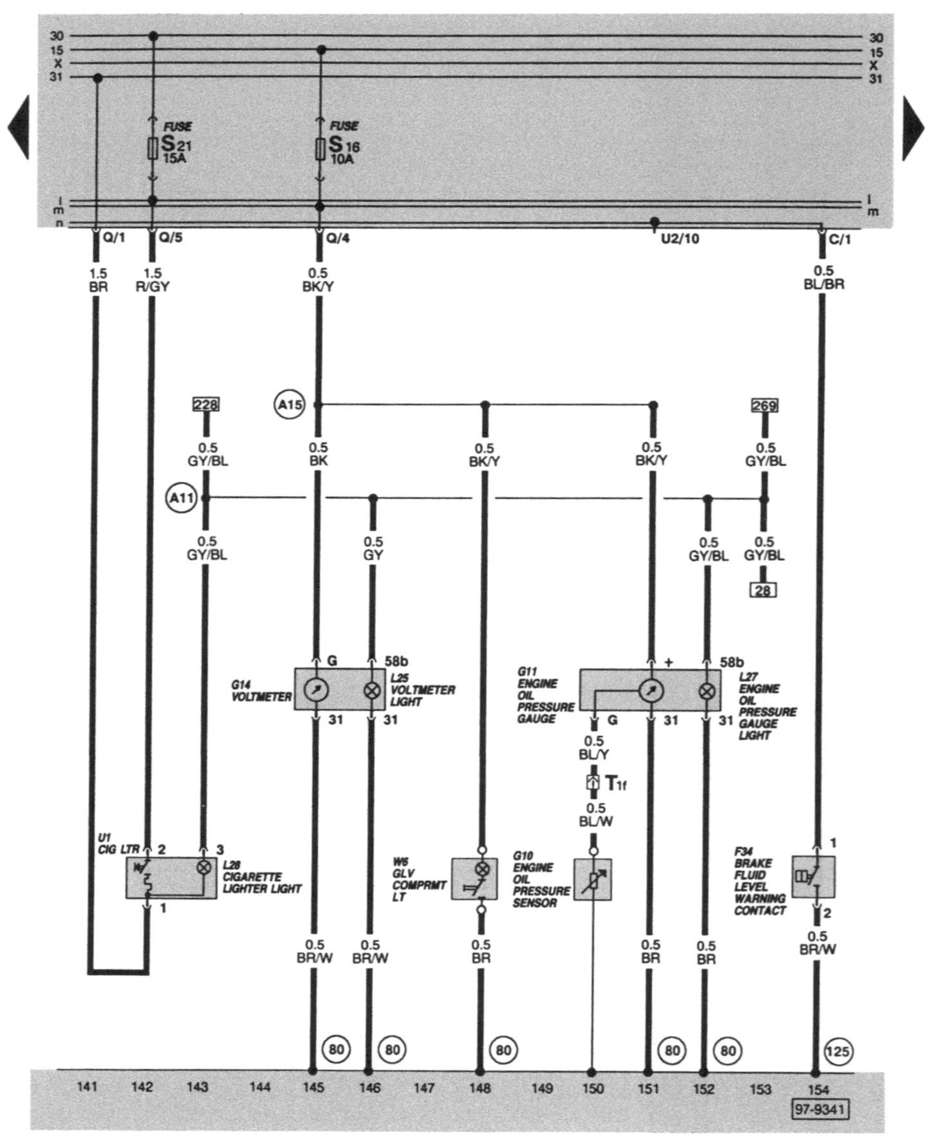

Corrado 1991–1992
Engine code – PG

Corrado 1991–1992
Engine code – PG

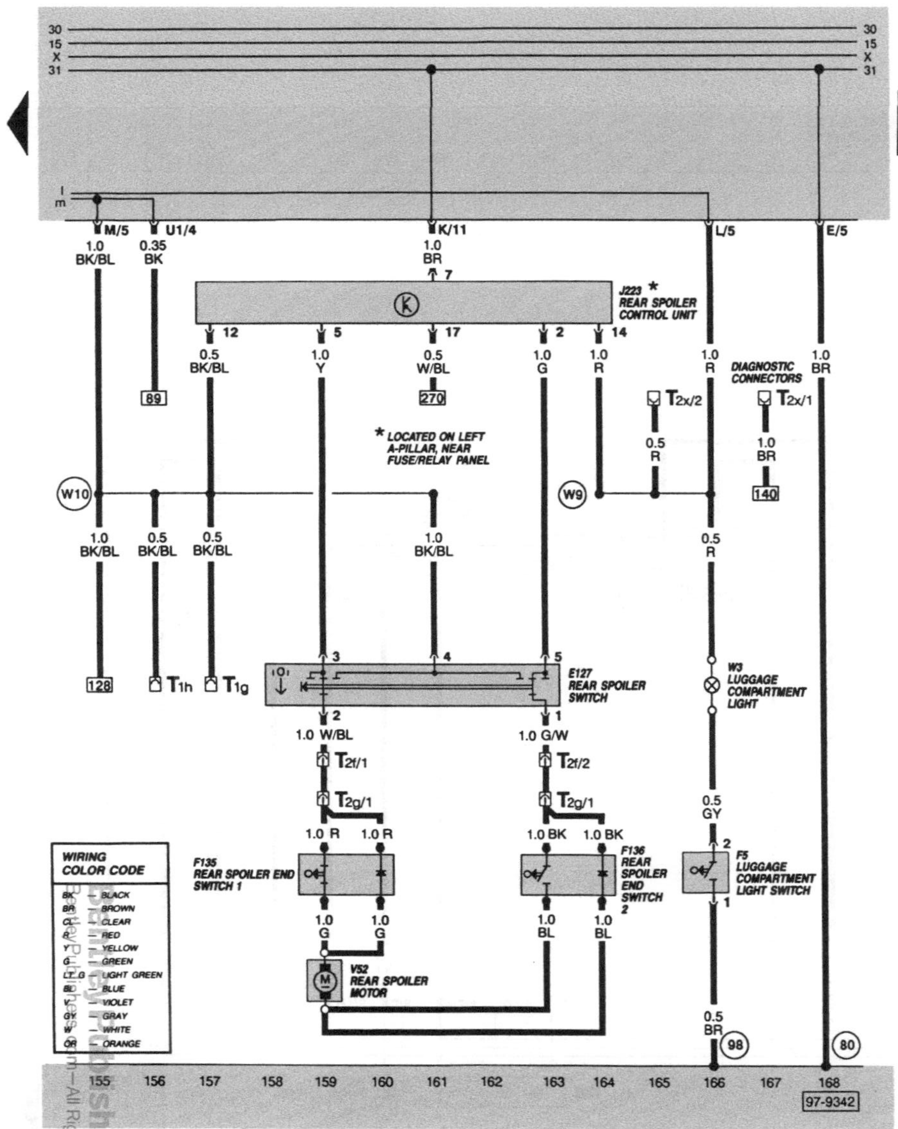

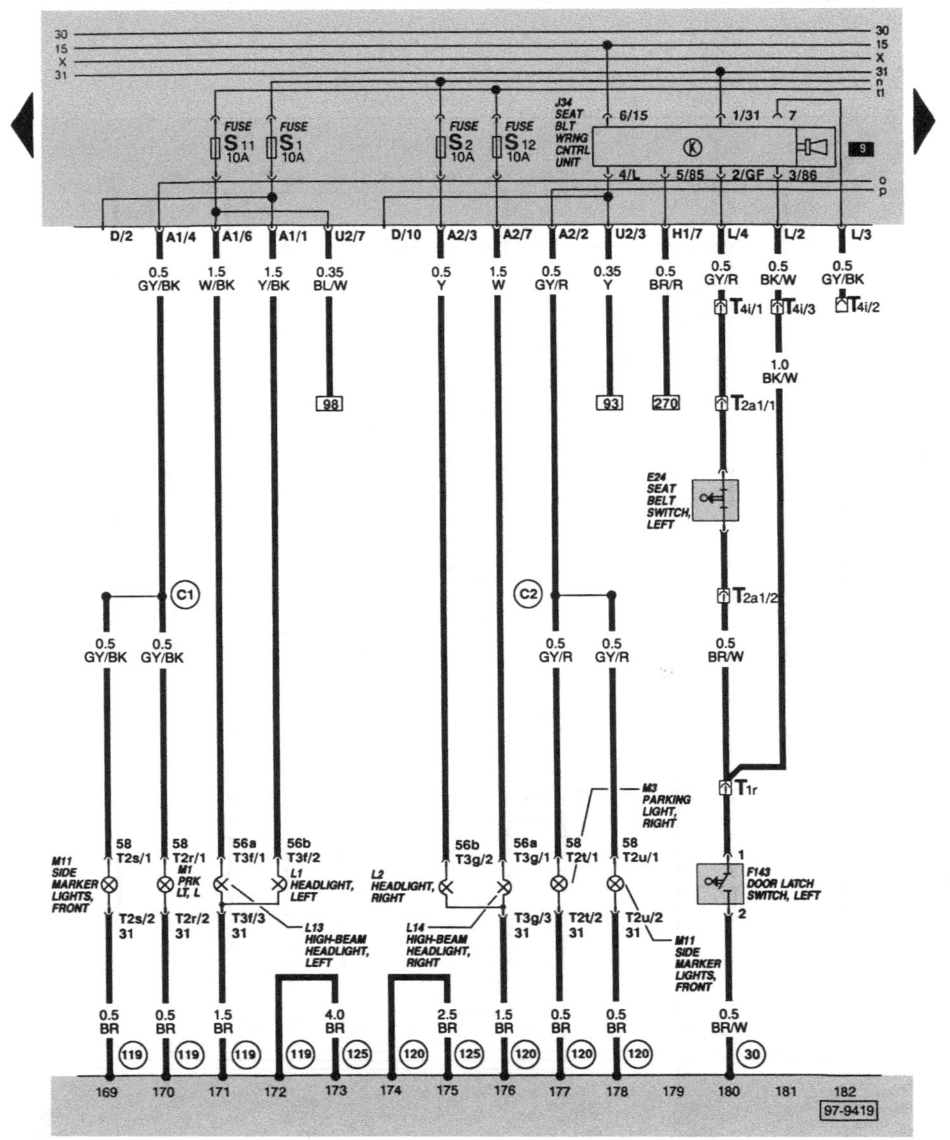

Corrado 1991–1992
Engine code – PG

Corrado 1991–1992
Engine code – PG

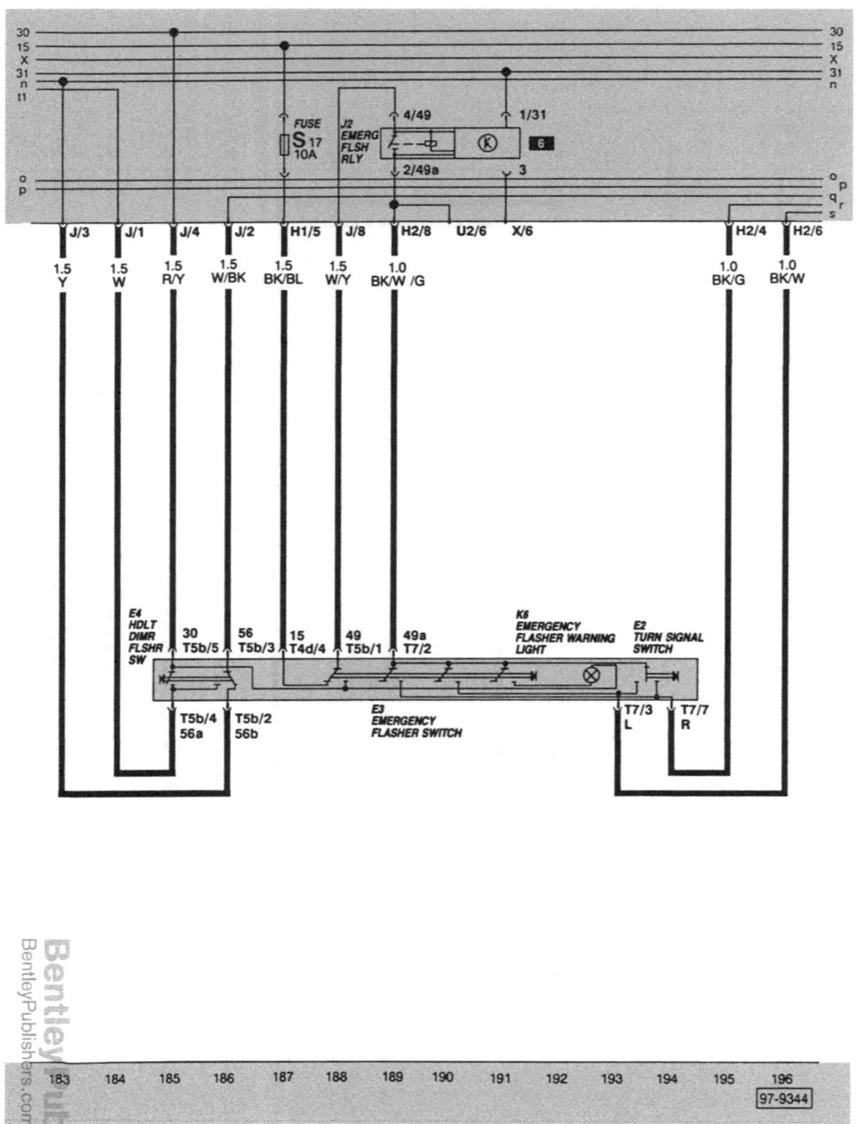
FUSE S17 10A
J2 EMERG FLSH RLY
4/49
2/49a
1/31
3
6
J/3
J/1
J/4
J/2
H1/5
J/8
H2/8
U2/6
X/6
H2/4
H2/6
1.5 Y
1.5 W
1.5 R/Y
1.5 W/BK
1.5 BK/BL
1.5 W/Y
1.0 BK/W /G
1.0 BK/G
1.0 BK/W
E4 HDLT DIMR FLSHR SW
30 T5b/5
56 T5b/3
15 T4d/4
49 T5b/1
49a T7/2
K6 EMERGENCY FLASHER WARNING LIGHT
E2 TURN SIGNAL SWITCH
T5b/4 56a
T5b/2 56b
E3 EMERGENCY FLASHER SWITCH
T7/3 L
T7/7 R
183 184 185 186 187 188 189 190 191 192 193 194 195 196
97-9344

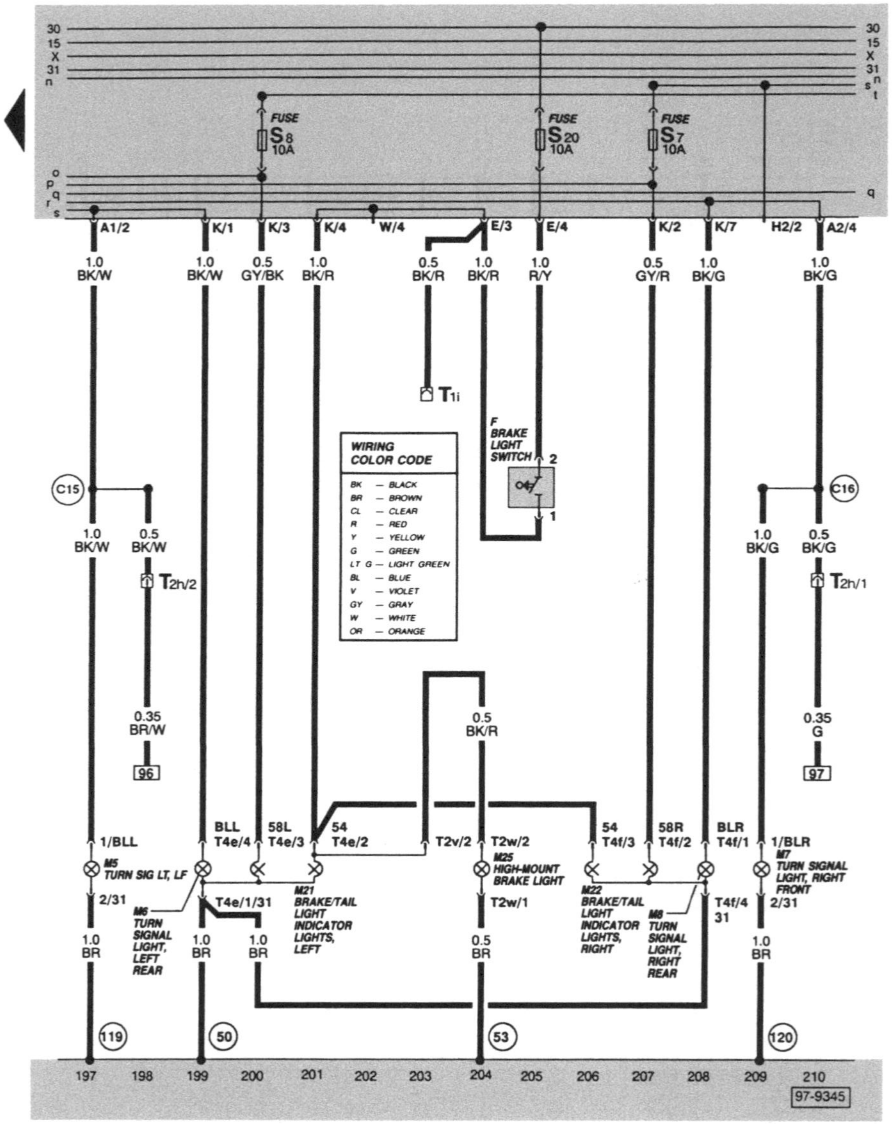
FUSE S8 10A
FUSE S20 10A
FUSE S7 10A
A1/2
K/1
K/3
K/4
W/4
E/3
E/4
K/2
K/7
H2/2
A2/4
1.0 BK/W
0.5 GY/BK
1.0 BK/R
0.5 BK/R
1.0 R/Y
0.5 GY/R
1.0 BK/G
T1i
F BRAKE LIGHT SWITCH
WIRING COLOR CODE
BK – BLACK
BR – BROWN
CL – CLEAR
R – RED
Y – YELLOW
G – GREEN
LT G – LIGHT GREEN
BL – BLUE
V – VIOLET
GY – GRAY
W – WHITE
OR – ORANGE
C15
C16
T2h/2
T2h/1
0.35 BR/W
0.35 G
96
97
1/BLL
BLL T4e/4
58L T4e/3
54 T4e/2
T2v/2
T2w/2
54 T4f/3
58R T4f/2
BLR T4f/1
1/BLR
M5 TURN SIG LT, LF
M6 TURN SIGNAL LIGHT, LEFT REAR
T4e/1/31
M21 BRAKE/TAIL LIGHT INDICATOR LIGHTS, LEFT
M25 HIGH-MOUNT BRAKE LIGHT
T2w/1
M22 BRAKE/TAIL LIGHT INDICATOR LIGHTS, RIGHT
M8 TURN SIGNAL LIGHT, RIGHT REAR
T4f/4 31
M7 TURN SIGNAL LIGHT, RIGHT FRONT
2/31
1.0 BR
0.5 BR
119
50
53
120
197 198 199 200 201 202 203 204 205 206 207 208 209 210
97-9345

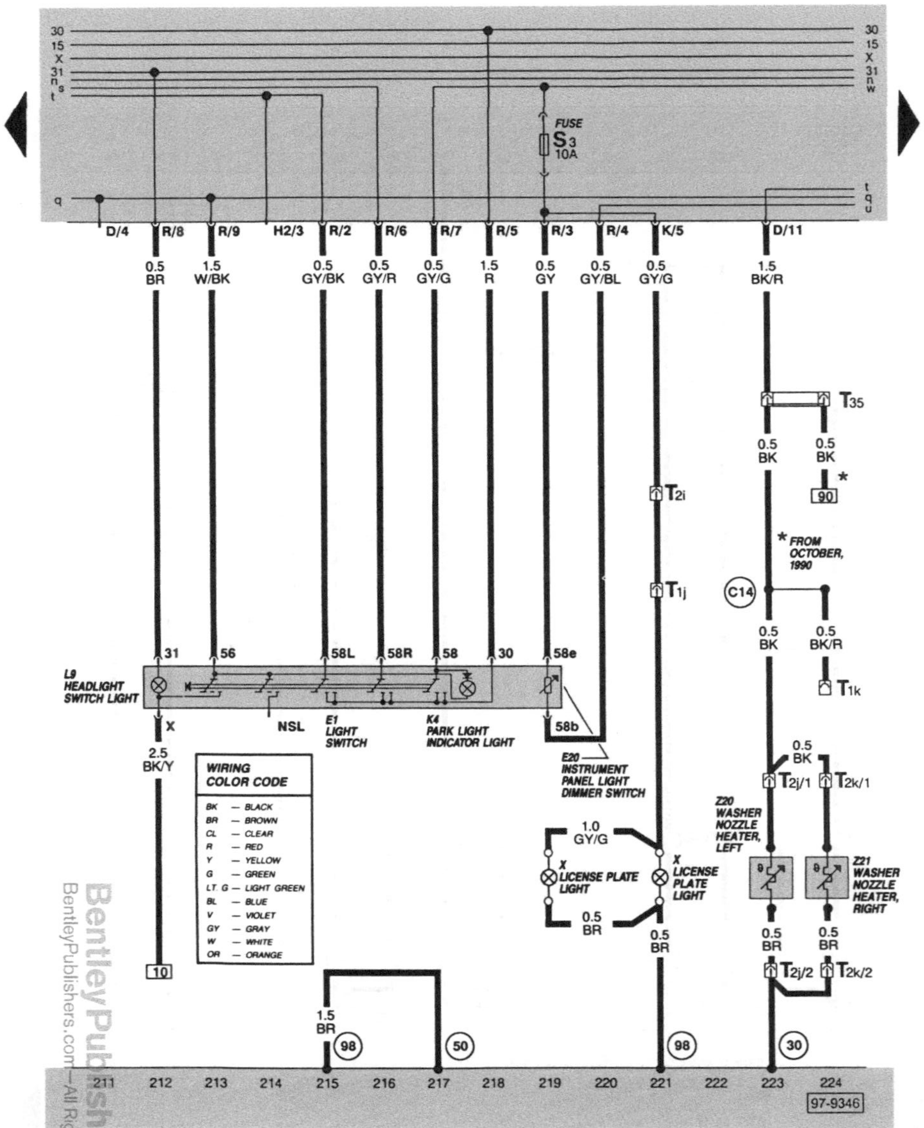
FUSE
S3
10A
D/4
R/8
R/9
H2/3
R/2
R/6
R/7
R/5
R/3
R/4
K/5
D/11
L9
HEADLIGHT
SWITCH LIGHT
E1
LIGHT
SWITCH
K4
PARK LIGHT
INDICATOR LIGHT
NSL
E20
INSTRUMENT
PANEL LIGHT
DIMMER SWITCH
X
LICENSE PLATE
LIGHT
Z20
WASHER
NOZZLE
HEATER,
LEFT
Z21
WASHER
NOZZLE
HEATER,
RIGHT
* FROM
OCTOBER,
1990
WIRING
COLOR CODE
BK — BLACK
BR — BROWN
CL — CLEAR
R — RED
Y — YELLOW
G — GREEN
LT G — LIGHT GREEN
BL — BLUE
V — VIOLET
GY — GRAY
W — WHITE
OR — ORANGE
211 212 213 214 215 216 217 218 219 220 221 222 223 224
97-9346

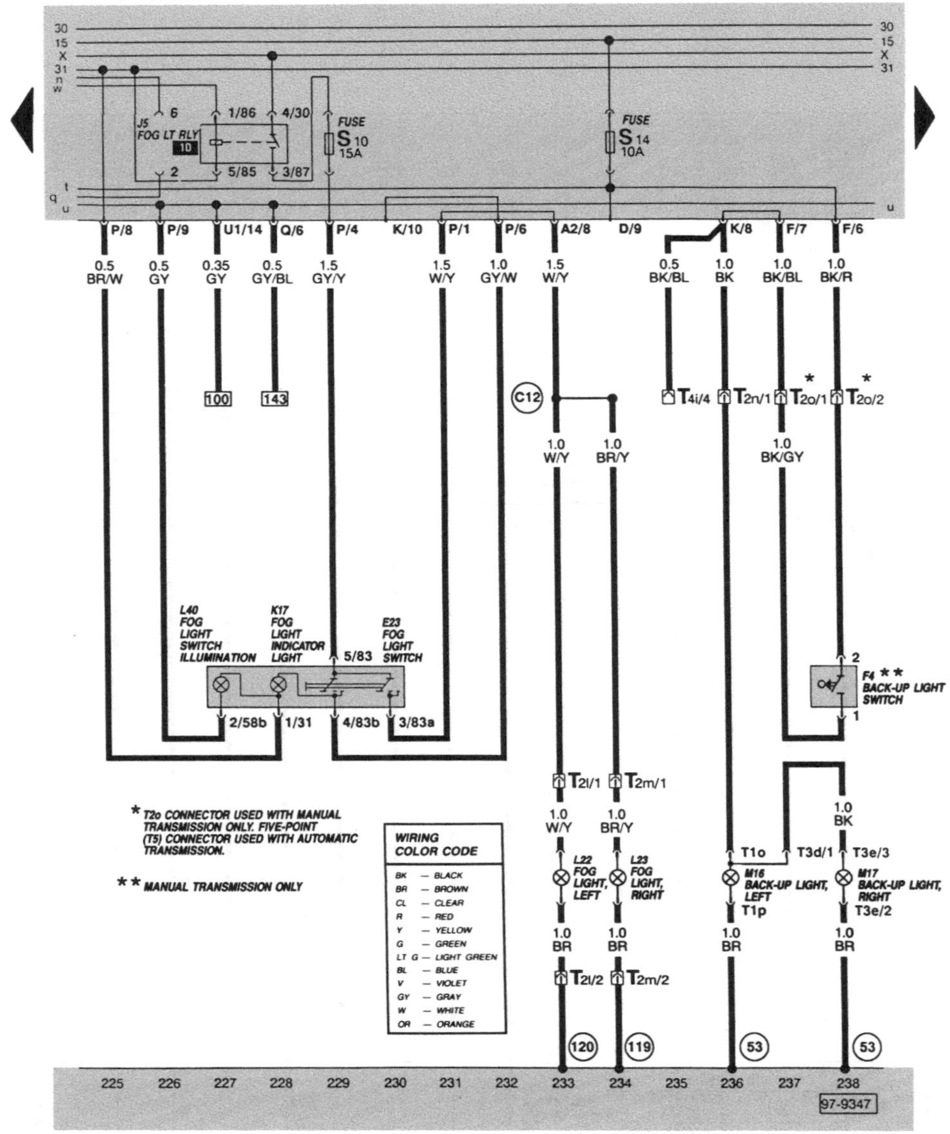
J5
FOG LT RLY
10
FUSE
S10
15A
FUSE
S14
10A
P/8
P/9
U1/14
Q/6
P/4
K/10
P/1
P/6
A2/8
D/9
K/8
F/7
F/6
L40
FOG
LIGHT
SWITCH
ILLUMINATION
K17
FOG
LIGHT
INDICATOR
LIGHT
E23
FOG
LIGHT
SWITCH
F4 **
BACK-UP LIGHT
SWITCH
L22
FOG
LIGHT,
LEFT
L23
FOG
LIGHT,
RIGHT
M16
BACK-UP LIGHT,
LEFT
M17
BACK-UP LIGHT,
RIGHT
* T2o CONNECTOR USED WITH MANUAL TRANSMISSION ONLY. FIVE-POINT (T5) CONNECTOR USED WITH AUTOMATIC TRANSMISSION.
** MANUAL TRANSMISSION ONLY
WIRING
COLOR CODE
BK — BLACK
BR — BROWN
CL — CLEAR
R — RED
Y — YELLOW
G — GREEN
LT G — LIGHT GREEN
BL — BLUE
V — VIOLET
GY — GRAY
W — WHITE
OR — ORANGE
225 226 227 228 229 230 231 232 233 234 235 236 237 238
97-9347

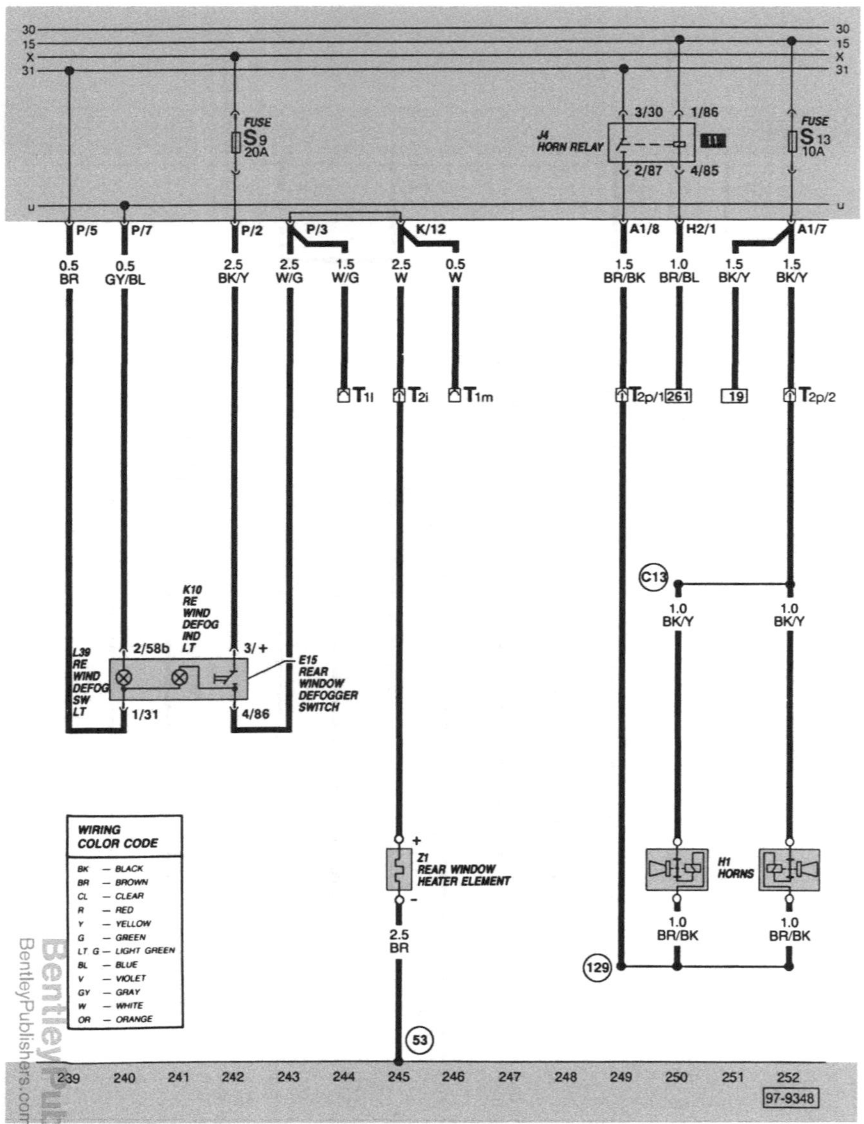
30
15
X
31
FUSE S9 20A
J4 HORN RELAY
FUSE S13 10A
K10 RE WIND DEFOG IND LT
L39 RE WIND DEFOG SW LT
E15 REAR WINDOW DEFOGGER SWITCH
Z1 REAR WINDOW HEATER ELEMENT
H1 HORNS
WIRING COLOR CODE
BK — BLACK
BR — BROWN
CL — CLEAR
R — RED
Y — YELLOW
G — GREEN
LT G — LIGHT GREEN
BL — BLUE
V — VIOLET
GY — GRAY
W — WHITE
OR — ORANGE
239 240 241 242 243 244 245 246 247 248 249 250 251 252
97-9348

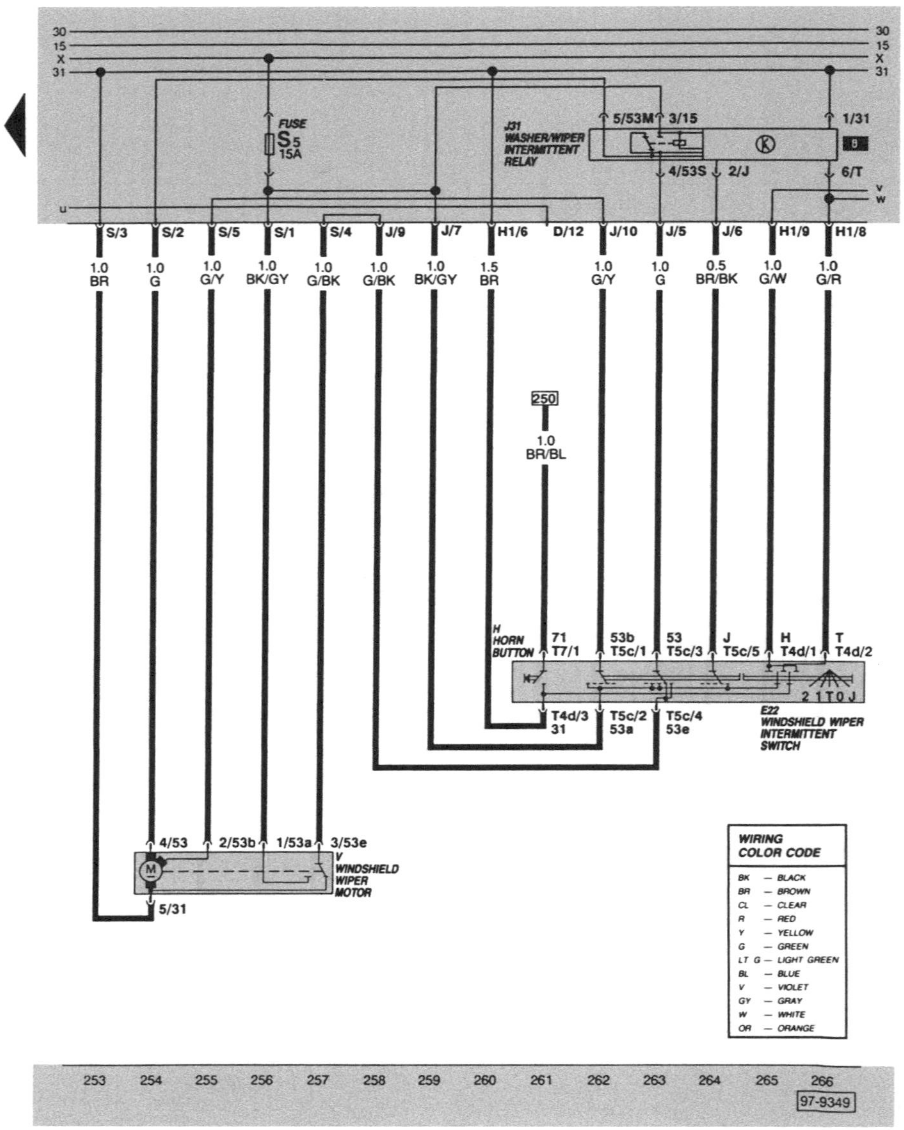
30
15
X
31
FUSE S5 15A
J31 WASHER/WIPER INTERMITTENT RELAY
H HORN BUTTON
E22 WINDSHIELD WIPER INTERMITTENT SWITCH
V WINDSHIELD WIPER MOTOR
WIRING COLOR CODE
BK — BLACK
BR — BROWN
CL — CLEAR
R — RED
Y — YELLOW
G — GREEN
LT G — LIGHT GREEN
BL — BLUE
V — VIOLET
GY — GRAY
W — WHITE
OR — ORANGE
253 254 255 256 257 258 259 260 261 262 263 264 265 266
97-9349

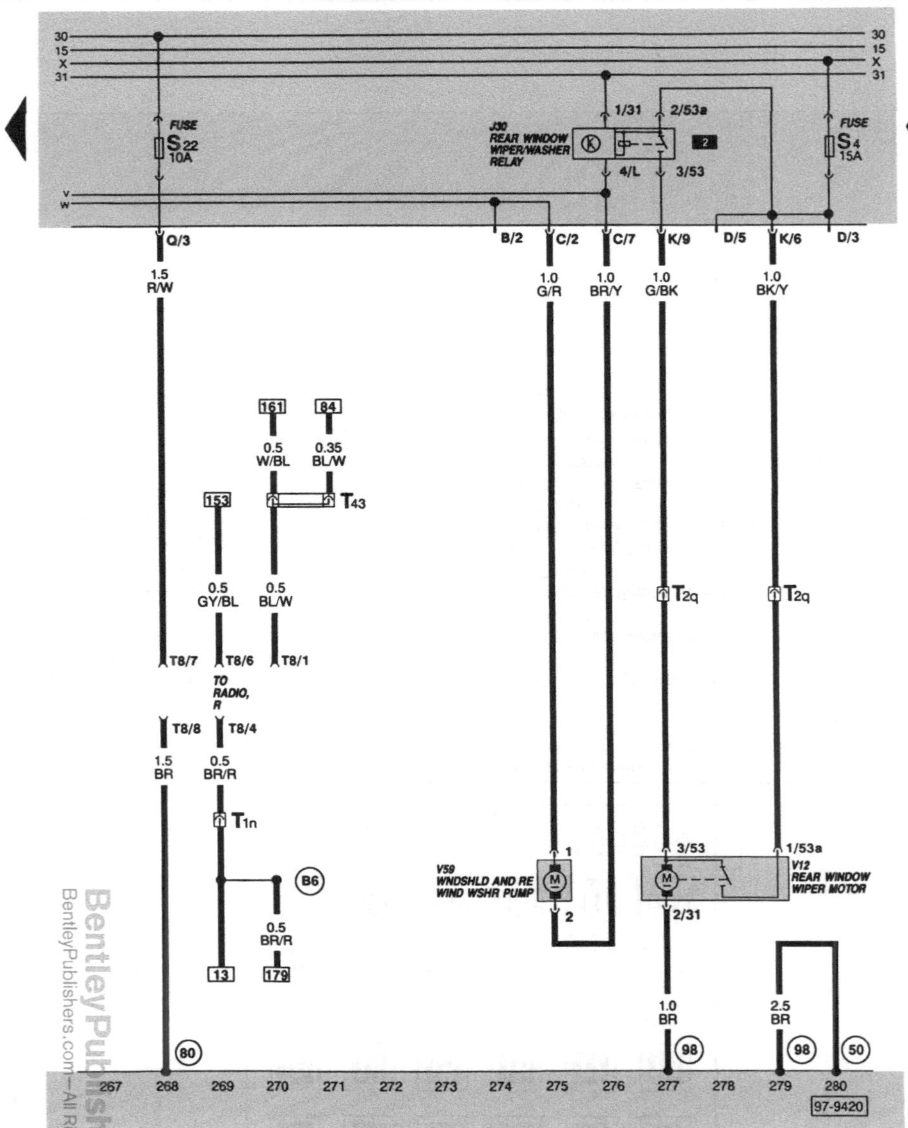

Corrado 1991–1992
Engine code – PG

THIS PAGE INTENTIONALLY LEFT
BLANK

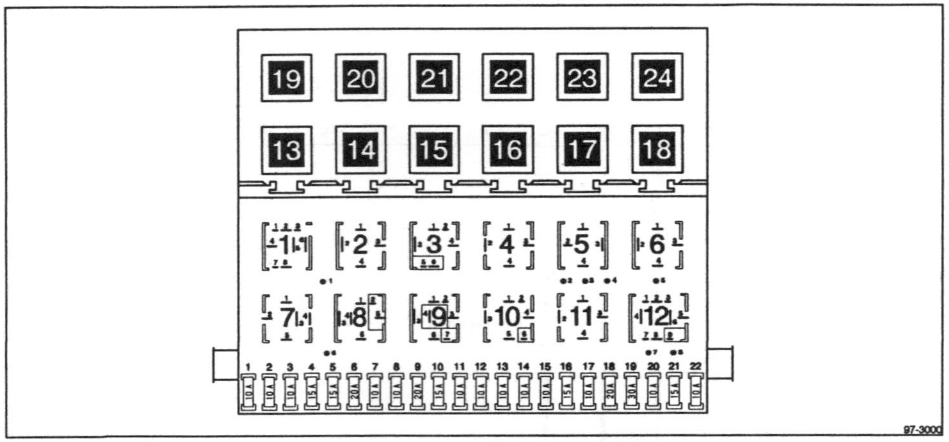

Fuse arrangement
according to the numbers on fuse panel from left to right:

		Amp.	Current track
1	Headlight Left, Terminal 56A	10	199
2	Headlight Right, Terminal 56A	10	204
3	Instrument Panel Light Dimmer Switch	10	247
4	Rear Window Wiper Motor	15	280
5	Windshield Wiper Motor	15	284
6	Fresh Air Blower Switch	20	23
7	Brake/Tail Lights, Right	10	235
8	Brake/Tail Lights, Left	10	228
9	Rear Window Defogger Switch	20	270
10	Open		
11	Headlight Left, Terminal 56B	10	199
12	Headlight Right, Terminal 56B	10	204
13	Horn	15	33

Fuse arrangement
according to the numbers on fuse panel from left to right:

		Amp.	Current track
14	Back-up Light Switch, Washer Nozzle Heater	10	262
15	Crankcase Ventilation Heating Element	10	97
16	Voltmeter, Engine Oil Pressure Gauge	15	173
17	Emergency Flashers Switch	10	215
18	HO_2S (Heated Oxygen Sensor) Control	20	53
19	Coolant FC (Fan Control) Thermoswitch	30	31
20	Brake Light Switch	10	233
21	Cigarette Lighter	15	170
22	Radio	10	296

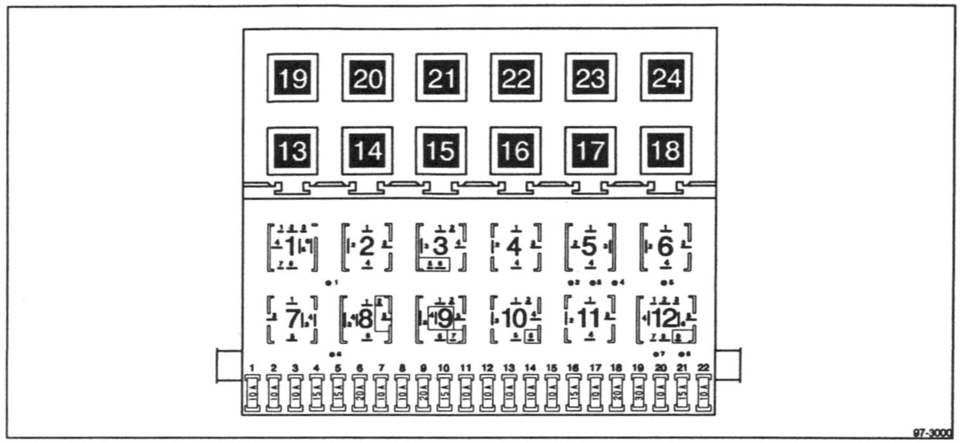

Relay location on fuse/relay panel

1 A/C Relay (13)*
2 Rear Window Wiper/Washer Relay (30)*
3 ECM (Engine Control Module) Power Supply Relay (109)*
4 Load Reduction Relay (18)*
5 Open
6 Emergency Flasher Relay (21)*
7 Headlight Washer System Relay (33)*
8 Washer/Wiper Intermittent Relay (19, 99)*
9 Seat Belt Warning Control Module (27, 4)*
10 Fog Light Relay (53)*
11 Dual Horn Relay (53)*
12 FP (Fuel Pump) Relay (67)*

Separate relays above fuse/relay panel

13 ABS Hydraulic Pump Relay (78)*
14 ABS Relay (79)*
15 Open
16 Open
17 Open
18 Power Seat Fuse
19 ABS Hydraulic Pump Fuse
20 ABS Valves Fuse
21 Open
22 Power Window Circuit Breaker
23 Fuse, S51
24 Open

*Numbers in parentheses indicate production control number stamped on relay housing.

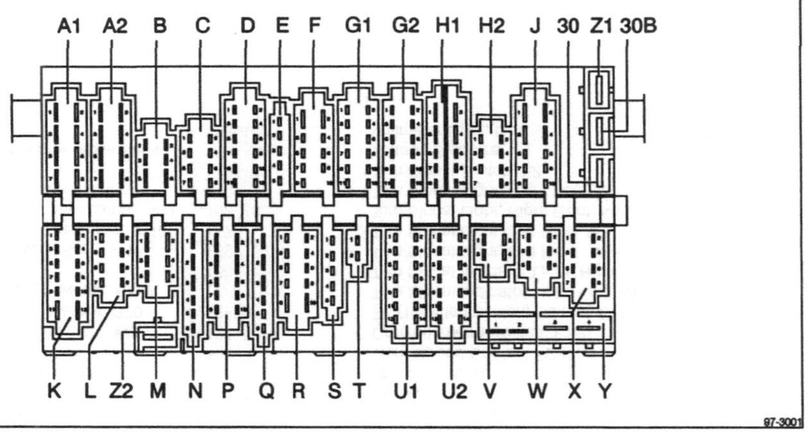

Connections and plugs on fuse/relay panel

- A1 – Headlight wiring harness (yellow), eight point
- A2 – Headlight wiring harness (yellow), eight point
- B – Open
- C – Headlight wiring harness (yellow), eight point
- D – Optional equipment wiring harness (green), twelve point
- E – Instrument wiring harness (green), five point
- F – Engine compartment wiring harness, right (white), ten point
- G1 – Engine compartment wiring harness, right (white), twelve point
- G2 – Engine compartment wiring harness, right (white), twelve point
- H1 – Steering column switch wiring harness (red), ten point
- H2 – Steering column switch wiring harness (red), eight point
- J – Steering column switch wiring harness (red), ten point
- K – Rear wiring harness (black), twelve point
- L – Rear wiring harness (black), eight point
- M – Rear wiring harness (black), six point

Connections and plugs on fuse/relay panel

- N – A/C wiring harness (green), six point
- P – Rear window defroster wiring harness (blue), ten point
- Q – Instrument wiring harness (blue), six point
- R – Light switch wiring harness (blue), ten point
- S – Engine compartment wiring harness, right (white), five point
- T – Connector (green), two point
- U1 – Instrument cluster wiring harness (blue), fourteen point
- U2 – Instrument cluster wiring harness (blue), fourteen point
- V – Multi-function indicator wiring harness (green), four point
- W – Connector (green), six point
- X – Warning lamp wiring harness (green), eight point
- Y – Four single connectors, teminal 30
- Z1 – Single connector
- Z2 – Single connector, teminal 31
- 30 – Single connector, terminal 30
- 30B – Single connector
- 31 – Single connector, terminal 31 (not shown)

Description	Current track
Back-Up Light Switch, F4	266
Back-Up Light, Left, M16	264
Back-Up Light, Right, M17	266
Battery, A	4
Brake and Parking Brake Indicator Light, K7	162–165
Brake Fluid Level Warning Switch, F34	182
Brake Light Switch, F	233
Brake/Tail Lights, Left, M21	228–229
Brake/Tail Lights, Right, M22	234–235
Central Locking System Switch, Left Front, F59	10–11
Cigarette Lighter Light, L28	171
Cigarette Lighter, U1	170
Control Module With Display Unit In Instrument Cluster, J285	104–144
Coolant Circulation Pump, V50	38
Coolant Fan, V7	34–36
Coolant FC (Fan Control) Control Module, J239	33–41
Coolant FC (Fan Control) Thermoswitch, F18	30–31
Coolant FC (Fan Control) Thermoswitch, F87	39
Crankcase Ventilation Heating Element, N79	97
Digital Display Light, L75	139,141
Door Contact Switch, Left Front, F2	155
Door Contact Switch, Right Front, F3	156
Door Latch Switch, Left, F143	208
Dual Horn Relay, J4	277–278
Dual Horns, H1	278,280
ECL (Engine Coolant Level) Sensor, G32	103
ECM (Engine Control Module) Power Supply Relay, J271	89–90
ECM (Engine Control Module), J220	47–90
ECT (Engine Coolant Temperature) Gauge, G3	108
ECT (Engine Coolant Temperature) Sensor, G2	40
ECT (Engine Coolant Temperature) Sensor, G62	63
ECT (Engine Coolant Temperature)/ ECL (Engine Coolant Level) Warning Light, K28	107
EGR Frequency Valve, N18	80
EGR Temperature Sensor, G98	67
Emergency Flasher Relay, J2	217–219

Description	Current track
Emergency Flasher Switch, E3	216–219
Emergency Flasher Warning Light, K6	220
Engine Oil Pressure Gauge Light, L27	180
Engine Oil Pressure Gauge, G11	178–179
Engine Oil Pressure Sensor, G10	178
Engine Oil Pressure Switch (0.3 Bar), F22	101
Engine Oil Pressure Switch, F1	102
Engine Oil Pressure Warning Light, K3	105
Engine Oil Temperature Sensor, G8	99
EVAP (Evaporative Emission) Frequency Valve, N80	91
Fog Light Relay, J5	254–256
FP (Fuel Pump) Relay, J17	50–51
FP (Fuel Pump), G6	151
Fresh Air Blower Series Resistance, N23	19–23
Fresh Air Blower Switch, E9	20–23
Fresh Air Blower, V2	18
Fresh Air Control Lever Light, L16	24
Fuel Gauge Sending Unit, G	150
Fuel Gauge, G1	109
Fuse Strip, S88	41
Fuse, S1, 10A	200
Fuse, S2, 10A	203
Fuse, S3, 10A	247
Fuse, S4, 15A	308
Fuse, S5, 15A	284
Fuse, S6, 20A	23
Fuse, S7, 10A	235
Fuse, S8, 10A	228
Fuse, S9, 20A	270
Fuse, S10, 15A	259
Fuse, S11, 10A	199
Fuse, S12, 10A	204
Fuse, S13, 10A	33
Fuse, S14, 10	262
Fuse, S15, 10A	97
Fuse, S16, 15A	173
Fuse, S17, 10A	215
Fuse, S18, 20A	53
Fuse, S19, 30A	31
Fuse, S20, 10A	223
Fuse, S21, 15A	170
Fuse, S22, 10A	296
GEN (Generator) Warning Light, K2	122
GEN (Generator), C	1–2
Glove Compartment Light, W6	176
Hall Generator, G40	47–49

Description	Current track
Headlight Dimmer/Flasher Switch, E4	213-214
Headlight High Beam Indicator Light, K1	126
Headlight Left, L1	200
Headlight Right, L2	203
Headlight Switch Light, L9	240
High Beam Headlight, Left, L13	199
High Beam Headlight, Right, L14	204
High-mount Brake Light, M25	232
HO_2S (Heated Oxygen Sensor) Control Module, J208	53-55
HO_2S (Heated Oxygen Sensor), G39	52-54
Horn Button, H	289
IAC (Idle Air Control) Valve, N71	77
IAT (Intake Air Temperature) Sensor, G42	66
Ignition Coil Power Output Stage, N157	45-46
Ignition Coil, N152	43
Ignition Distributor, O	43-45
Ignition/Starter Switch, D	9-13
Injector, Cyl. 1, N30	71
Injector, Cyl. 2, N31	72
Injector, Cyl. 3, N32	73
Injector, Cyl. 4, N33	74
Injector, Cyl. 5, N83	75
Injector, Cyl. 6, N84	76
Instrument Panel Light Dimmer Switch, E20	247
Instrument Panel Light, L10	130,131, 132
Interior Light With Delay Switch, W15	155
KS (Knock Sensor) 1, G61	57-59
KS (Knock Sensor) 2, G66	60-62
License Plate Light, X	247,249
Light Switch, E1	243-245
Luggage Compartment Light Switch, F5	194
Luggage Compartment Light, W3	194
MAF (Mass Air Flow) Sensor, G70	82-84
Make-Up Mirror Light Switch, Right, F148	160
Make-Up Mirror Light, Right, W14	159-160
Map/Reading Light, Right Front, W13	157
MIL (Malfunction Indicator Lamp), K83	162-165
Multi-Function Indicator Memory Switch, E109	146-149
Multi-Function Indicator Mode Select Switch, E86	146-149

Description	Current track
Multi-Function Indicator Vacuum Sensor, G55	127-129
Odometer Display, Y4	136-138
Outside Air (Ambient) Temperature Sensor, G17	153
Overheat Fuse, S24	21
Park Light Indicator Light, K4	246
Parking Brake Warning Light Switch (Ground), F9	196
Parking Light, Left, M1	198
Parking Light, Right, M3	205
Radio, R	296-298
Rear Spoiler Control Module, J223	184-192
Rear Spoiler End Switch 1, F135	189-190
Rear Spoiler End Switch 2, F136	192-193
Rear Spoiler Motor, V52	189
Rear Spoiler Switch, E127	185-187
Rear Window Defogger Indicator Light, K10	269
Rear Window Defogger Switch Light, L39	268
Rear Window Defogger Switch, E15	270
Rear Window Heat Element, Z1	273
Rear Window Wiper Motor, V12	305-307
Rear Window Wiper/Washer Relay, J30	304-305
RPM (Engine Speed) Sensor, G28	68-70
Seat Belt Switch, Left, E24	208
Seat Belt Warning Control Module, J34	206-209
Seat Belt Warning Light, K19	121
Side Marker Lights, Front, M11	197,206
Spark Plug Connectors, P	43-45
Spark Plugs, Q	43-45
Speedometer, G21	130-133
Starter, B	6-7
Tachometer, G5	113-116
Third Speed Coolant FC (Fan Control) Thermoswitch, F165	42
TP (Throttle Position) Sensor, G69	88-89
Turn Signal Indicator Light, Left, K65	124
Turn Signal Indicator Light, Right, K94	125
Turn Signal Light, Left Front, M5	225
Turn Signal Light, Left Rear, M6	227
Turn Signal Light, Right Front, M7	237
Turn Signal Light, Right Rear, M8	236
Turn Signal Switch, E2	222
Voltage Stabilizer, J6	118-120
Voltmeter Light, L25	174

Description	Current track
Voltmeter, G14	173
VR (Voltage Regulator), C1	1-2
VSS (Vehicle Speed Sensor), G22	145-146
Warning Buzzer, H3	109
Washer Nozzle Heater, Left, Z20	257
Washer Nozzle Heater, Right, Z21	258
Washer/Wiper Intermittent Relay, J31	290-294
Windshield and Rear Window Washer Pump, V59	303
Windshield Wiper Intermittent Switch, E22	290-294
Windshield Wiper Motor, V	282-285

Wire Connectors

- T1 - single, near starter
- T1a - single, behind relay panel
- T1c - single, behind relay panel
- T1d1 - single, behind relay panel
- T1e - single, on left A-pillar, near relay panel
- T1f - single, behind relay panel
- T1g - single, behind relay panel
- T1h - single, behind relay panel
- T1i - single, behind relay panel
- T1j - single, in rear lid
- T1l - single, behind relay panel
- T1m - single, behind relay panel
- T1n - single, behind relay panel
- T1o - single, on Back-Up Light, Left
- T1p - single, on Back-Up Light, Left
- T1r - single, behind relay panel
- T1s - single, behind relay panel
- T1y - single, behind relay panel
- T2 - double, above relay panel
- T2a - double, behind console
- T2a1 - double, below driver's seat
- T2a2 - double, in rear lid
- T2b - double, behind instrument panel, center
- T2c - double, behind instrument panel
- T2c - double, on Fresh Air Blower
- T2d - double, on left A-pillar, near relay panel
- T2e - double, behind relay panel
- T2f - double, on right A-pillar, upper part
- T2g - double, in luggage compartment, left
- T2h - double, behind relay panel
- T2i - double, in luggage compartment, left
- T2i - double, in luggage compartment, left
- T2j - double, on engine hood
- T2k - double, on engine hood
- T2l - double, in luggage compartment, left
- T2o - double, near headlight, left
- T2p - double, behind relay panel
- T2p - double, near headlight, right
- T2q - double, in luggage compartment, left
- T2r - double, on Parking Light, Left
- T2s - double, on Side Marker Lights, Front
- T2t - double, on Parking Light, Right
- T2u - double, on Side Marker Lights, Front
- T2v - double, on taillight, left
- T2w - double, on High-mount Brake Light
- T2x - double, black, diagnostic connector, below transmission shift lever cover
- T2z - double, white, diagnostic connector, below transmission shift lever cover
- T3a - three point, near starter
- T3b - three point, near exhaust manifold
- T3c - three point, near starter
- T3d - three point, on Ignition Coil Power Output Stage
- T3e - three point, on Back-Up Light, Right
- T3f - three point, on Headlight, Left
- T3g - three point, on Headlight, Right
- T4 - four point, on coolant fan control module
- T4a - four point, behind steering column switch cover
- T4b - four point, near exhaust manifold
- T4d - four point, behind steering column switch cover
- T4d - four point, behind steering column switch cover
- T4e - four point, on taillight, left
- T4f - four point, on taillight, right
- T4i - four point, behind relay panel
- T4l - four point, behind relay panel
- T5 - five point, on Fresh Air Blower Series Resistance
- T5b - five point, behind steering column switch cover
- T5c - five point, behind steering column switch cover
- T6 - six point, on Rear Spoiler Switch
- T7 - seven point, behind steering column switch cover
- T7 - seven point, behind steering column switch cover
- T8 - eight point, on radio
- T8a - eight point, behind relay panel
- T8d - eight point, on Brake and Parking Brake Indicator Light
- T8e - eight point, on coolant fan control module
- T10 - ten point, on Rear Spoiler Control Module
- T28 - twenty-eight point, on instrument cluster
- T42 - fourty-two point, on cylinder head, left
- T68 - sixty-eight point, on ECM (Engine Control Module)
- TV5 - Wire Connector, Terminal 15a, above relay panel

Welded wiring harness points

- (A11) – plus connection (58b), in instrument panel wiring harness
- (A15) – plus connection (15), in instrument wiring harness
- (B6) – wire connection (s), in steering column switch wiring harness
- (B8) – plus connection (58), in light switch wiring harness
- (C1) – plus connection (58L), in left front wiring harness
- (C2) – plus connection (58R), in left front wiring harness
- (C13) – plus connection, in dual tone horn wiring harness
- (C15) – wire connection –1–, in headlight wiring harness
- (C16) – wire connection –2–, in headlight wiring harness
- (E6) – plus connection –1– (15), in MFI (Multiport Fuel Injection) wiring
- (E7) – plus connection (87a), in MFI (Multiport Fuel Injection) wiring harness
- (G3) – plus connection, in injector wiring harness
- (K21) – wire connection –1–, in coolant fan wiring harness
- (K22) – wire connection –2–, in coolant fan wiring harness
- (K28) – plus connection (15), in coolant fan wiring harness
- (W9) – plus connection (30), in rear wiring harness
- (W10) – plus connection (15), in rear wiring harness

Ground connections

- (1) – ground strap, battery to body
- (2) – ground strap, transmission to body
- (15) – ground connection, on cylinder-head
- (18) – ground connection, on engine block
- (30) – ground connection, –1–, beside relay panel
- (43) – ground connection, on right A–pillar, lower part
- (50) – ground connection, in luggage compartment, left
- (53) – ground connection, in rear lid, right
- (80) – ground connection –1–, in instrument panel wiring harness
- (86) – ground connection –1–, in rear wiring harness
- (98) – ground connection, in rear lid wiring harness
- (119) – ground connection –1–, in headlight wiring harness
- (120) – ground connection –2–, in headlight wiring harness
- (125) – ground connection –3–, in headlight wiring harness
- (129) – ground connection, in dual tone horn wiring harness
- (137) – ground connection (end stage), in MFI (Multiport Fuel Injection)
- (138) – ground connection (control module), in MFI (Multiport Fuel Injection)
- (159) – ground connection –2–, in MFI (Multiport Fuel Injection) wiring harness
- (182) – ground connection –1–, in engine compartment wiring harness
- (193) – ground connection –1–, in coolant fan wiring harness

Wiring Color code	
BK	BLACK
BR	BROWN
CL	CLEAR
R	RED
Y	YELLOW
G	GREEN
LT. G	LIGHT GREEN
BL	BLUE
V	VIOLET
GY	GRAY
W	WHITE
OR	ORANGE

Wire Size

Wiring Diagrams identify wires by the metric wire size. Metric wire sizes indicate cross-sectional area in square millimeters (mm^2). The chart below lists metric wire sizes and their approximate equivalents in American Wire Gauge (AWG) sizes.

Metric size Cross-section (mm2)	American Wire Gauge size (AWG)
0.35	22
0.50	20
0.75	18
1.00	16
1.50	14
2.50	12
4.00	10
6.00	8
16.00	4
25.00	2
35.00	2

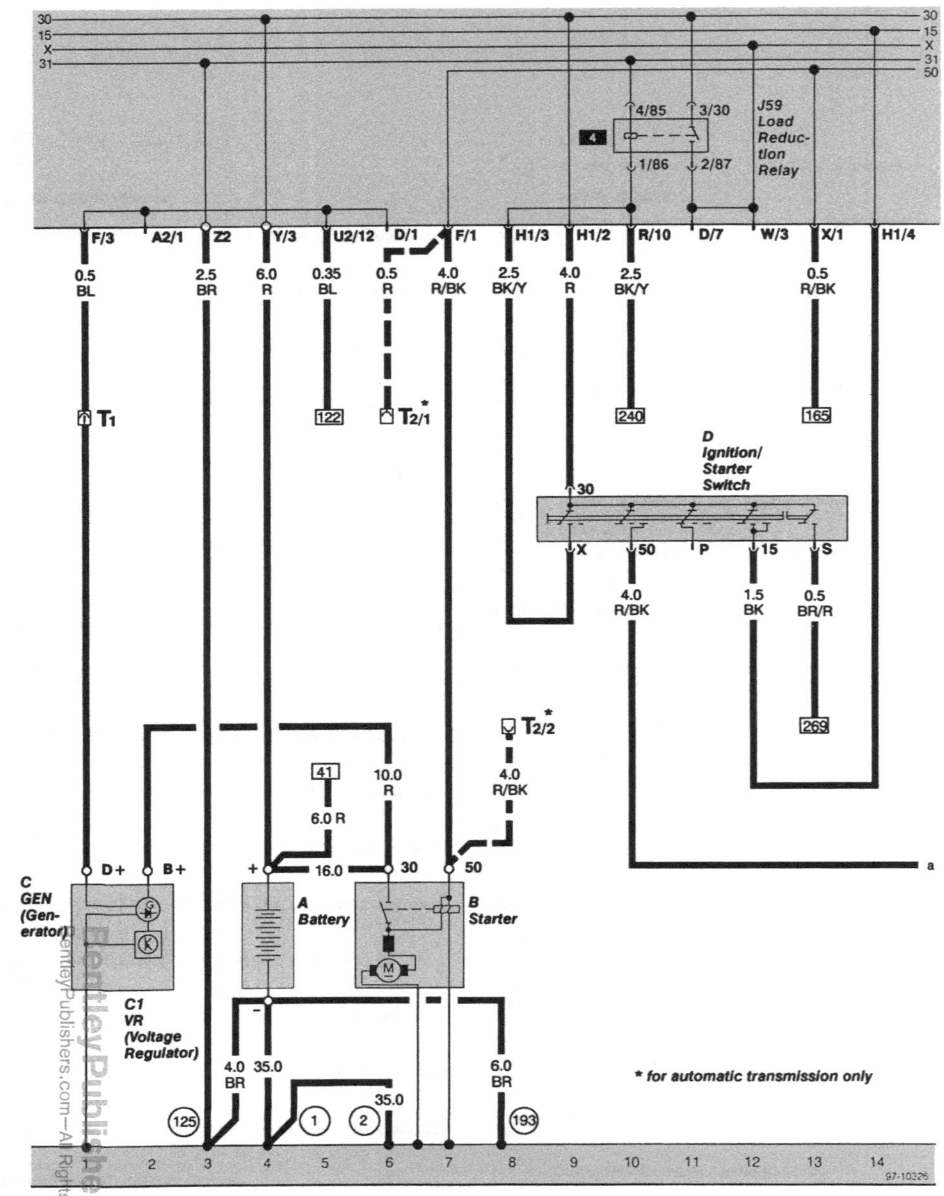

Corrado 1992–1993
Engine code – AAA
Production up to Jan. 1993

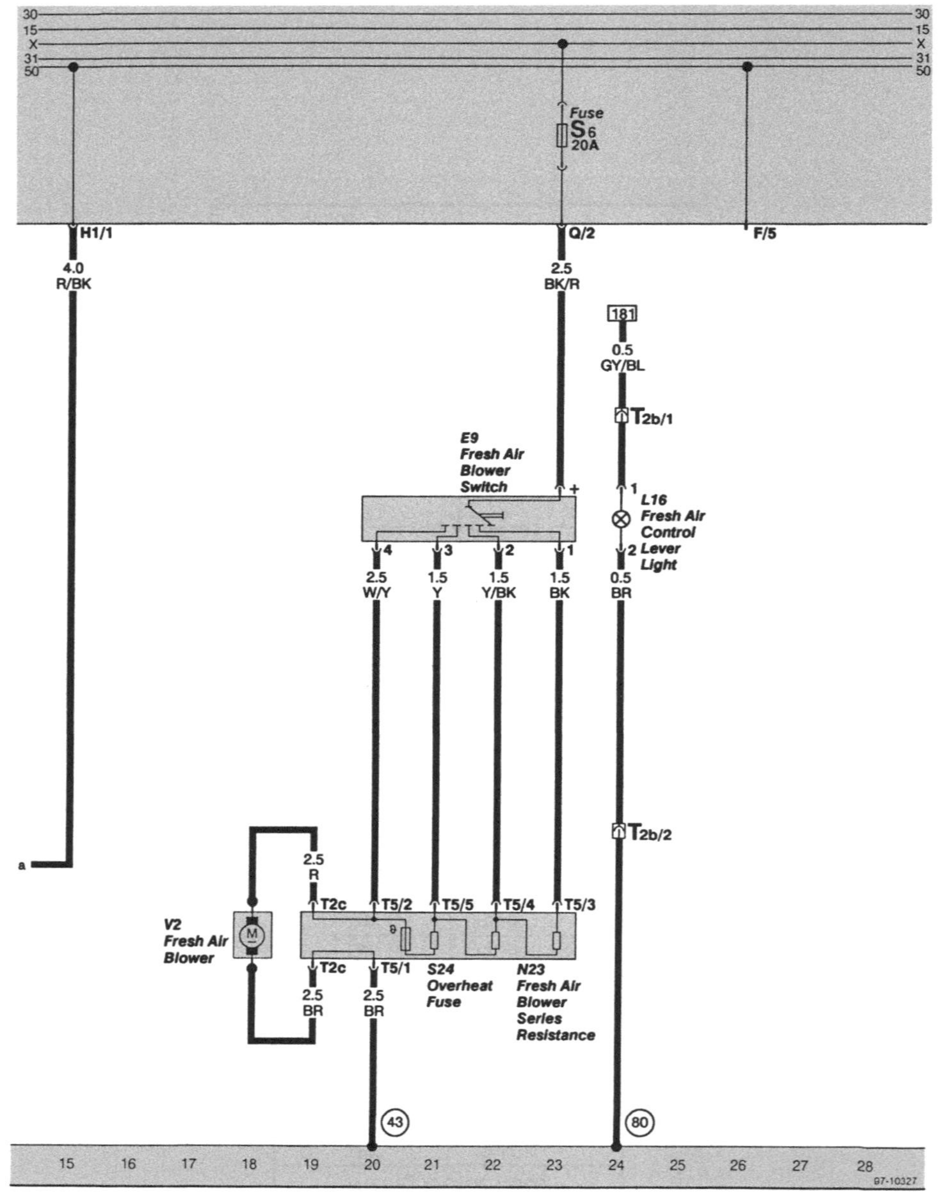

Corrado 1992–1993
Engine code – AAA
Production up to Jan. 1993

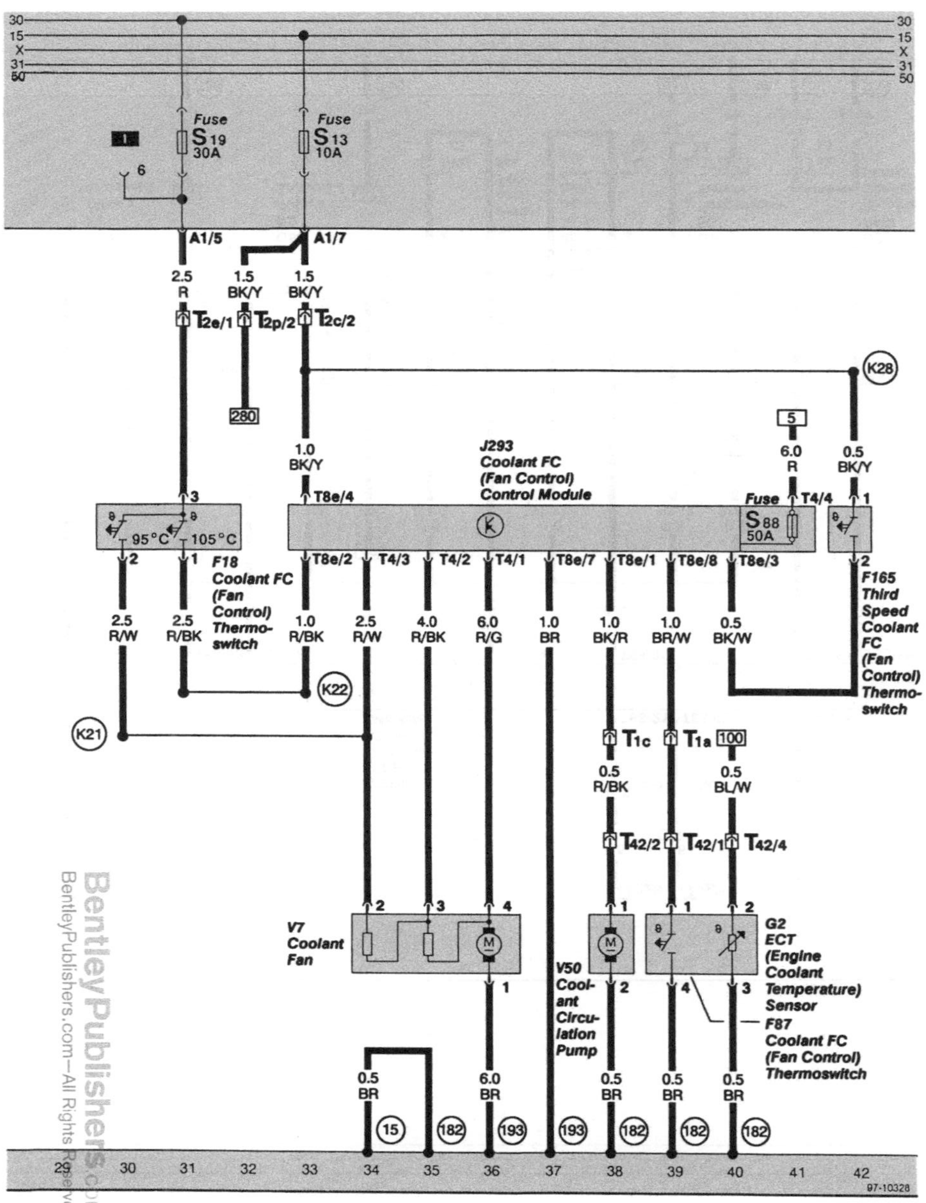

Corrado 1992–1993
Engine code – AAA
Production up to Jan. 1993

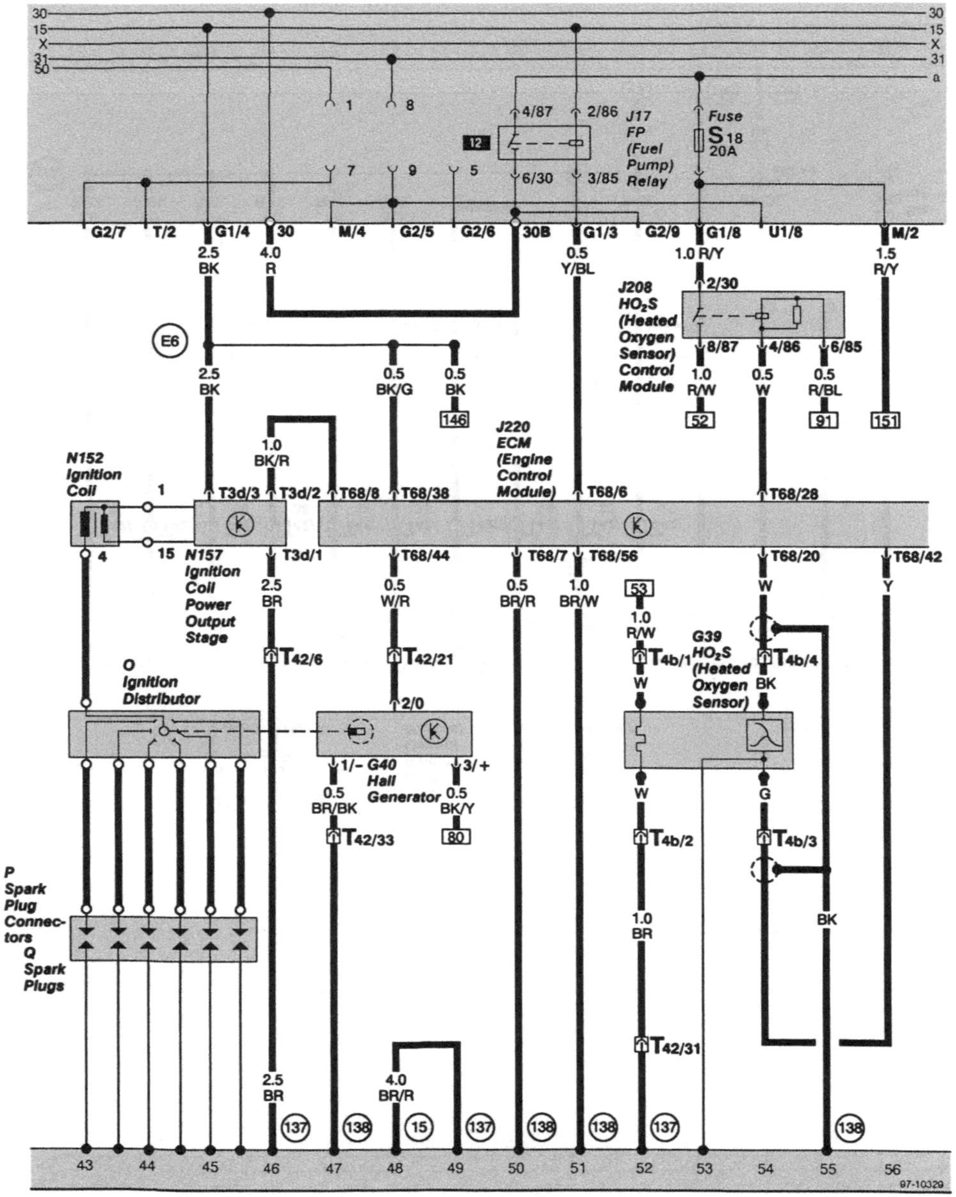

Corrado 1992–1993
Engine code – AAA
Production up to Jan. 1993

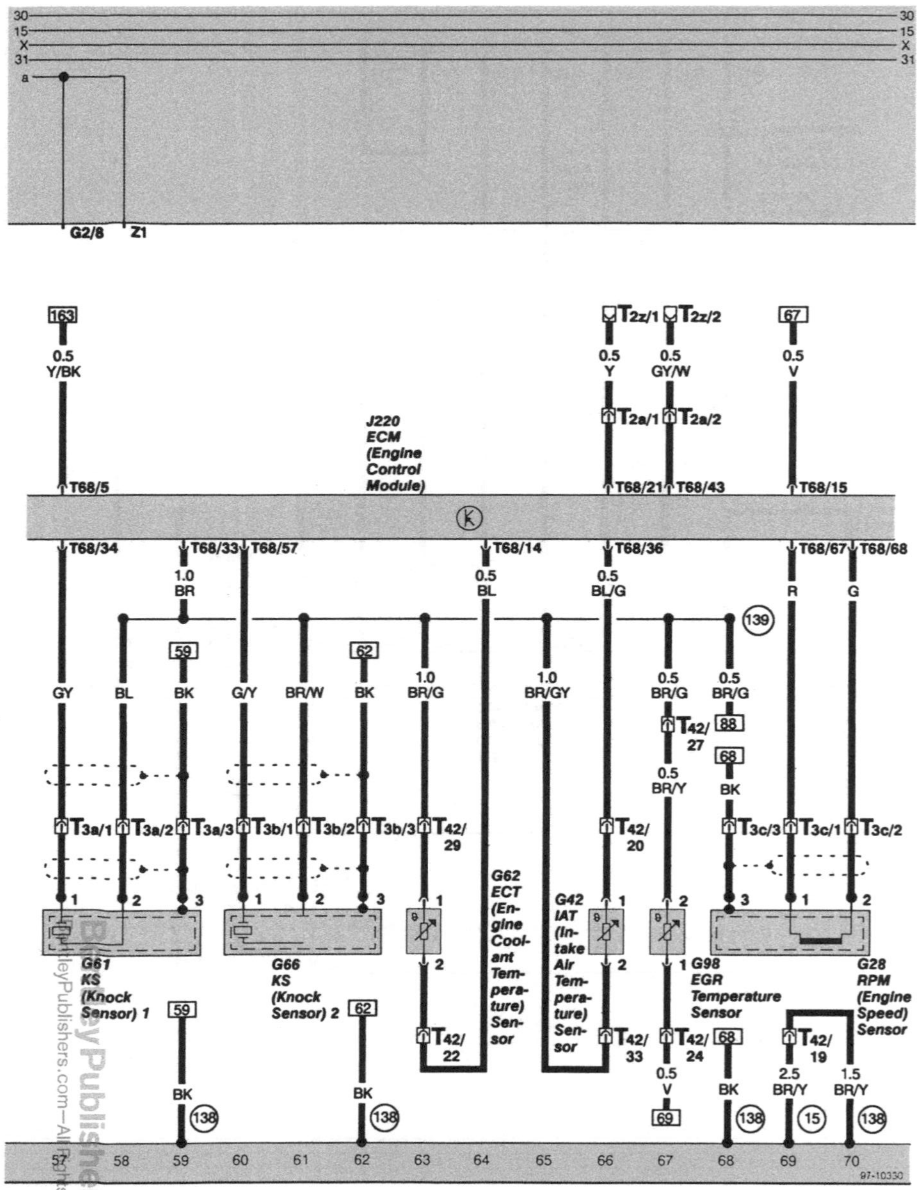

Corrado 1992–1993
Engine code – AAA
Production up to Jan. 1993

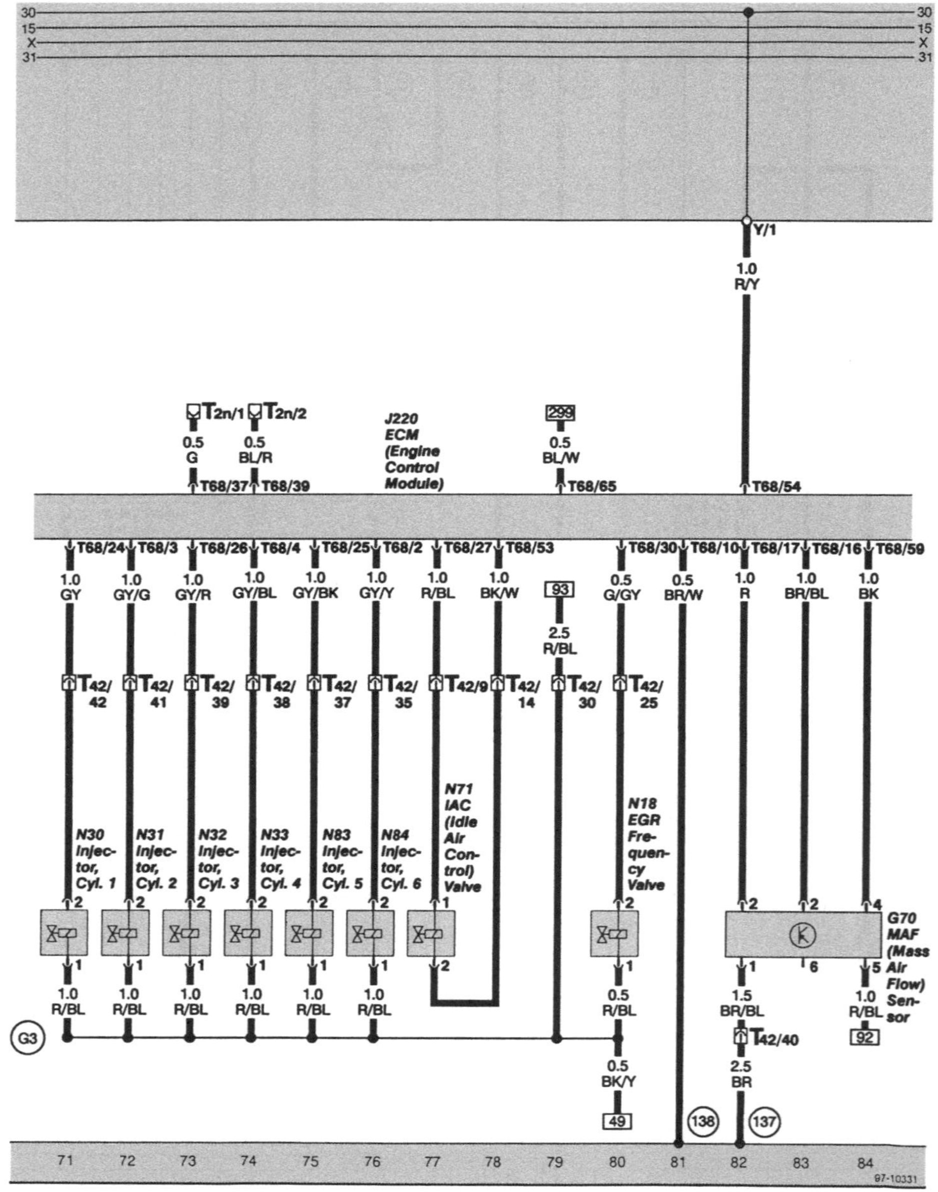

Corrado 1992–1993
Engine code – AAA
Production up to Jan. 1993

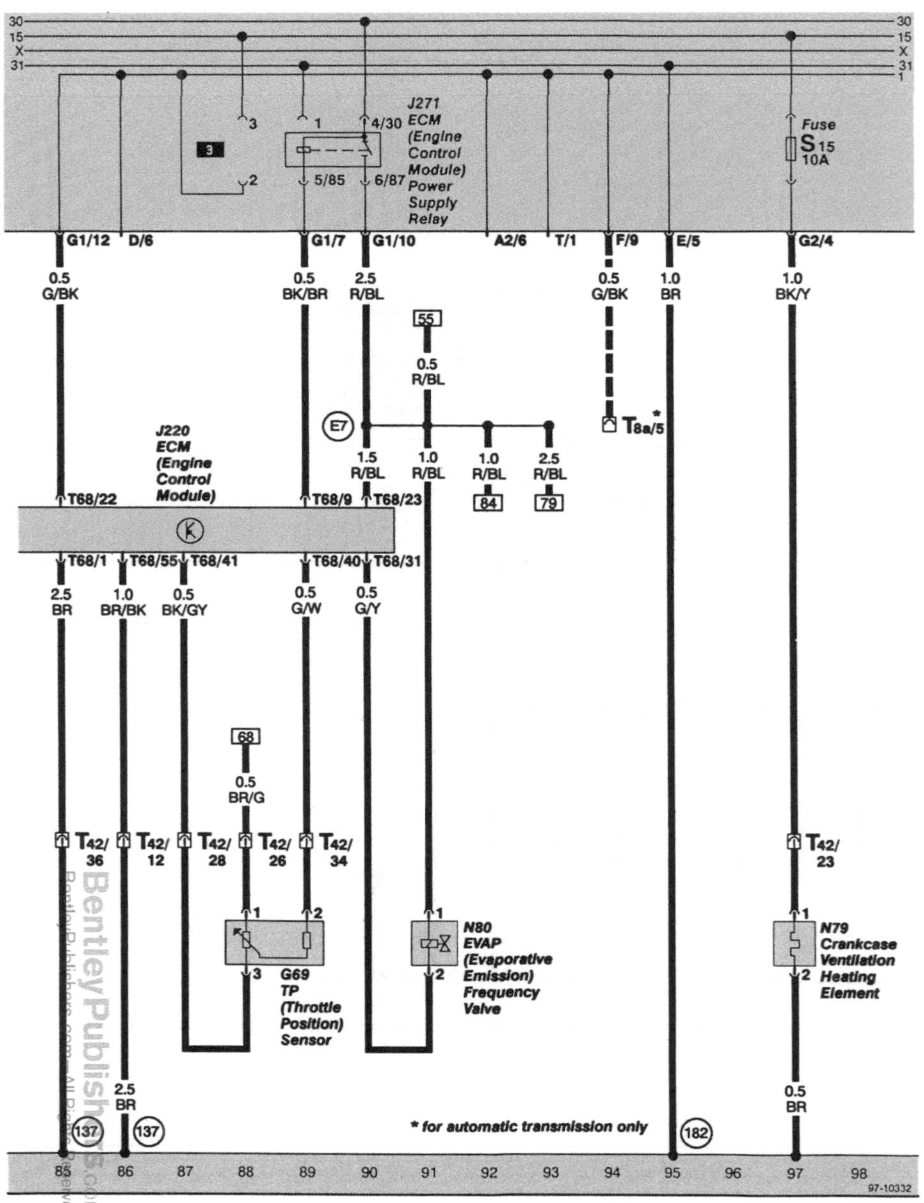

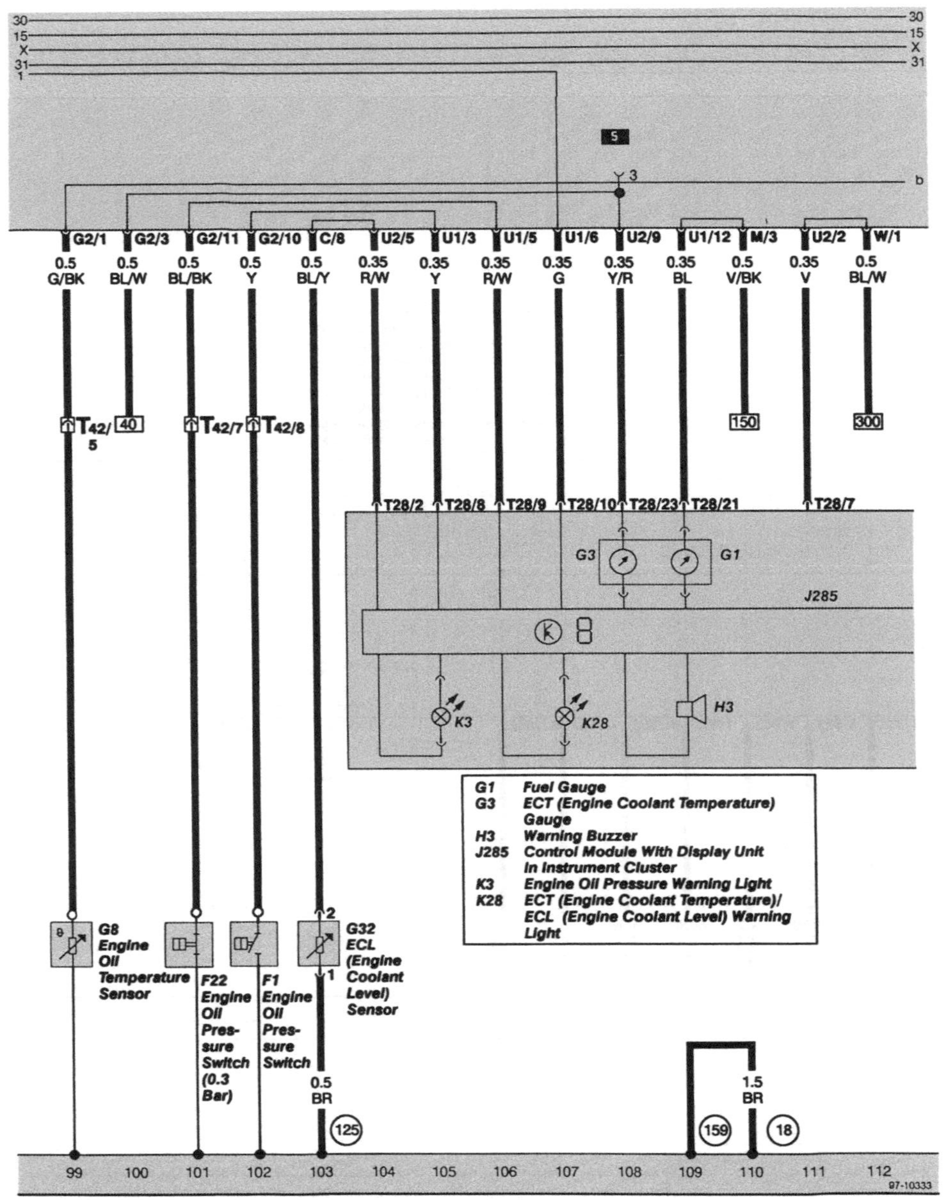

Corrado 1992–1993
Engine code – AAA
Production up to Jan. 1993

Corrado 1992–1993
Engine code – AAA
Production up to Jan. 1993

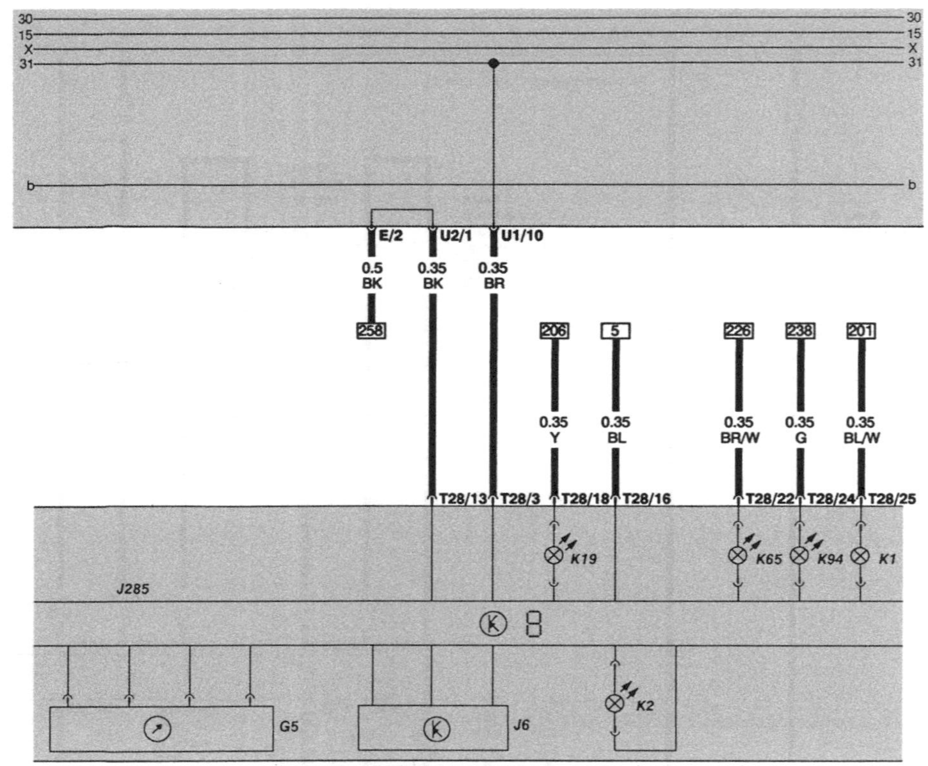

G5	Tachometer
J6	Voltage Stabilizer
J285	Control Module With Display Unit In Instrument Cluster
K1	Headlight High Beam Indicator Light
K2	GEN (Generator) Warning Light
K19	Seat Belt Warning Light
K65	Turn Signal Indicator Light, Left
K94	Turn Signal Indicator Light, Right

Corrado 1992–1993
Engine code – AAA
Production up to Jan. 1993

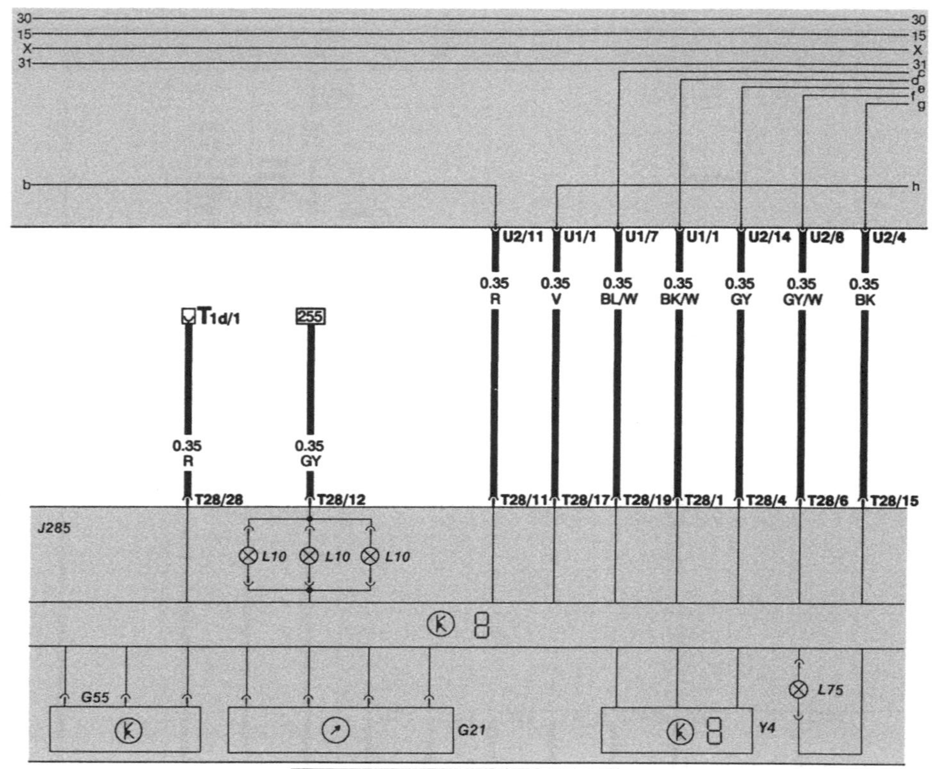

G21	Speedometer
G55	Multi-Function Indicator Vacuum Sensor
J285	Control Module With Display Unit In Instrument Cluster
L10	Instrument Panel Light
L75	Digital Display Light
Y4	Odometer Display

Corrado 1992–1993
Engine code – AAA
Production up to Jan. 1993

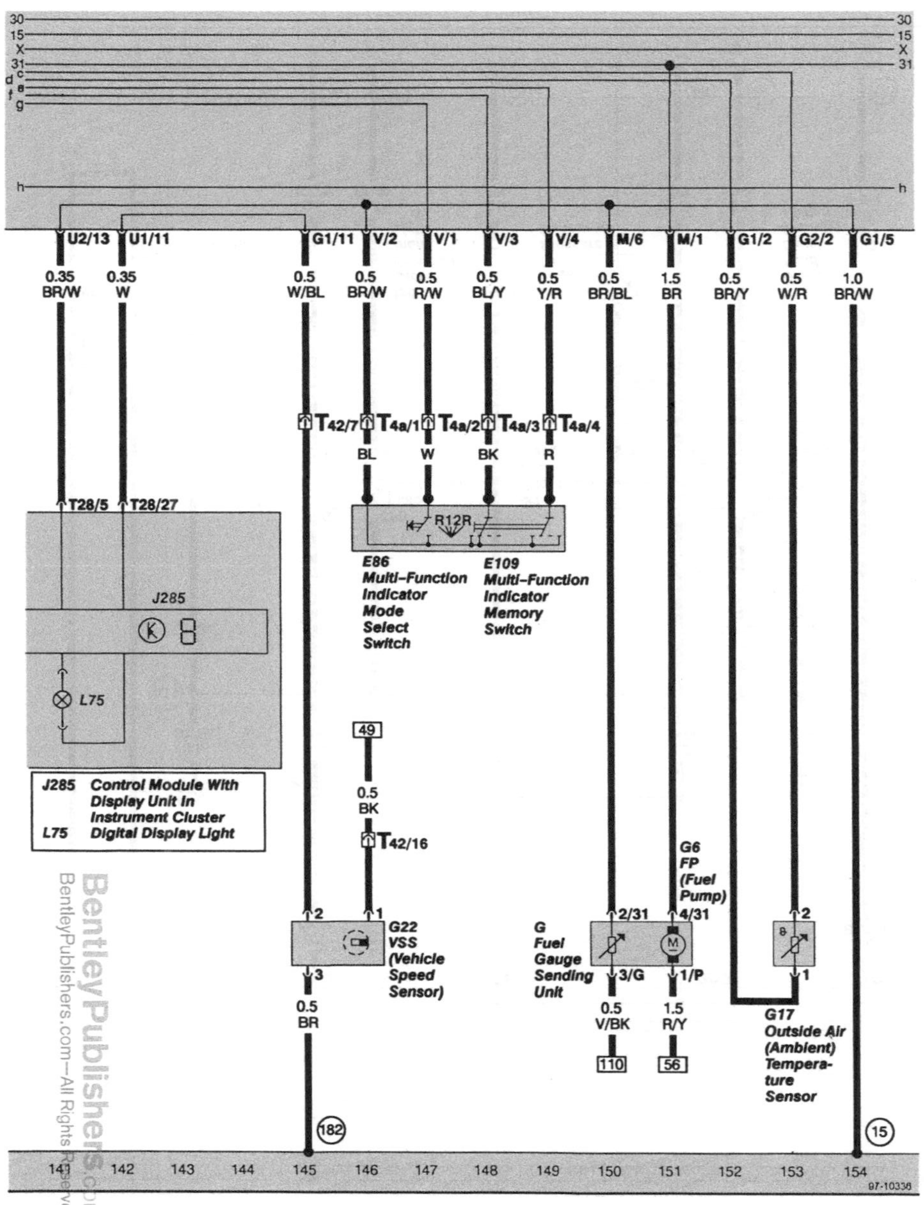

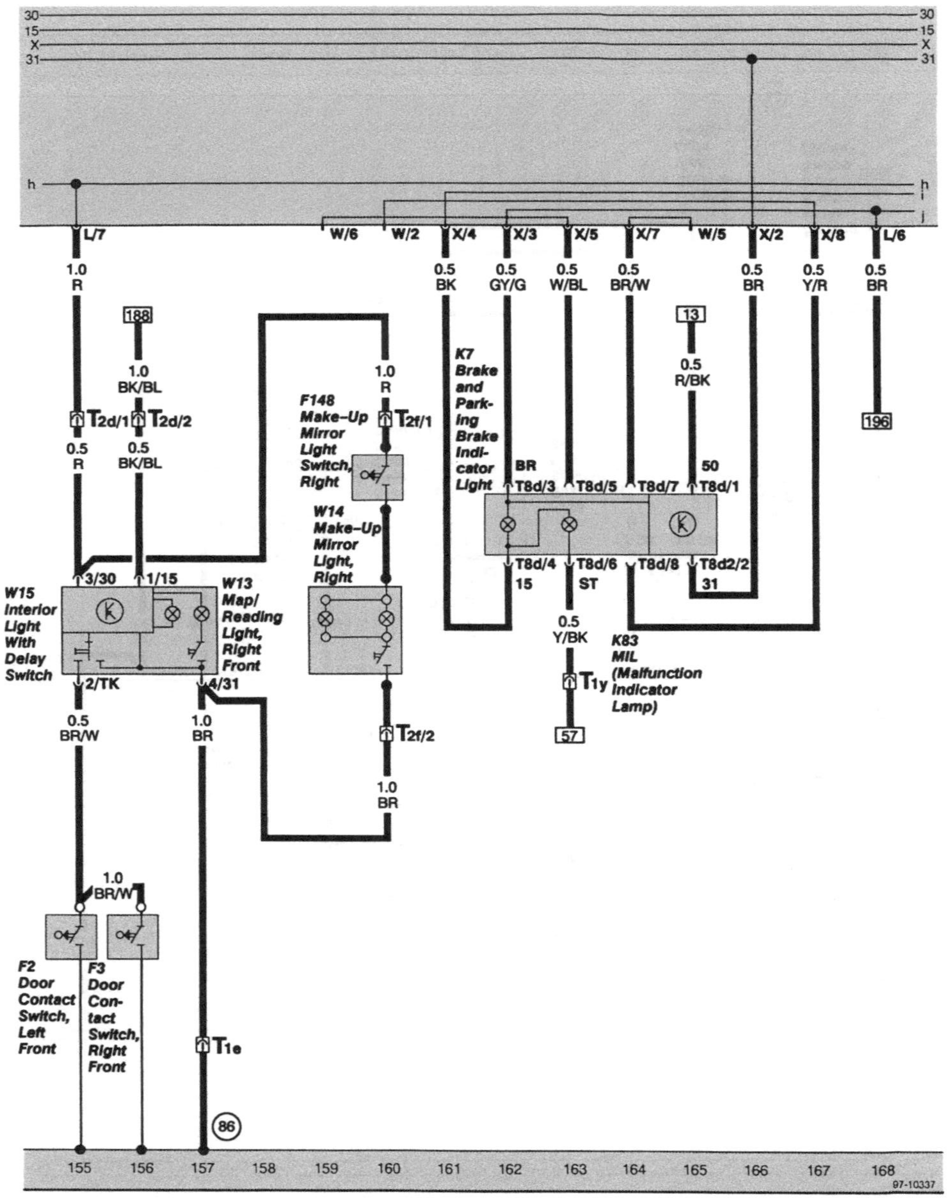

Corrado 1992–1993
Engine code – AAA
Production up to Jan. 1993

Corrado 1992–1993
Engine code – AAA
Production up to Jan. 1993

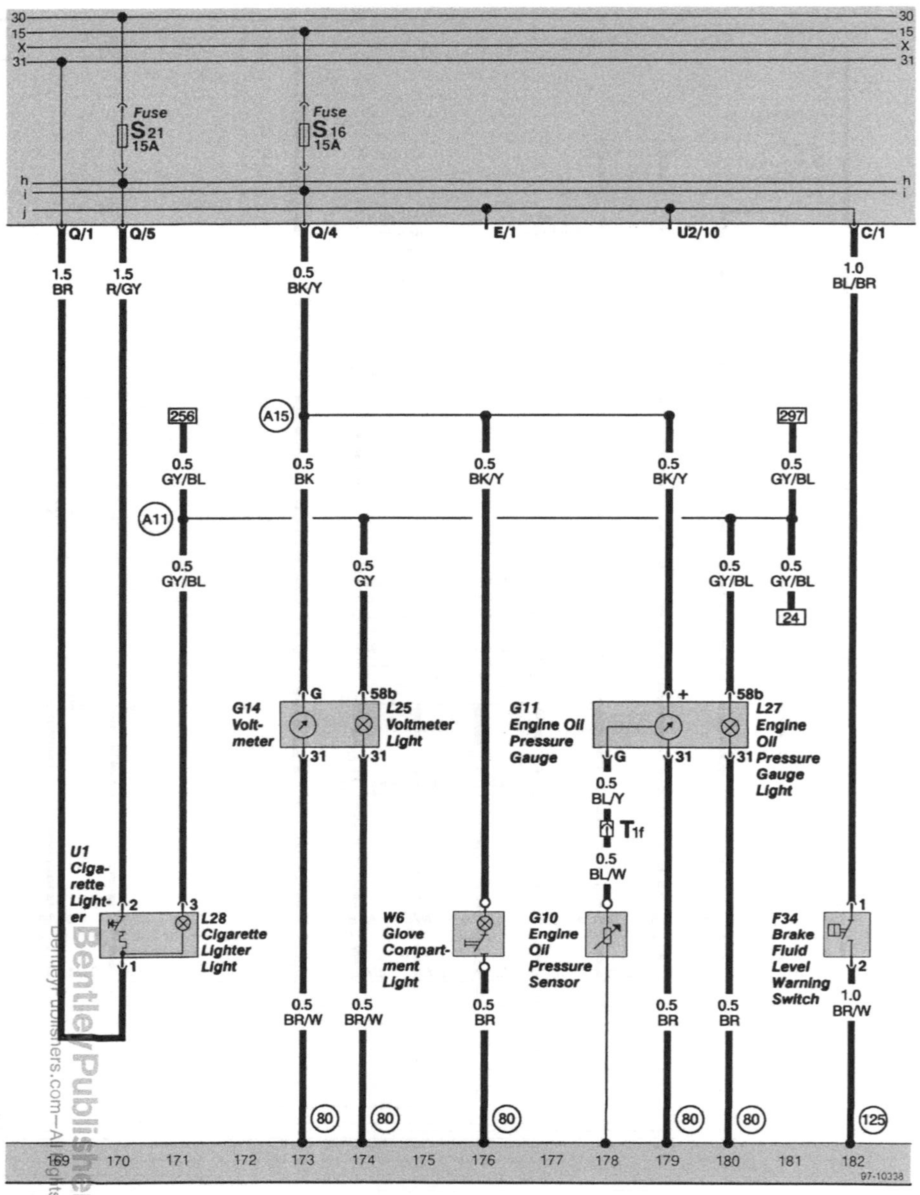

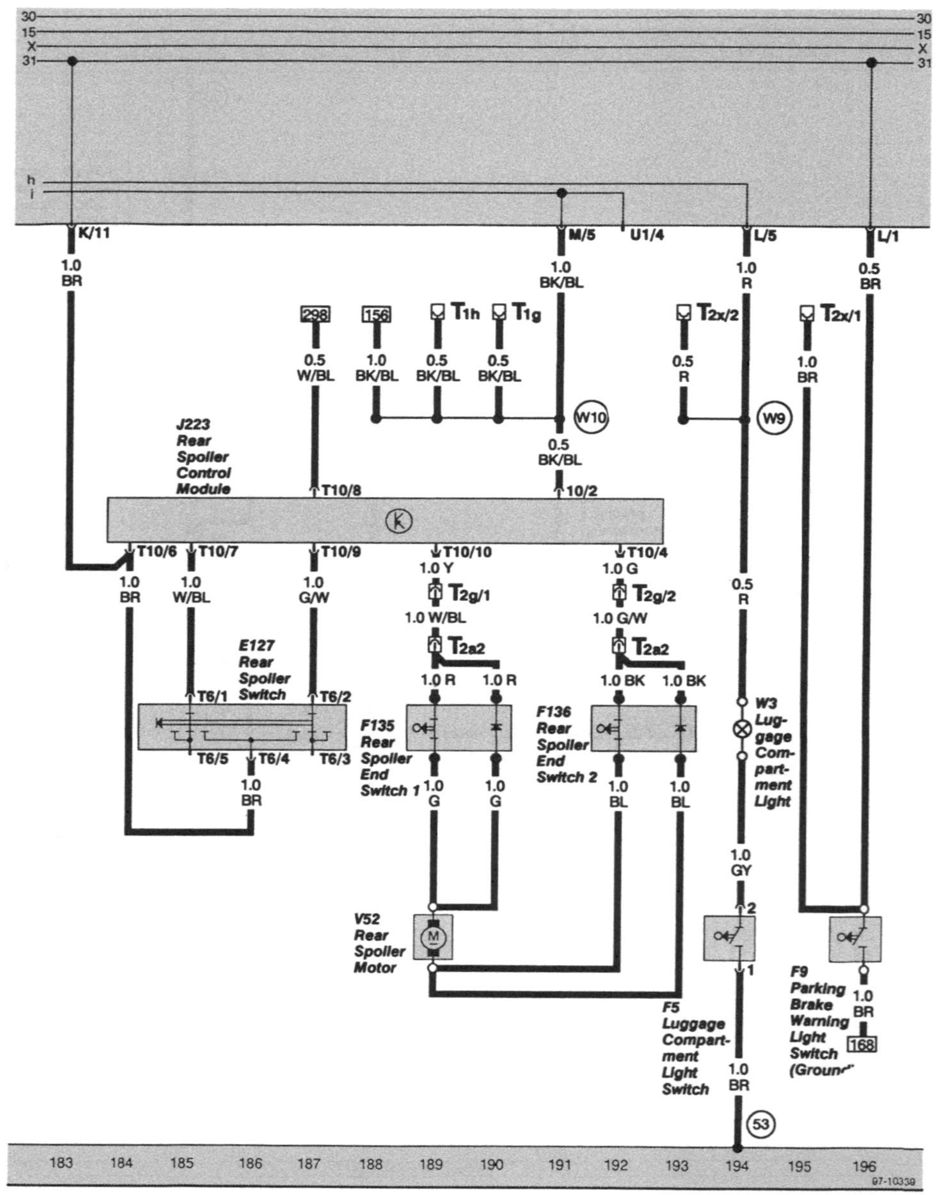

Corrado 1992–1993
Engine code – AAA
Production up to Jan. 1993

Corrado 1992–1993
Engine code – AAA
Production up to Jan. 1993

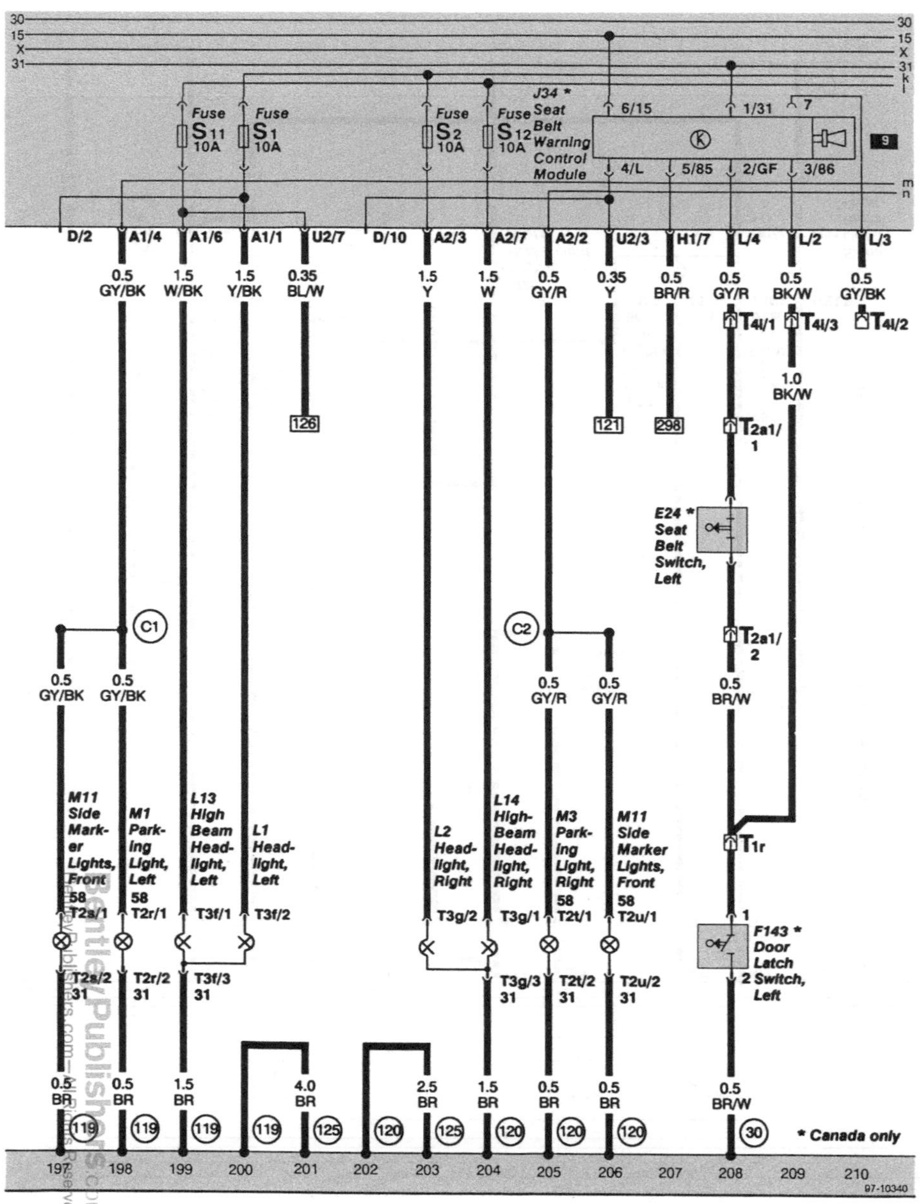

105 Headlights

Corrado 1992–1993
Engine code – AAA
Production up to Jan. 1993

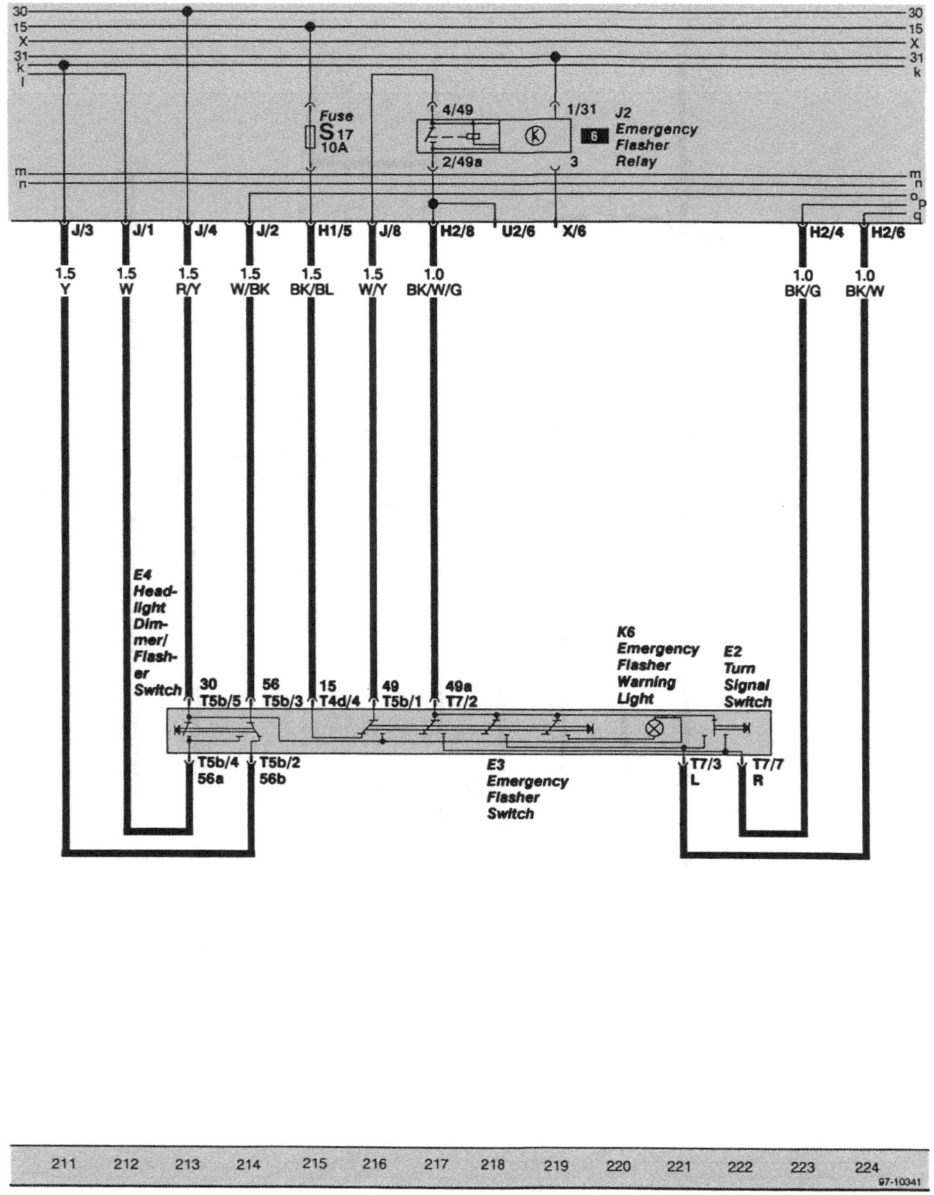

Corrado 1992–1993
Engine code – AAA
Production up to Jan. 1993

Emergency flasher switch
Turn signal switch 106

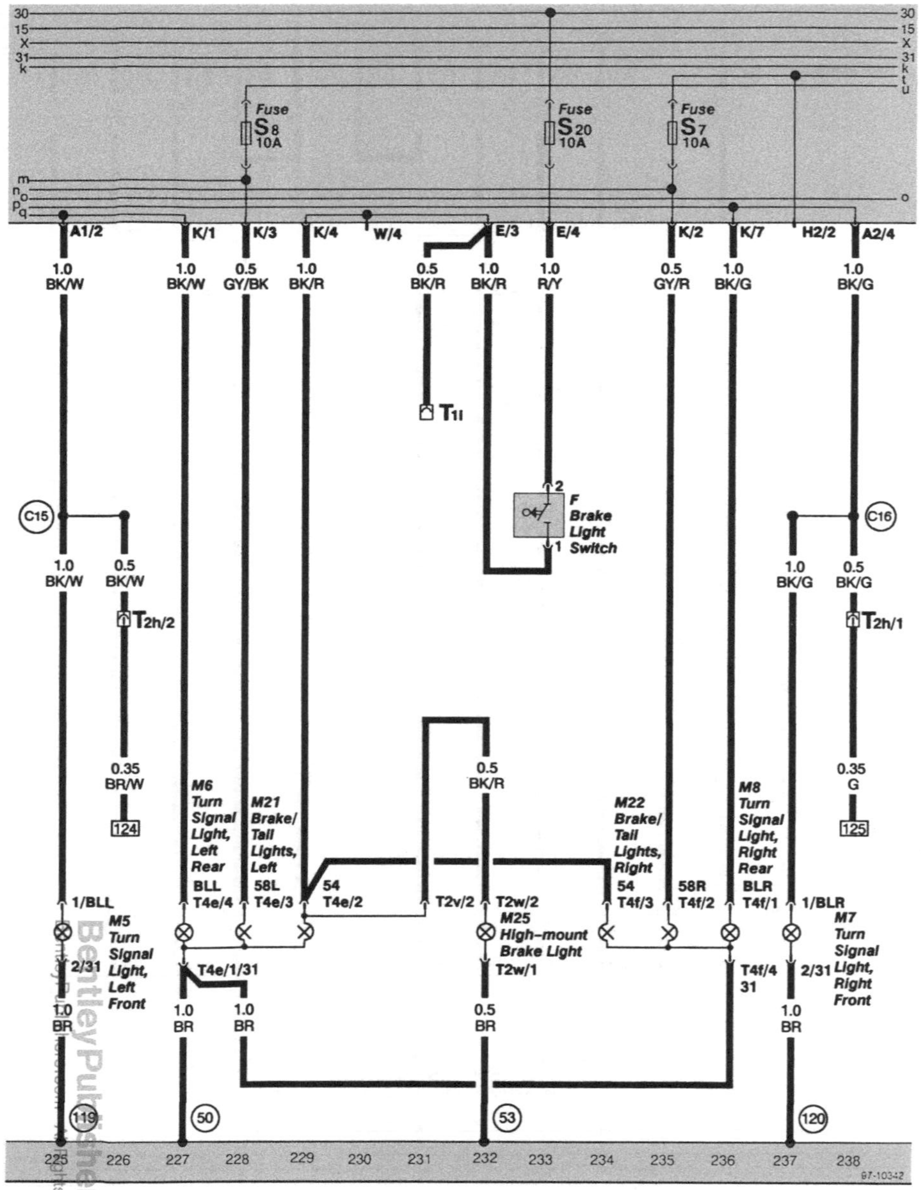

Corrado 1992–1993
Engine code – AAA
Production up to Jan. 1993

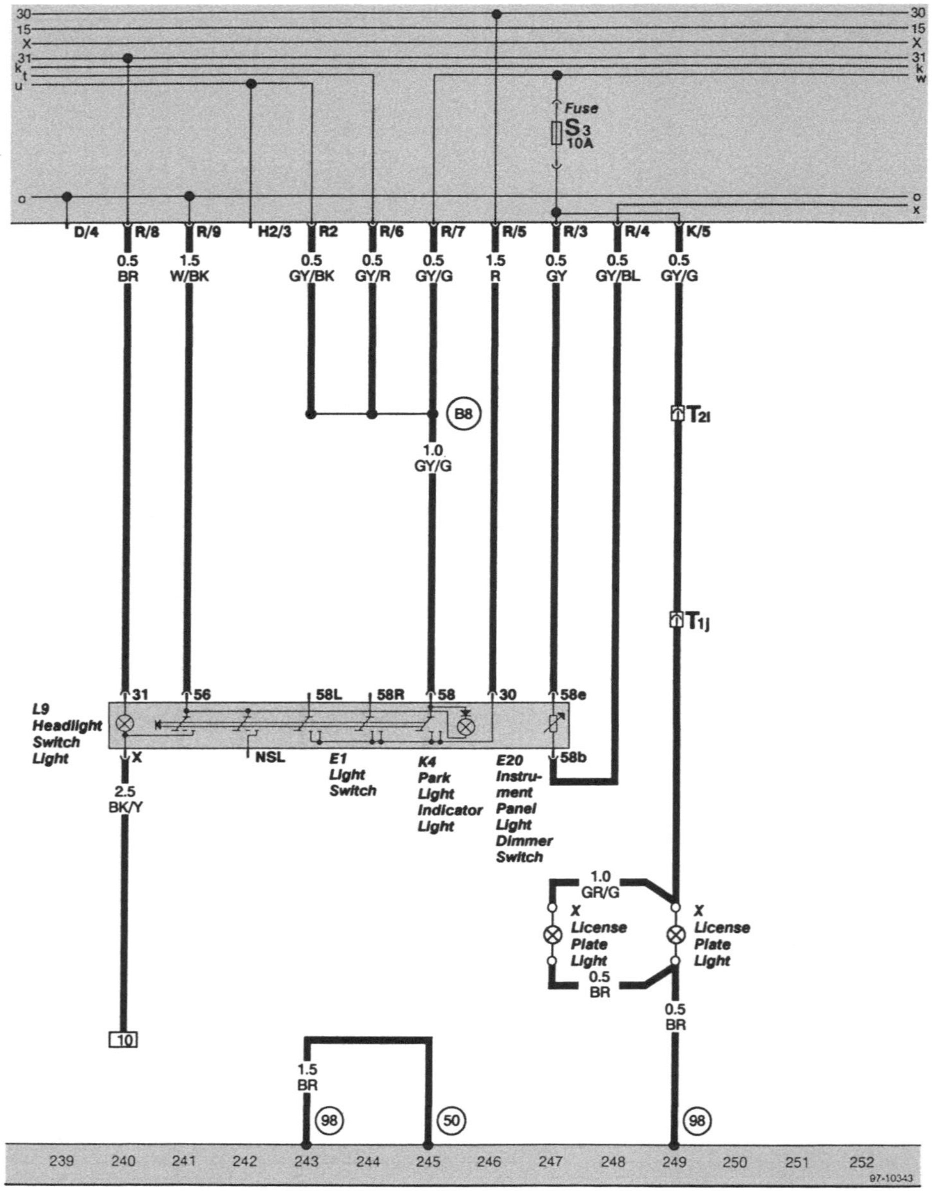

Corrado 1992–1993
Engine code – AAA
Production up to Jan. 1993

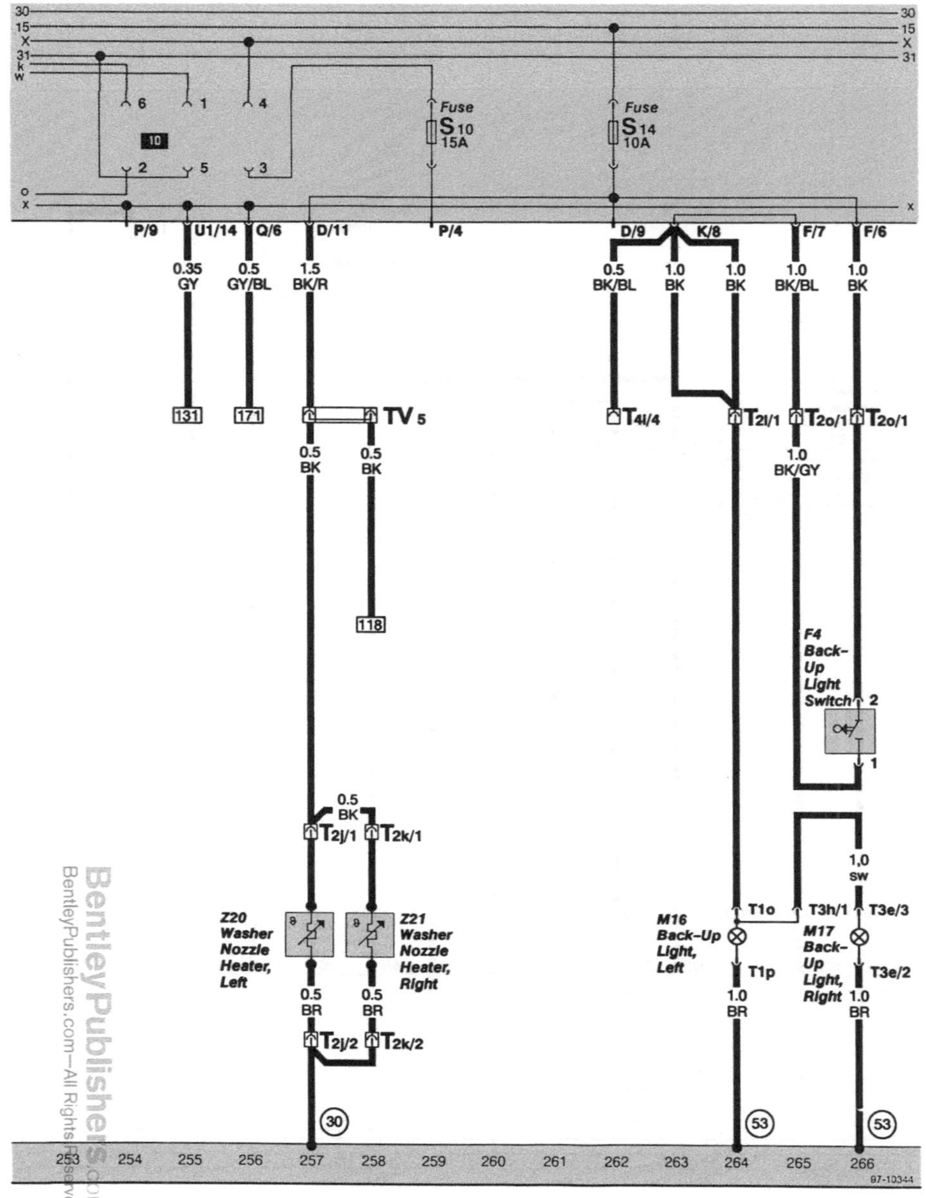

Corrado 1992–1993
Engine code – AAA
Production up to Jan. 1993

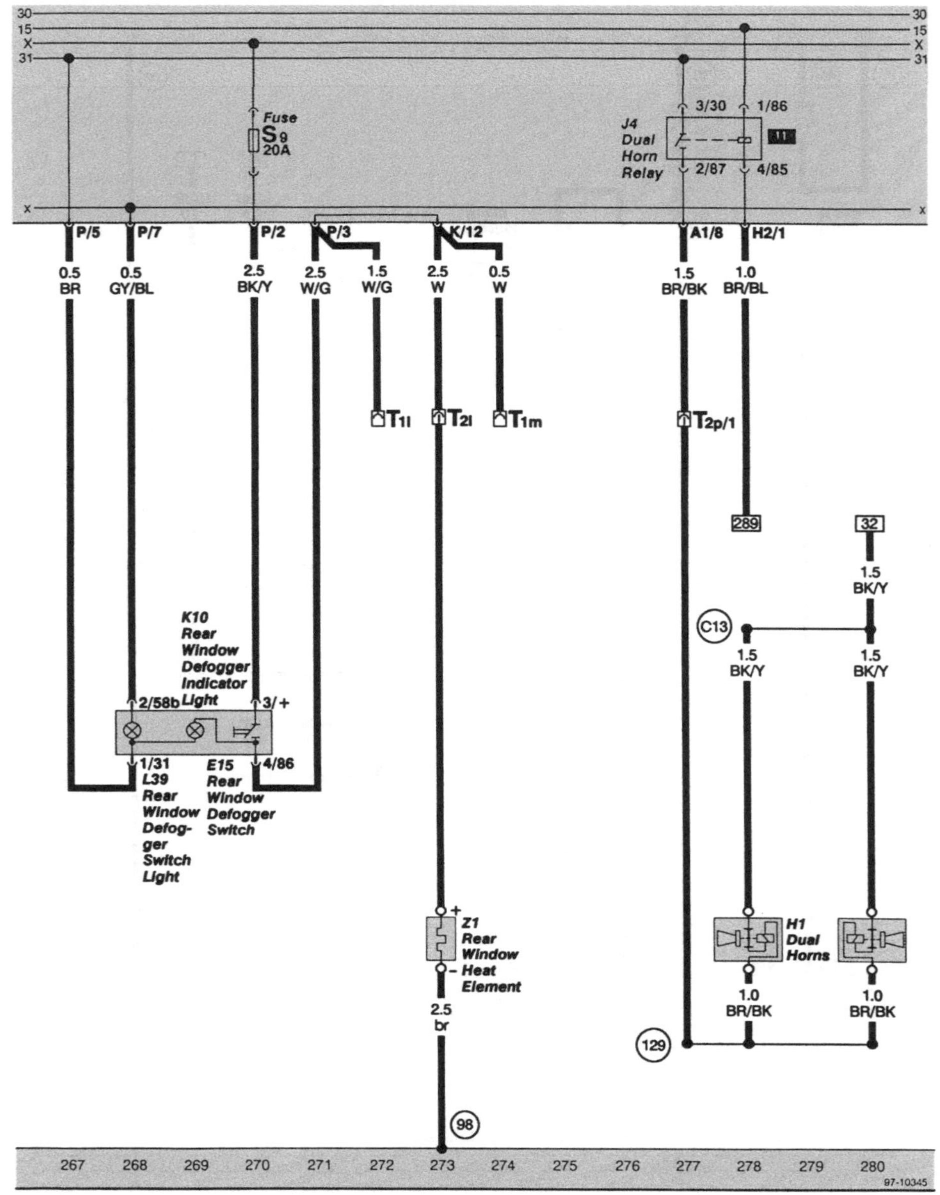

Corrado 1992–1993
Engine code – AAA
Production up to Jan. 1993

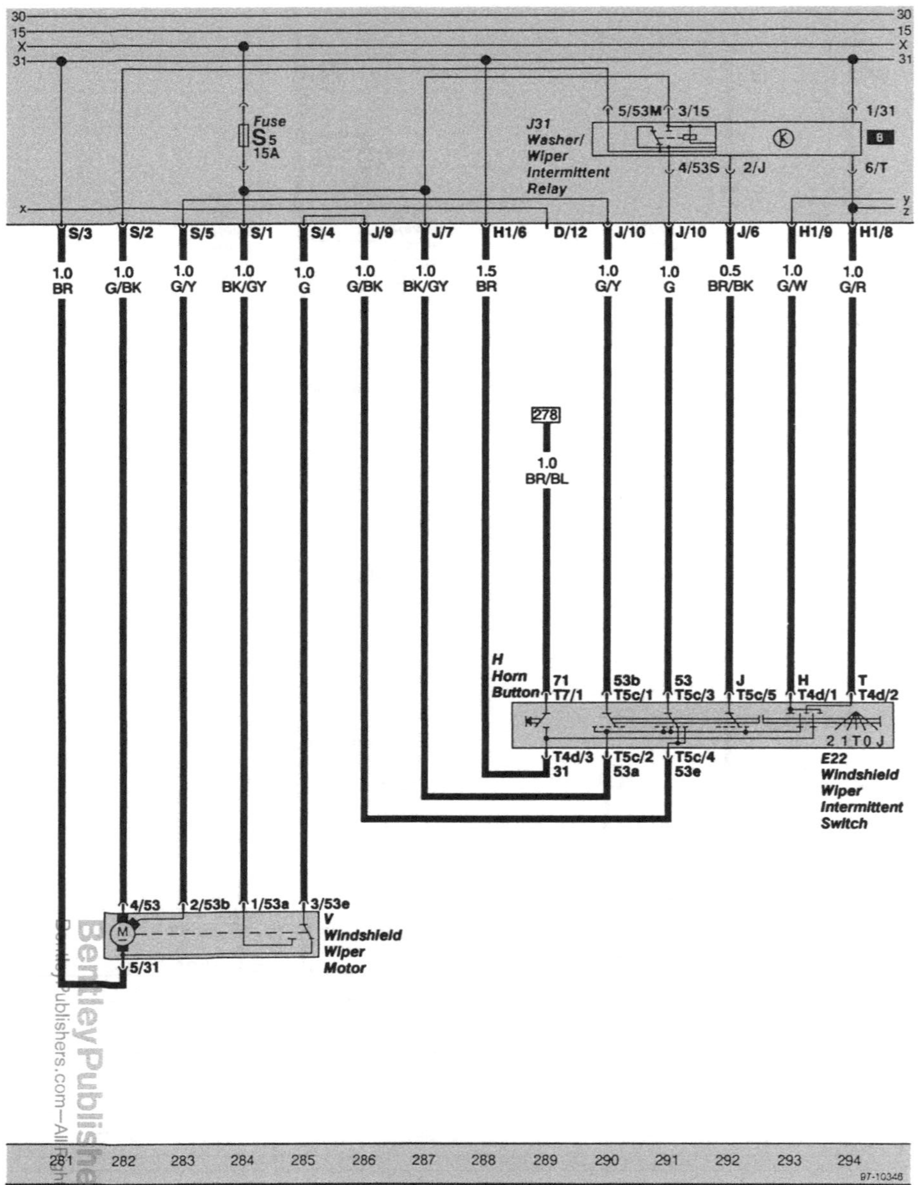

Corrado 1992–1993
Engine code – AAA
Production up to Jan. 1993

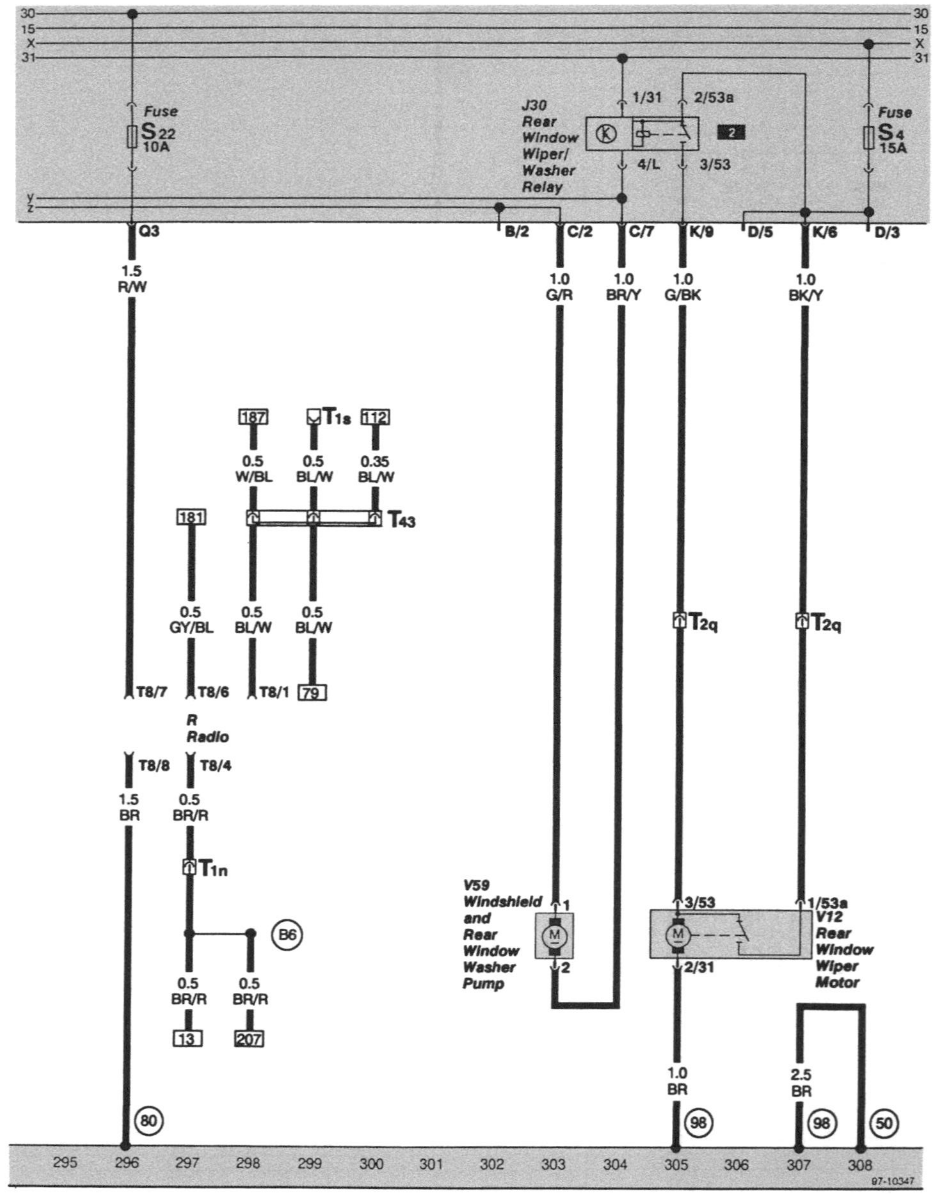

Corrado 1992–1993
Engine code – AAA
Production up to Jan. 1993

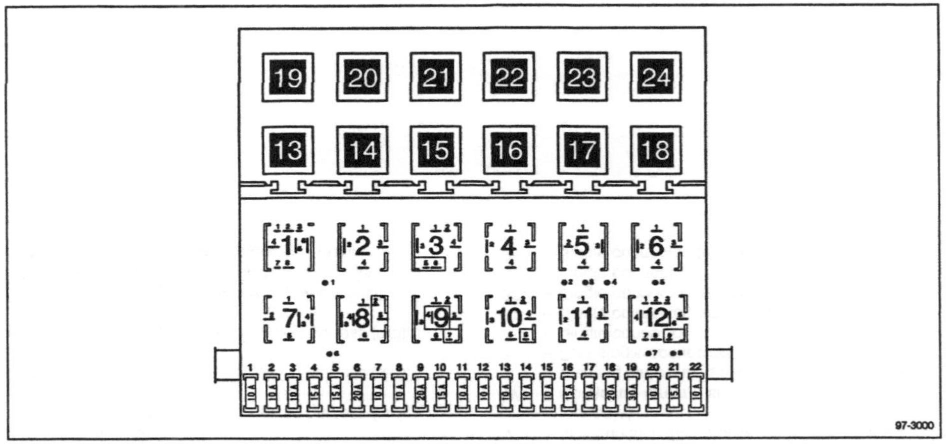

Fuse arrangement
according to the numbers on fuse panel from left to right:

		Amp.	Current track
1	Headlight, Low Beam, Left,	10	200
2	Headlight, Low Beam, Right	10	203
3	Instrument Panel Lights	10	247
4	Rear Window Wiper Motor	15	308
5	Windshield Wiper Motor	15	284
6	Fresh Air Blower	20	23
7	Brake/Tail Lights, Right	10	235
8	Brake/Tail Lights, Left	10	228
9	Rear Window Defogger	20	270
10	Open		
11	Headlight, High Beam, Left	10	199
12	Headlight, High Beam, Right	10	204
13	Horn	15	280
14	Back-up Lights, Washer Nozzle Heater	10	262
15	Positive Crankcase Ventilation (PCV) Heating Element	10	28
16	Voltmeter, Engine oil Pressure Gauge	15	177
17	Emergency Flashers	10	215
18	HO_2S (Heated Oxygen Sensor) Control	20	53
19	Coolant FC (Fan Control) Thermoswitch	30	31
20	Brake Lights	10	233
21	Cigarette Lighter	15	170
22	Radio	10	296

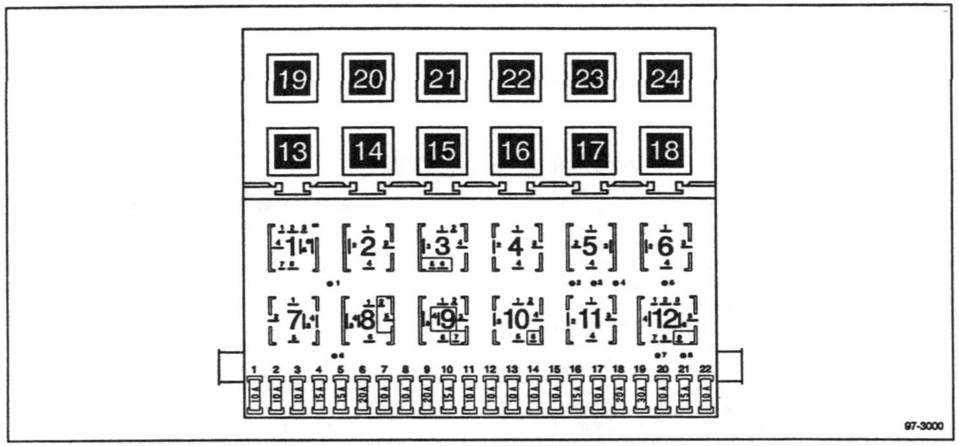

Relay location on fuse/relay panel

1 A/C Relay (13)*
2 Rear Window Wiper/Washer Relay (72)*
3 Motronic Engine Control Module (ECM) Power Supply Relay (32, 30)*
4 Load Reduction Relay (18)*
5 Engine Coolant Level (ECL) Control Module (43)*
6 Emergency Flasher Relay (21)*
7 Headlight Washer System Relay (33)*
8 Washer/Wiper Intermittent Relay (19, 99)*
9 Seat Belt Warning System Relay – Canada (4)*
10 Fog Light Relay (53)*
11 Dual Horn Relay (53)*
12 Fuel Pump (FP) Relay (67)*

Separate relays above fuse/relay panel

13 Back-Up Light Fuse (Auto. Trans.)
14 Coolant After-Run Control Module (31)*
15 ABS Hydraulic Pump Pelay (78)*
16 ABS Relay (79)*
17 Open
18 Open
19 A/C Relay (56)*
20 Park/Neutral Position (PNP) Relay (150)*
21 ABS Hydraulic Pump Fuse/Power Window Circuit Breaker
22 ABS Valves Fuse
23 Cruise Control Fuse/Automatic Shoulder Belt System Fuse
24 Open

*Numbers in parentheses indicate production control number stamped on relay housing.

Note: Locations of relays in positions 13–24 may vary from vehicle to vehicle depending on model and option content.

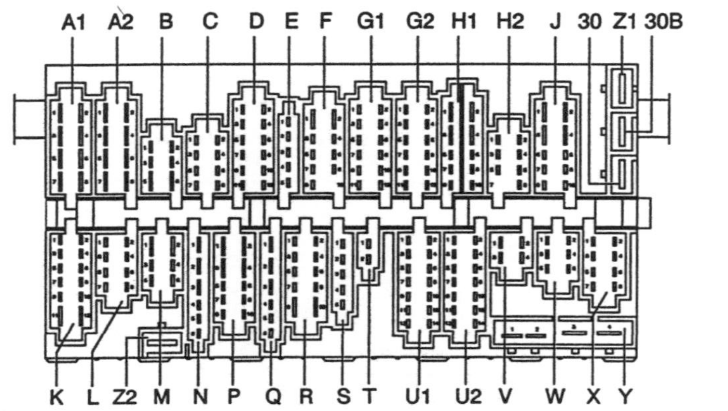

Connections and plugs on fuse/relay panel

A1 – Headlight wiring harness (yellow), eight point
A2 – Headlight wiring harness (yellow), eight point
B – Open
C – Headlight wiring harness (yellow), eight point
D – Optional equipment wiring harness (green), twelve point
E – Instrument wiring harness (green), five point
F – Engine compartment wiring harness, right (white), ten point
G1 – Engine compartment wiring harness, right (white), twelve point
G2 – Engine compartment wiring harness, right (white), twelve point
H1 – Steering column switch wiring harness (red), ten point
H2 – Steering column switch wiring harness (red), eight point
J – Steering column switch wiring harness (red), ten point
K – Rear wiring harness (black), twelve point
L – Rear wiring harness (black), eight point
M – Rear wiring harness (black), six point
N – A/C wiring harness (green), six point
P – Rear window defroster wiring harness (blue), ten point
Q – Instrument wiring harness (blue), six point
R – Light switch wiring harness (blue), ten point
S – Engine compartment wiring harness, right (white), five point
T – Connector (green), two point
U1 – Instrument cluster wiring harness (blue), fourteen point
U2 – Instrument cluster wiring harness (blue), fourteen point
V – Multi-function indicator wiring harness (green), four point
W – Connector (green), six point
X – Warning lamp wiring harness (green), eight point
Y – Four single connectors, terminal 30
Z1 – Single connector
Z2 – Single connector, terminal 31
30 – Single connector, terminal 30
30B – Single connector
31 – Single connector, terminal 31 (not shown)

Description	Current track
After-Run Coolant Fan Control (FC) Thermoswitch, F87	39
Back-Up Light Switch, F4	266
Back-Up Light, Left, M16	264
Back-Up Light, Right, M17	266
Battery, A	5
Brake and Parking Brake Indicator Light, K7	162-165
Brake Fluid Level Warning Switch, F34	182
Brake Light Switch, F	233
Brake/Tail Light, Left, M21	228-229
Brake/Tail Light, Right, M22	234-235
Cassette Storage Light, L66	180
Changeover Valve, N180	93
Cigarette Lighter Light, L28	171
Control Module With Display Unit In Instrument Cluster, J285	104-144
Coolant Fan Control (FC) Control Module, J293	33-41
Coolant Fan Control (FC) Thermoswitch, F18	30-31
Coolant Fan, V7	34-36
Digital Display Light, L75	139, 141
Door Contact Switch, Left Front, F2	155
Door Contact Switch, Right Front, F3	156
Door Latch Switch, Left, F143	208
Dual Horn Relay, J4	277-278
Dual Horns, H1	278, 280
EGR Temperature Sensor, G98	90
EGR Vacuum Regulator Solenoid Valve, N18	91
Emergency Flasher Relay, J2	217-219
Emergency Flasher Switch, E3	213-222
Emergency Flasher Warning Light, K6	221
Engine Coolant Level (ECL) Sensor, G32	103
Engine Coolant Level/Temperature (ECL/ECT) Warning Light, K28	107
Engine Coolant Temperature (ECT) Gauge, G3	108
Engine Coolant Temperature (ECT) Sensor, G2	40
Engine Coolant Temperature (ECT) Sensor, G62	63
Engine Speed (RPM) Sensor, G28	68-70
Evaporative Emission (EVAP) Canister Purge Regulator Valve, N80	98
Fresh Air Blower , V2	18

Description	Current track
Fresh Air Blower Series Resistance, N23	19-23
Fresh Air Blower Switch, E9	20-23
Fresh Air Control Lever Light, L16	24
Fuel Gauge, G1	109
Fuel Level Sensor, G	150
Fuel Pump (FP) Relay, J17	50-51
Fuel Pump (FP), G6	151
Fuse Strip, S88	41
Fuse, S1, 10A	200
Fuse, S2, 10A	203
Fuse, S3, 10A	247
Fuse, S4, 15A	308
Fuse, S5, 15A	284
Fuse, S6, 20A	23
Fuse, S7, 10A	235
Fuse, S8, 10A	228
Fuse, S9, 20A	270
Fuse, S10, 15A	259
Fuse, S11, 10A	199
Fuse, S12, 10A	204
Fuse, S13, 10A	180
Fuse, S14, 10A	262
Fuse, S15, 10A	28
Fuse, S16, 15A	177
Fuse, S17, 10A	215
Fuse, S18, 20A	53
Fuse, S19, 30A	31
Fuse, S20, 10A	233
Fuse, S21, 15A	170
Fuse, S22, 10A	296
Generator (GEN) Warning Light, K2	122
Generator (GEN), C	1-2
Glove Compartment Light, W6	174
Hall Sensor, G40	50-51
Headlight Dimmer/Flasher Switch, E4	213-222
Headlight High Beam Indicator Light, K1	126
Headlight Switch Light, L9	240
Headlight, Left, L1	200
Headlight, Right, L2	203
Heated Oxygen Sensor (HO2S) Control Module, J208	52-54
Heated Oxygen Sensor (HO2S), G39	52-54
Heater, A/C Clutch Fuse, S24	19-23
High Beam Headlight, Left, L13	199
High Beam Headlight, Right, L14	204
High-Mount Brake Light, M25	232
Horn Button, H	289

Description	Current track
Idle Air Control (IAC) Valve, N71	77
Ignition Coil, N152	43-49
Ignition/Starter Switch, D	10-14
Injector, Cyl. 1, N30	71
Injector, Cyl. 2, N31	72
Injector, Cyl. 3, N32	73
Injector, Cyl. 4, N33	74
Injector, Cyl. 5, N83	75
Injector, Cyl. 6, N84	76
Instrument Cluster Light, L10	130, 131, 132
Instrument Panel Light Dimmer Switch, E20	240-247
Intake Air Temperature (IAT) Sensor, G42	81
Interior Light With Delay Switch, W15	155-157
Knock Sensor (KS) I, G61	57-59
Knock Sensor (KS) II, G66	60-62
License Plate Light, X	247, 249
Light Switch, E1	240-247
Load Reduction Relay, J59	11-12
Luggage Compartment Light Switch, F5	194
Luggage Compartment Light, W3	194
Make-Up Mirror Light Switch, Right, F148	160
Make-Up Mirror Light, Right, W14	159-160
Malfunction Indicator Lamp (MIL), K83	162-165
Map/Reading Light, Right Front, W13	155-157
Mass Air Flow (MAF) Sensor, G70	82-84
Motronic Engine Control Module (ECM) Power Supply Relay, J271	94-95
Motronic Engine Control Module (ECM), J220	44-97
Multi-Function Indicator Memory Switch, E109	146-149
Multi-Function Indicator Mode Select Switch, E86	146-149
Odometer Display, Y4	136-138
Oil Pressure Gauge Light, L27	178
Oil Pressure Gauge, G11	175-178
Oil Pressure Sensor, G10	175
Oil Pressure Switch (0.3 Bar), F22	101
Oil Pressure Switch, F1	102
Oil Pressure Warning Light, K3	105
Oil Temperature Sensor, G8	99
Outside Air Temperature Sensor, G17	153
Park Light Indicator Light, K4	246
Parking Brake Warning Light Switch, F9	196

Description	Current track
Parking Light Right, M3	205
Parking Light, Left, M1	198
Positive Crankcase Ventilation (PCV) Heating Element, N79	28
Radio, R	296-298
Rear Spoiler Control Module, J223	184-192
Rear Spoiler End Switch 1, F135	189-190
Rear Spoiler End Switch 2, F136	192-193
Rear Spoiler Motor, V52	189
Rear Spoiler Switch, E127	184-187
Rear Window Defogger (Heat Element), Z1	273
Rear Window Defogger Indicator Light, K10	268-270
Rear Window Defogger Switch Light, L39	268-270
Rear Window Defogger Switch, E15	268-270
Rear Window Wiper Motor, V12	305-307
Rear Window Wiper/Washer Relay, J30	304-305
Seat Belt Switch, Left, E24	208
Seat Belt Warning Light, K19	121
Seat Belt Warning System Relay, J34	206-209
Secondary Air Injection (AIR) Pump Motor, V101	92
Secondary Air Injection (AIR) Relay, J316	92-96
Side Marker Lights, Front, M11	197, 206
Spark Plug Connector, P	44-49
Spark Plugs, Q	44-49
Speedometer, G21	130-133
Starter, B	7-8
Tachometer, G5	113-116
Third Speed Coolant Fan Control (FC) Thermoswitch, F165	42
Throttle Position (TP) Sensor, G69	65-66
Turn Signal Indicator Light, Left, K65	124
Turn Signal Indicator Light, Right, K94	125
Turn Signal Light, Left Front, M5	225
Turn Signal Light, Left Rear, M6	227
Turn Signal Light, Right Front, M7	237
Turn Signal Light, Right Rear, M8	236
Turn Signal Switch, E2	213-222
Vehicle Speed Sensor (VSS), speedometer, G22	145-146
Voltage Regulator (VR), C1	1-2
Voltage Stabilizer, J6	118-120
Voltmeter, G14	175-178
Warning Buzzer, H3	109

Description	Current track
Washer Nozzle Heater, Left, Z20	257
Washer Nozzle Heater, Right, Z21	258
Washer/Wiper Intermittent Relay, J31	290-294
Water Pump, V50	38
Windshield and Rear Window Washer Pump, V59	303
Windshield Wiper Intermittent Switch, E22	289-294
Windshield Wiper Motor, V	282-285

Wire connectors

T1	- Single Connector, near starter
T1a	- Single Connector, behind fuse relay/panel
T1b	- Single Connector, behind instrument panel, center
T1c	- Single Connector, behind fuse relay/panel
T1d	- Single Connector, behind fuse relay/panel
T1d1	- Single Connector, behind relay panel
T1e	- Single Connector, on left Apillar, near relay panel
T1f	- Single Connector, behind fuse relay/panel
T1g	- Single Connector, behind fuse relay/panel
T1h	- Single Connector, behind fuse relay/panel
T1i	- Single Connector, behind fuse relay/panel
T1j	- Single Connector, in rear lid
T1k	- Single Connector, behind fuse relay/panel
T1l	- Single Connector, behind fuse relay/panel
T1m	- Single Connector, behind fuse relay/panel
T1n	- Single Connector, behind fuse relay/panel
T1o	- Single Connector, on Back-Up Light, Left
T1p	- Single Connector, on Back-Up Light, Left
T1r	- Single Connector, behind fuse relay/panel
T1s	- Single Connector, behind fuse relay/panel
T1y	- Single Connector, behind fuse relay/panel
T2	- Double Connector, above relay panel
T2a	- Double Connector, behind fuse relay/panel
T2a1	- Double Connector, below driver's seat
T2a2	- Double Connector, in rear lid
T2b	- Double Connector, behind fuse relay/panel
T2b1	- Double Connector, behind instrument panel, center
T2c	- Double Connector, behind instrument panel
T2d	- Double Connector, on Fresh Air Blower
T2e	- Double Connector, behind fuse relay/panel
T2f	- Double Connector, on right A-pillar, upper part
T2g	- Double Connector, in luggage compartment, left
T2h	- Double Connector, behind fuse relay/panel
T2i	- Double Connector, in luggage compartment, left
T2j	- Double Connector, on engine hood
T2k	- Double Connector, on engine hood
T2l	- Double Connector, in luggage compartment, left
T2n	- Double Connector, behind fuse relay/panel
T2o	- Double Connector, near headlight, left
T2p	- Double Connector, near headlight, right
T2q	- Double Connector, in luggage compartment, left
T2r	- Double Connector, on Parking Light, Left
T2s	- Double Connector, on Side Marker Lights Front
T2t	- Double Connector, on Parking Light Right
T2u	- Double Connector, on Side Marker Lights, Front
T2v	- Double Connector, on taillight, left
T2w	- Double Connector, on High-Mount Brake Light
T2x	- Double Connector, black, Data Link connector, below transmission shift lever cover
T2z	- Double Connector, white, Data Link connector, below transmission shift lever cover
T3a	- 3-pin Connector, near starter
T3b	- 3-pin Connector, near exhaust manifold
T3c	- 3-pin Connector, near starter
T3e	- 3-pin Connector, on Back-Up Light, Right
T3f	- 3-pin Connector, on High Beam Headlight, Left
T3g	- 3-pin Connector, on High Beam Headlight, Right
T4	- 4-pin Connector, on Coolant Fan Control (FC) Control Module
T4a	- 4-pin Connector, behind steering column switch cover
T4b	- 4-pin Connector, near exhaust manifold
T4c	- 4-pin Connector, behind fuse/relay panel
T4d	- 4-pin Connector, behind steering column switch cover
T4e	- 4-pin Connector, on Turn Signal Light, Left Rear
T4e	- 4-pin Connector, on Turn Signal Light, Left Rear
T4f	- 4-pin Connector, on Turn Signal Light, Right Rear
T4i	- 4-pin Connector, behind fuse/relay panel
T5	- 5-pin Connector, on Fresh Air Blower Series Resistance
T5b	- 5-pin Connector, behind steering column switch cover
T5c	- 5-pin Connector, behind steering column switch cover

Wire connectors

- T6 – 6-pin Connector, on Rear Spoiler Switch
- T7 – 7-pin Connector, behind steering column switch cover
- T8 – 8-pin Connector, on Radio
- T8a – 8-pin Connector, behind fuse/relay panel
- T8d – 8-pin Connector, Brake and Parking Brake Indicator Light
- T8e – 8-pin Connector, on Coolant Fan Control (FC) Control Module
- T10 – 10-pin Connector, on Alarm System Control Module
- T10a – 10-pin Connector, on Rear Spoiler Control Module
- T28 – 28-pin Connector, on Instrument Cluster
- T42 – 42-pin Connector, on cylinder head, left
- T68 – 68-pin Connector, on Motronic Engine Control Module (ECM)
- TV4 – Wire Connector, Terminal 15, behind fuse/relay panel
- TV13 – Wire Connector, Vehicle Speed Signal, behind fuse/relay panel
- TV14 – Data Link Connector (DLC), behind fuse/relay panel

Welded wiring harness points

- (A11) – Plus connection (58b), in instrument panel wiring harness
- (A15) – Plus connection (15), in instrument wiring harness
- (B6) – Wire connection (s), in steering column switch wiring harness
- (B8) – Plus connection(58), in light switch wiring harness
- (C1) – Plus connection (58L), in left front wiring harness
- (C2) – Plus connection (58R), in left front wiring harness
- (C13) – Plus connection, in dual tone horn wiring harness
- (C15) – Wire connection –1–, in headlight wiring harness
- (C16) – Wire connection –2–, in headlight wiring harness
- (E6) – Plus connection –1– (15), in Multiport Fuel Injection (MFI) wiring harness
- (E14) – Wire connection –1–, in Multiport Fuel Injection (MFI) wiring harness
- (E20) – Plus connection (30), in Multiport Fuel Injection (MFI) wiring harness
- (E7) – Plus connection (87a), in Multiport Fuel Injection (MFI) wiring harness
- (G3) – Plus connection, in injector wiring harness
- (K21) – Wire connection –1–, in coolant fan wiring harness
- (K22) – Wire connection –2–, in coolant fan wiring harness
- (K28) – Plus connection (15), in coolant fan wiring harness
- (W9) – Plus connection (30), in rear wiring harness
- (W10) – Plus connection (15), in rear wiring harness

Ground connection

- (1) – Ground strap, battery to body
- (2) – Ground strap, transmission to body
- (15) – Ground connection, on cylinder head
- (18) – Ground connection, on engine block
- (30) – Ground connection, –1–, beside fuse/relay panel
- (43) – Ground connection, on right A–pillar, lower part
- (50) – Ground connection, in luggage compartment, left
- (53) – Ground connection, in rear lid, right
- (80) – Ground connection –1–, in instrument panel wiring harness

Ground connection

- (86) – Ground connection –1–, in rear wiring harness
- (98) – Ground connection, in rear lid wiring harness
- (119) – Ground connection –1–, in headlight wiring harness
- (120) – Ground connection –2–, in headlight wiring harness
- (125) – Ground connection –3–, in headlight wiring harness
- (129) – Ground connection, in dual–tone horn wiring harness
- (138) – Ground connection (sensor Ground), in Motronic Multiport Fuel Injection (MFI) wiring harness
- (139) – Ground connection (sensor Ground), in Motronic Multiport Fuel
- (159) – Ground connection –2–, in Digifant Multiport Fuel Injection (MFI) wiring harness
- (182) – Ground connection –1–, in engine compartment wiring harness (6 Cylinder)
- (193) – Ground connection –1–, in coolant fan wiring harness
- (210) – Ground connection –1–, in Motronic Multiport Fuel Injection (MFI) wiring harness

Wiring Color code	
BK	BLACK
BR	BROWN
CL	CLEAR
R	RED
Y	YELLOW
G	GREEN
LT. G	LIGHT GREEN
BL	BLUE
V	VIOLET
GY	GRAY
W	WHITE
OR	ORANGE

Wire Size

Wiring Diagrams identify wires by the metric wire size. Metric wire sizes indicate cross-sectional area in square millimeters (mm²). The chart below lists metric wire sizes and their approximate equivalents in American Wire Gauge (AWG) sizes.

Metric size Cross-section (mm2)	American Wire Gauge size (AWG)
0.35	22
0.50	20
0.75	18
1.00	16
1.50	14
2.50	12
4.00	10
6.00	8
16.00	4
25.00	2
35.00	2

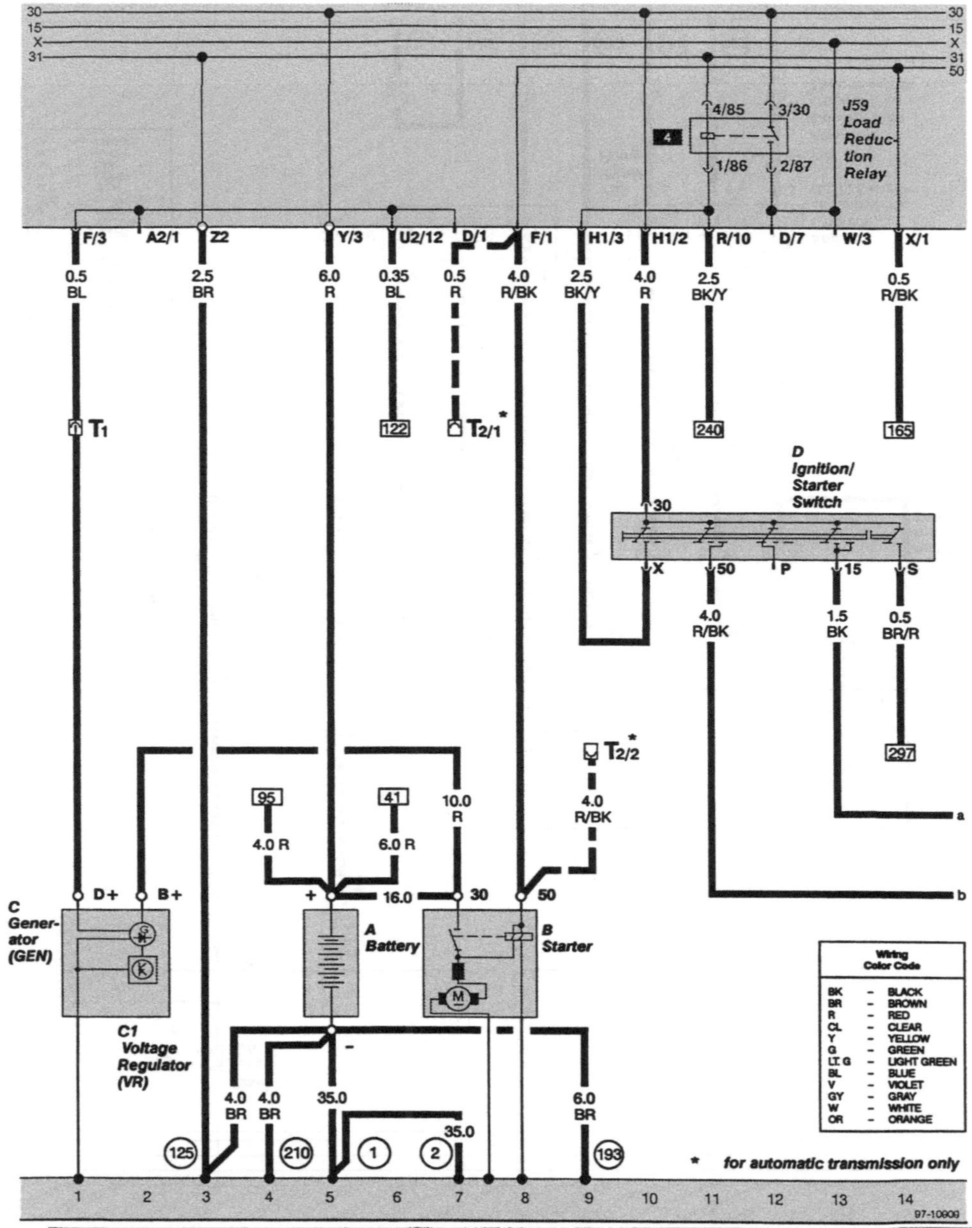

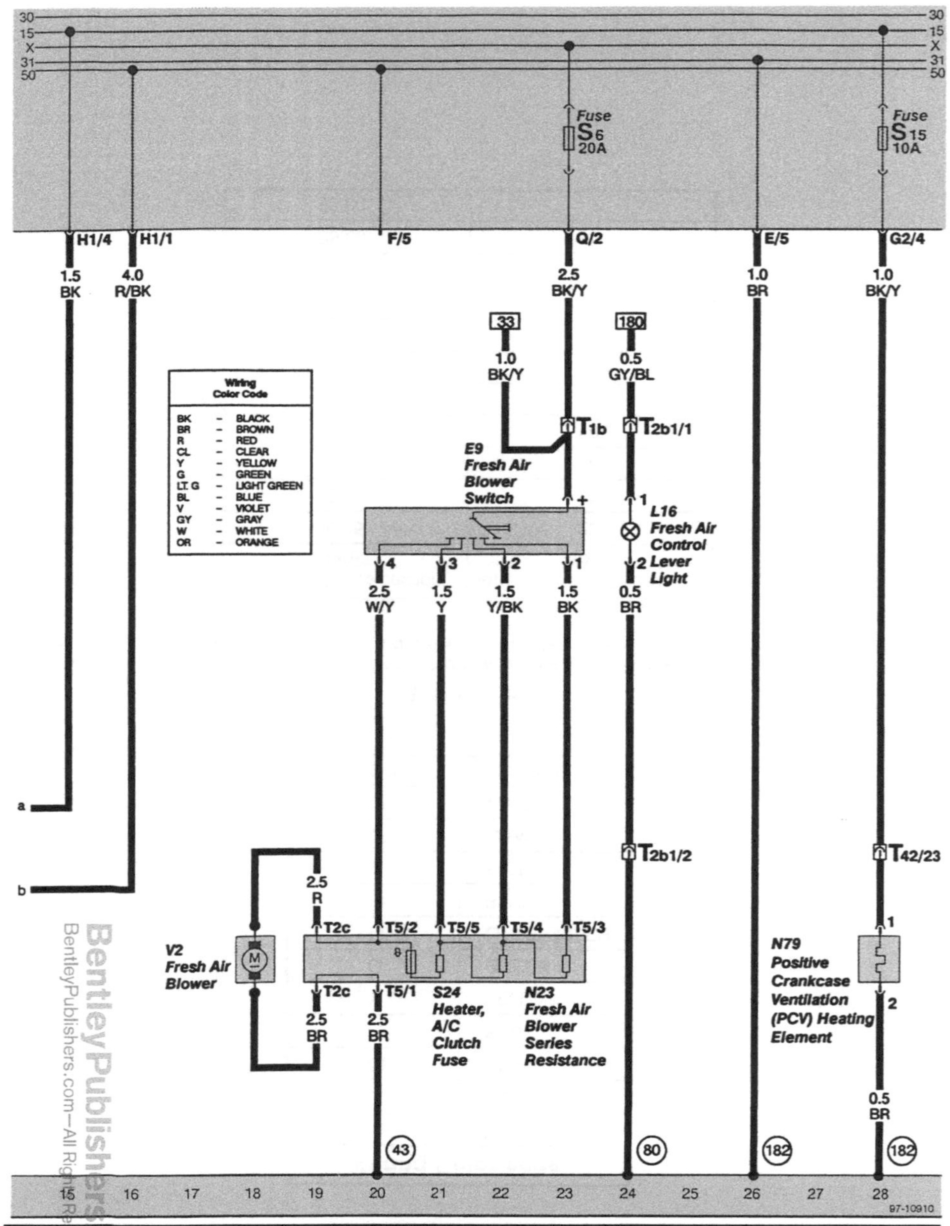

123 Fresh air blower

Corrado 1993
Production from Jan. 1993

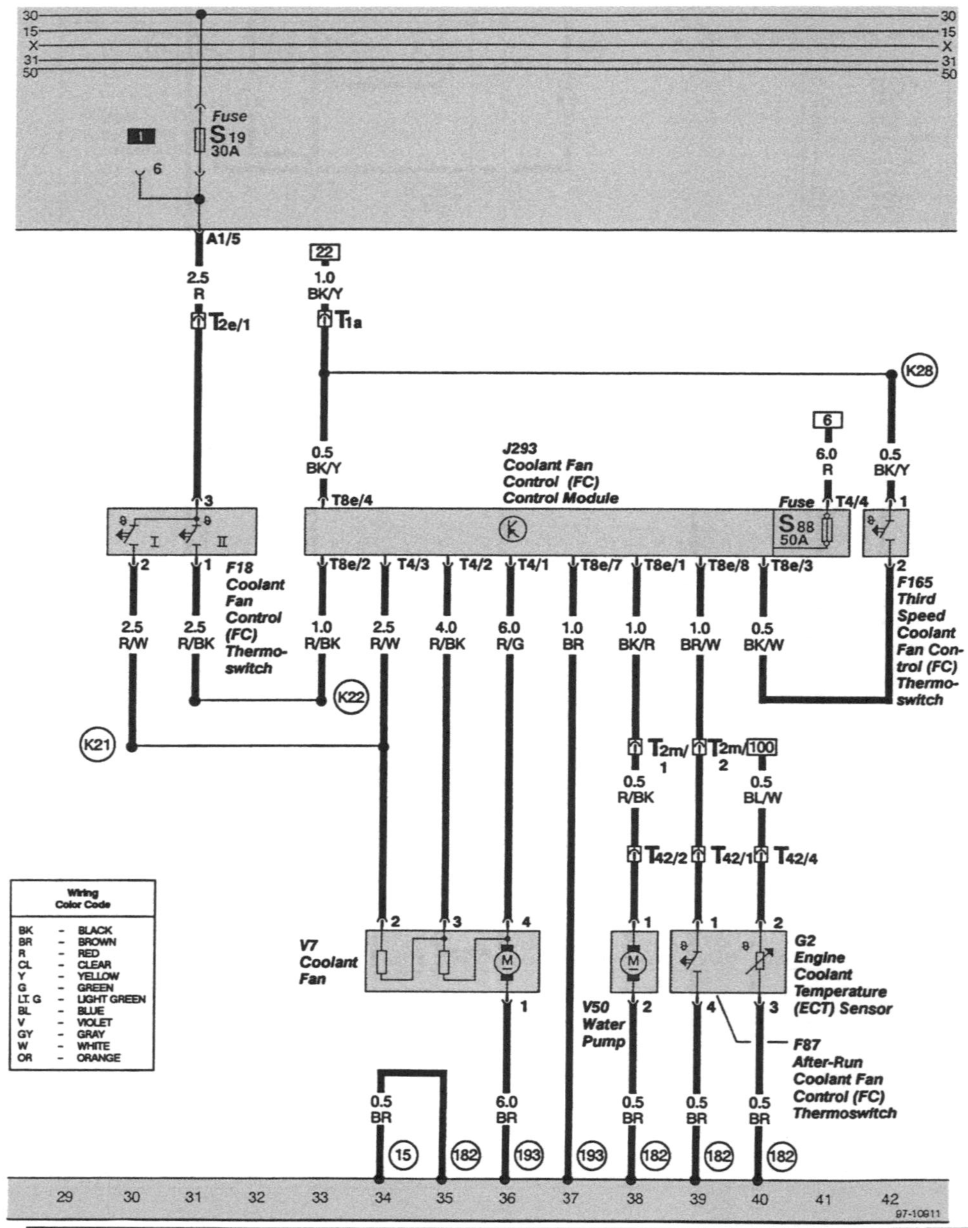

Corrado 1993
Production from Jan. 1993

Radiator cooling fan **124**

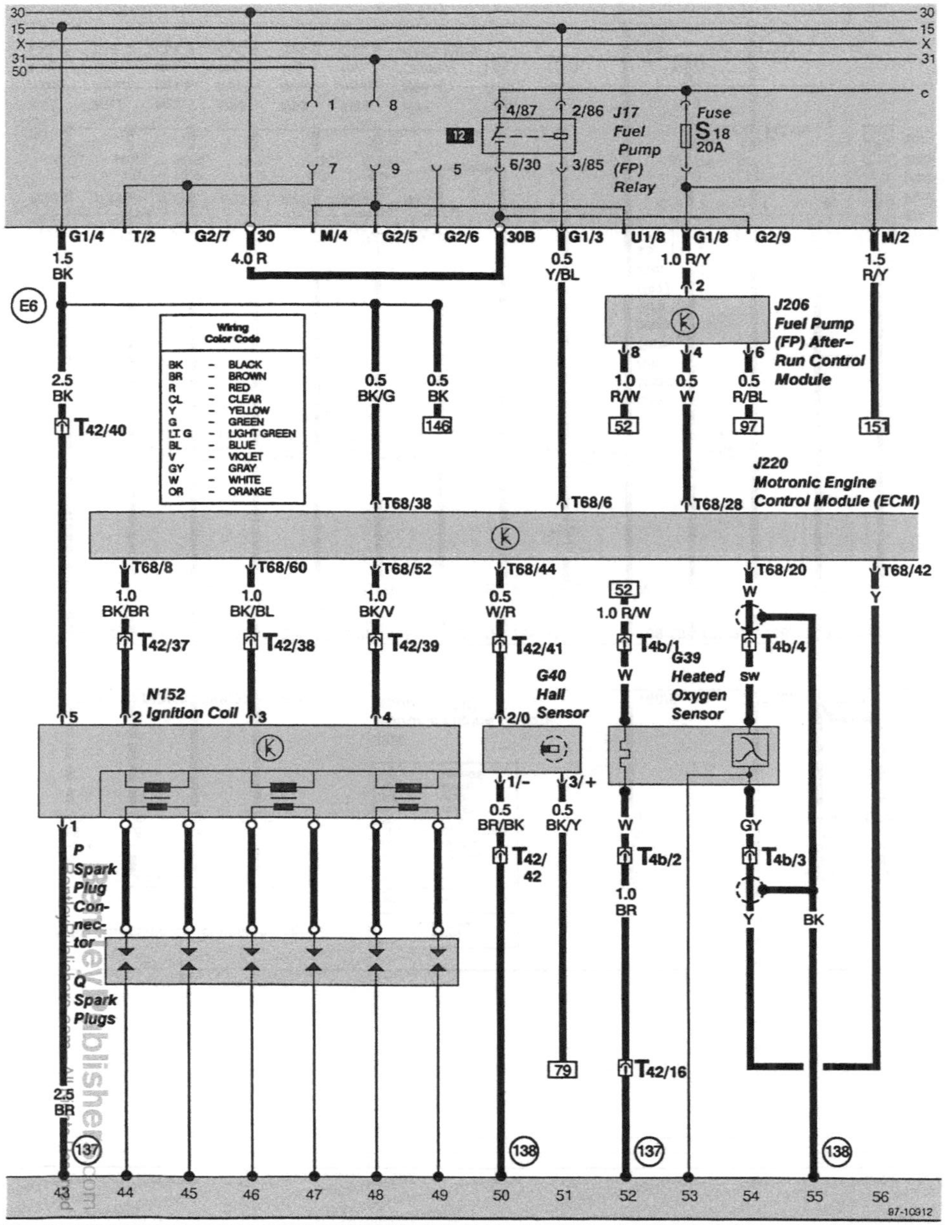

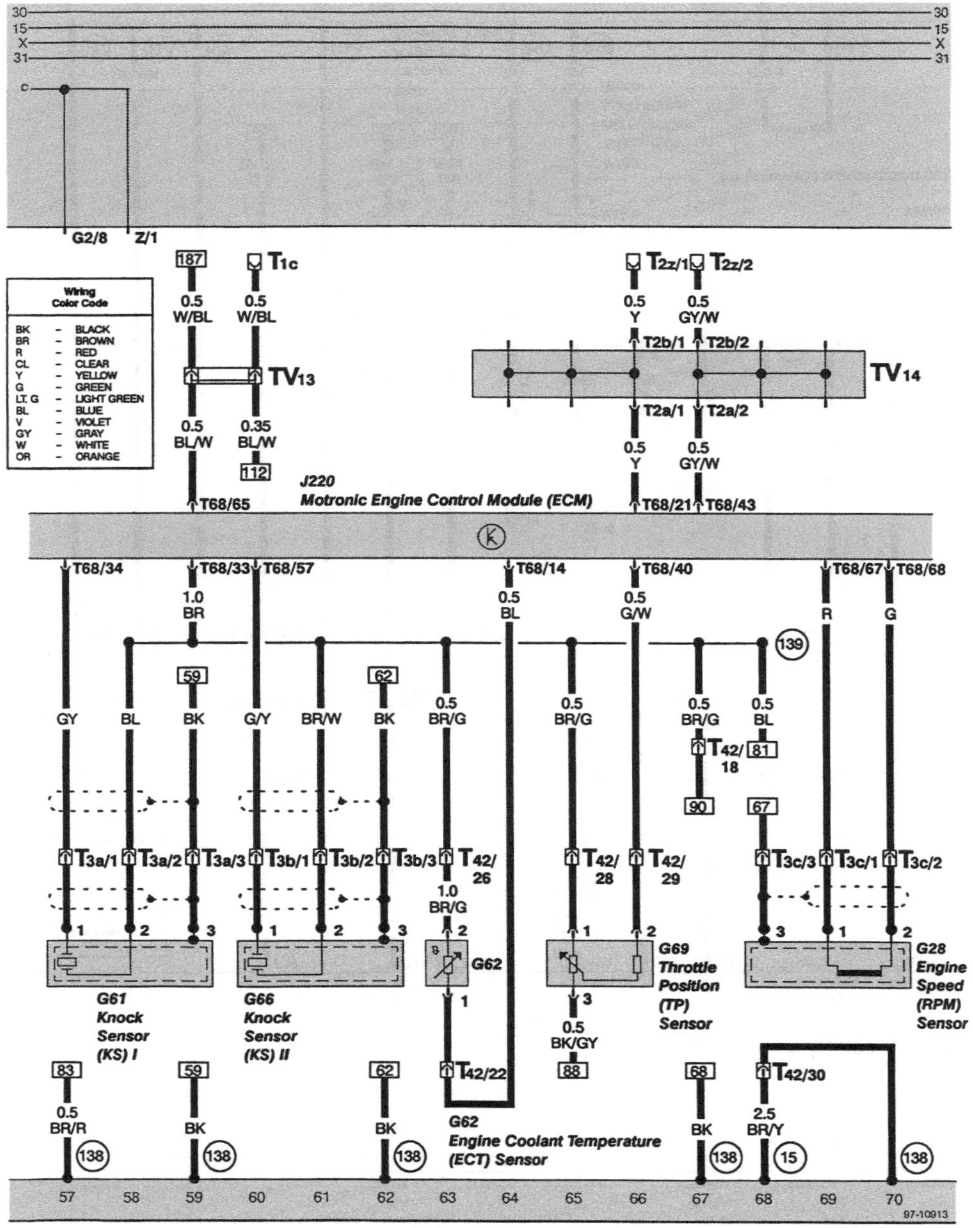

Corrado 1993
Production from Jan. 1993

Corrado 1993
Production from Jan. 1993

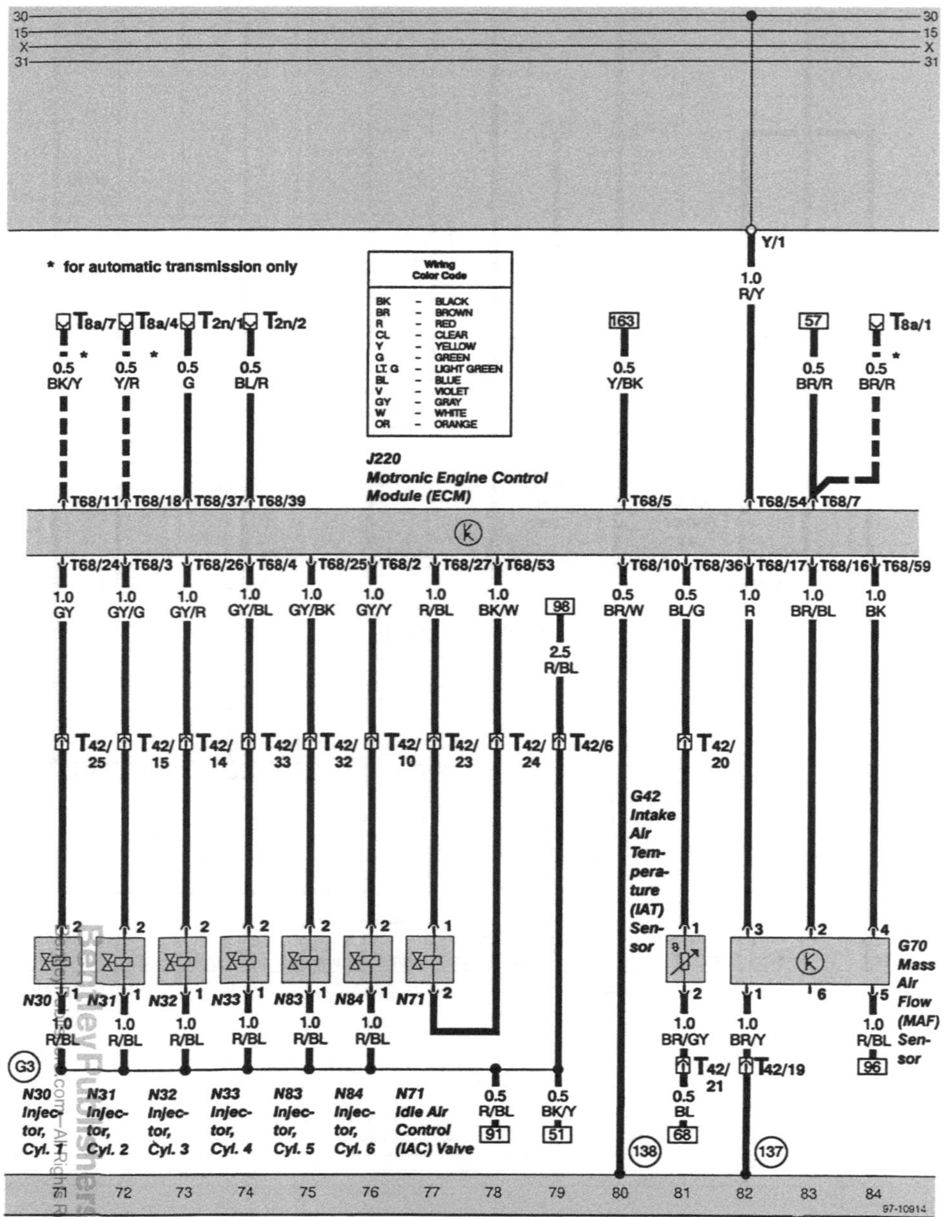

Corrado 1993
Production from Jan. 1993

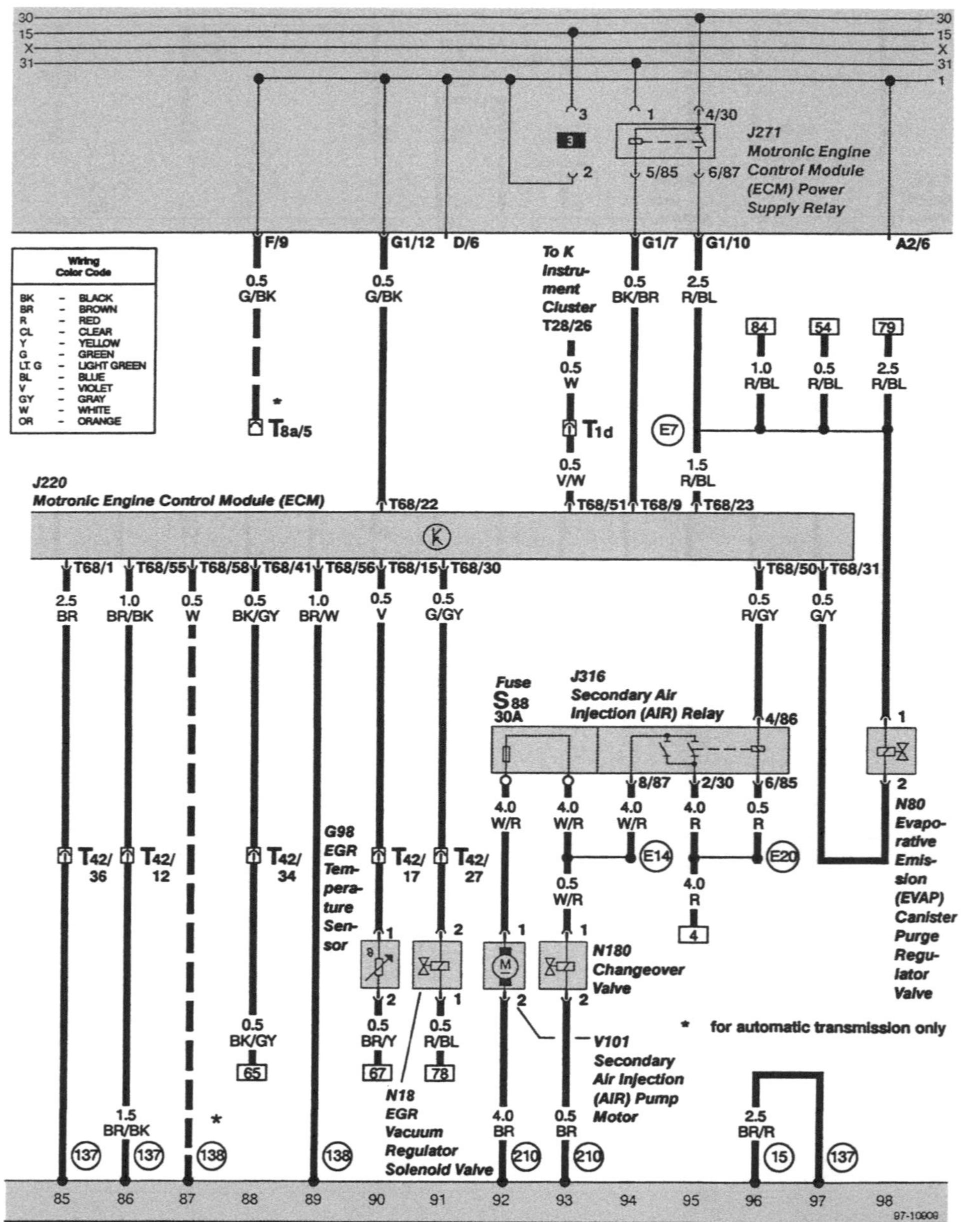

Corrado 1993
Production from Jan. 1993

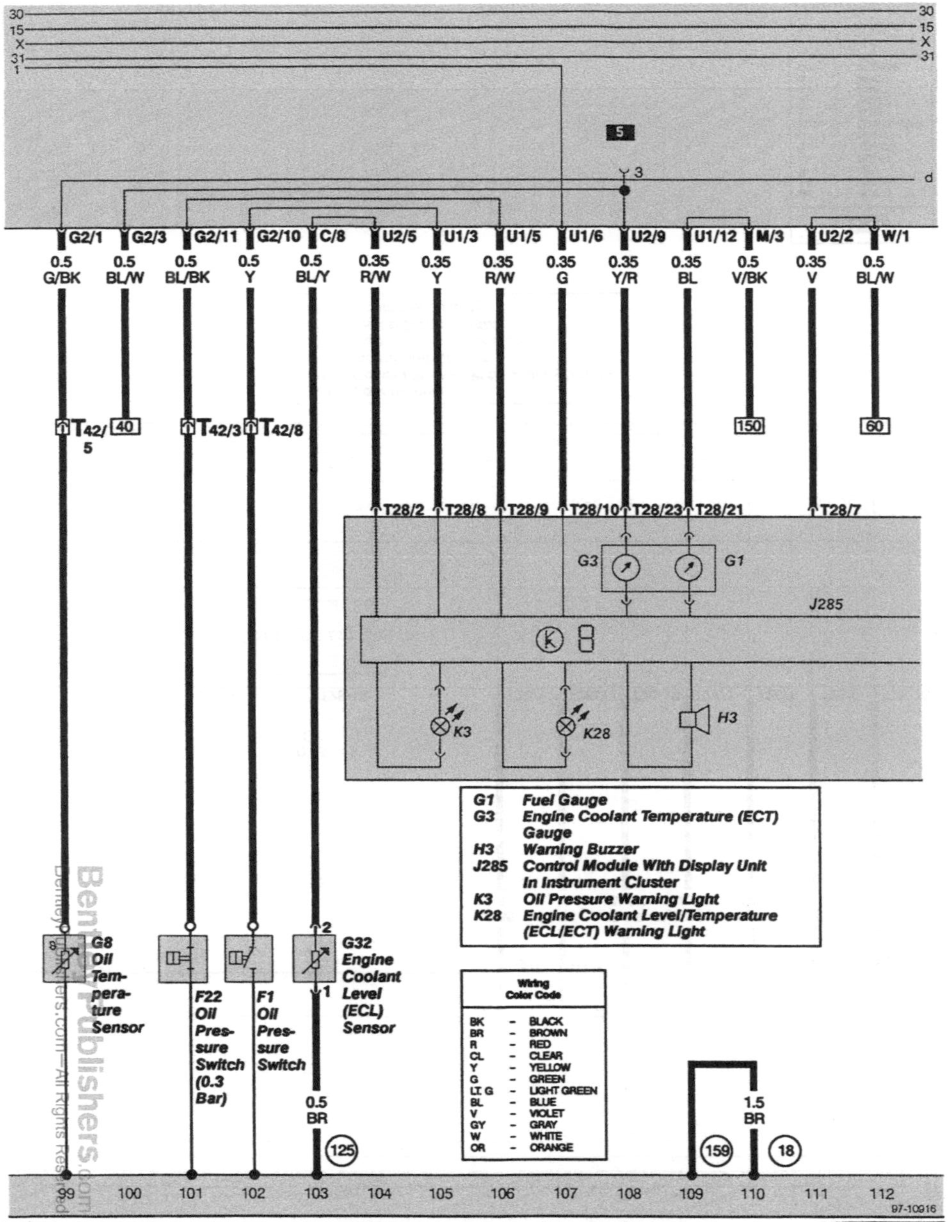

Corrado 1993
Production from Jan. 1993

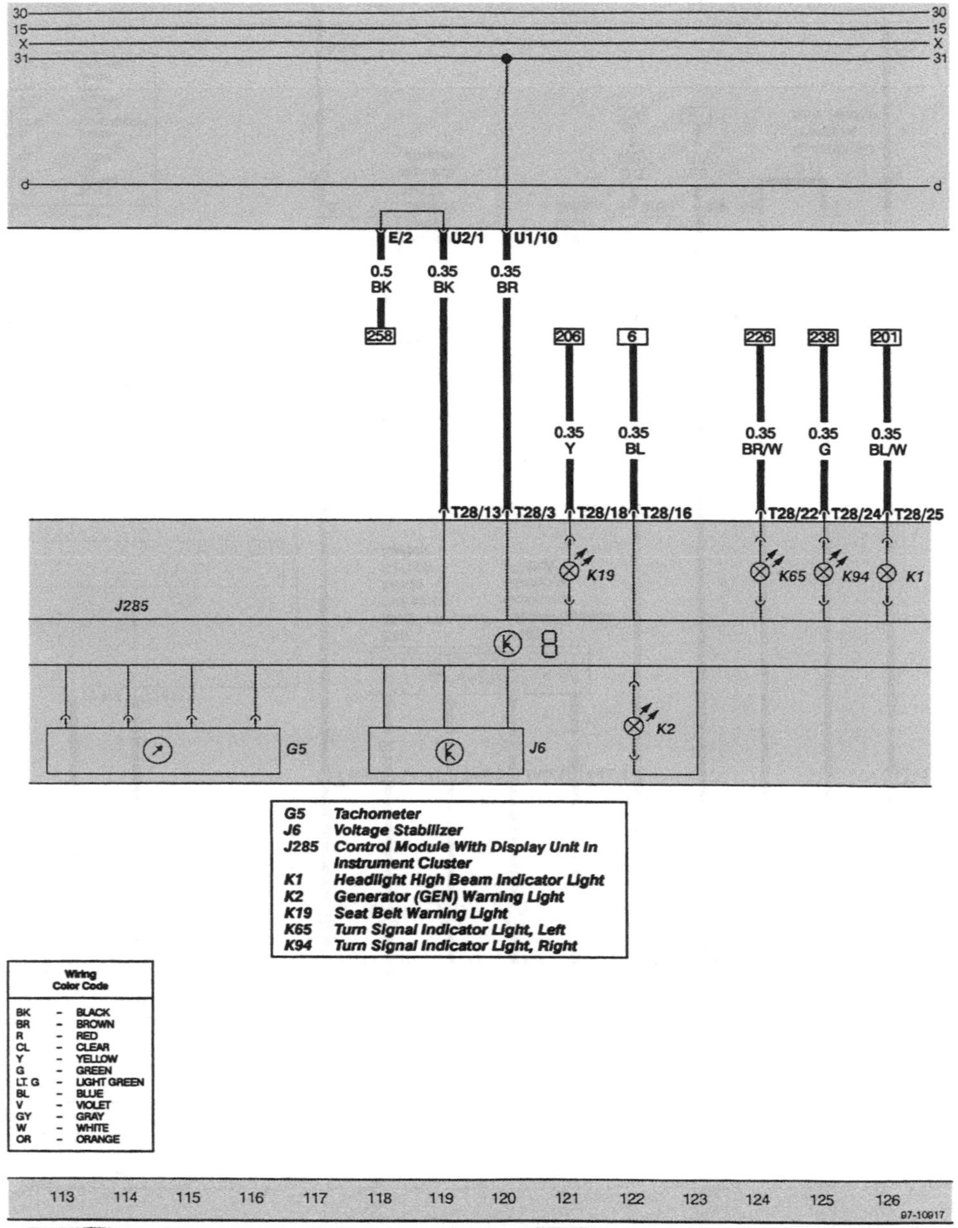

Corrado 1993
Production from Jan. 1993

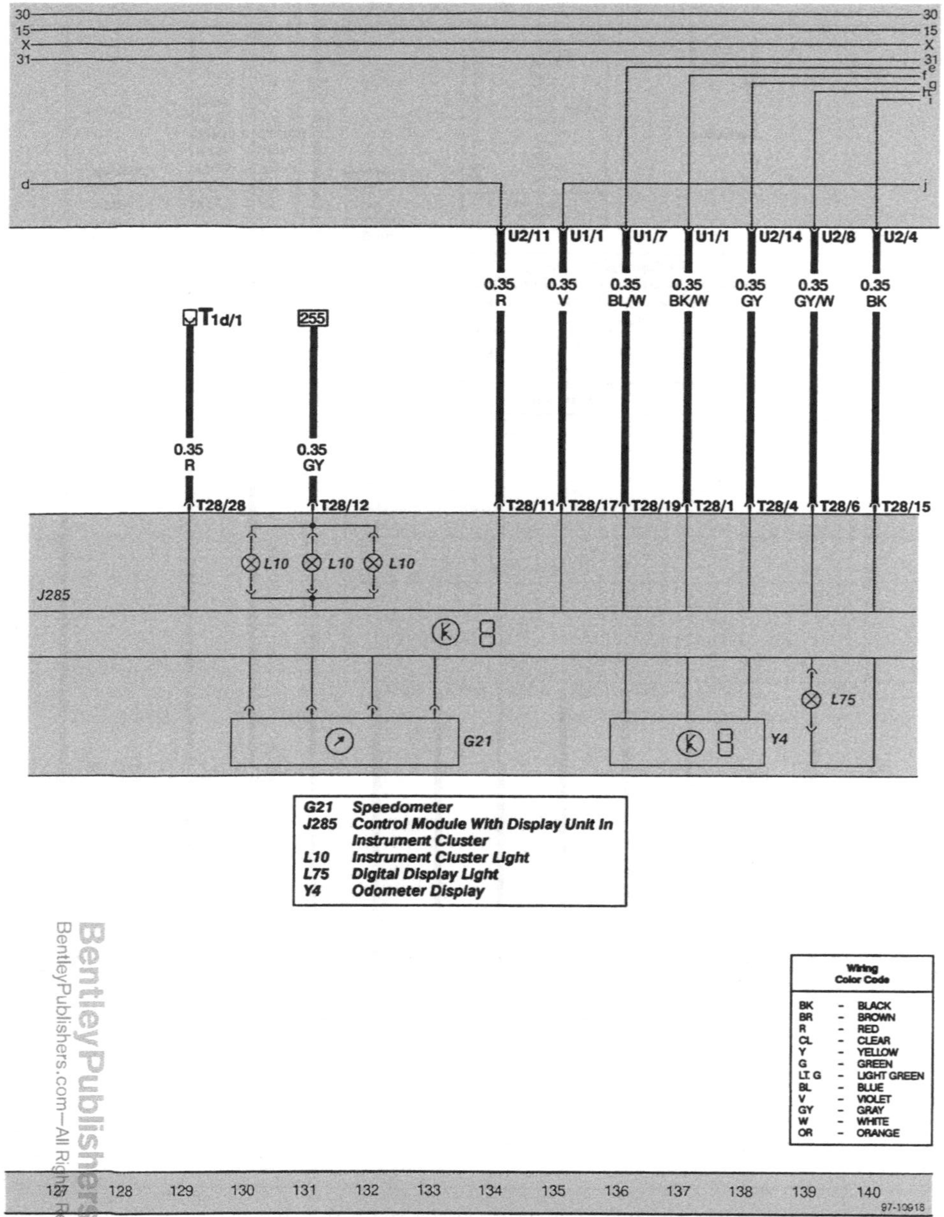

Corrado 1993
Production from Jan. 1993

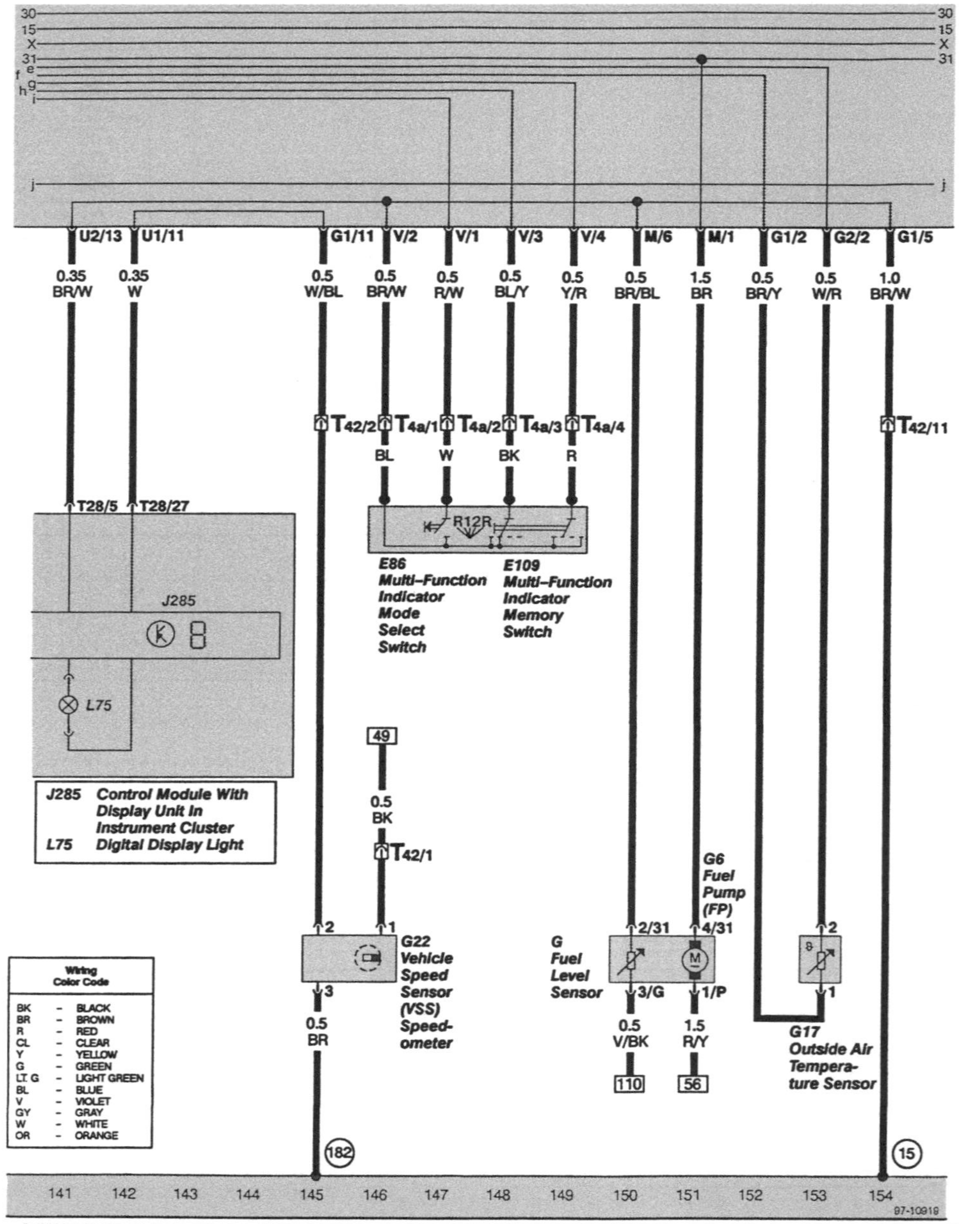

Corrado 1993
Production from Jan. 1993

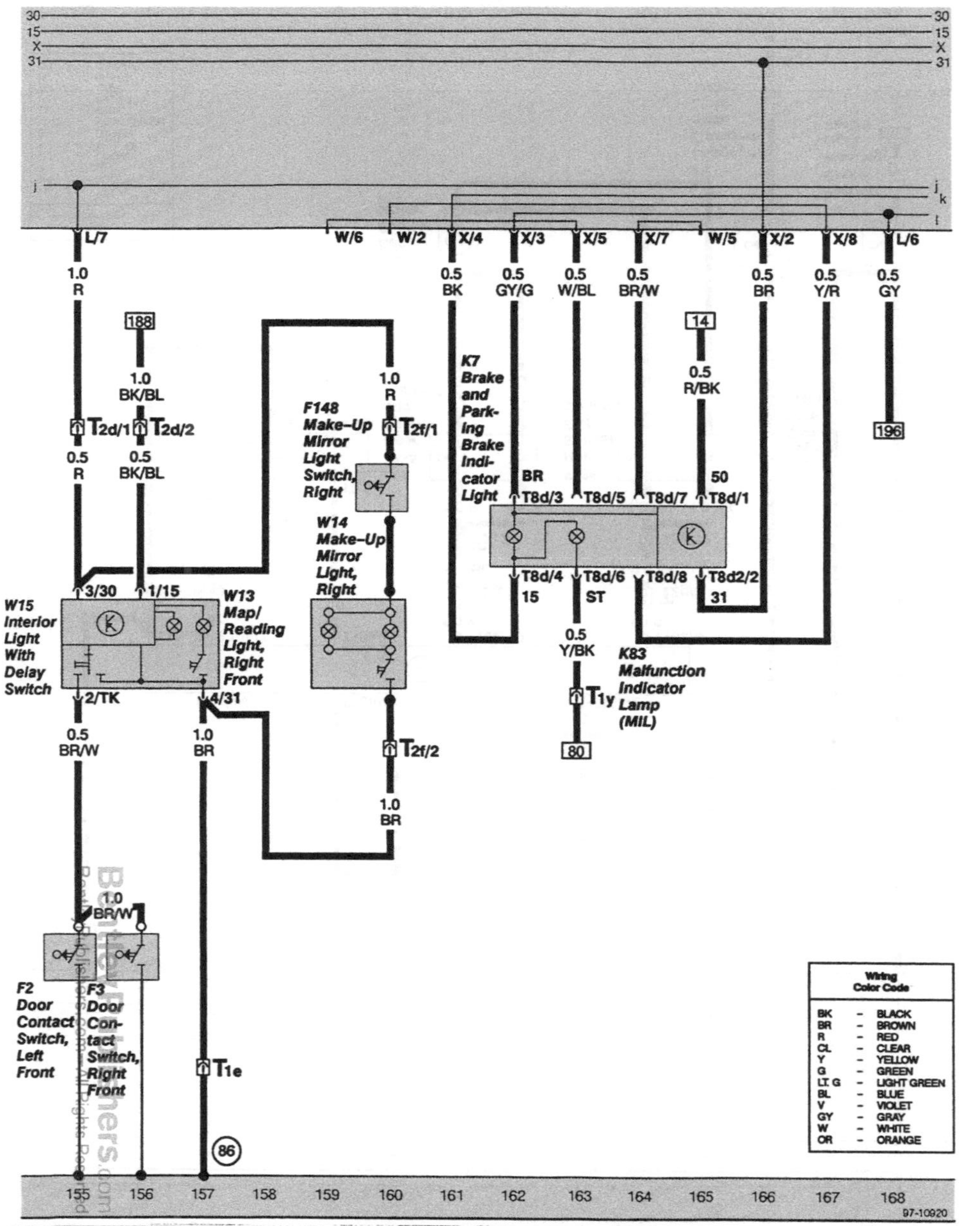

133 Interior lights
Brake warning lights

Corrado 1993
Production from Jan. 1993

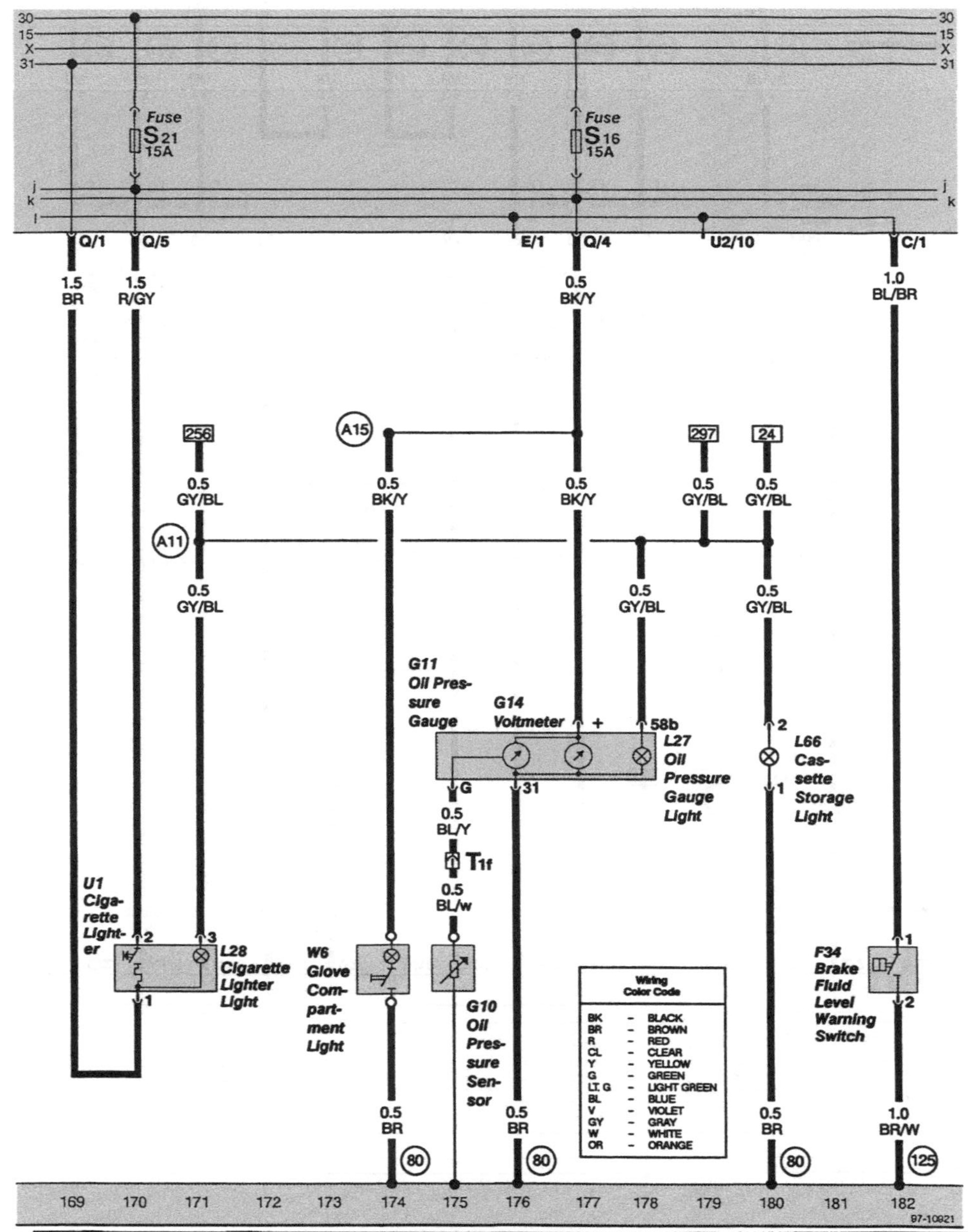

Corrado 1993
Production from Jan. 1993

Cigarette lighter **134**

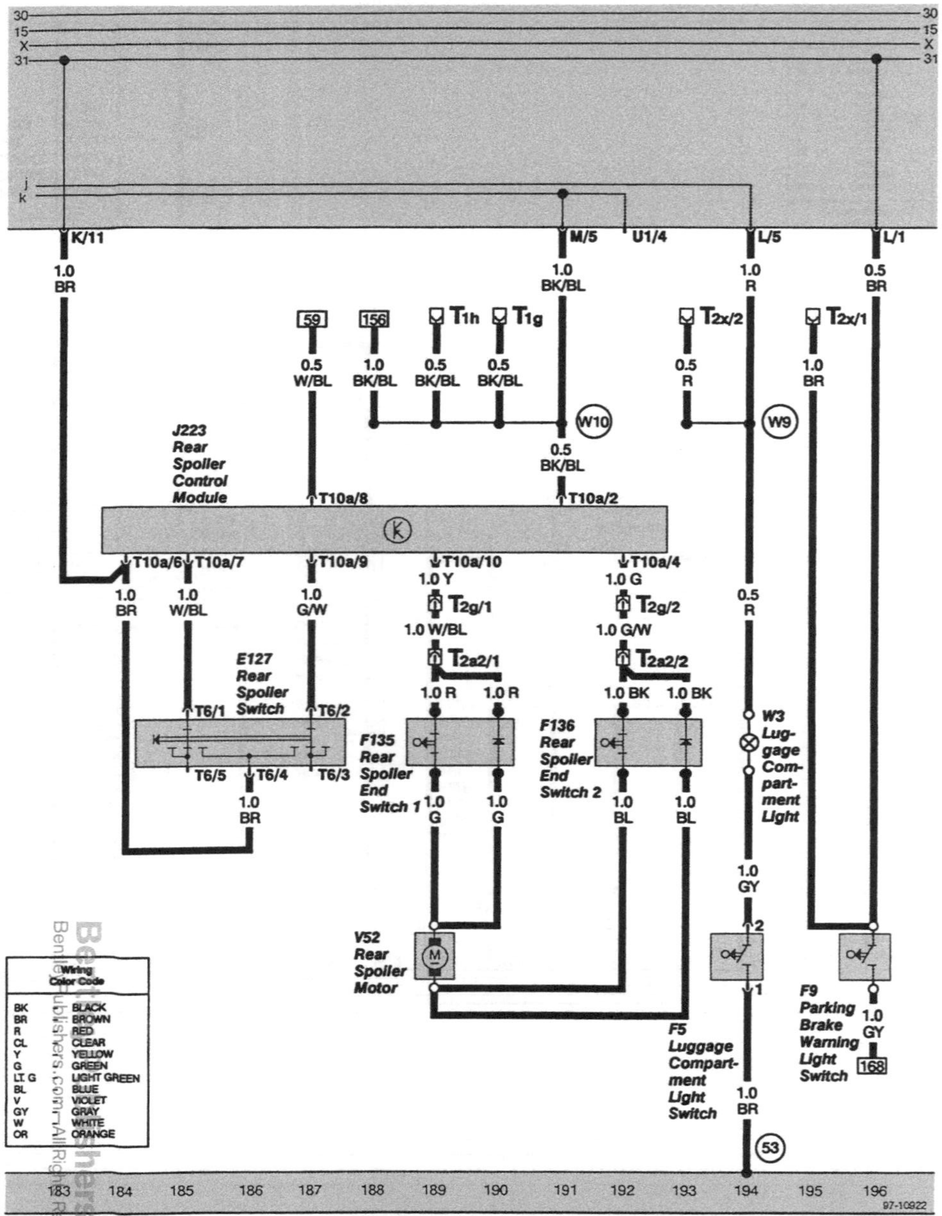

Corrado 1993
Production from Jan. 1993

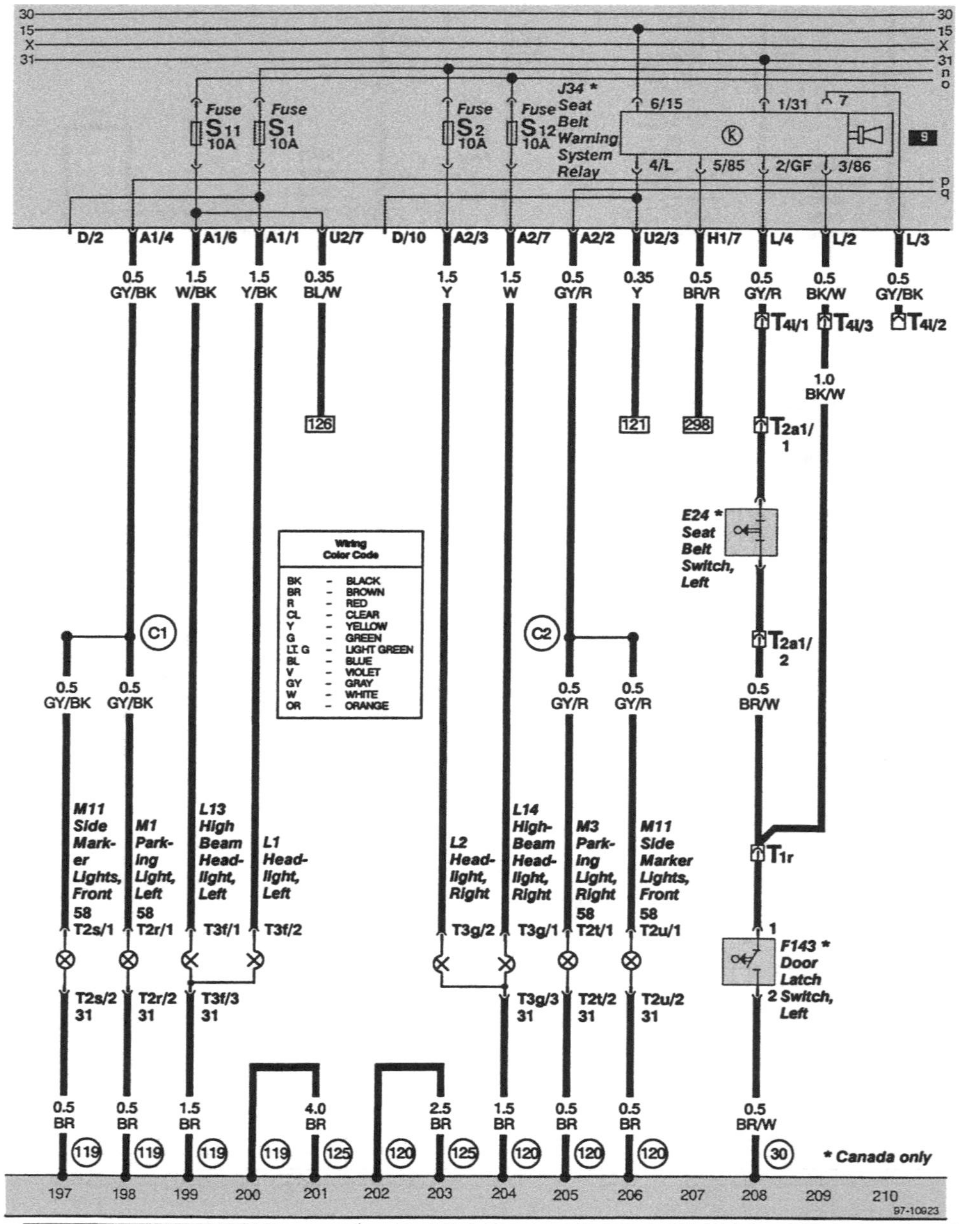

Corrado 1993
Production from Jan. 1993

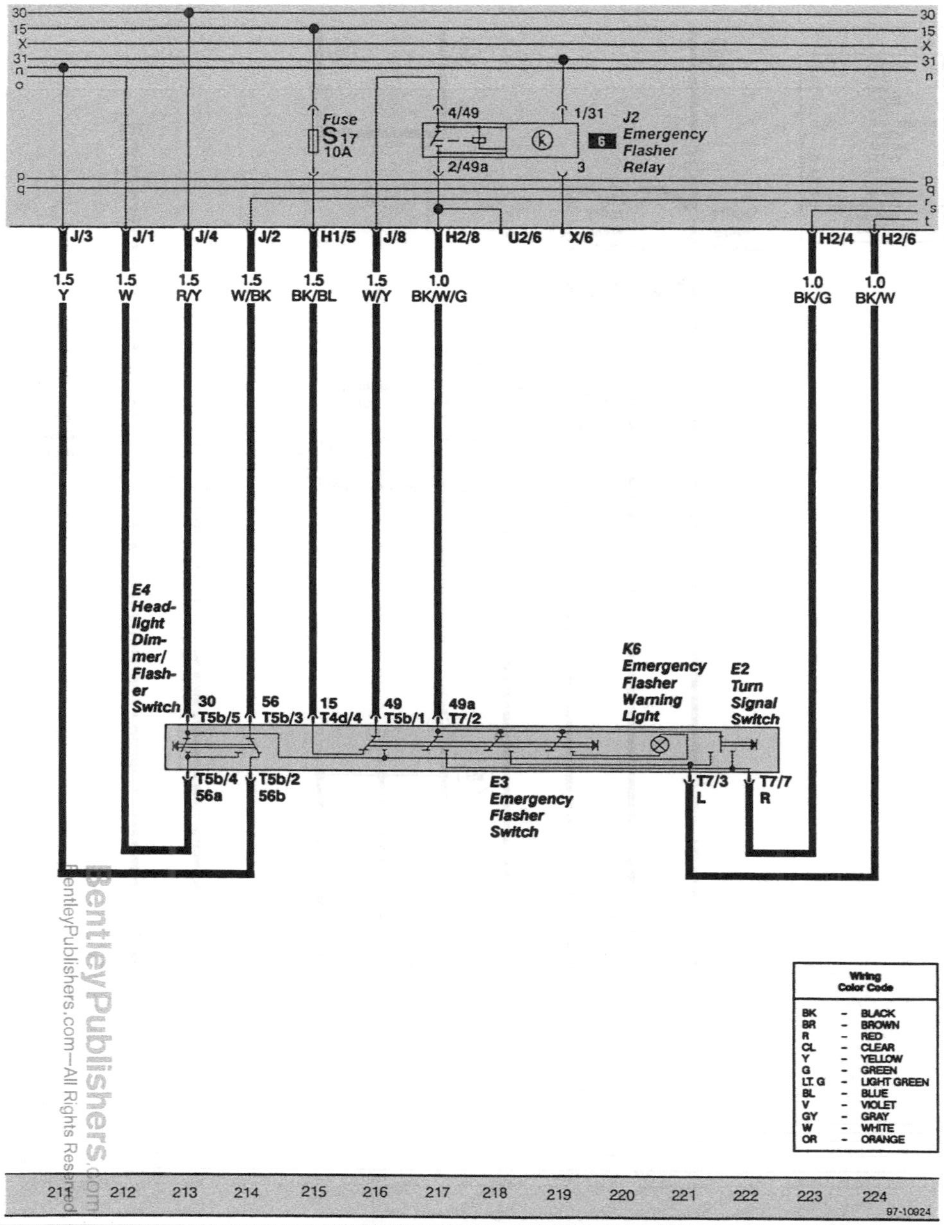

137 Emergency flasher switch
Turn signal switch

Corrado 1993
Production from Jan. 1993

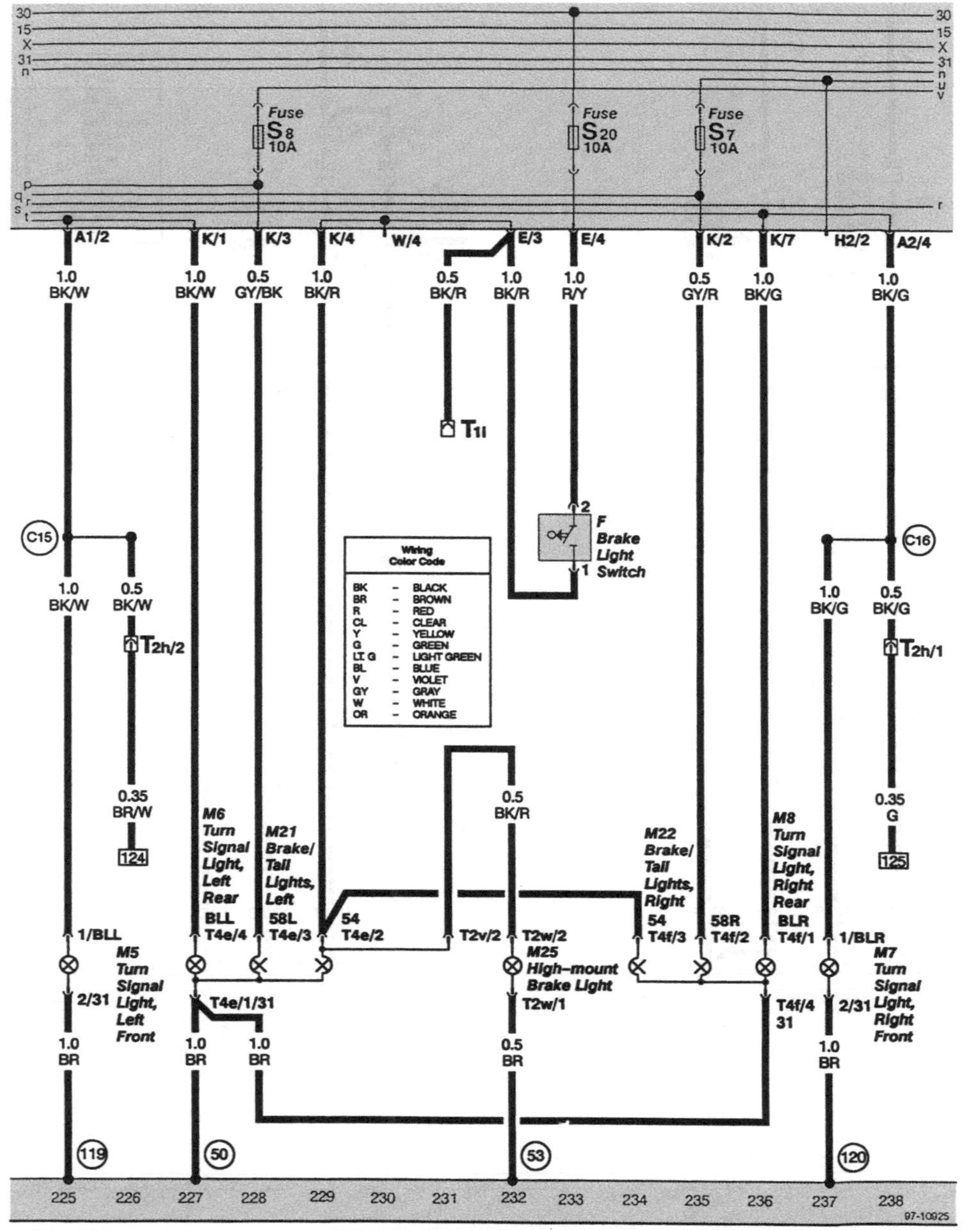

Corrado 1993
Production from Jan. 1993

Exterior lights **138**

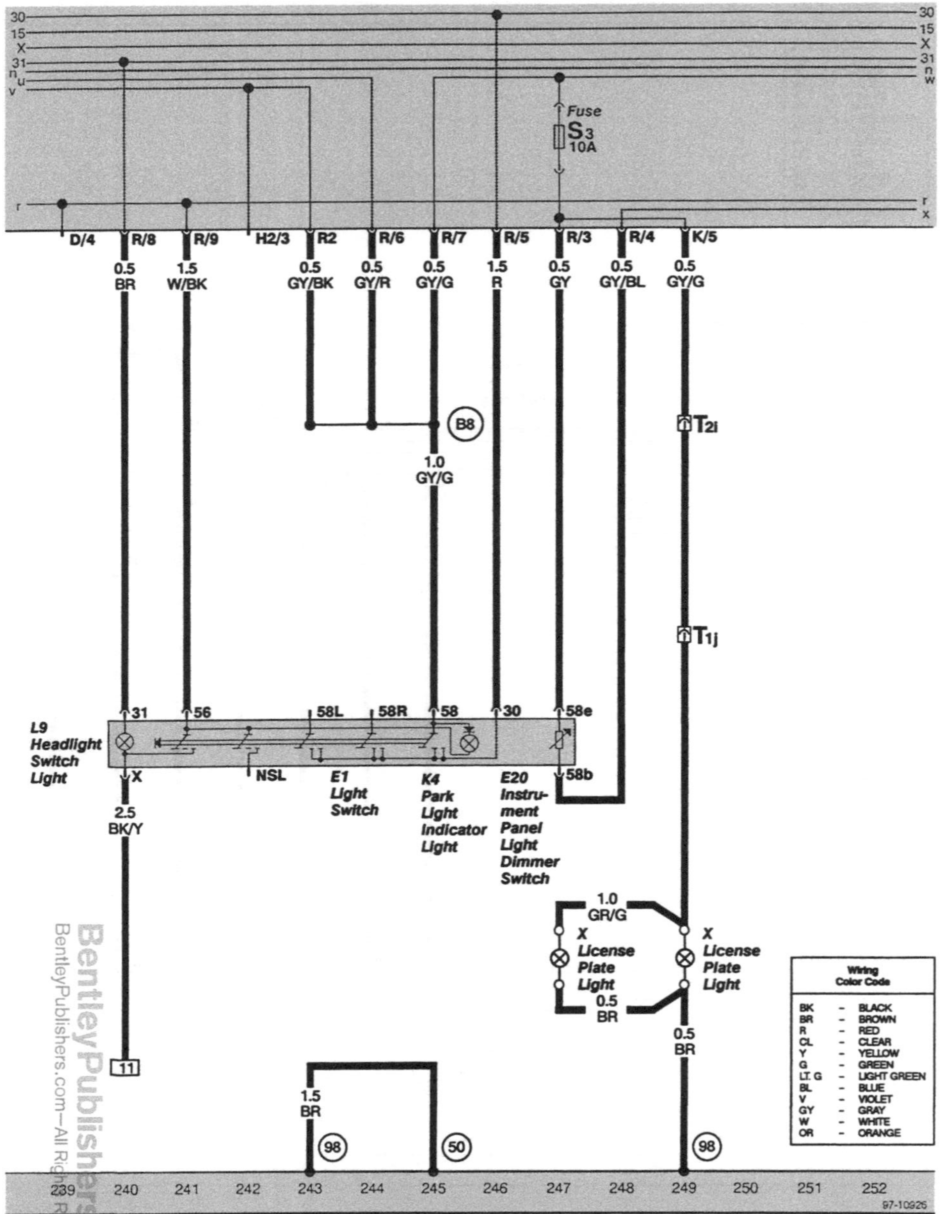

BK	-	BLACK
BR	-	BROWN
R	-	RED
CL	-	CLEAR
Y	-	YELLOW
G	-	GREEN
LT. G	-	LIGHT GREEN
BL	-	BLUE
V	-	VIOLET
GY	-	GRAY
W	-	WHITE
OR	-	ORANGE

Wiring Color Code

239 240 241 242 243 244 245 246 247 248 249 250 251 252

97-10926

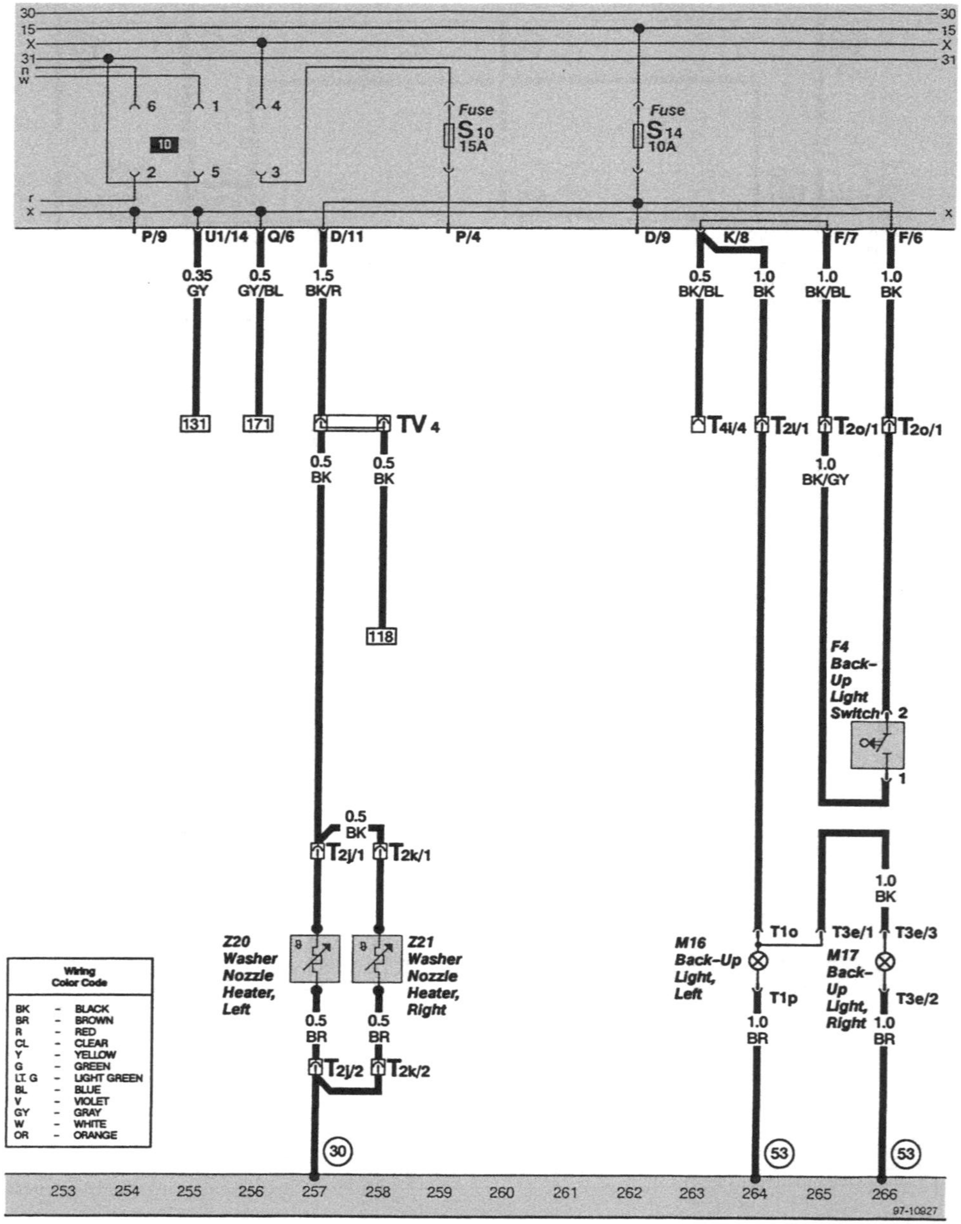

BK	-	BLACK
BR	-	BROWN
R	-	RED
CL	-	CLEAR
Y	-	YELLOW
G	-	GREEN
LT. G	-	LIGHT GREEN
BL	-	BLUE
V	-	VIOLET
GY	-	GRAY
W	-	WHITE
OR	-	ORANGE

Wiring Color Code

253 254 255 256 257 258 259 260 261 262 263 264 265 266

97-10927

Corrado 1993
Production from Jan. 1993

Corrado 1993
Production from Jan. 1993

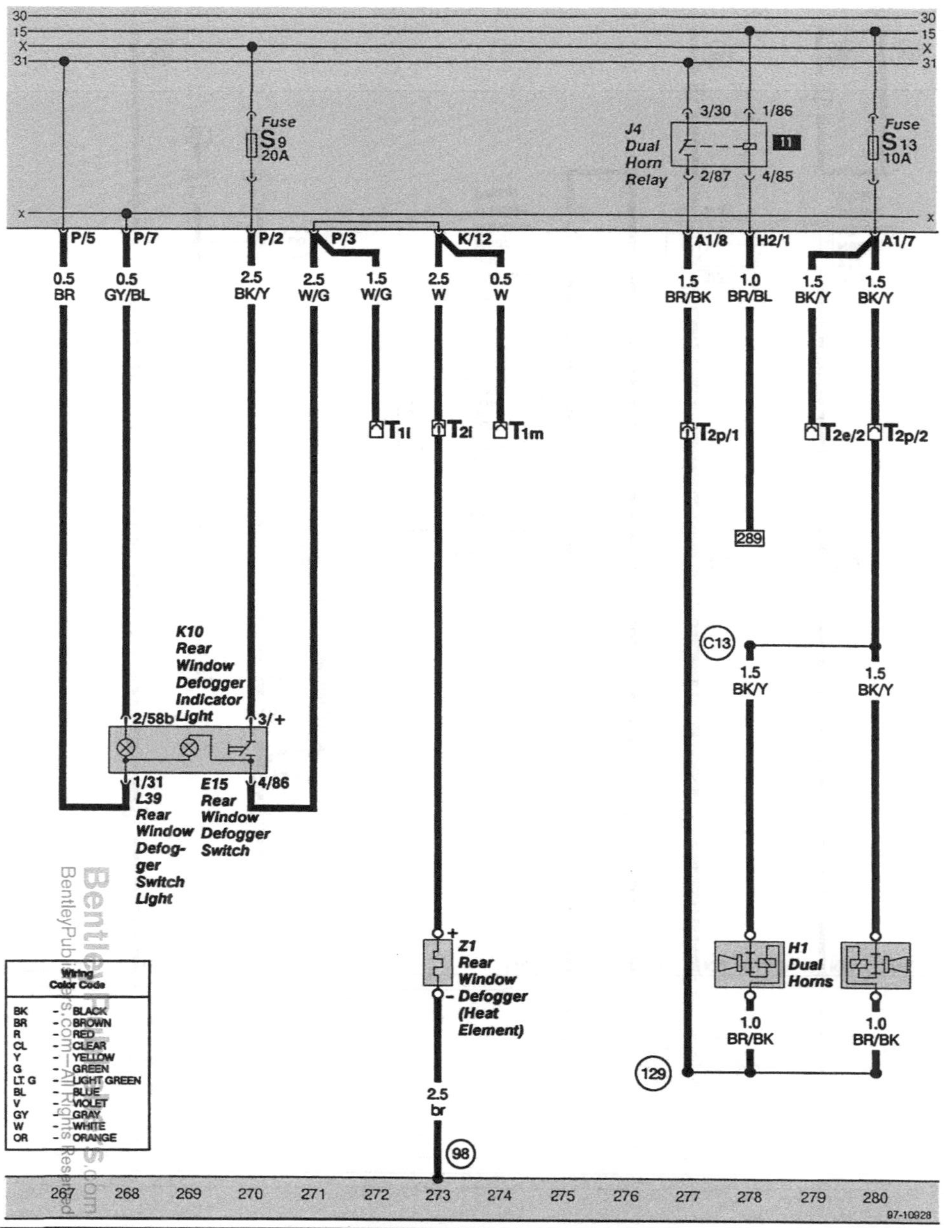

Corrado 1993
Production from Jan. 1993

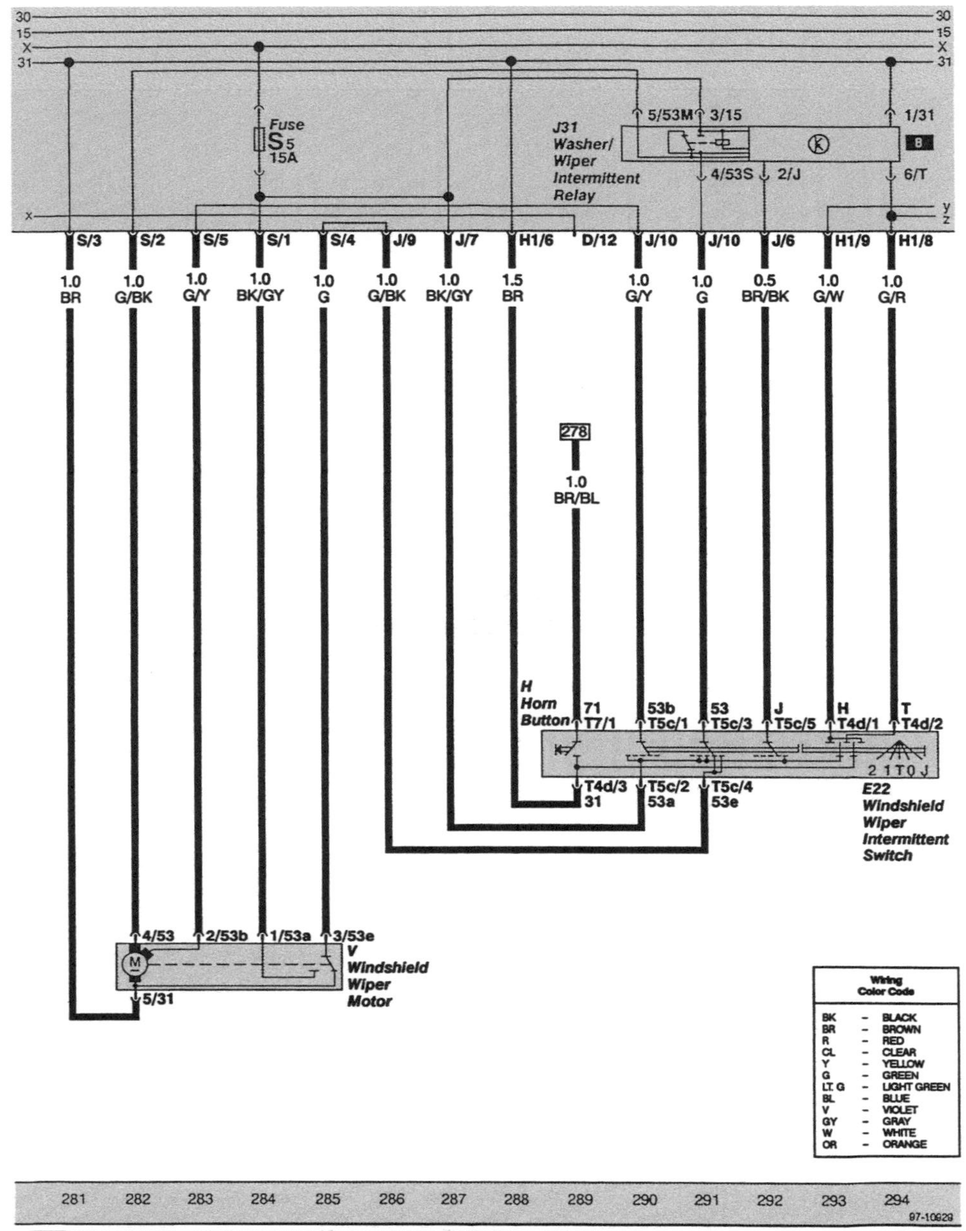

Corrado 1993
Production from Jan. 1993

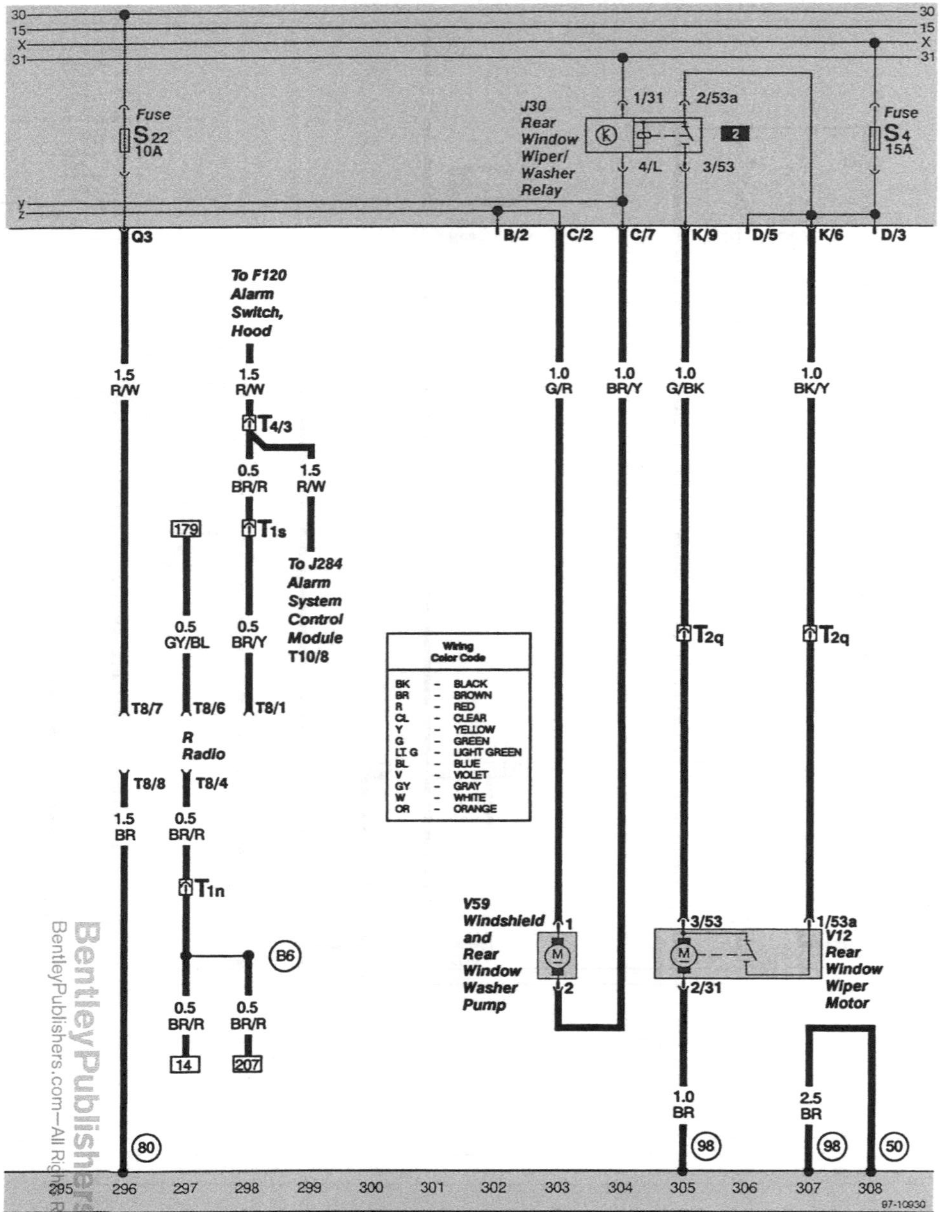

Corrado 1993
Production from Jan. 1993

THIS PAGE INTENTIONALLY LEFT
BLANK

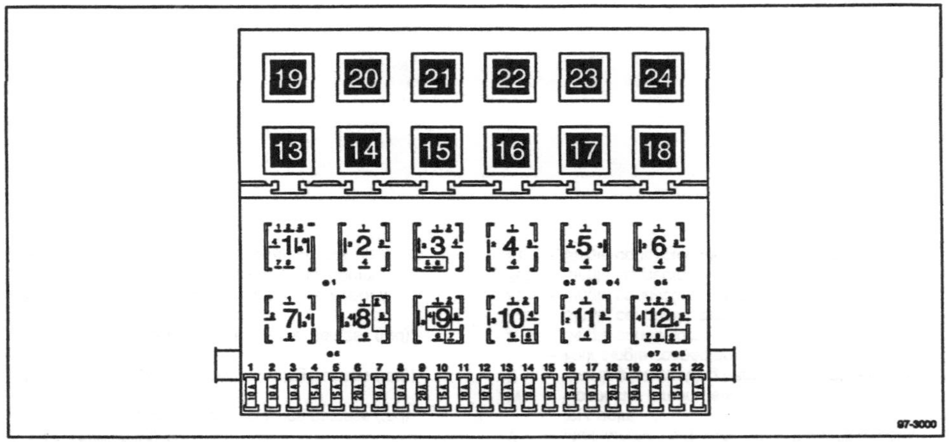

Fuse arrangement
according to the numbers on fuse panel from left to right:

		Amp.	Current track
1	Left Headlight	10	214
2	Right Headlight	10	217
3	Instrument Panel Light Dimmer Switch	10	261
4	Rear Window Wiper Motor	15	318
5	Windshield Wiper Motor	15	298
6	Fresh Air Blower Switch	20	23
7	Right Brake/Tail Light	10	249
8	Left Brake/Tail Light	10	242
9	Rear Window Defogger Switch Light	20	284
10	Open		
11	Headlight, High Beam, Left	10	213
12	Headlight, High Beam, Right	10	218
13	Horn	10	294
14	Back-up Lights Switch	10	276
15	Positive Crankcase Ventilation (PCV) Heating Element	10	28
16	Voltmeter, Oil Pressure Gauge	15	191
17	Emergency Flashers Switch	10	229
18	Heated Oxygen Sensor (HO_2S)	20	57
19	Coolant Fan Control (FC) Thermal Switch	30	31
20	Brake Light Switch	10	247
21	Cigarette Lighter	15	184
22	Radio	10	316

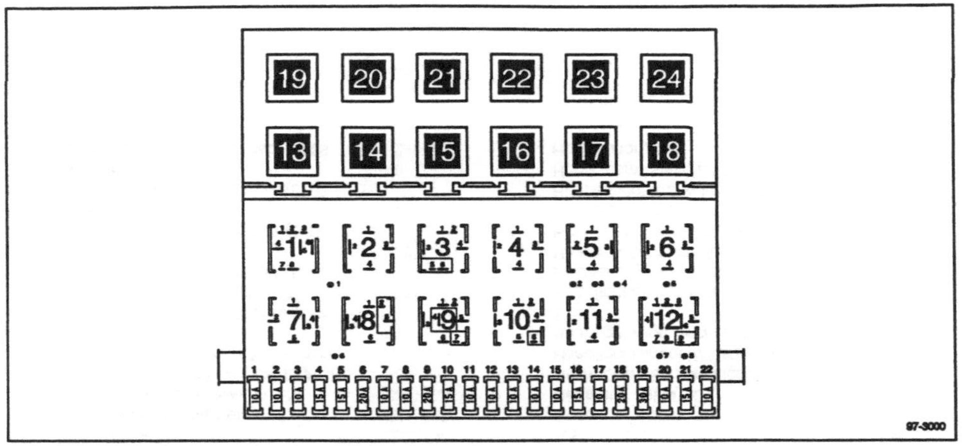

Relay location on fuse/relay panel

- 1 A/C Relay (13)*
- 2 Rear Window Wiper/Washer Relay (72)*
- 3 Motronic Engine Control Module (ECM) Power Supply Relay (32, 30)*
- 4 Load Reduction Relay (18)*
- 5 Engine Coolant Level (ECL) Control Module (43)*
- 6 Emergency Flasher Relay (21)*
- 7 Headlight Washer System Relay (33)*
- 8 Washer/Wiper Intermittent Relay (19, 99)*
- 9 Seat Belt Warning System Relay – Canada (4)*
- 10 Fog Light Relay (53)*
- 11 Dual Horn Relay (53)*
- 12 Fuel Pump (FP) Relay (67)*

Separate relays above fuse/relay panel

- 13 Back-Up Light Fuse (Auto. Trans.)
- 14 Coolant After-Run Control Module (31)*
- 15 ABS Hydraulic Pump Pelay (78)*
- 16 ABS Relay (79)*
- 17 Open
- 18 Open
- 19 A/C Relay (56)*
- 20 Park/Neutral Position (PNP) Relay (150)*
- 21 ABS Hydraulic Pump Fuse/Power Window Circuit Breaker
- 22 ABS Valves Fuse
- 23 Cruise Control Fuse/Automatic Shoulder Belt System Fuse
- 24 Open

*Numbers in parentheses indicate production control number stamped on relay housing.

Note: Locations of relays in positions 13–24 may vary from vehicle to vehicle depending on model and option content.

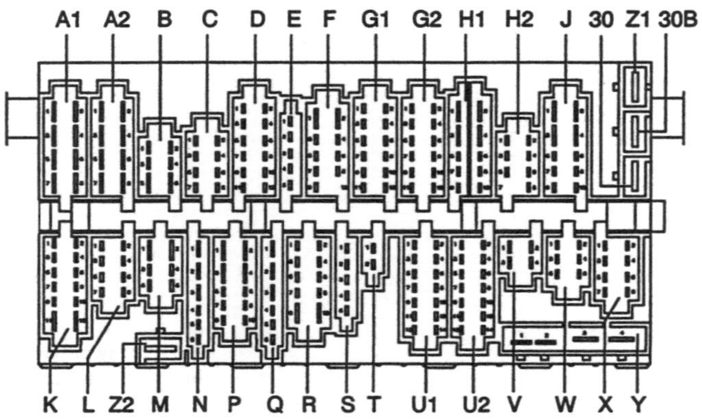

Connections and plugs on fuse/relay panel

- A1 – Headlight wiring harness (yellow), eight point
- A2 – Headlight wiring harness (yellow), eight point
- B – Open
- C – Headlight wiring harness (yellow), eight point
- D – Optional equipment wiring harness (green), twelve point
- E – Instrument wiring harness (green), five point
- F – Engine compartment wiring harness, right (white), ten point
- G1 – Engine compartment wiring harness, right (white), twelve point
- G2 – Engine compartment wiring harness, right (white), twelve point
- H1 – Steering column switch wiring harness (red), ten point
- H2 – Steering column switch wiring harness (red), eight point
- J – Steering column switch wiring harness (red), ten point
- K – Rear wiring harness (black), twelve point
- L – Rear wiring harness (black), eight point
- M – Rear wiring harness (black), six point
- N – A/C wiring harness (green), six point
- P – Rear window defroster wiring harness (blue), ten point
- Q – Instrument wiring harness (blue), six point
- R – Light switch wiring harness (blue), ten point
- S – Engine compartment wiring harness, right (white), five point
- T – Connector (green), two point
- U1 – Instrument cluster wiring harness (blue), fourteen point
- U2 – Instrument cluster wiring harness (blue), fourteen point
- V – Multi-function indicator wiring harness (green), four point
- W – Connector (green), six point
- X – Warning lamp wiring harness (green), eight point
- Y – Four single connectors, terminal 30
- Z1 – Single connector
- Z2 – Single connector, terminal 31
- 30 – Single connector, terminal 30
- 30B – Single connector
- 31 – Single connector, terminal 31 (not shown)

Corrado starting 1994
Engine code – AAA
Production from Mar. 1994

Description	Current track
ABS Warning Light, K47	176–179
Accelerator Sensor, G165	55–56
After-Run Coolant Fan Control (FC) Thermal Switch, F87	39
Alarm System Control Module, J284	317
Back-Up Light Switch, F4	280
Battery, A	5
Brake and Parking Brake Warning Light, K7	176–179
Brake Fluid Level Warning Switch, F34	196
Brake Light Switch, F	247
Camshaft Position (CMP) Sensor, G40	51–52
Cassette Storage Light, L66	194
Change-Over Valve, N180	105
Cigarette Lighter Light, L28	185
Cigarette Lighter, U1	184
Control Module With Display Unit In Instrument Cluster, J285	118–158
Coolant Fan (01018–3433), V7	34–36
Coolant Fan Control (FC) Thermal Switch, F18	30–31
Coolant FC (Fan Control) Control Module in engine compartment, left, J293	33–41
Cylinder 1 Fuel Injector, N30	85
Cylinder 2 Fuel Injector, N31	86
Cylinder 3 Fuel Injector, N32	87
Cylinder 4 Fuel Injector, N33	88
Cylinder 5 Fuel Injector, N83	89
Cylinder 6 Fuel Injector, N84	90
Digital Display Light, L75	153,155
Display, Y5	141–143
Dual Horn Relay, J4	291–292
Dual Horns, H1	292,294
Emergency Flasher Relay, J2	231–233
Emergency Flasher Switch, E3	230–233
Emergency Flasher Warning Light, K6	235
Engine Coolant Level (ECL) Sensor, G32	117
Engine Coolant Level/Temperature (ECL/ECT) Warning Light, K28	121
Engine Coolant Temperature (ECT) Gauge, G3	122
Engine Coolant Temperature (ECT) Sensor, G62	77
Engine Coolant Temperature (ECT) Sensor, G2	40
Engine Speed (RPM) Sensor , G28	82–84

Description	Current track
Evaporative Emission (EVAP) Canister Purge Regulator Valve, N80	112
Fresh Air Blower, V2	18
Fresh Air Blower Series Resistance, N23	19–23
Fresh Air Blower Switch, E9	20–23
Fresh Air Control Lever Light, L16	24
Front, Side Marker Lights, M11	211,220
Fuel Gauge, G1	123
Fuel Level Sensor, G	164
Fuel Pump (FP) Relay, J17	50–51
Fuel Pump (FP), G6	165
Fuse Strip, S88	41
Fuse Strip, S88	107
Fuse, S2, 10A	217
Fuse, S3, 10A	261
Fuse, S4, 15A	318
Fuse, S5, 15A	298
Fuse, S6, 20A	23
Fuse, S7, 10A	249
Fuse, S8, 10A	242
Fuse, S9, 20A	284
Fuse, S10, 15A	273
Fuse, S11, 10A	213
Fuse, S12, 10A	218
Fuse, S13, 10A	294
Fuse, S14, 10A	276
Fuse, S15, 10A	28
Fuse, S16, 15A	191
Fuse, S17, 10A	229
Fuse, S18, 20A	57
Fuse, S19, 30A	31
Fuse, S20, 10A	247
Fuse, S21, 15A	184
Fuse, S22, 10A	316
Generator (GEN) Warning Light, K2	136
Generator (GEN), C	1–2
Glove Compartment Light, W6	188
Headlight Dimmer/Flasher Switch, E4	227–228
Headlight High Beam Indicator Light, K1	140
Headlight Switch Light, L9	254
Heated Oxygen Sensor (HO2S), G39	59–61
Heater, A/C Clutch Fuse, S24	19–23
High Beam Headlight, Left, L13	213
High Beam Headlight, Right, L14	218
High-mount Brake Light, M25	246
Hood Alarm Switch, F120	318
Horn Button, H	203

Corrado starting 1994
Engine code – AAA
Production from Mar. 1994

Description	Current track
Idle Air Control (IAC) Valve, N71	91
Ignition Coil, N152	42-49
Ignition/Starter Switch, D	10-14
Instrument Cluster Light, L10	144-146
Instrument Panel Light Dimmer Switch, E20	261
Intake Air Temperature (IAT) Sensor, G42	95
Interior Light With Delay Switch, W15	169-171
Knock Sensor (KS) 1, G61	71-73
Knock Sensor (KS) II, G66	74-76
Left Back-Up Light, M16	278
Left Brake/Tail Light, M21	243-244
Left Door Latch Switch, F143	222
Left Front Door Contact Switch, F2	169
Left Front Turn Signal Light, M5	239
Left Headlight, L1	214
Left Parking Light, M1	212
Left Rear Turn Signal Light, M6	241
Left Seat Belt Switch, E24	222
Left Turn Signal Indicator Light, K65	138
Left Washer Nozzle Heater, Z20	271
License Plate Light, X	261,263
Light Switch, E1	257-259
Load Reduction Relay, J59	11-12
Luggage Compartment Light Switch, F5	208
Luggage Compartment Light, W3	208
Malfunction Indicator Lamp (MIL), K83	176-179
Manifold Absolute Pressure (MAP) Sensor, G71	53-54
Mass Air Flow (MAF) Sensor, G70	96-98
Motronic Engine Control Module (ECM) in plenum chamber, right, J220	44-111
Motronic Engine Control Module (ECM) PowerMulti-Function Indicator Memory Switch, E109	160-163
Multi-Function Indicator Mode Select Switch, E86	160-163
O.3bar Oil Pressure Switch, F22	115
Odometer Display, Y4	150-152
Oil Pressure Gauge Light, L27	192
Oil Pressure Gauge, G11	189-191
Oil Pressure Sensor,G10	189
Oil Pressure Switch, F1	116
Oil Pressure Warning Light, K3	119
Oil Temperature Sensor, G8	113
Outside Air Temperature Sensor, G17	167
Oxygen Sensor (O2S) II Behind Three Way Catalytic Converter (TWC), G130	66-68

Description	Current track
Park Light Indicator Light, K4	260
Parking Brake Warning Light Switch, F9	210
Positive Crankcase Ventilation (PCV) Heating Element, N79	28
Radio, R	316-318
Rear Spoiler Control Module,on left A-pillar, near relay panel, J223	198-206
Rear Spoiler End Switch 1, F135	203-204
Rear Spoiler End Switch 2, F136	206-207
Rear Spoiler Motor, V52	203
Rear Spoiler Switch, E127	199-201
Rear Window Defogger Indicator Light, K10	282-283
Rear Window Defogger Switch Light, L39	282-283
Rear Window Defogger Switch, E15	284
Rear Window Defogger/Heat Element, Z1	287
Rear Window Wiper Motor, V12	321-322
Rear Window Wiper/Washer Relay, J30	320-321
Right Back–Up Light, M17	280
Right Brake/Tail Light, M22	248-249
Right Front Door Contact Switch, F3	170
Right Front Map/Reading Light, W13	169-171
Right Front Turn Signal Light, M7	251
Right Headlight, L2	217
Right Make-Up Mirror Light Switch, F148	174
Right Make-up Mirror Light, W14	17-174
Right Parking Light, M3	219
Right Rear Turn Signal Light, M8	250
Right Turn Signal Indicator Light, K94	139
Right Washer Nozzle Heater, Z21	272
Seat Belt Warning Light, K19	135
Seat Belt Warning System Relay, J34	220-223
Secondary Air Injection (AIR) Pump Motor, V101	108
Secondary Air Injection (AIR) Relay, J316	107-110
Spark Plug Connectors, P	44-49
Spark Plugs, Q	44-49
Speedometer (00663), G21	144-147
Speedometer Vehicle Speed Sensor (VSS), G22	159-160
Starter, B	7-8
Supply Relay, J271	108-109
Tachometer, G5	127-130

Description	Current track
Third Speed Coolant Fan Control (FC) Thermal Switch, F165	42
Throttle Position (TP) Sensor, G69	79-80
Transmission Range (TR) Selector LeverTurn Signal Switch, E2	236
Voltage Regulator (VR), C1	1-2
Voltage Stabilizer, J6	132-134
Voltmeter, G14	189-191
Warning Buzzer, H3	123
Water Pump (00739), V50	38
Windshield and Rear Window Washer Pump, V59	319
Windshield Wiper Motor, V	296-299
Windshield Wiper/Washer Switch, E22	304-308
Wiper/Washer Intermittent Relay, J31	304-308

Wire connectors

T1 – Single Connector, near starter
T1a – Single Connector, behind fuse relay/panel
T1b – Single Connector, behind instrument panel, center
T1c – Single Connector, behind fuse relay/panel
T1d – Single Connector, behind fuse relay/panel
T1d1 – Single Connector, behind fuse/relay panel
T1e – Single Connector, on left Apillar, near relay panel
T1f – Single Connector, behind fuse relay/panel
T1g – Single Connector, behind fuse relay/panel
T1h – Single Connector, behind fuse relay/panel
T1i – Single Connector, behind fuse relay/panel
T1j – Single Connector, in rear lid
T1l – Single Connector, behind fuse relay/panel
T1m – Single Connector, behind fuse relay/panel
T1n – Single Connector, behind fuse relay/panel
T1o – Single Connectors, on Left Back-Up Light
T1p – Single Connectors, on Left Back-Up Light
T1r – Single Connector, behind fuse relay/panel
T1s – Single Connector, behind fuse relay/panel
T1t – Single Connector, behind fuse relay/panel
T1v – Single Connector
T1w – Single Connector
T1x – Single Connectors
T1y – Single Connectors
T2 – Double Connector, above relay panel
T2a – Double Connectors
T2a1 – Double Connector, below driver's seat
T2a2 – Double Connector, in rear lid
T2b – Double Connectors
T2b1 – Double Connector, behind instrument panel, center
T2c – Double Connector, behind instrument panel
T2c – Double Connectors
T2d – Double Connectors, behind fuse/relay panel
T2e – Double Connector, behind fuse relay/panel
T2f – Double Connector, on right A-pillar, upper part
T2g – Double Connector, in luggage compartment, left
T2h – Double Connector, behind fuse relay/panel
T2i – Double Connector, in luggage compartment, left
T2j – Double Connector, on engine hood
T2k – Double Connector, on engine hood
T2l – Double Connectors, behind fuse/relay panel
T2o – Double Connectors, behind fuse/relay panel
T2p – Double Connectors, behind fuse/relay panel
T2q – Double Connector, in luggage compartment, left
T2r – Double Connectors, on Left Parking Light
T2s – Double Connectors, on Front, Side Marker Lights
T2t – Double Connectors, on Right Parking Light
T2u – Double Connectors, on Front, Side Marker Lights
T2v – Double Connector, on taillight, left
T2w – Double Connector, on High-Mount Brake Light
T3a – 3-Pin Connector, near starter
T3b – 3-Pin Connector, near exhaust manifold
T3c – 3-Pin Connector, near starter
T3e – 3-Pin Connectors, on Right Back-Up Light
T3f – 3-Pin Connectors, on Left Headlight
T3g – 3-Pin Connectors, on Right Headlight
T4 – 4-Pin Connectors, on Coolant FC (Fan Control) Control Module
T4a – 4-Pin Connector, behind steering column switch cover
T4b – 4-Pin Connector, near exhaust manifold
T4c – 4-Pin Connector, behind fuse/relay panel
T4d – 4-Pin Connector, behind steering column switch cover
T4e – 4-Pin Connectors, on Left Rear Turn Signal Light
T4f – 4-Pin Connectors, on Right Rear Turn Signal Light
T4i – 4-Pin Connector, behind fuse/relay panel
T4r – 4-Pin Connectors, behind fuse/relay panel
T4r – 4-Pin Connectors, near exhaust manifold
T5 – 5-Pin Connector, on Fresh Air Blower Series Resistance
T5b – 5-Pin Connector, behind steering column switch cover
T5c – 5-Pin Connector, behind steering column switch cover
T6 – 6-Pin Connector, on Rear Spoiler Switch
T7 – 7-Pin Connector, behind steering column switch cover

Wire connectors

- T8 – 8-Pin Connector, on Radio
- T8a – 8-Pin Connector, behind fuse/relay panel
- T8d – 8-Pin Connector, on Brake and Parking Brake Indicator Light
- T10 – 10-Pin Connector, on Alarm System Control Module
- T10a – 10-Pin Connector, on Rear Spoiler Control Module
- T10b – 10-Pin Connectors, on Coolant FC (Fan Control) Control Module
- T16 – 16-Pin Connectors, On Board Diagnostic (OBD), near ignition/starter, right
- T28 – 28-Pin Connector, on Instrument Cluster
- T42 – 42-Pin Connector, on cylinder head, left
- T68 – 68-Pin Connector, on Motronic Engine Control Module (ECM)
- TV 4 – Terminal 15 Wire Connector, above fuse relay panel
- TV13 – Wire Connector, Vehicle Speed Signal, behind fuse/relay panel
- TV14 – Data Link Connector (DLC) wire Connector, behind console

Welded wiring harness points

- (A11) – plus connection (58b) in instrument panel wiring harness
- (A15) – plus connection (15) in instrument panel wiring harness
- (B6) – wire connection (s), in steering column switch wiring harness
- (B8) – plus connection(58), in light switch wiring harness
- (C1) – plus connection (58L), in left front wiring harness
- (C2) – plus connection (58R), in left front wiring harness
- (C13) – plus connection, in dual tone horn wiring harness
- (C15) – wire connection –1–, in headlight wiring harness
- (C16) – wire connection –2–, in headlight wiring harness
- (E6) – plus connection –1– (15), in Motronic Multiport Fuel Injection (MFI) wiring harness
- (E7) – plus connection (87a), in Motronic Multiport Fuel Injection (MFI) wiring harness
- (E17) – wire connection –2–, in Motronic Multiport Fuel Injection (MFI) wiring harness
- (E18) – wire connection –3–, in Motronic Multiport Fuel
- (G3) – plus connection, in injector wiring harness Injection (MFI) wiring harness
- (K21) – wire connection –1–, in coolant fan wiring harness
- (K28) – plus connection (15), in coolant fan wiring harness
- (W9) – plus connection (30), in rear wiring harness
- (W10) – plus connection (15), in rear wiring harness

Ground connection

- (1) – Ground strap, battery to body
- (2) – Ground strap, transmission to body
- (15) – Ground connection, on cylinder head
- (18) – Ground connection, on engine block
- (28) – Ground connection, on firewall
- (30) – Ground connection, –1–, beside fuse/relay panel
- (43) – Ground connection, on right A-pillar, lower part
- (50) – Ground connection, in luggage compartment, left

Ground connection

- (53) – Ground connection, in rear lid, right
- (80) – Ground connection –1–, in instrument panel wiring harness
- (86) – Ground connection –1–, in rear wiring harness
- (98) – Ground connection, in rear lid wiring harness
- (119) – Ground connection –1–, in headlight wiring harness
- (120) – Ground connection –2–, in headlight wiring harness
- (125) – Ground connection –3–, in headlight wiring harness
- (129) – Ground connection, in dual-tone horn wiring harness
- (138) – Ground connection (control module), in Motronic Multiport Fuel Injection (MFI) wiring harness
- (139) – Ground connection (sensor Ground), in Motronic Multiport Fuel Injection (MFI) wiring harness
- (159) – Ground connection –2–, in Digifant Multiport Fuel Injection (MFI) wiring harness
- (182) – Ground connection –1–, in engine compartment wiring harness (6 Cylinder)
- (193) – Ground connection –1–, in coolant fan wiring harness

Wiring Color code	
BK	BLACK
BR	BROWN
CL	CLEAR
R	RED
Y	YELLOW
G	GREEN
LT. G	LIGHT GREEN
BL	BLUE
V	VIOLET
GY	GRAY
W	WHITE
OR	ORANGE

Wire Size

Wiring Diagrams identify wires by the metric wire size. Metric wire sizes indicate cross-sectional area in square millimeters (mm²). The chart below lists metric wire sizes and their approximate equivalents in American Wire Gauge (AWG) sizes.

Metric size Cross-section (mm2)	American Wire Gauge size (AWG)
0.35	22
0.50	20
0.75	18
1.00	16
1.50	14
2.50	12
4.00	10
6.00	8
16.00	4
25.00	2
35.00	2

Corrado starting 1994
Engine code – AAA
Production from Mar. 1994

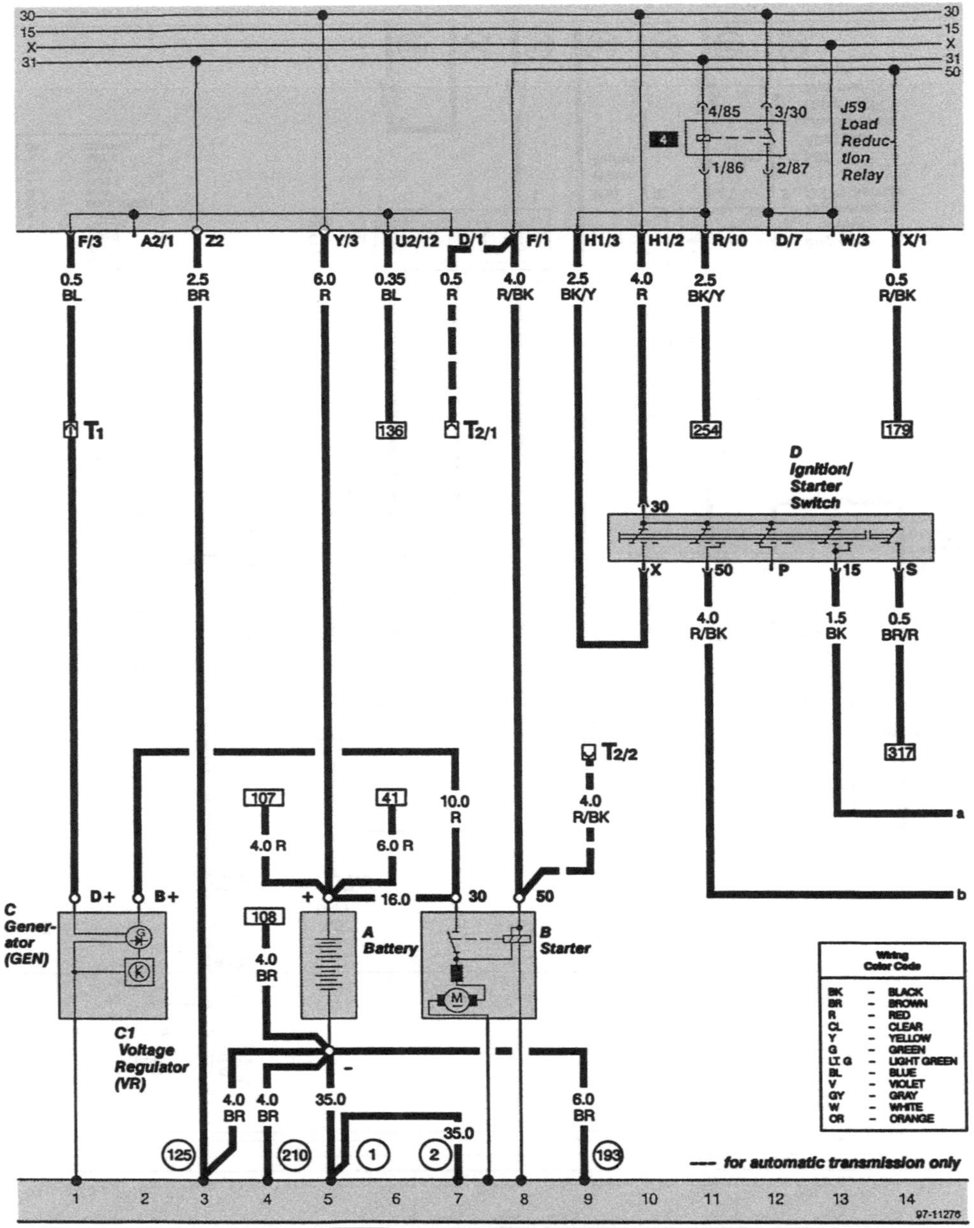

Corrado starting 1994
Engine code – AAA
Production from Mar. 1994

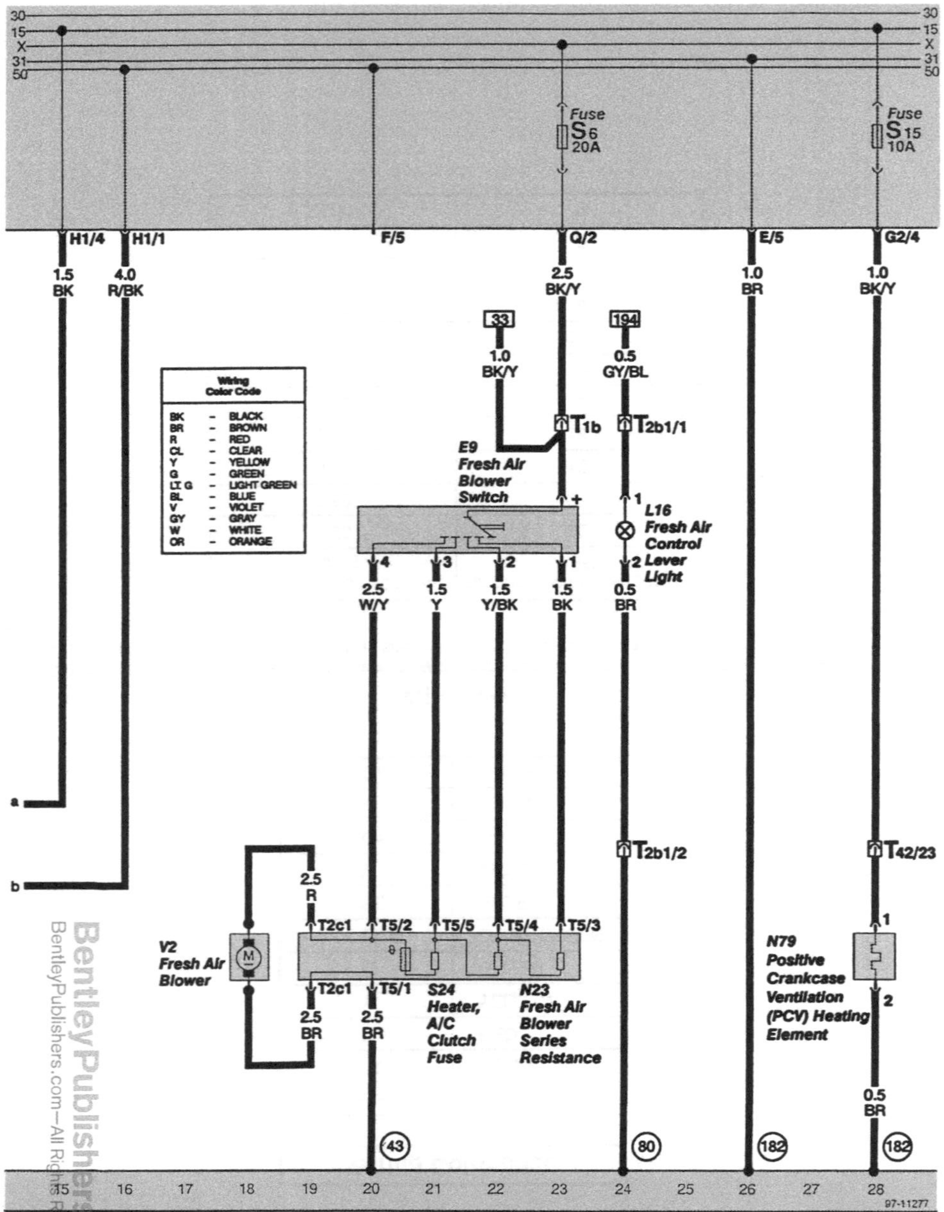

Corrado starting 1994
Engine code – AAA
Production from Mar. 1994

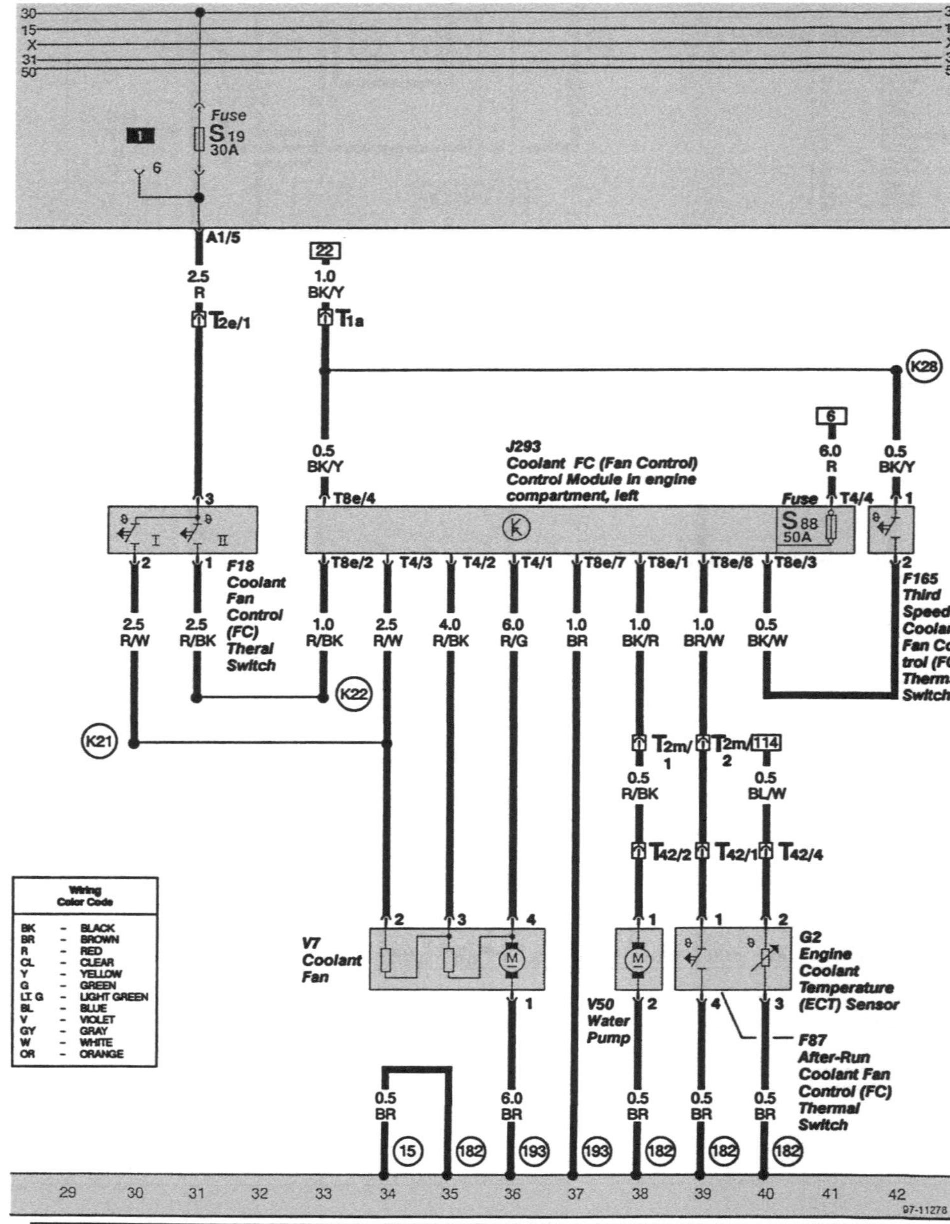

Corrado starting 1994
Engine code – AAA
Production from Mar. 1994

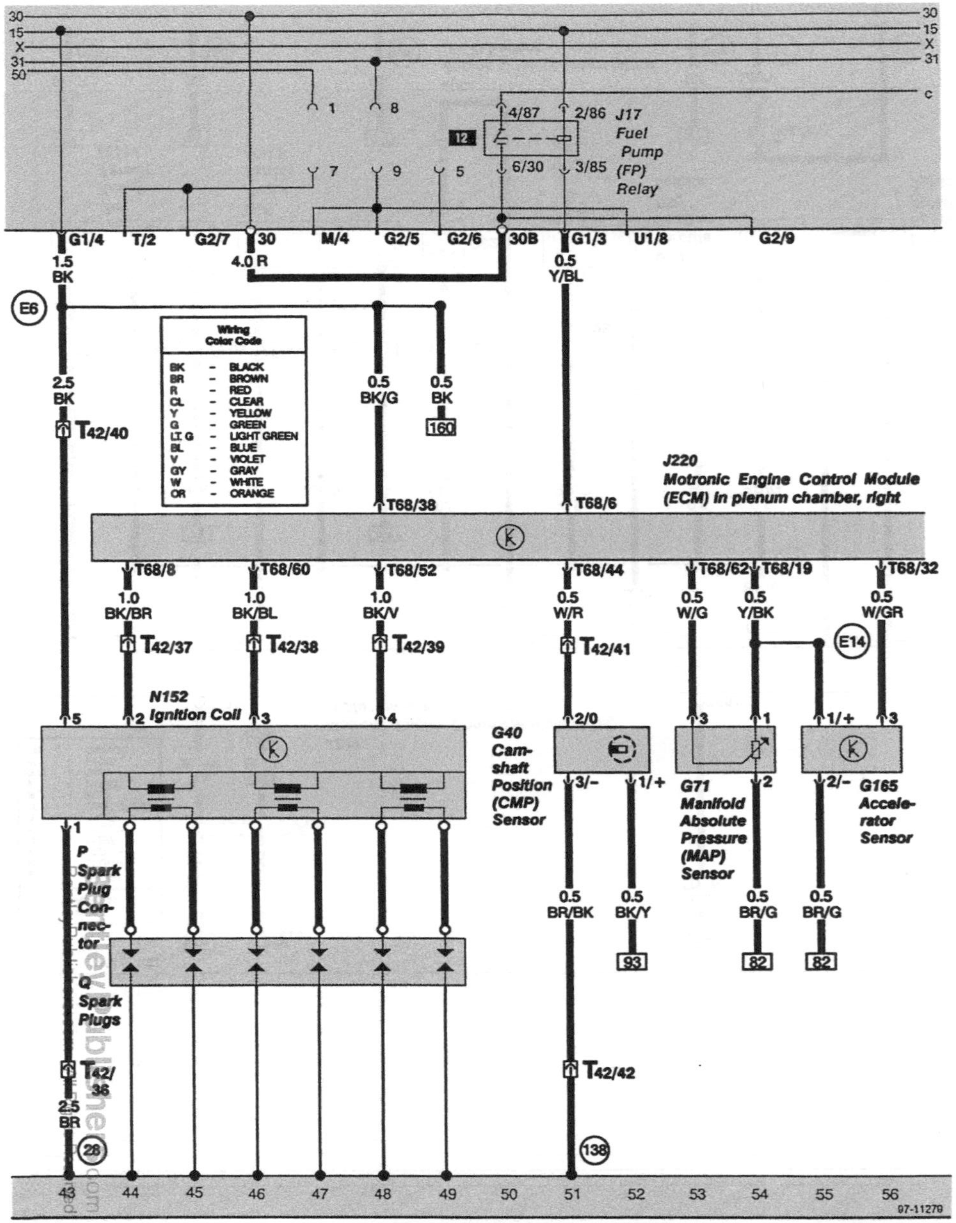

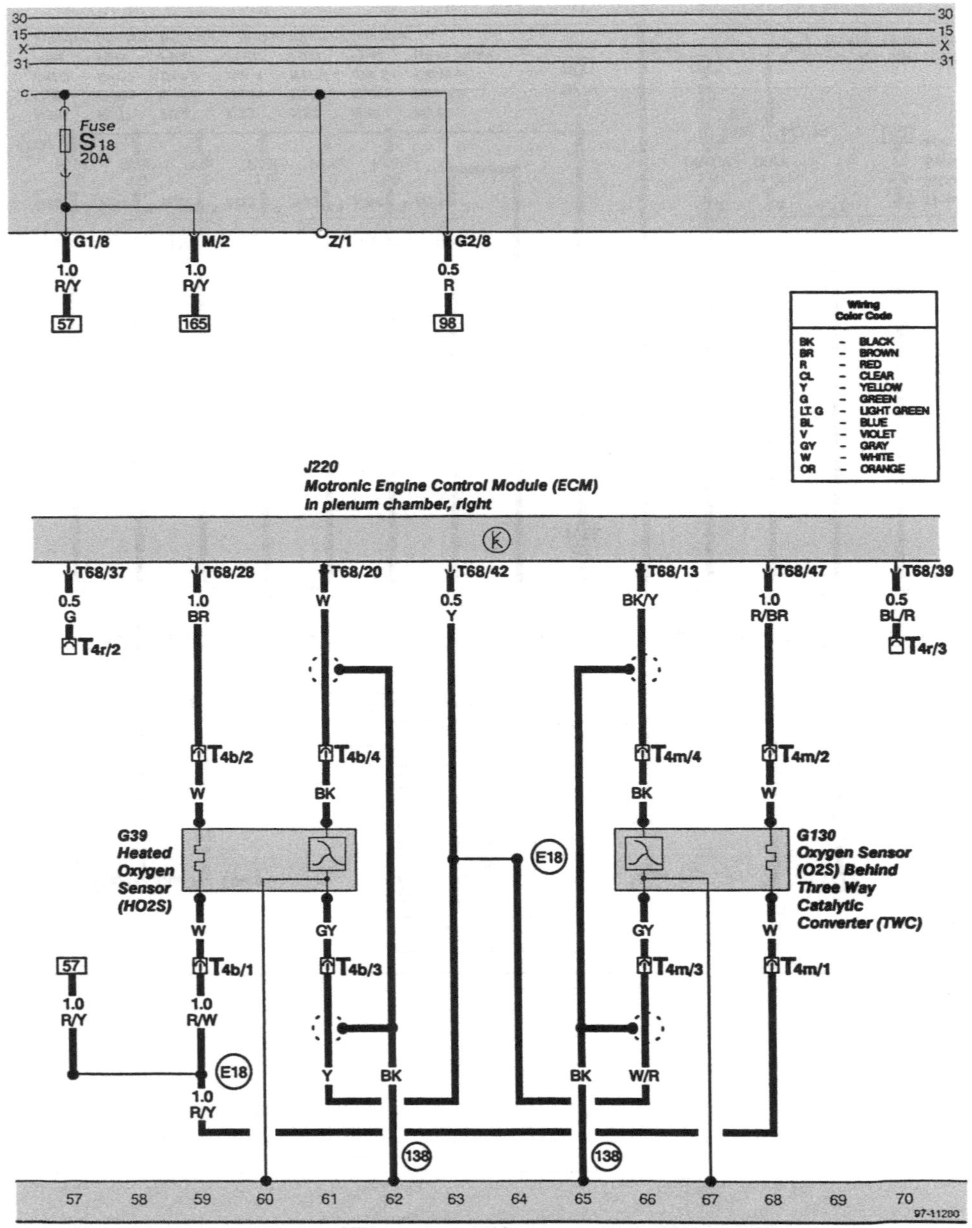

Corrado starting 1994
Engine code – AAA
Production from Mar. 1994

Corrado starting 1994
Engine code – AAA
Production from Mar. 1994

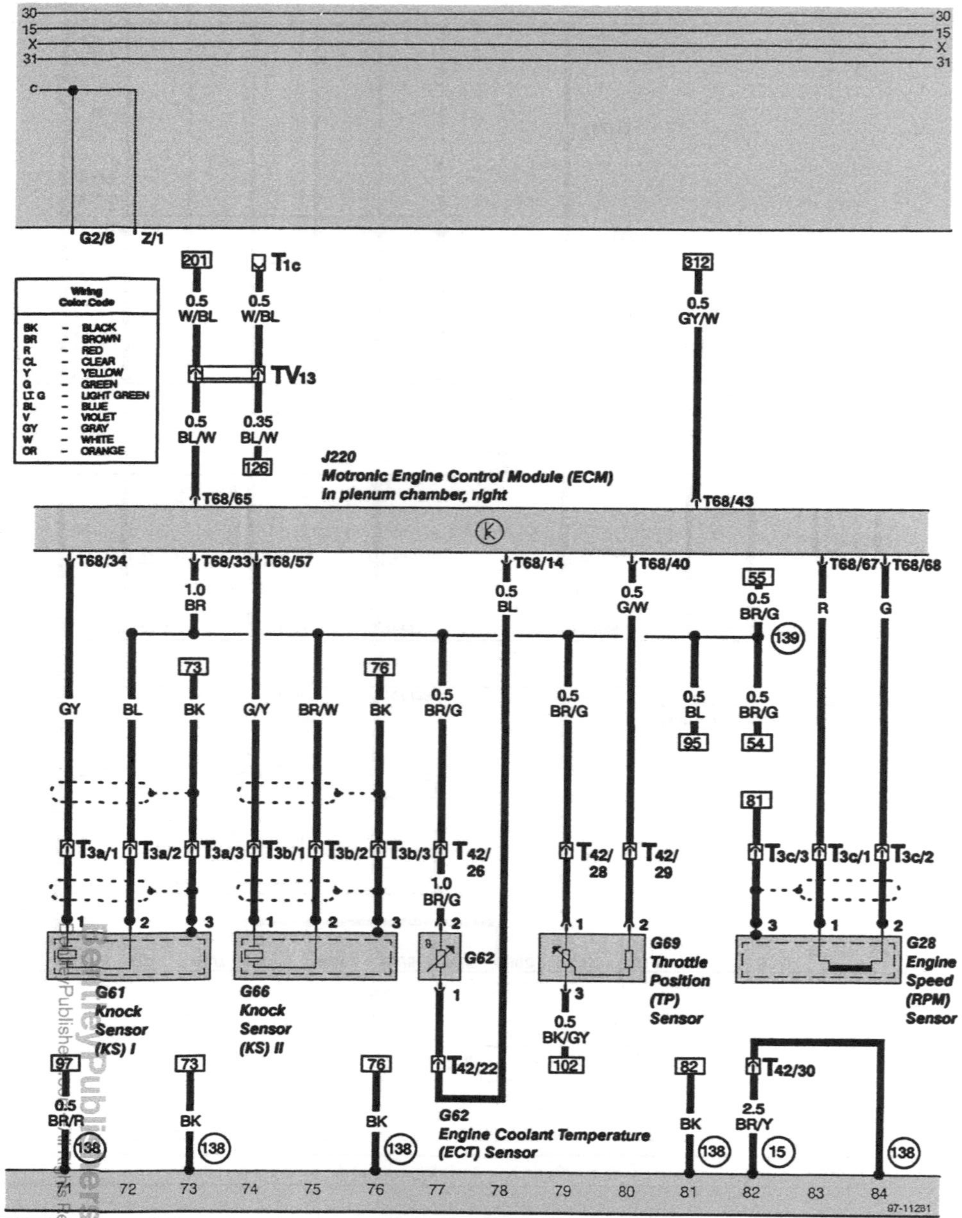

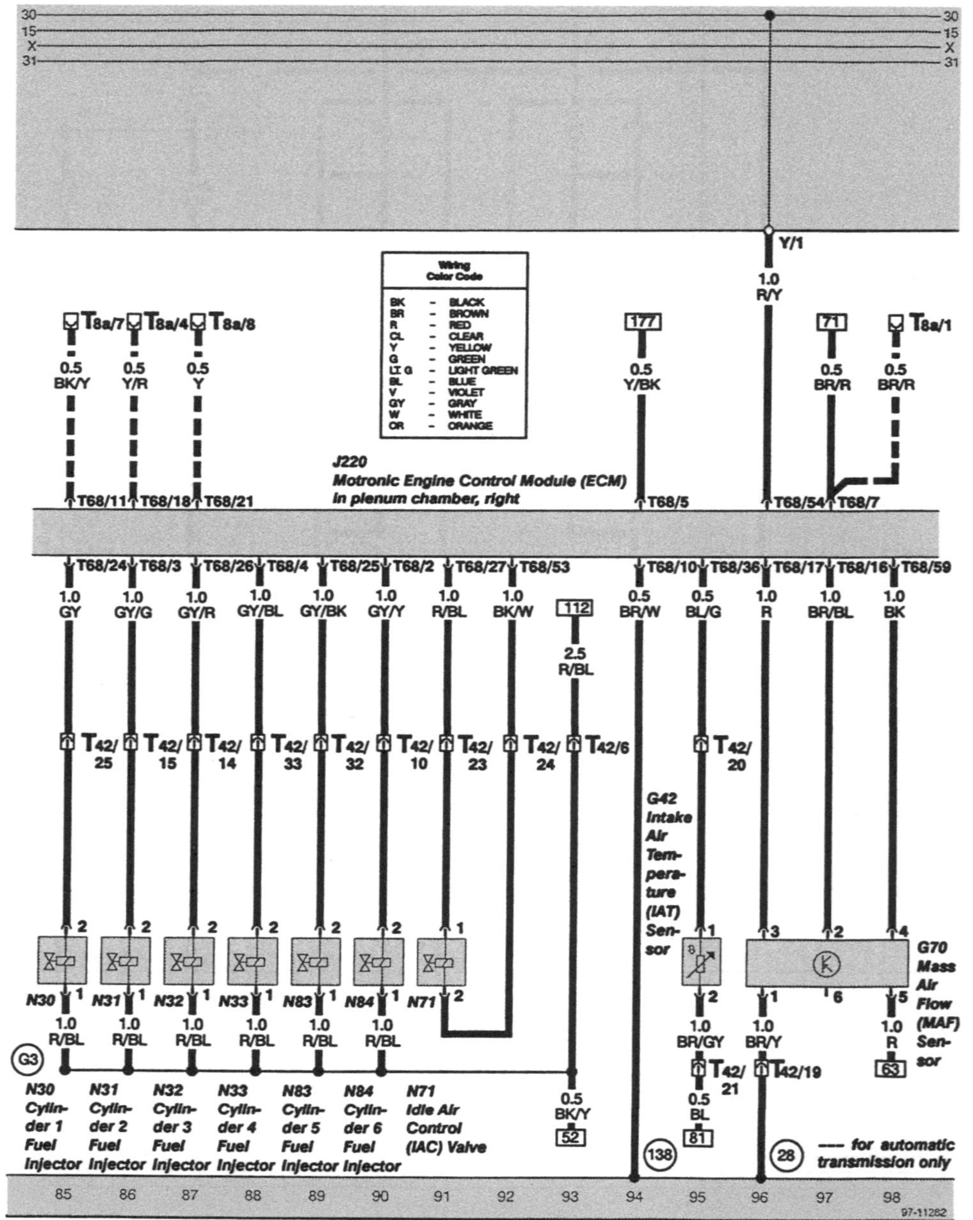

Corrado starting 1994
Engine code – AAA
Production from Mar. 1994

Corrado starting 1994
Engine code – AAA
Production from Mar. 1994

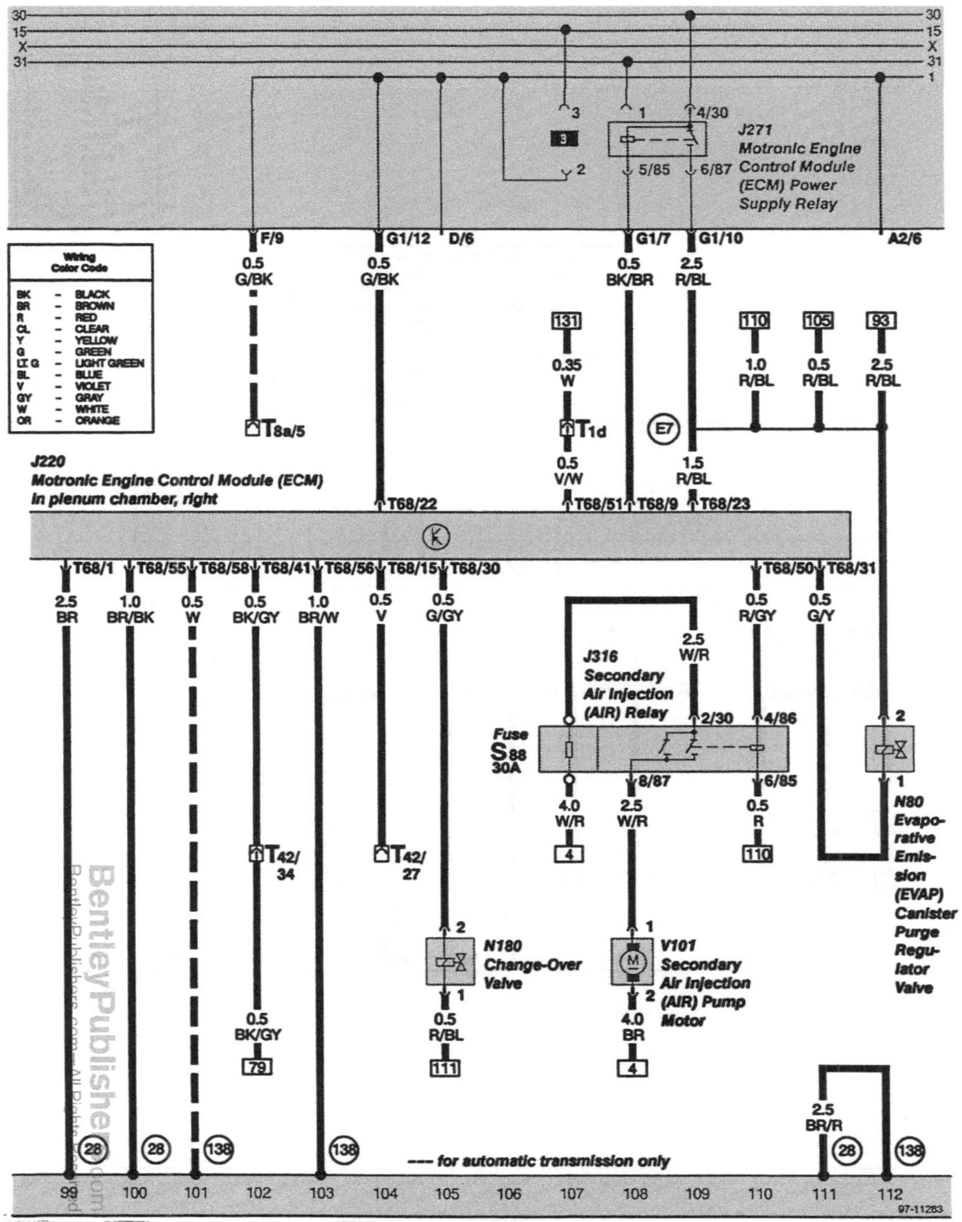

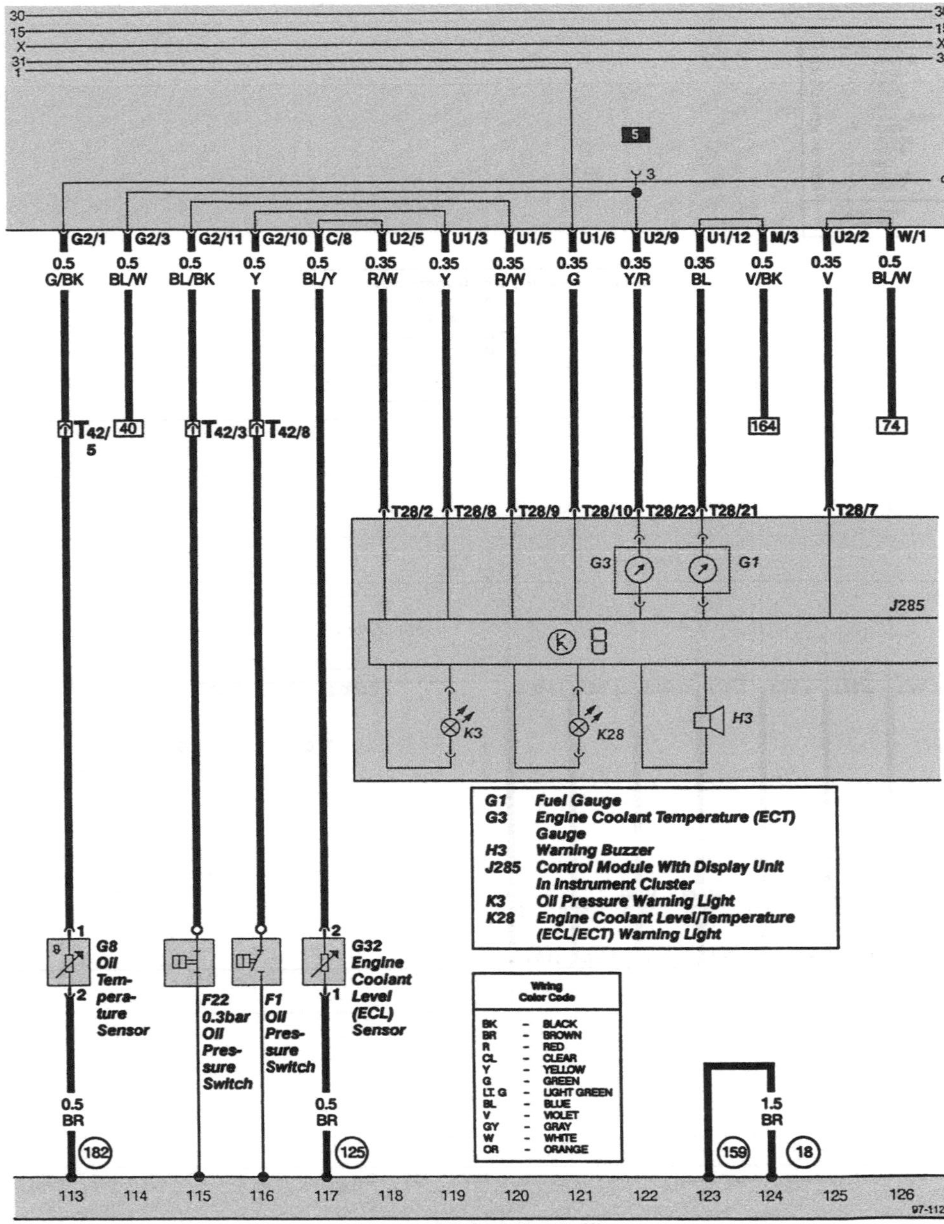

Corrado starting 1994
Engine code – AAA
Production from Mar. 1994

Corrado starting 1994
Engine code – AAA
Production from Mar. 1994

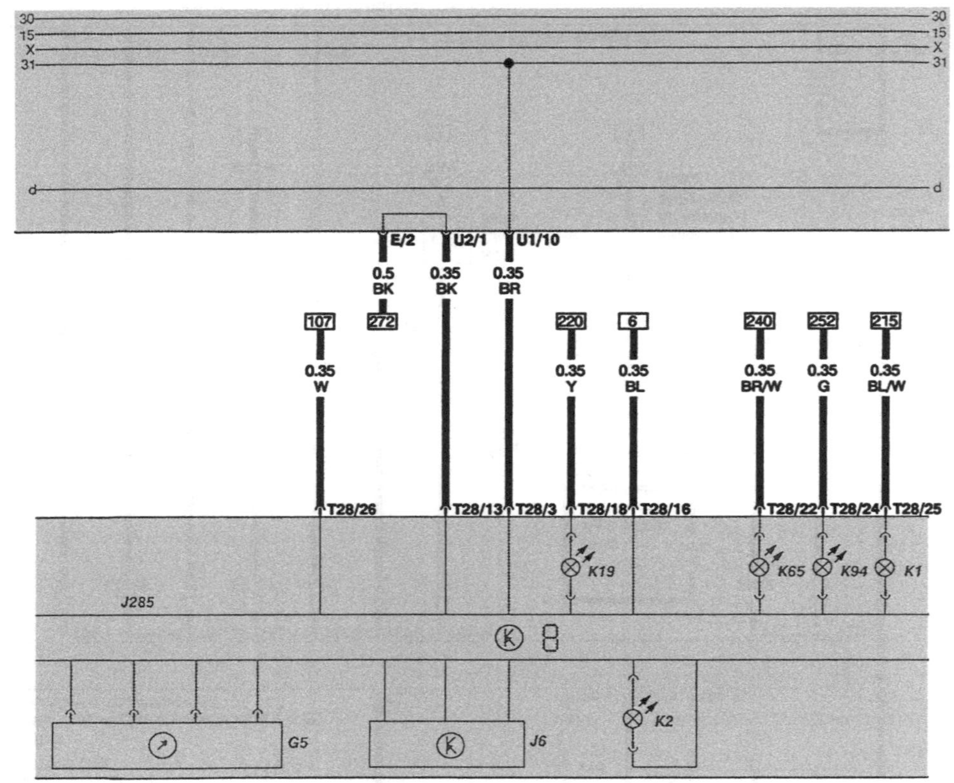

G5	Tachometer
J6	Voltage Stabilizer
J285	Control Module With Display Unit In Instrument Cluster
K1	Headlight High Beam Indicator Light
K2	Generator (GEN) Warning Light
K19	Seat Belt Warning Light
K65	Left Turn Signal Indicator Light
K94	Right Turn Signal Indicator Light

Wiring Color Code

BK	-	BLACK
BR	-	BROWN
R	-	RED
CL	-	CLEAR
Y	-	YELLOW
G	-	GREEN
LT G	-	LIGHT GREEN
BL	-	BLUE
V	-	VIOLET
GY	-	GRAY
W	-	WHITE
OR	-	ORANGE

127 128 129 130 131 132 133 134 135 136 137 138 139 140

97-11285

Corrado starting 1994
Engine code – AAA
Production from Mar. 1994

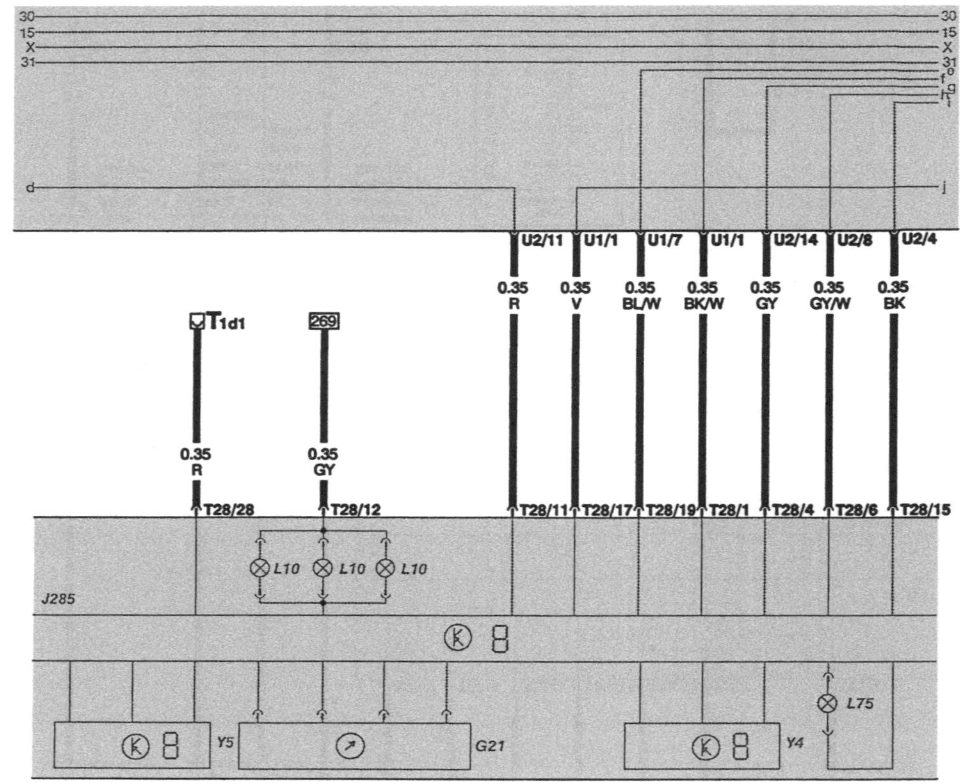

G21	Speedometer
J285	Control Module With Display Unit In Instrument Cluster
L10	Instrument Cluster Light
L75	Digital Display Light
Y4	Odometer Display
Y5	Transmission Range (TR) Selector Lever

Wiring Color Code

BK	-	BLACK
BR	-	BROWN
R	-	RED
CL	-	CLEAR
Y	-	YELLOW
G	-	GREEN
LT G	-	LIGHT GREEN
BL	-	BLUE
V	-	VIOLET
GY	-	GRAY
W	-	WHITE
OR	-	ORANGE

141 142 143 144 145 146 147 148 149 150 151 152 153 154

97-11286

Corrado starting 1994
Engine code – AAA
Production from Mar. 1994

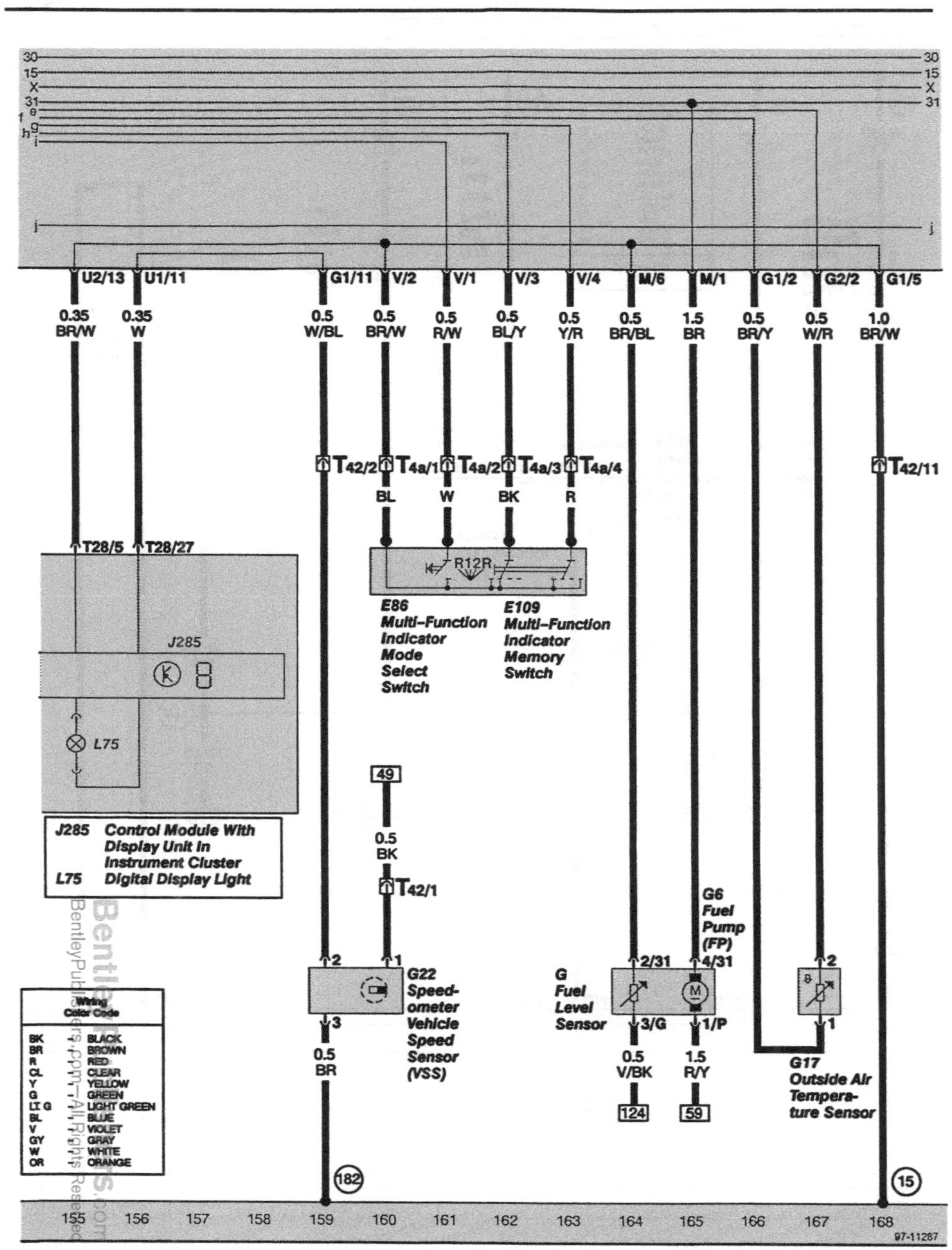

165 Instrument cluster

Corrado starting 1994
Engine code – AAA
Production from Mar. 1994

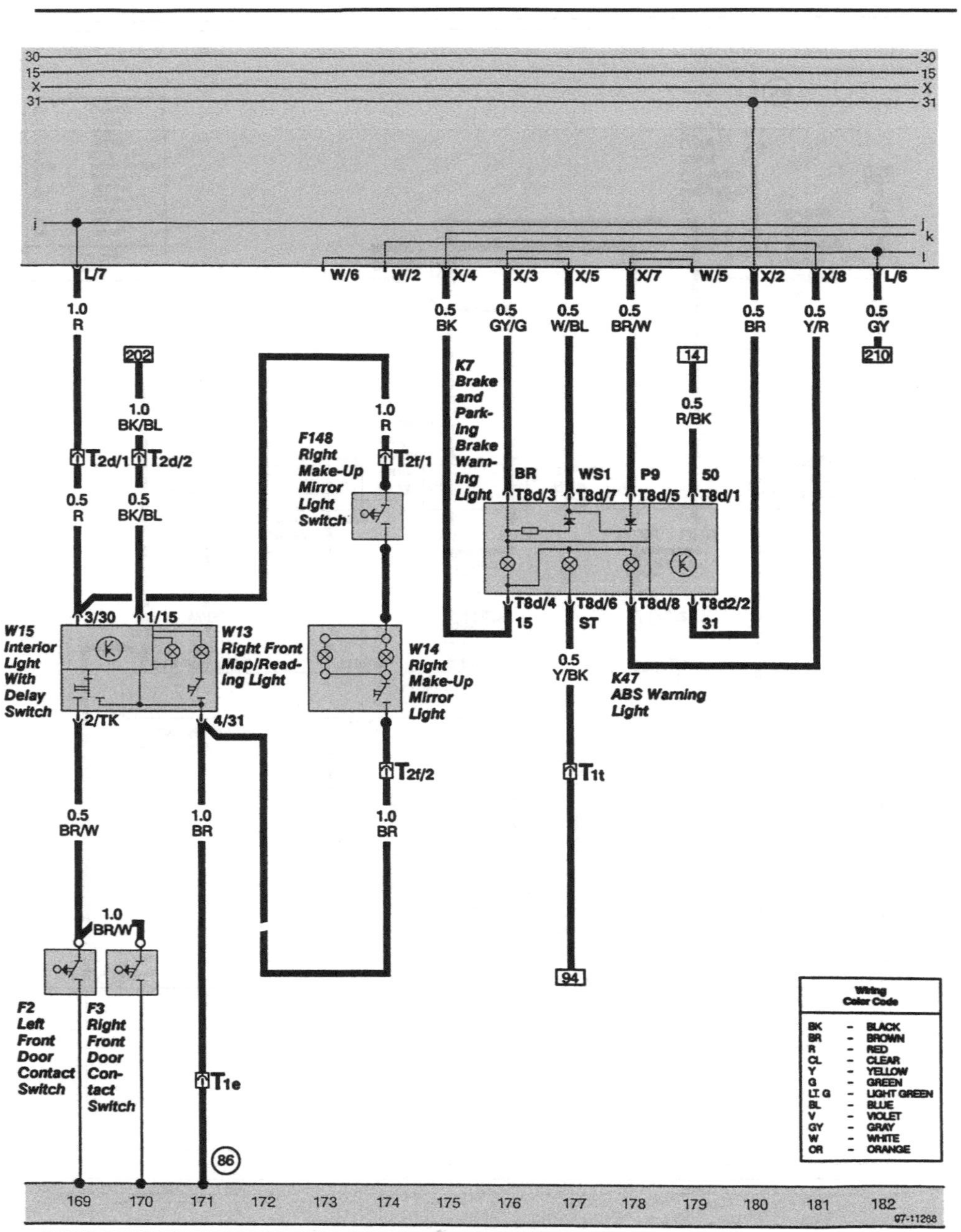

Corrado starting 1994
Engine code – AAA
Production from Mar. 1994

Interior lights Brake warning lights 166

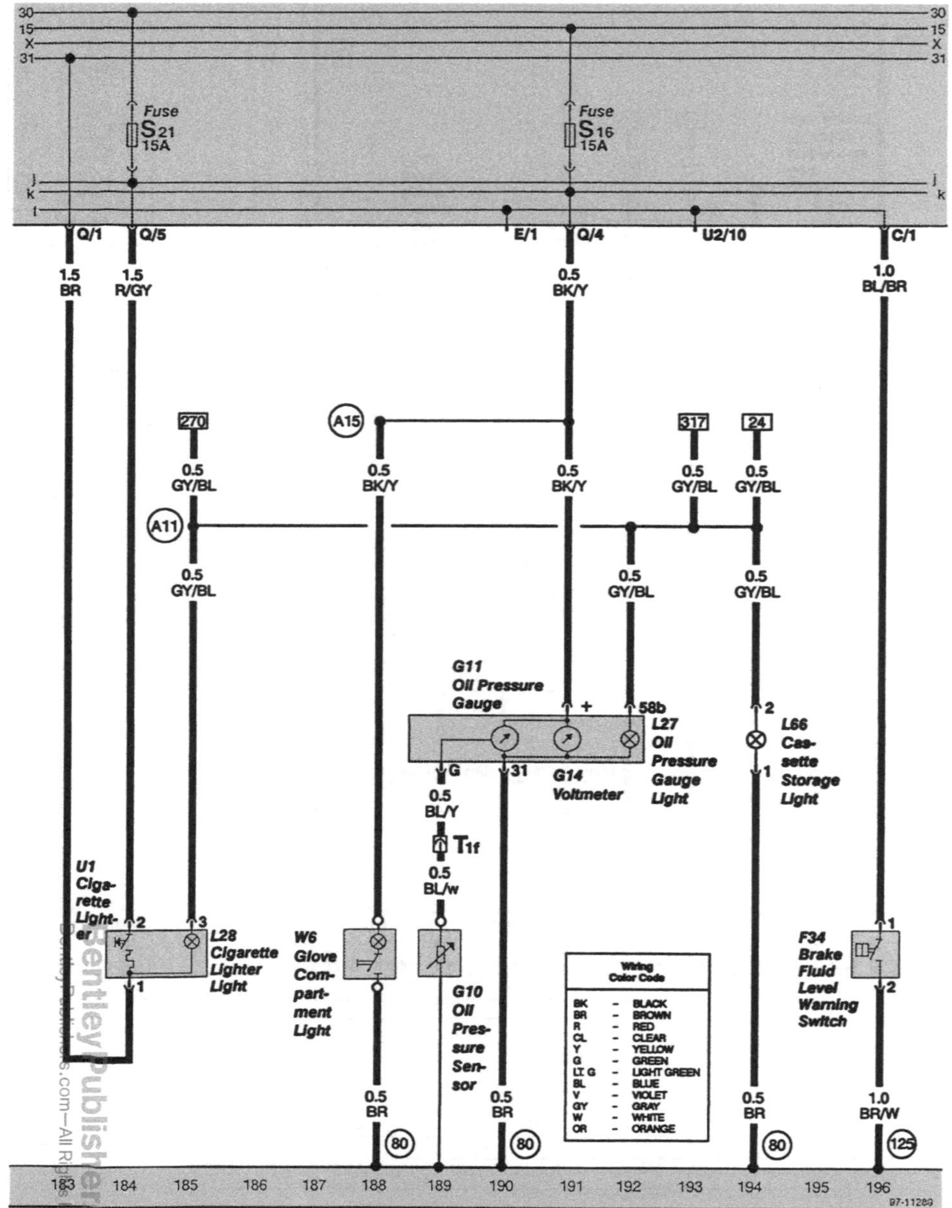

Corrado starting 1994
Engine code – AAA
Production from Mar. 1994

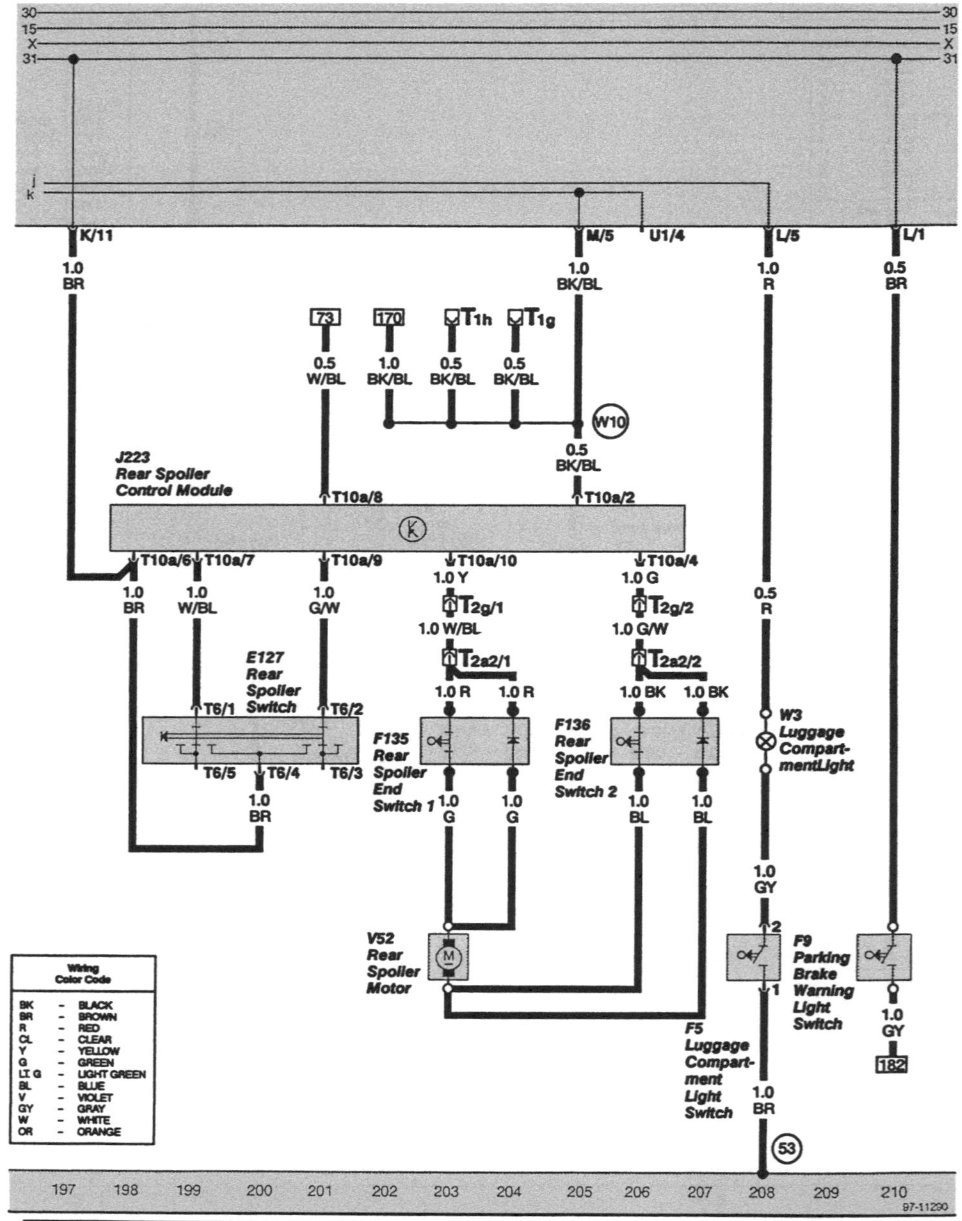

Corrado starting 1994
Engine code – AAA
Production from Mar. 1994

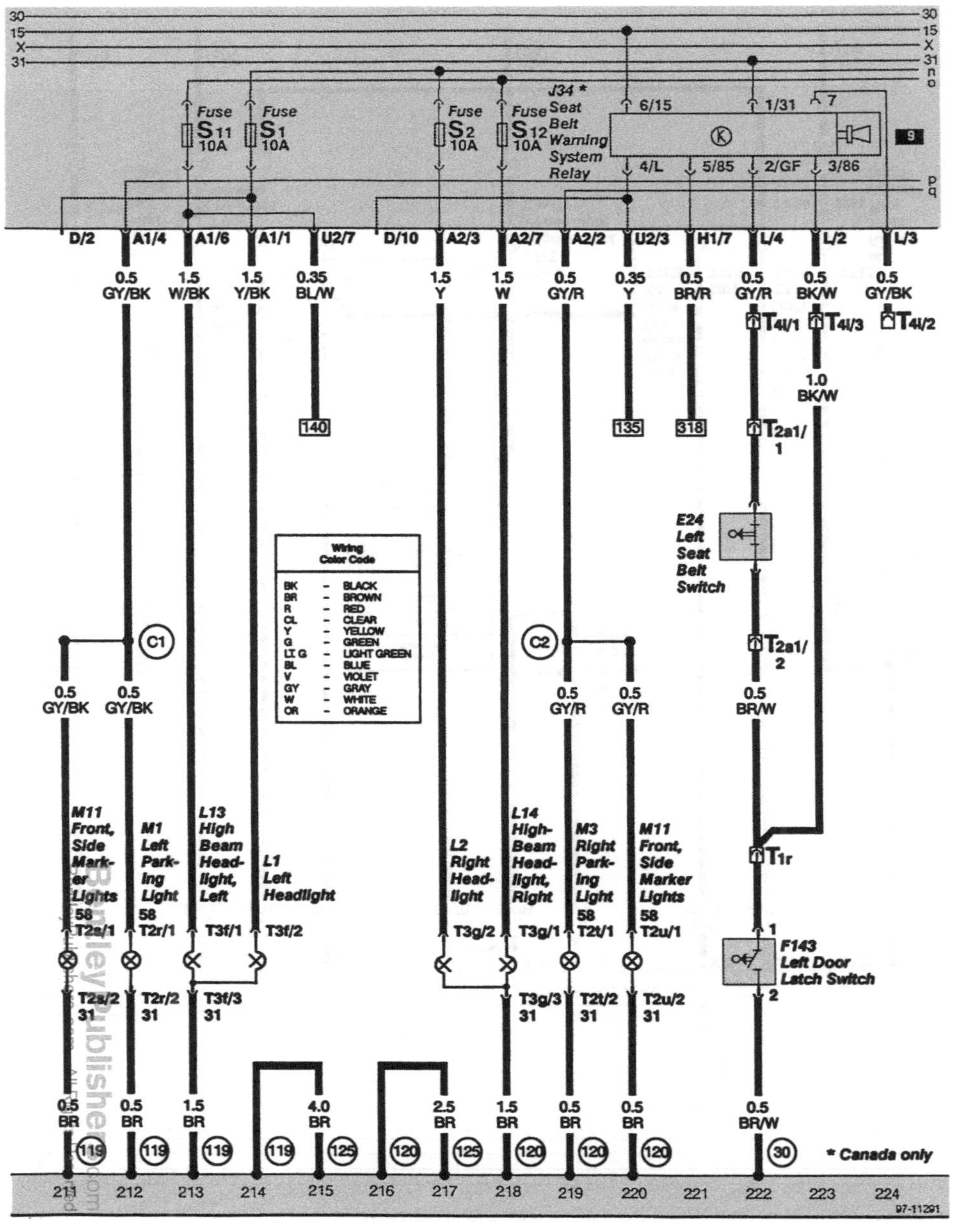

Corrado starting 1994
Engine code – AAA
Production from Mar. 1994

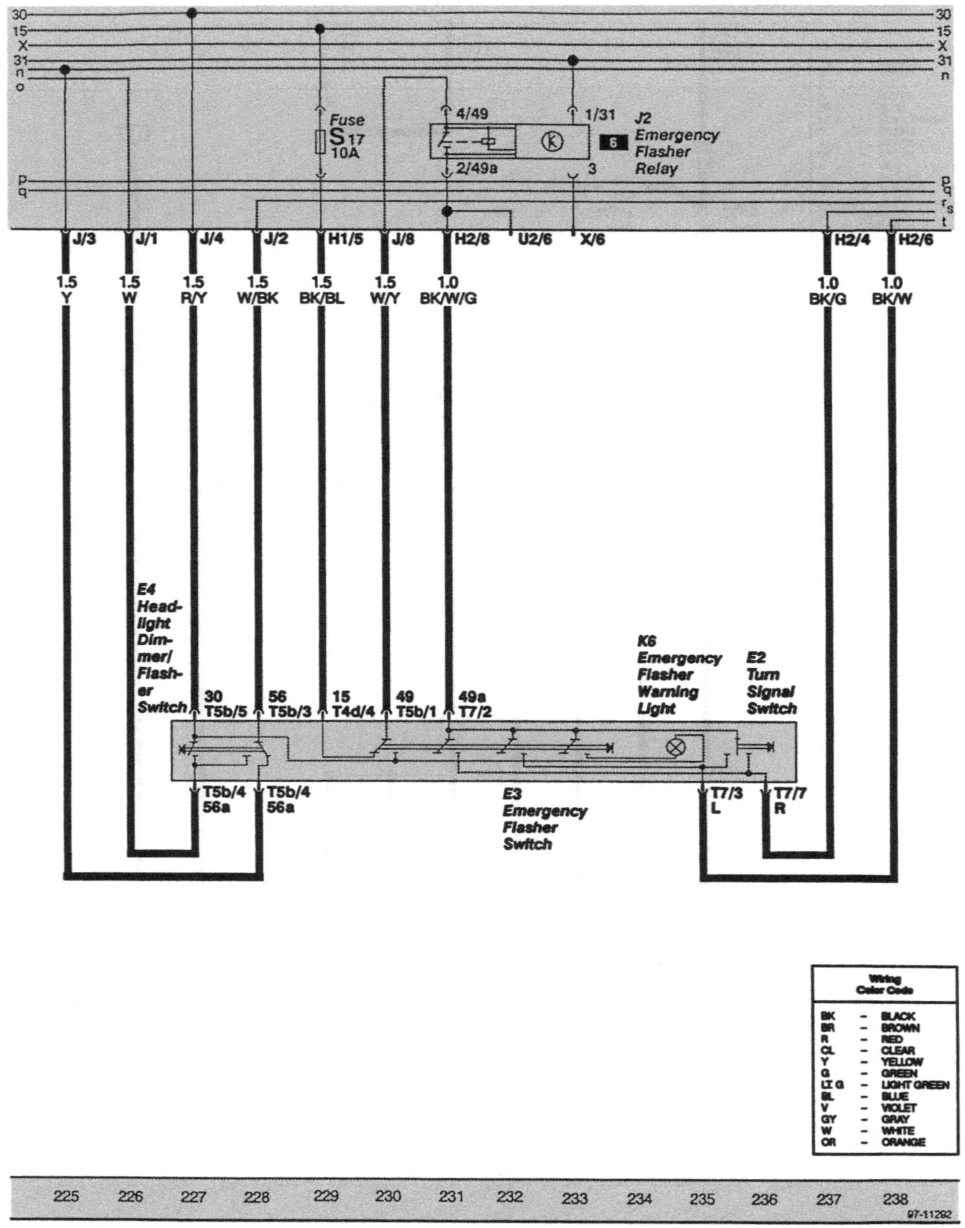

Corrado starting 1994
Engine code – AAA
Production from Mar. 1994

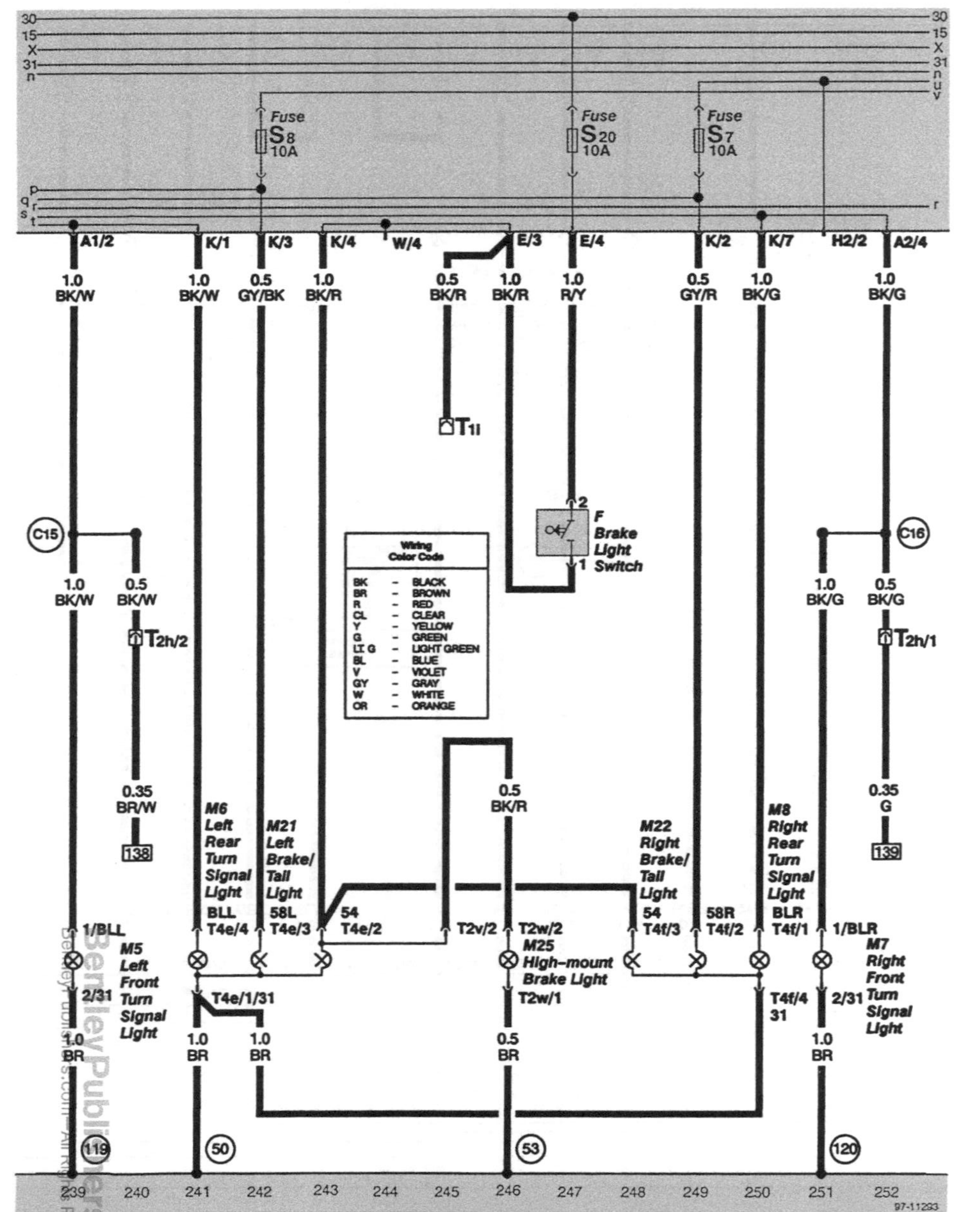

Corrado starting 1994
Engine code – AAA
Production from Mar. 1994

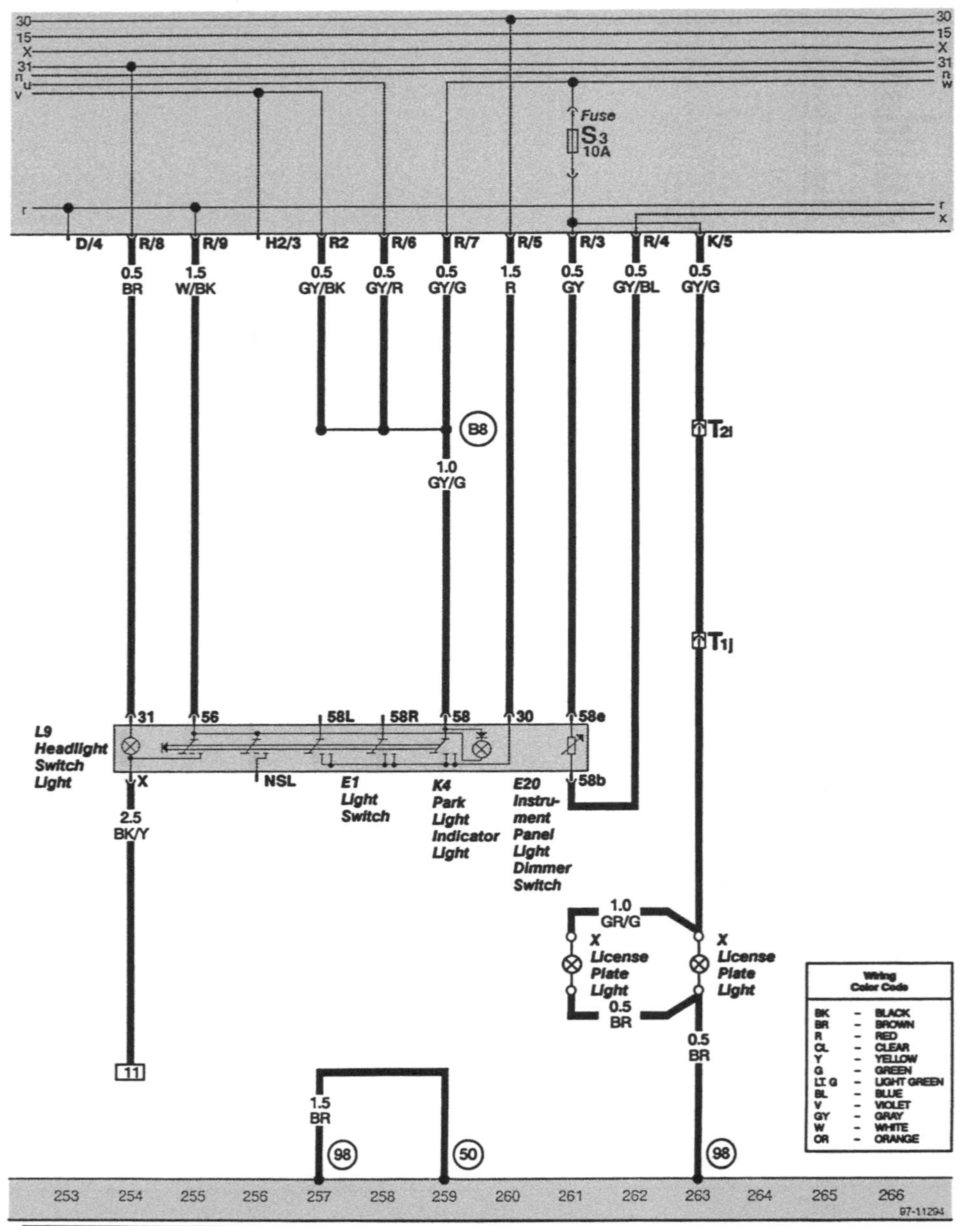

Corrado starting 1994
Engine code – AAA
Production from Mar. 1994

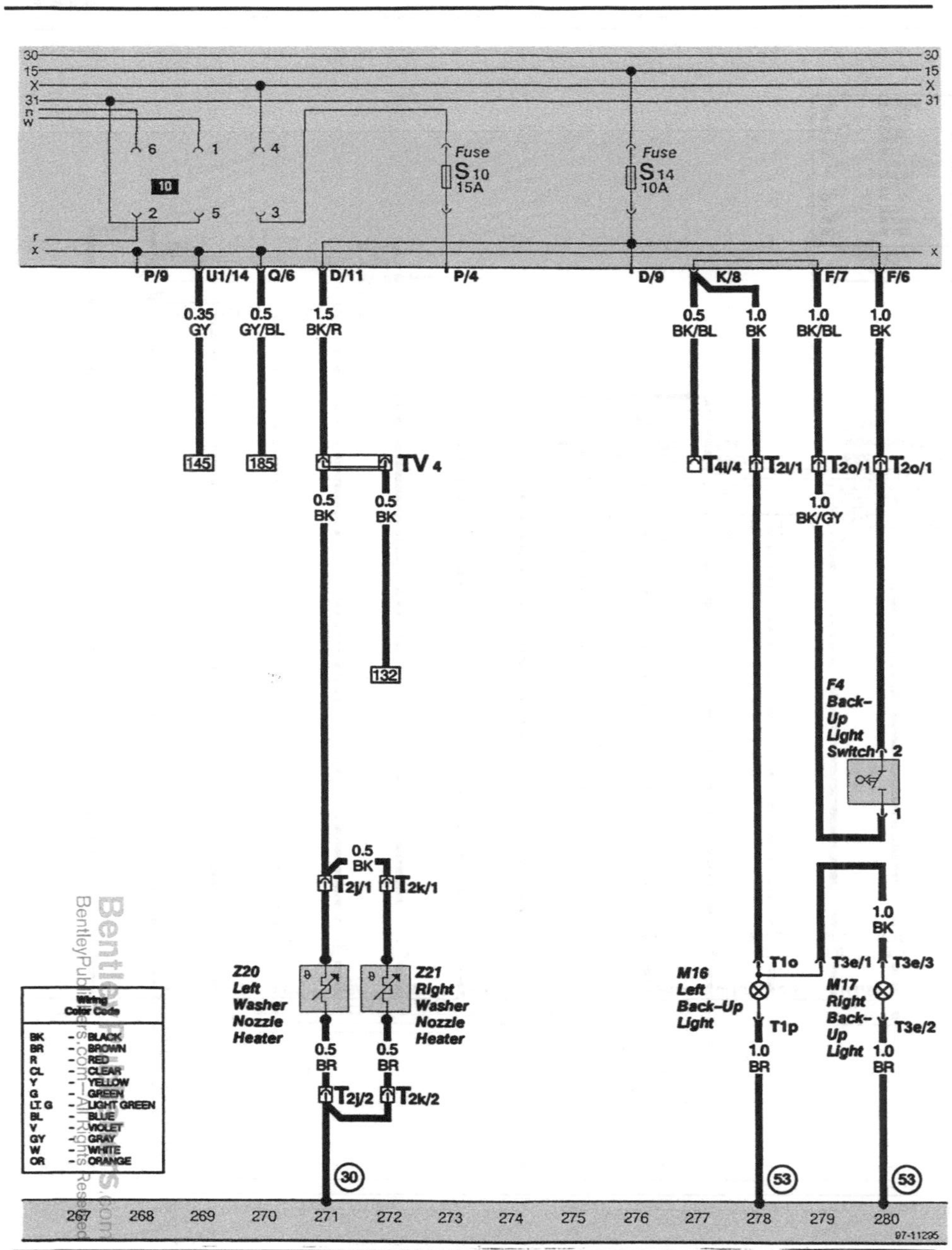

173 Washer nozzle heaters
Back-up lights

Corrado starting 1994
Engine code – AAA
Production from Mar. 1994

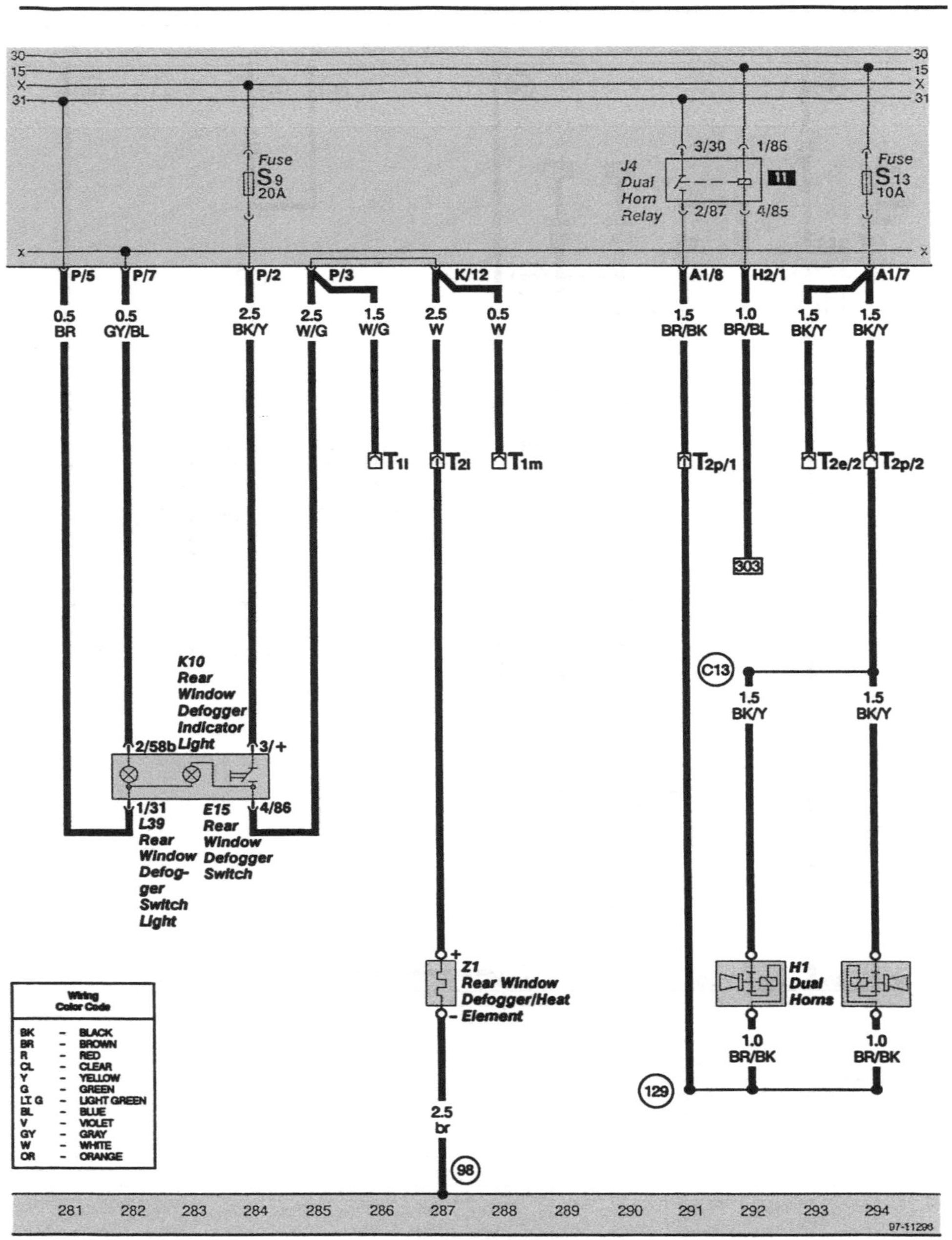

Corrado starting 1994
Engine code – AAA
Production from Mar. 1994

Rear window defogger
Horns **174**

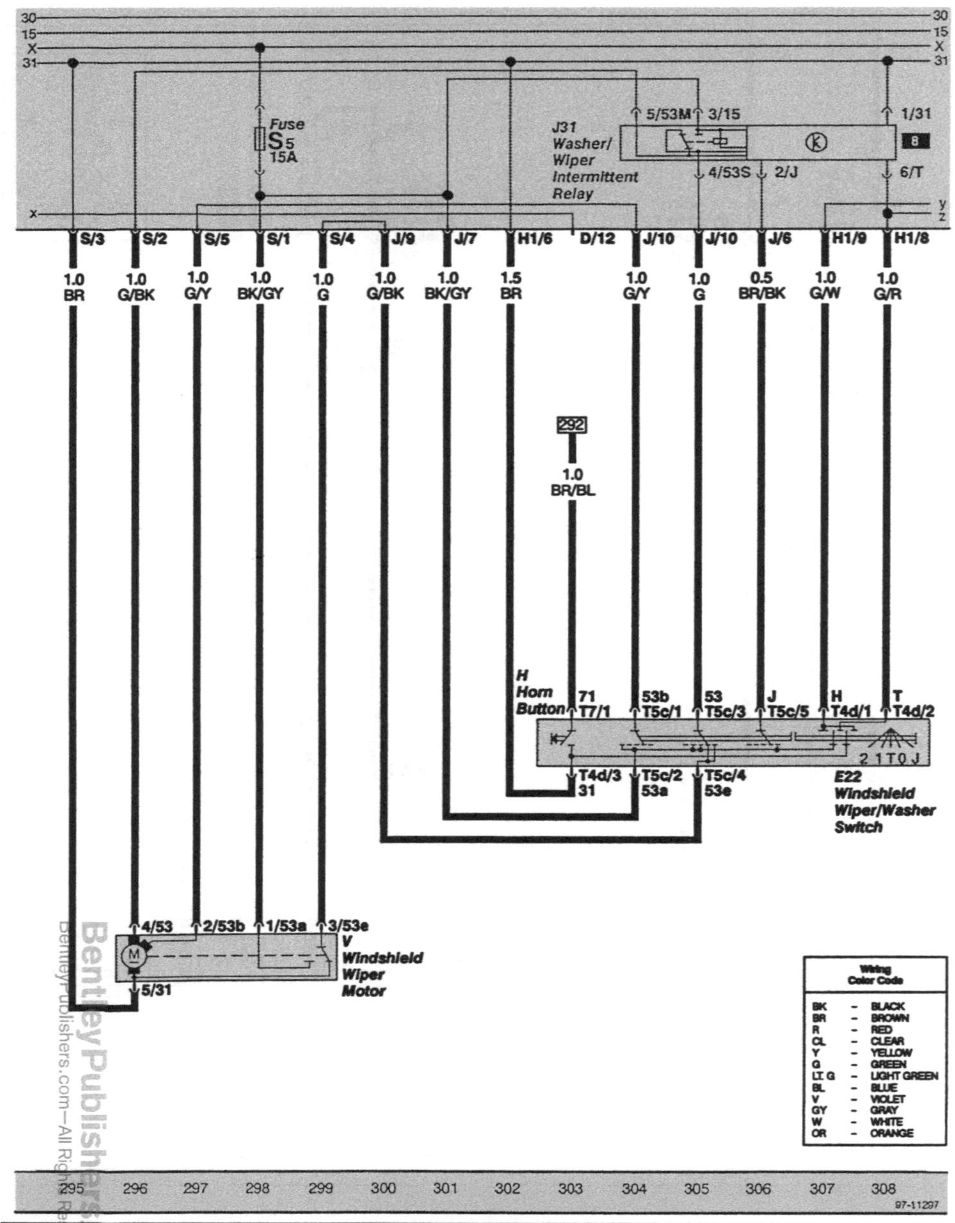

Corrado starting 1994
Engine code – AAA
Production from Mar. 1994

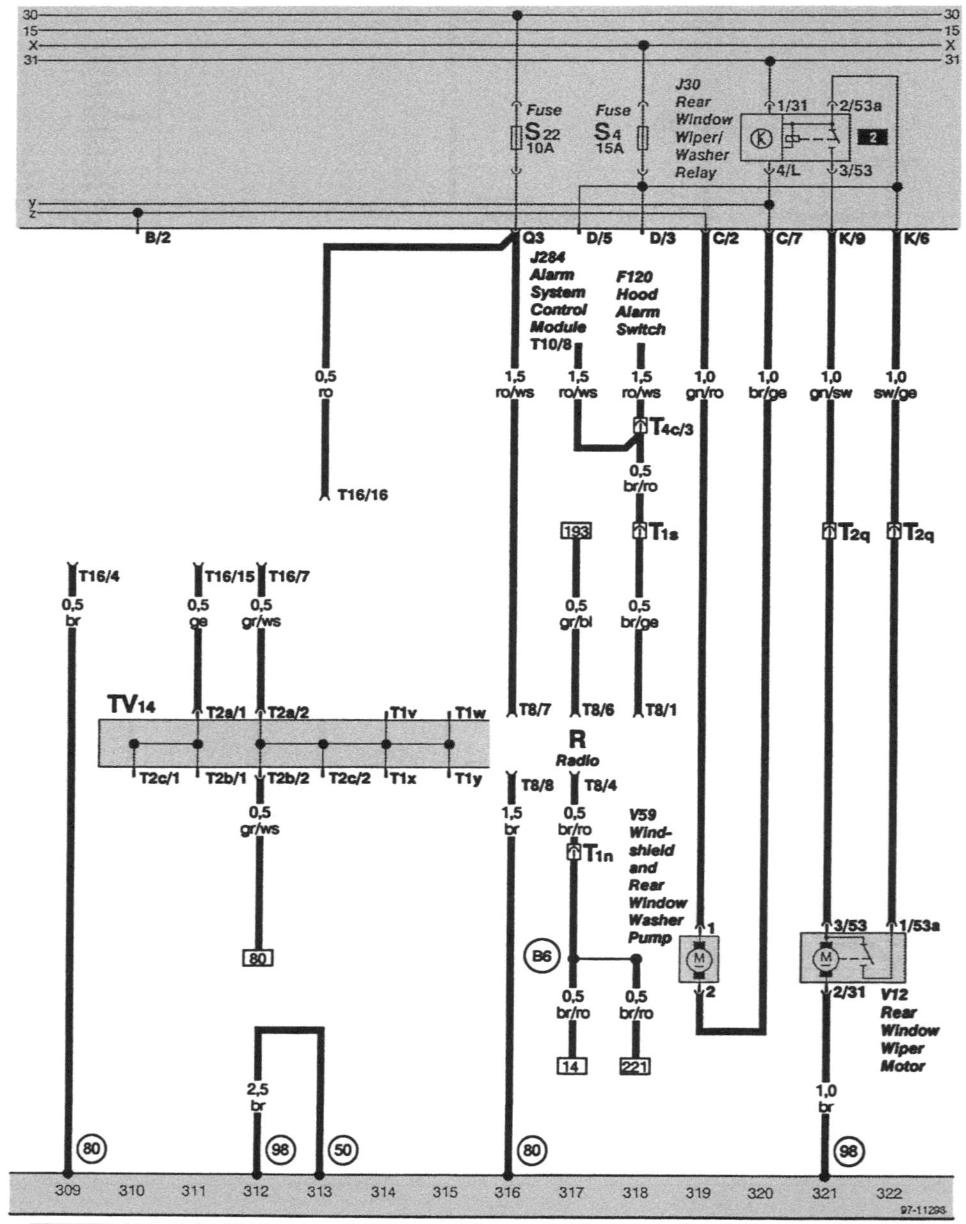

Corrado starting 1994
Engine code – AAA
Production from Mar. 1994

Description	Current track
A/C clutch relay, J44	16-17
A/C compressor clutch, N25	17
A/C pressure switch, F129	15-19
A/C relay, J32	9-13
A/C switch, E35	4-6
A/C thermostat, E33	12
Fresh air fan, V2	4
Fresh air fan series resistance, N23	1-4
Fresh air fan switch, E9	1-4
Fuse, S19	27
Fuse, S6	1
Fuse, S23	7
Overheat fuse, S24	4
Radiator cooling fan, V7	23-25
Radiator cooling fan relay (2nd stage), J101	21-23
Radiator cooling fan thermoswitch, F18	20-22

Wire connectors

T1 — single, on firewall
T1a — single, behind instrument panel
T1c — single, behind instrument panel, left
T2a — double, behind instrument panel
T2d — double, in engine compartment
T2l — double, in engine compartment
T3d — three-point, on headlight, left
T4a — four-point, in engine compartment
T5a — five-point, behind instrument panel
T5b — five-point, behind instrument panel

Ground connections

(43) — on lower right A-pillar
(97) — in A/C system wiring harness
(125) — in headlight wiring harness

Welded wiring harness points

(L8) — plus connection, in evaporator housing wiring harness
(L9) — plus connection, in A/C system wiring harness
(L10) — plus connection, in A/C system wiring harness

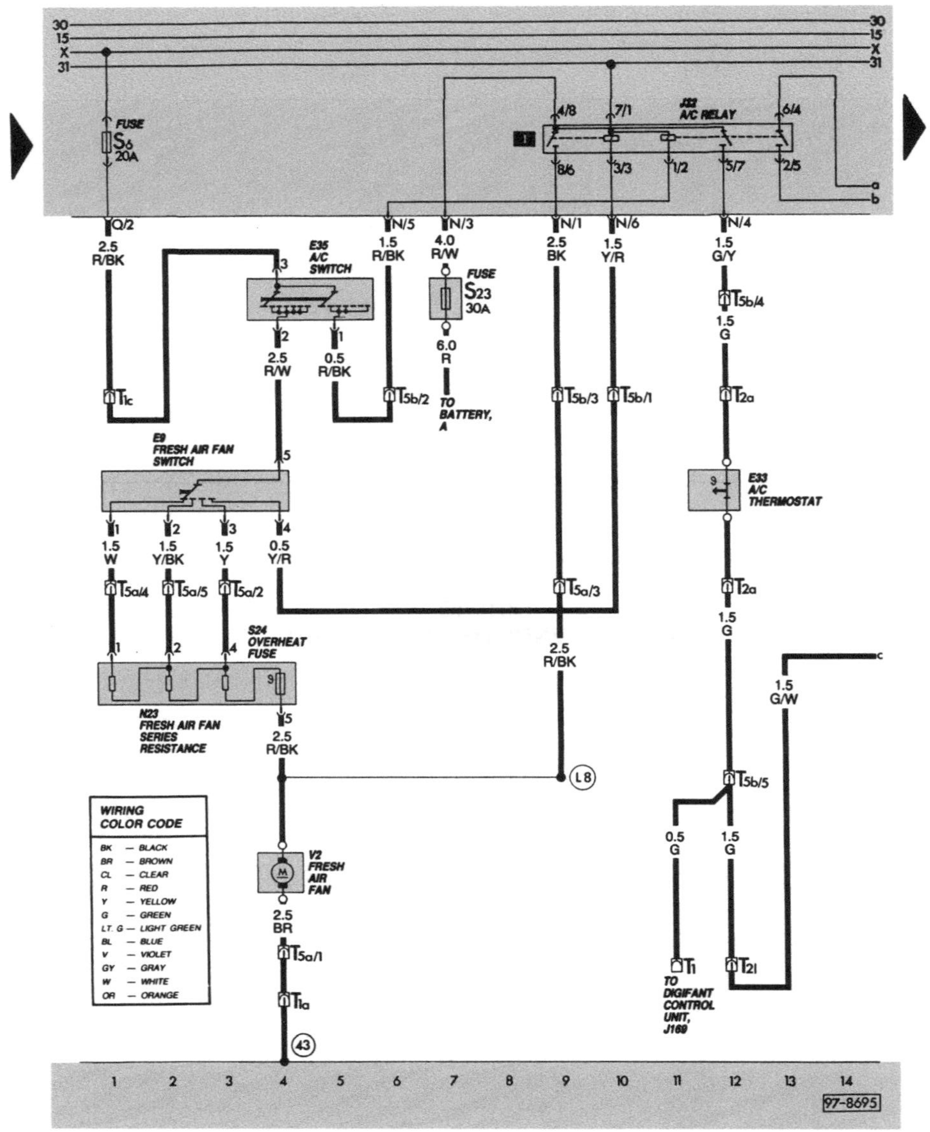

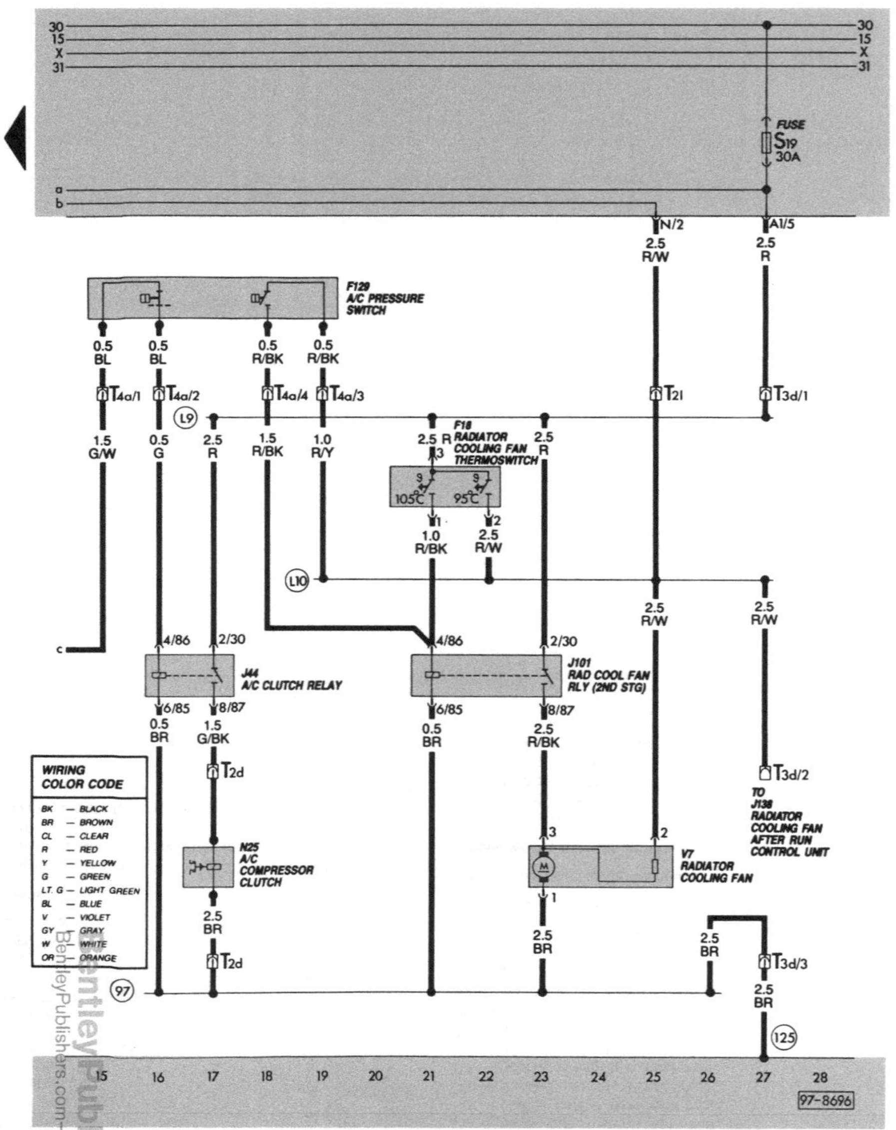

THIS PAGE INTENTIONALLY LEFT BLANK

Description	Current track
A/C compressor clutch, N25	22
Air conditioner pressure switch, F129	20–21
Air conditioner relay, J32	7–11
Air-conditioner evaporator temperature switch, E33	14
Air-conditioner switch, E35	3–5
Ambient temperature switch, F38	14
Battery, A	18
Coolant fan control unit, J293	18–26
Coolant fan thermoswitch, 3. speed, F165	21–22
Coolant fan thermoswitch, F87	21
Fresh air blower, V2	4
Fresh air fan switch, E9	1–4
Fuse, S6	1
Fuse, S13	28
Fuse, S19	26
Fuse, S23	17
Fuse, S88	18
Motronic control unit, J220	10,11
Pump for coolant circulation, V50	24
Radiator cooling fan thermoswitch, F18	26–27
Radiator fan, V7	25–27
Series resistance for fresh air blower with overheat fuse, N24	1–4

Wire connectors

- T1 – single, behind fuse/relay panel
- T1a – single, behind fuse/relay panel
- T1c – single, behind instrument panel
- T2a – double, behind instrument panel, left
- T2b – double, behind instrument panel, middle
- T2d – double, on compressor
- T2e – double, beside fuse/relay panel
- T2n – double, behind fuse/relay panel
- T4a – four point, behind instrument panel, left
- T5a – five point, behind instrument panel
- T42 – fourty-two point, on cylinder head, left

Welded wiring harness points

- (K20) – plus connection (30), in wiring harness coolant fan
- (K21) – wiring connection, in wiring harness coolant fan
- (K22) – wiring connection, in wiring harness coolant fan
- (K28) – plus connection (15), in wiring harness coolant fan
- (L5) – wiring connection, in wiring harness evaporator housing
- (L8) – wiring connection, in wiring harness evaporator housing
- (L12) – wiring connection (15), in wiring harness heating control (thermoswitch)

Ground connections

- (30) – beside fuse/relay panel
- (43) – on right pillar A, lower part
- (193) – in wiring harness fan for coolant

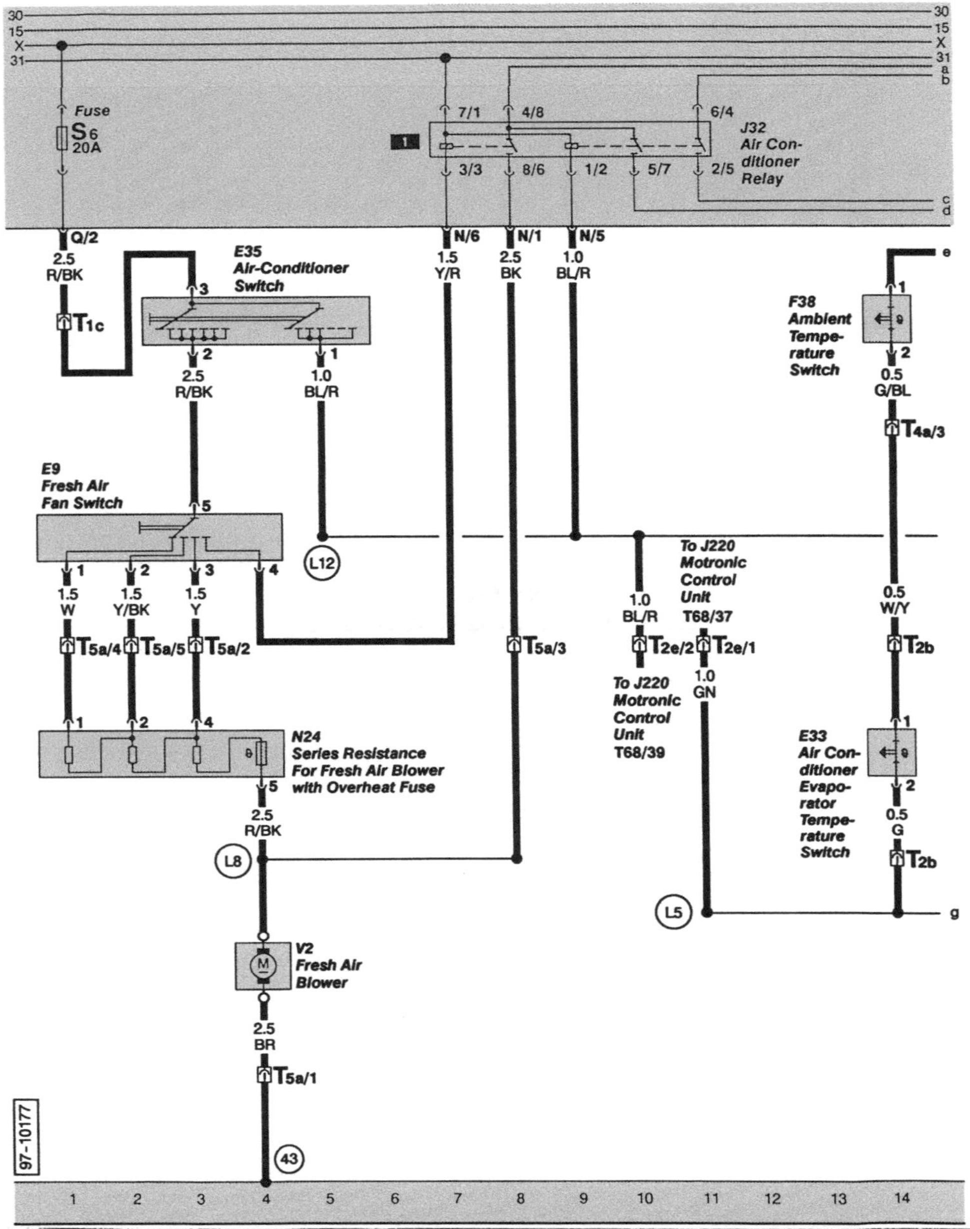

Corrado 1992–1993
Engine code – AAA
Production up to Aug. 1992

Corrado 1992–1993
Engine code – AAA
Production up to Aug. 1992

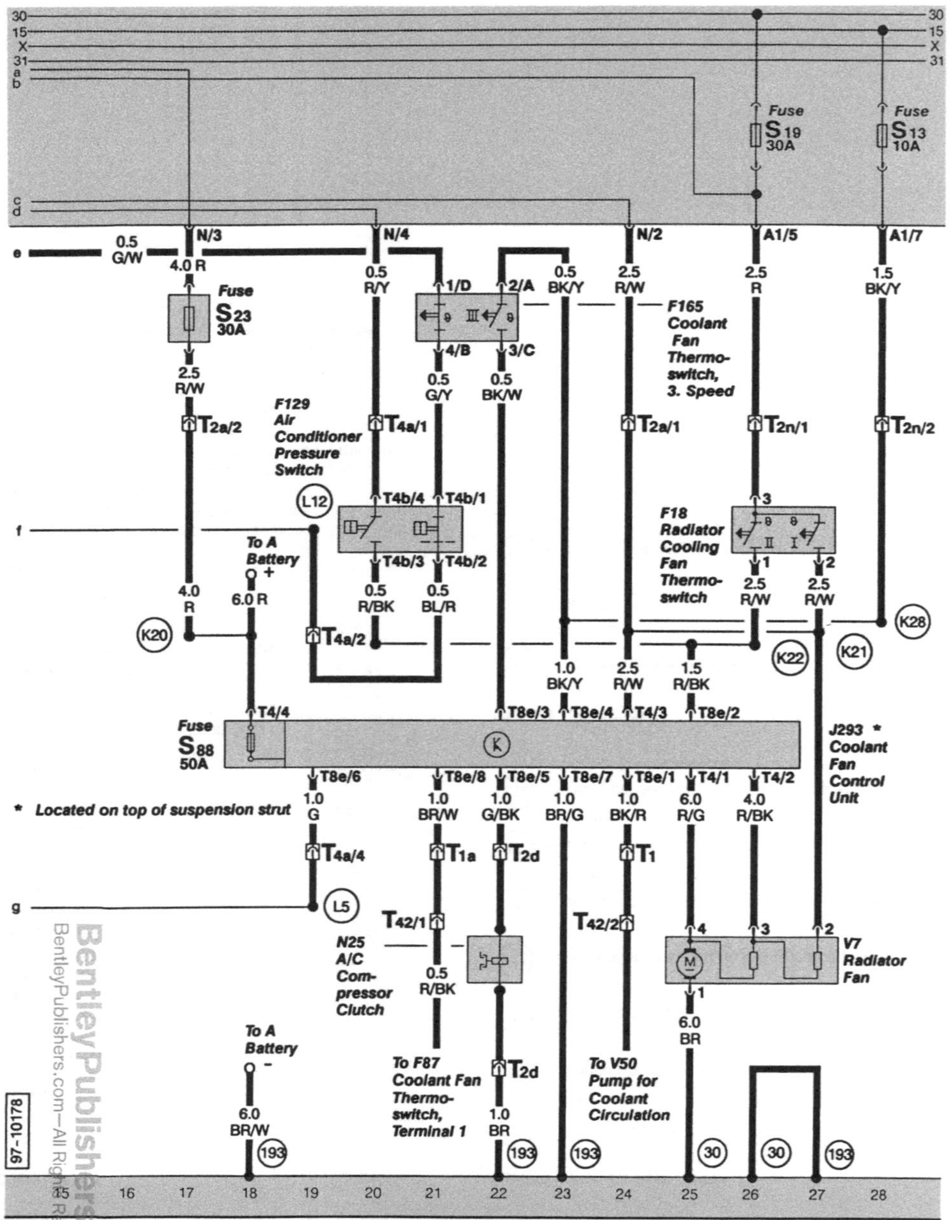

THIS PAGE INTENTIONALLY LEFT BLANK

Corrado 1992–1993
Engine code – AAA
Production up to Aug. 1992

Description	Current track
A/C Clutch Relay, behind Console, production control number (53), J44	17–19
A/C Clutch, N25	32
A/C Cut-Out Thermal Switch, F163	25
A/C Evaporator Temperature Switch, E33	28
A/C Indicator Light, K84	16
A/C Pressure Switch, F129	22–23
A/C Relay, production control number (133), J32	20–23
A/C Switch, E35	17–19
Ambient Temperature Switch, F38	23
Coolant Fan Control (FC) Control Module, J293	29–38
Coolant Fan Control (FC) Thermal Switch, F18	38–40
Coolant Fan, V7	36–40
Fresh Air Blower, V2	6
Fresh Air Blower Series Resistance With Fuse, N24	9–12
Fresh Air Blower Switch, E9	6–12
Fresh Air Control Lever Light, L16	5
Fresh Air/Recirculating Flap Two-Way Valve, N63	14
Fuse Strip, S88	29
Fuse, S6, 30A	7
Fuse, S19, 30A	38
Third Speed Coolant Fan Control (FC) Thermal Switch, F165	26

Wire connectors

T1 – Single Connector, behind fuse/relay panel
T2 – Double Connector, behind fuse/relay panel, for idle speed boost. Terminal 1: signal "on/off" – on T68/37 from Motronic Engine Control Module (ECM). Terminal 2: signal "ready" – on T68/39 from Motronic Engine Control Module (ECM).
T2a – Double Connector, behind instrument panel left
T2b – Double Connector, behind instrument panel, left
T2c – Double Connector, behind fuse/relay panel
T2d – Double Connector, at Compressor
T2e – Double Connector, on Fresh Air Blower Switch
T2n – Double Connector, behind fuse/relay panel
T4 – 4-pin Connector, on Coolant Fan Control (FC) Control Module
T4a – 4-pin Connector, behind instrument panel, left
T4b – 4-pin Connector, on A/C Pressure Switch
T6 – 6-pin Connector, on Fresh Air Blower Switch
T8 – 8-pin Connector, on Radio
T8e – 8-pin Connector, on Coolant Fan Control (FC) Control Module
T28 – 28-pin Connector, on Instrument Cluster
T42 – 42-pin Connector, on cylinder head, left
T68 – 68-pin Connector, on Motronic Engine Control Module (ECM)

Welded wiring harness point

(A11) – Plus connection (58b), in instrument panel wiring harness

(K21) – Wire connection –1–, in coolant fan wiring harness

(K24) – Wire connection –3–, in coolant fan wiring harness

(K25) – Wire connection –4–, in coolant fan wiring harness

(L1) – Plus connection, in A/C wiring harness

(L2) – Wire connection, in A/C wiring harness

(L5) – Wire connection –1–, in evaporator housing wiring harness

(L8) – Wire connection –4–, in evaporator housing wiring harness

Ground connections

(30) – Ground connection, –1–, beside fuse/relay panel

(43) – Ground connection, on right A-pillar, lower part

(80) – Ground connection –1–, in instrument panel wiring harness

(181) – Ground connection (for A/C), in engine compartment, left wiring harness

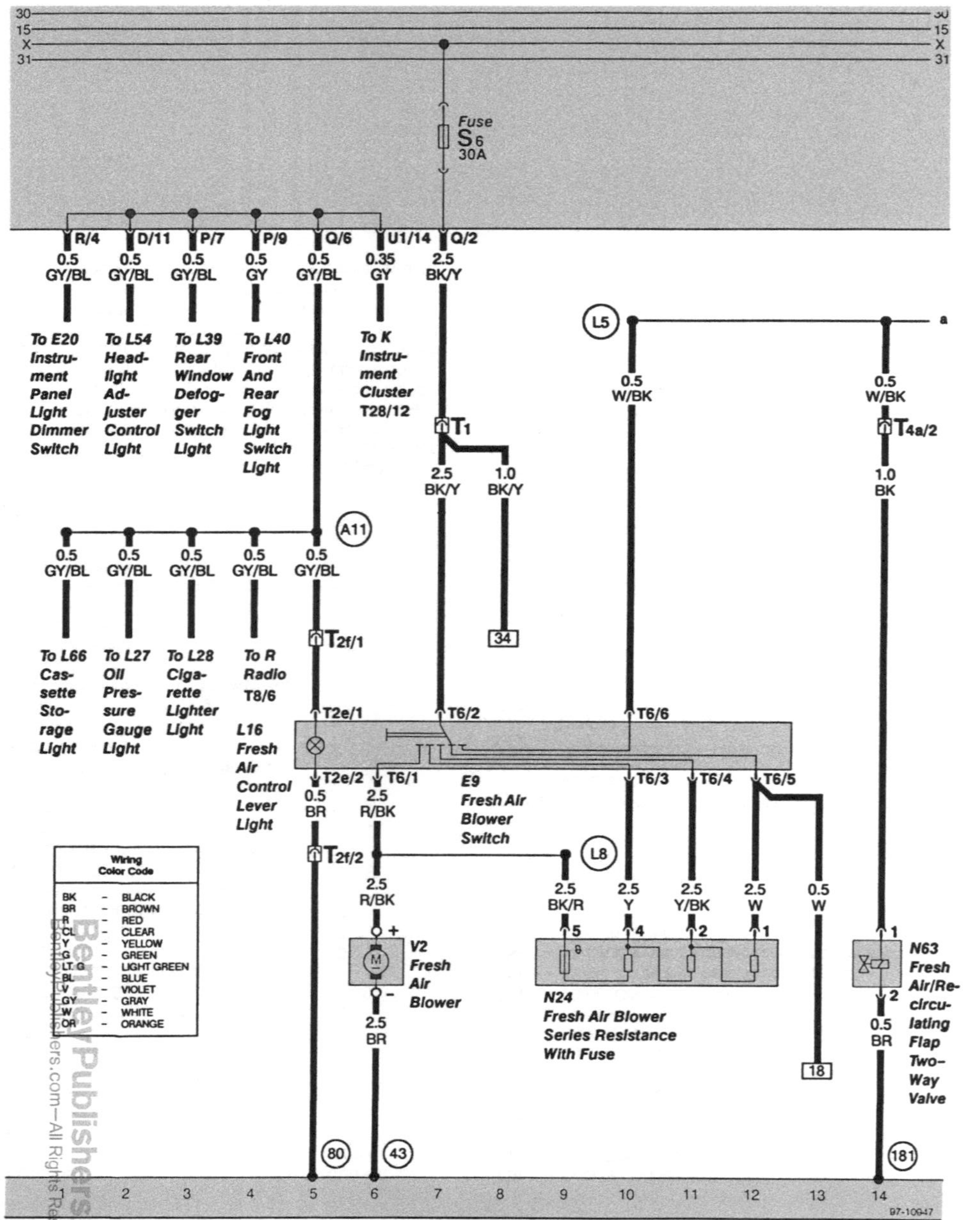

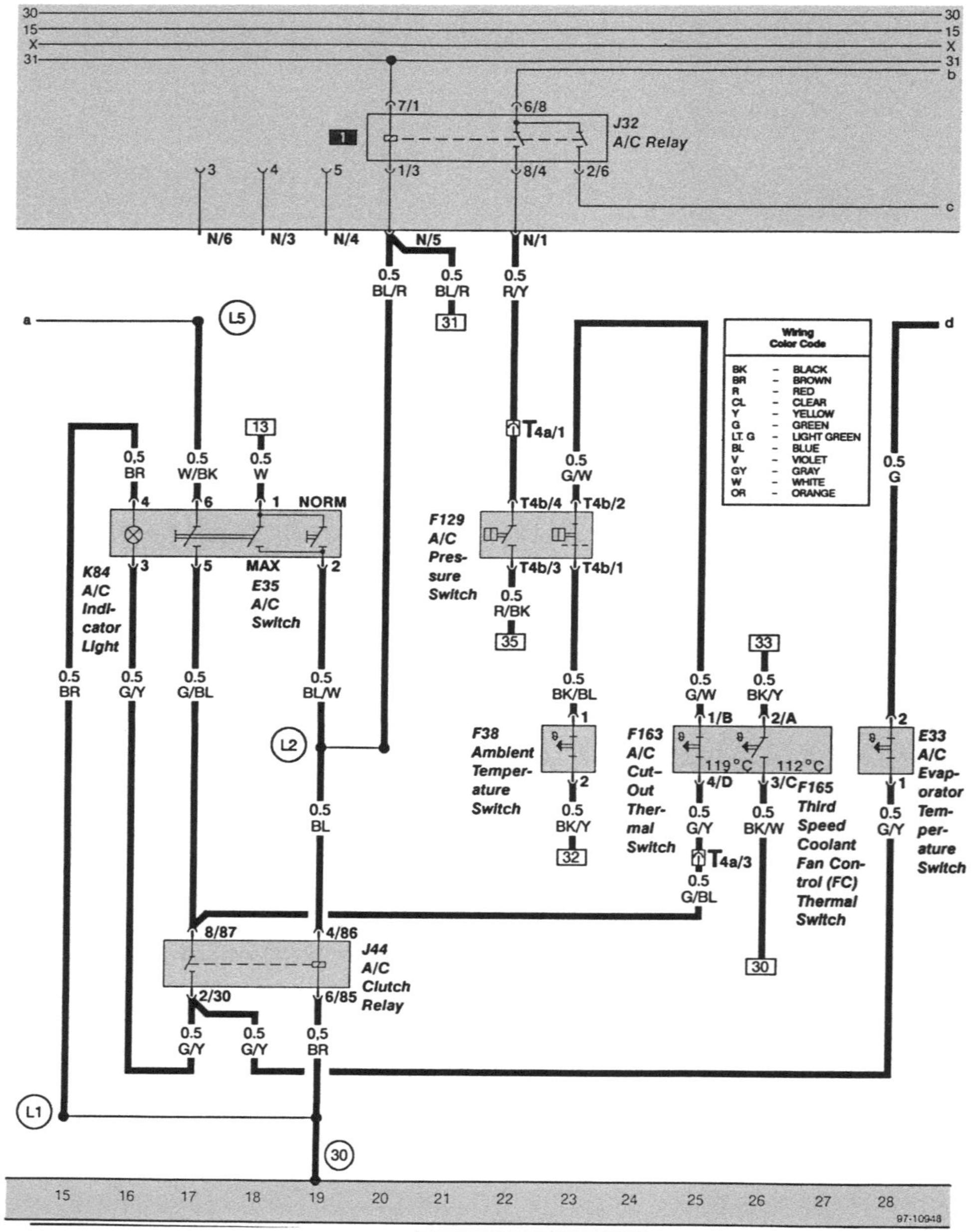

Corrado starting 1993
Engine code – AAA
Production from Aug. 1992

Corrado starting 1993
Engine code – AAA
Production from Aug. 1992

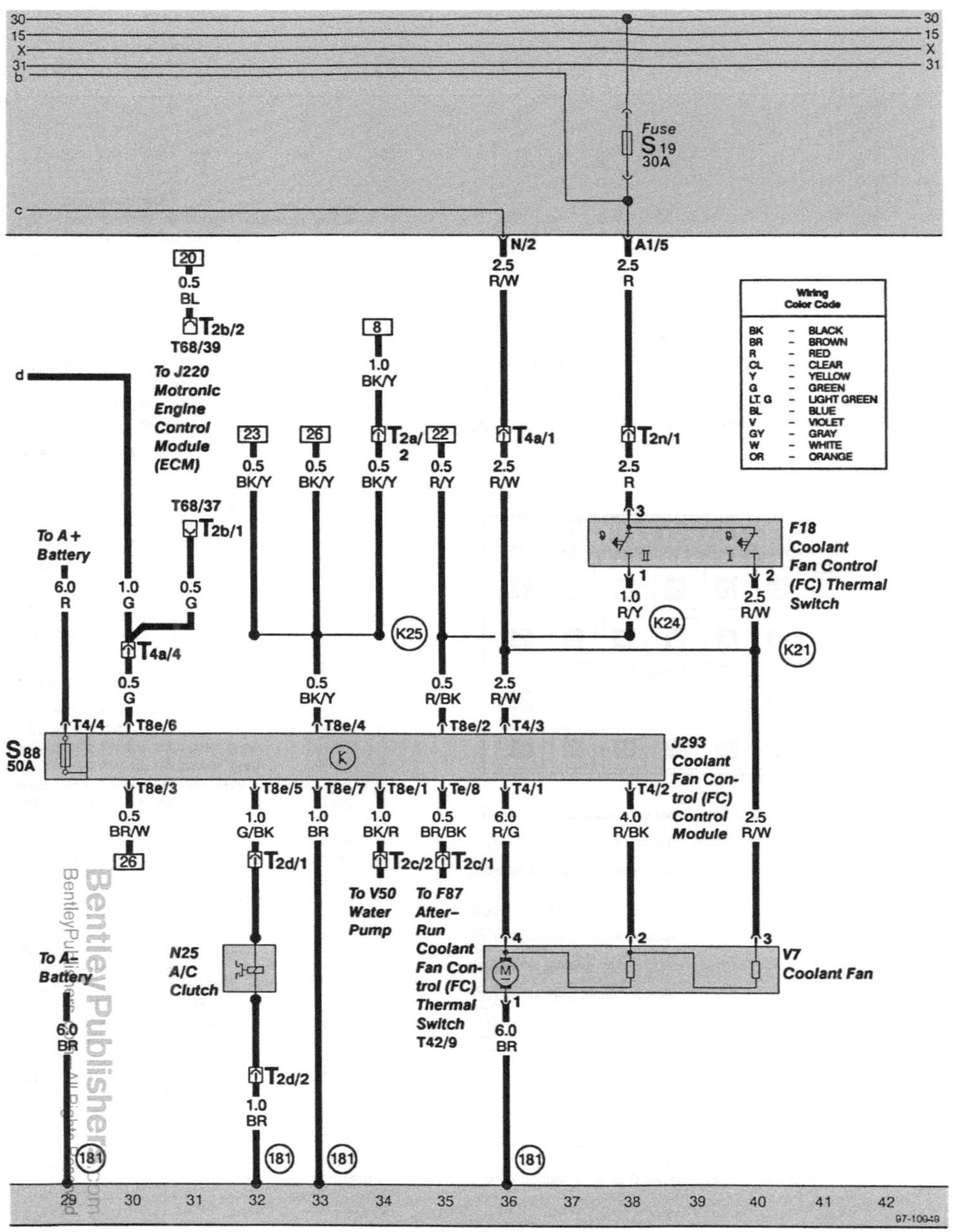

Corrado starting 1993
Engine code – AAA
Production from Aug. 1992

THIS PAGE INTENTIONALLY LEFT BLANK

Description	Current track
ABS brake fluid level warning switch, F117	18
ABS control unit, J104	18-36
ABS high pressure switch, F109	37
ABS hydraulic pump, V64	39
ABS hydraulic pump relay, J185	36-41
ABS hydraulic unit, N55	21-39
ABS inlet valve, left front, N101	23
ABS inlet valve, rear, N103	27
ABS inlet valve, right front, N99	32
ABS main valve, N105	21
ABS outlet valve, left front, N102	25
ABS outlet valve, rear, N104	29
ABS outlet valve, right front, N100	30
ABS pressure warning switch I, F115	36
ABS pressure warning switch II, F116	35
ABS relay, J102	9-13
ABS speed sensor condenser, C17	27-28, 31-32
ABS speed sensor, left front, G47	31-32
ABS speed sensor, left rear, G46	29-30
ABS speed sensor, right front, G45	27-28
ABS speed sensor, right rear, G44	24-25
ABS valves fuse, S54	12
ABS warning light, K47	1-5
Brake fluid level warning contact, F34	16
Diagnostic connector, T2y/1	4
Diagnostic connector, T2x/1	6
Diagnostic connector, T2x/2	6
Fuse, S16	4
Fuse, S21	6
Load reduction relay, J59	36
Luggage compartment light, W3	8
Luggage compartment light switch, F5	8

Wire connectors

T1a — single, behind fuse/relay panel
T1b — single, behind fuse/relay panel
T1c — single, behind fuse/relay panel
T2a — double, on ABS hydraulic unit
T2b — double, on ABS hydraulic unit
T2c — double, under rear seat, right
T2d — double, on firewall, right
T2e — double, on firewall, right
T2f — double, on firewall, left
T2g — double, on brake fluid reservoir
T2h — double, on brake fluid reservoir
T2x — double, black, diagnostic connector, below shift lever boot
T2y — double, blue or gray, diagnostic connector, below shift lever boot
T5 — five-point, on ABS hydraulic unit
T7 — seven-point, on ABS hydraulic unit
T28 — twenty eight-point, on instrument cluster

Ground connections

(14) — on transmission
(22) — on ABS hydraulic unit
(28) — on firewall
(30) — beside fuse/relay panel
(53) — on rear lid

Welded wiring harness points

(J3) — plus connection, in ABS wiring harness
(W1) — plus connection, in rear wiring harness

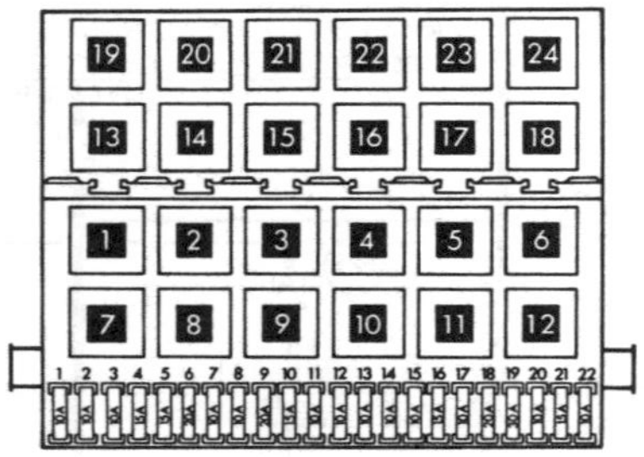

13 — ABS hydraulic pump relay, J 185
14 — ABS relay, J 102
19 — ABS hydraulic pump fuse, S 53
20 — ABS valves fuse, S 54

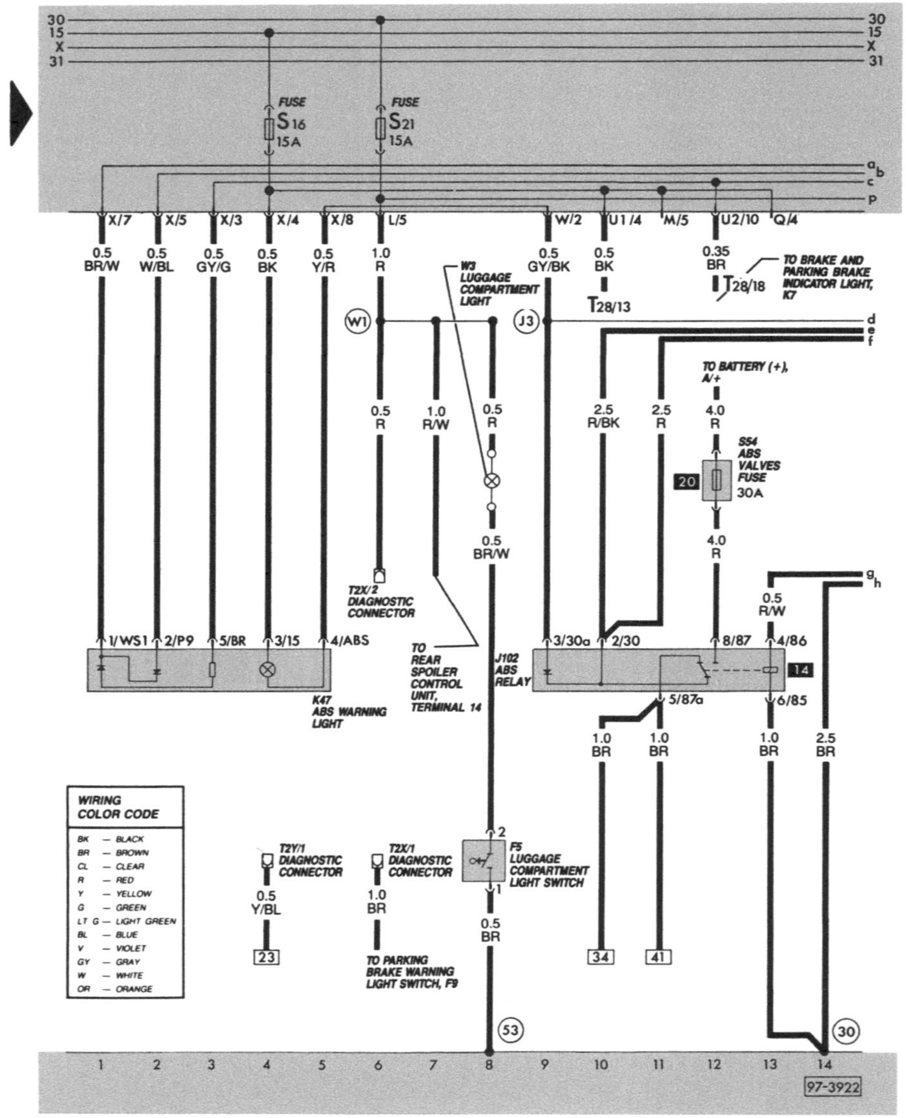

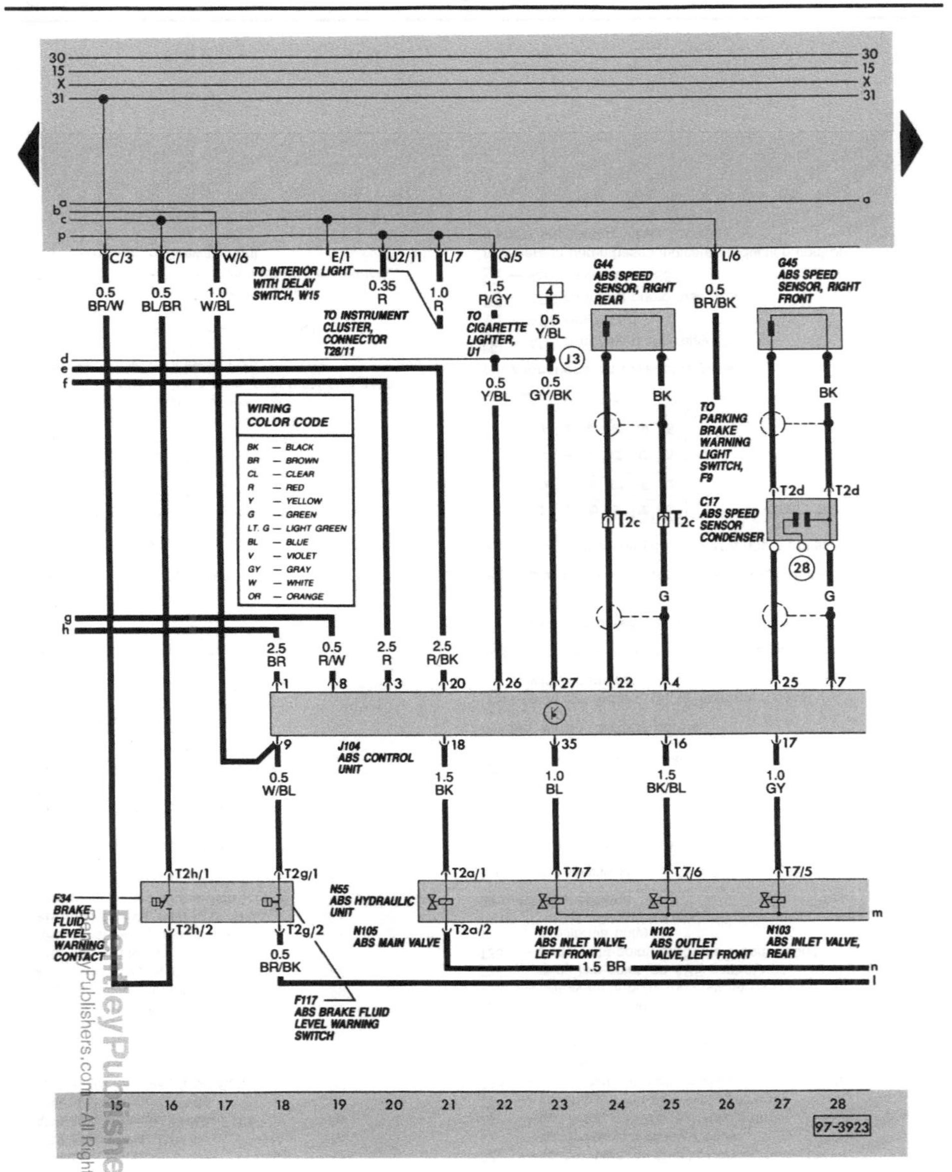

G44 ABS SPEED SENSOR, RIGHT REAR
G45 ABS SPEED SENSOR, RIGHT FRONT
C17 ABS SPEED SENSOR CONDENSER
J104 ABS CONTROL UNIT
N55 ABS HYDRAULIC UNIT
N105 ABS MAIN VALVE
N101 ABS INLET VALVE, LEFT FRONT
N102 ABS OUTLET VALVE, LEFT FRONT
N103 ABS INLET VALVE, REAR
F34 BRAKE FLUID LEVEL WARNING CONTACT
F117 ABS BRAKE FLUID LEVEL WARNING SWITCH
TO INTERIOR LIGHT WITH DELAY SWITCH, W15
TO INSTRUMENT CLUSTER, CONNECTOR T28/11
TO CIGARETTE LIGHTER, U1
TO PARKING BRAKE WARNING LIGHT SWITCH, F9
WIRING COLOR CODE
BK — BLACK
BR — BROWN
CL — CLEAR
R — RED
Y — YELLOW
G — GREEN
LT. G — LIGHT GREEN
BL — BLUE
V — VIOLET
GY — GRAY
W — WHITE
OR — ORANGE
97-3923

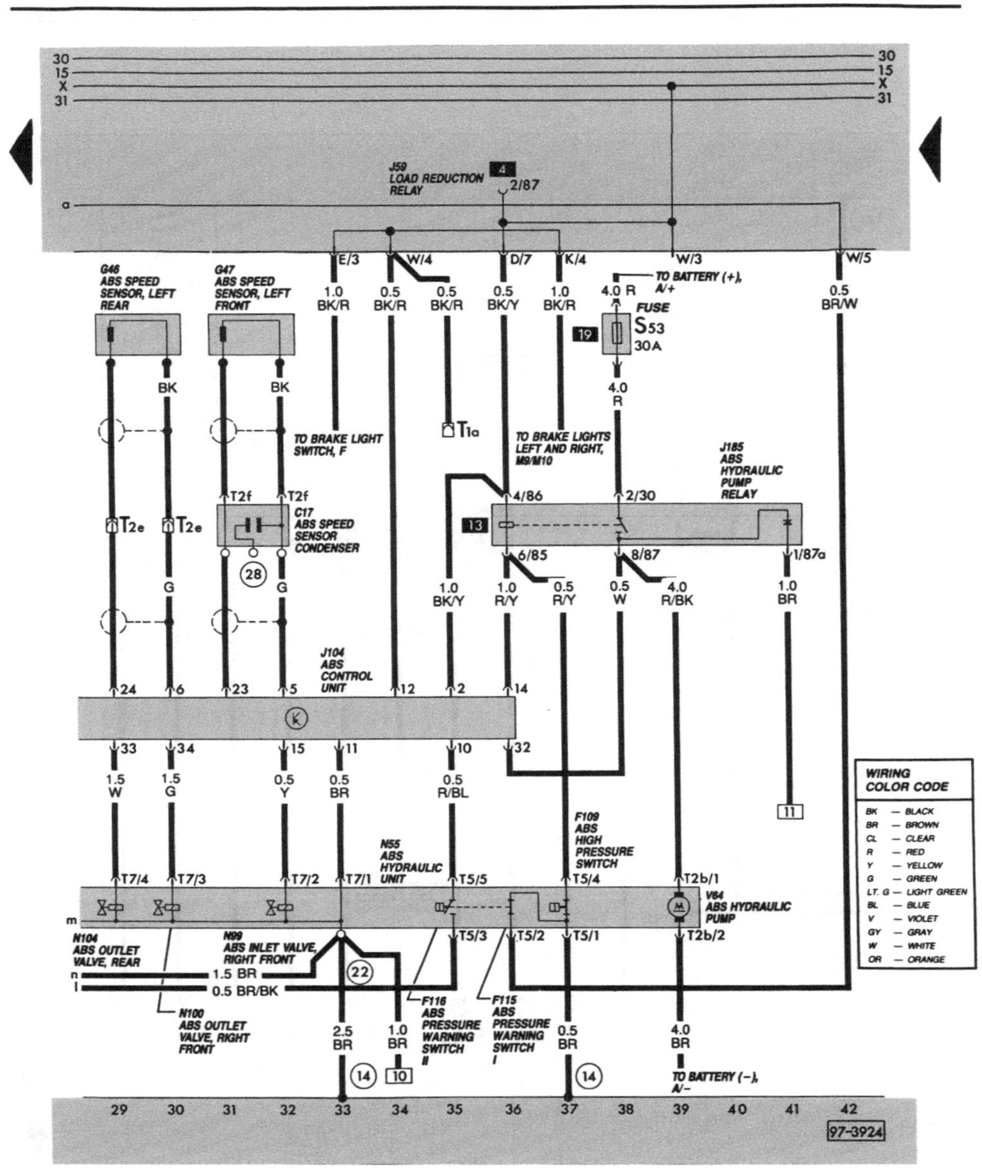

J59 LOAD REDUCTION RELAY
G46 ABS SPEED SENSOR, LEFT REAR
G47 ABS SPEED SENSOR, LEFT FRONT
C17 ABS SPEED SENSOR CONDENSER
TO BRAKE LIGHT SWITCH, F
TO BRAKE LIGHTS LEFT AND RIGHT, M9/M10
TO BATTERY (+), A/+
FUSE S53 30A
J185 ABS HYDRAULIC PUMP RELAY
J104 ABS CONTROL UNIT
N55 ABS HYDRAULIC UNIT
F109 ABS HIGH PRESSURE SWITCH
V64 ABS HYDRAULIC PUMP
N104 ABS OUTLET VALVE, REAR
N99 ABS INLET VALVE, RIGHT FRONT
N100 ABS OUTLET VALVE, RIGHT FRONT
F116 ABS PRESSURE WARNING SWITCH II
F115 ABS PRESSURE WARNING SWITCH I
TO BATTERY (–), A/–
WIRING COLOR CODE
BK — BLACK
BR — BROWN
CL — CLEAR
R — RED
Y — YELLOW
G — GREEN
LT. G — LIGHT GREEN
BL — BLUE
V — VIOLET
GY — GRAY
W — WHITE
OR — ORANGE
97-3924

Description	Current track
ABS brake fluid level warning switch, F117	18
ABS control unit, J104	18-35
ABS high pressure switch, F109	36
ABS hydraulic pump, V64	39
ABS hydraulic pump fuse, S53	38
ABS hydraulic pump relay, J185	35-40
ABS hydraulic unit, N55	23-39
ABS inlet valve, left front, N101	24
ABS inlet valve, rear, N103	26
ABS inlet valve, right front, N99	29
ABS main valve, N105	23
ABS outlet valve, left front, N102	25
ABS outlet valve, rear, N104	27
ABS outlet valve, right front, N100	28
ABS pressure warning switch I, F115	33
ABS pressure warning switch II, F116	31
ABS relay, J102	10-14
ABS speed sensor condenser, C17	27-28, 31-32
ABS speed sensor, left front, G47	31-32
ABS speed sensor, left rear, G46	29-30
ABS speed sensor, right front, G45	27-28
ABS speed sensor, right rear, G44	25-26
ABS valves fuse, S54	13
ABS warning light, K47	5
Brake and parking brake indicator light, K7	4
Brake fluid level warning contact, F34	16
Fuse, S16	4
Fuse, S21	42
OBD fault warning light, K83	6

Wire connectors

- T1a — single, behind fuse/relay panel
- T1b — single, behind fuse/relay panel
- T1e1 — single, behind fuse/relay panel (California only)
- T2a — double, on ABS hydraulic unit
- T2b — double, on ABS hydraulic unit
- T2c — double, under rear seat, right
- T2d — double, on left A-pillar, near fuse/relay panel
- T2e — double, on firewall, right
- T2f — double, on firewall, left
- T2g — double, on ABS brake fluid level warning switch
- T2g1 — double, above fuse/relay panel
- T2h — double, on brake fluid level warning switch
- T2j — double, on firewall, right

Wire connectors

- T2x — double, black, diagnostic connector, below transmission shift lever cover
- T2y — double, blue, diagnostic connector, below transmission shift lever cover
- T5 — five-point, on ABS hydraulic unit
- T7 — seven-point, on ABS hydraulic unit
- T8d — eight-point, on brake/parking brake warning light assembly
- T9 — nine-point, on starter interlock/back-up light relay
- T25 — twenty five-point, on Digifant control unit
- T28 — twenty eight-point, on instrument cluster
- T35 — thirty five-point, on ABS control unit
- T38 — thirty eight-point, on Digifant control unit (California only)

Ground connections

- (14) — on transmission
- (22) — on ABS hydraulic unit
- (28) — on firewall
- (39) — below rear seat, left
- (125) — in headlight wiring harness

Welded wiring harness points

- (A15) — plus connection (15), in instrument panel wiring harness
- (J3) — wire connection, in ABS wiring harness
- (J4) — wire connection (diagnosis), in ABS wiring harness
- (W9) — plus connection (30), in rear wiring harness
- (W10) — plus connection (15), in rear wiring harness

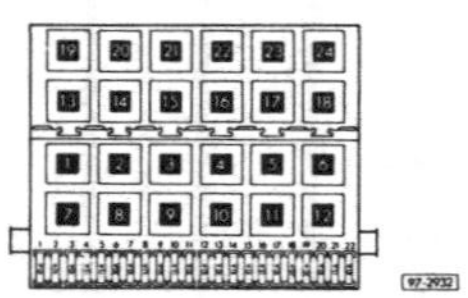

Relay locations on fuse/relay panel

- 13 — ABS hydraulic pump relay (78)*
- 14 — ABS relay (79)*
- 19 — ABS hydraulic pump fuse, S53
- 20 — ABS valves fuse, S54

*Numbers in parentheses indicate production control number stamped on relay housing

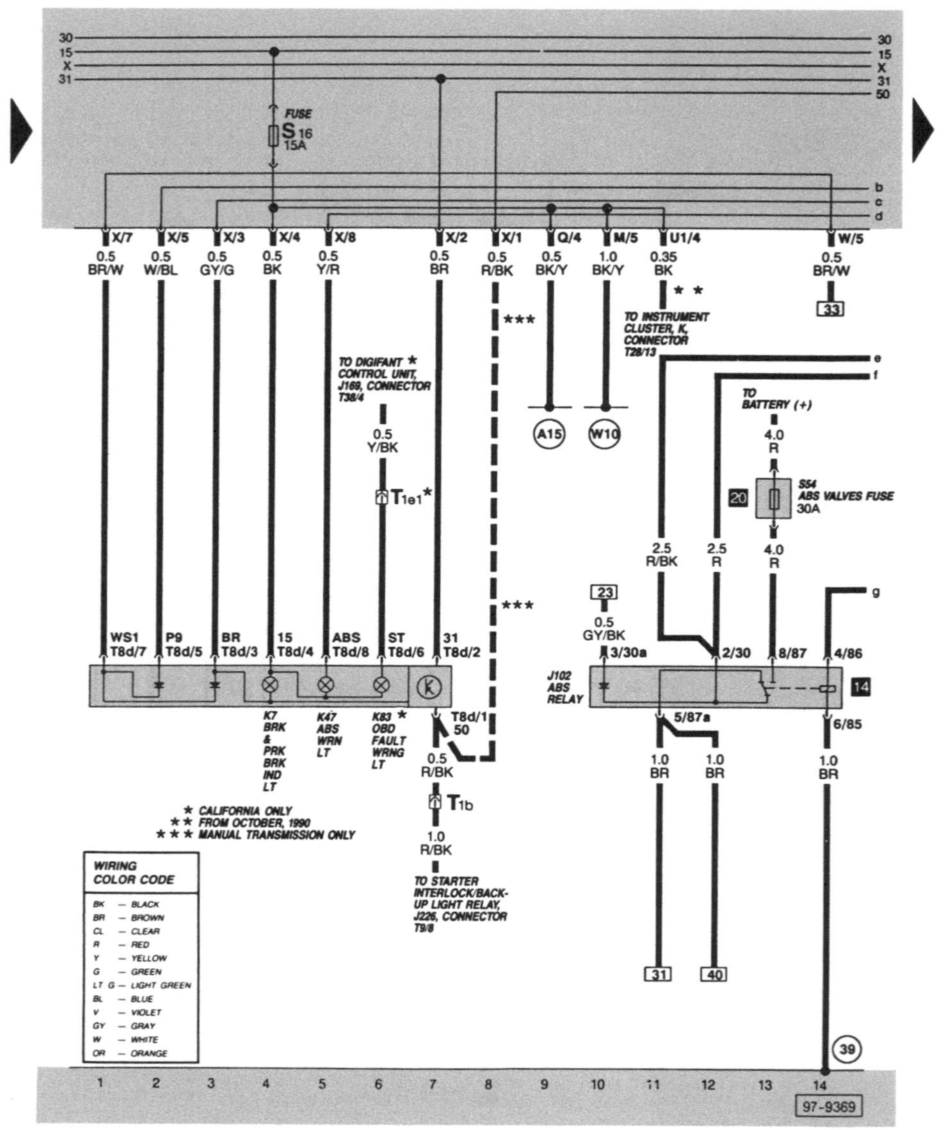

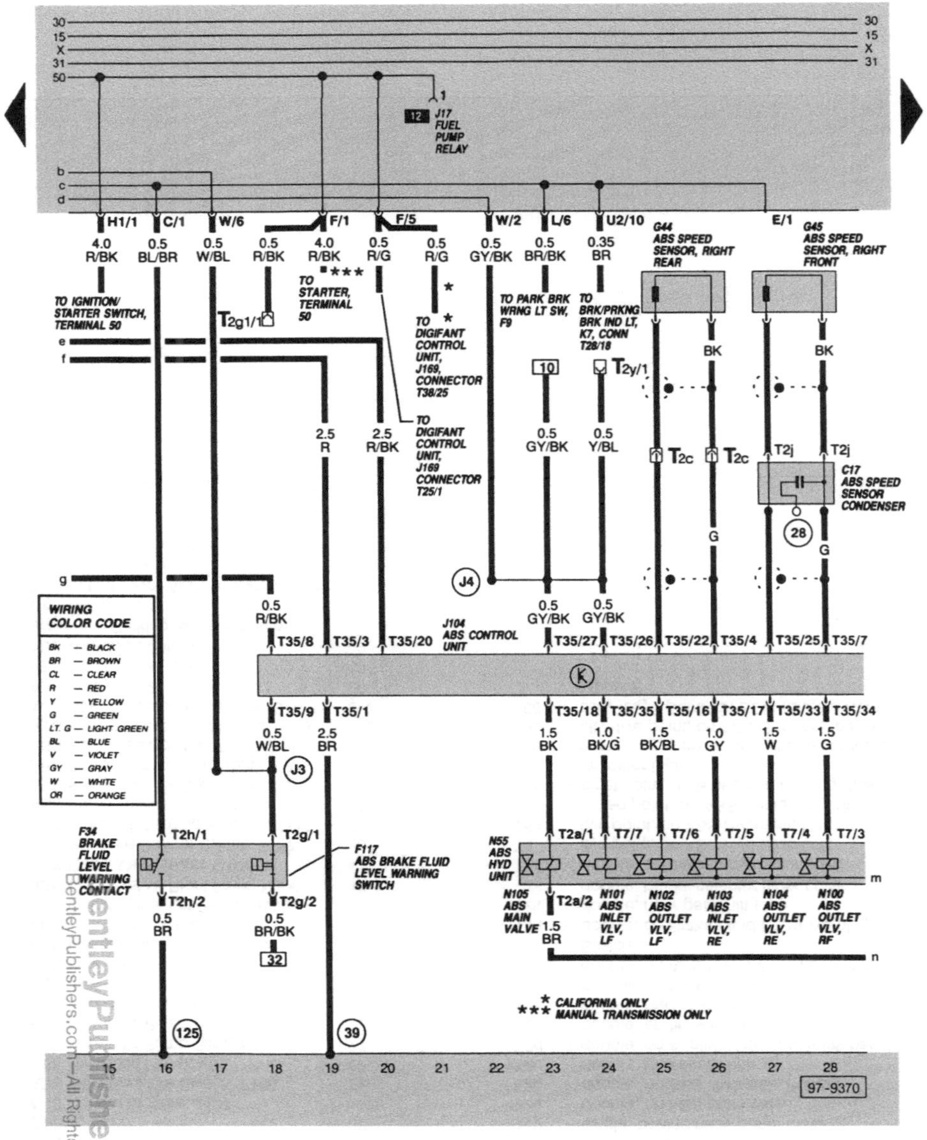
J17 FUEL PUMP RELAY
G44 ABS SPEED SENSOR, RIGHT REAR
G45 ABS SPEED SENSOR, RIGHT FRONT
C17 ABS SPEED SENSOR CONDENSER
TO IGNITION/ STARTER SWITCH, TERMINAL 50
TO STARTER, TERMINAL 50
TO DIGIFANT CONTROL UNIT, J169, CONNECTOR T38/25
TO DIGIFANT CONTROL UNIT, J169 CONNECTOR T25/1
TO PARK BRK WRNG LT SW, F9
TO BRK/PRKNG BRK IND LT, K7, CONN T28/18
J104 ABS CONTROL UNIT
F34 BRAKE FLUID LEVEL WARNING CONTACT
F117 ABS BRAKE FLUID LEVEL WARNING SWITCH
N55 ABS HYD UNIT
N105 ABS MAIN VALVE
N101 ABS INLET VLV, LF
N102 ABS OUTLET VLV, LF
N103 ABS INLET VLV, RE
N104 ABS OUTLET VLV, RE
N100 ABS OUTLET VLV, RF
* CALIFORNIA ONLY
*** MANUAL TRANSMISSION ONLY
WIRING COLOR CODE
BK — BLACK
BR — BROWN
CL — CLEAR
R — RED
Y — YELLOW
G — GREEN
LT. G — LIGHT GREEN
BL — BLUE
V — VIOLET
GY — GRAY
W — WHITE
OR — ORANGE
97-9370

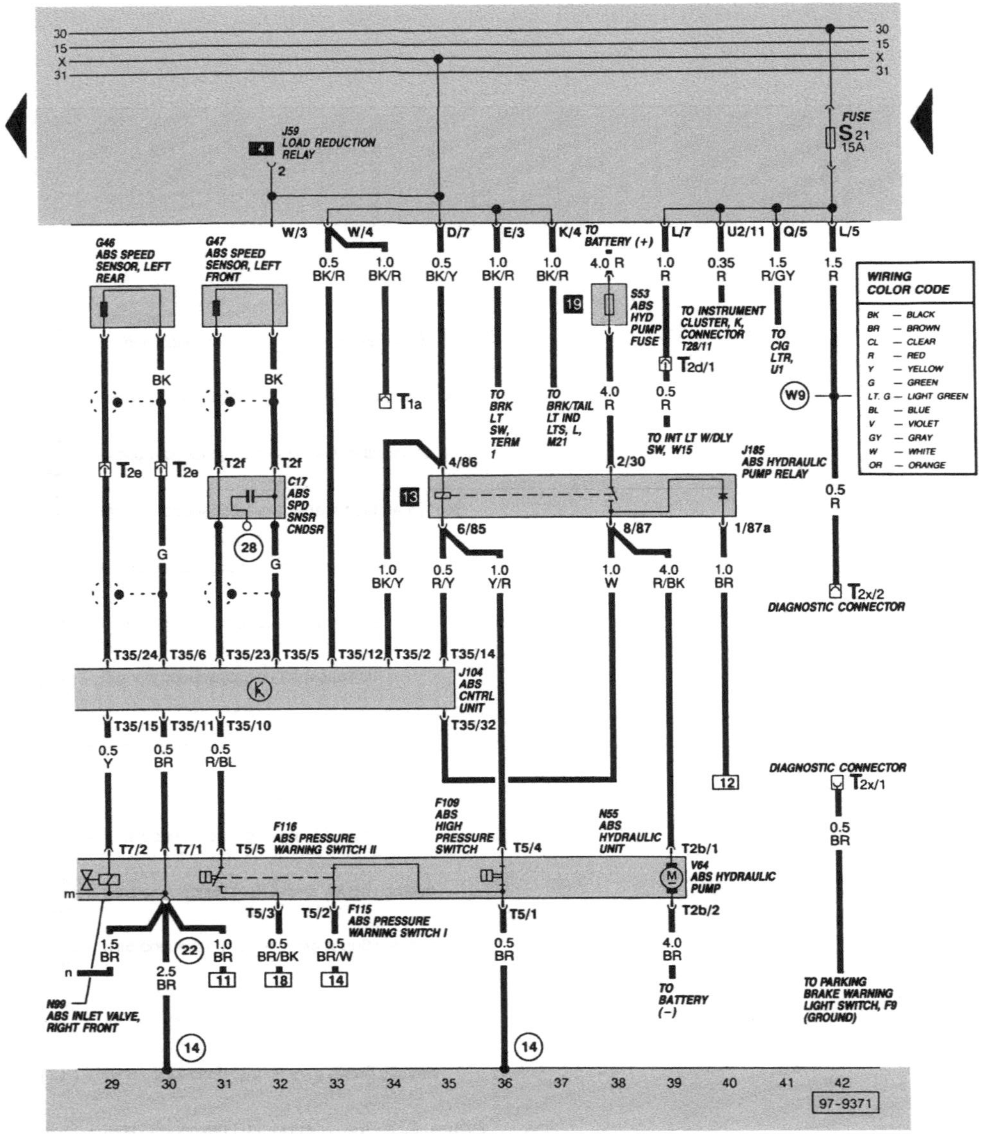
J59 LOAD REDUCTION RELAY
FUSE S21 15A
G46 ABS SPEED SENSOR, LEFT REAR
G47 ABS SPEED SENSOR, LEFT FRONT
C17 ABS SPD SNSR CNDSR
TO BATTERY (+)
S53 ABS HYD PUMP FUSE
TO INSTRUMENT CLUSTER, K, CONNECTOR T28/11
TO CIG LTR, U1
TO BRK LT SW, TERM 1
TO BRK/TAIL LT IND LTS, L, M21
TO INT LT W/DLY SW, W15
J185 ABS HYDRAULIC PUMP RELAY
DIAGNOSTIC CONNECTOR
J104 ABS CNTRL UNIT
F116 ABS PRESSURE WARNING SWITCH II
F109 ABS HIGH PRESSURE SWITCH
N55 ABS HYDRAULIC UNIT
V64 ABS HYDRAULIC PUMP
F115 ABS PRESSURE WARNING SWITCH I
N99 ABS INLET VALVE, RIGHT FRONT
TO BATTERY (–)
TO PARKING BRAKE WARNING LIGHT SWITCH, F9 (GROUND)
WIRING COLOR CODE
BK — BLACK
BR — BROWN
CL — CLEAR
R — RED
Y — YELLOW
G — GREEN
LT. G — LIGHT GREEN
BL — BLUE
V — VIOLET
GY — GRAY
W — WHITE
OR — ORANGE
97-9371

Description	Current track
ABS Brake Fluid Level Warning Switch, F117	36
ABS Control Module, J104	1-25
ABS High Pressure Switch, F109	26
ABS Hydraulic Pump, V64	27
ABS Hydraulic Pump Fuse, S53	27
ABS Hydraulic Pump Relay, J185	25-28
ABS Hydraulic Unit, N55	5-27
ABS Inlet Valve, Left Front, N101	8
ABS Inlet Valve, Rear, N103	10
ABS Inlet Valve, Right Front, N99	13
ABS Main Valve, N105	5
ABS Outlet Valve, Left Front, N102	9
ABS Outlet Valve, Right Front, N100	12
ABS Outlet Valve, Rear, N104	11
ABS Pressure Control Switch, F137	20
ABS Pressure Warning Switch I, F115	24
ABS Pressure Warning Switch II, F116	23
ABS Speed Sensor Condenser, C17	6-13
ABS Speed Sensor, Left Front , G47	12-13
ABS Speed Sensor, Left Rear, G46	10-11
ABS Speed Sensor, Right Front , G45	8-9
ABS Speed Sensor, Right Rear, G44	6-7
ABS Relay, J102	29-34
ABS Valves Fuse, S54	32
ABS Warning Light, K47	40
Brake and Parking Brake Indicator Light, K7	39
Brake Fluid Level Warning Switch, F34	35
Differential Lock Valve 1, N125	15
Differential Lock Valve 2, N126	16
Fuse, S16, 15A	39
Fuse, S21, 15A	15
MIL (Malfunction Indicator Lamp), K83	41

Wire connectors

T1i	- single, behind relay panel plate
T1y	- single, behind relay panel plate
T2a	- double, on ABS Main Valve
T2b	- double, on ABS Hydraulic Pump
T2c	- double, rear seat, right
T2d	- double, on A-pillar, left, near relay panel
T2i	- double, on firewall, right
T2e	- double, rear seat, left
T2f	- double, on firewall, left
T2g	- double, on ABS Brake Fluid Level Warning Switch
T2h	- double, on Brake Fluid Level Warning Switch
T2k	- double, near gearshift lever
T2x	- double, below selector lever cover
T2z	- double, below selector lever cover
T5	- five point, on ABS Hydraulic Unit
T5a	- five point, on ABS Hydraulic Unit
T7	- seven point, on ABS Hydraulic Unit
T8d	- eight point, on Brake and Parking Brake Indicator Light
T28	- twenty-eight point, on Instrument Cluster
T55	- fifty-five point, on ABS Control Module
T68	- sixty-eight point, on ECM (Engine Control Module)

Welded wiring harness points

(A15) - plus connection (15), in instrument wiring harness

(J1) - plus connection (30), in ABS wiring harness

(J4) - wire connection (diagnostic), in ABS wiring harness

(J5) - wire connection -2-, in ABS wiring harness

(W9) - Plus connection (30), in rear wiring harness

(W10) - Plus connection (15), in rear wiring harness

Ground Connections

(14) - ground connection, on transmission

(22) - ground connection, on hydraulic unit

(28) - ground connection on bulkhead

(39) - ground-connection, below rear seat bench, left

(40) - ground-connection, below rear seat bench, right

(86) - ground-connection -1-, in rear wiring harness

(125) - ground connection -3-, in headlight wiring harness

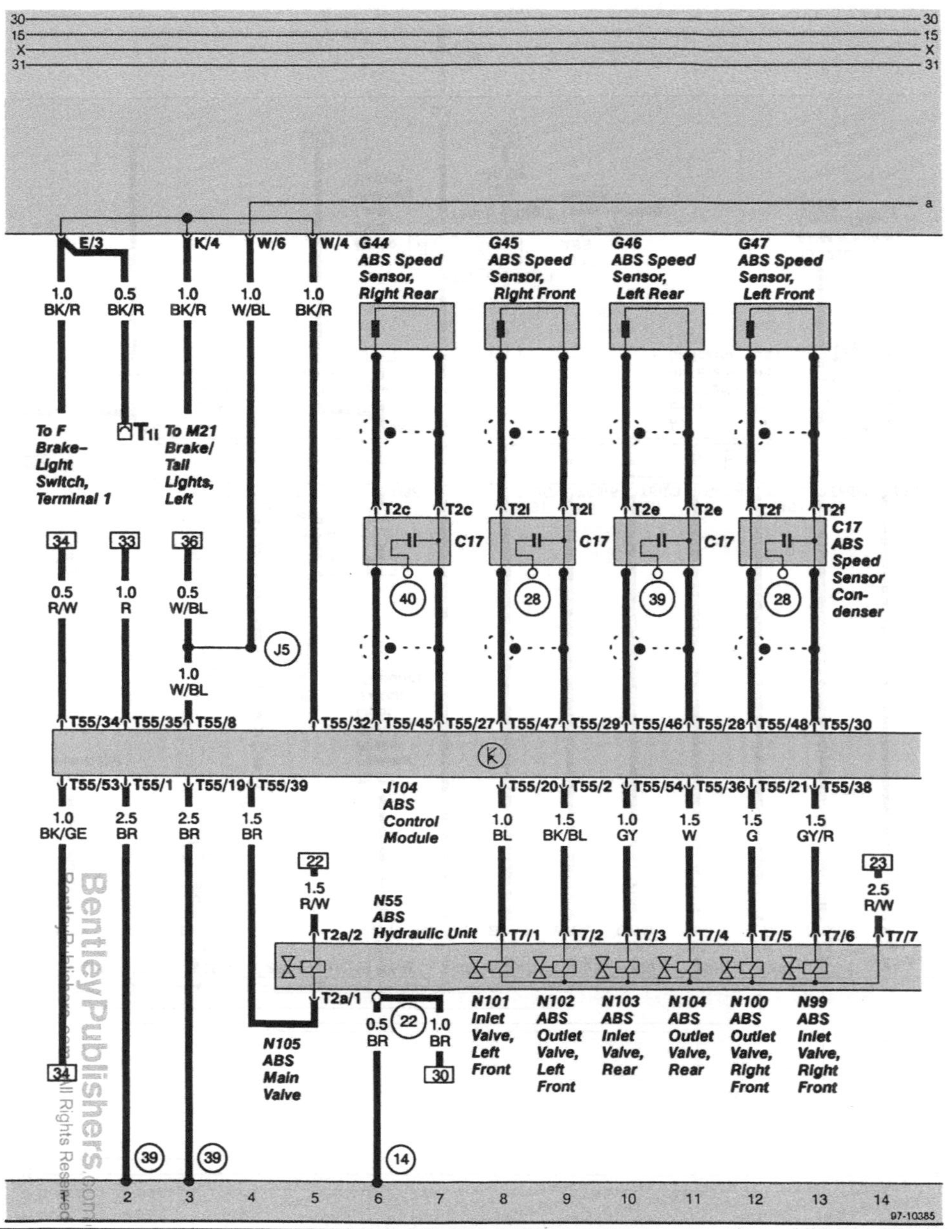

Corrado starting 1992
Engine code – AAA

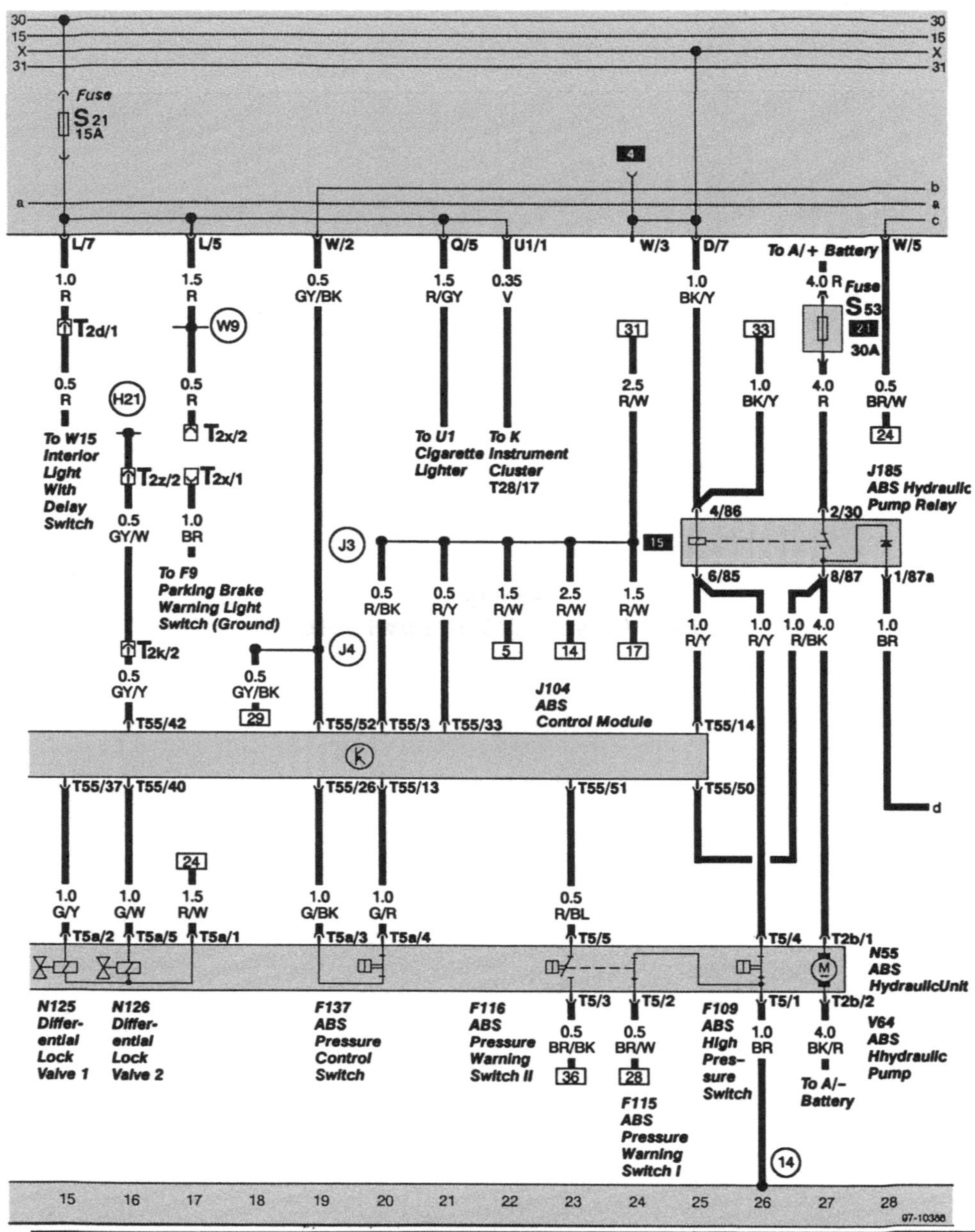

Corrado starting 1992
Engine code – AAA

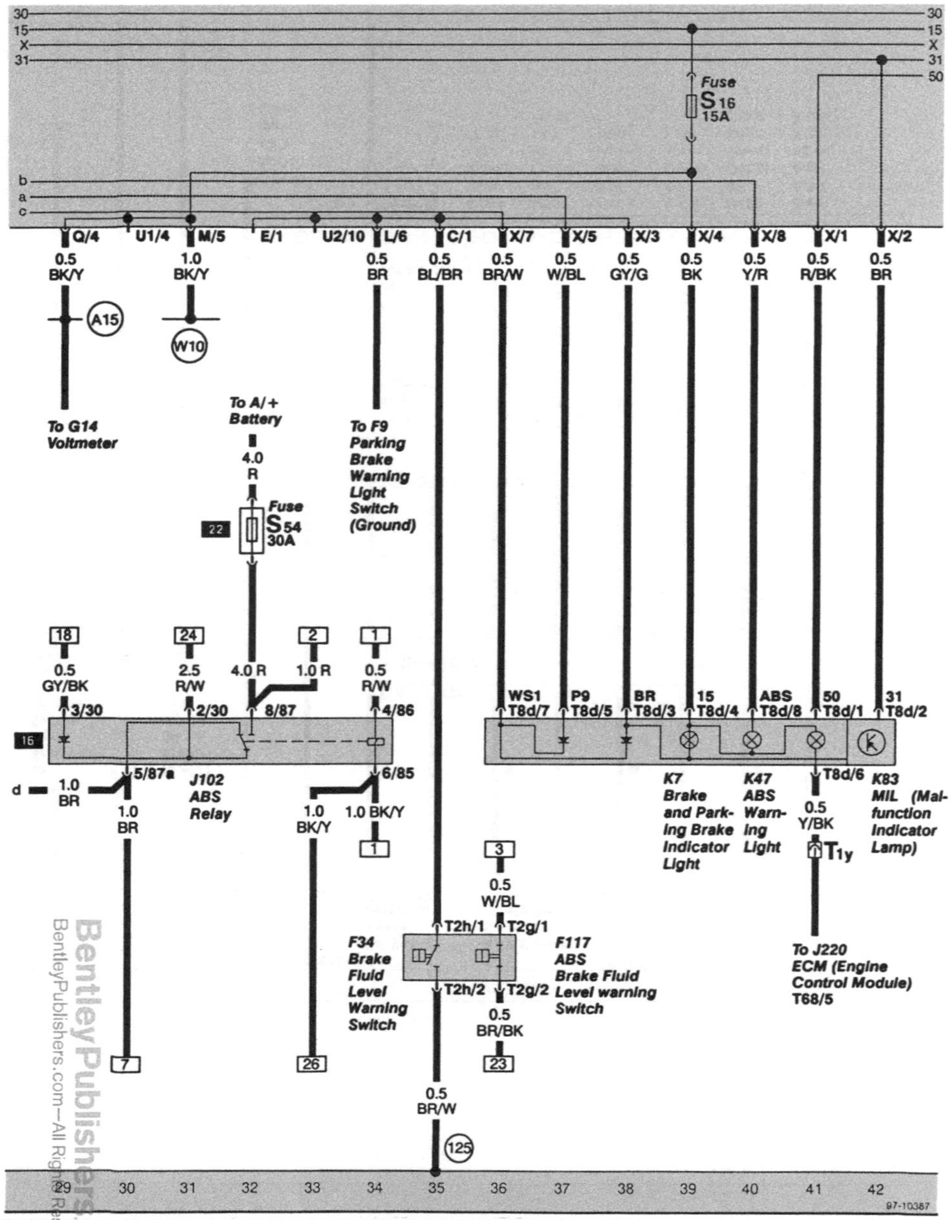

Corrado starting 1992
Engine code – AAA

THIS PAGE INTENTIONALLY LEFT BLANK

Description	Current track
Alarm trunk latch/luggage compartment light switch, F5	16
Alarm horn, H8	1
Alarm switch, hood, F120	2
Alarm switch, trunk lock, F124	6
Anti-theft alarm system control unit, J284	1-22
Central locking system switch (Passenger's door), F114	23-24
Door contact switch, left front, F2	9
Door contact switch, right front, F3	14
Fuel pump relay, J17	3

Wire connectors

- T1 — single, behind fuse/relay panel
- T1a — single, behind fuse/relay panel
- T1b — single, on left A-pillar, near fuse/relay panel
- T1c — single, on lower right A-pillar
- T1d — single, in rear lid
- T2a — double, on lower right A-pillar
- T2b — double, in luggage compartment, right
- T2c — double, behind fuse/relay panel
- T2d — double, behind fuse/relay panel
- T3 — three-point, on lower right A-pillar
- T3c — three-point, on lower right A-pillar
- T3d — three-point, on lower left A-pillar, near fuse/relay panel
- T3f — three-point, in passenger's door
- T3j — three-point, on left front turn signal
- T3k — three-point, on right front turn signal
- T4 — four-point, behind fuse/relay panel
- T6 — six-point, on anti-theft alarm system control unit
- T8a — eight-point, on power window control unit
- T10 — ten-point, on anti-theft alarm system control unit
- T12 — twelve-point, on central locking system pump and control unit
- T28 — twenty eight-point, on instrument cluster
- T36 — junction connector, terminal 30, behind fuse/relay panel

Ground connections

- (32) — behind instrument panel, left
- (52) — in rear lid, left
- (53) — in rear lid, right
- (106) — in anti-theft alarm system wiring harness

Welded wiring harness points

- (C15) — wire connection, in headlight wiring harness
- (C16) — wire connection, in headlight wiring harness
- (Q4) — plus (+) connection (30), in power window/central locking system wiring harness
- (Q22) — wire connection, in rear lid wiring harness
- (Q51) — plus (+) connection (30), in anti-theft alarm system wiring harness
- (Q53) — plus (+) connection (15), in anti-theft alarm system wiring harness
- (Q54) — wire connection (left flasher), in anti-theft alarm system wiring harness
- (Q55) — wire connection (right flasher), in anti-theft alarm system wiring harness
- (Q56) — wire connection (door locks open), in anti-theft alarm system wiring harness
- (Q57) — wire connection (door locks locked), in anti-theft alarm system wiring harness
- (S1) — plus (+) connection (30), in central locking system wiring harness
- (S4) — plus (+) connection (30), in central locking system wiring harness

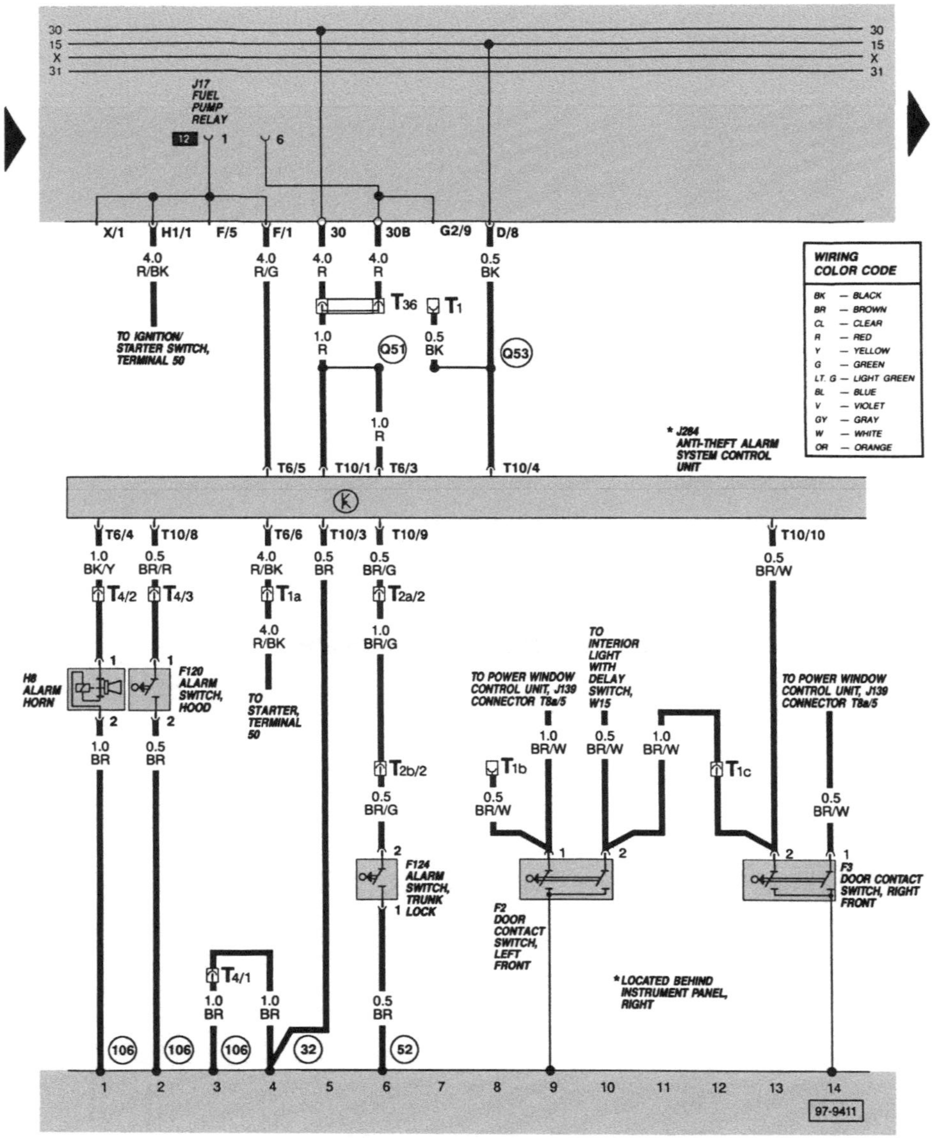

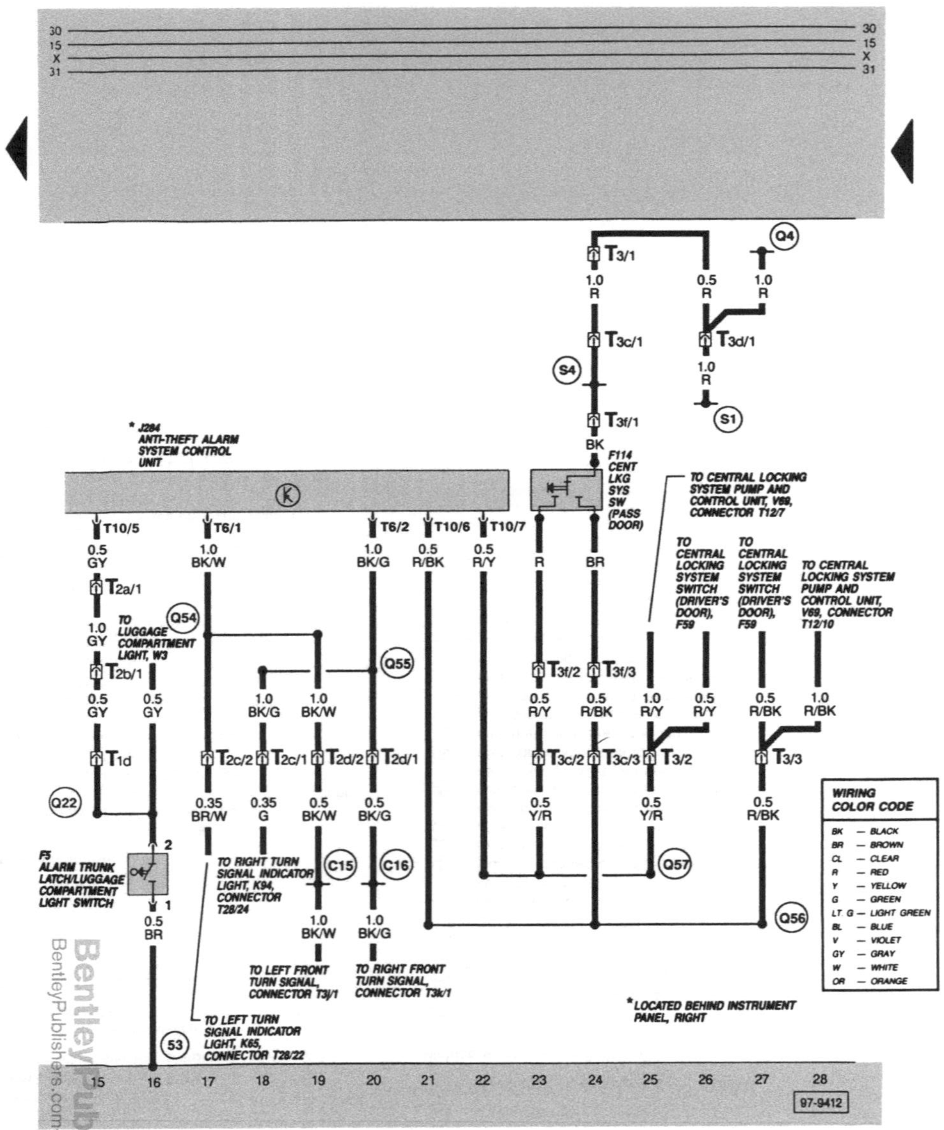

Corrado 1991–1993
Production up to Aug. 1992

THIS PAGE INTENTIONALLY LEFT BLANK

Description	Current track
Alarm Horn, H8	1
Alarm Switch Hood, F120	2
Alarm System Control Module Fuse, S91	6
Alarm System Control Module, J284	1-22
Alarm System Flasher Lights Fuse, S90	5
Alarm System Indicator Light, K95	8
Alarm/Central Locking Switch Trunk Lock, F124	6
Central Locking System Switch, Passenger's Door, F114	23-24
Door Contact Switch, Left Front, F2	10-11
Door Contact Switch, Right Front, F3	12-13
Luggage Compartment Light Switch, F5	16

Wire connectors

- T1 – single, behind relay panel plate
- T1a – single, behind relay panel plate
- T1b – single, on left A-pillar, near fuse/relay panel
- T1c – single, on lower right A-pillar
- T1d – single, in rear lid
- T1e – single, behind relay panel plate
- T1v – single, on lower left A-pillar
- T2a – double, on lower right A-pillar
- T2b – double, in luggage compartment, left
- T2d – double, behind relay panel plate
- T2e – double, on lower left A-pillar
- T2h – double, behind relay panel plate
- T3 – three point, on lower right A-pillar
- T3c – hree point, on lower right A-pillar
- T3d – three point, on left A-pillar, near fuse/relay panel
- T3f – hree point, in passenger's door
- T4 – four point, behind relay panel plate
- T6 – six point, on Alarm System Control Module
- T8a – eight point, on Power Window/Sunroof Control Module
- T10 – ten point, on Alarm System Control Module
- T12 – twelve point, on Central Locking System Pump and Control Module
- T28 – twenty-eight point, on Instrument Cluster

- TV2 – Wire Connector, Terminal 30 (B+), behind fuse/relay panel
- TV4 – Wire Connector, Terminal 15, behind fuse/relay panel

Welded wiring harness point

- (C15) – Wire connection -1-, in headlight wiring harness
- (C16) – Wire connection -2-, in headlight wiring harness
- (Q4) – Plus connection (30), in power window/central locking system and door contact switch wiring harness
- (Q8) – Wire connection (power window/rear cigarette lighter), in power window/central locking system and door contact switch wiring harness
- (Q22) – Wire connection -1-, in rear lid wiring harness
- (Q51) – Plus connection (30), in alarm system wiring harness
- (Q54) – Wire connection (left turn signal), in alarm wiring harness
- (Q55) – Wire connection (right turn signal), in alarm wiring harness
- (Q56) – Wire connection (open), in alarm system wiring harness
- (Q57) – Wire connection (closed),in alarm system wiring harness
- (S1) – Plus connection (30) -1-, in central locking sytem wiring harness
- (S4) – Plus connection (30) -2-, in central locking sytem wiring harness

Ground connections

- (32) – Ground connection, behind instrument panel, left
- (52) – Ground connection, in rear lid, left
- (53) – Ground connection, in rear lid, right
- (106) – Ground connection, in alarm system wiring harness

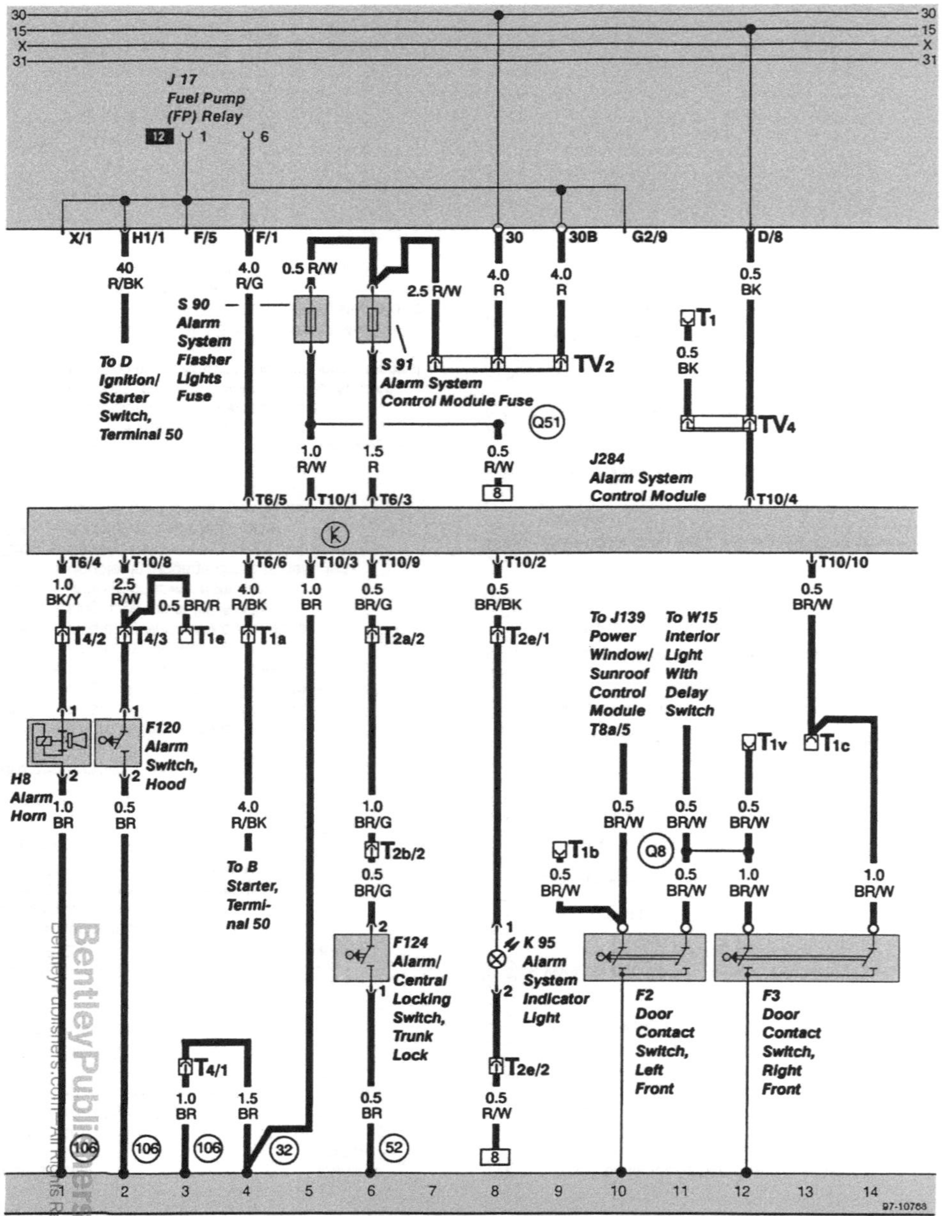

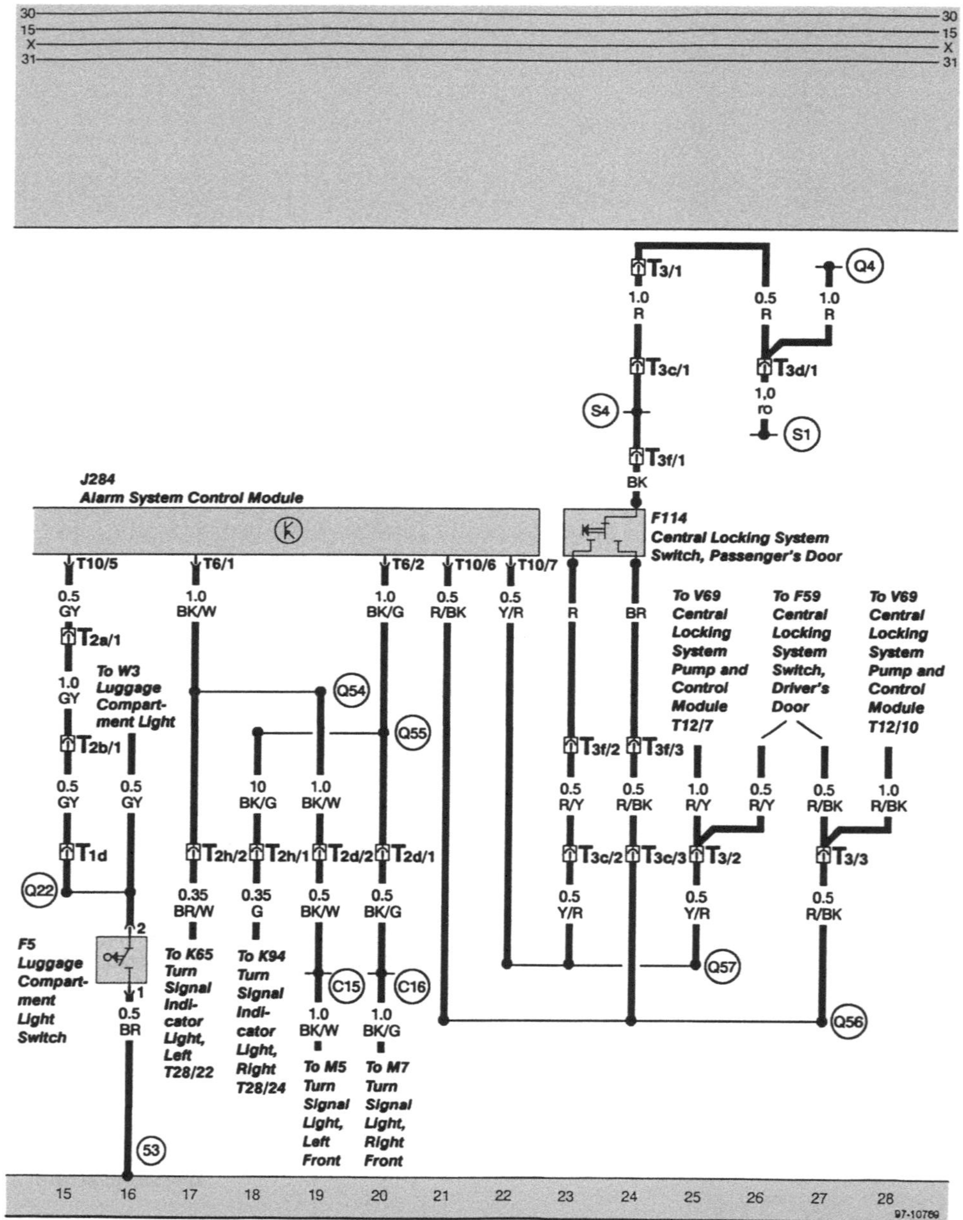

Corrado starting 1993
Production from Aug. 1992

Corrado starting 1993
Production from Aug. 1992

Description	Current track
Door latch switch, left, F143	27
Door latch switch, right, F144	7
Magnetic switch, left rear (B-pillar), F157	24
Microswitch, left front (A-pillar), F140	21
Microswitch, left rear (B-pillar), F145	23
Microswitch, right front (A-pillar), F141	5
Microswitch, right rear (B-pillar), F146	2
Motion/time delay sensor, G97	13-14
Seat belt control unit, left, J227	16-26
Seat belt control unit, right, J228	2-10
Seat belt motor, left, V74	25
Seat belt motor, right, V75	3
Seat belt warning system control unit, J34	4-7
Shoulder belt lock release solenoid, left, N129	15
Shoulder belt lock release solenoid, right, N130	12
Shoulder belt lock release switch, left (under seat), F149	15
Shoulder belt lock release switch, right (under seat), F150	12

Wire connectors

T4n — four-point, behind fuse/relay panel
T18 — eighteen-point, on seat belt control unit
T28 — *twenty eight-point, on instrument cluster*

Ground connections

(30) — beside fuse/relay panel
(141) — in seat belt wiring harness
(142) — in seat belt wiring harness

Welded wiring harness points

(Q60) — plus connection (30), in seat belt wiring harness

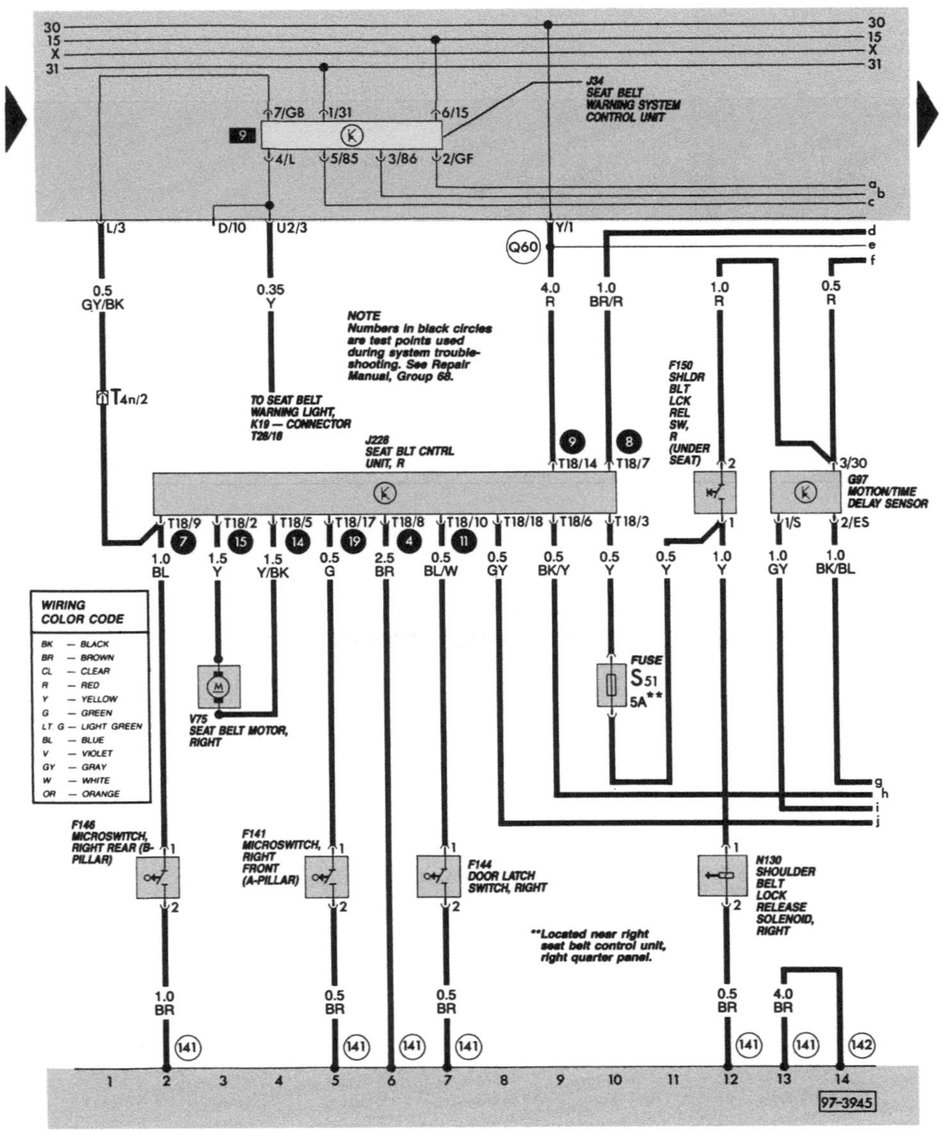

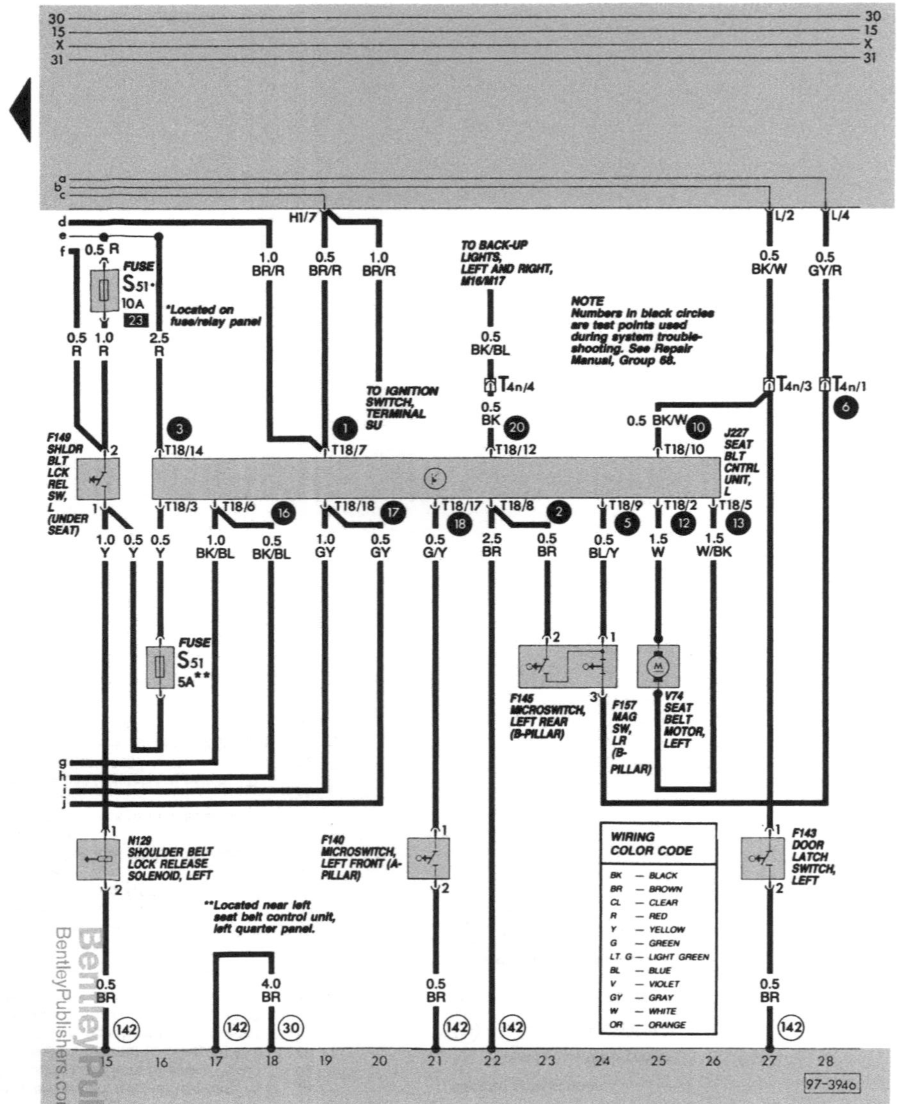

THIS PAGE INTENTIONALLY LEFT BLANK

Description	Current track
Door latch switch, left, F143	28
Door latch switch, right, F144	6
Fuse, S51	13
Fuse, S69	10
Fuse, S69	16
Magnetic switch, left rear (B-pillar), F157	26
Microswitch, left front (A-pillar), F140	21
Microswitch, left rear (B-pillar), F145	25
Microswitch, right front (A-pillar), F141	5
Microswitch, right rear (B-pillar), F146	2
Motion/time delay sensor, G97	13-14
Seat belt control unit, left, J227	16-26
Seat belt control unit, right, J228	2-10
Seat belt motor, left, V74	22
Seat belt motor, right, V75	3
Seat belt warning control unit, J34	2-7
Shoulder belt lock release solenoid, left, N129	15
Shoulder belt lock release solenoid, right, N130	12
Shoulder belt lock release switch, left (under seat), F149	15
Shoulder belt lock release switch, right (under seat), F150	12

Wire connectors

T1h — single, beind fuse/relay panel
T1x — single, behind fuse/relay panel
T2n — double, in luggage compartment, left
T2o — double, in engine compartment, left
T4i — four-point, behind fuse/relay panel
T8 — eight-point, on radio
T18 — eighteen-point, on seat belt control unit, right
T18a — eighteen-point, on seat belt control unit, left
T28 — twenty eight-point, on instrument cluster

Ground connections

(30) — beside fuse/relay panel
(141) — in automatic shoulder belt system wiring harness
(142) — in automatic shoulder belt system wiring harness

Welded wiring harness points

(B6) — wire connection (s), in steering column switch wiring harness
(Q60) — plus connection (30), in automatic shoulder belt system wiring harness
(Q66) — plus connection (30), in automatic shoulder belt system wiring harness
(Q67) — wire connection, in automatic shoulder belt system wiring harness
(Q68) — wire connection, in automatic shoulder belt system wiring harness

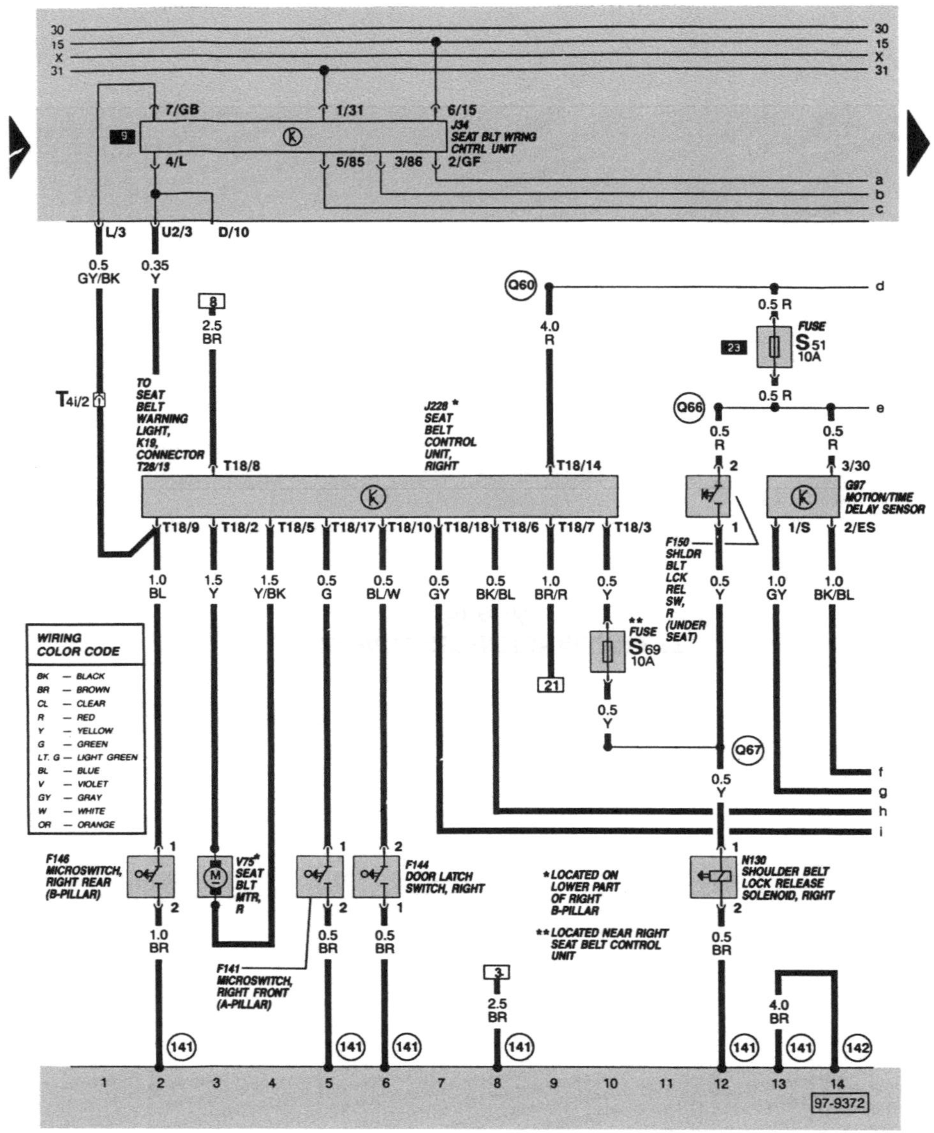

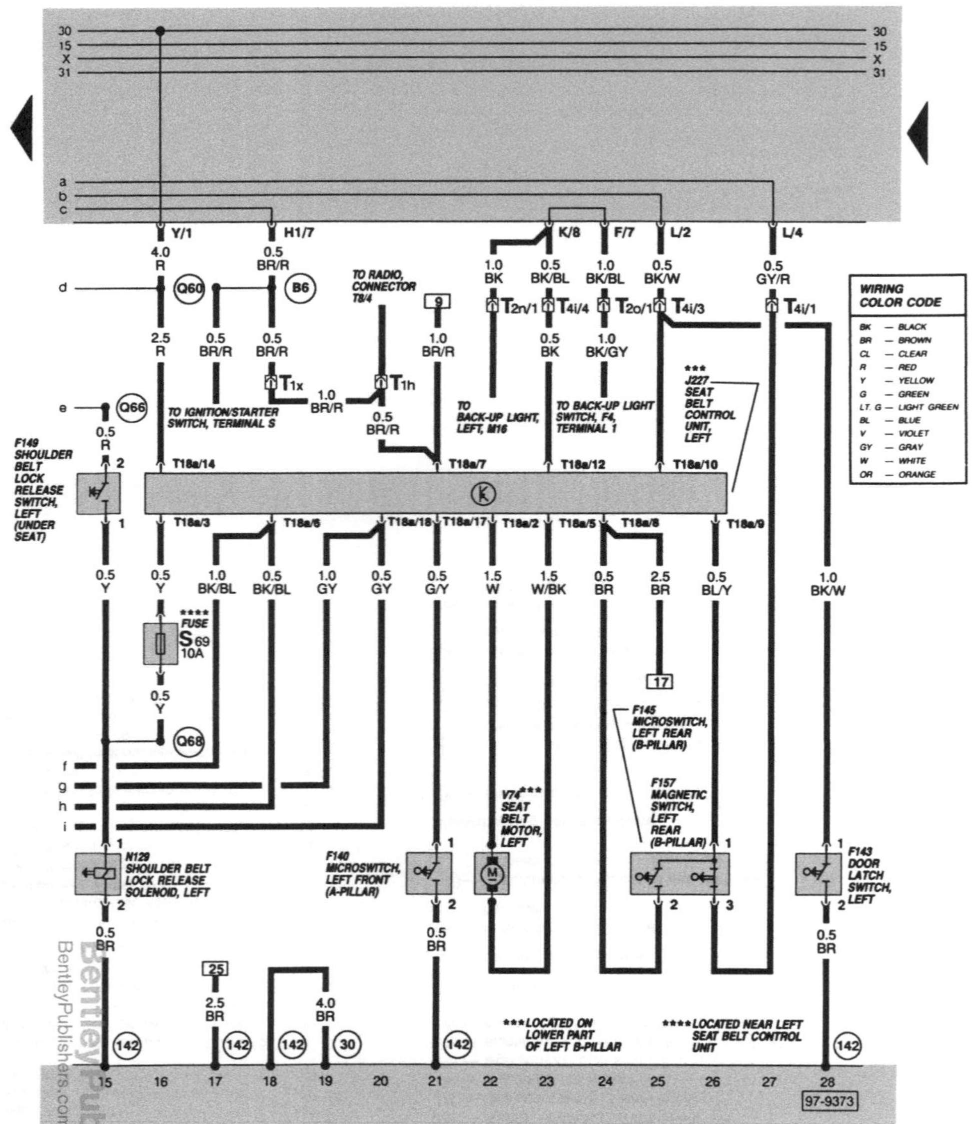

THIS PAGE INTENTIONALLY LEFT BLANK

Description	Current track
Door Latch Switch, Left, F143	28
Door Latch Switch, Right, F144	6
Fuse, S51,10A	13
Fuse, Seat Belt Lock Solenoid, Left, S69, 10A	16
Fuse, Seat Belt Lock Solenoid, Right, S69, 10A	10
Magnetic Switch, Left Rear (B-pillar), F157	26
Microswitch, Left Front (A-pillar), F140	21
Microswitch, Left Rear (B-pillar), F145	25
Microswitch, Right Front (A-pillar), F141	5
Microswitch, Right Rear (B-pillar), F146	2
Motion/Time Delay Sensor, G97	13–14
Seat Belt Control Module, Left, J227	16–26
Seat Belt Control Module, Right, J228	2–10
Seat Belt Motor, Left, V74	22
Seat Belt Motor, Right, V75	3
Seat Belt Warning Control Module, J34	2–7
Seat Belt Warning Light, K19	4
Shoulder Belt Lock Release Solenoid, Left, N129	15
Shoulder Belt Lock Release Solenoid, Right, N130	12
Shoulder Belt Lock Release Switch, Left (Under Seat), F149	15
Shoulder Belt Lock Release Switch, Right (Under Seat), F150	12

Wire connectors

- T1 - single, behind fuse/relay panel
- T1h - single, behind fuse/relay panel
- T1x - single, behind fuse/relay panel
- T2 - double, in left door
- T2a - double, in right door
- T2n - double, in luggage compartment, left
- T2o - double, in engine compartment, left
- T4l - four point, behind fuse/relay panel
- T8 - eight point, on Radio
- T18 - eighteen point, on Safety Belt Adjuster Control Unit, right
- T18a - eighteen point, on Safety Belt Adjuster Control Unit, left

Welded wiring harness points

- (B6) - wiring connection(s), in steering column switch wiring harness
- (Q60) - plus connection (30), in automatic shoulder belt system wiring harness
- (Q66) - plus connection (30), in automatic shoulder belt system wiring harness
- (Q67) - wiring connection, in automatic shoulder belt system wiring harness
- (Q68) - wiring connection, in automatic shoulder belt system wiring harness

Ground connections

- (30) - ground connection, -1-, beside relay panel
- (141) - ground connection -1-, in automatic shoulder belt system wiring harness
- (142) - ground connection -2-, in automatic shoulder belt system wiring

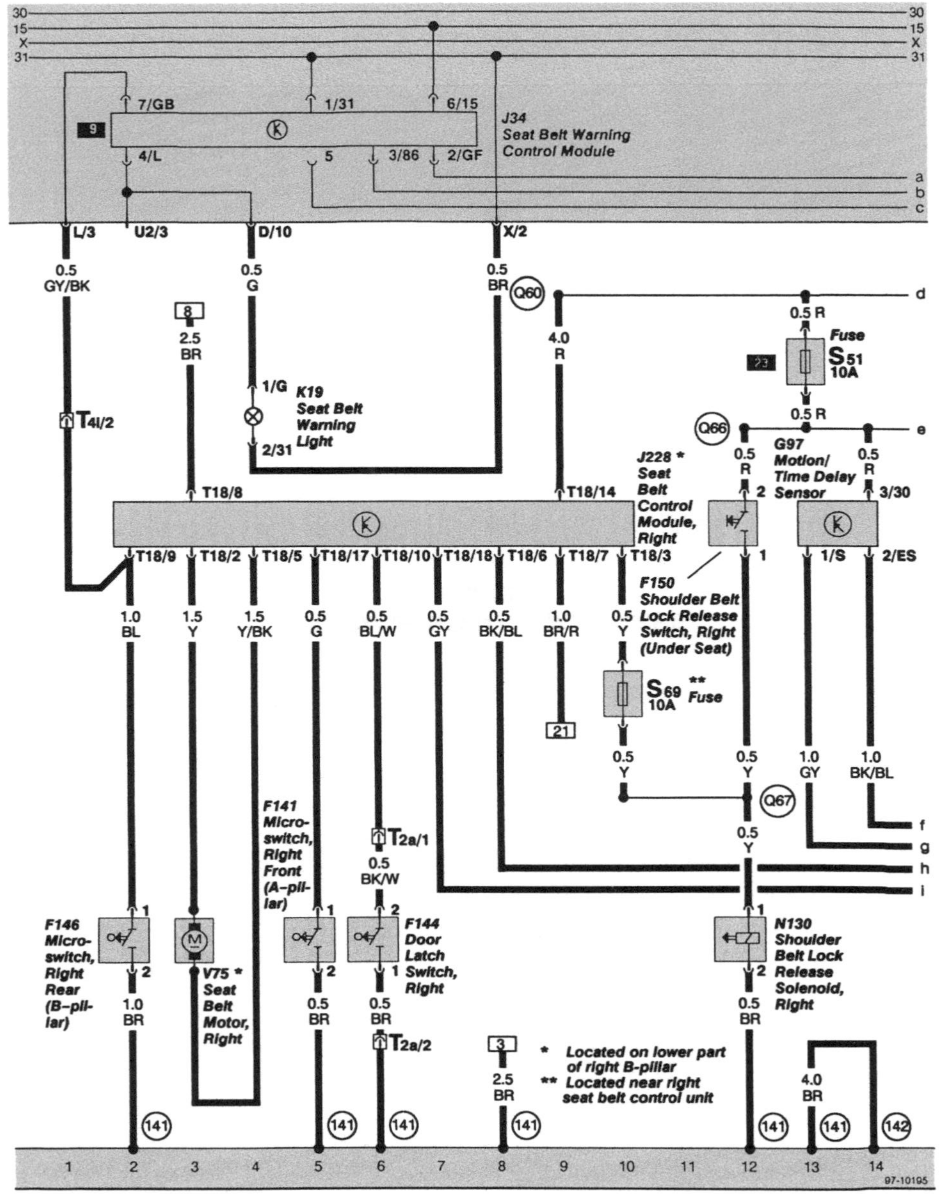

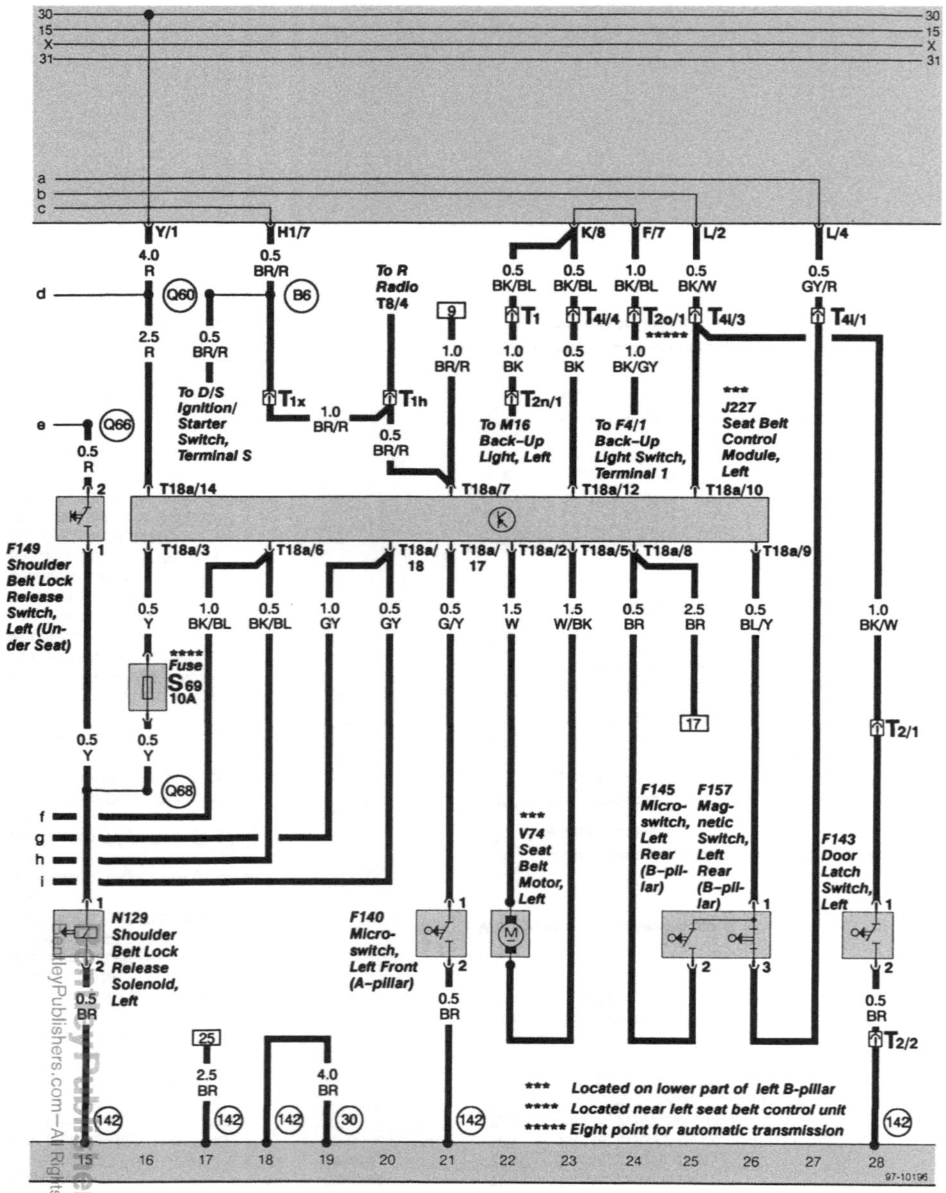

Corrado starting 1992
Engine code – AAA

THIS PAGE INTENTIONALLY LEFT BLANK

Description	Current track
Automatic transmission console light, L19	55
Automatic transmission control unit, J217	13-48
Back-up light, left, M16	3
Back-up light, right, M17	1
Full throttle switch, F81	34
Fuse, S14	5
Fuse, S21	21
Fuse, S16	53
Idle switch, F60	32
Kick-down switch, F8	33
Multifunction switch (Digimat), F125	36-39
OBD fault warning light, K83	47
Program switch (Auto Trans 4-ZF), E122	44
Program switch illumination, L70	45
Shift lock solenoid, N110	49
Solenoid valve 1, N88	21
Solenoid valve 2, N89	22
Solenoid valve 3, N90	23
Solenoid valve 4, N91	24
Solenoid valve 5, N92	25
Solenoid valve 6, N93	19
Solenoid valve 7, N94	26
Speed sensor (auto 4 speed), G68	31
Sports program indicator light (auto. trans.), K91	46
Starter interlock and back-up light relay, J226	4-10
Throttle valve potentiometer, G69	29-30, 37-42
Transmission temperature sensor, G93	27

Wire connectors

- T1b — single, behind fuse/relay panel
- T1g — single, behind fuse/relay panel
- T1h — single, behind fuse/relay panel
- T1i — single, behind fuse/relay panel
- T1o — single, on left back-up light
- T1p — single, on left back-up light
- T2 — double, above fuse/relay panel
- T2a — double, near transmission shift lever
- T2b — double, on speed sensor
- T2c — double, on kick-down switch
- T2d — double, on left A-pillar, near fuse/relay panel
- T2e — double, on automatic shift lock solenoid
- T2g — double, on automatic transmission console light

Wire connectors

- T2n — double, in luggage compartment, left
- T2x — double, black, diagnostic connector, below transmission shift lever cover
- T2z — double, white, diagnostic connector, below transmission shift lever cover
- T3b — three-point, near throttle valve housing
- T3d — three-point, on left back-up light
- T3e — three-point, on right back-up light
- T4 — four-point, near transmission shift lever
- T4c — four-point, near throttle valve housing
- T5 — five-point, on transmission program switch
- T5a — five-point, above fuse/relay panel
- T6 — six-point, in engine compartment, left
- T7 — seven-point, on multi-function switch
- T8d — eight-point, on brake/parking brake warning light assembly
- T9 — nine-point, on starter interlock/back-up light relay
- T12 — twelve-point, on transmission solenoid valve assembly
- T25 — twenty five-point, on Digifant control unit
- T28 — twenty eight-point, on instrument cluster
- T36 — junction, terminal 30, above fuse/relay panel
- T38 — thirty eight-point, on automatic transmission control unit
- T38a — thirty eight-point, on Digifant control unit

Ground connections

- (30) — beside fuse/relay panel
- (53) — in rear lid, right
- (94) — in Digifant wiring harness
- (150) — in automatic transmission wiring harness

Welded wiring harness points

- (A11) — plus connection (58b), in instrument panel wiring harness
- (A15) — plus connection (15), in instrument panel wiring harness
- (C14) — plus connection (15), in heated windshield washer nozzle wiring harness
- (U2) — wire connection (15), in automatic transmission wiring harness
- (U5) — wire connection (15), in automatic transmission wiring harness
- (W9) — plus connection (30), in rear wiring harness
- (W10) — plus connection (15), in rear wiring harness

Corrado 1991–1992
Engine code – PG

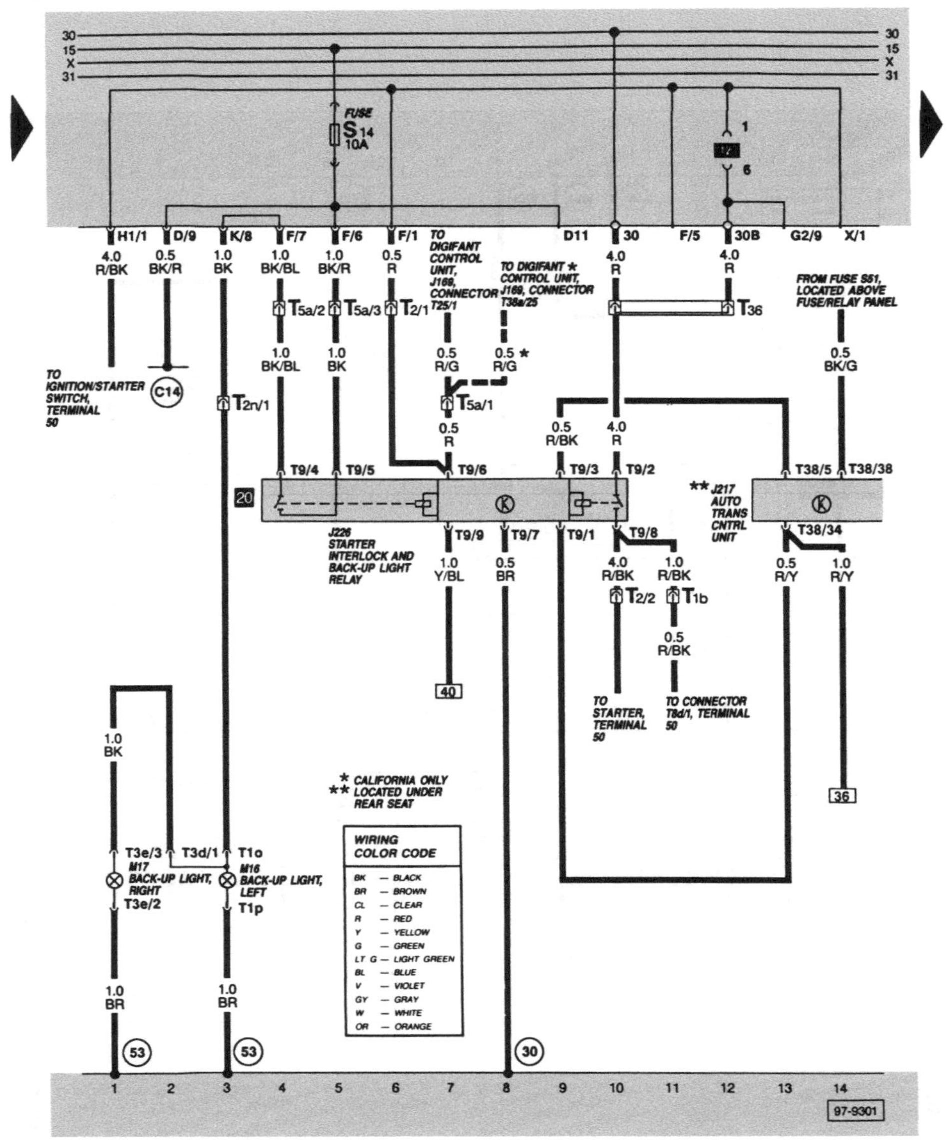

Corrado 1991–1992
Engine code – PG

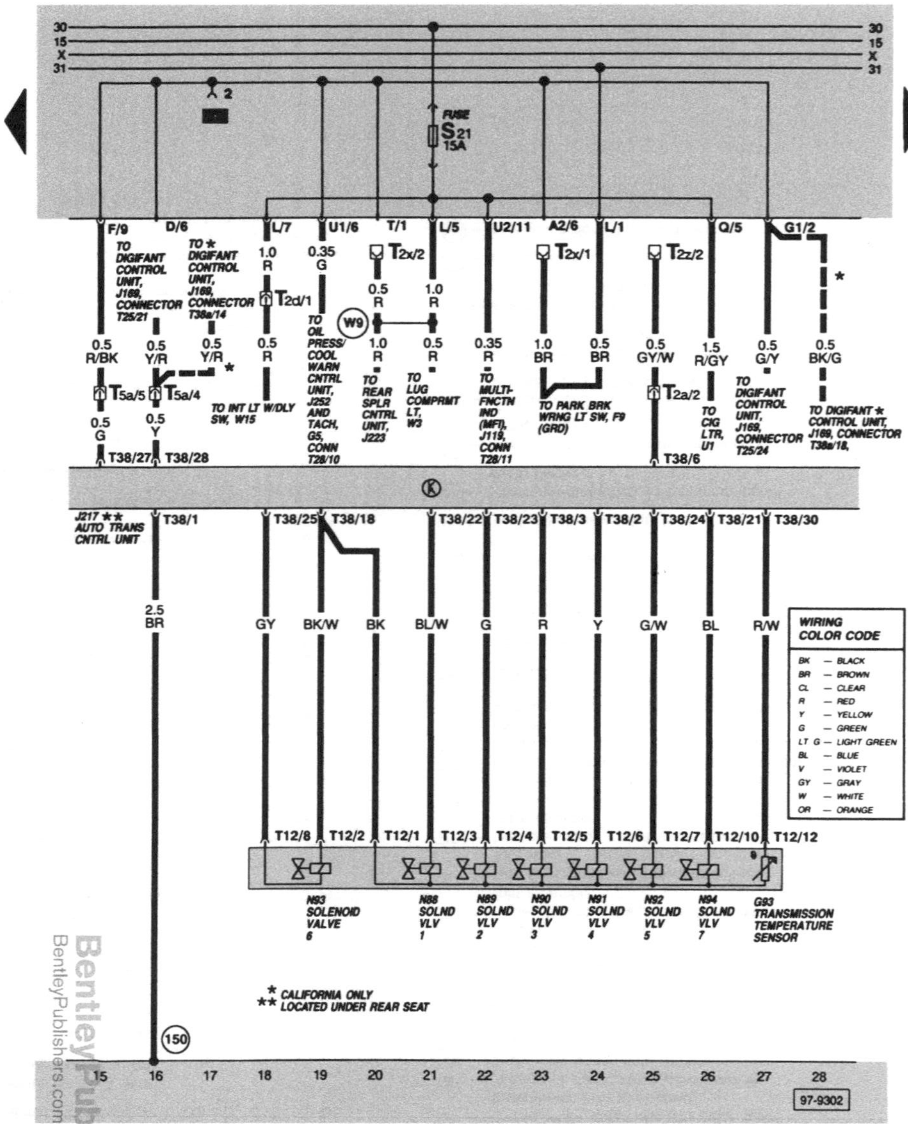
97-9302
WIRING COLOR CODE
* CALIFORNIA ONLY
** LOCATED UNDER REAR SEAT

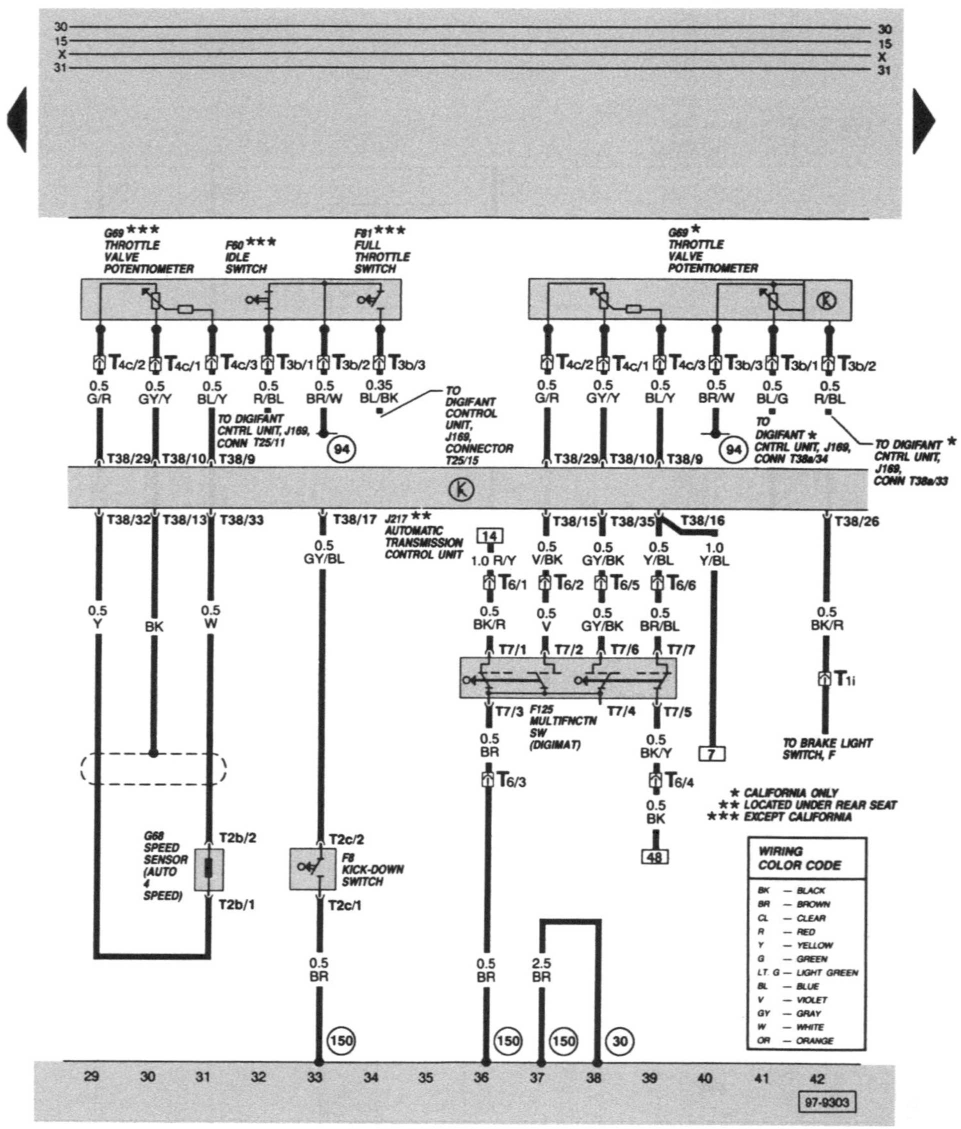
97-9303
WIRING COLOR CODE
* CALIFORNIA ONLY
** LOCATED UNDER REAR SEAT
*** EXCEPT CALIFORNIA

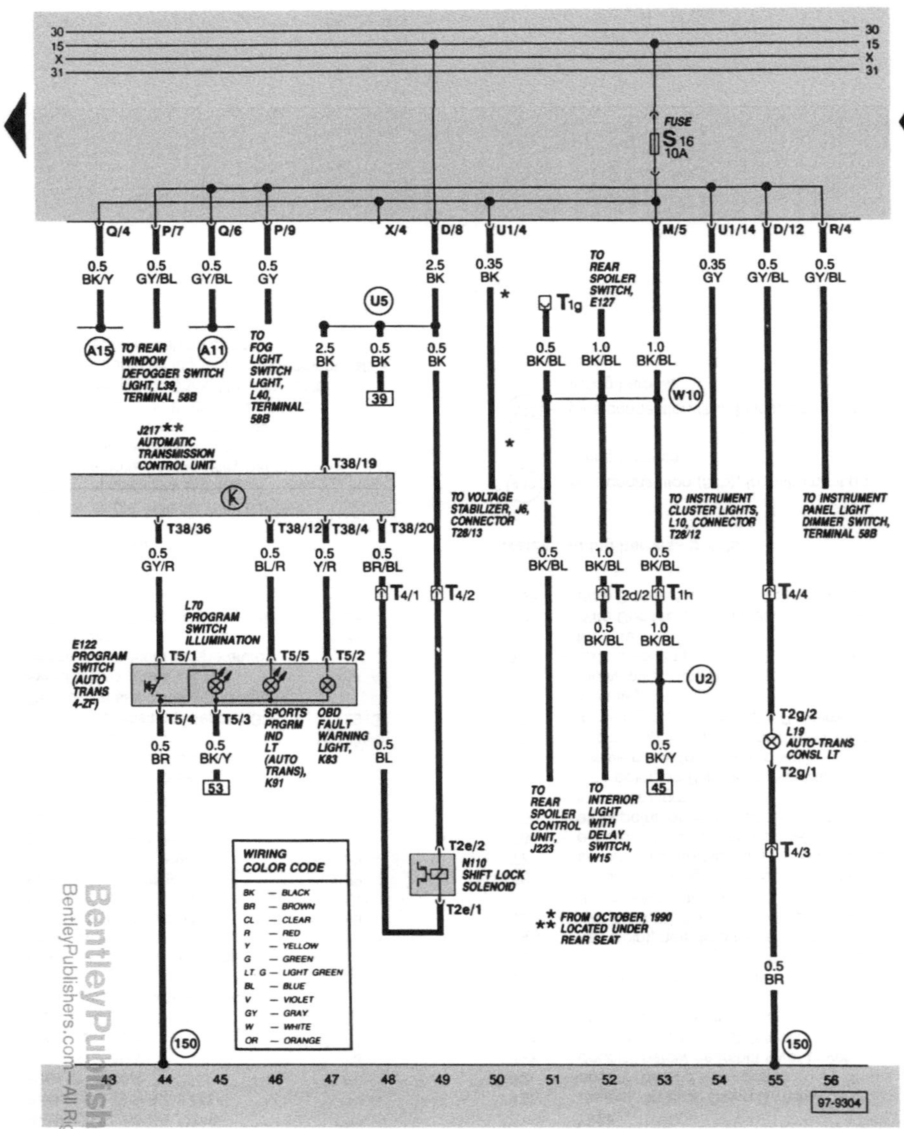

Corrado 1991–1992
Engine code – PG

THIS PAGE INTENTIONALLY LEFT BLANK

Description	Current track
Automatic Transmission Console Light, L19	55
Back-Up Light, Left, M16	3
Back-Up Light, Right,M17	1
Fuse, S14, 10A	5
Fuse, S16, 10A	53
Fuse, S21, 15A	21
Kick-Down Switch, F8	36
Multi-Function Switch (Digimat), F125	37-40
PNP (Park/Neutral Position) Relay, J226	4-10
Shift Lock Solenoid, N110	49
Solenoid Valve 1, N88	21
Solenoid Valve 2N89	22
Solenoid Valve 3, N90	23
Solenoid Valve 4, N91	24
Solenoid Valve 5, N92	25
Solenoid Valve 6, N93	19
Solenoid Valve 7, N94	26
TCM (Transmission Control Module), J217	13-48
TP (Throttle Position) Sensor, G69	29-34
TR (Transmission Range) Program Switch (Auto. Trans.), E122	44-47
Transmission Fluid Temperature Sensor, G93	27
VSS (Vehicle Speed Sensor), G68	36

Wire connectors

- T1 - single, behind relay panel
- T1b - above relay panel
- T1g - single, behind relay panel
- T1h - single, behind relay panel
- T1i - single, behind relay panel
- T1o - single, on Back-Up Light, Left
- T1p - single, on Back-Up Light, Left
- T2 - double, in luggage compartment, left
- T2a - double, behind console
- T2b - double, on VSS (Vehicle Speed Sensor)
- T2d - double, behind relay panel
- T2e - double, on Shift Lock Solenoid
- T2g - double, on Automatic Transmission Console Light
- T2h - double, on Kick-Down Switch
- T2l - double, behind relay panel
- T2x - double, below gearshift lever cover
- T2z - double, below gearshift lever cover
- T3a - three point, near throttle valve housing
- T3d - three point, on Back-Up Light, Left
- T3e - three point, on Back-Up Light, Right
- T4 - four point, below gearshift lever cover
- T4a - four point, near throttle valve housing
- T5 - five point, on TR (Transmission Range) Program Switch (Auto. Trans.)
- T6 - six point, in engine compartment, left
- T7 - seven point, on Multi-Function Switch (Digimat)
- T8a - eight point, above relay panel
- T8d - eight point, on Brake and Parking Brake Indicator Light
- T9 - nine point, on PNP (Park/Neutral Position) Relay
- T12 - twelve point, for Solenoid Valve
- T28 - twenty-eight point, on instrument cluster
- T38 - thirty-eight point, on TCM (Transmission Control Module)
- T42 - fourty-two point, on cylinder-head, left
- T68 - sixteen-eight point, on ECM (Engine Control Module)
- TV2 - Wire Connector, Terminal 30, above relay panel
- TV5 - Wire Connector, Terminal 15a, above relay panel

Welded wiring harness points

- (A11) - plus connection (58b), in instrument panel wiring harness
- (A15) - plus connection (15), in instrument panel wiring harness

Welded wiring harness points

- (U5) - wire connection -2- (15), in automatic transmission wiring harness
- (W9) - plus connection (30), in rear wiring harness
- (W10) - plus connection (15), in rear wiring harness

Ground Connections

- (30) - ground connection, -1-, beside relay panel
- (33) - ground connection, in rear lid, right
- (139) - ground connection (sensor ground), in MFI (Multiport Fuel Injection)
- (150) - ground connection, in Automatic Transmission (Digimat/AG4) wiring

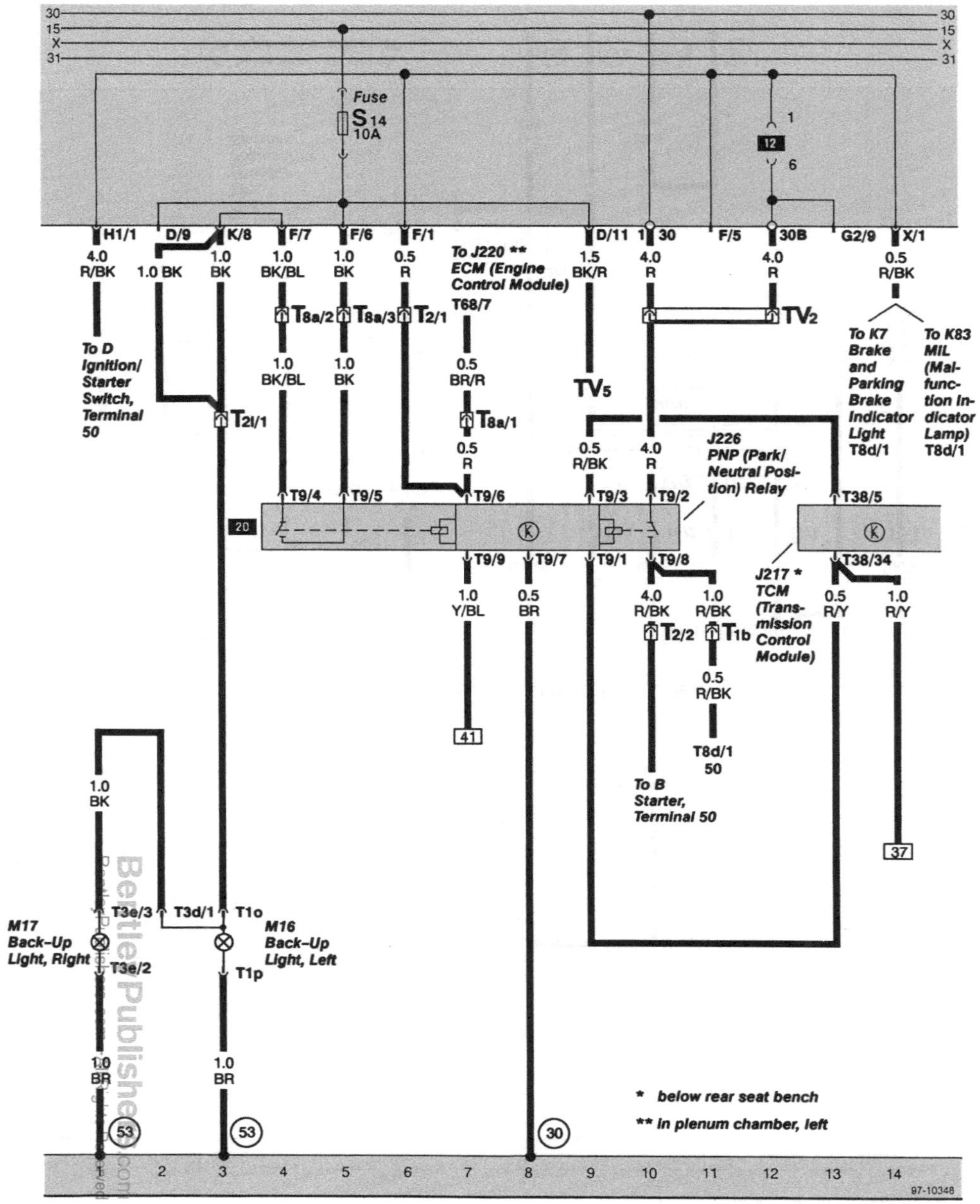

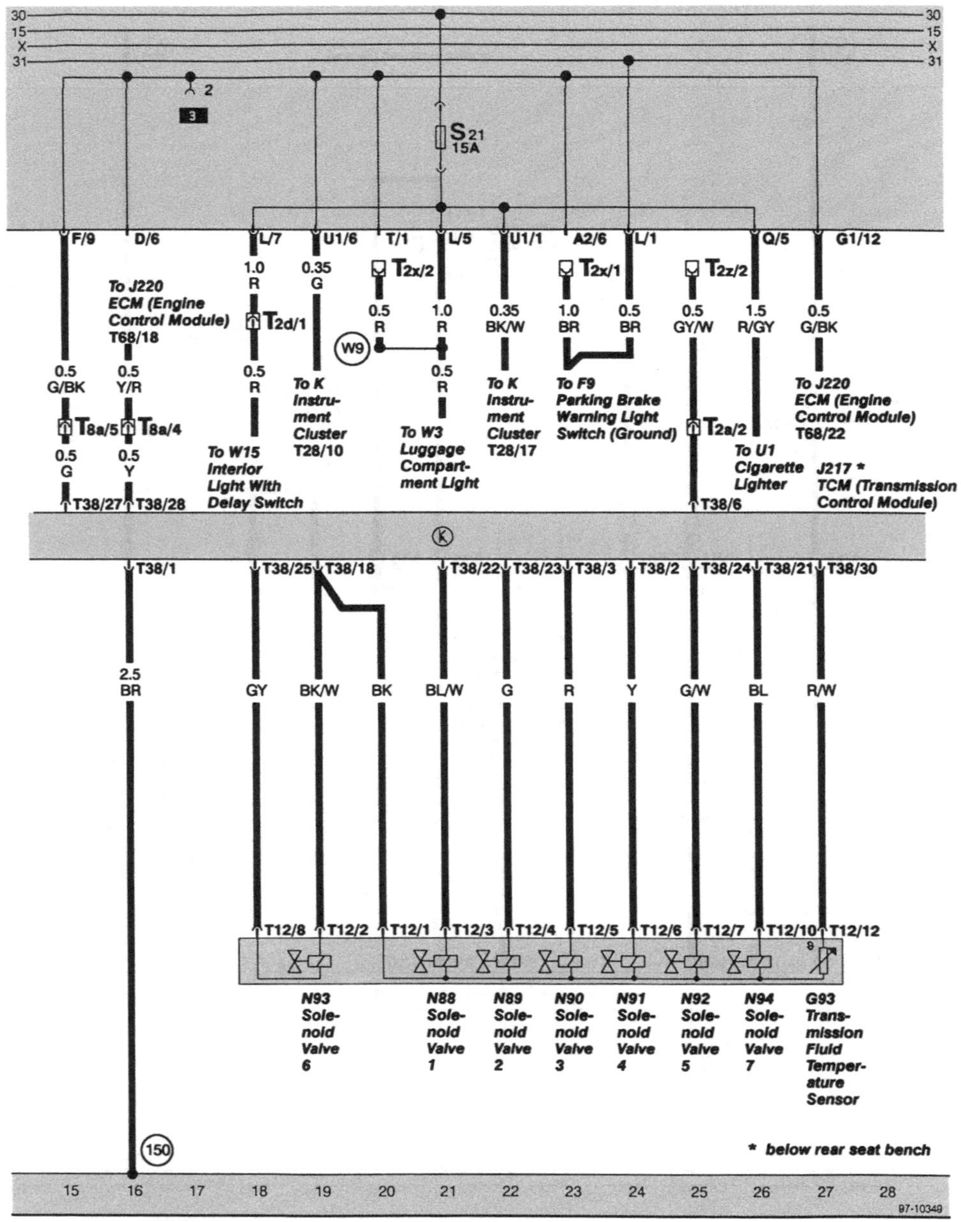

Corrado 1992–1993
Engine code – AAA
Production up to Jan. 1993

Corrado 1992–1993
Engine code – AAA
Production up to Jan. 1993

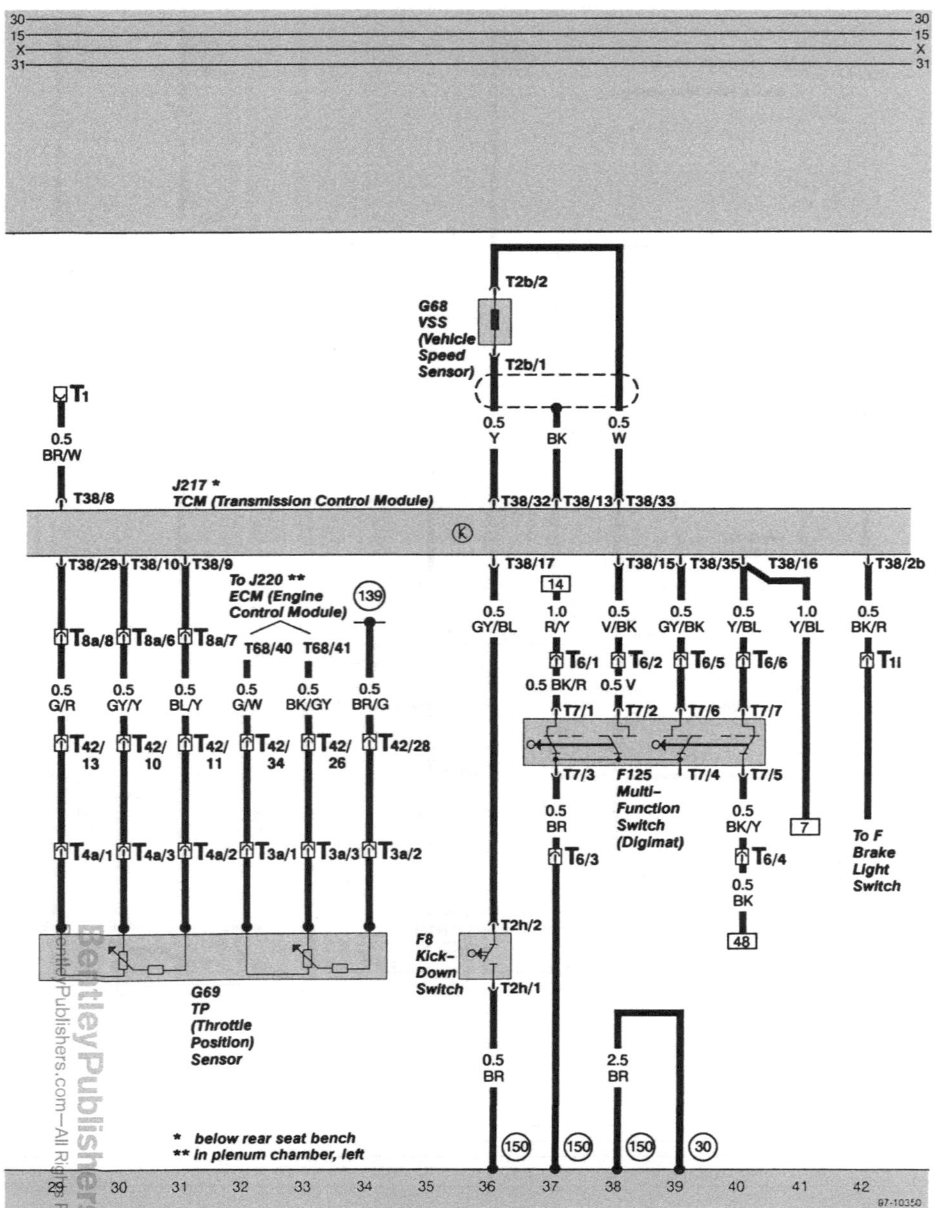

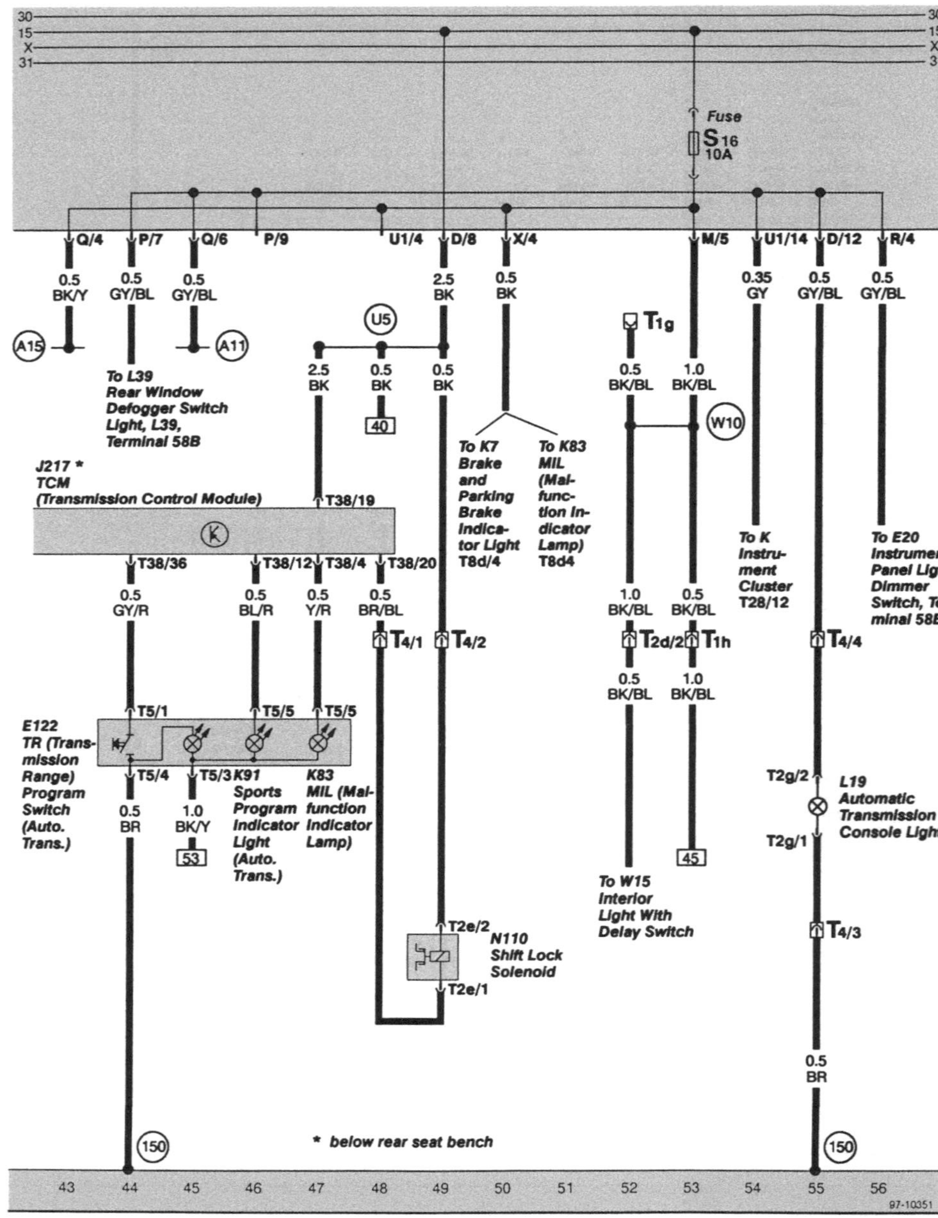

Corrado 1992–1993
Engine code – AAA
Production up to Jan. 1993

Corrado 1992–1993
Engine code – AAA
Production up to Jan. 1993

Description	Current track
Automatic Transmission Console Light, L19	55
Fuse, S14, 10A	5
Fuse, S16, 10A	53
Fuse, S21,.15A	21
Kick Down Switch, F8	36
Left Back-Up Light, M16	3
Multi-Function Transmission Range (TR) Switch, F125	37–40
Park/Neutral Position (PNP) Relay, J226	4–10
Right Back-Up Light, M17	1
Shift Lock Solenoid, N110	49
Solenoid Valve 1, N88	21
Solenoid Valve 2, N89	22
Solenoid Valve 3, N90	23
Solenoid Valve 4, N91	24
Solenoid Valve 5, N92	25
Solenoid Valve 6, N93	19
Solenoid Valve 7, N94	26
Transmission Control Module (TCM), J217	13–48
Transmission Fluid Temperature Sensor, G93	27
Vehicle Speed Sensor (VSS), G68	36

Wire connectors

- T1 – Single Connector, behind fuse/relay panel
- T1b – Single Connector, above relay panel
- T1g – Single Connector, behind fuse/relay panel
- T1h – Single Connector, behind fuse/relay panel
- T1i – Single Connector, behind fuse/relay panel
- T1o – Single Connector, on Left Back-Up Light
- T1p – Single Connector, on Left Back-Up Light
- T2 – Double Connectors, in luggage compartment, left
- T2a – Double Connectors, behind console, On Board Diagnostic (OBD)
- T2b – Double Connectors, on Vehicle Speed Sensor (VSS)
- T2d – Double Connectors, behind fuse/relay panel
- T2e – Double Connectors, on Shift Lock Solenoid
- T2g – Double Connectors, on Automatic Transmission Console Light
- T2h – Double Connectors, on Kick Down Switch
- T2l – Double Connectors, behind fuse/relay panel
- T2x – Double Connectors, below gearshift lever cover
- T2z – Double Connectors, below gearshift lever cover
- T3a – 3-Pin Connectors, near throttle valve housing
- T3d – 3-Pin Connectors, on Left Back-Up Light
- T3e – 3-Pin Connectors, on Right Back-Up Light
- T4 – 4-Pin Connectors, below gearshift lever cover
- T4a – 4-Pin Connectors, near throttle valve housing
- T5 – 5-Pin Connectors, on Transmission Range (TR) Program Switch (to December 1992)
- T6 – 6-Pin Connectors, in engine compartment, left
- T7 – 7-Pin Connectors, on Multi-Function Transmis sion Range (TR) Switch
- T8a – 8-Pin Connectors, above relay panel
- T8d – 8-Pin Connectors, on Brake and Parking Brake Warning Light
- T9 – 9-Pin Connectors, on Park/Neutral Position (PNP) Relay
- T12 – 12-Pin Connectors, for Solenoid Valve
- T28 – 28-Pin Connectors, on Instrument Cluster
- T38 – 38-Pin Connectors, on Transmission Control Module (TCM)
- T42 – 42-Pin Connectors, on cylinder head, left
- T68 – 68-Pin Connectors, on Motronic Engine Control Module (ECM)
- TV2 – Terminal 30 (B+) Wire Connector, above relay panel
- TV5 – Terminal 15a Wire Connector, above relay panel

Welded wiring harness points

- (A11) – plus connection (58b), in instrument panel wiring harness
- (A15) – plus connection (15), in instrument panel wiring harness

Welded wiring harness points

- (U5) – wire connection -2- (15), in automatic trans-mission wiring harness
- (W9) – plus connection (30), in rear wiring harness
- (W10) – plus connection (15), in rear wiring harness

Ground Connections

- (30) – Ground connection, -1-, beside fuse/relay panel
- (53) – Ground connection, in rear lid, right
- (139) – Ground connection (sensor Ground), in Motronic Multiport Fuel Injection (MFI) wiring harness
- (150) – Ground connection, in automatic transmission wiring harness

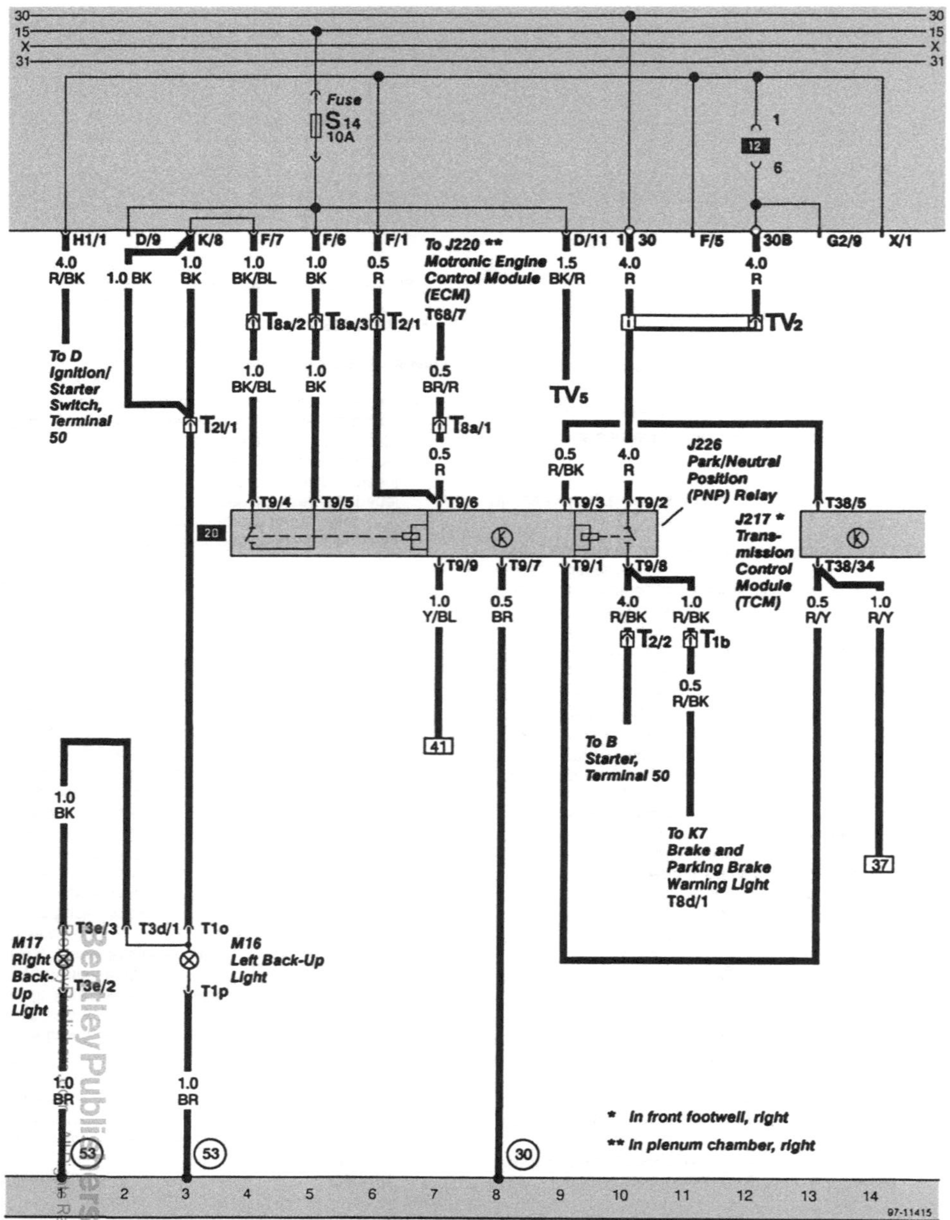

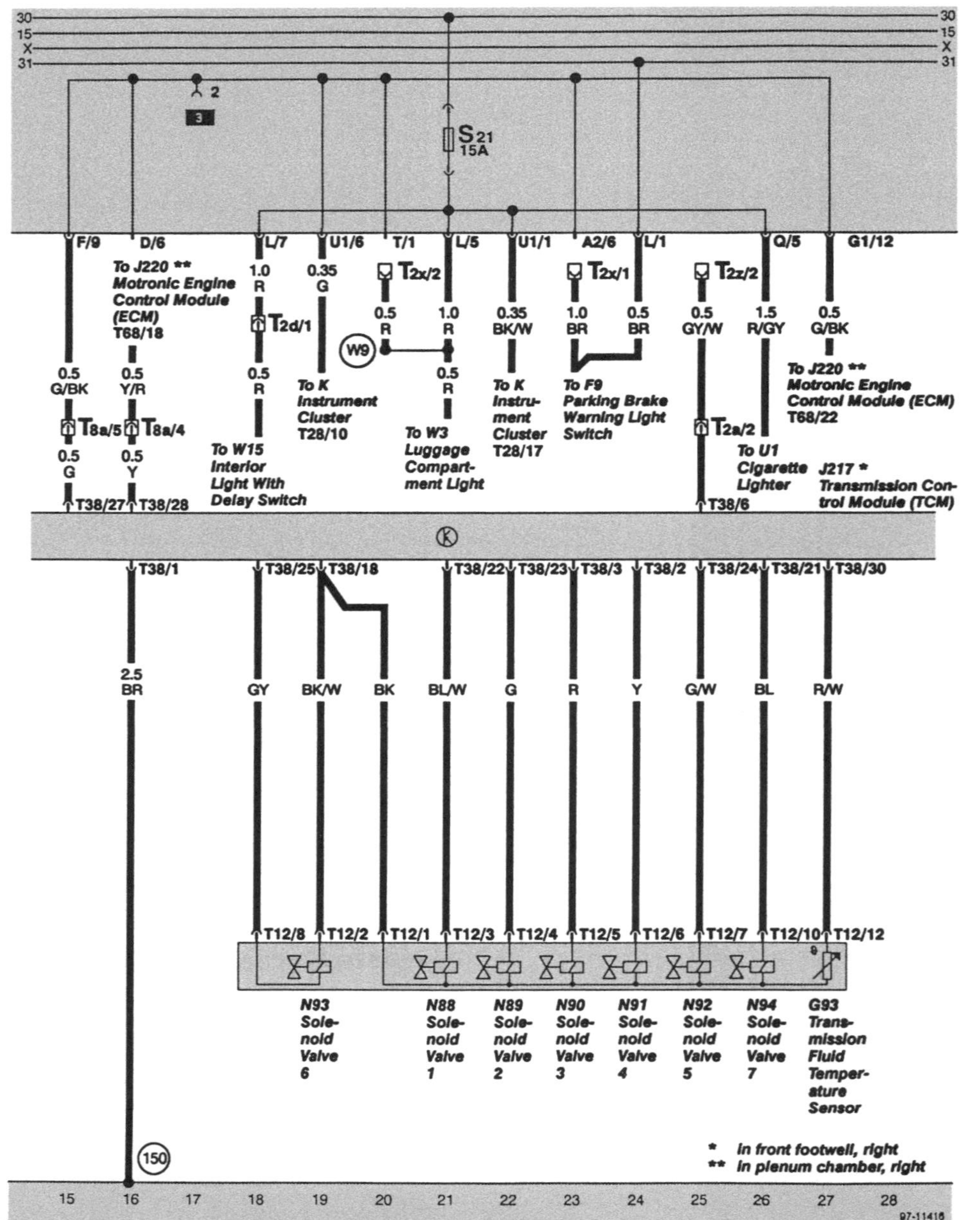

Corrado 1993–1994
Engine code – AAA
Production from Jan. 1993 to Mar. 1994

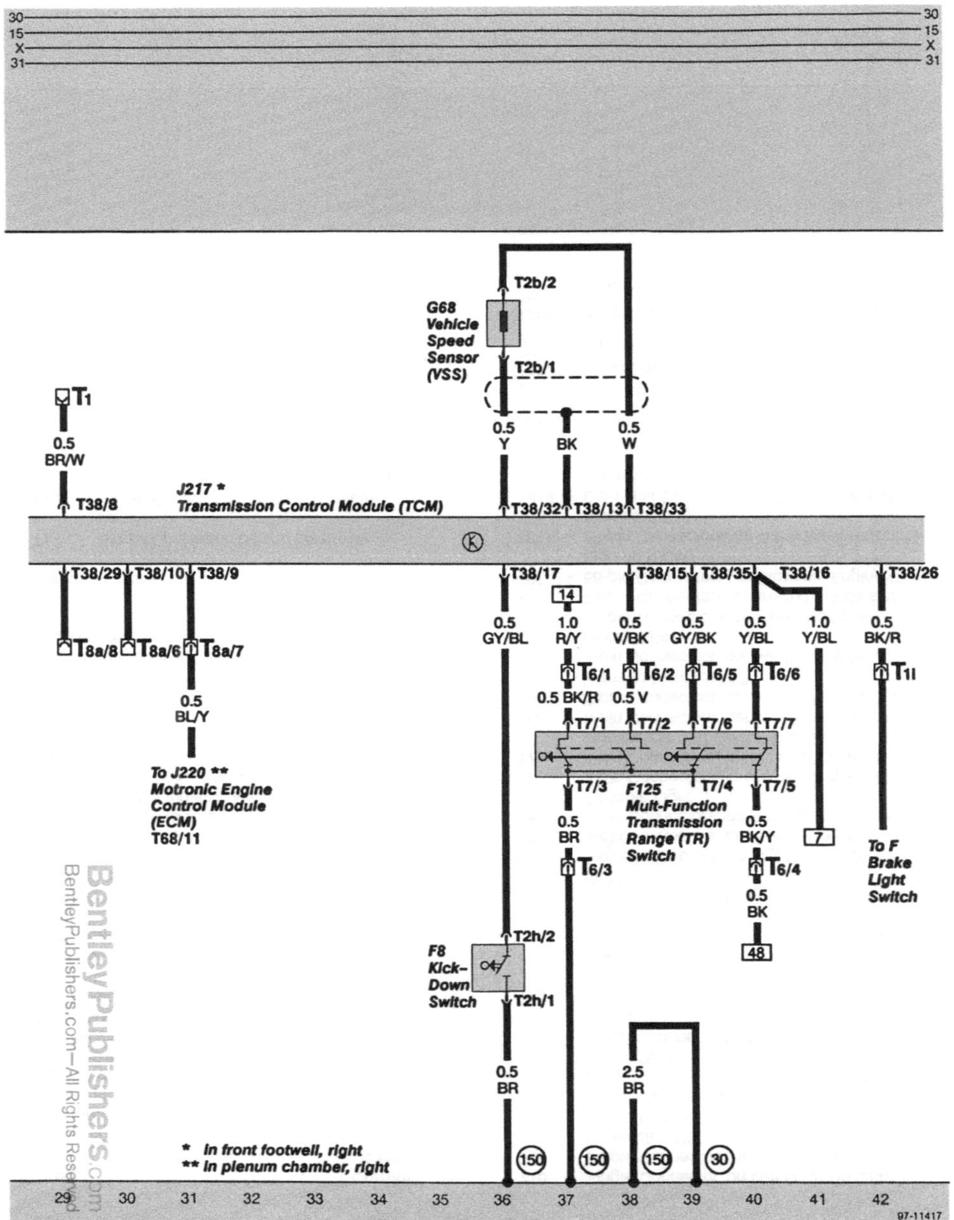

Corrado 1993–1994
Engine code – AAA
Production from Jan. 1993 to Mar. 1994

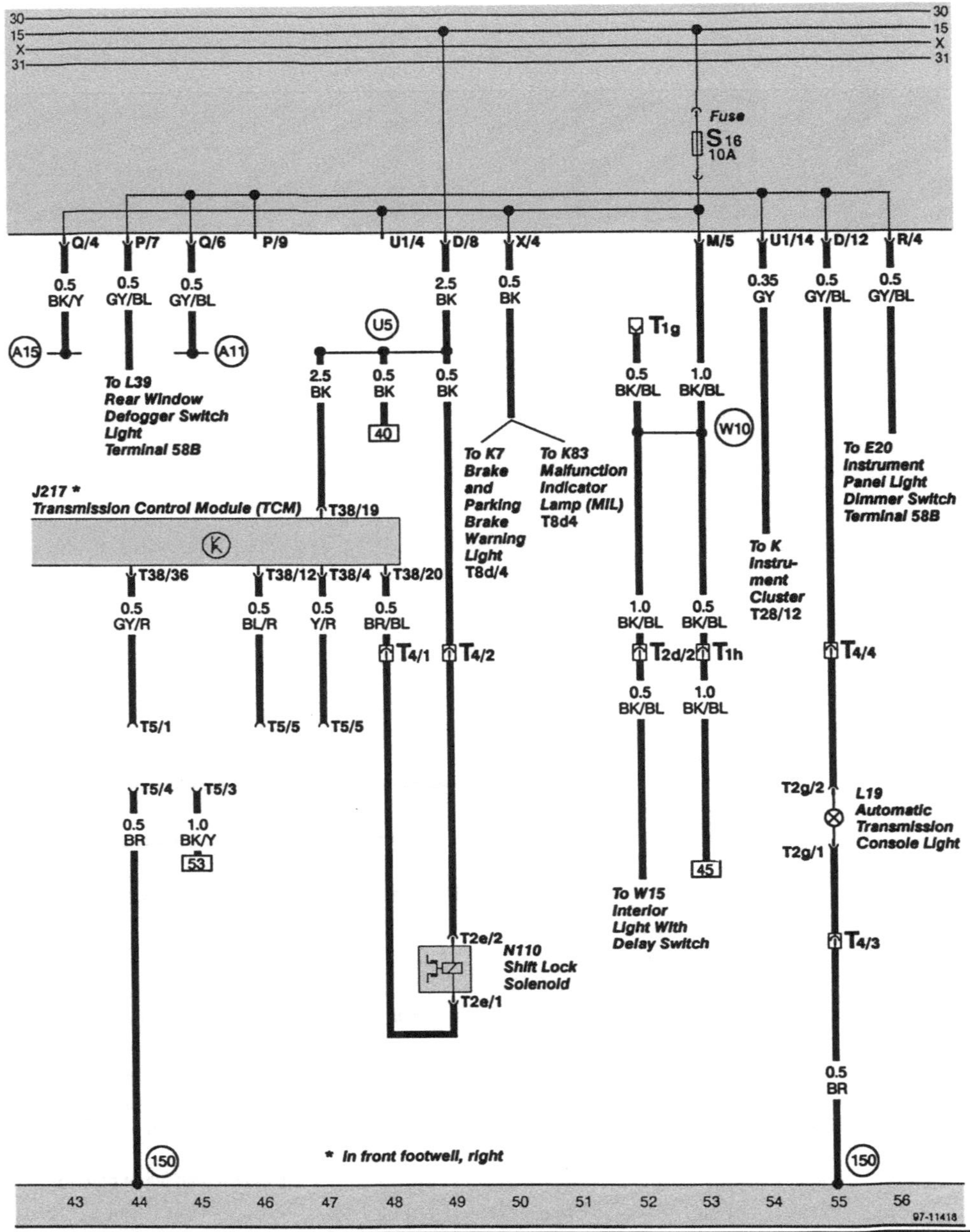

Corrado 1993–1994
Engine code – AAA
Production from Jan. 1993 to Mar. 1994

Description	Current track
Automatic Transmission Console Light, L19	56
Fuse, S14, 10A	6
Kick Down Switch (00660), F8	43
Left Back-Up Light, M16	1
Multi-Function Transmission Range (TR) (00293), F125	44-47
Park/Neutral Position (PNP) Relay, J226	4-10
Right Back-Up Light, M17	3
Shift Lock Solenoid (01236), N110	50
Solenoid Valve1 (00258), N88	20
Solenoid Valve 2 (00260), N89	21
Solenoid Valve 3 (00262), N90	22
Solenoid Valve 4 (00264), N91	23
Solenoid Valve 5 (00266), N92	24
Solenoid Valve 6 (00268), N93	18
Solenoid Valve 7 (00270), N94	25
Switch (00293), F125	44-47
Transmission Control Module (TCM), J217	13-49
Transmission Fluid Temperature Sensor (00300), G93	26
Vehicle Speed Sensor (VSS) (00281), G68	36

Wire connectors

- T1 - Single Connector, behind fuse/relay panel
- T1b - Single Connector, behind fuse/relay panel, on Speedometer Vehicle Speed Sensor (VSS)
- T1c - Single Connector, behind fuse/relay panel, on Transmission Range (TR) Selector Lever Display
- T1e - Single Connector, behind fuse/relay panel, on Air Conditioning
- T1h - Single Connector, behind fuse/relay panel
- T1o - Single Connector, on Left Back-Up Light
- T1p - Single Connector, on Left Back-Up Light
- T1v - Single Connector, behind console, on Data Link Connector (DLC) wire Connector
- T1y - Single Connector, behind fuse/relay panel
- T2 - Double Connectors, behind fuse/relay panel
- T2b - Double Connectors, on Vehicle Speed Sensor (VSS) (00281)
- T2e - Double Connectors, on Shift Lock Solenoid (01236)
- T2g - Double Connectors, on Automatic Transmission Console Light
- T2i - Double Connectors, in luggage compartment, left
- T2s - Double Connectors, on Kick Down Switch (00660)
- T3d - 3-pin Connectors, on Right Back-Up Light
- T3e - 3-pin Connectors, on Right Back-Up Light
- T4a - 4-pin Connectors, on Automatic Transmission Console Light
- T6 - 6-pin Connectors, in engine compartment, left
- T7 - 7-pin Connectors, on Multi-Function Transmission Range (TR) Switch (00293)
- T8a - 8-pin Connectors, behind fuse/relay panel
- T8d - 8-pin Connectors, Brake and Parking Brake Warning Light (01047)
- T9 - 9-pin Connectors, on Park/Neutral Position (PNP) Relay
- T12 - 12-pin Connectors, on Solenoid Valve
- T28 - 28-pin Connectors, on Instrument Cluster
- T68 - 68-pin Connectors, on Motronic Engine Control Module (ECM)
- T68a - 68-pin Connectors, on Transmission Control Module (TCM)
- TV14 - Data Link Connector (DLC) wire Connector, behind console

Welded wiring harness point

- (A11) - plus connection (58b) in instrument panel wiring harness
- (U3) - wire connection -1-, in automatic transmission wiring harness

Welded wiring harness point

- (U4) - wire connection (30), in automatic transmission wiring harness
- (U5) - wire connection -2- (15), in automatic transmission wiring harness
- (U6) - wire connection -2-, in automatic transmission wiring harness

Ground connections

- (30) - Ground connection, -1-, beside fuse/relay panel
- (53) - Ground connection, in rear lid, right
- (150) - Ground connection, in automatic transmission wiring harness

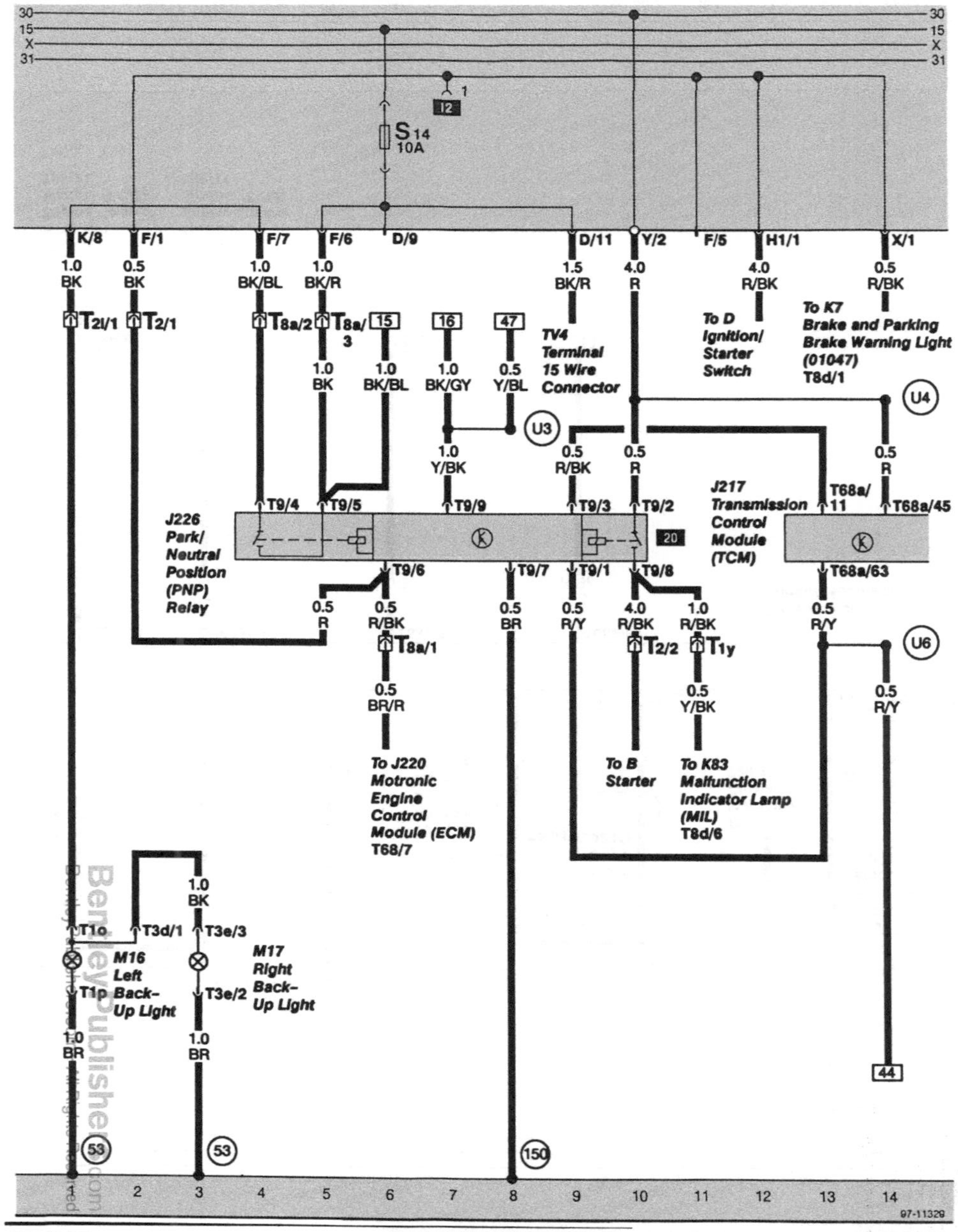

Corrado starting 1994
Engine code – AAA
Production from Mar. 1994

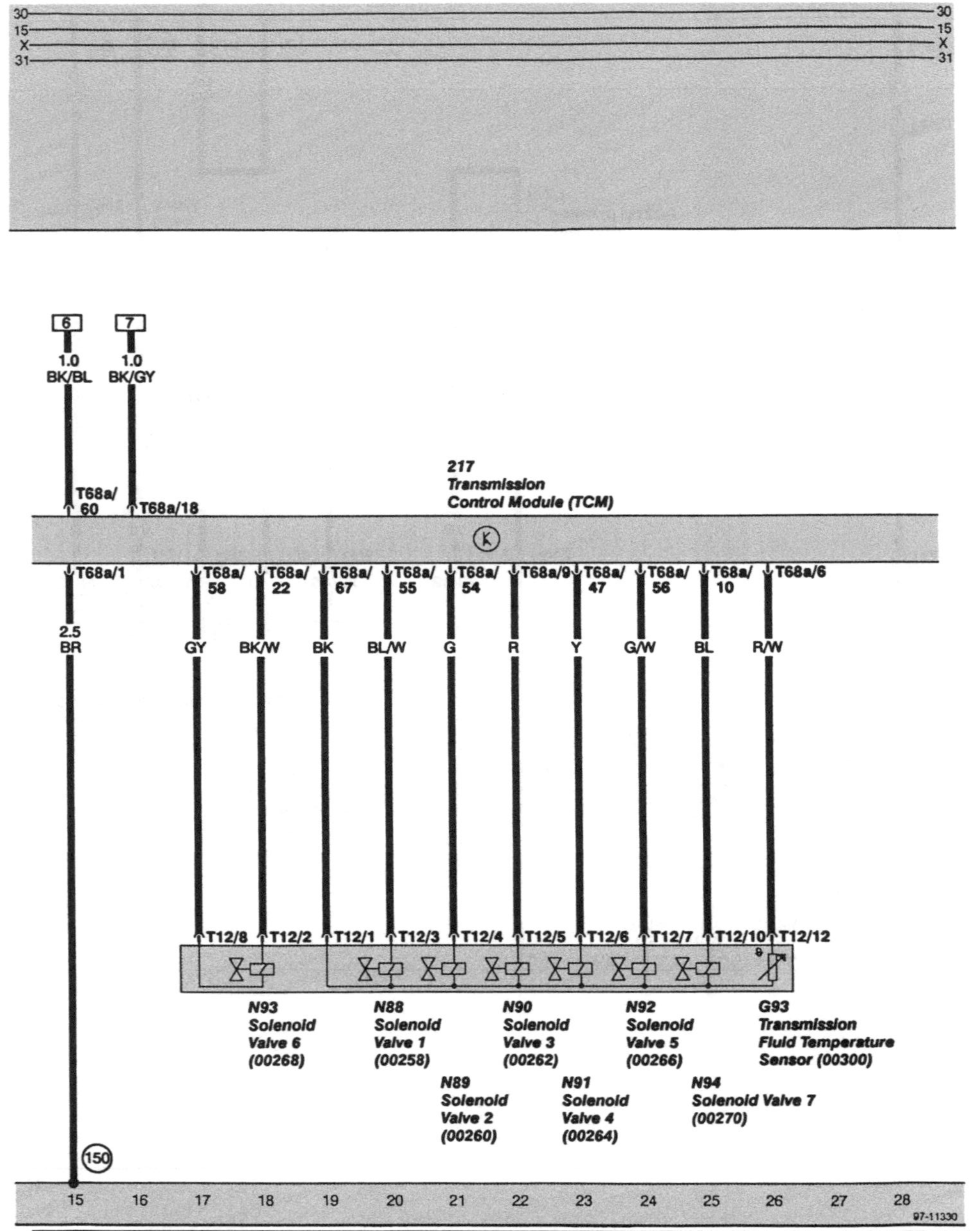

Corrado starting 1994
Engine code – AAA
Production from Mar. 1994

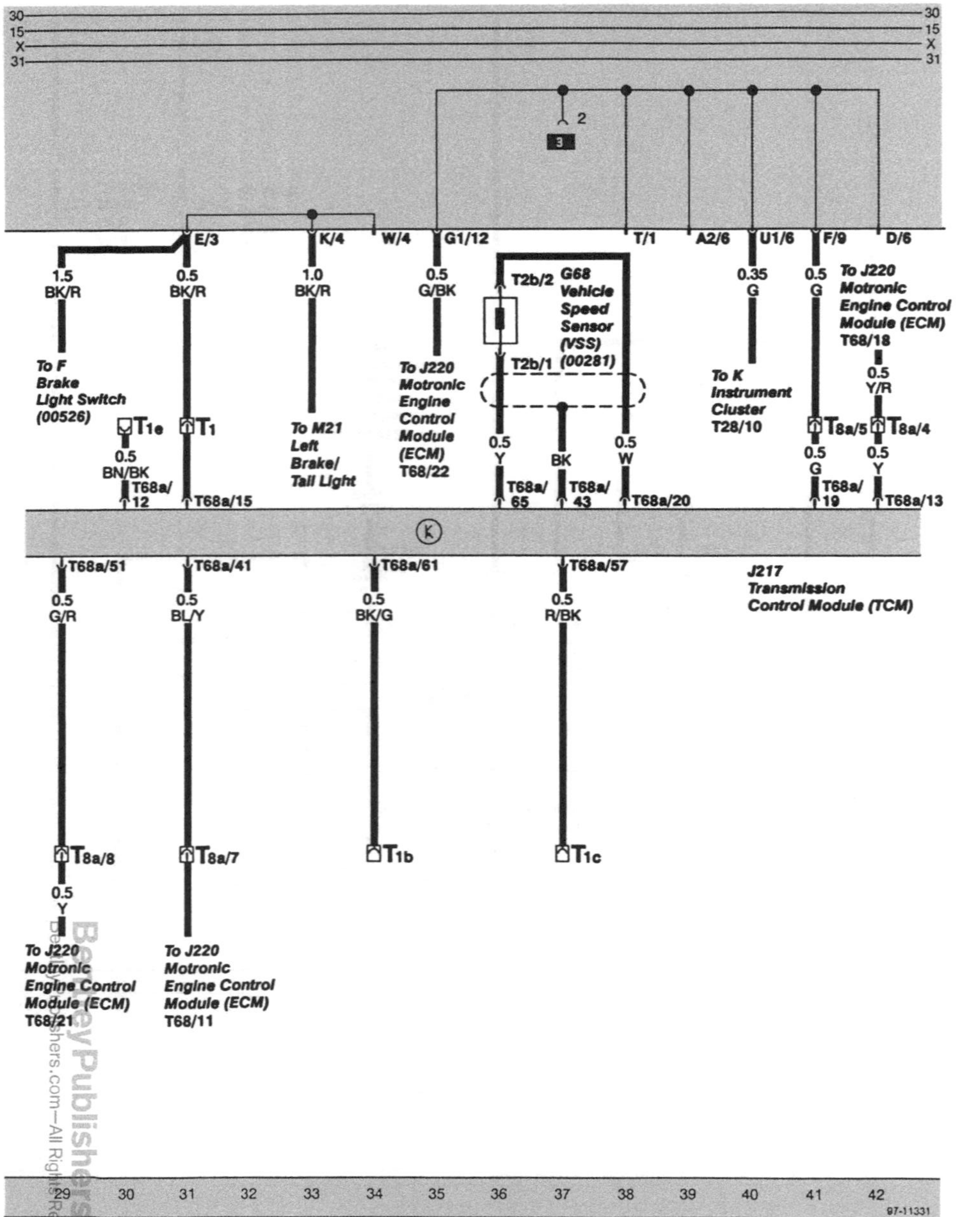

Corrado starting 1994
Engine code – AAA
Production from Mar. 1994

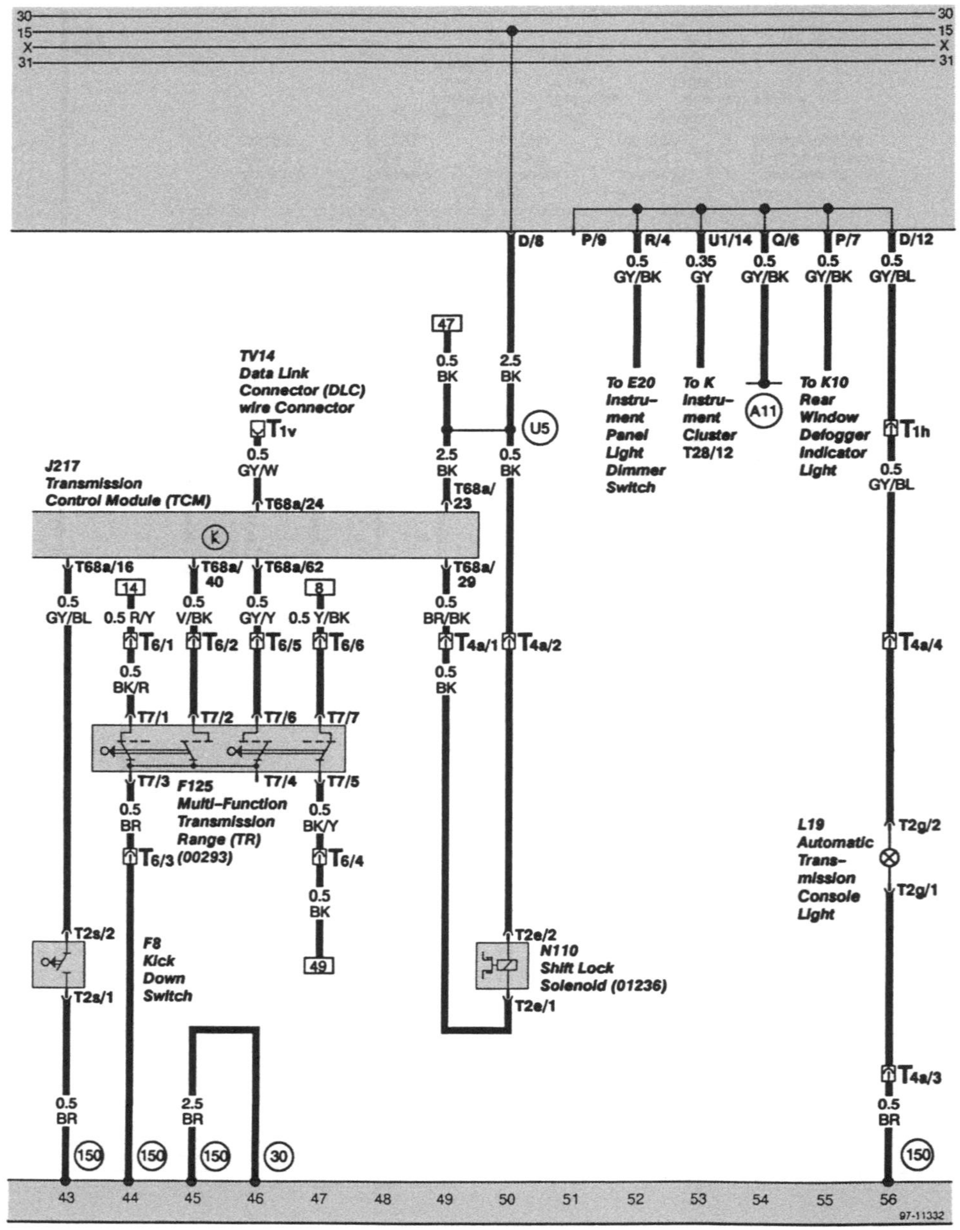

Corrado starting 1994
Engine code – AAA
Production from Mar. 1994

Description	Current track
Cruise control switch, E45	1-3
Cruise control, control unit, J213	4-13
Fuse, S14	3
Vacuum control motor, V18	8-10
Vacuum vent valve, brake, F47	13
Vacuum vent valve, clutch, F36	13

Wire connectors

T1g — single, behind fuse/relay panel
T1u — single, behind fuse/relay panel
T1v — single, behind fuse/relay panel
T1w — single, behind fuse/relay panel
T1x — single, behind fuse/relay panel
T1y — single, behind fuse/relay panel
T2 — double, on cruise control, control unit
T3 — three-point, behind fuse/relay panel
T4a — four-point, behind steering column switch cover
T4b — four-point, near steering column
T4h — four-point, under rear seat, center
T8 — eight-point, on cruise control, control unit
T28 — twenty eight-point, on instrument cluster

Ground connections

(30) — near fuse/relay panel

Welded wiring harness points

(A15) — plus connection (15), in instrument cluster wiring harness

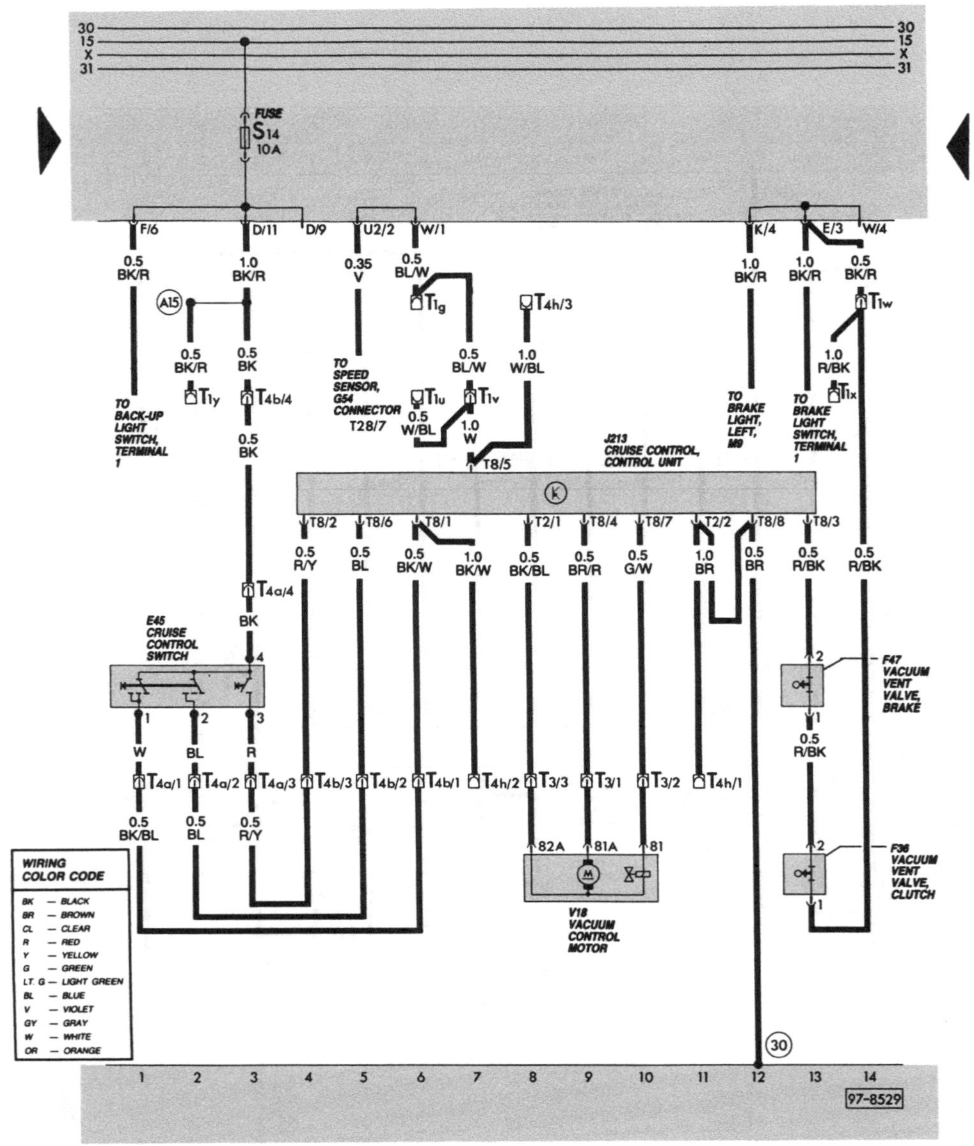

Description	Current track
Cruise control switch, E45	1-3
Cruise control vacuum pump, V18	8-10
Cruise control, control unit, J213	4-13
Fuse, S14	3
Fuse, S51	2
Vacuum vent valve, brake, F47	13
Vacuum vent valve, clutch, F36	13

Wire connectors

T1 — single, behind fuse/relay panel
T1a — single, behind fuse/relay panel
T1h — single, behind fuse/relay panel
T1i — single, behind fuse/relay panel
T2 — double, on control unit for cruise control
T3 — three-point, above fuse/relay panel
T4a — four-point, behind steering column switch cover
T4b — four-point, near steering column
T4h — four-point, rear seat, center
T8 — eight-point, on control unit for cruise control
T8a — eight-point, on radio
T28 — twenty eight-point, on instrument cluster
T38 — thirty eight-point, on automatic transmission control unit
T43 — wire junction for speed signal, behind fuse/relay panel

Ground connections

(30) — beside fuse/relay panel
(185) — in cruise control wiring harness

Welded wiring harness points

(A15) — plus connection (15), in instrument panel wiring harness

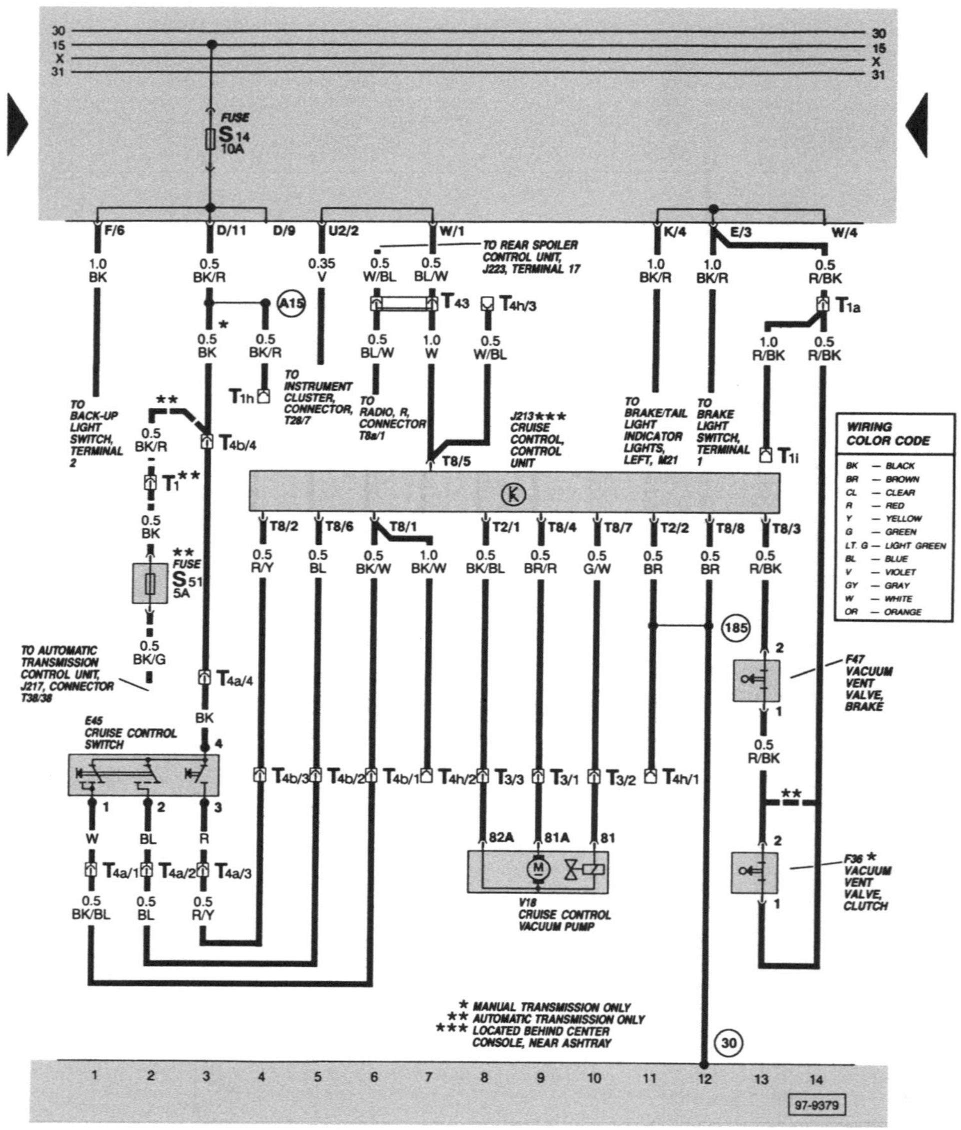

Description	Current track
Automatic cruise control vacuum pump, V18	8–10
Automatic transmission control unit, J217	2
Back-up light switch, F4	1
Brake pedal switch for cruise control/diesel injection system, F47	13
Brake-light switch, F	12
Brake/tail indicator lights, left, M21	11
Clutch pedal switch, F36	13
Cruise control switch, E45	1–3
Cruise control unit, J213	4–13
Fuse, S14	3
Fuse, S51	2
Instrument/panel insert combined instrument, K	5
Radio, R	6
Rear spoiler control module, J223	6

Wire connectors

- T1 – single, behind fuse/relay panel
- T1a – single, behind fuse/relay panel
- T1h – single, behind fuse/relay panel
- T1i – single, behind fuse/relay panel
- T2 – double, on control unit for cruise control
- T3 – three point, above fuse/relay panel
- T4a – four point, behind steering column switch cover
- T4b – four point, near steering column
- T4h – four point, behind console
- T8 – eight point, on radio
- T8a – eight point, on control unit for cruise control
- T28 – twenty-eight point, on instrument cluster
- T35 – wire connector, terminal 15, above fuse/relay panel
- T38 – thirty-eight point, on automatic transmission control unit
- T43 – wire connector (junction) for speed signal, behind fuse/relay panel

Welded wiring harness point

- (A15) – plus connection (15), in instrument panel wiring harness

Ground connection

- (30) – beside fuse/relay panel
- (185) – in cruise control wiring harness

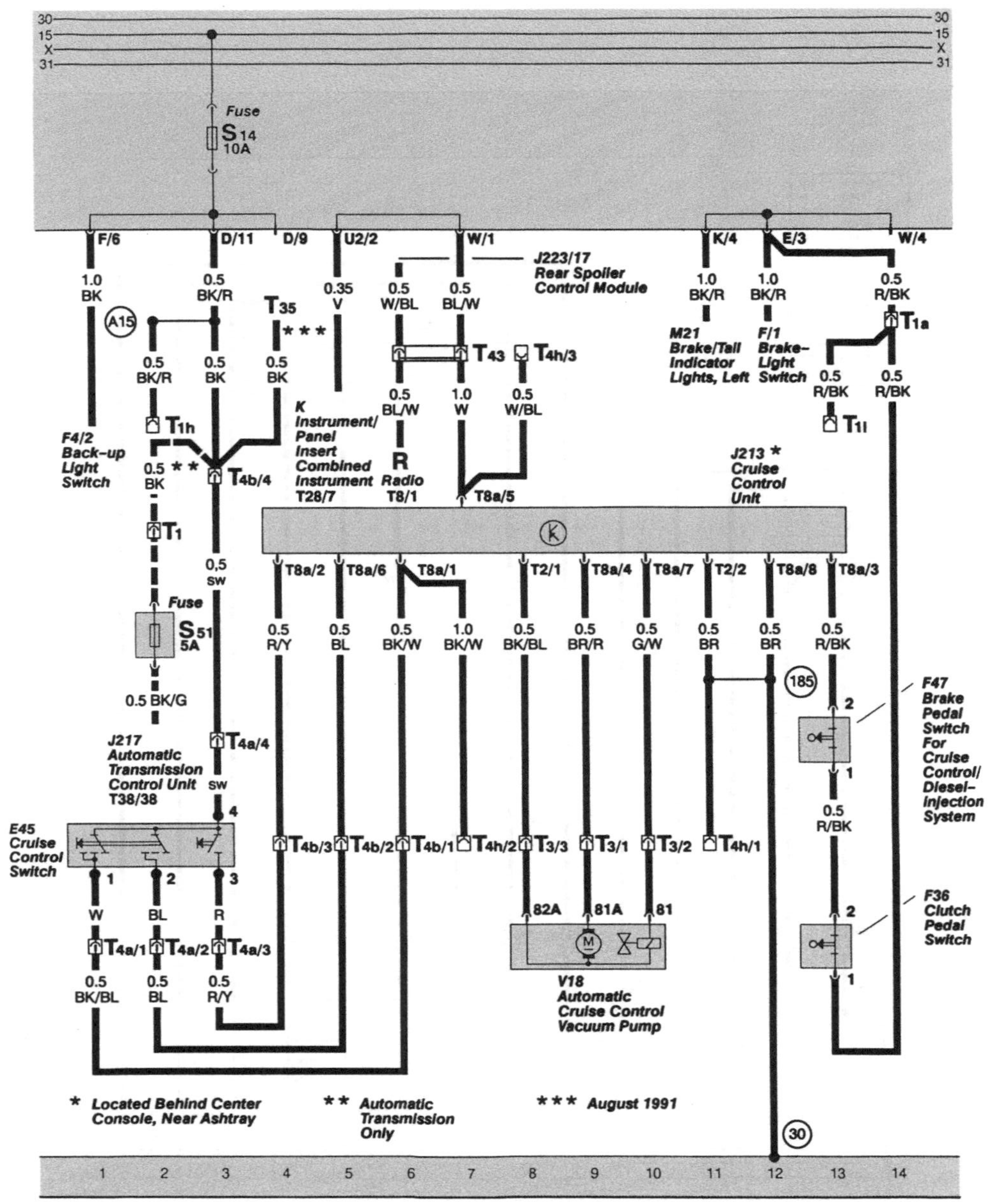

Description	Current track
Emergency flasher relay, J2	16, 17
Emergency flasher switch, E3	15-20
Emergency flasher warning light, K6	20
Fuse, S1	26
Fuse, S2	28
Fuse, S3	4
Fuse, S7	35
Fuse, S8	31
Fuse, S11	25
Fuse, S12	27
Fuse, S17	15
Headlight dimmer/flasher switch, E4	13-14
Headlight switch light, L9	5
Headlight, left, L1	26
Headlight, right, L2	28
High-beam headlight, left, L13	25
High-beam headlight, right, L14	27
Instrument panel light dimmer switch, E20	4
Light switch, E1	4-12
Load reduction relay, J59	21
Park light indicator light, K4	12
Park light switch, E19	23, 24
Parking light, left, M1	31
Parking light, right, M3	35
Series resistance wire, N74	10
Side marker lights, front, M11	30, 36
Tail light, left, M4	32
Tail light, right, M2	34
Turn signal switch, E2	21

Wire connectors

T1 — single, behind instrument panel
T1a — single, behind instrument panel
T1b — single, behind instrument panel
T4a — four-point, on left taillight
T4b — four-point, on right taillight
T4c — four-point, behind steering column switch cover
T5b — five-point, behind steering column switch cover
T7a — seven-point, behind steering column switch cover
T28 — twenty eight-point, on instrument cluster

Ground connections

(50) — in luggage compartment, left
(119) — in headlight wiring harness
(120) — in headlight wiring harness

Welded wiring harness points

(C1) — plus connection (58L), in left front wiring harness
(C2) — plus connection (58R), in left front wiring harness
(C12) — plus connection, in headlight wiring harness

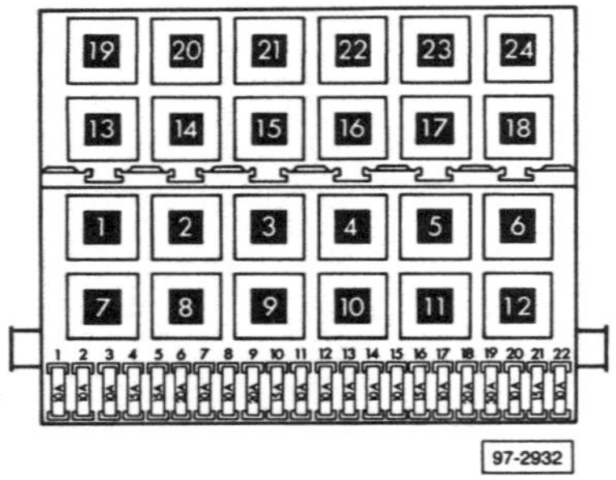

4 — Load reduction relay (18)*, J 59
6 — Hazard flasher relay (21)*, J 2
10 — Fog light relay

*Numbers in parenthesis indicate production control number stamped on relay housing.

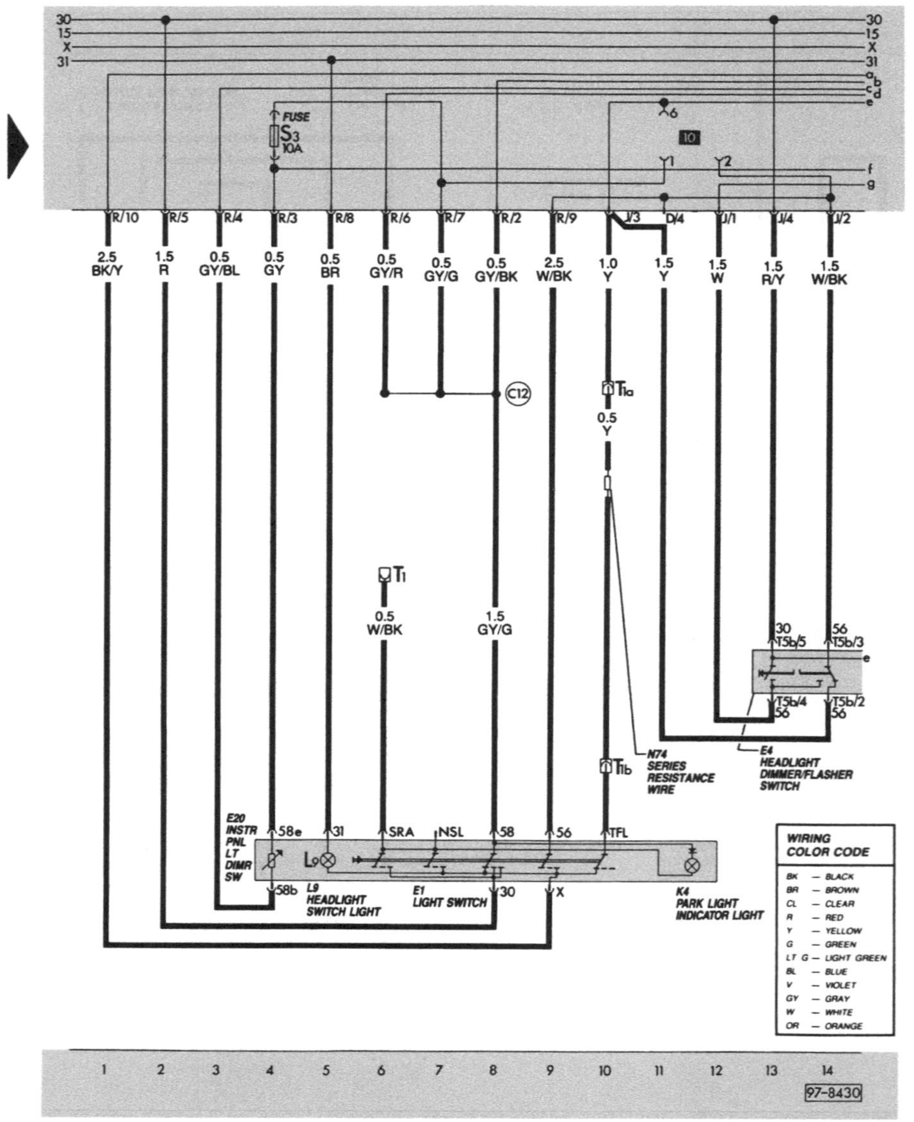

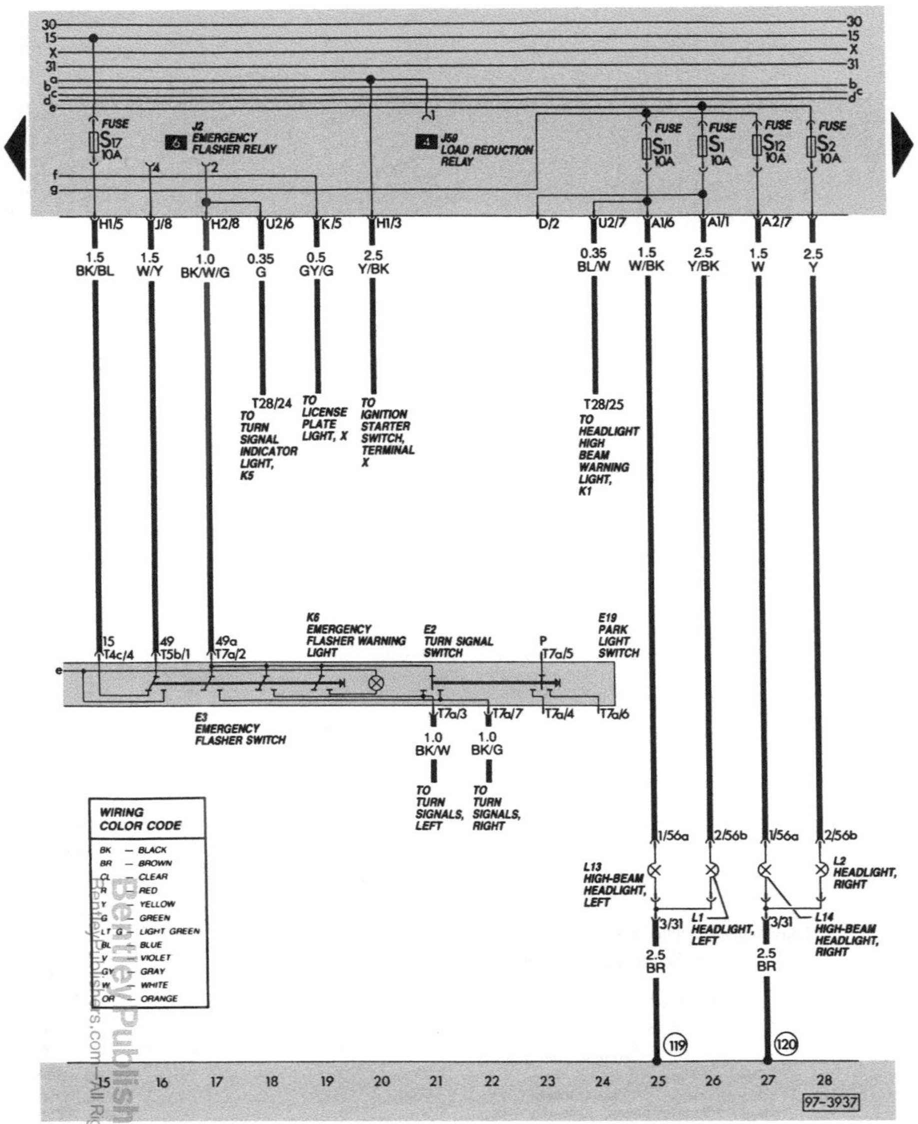

Corrado 1990
Canada only

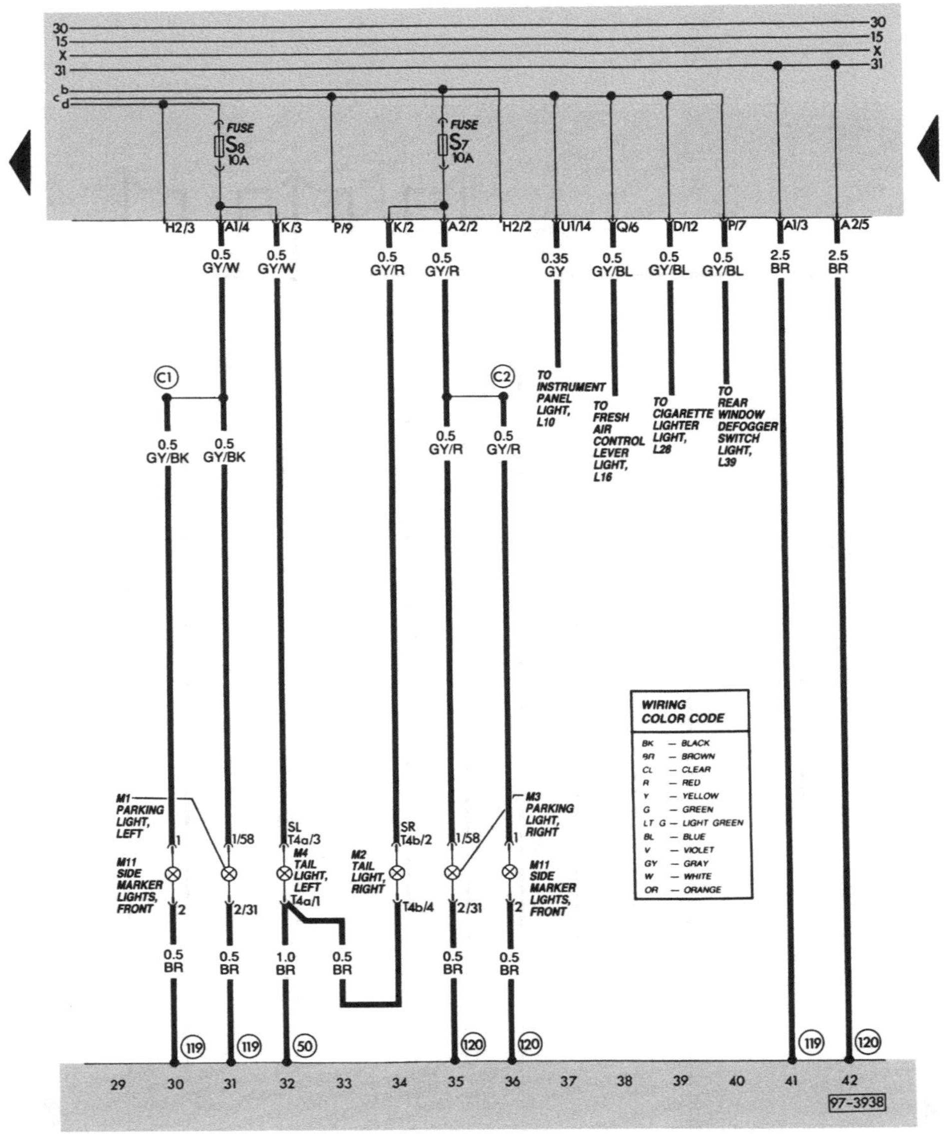

Corrado 1990
Canada only

Description	Current track
Fuse, S11	1
Fuse, S1	2
Fuse, S12	6
Fuse, S2	7
Fuse, S3	23
Headlight dimmer/flasher switch, E4	11-13
Headlight switch light, L9	22
Headlight, left, L1	2
Headlight, right, L2	7
High-beam headlight, left, L13	1
High-beam headlight, right, L14	6
Instrument panel light dimmer switch, E20	23
Light switch, E1	15-24
Park light indicator light, K4	19
Series resistance wire, N74	8

Wire connectors

T1 — single, behind fuse/relay panel
T1a — single, behind fuse/relay panel
T1b — single, behind fuse/relay panel
T5b — five-point, behind steering column switch cover

Ground connections

(119) — in headlight wiring harness
(120) — in headlight wiring harness

Welded wiring harness points

(C1) — plus connection (58L), in left front wiring harness

Corrado starting 1991
Canada only

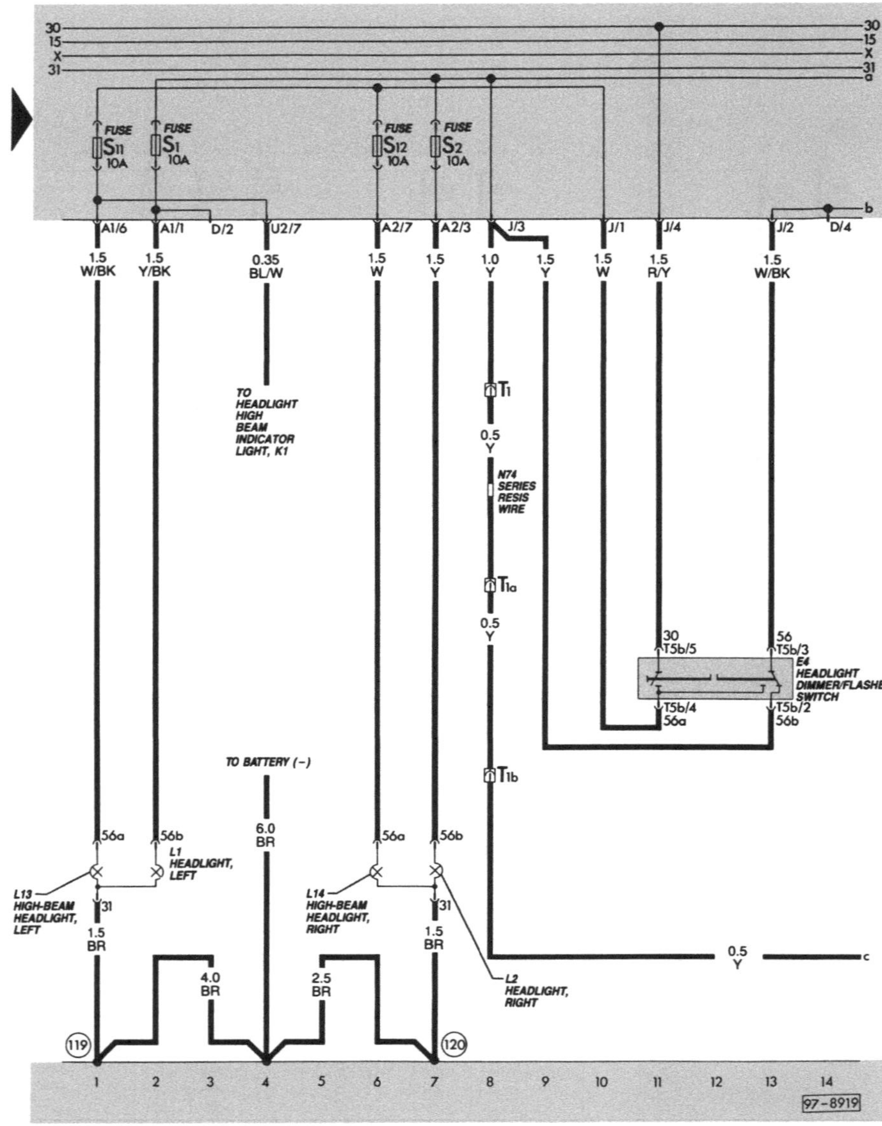

Corrado starting 1991
Canada only

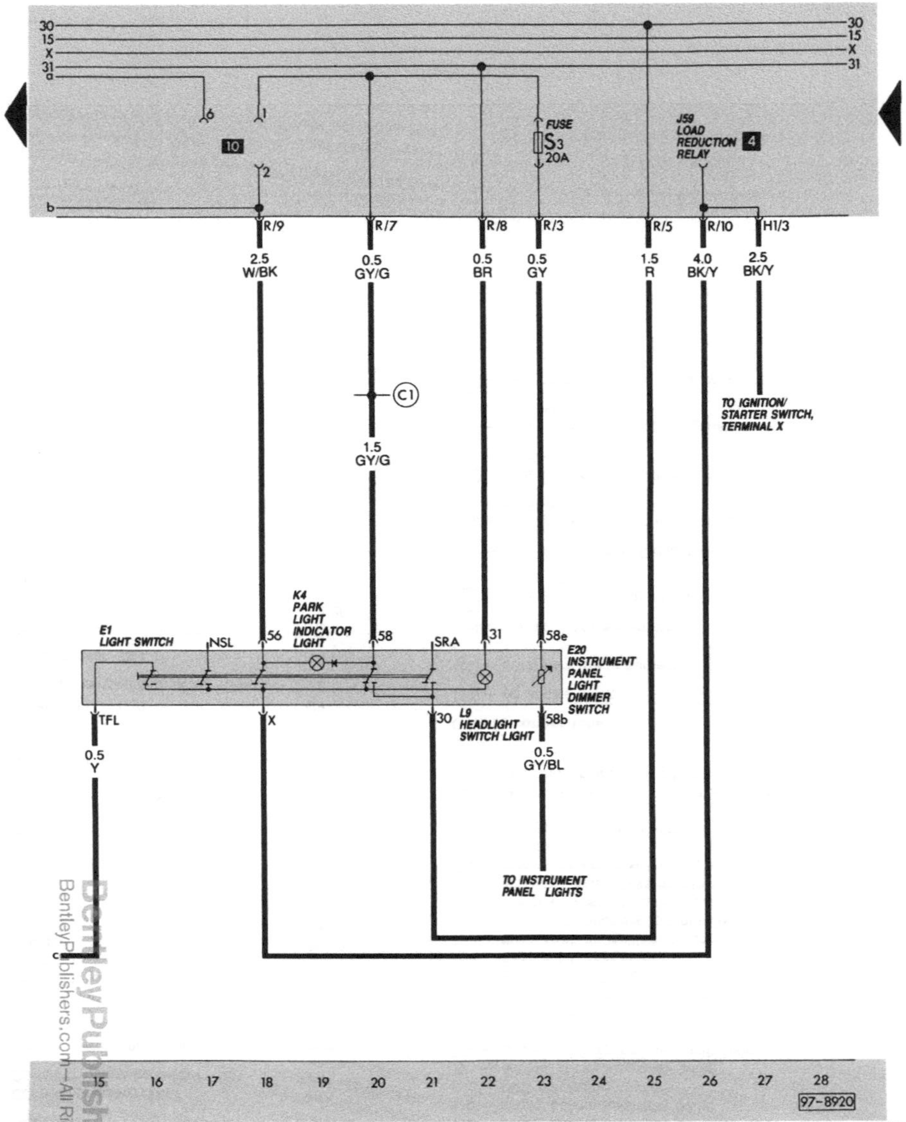

THIS PAGE INTENTIONALLY LEFT BLANK

Corrado starting 1991
Canada only

Description	Current track
Brake and parking brake indicator light, K7	44
CO potentiometer, G74	13
Coolant temperature sensor, G62	14
Digifant control unit, J169	5-34
Digifant control unit relay, J176	31-33
Fuel pump afterrunning control unit, J206	53-55
Fuel pump afterrunning pressure switch, F130	54
Fuel pump relay, J17	25-26
Fuse, S18	28
Fuse, S16	43
Fuse, S15	53
Hall generator, G40	5-7
Idle stabilizer valve, N71	34-35
Ignition coil, N152	1
Ignition coil power stage, N157	3-4
Ignition distributor, O	1-3
Injector, cyl. 1, N30	32
Injector, cyl. 2, N31	31
Injector, cyl. 3, N32	30
Injector, cyl. 4, N33	29
Intake air temperature sensor, G42	12
Knock sensor, G61	18-20
OBD fault warning light, K83	45
Oxygen sensor, G39	25-26
Spark plug connectors, P	1-3
Spark plugs, Q	1-3
Throttle valve potentiometer, G69	16-17

Wire connectors

T1b — single, behind fuse/relay panel
T1c — single, behind fuse/relay panel
T1e1— single, behind fuse/relay panel
T1g1— single, behind fuse/relay panel (A/C connector)
T2 — double, above fuse/relay panel
T2a — double, near transmission shift lever
T2c — double, near transmission shift lever
T2d — double, behind fuse/relay panel
T2z — double, white, diagnostic connector, below transmission shift lever cover
T3a — three-point, near starter
T3m— three-point, on ignition coil power stage
T3r — three-point, near exhaust manifold
T5a — five-point, above fuse/relay panel
T5h — five-point, near intake manifold
T8d — eight-point, on brake/parking brake warning light assembly
T9 — nine-point, on starter interlock/back-up light relay
T28 — twenty eight-point, on instrument cluster
T38 — thirty eight-point, on Digifant control unit

Ground connections

(18) — on engine block
(94) — in Digifant wiring harness

Welded wiring harness points

(A15) — plus connection (15), in instrument panel wiring harness
(G1) — plus connection, in Digifant wiring harness
(G3) — plus connection, in injector valves wiring harness
(H20) — wire connection, in self diagnosis wiring harness
(H21) — wire connection, in self diagnosis wiring harness
(W10) — plus connection (15), in rear wiring harness

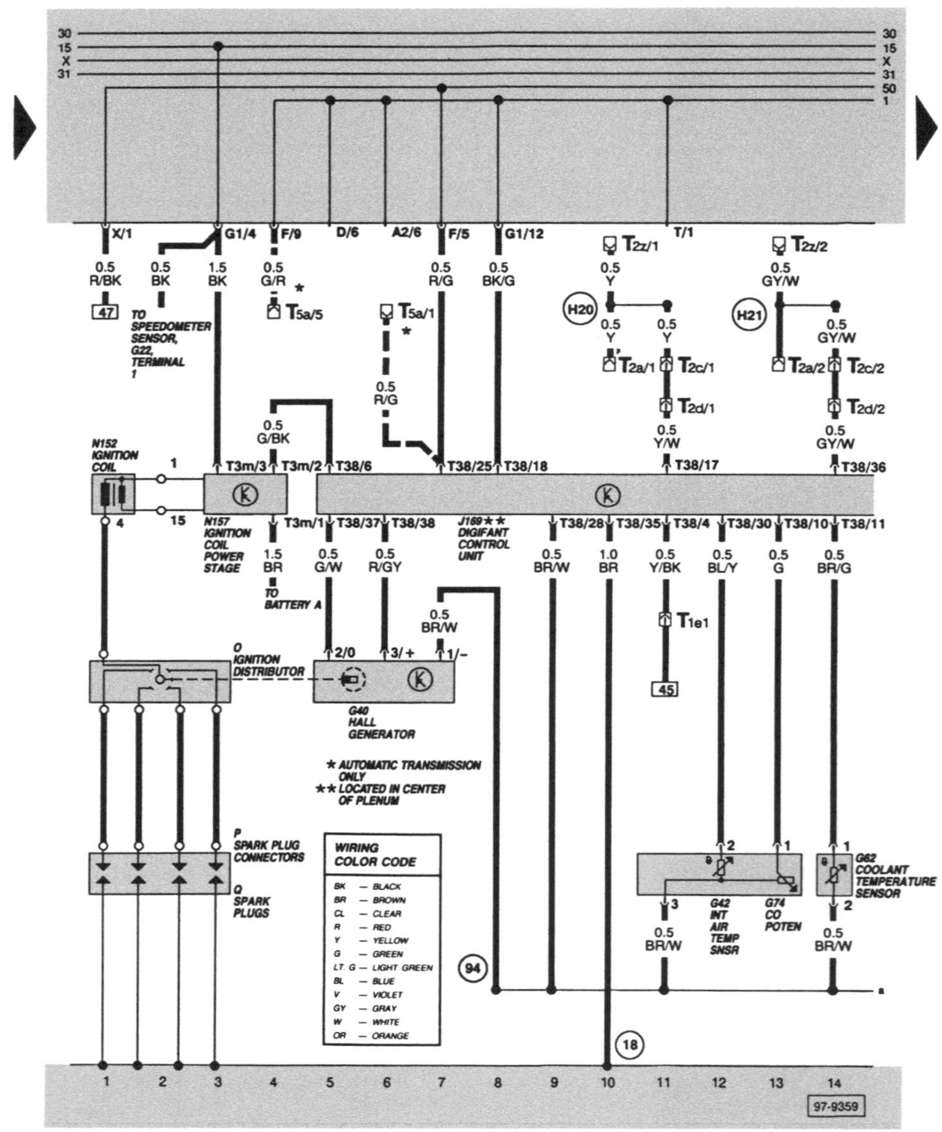

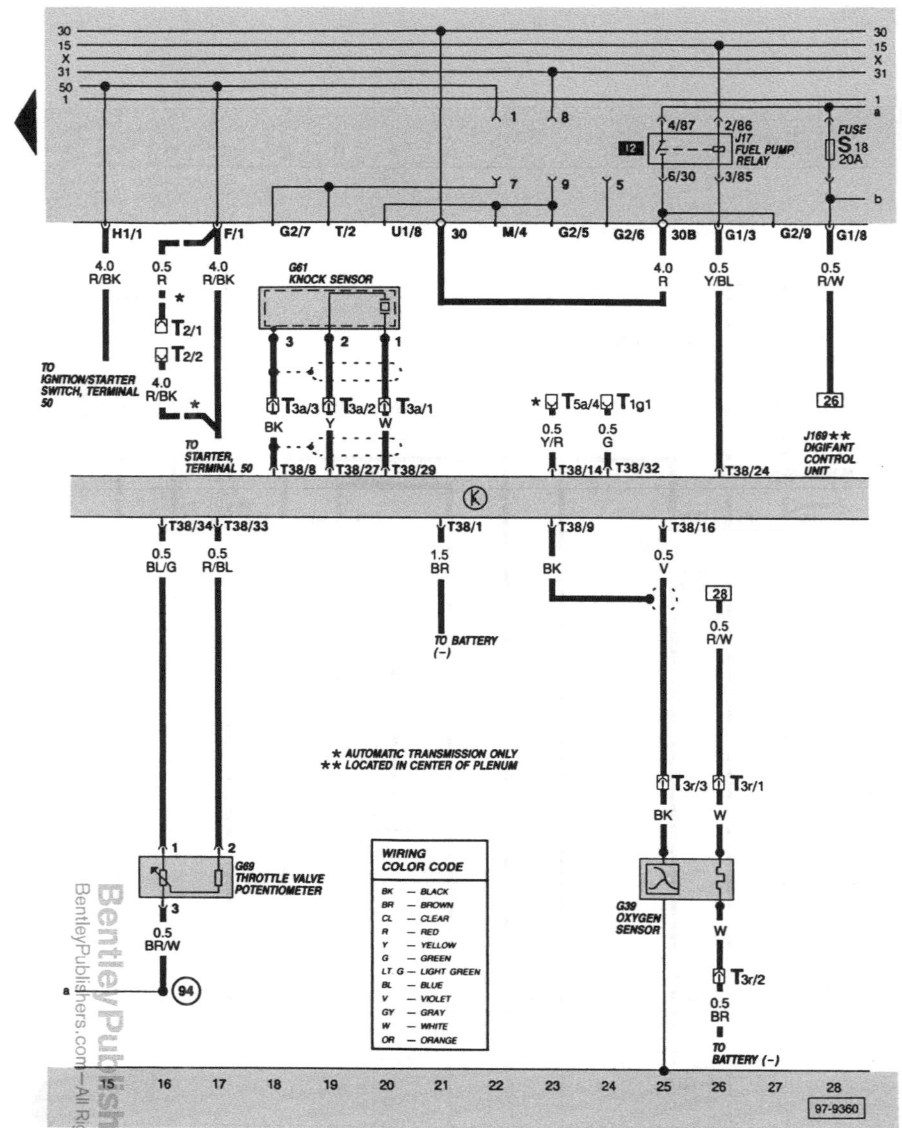

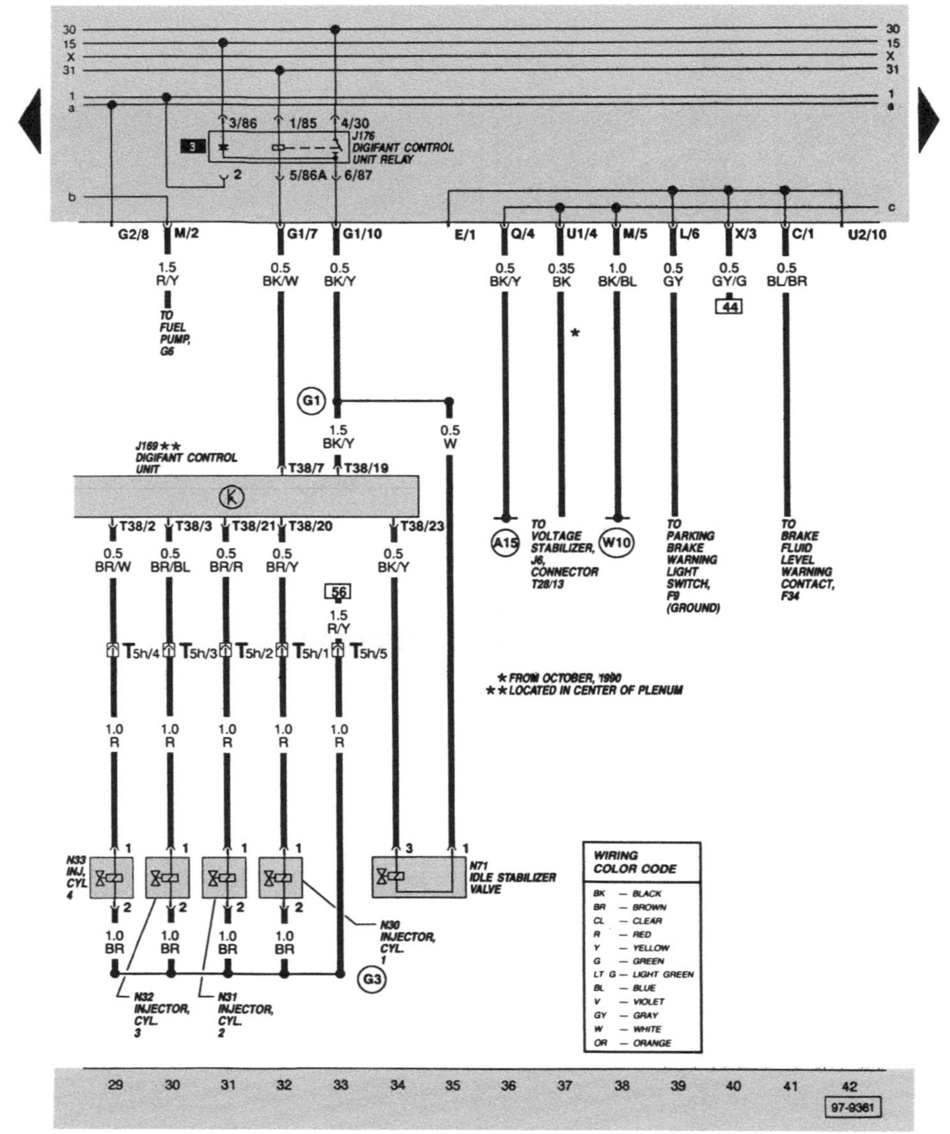

Corrado starting 1991
Engine code – PG
Digifant I, California only

Corrado starting 1991
Engine code – PG
Digifant I, California only

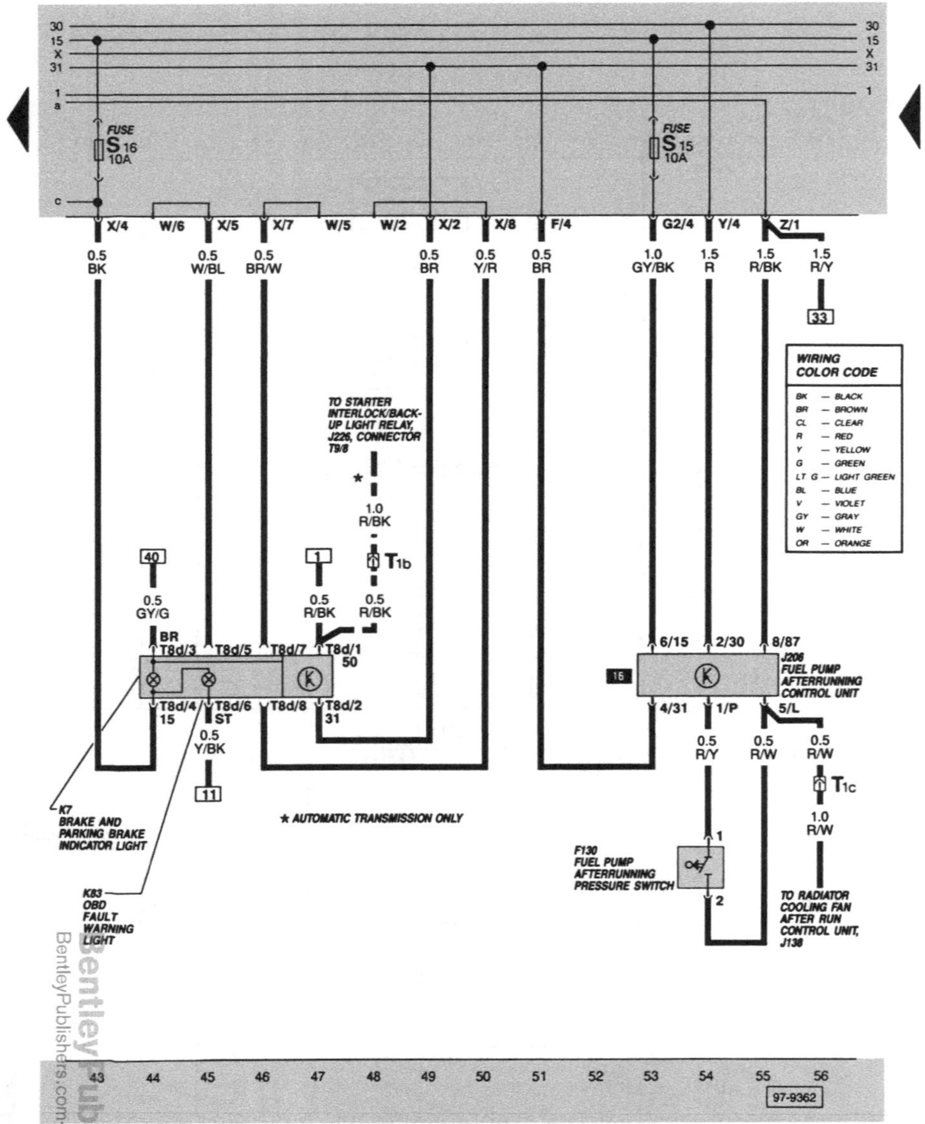

Corrado starting 1991
Engine code – PG
Digifant I, California only

THIS PAGE INTENTIONALLY LEFT BLANK

Description	Current track
Fuel gauge sending unit, G	4
Fuel pump afterrunning control unit, J206	9-13
Fuel pump afterrunning pressure switch, F130	11
Fuel pump relay, J17	3
Fuse, S18	1
Fuse, S15	10
Fuel pump, G6	3

Wire connectors

T1 — single, behind fuse/relay panel
T1h — single, behind fuse/relay panel

Ground connections

(14) — on transmission

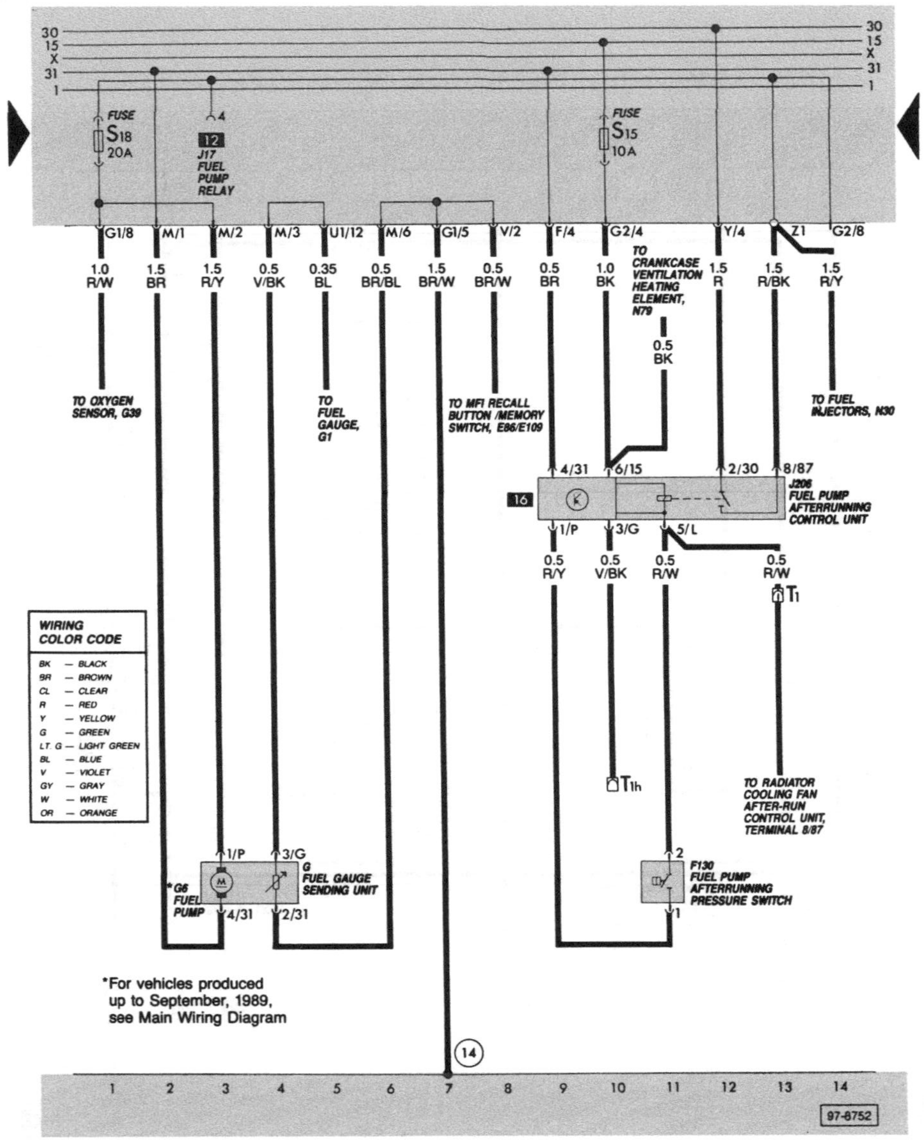

Description	Current track
Back-up light switch	3
Driver seat heater control unit	4
Driver seat temperature sensor	2
Heat element, driver backrest	1
Heat element, driver seat	1
Heat element, passenger backrest	14
Heat element, passenger seat	14
Heat regulating switch, driver seat	6
Heat regulating switch, passenger seat	10
Passenger seat heater control unit	12
Passenger seat temperature sensor	13
Seat heater switch indicator light	8

Wire connectors

T1j — single, behind relay panel
T1y — single, behind relay panel
T6d — six-point, below driver seat
T6f — six-point, below passenger seat
T6n — six-point, behind relay panel

Ground connections

(30) — beside relay panel
(96) — in heated seat wiring harness

Welded wiring points

(A11) — plus connection (58b), in instrument panel wiring harness
(02) — plus connection (15), in heated seat wiring harness
(03) — plus connection (58b), in heated seat wiring harness

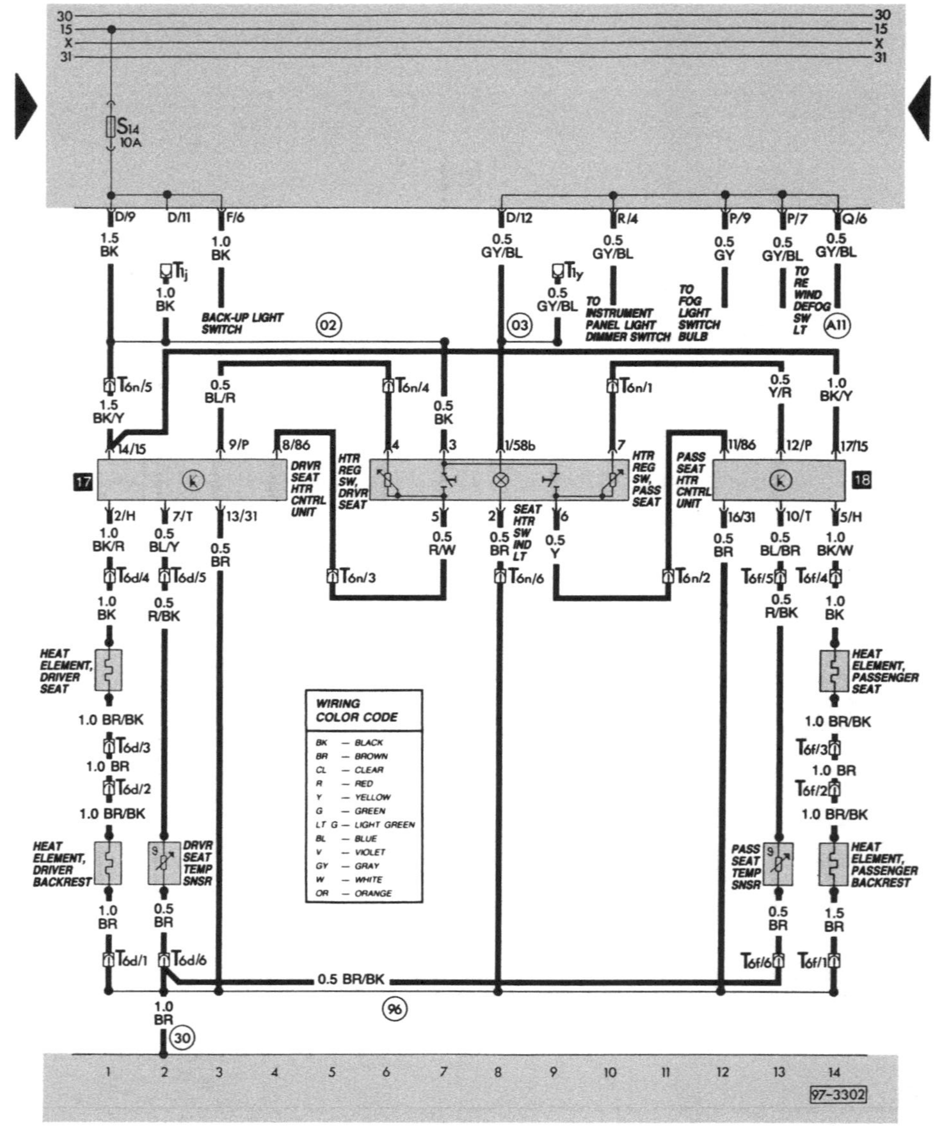

Description

Description	Current track
Back-up light switch, F4	5
Driver's seat heater control unit, J131	1-4
Driver's seat heater temperature sensor, G59	2
Driver's seat regulating heater switch, E94	7
Fog light switch illumination, L40	12
Fuse, S14	1
Heated driver's seat backrest, Z7	1
Heated driver's seat, Z6	1
Heated passenger's seat backrest, Z9	14
Heated passenger's seat, Z8	14
Instrument cluster, K	11
Instrument panel light dimmer switch, E20	10
Passenger's seat heater control unit, J132	12-14
Passenger's seat regulating heater switch, G60	13
Passenger's seat regulating heating switch, E95	9
Rear-window-defogger switch, L39	13
Seat heating indicator light, L44	8

Wire connectors

- T1j – single, behind fuse/relay panel
- T1y – single, behinde fuse/relay panel
- T6d – six point, below driver's seat
- T6f – six point, below passenger's seat
- T6n – six point, behind fuse/relay panel
- T28 – twenty-eight point, on instrument cluster
- TV5 – wire distributor for terminal 15a, behind fuse/relay panel
- TV8 – wire distributor for terminal 58b, behind fuse/relay panel

Welded wiring harness points

- (A11) – plus connection (58b), in instrument panel wiring harness
- (O2) – plus connection (15), in heated seats wiring harness
- (O3) – plus connection (58b), in heated seats wiring harness

Ground connections

- (30) – beside fuse relay/panel
- (96) – in wiring harness heated seats

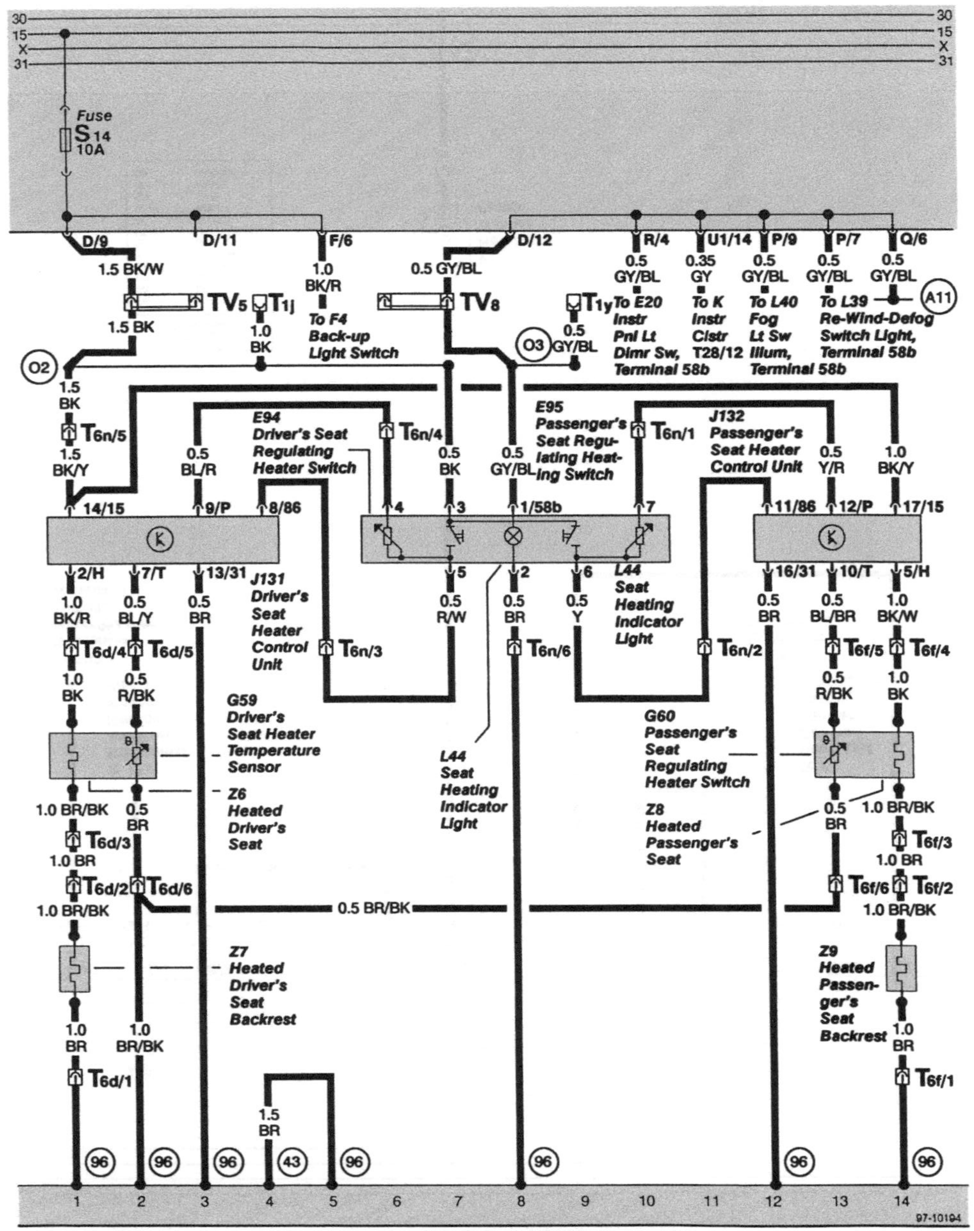

Description | **Current track**

Description	Current track
Headlight Washer Pump, V11	7
Headlight Washer System Relay, J39	7-9

Ground connections

(30) - ground connection, -1-, beside fuse/relay panel

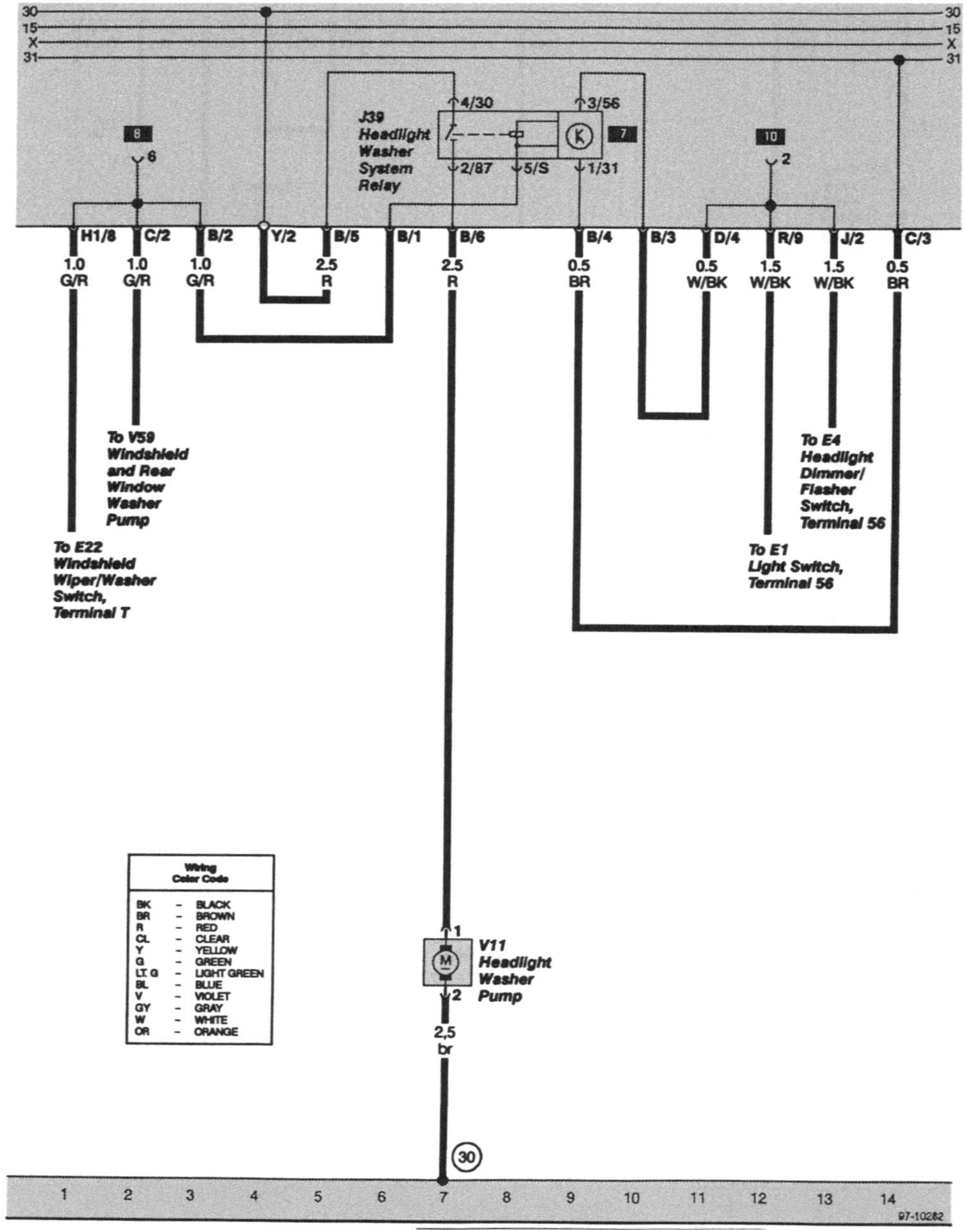

Description	Current track
Alternator warning light, K2	24
Control unit with display unit in instrument cluster, J285	9-46
Coolant low level sensor, G32	8
Coolant overheat warning light, K28	12
Coolant temperature gauge, G3	13
Coolant temperature sending unit, G2	5
Digital display light, L75	41, 43
Engine oil pressure switch (0.3 bar), F22	6
Engine oil pressure switch (1.8 bar), F1	7
Engine oil pressure warning light, K3	10
Engine oil temperature sensor, G8	53
Fuel gauge, G1	14
Fuse, S16	19
Fuse, S21	32
Headlight high beam indicator light, K1	28
Instrument panel light, L10	32-34
Multi-function indicator memory switch, E109	51
Multi-function indicator recall button, E86	49
Multi-function indicator vacuum sensor, G55	29-31
Odometer display, Y4	38-40
Outside air temperature sensor, G17	55
Seat belt warning light, K19	23
Speed sensor (speedometer), G22	48-49
Speedometer, G21	32-35
Tachometer, G5	15-19
Turn signal indicator light, left, K65	26
Turn signal indicator light, right, K94	27
Voltage stabilizer, J6	20-22
Warning buzzer, H3	14
Wire connection, terminal 15, T35	20
Wire connector, speed signal, T43	29

Wire connectors

T2d – double, behind fuse/relay panel
T2h – double, behind fuse/relay panel
T4a – four-point, behind steering column switch cover
T8a – eight point, above fuse/relay panel
T28 – twenty-eight point, on instrument cluster
T35 – wire connection, terminal 15, behind fuse/relay panel
T42 – forty-two point, on left side of cylinder head (6 cyl. only)
T43 – wire connection for speed signal, behind fuse/relay panel

Ground connections

(18) – on engine block
(125) – in headlight wiring harness
(159) – in Digifant/Motronic wiring harness

Welded wiring harness points

(A15) – plus connection (15), in instrument panel wiring harness
(C15) – wire connection, in headlight wiring harness
(C16) – wire connection, in headlight wiring harness
(E6) – plus connection (15), in Digifant/Motronic wiring harness
(W9) – plus connection (30), in rear wiring harness
(W10) – plus connection (15), in rear wiring harness

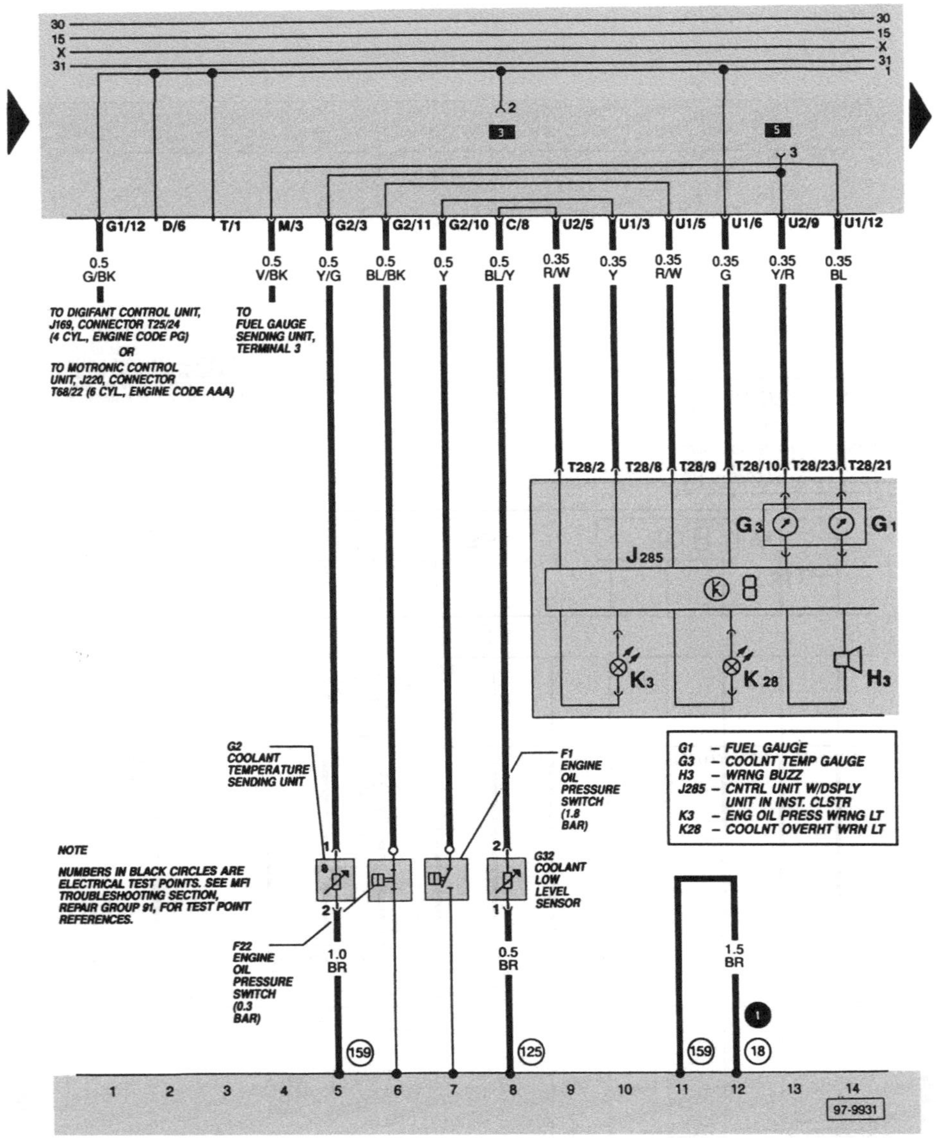

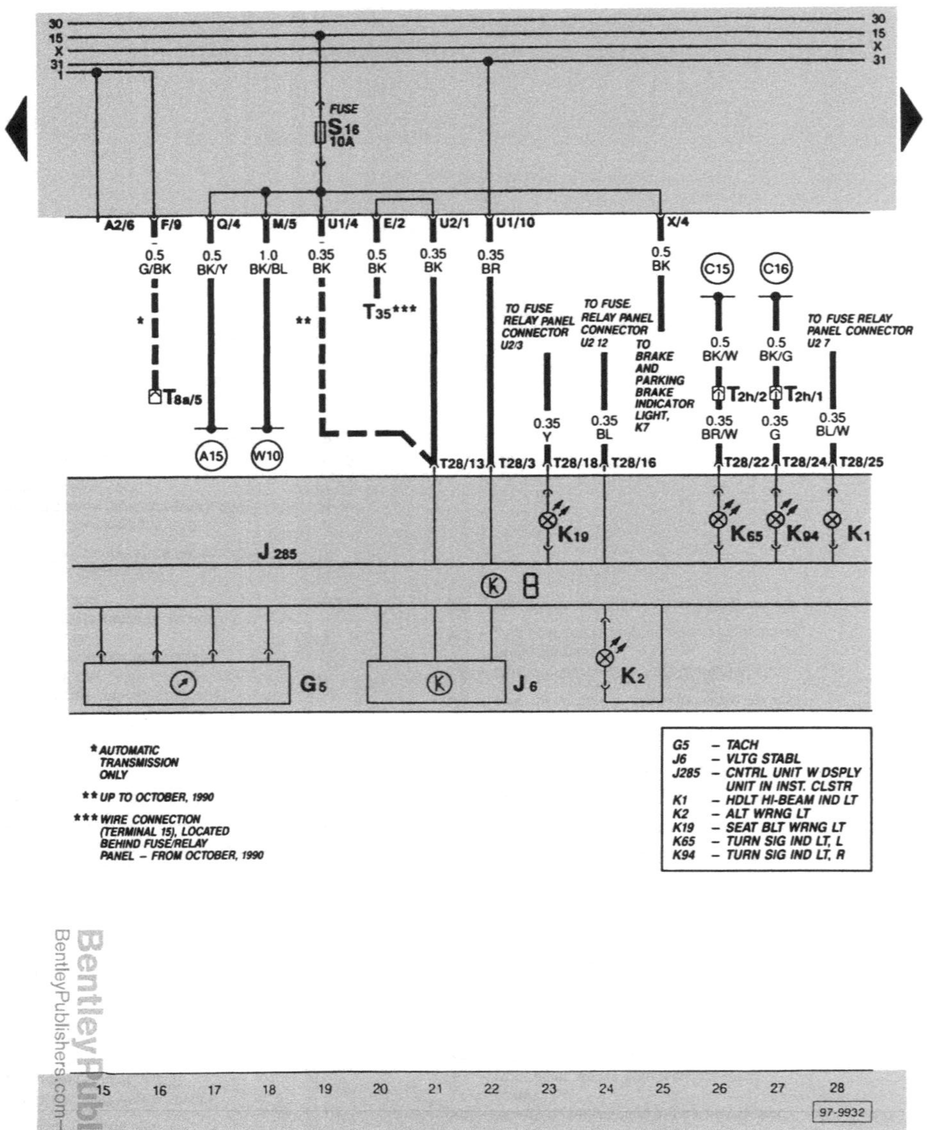

* AUTOMATIC TRANSMISSION ONLY

** UP TO OCTOBER, 1990

*** WIRE CONNECTION (TERMINAL 15), LOCATED BEHIND FUSE/RELAY PANEL – FROM OCTOBER, 1990

G5	–	TACH
J6	–	VLTG STABL
J285	–	CNTRL UNIT W DSPLY UNIT IN INST. CLSTR
K1	–	HDLT HI-BEAM IND LT
K2	–	ALT WRNG LT
K19	–	SEAT BLT WRNG LT
K65	–	TURN SIG IND LT, L
K94	–	TURN SIG IND LT, R

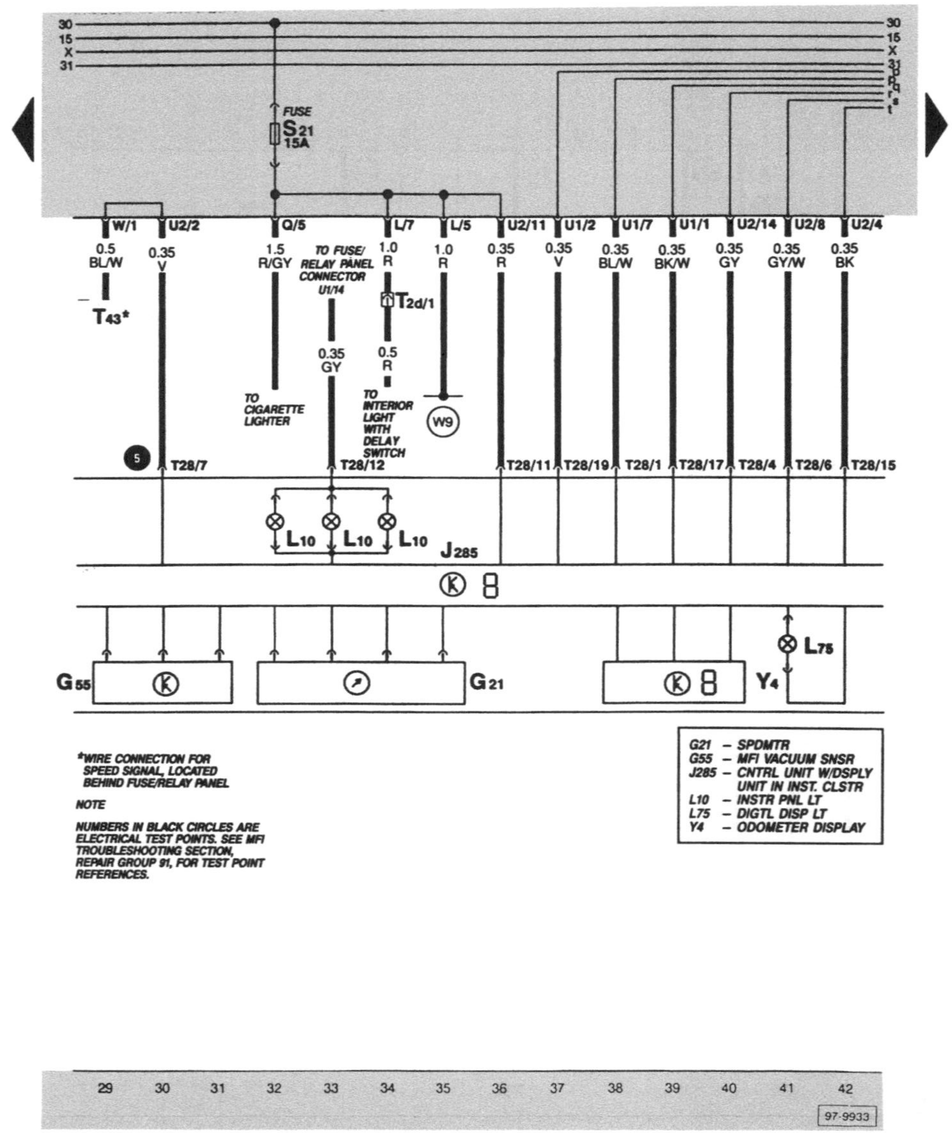

*WIRE CONNECTION FOR SPEED SIGNAL, LOCATED BEHIND FUSE/RELAY PANEL

NOTE

NUMBERS IN BLACK CIRCLES ARE ELECTRICAL TEST POINTS. SEE MFI TROUBLESHOOTING SECTION, REPAIR GROUP 91, FOR TEST POINT REFERENCES.

G21	–	SPDMTR
G55	–	MFI VACUUM SNSR
J285	–	CNTRL UNIT W/DSPLY UNIT IN INST. CLSTR
L10	–	INSTR PNL LT
L75	–	DIGTL DISP LT
Y4	–	ODOMETER DISPLAY

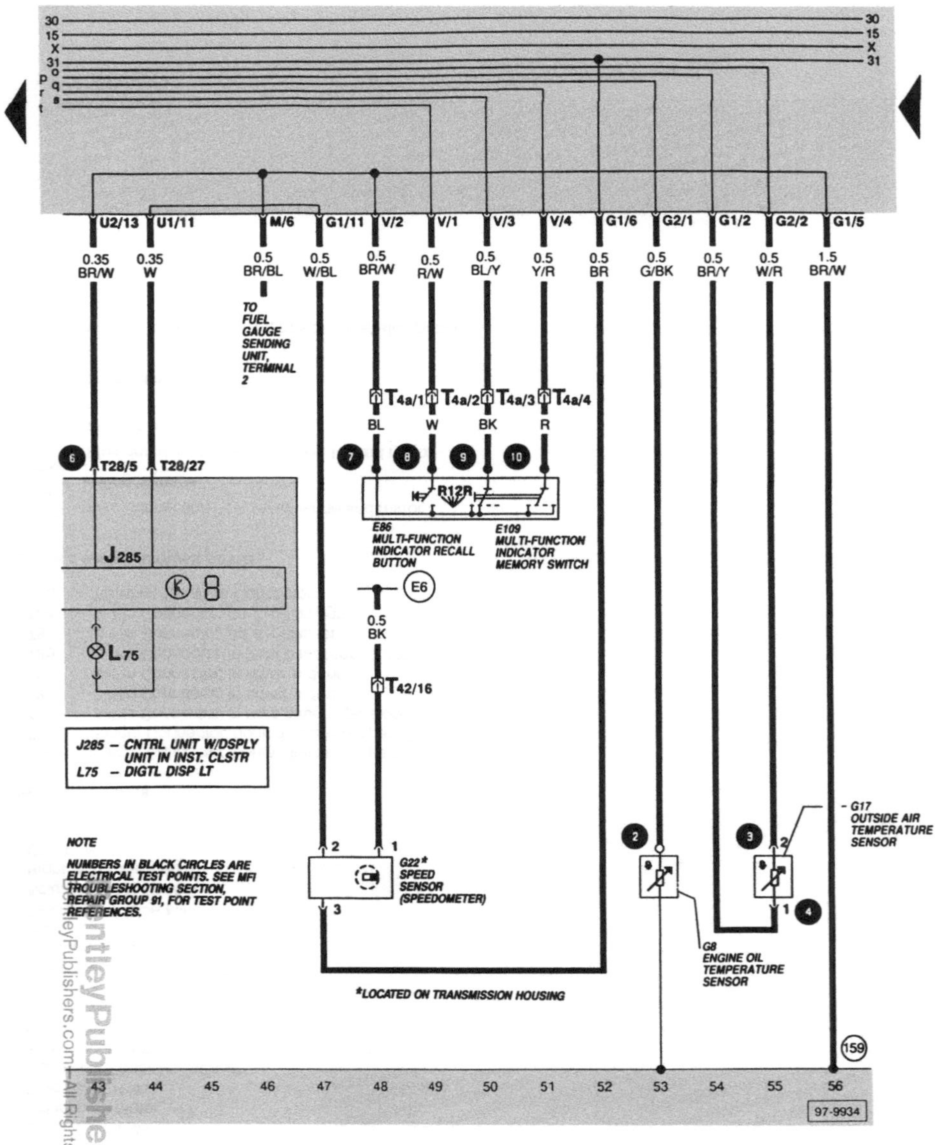

NOTE

NUMBERS IN BLACK CIRCLES ARE ELECTRICAL TEST POINTS. SEE MFI TROUBLESHOOTING SECTION, REPAIR GROUP 91, FOR TEST POINT REFERENCES.

THIS PAGE INTENTIONALLY LEFT BLANK

Description | **Current track**

Description	Current track
Driver's Side Heated Mirror, Z4	11
Driver's Side Mirror Adjustment Motor, V17	3
Fuse, S14, 10A	1
Instrument Panel/Adjustment Illumination, L87	7-9
Left Mirror Magnetic Clutch, N35	6
Mirror Adjustment Switch, E43	1-6
Mirror Selector Switch, E48	1
Passenger's Side Heated Mirror, Z5	10
Passenger's Side Mirror Adjustment Motor, V25	1
Right Mirror Magnetic Clutch, N42	7

Wire connectors

- T1l - Single Connector, behind fuse/relay panel
- T1m - Single Connector, behind fuse/relay panel
- T3m - 3-Pin Connector, in front passenger's door
- T3n - 3-Pin Connector, in driver's door
- T3o - 3-Pin Connector, in driver's door
- T3p - 3-Pin Connector, in front passenger's door
- T6 - 6-Pin Connector, on A-pillar, left
- T6a - 6-Pin Connector, on A-pillar, right
- TV5 - Terminal 15a Wire Connector

Welded wiring harness points

- (Z1) - Wire connection -1-, in power heated mirror wiring harness
- (Z4) - Plus connection (15a), in power heated mirror wiring harness

Ground connection

- (30) - Ground connection, -1-, beside fuse/relay panel

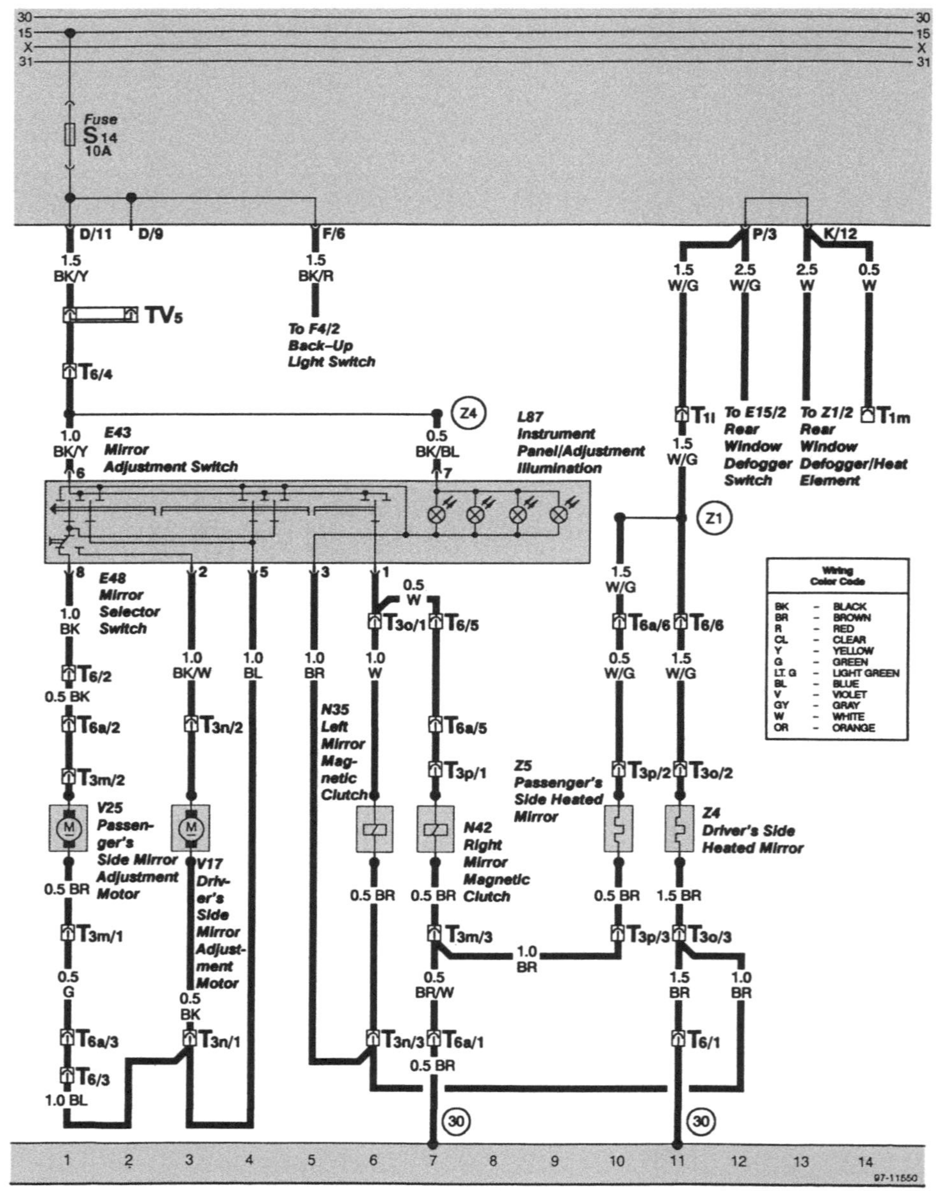

Description	Current track
Fuse, S4	7
Overheat fuse, S24	8
Rear window wiper/washer relay, J30	5
Sunroof stop switch (closed pos.), E56	7
Sunroof stop switch (open pos.), E57	5
Sunroof switch, E8	4-8
Sunroof motor, V1	4-9
Sunroof switch light, L65	8

Wire connectors

T2q — double, in luggage compartment, left
T3s — three-point, near interior dome light

Ground connections

(30) — beside fuse/relay panel
(128) — in interior light wiring harness

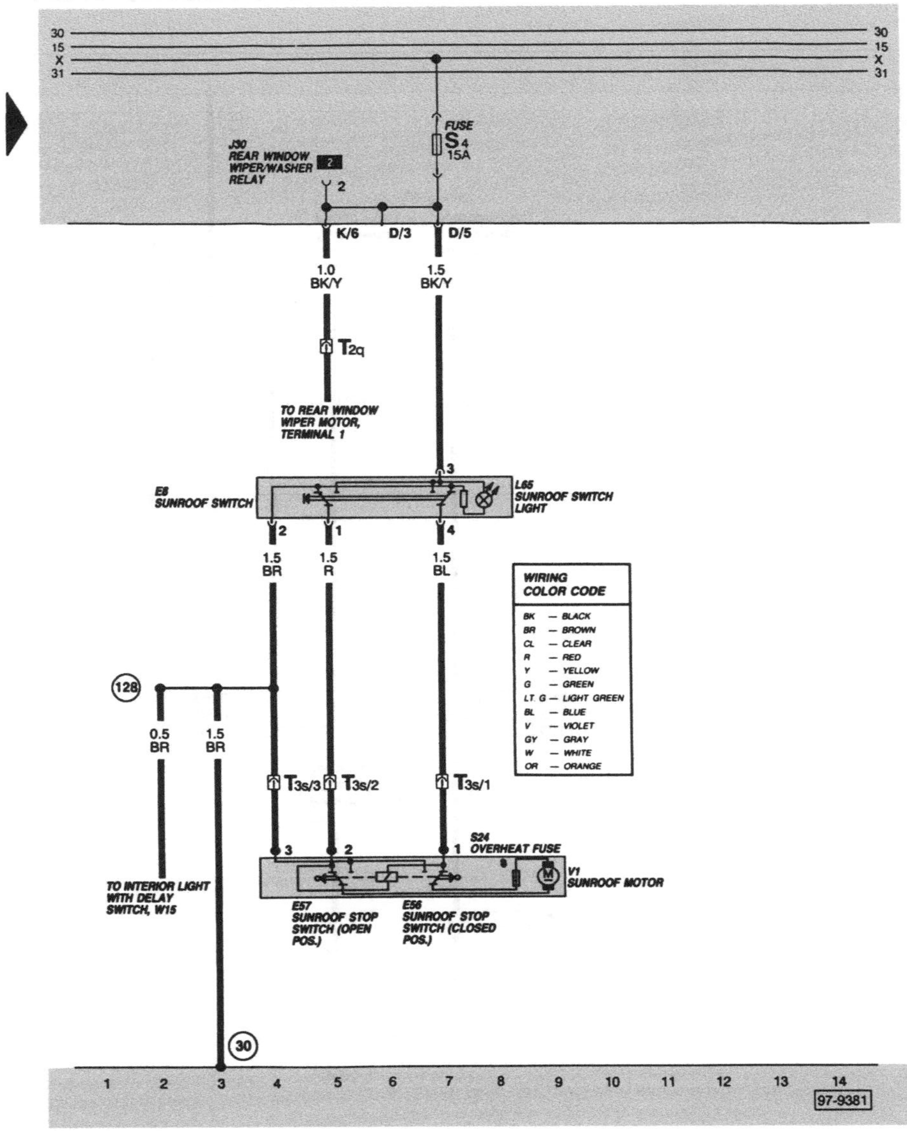

Description

Description	Current track
Fuse, S4, 15A	7
Overheat Fuse, S24	8
Sunroof Motor, V1	9
Sunroof Stop Switch (Closed Pos.), E56	7
Sunroof Stop Switch (Open Pos.), E57	5
Sunroof Switch Light, L65	8
Sunroof Switch, E8	4-7

Wire connectors

- T2q - double, in luggage compartment, left
- T3s - three point, near interior light
- TV6 - Wire Connector, Terminal X, behind relay panel plate

Ground connections

- (30) - ground connection, -1-, beside relay panel
- (128) - ground connection -1-, in interior light wiring harness

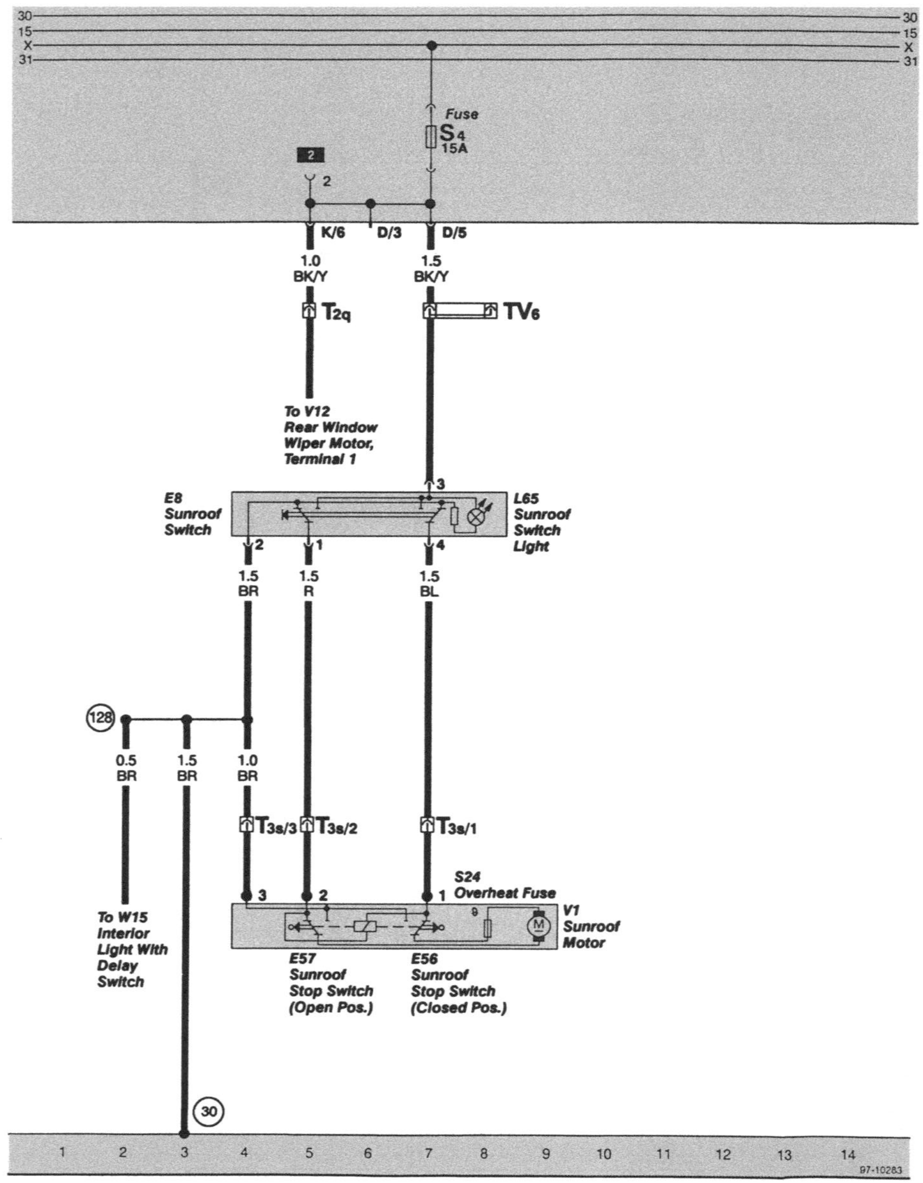

Description	Current track
Door Contact Switch, Left Front, F2	9-10
Door Contact Switch, Right Front, F3	11
Fuse, S16, 15A	4
Fuse, S21, 15A	10
Interior Light With Delay Switch, W15	10
Make-Up Mirror Light Switch, Right, F148	14
Make-Up Mirror Light, Right, W14	14
Map/Reading Light, Right Front, W13	12
Power Sunroof Control Module, J245	2-7
Sunroof Motor, V1	2-7
Sunroof Switch Light, L65	3-7
Sunroof Switch, E8	3-7

Wire connectors

- T1a – single, behind relay panel plate
- T1b – single, behind relay panel plate
- T1e – single, on the left A-pillar bottom
- T1v – single, on the left A-pillar bottom
- T2d – double, behind relay panel plate
- T2e – double, on the left A-pillar top
- T6 – six point, on Power Sunroof Control Module
- T8a – eight point, on Power Window/Sunroof Control Module
- T8c – eight point, on Power Window/Sunroof Control Module
- T28 – twenty-eight point, on Instrument Cluster

Welded wiring harness point

- (A15) – Plus connection (15), in instrument wiring harness
- (Q8) – Wire connection (power window/rear cigarette lighter), in power window/central locking system and door contact switch wiring harness
- (Q45) – Wire connection (comfort), in power window wiring harness

- (W9) – Plus connection (30), in rear wiring harness
- (W10) – Plus connection (15), in rear wiring harness
- (Y1) – Plus connection (30), in interior light wiring harness
- (Y3) – Plus connection (15), in interior light/sunroof wiring harness

Ground connections

- (86) – Ground connection –1–, in rear wiring harness
- (128) – Ground connection –1–, in interior light wiring harness

Corrado starting 1993
Production from Aug. 1992

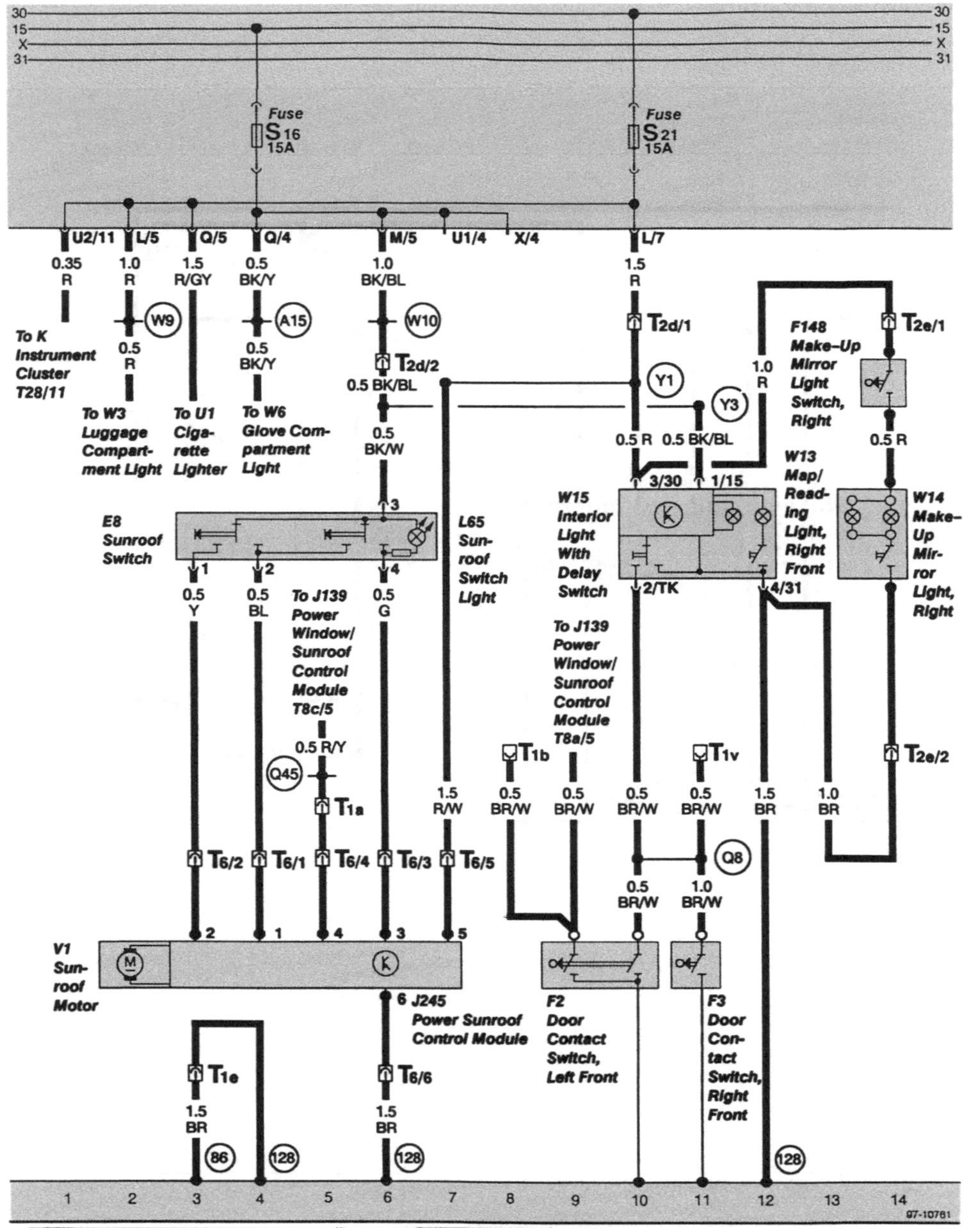

Corrado starting 1993
Production from Aug. 1992

Description	Current track
Central locking system door positioner, left front	22
Central locking system door positioner, right front	17
Central locking system pump and control unit	10
Central locking system switch	2
Central locking system switch, right door	5
Circuit breaker (power window)	24
Door contact switch, left front	26
Door contact switch, right front	28
Power window control unit	26,32
Power window switch light	43,48, 51
Window motor, left front door	30
Window motor, right front door	34
Window switch, left front	47
Window switch, RF (in LF door)	51
Window switch, right front	54

Wire connectors

T1 — single, behind relay panel
T2a — double, below left A-pillar near relay panel
T2b — double, below right A-pillar
T2c — double, in driver door
T2d — double, in passenger door
T3a — three-point, below left A-pillar near relay panel
T3b — three-point, below right A-pillar
T3c — three-point, below right A-pillar
T3d — three-point, below left A-pillar near relay panel
T3e — three-point, in driver door
T3f — three-point, in passenger door
T6h — six-point, below left A-pillar near relay panel
T6i — six-point, below right A-pillar
T8a — eight-point, on power window control unit
T8b — eight-point, on power window control unit
T8c — eight-point, on power window control unit
T12 — twelve-point, on central locking system pump

Ground connections

(30) — beside relay panel
(103) — in power window/central locking system wiring harness
(105) — in power window/central locking system wiring harness

Welded wiring harness points

(Q4) — plus connection (30), in power window/cetnral locking system wiring harness
(Q13) — plus connection (15), in power window/central locking system wiring harness

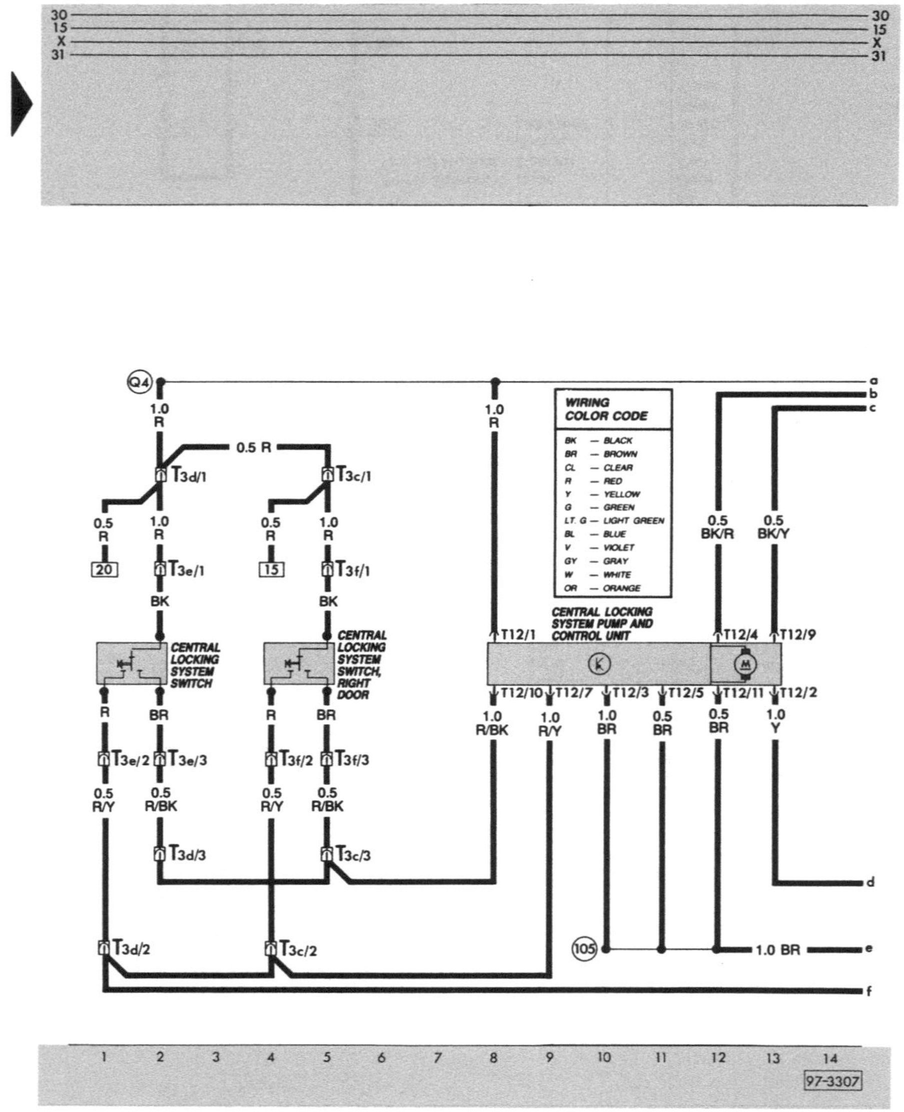

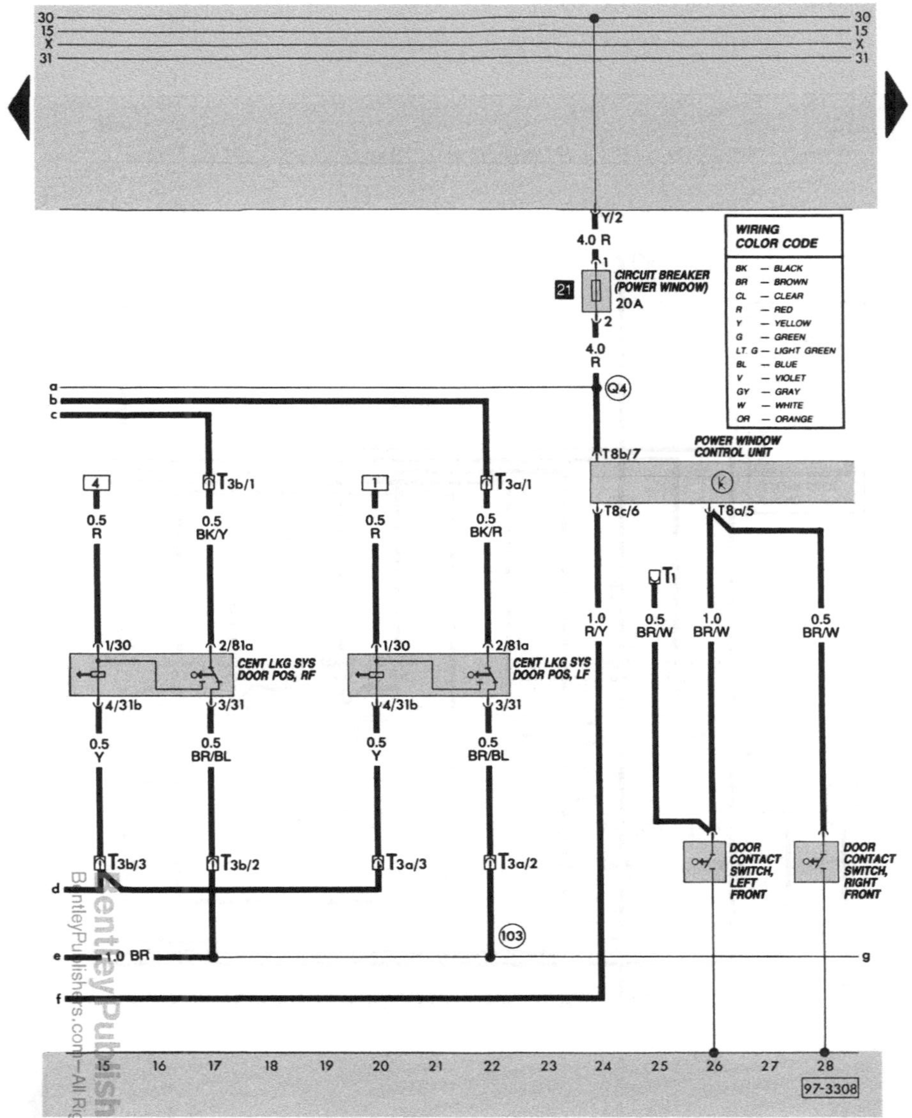

Y/2
4.0 R
CIRCUIT BREAKER (POWER WINDOW) 20A
21
4.0 R
Q4
T8b/7
T8c/6
T8a/5
POWER WINDOW CONTROL UNIT
WIRING COLOR CODE
BK — BLACK
BR — BROWN
CL — CLEAR
R — RED
Y — YELLOW
G — GREEN
LT. G — LIGHT GREEN
BL — BLUE
V — VIOLET
GY — GRAY
W — WHITE
OR — ORANGE
0.5 R
0.5 BK/Y
0.5 BK/R
T3b/1
T3a/1
1/30
2/81a
4/31b
3/31
CENT LKG SYS DOOR POS, RF
CENT LKG SYS DOOR POS, LF
0.5 Y
0.5 BR/BL
T3b/3
T3b/2
T3a/3
T3a/2
1.0 R/Y
0.5 BR/W
1.0 BR/W
T1
DOOR CONTACT SWITCH, LEFT FRONT
DOOR CONTACT SWITCH, RIGHT FRONT
103
1.0 BR
97-3308

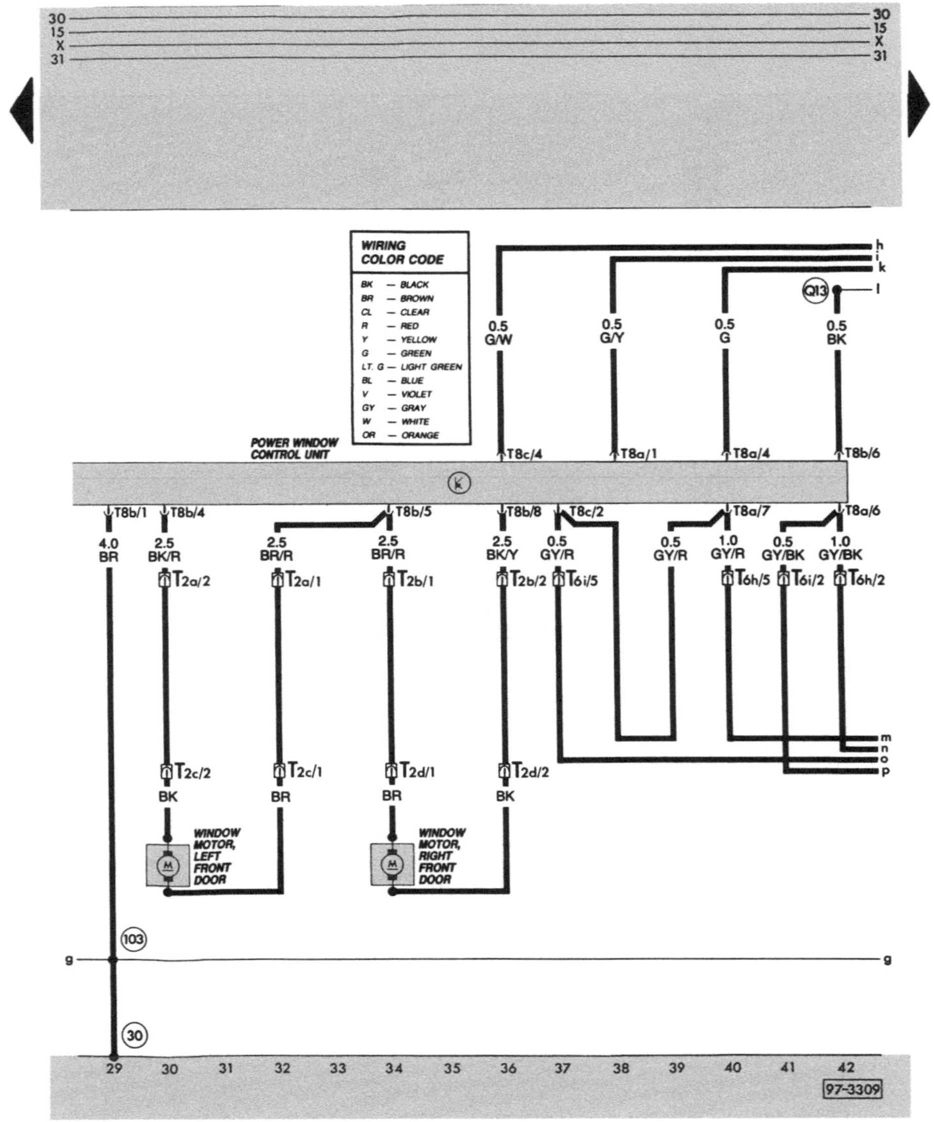

WIRING COLOR CODE
BK — BLACK
BR — BROWN
CL — CLEAR
R — RED
Y — YELLOW
G — GREEN
LT. G — LIGHT GREEN
BL — BLUE
V — VIOLET
GY — GRAY
W — WHITE
OR — ORANGE
POWER WINDOW CONTROL UNIT
0.5 G/W
0.5 G/Y
0.5 G
0.5 BK
Q13
T8c/4
T8a/1
T8a/4
T8b/6
T8b/1
T8b/4
T8b/5
T8b/8
T8c/2
T8a/7
T8a/6
4.0 BR
2.5 BK/R
2.5 BR/R
2.5 BK/Y
0.5 GY/R
1.0 GY/R
0.5 GY/BK
1.0 GY/BK
T2a/2
T2a/1
T2b/1
T2b/2
T6i/5
T6h/5
T6i/2
T6h/2
T2c/2
T2c/1
T2d/1
T2d/2
BK
BR
WINDOW MOTOR, LEFT FRONT DOOR
WINDOW MOTOR, RIGHT FRONT DOOR
103
30
97-3309

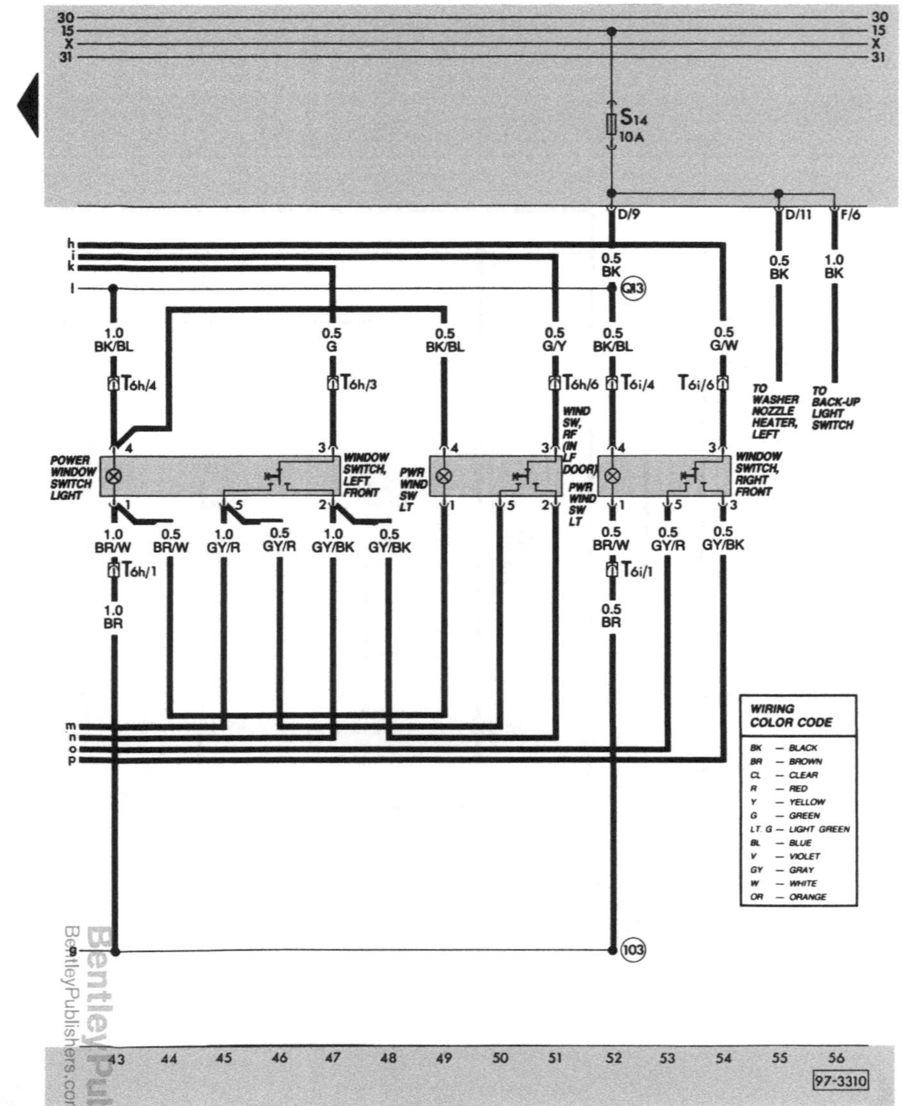

THIS PAGE INTENTIONALLY LEFT BLANK

Description	Current track
Central locking system door switch, left front, F131	52-54
Central locking system door switch, right front, F133	45-47
Central locking system fuse, S52	38
Central locking system pump and control unit, V69	36-41
Central locking system switch (driver's door), F59	30-31
Central locking system switch (passenger's door), F114	33-34
Door contact switch, left front, F2	2-3
Door contact switch, right front, F3	4-5
Fuse, S14	24
Fuse, S16	47
Power window control unit, J139	1-14
Power window switch light, L53	15, 21, 24
Window motor, left front door, V14	6
Window motor, right front door, V15	8
Window switch, left front, E40	15-19
Window switch, RF (in LF door), E81	21-23
Window switch, right front, E41	24-26

Wire connectors

T1 — single, on A-pillar, left, near fuse/relay panel
T1g — single, behind fuse/relay panel
T1h — single, behind fuse/relay panel
T2a — double, on A-pillar, left, near fuse/relay panel
T2b — double, on A-pillar, right, upper part
T2c — double, in driver's door
T2d — double, in passenger's door
T2e — double, behind fuse/relay panel
T2o — double, in engine compartment, left
T3a — three-point, on A-pillar, left, near fuse/relay panel
T3b — three-point, on A-pillar, right, upper part
T3c — three-point, on A-pillar, right, upper part
T3d — three-point, on A-pillar, left, near fuse/relay panel
T3e — three-point, in driver's door
T3f — three-point, in passenger's door
T6h — six-point, on A-pillar, left, near fuse/relay panel
T6i — six-point, on A-pillar, right, upper part
T8a — eight-point, brown, on power window control unit

Wire connectors

T8b — eight-point, white, on power window control unit
T8c — eight-point, black, on power window control unit
T12 — twelve-point, on central locking system pump and control unit
T28 — twenty eight-point, on instrument cluster

Ground connections

(30) — beside fuse/relay panel
(103) — in central locking system wiring harness
(105) — in central locking system wiring harness

Welded wiring harness points

(A15) — plus connection (15), in instrument panel wiring harness
(C14) — plus connection (15), in heated windshield washer nozzle wiring harness
(Q4) — plus connection (30), in power windows/ central locking system wiring harness
(Q13) — plus connection (15), in power windows/ central locking system wiring harness
(S1) — plus connection (30), in central locking system wiring harness
(S4) — plus connection (30), in central locking system wiring harness
(W10) — plus connection (15), in rear wiring harness

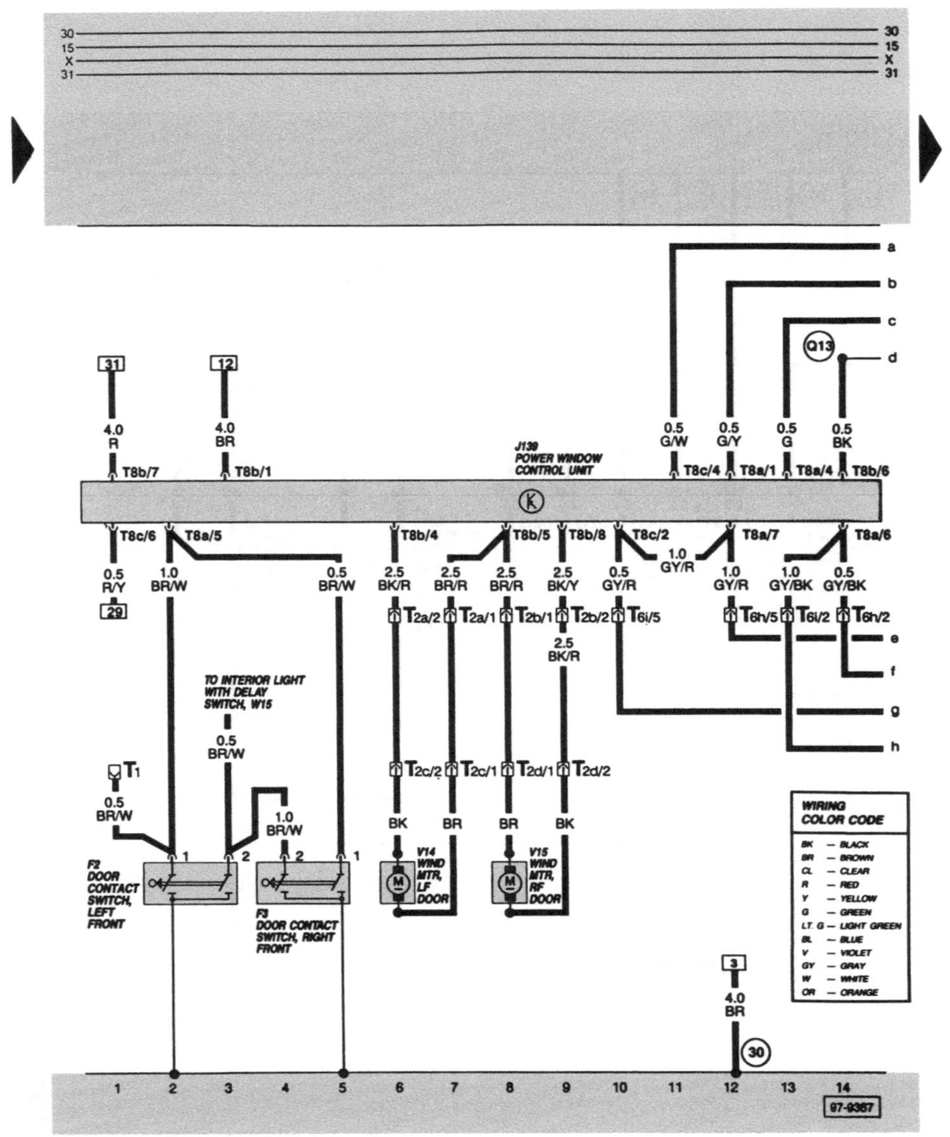

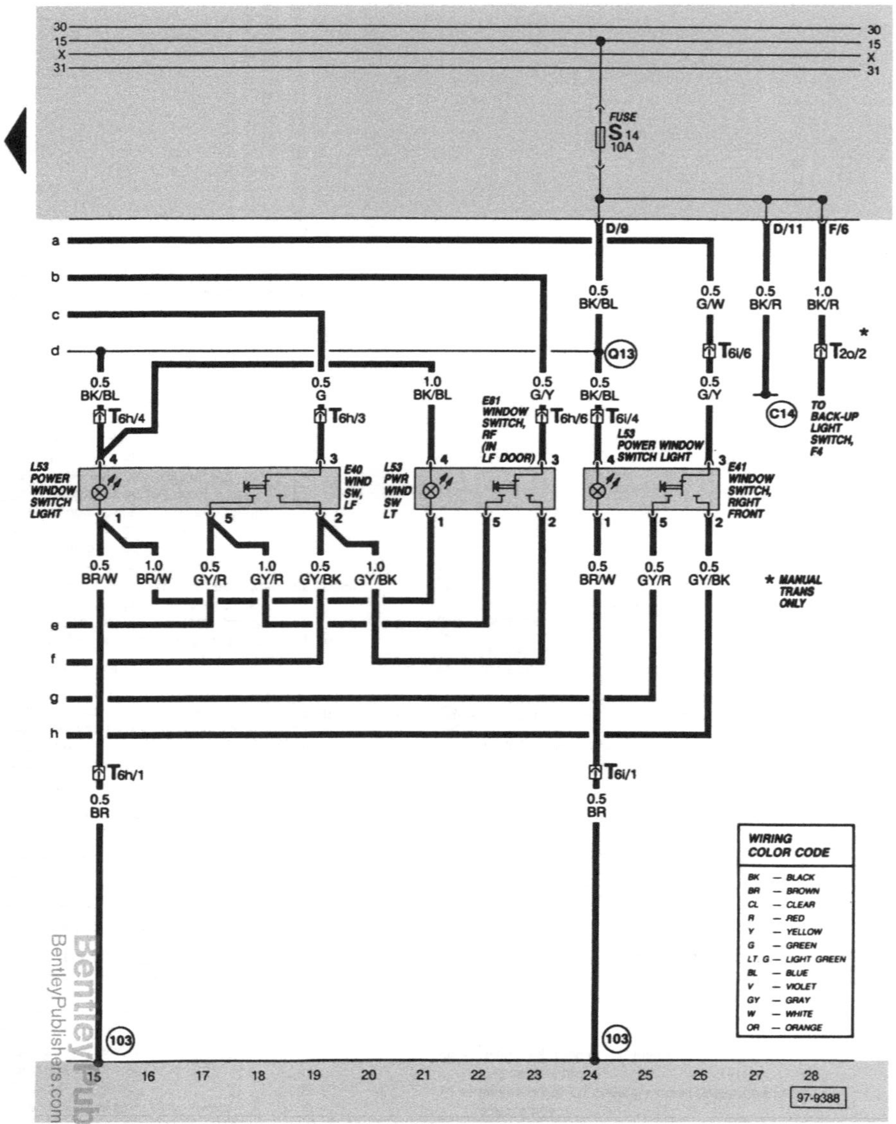
30
15
X
31
FUSE
S14
10A
D/9
D/11
F/6
a
b
c
d
e
f
g
h
0.5 BK/BL
0.5 G/W
0.5 BK/R
1.0 BK/R
Q13
T6i/6
T20/2
C14
TO BACK-UP LIGHT SWITCH, F4
0.5 G
1.0 BK/BL
0.5 G/Y
T6h/4
T6h/3
T6h/6
T6i/4
E81 WINDOW SWITCH, RF (IN LF DOOR)
L53 POWER WINDOW SWITCH LIGHT
L53 PWR WIND SW LT
E40 WIND SW, LF
E41 WINDOW SWITCH, RIGHT FRONT
0.5 BR/W
1.0 BR/W
0.5 GY/R
1.0 GY/R
0.5 GY/BK
1.0 GY/BK
* MANUAL TRANS ONLY
T6h/1
T6i/1
0.5 BR
103
WIRING COLOR CODE
BK — BLACK
BR — BROWN
CL — CLEAR
R — RED
Y — YELLOW
G — GREEN
LT G — LIGHT GREEN
BL — BLUE
V — VIOLET
GY — GRAY
W — WHITE
OR — ORANGE
15
16
17
18
19
20
21
22
23
24
25
26
27
28
97-9388

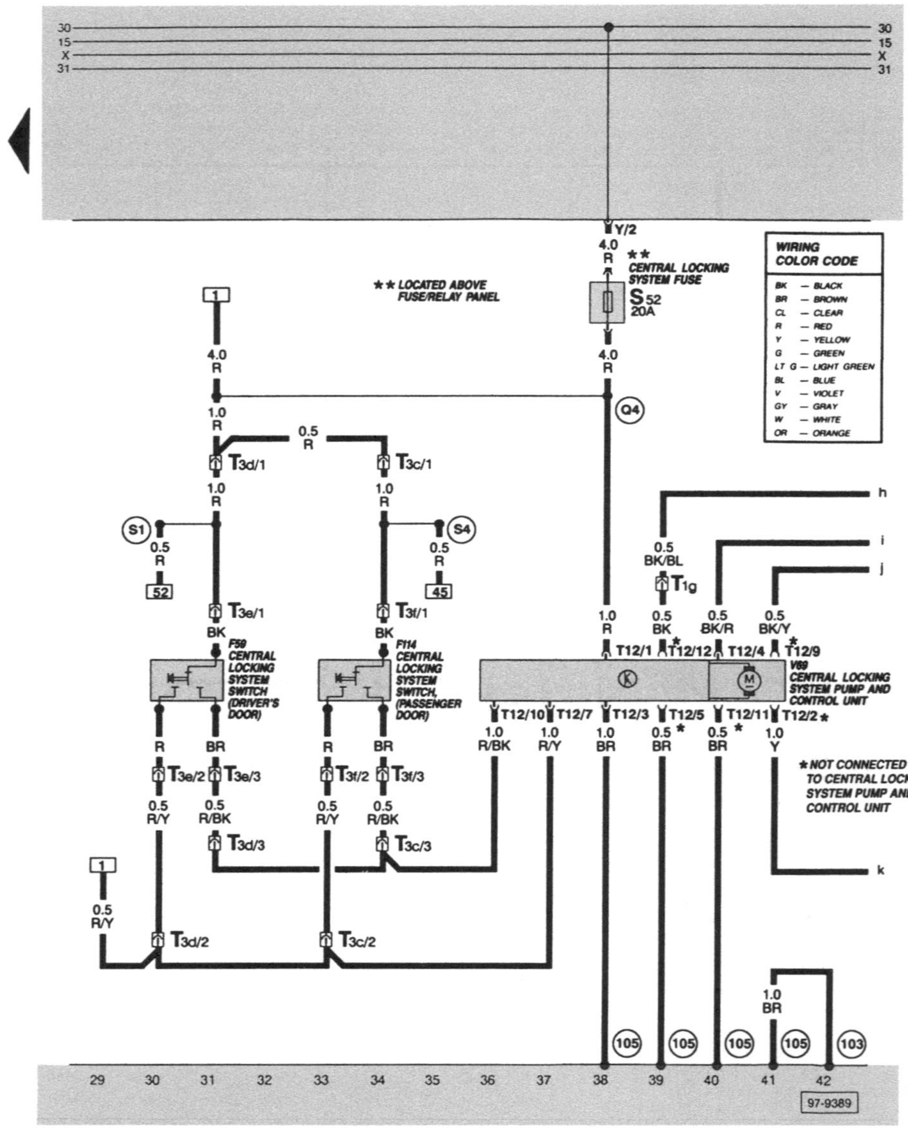
30
15
X
31
Y/2
4.0 R
** CENTRAL LOCKING SYSTEM FUSE
S52
20A
** LOCATED ABOVE FUSE/RELAY PANEL
WIRING COLOR CODE
BK — BLACK
BR — BROWN
CL — CLEAR
R — RED
Y — YELLOW
G — GREEN
LT G — LIGHT GREEN
BL — BLUE
V — VIOLET
GY — GRAY
W — WHITE
OR — ORANGE
Q4
S1
S4
52
45
T3d/1
T3c/1
T3e/1
T3f/1
F59 CENTRAL LOCKING SYSTEM SWITCH (DRIVER'S DOOR)
F114 CENTRAL LOCKING SYSTEM SWITCH, (PASSENGER DOOR)
T3e/2
T3e/3
T3f/2
T3f/3
T3d/3
T3c/3
T3d/2
T3c/2
0.5 R/Y
0.5 R/BK
T1g
0.5 BK/BL
0.5 BK/R
0.5 BK/Y
T12/1
T12/12
T12/4
T12/9
V69 CENTRAL LOCKING SYSTEM PUMP AND CONTROL UNIT
T12/10
T12/7
T12/3
T12/5
T12/11
T12/2
1.0 R/BK
1.0 R/Y
1.0 BR
0.5 BR
1.0 Y
* NOT CONNECTED TO CENTRAL LOCKING SYSTEM PUMP AND CONTROL UNIT
h
i
j
k
105
103
29
30
31
32
33
34
35
36
37
38
39
40
41
42
97-9389

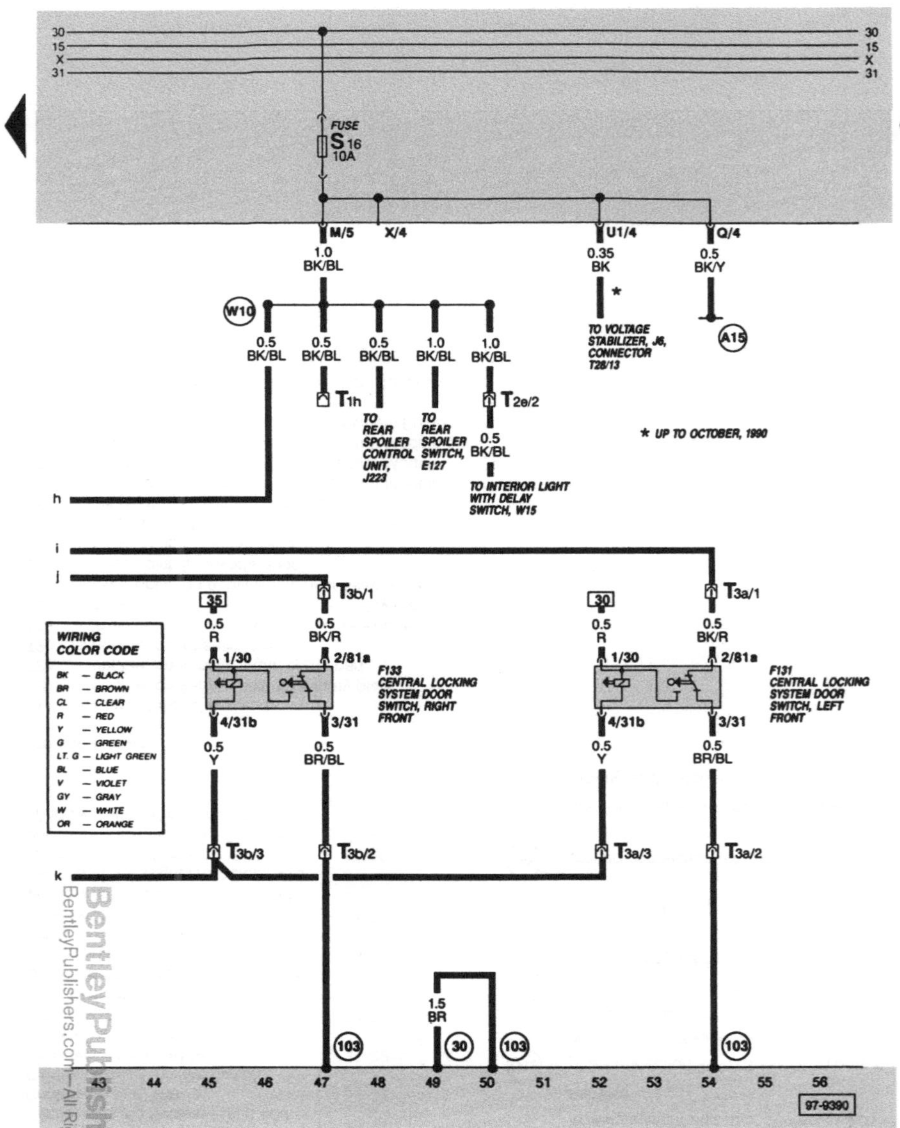

Corrado 1991–1992
Engine code – PG

THIS PAGE INTENTIONALLY LEFT BLANK

Description	Current track
Central Locking System Door Switch, Left Front, F131	40-42
Central Locking System Pump and Control Module, V69	33-38
Central Locking System Switch (Passenger Door), F114	31-32
Central Locking System Switch, Left Front, F59	29-30
Door Contact Switch, Left Front, F2	2-3
Door Contact Switch, Right Front, F3	4-5
Fuse, S52 (ABS Hydraulic Pump)	35
Power Window Control Module, J139	1-14
Power Window Switch Light, L53	15,21,24
Window Motor, Left Front Door, V14	7
Window Motor, Right Front Door, V15	9
Window Switch, Left Front, E40	17-19
Window Switch, RF (in LF Door), E81	21-23
Window Switch, Right Front, E41	24-26

Wire connectors

- T1 - single, fuse/relay panel
- T2a - double, on A-pillar, left, near relay plate
- T2b - double, on A-pillar, right, upper part
- T2o - double, near transmission
- T3a - three point, on A-pillar, left, near relay plate
- T3c - three point, on A-pillar, right, near relay plate
- T3d - three point, on A-pillar, left, near relay plate
- T3e - three point, in driver's door
- T3f - three point, in passenger's door
- T6h - six point, on A-pillar, left, near relay plate
- T6i - six piont, on A-pillar, right, upper part
- T8a - eight point, on Power Window Control Module
- T8b - eight point, on Power Window Control Module
- T8c - eight point, on Power Window Control Module
- T12 - twelve point, on Central Locking System Pump and Control Module
- TV5 - wire distrubtor for terminal 15a, above fuse/relay panel

Welded wiring harness points

- (Q4) - plus connection (30), in power window/central locking system and door contact switch wiring harness
- (Q13) - plus connection -1- (75 as/15), in power windows/power locks and door contact switch wiring harness
- (S1) - plus connection -1- (30), in central locking system wiring harness

Ground connections

- (30) - ground connection, -1-, beside relay panel
- (103) - ground connection, -2-, in power window/central locking system/door

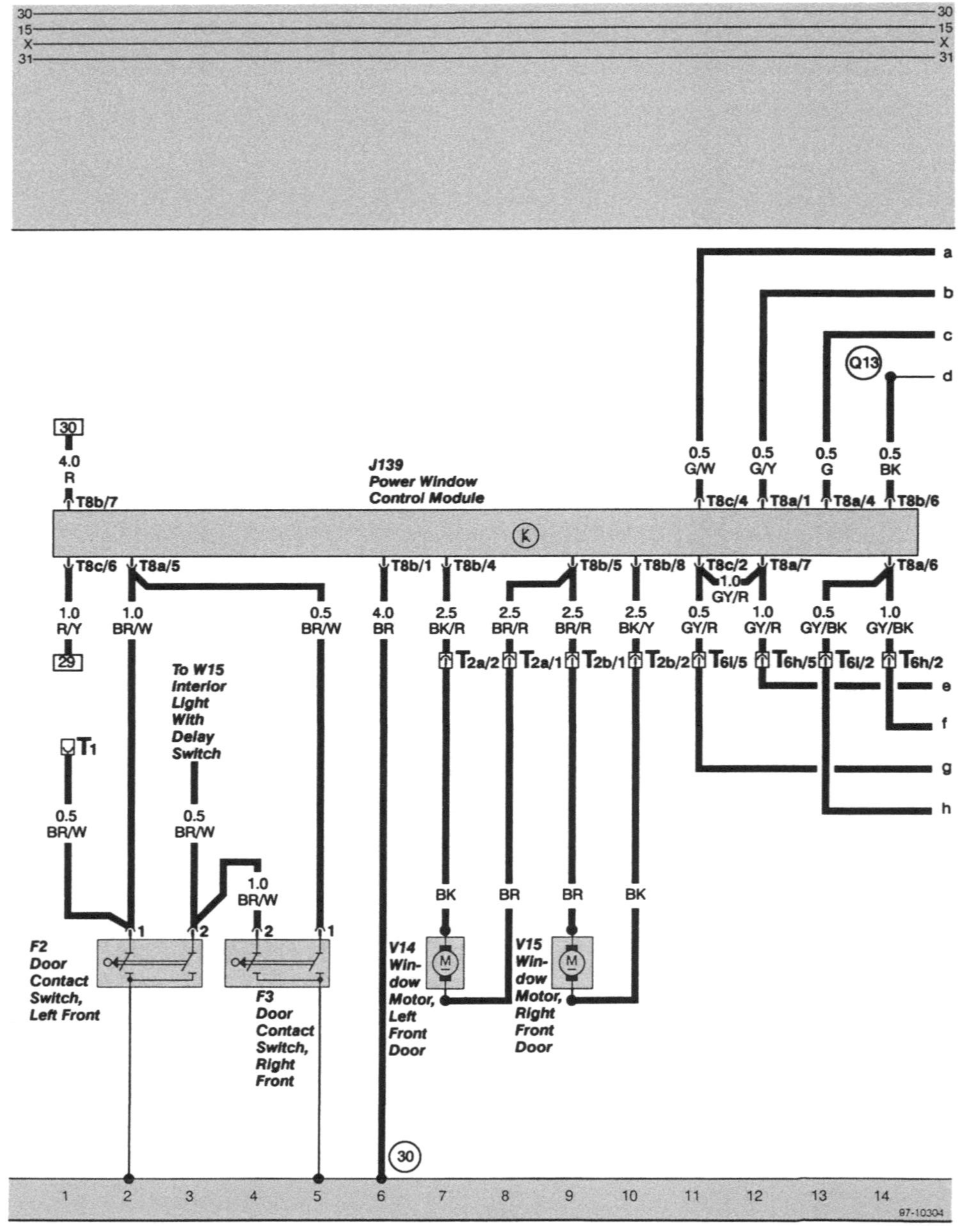

Corrado 1992–1993
Engine code – AAA
Production up to Aug. 1992

Corrado 1992–1993
Engine code – AAA
Production up to Aug. 1992

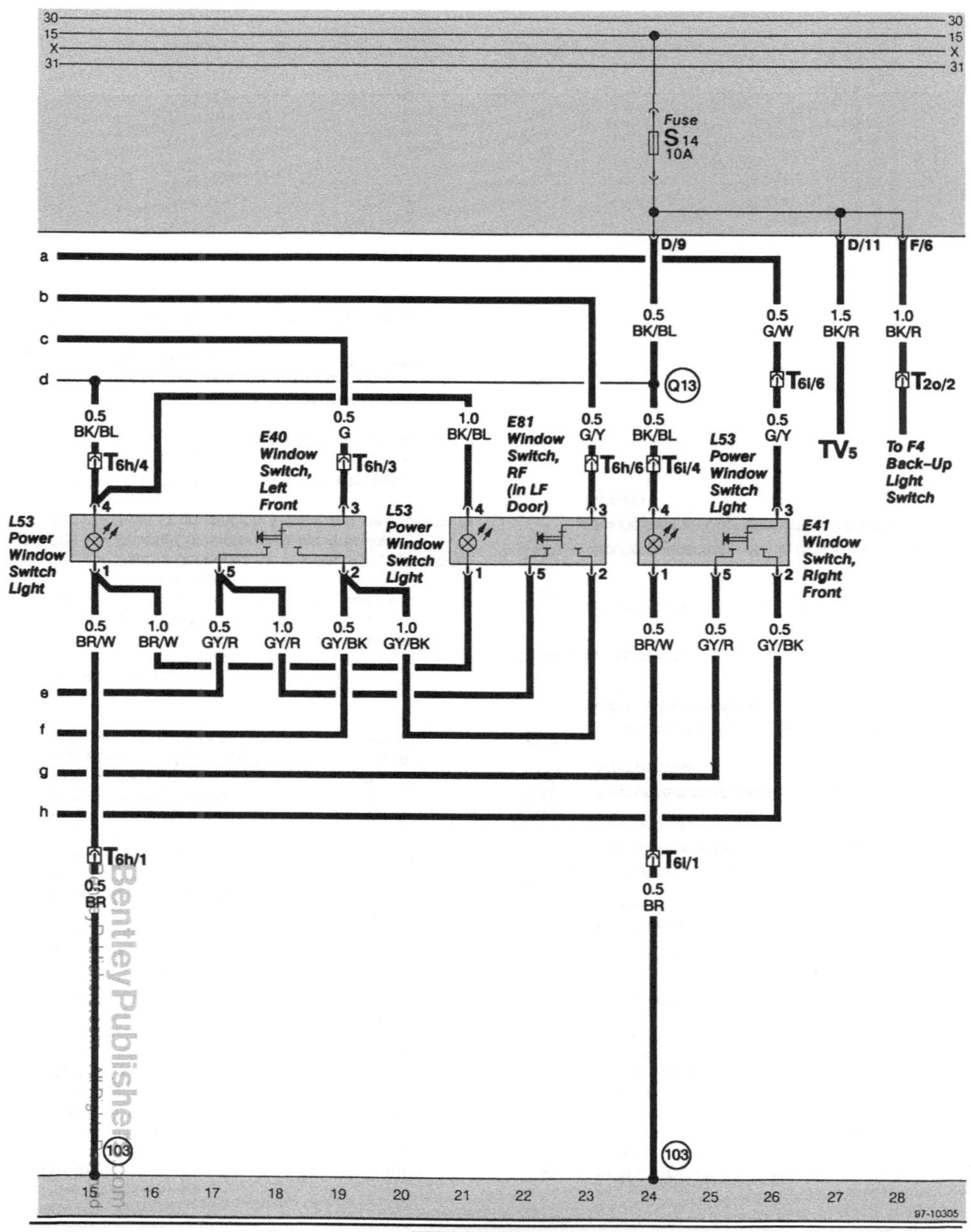

Corrado 1992–1993
Engine code – AAA
Production up to Aug. 1992

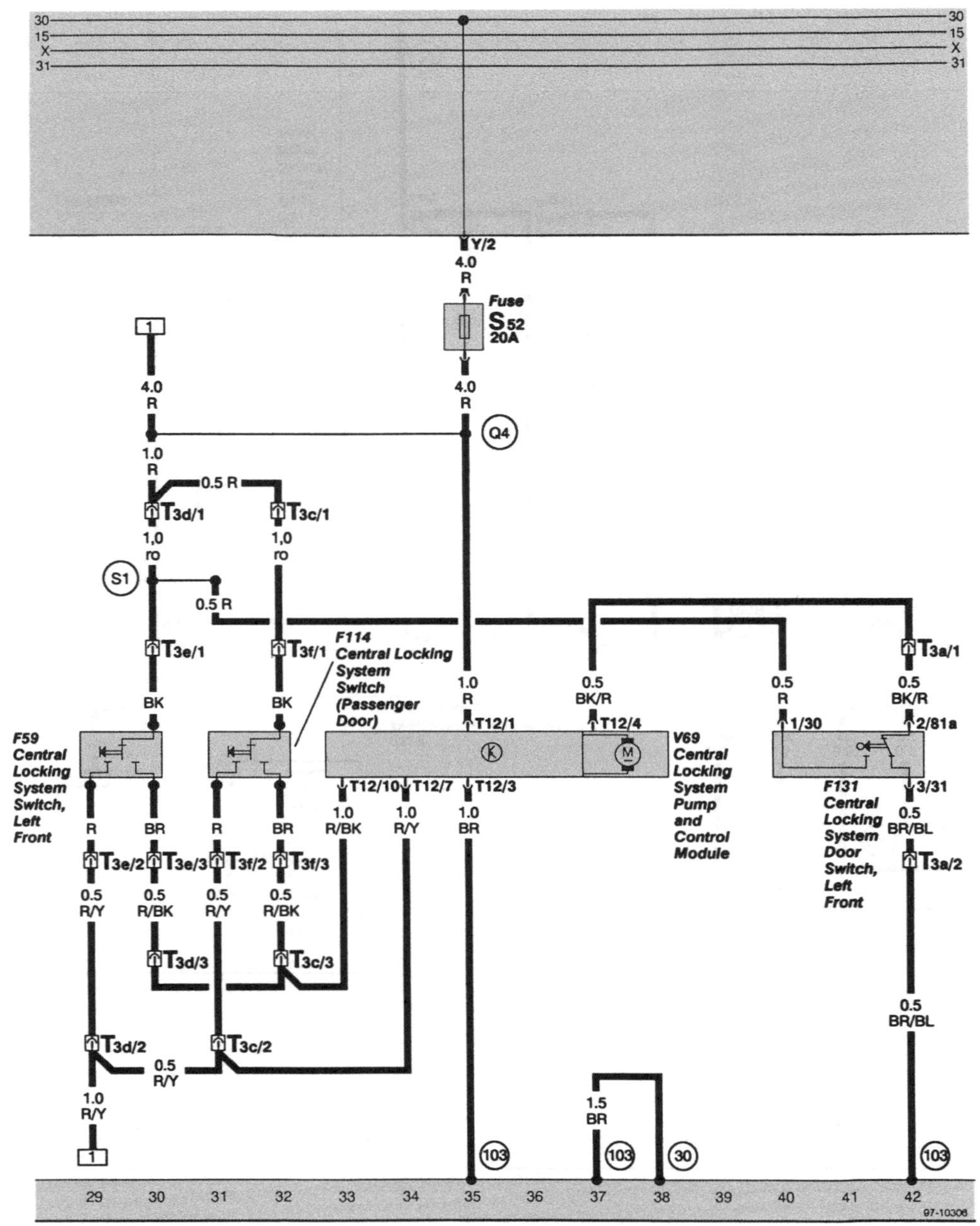

Corrado 1992–1993
Engine code – AAA
Production up to Aug. 1992

Description	Current track
Central Locking System Door Switch, Left Front, F131	40-42
Central Locking System, Fuse, S52	36
Central Locking System Pump and Control Module, V69	34-39
Central Locking System Switch, Passenger's Door, F114	32-33
Central Locking System Switch, Driver's Door, F59	30-31
Door Contact Switch, Left Front, F2	2-3
Door Contact Switch, Right Front, F3	4-5
Fuse, S14, 10A	24
Power Window, Sunroof Control Module, J139	1-14
Power Window Switch Light, L53	15,21,24
Right Front Window Switch (in Driver's Door), E81	21-23
Window Motor, Left, V14	7
Window Motor, Right, V15	9
Window Switch, Left Front, E40	17-19
Window Switch, Right Front, E41	24-26

Wire connectors

- T1 – Single Connector, behind fuse/relay panel
- T1a – Single Connector, behind fuse/relay panel
- T2a – Double Connector, on A-pillar, left, near fuse/relay panel
- T2b – Double Connector, on A-pillar, right, upper part
- T2o – Double Connector, near transmission
- T3a – 3-pin Connector, on A-pillar, left, near fuse/relay panel
- T3c – 3-pin Connector, on A-pillar, right, near fuse/relay panel
- T3d – 3-pin Connector, on A-pillar, left, near fuse/relay panel
- T3e – 3-pin Connector, in driver's door
- T3f – 3-pin Connector, in passenger's door
- T6h – 6-pin Connector, on A-pillar, left, near fuse/relay panel
- T6i – 6-pin Connector, on A-pillar, right, upper part
- T8a – 8-pin Connector, on Power Window Control Module
- T8b – 8-pin Connector, on Power Window Control Module
- T8c – 8-pin Connector, on Power Window Control Module
- T12 – 12-pin Connector, on Central Locking System Pump and Control Module
- TV5 – Wire Connector, Terminal 15a, above fuse/relay panel

Welded wiring harness points

- (Q4) – Plus connection (30), in power window/central locking system and door contact switch wiring harness
- (Q13) – Plus connection –1– (75 as/15), in power windows/central locking and door contact switch wiring harness
- (Q45) – Wire connection (comfort), in power window wiring harness
- (S1) – Plus connection (30) –1–, in central locking system wiring harness

Ground connections

- (30) – Ground connection, –1–, beside fuse/relay panel
- (103) – Ground connection, –2–, in power window/central locking system/door contact switch wiring harness

Corrado starting 1993
Production from Aug. 1992

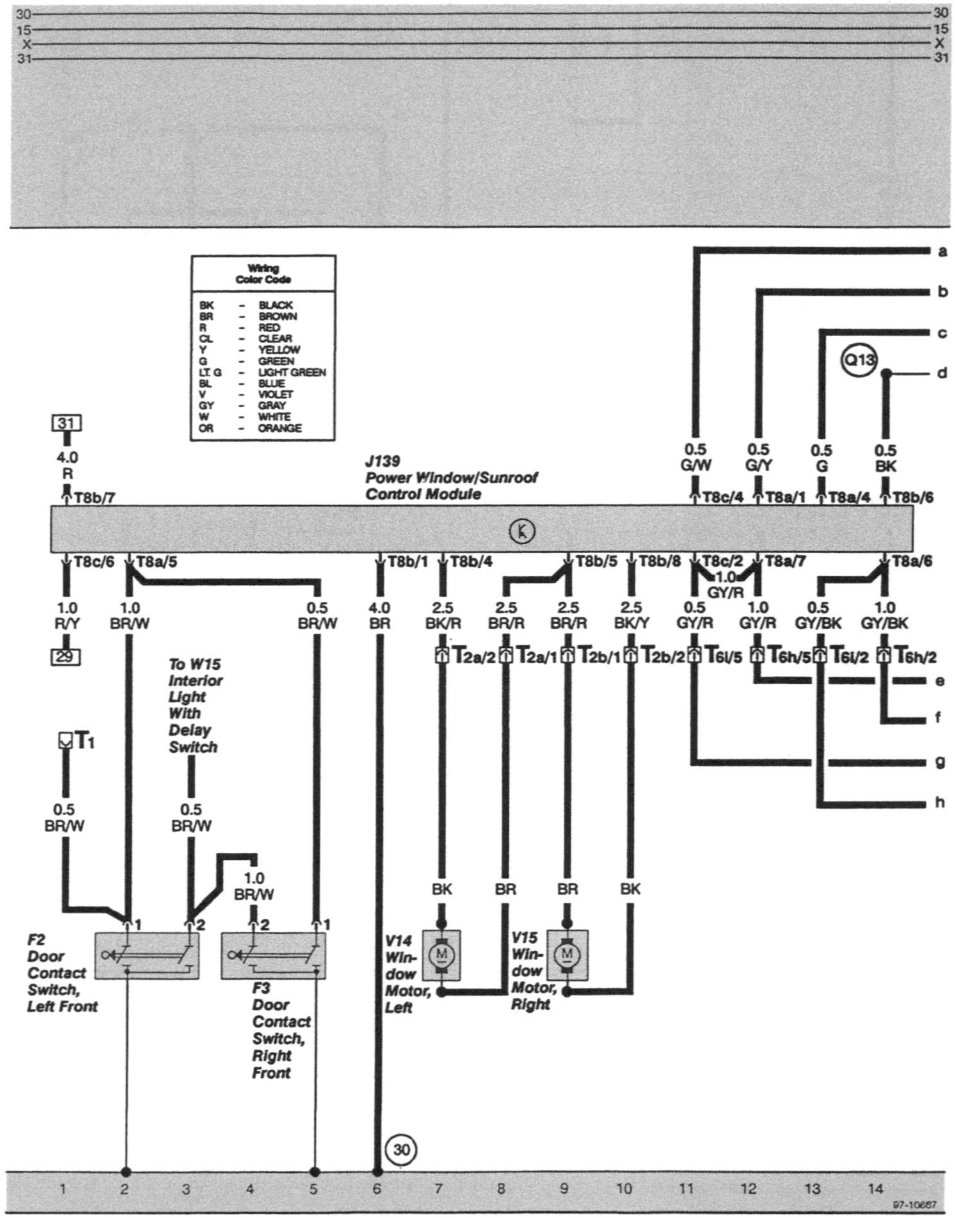

Corrado starting 1993
Production from Aug. 1992

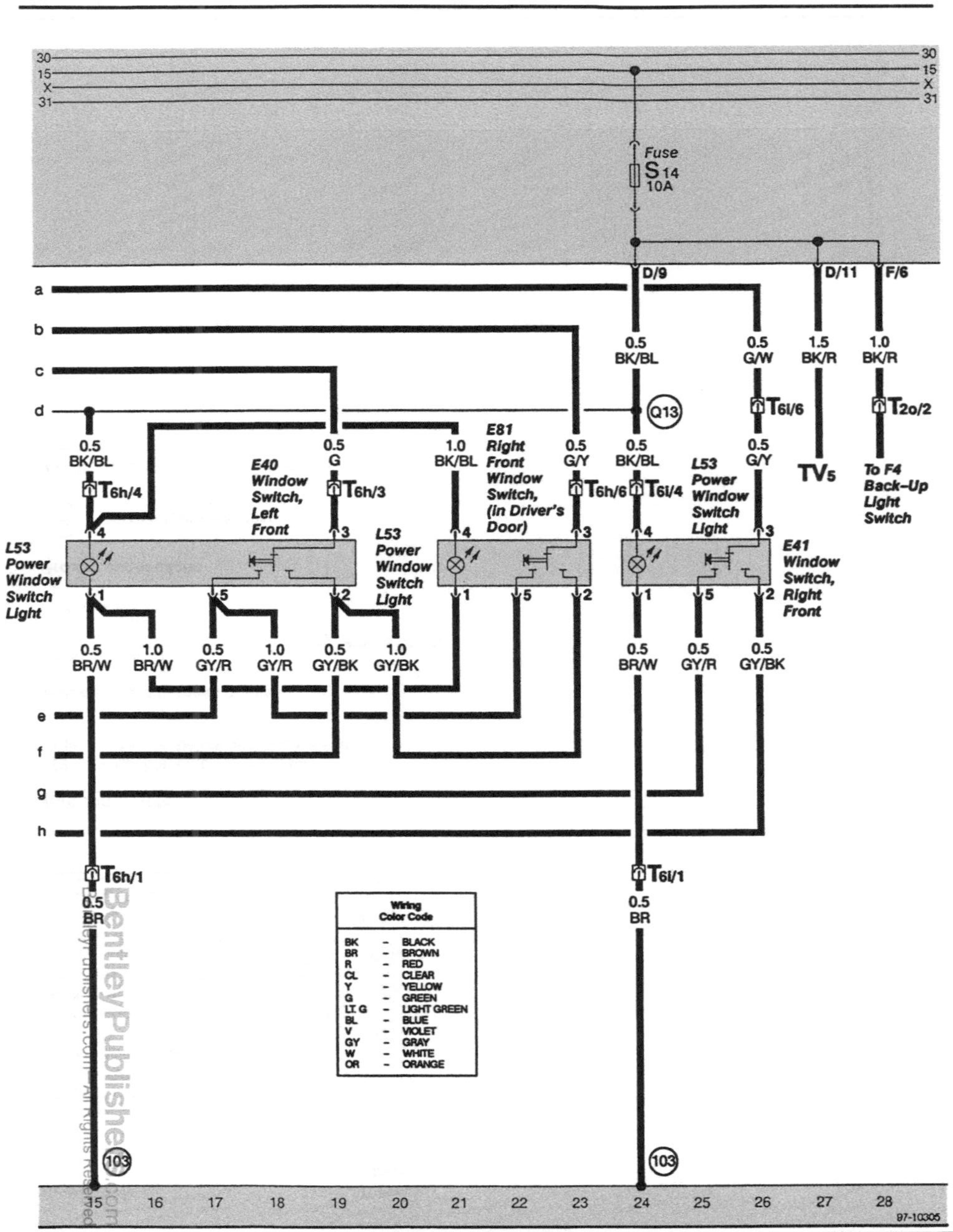

Corrado starting 1993
Production from Aug. 1992

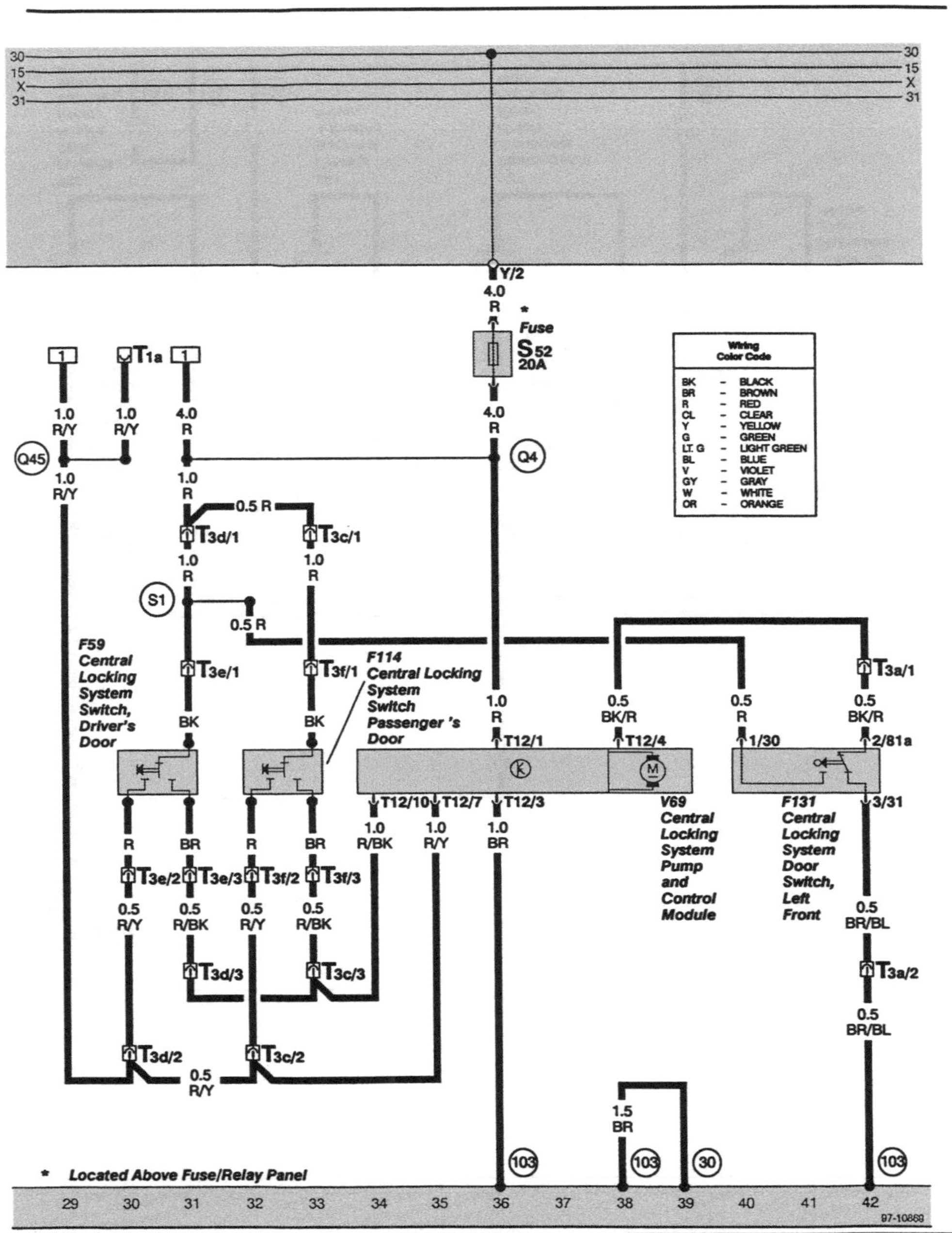

Corrado starting 1993
Production from Aug. 1992

Description	Current track
Circuit breaker, S45	1
Driver's backrest adjusting motor, V45	5
Driver's backrest switch, E96	5–6
Driver's seat height (rear) adjusting motor, V30	1
Driver–seat height adjusting switch, rear, E63	1–3
Front passenger seat height adjusting switch, rear, E66	8–10
Passenger seat backrest adjustment control switch, E98	12–13
Passenger's front–seat height (rear) adjusting motor, V33	8
Passenger's seat backrest adjustment electr. motor, V46	12

Wire connectors

T2 – double, below driver's seat
T2a – double, below passenger's seat

Welded wiring harness points

(M1) – plus connection (30), in wiring harness power seats

Ground connections

(30) – beside fuse/relay panel

(95) – in wiring harness seat-adjuster

Corrado starting 1992
Engine code – AAA
Canada only

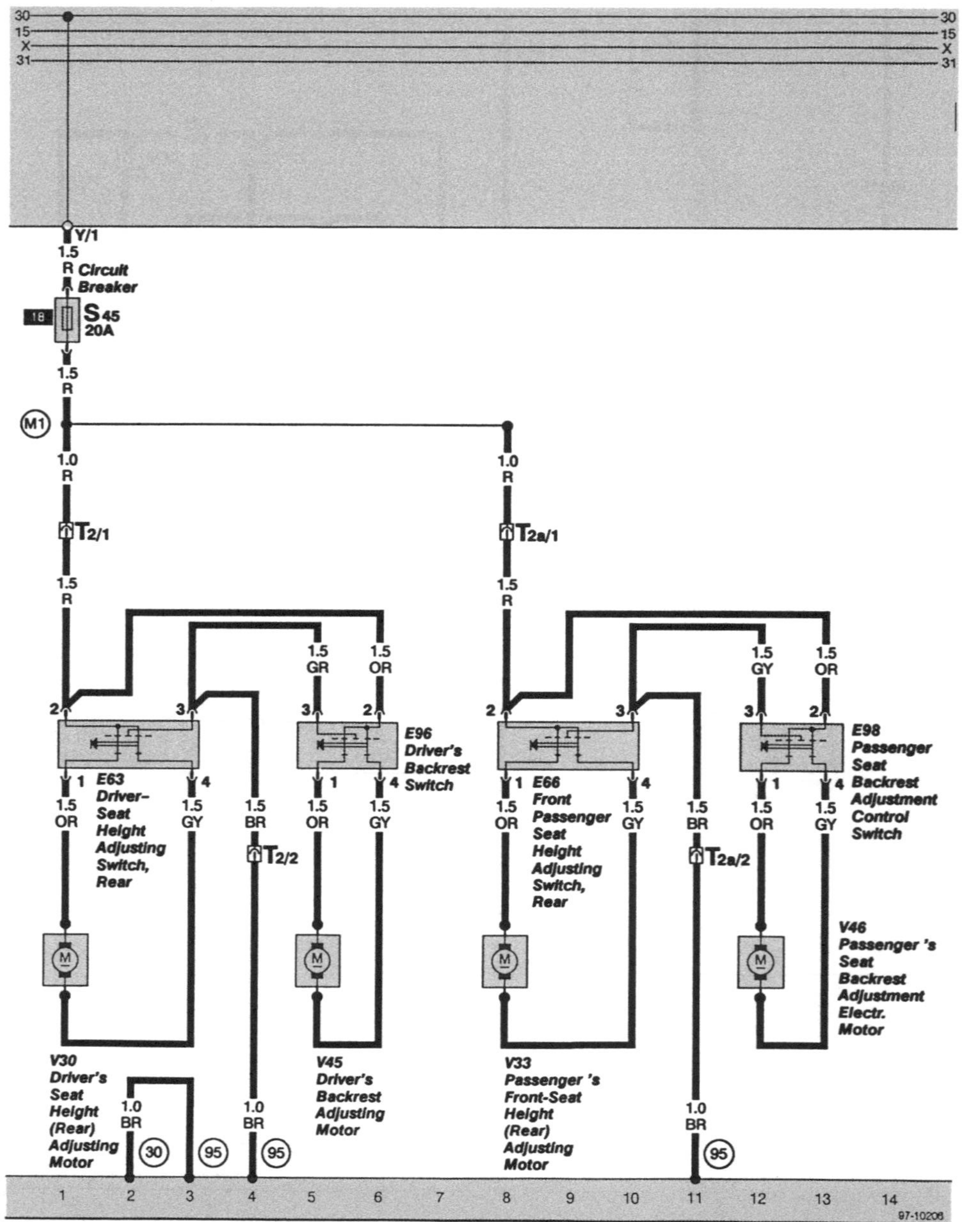

Corrado starting 1992
Engine code – AAA
Canada only

Description	Current track
Door contact switch, left front, F2	8-9
Door contact switch, right front, F3	6
Ignition/starter switch, D	3-6
Load reduction relay, J59	4-5
Seat belt switch, left, E24	11
Seat belt warning relay, J34	8-14

Wire connectors

T1a — single, behind fuse/relay panel
T2 — double, under driver's seat
T4 — *four-point, behind fuse/relay panel*
T28 — twenty eight-point, on instrument cluster

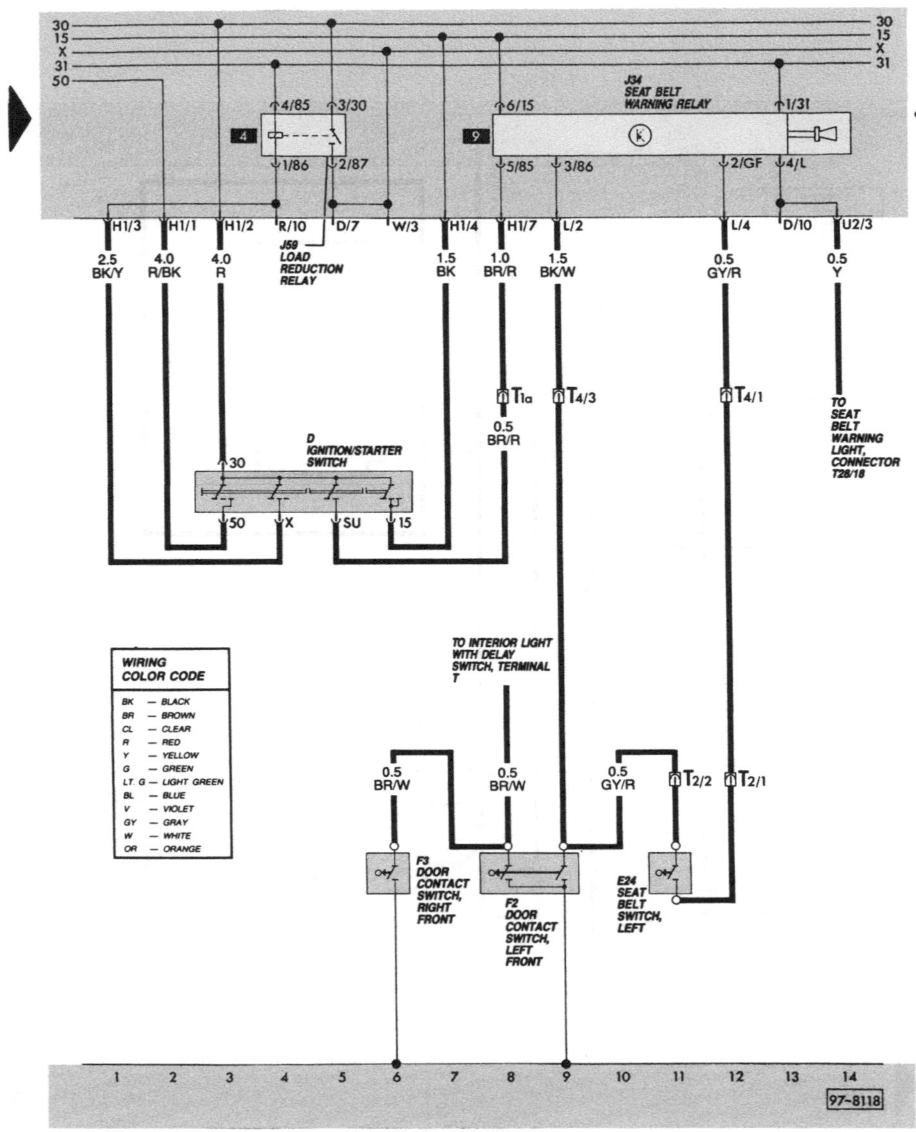

Description	Current track
Amplified speaker (bass), left, R32	6-12
Amplified speaker (bass), right, R33	15-21
Antenna, R11	15-16
Antenna booster, R24	15-16
Fuse, S22	7
Midrange speaker, left front, R26	4
Midrange speaker, left rear, R34	2
Midrange speaker, right front, R27	23
Midrange speaker, right rear, R35	25
Radio, R	9-27

Wire connectors

T1n — single, behind fuse/relay panel
T2b — double, behind instrument panel, center
T3b — three-point, behind instrument panel, center
T4 — four-point, behind instrument panel, right
T8 — eight-point, on radio
T10 — ten-point, on radio
T10a— ten-point, on A-pillar, left, lower part
T10b— ten-point, on A-pillar, right, lower part
T10c— ten-point, on amplified speaker (bass), left
T10d— ten-point, on amplified speaker (bass), right
T28 — twenty eight-point, on instrument cluster
T43 — wire junction for speed signal, behind fuse/relay panel

Ground connections

(43) — on lower right A-pillar

Welded wiring harness points

(A11) — plus connection (58b), in instrument panel wiring harness

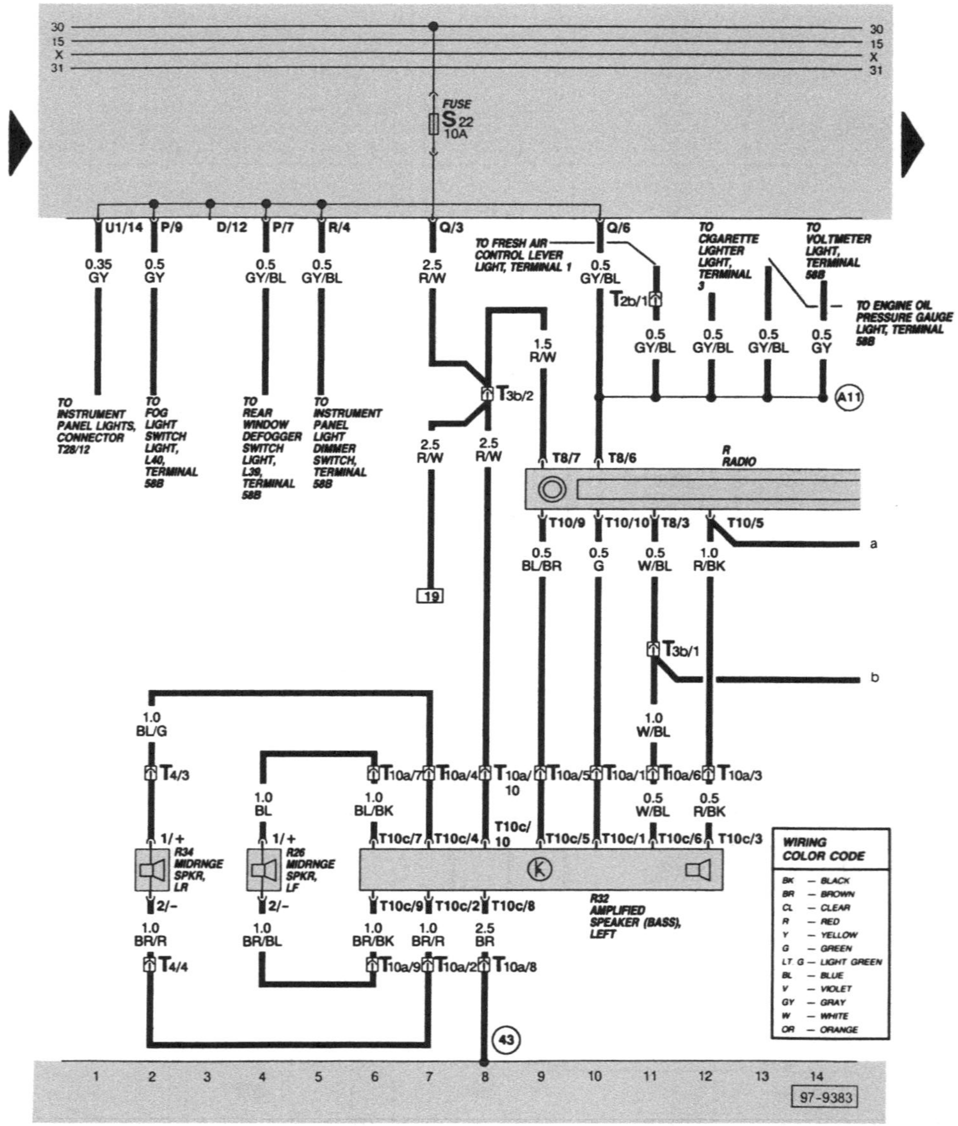

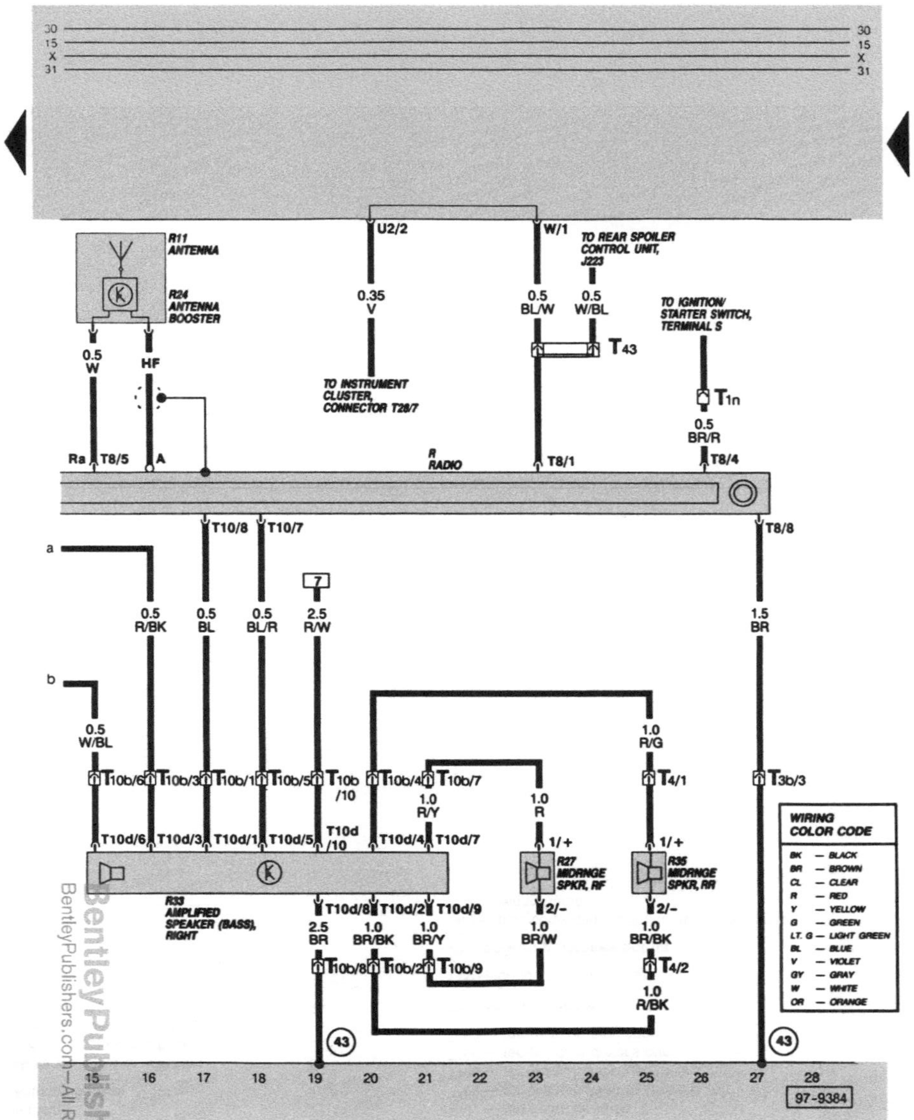

THIS PAGE INTENTIONALLY LEFT BLANK

Description	Current track
Amplified speaker (bass), left	7-13
Amplified speaker (bass), right	15-21
Antenna	17-18
Antenna booster	17-18
Midrange speaker, left front	5
Midrange speaker, left rear	3
Midrange speaker, right front	23
Midrange speaker, right rear	25
Radio	10-27

Wire connectors

T3b — three-point, behind dashboard, center
T4 — four-point, behind dashboard, right
T8 — eight-point, on radio
T10 — ten-point, on radio
T10a — ten-point, behind dashboard, left
T10b — ten-point, behind dashboard, right
T10c — ten-point, in driver's door
T10d — ten-point, in passenger's door

Ground connections

(43) — on lower right A-pillar

Welded wiring harness points

(A11) — plus connection (58b), in instrument panel wiring harness

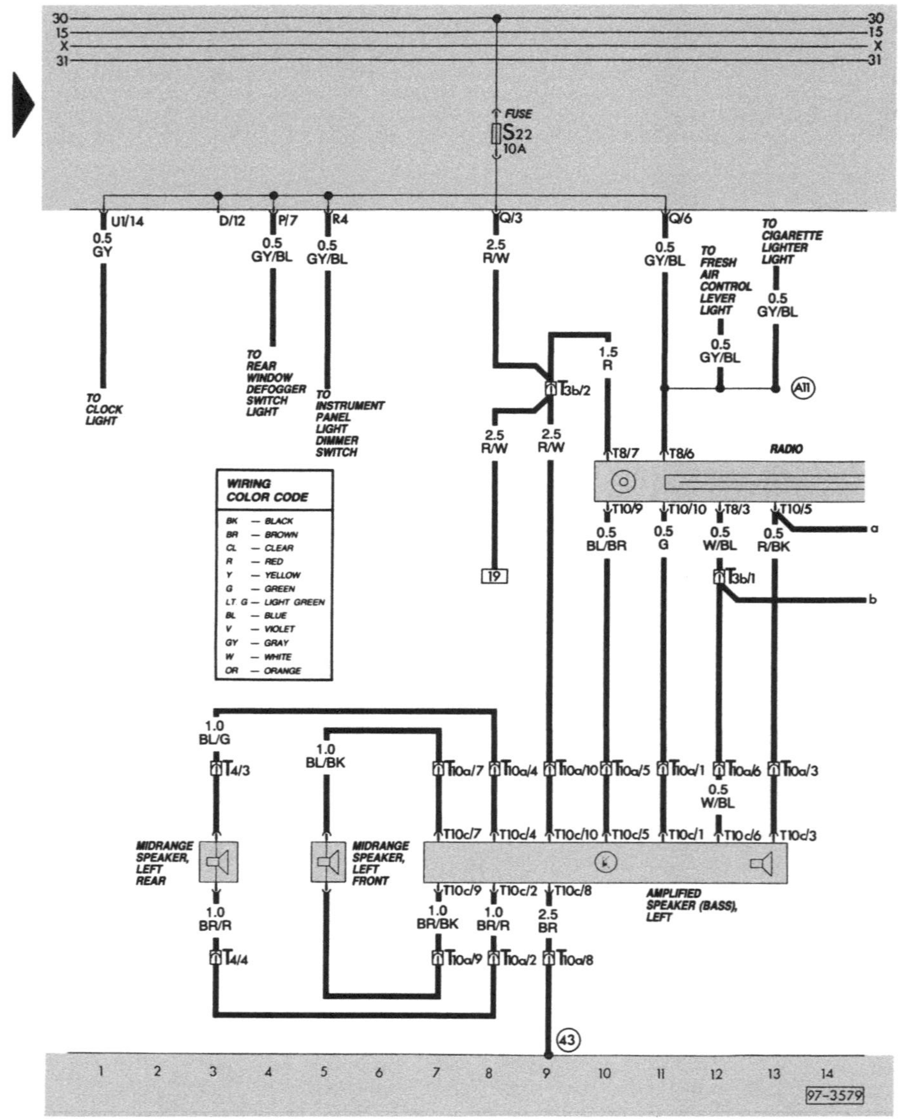

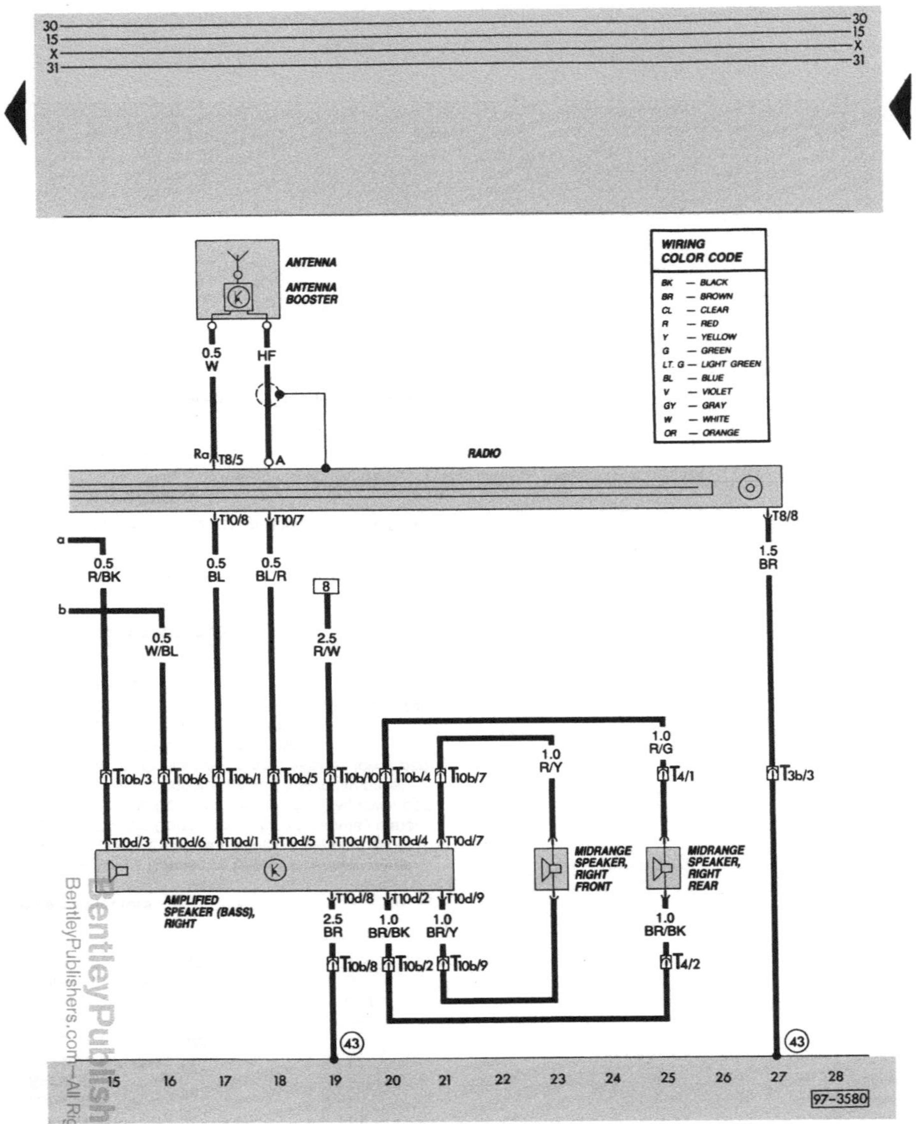

THIS PAGE INTENTIONALLY LEFT BLANK

Description	Current track
Amplifier, R12	10–11
Antenna, R11	10–11
Fuse, S22, 10A	1
Radio, R	1–14
Speaker, Left Rear, R4	12
Speaker, Right Rear, R5	13
Tweeter, Left Front, R20	2–4
Tweeter, Right Front, R22	8–9
Woofer, Left Front, R21	5
Woofer, Right Front, R23	7

Wire connectors

- T1e – Single Connector, behind fuse/relay panel
- T1n – Single Connector, behind fuse/relay panel
- T2 – Double Connector, behind fuse/relay panel
- T2a – Double Connector, on left A-pillar, lower part
- T4 – 4-pin Connector, behind fuse/relay panel
- T4h – 4-pin Connector, on right A-pillar, lower part
- T8 – 8-pin Connector, on Radio
- T8a – 8-pin Connector, on Radio
- T10 – 10-pin Connector, on Alarm System Control Module

Welded wiring harness point

- (A11) – Plus connection (58b), in instrument panel wiring harness
- (B6) – Wire connection (s), in steering column switch wiring harness

Ground connection

- (80) – Ground connection -1-, in instrument panel wiring harness

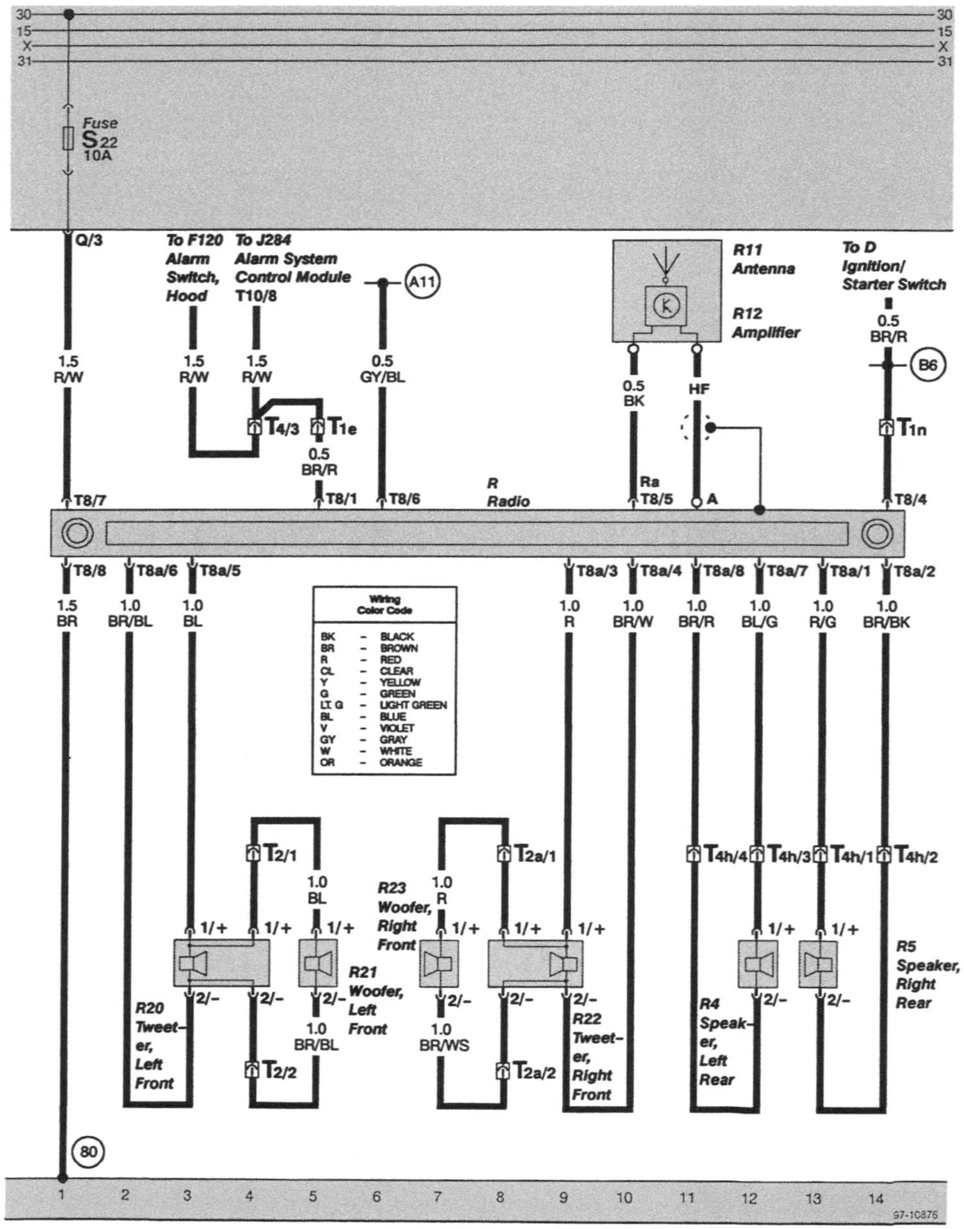

Description	Current track
Overheat fuse	6
Sun-roof stop switch (closed pos.)	5
Sun-roof stop switch (open pos.)	3
Sun-roof switch	2
Sunroof motor	7
Sunroof switch light	6

Wire connectors

T3s — three-point, near interior light

Ground connections

(30) — beside relay panel

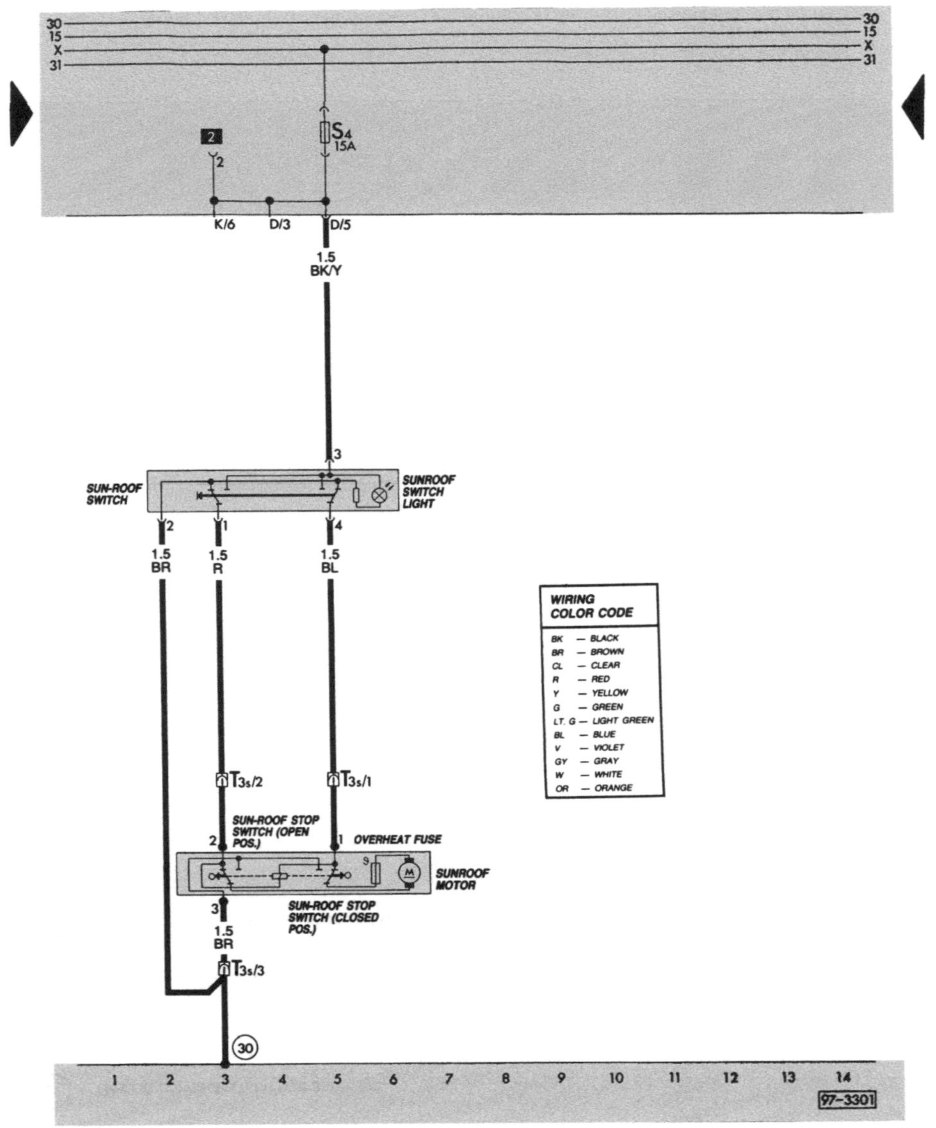

Description	Current track
ABS Brake Fluid Level Warning Switch, F117	17
ABS Control Module (w/EDL), J104	18–50
ABS High Pressure Switch, F109	52
ABS Hydraulic Pump Fuse, S53	53
ABS Hydraulic Pump Relay, J185	50–55
ABS Hydraulic Pump, V64	53
ABS Hydraulic Unit, N55	30–53
ABS Main Valve, N105	30
ABS Pressure Control Switch, F137	44
ABS Pressure Warning Switch I, F115	48
ABS Pressure Warning Switch II, F116	46
ABS Relay, J102	9–14
ABS Speed Sensor Capacitor, C17	35–36, 39–40
ABS Valves Fuse, S54	12
ABS Warning Light, K47	4
Brake and Parking Brake Warning Light, K7	2
Brake Fluid Level Warning Switch, F34	16
Differential Lock Valve 1, N125	41
Differential Lock Valve 2, N126	42
Left Front ABS Inlet Valve, N101	33
Left Front ABS Outlet Valve, N102	34
Left Front ABS Wheel Speed Sensor, G47	39–40
Left Rear ABS Wheel Speed Sensor, G46	37–38
Malfunction Indicator Lamp (MIL), K83	3
Rear ABS Inlet Valve, N103	35
Rear ABS Outlet Valve, N104	36
Right Front ABS Inlet Valve, N99	38
Right Front ABS Outlet Valve, N100	37
Right Front ABS Wheel Speed Sensor, G45	35–36
Right Rear ABS Wheel Speed Sensor, G44	33–34

Wire connectors

T1i – Single Connector, behind fuse/relay panel
T1t – Single Connector, behind fuse/relay panel

T1y – Single Connector
T2 – Double Connector, above fuse/relay panel
T2a – Double Connector, on ABS Hydraulic Unit
T2b – Double Connector, on ABS Hydraulic Unit
T2c – Double Connector, below rear seat, right
T2d – Double Connector, on firewall, right
T2e – Double Connector, below rear seat, left
T2f – Double Connector, on firewall, left
T2g – Double Connector, on ABS Brake Fluid Level Warning Switch
T2h – Double Connector, on Brake Fluid Level Warning Switch
T5 – 5-Pin Connector, on ABS Hydraulic Unit
T5a – 5-Pin Connector, on ABS Hydraulic Unit
T7 – 7-Pin Connector, on ABS Hydraulic Unit
T8d – 8-Pin Connector, on auto-check system monitor
T55 – 55-Pin Connector, on ABS Control Module (w/EDL)
T68 – 68-Pin Connector, on Motronic Engine Control Module (ECM)
TV14 – Data Link Connector (DLC) Wire Connector

Welded wiring harness points

(A15) – Plus connection (15), in instrument panel wiring harness
(J3) – Wire connection -1-, in ABS wiring harness
(J4) – Wire connection (diagnostic), in ABS wiring harness
(J5) – Wire connection -2-, in ABS wiring harness
(W9) – Plus connection (30), in rear wiring harness
(W10) – Plus connection (15), in rear wiring harness

Ground connections

(14) – Ground connection, on transmission
(22) – Ground connection, on hydraulic unit
(28) – Ground connection, on firewall
(39) – Ground connection, below rear seat bench, left
(125) – Ground connection -3-, in headlight wiring harness

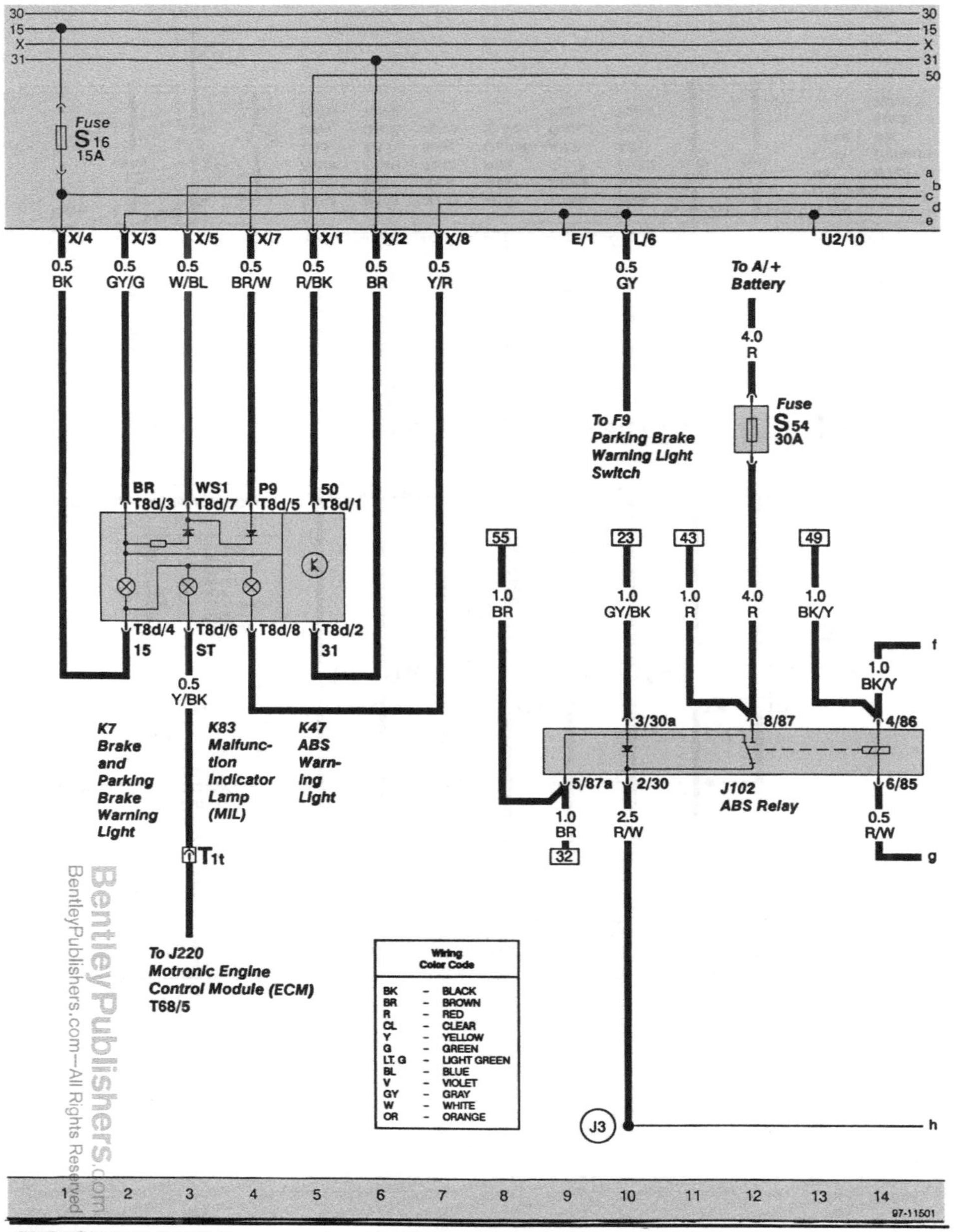

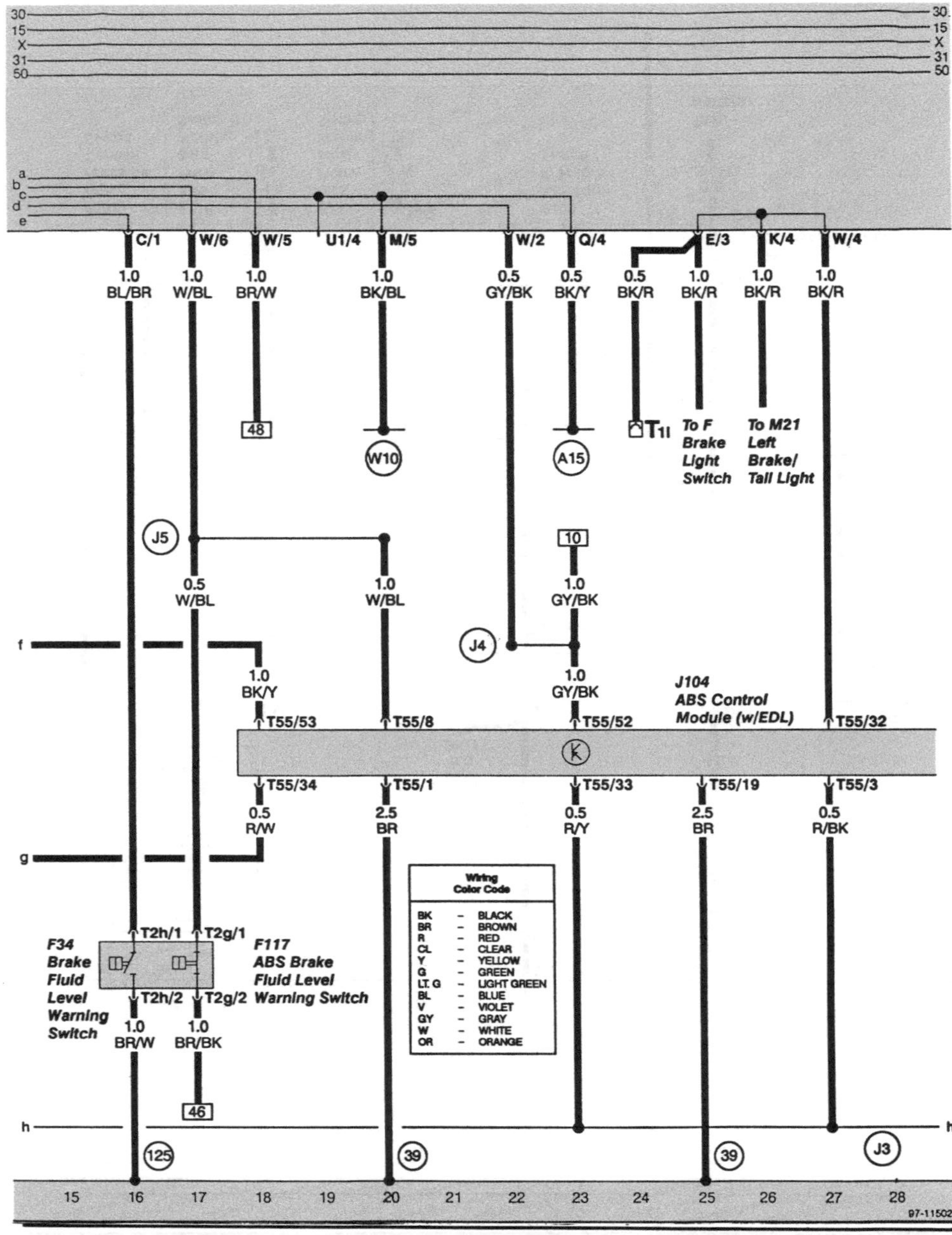

Corrado starting 1994

Corrado starting 1994

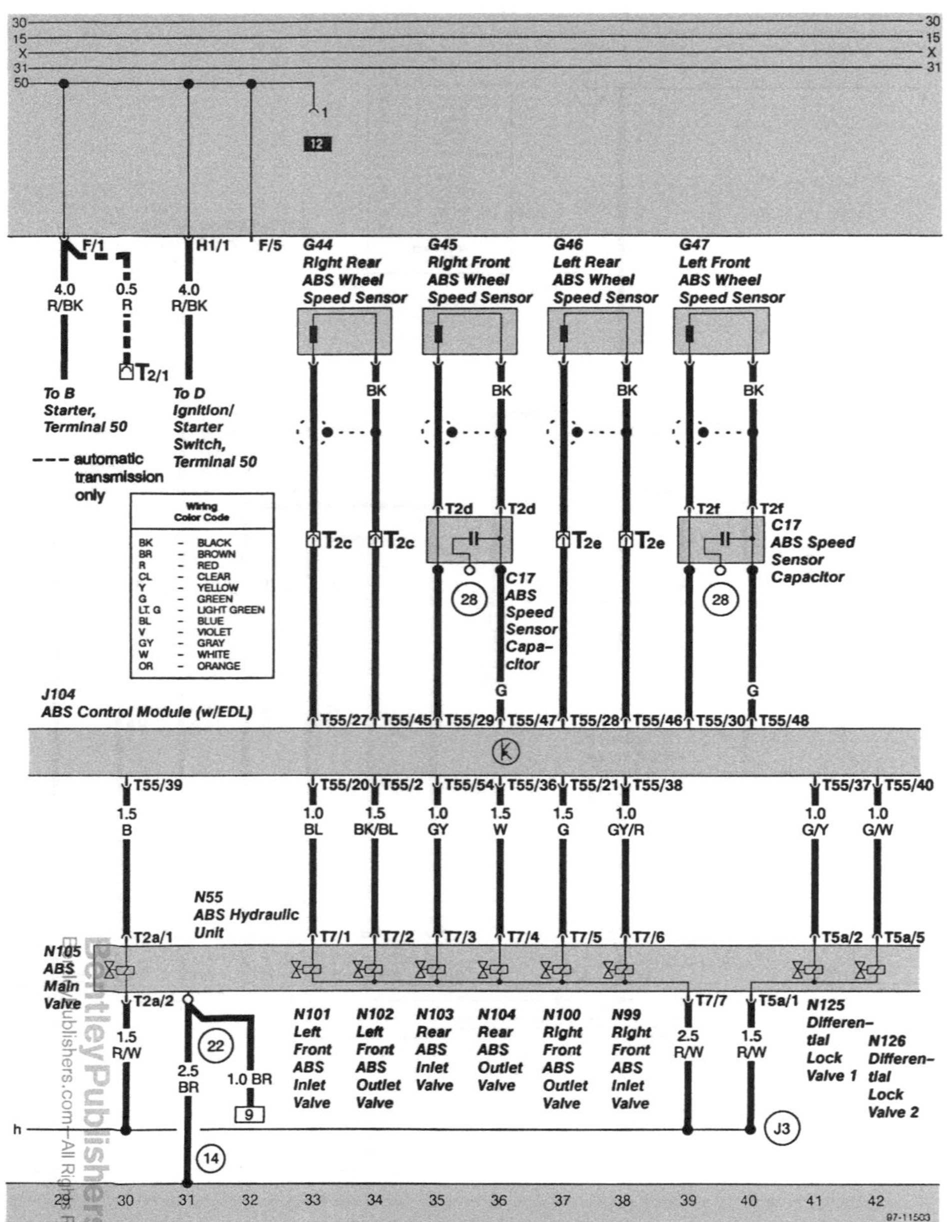

Corrado starting 1994

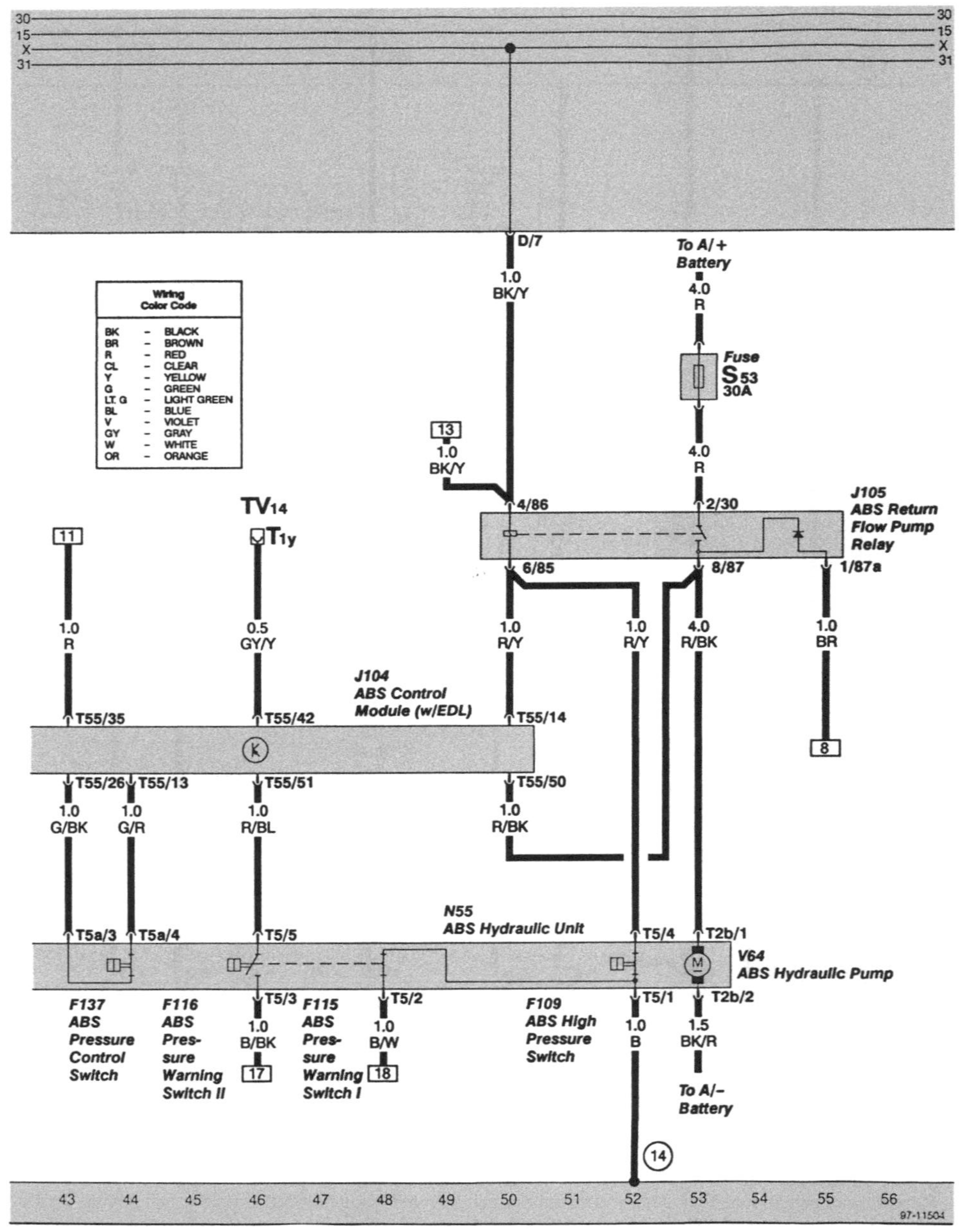

Corrado starting 1994

Note

For wiring pages 333 – 376 see Supplement at rear of book

01 (Fuel Injection and Ignition) On-Board Diagnostic

Engine code PG–Digifant II 01-1

Digifant II system troubleshooting, general information 01–1

Digifant II system, electrical checking 01–2

Engine Code PG–Digifant I 01-9

On-Board Diagnostic, general information (California models from 1991 01–9

Engine Control Module (ECM), technical features 01–12

VAG 1551 Scan Tool, connecting 01–13

Diagnostic Trouble Code (PTC) memory, reading erasing 01-15

Diagnostic Trouble Code (DTC), troubleshooting table 01–19

Function 08, measuring value block reading 01–23

Oxygen Sensor and Oxygen Sensor Control, checking 01–28

Acceleration/Full Throttle Enrichment and Deceleration Fuel Cutoff, checking .. 01–34

Engine Coolant Temperature sensor, checking 01–36

Intake Air Temperature (IAT) sensor, checking 01–40

CO Fuel Trim potentiometer, checking 01–44

Manifold Absolute Pressure (MAP) sensor, checking 01–49

Throttle Position Sensor, checking 01–51

Digifant system, electrical checking 01–55

Digifant I system, Test A electrical checking 01–56

Digifant I system, Test B electrical checking 01–62

Engine Code AAA 01-65

On-Board Diagnostic, general information 01–65

On-Board Diagnostic system, reading/erasing without VAG 1551 Scan Tool 01–67

VAG 1551 Scan Tool, connecting 01–68

VAG 1551 Scan Tool, selectable functions 01–71

Diagnostic Trouble Code (DTC) MEMORY,
reading/erasing using VAG 1551 Scan Tool 01–72

Diagnostic Trouble Code (DTC), Chart 01–77

Output DTM (Diagnostic Test Mode), checking 01–87

Measuring value block, reading 01–96

Scan Tool VAG 1551 voltage supply, checking 01–101

Electrical checks using VAG 1598 Test Box 01–103

Engine code PG | Digifant II

Digifant II system troubleshooting, general information

Engine Code PG applying to:

- 49 States and Canada, model years 1989 to present.
- California vehicles model years 1989 and 1990.

Note:

These vehicles do not have On Board Diagnostic capability. It is not possible to read a DTC memory on these vehicles regardless of method. This includes using the VAG 1551 Scan Tool and the VAG 1527B LED tester.

CAUTION!

A brown data link connector has been installed in the center console of some vehicles. Do not connect the VAG 1551 Scan Tool to this connector; it is not intended for use outside of the factory. When accessed, the codes that it provides will indicate faulty components in cases where this is not true.

Note:

Information for 1991 California vehicles (with Digifant I Engine Management System) starts on page 01-9.

Engine code PG

Digifant II system, electrical checking

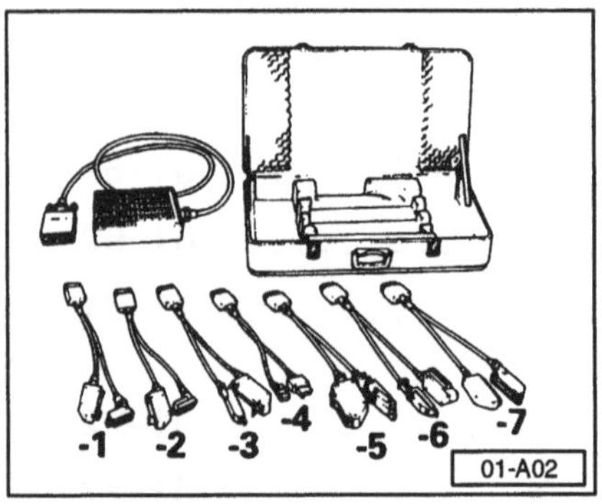

- use Test box VAG 1598 and adaptor cable VAG 1598 along with jumper leads from the VW 1594 adaptor kit to connect the test instruments to the Digifant Engine Control Module ECM
- the terminal numbers of the ECM harness connector and the numbers on the VAG 1598 Test box are identical

Requirements:

- battery voltage OK
- fuse 18 OK
- Digifant system ground wires OK (connections at engine, transmission, battery and body)
- fuel pump and fuel pump relay OK, checking, ⇒ Repair Group 20

Tests on Digifant ECM wiring harness connector (Test A)

♦ switch OFF ignition

♦ disconnect Digifant ECM (J 169) harness connector
  - ECM is installed in plenum

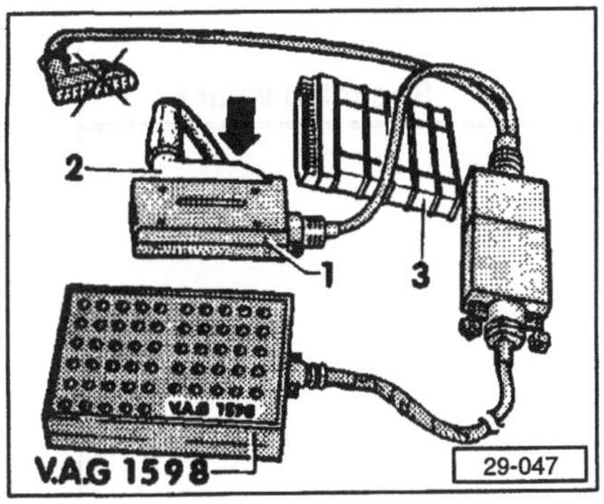

- connect VAG 1598 Test box 1 to ECM harness connector 2 using adaptor wire VAG 1598/2
  - control unit 3 remains disconnected
- switch multimeter to 20 Volt range
- connect multimeter to terminals of Test box as indicated

Test A

Test step	VAG 1598 terminals	To be tested	Test conditions or additional steps	Specified values
1	13 + 23 (starting m.y. 1990)	Power supply relay (J 176) for ECM (J 169)	Switch ON ignition	Approx. battery voltage
2	13 + 14 14 + 19	Voltage supply for ECM (J 169)	Switch ON ignition	Approx. battery voltage
3	13 + 25	Wire to terminal 1 of ignition coil	Switch ON ignition	Approx. battery voltage
4	3 + 14 bridged	Wiring to fuel pump relay	Switch ON ignition	Fuel pumps must run audibly
5	12 + 13	Wiring to fuel injectors (N 30, N 31, N 32, N 33)	Switch ON ignition Remove fuse 18 Bridge terminals 3 + 13	Approx. battery voltage
6	13 + 16	Signal from A/C	A/C system operation OK Switch ON ignition Switch ON A/C	Approx. battery voltage
7	For auto trans only: 1 + 14	Wiring to relay (J 226) for starter lockout	Switch ignition ON Selector lever in P then N Switch selector lever to 1, 2, 3, D and R	Approx. battery voltage No voltage
8	For manual trans only 1 + 13	Wires from starter terminal 50	Operate starter then switch OFF ignition	8 Volts minimum

Note:
Engine malfunctions that only occur at certain temperatures are due to an intermittent signal from the engine coolant temperature sensor. ⇒ Repair Group 24 for engine coolant temperature sensor checking.

Test step	VAG 1598 terminals	To be tested	Test conditions or additional steps	Specified values
♦ switch OFF ignition ♦ switch multimeter to resistance range				
9	6 + 10	Engine coolant temperature sensor (G 62)		
Note: Engine malfunctions that only occur at certain temperatures are due to an intermittent signal from the engine coolant temperature sensor. ⇒ Repair Group 24 for engine coolant temperature sensor checking.				

Test A continued

Test step	VAG 1598 terminals	To be tested	Test conditions or additional steps	Specified values
10	6 + 9	Intake air temp. sensor (G 42) in CO fuel trim potentiometer (G 74)	—	
11	5 + 6	CO fuel trim potentiometer (G 74)	—	0 to 2000 Ohms
12	14 + 22	Idle air control valve (N 71)	—	2 to 10 Ohms
13	6 + 11	Closed throttle position switch (F 6)	Throttle closed Throttle open	1.5 Ohms maximum Infinite Ohms (open)
14	6 + 15	Wide open throttle switch (F 81)	Throttle closed Throttle fully open	Infinite Ohms (open) 1.5 Ohms maximum (6 times)
15	6 + 8	Wiring to camshaft position sensor (G 40)	Disconnect camshaft position sensor and bridge all 3 terminals of harness connector	1.5 Ohms maximum
	6 + 18			1.5 Ohms maximum
16	4 + 7	Wiring to knock sensor (G 61)	Disconnect knock sensor and bridge all 3 terminals of harness connector	1.5 Ohms maximum
17	2 + 13	Wiring to Oxygen sensor	Disconnect oxygen sensor and connect black or lilac wire to ground	1.5 Ohms maximum
			Reconnect oxygen sensor	Infinite Ohms (open)

Engine code PG

With Digifant Engine Control Module (ECM) connected to VAG 1598 Test box (Test B)

Note:

The following checks can only be performed by using the VAG 1598 Test box and VAG 1598/2 adaptor cable.

> ***CAUTION!***
>
> ***When performing Test B:***
>
> - ***do NOT switch measuring ranges on the multimeter while the ignition is switched ON***
> - ***do NOT switch connections between terminals on the Test box while the ignition is switched ON***
>
> ***Ignoring these precautions could destroy the Digifant ECM.***

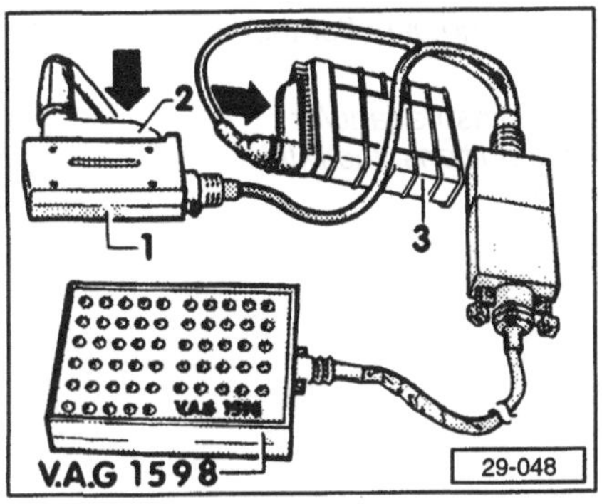

- switch OFF ignition
- connect VAG 1598 Test box using adaptor wire 1 which is connected to both harness connector 2 and Digifant ECM 3
- switch multimeter to 20 Volt range
- connect multimeter to terminals of Test box as indicated

Test B

Test step	VAG 1598 terminals	To be tested	Test conditions or additional steps	Specified values
♦ perform the following test step only if the engine will not start				
1	8 + 6	Voltage supply for camshaft positions sensor (G 40)	Switch ON ignition	10.0 Volts minimum
2	6 + 10	Resistance curve for engine coolant temp. sensor (G 62)	With engine cold, start and let warm up while observing multimeter	Indicated voltage must decrease as engine warms up
♦ perform the following test step only if the engine will not start ♦ switch OFF ignition ♦ use VAG 1527B LED tester with VW 1594 adaptors instead of the multimeter to perform the following checks ♦ disconnect harness connector from fuel rail (main fuel injector connector)				
3	6 + 18	Signal from camshaft position sensor (G 40)	Operate starter	LED tester must flicker
♦ perform the following test step only if the engine will not start				
4	3 + 25	Ignition signal function of Digifant ECM (J 169)	Operate starter	LED must flicker
5	19 + 24 (m.y. 1991)	RPM signal	Operate starter	LED must flicker

Engine code PG | Digifant I

On-Board Diagnostic, general information

Digifant I system (California)

The Digifant I control unit (J 169) for the ignition and fuel injection system is equipped with a Diagnostic Trouble Code (DTC) memory.

If malfunctions occur in the monitored sensors and/or components, they are stored in the DTC memory which also includes data regarding the type of malfunction.

After repairing a malfunction the permanent memory must be erased.

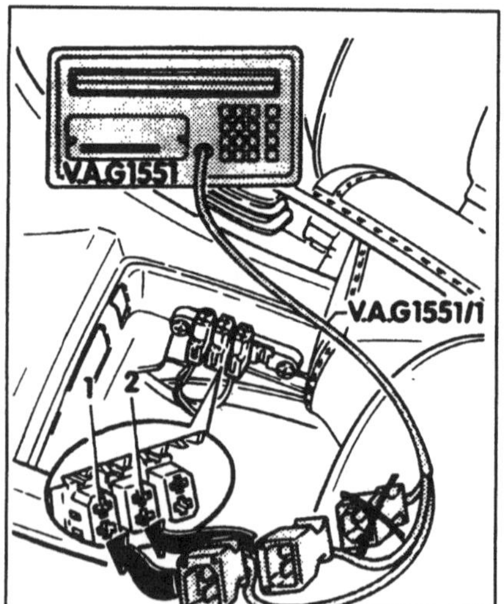

29-027

The VAG 1551 Scan Tool must be used to make full use of the wide range of capabilities of the Digifant On-Board Diagnostic system.

Note:

The Malfunction Indicator Lamp (MIL) must light up when the ignition is switched ON and go out after the engine starts.

If malfunctions occur at monitored sensors (and/or components) under the required conditions, the MIL will light up with the engine running.

DTC memory, reading

- switch ON ignition
- connect jumper cable 357 971 514E to data link connectors located in center console as follows:

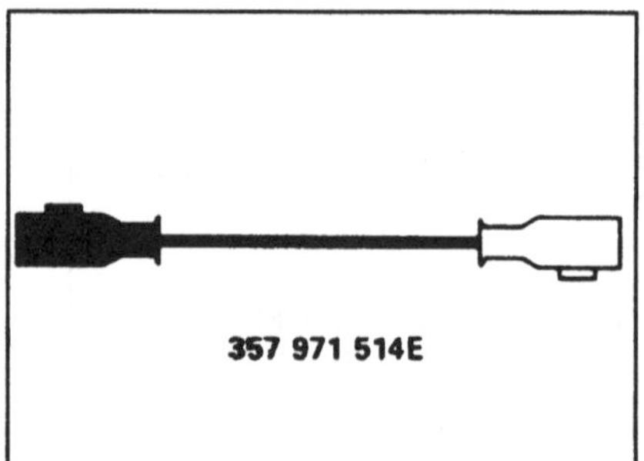

D2-A004

- connect black end of jumper to black data link connector 1
- connect white end of jumper to white data link connector 2
- maintain this connection for approx. 5 seconds
  - MIL light must begin blinking
- disconnect jumper cable
- repeat procedure until blink code 4444 or 0000 "end of output" appears
  - ⇒ DTC code troubleshooting table, page 01-19

> ***CAUTION!***
>
> ***Part numbers are for reference only. Always check with the Parts Department for the latest information.***

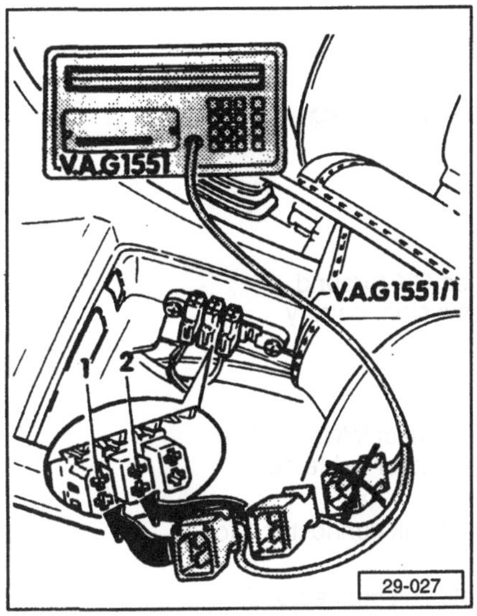

Erasing

- switch OFF ignition
- bridge data link connectors with jumper wire
- switch ON ignition
- remove jumper from data link connectors after approx. 5 seconds
  - MIL light must display blink code 4444 (no malfunctions indicated)
- switch OFF ignition

Engine code PG | Digifant I

Engine Control Module (ECM), technical features

Permanent memory	yes
Temporary memory	no
Data Output	
Rapid data transfer	yes
Malfunction indicator lamp	yes
(MIL light in instrument panel)	
Output Diagnostic Test Mode	no
Basic engine adjustment	no
Read measuring value block	yes
Read individual measured values	no
Engine Control Module, coding	no

Engine code PG | **Digifant I**

VAG 1551 Scan Tool, connecting

Requirements:

- battery voltage OK
- fuses 16, 18 and 21 OK
- Digifant system ground wires OK (ground connections: engine, transmission, battery and body)
- A/C switched OFF

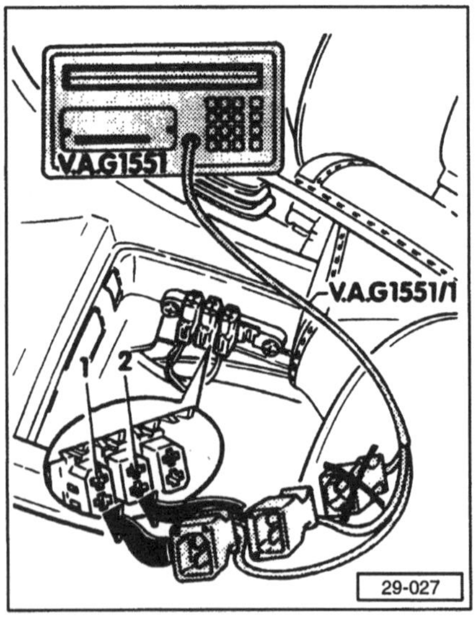

♦ remove shift lever cover and connect VAG 1551 Scan Tool using VAG 1551/1 adaptor cable as follows:

- black connector to black data link connector 1 for voltage supply to scan tool
- white connector to white data link connector 2
- blue connector not used

VAG-ON-BOARD DIAGNOSTIC	HELP
1 - Rapid data transfer *	

VAG-ON-BOARD DIAGNOSTIC	HELP
2 - blink code output *	

◄ The two displays will alternately appear as shown

Note:

♦ *press the HELP button to obtain additional operating information*

♦ *press the → button to step through the program sequence*

Engine code PG | **Digifant I**

Diagnostic Trouble Code (DTC) memory, reading/erasing

Note:

DTC memory cannot be erased until after it has first been read.

- ♦ connect VAG 1551 Scan Tool, ⇒ page 01-13
- ♦ switch ON ignition
- ♦ switch ON printer using print button
  - • LED in print button lights up

Rapid data transfer HELP Input address word XX	◄ press 1 button to select "Rapid data transfer" operating mode • display will appear as shown
Rapid data transfer Q 01 - engine electronics	◄ press 0 and 1 buttons to select "engine electronics" • display will appear as shown
Rapid data transfer Scan tool sends address word 01	◄ press Q button to enter input • displays will appear as shown in the sequence shown

01 (Fuel and Ignition) OBD

037906023 F DIGIFANT 0831 Coding 000 WSC 00000 00000	◄ Control unit identification, e.g.:
Rapid data transfer HELP ECM does not answer!	◄ press → button • display will appear as shown

Note:

By pressing the HELP button, a list of the possible malfunctions can be printed out.

After repairing

- ♦ press 0 and 1 buttons to select "engine electronics"
- ♦ press Q button to enter input

Rapid data transfer HELP Select function XX	◄ press → button • display will appear as shown
Rapid data transfer Q 02 - read DTC memory	◄ press 0 and 2 buttons to select Function "read DTC memory" • display will appear as shown

- ♦ press Q button to enter input
  - • the number of stored malfunctions and/or "No malfunctions recognized!" is displayed

2 malfunctions recognized!	◄ (Example only!)

or

No malfunctions recognized!	◄ (Example only!) • stored malfunctions (if there are any) are displayed and printed out in sequence

Rapid data transfer HELP
Select function XX

◄ press → button
- display will appear as shown

Rapid data transfer Q
05 - erase DTC memory

◄ press 0 and 5 buttons to select Function "erase DTC memory"
- display will appear as shown

Caution! DTC memory was not read!

◄ press Q button to enter input
- display will appear as shown

Note:

This display indicates that the DTC memory was not read. You must first read the DTC memory before you can erase it.

If the ignition was switched OFF between reading DTC memory and erasing; DTC memory will NOT be erased.

Rapid data transfer
DTC memory is erased!

◄ If the DTC memory is successfully erased, the display will appear as shown.

Rapid data transfer HELP
Select function XX

◄ press → button
- display will appear as shown

Rapid data transfer Q
06 - end output

◄ press 0 and 6 buttons to terminate the data flow between the ECM and the VAG 1551 Scan Tool
- display will appear as shown

♦ press Q button to enter input

♦ switch OFF ignition

♦ disconnect VAG 1551/1 adaptor cable from the data link connectors

♦ compare any malfunctions printed out with the DTC troubleshooting table starting on page 01-19, repair or replace as necessary

Note:

The diagnostic trouble codes are arranged in ascending order, based on the 5 digit rapid data transfer code.

♦ test drive vehicle for 5 minutes minimum until the following conditions are obtained
- coolant temperature must reach 80 C (176 F) minimum
- engine RPM must exceed 3,000 at least once
- accelerator pedal must be pushed to the floor completely at least once.

♦ read DTC memory again to ensure that all malfunctions have been corrected

Engine code PG | **Digifant I**

Diagnostic Trouble code (DTC), troubleshooting table

- the following table lists all of the possible malfunctions recognized by the Digifant (J 169) and displayed on the VAG 1551 Scan Tool (with printer switched on while DTC memory is being read)
- if malfunctioning components are indicated, check wiring for short or open circuits using wiring diagram before replacing the components.
- DTC code and blink code are displayed in operating mode 1 "Rapid data transfer" only when the VAG 1551 printer is switched ON

Example:
Rapid Data Transfer code 65535 — **Blink code** (if present) 1111

VAG 1551 printer display	Possible cause	Corrective step
00000 4444 No malfunction recognized!	—	—
00518 2212 Throttle position sensor – (**G69**)	(**G 69**) Malfunctioning	Check (**G 69**), ⇒ Repair Group 24
No signal	—	Break in wiring/short

Example:
Rapid Data Transfer code 65535 — **Blink code** (if present) 1111

VAG 1551 printer display	Possible cause	Corrective step
00519 2222 Manifold absolute pressure sensor (G 71) No signal	Vacuum line to Digifant ECM (J 169) Sensor (G 71) inside of ECM (J 169) malfunctioning	Check (G 71) Repair Group 24
00521 2242 CO Fuel Trim Potentiometer (G 74)	(G 74) malfunctioning	Check (G 74), Repair Group 24
No signal	Open circuit in wiring	
00522 2312 Engine coolant temperature sensor (G 62)	(G 62) malfunctioning	Check (G 62), ⇒ Repair Group 24
Break in wiring/short circuit to positive	Open circuit in wiring	—
00523 2322 Intake air temperature sensor (G 42)	(G 42) malfunctioning in CO fuel trim potentiometer	Check (G 42), ⇒ Repair Group 24
No signal	Open or short circuit in wiring	—
00525 2342 Oxygen sensor (G 39)	(G 39) malfunctioning	Check Oxygen sensor and Oxygen sensor control, ⇒ Repair Group 24
No signal	Open circuit in wiring	

Example:
Rapid Data Transfer code 65535
Blink code (if present) 1111

VAG 1551 printer display	Possible cause	Corrective step
00535 2141*		
First knock sensor regulator	Open circuit or short in the sensor wire	Check wires, page 01-61, test step 15
Regulator limit exceeded	Knock sensor (G 61) malfunctioning	Replace (G 61) ⇒ Repair Group 28
	No knock recognition from Digifant control unit (J 169)	Replace (J 169) ⇒ Repair Group 28
00537 2341		
Oxygen sensor control limit exceeded	Fuel tank empty	Erase DTC memory, ⇒ page 01-15
	CO content not within specified range	Check CO content, and idle speed, ⇒ Repair Group 24
	CO content too high	Check exhaust system sealing between exhaust manifold and Oxygen sensor ⇒ Repair Group 26
		Oxygen sensor short circuit to ground – Repair Group 24
		Check EVAP Canister Purge Regulator valve, ⇒ Repair Group 20
	CO content too low	Check intake system for sealing, ⇒ Repair Group 24
		Oxygen sensor short circuit to positive (+), ⇒ Repair Group 24
*Fault indicator lamp will display 2142 during blinkcode output		

Example:
Rapid Data Transfer code 65535
Blink code (if present) 1111

VAG 1551 printer display	Possible cause	Corrective step
01249 4411		Check wiring using Test step 6, page 01-58
Cylinder 1 injector (N 30) malfunctioning	Open circuit in wiring for fuel injector (N 30)	
	Fuel injector for cylinders 1, 2, 3 or 4 malfunctioning	Check injectors, ⇒ Repair Group 24
65535 1111		
ECM malfunctioning	—	Replace ECM (J 169), ⇒ Repair Group 24

Engine code PG | **Digifant I**

Function 08, measuring value block reading

- ♦ connect VAG 1551 Scan Tool, ⇒ page 01-13

If engine won't start

- ♦ switch ON ignition

If engine will start

- ♦ start engine and let idle

```
Rapid data transfer                HELP
Input address word XX
```

◄ press button 1 to select operating mode "Rapid data transfer"
- • display will appear as shown

```
Rapid data transfer                   Q
01 - engine electronics
```

◄ press 0 and 1 buttons to select Function 01 "engine electronics"
- • display will appear as shown

```
Rapid data transfer                   Q
Scan tool sends address word         01
```

◄ press Q button to enter input
- • displays will appear in sequence as shown

```
037906023 F DIGIFANT 0831
coding      000       WSC        00000
```

◄ ECM identification, e.g.:

```
Rapid data transfer                HELP
ECM does not respond!
```

◄ If display appears as shown

- ♦ press HELP button to print out list of possible DTC causes


```
Rapid data transfer                HELP
Select function XX
```

◄ After correcting:

- ♦ press 0 and 1 buttons again to select "engine electronics"
- ♦ press Q button to enter input
- ♦ press → button
  - • display will appear as shown

```
Rapid data transfer                   Q
08 - read measuring value block
```

◄ press buttons 0 and 8 to select Function "Read measuring value block"
- • display will appear as shown

- ♦ select desired display group number corresponding to the display group overview shown in the following texts
- ♦ press Q button to enter input for example: display group number 01

```
Read measuring value block            Q
Input display group number           01
```

◄ press 0 and 1 buttons to select "display group number 01"
- • display will appear as shown

```
Read measuring value block           1→
        1        2        3           4
```

◄ press Q button to enter input
- • channels will be displayed as shown.

Note:

- *when any of the display group numbers 01 to 05 are displayed, the group number is shown in the "read measuring value block" display without a zero*
- *when display group 00 is displayed, "read measuring value block" is shown by itself without the zeroes*
- *before selecting other display groups, first press the C button*

Display group overview

Display group number 01

Channel identification within display
1 = engine RPM
2 = manifold vacuum (load)* (via Throttle Position sensor)
3 = coolant temperature
4 = injection time

*Sensor (G 71) is inside of Digifant ECM (J 169)

Read measuring value block	2→
1 2 3 4	

- display for this group appears as shown

Display group number 02

Channel identification within display
1 = engine RPM
2 = not for customer service
3 = not for customer service
4 = injection time

Read measuring value block	2→
1 2 3 4	

- display for this group appears as shown

Display group number 03

Channel identification within display
1 = engine RPM
2 = coolant temperature
3 = not for customer service
4 = injection time

Read measuring value block	3→
1 2 3 4	

- display for this group appears as shown

Display group number 04

Channel identification within display
1 = engine RPM
2 = manifold vacuum (load)*
3 = not for customer service
4 = injection time

Read measuring value block	4→
1 2 3 4	

- display for this group appears as shown

Display group number 05

Channel identification within display
1 = engine RPM
2 = throttle position sensor
3 = not for customer service
4 = injection time

Read measuring value block	5→
1 2 3 4	

- display for this group appears as shown

Display group number	Channel identification within display
00	**1** = intake air temperature **2** = CO fuel trim–potentiometer **3** = coolant temperature **5** = oxygen sensor probe signal
Read measuring value block → 1 2 3 4 5 6 7 8 9 10	• display for this group appears as shown

Note:

The remaining display channels are not currently in use,

**Manifold Absolute Pressure Sensor (G 71) is inside of Digifant ECM (J 169)*

Engine code PG **Digifant I**

Oxygen Sensor and Oxygen Sensor Control, Checking

Requirements

- no leaks in exhaust system between catalytic converter and cylinder head
- voltage supply for Oxygen sensor heater OK

♦ connect VAG 1551 Scan Tool, ⇒ page 01-13 for additional information

♦ start engine and let idle

♦ select Function 08 "Read measuring value block", ⇒ page 01-23 for additional information

or

♦ while observing VAG 1551 display, select the following functions in sequence

1 – Rapid data transfer

01 – Engine electronics

08 – Read measuring value block

♦ press 0 key twice to select display group number 00

Read measuring value block	Q
Display group number input	00

- display will appear as shown

♦ press Q key to enter input

- following display appears

Measuring value block reading									
1	2	3	4	5	6	7	8	9	10

continue with checking only after:

- the displayed value in channel 3 is less than 11 (which corresponds to a coolant temperature of at least 80 C or 176 F)

♦ let engine idle for 2 more minutes

- read display value in channel 5 (Oxygen sensor signal)

Display value	Possible cause	Corrective step
Varies more than **14**	No fault, Oxygen sensor control **OK**	Checking complete
0 – 5	Short circuit to positive (+) in wiring between Oxygen sensor and control unit	Perform Test **A**
164 – 168	Open circuit in wiring between Oxygen sensor and control unit	Perform Test **B**
230 minimum	Short circuit to ground (–) in wiring between Oxygen sensor and control unit	Perform Test **C**

Test A

♦ disconnect Oxygen sensor/heater harness connector (on valve cover)

- displayed value must be 164 - 168

If YES

- Oxygen sensor control OK

♦ switch OFF ignition

♦ replace Oxygen sensor (G 39)

♦ read DTC memory and then erase, ⇒ page 01-15 for additional information

- displayed value must be 0 to 5

If NO

- Oxygen sensor control not OK

♦ switch OFF ignition

♦ check wiring for short circuit to ECM positive (+), repair or replace as necessary, ⇒ page 01-61, Test A, step 16

If wiring OK

♦ replace Digifant ECM (J 169)

♦ read DTC memory and then erase, ⇒ page 01-15 for additional information

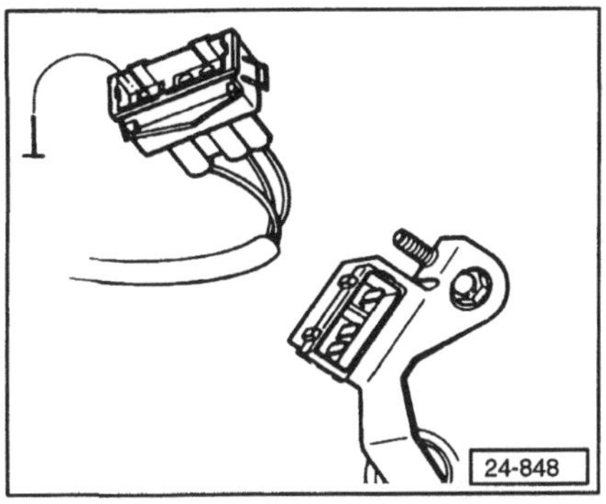

Test B

- ♦ disconnect Oxygen sensor/heater harness connector (on valve cover)
- ♦ touch violet wire (from Digifant ECM) to ground (-) using jumper from VW 1594 adaptor kit
  - displayed value must be 230 minimum

If YES
- Oxygen sensor control OK

- ♦ switch OFF ignition
- ♦ replace Oxygen sensor (G 39)
- ♦ read DTC memory and then erase, ⇒ page 01-15 for additional information
  - displayed value must be 164 - 168

If NO
- Oxygen sensor control NOT OK

- ♦ switch OFF ignition
- ♦ check wiring for open circuit to ECM using wiring diagram, repair or replace as necessary, also ⇒ page 01-61, Test A, step 16

If wiring OK

- ♦ replace Digifant ECM (J 169)
- ♦ read DTC memory then erase, ⇒ page 01-15 for additional information

Test C

- ♦ disconnect Oxygen sensor/heater harness connector (on valve cover)
  - displayed value must be 164 - 168

If YES
- Oxygen sensor control OK

- ♦ switch OFF ignition
- ♦ replace Oxygen sensor (G 39)
- ♦ read DTC memory and then erase, ⇒ page 01-15 for additional information
  - displayed value must be 230 minimum

If NO

- Oxygen sensor control NOT OK

♦ switch OFF ignition

♦ check wiring for short circuit between ECM and ground, repair or replace as necessary, also ⇒page 01-61, Test A, step 16

If wiring OK

♦ replace Digifant ECM (J 169)

♦ activate DTC memory and then erase, ⇒ page 01-15 for additional information

Engine code PG **Digifant I**

Acceleration/Full Throttle Enrichment and Deceleration Fuel Cutoff, Checking

Note:

Use the following procedure to check the deceleration fuel cutoff. If this check and the check for the throttle position sensor are OK, acceleration and full throttle enrichment are also OK.

♦ connect VAG 1551 Scan Tool, ⇒ page 01-13 for additional information

♦ start engine and let idle

♦ select Function 08 "read measuring block value", ⇒ page 01-23 for additional information

or:

♦ while observing VAG 1551 display, select the following functions in sequence

1 – Rapid data transfer

01 – Engine electronics

08 – Read measuring block value

♦ press 0 and 1 key to select display group number 00

Read measuring value block Q Display group number input 01

◄ • display will appear as shown

- ♦ press Q key to enter input
  - • display will appear as shown

Read measuring value block 1 1 2 3 4

◄ Continue with checking only if:
- • the engine coolant temperature displayed in channel 3 is 80 C (176 F) minimum

- ♦ while observing channel 1 displayed value, raise engine speed in excess of 3000 RPM three times

- ♦ let engine idle for 2 minutes

- ♦ observe displayed value in channel 4

- ♦ raise engine speed to a minimum of 3000 RPM by stepping on accelerator pedal then releasing
  - • the value displayed in channel 4 must increase in value; drop to 0 ms; then return to the previously observed value

If YES
- • acceleration/full throttle enrichment and deceleration fuel shutoff OK

If displayed value does not increase and does not return to 0 ms

- ♦ check throttle position sensor, ⇒ Repair Group 24

Engine code PG | Digifant I

Engine Coolant temperature sensor, checking

Requirement
- • engine cold

- ♦ connect VAG 1551 Scan Tool, ⇒ page 01-13 for additional information

- ♦ start engine and let idle

- ♦ select Function 08 “Read measuring value block”, ⇒ page 01-23 for additional information

or

- ♦ while observing VAG 1551 display, select the following functions in sequence

  1 – Rapid data transfer

  01 – Engine electronics

  08 – Read measuring value block

- ♦ press keys 0 and 1 to select “Display group number 01”

Read measuring value block Q
Display group number input 01

◄ • display will appear as shown

♦ press Q key to enter input

Read measuring value block 1
1 2 3 4

◄ • channels will be displayed as shown

♦ observe display value in channel 3

Display value	Possible cause	Corrective step
Must increase uniformly without interruption*	Coolant temperature sensor (G 62) and control OK	Checking complete
–40 C	Break in wiring, sensor wiring, or wiring to ECM	Perform Test A
80 C minimum **	Short circuit to negative (–) between coolant temperature sensor, wiring or ECM	Perform Test B

* If engine malfunctions occur during certain temperature ranges and the displayed value in channel 3 does NOT increase uniformly and without interruption, the problem is a faulty coolant temperature sensor.

If YES

♦ replace engine coolant temperature sensor (G 62)

** If the displayed value, 80 C (176 F) minimum does not correspond to the actual coolant temperature, e.g. actual coolant temperature under 60 C
♦ perform test A

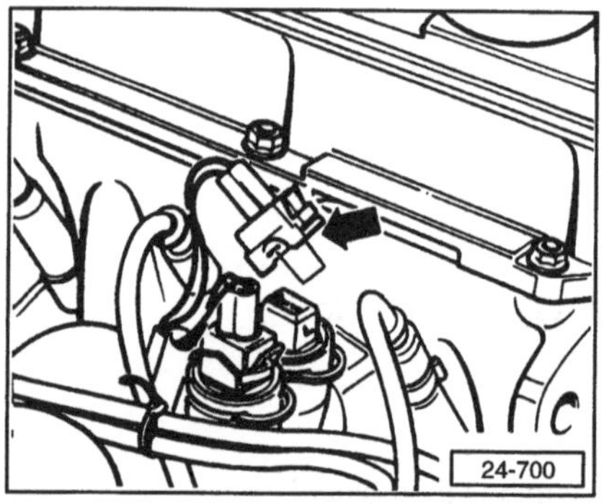

Test A

♦ disconnect engine coolant temperature sensor harness connector (blue) and bridge the connector terminals using jumpers from VW 1594 adaptor kit
  • displayed value must be 80 C (176 F) minimum

If YES
  • engine coolant temperature sensor control function OK

♦ switch OFF ignition

♦ replace engine coolant temperature sensor (G 62)

♦ read DTC memory and erase — ⇒ page 01-15 for additional information

Test B

♦ switch OFF ignition

♦ disconnect (blue) engine coolant temperature sensor harness connector

♦ switch Fluke 83 multimeter to resistance range

♦ connect multimeter between terminals of engine coolant temperature sensor using jumpers from VW 1594 adaptor kit

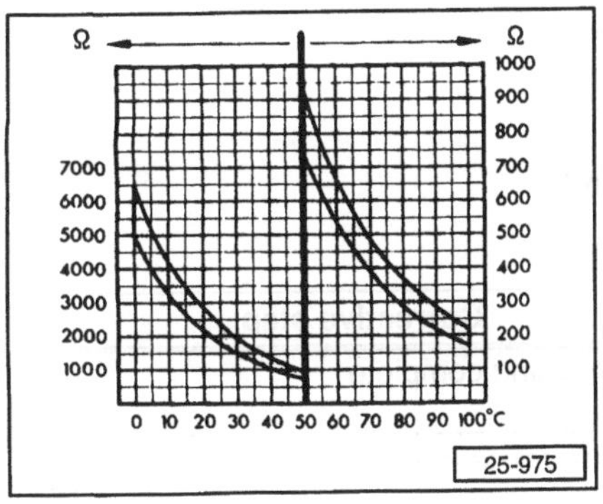

Specified Value: per chart

If OK

- re-connect engine coolant temperature sensor (G 62) harness connector
- check wiring for short circuit to ECM negative (–), repair or replace as necessary, ⇒ page 01-59 Test A, step 9

If wiring OK

- replace Digifant ECM (J 169)
- read DTC memory and then erase, ⇒ page 01-15 for additional information
- return ECM to basic setting, ⇒ Repair Group 01 (Transmission) OBD

If measured value is out of range

- replace engine coolant temperature sensor (G 62)

Engine code PG **Digifant I**

Intake Air Temperature (IAT) Sensor, checking

- connect VAG 1551 Scan Tool, ⇒ page 01-13 for additional information
- switch ON ignition
- select Function 08 "Read measuring value block", ⇒ page 01-23 for additional information

or

- while observing VAG 1551 display, select the following functions in sequence

  1 – Rapid data transfer

  01 – Engine electronics

  08 – Read measuring value block

- press 0 keys twice to select "Display group number 02"

Read measuring value block	Q
Display group number input	00

◄ • display will appear as shown

♦ press Q key to enter input

Read measuring value block 2
1 2 3 4 5 6 7 8 9 10

◄ • display will appear as shown

♦ read displayed value in channel 1

Display value	Possible cause	Corrective step
10 to 190	Engine coolant temperature sensor (G 62) and control OK	Checking complete
at least 250	Open circuit in wiring between sensor and ECM	Perform Test A
0	Short circuit to positive (+) in wiring between sensor and ECM	Perform Test B

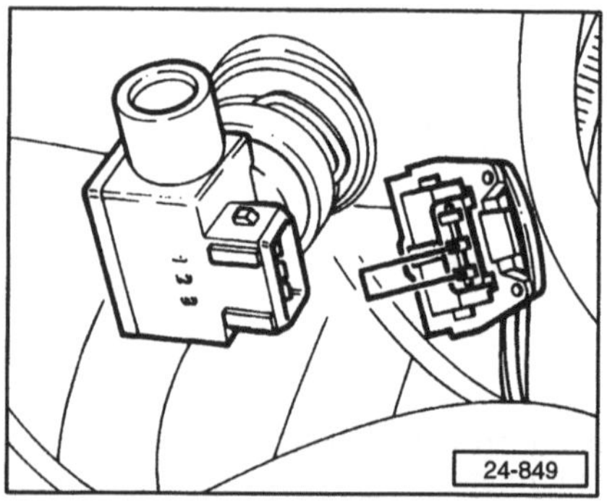

Test A

- ♦ disconnect CO fuel trim potentiometer harness connector (with sensor for intake air temperature and bridge terminals 2 and 3 of the connector using jumpers from VW 1594 adaptor kit
  - • displayed value must be 0

If YES

- • Intake air temperature sensor control OK
- ♦ switch OFF ignition
- ♦ replace CO fuel trim potentiometer (G 74) and intake air temperature sensor (G 42)
- ♦ read DTC memory and erase, ⇒ page 01-15 for additional information

Test B

- ♦ switch OFF ignition
- ♦ disconnect CO fuel trim potentiometer harness connector

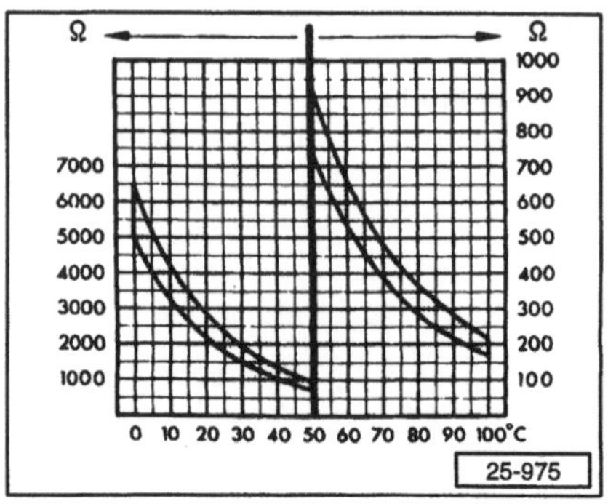

- ♦ measure sensor resistance across terminals 2 and 3 using Fluke 83 multimeter and jumper wires from VW 1594 adaptor kit
  - • measured value must fall within appropriate area of chart

If specified value obtained

- ♦ reconnect CO potentiometer harness connector
- ♦ check wiring for short circuit to ECM negative ([–]) using wiring diagram, repair or replace as necessary, also ⇒ page 01-59, Test A, step 10

If wiring OK

- ♦ replace Digifant ECM
- ♦ read DTC memory and then erase, ⇒ page 01-15 for additional information

If the specified value not obtained

- ♦ replace CO fuel trim potentiometer (G 74)
  - comes complete with with intake air temperature sensor (G 72)
- ♦ read DTC memory and then erase, ⇒ page 01-15 for additional information
- ♦ check idle speed and CO content, adjust if necessary, ⇒ Repair Group 24 for additional information

Engine code PG | Digifant I

CO Fuel Trim potentiometer, checking

Requirement
- idle speed OK
- coolant temperature 80 C (176 F) minimum

- ♦ connect VAG 1551 Scan Tool, ⇒ page 01-13 for additional information
- ♦ start engine
- ♦ select Function 08 "Read measuring value block", ⇒ page 01-15 for additional information

or

- ♦ while observing VAG 1551 display, select the following functions in sequence

1 – Rapid data transfer

01 – Engine electronics

08 – Read measuring value block

- ♦ press 0 keys twice to select "display group number 00"

Read measuring value block	Q
Display group number input	00

◄ • display will appear as shown

♦ press Q key to enter input

Read measuring value block 1
1 2 3 4 5 6 7 8 9 10

◄ • channels will be displayed as shown

♦ observe displayed value in channel 2

Display value	Possible cause	Corrective step
40 — 140	CO fuel trim potentiometer (G 74) and triggering OK	Checking complete
0 ... 2	Short negative (–): wire, CO fuel trim potentiometer, ECM	Perform Test A
at least 250	Open circuit in wiring between potentiometer and ECM	Perform Test B

* The displayed value depends on the CO content adjustment during idle adjustment.

Test A

♦ disconnect CO fuel trim potentiometer harness connector
- display value: 250 minimum

If YES
- triggering of CO potentiometer (G 74) OK

♦ switch OFF ignition

♦ replace CO fuel trim potentiometer assembly (G 74)
- includes intake air temperature sensor (G 42)

♦ read DTC memory and then erase, ⇒ page 01-9 for additional information

♦ check idle and CO values, adjust if necessary
- displayed value must be 0 to 2

If NO
- CO fuel trim potentiometer triggering (G 74) NOT OK

♦ switch OFF ignition

♦ reconnect CO fuel trim potentiometer (G 74) harness connector

♦ eliminate short in wiring between ECM and ground using wiring diagram, also ⇒ page 01-59, Test A, step 11

If wiring OK

- replace ECM (J 169)
- read DTC memory and then erase,
  ⇒ page 01-15 for additional information

Test B

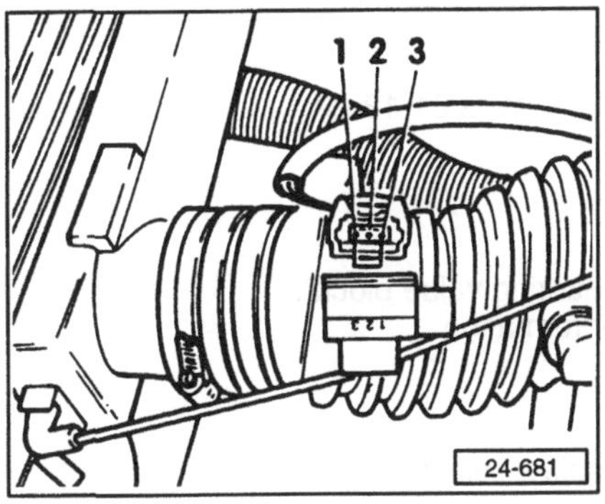

- disconnect CO fuel trim potentiometer harness connector (with intake air temperature sensor) and bridge terminals 1 and 3 of connector using jumper wires from VW 1594 adaptor kit
  - displayed value must be 0 to 2

If YES
  - ECM triggering OK

- switch OFF ignition
- replace CO fuel trim potentiometer (G 74) assembly
  - includes intake air temperature sensor (G 42)
- read DTC memory and then erase,
  ⇒ page 01-15 for additional information
- check idle speed and CO content, adjust if necessary
  - displayed value must be 250 minimum

If NO
  - ECM triggering not OK

- switch OFF ignition
- reconnect CO fuel trim potentiometer (G 74) harness connector
- determine open circuit and eliminate using wiring diagram, ⇒ page 01-60, Test A, step 11

If wiring OK

- replace Digifant ECM (J 169)
- read DTC memory and then erase,
  ⇒ page 01-15 for additional information

Engine code PG | Digifant I

Manifold Absolute Pressure (MAP) sensor, checking

Note:

The vacuum sensor (G 71) is installed inside the Digifant ECM (J 169).

- connect VAG 1551 Scan Tool, ⇒ page 01-13 for additional information
- start engine and let idle
- select Function 08 "Read measuring value block", ⇒ page 01-23 for additional information

or

- while observing VAG 1551 display, select the following functions in sequence

1 – Rapid data transfer

01 – Engine electronics

08 – Read measuring value block

- press 0 and 1 keys to select "display group number 01"

Read measuring value block	Q
Display group number input	00

- display will appear as shown
- press Q key to enter input

Read measuring value block	1
1 2 3 4	

- channels will be displayed as shown
- continue with the test only if the coolant temperature in channel 3 is 80 C minimum
- read display value in channel 2
  - displayed value must be 20 to 40% fluctuating

If YES

- vacuum line and vacuum sensor (G 71) inside Digifant ECM (J 169) OK

If displayed value 45% (minimum) and constant

- vacuum line and sensor (G 71) inside Digifant ECM (J 169) NOT OK

- check vacuum line between throttle body and ECM, replace if necessary

If vacuum line OK

- replace Digifant ECM
- read DTC memory and then erase, ⇒ page 01-15 for additional information

Throttle Position Sensor, checking

Note:

The throttle position sensor provides information to the Digifant ECM (J 169) for the following functions.

With throttle closed:

- injector shut off during fuel deceleration cutoff
- idle air control
- ignition adjustment during idle using idle air control function

with throttle partially opened:

- determines injection quantity for acceleration enrichment

with throttle fully opened:

- determines injection quantity during full throttle enrichment

Checking

- ♦ connect VAG 1551 Scan Tool, ⇒ page 01-13 for additional information
- ♦ switch ON ignition
- ♦ select Function 08 "Read measuring value block", ⇒ page 01-23 for additional information

or

- ♦ while observing VAG 1551 display, select the following functions in sequence

  1 – Rapid data transfer

  01 – Engine electronics

  08 – Read measuring value block

- ♦ press 0 and 5 keys to select "Display group number 05"

Read measuring value block	Q
Display group number input	05

◄ • display will appear as shown

- ♦ press Q key to enter input

Read measuring value block	5
1 2 3 4	

◄ • channels will be displayed as shown

- ♦ read displayed value in channel 2

Display value	Possible cause	Corrective step
Close throttle stop 5 — 50<	Throttle position sensor (G 69) and control OK	Checking complete
Wide open throttle stop* min, 90>	—	—
0.0>	Open circuit in wiring between throttle position sensor and ECM	Perform Test A
min. 90> at closed throttle stop	Short circuit to positive (+) in wiring between throttle position sensor and ECM	Perform Test A

- open throttle slowly and uniformly
  - displayed value in channel 2 must uniformly increase without interruption

Test A

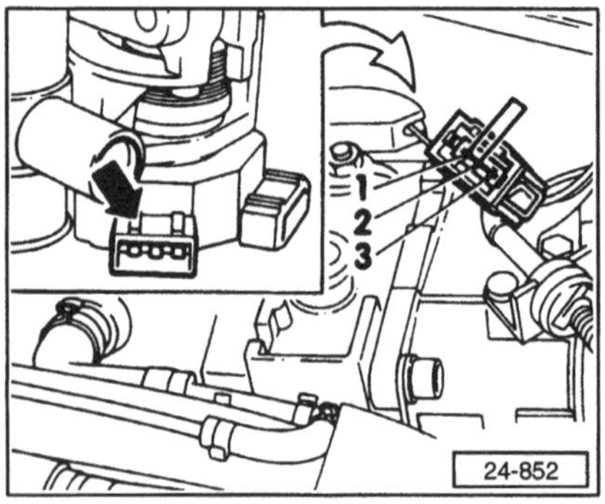

- disconnect 3-pin throttle position sensor (G 69) harness connector (arrow A)
  - displayed value must be 0.0 >
- bridge terminals 1 and 2 of the harness connector (arrow 2) using jumpers from VW 1594 adaptor kit
  - displayed value must be 90 > minimum

If YES

- throttle position sensor control OK

- switch OFF ignition
- replace throttle position sensor (G 69)

- read DTC memory and then erase, ⇒ page 01-15 for additional information

Vehicles with automatic transmission only:

- after removal/installation or replacement of throttle position sensor return ECM to basic adjustment, ⇒ 01 (Transmission) OBD

If NO

- throttle position sensor control NOT OK

- switch OFF ignition
- check wiring for open circuit or short to ECM positive (+), repair or replace as necessary, also ⇒ page 01-60, Test A, step 13 and/or Test B steps 1 and 2

If wiring OK

- replace Digifant ECM (J 169)
- read DTC memory and then erase, ⇒ page 01-15 for additional information

Engine code PG | Digifant I

Digifant system, electrical checking

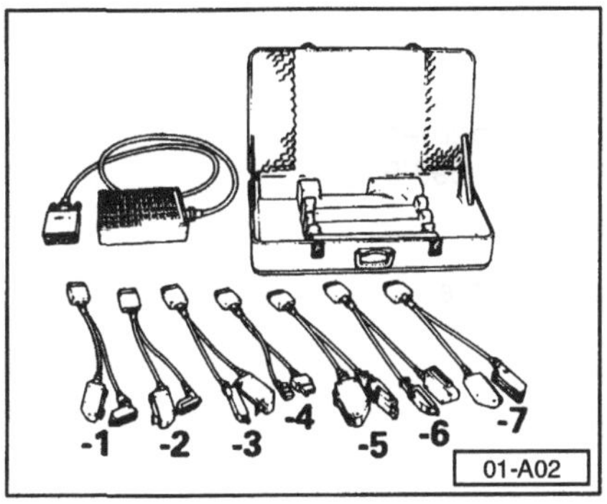

- use Test box VAG 1598 and adaptor cable VAG 1598/9 along with jumper leads from the VW 1594 adaptor kit to connect the test instruments to the Digifant ECM
- the terminal numbers of the ECM harness connector and the numbers on the VAG 1598 Test box are identical

Requirements:

- battery voltage OK
- fuses 16, 18 and 21 OK
- Digifant system ground wires OK (connections at engine, transmission, battery and body).
- fuel pump and fuel pump relay OK, checking, ⇒ Repair Group 20

Engine code PG | Digifant I

Digifant I system, Test A electrical checking

Tests on Digifant Engine Control Module (ECM) wiring harness connector

For vehicles with Automatic Transmission only:

♦ read DTC memory for transmission ECM (J 217) and check for any stored DTC's, ⇒ Repair Group 01 (Transmission) OBD

♦ repair or replace as necessary

♦ erase (J 217) DTC memory

All vehicles:

- ♦ switch OFF ignition
- ♦ disconnect Digifant ECM (J 169) harness connector
  - • ECM is installed in plenum

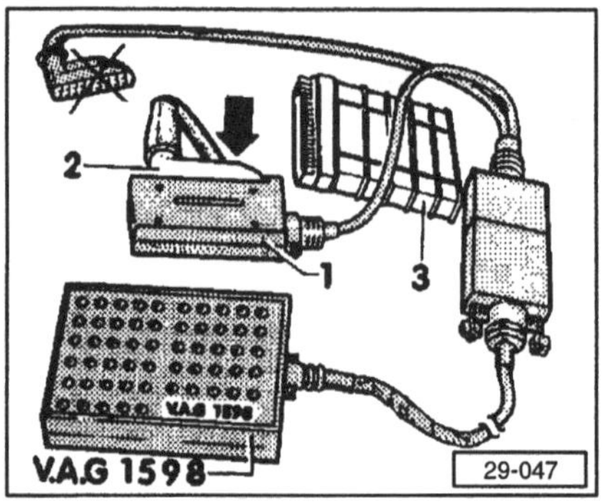

- ♦ connect VAG 1598 Test box to ECM harness connector 2 using adaptor wire VAG 1598/9
  - • ECM 3 remains disconnected
- ♦ perform checks using the following table

Test A

Test step	VAG 1598 terminals	To be tested	Test conditions or additional steps	Specified values
1	1 + 19 19 + 35	Voltage supply for Digifant ECM (J 169)	Ignition switched ON	Approx. battery voltage
2	1 + 6	Wiring to ignition coil (N 152)	Ignition switched ON	Approx. battery voltage
3	1 + 4	MIL lamp (K 83) (OBD light)	Ignition switched ON	Approx. battery voltage
4	1 + 23	Idle air control valve (N 71)	Ignition switched ON	Approx. battery voltage
5	1 + 24 bridged	Wiring to fuel pump relay (J 17)	Ignition switched ON	Fuel pump must run audibly
6	 1 + 20 1 + 21 1 + 3 1 + 2	Wiring to injectors (N 30 to N 33)* Cylinder 1 Cylinder 2 Cylinder 3 Cylinder 4	Ignition switched ON Remove fuse 18 Bridge terminals 1 and 24	Approx. battery voltage
*Check fuel injector resistance, ⇒ Repair Group 24				
7	For Auto trans only: 19 + 25	Wiring to starter lockout relay (J 226)	Ignition switched ON Selector lever in P or N Move selector lever to 1, 2, 3, D and R	Approx. battery voltage No voltage at any of these positions
8	For manual trans only: 1 + 25	Wiring from starter terminal 50	Operate starter and then switch OFF ignition	8 Volts minimum

Test A continued

Test step	VAG 1598 terminals	To be tested	Test conditions or additional steps	Specified values
♦ switch OFF ignition ♦ switch multimeter to resistance range				
9	10 + 28	Engine coolant temperature sensor (G 62)		
Note: Engine malfunctions that only occur at certain temperatures are due to an intermittent signal from the engine coolant temperature sensor. ⇒ Repair Group 24 for engine coolant temperature sensor checking.				
10	28 + 30	Intake air temp. sensor (G 42) in CO fuel trim potentiometer (G 74)		
11	17 + 36	Diagnostic connector (T 2)	Bridge terminals in white diagnostic connector	1.5 Ohms maximum
12	10 + 28	CO fuel trim potentiometer	—	0 to 2 Ohms

Test A continued

Test step	VAG 1598 terminals	To be tested	Test conditions or additional steps	Specified values
♦ switch OFF ignition ♦ switch multimeter to resistance range ♦ disconnect and perform the following check on removed harness connector for throttle Position Sensor				
13	34 + terminal 1 of connector 33 + terminal 2 of connector 28 + terminal 3 of connector	Throttle position sensor wiring (G 69)	Disconnect throttle position sensor	1.5 Ohms maximum
♦ disconnect and perform the following check on removed harness connector for Camshaft Position Sensor				
14	28 + terminal 1 of connector 37 + terminal 2 of connector 38 + terminal 3 of connector	Camshaft position sensor (G 40) wiring	Disconnect camshaft position sensor	1.5 Ohms maximum
♦ disconnect and perform the following check on removed harness connector for knock sensor				
15	29 + terminal 1 27 + terminal 2 8 + terminal 3	Wiring to knock sensor (G 61)	Disconnect knock sensor	1.5 Ohms maximum
♦ disconnect and perform the following check on removed harness connector for Oxygen sensor				

Test A continued

Test step	VAG 1598 terminals	To be tested	Test conditions or additional steps	Specified values
16	1 + terminal 2 of connector 16 + terminal 3 9 + terminal 3 1 + 16	Oxygen sensor (G 39) wiring	Disconnect Oxygen sensor Re–connect Oxygen sensor	1.5 Ohms maximum Infinite Ohms (open) Infinite Ohms (open)
17	Vehicle with A/C Only: 32 + 35	Wiring to A/C relay (J 32)	—	1 Ohm maximum

Note:
To check the wiring to the automatic transmission ECM (J 217) terminals 14 and 18; read DTC memory.

Resistance values for:

- intake air temperature sensor (G 42) in Volume Air Flow sensor Position Sensor (G 19)
- blue engine coolant temperature sensor (G 62)

Engine code PG | Digifant I

Digifant I system, Test B electrical checking

With Digifant I Engine Control Module (ECM) connected to VAG 1598 Test box

Note:

The following checks can only be performed by using the VAG 1598 Test box and VAG 1598/9 adaptor cable.

- switch OFF ignition

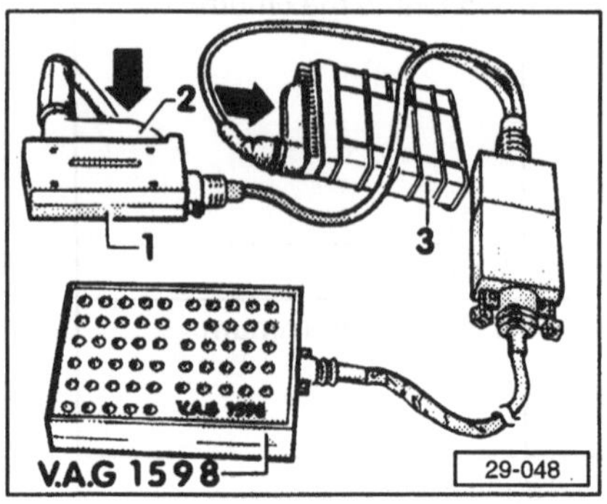

- connect VAG 1598 Test box using adaptor wire 1 which is connected to both harness connector 2 and Digifant ECM 3

CAUTION!

When performing the following tests:

- ***do NOT switch measuring ranges on the multimeter while the ignition is switched ON***
- ***do NOT switch connections between terminals on the Test box while the ignition is switched ON***

Either of these actions could destroy the Digifant ECM.

Test B

Test step	VAG 1598 terminals	To be tested	Test conditions or additional steps	Specified values
♦ read DTC memory and then erase				
1	33 + 34	Voltage supply for throttle position sensor (G 69)	Disconnect throttle position sensor	approx. 3.5 to 5.0 Volts
	28 + 34		Switch on ignition	4.5 . . . 5.0 Volts
♦ read DTC memory and then erase				
2	33 + 28	Throttle position sensor (G 69)	Switch ON ignition Re–connect throttle position sensor Throttle setting: ♦ idle: ♦ full throttle:	 0.3 — 1.7 Volts 3 — 5 Volts
♦ perform the following test step only if the engine will not start				
3	28 + 38	Voltage supply for camshaft position sensor (G 40)	Ignition switched ON	10 Volts minimum
♦ switch OFF ignition ♦ use VAG 1527B LED tester with VW 1594 adaptors instead of the multimeter to perform the following checks ♦ disconnect harness connector from fuel rail (main fuel injector connector) ♦ perform the following test step only if the engine will not start				
4	28 + 37	Signal from camshaft position sensor (G 40)	Operate starter	LED must flicker

Test B continued

Test step	VAG 1598 terminals	To be tested	Test conditions or additional steps	Specified values
♦ perform the following test step only if the engine will not start				
5	1 + 6	Ignition signal function of Digifant ECM (J 169)	Operate starter	LED must flicker
♦ perform the following test step only if the engine will not start				
6	18 + 28	RPM signal	Operate starter	LED must flicker

Engine code AAA

On-Board Diagnostic (OBD), general information

The Motronic engine control module (J 220) controls the fuel injection and ignition system. It is equipped with a DTC MEMORY (Diagnostic Trouble Code).

If malfunctions occur in monitored sensors or components a DTC (Diagnostic Trouble Code) indicating the type of malfunction is stored in the DTC MEMORY.

Stored DTC's (Diagnostic Trouble Codes) can be displayed by reading the DTC MEMORY, ⇒ pages 01-67 and 01-72.

The DTC MEMORY must be erased after the malfunctions have been eliminated.

To display DTC's relating to the 02S (Oxygen Sensor) control function you must drive the vehicle for a minimum of 10 minutes.

DTC's will also be stored which can be attributed to a temporary open circuit or loose terminal. These are displayed as "SP" or sporadic faults.

Vehicle with Malfunction Indicator lamp (MIL).

- the MIL lamp in the instrument panel must light up when the ignition is switched ON and then must switch OFF when the engine starts
- as long as interferences occur in the monitored sensors or component and the monitoring conditions are fulfilled, the MIL lamp will continue to remain lit, even with the engine running
- if the VAG 1551 Scan Tool is not available, DTC MEMORY can also be activated and erased using the adaptor cable stored in the glove compartment

Engine code AAA

On-Board Diagnostic system, reading/erasing without VAG 1551 Scan Tool

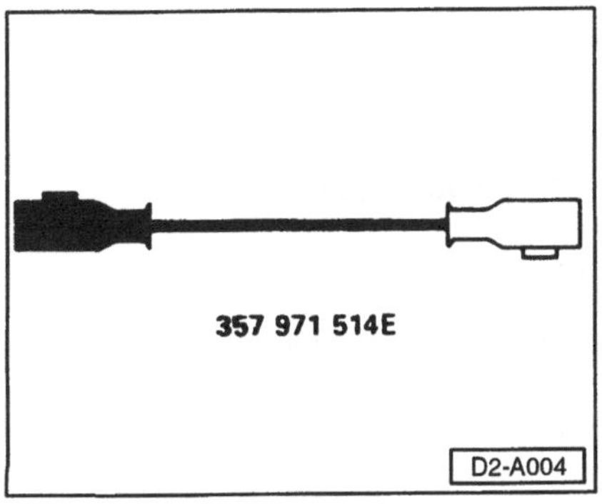

Requirement

- adaptor cable 357 971 514E (from glove box)

Activating

- ♦ switch ON ignition
- ♦ remove shift lever cover
- ♦ bridge the black and the white Data Link connectors using jumper wire for approx. 5 seconds
- ♦ repeat this procedure until blink code 4444 or 0000 "End of Output" appears
  - ⇒ starting on page 01-77 for the DTC coding chart

Erasing

- ♦ switch OFF ignition
- ♦ bridge Data Link connector
- ♦ switch ON ignition and after approx. 5 seconds remove adaptor wire from Data Link connectors
- ♦ after blink code 4444 appears, switch OFF ignition

Engine code AAA

VAG 1551 Scan Tool, connecting

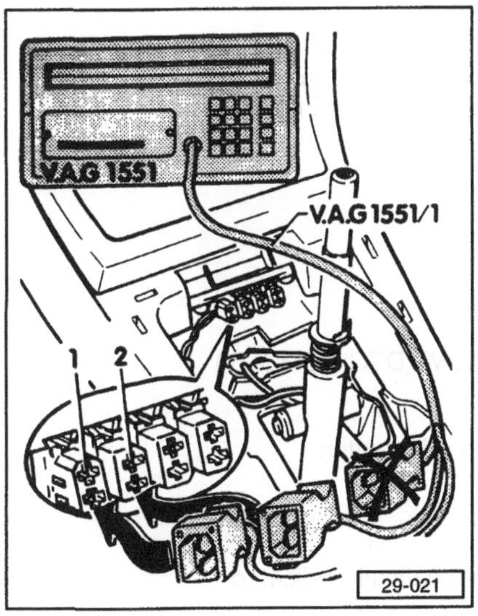

Requirements

- fuses 18 and 21 OK
- ground connections OK (engine ground connection: at cylinder head cover on left and beneath lower section of the intake manifold)

- ♦ connect Scan Tool VAG 1551 using adaptor cable VAG 1551/1 as follows:
- ♦ first connect black adaptor cable connector (voltage supply) to black Data Link connector 1

VAG On Board Diagnostic	HELP
1 - Rapid data transfer *	

VAG On Board Diagnostic	HELP
2 - Blink code output *	

- display will appear as shown

* These displays appear alternately

Note:

Do not connect the white adaptor cable connector if there is no display. Instead, check the voltage supply for Scan Tool VAG 1551, ⇒ page 01-101.

♦ connect white adaptor cable connector to the white/brown Data Link connector 2
  - the blue connector is not used

♦ depending on which function is required; either switch ON ignition or start engine and let idle, ⇒ Table, page 01-71

Note:

♦ *to obtain additional operating instructions press the HELP button*

♦ *to step through the program sequence, press the → button*

♦ press button 1 to select operating mode "Rapid data transfer"
  - display will appear as shown

Rapid data transfer	HELP
Enter address word XX	

♦ display 0 and 1 buttons to select address word 01 "Engine electronics"
  - display will appear as shown

Rapid data transfer	Q
01 - Engine electronics	

♦ press Q button to enter input

021906258 B ENGINE
Coding 1

  - the engine control module identification number is indicated via the display, for example: 021 906 258 B

Note:

Only during idle speed do correct specifications regarding the engine control module number and related coding occur.

♦ press → button

Rapid data transfer	HELP
ECM does not answer!	

  - if display appears as shown

♦ print out a list of possible causes by pressing the VAG 1551 HELP button

♦ after eliminating possible causes, press 0 and 1 buttons again to select address word 01 "Engine electronics"

♦ press Q button to enter input

♦ press → button

Rapid data transfer	HELP
Select function XX	

  - display will appear as shown

If function 02 "Check DTC memory" (DTC MEMORY) is to be selected:

♦ switch on the printer using the print button
  - lamp in print button lights

When printer is switched ON
  - stored malfunctions are automatically indicated and printed out in sequence

Engine code AAA

VAG 1551 Scan Tool, selectable functions

VAG 1551 Display	Ignition on	Engine idling
02 Check Diagnostic Trouble Code (DTC) Memory*	–	X
03 Output	X	–
04 Start basic setting	–	X
05 Erase DTC Memory***	–	X
06 End output	X	X
08 Read test value block	X	X

Engine code AAA

Diagnostic Trouble Code (DTC) MEMORY, reading/erasing using VAG 1551 Scan Tool

Requirement

- ground connections OK (engine ground connection: at left side cover and beneath the lower section of the intake manifold)

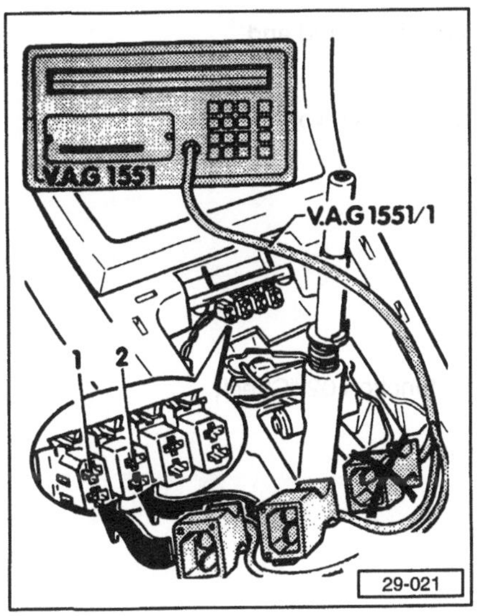

♦ connect Scan Tool VAG 1551 using adaptor cable VAG 1551/1 as follows:

♦ first connect black adaptor cable connector (voltage supply) to black data link connector 1

VAG On Board Diagnostic	HELP
1 - Rapid data transfer *	

- display will appear as shown

VAG On Board Diagnostic	HELP
2 - Blink code output *	

* These displays appear alternately

Note:

Do not connect the white adaptor cable connector if there is no display. Instead, check the voltage supply for Scan Tool VAG 1551, ⇒ page 01-101.

♦ connect white adaptor cable connector to the white/brown data link connector 2

- the blue connector is not required

♦ start engine and let idle

If engine does NOT start

♦ crank engine 4 to 6 seconds and leave ignition switched ON

♦ press 1 button to select operating mode “Rapid data transfer”

- display will appear as shown

Rapid data transfer	HELP
Enter address word XX	

♦ switch on the printer using the print button

- lamp in print button lights

When printer is switched ON

- stored DTC's are automatically indicated and printed out in sequence

♦ press 0 and 1 buttons to select address word 01 “Engine electronics”

- display will appear as shown

Rapid data transfer	Q
01 - Engine electronics	

♦ press Q button to enter input

♦ press → button

- display will appear as shown

Rapid data transfer	HELP
Select function XX	

♦ press 0 and 2 buttons to select function 02 “Check DTC memory”

- display will appear as shown

Rapid data transfer	HELP
Enter address word XX	

- press 0 and 2 buttons to select function 02 "Check DTC memory"
- press Q button to enter input
  - the number of stored DTC's or "No DTC recognized" is indicated on the display:

```
X DTC recognized
————————————
```

If no DTC has been recognized:

- press → button
  - display will appear as shown

```
Rapid data transfer                HELP
Select function XX
```

- press 0 and 5 buttons to select function 05 "Erase DTC memory"
- press Q button to enter input
  - display will appear as shown

```
Rapid data transfer
Fault Memory erased
```

Note:

If the engine was switched off between reading and erasing the DTC MEMORY, the DTC MEMORY is not erased.

- press → button
  - display will appear as shown

```
Rapid data transfer                HELP
Select function XX
```

- press 0 and 6 buttons to select function 06 "End output"
  - display will appear as shown

```
Rapid data transfer                   Q
06 - End output
```

- press Q button to enter input
- locate and eliminate any malfunctions printed out using the DTC table, ⇒ starting on page 01-77

Note:

The DTC (Diagnostic Trouble Code) chart is arranged in order of ascending sequence of the five digit code.

- perform a test drive of at least ten minutes during which:
  - coolant temperature must exceed 80°C (176°F) for 3 minutes minimum
  - the accelerator pedal must be completely floored briefly at a speed in excess of 4600 RPM
  - there must be acceleration at half throttle for approx. 4 seconds in 4th gear from a speed of roughly 30 mph
- read and erase DTC MEMORY again

Engine code AAA

Diagnostic Trouble Code (DTC), Chart

Requirements

- multi-function indicator and speedometer (G 21) OK
- before replacing components coded as faulty, first check the wiring and ground connections to these components using the proper wiring diagram. This is particularly relevant if the malfunction is displayed as being sporadic or "SP"
- the DTC (Diagnostic Trouble Code) and the blink code will only print out when the "Rapid data transfer" mode and then only if the Scan Tool printer is switched ON

Example:

	DTC (Diagnostic Trouble Codes)	
	5 digit (Rapid data transfer) 65535	4 digit (blink code) 1111

VAG 1551 display	Possible cause	Corrective step
No malfunction recognized	If a malfunction is present; it is not being recognized by OBD (On Board Diagnosis)	—
00281 1231 Vehicle speed sensor (G 68) Impossible signal No signal*	Multi-function indicator gets no signal from speedometer (G 21)	⇒ wiring diagrams, repair or replace as necessary
*To check the road speed signal; drive the vehicle at full load and raise the engine speed to 4500 RPM minimum.		
00513 2111 Engine speed sensor (G 28)	(G 28) malfunctioning	♦ check (G 28) and related wiring, using wiring diagram, repair or replace as necessary
No signal	Open circuit in wiring	⇒ page 01-109, Test step 26
Impossible signal*	Sensor wheel for engine speed sensor malfunctioning or loose	♦ check sensor wheel for secure fit ⇒ Repair Groups 13 and 24
	(G 28) loose	♦ check tightness of (G 28)
*It is unlikely that a malfunction due to abrupt stalling of the engine (sporadic) will be stored.		
00515 2113 Camshaft position sensor (G 40)	Camshaft position sensor (G 40) rotor twisted	♦ check distributor ♦ check control times
Open circuit/short to positive	Open circuit or short to positive in wiring (G 40)	♦ check wiring to (G 40) ⇒ page 01-108, Test step 19
Short to ground	Short to ground in wiring (G 40)	♦ check wiring to (G 40) ⇒ page 01-108, Test step 19

VAG 1551 display	Possible cause	Corrective step
00518 2212 Throttle position sensor (G 69) Short to positive Short to ground	Open circuit or short to positive in wiring (G 69) Short to ground in (G 69) wiring	♦ check wiring together with (G 69) ⇒ page 01-107, Test step 18 ♦ check wiring together with (G 69) for short to ground using wiring diagram, repair or replace as necessary
00522 2312 Engine coolant temperature sender (G 62)* Short to ground Open circuit/short to positive	(G 62) malfunctioning Short to ground in wiring (G 62) malfunctioning, or corrosion on connector Open circuit/short to positive in wiring	♦ check wiring and (G 62) for short ⇒ page 01-107, Test step 15
00524 2142 Knock sensor I (G 61) Open circuit/short to positive	(G 61) malfunctioning Open circuit/short circuit in sensor wiring Knocking not recognized by Motronic engine control module (J 220)	♦ replace (G 61) ♦ check wires to (G 61) ⇒ page 01-108, Test step 21 ♦ replace (J 220)
00525 2342 Oxygen sensor probe (G 39)* No signal Short to positive Short to ground	(G 39) malfunctioning Open circuit Open circuit or short to positive Short to ground in wiring or in (G 39) Fuel level too low	♦ replace (G 39) ♦ check wiring to (G 39) ⇒ page 01-107, Test step 17 ♦ fill fuel tank then erase DTC Memory

VAG 1551 display	Possible cause	Corrective step
00527 2412 Intake air temperature sensor (G 42) Short to ground Open circuit/short to positive	(G 42) malfunctioning Short to ground in wiring or (G 42) (G 72) malfunctioning	♦ replace (G 42) ♦ check wiring together with (G 42) for a short to ground ⇒ page 01-108, Test step 22
00532 2234 Supply voltage* Too high Too low	Battery voltage above 16.0 V Battery voltage below 10.0 V	♦ check generator (alternator) output ⇒ page 01-105, Test steps 1 + 2
00533 2231 Idle regulation adaption limit exceeded Adaption limit not met	Idle Air Control (IAC) valve (N 71) stuck or dirty Air leak between Mass Air Flow sensor (G 70) and engine	♦ check IAC valve, ⇒ Repair Group 24 ♦ check intake air system, ⇒ Repair Group 24 ♦ for vehicles with EGR, check EGR valve, ⇒ Repair Group 26
00535 2141 First knock regulation Impossible signal	Knock regulation in ECM (J 220) defective	♦ replace (J 220)
*The engine control module DOES NOT check voltage until the engine has been running for 3 minutes		

VAG 1551 display	Possible cause	Corrective step
00537 2341		
Oxygen sensor regulation*	O_2S (Oxygen Sensor) probe heater defective	♦ check voltage supply to probe heater, ⇒ page 01-106, Test step 4
Regulating limit exceeded		♦ check O_2S heater for continuity
	Exhaust system leaking between cylinder head and catalyst	♦ check exhaust system
	Misfiring	♦ check ignition system components
	Intake air system leaking (outside air)	♦ check intake air system
	Injector(s) (N 30, N 31, N 32, N 33, N 83, N 84) malfunctioning	♦ Check N 30, N 31, N 32, N 33, N 83, N 84) ⇒ Repair Group 24
	Fuel pressure too low	♦ check fuel pressure regulator
	Hot wire for MAF sensor (G 70) not burning off cleanly	♦ check (G 70)
	EVAP canister purge regulator valve (N 80) stuck (always open)	♦ perform Output DTM (Output checks)
Regulating limit not reached	Fuel pressure too high	♦ check fuel pressure regulator
	Injector(s) (N 30, N 31, N 32, N 33, N 83, N 84) malfunctioning, leaking	♦ check (N 30, N 31, N 32, N 33, N 83, N 84)
Signal to ground	Short to ground in O_2S (Oxygen Sensor) probe (G 39) or in wiring	♦ check wiring together with (G 39) for short to ground, ⇒ page 01-107, Test step 17

VAG 1551 display	Possible cause	Corrective step
00540 2144		
Knock sensor II (G 66)	(G 66) malfunctioning	♦ replace (G 66)
No signal	Open circuit/short circuit in sensor wiring	♦ check wires to (G 66) ⇒ page 01-109, Test step 24
	Knocking not recognized by Motronic engine control module (J 220)	♦ replace (J 220)
00543 2214		
Maximum engine speed exceeded	Incorrect switching (internal) causes control to exceed speed governor	
Signal too high	Signal from Engine RPM (speed) sensor (G 28) defective	♦ check wires together with (G 28) shield, ⇒ page 01-109, Test step 26
00545 2314		
Engine/transmission electrical connection	Wire has short to ground	♦ determine and eliminate short using wiring diagram, repair or replace as necessary
Short to ground		
00553 2324		
Mass air flow sensor (G 70)*	(G 70) malfunctioning	♦ check wiring to (G 70)
Signal too high	Short to positive in wiring	♦ replace (G 70), ⇒ page 01-109, Test step 25
Signal to low	(G 70) malfunctioning or short to ground in wiring	♦ check wiring to (G 70), if wiring OK, replace (G 70)
	Air leak between (G 70) and engine (outside air)	♦ eliminate leaks

VAG 1551 display	Possible cause	Corrective step
00561 2413		
Mixture adaption	Fuel pressure NOT OK	♦ check fuel pressure regulator
Adaption limit exceeded	Misfiring	♦ check ignition system components
or	Intake air system leaking (outside air)	♦ check intake air system
Adaption limit not reached	Injector(s) (N 30, N 31, N 32, N 33, N 83, N 84) defective, leaking	♦ check (N 30, N 31, N 32, N 33, N 83, N 84)
	Hot wire for MAF sensor (G 70) not burning off cleanly	♦ check (G 70), ⇒ page 01-109, Test step 25
	O_2S (Oxygen Sensor) regulation faulty	♦ check wire to O_2S (Oxygen Sensor) probe
	Exhaust system leaking between cylinder head and catalyst	♦ check exhaust system
	EVAP canister purge regulator valve (N 80) sticks in open position	♦ perform Output DTM (Output checks)
00585 2411		
EGR temperature sensor (G 98)*	Temperature sensor in EGR valve (G 98) malfunctioning or wire is shorted to ground	♦ check wires to (G 98), ⇒ page 01-110, Test step 29
Short to ground		
00586		
Controlling EGR system	EGR valve sticks or is malfunctioning	♦ check EGR valve, ⇒ Repair Group 26
Impossible signal		

VAG 1551 display	Possible cause	Corrective step
01025		
Triggering for On Board Diagnostic (OBD) lamp malfunctioning	Wiring has short to positive	♦ check wiring using wiring diagram, also ⇒ page 01-106, Test step 5
Short in positive		
Open circuit/short to ground	Malfunction indicator lamp malfunctioning	
01242 4332		
Output stages in engine control module	Loose terminal to one of the Output DTM components or to the Motronic engine control module (J 220)	♦ perform Output DTM (Output check) diagnosis
Mechanical malfunction		♦ check for loose terminal or connection in wiring to one of the Output DTM components or replace (J 220) if necessary
01247 4343		
EVAP canister purge regulator valve (N 80)	(N 80) malfunction or short to positive in wiring	♦ check wiring together with (N 80), ⇒ page 01-106, Test step 14
Open circuit/short to positive		
Open circuit/short to negative	(N 80) malfunction or short to negative in wiring	
01249 4411		
Injector for No. 1 cyl. (N 30)	(N 30) malfunction or short to positive positive	♦ check wiring together with (N 30) ⇒ page 01-106, Test step 8
Open circuit/short to positive		
Open circuit/short to negative	(N 30) malfunction or short to negative negative	♦ check wiring together with (N 30) ⇒ page 01-106, Test step 8

VAG 1551 display	Possible cause	Corrective step
01250 4412 Injector for No. 2 cyl. (N 31)	⇒ No. 1 cylinder (DTC Code 01249)	♦ check wiring and (N 31), ⇒ page 01-106, Test step 9
01251 4413 Injector for No. 3 cyl. (N 32)	⇒ No. 1 cylinder (DTC Code 01249)	♦ check wiring and (N 32), ⇒ page 01-106, Test step 10
01252 4414 Injector for No. 4 cyl. (N 33)	⇒ No. 1 cylinder (DTC Code 01249)	♦ check wiring and (N 33), ⇒ page 01-106, Test step 11
01253 4421 Injector for No. 5 cyl. (N 83)	⇒ No. 1 cylinder (DTC Code 01249)	♦ check wiring and (N 83), ⇒ page 01-106, Test step 12
01254 4422 Injector for No. 6 cyl. (N 84)	⇒ No. 1 cylinder (DTC Code 01249)	♦ check wiring and (N 84), ⇒ page 01-106, Test step 13
01257 4431 Idle air control valve (N 71) Signal to positive	(N 71) malfunction or short to positive	♦ check wires together with (N 71) ⇒ page 01-108, Test step 23
Signal to ground	(N 71) malfunction or short to negative in wiring	
Mechanical malfunction	(N 71) malfunction	♦ perform Output DTM (Output check) diagnosis
01259 4433 Fuel pump relay (J 17) Short to positive	(J 17) malfunction or wire has short to positive	♦ check wiring and (J 17), ⇒ section D2–200, Test step 3

VAG 1551 display	Possible cause	Corrective step
01265 4312 EGR vacuum regulator solenoid valve (N 18) Open circuit/short to ground	EGR valve (N 18) malfunction or open circuit or wiring is shorted to ground	♦ check wiring to (N 18), ⇒ page 01-110, Test step 28
Short to positive	(N 18) malfunction or wiring is shorted positive	♦ check wiring and (N 18), ⇒ pages 01-109, 01-110, Test steps 27 and 28
65535 1111 ECM malfunctioning	Components in the Motronic engine control module (J 220) are malfunctioning	♦ replace (J 220)

Output DTM (Diagnostic Test Mode), checking

The following components are activated by the Output DTM (Diagnostic Test Mode) in the sequence given:

1 - Fuel injector 1 (N 30)

2 - Fuel injector 2 (N 31)

3 - Fuel injector 3 (N 32)

4 - Fuel injector 4 (N 33)

5 - Fuel injector 5 (N 83)

6 - Fuel injector 6 (N 84)

7 - Idle Air Control (IAC) valve (N 71)

8 - EVAP Canister Purge Regulator valve (N 80)

9 - Heated Oxygen Sensor Relay (J 278)

10 - EGR Vacuum Regulator Solenoid valve (N 18) *

11 - MAP (Manifold Absolute Pressure) valve (N 156) **

* If engine has EGR; watch Scan Tool display

** Ignore display.

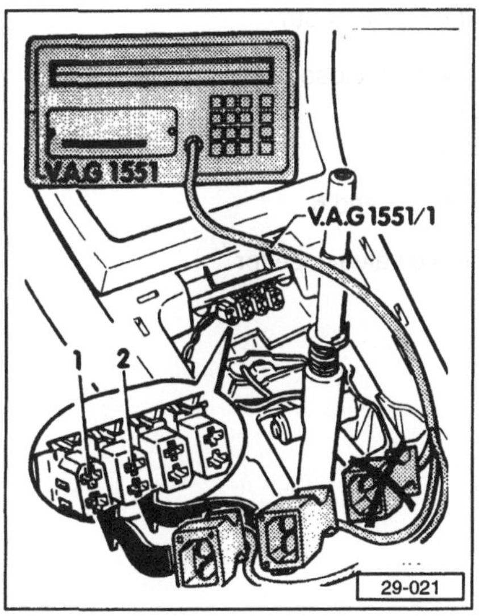

- connect Scan Tool VAG 1551 using adaptor cable VAG 1551/1 as follows:
- first connect black adaptor cable connector (voltage supply) to black Data Link connector 1

VAG On Board Diagnostic	HELP
1 - Rapid data transfer *	

- display will appear as shown

VAG On Board Diagnostic	HELP
2 - Blink code output*	

* These displays appear alternately

Note:

Do not connect the white adaptor cable connector if there is no display. Instead, check the voltage supply for Scan Tool VAG 1551, *⇒ page 01-101.*

♦ connect white adaptor cable connector to the white/brown Data Link connector 2
  • the blue connector is not used

♦ switch ON ignition

♦ press button 1 to select operating mode "Rapid data transfer"
  • display will appear as shown

Rapid data transfer	HELP
Enter address word XX	

♦ press 0 and 1 buttons to select address word 01 "Engine electronics"
  • display will appear as shown

Rapid data transfer	Q
01 - Engine electronics	

♦ press Q button to enter input

♦ press → button
  • display will appear as shown

Rapid data transfer	HELP
Select function XX	

♦ press 0 and 3 buttons to select function "Output DTM"
  • display will appear as shown

Rapid data transfer	Q
03 - Output DTM	

♦ press Q button to enter input
  • display will appear as shown

Output DTM
Injector - N 30

To activate fuel injector for cylinder No. 1 (N 30)

♦ manually open throttle valve and then close
  • fuel injector No. 1 must click
  • fuel injectors click very quietly

To activate fuel injectors for cylinders 2 and 6.

♦ re-open and then close the throttle valve once for each injector

To skip an individual check

♦ press → button

If one or more injectors does NOT click

- check activation using VAG 1527B LED tester

To activate IAC (Idle Air Control) valve (N 71)

Output DTM Idle Air Control valve (N 71)

- press → button
  - display will appear as shown

The IAC (Idling speed stabilization) valve must switch on and off until the next Output DTM (Output check) is selected by pressing → button.

If valve does not operate

- switch OFF ignition
- disconnect harness connector from IAC valve
- connect VAG 1527B LED tester to terminals 1 and 2 of harness connector using jumpers from VW 1594 adaptor kit
- switch ON ignition
- perform previous steps again to activate Idle Air Control valve (N 71)
  - LED tester must blink

If NO

- check wiring using wiring diagram, repair or replace as necessary

If wiring OK

- replace engine control module (J 220)

To activate EVAP Canister Purge Regulator valve (N 80)

Output DTM EVAP Canister Purge Regulator valve N 80

- press → button
  - display will appear as shown

Note:
The N80 valve (behind air cleaner housing on wheel housing) must click until Output DTM (Diagnostic Test Mode) is ended by pressing → button.

- disconnect hose from N 80 valve (coming from carbon canister)
- attach auxiliary hose to N 80 valve
- blow into hose (in direction of throttle housing) while Output DTM (Diagnostic Test Mode) is activated
  - N 80 valve must open and close

If solenoid valve does NOT click

- disconnect harness connector from N 80 valve
- connect VAG 1527B LED tester between terminals of harness connector using jumpers from VW 1594 adaptor kit
  - LED must blink

If YES

- replace EVAP Canister Purge Regulator valve (N 80)

If NO

- check wiring using wiring diagram, repair or replace as necessary

If wiring OK

- replace engine control module (J 220)

To activate Heated Oxygen Sensor relay (J 278)

Output DTM
Heated Oxygen Sensor relay - J 278

- press → button
  - display will appear as shown

Note:

The Heated Oxygen Sensor relay (mounted on the fuse/relay panel with a clip) must click until the Output DTM (Diagnostic Test Mode) is ended by pressing the → button.

If relay does NOT click

- check wiring using wiring diagram

To activate EGR Vacuum Regulator Solenoid valve (N 18)

Output DTM
EGR Vacuum Regulator Solenoid Valve N 18

- press → button
  - display will appear as shown

Note:

The N 18 valve must click until by pressing → button, Output DTM (Output check) ends.

If N 18 valve does NOT click

- disconnect harness connector from N 18 valve
- connect VAG 1527B LED tester to harness connector using jumpers from VW 1594 adaptor kit

If LED tester blinks

- ♦ replace EGR Vacuum Regulator Solenoid valve (N 18)

If LED tester does NOT blink

- ♦ check wiring using wiring diagram, repair or replace as necessary

If wiring OK

- ♦ replace Engine Control Module (J 220)
- ♦ press → button
- ♦ press 0 and 6 buttons to select function 06 "End output"
  - • display will appear as shown

Rapid data transfer	Q
06 - End output	

- ♦ press Q button to enter input

Engine code AAA

Measuring value block, reading

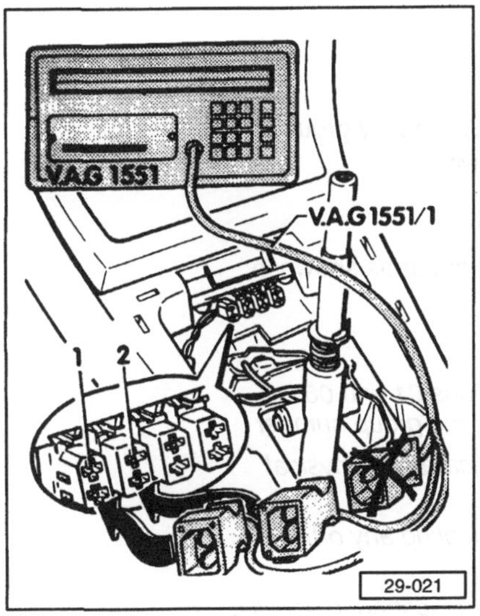

- ♦ connect Scan Tool VAG 1551 using adaptor cable VAG 1551/1 as follows:
- ♦ first connect black adaptor cable connector (voltage supply) to black Data Link connector 1

VAG On Board Diagnostic	HELP
1 - Rapid data transfer *	

VAG On Board Diagnostic	HELP
2 - Blink code output *	

- • display will appear as shown

* These displays appear alternately

Note:

Do not connect the white adaptor cable connector if there is no display. Instead, check the voltage supply for Scan Tool VAG 1551, ⇒ page 01-101.

- connect white adaptor cable connector to the white/brown Data Link connector 2
  - the blue Data Link connector is not used
- switch ON ignition
- press button 1 to select operating mode “Rapid data transfer”
  - display will appear as shown

```
Rapid data transfer                HELP
Enter address word XX
```

- press → button
- press 0 and 8 buttons to select function 08 “Measuring value block reading”
- press Q button to enter input


```
Read measuring value block         HELP
Enter display group no. XX
```

  - display will appear as shown
- enter desired display group number using the listings on page 01-99
- press Q button to enter input

Example

- press buttons 0 and 1 to select display group 01
  - display will appear as shown

```
Read measuring value block         HELP
Enter display group no. 01
```

- press Q button to enter input
  - channels 1 through 4 will appear as shown in display

```
Read measuring value block            1
1     2     3     4
```

Notes:

- *the VAG 1551 display for groups 01 and 06 is shown without the 0 (zero) in the group number*
- *the measured values are indicated as physical quantities (actual values)*
- *press the C button before selecting any other display groups*
- *keying in individual values is described in specific repair procedures ⇒ Repair Group 24*

- press → button
- press 0 and 6 buttons to select function 06 “End output”
  - display will appear as shown

```
Rapid data transfer                   Q
06 - Enter output
```

- press Q button to enter input

Display group number	Information in displayed channels
00	Only required for individual checks
01	**1** = Engine speed (RPM) **2** = Engine coolant temperature **3** = O_2S (Oxygen Sensor) factor **4** = Ignition angle
02	**1** = Engine speed (RPM) **2** = Injection time **3** = System VOLTAGE **4** = Exhaust gas recirculation temperature (if available)
03	**1** = Engine speed (RPM) **2** = Engine load **3** = Throttle valve angle **4** = Intake Air Temperature (at manifold)
04	**1** = Engine speed (RPM) **2** = Engine load **3** = Vehicle speed **4** = Switch positions* **10000** Acceleration enrichment **01000** Wide open throttle **00100** Partially open throttle **00010** Idle position **00001** Deceleration fuel shutoff **00000** Switch positions NOT OK (activate DTC Memory again if this numeric code appears)

* Switch positions can also be added, for example: acceleration from partial load = **10100**.
Note:: This display is only possible from Scan Tools equipped with program card 3 (or later).

Display group number	Information in displayed channels
05	**1** = Engine speed (RPM) **2** = IAC (Idle Air Control) valve **3** = IAC (Idle Air Control) valve (duty cycle) **4** = Switch positions: **10-00** A/C compressor **ON** **01-00** A/C **ON** **00-10** Transmission engagement (for automatic transmission ONLY) **00-01** Drive step (for automatic transmission ONLY)
06	**1** = O_2S (Oxygen Sensor) factor **2** = O_2S (Oxygen Sensor) adaptive learning (closed throttle) **3** = O_2S (Oxygen Sensor) adaptive learning (partially open throttle) **4** = O_2S (Oxygen Sensor) adaptive learning (full throttle)

* Switch positions can also be added, for example: acceleration from partial load = **10100**.
Note: This display is only possible from Scan Tools equipped with program card 3 (or later).

Engine code AAA

Scan Tool VAG 1551 voltage supply, checking

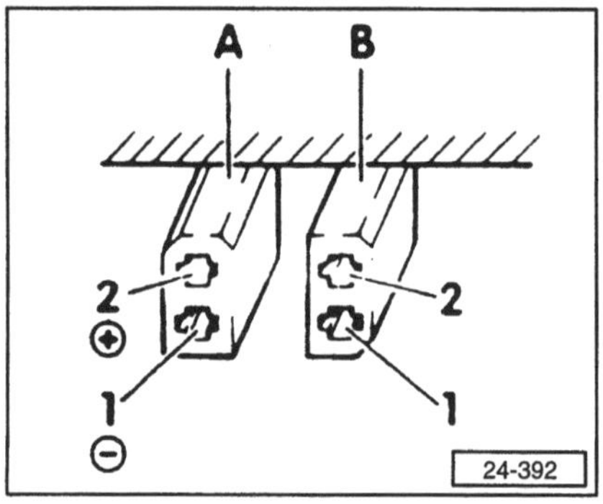

- check for voltage supply at black Data Link connector A in vehicle
  - observe polarity while performing this check:

    terminal 1 = Battery negative (–)

    terminal 2 = Battery positive (+)

If battery voltage is not obtained

- check wiring using wiring diagram, repair or replace as necessary
- switch Fluke 83 multimeter to resistance range
- connect multimeter to VAG 1551/1 adaptor cable and check for continuity

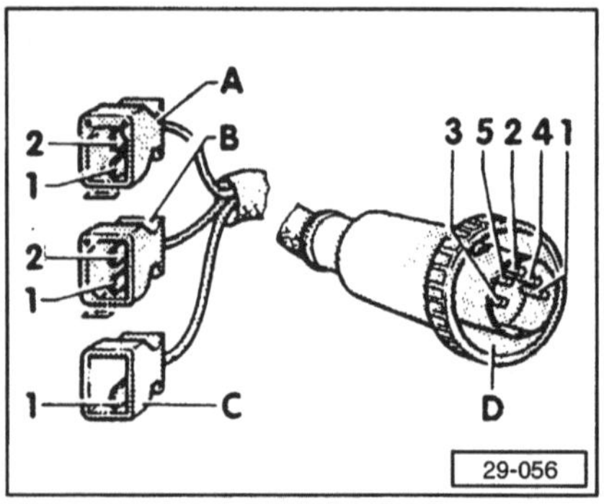

Terminal number at vehicle end of adaptor cable		Pin number in connector for Scan Tool VAG 1551	Connects to this point in the vehicle
Black A	**1**	3	Battery (–)
	2	2	Battery (+)
White B	**1**	4	L–wire
	2	1	K–wire
Blue C	**1**	5	MIL Lamp wire

Engine code AAA

Electrical checks using VAG 1598 Test Box

- the following checks use the VAG 1527B LED tester and the Fluke 83 multimeter in conjunction with the VAG 1598 Test Box
- use the VW 1594 adaptor kit to make connections between the Test Box, LED tester and multimeter
- the values given are valid within an ambient temperature range of 0 C to 40 C (32 F to 104 F)
- if measured values differ from the specifications; check the wiring to the component using the wiring diagram before replacing the component, especially for values specified at less than 10 Ohms in which case the resistance measurement for the component should be repeated
- if a measured value only deviates slightly from the specified value, clean the tester terminals, connectors and measuring wires and then repeat the test

CAUTION!

Switch multimeter to the appropriate measuring range BEFORE connecting the test leads to avoid damaging sensitive electronic components.

Requirements

- battery voltage OK
- fuses 15, 18 and 21 OK
- ground connections OK (engine ground connection: at left side of valve cover and beneath lower section of the intake manifold)
- fuel pump fuel pump relay OK

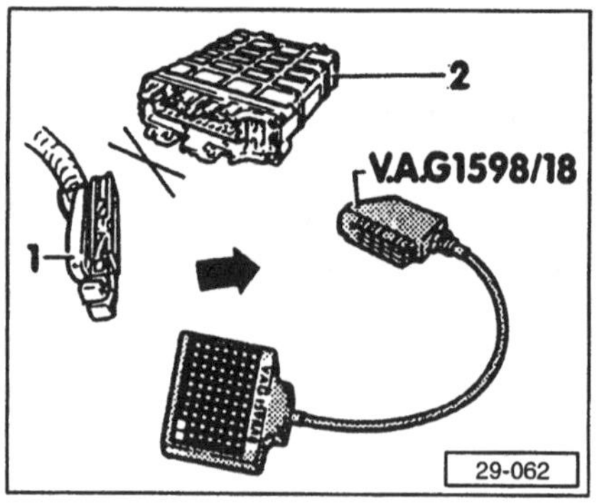

- connect adaptor cable VAG 1598/18 between the Test Box and engine control module harness connector 1
  - the engine control module 2 is left disconnected

CAUTION!

Ignition MUST be switched OFF before disconnecting engine control module. After switching OFF ignition allow 20 seconds for the hot wire in the MAF (Mass Air Flow) sensor to heat up and self clean.

Vehicles with automatic transmissions:

Performing electrical checks will generate and store malfunctions in the DTC MEMORY which must be erased, ⇒ Repair Group 01 (Automatic Transmission) for additional information.

- perform the electrical checks using the following charts

- switch multimeter to 20 Volt range

Test step	VAG 1598 terminals	To be tested	Test conditions or additional steps	Specified values
1	1 + 54	Voltage supply for DTC Memory	Ignition switched OFF	Approx. battery voltage
2	1 + 23	Voltage supply for ECM (J 220) via main relay (J 271)	Bridge terminals 9 + 55 of Test box	Approx. battery voltage
Vehicles with A/C ONLY:				

Test step	VAG 1598 terminals	To be tested	Test conditions or additional steps	Specified values
Vehicles with A/C ONLY:				
3	6 + 55 bridged	Wiring to fuel pump relay (J 17)	Ignition switched ON	Fuel pump must operate audibly
4	7 + 28 bridged	Heated O2S relay (J 278)	Ignition switched ON Bridge terminals 9 + 55 of Test box	Relay must briefly but audibly click
5	5 + 10 bridged	OBD lamp	Ignition switched ON Fuse 16 OK	MIL lamp must light up
Vehicles with A/C ONLY:				
6	55 + 39	Wiring for A/C	Ignition switched ON Switch on A/C	Approx. battery voltage
7	37 + 38	Wiring to A/C compressor	Ignition switched ON	A/C compressor clutch must briefly but audibly switch
♦ switch OFF ignition ♦ switch multimeter to resistance range				
8	24 + 23	Injector for cyl. 1 (N 30)	—	15 to 21.5 Ohms
9	3 + 23	Injector for cyl. 2 (N 31)	—	15 to 21.5 Ohms
10	26 + 23	Injector for cyl. 3 (N 32)	—	15 to 21.5 Ohms
11	4 + 23	Injector for cyl. 4 (N 33)	—	15 to 21.5 Ohms
12	25 + 23	Injector for cyl. 5 (N 83)	—	15 to 21.5 Ohms
13	2 + 23	Injector for cyl. 6 (N 84)	—	15 to 21.5 Ohms
14	23 + 31	EVAP Canister Purge Regulator valve (N 80)	—	40 to 80 Ohms

Test step	VAG 1598 terminals	To be tested	Test conditions or additional steps	Specified values
15	14 + 33	ECT sensor (G 62)		
If OBD (On Board Diagnosis) stores a DTC for this sensor which cannot be verified by these checks it may be due to an intermittent temperature signal. In this case, check the ECT sensor, ⇒ Repair Group 24.				
16	21 + terminal 1 of white diag. connector	Wiring to white diagnostic connector (beneath shift lever cover)	—	1.5 Ohms maximum
	43 + terminal 2 of white diag. connector	Wiring to white diagnostic connector (beneath shift lever cover)	—	1.5 Ohms maximum
17	20 + 42	Wiring to O_2S sensor (G 39)	Disconnect O_2S harness connector Bridge terminals 3 + 4 of harness connector	1.5 Ohms maximum
			Re–connect harness connector	Infinite Ohms (open)
18	33 + 41	Throttle position sensor	—	1750 Ohms
	33 + 40		Throttle valve closed Open throttle valve	1150 Ohms Resistance must increase
	40 + 41		Throttle valve closed Open throttle valve	2700 Ohms Resistance must decrease

Test step	VAG 1598 terminals	To be tested	Test conditions or additional steps	Specified values
19	7 + 44	Wiring to camshaft position sensor (G 40)	Disconnect harness connector from camshaft position sensor at distributor and bridge terminals 1 + 2 of harness connector	1.5 Ohms maximum
	7 + 23		Bridge terminals 1 + 3 of harness connector	1.5 Ohms maximum
20	8 + 55	Wiring to power output stage of ignition coil (N 157)	Disconnect power output stage harness connector and bridge terminals 1 + 2 of harness connector	1.5 Ohms maximum
	8 + 38		Bridge terminals 2 + 3 of harness connector	1.5 Ohms maximum
21	33 + 34	Wiring to knock sensor I (G 61)	Disconnect knock sensor harness connector	Infinite Ohms (open)
	7 + 34	3–pin connector at rear engine mount	Bridge terminals 1 + 3 of harness connector	1.5 Ohms maximum
22	33 + 36	IAT sensor (G 42)	Bridge terminals 1 + 2 of harness connector	1.5 Ohms maximum
23	27 + 53	IAC valve (N 71)	Bridge terminals 1 + 2 of harness connector	1.5 Ohms maximum

Test step	VAG 1598 terminals	To be tested	Test conditions or additional steps	Specified values
24	7 + 33	Wiring to knock sensor II (G 66)	Disconnect knock sensor harness connector	Infinite Ohms (open)
		3–pin connector (black) next to starter	Connect terminals 2 + 3 of harness connector	1.5 Ohms maximum
	56 + 57		Connect terminals 1 + 3 of harness connector	1.5 Ohms maximum
25	1 + 16	Wiring to MAF sensor (G 70)	Disconnect MAF sensor harness connector and bridge terminals 1 + 2 of harness connector	1.5 Ohms maximum
	1 + 17		Bridge terminals 1 + 3 of harness connector	1.5 Ohms maximum
	56 + 59		Bridge terminals 1 + 4 of harness connector	1.5 Ohms maximum
	1 + 23		Bridge terminals 1 + 5 of harness connector	1.5 Ohms maximum
26	67 + 68	Engine speed sensor (G 28) 3–pin connector (white) near starter	—	500 to 700 Ohms
	56 + 67	—	—	Infinite Ohms (open)
	56 + 68	—	—	Infinite Ohms (open)
27	09 + 56 bridged	EGR Vacuum Regulator Solenoid valve (N 18)	Bridge terminals 55 + 30 of Test box	Valve must briefly click

Test step	VAG 1598 terminals	To be tested	Test conditions or additional steps	Specified values
28	23 + 30	Wiring to EGR Vacuum Regulator Solenoid valve (N 18)	Disconnect N 18 valve harness connector and bridge terminals 1 + 2 of harness connector	1.5 Ohms maximum
29	15 + 33	EGR Temperature sensor (G 98)	Disconnect EGR temperature sensor harness connector and bridge terminals 1 + 2 of harness connector	1.5 Ohms maximum

01 (Automatic Transmission 096) On-Board Diagnostic

Automatic transmission 096 01–1

- Automatic transmission 096, On-Board Diagnostic 01–1
- Safety functions of the Transmission Control Module 01–2
- Return to the basic settings 01–2
- Automatic transmission 096, electronic components 01–3
- VAG 1551 Scan Tool, connecting 01–5
- Reading/erasing Diagnostic Trouble Code memory using VAG 1551 01–7
- Diagnostic Trouble Code (DTC) trouble shooting chart 01–10
- Diagnostic Trouble Code (DTC) memory erasing, using VAG 1551 Scan Tool . . . 01–17
- Automatic transmission 096, returning to basic setting 01–18
- Measuring value block, reading 01–19
- Individual measuring values, reading 01–36
- Individual measuring values 01–38
- Automatic transmission 096, troubleshooting using electrical test steps 01–40
- Transmission Control Module (J217) Harness Connector Terminals 01–42

Automatic transmission 096, On-Board Diagnostic

The Transmission Control Module (TCM) (J 217) has a permanent DTC memory. If problems occurat the monitored sensors or components of the transmission, the information is stored in the DTC memory.

Malfunctions that occur only occasionally are stored as sporadic malfunctions.

The automatic transmission TCM evaluates the data from 21 different messages and stores the DTC until the memory is read and its contents erased.

The On-Board Diagnostic must be initiated and the stored DTC's read using the VAG 1551 Scan Tool. The VAG 1551 Scan Tool can access the transmission control module memory via data link connectors beneath the center console in front of the shift lever.

Safety functions of the Transmission Control Module

If a critical failure relating to the transmission should occur during vehicle operation, the transmission will continue to operate, but only in the "limp home" mode. In the limp home mode the transmission will automatically engage 3rd gear hydraulically for any forward driving mode other than selector position 1. The vehicle can be operated in 1 and R while in the limp home mode.

During limp home mode operation, the transmission control module will not actuate the cruise control system.

If malfunctions that result in limp home mode operation are corrected, the transmission will remain in limp home mode until the ignition is switched OFF.

Malfunctions that result in the limp home mode are usually open circuits or short circuits in wiring, defective ECU or solenoids.

Returning to basic setting

Return system to basic setting after the following:

- engine replacement
- throttle replacement or cable adjustment
- throttle position sensor replacement or adjustment
- transmission control module replacement

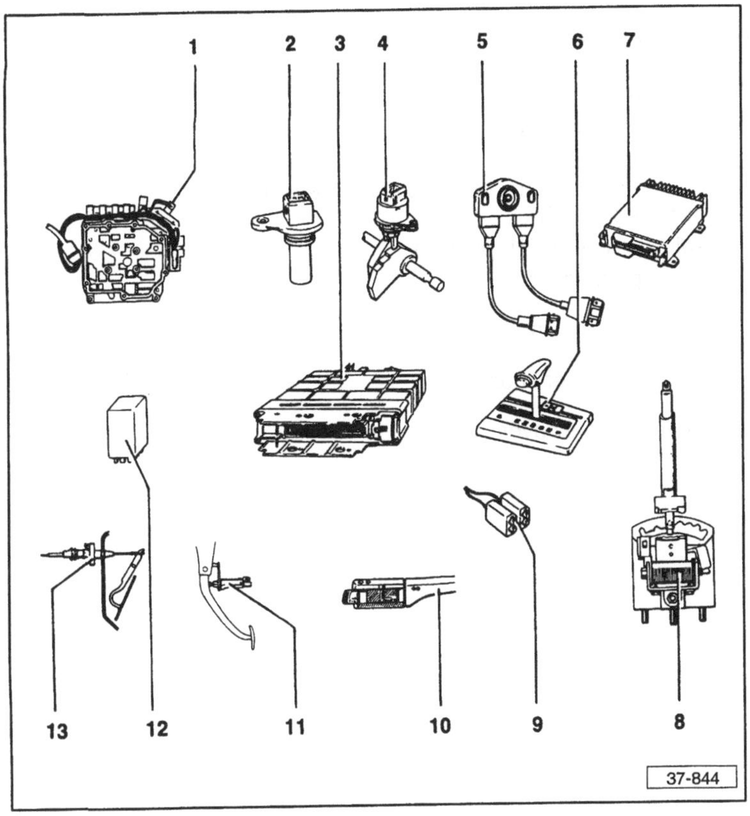

Automatic transmission 096, electronic components

1 – Valve body
- location: under oil pan solenoids (N 88, N 89, N 90, N 91, N 92, N 93, N 94) and Transmission Fluid temperature sensor (G 93) are mounted on valve body

2 – Vehicle speed sensor (G 68)
- location: on upper transmission case

3 – Transmission control module (TCM) (J 217)
- for automatic transmission location: at lower right "A" pillar area

4 – Multifunction transmission range switch (F 125)
- location: at rear of transmission case

5 – Throttle position sensor (G 69)
- location: on engine fuel distributor

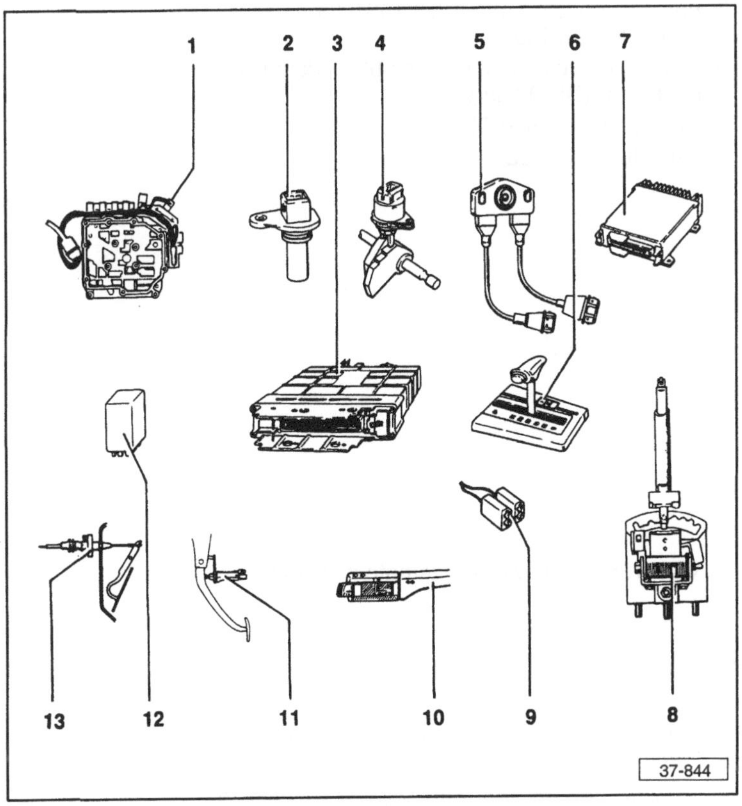

6 – Program switch
- location: in selector lever display on center console cover

7 – Engine control module
- location: ⇒ engine section

8 – Shift lock solenoid (N 110)
- location: on selector lever mount

9 – Data link connectors
- location: under center console cover in front of shift lever

10 – Cruise control transmission range switch (F 44)
- location: steering column switch

11 – Brake light switch (F)
- location: on pedal cluster

12 – Relay for starter interlock and backup light (J 226)
- location: relay panel (⇒ wiring diagram)

13 – Kick down switch (F8)
- location: integrated with accelerator cable in engine compartment

VAG 1551 Scan Tool, connecting

Requirements

- battery voltage OK
- fuses 14 and 21 OK
- ground connections on transmission and at left of relay panel OK
- selector lever in "N" position
- parking brake ON

♦ switch OFF ignition

♦ remove shift lever knob

♦ remove console cover

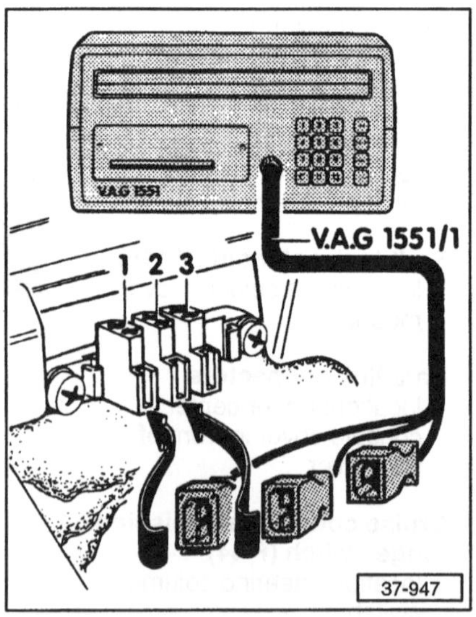

♦ connect diagnostic cable VAG 1551/1 for VAG 1551 Scan Tool:

- black plug (voltage supply) on black data link connector (DLC) 1 in vehicle
- white plug on white DLC 2
- blue DLC not used

VAG-On-Board Diagnostic	HELP
1 - Rapid data transfer	

◄ • VAG 1551 displays will alternate as shown:

VAG-On-Board Diagnostic	HELP
2 - Blink code output	

◄ and

Note:

For additional operating instructions push the HELP key.

If no display appears, check the voltage supply to the Scan Tool.

The → key advances the program to the next display.

Reading/erasing Diagnostic Trouble Code memory using VAG 1551

After connecting the tester the following operating modes appear alternately:

1 Rapid data transfer

and

2 Blink code output

Note:

Press the PRINT button for a complete printout of malfunctions.

- ♦ select operating mode 1, Rapid data transfer
- ♦ switch ignition ON

```
Rapid data transfer                    HELP
Input address word XX
```

◄ read display:

- ♦ press keys 0 and 2 (this selects program 02 for "transmission electronics")

```
Rapid data transfer                       Q
02 - Transmission electronics
```

◄ read display:

- ♦ press Q key to enter input

```
095927731 F Digimat                    0569
Code 127                          WSC131071
```

- ♦ read display showing transmission control module identification
- ♦ establish E-prom level of Digimat TCM installed in vehicle by noting four-digit number following word "Digimat"
- ♦ press → key


```
TCM does not respond!                  HELP

```

◄ If display appears as shown:

- ♦ press HELP key to print out a list of possible malfunctions (after eliminating cause, input 02 address word again)

Note:

If TCM does not respond! appears again:

- ♦ check for open circuits on data link connector — ⇒ wiring diagram
- ♦ ⇒ DTC chart under code 65535 — "TCM malfunctioning"

```
Rapid data transfer                    HELP
Select function XX
```

◄ read display:

- ♦ press 0 and 2 key (this selects program 02 for checking DTC memory)

```
Rapid data transfer                       Q
02 - Check DTC memory
```

◄ read display

- ♦ press Q key to enter input

```
2 DTC's recognized!
(Example)
```

◄ read display showing number of malfunctions

or

```
No DTC's recognized
(Example)
```

◄ The stored DTC's are indicated and printed out one after another.

- ♦ after last DTC has been displayed, press → key
- ♦ eliminate malfunctions using DTC chart, starting on page 01-10

Rapid data transfer	HELP
Select function XX	

◄ read display

♦ press 0 and 5 key (this selects program 05 for erasing DTC memory)

Rapid data transfer	Q
05 - Erase DTC memory	

◄ read display

♦ press Q key to enter input

Note:

If the engine is started or the ignition switched OFF between reading and erasing the DTC memory, no erasure of the memory will take place. Remember to read the DTC memory first, then erase.

Rapid data transfer
DTC memory is erased

◄ read display

- DTC memory is now erased

Note:

After reading and erasing the DTC memory, road test the vehicle and read the DTC memory again.

When reading the DTC memory again, "No DTC's recognized" must appear.

♦ press → key

Diagnostic Trouble Code (DTC) troubleshooting chart

Listed below are all the possible transmission DTC codes that can be recognized by the Transmission Control Module (TCM) (J 217) and displayed by the VAG 1551 Scan Tool.

If malfunctions occur only occasionally or if the DTC memory was not erased after elimination of the malfunction, it will be displayed as a sporadically occurring malfunction. Sporadic malfunctions are shown with an "SP/" at the right side of the display.

If malfunctioning components are found, also check the wiring to the components for short or open circuits using the wiring diagram.

VAG 1551 display Rapid Data code/Blink code	Possible causes	Corrective steps
0000 4444 No malfunction recognized	If after repair no malfunction appears, On Board Diagnostic has ended. If after repair, transmission does not function properly, conduct repairs per repair manual	
00258 1113 Solenoid valve 1 (N 88) open circuit short to ground	Open circuit or short circuit in wiring (K1 clutch, reverse gear brake B1) Malfunctioning solenoid valve 1	Check wiring, connections and solenoid Test step no. 6* Replace valve body
00260 1121 Solenoid valve 2 (N 89) open circuit short to ground	Open circuit or short circuit in wiring (B2 brake) Malfunctioning solenoid valve 2	Check wiring, connections and solenoid Test step no. 7* Replace valve body
00262 1123 Solenoid valve 3 (N 90) open circuit short to ground	Open circuit or short circuit in wiring (K3 clutch) Malfunctioning solenoid valve 3	Check wiring, connections and solenoid Test step no. 8* Replace valve body
00263 1124 Transmission 0107	Mechanical/hydraulic malfunction Malfunctioning clutch or valve body	Read measuring value block in whichever gear fault occurs Replace valve body or clutch
00264 1131 Solenoid valve 4 (N 91) open circuit short to ground	Open circuit or short circuit in wiring (K2 clutch) Malfunctioning solenoid valve 4	Check wiring, connections and solenoid Test step no. 9* Replace valve body

*⇒ page 01-40, Electrical testing

VAG 1551 display Rapid Data code/Blink code	Possible causes	Corrective steps
00266 1133 Solenoid valve 5 (N 92) open circuit short to ground	Open circuit or short circuit in wiring (B1 brake) Malfunctioning solenoid valve 5	Check wiring, connections and solenoid Test step no. 10* Replace valve body
00268 1141 Solenoid valve 6 (N 93) open circuit short to ground	Open circuit or short circuit in wiring (Modulator valve — main pressure in valve body) Malfunctioning solenoid valve 6	Check wiring, connections and solenoid Read measuring valve Test step no. 11* Replace valve body
00270 1143 Solenoid valve 7 N 94	Open circuit or short circuit in wiring (valve body pressure) Malfunctioning solenoid valve 7	Check wiring, connections and solenoid Test step no. 12* Replace valve body
00281 1231 Vehicle speed sensor G 68 No signal	Open circuit in wiring Malfunctioning vehicle speed sensor	Check wiring, connections and speed sensor Read measuring value Test step no. 15* Replace vehicle speed sensor
00293 1314 Multifunction transmission range switch F 125	Open circuit in wiring Malfunctioning multifunction switch	Check wiring and connections Read measuring value Test step no. 5* Replace multifunction switch

*⇒ page 01-40, Electrical testing

VAG 1551 display Rapid Data code/Blink code	Possible causes	Corrective steps
00296 **1323** Kickdown switch F 8	Open circuit in wiring Malfunctioning throttle valve position sensor Malfunctioning kickdown switch	Check wiring and connections Read measuring value Test step no. 14* and check throttle position sensor test step no. 2* Replace accelerator cable
00299 **1332** Transmission range program switch E 122 short circuit	Short circuit Malfunctioning program switch	Check wiring and connections Read measuring value Test step no. 16* Replace program switch
00300 **1333** Transmission fluid temperature sensor G 93 No DTC code recognized	Short circuit Malfunctioning temperature sensor	Check wiring and connections Read measuring value Test step no. 17* Replace valve body
00526 **2131** Brake light switch F No malfunction identified	Open circuit Malfunctioning brake light switch	Check wiring and connections Read measuring value block Test step no. 4* Replace brake light switch
00529 **2122** Engine speed signal missing	Open circuit	Check wiring and connections Check ECM
00518 **2212** Throttle position sensor G 69	Open circuit or short circuit Malfunctioning throttle position sensor	Check wiring and connections Read measuring value Test step no. 2* Replace throttle position sensor Return system to basic setting

*⇒ page 01-40, Electrical testing

VAG 1551 display Rapid Data code/Blink code	Possible causes	Corrective steps
00532 **2234** Supply voltage Voltage for all values too low	Malfunctioning battery Supply voltage NOT OK	Check battery voltage, replace if necessary Read individual measuring values Check supply voltage to TCM (J 217) Test step no. 1
00596 Short between solenoid wires	Break in wiring or short to ground	Check wiring and connections Perform Test Steps 6 through 12 and number 17
01236 **4314** Shift lock solenoid **N110** Open circuit* short to ground*	Open circuit or short circuit in wiring Malfunctioning interlock solenoid switch	Check wiring, connections and solenoid Test step no. 3* and 12* Replace shift interlock solenoid switch
00545 **2314** Engine/transmission electrical connection	Open circuit or short circuit in wiring No connection between ECM and TCM	Check wiring and connections Read individual measuring values If necessary, replace ECM Return system to basic setting

*⇒ page 01-40, Electrical testing

VAG 1551 display Rapid Data code/Blink code	Possible causes	Corrective steps
65535 1111 Transmission control module (TCM) (J 217)	Electrical interferences from outside sources or poor ground connection Malfunctioning TCM	Check wiring and connections Test step no. 1* Replace TCM if necessary Return system to basic setting
Note: *Replace the TCM and return system to basic setting only after determining and eliminating all possible mechanical and hydraulic malfunctions.*		
00638 N/A Engine/transmission electrical connection 2 Break Short to ground One of these indicators will appear in addition to display.	Break in wiring or short to ground Engine/transmission control module disconnected	Check wires and terminal connections Read test value block, ⇒ Repair Group D3 If necessary replace ECM, ⇒ Repair Group 24 Return system to basic setting, ⇒ Repair Group D3

*⇒ page 01-40, Electrical testing

VAG 1551 display Rapid Data code/Blink code	Possible causes	Corrective steps
00641 N/A ATF — Temperature Signal too large	Transmission fluid temperature exceeds max. of 148 C (299 F) Towing load of vehicle too high ATF-level not OK Note: If ATF — temperature too high, transmission will switch into next lowest gear	Check ATF level Read test value block and read ATF temperature
00652 N/A Gear monitoring Impossible sound	Electrical/hydraulic malfunction Clutch or valve body malfunctioning	Read test value block to determine in which gear the malfunction occurs Replace or adjust valve body, ⇒ Repair Group 38 Replace clutch, ⇒ Repair Group 37
00660 N/A Kickdown switch/throttle position sensor Impossible signal	Break in wiring Throttle position sensor (G 69) malfunctioning Kickdown switch (F8) malfunctioning	Check wiring and connectors Read test value block, ⇒ Repair Group D3, perform test step 14 and throttle position sensor test step 2 Adjust or replace accelerator cable, ⇒ Repair Group 20

*⇒ page 01-40, Electrical testing

Diagnostic Trouble Code (DTC) memory erasing, using VAG 1551 Scan Tool

Requirements

- DTC memory has been read and any malfunctions eliminated

Rapid data transfer Select function XX

◄ read display:

- ♦ press 0 and 5 key (this selects the program 05 for clearing the DTC memory)
- ♦ enter input with Q key

Rapid data transfer Erase DTC memory

◄ read display:

- ♦ if the ignition was switched OFF or the engine allowed to run between DTC memory reading and erasing, the display will read:

DTC memory must be read and erased again

Attention! DTC memory was not read

◄ after erasing DTC memory, test drive vehicle and repeat On Board Diagnostic procedure by reading DTC memory; display must appear as shown

No DTC recognized

◄ press → key

Automatic transmission 096, returning to basic setting

Rapid data transfer Select function XX	HELP

◄ with system in program 02 for transmission electronics, read display:

- ♦ press 0 and 4 key (this selects program 04 for "basic setting")

Rapid data transfer 04 Start basic setting	Q

◄ read display

- ♦ press Q key to enter input

Start basic setting Input display group number	HELP XX

◄ read display (appears with program card /2 only)

- ♦ press key 0 and 0
- ♦ push accelerator pedal to kickdown position and hold for three seconds
- ♦ release accelerator pedal and press Q key

System in basic setting	→

◄ read display

- ♦ press → key

The system is now returned to the basic setting.

Measuring value block, reading

Rapid data transfer HELP
Select function XX

◄ with system in program 02 for transmission electronics, read display

- press keys 0 and 8 (this selects function 08 for reading measuring value block)

Rapid data transfer Q
08 Read measuring value block

◄ read display

- press Q key to enter input

Read measuring value block HELP
Input display group number XX

◄ read display

- press keys 0 and 1 (this selects group number 01 and displays values corresponding to group 01 in following chart)

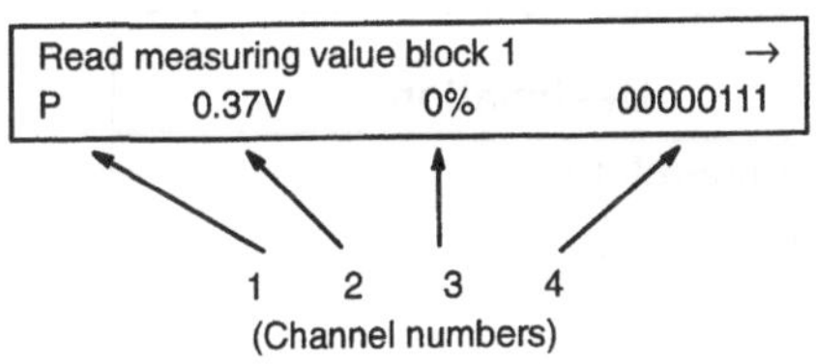

◄ press Q key to enter input

- read display
- press → key
- repeat group number entry for remaining groups (02 through 05) and press print key after each new display appears to print a record of diagnosis

Note:
Press keys 0 and 8 again before entering the next group number.

- *⇒ specified values on following pages*
- *if specified values appear on all channels press → key; the problem has not been found using this test sequence*
- *if specified values are not observed, conduct return to basic setting, page 01-18 and follow instructions in chart for further testing*

Measuring block values, overview

Examples

Read measuring value block 1
P 0.39V 0% 00100111

Read measuring value block 2
0.983A 0.996A 12.00V 2.48V

Read measuring value block 3
0 km/h 90/min 0 0%

Read measuring value block 4
1001 00 0 P 0km/h

Group number	Channel number	Designation
01	1	Selector lever position
	2	Throttle Position Sensor voltage
	3	Accelerator pedal
	4	Switch positions
02	1	Actual current of solenoid valve 6, (N 93)
	2	Specified current of solenoid valve 6, (N 93)
	3	Battery voltage
	4	Voltage at vehicle speed sensor, (G 68)
03	1	Vehicle speed
	2	Engine RPM
	3	Drive range
	4	Accelerator pedal value
04	1	Shift valves
	2	Drive range
	3	Selector lever position
	4	Vehicle speed

Read measuring value block 5
37°C 0011001 0 90/min

Group number	Channel number	Designation
05	1	ATF temperature
	2	Switch openings
	3	Drive range
	4	Engine RPM

Note:

Group numbers 01, 02, 03 and 05 are to be used for diagnosis.

Group number 04 is for reference only.

Measuring block values

Group number	Channel number	Test requirements		VAG 1551 display (specified value)	If test results NOT within specifications
01	1 Multi-function switch Ignition ON Engine OFF	vehicle stationary, selector lever in:	P	P	Check multifunction switch (F 125) VAG 1598 test step 5*
			R	R	
			N	N	
			D	D	
			3	3	
			2	2	
			1	1	

Group number	Channel number	Test requirements	VAG 1551 display (specified value)	If test results NOT within specifications
01	2 Throttle position sensor output Ignition ON Engine OFF	Vehicle Stationary Idle (Accelerator pedal not depressed) Full throttle (Accelerator pedal fully depressed)	 0.156 volts minimum 4.680 volts maximum	When accelerating from closed throttle to full throttle, the voltage value must increase.
	3 Throttle position sensor angular position Ignition ON Engine OFF	Vehicle Stationary Idle (Accelerator pedal not depressed) Full Throttle (Accelerator pedal fully depressed)	 0 to 1% 99 to 100%	Check throttle position sensor (G 69) The % value must increase when going from idle to full throttle. If adjusted or replaced, return system to basic setting

Group number	Channel number	Test requirements	VAG 1551 display (specified value)	If test results NOT within specifications
01	4 inputs to Digimat TCM Ignition ON Engine OFF	Brake [0] 0 0 0 0 0 0 0 This binary bit indicates when the brake is depressed. A "0" indicates that the brake pedal is not depressed. A "1" indicates that the brake pedal is depressed.		VAG 1598 test step 4
		Limited-slip control 0 [0] 0 0 0 0 0 0 This binary bit indicates the Limited-slip control, which is not applicable to this application and will always be set to "0".		Not applicable
		Program switch 0 0 [0] 0 0 0 0 0 This binary bit indicates the ECO/Sports program switch. A "0" indicates that the switch is set to Economy and a "1" indicates that the switch is set to Sport.		VAG 1598 test step 16
		Kickdown switch 0 0 0 [0] 0 0 0 0 This binary bit indicates the action of the Kickdown switch. When the Kickdown switch is actuated the bit will be "1" and when the switch is not actuated the bit will be "0".		VAG 1598 test step 14

Group number	Channel number	Test requirements	VAG 1551 display (specified value)	If test results NOT within specifications
01	4 inputs to Digimat TCM Ignition ON Engine OFF	Selector switch position 0 0 0 0 [0 0 0 0] **P** 0 1 1 1 **R** 1 1 1 1 **N** 1 0 1 1 **D** 1 0 1 0 **3** 1 0 0 0 **2** 1 1 0 0 **1** 0 1 0 0		Check multifunction switch (F 125) VAG 1598 test step 5

Group number	Channel number	Test requirements		VAG 1551 display (specified value)	If test results NOT within specifications
02 Modu-lator valve	1	vehicle stationary	full throttle	min. 0.0 Amps	VAG 1598 test step 11*
			idle	max. 1.1 Amps	
	2	vehicle stationary	full throttle	min 0.0 Amps	
			idle	max. 1.0 Amps	
	3	vehicle stationary		min. 10.8 Volts max. 16 Volts	Check battery, replace if necessary
	4	vehicle stationary		min. 2.20 Volts max. 2.52 Volts	VAG 1598 test step 15*

* page 01-40, Electrical test steps.

> ** ***CAUTION!***
> ***Use a second person to read displayed values when in the driving mode.***

Note:

When test conditions specify "in driving mode," the selector lever handle and cover assembly must be removed and the handle reinstalled. After testing reinstall the cover assembly and secure the handle set screw with AMV 200 000.

Group number	Channel number	Test requirements		VAG 1551 display (specified value)	If test results NOT within specifications
03 Inputs from TCM to valve body	1	in driving mode**		driven speed in km/h	Speedometer and VAG 1551 readings may deviate slightly
	2	engine running		engine RPM	Adjust per repair manual, ⇒ Engine section
	3	in driving mode**	neutral	O	Check solenoid valves VAG 1598 test steps 6* to 12*
			reverse	R	
			1 (hydraulic)	1	
			2 (hydraulic)	2	
			3 (hydraulic)	3H	
			3 (mechanical)	3M	
			4 (mechanical)	4	
	4	in driving mode**	idle	0 to 1	Check throttle Position Sensor (G 69) The % value must increase If adjusted or replaced return system to basic setting
			full throttle	99 to 100	

* page 01-40, Electrical test steps.

> ** ***CAUTION!***
> ***Use a second person to read displayed values when in the driving mode.***

Note:

When test conditions specify "in driving mode," the selector lever handle and cover assembly must be removed and the handle reinstalled. After testing reinstall the cover assembly and secure the handle set screw with AMV 200 000.

Group number	Channel number	Test requirements		VAG 1551 display (specified value)	If test results NOT within specifications
04	1 Valve actuation	in driving mode**			Individual solenoid valves will be switched depending on drive condition VAG 1598 test steps 6* to 10* and 12*
		(N 88)	switched	1	
			not switched	0	
		(N 89)	switched	1	
			not switched	0	
		(N 90)	switched	1	
			not switched	0	
		(N 91)	switched	1	
			not switched	0	
		(N 92)	switched	1	
			not switched	0	
		(N 94)	switched	1	
			not switched	0	

* page 01-40, Electrical test steps.

** ***CAUTION!***
Use a second person to read displayed values when in the driving mode.

Group number	Channel number	Test requirements		VAG 1551 display (specified value)	If test results NOT within specifications
04	2 Signal to solenoid valves	in driving mode **	neutral	O	Check solenoid valves VAG 1598 test steps 6* to 12*
			reverse	R	
			1 (hydraulic)	1	
			2 (hydraulic)	2	
			3 (hydraulic)	3H	
			3 (mechanical)	3M	
			4 (mechanical)	4	
	3 Multi-function switch	in driving mode ** selector lever in position:	P	P	Check multifunction switch (F 125) VAG 1598 test step 5*
			R	R	
			N	N	
			D	D	
			3	3	
			2	2	
			1	1	

* page 01-40, Electrical test steps.

** ***CAUTION!***
Use a second person to read displayed values when in the driving mode.

Group number	Channel number	Test requirements	VAG 1551 display (specified value)	If test results NOT within specifications
04	4	in driving mode **	driven speed in km/h	Speedometer and VAG 1551 readings may deviate slightly

* page 01-40, Electrical test steps.

** ***CAUTION!***
Use a second person to read displayed values when in the driving mode.

Group number	Channel number	Test requirements	VAG 1551 display (specified value)	If test results NOT within specifications
05	1	vehicle stationary with engine running	°C (display appears at approximately 60°C)	Check ATF level at a temperature of 60° to 80°C VAG 1598 test step 17*
	Air conditioning clutch	in driving mode** (1-7) 1-2 [0 0] 1 1 0 0 1 These two binary digits represent the signal level of the output of the Digimat TCM and the input to the CIS-E Motronic ECM that controls the firing angle retard when the transmission is shifting. These digits should always be the same. "1" when the transmission is shifting and "0" at all other times. This condition occurs very rapidly so when everything is working properly a change in level from 0 to 1 will probably not be seen. An error in this signal for a minimum of two seconds will cause a malfunction to be stored in the Digimat ECM.		Check harness routing per wiring diagram Replace engine or Transmission Control Module (J 217)
05	Shift lock solenoid	3-4 0 0 [1 1] 0 0 1 These two binary bits indicate the status of the signal to the shift lock solenoid. Both bits should be the same for correct operation of the system. A "1" indicates the lock is engaged and a "0" indicates disengagement. This signal will activate when the brake is depressed.		Check harness routing per wiring diagram Replace solenoid for shift lock VAG 1598 test step 3*

* page 01-40, Electrical test steps.
** ⇒ "CAUTION" note on previous page.

Group number	Channel number	Test requirements	VAG 1551 display (specified value)	If test results NOT within specifications
05	Cruise control power	5 0 0 [1 1] 0 0 1 This binary bit indicates when power is applied to the cruise control module. When the vehicle speed first exceeds 30 km the Digimat TCM turns on the power to the cruise control this is indicated by a "1". This bit will remain at "1" until the ignition switch is recycled off and on.		Check cruise control wiring harness routing per wiring diagram

* page 01-40, Electrical test steps.

** ⇒ "CAUTION" note on previous page.

Group number	Channel number	Test requirements	VAG 1551 display (specified value)	If test results NOT within specifications
5	Air conditioning clutch	0 0 1 1 0 [0] 1 This binary bit indicates when the Digimat TCM disengages the air conditioning compressor clutch. When the clutch is disengaged the bit will be a "1". This can be observed by depressing the pedal to kickdown while driving. This bit is normally a "0".		Check harness routing per wiring diagram Check air conditioning per repair manual
	Ⓐ ECO/Sport mode (1990) -or- Ⓑ Selector lever position (1991)	0 0 1 1 0 0 [1] Ⓐ ECO/Sport mode (model year 1990) This binary bit indicates the status of the ECO/Sport mode for model year 1990 only Ⓑ Shift lever position (model year 1991) This binary bit indicates the shift lever position in model year 1991 vehicles P, N = 1 R, D, 3, 2, 1 = 0		Check harness routing per wiring diagram Replace solenoid for shift lock VAG 1598 test step 3*

* page 01-40, Electrical test steps.

** ***CAUTION!***
Use a second person to read displayed values when in the driving mode.

Group number	Channel number	Test requirements	VAG 1551 display (specified value)	If test results NOT within specifications
	3 Signals to valve body	in driving mode **	O	If shifting does not occur, clutch or brake may be defective Replace transmission control module (J 217) VAG 1598 test step 12*
			R	
			1	
			2	
			3H	
			3M	
			4	
	4 Engine RPM	in driving mode **	Engine RPM	Check engine specifications

* page 01-40, Electrical test steps.

** ***CAUTION!***
Use a second person to read displayed values when in the driving mode.

Individual measuring values, reading

```
Rapid data transfer                HELP
Select function XX
```

◄ with system in program 02 for transmission electronics, read display:

- press keys 0 and 9 (this selects function 09 for reading individual test values)

```
Rapid data transfer                   Q
09 Read individual measuring value
```

◄ display reads:

- press Q key to enter input

```
Read individual measuring value
Feed in channel number XX
```

◄ display reads:

- press keys to input channel number of desired readout:
  - **00** test opening for solenoid valves (N 88, N 89, N 90) (not currently used)
  - **01** test opening for solenoid valves (N 91, N 92, N 94) (not currently used)
  - **02** voltage for transmission speed sender
  - **03** ATF temperature
  - **04** test opening for solenoid valve 6 (N 93)
  - **05** battery voltage
  - **06** not used
  - **07** voltage of throttle position sensor
- press Q key

Read individual measuring value
Channel 03 measuring value 2

◄ display reads:

Note:

Press keys 0 and 9 again before entering the next channel number.

- ⇒ specified values on following pages
- if specified values are obtained, press → button

Individual measuring values

Channel number	Designation	Conditions	VAG 1551 display value	If specified value not obtained
00	Test opening for solenoid valves (N 88, N 89, N 90)	not currently used	—	—
01	Test opening for solenoid valves (N 91, N 92, N 94)	not currently used	—	—
02	Vehicle speed sensor	vehicle stationary	minimum 112 maximum 133	VAG 1598 test step 15*
03	ATF temperature	vehicle stationary with engine running		The ATF temperature must be 60-80°C when checking the ATF level VAG 1598 test step 17*
		approximately 20°C	2	
		approximately 60°C	12	
		approximately 80°C	24	
04	Solenoid valve 6 (N 93) Modulator valve	vehicle stationary full throttle	minimum 0 maximum 10	VAG 1598 test step 11*
		idle	minimum 220 maximum 240	

* page 01-40, Electrical test steps.

Channel number	Designation	Conditions		VAG 1551 display value	If specified value not obtained
05	Battery voltage	vehicle stationary	maximum 16V	255	Check battery and replace if necessary
			minimum 10.8V	172	
06	Not used	—		—	—
07	Throttle position sensor (G 69)	vehicle stationary	full throttle	maximum 240	VAG 1598 test step 2*
			idle	minimum 8	

* page 01-40, Electrical test steps.

Automatic transmission 096, troubleshooting using electrical test steps

Testing equipment

- VAG 1598, Test box
- VAG 1598/9, adaptor
- Fluke 83, multi-meter

Testing

Perform only those recommended test steps from the Diagnostic Trouble Code chart, page 01-10 (indicated *)

Requirements

- battery voltage OK
- fuse 14 and 21 OK
- ground connections on transmission and at left of relay plate OK
- switch OFF ignition for all test steps
- disconnect multi-point connections from transmission ECM

voltage test — connect adaptor

VAG 1598/9 to ECM

resistance test — disconnect adaptor

VAG 1598/9 from ECM

- ECM ground connections OK

> ***CAUTION!***
>
> ***Switch multi-meter to appropriate range before connecting.***

Note:

The sockets on the Test box VAG 1598 are identical with the pin designations on the ECM (J 217).

Transmission Control Module (J 217) Harness Connector Terminals

1 – Ground (terminal 31)

2 – Solenoid valve 4, (N 91)

3 – Solenoid valve 3, (N 90)

4 – not used

5 – Park/neutral signal

6 – K-wire of diagnosis

7 – not used

8 – Kickdown for A/C

9 – Throttle Position Sensor (G 69)

10 – Throttle Position Sensor (G 69)

11 – not used

12 – Bulb for transmission range program switch (E 122)

13 – Vehicle speed sensor, (G 68), shielding

14 – L-wire of diagnosis

15 – Multi-function transmission range switch, (F 125)

16 – Multi-function transmission range switch, (F 125)

17 – Kickdown switch, (F 8)

18 – Supply voltage for solenoid valve

19 – Supply voltage (terminal 15)

20 – Shift lock solenoid, (N 110)

21 – Solenoid valve 7, (N 94)

22 – Solenoid valve 1, (N 88)

23 – Solenoid valve 2, (N 89)

24 – Solenoid valve 5, (N 92)

25 – Solenoid valve 6, (N 93)

26 – Brake light switch, signal voltage (F)

27 – Gas engine - engine speed sensor

28 – Gas engine - ignition timing reference

29 – Throttle position sensor, (G 69), ground

30 – ATF temperature

31 – Not used or selector lever display

32 – Vehicle speed sensor, (G 68)

33 – Vehicle speed sensor, (G 68)

34 – Multi-function transmission range switch, (F 125)

35 – Multi-function transmission range switch, (F 125)

36 – Program switch, (E 122)

37 – Closed throttle position switch

38 – Cruise control

Voltage tests

Test step	VAG 1598 terminals	Component to be tested	• Test conditions and additional steps	Specified value or results	If test results NOT within specs
Switch to 20 Volt range					
1	19 + 1	Transmission Control Module (J 217) voltage supply	• switch ignition ON	Battery voltage (approx.)	Check wire from terminal 1 to ground Check wire from terminal 19 for continuity with D/8 in relay panel
2	10 + 29	Throttle Position Sensor (G 69)	• switch ignition ON • disconnect throttle position sensor	4.6 to 5 Volts	Replace ECM Return system to basic setting
	9 + 29		throttle min. throttle max.	0.156 Volts 0.8 Volts	Calibration throttle position sensor; replace if necessary (⇒ engine section)
			wide open throttle	3.5 Volts to 4.68 Volts (max.)	Return system to basic setting
3	19 + 20	Shift Lock Solenoid (N 110)	• switch ignition ON • selector lever in P or N	Battery voltage (approx.)	Replace ECM Return system to basic setting
			brakes applied	0 Volts	Check signal from brake light switch (F) — ⇒ test step no. 4. Replace ECM if necessary Return system to basic setting

Test step	VAG 1598 terminals	Component to be tested	• Test conditions and additional steps	Specified value or results	If test results NOT within specs
4	26 + 1	Signal from brake light switch (F)	• switch ignition ON do NOT apply brakes	0 Volts	Check brake light switch and wiring per wiring diagram
			brakes applied	Battery voltage (approx.)	

Resistance tests

Test step	VAG 1598 terminals	Component to be tested	• Test conditions and additional steps	Specified value or results	If test results NOT within specs
5	34 + 1	Multi-function Transmission range switch (F 125)	• switch ignition ON move selector lever to positions N, D, and 3	4.5 to 5 Volts	Check wire routing per wiring diagram Replace multifunction TR switch
	15 + 1		move selector lever to position P and 1	0 to 0.8 Volts	
			move selector lever to position P, R, 2 and 1	4.5 to 5 Volts	
			move selector lever to position N, D and 3	0 to 0.8 Volts	
	35 + 1		move selector lever to position P, R, N and D	4.5 to 5 Volts	
			move selector lever to position 3, 2, and 1	0 to 0.8 Volts	
	16 + 1		move selector lever to position P, R, and N	approximate battery voltage	
			move selector lever to position D, 3, 2 and 1	0 to 0.8 Volts	

Resistance tests

Test step	VAG 1598 terminals	Component to be tested	• Test conditions and additional steps	Specified value or results	If test results NOT within specs
Switch to resistance measuring range 200 Ohms					
6	**22 + 18**	Solenoid valve 1 (N 88)	• switch ignition OFF	55-65 Ohms	Check harness per wiring diagram
	22 + 1		Transmission control module (J 217) removed	*∞ Ohms	Replace valve
7	**23 + 18**	Solenoid valve 2 (N 89)	• switch ignition OFF	55-65 Ohms	Check harness per wiring diagram
	23 + 1		Transmission control module (J 217) removed	*∞ Ohms	Replace valve
8	**3 + 18**	Solenoid valve 3 (N 90)	• switch ignition OFF	55-65 Ohms	Check harness routing per wiring diagram
	3 + 1		TCM (J 217) disconnected	∞ Ohms (open)	Replace valve
9	**2 + 18**	Solenoid valve 4 (N 91)	• switch ignition OFF	55-65 Ohms	Check harness routing per wiring diagram
	2 + 1		TCM (J 217) disconnected	∞ Ohms (open)	Replace valve
10	**24 + 18**	Solenoid valve 5 (N 92)	• switch ignition OFF	55-65 Ohms	Check harness routing per wiring diagram
	24 + 1		TCM (J 217) disconnected	∞ Ohms (open)	Replace valve
11	**25 + 18**	Solenoid valve 6 (N 93)	• switch ignition OFF	4.5-6.5 Ohms	Check harness routing per wiring diagram
	25 + 1		TCM (J 217) disconnected	∞ Ohms (open)	Replace valve

* switch multimeter to highest resistance range.

12	**21 + 18**	Solenoid valve 7 (N 94)	• switch ignition OFF	55-65 Ohms	Check harness routing per wiring diagram
	21 + 1		TCM (J 217) disconnected	∞ Ohms (open)	Replace valve

Resistance tests

Test step	VAG 1598 terminals	Component to be tested	• Test conditions and additional steps	Specified value or results	If test results NOT within specs
13	**19 + 20**	Shift lock solenoid (N 110)	• switch ignition OFF TCM (J 217) disconnected	14-25 Ohms	Check harness routing per wiring diagram Replace magnet for shift interlock
14	**1 + 17**	Kickdown switch (F 8)	• switch ignition OFF • TCM (J 217) do NOT press accelerator pedal	∞ Ohms (open)	Check harness routing per wiring diagram Adjust or replace accelerator cable
			depress accelerator fully	less than 1.5 Ohms	
Switch to measuring range 20 Ohms					
15	**32 + 33**	Vehicle speed sensor (G 68)	• switch ignition OFF • TCM (J 217) disconnected • set multimeter to 2 k Ohm scale	800 to 830 Ohms	Check harness routing per wiring diagram Replace vehicle speed sensor
16	**36 + 1**	Program switch (E 122)	• switch ignition OFF • TCM (J 217) disconnected TR Program switch not activated	*∞ Ohms (open)	Check harness routing per wiring diagram Replace TR program switch
			TR Program switch activated	less than 1.5 Ohms	

Resistance tests

Test step	VAG 1598 terminals	Component to be tested	• Test conditions and additional steps	Specified value or results	If test results NOT within specs
17	30 + 18	ATF temperature sensor (G 93)	• switch ignition OFF • TCM (J 217) disconnected set multimeter to 2 Meg Ohm scale ATF temperature (approx.) 20 C (68 F)	0.247 Meg Ohm	Check harness routing per wiring diagram Replace ATF temperature sensor
			Set multimeter to 200 k Ohm scale ATF temperature 60 C (140 F)	Approx 48,800 Ohms	
			ATF temperature 120 C (216 F)	Approx 7400 Ohms	

* switch multimeter to highest resistance range.

THIS PAGE INTENTIONALLY LEFT BLANK

01 (ABS) On-Board Diagnostic

Teves 02 ABS . . . 01–1

Distinguishing features of ABS and ABS/EDL . . . 01–1

Teves 02 ABS . . . 01–1

Teves 02 ABS/EDL . . . 01–2

Teves 02 ABS On-Board Diagnostic . . . 01–3

Function . . . 01–3

Technical data of On-Board Diagnostic . . . 01–5

Safety precautions, general troubleshooting . . . 01–6

Necessary technical information . . . 01–7

Test requirements for On-Board Diagnostic . . . 01–8

Connecting scan tool VAG 1551 and interrogating Diagnostic Trouble Code (DTC) memory . . . 01–9

Erasing Diagnostic Trouble Code (DTC) memory . . . 01–14

Diagnostic Trouble Code (DTC) table/Blink code output . . . 01–15

Teves 02 ABS Electrical Test . . . 01–24

Multi-pin connectors on Teves 02 ABS . . . 01–25

Test requirements . . . 01–27

Teves 02 ABS/EDL On-Board Diagnostic . . . 01–53

Function . . . 01–53

Technical data of On-Board Diagnostic . . . 01–55

Safety precautions, general troubleshooting . . . 01–57

Necessary technical information . . . 01–58

Test requirements for On-Board Diagnostic . . . 01–59

Connecting scan tool VAG 1551 . . . 01–61

Interrogating Diagnostic Trouble Code (DTC) memory (Rapid data transfer) . . . 01–66

Diagnostic Trouble Code (DTC) table (Rapid data transfer) . . . 01–69

Erasing/ending output Diagnostic Trouble Code memory (Rapid data transfer) . . 01–82

Interrogating Diagnostic Trouble Code memory (Blink code output) . . . 01–84

Erasing Diagnostic Trouble Code memory (Blink code output) . . . 01–88

Diagnostic Trouble Code (DTC) table (Blink code output) . . . 01–89

Electrical test of Teves 02 ABS/EDL . . . 01–100

Multi-pin connectors at Teves 02 ABS/EDL . . . 01–101

Test requirements . . . 01–104

Adjusting Brake light switch (F) . . . 01–138

On Board Diagnostic (OBD) of Teves ABS

Distinguishing features of Anti–Locking Brake Systems (ABS) and of anti–locking brake systems with electronic differential lock (ABS/EDL)

(Two different ABS systems are found on Corrado models.)

Teves 02 ABS

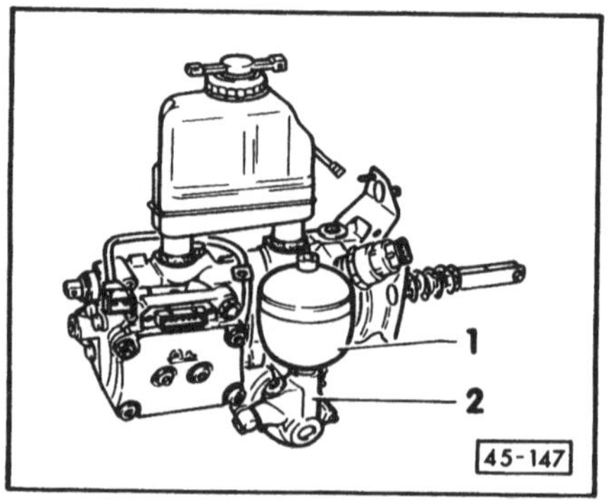

Hydraulic brake servo assistance

◄ ***Distinguishing features:***

- *Pressure accumulator –1–.*
- *Pump unit –2–.*
- *35–pin control module.*
- *OBD only in blink code output mode ⇒ page 01–5.*

Teves 02 ABS/EDL

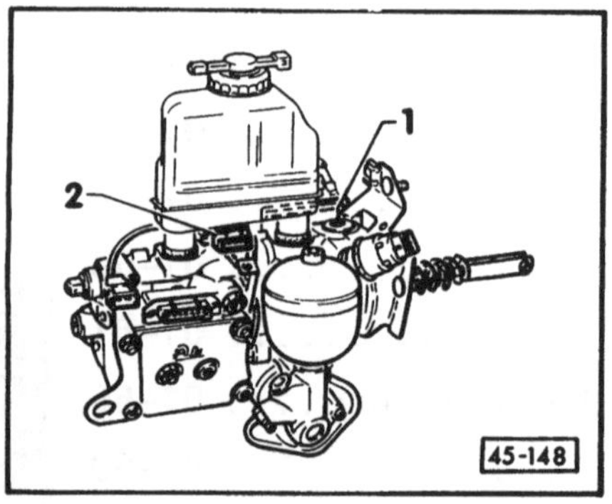

Hydraulic brake servo assistance

◄ ***Distinguishing features:***

- *EDL valve block –1–.*
- *5–pin connector below brake fluid reservoir –2–.*
- *55–pin control module.*
- *OBD in the blink code output and/or rapid data transfer modes ⇒ page 01–55.*

On Board Diagnostic (OBD) of Teves 02 ABS

Function

The ABS Control Module (J104) is located on the left below the rear seat. It is equipped with a diagnostic trouble code (DTC) memory. The connection for OBD is located below the boot of the gearshift lever.

The OBD relates to the electrical/electronic part of the ABS; in other words, only malfunctions via the electrical connection to the control module are recognized (e.g. interruption of a valve).

The control modules recognize malfunctions during ABS operation and store them in a permanent memory, the contents of which can be retrieved even after years despite the lack of any battery voltage.

Malfunctions which occur sporadically are likewise recognized and stored but, in contrast to other diagnostic compatible systems, without the suffix "SP" (sporadic).

Malfunctions are stored with differing priority: Solenoid valve malfunctions have the highest, while malfunctions which occur intermittently have the lowest priority.

After the ignition is switched on, the ABS Hydraulic Pump (V64) builds up the necessary operating pressure in the pressure accumulator within not more than 60 seconds provided the ABS is operating properly; the ABS Warning Light (K47) lights up for 2 ... 20 seconds depending on operating pressure and a test run is performed in the ABS Control Module (J104).

If an ABS malfunction is stored, the ABS is switched off – depending on the priority of the malfunction – and at the same time, the ABS Warning Light (K47) comes on.

OBD must be inititated at the start and the stored information should be retrieved with:

Scan Tool (ST) VAG 1551

The Diagnostic Trouble Code (DTC) information displayed is used in combination with a table of notes to determine the possible causes and to perform specific repair measures.

On Board Diagnostic (OBD), technical data

Control module

- 35–pin.

Diagnostic Trouble Code (DTC) memory

- Permanent memory–contents are retained even after the power supply is switched off.

Data output

- Blink code output (operating mode 2).
- ABS warning light in dash panel lights up if ABS is switched off.

On Board Diagnostic (OBD), safety precautions

General troubleshooting

- The ABS is a vehicle safety system; work on the system requires appropriate knowledge.
- Malfunctions in the ABS system are indicated by the ABS warning lamp lighting up; certain malfunctions are not recognized until the vehicle is traveling faster than a minimum speed of 40 km/h (performing road test).
- Do not drive with the connector unplugged from the ABS Control Module (J104).
- The Diagnostic Trouble Code (DTC) Memory must be interrogated before performing work on the ABS system in order to check complaints and to be able to perform specific OBD procedures.
- Do not separate any wiring connector when the ignition is switched on.
- Refer to the appropriate instructions regarding handling brake fluid ⇒ Repair Group 47.

Necessary technical information

- Design and operation of the anti–locking brake system ⇒ Service Training self–study program "Anti–Lock Brake System".
- Current flow diagram ⇒ Wiring Diagrams, Electrical Component Locations & Troubleshooting binder.

Test requirements for On Board Diagnostic (OBD)

- The tires fitted to all wheels must be of the specified type and same size. Inflate tires to the specified pressure.
- Conventional brake system with brake light switch and brake lights OK.
- No leaks at hydraulic connections and lines (visual inspection of hydraulic unit, brake cylinders).
- Wheel bearings and wheel bearing play OK.
- Connector at ABS Control Module (J104), on the left below the rear seat, correctly connected (lock is engaged).
- Ground connection close to ABS Control Module (J104) OK
- All fuses OK per wiring diagram.
- Supply voltage OK (at least 10.5V).

Diagnostic Trouble Code (DTC) Memory, connecting/interrogating with VAG 1551 Scan Tool

- Test requirements met ⇒ page 01–8.

– Unclip cover from gearshift lever.

– With ignition switched off, connect VAG 1551 with harness VAG 1551/1 to the Data Link Connectors (DLC) as follows:

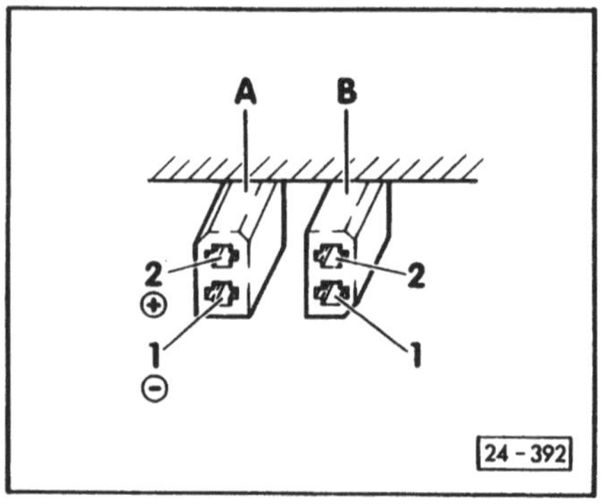

◄ – First of all plug black connector of VAG 1551/1 into black DLC –A– in vehicle (voltage supply).

VAG On Board Diagnostic	HELP
1 – Rapid data transfer*	
2 – Blink code output*	

◄ Readout in display:

*is displayed alternately.

Note:

If no readout appears in the display, do not plug in white connector! In this case, DLC (black connector) for OBD, electrical test ⇒ page 01–24, test step 38.

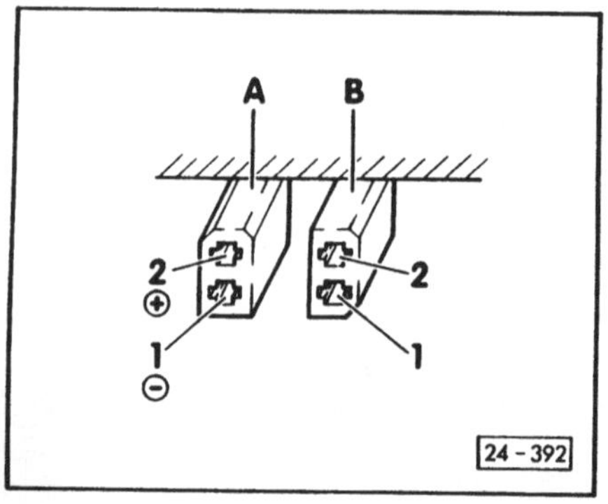

◄ – Plug white connector of VAG 1551/1 into grey (on early models, blue) DLC –B– in vehicle.
The blue connector of VAG 1551/1 is not required.

Notes:

- Depending on the program, additional operating information can be printed out by pressing the HELP button of VAG 1551.
- The → button is used for moving forward within the program.
- The PRINT button is used for switching on the printer (indicator lamp in button comes on).

– Press button –2– for "Check Diagnostic Trouble Code (DTC) Memory.

VAG On Board Diagnostic	HELP
Initiate with → key!	

◄ Readout in display:

– Press → button and hold.

Blink code output is initiated!

◄ The following readout appears first of all:

– Press → button and hold.

– If after about 2.5 seconds the readout

Blink code output →
Permanent ground on exciter wire!

appears, release → button.

– Switch on ignition. ABS Warning Light (K47) lights up.

Note:

If the ABS Warning Light (K47) does not light up, test the wiring connections for OBD and the warning light (K47) ⇒ page 01–24, test steps 38 ... 40.

– Briefly press → button.

The following readouts appear one after the other:

Blink code output is initiated!

Blink code XXXX

Blink code 1241
Right Front ABS Wheel Speed Sensor –G45

If a malfunction is stored, the DTC is indicated by means of the ABS Warning Light (K47) and appears in the display, e.g.

Notes:

Blink code 4444
No malfunction recognized!

◄ • *If this readout appears, OBD is ended.*

– *Switch off ignition.*

If the malfunction has not been corrected, continue with Electrical Testing ⇒ page 01–24, test steps 1 ... 37.

Blink code
No blink code received!

◄ • *If this readout appears, test the Data Link Connectors (DLC) for OBD ⇒ page 01–24, test steps 38 ... 40.*

• *If a malfunction is displayed (e.g. inlet or outlet valve) and if it is then not possible to retrieve any further DTC's, the malfunction found has caused the control module to switch off. After repairing this malfunction, repeat OBD.*

– Briefly press ⇒ button in order to interrogate the next DTC.

– DTC interrogation is continued until the blink code 0000 appears:

Blink code 0000
End output

Blink code output
is ended!

and then

- Switch off ignition.
- With the DTC last determined go to the DTC table ⇒ page 01–15 and correct the malfunction.
- After completing repairs, switch on ignition for about 4 seconds (self–test):

- If the ABS Warning Light (K47) goes out, there is no malfunction in the ABS system and the DTC memory in the control module must be erased ⇒ page 01–14.
- If the ABS Warning Light (K47) does not go out, once again interrogate the DTC memory. When this is done, not only the next DTC but also once again the previous DTC is displayed (although this malfunction has meanwhile been corrected).

Diagnostic Trouble Code (DTC) Memory, erasing

Requirements:

- 1. DTC Memory has been interrogated until DTC 0000 has appeared.
- 2. Ignition is switched off and repair of the malfunction already displayed has been performed.

- Carry out road test and briefly exceed a speed of 25 mph during test. This erases the DTC memory.

Important!

During the road test, neither the Scan Tool VAG. 1551 nor any harness adapters must be connected.

The cover for the gearshift lever should be installed again before driving.

- After erasing the DTC Memory, repeat OBD! OBD is not ended until the DTC 4444 appears.

Diagnostic Trouble Code (DTC) Table

Notes:

- *All the possible malfunctions which can be recognized by the ABS Control Module (J104) and printed out on VAG 1551 are listed below according to the 4–digit DTC number.*
- *After performing repairs, the DTC Memory must always be interrogated once again with the Scan Tool VAG 1551 and erased, and a road test (with vehicle driven at more than 25 mph) also performed.*
- *Repeat the OBD after the road test.*

Output at printer of VAG 1551	Possible cause of malfunction	Correction
Blink code 0000 End output	If "No malfunction recognized' appears after completing repairs, OBD is ended. Should the ABS system not operate properly although "No malfunction recognized" appears, continue troubleshooting without OBD and work through the electrical test complete ⇒ page 01–26.	
Blink code 1111 Control Module (J104) faulty	Electrical malfunctions caused by outside sources of interference or poor ground (GND) connections.	– Test wiring, wiring connections and GND connections: ⇒ Test step [1] 1, 2, and 3
Blink code 1112 Left Front ABS Inlet Valve –N101	Faulty wiring, wiring connector or valve coil.	– Test wiring, wiring connections and GND connections: ⇒ Test step [1] 23, 29, and 35

[1] ⇒ page 01–24, Electrical test

01 (ABS) OBD

Output at printer of VAG 1551	Possible cause of malfunction	Correction
Blink code 1114 Right Front ABS Inlet Valve –N99	Faulty wiring, wiring connection or valve coil.	– Test wiring, wiring connections and GND connections ⇒ Test step 1) 22, 29 and 36
Blink code 1122 Rear ABS Inlet Valve –N103	Faulty wiring, wiring connection or valve coil.	– Test wiring, wiring connections and GND connections ⇒ Test step 1) 21, 29 and 36
Blink code 1132 Left Front ABS Outlet Valve –N102.	Faulty wiring, wiring connection or valve coil.	– Test wiring, wiring connections and GND connections: ⇒ Test step 1) 26, 29 and 35

1) ⇒ page 01–24, Electrical test

Output at printer of VAG 1551	Possible cause of malfunction	Correction
Blink code 1134 Right Front ABS Outlet Valve –N100	Faulty wiring, wiring connection or valve coil.	– Test wiring, wiring connections and GND connections ⇒ Test step 1) 25, 29 and 36
Blink code 1122 Rear ABS Outlet Valve –N104	Faulty wiring, wiring connection or valve coil.	– Test wiring, wiring connections and GND connections ⇒ Test step 1) 24, 29 and 37
Blink code 1142 ABS Main Valve –N105	Faulty wiring, wiring connection or valve coil.	– Test wiring, wiring connections at main valve and also wiring connector T7 and valve coil: ⇒ Test step 1) 20, 29 and 33

1) ⇒ page 01–24, Electrical test

Output at printer of VAG 1551	Possible cause of malfunction	Correction
Blink code 1233 Left Front ABS Wheel Speed Sensor –G47	Open circuit, short circuit or loose contact in: • Speed sensor cable. • Wiring connection. • Speed sensor Excessive or insufficient gap between speed sensor and pulse wheel (signal not in order). Pulse wheel and speed sensor dirty or damaged.	– Test wiring, wiring connections and speed sensor, GND cables to control module as well as capacitor for correct GND conn–ections: ⇒ Test step [1] 9, 11 and 15 – Check installation of speed sensor and pulse wheel, clean if necessary.

[1] ⇒ page 01–24, Electrical test

Output at printer of VAG 1551	Possible cause of malfunction	Correction
DTC 1241 Right Front ABS Wheel Speed Sensor –G45	Open circuit, short circuit or loose contact in: • Speed sensor cable. • Plug connection. • Speed sensor Excessive or insufficient gap between speed sensor and pulse wheel (signal not in order). Pulse wheel and speed sensor dirty or damaged.	– Test wiring, wiring connections and speed sensor, GND cables to control module as well as capacitor for correct GND connec–tions: ⇒ Test step [1] 8, 10 and 14 – Check installation of speed sensor and pulse wheel, clean if necessary.

[1] ⇒ page 01–24, Electrical test

Output at printer of VAG 1551	Possible cause of malfunction	Correction
DTC 1243 Right Rear ABS Wheel Speed Sensor –G44	Open circuit, short circuit or loose contact in: • Speed sensor cable. • Plug connection. • Speed sensor Excessive or insufficient gap between speed sensor and pulse wheel (signal not in order). Pulse wheel and speed sensor dirty or damaged.	– Test wiring, wiring connections and speed sensor, GND cables to control module as well as capacitor for correct GND connections: ⇒ Test step [1] 6, 12 and 16 – Check installation of speed sensor and pulse wheel, clean if necessary.

[1] ⇒ page 01–24, Electrical test

Output at printer of VAG 1551	Possible cause of malfunction	Correction
DTC 1311 Left Rear ABS Wheel Speed Sensor –G46	Open circuit, short circuit or loose contact in: • Speed sensor cable. • Plug connection. • Speed sensor Excessive or insufficient gap between speed sensor and pulse wheel (signal not in order). Pulse wheel and speed sensor dirty or damaged.	– Test wiring, wiring connections and speed sensor, GND cables to control module as well as capacitor for correct GND connections: ⇒ Test step [1] 7, 13 and 17 – Check installation of speed sensor and pulse wheel, clean if necessary.

[1] ⇒ page 01–24, Electrical test

Output at printer of VAG 1551	Possible cause of malfunction	Correction
DTC 1312 ABS Pressure Warning Switch II F116 ABS Brake Fluid Warning Switch –F117	Wiring between terminal 9 and terminal 10 of ABS Control Module (J104) at components F116 and F117 has connection to Battery Positive (B+).	– Test cable between contact 9 and contact 10: ⇒ Test step [1] 31 and 32 – Unplug multi–pin connector from ABS Control Module (J104), connect to test box VAG 1598 and adapter VAG. 1598/3. – Use digital multimeter Fluke 83 (US 1119) to perform voltage measurement (20 V range) between contacts: 1 and 9 1 and 10 Specification: max. 0.1 V.

[1] ⇒ page 01–24, Electrical test

Electrical test of Teves 02 ABS

The test steps on the pages which follow apply to:

- Vehicles without On–Board Diagnostic (OBD).

– Perform test steps 1 ... 37.

- Vehicles on which OBD does not provide any indication of source of malfunction:

– Perform test steps 1 ... 37.

- Vehicles on which OBD provides a direct indication of source of malfunction:

– Perform only the test steps recommended from the DTC table (targeted test initiation).

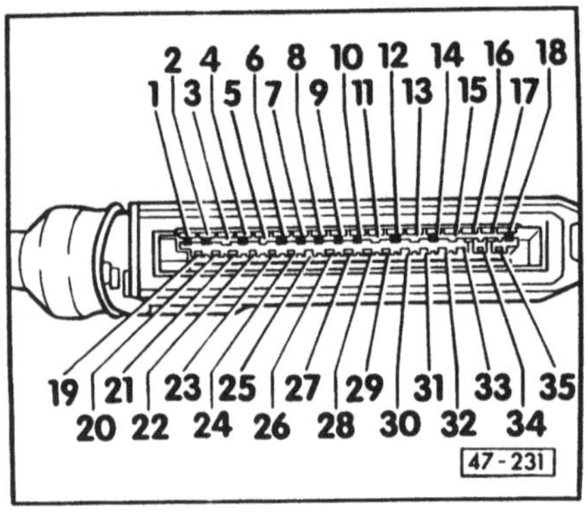

Multi–pin connectors on Teves 02 ABS

◄ 35–pin connector for ABS Control Module (J104).

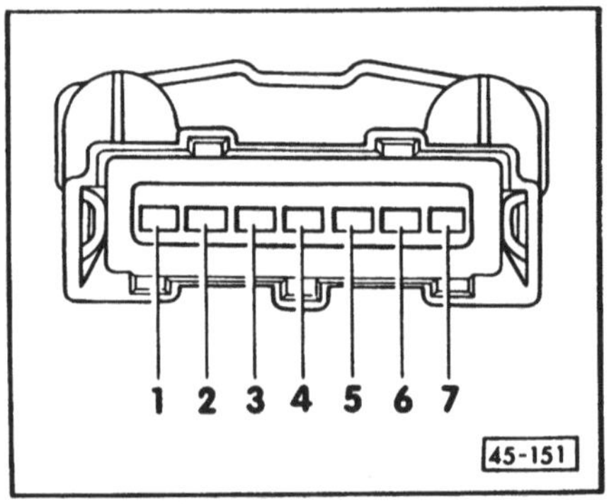

◄ 7–pin connector for valve block.

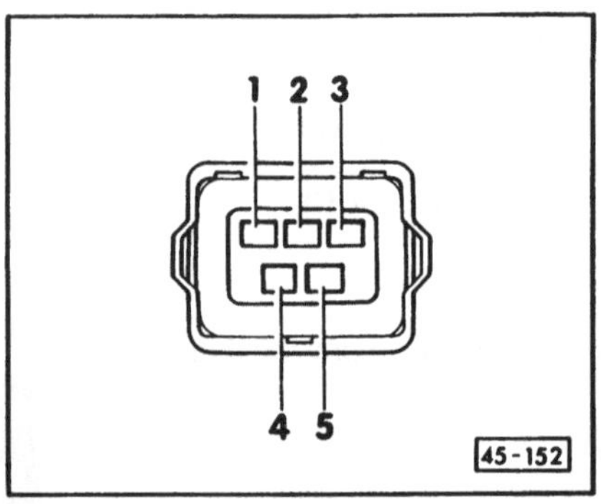

◄ 5–pin connector for Pressure Warning Switch.

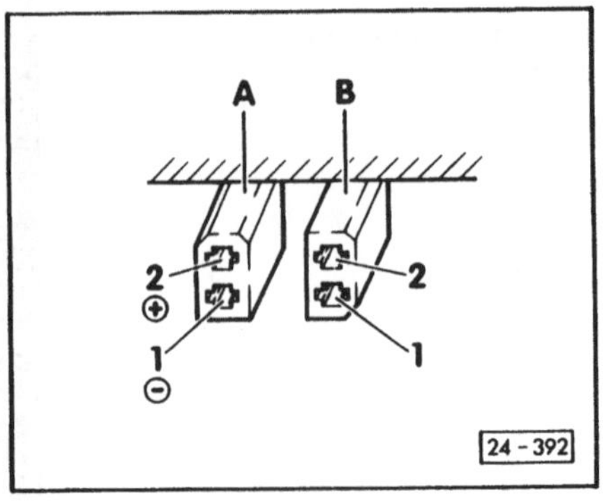

◄ Contact assignment of connectors for voltage supply and OBD with VAG 1551 Scan Tool.

Connector –A– (black) – voltage supply
Terminal 1 = Ground (GND)
Terminal 2 = Battery positive (B+)

Connector –B– (blue or grey) – OBD
Terminal 1 = Connection to contact 26 of ABS Control Module.

Measuring equipment required:

- Test box VAG 1598 with adapter VAG 1598/3.
- Adapter cable set VAG 1594.
- Digital multimeter Fluke 83 (US 1119).
- Raise vehicle with lifting platform for test steps 6 ... 9 and 35 ... 37.

Test requirements

- Fuses S16, S53, S54 in order.
- Ground (GND) cables close to ABS Control Module (J104) OK.
- Before starting test, switch off ignition and electrical accessories (headlamps, lighting, fan ...).
- The specifications are matched to Fluke 83 (US 1119) and do not necessarily apply to other test equipment.
- Disconnect multi–pin connector from ABS Control Module (J104) and connect test box VAG 1598 with adapter VAG 1598/3 to the wiring harness connector.

- The second connection of adapter VAG 1598/10 remains free!
  The control module is not connected.

Notes:

- *The socket designations of test box VAG 1598 are identical to the contact designations on the ABS Control Module (J104) ⇒ wiring diagrams, troubleshooting and component locations binder.*
- *If the readings obtained differ from the specifications, perform correction measures in the right–hand part of the table ⇒ wiring diagrams, troubleshooting and component locations binder.*
- *Before replacing the particular components, test wiring and connections and also, particularly for specifications of less than 10 Ω, repeat resistance measurement at component. If the readings obtained differ only slightly from the specifications, clean sockets and connectors of test equipment and also of test harnesses (with contact spray G 000 700 04 or equivalent) and repeat test.*

Switch on measuring range: Voltage measurement (20 V ≈)					
Test step [1]	**VAG 1598 sockets**	**Test of**	**• Test conditions – Additional operations**	**Specification**	**Correction if different from specification**
1	2 + 1	Voltage supply (terminal 15) at ABS Control Module (J104)	• Ignition switched on.	approx. battery voltage	– Test wiring from contact 1 to ground (GND). – Test wiring from contact 2 to contact D/7 (relay plate).
2	3 + 1	Function of ABS Relay (J102) Voltage supply at ABS Control Module (J104)	• Ignition switch off – Remove fuse S16. – Bridge sockets 2 and 8 at VAG 1598. – Switch on ignition: ABS Relay (J102) clicks.	approx. battery voltage	– Test wiring from contact 1 to GND. – Test wiring from contact 3 via J102 and S54 to Battery Positive (B+). – If J102 does not click, test actuation: Test wiring from contact 8 to J102/4 and relay J102.

[1] ⇒ page 01–27

Switch on measuring range: Voltage measurement (20 V ≈)					
Test step [1]	**V.A.G. 1598 sockets**	**Test of**	**• Test conditions – Additional operations**	**Specification**	**Correction if different from specification**
3	20 + 1	Function of ABS Relay (J102) (reference voltage)	• Ignition switched on. – Remove bridge after measurement – Insert fuse S16.	approx. battery voltage	– Test wiring from contact 20 via J102 and S54 to Battery Positive (B+). – If J102 does not click, test actuation: Test wiring from contact 8 to J102/4 and relay J102.

[1] ⇒ page 01–27

Switch on measuring range: Voltage measurement (20 V ≈)					
Test step 1)	**VAG 1598 sockets**	**Test of**	• **Test conditions** – **Additional operations**	**Specification**	**Correction if different from specification**
4	12 + 1	Function of Brake Light switch (F)	• Ignition switched off. – Depress brake pedal and hold.	After operating brake light switch: approx. battery voltage	– Test fuse S20 and Brake Light Switch (F). – Test wiring from contact 12 to contact W/4 (relay plate). – Test wiring from E/3 (relay plate) to Brake Light Switch (F).

1) ⇒ page 01–27, Electrical test

Switch on measuring range: Voltage measurement (20 V ≈)					
Test step 1)	**VAG 1598 sockets**	**Test of**	• **Test conditions** – **Additional operations**	**Specification**	**Correction if different from specification**
5	32 + 1	Operation of ABS Hydraulic Pump Relay (J185)	• Ignition switched off. – Unplug connector T2 at ABS Hydraulic Pump (V64). – Depress brake pedal 20 times as far as stop (discharge accumulator). – Switch on ignition. – After measurement, plug in connector T2 again.	approx. battery voltage	– Test wiring from contact 1 to GND. – Test wiring from contact 32 via J185 and S53 to Battery Positive (B+).

1) ⇒ page 01–27

Switch on measuring range: Voltage measurement (2 V ≈)					
Test step [1]	**VAG 1598 sockets**	**Test of**	• **Test conditions** – **Additional operations**	**Specification**	**Correction if different from specification**
6	4 + 22	Voltage signal of Right Rear ABS Wheel Speed Sensor (G44)	• Vehicle raised. • Ignition switch off. – Rotate right rear wheel about 1 rev./sec.	min. 65 mV alternating voltage	– Check T2 wiring connection, installation of speed sensor and pulse wheel. – Check whether speed sensor incorrectly connected.
7	6 + 24	Voltage signal of Left Rear ABS Wheel Speed Sensor (G46)	• Vehicle raised. • Ignition switched off. – Rotate left rear wheel about 1 rev./sec.	min. 65 mV alternating voltage	– Check T2 wiring connection, installation of speed sensor and pulse wheel. – Check whether speed sensor incorrectly connected.

1) ⇒ page 01–27

Switch on measuring range: Voltage measurement (2 V ≈)					
Test step [1]	**VAG 1598 sockets**	**Test of**	• **Test conditions** – **Additional operations**	**Specification**	**Correction if different from specification**
8	7 + 25	Voltage signal of Right Front ABS Wheel Speed Sensor (G45)	• Vehicle raised. • Ignition switch off. – Rotate right front wheel about 1 rev./sec.	min. 65 mV alternating voltage	– Check T2 wiring connection, installation of speed sensor and pulse wheel. – Check whether speed sensor incorrectly connected.
9	5 + 23	Voltage signal of Left Rear ABS Wheel Speed Sensor (G46)	• Vehicle raised. • Ignition switched off. – Rotate left rear wheel about 1 rev./sec.	min. 65 mV alternating voltage	– Check T2 wiring connection, installation of speed sensor and pulse wheel. – Check whether speed sensor incorrectly connected.

1) ⇒ page 01–27

Switch on measuring range:
Resistance measurement (2 kΩ)

Test step [1]	VAG 1598 sockets	Test of	• Test conditions – Additional operations	Specification	Correction if different from specification
10	7 + 25	Resistance of Right Front ABS Wheel Speed Sensor (G45)	• Ignition switched off.	1.0 ... 1.3 kΩ	– Test wiring connection T2. – Test resistance of speed sensor. (1.0 ... 1.2 kΩ). – Test wiring to speed sensor; move wiring during test (loose contact).
11	5 + 23	Resistance of Left Front ABS Wheel Speed Sensor (G47)	• Ignition switched off.	1.0 ... 1.3 kΩ	– Test wiring connection T2. – Test resistance of speed sensor. (1.0 ... 1.2 kΩ). – Test wiring to speed sensor; move wiring during test (loose contact).

[1] ⇒ page 01–27

Switch on measuring range:
Resistance measurement (2 kΩ)

Test step [1]	VAG 1598 sockets	Test of	• Test conditions – Additional operations	Specification	Correction if different from specification
12	4 + 22	Resistance of Right Rear ABS Wheel Speed Sensor (G44)	• Ignition switched off.	1.0 ... 1.3 kΩ	– Test wiring connection T2. – Test resistance of speed sensor. (1.0 ... 1.2 kΩ). – Test wiring to speed sensor; move wiring during test (loose contact).
13	6 + 24	Resistance of Left Rear ABS Wheel Speed Sensor (G46)	• Ignition switched off.	1.0 ... 1.3 kΩ	– Test wiring connection T2. – Test resistance of speed sensor. (1.0 ... 1.2 kΩ). – Test wiring to speed sensor; move wiring during test (loose contact).

[1] ⇒ page 01–27

Switch on measuring range: Resistance measurement (2 MΩ)					
Test step 1)	**V.A.G. 1598 sockets**	**Test of**	**• Test conditions – Additional operations**	**Specification**	**Correction if different from specification**
14	1 + 7	Insulation resistance of screened wire to Right Front ABS Wheel Speed Sensor (G45)	• Ignition switched off.	min. 2 MΩ	– Check wiring for signs of damage (chafing); move wiring while checking (loose contact).
15	1 + 5	Insulation resistance of screened wire to Left Front ABS Wheel Speed Sensor (G47)	• Ignition switched off.	min. 2 MΩ	– Check wiringfor signs of damage (chafing); move wiring while checking (loose contact).

1) ⇒ page 01–27

Switch on measuring range: Resistance measurement (2 MΩ)					
Test step 1)	**VAG 1598 sockets**	**Test of**	**• Test conditions – Additional operations**	**Specification**	**Correction if different from specification**
16	1 + 4	Insulation resistance of screened wire to Right Rear ABS Wheel Speed Sensor (G44)	• Ignition switched off.	min. 2 MΩ	– Check wiring for signs of damage (chafing); move wiring while checking (loose contact).
17	1 + 6	Insulation resistance of screened wire to Left Rear ABS Wheel Speed Sensor (G46)	• Ignition switched off.	min. 2 MΩ	– Check wiring for signs of damage (chafing); move wiring while checking (loose contact).

1) ⇒ page 01–27

Switch on measuring range:
Resistance measurement (200 Ω)

Test step [1]	V.A.G. 1598 sockets	Test of	• Test conditions – Additional operations	Specification	Correction if different from specification
18	1 + 8	Coil resistance of ABS Relay (J102)	• Ignition switched off.	50 ... 100 Ω	– Test wiring from contact 8 via J102 to GND. – Test coil resistance of J102 (50 ... 100 Ω); replace J102, if necessary.
19	2 + 14	Coil resistance of ABS Hydraulic Pump Relay (J185)	• Ignition switched off.	50 ... 100 Ω	– Test wiring from contact 2 via J185 to contact 14. – Test coil resistance of J185 (50 ... 100 Ω); replace J185, if necessary.
20	1 + 18	Resistance of ABS Main Valve (N105)	• Ignition switched off.	2 ... 7 Ω	– Test wiring from contact 18 via N105 to GND. – Test coil resistance of N105 (2 ... 5 Ω); replace servo unit with tandem brake master cylinder ⇒ Suspension, Wheels, Brakes, Steering.

[1] ⇒ page 01–27

Switch on measuring range:
Resistance measurement (200 Ω)

Test step [1]	VAG 1598 sockets	Test of	• Test conditions – Additional operations	Specification	Correction if different from specification
21	11 + 17	Resistance of Rear ABS Inlet Valve (N103)	• Ignition switched off.	5 ... 9 Ω	– Test wiring from contact 17 via N103 to GND. – Test resistance of N103 (5 ... 7 Ω); replace valve block if necessary ⇒ Suspension, Wheels, Brakes, Steering.
22	11 + 15	Resistance of Right Front ABS Inlet Valve (N99)	• Ignition switched off.	5 ... 9 Ω	– Test wiring from contact 15 via N99 to earth – Test resistance of N99 (5 ... 7 Ω); replace valve block if necessary ⇒ Suspension, Wheels, Brakes, Steering.
23	11 + 35	Resistance of Left Front ABS Inlet Valve (N101).	• Ignition switched off.	5 ... 9 Ω	– Test wiring from contact 35 via N101 to GND. – Test resistance of N101 (5 ... 7 Ω); replace valve block if necessary ⇒ Suspension, Wheels, Brakes, Steering.

[1] ⇒ page 01–27

Switch on measuring range:
Resistance measurement (200 Ω)

Test step [1]	VAG 1598 sockets	Test of	• Test conditions – Additional operations	Specification	Correction if different from specification
24	11 + 33	Resistance of Rear ABS Outlet Valve (N104)	• Ignition switched off.	3 ... 7 Ω	– Test wiring from contact 33 via N104 to GND. – Test resistance of N104 (3 ... 5 Ω); replace valve block if necessary ⇒ Suspension, Wheels, Brakes, Steering.
25	11 + 34	Resistance of Right Front ABS Outlet Valve (N100)	• Ignition switched off.	3 ... 7 Ω	– Test wiring from contact 34 via N100 to GND. – Test resistance of N100 (3 ... 5 Ω); replace valve block if necessary ⇒ Suspension, Wheels, Brakes, Steering.
26	11 + 16	Resistance of Left Front ABS Outlet Valve (N102).	• Ignition switched off.	3 ... 7 Ω	– Test wiring from contact 16 via N102 to GND. – Test resistance of N102 (3 ... 5 Ω); replace valve block if necessary ⇒ Suspension, Wheels, Brakes, Steering.

[1] ⇒ page 01–27

Switch on measuring range:
Resistance measurement (200 Ω)

Test step [1]	VAG 1598 sockets	Test of	• Test conditions – Additional operations	Specification	Correction if different from specification
27	1 + 3	Continuity of ABS Relay (J102)	• Ignition switched off.	max. 1.5 Ω	– Test wiring from contact 3 via J102 to GND at hydraulic unit. – Test wiring from contact 1 to GND close to control module. – If no contact resistance/open circuit exists, replace J102.
28	1 + 20	Continuity of ABS Relay (J102)	• Ignition switched off.	max. 1.5 Ω	– Test wiring from contact 20 via J102 to GND point at hydraulic unit. – Test cable from contact 1 to GND close to control module. – If no contact resistance/open circuit exists, replace J102.

[1] ⇒ page 01–27

Switch on measuring range: Resistance measurement (200 Ω)					
Test step [1]	**VAG 1598 sockets**	**Test of**	**• Test conditions – Additional operations**	**Specification**	**Correction if different from specification**
29	1 + 11	GND connection for inlet and outlet valves (N99...N104)	• Ignition switched off.	max. 1.5 Ω	– Test wiring from contact 11 via GND at hydraulic unit to GND at transmission. – Test wiring from contact 1 to GND close to control module.
30	1 + 14	Continuity of ABS High–Pressure Switch (F109)	• Ignition switched off. – Depress brake pedal 20 times as far as stop (discharge accumulator).	max. 1.5 Ω	– Test wiring from contact 14 via J185/6 and F109 to GND at transmission. – Test wiring from contact 1 to GND close to control module. – Test contact resistance of F109 (max. 1.5 Ω); replace if necessary

[1] ⇒ page 01–27

Switch on measuring range: Resistance measurement (200 Ω)					
Test step [1]	**VAG 1598 sockets**	**Test of**	**• Test conditions – Additional operations**	**Specification**	**Correction if different from specification**
31	9 + 10	Continuity of ABS Pressure Warning Switch II (F116) and ABS Brake Fluid Level Warning Switch (F117)	• Brake fluid level OK. – Switch on ignition. – Wait until ABS Hydraulic Pump (V64) switches off (accumulator dicharged).	After pump has run: max. 1.5 Ω	– Test wiring from contact 9 via F116 and F117 to contact 10.

[1] ⇒ page 01–27

Switch on measuring range: Resistance measurement (200 Ω)					
Test step 1)	**VAG 1598 sockets**	**Test of**	**• Test conditions** **– Additional operations**	**Specification**	**Correction if different from specification**
32	9 + 10	Operation of ABS Brake Fluid Level Warning Switch (F117)	– Switch off ignition. – Switch over range of VAG 1526 to 2 MΩ. – Withdraw warning contact for ABS Brake Fluid Level Warn ing Switch (F117) from reservoir. – After test, insert warning contact for (F117).	If warning contact is removed from fluid (corresponds to fluid level below min.) 2 MΩ	– If F117 has continuity outside of brake fluid, replace F117 ⇒ Suspension, Wheels, Brakes, Steering.
		Operation of ABS Pressure Warning Switch II (F116)	– Depress brake pedal 20 times as far as stop (discharge accumulator).	After depressing: 2 MΩ	– If F116 has continuity between contacts 3 and 5, replace F116 ⇒ Suspension, Wheels, Brakes, Steering.

1) ⇒ page 01–27

Operational test: Main valve (N105)					
Test step 1)	**VAG 1598 sockets**	**Test of**	**• Test conditions** **– Additional operations**	**Specification**	**Correction if different from specification**
33	Bridge 2 = 18	Operation of ABS Main Valve (N105)	• Ignition switched off. – Depress brake pedal as far as stop and hold. – Switch on ignition for not more than 30 seconds.	When ignition is switched on, brake pedal pushes against foot pressure	– N105 faulty, replace servo unit with tandem brake master cylinder ⇒ Suspension, Wheels, Brakes, Steering.

1) ⇒ page 01–27

Operational test: Main valve (N105)					
Test step 1)	VAG1598 sockets	Test of	• Test conditions – Additional operations	Specification	Correction if different from specification
34	—	Operation of ABS Hydraulic Pump (V64)	• Ignition switched off. – Depress brake pedal 20 times (discharge accumulator). – Mark fluid level on reservoir. – Switch on ignition.	Fluid level in reservoir drops by about 1.0 cm After switching off hydraulic pump	– Test wiring from Battery Positive (B+) via fuse S53,J185 and V64 to battery. – If no open circuit exists, replace V64 ⇒ Suspension, Wheels, Brakes, Steering.

1) ⇒ page 01–27

Operational test: Front left inlet and outlet valves (N101/N102)					
Test step 1)	VAG1598 sockets	Test of	• Test conditions – Additional operations	Specification	Correction if different from specification
35	Bridge 2 + 16 2 = 35	Operation of Left Front ABS Inlet and Outlet Valves (N101/N102)	• Vehicle raised. • Ignition switched off. – Depress brake pedal and hold. – Switch on ignition for not more than 30 seconds. – Depress brake pedal and hold. – After test, remove bridge.	Left front wheel locks Left front wheel must rotate freely; pedal must not give	– Check whether brake lines are properly connected. – Valve block faulty, replace ⇒ Suspension, Wheels, Brakes, Steering.

1) ⇒ page 01–27

Operational test: Front right inlet and outlet valves (N99/N100)					
Test step 1)	**VAG1598 sockets**	**Test of**	**• Test conditions – Additional operations**	**Specification**	**Correction if different from specification**
36	Bridge 2 + 15 2 = 34	Operation of front right inlet and outlet valves (N99/N100)	• Vehicle raised. • Ignition switched off. – Depress brake pedal and hold. – Switch on ignition for not more than 30 seconds. – Depress brake pedal and hold. – After test, remove bridge.	 Right front wheel locks Right front wheel must rotate freely; pedal must not give	– Check whether brake lines are properly connected. – Valve block faulty, replace ⇒ Suspension, Wheels Brakes, Steering.

1) ⇒ page 01–27

Operational test: Rear inlet and outlet valves (N103/N104)					
Test step 1)	**VAG 1598 sockets**	**Test of**	**• Test conditions – Additional operations**	**Specification**	**Correction if different from specification**
37	Bridge 2 + 17 2 = 33	Operation of Rear ABS Inlet and Outlet Valves (N103/N104)	• Vehicle raised. • Ignition switched off. – Depress brake pedal and hold. – Switch on ignition for not more than 30 seconds. – Depress brake pedal and hold. – After test, remove bridge.	 Rear wheels locked Rear wheels must rotate freely; pedal must not give	– Check whether brake lines are properly connected. – Valve block faulty, replace ⇒ Suspension, Wheels Brakes, Steering.

1) ⇒ page 01–27

Switch on measuring range: Voltage measurement (20V=) in test step 38, resistance measurement (200 Ω) in test step 39					
Test step 1)	**VAG1598 sockets**	**Test of**	**• Test conditions – Additional operations**	**Specification**	**Correction if different from specification**
38	—	Voltage supply for VAG 1551, connector T2x	Connect multimeter Fluke 83 (US 1119) with adapter VAG 1594 to T2x 2) (black): Contact 1 – Ground (GND) Contact 2 – positive	approx. battery voltage	– Test wiring from T2x/1 to GND. – Test wiring from T2x/2 via S21 to terminal 30.
39	—	Resistance of wiring for OBD, connector T2y/1	• Ignition switched off. – Connect socket 26 of VAG 1598 and Connector T2y, contact 1 with adapter VAG 1594 to digital multimeter Fluke 83 (US 1119) in the 200 Ω range.	max. 1.5 Ω	– Test wiring from T2y/1 to contact 26.

1) ⇒ page 01–27
2) ⇒ page 01–25

Operational test : ABS warning lamp (K47)					
Test step 1)	**VAG1598 sockets**	**Test of**	**• Test conditions – Additional operations**	**Specification**	**Correction if different from specification**
40	—	Operation of ABS Warning Light (K47)	• Ignition switched off. – Switch on ignition.	ABS Warning Light (K47) lights up	– Test wiring from contact W/2 (relay plate) via relay J102 (diode) to GND – Test wiring from contact X/8 (relay plate) via K47 to contact X/4 (relay plate). Replace K47 if necessary. – Test wiring from contact 52 to contact W/2 (relay plate).
	2 = 8		– Bridge sockets 2 and 8 at VAG 1598 (J102 clicks):	K47 goes out	
	1 = 26		– Bridge sockets 1 and 26 at VAG 1598. – After test, remove bridges.	ABS Warning Light (K47) lights up	

1) ⇒ page 01–27

On Board Diagnosis (OBD) of Teves 02 ABS/EDL

Function

The ABS control module (J104) is located on the left below the rear seat. It is equipped with a Diagnostic Trouble Code Memory. The connection for On Board Diagnosis (OBD) is located below the boot of the gearshift lever.

The term On Board Diagnostic relates to the electrical/electronic part of the ABS, i.e. only malfunctions which produce electrical signals are recognized.

The control modules recognize malfunctions during ABS/EDL operation or during diagnostic and store them in a permanent memory, the contents of which can be retrieved even after years despite the lack of any battery voltage.

Malfunctions which occur sporadically are likewise recognized and stored but, in contrast to other diagnostic compatible systems, without the suffix "SP" (sporadic).

Malfunctions are stored with differing priority: Solenoid valve malfunctions have the highest, while malfunctions which occur intermittently have the lowest priority.

After the ignition is switched on, the ABS Hydraulic Pump (V64) builds up the necessary operating pressure in the pressure accumulator within not more than 60 seconds provided the ABS is operating properly; the ABS Warning Light (K47) lights up for 2 ... 20 seconds depending on operating pressure and a test run is performed in the ABS/EDL Control Module (J104).

If an ABS malfunction is stored, the ABS is partially or fully switched off – depending on the priority of the malfunction (e.g. valve); at the same time the ABS Warning Light (K47) remains switched on.

If an EDL malfunction is stored, the EDL system is switched off. No warning is given by means of the ABS Warning Light (K47).

Diagnostic Trouble Code (DTC) memory interrogation must be initiated at the start of OBD and the stored information should be retrieved with the

Scan Tool VAG 1551

The DTC information displayed is used in combination with a DTC table with notes to determine the possible causes and to perform specific repair measures.

On Board Diagnostic (OBD), technical data

Control module

- 55–pin.

Diagnostic Trouble Code (DTC) Memory

- Permanent memory–contents are retained even after the power supply is switched off.

Data output

- Rapid data transfer (operating mode 1) as from 08/91.
- Blink code output (operating mode 2) up to 08/91.
- ABS Warning Light in dash panel lights up if ABS is switched off.

Functions

- Interrogate control module version (only rapid data transfer mode)
- Interrogate DTC Memory.
- Erase DTC Memory.

Note:

If a control module with blink code output is replaced when performing repairs, only a control module with rapid data transfer is available as a replacement part.

The vehicle is then automatically equipped with rapid data transfer.

When performing OBD, it is therefore necessary to always begin with the rapid transfer mode.

On Board Diagnostic (OBD), safety precautions

General

- The ABS is a vehicle safety system; work on the system requires appropriate knowledge.
- Do not drive with the connector disconnected from the ABS Control Module (J104).
- The DTC Memory must be interrogated before performing work on the ABS system in order to check complaints and to be able to perform specific diagnosis.
- Do not separate any wiring connection when the ignition is switched on.
- Refer to the appropriate instructions regarding handling brake fluid ⇒ Repair Group 47.
- Malfunctions in the ABS system are indicated by the ABS warning light going on; certain malfunctions are not recognized until the vehicle is traveling faster than a minimum speed of 25 mph (performing road test).

Necessary technical information

- Design and operation of the anti–locking brake system ⇒ Service Training Self–study program "Anti–Lock Brake System"
- Current flow diagram ⇒ Wiring diagrams, Electrical trouble shooting and Component locations binder.

Test requirements for On Board Diagnostic (OBD)

- The tires fitted to all wheels must be of the specified and same size. Inflate tires to the specified pressure.
- Conventional brake system with brake light switch and brake lights OK.
- No leaks at hydraulic connections and lines (visual inspection of hydraulic unit, brake cylinders).
- Wheel bearings and wheel bearing play OK.
- Connector at ABS Control Module (J104), on the left below the rear seat, correctly connected (lock is engaged).
- Ground (GND) connection close to ABS Control Module (J104) OK.
- All fuses OK per wiring diagram.
- Supply voltage OK (at least 10.5V).

Important:

In Corrado models with ABS/EDL control modules with different DTC Memory, interrogation is used.

Up to 07/91 the DTC Memory was read in the blink code output mode.

From 08/91 (indefinite production breakpoint) the DTC Memory is read in the rapid data transfer mode.

If a control module with blink code output is replaced when performing repairs, only a control module with rapid data transfer is available as a replacement part.

In this case the vehicle is automatically equipped with rapid data transfer.

In view of the DTC different procedures for interrogating and erasing the Memory (resulting from different control modules), it is always necessary to begin OBD with the rapid data transfer mode.

Scan Tool V.A.G. 1551, connecting

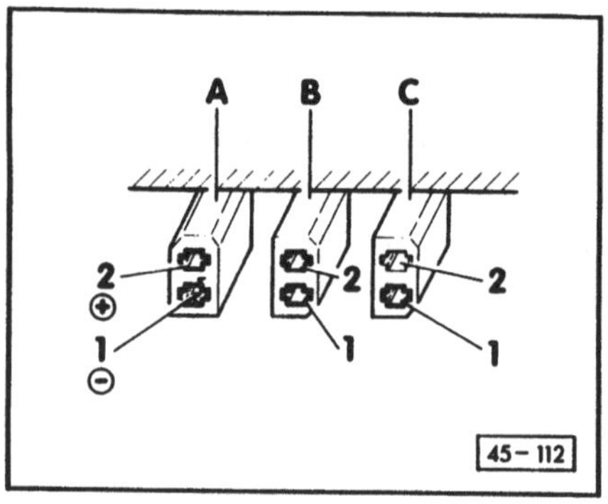

- Test requirements met ⇒ page 01–59.

– Unclip cover from gearshift lever.

– With ignition switched off, connect VAG 1551 with cable VAG 1551/1 to the Data Link Connectors (DLC) as follows:

◄ – First of all plug black connector of VAG 1551/1 into black DLC –A– in vehicle (voltage supply).

```
VAG On Board Diagnosis                HELP
1 – Rapid data transfer*
2 – Blink code output*
```

◄ Readout display:

* is displayed alternately.

Note:

If no readout appears in the display, do not connect white or blue connector! In this case, test connection (black connector) for OBD, Electrical test ⇒ page 01–100, test step 42.

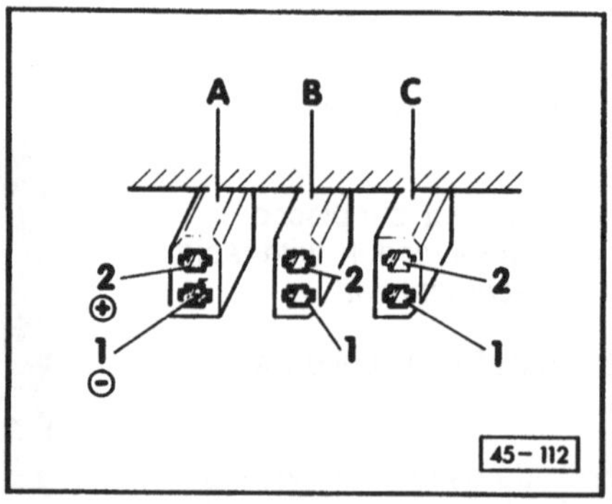

◄ – Plug white connector of VAG 1551/1 into white DLC –B– in vehicle.

– Plug blue connector of VAG 1551/1 into blue DLC –C– in vehicle (if fitted in vehicle).

Notes:

- *Depending on the program, additional operating information can be printed out by pressing the HELP button of VAG 1551.*
- The → button is used for moving forward within the program.
- The PRINT button is used for switching on the printer (indicator lamp in button comes on)

– Switch on ignition.

– Press button –1– for "Rapid data transfer" mode.

Rapid data transfer Enter address word XX	HELP

Readout display:

– Press buttons–0– and –3–;
the address word of the vehicle system to be tested "Brakes electronics" is entered with 03.

Rapid data transfer 03 Brake electronics	Q

Readout in display:

– Confirm entry with Q button

535 907 379 E	ABS + EDL

and then the following readout appears:

The identification number of the control module is displayed.

- Allocation of control module for ABS/EDL ⇒ Parts program.

Rapid data transfer Control module does not answer!	HELP

If this readout appears, continue OBD in the blink code output mode ⇒ page 01–84.

Notes:

Rapid data transfer Control module does not answer!	HELP

If one of the following messages appears in the display, the possible causes of the malfunction can be printed out by pressing HELP button.

Rapid data transfer K wire not switching to positive!	HELP

(The ignition must be switched on!)

Rapid data transfer No signal from control module!	→

- *Malfunctions have occurred at the start or during the program (external sources of malfunctions?).*

Rapid data transfer Fault in communication build–up	→

– Test diagnostic cables and also voltage supply and Ground (GND) connection close to ABS Control Module (J104), Electrical test ⇒ page 01–100 test step 42 ... 46.

– After correcting the possible causes of the malfunction, once again enter the address word 03 for "Brakes electronics" and confirm with Q button.

01 (ABS) OBD

Rapid data transfer
Tester sends address word 03

◄ Readout in display after entering the address word 03

535 907 379 E ABS + EDL

◄ and then the following readout appears:

The identification number of the control module is displayed.

– Press → button.

Rapid data transfer HELP
Select function XX

◄ Readout in display
(Function selection, e.g. 02 – Interrogate DTC Memory)

Note:

A list of the possible functions is printed out after pressing the HELP button.

After pressing the → button, the program of VAG 1551 returns to the starting position.

Rapid data transfer HELP
Select function XX

◄ Readout in display
(Function selection, e.g. 02 – Interrogate DTC Memory)

Diagnostic Trouble Code (DTC) Memory interrogating (Rapid Data Transfer)

- Scan tool VAG 1551 connected ⇒ page 01–61.
- Ignition switched on.

– Press → button.

Rapid data transfer HELP
Select function XX

◄ Readout display
(Function selection)

– Press buttons –0– and –2–;
the function "Interrogate DTC Memory" is entered with 02.

X malfunctions recognized!

◄ Readout display:

– Confirm entry with Q button.

No malfunction recognized! →

◄ The number of stored malfunctions or "No malfunction recognized" appears in the display.

– Press → button

The stored malfunctions are displayed and printed out one after the other.

As with "No malfunctions recognized", the program returns to the starting position after the → button is pressed:

- Move with the last message determined into the DTC table/rapid data transfer ⇒ page 01–69 and correct malfunctions.

Rapid data transfer Select function XX	HELP

◄ Readout in display (Function selection)

- End output (function 06) ⇒ page 01–82.
- Switch off ignition and disconnect Data Link Connectors (DLC)

Notes:

- *If a malfunction was recognized*
  1. Correct malfunction.
  2. Interrogate DTC Memory (function 02).
  3. Erase DTC Memory (function 05).
- *If a malfunction still exists without having been recognized by the OBD, in this case continue without OBD and work through the electrical test complete.*

Diagnostic Trouble Code (DTC) table (Rapid data transfer)

Notes:

- *All the possible malfunctions which can be recognized by the ABS Control Module (J104) and printed out on VAG 1551, are listed below according to the 5–digit DTC.*
- *DTC and blink code appear (in the "Rapid data transfer" mode) only in the printout.*

  Example:

DTC	*Blink code*
5–digit	*4–digit*
00277	*1221*

- *Entries marked with 2) in the DTC table provide additional information regarding locating malfunctions.*
- *Reference is made to individual test steps of the electrical test in the column "Correction".*

- *Before replacing components indicated as being faulty, test all the appropriate connections, Ground (GND) wiring and connections per the wiring diagram.*
- *After performing repairs, always once again interrogate and erase the ABS DTC Memory with the Scan Tool VAG 1551 and also perform a road test (briefly exceeding a speed of 25 mph).*
- *Repeat the OBD after the road test.*

Output at printer of VAG 1551	Possible cause of malfunction	Correction
4444 No malfunction recognized	If "No malfunction recognized" appears after performing repairs, OBD is ended. If despite "No malfunction recognized" appearing in the display, the ABS system does not operate properly, continue without OBD and work through the electrical test complete ⇒ page 01–100.	
1111 Control Module (J104) faulty	Electrical malfunctions caused by external sources or poor ground (GND) connections.	– Test wiring, wiring connections and GND connections to control module: – Test voltage supply: ⇒ Test step [1] 1 ... 4
00257 1112 Left Front ABS Inlet Valve –N101	Faulty wiring, wiring connection or valve coil. Malfunction at contact 3 of ABS Control Module (J104).	– Test wiring, wiring connections and valve coil: ⇒ Test step [1] 22 and 40 ⇒ Test step [1] 4

1) ⇒ page 01–100, Electrical test

Output at printer VAG 1551	Possible cause of malfunction	Correction
00259 1114 Right Front ABS Inlet Valve –N99	Faulty wiring, wiring connection or valve coil.	– Test wiring, wiring connections and valve coil: ⇒ Test step [1)] 21 and 39
00261 1122 Rear ABS Inlet Valve –N103	Faulty wiring, wiring connection or valve coil.	– Test wiring, wiring connections and valve coil: ⇒ Test step [1)] 20 and 37
00265 1132 Left Front ABS Outlet Valve –N102	Faulty wiring, wiring connection or valve coil.	– Test wiring, wiring connections and valve coil: ⇒ Test step [1)] 25 and 39

1) ⇒ page 01–100, Electrical test

Output at printer VAG 1551	Possible cause of malfunction	Correction
00267 1134 Right Front ABS Outlet Valve –N100	Faulty wiring, wiring connection or valve coil.	– Test wiring, wiring connections and valve coil: ⇒ Test step [1] 24 and 39
00269 1142 Rear ABS Outlet Valve –N104	Faulty wiring, wiring connection or valve coil.	– Test wiring, wiring connections and valve coil: ⇒ Test step [1] 23 and 37
00278 1122 ABS Main Valve –N105	Faulty wiring, wiring connection or valve coil.	– Test wiring, wiring connections of main valve ⇒ Test step [1] 19 and 35

1) ⇒ page 01–100, Electrical test

Output at printer VAG 1551	Possible cause of malfunction	Correction
00279 1223 Differential Lock Valve 1 –N125	Faulty wiring, wiring connection or valve coil.	– Test wiring, wiring connections and valve coil: ⇒ Test step [1] 26 and 40
00280 1224 Differential Lock Valve 2 –N126	Faulty wiring, wiring connection or valve coil.	– Test wiring, wiring connections and valve coil: ⇒ Test step [1] 27 and 41

1) ⇒ page 01–100, Electrical test

Output at printer of VAG 1551	Possible cause of malfunction	Correction
00283 1233 Front left speed sensor –G47 Mechanical malfunction [2] Signal not to tolerance [2]	Open circuit, short circuit or loose contact in: • Speed sensor cable. • Plug connection. • Speed sensor. Excessive or insufficient gap between speed sensor and pulse wheel (signal not OK). Pulse wheel and speed sensor dirty or damaged.	– Test wiring, plug connections, speed sensor, GND cables to control module and also capacitor for correct GND connection: ⇒ Test step [1] 10, 12 and 16 – Check installation of speed sensor and pulse wheel; clean if necessary.

1) ⇒ page 01–100, Electrical test
2) ⇒ Type of malfunction

Output at printer of VAG 1551	Possible cause of malfunction	Correction
00285 1241 Right Front ABS Wheel Speed Sensor –G45 Mechanical malfunction [2] Signal not to tolerance [2]	Open circuit, short circuit or loose contact in: • Speed sensor wiring. • Wiring connection. • Speed sensor. Excessive or insufficient gap between speed sensor and pulse wheel (signal not OK). Pulse wheel and speed sensor dirty or damaged.	– Test wiring, wiring connections and speed sensor, GND to control module as well as capacitor for correct GND connection: ⇒ Test step [1] 9, 11 and 15 – Check installation of speed sensor and pulse wheel; clean if necessary.

1) ⇒ page 01–100, Electrical test
2) ⇒ Type of malfunction

Output at printer of VAG 1551	Possible cause of malfunction	Correction
00287 1243 Right Rear ABS Wheel Speed Sensor –G44	Open circuit, short circuit or loose contact in: • Speed sensor wiring. • Wiring connection. • Speed sensor.	– Test wiring, wiring connections and speed sensor, GND to control module as well as capacitor for correct GND connection: ⇒ Test step 1) 7, 13 and 17 – Check installation of speed sensor and pulse wheel; clean if necessary.
Mechanical malfunction 2)	Excessive or insufficient gap between speed sensor and pulse wheel (signal not OK).	
Signal not to tolerance 2)	Pulse wheel and speed sensor dirty or damaged.	

1) ⇒ page 01–100, Electrical test
2) ⇒ Type of malfunction

Output at printer of VAG 1551	Possible cause of malfunction	Correction
00290 1311 Left Rear ABS Wheel Speed Sensor –G46	Open circuit, short circuit or loose contact in: • Speed sensor wiring. • Wiring connection. • Speed sensor.	– Test wiring, wiring connections and speed sensor, GND to control module as well as capacitor for correct GND connec tion: ⇒ Test step 1) 8, 14 and 18 – Check installation of speed sensor and pulse wheel; clean if necessary.
Mechanical malfunction 2)	Excessive or insufficient gap between speed sensor and pulse wheel (signal not OK).	
Signal not to tolerance 2)	Pulse wheel and speed sensor dirty or damaged.	

1) ⇒ page 01–100, Electrical test
2) ⇒ Type of malfunction

Output at printer of V.A.G. 1551	Possible cause of malfunction	Correction
00291 1312 ABS Pressure Warning Switch II –F116 ABS Brake Fluid Level Warning Switch –F117 Short circuit to positive [2]	Wiring between contact 8 and contact 51 of ABS Control Module (J104) and components F116 and F117 has connection to Battery Positive (B+).	– Visual inspection of harness routing between contact 8 and contact 51 ⇒ Wiring Diagram – Disconnectmulti–pin connector from control module, connect to test box V.A.G. 1598 and measure voltage with digital multimeter Fluke 83 (US 1119) between contacts: 8 and 1: max 0.1 V 51 and 1: max 0.1 V
00532 2234 Supply voltage	Faulty wiring, wiring connections, fuse S54 or ABS Relay (J102).	Test wiring, wiring connetions and GND cables to control module: Test supply voltage: ⇒ Test step [1]1... 4

[1] ⇒ page 01–100, Electrical test
[2] Type of malfunction

Output at printer of V.A.G. 1551	Possible cause of malfunction	Correction
00547 2133 ABS Pressure Control Switch –F137	Wiring between contact 13 and contact 26 of ABS Control Module (J104) with component F137 has connection to Battery Positive (B+).	– Visual inspection of harness routing between contact 13 and contact 26 ⇒ Wiring diagram. – Disconnect multi–pin connector from control module,connect to test box V.A.G. 1598 and measure voltage with digital multimeter Fluke 83 (US 1119) between contacts: 13 and 1: max 0.1 V 26 and 1: max 0.1 V Further test step [1] 32

[1] ⇒ page 01–100, Electrical test

Output at printer of V.A.G. 1551	Possible cause of fault	Rectifying fault
00599 Plausibility of pressure/brake light switch Mechanical malfunction [2]	Faulty cable, wiring connection, fuse S20 or Brake Light Switch (F).	– Test wiring from contact 32 at ABS Control Module (J104) to contact W/4 (relay plate). – Test wiring from E/3 (relay plate) to Brake Light Switch (F). – Test fuse S20 and Brake Light Switch (F); adjust if necessary ⇒ page 01-138 ⇒ Test step [1] 6

[1] ⇒ page 01–100, Electrical test
[2] Type of malfunction

Diagnostic Trouble Code (DTC) Memory, erasing/ending output (Rapid data transfer)

Requirements:

- DTC Memory has been interrogated.

1. Erasing DTC Memory

– Press → button

Rapid data transfer HELP Select function XX	◄ Readout in display (Clear for function selection)

– Press buttons –0– and –5–;
the function "Erase DTC memory" is entered with 05.

Rapid data transfer Q 05 Erase DTC Memory	◄ Readout in display:

– Confirm entry with Q button

Rapid data transfer DTC Memory is erased!	◄ Readout display:

– Press → button.

Rapid data transfer HELP Select function XX	◄ Readout in display (Clear for function selection)

2. Ending output

– Press buttons–0– and –6–. This ends the output.

Rapid data transfer	Q
06 End output	

◄ Readout display:

– Confirm entry with Q button.

Rapid data transfer	HELP
Enter address word XX	

◄ Readout display:

– Switch off ignition.

– Separate DLC to Scan Tool VAG 1551.

– Switch on ignition.

ABS Warning Light (K47) must go out after about 2 seconds.

Diagnostic Trouble Code (DTC) Memory, interrogating (Blink code output)

- Scan Tool VAG 1551 connected ⇒ page 01–61.
- Ignition switched off.

VAG On Board Diagnostic	HELP
1 – Rapid data transfer*	
2 – Blink code output*	

◄ Readout display:

* appears alternately.

– Press key –2– for "Blink code output" mode.

Blink code output	HELP
Initiate with the → key	

◄ Readout display:

– Depress ⇒ key and hold.

Blink code output is initiated!

◄ The following readout first of all appears:

Blink code output	→
Permanent ground on exciter wire!	

◄ If the following readout appears after about 2.5 seconds:

– Release ⇒ key.

– Switch on ignition. ABS warning lamp (K47) lights up.

01 (ABS) OBD

Note:

If the ABS Warning Light (K47) does not light up, test the connectors for OBD and the ABS Warning Light (K47) ⇒ page 01–100, test steps 42 46.

– Briefly press → button

The following readouts appear one after the other:

Blink code output is intiatied!

Blink code XXXX

If a malfunction is stored, the DTC is indicated via the ABS Warning Light (K47) and appears in the display, e.g.:

Blink code 1241 Right front ABS wheel speed sensor –G45

– Briefly press → button in order to interrogate the next blink code.

– Blink code interrogation is continued until the blink code 0000 appears.

Blink code 0000 End output

and then

Blink code output is ended

– Switch off ignition

- If a malfunction is displayed (e.g. inlet or outlet valve) and it is then not possible to retrieve any further blink codes, the malfunction determined has caused the control module to switch off. After repairing this malfunction, repeat OBD.

– Move with the blink code last determined into the DTC table/blink code output ⇒ page 01–89 and correct malfunction.

– After completing repairs, switch on ignition for approx. 4 seconds (self–test):

- If the ABS Warning Light (K47) goes out, there is no malfunction in the ABS system and the DTC Memory in the control module must be erased ⇒ page 01–88.

- If the ABS Warning Light (K47) does not go out, the DTC Memory should once again be interrogated. When doing this, not only the next DTC but also the previous DTC is once again displayed (although this malfunction has meanwhile been corrected).

Notes:

Blink code 4444 No malfunction recognized

◄ *If this readout appears, OBD is ended.*

If the malfunction has not been corrected, continue troubleshooting without OBD ⇒ "Electrical test", page 01–100.

Blink code No blink code received!

◄ If this readout appears, test the connections for OBD ⇒ page 01–100, test steps 42 ... 46.

Diagnostic Trouble Code (DTC) Memory, erasing (Blink code output)

Requirements:

- 1. DTC Memory has been interrogated until blink code 0000 has appeared.
- 2. Ignition is switched off and repair of the malfunction already displayed has been performed.

– Carry out road test and briefly exceed a speed of 25 mph during test. This erases the DTC memory.

> ***Important!***
>
> ***During the road test, neither the Scan Tool VAG 1551 nor any harness adapters must be connected.***
>
> ***The cover for the gearshift lever should be installed again before driving.***

– After erasing the DTC Memory, repeat OBD! OBD is not ended until the blink code 4444 appears.

– After performing repairs, switch on ignition for approx. 4 seconds (self–test).

01 (ABS) OBD

Diagnostic Trouble Code (DTC) table/Blink code output

Notes:

- *All the possible malfunctions which can be recognized by the ABS Control Module (J104) and printed out by the VAG 1551 Scan Tool are listed below according to the 4–digit blink code number.*
- *Reference is made to individual test steps of the electrical test in the column "Correction".*
- *Before replacing the components indicated as being faulty, test all the appropriate connectors, wiring and Ground (GND) connections per the wiring diagram.*
- *After performing repairs, the DTC Memory must always be interrogated once again with the Scan Tool VAG 1551 and erased, and a road test (with vehicle driven at more than 25 mph) also performed.*
- *Repeat the OBD after the road test.*

Output at printer of VAG 1551	Possible cause of malfunction	Correction
Blink code: 0000 End output		Only if blink code 0000 appears when first interrogated and ABS Warning Light (K47) shows a steady light: test voltage supply.
Blink code: 1111 Control Module (J104) faulty	Electrical malfunctions caused by outside sources of interference or poor Ground (GND) connections.	– Test wiring, wiring connections and Ground (GND) connections to control module: – Test voltage supply ⇒ Test step [1] 1 ... 4
Blink code: 1112 Left Front ABS Inlet Valve –N101	Faulty cable, wiring connection or valve coil. Malfunction at contact 3 of control module (J104).	– Test wiring, wiring connections and valve coil: ⇒ Test step [1]22 and 39 ⇒ Test step [1] 4

1) ⇒ page 01–100, Electrical test

Output at printer of VAG 1551	Possible cause of malfunction	Correction
Blink code: 1114 Right Front ABS Inlet Valve –N99	Faulty wiring, wiring connection or valve coil.	– Test wiring, wiring connections and valve coil: ⇒ Test step [1] 12 and 39
Blink code: 1122 Rear ABS Inlet Valve –N103	Faulty wiring, wiring connection or valve coil.	– Test wiring, wiring connections and valve coil: ⇒ Test step [1] 20 … 37
Blink code: 1132 Left Front ABS Outlet Valve –N102	Faulty wiring, wiring connection or valve coil.	– Test wiring, wiring connections and valve coil: ⇒ Test step [1] 25 and 39

1) ⇒ page 01–100, Electrical test

Output at printer of VAG 1551	Possible cause of malfunction	Correction
Blink code: 1134 Right Front ABS Outlet Valve –N100	Faulty wiring, wiring connection or valve coil.	– Test wiring, wiring connections and valve coil: ⇒ Test step [1] 24 and 39
Blink code: 1142 Rear ABS Outlet Valve –N104	Faulty wiring, wiring connection or valve coil.	– Test wiring, wiring connections and valve coil: ⇒ Test step [1] 23 … 37
Blink code: 1222 ABS Main Valve –N105	Faulty wiring, wiring connection or valve coil.	– Test wiring, wiring connections at main valve and also valve coil: ⇒ Test step [1] 3, 19 and 35

1) ⇒ page 01–100, Electrical test

Output at printer of VAG 1551	Possible cause of malfunction	Correction
Blink code: 1223 Differential Lock Valve 1 –N125	Faulty wiring, wiring connection or valve coil.	– Test wiring, wiring connections and valve coil: ⇒ Test step 1) 26 and 40
Blink code: 1224 Differential Lock Valve 2 –N126	Faulty wiring, wiring connection or valve coil.	– Test wiring, wiring connections and valve coil: ⇒ Test step 1) 27 ... 41

1) ⇒ page 01–100, Electrical test

Output at printer of VAG 1551	Possible cause of malfunction	Correction
Blink code: 1233 Left Front ABS Wheel Speed Sensor –G47	Open circuit, short circuit or loose contact in: • Speed sensor harness • Wiring connection. • Speed sensor. Excessive or insufficient gap between speed sensor and pulse wheel (signal not in order). Pulse wheel and speed sensor dirty or damaged.	– Test wiring, wiring connections and speed sensor, GND cables to control module as well as capacitor for correct GND connection: ⇒ Test step 1) 10, 12 and 16 – Check installation of speed sensor and pulse wheel, clean if necessary.

1) ⇒ page 01–100, Electrical test

Output at printer of VAG 1551	Possible cause of malfunction	Correction
Blink code: 1241 Right Front ABS Wheel Speed Sensor –G45	Open circuit, short circuit or loose contact in: • Speed sensor harness • Wiring connection. • Speed sensor. Excessive or insufficient gap between speed sensor and pulse wheel (signal not in order). Pulse wheel and speed sensor dirty or damaged.	– Test wiring, wiring connections and speed sensor, GND cables to control module as well as capacitor for correct GND connection: ⇒ Test step [1] 9, 11 and 15 – Check installation of speed sensor and pulse wheel, clean if necessary.

[1] ⇒ page 01–100, Electrical test

Output at printer of VAG 1551	Possible cause of malfunction	Correction
Blink code: 1243 Right Rear ABS Wheel Speed Sensor –G44	Open circuit, short circuit or loose contact in: • Speed sensor harness • Wiringconnection. • Speed sensor. Excessive or insufficient gap between speed sensor and pulse wheel (signal not in order). Pulse wheel and speed sensor dirty or damaged.	– Test wiring, wiring connections and speed sensor, GND cables to control module as well as capacitor for correct GND connection: ⇒ Test step [1] 7, 13 and 17 – Check installation of speed sensor and pulse wheel, clean if necessary.

[1] ⇒ page 01–100, Electrical test

Output at printer of VAG 1551	Possible cause of malfunction	Correction
Blink code: 1311 Left Rear ABS Wheel Speed Sensor -G46	Open circuit, short circuit or loose contact in: • Speed sensor harness. • Wiring connection. • Speed sensor. Excessive or insufficient gap between speed sensor and pulse wheel (signal not in order). Pulse wheel and speed sensor dirty or damaged.	– Test wiring, wiring connections and speed sensor, GND cables to control module as well as capacitor for correct GND connection: ⇒ Test step [1] 8, 14 and 18 – Check installation of speed sensor and pulse wheel, clean if necessary.

[1] ⇒ page 01–100, Electrical test

Output at printer of VAG 1551	Possible cause of malfunction	Correction
Blink code: 1312 ABS Pressure Warning Switch II -F116 ABS Brake Fluid Level Warning Switch -F117	Wiring between contact 8 and contact 51 of ABS Control Module (J104) at components F116 and F117 has connection to Battery Positive (B+).	– Visual inspection of harness routing between contact 8 and contact 51 ⇒ Wiring diagram. – Unplug multi-pin connector from control module, connect to tester VAG 1598 and measure voltage with digital multimeter Fluke 83 (US 1119) between the contacts. – Unplug multi -pin connector from control module, connect to test box VAG 1598 and measure voltage with digital multimeter Fluke 83 (US 1119) between the contacts: 8 and 1: max. 01. V 51 and 1: max. 0.1 V

[1] ⇒ page 01–100, Electrical test

Output at printer of VAG 1551	Possible cause of malfunction	Correction
Blink code: 2133 ABS Pressure Control Switch -F137	Wiring between contact 13 and contact 26 Control Module (J104) at components F137 has connection to Battery Positive (B +).	– Visual inspection of harness routing between contact 13 and contact 26 ⇒ Wiring diagram. – Unplug multi-pin connector from control module, connect to tester VAG 1598 and measure voltage with digital multimeter Fluke 83 (US 1119) between the contacts: 13 and 1: max. 0.1 V 26 and 1: max. 0.1 V ⇒ Further test step [1] 32
Blink code 4444 No DTC recognized!	If blink code 4444 appears after performing repairs, OBD is ended. If, despite blink code 4444 appearing, the ABS does not operate properly, continue trouble-shooting without OBD and work through the electrical test complete page 01-100.	

[1] ⇒ page 01-100, Electrical test

Electrical test of Teves 02 ABS/EDL

The test steps on the pages which follow apply to:

- Vehicles on which the ABS Warning Light (K47) and/or the Scan Tool VAG 1551 did ot react after an attempt to initiate OBD.

– Perform test steps 42 ... 46 and repeat OBD.

- Vehicles on which OBD does not provide any indicationof the source of the malfunction:

– Perform test steps 1 ... 41 one after the other.

- Vehicles on which OBD provides an indication of the source of the malfunction:

– Perform only the test steps recommended from the Diagnostic Trouble Code (DTC) table (targeted test initiation).

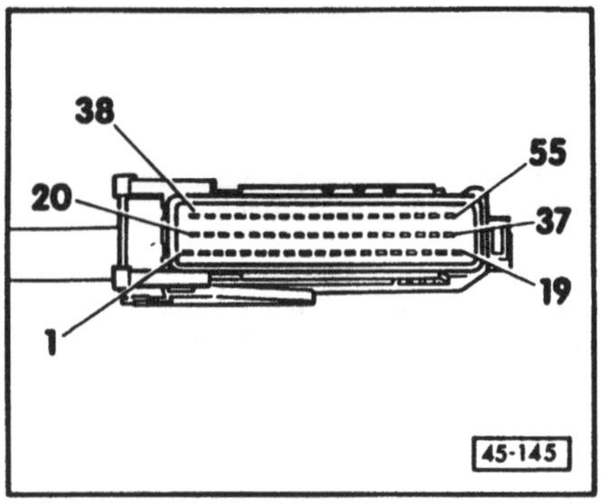

Multi–pin connectors at Teves 02 ABS/EDL

◄ 55–pin connector for ABS Control Module with EDL (J104).

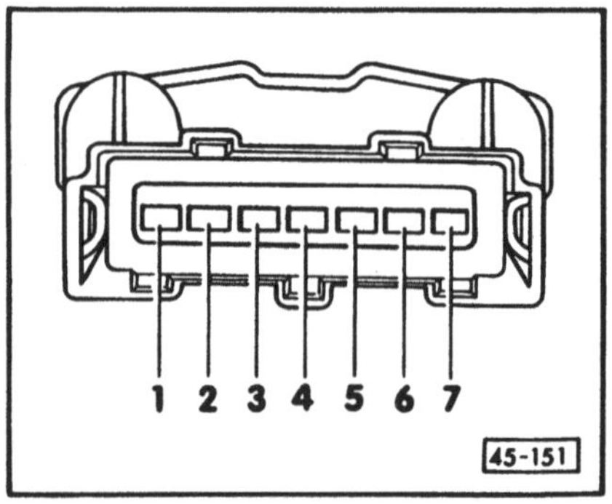

◄ 7–pin connector for valve block.

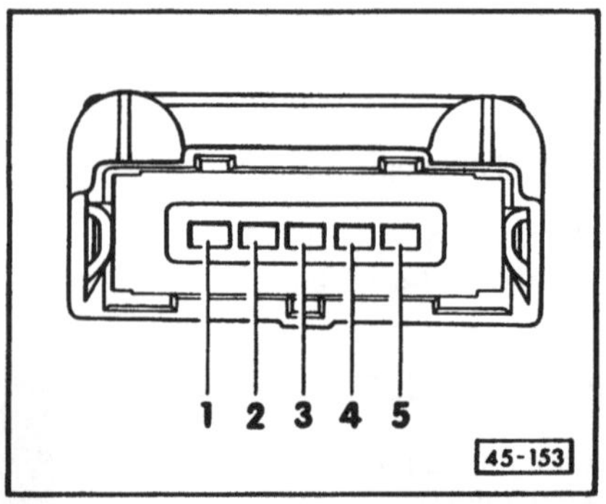

◄ 5–pin connector for EDL valve block.

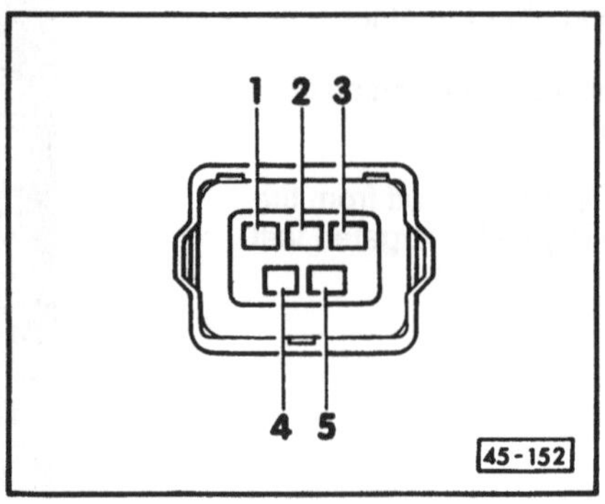

◄ 5–pin connector for ABS Pressure Warning Switch.

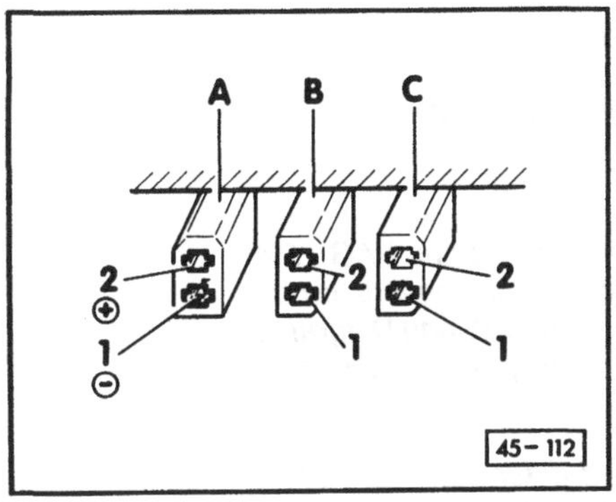

◄ Contact assignment of connectors for voltage supply and OBD with VAG 1551.

- Connector –A– (black) – voltage supply VAG 1551
  Terminal 1 = Ground (GND)
  Terminal 2 = Positive
- Connector –B– (white) – diagnostic cable
  Terminal 1 – L wire
  Terminal 2 = K wire
- Connector –C– (blue) diagnostic cable
  Terminal 1 = Lamp wire

Measuring equipment and aids required:

- Test box VAG 1598 with adapter VAG 1598/10.
- Adapter harness set VAG 1594.
- Digital multimeter Fluke 83 (US 1119).
- Raise vehicle with lifting platform for test steps 7 ... 10 and 37 ... 39.

Test requirements

- Fuses S16, S21, S53, S54 OK.
- Ground (GND) cables close to ABS Control Module (J104) OK.
- Before starting test, switch off ignition and electrical accessories (headlamps, lighting, fan ...).
- The specifications are matched to US 1119 and do not necessarily apply to other test equipment.
- Unplug multi–pin connector from ABS Control Module (J104) and connect test box VAG 1598 with adapter VAG 1598/10 to the wiring harness connector.

The second connection of adapter V.A.G. 1598/10 remains free!
The control module is not connected!

Notes:

- *The socket designations of test box VAG 1598 are identical to the contact designations on the ABS Control Module (J104) ⇒ Wiring Diagrams, Troubleshooting and Component Locations binder.*

- *If the readings obtained differ from the specifiations, perform corrective measures in the right-hand part of the table ⇒ Wiring Diagrams, Troubleshooting and Component Locations binder.*

- *Before replacing the particular components, test wiring and connections, particularly for specifications of less than 10Ω, repeat resistance measurement at component. If the readings obtained differ only slightly from the specifications, clean sockets and connectors of test equipment and also of test harnesses (with contact spray G 000 700 04) and repeat test.*

Switch on measuring range: Voltage measurement (20 V =)					
Test step[1)]	**VAG 1598 sockets**	**Test of**	**• Test conditions – Additional operations**	**Specification**	**Measures if different from specification**
1	35 + 19	Voltage supply (terminal 30) at ABS Control Module (J104)	• Ignition switched off.	approx. battery voltage	– Test wiring from contact 19 to Ground (GND). – Test wiring from contact 35 via J102 and S54 to Battery Positive (B+).
2	53 + 1	Voltage supply (terminal 15) at ABS Control Module (J104)	• Ignition switched on.	After switching on ignition: approx. battery voltage	– Test wiring from contact 1 to GND – Test wiring from contact 53 to contact D/7 (relay plate).

1) ⇒ page 01–104

Switch on measuring range: Voltage measurement (20 V =)					
Test step[1)]	**VAG 1598 sockets**	**Test of**	**• Test conditions – Additional operations**	**Specification**	**Measures if different from specification**
3	33 + 1	Operation of ABS Relay (J102)	• Ignition switched off. – Fuse S16 removed. – Bridge sockets 1 and 34 at VAG 1598. – Switch on ignition. – After measurement, remove bridge of sockets 1 and 34 – Insert fuse S16.	approx. battery voltage	– Test wiring from contact 1 to GND. – Test wiring from contact 33 via J102 and S54 to Battery Positive (B+). – If J102 does not click, test actuation: Test wiring from contact 34 to J102/4 and relay J102.

1) ⇒ page 01–104

Switch on measuring range: Voltage measurement (20 V =)					
Test step[1)]	**VAG 1598 sockets**	**Test of**	**• Test conditions – Additional operations**	**Specification**	**Measures if different from specification**
4	3 + 1	Operation of ABS Relay (J102) (reference voltage)	• Ignition switched off. – Fuse S16 removed. – Bridge sockets 1 and 34 at VAG 1598. – Switch on ignition. – After measurement, remove bridge of sockets 1 and 34 – Insert fuse S16.	approx. battery voltage	– Test wiring from contact 1 to GND. – Test wiring from contact 3 via J102 and S54 to Battery Positive (B+). – If J102 does not click, test actuation: Test wiring from contact 34 to J102/4 and relay J102.

1) ⇒ page 01–104

Switch on measuring range: Voltage measurement (20 V =)					
Test step [1]	**VAG 1598 sockets**	**Test of**	**• Test conditions** **– Additional operations**	**Specification**	**Measures if different from specification**
5	50 + 1	Operation of ABS Hydraulic Pump Relay (J185)	• Ignition switched off.		– Test wiring from contact 50 via J185 and S53 to Battery Positive (B+).
			– Unplug connector T2 at ABS Hydraulic Pump V 64.		
			– Bridge sockets 1 and 14 at V.A.G. 1598.		– Test wiring from contact 1 to GND.
			– Switch on ignition (J185 picks up).	approx. battery voltage	– If J185 does not click, test actuation: Test wiring from contact 14 to J185/6 and also relay J185.
			– After measurement, remove bridge at sockets 1 and 14 and plug in connector T2 again.		

[1] ⇒ page 01–104

Switch on measuring range: Voltage measurement (20 V =)					
Test step [1]	**VAG 1598 sockets**	**Test of**	**• Test conditions** **– Additional operations**	**Specification**	**Measures if different from specification**
6	32 + 1	Operation of Brake Light Switch (F)	• Ignition switched off.	After operating brake light switch:	– Test fuse S20 and Brake Light Switch (F)
			– Depress brake pedal and hold.	approx. battery voltage	– Test wiring from contact 32 to contact W/4 (relay plate)

[1] ⇒ page 01–104

Switch on measuring range: Voltage measurement (2 V ≅)					
Test step[1]	**VAG 1598 sockets**	**Test of**	**• Test conditions – Additional operations**	**Specification**	**Measures if different from specification**
7	45 + 27	Voltage signal of Right Rear ABS Wheel Speed Sensor (G44)	• Vehicle raised. • Ignition switched off. – Rotate right rear wheel approx. 1 rev./sec.	min. 65 mV alternating voltage	– Check wiring connector T2, installation of speed sensor and pulse wheel. – Check whether speed sensor incorrectly connected.
8	46 + 28	Voltage signal of Left Rear ABS Wheel Speed Sensor (G46)	• Vehicle raised. • Ignition switched off. – Rotate left rear wheel approx. 1 rev./sec.	min. 65 mV alternating voltage	– Check wiring connector T2, installation of speed sensor and pulse wheel. – Check whether speed sensor incorrectly connected.

1) ⇒ page 01–104

Switch on measuring range: Voltage measurement (2 V ≅)					
Test step[1]	**VAG 1598 sockets**	**Test of**	**• Test conditions – Additional operations**	**Specification**	**Measures if different from specification**
9	47 + 29	Voltage signal of Right Front ABS Wheel Speed Sensor (G45)	• Vehicle raised. • Ignition switched off. – Rotate right rear wheel approx. 1 rev./sec.	min. 65 mV alternating voltage	– Check wiring connector T2, installation of speed sensor and pulse wheel. – Check whether speed sensor incorrectly connected.
10	46 + 30	Voltage signal of Left Rear ABS Wheel Speed Sensor (G46)	• Vehicle raised. • Ignition switched off. – Rotate left rear wheel approx. 1 rev./sec.	min. 65 mV alternating voltage	– Check wiring connector T2, installation of speed sensor and pulse wheel. – Check whether speed sensor incorrectly connected.

1) ⇒ page 01–104

Switch on measuring range: Resistance measurement (2 kΩ)					
Test step[1]	**VAG 1598 sockets**	**Test of**	**• Test conditions – Additional operations**	**Specification**	**Measures if different from specification**
11	47 + 29	Resistance of Right Front ABS Wheel Speed Sensor (G45)	• Ignition switched off.	1.0 ... 1.3 kΩ	– Check wiring connector T2 – Test resistance of speed sensor (1.0 ... 1.2 kΩ). – Check wiring to speed sensor; move wiring when checking (loose contact).
12	48 + 30	Resistance of Left Front ABS Wheel Speed Sensor (G47)	• Ignition switched off.	1.0 ... 1.3 kΩ	– Check wiring connector T2. – Test resistance of speed sensor (1.0 ... 1.2 kΩ). – Check wiring to speed sensor; move wiring when checking (loose contact).

1) ⇒ page 01–104

Switch on measuring range: Resistance measurement (2 kΩ)					
Test step[1]	**VAG 1598 sockets**	**Test of**	**• Test conditions – Additional operations**	**Specification**	**Measures if different from specification**
13	45 + 27	Resistance of Right Rear ABS Wheel Speed Sensor (G44)	• Ignition switched off.	1.0 ... 1.3 kΩ	– Check wiring connector T2 – Test resistance of speed sensor (1.0 ... 1.2 kΩ). – Check wiring to speed sensor; move wiring when checking (loose contact).
14	46 + 28	Resistance of Left Rear ABS Wheel Speed Sensor (G46)	• Ignition switched off.	1.0 ... 1.3 kΩ	– Check wiring connector T2. – Test resistance of speed sensor (1.0 ... 1.2 kΩ). – Check wiring to speed sensor; move wiring when checking (loose contact).

1) ⇒ page 01–104

Switch on measuring range: Resistance measurement (2 MΩ)					
Test step[1)]	**VAG 1598 sockets**	**Test of**	**• Test conditions – Additional operations**	**Specification**	**Measures if different from specification**
15	1 + 29	Insulation resistance of screened wire to Right Front ABS Wheel Speed Sensor (G45)	• Ignition switched off.	min. 2 MΩ	– Check wiring for damage (chafing); move wiring when checking (loose contact).
16	1 + 30	Insulation resistance of screened wire to Left Front ABS Wheel Speed Sensor (G47)	• Ignition switched off.	min. 2 MΩ	– Check wiring for damage (chafing); move wiring when checking (loose contact).

1) ⇒ page 01–104

Switch on measuring range: Resistance measurement (2 MΩ)					
Test step[1)]	**VAG 1598 sockets**	**Test of**	**• Test conditions – Additional operations**	**Specification**	**Measures if different from specification**
17	1 + 27	Insulation resistance of screened wire to Right Rear ABS Wheel Speed Sensor (G44)	• Ignition switched off.	min. 2 MΩ	– Check wiring for damage (chafing); move wiring when checking (loose contact).
18	1 + 28	Insulation resistance of screened wire to Left Rear ABS Wheel Speed Sensor (G46)	• Ignition switched off.	min. 2 MΩ	– Check wiring for damage (chafing); move wiring when checking (loose contact).

1) ⇒ page 01–104

Switch on measuring range: Resistance measurement (200 Ω)					
Test step[1)]	**VAG 1598 sockets**	**Test of**	**• Test conditions – Additional operations**	**Specification**	**Measures if different from specification**
19	39 + 3	Resistance of ABS Main Valve (N105)	• Ignition switched off.	3 ... 7 Ω	– Test wiring from contact 39 via N105 to contact 3. – Test coil resistance of N105 (2 ... 5 Ω); replace servo unit with tandem brake master cylinder, if necessary ⇒ Suspension, Wheels, Brakes, Steering
20	54 + 3	Resistance of Rear ABS Inlet Valve (N103)	• Ignition switched off.	5 ... 9 Ω	– Test wiring from contact 54 via N103 to contact 3. – Test resistance of N103 (5 ... 7 Ω); replace valve block if necessary

1) ⇒ page 01–104

Switch on measuring range: Resistance measurement (200 MΩ)					
Test step[1)]	**VAG 1598 sockets**	**Test of**	**• Test conditions – Additional operations**	**Specification**	**Measures if different from specification**
21	38 + 3	Resistance of Right Front ABS Inlet Valve (N99)	• Ignition switched off.	5 ... 9 Ω	– Test wiring from contact 38 via N99 to contact 3. – Test coil resistance of N99 (5 ... 7 Ω); replace valve block if necessary ⇒ Suspension, Wheels, Brakes, Steering
22	20 + 3	Resistance of Left Front ABS Inlet Valve (N101)	• Ignition switched off.	5 ... 9 Ω	– Test wiring from contact 20 via N101 to contact 3. – Test resistance of N101 (5 ... 7 Ω); replace servo unit with tandem brake master cylinder if necessary ⇒ Suspension, Wheels, Brakes, Steering

1) ⇒ page 01–104

Switch on measuring range: Resistance measurement (200 MΩ)					
Test step[1]	**VAG 1598 sockets**	**Test of**	**• Test conditions** **– Additional operations**	**Specification**	**Measures if different from specification**
23	36 + 3	Resistance of Rear ABS Outlet Valve (N104)	• Ignition switched off.	3 ... 7 Ω	– Test wiring from contact 36 via N104 to contact 3. – Test resistance of N204 (3 ... 5 Ω); replace valve block if necessary ⇒ Suspension, Wheels, Brakes, Steering
24	21 + 3	Resistance of Right Front ABS Outlet Valve (N100)	• Ignition switched off.	3 ... 7 Ω	– Test wiring from contact 21 via N100 to contact 3. – Test resistance of N100 (3 ... 5 Ω); replace valve block if necessary ⇒ Suspension, Wheels, Brakes, Steering

1) ⇒ page 01–104

Switch on measuring range: Resistance measurement (200 Ω)					
Test step[1]	**VAG 1598 sockets**	**Test of**	**• Test conditions** **– Additional operations**	**Specification**	**Measures if different from specification**
25	2+ 3	Resistance of Left Front ABS Outlet Valve (N102)	• Ignition switched off.	3 ... 7 Ω	– Test wiring from contact 2 via N102 to contact 3. – Test resistance of N102 (3 ... 5 Ω); replace valve block if necessary ⇒ Suspension, Wheels, Brakes, Steering
26	37 + 3	Resistance of Differential Lock Valve 1 (N125)	• Ignition switched off.	6 ... 10 Ω	– Test wiring from contact 37 via N125 to contact 3. – Test resistance of N125 (6 ... 8); replace servo unit with tandem brake master cylinder if necessary ⇒ Suspension, Wheels, Brakes, Steering

1) ⇒ page 01–104

Switch on measuring range: Resistance measurement (200 Ω)					
Test step[1)]	**VAG 1598 sockets**	**Test of**	**• Test conditions – Additional operations**	**Specification**	**Measures if different from specification**
27	40 + 3	Resistance of Differential Lock Valve 2 (N126)	• Ignition switched off.	6 ... 10 Ω	– Test wiring from contact 40 via N125 to contact 3. – Test resistance of N126 (6 ... 8 Ω); replace servo unit with tandem brake master cylinder if necessary ⇒ Suspension, Wheels, Brakes, Steering
28	53 + 34	Coil resistance of ABS Relay (J102)	• Ignition switched off.	50 ... 100 Ω	– Test wiring from contact 53 via J102 (50 ... 100 Ω) to contact 34. – Test coil resistance of J102 (50 ... 100 Ω); replace J102 if necessary.
29	53 + 14	Coil resis tance of ABS Hydraulic Pump Relay (J185)	• Ignition switched off.	50 ... 100 Ω	– Test wiring from contact 53 via J185 to con tact 14. – Test coil resistance of J185 (50 ... 100 Ω); replace J185 if necessary.

1) ⇒ page 01–104

Switch on measuring range: Resistance measurement (200 Ω)					
Test step[1)]	**VAG 1598 sockets**	**Test of**	**• Test conditions – Additional operations**	**Specification**	**Measures if difference from specification**
30	1 + 3 1 + 33 1 + 19	Continuity of ABS Relay (J102) Ground (GND) connection	• Ignition switched off.	max ... 1.5 Ω	– Test wiring from contact 33 and 3 via J102 to GND at transmission. – Test wiring from contact 1 and 19 to GND close to control module.

1) ⇒ page 01–104

Switch on measuring range: Resistance measurement (200 Ω in test step 32 also 2 MΩ)					
Test step[1)]	**VAG 1598 sockets**	**Test of**	**• Test conditions – Additional operations**	**Specification**	**Measures if difference from specification**
31	1 + 14	Continuity via ABS High Pressure switch (F109)	• Ignition switched off. – Depress brake pedal 20 times as far as stop (discharge accumulator).	max. 1.5 Ω	– Test wiring from contact 14 to J185/6 via F109 to GND. – Test continuity of F109; replace if necessary ⇒ Suspension, Wheels, Brakes, Steering
32	13 + 26	Operation of ABS Pressure Control Switch (F137)	• Accumulator charged. • Ignition switched off. • Brake pedal not depressed. – Switchover range of US 1119 to 2 MΩ. – Depress brake pedal as far as stop and hold.	max. 1.5 Ω max. 2 MΩ	– Test wiring from contact 13 via F137 to contact 26. – If the specifications are not achieved, replace servo unit with tandem brake master cylinder if necessary ⇒ Suspension, Wheels, Brakes, Steering

1) ⇒ page 01–104

Switch on measuring range: Resistance measurement (200 Ω)					
Test step[1)]	**VAG 1598 sockets**	**Test of**	**• Test conditions – Additional operations**	**Specification**	**Measures if difference from specification**
33	8 + 51	Continuity of ABS Pressure Warning Switch II (F116) and ABS Brake Fluid Level Warning Switch (F117)	• Brake fluid level OK. – Switch on ignition. – Wait until ABS Hydraulic Pump (V64) switches off (accumulator charged).	After pump has run: max. 1.5 Ω	– Test wiring from contact 8 via F116 and F117 to contact 51.

1) ⇒ page 01–104

Switch on measuring range: Resistance measurement (200 Ω)					
Test step[1]	**VAG 1598 sockets**	**Test of**	**• Test conditions – Additional operations**	**Specification**	**Measures if different from specification**
34	8 + 51	Operation of ABS Brake Fluid Level Warning Switch (F117)	• Ignition switched off. – Remove warning contact for ABS Brake Fluid Level Warning Switch (F117) from reservoir. – After test, install warning contact (F117).	Warning contact is removed from fluid (corresponds to fluid level below min.) 2 MΩ	– If F117 has continuity outside of brake fluid, replace F117 ⇒ Suspension, Wheels, Brakes, Steering
		Operation of ABS Pressure Warning Switch II (F116)	– Depress brake pedal 20 times as far as stop (discharge accumulator).	After depressing: 2 MΩ	– If F116 has continuity between contacts 3 and 5, replace F116 ⇒ Suspension, Wheels, Brakes, Steering

[1] ⇒ page 01–104

Operational test: ABS Main Valve (N105)					
Test step[1]	**VAG 1598 sockets**	**Test of**	**• Test conditions – Additional operations**	**Specification**	**Measures if difference from specification**
35	Bridge 1 = 34 1 = 39	Operation of ABS Main Valve (N105)	• Ignition switched off. – Depress brake pedal and hold. – Switch on ignition for not more than 30 seconds. – After test, remove bridge.	When ignition is switched on, brake pedal pushes agains foot pressure.	– N105 faulty; replace servo unit with tandem brake master cylinder ⇒ Suspension, Wheels, Brakes, Steering

[1] ⇒ page 01–104

Operational test: ABS Hydraulic Pump (V64)

Test step[1)]	VAG 1598 sockets	Test of	• Test conditions – Additional operations	Specification	Measures if difference from specification
36	—	Operation of ABS Hydraulic Pump (V64)	• Ignition switched off. – Depress brake pedal 20 times (discharge accumulator). – Mark fluid level on reservoir. – Switch on ignition.	After switching on ignition: Fluid level in reservoir drops by about 1.0 cm	– Test wiring from Battery Positive (B+) via fuse S53, J185 and V64 to battery –. – If no open circuit exists, replace V64 ⇒ Suspension, Wheels, Brakes, Steering

1) ⇒ page 01–104

Operational test: Rear ABS Inlet/Outlet Valves (N103/N104)

Test step[1)]	VAG 1598 sockets	Test of	• Test conditions – Additional operations	Specification	Measures if difference from specification
37	Bridge 1 = 34 1 = 36 1 = 54	Operation of Rear ABS Inlet/Outlet Valves (N103/N104)	• Vehicle raised. • Ignition switched off. – Depress brake pedal and hold. – Switch on ignition for not more than 30 seconds – Depress brake pedal and hold. – After test, remove bridge.	Rear wheels lock Rear wheels must rotate freely; pedal must not give	– Check if brake lines correctly connected. – Valve block faulty, replace ⇒ Suspension, Wheels, Brakes, Steering

1) ⇒ page 01–104

Operational test: Right Front ABS Inlet/Outlet Valves (N99/N100)					
Test step[1)]	**VAG 1598 sockets**	**Test of**	**• Test conditions – Additional operations**	**Specification**	**Measures if difference from specification**
38	Bridge 1 = 34 1 = 21 1 = 38	Operation of Right Front ABS Inlet/ Outlet Valves (N99/N100)	• Vehicle raised. • Ignition switched off. – Depress brake pedal and hold. – Switch on ignition for not more than 30 seconds – Depress brake pedal and hold. – After test, remove bridge.	Front right wheel locks Front right wheel must rotate freely; pedal must not give	– Check if brake lines correctly connected. – Valve block faulty, replace ⇒ Suspension, Wheels, Brakes, Steering

1) ⇒ page 01–104

Operational test: Left Front ABS Inlet/Outlet Valves (N101/N102)					
Test step[1)]	**VAG 1598 sockets**	**Test of**	**• Test conditions – Additional operations**	**Specification**	**Measures if difference from specification**
39	Bridge 1 = 34 1 = 20 1 = 2	Operation of Left Front ABS Inlet/ Outlet Valves (N101/N102)	• Vehicle raised. • Ignition switched off. – Depress brake pedal and hold. – Switch on ignition for not more than 30 seconds – Depress brake pedal and hold. – After test, remove bridge.	Front left wheel locks Front left wheel must rotate freely; pedal must not give	– Check if brake lines correctly connected. – Valve block faulty, replace ⇒ Suspension, Wheels, Brakes, Steering

1) ⇒ page 01–104

Operational test: Differential Lock Valve 1 (N125)					
Test step[1]	**VAG 1598 sockets**	**Test of**	**• Test conditions – Additional operations**	**Specification**	**Measures if different from specification**
40	Bridge 1 = 34 1 = 40 1 = 37	Operation of Differential Lock Valve 1 (N125)	• Ignition switched off	After switching on ignition:	
			– Switch on ignition for not more than 30 seconds. – Briefly unplug connector 2 ... 3 times from socket 37.	Switching noises heard from Differential Lock Valve 1 (N125) and ABS Hydraulic Pump (V64) runs until accumulator is filled.	– Test wiring from contact 37 via N125, ABS Relay J102 and fuse S54 to Battery Positive (B+). – Differential Lock Valve 1 (N125) faulty; replace servo unit with tandem brake master cylinder ⇒ Suspension, Wheels, Brakes, Steering

1) ⇒ page 01–104

Operational test: Differential Lock Valve 2 (N126)					
Test step[1]	**VAG 1598 sockets**	**Test of**	**• Test conditions – Additional operations**	**Specification**	**Measures if different from specification**
41	Bridge 1 = 34 1 = 40 1 = 37	Operation of Differential Lock Valve 2 (N126)	• Ignition switched off	After switching on ignition:	
			– Switch on ignition for not more than 30 seconds. – Briefly unplug connector 2 ... 3 times from socket 40. – After checking, remove bridge.	Switching noises heard from Differential Lock Valve 2 (N126) and ABS Hydraulic Pump (V64) runs until accumulator is filled.	– Test wiring from contact 40 via N126, ABS Relay J102 and fuse S54 to Battery Positive (B+). – Differential Lock Valve 2 (N126) faulty; replace servo unit with tandem brake master cylinder ⇒ Suspension, Wheels, Brakes, Steering

1) ⇒ page 01–104

Switch on measuring range:
Voltage measurement (20V ≅)

Test step[1]	VAG 1598 sockets	Test of	• Test conditions – Additional operations	Specification	Measures if different from specification
42	—	Voltage supply for VAG 1551, connector T2x	– Connect multimeter Fluke 83 (US 1119) with adapter harness set VAG 1594 to T2x [2] (black): Contact 1 – Ground (GND) Contact 2 – positive	approx. battery voltage	– Test cable from T2x/1 to GND. – Test cable from T2x/2 via S21 to terminal 30.

[1] ⇒ page 01–104
[2] ⇒ page 01–101

Switch on measuring range:
Resistance measurement (200 Ω)

Test step[1]	VAG 1598 sockets	Test of	• Test conditions – Additional operations	Specification	Measures if different from specification
43	—	Resistance of K wire for OBD, wiring connector T2z/2	• Ignition switched off. – Connect digital multimeter Fluke 83 (US 1119) to: – Socket 42 of VAG 1598 – and with adapter harness set VAG 1594 to wiring connector T2z (white), contact 2.	max. 1.5 Ω	– Test wiring from T2z/2 to contact 42.

[1] ⇒ page 01–104
[2] ⇒ page 01–101

Switch on measuring range: Resistance measurement (200 Ω)

Test step[1]	VAG 1598 sockets	Test of	• Test conditions – Additional operations	Specification	Measures if different from specification
44	—	Resistance of L wire for OBD, wiring connector T2z/1	• Ignition switched off. – Connect digital multi–meter Fluke 83 (US 1119) to: – Socket 23 of VAG 1598 – and with adapter harness set VAG 1594 to wiring connector T2z (white), contact 1.	max. 1.5 Ω	– Test wiring from T2z/2 to contact 42.

1) ⇒ page 01–104
2) ⇒ page 01–101

Operational test: OBD connections

Test step[1]	VAG 1598 sockets	Test of	• Test conditions – Additional operations	Specification	Measures if different from specification
45	—	Resistance of lamp wire for OBD, wiring connector T2y/1	• Ignition switched off. – Connect digital multimeter Fluke 83 (US 1119) to: – Socket 52 of VAG 1598 – and with adapter cable set VAG 1594 to wiring connector T2y (blue), contact 1.	max. 1.5 Ω	– Test wiring from T2y/2 to contact 52.

1) ⇒ page 01–104
2) ⇒ page 01–101

Operational test: ABS Warning Light (K47)					
Test step[1]	**VAG 1598 sockets**	**Test of**	**• Test conditions – Additional operations**	**Specification**	**Measures if different from specification**
46	—	Operation of ABS Warning Light (K47)	• Ignition switched off.		– Test wiring from contact W/2 (relay plate) via ABS relay J102 (diode) to GND
			– Switch on ignition.	ABS Warning Light (K47) lights up	– Test wiring from contact X/8 (relay plate) via K47 to contact X/4 (relay plate); replace K47 if necessary.
	1 = 34		– Bridge sockets 1 and 34 at VAG 1598: J102 picks up.	ABS Warning Light (K47) goes out	– Test wiring from contact 52 to contact W/2 (relay plate).
	1 = 52		– Bridge sockets 1 and 52 at VAG 1598.		
			– After test, remove bridges.	ABS Warning Light (K47) lights up	

1) ⇒ page 01–104
2) ⇒ page 01–101

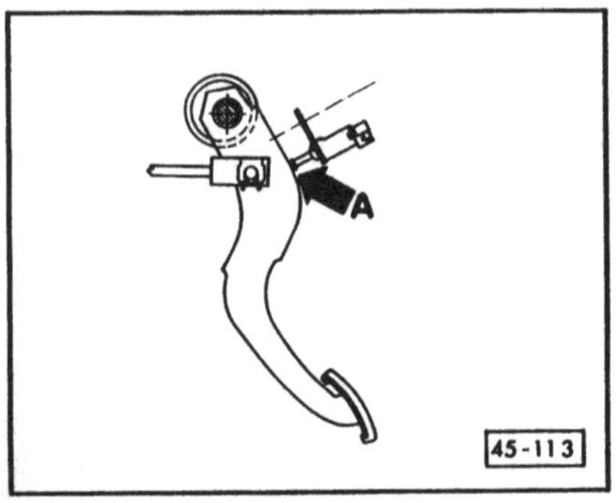

Brake Light Switch (F), adjusting

- Depress brake pedal slightly by hand.
- ◄ Hold 1 mm thick feeler gauge between brake pedal and brake light switch (arrow).
- Pull brake pedal back by hand as far as stop (brake light switch is pushed back by one further detent).
- Check operation of brake lights.

Supplement
July 1994

Volkswagen Corrado

Repair Manual Supplement — July, 1994

This Supplement includes new information for the 1993 and 1994 Corrado that was unavailable when this Official Factory Repair Manual was first published. It covers the revised VR6 ignition system, the automatic transmission shiftlock, a new air conditioning system, and additional wiring diagrams.

> ***WARNING!***
> ***When using the information in this Supplement, please remember to read all the Cautions and Warnings listed at the front of the manual, as well as any Cautions and Warnings that may accompany a specific procedure or other service information in this Supplement.***

Repair Group	Subject	Page

28 Motronic Ignition System—VR6 engine code AAA

This Supplement section covers the "distributorless" ignition system, distinguished by an ignition coil with six separate high–voltage output terminals near the front of the engine, and the lack of the more familiar camshaft–driven distributor.

General information
- ignition system, servicing 28-1
- ignition system components 28-2

Motronic ignition system, components
- Camshaft Position (CMP) Sensor, checking 28-7
- Ignition Coil (N152), checking 28-9

Safety rules 28-5

Technical data, spark plugs 28-6

37 Automatic Transmission 096 Shiftlock III

This Supplement section covers "Shiftlock III," the gear selector lever and ignition switch locking mechanism for all 1993 and 1994 Corrados with automatic transmission.

Ignition lock mechanism, checking 37-4

Lock cable
- adjusting 37-24
- removing and installing 37-20

Selector lever/cable assembly 37-5

Selector lever, disassembling and assembling 37-13

System operation 37-1

Repair Group	Subject	Page

87 Air Conditioning—systems using R-134a refrigerant

This Supplement section covers the air conditioning system for 1993 and 1994 vehicles that uses the new CFC-free refrigerant R-134a. This system is completely new. None of the procedures and specifications are carried over from the older systems using Freon® R-12 refrigerant.

A/C Clutch (N25), assembly 87-52

A/C refrigerant system (R-134a), general information
- basics when working on refrigerant system (R-134a) 87-14
- component functions 87-18
- refrigerant R-134a, properties 87-8
- safety measures 87-1
- system identification 87-17

A/C refrigerant system (R-134a), system components
- components, body 87-22
- components, engine compartment 87-30
- fuses 87-39
- valves and switches 87-40

A/C refrigerant system (R-134a), testing and servicing
- capacities 87-64
- checking for leaks 87-68
- checking pressures and temperatures 87-66
- components, replacing 87-79
- discharging 87-71
- evacuating and recharging 87-77
- flushing with compressed air and nitrogen 87-74
- pressure gauges, testing with 87-69
- pressures, checking 87-88

A/C refrigerant system (R-134a), troubleshooting 87-86
- A/C does not cool 87-101
- fresh air cannot be shut off 87-94
- fresh air operation (ventilation) cannot be adjusted; no fresh air flows from instrument panel and footwell outlets 87-95
- insufficient A/C cooling output 87-104
- insufficient or no heat output 87-97
- windows do not clear in defrost mode 87-99

Air distribution and evaporator housing, assembly 87-48

Compressor mounting bracket, removing/installing 87-83

Condenser, removing/installing 87-81

Control cables, installing/adjusting 87-59

Control head, assembly 87-58

Heater/air conditioner air ducts, assembly 87-42

Safety measures 87-1

Vacuum hose layout 87-63

Wiring Diagrams

Air conditioning — from August, 1993 333

Main wiring diagram — September, 1993 to March, 1994 339

Radio, stereo 371

Radio with CD changer 373

Motronic ignition system, general information

Ignition system, servicing

For information on servicing and repairing the Multiport Fuel Injection (MFI) part of the Motronic system ⇒ Repair Group 24.

Notes:

- *Only the components which specifically relate to the ignition system are dealt with here. For the other components of the injection and ignition system ⇒ Repair Group 24.*
- *The Motronic Engine Control Module (J220) for ignition and fuel injection is equipped with Diagnostic Trouble Code (DTC) Memory.*
- *Components marked with an asterisk (*) are checked by the On Board Diagnostic (OBD) program.*
- *Electrical check of Motronic Multiport Fuel Injection (MFI) and ignition system ⇒ Repair Group 01.*
- *Before carrying out repairs or troubleshooting, initiate the On Board Diagnostic program and check for malfunctions stored in the Diagnostic Trouble Code (DTC) Memory, ⇒ Repair Group 01.*

Safety rules ⇒ page 28-5

Technical data, spark plugs ⇒ page 28-6

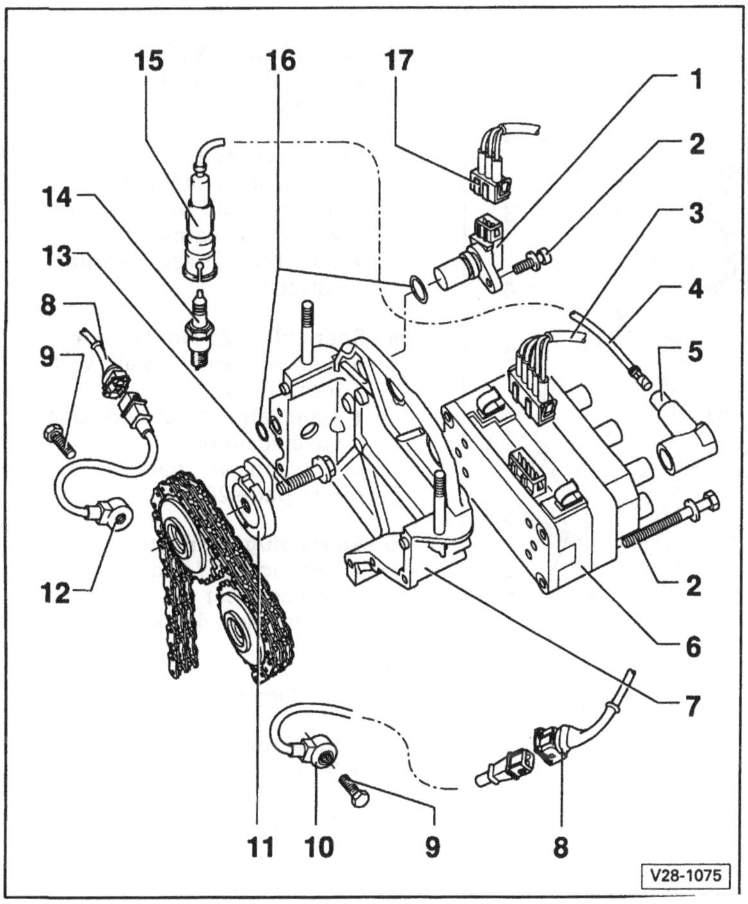

Motronic ignition system, components

Note:

Motronic Engine Control Module (J220) with connector* ⇒ Repair Group 24.

1 - **Camshaft Position (CMP) Sensor (G40)***
- Checking ⇒ Page 28-7
- Note spacer rings

2 - **10 Nm (7 ft lb)**

3 - **Connector**
- 5-pin
- For Ignition Coil (N152)

4 - **Ignition cable**
- Check for continuity

5 - **Suppressor**
- Resistance: 0.6–1.4 kΩ

6 - **Ignition Coil (N152)***
- With ignition cable identification, do not interchange
- Checking ⇒ page 28-9

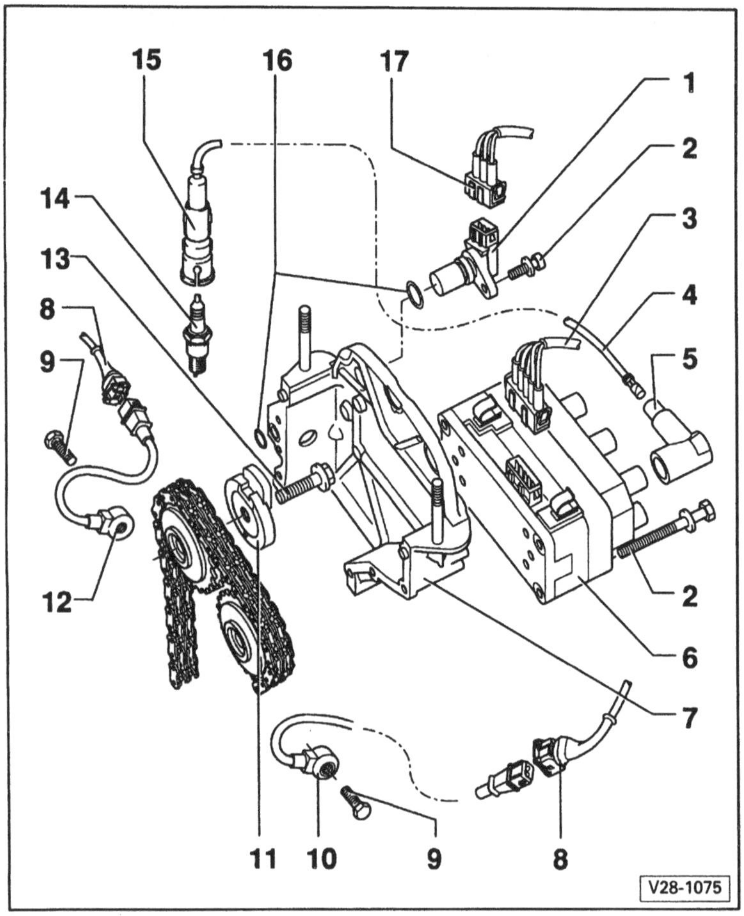

7 - Camshaft sprocket cover
- Removing and installing ⇒ Repair Manual, *2.8 Liter General, Engine,* Repair Group 15

8 - Connector
- 3-pin

9 - 20 Nm (15 ft lb)
- Tightening torque influences Knock Sensor function

10 - Knock Sensor 2 (G66)*
- Mounting location: cylinder block, intake side

11 - Sensor wheel
- For Camshaft Position (CMP) Sensor (G40)
- Contact surface on camshaft sprocket must be dry
- If sensor wheel has been removed, check timing after installing

⇒ Repair Manual, *2.8 Liter General, Engine,* Repair Group 13

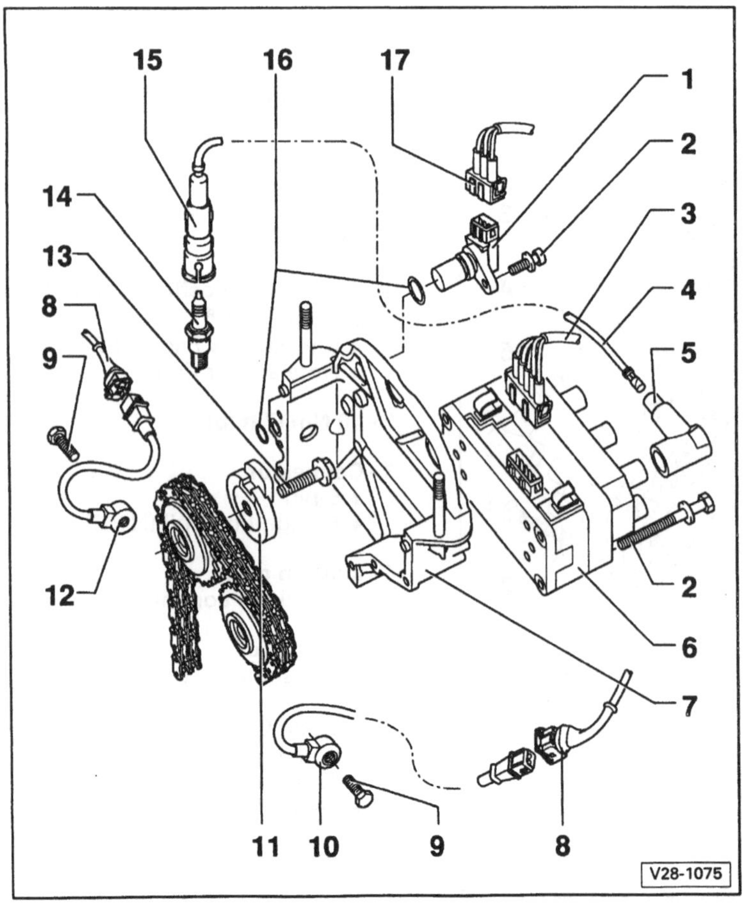

12 - Knock Sensor 1 (G61)*
- Mounting location: Cylinder block, exhaust side

13 - 100 Nm (74 ft lb)
- Removing/installing counter-hold with 24 mm wrench on camshaft
- Oil bolt head contact surface

14 - Spark plug, 25 Nm (18 ft lb)
- Remove and install with tool 3122B
- Type and electrode gap specifications

⇒ *Technical data, spark plugs,* page 28-6

15 - Spark plug connector
- Use tool 3277 to disconnect and connect
- Resistance: 4–6 kΩ

16 - O ring
- Replace if damaged

17 - Connector
- 3-pin
- For Camshaft Position (CMP) Sensor (G40) -1-

Safety rules

> ***CAUTION!***
>
> **To prevent personal injury or damage to sensitive electrical components:**
>
> - ***Switch ignition OFF before connecting or disconnecting components or test equipment.***
> - ***DO NOT touch or disconnect ignition wiring while the engine is running or being cranked by the starter.***
> - ***If engine must be cranked but not started (for compression testing etc.) disconnect the 5-pin connector from the Ignition Coil (N152).***
> - ***DO NOT use battery booster for longer than one minute and DO NOT exceed 16.5 Volts.***
> - ***DO NOT wash the engine unless the ignition has been switched OFF.***
> - ***Switch ignition OFF when arc or spot welding.***
> - ***Connect or disconnect Battery only with ignition switched OFF to prevent damage to the Motronic Engine Control Module (ECM).***
> - ***Before disconnecting the Battery, ALWAYS ask for the correct radio code (vehicles equipped with anti-theft radio).***

Note:

A variety of electrical connectors are used on this vehicle. Always use VW 1594 adaptor kit to connect test instruments to these connectors.

Technical data, spark plugs

Engine code	**AAA**
Ignition timing	Not adjustable
Checking	⇒ Repair Group 24
Firing order	1–5–3–6–2–4
Spark plugs[1]	
VW/Audi Part No.	101 000 035 AB
Manufacturer's designation	BKR 5 EKU
Electrode gap	0.7 mm (0.028 in.) [2]
Tightening torque	25 Nm (18 ft lb)

[1] Use tool 3277 to connect and disconnect Spark Plug connector. Remove and install spark plugs with tool 3122B

[2] Gap between Ground (GND) electrode and center electrode

Camshaft Position (CMP) Sensor, checking

- "Camshaft Position Sensor-G40" displayed by VAG 1551 On Board Diagnostic (OBD) program (function 02, Check DTC Memory).

- Carry out electrical testing, test steps 2 and 23 ⇒ Repair Group 01.

CMP Sensor voltage supply, checking

- Disconnect 3-pin connector from Camshaft Position Sensor (G40).

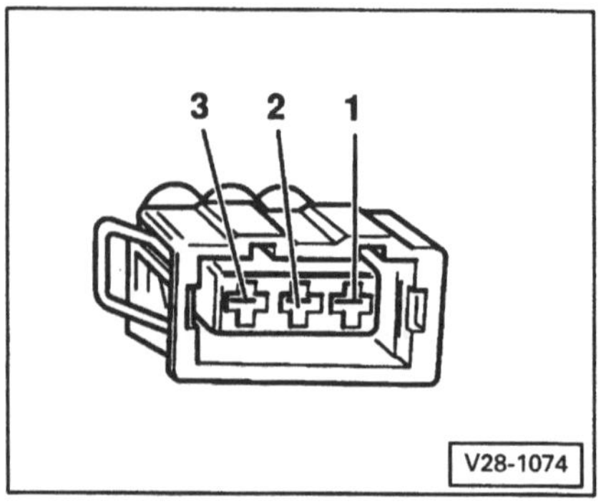

◄ - Connect hand multimeter (Fluke 83, US 1119 or equivalent) to measure voltage across disconnected connector terminals -1- and -3-, using jumper cables from VAG 1594 adapter kit.

- Switch ignition ON.

  ♦ Multimeter should read 10 V (minimum).

- Switch ignition OFF.

If voltage is OK:

- Replace Camshaft Position (CMP) Sensor (G40) ⇒ page 28-2, item -1-.

If no voltage is present:

- Replace Motronic Engine Control Module (ECM) (J220) ⇒ Repair Group 24.

Functional check

- Remove fuse 18 (S18).

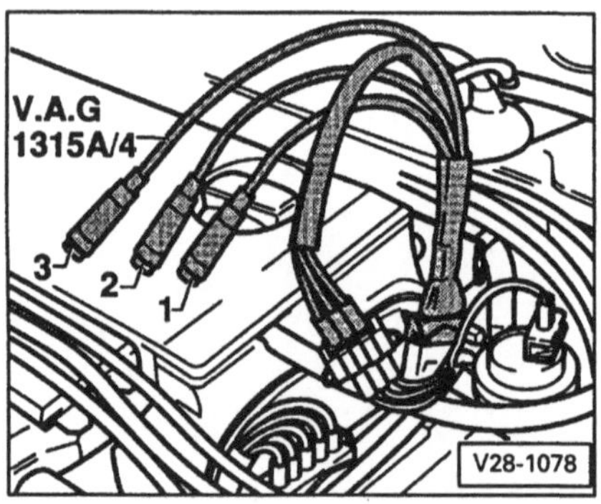

◄ - Bridge connector and Camshaft Position (CMP) Sensor (G40) using VAG 1315A/4 adapter cable.

- Connect VAG 1527B voltage tester (LED test light) to test connectors -2- and -3-, using cables from VAG 1594 adapter kit.

- Operate starter.

  ♦ LED must flicker

If necessary replace Camshaft Position (CMP) Sensor (G40) ⇒ page 28-2, item -1-.

Ignition Coil (N152), checking

- "Ignition Output 1," "Ignition Output 2" or "Ignition Output 3" displayed by VAG 1551 On Board Diagnostic (OBD) program (function 02, Check DTC Memory).
- Camshaft Position (CMP) Sensor OK (checking ⇒ page 28-7).

– Carry out electrical check, test steps 2 and 24 ⇒ Repair Group 01.

Only if the engine will not start:

Ignition Coil voltage supply, checking

– Pull 5-pin connector off Ignition Coil (N152).

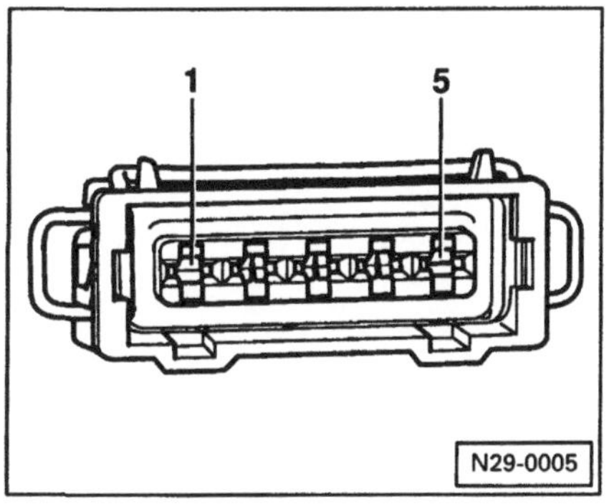

◄ – Connect Fluke 83 hand multimeter (US 1119 or equivalent) to measure voltage across connector terminals -1- and -5-, using cables from VAG 1594 adapter kit.

– Switch ignition ON.

♦ Multimeter should read 10 V (minimum).

– Switch ignition OFF.

Checking activation

– Remove fuse no. 18 (S18).

– Connect VAG 1527B voltage tester (LED test light) to connector terminals -5- and -2-, using cables from VAG 1594 adapter kit.

– Operate starter and check ignition signal from Motronic Engine Control Module (ECM).

♦ LED must flicker.

– Repeat the test twice more, with the voltage tester connected to terminals -5- and -3-, then -5- and -4-.

If LED does NOT flicker:

– Replace Motronic Engine Control Module (ECM) (J220) ⇒ Repair Group 24.

Shiftlock III, system operation

> ***CAUTION***
> ***Before working on a running engine, always:***
> ***♦ place selector lever in P***
> ***♦ engage parking brake***

Checking with selector lever in P

Note:

During the checking procedure, DO NOT operate the brake pedal until instructed to do so.

– place selector lever in **P**

– turn ignition **ON**

– try shifting out of **P**

♦ it must not be possible to shift out of **P**

If selector lever can be shifted out of P

♦ shiftlock mechanism is malfunctioning: solenoid is NOT triggering a "blocking" action at the selector lever

If selector lever cannot be shifted out of **P**

♦ shiftlock mechanism is functioning: solenoid is HAS triggered a "blocking" action at the selector lever

– apply brake pedal and hold

– shift selector lever out of **P**

If selector lever cannot be shifted out of **P**

♦ system is malfunctioning: solenoid is NOT being deactivated to release selector lever

If selector lever can be shifted out of **P**

♦ system is functioning: solenoid, IS deactivated to release selector lever

Checking with selector lever in N

Note:

Repeat checking procedure, this time with selector lever in ***N*** *(ignition* ***ON****). As before, DO NOT operate the brake pedal until instructed to do so. The results must be identical to those specified for checking with the selector lever in* **P**.

Checking with the vehicle in motion

> ***Warning:***
>
> ***Do not perform the following checks on a public road in traffic. Test drive vehicle in a parking lot or other area away from traffic.***

- operate vehicle at speeds OVER 5 km/h (3 mph)
- shift selector lever into **N**
  - ♦ Shift lock solenoid must NOT block selector lever (e.g. it must be possible to shift back into **D**.)
- operate vehicle at speeds UNDER 5 km/h (3 mph)
- shift selector lever into **N**
  - ♦ Shift lock solenoid must block selector lever after one second (e.g. it must NOT be possible to shift back into **D**)

Note:

Once the brake pedal is applied, however, the selector lever can again be shifted out of the ***N*** *postion.*

Additional function check

- Check that the vehicle will NOT start with the selector lever in positions **R, D, 3, 2,** or **1.**

Ignition lock mechanism, checking

- place selector lever in **P** position
- turn ignition key to OFF position
- remove key
  - ♦ it must NOT be possible to remove key with ignition switch in any ony position but OFF
- attempt to shift selector lever (with key removed)
  - ♦ it must NOT be possible to shift selector lever out of **P**

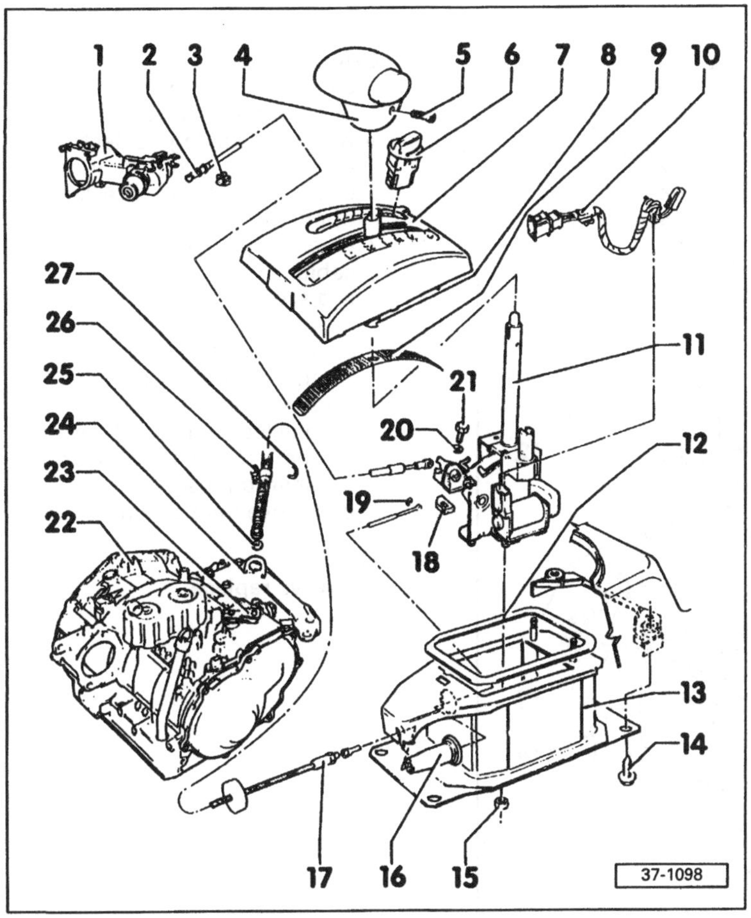

Selector lever/cable assembly

1 - **Ignition switch with housing**

2 - **Lock cable**
- do not bend or kink
- adjusting ⇒ page 37-24
- removing and installing ⇒ page 37-20

3 - **Clip**
- for securing lock cable
- always replace

4 - **Handle**

5 - **Shaft bolt**
- install with locking compound AMV 200 000

6 - **Transmission Range (TR) Program Switch -E122-**
- discontinued on vehicles built from 01/93
- checking, ⇒ *Measuring value block test tables,* Repair Group 01

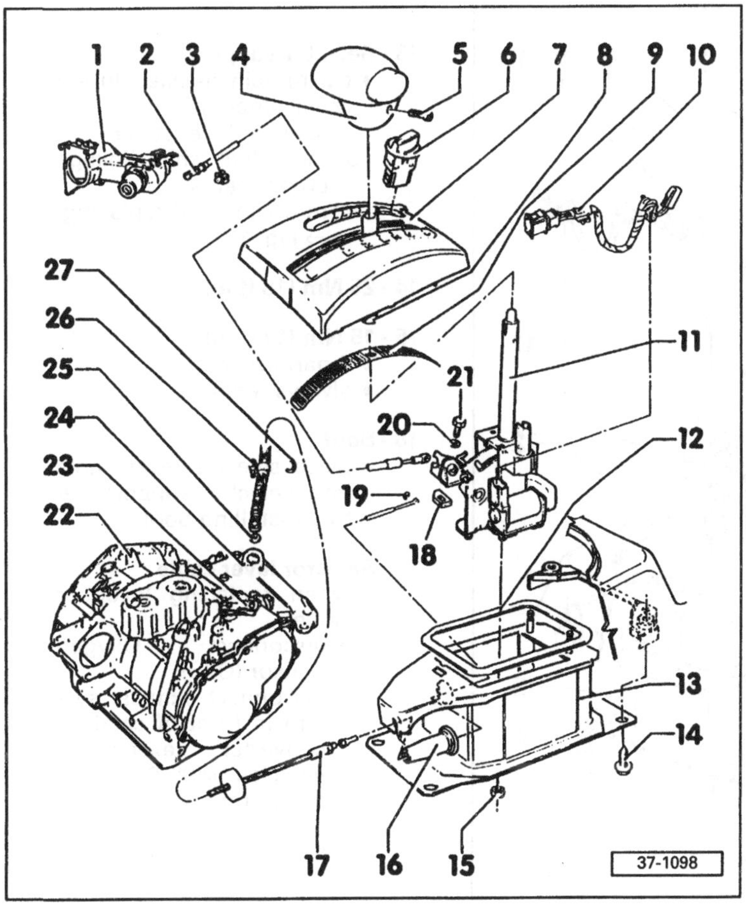

7 - **Cover**
- with gear selector scale

8 - **Protective sleeve for selector lever**

9 - **Cover strip**
- insert in cover, item -7-

10 - **Wiring harness**
- for shiftlock solenoid (N110). On some vehicles harness may include solenoid
- For illumination of selector scale
- fastening, ⇒ Fig. 2

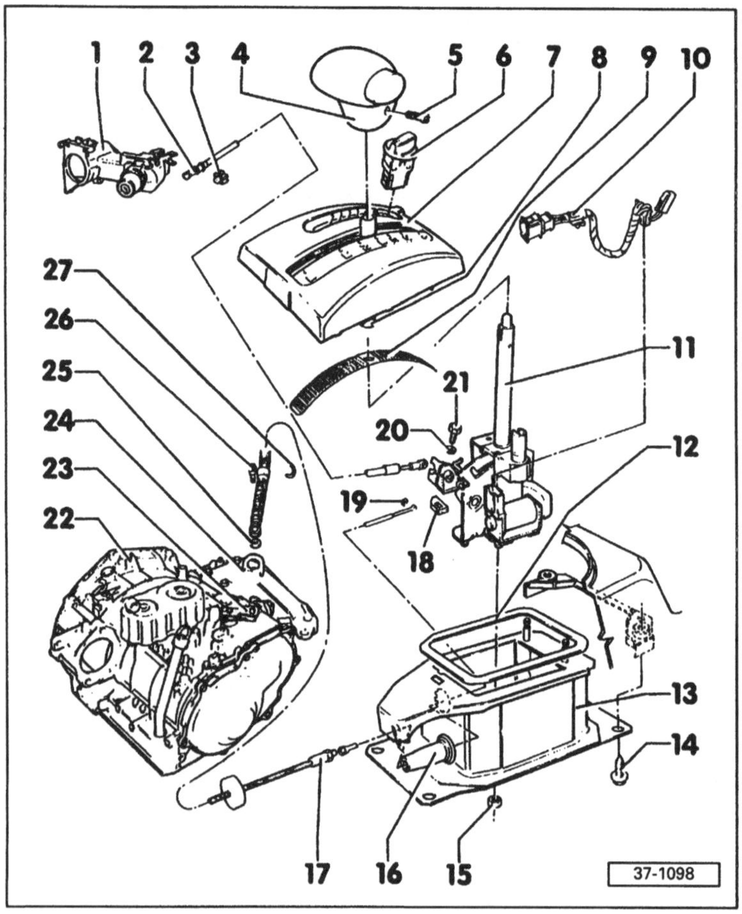

11 - Selector lever

- disassembling and assembling ⇒ page 37-13
- when installing selector lever, place anchor for lock cable in shift bracket opening
- when replacing lever or cable:
  - a. disconnect Oxygen Sensor (O2S)
  - b. disconnect exhaust system in front of catalytic converter (exception–vehicles with 6-cyl. engine: remove catalytic converter also)
  - c. remove heat shield

12 - Gasket

- always replace

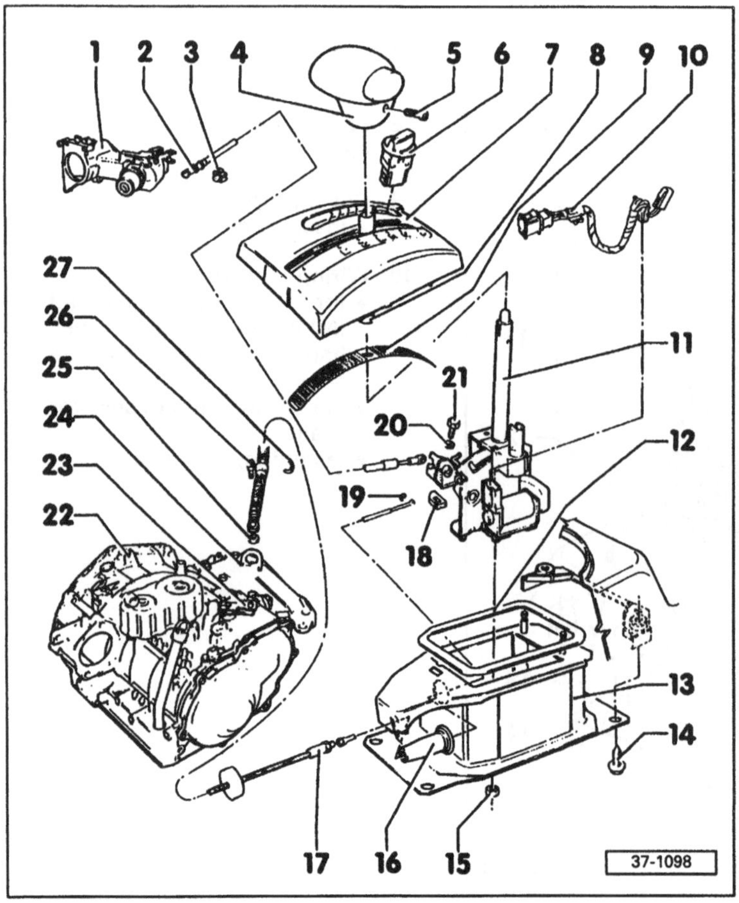

13 - Selector support

- can remain installed during replacement of:
  - a. selector lever cable
  - b. individual shift linkage components
  - c. removing and replacing, ⇒ Fig. 1

14 - 25 Nm (18 ft lb)

15 - 15 Nm (11 ft lb)

- 3-part assembly
- always replace

16 - Boot

- place selector lever cable into selector support before installing boot

17 - Selector lever cable

- when replacing lever or cable:
  - a. disconnect Oxygen Sensor (O2S)
  - b. disconnect exhaust system in front of catalytic converter (exception–vehicles with 6-cyl. engine: remove catalytic converter also)
  - c. remove heat shield

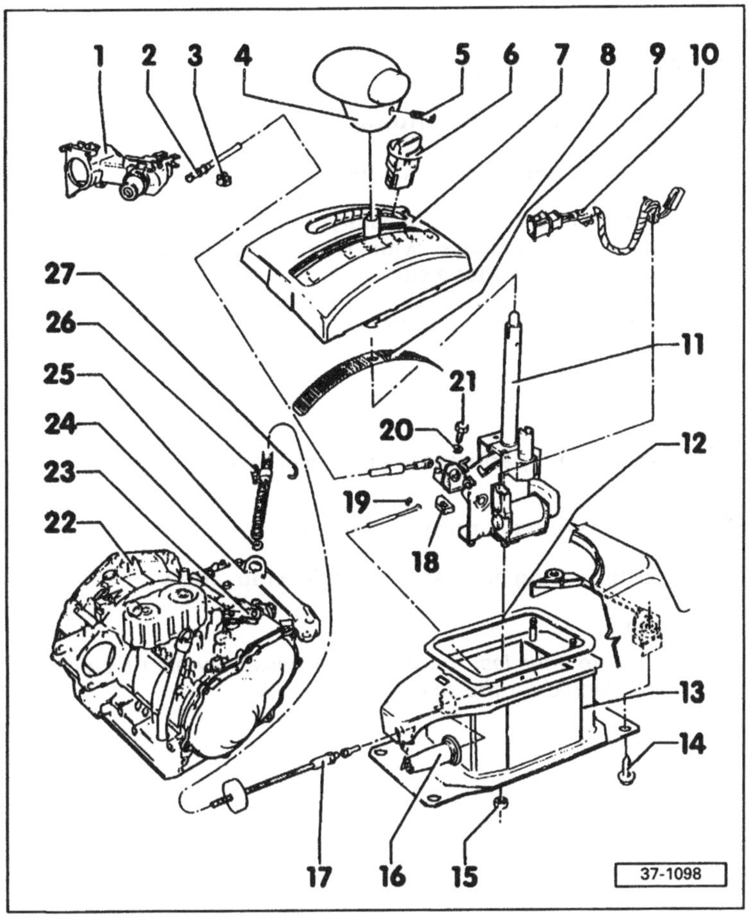

- do not bend or kink cable eye
- lubricate cable eye and ends before installing
- tighten selector lever cable at selector support to 20 Nm (15 ft lb)
- check guide tube for correct fit
- check protective sleeve between bracket, item -24- and lever, item -23-
- fasten selector lever cable to selector lever ⇒ Fig. 3
- Adjusting:
  a. position selector lever in **P**
  b. loosen shoulder bolt, item -26-
  c. move lever, item -23- into **P** (detent lever must engage in parking lock gear, locking both front wheels)
  d. in this position, tighten shoulder bolt to 25 Nm (18 ft lb)

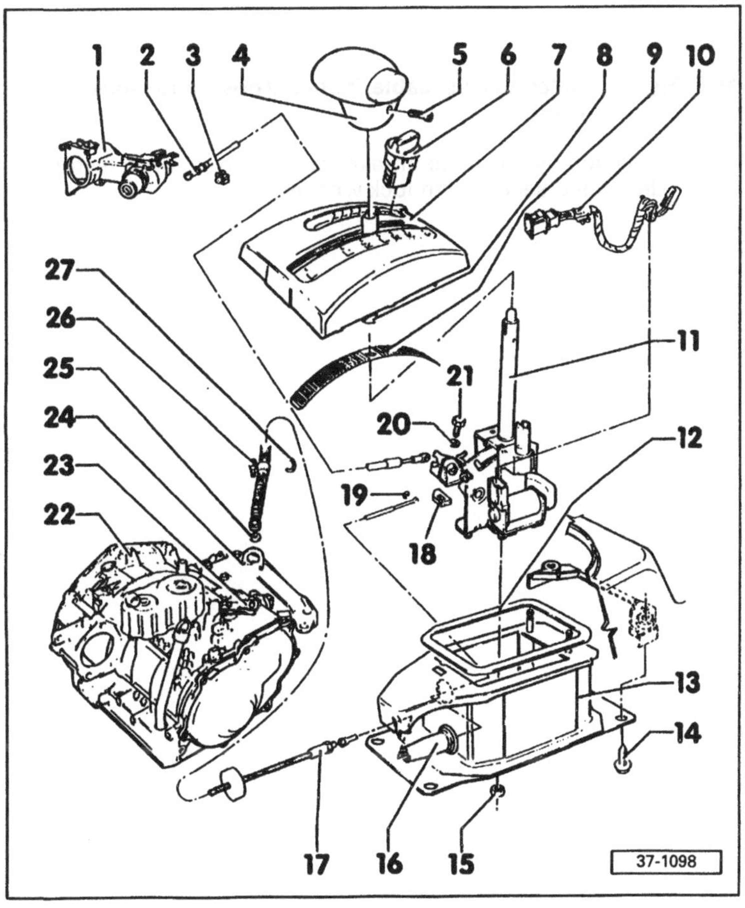

18 - Seal
- bonded to cable anchor
- replace if damaged

19 - Circlip
- insert after installing selector lever cable ⇒ Fig. 3

20 - Washer

21 - 6 Nm (53 in lb or 61 cm kg)

22 - Transmission

23 - Gear lever

24 - Bracket
- With anchor for selector lever cable

25 - Washer

26 - Shoulder bolt – 25 Nm (18 ft lb)

27 - Circlip
- for selector lever cable

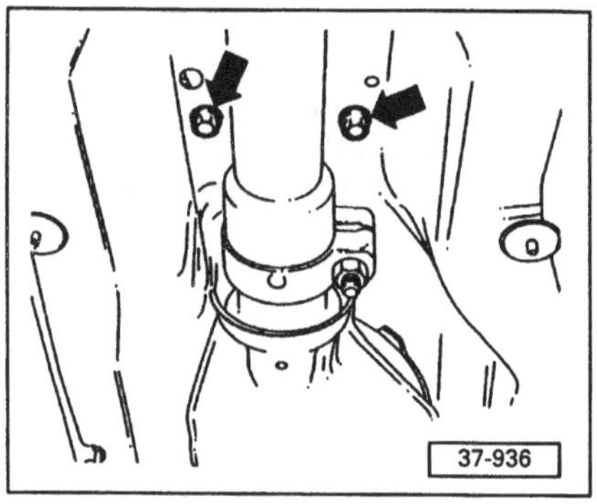

Fig. 1 Selector support, removing and installing

- remove front bolts -arrows-
- remove connector for Oxygen Sensor (O2S)
- disconnect front end of catalytic converter (exception–vehicles with 6-cyl. engine: disconnect catalytic converter also)
- remove front heat shield
- remove rear bolts

Note:

Installation is performed in the reverse order.

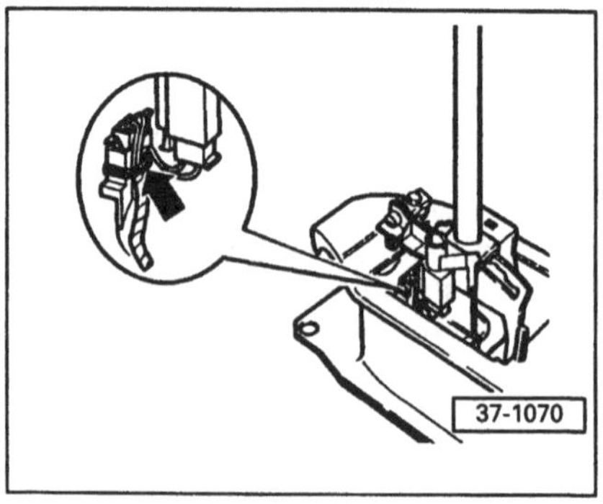

Fig. 2 Wiring harness, fastening

- fasten wiring harness with cable wrap -arrow-

Note:

On some vehicles, the wiring harness may also include the solenoid for the selector lever lock.

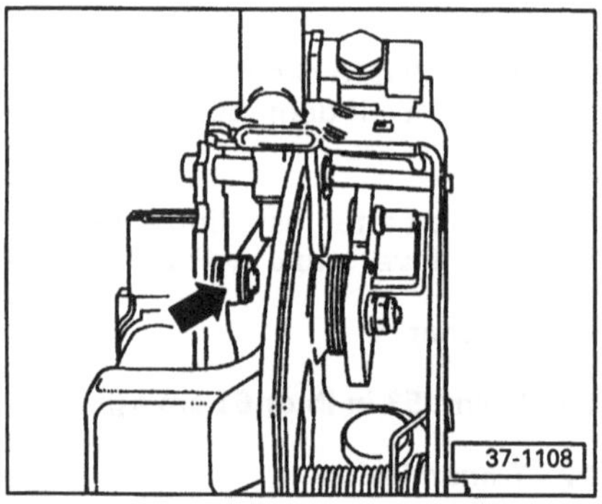

Fig. 3 Selector lever cable, installed position at selector lever

- position eye of selector lever cable -arrow- on selector lever and secure with lock washer.

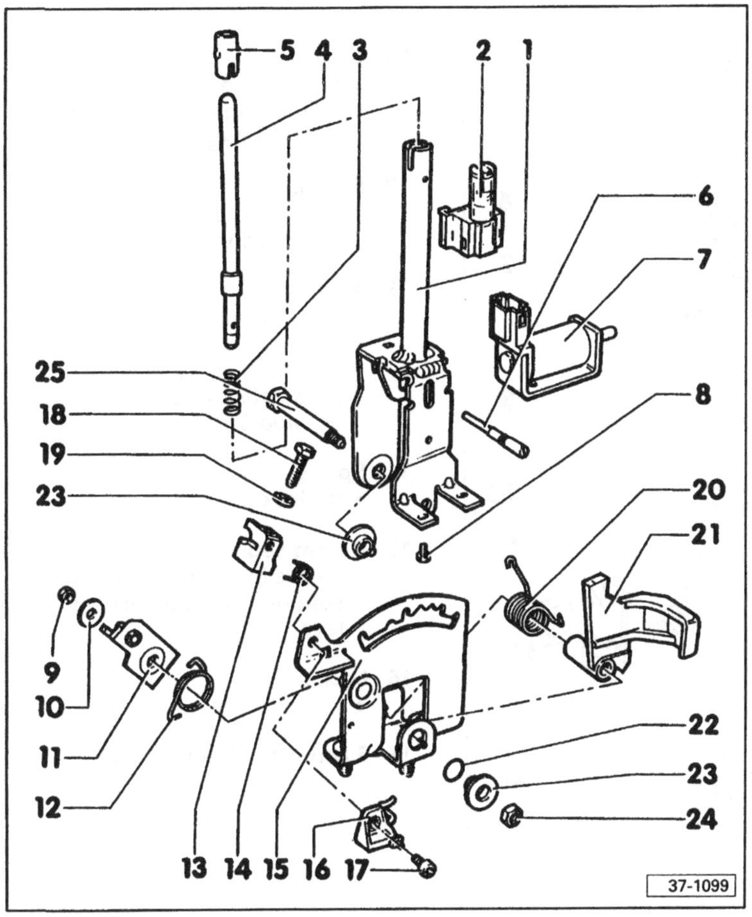

Selector lever, disassembling and assembling

Note:

In order to replace an individual selector lever component, the lever itself must be removed. See *Selector lever/cable assembly,* *⇒ page 37-5.*

1 - Selector lever
- ♦ assembling ⇒ Fig. 1

2 - Selector scale light

3 - Spring

4 - Pressure rod
- ♦ apply grease to contact surfaces

5 - Guide sleeve
- ♦ install pressure rod first, before installing sleeve
- ♦ install with locking compound AMV 200 000

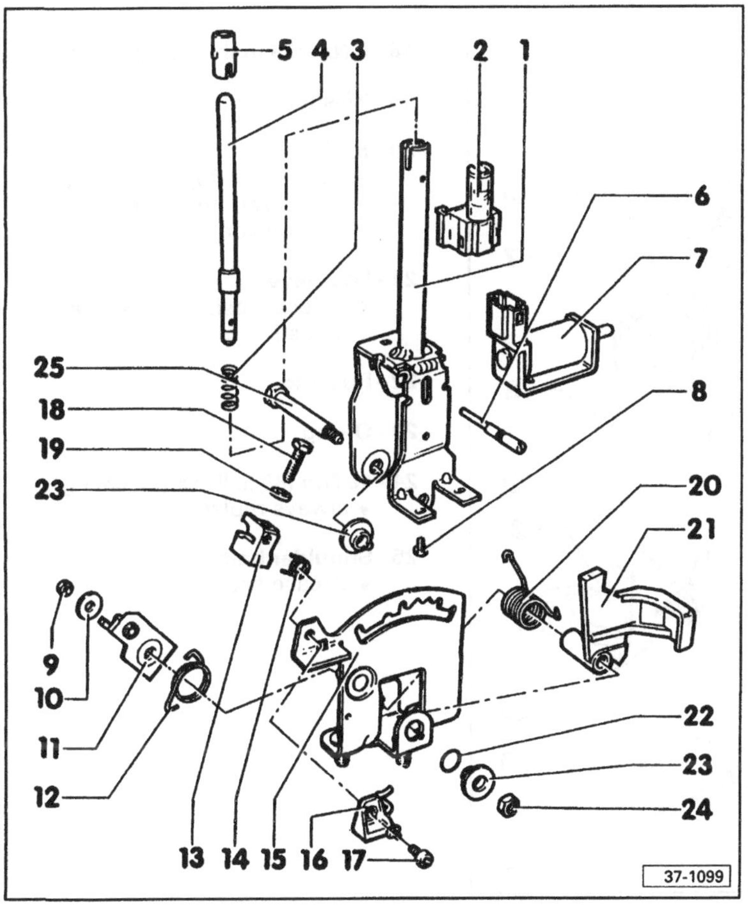

6 - Shaft bolt
- ♦ always replace
- ♦ in order to loosen shaft bolt, first remove solenoid for selector lever lock and selector scale light; clamp shaft bolt in vice and loosen by turning selector lever

7 - Solenoid for selector lever lock -N110-
- ♦ on some vehicles, wiring harness may include solenoid
- ♦ adjusting ⇒ Fig. 2
- ♦ checking function ⇒ Fig. 3
- ♦ checking in-vehicle, through, ⇒ *On-Board Diagnostic (OBD),* Repair Group 01

8 - Bolt assembly – 3 Nm (27 in lb or 31 cm kg)
- ♦ install with locking compound AMV 200 000

9 - 5 Nm (44 in lb or 51 cm kg)
- ♦ always replace

10 - Washer

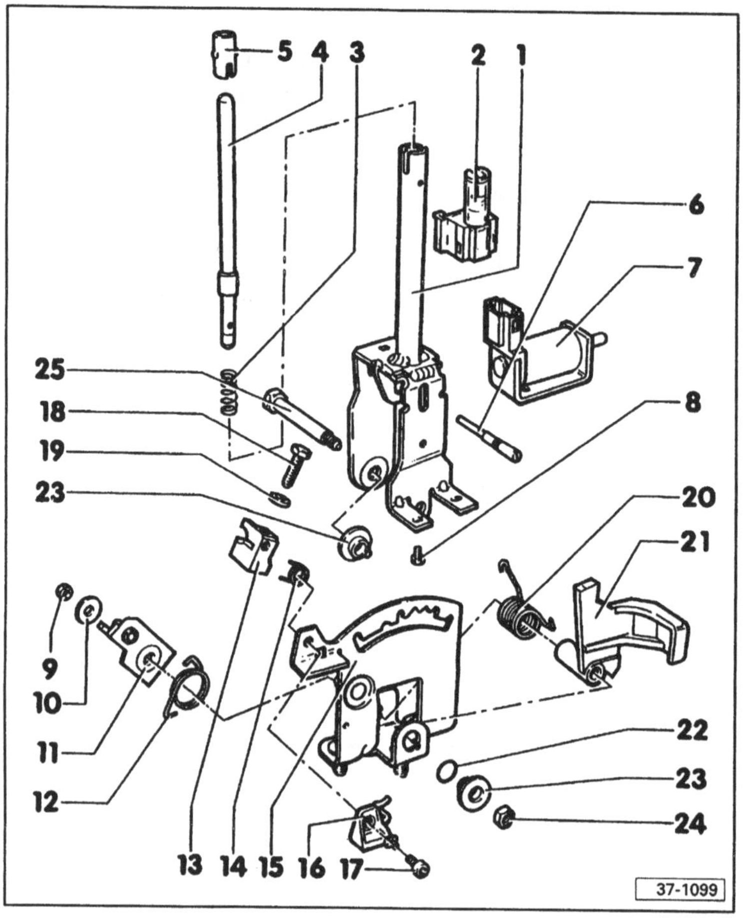

11 - Lock lever
- for ignition starter lock
- adjust together with lock cable ⇒ page 37-24

12 - Spring
- Hook into opening in detent segment, item -15-

13 - Anchor
- For lock cable
- *Lock cable, removing and installing* ⇒ page 37-20
- *Lock cable, adjusting,* ⇒ page 37-24

14 - Spring
- for lock pin

15 - Detent segment
- before installing in selector support install selector lever

16 - Lock pin
- with bracket

17 - 6 Nm (53 in lb or 61 cm kg)

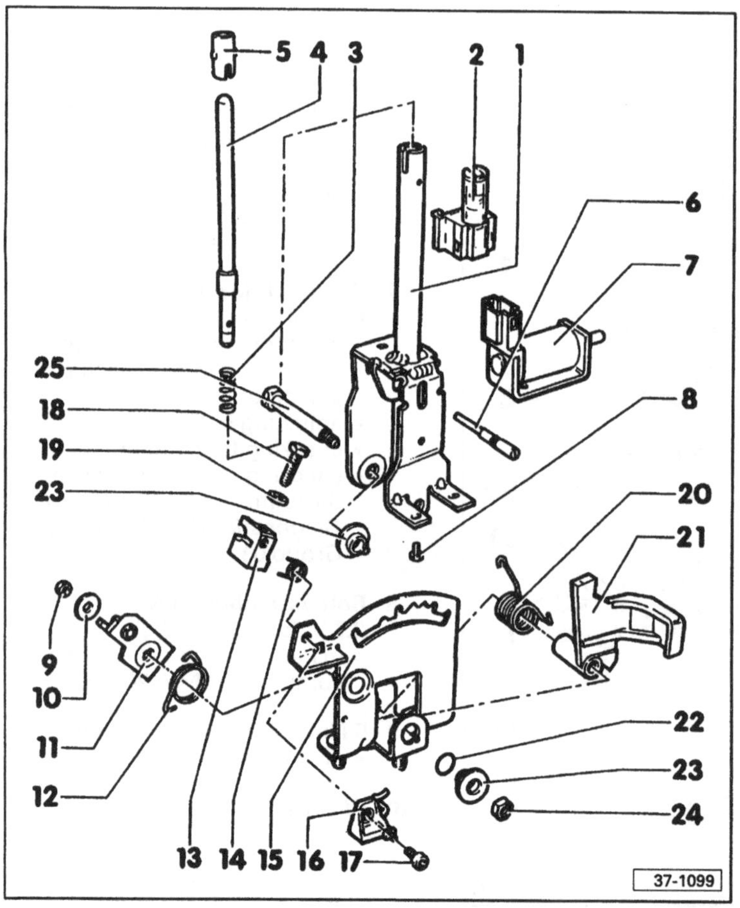

18 - 6 Nm (53 in lb or 61 cm kg)

19 - Washer

20 - Spring
- assemble together with lock lever, item -21- in selector lever

21 - Lock lever
- to lock, position in **N** or **P**
- adjusting ⇒ Fig. 2

22 - Bushing

23 - O-ring

24 - 10 Nm (89 in lb or 102 cm kg)
- always replace

25 - Shoulder bolt
- Grease shaft

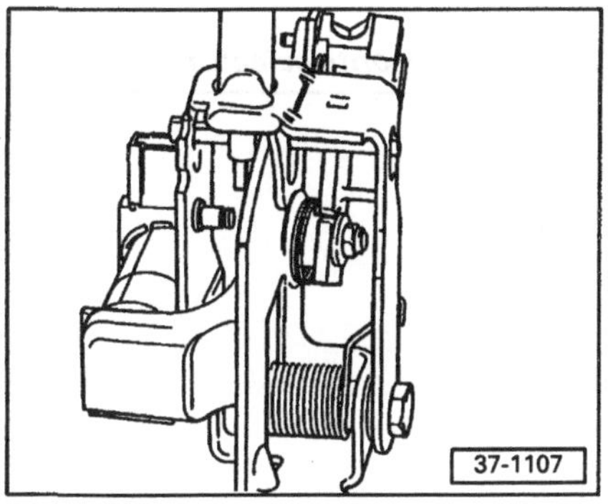

Fig. 1 Selector lever, assembling before installing

Note:

Before installing selector lever in selector support, be sure:

- *shiftlock solenoid (N110) is adjusted* ⇒ Fig. 2
- *functional test* ⇒ Fig. 3 has been performed

- insert shaft bolt, item -6- ⇒ page 37-14 and shiftlock solenoid bolts with locking compound AMV 200 000.

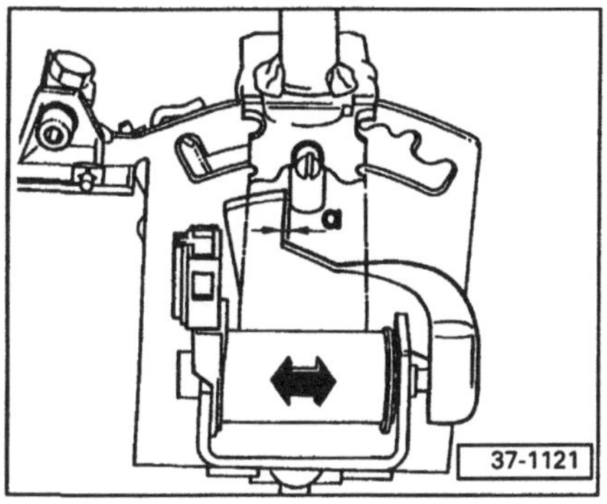

Fig. 2 Shiftlock solenoid -N110-, adjusting

- adjust solenoid by sliding in slots -arrows- so that a free play of 0.3mm (0.012 in.) exists between pressure rod and lever ⇒ Fig. 3.

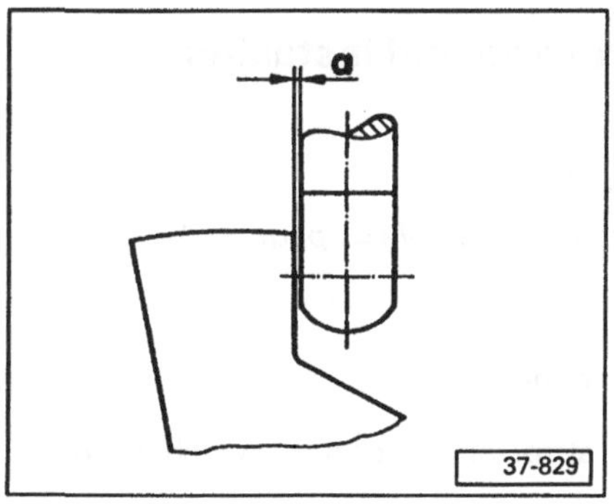

Fig. 3 Free-play between pressure rod and lever, adjusting

Dimension -a- = 0.3 mm (0.012 in.)

Functional test

- place selector lever in **P**
- apply 12 Volts to solenoid
  - must not be possible to move selector lever into a drive gear
- interrupt solenoid operation.
  - selector can now be moved into a drive gear

CAUTION
Be sure pressure rod does not hook onto selector lever.

- place selector lever in **N**
- apply 12 Volts to solenoid
  - must not be possible to move selector lever into a drive gear
- interrupt solenoid operation.
  - selector can now be moved into a drive gear

CAUTION

Be sure pressure rod does not hook onto selector lever.

- adjust play
- install selector lever completely

Locking cable, removing and installing

Note:

- *lock cable must not be kinked*
- *after installing, adjust locking cable ⇒ page 37-24*

Removing

- remove selector lever handle
- remove cover, cover strip, and program switch if so equipped
- loosen bolt for cable anchor, withdraw lock cable from lock lever and pull out of anchor
- remove trim from center console left side, in floor outlet, ⇒ Repair Group 70
- remove trim from ignition switch
- remove lock cable from clip -arrow-
- press out lock cable clip from ignition switch and remove lock cable from cable head
- pull out cable head

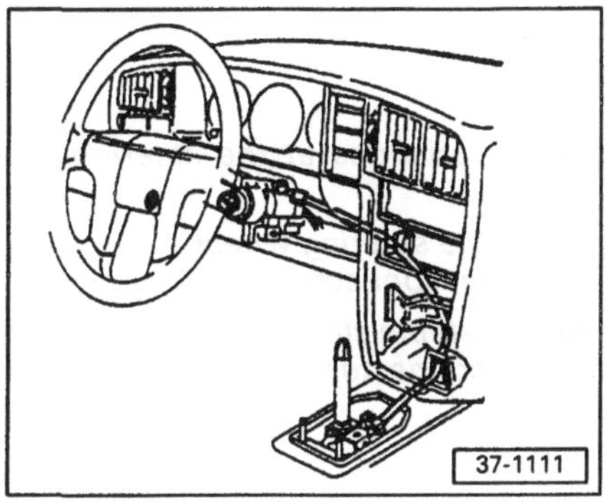

Installing

> ***CAUTION***
> ***During installation, be sure that cable:***
> ***♦ is routed correctly***
> ***♦ does not bend or kink***

◄ – lock cable, installed position

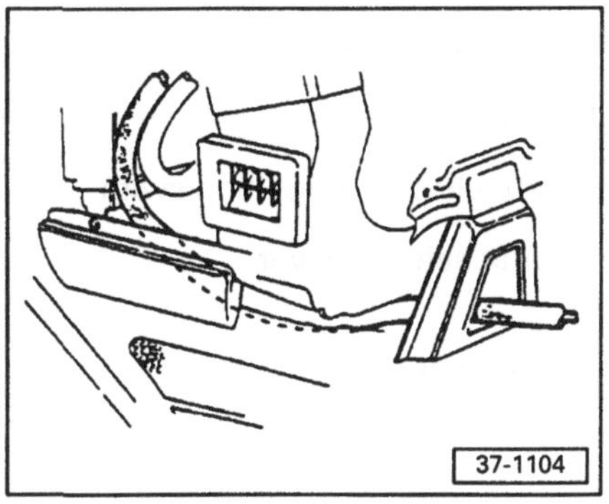

◄ – route cable at center console next to floor outlet, as shown

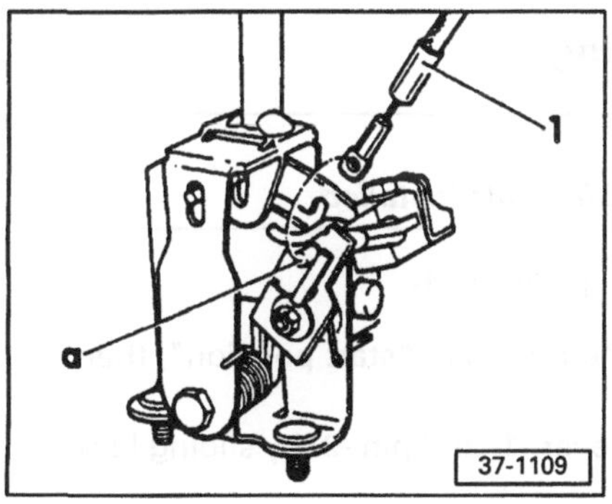

◄ – position eye of lock cable -1- on lock lever -a-

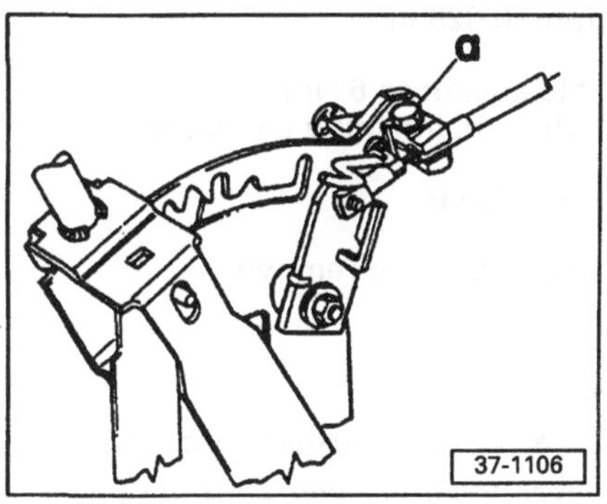

◄ – position lock cable in cable anchor, as shown

- ♦ do not tighten bolt -a- for anchor until lock cable has been adjusted in installed position ⇒ page 37-24

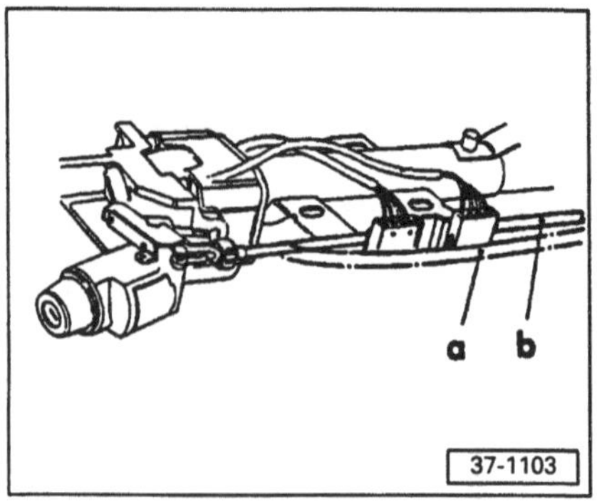

Note:

There are two cable routing variations at the steering wheel:

*a. vehicles **with** height-adjustable steering wheel – route cable in front of multi-pin connector*

*B. vehicles **without** height-adjustable steering wheel – route cable behind multi-pin connector*

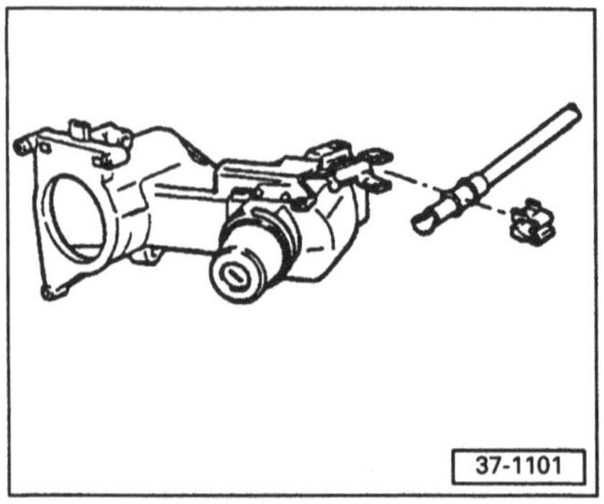

◄ – attach lock cable clip to lock cable first

– press lock cable onto ball head

– push lock cable with clip attached, into ignition swith

– adjust lock cable in anchor at selector lever ⇒ page 37-24

– install cover with gear selector scale

– install gearshift handle

– install ignition switch trim

– install trim for center console

Lock cable, adjusting

CAUTION
Always adjust lock cable while installed.

– place selector lever in position -1-

– turn ignition key to the right, i.e. "start position", then release

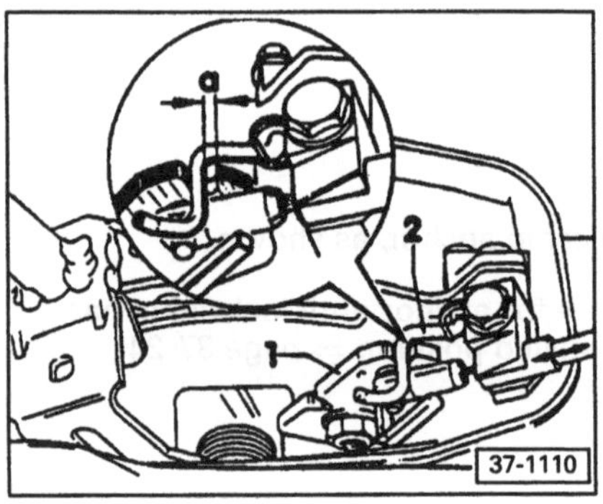

◄ – adjust play between lever -1- and pin -2- by sliding lock cable sleeve

♦ dimension -a- = 0.7 mm (0.028 in.)

– tighten bolt for lock cable anchor to 6 Nm (53 in lb or 61 cm kg), while maintaining dimension -a-

– place selector lever in position **P**

– turn igntion key to last position, and remove

Note:

It must NOT be possible:

♦ *to remove key in any selector lever position except **P***

♦ *to shift out of selector lever position **P** with key removed*

A/C refrigerant system (R-134a), safety measures

The subassemblies and lines of the air conditioner are filled with Tetrafluoroethane also known as refrigerant R-134a.

Refrigerant R-134a replaces the refrigerant R-12 which is reported to have harmful effects on the earth's ozone layer.

R-134a and R-12 systems are similar in design but there are very important differences in the refrigerant, lubricating oils, service equipment and A/C system components.

WARNING

*– Always use an Underwriter's Laboratory (UL) approved refrigerant recovery/ recycling/recharging unit such as **Kent-Moore ACR4**, or equivalent, whenever discharging an R-134a A/C system.*

*– As of January 1, 1992 any person who services a motor vehicle air conditioner **MUST**, by law, be properly trained and certified and use approve refrigerant recycling equipment. Technicians must complete an EPA approved recycling course to be certified.*

– State and local governments may have additional requirements regarding air conditioning servicing. Always comply with state and local laws.

WARNING

– A/C system is filled with refrigerant gas which is under pressure.

– Always be careful that refrigerant does not come in contact with your skin.

– If liquid refrigerant has come in contact with your skin or eyes:

- *do not rub skin or eyes*

immediately flush with cool water for 15 minutes

- *rush to a doctor or hospital*
- *do not attempt to treat yourself*

– Work in a well ventilated area because refrigerant gases are heavier than air, displace oxygen and may cause suffocation in areas of poor air circulation, like under the car.

– Switch on existing exhaust/ventilation systems when working on the refrigerant system.

– Avoid breathing refrigerant vapors. Exposure may irritate eyes, nose and throat.

– Always wear hand and eye protection (gloves and goggles) when working around the A/C system.

– Do not expose any component of the A/C system to high temperatures (above 176°F/ 80°C) or open flames). Excessive heat will cause a pressure increase which could burst the system.

WARNING

– Keep refrigerant containers stored below 50°C (122°F) and never drop from high places. ***DO NOT*** *warm refrigerant containers with an open flame. If refrigerant needs to be warmed, place bottom of tank in warm water.*

– Keep refrigerant away from open flames because poisonous gas will be produced if it burns. Do not smoke when refrigerant gases are present for the same reason.

– Electric welding near refrigerant hoses causes R-134a to decompose from ultraviolet light. Discharge system before electric welding.

– Pressurized R-134a refrigerant in the presence of oxygen may form a combustible mixture. Never introduce compressed air into any R-134a container (full or empty), capped off A/C component or piece of service equipment.

– ***DO NOT*** *exceed maximum rated capacity of refrigerant containers. Never fill a container to more than 60% of its gross weight rating (for example, 18 lb. in a 30 lb. container.) Without sufficient room for expansion of R-134a (gas cushion), the container could explode when the temperature rises causing serious injury.*

– Do not steam clean condensers or evaporators. Use only cold water or compressed air.

CAUTION

– R-12 and R-134a refrigerant are ***NOT*** *compatible. Never add R-12 refrigerant to an R-134a system or R-134a refrigerant to an R-12 system. If the refrigerants are mixed, total system contamination will occur and compressor failure may result.*

– Refrigerant oils used for the R-134a system and R-12 system are ***NOT*** *compatible. Use only the specified synthetic oil (Polyalkylene Glycol/PAG) for the R-134a refrigerant system.* ***DO NOT*** *use R-12 system oil in an R-134a system or R-134a system oil in an R-12 system. If the refrigerant oils are mixed, system contamination will occur and compressor failure may result.*

– R-134a refrigerant system oil (PAG oil) absorbs moisture very rapidly. Moisture combines with the refrigerant to form acids which will damage the system. Use only the specified oil from a sealed container and ***ALWAYS*** *reseal oil container immediately after use.* ***DO NOT*** *use oil if it has become contaminated with moisture or if container has been left open.*

– Immediately plug open connections on A/C components to prevent dirt and moisture contamination. Likewise, ***DO NOT*** *remove new component from packaging until ready to install. Immediately tighten component connections after installation.*

CAUTION

*– Always use separate refrigerant recovery/recycling/recharging servicing equipment for R-12 and R-134a systems. **DO NOT** use one piece of equipment for both R-12 and R-134a systems. The residual traces of refrigerant will contaminate and damage the equipment. Servicing equipment includes recovery/ recycling/recharging unit, charging station, vacuum pump, manifold gauges, etc.*

*– **DO NOT** use R-12 servicing equipment on R-134a systems or R-134a equipment on R-12 systems or damage to both the vehicle A/C system and servicing equipment may result. Use only equipment designed to meet Society of Automotive Engineers (SAE) standards.*

*– R-134a and R-12 systems use different size service fittings. **NEVER** use adaptors to convert an R-12 fitting to R-134a size or R-134a fitting to R-12 size.*

*– R-134a and R-12 A/C components including compressor, hoses, O-rings, evaporator, condenser, receiver–drier, etc. are **NOT** interchangeable. Components of the R-134a system are identified by lettering "R-134a" or by a green label (or stripe). In addition, a label on the upper radiator support in the engine compartment identifies which type refrigerant is used. Use only the correct system component for each refrigerant type.*

CAUTION

– Flush refrigerant system with compressed air and nitrogen if:

• moisture or dirt have entered the refrigerant system (i.e. following an accident)

• refrigerant oil is dark and viscous (thick)

• too much refrigerant oil is in the system following compressor replacement

• the compressor is replaced due to internal damage (i.e. noises)

– During flushing with compressed air and nitrogen, use existing exhaust/ventilation systems to draw off the gas mixture escaping from the A/C system.

– Always replace damaged and/or leaking A/C system components. Do not attempt repair by soldering, brazing or welding.

– Work area must be extremely clean when working on A/C system components.

– Use only tools, equipment and parts specified for use with R-134a.

*– Discharge A/C system using refrigerant recovery/ recycling/ recharging unit **Kent-Moore ACR[4]**, or equivalent, before removing any A/C system component.*

CAUTION

*– Always replace O-rings, **DO NOT** reuse. Use only the correct size and type of O-rings specified for use with R-134a refrigerant. Lubricate O-ring with refrigerant oil before installing.*

– Always reinstall cap(s) over A/C service valve(s).

– Before working on the electrical system, obtain radio security code and disconnect battery Ground (GND) strap.

A/C refrigerant R-134a, properties

Commercial designation

The following designations are used and appear on component labels and factory tags:

- R-134a
- Tetrafluoroethane
- CH_2 F CF_3
- H–FKW 134a
- SUVA® TRANS A/C
- ARCTON® 134a

Note

R-134a refrigerant is packaged in different containers. One variation is SUVA® COLD MP used only for commercial applications. This type of R-134a is sold in cylinders using a 1/4 flare fitting which will not connect to the 1/2 in.-16 ACME fittings used on the vehicle and servicing equipment. Use only R-134a which comes in containers having the correct type of service fitting.

Color

Refrigerant R-134a is colorless and is invisible as a gas. R-134a when viewed through the sight glass may appear milky due to the mixture of refrigerant and lubricating oil (PAG oil).

Temperature/pressure relationship

Similar to R-12 refrigerant, R-134a in an enclosed container will have a specific temperature/pressure relationship as follows:

Temperature in °C (°F)	Pressure in bar (psi)
–30 (–22)	0.0 (0.0)
–20 (–4)	0.3 (4.4)
–10 (14)	1.0 (14.5)
0 (32)	1.9 (27.5)
10 (50)	3.1 (45.0)
20 (68)	4.7 (68.2)
30 (86)	6.7 (97.2)
40 (104)	9.1 (132.0)
50 (122)	12.2 (177.0)
60 (140)	15.8 (229.2)
70 (158)	20.2 (293.0)

Refrigerant oil

A special Polyalkylene Glycol (PAG) synthetic oil is used in R-134a systems. This oil is **NOT** compatible with mineral based oils used in R-12 systems. See safety measures, page 87-1, for Warnings and Cautions regarding proper oil usage.

Airborne properties

Escaped refrigerant gases are heavier than air and will gather first in low places, like under the car. R-134a refrigerant gas displaces oxygen and may cause suffocation in low areas of poor air circulation.

Effects on the environment

Because R-134a does not contain chlorine atoms, the major catalyst in ozone depletion, R-134a has no ozone depletion potential.

R-134a does not cause smog and has a considerably lower global warming potential (greenhouse effect) than R-12 refrigerant. Nonetheless, do not release R-134a into the environment and always recover/recycle using approved service equipment.

Effects on plastics

Refrigerant R-134a will deteriorate some plastics. When system temperature drops, these plastics can be deposited at the expansion valve or restrictor causing blockage. Therefore, when making system repairs, use only genuine VW replacement parts which are specified for use with R-134a refrigerant.

Effects on metal

In its pure state, refrigerant R-134a is chemically stable and will not attack iron or aluminum. However, the mixture of R-134a and PAG oil may deteriorate certain metals (copper) which could lead to blockage, leaks or deposits on the compressor piston. Therefore, when making system repairs, use only genuine VW replacement parts which are specified for use with R-134a refrigerant.

Water solubility

Liquid R-134a refrigerant will absorb only very minute quantities of moisture. However, R-134a vapor can absorb large amounts of moisture.

Water in the system forms ice in the expansion valve or restrictor causing the system to stop cooling. In addition, moisture combines with the refrigerant to form acids which will damage the system.

Toxicity

Refrigerant R-134a is non-toxic up to a temperature of 101°C (214°F) and is safe when handled and used properly.

Above 101°C (214°F), gas pressure is above 39.5 bar (573 psi) and R-134a begins to decompose into poisonous gases (hydrogen fluoride, carbonyl fluoride and fluorine). These gases give adequate warning of their presence since they cause irritation of the mucous membrane (throat).

Refrigerant R-134a gas has no effect on foodstuffs; it does not poison food or make it inedible.

Flammability

R-134a refrigerant is not flammable. In fact, it tends to inhibit combustion and has a fire extinguishing effect. However, a flame or glowing hot surface, as well as ultraviolet light (from electric welding), will decompose the refrigerant into poisonous gases (hydrogen fluoride, carbonyl fluoride and fluorine). These gases give adequate warning of their presence since they cause irritation of the mucous membrane (throat).

R-134a can become flammable at pressures above ambient pressure in conjuction with air concentrations greater than 60% by volume.

Refrigerant containers

DO NOT exceed maximum rated capacity of refrigerant containers. Never fill a container to more than 60% of its gross weight rating (for example, 18 lb in a 30 lb container.)

Without sufficient room for expansion of R-134a (gas cushion), the container could explode when the temperature rises causing serious injury.

Leak detection

Use halogen leak detector **Hitec HI400A-TEL**, or equivalent, to check for R-134a system leaks. This tool can also be used to detect leaks in R-12 systems. Many currently available R-12 leak detectors cannot detect R-134a refrigerant leaks.

R-134a refrigerant oil

A special Polyalkylene Glycol (PAG) synthetic oil is used in R-134a systems. This oil is **NOT** compatible with mineral based oils used in R-12 systems.

> ***CAUTION***
>
> *– Refrigerant oils used for the R-134a system and R-12 system are **NOT** compatible. Use only the specified synthetic oil for the R-134a refrigerant system. **DO NOT** use R-12 system oil in an R-134a system or R-134a system oil in an R-12 system. If the refrigerant oils are mixed, system contamination will occur and compressor failure may result.*
>
> *– R-134a refrigerant system oil (PAG oil) absorbs moisture very rapidly. Moisture combines with the refrigerant to form acids which will damage the system. Use only the specified oil from a sealed container and **ALWAYS** reseal oil container immediately after use. **DO NOT** use oil if it has become contaminated with moisture or if container has been left open.*
>
> *– Do not allow refrigerant oil to come in contact with engine oil or transmission fluid (for example when disposing). Dispose of used PAG refrigerant oil separately following laws governing hazardous waste disposal.*

Basics when working on refrigerant system (R-134a)

> ***CAUTION***
>
> *– Keep work area and all tools clean.*
>
> *– Use only tools, equipment and parts specified for use with R-134a.*
>
> *– Always wear protective glasses and gloves when working on A/C refrigerant system.*
>
> *– Switch on existing exhaust/ventilation systems when working on the refrigerant system.*
>
> *– Discharge A/C system before removing any A/C refrigerant system component.*
>
> *– Always use an Underwriter's Laboratory (UL) approved refrigerant recovery/recycling/recharging unit such as **Kent-Moore ACR4**, or equivalent, whenever discharging an R-134a A/C system.*
>
> *– Immediately plug open connections on A/C components to prevent dirt and moisture contamination. Likewise, **DO NOT** remove new component from packaging until ready to install. Immediately tighten component connections after installation.*
>
> *– **DO NOT** top-off a charged refrigerant system. Instead, discharge, evacuate and recharge system.*

CAUTION

*– Always replace O-rings, **DO NOT** reuse. Use only the correct size and type of O-rings specified for use with R-134a refrigerant. Lubricate O-ring with refrigerant oil before installing.*

*– Use only the specified PAG oil from a sealed container and **ALWAYS** reseal oil container immediately after use. **DO NOT** use oil if it has become contaminated with moisture or if oil container has been left open.*

– Always reinstall cap(s) over A/C service valve(s).

– Flush refrigerant system with compressed air and nitrogen (available locally) if:

- *moisture or dirt have entered the refrigerant system (i.e. following an accident)*
- *refrigerant oil is dark and thick*
- *too much refrigerant oil is in the system following compressor replacement*
- *the compressor is replaced due to internal damage (i.e. noises)*

*– **DO NOT** flush refrigerant system with R-11.*

– Evacuate refrigerant system for a minimum of 30 minutes.

*– **DO NOT** top-off a charged refrigerant system. Instead, discharge, evacuate and recharge system.*

CAUTION

– After system recharge, manually rotate A/C compressor approx. 10 turns before starting engine. Start engine with A/C OFF. After idle speed has stabilized, switch A/C ON and let engine idle (compressor running) for a minimum of two minutes before raising engine speed.

Note

- *For information on A/C wiring see applicable wiring diagram.*
- *See page 87-1 for A/C refrigerant system safety measures.*
- *Green label in engine compartment indicates A/C system is charged with refrigerant R-134a.*

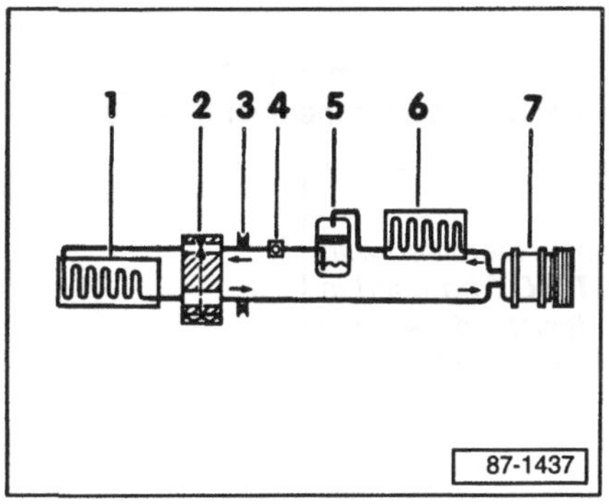

87-1437

A/C refrigerant system (R-134a), system identification

Refrigerant circuit

1 – Evaporator

2 – Expansion valve

3 – High pressure service valve

4 – Sight glass

5 – Receiver drier

6 – Condenser

7 – Compressor

Note

Arrows indicate flow direction of refrigerant.

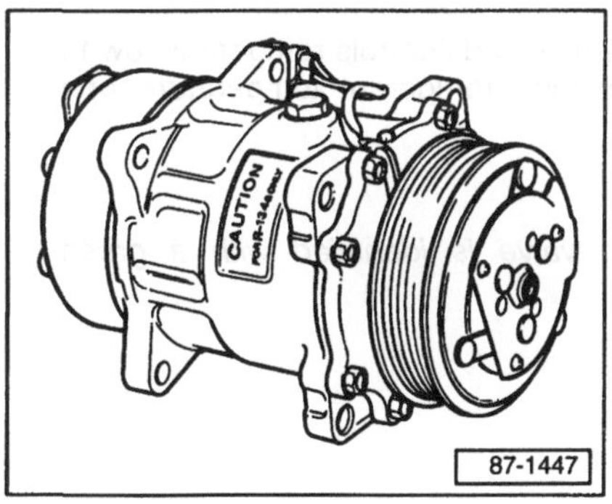

87-1447

A/C refrigerant system (R-134a), component functions

Compressor

The compressor is driven via a belt on the engine when the A/C clutch engages (A/C ON).

Low pressure refrigerant gas from the evaporator is compressed by the compressor. After compression, the refrigerant gas (now high pressure) flows to the condenser.

Note

- *The compressor contains refrigerant oil (PAG oil) that is mixable under all temperatures with R-134a refrigerant.*
- *A label on the compressor indicates that compressor is for R-134a systems only.*

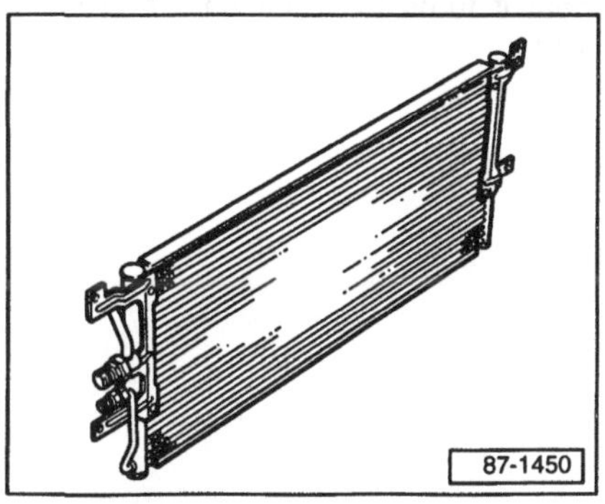
87-1450

Condenser

The condenser transfers heat from the compressed refrigerant gas to the outside air which causes the refrigerant to change state from a gas to a liquid.

Note

The condenser for the R-134a refrigerant system is identified with a green label.

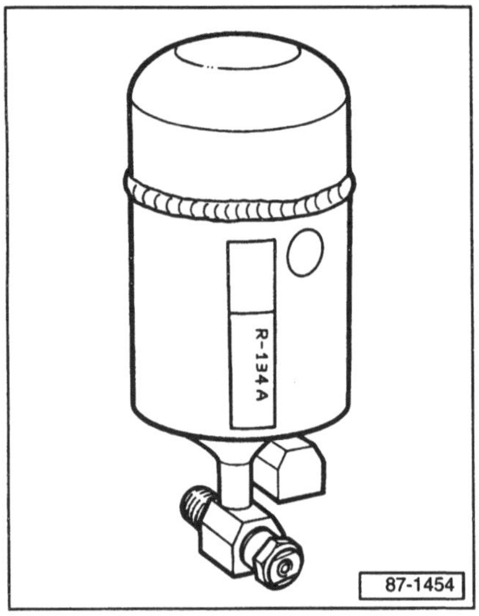

87-1454

Receiver drier

The receiver drier acts as a refrigerant reservoir for the system. Any moisture in the system is absorbed in the drier desiccant.

> ***CAUTION***
>
> *The R-134a receiver drier is identified with a green label and **MUST NOT** be used in R-12 systems (drier desiccant is only compatible with R-134a refrigerant). Likewise, **NEVER** use an R-12 receiver drier in an R-134a system.*

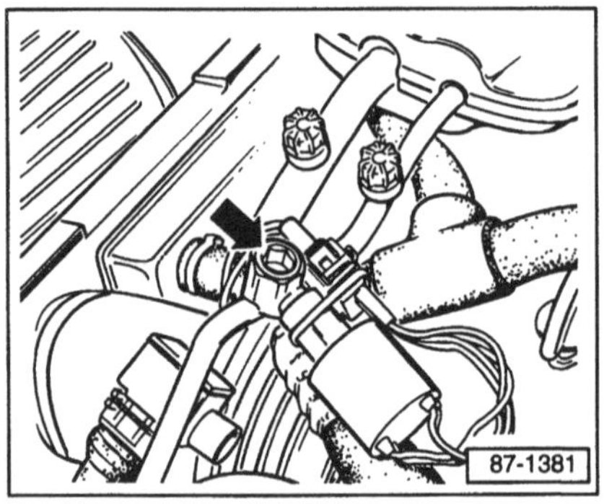

87-1381

Sight glass

With a properly charged A/C system and A/C clutch engaged (engine running), the sight glass -arrow- should remain clear and bubble-free.

Note

The mixture of R-134a refrigerant and refrigerant oil (PAG oil) may appear milky in the sight glass.

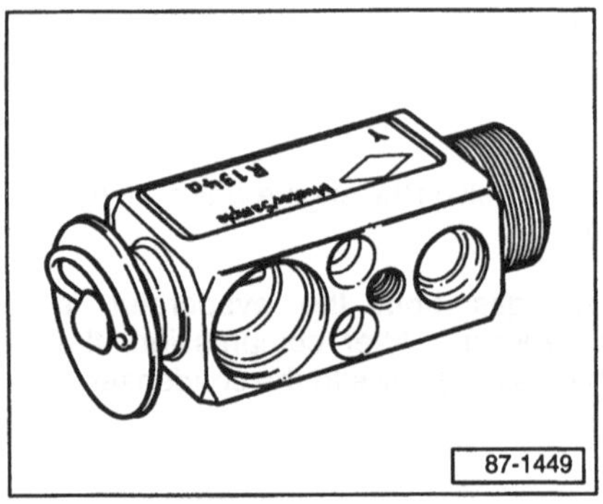

87-1449

Expansion valve

The expansion valve restricts and controls refrigerant flow to the evaporator thus lowering refrigerant temperature and pressure.

Note

The R-134a expansion valve is identified with a green adhesive label.

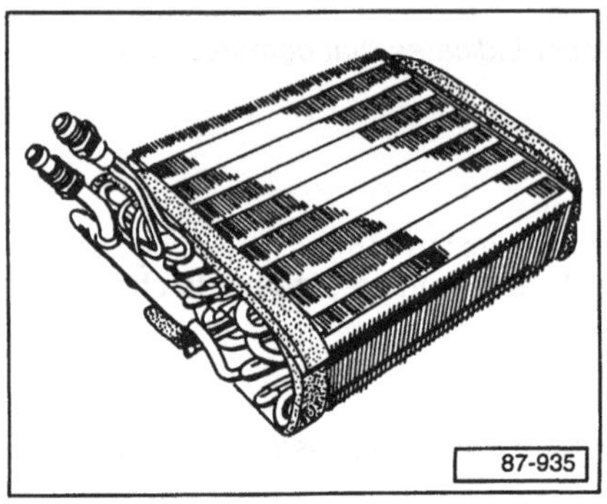

87-935

Evaporator

Liquid refrigerant entering the evaporator absorbs heat from air passing through the evaporator fins and cools the air. As the refrigerant absorbs heat it turns to vapor and then is suctioned by the compressor.

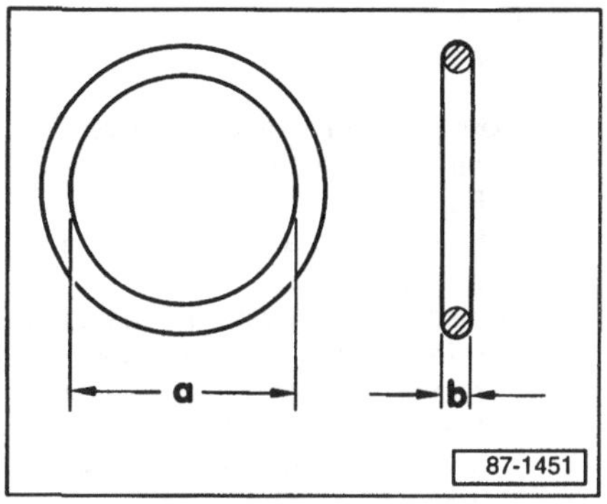

O-rings

O-rings seal the connections between components of the A/C system.

Use only new red or green O-rings which are compatible with R-134a refrigerant and refrigerant (PAG) oil.

Always use correct size O-rings (dimensions -a- and -b-).

Note

- *Do not reuse O-rings, always replace.*
- *Lubricate O-rings with refrigerant oil before installing.*

Pressure relief valve

The pressure relief valve is mounted on the compressor or the receiver drier. At pressures above 40 bar (580 psi), the pressure relief valve opens to vent excessive pressure. When system pressure is reduced, the valve closes to prevent total refrigerant loss.

A cap on the pressure relief valve will pop out if the valve has opened.

A/C system hoses and lines

The mixture of refrigerant oil (PAG oil) and refrigerant R-134a attacks some metals and alloys (for example, copper) and breaks-down certain hose material. Use only hoses and lines which are identified with a green mark (stripe) or the lettering "R-134a".

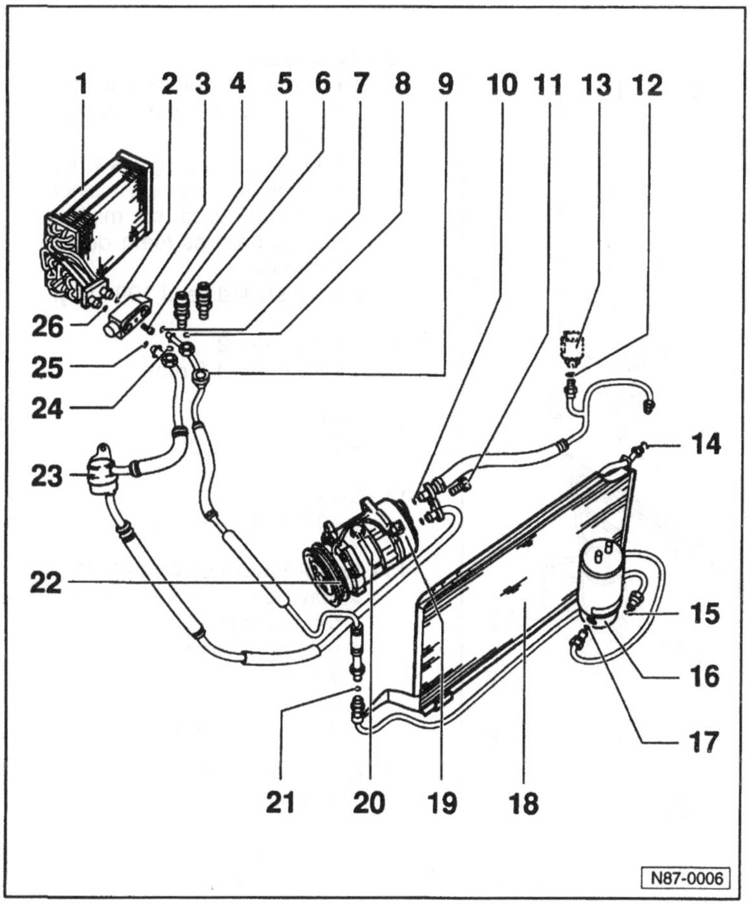

A/C refrigerant system (R-134a), components

Note

- *For A/C refrigerant system components, replacing, see page 87-79*
- *For A/C refrigerant system capacities, see page 87-64*
- *A/C components specified for use with R-134a are identified by a green label or stripe.*
- *O-rings specified for use with R-134a are red or black*
- *The A/C pressure switch -F129 – shown as a dotted line – can be removed and installed without discharging the A/C system (valve under switch).*

1 - Evaporator

2 - O-ring
- 10.8 mm x 1.8 mm

3 - Expansion valve
- identified with a green label
- cooling output of evaporator is set and cannot be changed
- insulator, removing/ installing, see page 87-35

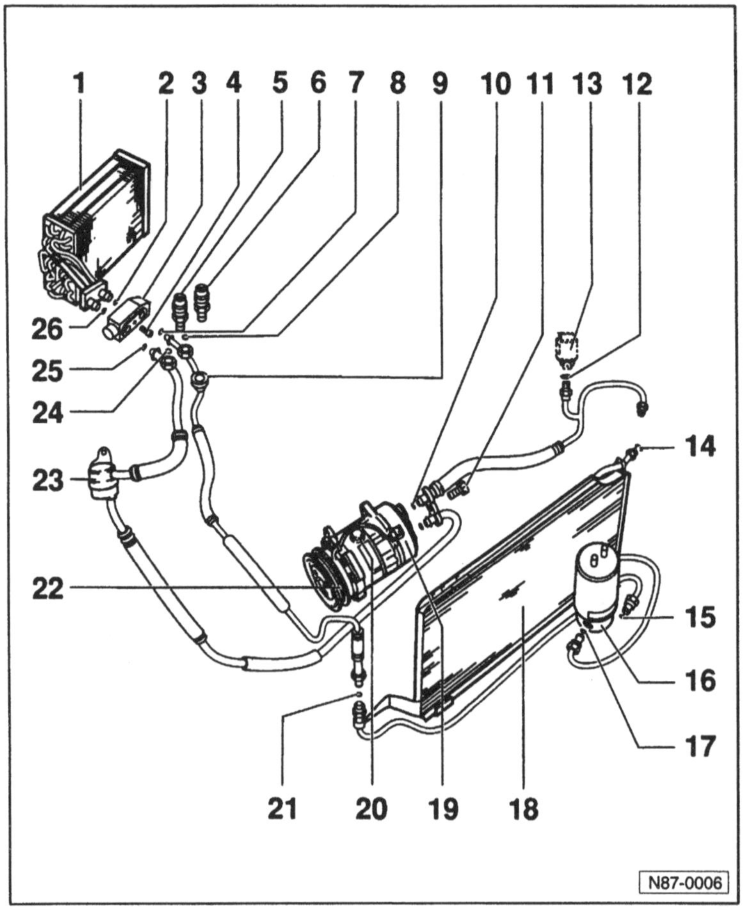

4 - Bolt
- 7 Nm (62 in. lb)

5 - Low pressure service valve
- removing/installing, see Fig.1
- for charging and discharging system
- discharge system using refrigerant recovery/ recycling/ recharging unit **Kent-Moore ACR4** or equivalent; do not release refrigerant into the air

6 - High pressure service valve
- removing/installing, see Fig. 2
- for charging and discharging system
- discharge system using refrigerant recovery/ recycling/ recharging unit **Kent-Moore ACR4** or equivalent; do not release refrigerant into the air

7 - O-ring
- 7.6 mm x 1.8 mm

8 - O-ring
- 10.8 mm x 1.8 mm

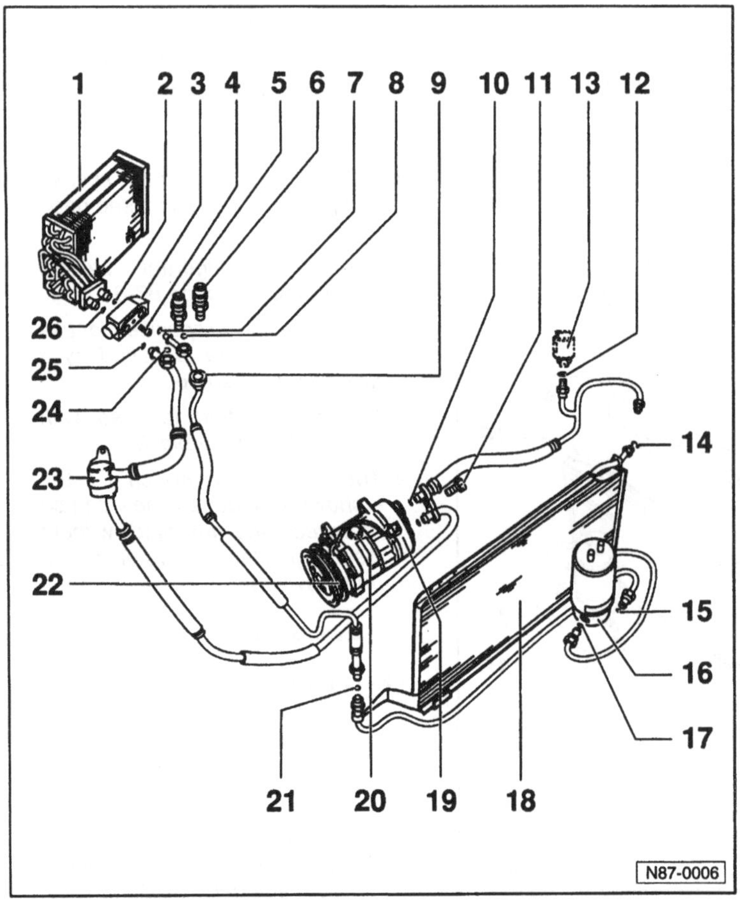

9 - Sight glass
- mixture of refrigerant and refrigerant oil may appear milky
- do not use sight glass when recharging system since refrigerant and oil mixture may appear as foam during recharge
- not installed as of 1994 m.y.

10 - O-ring, Qty: 2
- 20.5 mm x 1.8 mm

11 - Bolt
- 30 Nm (22 ft lb)

12 - O-ring
- 10.8 mm x 1.8 mm

13 - A/C pressure switch -F129
- may be removed without discharging A/C system
- ensure O-ring is properly seated in groove
- 8 Nm (71 in. lb)
- function, see page 87-40

14 - O-ring, in 3/4 in. threaded connection
- 10.8 mm x 1.8 mm
- threaded connection torque: 30 Nm (22 ft lb)

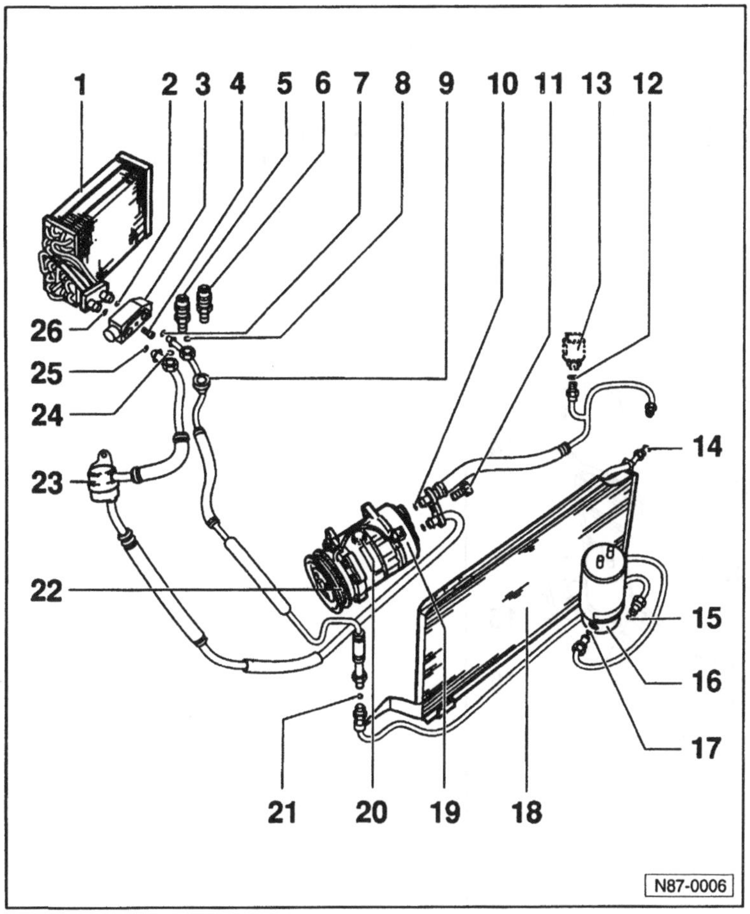

15 - O-ring, in 5/8 in. threaded connection
- 7.6 mm x 1.8 mm
- threaded connection torque: 15 Nm (11 ft lb)

16 - Receiver drier
- replace whenever A/C system is discharged
- to replace, remove bumper and upper radiator support

17 - O-ring, in 3/4 in. threaded connection
- 10.8 mm x 1.8 mm
- threaded connection torque: 30 Nm (22 ft lb)

18 - Condenser
- to replace, remove bumper and upper radiator support
- removing/installing, see page 87-81

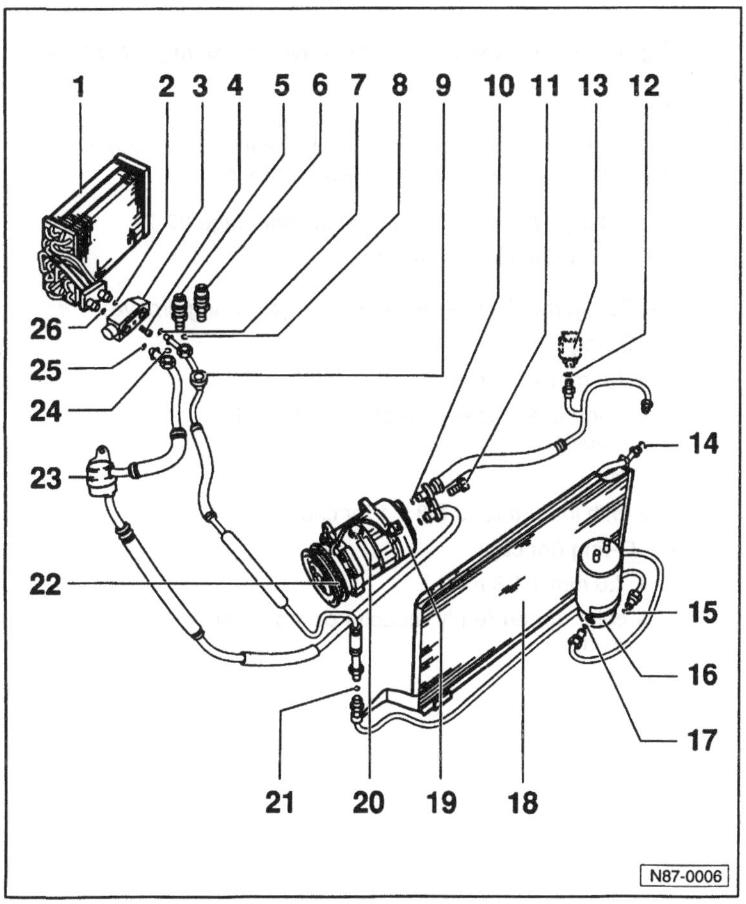

19 - Compressor (Sanden SD7-V16, SD7-V16L)
- variable displacement compressor maintains suction (low) pressure at 2 bar (29 psi) in the evaporator
- if replacing, note directional rotation of compressor shaft and ensure replacement compressor is the same
- engine code AAA, 2.8L VR6 engine: compressor rotation direction (clockwise) as noted on compressor label

20 - Oil drain plug
- oil capacity, 115 ± 15 cc see page 87-64

21 - O-ring, in 5/8 in. threaded connection
- 7.6 mm x 1.8 mm
- threaded connection torque: 15 Nm (11 ft lb)

22 - A/C Clutch -N 25
- A/C clutch, assembly (same as on Sanden compressor SD7-H15), see page 87-52

23 - Damper

24 - O-ring
- 7.6 mm x 1.8 mm

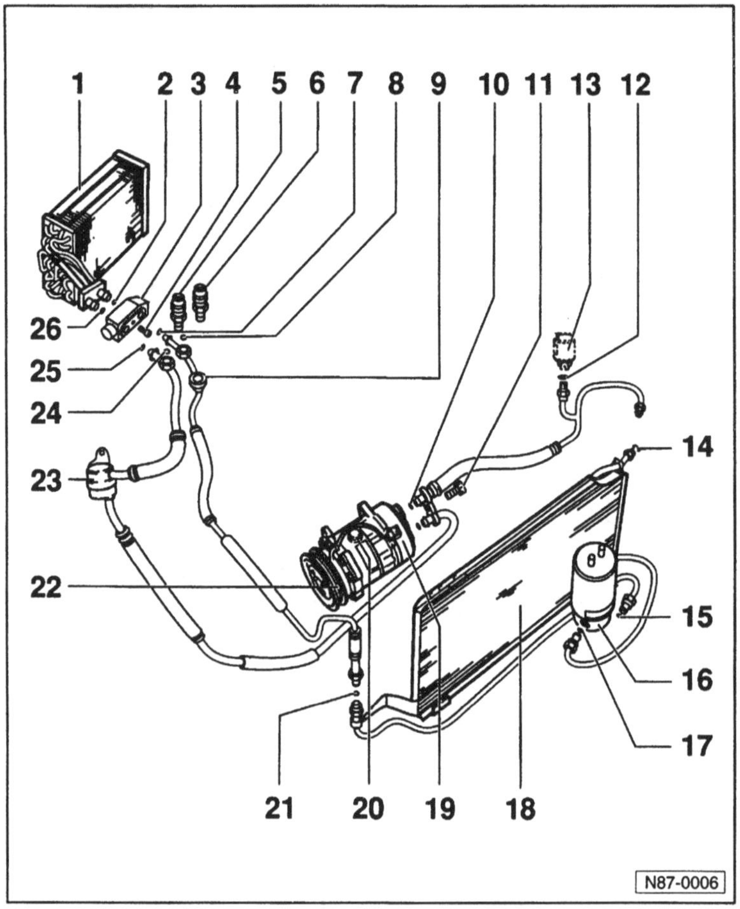

25 - O-ring

♦ 10.7 mm x 1.8 mm

26 - O-ring

♦ 14.0 mm x 1.8 mm

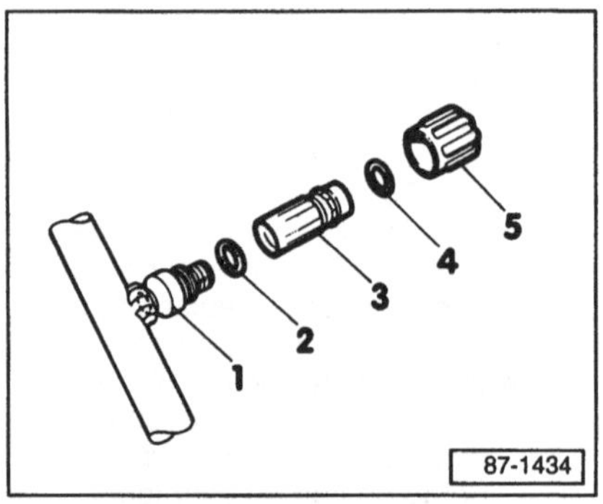

Fig. 1 Low pressure service valve, removing/installing

Note

Low pressure and high pressure service valve connections used on R-134a systems are a different size.

– Discharge A/C refrigerant system, see page 87-71

– loosen valve -3- and remove

1 – Refrigerant line valve connection (w/ O-ring groove)

2 – O-ring

♦ 7.6 mm x 1.8 mm

♦ color: green (early models) or red (later models)

3 – Valve

♦ with inner thread M8 x 1 for cap

4 – O-ring (for cap)

♦ 7.6 mm x 1.8 mm

♦ color: green (early models) or red (later models)

5 – Cap

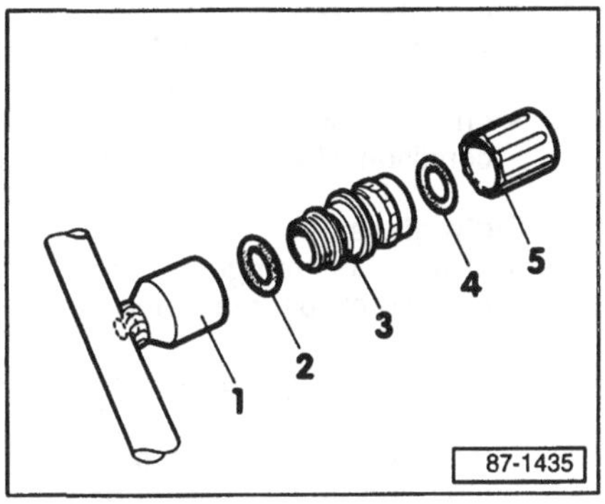

Fig. 2 High pressure service valve, removing/installing

Note

High pressure and low pressure service valve connections used on R-134a systems are a different size.

– discharge A/C refrigerant system, see page 87-71

– loosen valve -3- and remove

1 – Refrigerant line valve connection (socket) with inner thread

2 – O-ring

- ♦ 10.8 mm x 1.8 mm
- ♦ color: green (early models) or red (later models)

3 – Valve

- ♦ with inner thread M 8 x 1 for cap and O-ring groove

4 – O-ring

- ♦ 10.8 mm x 1.8 mm
- ♦ color: green (early models) or red (later models)

5 – Cap

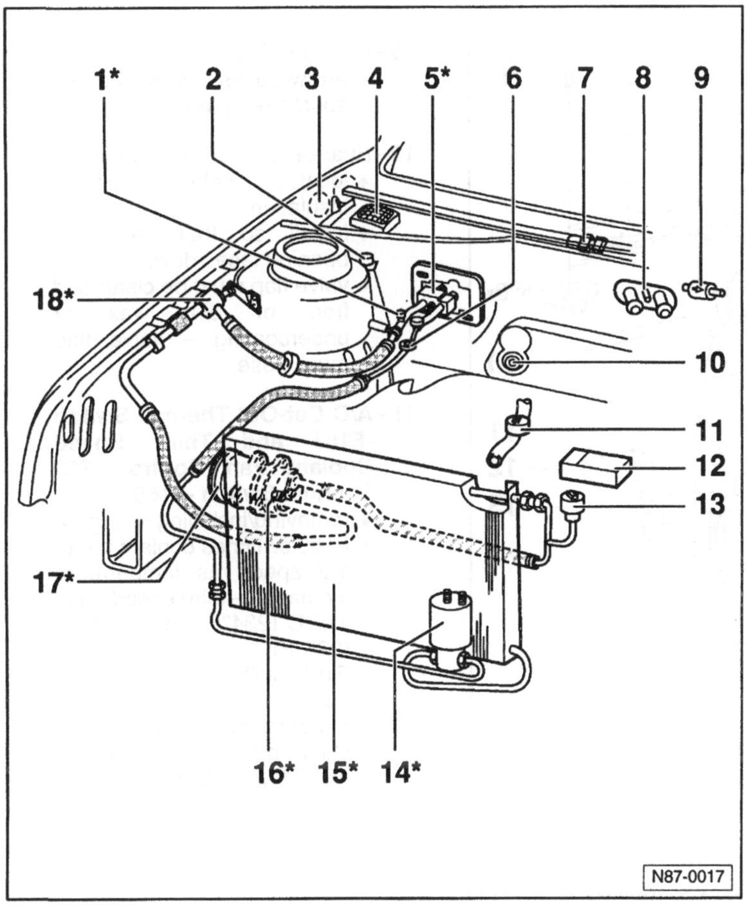

A/C refrigerant system (R-134a) components, in engine compartment

Note

Discharge A/C system before removing components marked with an asterisk (*).

1 - Low pressure service valve*
- ♦ always reinstall cap

2 - Fresh air/recirculating flap two–way valve -N63
- ♦ checking ⇒ Fig. 1
- ♦ valve opens (allows vacuum to servo) when voltage is applied
- ♦ vacuum hose layout ⇒ Page 87-63

3 - Vacuum reservoir
- ♦ insert vacuum hose 30 mm (1 3/16 in.) into reservoir
- ♦ removing/installing ⇒ Fig. 2

4 - Air intake grille
- ♦ with water tight seal
- ♦ dust and pollen filter not available due to limited space

* Discharge A/C system before removing this component.

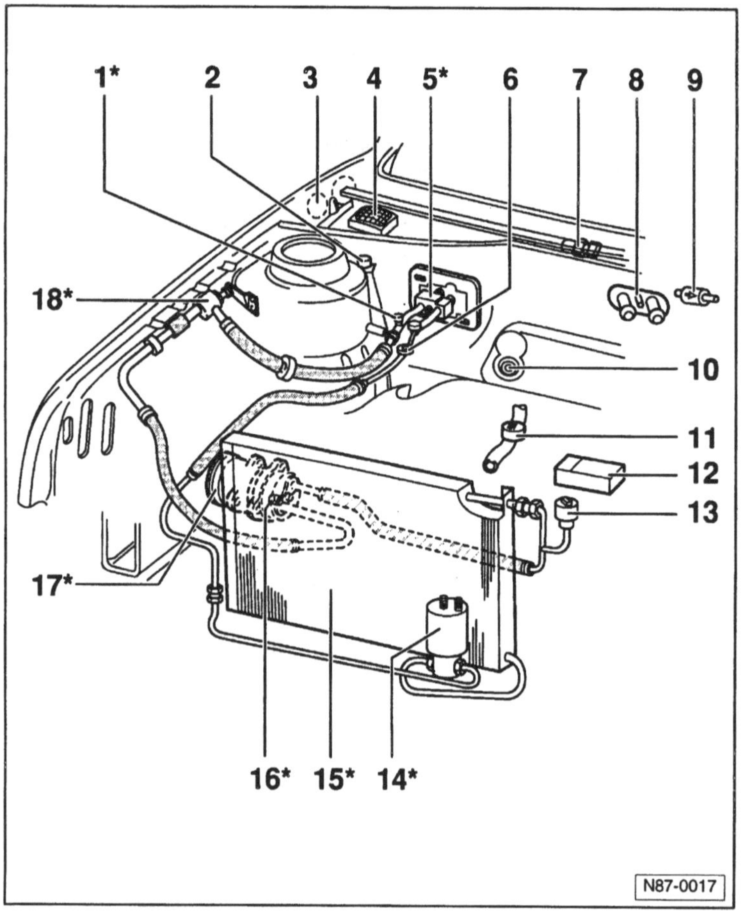

5 - **Expansion valve***
- ensure valve is protected from rain and splash water
- ensure insulator is in good condition, checking ⇒ Fig. 3

6 - **Sight glass**
- mixture of refrigerant and refrigerant oil may appear milky
- do not use sight glass when recharging system since refrigerant and oil mixture may appear as foam during recharge

7 - **Ambient temperature switch -F38**
- switches A/C clutch -N 25 off at low ambient (outside) temperature – A/C clutch off at –1°C (30°F), on at 7°C (45°F)

8 - **Heater core connections and vacuum hose connection**
- vacuum hose layout ⇒ Page 87-63

* Discharge A/C system before removing this component.

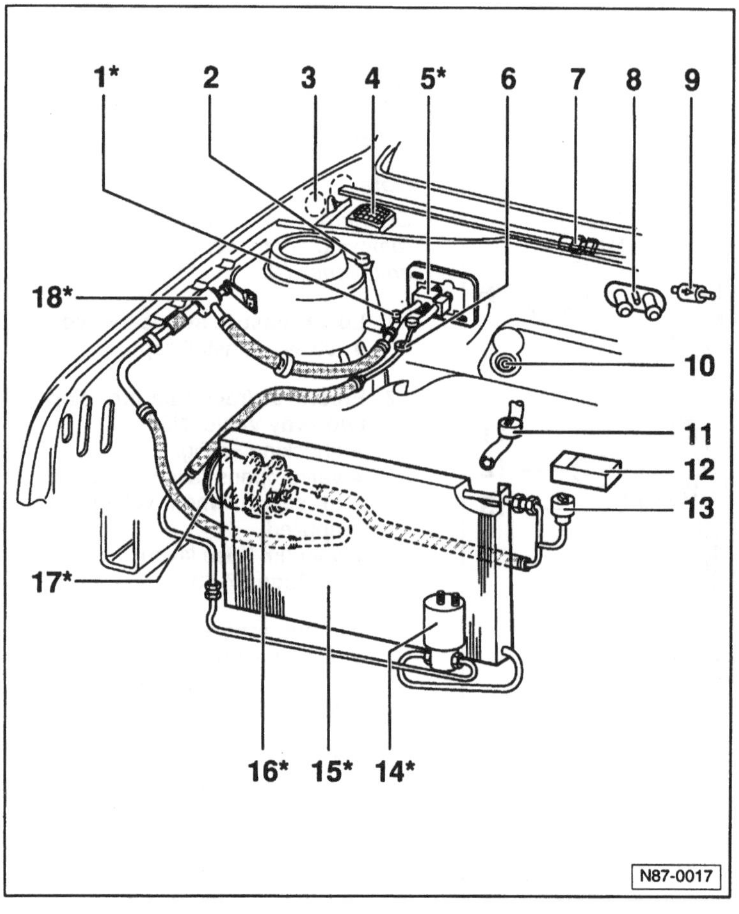

9 - **Check valve**
- arrow points toward vacuum source (engine)

10 - **Evaporator water drain valve**
- located behind firewall insulation
- install so that valve lip opening faces down
- valve flap must be clean and free of dirt, wax or undercoating – valve flap must close

11 - **A/C Cut-Out Thermal Switch -F163 and Third Speed Coolant Fan Control (FC) Thermal Switch -F165**
- removing/installing ⇒ Fig. 4
- F165 switches coolant fan to 3rd speed as temperature increases – third speed on at 112°C (234°F), off at 108°C (226°F)
- F163 switches A/C clutch -N25 off if coolant temperature is too high – A/C clutch off at 119°C (246°F), on at 112°C (234°F)

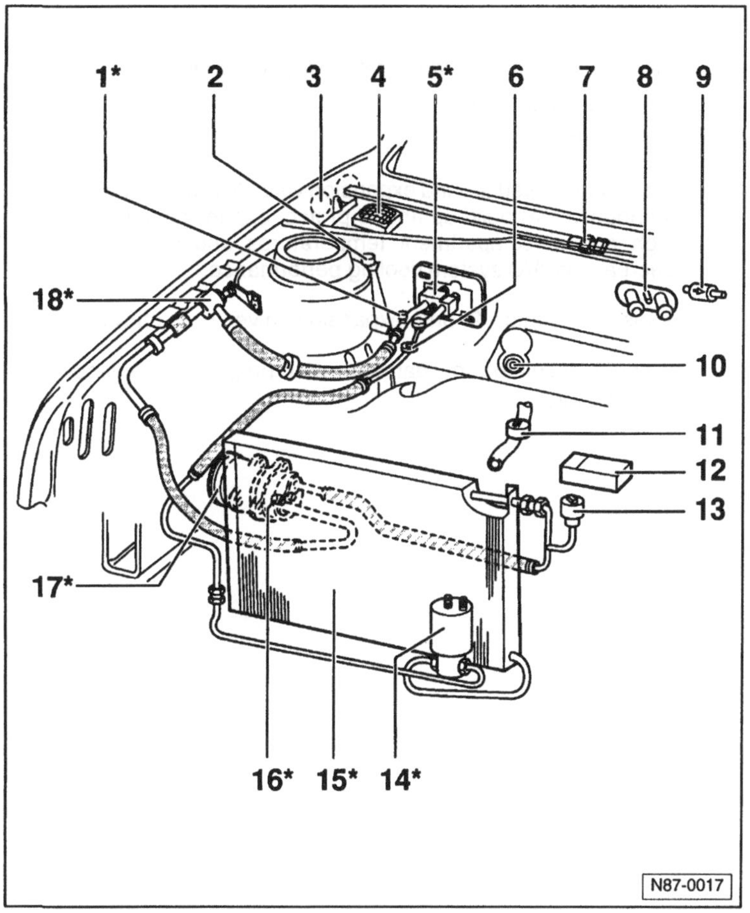

12 - **Coolant Fan Control (FC) Control Module -J293**
- with fuse -S88
- terminal identification, checking ⇒ Fig. 6

13 - **A/C pressure switch -F129**
- functions ⇒ page 87-40
- checking ⇒ Fig. 5
- removing/installing:
  - torque to: 8 Nm (71 in. lb)
  - always replace O-ring (Note Part No.).

14 - **Receiver drier***

15 - **Condenser***

16 - **Pressure relief valve***

17 - **A/C clutch -N 25***

18 - **Damper***

* Discharge A/C system before removing this component.

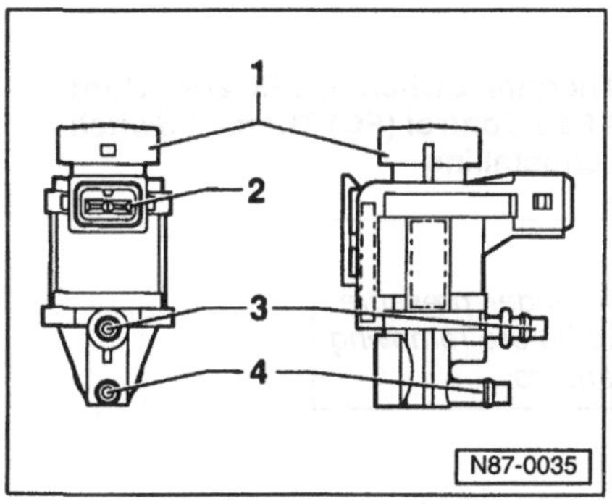

Fig. 1 Fresh air/recirculating flap two–way valve -N 63, checking

1 - Cap, with air vent
Check air vent slot – must not be plugged

2 - 12V (B+) connection

3 - Vacuum hose connection with yellow ring

4 - Vacuum hose connection

To check,

– without voltage (B+) applied: vacuum passage is routed from cap -1- to hose connection -3-

– with voltage (B+) applied: vacuum passage is routed from hose connection -3- to hose connection-4-

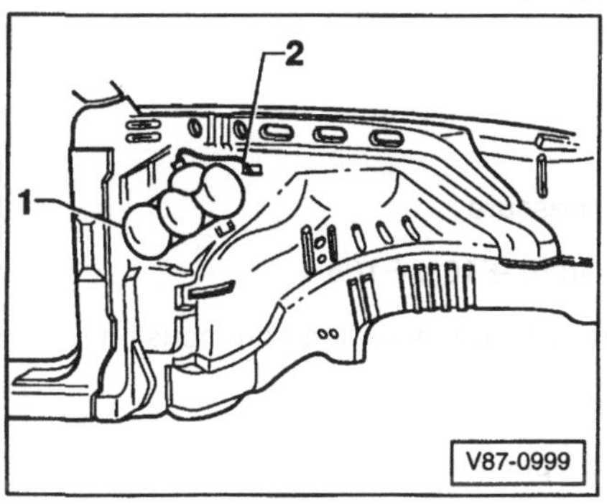

Fig. 2 Vacuum reservoir, removing/installing

1 - Vacuum reservoir

2 - Vacuum hose
insert into reservoir 30 mm (1 3/16 in.)

– remove right wheel housing liner, see Repair Group 66

– remove vacuum hose -2- and and remove reservoir -1-

– install in reverse order of removal

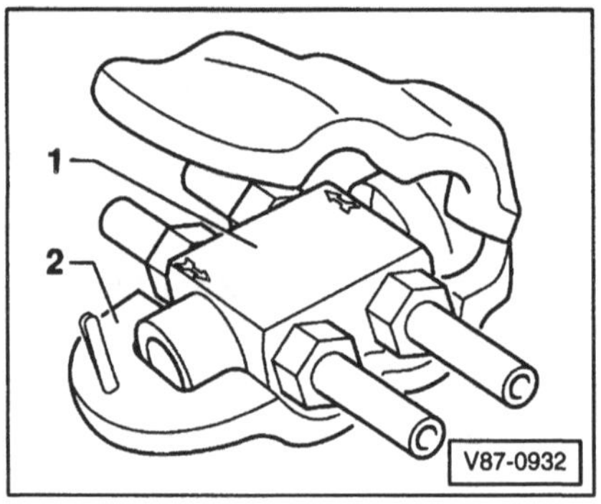

Fig. 3 Expansion valve insulator, checking

1 - Expansion valve

2 - Insulator

Insulation is used around expansion valve to prevent moisture (condensation) build-up and to protect expansion valve from high engine compartment temperatures and the resulting decrease in A/C system cooling performance.

– install insulator -2- onto expansion valve -1- as shown

– check insulator for damage and proper fit, replace if necessary

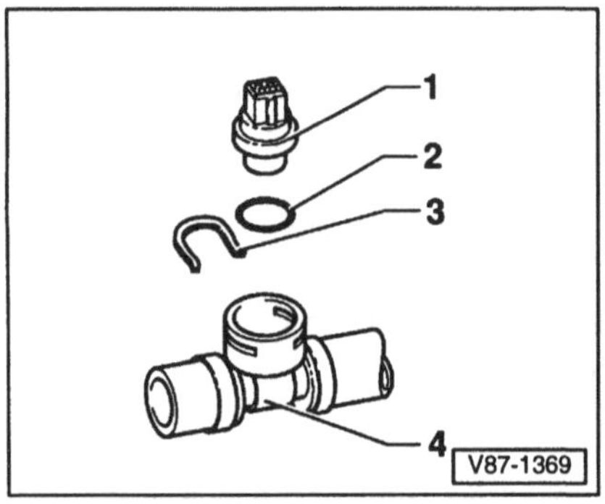

Fig. 4 A/C Cut-Out Thermal Switch -F163 and Third Speed Coolant Fan Control (FC) Thermal Switch -F165, removing/installing

CAUTION

Cooling system may be under pressure. Relieve system pressure before removing any thermal switches/sensors.

Note

Refill engine coolant to proper level after installing thermal switch/sensor.

1 - Thermal Switch -F163/-F165

2 - O-ring

3 - Clip

4 - Coolant hose

– relieve cooling system pressure

– remove wire connector from -F163/-F165

– remove clip -3- and remove thermal switch from housing

– install in reverse order of removal

– refill engine coolant to proper level

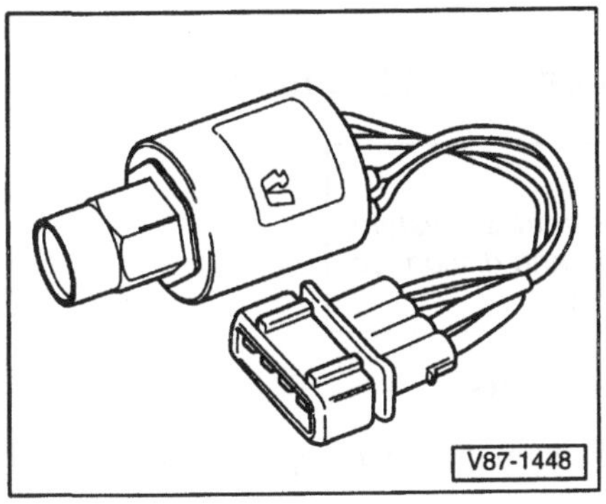

Fig. 5 A/C pressure switch -F129-, checking

- ensure O-ring (color: red, 10.8 mm x 1.8 mm) fits properly in groove
- to remove, unscrew switch from Schrader valve; tightening torque: 8 Nm (71 in. lb)

Functions:

– switches A/C clutch -N 25 OFF if system pressure is too low (low refrigerant charge – switching occurs between terminals 1 (T4b/1) and 2 (T4b/2)

- switch opens (compressor OFF): below 1.2 bar (17.4 psi), switch closes (compressor ON): above 2.4 bar (34.8 psi)

– switches A/C clutch -N 25 OFF if system pressure is too high – switching occurs between terminals 1 (T4b/1) and 2 (T4b/2)

- switch opens (compressor OFF): above 32.0 bar (464.1 psi), switch closes (compressor ON): below 24.0 bar (348.1 psi)

– switches coolant fan -V 7 to second speed when A/C system pressure increases – switching occurs between terminals 3 (T4b/3) and 4 (T4b/4)

- switch closes (2nd speed ON): above 16.0 bar (232.0 psi), switch opens (2nd speed OFF): below 12.5 bar (181.3 psi)

If A/C clutch -N 25 does not engage with A/C ON,

– start engine

– using a jumper wire, briefly bridge terminals1 (T4b/1) and 2 (T4b/2) of A/C pressure switch connector

If A/C clutch -N25 now engages, refrigerant system is discharged (empty),

– check A/C system for leaks, repair as necessary

Note

With A/C Pressure Switch -F129 removed and connector attached, A/C clutch should not engage and A/C should not cool.

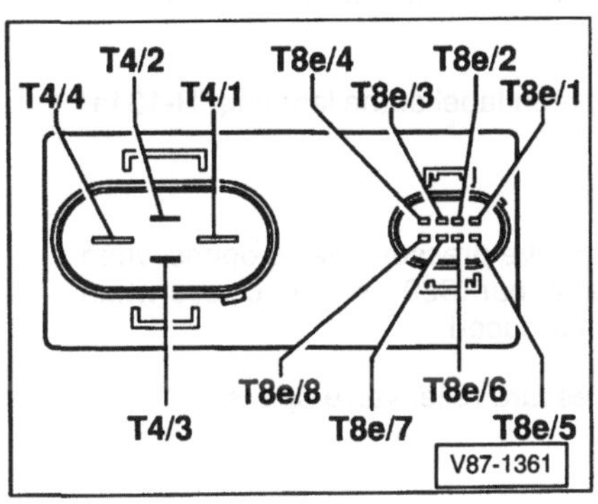

Fig. 6 Coolant Fan Control (FC) Control Module -J293, terminal identification

A/C refrigerant system (R-134a), fuses

> ***CAUTION***
> *Before working on the electrical system, obtain radio security code and disconnect battery Ground (GND) strap.*

Fuses for electrical components

For exact fuse and wiring information, see applicable wiring diagram.

A/C system single fuses in fuse/relay panel:

- S6 – 30A
- S19 – 30A

Coolant fan fuse:

- S88 – Strip fuse, in Coolant Fan Control (FC) Control Module -J 293

  Note Part No. of -J 293 for correct replacement.

A/C refrigerant system (R-134a), valves/switches

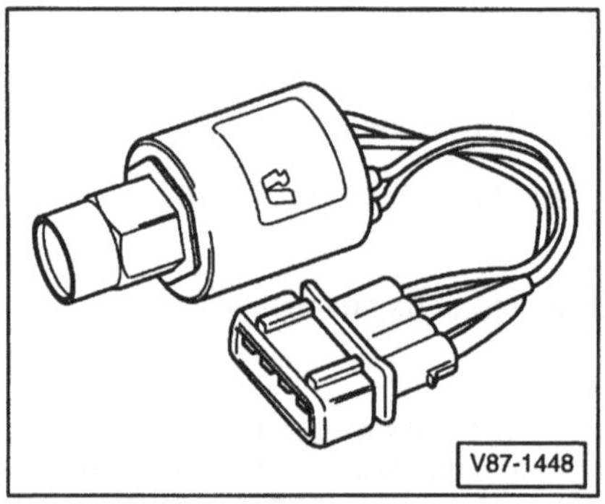

V87-1448

A/C Pressure Switch -F129

Functions:

- switches A/C clutch -N 25 OFF if system pressure is too low (low refrigerant charge) – switching occurs between terminals 1 (T4a/1) and 2 (T4a/2)
- switches A/C clutch -N 25 OFF if system pressure is too high – switching occurs between terminals 1 (T4a/1) and 2 (T4a/2)
- switches coolant fan -V 7 to second speed when A/C system pressure increases – switching occurs between terminals 3 (T4a/3) and 4 (T4a/4)
- switch identified with a green label or the lettering "R-134a"

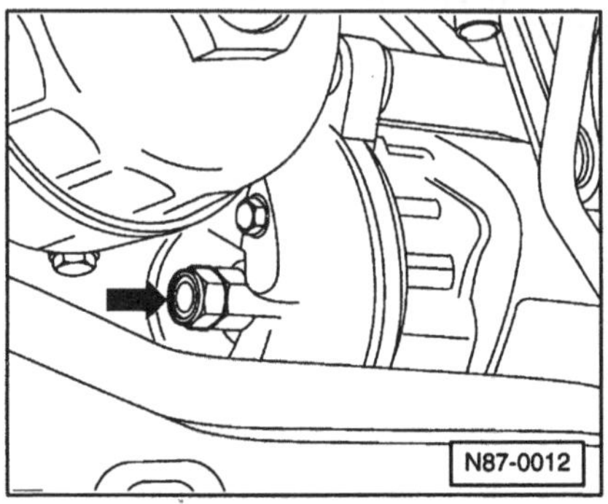

N87-0012

Pressure relief valve

Protects system from excessive pressure; valve opens when system pressure is above 40 bar (580 psi) and closes when system pressure has been reduced.

Cap -arrow- pops out if pressure relief valve opens.

Note

If the cap on the pressure relief valve is pushed out, check system for cause of excessive pressure.

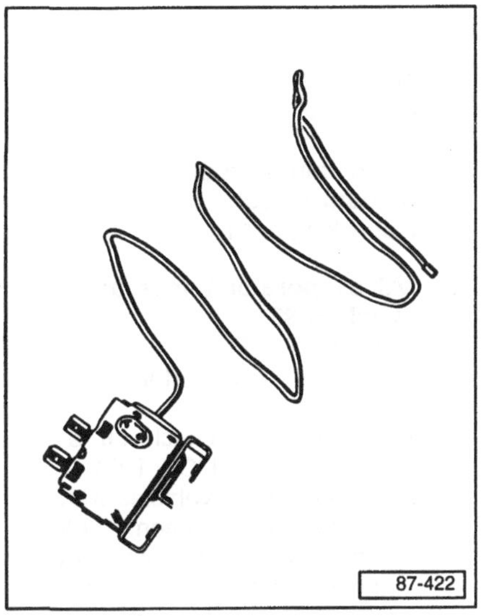

A/C evaporator temperature switch -E33-

Senses evaporator temperature and switches A/C clutch off when evaporator temperature drops to 0°C (32°F) to prevent evaporator icing.

Not required on vehicles with variable displacement, self regulating compressors (these compressors will not allow evaporator temperatures to drop below freezing).

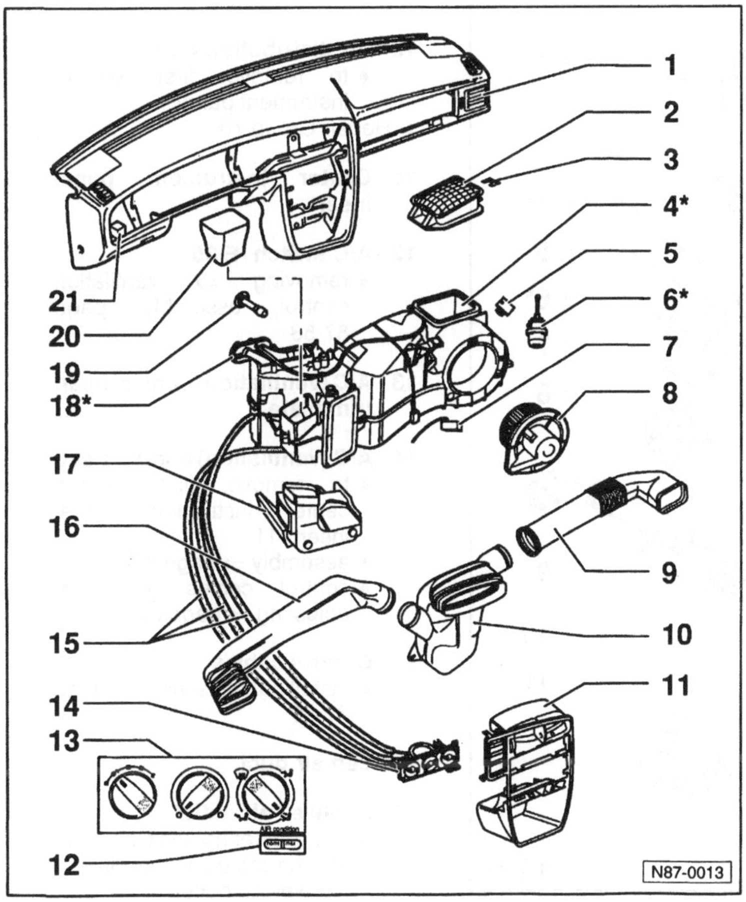

Heater/air conditioner air ducts, assembly

Note

Discharge A/C system before removing components marked with an asterisk (*).

1 - **Instrument panel**
- removing/installing,

⇒ General body repairs, Repair group 70, Instrument panel removing/installing; first remove locking cable from steering lock housing if applicable

2 - **Air intake grille**
- with water tight seal
- dust and pollen filter not available due to limited space

3 - **Air intake grille clamp**

4 - **Air distribution and evaporator housing***
- remove together with cables and controls
- install so that cables are not kinked or sharply bent otherwise control knobs will be difficult to turn

* Discharge A/C system before removing this component.

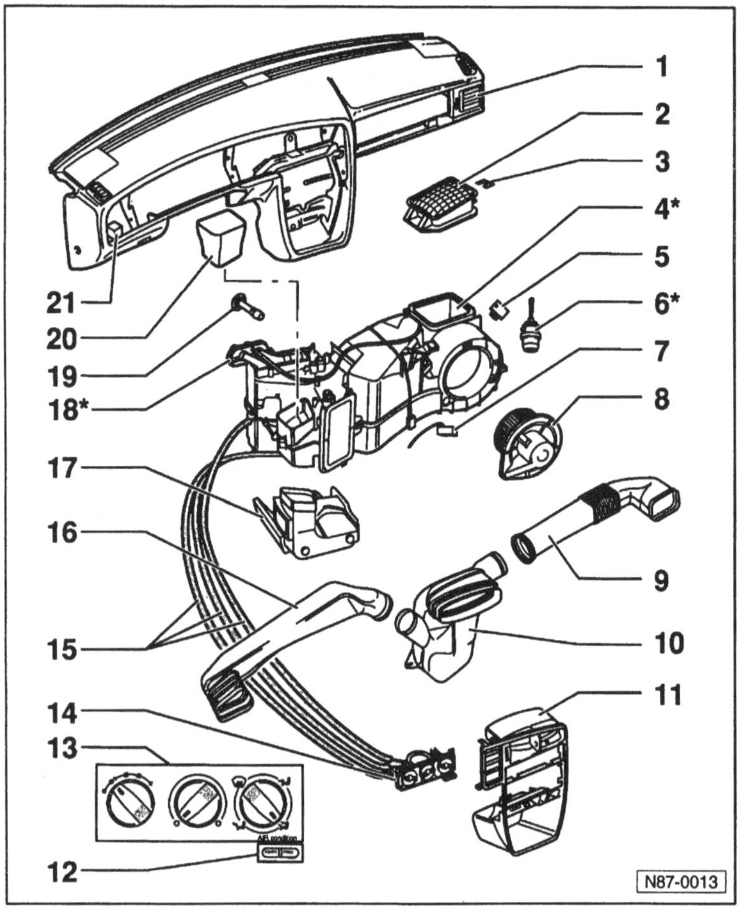

5 - Fresh Air Blower Series Resistance -N 24
- replacing ⇒ Fig. 1

6 - Fresh air/recirculating flap vacuum servo*
- checking ⇒ Fig. 2

7 - A/C evaporator temperature switch -E 33
- checking ⇒ Fig. 3
- prevents evaporator from icing up
- switches A/C clutch off at 34°F ± 2°F (1.1°C ± 1°C)
- not needed on vehicles with variable displacement, self regulating compressors (these compressors will not allow evaporator temperatures to drop below freezing)

8 - Fresh air blower -V 2
- removing/installing ⇒ Fig. 4

9 - Right air duct

* Discharge A/C system before removing this component.

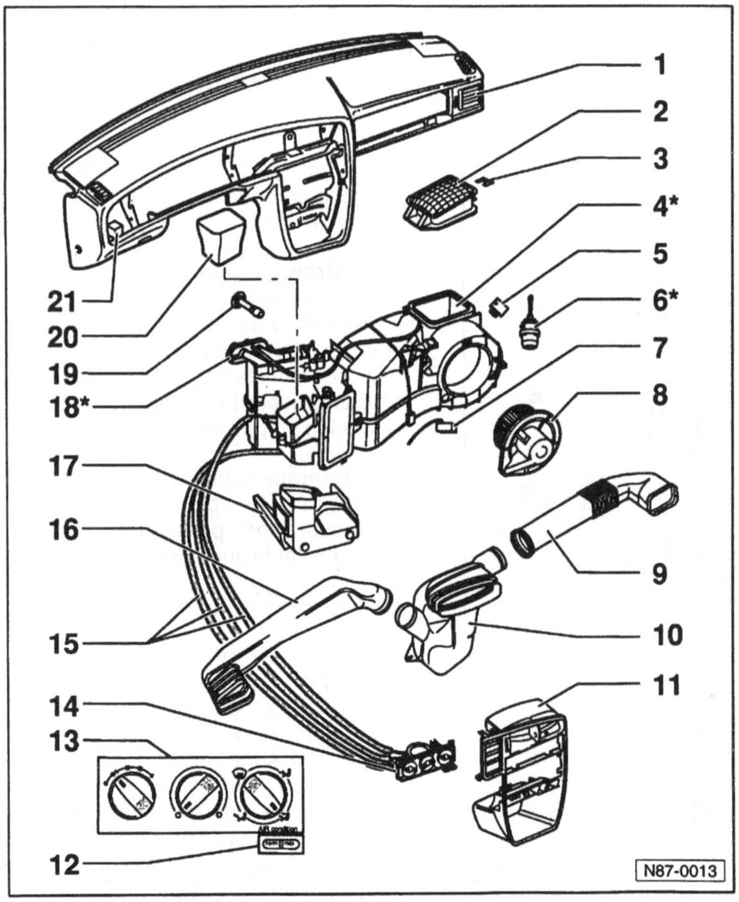

10 - Air distribution duct
- to remove: first remove instrument panel,

⇒ Repair Group 70

11 - Center instrument panel insert

12 - A/C switch -E 35
- removing ⇒ ventilation control, assembly, page 87-58

13 - A/C (ventilation) control head trim plate

14 - A/C (ventilation) control head
- to remove, first remove center instrument panel insert -11-
- assembly ⇒ page 87-58
- control cables, installing/ adjusting ⇒ page 87-59

15 - Control cables
- installing/adjusting ⇒ page 87-59

16 - Left air duct

17 - Footwell air outlet
- route steering lock housing locking cable on tunnel below footwell air outlet

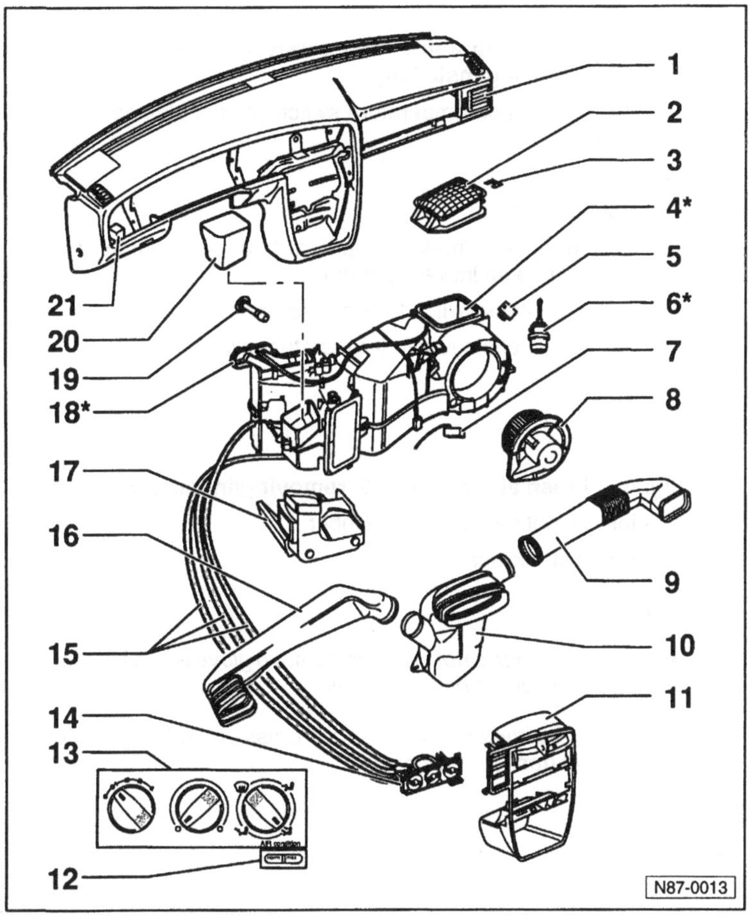

18 - Heater core*
- when removing heater core, drain entire cooling system and refill with new coolant.

19 - Evaporator water drain valve
- to remove, pull from housing
- install so that valve lip opening faces down

20 - Connecting duct

21 - A/C clutch relay -J 44
- stamped with production control number 53

* Discharge A/C system before removing this component.

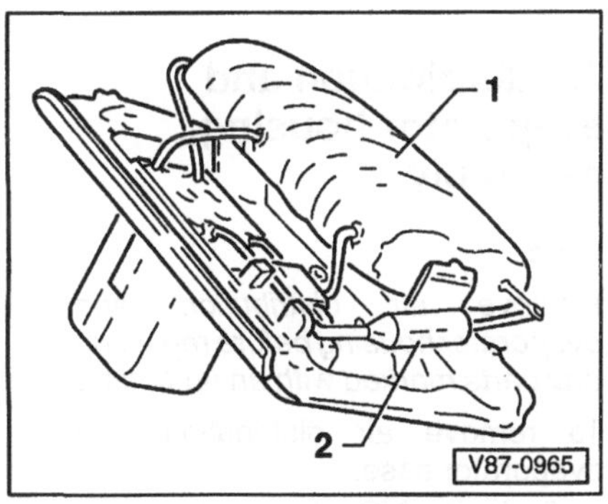

Fig. 1 Fresh air blower series resistance -N 24 with overheat fuse -S 24, removing/installing

1 – Fresh air blower series resistance -N 24

2 – Overheat fuse -S 24

– remove glove compartment

– remove connector and series resistance

– install in reverse order of removal

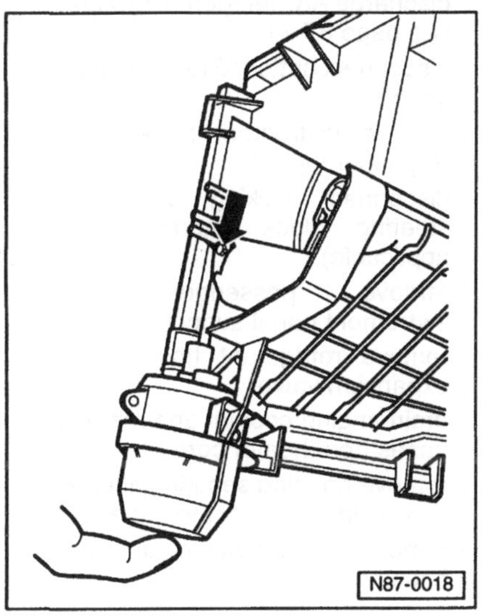

Fig. 2 Fresh air/recirculating flap vacuum servo, checking

Note

For clarity, illustration shows vacuum servo with evaporator housing removed. Vacuum servo can be checked with evaporator housing installed.

– remove vacuum hose from vacuum servo

– push rod into vacuum servo in direction of -arrow- to stop and hold

– using a finger, plug off vacuum servo vacuum line inlet

– release rod

● rod must remain in servo and not extend

If rod extends,

– vacuum servo leaks; replace servo

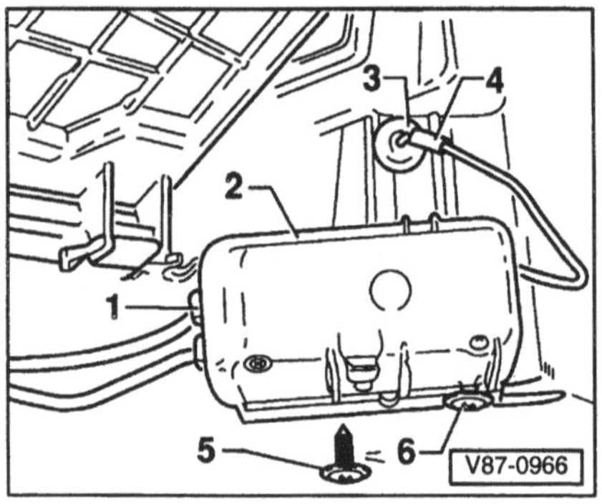

Fig. 3 A/C evaporator temperature switch -E33, removing/installing

1 - A/C evaporator temperature switch (do not kink sensing tube)

2 - Cover

3 - Rubber grommet

4 - Sensing tube – mark with tape at 330 mm (13.0 in.) for insertion depth into evaporator

5 - Screw (fastens switch to evaportator housing)

6 - Screw (fastens cover to temperature switch)

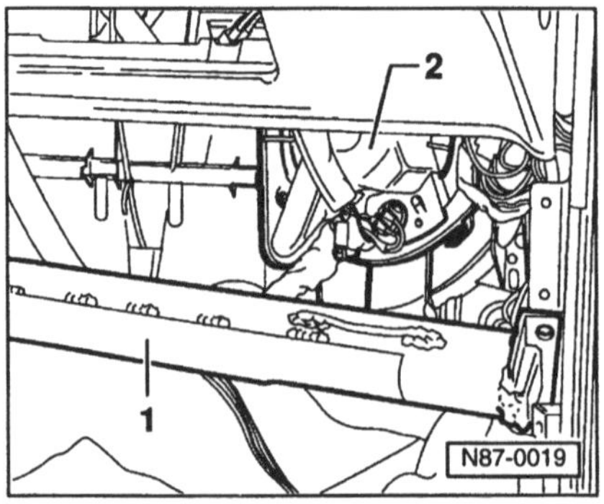

Fig. 4 Fresh air blower -V 2, removing/installing

1 - Instrument panel cross member

2 - Fresh air blower -V2

Note

- *If cross member is not installed, remove glove compartment before removing fresh air blower.*
- *Loosen instrument panel and cross member before removing fresh air blower.*

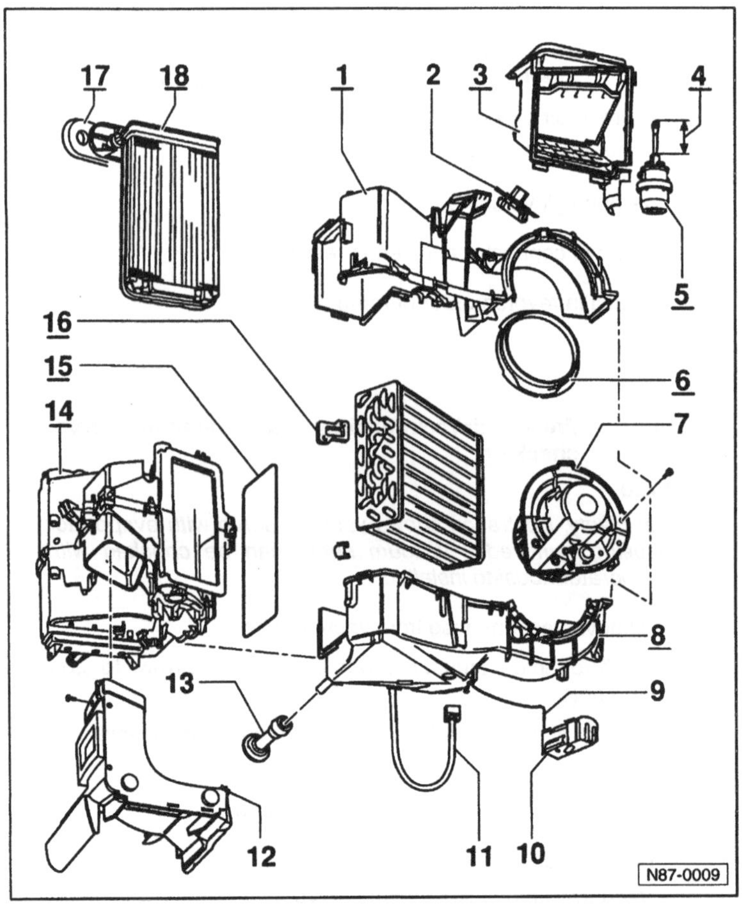

Air distribution and evaporator housing, assembly

Note

Remove air distribution and evaporator housing before replacing any parts marked with an underline.

To remove air distribution and evaporator case:

- discharge A/C refrigerant system, see page 87-71
- disconnect battery Ground (GND) strap
- remove instrument panel, see Repair Group 70
- disconnect locking cable on steering lock housing (if applicable)
- remove passenger side instrument panel support bracket
- remove refrigerant hoses from expansion valve
- drain engine coolant and remove hoses from heater core
- remove housing securing nuts on firewall in engine compartment
- remove air distribution/evaporator housing

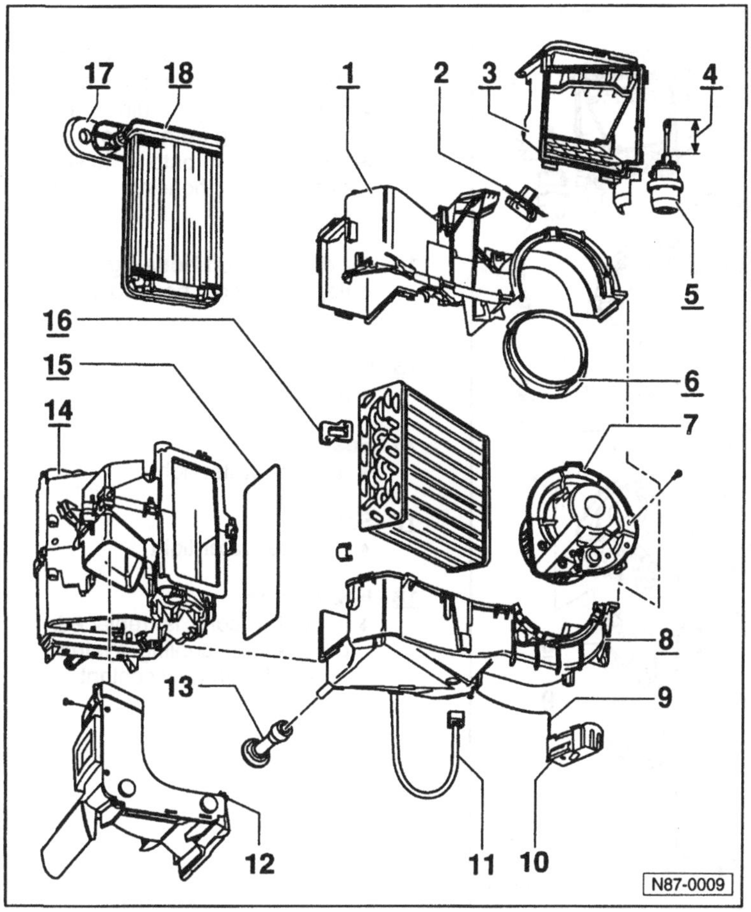

> **CAUTION**
>
> Part numbers are for reference only. Always check with your Parts Department for latest parts information.

1 - Upper evaporator housing
- seal grooves and contact surfaces with silicone rubber sealant Part No. AKD 476 KD5 05 or equivalent

2 - Fresh air blower series resistance with fuse -N 24

3 - Intake duct
- glue seal to ensure water tight connection

4 - Fresh air/recirculation flap vacuum servo rod
- adjust rod to 56 mm (2 13/64 in.)

5 - Fresh air/recirculation flap vacuum servo

6 - Air intake ring

7 - Fresh air blower -V 2

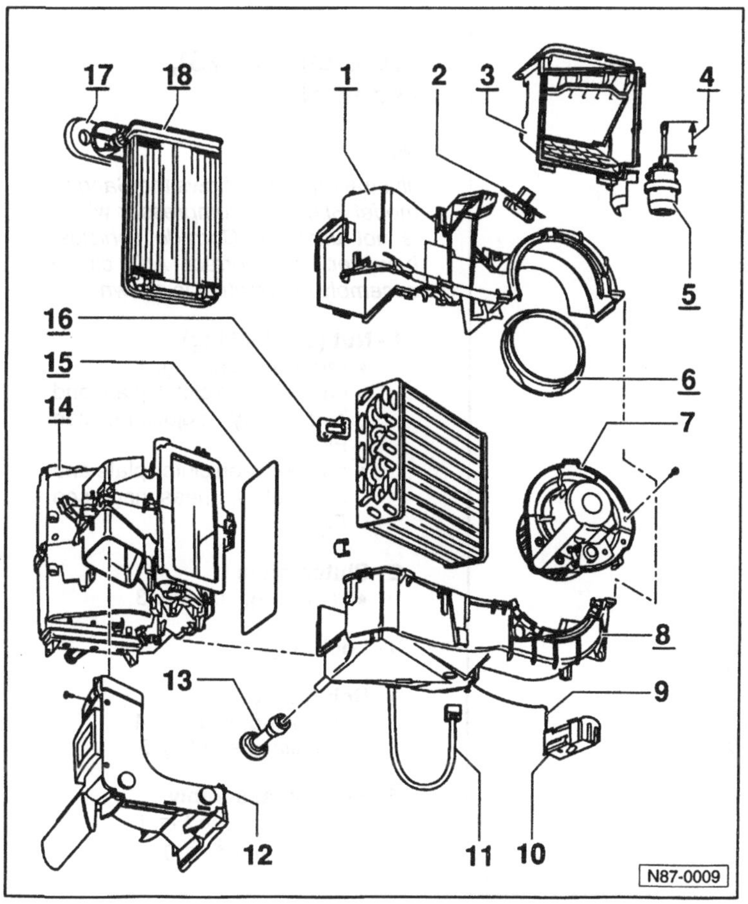

8 - Lower evaporator housing
- seal grooves and contact surfaces with silicone rubber sealant Part No. AKD 476 KD5 05 or equivalent

9 - A/C evaporator temperature switch -E 33 sensing tube

10 - A/C evaporator temperature switch -E 33
- insert sensing tube of temperature switch into evaporator 330 mm (13.0 in.)
- switches A/C clutch off at 34°F ± 2°F (1°C ± 1°C)
- replaceable only with cover and insulation
- prevents evaporator from freezing
- not needed on vehicles with variable displacement, self regulating compressors (these compressors will not allow evaporator temperatures to drop below freezing)

11 - Retaining strap
- replace with Part No. 191 820 845

12 - Footwell vent

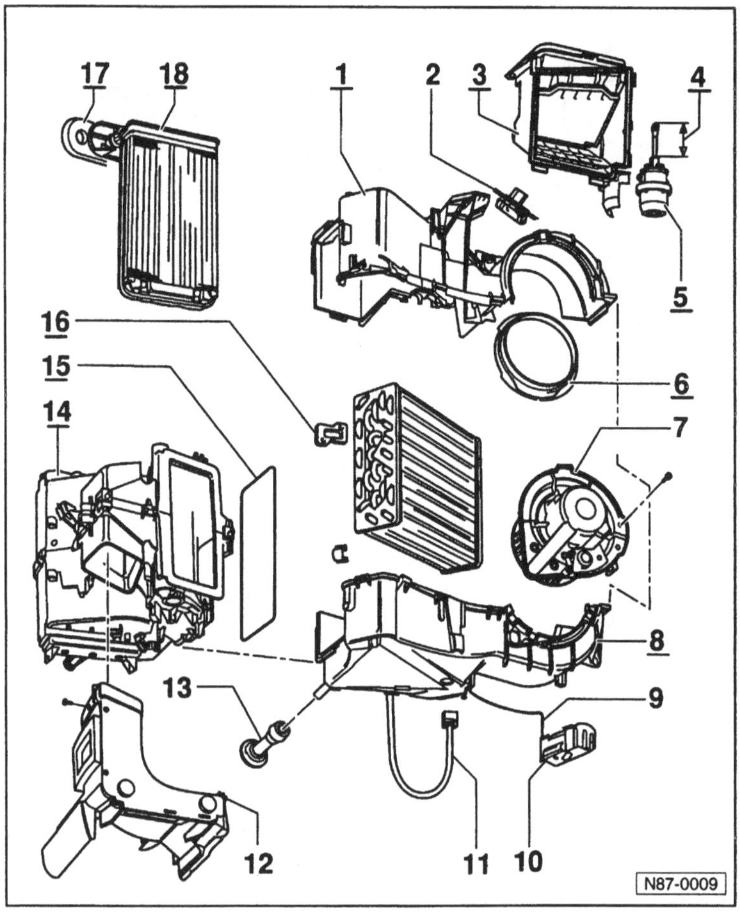

13 - Evaporator water drain valve
- ♦ install with lip facing down
- ♦ remove any dirt or undercoating from valve
- ♦ press fit onto lower evaporator housing (not glued)

14 - Air distribution housing
- ♦ flaps cannot be replaced individually

15 - Seal (round)
- ♦ must be water tight

16 - Seal
- ♦ must be water tight

17 - Seal
- ♦ must be water tight

18 - Heater core
- ♦ if heater core is replaced, drain entire cooling system and refill with new coolant

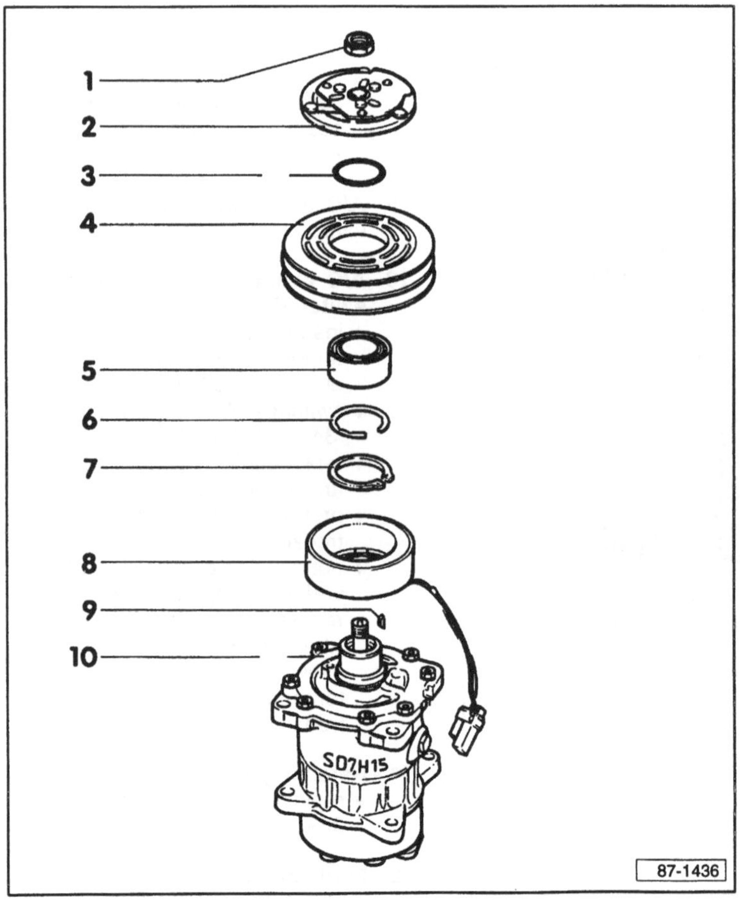

A/C clutch (N 25), assembly

Note

Illustration at left shows the Sanden model SD7 H15 compressor which is not used on Corrado vehicles; however, the correct A/C clutch assembly information is shown.

1 - Nut (self locking)
- ♦ removing, see Fig. 1
- ♦ gap between clutch plate and belt pulley, adjusting, see Fig.1
- ♦ gap between clutch plate and belt pulley, measuring, see Fig. 2

2 - Clutch plate
- ♦ removing, see Fig. 3

3 - Lock ring

4 - Belt pulley
- ♦ removing, see Fig. 4
- ♦ installing, see Fig. 5

5 - Bearing assembly
- ♦ removing, see Fig. 6
- ♦ installing, see Fig. 7

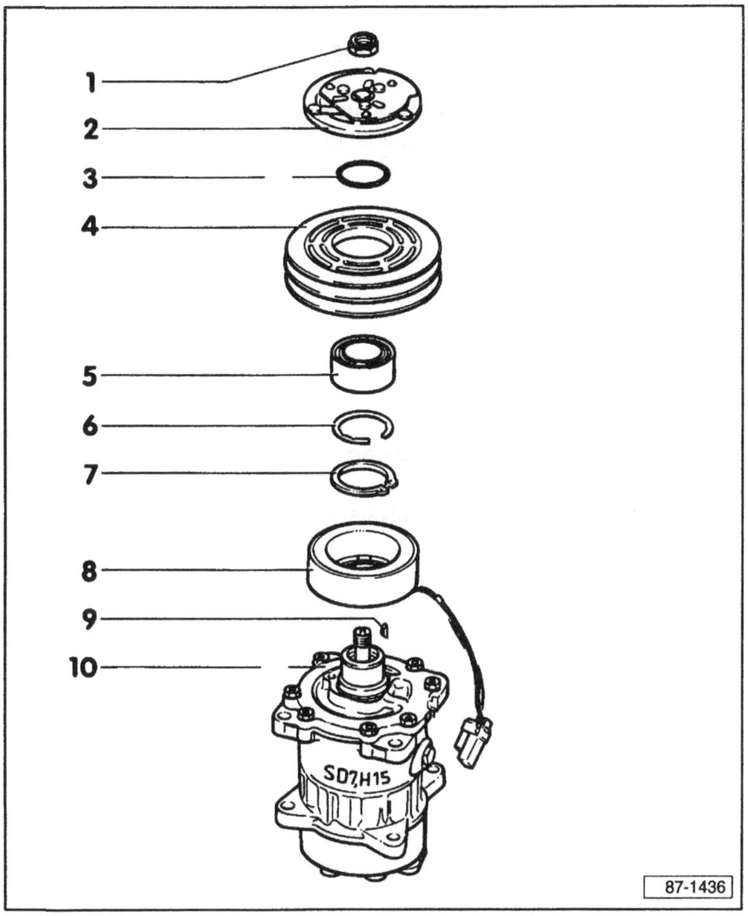

6 - **Snap ring**

7 - **Snap ring**

8 - **A/C Clutch coil**

9 - **Key**

10 - **A/C compressor**
- ♦ refrigerant oil capacity: 115 cc (3.9 oz)

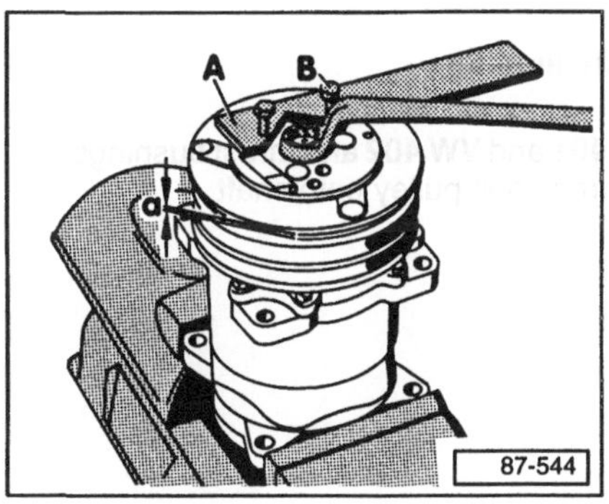

Fig. 1 Clutch plate nut, removing/installing and gap between clutch plate and belt pulley, adjusting

– hold clutch plate using spanner wrench -A- (Mastercool tool No. 90901 or equivalent) and remove clutch plate nut

A - Mastercool tool No. 90901 or equivalent

B - Hex head bolt, 1/4 in. (for tool)

To adjust gap between clutch plate and belt pulley,

– hold clutch plate and tighten clutch plate nut (self-locking) until gap -a- = 0.7 mm (0.028 in. or approx. 1/32 in.)

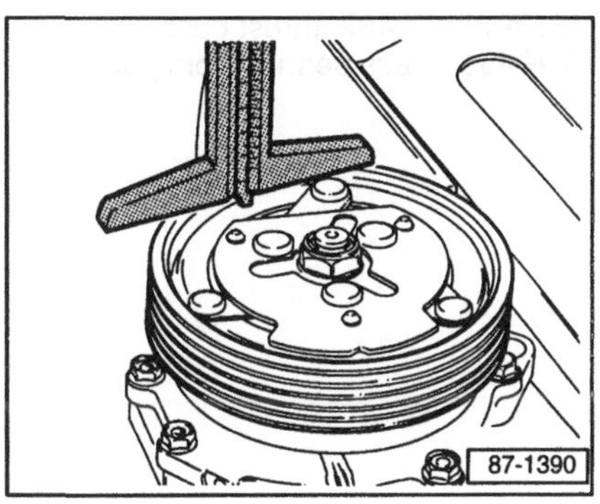

Fig. 2 Gap between clutch plate and belt pulley, measuring

– measure distance between clutch plate and belt pulley as shown

– using a fused 12 Volt power supply, connect 12 V (power (B+) and Ground) to A/C clutch connector

– with clutch engaged, measure distance between clutch plate and belt pulley once again

– difference between first and second measurement is gap

● must be 0.7 mm (0.028 in. or approx. 1/32 in.)

– to adjust gap, see Fig. 1

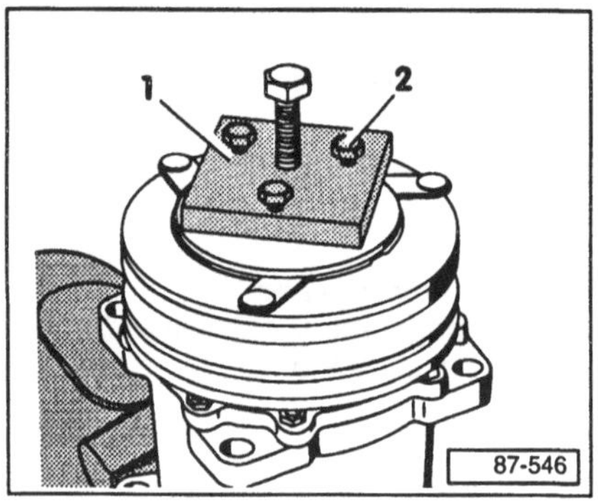

Fig. 3 Clutch plate, removing

- remove front clutch plate using clutch plate puller -1- (Mastercool tool No. 90902 or equivalent – similar tool shown) and 1/4 in. bolts -2-

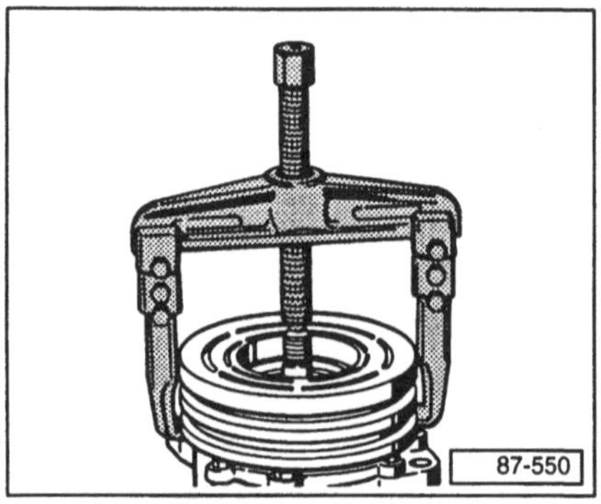

Fig. 4 Belt pulley, removing

- using puller **20/10** or equivalent, remove belt pulley

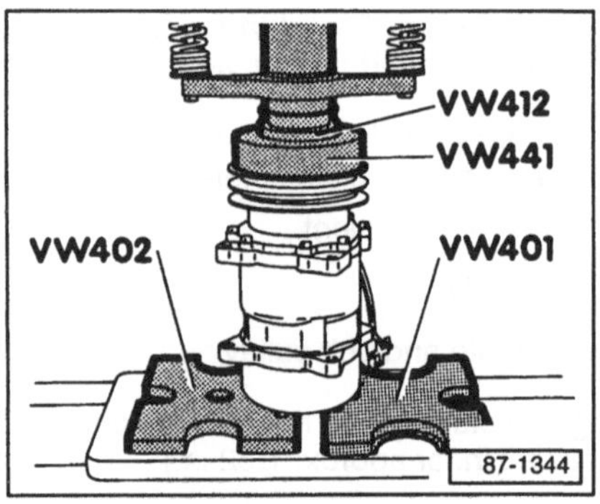

Fig. 5 Belt pulley, installing

- using thrust plates **VW 401** and **VW 402** and thrust bushings **VW 412** and **VW 441** press belt pulley onto shaft

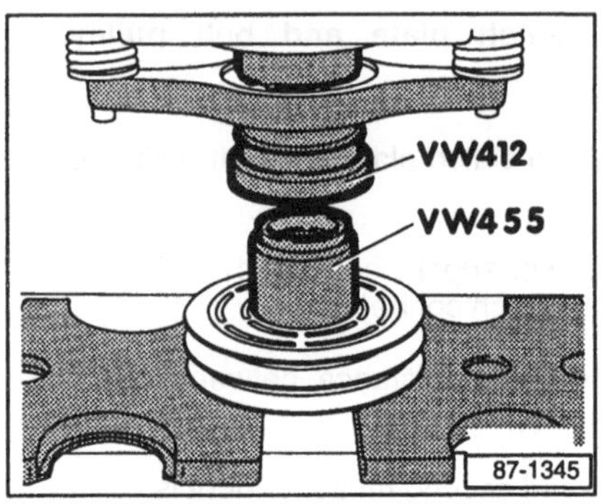

Fig. 6 Bearing assembly, removing

- using thrust plates **VW 401** and **VW 402,** thrust bushing **VW 412** and thrust tube **VW 455**, press bearing from pulley

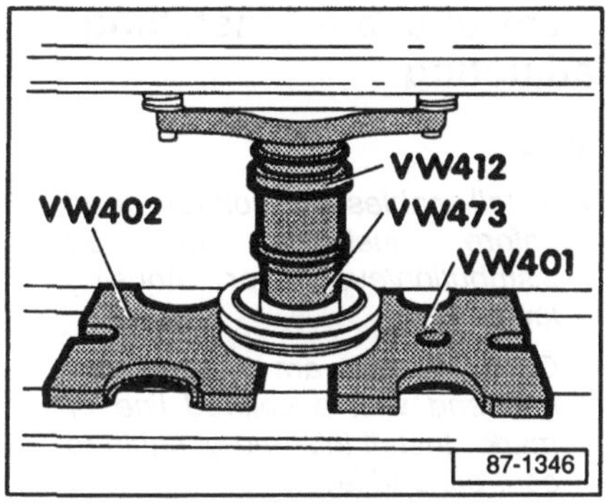

Fig. 7 Bearing assembly, installing

– using thrust plates **VW 401** and **VW 402** and thrust bushings **VW 412** and **VW 473**, press bearing onto pulley until fully seated

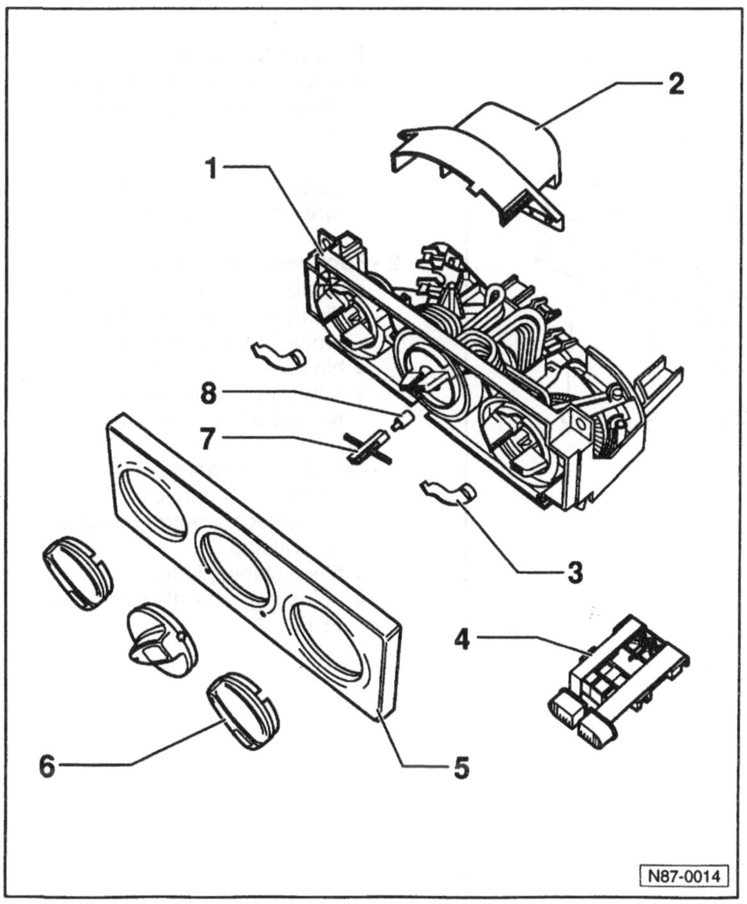

A/C (ventilation) control head, assembly

1 - Control assembly
- with fresh air blower switch -E9

2 - Cover

3 - Light conductor for knob

4 - A/C switch -E35
- to remove, push switch down approx. 0.5 mm (1/64 in.) away from controls and simultaneously slide out

5 - A/C (ventilation) control head trim plate

6 - Control knobs
- to remove, pull off

7 - Light conductor with bulb holder

8 - Fresh air (ventilation) controls illumination bulb -L16

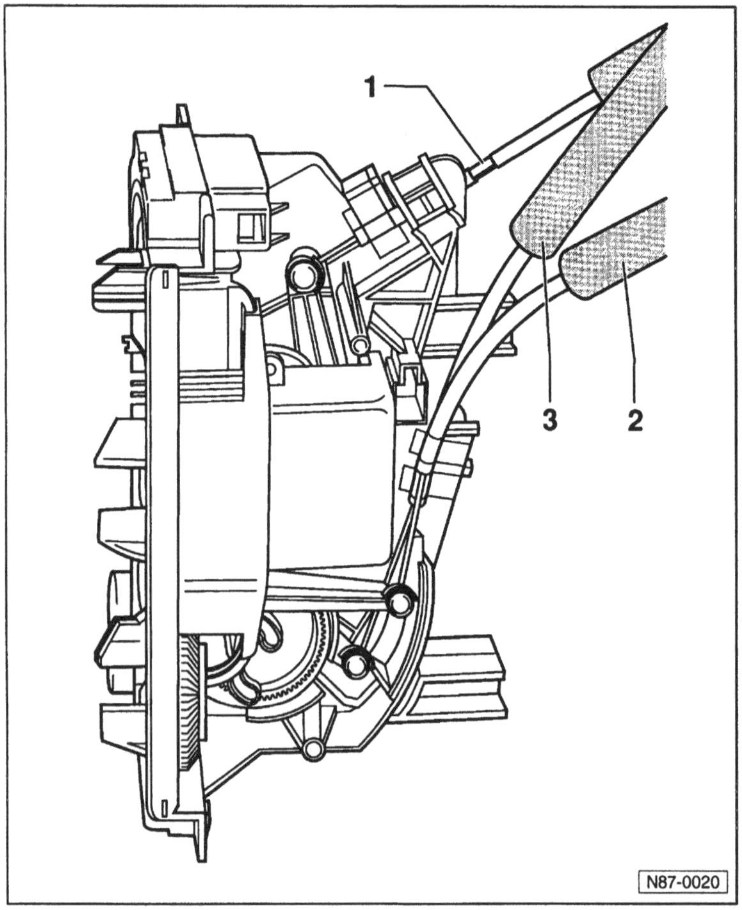

Control cables, installing/ adjusting

Note

- *Install cables to control head before installing on air distribution/evaporator housing levers.*
- *Cable sleeves are identified on one end with a colored line or mark. Install this end of cable to control head lever.*
- *Push cable sleeve against control head stop then secure with clip.*
- *Adjust cables at flap levers with control head installed.*
- *All flaps must audibly contact stops when controls knobs are turned.*

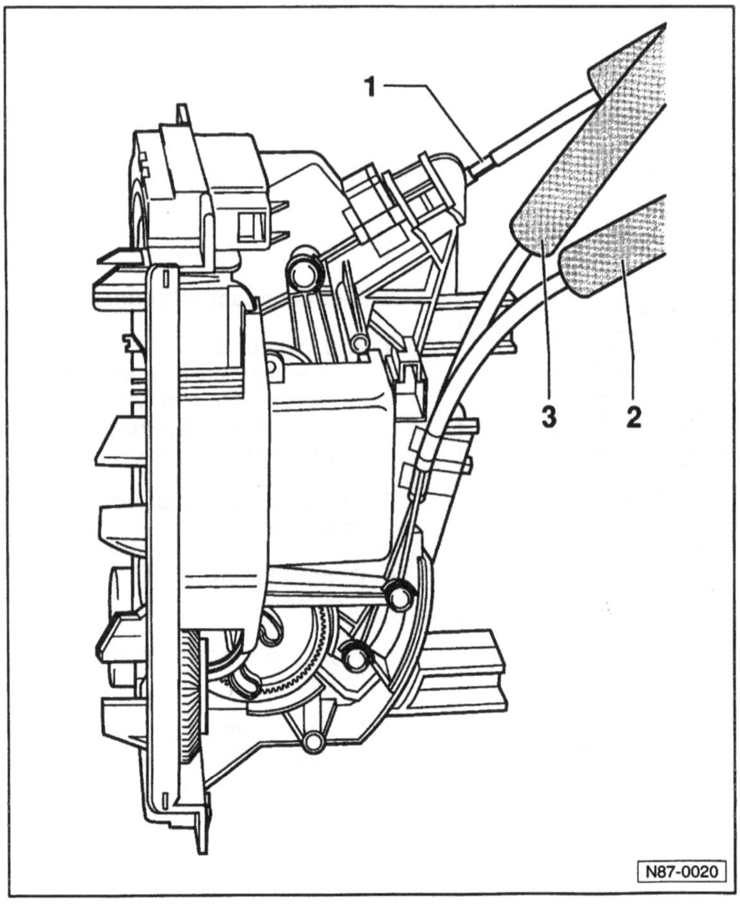

1 - Temperature flap cable
- cable connects temperature control knob to temperature control flap
- blue sleeve
- adjusting cable at temperature control flap ⇒ Fig. 1

2 - Footwell/defroster flap cable
- cable connects air distribution control knob to footwell/defroster flap
- black sleeve, short
- adjusting cable at footwell/ defroster flap ⇒ Fig. 2

2 - Central flap cable
- cable connects air distribution control knob to central flap
- black sleeve, long
- adjusting cable at central flap ⇒ Fig. 3

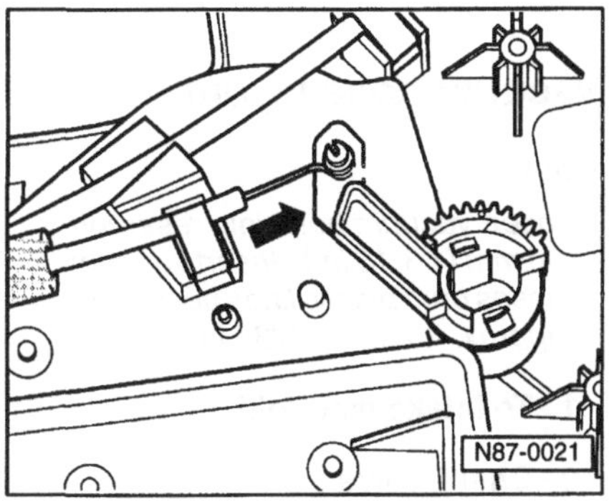

Fig. 1 Temperature flap cable, adjusting

- install A/C (ventilation) control head
- adjust temperature control knob to full cold (against stop)
- connect temperature flap cable (blue sleeve) to temperature flap lever
- push lever in direction of -arrow- to stop
- hold lever in this position and secure cable with cable retaining clip

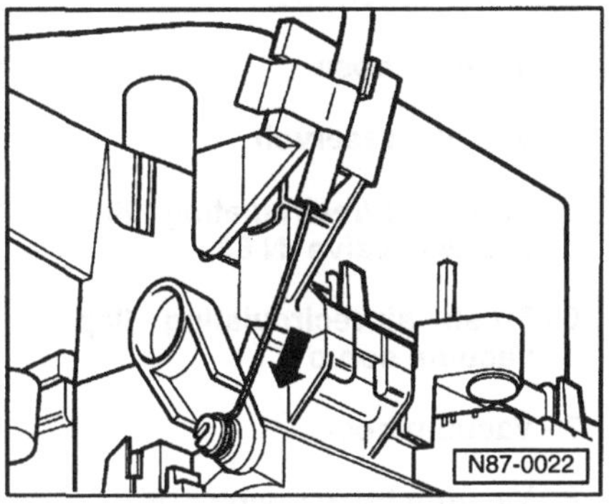

Fig. 2 Footwell/defroster flap cable, adjusting

- install A/C (ventilation) control head
- adjust air distribution control knob to "DEFROST" position (against stop)
- connect footwell/defroster flap cable (black sleeve) to footwell/defroster flap lever
- push lever in direction of -arrow- to stop
- hold lever in this position and secure cable with cable retaining clip

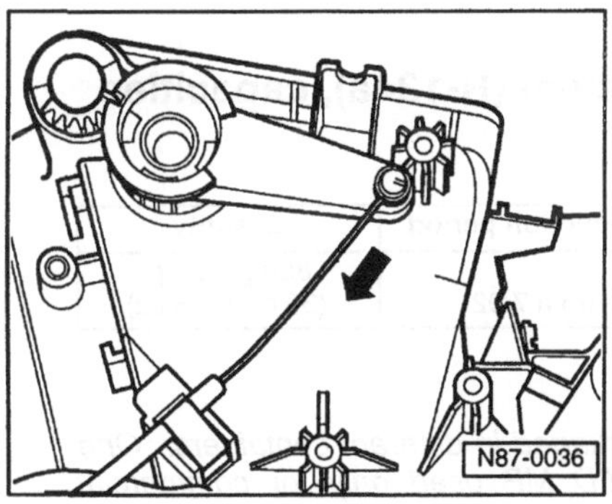

Fig. 3 Central flap cable, adjusting

- install A/C (ventilation) control head
- adjust air distribution control knob to "DEFROST" position (against stop)
- connect central flap cable (black sleeve) to central flap lever
- push lever in direction of -arrow- to stop
- hold lever in this position and secure cable with cable retaining clip

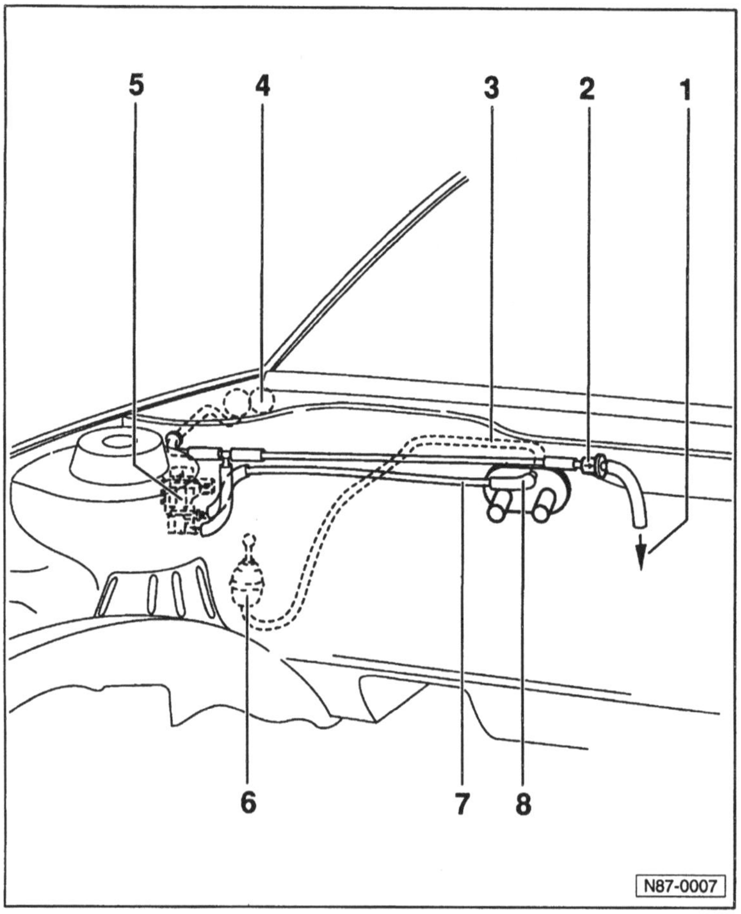

Vacuum hose layout

Note

- *Insert vacuum hose into reservoir 30 mm (1 3/16 in.). Insert vacuum hoses onto plastic lines up to stop, approx. 15 mm (19/32 in.).*

1 - To intake manifold

2 - Check valve

- arrow points toward vacuum source (engine)

3 - Vacuum hose

4 - Vacuum reservoir

5 - Fresh air/recirculating flap two-way valve -N 63

6 - Fresh air/recirculating flap vacuum servo

7 - Vacuum hose

8 - Vacuum hose connector

A/C refrigerant system (R-134a), capacities

Refrigerant R-134a

Model	Production period	Quantity
Corrado	from 7/92	850 g + 50 g (30 oz + 1.8 oz)

Note

R-134a refrigerant is packaged in different containers. One variation is SUVA® COLD MP used only for commercial applications. This type of R-134a is sold in cylinders using a 1/4 flare fitting which will not connect to the 1/2 in.-16 ACME fittings used on the vehicle and servicing equipment. Use only R-134a which comes in containers having the correct type of service fitting.

Refrigerant oil

> ***CAUTION***
>
> *Part numbers are for reference only. Always check with your Parts Department for latest parts information.*

Use only SP-10 refrigerant oil (PAG oil), Part No. G 052 154 A2, or equivalent.

Model	Production period	Quantity
Corrado	from 7/92	115 cc (3.9 fl oz)

Note

Replacement A/C compressors supplied by the Parts Department are filled with 115 cc (3.9 fl oz) of refrigerant oil. This is the total A/C system refrigerant oil capacity.

Refrigerant oil distribution

With the A/C system operating, the refrigerant (PAG) oil is distributed throughout the A/C system as follows:

Compressor	approx. 50% (57.5 cc/1.9 fl oz)
Condenser	approx. 10% (11.5 cc/0.4 fl oz)
Low pressure hose	approx. 10% (11.5 cc/0.4 fl oz)
High pressure hose	---
Evaporator	approx. 20% (23 cc/0.8 fl oz)
Receiver drier	approx. 10% (11.5 cc/0.4 fl oz)

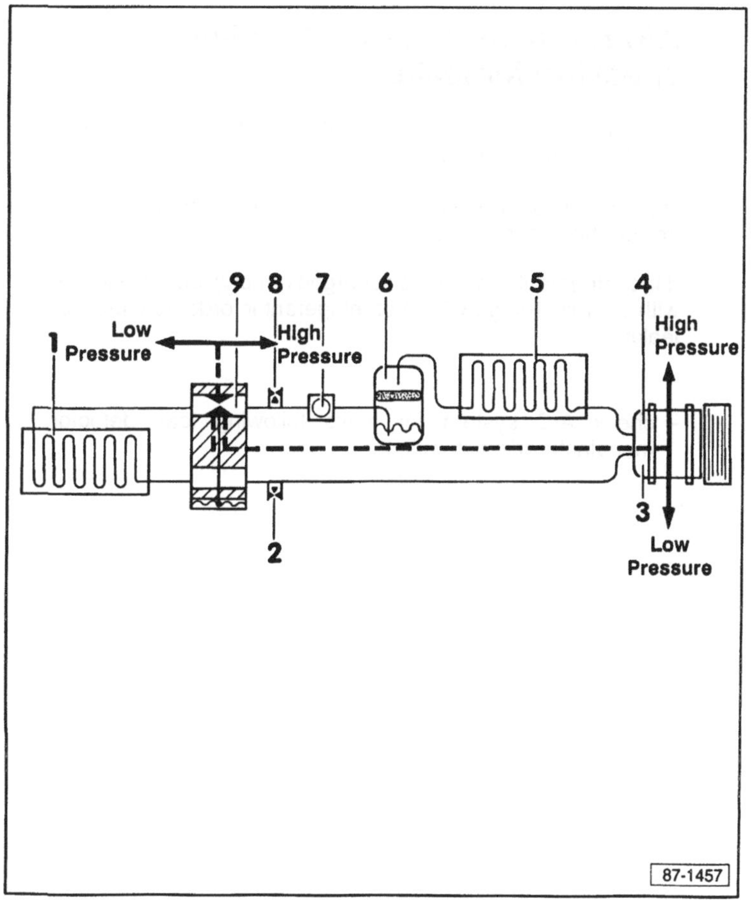

A/C refrigerant system (R-134a) pressures and temperatures, checking

The pressures and temperatures in the A/C system will vary depending on Engine Speed (RPM), coolant fan speed, engine temperature, whether A/C clutch is engaged or not, outside temperature, humidity, etc.

The pressure and temperature specifications given on the next page are based on the following:

- Engine Speed (RPM) at 1500
- fresh air blower on high speed
- A/C switch adjusted to Max. cooling

1 - Evaporator

2 - Low pressure service valve

3 - Compressor, low pressure side

4 - Compressor, high pressure side

5 - Condenser

6 - Receiver drier

7 - Sight glass

8 - High pressure service valve

9 - Expansion valve

Refer to illustration 87-1457 on previous page for component identification.

Component	Refrigerant State	Approximate Pressure (bar/psi)	Approximate Temperature (°C/°F)
1 – Evaporator, inlet to outlet	Vapor	1.2 bar (17.4 psi) *	–7°C (19°F) **
2,3 – Compressor, low pressure side	Gas	1.2 bar (17.4 psi)	–1°C (30°F)
4 – Compressor, high pressure side	Gas	14 bar (203 psi)	65°C (149°F)
5 – Condenser	Gas Vapor Liquid	14 bar (203 psi)	at outlet, 55°C (131°F)
6 – Receiver drier	Liquid	14 bar (203 psi)	55°C (131°F)
7 – Sight glass	Liquid	14 bar (203 psi)	55°C (131°F)
8 – Service valve, high pressure	Liquid	14 bar (203 psi)	55°C (131°F)
9 – Expansion valve	Liquid, turns to vapor	inlet, 14 bar (203 psi) outlet, 1.2 bar (17.4 psi)	inlet, 55°C (131°F) outlet, –7°C (19°F)

* This pressure is maintained in the refrigerant system by the variable displacement compressor despite changing conditions and varying Engine Speeds (RPM).

** This temperature is maintained in the refrigerant system by the variable displacement compressor despite changing conditions and varying Engine Speeds (RPM).

A/C refrigerant system (R-134a), checking for leaks

Use halogen leak detector **Hitec HI400A-TEL** or equivalent to check for refrigerant leaks.

Refrigerant gas dissipates very quickly. Avoid drafty or windy areas when checking for leaks.

If the refrigerant system is discharged (empty), charge system with approx. 100 g (3.5 oz) of refrigerant in order to check for leaks.

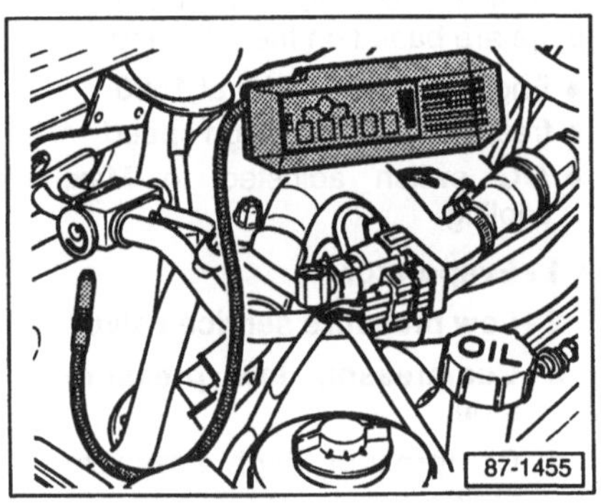

- check A/C system for leaks following leak detector manufacturer's instructions

A/C refrigerant system, testing with pressure gauges

Due to the constant temperature/pressure relationship of R-134a, approximate high side system temperature can be determined based on system pressure.

Pressure gauges may have one or more temperature scales in addition to the pressure scale. Since various refrigerants have different temperature/pressure relationships, each gauge is specific for a particular refrigerant.

Measuring pressure and temperature in a "switched-on" refrigerant system

Use the A/C refrigerant high pressure gauge to measure high pressure between the compressor and restrictor (including condenser) with the A/C switched on.

Use the A/C refrigerant low pressure gauge to measure low pressure between the expansion valve and the compressor inlet (including evaporator) with the A/C switched on.

The temperature displays of the low and high pressure gauges apply only to a portion of the low and high pressure sides. The low pressure gauge displays approximate temperature between the expansion valve and evaporator outlet. The high pressure gauge displays approximate temperature between the compressor outlet and restrictor.

Note

The temperature/pressure relationship only holds true when the refrigerant system contains liquid and vapor, but not when it contains only gas. In the gaseous state, the temperature is approx. 10-30°C (18-54°F) higher than indicated on the pressure gauge.

Measuring pressure and temperature in a closed container or "switched-off" refrigerant system

Due to the constant temperature/pressure relationship of R-134a, approximate temperature of refrigerant in a closed container or in a switched-off A/C system can be determined based on its pressure.

If the temperature displayed on a gauge is lower than the actual temperature of the refrigerant, the refrigerant container or system is empty (discharged).

Note

The temperature/pressure relationship only holds true when the refrigerant system contains liquid and vapor, but not when it contains only gas. In the gaseous state, the temperature is approx. 10-30°C (18-54°F) higher than indicated on the pressure gauge.

A/C refrigerant system (R-134a), discharging

> ***WARNING***
>
> *– Always use an Underwriter's Laboratory (UL) approved refrigerant recovery/ recycling/recharging unit such as* ***Kent-Moore ACR4****, or equivalent, whenever discharging an R-134a A/C system.*
>
> *– Any person who services a motor vehicle air conditioner MUST, by law, be properly trained and certified and use approved refrigerant recycling equipment. Technicians must complete an EPA approved recycling course to be certified.*
>
> *– State and local governments may have additional requirements regarding air conditioning servicing. Always comply with state and local laws.*
>
> *– Always wear safety goggles when charging or discharging an A/C system.*

> ***CAUTION***
>
> *– Always use separate refrigerant recovery/recycling/recharging servicing equipment for R-12 and R-134a systems.* ***DO NOT*** *use one piece of equipment for both R-12 and R-134a systems. The residual traces of refrigerant inside the equipment will contaminate and damage the equipment. Servicing equipment includes recovery/recycling/recharging unit, charging station, vacuum pump, manifold gauges, etc.*
>
> *–* ***DO NOT*** *use R-12 servicing equipment on R-134a systems or R-134a equipment on R-12 systems or damage to both the vehicle A/C system and servicing equipment may result. Use only equipment designed to meet Society of Automotive Engineers (SAE) standards.*
>
> *– R-134a and R-12 systems use different size service fittings.* ***NEVER*** *use adaptors to convert an R-12 fitting to R-134a size or R-134a fitting to R-12 size.*

Note

Follow all safety measures listed on page 87-1 when discharging or charging A/C refrigerant system.

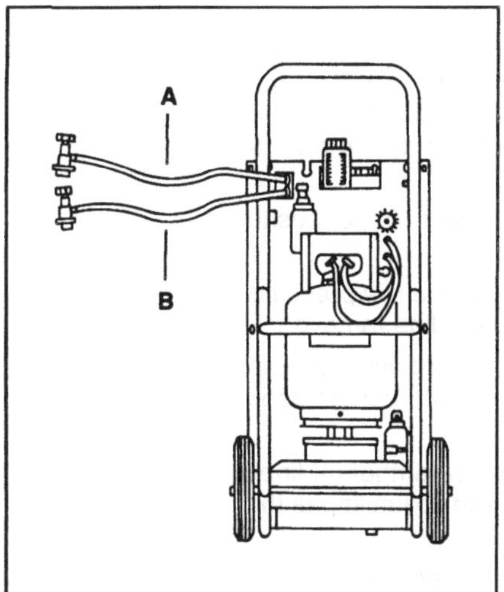

87-A197

Note

Ensure initial set-up has been performed on the refrigerant recovery/recycling/recharging unit before discharging A/C system.

- connect red high pressure hose -A- of refrigerant recovery/recycling/recharging unit to high side fitting on vehicle and open coupler valve

- connect blue low pressure hose -B- of refrigerant recovery/recycling/recharging unit to low side fitting on vehicle and open coupler valve

CAUTION

Always follow manufacturer's instructions when using a refrigerant/recovery/ recycling/recharging unit.

- following refrigerant recovery/recycling/recharging unit manufacturer's instructions, discharge A/C system into refrigerant recovery/recycling/recharging unit

- disconnect power supply from A/C clutch to prevent accidental compressor operation with A/C system discharged

A/C refrigerant system, flushing with compressed air and nitrogen

CAUTION

When using compressed nitrogen always use a pressure regulator and the proper adaptor hoses and fittings (available locally). During flushing, use existing exhaust/ventilation systems to draw off the gas mixture escaping from the A/C system.

CAUTION

DO NOT *flush R-134a refrigerant system with R-11. R-11 is not compatible with R-134a refrigerant and PAG oil and will cause total system contamination.*

- Use compressed air and nitrogen (available locally) to remove moisture, impurities and old refrigerant oil from A/C refrigerant system.
- First blow out old refrigerant oil and dirt with compressed air, then dry components with nitrogen.
- **DO NOT** blow compressed air and nitrogen through the compressor or expansion valve. Only blow compressed air and nitrogen through disconnected, free flowing components (i.e. disconnected hose, condenser, evaporator, etc.)

- **DO NOT** blow compressed air and nitrogen into a capped off A/C component. Pressurized R-134a refrigerant in the presence of oxygen may form a combustible mixture.
- Always flush components in opposite direction of refrigerant flow.
- Flush evaporator through the low pressure line with the high pressure line removed.
- If any component has dark thick deposits that cannot be removed with compressed air, replace component.
- Thin light gray deposits in refrigerant lines and hoses are normal and do not impair the function of the system.
- Always replace receiver drier and expansion valve after flushing.
- Dispose of contaminated refrigerant (PAG) oil following laws governing hazardous waste disposal. Do not combine PAG oil with any other old oils such as engine oil or transmission fluid.
- Flush refrigerant system with compressed air and nitrogen if:
    - refrigerant oil is dark and viscous (thick)
    - too much refrigerant oil is in the system following compressor replacement
    - unclear or do not know how much refrigerant oil is in the system

    - moisture, dirt or other impurities have entered the refrigerant system (i.e. following an accident)
    - unable to pull a constant vacuum during evacuation of a leak-free system due to excessive moisture in the system
    - refrigerant system has been open longer than the time required for normal repairs (i.e. following an accident)
    - based on temperature and pressure measurements, system is diagnosed with moisture contamination
    - compressor is replaced due to noises or internal damage
    - required after replacing certain components in certain situations, see A/C refrigerant system, components, replacing, page 87-79

A/C refrigerant system (R-134a), evacuating and recharging

> ***WARNING***
>
> *– Always use an Underwriter's Laboratory (UL) approved refrigerant recovery/ recycling/recharging unit such as* ***Kent-Moore ACR4****, or equivalent, when evacuating and recharging an R-134a A/C system.*
>
> *– Any person who services a motor vehicle air conditioner* ***MUST****, by law, be properly trained and certified and use approved refrigerant recycling equipment. Technicians must complete an EPA approved recycling course to be certified.*
>
> *– State and local governments may have additional requirements regarding air conditioning servicing. Always comply with state and local laws.*
>
> *– Always wear safety goggles when discharging, evacuating and recharging an A/C system.*

> ***CAUTION***
>
> *Always follow manufacturer's instructions when using a refrigerant/recovery/ recycling/recharging unit.*

Note

Follow all safety measures starting on page 87-1 when evacuating and recharging the A/C refrigerant system.

- follow refrigerant recovery/recycling/recharging unit manufacturer's instructions for evacuating and recharging A/C system
- evacuate refrigerant system for a minimum of 30 minutes
- when recharging A/C system, add correct amount of refrigerant and PAG oil to system, see A/C refrigerant system capacities, page 87-64
- after system recharge, manually rotate A/C compressor approx. 10 turns before starting engine. Start engine with A/C OFF. After idle speed has stabilized, switch A/C ON and let engine idle (compressor running) for a minimum of two minutes before raising engine speed.

A/C refrigerant system (R-134a) components, replacing

Replacement A/C compressors, evaporators and condensers supplied by the Parts Dept. are filled with Nitrogen. If gas (Nitrogen) does not escape when component is first opened, the component may be faulty (leaking), do not install.

Replacement A/C compressors are filled with the total refrigerant oil (PAG oil) quantity needed for the entire refrigerant system.

Always replace the receiver drier whenever the refrigerant system has been opened. Install receiver drier immediately after opening to prevent moisture contamination of drier dessicant. Moreover, keep refrigerant system and other replacement components sealed for as long as possible to minimize the chance of dirt and moisture contamination.

Always plug open refrigerant line connections to prevent dirt and moisture contamination.

If the system has been discharged due to a damaged or leaking component (refrigerant hose/line, compressor, evaporator, condenser, etc.), flush the refrigerant system first with compressed air, then with nitrogen (available locally) and collect the oil that runs out. This will remove the refrigerant oil which may be saturated with moisture.

> ***CAUTION***
>
> ***DO NOT*** *flush R-134a refrigerant system with R-11. R-11 is not compatible with R-134a refrigerant and PAG oil and will cause total system contamination.*

After flushing the system, if the compressor is not replaced new, fill compressor with the correct type and quantity of refrigerant (PAG) oil as specified on page 87-64. Fill compressor with the total system oil capacity.

If the compressor is replaced new, do not add any additional oil as the total amount of oil required is already in the compressor.

Dispose of contaminated refrigerant (PAG) oil following laws governing hazardous waste disposal. Do not combine PAG oil with any other old oils such as engine oil or transmission fluid.

Follow all Warnings and Cautions starting on page 87-1 when working on the R-134a A/C system

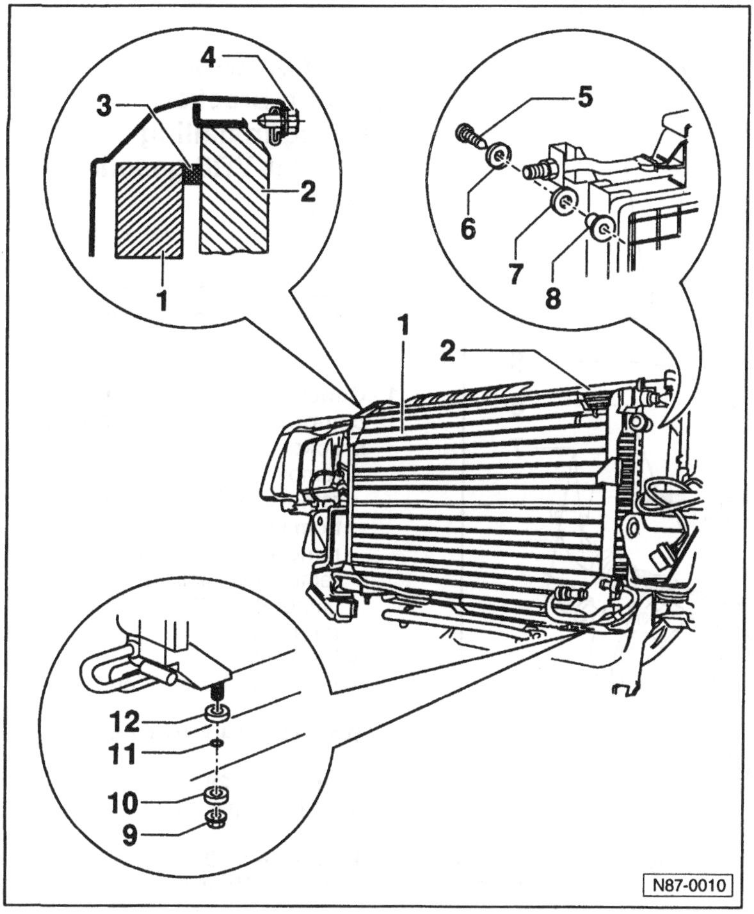

Condenser, removing/installing

Removing

– obtain radio security code and disconnect battery Ground (GND) strap

– remove battery

– discharge A/C refrigerant system, see page 87-71

– remove front lock support and grille

– remove front bumper

– remove air deflectors

– remove refrigerant lines from condenser -1- and plug all open line connections

– remove condenser

– install in reverse order of removal

1 - Condenser

2 - Radiator

3 - Seal

♦ install seal around entire circumference of condenser between condenser -1- and radiator -2-

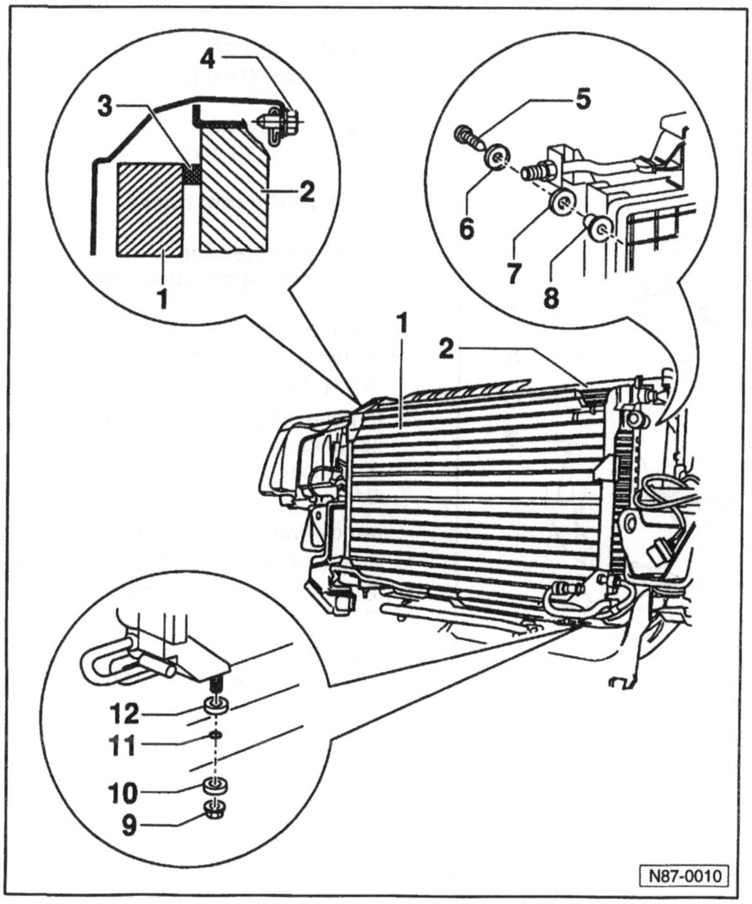

4 - Bolt – M8

♦ 6 Nm (53 in. lb)

5 - Self tapping screw 6.3 X 27

♦ 6 Nm (53 in. lb)

6 - Dished washer

7 - Rubber washer

♦ secured in evaporator bracket

8 - Shouldered sleeve

9 - Lock nut – M6

♦ 6 Nm (53 in. lb)

10 - Rubber washer

11 - Condenser mounting hole

12 - Rubber washer

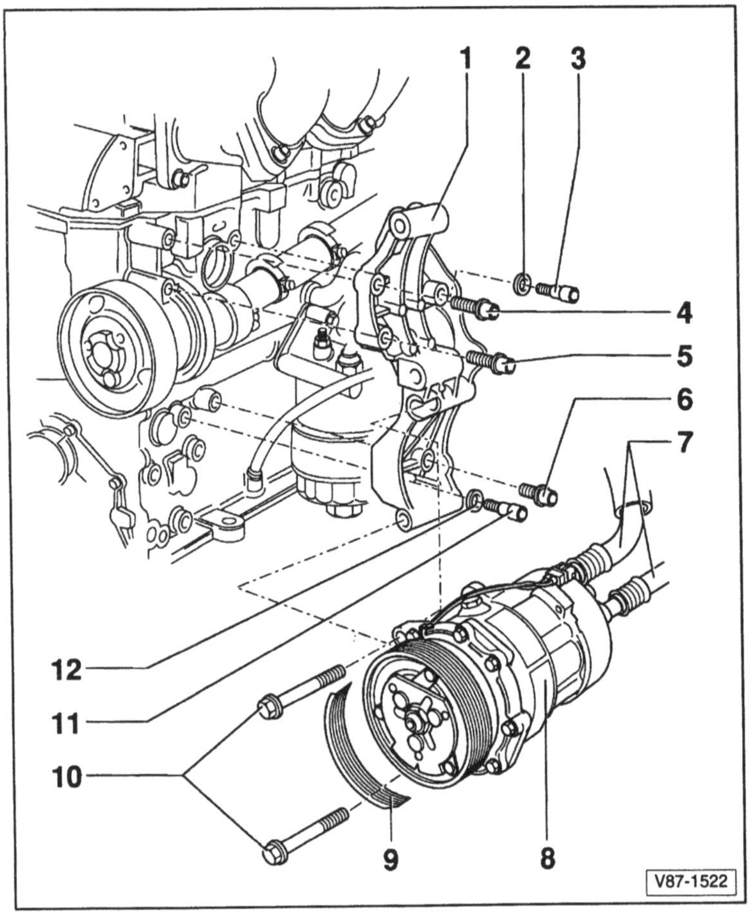

A/C compressor mounting bracket, removing/installing (6-cylinder VR6 gasoline engine)

Note

The A/C compressor mounting bracket can be removed/installed without discharging the A/C system.

1 - Generator (GEN) and A/C compressor mounting bracket
- ♦ removing:
- – remove A/C compressor -8- from bracket
- – remove Generator (GEN)
- – removing/installing Generator (GEN), ⇒ Repair group 27
- ♦ installing:
- – first install bolts -3- and -11-

2 - Washer – 8.4 x 16 x 1.6

3 - Bolt – M8 x 28
- ♦ 25 Nm (18 ft lb)

4 - Bolt – M8 x 30
- ♦ 25 Nm (18 ft lb)

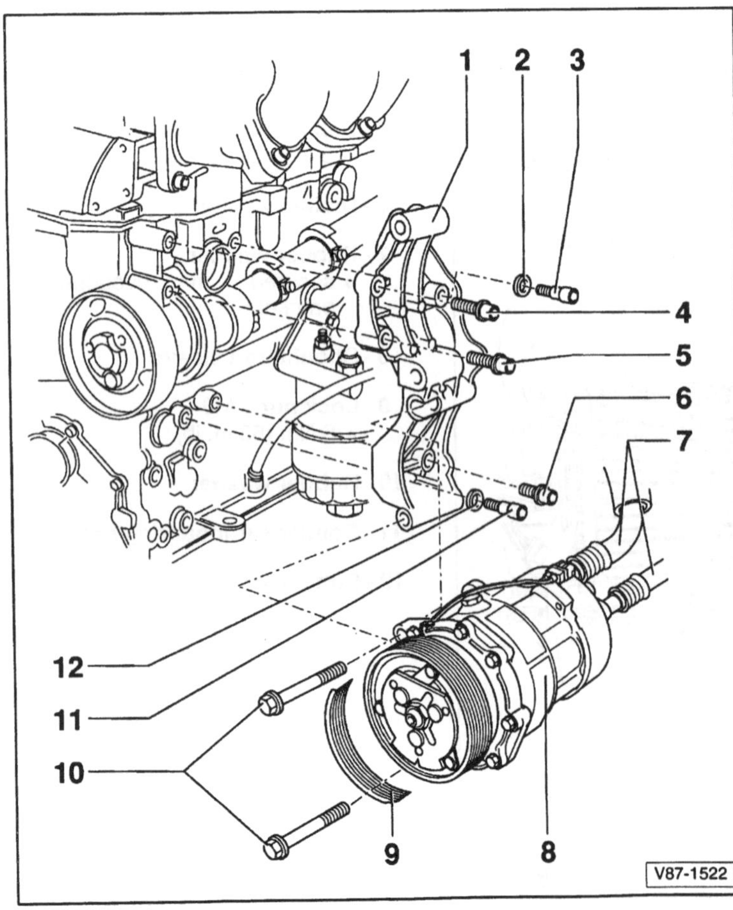

5 - Bolt – M8 x 38
- ♦ 25 Nm (18 ft lb)

6 - Bolt – M8 x 30
- ♦ 25 Nm (18 ft lb)

7 - A/C refrigerant hoses

8 - A/C compressor
- ♦ removing from bracket:

- – remove front bumper,
  ⇒ General body repairs, Repair Group 63
- – remove right front headlight housing,
  ⇒ Electrical system, Repair Group 94
- – remove ribbed drive belt -9-
- – remove Generator (GEN),
  ⇒ Electrical system, Repair Group 27
- – loosen A/C compressor bolts -10- and push back threaded bushings
- – remove bolts -10- and remove compressor

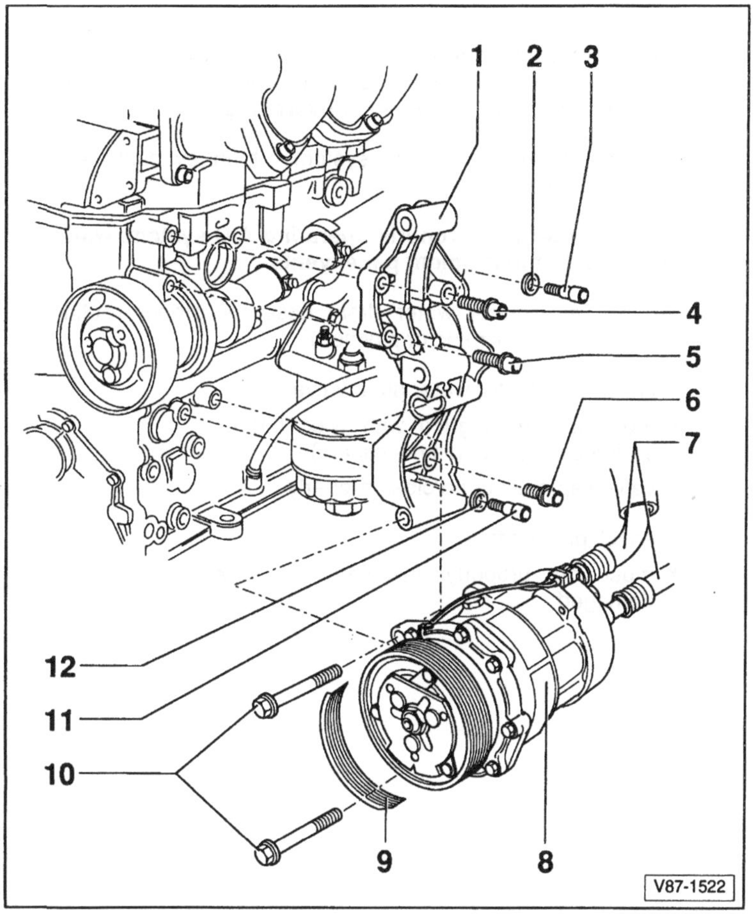

9 - **Ribbed drive belt**
- removing/installing, ⇒ 6-cylinder VR6 engine mechanical, Repair Group 13

10 - **Bolt – M10 x 112**
- 45 Nm (33 ft lb)

11 - **Bolt – M8 x 20**
- 25 Nm (18 ft lb)

12 - **Washer – 8.4 x 16 x 1.6**

A/C refrigerant system (R-134a), troubleshooting

Test requirements

A/C and heater electrical, vacuum and air distribution systems OK; correct/repair as necessary berfore proceeding.

– If one of the following conditions exists, follow recommended service:

A/C does not cool (cooling has stopped completely)

– check A/C system pressures, see page 87-88

Insufficient cooling during all driving speeds or Engine Speeds (RPM)

– check A/C system pressures, see page 87-88

No cooling or insufficient cooling after driving several miles

– check A/C system pressures, see page 87-88

A/C compressor noisy

– tighten A/C compressor and compressor bracket mounting bolts using torque wrench

– check routing of A/C refrigerant hoses and lines; hoses and lines must not contact other components and must be installed without tension or kinks, reposition or realign as necessary

A/C compressor noisy immediately after switching A/C ON and/or when driving around corners or braking (refrigerant shock) – system is overcharged

– discharge A/C refrigerant system, then evacuate and recharge, see page 87-77

With an otherwise properly operating A/C system, water sprays out of instrument panel vents

– check evaporator water drain hose for proper routing (should not be kinked or pinched)

– check evaporator water drain valve

- must not be plugged with wax, dirt or undercoating (clean if necessary)
- valve flap must open/close properly

– check plenum tray

- must be properly installed and not damaged

A/C refrigerant system (R-134a) pressures, checking

Check A/C refrigerant system pressures using these three tests:

1 - Air temperature drop from center instrument panel vent (A/C system cooling performance)

2 - A/C system high pressure

3 - A/C system low pressure

The combined results of all three steps determine which part of the A/C system is causing the problem, see table on page 87-92.

Test conditions

- A/C refrigerant system fully charged; discharge, evacuate and recharge system if necessary
- condenser and radiator clean and free of obstructions (spray clean if necessary)
- air distribution can be adjusted correctly using control knobs (all air distribution flaps reach end positions)
- no wiring problems as per wiring diagram
- outside (ambient) air temperature between 68 – 86°F (20 – 30°C)
- drive belts for A/C compressor and Generator in good condition and properly tensioned

Test 1 - Air temperature drop from center instrument panel vent (A/C system cooling performance, checking)

- start engine
- adjust air distribution so air flows from instrument panel vents
- set temperature control knob to full "cold" position (against stop)
- set fresh air blower to second speed (posiiton 2)
- insert thermometer into center instrument panel vent
- switch A/C on by pressing A/C Norm button
- raise Engine Speed (RPM) to 2000

Specified result

With humidity normal and outside (ambient) temperature between 68 – 77°F (20 – 25°C):

- system is cooling satisfactorily if air temperature from center instrument panel vent drops below 50°F (+10°C) within one minute

For higher ambient temperatures and/or higher humidity, specified air temperature from center instrument panel vent can be slightly higher.

If specified reading is not obtained, perform tests 2 and 3, then compare results of all three tests with table on page 87-92.

Test 2 – A/C system high pressure, checking

- connect high and low pressure hoses of refrigerant recovery/ recycling/recharging unit **Kent-Moore ACR4**, or equivalent, to high and low pressure service valves
- disconnect coolant fan connector
- adjust air distribution knob to footwell outlets
- adjust temperature control knob to full hot
- set fresh air blower to high speed (posiiton 4)
- start engine
- switch Max. A/C on by pressing A/C Max. button
- raise Engine Speed (RPM) to approx. 1500

Specified result

- system high pressure is normal if high pressure gauge reads 232 psi (16 bar) within 30 seconds

If specified reading is not obtained, also perform test 3 and compare results of all three tests with table on page 87-92.

Test 3 – A/C system low pressure, checking

- connect high and low pressure hoses of refrigerant recovery/ recycling/recharging unit **Kent-Moore ACR[4]**, or equivalent, to high and low pressure service valves
- adjust air distribution knob to instrument panel outlets
- adjust temperature control knob to full cold
- set fresh air blower to low speed (position 1)
- start engine
- switch A/C on by pressing A/C Norm button
- raise Engine Speed (RPM) to approx. 1500

Specified result

- system low pressure is normal if low pressure gauge reads 22 – 36 psi (1.5 – 2.5 bar) within 30 seconds

If specified reading is not obtained, compare results of all three tests with table on page 87-92.

Test 1 – Temperature from center air vent*	Test 2 – High pressure**	Test 3 – Low pressure**	Possible causes of incorrect readings	Corrective measures
Normal	Normal	Normal	None	---
Too high	Normal	Normal	Temperature flap position incorrect	Adjust temperature flap cable
Too high	Too low	Normal	Compressor	Replace compressor
Normal	Too low	Normal	Compressor	Replace compressor
Normal	Normal	Too high or too low	Expansion valve or compressor	Replace expansion valve or compressor
Too high	Normal	Too high or too low	Expansion valve or compressor	Replace expansion valve or compressor
Normal	Too high or too low	Too high or too low	Expansion valve or compressor	Replace expansion valve or compressor

* Normal air outlet temperature approx. 43°F (6°C).

** For normal system temperatures and pressures, see page 87-66.

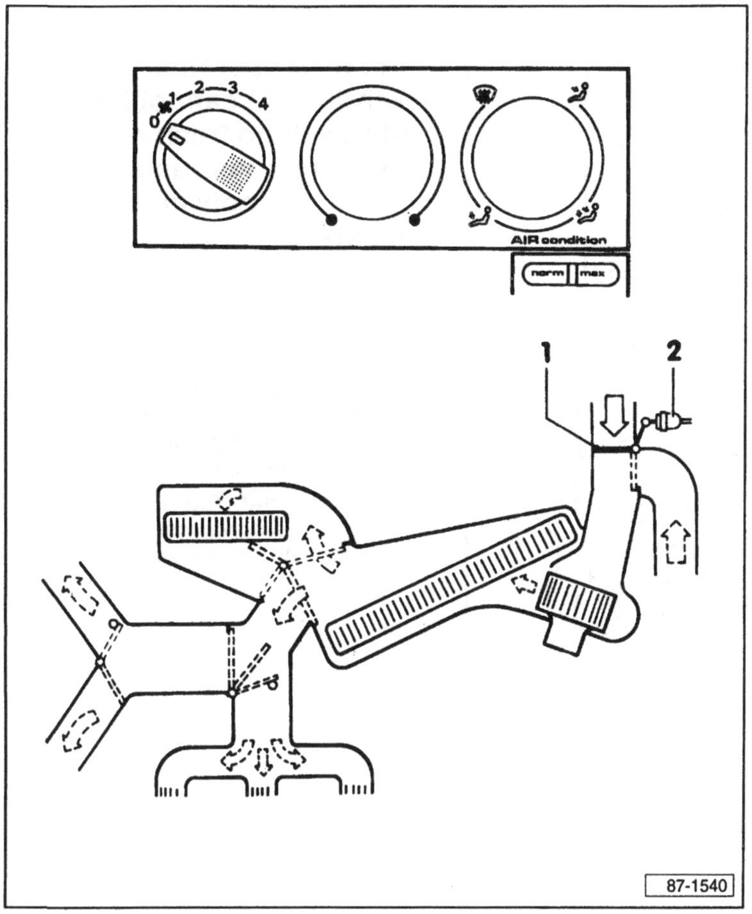

Fresh air cannot be shut off

Note

- *See next page for troubleshooting procedure.*
- *The A/C switch "norm" and "max" buttons have changed and do not appear on car as shown. The buttons are now identified with "A/C" and ↻. The ↻ symbol designates the recirculation mode for max. A/C operation.*

Fresh air cannot be shut off

Refer to illustration 87-1540.

Test condtion

- engine running, vacuum is present

Test procedure

– turn fresh air blower switch to "0" (OFF) position

Note

The position of the other two ventilation control knobs does not affect fresh air cut-off.

– check components and specified functions using following table:

Components	Specified function and check	Remedy if specified function not obtained
1 - Fresh air/recirculation flap	fresh air inlet -arrow- closed to outside air vacuum servo -2- retracted (vacuum applied)	– check fresh air recirculation flap two-way valve -N63 according to wiring diagram – check fresh air/recirculation flap two-way valve -N63 breather cap for an opening/air bleed. If cap does not have air bleed, remove cap and discard – check vacuum hoses and connections according to vacuum hose layout, see page 87-63

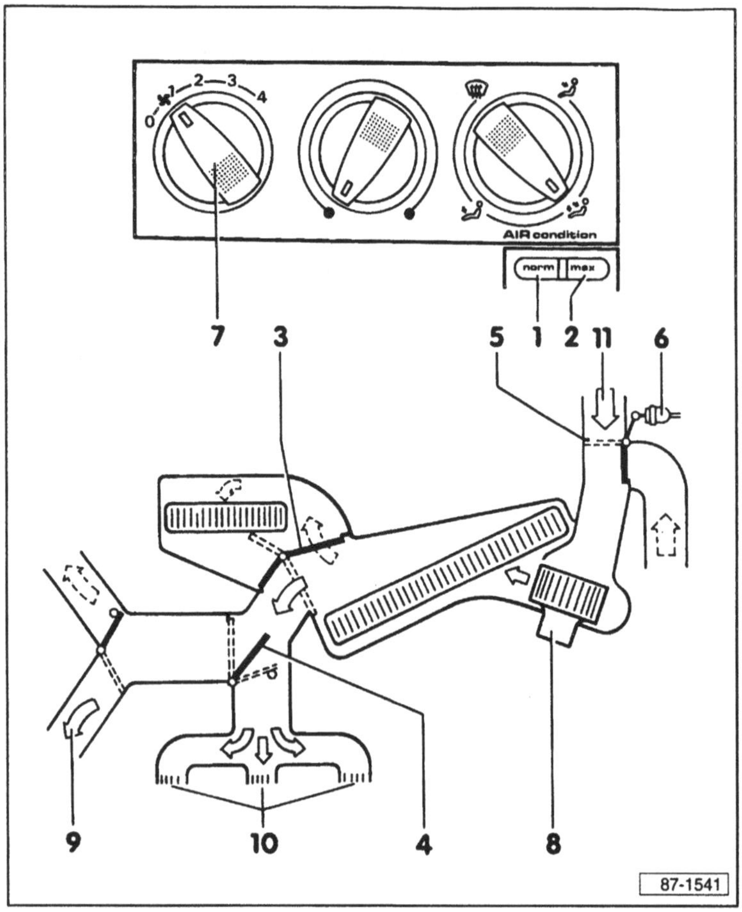

Fresh air operation (ventilation) cannot be adjusted; no fresh air flows from instrument panel and footwell outlets

Note

- *See next page for troubleshooting procedure.*
- *The A/C switch "norm" and "max" buttons have changed and do not appear on car as shown. The buttons are now identified with "A/C" and ⟲. The ⟲ symbol designates the recirculation mode for max. A/C operation.*

Fresh air operation (ventilation) cannot be adjusted; no fresh air flows from instrument panel and footwell outlets

Refer to illustration 87-1541.

Test condtion

- fresh air blower -8- runs in all four speeds

Test procedure

– set ventilation control knobs as shown in illustration 87-1541
– A/C switch -E35 OFF (Norm button -1- or Max button -2- not pressed)
– check components and specified functions using following table:

Components	Specified function and check	Remedy if specified function not obtained
3 - Temperature flap	temperature flap -3- must make audible contact when temperature control knob (center) is turned back and forth to full "cold" position	– adjust temperature flap control cable
4 - Central flap	air flows from instrument panel and footwell vents (-9-, -10-) (residual air should flow to windshield	– adjust central flap control cable
5 - Fresh air/recirculation flap	fresh air inlet -11- open (arrow) vacuum servo -6- extended (no vacuum applied)	– check vacuum system – check fresh air recirculation flap two-way valve -N63 according to wiring diagram – check fresh air/recirculation flap two-way valve -N63 *breather cap for an opening/air bleed.* If cap does not have air bleed, remove cap and discard – check fresh air blower switch -7- (-E 9) according to wiring diagram

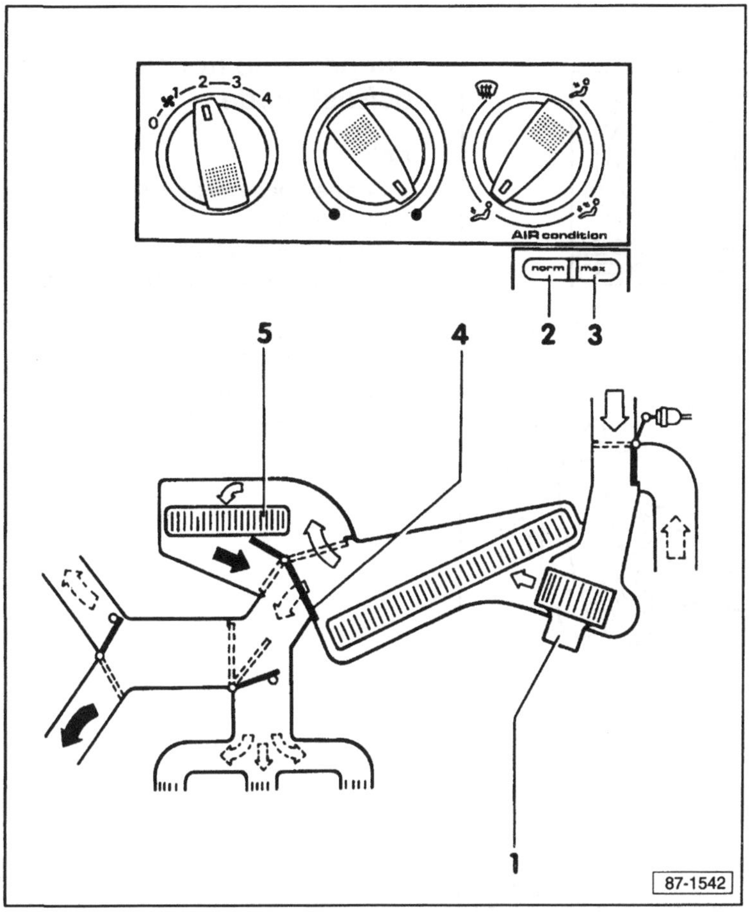

Insufficient or no heat output

Note

- *See next page for troubleshooting procedure.*
- *The A/C switch "norm" and "max" buttons have changed and do not appear on car as shown. The buttons are now identified with "A/C" and ↻. The ↻ symbol designates the recirculation mode for max. A/C operation.*

Insufficient or no heat output

Refer to illustration 87-1542.

Test condtion

- engine running, at normal operating temperature
- fresh air blower -1- runs in all four speeds

Test procedure

- set ventilation control knobs as shown in illustration 87-1542
- A/C switch -E35 OFF (Norm button -2- or Max button -3- not pressed)
- check components and specified functions using following table:

Components	Specified function and check	Remedy if specified function not obtained
4 - Temperature flap	temperature flap -4- must make audible contact when temperature control knob (center) is turned back and forth to full "cold" position	– adjust temperature flap control cable
5 - Heater core	warm	– check thermostat, see Repair Group 19

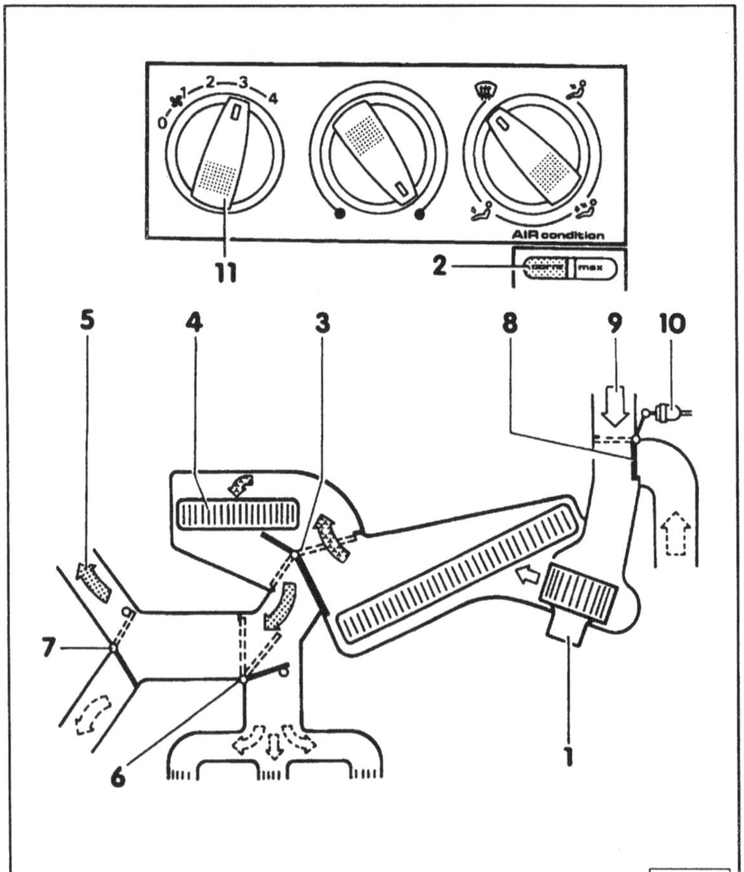

87-1543

Windows do not clear in defrost mode

Note

- *See next page for troubleshooting procedure.*
- *The A/C switch "norm" and "max" buttons have changed and do not appear on car as shown. The buttons are now identified with "A/C" and ↻. The ↻ symbol designates the recirculation mode for max. A/C operation.*

Windows do not clear in defrost mode

Refer to illustration 87-1543.

Test condtions

- outside (ambient) temperature above 45°F (+7°C)
- A/C cools with full output
- fresh air blower -1- runs in all four speeds

Test procedure

- start engine
- set ventilation control knobs as shown in illustration 87-1543
- switch A/C ON by pressing Norm button -2-
- check components and specified functions using following table:

Components	Specified function and check	Remedy if specified function not obtained
3 - Temperature flap	cooled air flows through heater core -4-. Temperature flap -3- must make audible contact when temperature control knob (center) is turned back and forth to full "cold" position	– adjust temperature flap control cable
4 - Heater core	coolant hoses warm	– check thermostat, see Repair Group 19
5 - Defroster vents	luke-warm air flows from defroster vents	– adjust central flap control cable and footwell/defroster flap control cable
6 - Fresh air/recirculation flap	fresh air inlet -9- open (arrow) vacuum servo -10- extended (no vacuum applied)	– check vacuum system – check fresh air recirculation flap two-way valve -N63 according to wiring diagram – check fresh air/recirculation flap two-way valve -N63 breather cap for an opening/air bleed. If cap does not have air bleed, remove cap and discard – check fresh air blower switch -11- (-E 9) according to wiring diagram

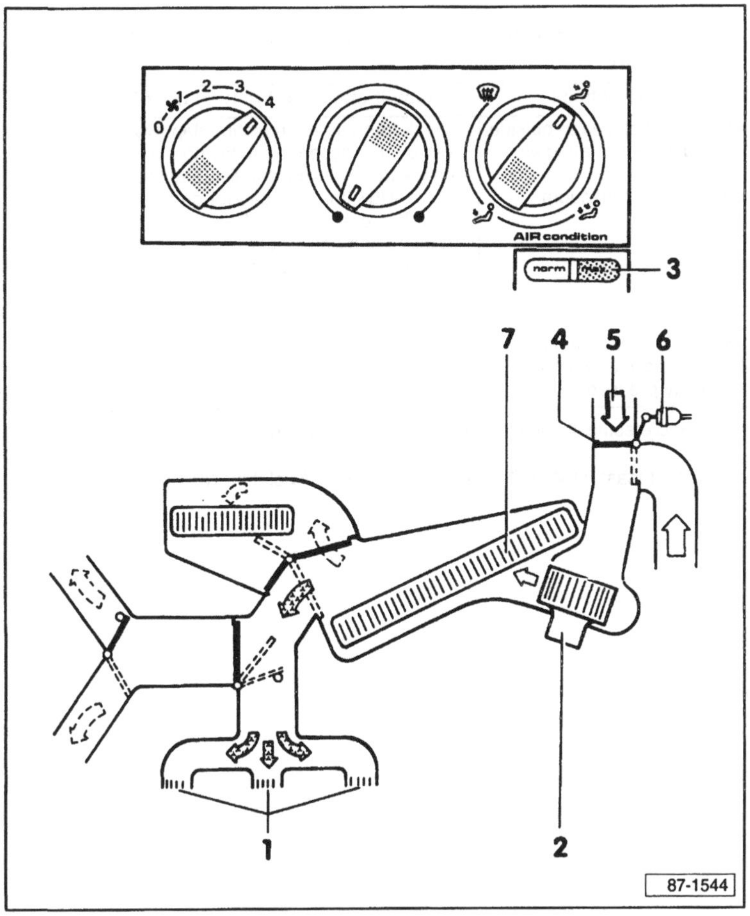

A/C does not cool

Note

- *See next page for troubleshooting procedure.*
- *The A/C switch "norm" and "max" buttons have changed and do not appear on car as shown. The buttons are now identified with "A/C" and ⟲. The ⟲ symbol designates the recirculation mode for max. A/C operation.*

A/C does not cool

(Evaporator -7- A/C lines at room temperature)

Refer to illustration 87-1544.

Test condtion

- check applicable A/C system fuses, see page 87-39
- A/C compressor drive belt in good condition and properly tensioned
- fresh air blower -2- runs in all four speeds

Note

The A/C system charge cannot be determined using the sight glass as the mixture of refrigerant R-134a and refrigerant (PAG) oil may appear foamy.

Test procedure

- open instrument panel vents -1-
- set ventilation control knobs as shown in illustration 87-1544
- switch A/C ON by pressing Max button -3-

– check components and specified functions using following table:

Components	Specified function and check	Remedy if specified function not obtained
4- Fresh air/recirculation flap	fresh air inlet -5- (arrow) closed to outside air (system operating in recirculating air mode) vacuum servo -6- retracted (vacuum applied)	– check fresh air recirculation flap two-way valve -N63 according to wiring diagram – check fresh air/recirculation flap two-way valve -N63 breather cap for an opening/air bleed. If cap does not have air bleed, remove cap and discard – check vacuum hoses and connections according to vacuum hose layout, see page 87-63
3 - A/C switch -E35 (Max button -3-)	with engine switched off and ignition on, A/C clutch -N25 audibly clicks when Max button -3- is repeatedly pressed*	– locate malfunction using wiring diagram*

* Remove 4-pin connector from A/C pressure switch -F129 and electrically bridge terminals 1 and 2 of connector. If A/C clutch -N25 clicks on, system is empty (no refrigerant).
If A/C clutch -N25 does not click on and voltage is present (at least 9.5V) at A/C clutch connector, A/C clutch -N25 is faulty.

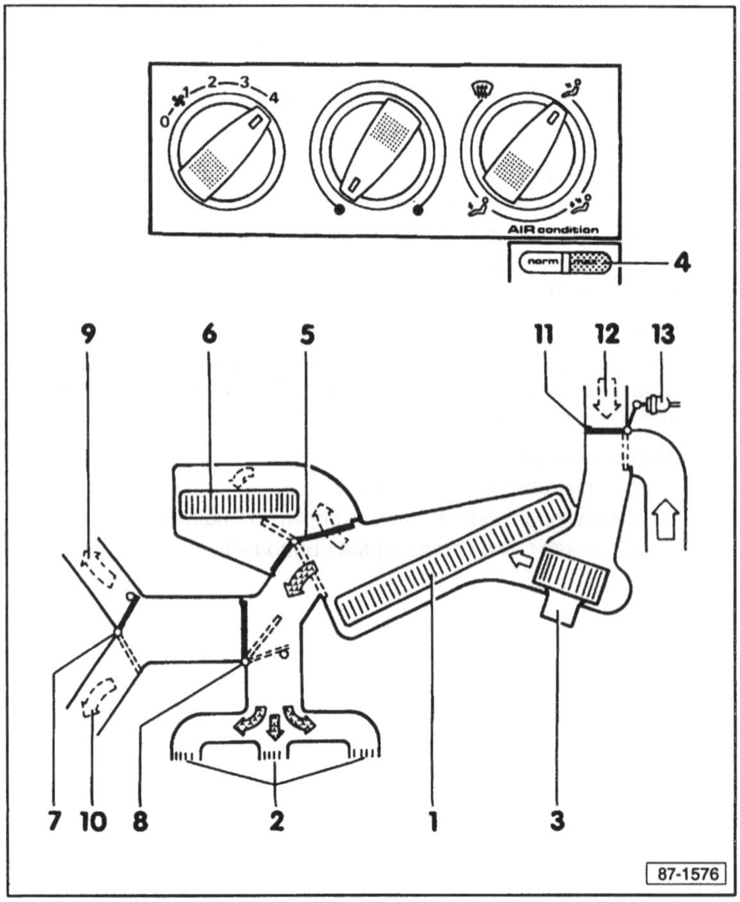

Insufficient A/C cooling output

Note

- *See next page for troubleshooting procedure.*
- *The A/C switch "norm" and "max" buttons have changed and do not appear on car as shown. The buttons are now identified with "A/C" and ↺. The ↺ symbol designates the recirculation mode for max. A/C operation.*

Insufficient A/C cooling output

Refer to illustration 87-1576.

Note

- *If evaporator -1- ices up, air flow is reduced after driving a few miles. If this occurs, check insertion depth of A/C evaporator temperature switch -E 33 sensing tube, replace temperature switch is necessary.*
- *If water sprays from air outlets -2-, check evaporator water drain valve.*
- *The A/C system charge cannot be determined using the sight glass as the mixture of refrigerant R-134a and refrigerant (PAG) oil may appear foamy.*

Test condtions

- condenser and radiator are clean, spray clean if necessary
- air ducts for condenser and radiator are properly installed
- do not expose vehicle to direct sunlight when performing A/C system checks and measurements
- fresh air blower -3- runs in all four speeds
- adjust Engine Speed (RPM) to approx. 2000

Test procedure

– open instrument panel air vents -2-

– insert thermometer into left air vent -2-

– switch A/C ON by pressing Max button -4- (compressor must run)

– set ventilation control knobs as shown in illustration 87-1576

– check components and specified functions using following table:

Components	Specified function and check	Remedy if specified function not obtained
5 - Temperature flap	Temperature flap -5- must make audible contact when temperature control knob (center) is turned back and forth to full "cold" position (air flow through heater core -6- blocked)	– adjust temperature flap control cable
7 - Footwell/defroster flap	no air flow from defroster or footwell vents (-9,-10)	– adjust control cables – pinion gear on A/C (ventilation) control head faulty, replace A/C (ventilation) control head
8 - Central flap	no air flow from defroster vents -9- no air flow from footwell vents -10-	– adjust control cables – pinion gear on A/C (ventilation) control head faulty, replace A/C (ventilation) control head
11- Fresh air/recirculation flap	fresh air inlet -12- (arrow) closed to outside air (system operating in recirculating air mode) vacuum servo -13- retracted (vacuum applied)	– check fresh air recirculation flap two-way valve -N63 according to wiring diagram – check fresh air/recirculation flap two-way valve -N63 breather cap for an opening/air bleed. If cap does not have air bleed, remove cap and discard – check vacuum hoses and connections according to vacuum hose layout, see page 87-63

A/C system cooling performance, checking

- start engine
- set fresh air blower to second speed (posiiton 2)
- set temperature control knob to full "cold" position (against stop)
- adjust air distribution so air flows from instrument panel vents
- switch A/C on by pressing A/C Norm button
- insert thermometer into center instrument panel outlet
- raise Engine Speed (RPM) to 2000

Specified result

With normal humidity and outside (ambient) temperature between 68 – 77°F (20 – 25°C):

- system is sufficiently charged if air temperature from center instrument panel vent drops below 50°F (+10°C) within one minute

For higher ambient temperatures and/or higher humidity, specified air temperature from center instrument panel vent can be slightly higher.

Description	Current track
A/C Clutch, N25	31
A/C Cut-Out Thermal Switch, F163	25
A/C Indicator Light, K84	16
A/C Pressure Switch, F129	22-23
A/C Switch, E35	17-19
After-Run Coolant Fan Control (FC) Thermal Switch, F87	35
Ambient Temperature Switch, F38	23
Coolant Fan Control (FC) Control Module, J293	29-41
Coolant Fan Control (FC) Thermal Switch, F18	40-42
Coolant Fan First Speed and A/C Clutch Fuse, S108, 20A	30
Coolant Fan Second and Third Speed Fuse, S107, 50A	29
Coolant Fan, V7	40-42
Coolant Pump, V50	38
Engine Coolant Pump Circuit Breaker, S78, 5A	38
Engine Coolant Temperature (ECT) Sensor, G2	34
Fresh Air Blower Series Resistance With Fuse, N24	9-12
Fresh Air Blower Switch, E9	6-12
Fresh Air Blower, V2	6
Fresh Air Control Lever Light, L16	5
Fresh Air/Recirculating Flap Two-Way Valve, N63	17
Fuse, S6, 30A	7
Fuse, S19, 30A	42
Third Speed Coolant Fan Control (FC) Thermal Switch, F165	26

Wire connectors

- T1 - Single Connector, behind fuse/relay panel
- T1a - Single Connector, on evaporator housing
- T1b - Single Connector, on evaporator housing
- T2b - Double Connector, behind instrument panel, left
- T2d - Double Connector, on compressor
- T2e - Double Connector, on Fresh Air Control Lever Light
- T2f - Double Connector, behind fuse/relay panel
- T4 - 4-Pin Connector, on Coolant Fan Control (FC) Control Module
- T4a - 4-Pin Connector, behind instrument panel, left
- T4b - 4-Pin Connector, on evaporator housing
- T4c - 4-Pin Connector, behind instrument panel, left
- T6 - 6-Pin Connector, on Fresh Air Blower Switch
- T8 - 8-Pin Connector, behind instrument panel, center
- T10e - 10-Pin Connector, on Coolant Fan Control (FC) Control Module
- T28 - 28-Pin Connector, on instrument cluster
- T42 - 42-Pin Connector, near oil filter
- T68 - 68-Pin Connector, on Motronic Engine Control Module (ECM)

Welded wiring harness points

- (A11) - Plus connection (58b), in instrument panel wiring harness
- (K21) - Wire connection -1-, in coolant fan wiring harness
- (K22) - Wire connection -2-, in coolant fan wiring harness
- (K24) - Wire connection -3-, in coolant fan wiring harness
- (K25) - Wire connection -4-, in coolant fan wiring harness
- (L2) - Wire connection, in A/C wiring harness
- (L5) - Wire connection -1-, in evaporator housing wiring harness
- (L8) - Wire connection -4-, in evaporator housing wiring harness

Ground connections

- (43) - Ground connection, on right A-pillar, lower part
- (80) - Ground connection -1-, in instrument panel wiring harness
- (182) - Ground connection -1-, in engine compartment wiring harness (6 Cylinder)

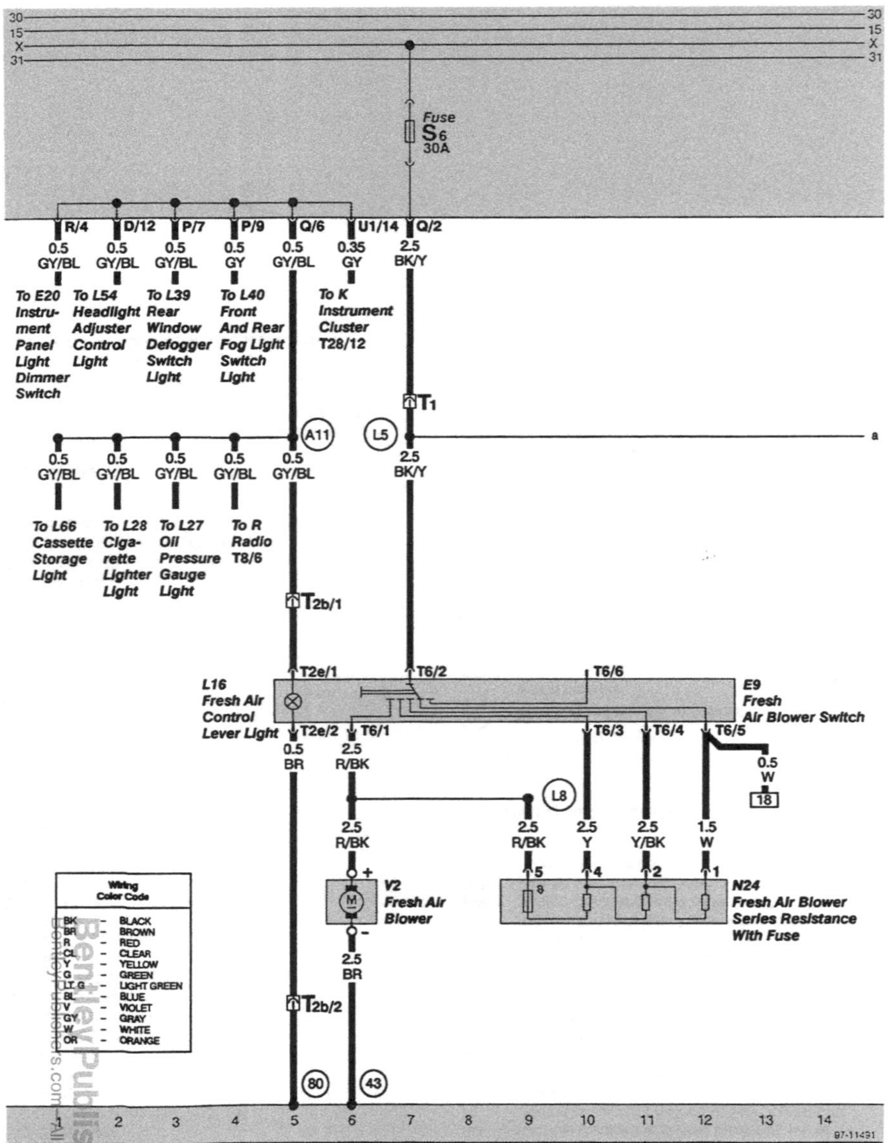
Fuse S6 30A
R/4 D/12 P/7 P/9 Q/6 U1/14 Q/2
To E20 Instrument Panel Light Dimmer Switch
To L54 Headlight Adjuster Control Light
To L39 Rear Window Defogger Switch Light
To L40 Front And Rear Fog Light Switch Light
To K Instrument Cluster T28/12
To L66 Cassette Storage Light
To L28 Cigarette Lighter Light
To L27 Oil Pressure Gauge Light
To R Radio T8/6
L16 Fresh Air Control Lever Light
E9 Fresh Air Blower Switch
V2 Fresh Air Blower
N24 Fresh Air Blower Series Resistance With Fuse
Wiring Color Code
BK - BLACK
BR - BROWN
R - RED
CL - CLEAR
Y - YELLOW
G - GREEN
LT G - LIGHT GREEN
BL - BLUE
V - VIOLET
GY - GRAY
W - WHITE
OR - ORANGE
97-11491
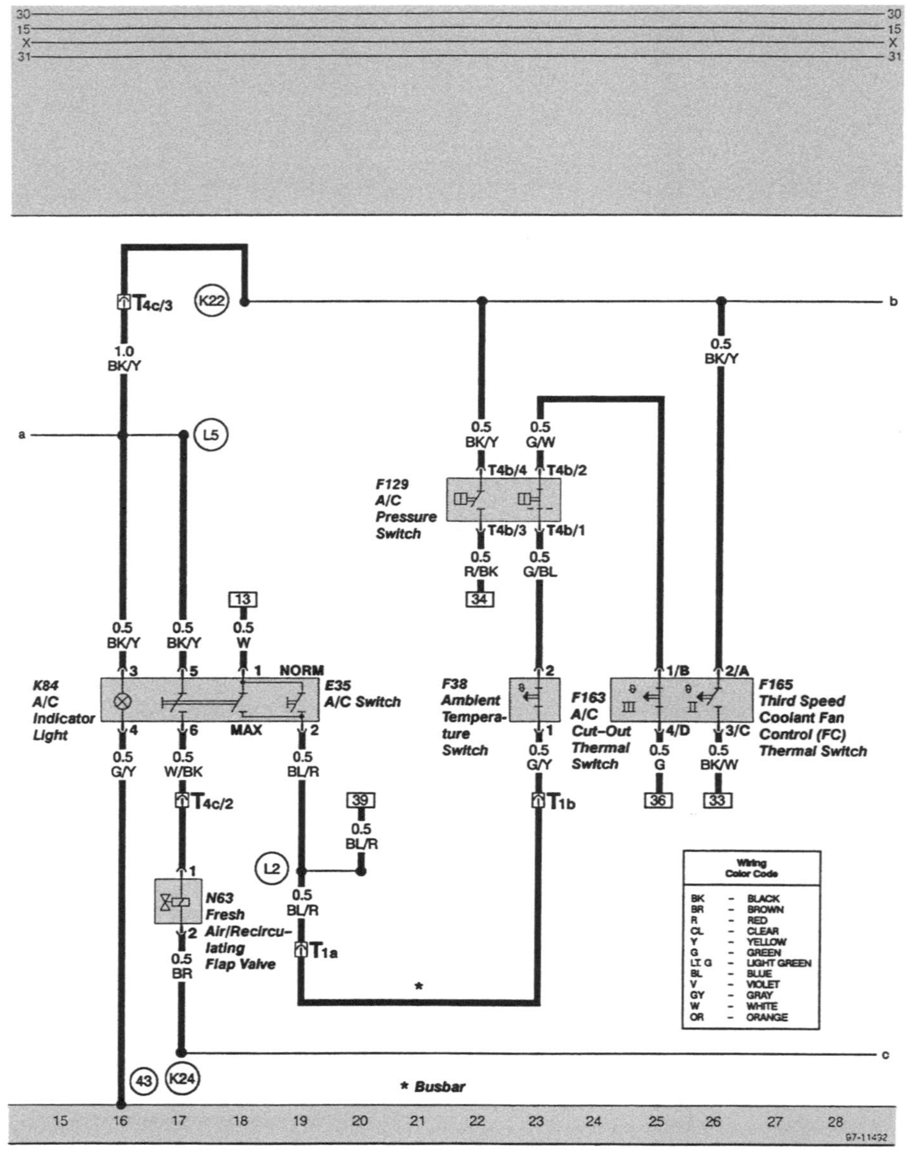
K84 A/C Indicator Light
E35 A/C Switch
NORM
MAX
F129 A/C Pressure Switch
F38 Ambient Temperature Switch
F163 A/C Cut-Out Thermal Switch
F165 Third Speed Coolant Fan Control (FC) Thermal Switch
N63 Fresh Air/Recirculating Flap Valve
* Busbar
Wiring Color Code
BK - BLACK
BR - BROWN
R - RED
CL - CLEAR
Y - YELLOW
G - GREEN
LT G - LIGHT GREEN
BL - BLUE
V - VIOLET
GY - GRAY
W - WHITE
OR - ORANGE
97-11492

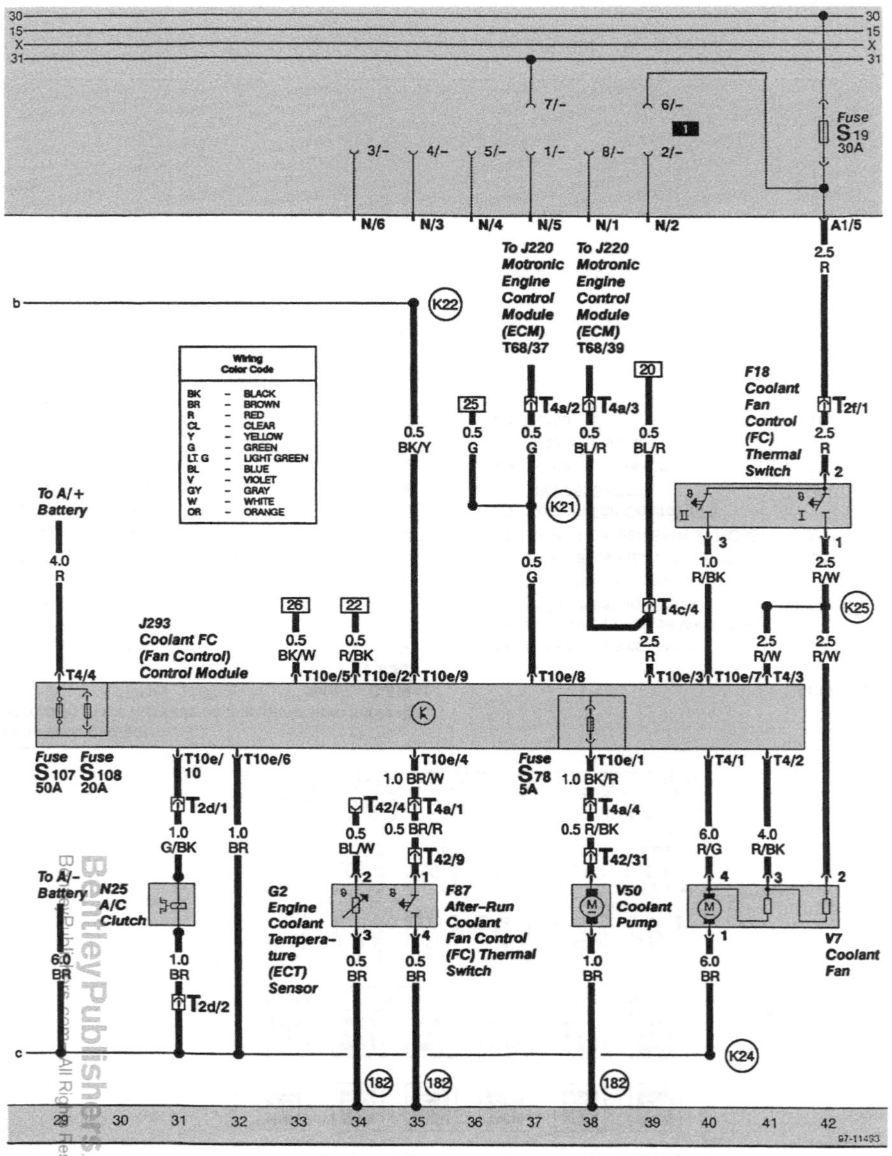

THIS PAGE INTENTIONALLY LEFT BLANK

Corrado VR6 (Canada) 1994
Corrado SLC (USA) 1994

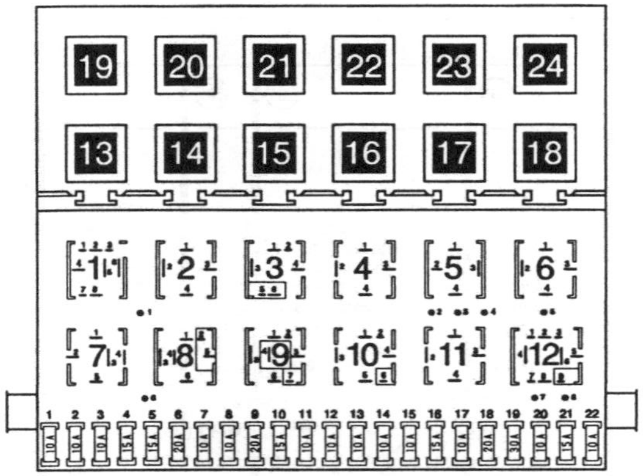

Fuse arrangement
according to the numbers on fuse panel from left to right:

		Amp.	Current track
1	Left Headlight	10	200
2	Right Headlight	10	203
3	Instrument Panel Light Dimmer Switch	10	247
4	Rear Window Wiper Motor	15	304
5	Windshield Wiper Motor	15	284
6	Fresh Air Blower Switch	20	23
7	Right Brake/Tail Light	10	235
8	Left Brake/Tail Light	10	228
9	Rear Window Defogger Switch Light	20	270
10	Open		
11	Headlight, High Beam, Left	10	199
12	Headlight, High Beam, Right	10	204
13	Horn	10	280
14	Back-Up Lights Switch	10	262
15	Positive Crankcase Ventilation (PCV) Heating Element	10	28
16	Voltmeter, Oil Pressure Gauge	15	177
17	Emergency Flashers Switch	10	215
18	Heated Oxygen Sensor (HO_2S)	20	53
19	Coolant Fan Control (FC) Thermal Switch	30	31
20	Brake Light Switch	10	233
21	Cigarette Lighter	15	170
22	Radio	10	302

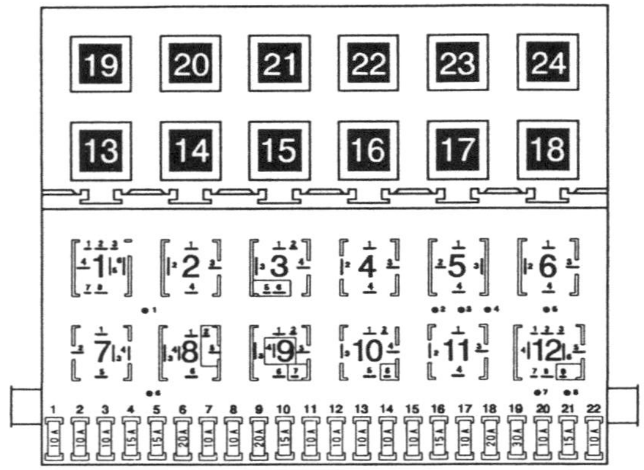

Relay location on fuse/relay panel

1. A/C Relay (13)*
2. Rear Window Wiper/Washer Relay (72)*
3. Motronic Engine Control Module (ECM) Power Supply Relay (32, 30)*
4. Load Reduction Relay (18)*
5. Engine Coolant Level (ECL) Control Module (43)*
6. Emergency Flasher Relay (21)*
7. Headlight Washer System Relay (33)*
8. Washer/Wiper Intermittent Relay (19, 99)*
9. Seat Belt Warning System Relay – Canada (4)*
10. Fog Light Relay (53)*
11. Dual Horn Relay (53)*
12. Fuel Pump (FP) Relay (67)*

Separate relays above fuse/relay panel

13. Back-Up Light Fuse (Auto. Trans.)
14. Coolant After-Run Control Module (31)*
15. ABS Hydraulic Pump Pelay (78)*
16. ABS Relay (79)*
17. Open
18. Open
19. A/C Relay (56)*
20. Park/Neutral Position (PNP) Relay (150)*
21. ABS Hydraulic Pump Fuse/Power Window Circuit Breaker
22. ABS Valves Fuse
23. Cruise Control Fuse/Automatic Shoulder Belt System Fuse
24. Open

*Numbers in parentheses indicate production control number stamped on relay housing.

Note: Locations of relays in positions 13–24 may vary from vehicle to vehicle depending on model and option content.

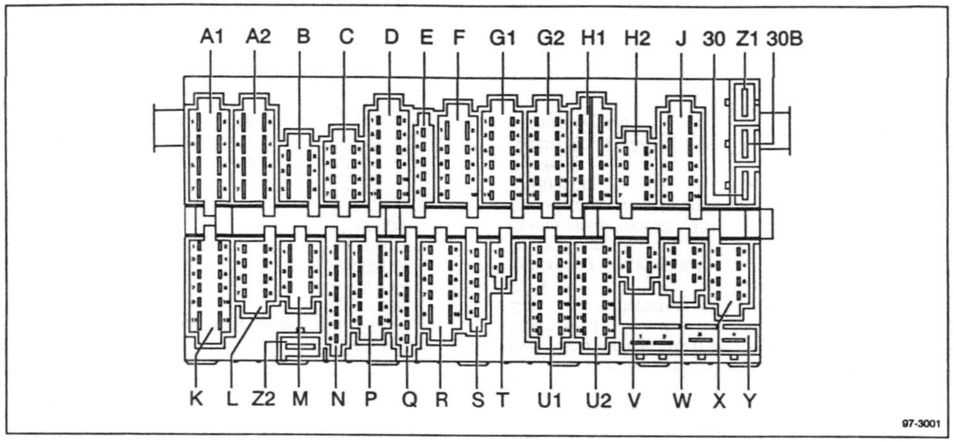

Connections and plugs on fuse/relay panel

A1 – Headlight wiring harness (yellow), eight point
A2 – Headlight wiring harness (yellow), eight point
B – Open
C – Headlight wiring harness (yellow), eight point
D – Optional equipment wiring harness (green), twelve point
E – Instrument wiring harness (green), five point
F – Engine compartment wiring harness, right (white), ten point
G1 – Engine compartment wiring harness, right (white), twelve point
G2 – Engine compartment wiring harness, right (white), twelve point
H1 – Steering column switch wiring harness (red), ten point
H2 – Steering column switch wiring harness (red), eight point
J – Steering column switch wiring harness (red), ten point
K – Rear wiring harness (black), twelve point
L – Rear wiring harness (black), eight point
M – Rear wiring harness (black), six point
N – A/C wiring harness (green), six point
P – Rear window defroster wiring harness (blue), ten point
Q – Instrument wiring harness (blue), six point
R – Light switch wiring harness (blue), ten point
S – Engine compartment wiring harness, right (white), five point
T – Connector (green), two point
U1 – Instrument cluster wiring harness (blue), fourteen point
U2 – Instrument cluster wiring harness (blue), fourteen point
V – Multi-function indicator wiring harness (green), four point
W – Connector (green), six point
X – Warning lamp wiring harness (green), eight point
Y – Four single connectors, terminal 30
Z1 – Single connector
Z2 – Single connector, terminal 31
30 – Single connector, terminal 30
30B – Single connector
31 – Single connector, terminal 31 (not shown)

Description	Current track
ABS Warning Light, K47	162-165
After-Run Coolant Fan Control (FC) Thermal Switch, F87	39
Alarm System Control Module, J284	298
Back-Up Light Switch, F4	266
Battery, A	5
Brake and Parking Brake Warning Light, K7	162-165
Brake Fluid Level Warning Switch, F34	182
Brake Light Switch, F	233
Camshaft Position (CMP) Sensor, G40	50-51
Cassette Storage Light, L66	180
Change-Over Valve, N180	93
Cigarette Lighter Light, L28	171
Cigarette Lighter, U1	170
Control Module With Display Unit In Instrument Cluster, J285	104-144
Coolant Fan (01018-3433), V7	34-36
Coolant Fan Control (FC) Thermal Switch, F18	30-31
Coolant FC (Fan Control) Control Module in engine compartment, left, J293	33-41
Cylinder 1 Fuel Injector, N30	71
Cylinder 2 Fuel Injector, N31	72
Cylinder 3 Fuel Injector, N32	73
Cylinder 4 Fuel Injector, N33	74
Cylinder 5 Fuel Injector, N83	75
Cylinder 6 Fuel Injector, N84	76
Digital Display Light, L75	139,141
Display, Y5	127-129
Dual Horn Relay, J4	277-278
Dual Horns, H1	278,280
EGR Temperature Sensor, G98	90
EGR Vacuum Regulator Soleniod Valve, N18	91
Emergency Flasher Relay, J2	217-219
Emergency Flasher Switch, E3	216-219
Emergency Flasher Warning Light, K6	221
Engine Coolant Level (ECL) Sensor, G32	103
Engine Coolant Level/Temperature (ECL/ECT) Warning Light, K28	107
Engine Coolant Temperature (ECT) Gauge, G3	108
Engine Coolant Temperature (ECT) Sensor, G62	63
Engine Coolant Temperature (ECT) Sensor, G2	40

Description	Current track
Engine Speed (RPM) Sensor , G28	68-70
Evaporative Emission (EVAP) Canister Purge Regulator Valve, N80	98
Fresh Air Blower, V2	18
Fresh Air Blower Series Resistance, N23	19-23
Fresh Air Blower Switch, E9	20-23
Fresh Air Control Lever Light, L16	24
Front, Side Marker Lights, M11	197,206
Fuel Gauge, G1	109
Fuel Pump (FP) After-Run Control Module, J206	52-54
Fuel Level Sensor, G	150
Fuel Pump (FP) Relay, J17	50-51
Fuel Pump (FP), G6	151
Fuse Strip, S88	41
Fuse Strip, S88	92
Fuse, S2, 10A	203
Fuse, S3, 10A	247
Fuse, S4, 15A	304
Fuse, S5, 15A	284
Fuse, S6, 20A	23
Fuse, S7, 10A	235
Fuse, S8, 10A	228
Fuse, S9, 20A	270
Fuse, S10, 15A	259
Fuse, S11, 10A	199
Fuse, S12, 10A	204
Fuse, S13, 10A	280
Fuse, S14, 10A	262
Fuse, S15, 10A	28
Fuse, S16, 15A	177
Fuse, S17, 10A	215
Fuse, S18, 20A	53
Fuse, S19, 30A	31
Fuse, S20, 10A	233
Fuse, S21, 15A	170
Fuse, S22, 10A	302
Generator (GEN) Warning Light, K2	122
Generator (GEN), C	1-2
Glove Compartment Light, W6	174
Headlight Dimmer/Flasher Switch, E4	213-214
Headlight High Beam Indicator Light, K1	126
Headlight Switch Light, L9	240
Heated Oxygen Sensor (HO2S), G39	52-54
Heater, A/C Clutch Fuse, S24	19-23

Description	Current track
High Beam Headlight, Left, L13	199
High Beam Headlight, Right, L14	204
High-mount Brake Light, M25	232
Hood Alarm Switch, F120	298
Horn Button, H	189
Idle Air Control (IAC) Valve, N71	77
Ignition Coil, N152	43-49
Ignition/Starter Switch, D	10-14
Instrument Cluster Light, L10	130-132
Instrument Panel Light Dimmer Switch, E20	247
Intake Air Temperature (IAT) Sensor, G42	81
Interior Light With Delay Switch, W15	155-157
Knock Sensor (KS) 1, G61	57-59
Knock Sensor (KS) II, G66	60-62
Left Back-Up Light, M16	264
Left Brake/Tail Light, M21	229-230
Left Door Latch Switch, F143	208
Left Front Door Contact Switch, F2	155
Left Front Turn Signal Light, M5	225
Left Headlight, L1	200
Left Parking Light, M1	198
Left Rear Turn Signal Light, M6	227
Left Seat Belt Switch, E24	208
Left Turn Signal Indicator Light, K65	124
Left Washer Nozzle Heater, Z20	257
License Plate Light, X	247,249
Light Switch, E1	243-245
Load Reduction Relay, J59	11-12
Luggage Compartment Light Switch, F5	194
Luggage Compartment Light, W3	194
Malfunction Indicator Lamp (MIL), K83	162-165
Mass Air Flow (MAF) Sensor, G70	82-84
Motronic Engine Control Module (ECM) in plenum chamber, right, J220	44-97
Motronic Engine Control Module (ECM) PowerMulti-Function Indicator Memory Switch, E109	146-149
Multi-Function Indicator Mode Select Switch, E86	146-149
O.3bar Oil Pressure Switch, F22	101
Odometer Display, Y4	136-138
Oil Pressure Gauge Light, L27	178
Oil Pressure Gauge, G11	175-177
Oil Pressure Sensor,G10	175
Oil Pressure Switch, F1	102
Oil Pressure Warning Light, K3	105
Oil Temperature Sensor, G8	99
Outside Air Temperature Sensor, G17	153

Description	Current track
Park Light Indicator Light, K4	246
Parking Brake Warning Light Switch, F9	196
Positive Crankcase Ventilation (PCV) Heating Element, N79	28
Radio, R	302-304
Rear Spoiler Control Module,on left A-pillar, near relay panel, J223	184-192
Rear Spoiler End Switch 1, F135	192-193
Rear Spoiler End Switch 2, F136	192-193
Rear Spoiler Motor, V52	189
Rear Spoiler Switch, E127	185-187
Rear Window Defogger Indicator Light, K10	268-269
Rear Window Defogger Switch Light, L39	268-269
Rear Window Defogger Switch, E15	270
Rear Window Defogger/Heat Element, Z1	273
Rear Window Wiper Motor, V12	307-308
Rear Window Wiper/Washer Relay, J30	306-307
Right Back-Up Light, M17	266
Right Brake/Tail Light, M22	234-235
Right Front Door Contact Switch, F3	156
Right Front Map/Reading Light, W13	155-157
Right Front Turn Signal Light, M7	237
Right Headlight, L2	203
Right Make-Up Mirror Light Switch, F148	160
Right Make-up Mirror Light, W14	159-160
Right Parking Light, M3	205
Right Rear Turn Signal Light, M8	236
Right Turn Signal Indicator Light, K94	125
Right Washer Nozzle Heater, Z21	268
Seat Belt Warning Light, K19	121
Seat Belt Warning System Relay, J34	206-209
Secondary Air Injection (AIR) Pump Motor, V101	92
Secondary Air Injection (AIR) Relay, J316	92-96
Spark Plug Connectors, P	44-49
Spark Plugs, Q	44-49
Speedometer (00663), G21	130-133
Speedometer Vehicle Speed Sensor (VSS), G22	145-146
Starter, B	7-8
Supply Relay, J271	94-95
Tachometer, G5	113-116

Description	Current track
Third Speed Coolant Fan Control (FC) Thermal Switch, F165	42
Throttle Position (TP) Sensor, G69	65-66
Transmission Range (TR) Selector LeverTurn Signal Switch, E2	222
Voltage Regulator (VR), C1	1-2
Voltage Stabilizer, J6	118-120
Voltmeter, G14	175-177
Warning Buzzer, H3	109
Water Pump (00739), V50	38
Windshield and Rear Window Washer Pump, V59	305
Windshield Wiper Motor, V	282-285
Windshield Wiper/Washer Switch, E22	290-294
Wiper/Washer Intermittent Relay, J31	290-294

Wire connectors

T1 – Single Connector, near starter
T1a – Single Connector, behind fuse relay/panel
T1b – Single Connector, behind instrument panel, center
T1c – Single Connector, behind fuse relay/panel
T1d – Single Connector, behind fuse relay/panel
T1d1 – Single Connector, behind fuse/relay panel
T1e – Single Connector, on left Apillar, near relay panel
T1f – Single Connector, behind fuse relay/panel
T1g – Single Connector, behind fuse relay/panel
T1h – Single Connector, behind fuse relay/panel
T1i – Single Connector, behind fuse relay/panel
T1j – Single Connector, in rear lid
T1l – Single Connector, behind fuse relay/panel
T1m – Single Connector, behind fuse relay/panel
T1n – Single Connector, behind fuse relay/panel
T1o – Single Connector, on Left Back-Up Light
T1p – Single Connector, on Left Back-Up Light
T1r – Single Connector, behind fuse relay/panel
T1s – Single Connector, behind fuse relay/panel
T1t – Single Connector, behind fuse relay/panel
T1v – Single Connector
T1w – Single Connector
T1x – Single Connector
T1y – Single Connector
T2 – Double Connector, above fuse/relay panel
T2a – Double Connector
T2a1 – Double Connector, below driver's seat
T2a2 – Double Connector, in rear lid
T2b – Double Connector
T2b1 – Double Connector, behind instrument panel, center
T2c – Double Connector
T2c1 – Double Connector, behind instrument panel
T2d – Double Connector, behind fuse/relay panel
T2e – Double Connector, behind fuse relay/panel
T2f – Double Connector, on right A-pillar, upper part
T2g – Double Connector, in luggage compartment, left
T2h – Double Connector, behind fuse relay/panel
T2i – Double Connector, in luggage compartment, left
T2j – Double Connector, on engine hood
T2k – Double Connector, on engine hood
T2l – Double Connector, behind fuse/relay panel
T2n – Double Connector, behind fuse/relay panel
T2o – Double Connector, behind fuse/relay panel
T2p – Double Connector, behind fuse/relay panel
T2q – Double Connector, in luggage compartment, left
T2r – Double Connector, on Left Parking Light
T2s – Double Connector, on Front, Side Marker Lights
T2t – Double Connector, on Right Parking Light
T2u – Double Connector, on Front, Side Marker Lights
T2v – Double Connector, on taillight, left
T2w – Double Connector, on High-Mount Brake Light
T3a – 3-Pin Connector, near starter
T3b – 3-Pin Connector, near exhaust manifold
T3c – 3-Pin Connector, near starter
T3e – 3-Pin Connector, on Right Back-Up Light
T3f – 3-Pin Connector, on Left Headlight
T3g – 3-Pin Connector, on Right Headlight
T4 – 4-Pin Connector, on Coolant FC (Fan Control) Control Module
T4a – 4-Pin Connector, behind steering column switch cover
T4b – 4-Pin Connector, near exhaust manifold
T4c – 4-Pin Connector, behind fuse/relay panel
T4d – 4-Pin Connector, behind steering column switch cover
T4e – 4-Pin Connector, on Left Rear Turn Signal Light
T4f – 4-Pin Connector, on Right Rear Turn Signal Light
T4i – 4-Pin Connector, behind fuse/relay panel
T5 – 5-Pin Connector, on Fresh Air Blower Series Resistance
T5b – 5-Pin Connector, behind steering column switch cover
T5c – 5-Pin Connector, behind steering column switch cover
T6 – 6-Pin Connector, on Rear Spoiler Switch
T7 – 7-Pin Connector, behind steering column switch cover

Wire connectors

T8 – 8-Pin Connector, on Radio
T8a – 8-Pin Connector, behind fuse/relay panel
T8d – 8-Pin Connector, on Brake and Parking Brake Indicator Light
T10 – 10-Pin Connector, on Alarm System Control Module
T10a – 10-Pin Connector, on Rear Spoiler Control Module
T10b – 10-Pin Connectors, on Coolant FC (Fan Control) Control Module
T28 – 28-Pin Connector, on Instrument Cluster
T42 – 42-Pin Connector, on cylinder head, left
T68 – 68-Pin Connector, on Motronic Engine Control Module (ECM)
TV 4 – Terminal 15 Wire Connector, above fuse relay panel
TV13 – Wire Connector, Vehicle Speed Signal, behind fuse/relay panel
TV14 – Data Link Connector (DLC) wire Connector, behind console
T16 – 16-Pin Connector, on board diagnostic (OBD), near ignition/starter switch, right

Welded wiring harness points

(A11) – plus connection (58b) in instrument panel wiring harness
(A15) – plus connection (15) in instrument panel wiring harness
(B6) – wire connection (s), in steering column switch wiring harness
(B8) – plus connection(58), in light switch wiring harness
(C1) – plus connection (58L), in left front wiring harness
(C2) – plus connection (58R), in left front wiring harness
(C13) – plus connection, in dual tone horn wiring harness
(C15) – wire connection -1-, in headlight wiring harness
(C16) – wire connection -2-, in headlight wiring harness

(E6) – plus connection -1- (15), in Motronic Multiport Fuel Injection (MFI) wiring harness
(E7) – plus connection (87a), in Motronic Multiport Fuel Injection (MFI) wiring harness
(G3) – plus connection, in injector wiring harness Injection (MFI) wiring harness
(K21) – wire connection -1-, in coolant fan wiring harness
(K28) – plus connection (15), in coolant fan wiring harness
(W9) – plus connection (30), in rear wiring harness
(W10) – plus connection (15), in rear wiring harness

Ground connections

(1) – Ground strap, battery to body
(2) – Ground strap, transmission to body
(15) – Ground connection, on cylinder head
(18) – Ground connection, on engine block
(30) – Ground connection, -1-, beside fuse/relay panel
(43) – Ground connection, on right A-pillar, lower part
(50) – Ground connection, in luggage compartment, left
(53) – Ground connection, in rear lid, right
(80) – Ground connection -1-, in instrument panel wiring harness
(86) – Ground connection -1-, in rear wiring harness

Ground connections

(98) – Ground connection, in rear lid wiring harness
(119) – Ground connection -1-, in headlight wiring harness
(120) – Ground connection -2-, in headlight wiring harness
(125) – Ground connection -3-, in headlight wiring harness
(129) – Ground connection, in dual-tone horn wiring harness
(137) – Ground connection (end stage), in Motronic Multiport Fuel Injection (MFI) wiring harness
(138) – Ground connection (control module), in Motronic Multiport Fuel Injection (MFI) wiring harness
(139) – Ground connection (sensor Ground), in Motronic Multiport Fuel Injection (MFI) wiring harness
(159) – Ground connection -2-, in Digifant Multiport Fuel Injection (MFI) wiring harness
(182) – Ground connection -1-, in engine compartment wiring harness (6 Cylinder)
(193) – Ground connection -1-, in coolant fan wiring harness

Wiring Color Code	
BK	BLACK
BR	BROWN
CL	CLEAR
R	RED
Y	YELLOW
G	GREEN
LT. G	LIGHT GREEN
BL	BLUE
V	VIOLET
GY	GRAY
W	WHITE
OR	ORANGE

Wire Size

Wiring Diagrams identify wires by the metric wire size. Metric wire sizes indicate cross-sectional area in square millimeters (mm^2). The chart below lists metric wire sizes and their approximate equivalents in American Wire Gauge (AWG) sizes.

Metric size Cross-section (mm^2)	American Wire Gauge size (AWG)
0.35	22
0.50	20
0.75	18
1.00	16
1.50	14
2.50	12
4.00	10
6.00	8
16.00	4
25.00	2
35.00	2

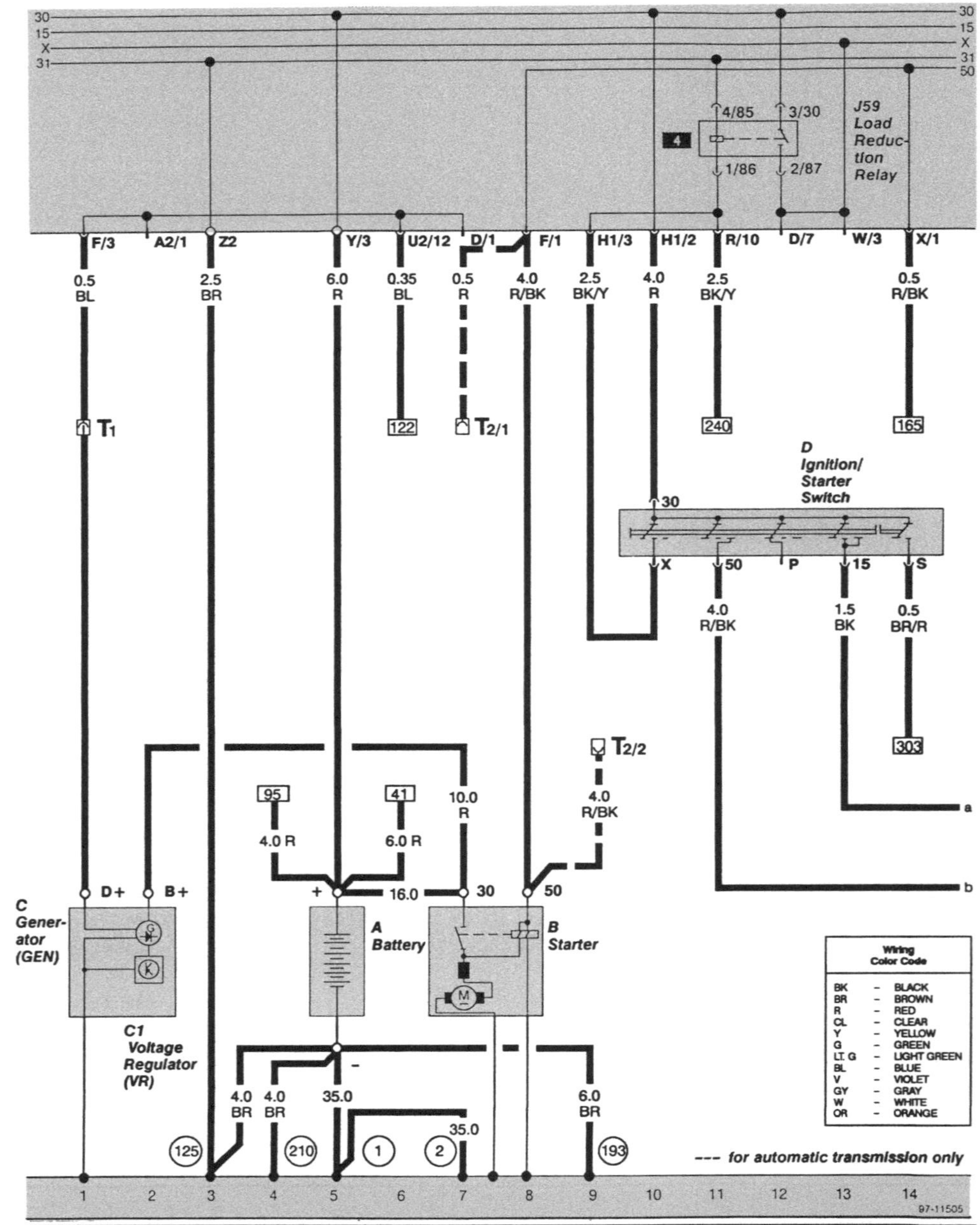

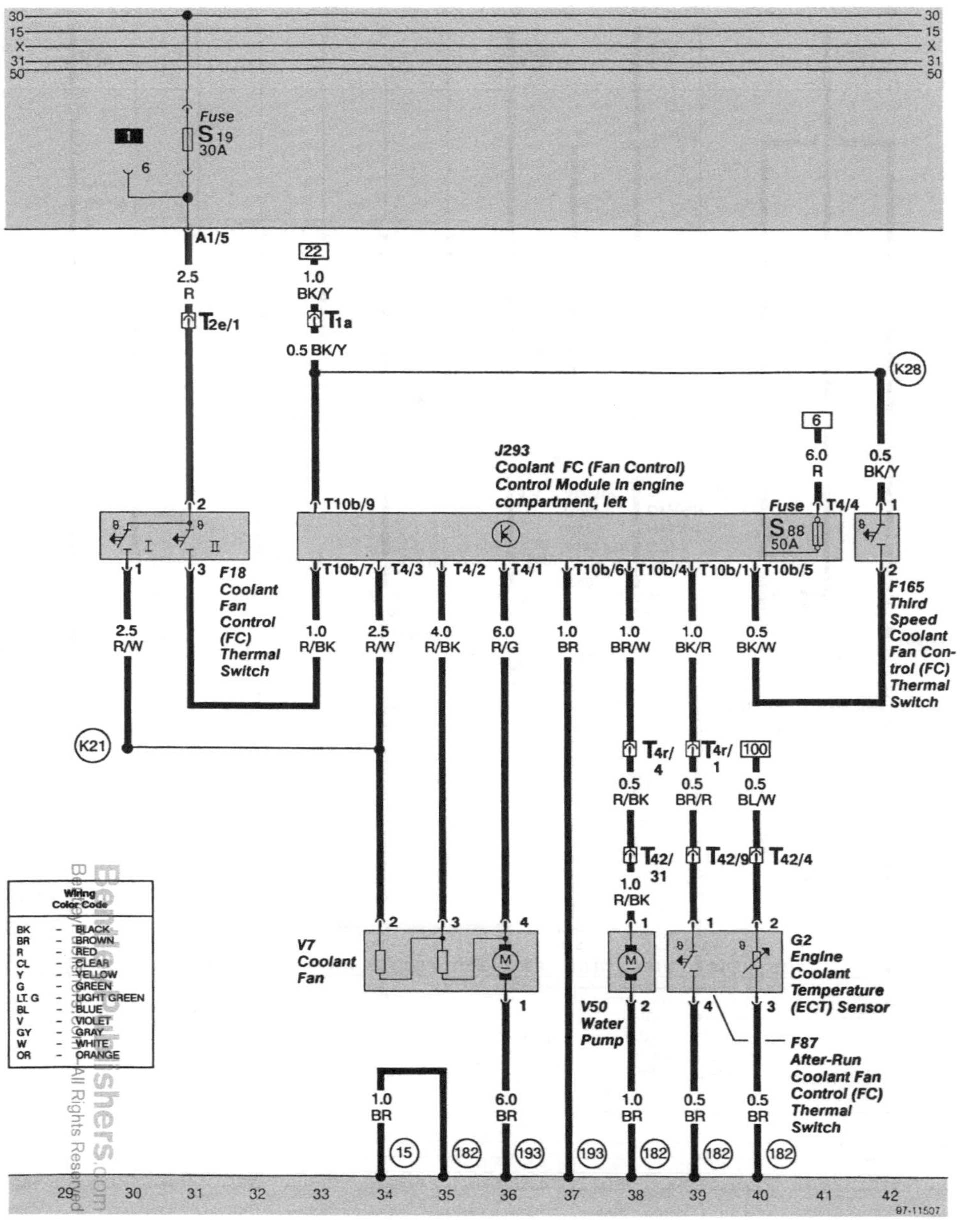

Wiring Color Code

BK	-	BLACK
BR	-	BROWN
R	-	RED
CL	-	CLEAR
Y	-	YELLOW
G	-	GREEN
LT. G	-	LIGHT GREEN
BL	-	BLUE
V	-	VIOLET
GY	-	GRAY
W	-	WHITE
OR	-	ORANGE

Corrado VR6 (Canada) 1994
Corrado SLC (USA) 1994

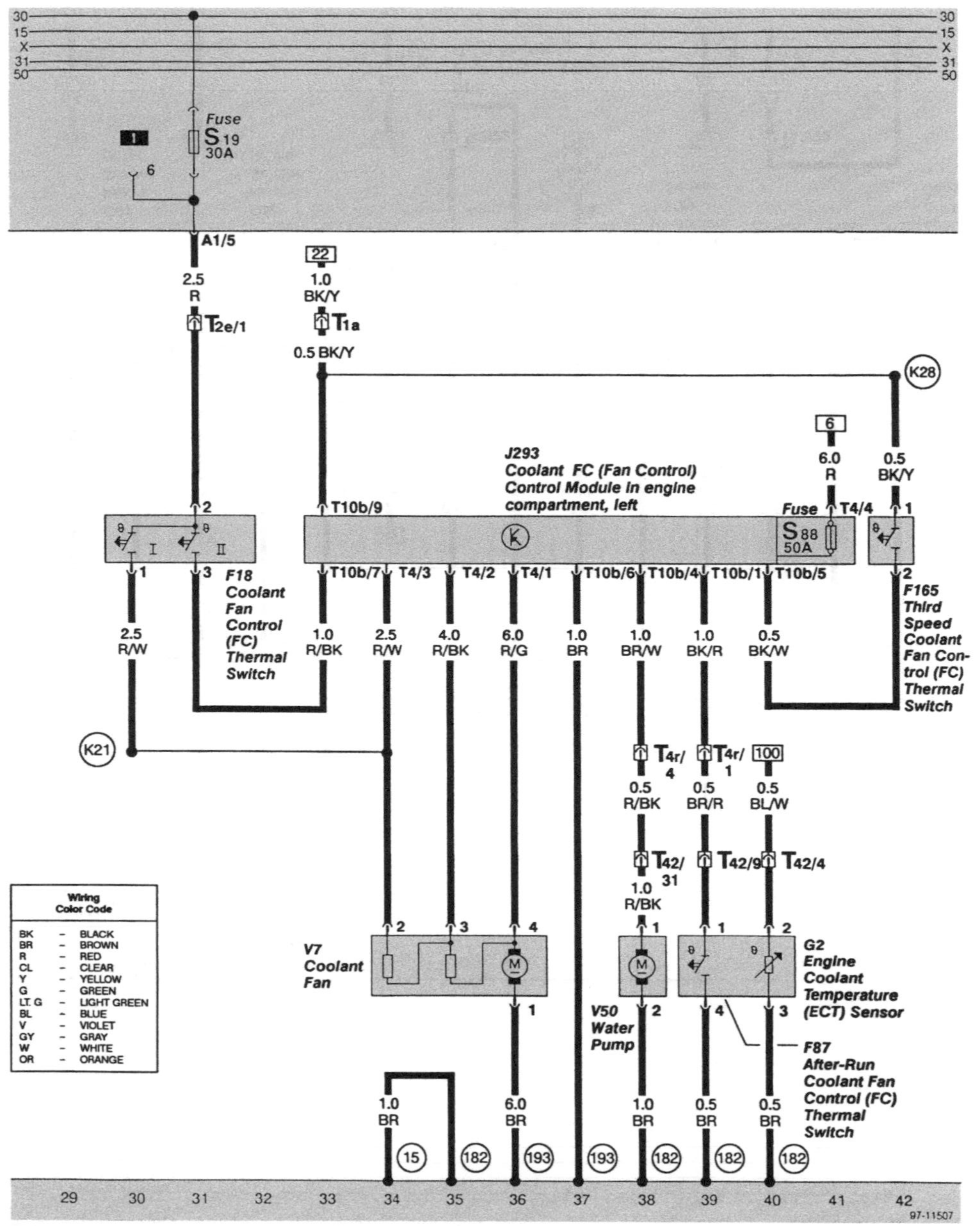

Wiring Color Code

BK	-	BLACK
BR	-	BROWN
R	-	RED
CL	-	CLEAR
Y	-	YELLOW
G	-	GREEN
LT. G	-	LIGHT GREEN
BL	-	BLUE
V	-	VIOLET
GY	-	GRAY
W	-	WHITE
OR	-	ORANGE

Corrado VR6 (Canada) 1994
Corrado SLC (USA) 1994

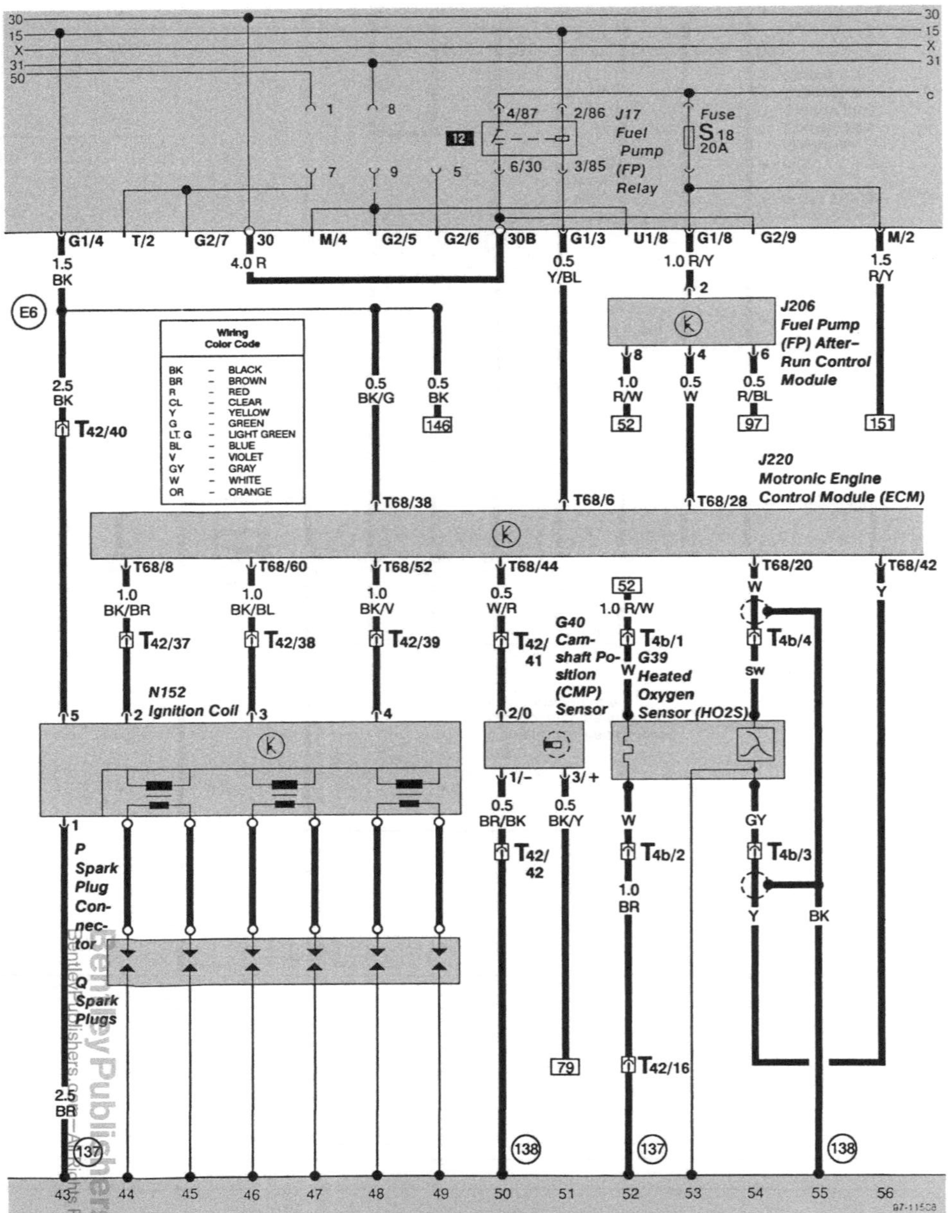

Corrado VR6 (Canada) 1994
Corrado SLC (USA) 1994

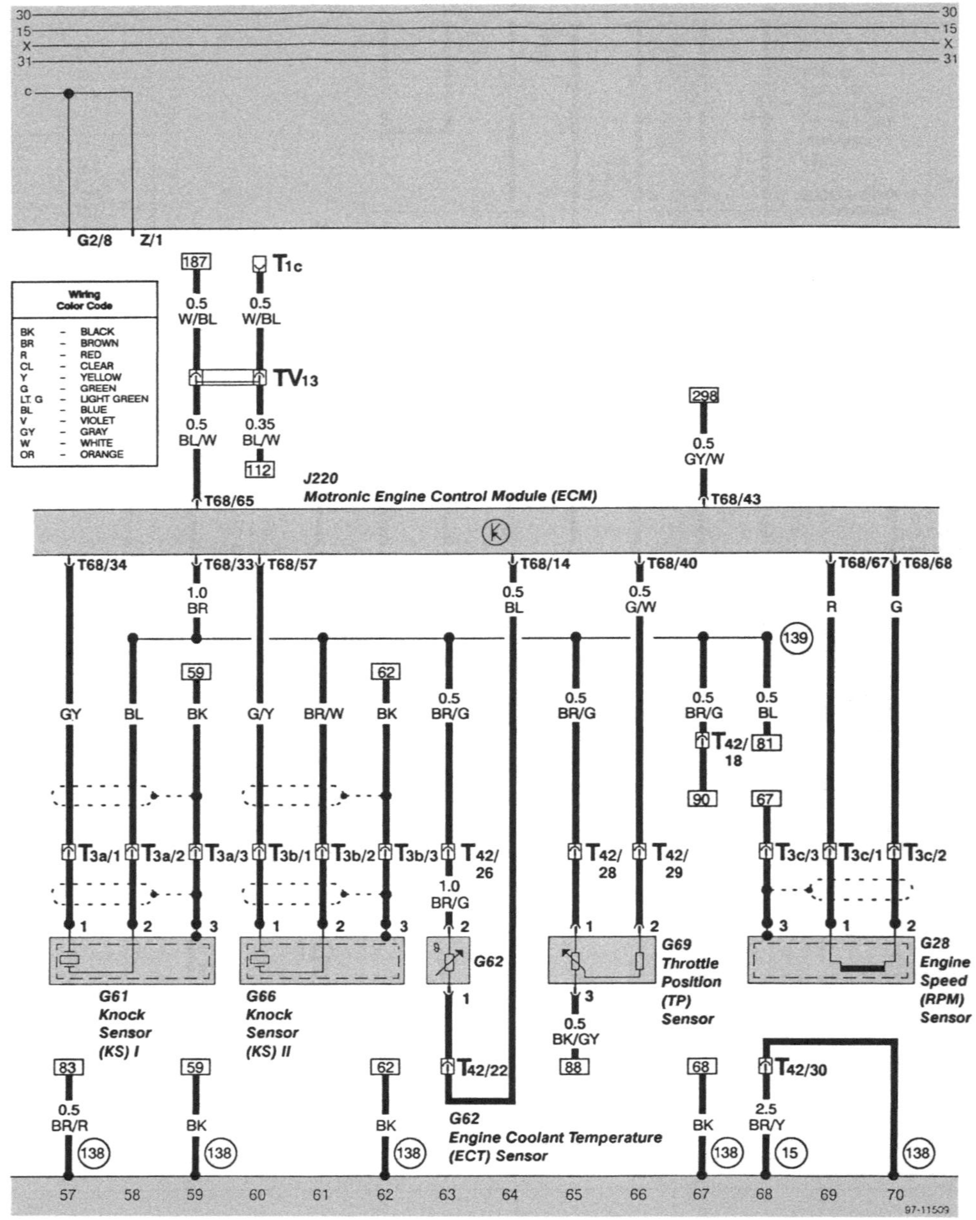

Corrado VR6 (Canada) 1994
Corrado SLC (USA) 1994

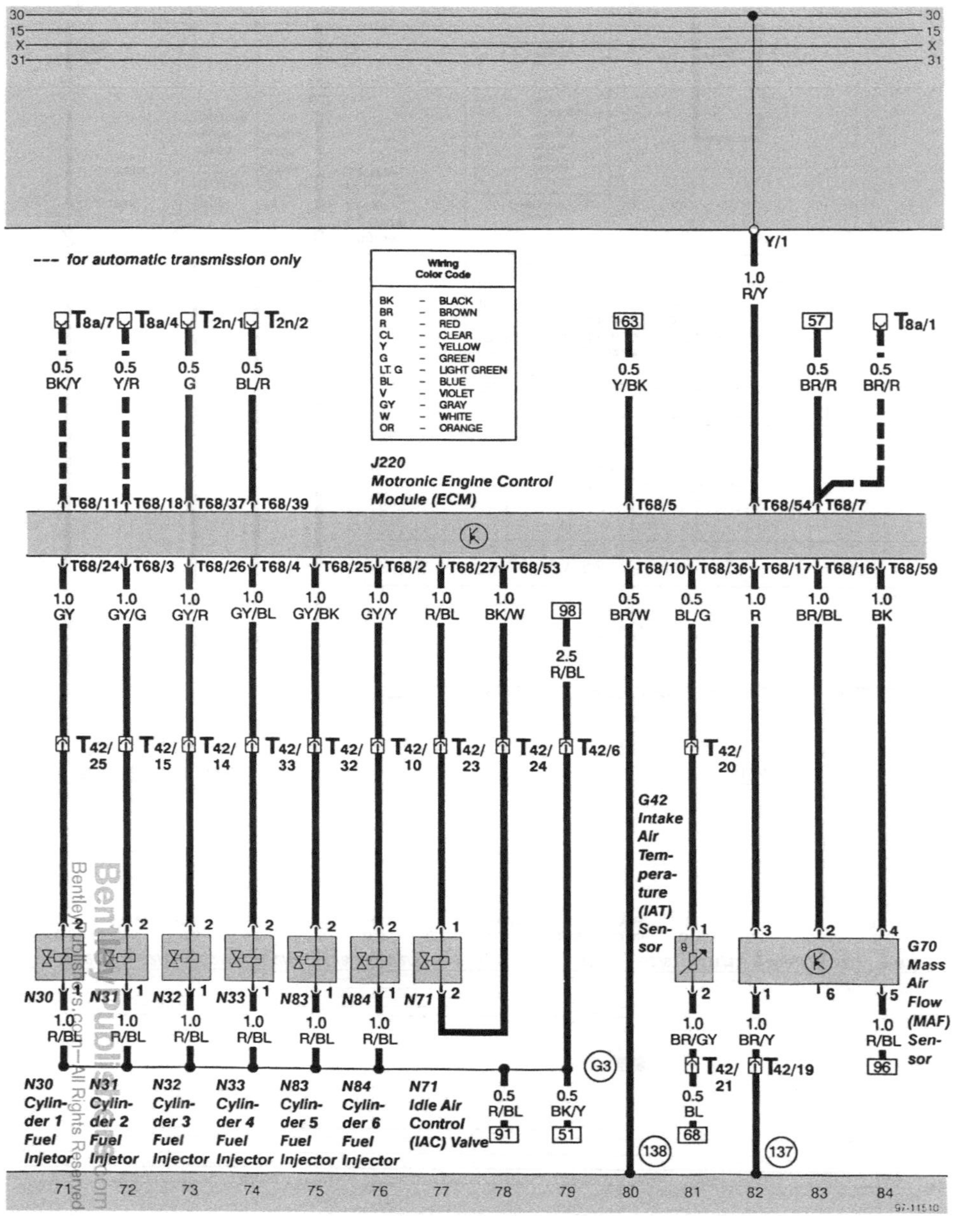

Corrado VR6 (Canada) 1994
Corrado SLC (USA) 1994

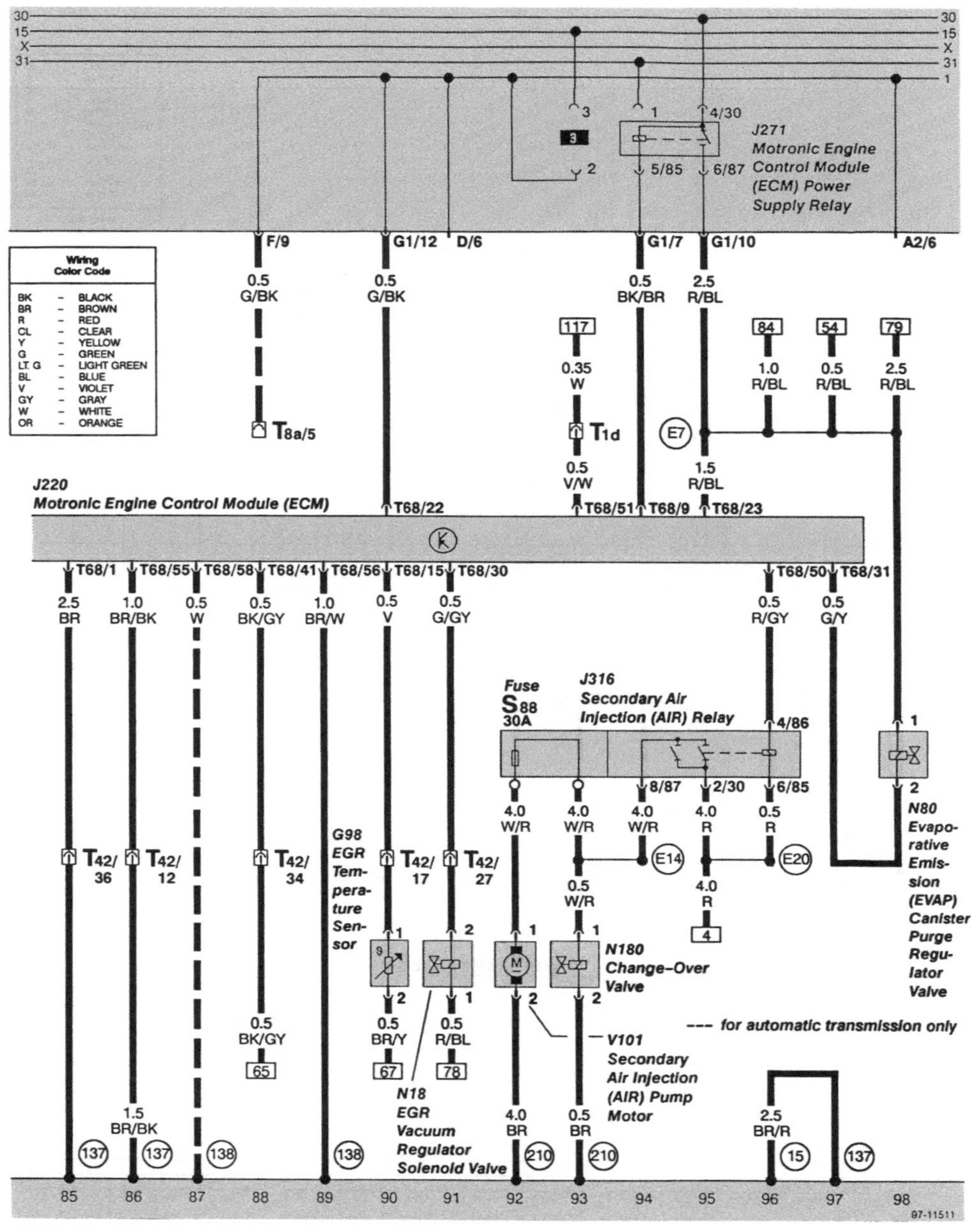

Corrado VR6 (Canada) 1994
Corrado SLC (USA) 1994

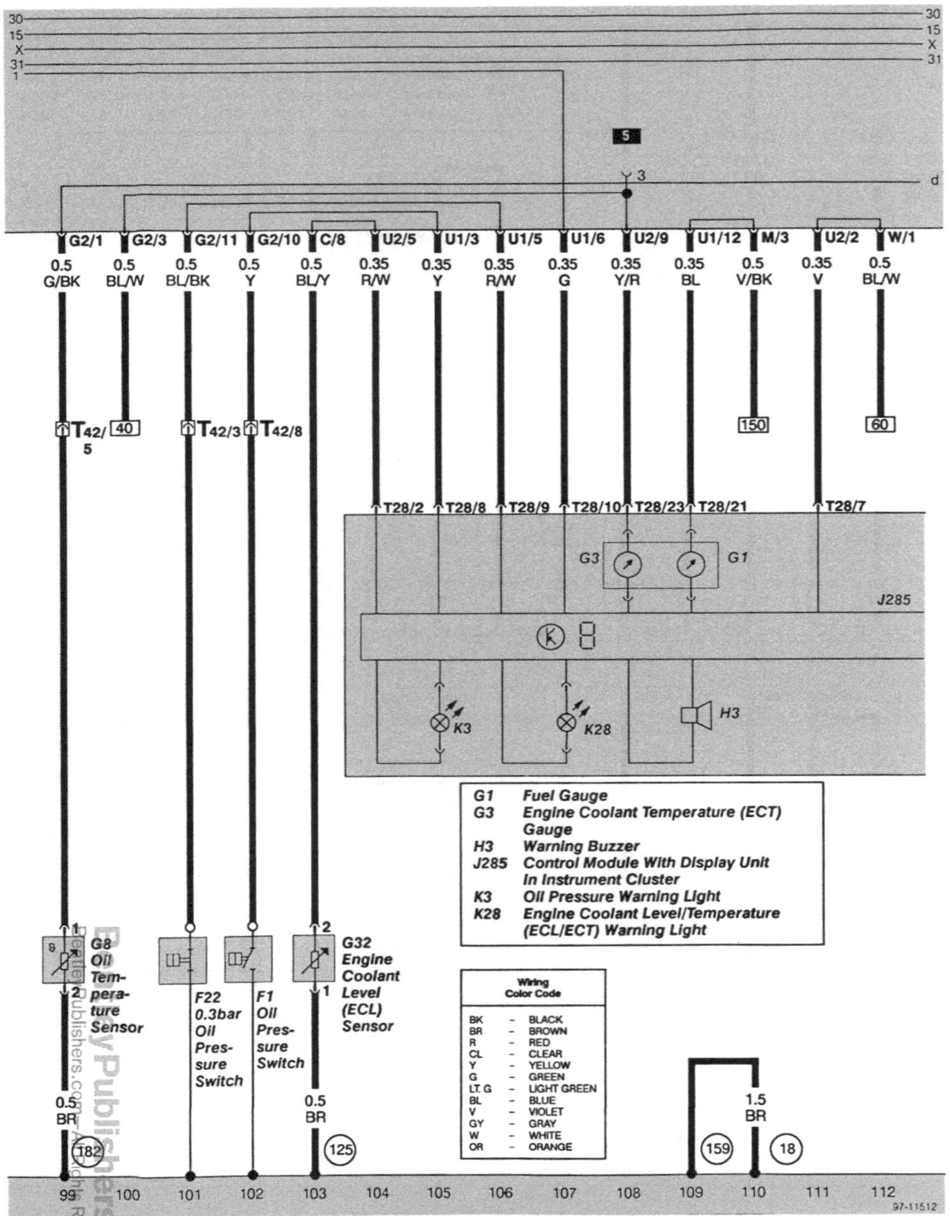

G1	Fuel Gauge
G3	Engine Coolant Temperature (ECT) Gauge
H3	Warning Buzzer
J285	Control Module With Display Unit In Instrument Cluster
K3	Oil Pressure Warning Light
K28	Engine Coolant Level/Temperature (ECL/ECT) Warning Light

Wiring Color Code

BK	-	BLACK
BR	-	BROWN
R	-	RED
CL	-	CLEAR
Y	-	YELLOW
G	-	GREEN
LT. G	-	LIGHT GREEN
BL	-	BLUE
V	-	VIOLET
GY	-	GRAY
W	-	WHITE
OR	-	ORANGE

Corrado VR6 (Canada) 1994
Corrado SLC (USA) 1994

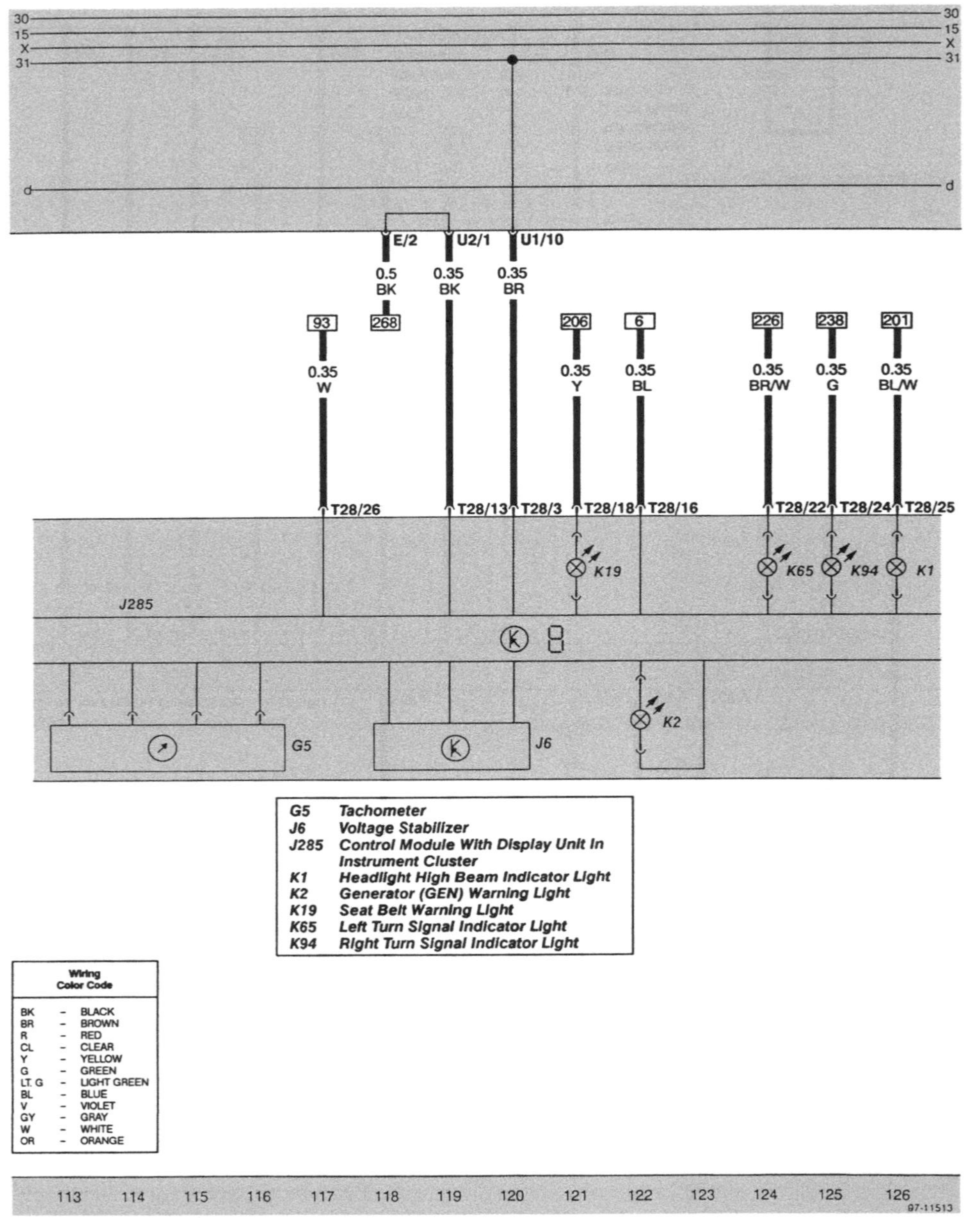

G5	Tachometer
J6	Voltage Stabilizer
J285	Control Module With Display Unit In Instrument Cluster
K1	Headlight High Beam Indicator Light
K2	Generator (GEN) Warning Light
K19	Seat Belt Warning Light
K65	Left Turn Signal Indicator Light
K94	Right Turn Signal Indicator Light

Wiring Color Code

BK	-	BLACK
BR	-	BROWN
R	-	RED
CL	-	CLEAR
Y	-	YELLOW
G	-	GREEN
LT. G	-	LIGHT GREEN
BL	-	BLUE
V	-	VIOLET
GY	-	GRAY
W	-	WHITE
OR	-	ORANGE

Corrado VR6 (Canada) 1994
Corrado SLC (USA) 1994

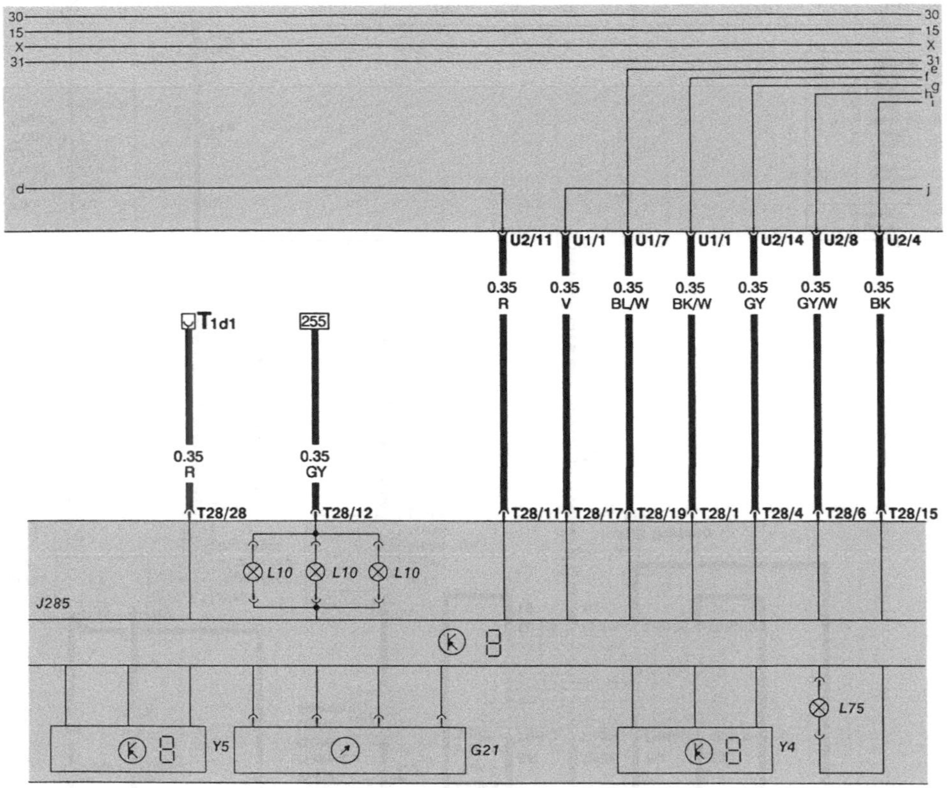

G21	Speedometer
J285	Control Module With Display Unit In Instrument Cluster
L10	Instrument Cluster Light
L75	Digital Display Light
Y4	Odometer Display
Y5	Transmission Range (TR) Selector Lever

Wiring Color Code

BK	-	BLACK
BR	-	BROWN
R	-	RED
CL	-	CLEAR
Y	-	YELLOW
G	-	GREEN
LT. G	-	LIGHT GREEN
BL	-	BLUE
V	-	VIOLET
GY	-	GRAY
W	-	WHITE
OR	-	ORANGE

127 128 129 130 131 132 133 134 135 136 137 138 139 140

97-11514

Corrado VR6 (Canada) 1994
Corrado SLC (USA) 1994

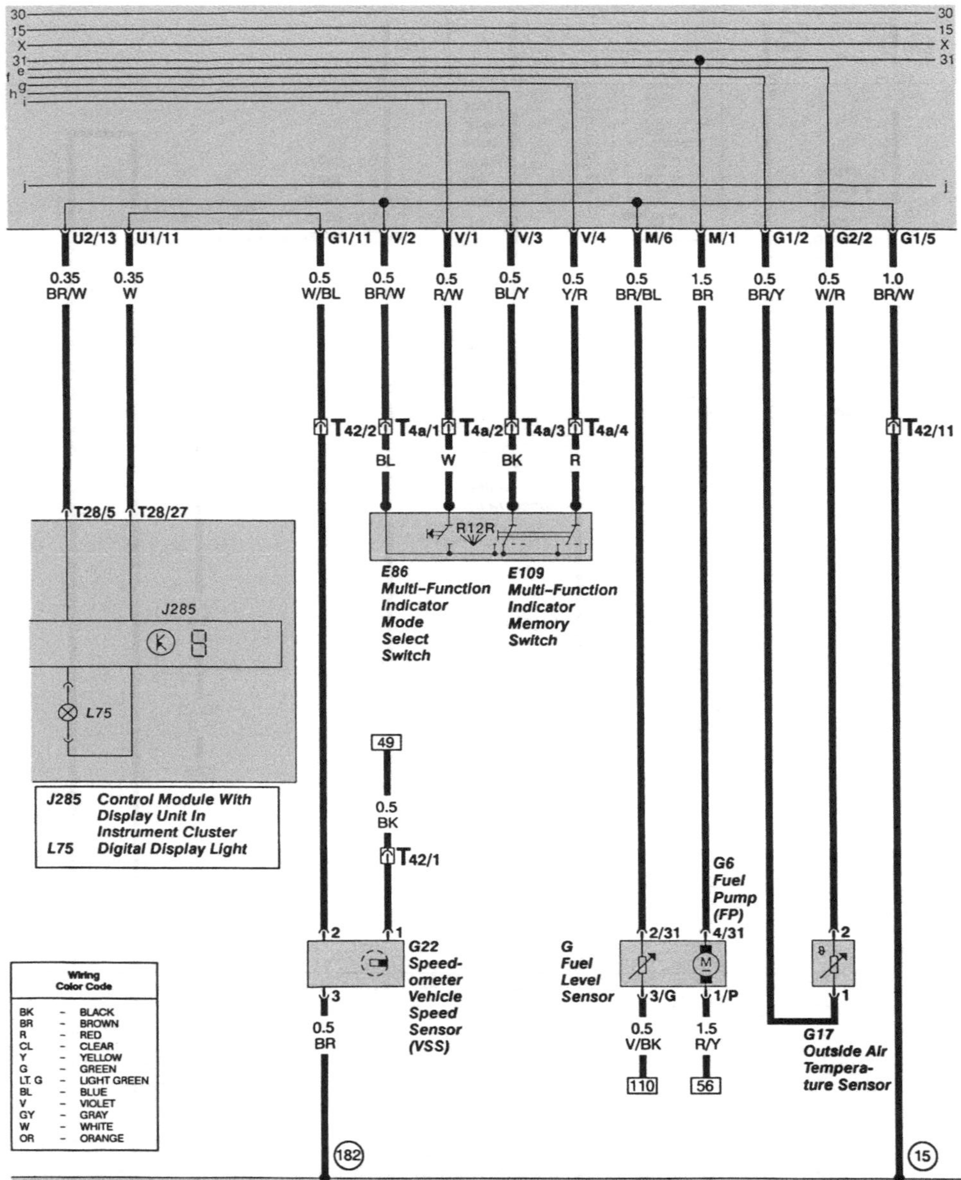

J285	Control Module With Display Unit In Instrument Cluster
L75	Digital Display Light

Wiring Color Code

BK	-	BLACK
BR	-	BROWN
R	-	RED
CL	-	CLEAR
Y	-	YELLOW
G	-	GREEN
LT. G	-	LIGHT GREEN
BL	-	BLUE
V	-	VIOLET
GY	-	GRAY
W	-	WHITE
OR	-	ORANGE

141 142 143 144 145 146 147 148 149 150 151 152 153 154

97-11515

Corrado VR6 (Canada) 1994
Corrado SLC (USA) 1994

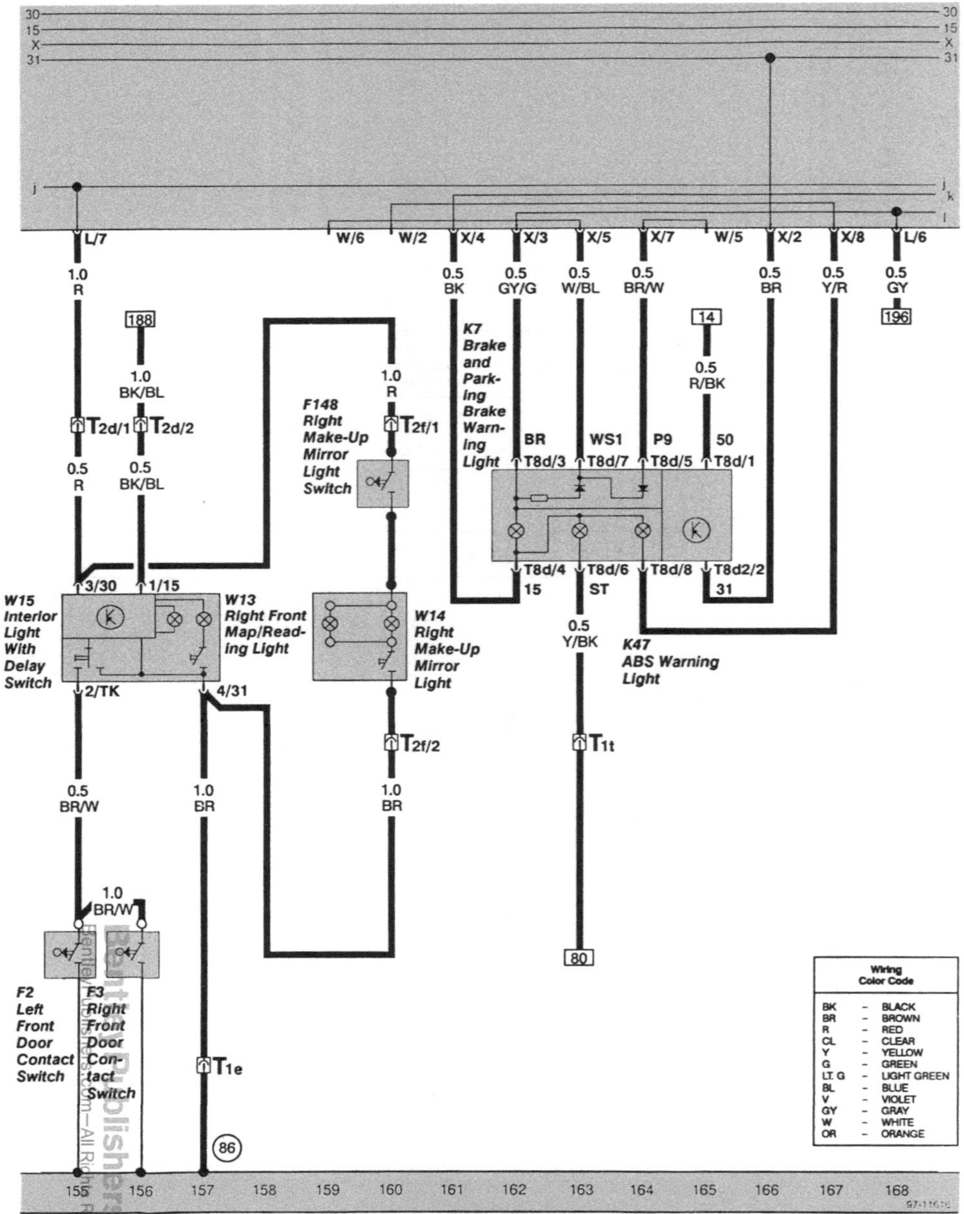

Wiring Color Code

BK	-	BLACK
BR	-	BROWN
R	-	RED
CL	-	CLEAR
Y	-	YELLOW
G	-	GREEN
LT. G	-	LIGHT GREEN
BL	-	BLUE
V	-	VIOLET
GY	-	GRAY
W	-	WHITE
OR	-	ORANGE

Corrado VR6 (Canada) 1994
Corrado SLC (USA) 1994

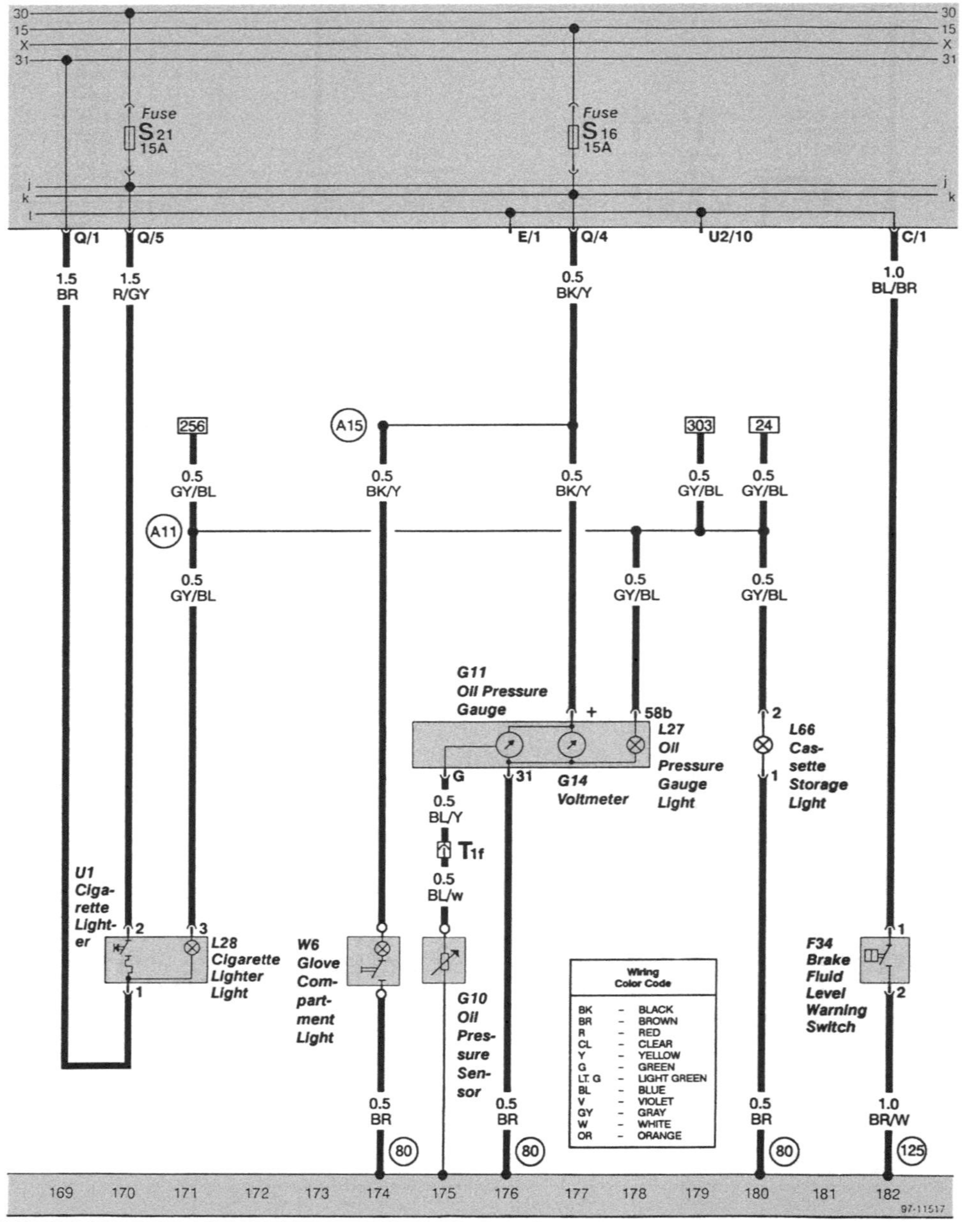

Wiring Color Code

BK	-	BLACK
BR	-	BROWN
R	-	RED
CL	-	CLEAR
Y	-	YELLOW
G	-	GREEN
LT. G	-	LIGHT GREEN
BL	-	BLUE
V	-	VIOLET
GY	-	GRAY
W	-	WHITE
OR	-	ORANGE

Corrado VR6 (Canada) 1994
Corrado SLC (USA) 1994

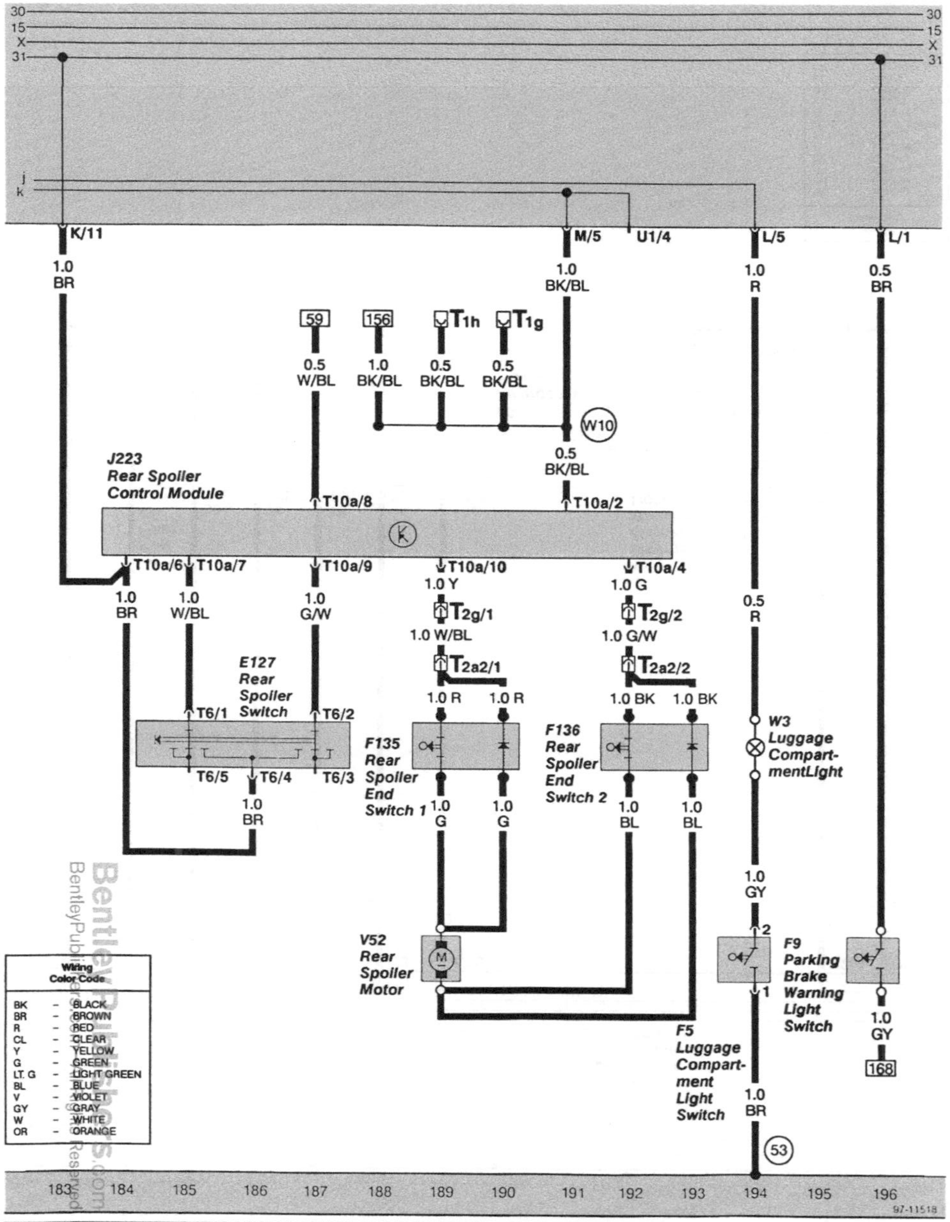

Corrado VR6 (Canada) 1994
Corrado SLC (USA) 1994

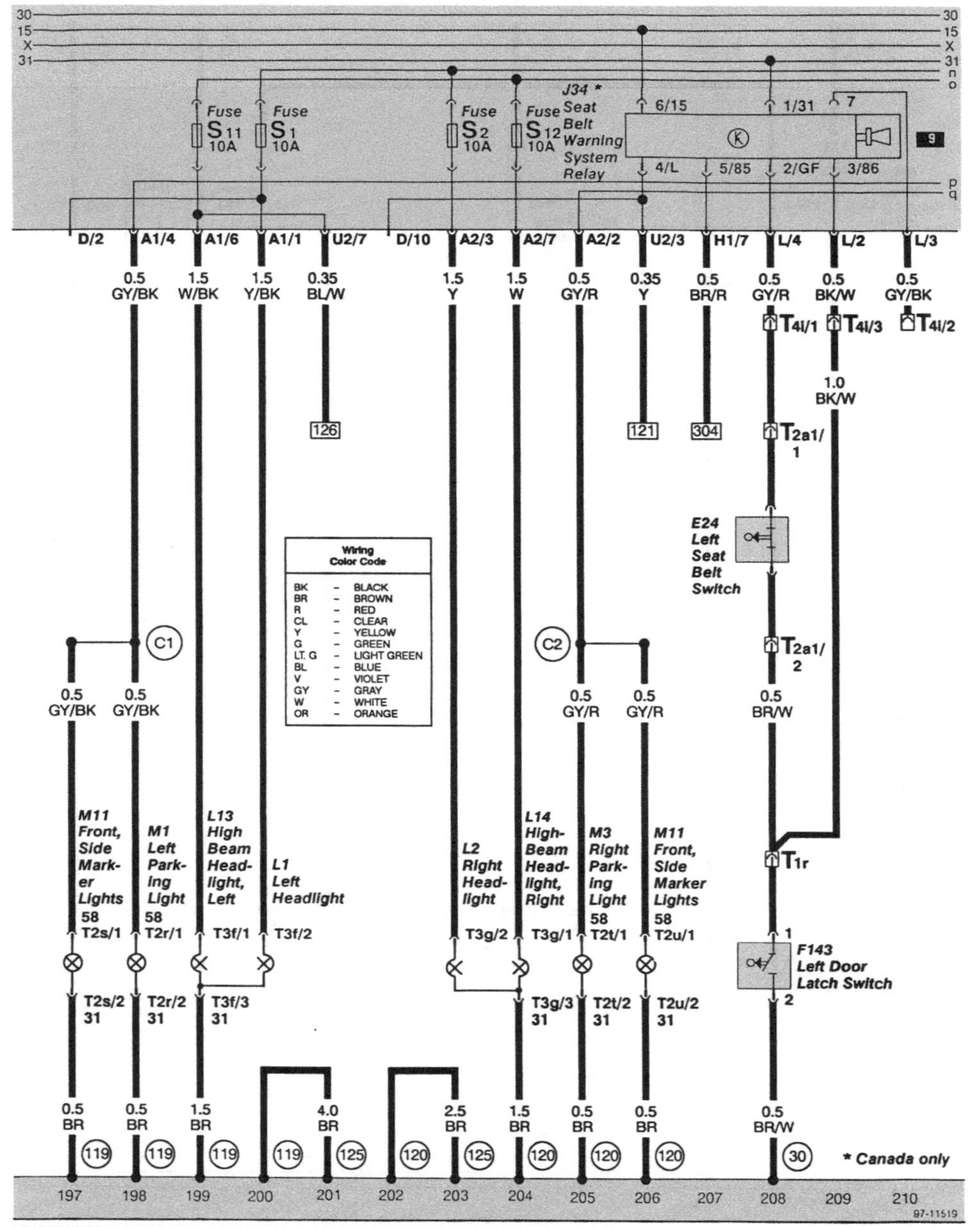

Corrado VR6 (Canada) 1994
Corrado SLC (USA) 1994

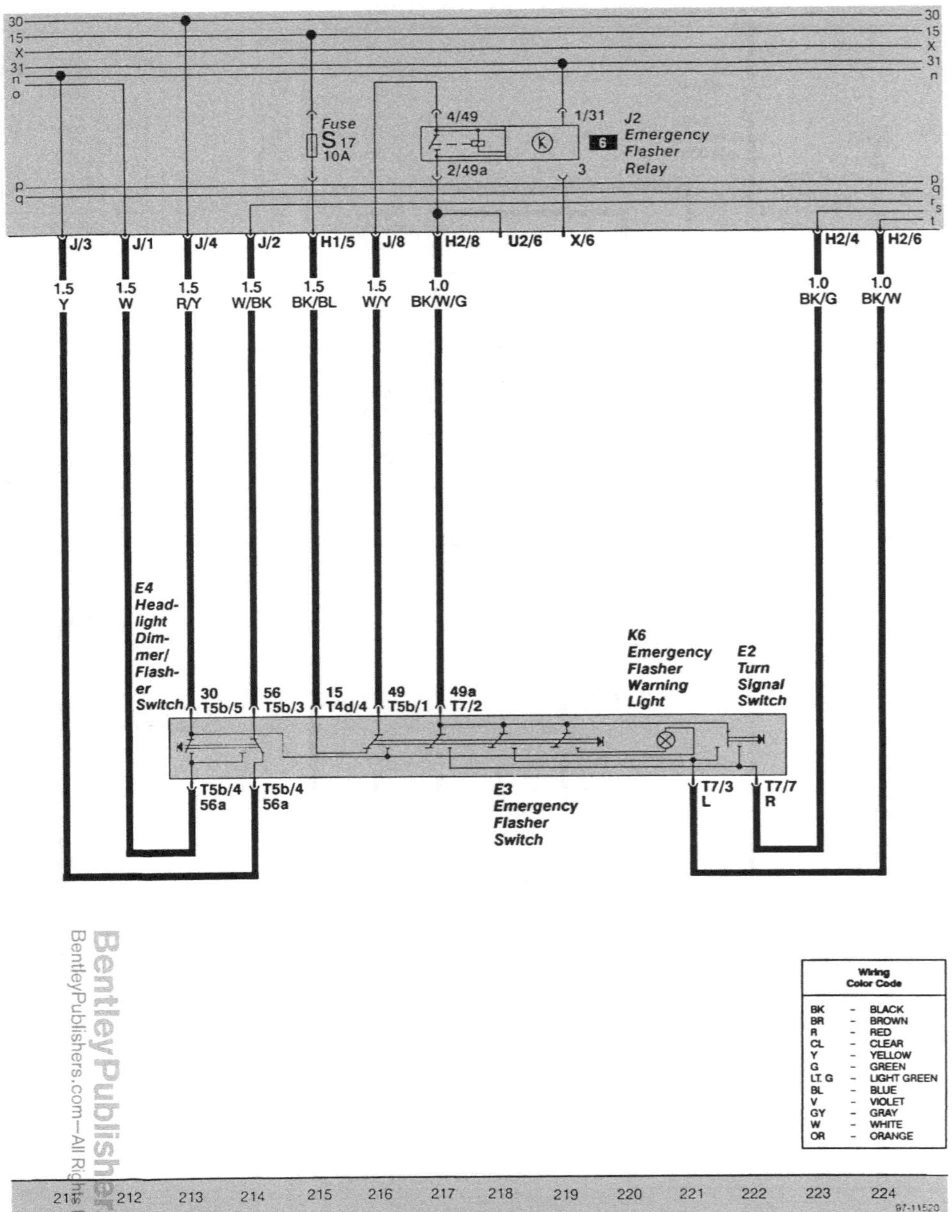

Corrado VR6 (Canada) 1994
Corrado SLC (USA) 1994

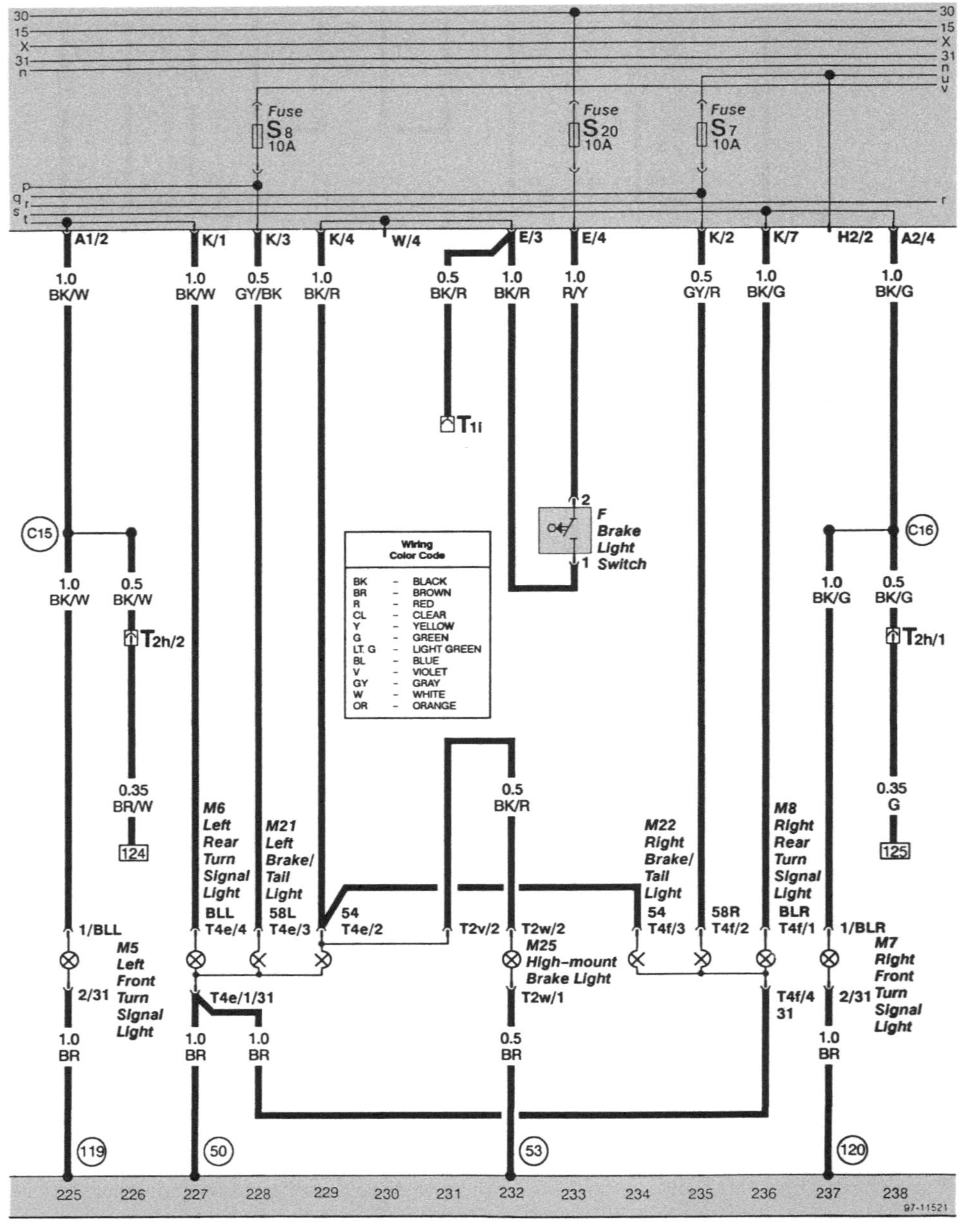

Corrado VR6 (Canada) 1994
Corrado SLC (USA) 1994

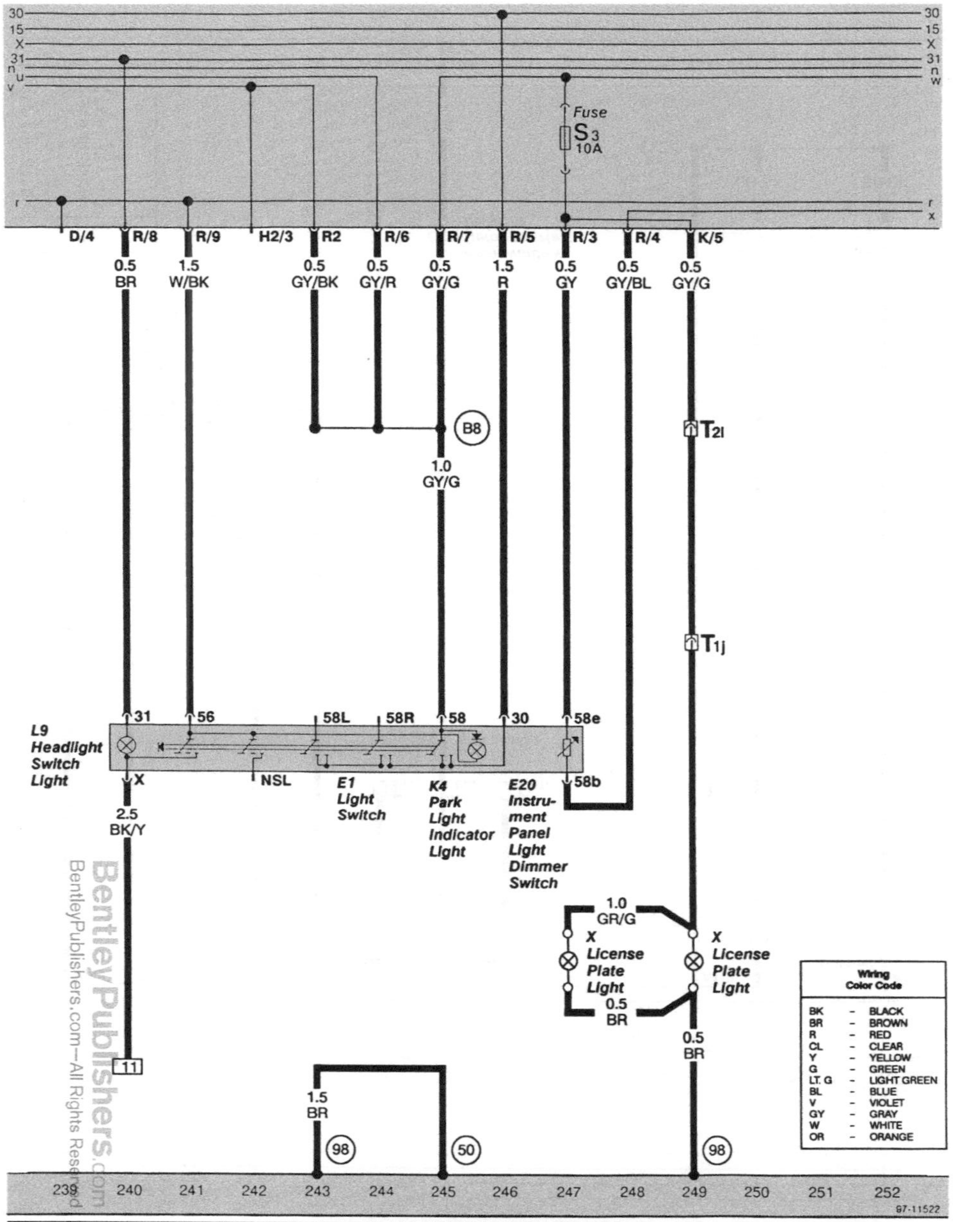
Fuse S3 10A
D/4 R/8 R/9 H2/3 R2 R/6 R/7 R/5 R/3 R/4 K/5
L9 Headlight Switch Light
E1 Light Switch
K4 Park Light Indicator Light
E20 Instrument Panel Light Dimmer Switch
X License Plate Light
X License Plate Light
T2l
T1j
B8
Wiring Color Code
BK - BLACK
BR - BROWN
R - RED
CL - CLEAR
Y - YELLOW
G - GREEN
LT. G - LIGHT GREEN
BL - BLUE
V - VIOLET
GY - GRAY
W - WHITE
OR - ORANGE
239 240 241 242 243 244 245 246 247 248 249 250 251 252
97-11522

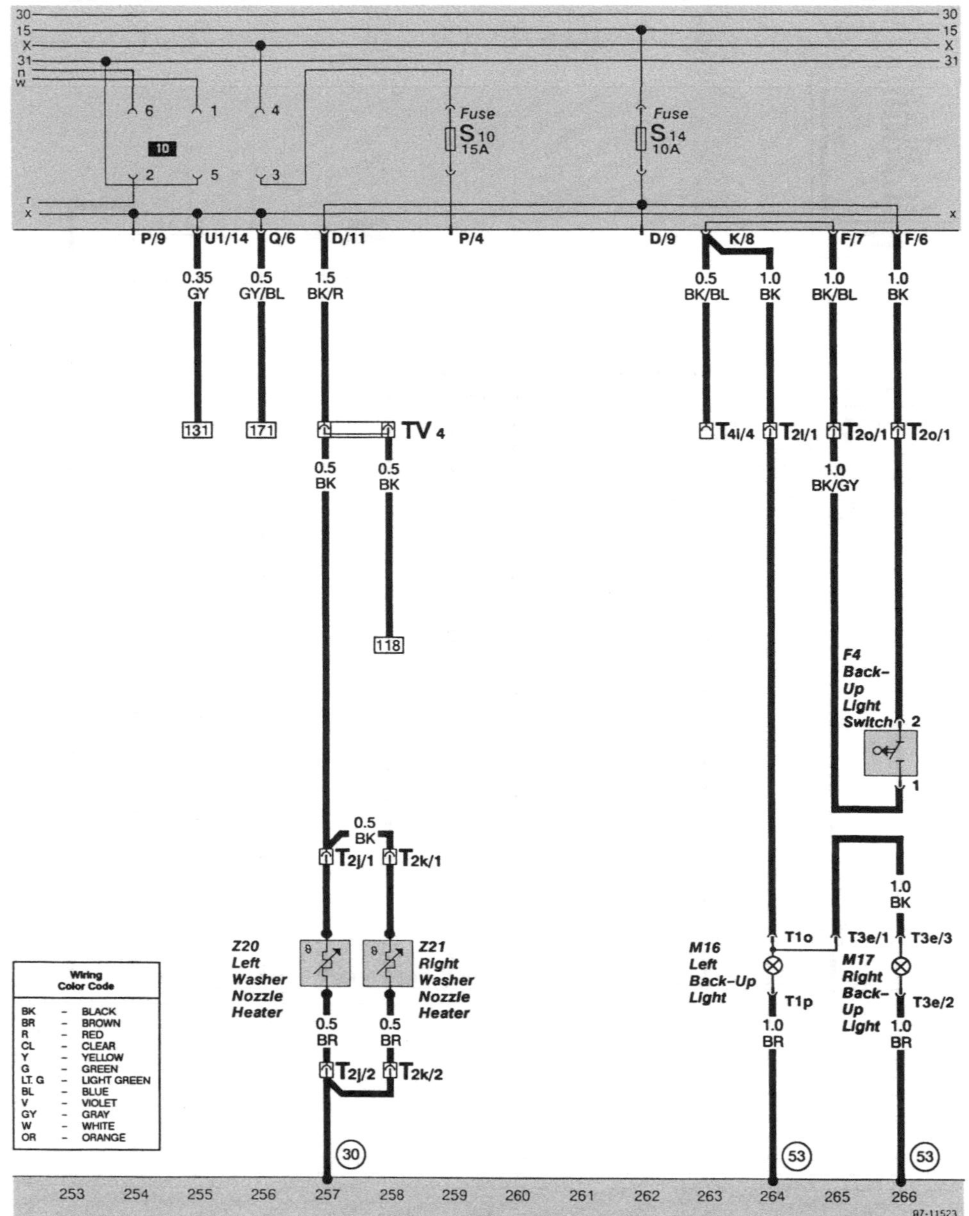
Fuse S10 15A
Fuse S14 10A
P/9 U1/14 Q/6 D/11 P/4 D/9 K/8 F/7 F/6
TV 4
T4l/4 T2l/1 T2o/1 T2o/1
F4 Back-Up Light Switch
Z20 Left Washer Nozzle Heater
Z21 Right Washer Nozzle Heater
M16 Left Back-Up Light
M17 Right Back-Up Light
Wiring Color Code
BK - BLACK
BR - BROWN
R - RED
CL - CLEAR
Y - YELLOW
G - GREEN
LT. G - LIGHT GREEN
BL - BLUE
V - VIOLET
GY - GRAY
W - WHITE
OR - ORANGE
253 254 255 256 257 258 259 260 261 262 263 264 265 266
97-11523

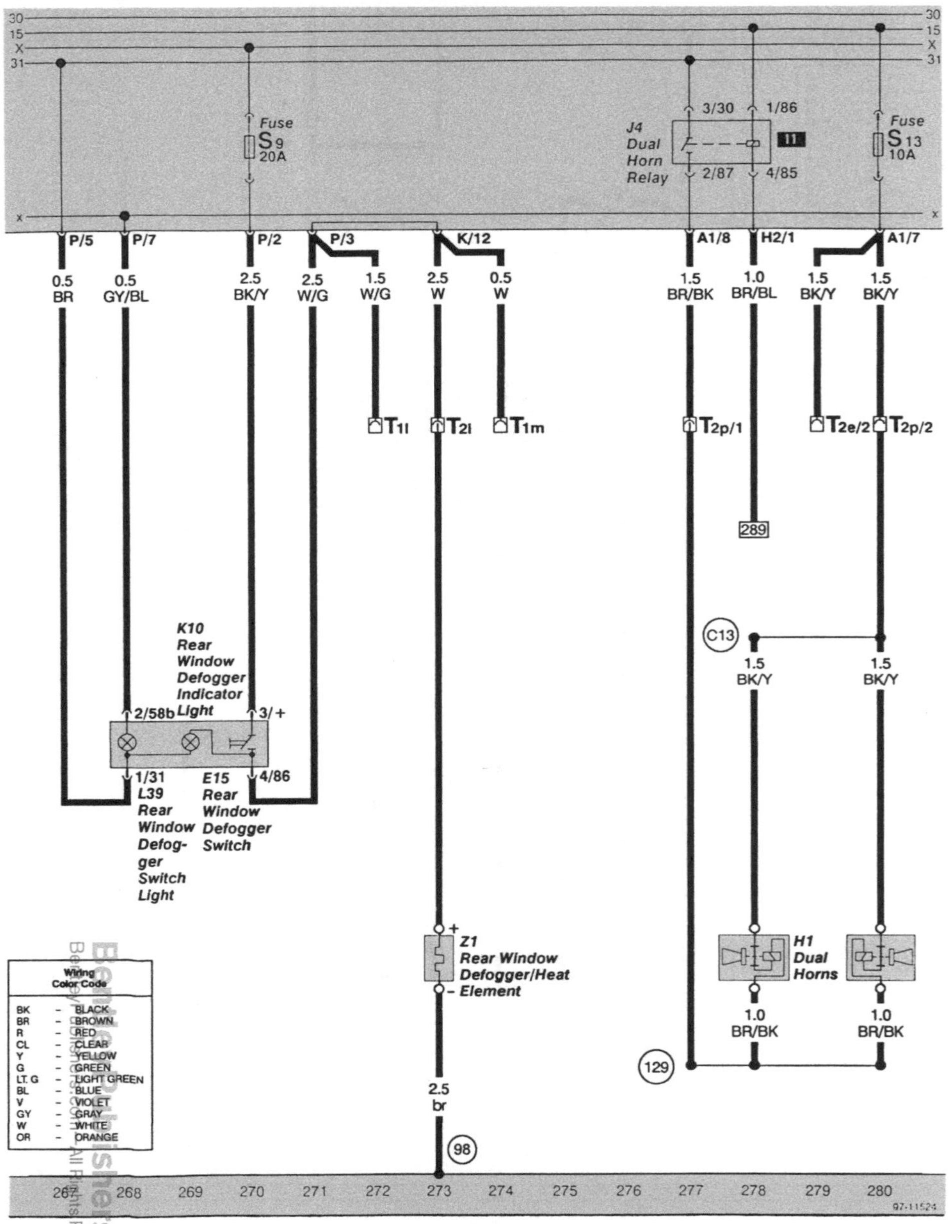

Corrado VR6 (Canada) 1994
Corrado SLC (USA) 1994

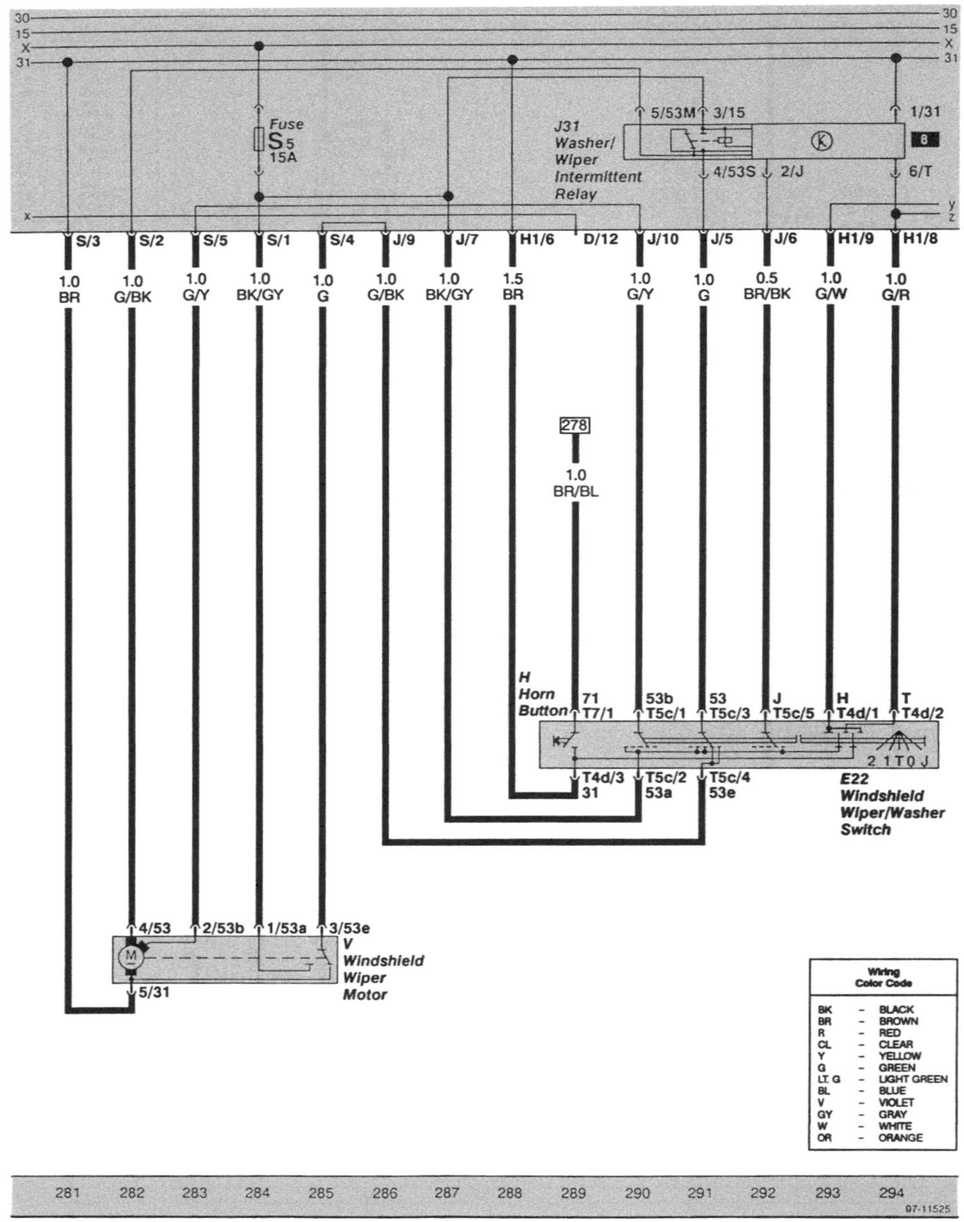

Corrado VR6 (Canada) 1994
Corrado SLC (USA) 1994

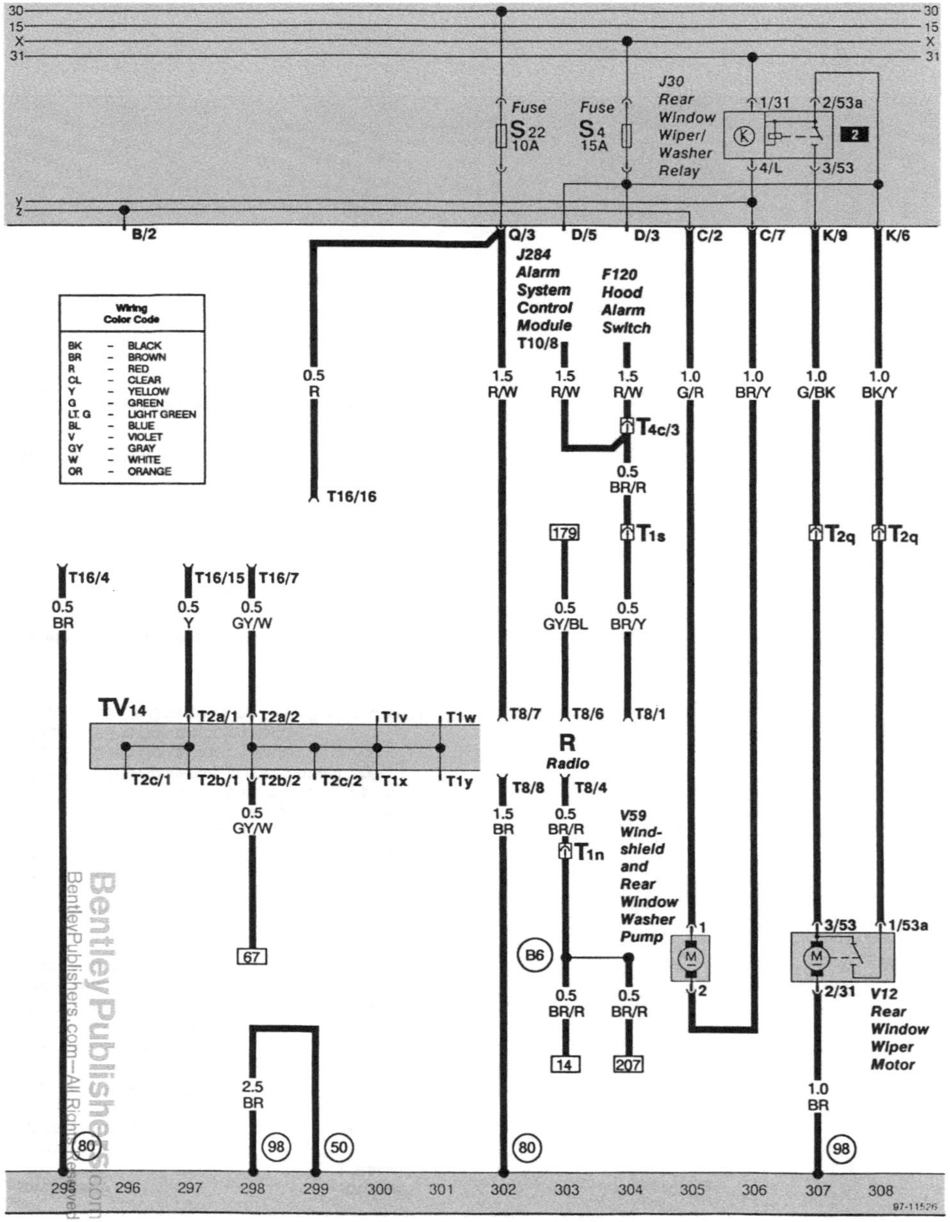

Corrado VR6 (Canada) 1994
Corrado SLC (USA) 1994

THIS PAGE INTENTIONALLY LEFT BLANK

Description | **Current track**

Description	Current track
Antenna , R11	10
Antenna Amplifier, R24	10
Fuse, S22, 10A	1
Left Front Tweeter, R20	3-4
Left Front Woofer, R21	5
Left Rear Speaker, R4	12
Radio, R	1-14
Right Front Tweeter, R22	8-9
Right Front Woofer, R23	7
Right Rear Speaker, R5	13

Wire connectors

- T1c – Single Connector, behind fuse/relay panel
- T1n – Single Connector, behind fuse/relay panel
- T2 – Double Connector, on left A-pillar, lower part
- T2a – Double Connector, on right A-pillar, lower part
- T4 – 4-Pin Connector, behind fuse/relay panel
- T4h – 4-Pin Connector, in front footwell, right
- T8 – 8-Pin Connector, on Radio
- T8a – 8-Pin Connector, on Radio
- T10 – 10-Pin Connector, on Alarm System Control Module

Welded wiring harness points

(A11) – Plus connection (58b), in instrument panel wiring harness

(B6) – Wire connection (s), in steering column switch wiring harness

Ground connections

(80) – Ground connection -1-, in instrument panel wiring harness

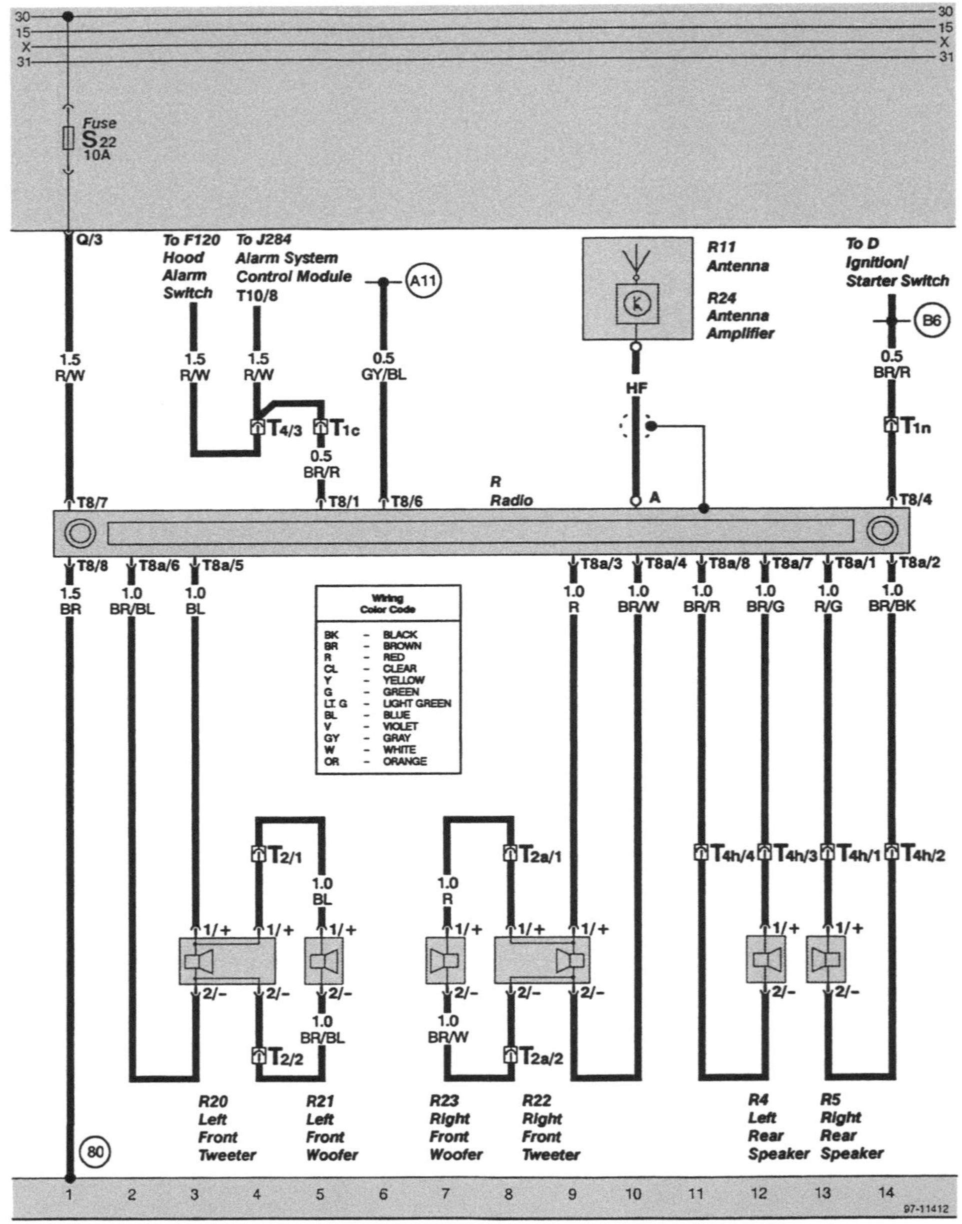

Description | **Current track**

Description	Current track
Antenna Amplifier, R24	20
Antenna, R11	20
CD Changer Unit, R41	17-26
Fuse, S22, 10A	1
Left Front Tweeter, R20	3-4
Left Front Woofer, R21	5
Left Rear Speaker, R4	12
Radio, R	1-28
Right Front Tweeter, R22	8-9
Right Front Woofer, R23	7
Right Rear Speaker, R5	13

Wire connectors

- T1e – Single Connector, behind fuse/relay panel
- T1n – Single Connector, behind fuse/relay panel
- T2 – Double Connector, on A-pillar, left
- T2a – Double Connector, on A-pillar, right
- T4 – 4-Pin Connector, behind fuse /relay panel
- T4h – 4-Pin Connector, in front footwell, right
- T8 – 8-Pin Connector, on Radio
- T8a – 8-Pin Connector, on Radio
- T8b – 8-Pin Connector, on CD Changer Unit
- T10 – 10-Pin Connector, on Alarm System Control Module
- T10a – 10-Pin Connector, on Radio

Welded wiring harness points

- (A11) – Plus connection (58b), in instrument panel wiring harness
- (B6) – Wire connection (s), in steering column switch wiring harness

Ground connection

- (80) – Ground connection -1-, in instrument panel wiring harness

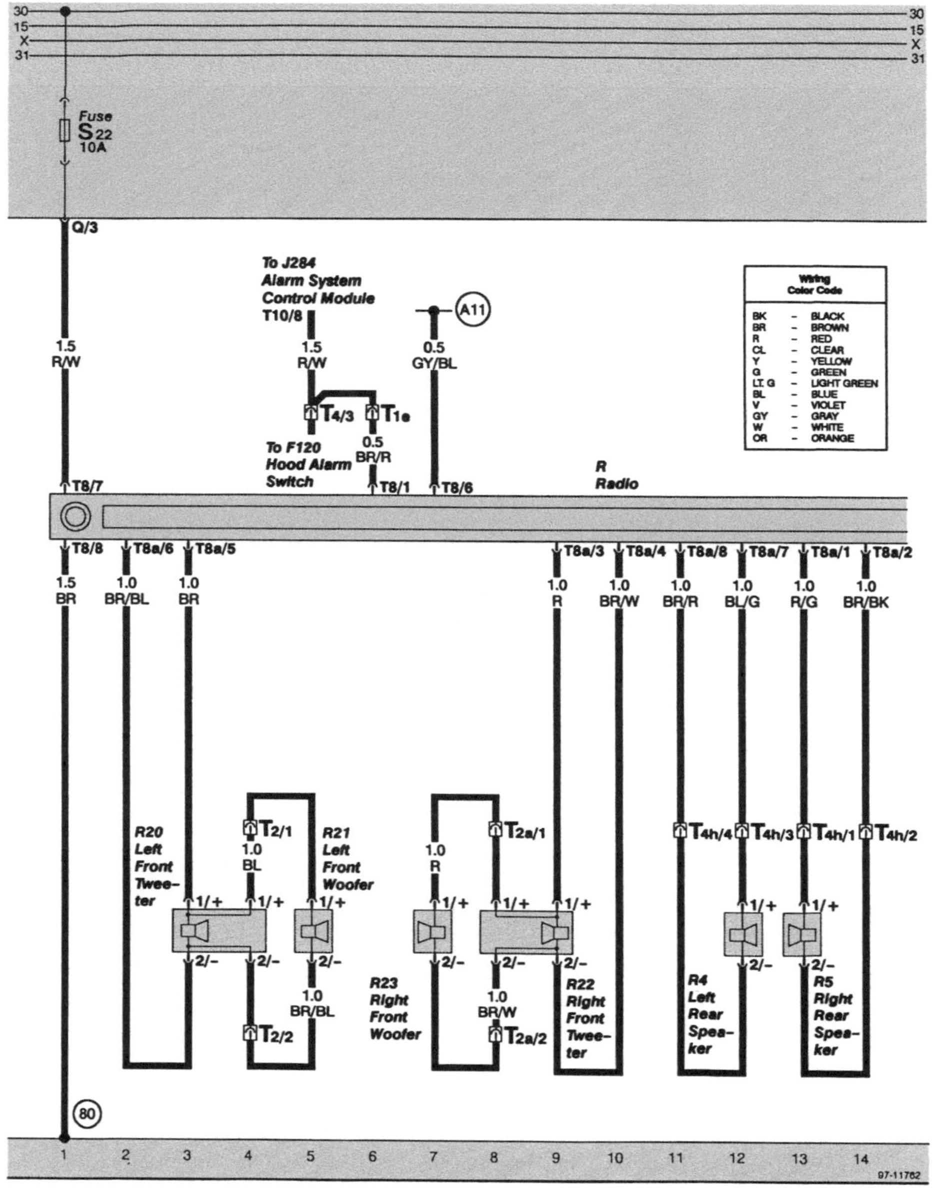

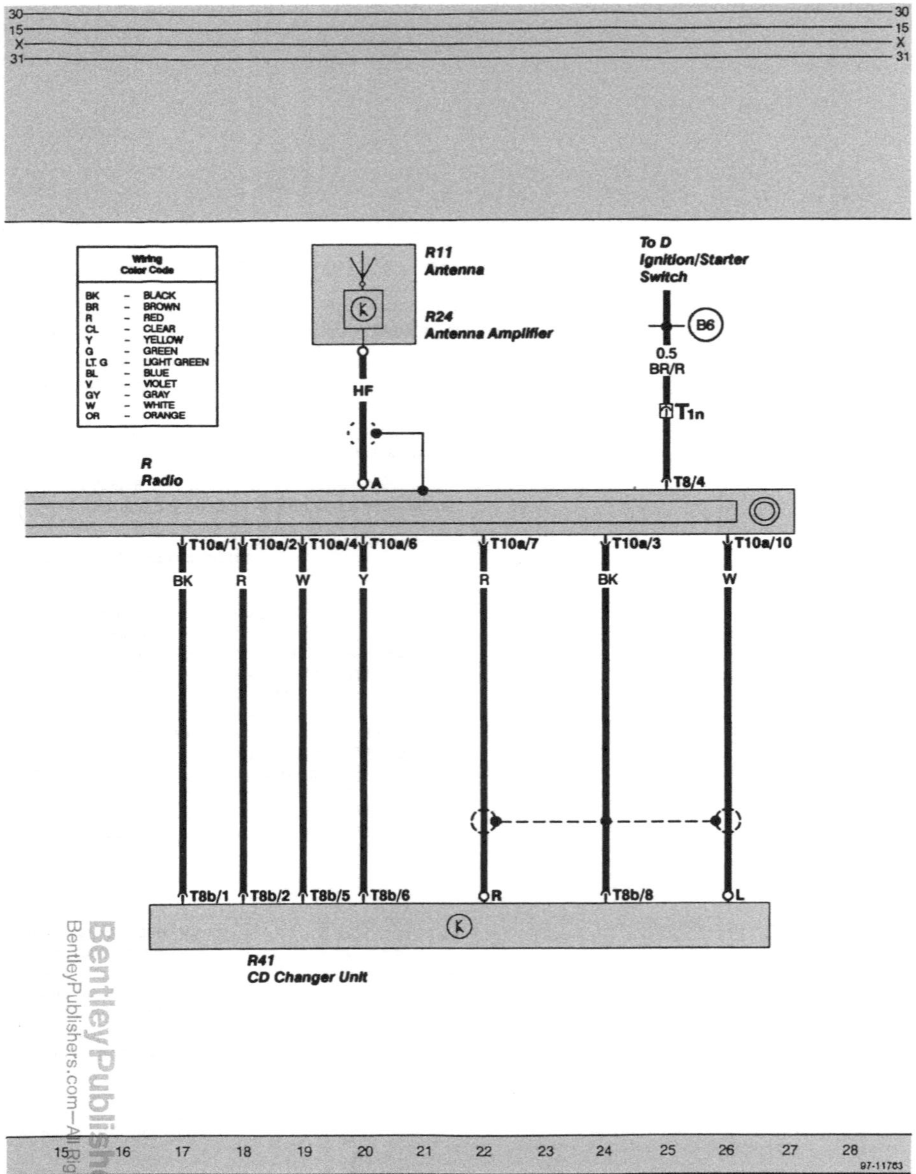

THIS PAGE INTENTIONALLY LEFT BLANK